Biology

FOURTH EDITION

ELDRA PEARL SOLOMON
University of South Florida

LINDA R. BERG
St. Petersburg Junior College

DIANA W. MARTIN
Rutgers University

CLAUDE VILLEE
Emeritus, Harvard University

SAUNDERS COLLEGE PUBLISHING
Harcourt Brace College Publishers

Fort Worth Philadelphia San Diego New York Orlando Austin San Antonio
Toronto Montreal London Sydney Tokyo

To our families, friends, and colleagues
who gave freely of their love, support, knowledge, and time
as we labored over this revision of BIOLOGY . . .

Especially, to . . .
Rabbi Theodore and Freda Brod
Kathleen M. Heide, Ph.D.
Alan Berg
Jennifer Berg
Margaret Martin
Dorothy Villee, M.D.

Text Typeface: Palatino
Compositor: York Graphic Services
Acquisitions Editor: Julie Levin Alexander
Developmental Editors: Rebecca Gruliow and Christine Rickoff
Managing Editor: Carol Field
Project Editor: Nancy Lubars
Copy Editor: Judy Patton
Manager of Art and Design: Carol Bleistine
Art Director: Joan Wendt
Illustration Supervisor: Sue Kinney
Art Editor: Ray Tschoepe
Photo Editor: Robin Bonner
Text Designer and Layout Artist: Rebecca Lloyd Lemna
Cover Designer: Lawrence R. Didona
Text Artwork: Rolin Graphics
Director of EDP: Tim Frelick
Production Manager: Joanne Cassetti
Product Manager: Sue Westmoreland
Director of Marketing: Marjorie Waldron
Cover Credit: © Don Riepe/Peter Arnold, Inc.

Printed in the United States of America

Biology, Fourth Edition

ISBN 0-03-0103533
Library of Congress Catalog Card Number: 95-070696

5678901234 032 3210987654

ABOUT THE COVER

A flower spider (*Misumena vatia*) blends into the white petals of a rose (*Rosa rugosa*) as it waits to ambush unwary insects that visit the flower. The camouflage of the flower spider, which is white or yellow depending on the color of the flower on which it hides, enables it to catch insect prey more effectively. The requirement that certain animals consume other animals in order to survive has resulted in the evolution of very efficient ways to catch prey.

Preface

As we enter the 21st century, we reflect on the important contributions of the biological sciences to modern society. As biologists have studied life, we have gained greater understanding of human life processes and have become more aware of our interdependence with the vast diversity of organisms with which we share our planet. With new advances in biological research, our lives have become healthier, safer, more comfortable, and also more challenging.

One of our principal goals in developing *Biology,* fourth edition, has been to share with beginning biology students our sense of excitement about the biological sciences. We want students to understand and appreciate the diverse organisms on Earth, their remarkable adaptations to the environment, and their evolutionary and ecological relationships. Special emphasis is placed on the basic unity of life and the fundamental similarities of the challenges that have been faced and solved by all living organisms. We are very aware of our responsibility to impress upon our readers our interdependence with the many life forms with which we share planet Earth.

IMPORTANT FEATURES OF THE FOURTH EDITION

The evolution of *Biology* through its editions reflects the advances in the biological sciences and in biological education. Every effort has been made to update its content and pedagogy so that this book accurately presents modern biology.

Student Focus Groups

A significant effort was made in the preparation of this edition to speak directly to students and determine how they use the text and what might be done to enhance its efficacy in teaching biology. To that end, student focus groups were conducted at a number of colleges and universities to gauge how students felt we could improve the utility of the text, artwork, and pedagogy. Students at Delaware County Community College, Montgomery County Community College, Ohio University, Orange Coast Community College, and the University of Delaware met with our editors and told them how they study, which pedagogical elements are most helpful, and which pieces of art are effective in helping them learn the subject matter.

Our student focus group participants told us that illustrations paired with electron micrographs are a very effective visual teaching aid, so we've added more to this edition. They told us that the energy chapters represent some of the most challenging topics for them to comprehend. In response, we made revision of the energy chapters a priority, and applied a step by step approach to build the key concepts of energy transfer through living systems. Students suggested that a key terminology list would help them focus on the important terms they must know. A **Selected Key Terms** list is now featured in every chapter.

Additionally, focus group participants told us that analogies linking the biology presented in the text to real-life examples help them synthesize difficult material. Thus an effort was made to increase such analogies throughout the text. Finally, students were concerned about the depth and quality of the glossary. The glossary has been completely revised and expanded in this edition. Overall, our student focus groups were a key factor in determining how we revised the fourth edition, and we thank the students who participated for their thoughtful suggestions.

Themes

Throughout the book, we emphasize three basic themes of biology—transmission of information, evolution of life, and flow of energy through living systems. As we introduce the concepts of modern biology, we explain how these three themes are connected and how life depends upon them. In this new edition, we also emphasize the process of science and the role of the many scientists who have contributed to our current understanding of biology.

The Author Team

Rapid advances in the biological sciences require a multi-author team who specialize in particular areas of biology. The author team of the fourth edition includes Dr. Eldra Solomon, zoologist and physiologist; botanist Dr. Linda Berg; and cell biologist/geneticist, Dr. Diana Martin. We also acknowledge the legacy of Dr. Claude A. Villee, Professor Emeritus, Harvard University, who contributed to previous editions. All of the authors are experienced college biology teachers.

Tools for Learning

Learning the principles of biology is a challenging endeavor. A variety of learning aids are included within the textbook to help the student achieve mastery of the concepts presented.

1. **Learning Objectives** at the beginning of each chapter indicate what the student must be able to do in order to demonstrate mastery of the material in the chapter.
2. **Concept-statement heads** introduce each section, previewing and summarizing the key idea that will be discussed in that section.
3. **Making the Connection boxes** and **Focus boxes** facilitate integration of concepts and spark interest. For example, in Chapter 12, Making the Connection: "Split Genes and Evolution" relates the discussion of interrupted coding sequences in DNA to the evolution of eukaryotes. This box emphasizes the scientific process.
4. **On the Cutting Edge boxes** present exciting research areas that are currently being explored.
5. **Career Visions** present a variety of professional possibilities in the biological sciences for students to explore. An interview with a professional who majored in biology is presented in each part of the book. Those interviewed discuss how they decided on and prepared for their career, and what they do professionally. Careers new to the fourth edition include science journalism and bioremediation specialist.
6. Numerous **tables,** many of them illustrated, summarize and organize material presented in the text.
7. Carefully rendered **illustrations,** many of them new in this edition, support concepts covered in the text. Many of the illustrations are sequential, with close-ups "exploded" to reveal greater detail. Composite pieces of line art and photographs help students interpret electron micrographs. **Scale bars** accompany micrographs to provide information regarding magnification.
8. **Sequence Summaries** review sequential material discussed in the text.
9. **Boldface terms** facilitate identification of key terms and their definitions and also provide emphasis.
10. A **Chapter Summary** in outline form at the end of each chapter provides a review of the material presented.
11. **Selected Key Terms** at the end of each chapter provides the student with an alphabetical list of many of the important terms defined in the chapter.
12. A **Post Test,** which tests knowledge of the Key Terms, provides the opportunity to evaluate mastery of the material within the chapter; answers are provided in an Appendix.
13. **Review questions** test knowledge of important concepts and applications. They are designed to help students test their mastery of the chapter learning objectives.
14. Many **You Make the Connection** questions challenge the student to relevant principles in other chapters. Others require the student to apply concepts to new situations. These questions can be used for class discussions or essay assignments.
15. A list of **Recommended Readings** at the end of each chapter provides references for further learning.
16. A separate **Glossary,** completely revised and expanded for the fourth edition, facilitates rapid definition of terms.
17. **Appendices** provide help in understanding biological terms, measurement, career information, and biological classification.

THE ORGANIZATION OF *BIOLOGY,* FOURTH EDITION

Educators present the major topics of an introductory biology course in a variety of orders. A lack of consensus regarding sequence of topics is understandable, because reasonable arguments can be advanced for each of the many possible combinations and permutations. All aspects of biology are intimately related, and each could be grasped much more readily if all other topics had been mastered previously. Because this feat cannot be accomplished, each instructor must select the topic sequence that seems most reasonable. For this reason, we have carefully designed each of the eight parts so that they do not depend heavily on preceding chapters and parts. The eight parts and their chapters can be presented in any number of sequences with pedagogic success.

PART 1: THE ORGANIZATION OF LIFE

Chapter 1, *A View of Life: Basic Concepts of Biology* introduces several major concepts of biology, including the fundamental similarities of all living things; the organization of life on individual and ecological levels; the transfer of information; the evolution of life on our planet; the diversity of life and how biologists classify organisms; energy transfer among organisms; and how science works. Chapters 2 and 3 focus on the molecular level of organization and lay the foundations in chemistry needed for an understanding of biological processes. Chapters 4 and 5 focus on the cellular level of organization, with emphasis on recent advances in cell biology.

PART 2: THE ENERGY OF LIFE

Part 2, which focuses on the metabolism and energy transactions involved in life processes, has been thoroughly revised for the fourth edition. Chapter 6 introduces energy in cells and organisms. Chapters 7 and 8 discuss the metabolic adaptations by which organisms obtain and use energy through photosynthesis and cellular respiration. Chapters 7 and 8 have been rewritten so that they can be taught in either order.

Part 3: The Continuity of Life: Genetics

This unit begins with a discussion of mitosis and meiosis in Chapter 9. Chapter 10 describes Mendelian genetics and related patterns of inheritance. Chapter 11 discusses the structure and replication of DNA and Chapter 12 presents RNA and protein synthesis. Gene regulation is discussed in Chapter 13. In Chapter 14, we focus on genetic engineering and in Chapter 15 we focus on human genetics. In Chapter 16, we introduce the role of genes in development, including the latest findings in this exciting and rapidly changing area of biology.

Part 4: The Continuity of Life: Evolution

The unit on evolution has been revised for this edition. Chapter 17 introduces Darwinian evolution and presents scientific evidence for evolution. In Chapter 18, we examine evolution at the population level. Chapter 19 describes the evolution of new species and discusses aspects of macroevolution. Chapter 20 summarizes the evolutionary history of life on Earth. In Chapter 21 we recount the evolution of the primates, including humans.

Part 5: The Diversity of Life

An evolutionary framework is used in our survey of the kingdoms of organisms. In Chapter 22 we discuss why and how organisms are classified. Chapter 23, devoted to the viruses and to Kingdom Prokaryotae, has been revised for this edition. Chapter 25 describes the fungi. Chapters 26 and 27 present the members of the plant kingdom. Chapters 28 through 30 focus on the diversity of animals. The discussion of each group of organisms focuses on their evolutionary relationships and on their structural and functional adaptations. Several new, illustrated tables summarize groups of organisms, such as the bacteria in Chapter 23 and the orders of insects in Chapter 29.

Part 6: Structure and Life Processes in Plants

This part integrates plant structure and function, beginning in Chapter 31 with a discussion of plant structure, growth, and differentiation. New tables, complete with labeled micrographs, have been added in this edition. Chapters 32 through 34 discuss the structural and physiological adaptations of leaves, stems, and roots. Chapter 35 describes reproduction in flowering plants, including asexual reproduction, flowers, fruits, and seeds. Chapter 36 focuses on growth responses and regulation of growth. New topics include thigmomorphogenesis, genetic regulation of auxin, circadian clock mutants, and new chemical regulators.

Part 7: Structure and Life Processes in Animals

This part emphasizes the structural, functional, and behavioral adaptations that animals have evolved to meet environmental challenges. As each system of the animal body is discussed, a comparative approach is used to examine how various animal groups have solved similar and diverse problems. Chapter 37 is devoted to the architecture of the animal body, emphasizing the various tissues and organ systems. Then Chapters 38 through 49 present animal life processes. After a comparison of how different animal groups carry on a particular process—digestion, gas exchange, internal transport, etc.—each chapter considers the human adaptations for that process. The unit ends with a discussion of behavioral adaptations in Chapter 50.

Part 8: The Interactions of Life: Ecology

The ecology unit has been updated for this edition. Chapters 51 through 54 provide the foundations of ecology with the final chapter (55) focusing on environmental problems caused by humans. New topics include coral reef ecology, chaos theory, scramble and contest competition, new human population data, ENSO, keystone species and conservation, and expanded coverage of predation.

Supplements

To further facilitate learning and teaching, a supplement package has been carefully designed for the student and instructor. It includes a **Study Guide, Instructor's Resource Manual, Test Bank, Computerized Test Bank** (available for the IBM PC and Apple Macintosh series), and **BIOXL** (available for both IBM and Macintosh formats). Other important components of *Biology's* supplement package are a set of 250 **Overhead Transparencies** based on diagrams in the book; a set of 150 **Electron Micrograph Overhead Transparencies; BioArt,** which is composed of 100 black-and-white unlabeled line drawings from the text; and 50 **General Sequence Overhead Transparencies,** which contain topics displayed in a series of stages or layers.

A **Laboratory Manual** written by Russell V. Skavaril, Mary M. Finnen, and Steven M. Lawton, all of Ohio State University, and an accompanying **Laboratory Instructor's Manual** are available. Also available is a **Laboratory Manual** written by Carolyn Eberhard of Cornell University. Additionally, a **Custom Publication** service is available from which a wide variety of individual laboratory exercises may be selected and combined in a single volume. Two supplementary texts are available by Randy Moore of the University of Akron. *Writing to Learn Science* and *Classic and Modern Readings in Biology* provide interesting articles and numerous exercises that will enhance understanding of science and biological concepts.

Multi Media Offerings

The **Saunders General Biology Videodisc** has been prepared to enhance lecture or laboratory presentation of material that is difficult to visualize. The 60-minute videodisc contains more than 1500 still images and a collection of video clips from *Encyclopedia Britannica* and other sources, in addition to animated figures from the text. The videodisc is accompanied by the **Saunders General Biology Videodisc Directory,** which contains complete descriptions, barcode labels, reference numbers, and instructions for using the videodisc, and by LectureActive™, a software interface that enables instructors to customize the videodisc for lectures as well as enabling students to use the videodisc for self-directed study.

ACKNOWLEDGMENTS

The development and production of this new edition of *Biology* required extensive interaction and cooperation among the authors and many individuals in our home and professional environments. We appreciate the valuable input and support from editors, colleagues, students, family, and friends.

We are grateful to the editorial and production staffs at Saunders College Publishing for their help and support throughout this project. We thank our publisher Elizabeth Widdicombe and our Acquisitions Editor Julie Alexander for their support, enthusiasm, and ideas. Our Developmental Editors Becca Gruliow and Christine Rickoff efficiently guided the project through the revision process, providing us with many thoughtful reviews and useful suggestions.

We are grateful to our very talented Art Editor Ray Tschoepe who always contributes new and exciting ideas for reconceptualizing and improving the art. We thank Photo Editor Robin Bonner for helping us find outstanding photographs that enhance the text and for her valuable input on the illustration program.

We appreciate the help of our Project Editor Nancy Lubars who expertly guided the project through the complexities of production. We thank Art Director Joan Wendt and Illustrations Supervisor Sue Kinney for coordinating the art program and design. We appreciate the efforts of Sue Westmoreland, Product Manager, in marketing our book. All of these dedicated professionals and many others at Saunders provided the skill and attention needed to produce *Biology,* fourth edition. We thank them for their help and support throughout this project.

We thank our families and friends for their understanding, support, and encouragement as we struggled through many revisions and deadlines. We especially thank Dr. A. Orson Brod, Mical Solomon, Dr. Amy Solomon, Belicia Efros, Dr. Kathleen M. Heide, Alan and Jennifer Berg, and Dr. Charles Martin and Margaret Martin for their input and support.

Our colleagues and students who have used our book have provided valuable input by sharing their responses to the third edition of *Biology* with us. We thank them and ask again for their comments and suggestions as they use this new edition. We can be reached through our editors at Saunders College Publishing. We express our thanks to the many biologists who have read the manuscript during various stages of its preparation and provided us with valuable suggestions for improving it. Their input has contributed greatly to our final product. Fourth edition reviewers include:

Jane Aloi, Saddleback College
Marvin Alvarez, University of South Florida
Sonya Baird, University of Georgia
Ed Bedecarrax, City College of San Francisco
Todd Bennethum, Purdue University
Charles Biggers, University of Memphis
George Boyajian, University of Pennsylvania
Gary Brusca, Humboldt State University
David Benner, Eastern Tennessee State University
Vicki Cameron, Ithaca College
Gary Cole, University of Texas at Austin
Bruce Condon, Seattle Pacific University
Warren Dolphin, Iowa State University
Sharon Eversman, Montana State University
Guy Farish, Adams State College
David Goldstein, Wright State University
Floyd Grimm, Harford Community College
Alexander Harcourt, University of California at Davis
Ricky Hirschorn, Hood College
Dan Hoffman, Bucknell University
Rebecca Holburton, University of Mississippi
Dan Johnson, Eastern Tennessee State University
Tasneem Khaleel, Montana State University
Ross Koning, Eastern Connecticut State University
Dan E. Krane, Wright State University
William Kroen, Wesley College
Virginia Latta, Jefferson State College
Jeffrey May, Marshall University
Miriam Lobstein, Northern Virginia Community College

Charles Mallery, University of Miami
Joseph Michalewicz, Holy Family College
Ann Mickle, LaSalle University
Lillian Miller, Florida Community College
Marion Monahan, Immaculata College
Jim Morrone, Louisiana State University
Anthony G. Moss, Auburn University
Carolyn Ogren, Parkland College
John Olsen, Rhodes College
Dave Polcyn, California State University at San Bernadino
Susan Pross, University of South Florida
Mary Colavito Shepanski, Santa Monica College
Lisa Shimeld, Crafton Hills College
Philip Snider, University of Houston
Robert Stockhouse, Pacific University
Gerald Summers, University of Missouri
Marshall Sundberg, Louisiana State University
Kenneth Thomulka, Philadelphia College of Pharmacy
John Utley, University of New Orleans
Darrel Vodopich, Baylor University
Eileen Walsh, Westchester Community College
Fred Wasserman, Boston University
Jacqueline F. Webb, Villanova University
David Wilson, University of Miami
Roger Young, Texas A & M

Student Reviewers

Sreddevi Chittineni, University of Delaware
Beverly Cimino, Montgomery County Community College
Christopher David Colson, Ohio University
Kimberly Dunham, University of Delaware
Michele France, University of Delaware
Beth Glaze, Orange Coast College
Cory Hinchman, Orange Coast College
Stacy Hirth, Ohio University
Jeffrey L. Kacsandi, Ohio University
Michael Kane, Delaware County Community College
Elizabeth Kucera, Ohio University
Mike Tien Minh Le, Orange Coast College
Nancy Lee, Orange Coast College
Nasser Mahaud, Delaware County Community College
Emedio Marchozzi, Montgomery County Community College
Joe Matthews, Delaware County Community College
Glenda McCourt, Delaware County Community College
Marianna J. McSweeney, University of Delaware
Jami Miller, Ohio University
Rick Poce, Delaware County Community College
Tanga M. Ray, University of Delaware
Heather Slater, Montgomery County Community College
Katherine Strafford, Ohio University
Theresa Tidd, Delaware County Community College
Travis Vaughn, University of Delaware
Irene Wedderien, Orange Coast College
Alex M. Zadeh, Orange Coast College

We would also like to thank the Introductory Biology Students at Ohio University and Montgomery County Community College.

To The Student

Biology is one of the most varied subjects one can study. It is therefore not surprising that biologists are a diverse group, with different interests, talents, and personalities. Almost anyone who has a desire to understand living things can find a suitable niche in the field of biology.

The thousands of students we have taught have differed in their life goals and learning styles. Some have had excellent backgrounds in science, others poor ones. Regardless of their backgrounds, it is common for students taking their first college biology course to find they must work harder than they expected. You can make the task easier by using approaches to learning that are usually successful for a broad range of students.

Many students "study" passively. An active learner always has questions in mind and is constantly making connections. For example, in biology there are many processes that must be understood. Do not try to blindly memorize these; instead think about causes and effects, so that every process becomes a story. Eventually you will see that many processes are connected by common elements.

Active learning is facilitated if you do some of your studying in a small group. In a study group the roles of teacher and learner should become interchangeable, for the best way to make sure you understand is to teach. A study group allows you to be challenged in a nonthreatening environment and can provide some emotional support.

One stumbling block for many students is the necessity to learn a great deal of terminology. In fact, it would be much more difficult to learn and communicate if we did not have this terminology, for words are really "tools for thinking." Learning terminology generally becomes easier because most biological terms are modular. They are composed of mostly Latin and Greek roots, and once you learn many of these you will find you may have a good idea of the meaning of a new word even before it is defined. For this reason we have included Appendix A, Understanding Biological Terms. Of course, to make sure you understand the precise definition, you will want to use the Index and Glossary. The more you use biological terms, in both speech and writing, the more comfortable you will be.

Although biology is a demanding subject, the time and effort you spend studying will be well spent, because this is a very exciting time to be a biologist. Today we have the tools to study living things in ways that were only a dream in the not too distant past. As we gain new information, our concepts are constantly evolving. We find this to be one of the most exhilarating aspects of biology, and we hope you will too!

INTRODUCING *BIOLOGY*, FOURTH EDITION

Many features of ***Biology*, fourth edition** have been designed, or increased in number from the third edition, to address specific needs of students, as revealed in student focus groups.

Making the Connection boxes—now in every chapter—encourage students to integrate concepts from various chapters.

180 *Part 2 Energy Transfer Through Living Systems*

Making the Connection

Electron Transport and Heat

What is the source of our body heat? Essentially, it is a byproduct of various exergonic reactions, especially those involving the electron transport chains in our mitochondria. Some organisms are able to produce unusually large amounts of heat by uncoupling electron transport from ATP production. As a consequence, most of the energy of glucose is converted to heat rather than to ATP energy. Even some plants (which we don't generally consider to be "warm" organisms) have this capability. For example, skunk cabbage, which lives in North American swamps and wet woodlands, generally flowers when the ground is still covered with snow. Its uncoupled mitochondria generate large amounts of heat, enabling the plant to melt the snow and attract pollinators. ▲

Skunk cabbage *(Symplocarpus foetidus).* (Earth Scenes © 1996 Leonard Lee Rue III)

ing energy as they go. Finally, the last cytochrome in the chain, cytochrome a_3, passes two electrons to oxygen. The electrons simultaneously unite with protons from the surrounding medium to form hydrogen; the chemical reaction between hydrogen and oxygen produces water.

Because oxygen is the final electron acceptor in the electron transport system, organisms that respire aerobically require oxygen. What happens when cells that are strict aerobes are deprived of oxygen? When no oxygen is available to accept them, the last cytochrome in the chain is stuck with its electrons. When that occurs, each acceptor molecule in the chain remains stuck with electrons (i.e., is reduced), and the entire system is blocked all the way back to NADH. Because phosphorylation is coupled to electron transport, no further ATPs are produced by way of the electron transport system. Most cells of complex organisms cannot live long without oxygen because the amount of energy they produce in its absence is insufficient to sustain life processes.

Lack of oxygen is not the only factor that interferes with the electron transport system. Some poisons, including cyanide, inhibit the normal activity of the cytochrome system. Cyanide binds tightly to the iron in cytochrome a_3, making it unable to transport electrons on to oxygen. This blocks the further passage of electrons through the chain, halting ATP production.

The chemiosmotic model explains the coupling of ATP synthesis to electron transport

The flow of electrons in electron transport is usually tightly coupled to the production of ATP and does not occur unless the phosphorylation of ADP can also proceed. This prevents a waste of energy, because high-energy electrons do not flow unless ATP can be produced. If electron flow were uncoupled from the phosphorylation of ADP, there would be no production of ATP, and the energy of the electrons would be converted to heat (see *Making the Connection: Electron Transport and Heat*).

For decades scientists were aware that oxidative phosphorylation occurs in mitochondria, and many experiments had shown that the transfer of two electrons from each NADH to oxygen (via the electron transport chain) usually results in the production of up to three ATP molecules. However, for a long time, just *how* ATP synthesis is related to electron transport remained a mystery. Then, in 1961 Peter Mitchell proposed the **chemiosmotic model,** based on his experiments with bacteria. His model was

High-interest *Focus On* boxes draw attention to interesting biological phenomena and current research.

Chapter 18 Population Genetics 425

Focus On

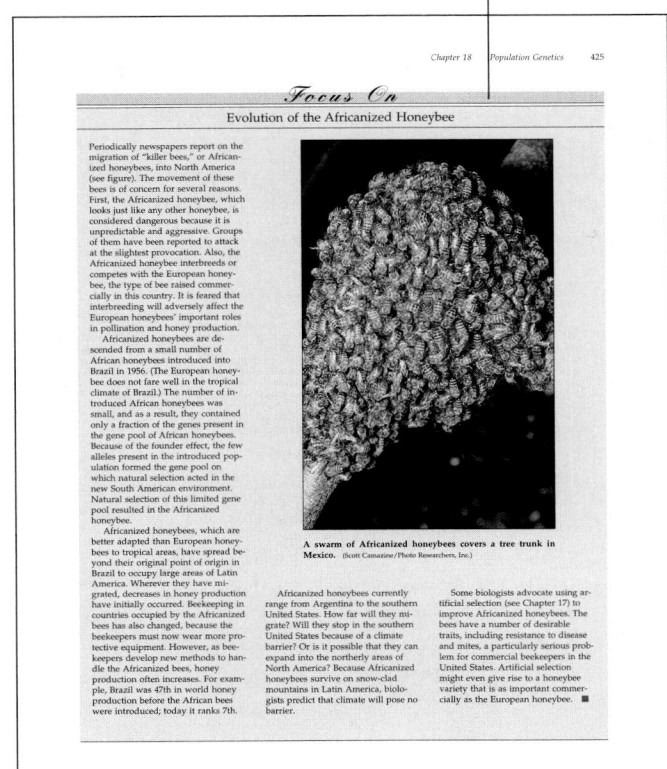

Evolution of the Africanized Honeybee

Periodically newspapers report on the migration of "killer bees," or Africanized honeybees, into North America (see figure). The movement of these bees is of concern for several reasons. First, the Africanized honeybee, which looks just like any other honeybee, is considered dangerous because it is unpredictable and aggressive. Groups of them have been reported to attack at the slightest provocation. Also, the Africanized honeybee interbreeds or competes with the European honeybee, the type of bee raised commercially in this country. It is feared that interbreeding will adversely affect the European honeybees' important roles in pollination and honey production.

Africanized honeybees are descended from a small number of African honeybees introduced into Brazil in 1956. (The European honeybee does not fare well in the tropical climate of Brazil.) The number of introduced African honeybees was small, and as a result, they contained only a fraction of the genes present in the gene pool of African honeybees. Because of the founder effect, the few alleles present in the introduced population formed the gene pool on which natural selection acted in the new South American environment. Natural selection of this limited gene pool resulted in the Africanized honeybee.

Africanized honeybees, which are better adapted than European honeybees to tropical areas, have spread beyond their original point of origin in Brazil to occupy large areas of Latin America. Wherever they have migrated, decreases in honey production have initially occurred. Beekeeping in countries occupied by the Africanized bees has also changed, because the beekeepers must now wear more protective equipment. However, as beekeepers develop new methods to handle the Africanized bees, honey production often increases. For example, Brazil was 47th in world honey production before the African bees were introduced; today it ranks 7th.

A swarm of Africanized honeybees covers a tree trunk in Mexico. (Scott Camazine/Photo Researchers, Inc.)

Africanized honeybees currently range from Argentina to the southern United States. How far will they migrate? Will they stop in the southern United States because of a climate barrier? Or is it possible that they can expand into the northerly areas of North America? Because Africanized honeybees survive on snow-clad mountains in Latin America, biologists predict that climate will pose no barrier.

Some biologists advocate using artificial selection (see Chapter 17) to improve Africanized honeybees. The bees have a number of desirable traits, including resistance to disease and mites, a particularly serious problem for commercial beekeepers in the United States. Artificial selection might even give rise to a honeybee variety that is as important commercially as the European honeybee. ■

The evolution of ***Biology,* fourth edition** is particularly apparent in the incorporation of superb artwork—much of it revised or new. Students and instructors will respond enthusiastically to this visual enhancement, which includes

- **Sequential art** that incorporates the use of close-ups "exploded" to reveal greater detail
- **Composite pieces of line art and photographs** to help students interpret electron micrographs
- **Scale bars** accompanying the electron micrographs, providing a guide that clarifies size
- **Sequence summaries** show biological processes at a glance and help students review for exams

Chapter 5 Biological Membranes 135

particle binds to receptor proteins on plasma membrane → coated pit forms → coated vesicle forms by endocytosis → "uncoating" of vesicle → vesicles fuse to form endosome → division of endosome:

→ some vesicles return receptors to plasma membrane
→ other vesicles fuse with lysosomes and contents are digested and processed

Cholesterol in the blood is taken up by animal cells by receptor-mediated endocytosis. Much of the receptor-mediated endocytosis pathway was detailed through studies of the receptor for low-density lipoprotein (LDL—a primary cholesterol carrier in blood) by M. Brown and J. Goldstein. In 1986 these investigators were awarded the Nobel Prize for their pioneering work. Their findings have important medical implications, for cholesterol that remains in the blood instead of entering the cells can become deposited on the artery walls and increase the risk of heart attack.

The recycling of the LDL receptor to the plasma membrane through vesicles illustrates a problem common to all cells that employ endocytotic and exocytotic mechanisms. In cells that are constantly involved in secretion, an equivalent amount of membrane must be returned to the interior of the cell for each vesicle that fuses with the plasma membrane; if it is not, the cell surface will keep expanding even though the growth of the cell itself may be arrested. A similar situation exists for cells that use endocytosis. A macrophage, for example, ingests the equivalent of its entire surface membrane in about 30 minutes, requiring an equivalent amount of recycling or new membrane synthesis for the cell to maintain its surface area.

JUNCTIONS ARE SPECIALIZED CONTACTS BETWEEN CELLS

Cells in close contact with each other may develop specialized intercellular junctions that involve their plasma membranes as well as other components. These structures may allow neighboring cells to form strong connections with each other, prevent passage of materials, or establish rapid communications between adjacent cells. In animals there are three common types of intercellular contacts: desmosomes, tight junctions, and gap junctions. Plant cells are connected by plasmodesmata.

Desmosomes are points of attachment between some animal cells

Adjacent epithelial cells, such as those found in the upper layer of the skin, are so tightly bound to each other that strong mechanical forces are required to separate them. They are held together by structures called **desmosomes** (Fig. 5–23). Each desmosome includes parts of two adjacent cells. It is made up of regions of dense material associated with the cytosolic sides of the two plasma membranes, plus protein filaments that cross the 24 nanometer-wide intercellular space between them.

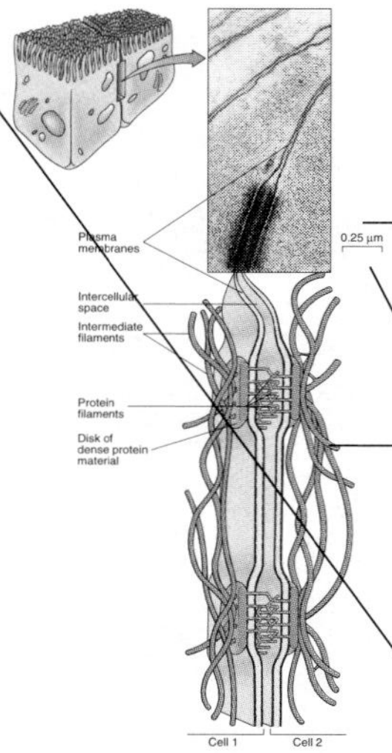

FIGURE 5–23 Desmosomes may connect two animal cells. The dense structure in the transmission electron micrograph is a desmosome. As can be seen in the interpretive drawing, each desmosome consists of a pair of button-like discs associated with the plasma membranes of adjacent cells, plus the intercellular protein filaments that connect them. Intracellular intermediate filaments attached to the discs are connected to other desmosomes. (D.W. Fawcett)

Learning Objectives at the beginning of each chapter indicate exactly what the student must be able to do in order to demonstrate mastery of the material in the chapter.

Concept statement heads formulated as complete statements, not as "disembodied" terminology, provide a clear preview and summary of conceptual discussion to follow. An innovation in the third edition, full-statement headings have remained popular with students.

Numerous **tables,** many of them illustrated, summarize and organize material presented in the text.

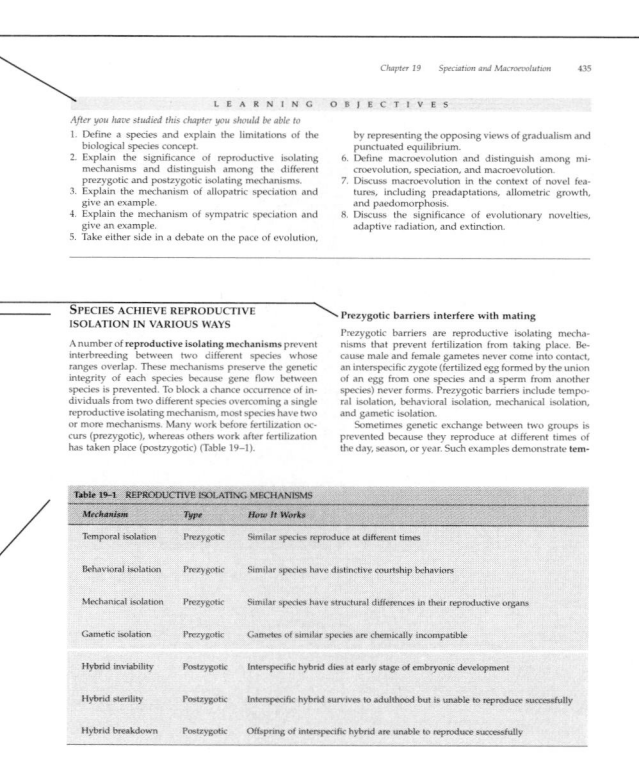

Chapter 19 Speciation and Macroevolution 435

L E A R N I N G O B J E C T I V E S

After you have studied this chapter you should be able to

1. Define a species and explain the limitations of the biological species concept.
2. Explain the significance of reproductive isolating mechanisms and distinguish among the different prezygotic and postzygotic isolating mechanisms.
3. Explain the mechanism of allopatric speciation and give an example.
4. Explain the mechanism of sympatric speciation and give an example.
5. Take either side in a debate on the pace of evolution, by representing the opposing views of gradualism and punctuated equilibrium.
6. Define macroevolution and distinguish among microevolution, speciation, and macroevolution.
7. Discuss macroevolution in the context of novel features, including preadaptations, allometric growth, and paedomorphosis.
8. Discuss the significance of evolutionary novelties, adaptive radiation, and extinction.

SPECIES ACHIEVE REPRODUCTIVE ISOLATION IN VARIOUS WAYS

A number of **reproductive isolating mechanisms** prevent interbreeding between two different species whose ranges overlap. These mechanisms preserve the genetic integrity of each species because gene flow between species is prevented. To block a chance occurrence of individuals from two different species overcoming a single reproductive isolating mechanism, most species have two or more mechanisms. Many work before fertilization occurs (prezygotic), whereas others work after fertilization has taken place (postzygotic) (Table 19–1).

Prezygotic barriers interfere with mating

Prezygotic barriers are reproductive isolating mechanisms that prevent fertilization from taking place. Because male and female gametes never come into contact, an interspecific zygote (fertilized egg formed by the union of an egg from one species and a sperm from another species) never forms. Prezygotic barriers include temporal isolation, behavioral isolation, mechanical isolation, and gametic isolation.

Sometimes genetic exchange between two groups is prevented because they reproduce at different times of the day, season, or year. Such examples demonstrate **tem-**

Table 19–1 REPRODUCTIVE ISOLATING MECHANISMS

Mechanism	*Type*	*How It Works*
Temporal isolation	Prezygotic	Similar species reproduce at different times
Behavioral isolation	Prezygotic	Similar species have distinctive courtship behaviors
Mechanical isolation	Prezygotic	Similar species have structural differences in their reproductive organs
Gametic isolation	Prezygotic	Gametes of similar species are chemically incompatible
Hybrid inviability	Postzygotic	Interspecific hybrid dies at early stage of embryonic development
Hybrid sterility	Postzygotic	Interspecific hybrid survives to adulthood but is unable to reproduce successfully
Hybrid breakdown	Postzygotic	Offspring of interspecific hybrid are unable to reproduce successfully

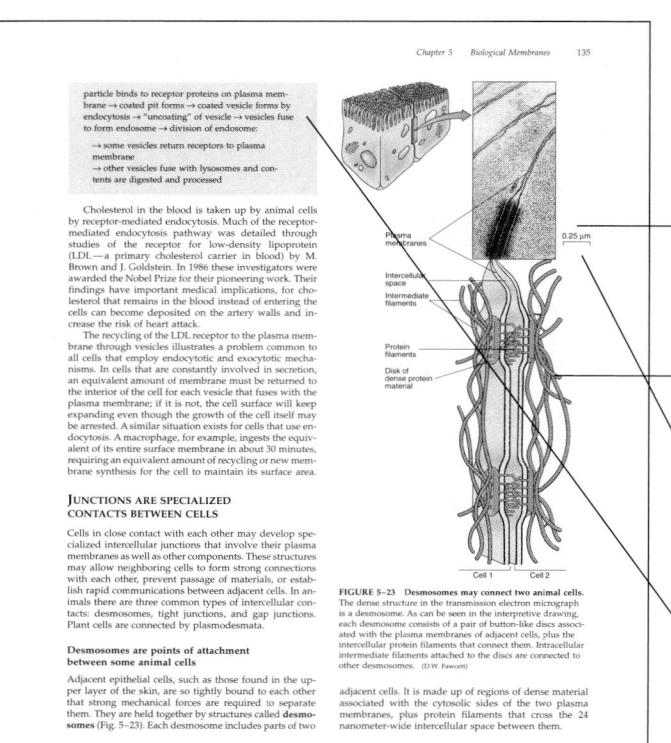

On the Cutting Edge boxes highlight current, often controversial research; this feature complements the text's emphasis on the process of science and the influence of research applications on new thinking and future research.

ON THE CUTTING EDGE

Test-Tube Evolution

Objective: To study bacterial evolution in the laboratory.

Method: Expose starving bacteria to a new food that they cannot metabolize.

Results: Mutations appear more frequently than would be expected in the starving cells, and some of these mutations are adaptive, allowing the bacteria to metabolize the new food. As a result, the bacteria survive and pass the ability to metabolize the new food on to their offspring.

Conclusion: Under conditions of starvation, adaptive mutation may occur in bacteria.

In 1988 a well-known British biologist, John Cairns, published the results of a study that indicated a different aspect of evolution known as **adaptive mutation.** By starving bacteria, Cairns found that they mutate in a nonrandom way and thus obtain the mutation they need to survive. His experiment startled many biologists because it seemed to support Lamarck's discredited idea that evolution is directed and purposeful. Like Lamarck's classical example of giraffes needing—and therefore evolving—longer necks to stretch into treetops, Cairn's starving bacteria needed—and seemed to evolve—a certain mutation that allowed them to use lactose for food.

The mutations produced by these bacteria did not appear to be the random mutations (that is, produced without regard for usefulness) characteristic of neo-Darwinian evolution. Adaptive mutation allowed the bacteria to obtain the needed mutation in a fraction of the number of cell divisions that "normal" random mutations would require.

Several other biologists repeated Cairn's experiment and reproduced his results, but the mechanism or mechanisms responsible for such apparently directed evolution remained unknown. Then in 1994 a group of researchers headed by Susan Rosenberg from the University of Alberta published two papers in the journal *Science** that reported a probable molecular mechanism for the variation produced by adaptive mutation. Certain proteins used during the recombination of bacterial DNA are necessary for adaptive mutation. The absence or inactivation of the genes encoding these proteins results in a strain of bacteria that cannot undergo rapid mutation when they are starving.

Many questions remain unanswered. Are these mutations occurring *because* they are needed? Why do the recombination genes cause adaptive mutation in starving bacteria but not in well-fed, rapidly growing cells? How important is adaptive mutation in bacterial evolution? Certain eukaryotic cells (yeasts) appear to have a process like adaptive mutation. Does it use a similar mechanism? Biologists disagree about the evolutionary implications of adaptive mutation, which will likely make this controversial phenomenon the target of investigation for many years.

*1. Harris, R.S., S. Longerich, and S.M. Rosenberg. "Recombination in adaptive mutation." *Science*, Vol. 264, 8 April 1994. 2. Rosenberg, S.M., S. Longerich, P. Gee, and R.S. Harris. "Adaptive mutation by deletions in small mononucleotide repeats." *Science*, Vol. 265, 15 July 1994.

reasoning, Lamarck suggested that the long neck of the giraffe developed when a short-necked ancestor began browsing on the leaves of trees instead of on grass. Lamarck speculated that the ancestral giraffe, by reaching up, stretched and elongated its neck. Its offspring inherited the longer neck, which stretched still further. This process, repeated over many generations, supposedly resulted in the long necks of modern giraffes.

The proposed mechanism for Lamarckian evolution was an "inner drive" for self-improvement, a notion that was discredited when the actual basis of heredity was later discovered. (However, see *On the Cutting Edge: Test-Tube Evolution* for a possible example of "Lamarckian" evolution in bacteria.) Lamarck's contribution to science is important because he was the first to propose that organisms undergo change over time as a result of some natural phenomenon rather than divine intervention. It remained for Charles Darwin to discover the mechanism of evolution by natural selection.

DARWIN WAS INFLUENCED BY HIS CONTEMPORARIES

Charles Darwin, the son of a prominent physician, was sent at the age of 15 to study medicine at the University of Edinburgh. Finding himself unsuited for medicine, he transferred to Cambridge University to study theology. Shortly after receiving his degree, Darwin embarked as a naturalist on the *H.M.S. Beagle*, which was taking a five-year exploratory cruise around the world to prepare navigation charts for the British Navy.

Career Visions

Tissue Bank Director

MARTHA ANDERSON

Martha Anderson originally planned to attend medical school after earning a biology degree at Colorado State University. But by the time she transferred to complete her biology studies at the University of Colorado, her focus had changed. Upon graduating, she began a career that enabled her to help many more patients and their families than a single doctor ever could. Martha is Vice President of Donor Services for the Musculoskeletal Transplant Foundation, a non-profit organization that "banks" human tissue for transplant use and conducts research to further develop transplantation techniques. The Foundation stores tissues and bone that are used in a variety of procedures, from periodontal reconstructive surgery to replacement of cancerous tissue. The Foundation is a membership organization composed of medical and academic institutions as well as other organ and tissue procurement agencies.

You had planned on going to medical school and decided against that profession. What did you do after college?

I went to work in one of Denver's trauma hospitals where I took an administrative position. Eventually I was promoted into what was probably the best job of my life as a patient representative at St. Anthony's Hospital.

What does a patient representative do?

A patient representative acts as an advocate for patients. You resolve a lot of patient care related issues and are responsible for crisis intervention with many families. Sometimes you serve as a person to talk to for the patient who doesn't have any family or friends around. We had the first helicopter rescue system in the country at St. Anthony's, and as a result, we had a great deal of trauma victims. Often these patients arrived unconscious and without any identification. It was my job to find out who they were and contact their families. It was a very stressful job, but it was rewarding because we could help people negotiate the health care system. Many people don't know anything about health care, and if they are incapacitated by their illness or by grief it's difficult for them to know what questions to ask so they can get the help that they need. Through my work as a patient representative I became involved with the organ and tissue donation process.

And that's how you got into your current position?

Yes, I helped set up the largest donor program in Colorado. We had more donors from St. Anthony's in Denver than from any other hospital in the state.

How did you do that?

We were very pro-active in identifying potential donors and offering their families the opportunity to talk about donation and find out if they wanted to be donors.

I suppose there are other organizations that "bank" bone or tissue. What's different about your organization?

There are probably about 100 or 150 tissue banks in the country. Ours is the largest and we have a unique structure. Many organ and tissue banks are regional or local, but we're a national consortium of medical schools and other recovery agencies. When someone has the potential of being an organ or tissue donor the local organ procurement agency is usually contacted. There are over 25 locations around the country that recover tissue for us—the bones, tendons and ligaments—which we then provide to surgeons and hospitals throughout the U.S.

Then you process them for storage so they can be transplanted?

Yes, we have them processed in our New Jersey facility. The tissue is formed into different sorts of shapes depending on whatever the local need is. For example, there may be a hospital that does a lot of spine surgery so we would cut the bone into sizes for spine surgeons to use. Or there may be a need for an entire femur which would probably be used for someone who has cancer. In the past the only option for removing a cancerous tumor in a femur was to amputate the leg. Now we can take out the portion of bone that has the tumor in it and replace it with another that's the right size. Such a person can now keep their leg, and maintain an active lifestyle. This procedure is called limb salvage.

Do you feel that you use your biology degree in your current job? And, if so, how do you use it?

I used my biology degree when I was a patient representative because my studies focused on human physiology and anatomy. I had a very good basic understanding of how the human body works and that helped me a lot. One of the things that I do use every day are the critical thinking skills I developed as an undergraduate. I have to figure out answers to problems that have a basis in science. One of the things you learn in basic biology is that you have to think and figure out how things work. These are skills I use quite frequently.

Is there a specific kind of science or education background required to work in your field?

A lot of the people who work for me are either nurses or operating room technicians. A few people have bachelors degrees in biology. Having only a bachelor's degree in biology may not be enough anymore, and I think a graduate degree is probably required to advance in the field.

Career Visions, interviews with former biology majors who describe what they now do professionally, convey persuasive, practical information on the merits of studying biology, and the variety of opportunities open to biology majors.

Making the Connection

Evolution and Genetics

Can biologists make use of the fact that mutations in DNA sequences have been occurring throughout the course of evolution? Within a given taxonomic group, mutations appear to have occurred at a uniform rate over millions of years. From the number of alterations in the DNA nucleotide sequence of one organism compared with another, we can develop a **molecular clock** to estimate the time of divergence between two closely related species or higher taxonomic groups. Such molecular clocks can be used to complement geological estimates of the divergence of species or to assign tentative dates to evolutionary events that lack fossil evidence.

Molecular clocks must be developed and interpreted with care. The rates of mutation appear to vary among different genes and among different taxonomic groups. Therefore, although some mutations occur at a fairly uniform rate, a single molecular clock for all genes in all organisms cannot be elucidated. ▲

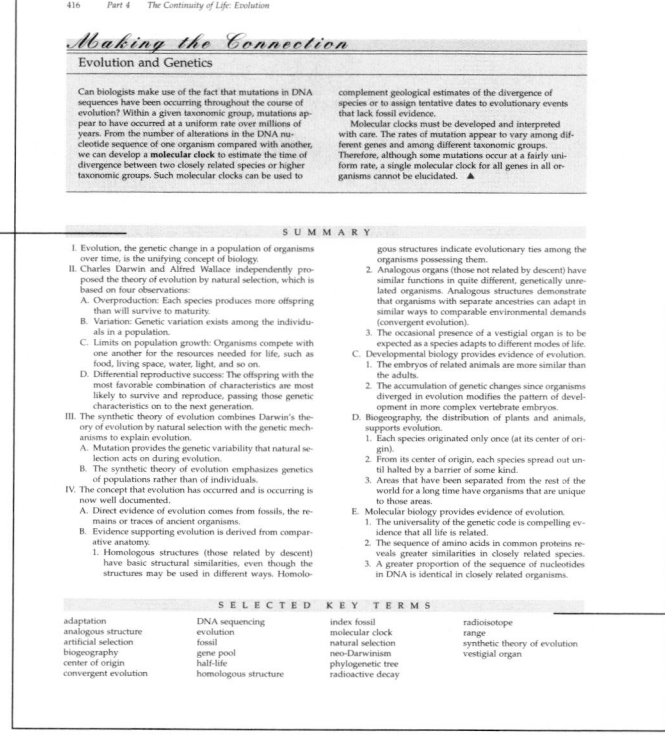

A **Chapter Summary** in outline form at the end of each chapter provides a review of the material presented.

SUMMARY

I. Evolution, the genetic change in a population of organisms over time, is the unifying concept of biology.
II. Charles Darwin and Alfred Wallace independently proposed the theory of evolution by natural selection, which is based on four observations:
   A. Overproduction: Each species produces more offspring than will survive to maturity.
   B. Variation: Genetic variation exists among the individuals in a population.
   C. Limits on population growth: Organisms compete with one another for the resources needed for life, such as food, living space, water, light, and so on.
   D. Differential reproductive success: The offspring with the most favorable combination of characteristics are most likely to survive and reproduce, passing those genetic characteristics on to the next generation.
III. The synthetic theory of evolution combines Darwin's theory of evolution by natural selection with the genetic mechanisms to explain evolution.
   A. Mutation provides the genetic variability that natural selection acts on during evolution.
   B. The synthetic theory of evolution emphasizes genetics of populations rather than of individuals.
IV. The concept that evolution has occurred and is occurring is now well documented.
   A. Direct evidence of evolution comes from fossils, the remains or traces of ancient organisms.
   B. Evidence supporting evolution is derived from comparative anatomy.
      1. Homologous structures (those related by descent) have basic structural similarities, even though the structures may be used in different ways. Homologous structures indicate evolutionary ties among the organisms possessing them.
      2. Analogous organs (those not related by descent) have similar functions in quite different, genetically unrelated organisms. Analogous structures demonstrate that organisms with separate ancestries can adapt in similar ways to comparable environmental demands (convergent evolution).
      3. The occasional presence of a vestigial organ is to be expected as a species adapts to different modes of life.
   C. Developmental biology provides evidence of evolution.
      1. The embryos of related animals are more similar than the adults.
      2. The accumulation of genetic changes since organisms diverged in evolution modifies the pattern of development in more complex vertebrate embryos.
   D. Biogeography, the distribution of plants and animals, supports evolution.
      1. Each species originated only once (at its center of origin).
      2. From its center of origin, each species spread out until halted by a barrier of some kind.
      3. Areas that have been separated from the rest of the world for a long time have organisms that are unique to those areas.
   E. Molecular biology provides evidence of evolution.
      1. The universality of the genetic code is compelling evidence that all life is related.
      2. The sequence of amino acids in common proteins reveals greater similarities in closely related species.
      3. A greater proportion of the sequence of nucleotides in DNA is identical in closely related organisms.

SELECTED KEY TERMS

adaptation	DNA sequencing	index fossil	radioisotope
analogous structure	evolution	molecular clock	range
artificial selection	fossil	natural selection	synthetic theory of evolution
biogeography	gene pool	neo-Darwinism	vestigial organ
center of origin	half-life	phylogenetic tree	
convergent evolution	homologous structure	radioactive decay	

Selected Key Terms at the end of every chapter direct students to important vocabulary explained in the chapter.

A **Post-Test** provides the opportunity to evaluate mastery of the material using new key terms from within the chapter; answers are provided.

POST-TEST

1. The fact that all species developed from earlier forms by the accumulation of genetic changes over many successive generations is known as the theory of ________ .
2. The genetic constitution of an entire population of a given species is termed its ________ ________ .
3. Darwin proposed ________ ________ as the mechanism by which evolutionary change takes place.
4. The four premises of ________ ________ are overproduction, variation, limits on population growth, and differential reproductive success.
5. In natural selection, the selecting agent is the environment, whereas in ________ ________ , the selecting agent is humans.
6. The modern theory of evolution in which Darwin's observation of variation is explained by mutation is known as the ________ ________ of evolution.
7. The synthetic theory of evolution is also called ________ .
8. Geologists can identify specific layers of rock by the presence of certain key fossils, known as ________ ________ .
9. Unstable isotopes that spontaneously emit radiation, known as ________ , are used to date the rocks in which fossils appear.
10. An organ that appears to have little or no function, and is smaller than a similar, fully functional equivalent in the organism's ancestor or relatives, is known as a (an) ________ organ.
11. The wings of butterflies and bats have similar functions but are quite different in structure. This is an example of ________ structures.
12. The independent evolution of similar structures in two unrelated organisms is known as ________ ________ .
13. The study of the distribution of plants and animals is called ________ .
14. The portion of Earth over which a given species is found is its ________ .
15. One type of molecular evidence for evolution is ________ ________ , in which the order of nucleotide bases in a strand of DNA that codes for a gene shared by several organisms is determined.

REVIEW QUESTIONS

1. Explain briefly the concept of evolution by natural selection.
2. Why are only inherited variations important in the evolutionary process?
3. What part cf Darwin's theory was he unable to explain? How does the synthetic theory of evolution explain this?
4. How do scientists date fossils? How do fossils provide evidence of evolution?
5. Distinguish among homologous, analogous, and vestigial structures. How does each provide evidence of evolution?
6. How does developmental biology provide evidence of a common ancestry for vertebrates as diverse as reptiles, birds, pigs, and humans?
7. Explain why marsupials are widespread in Australia and almost nonexistent elsewhere.
8. What is indicated if the DNA from two species is found to be almost identical?

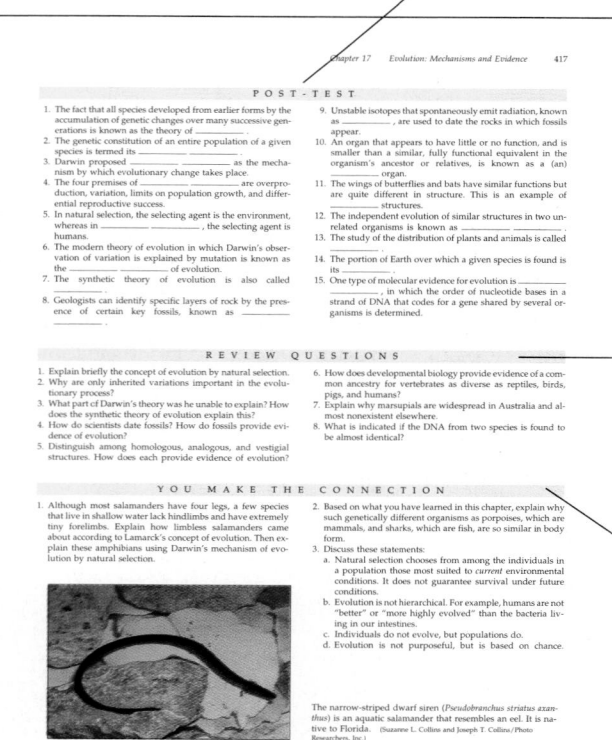

Review Questions focus on mastery of chapter Learning Objectives.

YOU MAKE THE CONNECTION

1. Although most salamanders have four legs, a few species that live in shallow water lack hindlimbs and have extremely tiny forelimbs. Explain how limbless salamanders came about according to Lamarck's concept of evolution. Then explain these amphibians using Darwin's mechanism of evolution by natural selection.
2. Based on what you have learned in this chapter, explain why such genetically different organisms as porpoises, which are mammals, and sharks, which are fish, are so similar in body form.
3. Discuss these statements:
   a. Natural selection chooses from among the individuals in a population those most suited to *current* environmental conditions. It does not guarantee survival under future conditions.
   b. Evolution is not hierarchical. For example, humans are not "better" or "more highly evolved" than the bacteria living in our intestines.
   c. Individuals do not evolve, but populations do.
   d. Evolution is not purposeful, but is based on chance.

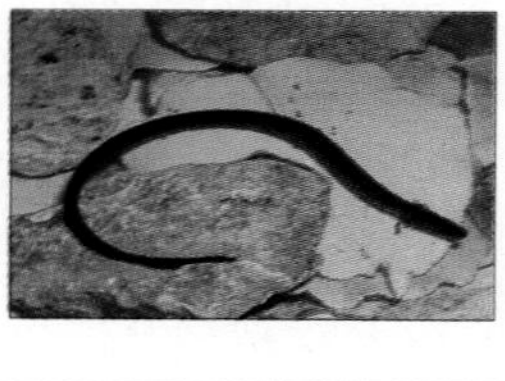

The narrow-striped dwarf siren (*Pseudobranchus striatus axanthus*) is an aquatic salamander that resembles an eel. It is native to Florida. (Suzanne L. Collins and Joseph T. Collins/Photo Researchers, Inc.)

You Make the Connection questions, after every chapter, challenge students to synthesize material by applying concepts in the chapter to new situations.

Contents Overview

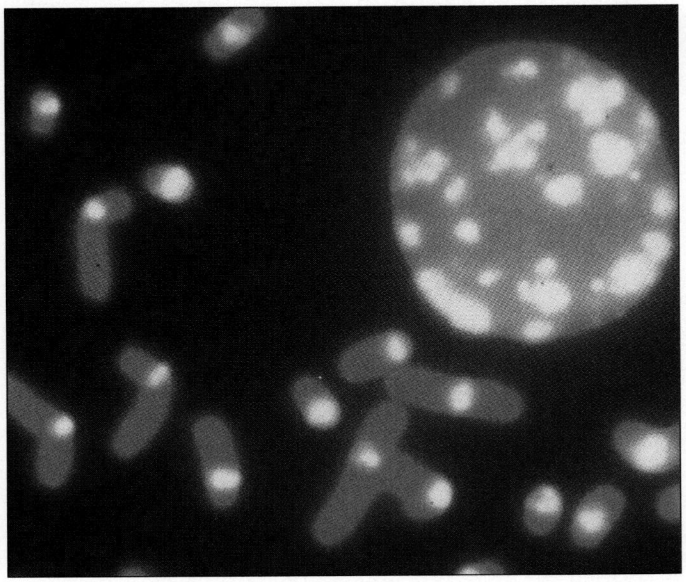

PART 5

THE DIVERSITY OF LIFE 489

PART 6

STRUCTURE AND LIFE PROCESSES IN PLANTS 687

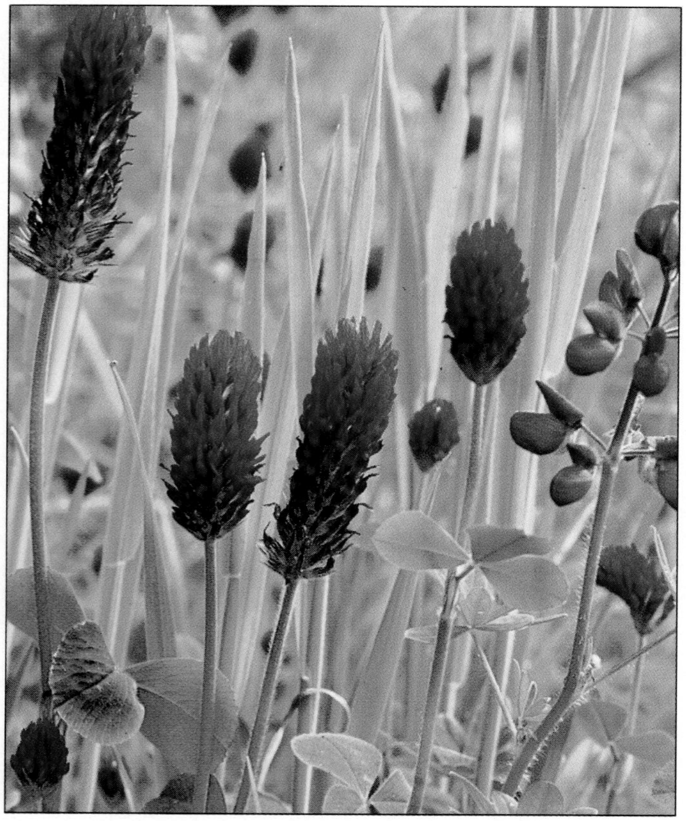

PART 7

STRUCTURE AND LIFE PROCESSES IN ANIMALS 799

PART 8

THE INTERACTIONS OF LIFE: ECOLOGY 1133

Contents

PART 2

ENERGY TRANSFER THROUGH LIVING SYSTEMS 143

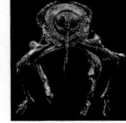

PART 1

The Organization of Life

Biologists study the details of living organisms and all of their interactions. The Louisiana heron *(Hydranassa tricolor)* and the turtle *(Chrysemys nelsoni)* on which it has perched are complex organisms that make their home in the Florida cypress swamp. (Mark J. Thomas/Dembinsky Photo Associates)

C H A P T E R 1

A View of Life

This frog *(Rhacophorus pardalis)* looks out at life—perhaps searching visually for food—from the safety of a mushroom. (Photographed in Borneo; Frans Lanting/Minden Pictures)

As we prepare to enter the 21st century, we face many global challenges whose solutions depend on an understanding of the biological sciences. Expanding human population and increased consumer demand for energy, homes, highways, and manufactured goods are causing the rapid destruction of vital ecosystems such as oceans, lakes, wetlands, rain forests, and coral reefs. We struggle with the issues of overpopulation, world hunger and malnutrition, diminishing natural resources, effects of pesticides, air and water pollution, declining biodiversity and endangered species, and diseases such as AIDS, cancer, and heart disease. Resolution of these problems will require the combined efforts of biologists and other scientists, politicians, and biologically informed citizens. Whatever your college major or career goals, a knowledge of biological concepts is a vital tool for understanding and meeting the pressing challenges that confront us. More than any other discipline, **biology,** the science of life, helps us understand ourselves and our planet.

During the past few decades research in the science of biology has yielded amazing knowledge about our human species and about the millions of other diverse life forms with which we share our planet. Applications of this basic research have provided us with the technology to transplant hearts, manipulate genes, control many diseases, and increase world food production. Research in molecular biology and genetics has led to new insights into disease processes, leading to the new science of gene therapy. New discoveries in biology have enriched the quality of life that many of us enjoy.

As biologists continue to study interrelationships of the living systems, or **organisms,** that inhabit our planet, they enhance our awareness of our own impact on other organisms and on the environment (Fig. 1–1). This book is a starting point for your exploration of biology. It will provide you with tools to become a part of this fascinating science and a more informed inhabitant of our planet. Perhaps you will decide to become a research biologist and help unravel the complexities of the human brain, discover new hormones that cause plants to flower, identify new species of animals or bacteria, or develop new treatments for diseases such as cancer. Or perhaps you will choose to enter an applied field of biology such as forestry, dentistry, medicine, or veterinary medicine. Even if you are not planning a career in one of the biological sciences, learning about this exciting science will increase your understanding of yourself, your environment, and the organisms with which you share your planet.

In this first chapter we will introduce three basic themes of biology: evolution of life, transmission of information, and flow of energy through living systems. How evolution contributes to the organization of life is a major theme of this book. Scientists have accumulated a wealth of evidence showing that the life forms on our planet are related and that complex organisms have evolved through time from simpler life forms.

Organization also depends on the precise, orderly transmission of information. Instructions for organizing each living organism and each new generation are encoded in the DNA molecules that make up the genes.

Energy is required to maintain the precise order that characterizes living systems. Maintaining the chemical transactions and cellular organization essential to life requires a continuous input of energy. We begin our study of biology by developing a more precise understanding of what life is.

LEARNING OBJECTIVES

After you have studied this chapter you should be able to

1. Define biology and discuss its applications to human life and society.
2. Distinguish between living organisms and nonliving things by describing the features that characterize living organisms.
3. Summarize the importance of information transfer to living systems, giving specific examples.
4. Give a brief overview of the theory of evolution and explain why it is the principal unifying concept in biology.
5. Apply the theory of natural selection to any given adaptation, suggesting a logical explanation of how the adaptation may have evolved.
6. Construct a hierarchy of biological organization, including individual and ecological levels.
7. Demonstrate the binomial system of nomenclature using several specific examples, and classify an organism (a human, for example) according to kingdom, phylum, class, order, family, genus, and species.
8. Contrast the five kingdoms of living organisms and cite examples of each group.
9. Relate metabolism and homeostasis and give specific examples of these life processes.
10. Contrast the roles of producers, consumers, and decomposers, and cite examples of their interdependence.
11. Design an experiment to test a given hypothesis using the procedure and terminology of the scientific method.

(a)

(b)

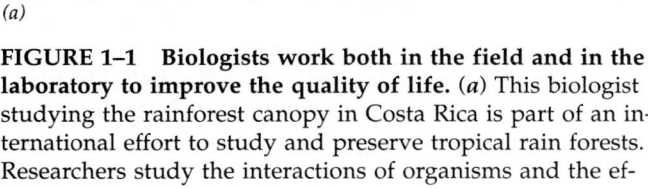

FIGURE 1–1 Biologists work both in the field and in the laboratory to improve the quality of life. (*a*) This biologist studying the rainforest canopy in Costa Rica is part of an international effort to study and preserve tropical rain forests. Researchers study the interactions of organisms and the effects of human activities on the rain forests. (*b*) These biologists, doing research in developing an AIDS vaccine, are examining a DNA sequencing autoradiogram over a light box. (*a*, Mark Moffett/Minden Pictures; *b*, Hank Morgan/Photo Researchers, Inc.)

LIFE CAN BE DEFINED IN TERMS OF THE CHARACTERISTICS OF ORGANISMS

We can easily determine that a human, an oak tree, and a butterfly are living, whereas a rock is not. Despite their diversity, the living organisms that inhabit our planet share a common set of characteristics that distinguish them from nonliving things. These features include a precise kind of organization, a variety of chemical reactions that we refer to as metabolism, the ability to maintain an appropriate internal environment even when the external environment changes (a process referred to as *homeostasis*), movement, responsiveness, growth and development, reproduction, and adaptation to environmental change. We consider each of these characteristics in the following sections.

Organisms are composed of cells

The cell theory, one of the fundamental unifying concepts of biology, states that all living organisms are composed of basic units called **cells** and of substances produced by cells. Although they vary greatly in size and appearance, all organisms[1] are composed of these small building blocks. Some of the simplest life forms, such as bacteria, are *unicellular:* they consist of a single cell. In contrast, the body of a human or a maple tree is made of billions of cells. In these complex *multicellular* organisms, life processes depend on the coordinated functions of the component cells (Fig. 1–2).

[1]As we will see in Chapter 23, viruses are not considered organisms. They can carry on life activities and reproduce only by using the metabolic machinery of *the cells they parasitize,* and so are said to be on the borderline between living and nonliving things.

Living organisms grow and develop

Some nonliving things appear to grow. Crystals may form in a supersaturated solution of a salt; as more of the salt comes out of solution, the crystals may enlarge. However, this is not growth in the biological sense. Biologists define *growth* as an increase in the amount of living substance in the organism. **Growth** can result from an increase in the *size* of the individual cells, the *number* of cells, or both (Fig. 1–3). Growth may be uniform in the various parts of an organism, or it may be greater in some parts than in others, causing the body proportions to change as growth occurs.

Some organisms — most trees, for example — continue to grow indefinitely. Many animals have a defined growth period that terminates when a characteristic adult size is reached. One of the remarkable aspects of the growth process is that each part of the organism continues to function as it grows.

Living organisms develop as well as grow. **Development** includes all the changes that take place during the life of an organism. Humans and many other organisms begin life as a fertilized egg, which then grows and develops specialized structures and body form.

Metabolism includes the chemical processes essential to growth, repair, and reproduction

In all living organisms, chemical reactions and energy transformations take place that are essential to nutrition, growth and repair of cells, and conversion of energy into usable forms. The sum of all the chemical activities of the organism is its **metabolism.** Metabolic reactions occur continuously in every living organism, and they must be carefully regulated to maintain a balanced internal state.

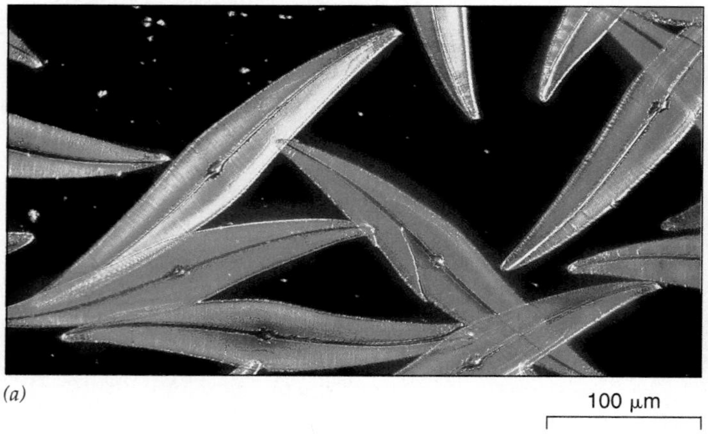

(a)

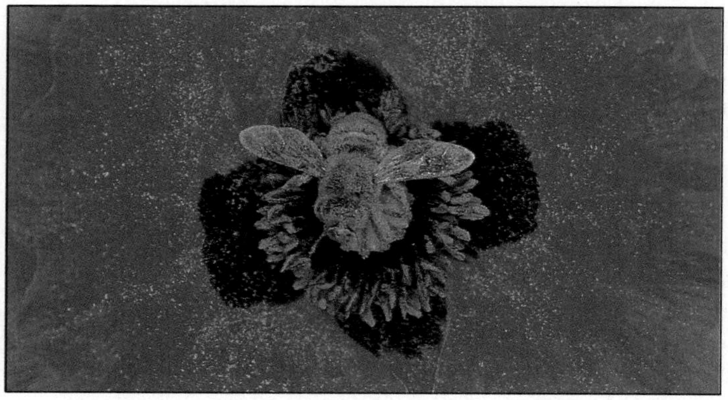
(b)

FIGURE 1–2 Some life forms are single-celled, others are multicellular. (*a*) Single-celled organisms are generally smaller and consist of one intricate cell that performs all of the functions essential to life. The pennate diatoms *(Pleurosigma angulatum),* members of the protist kingdom, have a glasslike cell wall. Diatoms account for a large part of the plankton in marine and freshwater food webs. The blue color is a diffraction effect. (× approximately 150) (*b*) Both the honeybee *(Apis mellifera)* and the field poppy *(Papaver rhoeas)* it is pollinating are complex multicellular organisms. Each consists of thousands of specialized cells that carry on specific tasks. (*a*, Alfred Pasieka/Science Photo Library/Photo Researchers, Inc.; *b*, Dr. Jeremy Burgess/Science Photo Library/Photo Researchers, Inc.)

FIGURE 1–3 An organism grows by using raw materials to increase its size. The young olive baboon shown with its mother will eat and grow until it reaches adult size. These animals were photographed in Lake Manyara NP, Tanzania. (Stan Osolinski/Dembinsky Photo Associates)

The tendency of organisms to maintain a relatively constant internal environment is termed **homeostasis,** and the mechanisms that accomplish the task are known as **homeostatic mechanisms.** Metabolism and homeostasis are discussed later in this chapter and in other sections of this book.

Movement is a basic property of cells

Movement, although not necessarily locomotion (moving from one place to another), is another characteristic of living organisms. The living material within cells is in continuous motion, and organisms move as they interact with the environment.

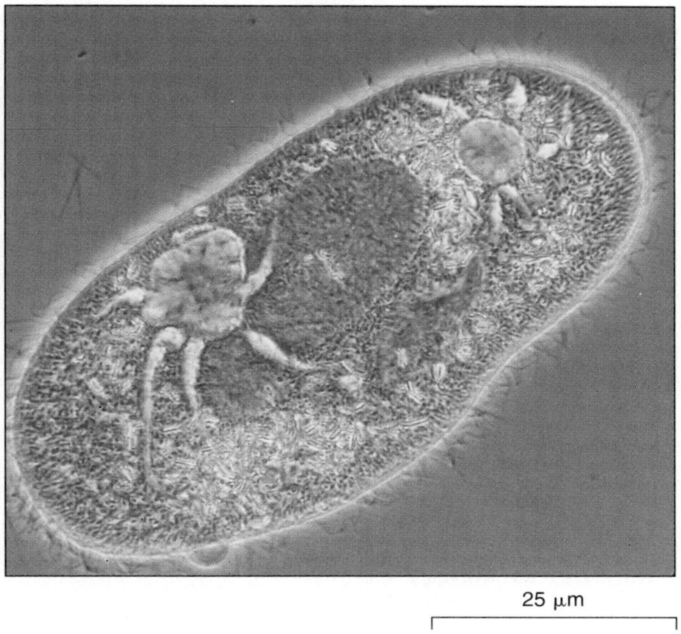

FIGURE 1–4 Movement is characteristic of all living things. Ciliates, such as this *Paramecium,* move about by beating their hair-like cilia. (Roland Birke/Peter Arnold, Inc.)

Most animals move very obviously; they wiggle, crawl, swim, run, or fly. Locomotion may result from the slow oozing of the cell (a process called *amoeboid motion*), from the beating of tiny hairlike extensions of the cell called **cilia** or longer structures known as **flagella** (Fig. 1–4), or from the contraction of muscles. A few animals, such as sponges, corals, and oysters, have free-swimming larval stages but do not move from place to place as adults. Even though these adults, described as **sessile,** remain firmly attached to some surface, they may have cilia or flagella. These structures beat rhythmically, moving the surrounding water that brings food and other necessities to the organism.

Although plants do not move about in the ways we associate with animals, they do move. For example, plants orient their leaves to the sun and grow toward light. In some plants, for example the Venus flytrap, movement is obvious, even dramatic; this phenomenon is described in the next section.

Organisms respond to stimuli

All forms of life respond to **stimuli,** physical or chemical changes in their internal or external environment. Stimuli that evoke a response in most organisms are changes in the color, intensity, or direction of light; changes in temperature, pressure, or sound; and changes in the chemical composition of the surrounding soil, air, or water. In simple organisms, the entire organism may be sensitive to stimuli. Certain single-celled organisms, for example, respond to bright light by retreating. In complex animals

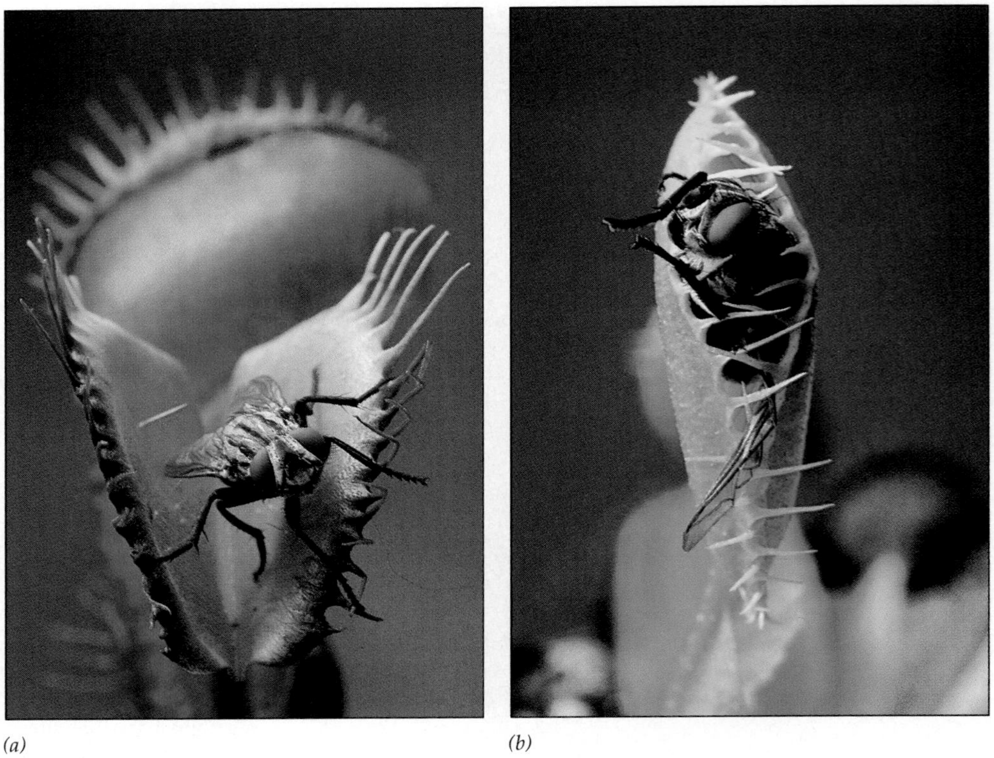

FIGURE 1–5 A few plants, such as the Venus flytrap *(Dionaea muscipula)*, can respond to the touch of an insect by trapping it. (*a*) Here a fly lights on a leaf of the Venus flytrap. The leaves of this plant have a scent that attracts insects. Hairs on the leaf surface detect the presence of the insect. When triggered, the leaf, which is hinged along its midrib, folds. (*b*) The edges come together and interlock, preventing the fly's escape. The leaf then secretes enzymes that kill and digest the insect. (David M. Dennis/Tom Stack & Associates)

such as polar bears or humans, certain cells of the body are highly specialized to respond to certain types of stimuli. For example, cells in the retina of the eye respond to light.

Although their responses may not be as obvious as those of animals, plants do respond to light, gravity, water, touch, and other stimuli. The streaming motion of the cytoplasm in plant cells may be speeded up or stopped by changes in the amount of light. Many plant responses are carried out by different rates of growth of various parts of the plant body. A few plants, such as the Venus flytrap of the Carolina swamps (Fig. 1–5), are remarkably sensitive to touch and can catch insects. Their leaves are hinged along the midrib and they have a scent that attracts insects. The presence of an insect on the leaf, detected by trigger hairs on the leaf surface, stimulates the leaf to fold. When the edges come together, the hairs interlock to prevent escape of the prey. The leaf then secretes enzymes that kill and digest the insect. The Venus flytrap is usually found in soil that is deficient in nitrogen. The plant obtains part of the nitrogen required for its growth from the insect prey it "eats."

Organisms reproduce

Although at one time worms were thought to arise spontaneously from horsehairs in a water trough, maggots from decaying meat, and frogs from the mud of the Nile, we now know that each can come only from previously existing organisms. One of the fundamental principles of biology is that "all life comes only from living things." If any one characteristic can be said to be the very essence of life, it is the ability of an organism to reproduce its kind.

In simple organisms such as the amoeba, reproduction may be **asexual**—that is, without sex (Fig. 1–6). When an amoeba has grown to a certain size it reproduces by splitting in half to form two new amoebas. Before it divides, an amoeba makes a duplicate copy of its hereditary material (genes) and distributes one complete set to each new cell. Except for size, each new amoeba is similar to the parent cell. (The new amoebas may not be identical to each other or to the parent cell because mutations may occur.)

In most plants and animals, **sexual reproduction** is carried out by the production of specialized egg and sperm cells that fuse to form a fertilized egg. The new organism develops from the fertilized egg. Offspring produced by sexual reproduction are the product of the interaction of various genes contributed by both the mother and the father. Such genetic variation provides raw material for the vital processes of evolution and adaptation.

Populations evolve and become adapted to the environment

The ability of a population to evolve (change) and adapt to its environment enables it to survive in a changing

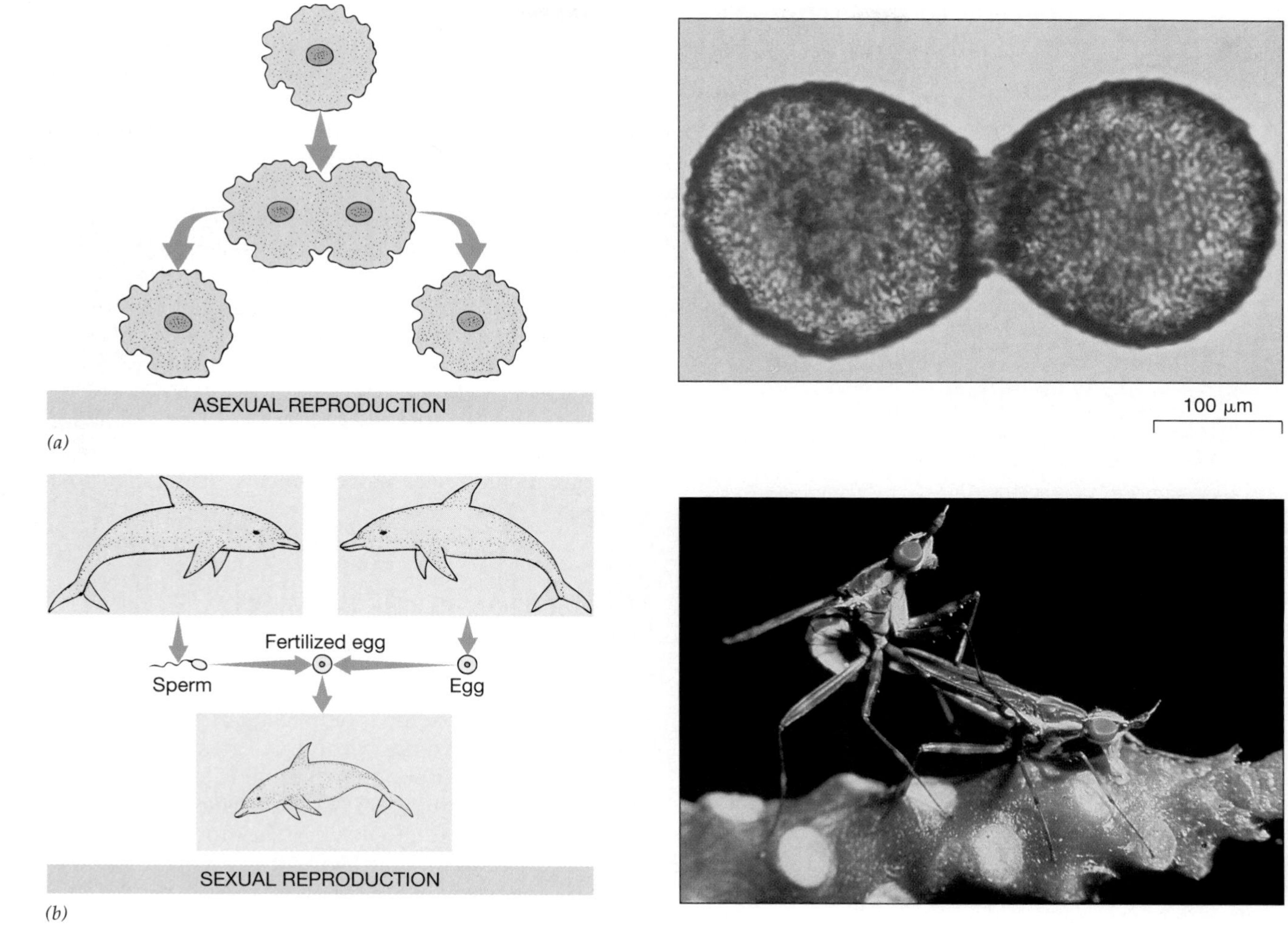

FIGURE 1–6 Two approaches to reproduction are asexual and sexual. (***a***) Asexual reproduction in *Difflugia,* a one-celled organism. In asexual reproduction, one individual gives rise to two or more offspring—all similar to the parent. (***b***) In sexual reproduction, typically two parents each contribute a gamete (sperm or egg). Gametes join to produce the offspring, which is a combination of the traits of both parents. A pair of tropical flies mating. (*a,* Visuals Unlimited/Cabisco; *b,* L.E. Gilbert, University of Texas at Austin/Biological Photo Service)

world. **Adaptations** are traits that enhance an organism's ability to survive in a particular environment. They may be structural, physiological, behavioral, or a combination of all three. The long, flexible tongue of the frog is an adaptation for catching insects, and the thick fur coat of the polar bear is an adaptation for surviving frigid temperatures. Every biologically successful organism is a complex collection of coordinated adaptations produced through evolutionary processes.

INFORMATION MUST BE TRANSMITTED WITHIN AND BETWEEN INDIVIDUALS

In order for a living organism to grow, develop, carry on self-regulated metabolism, move, respond, and reproduce, it must have precise instructions. The information an organism needs to carry on all of these processes is coded and delivered in the form of chemical substances and electrical impulses.

DNA transmits information from one generation to the next

Humans give birth only to human babies, not to giraffes or rose bushes. In organisms that reproduce sexually, each offspring is a combination of the traits of its parents. In 1953, James Watson and Francis Crick worked out the structure of deoxyribonucleic acid, more simply known as **DNA.** This chemical substance makes up the **genes,** the units of hereditary material. The work of Watson and Crick led to the understanding of the genetic code that transmits information from generation to generation. This code works somewhat like our alphabet; it can spell an

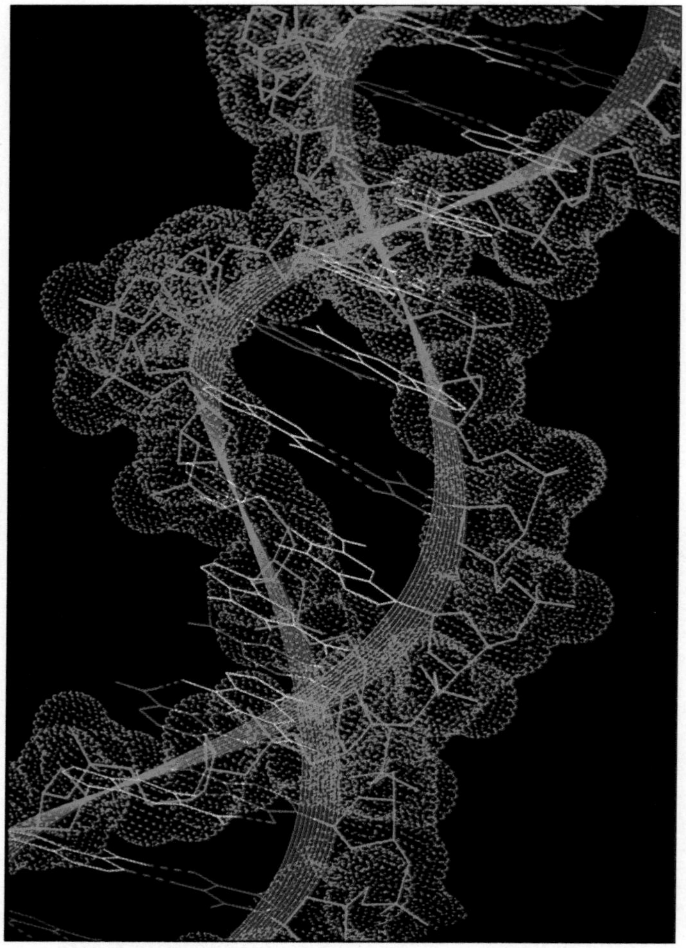

FIGURE 1–7 An organism's ability to transmit information from one generation to the next is essential to the continuity of life. In all organisms, the hereditary material is DNA. This computer-generated image shows the double-helix configuration of DNA. (Will & Deni McIntyre/Photo Researchers, Inc.)

amazing variety of instructions for making organisms as diverse as bacteria, frogs, and redwood trees. The same genetic code is used to specify instructions for making every living organism, a dramatic example of the unity of life (Fig. 1–7).

Information is transmitted by many types of molecules and by nervous systems

The genes are also responsible for controlling the development and functioning of each individual organism. DNA contains the "recipes" for making all of the **proteins** needed by the organism. Proteins are very large molecules that are slightly different in each type of organism and in each individual. For example, brain cells are different from muscle cells, in large part because they have different types of proteins.

Hormones are molecules that function as chemical messengers that transmit information from one part of an organism to another. A hormone can signal cells to produce or secrete a certain protein or other substance. As you proceed in your study of biology, you will learn about many types of molecules that code or transmit information.

Many living organisms use electrical signals to transmit information. Most animals have nervous systems that transmit information by way of both electrical impulses and chemical compounds known as neurotransmitters. Information conveyed by the nervous system keeps the individual informed of changes, both in the outside world and within the body.

Information must also be transmitted from one organism to another. Mechanisms for this type of communication include release of chemicals, visual displays, and sounds.

EVOLUTION IS THE PRIMARY UNIFYING CONCEPT OF BIOLOGY

How populations of organisms have changed, or **evolved,** over time is a major focus of investigation and debate. The theory of evolution has become the greatest unifying concept of biology. Although evolution is discussed in depth in Chapters 17 through 21, we present a brief overview here to give you the tools necessary to understand other aspects of biology. Some element of an evolutionary perspective is present in almost every specialized field within biology. Biologists in almost every subdiscipline try to understand the features and functions of organisms, and their constituent cells and parts, by considering them in light of the long, continuing process of evolution.

Species adapt in response to changes in the environment

Every organism is the product of complex interactions between environmental conditions and the genes of its ancestors. If every organism of a species[2] were exactly like every other, any change in the environment might be disastrous to all, and the species would become extinct. Adaptation to changes in the environment involves changes in populations rather than in individual organisms. Such adaptations are the result of evolutionary processes that occur over long periods of time and involve many generations.

---

[2]A species is a group of organisms with similar structure and function; in nature they play a similar role, breed only with each other, and share a common ancestry. Members of a species have a common gene pool.

Natural selection is an important mechanism by which evolution proceeds

Although the concept of evolution had been discussed by philosophers and naturalists through the ages, Charles Darwin and Alfred Wallace first brought the theory of evolution to general attention and suggested a plausible mechanism—natural selection—to explain it. In his book *On the Origin of Species by Means of Natural Selection,* published in 1859, Darwin synthesized many new findings in geology and biology. In his **theory of organic evolution** he presented a wealth of evidence that the present forms of life on Earth descended with modifications from previously existing forms. Darwin's book raised a storm of controversy in both religion and science, some of which still lingers.

Darwin's theory of evolution has helped shape the biological sciences to the present day. His work generated a great wave of scientific observation and research that has provided much additional evidence that evolution is responsible for the great diversity of organisms present on our planet.

Darwin based his theory of natural selection on the following four observations: (1) Individual members of a species show some variation from one another. (2) Many more organisms are produced than survive to reproduce (Fig. 1–8). (3) Competition for necessary resources like food, sunlight, and space takes place among the many individuals produced. Individuals who happen to have characteristics that give them some advantage are more likely to survive. (4) The survivors live to reproduce and pass their genetic "recipe" for survival on to their offspring. Thus, the best adapted individuals of a population leave, on average, more offspring than do other individuals. Because of this differential reproduction, a greater proportion of the population becomes adapted to the prevailing environmental conditions. The environment *"selects"* the best adapted organisms for survival.

Darwin did not know about DNA or understand the mechanisms of inheritance. We now understand that the variations among organisms are a result of different varieties of genes that code for each characteristic. The ultimate sources of these variations are random **mutations,** chemical changes in DNA that persist and can be inherited. Mutations modify genes and provide the raw material for evolution.

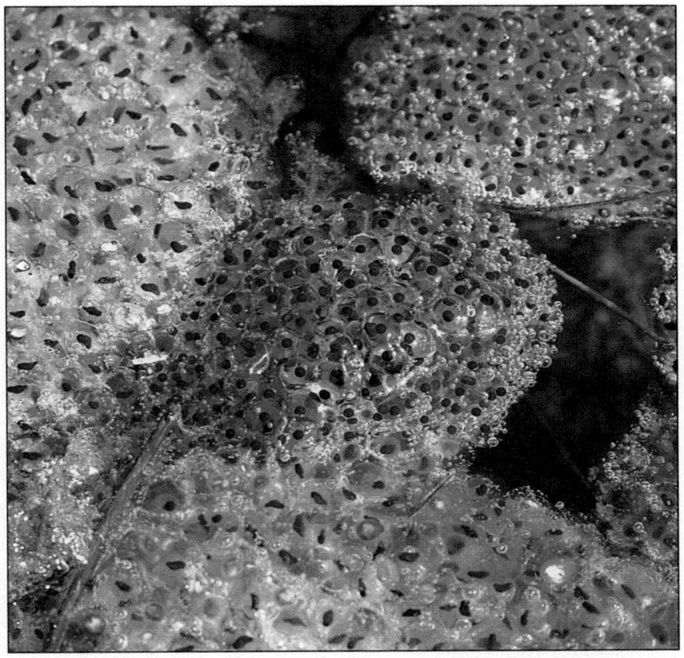

FIGURE 1–8 Egg masses of the wood frog *(Rana sylvatica)* were fertilized in a pond. Many more eggs are produced than can possibly develop into adult frogs. Random events are largely responsible for determining which of these developing frogs will hatch, reach adulthood, and reproduce. However, certain traits possessed by each organism will also contribute to its probability for success in its environment. Although not all organisms are as prolific as the frog, the generalization that more organisms are produced than survive is true throughout the living world. (J. Serrao/Photo Researchers, Inc.)

Populations evolve as a result of selective pressures from changes in the environment

All the genes present in a population make up its gene pool. By virtue of its gene pool, a population is a reservoir of variation. Natural selection acts on individuals within a population. Selection favors individuals with genes specifying traits that enable them to effectively cope with pressures exerted by the environment. These organisms are most likely to survive and produce offspring. As these successful organisms pass on their genetic recipe for survival, their traits become more widely distributed in the population. Over long periods of time, as organisms continue to change (and as the environment itself changes, bringing different selective pressures), the members of the population become increasingly unlike their ancestors (see *Focus On: Evolution in Action*).

A successful organism is adapted to its environment. Adaptations, and thus well-adapted organisms, are the products of evolution. The long neck of the giraffe, for example, is an adaptation for reaching leaves on trees (Fig. 1–9). The antelope-like ancestors of modern-day giraffes did not have long necks. As with other traits, there was a standard, bell-shaped distribution of neck heights: while most of the ancestors of modern giraffes had necks that were about the same length, a few had relatively short necks and others had relatively long ones. The ancestral giraffes with the shortest necks could not compete

(a) *(b)*

FIGURE 1–9 A successful organism is adapted to its environment. (***a***) The long neck of the giraffe is an adaptation for reaching leaves high on trees. The giraffe shown here is browsing on *Acacia.* (***b***) The scorpion fish, photographed off the coast of Maine, blends with its background so well that it looks like a rock on the ocean floor. It is well adapted to make dinner of any small organism that unwarily swims by. (*a,* Visuals Unlimited/Walt Anderson; *b,* Robert Shupak)

effectively for the leaves on the trees of the African savanna. They were less likely to survive and reproduce. Those with longer than average necks were best able to reach the leaves. These giraffes survived and reproduced, passing on their genes (DNA) for long necks. Through thousands of generations, the giraffe neck became longer and longer.

Biological organization reflects the course of evolution

Whether we study an individual organism or the world of life as a whole, we can identify a pattern of increasing complexity (Fig. 1–10).

Organisms have several levels of organization

The **chemical level** is the most basic level of organization. It includes the primary particles of all matter: atoms and combinations of atoms called molecules. An **atom** is the smallest unit of a chemical element (fundamental substance) that retains the characteristic properties of that element. For example, an atom of iron is the smallest possible amount of iron. Atoms combine chemically to form **molecules.** Two atoms of hydrogen combine with one atom of oxygen, to form one molecule of water.

At the **cellular level** we find that many diverse molecules may associate with one another to form complex and highly specialized structures, called **organelles,** within cells. Many organelles are suspended within the jelly-like cytoplasm of the cell. The **cell** itself is the basic structural and functional unit of life, the simplest component of living matter that can carry on all of the activities necessary for life. The **plasma membrane** surrounding the cell and the **nucleus** containing the hereditary material are examples of organelles.

In most multicellular organisms, cells associate to form **tissues,** such as muscle tissue in animals, or epidermis (a tissue that serves as a protective covering) in plants. Tissues, in turn, are arranged into functional structures, called **organs,** such as the heart and stomach in animals and the roots and leaves in plants. In animals, each major group of biological functions is performed by a coordinated group of tissues and organs called an **organ system.** The circulatory and digestive systems are examples of organ systems. Functioning together with great precision, the organ systems make up the complex multicellular **organism.**

Focus On

Evolution in Action: The Case of the Peppered Moth

An interesting case of evolution in action has been documented in England since 1850. The tree trunks in a certain region of England were once white because of a type of fungus, a lichen, that grew on them. The common peppered moth was beautifully adapted for landing upon these white tree trunks because its light color blended with the trunks and protected it from predacious birds (see figure). At that time black moths were rare.

Then humans changed the environment. They built industries that polluted the air with soot, killing the lichens and coloring the tree trunks black. The light-colored moths became easy prey to the birds. Now the black moths blended with the dark trunks and escaped the sharp eyes of predators. In these new surroundings, the dark moths were better adapted and were selected for survival. Eventually, more than 90% of the peppered moths in the industrial areas of England were dark. Interestingly, with recent efforts to control air pollution, there has been an increase in the population of the light-colored moths.

Adaptation of the peppered moth was studied in the 1950s by H. B. D. Kettlewell of Oxford, who marked hundreds of male moths with a spot of paint under their wings and then released them in both rural and industrial areas. Observers reported that birds preyed on the moths that were more visible. After a period of time, surviving moths were recaptured by attracting them with light or females. Based on observation and on the percentage of each type of moth recaptured, these studies confirmed that significantly more dark moths survived in industrial areas and more light moths survived in rural areas. ■

Both a dark and a light peppered moth can be seen on the tree trunk photographed here. Which is most likely to become food for the bird? Note that the tree trunk is covered by lichens. (Visuals Unlimited/John D. Cunningham)

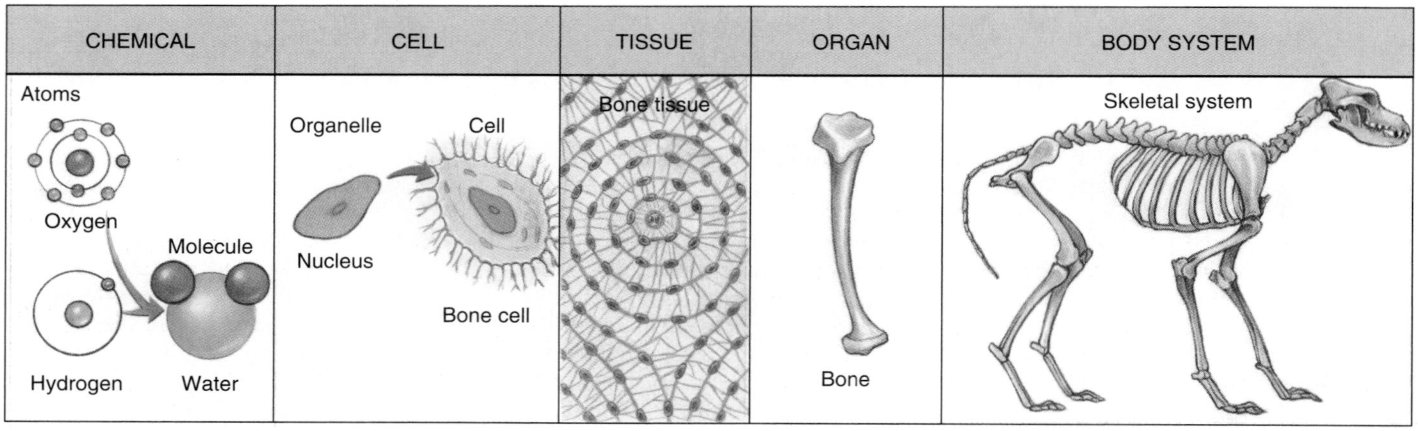

Several levels of ecological organization can be identified

Organisms interact to form still more complex levels of biological organization. All of the members of one species that live in the same geographic area at the same time make up a **population.** The populations of organisms that inhabit a particular area and interact with one another form a **community.** A community can consist of hundreds of different types of organisms. All communities of living organisms on Earth are collectively referred to as the **biosphere.**

A community together with its nonliving environment is referred to as an **ecosystem.** An ecosystem can be as small as a pond (or even a puddle) or as vast as the Great Plains of North America or the Arctic tundra. The largest ecosystem is the planet Earth with all its inhabitants: the **ecosphere.** The ecosphere encompasses all of the interactions among the biosphere, the Earth's atmosphere, the hydrosphere (water in any form), and the lithosphere (Earth's crust). The study of how organisms of a community relate to one another and to their physical environment is called **ecology** (derived from the Greek *oikos,* meaning "house").

Millions of kinds of organisms have evolved on our planet

The variety of living organisms on our planet challenges the imagination. In order to study life, we need a system for organizing, naming, and classifying its myriad forms.

Biologists use a binomial system for classifying organisms

To facilitate effective communication with one another, biologists have developed a formal system of classifying and naming organisms, called **taxonomy;** the biologists who specialize in classification are **taxonomists.**

In the 18th century Carolus Linnaeus, a Swedish botanist, developed a system of classification that, with some modification, is still used today. The basic unit of classification is the **species.** Closely related species may be grouped together in the next higher unit of classification, the **genus** (pl. *genera*).

The Linnaean system of naming species is referred to as the **binomial system of nomenclature** because each species is assigned a two-part name. The first part of the name is the genus, and the second part, the **specific ep-**

Table 1–1 CLASSIFICATION OF DOMESTIC CAT, HUMAN, AND WHITE OAK

Category	*Classification of Cat*	*Classification of Human*	*Classification of White Oak*
Kingdom	Animalia	Animalia	Plantae
Phylum	Chordata	Chordata	Anthophyta
Subphylum	Vertebrata	Vertebrata	None
Class	Mammalia	Mammalia	Dicotyledones
Order	Carnivora	Primates	Fagales
Family	Felidae	Hominidae	Fagaceae
Genus and species	*Felis domestica*	*Homo sapiens*	*Quercus alba*

FIGURE 1–10 Biological organization is hierarchical.
(Earth viewed from space: NASA)

ithet, designates a particular species belonging to that genus. The specific epithet is often a descriptive word expressing some quality of the organism. It is always used together with the full or abbreviated generic name preceding it. The generic name is always capitalized; the specific epithet is never capitalized. Both names are always italicized or underlined. For example, the dog, *Canis familiaris* (sometimes abbreviated *C. familiaris*), and the timber wolf, *Canis lupus* (*C. lupus*), belong to the same genus. The cat, *Felis domestica,* belongs to a different genus. The scientific name of the American white oak is *Quercus alba,* whereas the name of the European white oak is *Quercus robur.* Another tree, the white willow, *Salix alba,* belongs to a different genus. The scientific name for our own species is *Homo sapiens.*

Taxonomic classification is hierarchical

Just as species may be grouped together in a common genus, a number of related genera constitute a **family** (Table 1–1). In turn, families may be grouped into **orders,** orders into **classes,** and classes into **phyla.** Plants and fungi were traditionally classified into divisions rather than phyla. The International Botanical Congress, however, recently approved the phylum designation for plants and fungi. The family Canidae, which includes all doglike carnivores (animals that eat mainly meat), consists of 12 genera and about 34 living species. Family Canidae, along with family Ursidae (bears), family Felidae (catlike animals), and several other families that eat mainly meat, is placed in order Carnivora. Order Carnivora, order Primates (the order to which humans belong), and several other orders belong to class Mammalia (mammals). Class Mammalia, class Aves (birds), class Reptilia (reptiles), and four other classes are grouped together as subphylum Vertebrata. The vertebrates belong to phylum Chordata, which is part of kingdom Animalia.

FIGURE 1–11 A survey of the kingdoms of life. (***a***) Kingdom Prokaryotae is made up of bacteria. False-color scanning electron micrograph shows the large rod-shaped bacteria *Bacillus anthracis,* the causative agent of anthrax, a disease of cattle and sheep that can infect humans. (***b***) Member of the kingdom Protista. These one-celled, stalked organisms are ciliate protozoa *(Vorticella sp.;* × approximately 100). (***c***) Mushrooms, such as these fly agaric mushrooms *(Amanita muscaria)* belong to kingdom Fungi. (***d***) The plant kingdom claims many beautiful and diverse forms. Shown here is the blue flag iris *(Iris versicolor).* (***e***) Mountain lion mother with cubs *(Felis concolor).* Among the fiercest members of the animal kingdom, mountain lions are also among the most sociable. They live peaceably in prides (groups) of as many as 35. (*a,* CNRI/Science Photo Library/Photo Researchers, Inc.; *b,* Visuals Unlimited/Cabisco; *c,* Ulf Sjostedt/FPG International; *d,* Skip Moody/Dembinsky Photo Associates; *e,* Sharon Cummings/Dembinsky Photo Associates)

Most biologists recognize five kingdoms

From the time of Aristotle until recently, biologists divided the living world into two **kingdoms,** Plantae and Animalia. After microscopes were developed it became increasingly evident that many organisms could not easily be assigned to either the plant or animal kingdom.

In the system of classification used in this book, every organism is assigned to one of five kingdoms: Prokaryotae (formerly known as Monera), Protista, Fungi, Plantae, or Animalia (Fig. 1–11). The members of kingdom Plantae, the plants, and of kingdom Animalia, the animals, are the organisms most familiar to us. Some biologists now recognize six kingdoms. They divide kingdom Prokaryotae into two kingdoms: Eubacteria and Archaebacteria.

In the five kingdom system, all bacteria are assigned to kingdom **Prokaryotae.** Bacteria are single-celled and differ from all other organisms in that they lack a discrete nucleus and most other cellular organelles. Kingdom **Protista** consists of protozoa, algae, water molds, and slime molds. These are single-celled or simple multicellular organisms. Some protists are adapted to carry out **photosynthesis,** the process in which light energy is converted to the chemical energy of food molecules.

Kingdom **Fungi** is composed of the yeasts, mildews, molds, and mushrooms. These organisms do not carry on photosynthesis. They obtain their nutrients by secreting digestive enzymes into food and then absorbing the predigested food.

Plants are complex multicellular organisms adapted to carry out photosynthesis. Among the characteristic plant features are the *cuticle* (a waxy covering over aerial parts that reduces water loss); *stomata* (tiny openings in stems and leaves for gas exchange); and multicellular *gametangia* (organs that protect developing reproductive cells). The kingdom Plantae includes both nonvascular plants (mosses) and vascular plants (ferns, conifers, and flowering plants).

Animals are multicellular organisms that must eat other organisms for nourishment. Complex animals have high degrees of tissue specialization and body organization; these have evolved along with motility, complex sense organs, nervous systems, and muscular systems.

A more detailed presentation of the kingdoms can be found in Chapters 22 through 30, and classification of living organisms is summarized in Appendix A. We refer to these groups repeatedly throughout this book as we consider the many kinds of problems faced by living organisms and the various adaptations that have evolved in response to these problems.

LIFE DEPENDS ON CONTINUOUS INPUT OF ENERGY

Life on Earth depends on a continuous input of energy from the sun. Every activity of a living cell or organism requires energy. Whenever energy is used to perform biological work, some is converted to heat and dispersed into the environment.

Energy flows through individual cells and organisms

Recall that all of the energy and chemical processes that occur within an organism are referred to as metabolism. Energy is necessary in order to carry on the metabolic activities essential for growth, repair, and maintenance. Each cell of an organism requires nutrients. Some nutrients are used as "fuel" for **cellular respiration,** a process during which some of the energy (stored in nutrient molecules) is released for use by the cells (Fig. 1–12). This energy can be used for cellular work or for synthesis of needed materials such as new cellular components.

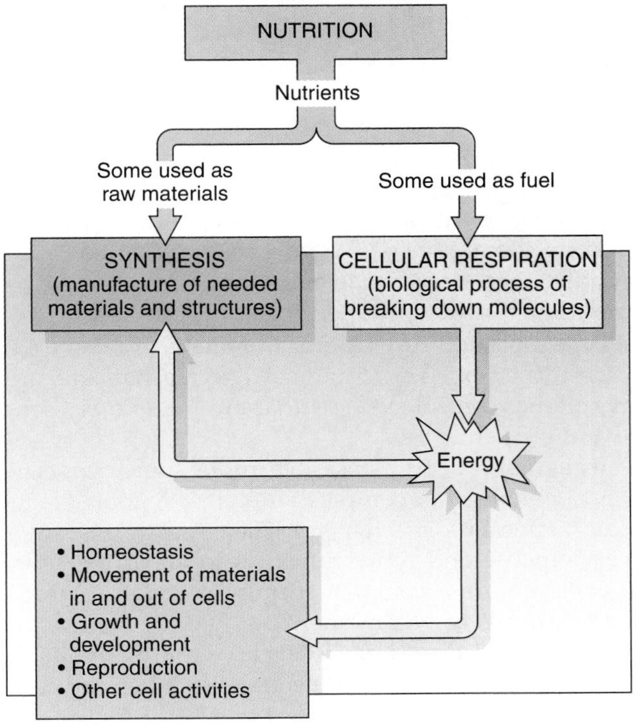

FIGURE 1–12 Metabolic reactions occur continuously in every living organism. Relationships of some metabolic activities in animals and other consumers. Some of the nutrients in food are used to synthesize needed materials and cell parts. Other nutrients are used as fuel for cellular respiration, a process that releases energy stored in food. This energy is needed for synthesis and for other forms of cellular work. Cellular respiration also requires oxygen, which is provided by gas exchange. Wastes from the cells such as carbon dioxide and water must be excreted.

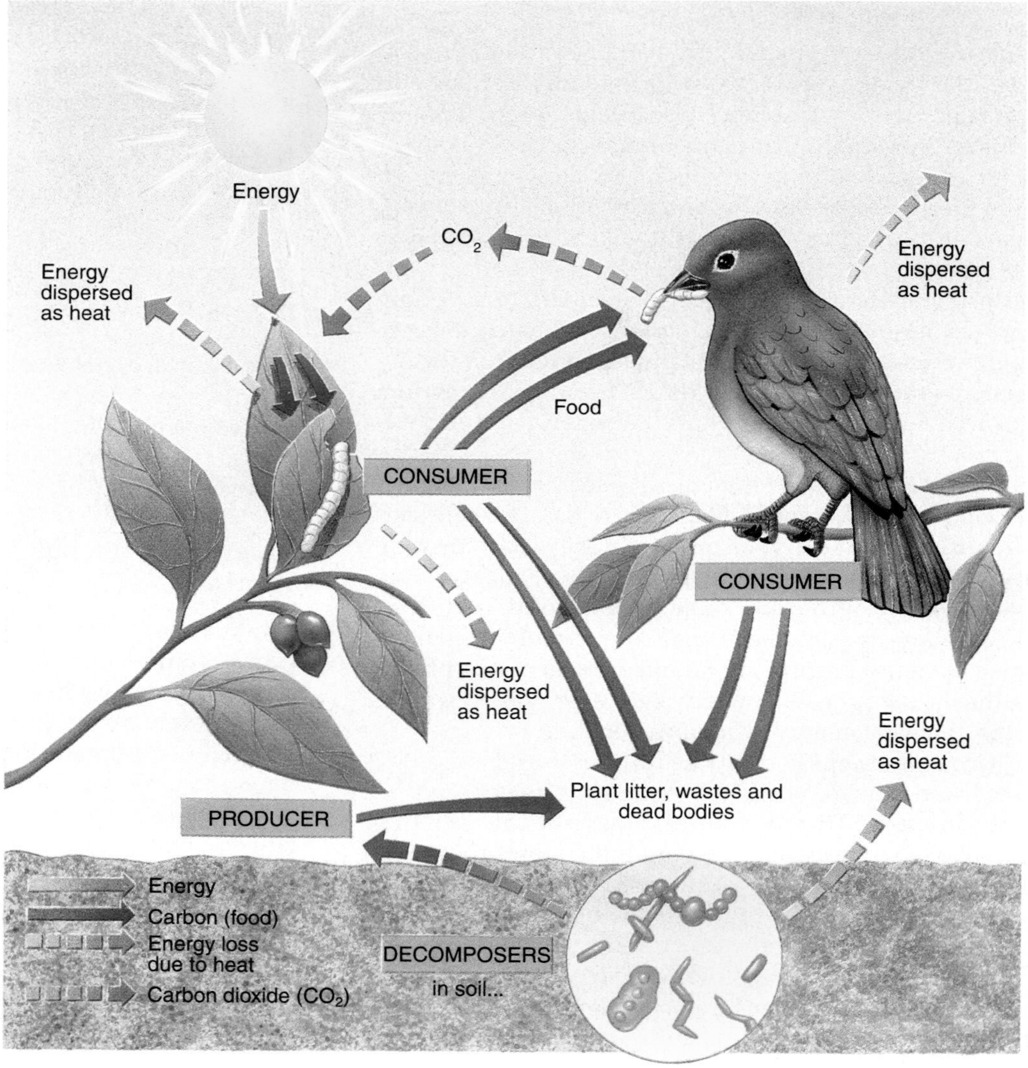

FIGURE 1–13 Energy flows through the ecosphere. Continuous energy input from the sun is required to keep the ecosphere in operation. Some energy is dispersed as heat during every energy transaction. During photosynthesis, producers use the energy from sunlight to make complex molecules from carbon dioxide and water. Consumers obtain energy, carbon, and other needed materials when they eat producers. Wastes and dead organic material supply decomposers with energy.

In all organisms, metabolic processes must be carefully and constantly regulated to maintain a balanced internal state (homeostasis). When enough of some cellular product has been made, its manufacture must be decreased or turned off. When the supply of energy declines, appropriate processes for obtaining more energy must be turned on. These homeostatic mechanisms are self-regulating control systems that are remarkably sensitive and efficient.

The regulation of glucose (a simple sugar) concentration in the blood of complex animals is a good example of a homeostatic mechanism. The circulatory system delivers glucose and other nutrients to all of the cells. Most cells require a constant supply of glucose, which they break down to obtain energy. When the concentration of glucose in the blood rises above normal limits, it is stored in the liver and in muscle cells. When the concentration begins to fall (between meals), stored nutrients are converted to glucose so that the concentration in the blood returns to normal levels. When glucose becomes depleted, we also feel hungry and restore nutrients by eating.

Energy flows through ecosystems

Like individual organisms, ecosystems depend on a continuous input of energy. A self-sufficient ecosystem contains three types of organisms—producers, consumers, and decomposers—and has a physical environment appropriate for their survival. These organisms depend on each other and on the environment for nutrients, energy, oxygen, and carbon dioxide. However, there is a one-way flow of energy through ecosystems. Organisms can neither create energy nor use it with complete efficiency. During every energy transaction, some energy is dispersed to the environment as heat (Fig. 1–13).

Producers manufacture their own food

Producers, or **autotrophs,** are plants, algae, and certain bacteria that can produce their own food from simple raw materials. Most of these organisms use sunlight as an energy source and carry out photosynthesis.

During photosynthesis, the energy from sunlight is used to synthesize complex molecules from carbon dioxide and water. The light energy is transformed into chemical energy, which is stored within the chemical bonds of the food molecules produced. Oxygen, which is required not only by plant cells but also by the cells of most other organisms, is produced as a byproduct of photosynthesis.

$$\text{Carbon dioxide} + \text{Water} + \text{Energy} \longrightarrow \text{Food} + \text{Oxygen}$$

Consumers obtain energy by eating producers

Animals are **consumers,** or **heterotrophs,** organisms that depend on producers for food, energy, and oxygen. They obtain energy by breaking down food molecules originally produced during photosynthesis. Recall that the biological process of breaking down "fuel" molecules is known as cellular respiration. When chemical bonds are broken during cellular respiration, their stored energy is made available for life processes.

$$\text{Food} + \text{Oxygen} \longrightarrow \text{Carbon dioxide} + \text{Water} + \text{Energy}$$

Gas exchange between producers and consumers by way of the nonliving environment helps maintain the life-sustaining mixture of gases in the atmosphere. Thus, consumers also contribute to the balance of the ecosystem.

Decomposers obtain energy from wastes and dead organisms

Decomposers—the bacteria and fungi—are heterotrophs that obtain nutrients by breaking down wastes and the bodies of dead organisms. In their process of obtaining energy, decomposers make the components of wastes and dead organisms available for reuse. If decomposers did not exist, nutrients would remain locked up in dead bodies, and the supply of elements required by living systems would soon be exhausted.

BIOLOGY IS STUDIED USING THE SCIENTIFIC METHOD

This book is about the **science** of biology. The word *science* comes from a Latin word meaning "to know." Science is a way of thinking and a method for investigating the world around us in a systematic manner. Science enables us to uncover ever more about the world we live in and leads us to an expanded appreciation of our universe.

The *process* of science is creative and dynamic, changing over time and influenced by cultural, social, and historical settings as well as by the personalities of scientists themselves. The process of science may be different for every individual scientist. The observations made, the range of questions posed, and the design of experiments depend on the creativity of the scientist. In contrast, the **scientific method** involves a series of ordered steps and is a tool used by all successful scientists.

Using the scientific method, scientists make careful observations, recognize and state problems, develop hypotheses (educated guesses), make predictions that can be tested, and design experiments to test their predictions (Fig. 1–14). They study the results of their experiments and draw conclusions from them. Even results that do not support the hypothesis may be valuable and may lead to new hypotheses. If the results support the hypothesis, the scientist may use them to generate related hypotheses.

Science is systematic because of the attention it gives to organizing knowledge, making it readily accessible to all who wish to build on its foundation. In this way science is both a personal and a social endeavor. Science is not mysterious. Anyone who understands its rules and procedures can take on its challenges. What distinguishes science is its insistence on rigorous methods to examine a problem. Science seeks to give us precise knowledge about those aspects of the world that are accessible to its methods of inquiry. It is not a replacement for philosophy, religion, or art. Being a scientist does not prevent one from participating in other fields of human endeavor, just as being an artist does not prevent one from practicing science.

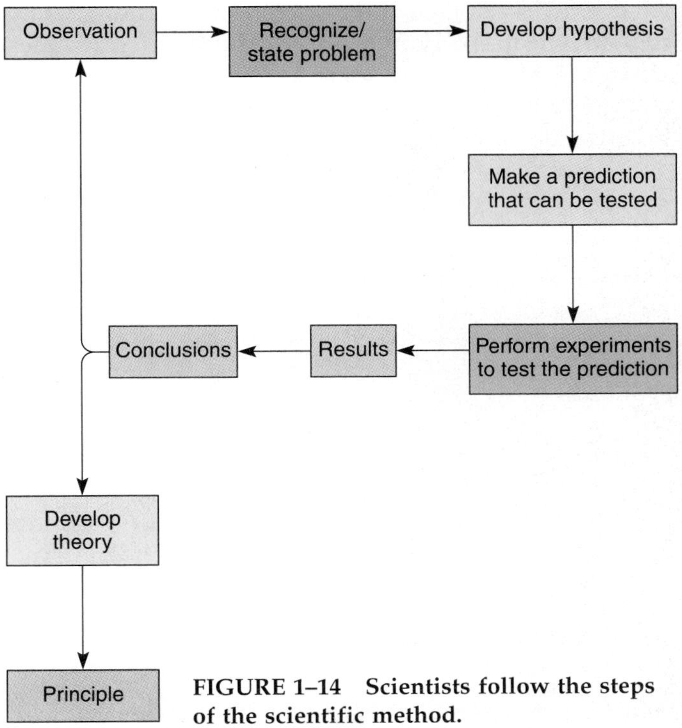

FIGURE 1–14 Scientists follow the steps of the scientific method.

Science is based on systematic thought processes

The systematic thought processes on which science is based generally fall into two categories: deduction and induction. With **deductive reasoning,** we begin with supplied information, called *premises,* and draw conclusions on the basis of that information. Deduction proceeds from general principles to specific conclusions. For example, if we accept the premise that all birds have wings, and the second premise that sparrows are birds, we can conclude deductively that sparrows have wings (Fig. 1–15).

Although undiscovered relationships may exist among the facts that we already know, scientists often focus on discovering *new* general principles. For example, we know that if we hold up an apple and then let go, it will fall to the ground. When oranges, rocks, and trees are released from support they also fall to the ground. By a reasoning process that is the opposite of deduction, we can conclude that when we release any object from its support, it will fall. We may further conclude that some force (gravity) acting on these objects attracts them to the ground.

With **inductive reasoning,** we begin with specific observations (e.g. that apples and oranges fall when released from support), and we draw a conclusion or discover a general principle. The inductive method can be used to organize raw data into manageable categories by answering the question, "What do all these facts have in common?"

A weakness of inductive reasoning is that conclusions generalize the facts to all possible examples. We go from many observed examples to all possible examples when we formulate the general principle. This is known as the **inductive leap.** Without it, we could not arrive at generalizations. However, we must be sensitive to exceptions (for example, helium balloons may not immediately fall to the ground) and also to the possibility that the conclusion is not valid. The generalizations that inductive conclusions contain come from the creative insight of the human mind, and creativity, however admirable, is not infallible.

FIGURE 1–15 Why birds have wings. A diagrammatic example of a syllogism, the classic form of deductive reasoning.

Scientists make careful observations and recognize problems

Significant discoveries are usually made by those who are in the habit of looking critically at nature. Chance and luck are often involved in recognizing a problem. In 1928 the British bacteriologist Alexander Fleming *observed* that one of his bacterial cultures had become invaded by a blue mold. He almost discarded it but before he did, he *noticed* that the area contaminated by the mold was surrounded by a zone where bacterial colonies did not grow well.

The bacteria were disease organisms of the genus *Staphylococcus,* which can cause boils and skin infections. Anything that could kill them was interesting! Fleming saved the mold, a variety of *Penicillium* (blue bread mold). It was subsequently discovered that the mold produced a substance that slowed reproduction of the bacterial population but was usually harmless to laboratory animals and humans. The substance was penicillin, one of the first antibiotics.

We may wonder how often the same mold grew on the cultures of other bacteriologists who failed to make the connection and just threw away their contaminated cultures. Fleming benefited from chance, but his mind was prepared to make observations and his pen to publish them. It was left to others, however, to develop the practical applications. Though Fleming recognized the potential practical benefit of penicillin, he did not vigorously promote it, and more than ten years passed before the drug was put to significant use (see *Making the Connection: How Do Basic Science and Technology Interact?*).

A hypothesis is a proposed explanation

A hypothesis is an educated guess. In the early stages of an investigation, a scientist usually thinks of many possible explanations and hopes that the right one is among them. He or she then decides which, if any, could and should be subjected to experimental test. Why not test them all? Time and money are important considerations in conducting research. We must establish priority among the hypotheses in order to decide which to test first. Fortunately, some guidelines do exist. A good hypothesis exhibits the following:

1. It is reasonably consistent with well-established facts.
2. It is capable of being tested; that is, it should generate definite predictions, whether the results are positive or negative. Test results should also be repeatable by independent observers.
3. It is falsifiable, which means it can be proved false.

Making the Connection

How Do Basic Science and Technology Interact?

Why is research in basic science important? For many people, the chief justification for scientific inquiry is the practical, everyday applications that may result. Oddly, though, discoveries that prove to have the greatest practical value often come from basic (often purely abstract) research. In fact, some very important discoveries have occurred by accident during the course of general investigation. Science and technology tend to promote one another.

The 17th- and 18th-century microscopists such as Leeuwenhoek and Hooke could not have seen cells at all without the use of lens-making techniques that were then state-of-the-art (see Chapter 4). Our modern knowledge of cell structure could not have been obtained without the electron microscope, first developed in the late 1930s with what was then the latest in electronic technology. The electron microscope uses a focused electron beam to form an image.

Almost everything we know about photosynthesis and, to a somewhat lesser degree, cellular respiration, has been discovered using radioactively labelled chemical substances. Without nuclear technology, we would probably still be wondering whether the hereditary material of the cell was protein, nucleic acid, or something else. Thus, advances in the physical sciences have led to further advances in the biological sciences.

The lensmakers, electron-tube designers, and nuclear physicists who made modern biology possible probably never dreamt of the further scientific advances that would occur because of *their* discoveries. Biological technology also advances biological research, for example, genetic engineering. Biological technology has applications in other fields, as well. For example, recently DNA molecules have been used to solve a problem in computing, leading to the speculation that one day DNA-based computers will be developed. Science nourishes itself, particularly when scientists from different areas of study interact. ▲

A hypothesis cannot really be proved true, but in theory (though not necessarily in practice) a well-stated hypothesis can be proved false. If one believes in an unfalsifiable hypothesis (e.g. the existence of invisible and undetectable angels), it must be on grounds other than scientific ones.

Consider the following hypothesis: All female mammals (animals that have hair and produce milk for their young) bear live young. Consider further that a female dog named Princess is a mammal. Therefore, we can predict that Princess should bear live young. When Princess has a litter of puppies, this supports the hypothesis. Yet it does not really prove the hypothesis.

Before the Southern Hemisphere was explored, most individuals would probably have believed that hypothesis without question, because all known furry, milk-giving animals did, in fact, bear live young. But it was discovered that two Australian animals (the duck-billed platypus and the spiny anteater) had fur, produced milk for their young, but laid eggs (Fig. 1–16).

The hypothesis, as stated, was false no matter how many times it had previously been supported. As a result, biologists had to either consider the platypus and the spiny anteater as nonmammals, or had to broaden their definition of mammals to include them (they chose the latter).

A hypothesis is not true just because some of its predictions (the ones we happen to have thought of or have thus far been able to test) have come true. After all, they could be true by coincidence. Failure to observe a prediction does not make a hypothesis false, but it does not show that the hypothesis is true, either.

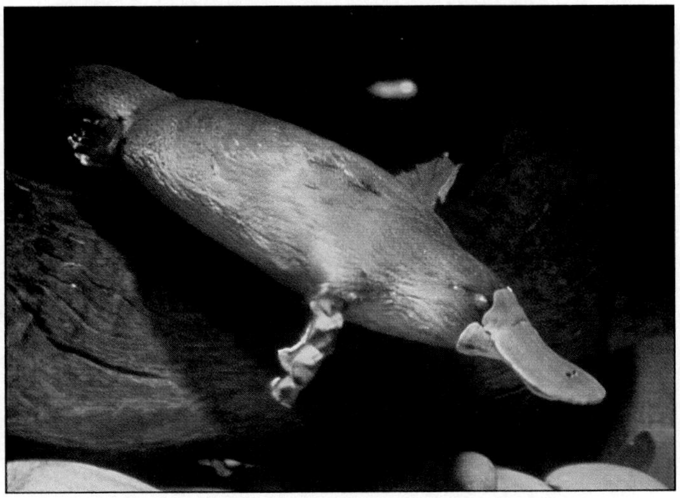

FIGURE 1–16 The duck-billed platypus is classified as a mammal because it has fur and produces milk for its young. However, unlike most mammals, it lays eggs. (Tom McHugh/Photo Researchers, Inc.)

A prediction is a logical consequence of a hypothesis

A hypothesis is an abstract idea, so there is no way to test it directly. But hypotheses suggest certain logical consequences, that is, observable things that cannot be false if the hypothesis is true. On the other hand, if the hypothesis is, in fact, false, *other* definite predictions should disclose that. As used here, then, a **prediction** is a deductive logical consequence of a hypothesis. It does not have to be a future event.

Predictions can be tested by experiment

A prediction can be tested by controlled experiments. We will consider two examples. Early biologists observed that the nucleus was the most prominent part of the cell, and they hypothesized that it might be essential for the well-being of the cell. Experiments were performed in which the nucleus of a single-celled amoeba was removed surgically with a micro-loop. After this surgery, the amoeba continued to live and move but it did not grow and after a few days it died. These results suggested that the nucleus is necessary for the metabolic processes that provide for growth and cell reproduction (Fig. 1–17).

But, the investigators asked, what if the operation itself and not the loss of the nucleus caused the amoeba to die? They performed a *controlled experiment* in which two groups of amoebas were subjected to the same operative

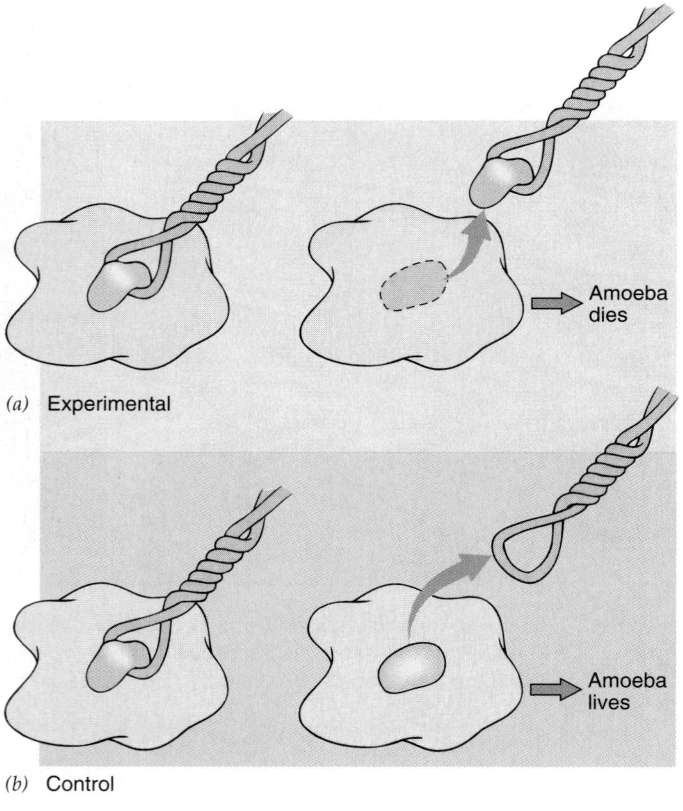

FIGURE 1–17 A controlled experiment demonstrating that the nucleus is essential for the well-being of the cell. (*a*) When its nucleus is surgically removed with a micro-loop, the amoeba dies. (*b*) Control amoeba subjected to similar surgical procedures (including insertion of a micro-loop) but without actual removal of the nucleus does not die.

trauma. However, in the **experimental group** the nucleus was removed, whereas in the **control group** it was not. An experimental group ideally differs from a control group only with respect to the variable being studied. In the control group, a micro-loop was inserted into each amoeba and pushed around inside the cell to simulate the removal of the nucleus; then the needle was withdrawn, leaving the nucleus inside. Amoebas treated with such a sham operation recovered and subsequently grew and divided, but the amoebas without nuclei died. This confirmed the hypothesis that it was the removal of the nucleus and not simply the operation that caused the death of the amoebas.

Let us consider another example of an experiment. Suppose a pharmaceutical company wants to determine whether a new drug will improve memory in elderly patients with memory problems. To test the drug, the company solicits the cooperation of physicians who work with such patients. The physicians administer a memory test and then prescribe the drug to 500 patients for a period of 2 months. They then administer another memory test and find that the patients demonstrate a 20% increase in their ability to remember things. Can the drug company legitimately conclude that its hypothesis is correct, that the drug does indeed improve memory in elderly patients? Alternative explanations might be possible. For instance, the attention paid to the patients might in itself have stimulated them to be more attentive.

To avoid such objections, the experiment must have a control. A second similar group of patients must be given a **placebo,** a harmless starch pill similar in size, shape, color, and taste to the pill being tested. Neither group of patients is told which pill—the drug or the placebo—has been given. In fact, to prevent bias, most medical experiments today are carried out in **"double-blind"** fashion: Neither the patient nor the physician knows who is getting the experimental compound and who is getting the placebo. The pills or treatments are coded in some way, and only after the experiment is over and the results are in is the code broken.

Not all experiments can be so neatly designed; for one thing, it is often difficult to establish appropriate controls. For example, we know that the carbon dioxide content of the Earth's atmosphere is increasing because of the combustion of fossil fuels and because of widespread clearing and burning of forests. This increased carbon dioxide may be producing a "greenhouse effect," by trapping heat from solar radiation. Some scientists have warned that this thermal blanket around the globe may increase the average temperature of the Earth and alter its climate. Even if the temperature does increase, however, how can we be certain that the change has resulted from human activities?

This raises an important practical question. Obviously we do not have a second, unindustrialized Earth whose climate could be compared with our own. Without such a control, scientists have had to base their predictions of the future climate on mathematical modeling techniques that fall short of perfection. Should we postpone action pending the development of a perfectly predictive model? Clearly, that would involve a long wait, and by then it might be impossible to act effectively.

Scientists draw conclusions from the results of experiments

Scientists gather data in an experiment, study their results, and then formulate conclusions. One reason for inaccurate conclusions is **sampling error.** Since *all* cases of what is being studied cannot be observed or tested, we must be content with a sample, or subset, of them. Yet how can we know whether that sample is truly representative of whatever we are studying? In the first place, the sample may be too small, so that it is likely to be different because of random factors. This problem can usually be solved by applying the mathematics of statistical analysis. In the second place, the sample may not be typical of the group that we intend to study. Again, statistical techniques to ensure that there is no consistent bias in the way that experimental samples are chosen can be employed (Fig. 1–18).

Even if a conclusion is based on results from a carefully designed experiment, it is still possible that new observations or results from other experiments can challenge the conclusion. If we test a large number of cases, we are more likely to draw accurate scientific conclusions. The scientist seeks to state with confidence that any specific conclusion has a certain statistical probability of being correct.

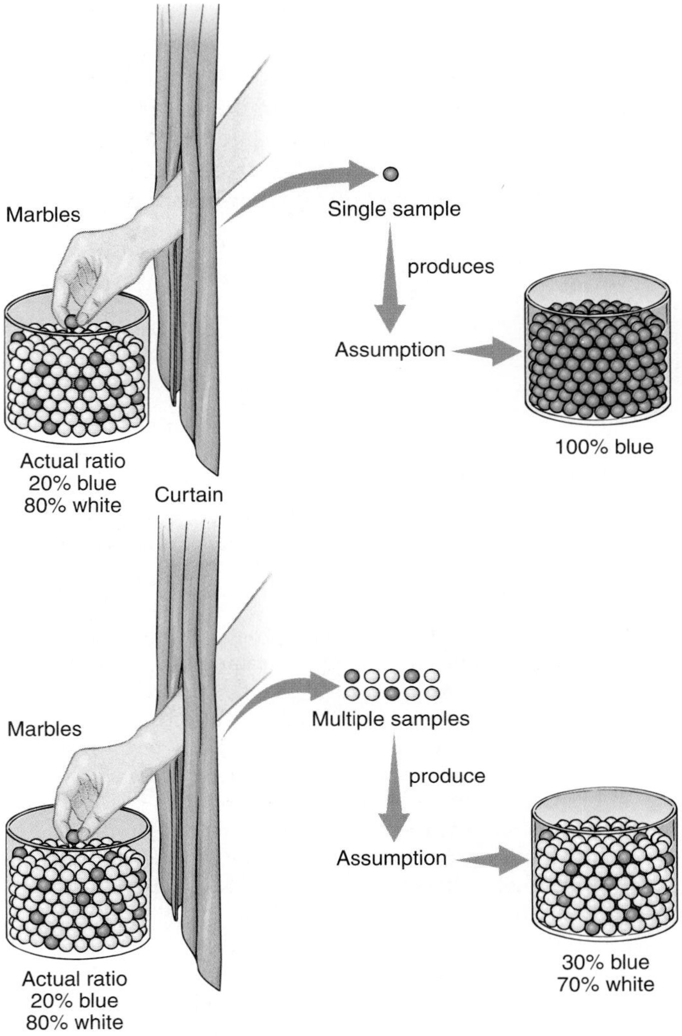

FIGURE 1–18 Taking a single sample can result in sampling error. The greater the number of samples we take of an unknown, the more likely we can make valid assumptions about it.

A well-supported hypothesis may lead to a theory

Nonscientists often use the word *theory* incorrectly when they mean to refer to a hypothesis. A **theory** can be developed only when a hypothesis has been supported by consistent results from many observations or experiments. A good theory relates facts that previously appeared to be unrelated. A good theory grows. It relates additional facts as they become known. It predicts new facts and suggests new relationships among phenomena. It may even suggest practical applications.

A good theory, by showing the relationships among classes of facts, simplifies and clarifies our understanding of natural phenomena. As Einstein wrote, "In the whole history of science from Greek philosophy to modern physics, there have been constant attempts to reduce the apparent complexity of natural phenomena to simple, fundamental ideas and relations."

A theory that, over a long period of time, has yielded true predictions and is thus almost universally accepted is referred to as a scientific **principle.** The term **law** is sometimes used for a principle judged to be of great basic importance—the law of gravity, for instance.

Science has ethical dimensions

Researchers who publish their work in scientific journals describe their experiments in sufficient detail to be independently performed by others. This permits objective observers to detect errors or bias in the original study, and helps to guard against the occasional odd result caused by random or uncontrolled factors, as well as those tainted by dishonesty on the part of the original researcher.

Scientific investigation depends on commitment to such practical ideals as truthfulness and the obligation to communicate results. Honesty is particularly important in science. Consider the great (though temporary) damage done whenever an unprincipled or even desperate researcher, whose career might depend on publication of a research study, knowingly disseminates false data. Until the deception is uncovered, researchers might devote many thousands of dollars or hours of precious professional labor to futile lines of research inspired by untrue reports.

Fortunately, science tends to be self-correcting by the consistent use of the scientific process itself. Sooner or later, someone's experimental results are bound to cast doubt on dishonest data.

Scientists face many important ethical issues, such as human and animal experimentation, research on human embryos, and applications of genetic engineering. Many of these concerns will be discussed in this book.

SUMMARY

I. A living organism is able to grow and develop, carry on metabolism, maintain homeostasis, move, respond to stimuli, and reproduce. In addition, species evolve and adapt to the environment.
   A. All living organisms are composed of cells.
   B. Living organisms grow by increasing the size and number of their cells.
   C. Metabolism refers to all the chemical activities that take place in the organism, including the chemical reactions essential to nutrition, growth and repair, and the conversion of energy to usable forms.
   D. Homeostasis is the tendency of organisms to maintain a constant internal environment.
   E. Movement, although not necessarily locomotion, is characteristic of living organisms.
   F. Living organisms respond to physical and chemical changes in their external or internal environment.
   G. Reproduction may be asexual, in which the offspring may be similar to the parent, or sexual, in which the offspring are the product of genes contributed by two parents.

II. Information encoded in DNA is transmitted from one generation to the next. DNA, proteins, hormones, and nervous systems transmit information within individuals.

III. Populations of organisms evolve over time in response to changes in the environment.
   A. Natural selection favors organisms with traits that enable them to cope with environmental changes. These organisms are most likely to survive and produce offspring.
   B. As successful organisms pass on their genes for survival, their traits become more widely distributed in the population.
   C. Biological organization reflects the course of evolution.
      1. A complex organism is organized at the chemical, cellular, tissue, organ, and organ system levels.
      2. The basic unit of ecological organization is the population. Various populations form communities; a community and its physical environment are referred to as an ecosystem. Planet Earth and all of its inhabitants make up the ecosphere.
   D. Millions of kinds of organisms have evolved on our planet.
      1. Biologists use a binomial system of nomenclature in which the name of each kind of organism includes a genus name and a specific epithet.
      2. Taxonomic classification is hierarchical; it includes species, genus, family, order, class, phylum, and kingdom.
      3. Living organisms can be classified into five kingdoms: Prokaryotae (bacteria), Protista (protozoa, algae, water molds, and slime molds), Fungi (molds and yeasts), Plantae, and Animalia.

IV. Life depends on continuous energy input from the sun. Activities of living cells require energy.
   A. Plants and many other types of producers use the energy of sunlight to synthesize complex molecules from carbon dioxide and water; this process is called photosynthesis.
   B. During cellular respiration, cells capture the energy stored in nutrients by producers. Some of that energy is then used to synthesize needed materials or to carry on other cell activities.
   C. A self-sufficient ecosystem includes producers that make their own food, consumers that depend on producers for food and energy, and decomposers that obtain energy by breaking down wastes and dead organisms.

V. Scientific method is a system of observing, recognizing a problem, developing a hypothesis, making a prediction that can be tested, performing experiments, and drawing conclusions from the results that support or falsify the hypothesis.
   A. Deductive reasoning and inductive reasoning are two categories of systematic thought processes used in the scientific method.
   B. A hypothesis is a trial idea about the nature of an observation or relationship.
   C. A properly designed scientific experiment has a control and must be as free as possible from bias.
   D. A hypothesis supported by conclusions from many experiments may lead to a theory.
   E. Science has important ethical dimensions.

SELECTED KEY TERMS

adaptation
atom
binomial system of nomenclature
biosphere
cell
cellular level
cellular respiration
chemical level
cilia
community
consumer
control group
DNA
decomposer
deductive reasoning
ecosphere
ecosystem
evolution
experiment
family
Fungi
genus
homeostasis
hormone
hypothesis
inductive reasoning
kingdom
metabolism
molecule
mutation
natural selection
order
organism
photosynthesis
phylum
population
producer
Prokaryotae
Protista
reproduction
species
stimulus (stimuli)
taxonomy
theory
tissue

POST-TEST

1. The sum of all the chemical activities of the organism is its ________ .
2. The tendency of organisms to maintain a constant internal environment is termed ________ .
3. A(n) ________ is a physical or chemical change in the internal or external environment that causes a response in an organism.
4. ________ and flagella are used by some organisms for locomotion.
5. The splitting of an amoeba into two is an example of ________ ________ .
6. A population must be able to ________ to changes in its environment in order to survive.
7. The building blocks of living organisms are called ________ .
8. At the chemical level, atoms join to form ________ .
9. Information is transmitted from one generation to the next encoded in the molecule known as ________ .
10. ________ are chemical messengers that transmit information from one part of an organism to another.
11. A(n) ________ consists of similar cells that associate and carry out a specific function.
12. In an ecosystem we can distinguish producers, consumers, and decomposers. Plants are ________ ; fungi and most bacteria are ________ ; animals are ________ .
13. During ________ ________, energy stored in nutrients is released for use by cells.
14. In the binomial system of nomenclature, the first part of an organism's name designates the ________ .
15. The yeasts and molds are assigned to kingdom ________ .
16. Bacteria are assigned to kingdom ________ .
17. ________ are chemical changes in DNA.
18. Darwin proposed that evolution takes place by the mechanism of ________ ________ .
19. An educated guess or trial explanation is called a ________ .
20. A(n) ________ may be developed when a well-tested hypothesis has been supported by results from many experiments.

REVIEW QUESTIONS

1. Contrast a living organism with a nonliving object.
2. In what ways might the metabolisms of an oak tree and a tiger be similar? Relate these similarities to the biological themes of transmission of information, energy, and evolution.
3. What would be the consequences if an organism's homeostatic mechanisms failed? Explain your answer.
4. What components do you think might be present in a balanced forest ecosystem? In what ways are consumers dependent on producers? On decomposers? Include energy considerations in your answer.
5. Why do you suppose that the binomial system of nomenclature has survived for more than 200 years and is still used by biologists?
6. How might you explain the sharp claws and teeth of tigers in terms of natural selection?
7. What is meant by a "controlled" experiment?
8. Make a prediction and devise a suitably controlled experiment to test each of the following hypotheses: (a) A strain of mold found in your garden does not produce an effective antibiotic (a chemical that inhibits the growth of bacteria). (b) The rate of growth of a bean seedling is affected by temperature. (c) Beriberi is not caused by a deficiency of the vitamin thiamine.

YOU MAKE THE CONNECTION

1. Discuss how your study of biology could help you develop effective strategies for managing some of the complex issues that confront you in everyday life. How could learning more about biology help you function as a more enlightened citizen?
2. How might a firm understanding of evolutionary processes be helpful to a biologist who is doing research in (a) animal behavior, (b) ecology, and (c) the development of a vaccine against HIV, the virus that causes AIDS?

RECOMMENDED READINGS

Moore, J.A. *Science as a Way of Knowing: The Foundations of Modern Biology.* Harvard University Press, Cambridge, MA, 1993. An account of scientific thought as related to the history of modern biology.

Scientific American, September 1989, Vol. 261, No. 3. Special Issue: Managing Planet Earth. Eleven articles on various environmental issues focus on the human impact on our planet and strategies for sustaining our world. See especially "Threats to the Biodiversity," by Edward O. Wilson.

Scientific American, October 1994, Vol. 271, No. 4. Special Issue: Life in the Universe. Ten articles on the origin and evolution of the universe and of planet Earth written by authors such as Carl Sagan and Stephen Jay Gould.

Scientific American, September 1995, Vol. 273, No. 3. Special Issue: Key Technologies for the 21st Century. This issue includes several interesting articles on medical technology and on "Energy and the Environment."

C H A P T E R 2

Atoms and Molecules: The Chemical Basis of Life

This jaguar and the plants of the rain forest share fundamental chemical similarities, including the need for water. (Frans Lanting/Minden Pictures)

Because every organism, as well as its environment, is made up of atoms and molecules, biologists use their knowledge of chemistry every day. In fact, much of modern biology is concerned with **molecular biology**—the chemistry and physics of the molecules that constitute living things. A molecular biologist might study how the structure of a DNA molecule allows it to store and transmit genetic information, or might investigate the precise interactions among a cell's atoms and molecules that maintain the energy flow essential to life. However, an understanding of chemistry is essential to *all* biologists. An evolutionary biologist might study evolutionary relationships by comparing proteins produced by different types of organisms. An ecologist might study the biological effects of changes in the salinity of the water in an estuary. A botanist might be a "chemical prospector," seeking new sources of medicines from plants.

In recent years biologists have learned that, although organisms are strikingly diverse, their chemical composition and fundamental metabolic processes are remarkably alike (Fig. 2–1). These chemical similarities provide strong evidence for the evolution of all organisms from a common ancestor, and explain why much of what biologists learn from studying bacteria or rats in laboratories can be applied to other organisms, including humans. Furthermore, the physical and chemical principles governing organisms are not unique. They are the same as those governing nonliving systems.

In this chapter we lay a foundation for understanding how the structure of atoms determines the way they form chemical bonds to produce compounds of increasing complexity. Most of our discussion will center around small, simple substances known as **inorganic compounds.** Among the biologically important groups of inorganic compounds are water, simple acids and bases, and simple salts. We pay particular attention to water, the most abundant substance on Earth's surface, and examine how its unique properties affect organisms as well as their nonliving environment.

In Chapter 3 we extend our discussion to **organic compounds,** which are generally large and complex and which always contain carbon atoms joined together to form the backbone, or skeleton, of the molecule.

LEARNING OBJECTIVES

After you have studied this chapter you should be able to

1. Name the principal chemical elements in living things and give an important function of each.
2. Compare the physical properties (mass and charge) and the locations of electrons, protons, and neutrons.
3. Distinguish between the *atomic number* and the *mass number* of an element.
4. Define the term *electron orbital,* and relate orbitals to energy levels.
5. Explain how the number of valence electrons of an atom is related to its chemical properties.
6. Distinguish among covalent bonds, hydrogen bonds, and ionic bonds. Compare them in terms of the mechanisms by which they are formed and their relative bond strengths.
7. Explain how *cations* and *anions* can be formed and how they interact.
8. Distinguish between and apply the terms *oxidation* and *reduction,* and relate these processes to the transfer of energy.
9. Sketch a water molecule. Show how hydrogen bonds are responsible for many of the properties of water.
10. Contrast acids and bases, and discuss their properties.
11. Use the pH scale to express the hydrogen ion concentration of a solution. Describe how buffers help minimize changes in pH.
12. Describe the composition of a salt, and explain why salts are important in organisms.

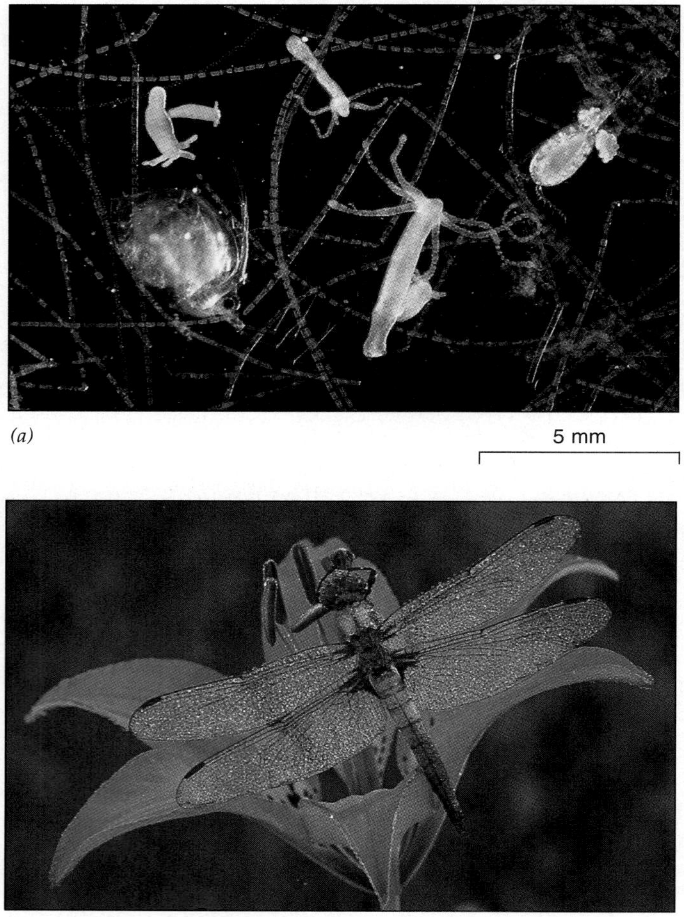

FIGURE 2–1 The chemical composition and metabolic processes of all living things are remarkably similar. The metabolic reactions that take place in these small organisms, including hydra, daphnia, and filamentous algae (***a***) are similar to those carried out within cells of larger and more complex organisms such as this dragonfly (*Aeshna sp.*) and the wood lily (*Lilium philadelphicum*) on which it is resting (***b***). (*a*, Dwight Kuhn; *b*, Visuals Unlimited/John Gerlach)

ELEMENTS ARE NOT CHANGED IN CHEMICAL REACTIONS

Elements are substances that cannot be broken down into simpler substances by chemical reactions. The matter of the universe is composed of 92 naturally occurring elements, ranging from hydrogen, the lightest, to uranium, the heaviest. Just four elements—oxygen, carbon, hydrogen, and nitrogen—are responsible for over 96% of the mass of most organisms.[1] Others, such as calcium, phosphorus, potassium, and magnesium, are also consistently present but in smaller quantities. Some elements, such as iodine and copper, are known as **trace elements** because they are present only in minute amounts. Chemical elements cycle between organisms and the nonliving environment.

Scientists have assigned each element a **chemical symbol:** usually the first letter or first and second letters of the English or Latin name of the element. For example, O is the symbol for oxygen, C for carbon, H for hydrogen, N for nitrogen, and Na for sodium (Latin *natrium*). Table 2–1 lists the elements that make up two representive organisms, a human and a typical nonwoody plant, and briefly explains why each is important.

ATOMS ARE THE BASIC PARTICLES OF ELEMENTS

An **atom** is the smallest portion of an element that retains its chemical properties. Atoms are much smaller than the tiniest particle visible under a light microscope. By spe-

[1]For convenience we consider mass and weight to be equivalent, although this is not always true. Mass does not depend on the force of gravity; weight does. Thus, you would have the same mass on the moon as you do on Earth, but your weight would be less on the moon because of its weaker gravity.

Table 2–1 ELEMENTS THAT MAKE UP SOME REPRESENTATIVE ORGANISMS

Element and Chemical Symbol	*Approximate % of Total Mass of Human Body*	*Approximate % of Total Mass of Nonwoody Plant*	*Importance or Functions*
Oxygen (O)	65	78	Required for cellular respiration; present in most organic compounds; component of water
Carbon (C)	18	11	Forms backbone of organic molecules; can form four bonds with other atoms
Hydrogen (H)	10	9	Present in most organic compounds; component of water; hydrogen ion (H^+) is involved in energy transformations
Nitrogen (N)	3	*	Component of proteins and nucleic acids; component of chlorophyll in plants
Calcium (Ca)	1.5	*	Structural component of bones and teeth; calcium ion (Ca^{2+}) is important in muscle contraction, conduction of nerve impulses, and blood clotting; associated with plant cell wall
Phosphorus (P)	1	*	Component of nucleic acids and of phospholipids in membranes; important in energy transfer reactions; structural component of bone
Potassium (K)	*	*	Potassium ion (K^+) is principal positive ion (cation) in interstitial (tissue) fluid of animals; important in nerve function; affects muscle contraction; controls opening of stomata in plants
Sulfur (S)	*	*	Component of most proteins
Sodium (Na)	*	*	Sodium ion (Na^+) is principal positive ion (cation) in interstitial (tissue) fluid of animals; important in fluid balance; essential for conduction of nerve impulses; not essential in most plants
Magnesium (Mg)	*	*	Needed in blood and other tissues of animals; activates many enzymes; component of chlorophyll in plants
Chlorine (Cl)	*	*	Chloride ion (Cl^-) is principal negative ion (anion) in interstitial (tissue) fluid of animals; important in water balance; essential for photosynthesis
Iron (Fe)	*	*	Component of hemoglobin in animals; component of cytochromes; activates certain enzymes

*The asterisk indicates that these elements represent less than 1% of the total mass. Other elements found in very small (trace) amounts in animals, plants, or both include iodine (I), manganese (Mn), copper (Cu), zinc (Zn), cobalt (Co), fluorine (F), molybdenum (Mo), selenium (Se), boron (B), silicon (Si), and a few others.

cial scanning electron microscopy (see Chapter 4), with magnification as high as × 5 million, researchers have been able to photograph the positions of some large atoms in molecules.

ATOMS CONTAIN PROTONS, NEUTRONS, AND ELECTRONS

Physicists have discovered a number of subatomic particles, but for our purposes we need consider only three: protons, neutrons, and electrons. Each **proton** has one unit of a positive electrical charge; a **neutron** is an uncharged particle with about the same mass as a proton. The mass of a proton or neutron is exceedingly small, much too small to be conveniently expressed in terms of grams or even micrograms. Such masses are expressed in terms of the **atomic mass unit** (**amu**), also called the **dalton** in honor of John Dalton, who formulated an atomic theory in the early 1800s. An amu is equal to the *approximate* mass of a proton or neutron. Protons and neutrons make up almost all of the mass of an atom and are concentrated in the **atomic nucleus.**

Each **electron** has one unit of negative electrical charge and an extremely small mass (only about 1/1800 of the mass of a proton). The number of electrons in an electrically neutral atom equals the number of protons. The electrons do not have fixed locations, but are moving rapidly in the space outside the atomic nucleus (Fig. 2–2).

The characteristics of protons, electrons, and neutrons are summarized below:

Particle	*Charge*	*Mass*	*Location*
Proton	positive	1 amu	nucleus
Neutron	neutral	1 amu	nucleus
Electron	negative	negligible	outside nucleus

An atom is uniquely identified by its number of protons

Each kind of element has a fixed number of protons in the atomic nucleus. This number, called the **atomic number,** is written as a subscript to the left of the chemical symbol. Thus $_1H$ indicates that the hydrogen nucleus contains one proton, and $_8O$ that the oxygen nucleus contains eight protons. It is the atomic number, the number of protons in its nucleus, that determines an atom's identity.

Protons plus neutrons determine atomic mass

The **atomic mass** of an element is a number that indicates how massive an atom of that element is compared with an atom of another element. This value is determined by adding the number of protons to the number of neutrons and expressing the result in atomic mass units or daltons. The mass of the electrons is ignored because it is so small. The atomic mass is indicated by a superscript to the left of the chemical symbol. The common form of the oxygen atom, with eight protons and eight neutrons in its nucleus, has an atomic number of 8 and a mass number of 16 atomic mass units. It is indicated by the symbol $^{16}_{8}O$.

Scientists commonly refer to the **atomic weight** of an element. This is calculated as the *ratio* of its atomic mass compared with the atomic mass of the common form of carbon (12 amu). When this ratio is calculated, the atomic mass units cancel out. For this reason, atomic weight and atomic mass are numerically equal, but atomic weight is dimensionless. Therefore, while the atomic mass of the common form of oxygen is properly expressed as 16 atomic mass units, the atomic weight is simply 16.

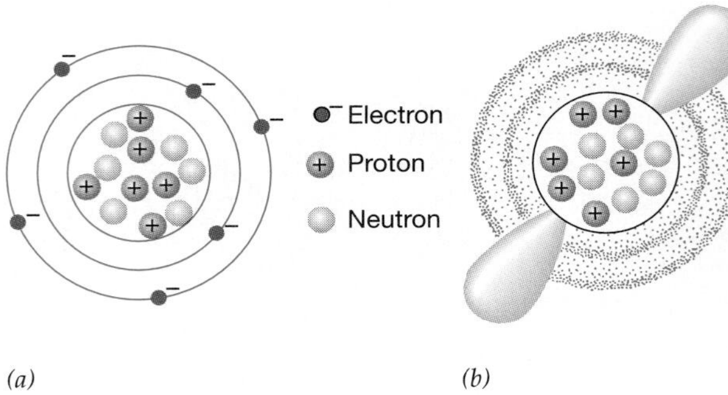

FIGURE 2–2 The electron configuration of an atom can be represented in more than one way. (***a***) The Bohr model of a carbon atom. Although the Bohr model does not depict electron configuration accurately, it is commonly used because of its simplicity and convenience. (***b***) Electron clouds. The density of the dots represents the probability that an electron is in that particular location at any given moment.

Isotopes differ in number of neutrons

Most elements consist of a mixture of atoms with different numbers of neutrons and thus different masses. Such atoms are called **isotopes.** Isotopes of the same element have the same number of protons and electrons; only the number of neutrons varies. The three isotopes of hydrogen, $^{1}_{1}H$ (ordinary hydrogen), $^{2}_{1}H$ (deuterium), and $^{3}_{1}H$ (tritium), contain zero, one, and two neutrons, respectively. Two isotopes of carbon, ^{12}C and ^{14}C, are illustrated in Figure 2–3.

The standard for comparing elements is based on assignment of an atomic mass of exactly 12 to $^{12}_{6}C$, the most common isotope of carbon. The atomic mass for an element reflects the masses of the mixtures of isotopes that occur in nature. For example, although more than 99% of the hydrogen atoms in a naturally occurring sample have

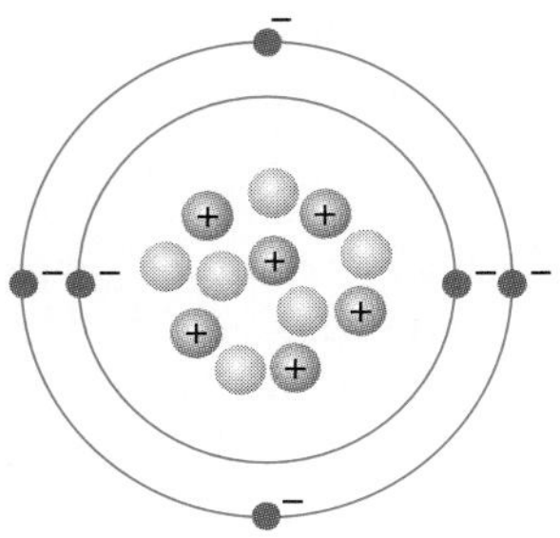

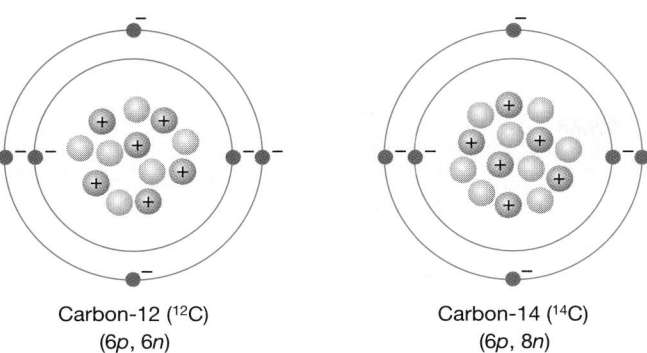

FIGURE 2–3 Isotopes differ in atomic mass. Carbon-12 (^{12}C) is the most common isotope of carbon. Its nucleus contains six protons and six neutrons, so its atomic mass is 12. Carbon-14 (^{14}C) is a rare radioactive isotope of carbon. Because it contains eight neutrons in its nucleus rather than six, its atomic mass is 14. Both types of carbon have six protons, so both have atomic number 6. Carbon-14 is often used in research to trace the fate of carbon atoms in metabolic processes.

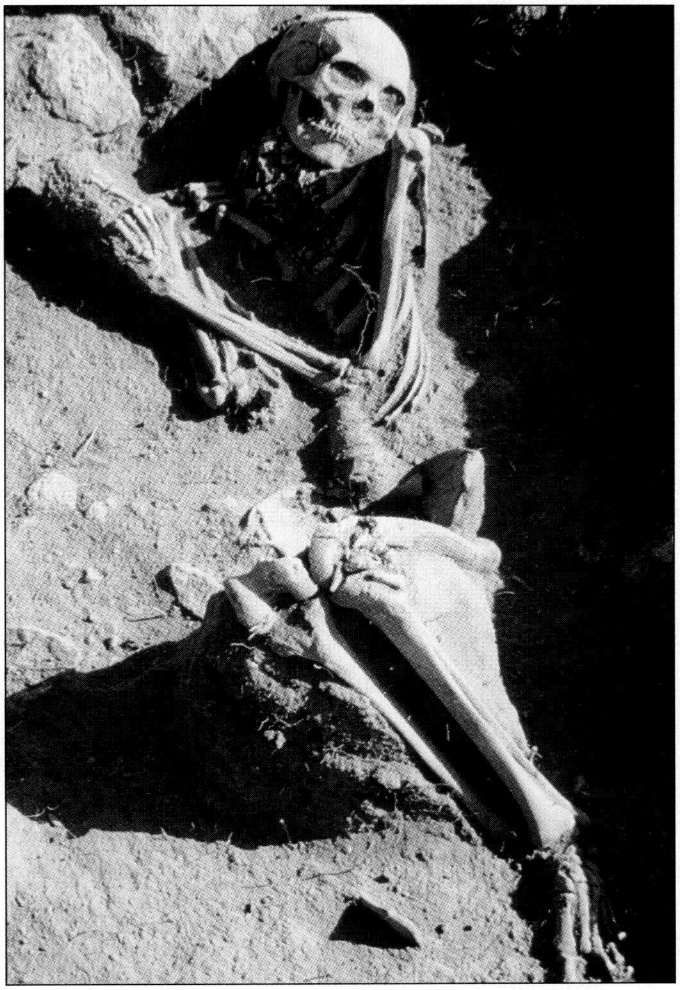

FIGURE 2–4 Anthropologists use radioisotope content to date and study fossils. This skeleton of an 11th-century inhabitant of a South African Iron Age village posed an anthropological puzzle. Physically the man's skeleton was different from those of the other villagers, suggesting that he was not a native of the area. However, when the skeleton was analyzed, the ratio of carbon isotopes was found to be similar to that of other skeletons from the same village. Because different kinds of plants incorporate different proportions of isotopes into the food produced from them, this similarity of isotope content indicates that these individuals all ate the same foods. Thus, anthropologists concluded that this man had probably spent most of his life in the village after migrating there from a distant region. (Nicholas J. van der Merwe)

an atomic mass of 1 amu (to be precise, 1.0000078 amu on the carbon-12 scale), the atomic mass of hydrogen is 1.0079 amu. Likewise, its atomic weight is 1.0079. This reflects the fact that a small amount of deuterium, $^{2}_{1}H$ (mass number 2), and an even smaller amount of tritium, $^{3}_{1}H$ (mass number 3), occur along with the common form of hydrogen, $^{1}_{1}H$.

All the isotopes of a given element have essentially the same chemical characteristics. However, some isotopes with an excess of neutrons are unstable and tend to break down, or decay, to a more stable isotope (usually becoming a different element). Such isotopes are termed **radioisotopes** because they emit radiation when they decay. Sophisticated instruments allow scientists to detect and measure this radiation.

Radioisotopes such as ^{3}H (tritium), ^{14}C, and ^{32}P have been extremely valuable research tools, useful in studies ranging from determining the age of fossils (Fig. 2–4) to DNA synthesis to sugar transport in plants. This is because, despite the difference in the number of neutrons, the organism treats all isotopes of a given element in a similar way.

In medicine, radioisotopes are used for both diagnosis and treatment. The reactions of a sugar, hormone, or drug can be followed in the body by labeling the substance with a radioisotope such as carbon-14 or tritium. For example, the active component in marijuana (tetrahydrocannabinol) can be labeled and administered intravenously. Then the amount of radioactivity in the blood and urine can be measured at successive intervals. Results of such measurements have determined that this compound remains in the blood for several weeks, and products of its metabolism can be detected in the urine. Radioisotopes are also used to test thyroid gland function, to measure the rate of red blood cell production, and to study many other aspects of body function and chemistry.

Because radiation can interfere with cell division, radioisotopes have been used in the treatment of cancer (a disease often characterized by rapidly dividing cells).

Electrons can be involved in chemical changes

Although the mass of electrons makes only a negligible contribution to the mass of an atom, electrons carry an electrical charge that profoundly affects the chemical properties of the atom. Each electron bears a charge of −1, exactly equal but opposite to the charge on a proton. The electrons are attracted by the positive charges of the protons. The atom as a whole has no net charge. The positive charges of the protons equal the negative charges of the electrons.

Although the number of protons in an atom does not change during a chemical reaction, the number and relative positions of the electrons usually do change.

Electrons occupy orbitals corresponding to energy levels

Electrons move rapidly through characteristic regions of space termed **orbitals.** Each orbital contains a maximum of two electrons. It is impossible to know an electron's position at any given time. For this reason each orbital represents the region in which the electrons that occupy it are most probably found. This probability can be depicted as a shaded area whose density is proportional to the probability that an electron is present there at any given instant. Such a representation of an orbital is referred to as an "electron cloud" (see Fig. 2–2*b*).

The energy of an electron depends on the orbital it occupies. The lowest energy orbital, the 1*s* orbital, is nearest the nucleus and is spherical in shape (Fig. 2–5*a*). The next, 2*s*, is also spherical. The electron clouds of the 2*p* orbitals are dumbbell-shaped. These occur in groups of three, arranged at right angles to each other (Fig. 2–5*b*). Other, higher energy electron orbitals are either spherical or dumbbell-shaped or are represented by more complex three-dimensional coordinates.

Electrons in orbitals with similar energies, said to be at the same **principal energy level,** make up an **electron shell.** In general, electrons in a shell distant from the nucleus have greater energy than those in a shell close to the nucleus. For example, the lowest energy electrons, the 1*s* electrons, occupy the first (innermost) electron shell. Although each orbital may contain no more than two electrons, there may be several orbitals within a given electron shell. The 2*s* electrons and the 2*p* electrons have similar energies and occupy the second electron shell. The number of electrons in the highest energy level (outermost electron shell) determines the chemical properties of an atom.

The way that electrons are arranged around an atom is referred to as the **electron configuration** of that atom. The electron configuration can be conveniently represented by a series of concentric circles around the nucleus, each circle representing an electron shell, as in Figures 2–2*a* and 2–3. It is important to remember, however, that electrons do *not* circle the nucleus in fixed concentric pathways.

Electrons always fill the lower energy orbitals before occupying the higher energy ones. The maximum number of electrons in the innermost shell (1*s*, which is a sin-

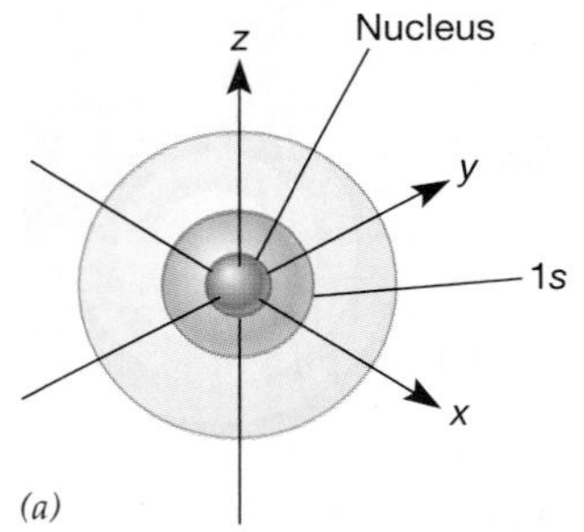

(*a*)

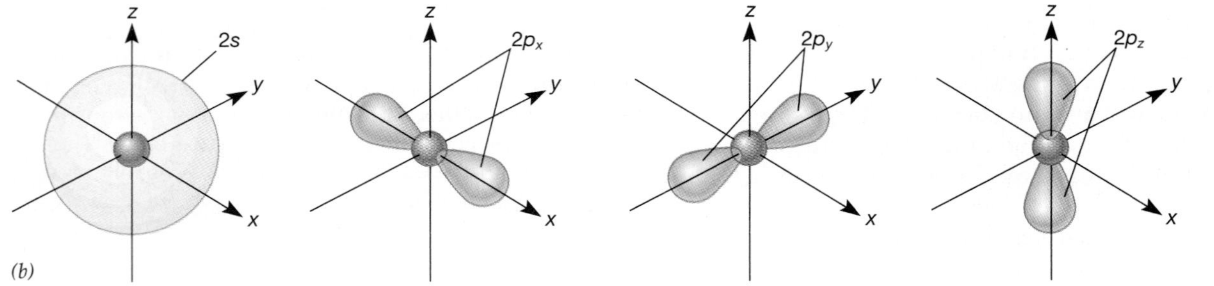

(*b*)

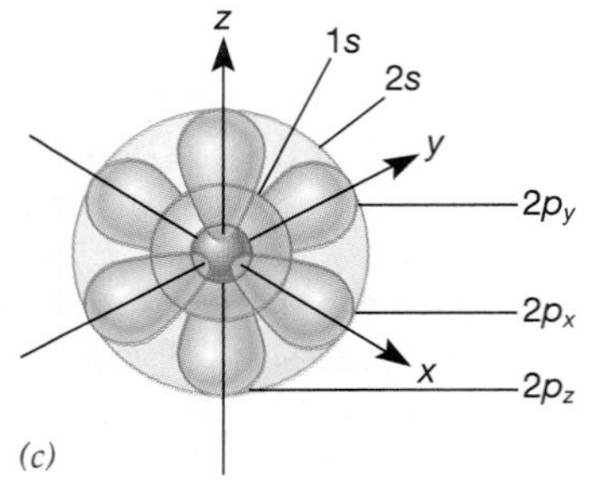

(*c*)

FIGURE 2–5 Atomic orbitals can be represented as probability densities ("electron clouds"). (***a***) The first principal energy level is a single spherical orbital (designated 1*s*) that can hold a maximum of two electrons. The electrons depicted in the diagram could be present anywhere within the deep blue area. (***b***) The second principal energy level has four orbitals, one spherical (2*s*) and three dumbbell-shaped (2*p*). (***c***) Orbitals of the first and second principal energy levels are shown superimposed on one another.

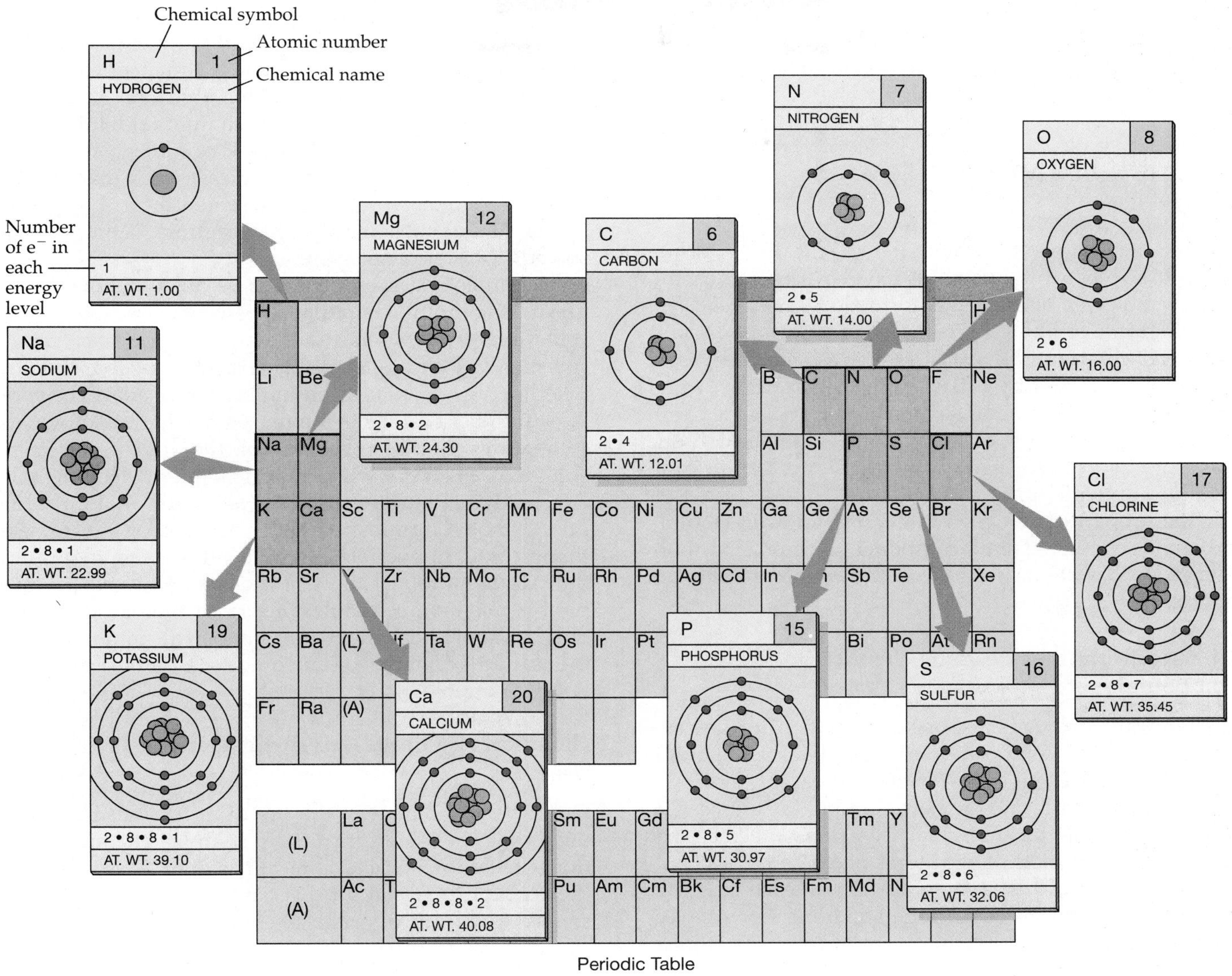

FIGURE 2–6 This representation of the periodic table shows Bohr models of some biologically important atoms.

gle spherical orbital) is two. The second shell has four orbitals (2*s*, which is spherical, plus the three dumbbell-shaped 2*p* orbitals) and thus can contain a maximum of eight electrons (Fig. 2–5). The third shell has a maximum of 18 electrons arranged in nine orbitals, and the fourth has 32 electrons in 16 orbitals.

Although the third and outer shells can each contain more than eight electrons, they are stable when only eight are present. We may consider the first shell to be complete when it contains two electrons, and the other shells to be complete when they each contain eight electrons. The electron configurations of some elements important in organisms—including carbon, hydrogen, oxygen, and nitrogen—are shown in Figure 2–6.

Remember that each atom is largely empty space. The distance from an electron to the protons and neutrons in the central nucleus may be thousands of times greater than the diameter of the nucleus itself. The tendency of the negatively charged electrons to fly off into space is countered by their attraction to the positive charges of the protons in the nucleus.

Energy is required to move a negatively charged electron farther away from the positively charged nucleus. An electron can move to an orbital farther from the nucleus by receiving more energy, or it can give up energy and sink to a lower energy level in an orbital nearer the nucleus.

When energy is added, an electron can jump from one energy level to the next, *but it cannot stop at some intermediate energy level.* To move an electron from one level to the next, the atom must absorb a tiny discrete packet of energy known as a **quantum** (pl. *quanta*), which contains just the right amount of energy for the transition—no more and no less. The term *quantum leap* is used in everyday language to indicate a sudden discontinuous move from one level to another. Changes in electron energy levels are important in energy conversions in organisms. For example, in photosynthesis (see Chapter 8) quanta of light energy absorbed by chlorophyll molecules cause electrons to move to a higher energy level.

ATOMS FORM MOLECULES AND COMPOUNDS

Two or more atoms may combine chemically to form a **molecule.** When two atoms of oxygen combine chemically, a molecule of oxygen is formed. Atoms of *different* elements can combine to form chemical compounds. A **chemical compound** consists of two or more different elements combined in a fixed ratio. For example, water is a chemical compound consisting of molecules produced when two atoms of hydrogen combine with one atom of oxygen. However, as we shall see, not all compounds are made up of molecules.

The properties of a chemical compound can be quite different from those of its component elements: at room temperature, water is usually a liquid, whereas hydrogen and oxygen are gases.

Chemical formulas describe chemical compounds

A **chemical formula** is a shorthand method for describing the chemical composition of a compound. Chemical symbols indicate the types of atoms present, and subscript numbers indicate the ratios among the atoms. The chemical formula for molecular oxygen, O_2, tells us that this molecule consists of two atoms of oxygen. The chemical formula for water, H_2O, indicates that each molecule consists of two atoms of hydrogen and one atom of oxygen. (Note that when a single atom of one type is present, it is not necessary to write 1; we do *not* write H_2O_1.)

Another type of formula is the **structural formula,** which shows not only the types and numbers of atoms in a compound but also their arrangement in a molecule. From the chemical formula for water, H_2O, you would not know whether the atoms were arranged H—H—O or H—O—H. The structural formula, H—O—H, settles the matter, indicating that the two hydrogen atoms are attached to the oxygen atom.

One mole of any substance contains the same number of units

The **molecular mass** of a compound is the sum of the atomic masses of the component atoms of a single molecule; thus, the molecular mass of water, H_2O, is (2 × 1 amu) + (16 amu), or 18 amu. (Owing to the presence of isotopes, atomic mass values are not whole numbers. However, for our purposes each atomic mass value has been rounded off to a whole number.) The molecular mass of the simple sugar glucose, $C_6H_{12}O_6$, which is a key compound in cellular metabolism, is (6 × 12 amu) + (12 × 1 amu) + (6 × 16 amu), or 180 amu. Scientists commonly refer to the **molecular weight** of a compound. Like atomic weight, molecular weight is dimensionless. The molecular weight of glucose is therefore 180.

The amount of an element or compound whose mass in grams is equivalent to its atomic or molecular mass is termed 1 **mole.** Thus 1 mole of glucose has a mass of 180 grams. A 1-molar solution, represented by 1 *M*, contains 1 mole of that substance (e.g., 180 grams of glucose) dissolved in 1 liter of solution.

The mole is a useful concept because it allows us to make meaningful comparisons between atoms and molecules of very different mass. This is because *one mole of any substance always has exactly the same number of units,* whether they be small atoms or large molecules. The very large number of units in a mole, 6.02×10^{23}, is known as **Avogadro's number,** named for the Italian physicist, Amadeo Avogadro, who first calculated it. Thus 1 mole (180 grams) of glucose contains 6.02×10^{23} molecules, as does 1 mole (2 grams) of molecular hydrogen (H_2). Although it is impossible to count atoms and molecules individually, this fact allows a scientist to count them simply by weighing a sample. Molecular biologists usually deal with smaller values—millimoles (a mmole is one thousandth of a mole) or micromoles (a μmole is one millionth of a mole).

Chemical equations describe chemical reactions

During any moment in the life of an organism, be it a mushroom or a butterfly, many complex chemical reactions are taking place. Chemical reactions—for example, the reaction between glucose and oxygen—can be described by means of chemical equations:

$$\underset{\text{Glucose}}{C_6H_{12}O_6} + \underset{\text{Oxygen}}{6\,O_2} \longrightarrow \underset{\text{Carbon dioxide}}{6\,CO_2} + \underset{\text{Water}}{6\,H_2O} + \text{Energy}$$

In a *chemical equation,* the **reactants** (the substances that participate in the reaction) are generally written on the left side, and the **products** (the substances formed by the reaction) are written on the right side. The arrow means "yields" and indicates the direction in which the reaction tends to proceed.

Chemical compounds react with each other in quantitatively precise ways. The numbers preceding the chemical symbols or formulas indicate the relative number of atoms or molecules reacting. For example, 1 mole of glucose burned in a fire or metabolized in a cell reacts with 6 moles of oxygen to form 6 moles of carbon dioxide (CO_2) and 6 moles of water.

Reactions may proceed in the reverse direction (to the left) as well as forward (to the right); at **equilibrium** the rates of the forward and reverse reactions are equal. Reversible reactions are indicated by double arrows:

$$\underset{\text{Carbon dioxide}}{CO_2} + \underset{\text{Water}}{H_2O} \rightleftarrows \underset{\text{Carbonic acid}}{H_2CO_3}$$

In this example, the arrows are drawn different lengths to indicate that when the reaction reaches equilibrium there will be more reactant(s) than product.

ATOMS ARE JOINED BY CHEMICAL BONDS

The chemical behavior of an atom is determined primarily by the number and arrangement of electrons in the highest principal energy level (outermost electron shell). In a few elements, called the "noble gases," the outermost shell is filled. These elements are chemically inert, meaning that they do not readily combine with other elements. Two such elements are helium, with two electrons (a complete shell), and neon, with ten electrons (a complete inner shell of two and a complete second shell of eight).

The electrons in the outermost shell of an atom are referred to as **valence electrons.** The outer electron shell, called the **valence shell,** of hydrogen or helium is full when it contains two electrons. The valence shell of any other atom is full when it contains eight electrons. When the valence shell is not full, the atom tends to lose, gain, or share electrons to achieve a full outer shell.

The atoms of a compound are held together by forces of attraction called **chemical bonds.** Each bond represents a certain amount of chemical energy. **Bond energy** is the energy necessary to break a bond. The valence electrons dictate how many bonds an atom can participate in. The two principal types of chemical bonds are covalent bonds and ionic bonds.

Electrons are shared in covalent bonds

Covalent bonds involve the sharing of electrons between atoms in a way that results in each having a filled valence shell. A compound consisting mainly of covalent bonds is called a **covalent compound.** A simple example of a covalent bond is the one joining two hydrogen atoms in a molecule of hydrogen gas, H_2 (Fig. 2–7*a*). Each atom of

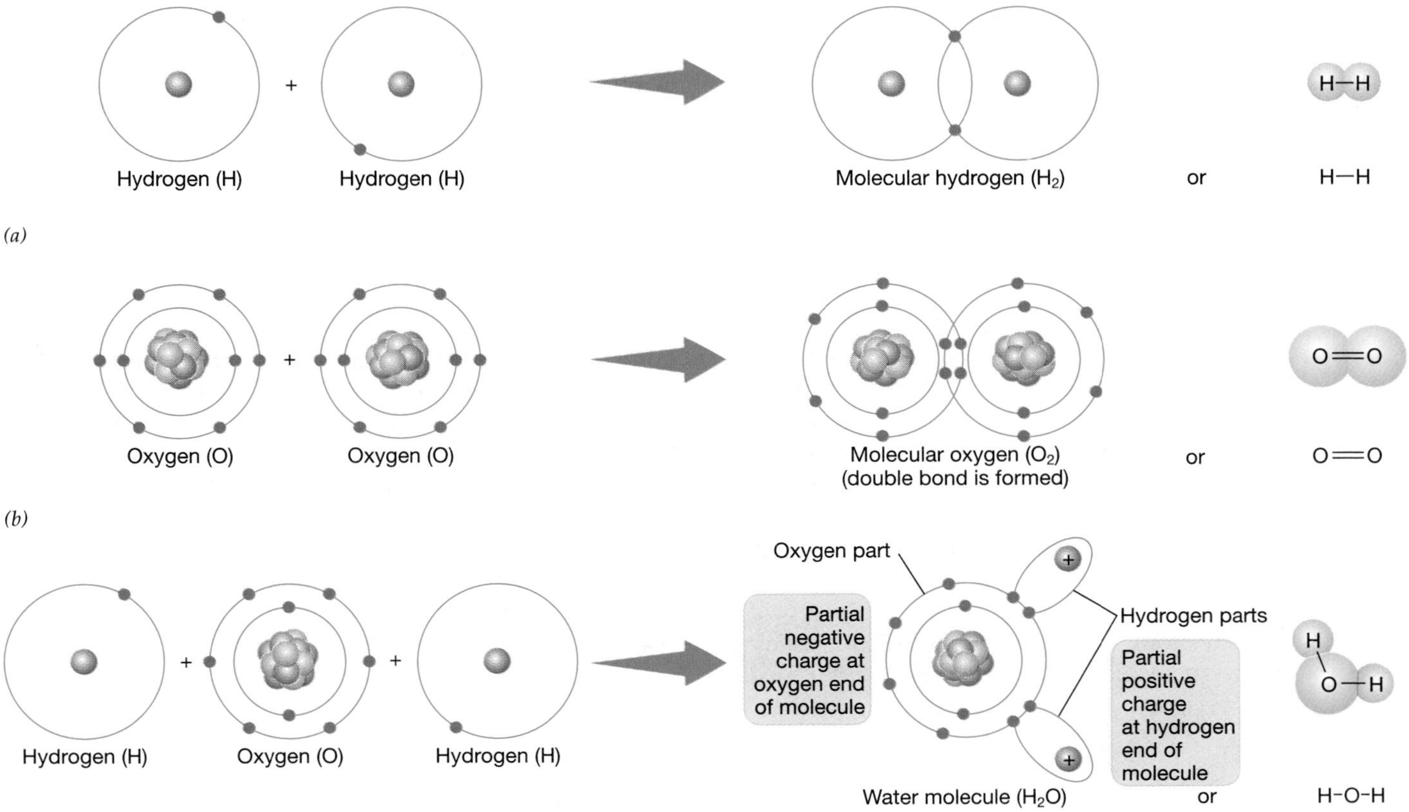

FIGURE 2–7 Covalent compounds are formed by sharing of electrons. (*a*) Two hydrogen atoms achieve stability by sharing electrons, thereby forming a molecule of hydrogen. The structural formula shown on the right is a simpler way of representing molecular hydrogen. The straight line between the hydrogen atoms represents a single covalent bond. (*b*) Two oxygen atoms share two pairs of electrons to form molecular oxygen. Note the double covalent bond. (*c*) When two hydrogen atoms share electrons with an oxygen atom, the result is a molecule of water. Note that the electrons tend to stay closer to the nucleus of the oxygen atom than to the hydrogen nuclei. This results in a partial negative charge on the oxygen portion of the molecule and a partial positive charge at the hydrogen end. Although the water molecule as a whole is electrically neutral, it is a polar covalent compound.

hydrogen has one electron, but two electrons are required to complete the first energy level (the valence shell for hydrogen). The hydrogen atoms have equal capacities to attract electrons, so neither donates an electron to the other. Instead, the two hydrogen atoms share their single electrons so that each of the two electrons is attracted simultaneously to the two protons in the two hydrogen nuclei. The two electrons thus whirl around *both* atomic nuclei, joining the two atoms.

A simple way of representing the electrons in the valence shell of an atom is to use dots placed around the chemical symbol of the element to represent the electrons. Such a representation is referred to as the *Lewis* structure of the atom. In a water molecule, two hydrogen atoms are covalently bonded to an oxygen atom:

$$\mathrm{H\cdot + H\cdot + \cdot\overset{\cdot\cdot}{\underset{\cdot\cdot}{O}}\cdot \longrightarrow H:\overset{\cdot\cdot}{\underset{\cdot\cdot}{O}}:H}$$

Oxygen has six valence electrons; by sharing electrons with two hydrogen atoms, it completes its outer shell of eight. At the same time each hydrogen atom obtains a complete outer shell of two. (Note that in the structural formula H—O—H each pair of shared electrons constitutes a covalent bond, represented by a solid line. Unshared electrons are usually omitted in a structural formula.)

The carbon atom has four electrons in its valence shell. These four electrons are available for covalent bonding:

$$\mathrm{\cdot\overset{\cdot}{\underset{\cdot}{C}}\cdot}$$

When one carbon and four hydrogen atoms share electrons, a molecule of methane, CH_4, is formed:

$$\begin{array}{c} \mathrm{H} \\ \mathrm{H : \overset{\cdot\cdot}{\underset{\cdot\cdot}{C}} : H} \\ \mathrm{H} \end{array} \qquad \text{or} \qquad \begin{array}{c} \mathrm{H} \\ | \\ \mathrm{H—C—H} \\ | \\ \mathrm{H} \end{array}$$

Lewis structure Structural formula

Each atom shares its outer shell electrons with the other, thereby completing the first electron shell of each hydrogen atom and the second shell of the carbon atom.

The nitrogen atom has five electrons in its outer shell:

$$\mathrm{\cdot\overset{\cdot\cdot}{\underset{\cdot}{N}}\cdot}$$

When a nitrogen atom shares electrons with three hydrogen atoms, a molecule of ammonia, NH_3, is formed:

$$\begin{array}{c} \mathrm{H : \overset{\cdot\cdot}{\underset{\cdot\cdot}{N}} : H} \\ \mathrm{H} \end{array} \qquad \text{or} \qquad \begin{array}{c} \mathrm{H—N—H} \\ | \\ \mathrm{H} \end{array}$$

Lewis structure Structural formula

When an electron pair is shared between two atoms, the covalent bond is referred to as a **single covalent bond.** Two oxygen atoms may achieve stability by forming covalent bonds with one another. Each oxygen atom has six electrons in its outer shell. To become stable, the two atoms share two pairs of electrons, forming molecular oxygen (Fig. 2–7*b*). When two pairs of electrons are shared in this way, the covalent bond is called a **double covalent bond** and is represented by two parallel solid lines. Similarly, a **triple covalent bond** (represented by three parallel solid lines) is formed when three pairs of electrons are shared.

The number of covalent bonds usually formed by the atoms commonly present in biologically important molecules is summarized below:

Atom	*Symbol*	*Covalent Bonds*
Hydrogen	H	1
Oxygen	O	2
Carbon	C	4
Nitrogen	N	3
Phosphorus	P	5
Sulfur	S	2

Orbitals may change shape when covalent bonds form

Each kind of molecule has a characteristic size and a general overall shape. Although the shape of a molecule may change (within certain limits), the functions of molecules in living cells are largely dictated by their geometric shapes. A molecule that consists of two atoms, for example, has a linear shape. Molecules composed of more than two atoms may have more complicated shapes. The geometric shape of a molecule provides the optimal distance between the atoms to counteract the repulsion of electron pairs.

When an atom forms covalent bonds with other atoms, the orbitals in the outer shell may become rearranged, or **hybridized.** Atoms with valence electrons in both the *s* and *p* orbitals undergo a rearrangement such that the *s* and *p* orbitals hybridize to form four new molecular orbitals; these extend outward from the nucleus. When the ends of the orbitals are connected, the result is a three-dimensional structure called a *tetrahedron.* When four hydrogen atoms combine with a carbon atom to form a molecule of methane, one hydrogen atom is present at each of the four corners of the tetrahedron. Each hydrogen shares a pair of electrons in the hybrid orbital with the carbon atom (Fig. 2–8).

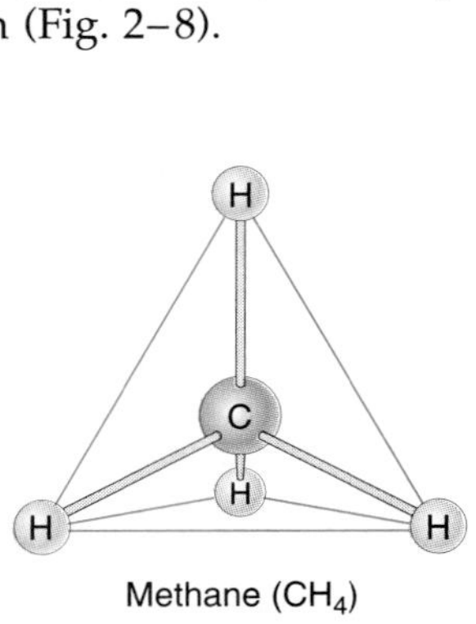

FIGURE 2–8 The four hydrogens of a methane molecule are located at the corners of a tetrahedron.

Covalent bonds can be nonpolar or polar

The atoms of each element have a characteristic affinity for electrons. **Electronegativity** is a measure of an atom's attraction for electrons in chemical bonds. Very electronegative atoms are sometimes called "electron-greedy." When covalently bound atoms have similar electronegativities, the electrons are shared equally and the covalent bond is described as **nonpolar.** The covalent bond of the hydrogen molecule is nonpolar; so are the covalent bonds of molecular oxygen and methane.

In a covalent bond between two different elements, such as oxygen and hydrogen, the electronegativities of the atoms may be different. If so, electrons are pulled closer to the atomic nucleus of the element with the greater electron affinity (in this case, oxygen). A covalent bond between atoms that differ in electronegativity is called a **polar covalent bond.** Such a bond has two dissimilar ends (or "poles"), one with a partial positive charge and the other with a partial negative charge. Each of the two covalent bonds in water is polar because there is a partial positive charge at the hydrogen end of the bond and a partial negative charge at the oxygen end, where the "shared" electrons are more likely to be found.

Covalent bonds differ in their degree of polarity, ranging from those in which the electrons are exactly shared (as in the hydrogen molecule) to those in which the electrons are much closer to one atom than to the other (as in water). Oxygen is quite electronegative and forms polar covalent bonds with carbon, hydrogen, and many other atoms. Nitrogen is also relatively electronegative, although less so than oxygen.

A molecule with one or more polar covalent bonds is electrically neutral as a whole. However, because of the geometric arrangement of the bonds, some such molecules are themselves polar. A **polar molecule** has one end with a partial positive charge and another end with a partial negative charge. One example is water. Each hydrogen atom has a partial positive charge and the oxygen atom has a partial negative charge (Fig. 2–7*c*). Because the atoms are arranged in a "V" shape, rather than linearly, the oxygen end constitutes the negative pole and the end with the two hydrogens constitutes the positive pole.

Ionic bonds form between cations and anions

Some atoms or groups of atoms are not electrically neutral. A particle with one or more units of electrical charge is called an **ion.**

An atom becomes an ion if it gains or loses one or more electrons. An atom with one, two, or three electrons in its valence shell tends to lose electrons to other atoms. When such an atom loses electrons, it becomes positively charged because its nucleus contains excess protons. Positively charged ions are termed **cations.** Atoms with five, six, or seven valence electrons tend to gain electrons from other atoms and become negatively charged **anions.**

It is important to understand that ions have properties that are quite different from those of the electrically neutral atoms from which they were derived. For example, although chlorine is a poison, chloride ions (Cl^-) are essential to life. Because their electrical charges provide a basis for many interactions, cations and anions are involved in energy transformations within the cell, the transmission of nerve impulses, muscle contraction, and many other life processes (Fig. 2–9).

Molecules can also become ions. Unlike atoms, molecules can lose or gain protons (derived from hydrogen atoms) as well as electrons. Therefore, a molecule can become a cation if it loses one or more electrons or gains one or more protons. A molecule becomes an anion if it gains one or more electrons, or loses one or more protons.

An **ionic bond** forms as a consequence of the attraction between the positive charge of a cation and the negative charge of an anion. An **ionic compound** is a substance consisting of anions and cations bonded together by their opposite charges.

A good example of how ionic bonds are formed is the attraction between sodium ions and chloride ions. A sodium atom, with atomic number 11, has two electrons in its inner shell, eight in the second, and one in the third. A sodium atom cannot fill its third shell by obtaining seven electrons from other atoms, for it would then have a very large unbalanced negative charge. Instead, it gives

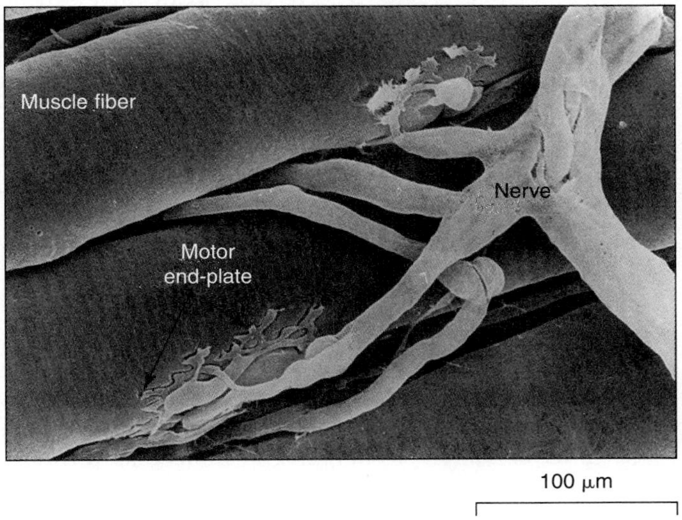

FIGURE 2–9 Ions play important roles in biological processes. Sodium, potassium, and chloride ions are among the ions essential in the conduction of nerve impulses. This scanning electron micrograph shows a nerve fiber communicating with several muscle cells. The nerve fiber transmits impulses to the muscle cells, stimulating them to contract. The muscle cells are rich in calcium ions, which are required for muscle contraction. (D. W. Fawcett)

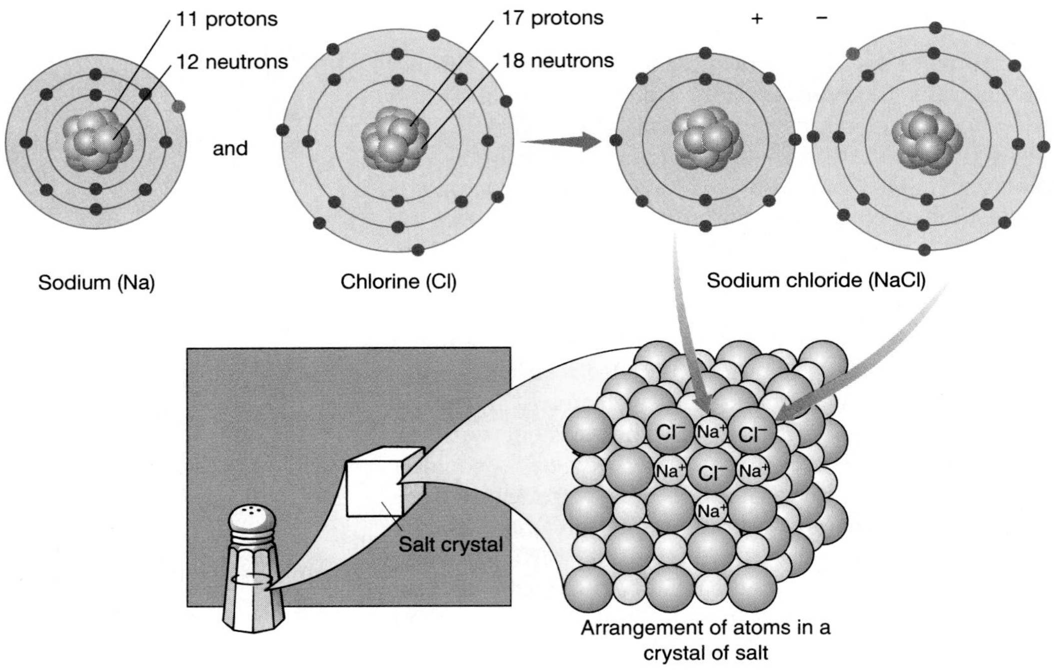

FIGURE 2–10 Sodium chloride is an ionic compound. Sodium becomes a positively charged sodium ion when it donates its single valence electron to chlorine, which has seven valence electrons. With this additional electron, chlorine completes its outer shell and becomes a negatively charged chloride ion. These ions are attracted to one another by their unlike electrical charges, forming the ionic compound sodium chloride. The force of attraction holding these ions together is called an ionic bond.

up the single electron in its third shell to an electron acceptor, leaving the second shell as the complete outer shell (Fig. 2–10). A chlorine atom is very electronegative. With atomic number 17, it has 17 protons in its nucleus, two electrons in its inner shell, eight in the second shell, and seven in the third shell. The chlorine atom achieves a complete outer shell not by losing the seven electrons in its third shell, for it would then have a vast positive charge, but by stripping an electron from an electron donor (such as sodium) to complete its outer third shell.

When sodium reacts with chlorine, its outermost electron is transferred completely to chlorine. The resulting sodium ion now has 11 protons in its nucleus and 10 electrons circling the nucleus, and one unit of positive charge. The resulting chloride ion has 17 protons in its nucleus, 18 electrons circling the nucleus, and one unit of negative charge. These ions attract each other as a result of their opposite charges. They are held together by this electrical attraction in ionic bonds to form sodium chloride,[2] common table salt.

The term *molecule* does not adequately explain the properties of ionic compounds such as NaCl. When NaCl is in its solid crystal state, each ion is actually surrounded by six ions of opposite charge. The molecular formula NaCl indicates that sodium ions and chloride ions are present in a one-to-one ratio, but in the actual crystal, no discrete molecules composed of one Na^+ ion and one Cl^- ion are present.

Compounds joined by ionic bonds, such as sodium chloride, have a tendency to *dissociate* (separate) into their individual ions when placed in water. In the solid form of an ionic compound, the constituent ions require considerable energy to be pulled apart. Water, however, is an excellent **solvent;** as a liquid it is capable of dissolving many substances, particularly those that are polar or ionic. This is because of the polarity of water molecules.

The localized partial positive charges (on the hydrogen atoms) and partial negative charges (on the oxygen atom) on each water molecule attract the anions and cations on the surface of an ionic solid. As a result, the solid dissolves. A dissolved substance is referred to as a **solute.** In solution, each cation and anion of the ionic compound is surrounded by oppositely charged ends of the water molecules (Fig. 2–11). This process is known as **hydration.** Hydrated ions still interact with each other to some extent, but the transient ionic bonds formed are much weaker than those in a solid crystal.

$$\underset{\text{Sodium chloride}}{NaCl} \xrightarrow{\text{in } H_2O} \underset{\text{Sodium ion}}{Na^+} + \underset{\text{Chloride ion}}{Cl^-}$$

Hydrogen bonds are weak attractions involving partially charged hydrogen atoms

Another type of bond important in organisms is the **hydrogen bond.** When hydrogen is combined with oxygen (or with another relatively electronegative atom such as nitrogen), it has a partial positive charge because its electron spends more time closer to the electronegative atom.

---

[2]In both covalent and ionic binary compounds (*binary* denotes compounds consisting of two elements), the element having the greater attraction for electrons is named second, and an *-ide* ending is added to the stem name — e.g., sodium chloride, hydrogen fluoride. The *-ide* ending is also used to indicate an anion, as in chloride (Cl^-) and hydroxide (OH^-).

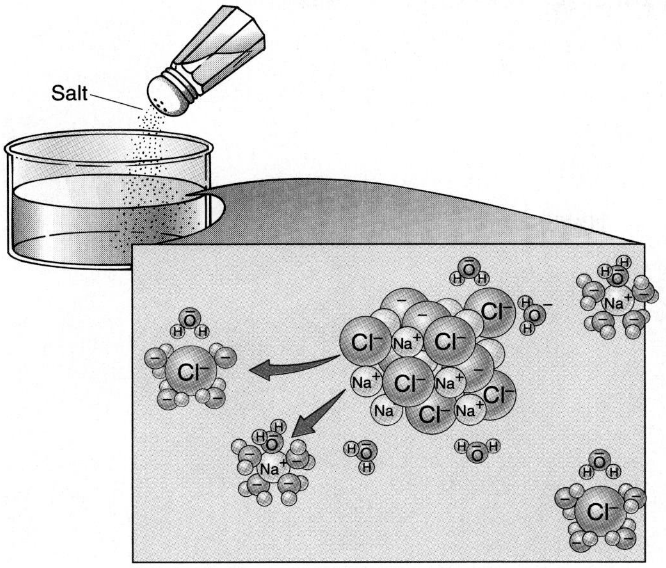

FIGURE 2–11 An ionic compound becomes hydrated through interaction with water molecules. The crystal of NaCl consists of regularly spaced ionic bonds between the Na^+ and Cl^-. When NaCl is added to water, the partial negative ends of the water molecules are attracted to the positive sodium ions and tend to pull them away from the chloride ions. At the same time, the partial positive ends of the water molecules are attracted to the negative chloride ions, separating them from the sodium ions. When the NaCl is dissolved, each of the sodium and chloride ions is surrounded by water molecules electrically attracted to it.

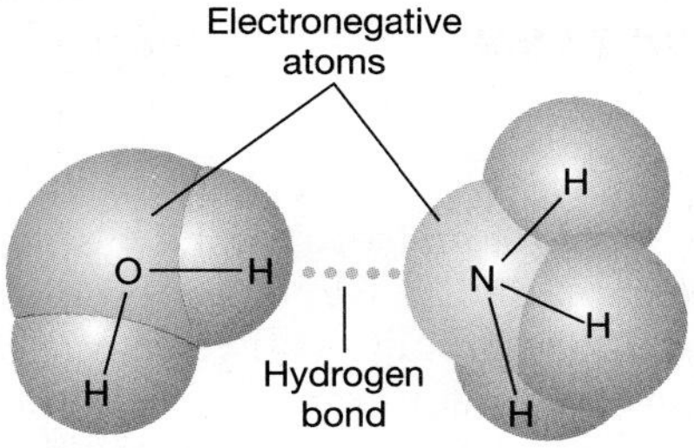

FIGURE 2–12 A hydrogen bond is formed between a hydrogen atom with a partial positive charge and an electronegative atom with a partial negative charge. A hydrogen atom in a water molecule (H_2O) has a partial positive charge because of its polar covalent bond with oxygen. Nitrogen is relatively electronegative and, in molecules like ammonia (NH_3), has a partial negative charge because of its polar covalent bonds with hydrogen. In this example, the nitrogen atom of the ammonia molecule is joined by a hydrogen bond to a hydrogen atom of a water molecule. A hydrogen bond is generally indicated by a dotted line.

Hydrogen bonds tend to form between an atom with a partial negative charge and a hydrogen atom that is covalently bonded to oxygen or nitrogen (Fig. 2–12). The atoms involved may be in two parts of the same large molecule or in two different molecules. Water molecules interact with each other extensively through hydrogen bond formation.

Hydrogen bonds are weak and are readily formed and broken. They have a specific length and orientation; this feature is very important in their role of helping determine the three-dimensional structure of large molecules such as DNA and proteins. Although individual hydrogen bonds are relatively weak, they are collectively strong when present in large numbers.

Molecules may interact through van der Waals forces and hydrophobic forces

Even electrically neutral, nonpolar molecules can develop transient regions of weak positive and negative charge. These slight charges develop as a consequence of the fact that electrons are in constant motion. A region with a temporary excess of electrons will have a weak negative charge, while one with an electron deficit will have a weak positive charge. Adjacent molecules may interact in regions of slight opposite charge. These attractive forces, called **van der Waals forces,** operate over very short distances and are weaker and less specific than the other types of interactions we have considered. They are most important when they occur in large numbers and when the shapes of the molecules involved permit close contact between the atoms. As with hydrogen bonds, the bonding force of a single interaction is very weak. In molecules and structures with a large number of these interactions working together, however, the binding force can be significant.

Hydrophobic ("water-hating") **interactions** occur between groups of nonpolar molecules. Such molecules are insoluble in water and tend to cluster together. This is not due to formation of bonds between the nonpolar molecules, but rather to the fact that the hydrogen-bonded water molecules exclude them and in a sense "drive them together." Hydrophobic interactions explain why oil tends to form globules when added to water.

OXIDATION INVOLVES THE LOSS OF ELECTRONS; REDUCTION INVOLVES THE GAIN OF ELECTRONS

Rusting—the combination of iron (symbol Fe) with oxygen—is a familiar example of oxidation and reduction:

$$4\ Fe + 3\ O_2 \longrightarrow 2\ Fe_2O_3$$

Oxidation and reduction always occur together, but initially we will discuss them separately. **Oxidation** is a chemical process in which an atom, ion, or molecule loses electrons. In rusting, each iron atom becomes oxidized as it loses three electrons.

$$4\ Fe \longrightarrow 4\ Fe^{3+} + 12\ e^-$$

The e^- is a symbol for an electron; the + superscript represents an electron deficit. (When an atom loses an electron, it acquires one unit of positive charge from the excess of one proton. In our example, each iron atom loses three electrons and aquires three units of positive charge.)

You will recall that the oxygen atom is very electronegative, able to remove electrons from other atoms. In this reaction, oxygen gains electrons from iron.

$$3\ O_2 + 12\ e^- \longrightarrow 6\ O^{2-}$$

Oxygen becomes reduced when it accepts electrons from the iron. **Reduction** is a chemical process in which an atom, ion, or molecule *gains* electrons. (The term reduction refers to the fact that the gain of an electron results in the *reduction* of any positive charge that might be present).

Oxidation and reduction reactions occur simultaneously because one substance must accept the electrons that are removed from the other. Oxidation-reduction reactions are sometimes referred to as **redox reactions.** In a redox reaction, one component, the *oxidizing agent,* accepts one or more electrons and becomes reduced. Oxidizing agents other than oxygen are known, but oxygen is such a common one that its name was given to the process. Another reaction component, the *reducing agent,* gives up one or more electrons and becomes oxidized.

In our example there was a complete transfer of electrons from iron to oxygen. Some redox reactions are not so obvious, however. In these, electrons simply move farther from the reducing agent and closer to the oxidizing agent.

Electrons are not easily removed from covalent compounds unless an entire atom is removed. In cells, oxidation often involves the removal of a hydrogen *atom* (an electron plus a proton that "goes along for the ride") from a compound; reduction often involves the addition of hydrogen (see Chapter 6).

Redox reactions are central to many of the energy conversions that go on in a cell. This is because the transfer of an electron also involves the transfer of the energy of that electron. Therefore, a reduced substance has gained not only one or more electrons, but also energy. Both cellular respiration (Chapter 7) and photosynthesis (Chapter 8) are essentially redox processes.

WATER IS ESSENTIAL TO LIFE

A large part of the mass of most organisms is water. In human tissues the percentage of water ranges from 20% in bones to 85% in brain cells. About 70% of our total body weight is water; as much as 95% of a jellyfish and certain plants is water. Water is the source, through photosynthesis (see Chapter 8), of the oxygen in the air we breathe, and its hydrogen atoms become incorporated into many organic compounds. Water is also the solvent for most biological reactions and a reactant or product in many chemical reactions.

FIGURE 2–13 Earth is sometimes referred to as the *water planet* because most of its surface is covered with water. Here, Earth is seen from Apollo II, about 182,000 km away. (NASA)

Water is not only important inside organisms but also is one of the principal environmental factors affecting them. Many organisms live within the sea or in freshwater rivers, lakes, or puddles. Water's unique combination of physical and chemical properties has permitted living things to originate, survive, and evolve on Earth (Fig. 2–13).

Water molecules are polar

As discussed previously, water molecules are polar, that is, one end of each molecule bears a partial positive charge and the other a partial negative charge (see Fig. 2–7*c*). The water molecules in liquid water and in ice associate by hydrogen bonds. The hydrogen atom of one water molecule, with its partial positive charge, is attracted to the oxygen atom of a neighboring water molecule, with its partial negative charge, forming a hydrogen bond. Each water molecule can form hydrogen bonds with a maximum of four neighboring water molecules (Fig. 2–14).

Water is the principal solvent in organisms

Because its molecules are polar, water is an excellent solvent, a liquid capable of dissolving many different kinds

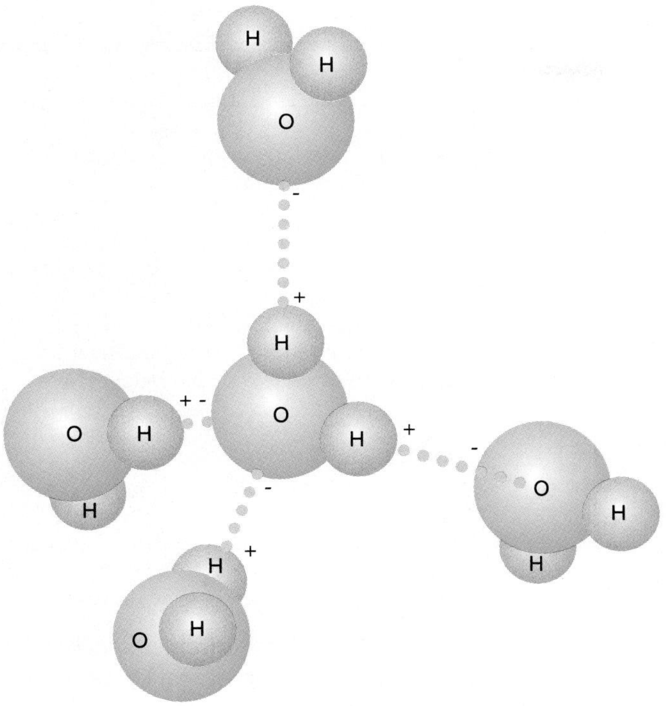

FIGURE 2–14 Each water molecule can form hydrogen bonds with up to four neighboring water molecules. The hydrogen bonds are indicated by dotted lines.

of substances, especially polar and ionic compounds. Previously in this chapter, we discussed how polar water molecules pull the ions of ionic compounds apart so that they dissociate (see Fig. 2–11). Because of its solvent properties and the tendency of the atoms in certain compounds to form ions when in solution, water plays an important role in facilitating chemical reactions. Substances that interact readily with water are said to be **hydrophilic** ("water-loving"). Not all substances in organisms are hydrophilic, however. Many **hydrophobic** ("water-hating") substances found in living things are especially important because of their ability to form structures that are not dissolved by water.

Hydrogen-bonding makes water cohesive and adhesive

Water molecules have a very strong tendency to stick to each other; that is, they are **cohesive.** This is due to the hydrogen bonds among the molecules. Water molecules also stick to many other kinds of substances, most notably those with charged groups of atoms or molecules on their surfaces. These **adhesive forces** explain how water makes things wet.

A combination of adhesive and cohesive forces accounts for the tendency, termed **capillary action,** of water to move in narrow tubes, even against the force of gravity (Fig. 2–15). For example, water moves through the microscopic spaces between soil particles to the roots

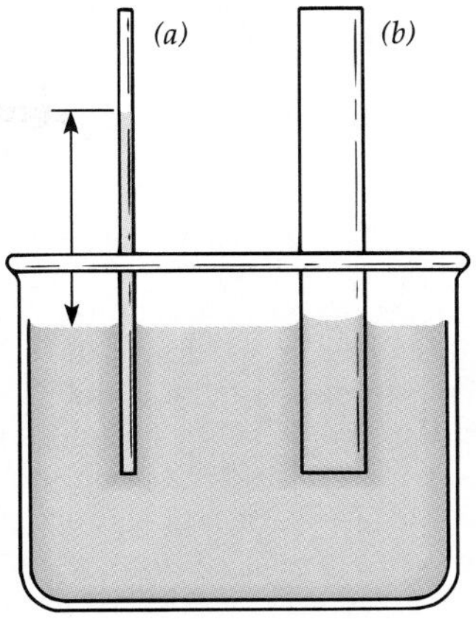

FIGURE 2–15 The cohesive and adhesive properties of water account for capillary action. (*a*) In the smaller tube, adhesive forces attract water molecules to charged groups on the surfaces of the tube. Other water molecules inside the tube are then "pulled along" by cohesive forces, which are actually due to hydrogen bonds between the water molecules. (*b*) In the large-diameter tube, a smaller percentage of the water molecules line the glass. Because of this, the adhesive forces are not strong enough to overcome the cohesive forces of the water beneath the surface level of the container, and water in the tube rises only slightly.

of plants by capillary action. Because of the cohesive nature of water molecules, any force exerted on part of a column of water will be transmitted to the column as a whole. The major mechanism of water movement in plants (see Chapter 33) depends on this fact.

Water has a high degree of **surface tension** because of the cohesiveness of its molecules, which have a much greater attraction for each other than for molecules in the air. Thus, water molecules at the surface crowd together, producing a strong layer as they are pulled downward by the attraction of other water molecules beneath them (Fig. 2–16).

FIGURE 2–16 Water has a very high surface tension because of the strength of its hydrogen bonds. These water striders, although more dense than water, can walk on the surface of a pond. Fine hairs at the ends of their legs spread their weight over a large area, allowing their bodies to be supported by the surface tension of the water. (Dennis Drenner)

Making the Connection

Hydrogen Bonding and the Environment

Why does ice float? This is because hydrogen bonds contribute another important property of water. Liquid water expands as it freezes, due to the fact that the hydrogen bonds joining the water molecules in the crystalline lattice keep the molecules far enough apart to give ice a density about 10% less than the density of liquid water (see Fig. 2–17). When ice has been heated enough to raise its temperature above 0°C (32°F), hydrogen bonds between the water molecules are broken, freeing the molecules to slip closer together. The density of water is greatest at 4°C, above which water begins to expand again as the speed of its molecules increases. As a result, ice floats on the denser cold water.

This unusual property of water has been important in enabling life as we know it to appear, survive, and evolve on the Earth. If ice had a greater density than water, it would sink; eventually all ponds, lakes, and even oceans would freeze solid from the bottom to the surface, making life impossible. When a body of deep water cools, it becomes covered with floating ice. The ice insulates the liquid water below it, preventing it from freezing and permitting a variety of animals and plants to survive below the icy surface. ▲

Water helps maintain a stable temperature

Raising the temperature of a substance involves adding heat energy to make its molecules move faster—that is, to increase the **kinetic energy** (energy of motion) of the molecules (see Chapter 6). The term **heat** refers to the *total* amount of kinetic energy in a sample of a substance; **temperature** refers to the *average* kinetic energy of the particles. Water has a high **specific heat;** that is, the amount of energy required to raise the temperature of water is quite large. A **calorie** is a unit of heat energy (equivalent to 4.184 joules) that equals the amount of heat required to raise the temperature of 1 gram of water 1 degree Celsius. The specific heat of water is therefore 1 calorie per gram of water per degree Celsius. Most other common substances have much lower specific heat values.

The high specific heat of water results from the hydrogen-bonding of its molecules. Some of the hydrogen bonds holding the water molecules together must first be broken to permit the molecules to move more freely. Much of the energy added to the system is used up in breaking the hydrogen bonds, and only a portion of the heat energy is available to speed the movement of the water molecules (thereby increasing the temperature of the water). Conversely, when liquid water changes to ice, additional hydrogen bonds must be formed, liberating a great deal of heat into the environment.

Because so much heat input is required to raise the temperature of water (and so much heat is lost when the temperature is lowered), oceans and other large bodies of water have relatively constant temperatures. Thus, many organisms living in the oceans are provided with a relatively constant environmental temperature. The properties of water are crucial in stabilizing temperatures on the surface of the Earth. Although surface water is only a thin film relative to Earth's volume, the quantity is enormous compared to the exposed land mass. This relatively large mass of water resists both the warming effect of heat and the cooling effect of low temperatures. In addition, hydrogen-bonding gives ice unique properties that have important environmental consequences (see *Making the Connection: Hydrogen Bonding and the Environment*).

The high water content of organisms helps them maintain relatively constant internal temperatures. Such minimizing of temperature fluctuations is important because biological reactions can take place only within a relatively narrow temperature range.

Because its molecules are held together by hydrogen bonds, water has a high **heat of vaporization.** To change 1 gram of liquid water into 1 gram of water vapor, 540 calories of heat are required. The heat of vaporization of most other common substances is much less. As a sample of water is heated, some molecules are moving much faster than others (that is, they have more heat energy). These faster-moving molecules are more likely to escape the liquid phase and enter the vapor phase (Fig. 2–17). When they do, they take their heat energy with them (thus lowering the temperature of the sample). For this reason the human body can dissipate excess heat as sweat evaporates from the skin, and a leaf can keep cool in the bright sunlight as water evaporates from its surface.

ACIDS ARE PROTON DONORS; BASES ARE PROTON ACCEPTORS

Water molecules have a slight tendency to **ionize**—that is, to dissociate into hydrogen ions (H^+) and hydroxide ions (OH^-).[3] In pure water, a very small number of wa-

[3]The H^+ immediately combines with a negatively charged region of a water molecule, forming a hydronium ion (H_3O^+). However, by convention H^+, rather than the more accurate H_3O^+, is used.

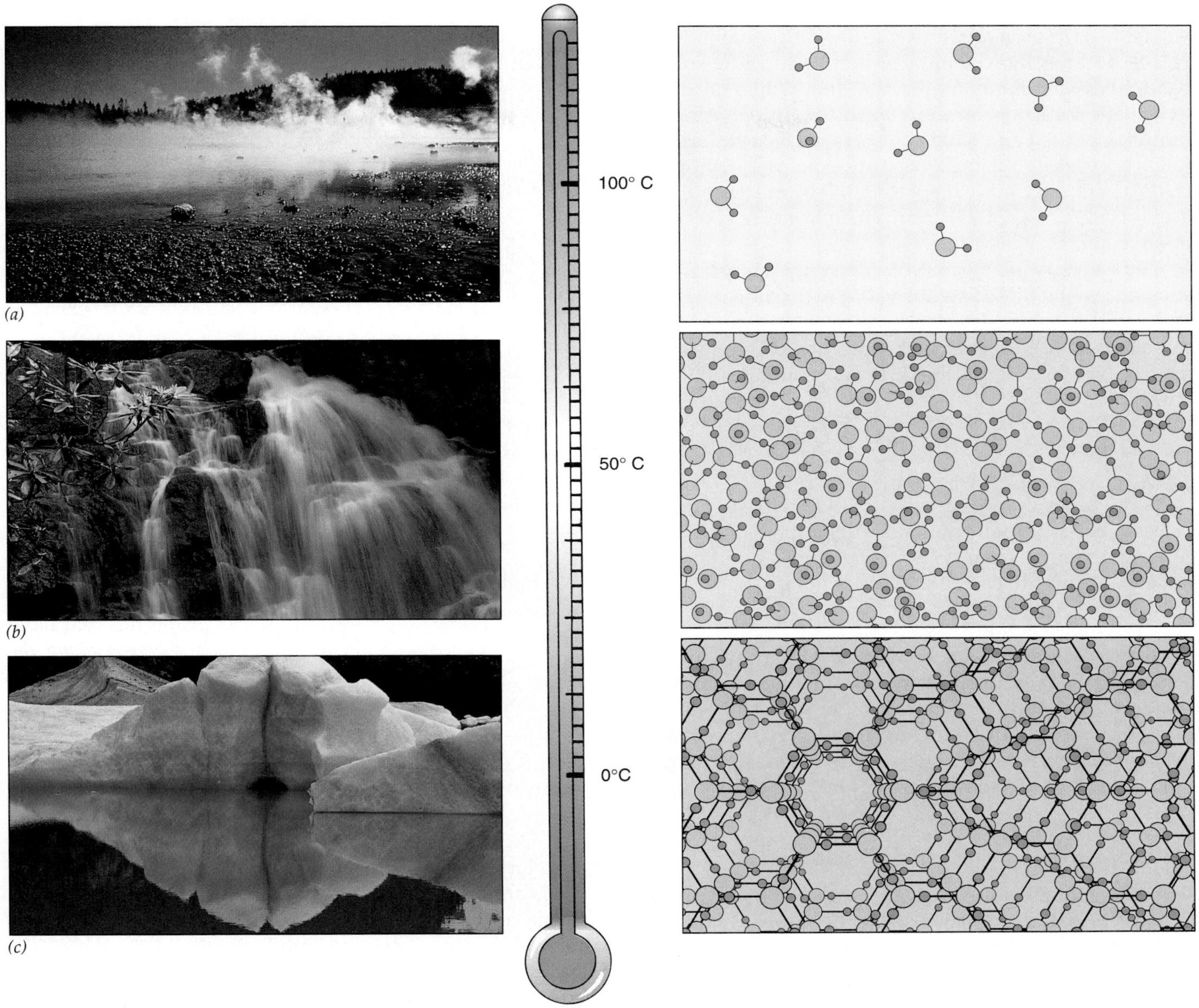

FIGURE 2–17 Water can exist as a gas (water vapor), a liquid, or a solid (ice). These three forms differ in the extent of hydrogen bond formation. (***a***) When water boils, as in this hot spring at Yellowstone National Park, hydrogen bonds are broken. The water molecules in the steam that forms are far apart and move freely and rapidly. (***b***) Water molecules in a liquid state continually form, break, and re-form hydrogen bonds with each other. (***c***) In ice, each water molecule participates in four hydrogen bonds with adjacent molecules, resulting in a regular, evenly distanced crystalline lattice structure. The water molecules move apart slightly as the hydrogen bonds form. For this reason, water expands as it freezes, making ice one of the very few substances that is less dense in its solid form than in its liquid form. Thus ice, as in these icebergs at Portage Glacier, Alaska, floats on water instead of accumulating on the bottom. Note the regular, evenly distanced hydrogen bonds in the superstructure of ice.
(*a,* Woodbridge Wilson/National Park Service; *b,* Gary R. Bonner; *c,* Barbara O'Donnell/Biological Photo Service)

ter molecules ionize. This slight tendency of water to dissociate is reversible as hydrogen ions and hydroxide ions reunite to form water:

$$HOH \rightleftarrows H^+ + OH^-$$

Because each water molecule splits into one hydrogen ion and one hydroxide ion, the concentrations of hydrogen ions and hydroxide ions in pure water are exactly equal (0.0000001 or 10^{-7} moles per liter for each ion). Such a solution is said to be **neutral,** neither acidic nor basic (alkaline).

An **acid** is a substance that dissociates in solution to yield hydrogen ions (H^+) and an anion.

$$\text{Acid} \longrightarrow H^+ + \text{anion}$$

An acid is a proton *donor.* (Recall that a hydrogen ion, or H^+, is nothing more than a proton.) An acidic solution has a hydrogen ion concentration that is higher than its hydroxide ion concentration. Acidic solutions turn blue litmus paper red and have a sour taste. Hydrochloric acid (HCl) and sulfuric acid (H_2SO_4) are examples of inorganic acids. Lactic acid ($CH_3CHOHCOOH$) from sour milk and acetic acid (CH_3COOH) from vinegar are two common organic acids.

A **base** is defined as a proton *acceptor.* Most bases are substances that dissociate to yield a hydroxide ion (OH^-) and a cation when dissolved in water. A hydroxide ion can act as a base by accepting a proton (H^+) to form water. Sodium hydroxide (NaOH) is a common inorganic base.

$$NaOH \longrightarrow Na^+ + OH^-$$
$$OH^- + H^+ \longrightarrow H_2O$$

Some bases do not dissociate to yield hydroxide ions directly. For example, ammonia (NH_3) acts as base by accepting a proton from water, producing an ammonium ion ($NH_4{}^+$) and releasing a hydroxide ion.

$$NH_3 + H_2O \longrightarrow NH_4^+ + OH^-$$

A basic solution is one in which the hydrogen ion concentration is lower than the hydroxide ion concentration. Basic solutions turn red litmus paper blue and feel slippery to the touch. In later chapters we shall encounter a number of organic bases such as the purine and pyrimidine bases that are components of nucleic acids.

pH is a convenient measure of acidity

The degree of a solution's acidity is generally expressed in terms of **pH,** defined as the *negative logarithm (base 10) of the hydrogen ion concentration (expressed in moles per liter).*

$$pH = -\log_{10} [H^+]$$

The brackets refer to concentration; therefore the term $[H^+]$ means "the concentration of hydrogen ions," which is expressed in moles per liter because we are interested in the *number* of hydrogen ions per liter. Because the range of possible pH values is very broad, a logarithmic scale (with a ten-fold difference between successive units) is more convenient than a linear scale.

Hydrogen ion concentrations are nearly always less than 1 mole per liter. One gram of hydrogen ions dissolved in 1 liter of water (a 1-molar solution) may not sound very impressive, but such a solution would be extremely acidic. The logarithm of a number less than one is a negative number; thus the *negative* logarithm corresponds to a *positive* pH value.

Whole number pH values are easy to calculate (Table 2–2). For instance, consider our example of pure water, which has a hydrogen ion concentration of 0.0000001 (10^{-7}) moles per liter. The logarithm is −7. The negative logarithm is 7; therefore the pH is 7.

Table 2–2 THE RELATION OF pH TO HYDROGEN ION CONCENTRATION

Substance	$[H^+]$*	log $[H^+]$	pH
Gastric juice	0.01, 10^{-2}	−2	2
Pure water, neutral solution	0.0000001, 10^{-7}	−7	7
Household ammonia	0.00000000001, 10^{-11}	−11	11

*$[H^+]$ = hydrogen ion concentration (moles/L)

If the hydrogen ion concentration of a solution is known, the hydroxide ion concentration can be easily calculated. The product of the hydrogen ion concentration and the hydroxide ion concentration is 1×10^{-14}.

$$[H^+]\,[OH^-] = 1 \times 10^{-14}$$

In pure (freshly distilled) water, the hydrogen ion concentration is 10^{-7}; therefore the hydroxide concentration is also 10^{-7}. Such a solution, in which the concentrations are equal, is said to be neutral. Acidic solutions (those with an excess of hydrogen ions) have pH values smaller than 7; basic solutions (those with an excess of hydroxide ions) have pH values greater than 7. A more acidic solution has a lower pH value. The hydrogen ion concentration of a solution with pH 1 is ten times that of a solution with pH 2.

The pH values of some common substances are shown in Figure 2–18. The contents of most animal and plant cells are neither strongly acidic nor strongly basic but instead are an essentially neutral mixture of acidic and basic substances. Any substantial change in the pH of the cell is incompatible with life (Fig. 2–19). The pH of living cells ordinarily ranges from around 7.2 to 7.4.

Buffers minimize pH change

Many homeostatic mechanisms operate to maintain appropriate pH values. For example, the pH of human blood is about 7.4 and must be maintained within very narrow limits. Should the blood become too acidic (for example, as a result of respiratory disease), coma and death may result. Excessive alkalinity can result in overexcitability of the nervous system and even convulsions.

Organisms contain many natural buffers. A **buffer** is a substance or combination of substances that resists changes in pH when an acid or base is added. A buffering system includes a weak acid or a weak base (Fig. 2–20). A weak acid or weak base does not ionize completely. That is, at any given instant only a fraction of the molecules are ionized; most are undissociated.

One of the most common buffering systems is important in human blood (see Chapter 44). Carbon diox-

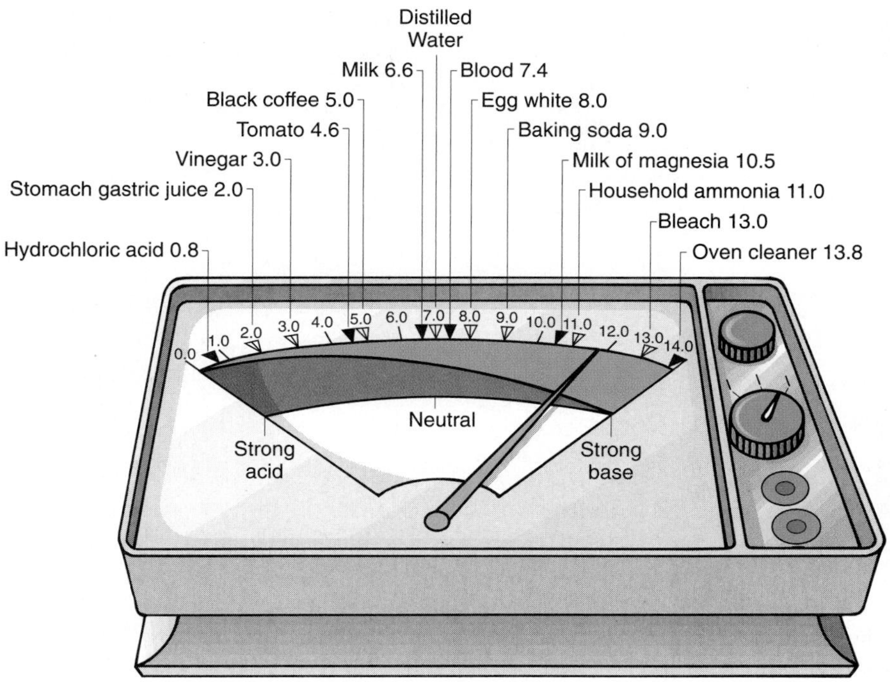

FIGURE 2–18 A pH meter is an electronic device used to measure the acidity of a solution. A solution with a pH of 7 is neutral because the concentrations of H^+ and OH^- are equal. As the pH value falls below 7, the H^+ concentration increases, and the solution becomes more acidic. As the pH rises above 7, the concentration of H^+ ions decreases and the concentration of OH^- increases, making the solution less acidic (more basic).

ide, produced as a waste product of cellular metabolism, enters the blood, the main constituent of which is water. The carbon dioxide reacts with the water to form carbonic acid, a weak acid that dissociates to yield a hydrogen ion and a bicarbonate ion. The buffering system is described by the following expression:

$$\underset{\text{Carbon dioxide}}{CO_2} + \underset{\text{Water}}{H_2O} \rightleftharpoons \underset{\text{Carbonic acid}}{H_2CO_3} \rightleftharpoons H^+ + \underset{\text{Bicarbonate ion}}{HCO_3^-}$$

Note that the expression is not an equation and does not need to be "balanced." As indicated by the arrows, all the reactions are reversible. Because carbonic acid is a weak acid, undissociated molecules are always present, as are all the other components of the system. The expression describes the system when it is at equilibrium, when the rates of the forward and reverse reactions are equal and the relative concentrations of the components are not changing. If a system is at equilibrium, it can be "shifted to the right" by adding reactants or removing

FIGURE 2–19 The trees shown here (photographed in the Black Forest, Germany) are casualties of acid rain. Sulfur oxides, emitted from fossil fuel plants and industry, and nitrogen oxides, mainly from automobile exhaust, are converted in the moist atmosphere into acids of, respectively, sulfur and nitrogen, such as sulfurous and nitrous acid. These acids are dispersed over wide areas by airflow patterns in the atmosphere. Whereas the pH of unpolluted rain averages 5.6, the pH of acid rain has been measured at 4.2 and even lower. (Spencer Swanger/Tom Stack and Associates).

FIGURE 2–20 Buffering is sometimes used as a remedy for excess stomach acid. The bubbles are CO_2 from the reaction between an acid (citric acid) and the bicarbonate ion (HCO_3^-) from sodium bicarbonate. (Charles D. Winters)

Table 2–3 SOME BIOLOGICALLY IMPORTANT IONS

Name	*Formula*	*Charge*
Sodium	Na^+	1+
Potassium	K^+	1+
Hydrogen	H^+	1+
Magnesium	Mg^{2+}	2+
Calcium	Ca^{2+}	2+
Iron	Fe^{2+} or Fe^{3+}	2+ [iron(II)] or 3+ [iron(III)]
Ammonium	NH_4^+	1+
Chloride	Cl^-	1−
Iodide	I^-	1−
Carbonate	CO_3^{2-}	2−
Bicarbonate	HCO_3^-	1−
Phosphate	PO_4^{3-}	3−
Acetate	CH_3COO^-	1−
Sulfate	SO_4^{2-}	2−
Hydroxide	OH^-	1−
Nitrate	NO_3^-	1−
Nitrite	NO_2^-	1−

products. Conversely, it can be "shifted to the left" by adding products or removing reactants. Hydrogen ions are the important products to consider in this system.

The addition of excess hydrogen ions has the effect of temporarily shifting the system to the left, as they combine with the bicarbonate ions to form carbonic acid. Eventually a new equilibrium is established; at this point the hydrogen ion concentration is similar to the original concentration.

If hydroxide ions are added, they combine with the hydrogen ions to form water, effectively removing a product and thus shifting the system to the right. More carbonic acid then ionizes, replacing the hydrogen ions that were removed.

Organisms contain many weak acids and weak bases, thus maintaining an essential reserve of buffering capacity and avoiding pH extremes.

An acid and a base react to form a salt

When an acid and a base are mixed together, the H^+ of the acid unites with the OH^- of the base to form a molecule of water. The remainder of the acid (an anion) combines with the remainder of the base (a cation) to form a salt. For example, hydrochloric acid reacts with sodium hydroxide to form water and sodium chloride:

$$HCl + NaOH \longrightarrow H_2O + NaCl$$

A **salt** is a compound in which the hydrogen ion of an acid is replaced by some other cation. Sodium chloride, NaCl, is a compound in which the hydrogen ion of HCl has been replaced by the cation Na^+.

When a salt, an acid, or a base is dissolved in water, its dissociated ions can conduct an electrical current; these substances are called **electrolytes.** Sugars, alcohols, and many other substances do not form ions when dissolved in water; they do not conduct an electrical current and are referred to as **nonelectrolytes.**

Cells and extracellular fluids (such as blood) of animals and plants contain a variety of dissolved salts that are the source of the many important mineral ions essential for fluid balance and acid-base balance. Nitrates and ammonium from the soil are the important nitrogen sources for plants. In animals, nerve and muscle function, blood clotting, bone formation, and many other aspects of body function depend on ions. Sodium, potassium, calcium, and magnesium are the chief cations present; chloride, bicarbonate, phosphate, and sulfate are important anions (Table 2–3).

The body fluids of terrestrial animals differ considerably from sea water in their total salt content. However, they tend to resemble sea water in the kinds of salts present and in their relative abundances. The total concentration of salts in the body fluids of most invertebrate marine animals is equivalent to that in sea water, about 3.4%. Vertebrates, whether terrestrial, freshwater, or marine, have less than 1% salt in their body fluids.

Although the concentration of salts in cells and body fluids of plants and animals is small, the concentrations and relative amounts of the respective cations and anions are kept remarkably constant. Any marked change results in impaired cellular functions and may lead to death.

SUMMARY

I. The chemical composition and metabolic processes of all organisms are very similar; the physical and chemical principles that govern nonliving things also govern organisms.

II. Organisms are made up of small, simple, inorganic compounds as well as large, complex, carbon-containing organic compounds.

III. An element is a substance that cannot be decomposed into simpler substances by chemical reactions.
   A. The matter of the universe is composed of 92 elements, ranging from hydrogen, the lightest, to uranium, the heaviest.
   B. Four elements—carbon, hydrogen, oxygen, and nitrogen—make up 96% or more of an organism's mass.

IV. Atoms are composed of a nucleus containing protons and neutrons. In the space outside the nucleus, electrons move rapidly in orbitals that correspond to energy levels.
   A. An atom is identified by its number of protons (atomic number). Atoms of the same element with different numbers of neutrons (different mass numbers) are isotopes.
   B. In an atom, the number of protons equals the number of electrons.

V. Different atoms are joined by chemical bonds to form compounds.
   A. One mole (the atomic or molecular mass in grams) of any substance contains 6.02×10^{23} units (atoms or molecules).
   B. Covalent bonds are strong, stable bonds formed when atoms share electrons, forming molecules.
      1. Covalent bonds are nonpolar if the electrons are shared equally between the two atoms.
      2. Covalent bonds are polar if one atom is more electronegative (has a greater affinity for electrons) than the other.
   C. An ionic bond is formed between a positively charged cation and a negatively charged anion.
   D. Hydrogen bonds are relatively weak bonds formed when a hydrogen atom with a partial positive charge is attracted to an atom (usually oxygen or nitrogen) with a partial negative charge in another molecule or in another part of the same molecule.

VI. Oxidation and reduction are chemical processes that occur simultaneously. Electrons (and their energy) are transferred from the substance that becomes oxidized to the substance that becomes reduced.

VII. Water accounts for a large part of the mass of most organisms, is important in many chemical reactions within living things, and has unique properties that also affect the environment.
   A. Because its molecules are polar, water is an excellent solvent for ionic or polar solutes.
   B. Water molecules are cohesive because of the hydrogen-bonding between the molecules; water molecules also adhere to substances with ionic or polar regions.
   C. Water has a high specific heat, which helps organisms maintain a relatively constant internal temperature; this property also helps keep the oceans and other large bodies of water at a constant temperature.
   D. Water has a high heat of vaporization. Molecules entering the vapor phase carry a great deal of heat, which accounts for evaporative cooling.
   E. The fact that ice is less dense than liquid water makes the environment less extreme.
   F. Water has a slight tendency to dissociate to form hydrogen ions, or protons (H^+), and hydroxide ions (OH^-).

VIII. Acids are proton (H^+) donors; bases are proton acceptors. Many bases dissociate in solution to yield hydroxide ions, which then accept protons.
   A. The pH scale is a logarithmic expression of the hydrogen ion concentration of a solution. As a solution becomes more acidic, its pH falls below 7 (neutrality). As a solution becomes more basic (alkaline), its pH rises above 7.
   B. A buffering system is based on a weak acid or a weak base. Buffers resist changes in the pH of a solution when acids or bases are added.
   C. A salt is a compound in which the hydrogen atom of an acid is replaced by some other cation. Salts provide the many mineral ions essential for life functions.

SELECTED KEY TERMS

acid	covalent bond	ionic bond	polar covalent bond
anion	electrolyte	isotope	proton
atom	electron	mole	quantum
atomic mass	electronegativity	molecular mass	radioisotope
atomic number	element	molecular weight	redox reaction
atomic weight	heat of vaporization	molecule	reduction
base	hydrogen bond	neutron	salt
buffer	hydrophilic	orbital	valence electrons
calorie	hydrophobic	oxidation	van der Waals forces
cation	ion	pH	

POST-TEST

1. Carbon, oxygen, hydrogen, and nitrogen are the __________ that make up about 96% of the mass of most organisms.
2. Subatomic particles with a negative electric charge and an extremely small mass are __________ .
3. The number of protons in the nucleus, called the __________ __________, is written as a subscript to the left of the chemical symbol.
4. The sum of the protons and the neutrons in the nucleus of the atom, termed the __________ __________, is indicated by a superscript to the left of the chemical symbol.
5. Atoms of the same element containing the same number of protons but different numbers of neutrons are __________ .
6. Each __________ may contain a maximum of two electrons.
7. Particles with one or more units of electrical charge are called __________ ; __________ are positively charged and __________ are negatively charged.
8. Substances that interact with water, such as ions and polar compounds, are __________ ; nonpolar compounds are __________ .
9. __________ electrons in the outer electron shell determine how many electrons an atom can donate, receive, or share.
10. A(an) __________ is composed of two or more atoms joined by covalent chemical bonds.
11. A(an) __________ bond results from the attraction between an anion and a cation.
12. Atoms share electrons in a(an) __________ bond.
13. Water molecules associate with each other by __________ bonds.
14. The amount of energy required to change 1 gram of liquid water to 1 gram of water vapor is termed the __________ __________ __________ of water.
15. A(an) __________ is a proton donor; a(an) __________ is a proton acceptor.
16. An acidic solution has a(an) __________ less than 7.
17. A(an) __________ is a substance that resists change in pH when an acid or base is added.
18. Substances that conduct an electrical current when dissolved in water are known as __________ .

REVIEW QUESTIONS

1. What is the relationship between molecules and compounds? Are all compounds composed of molecules?
2. What are the ways an atom or molecule can become an anion or a cation?
3. What is a radioisotope? Why is it able to substitute for an ordinary (nonradioactive) atom of the same element in a molecule? What are some of the ways radioisotopes are used in biological research?
4. Element *a* has two electrons in its valence shell (which is complete when it contains eight electrons). Would you expect element *a* to share, donate, or accept electrons? What would you expect of element *b*, which has four valence electrons, and element *c*, which has seven?
5. How do ionic and covalent bonds differ?
6. Why does water form hydrogen bonds? Enumerate some of the properties of water that result from hydrogen bonding. How do these properties contribute to the role of water as an essential component of organisms?
7. How can weak forces, such as hydrogen bonds and van der Waals interactions, have significant effects in organisms?
8. A solution has a hydrogen ion concentration of 0.01 moles/liter. What is its pH? What is its hydroxide ion concentration? How would this solution differ from one with a pH of 1?
9. Why are buffers important in organisms? Give a specific example of how a buffering system works.
10. Differentiate clearly among acids, bases, and salts.
11. Why must oxidation and reduction occur simultaneously? Why are redox reactions important in some energy transfers?
12. Describe a reversible reaction that is at equilibrium. What would be the consequences of adding or removing a reactant or a product?

YOU MAKE THE CONNECTION

1. A hydrogen bond formed between two water molecules is only about one-twentieth as strong as a covalent bond between hydrogen and oxygen. In what ways would the physical properties of water be different if these hydrogen bonds were stronger (e.g., one-tenth the strength of covalent bonds)?
2. Consider the following reaction (in water).

$$HCl \longrightarrow H^+ + Cl^-$$

Name the reactant(s) and product(s). Does the expression indicate that the reaction is reversible? Could HCl be used as a buffer?

RECOMMENDED READINGS

Bettelheim, F.A., and J. March. *Introduction to General, Organic and Biochemistry,* 4th ed. Saunders College Publishing, Philadelphia, 1994. A very readable reference text for those who would like to know more about the chemistry basic to life.

Gorham, E. "Neutralizing Acid Rain." *Nature,* Vol. 367, January 27, 1994. A discussion of the current status of acid precipitation.

Zimmer, C. "Wet, Wild and Weird." *Discover,* December 1992. Computer simulations illustrate the many ways water molecules can interact through hydrogen bonding.

CHAPTER 3

The Chemistry of Life: Organic Compounds

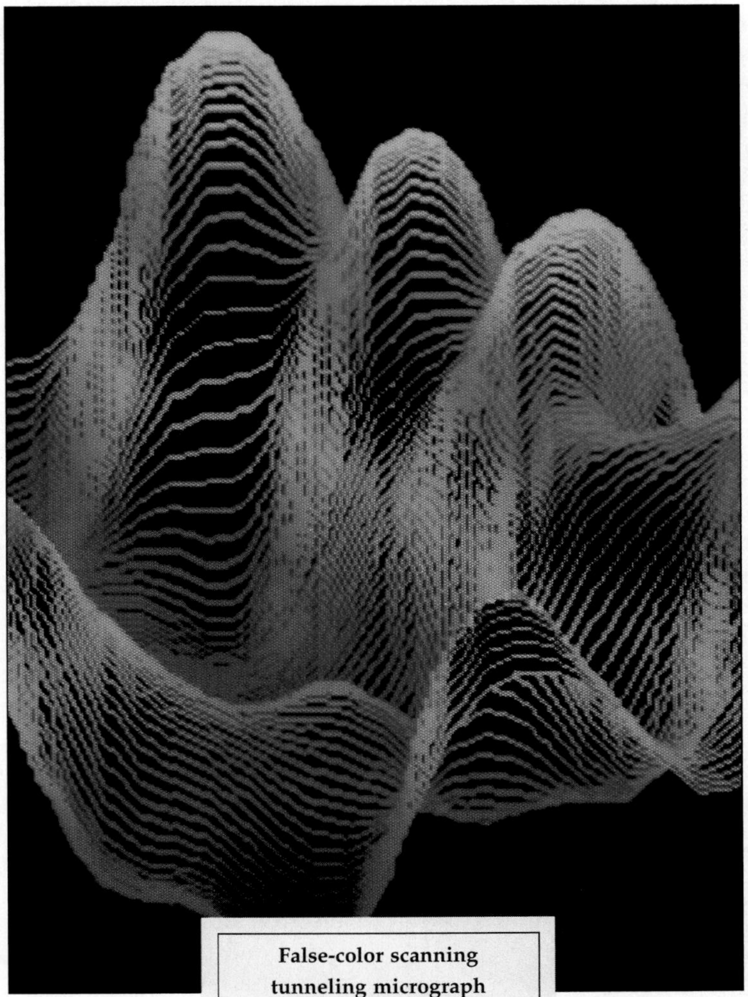

False-color scanning tunneling micrograph (STM) of DNA. An image is formed by scanning a fine point just above the specimen surface and electronically recording the height of the point as it moves. This image shows a section of a double-stranded DNA molecule with the helix coils at the center (row of orange/yellow peaks). (Department of Energy Library/Photo Researchers, Inc.)

Most chemical compounds present in living organisms contain backbones of covalently bonded carbon. These biological molecules are known as **organic compounds** because at one time they were thought to be produced only by living (organic) organisms. In 1928 the German chemist Friedrich Wöhler synthesized urea, a metabolic waste product. Since that time, scientists have learned to synthesize many organic molecules and have discovered organic compounds not found in organisms.

More than five million organic compounds have been identified. Perhaps because it can form a greater variety of molecules than any other element, carbon is the central component of all organic compounds. The carbon atom can form bonds with a greater number of different elements than any other type of atom. Organic compounds consisting only of carbon and hydrogen are known as **hydrocarbons.** Living organisms use hydrocarbon skeletons to build diverse organic compounds. The addition of chemical groups containing other atoms, especially oxygen, nitrogen, phosphorus, and sulfur, can profoundly change the properties of an organic molecule.

Organic compounds are the main structural components of cells and tissues. They participate in and regulate metabolic reactions, transmit information, and provide energy for life processes. Evolution involves chemical changes in the organic compounds produced by organisms. In this chapter we focus on some of the major groups of organic compounds important in organisms, including carbohydrates, lipids, proteins, and nucleic acids (DNA and RNA). Most of these compounds are constructed in the cell from modular subunits. For example, protein molecules are built from smaller compounds called amino acids.

LEARNING OBJECTIVES

After you have studied this chapter you should be able to

1. Distinguish between organic and inorganic compounds.
2. Describe the properties of carbon that make it the central component of organic compounds.
3. Distinguish among the three principal types of isomers.
4. Identify the major functional groups present in organic compounds, and describe their properties.
5. Compare the functions and chemical compositions of the major groups of organic compounds: carbohydrates, fats, proteins, and nucleic acids.
6. Distinguish among monosaccharides, disaccharides, and polysaccharides. Compare storage polysaccharides with structural polysaccharides.
7. Distinguish among neutral fats, phospholipids, and steroids, and describe the compositions, characteristics, and biological functions of each.
8. Sketch the structure of an amino acid. Explain how amino acids are grouped into classes based on the characteristics of their side chains.
9. Distinguish among the levels of organization of protein molecules.
10. Describe the components of a nucleotide. Name some nucleic acids, and discuss the importance of these compounds in living organisms.

CARBON ATOMS CAN FORM AN ENORMOUS VARIETY OF STRUCTURES

Although carbon can exist in simple inorganic form, it has unique properties that permit formation of the large, complex molecules essential to life (Fig. 3–1). A carbon atom has a total of six electrons—two in its lowest energy level and four valence electrons in its highest energy level. This means that a carbon atom can complete its outer shell by forming a total of four covalent bonds. Each bond can link it to another carbon atom or to an atom of a different element. Carbon is particularly well suited to serve as the backbone of a large molecule because carbon-to-carbon bonds are strong and not easily broken.

FIGURE 3–1 Carbon is the basis of organic compounds, of which all living things are made and also occurs in inorganic forms. The human, the plant, the paper, and the wood of the desk all contain organic molecules. Inorganic carbon is represented by the diamond and the pencil "lead," which is a form of elemental carbon called graphite. (Kenneth Knott/Fine Light Photography)

However, they are not so strong that an unreasonable amount of energy is required to break them. Carbon-to-carbon bonds are not limited to single bonds (based on sharing of one electron pair). Two carbon atoms can share two electron pairs with each other, forming double bonds $\left(\rangle C{=}C\langle\right)$. In some compounds, triple carbon-to-carbon bonds $\left(—C{\equiv}C—\right)$ are formed. Carbon chains can be unbranched or branched, and carbon atoms can also be joined into rings (Fig. 3–2). Rings and chains are joined in some compounds.

The shape of a molecule is important in determining its biological properties and function. A carbon-containing molecule has a three-dimensional structure due to the tetrahedral nature of its bond angles. When a carbon atom forms four covalent single bonds with other atoms, the electron orbitals in its outer energy level become elongated and project from the carbon atom toward the corners of a tetrahedron (Fig. 3–3). The angle between any two of these bonds is about 109.5 degrees. This bond angle is similar in diverse organic compounds. Keep in mind that, for simplicity, many of the figures in this book are drawn as two-dimensional graphic representations of three-dimensional molecules. For example, hydrocarbon chains, such as those seen in Figure 3–2, are not actually straight, but have a three-dimensional zigzag structure.

Generally, there is freedom of rotation around each carbon-to-carbon single bond. This property permits organic molecules to be flexible and assume a variety of shapes, depending on the extent to which each single bond is rotated. Double and triple bonds do not permit rotation, so regions of a molecule with such bonds tend to be inflexible.

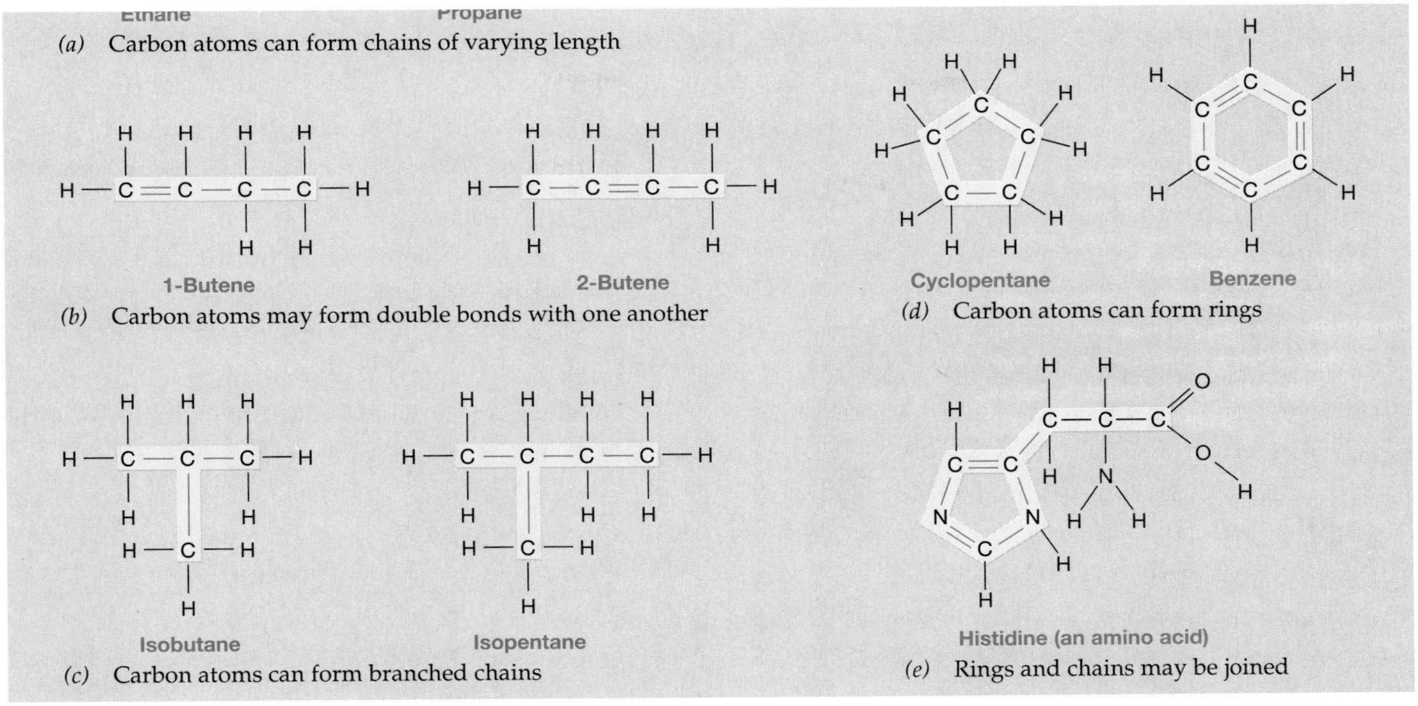

FIGURE 3–2 These structural formulas illustrate common variations in the architecture of organic molecules. Note that each carbon atom has four covalent bonds.

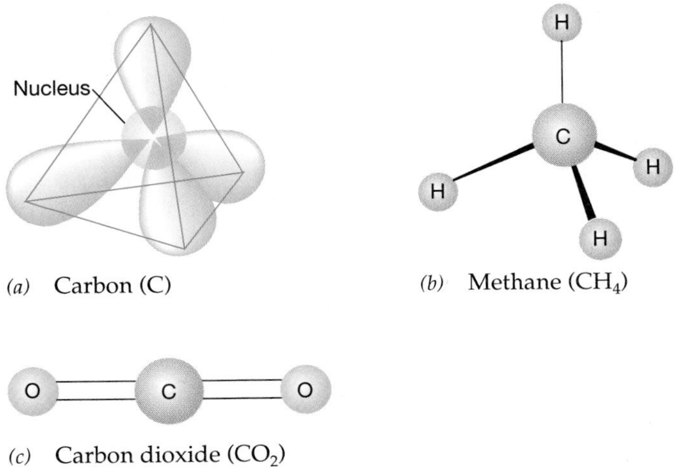

FIGURE 3–3 A carbon atom can form four covalent bonds in three-dimensional space. (*a*) The bonds of a carbon atom point to the four corners of a tetrahedron. This arrangement maximizes the distance between the atoms bonded to the carbon atom. (*b*) Methane consists of a single carbon atom bonded to four hydrogen atoms. The hydrogens are bonded symmetrically around the carbon at the points of a tetrahedron. (*c*) In carbon dioxide, each oxygen atom is connected to the carbon atom by a double bond. Each bond is polar, with the carbon having a partial positive charge and each oxygen having a partial negative cha rge. However, in this linear molecule the bonds point in opposite directions, and the molecule as a whole is nonpolar.

ISOMERS HAVE THE SAME MOLECULAR FORMULA, BUT DIFFER IN STRUCTURE

Compounds with the same molecular formulas but different structures and thus different properties are called **isomers.** Isomers do not have identical physical or chemical properties and may have different common names. Cells can distinguish between isomers. Usually, one isomer is biologically active and the other is not. Three types of isomers are structural isomers, geometric isomers, and enantiomers.

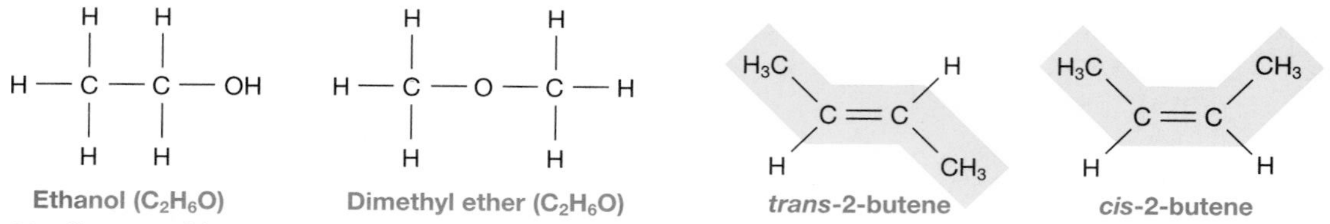

(a) Structural isomers *(b)* Geometric isomers

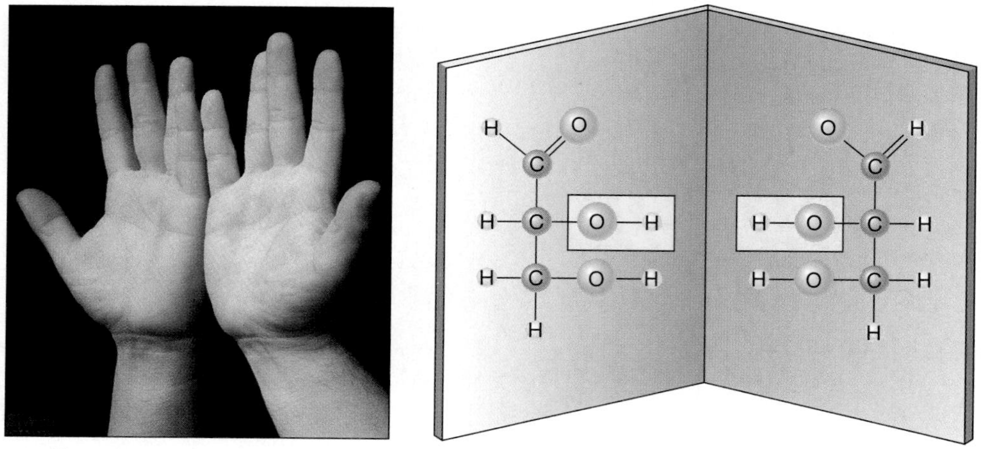

(c) Enantiomers are mirror images *(d)* Enantiomers of glyceraldehyde

FIGURE 3–4 Isomers have the same molecular formula, but their atoms are arranged differently. (***a***) Structural isomers differ in the covalent arrangement of their atoms. (***b***) Geometric, or *cis-trans,* isomers have identical covalent bonds, but differ in the order in which groups of atoms are arranged in space. (***c*** and ***d***) Enantiomers are isomers that are mirror images of one another. (*c,* Dennis Drenner)

Structural isomers are compounds that differ in the covalent arrangements of their atoms (Fig. 3–4*a*). For example, there are two structural isomers of the four-carbon hydrocarbon butane, one with a straight chain and the other with a branched chain (isobutane). Large compounds have more possible structural isomers. There are only two structural isomers of butane, but there can be up to 366,319 isomers of $C_{20}H_{42}$.

Geometric isomers are compounds that are identical in the arrangement of their covalent bonds, but different in the spatial arrangement of groups of atoms. Geometric isomers, also called *cis-trans* isomers, are present in some compounds with carbon-to-carbon double bonds. Because double bonds are not flexible like single bonds, atoms joined to the carbons of a double bond cannot rotate freely about the axis of the bonds. The *cis-trans* isomers may be drawn as shown in Figure 3–4*b*. The designation *cis* indicates that the two larger components are on the same side of the double bond. If they are on opposite sides of the double bond, the compound is designated a *trans* isomer.

Enantiomers are molecules that are mirror images of one another. Recall that the four groups bonded to a single carbon atom are arranged at the vertices of a tetrahedron. If the four bonded groups are all different, the central carbon is described as asymmetric. Figure 3–4*d* illustrates that the four groups can be arranged about the asymmetric carbon in two different ways that are mirror images of each other. The two molecules are enantiomers if they cannot be superimposed on one another no matter how they are rotated in space. Although enantiomers have similar chemical properties and identical physical properties, cells distinguish between them, and usually only one form is found in organisms.

FUNCTIONAL GROUPS CHANGE THE PROPERTIES OF ORGANIC MOLECULES

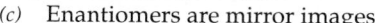

Covalent bonds between hydrogen and carbon are nonpolar. Hydrocarbons therefore lack distinct charged regions. They are quite hydrophobic and do not interact readily with other compounds. However, the characteristics of a molecule can be changed dramatically by replacing one of the hydrogens with a group of atoms known as a **functional group.** Functional groups help determine the types of chemical reactions in which the compound participates. Most functional groups readily form associations, such as ionic and hydrogen bonds, with other molecules. Polar and ionic functional groups are hydrophilic because they associate strongly with polar water molecules.

As illustrated in Table 3–1, each class of organic compounds is characterized by the presence of one or more specific functional groups. Note that the symbol *R* is used to represent the *remainder* of the molecule of which the functional group is a part.

Most compounds present in cells contain two or more different functional groups. When we know what kinds of functional groups are present in an organic compound, we can predict its chemical behavior.

The **hydroxyl group** (abbreviated R—OH) must not be confused with the hydroxide ion (OH^-) discussed in Chapter 2. The hydroxyl group is polar due to the presence of a strongly electronegative oxygen atom. If a hydroxyl group replaces one of the hydrogens of a hydrocarbon, the resulting molecule can have significantly altered properties. For example, ethane (Fig. 3–2*a*) is a hydrocarbon that takes the form of a gas at room temperature. If a hydrogen is replaced by a hydroxyl group, the resulting molecule is ethyl alcohol, or ethanol (Fig. 3–4*a*). Ethanol is somewhat cohesive because the polar hydroxyl groups of adjacent molecules interact; it is therefore liquid at room temperature. Unlike ethane, ethyl alcohol can dissolve in water because the polar hydroxyl groups interact with the polar water molecules.

The **carbonyl group** consists of a carbon atom that has a double covalent bond with an oxygen atom. This double bond is polar because of the electronegativity of the oxygen; thus the group is hydrophilic. The position of the carbonyl group in the molecule determines the class to which the molecule belongs. An **aldehyde** has a carbonyl group positioned at the end of the carbon skeleton (abbreviated R—CO); a **ketone** has an internal carbonyl group (abbreviated R—CO—R).

The **carboxyl group** (abbreviated R—COOH) consists of a carbon atom joined by a double covalent bond to an oxygen atom, and by a single covalent bond to another oxygen, which is in turn bonded to a hydrogen atom. Two electronegative oxygen atoms in such close proximity establish an extremely polarized condition, which can cause the hydrogen atom to be stripped of its electron and released as a hydrogen ion (H^+). The carboxyl group then has one unit of negative charge (R—COO^-).

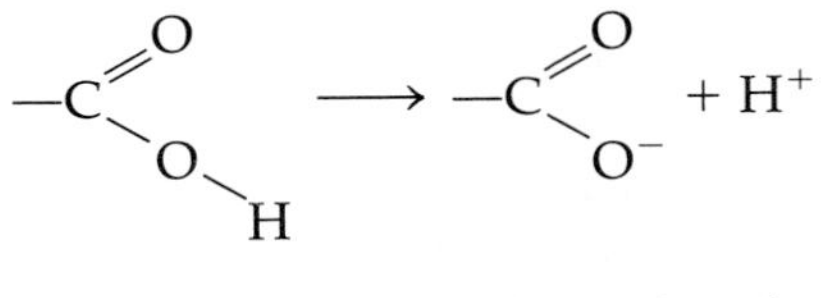

Carboxyl groups are weakly acidic; only a fraction of the molecules ionize in this way. This group can therefore exist in one of two hydrophilic states: ionic or polar. Carboxyl groups are essential constituents of amino acids.

An **amino group** (abbreviated R—NH_2) includes a nitrogen atom covalently bonded to two hydrogen atoms. Amino groups are weakly basic; some of the molecules accept a hydrogen ion (proton), thus acquiring a unit of positive charge. Amino groups are components of amino acids and of nucleic acids.

A **phosphate group** (abbreviated R—PO_4H_2) is weakly acidic. It can release one or two hydrogen ions, resulting in ionized forms with one or two units of negative charge. Phosphates are constituents of nucleic acids and certain lipids. They are very significant in many energy transfer reactions that take place in the cell (see Chapter 6).

The **sulfhydryl group** (abbreviated R—SH), consisting of an atom of sulfur covalently bonded to a hydrogen atom, is found in molecules called *thiols*. As we shall see, amino acids that contain a sulfhydryl group can make important contributions to the structure of proteins.

MANY BIOLOGICAL MOLECULES ARE POLYMERS

Many biological molecules such as proteins and nucleic acids are very large, consisting of thousands of atoms. Such giant molecules are known as **macromolecules.** Most macromolecules are **polymers,** produced by linking small organic compounds called **monomers** (Fig. 3–5). Just as all the words in this book have been written by arranging the 26 letters of the alphabet in various combinations, monomers can be grouped together to form an almost infinite variety of larger molecules. Just as we use different words to convey information, cells use different molecules to convey information. The thousands of different complex organic compounds present in organisms are constructed from about 40 small, simple monomers. For example, the 20 common types of amino acid monomers can be linked end-to-end in countless ways to form the polymers we know as proteins.

Each organism is unique because of differences in the monomer sequence within its DNA, the polymer that constitutes the information in the genes. Cells and tissues within the same organism are also different, due to variations in their component polymers. Muscle tissue and brain tissue differ in large part because of differences in the types and sequences of amino acids in proteins. Ultimately this protein structure is dictated by the sequence of monomers within the DNA of the organism, which is expressed somewhat differently in each cell type.

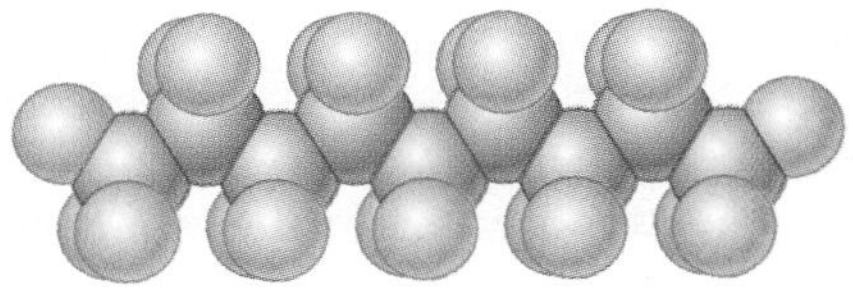

FIGURE 3–5 Monomers may be linked to form polymers. This small polymer of polyethylene is formed by linking two-carbon ethylene (C_2H_4) monomers.

Table 3–1 SOME BIOLOGICALLY IMPORTANT FUNCTIONAL GROUPS

Functional group	*Structural formula*	*Class of compounds characterized by group*	*Description*
Hydroxyl	R—OH	Alcohols 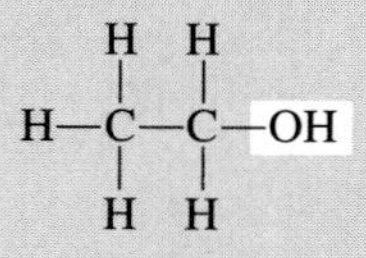 Ethanol	Polar because electronegative oxygen attracts covalent electrons
Amino	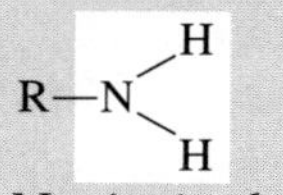 Nonionized 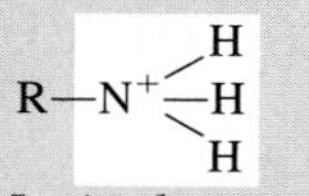 Ionized	Amines 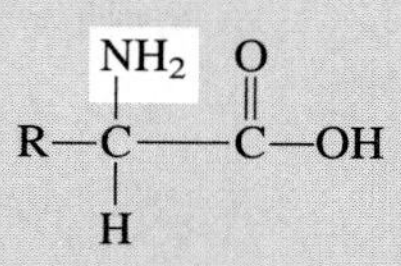 Amino acid	Weakly basic; can accept an H^+ ion
Carbonyl	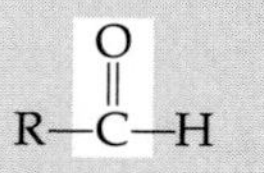	Aldehydes 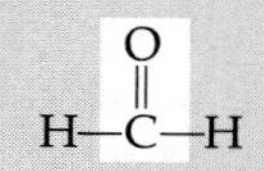 Formaldehyde	Carbonyl group carbon is bonded to at least one H atom; polar because electronegative oxygen attracts covalent electrons
		Ketones 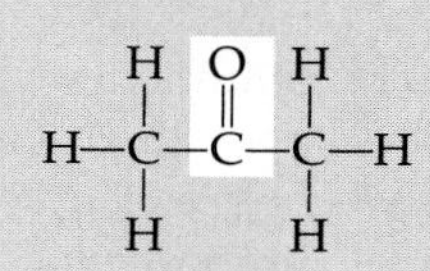 Acetone	Carbonyl group carbon is bonded to two other carbons; polar because electronegative oxygen attracts covalent electrons
Carboxyl	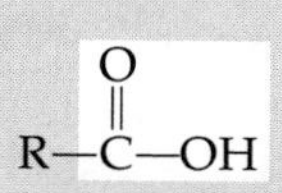 Nonionized 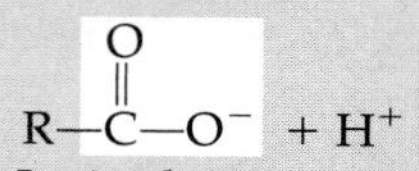 Ionized	Carboxylic acids (organic acids) 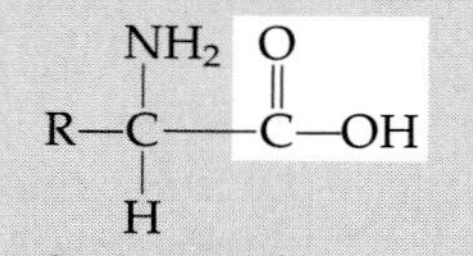 Amino acid	Weakly acidic; can release an H^+ ion

Polymers can be degraded to their component monomers by **hydrolysis** (which means "to break with water"). In a reaction regulated by a specific enzyme,[1] a hydrogen from a water molecule attaches to one monomer, and a hydroxyl from water attaches to the adjacent monomer.

The synthetic process by which monomers are covalently linked is called **condensation.** Because the *equivalent* of a molecule of water is removed during the reactions that combine monomers, the term *dehydration synthesis* is sometimes used to describe the process. However, in biological systems the synthesis of a polymer is not simply the reverse of hydrolysis, even though the net effect is the opposite of hydrolysis. Synthetic processes require energy and are regulated by different enzymes.

[1]An enzyme (see Chapter 6) is a protein catalyst that accelerates a specific chemical reaction.

Table 3–1 continued

Functional group	*Structural formula*	*Class of compounds characterized by group*	*Description*
Methyl	R—CH_3	Component of many organic compounds 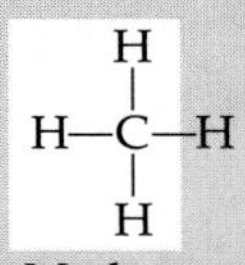 Methane	Hydrocarbon; nonpolar
Phosphate	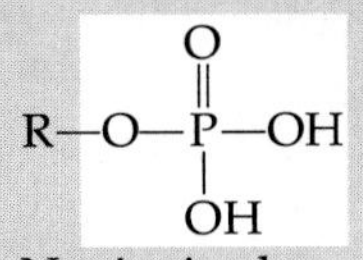 Nonionized 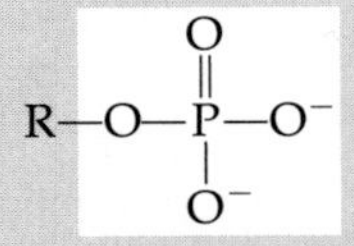 + 2 H^+ Ionized	Organic phosphates 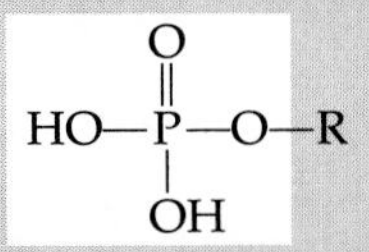 Phosphate ester (as found in ATP)	Weakly acidic; one or two H^+ ions can be released
Sulfhydryl	R—SH	Thiols 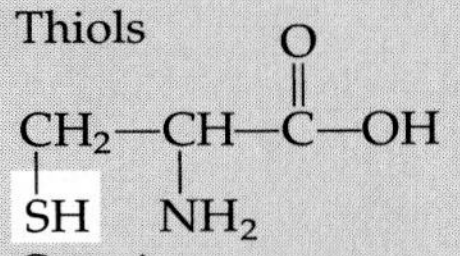 Cysteine	Help stabilize internal structure of proteins

Specific examples of condensation and hydrolysis reactions are presented as we discuss the groups of organic compounds in more detail. The principal groups of biologically important organic compounds are summarized in Table 3–2.

CARBOHYDRATES INCLUDE SUGARS, STARCHES, AND CELLULOSE

Sugars, starches, and cellulose are **carbohydrates.** Sugars and **starches** serve as energy sources for cells; **cellulose** is the main structural component of the walls that surround plant cells. Carbohydrates contain carbon, hydrogen, and oxygen atoms in a ratio of approximately one carbon to two hydrogens to one oxygen $(CH_2O)_n$. The term *carbohydrate,* meaning "hydrate (water) of carbon," reflects the 2:1 ratio of hydrogen to oxygen, the same ratio found in water (H_2O). Carbohydrates contain one sugar unit (*mono*saccharides), two sugar units (*di*saccharides), or many sugar units (*poly*saccharides).

Monosaccharides are simple sugars

Monosaccharides typically contain from three to seven carbon atoms. The simplest carbohydrates are the two three-carbon sugars (trioses): glyceraldehyde and dihydroxyacetone (Fig. 3–6). Ribose and deoxyribose are common pentoses, sugars that contain five carbons; they are components of nucleic acids (DNA, RNA, and related compounds). Glucose, fructose, galactose, and other six-carbon sugars are called **hexoses.** (Note that the names of carbohydrates typically end in *-ose.*)

In a monosaccharide, a hydroxyl group is bonded to each carbon except one; that carbon is double-bonded to an oxygen atom, forming a carbonyl group. If the carbonyl group is at the end of the chain, the monosaccharide is an aldehyde; if the carbonyl group is at any other position, the monosaccharide is a ketone. (By convention, the numbering of the carbon skeleton of a sugar begins with the carbon at or nearest the carbonyl end of the open chain.) The large number of polar hydroxyl groups, plus the carbonyl group, give a monosaccharide hydrophilic properties.

(Text continues on page 56)

Table 3–2 SOME OF THE GROUPS OF BIOLOGICALLY IMPORTANT ORGANIC COMPOUNDS

Class of Compounds	*Component Elements*	*Description*	*How to Recognize*	*Principal Function in Living Systems*
Carbohydrates	C, H, O	Contain approximately 1 C:2 H:1 O (but make allowance for loss of oxygen and hydrogen when sugar units are linked)	Count the carbons, hydrogens, and oxygens.	Cellular fuel; energy storage; structural component of plant cell walls; component of other compounds such as nucleic acids and glycoproteins
		1. Monosaccharides (simple sugars). Mainly five-carbon (pentose) molecules such as ribose or six-carbon (hexose) molecules such as glucose and fructose	Look for the ring shapes: 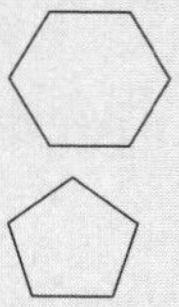	Cellular fuel; components of other compounds
		2. Disaccharides. Two sugar units linked by a glycosidic bond, e.g., maltose, sucrose	Count sugar units.	Components of other compounds; form of sugar transported in plants
		3. Polysaccharides. Many sugar units linked by glycosidic bonds, e.g., glycogen, cellulose	Count sugar units.	Energy storage; structural components of plant cell walls
Lipids	C, H, O	Contain much less oxygen relative to carbon and hydrogen than do carbohydrates		Energy storage; cellullar fuel, structural components of cells; thermal insulation
		1. Neutral fats. Combination of glycerol with one to three fatty acids. Monoacylglycerol contains one fatty acid; diacylglycerol contains two fatty acids; triacylglycerol contains three fatty acids. If fatty acids contain double carbon-to-carbon linkages (C=C), they are unsaturated; otherwise they are saturated	Look for glycerol at one end of molecule: 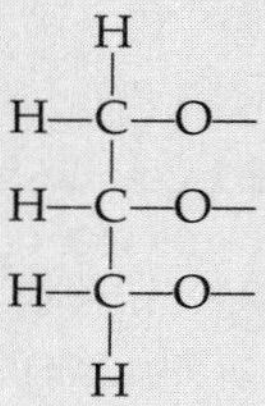	Cellular fuel; energy storage
		2. Phospholipids. Composed of glycerol attached to one or two fatty acids and to an organic base containing phosphorus	Look for glycerol and side chain containing phosphorus and nitrogen.	Components of cell membranes
		3. Steroids. Complex molecules containing carbon atoms arranged in four attached rings (three rings contain six carbon atoms each and the fourth ring contains five)	Look for four attached rings: 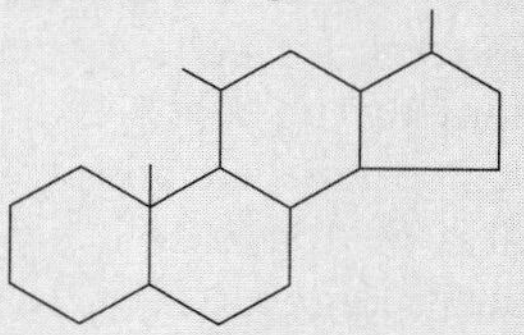	Some are hormones, others include cholesterol, bile salts, vitamin D, components of cell membranes
		4. Carotenoids. Orange and yellow pigments; consist of isoprene units	Look for isoprene units.	Retinal (important in photoreception) and vitamin A are formed from carotenoids.

Table 3–2 continued

Class of Compounds	*Component Elements*	*Description*	*How to Recognize*	*Principal Function in Living Systems*
Proteins	C, H, O, N, usually S	One or more polypeptides (chains of amino acids) coiled or folded in characteristic shapes	Look for amino acid units joined by C—N bonds.	Serve as enzymes; structural components; muscle proteins; hemoglobin
Nucleic acids	C, H, O, N, P	Backbone composed of alternating pentose and phosphate groups, from which nitrogenous bases project. DNA contains the sugar deoxyribose and the bases guanine, cytosine, adenine, and thymine. RNA contains the sugar ribose and the bases guanine, cytosine, adenine, and uracil. Each molecular subunit, called a *nucleotide,* consists of a pentose, a phosphate, and a nitrogenous base.	Look for a pentose-phosphate backbone. DNA forms a double helix.	Storage, transmission, and expression of genetic information

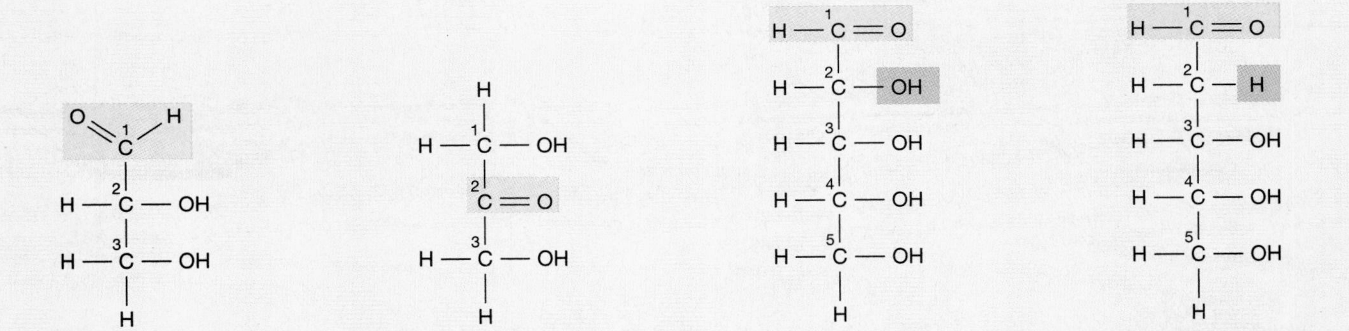

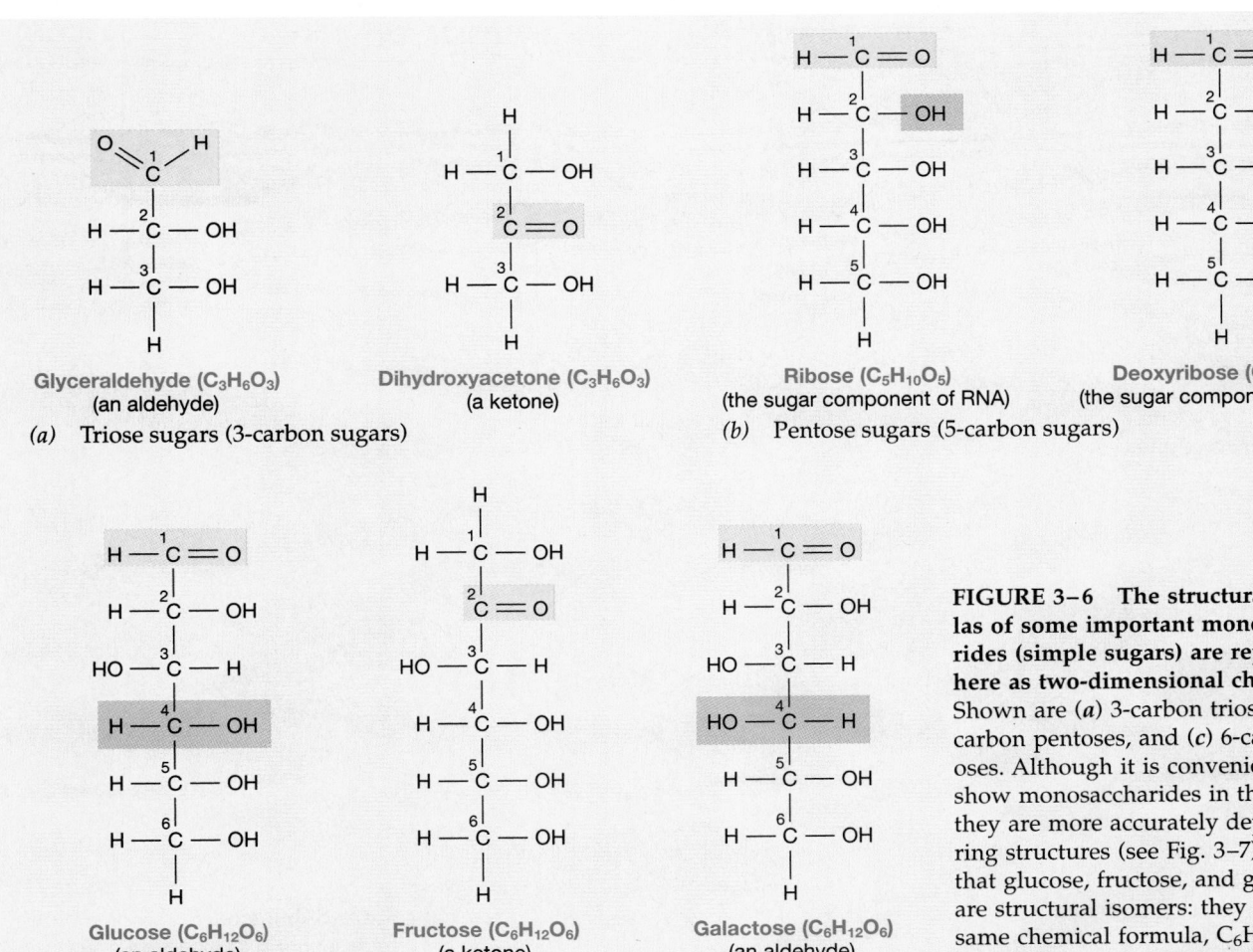

Glyceraldehyde ($C_3H_6O_3$) (an aldehyde)

Dihydroxyacetone ($C_3H_6O_3$) (a ketone)

(*a*) Triose sugars (3-carbon sugars)

Ribose ($C_5H_{10}O_5$) (the sugar component of RNA)

Deoxyribose ($C_5H_{10}O_4$) (the sugar component of DNA)

(*b*) Pentose sugars (5-carbon sugars)

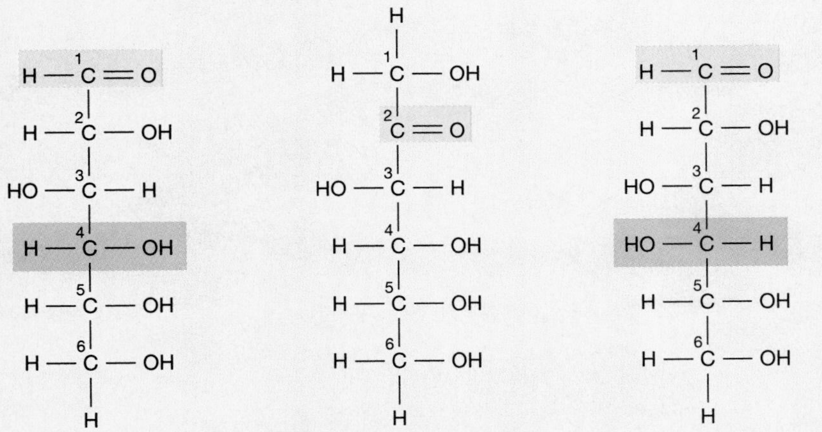

Glucose ($C_6H_{12}O_6$) (an aldehyde)

Fructose ($C_6H_{12}O_6$) (a ketone)

Galactose ($C_6H_{12}O_6$) (an aldehyde)

(*c*) Hexose sugars (6-carbon sugars)

FIGURE 3–6 The structural formulas of some important monosaccharides (simple sugars) are represented here as two-dimensional chains. Shown are (***a***) 3-carbon trioses, (***b***) 5-carbon pentoses, and (***c***) 6-carbon hexoses. Although it is convenient to show monosaccharides in this form, they are more accurately depicted as ring structures (see Fig. 3–7). Note that glucose, fructose, and galactose are structural isomers: they have the same chemical formula, $C_6H_{12}O_6$, but their atoms are arranged differently.

Glucose ($C_6H_{12}O_6$), the most abundant monosaccharide, is extremely important in life processes. During photosynthesis, algae and some plant cells produce glucose from carbon dioxide and water, using sunlight as an energy source. Then, during cellular respiration, cells oxidize the glucose molecule, converting the stored energy to a form that can be readily used for cellular work. Glucose is also used as a component in the synthesis of other types of compounds such as amino acids and fatty acids. So important is glucose in metabolism that its concentration is carefully kept at a homeostatic (relatively constant) level in the blood of humans and other complex animals.

Glucose and fructose are structural isomers: they have identical molecular formulas, but their atoms are arranged differently. In fructose (a ketone) the double-bonded oxygen is linked to a carbon within the chain rather than to a terminal carbon as in glucose (which is an aldehyde). Because of these differences, the two sugars have different properties. For example, fructose tastes sweeter than glucose.

Glucose and galactose differ from each other in another way. Both are hexoses and both are aldehydes, but they differ in the arrangement of their atoms around carbon atom 4. They are mirror images, or enantiomers.

The "stick" formulas in Figure 3–6 give a clear but somewhat unrealistic picture of the structures of some common monosaccharides. As has been discussed, molecules are not the simple two-dimensional structures depicted on a printed page. In fact, the properties of each compound depend in part on its three-dimensional structure. Thus, three-dimensional formulas are helpful in understanding the relationship between molecular structure and biological function. Molecules of glucose and other monosaccharides in solution are actually rings, rather than the extended straight carbon chains shown in Figure 3–6.

Glucose in solution typically exists as a ring of five carbons and one oxygen. It assumes this configuration when a covalent bond connects carbon 1 to the oxygen attached to carbon 5 (Fig. 3–7). When glucose forms a ring,

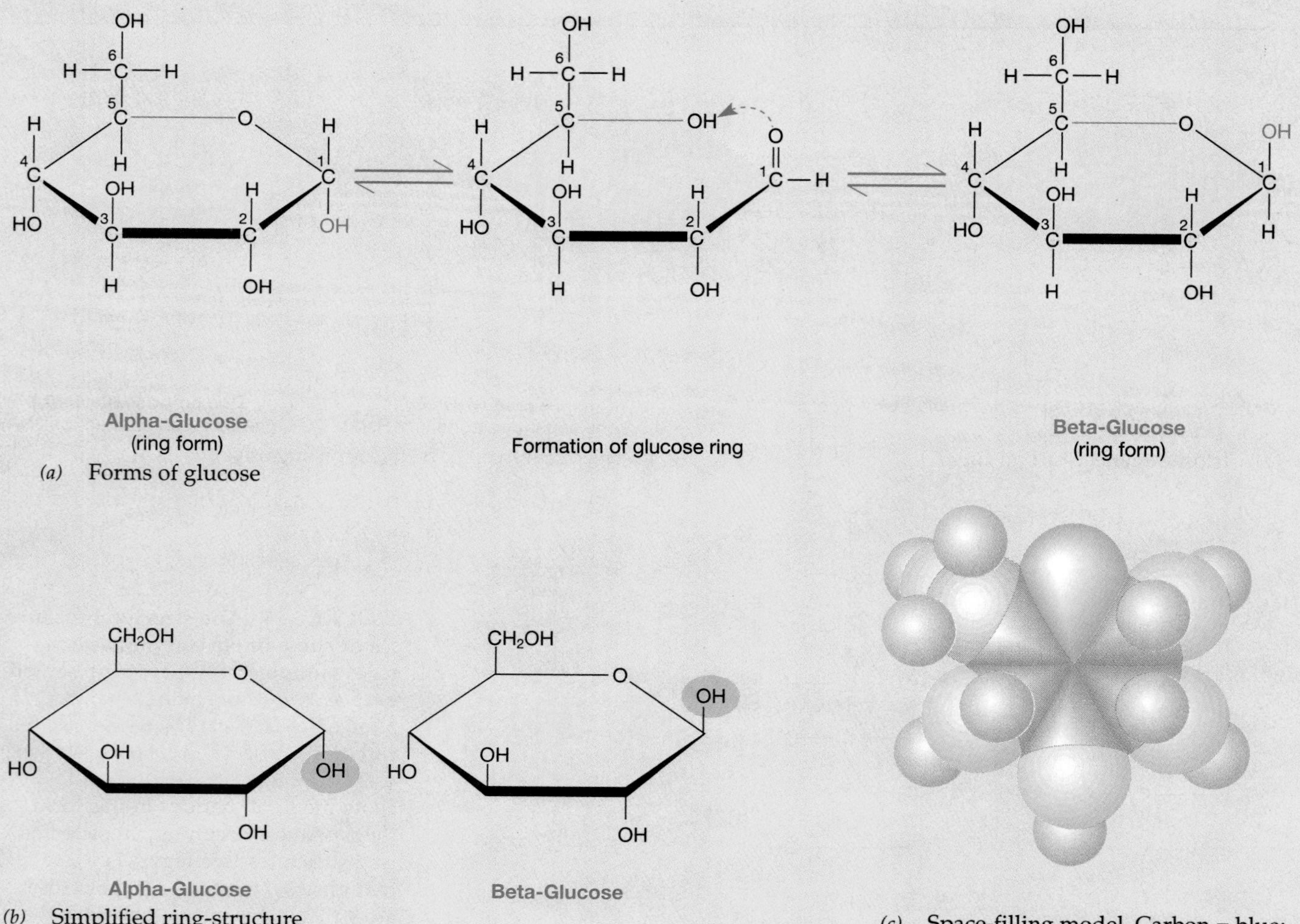

(a) Forms of glucose

(b) Simplified ring-structure

(c) Space-filling model. Carbon = blue; hydrogen = red; oxygen = green.

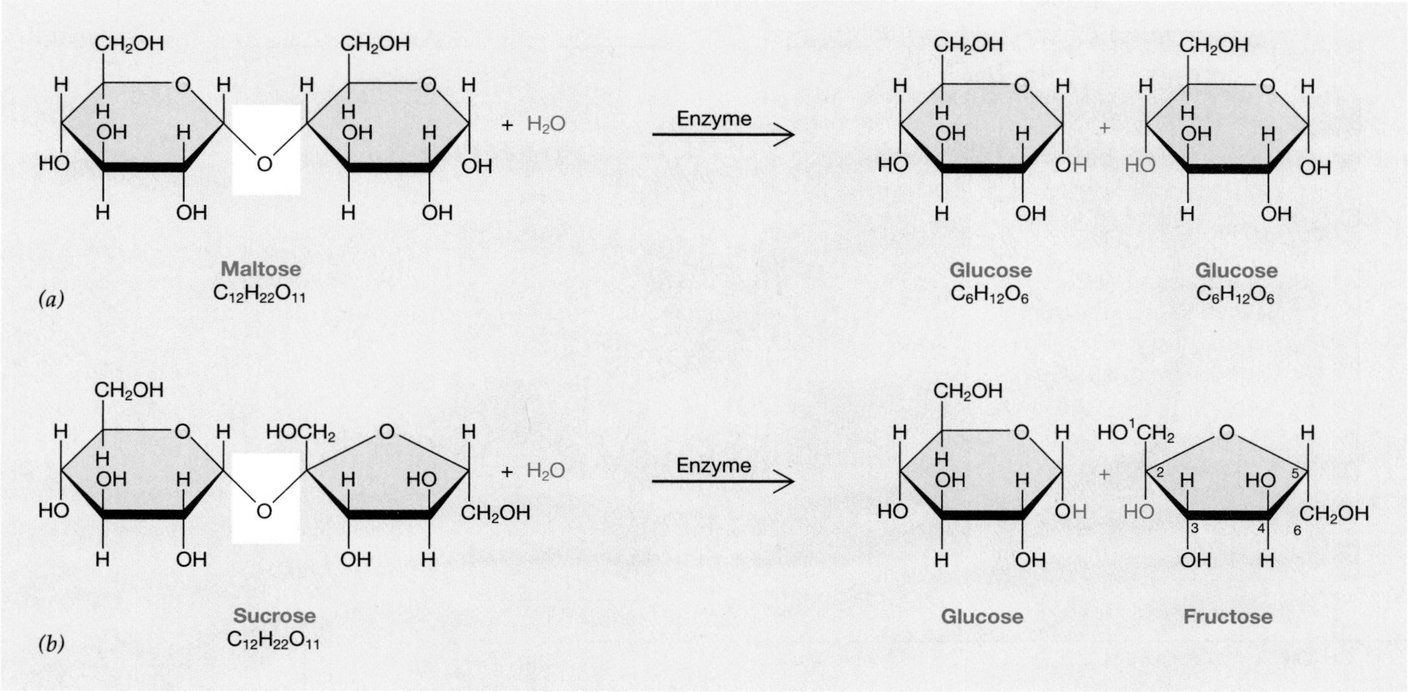

FIGURE 3–8 A disaccharide can be cleaved to yield two monosaccharide units. (*a*) Maltose may be broken down (as it is during digestion) to form two molecules of glucose. The glycosidic linkage is broken in a hydrolysis reaction, which requires the addition of water. (*b*) Sucrose can be hydrolyzed to yield a molecule of glucose and a molecule of fructose. Note that an enzyme (a protein catalyst) is needed to promote these reactions.

two isomeric forms are possible, differing only in the orientation of the hydroxyl (—OH) group attached to carbon 1. When this hydroxyl group is on the same side of the plane of the ring as the $—CH_2OH$ side group, the glucose is designated *β-glucose.* When it is on the side (with respect to the plane of the ring) opposite the $—CH_2OH$ side group, the compound is designated *α-glucose.*

◀ **FIGURE 3–7 Glucose can form an α-ring or a β-ring.** (*a*) When glucose (center) dissolves in water, the molecule bends so that the —OH group on carbon 5 comes close to the —O on carbon 1. The hydrogen moves from one oxygen to the other. This permits carbon 1 to bond with the oxygen on carbon 5, producing a ring structure. Two isomeric forms that differ in the orientation of the —OH group on carbon 1 are possible. In α-glucose, the —OH is on the side of the plane of the ring opposite the $—CH_2OH$ side chain; in β-glucose, the two groups are on the same side of the ring. Although the drawing does not attempt to show the complete three-dimensional structure, the thick, tapered bonds in the lower portion of each ring represent the part of the molecule that would project out of the page toward you. (*b*) A drawing of a simplified structure of glucose, which makes the essential features of the molecule more readily apparent. By convention, a carbon atom is assumed to be present at each angle in the ring unless another atom is shown. Most hydrogen atoms have been omitted. (*c*) A space-filling model of a glucose molecule.

Disaccharides consist of two monosaccharide units

A **disaccharide** (two sugars) contains two monosaccharides joined by a **glycosidic linkage,** consisting of a central oxygen covalently bonded to two carbons, one in each ring. The glycosidic linkage of a disaccharide generally forms between carbon 1 of one molecule and carbon 4 of the other molecule. The disaccharide maltose (malt sugar) consists of two covalently linked α-glucose units. Sucrose, common table sugar, consists of a glucose unit combined with a fructose unit. Lactose (the sugar present in milk) is composed of one molecule of glucose and one of galactose.

A disaccharide can be hydrolyzed—that is, split by the addition of water—into two monosaccharide units. During digestion, maltose is hydrolyzed to form two molecules of glucose:

$$\text{Maltose} + \text{Water} \longrightarrow \text{Glucose} + \text{Glucose}$$

Similarly, sucrose is hydrolyzed to form glucose and fructose:

$$\text{Sucrose} + \text{Water} \longrightarrow \text{Glucose} + \text{Fructose}$$

Structural formulas for the compounds in these reactions are shown in Figure 3–8.

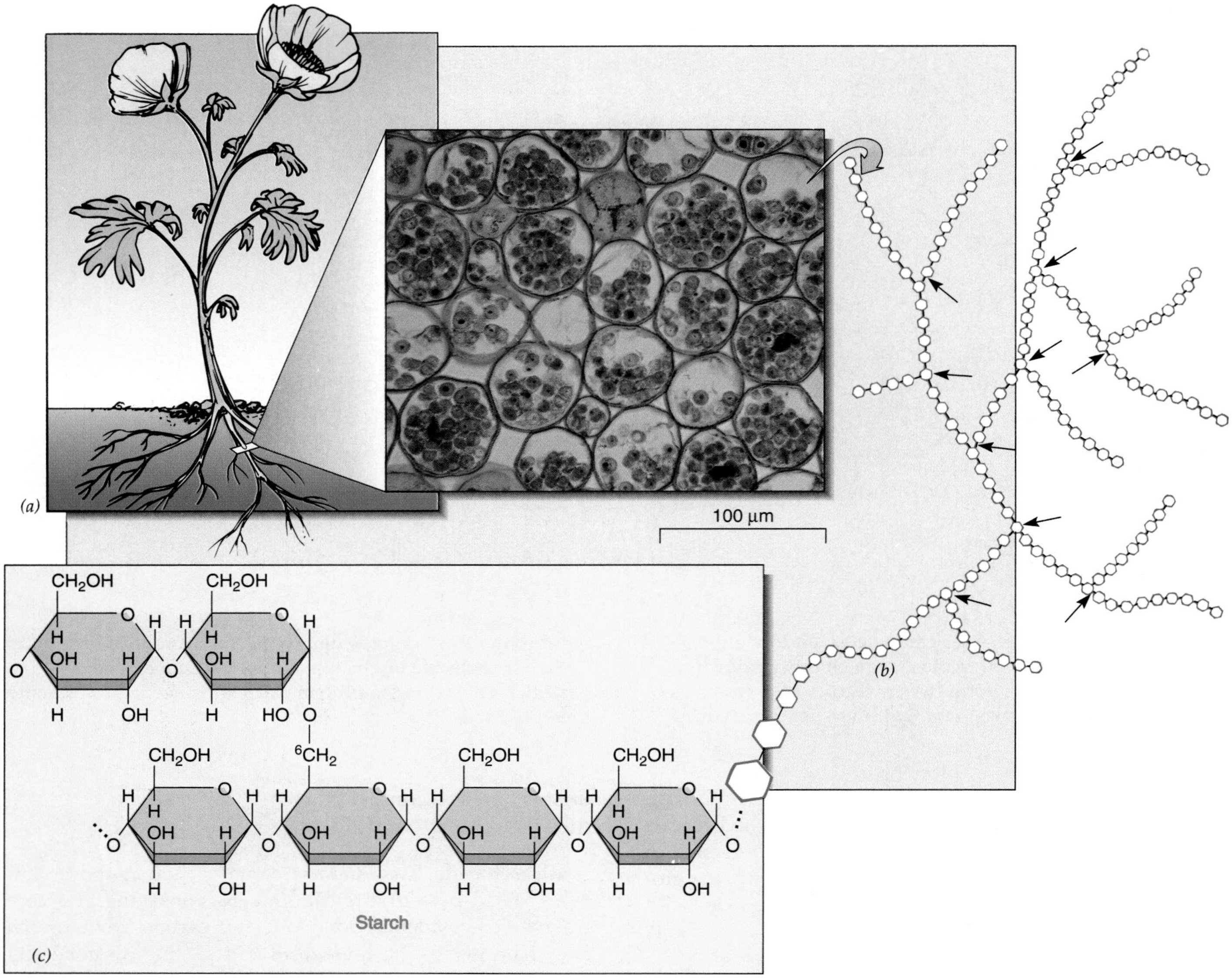

FIGURE 3–9 Starch is a storage polysaccharide. (*a*) Starch (stained purple) is stored in specialized organelles, called *amyloplasts*, in these cells of a buttercup root. (*b*) In this diagrammatic represention of starch, the arrows represent the branch points. The drawing is simplified; each chain is actually in the form of a coil or helix, stabilized by hydrogen bonds between the glucose subunits. (*c*) Starch is a branched polysaccharide composed of α-glucose molecules joined by glycosidic bonds. At the branch points are bonds between carbon 6 of the glucose in the straight chain and carbon 1 of the glucose in the branching chain. Glycogen, or "animal starch" has a similar structure but is more highly branched. (*a*, Ed Reschke)

Polysaccharides can store energy or provide structure

The most abundant carbohydrates are the **polysaccharides,** a group that includes starches, glycogen, and cellulose. A polysaccharide is a macromolecule consisting of repeating units of simple sugars, usually glucose. Although the precise number of sugar units varies, thousands of units are typically present in a single molecule. The polysaccharide may be a single long chain or a branched chain. Because they are composed of different isomers and because the units may be arranged differently, polysaccharides vary in their properties.

Starch, the typical form of carbohydrate used for energy storage in plants, is a polymer consisting of α-glucose subunits. The monomers are joined by α 1—4 linkages, which means that carbon 1 of one glucose is linked to carbon 4 of the next glucose in the chain (Fig. 3–9). Starch occurs in two forms, amylose and amylopectin. Amylose, the simpler form, is unbranched. Amylopectin, the more common form, usually consists of about 1000 units in a branched chain. Branching takes place at about every 20 to 25 units and involves a C-1 to C-6 glycosidic linkage.

Plant cells store starch as granules within specialized organelles called **plastids** (Fig. 3–9*a*). When energy is needed for cellular work, the plant can hydrolyze the starch, releasing the glucose subunits. Humans and other animals that eat plant foods have enzymes to hydrolyze starch.

Glycogen (sometimes referred to as *animal starch*) is the form in which glucose is stored as an energy source in animal tissues. This polysaccharide is highly branched and more water-soluble than plant starch. Glycogen is stored mainly in liver and muscle cells.

Carbohydrates are the most abundant group of organic compounds on Earth, and **cellulose** is the most abundant carbohydrate; it accounts for 50% or more of all the carbon in plants (Fig. 3–10). Cellulose is a structural carbohydrate. Wood is about half cellulose, and cotton is at least 90% cellulose. Plant cells are surrounded by strong supporting cell walls consisting mainly of cellulose.

Cellulose is an insoluble polysaccharide composed of many glucose molecules joined together. The bonds joining these sugar units are different from those in starch. Recall that starch is composed of α-glucose subunits, joined by α 1—4 glycosidic linkages. Cellulose contains β-glucose monomers joined by β 1—4 linkages. These bonds are not split by the enzymes that hydrolyze the alpha linkages in starch. Humans, like most organisms, do not have enzymes that can digest cellulose and therefore cannot use it as a nutrient. However, as discussed in Chapter 45, cellulose is an important component of dietary fiber and helps keep the digestive tract functioning properly.

Some microorganisms can digest cellulose to glucose. In fact, cellulose-digesting bacteria live in the digestive systems of cows and sheep, enabling these grass-eating animals to obtain nourishment from cellulose. Similarly, the digestive systems of termites contain microorganisms that digest cellulose.

Cellulose molecules have characteristics that make them well-suited for a structural role. The β-glucose sub-

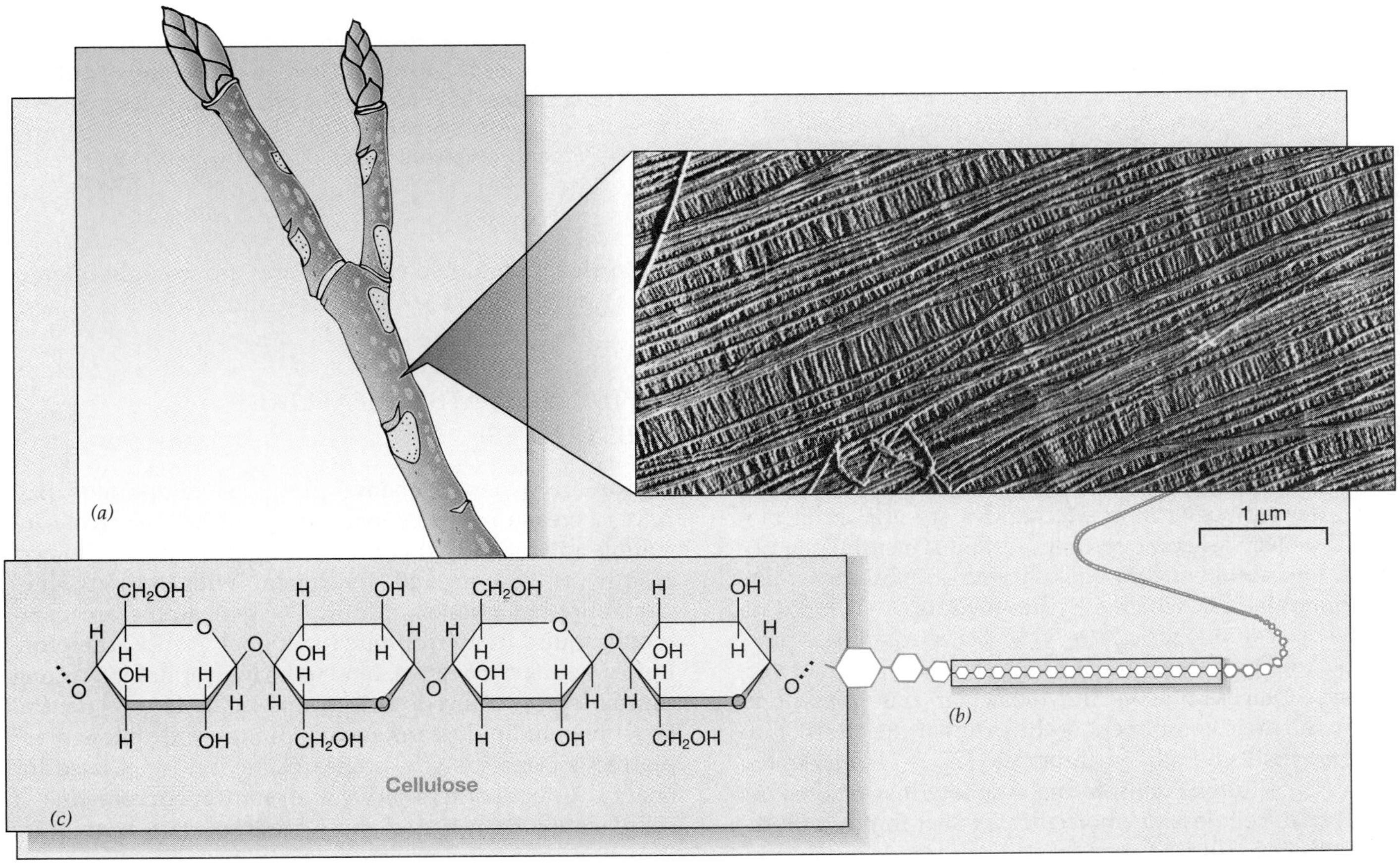

FIGURE 3–10 Cellulose, a structural polysaccharide, is an important component of plant cell walls. (*a*) An electron micrograph of cellulose fibers from a cell wall. The fibers visible in the photograph consist of bundles of cellulose molecules, interacting through hydrogen bonds. (***b*** and ***c***) The cellulose molecule is an unbranched polysaccharide composed of approximately 10,000 β-glucose units joined by glycosidic bonds. (*a*, Omikron/Photo Researchers, Inc.)

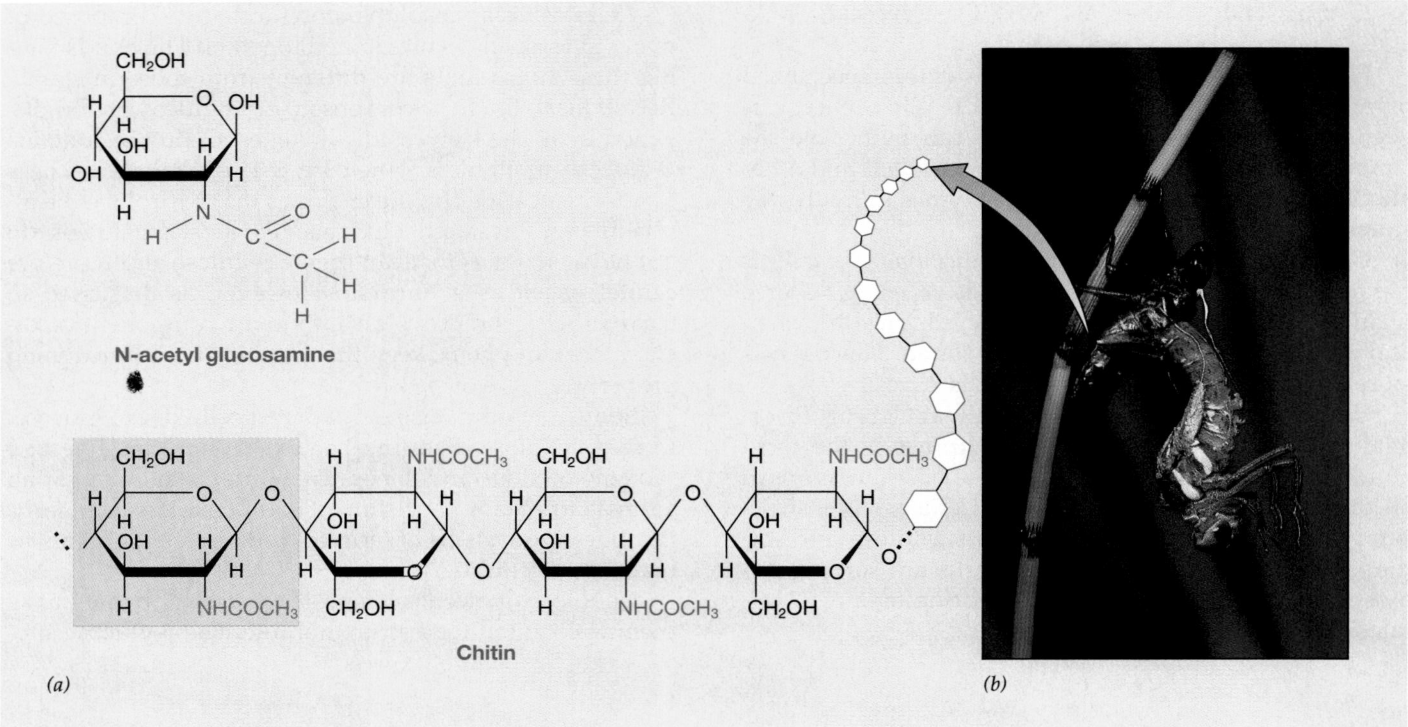

FIGURE 3–11 Chitin is a structural polysaccharide that rivals cellulose as the most abundant carbohydrate on Earth. (*a*) Chitin is a polymer composed of subunits of the amino sugar *N*-acetyl glucosamine (NAG), joined by glycosidic bonds. As in cellulose, the structural strength of this material is a consequence of the fact that the molecules aggregate through hydrogen-bonding. (*b*) Chitin is an important component of the armor-like exoskeleton (outer covering) of arthropods such as this dragonfly. In the process of molting, shown here, the dragonfly sheds its exoskeleton so that it can grow. A new, larger exoskeleton develops. (*b*, Dwight R. Kuhn)

units are joined in a way that allows extensive hydrogen bonding between different cellulose molecules. Thus, cellulose molecules aggregate in long bundles of fibers, as shown in Figure 3–10*a*.

Some modified and complex carbohydrates have special roles

Many derivatives of monosaccharides are important biological molecules. Some form important structural components. The amino sugars glucosamine and galactosamine are compounds in which a hydroxyl group (—OH) is replaced by an amino group (—NH_2). Galactosamine is present in cartilage, a constituent of the skeletal system of vertebrates. Glucosamine is the molecular unit present in **chitin,** the main component of the external skeletons of insects, crayfish, and other arthropods (Fig. 3–11), and of the cell walls of fungi. Chitin forms very tough structures because, as in cellulose, its molecules interact through multiple hydrogen bonds.

Carbohydrates may also be combined with proteins to form **glycoproteins,** compounds present on the outer surface of many eukaryotic cells. Most proteins secreted by cells are glycoproteins. Carbohydrates can combine with lipids to form **glycolipids,** compounds present on the surfaces of animal cells that are important in interactions among cells (see Chapters 4 and 5).

LIPIDS ARE FATS OR FATLIKE SUBSTANCES

Lipids are a heterogeneous group of compounds that have a greasy or oily consistency and are relatively insoluble in water. This is because lipid molecules consist mainly of carbon and hydrogen, with few oxygen-containing functional groups. Oxygen atoms are commonly found in hydrophilic functional groups; therefore lipids, with little oxygen, tend to be hydrophobic. Among the biologically important groups of lipids are neutral fats, phospholipids, carotenoids (orange and yellow plant pigments), steroids, and waxes. Some lipids are used for energy storage, others serve as structural components of cellular membranes, and some are important hormones.

Neutral fats contain glycerol and fatty acids

The most abundant lipids in living organisms are the **neutral fats.** These compounds are an economical form of reserve fuel storage because they yield more than twice as

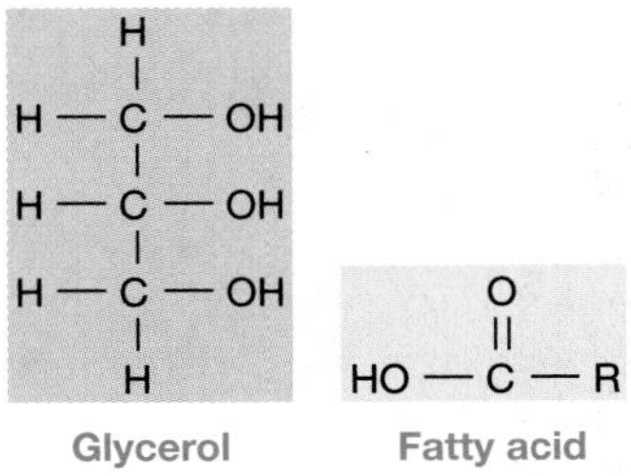

(a)

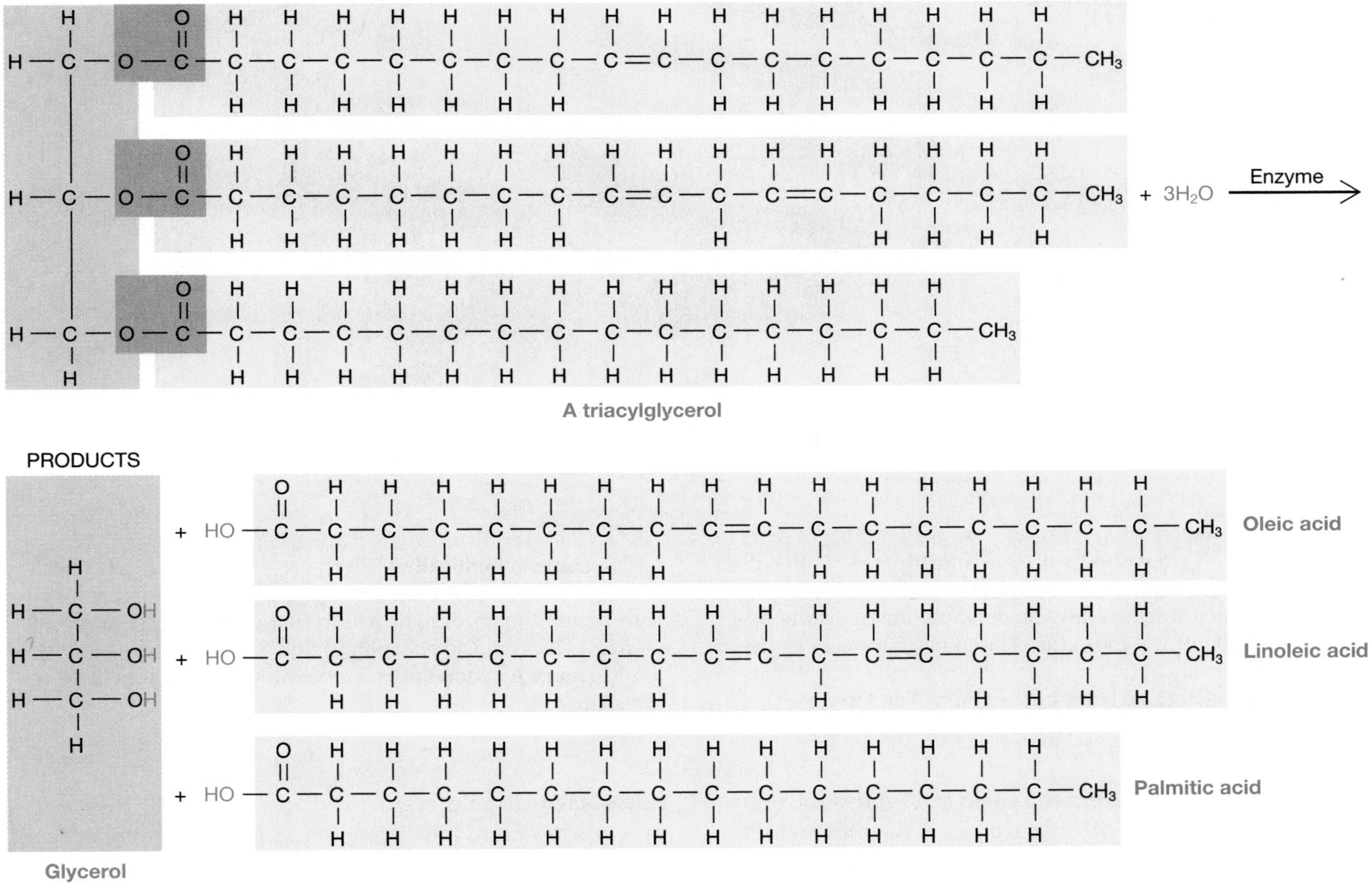

(b)

(c)

FIGURE 3–12 Neutral fats are formed by the reaction of glycerol with one to three fatty acids. (*a*) Structures of glycerol and of a fatty acid. The carboxyl (—COOH) group is present in all fatty acids. The R represents the remainder of the molecule, a long hydrocarbon that varies with each type of fatty acid. (*b*) The glycerol is attached to each fatty acid by an ester linkage, shown in green. Hydrolysis of a triacylglycerol yields glycerol plus three fatty acids. Note that the triacylglycerol is an unsaturated fat; two of its fatty acid components contain double bonds between carbon atoms. For simplicity these unsaturated fatty acids are drawn as straight chains, but the molecules of oleic and linoleic acid are actually bent or kinked wherever a carbon-to-carbon double bond appears. (*c*) Commonly used cooking fats, such as butter, olive oil, solid vegetable shortening, etc. contain triacylglycerols. (*c*, Kenneth Knott/Fine Light Photography)

much energy per gram as do carbohydrates. Carbohydrates and proteins can be transformed by enzymes into fats and stored within the cells of adipose (fat) tissue.

A neutral fat consists of glycerol joined to one, two, or three fatty acids. **Glycerol** is a three-carbon alcohol that contains three hydroxyl (—OH) groups (Fig. 3–12). A **fatty acid** is a long, unbranched hydrocarbon chain with a carboxyl group (—COOH) at one end. About 30 different fatty acids are commonly found in lipids, and they typically have an even number of carbon atoms. For ex-

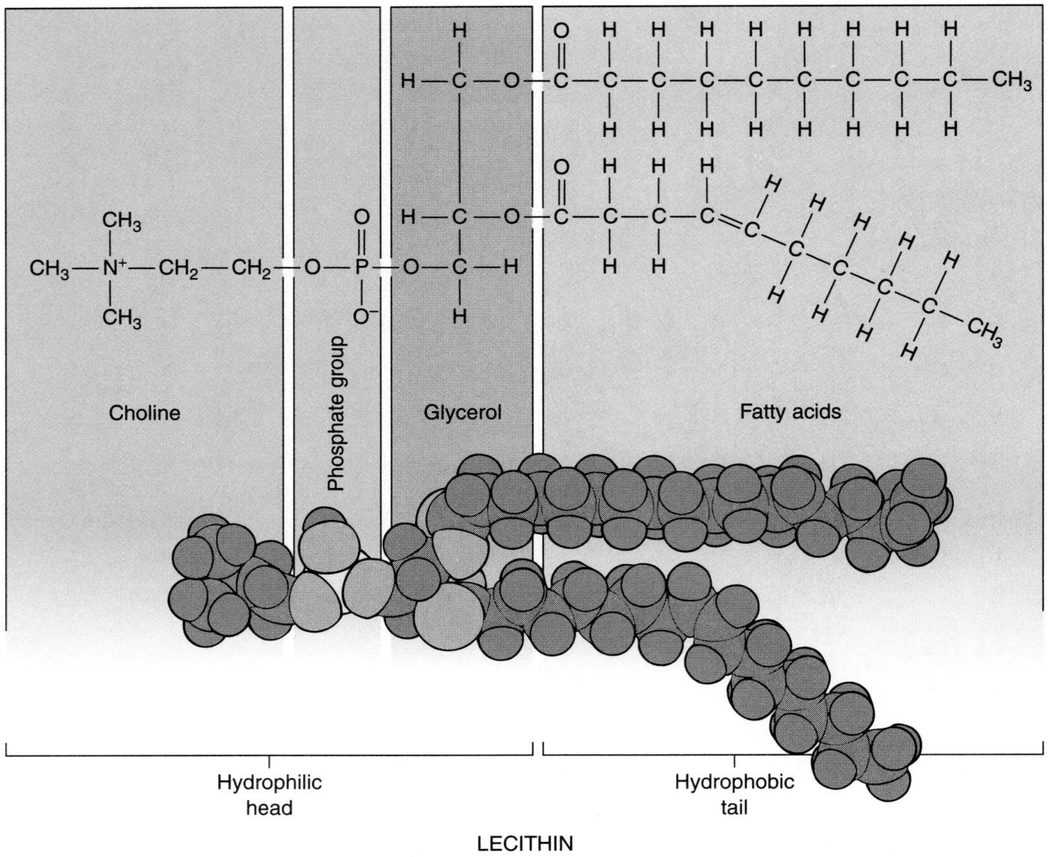

FIGURE 3–13 A phospholipid is an amphipathic molecule, having a hydrophilic region and a hydrophobic region. A phospholipid includes two fatty acids; these make up the hydrophobic region of the molecule. The lower fatty acid in the figure is monounsaturated; it contains one double bond that produces a characteristic bend in the chain. The fatty acids are combined chemically with the hydrophilic region, which includes a glycerol bonded to a phosphate group, which is in turn bonded to an organic group that can vary. The molecule shown is lecithin (or phosphatidylcholine); choline is its organic group. A space-filling model of lecithin is shown below the structural formula.

ample, butyric acid, present in rancid butter, has four carbon atoms. Oleic acid, with 18 carbons, is the most widely distributed fatty acid in nature and is found in most animal and plant fats.

Saturated fatty acids contain the maximum possible number of hydrogen atoms. Fats high in saturated fatty acids tend to be solid at room temperature. This is because extensive van der Waals interactions (see Chapter 2) can occur among the long hydrocarbon chains. Animal fat and solid vegetable shortening are examples of saturated fats.

Unsaturated fatty acids include one or more adjacent pairs of carbon atoms joined by a double bond. They therefore are not fully saturated with hydrogen. Fatty acids with one double bond are called **monounsaturated fatty acids,** while those with more than one double bond are **polyunsaturated fatty acids.** Fats containing a high proportion of monounsaturated or polyunsaturated fatty acids tend to be liquid at room temperature. This is because each double bond produces a "bend" in the hydrocarbon chain that prevents it from aligning closely with an adjacent chain, thereby limiting van der Waals interactions (Fig. 3–13).

At least two unsaturated fatty acids (linoleic acid and arachidonic acid) are essential nutrients that must be obtained from food because the human body cannot synthesize them. However, the amounts required are small, and deficiencies are rarely seen. There is no dietary requirement for saturated fatty acids.

When a glycerol molecule combines chemically with one fatty acid, a **monoacylglycerol** (sometimes called *monoglyceride*) is formed. **Diacylglycerols** (or *diglycerides*) and **triacylglycerols** (or *triglycerides*) contain two or three fatty acids, respectively. In each reaction, the equivalent of a water molecule is removed as one of the glycerol's hydroxyl groups reacts with the carboxyl group of a fatty acid. Such a reaction between a hydroxyl group and a carboxyl group results in the formation of a covalent link-

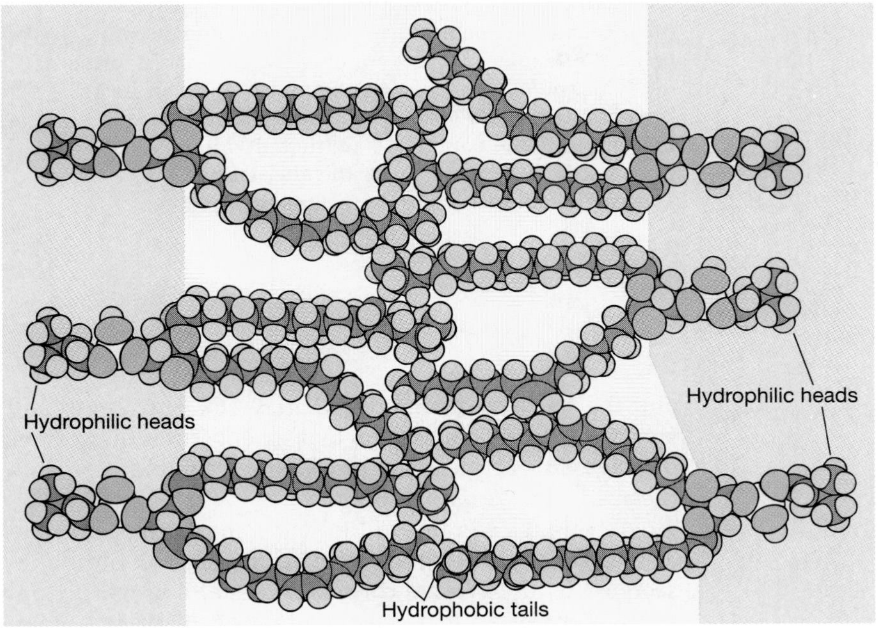

(a)

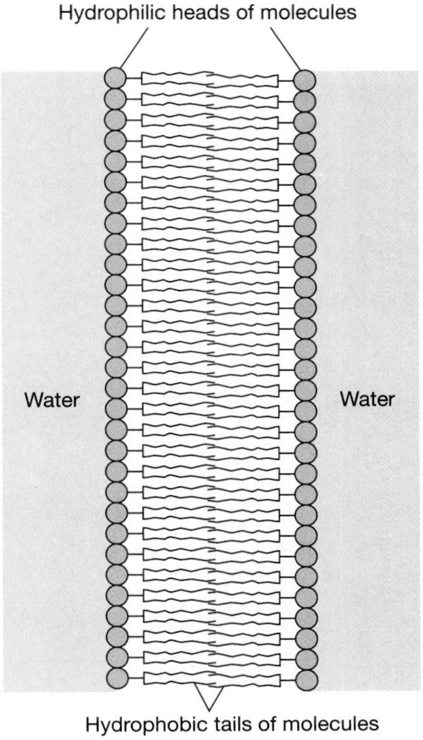

(b)

FIGURE 3–14 In the presence of water, phospholipid molecules form a bilayer, with their hydrophilic water-soluble heads facing outward toward the surrounding water and their hydrophobic tails filling the middle of the bilayer. (***a***) Space-filling model of phospholipids in a bilayer. (***b***) Diagram of a lipid bilayer, such as that found in cell membranes.

age known as an **ester linkage** (Fig. 3–12*b*). During digestion, the neutral fats are hydrolyzed to produce fatty acids and glycerol (see Chapter 45).

Phospholipids are components of cellular membranes

Phospholipids belong to a group of lipids, called **amphipathic lipids,** in which one end of each molecule is hydrophilic and the other end is hydrophobic. The properties of these molecules make them uniquely suited to form structures known as cellular membranes (see Chapter 5). Every cell is enclosed by a cellular membrane, and complex cells contain many internal membranes as well. A phospholipid consists of a glycerol molecule attached to two fatty acids and to a phosphate group linked to an organic compound such as choline. The organic compound usually contains nitrogen (Fig. 3–13). (Note that phosphorus and nitrogen are absent in the neutral fats.)

The two ends of a phospholipid molecule differ physically as well as chemically. The fatty acid portion of the molecule (containing the two hydrocarbon "tails") is hydrophobic and not soluble in water. However, the portion composed of glycerol, phosphate, and the organic base (the "head" of the molecule) is ionized and readily water-soluble. The amphipathic properties of these lipid molecules cause them to assume a predictable configuration, known as a *lipid bilayer* (two layers of phospholipid molecules), in the presence of water. The fundamental structure of a cellular membrane is a phospholipid bilayer in which the hydrophilic heads face toward the surrounding water and the hydrophobic tails meet in the middle (Fig. 3–14).

Carotenoid plant pigments are derived from isoprene units

The orange and yellow plant pigments called **carotenoids** are classified with the lipids because they are insoluble in water and have an oily consistency. These pigments, found in the cells of all plants, play a role in photosynthesis. The carotenoid molecule consists of five-carbon hydrocarbon monomers known as *isoprene units* (Fig. 3–15*a*). Many important pigments are derived from isoprene units (see *Making the Connection: Molecules that Absorb Light*).

Steroids contain four rings of carbon atoms

A **steroid** consists of carbon atoms arranged in four attached rings; three of the rings contain six carbon atoms,

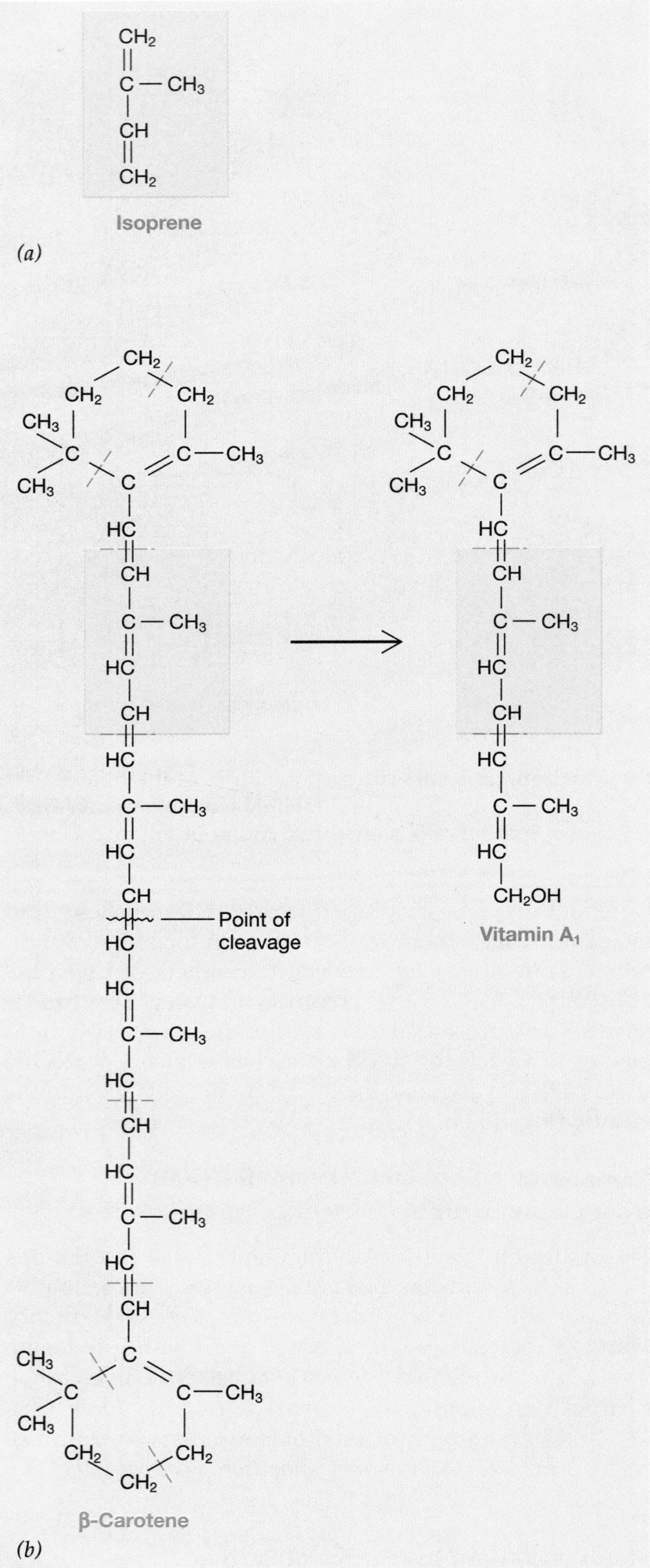

FIGURE 3–15 Carotenoids are formed from isoprene units. (*a*) An isoprene monomer. (*b*) β-carotene, a yellow pigment that gives carrots, sweet potatoes, and other orange vegetables their color. Most animals can break carotenoids at the point of cleavage marked in the diagram and convert them to vitamin A. Vitamin A can be converted to the visual pigment retinal. The dashed lines indicate the boundaries of the individual isoprene units within β-carotene. Note the alternating double and single bonds that are characteristic of pigments.

and the fourth contains five (Fig. 3–16). The length and structure of the side chains that extend from these rings distinguish one steroid from another. Steroids are synthesized from isoprene units.

Among the steroids of biological importance are cholesterol, bile salts, reproductive hormones, and hormones secreted by the adrenal cortex. Cholesterol is a structural component of animal cell membranes; plant cell membranes contain molecules similar to cholesterol. Bile salts emulsify fats in the intestine so that they can be enzymatically hydrolyzed. Steroid hormones regulate certain aspects of metabolism in a variety of animals, including vertebrates, insects, and crabs.

PROTEINS ARE MACROMOLECULES FORMED FROM AMINO ACIDS

Proteins are of central importance in the chemistry of life. These macromolecules serve as structural components of cells and tissues; growth and repair, as well as maintenance of the organism, depend on an adequate supply of these compounds. Many proteins serve as **enzymes,** molecules that speed up the thousands of different chemical reactions that take place in an organism.

The protein constituents of a cell are the clues to its lifestyle. Each cell type has characteristic types, distributions, and amounts of protein that determine what the cell looks like and how it functions. A muscle cell differs from other cell types by virtue of its large content of the proteins myosin and actin, which are largely responsible for its appearance as well as for its ability to contract. The protein hemoglobin, found in red blood cells, is responsible for the specialized function of oxygen transport.

Although carbohydrates and lipids tend to have the same structures among different species, most proteins are species-specific; that is, their structures vary from species to species. The specific proteins present (determined by the instructions in the genes) are largely responsible for differences among species. Thus, the proteins in the cells of a dog vary somewhat from those of a

Making the Connection

Molecules that Absorb Light

What do some biologically important pigments have in common? Splitting a molecule of the yellow plant pigment carotene in half yields a molecule of vitamin A, or retinol (Fig. 3–15*b*). Retinal, the light-sensitive chemical present in the retina of the eye (see Chapter 41), is a derivative of vitamin A. Interestingly, eyes have evolved independently in three different lines of animals: the mollusks, insects, and vertebrates. These animals have no common evolutionary ancestor equipped with eyes, yet the eyes of each have the same compound—retinal—involved in the process of light reception. That retinal is present in all three types of eyes is evidence of the unique fitness of this kind of molecule for light reception.

Notice that the carotenoid and vitamin A molecules shown in Figure 3–15 have a pattern of double bonds alternating with single bonds. The electrons that make up these bonds can move about relatively easily when light strikes the molecule. Such molecules tend to be highly colored pigments, because they strongly absorb light of certain wavelengths and reflect other wavelengths. ▲

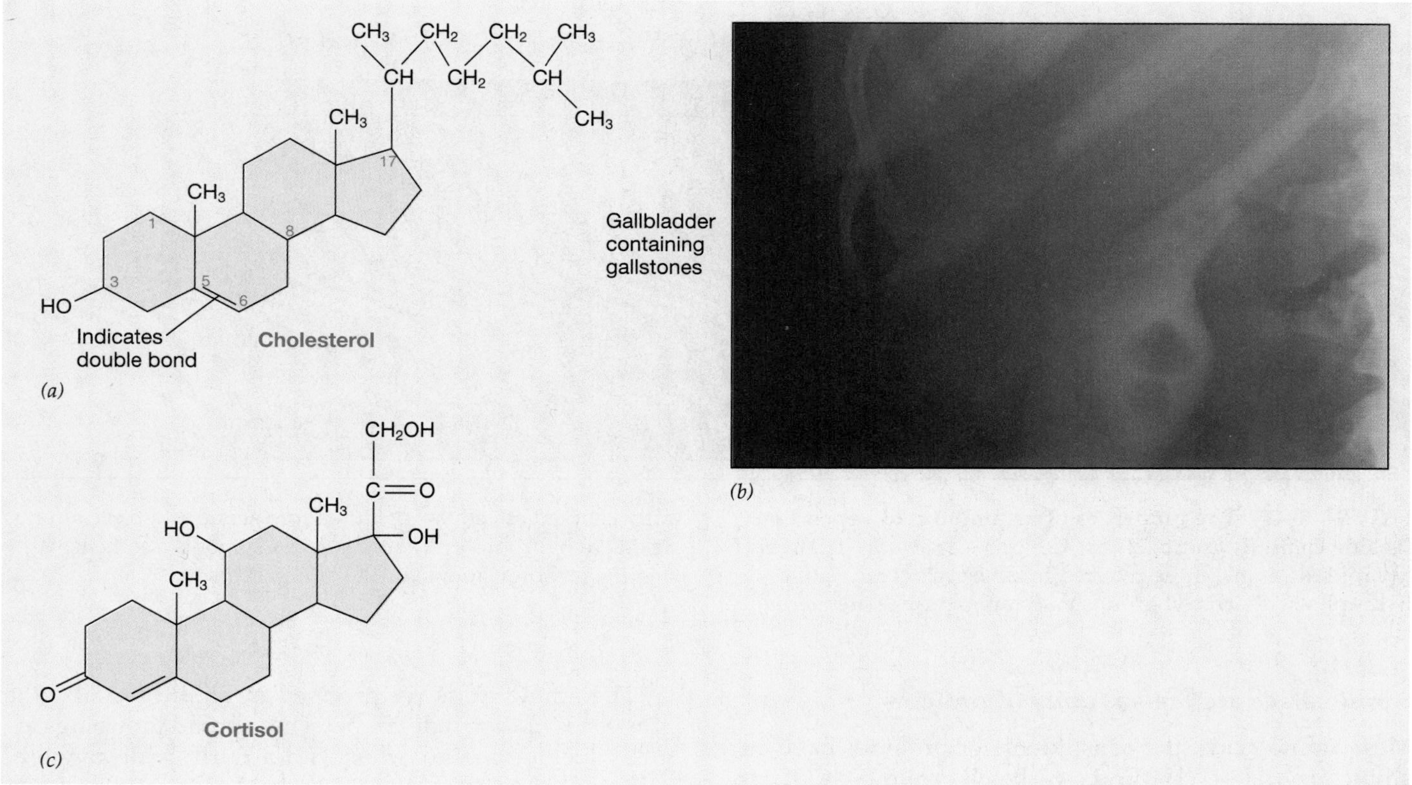

FIGURE 3–16 A steroid contains four attached rings: three six-carbon rings and one with five carbons. Note that some carbons are "shared" by two rings. In these simplified structures, a carbon atom is present at each angle of a ring; the hydrogen atoms attached directly to the ring have not been drawn. (***a***) Cholesterol is an essential component of animal cell membranes. (***b***) Gallstones, such as those present in the gallbladder shown here, may form from cholesterol. (***c***) Cortisol is a steroid hormone secreted by the adrenal glands. Notice that it differs from cholesterol in its attached functional groups. (*b,* Science Photo Library/Photo Researchers Inc.)

fox and even more from those of an oak tree. The degree of difference in the proteins of two species is thought to depend on evolutionary relationships. Distantly related organisms have proteins that differ more markedly than those of closely related forms.

Some proteins differ slightly even among individuals of the same species; hence most individuals are biochemically unique. Only genetically identical organisms—identical twins or members of closely inbred strains of organisms—have identical proteins.

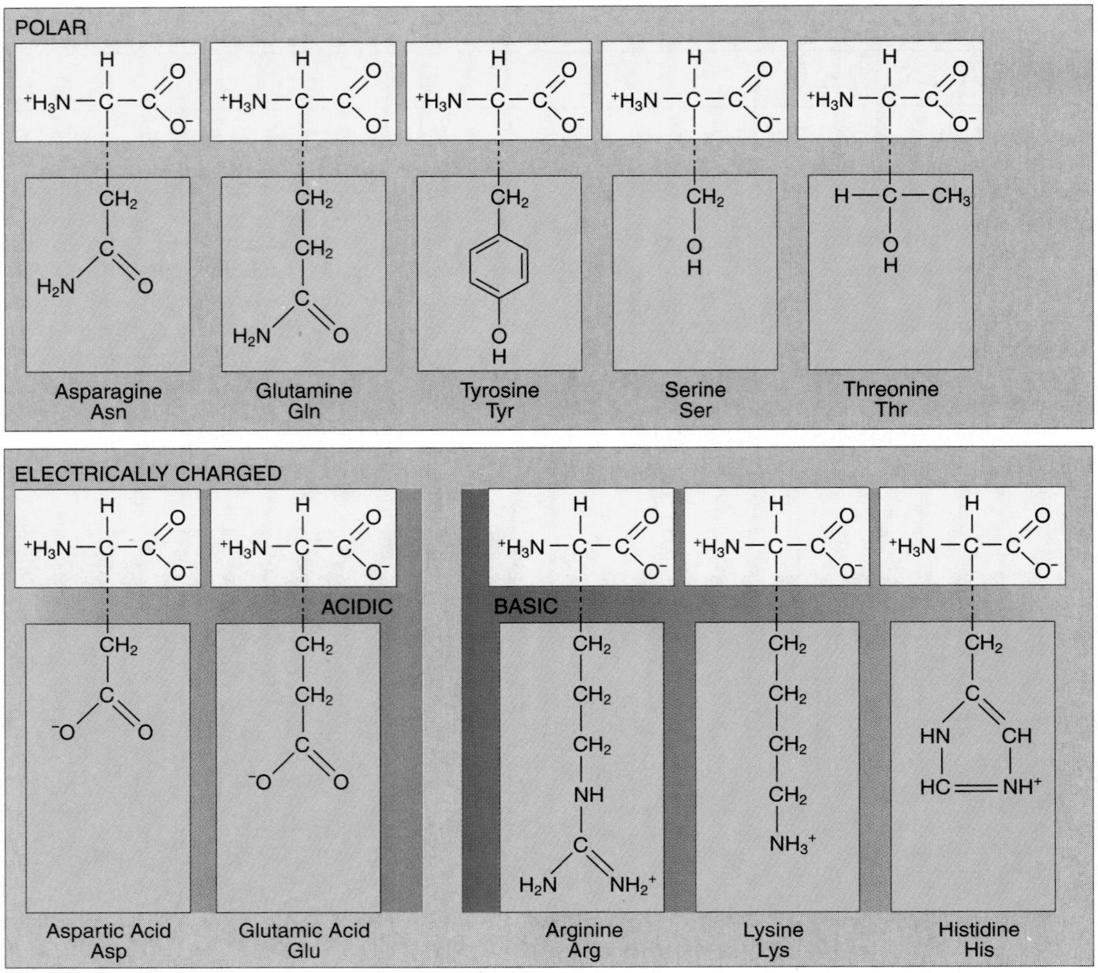

FIGURE 3–17 The properties of an amino acid depend on its side chain (R group). Those designated *polar* are relatively hydrophilic, while those referred to as *nonpolar* are relatively hydrophobic. Carboxyl groups and amino groups are electrically charged at cellular pH; therefore acidic and basic amino acids are hydrophilic. The three-letter symbols are the conventional abbreviations for the amino acids.

Amino acids are the subunits of proteins

Most amino acids, the constituents of proteins, have an amino group ($-NH_2$) and a carboxyl group ($-COOH$) bonded to the same asymmetric carbon atom, the **alpha carbon.** There are about 20 amino acids commonly found in proteins, each uniquely identified by the variable group (R group) bonded to the alpha carbon (Fig. 3–17). Glycine, the simplest amino acid, has a hydrogen atom as its R group; alanine has a methyl ($-CH_3$) group.

Amino acids in solution at neutral pH are mainly dipolar ions. This is generally how amino acids exist at cellular pH. Each carboxyl group ($-COOH$) donates a proton and becomes dissociated ($-COO^-$), while each amino group ($-NH_2$) accepts a proton and becomes $-NH_3^+$ (Fig. 3–18). Because of their amino and carboxyl groups, amino acids in solution resist changes in acidity and alkalinity and so are important biological buffers.

The amino acids are grouped in Figure 3–17 by the properties of their side chains. These broad groupings actually include amino acids with a fairly wide range of properties. Amino acids classified as having *nonpolar* side chains tend to have hydrophobic properties, whereas those classified as *polar* are more hydrophilic. An acidic amino acid has a side chain that contains a carboxyl group. At cellular pH the carboxyl group is dissociated,

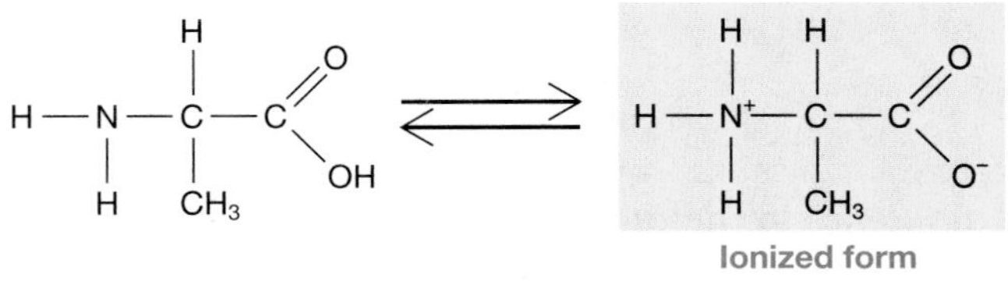

FIGURE 3–18 At the pH of living cells, amino acids exist mainly as dipolar ions.

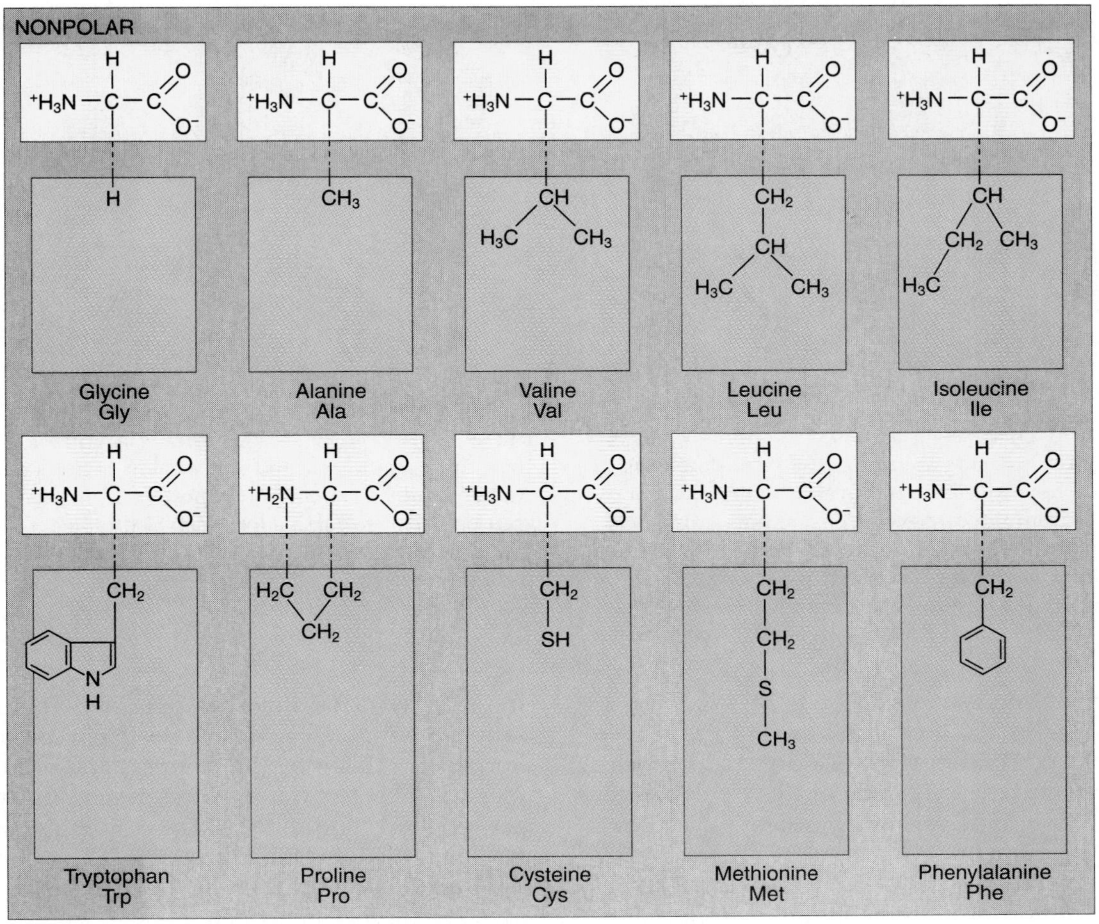

giving the R group a negative charge. A basic amino acid becomes positively charged when the amino group in its side chain accepts a hydrogen ion. Acidic and basic side chains are ionic at cellular pH and therefore hydrophilic.

In addition to the 20 common amino acids, some proteins have unusual ones. These rare amino acids are produced by the modification of common ones after they have become part of a protein. For example, lysine and proline may be converted to hydroxylysine and hydroxyproline after they have been incorporated into collagen. These amino acids can form cross links between the peptide chains that make up collagen. Such cross links are responsible for the firmness and great strength of the collagen molecule, which is a major component of cartilage, bone, and other connective tissues.

With some exceptions, bacteria and plants can synthesize all their needed amino acids from simpler substances. If the proper raw materials are available, the cells of humans and animals can manufacture some, but not all, of the biologically significant amino acids. Those that animals cannot synthesize and so must obtain from the diet are known as **essential amino acids.** Animals differ in their biosynthetic capacities; what is an essential amino acid for one species may not be for another. The essential amino acids for humans include isoleucine, leucine, lysine, methionine, phenylalanine, threonine, tryptophan, valine, histidine, and (in children) arginine.

Peptide bonds join amino acids

Amino acids combine chemically with one another by bonding the carboxyl carbon of one molecule to the amino nitrogen of another (Fig. 3–19). The covalent carbon-to-nitrogen bond linking two amino acids together is called a **peptide bond.** When two amino acids combine, a **dipeptide** is formed; a longer chain of amino acids is a **polypeptide.** A polypeptide chain has direction, with a free amino group at one end and a free carboxyl group (belonging to the last amino acid added to the chain) at the opposite end. The other amino and carboxyl groups of the amino acid monomers (except those in side chains) are part of the peptide bonds. The complex process by which polypeptides are synthesized is discussed in Chapter 12.

A polypeptide may contain hundreds of amino acids joined in a specific linear order. The backbone of the chain

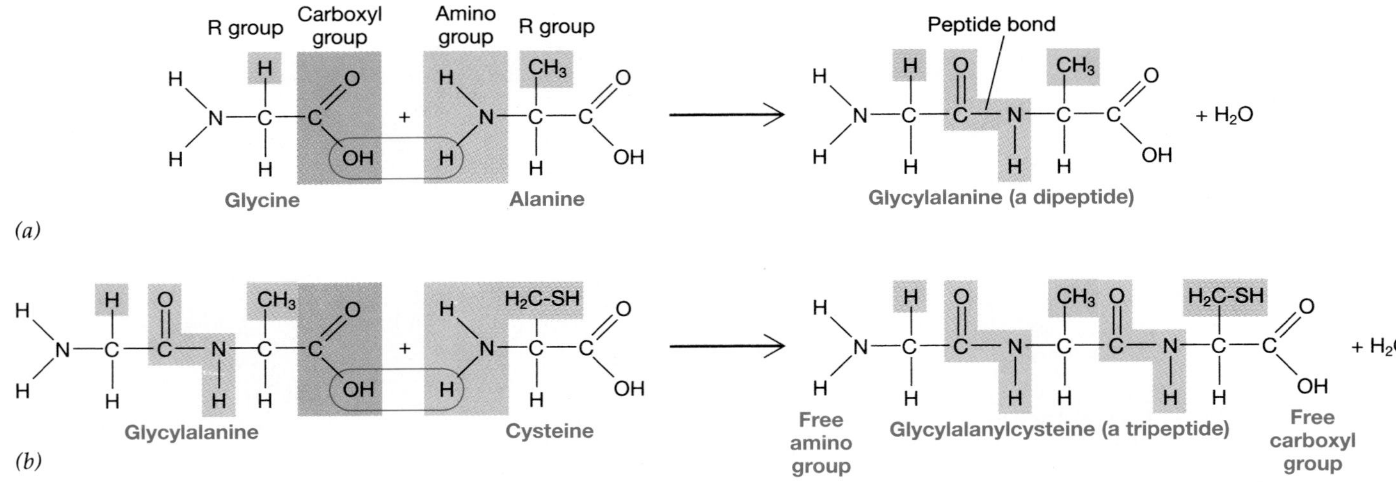

FIGURE 3–19 Amino acids are linked by peptide bonds. (*a*) A dipeptide is formed by the removal of the equivalent of a water molecule from the carboxyl group of one amino acid and the amino group of another amino acid. The resulting peptide bond is a covalent carbon-to-nitrogen bond. (*b*) The carboxyl group of the dipeptide reacts with the amino group of a third amino acid to form a chain of three amino acids (a tripeptide, or small polypeptide). Additional amino acids can be added to form a long polypeptide chain with a free amino group at one end and a free carboxyl group at the other.

includes the repeating sequence —C—N—C—C—N —C—C—N— plus all other atoms *except those in the R groups*. The R groups of the amino acids extend from this backbone. A protein consists of one or more polypeptide chains. An almost infinite variety of protein molecules is possible, differing from one another in the number, types, and sequences of amino acids they contain. The 20 types of amino acids found in proteins may be thought of as letters of a protein alphabet; each protein is a very long sentence made up of amino acid letters.

Proteins have four levels of organization

The polypeptide chains making up a protein are twisted or folded to form a macromolecule with a specific *conformation* or three-dimensional shape. Some polypeptide chains form long fibers. *Globular* proteins are tightly folded into compact, roughly spherical shapes. The conformation of a protein determines its function. For example, a typical enzyme is a globular protein with a unique shape that permits it to catalyze a specific reaction. Similarly, the shape of a protein hormone enables it to combine with receptors on its target cell (the cell the hormone acts upon).

Four main levels of protein organization can be recognized: primary, secondary, tertiary, and quaternary (Fig. 3–20).

Primary structure is the amino acid sequence

The sequence of amino acids, joined by peptide bonds, is the **primary structure** of a polypeptide chain. This sequence, discussed in Chapter 12, is specified by the instructions in a gene. Using analytical methods developed in the early 1950s, investigators can determine the exact sequence of amino acids in a protein molecule. Insulin, a hormone secreted by the pancreas and used in the treatment of diabetes, was the first protein for which the exact sequence of amino acids in the polypeptide chains was identified. Insulin is a very small protein, consisting of 51 amino acid units in two linked chains, each with its own primary sequence (Fig. 3–21).

Secondary structure results from hydrogen-bonding involving the backbone

Some regions of a polypeptide exhibit **secondary structure,** which is highly regular because it is maintained by

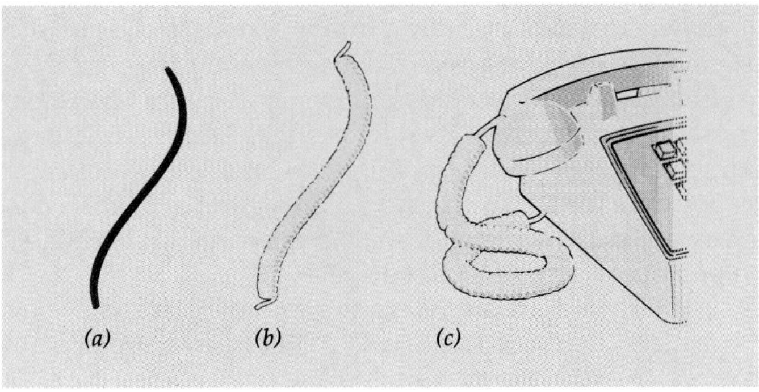

FIGURE 3–20 A telephone cord provides a familiar analogy that illustrates the levels of structure in a polypeptide. (*a*) Primary structure, which can also be thought of as "beads on a string." (*b*) Secondary structure, in this case an α-helix. (*c*) Tertiary structure, in which interactions among side chains cause the molecule to fold back on itself.

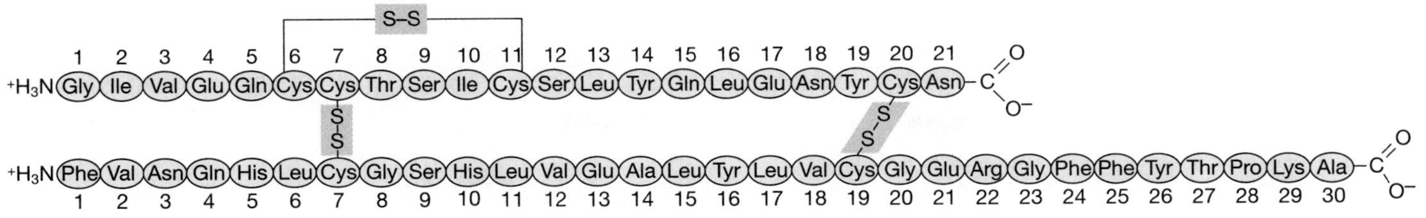

FIGURE 3–21 The primary structure of a polypeptide is its linear sequence of amino acids. Insulin is a very small protein made up of two polypeptides, each with its own primary structure. The amino acid sequence is indicated by ovals containing the abbreviated names of amino acids (see Fig. 3–17).

hydrogen bonds between certain atoms of the polypeptide chain's uniform backbone.

A common secondary structure in protein molecules is the **α-helix,** a region where a polypeptide chain forms a uniform spiral coil (Fig. 3–22*a*). The helical structure is determined and maintained by the formation of hydrogen bonds between the backbones of the amino acids in successive turns of the spiral coil. Each hydrogen bond forms between an oxygen with a partial negative charge and a hydrogen with a partial positive charge. The oxygen is part of the remnant of the carboxyl group of one amino acid; the hydrogen is part of the remnant of the amino group of the fourth amino acid down the chain. Thus 3.6 amino acids are included in each complete turn of the helix. Every amino acid in an α-helix is hydrogen-bonded in this way.

The α-helix is the basic structural unit of some fibrous proteins that make up wool, hair, skin, and nails. The fiber is elastic because the hydrogen bonds can be broken and then reformed. This is why human hairs can stretch to some extent and then snap back to their original length.

A second type of secondary structure is the **β-pleated sheet**[2] (Fig. 3–22*b*). The hydrogen-bonding in a β-pleated sheet takes place between different polypeptide chains, or different regions of a polypeptide chain that has turned back on itself. Each chain is fully extended, but because each has a zigzag structure, the resulting "sheet" has an overall pleated conformation (much like a sheet of paper that has been folded to make a fan.) A pleated sheet is flexible rather than elastic. Fibroin, the protein of silk, is characterized by a β-pleated sheet structure, and the cores of many globular proteins consist of β-sheets.

Tertiary structure depends on interactions among side chains

The **tertiary structure** of a protein molecule is the overall shape assumed by each individual polypeptide chain (Fig. 3–23). This three-dimensional structure is determined by four main factors that involve *interactions among R groups (side chains) belonging to the same polypeptide chain.*

1. Hydrogen bonds form between R groups of certain amino acid subunits.
2. Ionic attraction can occur between an R group with a unit of positive charge and one with a unit of negative charge.
3. Hydrophobic interactions result from the tendency of nonpolar R groups to be excluded by the surrounding water and to therefore associate in the interior of the globular structure.
4. Covalent bonds known as disulfide bonds or disulfide bridges (—S—S—) may link the sulfur atoms of two cysteine subunits. A disulfide bridge forms when the sulfhydryl groups of two cysteines react; the two hydrogens are removed and the two sulfur atoms that remain become covalently linked.

Quaternary structure results from interactions among polypeptides

If a protein is composed of two or more polypeptide chains, these interact in specific ways to form the biologically active protein molecule. **Quaternary structure** is the resulting arrangement of the polypeptide chains (each with its own primary, secondary, and tertiary structure). Quaternary structure does not involve additional forces; hydrogen-bonding, ionic bonding, hydrophobic interactions, and disulfide bridges can all contribute to quaternary structure. For example, the two polypeptide chains that make up the insulin molecule shown in Figure 3–21 are joined by two disulfide bridges.

Hemoglobin, the protein in red blood cells that is responsible for oxygen transport, is an example of a globular protein with quaternary structure (Figure 3–24). Hemoglobin consists of 574 amino acids arranged in four polypeptide chains: two identical chains called alpha chains and two identical chains called beta chains.

Protein structure determines function

The structure of a protein helps determine its biological activity. A single protein may have more than one distinct structural region, each with its own function. Many pro-

[2]Note that the designations α and β refer simply to the order in which these two types of secondary structures were discovered.

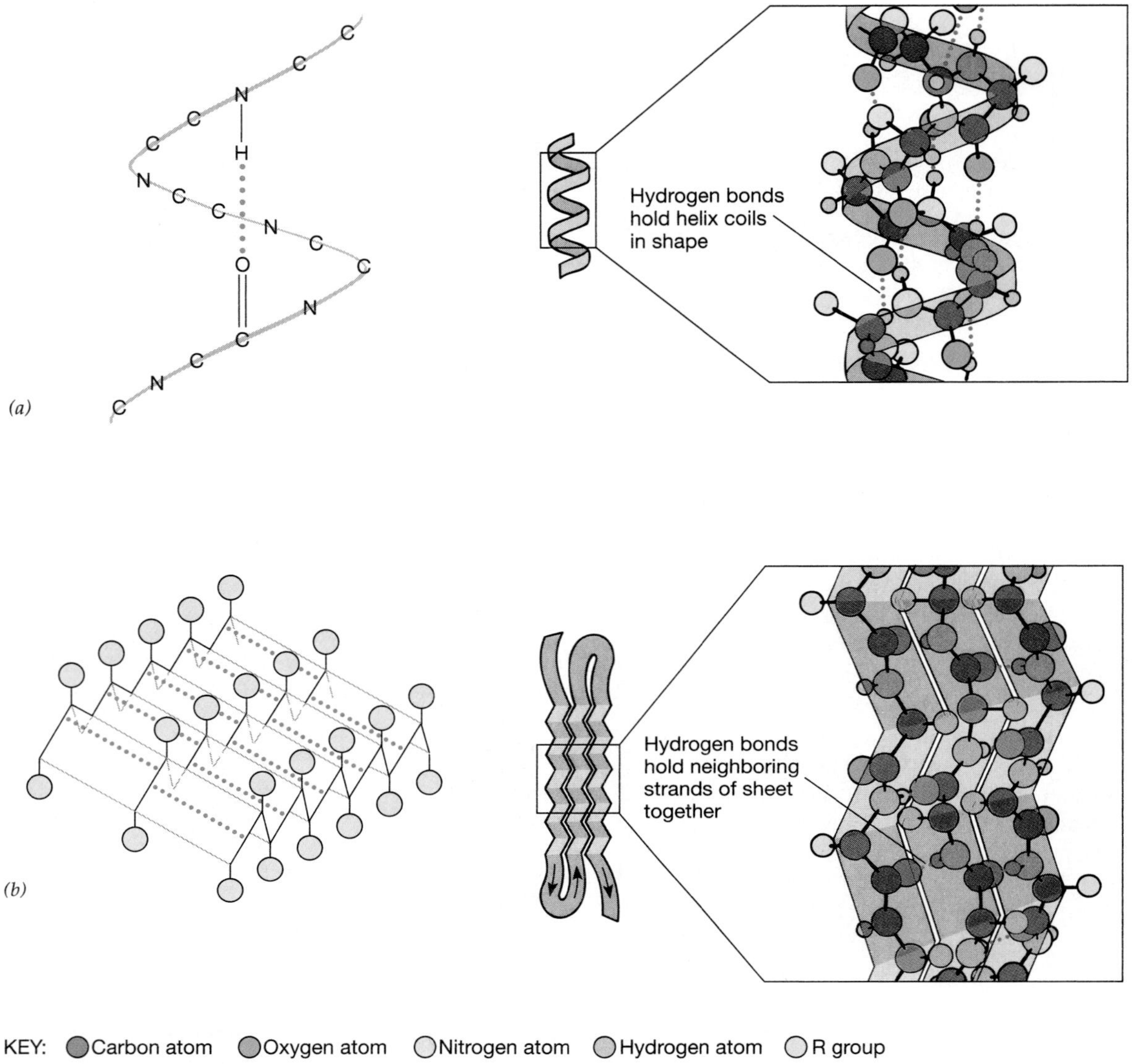

FIGURE 3–22 The secondary structure of a protein is very regular because it depends on the hydrogen-bonding of uniform backbone elements. (***a***) The coils of an α-helix are held together by hydrogen bonds between oxygen and hydrogen atoms. Each bond involves two amino acids separated by three intervening amino acids. Note that the R groups project out from the sides of the helix. (The R groups have been omitted in the simplified diagram at left.) (***b***) In a β-pleated sheet the backbone of the polypeptide chain is stretched out into a zigzag shape. Hydrogen bonds can form between regions of the same polypeptide chain or between two or more different chains. Note that half the R groups project above the sheet and the other half project below it. (***c***) The silk used by this spider to spin its web and wrap its prey is an extremely strong and flexible protein, fibroin, that exhibits a β-pleated sheet structure. The silk fibers harden as they are spun from the glands in the spider's abdomen. (*c,* Skip Moody/Dembinsky Photo Associates)

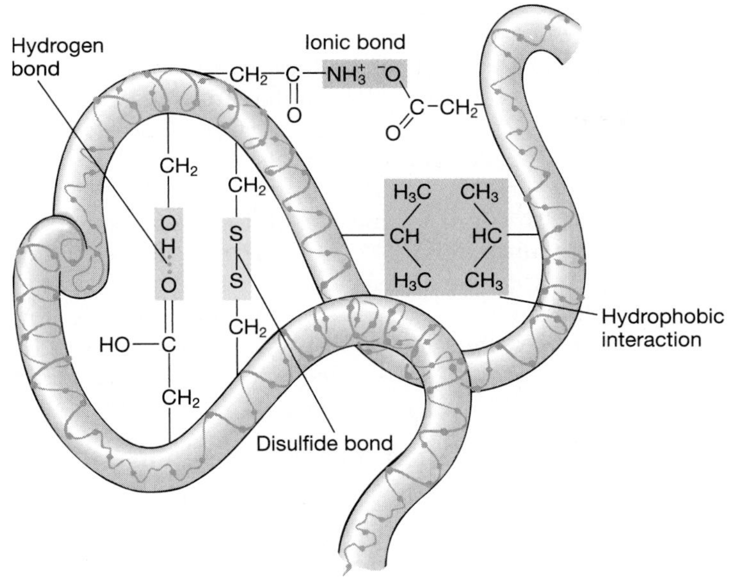

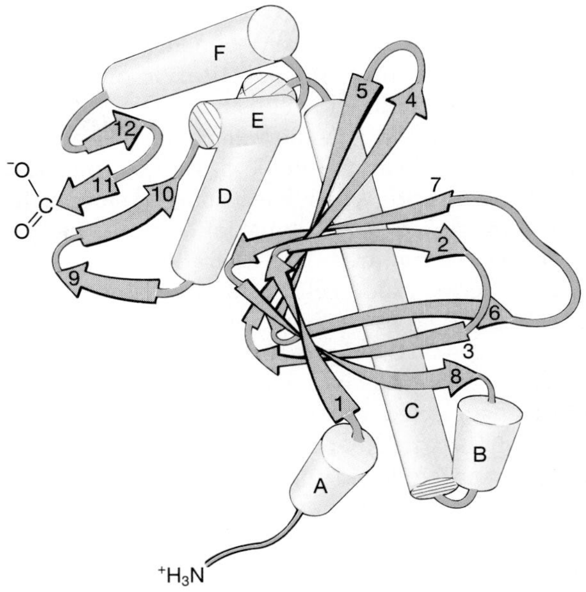

FIGURE 3–23 The tertiary structure of a protein results from interactions among R groups that cause an α-helix (or other secondary structure) to coil and fold into an overall globular or other shape. (***a***) Hydrogen bonds, hydrophobic interactions, and ionic attractions between R groups are among the forces that hold the parts of the molecule in the designated shape. Disulfide bonds, shown here and in Figure 3–21, are covalent bonds between the sulfur atoms of two cysteines. (***b***) Schematic drawing of the tertiary structure of a polypeptide that also has both α-helical and β-sheet secondary structure. The polypeptide is a subunit of a DNA-binding protein (CAP) from the bacterium *Escherichia coli.* The regions of the polypeptide that are in α-helical conformation are represented as blue tubes lettered A through F. Regions in beta-conformation are represented as gray arrows numbered 1 through 12. Green lines represent connecting regions. Notice that, although the molecule seems very complicated, it is a single polypeptide chain, starting at the amino end (bottom left) and terminating at the carboxyl end. Most of the bends and foldbacks that give the molecule its overall conformation (tertiary structure) are stabilized by R-group interactions.

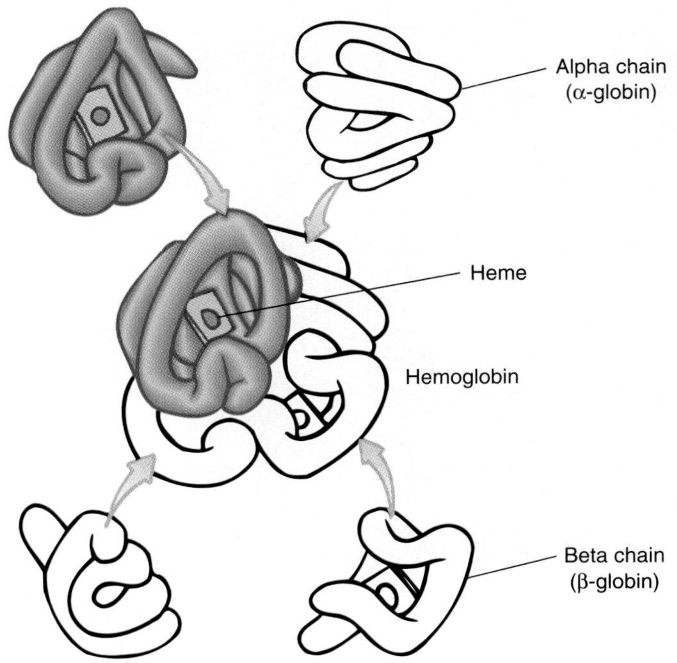

FIGURE 3–24 Hemoglobin has quaternary structure because it consists of more than one polypeptide. All four of its polypeptides must associate in the proper conformation for hemoglobin to function in oxygen transport. Each polypeptide is joined to an iron-containing molecule, a heme, but only two of the hemes can be seen in this view of the molecule.

teins are modular, consisting of two or more globular regions, called *domains,* connected by less compact regions of the polypeptide chain. Each domain may have a different function.

Conformation is determined by the primary structure of the polypeptide. Predicting the structure of a protein from its primary sequence of amino acids, however, is quite difficult because of the many possible combinations of folding patterns. Very complex computer programs are being developed to predict the three-dimensional shape of a protein from its amino acid sequence (see Fig. 3–23*b*).

The biological activity of a protein can be disrupted by a change in amino acid sequence that results in a change in conformation. For example, the genetic disease known as *sickle cell anemia* (see Chapter 15) is due to a mutation that causes the substitution of the amino acid valine for glutamic acid at position 6 (the sixth amino acid from the amino end) in the beta chain of hemoglobin. The substitution of valine (which has a nonpolar side chain) for glutamic acid (which has a charged side chain) makes the hemoglobin less soluble and more likely to form crystal-like structures. This alteration of the hemoglobin affects the red blood cells, changing them to the crescent or "sickle" shapes that characterize this disease.

Changes in the three-dimensional structure of a protein also disrupt its biological activity. When a protein is heated, subjected to significant pH changes, or treated with any of a number of chemicals, its structure can become disordered and the coiled peptide chains can unfold to give a more random conformation. This unfolding, which is mainly due to the disruption of hydrogen bonds and ionic bonds, is typically accompanied by a loss of the protein's biological activity—for example, its ability to act as an enzyme. Such changes in shape and the accompanying loss of biological activity are termed **denaturation** of the protein. Denaturation generally cannot be reversed. However, under certain conditions, some proteins that have been denatured return to their original shape and biological activity when normal environmental conditions are restored.

DNA AND RNA ARE NUCLEIC ACIDS

Nucleic acids transmit hereditary information and determine what proteins a cell manufactures. There are two classes of nucleic acids found in cells: **ribonucleic acids (RNA)** and **deoxyribonucleic acids (DNA).** DNA comprises the genes, the hereditary material of the cell, and contains instructions for making all the proteins needed by the organism (Fig. 3–25). RNA participates in the process of protein synthesis. Like proteins, nucleic acids are large, complex molecules. They were first isolated by Friederich Miescher in 1870 from the nuclei of pus cells; the name *nucleic acid* reflects the fact that they are acidic and were first identified in nuclei.

Nucleic acids consist of nucleotide subunits

Nucleic acids are polymers of **nucleotides,** molecular units that consist of (1) a five-carbon sugar, either ribose or deoxyribose, (2) a phosphate group, which makes the molecule acidic, and (3) a nitrogenous base, a ring compound that contains nitrogen. The nitrogenous base may be either a double-ringed purine or a single-ringed pyrimidine (Fig. 3–26). DNA commonly contains the purines adenine (A) and guanine (G), the pyrimidines cytosine (C) and thymine (T), the sugar deoxyribose, and phosphate. RNA contains the purines adenine and guanine, and the pyrimidines cytosine and uracil (U), to-

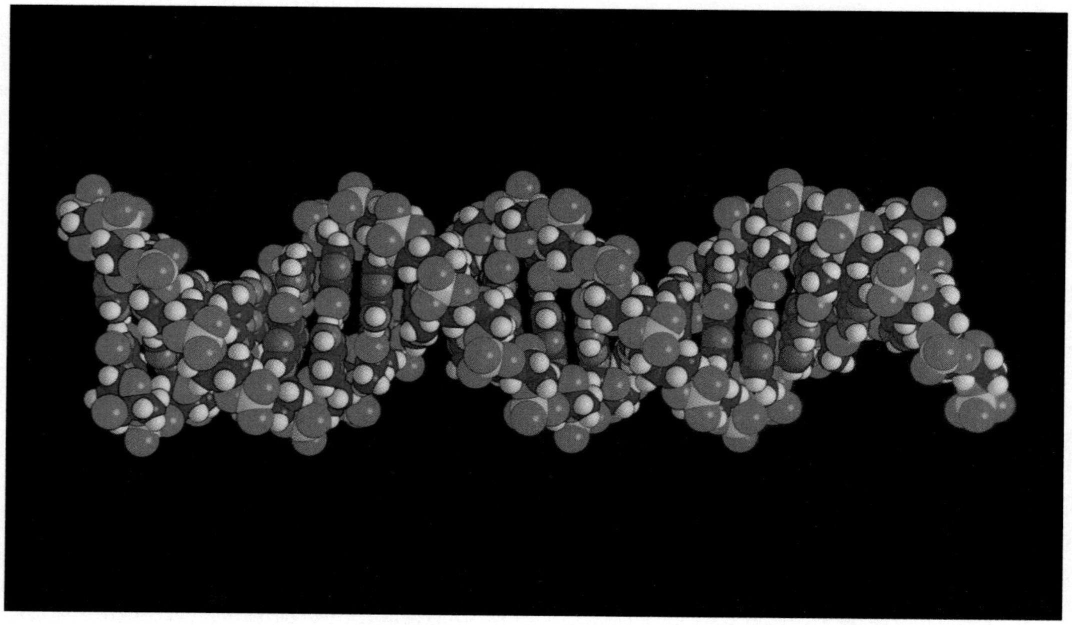

FIGURE 3–25 DNA is a long double helix. The photograph is a computer-generated simulation of the colored plastic, space-filling molecular models of DNA used by chemists. Twenty base pairs are shown in the crystalline B form first studied by Watson and Crick. *Red,* oxygen; *blue,* nitrogen; *dark blue,* carbon; *yellow,* phosphorus; *white,* hydrogen. (N.L. Max, University of California/Biological Photo Service)

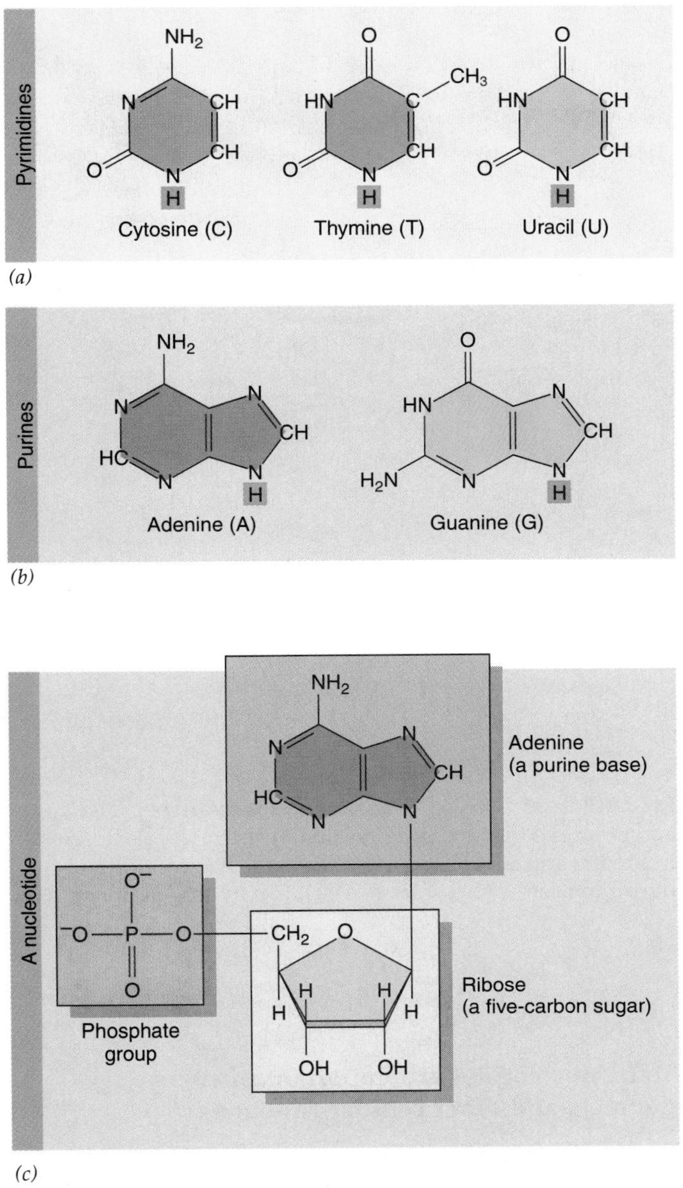

FIGURE 3–26 A nucleic acid consists of subunits called nucleotides. Each nucleotide consists of (1) a nitrogenous base, which may be either a purine or a pyrimidine; (2) a five-carbon sugar, either ribose (in RNA) or deoxyribose (in DNA); and (3) a phosphate group. (*a*) The three major pyrimidine bases found in nucleotides are cytosine, thymine (in DNA only), and uracil (in RNA only). (*b*) The two major purine bases found in nucleotides are adenine and guanine. The hydrogens indicated by the boxes are removed when the base is attached to a sugar. (*c*) A nucleotide, adenosine monophosphate (AMP).

gether with the sugar ribose, and phosphate. The removal of the phosphate group from a nucleotide yields a compound, termed a *nucleoside,* composed of the base and a sugar.

The molecules of nucleic acids are made of linear chains of nucleotides, which are joined by **phosphodiester linkages,** each consisting of a phosphate group and the covalent bonds that attach it to the sugars of adjacent nucleotides (Fig. 3–27). The specific information of the nucleic acid is coded in the unique sequence of the four kinds of nucleotides (each of which includes a specific base) present in the chain (see Chapter 12). DNA is composed of two nucleotide chains held together by hydrogen bonds and entwined around each other in a double helix.

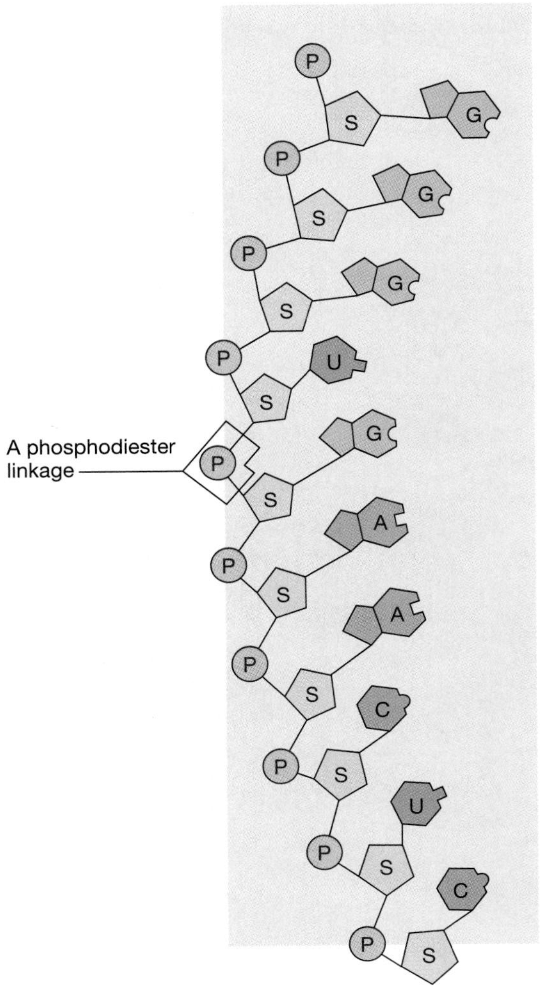

FIGURE 3–27 The nucleotides of a nucleic acid, such as the RNA molecule shown here, are arranged in a specific linear sequence. Each nucleotide includes a specific base, so the bases are also arranged in a specific sequence. P, phosphate; S, sugar; G, guanine; C, cytosine; A, adenine; U, uracil. Each phosphodiester linkage includes a phosphate group and the covalent bonds that link it to the sugars of adjacent nucleotides.

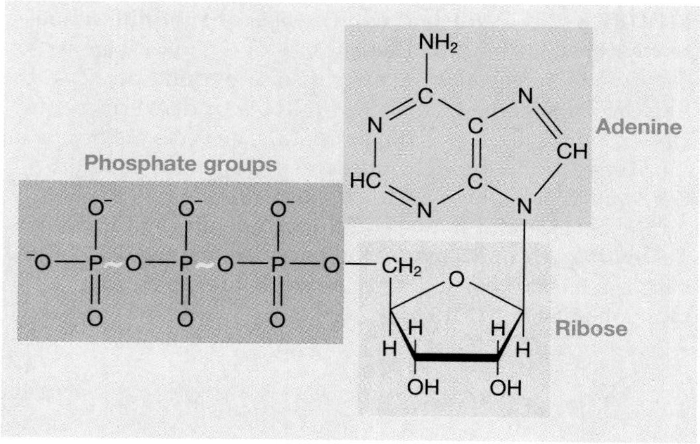

FIGURE 3–28 ATP (adenosine triphosphate) consists of a nucleotide joined to two terminal phosphate groups by unstable bonds (indicated by wavy lines).

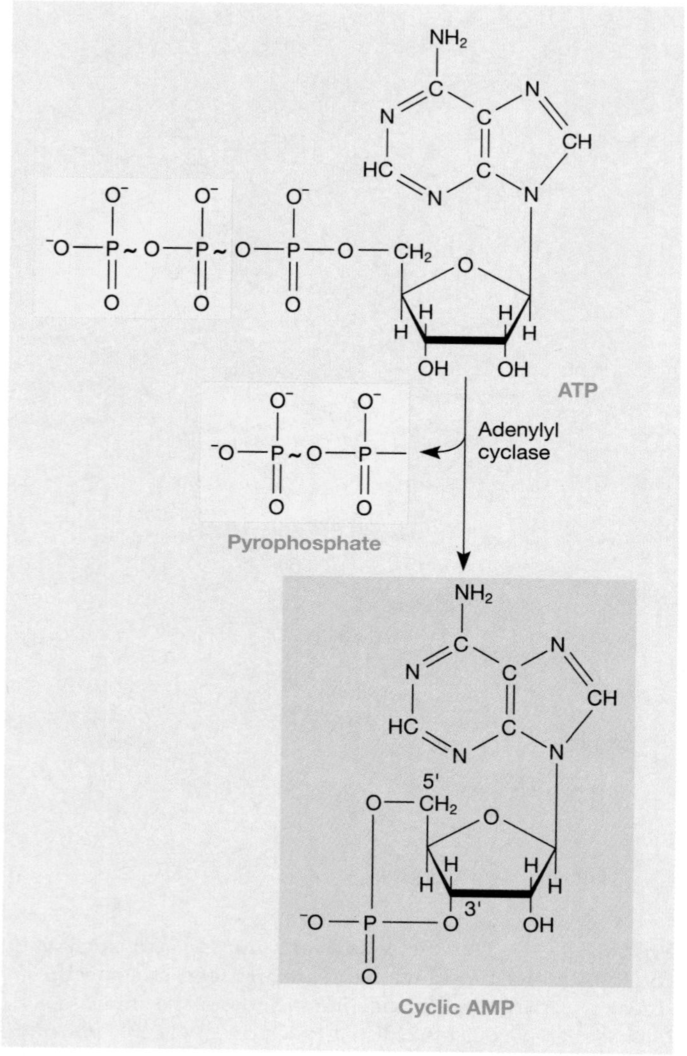

FIGURE 3–29 ATP is converted to cyclic AMP by adenylyl cyclase. The two terminal phosphates are released as pyrophosphate. The remaining phosphate becomes part of a ring connecting two regions of the ribose.

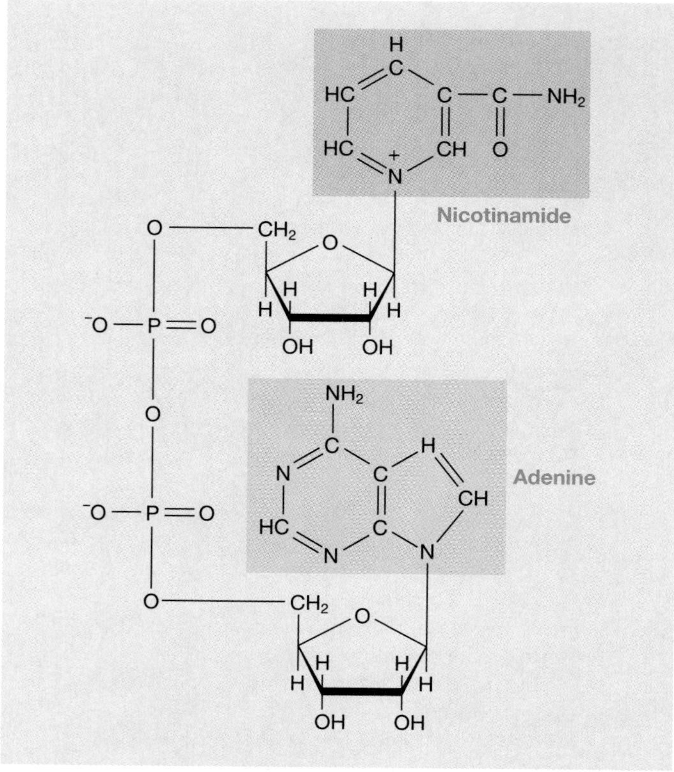

FIGURE 3–30 NAD^+ is an important hydrogen (electron) acceptor. The nicotinamide portion of the molecule accepts hydrogen and is reduced in the process. The resulting reduced molecule, known as NADH, is a hydrogen donor.

Some nucleotides are important in energy transfers and other cellular functions

Besides their importance as subunits of DNA and RNA, nucleotides serve other vital functions in living cells. **Adenosine triphosphate (ATP),** composed of adenine, ribose, and three phosphates (Fig. 3–28), is of major importance as the primary energy currency of all cells (see Chapter 6). The two terminal phosphate groups are joined to the nucleotide by unstable bonds, traditionally indicated by wavy lines. ATP can transfer a phosphate group to another molecule, making that molecule unstable and more reactive. In this way ATP is able to donate some of its chemical energy. Most of the readily available chemical energy of the cell is associated with the phosphate groups of ATP, which are ready to be transferred to other molecules.

A nucleotide may be converted to a cyclic form by an enzyme called a *cyclase.* ATP, for example, is converted to cyclic AMP (cyclic adenosine monophosphate) by the enzyme adenylyl cyclase (Fig. 3–29). Cyclic nucleotides mediate the effects of some hormones and regulate certain aspects of cellular function (see Chapters 13 and 47).

Cells contain several dinucleotides, which are of great importance in metabolic processes. For example, as discussed in Chapter 6, **nicotinamide adenine dinucleotide** (Fig. 3–30) has a primary role in biological oxidations and reductions within cells. It can exist in an oxidized form **(NAD^+)**, which is converted to a reduced form **(NADH)** when it accepts hydrogen. The electrons of these hydrogens, along with their energy, can be transferred to other molecules.

SUMMARY

I. The major groups of biologically important organic compounds are carbohydrates, lipids, proteins, and nucleic acids.

II. Carbon atoms form the backbone of a large variety of organic compounds essential to life.
   A. Each carbon atom can form four covalent bonds with four other atoms; these can be single, double, or triple bonds.
   B. Carbon forms covalent bonds with a greater number of different elements than does any other type of atom. Carbon atoms can form straight or branched chains, or can be joined into rings.

III. Isomers are compounds with the same molecular formula but different structures.
   A. Structural isomers differ in the covalent arrangements of their atoms.
   B. Geometric isomers, or *cis-trans* isomers, differ in the spatial arrangements of their atoms.
   C. Enantiomers are isomers that are mirror images of each other. Cells can distinguish between these configurations.

IV. Organic compounds are made up of specific functional groups with characteristic properties.
   A. Polar and ionic functional groups interact each other and dissolve in water.
   B. Partial charges on atoms at opposite ends of a bond are responsible for the polar property of a functional group. Hydroxyl and carbonyl groups are polar.
   C. Carboxyl and phosphate groups are acidic, becoming negatively charged when they release hydrogen ions. The amino group is basic, becoming positively charged when it accepts a hydrogen ion.

V. Long chains of similar organic compounds linked together are called polymers. Large polymers such as proteins and DNA are referred to as macromolecules.

VI. Carbohydrates contain carbon, hydrogen, and oxygen in a ratio of approximately one carbon to two hydrogens to one oxygen.
   A. Monosaccharides are simple sugars such as glucose, fructose, and ribose.
   B. Two monosaccharides can be joined by a glycosidic linkage, forming a disaccharide such as maltose or sucrose.
   C. Most carbohydrates are polysaccharides, long chains of repeating units of a simple sugar.
      1. Carbohydrates are typically stored in plants as starch and in animals as glycogen.
      2. The cell walls of plants are composed mainly of the polysaccharide cellulose.

VII. Lipid molecules are composed mainly of hydrocarbon-containing regions, with few oxygen-containing (polar or ionic) functional groups. Lipids have a greasy or oily consistency and are relatively insoluble in water.
   A. Neutral fats are used for fuel storage. A fat consists of a molecule of glycerol combined with one to three fatty acids.
      1. Monoacylglycerols, diacylglycerols, and triacylglycerols are neutral fats containing one, two, and three fatty acids, respectively.
      2. A fatty acid can be either saturated with hydrogen, or unsaturated.
   B. Phospholipids are structural components of cellular membranes.
   C. Steroid molecules contain carbon atoms arranged in four attached rings. Cholesterol, bile salts, and certain hormones are important steroids.

VIII. Proteins are large, complex molecules made of simpler subunits, called amino acids, joined by peptide bonds.
   A. Proteins are important structural components of cells and tissues. Many serve as enzymes. Most proteins are species-specific.
   B. Proteins are composed of various linear sequences of 20 different amino acids. Two amino acids combine to form a dipeptide. A longer chain of amino acids is a polypeptide.
      1. All amino acids contain an amino group and a carboxyl group, but vary in their side chains. The side chains of amino acids dictate their chemical properties.
      2. Amino acids generally exist as dipolar ions at cellular pH and serve as important biological buffers.
   C. Four levels of organization can be distinguished in protein molecules.
      1. Primary structure is the linear sequence of amino acids in the polypeptide chain.
      2. Secondary structure is a regular conformation, such as an α-helix or a β-pleated sheet, due to hydrogen-bonding between elements of the uniform backbone of the polypeptide.
      3. Tertiary structure is the overall shape of the polypeptide chains as dictated by chemical properties and interactions of the side chains of specific amino acids. Hydrogen-bonding, ionic bonds, hydrophobic interactions, and disulfide bridges contribute to tertiary structure.
      4. Quaternary structure is the association of two or more polypeptide chains.

IX. The nucleic acids DNA and RNA store and transfer information that governs the sequence of amino acids in proteins and ultimately the structure and function of the organism.
   A. Nucleic acids are composed of long chains of nucleotide subunits, each composed of a purine or pyrim-

idine nitrogenous base, a five-carbon sugar (ribose or deoxyribose), and a phosphate group.

B. ATP is a nucleotide of special significance in energy metabolism. NAD^+ is also involved in energy metabolism through its role as an electron and hydrogen acceptor in biological oxidations.

SELECTED KEY TERMS

adenosine triphosphate (ATP)
aldehyde
amino acid
amino group
amphipathic
carbohydrate
carbonyl group
carboxyl group
carotenoid
cellulose
condensation synthesis
cyclic AMP
deoxyribonucleic acid (DNA)
diacylglycerol (diglyceride)
dipeptide
disaccharide
enantiomer
fatty acid (saturated, unsaturated, polyunsaturated)
geometric isomer
glucose
glycerol
glycogen
hydrocarbon
hydrolysis
hydroxyl group
isomer
ketone
lipid
macromolecule
monoacylglycerol (monoglyceride)
monosaccharide
neutral fat
nicotinamide adenine dinucleotide
nucleotide
peptide bond
phosphate group
phospholipid
polypeptide
polysaccharide
primary structure
protein
quaternary structure
ribonucleic acid (RNA)
secondary structure
starch
steroid
structural isomer
sulfhydryl group
tertiary structure
triacylglycerol (triglyceride)

POST-TEST

1. Polysaccharides are composed of __________ subunits.
2. Neutral fats contain glycerol and one to three __________ __________ .
3. The primary structure of a protein refers to the sequences of __________ __________, linked by __________ bonds.
4. __________ is a polysaccharide that is an important component of the cell walls of plants.
5. A phospholipid, with its hydrophobic region and its hydrophilic region, is said to be a(an) __________ lipid.
6. The three functional groups that make up an amino acid are the __________ group, the __________ group, and the R group.
7. Animals store glucose in the form of __________ .
8. The three components of a nucleotide are a base, a(an) __________ and a(an) __________ .
9. In biological systems, monomers are most commonly linked together by __________ , the removal of the equivalent of a water molecule.
10. Two molecules that are mirror images are called __________ .
11. A fatty acid containing two or more double carbon-to-carbon bonds is said to be a(an) __________ fatty acid.
12. Interactions among side chains of the amino acids of a single polypeptide are responsible for the __________ structure of a protein.
13. A monosaccharide with a carbonyl group that includes a terminal carbon (when drawn in open chain form) is referred to as a(an) __________ sugar.
14. A regular α-helix is an example of __________ protein structure.
15. __________ is able to make another molecule more reactive by transferring a phosphate group to it.

REVIEW QUESTIONS

1. Marble is composed of the carbon-containing compound calcium carbonate ($CaCO_3$). Should calcium carbonate be considered an organic molecule? Why or why not?
2. What are some of the ways that the features of carbon-to-carbon bonds influence the stability and three-dimensional structure of molecules with carbon backbones?
3. Draw pairs of simple sketches comparing two (a) structural isomers, (b) geometric isomers, and (c) enantiomers.
4. Sketch the following functional groups: methyl, amino, carbonyl, hydroxyl, carboxyl, and phosphate. Classify each as either nonpolar, polar, acidic, or basic.
5. What features related to hydrogen-bonding give storage polysaccharides, such as starch and glycogen, very different properties than structural polysaccharides, such as cellulose and chitin?
6. Draw a structural formula of a simple amino acid, and identify the carboxyl group, amino group, and R group.
7. How does the primary structure of a polypeptide influence its secondary and tertiary structures? How can the conformation of a protein be disrupted?
8. Compare the functions of proteins and nucleic acids.

YOU MAKE THE CONNECTION

1. Describe some of the ways the electronegativity of oxygen affects the properties of water, and of the biological molecules of which oxygen is a part.
2. How do the differences in the structures of neutral fats and phospholipids determine their roles in the cell?

RECOMMENDED READINGS

Atkins, P.W. *Molecules,* Scientific American Library, W.H. Freeman and Co., New York, 1987. A fascinating account of the relationship between the properties and the molecular structures of various substances.

Bettelheim, F.A. and J. March. *Introduction to General, Organic and Biochemistry,* 4th ed. Saunders College Publishing, Philadelphia, 1995. A very readable reference text for those who would like to know more about the chemistry basic to life.

Garrett, R.H. and C.M. Grisham. *Biochemistry.* Saunders College Publishing, Philadelphia, 1995. A comprehensive advanced biochemistry text.

Olsen, A.J. and D.S. Goodsell. "Visualizing Biological Molecules." *Scientific American,* November 1992, Vol. 267, No. 5. A discussion of the use of computer-generated images to understand the structure of macromolecules.

Richards, F.M. "The Protein Folding Problem." *Scientific American,* January 1991, Vol. 264, No. 1. A discussion of the mechanisms involved when a protein folds into its biologically active shape.

CHAPTER 4

Cellular Organization

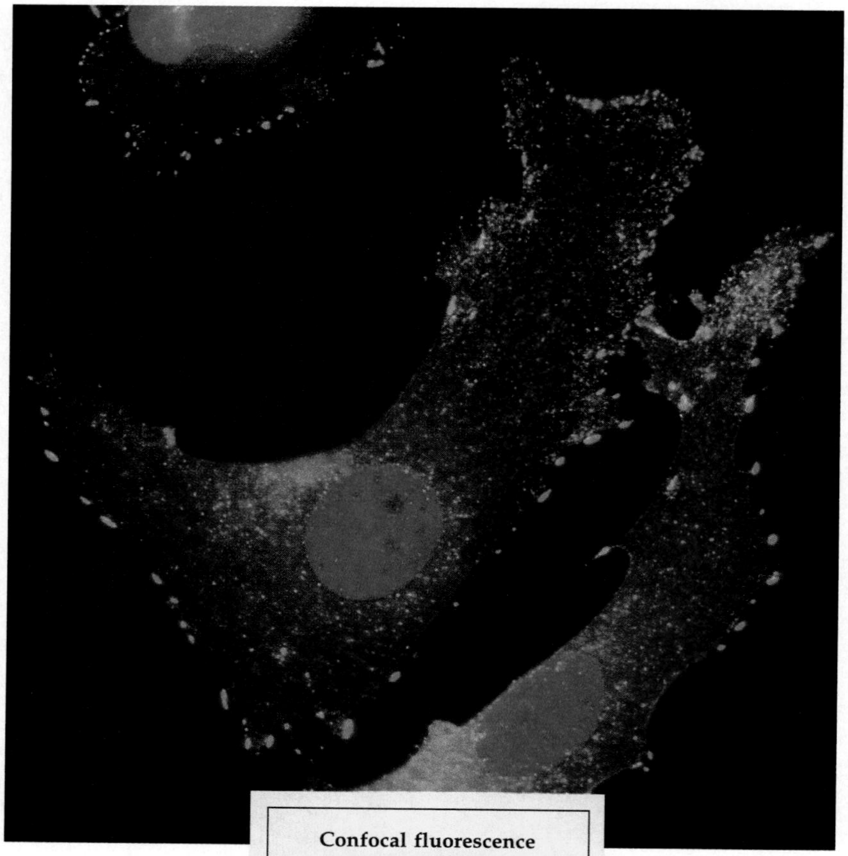

Confocal fluorescence micrograph of human breast cells. (Nancy Kedersha/ImmunoGen, Inc.)

A cell is a virtual microcosm of life, for it is the smallest unit that can carry out all life activities. Although some are more complex than others, most cells have all the physical and chemical components needed for their own maintenance, growth, and division. Cells convert energy from one form to another and use that energy to carry out various activities, ranging from mechanical work to chemical synthesis. Cells store genetic information in DNA molecules, which are faithfully replicated and passed on to the progeny during cell division. Cells use the information in the DNA to control their metabolisms and specify their structures. Of course a cell requires exchange of materials and energy with the environment, but when provided with essential nutrients and an appropriate environment, some cells can be kept alive and growing in the laboratory for many years. By contrast, no isolated cell part is capable of sustained survival.

Cells are the building blocks of even the most complex multicellular organisms. They are extraordinarily versatile modules that can be modified in a variety of ways to carry out specialized functions.

Cells provide dramatic testimony for the underlying unity of all living things. When we examine a wide range of seemingly very diverse organisms, ranging from simple bacteria to the most complex plants and animals, we find striking similarities at the cellular level. Careful studies of shared cellular features help us trace the evolutionary history of various groups of organisms, and furnish powerful evidence that all organisms alive today had a common origin.

LEARNING OBJECTIVES

After you have studied this chapter you should be able to

1. Point out the basic needs of all living things, and explain how a cell is able to meet these needs. Justify why the cell is considered the basic unit of life, and explain some of the ramifications of the cell theory.
2. Discuss the general characteristics of prokaryotic and eukaryotic cells.
3. Evaluate size relationships among different cells and cell structures.
4. Explain why the relationship between surface area and volume of a cell is important in determining cell size limits.
5. Describe the structure of the nucleus and relate this information to its function in eukaryotic cells.
6. Distinguish between smooth and rough endoplasmic reticulum in terms of both structure and function, and discuss the relationship between the endoplasmic reticulum and other internal membranes in the cell.
7. Trace the path of certain proteins synthesized in the rough endoplasmic reticulum as they are subsequently processed, modified, and sorted by the Golgi complex.
8. Describe the functions of lysosomes. Explain what can happen when they fail to carry out their functions or when they leak.
9. Distinguish between the functions of chloroplasts and mitochondria, and explain why both organelles synthesize ATP.
10. Describe the structures of the major types of fibers that make up the cytoskeleton. Explain the importance of the cytoskeleton to the cell.
11. Relate the structural features of cilia and flagella to the way in which these organelles are able to move.

THE CELL IS THE SMALLEST UNIT OF LIFE

The **cell theory** states that cells are the fundamental units of all organisms. Two German scientists, botanist Matthias Schleiden in 1838 and zoologist Theodore Schwann in 1839, were the first to point out that plants and animals are composed of groups of cells, and to put forth the unifying concept that the cell is the basic unit of living organisms.

The cell theory was extended in 1855 by Rudolph Virchow, who stated that new cells are formed only by the division of previously existing cells. In other words, cells do not arise by spontaneous generation from nonliving matter (an idea that was rooted in the writings of Aristotle and persisted over many centuries). About 1880 another famous biologist, August Weismann, added an important corollary to Virchow's statement by pointing out that the ancestry of all the cells alive today can be traced back to ancient times. Evidence that all presently living cells have a common origin is provided by the basic similarities in their structures and in the molecules of which they are made.

CELLS SHARE MANY ATTRIBUTES

Although cells may appear to be very diverse, their fundamental features are remarkably similar. This is a reflection of their evolution from a common ancestor, as well as the fact that they have many common needs.

First and foremost, every cell must be able to keep its contents together and separated from the external environment. For this reason, all cells, from bacteria to human cells, are enclosed by a surface membrane, commonly known as the **plasma membrane.**[1] Cells must also be able to accumulate materials and energy stores and to exchange materials with the environment, usually in a highly regulated fashion. Therefore the plasma membrane must serve as an extremely selective barrier, making the interior of the cell an enclosed compartment with a chemical composition quite different from that outside.

All living cells need one or more sources of energy, but a cell rarely obtains energy in a form that is immediately usable. All cells must therefore have the ability to convert energy to a convenient form, usually ATP (adenosine triphosphate; see Chapter 3). Although the specifics vary, the basic strategies cells use for energy conversion are very similar. The chemical reactions that convert energy from one form to another are essentially the same in all cells, from bacteria to those of complex plants and animals.

Every cell needs to control its activities and specify its structure. Genetic information has the central role in directing cellular activities (see *Making the Connection: Biological Molecules and Cellular Control*). The genetic information pathway (Fig. 4–1) interacts with many other kinds of information from a wide variety of sources.

[1]The term *membrane* is widely used in biology to refer to any structure that is like a thin sheet. However, the cellular membranes discussed in this chapter have a unique structure consisting of a lipid bilayer and other molecular components (see Chapters 3 and 5).

Making the Connection

Biological Molecules and Cellular Control

How does a cell store and use genetic information? Recall from Chapter 3 that a cell stores information in the form of **deoxyribonucleic acid (DNA).** The DNA molecule contains a linear sequence of components called nucleotides. In all cells this sequence of nucleotides serves as a code that specifies the amino acid sequence (primary structure) in proteins. Proteins are the molecules that carry out most cellular functions.

Many proteins are **enzymes,** special molecules that catalyze virtually every chemical reaction that takes place in organisms. By acting as catalysts, enzymes dramatically speed up specific chemical reactions to rates compatible with life. Thus, by specifying the structure of enzymes and other proteins, which in turn direct both the synthesis and the breakdown of all biological molecules (including lipids, carbohydrates, and nucleic acids), the DNA molecule directs the metabolism of the cell.

DNA uses a second nucleic acid, **ribonucleic acid (RNA),** as an intermediary. The sequence of bases in DNA that encodes a protein is copied as a sequence of bases in RNA through a process known as **transcription.** A particular kind of RNA called **messenger RNA (mRNA)** is responsible for carrying the information needed for synthesis of proteins (a process known as **translation**). The translation process itself is complex, requiring complicated machinery (see Chapter 12). ▲

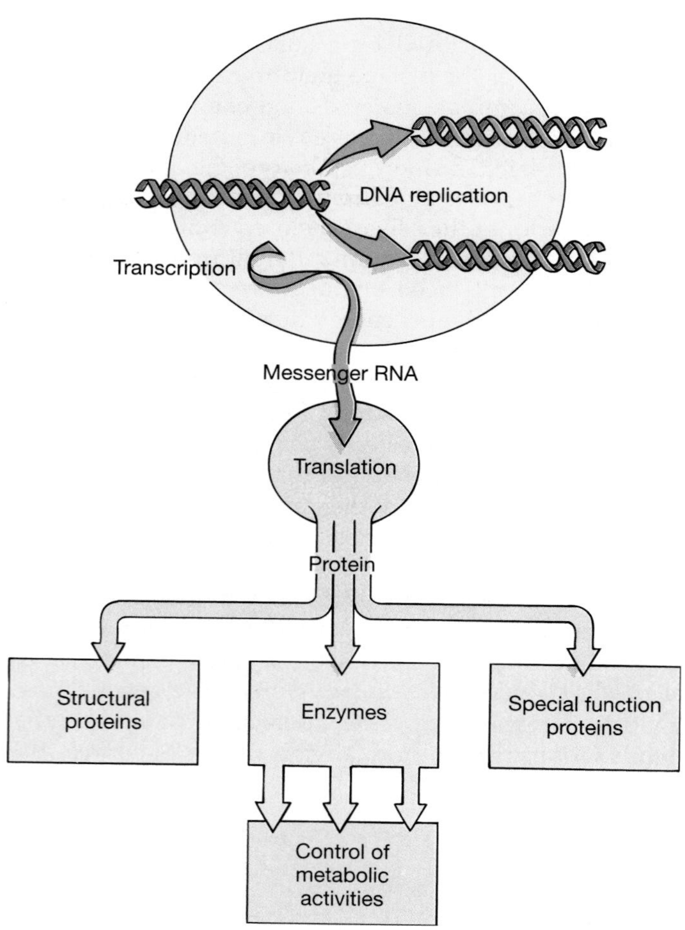

FIGURE 4–1 Genetic information generally flows from DNA to RNA to protein.

CELL SIZE IS LIMITED

Although their sizes vary over a wide range (Fig. 4–2), most cells are microscopically small. We therefore require very small units in order to measure cells and their internal structures. The basic unit of linear measurement in the metric system (see inside back cover) is the meter (m), which is just a little longer than a yard. A millimeter (mm) is 1/1000 of a meter and is about as long as the bar enclosed in parentheses (-). The micrometer (μm) is the most convenient unit for measuring cells. A bar 1 μm long is far too short to be seen with the unaided eye, for it is 1/1,000,000 (one millionth) of a meter, or 1/1000 of a millimeter, long. Most of us have difficulty thinking about units that are too small to see, but it is very helpful to remember that a micrometer has the same relationship to a millimeter that a millimeter has to a meter (1/1000). As small as it is, the micrometer is actually too large to measure most subcellular structures. For these purposes we use the nanometer (nm), which is 1/1,000,000,000 (one billionth) of a meter, or 1/1000 of a micrometer. To mentally move down to the world of the nanometer, we make a now-familiar transition. Just as a millimeter is one thousandth of a meter, a micrometer is one thousandth of a millimeter, and a nanometer is one thousandth of a micrometer.

A good light microscope allows us to see most types of bacterial cells, and some specialized animal cells are large enough to be seen with the naked eye. The human egg cell, for example, is about 130 μm in diameter, or approximately the size of the period at the end of this sentence. The largest cells are birds' eggs, but they are atypical because both the yolk and the "egg white" consist of food reserves. The functioning part of the cell is a microscopic mass on the surface of the yolk.

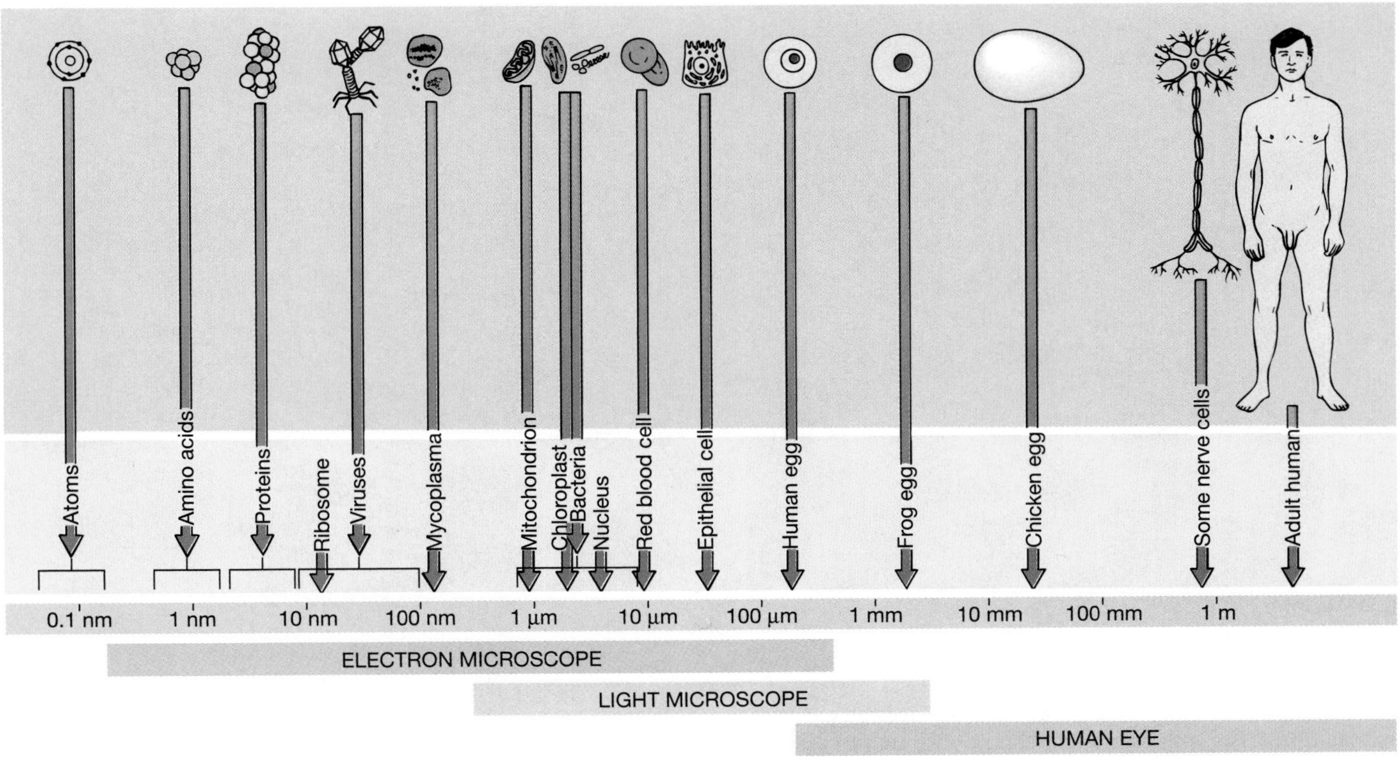

FIGURE 4–2 The relative lengths of some well-known cells, their organelles, and other components are most conveniently compared using a logarithmic scale. The figure is relatively compact because a logarithmic scale has a tenfold difference between successive units. If a linear scale had been used, the figure would be more than 10,000 kilometers wide! Prokaryotic cells (including bacteria) vary in size from about 1 to 10 μm long; eukaryotic cells (cells of plants and animals) generally fall within the range of 10 to 100 μm long; the majority are between 10 and 30 μm. While there is actually some overlap between the largest prokaryotic cells and the smallest eukaryotic cells, an average prokaryotic cell is about 1/10 the length of an average eukaryotic cell. The nuclei of animal and plants cells range from about 3 to 10 μm in diameter. Mitochondria are about the size of small bacteria, whereas chloroplasts are usually larger, about 5 μm long.

The sizes and shapes of cells are related to the functions they perform (Fig. 4–3). Some cells, such as the amoeba and the white blood cell, can change their shape as they move about. Sperm cells have long, whiplike tails, called flagella, for locomotion. Nerve cells possess long, thin extensions that permit them to transmit messages over great distances. The extensions on some nerve cells in the human body may be as long as 1 m. Other cells, such as epithelial cells, are almost rectangular in shape and are stacked much like building blocks to form sheet-like structures.

Why are most cells so small? If you consider what a cell must do to grow and survive, it may be easier to understand the reasons for its small size. A cell must take in food and other materials through its plasma membrane. Once inside, molecules of these substances must be transported to the locations where they are converted into other forms. Also, waste byproducts generated by various metabolic reactions must rapidly move out of the cell before they accumulate to toxic concentrations. Because cells are small, the distances molecules travel within them are relatively short, which speeds up many cellular activities. In addition, because essential molecules and waste products must all pass through the plasma membrane, the more surface area the cell has, the faster a given quantity of molecules can pass through it. This means that a critical factor in determining cell size is the ratio of its surface area to its volume (Fig. 4–4).

The fact that as a cell becomes larger its volume increases more rapidly than its surface area places an upper limit on cell size. Above some critical size, the number of molecules required by the cell could not be transported into the cell fast enough to sustain its needs. Of course, not all cells are spherical or cuboidal in shape. Some very large cells have relatively favorable ratios of surface area to volume because of their shapes. In fact, much of the variation in cell shape represents different ways of increasing the ratio of surface area to volume. For example, some cells such as the epithelial cells pictured in Figure 4–3*e* have finger-like projections of the plasma membrane, called *microvilli,* that significantly increase the surface area used to absorb nutrients.

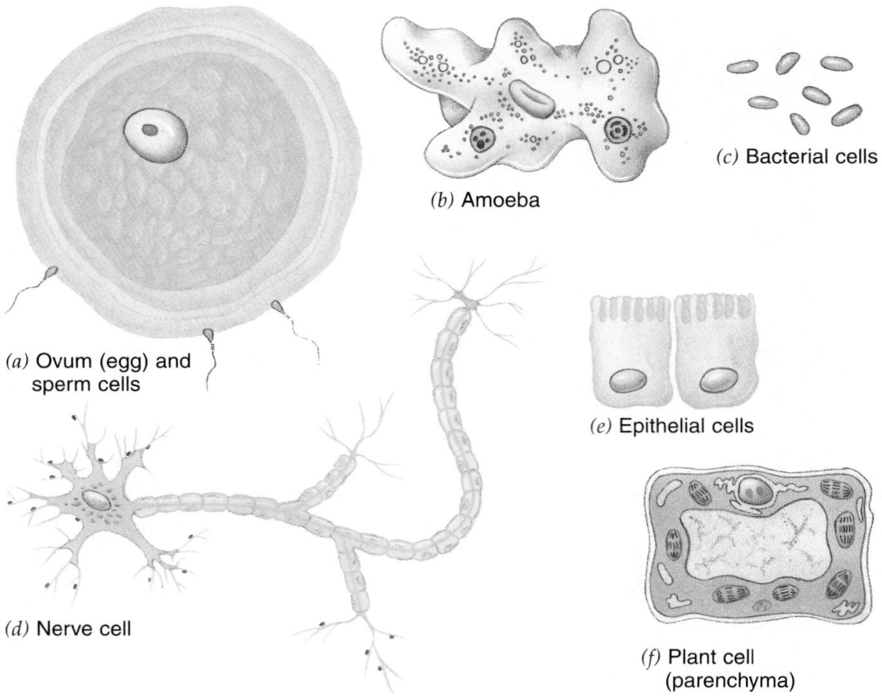

FIGURE 4–3 The size and shape of a cell are related to its functions. The cells shown here are not drawn to scale. (***a***) An ovum (egg cell) and sperm cells. Ova are among the largest cells; sperm cells are comparatively tiny. Note the long tail (flagellum) used by the sperm cell in locomotion. By whipping its flagellum, the sperm can move toward the egg. (***b***) The amoeba changes its shape as it moves from place to place. (***c***) The small size of bacterial cells enables them to grow and divide rapidly. (***d***) Nerve cells are specialized to transmit messages from one part of the body to another. (***e***) Epithelial cells join to form tissues that cover body surfaces and line body cavities. (***f***) The bulk of the organs of most young plants consists of relatively unspecialized cells called parenchyma cells.

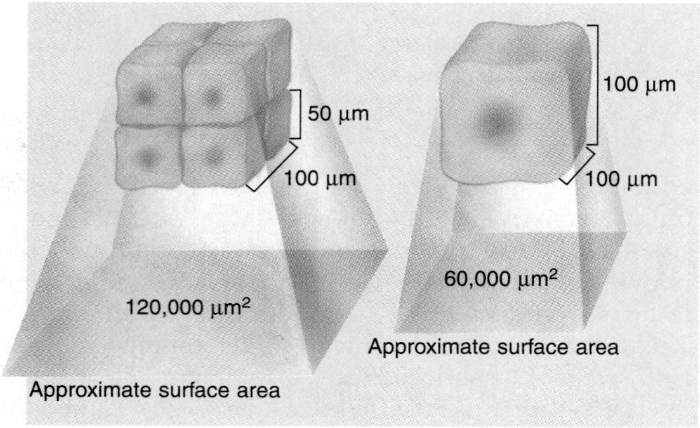

FIGURE 4–4 Eight small cells have a much greater surface area (plasma membrane) in relation to their total volume than does one large cell. The volume of a cube is equal to the cube of the length of one of its sides. The volume of the large cube on the right, which is equal to the total volume of the eight small cubes on the left, is $(100\ \mu m)^3 = 1{,}000{,}000\ \mu m^3$. The surface area of a cube is equal to the square of the length of one of its sides, multiplied by 6 (the number of sides). The surface area of the large cube on the right is $(100\ \mu m)^2 \times 6 = 60{,}000\ \mu m^2$. The ratio of its surface area to its volume (surface area:volume) is 0.06. The total surface area of the eight small cubes on the left is $(50\ \mu m)^2 \times 6 \times 8 = 120{,}000\ \mu m^2$. The ratio of their total surface area to their total volume is 0.12, which is *double* the surface-to-volume ratio of the single large cube.

CELLS ARE STUDIED BY A COMBINATION OF METHODS

Because cells are so small, scientists have had to be extremely clever in developing methods for studying them. Traditionally, one of the most important tools used to study cell structures has been the microscope. In fact, cells were not described until 1665, when Robert Hooke examined a piece of cork using a microscope he had made. Hooke did not actually see cells in the cork; rather, he saw the walls of dead cork cells (Fig. 4–5). Not until much later was it realized that the interior enclosed by the walls is the important part of living cork cells.

By the early 20th century, refined versions of the *light microscope* (Fig. 4–6), as well as certain organic chemicals that specifically stain different cellular structures, became available. These enabled biologists to discover that cells contain a number of different internal structures called **organelles** (literally, "little organs"). We now know that each type of organelle performs specific functions required for the cell's existence. The contribution of organic chemists in the development of biological stains was essential to this understanding, because the interior of many cells is transparent. Most of the methods used to prepare and stain cells for observation, however, also kill them in the process. More recently, sophisticated types of

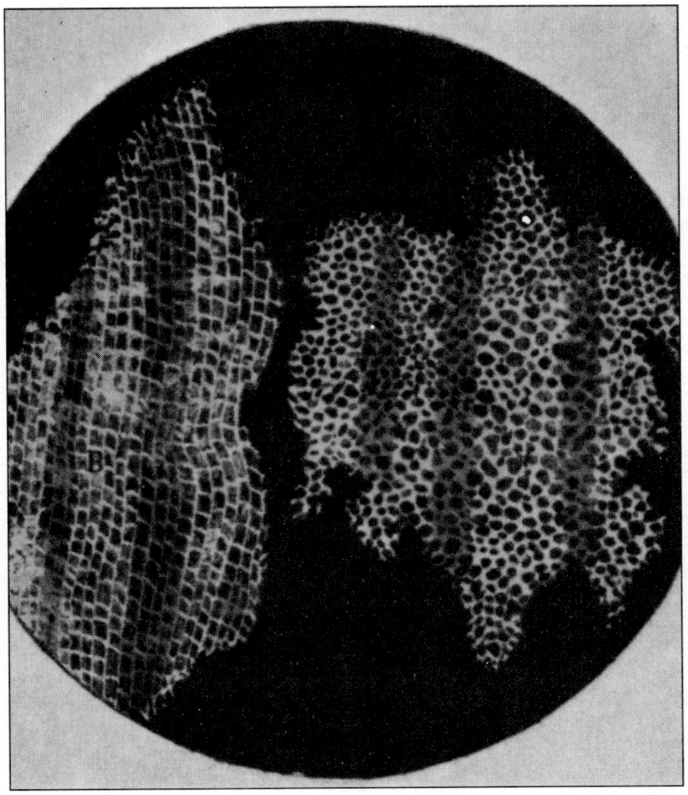

FIGURE 4–5 This drawing of the microscopic structure of a thin slice of cork was made by Robert Hooke. Hooke was the first to describe cells, basing his observations on the cell walls of these dead cork cells. Hooke used the term *cell* because the tissue reminded him of the small rooms that monks lived in during that period. (From the book *Micrographica*, published in 1665, in which Hooke described many of the objects he had viewed through the compound microscope he himself had constructed).

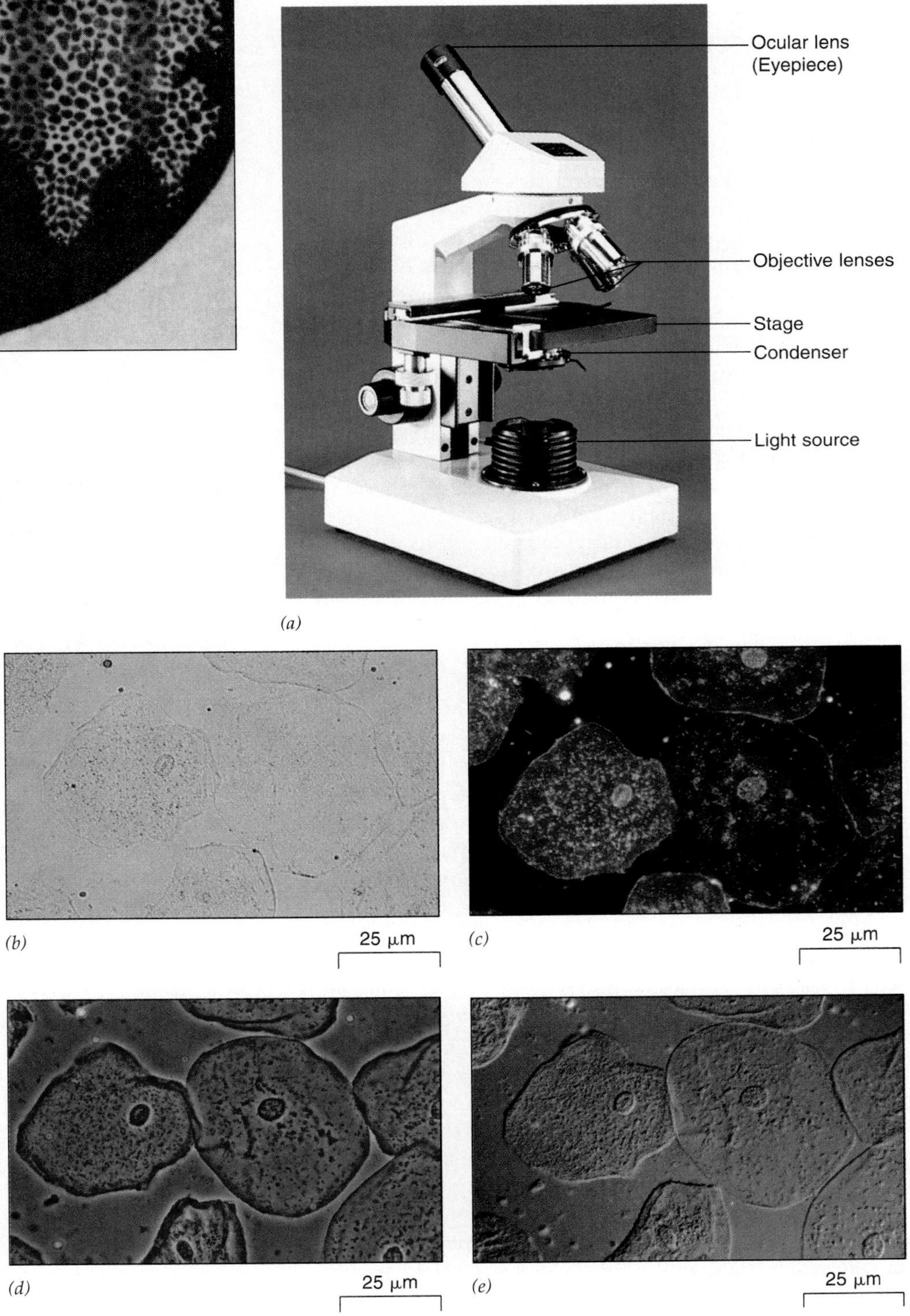

FIGURE 4–6 Several kinds of light microscopes are available to biologists today. (*a*) A student light microscope. Unstained epithelial cells photographed using (*b*) bright field (transmitted light) (*c*) dark field, (*d*) phase contrast, and (*e*) Nomarski differential interference microscopy. The phase contrast and differential interference microscopes enhance detail by increasing the differences in optical density in different regions of the cells. (*a*, Carolina Biological Supply Company/Phototake–NYC; *b–e*, Jim Solliday/Biological Photo Service)

light microscopes have been developed that use interfering waves of light to enhance the internal structures of cells. With *phase contrast* and *Nomarski differential interference microscopes,* some internal structures can be seen in unstained living cells (Fig. 4–6*d* and *e*). One of the most striking things that can be observed with these microscopes is that living cells contain numerous internal structures that are constantly changing shape and location.

Fluorescence microscopes are used to detect the locations of specific molecules in cells. Fluorescent stains (like paints that glow under black light) are molecules that absorb light energy of one wavelength and then release some of that energy as light of a longer wavelength. One such stain binds specifically to DNA molecules and emits green light after absorbing ultraviolet light. Cells can be stained, and the location of the DNA can be determined, by observing the source of the green fluorescent light within the cell. Some fluorescent stains can be chemically bonded to *antibodies,* special protein molecules that can then bind to a highly specific region of a molecule in the cell (Fig. 4–7*a*; see also Chapter 43). A single type of antibody molecule can bind to only one type of structure, such as a part of a specific protein or some of the sugars in a specific polysaccharide. Purified fluorescent antibodies known to bind to a specific protein isolated from a cell can be used to determine where that protein is located. Recently, powerful new computer imaging methods have allowed the development of the *confocal fluorescence microscope,* which greatly improves the resolution of structures labeled by fluorescent dyes (Fig. 4–7*b*).

Cells and their components are so small that ordinary light microscopes can distinguish only the gross details of many cell parts. In most cases all that can be seen clearly is the outline of a structure and its ability to be stained by some dyes and not by others. Not until the development of the **electron microscope (EM),** which came into wide use in the 1950s, were researchers able to study the fine details, or **ultrastructure,** of cells.

Two features of a microscope determine how clearly a small object can be viewed. The *magnification* of the instrument is the ratio of the size of the image seen with the microscope to the actual size of the object. The best light microscopes usually magnify an object no more than 1000 times, whereas the electron microscope can magnify it 250,000 times or more. The other, even more important, feature of a microscope is its capacity to see fine detail, called its **resolving power.** This is defined as the minimum distance between two points at which they can both be seen separately rather than as a single blurred point. Resolving power depends on the quality of the lenses and the *wavelength* of the illuminating light; as the wavelength decreases, the resolution increases. The visible light used by light microscopes has wavelengths ranging from about 400 nm (violet) to 700 nm (red); this limits the resolution (resolving power) of the light microscope to details no smaller than the diameter of a small bacterial cell.

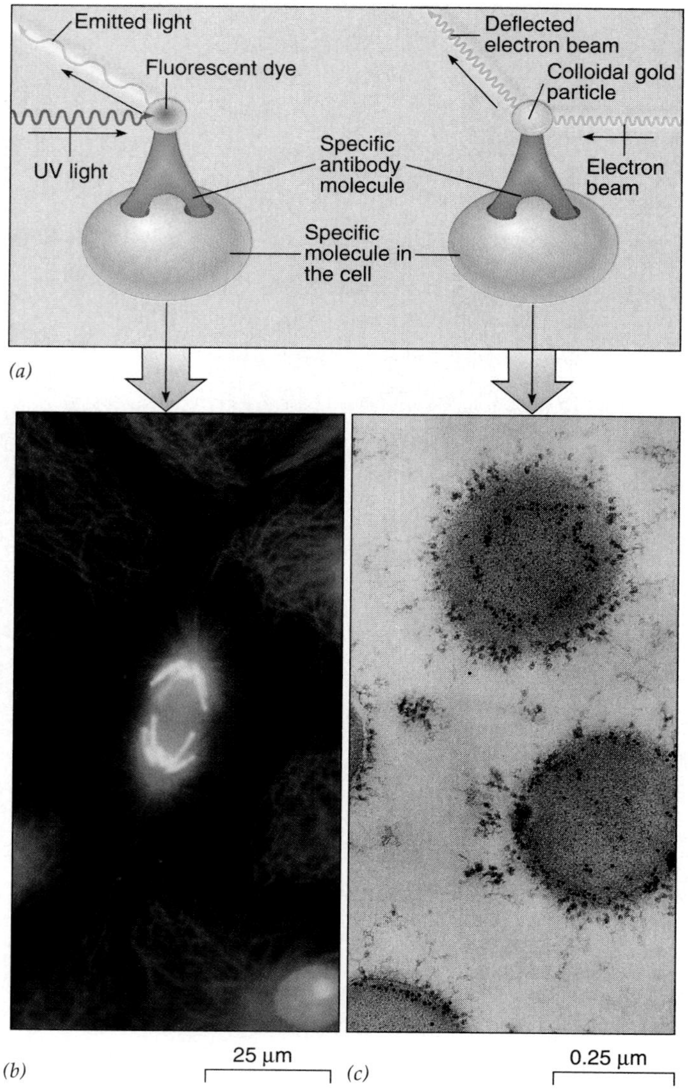

FIGURE 4–7 A specific antibody can be used to determine the cellular location of a particular molecule. (***a***) The antibody molecule on the left has been linked to a fluorescent dye, which emits visible light when illuminated with ultraviolet light. The antibody molecule on the right has been linked to a gold particle, which can be detected by electron microscopy because it deflects the electron beam. (***b***) Confocal fluorescence micrograph (see text) of cultured animal cells. The cell in the center is dividing. The DNA of the nuclei and chromosomes is yellow; the microtubules are red. (***c***) Transmission electron micrograph of part of a rat adrenal medullary cell. Tiny gold particles (known as colloidal gold) were linked to antibody molecules; the resulting complex binds specifically to the membranes of vesicles within the cell. (*b*, courtesy of Dr. John M. Murray, Department of Cell and Developmental Biology, University of Pennsylvania; *c*, courtesy of Carl Zeiss, Inc.)

Whereas the best light microscopes have about 500 times more resolving power than the human eye, the electron microscope multiplies the resolving power by more than 10,000 (Fig. 4–8). This is because electrons

(Text continues on page 86)

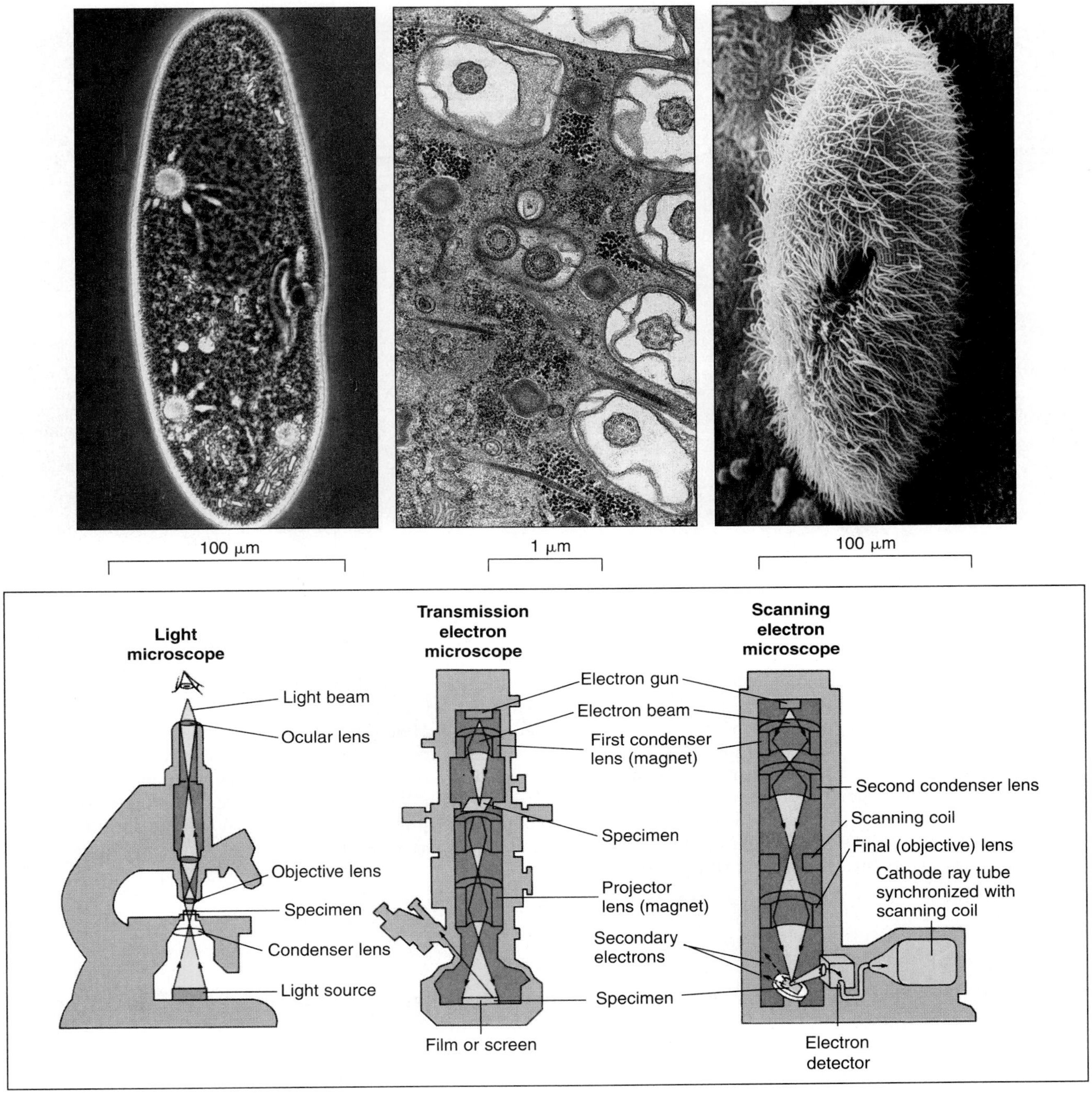

FIGURE 4–8 Distinctive images of cells, such as the protist *Paramecium* shown in the top row of photographs, are provided by a phase-contrast light microscope (left), a transmission electron microscope (center), and a scanning electron microscope (right). A light microscope can be used to view stained or living cells, but at relatively low resolution. The transmission electron microscope (TEM) produces a high resolution image that can be greatly magnified. Because of the high magnification, only a small part of the *Paramecium* cell is shown in the photograph. The scanning electron microscope (SEM) is used to provide a clear view of surface features. All three microscopes are focused by similar principles. A beam of light or an electron beam is directed by the condenser lens onto the specimen and is magnified by the objective lens and the projector lens in the TEM, or by the objective lens and the eyepiece in the light microscope. The TEM image is focused onto a fluorescent screen, and the SEM image is viewed on a type of "television" screen. Lenses in the electron microscopes are actually magnets that bend the beam of electrons. (Photos courtesy of T.K. Maugel/University of Maryland)

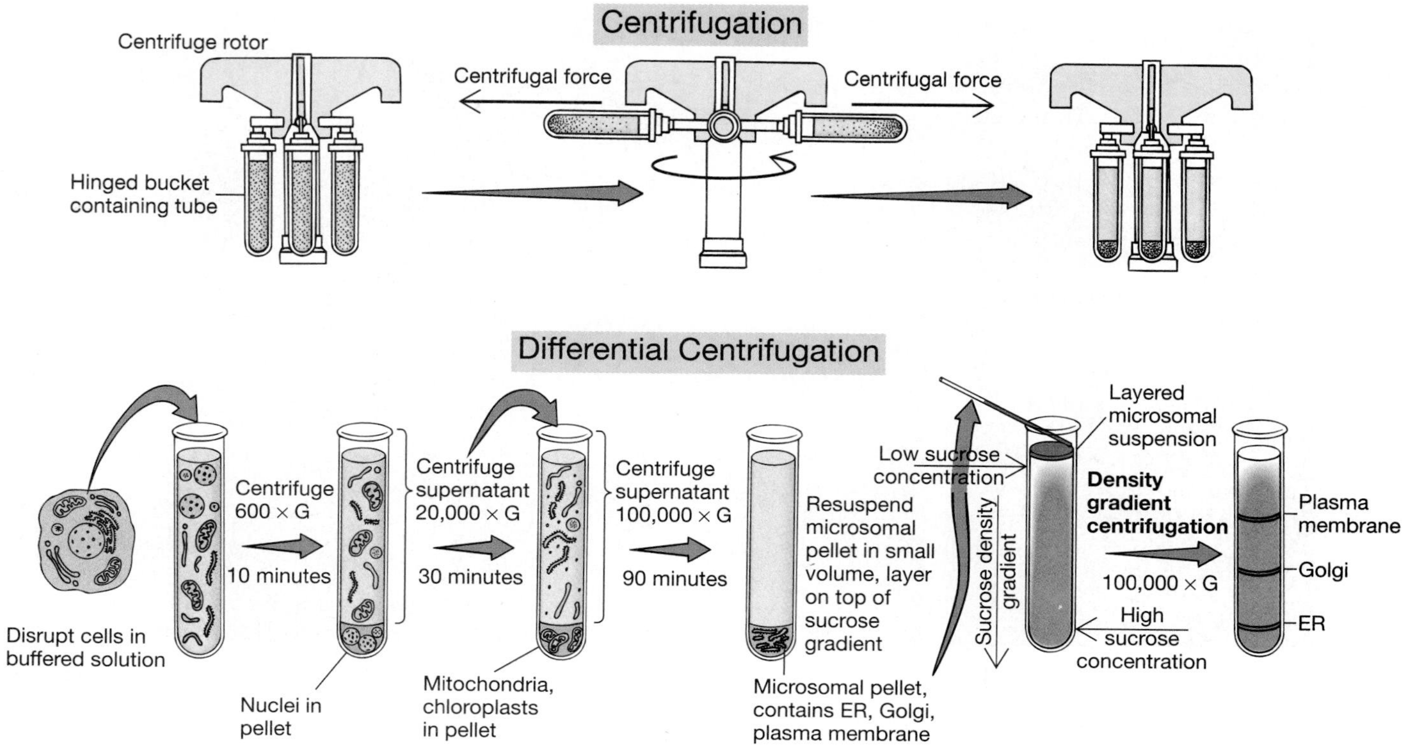

FIGURE 4–9 Cell fractionation allows separation of cellular components. Cell membranes and organelles are usually separated by centrifuges, machines that can spin tubes that are much like test tubes. Spinning the tubes exerts a centrifugal force on the contents; this force causes particles suspended in solution to sediment. Some of these particles, such as mitochondria, are entire organelles. Others are small closed vesicles, known as *microsomes,* made up of membranes from the ER, Golgi, and plasma membrane. After centrifugation some of the particles form a pellet at the bottom of the tube. Different cell parts have different densities; this allows them to be separated into various cell fractions by centrifuging the suspension at increasing revolutions per minute **(differential centrifugation).** Membranes and the organelles from the resuspended pellets can then be further purified by **density gradient centrifugation,** shown as the last step in the figure. The resuspended pellet is placed in a layer on top of a **density gradient,** usually made up of a solution of sucrose and water. Because the densities of the organelles and membranes differ, each will migrate during centrifugation and form a band at a different position in the gradient.

have very short wavelengths, on the order of about 0.1 to 0.2 nm. Although this implies that the limit of resolution in the electron microscope comes close to the diameter of a water molecule, such resolution is difficult to achieve with biological material. It can be approached, however, when isolated molecules such as proteins or DNA are examined.

The image formed by the electron microscope cannot be seen directly. The electron beam itself consists of energized electrons, which, because of their negative charge, can be focused by electromagnets just as images are focused by glass lenses in a light microscope (Fig. 4–8). For **transmission electron microscopy (TEM),** the specimen is embedded in plastic and then cut into extraordinarily thin sections (50 to 100 nm thick) with a glass or diamond knife. A section is then placed on a small metal grid. The electron beam passes through the specimen and then falls onto a photographic plate or a fluorescent screen that works much like a television screen. When you look at transmission electron microscope photographs in this chapter (and elsewhere), keep in mind that each represents only a thin cross section of a cell.

To reconstruct how something inside the cell looks in three dimensions, it is necessary to study many consecutive cross-sectional views (called *serial sections*) through the object. To understand the enormity of such a task, try imagining what it would be like to reconstruct the contents of your home from a set of hundreds of consecutive 5-cm sections.

Special methods allow the detection of specific molecules in electron microscope images, using antibody molecules that have very tiny gold particles bound to them. The dense gold particles block the electron beam and identify the location of the proteins recognized by the antibodies as precise black spots on the electron micrograph (Fig. 4–7*a* and *c*).

In another type of electron microscope, the **scanning electron microscope (SEM),** the electron beam does not pass through the specimen. Instead, the specimen is coated with a thin gold film. When the electron beam strikes various points on the surface of the specimen, secondary electrons are emitted whose intensity varies with the contour of the surface. The recorded emission patterns of the secondary electrons give a three-dimensional picture of the surface (Fig. 4–8). This special kind of micrograph provides information about the shape and external features of the specimen that cannot be obtained with the transmission electron microscope.

The electron microscope is a powerful tool for studying cell structure, but it has limitations. The methods used to prepare cells for electron microscopy kill them and may even alter their structure. Furthermore, electron microscopy usually provides only clues about the *functions* of organelles and other cell components. To determine what organelles actually do, researchers had to be able to purify different parts of cells so that they could be studied by physical and chemical methods. **Cell fractionation** procedures are methods for purifying organelles. Generally, cells are broken apart as gently as possible, and the mixture is subjected to centrifugal force by spinning in a device called a **centrifuge.** As the number of rpm's (revolutions per minute) increases, so does the centrifugal force (measured as *G,* which is equal to the force of gravity). This permits various cell components to be separated on the basis of their different densities (Fig. 4–9). The purified organelles can then be examined to determine what kinds of proteins and other molecules they might contain, as well as the nature of the chemical reactions that take place within them. Today, cell biologists often use a combination of experimental approaches to understand the functions of cellular structures.

EUKARYOTIC CELLS ARE MUCH MORE COMPLEX THAN PROKARYOTIC CELLS

Organisms can be placed into two major groups according to the structure and complexity of their cells. **Eukaryotes** are organisms whose cells contain highly organized membrane-bounded organelles. The most prominent of these is the *nucleus,* which serves to localize the hereditary material, DNA. In fact, the name eukaryote means "true nucleus." The DNA of **prokaryotes** (meaning "before the nucleus") is not enclosed in a nucleus, and other membranous organelles are also lacking. These cells, which include bacteria, are generally much smaller than eukaryotic cells. In fact, the average prokaryotic cell is only about one-tenth the diameter of the average eukaryotic cell (see Fig.4–2).

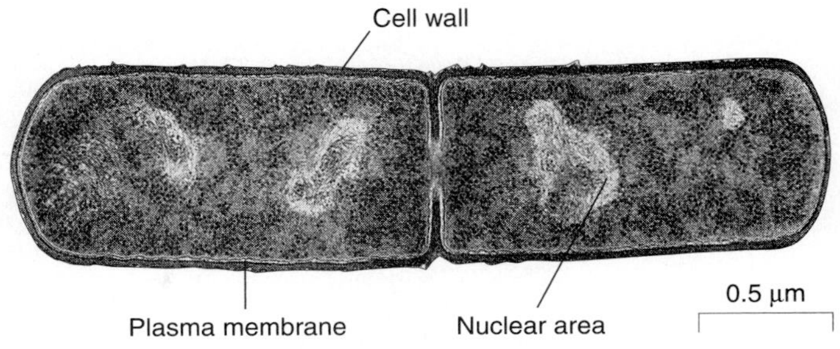

FIGURE 4–10 The prokaryotic cell shown in this transmission electron micrograph is about to complete cell division. This bacterium, *Bacillus subtilis,* has a prominent cell wall surrounding the plasma membrane. The nuclear areas are clearly visible. (Courtesy of A. Ryter)

Table 4–1 summarizes the types of organelles typically found in eukaryotic cells. Some organelles may be found only in specific cells. For example, chloroplasts, structures that trap sunlight for energy conversion, are found only in cells that carry on photosynthesis, such as certain plant cells. The many specialized organelles of eukaryotic cells allow them to overcome some of the problems associated with large size, so eukaryotic cells can be considerably larger than prokaryotic cells.

Prokaryotes are single-celled organisms that belong to the kingdom Prokaryotae, which includes the bacteria. The DNA in prokaryotic cells is usually concentrated in one or more regions; each is called a **nuclear area** (Fig. 4–10). A nuclear area is not enclosed by a separate membrane.

Like eukaryotic cells, prokaryotic cells have a surface **plasma membrane** which confines the contents of the cell to an internal compartment, but they do not have distinct internal membranous organelles. In some prokaryotic cells the plasma membrane may be folded inward to form a complex of membranes along which the cell's energy-transforming reactions are thought to take place. Some prokaryotic cells may also have **cell walls** or **outer membranes,** structures that enclose the entire cell, including the plasma membrane. Prokaryotic cells are discussed in more detail in Chapter 23.

EUKARYOTIC CELLS CONTAIN SPECIALIZED ORGANELLES

Early biologists thought that the cell consisted of a homogeneous jelly, which they called *protoplasm.* With the electron microscope and other modern research tools, perception of the environment within the cell has been greatly expanded. We now know that the cell is highly

Table 4–1 EUKARYOTIC CELL STRUCTURES AND THEIR FUNCTIONS

Structure	*Description*	*Function*
The Cell Nucleus		
Nucleus	Large structure surrounded by double membrane; contains nucleolus and chromosomes	Information in DNA is transcribed in RNA synthesis; specifies cellular proteins
Nucleolus	Granular body within nucleus; consists of RNA and protein	Site of ribosomal RNA synthesis; ribosome subunit assembly
Chromosomes	Composed of a complex of DNA and protein known as chromatin; become visible as rodlike structures when the cell divides	Contain genes (units of hereditary information that govern structure and activity of cell)
Cytoplasmic Organelles		
Plasma membrane	Membrane boundary of living cell	Encloses cellular contents; regulates movement of materials in and out of cell; helps maintain cell shape; communicates with other cells (also present in prokaryotes)
Endoplasmic reticulum (ER)	Network of internal membranes extending through cytoplasm	Synthetic site of membrane lipids and many membrane proteins; origin of intracellular transport vesicles carrying proteins to be secreted
Smooth	Lacks ribosomes on outer surface	Lipid biosynthesis; drug detoxification
Rough	Ribosomes stud outer surface	Manufacture of many proteins destined for secretion or for incorporation into membranes
Ribosomes	Granules composed of RNA and protein; some attached to ER, some free in cytosol	Synthesize polypeptides in both prokaryotes and eukaryotes
Golgi complex	Stacks of flattened membrane sacs	Modifies proteins; packages secreted proteins; sorts other proteins to vacuoles and other organelles
Lysosomes	Membranous sacs (in animals)	Contain enzymes to break down ingested materials, secretions, wastes
Vacuoles	Membranous sacs (mostly in plants, fungi, algae)	Transport and store materials, wastes, water

organized and complex (Figs. 4–11 to 4–14). It has its own control center, internal transportation system, power plants, factories for making needed materials, packaging plants, and even a "self-destruct" system. Today the word *protoplasm,* if used at all, is applied in a very general way. Specifically, the portion of the protoplasm outside the nucleus is called the **cytoplasm,** and the corresponding material within the nucleus is termed the **nucleoplasm.** Various organelles are suspended within the fluid component of the cytoplasm, which is generally referred to as the **cytosol.** Therefore, the term *cytoplasm* includes both the cytosol and all the organelles other than the nucleus.

Membranous organelles carry out specific functions

Membranes have unique properties that enable membranous organelles to carry out a wide variety of functions. For example, cellular membranes never have free ends; therefore a membranous organelle always contains at least one enclosed internal space or compartment. These membrane-bounded compartments allow certain cellular activities to be localized within specific enclosed regions of the cell. Reactants that are concentrated in only a small part of the total cell volume are far more likely to come in contact, and the rate of the reaction can be dramatically increased. Membrane-bounded compartments also keep certain reactive compounds away from other parts of the cell that might be adversely affected by them.

Membranes also allow the storage of energy. The membrane provides a barrier that is analogous to a dam on a river. A difference in the concentration of some substance on the two sides of a membrane is a form of stored energy or *potential energy* (see Chapter 6). As particles of the substance move across the membrane from the side of high concentration to the side of low concentration, the cell can convert some of their energy to the chemical energy of ATP molecules. This process of energy conversion

Table 4–1 continued

Structure	Description	Function
Microbodies (e.g., peroxisomes)	Membranous sacs containing a variety of enzymes	Sites of many diverse metabolic reactions
Mitochondria	Sacs consisting of two membranes; inner membrane is folded to form cristae and encloses matrix	Site of most reactions of cellular respiration; transformation of energy originating from glucose or lipids into ATP energy
Plastids (e.g., chloroplasts)	Double membrane structure enclosing internal thylakoid membranes; chloroplasts contain chlorophyll in thylakoid membranes	Chlorophyll captures light energy; ATP and other energy-rich compounds are formed and then used to convert CO_2 to glucose
The Cytoskeleton		
Microtubules	Hollow tubes made of subunits of tubulin protein	Provide structural support; have role in cell and organelle movement and cell division; components of cilia, flagella, centrioles, basal bodies
Microfilaments	Solid, rodlike structures consisting of actin protein	Provide structural support; play role in cell and organelle movement and cell division
Centrioles	Pair of hollow cylinders located near center of cell; each centriole consists of nine microtubule triplets (9×3 structure)	Mitotic spindle forms between centrioles during animal cell division; may anchor and organize microtubule formation in animal cells; absent in most plants
Cilia	Relatively short projections extending from surface of cell; covered by plasma membrane; made of two central and nine peripheral microtubules ($9 + 2$ structure)	Movement of some single-celled organisms; used to move materials on surface of some tissues
Flagella	Long projections made of two central and nine peripheral microtubules ($9 + 2$ structure); extend from surface of cell; covered by plasma membrane	Cellular locomotion by sperm cells and some single-celled eukaryotes

(discussed in Chapters 7 and 8) is a basic mechanism that cells use to capture and convert energy to sustain life.

Membranes also serve as important cellular work surfaces. For example, a number of chemical reactions in cells are carried out by enzymes that are bound to membranes. By organizing the enzymes that carry out successive steps of a series of reactions close together on a membrane surface, certain series of chemical reactions can occur more rapidly.

The cell nucleus contains DNA

The most prominent organelle is usually the **nucleus.** In most cases it is spherical or oval and averages 5 μm in diameter. Owing to its size and the fact that it often occupies a relatively fixed position near the center of the cell, some early investigators guessed long before experimental evidence was available that the nucleus served as the control center of the cell (see *Focus On: Acetabularia*). Most cells have one nucleus, although there are exceptions.

The nuclear envelope controls traffic between the nucleus and the cytoplasm

The **nuclear envelope** consists of two concentric membranes that separate the nuclear contents from the surrounding cytoplasm (Fig. 4–15). These membranes are fused at intervals, forming **nuclear pores.** Nuclear pores appear to allow the passage of materials to the cytoplasm from the interior of the nucleus and vice versa, but the process is highly selective, permitting only specific molecules to pass through these openings.

Chromosomes consist of chromatin, a DNA/protein complex

Almost all the DNA in a cell is located in the interior of the nucleus. The DNA molecules make up the **genes,** which contain the chemically coded instructions for producing virtually all the proteins needed by the cell. The

(Text continues on page 96)

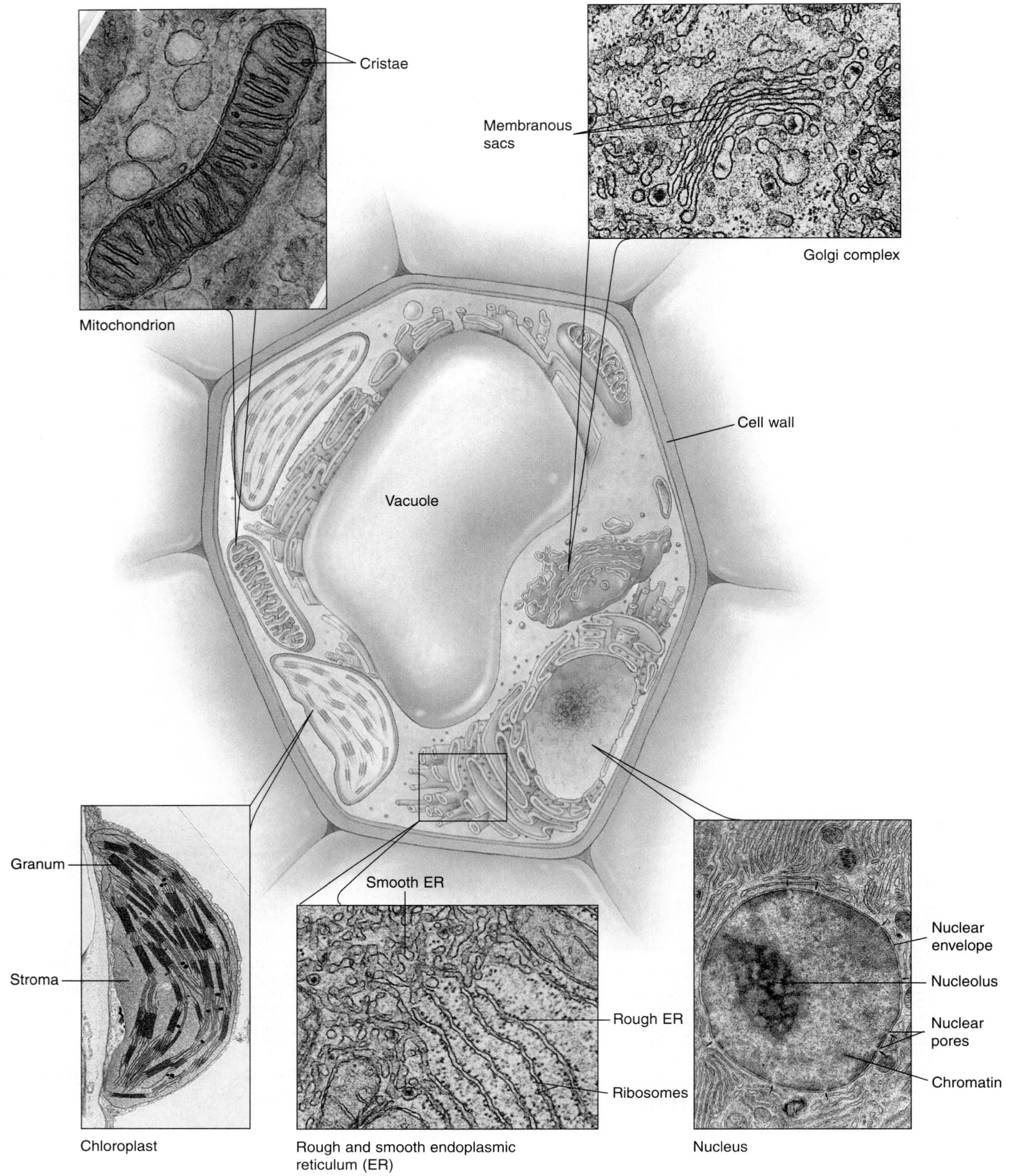

FIGURE 4–11 This composite diagram includes organelles typically found in plant cells. Some plant cells do not have all the organelles shown. Cells from photosynthetic tissues contain chloroplasts, for example, whereas root cells do not. Chloroplasts or other plastids, a cell wall, and prominent vacuoles are characteristic of plant cells. Many of the other components, such as the nucleus, the mitochondria, and the endoplasmic reticulum, are also found in animal cells. (Clockwise from top left: D.W. Fawcett; D.W. Fawcett and R. Bolender; D.W. Fawcett; Visuals Unlimited/R. Bolender and D.W. Fawcett; E.H. Newcomb and W.P. Wergin, University of Wisconsin/Biological Photo Service)

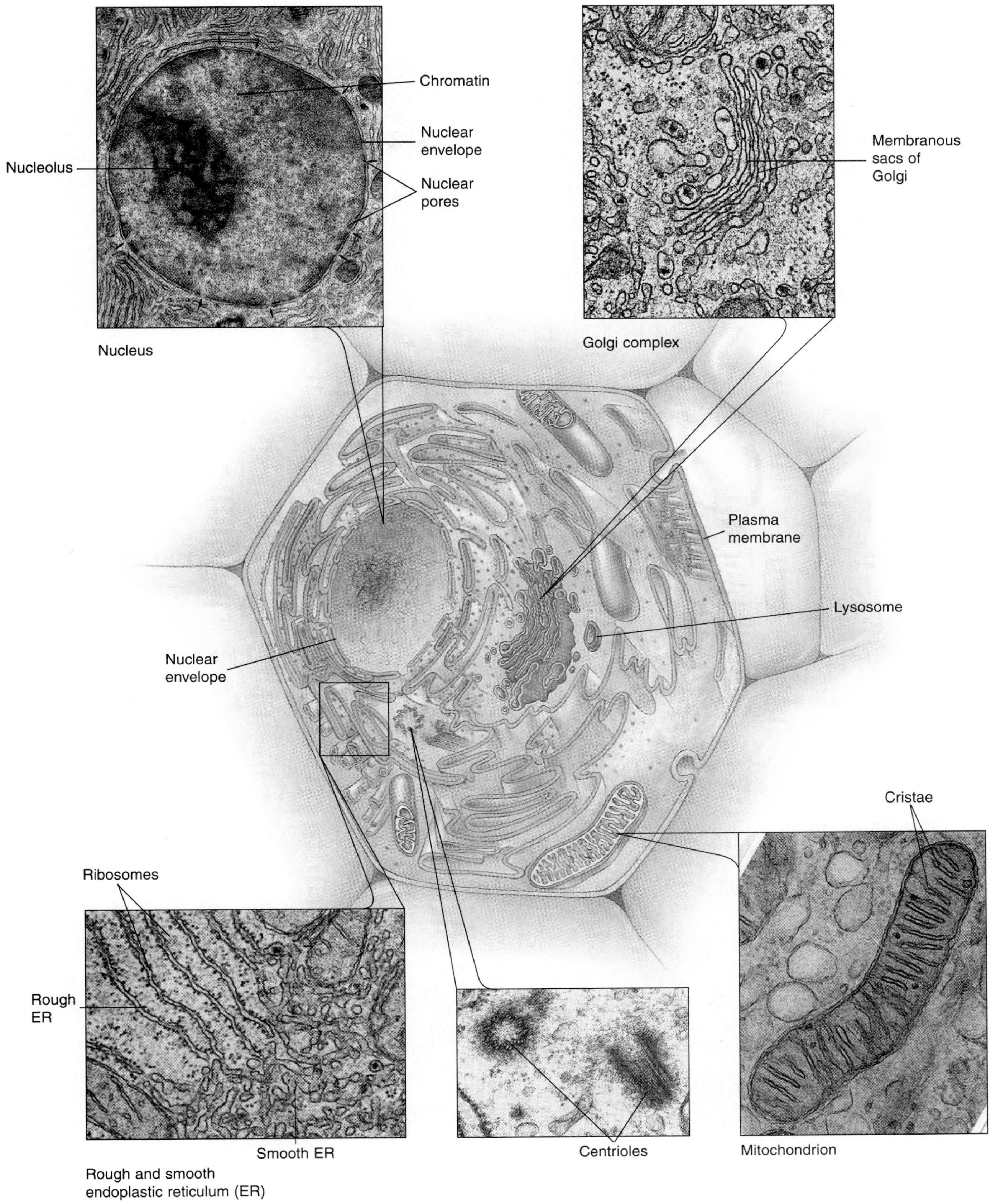

FIGURE 4–12 This composite diagram includes organelles present in most animal cells. This generalized animal cell is shown in realistic context surrounded by adjacent cells that cause it to be slightly compressed. Depending on the cell type, certain organelles may be more or less prominent. Note that chloroplasts, a cell wall, and large vacuoles are lacking. (Clockwise from top left: D.W. Fawcett; D.W. Fawcett and R. Bolender; D.W. Fawcett; B.F. King, School of Medicine, University of California, Davis/Biological Photo Service; Visuals Unlimited/R. Bolender and D.W. Fawcett)

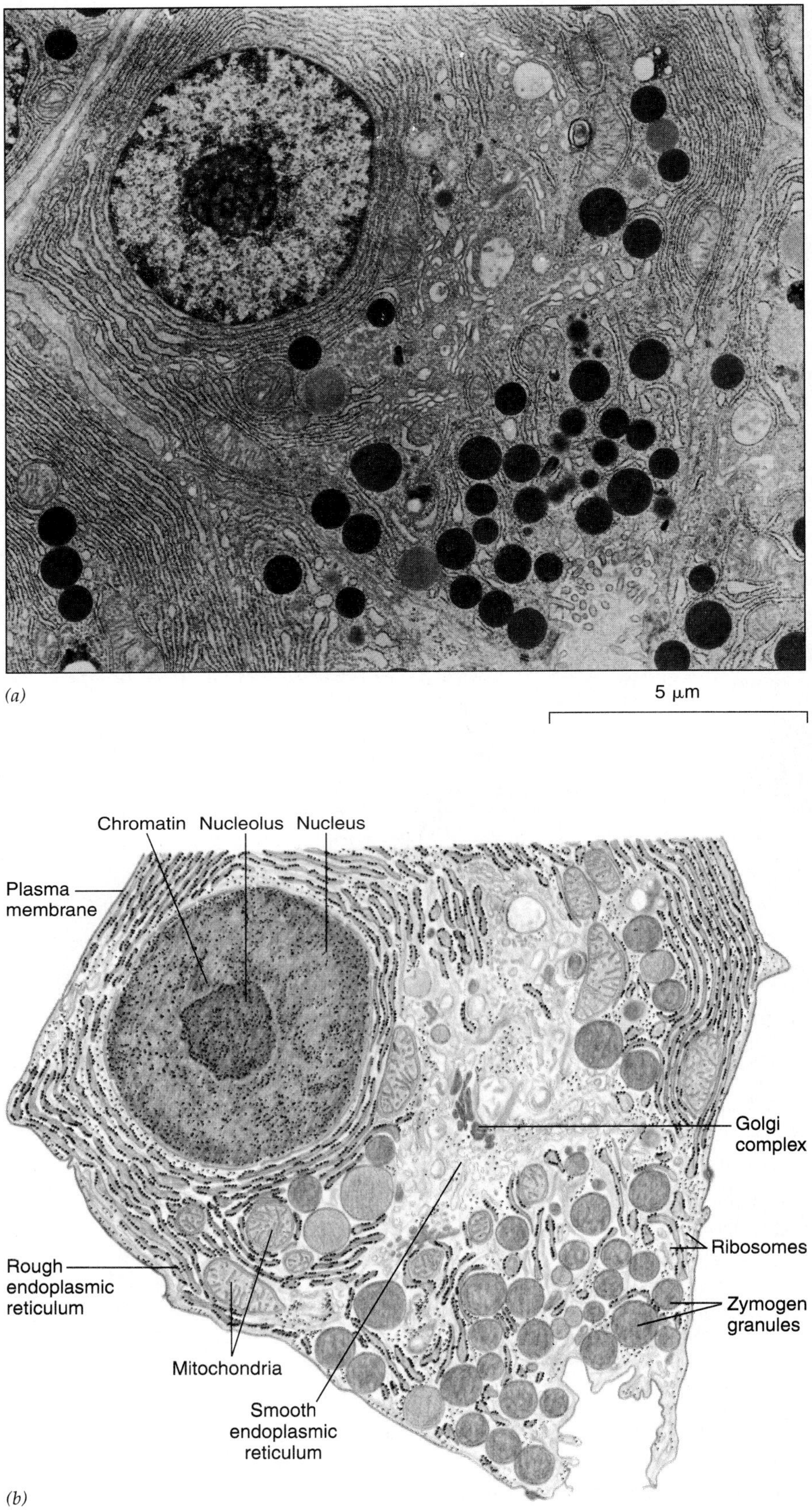

FIGURE 4–13 A transmission electron micrograph of a human pancreas cell (*a*), which is specialized to secrete large amounts of a protein, is paired with an interpretive drawing (*b*). Most of the structures of a typical animal cell are present. However, like most animal cells, this one has certain features associated with its specialized functions. Most of its membranes are part of the rough endoplasmic reticulum, because that is the site where the secreted protein is synthesized. The large, dark, circular bodies are zymogen granules containing inactive enzymes. When released from the cell, they catalyze chemical reactions such as the breakdown of peptide bonds of ingested proteins in the intestine. A different cell type might have many of the same organelles, but in different proportions. Heart muscle cells, for example, are packed with mitochondria and have very little endoplasmic reticulum.
(*a*, Dr. Susumu Ito, Harvard Medical School)

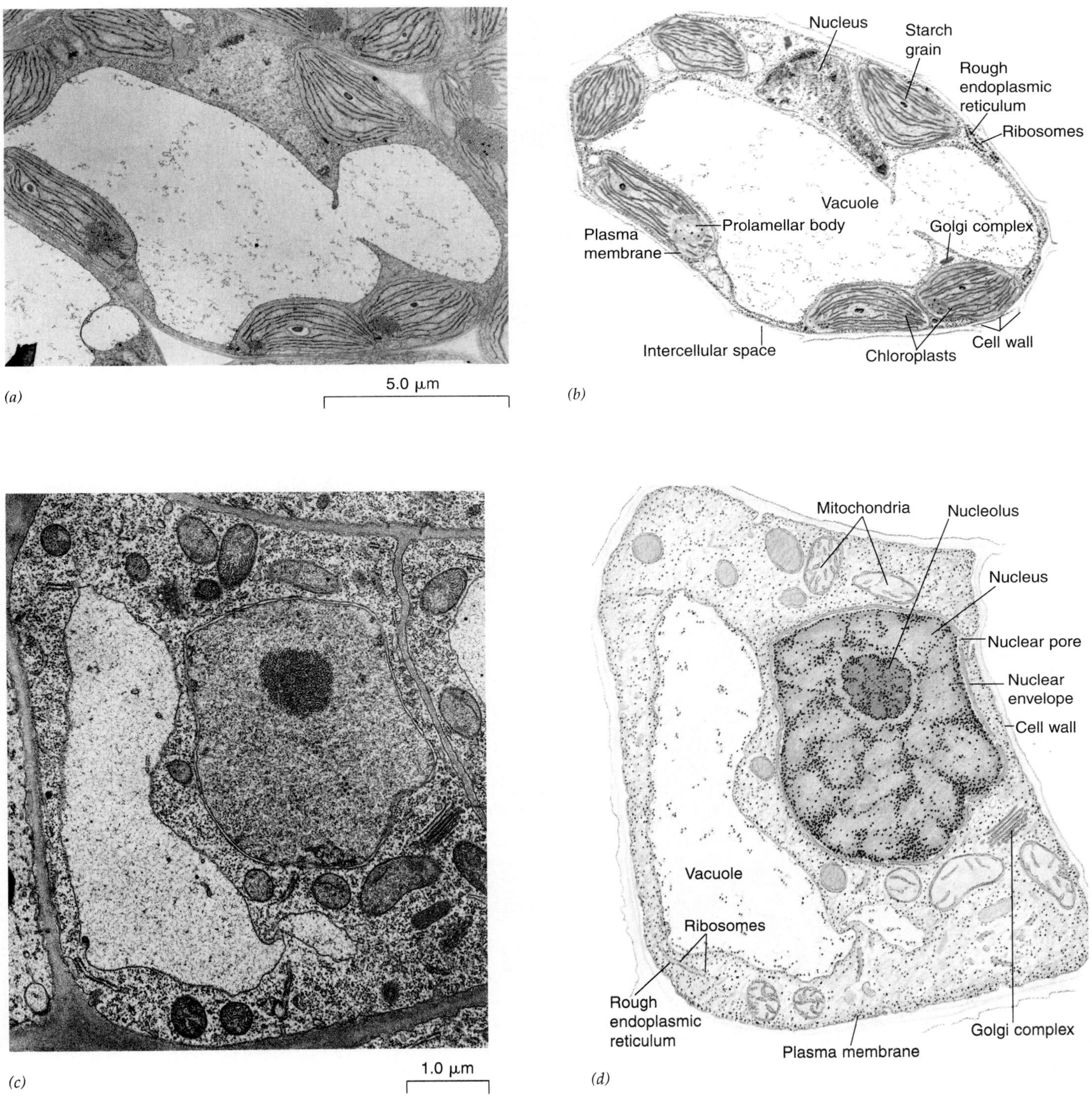

FIGURE 4–14 Transmission electron micrographs of two different types of plant cells are paired with interpretive drawings. (***a***) and (***b***) Most of this cross section of a cell from the leaf of a young bean plant, *Phaseolus vulgaris*, is dominated by the vacuole. Prolamellar bodies are membranous regions typically seen in developing chloroplasts. (***c***) and (***d***) This cross section of a root cell from *Arabidopsis thailiana* lacks chloroplasts. A number of mitochondria are evident. (*a*, courtesy of Dr. Kenneth Miller, Brown University; *c*, Biophoto Associates)

Focus On

Acetabularia: The Mermaid's Wineglass and the Control of Cellular Activities

As discussed in Chapter 1, we can learn something about the role of the nucleus by removing it from a cell and examining the consequences. When the nucleus of a single-celled amoeba is removed with a microneedle, the amoeba continues to live and move, but it ceases to grow and dies after a few days. A control amoeba, subjected to similar trauma but without removal of the nucleus, does not die. We conclude that the nucleus is necessary for the metabolic processes that provide for growth and cell reproduction.

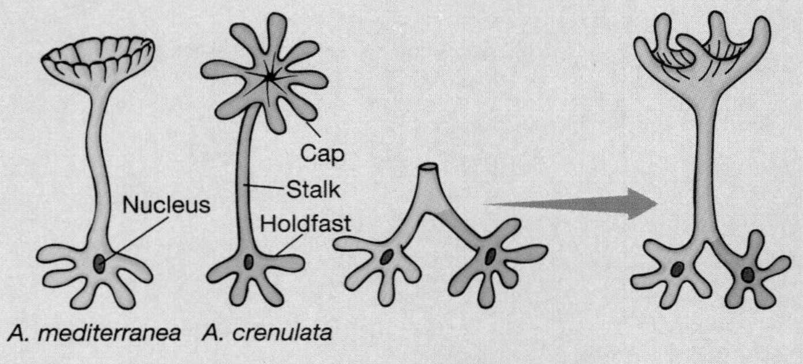

Acetabularia Is Easy to Manipulate Because It Is a Single Giant Cell

To the romantically inclined, the little seaweed *Acetabularia* resembles a mermaid's wineglass, although the literal translation of its name, "vinegar cup," is somewhat less elegant.

In the 19th century, biologists discovered that this marine eukaryotic alga consists of a single giant cell. At about 5 cm in length, *Acetabularia* is small for a seaweed but gigantic for a cell. It consists of a rootlike **holdfast,** a long cylindrical **stalk,** and a cuplike **cap.** The nucleus is found in the holdfast, about as far away from the cap as it can be.

Regeneration Experiments Demonstrated That the Cap Shape Is under the Control of Something in the Stalk or the Holdfast

If the cap of *Acetabularia* is removed experimentally, another one grows after a few weeks. Such behavior, common among simple organisms, is called **regeneration.** This fact attracted the attention of investigators, especially J. Hämmerling and J. Brachet, who became interested in whether a relationship exists between the nucleus and the physical characteristics of the alga. Because of its great size, *Acetabularia* could be subjected to surgery that would be impossible with smaller cells. During the 1930s and 1940s these researchers performed brilliant experiments that in many ways laid the foundation for much of our modern knowledge of the nucleus. Two species were used for most experiments, *A. mediterranea,* which has a smooth cap, and *A. crenulata,* which has a cap broken up into a series of finger-like projections.

The kind of cap that is regenerated depends on the species of *Acetabularia* used in the experiment. As you might expect, *A. crenulata* regenerates a "cren" cap, and *A. mediterranea* regenerates a "med" cap. But it is possible to graft together two capless algae of different species. Through this union, they regenerate a common cap that has characteristics intermediate between those of the two species involved. Thus, it is clear that something about the lower part of the cell controls cap shape.

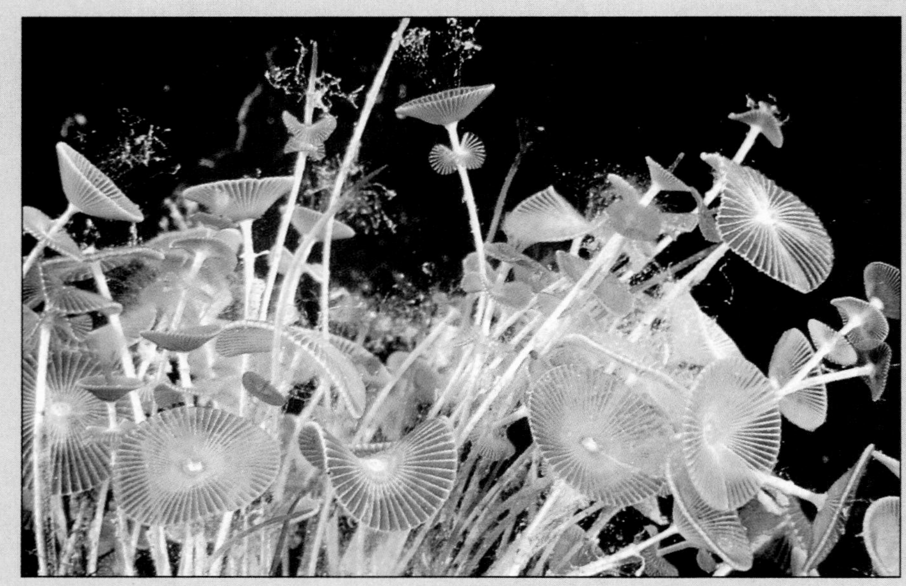

Acetabularia. (Visuals Unlimited/L. Sims)

Stalk Exchange Experiments Indicated That Short-Term Control Can Be Exerted by the Stalk, but Long-Term Control Is in the Holdfast

It is possible to attach a section of *Acetabularia* to a holdfast that is not its own by telescoping the cell walls of the two into one another. In this way the stalks and holdfasts of different species may be intermixed.

First, we take *A. mediterranea* and *A. crenulata* and remove their caps. Then we sever the stalks from the holdfasts. Finally, we exchange the parts. What happens? Not, perhaps, what you would expect! The caps that regenerate are characteristic not of the species donating the holdfasts but of those donating the stalks!

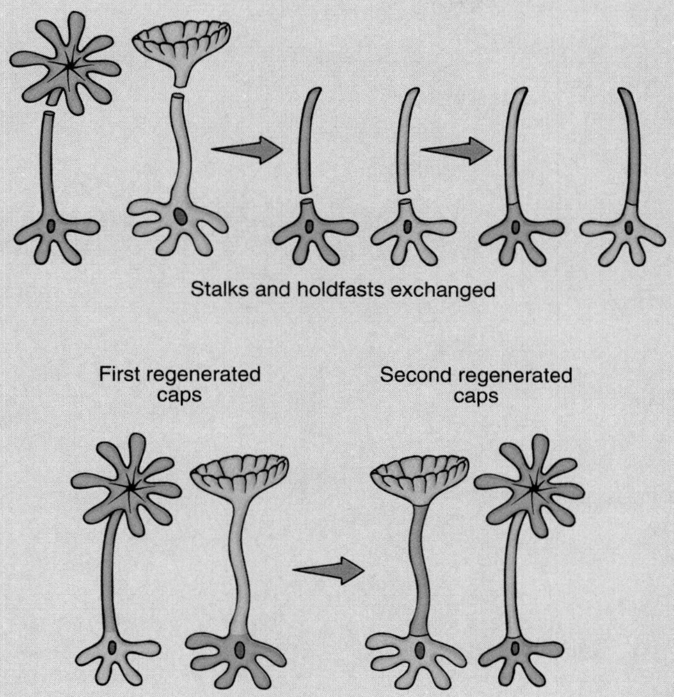
Stalks and holdfasts exchanged
First regenerated caps
Second regenerated caps

However, if the caps are removed once again, this time the caps that regenerate are characteristic of the species that donated the holdfasts. This continues to be the case no matter how many more times the regenerated caps are removed.

From all these results Hämmerling and Brachet deduced that the ultimate control of the *Acetabularia* cell is associated with the holdfast. Because there is a time lag before the holdfast appears to take over, they hypothesized that it produces some temporary cytoplasmic messenger substance whereby it exerts its control. They further hypothesized that initially the grafted stalks still contain enough of that substance from their former holdfasts to regenerate a cap of the former shape. But this still leaves us with the question of what it is about the holdfast that accounts for its apparent control. An obvious suspect is the nucleus.

Nuclear Exchange Experiments Demonstrated That the Nucleus Is the Ultimate Source of Information for the Control of Cellular Activities

If the nucleus is removed and the cap cut off, a new cap regenerates. *Acetabularia,* however, can usually regenerate only once without a nucleus.

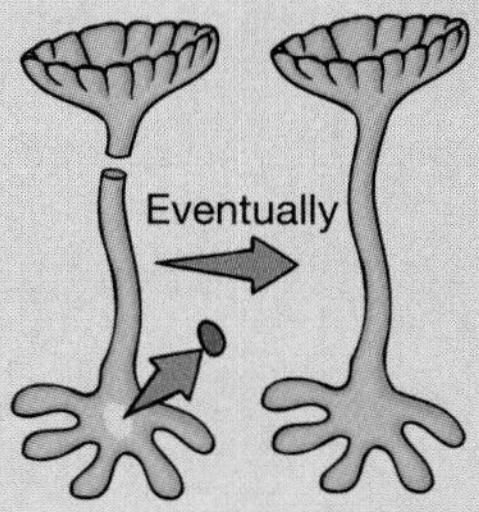
Eventually

If the nucleus of an alien species is now inserted, and the cap is cut off once again, a new cap regenerates that is characteristic of the species of the nucleus! If two kinds of nuclei are inserted, the regenerated cap is intermediate in shape between those of the species that donated the nuclei.

As a result of these and other experiments, biologists began to accept

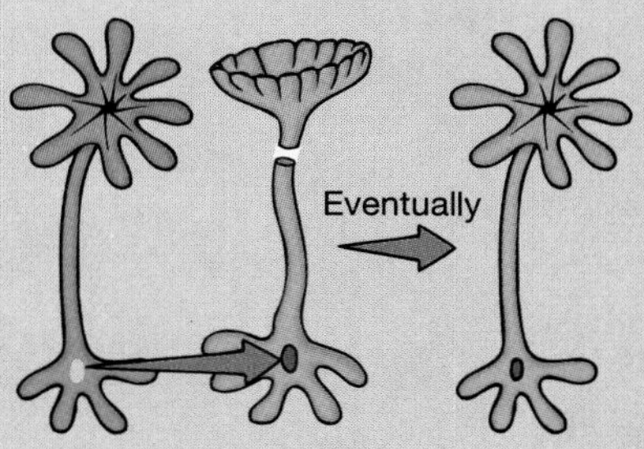
Eventually

certain basic ideas. The control of the cell exerted by the holdfast is attributable to the nucleus that is located there. Further, the nucleus is the apparent source of some "messenger substance" that can temporarily exert control but is limited in quantity and cannot be produced without the nucleus. This information helped provide a starting point for research on the role of the nucleic acids in the control of all cells.

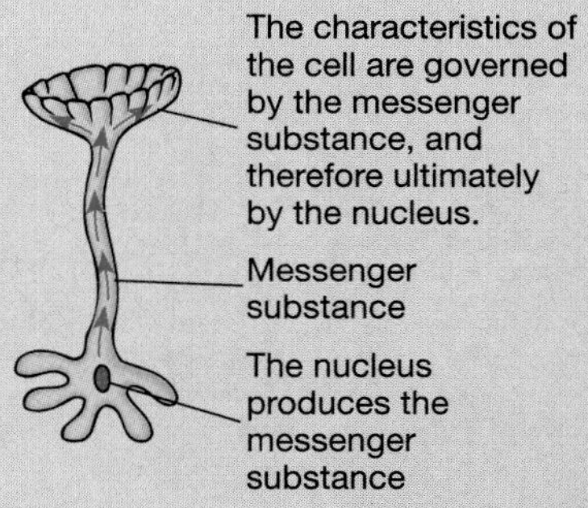
The characteristics of the cell are governed by the messenger substance, and therefore ultimately by the nucleus.
Messenger substance
The nucleus produces the messenger substance

Today we see these ideas extended in our modern view of information flow and control in the cell (see Fig. 4–1). We now know that the nucleus of eukaryotes controls the cell's activities because it contains DNA (deoxyribonucleic acid), the ultimate source of biological information. DNA can pass on its information to successive generations because it is able to precisely duplicate or replicate itself. The information in the DNA is used to specify the sequence of amino acids in all the proteins of the cell. To carry out its mission, DNA uses ribonucleic acid (RNA) as the cytoplasmic messenger substance. ■

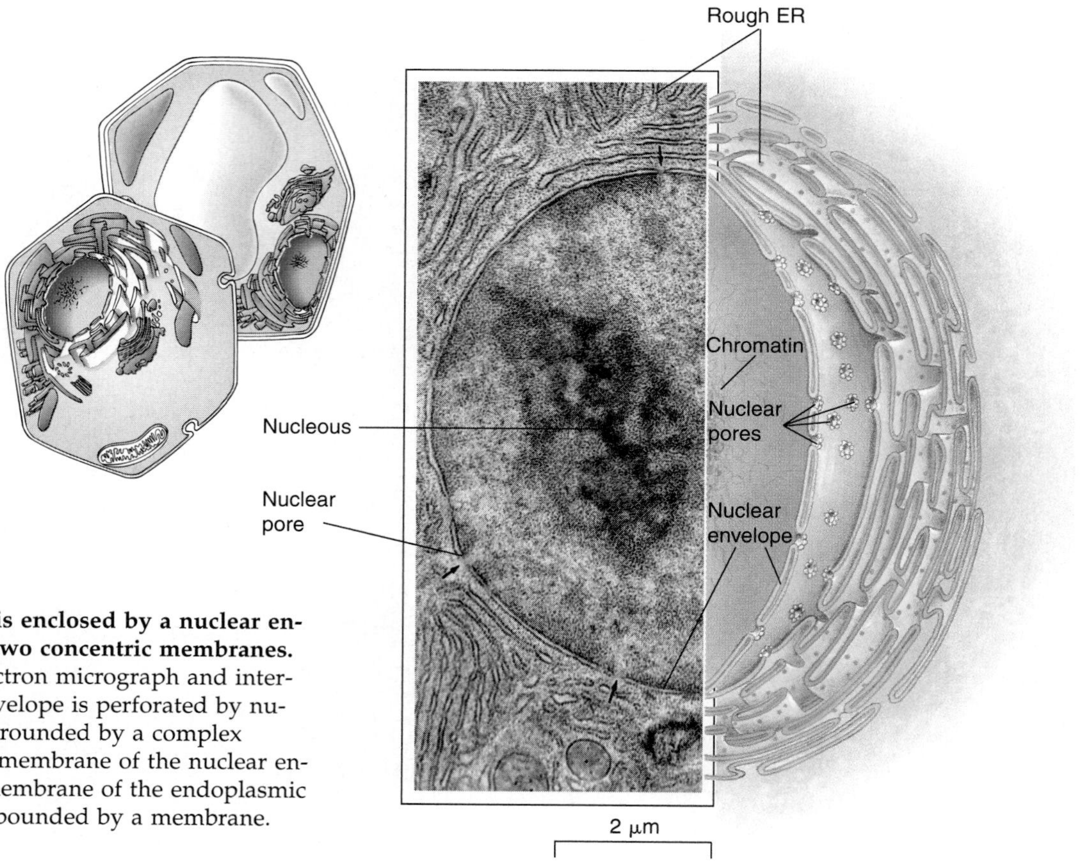

FIGURE 4–15 A cell nucleus is enclosed by a nuclear envelope, which is composed of two concentric membranes. As seen in this transmission electron micrograph and interpretive drawing, the nuclear envelope is perforated by nuclear pores, each of which is surrounded by a complex made up of proteins. The outer membrane of the nuclear envelope is continuous with the membrane of the endoplasmic reticulum. The nucleolus is not bounded by a membrane. (D.W. Fawcett)

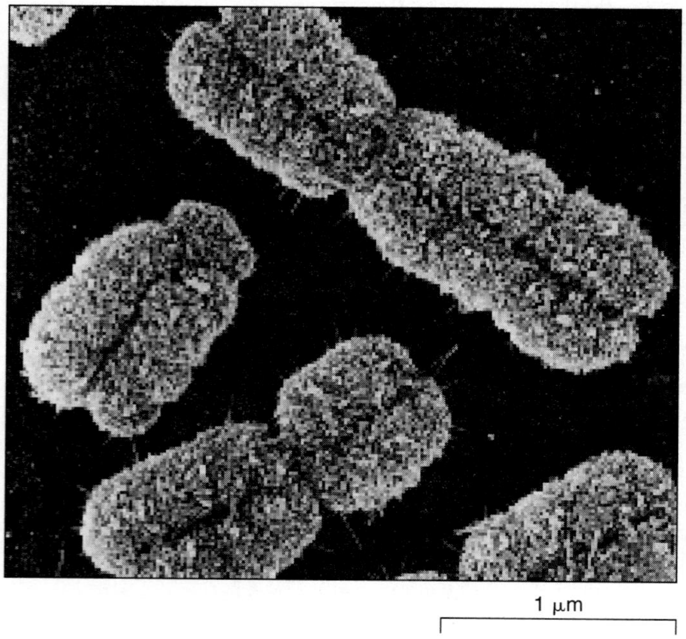

FIGURE 4–16 This scanning electron micrograph shows the structure of human chromosomes that have been isolated from a dividing cell. The knotted coils you see here are made up of DNA and associated proteins. (Biophoto Associates)

nucleus controls protein synthesis (which takes place in the cytoplasm) by sending messenger RNA (mRNA) molecules (copies of the parts of genes that code for proteins; see Chapter 12) through the nuclear envelope to the cytoplasm.

DNA is associated with proteins to form a complex known as **chromatin,** which appears to be an irregular network of granules and strands in cells that are not dividing. Although chromatin appears disorganized, it is not. Because DNA molecules are extremely long and thin, they must be packed inside the nucleus in a very regular fashion. The chromatin is arranged into structures called **chromosomes.** As a cell divides, the chromosomes must be duplicated within the nucleus, and the two copies must be separated in such a way that no portion of either is lost or ends up in the wrong place. As the cell prepares to divide, the DNA and proteins that form each chromosome become even more tightly coiled than usual, so that the chromosomes get shorter and thicker and ultimately become visible with the microscope (Fig. 4–16).

Ribosomal subunits are assembled in the nucleolus

In all cells, whether prokaryotic or eukaryotic, messenger RNA must become attached to small but complex

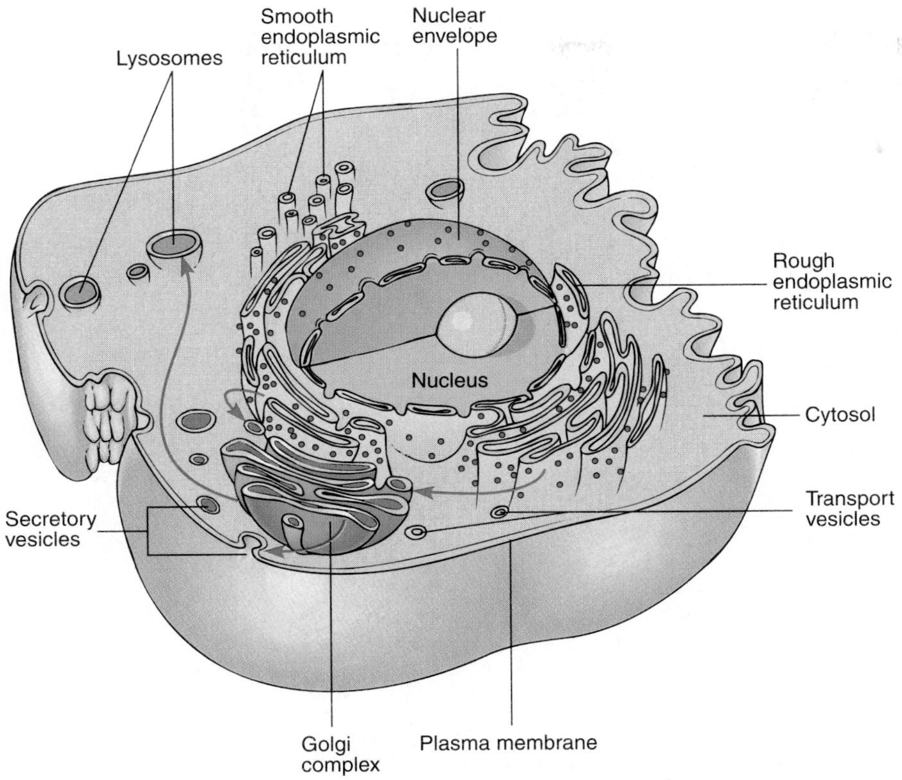

FIGURE 4–17 The internal membrane, or endomembrane, system consists of functionally different membranous structures that communicate with each other. Some membranes are physically connected and others interact through transport vesicles, which bud from one membrane and fuse with another membrane in the system. Many of the membrane components originate in the ER and then progress to the cell surface or to other organelles by way of the Golgi complex. A molecule in the lumen of the ER might move via vesicles through several other compartments in the system and then pass through the plasma membrane to the outside by way of a secretory vesicle. The compartments enclosed by these internal membranes can thus be considered equivalent to the exterior of the cell.

beadlike structures called **ribosomes** in order to carry out protein synthesis (see Chapter 12). Prokaryotic ribosomes are similar, but not identical, to those of eukaryotes. All ribosomes consist of two main parts: a large subunit and a small subunit. Each contains a special type of RNA, known as ribosomal RNA or rRNA, and several ribosomal proteins. In eukaryotes, the subunits of the ribosomes are assembled in a specialized region of the nucleus called the **nucleolus** (pl., *nucleoli*). The nucleolus, a compact body that is *not* membrane-bounded, usually stains differently from the surrounding chromatin. Ribosomal RNA is synthesized in the nucleolus. The ribosomal proteins are synthesized in the cytoplasm and imported into the nucleolus. These components are assembled into large and small ribosomal subunits. The subunits leave the nucleus through the nuclear pores and enter the cytoplasm. There they associate with each other to form complete ribosomes.

Organelles of the internal membrane system interact extensively

Certain organelles of a eukaryotic cell cooperate in a variety of ways; collectively these make up the **internal membrane system,** or **endomembrane system,** of the cell (Fig. 4–17). Some organelles have direct connections between their membranes and compartments. Others interact by means of **vesicles,** small membrane-bounded sacs that can carry materials from one cellular compartment to another. Through a complex series of steps, a vesicle can form as a "bud" from one membrane and then be transported to another membrane to which it fuses, thus delivering its contents. A number of structures are generally considered to be part of the internal membrane system, including the endoplasmic reticulum, the outer membrane of the nuclear envelope, the Golgi complexes, the lysosomes, and the vacuoles. Although it is not internal, the plasma membrane is also included because of its participation in the activities of the internal membrane system.

The endoplasmic reticulum is a major manufacturing center

One of the most prominent features in the electron micrograph in Figure 4–13 is a maze of parallel internal membranes that encircle the nucleus and extend into many regions of the cytoplasm. This complex of membranes is the **endoplasmic reticulum (ER),** which can form a significant part of the total volume of the cytoplasm in certain types of cells. A higher-magnification micrograph of the ER is shown in Figure 4–18. Remember that the electron micrograph represents only a thin cross section of the cell, so there is a tendency to interpret the photographs as depicting a series of tubes. In fact, these

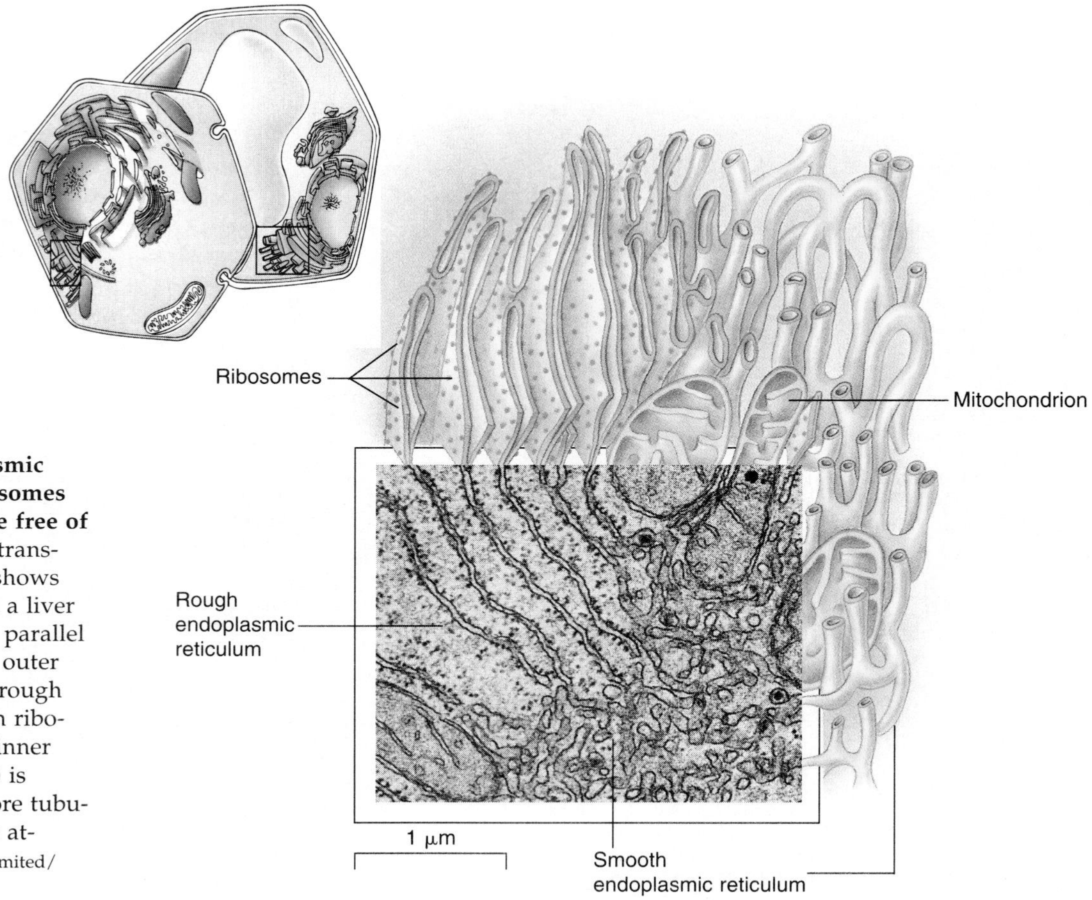

FIGURE 4–18 The endoplasmic reticulum (ER) can have ribosomes attached (rough ER) or can be free of ribosomes (smooth ER). The transmission electron micrograph shows both rough and smooth ER in a liver cell. The rough ER consists of parallel arrays of broad, flat sacs. The outer surface (cytosolic side) of the rough ER membrane is studded with ribosomes; the surface facing the inner compartment (ER lumen side) is smooth. The smooth ER is more tubular in form and does not have attached ribosomes. (Visuals Unlimited/ R. Bolender and D.W. Fawcett)

membranes usually consist of a series of tightly packed and flattened saclike structures that form interconnected compartments within the cytoplasm.

The internal space enclosed by the membranes is called the ER **lumen.** In most cells the ER lumen forms a single internal compartment. Evidence also suggests that the ER membrane is continuous with the outer membrane of the nuclear envelope (Fig. 4–15) so that the compartment formed between the two nuclear membranes is connected to the ER lumen. The membranes of other organelles are not directly connected to the ER and appear to form distinct and separate compartments within the cytoplasm.

The ER membranes and lumen contain a large variety of enzymes that catalyze many different types of chemical reactions. In some cases the membranes serve as a framework for systems of enzymes that carry out sequential biochemical reactions. The two surfaces of the membrane contain different sets of enzymes and represent regions of the cell with different synthetic capabilities, just as different regions of a factory are used to make different parts of a particular product. Other ER enzymes are located within the ER lumen.

Notice that in both Figures 4–17 and 4–18, one membrane face (the lumen side) appears to be bare, while the other membrane face (the cytosolic side) is studded with dark particles; these are ribosomes. As stated earlier, ribosomes are a major component of the protein-synthesizing machinery of the cell. Many of the ribosomes found in the cell at any one time are bound to the ER surface. However, not all proteins are synthesized on the surface of the ER membranes; some proteins are synthesized on ribosomes found free within the cytoplasm.

The ER plays a central role in the synthesis and assembly of proteins. Many proteins that are exported from the cell (such as digestive enzymes) and those destined for other organelles are synthesized on ribosomes attached to the ER membrane. These proteins are transported across the membrane into the ER lumen, where they may be modified by enzymes that add complex carbohydrates or lipids to them. Other enzymes in the ER lumen may be involved in assisting proteins to fold into their proper conformations. The proteins are then transferred to other membranes by small **transport vesicles,** which bud off of the ER membrane and are then inserted into the target membrane.

Two distinct regions of the ER can be seen in electron micrographs. Although these regions have different functions, their membranes are connected and their internal spaces are continuous. **Rough ER,** as just discussed, has ribosomes attached to it and consequently appears rough in electron micrographs. **Smooth ER** is more tubular in

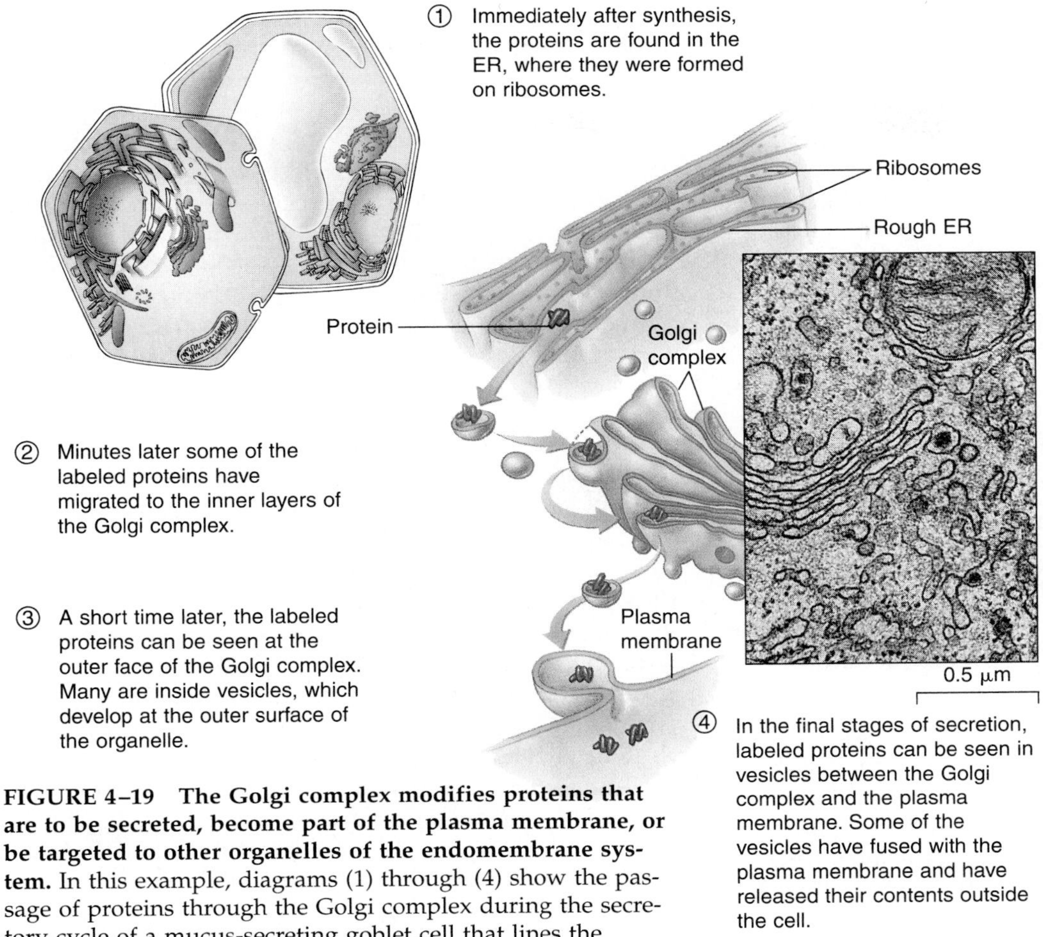

① Immediately after synthesis, the proteins are found in the ER, where they were formed on ribosomes.
Ribosomes
Rough ER
Protein
Golgi complex
② Minutes later some of the labeled proteins have migrated to the inner layers of the Golgi complex.
③ A short time later, the labeled proteins can be seen at the outer face of the Golgi complex. Many are inside vesicles, which develop at the outer surface of the organelle.
Plasma membrane
0.5 μm
④ In the final stages of secretion, labeled proteins can be seen in vesicles between the Golgi complex and the plasma membrane. Some of the vesicles have fused with the plasma membrane and have released their contents outside the cell.

FIGURE 4–19 The Golgi complex modifies proteins that are to be secreted, become part of the plasma membrane, or be targeted to other organelles of the endomembrane system. In this example, diagrams (1) through (4) show the passage of proteins through the Golgi complex during the secretory cycle of a mucus-secreting goblet cell that lines the intestine. Mucus is a complex mixture of covalently linked proteins and carbohydrates. By labeling newly synthesized proteins briefly with radioactive amino acids, it is possible to follow their movement in the cell at different times after their synthesis. (D.W. Fawcett and R. Bolender)

nature and does not have ribosomes bound to it, so its outer membrane surfaces have a smooth appearance. The smooth ER is the primary site of phospholipid, steroid, and fatty acid metabolism. Smooth ER also serves an important function by localizing detoxifying enzymes that break down chemicals such as carcinogens (cancer-causing agents). These chemicals are then converted to water-soluble products that can be excreted from the body. Certain types of cells contain extensive amounts of smooth ER. An example is the cells of the human liver, which synthesize and process much of the cholesterol and other lipids and serve as the major detoxification site of the body. The smooth ER may be a minor membrane component in other cells.

The Golgi complex processes and packages proteins

The **Golgi complex** (also known as the *Golgi body* or *Golgi apparatus*) was first described in 1898 by the Italian microscopist Camillo Golgi, who found a way to specifically stain that organelle. In many cells the Golgi complex consists of stacks of flattened membranous sacs, which may be distended in certain regions because they are filled with cellular products (Fig. 4–19). Each of the flattened sacs has an internal space, or lumen. However, unlike the endoplasmic reticulum, most of these internal spaces of the Golgi complex and the membranes that form them are not continuous. Hence a Golgi complex contains a number of separate compartments, as well as some that are interconnected. In a cross-sectional view like that in Figure 4–19, many of the ends of the sheetlike layers of Golgi membranes are distended. The arrangement of the membranes in that figure is characteristic of well-developed Golgi complexes in many types of cells. In some animal cells the Golgi complex is often located at one side of the nucleus; in other animal and plant cells there are many Golgi bodies, usually consisting of separate stacks of membranes dispersed throughout the cell.

The Golgi complex functions principally as an apparatus for processing, sorting, and modifying proteins. Most proteins that pass through the Golgi complex either are secreted from the cell, become a part of the plasma membrane, or are routed to other organelles of the internal membrane system. After those proteins have been synthesized on ribosomes attached to the rough ER, they are transported to the Golgi complex enclosed in small transport vesicles formed from the ER membrane. These vesicles then fuse with the membranes of the complex that are closest to the nucleus (Fig. 4–19; see also Chapter 5). The proteins then pass through the separate layers of the Golgi complex by way of membrane transport vesicles.

While moving through the Golgi complex, the proteins are modified in different ways, resulting in the formation of complex biological molecules. Often sugars that were added to proteins in the rough ER become further modified in the Golgi. The resulting **glycoproteins** are proteins with complex branched-chain polysaccharides attached to a number of different amino acids. Each type of protein is modified in a different way, although the purpose of such modifications is poorly understood. In a few cases the carbohydrate added to the protein may be used as a "sorting signal," allowing the Golgi complex to route the protein to a specific organelle. In addition to adding carbohydrates to proteins, the Golgi complexes of plant cells produce extracellular polysaccharides that are used as components of the cell wall.

Lysosomes are compartments for digestion

Small sacs of digestive enzymes called **lysosomes** are dispersed in the cytoplasm of animal cells (Fig. 4–20). The enzymes in these organelles break down complex molecules that originate inside or outside the cell. These include lipids, proteins, carbohydrates, and nucleic acids. About 40 different hydrolytic enzymes have been identified in lysosomes; most are active under rather acidic conditions (pH 5). These enzymes originate in the Golgi complex, where they are identified and sorted to the lysosomes by unique carbohydrate signals attached to the proteins. *Primary lysosomes* are formed by budding from the Golgi complex.

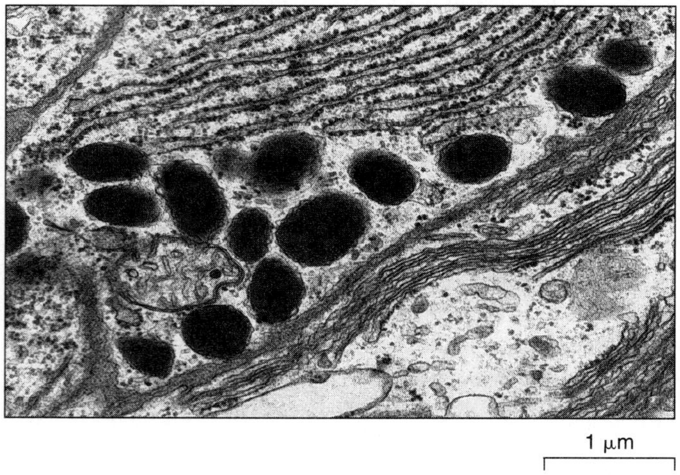

FIGURE 4–20 The dark vesicles in this transmission electron micrograph are lysosomes. A mitochondrion and rough and smooth endoplasmic reticulum are also visible in these cells of the flatworm *Polystylitera.* (Dwight Kuhn)

In a cell that is short of fuel, lysosomes may break down organelles so that their components can be used as an energy source. Lysosomes are also used to degrade foreign molecules ingested by cells. When a white blood cell or a scavenger cell ingests a bacterium or debris from dead cells, the foreign matter is enclosed in a vesicle formed from part of the plasma membrane. One or more primary lysosomes then fuse with the vesicle containing foreign matter to form a larger vesicle called a *secondary lysosome.* In the secondary lysosome the powerful enzymes come in contact with the foreign molecules and degrade them into their components (see Chapter 5).

Lysosomal enzymes are apparently released into the cell in some normal processes, such as the resorption of the tail of a tadpole undergoing metamorphosis. However, some forms of tissue damage may be related to "leaky" lysosomes. Rheumatoid arthritis is thought to result in part from damage done to cartilage cells in the joints by enzymes released from lysosomes.

In certain genetic diseases of humans, known as lysosomal storage diseases, one of the normally present hydrolytic enzymes is lacking. The substrate of that enzyme accumulates in the lysosomes, ultimately interfering with cellular activities. One such disease is Tay-Sachs disease (see Chapter 15), in which a normal lipid fails to break down in brain cells, resulting in mental retardation and death.

Vacuoles are large, fluid-filled sacs

Although lysosomes have been identified in almost all kinds of animal cells, their occurrence in plant and fungal cells is open to debate. Many of the functions carried out in animal cells by lysosomes are performed in plant and fungal cells by a large, single, membrane-bounded sac referred to as the **vacuole** (see Fig. 4–11). The vacuolar membrane is referred to as the **tonoplast.** Although the terms *vacuole* and *vesicle* are sometimes used interchangeably, vacuoles are usually larger structures, sometimes produced by the merging of many vesicles.

As much as 90% of the volume of a plant cell may be occupied by a large central vacuole containing water, stored food, salts, pigments, and wastes. The term vacuole, which means "empty," refers to the fact that these organelles have no internal structure (with the exception of crystals that are sometimes present). Plants lack systems of organs to dispose of toxic metabolic waste products; these may be recycled in the vacuole or they may aggregate and form small crystals inside the vacuole.

The vacuole may also serve as a storage compartment for inorganic compounds in plant cells and for molecules such as proteins in seeds. Compounds that are noxious to herbivores (animals that eat plants) may also be stored in some plant vacuoles as a means of defense. Plant vac-

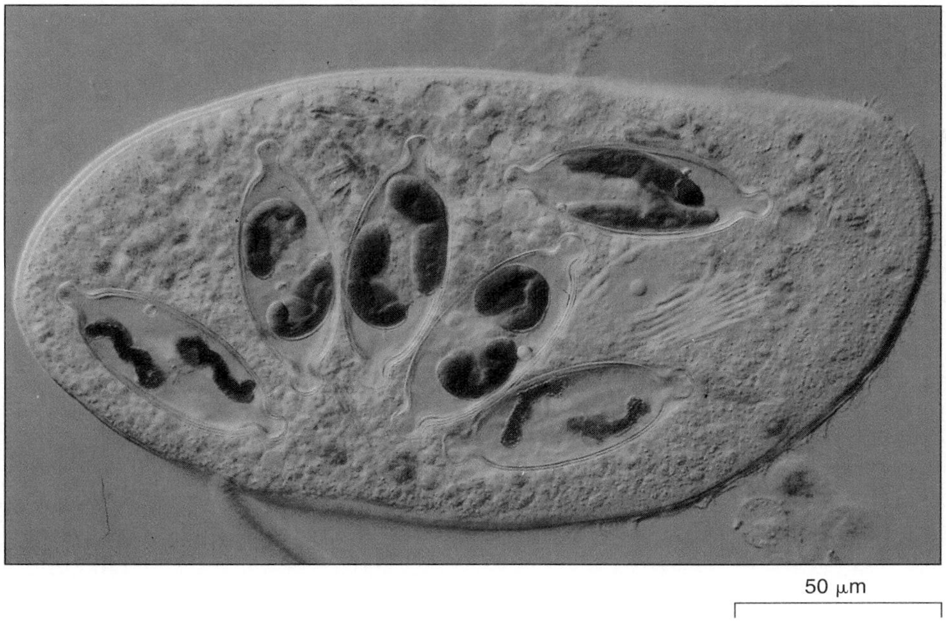

FIGURE 4–21 Within this protozoon, *Chilodonella,* are many food vacuoles containing ingested diatoms (small, photosynthetic protists). From the number of diatoms scattered about its insides, one might judge that *Chilodonella* has a rather voracious appetite. (M.I. Walker/Photo Researchers, Inc.)

uoles are lysosome-like in their ability to break down unneeded organelles and other cellular components.

Vacuoles also play a significant role in plant growth and development. Immature plant cells are generally small and contain numerous small vacuoles. As water accumulates in these vacuoles, they tend to coalesce, forming a large central vacuole. A plant cell increases in size mainly by adding water to this central vacuole.

Vacuoles can have numerous other functions and are also present in many types of animal cells and in single-celled protists. Most protozoa have food or digestion vacuoles, which fuse with lysosomes so that the food they contain can be digested (Fig. 4–21); many also have contractile vacuoles, which remove excess water from the cell (see Chapter 24).

Mitochondria and chloroplasts are energy-converting organelles

When a cell obtains energy from its environment, it is usually in the form of chemical energy in food molecules (such as glucose) or in the form of light energy. These types of energy must be converted to forms that can be used more conveniently by cells. Some energy conversions go on in the cytosol, but other types take place in mitochondria and chloroplasts, organelles that are specialized to facilitate conversion of energy from one form to another. Energy is most commonly converted to ATP. You will recall from Chapter 3 that the chemical energy of ATP can be used to drive a variety of chemical reactions in the cell. Figure 4–22 summarizes the main activ-

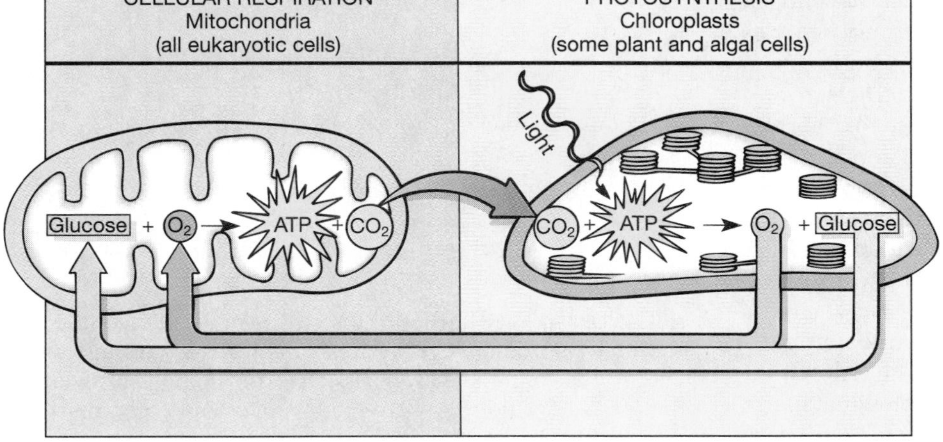

FIGURE 4–22 Cellular respiration and photosynthesis can be thought of as complementary processes. In cellular respiration, which takes place in the mitochondria of virtually all eukaryotic cells, the chemical energy of glucose is converted to chemical energy in the form of ATP. Photosynthesis, which is carried out in chloroplasts in some plant and algal cells, converts light energy to ATP and other forms of chemical energy. This energy is used to synthesize glucose from carbon dioxide.

Making the Connection

Mitochondria, Chloroplasts, and Cellular Evolution

What is the evolutionary relationship between simple prokaryotic cells and the complex cells of eukaryotes? Mitochondria and chloroplasts have provided valuable insights because these organelles have been shown to have many prokaryote-like features. For example, although most of the DNA in eukaryotic cells resides in the nucleus, both mitochondria and chloroplasts (as well as proplastids and other plastids) have DNA molecules in their inner compartments. These DNA molecules code for a small number of the proteins found in these organelles. These proteins are synthesized on mitochondrial or chloroplast ribosomes, which are similar to the ribosomes of prokaryotes. The majority of the mitochondrial and chloroplast proteins, however, are coded for by nuclear genes, made on free ribosomes outside the organelles, and then transported to their appropriate locations within. The existence of a separate set of ribosomes and DNA molecules in mitochondria and chloroplasts, along with other prokaryote-like characteristics, provide support for the view that these organelles evolved from prokaryotic organisms that originally lived inside larger cells, eventually losing the ability to function as autonomous organisms. This idea has become a major part of one theory, called the **endosymbiont theory,** of how eukaryotic organisms evolved from prokaryotic ancestors (see Chapter 20). ▲

ities that take place in mitochondria, found in almost all eukaryotic cells, and in chloroplasts, found only in algae and certain plant cells.

In addition to their central roles in energy metabolism, mitochondria and chloroplasts provide important clues about the evolution of eukaryotic cells (see *Making the Connection: Mitochondria, Chloroplasts, and Cellular Evolution*).

Mitochondria make ATP through cellular respiration

Virtually all eukaryotic cells (plant, animal, fungal, and protist) contain complex organelles called **mitochondria** (sing., *mitochondrion*). These organelles are the site of **aerobic cellular respiration** (see Chapter 7), a process that includes most of the reactions that convert the chemical energy present in certain foods to ATP. Aerobic cellular respiration requires oxygen and results in the release of carbon atoms from food molecules as carbon dioxide (a waste product). Mitochondria are most numerous in cells that are very active and therefore have high energy requirements. More than 1000 mitochondria have been counted in a single liver cell, but the number varies among cell types. Mitochondria vary in size, ranging from 2 to 8 μm in length, and are capable of changing size and shape rapidly. Mitochondria usually give rise to other mitochondria by growth and subsequent division.

Each mitochondrion is bounded by a double membrane, which forms two *different* compartments within the organelle (Fig. 4–23; see Chapter 7 for more detailed descriptions of mitochondrial structure). The **intermembrane space** is the compartment formed between the outer and inner membranes; the **matrix** is the compartment enclosed by the inner membrane. The **mitochondrial outer membrane** is smooth and somewhat like a sieve in that it allows many small molecules to pass

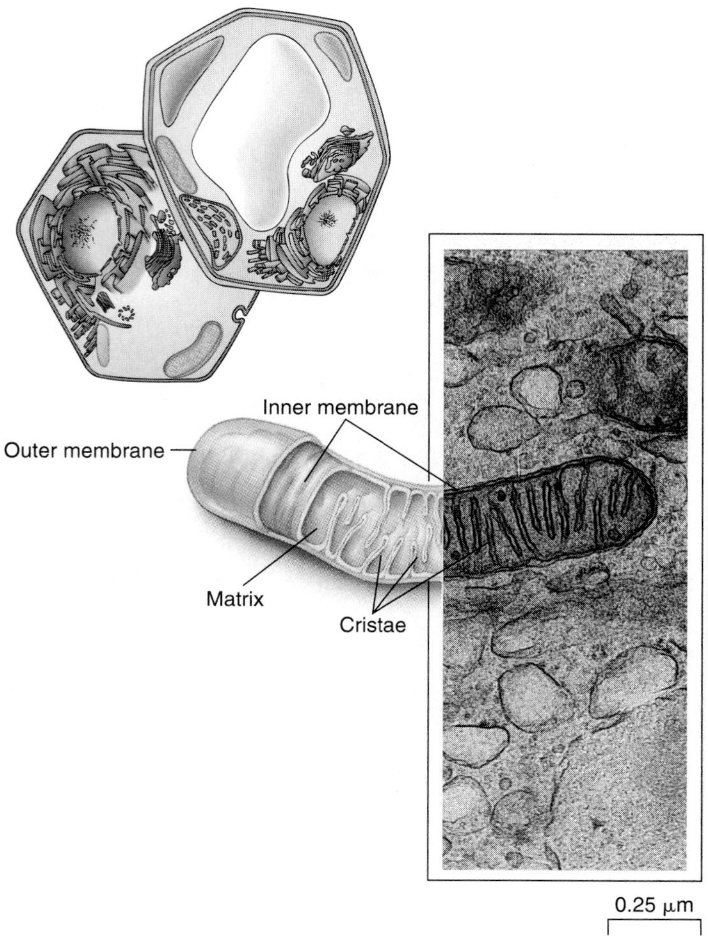

FIGURE 4–23 Mitochondria are the centers for aerobic cellular respiration. Cristae are evident in the transmission electron micrograph as well as in the drawing; the drawing reveals the relationship between the inner and outer membranes. (D.W. Fawcett)

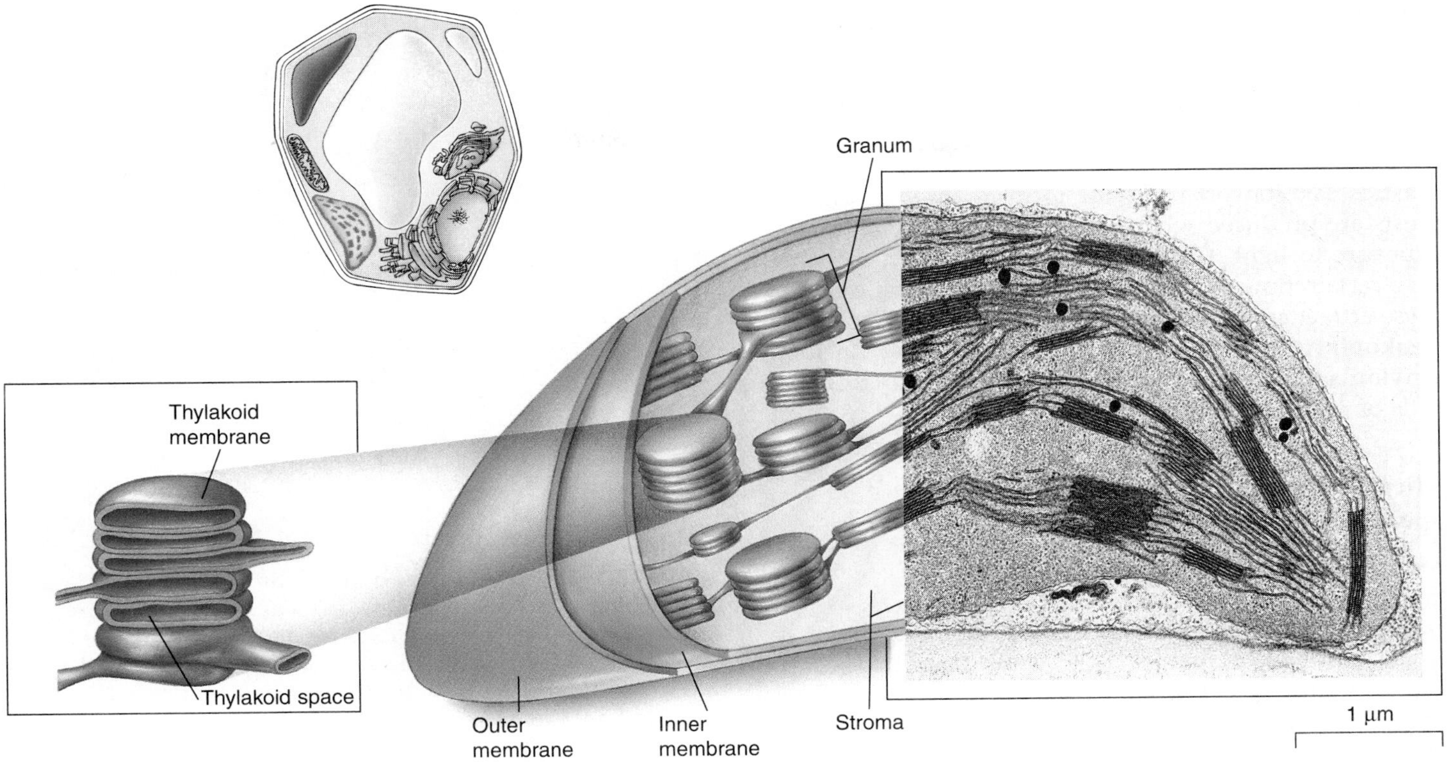

FIGURE 4–24 Photosynthesis takes place in chloroplasts. The transmission electron micrograph shows a chloroplast from a corn leaf cell. In the drawing the thylakoids are seen to be an interconnected set of flat, disclike sacs, each containing an internal space (the thylakoid space). The thylakoids are arranged in stacks called grana. The membranous interconnections between thylakoids can only be suggested in this diagram. (E.H. Newcomb and W.P. Wergin, University of Wisconsin/Biological Photo Service)

through it. By contrast, the **mitochondrial inner membrane** strictly regulates the types of molecules that can move across it. The inner membrane is folded repeatedly into projections, called **cristae** (sing., *crista*), that serve to increase its surface area. The matrix compartment contains enzymes that break down food molecules and convert their energy to other forms of chemical energy. The inner membrane contains a complex series of enzymes and other proteins involved in transforming the chemical energy in food molecules into that present in ATP.

Chloroplasts convert light energy to chemical energy through photosynthesis

Certain plant cells and algal cells carry out a complex set of energy conversion reactions known as *photosynthesis* (see Chapter 8). Organelles known as **chloroplasts** contain the green pigments **chlorophyll *a*** and ***b*,** which trap light energy for photosynthesis. Chloroplasts also contain a variety of yellow and orange light-absorbing pigments known as **carotenoids** (see Chapter 3).

A unicellular alga may have only a single large chloroplast, whereas a plant leaf cell may have as many as 20 to 100. Chloroplasts tend to be somewhat larger than mitochondria, with lengths typically ranging from about 5 to 10 μm or longer.

Chloroplasts are typically complex disc-shaped structures bound by an inner and an outer membrane (Fig. 4–24; see Chapter 8 for more detailed descriptions of structure). The inner membrane encloses a fluid-filled space called the **stroma,** which contains enzymes responsible for producing glucose from carbon dioxide and water using energy trapped from sunlight. The inner chloroplast membrane also encloses a third system of membranes, consisting of an interconnected set of flat, disclike sacs called **thylakoids.** The thylakoids are arranged in stacks called **grana** (sing., *granum*).

The thylakoid membranes enclose a third, innermost compartment within the chloroplast called the **thylakoid space.** The thylakoid membranes, in which chlorophyll is found, are similar to the inner membranes of the mitochondria in that they are involved in the formation of ATP. Energy trapped from sunlight by the chlorophyll molecules is used to excite electrons; the energy in these excited electrons is then used to form ATP and other molecules that can transfer chemical energy. This chemical energy is then used to form glucose and other substances from carbon dioxide and water in the stroma.

Chloroplasts belong to a group of organelles known as **plastids,** which produce and store food materials in cells of plants and algae. All plastids develop from **pro-**

plastids, precursor organelles found in less specialized plant cells, particularly in growing, undeveloped tissues. Depending on the special functions a cell will eventually have, its proplastids can mature into a variety of specialized mature plastids. These are extremely versatile organelles; in fact, under certain conditions even mature plastids can convert from one form to another. Chloroplasts are produced when proplastids are stimulated by exposure to light. **Chromoplasts** contain pigments that give certain flowers and fruits their characteristic colors; these attract animals as pollinators or as seed dispersers. **Leukoplasts** are unpigmented plastids; they include **amyloplasts** (see Chapter 3), which store starch in the cells of roots and tubers.

Microbodies are compartments for specialized chemical reactions

Microbodies are diverse membrane-bounded organelles containing a variety of enzymes that catalyze an assortment of metabolic reactions. During the breakdown of lipids, hydrogen peroxide (H_2O_2), a substance toxic to the cell, is produced. **Peroxisomes,** the microbodies in which these reactions occur, contain enzymes that split hydrogen peroxide, rendering it harmless. Peroxisomes in liver and kidney cells may be important in detoxifying certain compounds such as ethanol, the alcohol in alcoholic beverages.

Plant cells contain two main types of microbodies. A type of peroxisome found in leaf cells (Fig. 4–25) plays a part in photosynthesis (see Chapter 8). Another type of microbody, the **glyoxysome,** contains enzymes used to convert stored fats in plant seeds to sugars. The sugars are used by the young plant as an energy source and as a component needed to synthesize other compounds. Animal cells lack glyoxysomes and cannot convert fatty acids into sugars.

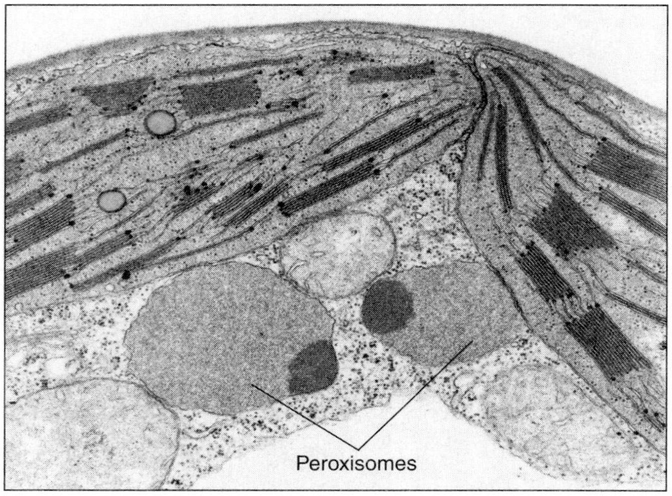

FIGURE 4–25 Peroxisomes of leaf cells generally associate with chloroplasts because they carry out reactions that are connected to some aspects of photosynthesis. Portions of two chloroplasts, as well as two mitochondria, are seen adjacent to the peroxisomes in this transmission electron micrograph of a tobacco leaf cell. (E.H. Newcomb and S.E. Frederick, University of Wisconsin/Biological Photo Service)

All eukaryotic cells contain a cytoskeleton

A close examination of cells from different animal tissues reveals striking and characteristic differences in cell shapes. When we watch these cells growing in the laboratory, we see that they can change shape and in many cases move about. The shapes of these cells and their ability to move are determined in large part by a complex network of protein fibers found within all eukaryotic cells called the **cytoskeleton.** The term is somewhat misleading because it implies a static structure, whereas the cytoskeleton as a whole is highly dynamic and constantly changing.

The protein filaments that make up the cytoskeletal framework were originally classified on the basis of their relative sizes. The two major types of filaments that make up the cytoskeleton in all eukaryotic cells are **microfilaments** (also known as **actin filaments**), which are 7 nm in diameter, and **microtubules,** 25 nm in diameter. Both microfilaments and microtubules are fibers formed from beadlike *globular* protein subunits, which can be rapidly assembled and disassembled. Although both types of fibers are major components of the cytoskeleton, they also play a role in forming other structures involved in cellular movement and organization.

In many animal cells there is also a third class of filaments, intermediate between the other two in size. These **intermediate filaments** have diameters from 8 to 10 nm. Intermediate filaments are made from *fibrous* protein subunits and are more stable than microtubules and microfilaments.

Figure 4–26 depicts some of the possible relationships among various cytoskeletal elements, including intermediate filaments, microtubules, and microfilaments.

Microtubules are hollow cylinders

A microtubule grows by the addition of *dimers*[2] of protein subunits called **tubulins** (Fig. 4–27). Each dimer is made up of two very similar subunits, α-tubulin and β-tubulin. Dimers are added preferentially to one end (referred to as the "plus" or fast-growing end) of the cylinder. The opposite, slow-growing end of a microtubule is referred to as the "minus" end. Microtubules can be disassembled by the removal of dimers, which can then be recycled to form microtubules in other parts of the cell.

[2]A dimer is a structure formed by the association of two monomers (similar, simpler units).

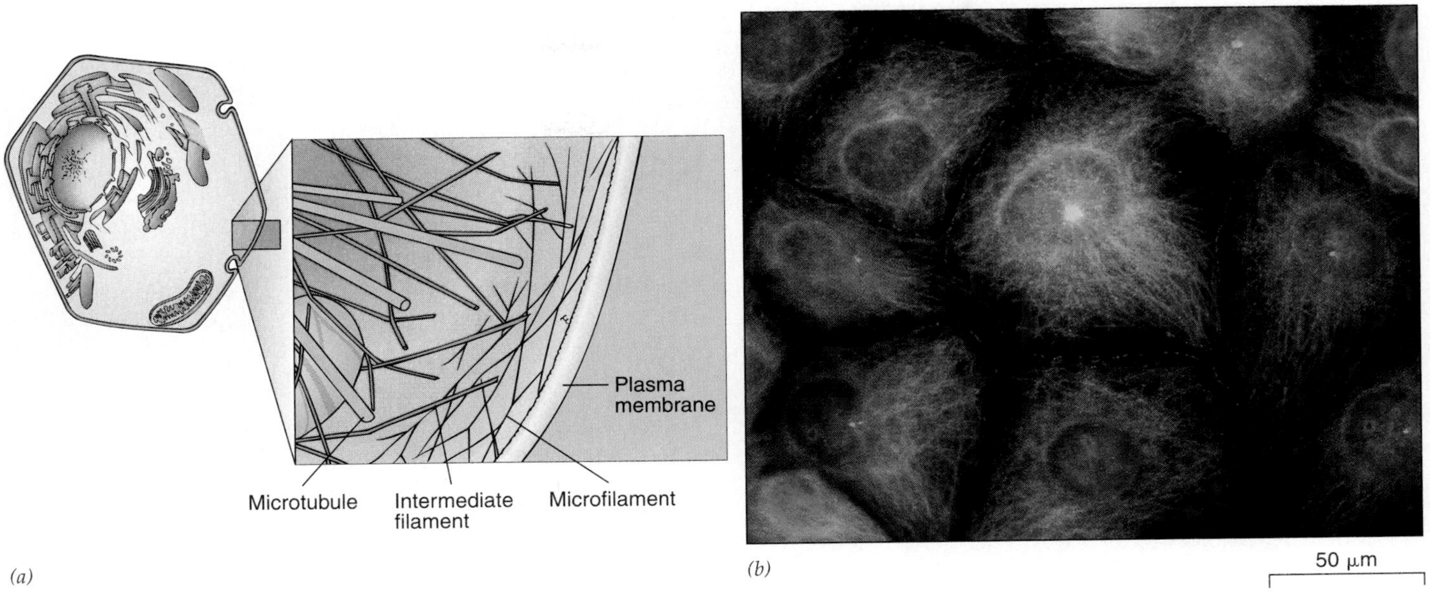

FIGURE 4–26 All eukaryotic cells contain a dynamic complex of protein filaments known as the cytoskeleton. This diagram illustrates some cytoskeletal elements that might be present in a typical animal cell. (***a***) The cytoskeleton consists of networks of several types of fibers, including microtubules, microfilaments, and intermediate filaments. The cytoskeleton contributes to the shape of the cell, anchors organelles, and sometimes rapidly changes shape during cellular locomotion. (***b***) Confocal fluorescence micrograph showing microtubules in green. A microtubule organizing center (pink dot) is visible near most of the cell nuclei (blue). (*b*, Nancy Kedersha/ImmunoGen, Inc.)

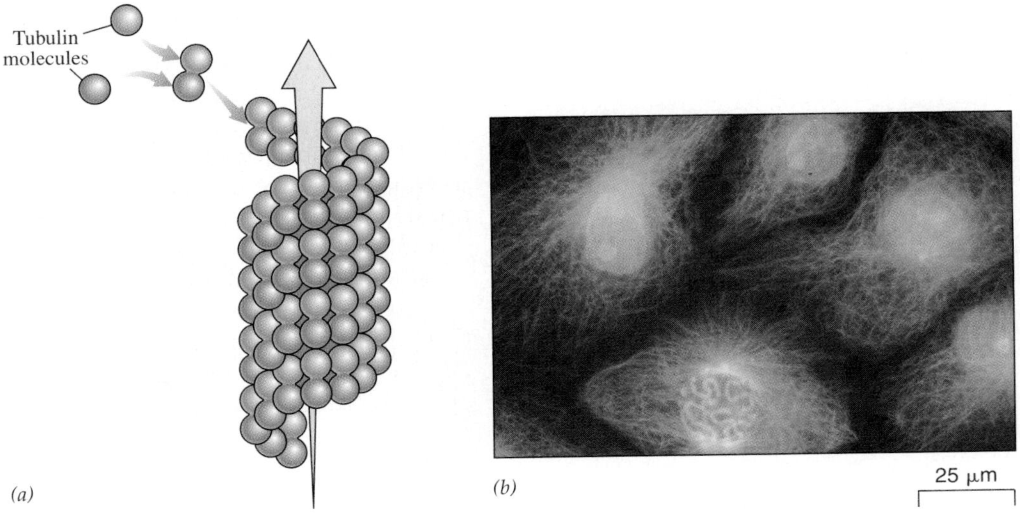

FIGURE 4–27 Microtubules are versatile cytoskeletal components. (***a***) Microtubules are constructed by adding dimers of α-tubulin and β-tubulin to an end of the hollow cylinder. Notice that the cylinder has polarity. The end shown at the top of the figure is the fast-growing or "plus" end; the opposite end is the slow-growing "minus" end. Each turn of the spiral requires 13 dimers. Disassembly occurs by removal of subunits from the ends of the filaments. (***b***) Confocal fluorescence micrograph showing an extensive distribution of microtubules. These cells were stained with fluorescent antibodies that bound to the tubulin, permitting the microtubules to be viewed (green). Different fluorescent antibodies were used to stain the DNA (orange). (*b*, courtesy of Dr. John M. Murray, Department of Cell and Developmental Biology, University of Pennsylvania)

Microtubules are extremely adaptable structures. In addition to playing a structural role in the formation of the cytoskeleton, they are involved in the movement of chromosomes during cell division, form "highways" for several other kinds of intracellular movement, and are the major structural components of cilia and flagella, special structures used in some cell movements.

For microtubules to act as a structural framework or participate in cell movement, they must be anchored to other parts of the cell. In nondividing cells, the micro-

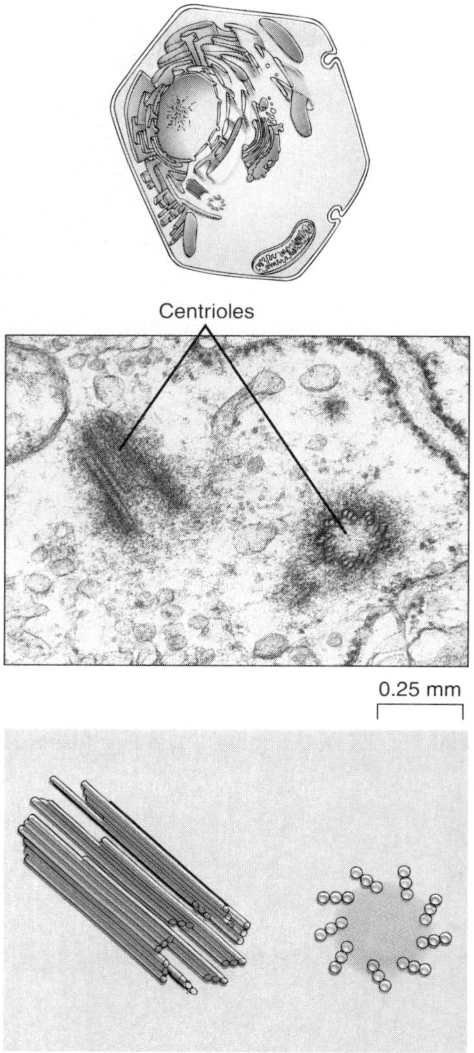

FIGURE 4–28 Centrioles are positioned at right angles to each other, near the nucleus of a nondividing animal cell. The transmission electron micrograph is paired with an interpretive drawing. Note the 9 × 3 arrangement of microtubules. The centriole on the left has been cut longitudinally and the one on the right transversely. (Photo, B.F. King, School of Medicine, University of California, Davis/Biological Photo Service)

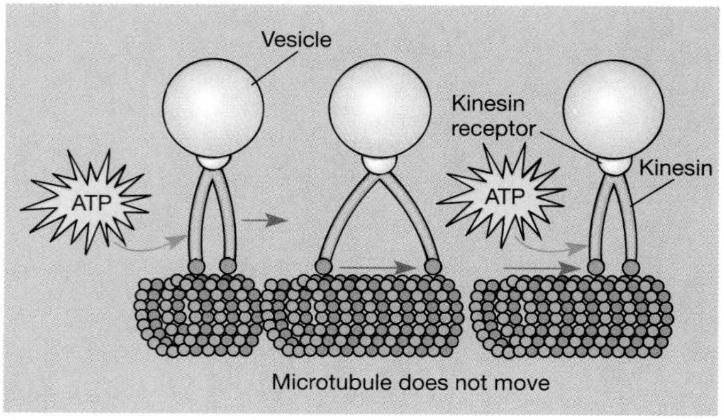

FIGURE 4–29 The drawing illustrates a hypothetical model of a kinesin motor. A kinesin molecule attaches to a specific receptor on the vesicle. It has been postulated that ATP energy allows the kinesin molecule to change its conformation and "walk" along the microtubule, carrying the vesicle along.

tubules appear to extend from a region called the **cell center** or **microtubule-organizing center (MTOC).**

Associated with the microtubule-organizing center of almost all animal cells are two structures called **centrioles** (Fig. 4–28). These structures, referred to as *9 × 3 structures,* are at right angles to each other and are made of nine sets of three microtubules arranged to form a hollow cylinder. The centrioles are duplicated before cell division and may play a role in some types of microtubule assembly, although their specific function is unknown. Most plant cells have a microtubule-organizing center, but lack centrioles. This suggests either that centrioles are not essential to most microtubule assembly processes or that alternative assembly mechanisms are possible.

The ability of microtubules to be assembled and disassembled rapidly can be seen during cell division (see Chapter 9), when much of the cytoskeletal apparatus in cells appears to break down. Many of the tubulin subunits are then reassembled into a structure called the **spindle,** which serves as a framework for the orderly distribution of chromosomes when the cell divides.

Microtubules also serve as tracks along which organelles can be moved to different cellular locations. Mitochondria, transport and secretory vesicles, and other organelles become attached to microtubules. They are then transported to various parts of the cell along the microtubule network by ATP-requiring proteins, which act as "motors." One such motor protein, named *kinesin,* has been isolated and can be shown to direct the movement of isolated organelles along purified microtubules from their "minus" ends to their "plus" ends (Fig. 4–29). *Dynein,* another motor protein, has been shown to transport vesicles in the opposite direction, from the "plus" end to the "minus" end.

Cilia and flagella are microtubule-containing structures used in cell movements

Projecting from surfaces of many cells are thin, movable structures that exhibit a beating motion. If a cell has one, or only a few, of these appendages and if they are relatively long in proportion to the size of the cell, they are called **flagella** (sing., *flagellum*). If the cell has many short appendages, they are called **cilia** (sing., *cilium*). Both cilia and flagella are used by cells to move through a watery environment or to pass liquids and particles across the cell surface. These structures are commonly found on unicellular and small multicellular organisms. In animals and certain plants, flagella serve as the tails of sperm cells. Cilia commonly occur on the surfaces of cells of an-

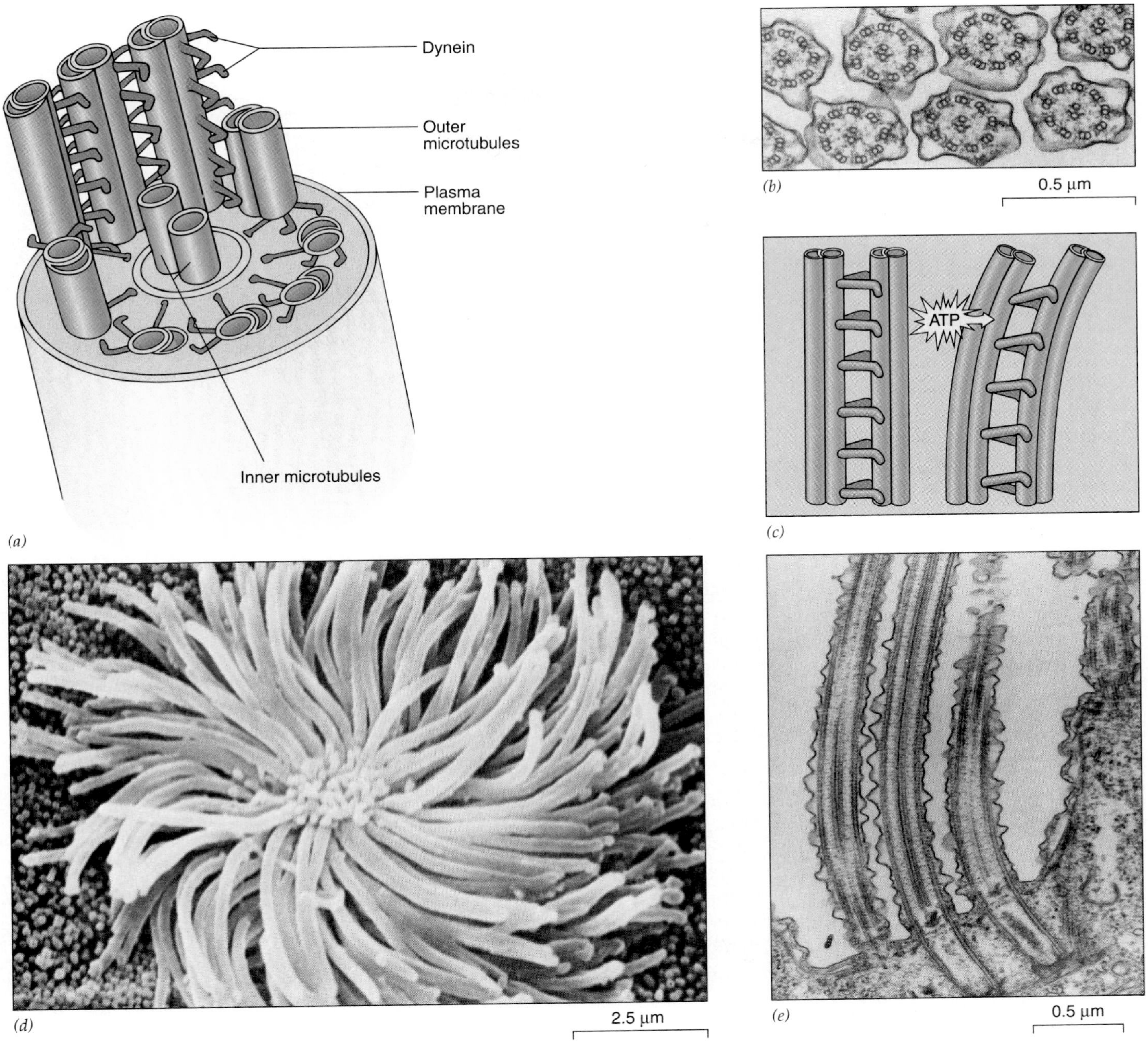

FIGURE 4–30 A cilium contains microtubules in a 9 + 2 arrangement. (*a*) This three-dimensional representation shows nine attached microtubule pairs (doublets) arranged in a cylinder, with two unattached microtubules in the center. The "arms" shown in the figure are made of dynein, a force-generating protein that uses energy from ATP to bend the cilia by "walking" up and down the neighboring pair of microtubules. The dynein arms, shown widely spaced for clarity, are actually much closer together along the longitudinal axis. (*b*) Transmission electron micrograph of cross sections through cilia showing the 9 + 2 arrangement of microtubules. (*c*) The dynein arms move the microtubules by forming and breaking cross bridges on the adjacent microtubules, so that one tubule "walks" along its neighbor. (*d*) Scanning electron micrograph of a group of cilia projecting from the surface of a cell in the lining of a rat's trachea ("windpipe"). In the center of the radiating cilia is a group of finger-like microvilli. (*e*) Transmission electron micrograph of a longitudinal section of the cilia of the protist *Tetrahymena,* an organism often used in genetic research. Some of the interior microtubules can be clearly seen. (*b, e,* W.L. Dentler, University of Kansas/Biological Photo Service; *d,* courtesy of Drs. James A. Popp and Joseph T. Martin, *American Journal of Anatomy,* 169, 1984)

imals that line internal ducts of the body (e.g., respiratory passageways).

Eukaryotic cilia and flagella are structurally alike. Each consists of a slender, cylindrical stalk covered by an extension of the plasma membrane. The core of the stalk contains a group of microtubules arranged so that there are nine attached pairs of tubules around the circumference and two unpaired microtubules in the center (Fig. 4–30). This *9 + 2* arrangement of microtubules is characteristic of virtually all eukaryotic cilia and flagella. The microtubules move by sliding in pairs past each other. The sliding force is generated by dynein proteins, which are attached to the microtubules like small arms. ATP energy causes the arms on one pair of tubules to change their conformation and "walk" along the adjacent pair. Thus, the microtubules on one side of a cilium or a flagellum extend farther toward the tip than those on the other side. Other components of the system exert constraints that cause the relative tip-to-base movement of the microtubules to be translated into a back-and-forth bending motion (Fig. 4–30*c*).

At the base of each cilium or flagellum is a **basal body,** which has nine sets of three microtubules in a cylindrical array. Both basal bodies and centrioles (see Fig. 4–28) are therefore referred to as *9 × 3 structures.* The basal body appears to be the organizing structure for the cilium or flagellum when it first begins to form. However, experiments have shown that as growth proceeds, the tubulin subunits are added much faster to the tips of the microtubules than to the base. Basal bodies and centrioles appear to be functionally related as well as structurally similar. In fact, centrioles are typically found in the cells of organisms that are capable of producing flagellated or ciliated cells; these include animals, certain protists, and a few plants.

Microfilaments consist of intertwined strings of actin

Microfilaments are solid fibers composed of the globular protein **actin** and actin-associated proteins (Fig. 4–31). In muscle cells actin is associated with another protein, myosin, to form fibers that generate the forces involved in muscle contraction (see Chapter 38). Actin microfilaments perform two different types of functions in nonmuscle cells. When actin is associated with myosin, it can form contractile structures that are involved in various cell movements. Actin can also be cross-linked with other proteins to form bundles of fibers that provide mechanical support for various cell structures. Actin fibers themselves cannot contract, but they can generate movement by rapidly assembling and disassembling.

Many types of cells have finger-like *microvilli* projecting from their surfaces (Fig. 4–30*d*). These structures can extend and retract as a result of the polymerization and depolymerization of actin fibers within the microvilli.

Actin microfilaments associated with myosin are involved in transient functions, such as cell division in animals, in which contraction of a ring of actin associated with myosin causes the constriction of the cell to form two daughter cells. This occurs after microtubules act to separate duplicated chromosomes (see Chapter 9).

Intermediate filaments help stabilize cell shape

Intermediate filaments are very stable, tough fibers made of polypeptides that can vary widely in size among different cell types and different species of animals. These fibers are thought to help strengthen the cytoskeleton and are abundant in parts of a cell that may be subject to mechanical stress. The assembly of these filaments is probably irreversible; unpolymerized subunits are not abundant in cells. Cells may be able to regulate the length of intermediate filaments, however, by use of enzymes that break down their polypeptides into smaller fragments. It is not clear whether intermediate filaments are involved in cellular functions beyond their structural role.

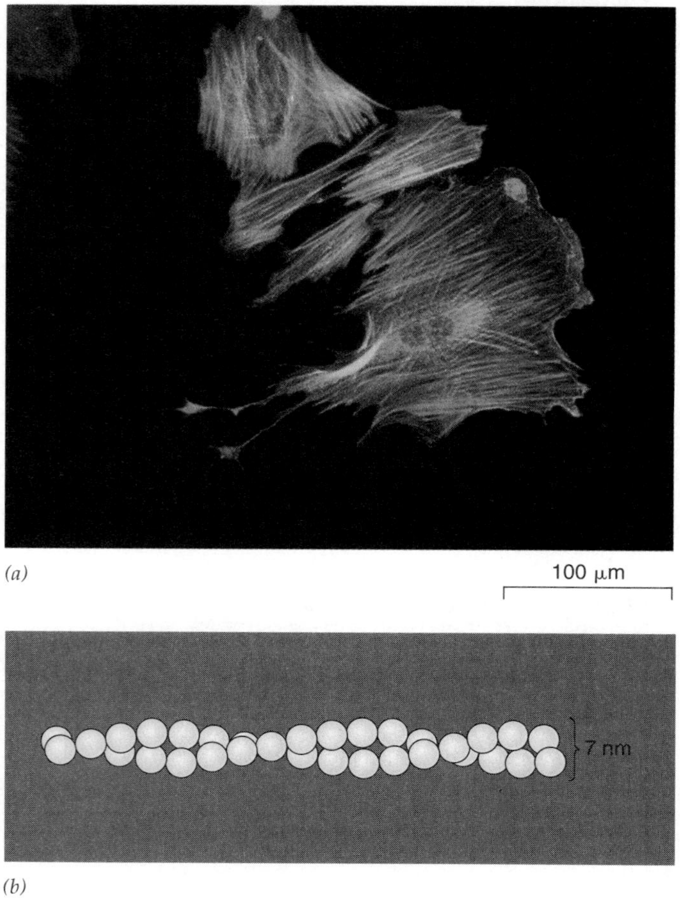

FIGURE 4–31 Actin microfilaments generally form bundles or networks in the cytoplasm. (***a***) Many bundles of aggregated microfilaments (green) are evident in this confocal fluorescence micrograph of fibroblast cells. (***b***) An individual microfilament consists of two intertwined strings of beadlike actin protein molecules. (*a*, Nancy Kedersha/ImmunoGen, Inc.)

An extracellular matrix surrounds most cells

Although the contents of the cell are effectively contained by the plasma membrane, most cells are also surrounded by some type of secreted coating that extends beyond the cell surface. Plant cells are surrounded by thick **cell walls** that contain multiple layers of the polysaccharide **cellu-**

lose. These molecules are formed into bundles of fibers that make up the bulk of the cell wall (see Fig. 3–10). Other polysaccharides in the cell wall form cross links between the cellulose fibers. Each cellulose fiber layer in a plant cell wall runs in a different direction from the adjacent layer, giving the structure great mechanical strength. A growing plant cell secretes a thin *primary cell wall,* which can stretch and expand as the cell increases its size (Fig. 4–32). After the cell stops growing, either new wall material is secreted that thickens and solidifies the primary wall, or multiple layers of a *secondary cell wall* with a different composition are formed between the primary wall and the plasma membrane. Between the primary cell walls of adjacent cells is the **middle lamella,** a layer of gluelike polysaccharides called *pectins.* The middle lamella causes the cells to adhere tightly to one another.

Animal cells do not have rigid cell walls, although many secrete proteins and polysaccharides that are bound to their outer surfaces and fill spaces between cells in tissues. The **glycocalyx** is a coat formed by polysaccharide side chains of lipids and proteins that are a part of the plasma membrane. Many of these molecules contain negatively charged regions, giving a negative charge to the surfaces of most cells. In many cases these coatings play a role in cellular contact and recognition, in addition to increasing the mechanical strength of multicellular tissues.

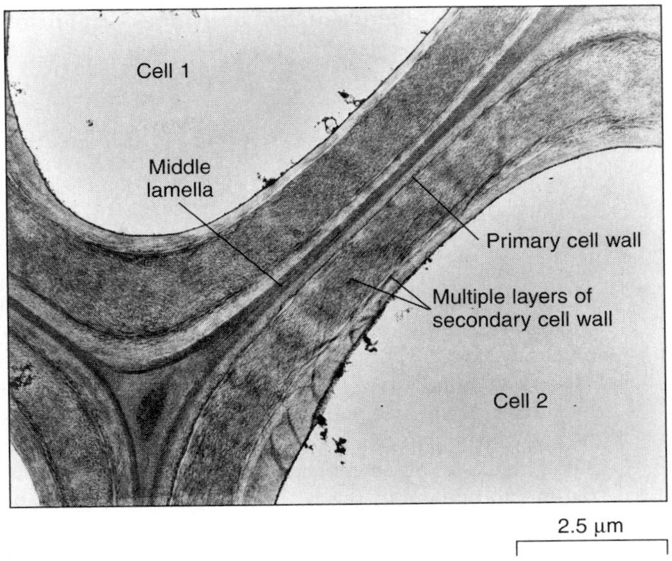

FIGURE 4–32 Plant cell walls include multiple layers. The cell walls of two adjacent plant cells are labelled in this transmission electron micrograph. The cells are cemented together by the middle lamella, a layer of gluelike polysaccharides called pectins. A growing plant cell first secretes a thin primary wall that is flexible and can stretch as the cell grows. The thicker layers of the secondary wall are secreted *inside* the primary wall after the cell stops growing. (Biophoto Associates)

SUMMARY

I. The cell is considered the basic unit of life because it is the smallest self-sufficient unit of living material.

II. Modern cell theory states that organisms are composed of cells and products of cells. All cells arise by division of preexisting cells.

III. Biologists have learned much about cellular structure by studying cells with light and electron microscopes. The electron microscope has superior resolving power, enabling investigators to see details of cell structures not observable with conventional microscopes. Information about the function of cellular structures requires the use of cell fractionation and biochemical methods in addition to microscopic observations.

IV. Every cell is surrounded by a plasma membrane that forms a cytoplasmic compartment, inside of which are found the contents of the cell.
   A. Because cells are small, the ratio of surface area to volume is favorable for rapid transport of molecules into or out of the cell.
   B. Prokaryotic cells are bounded by a plasma membrane but lack a nucleus and have little or no internal membrane organization.
   C. Eukaryotic cells have a nucleus and cytoplasm, which is organized into organelles that form membrane-bounded compartments. Plant cells differ from animal cells in that they possess rigid cell walls, plastids, and large vacuoles; cells of most plants lack centrioles.

V. The organelles of eukaryotic cells assume many diverse functions.
   A. Membrane-bounded compartments serve as a system of energy storage, allow cells to conduct specialized activities within small areas of the cytoplasm, concentrate molecules, and organize metabolic reactions within the cells.
   B. The nucleus, the control center of the cell, contains genetic information in the form of genes in the chromosomes.
      1. The nucleus is bounded by a double-membrane system with pores that communicate with the cytoplasm.
      2. Genetic information in the nucleus is carried by the DNA, which associates with protein to form chromatin. Chromatin complexes are organized into chromosomes, which become visible when the cell divides.
      3. The nucleolus is a region in the nucleus that is the site of ribosomal RNA synthesis and ribosome assembly.
   C. The endoplasmic reticulum (ER) is a series of folded internal membranes with many functions.
      1. Rough ER is studded along its outer walls with ribosomes, which manufacture proteins.

2. Smooth ER is the site of lipid biosynthesis and detoxifying enzymes.
3. Proteins synthesized on rough ER can be transferred to other membranes or secreted from the cells by transport vesicles, which are formed by membrane budding and then targeted to different cellular membrane locations.

D. The Golgi complex is a series of flattened membrane sacs that process, sort, and modify proteins synthesized on the ER. It adds carbohydrates and lipids to proteins and can route proteins (by way of vesicles) to the plasma membrane, to the outside of the cell, and to the lysosomes and possibly other membrane systems.

E. Lysosomes function in intracellular digestion; they contain degradative enzymes that break down both worn-out cell structures and substances taken into cells.

F. Mitochondria, the sites of aerobic respiration, are double-membrane organelles in which the inner membrane is folded to form cristae.

G. Cells of algae and plants contain plastids; chloroplasts, the sites of photosynthesis, are double-membrane structures enclosing internal thylakoid membranes, which are organized as stacks of flat, disclike structures called grana.

H. Microbodies are membrane-bounded sacs that can contain enzymes with diverse functions. Peroxisomes are microbodies that break down hydrogen peroxide.

I. The cytoskeleton is an internal framework made of at least three types of fibers.
1. Microtubules are hollow cylinders assembled from subunits of the protein tubulin.
2. Microfilaments, rapidly assembled and disassembled filaments with a smaller diameter than microtubules, are formed from subunits of the protein actin.
3. Intermediate filaments are stable structures formed from several different types of protein.

J. Cilia and flagella are structures that project from the cell surface and are used for cell movement. They are 9 + 2 structures formed from microtubules and covered by the plasma membrane.

K. Microtubules are components of centrioles and basal bodies (9 × 3 structures), which appear to be associated with organizing centers for microtubule formation in animal cells.

L. Plant cells secrete cellulose and other polysaccharides to form rigid cell walls.

M. Some animal cells are covered by a glycocalyx, a coating formed from carbohydrate regions of glycoproteins and glycolipids on the cell surface.

SELECTED KEY TERMS

actin
basal body
cell fractionation
cell theory
centriole
chlorophyll
chloroplast
chromatin
chromosome
cilium (cilia)
cytoplasm
cytoskeleton
cytosol
deoxyribonucleic acid (DNA)
electron microscope
endoplasmic reticulum (ER)
endosymbiont theory
eukaryote
flagellum (flagella)
glycocalyx
glycoprotein
glyoxysome
Golgi complex
granum (grana)
intermediate filament
lysosome
microbody
microfilament
microtubule
microtubule-organizing center (MTOC)
mitochondrion (mitochondria)
nuclear envelope
nuclear pores
nucleolus
nucleoplasm
nucleus
organelle
peroxisome
photosynthesis
plasma membrane
plastid
prokaryote
resolving power
ribosome
secretory vesicle
stroma
thylakoid
vacuole
vesicle

POST-TEST

1. The ability of a microscope to reveal fine detail is known as __________ __________.
2. Proteins to be secreted from the cell are synthesized by ribosomes bound to the __________ __________.
3. In eukaryotic cells DNA is associated with proteins to form __________, which makes up the chromosomes.
4. Powerful hydrolytic enzymes contained in the __________ are released when the cell dies; they digest the cellular remains.
5. Membrane-bounded organelles that break down H_2O_2 are termed __________.
6. __________ are the sites of aerobic cellular respiration.
7. The __________ are organelles involved in the synthesis and storage of carbohydrates.
8. Chlorophyll, which is located in the __________ membranes of chloroplasts, is used to trap energy from sunlight for use in __________.
9. Cylindrical, hollow cytoplasmic rods called __________ play a role in controlling the shape and movement of cells.
10. The flexible framework in the cytoplasm of the cell called the cytoskeleton is mainly composed of __________, __________, and __________ filaments.
11. __________ and __________ are movable, whiplike structures projecting from the cell surface. These are used to move the cell through surrounding liquid or to move liquid across the surface of the cell. The core of each is composed of a 9 + 2 arrangement of __________.
12. In addition to having a plasma membrane, plant cells are surrounded by a __________ __________, formed primarily from fibers of the polysaccharide cellulose.

13. __________ are the membrane compartments in plant cells that are used for the storage of water and waste products. They may also have functions similar to __________ of animal cells.
14. Ribosomes are assembled in the __________ region of the __________ in eukaryotic cells.
15. The Golgi complex modifies proteins by adding complex carbohydrates to certain amino acids in the polypeptide chains to form __________ .

REVIEW QUESTIONS

1. Trace the development of the cell theory. Why is this theory important to an understanding of how living things function?
2. What are the main differences between prokaryotic and eukaryotic cells?
3. Draw diagrams of a prokaryotic cell, a plant cell, and an animal cell. Label the organelles. Which organelles might be found in a plant cell but not an animal cell (and vice versa)?
4. Sketch the membranes of chloroplasts and mitochondria. Label the membranes and their compartments.
5. What are the functions of each of the following?
   a. ribosomes
   b. endoplasmic reticulum
   c. Golgi complex
   d. lysosomes
6. Trace the path of a protein from its site of synthesis to its final destination for the following:
   a. a secreted protein
   b. a protein found inside a lysosome
   c. a protein associated with the plasma membrane
7. Describe the differences between microfilaments and microtubules. Compare their structures and the different roles they play in cell structure and function.
8. Why are lysosomes sometimes referred to as the "self-destruct system" of the cell?
9. Describe plant cell walls. How are they formed?
10. Label the diagram of the animal cell and the plant cell. What are the fundamental differences between the two cell types?

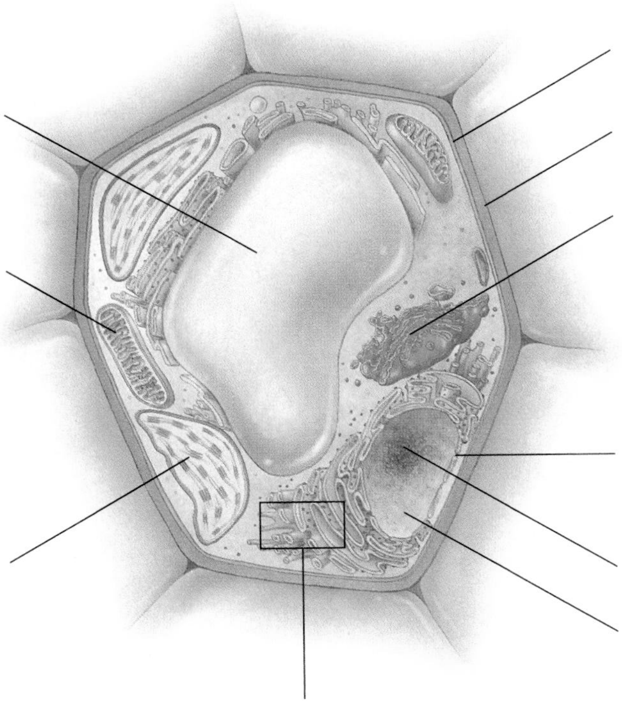

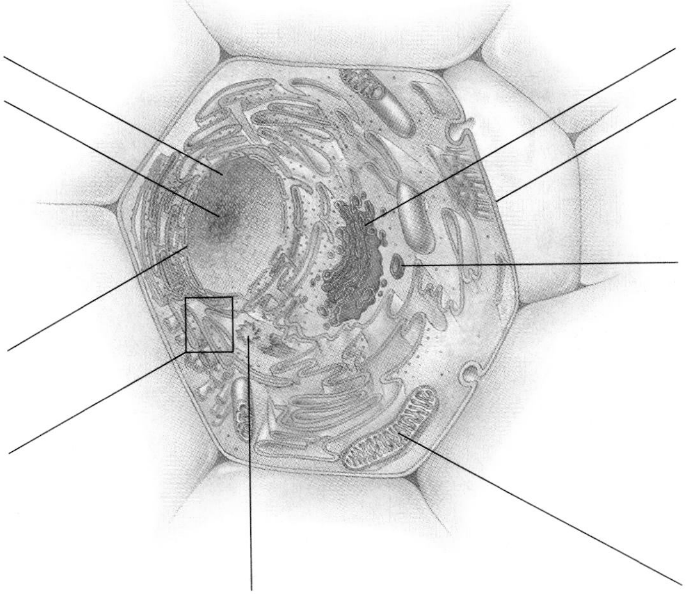

YOU MAKE THE CONNECTION

1. Why does a eukaryotic cell need both membranous organelles and fibrous cytoskeletal components?
2. The *Acetabularia* experiments described in this chapter suggest that DNA is much more stable in the cell than messenger RNA. Is this advantageous or disadvantageous to the cell? Why?

RECOMMENDED READINGS

Alberts, B., D. Bray, J. Lewis, M. Raff, K. Roberts, and J.D. Watson. *Molecular Biology of the Cell.* 3rd edition. Garland, New York, 1994. A comprehensive, well-written, and well-illustrated text on cell biology.

de Duve, C. *A Guided Tour of the Living Cell.* Scientific American Library, New York, 1984. An engrossing, beautifully illustrated tour of the cell in the form of journeys through different membrane and organelle systems.

Goodsell, D.S. "A Look Inside the Living Cell." *American Scientist,* Vol. 80, No. 5, 1992. A fascinating effort to visualize how biological molecules actually look in their natural environment, the cell.

Loewy, A., P. Siekovitz, J. Menninger, and J. Gallant. *Cell Structure and Function: An Integrated Approach.* Saunders College Publishing, Philadelphia, 1991. An introductory cell biology text.

Taylor, D.L., M. Nederlof, F. Lanni, and A.S. Waggoner. "A New Vision of Light Microscopy." *American Scientist,* Vol. 80, No. 4, 1992. A description of how new technology is allowing scientists to use light microscopes to study living cells in sophisticated ways.

CHAPTER 5

Biological Membranes

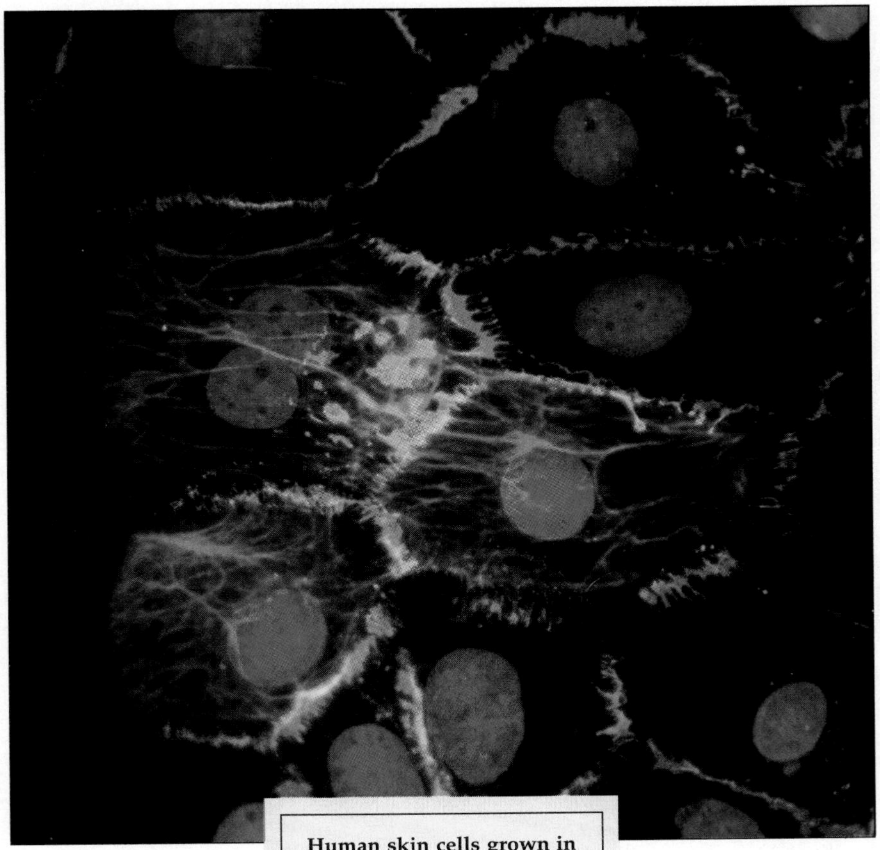

Human skin cells grown in culture and stained with fluorescent antibodies.
(Nancy Kedersha, ImmunoGen, Inc.)

To carry out the many chemical reactions necessary to sustain life, a cell must maintain an appropriate internal environment. This is possible because all cells are physically separated from the outside world by a limiting plasma membrane, which defines the cell as a distinct entity. Many biologists, in fact, view the evolution of biological membranes as an essential step in the origin of life. One can argue further that membranes made the evolution of complex cells possible, because the extensive internal membranes of eukaryotes form multiple compartments with unique environments for highly specialized activities.

Cellular membranes are not inanimate walls; they are complex and dynamic structures made from lipid and protein molecules that are in constant motion. The unusual properties of membranes allow them to perform many functions. These include serving as work surfaces for many chemical reactions, regulating movement of materials in and out of the cell, transmitting signals and information between the environment and the interior of the cell, and acting as an essential part of an energy transfer and storage system (see Chapters 7 and 8).

To understand how membranes do these things, we first consider what is known about their composition and structure. This chapter then surveys how various materials, ranging from simple to complex molecules, and even particles, are able to move across membranes. It also considers how information can cross the plasma membrane through a signal relay system. Finally, specialized structures that permit interactions between membranes of different cells are examined. Although most of our discussion centers on the structure and functions of plasma membranes, most of the concepts are also applicable to internal membrane systems.

LEARNING OBJECTIVES

After you have studied this chapter you should be able to

1. Evaluate the importance of membranes to the cell, emphasizing their various functions.
2. Make a detailed sketch of the fluid mosaic model of cell membrane structure.
3. Explain how the properties of the lipid bilayer are responsible for many of the physical properties of a cell membrane.
4. Explain how the various classes of membrane proteins associate with the lipid bilayer, and discuss the different roles that membrane proteins assume.
5. Contrast the physical processes of simple diffusion and osmosis with the carrier-mediated physiological processes by which materials are transported across cell membranes.
6. Solve simple problems involving osmosis; for example, predict whether cells will swell or shrink under various osmotic conditions.
7. Summarize the main ways that small hydrophilic molecules can move across membranes.
8. Differentiate between the processes of facilitated diffusion and active transport, and discuss the ways in which energy is supplied to active transport systems.
9. Compare endocytotic and exocytotic transport mechanisms.
10. Describe the structures and compare the functions of desmosomes, tight junctions, gap junctions, and plasmodesmata.

BIOLOGICAL MEMBRANES ARE LIPID BILAYERS WITH ASSOCIATED PROTEINS

When you examine an electron micrograph and compare the sizes of different cell structures, one of the most striking observations is how exceedingly uniform and thin the membranes are (Fig. 5–1). Cellular membranes are no more than 10 nanometers thick.

Long before the development of the electron microscope, it was known that membranes are composed of both lipids and proteins. Work by researchers in the 1920s and 1930s had provided clues that the core of the cell membrane is composed of lipids, mostly phospholipids (see Chapter 3). Furthermore, by examining the membrane of the mammalian red blood cell (which has only a plasma membrane) and comparing the surface area of the membrane with the total number of lipid molecules per cell, investigators were able to calculate that the phospholipids are probably arranged so that the membrane is no more than two phospholipid molecules thick.

These findings, together with other information, led H. Davson and J.F. Danielli in 1935 to propose a model in which they envisioned a membrane as a kind of "sandwich" consisting of a *lipid bilayer* (a double layer of lipid)

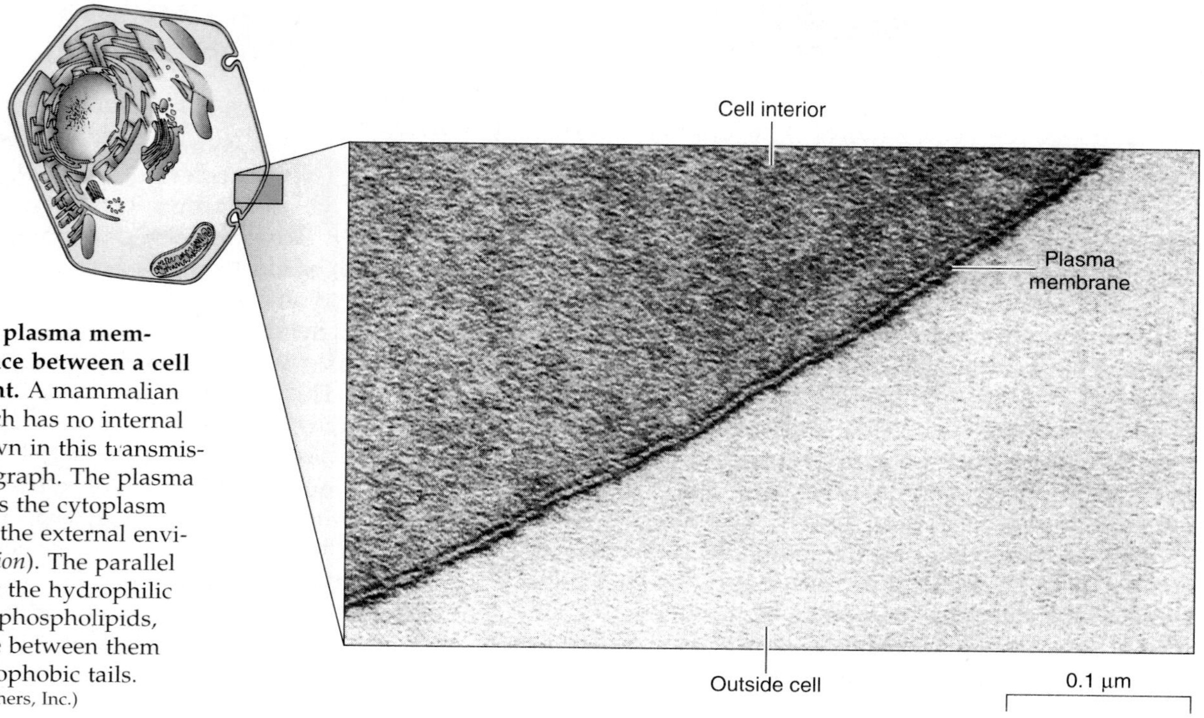

FIGURE 5–1 The plasma membrane is the interface between a cell and its environment. A mammalian red blood cell, which has no internal membranes, is shown in this transmission electron micrograph. The plasma membrane separates the cytoplasm (*darker region*) from the external environment (*lighter region*). The parallel dark lines represent the hydrophilic head groups of the phospholipids, while the light zone between them represents the hydrophobic tails. (Omikron/Photo Researchers, Inc.)

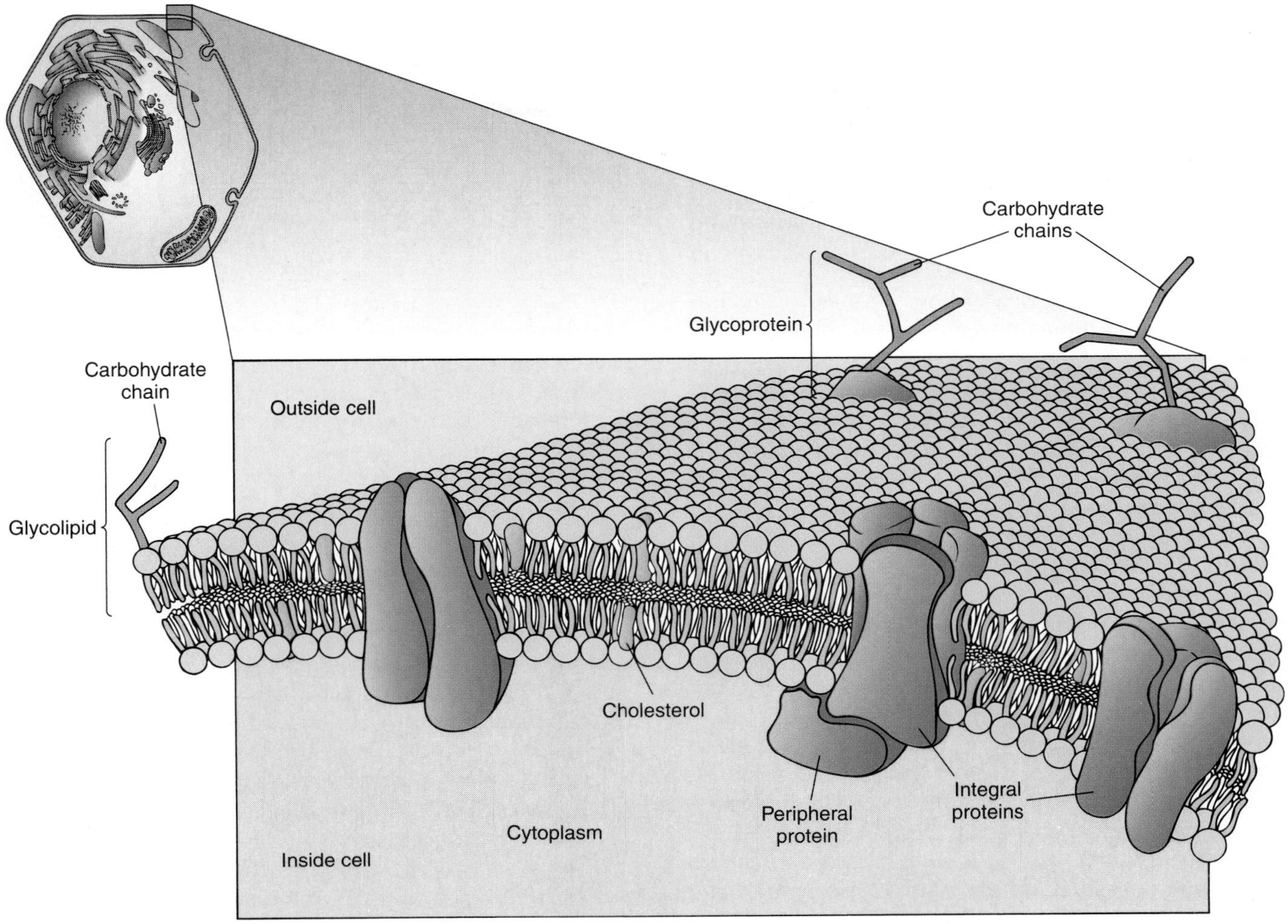

FIGURE 5–2 According to the fluid mosaic model, a cellular membrane is made up of a fluid lipid bilayer with a constantly changing "mosaic pattern" of associated proteins. The diagram illustrates the plasma membrane of a eukaryotic cell, in this case an animal cell. The lipid bilayer, which is about 5 nanometers thick, consists mainly of phospholipids. Other lipids include steroids such as cholesterol. Peripheral proteins are loosely associated with the bilayer, while integral proteins are tightly bound. Transmembrane proteins are integral proteins that extend through the bilayer. Carbohydrates attached to lipids (glycolipids) and to proteins (glycoproteins) are exposed on the extracellular surface; these appear to provide some mechanical protection and to have roles in cell recognition and adhesion. Some membrane proteins are attached to cytoskeletal elements (not shown).

between two protein layers. This very useful model (as well as some related models) had a great influence on the direction of membrane research for over 20 years. Models are very important in biological research; good ones not only explain the available data, but are *testable.* That is, scientists can use the model to help them develop hypotheses that can be tested experimentally (see Chapter 1).

As data accumulated, the evidence supporting the idea that membrane lipids are associated as a bilayer became much stronger. However, a major paradox emerged regarding arrangement of the proteins. Many membrane proteins were found to have diameters greater than 10 nanometers. How could these proteins be arranged to fit in a membrane less than 10 nanometers thick? Later modifications of the model addressed this objection by hypothesizing that the proteins on the membrane surfaces were a flattened, extended form, perhaps a β-pleated sheet (see Chapter 3).

In 1972, S.J. Singer and G.L. Nicolson proposed a membrane structure model that represented a synthesis of the known properties of biological membranes (discussed later in this chapter). According to their **fluid mosaic model,** a cellular membrane consists of a fluid bilayer of lipid molecules in which the proteins are embedded or otherwise associated, much like the tiles in a mosaic picture. This mosaic pattern is not static, however, because the positions of the proteins are constantly changing as they move about like icebergs in a fluid sea of lipids. This model has provided a great impetus to research; it has been repeatedly tested and has been shown to accurately predict the properties of many kinds of cellular membranes. Figure 5–2 depicts the plasma membrane of a eukaryotic cell; prokaryotic plasma membranes are discussed in Chapter 23.

We now know that lipids are primarily responsible for the physical properties of biological membranes. This is because certain lipids have unique attributes, including features that allow them to form bilayered structures.

Phospholipids form bilayers in water

A phospholipid contains two fatty acid chains linked to two of the three carbons of a glycerol molecule. The fatty acid chains make up the nonpolar, hydrophobic ("water-hating") portion of the phospholipid. Bonded to the third carbon of the glycerol is a negatively charged, hydrophilic ("water-loving") phosphate group, which in turn is linked to a polar, hydrophilic organic group. Molecules of this type, which have distinct hydrophobic and hydrophilic regions, are called **amphipathic** molecules (see Chapter 3). All lipids that make up the core of biological membranes have amphipathic characteristics.

Because one end of each phospholipid associates freely with water and the opposite end does not, the most favorable orientation for them to assume in water results in the formation of a bilayer structure (Fig. 5–3). This arrangement allows the hydrophilic headgroups of the phospholipids to be in contact with the aqueous medium, while the hydrophobic fatty acid chains are buried in the interior of the structure away from the water molecules.

Amphipathic properties alone do not predict the ability of lipids to associate as a bilayer. Shape is also important. For example, many common detergents are amphipathic molecules, each containing a single hydrocarbon chain (like a fatty acid) at one end and a hydrophilic region at the other. These molecules are roughly cone-shaped, with the hydrophilic end forming the broad base

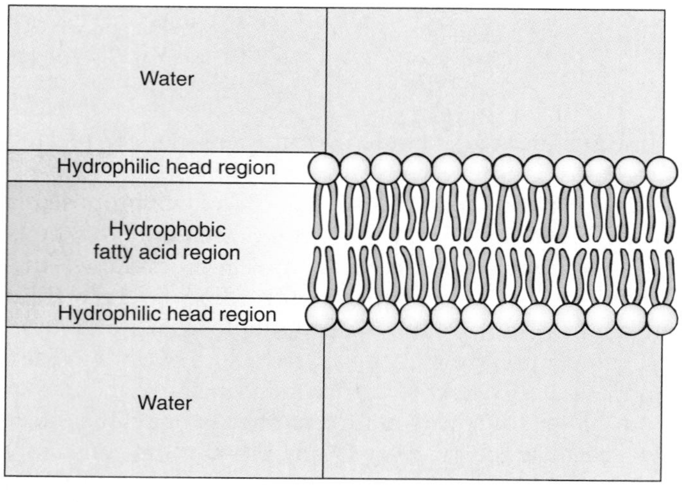

FIGURE 5–3 Phospholipids form bilayers in water. The hydrophobic fatty acid chains associate with each other and are not exposed to the water. The hydrophilic phospholipid headgroups are in contact with the aqueous medium.

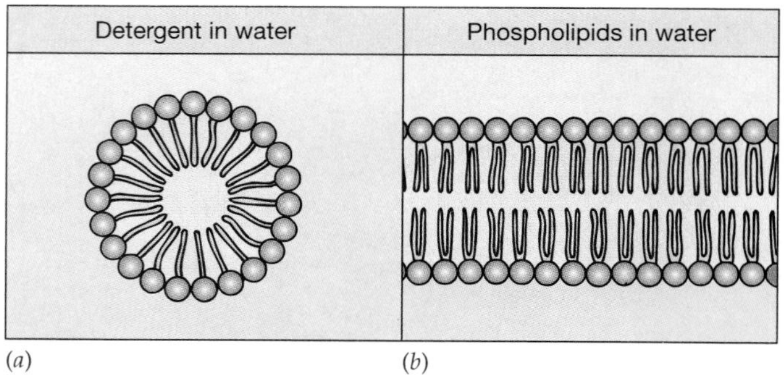

FIGURE 5–4 The ability of lipids to form bilayers in water depends on their amphipathic properties and their shapes. (***a***) Detergent molecules are roughly cone-shaped amphipathic molecules that associate in water as spherical structures. (***b***) Phospholipids associate as bilayers in water because they are roughly cylindrical amphipathic molecules.

and the hydrocarbon tail leading to the point. Because of their shapes, these molecules do not associate as bilayers, but instead tend to form spherical structures in water (Fig. 5–4*a*). Detergents are able to "solubilize" oil because the oil molecules associate with the hydrophobic interiors of the spheres.

Phospholipids tend to have uniform widths; their roughly cylindrical shapes, together with their amphipathic properties, are responsible for bilayer formation (Fig. 5–4*b*). Thus, phospholipids regularly form bilayers because the molecules have (1) two distinct regions, one strongly hydrophobic and the other strongly hydrophilic (making them strongly amphipathic), and (2) cylindrical shapes that allow them to associate with water most favorably as a bilayer structure.

Biological membranes are two-dimensional fluids

An important physical property of phospholipid bilayers is that they behave as *liquid crystals* (Fig. 5–5). The bilayers are crystal-like in that the lipid molecules form an or-

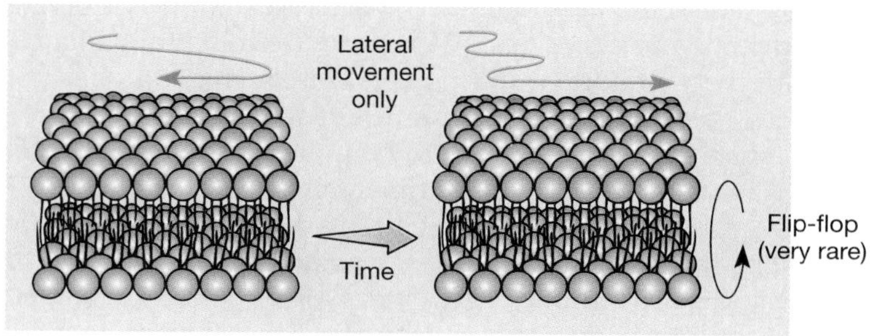

FIGURE 5–5 The phospholipid bilayers of cellular membranes are in a liquid-crystalline state. Although the phospholipid molecules have specific orientations in the bilayer, their hydrocarbon chains are in constant motion, allowing each molecule to move laterally on the same side of the bilayer.

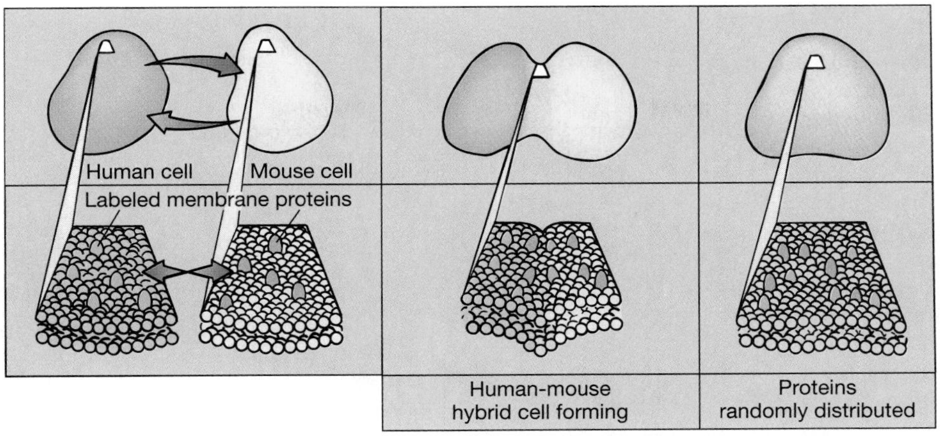

FIGURE 5–6 The elegant series of experiments by Frye and Ediden demonstrated the mobility of membrane proteins. Membrane proteins of mouse cells and human cells were labeled with fluorescent dye markers in two different colors. When the plasma membranes of a mouse cell and a human cell were then fused, mouse proteins were observed migrating to the human side and human proteins to the mouse side. After a short time mouse and human proteins became randomly distributed on the cell surface. This demonstration was convincing evidence that at least some membrane proteins are highly mobile entities in a two-dimensional fluid.

dered array with the headgroups on the outside and fatty acid chains on the inside; they are liquid-like in that, despite the orderly arrangement of the molecules, their hydrocarbon chains are in constant motion. Thus molecules are free to rotate and can move laterally within their single layer. Such movement gives the bilayer the property of a *two-dimensional fluid.* Under normal conditions this means that a single phospholipid molecule can travel across the surface of a eukaryotic cell in seconds.

The fluid-like qualities of lipid bilayers also allow molecules embedded in them to move along the plane of the membrane (as long as they are not anchored in some way). This was elegantly demonstrated by David Frye and Michael Ediden in 1970. They conducted experiments in which they followed the movement of membrane proteins on the surface of two cells that had been joined together (Fig. 5–6). When the plasma membranes of a mouse cell and a human cell are fused, within minutes at least some of the membrane proteins from each cell migrate and become randomly distributed over the single continuous plasma membrane that surrounds the joined cells.

If a membrane is to function properly, its lipids must be in a state of optimal fluidity. The structure of a membrane is weakened if its lipids are too fluid. On the other hand, it has been shown that many membrane functions, such as the transport of certain substances, are inhibited or cease if the lipid bilayer is too rigid.

Certain properties of membrane lipids have significant effects on the fluidity of the bilayer. Recall from Chapter 3 that molecules are free to rotate around single carbon-to-carbon covalent bonds. Because most of the bonds in hydrocarbon chains are single bonds, the chains themselves can undergo very rapid twisting motions that increase as the temperature increases. Although most biological membranes are in the liquid-crystalline state in living cells, the motion of the fatty acid chains is slowed at low temperatures.

Van der Waals interactions (see Chapters 2 and 3) can take place between hydrocarbon chains if they are lined up close to each other. When this happens, the phospholipid bilayer is converted to a solid gel state. You may be familiar with a similar situation in cooking fats: some fats are solid at room temperature, whereas others are liquid. One of the major differences between these two types of fat is the number of double bonds in the hydrocarbon chains of their fatty acids. If a fatty acid is saturated, it has no double bonds; if it is unsaturated, it has one (monounsaturated) or two or more (polyunsaturated). Double bonds produce "bends" in the molecules that prevent the hydrocarbon chains from coming close enough together to form van der Waals contacts, thus effectively lowering the temperature at which the oil or the membrane lipids crystallize (see Fig. 3–13).

Many organisms have regulatory mechanisms that allow them to maintain their membranes in an optimally fluid state. For example, many organisms that are unable to maintain a constant internal temperature can compensate for temperature changes by altering the fatty acid content of their membrane lipids. When grown at colder temperatures, such organisms are found to have relatively high proportions of unsaturated fatty acids in their membrane lipids.

Some membrane lipids have the ability to help stabilize membrane fluidity within certain limits. One such "fluidity buffer" is cholesterol, a steroid found in animal cell membranes. A cholesterol molecule is largely hydrophobic, but is slightly amphipathic due to the presence of a single hydroxyl group (see Fig. 3–16*a*). This hydroxyl group associates with the hydrophilic headgroups of the

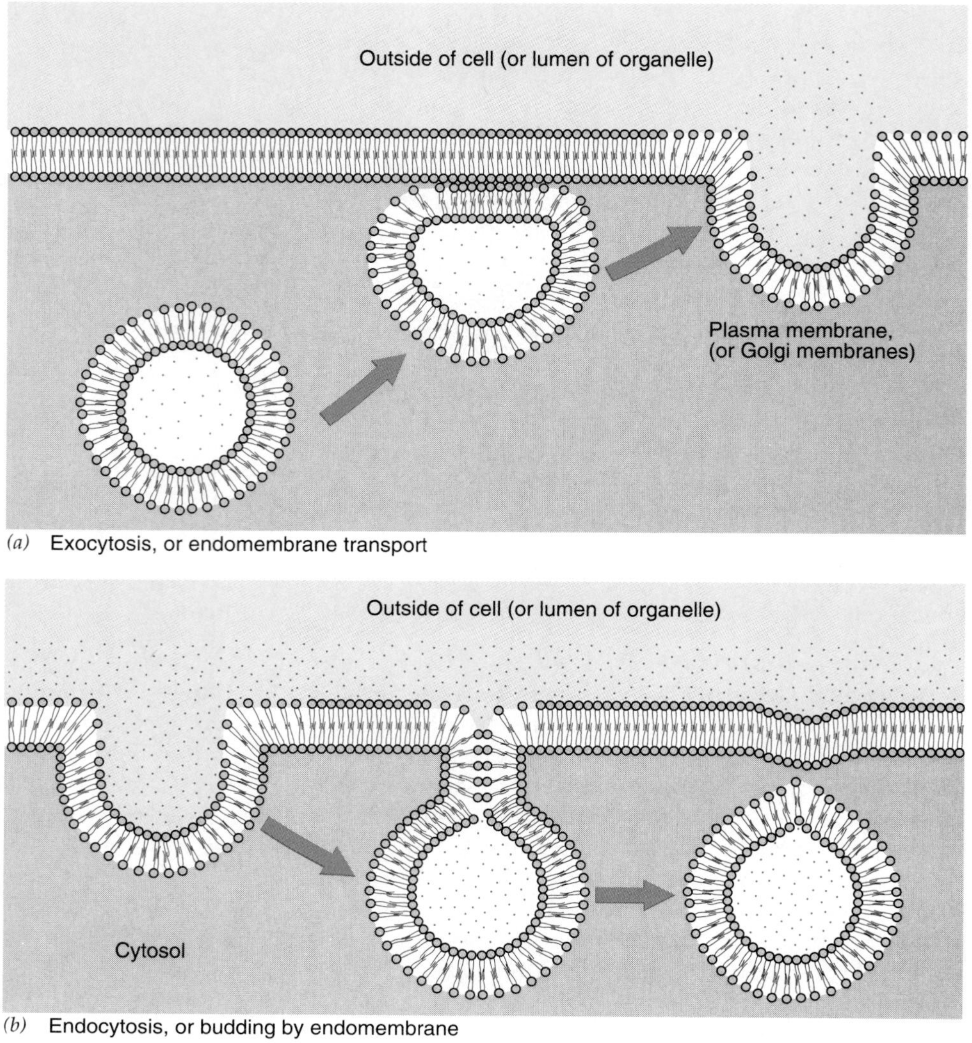

FIGURE 5–7 Endocytosis and exocytosis, as well as intracellular transport by vesicles, involve the fusion of lipid bilayers. It has been shown experimentally that, at least in certain well-documented cases, elaborate protein complexes referred to as "budding machines" and "fusion machines" are required in addition to the lipid bilayers shown here. (***a***) Both exocytosis and movement of molecules between organelles involve the contact and fusion of a vesicle with a membrane, releasing the vesicle's contents into another compartment or to the outside of the cell. (***b***) Endocytosis results from invagination of the plasma membrane, while the formation of an intracellular transport vesicle involves the formation of a "bud" from an internal membrane. These events are followed by the fusion of two regions of the membrane that come in contact with each other. Note that in exocytosis the two cytoplasmic sides of the membrane make contact with each other to initiate membrane fusion, while in endocytosis the two noncytoplasmic layers make the first contacts. This means that, even if we consider only the behavior of the lipids, endocytosis and exocytosis are not exactly the reverse of each other.

phospholipids; the hydrophobic remainder of the cholesterol molecule fits between the fatty acid hydrocarbon chains (see Fig. 5–2).

At low temperatures the cholesterol molecules act as "spacers" between the hydrocarbon chains, restricting van der Waals interactions that would promote crystallization. Cholesterol also helps prevent the membrane from becoming weakened or unstable at higher temperatures. This is because the cholesterol molecules interact strongly with the portions of the hydrocarbon chains closest to the phospholipid head groups, thereby restricting motion in these regions. Plant cells have steroids other than cholesterol that carry out similar functions.

Biological membranes fuse and form closed vesicles

Lipid bilayers, particularly those in the liquid-crystalline state, have additional important physical properties. Bilayers tend to resist forming free ends; as a result, they are self-sealing and under most conditions spontaneously round up to form closed vesicles. Fluid bilayers also are flexible, allowing cell membranes to change shape without breaking. Finally, under appropriate conditions lipid bilayers have the ability to fuse with other bilayers.

Membrane fusion is an important cellular phenomenon (Fig. 5–7). When a vesicle fuses with another membrane, both membrane bilayers and their compartments become continuous. This allows materials to be transferred from one compartment to another or to move from a secretory vesicle to the outside of a cell by a process known as *exocytosis.* In a similar but reverse process, *endocytosis,* large molecules are brought into the cell from the outside by the formation of vesicles from a section of membrane. Both endocytosis and exocytosis are discussed later in this chapter.

Membrane proteins may be integral or peripheral

It was not always clear how proteins might be associated with membranes. Early investigators, in fact, found it difficult to accept the idea that proteins could associate with any part of membranes other than their surfaces. It was widely assumed that membrane proteins must be very uniform and must have shapes that would allow them to lie like thin sheets on the membrane surface. However, several lines of evidence eventually argued against these ideas, and influenced Singer and Nicolson in their development of the fluid-mosaic model.

Proteins from membranes purified by cell fractionation were found to be far from uniform; in fact, they varied widely in size and composition. Other studies provided evidence that instead of having sheetlike structures, many membrane proteins are rounded in shape, or *globular.* Their diameters are so large that the membrane would have to be much thicker than 10 nanometers if the proteins were located only on the surface. Studies of a number of individual membrane proteins showed that one region (or domain) of the molecule could always be found on one side of the bilayer, while another part of the protein might be located on the opposite side. It appeared that, rather than forming a thin surface layer, many membrane proteins extend completely through the lipid bilayer. Thus, membranes appear to contain many different types of proteins of different shapes and sizes that are associated with the bilayer in a mosaic pattern. However, this pattern is far from static, for, as shown by the experiments of Frye and Ediden, the fluidity of the lipids allows many of the proteins to move in the plane of the bilayer, producing an ever changing configuration.

The two major classes of membrane proteins, integral proteins and peripheral proteins, are defined by how tightly they are associated with the lipid bilayer (see Fig. 5–2). **Integral membrane proteins** are firmly bound to the membrane; usually they can be released only by disrupting the bilayer with detergents. Some integral proteins are not embedded in the membrane, but are covalently bound to lipids that are part of the bilayer. Available evidence suggests that a few integral proteins may be partially embedded in the membrane.

Most integral proteins actually span the membrane through regions of their polypeptide chains that pass through the hydrophobic interior of the lipid bilayer. Some of these **transmembrane proteins** span the membrane only once, while others pass back and forth as many as 24 times.

Transmembrane proteins are able to insert into the lipid bilayer because the regions of the molecules that are within the membrane have hydrophobic surfaces, compatible with the interior of the bilayer. If a membrane protein has a region with a hydrophilic surface, it is usually found protruding from the membrane, in contact with the aqueous medium. The most common kind of membrane-spanning region is an α-helix (see Chapter 3), with hydrophobic amino acid side chains projecting out from the helix (Fig. 5–8).

Peripheral membrane proteins can be easily removed from the membrane without disrupting the structure of the bilayer. They usually bind to exposed regions of integral membrane proteins and are held there by noncovalent interactions.

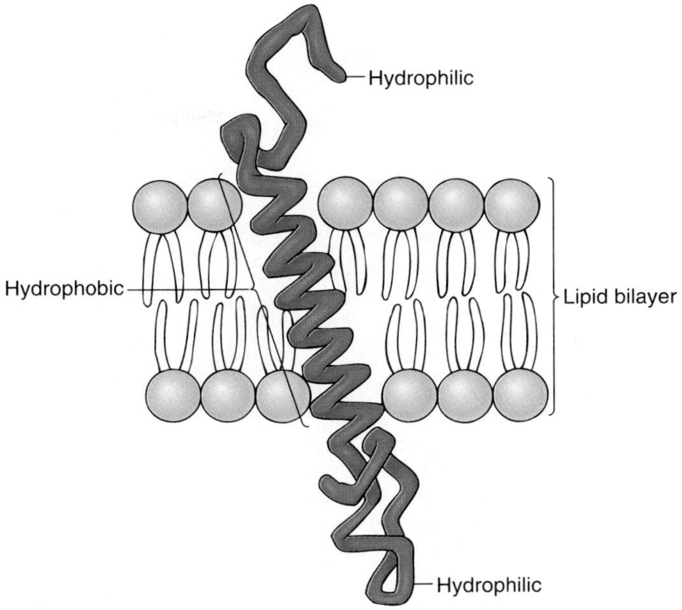

FIGURE 5–8 A transmembrane protein has one or more hydrophobic regions that cross the lipid bilayer. The protein illustrated has hydrophilic regions on both sides of the bilayer, connected by a membrane-spanning α-helix with hydrophobic amino acid side chains (not shown).

Proteins are oriented asymmetrically across the bilayer

One of the most remarkable demonstrations that proteins are actually embedded in the lipid bilayer comes from freeze-fracture electron microscopy (Fig. 5–9), which enables investigators to literally see the membrane from "inside out." When cellular membranes are examined in this way, numerous particles are observed on the fracture faces. These particles are clearly integral membrane proteins because they are never seen in freeze-fractured artificial lipid bilayers. These findings profoundly influenced Singer and Nicolson in their development of the fluid mosaic model.

When the two sides of a membrane are compared by this method (as in Fig. 5–9), large numbers of particles are found on one side and very few on the other. This does not necessarily mean that there are more proteins on one side of the membrane than on the other, but rather that most are more firmly attached to a given side. Thus, the protein molecules are *asymmetrically oriented.* Each side of a membrane has different characteristics because each type of protein is oriented in the bilayer in only one way. Proteins are not randomly placed into membranes; asymmetry is produced by the highly specific way in which each protein is inserted into the bilayer.

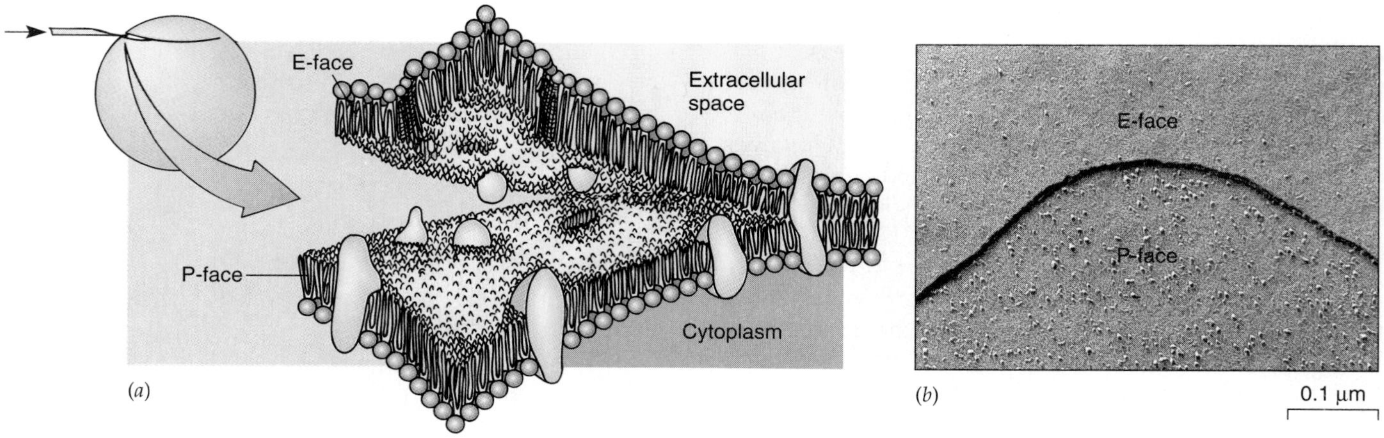

FIGURE 5–9 Freeze-fracture electron microscopy reveals the asymmetry of the plasma membrane. (***a***) In the freeze-fracture method, the path of membrane cleavage is along the hydrophobic interior of the lipid bilayer, resulting in two complementary fracture faces: (1) an inner half-membrane presenting the P-face (or *protoplasmic* face), from which project the majority of the membrane proteins, and (2) a relatively smooth, outer half-membrane presenting the E-face (or *external* face), which shows occasional protein particles. In a good fracture, particles are visible on both of the inside faces of the fractured membrane, as shown here. These particles are transmembrane proteins inserted into the lipid bilayer. Freeze-fractured bilayers of lipids alone do not have particles on the fracture planes. (***b***) A freeze-fracture electron micrograph. Notice the greater number of proteins on the P-face of the membrane. (*b*, D.W. Fawcett)

As an example, look again at Figure 5–2 and notice the different ways that protein molecules are oriented in the plasma membrane. Note, for example, that carbohydrates are attached to the parts of the proteins exposed on the surface of the cell, but not to the parts exposed to the cytosol.

As you recall from Chapter 4, plasma membrane proteins are initially formed by ribosomes on the rough endoplasmic reticulum (ER) and are inserted through the ER membrane as they are synthesized. Only a part of each protein passes through the membrane, so each completed protein has some regions that are located in the ER lumen and other regions that remain in the cytosol. Enzymes that attach the sugars to certain amino acids on the protein are found only in the lumen of the ER. Thus, carbohydrates can be added only to the parts of proteins that are located in that compartment.

If you follow the vesicle budding and membrane fusion events that are part of the transport process (Fig. 5–10), you can see that the same region of the protein that protruded into the ER lumen is also transferred to the lumen of the Golgi complex. There additional enzymes that can further modify the carbohydrate groups are located. That region of the protein remains inside a membrane compartment of a secretory vesicle as it buds from the Golgi complex. When the secretory vesicle fuses with the plasma membrane, the carbohydrate-containing part of the protein that was formerly located on the inside of the vesicle becomes the part of the membrane protein that is exposed on the cell surface.

ER lumen → transport vesicle → Golgi lumen (successive compartments) → secretory vesicle → cell surface

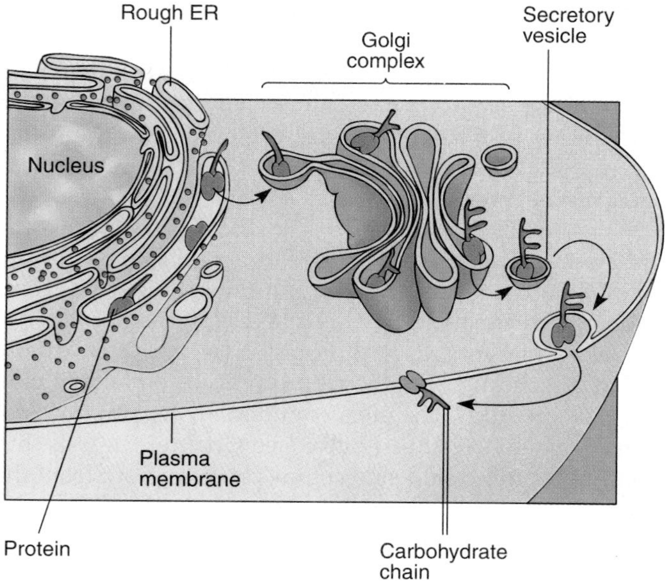

FIGURE 5–10 The orientation of a protein in the plasma membrane is a consequence of the pathway of synthesis and transport in the cell. The regions of a protein found on the extracellular surface of the plasma membrane originated in the lumen of the rough ER and were then passed through the compartments of the Golgi complex and secretory vesicles. Carbohydrates were added to these protein regions in the ER lumen and then modified in the Golgi complex.

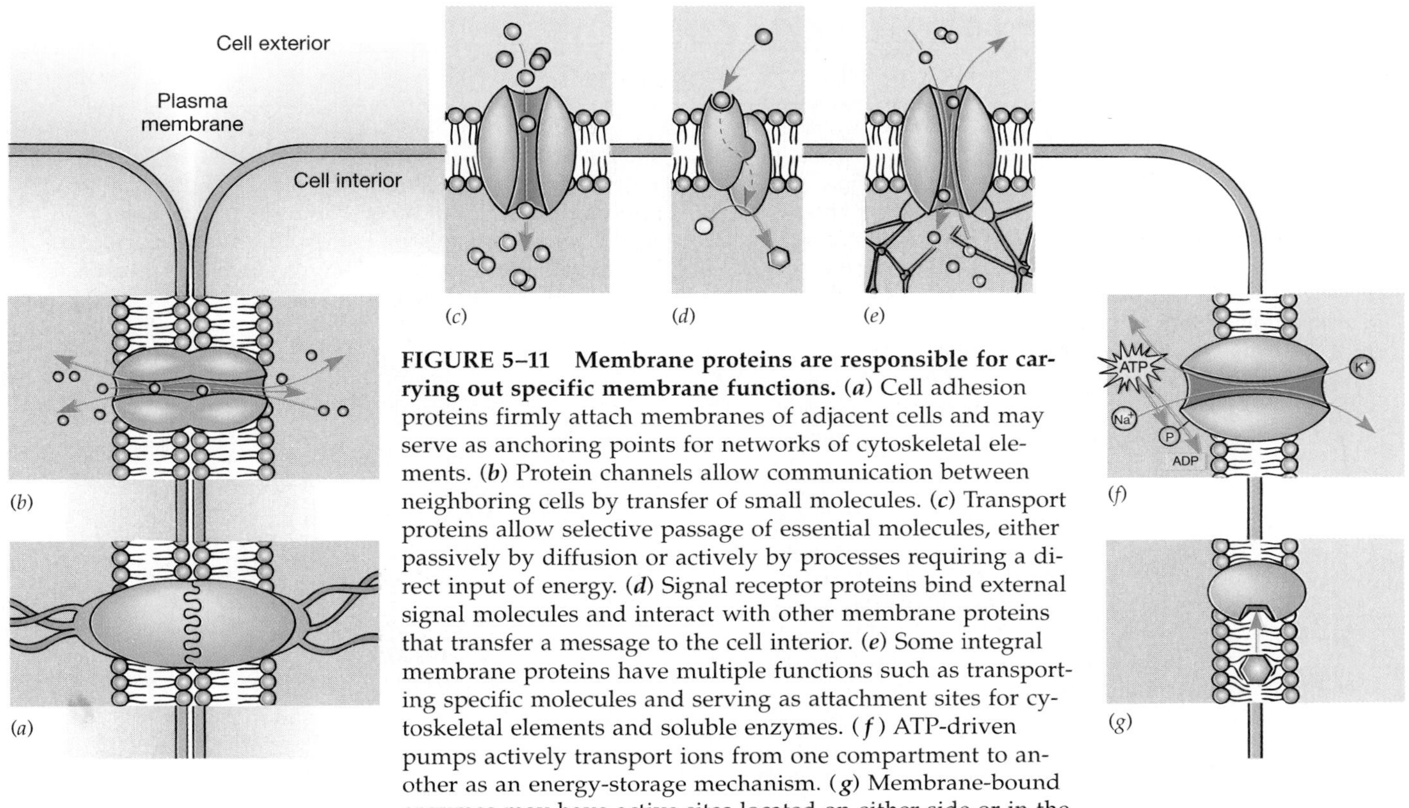

FIGURE 5–11 Membrane proteins are responsible for carrying out specific membrane functions. (*a*) Cell adhesion proteins firmly attach membranes of adjacent cells and may serve as anchoring points for networks of cytoskeletal elements. (***b***) Protein channels allow communication between neighboring cells by transfer of small molecules. (***c***) Transport proteins allow selective passage of essential molecules, either passively by diffusion or actively by processes requiring a direct input of energy. (***d***) Signal receptor proteins bind external signal molecules and interact with other membrane proteins that transfer a message to the cell interior. (*e*) Some integral membrane proteins have multiple functions such as transporting specific molecules and serving as attachment sites for cytoskeletal elements and soluble enzymes. (*f*) ATP-driven pumps actively transport ions from one compartment to another as an energy-storage mechanism. (***g***) Membrane-bound enzymes may have active sites located on either side or in the interior of the membrane.

Membrane proteins have specific functions

Why should a membrane such as the plasma membrane illustrated in Figure 5–2 require so many different proteins? This diversity is a reflection of the number of activities that take place in or on the membrane.

Generally, plasma membrane proteins fall into several broad functional groups (Fig. 5–11). A number of membrane proteins are involved in the *membrane transport* of small molecules. *Enzymes* that modify molecules needed near the cell surface can also be found associated with both sides of the plasma membrane. Many membrane-bound *receptor proteins* receive information from the environment and transmit it to the cell interior (see *Making the Connection: Information Transfer Across the Plasma Membrane*).

Other integral or peripheral proteins may be parts of specialized structures that link cells together, attach to cytoskeletal elements, or transmit signals to neighboring cells. The remainder of this chapter examines this functional diversity of membrane proteins.

Cellular Membranes Are Selectively Permeable

Whether a membrane permits a substance to pass through it depends on the size and charge of its particles and on the composition of the membrane. A membrane is said to be *permeable* to a given substance if it permits it to pass through, and *impermeable* if it does not. A *selectively permeable* membrane allows some but not other substances to pass through it readily. In general, biological membranes are most permeable to small molecules and to lipid-soluble substances able to cross the hydrophobic interior of the bilayer.

Although they are polar (and therefore not lipid-soluble) water molecules can rapidly cross a fluid lipid bilayer because they are small enough to pass through gaps that occur as a fatty acid chain momentarily moves out of the way. Gases such as oxygen, carbon dioxide, and nitrogen; small polar molecules like glycerol; plus larger, nonpolar (hydrophobic) substances such as hydrocarbons can also freely traverse a lipid bilayer. Slightly larger polar molecules, such as glucose, and charged ions of any size do not pass freely through the bilayer, either because of their size or because they are repulsed by a layer of electrical charges on the surface of the membrane (Table 5–1).

Although the bilayer is relatively impermeable to ions, cells must be able to move ions and large and small polar molecules such as amino acids and sugars across membranes. The permeability of membranes to those substances is due primarily to the activities of specialized membrane proteins. All the biological membranes surrounding cells, nuclei, vacuoles, mitochondria, chloro-

Making the Connection

Information Transfer Across the Plasma Membrane

Can an extracellular signal molecule transmit information to the cell interior without physically crossing the plasma membrane? In fact, most signal molecules never enter the cell. They instead rely on systems of interacting integral membrane proteins to transmit the signal by a process known as **signal transduction.** Each component of a signal transduction system acts as a relay "switch," which can be in an activated ("on") state or an inactive ("off") state. The first component is a *receptor,* a transmembrane protein with a domain exposed on the extracellular surface. In a typical sequence of events, the binding of the external signal activates the receptor by changing its conformation. The activated receptor then changes the conformation of a second protein, which then becomes activated. Ultimately these interactions result in the activation of a specific membrane-bound enzyme, which may activate intracellular enzymes or catalyze the production of large numbers of intracellular signal molecules. In this way the original signal received by the receptor protein is amplified many times, and the metabolism of the cell may be profoundly altered.

Some activated receptors are enzymes themselves, or they interact directly with enzymes (see *Focus on Oncogenes and Cancer,* Chapter 16). Others regulate enzymes indirectly by using certain integral membrane proteins, referred to as **G proteins,** as intermediaries. In 1994 Dr. Alfred G. Gilman and Dr. Martin Rodbell were awarded a Nobel Prize for their research on G proteins, so-named because the active form is bound to GTP, or guanosine triphosphate, a molecule similar to ATP but containing the base guanine instead of adenine. G proteins are now known to be involved in a number of important signal transductions, including the action of many hormones (see Chapter 47). Some G proteins regulate channels that allow ions to cross the plasma membrane and still others play important roles in the senses of sight, smell, and taste (see Chapter 41). ▲

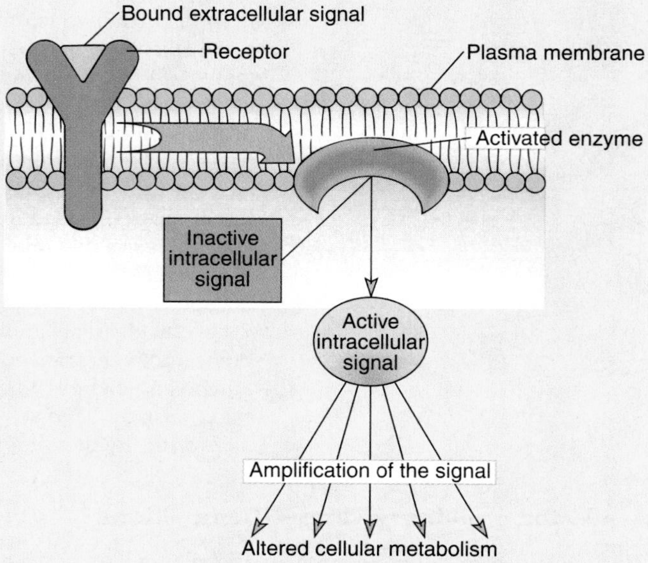

plasts, and the other subcellular organelles are selectively permeable to different types of molecules.

In response to varying environmental conditions or cellular needs, a plasma membrane may be a barrier to a particular substance at one time and actively promote its passage at another. By regulating chemical traffic in this way, a cell can exert some control over its own internal ionic and molecular composition, which can be very different from that on the outside. In the nonliving world, materials move passively by physical processes such as diffusion; in living organisms, some particles can diffuse across the bilayer, while other materials can be moved very rapidly by physiological processes such as active transport, exocytosis, and endocytosis. These processes (discussed later in this chapter) require a direct expenditure of metabolic energy by the cell.

Table 5–1 PERMEABILITY OF THE LIPID BILAYER TO DIFFERENT SUBSTANCES

Type of Molecule	*Example*	*Permeability*
Gases	N_2, O_2, CO_2	Freely permeable
Hydrophobic	Hydrocarbons	Freely permeable
Small polar	H_2O, glycerol, urea	Freely permeable
Large polar	Glucose, other uncharged monosaccharides, disaccharides	Not permeable
Ions/charged molecules	Amino acids, H^+, HCO_3^-, Na^+, K^+, Ca^{2+}, Cl^-, Mg^{2+}	Not permeable

Making the Connection

Diffusion, Time, and Distance

Under what conditions is diffusion important in biology? The overall rate of diffusion depends on many factors, including the temperature, the sizes of the particles, and their charges. However, distance is a primary consideration because the time required for diffusion to take place is proportional to the *square* of the distance involved. As diffusion occurs, each individual particle may move as fast as several hundred meters per second in a straight line. However, it typically moves only a fraction of a nanometer before it collides with another particle and rebounds. As the particles engage in this "random walk," their overall progress in any particular direction is extremely slow.

Diffusion can be very rapid in the microscopic world of the cell. When distances are measured in micrometers or nanometers, events occur in milliseconds (a millisecond is a thousandth of a second) or fractions of milliseconds. For example, oxygen diffuses very rapidly across the 10-nanometer width of the plasma membrane.

We rarely observe diffusion in our familiar macroscopic world because diffusion of even very small particles becomes excruciatingly slow over distances greater than about 1 millimeter. Events commonly attributed to diffusion, such as sugar dissolving in water, are mainly consequences of other forces, such as convection currents. For these reasons, the evolution of special transport systems has been a major trend in the history of complex plants and animals (see Chapters 33, 42, and 44). ▲

Random motion of particles leads to diffusion

Some substances pass into or out of cells and move about within cells by simple diffusion, a physical process based on random motion. All atoms and molecules possess kinetic energy, or energy of motion, at temperatures above absolute zero (0 Kelvin, −273° Celsius, or −459.4° F). Matter may exist as a solid, liquid, or gas, depending on the freedom of movement of its constituent particles. The particles of a solid are closely packed, and the forces of attraction between them allow them to vibrate but not to move around. In a liquid the particles are farther apart; the attractions are weaker, and the particles move about with considerable freedom. In a gas the particles are so far apart that intermolecular forces are negligible; molecular movement is restricted only by the walls of the container that encloses the gas. This means that atoms and molecules in liquids and gases move in a kind of "random walk," changing directions as they collide.

Although the movement of the individual particles is undirected and unpredictable, we can nevertheless make predictions about the behavior of groups of particles. If the particles (atoms, ions, or molecules) are not evenly distributed, then at least two regions exist, one with a higher concentration of particles and the other with a lower concentration. Such a difference in the concentration of a substance from one place to another is a **concentration gradient.**

In the phenomenon of **diffusion,** the random motion of particles results in their *net* movement "down" their own concentration gradient (from the region of higher concentration to the one of lower concentration). This does not mean that individual particles are prohibited from moving "against" the gradient. However, because there are initially more particles in the region of high concentration, it logically follows that more particles move randomly from there into the low-concentration region than vice versa.

Diffusion can occur rapidly over very short distances (see *Making the Connection: Diffusion, Time, and Distance*). The rate of diffusion is determined by the movement of the particles, which in turn is a function of their size and shape, their electrical charges, and the temperature. As the temperature rises, the particles move faster and the rate of diffusion increases.

Particles of any number of different substances in a mixture diffuse independently of each other. If particles are not added to or removed from the system, a state of equilibrium (condition of no net change in the system) is ultimately reached. At equilibrium the particles are uniformly distributed.

More commonly in organisms, equilibrium is never attained. For example, carbon dioxide is continually formed within a human cell as sugars and other molecules are metabolized during the process of aerobic cellular respiration. Carbon dioxide readily diffuses across the plasma membrane but then is rapidly removed by the blood. This limits the opportunity for the molecules to reenter the cell, so a sharp concentration gradient of carbon dioxide molecules always exists across the membrane.

Dialysis is the diffusion of a solute across a selectively permeable membrane

To demonstrate **dialysis,** one can fill a cellophane bag[1] with a sugar solution and immerse it in a beaker of pure

[1]Cellophane is often used as an "artificial membrane." It is made from cellulose and can be formed into a thin sheet that allows the passage of water molecules. Such membranes can be constructed with varying permeability to different solutes, and can be very different from biological membranes in their permeability. (When cellophane is used to package foods, it is coated to make it impermeable to air and water.)

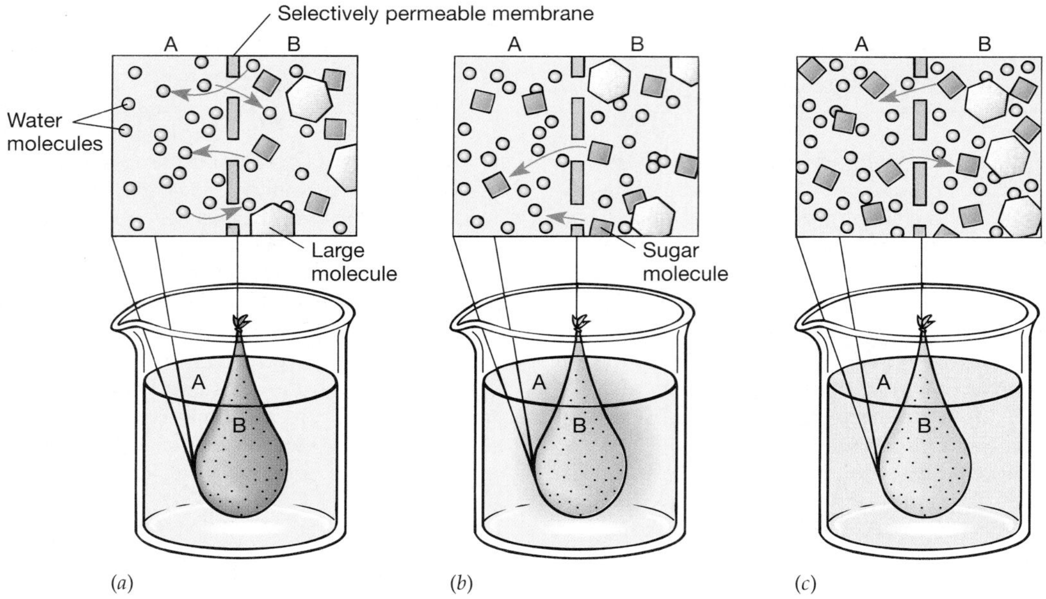

FIGURE 5–12 Dialysis can be used to separate molecules on the basis of size. In the magnified (upper) views, spheres represent water molecules, squares represent sugar molecules, and hexagons represent large molecules such as proteins. In the lower views, dots represent large molecules and purple indicates sugar and water. (***a***) A cellophane bag, filled with a mixture of sugar, water, and large molecules such as proteins, is immersed in a beaker of water. The cellophane acts as a selectively permeable membrane, permitting passage of the sugar and water molecules (*arrows*), but preventing passage of larger molecules. (***b***) The arrows indicate net movement of sugar molecules through the membrane into the water of the beaker. (***c***) Eventually the sugar becomes distributed equally between the two compartments. Although sugar and water molecules continue to diffuse back and forth (*arrows*), net movement is zero.

water (Fig. 5–12). If the cellophane membrane is permeable to sugar as well as to water, the sugar molecules will pass through it, and the concentrations of sugar molecules in the water on the two sides of the membrane will eventually become equal. Subsequently, both solute and water molecules will continue to cross the membrane, but there will be no net change in their concentrations.

The principle of dialysis has many practical applications. A dialysis machine can be used to cleanse the blood of wastes if the kidneys do not function properly. Waste products in the form of small molecules diffuse readily across the artificial membrane in the dialysis apparatus. Wastes can thus be removed from the blood, while blood cells, blood proteins, and other large molecules are retained.

Osmosis is the diffusion of water (solvent) across a selectively permeable membrane

The selective permeability of cell membranes results in a special kind of diffusion called **osmosis,** which involves the movement of *solvent* (in this case, water) molecules through a selectively permeable membrane. The water molecules pass freely in both directions, but, as in all types of diffusion, *net* movement is from the region where the water molecules are more concentrated to the region where they are less concentrated. Most solute molecules cannot diffuse freely through selectively permeable membranes of the cell (see Table 5–1).

The principles involved in osmosis can be illustrated using an apparatus called a U-tube (Fig. 5–13). The U-tube is divided into two sections by a selectively permeable membrane that allows solvent (water) molecules to pass freely but excludes solute molecules (e.g., sugar, salt). A water/solute solution is placed on one side, and pure water is placed on the other. The side containing the solute dissolved in the water has a lower effective concentration of water than the pure water side. This is because the solute particles, which are charged (ionic) or polar, interact with the partial electrical charges on the polar water molecules. Many of the water molecules are thus "bound up" and no longer free to diffuse across the membrane.

Because of the difference in effective water concentration, there is net movement of water molecules from the pure water side (with a high effective concentration of water) to the water/solute side (with a lower effective concentration of water). As a result, the fluid level drops on the pure water side and rises on the water/solute side. Because the solute molecules do not diffuse across the

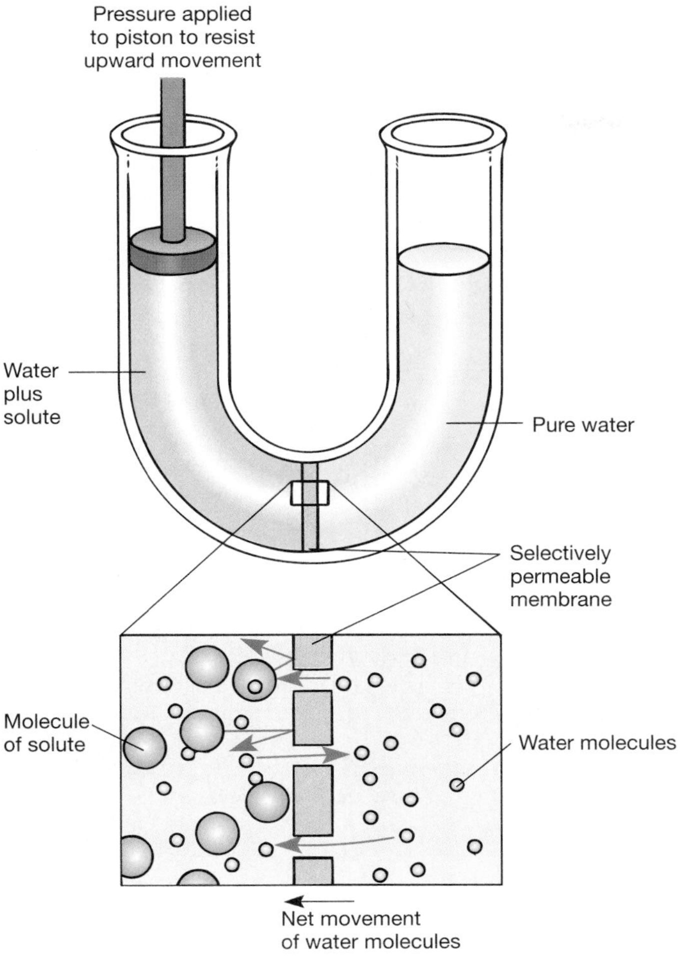

FIGURE 5–13 A U-tube demonstrates the principles of osmosis. The U-tube contains pure water on the right and water plus a solute on the left, separated by a selectively permeable membrane. The red arrows indicate that the water molecules are able to cross the membrane in both directions. Solute molecules are unable to cross (*green arrows*). The fluid level is expected to rise on the left and fall on the right because *net* movement of water (*black arrow*) is to the left. The force that must be exerted by the piston in order to prevent the rise in fluid level is equal to the osmotic pressure of the solution.

membrane, equilibrium is never attained. Net movement of water continues, and the fluid level continues to rise on the side containing the solute. Under weightless conditions this process could go on indefinitely, but on Earth the weight of the rising column of fluid eventually exerts enough pressure to stop further changes in fluid levels, although water molecules continue to pass through the selectively permeable membrane in both directions.

We define the **osmotic pressure** of a solution as the tendency of water to move into that solution by osmosis. In our U-tube example, we could measure the osmotic pressure by inserting a piston on the water/solute side of the tube and measuring how much pressure must be exerted by the piston to prevent the rise of fluid on that side of the tube. A solution with a high solute concentration has a low effective water concentration and a high osmotic pressure; conversely, a solution with a low solute concentration has a high effective concentration of water and a low osmotic pressure.

Two solutions may be isotonic to each other, or one may be relatively hypertonic and the other relatively hypotonic Dissolved in the fluid compartment of every living cell are salts, sugars, and other substances that give that fluid a certain osmotic pressure. When a cell is placed in a fluid with exactly the same osmotic pressure, no net movement of water molecules occurs, either into or out of the cell; the cell neither swells nor shrinks. Such a fluid is said to be **isotonic** or **isosmotic** (i.e., of equal osmotic pressure) to the fluid within the cell (Table 5–2). Normally, our blood plasma (the fluid component of blood) and all our other body fluids are isotonic to our cells; they contain a concentration of water equal to that in the cells.

Table 5–2 OSMOTIC TERMINOLOGY

Solute Concentration in Solution A	*Solute Concentration in Solution B*	*Tonicity*	*Direction of Net Movement of Water*
Greater	Less	A hypertonic to B B hypotonic to A	B to A
Less	Greater	B hypertonic to A A hypotonic to B	A to B
Equal	Equal	A and B are isotonic to each other	No net movement

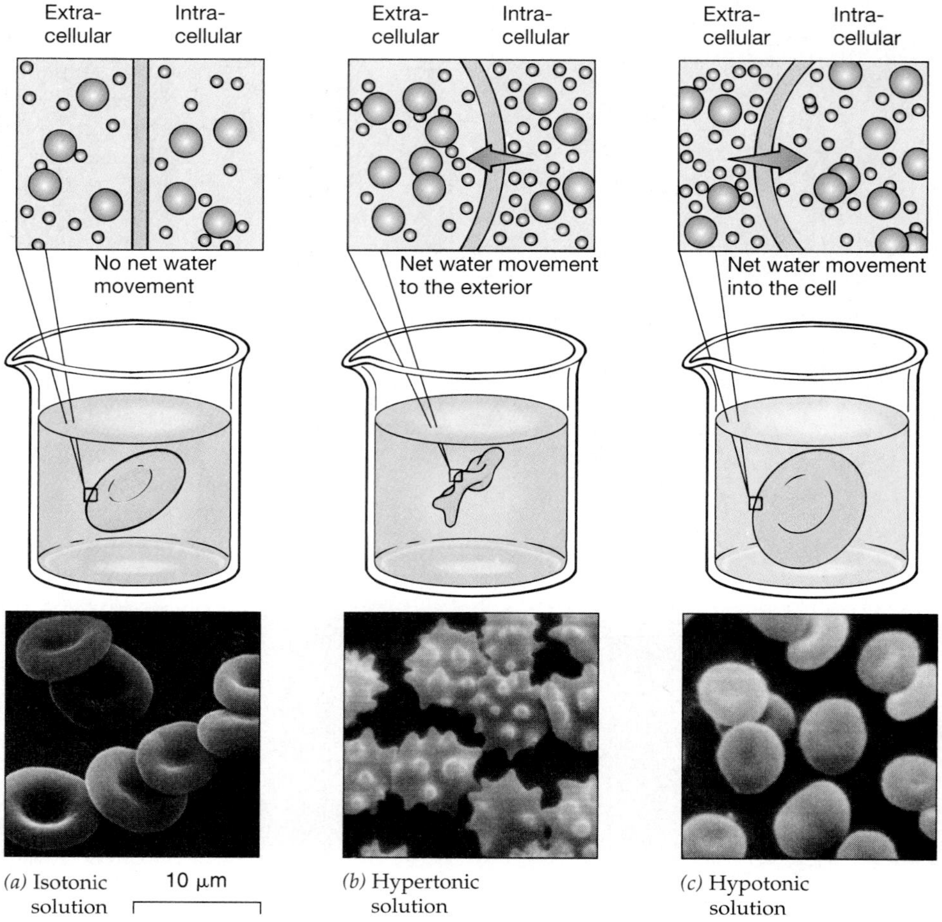

FIGURE 5–14 Living cells respond to osmotic pressure differences. (*a*) A cell is placed in an isotonic solution. Because the concentration of solutes (and thus the effective concentration of water molecules) is the same in the solution as in the cell, water can pass in and out of the cell, but the net movement is zero. (*b*) A cell is placed in a hypertonic solution. This solution has a greater solute concentration (and thus a lower effective water concentration) than does the cell. This results in a net movement of water out of the cell (*arrow*), and the cell becomes dehydrated, crenated (shrunken), and may die. (*c*) A cell is placed in a hypotonic solution. The solution has a lower solute concentration (and thus a greater effective water concentration) than does the cell. The cell contents therefore have higher osmotic pressure than the solution. There is a net movement of water molecules into the cell (*arrow*), causing the cell to swell or even burst. (Micrographs of human red blood cells courtesy of Dr. R. F. Baker, University of Southern California Medical School)

A solution of 0.9% sodium chloride (sometimes called *physiological saline*) is isotonic to the cells of humans and other mammals. Human red blood cells placed in 0.9% sodium chloride neither shrink nor swell (Fig. 5–14*a*).

If the surrounding fluid has a concentration of dissolved substances greater than the concentration within the cell, it has a higher osmotic pressure than the cell and is said to be **hypertonic (hyperosmotic)** to the cell. Because the hypertonic solution has a lower effective water concentration, a cell placed in such a solution shrinks as it loses water by osmosis. Human red blood cells placed in a solution of 1.3% sodium chloride shrink and are said to be *crenated* (Fig. 5–14*b*). If a cell that has a cell wall is placed in a hypertonic medium, it loses water to its surroundings, and its contents shrink away from the wall; this process is called **plasmolysis** (Fig. 5–15*b, c*). Plasmolysis occurs in plants when the soil or water around them contains high concentrations of salts or fertilizers.

If the surrounding fluid contains a lower concentration of dissolved materials than does the cell, it has a lower osmotic pressure and is said to be **hypotonic (hypoosmotic)** to the cell; water then enters the cell and causes it to swell. Red blood cells placed in a solution of 0.6% sodium chloride gain water, swell (Fig. 5–14*c*), and may eventually burst. Many cells that normally live in hypotonic environments have adaptations to prevent excessive water accumulation. For example, certain protists have a contractile vacuole that they use to expel excess water.

Turgor pressure is the internal hydrostatic pressure usually present in walled cells The relatively rigid cell walls of plant cells, algae, bacteria, and fungi enable these cells to withstand, without bursting, an external medium that is very dilute, containing only a very low concentration of solutes. Because of the substances dissolved in the cytoplasm, the cells are hypertonic to the outside medium (conversely, the outside medium is hypotonic to the cytoplasm). Water moves into the cells by osmosis, filling their central vacuoles and distending the cells. The cells swell, building up a pressure, termed **turgor pressure,** against the rigid cellulose cell walls (Fig. 5–15*a*). The cell walls can be stretched only very slightly, and a steady state is reached when their resistance to stretching prevents any further increase in cell size and thereby halts the net movement of water molecules into the cells (although, of course, molecules continue to move back and forth across the plasma membrane). Turgor pressure in

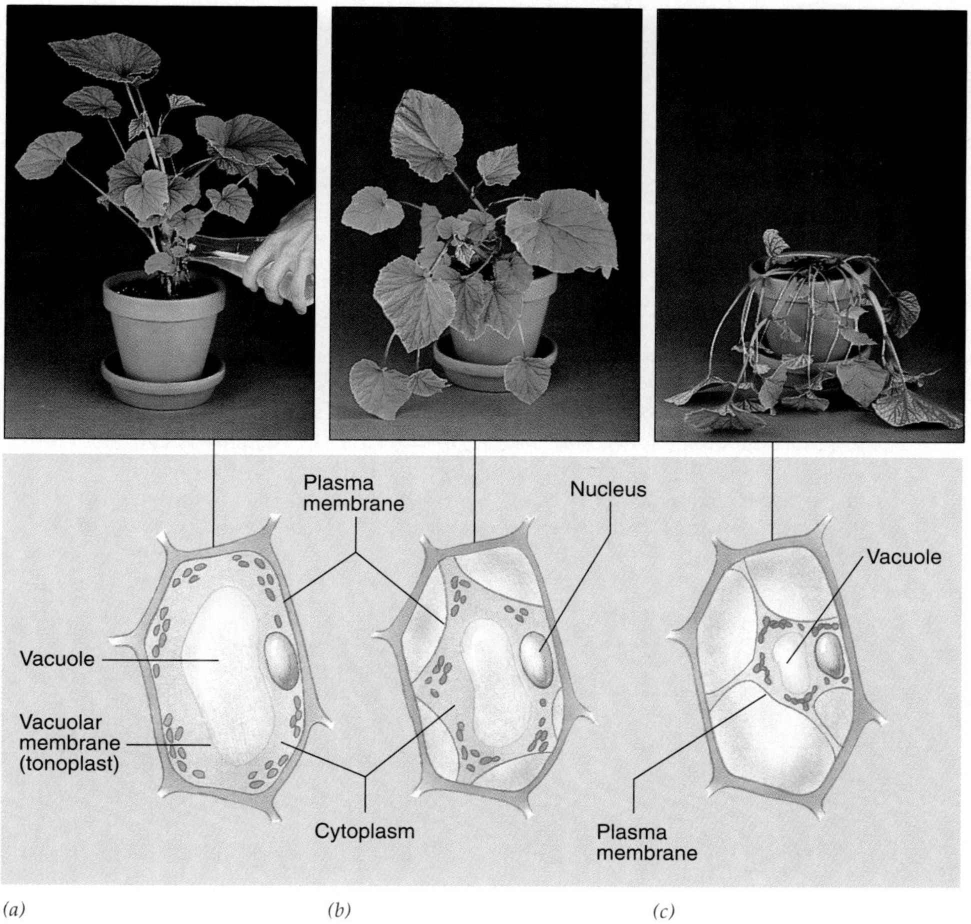

FIGURE 5–15 Healthy plant cells are usually turgid. (*a*) A walled cell in hypotonic surroundings becomes turgid as water enters by osmosis. The cells of this healthy begonia plant are turgid. (*b*), (*c*) If the cell is placed in a hypertonic medium, it becomes plasmolyzed as it loses water. When the begonia plant is exposed to a hypertonic salt solution, it wilts and eventually dies as its cells become plasmolyzed. (Dennis Drenner)

the cells is an important factor in providing support for the body of nonwoody plants. Thus, lettuce becomes limp in a salty salad dressing and a picked flower wilts due to lack of water.

Carrier-mediated transport of solutes requires special integral membrane proteins

A lipid bilayer is relatively impermeable to most of the larger polar molecules (see Table 5–1). This is advantageous to cells for a number of reasons. Most of the compounds required in metabolism are polar, and the impermeability of the cell membrane prevents their loss by diffusion. A lipid bilayer is also impermeable to ions, which play important roles in many physiological processes. Some ions, such as calcium ions, are used as intracellular signals, and changes in their cytoplasmic concentration trigger changes in a number of cellular processes (such as muscle contraction; see Chapter 38). As a cell controls the influx and efflux of ions, it is able to directly or indirectly control many metabolic activities (see *Focus On: How the Patch Clamp Technique Has Revolutionized the Study of Ion Channels*).

Cells also must continually acquire essential polar nutrient molecules such as glucose and amino acids. To transport ions and nutrients through membranes, systems of carrier proteins apparently evolved very early in the origin of cells. This transfer of solutes by proteins located within the membrane is termed **carrier-mediated transport.** The two forms of carrier-mediated transport, facilitated diffusion and carrier-mediated active transport, differ in their capabilities and energy sources.

Facilitated diffusion occurs down a concentration gradient

In all processes in which substances move across membranes by passive diffusion, the net transfer of those molecules from one side to the other occurs as a result of a concentration gradient. If the membrane is permeable to a substance, there is net movement from the side of the membrane where it is more highly concentrated to the side where it is less concentrated. Such a gradient across the membrane is actually a form of stored energy. A concentration gradient can be established as a result of certain processes taking place in the cell. The stored energy

Focus On

How the Patch Clamp Technique Has Revolutionized the Study of Ion Channels

Because of their electrical charges, which provide a basis for various interactions, ions play an essential role in most cellular processes. Their electrical charges, however, prevent them from crossing a lipid bilayer by simple diffusion. For this reason, every membrane of every cell contains numerous ion carriers, or *ion channels.* Some ion channels, referred to as "ion pumps," require a direct input of metabolic energy, while others provide for facilitated diffusion.

Movement of ions across a membrane can result in a charge difference (electrochemical gradient). If the cell is large enough, this charge difference (usually expressed in millivolts, mV) can be measured by using two microelectrodes connected to an extremely sensitive oscilloscope or voltmeter. One of the microelectrodes is inserted into the cell and the other is placed just outside the plasma membrane. Although valuable, these techniques have serious limitations, for they cannot be used on smaller cells and do not provide information on the function of individual ion channels.

In the mid-1970s Erwin Neher and Bert Sakmann developed a method, known as the **patch clamp technique,** which allows researchers to study single ion channels of very small cells. In this technique, a micropipette is tightly sealed to a patch of membrane so small that it generally contains only a single ion channel. The flow of ions through the channel can be measured using an extremely sensitive recording device. This basic technique has been modified in many ways and has been applied to studies of the roles of ion channels in a wide range of cellular processes in both plants and animals. For example, studies of single ion channels enabled researchers to demonstrate that the genetic disease cystic fibrosis (see Chapter 15) is caused by a defect in a specific type of chloride ion channel. Because of the far-reaching implications of their work, Neher and Sakmann were awarded a Nobel Prize in 1991. ■

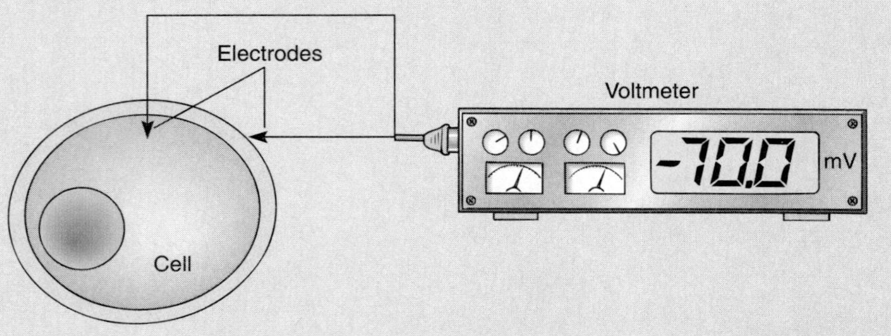

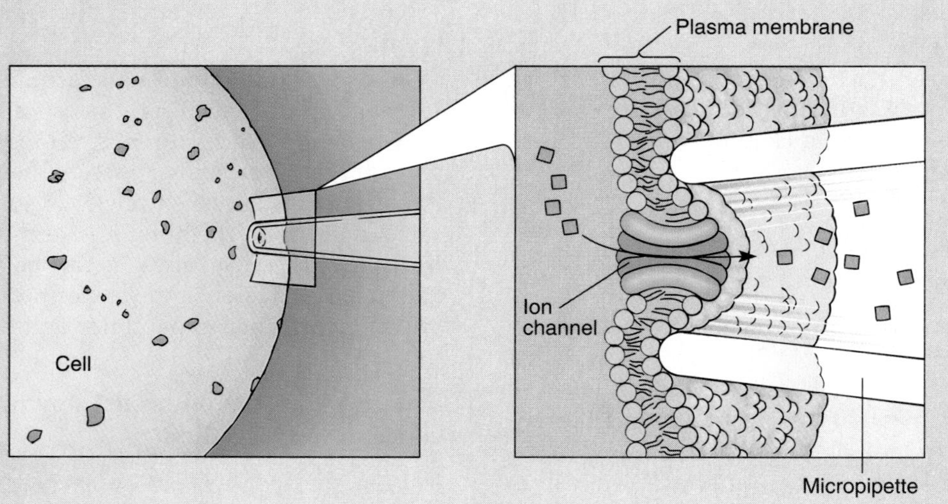

of the concentration gradient is released when molecules move from a region of high concentration to one of low concentration; movement down a concentration gradient is therefore spontaneous. (These types of energy and spontaneous processes are discussed in greater detail in Chapter 6.)

In the type of transport known as **facilitated diffusion,** the membrane may be made permeable to a solute, such as an ion or a polar molecule, by a specific *carrier* or *transport protein* that combines temporarily with the solute particle and permits it to move through the membrane. The carrier protein is not changed by this action;

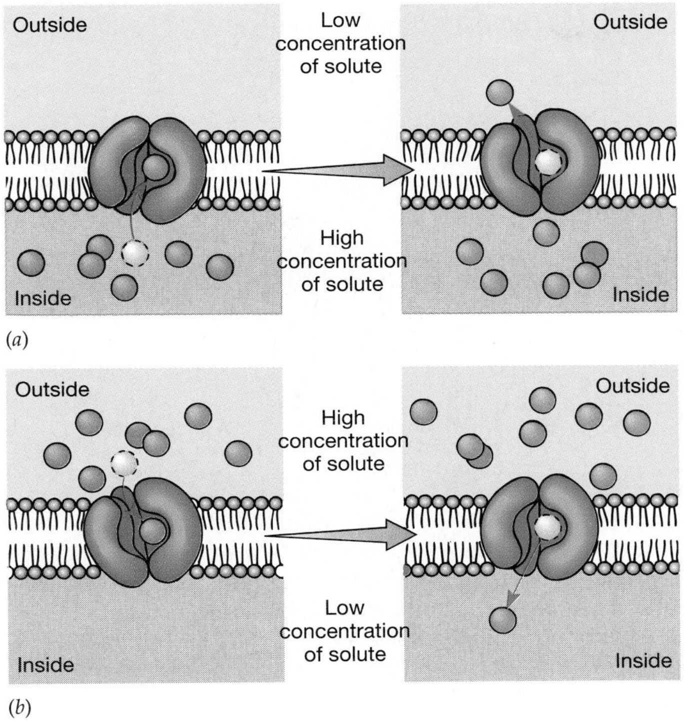

FIGURE 5–16 Facilitated diffusion requires a carrier or transport protein in the membrane, as well as the potential energy of a concentration gradient. The transport protein is capable of binding a solute particle on one side of the plasma membrane and then changing its shape so that a channel is opened to the other side. A specific solute can be transported from the inside of the cell to the outside (*a*), or from the outside to the inside (*b*), but net movement is always from a region of higher solute concentration to a region of lower concentration.

after it transports a solute particle, it is free to bind with another (Fig. 5–16).

An important example of a carrier that works by facilitated diffusion is the glucose transporter in red blood cells. These cells keep the internal concentration of glucose low by immediately adding a phosphate group to entering glucose molecules, converting them to highly charged glucose phosphates that cannot pass back through the membrane. Because glucose phosphate is a different molecule, it does not contribute to the glucose gradient. Thus a steep concentration gradient for glucose is continually maintained, and glucose rapidly diffuses into the cell, only to be immediately changed to the phosphorylated form.

The mechanism of facilitated diffusion of glucose is not entirely clear. It appears that the carrier protein does not form a "hole" in the membrane for glucose to pass through; if that were the case, related molecules or molecules smaller than glucose could also pass through the pore. It is more likely that glucose binds specifically to a region of the protein that is exposed to the outside of the cell; this binding changes the shape of the protein, allowing the glucose molecule to be released on the inside. According to this model, when the glucose is released into the cytoplasm, the protein reverts to its original structure and is available to bind the next glucose molecule on the outside of the cell.

Some carrier-mediated active transport systems "pump" substances against their concentration gradients

Although adequate amounts of some substances can be transported across cellular membranes by diffusion, a cell often needs to move solutes against a concentration gradient. Many substances are required by the cell in concentrations higher than those outside the cell. These molecules are moved across cellular membranes by **carrier-mediated active transport** mechanisms. Because active transport requires that particles be "pumped" from a region of low concentration to a region of high concentration (i.e., *against a concentration gradient*), the energy inherent in the gradient is unavailable and hence a different energy source (often ATP) is required.

One of the most striking examples of an active transport mechanism is the **sodium-potassium pump** found in virtually all animal cells (Fig. 5–17). The pump is a group of specific proteins in the plasma membrane that uses energy in the form of ATP to exchange sodium ions on the inside of the cell for potassium ions on the outside of the cell. The exchange is unequal, so that usually only two potassium ions are imported inside for every three sodium ions exported. Because these particular concentration gradients involve ions, an electrical potential (separation of electrical charges) is generated across the membrane, and we say that the membrane is polarized.

Both sodium and potassium ions are positively charged, but because there are fewer potassium ions inside relative to the sodium ions outside, the inside of the cell is negatively charged relative to the outside. We refer to such a gradient as an *electrochemical* gradient because it involves not only a concentration difference on the two sides of the membrane, but also a charge difference. These gradients are also a form of energy storage (like water stored behind a dam), which can be used to drive other transport systems. So important is the electrochemical gradient produced by these pumps that some cells (e.g., nerve cells) expend 70% of their total energy just to power this one transport system.

Sodium-potassium pumps (as well as all other ATP-driven pumps) are transmembrane proteins that extend entirely through the membrane. By undergoing a series of conformational changes, the pumps are able to exchange sodium for potassium across the plasma membrane. Unlike facilitated diffusion, at least one of the conformational changes in the pump cycle requires energy, which is provided by ATP. The shape of the pump protein changes as

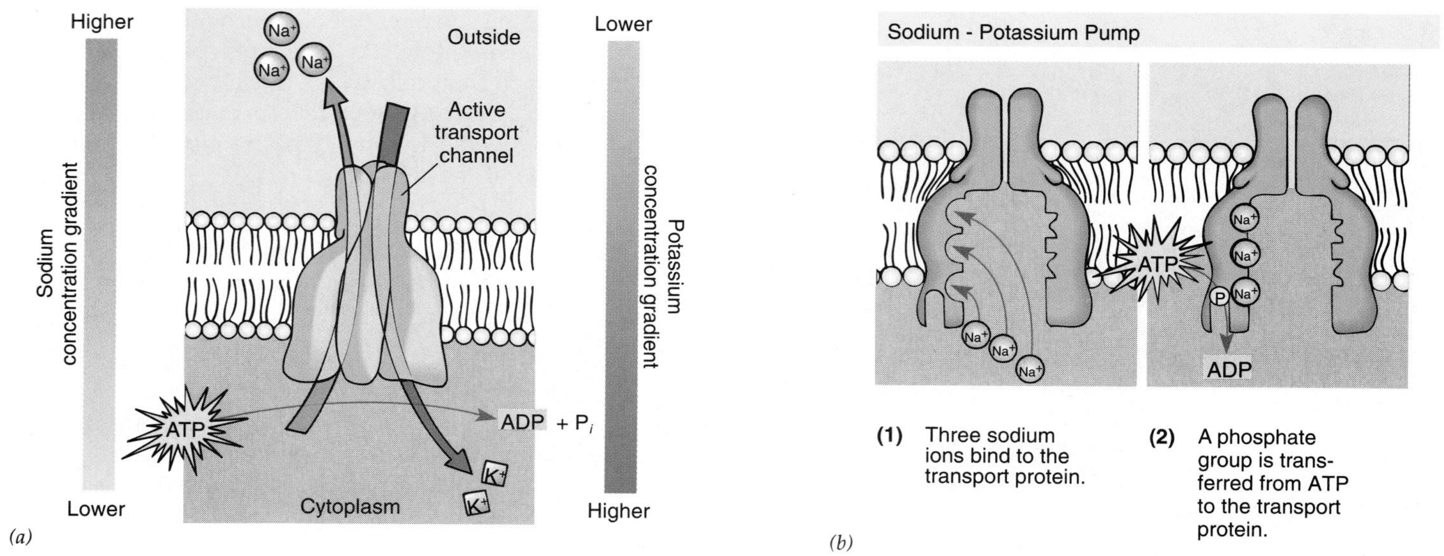

FIGURE 5–17 The sodium-potassium pump is an ATP-driven active transport system. (*a*) A diagram of the sodium-potassium ATPase in the plasma membrane. Each complete pumping cycle uses one molecule of ATP and results in the export of three sodium ions and the import of two potassium ions. (*b*) A model illustrating the 7 steps of the sodium-potassium pumping cycle.

a phosphate group (from ATP) first binds to it and is subsequently removed later in the pump cycle.

The use of electrochemical potentials for energy storage is not confined to the plasma membrane of animal cells. Plant and fungal cells use ATP-driven plasma membrane pumps to transfer protons from the cytoplasm of their cells to the outside. Removal of positively charged protons from the cytoplasm of these cells results in a large difference in the concentration of protons, such that the outside of the cells is relatively positively charged and the inside of the plasma membrane is relatively negatively charged. The energy stored in these electrochemical gradients can be made available to do certain kinds of cellular work.

Other proton pumps can be used in "reverse" to synthesize ATP. As we shall see in Chapters 7 and 8, bacteria, mitochondria, and chloroplasts use energy from food or from light to establish proton concentration gradients. When the protons diffuse through the proton carriers from a region of high proton concentration to one of low concentration, ATP is synthesized. These electrochemical gradients form the basis for the major energy-conversion system in virtually all cells.

Ion pumps have other important roles. For example, they are instrumental in the ability of an animal cell to equalize the osmotic pressures of its cytoplasm and its external environment. If an animal cell does not control its internal osmotic pressure, its contents will become hypertonic relative to the exterior. Water will enter the cell by osmosis, causing it to swell and possibly burst (see Fig. 5–14*c*). By controlling the ion distribution across the membrane, the cell is able to indirectly control the movement of water, for when ions are pumped out of the cell, water leaves by osmosis.

Linked cotransport systems indirectly provide energy for active transport

The electrochemical concentration gradients generated by the sodium-potassium pump also provide sufficient energy to power the active transport of a number of other essential substances. In these systems a transport protein can **cotransport** the required molecules *against* their concentration gradient, while transporting sodium, potassium, or hydrogen ions *down* their gradient. Energy from ATP is used indirectly in this process, for it produces the ion gradient; the energy of this gradient is then used to drive the active transport of a required substance against its gradient.

Integrated multiple transport systems use indirect linkages between active transport and facilitated diffusion

In some cells, more than one system may work to transport a given substance. For example, the transport of glucose from the intestine to the blood occurs through a thin sheet of epithelial cells that line the intestine (Fig. 5–18) and have highly specialized regions on their plasma membranes. The surface that is exposed to the intestine has many **microvilli** (sing., *microvillus*), finger-like extensions that effectively increase the surface area of the membrane available for absorption. The glucose transporter protein on that region of the cell surface is part of an active transport system for glucose that is "driven" by

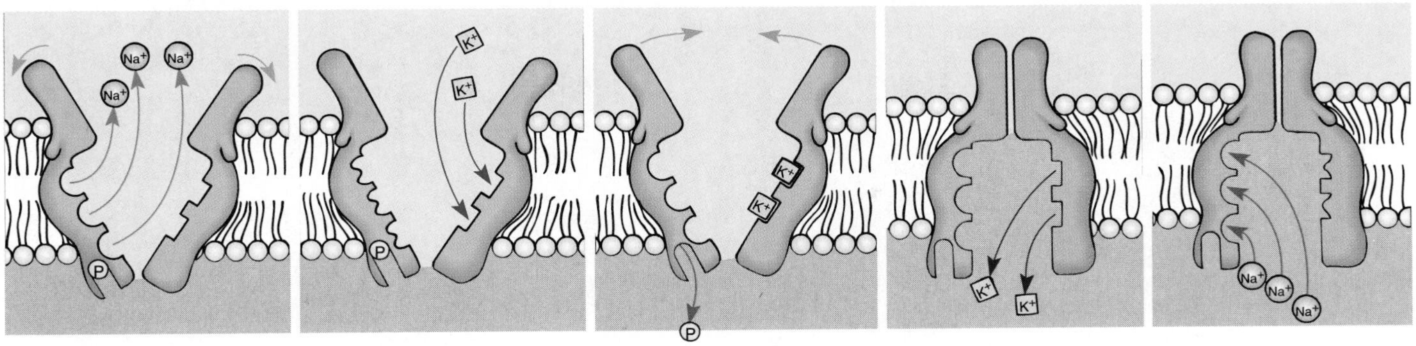

(3) The transport protein undergoes a conformational change, releasing three sodium ions outside the cell.

(4) Two potassium ions bind to the transport protein.

(5) The phosphate is released.

(6) The transport protein returns to its original shape; two potassium ions are released inside the cell.

(7) The process begins again with the binding of sodium ions.

the cotransport of sodium. The sodium concentration inside the cell is kept low by an ATP-requiring sodium-potassium pump that transports sodium out of the cell and into the blood. Because of its high concentration inside the cell (relative to the blood), glucose can be transported to the blood by facilitated diffusion.

Some of the current research in cell biology concerns understanding mechanisms such as those that allow the cell to place different transport proteins in two separate regions of the same plasma membrane. What are the signals that target each protein to its appropriate region? If the cell did not have a specific mechanism for handling this problem, proteins might be inserted randomly on both sides of the cell, leading to no net transport of glucose.

Facilitated diffusion is powered by a concentration gradient; active transport requires another energy source

It is a common misconception that diffusion, whether simple or facilitated, is somehow "free of cost" and that

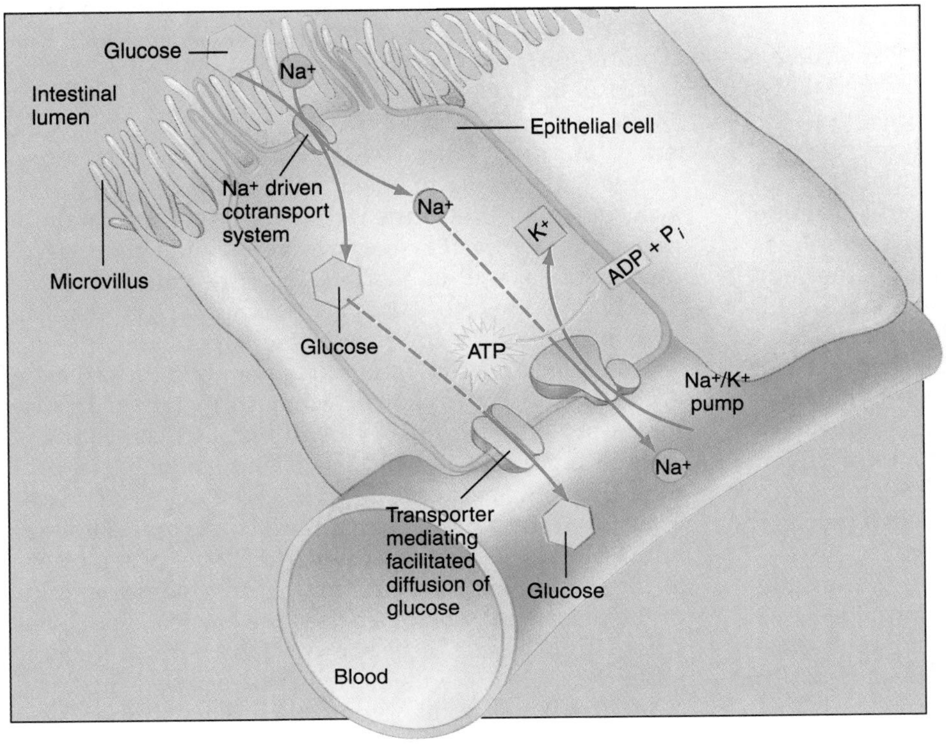

FIGURE 5–18 Integrated multiple transport systems allow glucose to be transported from the intestine to the blood through an intestinal epithelial cell. Glucose is actively transported into the cell by a sodium ion–driven cotransport system located only on the part of the plasma membrane in contact with the intestinal lumen. The sodium ion gradient across the plasma membrane is maintained by sodium-potassium pumps, which keep the intracellular sodium ion concentration low by actively transporting it from the cytoplasm into the blood. The active transport of glucose keeps its intracellular concentration high, so that it can enter the blood by facilitated diffusion. The carrier proteins responsible for the facilitated diffusion of glucose are located only on the regions of the plasma membrane in contact with the capillary.

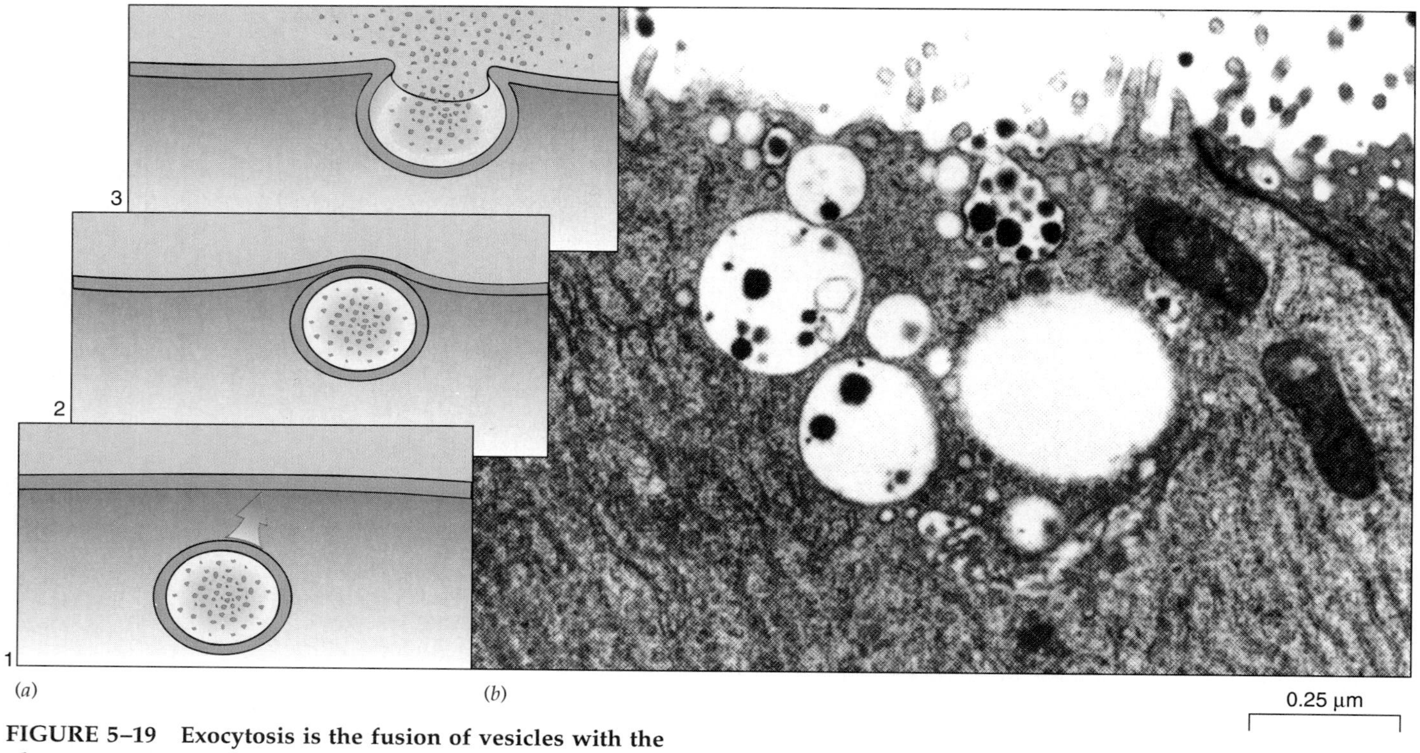

FIGURE 5–19 Exocytosis is the fusion of vesicles with the plasma membrane, resulting in the export of large particles. (*a*) The diagram shows a vesicle approaching the plasma membrane, fusing with it, and releasing its contents outside the cell. (*b*) Transmission electron micrograph showing exocytosis of the protein components of milk by a mammary gland cell. (*b*, A. Ichikawa/from D.W. Fawcett)

only active transport mechanisms require energy. Because diffusion always involves net movement of a substance down its concentration gradient, we say that the concentration gradient "powers" the process. However, energy is required to do the work of establishing and maintaining the gradient. Think back to the example of facilitated diffusion of glucose. The cell maintains a steep gradient (high outside, low inside) by phosphorylating the glucose molecules once they enter the cell. An ATP molecule is spent for every glucose molecule phosphorylated (not to mention such additional costs as the energy required to make the enzymes that carry out the reaction).

An active transport system can work *against* a concentration gradient (pumping materials from a region of low concentration to a region of high concentration). The energy stored in the concentration gradient is not only unavailable to the system, but actually works against it. For this reason some other source of energy must be provided. As we have seen, in many cases ATP energy is used directly. In a cotransport system, energy is provided by a concentration gradient for some other substance (e.g., an ion). Of course, ATP energy is required indirectly to power the pump that produces the ion gradient. As we shall see in Chapters 7 and 8, energy for active transport can come from other sources as well.

To summarize, both diffusion and active transport require energy. The energy for diffusion is provided by a concentration gradient *for the substance being transported*. Active transport requires some other, usually more direct, expenditure of metabolic energy.

In exocytosis and endocytosis large particles are transported by vesicles or vacuoles

In both simple and facilitated diffusion, and in carrier-mediated active transport, individual molecules and ions pass through the plasma membrane. Larger quantities of material, such as particles of food or even whole cells, must also be moved into or out of cells. Such cellular work requires that cells expend energy directly (making it a form of active transport).

In **exocytosis,** a cell ejects waste products or specific secretion products such as hormones by the fusion of a vesicle with the plasma membrane of the cell (Fig. 5–19, also see 5–7*a*). Exocytosis results in the incorporation of the membrane of the secretory vesicle into the plasma membrane as well as the release of the contents of the vesicle from the cell. This is also the primary mechanism by which plasma membranes grow larger.

In **endocytosis,** materials are taken into the cell. Several types of endocytotic mechanisms operate in biological systems. In **phagocytosis** (literally, "cell-eating"), the cell ingests large solid particles such as bacteria or food (Fig. 5–20, also see 5–7*b*). Phagocytosis is a mechanism used by certain protists and by several classes of white blood cells to ingest particles, some of which are as large as an entire bacterium. During ingestion, folds of the plasma membrane enclose the particle, which has bound to the surface of the cell, and form a vacuole around it. When the membrane has encircled the particle, it fuses at the point of contact. The vacuole then fuses

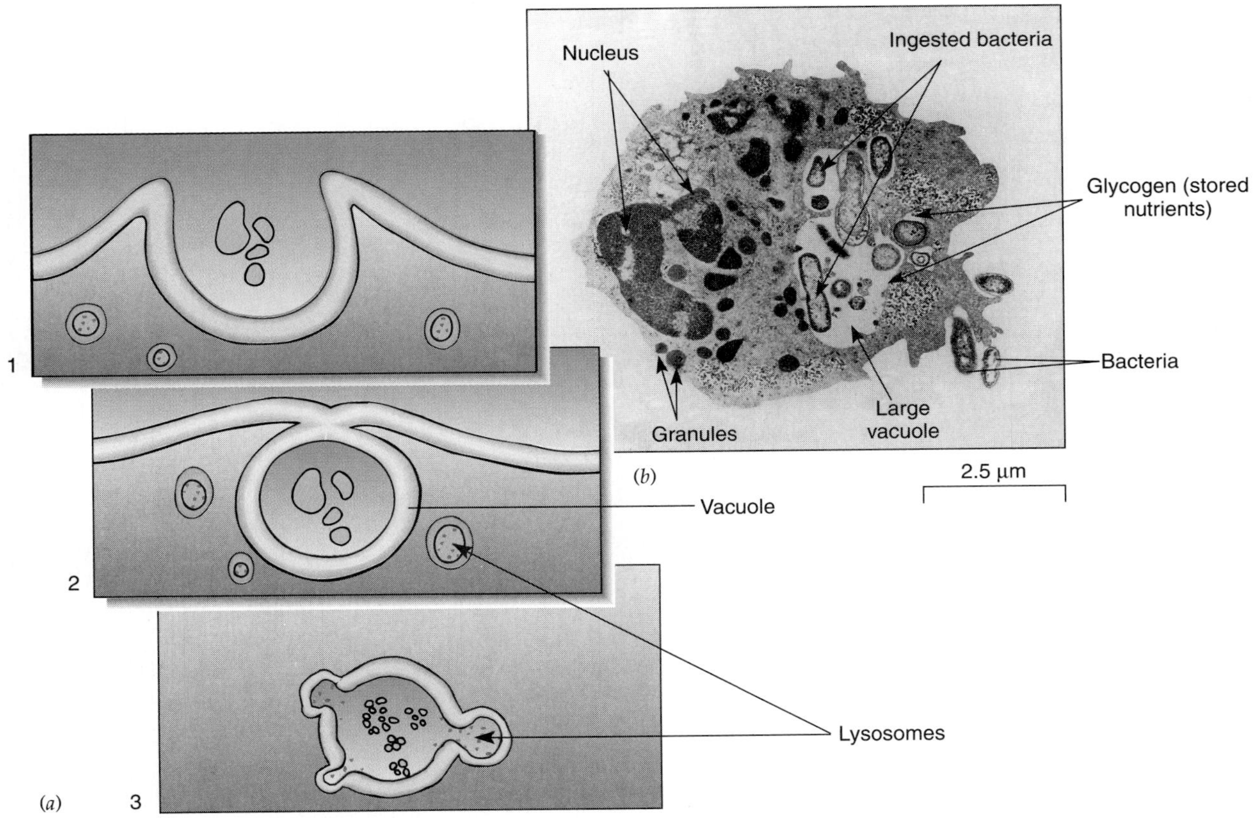

FIGURE 5–20 Phagocytosis is a form of endocytosis whereby a cell ingests relatively large solid particles. (*a*) A diagram showing the steps of endocytosis: (1) Folds of the plasma membrane surround the particle to be ingested, forming a small vacuole around it. (2) The vacuole then pinches off inside the cell. (3) Lysosomes may fuse with the vacuole and pour their potent digestive enzymes onto the ingested material. (*b*) The white blood cell (known as a *neutrophil*) shown in this transmission electron micrograph is phagocytizing bacteria. The vacuoles contain bacteria that have already been phagocytized, while other bacteria are still outside the cell. The granules in the cytoplasm contain digestive enzymes, which have already partially digested the white blood cell's own nucleus. (D.W. Fawcett)

with lysosomes, and the ingested material is degraded.

In the form of endocytosis known as **pinocytosis** ("cell-drinking"), the cell takes in dissolved materials. Tiny droplets of fluid are trapped by folds in the plasma membrane (Fig. 5–21), which pinch off into the cytoplasm as tiny vesicles. The liquid contents of these vesicles are then slowly transferred into the cytoplasm; the vesicles may become progressively smaller, to the point that they appear to vanish.

In a third type of endocytosis, called **receptor-mediated endocytosis,** specific proteins or particles combine with *receptor proteins* embedded in the plasma membrane of the cell. The receptor-bound molecules then migrate into *coated pits,* which are regions on the cytoplasmic

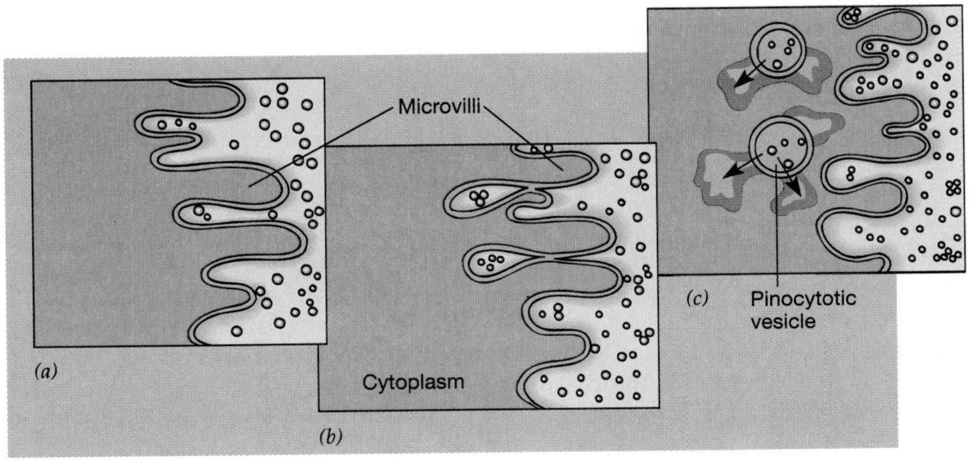

FIGURE 5–21 Pinocytosis is a form of endocytosis commonly known as "cell-drinking." (*a*) Tiny droplets of fluid are trapped by folds of the plasma membrane. These pinch off (*b*) into the cytoplasm as small fluid-filled vesicles (*c*). The content of these vesicles is then slowly transferred to the cytoplasm across their membrane linings.

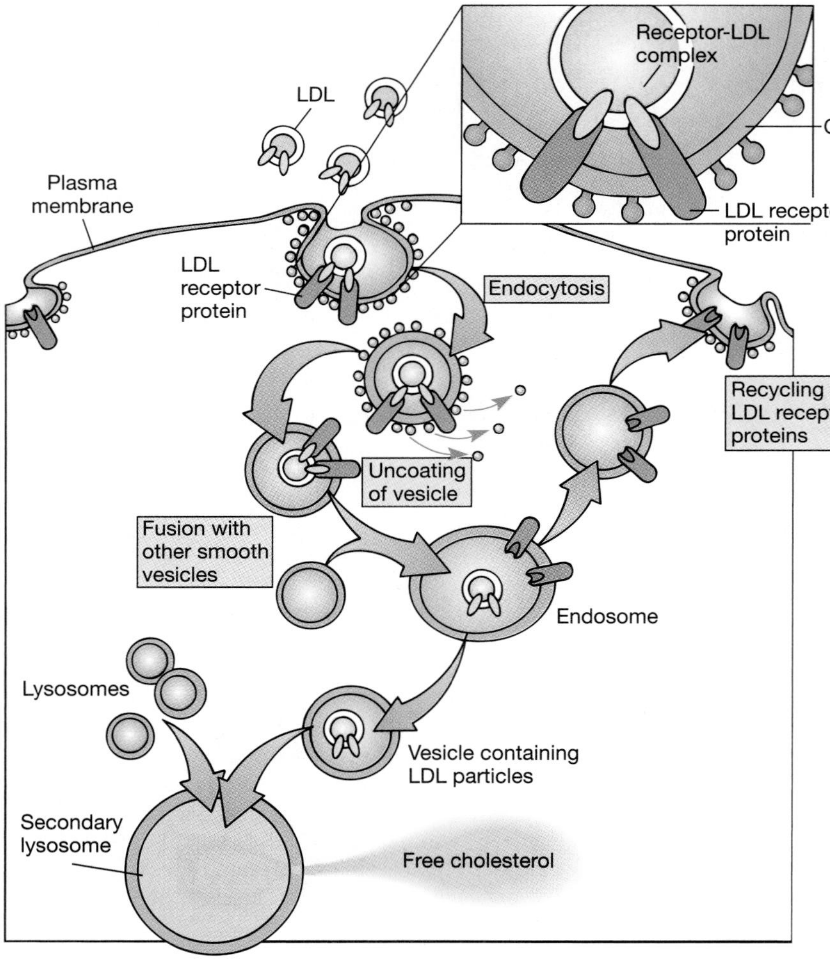

(*a*)

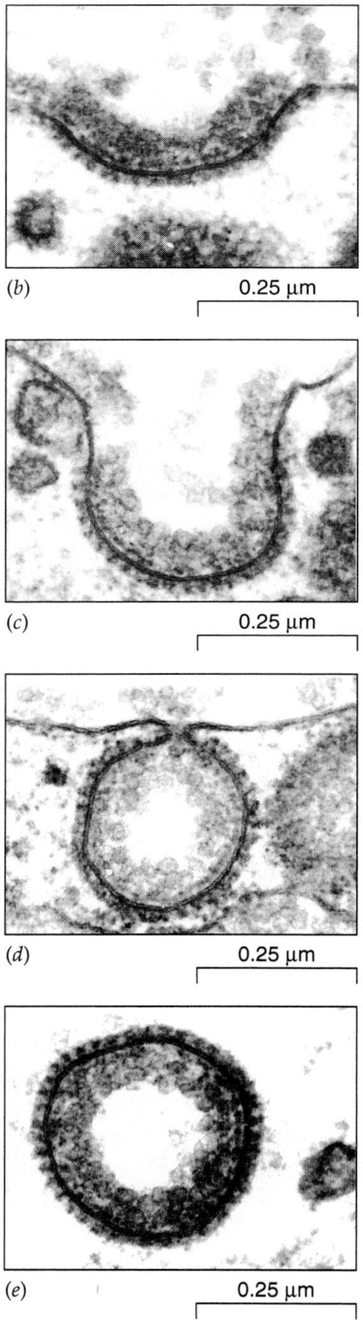

FIGURE 5–22 Receptor-mediated endocytosis provides for the efficient transport of a specific substance. (***a***) Low-density lipoprotein (LDL) particles, which transport cholesterol in the blood, attach to specific receptor proteins on the plasma membrane. The receptor-LDL complexes move along the surface of the fluid membrane and cluster in coated pit regions of the membrane surface. Endocytosis of the coated pit results in the formation of a coated vesicle in the cytoplasm. Seconds later the coat is removed, and the vesicles fuse with their counterparts to form large, smooth vesicles called endosomes. The receptors and the LDL particles dissociate in the endosomes and move to different regions of the vesicles. New vesicles form from the endosomes. The vesicles containing the receptors move to the surface and fuse with the plasma membrane, recycling the receptors to the cell surface. Vesicles containing the LDL particles fuse with lysosomes to form a secondary lysosome. Hydrolytic enzymes then release the cholesterol from the particles for use by the cell. ***(b–e)*** Series of transmission electron micrographs showing the formation of a coated vesicle from a coated pit. (*b–e,* From Perry, M.M., and A.B. Gilbert, *J. Cell. Sci.* 39:257–272, 1979)

surface of the membrane coated with whisker-like structures. These coated pits form *coated vesicles* (Fig. 5–22) by endocytosis. The coating on a vesicle consists of molecules of a protein known as clathrin, which momentarily forms a basket-like structure around it. Seconds after the vesicles are released, however, the coating dissociates from them, leaving the vesicles free in the cytoplasm. The vesicles then fuse with other similar vesicles to form *endosomes,* larger vesicles in which the materials being transported are free inside and no longer attached to the membrane receptors.

An endosome can divide to form two kinds of vesicles. One kind contains the receptors and can be returned to the plasma membrane; the other, which contains the ingested particles, fuses with lysosomes and is then processed by the cell.

particle binds to receptor proteins on plasma membrane → coated pit forms → coated vesicle forms by endocytosis → "uncoating" of vesicle → vesicles fuse to form endosome → division of endosome:

→ some vesicles return receptors to plasma membrane
→ other vesicles fuse with lysosomes and contents are digested and processed

Cholesterol in the blood is taken up by animal cells by receptor-mediated endocytosis. Much of the receptor-mediated endocytosis pathway was detailed through studies of the receptor for low-density lipoprotein (LDL—a primary cholesterol carrier in blood) by M. Brown and J. Goldstein. In 1986 these investigators were awarded the Nobel Prize for their pioneering work. Their findings have important medical implications, for cholesterol that remains in the blood instead of entering the cells can become deposited on the artery walls and increase the risk of heart attack.

The recycling of the LDL receptor to the plasma membrane through vesicles illustrates a problem common to all cells that employ endocytotic and exocytotic mechanisms. In cells that are constantly involved in secretion, an equivalent amount of membrane must be returned to the interior of the cell for each vesicle that fuses with the plasma membrane; if it is not, the cell surface will keep expanding even though the growth of the cell itself may be arrested. A similar situation exists for cells that use endocytosis. A macrophage, for example, ingests the equivalent of its entire surface membrane in about 30 minutes, requiring an equivalent amount of recycling or new membrane synthesis for the cell to maintain its surface area.

Junctions are specialized contacts between cells

Cells in close contact with each other may develop specialized intercellular junctions that involve their plasma membranes as well as other components. These structures may allow neighboring cells to form strong connections with each other, prevent passage of materials, or establish rapid communications between adjacent cells. In animals there are three common types of intercellular contacts: desmosomes, tight junctions, and gap junctions. Plant cells are connected by plasmodesmata.

Desmosomes are points of attachment between some animal cells

Adjacent epithelial cells, such as those found in the upper layer of the skin, are so tightly bound to each other that strong mechanical forces are required to separate them. They are held together by structures called **desmosomes** (Fig. 5–23). Each desmosome includes parts of two adjacent cells. It is made up of regions of dense material associated with the cytosolic sides of the two plasma membranes, plus protein filaments that cross the 24 nanometer-wide intercellular space between them.

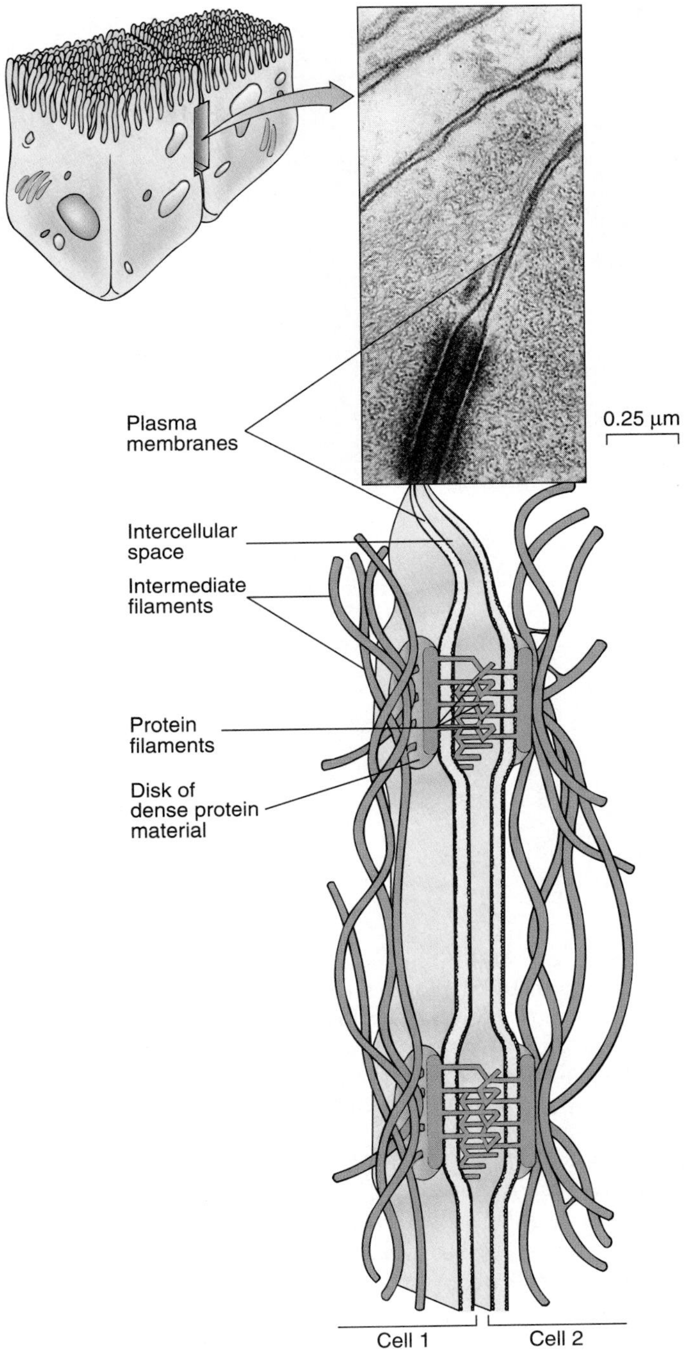

FIGURE 5–23 Desmosomes may connect two animal cells. The dense structure in the transmission electron micrograph is a desmosome. As can be seen in the interpretive drawing, each desmosome consists of a pair of button-like discs associated with the plasma membranes of adjacent cells, plus the intercellular protein filaments that connect them. Intracellular intermediate filaments attached to the discs are connected to other desmosomes. (D.W. Fawcett)

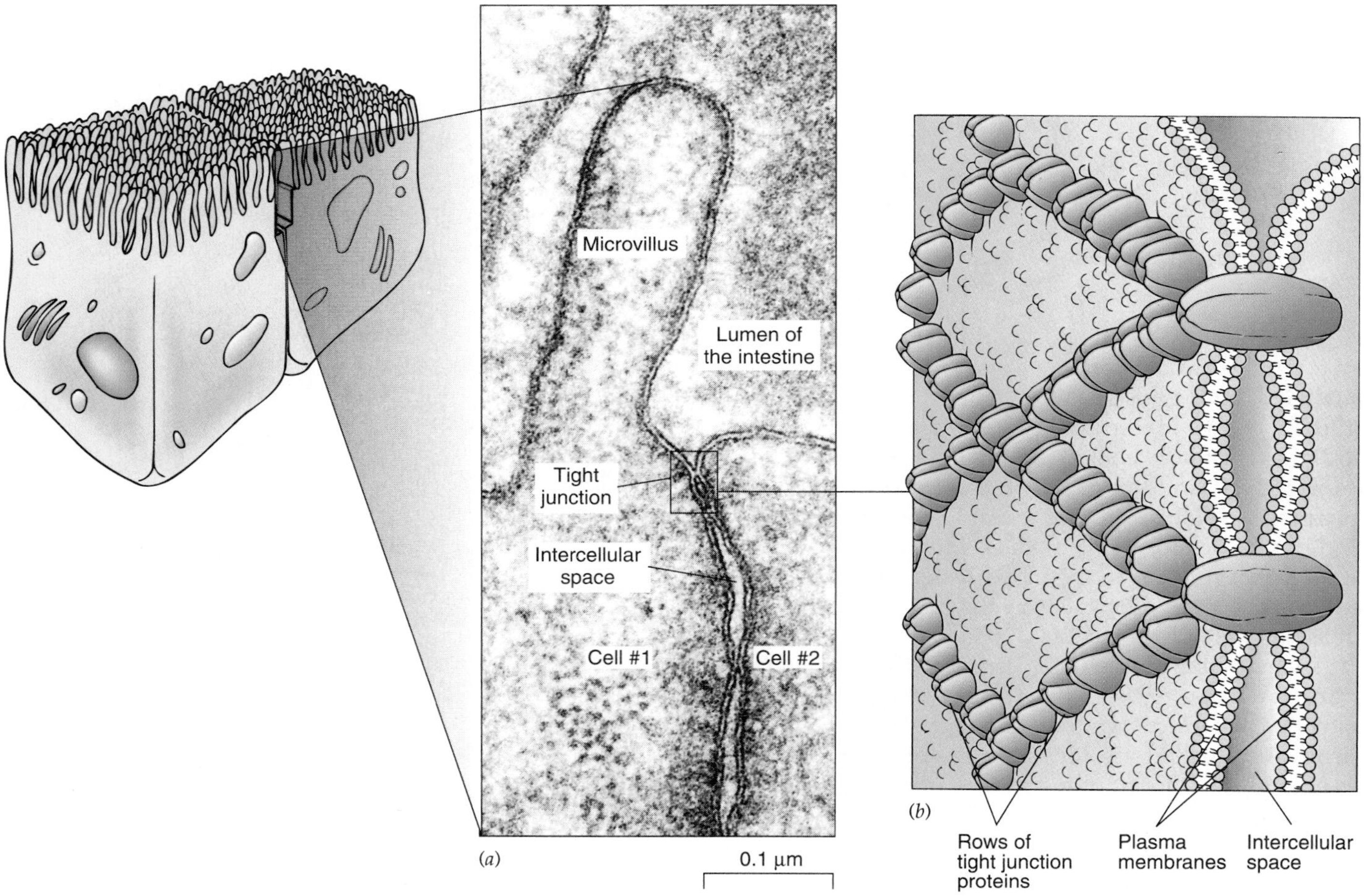

FIGURE 5–24 Tight junctions prevent passage of materials through intercellular spaces between cells. The tight junctions occur at the points of contact between two cells and would extend completely around the cells in a three-dimensional view. (***a***) This transmission electron micrograph shows points of fusion between the plasma membranes of adjacent cells lining the intestine. One tight junction is marked by the box. (***b***) This interpretive diagram shows that a tight junction is formed by linkages between rows of proteins of adjacent cells. These proteins are tightly packed in rows that seal off the intercellular space. (*a*, G. E. Palade)

Desmosomes are anchored to systems of intermediate filaments inside the cells. Thus the intermediate filament networks of adjacent cells are connected so that mechanical stresses are distributed throughout the tissue. The function of the desmosomes appears to be purely mechanical; they hold cells together at one point like a rivet or a spot weld. As a result, cells can form strong sheets, but substances can still pass freely through the spaces between the plasma membranes.

Tight junctions seal off intercellular spaces between some animal cells

Tight junctions are literally areas of tight connections between the membranes of adjacent cells. These connections are so tight that no spaces remain around the cells. Certain substances can thus be prevented from passing through the layer of cells. Electron micrographs of tight junctions show that in the region of the junction the membranes of the two cells are in actual contact with each other (Fig. 5–24), held together by proteins linking the two cells.

Cells connected by tight junctions seal off body cavities. For example, tight junctions between cells lining the intestine prevent substances in the intestine from entering the body or the blood by passing around the cells. The sheet of cells thus acts as a selective barrier; food substances required by the body must be transported across the plasma membranes of the intestinal cells before they enter the blood, as we saw in Figure 5–18. In this way some toxins and other unwanted materials are prevented from entering the blood.

Gap junctions permit transfer of small molecules and ions between some animal cells

A third type of intercellular connection in animal cells, the **gap junction,** is like the desmosome in that it bridges the space between cells; however, the space it spans is

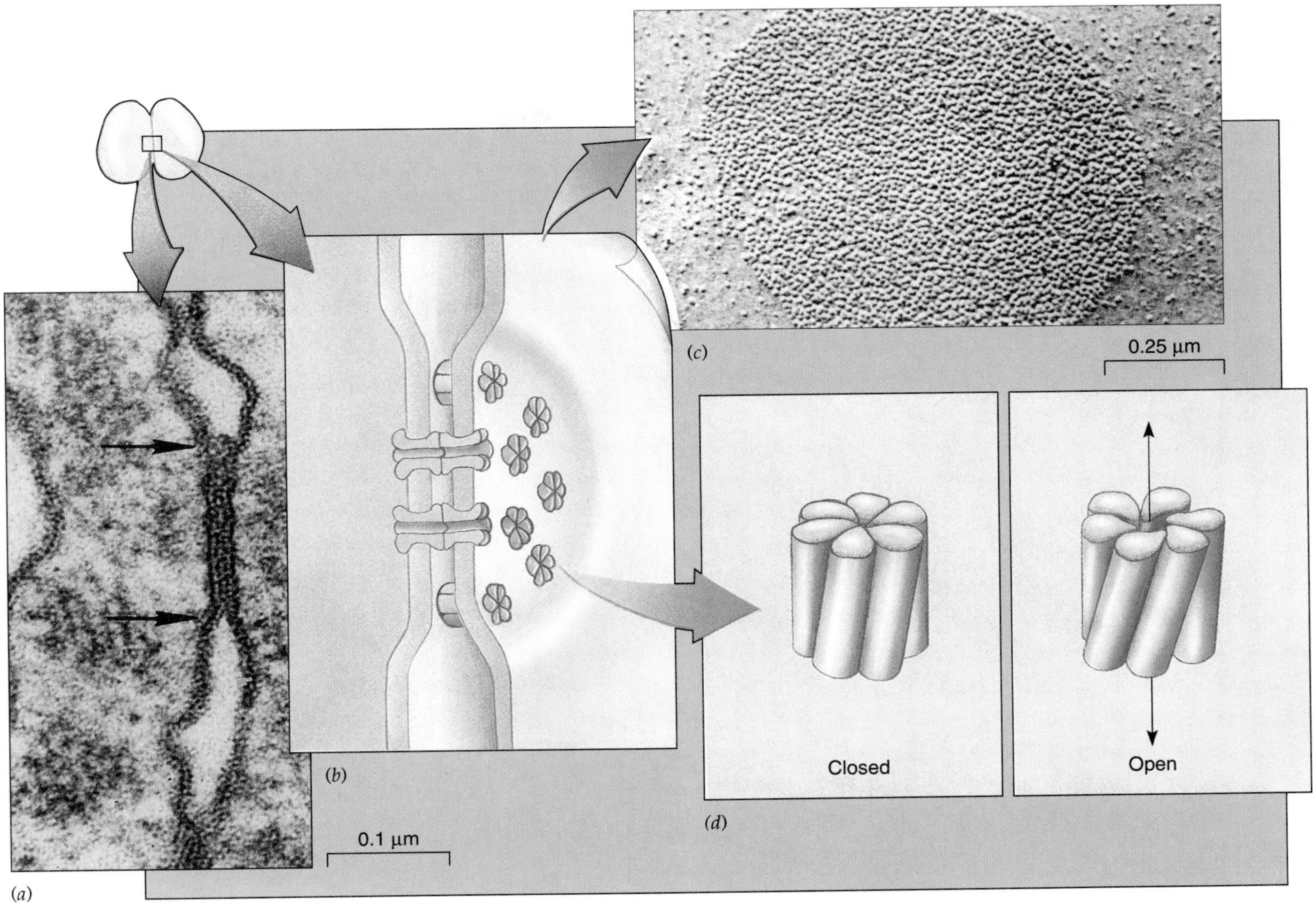

FIGURE 5–25 Gap junctions permit transfer of small molecules and ions between adjacent cells. (***a***) A transmission electron micrograph of a gap junction (*between the arrows*). (***b***) Model of a gap junction based on electron-microscopic and x-ray diffraction data. The two membranes contain cylinders composed of six protein subunits arranged to form a pore. Two cylinders from opposite membranes are joined to form a pore about 1.5 to 2.0 nanometers in diameter connecting the cytoplasmic compartments of the two cells. (***c***) Freeze-fracture replica of the P-face of a gap junction between two ovarian cells of a mouse, showing the numerous protein particles present. (***d***) Model illustrating how a gap junction pore might open and close. (*a*, D.W. Fawcett; *c*, E. Anderson, *J. Morph.* 156:339–366, 1978)

somewhat narrower (Fig. 5–25). Gap junctions also differ in that they not only connect the membranes, but also contain pores connecting the cytoplasm of adjacent cells. A gap junction consists of a hexagonal array of proteins forming a cluster of pores, each of which is about 1 to 2 nanometers in diameter. Small inorganic molecules (e.g., ions) and some biological molecules (e.g., derivatives of ATP) can pass through the pores, but larger molecules are excluded. When appropriate marker substances are injected into one of a group of cells connected by gap junctions, the marker passes rapidly into the adjacent cells but does not enter the space between the cells.

Gap junctions provide for rapid chemical and electrical communications between cells. Cells in the pancreas, for example, are linked together by gap junctions in such a way that if one of a group of cells is stimulated to secrete insulin, the signal is passed through the junctions to the other cells in the cluster, ensuring a coordinated response to the initial signal. Gap junctions allow some nerve cells to be electrically coupled. Heart muscle cells are linked by gap junctions that provide for synchronization of their contractions.

Cells are able to control the passage of materials through gap junctions by opening and closing the pores (Fig. 5–25*d*). There is evidence that the open and closed states are regulated mainly by the intracellular concentrations of certain ions.

Plasmodesmata allow movement of certain molecules and ions between plant cells

Plant cells do not need desmosomes for strength because they have cell walls. However, these same walls would isolate the cells, preventing them from communicating.

For this reason, plant cells require connections that are functionally equivalent to the gap junctions of some animal cells. **Plasmodesmata** (sing., *plasmodesma*) are 20 to 40 nanometer-wide channels through adjacent cell walls connecting the cytoplasm of neighboring cells (Fig. 5–26). The plasma membranes of adjacent cells are therefore continuous with each other through the plasmodesmata. Most plasmodesmata contain a cylindrical membranous structure, called the *desmotubule,* which also runs through the opening and connects the endoplasmic reticulum of the two adjacent cells. Plasmodesmata generally allow molecules and ions, but not organelles, to pass through the openings from cell to cell. The movement of ions through the plasmodesmata allows for a very slow type of electrical signalling in plants.

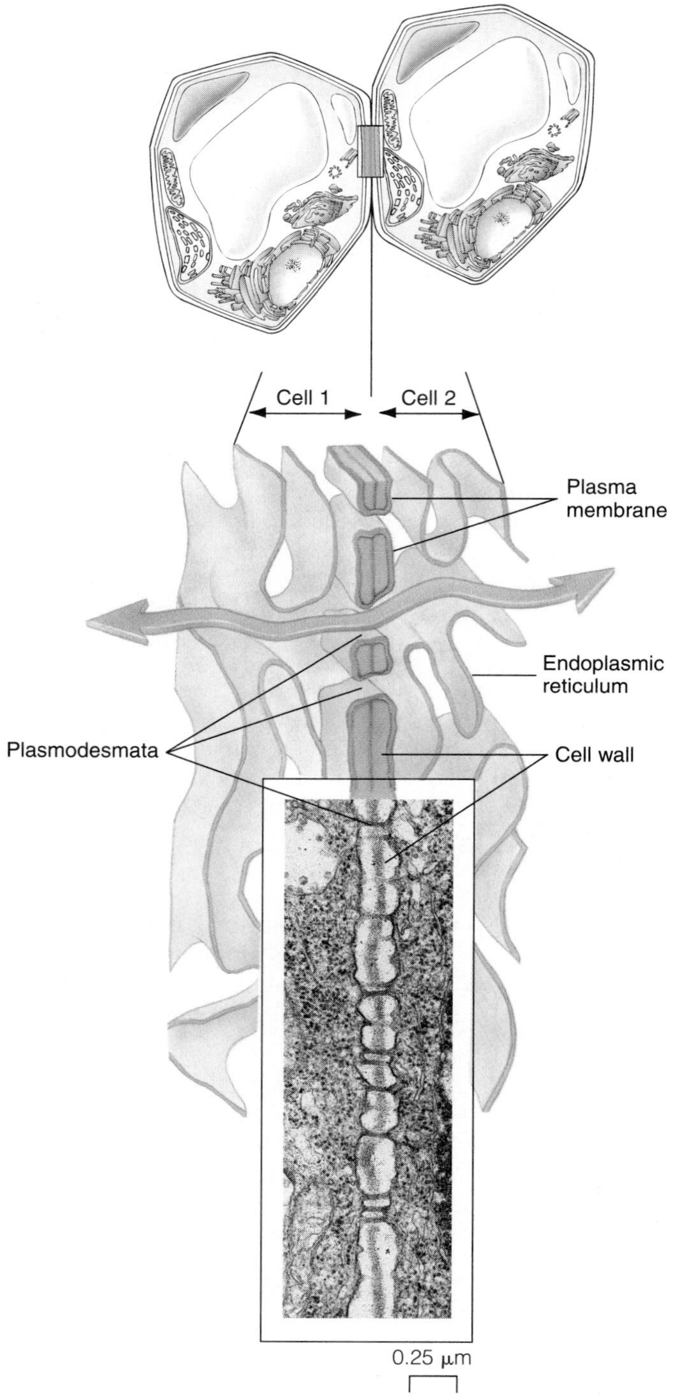

FIGURE 5–26 Plasmodesmata are cytoplasmic channels that pass through the cell walls of adjacent plant cells. The transmission electron micrograph and interpretive drawing show adjacent cells connected by plasmodesmata, which allow the passage of water, ions, and small molecules (*arrows*). The channels are lined with the fused plasma membranes of the two adjacent cells. The desmotubules are not shown. (E. H. Newcomb, University of Wisconsin/Biological Photo Service)

SUMMARY

I. Cellular membranes are complex structures that (1) physically separate the interior of the cell from the extracellular environment and (2) form compartments within the cells of eukaryotes that allow them to perform complex functions.
   A. Membranes have many different structural and functional roles.
      1. They regulate the passage of materials.
      2. They receive information that permits the cell to sense changes in its environment and respond to them.
      3. They contain specialized structures that allow specific contacts and communications with other cells.
      4. They serve as work surfaces for various biochemical reactions.
   B. According to the fluid mosaic model of membrane structure, membranes consist of a fluid lipid bilayer in which a variety of proteins are embedded.
      1. The lipid bilayer is arranged in such a way that the hydrophilic headgroups of the phospholipids are at the two surfaces of the bilayer and their hydrophobic fatty acid chains are in the interior.
      2. In almost all biological membranes, the lipids of the bilayer are in a fluid or liquid-crystalline state, which allows the molecules to move rapidly in the plane of the membrane.
      3. Integral membrane proteins are embedded in the bilayer in such a way that their hydrophilic surfaces are exposed to the aqueous environment and their hydrophobic surfaces are in contact with the hydrophobic interior of the bilayer.
      4. Peripheral membrane proteins are associated with the surface of the bilayer and are easily removed without disrupting the structure of the membrane.
      5. Membrane proteins, lipids, and carbohydrates are asymmetrically positioned with respect to the bilayer so that one side of the membrane has a different composition and structure from the other.

II. Biological membranes are selectively permeable; that is, they allow the passage of some substances but not others.
   A. Some molecules pass through membranes by simple diffusion.
      1. Diffusion is the net movement of a substance down its concentration gradient (from a region of high concentration to one of low concentration).
      2. Dialysis is the diffusion of a solute across a membrane.
      3. Osmosis is a kind of diffusion in which molecules of water pass through a selectively permeable membrane from a region where water has a higher effective concentration to a region where its effective concentration is lower.
      4. The osmotic pressure of a solution is determined by its concentration of dissolved substances (solutes). Cells regulate their internal osmotic pressure to prevent shrinking or bursting. Plant cells can withstand high internal hydrostatic pressure because their cell walls prevent them from expanding and bursting.
   B. Some substances pass through membranes by facilitated diffusion, a form of carrier-mediated transport in which a carrier protein helps a molecule move through the membrane. Facilitated diffusion uses the energy of a concentration gradient for the substance being transported and cannot work against the gradient.
   C. In carrier-mediated active transport the cell expends metabolic energy to move ions or molecules against a concentration gradient.
   D. In endocytosis (phagocytosis, pinocytosis, and receptor-mediated endocytosis) materials such as food may be moved into the cell; a portion of the plasma membrane envelops the material, enclosing it in a vacuole, or small vesicle, that is then released inside the cell.
   E. In exocytosis, the cell ejects waste products or secretes substances such as mucus.

III. Plasma membranes of animal and plant cells contain specialized structures that allow them to have contact and communication with adjacent cells.
   A. Desmosomes, tight junctions, and gap junctions are specialized structures associated with the plasma membranes of animal cells.
      1. Desmosomes weld cells together to form strong tissues.
      2. Tight junctions seal membranes of adjacent cells together to prevent substances from passing between cells.
      3. Gap junctions are protein pores in membranes that allow communication between the cytoplasm of adjacent cells.
   B. Plasmodesmata are channels connecting adjacent plant cells. Openings in the cell walls allow the plasma membranes and cytoplasm to be continuous, thus permitting certain molecules and ions to pass from cell to cell.

SELECTED KEY TERMS

active transport
amphipathic
carrier-mediated active transport
carrier-mediated transport
concentration gradient
cotransport
desmosome
dialysis
diffusion
endocytosis
exocytosis
facilitated diffusion
fluid mosaic model
gap junction
G protein
hypertonic (hyperosmotic)
hypotonic (hypoosmotic)
integral membrane protein
isotonic (isosmotic)
microvilli
osmosis
osmotic pressure
patch clamp technique
peripheral membrane protein
phagocytosis
pinocytosis
plasmodesma (plasmodesmata)
plasmolysis
receptor-mediated endocytosis
selectively permeable
signal transduction
sodium-potassium pump
tight junction
transmembrane protein
turgor pressure

POST-TEST

1. According to the ________ ________ model of membrane structure, a cellular membrane consists of a lipid bilayer with embedded proteins.
2. Bilayers are formed from ________ lipids—that is, lipids with prominent hydrophobic and hydrophilic regions.
3. ________ membrane proteins are tightly associated with the bilayer. ________ membrane proteins are associated with the bilayer through interactions with other membrane components on the surface of the membrane.
4. Net movement of particles from a region where they are more concentrated to a region where they are less concentrated is called ________ .
5. Diffusion of water through a selectively permeable membrane down its own concentration gradient is called ________ .
6. A solution with an equivalent concentration of water and hence the same osmotic pressure as the fluid inside the cell is said to be ________ to the cell's contents.
7. When a plant cell is placed in solution A, the contents inside the plasma membrane can be seen to shrink away from the cell wall. Solution A is ________ to the interior of the plant cell.
8. When red blood cells are placed in solution B, the cells are seen to expand, and many of them burst. Solution B is ________ compared to the interior of the red blood cells.
9. When glucose enters cells by moving down its concentration gradient with the help of a carrier protein, but with no requirement for an additional source of energy, it does so by ________ ________ .
10. If the glucose concentration outside the cell is lower than the concentration inside the cell, glucose must be moved into the cell by ________ ________ .
11. The sodium-potassium pump is a carrier-mediated ________ transport system that pumps sodium ions from the inside of the cell to the outside and pumps potassium ions from the outside of the cell to the inside.
12. Active transport systems that derive required energy from transport of a second molecule down its concentration gradient are called ________ systems.
13. Large molecules and large particles are transported into cells by an endocytotic mechanism called ________ .
14. Molecules that bind to specific molecules on the plasma membrane and then move into the interior of the cell through small vesicles do so by the process of ________-________ endocytosis.
15. ________ are specialized regions of the plasma membrane in animal cells that "spot weld" adjacent cells together.
16. ________ ________ contain specialized protein pores in the plasma membranes of animal cells that allow small molecules to pass from the cytoplasm of one cell to the cytoplasm of its neighbor.
17. ________ ________ are specialized regions of plasma membranes in animal cells that prevent substances from passing around the outsides of cells in a tissue.
18. ________ provide for the transfer of water, ions, and small molecules between adjacent plant cells.

REVIEW QUESTIONS

1. What molecules are primarily responsible for the physical properties of a cellular membrane?
2. Illustrate how a transmembrane protein might be positioned in a lipid bilayer. How do the hydrophilic and hydrophobic regions of the protein affect its orientation?
3. Describe the pathway used by cells to place carbohydrates on plasma membrane proteins. Explain why this results in the carbohydrate groups being on only one side of the lipid bilayer.
4. What is the source of energy for diffusion? State a rule for predicting the movement of particles along their concentration gradient. Is the rule different for facilitated diffusion compared with simple diffusion?
5. Distinguish clearly between osmosis and dialysis.
6. Predict the consequences if a plant cell were to be placed in a relatively (a) isotonic, (b) hypertonic, or (c) hypotonic environment. Would you have to modify any of your predictions for an animal cell?
7. What are some of the functions of the plasma membrane? Discuss the nature of the proteins that carry out those functions, and explain how their properties make them especially adapted for their functions.
8. What is the most common energy source for active transport? In what ways are facilitated diffusion and carrier-mediated active transport similar? In what ways do they differ?
9. Draw a diagram illustrating how membrane lipid bilayers fuse during the processes of exocytosis and endocytosis. Is one the exact reverse of the other? Why or why not?
10. Discriminate between the processes of phagocytosis and pinocytosis.
11. How are desmosomes and tight junctions functionally similar? How do they differ? Do they share any structural similarities?
12. What is the justification for considering gap junctions and plasmodesmata to be functionally similar? How do they differ structurally?
13. Label the diagram (at the top of page 141) of a typical plasma membrane.

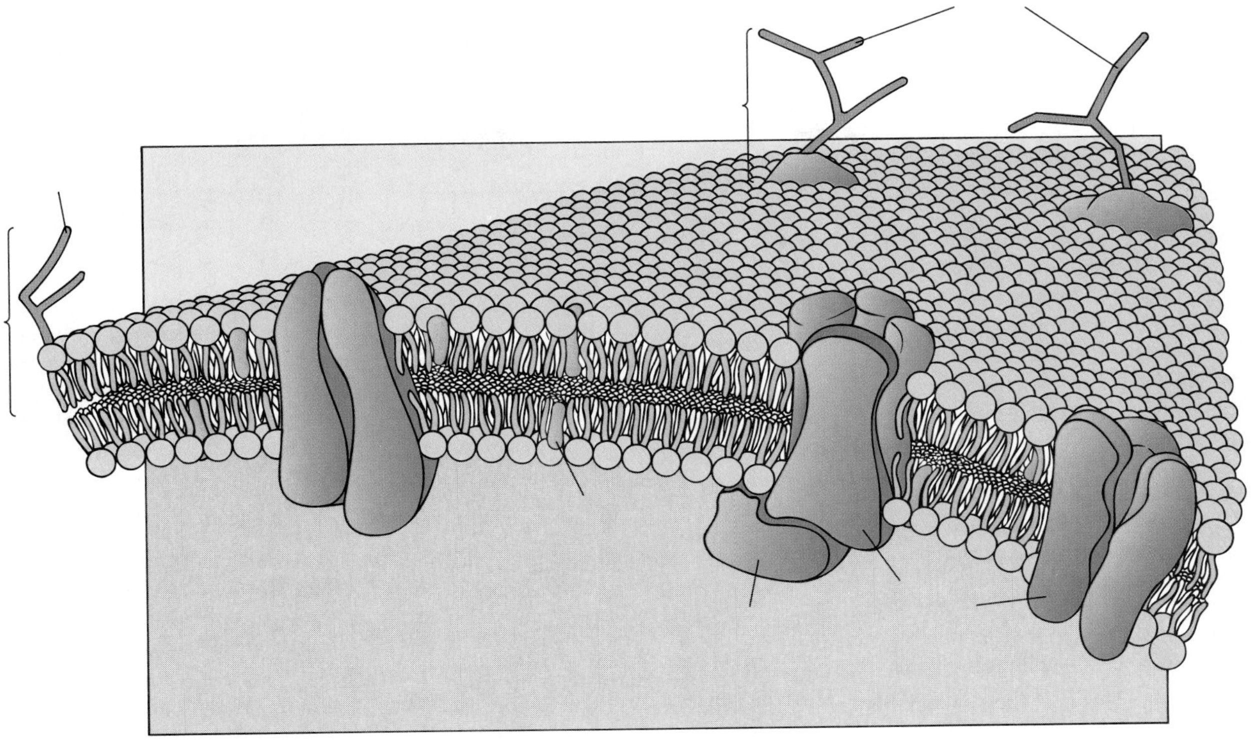

YOU MAKE THE CONNECTION

1. Why can't larger polar molecules and ions cross a lipid bilayer? Would it be advantageous to the cell if they could?
2. Most cells do not actively transport water, yet water is essential to life. How, then, are cells able to control their water content?
3. You prepare a salad with dressing in the morning, but find that it is limp and unappetizing by lunch time. Why?
4. Most adjacent living cells in a plant are connected by plasmodesmata. On the other hand, only certain adjacent animal cells are associated through gap junctions. Why?

RECOMMENDED READINGS

Alberts, B.,D. Bray, J. Lewis, M. Raff, K. Roberts, and J.D. Watson. *Molecular Biology of the Cell,* 2nd ed. Garland Publishing, New York, 1994, Chapter 10. A discussion of the structure and functions of the plasma membrane.

Brown, M.S., and J.L. Goldstein. "How LDL Receptors Influence Cholesterol and Atherosclerosis." *Scientific American,* November, 1984. Nobel prize–winning authors explain how improper functioning of high density lipoprotein (LDL) in the blood and LDL receptors on cells can lead to cardiovascular disease.

Linder, M.E., and A. Gilman. "G Proteins." *Scientific American,* Vol. 267, No. 1, July 1992. Pioneers in cell-signalling research discuss the many roles of G proteins.

Nehr, E., and B. Sakmann. "The Patch Clamp Technique." *Scientific American,* March 1992. The developers of the patch clamp technique discuss its varied applications.

Vogel, S. "Dealing Honestly with Diffusion." *The American Biology Teacher,* October, 1994. An explanation of why most macroscopic phenomena attributed to diffusion actually have other explanations. This article emphasizes the fact that diffusion is rapid only over extremely short distances.

Wayne, R. "Excitability in Plant Cells." *American Scientist,* 81, March–April 1993. A discussion of how modern technology, including the patch clamp technique, is being used to study ion movements in plants.

Career Visions

Bioremediation Specialist

ANDREA LEESON

Andrea Leeson began her college career at Eastern Kentucky State University with an interest in working toward a degree in environmental resources. As a bioremediation specialist she applies her combined education in biology, mathematics, and engineering to cleaning up hazardous waste sites for the Air Force and the Environmental Protection Agency. Since 1991, she has worked at the Battelle Memorial Institute in Columbus, Ohio—a non-profit organization involved in industrial, medical, and environmental research and its practical applications to technology. Andrea works in field research to develop methods that will more effectively use microbes to eliminate toxic wastes.

Did you major in biology?

I originally started out in an environmental resources program because I wanted to do some sort of environmental work, but I switched to biology after about a year.

How did your undergraduate degree lead you to your career?

I guess what really pushed me toward my career was the math that I took for the biology major. I took calculus in my senior year of college and had an excellent teacher—Dr. King at Eastern Kentucky State University. This was probably one of my favorite classes and it made me think seriously about combining my interest in environmental work with mathematics. I had also spent a semester doing research at a national laboratory. The research experience also helped me decide that I really wanted to combine the two fields. The perfect combination seemed to be environmental engineering, since engineering requires a fair amount of math.

You then attended Johns Hopkins University. Did they have a particular program that met the criteria you were interested in?

Yes, at Johns Hopkins I could specialize in bioremediation, a field that some graduate programs in environmental engineering don't offer.

What was different about the program at Johns Hopkins?

A lot of the environmental engineering programs are incorporated into civil engineering departments. Hopkins was unique in that they have their own department of environmental engineering—actually geography and environmental engineering. The department was pretty strong in areas like wastewater treatment, drinking water treatment, and hazardous waste treatment. Those are the three main areas for an environmental engineer.

What is bioremediation? What does your job entail?

Essentially bioremediation is using any kind of microorganism to clean up hazardous waste sites. We do field research and we are always looking at new ways to enhance the natural microbial activity that's already there to clean up a site. Primarily I work on petroleum contaminated sites now, but I work with chlorinated compounds as well.

How do you use bacteria and other microorganisms to clear up a site? How is your biology background used?

We use bacteria at a site by exploiting their natural ability to use contaminants as a source of carbon and/or energy. We supply them with whatever they need—usually oxygen—to enable them to continue to break down contaminants. My biological background is necessary in order to understand what is needed for the bacteria to grow, while the engineering background is necessary to understand how to supply the needed nutrients. The major technology I use is bioventing which is injecting air into the ground. Therefore, you have to be able to design an air injection system. You have to know something about well construction.

You mentioned working with chlorinated compounds. What does that involve?

Petroleum-contaminated sites are the most common kind of contamination there is, but after that there are chlorinated solvents. A lot of military installations use chlorinated solvents as degreasers. In the past, equipment may have been sprayed down with a solvent, generating runoff that contaminated the soil. Dry cleaners also use chlorinated solvents. It's a pretty common problem.

What would be the best way for a student interested in bioremediation to get the appropriate skills to do this kind of work?

People enter graduate training in environmental engineering with a mix of backgrounds. I really think it was good having a major like biology for my bachelors degree, but to do what I'm doing now you really need some engineering so I'd recommend going to graduate school in an engineering program.

PART 2

Energy Transfer Through Living Systems

Almost all life utimately depends on solar energy. (Adam Jones/Dembinsky Photo Associates)

CHAPTER 6

Energy and Metabolism

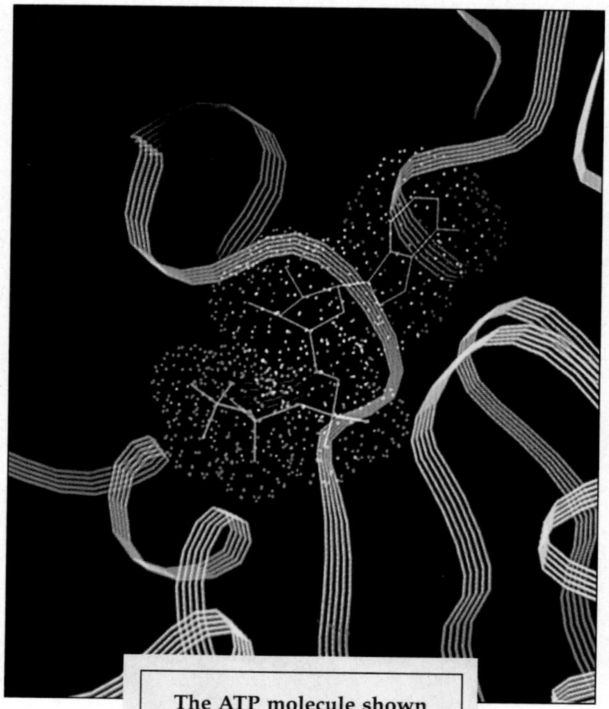

The ATP molecule shown in this computer graphics close-up is bound to phosphoglycerate kinase (*green ribbons*), an enzyme that catalyzes the transfer of a phosphate from ATP to another compound. (Oxford Molecular Biophysics Laboratory/Science Photo Library/Photo Researchers, Inc.)

All living things require energy because life processes involve work. It may seem obvious that cells need energy to grow and reproduce, but even non-growing cells need energy simply to maintain themselves.

The sun is the ultimate source of almost all the energy that powers life. Plants and other photosynthetic organisms capture a tiny portion of the sun's energy and, in the process of photosynthesis, convert it to chemical energy in organic molecules. When plants, animals, or other organisms need the energy stored in these organic molecules, they commonly use the process of cellular respiration to break them apart and convert their energy to more immediately usable forms.

Because energy can be neither created nor destroyed, cells have no way of producing new energy. Energy is captured from the environment, temporarily stored, and then used to perform biological work. However, not all of the captured energy can be used; at every step some inevitably becomes converted to heat and is dispersed back into the environment.

Cells obtain energy in many forms, but seldom can that energy be used directly to power cellular processes. For this reason metabolic mechanisms have evolved that enable cells to convert energy from one form to another. Because most of the components of these energy conversion systems evolved very early, the most fundamental aspects of energy metabolism tend to be very similar in a wide range of different organisms.

This chapter focuses on some of the basic principles that govern how cells capture, transfer, store, and use energy. We discuss the functions of ATP and other molecules used in energy conversions, including those that transfer electrons in redox reactions. We also pay particular attention to the essential role of enzymes in cellular energy dynamics. In Chapter 7 we will explore some of the main metabolic pathways used in cellular respiration, and in Chapter 8 we will discuss the chemical and energy transformations of photosynthesis. The flow of energy in ecosystems is discussed in Chapter 54.

LEARNING OBJECTIVES

After you have studied this chapter you should be able to

1. Define the term *energy*, emphasizing how it is related to work and to heat.
2. Use examples to contrast potential energy and kinetic energy.
3. State the first and second laws of thermodynamics, and discuss the implications of these laws as they relate to organisms and to the ecosphere.
4. Discuss how changes in free energy in a reaction are related to changes in entropy and enthalpy.
5. Compare the energy dynamics of a reaction at equilibrium with the dynamics of a reaction not at equilibrium.
6. Distinguish between endergonic and exergonic reactions, and give examples of how they may be coupled. Explain why energy coupling does not violate the laws of thermodynamics.
7. Explain how the chemical structure of ATP allows it to transfer a phosphate group. Discuss the central role of ATP in the overall energy metabolism of the cell.
8. Relate the transfer of electrons (or hydrogens) to the transfer of energy.
9. Explain how an enzyme lowers the required energy of activation for a reaction.
10. Describe how factors such as pH and temperature influence how an enzyme functions. Describe some of the ways an enzyme is regulated.

BIOLOGICAL WORK REQUIRES ENERGY

Energy may seem to be an abstract concept, but it helps to remember that energy can be understood in the context of **matter** (anything that has mass and takes up space). This is because energy can be defined as the capacity to do **work,** which is any change in the state or motion of matter (Fig. 6–1). Biologists generally express energy in units of work (**kilojoules, kJ**) or units of heat energy (**kilocalories, kcal**). One kilocalorie equals 4.184 kilojoules. Because heat energy cannot do cellular work (see *Making the Connection: Energy, Work, and Heat*), the kilojoule is the unit preferred by most biologists today. However, we will use both because references to the kilocalorie are common in the scientific literature.

Much of the work an organism does is mechanical work. At this very moment you are expending considerable energy to carry out such activities as breathing and circulating your blood. In these examples, it is evident that the state or motion of matter is being changed in some way. However, all these rather obvious forms of mechanical work are the consequence of work being performed on a microscopic scale by your individual cells. For example, the cells of the heart muscle use a great deal of energy to contract, thereby pumping the blood through your body. As we shall see, however, not all of the work of cells is mechanical. A great deal of it is chemical work. For example, heart muscle cells expend energy to synthesize the proteins required for contraction. Energy can be converted to many different forms, including not only mechanical and chemical energy, but also heat energy and radiant energy (the energy of electromagnetic waves, such as radio waves, visible light, x rays, and gamma rays).

FIGURE 6–1 This bobcat is doing work (expending energy) in an effort to capture the snowshoe hare. If caught and eaten, the hare will provide nutrients containing energy for future activity. For its part, the hare is expending a great deal of energy in its effort to escape becoming an energy source for the bobcat. However, even if these animals were both at rest, they would still require significant amounts of energy to maintain their life functions. (Sharon Cummings/Dembinsky Photo Associates)

Making the Connection

Energy, Work, and Heat

Why can we express energy both in units of work (kilojoules) and in units of heat energy (kilocalories)? We can because of a conceptual breakthrough that came about in the 1800s, after the invention of the steam engine. Scientists studying the connection between the heat energy that powered the engine, the mechanical work that the engine was able to perform, and the heat that was transferred to the environment were able to demonstrate that all these forms of energy are interconvertable.

Mechanical work depends on the force applied to matter (which is related to its mass) and the distance that the matter is displaced by that force. Thus we can calculate work as the product of a force applied to matter, times the distance over which matter is displaced.

$$\text{Work (in kilojoules)} = \text{Force} \times \text{Distance}$$

Today we know that not only mechanical work but all forms of energy can be converted to heat. In fact, the study of energy and its transformations has been named **thermodynamics** — that is, heat dynamics. (Recall from Chapter 2 that the term *heat* refers to the total amount of kinetic energy in a sample of a substance, whereas *temperature* refers to the average kinetic energy of the particles.) **Heat energy** is energy that can flow from an object with a higher temperature (known as the heat source) to an object with a lower temperature (the heat sink).

Cells cannot work as heat engines because they are isothermal; they are too small to have regions that differ in temperature. Therefore, *heat cannot be used to do biological work.* Nevertheless, the fact that all forms of energy can be converted to heat is useful to scientists because heat energy is particularly convenient to measure.

Nutritionists use the kilocalorie to express the potential energy of foods and usually refer to it as a *Calorie* (with a capital C; see Chapter 45). For example, the energy content of 10 grams (about 2 teaspoons) of table sugar (sucrose) is about 36 Calories (36 kcal or 151 kJ). A person weighing 58 kilograms (130 pounds) uses about 1 kcal (4.184 kJ) per minute to maintain the body while sleeping, and up to 10 kcal (41.84 kJ) per minute when engaged in strenuous activity. ▲

Organisms carry out conversions between potential energy and kinetic energy

When an archer draws a bow (Fig. 6–2), **kinetic energy,** which is energy of motion, is used; work is done because a force is exerted over a distance. The resulting tension in the bow and string represents stored energy, or **potential energy.** Potential energy has the capacity to do work owing to its position or state. When the string is released, this potential energy is converted to kinetic energy (the motion of the bow, which propels the arrow).

Most of the actions of an organism involve a complex series of energy transformations that occur as kinetic energy is converted to potential energy (which may again become kinetic energy). For example, potential energy (derived from chemical energy of food molecules) is converted to kinetic energy in the muscles of the archer.

TWO LAWS OF THERMODYNAMICS GOVERN ENERGY TRANSFORMATIONS

All the activities of our universe — from the life and death of cells to the life and death of stars — are governed by the laws of thermodynamics. Each of these laws may be stated in several ways and has a number of implications.

The total energy in the universe does not change

*According to the **first law of thermodynamics,** known also as the law of conservation of energy, the total energy of any **closed system** — that is, of any object and its surroundings — remains constant.* Because the term *surroundings* refers to the rest of the universe, which is thought to be a closed system, the total energy of the universe is defined as constant. The first law of thermodynamics thus holds that energy can be neither created nor destroyed. As we have seen, however, energy *can* be transferred or changed in form.

An organism is an **open system;** it can exchange matter and energy with its surroundings. Any change in its energy content must be balanced by a corresponding change in the energy content of the surroundings.

Because an organism can neither make nor destroy energy, it must have adaptations that allow it to capture energy from its environment, convert it to other forms, and use it for its own needs. During photosynthesis, plant cells transform light energy to electrical energy and then to chemical energy stored in chemical bonds. Some of that chemical energy may later be used to carry out the life functions of the plant cells, or it may be transformed by an animal that eats the plant to the mechanical energy of muscle contraction or some other needed form (Fig. 6–3).

(Text continues on page 148)

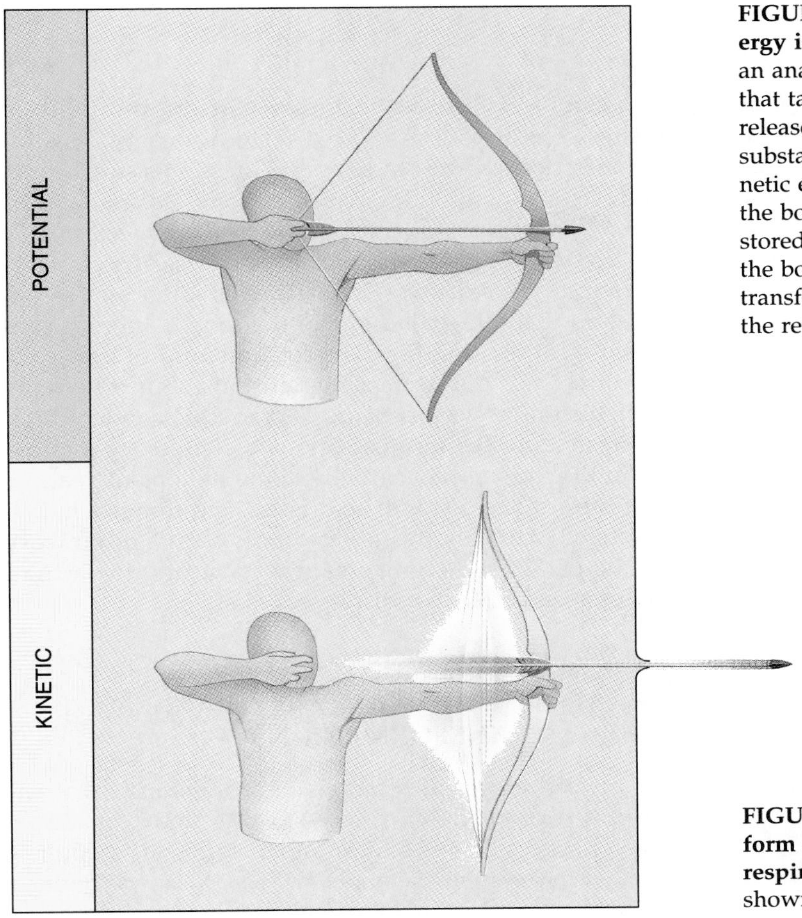

FIGURE 6–2 Potential energy is stored energy; kinetic energy is the energy of motion. Shooting a bow and arrow is an analogy that illustrates some of the energy transformations that take place in organisms. The potential chemical energy released by cellular respiration and held temporarily in the substance adenosine triphosphate (ATP) is converted to kinetic energy in the muscles, which do the work of drawing the bow, and to waste heat (not shown). The potential energy stored in the drawn bow is transformed into kinetic energy as the bowstring pushes the arrow toward its target. In all these transformations, energy is neither created nor destroyed, but the remaining total *useful* energy decreases at each step.

FIGURE 6–3 Three major types of processes that transform energy in the world of life are photosynthesis, cellular respiration, and biological work. (The chemical reactions shown are simplified. See Chapters 7 and 8 for more detail.)

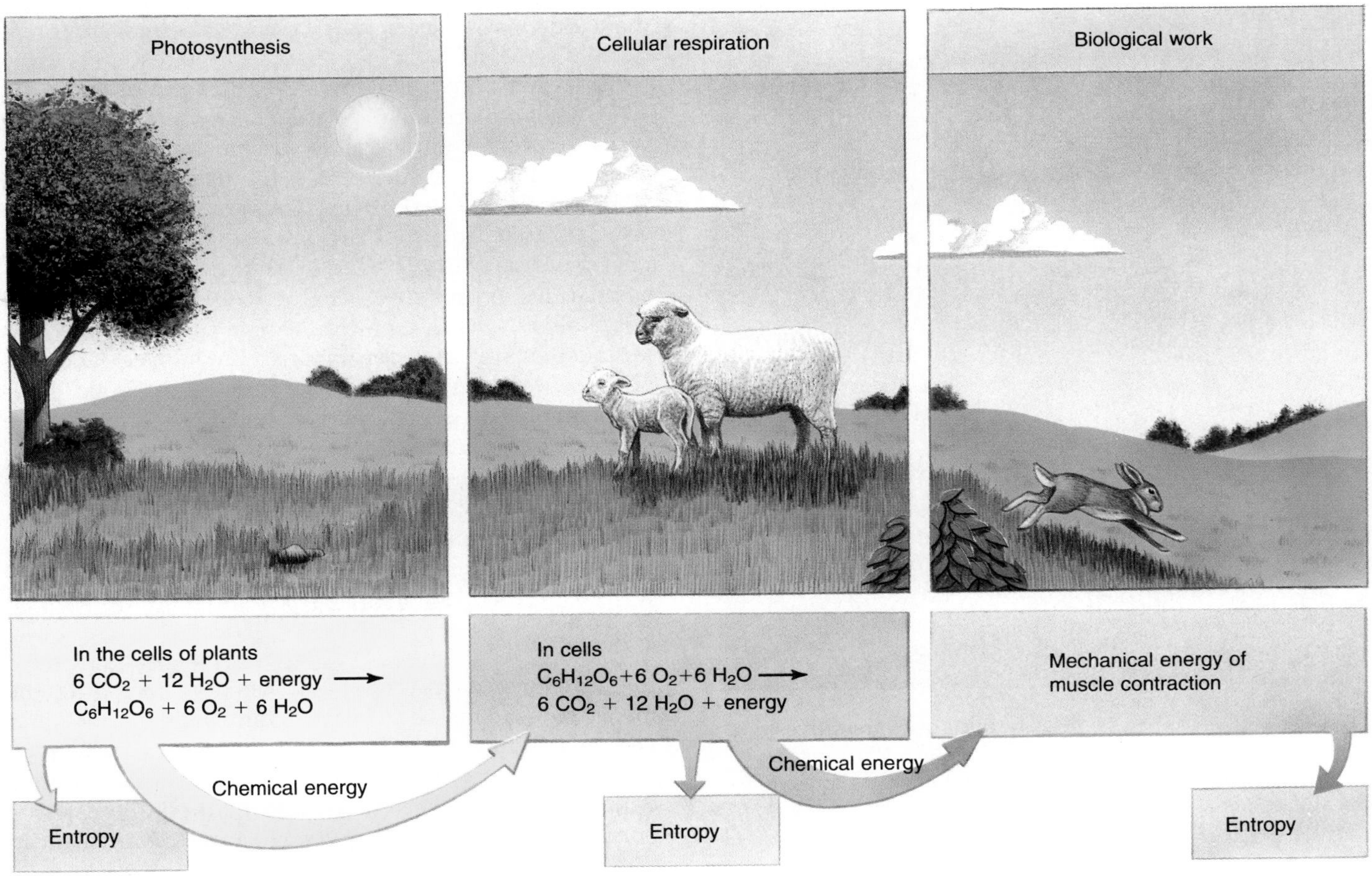

As these many transformations take place, some of the energy is converted to heat and dissipated into the environment. Although this energy can never again be used by the organism, it is not destroyed; it still exists in the surroundings.

The entropy (disorder) of the universe is increasing

In almost all energy transformations, some energy is lost in the form of heat to the surroundings (Fig. 6–4). This energy is not destroyed, but its capacity to do work is diminished because heat can do work only if heat energy flows from a region of higher temperature to a region of lower temperature.

Heat is the random movement of atoms and molecules; thus an increase in heat makes the system more disordered. Because there is a relationship between heat and the amount of disorder in the system, we can think of heat as the most disorganized form of energy. This increase in disorder or randomness can be measured as an increase in a quantity known as **entropy.** *According to the **second law of thermodynamics,** the total amount of entropy (randomized, disordered energy that is unavailable to do work) increases in the universe* (Fig. 6–5). It is important to understand that the second law of thermodynamics is consistent with the first law. The total amount of energy in the universe is not decreasing with time, but the energy available to do work is being degraded to random molecular motion. Because entropy is continuously increasing in the universe, eventually, some billions of years in the future, all energy will be random and uniform in distribution. With only this useless form of energy, no work will be possible; the universe will have run down from an organized state to a disorganized state.

FIGURE 6–4 As she perspires (a form of evaporative cooling) this athlete is losing energy as heat. That energy is gained by the surroundings, so the net energy change for the athlete plus her surroundings is zero. (Gerard Vandystadt/ Photo Researchers, Inc.)

METABOLIC REACTIONS INVOLVE ENERGY TRANSFORMATIONS

The myriad chemical reactions of an organism that enable it to carry on its activities—to grow, move, maintain and repair itself, reproduce, and respond to stimuli—together make up its metabolism. *Metabolism* was defined in Chapter 1 as the sum total of all the chemical and physical changes that take place in an organism, interacting to make up what has been called the **metabolic web.** An organism's metabolic web consists of many intersecting series of reactions, or pathways, which are of two main types. **Anabolism** refers to the various metabolic pathways in which complex molecules are synthesized from simpler substances, such as the linking of amino acids to form proteins. **Catabolism** includes the pathways in which larger molecules are broken down into smaller ones, such as the degradation of starch to form monosaccharides. As we shall see, these changes not only involve alterations in the arrangement of atoms, but also various energy transformations. Catabolism and anabolism are complementary processes; catabolic pathways involve an overall release of energy, some of which is used to power the anabolic pathways, which have an overall energy requirement. In the following sections we will discuss how to predict whether a particular chemical reaction requires energy or releases it.

Enthalpy (H) is the total potential energy of a system

In the course of any chemical reaction, including the metabolic reactions of a cell, chemical bonds break and new and different bonds may form. Every specific type of chemical bond has a certain amount of **bond energy,** defined by chemists as the energy required to break that bond. The total bond energy is essentially equivalent to

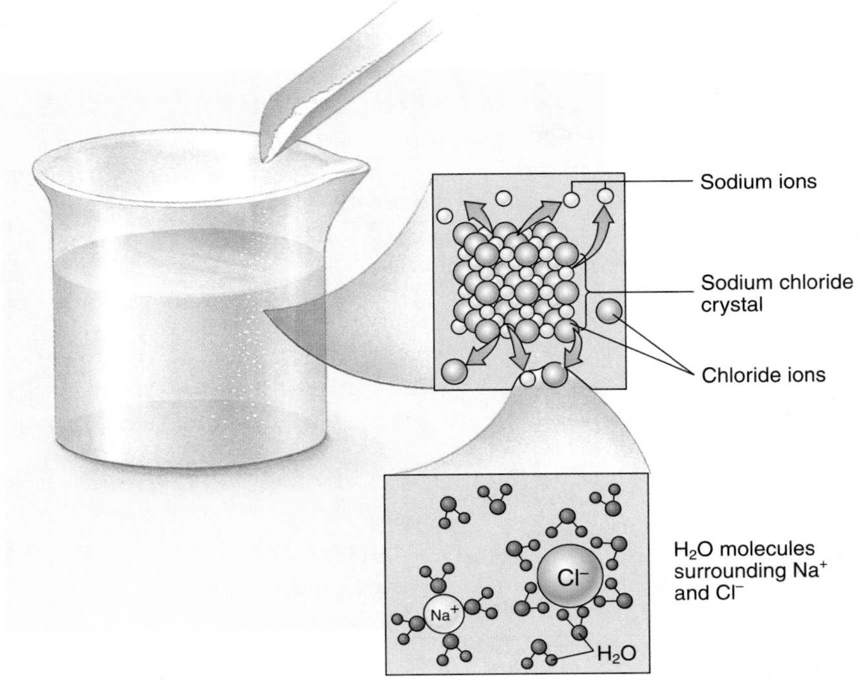

(a)

(b)

FIGURE 6–5 Entropy is a measure of disorder. (***a***) As particles leave a crystal to go into a solution, they become more disordered. The entropy (randomness) of this system increases during the process. (***b***) Imagine that you shake a beaker containing two colors of marbles. The number of possible disorderly arrangements of the marbles is much greater than the number of possible orderly arrangements. Therefore a disordered arrangement, such as the one on the right, is much more probable than the ordered arrangement, shown on the left, in which all marbles of the same color remain together. (*b*, Charles Steele)

the total potential energy of the system, a quantity known as **enthalpy, *H*.** Because energy can be conveniently measured as heat, enthalpy is often referred to as the *heat content* of the system.

Free energy is available to do cellular work

Entropy and enthalpy are related by a third dimension of energy, termed **free energy,** which can be expressed in kilojoules or kilcalories per mole. Free energy is *available energy;* it is the component of the total energy of a system that is available to do work under defined conditions of constant temperature and pressure. Changes in pressure are not generally important in biochemistry, but changes in temperature are. For our purposes, then, we can consider the free energy of a system to be the *maximum amount of energy that is available to do work, without a change in temperature.* The requirement that there be no change in temperature is important because heat energy cannot do cellular work. Free energy, the only kind of energy that can do cellular work, is the aspect of thermodynamics of greatest interest to a biologist.

Entropy, represented by the letter *S*, and free energy, represented by *G*, are related inversely; as entropy increases, the amount of free energy decreases. The two are related by the following equation:

$$G = H - TS$$

in which *H* is the enthalpy of the system, *T* is the absolute temperature (expressed in degrees Kelvin; K = °C + 273), and *S* is entropy. If we assume that the entropy is zero, the free energy is simply equal to the total potential energy (enthalpy); entropy reduces the free energy. What, then, is the significance of the temperature (*T*)? Remember that as the temperature increases, there is an increase in random molecular motion that contributes to disorder and multiplies the effect of the entropy term.

Chemical reactions involve changes in free energy

Biologists need ways of analyzing the role of energy in the many reactions that make up cellular metabolism. Although the total free energy of a system (*G*) cannot be effectively measured, the equation $G = H - TS$ is nevertheless useful because it can be extended to predict whether any particular chemical reaction will release en-

Making the Connection

Energy and Diffusion

What is the source of energy for diffusion? In Chapter 5 we saw that randomly moving particles can diffuse *down their own concentration gradient*. That is, although the movements of the individual particles are random, net movement of the group of particles seems to be directional. What provides the energy for this seemingly directed process? A concentration gradient, with a region of higher concentration and another region of lower concentration, is an orderly state.

A cell must expend energy to produce a concentration gradient. Because work must be done to produce this order, the concentration gradient is a form of potential energy. As the particles move about randomly, disorder increases. Although there is no change in enthalpy, entropy increases. The process is spontaneous because there is an overall decrease in free energy (negative ΔG); diffusion is paid for by an increase in entropy. In Chapters 7 and 8 we shall see that the potential energy stored in a concentration gradient can be transformed into chemical energy in ATP. ▲

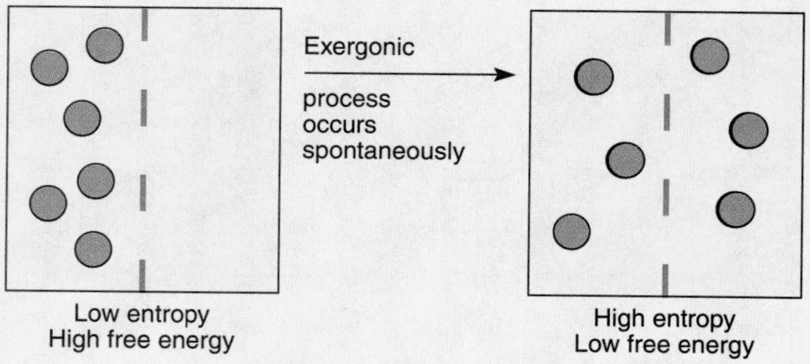

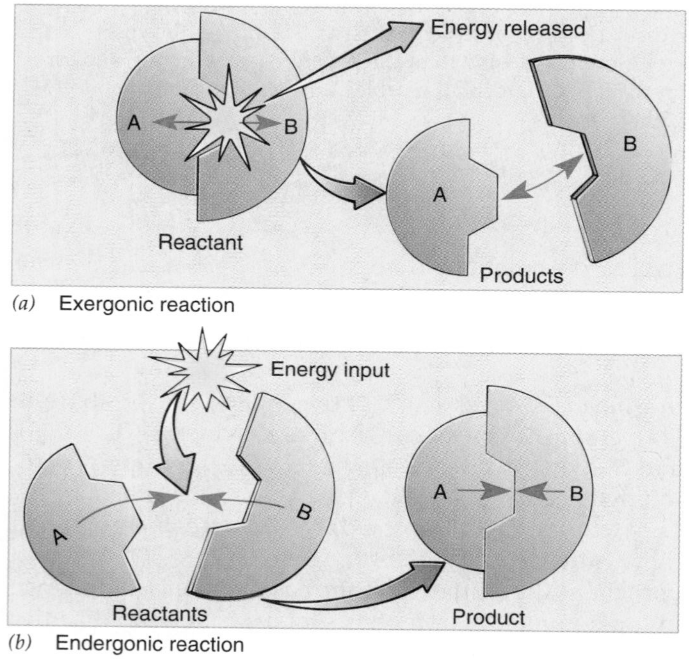

FIGURE 6–6 An exergonic reaction involves a loss of free energy; free energy is gained in an endergonic reaction. The reaction shown is exergonic in one direction and endergonic in the opposite direction. (***a***) In an exergonic reaction there is a net loss of free energy; the products have less free energy than was present in the reactants. (***b***) In endergonic reactions there is a net gain in free energy; the products have more free energy than was present in the reactants. An endergonic reaction occurs only if energy lost from some other system is fed into the reaction.

ergy or require an input of energy. This is because *changes* in free energy can be measured.

We use the Greek letter delta (Δ) to denote any change that occurs in the system between its *initial state* (the state before the reaction) and its *final state* (its state after the reaction). To express what happens with respect to energy in a chemical reaction, the equation becomes:

$$\Delta G = \Delta H - T\Delta S$$

Notice that the temperature does not change; it is held constant during the reaction. Thus, the change in free energy (ΔG) during the reaction is equal to the change in enthalpy (ΔH) minus the product of the absolute temperature (T) multiplied by the change in entropy (ΔS). ΔG and ΔH are expressed in kilojoules or kilocalories per mole; ΔS is expressed in kilojoules per degree or in kilocalories per degree.

Exergonic reactions do not require outside energy

In accordance with the second law of thermodynamics, no chemical reaction is 100% efficient. No reaction can take place without a decrease in enthalpy, an increase in entropy, or both (see *Making the Connection: Energy and Diffusion*). For this reason, the total free energy of the system in its final state is always less than the total free energy of the system in its initial state. When calculated in this way, ΔG, whether expressed in kilojoules per mole or kilocalories per mole, is a negative number. Such a reaction, with a negative value of ΔG ($-\Delta G$), is referred to as an **exergonic reaction** (Figs. 6–6*a* and 6–7*a*).

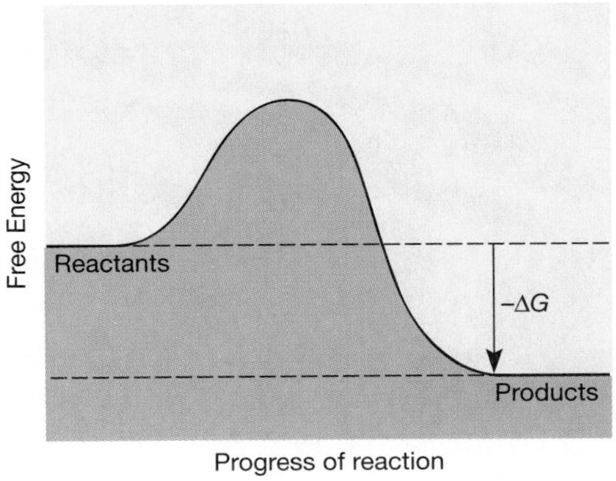

(*a*)

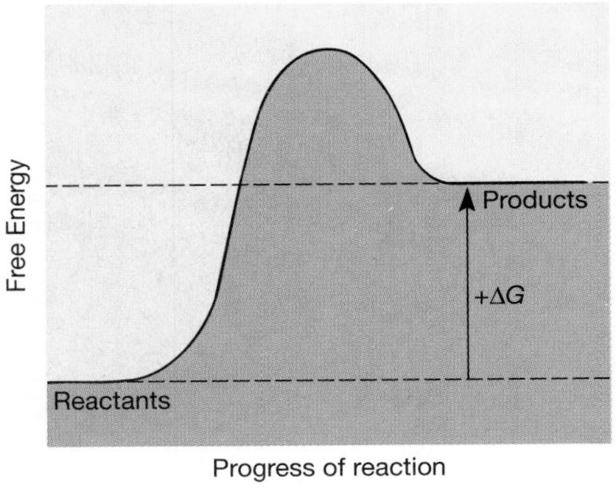

(*b*)

FIGURE 6–7 Exergonic and endergonic reactions can be distinguished by comparing the free energy change that takes place during the reaction. (***a***) In exergonic reactions free energy is released, and the product has less energy than the reactants. (***b***) In endergonic reactions there is a net input of free energy, so the products contain more energy than the reactants. Note that even the exergonic reaction requires some input of energy to get started. This initial investment energy is termed *activation energy*.

An exergonic reaction releases energy and is said to be a **spontaneous** or a "downhill" reaction. The term *spontaneous* may give the false impression that such reactions are always instantaneous. In fact, spontaneous reactions do not necessarily occur readily; some are extremely slow. This is because energy, known as activation energy, is required to initiate every reaction, even a spontaneous one.

All reactions have a required energy of activation

The energy necessary to begin a reaction is the required **energy of activation** or **activation energy.** This is the amount of energy needed to break the existing chemical bonds and initiate the reaction. The activation energy may be supplied as heat, which raises the temperature.

For bonds to break, the reacting molecules must come close together and be in the right orientation. The molecules have a certain average kinetic energy that depends on the temperature. As the temperature increases, random motions of the molecules increase. Furthermore, the internal vibrations in each molecule increase, causing the bonds to become less stable and more reactive. The reactants move faster and collide with one another more frequently and more forcefully. Some of these collisions are in the proper orientation and occur with a force sufficient to break the weakened chemical bonds.

For example, molecular hydrogen and molecular oxygen can react explosively to form water:

$$2\ H_2 + O_2 \longrightarrow 2\ H_2O$$

This reaction is spontaneous (exergonic), yet hydrogen and oxygen can be safely mixed as long as all sparks are kept away. This is because the required energy of activation for this particular reaction is relatively high. A tiny spark provides the activation energy that allows a few molecules to react. Their reaction liberates so much heat that the rest react, producing an explosion (Fig. 6–8).

Later in this chapter we will see how cells use enzymes to lower the activation energy barrier for specific reactions and thereby regulate their metabolism.

An endergonic reaction requires an energy source

It is possible to write an equation for an **endergonic reaction**—that is, a reaction in which there is a gain of free energy. Because the free energy of the products is greater than the free energy of the reactants, ΔG has a positive value. Such a reaction (Figs. 6–6*b* and 6–7*b*), cannot actually take place in isolation. Instead, it must occur in such a way that energy can be supplied from the surroundings. Of course, many energy-requiring reactions take place in cells, and, as we shall see, metabolic mechanisms have evolved that supply the energy needed to "drive" these nonspontaneous cellular reactions in a particular direction.

Actual free energy changes depend on the concentrations of reactants and products

In order to compare different reactions, chemists measure free energy changes under a defined set of standard conditions that would never exist in an organism. This standard free energy change, which can be thought of as a measure of the intrinsic free energy difference between the reactants and the products, depends mainly on the difference in their bond energies (enthalpy, *H*). The actual free energy change in a cellular reaction depends only partly on the standard free energy change; it is also af-

FIGURE 6–8 Hydrogen and oxygen combine explosively to form water if the required energy of activation is supplied. The Hindenburg was an airship that used hydrogen gas, which is lighter than air, for buoyancy. On its initial transatlantic voyage in 1937 it exploded upon landing, probably because a small spark overcame the activation energy for the reaction of hydrogen with oxygen from the air. The disaster resulted in the tragic deaths of 35 of the 97 people on board and an additional person on the ground. (Archive Photos)

fected by the relative concentrations of reactants and products.

In most biochemical reactions there is little standard free energy difference between reactants and products. Such reactions are reversible, a fact that is indicated by drawing double arrows ($\rightleftharpoons$) between the reactants and the products.

At the beginning of a reaction, only the reactant molecules may be present. These molecules move about and collide with one another with sufficient energy and in the right orientation to react. Thus the concentration of the reactant molecules decreases while the concentration of the product molecules increases. The product molecules collide more frequently as their concentration increases, and some have sufficient energy to initiate the reverse reaction. The reaction thus proceeds in both directions simultaneously; if undisturbed it could eventually reach a state of *dynamic equilibrium,* in which the rate of the reverse reaction is equal to the rate of the forward reaction. At equilibrium there is no net change in the system; every forward reaction is balanced by a reverse reaction.

Keep in mind that the knowledge that a system is at equilibrium tells us nothing about the relative concentrations of reactants and products. If the reactants have much greater intrinsic free energy than the products, the reaction goes almost to completion; that is, it reaches equilibrium at a point at which most of the reactants have been converted to products. This fact is indicated by drawing the arrows unequally, with the longer arrow pointing to the products ($A \overset{\longrightarrow}{\leftarrow} B$). Reactions in which the reactants have much less intrinsic free energy than the products reach equilibrium at a point where very few of the reactant molecules have been converted to products ($A \overset{\rightarrow}{\longleftarrow} B$). If the arrows are equal ($A \rightleftharpoons B$), we assume that there is very little standard free energy difference between the reactants and the products and that their concentrations are therefore about equal at equilibrium.

When a reaction is at equilibrium, even if the standard free energy difference between the reactants and products is great, the actual free energy difference is zero. Therefore, *a system at equilibrium can do no work.* The system will stay at equilibrium unless an external stress is applied to it, at which time the system reacts to partially alleviate that stress. According to this rule, known as **Le Chatelier's principle,** a change that affects the reacting system may cause the equilibrium to shift temporarily until a new equilibrium is established.

When little intrinsic free energy difference exists between the reactants and the products ($A \rightleftharpoons B$), the overall direction of the reaction is determined mainly by their relative initial concentrations. This concept, which is actually part of Le Chatelier's principle, is sometimes referred to as the **Law of Mass Action.**

If we increase the initial concentration of A, then the equilibrium will "shift to the right" and more A will be converted to B. A similar effect can be obtained if B is removed from the reaction mixture. An opposite effect (a

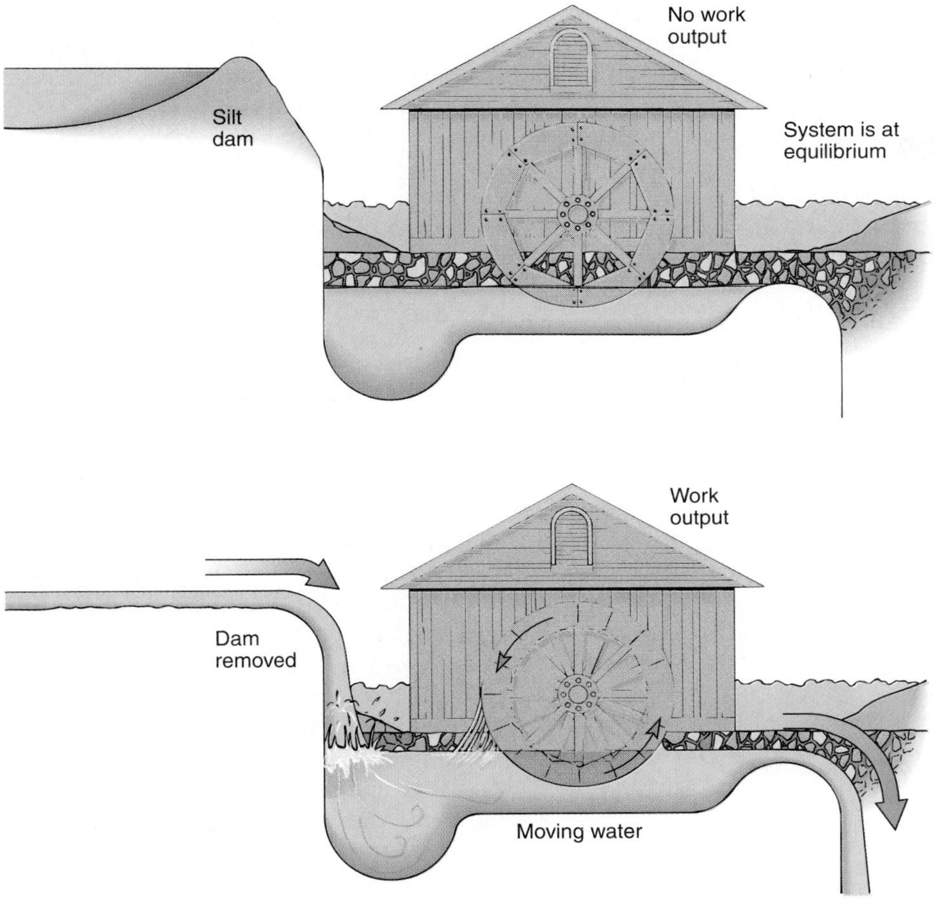

FIGURE 6–9 A system at equilibrium can do no work.

"shift to the left") occurs if the concentration of B is increased, or if A is removed. The actual free energy change that occurs during a reaction is defined mathematically to include these mass action effects, which are a consequence of the relative initial concentrations of reactants and products.

A cell maintains its reactions far from equilibrium

Mass action effects are of profound importance in biology. The cell uses its ordered components and energy derived from the surroundings to manipulate the relative concentrations of reactants and products of almost every reaction. Cellular reactions are virtually never at equilibrium. *By displacing its reactions far from equilibrium, a cell is able to supply energy to endergonic reactions and direct its metabolism in accordance with its needs* (Fig. 6–9).

Cells drive endergonic reactions by coupling them to exergonic reactions

Many metabolic reactions in a cell—protein synthesis, for example—are anabolic and endergonic. Because an endergonic reaction cannot take place without an input of energy, endergonic cellular reactions are **coupled** to exergonic cellular reactions. In energy coupling, the thermodynamically favorable exergonic reaction provides the energy required to drive the thermodynamically unfavorable endergonic reaction. The endergonic reaction can proceed only if it absorbs less free energy than is released by the exergonic reaction to which it is coupled.

Mass action effects play important roles in energy coupling

Consider the standard free energy changes, ΔG, in the following reactions:

$$(1)\ A \longrightarrow B \qquad \Delta G = +16.7\ \text{kJ/mole}\ (+4\ \text{kcal/mole})$$
$$(2)\ B \longrightarrow C \qquad \Delta G = -41.8\ \text{kJ/mole}\ (-10\ \text{kcal/mole})$$

Reaction 1, with a positive value of ΔG, is endergonic. Its product, B, is the reactant in Reaction 2. Reaction 2 has a negative value of ΔG and is exergonic.

In analyzing free energy changes we consider the initial and final states of reactions. If we combine reactions 1 and 2 we obtain:

$$(1)\ A \longrightarrow B \qquad \Delta G = +16.7\ \text{kJ/mole}\ (+4\ \text{kcal/mole})$$
$$(2)\ B \longrightarrow C \qquad \Delta G = -41.8\ \text{kJ/mole}\ (-10\ \text{kcal/mole})$$
$$A \longrightarrow C \qquad \Delta G = -25.1\ \text{kJ/mole}\ (-6\ \text{kcal/mole})$$

The pathway for the conversion of Reactant A to the final product, C, is exergonic overall. Keep in mind that the relative initial concentrations of reactants and products can have a significant effect on the actual free energy change. Reaction 2, which is highly exergonic, can, under the right conditions, dramatically lower the concentration of Product B of Reaction 1, thus shifting that reaction to the right. This mass action effect can be sufficient to make the actual free energy change of Reaction 1 negative ($-\Delta G$), such that it becomes thermodynamically favorable.

Reactions can be coupled through an "energized intermediate"

Consider the standard free energy change, ΔG, in the following reaction:

$$(3)\ A \longrightarrow B \qquad \Delta G = +20.9 \text{ kJ/mole } (+5 \text{ kcal/mole})$$

Because ΔG has a positive value, we know that the product of this reaction has more free energy than the reactant. This is an endergonic reaction. It is not spontaneous and does not take place without an energy source.

By contrast, consider the following reaction:

$$(4)\ C \longrightarrow D \qquad \Delta G = -33.5 \text{ kJ/mole } (-8 \text{ kcal/mole})$$

The negative value of ΔG tells us that the standard free energy of the reactant is greater than the free energy of the product. This exergonic reaction can proceed spontaneously.

We can sum up Reactions 3 and 4 as follows:

$$\begin{array}{ll} (3)\ A \longrightarrow B & \Delta G = +20.9 \text{ kJ/mole } (+5 \text{ kcal/mole}) \\ (4)\ C \longrightarrow D & \Delta G = -33.5 \text{ kJ/mole } (-8 \text{ kcal/mole}) \\ \hline \quad \text{Overall} & \Delta G = -12.6 \text{ kJ/mole } (-3 \text{ kcal/mole}) \end{array}$$

Because thermodynamics considers the overall changes in these two reactions, which show a net negative value of ΔG, the two reactions taken together are exergonic.

The fact that we can write reactions this way is a useful bookkeeping device, but it does not mean that an exergonic reaction can mysteriously transfer energy to an endergonic "bystander" reaction. However, these reactions can be coupled if their pathways are altered such that they are linked by a common intermediate. It is important to remember that the free energy change in a reaction depends only on the difference in free energy between the reactant(s) and the product(s). For this reason, changing the reaction pathway does not alter the overall change in free energy. Reactions 3 and 4 might be coupled in the following way:

$$\begin{array}{ll} (5)\ A + C \longrightarrow I & \Delta G = -8.4 \text{ kJ/mole } (-2 \text{ kcal/mole}) \\ (6)\ I \longrightarrow B + D & \Delta G = -4.2 \text{ kJ/mole } (-1 \text{ kcal/mole}) \\ \hline \quad \text{Overall} & \Delta G = -12.6 \text{ kJ/mole } (-3 \text{ kcal/mole}) \end{array}$$

"I" refers to a transitory common intermediate. It can be called an "energized" intermediate because it is unstable and can enter into reactions that would be impossible for Reactant A. Note that Reactions 5 and 6 are sequential. Thus the reaction pathways have changed, but overall the reactants and products are the same and the overall free energy change is the same.

Generally, for each endergonic reaction occurring in a living cell, there is a coupled exergonic reaction to drive it. Often, the exergonic chemical reaction involves the breakdown of adenosine triphosphate (ATP). We will examine specific examples of the role of ATP in energy coupling in the following section.

ADENOSINE TRIPHOSPHATE (ATP) IS THE ENERGY CURRENCY OF THE CELL

In all living cells, energy is temporarily packaged within a remarkable chemical compound called **adenosine triphosphate (ATP),** which holds readily available energy for very short periods of time. We may think of ATP as the energy currency of the cell. When you work to earn money, you might say that your energy is symbolically stored in the money you earn. The energy the cell requires for immediate use is temporarily stored in ATP, which is like cash. When you earn extra money, you might deposit some in the bank; similarly, a cell might deposit energy in the chemical bonds of lipids, starch, or glycogen. Moreover, just as you dare not make less money than you spend, so too the cell must avoid energy bankruptcy, which would mean its death. Finally, just as you (alas) do not keep what you make very long, so too the cell continuously spends its ATP, which must be replaced immediately.

The ATP molecule has three main parts

ATP is a nucleotide consisting of three main parts (Fig. 6–10): (1) a nitrogen-containing organic base, adenine; (2) ribose, a five-carbon sugar; and (3) three phosphate groups, identifiable as phosphorus atoms surrounded by oxygen atoms. Notice that the phosphate groups are bonded to the end of the molecule in a series, rather like three cars behind a locomotive, and, like the cars of a train, they can be attached and detached.

ATP can donate energy through the transfer of a phosphate group

The bonds linking the phosphate groups of ATP can be broken by hydrolysis. This is due to the fact that the three phosphate groups tend to repel one another because they are negatively charged at cellular pH. The special nature of these particular phosphate bonds is commonly signified by drawing each as a wavy line (Fig. 6–10). Traditionally such bonds were referred to as "high-energy" or "energy-rich" bonds, but these terms are misleading. As discussed earlier, the energy of a bond is the energy re-

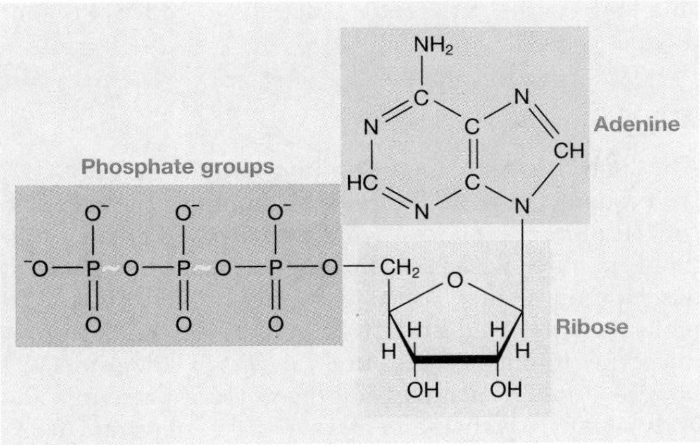

FIGURE 6–10 ATP, the energy currency of all living things, is composed of adenine, ribose, and three phosphate groups. The bonds attaching the two terminal phosphate groups of ATP, referred to as *phosphate bonds,* are shown as wavy yellow lines. When a phosphate bond is hydrolyzed (an exergonic reaction), the phosphate group can be readily transferred to another molecule, together with some of the energy of the bond.

quired to break it; these bonds are easily broken. ATP itself is sometimes called a "high-energy" compound because the hydrolysis reaction that releases a phosphate has a relatively large negative value of ΔG. More properly, the ATP molecule can be said to have a *high group transfer potential* because its terminal phosphate group (and some of its energy) is readily transferred to other molecules.

When the terminal phosphate is removed from ATP, the remaining molecule is **adenosine diphosphate (ADP).** If the phosphate group is not transferred to another molecule, it is released as inorganic phosphate (P_i). This is an exergonic reaction:

(7) $ATP + H_2O \longrightarrow ADP + P_i$
$\Delta G = -32$ kJ/mole (or -7.6 kcal/mole)[1]

The free energy released in Reaction 7 is "wasted" as heat. In energy coupling, however, some of this energy can be transferred and used to drive an endergonic reaction.

Consider the following endergonic reaction, in which the disaccharide sucrose is formed from two monosaccharides, glucose and fructose (see Chapter 3).

(8) Glucose + Fructose $\longrightarrow$ Sucrose + H_2O
$\Delta G = +27$ kJ/mole (or $+6.5$ kcal/mole)

With a free energy change of -32 kJ/mole (-7.6 kcal/mole), the hydrolysis of ATP in Reaction 7 can drive Reaction 8, but only if the reactions can be coupled through a common intermediate, with the help of one or more specific enzymes. The following series of reactions is a simplified version of an alternative pathway used by some bacteria. (Note: eukaryotes do not produce sucrose in this way.)

(9) Glucose + ATP $\longrightarrow$ Glucose-P + ADP
(10) Glucose-P + Fructose $\longrightarrow$ Sucrose + P_i

Reaction 9 is a **phosphorylation reaction,** one in which a phosphate group is transferred to some other compound. Glucose is phosphorylated to form glucose-P, which stands for glucose phosphate, the intermediate that links the two reactions. Glucose-P, which corresponds to "I" in Reactions 5 and 6, reacts exergonically with fructose to form sucrose. For energy coupling to work in this way, Reactions 9 and 10 must occur in sequence. It is convenient to summarize the reactions in the following way:

(11) Glucose + Fructose + ATP $\longrightarrow$ ADP + P_i + Sucrose
$\Delta G = -5$ kJ/mole (-1.2 kcal/mole)

When encountering an equation written in this way, it is important to keep in mind that it is actually a summary of a series of reactions, and that transitory intermediate products are sometimes not shown. A generalized example of a phosphorylation reaction is illustrated in Figure 6–11.

ATP links exergonic and endergonic reactions

Phosphorylation, the transfer of a phosphate group from ATP to some other compound, can be coupled to ender-

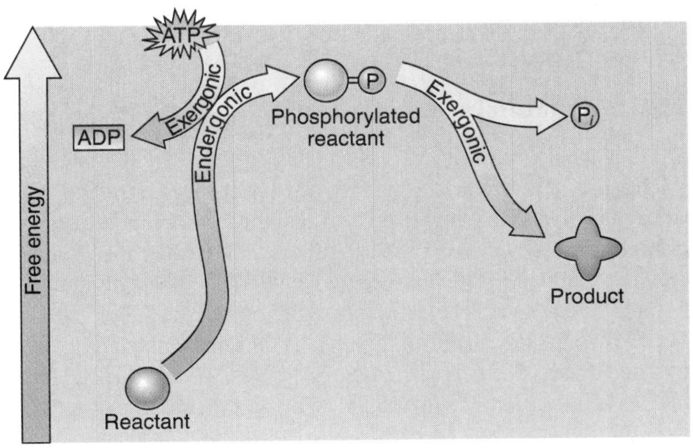

FIGURE 6–11 ATP transfers energy by transferring a phosphate group to a reactant. The reaction sequence is catalyzed by one or more specific enzymes. For simplicity, not all participants in the reaction are shown.

[1]Calculations of the standard free energy of ATP hydrolysis vary somewhat, but range between about -28 and -37 kJ/mole (-6.8 to -8.7 kcal/mole).

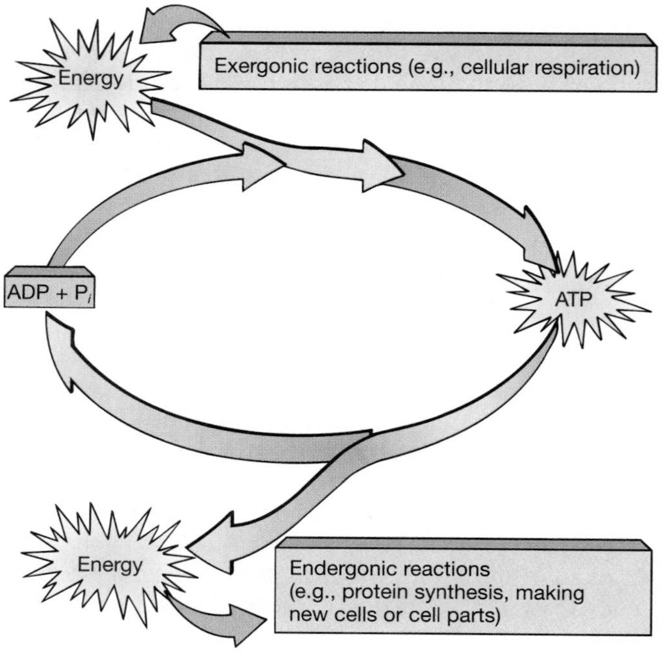

FIGURE 6–12 Because ATP is responsible for coupling many endergonic and exergonic reactions, it is an important link between anabolism and catabolism in living cells.

gonic processes in the cell with the help of specific enzymes. Conversely, energy is required to add a phosphate group to AMP (forming ADP) or to ADP (forming ATP).

$$\text{AMP} + \text{P}_i + \text{Energy} \longrightarrow \text{ADP}$$
$$\text{ADP} + \text{P}_i + \text{Energy} \longrightarrow \text{ATP}$$

These endergonic reactions take place only if they are coupled to certain strongly exergonic reactions. Thus ATP occupies an intermediate position in the energy metabolism of the cell and is an important link between exergonic reactions, which are generally components of catabolic pathways, and endergonic reactions, which are generally part of anabolic pathways (Fig. 6–12).

The cell maintains a very high ratio of ATP to ADP

Although the standard free energy change for the hydrolysis of ATP is relatively high, the actual free energy change under cellular conditions is even higher (actual $\Delta G = -42$ to -50 kJ/mole or -10 to -12 kcal/mole). This is because the cell maintains the ratio of ATP to ADP far from the equilibrium point. ATP is constantly being formed from ADP and inorganic phosphate as nutrients are oxidized (see Chapter 7) or as the radiant energy of sunlight is trapped in photosynthesis (see Chapter 8). At any point in time a typical cell contains more than 10 ATP molecules for every ADP molecule. The fact that the cell maintains the ATP concentration at such a high level (relative to the concentration of ADP) makes its hydrolysis reaction even more strongly exergonic, and more able to drive the endergonic reactions to which it is coupled.

ATP cannot be stockpiled

Although the cell maintains a high ratio of ATP to ADP, large quantities of ATP cannot be stored in the cell. The concentration of ATP is always very low, less than 1 millimole per liter. In fact, studies suggest that a bacterial cell has no more than a 1-second supply of ATP. Thus, ATP molecules are used almost as quickly as they are produced. A human at rest uses about 45 kilograms (99 pounds) of ATP each day, but the amount present in the body at any given moment is less than 1 gram (0.035 ounces). Every second in every cell an estimated 10 million molecules of ATP are made from ADP and phosphate, and an equal number transfer their phosphate groups (and thereby their energy) to whatever life processes may require them (Fig. 6–13).

The extensive and rapid formation of ATP is coupled to the exergonic breakdown of the primary long-term

FIGURE 6–13 The chemical energy of ATP is converted to light energy in the light organs of this black sea dragon, *Idioacanthus antrostomus*. The luminous "fishing lure" attracts prey. (Norbert Wu/Peter Arnold, Inc.)

storage molecules of a cell, the lipids and polysaccharides.

CELLS TRANSFER ENERGY BY REDOX REACTIONS

We have seen that cells can transfer energy through the transfer of a phosphate group from ATP. Energy can also be transferred through the transfer of electrons. As discussed in Chapter 2, *oxidation* is the chemical process in which a substance loses electrons, whereas *reduction* is the complementary process in which a substance gains electrons. Because electrons released during an oxidation reaction cannot exist in the free state in living cells, every oxidation reaction must be accompanied by a reduction reaction, in which the electrons are accepted by another atom, ion, or molecule. Because oxidation and reduction reactions are simultaneous, they are often called *redox reactions.* The substance that becomes oxidized gives up energy as it releases electrons, and the substance that becomes reduced receives energy as it gains electrons.

A series of redox reactions takes place in cells as electrons are transferred from one compound to another. These electron transfers, which are equivalent to energy transfers, are an essential part of cellular respiration, photosynthesis, and many other chemical reactions.

Most electron carriers carry hydrogen atoms

Generally it is not easy to remove one or more electrons from a covalent compound; it is much easier to remove a whole atom. For this reason, biological redox reactions usually involve the transfer of a hydrogen atom rather than just an electron. A hydrogen atom contains a proton (which does not participate in the oxidation/reduction) and an electron. (*Note*: Redox reactions can involve hydrogen *atoms*; hydrogen ions do not contain electrons and are not involved in redox reactions.)

Electron carriers transfer energy

When an electron, either singly or as part of a hydrogen atom, is removed from an organic compound, it takes with it some of the energy stored in the chemical bond of which it was a part. That electron, along with its energy, is transferred to an acceptor molecule. An electron progressively loses free energy as it is transferred from one acceptor to another.

One of the most frequently encountered hydrogen acceptors is **nicotinamide adenine dinucleotide,** more conveniently referred to as **NAD^+.** When it becomes reduced, it temporarily stores large amounts of free energy. Here is a generalized equation showing the transfer of hydrogen from a compound we call *X* to NAD^+:

$$XH_2 + NAD^+ \longrightarrow \underset{\text{Oxidized}}{X} + \underset{\text{Reduced}}{NADH} + H^+$$

Note that the NAD^+ becomes reduced when it combines with hydrogen. NAD^+ is an ion with a net charge of +1. When two electrons and one proton are added, the charge is neutralized and the reduced form of the compound, **NADH,**[2] is produced (Fig. 6–14). Some of the energy stored in the bonds holding the hydrogens to molecule X has been transferred by this redox reaction and is temporarily held by NADH. When NADH transfers the electrons to some other molecule, some of their energy is transferred as well (Fig. 6–15). This energy is usually transferred through a complex series of reactions that result in the formation of ATP (see Chapter 7).

Nicotine adenine dinucleotide phosphate ($NADP^+$) is a hydrogen acceptor that is chemically similar to NAD^+ but with an extra phosphate, which is not directly involved in energy transfers. Unlike NADH, the reduced form of $NADP^+$ (abbreviated **NADPH**) is not involved in ATP synthesis. Instead, its electrons are used more directly to provide energy for certain reactions, including certain essential reactions of photosynthesis (see Chapter 8).

Other important hydrogen acceptors or electron acceptors include **flavin adenine dinucleotide (FAD)** and the **cytochromes.** FAD is a nucleotide that accepts hydrogens and their electrons; its reduced form is **$FADH_2$.** The cytochromes are proteins that contain iron; the iron component accepts electrons from hydrogens and then transfers these electrons to some other compound. Like NAD^+ and $NADP^+$, FAD and the cytochromes are electron transfer agents. Each can exist in a reduced state in which it has more free energy, or in an oxidized state in which it has less. Each is an essential component of many cellular redox reaction sequences.

ENZYMES ARE CHEMICAL REGULATORS

The principles of thermodynamics help us predict whether or not a reaction can occur, but they tell us nothing about the speed of the reaction. The breakdown of glucose is an exergonic reaction, yet a glucose solution keeps virtually indefinitely in a bottle if kept free of bacteria and molds and not subjected to high temperature or strong acids or bases. Cells cannot wait for centuries for glucose to break down, nor can they use extreme conditions to cleave glucose molecules. Cells regulate the rates of chemical reactions with **enzymes,** protein **catalysts**[3] that affect the speed of a chemical reaction without being consumed by the reaction.

---

[2]Although the correct way to write the reduced form of NAD^+ is $NADH + H^+$, for simplicity we will present the reduced form as NADH in this and succeeding chapters.

[3]In recent years scientists have learned that protein enzymes are not the only cellular catalysts; some types of RNA molecules have catalytic activity as well (see Chapter 12).

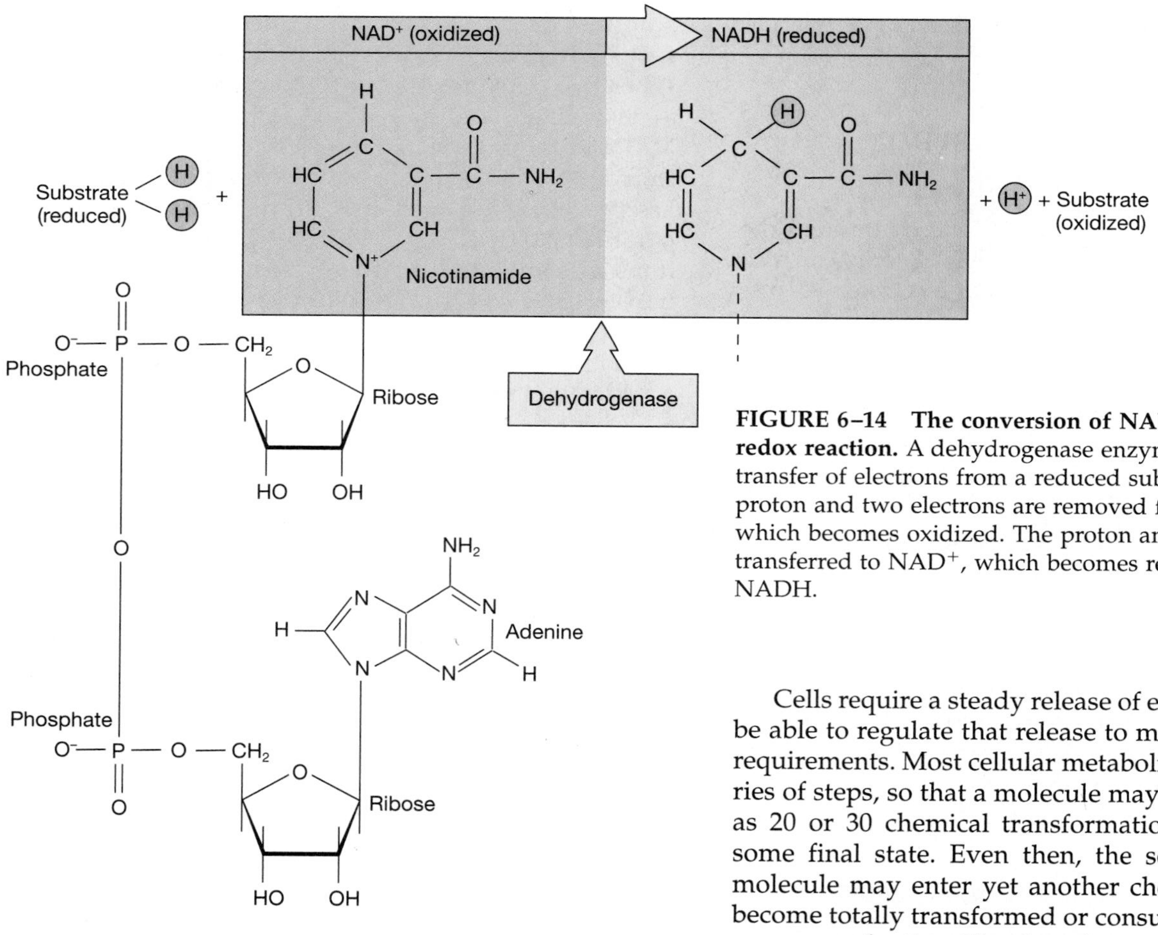

FIGURE 6–14 The conversion of NAD$^+$ to NADH is a redox reaction. A dehydrogenase enzyme is required for the transfer of electrons from a reduced substrate to NAD$^+$. A proton and two electrons are removed from the substrate, which becomes oxidized. The proton and two electrons are transferred to NAD$^+$, which becomes reduced to form NADH.

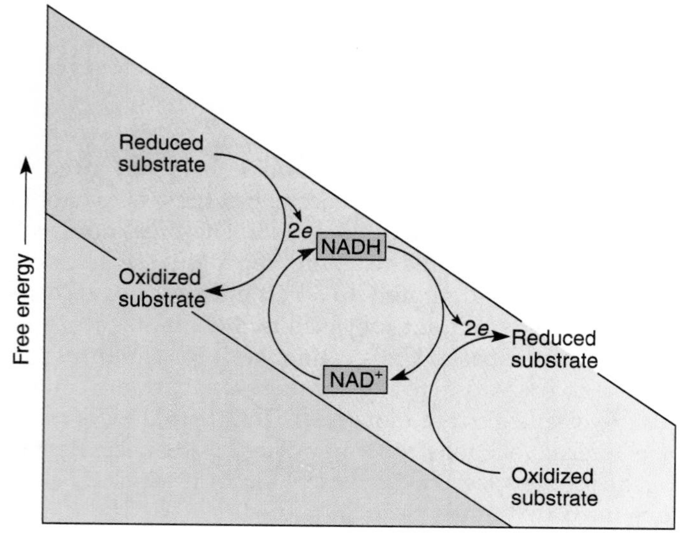

FIGURE 6–15 This is a short redox chain involving NAD$^+$ and NADH. Each component of the chain can exist in a reduced form, with more free energy, and an oxidized form, with less free energy. The free energy of the electrons decreases with each successive transfer. The proton that is transferred as part of a hydrogen, but is not involved in the redox reaction, is not shown.

Cells require a steady release of energy, and they must be able to regulate that release to meet metabolic energy requirements. Most cellular metabolism proceeds by a series of steps, so that a molecule may go through as many as 20 or 30 chemical transformations before it reaches some final state. Even then, the seemingly completed molecule may enter yet another chemical pathway and become totally transformed or consumed in the course of energy production. The changing needs of the cell require a system of flexible metabolic chemical control. The key directors of this control system are the remarkable enzymes (see *Making the Connection: Energy and Information*).

An enzyme lowers the activation energy needed to initiate a chemical reaction

As do all catalysts, enzymes affect the rate of a reaction by lowering the energy needed to initiate the reaction. As discussed previously, even a strongly exergonic reaction, one that releases substantial quantities of energy as it proceeds, may be prevented by an energy barrier. This barrier, known as the *activation energy,* is the energy required to break the existing bonds and begin the reaction. An enzyme greatly reduces the activation energy necessary to initiate a chemical reaction (Fig. 6–16).

In a population of molecules of any kind, some have a relatively high energy content, while others have a lower energy content. The energy content of the entire population of molecules conforms to a bell-shaped curve of normal distribution (Fig. 6–17). Only molecules with a relatively high energy content are likely to react to form the product. If molecules need less energy to react (i.e., if the activation barrier is lowered), a larger fraction of the

(Text continues on page 160)

Making the Connection

Energy and Information

What makes cellular energy transformations possible? As the fundamental unit of life, a cell transforms energy obtained from the environment to useful forms, and then uses that transformed energy to displace its chemical reactions far from equilibrium. This is essential because a chemical reaction at equilibrium can do no work; a biochemical equilibrium is the equivalent of death. The ordered components of the cell provide the informational system that directs the kinds of energy transformations that occur. For example, the ordered structure of the active site of an enzyme controls the way in which the catalyzed reaction takes place. By greatly speeding up certain chemical reactions (and "ignoring" others), enzymes express the information that dictates which reactions dominate the metabolism of the cell. As long as suitable external energy sources are available, the information inherent in the order of the cell allows the cell to make energy available when and where it is needed.

But, of course, we have come full circle. Energy is required to do the work of providing the entire informational system of the cell. The ordered structures of the cell represent an incredibly improbable state of matter; much of the energy obtained by a cell is used to produce and maintain this improbable state. Thus a cell maintains its low entropy state at the expense of externally supplied energy. ▲

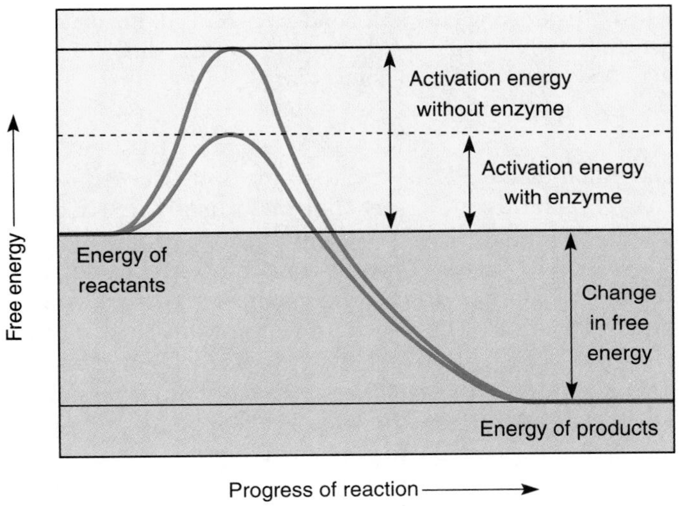

FIGURE 6–16 A catalyst such as an enzyme speeds up a reaction by lowering its activation energy. A catalyzed reaction proceeds more quickly than an uncatalyzed reaction because it has a lower activation energy barrier to overcome.

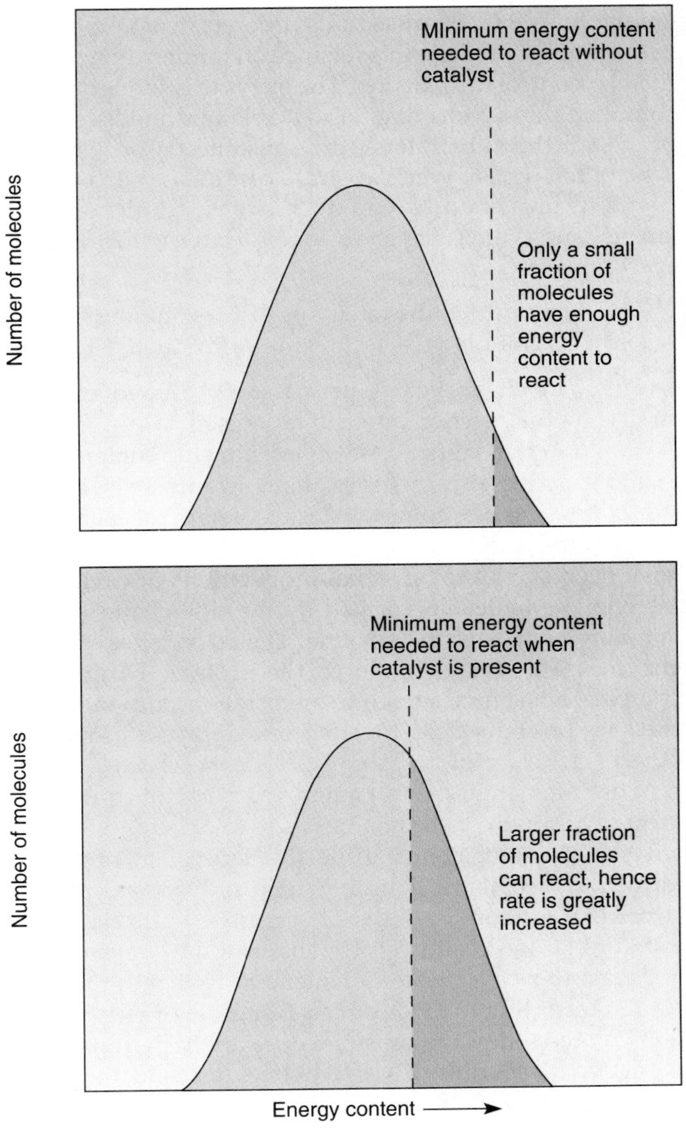

FIGURE 6–17 The kinetic energy in a population of molecules follows a normal distribution. (*a*) If the reaction is uncatalyzed, only a small fraction of the molecules have sufficient energy to react. (*b*) A catalyst lowers the activation energy and increases the fraction of the molecules that can react.

reactant population reacts at any one time. An enzyme lowers the requirement; it reduces the required activation energy for the particular reaction, allowing it to proceed more quickly.

An enzyme has no effect on free energy change

As can be seen in Figure 6–16, an enzyme lowers the activation energy for a reaction but has no effect on the overall free energy change. An enzyme can only promote a chemical reaction that could proceed without it. No catalyst can cause a reaction to proceed in a thermodynamically unfavorable direction, or influence the final concentrations of reactants and products if the reaction goes to equilibrium. Enzymes simply speed up reaction rates.

An enzyme works by forming an enzyme-substrate complex

An uncatalyzed reaction depends on random collisions among reactants. Because of its ordered structure, an enzyme is able to reduce this reliance on random events and thereby control the reaction. The enzyme is thought to accomplish this by forming an unstable intermediate complex with the **substrate(s),** the substance(s) on which it acts. When the **enzyme-substrate complex** (called the ES complex) breaks up, the product is released; the original enzyme molecule is regenerated and is free to form a new ES complex.

$$\text{Enzyme} + \text{Substrate(s)} \longrightarrow \text{ES complex}$$
$$\text{ES complex} \longrightarrow \text{Enzyme} + \text{Product(s)}$$

The enzyme itself is not permanently altered or consumed by the reaction and can be reused.

As shown in Figure 6–18, every enzyme contains one or more **active sites,** regions that can interact with the substrate. The ES complex forms when the substrate binds to the active site. The active sites of some enzymes have been shown to be actual indentations or cavities in the enzyme molecule, formed by the side chains of certain amino acids of the enzyme. The active sites of most enzymes are located close to the surface. During the course of a reaction, substrate molecules occupying these sites are brought close together and react with one another.

The interaction of an enzyme and its substrate was formerly thought to be analogous to a lock-and-key. The enzyme was envisioned to be the equivalent of a preformed molecular lock into which only a specifically shaped molecular key—the substrate—would fit. However, unlike a lock and key, the shape of the enzyme does not seem to be exactly complementary to that of the substrate. According to the **induced-fit model** of enzyme action favored today, the substrate does not fit perfectly into the active site. Because the active site is not rigid, the binding of the substrate induces a change in the shape of the enzyme molecule. Usually the shape of the substrate also changes slightly, in a way that may distort its chemical bonds (Fig. 6–18).

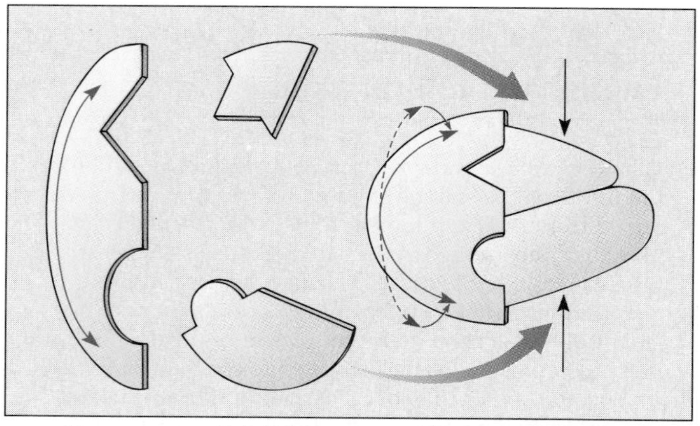

FIGURE 6–18 The interaction between the active site of an enzyme and its substrate is best described by the induced-fit model. Chemical reactions are favored when substrate molecules get close enough to one another to react, when they are presented to each other in the right orientation, and when their existing chemical bonds are strained. Enzymes generally facilitate all three of these processes. Thus the existing bonds break and new bonds form.

Why does the enzyme-substrate complex break up into chemical products different from those that participated in its formation? The substrate is bound in close *proximity* to the atoms with which it can now react. Furthermore, the highly ordered *orientations* of the reactants make them likely to react in a specific way. The proximity and orientation of the reactants, together with strains in their critical bonds, facilitate the breakage of old bonds and the formation of new ones. Thus the substrate is changed into product, which randomly moves away from the enzyme. The enzyme is then free to catalyze the reaction of more substrate molecules to form more product.

Enzymes are very efficient catalysts

The catalytic ability of some enzymes is truly remarkable. For example, hydrogen peroxide (H_2O_2) breaks down extremely slowly if the reaction is uncatalyzed, but a single molecule of the enzyme **catalase** brings about the decomposition of 5 million molecules of hydrogen peroxide per minute at 0°C! Catalase protects cells because hydrogen peroxide is a poisonous substance produced as a byproduct of a number of cellular reactions (Fig. 6–19).

Most enzyme names end in *-ase*

Enzymes are usually named by the addition of the suffix *-ase* to the name of the substance acted upon. For exam-

FIGURE 6–19 A bombardier beetle uses the catalyzed decomposition of hydrogen peroxide as a defense mechanism. The oxygen gas formed in the decomposition ejects water and other chemicals with explosive force. Because the reaction releases a great deal of heat, the water comes out as steam. The beetle shown in this photo is immobilized by a wire attached to its back by a drop of adhesive. (Thomas Eisner and Daniel Aneshansley/Cornell University)

ple, sucrose is split into glucose and fructose by the enzyme **sucrase.** Note that a few enzymes retain traditional names that do not end in *-ase.* Some of these end in *-zyme;* lysozyme, for example, an enzyme found in tears and saliva, breaks down bacterial cell walls. Other examples of enzymes with traditional names include pepsin and trypsin, which break internal peptide bonds in proteins.

Enzymes are specific

Virtually every chemical reaction that takes place in an organism is catalyzed by an enzyme. Because there is a close relationship between the shape of the active site and the shape of the substrate, the majority of enzymes are highly specific. Most are capable of catalyzing only a few closely related chemical reactions or, in many cases, only one particular reaction. For example, the enzyme urease, which decomposes urea to ammonia and carbon dioxide, attacks no other substrate. The enzyme sucrase splits only sucrose; it does not act on maltose or lactose.

A few enzymes are specific only to the extent that they require the substrate to have a certain kind of chemical bond. For example, the lipase secreted by the pancreas splits the ester linkages connecting the glycerol and fatty acids of a wide variety of fats.

Enzymes that catalyze similar reactions are classified into groups, although each particular enzyme in the group may catalyze only one specific reaction. Some of the important classes of enzymes and their roles are listed in Table 6–1. Each class is divided into many subclasses.

Table 6–1 SOME IMPORTANT CLASSES OF ENZYMES

Enzyme Class	*Function*
Oxidoreductases	Catalyze oxidation-reduction reactions
Transferases	Catalyze the transfer of a functional group from a donor molecule to an acceptor molecule
Hydrolases	Catalyze hydrolysis reactions
Isomerases	Catalyze conversion of a molecule from one isomeric form to another
Ligases	Catalyze certain reactions in which two molecules are joined
Lyases	Catalyze certain reactions in which double bonds are formed or broken.

For example, sucrase, mentioned above, is referred to as a glycosidase because it cleaves a glycosidic linkage. Glycosidases are a subclass of the hydrolases.

Many enzymes require cofactors

Some enzymes—for example, pepsin, secreted by the stomach—consist only of protein. Others have two components: a protein referred to as the **apoenzyme** and an additional chemical component called a **cofactor.** Neither the apoenzyme nor the cofactor alone has catalytic activity; only when the two are combined does the enzyme function. A cofactor may be inorganic, or it may be an organic molecule.

Some enzymes require a specific metal ion as a cofactor. Two very common inorganic cofactors are magnesium ions and calcium ions. Most of the trace elements—such as iron, copper, zinc, and manganese, which are required in very small amounts—function as cofactors.

An organic, nonpolypeptide compound that binds to the apoenzyme and serves as a cofactor is called a **coenzyme.** Most coenzymes can be categorized as *transfer agents*—that is, agents that transfer some component from one molecule to another. Some examples of coenzymes have already been introduced in this chapter. NADH, NADPH, and $FADH_2$ are coenzymes; they transfer electrons. ATP functions as a coenzyme; it is responsible for transferring phosphate groups. Yet another coenzyme, coenzyme A, is involved in the transfer of groups derived from organic acids. Most vitamins (organic compounds that an organism requires in small amounts but cannot synthesize itself) are coenzymes or components of coenzymes (see Table 45–4).

Enzymes are most effective at optimal conditions

Enzymes generally work best under certain narrowly defined conditions referred to as *optima*. These include appropriate temperature, pH, and ion concentration. Any departure from optimal conditions adversely affects enzyme activity.

Temperature affects enzyme activity

Most enzymes have an optimal temperature at which the rate of reaction is fastest. For human enzymes, the temperature optima are near body temperature (35° to 40°C). Enzymatic reactions occur slowly or not at all at low temperatures. The rates of most enzyme-controlled reactions increase as the temperature increases, within limits (Fig. 6–20*a*). High temperatures rapidly inactivate most enzymes. The molecular conformation of the protein becomes altered as the hydrogen bonds responsible for its secondary, tertiary, and quaternary structures are broken. This inactivation is usually not reversible; that is, activity is not regained when the enzyme is cooled.

Most organisms are killed by even short exposure to high temperature; their enzymes are inactivated, and they are unable to continue metabolism. A few remarkable exceptions to this rule exist: certain species of bacteria can survive in the waters of hot springs, such as those in Yellowstone Park, where the temperature is almost 100°C; these organisms are responsible for the brilliant colors in the terraces of the hot springs. Still other bacteria live at temperatures much above that of boiling water near deep-sea vents, where the extreme pressure keeps water in its liquid rather than its steam form (see Chapter 23 and *Focus On: Life Without the Sun* in Chapter 53).

Each enzyme has an optimal pH

Most enzymes have an optimal pH and are active only over a narrow pH range. The optimal pH for most human enzymes is between 6 and 8 (Fig. 6–20*b*). Pepsin, a protein-digesting enzyme secreted by cells lining the stomach, is remarkable in that it works only in a very acid medium—optimally at pH 2. In contrast, trypsin, the protein-splitting enzyme secreted by the pancreas, functions best under slightly basic conditions.

The activity of an enzyme may be markedly changed by any alteration in pH, which in turn alters charges on the enzyme. Changes in charge affect the ionic bonds that contribute to tertiary and quaternary structure, thus changing the protein's conformation and activity. Many enzymes become inactive, and usually irreversibly denatured, when the medium is made very acidic or very basic.

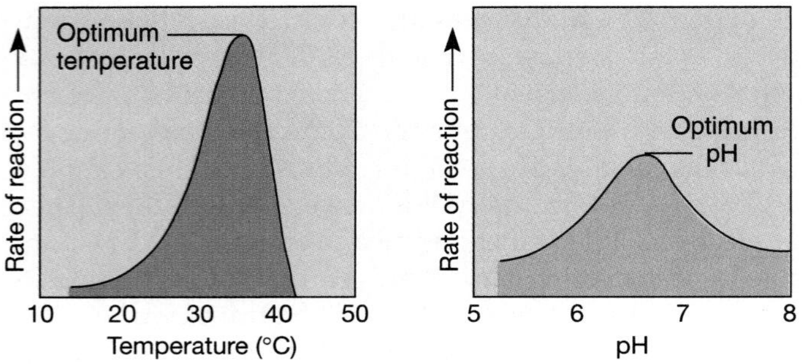

(*a*) (*b*)

FIGURE 6–20 Every enzyme has an optimal temperature (*a*) and an optimal pH (*b*). Substrate and enzyme concentrations are held constant in the reactions illustrated.

Enzymes are organized into teams in metabolic pathways

Enzymes play an essential role in energy coupling because they usually work in sequence, with the product of one enzyme-controlled reaction serving as the substrate for the next. We can picture the inside of a cell as a factory with many different assembly (and disassembly) lines operating simultaneously. An assembly line is composed of a number of enzymes. Each enzyme carries out one step, such as changing molecule A into molecule B. Then molecule B is passed along to the next enzyme, which converts it into molecule C, and so on. Such a series of reactions is referred to as a **metabolic pathway.**

$$A \xrightarrow{\text{Enzyme 1}} B \xrightarrow{\text{Enzyme 2}} C$$

Each of these reactions is theoretically reversible, and the fact that it is catalyzed by an enzyme does not change that fact. An enzyme does not itself determine the direction of the reaction it catalyzes. However, the overall reaction sequence is portrayed as proceeding from left to right. You will recall that if there is little intrinsic free energy difference between the reactants and products for a particular reaction, its direction will be determined mainly by the relative concentrations of reactants and products.

In biological pathways, both intermediate and final products are often removed and converted to other chemical compounds. Such removal drives the sequence of reactions in a particular direction. Let us assume that Reactant A is being constantly supplied and that its concentration remains constant. Enzyme 1 converts Reactant A to Product B. The concentration of B is always lower than the concentration of A, because B is removed as it is converted to C in the reaction catalyzed by enzyme 2. If C is removed as quickly as it is formed (perhaps by leaving the cell) the entire pathway is "pulled" toward C.

The cell regulates enzymatic activity

Enzymes regulate the chemistry of the cell, but what controls the enzymes? One mechanism depends simply on controlling the amount of enzyme produced. The syn-

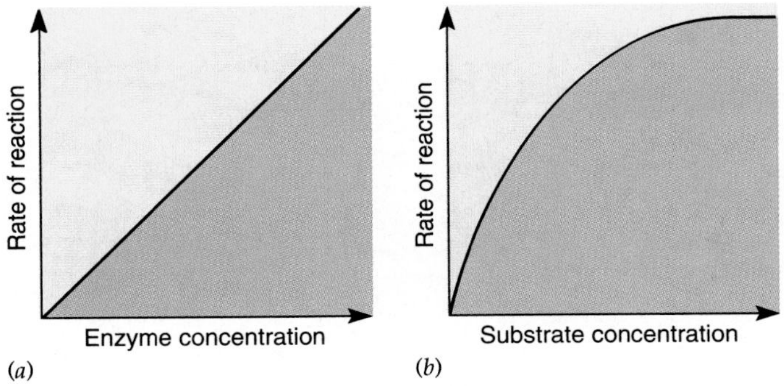

FIGURE 6–21 The rate of a reaction catalyzed by an enzyme is influenced by enzyme concentration or by substrate concentration. (*a*) Reaction rate as a function of the enzyme concentration. In this example it is assumed that an excess of substrate is present at all times. The rate of the reaction is therefore directly proportional to the enzyme concentration. (*b*) Reaction rate as a function of the concentration of substrate. The enzyme concentration is constant in this example. If the substrate concentration is relatively low, then the reaction rate is directly proportional to substrate concentration. However, higher substrate concentrations do not increase the reaction rate. The curve levels off because the enzyme molecules become saturated with substrate.

thesis of each type of enzyme is directed by a specific gene. The gene, in turn, may be switched on by a signal from a hormone or by some other type of cellular product (see Chapters 13 and 47). When the gene is switched on, the enzyme is synthesized. The amount of enzyme present then influences the rate of the reaction.

If the pH and temperature are kept constant, the rate of the reaction can be affected by the substrate concentration or by the enzyme concentration. If an excess of substrate is present, the enzyme concentration is the rate-limiting factor. The initial rate of the reaction is then directly proportional to the concentration of enzyme present (Fig. 6–21*a*).

If the enzyme concentration is kept constant, the initial rate of an enzymatic reaction is proportional to the concentration of substrate present. Substrate concentration is the rate-limiting factor at lower concentrations; the rate of the reaction is therefore directly proportional to the substrate concentration. However, at higher substrate concentrations the enzyme molecules become saturated with substrate, and increasing the substrate concentration does not increase the reaction rate (Fig. 6–21*b*).

The product of one enzymatic reaction may control the activity of another enzyme, especially in a complex sequence of enzymatic reactions. For example, in the following system,

$$A \xrightarrow{\text{Enzyme 1}} B \xrightarrow{\text{Enzyme 2}} C \xrightarrow{\text{Enzyme 3}} D \xrightarrow{\text{Enzyme 4}} E$$

each step is catalyzed by a different enzyme. The final product, E, may inhibit the activity of Enzyme 1. When the concentration of E is low, the sequence of reactions proceeds rapidly. However, an increasing concentration of E serves as a signal for Enzyme 1 to slow down and eventually to stop functioning. Inhibition of Enzyme 1 stops this entire sequence of reactions. This type of enzyme regulation, in which the formation of a product inhibits an earlier reaction in the sequence, is called **feedback inhibition.**

Another important method of enzymatic control depends on the activation of enzyme molecules. In their inactive form the active sites of the enzyme are inappropriately shaped, so that the substrates do not fit. Among the factors that influence the shape (conformation) of the enzyme are pH, the concentration of certain ions, and the addition of phosphate groups to certain amino acids in the enzyme.

Some enzymes, known as **allosteric enzymes,** possess a receptor site, called an **allosteric site,** on some region of the enzyme molecule other than the active site. (The word *allosteric* means "another space.") Substances that affect enzyme activity by binding to allosteric sites are called **allosteric regulators.** Some regulators are *inhibitors* that keep the enzyme in its inactive shape. Others are *activators* that stabilize the active shape of the enzyme (the shape with a functional active site).

The enzyme *cyclic AMP-dependent protein kinase* is an allosteric enzyme with a regulator that is an inhibitory protein that binds reversibly to the allosteric site and inactivates the enzyme. Protein kinase is in this inactive form most of the time (Fig. 6–22). When protein kinase activity is needed, the compound cyclic AMP (cAMP; see Fig. 3–29) contacts the enzyme-inhibitor complex and removes the inhibitor, thereby activating the protein kinase. Activation of protein kinases by cyclic AMP is an important aspect of the mechanism of action of certain hormones (see Chapter 47).

Enzymes can be inhibited by certain chemical agents

Most enzymes may be inhibited (so that their activity is decreased) or even destroyed by certain chemical agents. Enzyme inhibition may be reversible or irreversible.

Reversible inhibition occurs when an inhibitor forms weak chemical bonds with the enzyme. Reversible inhibition can be competitive or noncompetitive. In **competitive inhibition,** the inhibitor competes with the normal substrate for binding to the active site of the enzyme (Fig. 6–23). Usually a competitive inhibitor is structurally similar to the normal substrate and so fits into the active site and combines with the enzyme. However, it is not similar enough to substitute fully for the normal substrate in the chemical reaction, and the enzyme cannot attack it to form reaction products. A competitive inhibitor occupies the active site only temporarily and does not permanently damage the enzyme. In competitive inhibition, an active site is occupied by the inhibitor part of the time and by

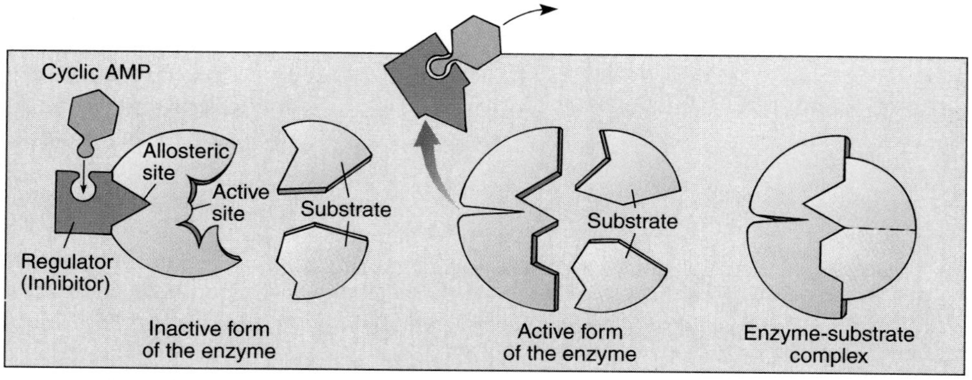

FIGURE 6–22 The enzyme protein kinase is inhibited by a regulatory protein that binds reversibly to its allosteric site. When the enzyme is in this inactive form, the shape of the active site is modified so that the substrate cannot combine with it. Cyclic AMP removes the allosteric inhibitor and activates the enzyme. The substrate can then combine with the active site.

the normal substrate part of the time. If the concentration of the substrate is increased relative to the concentration of the inhibitor, the active site will usually be occupied by the substrate. Competitive inhibition can be recognized experimentally by the fact that it can be reversed by increasing the substrate concentration.

In **noncompetitive inhibition,** the inhibitor binds with the enzyme at a site other than the active site. Such an inhibitor inactivates the enzyme by altering its shape so that the active site cannot bind with the substrate. Many important noncompetitive inhibitors are metabolic substances that regulate enzyme activity by combining reversibly with the enzyme. Noncompetitive inhibition has some features in common with allosteric inhibition discussed previously.

In **irreversible inhibition,** an inhibitor permanently inactivates or destroys an enzyme when it combines with one of its functional groups. Many poisons are irreversible enzyme inhibitors. For example, heavy metals such as mercury and lead bind irreversibly to and denature many proteins, including enzymes. Certain nerve gases poison the enzyme acetylcholinesterase, which is

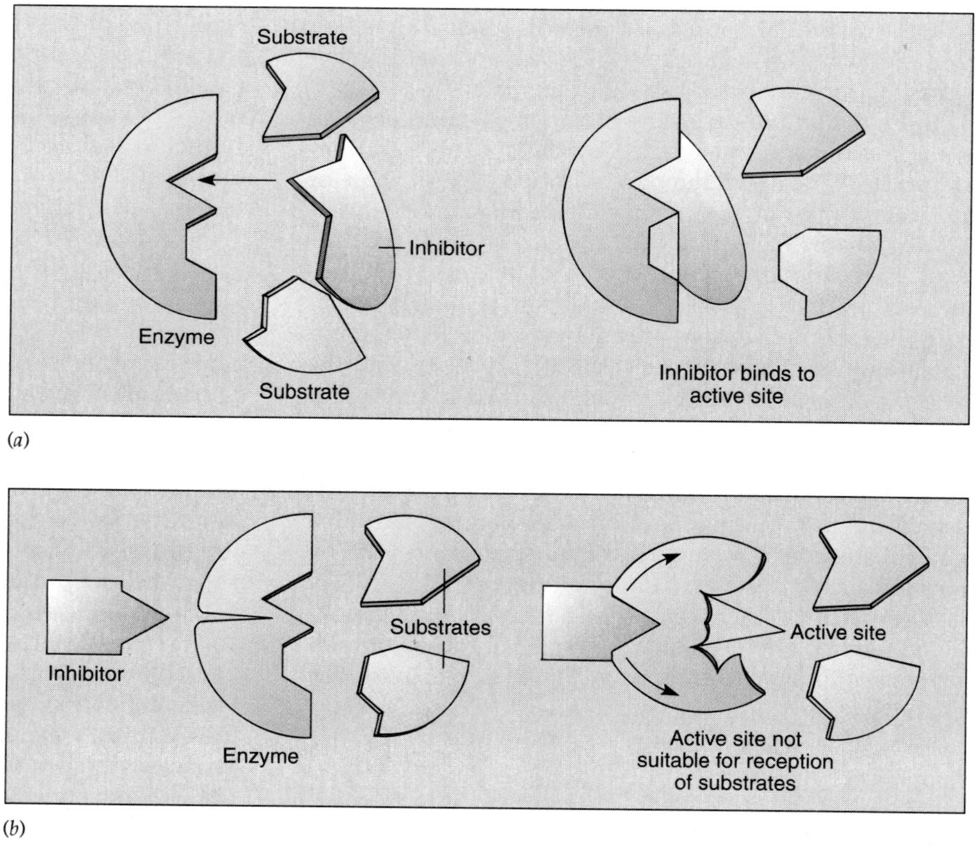

FIGURE 6–23 The distinction between competitive and noncompetitive inhibition rests on whether or not the active site of the enzyme is involved. (*a*) In competitive inhibition, the inhibitor competes with the normal substrate for the active site of the enzyme. A competitive inhibitor occupies the active site only temporarily. (*b*) In noncompetitive inhibition, the inhibitor binds with the enzyme at a site other than the active site, altering the shape of the enzyme and thereby inactivating it. Noncompetitive inhibition may be reversible. Allosteric regulation, used by cells to control enzyme action, is a somewhat similar process (see Fig. 6–22).

Focus On

Enzyme Inhibition and Antibacterial Drugs

Many bacterial infections are treated with drugs that directly or indirectly inhibit bacterial enzyme activity. For example, sulfa drugs have a chemical structure similar to that of the nutrient para-aminobenzoic acid (PABA). When PABA is available, microorganisms can synthesize the vitamin folic acid. Humans do not synthesize folic acid from PABA, and that is why sulfa drugs selectively affect bacteria.

When a sulfa drug is present, competitive inhibition occurs within the bacterium: the drug competes with PABA for the active site of the bacterial enzyme. When the bacteria use the sulfa drug instead of PABA, they synthesize a compound that has a structure somewhat similar to that of folic acid, a coenzyme. However, this imposter folic acid does not work as a coenzyme. Instead, it competitively inhibits the enzyme's action so that the bacteria are unable to make needed amino acids and nucleotides.

Penicillin and related antibiotics irreversibly inhibit a bacterial enzyme called transpeptidase. This enzyme is responsible for establishing some of the chemical linkages in the bacterial cell wall (see Chapter 23). Unable to produce properly constructed cell walls, cytoplasm spills out and susceptible bacteria are prevented from multiplying effectively (see figure). Human cells do not have cell walls and do not employ this enzyme. Thus, except for individuals allergic to it, penicillin is harmless to humans. Unfortunately, during the years since it was introduced, resistance to penicillin has evolved in many bacterial strains. The resistant bacteria fight back with an enzyme of their own, penicillinase, which breaks down the penicillin and renders it ineffective. Because bacteria evolve at such a high rate, drug resistance is a growing problem in medical practice. Although new antibacterial drugs are constantly under development, certain serious infections, such as tuberculosis, are becoming increasingly difficult to treat. ■

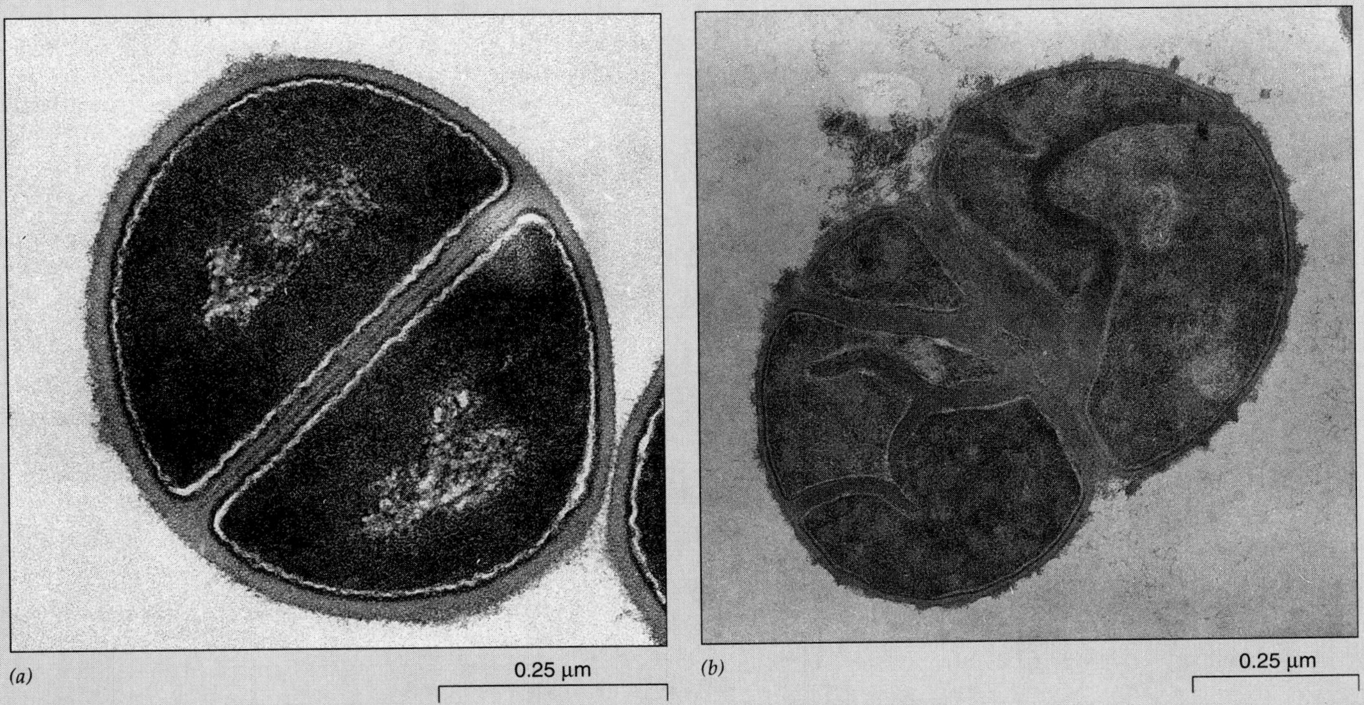

Penicillin is an irreversible enzyme inhibitor. (*a*) Normal bacteria, showing the new cell wall laid down between daughter cells during cell division. (*b*) Penicillin has damaged these bacterial cell walls. (Courtesy of Drs. Victor Lorian and Barbara Atkinson, with permission of *The American Journal of Clinical Pathology*)

important to the function of nerves and muscles. Cytochrome oxidase, one of the enzymes that transports electrons in cellular respiration, is especially sensitive to cyanide. Death results from cyanide poisoning because cytochrome oxidase is irreversibly inhibited and can no longer transfer electrons from substrate to oxygen. A number of insecticides and drugs are irreversible enzyme inhibitors (see *Focus on: Enzyme Inhibition and Antibacterial Drugs*). Irreversible inhibition may also occur if a protein is denatured by heat or organic solvents.

Enzymes can do damage if they function inappropriately

Enzymes themselves can act as poisons if they are active in the wrong compartment of a cell or of the body. As little as 1 milligram of trypsin can kill a rat if injected intravenously. Some enzymes are synthesized as inactive **precursors;** they are then converted to their active form after being transported to a location where they can function safely. Trypsin and chymotrypsin, proteolytic digestive enzymes produced by the pancreas, are synthesized as precursors that are somewhat larger than the active enzyme. These precursors are packaged in granules and secreted into the duct of the pancreas. Because the precursors are inactive, the pancreas is not digested by the enzymes it synthesizes. The inactive enzymes are made active by other enzymes that cleave off a portion of the precursor molecule to yield the active enzyme (see Chapter 45). Acute pancreatitis, a serious, sometimes fatal disease, occurs when the proteolytic enzymes become active while still within the pancreas and digest cells and blood vessels.

SUMMARY

I. All life depends on a continuous input of energy. Most producers capture energy during photosynthesis and incorporate some of it into the chemical bonds of organic compounds. Some of this chemical energy then becomes available to consumers and decomposers.

II. All forms of energy are interconvertable.
   A. Energy can be defined as the capacity to do work (expressed in kilojoules, kJ). Cells do mainly mechanical, chemical, and transport work.
   B. Potential energy is stored energy; kinetic energy is energy of motion.
   C. Energy can be conveniently measured as heat energy; the unit of heat energy is the kilocalorie (kcal), which is equal to 4.184 kilojoules. Heat energy cannot do cellular work.

III. The first law of thermodynamics states that energy can be neither created nor destroyed but can be transferred and changed in form. The second law of thermodynamics states that disorder (entropy) in the universe is continuously increasing.
   A. The first law explains why organisms cannot produce energy but must continuously capture it from the surroundings.
   B. The second law explains why no process requiring energy is ever 100% efficient; in every energy transaction, some energy is dissipated as heat, which contributes to entropy.

IV. When a chemical reaction is at equilibrium, the rate of change in one direction is exactly the same as the rate of change in the opposite direction; the system can do no work because the free energy difference between the reactants and products is zero.

V. Reactions that release free energy and can perform work are said to be spontaneous.
   A. An exergonic reaction releases free energy and is spontaneous.
   B. An endergonic reaction requires an input of free energy, which may be supplied by coupling it to an exergonic reaction, often through a common intermediate.

VI. ATP is the immediate energy currency of the cell; it generally transfers energy through the transfer of its terminal phosphate group to acceptor molecules.
   A. ATP is formed by the phosphorylation of ADP, a process that requires an input of energy.
   B. ATP is the common cellular link between exergonic and endergonic reactions, and between catabolism and anabolism.

VII. Energy can be transferred in oxidation-reduction (redox) reactions.
   A. A substance that becomes oxidized gives up one or more electrons (and energy) to a substance that becomes reduced. Electrons are typically transferred as part of hydrogen atoms.
   B. NAD^+ and $NADP^+$ accept electrons and become reduced to form NADH and NADPH, respectively; these electrons (along with some of their energy) can be transferred to other acceptors.

VIII. An enzyme is a biological catalyst; it greatly increases the speed of a chemical reaction without being consumed.
   A. An enzyme lowers the activation energy necessary to get a reaction going.
   B. An active site of an enzyme is a three-dimensional region where substrates come into close contact and thereby react more readily.
   C. Some enzymes consist of an apoenzyme and a cofactor.
      1. Most inorganic cofactors are metal ions.
      2. A coenzyme is an organic cofactor; most coenzymes are agents such as ATP, NADH, and NADPH that transfer a component from one molecule to another.
   D. Enzymes work best at specific temperature and pH optima.
   E. A cell can regulate enzymatic activity by controlling the amount of enzyme produced and by regulating metabolic conditions that influence the shape of the enzyme.
   F. Most enzymes can be inhibited by certain chemical substances. Reversible inhibition may be competitive or noncompetitive.

SELECTED KEY TERMS

activation energy
active site
allosteric site
anabolism
apoenzyme
bond energy
catabolism
catalyst
coenzyme
cofactor
competitive inhibition
coupled reactions
endergonic reaction
energy
enthalpy
entropy
enzyme
exergonic reaction
feedback inhibition
first law of thermodynamics
free energy
heat energy
induced-fit model
irreversible inhibition
kinetic energy
matter
noncompetitive inhibition
potential energy
redox reaction
second law of thermodynamics
spontaneous reaction
substrate
thermodynamics
work

POST-TEST

1. ________ is a change in the state or motion of matter.
2. A particle in motion possesses ________ energy.
3. ________ is the branch of science that deals with energy and its transformations.
4. "Energy may be changed from one form to another but is neither created nor destroyed" is a statement of the ________________ .
5. In thermodynamics, ________ is a measure of amount of disorder in the system.
6. "Physical and chemical processes proceed in such a way that the entropy of the system becomes maximal" is a statement of the ________________ .
7. The ________ energy of a system is that part of the total energy available to do cellular work.
8. A reaction that requires an input of free energy is described as ________ .
9. A reaction that releases energy is ________ .
10. A spontaneous reaction is one in which the change in ________ energy has a negative value.
11. To drive a reaction that requires an input of energy, some reaction that yields energy must be ________ to it.
12. The energy required to initiate a reaction is called ________ energy.
13. A substance that affects the rate of a chemical reaction without being consumed by the reaction is a(an) ________ .
14. ________ are biological catalysts produced by cells.
15. Enzymes and their substrates combine temporarily to form a(an) ________-________ complex.
16. The region of an enzyme molecule that combines with the substrate is the ________ ________ .
17. An organic cofactor is a(an) ________ .
18. A(an) ________ inhibitor binds to the active site of an enzyme.

REVIEW QUESTIONS

1. Why can we express energy either in kilojoules or in kilocalories? Which is more convenient to measure? Which has more meaning in biology?
2. You exert tension on a spring and then release it. Explain how these actions relate to work, potential energy, and kinetic energy.
3. Life is sometimes described as a constant struggle against the second law of thermodynamics. How do organisms succeed in this struggle without violating the second law?
4. Consider the standard free energy change in a reaction in which enthalpy decreases and entropy increases. Is ΔG zero, or does it have a positive value or a negative value? Is the reaction endergonic or exergonic?
5. Why do coupled reactions typically have common intermediates? Give a generalized example involving ATP, and another involving NADH.
6. Why is ATP able to serve as an important link between exergonic and endergonic reactions?
7. What is activation energy? What effect does a catalyst have on the required activation energy?
8. Give the function of each of the following: (a) active site of an enzyme; (b) coenzyme; (c) allosteric site.
9. Describe three factors that influence how an enzyme functions.

YOU MAKE THE CONNECTION

1. Reaction 1 and Reaction 2 happen to have the same standard free energy change: $\Delta G = -41.8$ kJ/mole (-10 kcal/mole). Reaction 1 is at equilibrium, but Reaction 2 is far from equilibrium. Is either reaction capable of performing work? If so, which one?
2. Consider the following reaction:

$$A \rightleftharpoons B$$

Suppose you have a limited amount of Substance A, but you want to convert virtually all of it to Substance B. How might you accomplish this?

3. You are doing an experiment in which you are measuring the rate at which succinic acid is converted to fumaric acid by the enzyme succinic dehydrogenase. You decide to add a little malonic acid to make things interesting. You observe that the reaction rate slows markedly and conclude that malonic acid must be acting as an inhibitor. Design an experiment that will help you decide if malonic acid is acting as a competitive inhibitor or a noncompetitive inhibitor.

RECOMMENDED READINGS

Alberts, B., D. Bray, J. Lewis, M. Raff, K. Roberts, and J.D. Watson. *Molecular Biology of the Cell,* 3rd ed. Garland Publishing, New York, 1994. A detailed, well-written account of cellular energy conversions.

Atkins, P.W. *The Second Law.* W.H. Freeman, San Francisco, 1984. A basic, understandable introduction to thermodynamics with an extensive section devoted to its biological implications.

Garrett, R.H., and C.M. Grisham. *Biochemistry.* Saunders College Publishing, 1995. A comprehensive biochemistry text with good coverage of cellular energetics.

Lodish, H., D. Baltimore, A. Berk, S.L. Zipursky, P. Matsudaira, and J. Darnell. *Molecular Cell Biology,* 3rd ed. Scientific American Books, New York, 1995. A readable and detailed account of energy transformations within the cell.

CHAPTER 7

Energy-Releasing Pathways and Biosynthesis

This American goldfinch will obtain energy from the organic molecules stored in the seeds of the thistle. (George E. Stewart/Dembinsky Photo Associates)

Cells are tiny factories that process materials on the molecular level, through thousands of metabolic reactions. As discussed in Chapter 6, metabolism has two complementary components: **catabolism,** which is the splitting of molecules into smaller components, and **anabolism,** the synthesis of complex molecules from simpler building blocks. Cells carry out many anabolic reactions to produce useful substances that help to maintain the life of the cell or the organism of which it is a part. These materials are required for all life functions, including growth; reproduction; movement; response to stimuli; and homeostasis, the maintenance of a constant internal environment. Many anabolic reactions are endergonic and require an energy source (usually ATP) to drive them.

Thus, cells must continually obtain energy from the environment and use it to synthesize ATP. For example, in photosynthesis, light energy is converted to chemical energy in ATP and other molecules used in energy transfer reactions. Because these molecules are not suitable for long-term energy storage, the cell uses their energy to power the anabolic pathway by which it converts carbon dioxide and water to carbohydrates that can be stored. Thus a cell carrying out photosynthesis ultimately converts light energy to a form that can be stored: the chemical energy in carbohydrates or other organic molecules.

How can the chemical energy stored in an organic molecule be made available to plant cells, or to the cells of consumers such as animals and decomposers? Certain catabolic reactions are the key. Cells use enzymes to catabolize glucose and other organic molecules in a series of controlled steps, thereby converting some of their free energy to free energy in ATP. Cells use three main catabolic pathways to obtain free energy from nutrients: aerobic respiration (which is by far the most common pathway and the main subject of this chapter), anaerobic respiration, and fermentation.

LEARNING OBJECTIVES

After you have studied this chapter you should be able to

1. Write a summary reaction for aerobic respiration, showing which reactant becomes oxidized and which becomes reduced.
2. List and give a brief overview of the four stages of aerobic respiration, indicate where each stage takes place in a eukaryotic cell, and add up the energy captured (as ATP, NADH, and $FADH_2$) in each stage.
3. Draw a diagram illustrating chemiosmosis and explain (1) how a gradient of protons is established across the inner mitochondrial membrane, and (2) the process by which the proton gradient drives ATP synthesis.
4. Summarize how the products of protein and lipid metabolism feed into the same metabolic pathway that oxidizes glucose.
5. Compare and contrast aerobic and anaerobic pathways used by cells to extract free energy from nutrients; include the mechanism of ATP formation, the final electron acceptor, and the end products.
6. Summarize the basic similarities of alcohol and lactate fermentation.
7. Explain why both anabolic and catabolic reactions are essential to cells.

CELLS USE REDOX REACTIONS TO EXTRACT ENERGY FROM NUTRIENTS

A glucose molecule (or some other nutrient) contains a great deal of energy in the electrons that make up its chemical bonds. This energy can be converted to ATP energy in a cell of the plant that made it, or in a cell of an animal or other consumer in a process known as **respiration** or **cellular respiration**[1] (Fig. 7–1). Most eukaryotes and prokaryotes use a form of respiration requiring oxygen and hence carry out **aerobic respiration,** a process that is the major focus of this chapter. Energy-yielding processes that do not require oxygen are discussed later in the chapter.

[1]The term *cellular respiration* is used to distinguish these cellular processes from *organismic respiration,* the exchange of oxygen and carbon dioxide with the environment by relatively complex animals that have special modifications, such as lungs or gills, for gas exchange.

Aerobic respiration is a redox process

During aerobic respiration, nutrients are catabolized to carbon dioxide and water. Most cells of plants, animals, protists, fungi, and bacteria use aerobic respiration to obtain energy from glucose. The overall reaction pathway for the aerobic respiration of glucose is summarized as follows:

$$C_6H_{12}O_6 + 6\ O_2 + 6\ H_2O \dashrightarrow 6\ CO_2 + 12\ H_2O + \text{Energy (as ATP)}$$

Note that water is shown on both sides of the equation; this is because it is a reactant in some reactions, and

FIGURE 7–1 This red-eyed tree frog, its grasshopper prey, as well as the plants (some of which may be eaten by the grasshopper) all carry out aerobic respiration. It is obvious that the frog needs energy for its spectacular leap, but in fact, all of these organisms need ATP produced during aerobic respiration to carry out their life functions (Animals Animals © 1996 Stephen Dalton.)

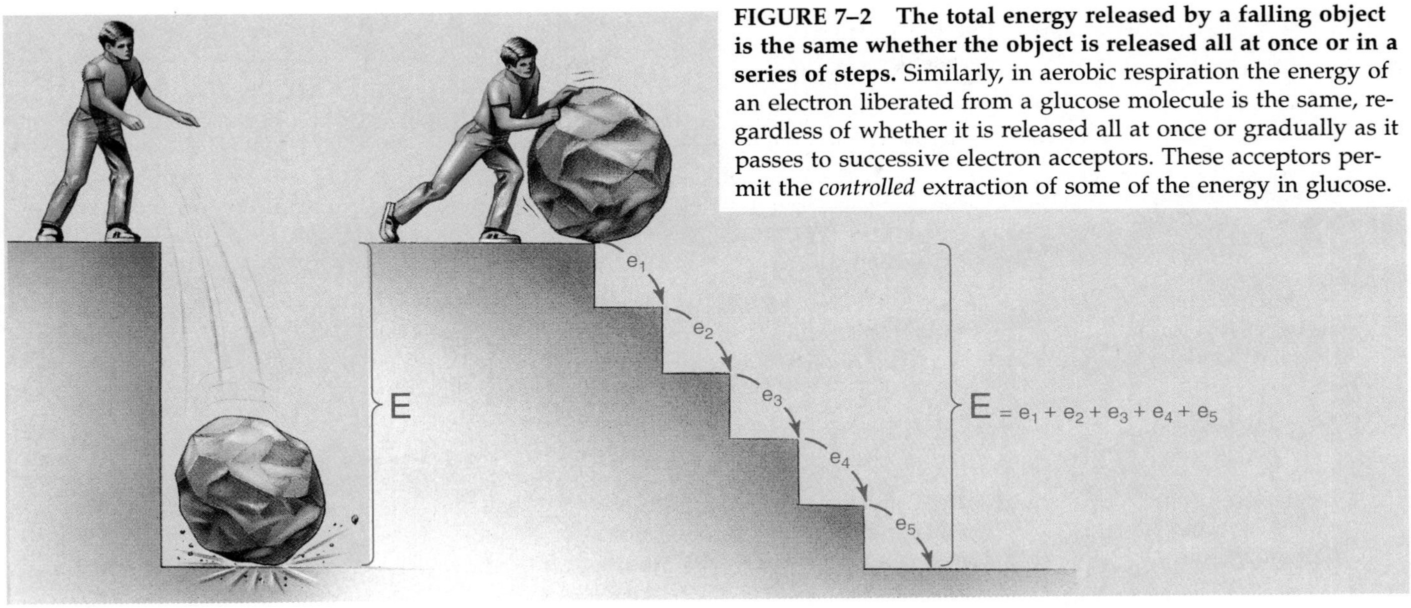

FIGURE 7–2 The total energy released by a falling object is the same whether the object is released all at once or in a series of steps. Similarly, in aerobic respiration the energy of an electron liberated from a glucose molecule is the same, regardless of whether it is released all at once or gradually as it passes to successive electron acceptors. These acceptors permit the *controlled* extraction of some of the energy in glucose.

the product of others. For purposes of discussion, the equation can be simplified to indicate that there is a net yield of water:

oxidation

$$C_6H_{12}O_6 + 6\,O_2 \dashrightarrow 6\,CO_2 + 6\,H_2O + \text{Energy (as ATP)}$$

reduction

If we analyze this summary reaction, it appears that carbon dioxide is produced by the removal of hydrogen atoms from glucose. Conversely, water appears to be formed as the hydrogen atoms are accepted by oxygen. Because the transfer of hydrogen atoms is equivalent to the transfer of electrons, this is a redox process (see Chapter 6).

The products of the reaction would be the same if the glucose were simply burned in the presence of oxygen. However, if cells were to burn glucose, its energy would be released as heat, which would not only be unavailable to the cell but would actually destroy it. For this reason, cells do not transfer hydrogens directly from glucose to oxygen. Aerobic respiration is a redox process in which electrons associated with the hydrogen atoms in glucose are transferred to oxygen in a series of steps. The complete catabolism of glucose requires a long sequence of enzymatic reactions (Fig. 7–2). During this process, the free energy of the electrons is used for ATP synthesis.

AEROBIC RESPIRATION HAS FOUR STAGES

The chemical reactions of the aerobic respiration of glucose can be grouped into four stages (Fig. 7–3; Table 7–1). In eukaryotes, the first stage (glycolysis) takes place in the cytosol; the rest take place inside the mitochondria. Most bacteria also carry out these processes, but because these cells lack mitochondria, all the events take place either in the cytoplasm or in association with the plasma membrane.

1. **Glycolysis.** A six-carbon glucose molecule is converted to two three-carbon molecules of pyruvate,[2] and ATP and NADH are formed.
2. **Formation of acetyl coenzyme A.** Each pyruvate enters a mitochondrion and is oxidized to a two-carbon group (acetate) that combines with coenzyme A, forming acetyl coenzyme A; NADH is produced and carbon dioxide is released as a waste product.
3. **The citric acid cycle.** The acetate of acetyl coenzyme A combines with a four-carbon molecule (oxaloacetate) to form a six-carbon molecule (citrate). In the course of the cycle, citrate is eventually converted to oxaloacetate and carbon dioxide is released as a waste product. Energy is captured as ATP and the reduced, high-energy compounds NADH and $FADH_2$.
4. **The electron transport system and chemiosmosis.** The electrons removed from glucose during the preceding stages are transferred from NADH and $FADH_2$ to a chain of electron acceptor compounds. As the electrons are passed from one electron acceptor to another, some of their energy is used to pump hydrogen ions (protons) across the inner mitochondrial membrane, forming a proton gradient. In a process known as

[2]Pyruvate and many other compounds in glycolysis and the citric acid cycle exist as anions at the pH found in the cell. They sometimes associate with H^+ to form acids. For example, pyruvate forms pyruvic acid. In some textbooks these compounds are presented in the acid form.

Table 7–1 SUMMARY OF AEROBIC RESPIRATION

Phase	*Summary*	*Some Starting Materials*	*Some End Products*
1. Glycolysis (in cytosol)	Series of about ten reactions in which glucose is degraded to pyruvate; net profit of 2 ATPs; hydrogens are transferred to carriers; can proceed anaerobically	Glucose, ATP, NAD^+ ADP, Pi	Pyruvate, ATP, NADH
2. Formation of acetyl CoA (in mitochondria)	Pyruvate is degraded and combined with coenzyme A to form acetyl CoA; hydrogens are transferred to carriers; CO_2 is released	Pyruvate, coenzyme A	Acetyl CoA, CO_2, NADH
3. Citric acid cycle (in mitochondria)	Series of reactions in which the acetyl portion of acetyl CoA is degraded to CO_2; hydrogens are transferred to carriers	Acetyl CoA, H_2O	CO_2, NADH, $FADH_2$, ATP
4. Electron transport and chemiosmosis (in mitochondria)	Chain of several electron transport molecules; H's (or their electrons) are passed along chain; energy released is used to form proton gradient; ATP is synthesized as protons diffuse down the gradient; oxygen is final electron acceptor	NADH, $FADH_2$, oxygen	ATP, H_2O

chemiosmosis, to be described later, the energy of this proton gradient is used to produce ATP.

Most reactions involved in aerobic respiration are one of three types: dehydrogenations, decarboxylations, and those that we will informally categorize as preparation reactions. **Dehydrogenations** are reactions in which two hydrogens (actually, two electrons plus two protons) are removed from the substrate and transferred to a coenzyme such as NAD^+ or FAD, which acts as a primary acceptor. **Decarboxylations** are reactions in which part of a carboxyl group (—COOH) is removed from the substrate as a molecule of CO_2. The carbon dioxide we exhale with each breath is derived from decarboxylations that occur in our cells. The rest of the reactions are preparation reactions in which molecules undergo rearrangements and other changes so that they can subsequently undergo further dehydrogenations or decarboxylations. As we examine the individual reactions of aerobic respiration, we will encounter many examples of these three basic types.

In following the reactions of aerobic respiration, it helps to do some bookkeeping as you go along. Because

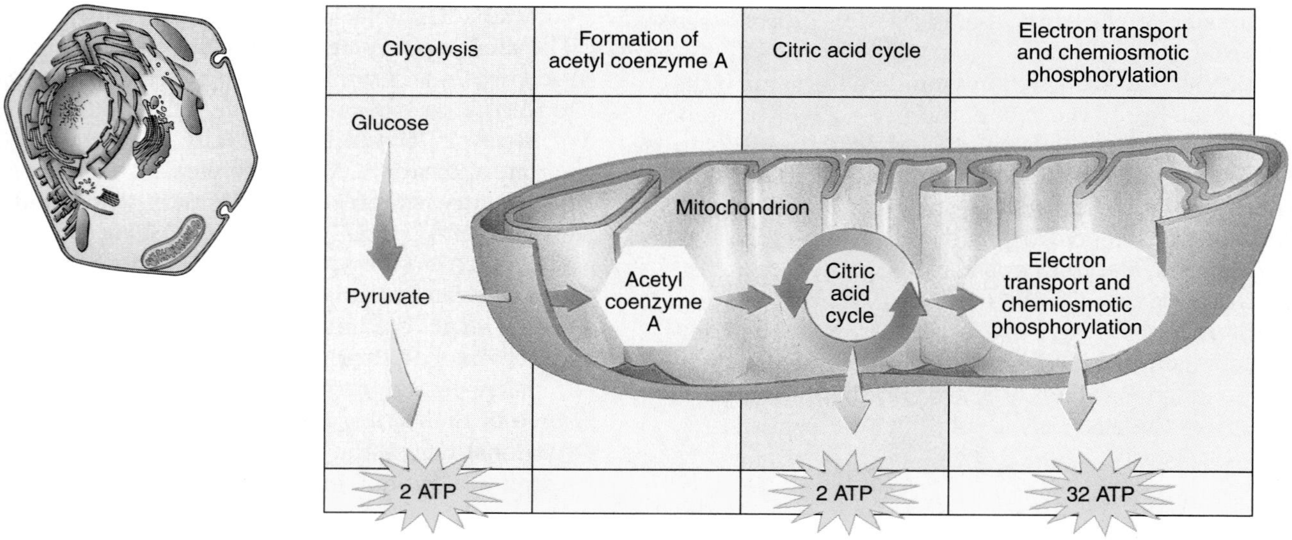

FIGURE 7–3 The four main phases in aerobic respiration are (1) glycolysis, (2) the formation of acetyl coenzyme A from pyruvate, (3) the citric acid cycle, and (4) the electron transport system and chemiosmosis. Glycolysis occurs in the cytosol. Pyruvate, the product of glycolysis, enters a mitochondrion, where cellular respiration continues. Most ATP is synthesized by chemiosmosis.

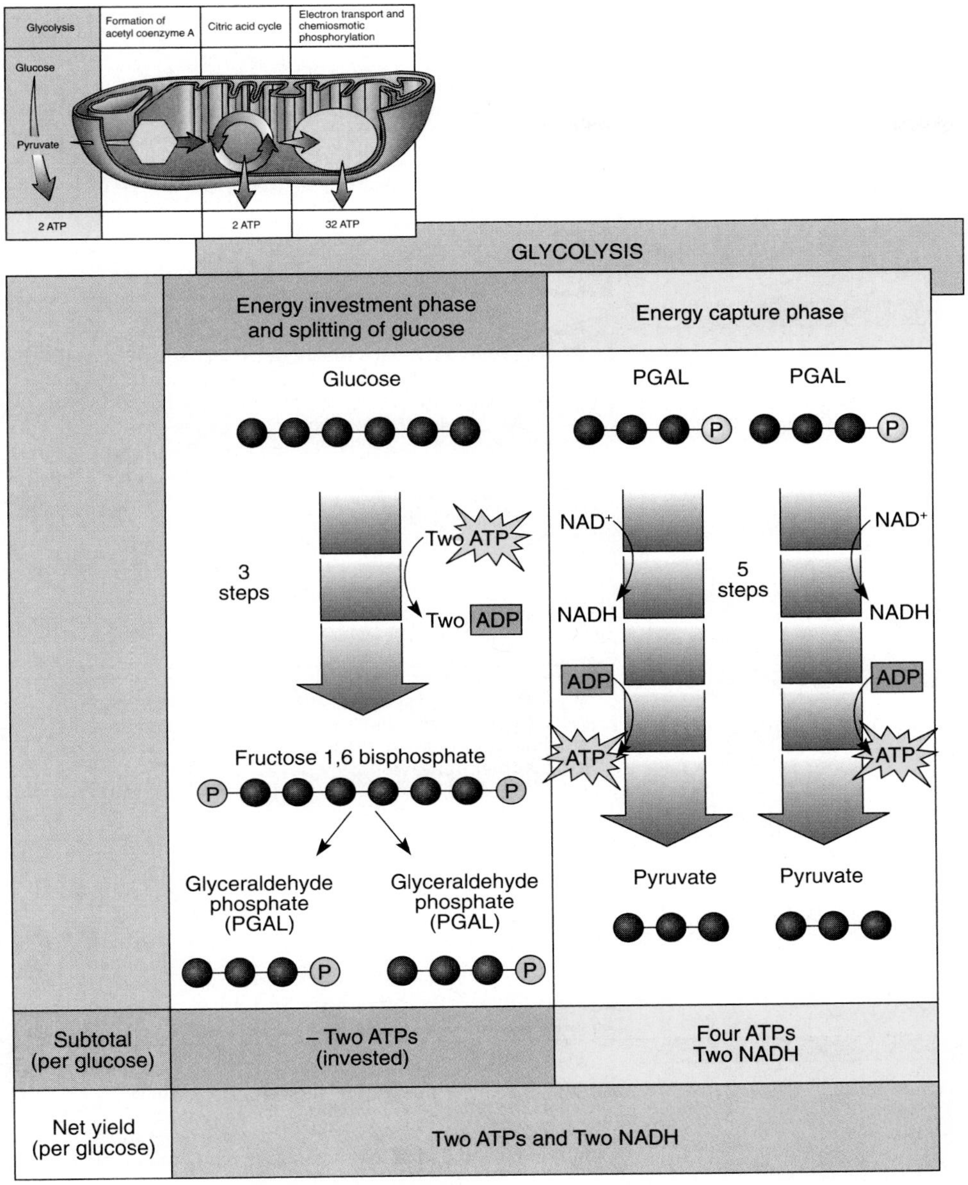

FIGURE 7–4 The energy investment phase of glycolysis leads to the splitting of sugar; ATP and NADH are produced during the energy capture phase. A glucose is converted to two pyruvates, with a net yield of two ATP molecules and two NADH molecules.

glucose is the starting material, it is useful to express changes on a per glucose basis. We will be paying particular attention to changes in the number of carbon atoms per molecule and to steps where some type of energy transfer takes place.

In glycolysis, glucose is converted to pyruvate

A simple summary of **glycolysis** (literally, "splitting sugar") is given in Figure 7–4. A glucose molecule (a six-carbon compound) is converted to two molecules of pyruvate (a three-carbon compound). Some of the energy in the glucose is captured; there is a net yield of two ATP molecules and two NADH molecules. The reactions of glycolysis take place in the cytosol, where the necessary ingredients, such as ADP, NAD^+, and inorganic phosphates, float freely and are used as needed. Glycolysis does not require oxygen and can proceed under aerobic or anaerobic conditions.

The glycolysis pathway includes ten separate reactions, each of which is catalyzed by a specific enzyme (Fig. 7–5). Glycolysis is divided into two major phases: the first includes endergonic reactions that require ATP, while the second includes exergonic reactions that yield ATP and NADH.

Splitting glucose requires an initial investment of ATP

The first phase of glycolysis, sometimes referred to as the "energy investment" phase, includes the first three steps

(Text continues on page 176)

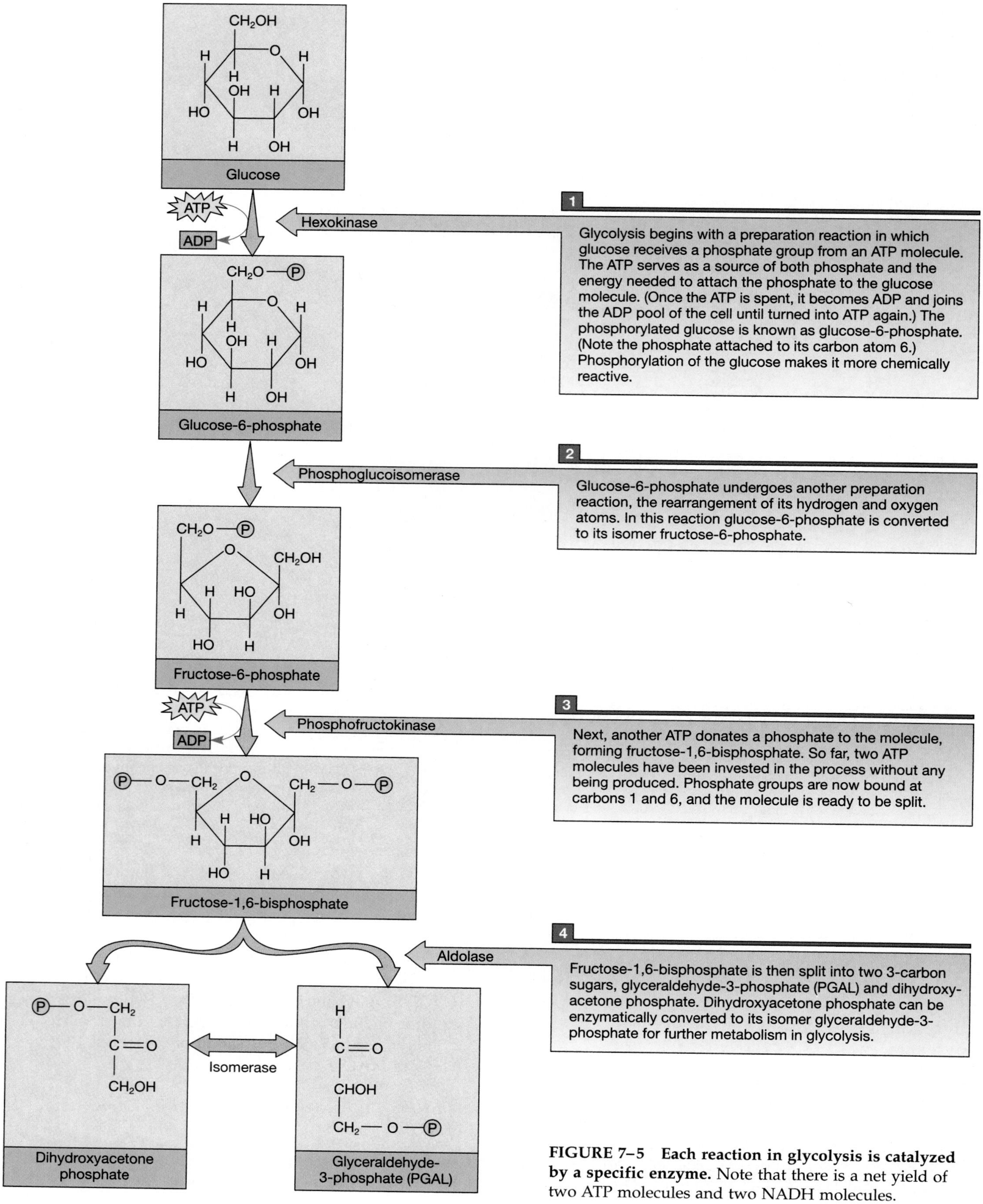

Glucose
ATP
ADP
Hexokinase
1
Glycolysis begins with a preparation reaction in which glucose receives a phosphate group from an ATP molecule. The ATP serves as a source of both phosphate and the energy needed to attach the phosphate to the glucose molecule. (Once the ATP is spent, it becomes ADP and joins the ADP pool of the cell until turned into ATP again.) The phosphorylated glucose is known as glucose-6-phosphate. (Note the phosphate attached to its carbon atom 6.) Phosphorylation of the glucose makes it more chemically reactive.
Glucose-6-phosphate
Phosphoglucoisomerase
2
Glucose-6-phosphate undergoes another preparation reaction, the rearrangement of its hydrogen and oxygen atoms. In this reaction glucose-6-phosphate is converted to its isomer fructose-6-phosphate.
Fructose-6-phosphate
ATP
ADP
Phosphofructokinase
3
Next, another ATP donates a phosphate to the molecule, forming fructose-1,6-bisphosphate. So far, two ATP molecules have been invested in the process without any being produced. Phosphate groups are now bound at carbons 1 and 6, and the molecule is ready to be split.
Fructose-1,6-bisphosphate
Aldolase
4
Fructose-1,6-bisphosphate is then split into two 3-carbon sugars, glyceraldehyde-3-phosphate (PGAL) and dihydroxyacetone phosphate. Dihydroxyacetone phosphate can be enzymatically converted to its isomer glyceraldehyde-3-phosphate for further metabolism in glycolysis.
Isomerase
Dihydroxyacetone phosphate
Glyceraldehyde-3-phosphate (PGAL)

FIGURE 7–5 Each reaction in glycolysis is catalyzed by a specific enzyme. Note that there is a net yield of two ATP molecules and two NADH molecules.

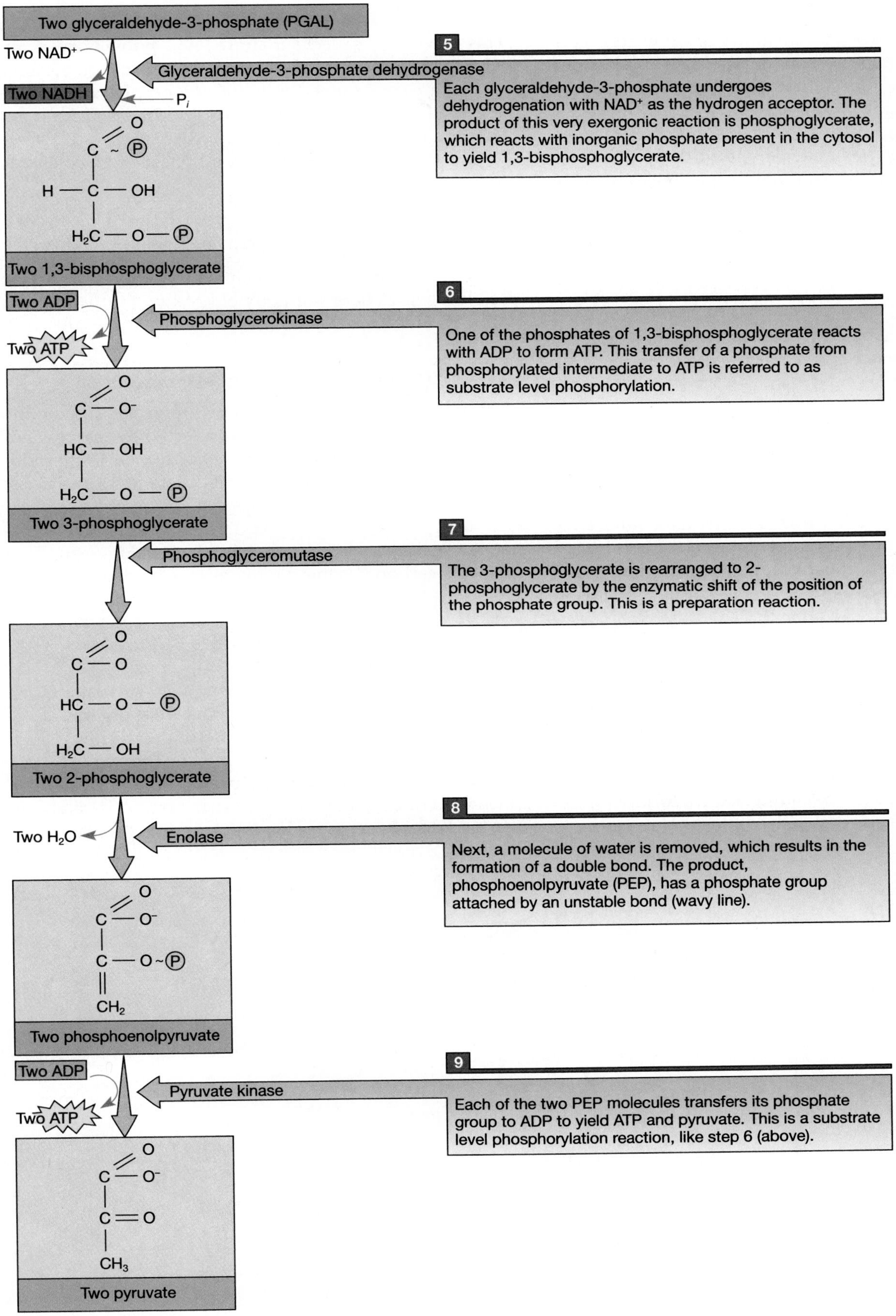
Two glyceraldehyde-3-phosphate (PGAL)
Two NAD+
Two NADH
Pi
5
Glyceraldehyde-3-phosphate dehydrogenase
Each glyceraldehyde-3-phosphate undergoes dehydrogenation with NAD+ as the hydrogen acceptor. The product of this very exergonic reaction is phosphoglycerate, which reacts with inorganic phosphate present in the cytosol to yield 1,3-bisphosphoglycerate.
Two 1,3-bisphosphoglycerate
Two ADP
Two ATP
6
Phosphoglycerokinase
One of the phosphates of 1,3-bisphosphoglycerate reacts with ADP to form ATP. This transfer of a phosphate from phosphorylated intermediate to ATP is referred to as substrate level phosphorylation.
Two 3-phosphoglycerate
7
Phosphoglyceromutase
The 3-phosphoglycerate is rearranged to 2-phosphoglycerate by the enzymatic shift of the position of the phosphate group. This is a preparation reaction.
Two 2-phosphoglycerate
Two H2O
8
Enolase
Next, a molecule of water is removed, which results in the formation of a double bond. The product, phosphoenolpyruvate (PEP), has a phosphate group attached by an unstable bond (wavy line).
Two phosphoenolpyruvate
Two ADP
Two ATP
9
Pyruvate kinase
Each of the two PEP molecules transfers its phosphate group to ADP to yield ATP and pyruvate. This is a substrate level phosphorylation reaction, like step 6 (above).
Two pyruvate

illustrated in Figure 7–5 and ends with the splitting of glucose into two three-carbon molecules. Glucose is a relatively stable molecule and is not easily broken. In two separate **phosphorylation** reactions, a phosphate group is transferred from ATP to the sugar. The resulting phosphorylated sugar (fructose-1,6-bisphosphate) is less stable and is broken enzymatically; ultimately two molecules of a three-carbon compound, glyceraldehyde-3-phosphate (PGAL) are formed. We may summarize this portion of glycolysis as follows:

$$\underset{\text{Six-carbon compound}}{\text{Glucose}} + 2\ \text{ATP} \dashrightarrow \underset{\text{Three-carbon compound}}{2\ \text{PGAL}} + 2\ \text{ADP}$$

The dashed arrow is used to indicate that the equation summarizes a sequence of several reactions.

The second phase of glycolysis yields NADH and ATP

In the steps shown on the right-hand side of Figure 7–5, each PGAL is converted to pyruvate. In the first step of this process each PGAL is oxidized by the removal of two electrons (as part of two hydrogen atoms). These immediately combine with the hydrogen carrier molecule, NAD^+:

$$\underset{\text{Oxidized}}{NAD^+} + \underset{\text{(from PGAL)}}{2\ H} \longrightarrow \underset{\text{Reduced}}{NADH + H^+}$$

Because there are two PGAL molecules for every glucose, two NADH are formed. The energy of the electrons carried by NADH can be used to form ATP. The process by which this is accomplished is discussed in conjunction with the electron transport system.

In two of the reactions leading to the formation of pyruvate, ATP is formed when a phosphate group is transferred to ADP from a phosphorylated intermediate. This process is called **substrate-level phosphorylation.** Note that in the first phase of glycolysis two molecules of ATP are consumed, but in the second phase four molecules of ATP are *produced.* Thus, glycolysis *yields* a net energy profit of two ATPs per glucose.

We may summarize the second phase of glycolysis as follows:

$$2\ \text{PGAL} + 2\ NAD^+ + 4\ \text{ADP} \dashrightarrow 2\ \text{Pyruvate} + 2\ \text{NADH} + 4\ \text{ATP}$$

Refer to Figure 7–4 for a summary of the two phases of glycolysis. There is a net yield of two pyruvates, two NADH, and two ATPs for every glucose.

Each pyruvate is converted to acetyl CoA

The pyruvate molecules formed in glycolysis enter the mitochondria, where they are converted to **acetyl coenzyme A (acetyl CoA).** (These reactions occur in the cytoplasm of aerobic prokaryotes.) In this series of reactions, pyruvate undergoes a process known as oxidative decarboxylation. First, a carboxyl group is removed as carbon dioxide, which diffuses out of the cell (Fig. 7–6). Then the two-carbon fragment remaining is oxidized, and the electrons that were removed during the oxidation are accepted by NAD^+. Finally, the oxidized two-carbon fragment, an acetyl group, becomes attached to **coenzyme A,** yielding acetyl coenzyme A.

Recall from Chapter 6 that coenzyme A acts as a transfer agent; it transfers groups derived from organic acids. In this case, it transfers an acetyl group, which is related to acetic acid. Coenzyme A is manufactured in the cell from one of the B vitamins, pantothenic acid. The attachment of an acetyl group to coenzyme A is catalyzed by a multienzyme complex that contains several copies

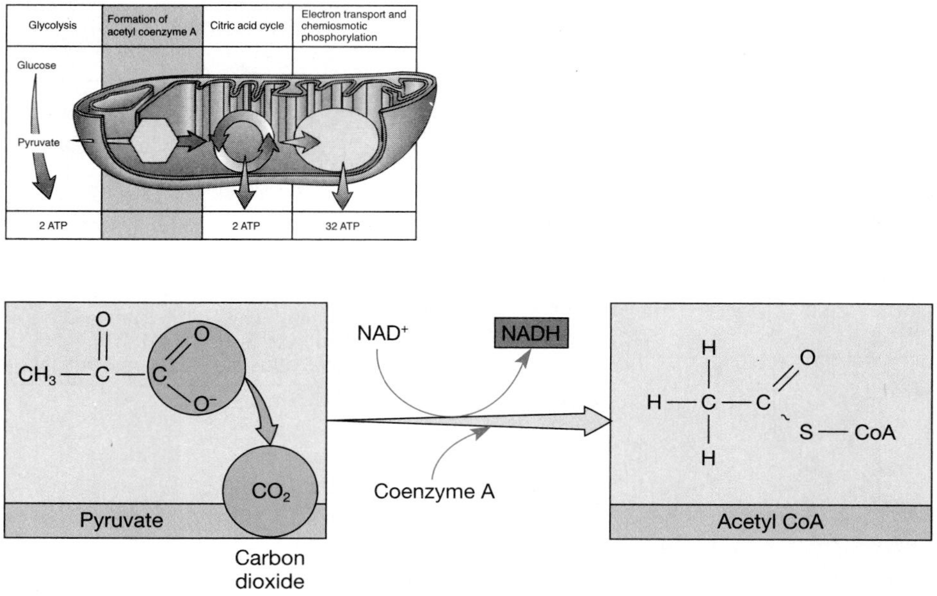

FIGURE 7–6 Pyruvate is converted to acetyl CoA in the mitochondrion. Pyruvate, the end product of glycolysis, undergoes oxidative decarboxylation. First, the carboxyl group is split off as carbon dioxide. Then the remaining two-carbon fragment is oxidized and its electrons are transferred to NAD^+. (The oxidation step is not apparent in the figure because it involves an intermediate compound that is not shown.) Finally, the oxidized two-carbon group, an acetyl group, is attached to coenzyme A. Coenzyme A has a sulfur atom that forms a very unstable bond, shown as a wavy line, with the acetyl group.

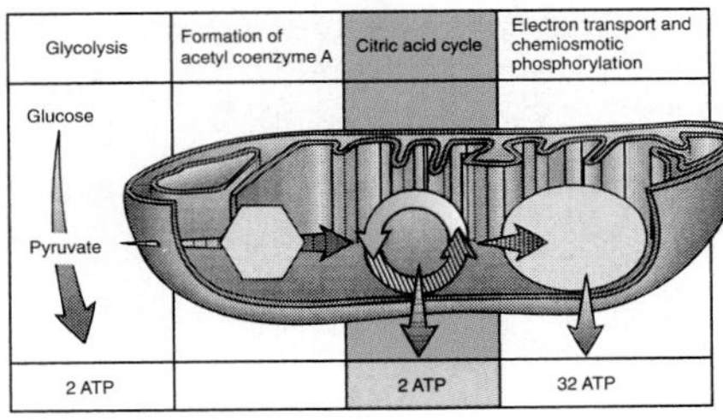

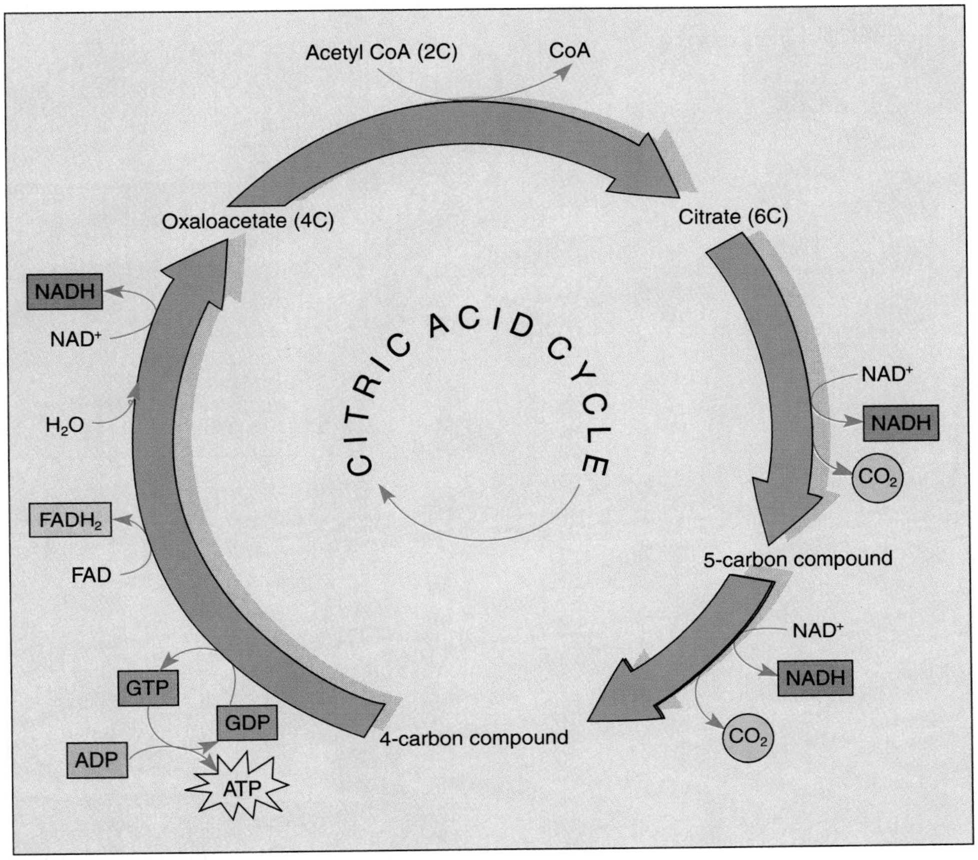

FIGURE 7–7 The products of the citric acid cycle are ATP, NADH, $FADH_2$, and carbon dioxide. Two acetyl groups enter the citric acid cycle for every glucose. Each two-carbon acetyl group combines with the four-carbon compound oxaloacetate to form the six-carbon compound citrate. The cycle includes two decarboxylations, four dehydrogenations, and several preparation reactions. Energy is captured as ATP, NADH, and $FADH_2$. The four-carbon oxaloacetate is regenerated, and the cycle begins anew. Note that the two CO_2 molecules produced account for the two carbons that entered the cycle as part of one acetyl CoA molecule. Energy is captured as one ATP, three NADH, and one $FADH_2$ per acetyl group (or two ATPs, six NADH, and two $FADH_2$ per glucose).

of each of three different enzymes. The overall reaction for the formation of acetyl coenzyme A is the following:

$$2 \text{ Pyruvate} + 2\ NAD^+ + 2CoA \longrightarrow 2 \text{ Acetyl CoA} + 2\ NADH + 2\ CO_2$$

Note that the original glucose molecule has now been partially oxidized, yielding two acetyl groups and two CO_2 molecules. The electrons removed have reduced NAD^+ to NADH. At this point in aerobic respiration, four NADH molecules have been formed as a result of the catabolism of a single glucose molecule: two during glycolysis and two during the formation of acetyl CoA from pyruvate.

The citric acid cycle oxidizes acetyl CoA

The **citric acid cycle** is also known as the **tricarboxylic acid (TCA) cycle,** and as the **Krebs cycle** after Sir Hans Krebs, who worked out its details in the 1930s. The citric acid cycle takes place in the mitochondria and consists of eight steps, illustrated and described in Figures 7–7 and 7–8. Each reaction is catalyzed by a specific enzyme.

The first reaction of the cycle occurs when acetyl CoA transfers its two-carbon acetyl group to the four-carbon compound **oxaloacetate,** forming **citrate,** a six-carbon compound:

$$\underset{\text{Four-carbon compound}}{\text{Oxaloacetate}} + \underset{\text{Two-carbon compound}}{\text{Acetyl CoA}} \longrightarrow \underset{\text{Six-carbon compound}}{\text{Citrate}} + \text{CoA}$$

The citrate then goes through a series of chemical transformations, losing first one and then a second carboxyl group as CO_2. Most of the energy made available by the oxidative steps of the cycle is transferred as energy-rich electrons to NAD^+, forming NADH. For each acetyl

(Text continues on page 179)

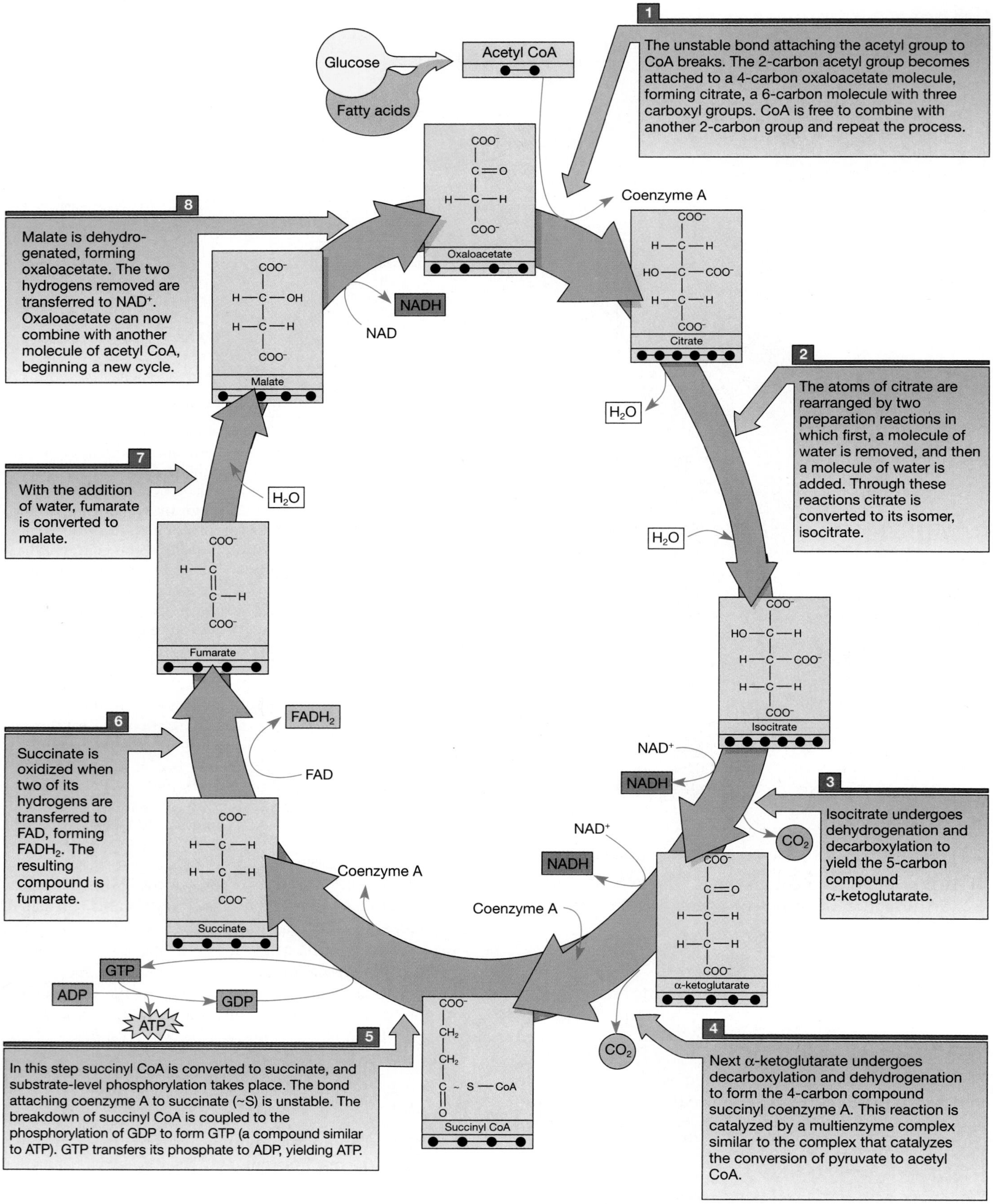
Glucose
Fatty acids
Acetyl CoA
1
The unstable bond attaching the acetyl group to CoA breaks. The 2-carbon acetyl group becomes attached to a 4-carbon oxaloacetate molecule, forming citrate, a 6-carbon molecule with three carboxyl groups. CoA is free to combine with another 2-carbon group and repeat the process.
Oxaloacetate
Coenzyme A
Citrate
2
The atoms of citrate are rearranged by two preparation reactions in which first, a molecule of water is removed, and then a molecule of water is added. Through these reactions citrate is converted to its isomer, isocitrate.
H_2O
H_2O
Isocitrate
NAD^+
NADH
CO_2
3
Isocitrate undergoes dehydrogenation and decarboxylation to yield the 5-carbon compound α-ketoglutarate.
α-ketoglutarate
NAD^+
NADH
Coenzyme A
CO_2
4
Next α-ketoglutarate undergoes decarboxylation and dehydrogenation to form the 4-carbon compound succinyl coenzyme A. This reaction is catalyzed by a multienzyme complex similar to the complex that catalyzes the conversion of pyruvate to acetyl CoA.
Succinyl CoA
Coenzyme A
GTP
ADP
GDP
ATP
5
In this step succinyl CoA is converted to succinate, and substrate-level phosphorylation takes place. The bond attaching coenzyme A to succinate (~S) is unstable. The breakdown of succinyl CoA is coupled to the phosphorylation of GDP to form GTP (a compound similar to ATP). GTP transfers its phosphate to ADP, yielding ATP.
Succinate
$FADH_2$
FAD
6
Succinate is oxidized when two of its hydrogens are transferred to FAD, forming $FADH_2$. The resulting compound is fumarate.
Fumarate
H_2O
7
With the addition of water, fumarate is converted to malate.
Malate
NADH
NAD
8
Malate is dehydrogenated, forming oxaloacetate. The two hydrogens removed are transferred to NAD^+. Oxaloacetate can now combine with another molecule of acetyl CoA, beginning a new cycle.

FIGURE 7–8 In the course of the citric acid cycle, energy is captured and the entry of a two-carbon acetyl group is balanced by the release of two molecules of carbon dioxide. Electrons are transferred to NAD^+ or FAD, yielding NADH and $FADH_2$, respectively, and ATP is formed by substrate-level phosphorylation.

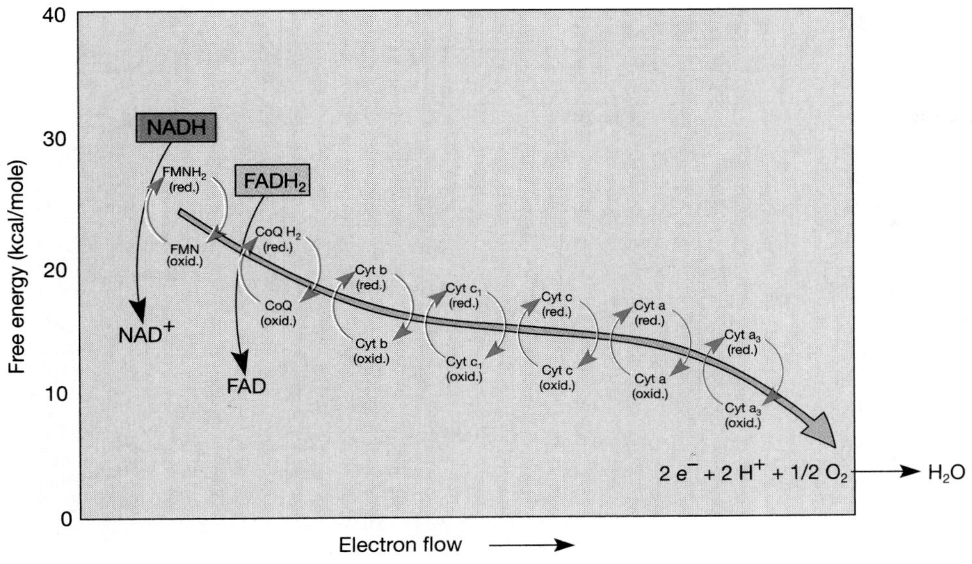

FIGURE 7–9 Electrons fall to successively lower energy levels as they are passed down the electron transport chain located in the mitochondrial inner membrane. The carriers are alternately reduced and oxidized as they accept and donate electrons (or electrons that are part of hydrogen atoms). The terminal acceptor is oxygen. Because oxygen cannot exist in atomic form, one of the two atoms of an oxygen molecule (written $\frac{1}{2}$ O_2) accepts two electrons, which are added to two protons from the surrounding medium to produce water. The energy released as electrons move to lower energy levels is used to make ATP by chemiosmosis.

group that enters the citric acid cycle, three molecules of NADH are produced. In Step 6 (Fig. 7–8), electrons are transferred to the electron acceptor FAD (forming $FADH_2$) rather than to NAD^+.

In the course of the citric acid cycle, two molecules of CO_2 and the equivalent of eight hydrogen atoms (eight protons and eight electrons) are removed, forming three NADH and one $FADH_2$. You may wonder why more hydrogen is generated by these reactions than entered the cycle with the acetyl CoA molecule. These hydrogens come from water molecules that are added during the reactions of the cycle. The CO_2 produced accounts for the two carbon atoms of the acetyl group that entered the citric acid cycle. At the end of each cycle, the four-carbon oxaloacetate has been regenerated, and a new cycle can begin.

Because two acetyl CoA molecules are produced from each glucose molecule, two cycles are required per glucose. After two turns of the cycle, the original glucose has lost all of its carbons and may be regarded as having been completely consumed. To summarize, the citric acid cycle yields (per glucose):

$$4\ CO_2 + 6\ NADH + 2\ FADH_2 + 2\ ATP$$

At the end of the citric acid cycle, glucose has been completely catabolized. Only four molecules of ATP have been formed by substrate-level phosphorylation: two during glycolysis and two during the citric acid cycle. Most of the energy of the glucose is in the form of high-energy electrons in NADH and $FADH_2$. Their energy will be used to power the synthesis of additional ATP through the electron transport system and chemiosmosis.

The electron transport system is coupled to ATP synthesis

Let us consider the fate of all the electrons removed from a molecule of glucose during glycolysis, acetyl CoA formation, and the citric acid cycle. Recall that these electrons were transferred (as part of hydrogen atoms) to the primary hydrogen acceptors NAD^+ and FAD, forming NADH and $FADH_2$. These reduced compounds now enter the **electron transport chain,** where the high-energy electrons of their hydrogens are shuttled from one acceptor to another. As the electrons are passed along in a series of exergonic redox reactions, some of their energy is used to drive the synthesis of ATP (an endergonic process). Because ATP synthesis (by phosphorylation of ADP) is coupled to the redox reactions in the electron transport chain, the entire process is known as **oxidative phosphorylation.**

The electron transport chain transfers electrons from NADH and $FADH_2$ to oxygen

The electron transport system is a chain of electron carriers embedded in the inner membrane of the mitochondrion of eukaryotes and in the plasma membrane of aerobic prokaryotes. Like NADH and $FADH_2$, each carrier can exist in an oxidized form or a reduced form. Electrons pass down the electron transport chain in a series of redox reactions. The electrons entering the electron transport system have a relatively high energy content. They lose some of their energy at each step as they pass along the chain of electron carriers.

Some of the electron carriers accept electrons that are part of hydrogen atoms, while others accept electrons only. For example, hydrogens are passed from NADH to flavin mononucleotide (FMN), the first acceptor in the chain (Fig. 7–9). The other members of the electron transport chain include ubiquinone (CoQ) and a group of closely related proteins called **cytochromes.** Cytochromes accept only the electron from each hydrogen, not the entire atom. The several types of cytochromes hold electrons at slightly different energy levels. Electrons are passed along from one cytochrome to the next in the chain, los-

Making the Connection

Electron Transport and Heat

What is the source of our body heat? Essentially, it is a byproduct of various exergonic reactions, especially those involving the electron transport chains in our mitochondria. Some organisms are able to produce unusually large amounts of heat by uncoupling electron transport from ATP production. As a consequence, most of the energy of glucose is converted to heat rather than to ATP energy. Even some plants (which we don't generally consider to be "warm" organisms) have this capability. For example, skunk cabbage, which lives in North American swamps and wet woodlands, generally flowers when the ground is still covered with snow. Its uncoupled mitochondria generate large amounts of heat, enabling the plant to melt the snow and attract pollinators. ▲

Skunk cabbage *(Symplocarpus foetidus).* (Earth Scenes © 1996 Leonard Lee Rue III)

ing energy as they go. Finally, the last cytochrome in the chain, cytochrome a_3, passes two electrons to oxygen. The electrons simultaneously unite with protons from the surrounding medium to form hydrogen; the chemical reaction between hydrogen and oxygen produces water.

Because oxygen is the final electron acceptor in the electron transport system, organisms that respire aerobically require oxygen. What happens when cells that are strict aerobes are deprived of oxygen? When no oxygen is available to accept them, the last cytochrome in the chain is stuck with its electrons. When that occurs, each acceptor molecule in the chain remains stuck with electrons (i.e., is reduced), and the entire system is blocked all the way back to NADH. Because phosphorylation is coupled to electron transport, no further ATPs are produced by way of the electron transport system. Most cells of complex organisms cannot live long without oxygen because the amount of energy they produce in its absence is insufficient to sustain life processes.

Lack of oxygen is not the only factor that interferes with the electron transport system. Some poisons, including cyanide, inhibit the normal activity of the cytochrome system. Cyanide binds tightly to the iron in cytochrome a_3, making it unable to transport electrons on to oxygen. This blocks the further passage of electrons through the chain, halting ATP production.

The chemiosmotic model explains the coupling of ATP synthesis to electron transport

The flow of electrons in electron transport is usually tightly coupled to the production of ATP and does not occur unless the phosphorylation of ADP can also proceed. This prevents a waste of energy, because high-energy electrons do not flow unless ATP can be produced. If electron flow were uncoupled from the phosphorylation of ADP, there would be no production of ATP, and the energy of the electrons would be converted to heat (see *Making the Connection: Electron Transport and Heat*).

For decades scientists were aware that oxidative phosphorylation occurs in mitochondria, and many experiments had shown that the transfer of two electrons from each NADH to oxygen (via the electron transport chain) usually results in the production of up to three ATP molecules. However, for a long time, just *how* ATP synthesis is related to electron transport remained a mystery. Then, in 1961 Peter Mitchell proposed the **chemiosmotic model,** based on his experiments with bacteria. His model was so radical that it was not accepted immediately, but by 1978 so much supporting evidence had accumulated that Peter Mitchell was awarded a Nobel Prize.

Because the respiratory electron transport chain is located in the plasma membrane of an aerobic bacterial cell, the bacterial plasma membrane can be considered comparable to the inner mitochondrial membrane. Mitchell was able to show that if bacterial cells were placed in an acidic environment (i.e., an environment with a high hydrogen ion, or proton, concentration), the cells would synthesize ATP even if no electron transport were taking place. On the basis of these and other experiments, Mitchell proposed that electron transport and ATP synthesis are coupled by means of a proton gradient across the inner mitochondrial membrane in eukaryotes (or across the plasma membrane in bacteria).

The proton gradient is established by the electron transport system; some of the energy released as electrons pass down the electron transport chain is used to pump protons across a membrane. In eukaryotes the protons are pumped across the inner mitochondrial membrane into the intermembrane space (the space between the inner and outer mitochondrial membranes; Fig. 7–10*a*); in prokaryotes the protons are pumped out of the cell into the surrounding growth medium. Hence the inner mitochondrial membrane separates a space with a higher con-

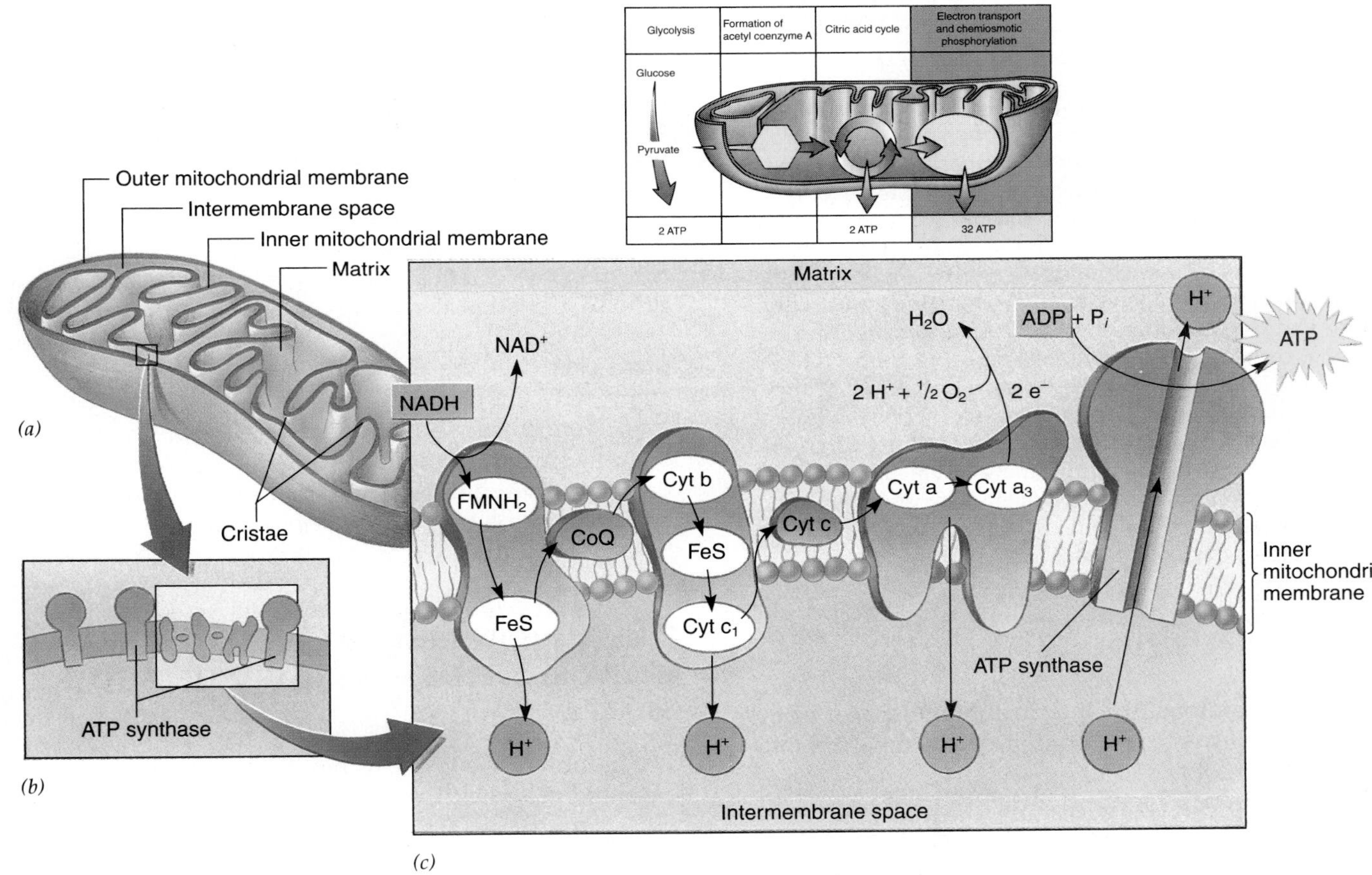

FIGURE 7–10 ATP is produced by chemiosmosis. (*a*) The mitochondrial inner membrane separates the innermost space (the matrix) from the intermembrane space. (*b* and *c*) According to the chemiosmotic model, the electron transport chain in the inner mitochondrial membrane includes three proton pumps. The electron acceptors in the membrane are located in three main complexes. FMN, which oxidizes NADH, is located in the first complex. The cytochrome b-c_1 complex consists of two cytochromes and some additional electron acceptors. The third complex includes cytochromes a and a_3. Coenzyme Q and cytochrome c are mobile carriers that transfer electrons between the complexes. At three sites in the chain, the energy released during electron transport is used to transport protons (H^+) from the mitochondrial matrix to the intermembrane space, where a high concentration of protons accumulates. The protons are prevented from diffusing back into the matrix through the inner membrane except through special channels in ATP synthase in the membrane. The flow of the protons through ATP synthase generates ATP. Phosphorylation is "paid for" by the decline in the free energy of the system (due to an increase in entropy) as the protons diffuse from a region of high concentration to a region of lower concentration.

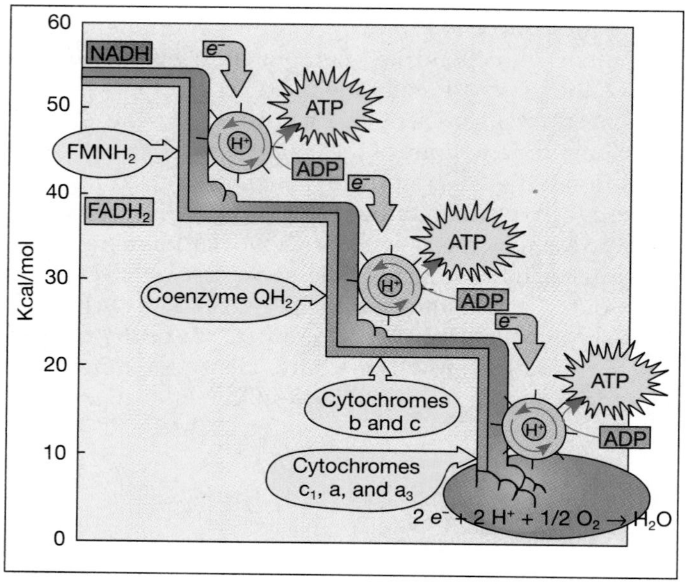

FIGURE 7–11 Oxidative phosphorylation requires the coupling of electron transport (a redox process) and ATP production (phosphorylation). Electron transport can be compared with a stream of water (electrons) that has three waterfalls. The flow of water drives water wheels (the proton gradient). There are three points in the electron transport system where protons are pumped. The electrons end up in the pond at the bottom of the waterfalls, where they unite with protons and oxygen to yield H_2O. Electron transfer from NADH to oxygen is very exergonic, releasing about 222 kJ/mole (53 kcal/mole). If released all at once, most of this energy would be lost as heat. Instead, the energy is released slowly in a series of steps, as shown here, and used to transport protons across the inner mitochondrial membrane. The potential energy of the proton gradient is the source of energy needed to synthesize ATP. For each pair of electrons that enters this pathway, a maximum of three ATP molecules is produced.

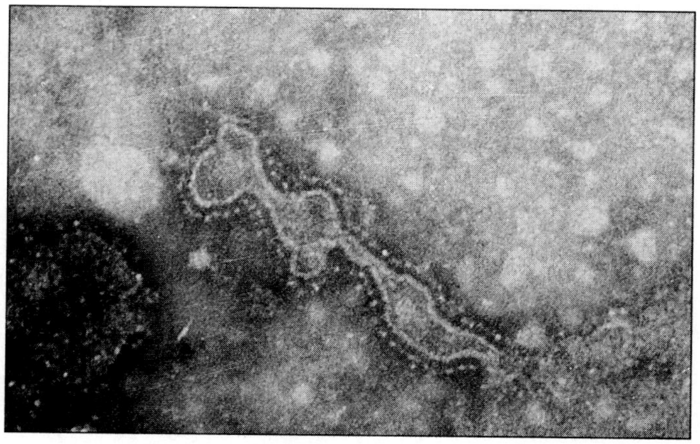

FIGURE 7–12 The lollipop-like structures shown in this electron micrograph are composed of complexes of the enzyme ATP synthase. Each complex provides a channel through the inner mitochondrial membrane and catalyzes the formation of ATP, which is released into the mitochondrial matrix. (Visuals Unlimited/R. Bhatnagar)

centration of protons (the intermembrane space) from a space with a lower concentration of protons (the mitochondrial matrix).

Protons are pumped across the inner mitochondrial membrane by three electron transfer complexes, each associated with particular steps in the electron transport system (Fig. 7–10*b* and *c*). The result is a proton gradient across the inner mitochondrial membrane. This gradient is a form of potential energy, like water behind a dam; it can be harnessed to provide the energy for ATP synthesis (Fig. 7–11).

Diffusion of protons through the inner mitochondrial membrane is limited to certain specific channels in complexes made up of the enzyme **ATP synthase,** a transmembrane protein. Portions of these complexes project from the inner surface of the membrane (the surface that faces the matrix) and are visible by electron microscopy (Fig. 7–12). Diffusion of the protons down their gradient, through the ATP synthase complex, is exergonic because the entropy of the system increases. This exergonic process provides the energy for ATP production (Fig. 7–10*c*), although the exact mechanism by which ATP synthase catalyzes the phosphorylation of ADP is still incompletely understood.

Chemiosmosis is a fundamental mechanism of energy coupling in cells; it allows exergonic redox processes to drive the endergonic reaction in which ATP is produced by phosphorylating ADP. In photosynthesis (see Chapter 8), ATP is produced by a comparable process.

THE AEROBIC RESPIRATION OF ONE GLUCOSE YIELDS A MAXIMUM OF 36 TO 38 ATPS

Let us now review where biologically useful energy is captured in aerobic respiration and calculate the total energy yield from the complete oxidation of glucose. Table 7–2 summarizes the arithmetic involved.

(1) In glycolysis, glucose is activated by the addition of phosphates from two ATP molecules and converted ultimately to two pyruvates + two NADH + four ATPs, yielding a net profit of two ATPs.
(2) The two pyruvates are metabolized to two acetyl CoA + two CO_2 + two NADH.
(3) In the citric acid cycle the two acetyl CoA molecules are metabolized to four CO_2 + six NADH + two $FADH_2$ + two ATPs.

Because the oxidation of NADH in the electron transport system yields up to three ATPs per mole-

Focus On

Shuttles Across the Mitochondrial Membrane

The inner mitochondrial membrane is not permeable to NADH, which is a large molecule. Therefore the NADH molecules produced in the cytosol during glycolysis cannot diffuse into the mitochondria to transfer their electrons to the electron transport system. Unlike ATP and ADP, NADH does not have a carrier protein to transport it across the membrane. Instead, several systems have evolved to transfer the *electrons* of NADH (although not the NADH molecules themselves) into the mitochondria.

In liver, kidney, and heart cells, a special shuttle system transfers the electrons from NADH through the inner mitochondrial membrane to an NAD^+ molecule in the matrix. These electrons are transferred to the electron transport system in the inner mitochondrial membrane, and up to three molecules of ATP are produced per pair of electrons.

In skeletal muscle, brain, and some other types of cells, another type of shuttle operates. Because this shuttle requires more energy than the shuttle in liver, kidney, and heart cells, the electrons are at a lower energy level when they enter the electron transport chain. They are accepted by coenzyme Q rather than by NAD+ and so generate a maximum of two ATP molecules per pair of electrons. This is why the number of ATPs produced by aerobic respiration of a molecule of glucose in skeletal muscle cells is 36 rather than 38. ■

cule, the ten NADH molecules can yield up to 30 ATPs. The two NADH molecules from glycolysis, however, yield either two or three ATPs each. This is because certain types of eukaryotic cells must expend energy to shuttle the NADH produced by glycolysis across the mitochondrial membrane (see *Focus On: Shuttles Across the Mitochondrial Membrane*). Prokaryotic cells lack mitochondria; hence they have no need to shuttle NADH molecules. For this reason, bacteria are able to generate three ATPs for every NADH, even those produced during glycolysis.

Thus, the maximum number of ATPs formed using the energy from NADH is 28 to 30. The oxidation of $FADH_2$ yields two ATPs per molecule, so the two $FADH_2$ molecules produced in the citric acid cycle yield four ATPs.

(4) Summing all the ATPs (two from glycolysis, two from the citric acid cycle, and 32 to 34 from electron transport and chemiosmotic phosphorylation), we see that the complete aerobic metabolism of one molecule of glucose yields a maximum of 36 to 38 ATPs. Note that all but two ATPs are generated by reactions taking place in the mitochondria. Of the 34 to 36 ATPs generated in the mitochondria, 32 to 34 are produced by the electron transport system and chemiosmosis. Only two ATPs are formed by substrate-level phosphorylation in the citric acid cycle.

We can analyze the efficiency of the overall process of aerobic respiration by comparing the free energy captured as ATP to the total free energy in a glucose molecule. You will recall from Chapter 6 that, although heat

Table 7–2 ENERGY YIELD FROM THE COMPLETE OXIDATION OF GLUCOSE

1. Net ATP profit from glycolysis		2 ATP* (substrate-level phosphorylation)
Also from glycolysis:	2 NADH --->	4–6 ATP (oxidative phosphorylation)
2. 2 pyruvate to 2 acetyl CoA	2 NADH --->	6 ATP (oxidative phosphorylation)
3. 2 acetyl CoA through citric acid cycle and electron transport system		2 ATP (substrate-level phosphorylation)
	6 NADH --->	18 ATP (oxidative phosphorylation)
	2 $FADH_2$ --->	4 ATP (oxidative phosphorylation)
Total ATP profit		36–38 ATP

*These are the only two ATPs that can be generated anaerobically; production of all other ATPs depends on the presence of oxygen.

energy cannot power cellular reactions, it is convenient to measure energy as heat. This can be done through the use of a calorimeter, an instrument that measures the heat of a reaction. A sample is placed in a compartment surrounded by a chamber of water. As the sample burns (becomes oxidized), the temperature of the water rises, providing a measure of the heat released during the reaction.

When one mole of glucose is burned in a calorimeter, some 686 kcal are released as heat. (If this heat energy could be used to do work, it would be equivalent to 2870 kJ/mole.) The free energy temporarily held in the phosphate bonds of ATP is about 7.6 kcal per mole; when 36 to 38 ATPs are generated during the aerobic respiration of glucose, the free energy momentarily trapped in ATP amounts to 7.6 kcal per mole × 36, or about 274 kcal per mole. Thus, the efficiency of aerobic respiration is 274/686, or about 40%. (By comparison, a steam power plant has an efficiency of 35% to 36% in converting its fuel energy into electricity.) The remainder of the energy in the glucose is released as heat.

NUTRIENTS OTHER THAN GLUCOSE ALSO PROVIDE ENERGY

Many organisms depend on nutrients other than glucose (or in addition to glucose) as a source of energy. Humans and many other animals usually obtain more of their energy by oxidizing fatty acids than by oxidizing glucose. Amino acids from protein digestion are also used as fuel molecules. Such nutrients are transformed into one of the metabolic intermediates that are fed into glycolysis or the citric acid cycle (Fig. 7–13).

Amino acids are metabolized by reactions in which the amino group is first removed, a process called **deamination.** In mammals and some other animals, the amino group is converted to urea and excreted, but the carbon chain is metabolized and eventually enters the citric acid cycle. The amino acid alanine, for example, undergoes deamination to become pyruvate, the amino acid glutamate is converted to α-ketoglutarate, and the amino acid aspartate yields oxaloacetate. Ultimately, the carbon chains of all the amino acids are metabolized in this way.

Each gram of lipid in the diet contains 9 kcal (38 kJ), more than twice as much energy as is found in glucose or amino acids, which have about 4 kcal (17 kJ) per gram. Lipids are rich in energy because they are highly reduced; that is, they have many hydrogen atoms and few oxygen atoms. When completely oxidized in aerobic respiration, a molecule of a six-carbon fatty acid generates up to 44 ATPs (compared with 36 to 38 ATPs for a molecule of glucose, which also has six carbons).

Both the glycerol and fatty acid components of a neutral fat (see Chapter 3) are used as fuel; phosphate is added to glycerol, converting it to PGAL or another com-

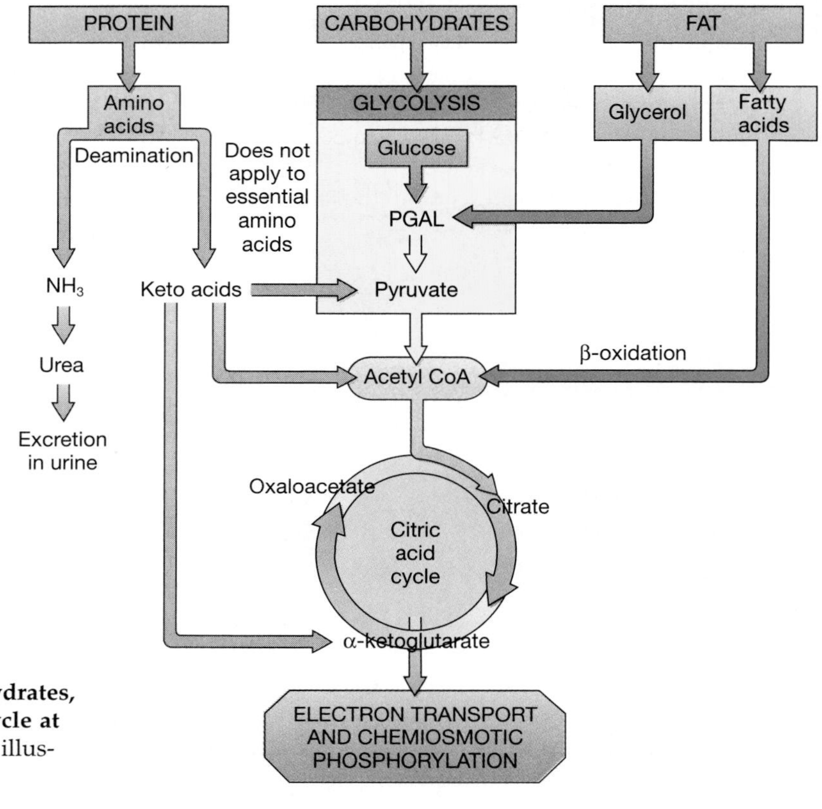

FIGURE 7–13 Products of the catabolism of carbohydrates, proteins, and fats enter glycolysis or the citric acid cycle at various points. This diagram is greatly simplified and illustrates only a few of the principal pathways.

pound that enters glycolysis. Fatty acids are oxidized and split enzymatically into two-carbon acetyl groups that are bound to coenzyme A; that is, fatty acids are converted to acetyl CoA. This process, which occurs in the mitochondrial matrix, is called ***β*-oxidation.** Acetyl CoA molecules formed by β-oxidation enter the citric acid cycle.

As excess ATP is used by the cell, ADP (and AMP) is produced. Now the inhibitor site of phosphofructokinase is no longer occupied by ATP. Instead, the activator sites are filled with ADP and AMP. Thus the enzyme becomes activated and binds ATP to its active site; respiration proceeds, generating more ATP.

Cells regulate aerobic respiration

Aerobic respiration requires a steady input of fuel molecules and oxygen. Under normal conditions these materials are adequately provided and do not affect the rate of respiration. Instead, the rate of aerobic respiration is regulated by how much ADP and phosphate are available. In a resting muscle cell, for example, ATP synthesis continues until most of the ADP has been converted to ATP. At this point oxidative phosphorylation slows considerably. Because electron flow is tightly coupled to oxidative phosphorylation, the flow of electrons also slows, which in turn slows down the citric acid cycle.

When ATP transfers energy to power an energy-requiring process like muscle contraction, many molecules of ATP are hydrolyzed. The ADP molecules produced can then accept phosphate to become ATP once again; aerobic respiration speeds up until most of the ADP has again been converted to ATP.

The control of most metabolic pathways is exerted on a particular enzyme that catalyzes a reaction early in the pathway (see discussion of feedback inhibition in Chapter 6). The regulated enzyme is usually inhibited by the presence of the end product of the pathway. One of the important control points in aerobic respiration in mammals is phosphofructokinase, an enzyme that catalyzes an early reaction of glycolysis (see Fig. 7–5). The active site of phosphofructokinase binds ATP and fructose 6-phosphate. However, the enzyme is inhibited by the presence of very high levels of ATP and activated by the presence of ADP and AMP (adenosine monophosphate, a molecule formed when two phosphates are removed from ATP). Therefore, this enzyme is inactivated when ATP levels are high and activated when they are low.[3]

Phosphofructokinase possesses allosteric receptor sites (see Chapter 6) for both enzyme inhibitors (in this case, ATP) and enzyme activators (in this case, ADP and AMP). When respiration produces more ATP than the cell currently needs, some of the excess ATP binds to phosphofructokinase, changing its conformation so that it is no longer active. Thus, glycolysis and aerobic respiration slow down and less ATP is produced.

[3]Other materials, including citrate, also affect the activity of phosphofructokinase.

Anaerobic respiration and fermentation do not require oxygen

As stated previously, cells use three types of pathways to extract free energy from nutrients: aerobic respiration, anaerobic respiration, and fermentation, which is also anaerobic (Fig. 7–14). In all three processes, glucose or other nutrients are oxidized, and their high-energy electrons are transferred to NAD^+, which becomes reduced to NADH. Table 7–3 compares these pathways with respect to the fates of the NADH electrons and the mechanism of ATP synthesis.

In both aerobic and anaerobic respiration, some ATP is synthesized at the substrate level. However, most phosphorylation occurs by chemiosmosis. NADH transfers its electrons to an electron transport chain, and the passage of these electrons down the chain is coupled to chemiosmotic ATP synthesis.

We have seen that molecular oxygen is the final electron acceptor for the electron transport chain in aerobic respiration. Some types of bacteria that live in soil or stagnant ponds where molecular oxygen is in short supply engage solely in anaerobic respiration. **Anaerobic respiration** is similar to aerobic respiration in the sense that electrons are transferred from glucose to NADH; they then pass down an electron transport chain that is coupled to ATP synthesis. However, an inorganic substance such as nitrate (NO_3^-) or sulfate (SO_4^{2-}) replaces molecular oxygen as the terminal electron acceptor. The end products of this type of anaerobic respiration are carbon dioxide, one or more reduced inorganic substances, and ATP. One representative type of anaerobic respiration, summarized below, is part of the biogeochemical cycle known as the nitrogen cycle (see Chapter 54).

$$C_6H_{12}O_6 + \underset{\text{potassium nitrate}}{12\ KNO_3} \dashrightarrow 6\ CO_2 + 6\ H_2O + \underset{\text{potassium nitrite}}{12\ KNO_2} + \text{Energy (as ATP)}$$

Certain other bacteria, as well as some fungi, regularly use **fermentation,** an anaerobic pathway that does not involve an electron transport chain. All ATP formed during fermentation is produced by phosphorylation at the substrate level during glycolysis (only two ATPs per

Table 7–3 A COMPARISON OF AEROBIC RESPIRATION, ANAEROBIC RESPIRATION, AND FERMENTATION

	Aerobic Respiration	*Anaerobic Respiration*	*Fermentation*
Immediate Fate of Electrons in NADH	Transferred to an electron transport chain	Transferred to an electron transport chain	Transferred to an organic molecule
Terminal Electron Acceptor of Electron Transport Chain	O_2	Inorganic substances such as NO_3^- or SO_4^{2-}	(No electron transport chain)
Reduced Product(s) Formed	Water	Relatively reduced inorganic substances	Relatively reduced organic compounds (commonly alcohol or lactate)
Mechanism of ATP Synthesis	Chemiosmosis; also substrate-level phosphorylation	Chemiosmosis; also substrate-level phosphorylation	Substrate-level phosphorylation only (during glycolysis)

glucose), not by chemiosmosis. One might expect that a cell that obtains energy from glycolysis would produce pyruvate (the end product of glycolysis). However, this cannot happen because every cell has a limited supply of NAD^+. If virtually all NAD^+ becomes reduced to NADH during glycolysis, then glycolysis stops and no more ATP can be produced. In fermentation, NADH molecules transfer their hydrogens to organic molecules, thus regenerating the NAD^+ needed to keep glycolysis going. The relatively reduced organic molecules produced (commonly alcohol or lactic acid) tend to be toxic to the cells and are essentially waste products.

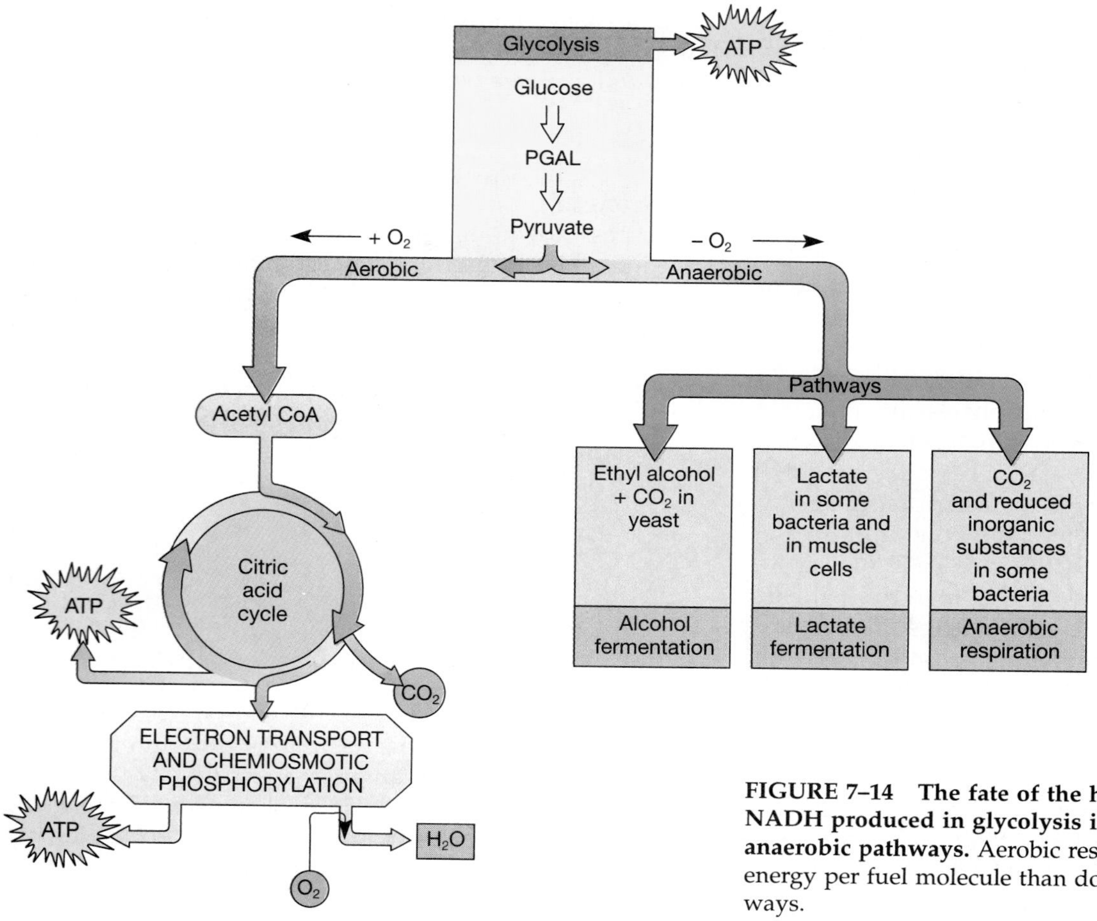

FIGURE 7–14 The fate of the high-energy electrons in the NADH produced in glycolysis is different for aerobic and anaerobic pathways. Aerobic respiration yields much more energy per fuel molecule than does any of the anaerobic pathways.

FIGURE 7–15 Both alcohol fermentation and lactate fermentation regenerate NAD$^+$ required for glycolysis. Both processes yield only two ATPs per glucose. Alcohol fermentation *(top)* is a two-step process. Pyruvate is decarboxylated, yielding carbon dioxide and acetaldehyde, which is converted to ethyl alcohol. In lactic acid fermentation *(bottom)*, pyruvate is converted to lactate.

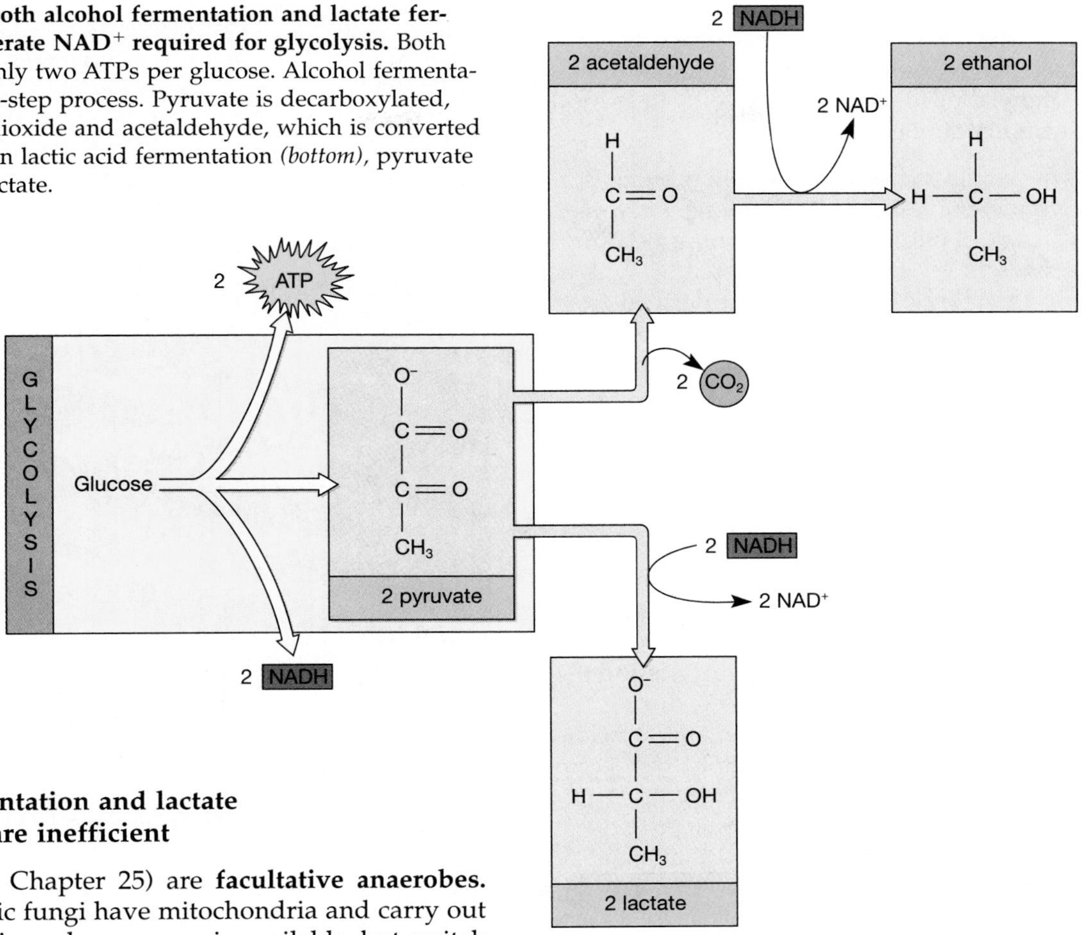

Alcohol fermentation and lactate fermentation are inefficient

Yeast cells (see Chapter 25) are **facultative anaerobes.** These eukaryotic fungi have mitochondria and carry out aerobic respiration when oxygen is available, but switch to **alcohol fermentation** when deprived of oxygen. They have enzymes that decarboxylate pyruvate, releasing carbon dioxide and forming a two-carbon compound called acetaldehyde (Fig. 7–15). NADH produced during glycolysis transfers hydrogens to acetaldehyde, reducing it to form **ethyl alcohol.** Alcohol fermentation is the basis for the production of beer, wine, and other alcoholic beverages. Yeast cells are also used in baking to produce the carbon dioxide that causes dough to rise (the alcohol evaporates during baking).

Certain fungi and bacteria carry on **lactate (lactic acid) fermentation.** In this alternative pathway, NADH produced during glycolysis transfers hydrogens to pyruvate, reducing it to form **lactate** (see Fig. 7–15). The ability of some bacteria to produce lactate is exploited by humans, who use them to make yogurt and to ferment cabbage for sauerkraut. Lactate is also produced during strenuous activity in the muscle cells of humans and other complex animals. If the amount of oxygen delivered to muscle cells is insufficient to support aerobic respiration, the cells shift temporarily to lactate fermentation. As lactate accumulates in muscle cells, it contributes to muscle fatigue.

Both alcohol fermentation and lactate fermentation are very inefficient because the fuel is only partially oxidized. Alcohol, the end product of fermentation by yeast cells, can be burned and can even be used as automobile fuel; obviously, it contains a great deal of energy that the yeast cells are unable to extract using anaerobic methods. Lactate, a three-carbon compound, contains even more energy than the two-carbon alcohol. In contrast, all available energy is removed during aerobic respiration because the fuel molecules become completely oxidized to CO_2. A net profit of only two ATPs is produced by the fermentation of one molecule of glucose, compared with up to 36 to 38 ATPs when oxygen is available.

The inefficiency of anaerobic metabolism necessitates a large supply of fuel. For example, skeletal muscle cells, which often metabolize anaerobically for short periods, store large quantities of glucose in the form of glycogen (see Chapter 38). To perform the same amount of work, a cell engaged in fermentation must consume up to 20 times more glucose or other carbohydrate per second than a cell using aerobic respiration.

ANABOLIC REACTIONS ARE PART OF BIOSYNTHETIC PROCESSES

Our discussion thus far has focused on catabolic reactions, processes that break down organic molecules and conserve their energy in the biologically useful form of

ATP. Cells also possess a remarkable array of enzymes that catalyze a variety of biosynthetic processes. This anabolic (building up, or synthesizing, of complex molecules) aspect of metabolism is quite complex, but several basic generalizations can be made:

1. Each cell type usually synthesizes its own proteins, nucleic acids, lipids, polysaccharides, and other complex molecules and does not receive them preformed from other cells. Muscle glycogen, for example, is synthesized within the muscle cell and is not derived from liver glycogen.
2. Each step in the biosynthesis of a molecule is catalyzed by a separate enzyme.
3. Although certain steps in a biosynthetic sequence may proceed without the use of ATP, the overall synthesis of complex molecules requires that chemical energy be transferred from ATP and similar compounds at various points along the way.
4. Anabolic processes use relatively few substances as raw materials; among these are acetyl CoA, glycine, succinyl CoA, ribose, pyruvate, and glycerol. Many of these molecules are intermediates in glycolysis and the citric acid cycle. Some others are formed during the catabolism of carbohydrates, lipids, or amino acids.
5. In general, anabolic processes are not simply the reverse of catabolic processes in which the molecule is degraded, but include one or more separate steps that differ from any step in catabolism. The fact that anabolic and catabolic processes are catalyzed by different enzymes permits separate control mechanisms to govern the synthesis and degradation of complex molecules.
6. Each cell's constituent molecules are in a dynamic state; that is, some molecules are being degraded while others are being synthesized. Both anabolism and catabolism are continually occurring.

Even a cell that is not growing or increasing in mass uses a considerable portion of its total energy for the chemical work of biosynthesis. A cell that is growing rapidly allocates a correspondingly larger fraction of its total energy to biosynthetic processes, especially to the biosynthesis of protein (Fig. 7–16). For example, a rapidly growing bacterial cell may use as much as 90% of its total energy for protein synthesis.

FIGURE 7–16 The generalizations about cellular metabolism also apply to the metabolism of a multicellular organism. An organism carries out catabolic reactions to provide the energy required to do various kinds of work, including the work of biosynthesis. An actively growing organism, such as the juvenile flamingo, obviously has a high rate of synthesis of various needed components, but even if growth has ceased there is a continuous turnover of organic molecules. (M.P. Kahl/VIREO, Academy of Natural Sciences)

SUMMARY

I. Cells use three different types of catabolic pathways to extract free energy from nutrients: aerobic respiration, anaerobic respiration, and fermentation.

II. During aerobic respiration, a fuel molecule such as glucose is oxidized to form carbon dioxide and water. Energy is captured through the formation of up to 36 to 38 ATPs per molecule of glucose.

III. Aerobic respiration is a redox process in which electrons are transferred from glucose (which becomes oxidized) to oxygen (which becomes reduced).

IV. The chemical reactions of aerobic respiration occur in four stages: glycolysis, formation of acetyl CoA, the citric acid cycle, and the electron transport system/chemiosmosis.
   A. During glycolysis, which occurs in the cytosol, a molecule of glucose is degraded to form two molecules of pyruvate.
      1. Two ATP molecules (net) are produced during glycolysis.
      2. Four hydrogen atoms are removed from the fuel molecule (as two NADH).

B. The two pyruvate molecules each lose a molecule of carbon dioxide, and the remaining acetyl groups each combine with coenzyme A, producing two molecules of acetyl CoA. One NADH is formed as each pyruvate is converted to acetyl CoA.

C. Each acetyl CoA enters the citric acid cycle by combining with a four-carbon compound, oxaloacetate, to form citrate, a six-carbon compound. Two acetyl CoA enter the cycle for every glucose.

1. For every two carbons that enter the cycle as acetyl CoA, two leave as carbon dioxide.
2. For every acetyl CoA, hydrogens are transferred to three NAD^+ and one FAD; only one ATP is produced by substrate-level phosphorylation.

D. Hydrogen atoms (or their electrons) removed from fuel molecules are transferred from one electron acceptor to another down an electron transport system located in the mitochondrial inner membrane.

1. Water is formed when oxygen combines with H^+ and with electrons from the electron transport chain.
2. According to the chemiosmotic theory, some of the energy of the electrons in the electron transport chain is used to establish a proton gradient across the inner mitochondrial membrane.
3. The diffusion of protons back through the membrane from the intermembrane space to the mitochondrial matrix (by way of the enzyme ATP synthase) provides the energy needed to synthesize ATP. Some of the protons from the matrix combine with oxygen and electrons from the electron transport chain to form water.

V. Organic nutrients other than glucose are converted to appropriate compounds and fed into glycolysis or the citric acid cycle.

A. Amino acids are deaminated and their carbon skeletons are converted to metabolic intermediates such as pyruvate.

B. Both the glycerol and fatty acid components of lipids are oxidized as fuel. Fatty acids are converted to acetyl coenzyme A molecules by the process of β-oxidation.

VI. In anaerobic respiration, electrons are transferred from fuel molecules to an electron transport chain; the final electron acceptor is nitrate or sulfate (not molecular oxygen).

VII. Fermentation is an anaerobic process that does not use an electron transport chain. There is a net gain of only two ATPs per glucose (produced during glycolysis). To maintain the supply of NAD^+ essential for glycolysis, hydrogens are transferred from NADH to an organic compound derived from the initial nutrient.

A. Yeast cells carry on alcohol fermentation, in which ethyl alcohol and carbon dioxide are the final (waste) products.

B. Certain fungi, certain bacteria, and certain animal cells (in the absence of sufficient oxygen) carry on lactate fermentation, in which hydrogen atoms are added to pyruvate to form lactate (a waste).

VIII. The cells of living organisms exist in a dynamic state and are continuously building up and breaking down the many different cell constituents.

A. Each cell usually synthesizes its own complex macromolecules, and each step in the process is catalyzed by a separate enzyme.

B. Anabolic pathways include many endergonic reactions that require ATP or some other energy source to drive them.

Summary reactions for aerobic respiration

Summary reaction for the complete oxidation of glucose:

$$C_6H_{12}O_6 + 6\,O_2 + 6\,H_2O \dashrightarrow 6\,CO_2 + 12\,H_2O + \text{Energy (as ATP)}$$

Summary reaction for glycolysis:

$$C_6H_{12}O_6 + 2\,ATP + 2\,ADP + 2\,P_i + 2\,NAD^+ \dashrightarrow 2\,\text{Pyruvate} + 4\,ATP + 2\,NADH + H_2O$$

Summary reaction for the conversion of pyruvate to acetyl CoA:

$$2\,\text{Pyruvate} + 2\,\text{Coenzyme A} + 2\,NAD^+ \dashrightarrow 2\,\text{Acetyl CoA} + 2\,CO_2 + 2\,NADH$$

Summary reaction for the citric acid cycle:

$$2\,\text{Acetyl CoA} + 6\,NAD^+ + 2\,FAD + 2\,ADP + 2\,P_i + 2\,H_2O \dashrightarrow 4\,CO_2 + 6\,NADH + 2\,FADH_2 + 2\,ATP + 2\,CoA$$

Summary reactions for the processing of the hydrogens of NADH and $FADH_2$ in the electron transport system:

$$NADH + 3\,ADP + 3\,P_i + \tfrac{1}{2}\,O_2 \dashrightarrow NAD^+ + 3\,ATP + H_2O$$

$$FADH_2 + 2\,ADP + 2\,P_i + \tfrac{1}{2}\,O_2 \dashrightarrow FAD + 2\,ATP + H_2O$$

Summary reactions for fermentation (beginning with pyruvate)

$$\text{Pyruvate} + NADH \longrightarrow \text{Lactate} + NAD^+$$

$$\text{Pyruvate} \longrightarrow CO_2 + \text{Acetaldehyde}$$

$$\text{Acetaldehyde} + NADH \longrightarrow \text{Ethyl alcohol}$$

SELECTED KEY TERMS

acetyl coenzyme A (acetyl CoA)
acetyl group
aerobe
aerobic respiration
anabolism
anaerobic respiration
ATP synthase
β-oxidation
catabolism
chemiosmosis
citrate (citric acid)
citric acid cycle (tricarboxylic acid cycle; Krebs cycle)
coenzyme A
cytochromes
deamination
decarboxylation
dehydrogenation
electron transport chain
ethyl alcohol
facultative anaerobe
FAD/$FADH_2$ (flavin adenine dinucleotide)
fermentation
glycolysis
lactate (lactic acid)
NAD^+/NADH (nicotinamide adenine dinucleotide)
oxidation
oxidative phosphorylation
phosphorylation
pyruvate (pyruvic acid)
reduction

POST-TEST

1. The process of splitting larger molecules into smaller ones is an aspect of metabolism called ________ .
2. The synthetic aspect of metabolism is referred to as ________ .
3. A chemical process during which a substance gains electrons is called ________ .
4. The pathway through which glucose is degraded to pyruvate is referred to as ________ .
5. The reactions of ________ take place within the cytosol of eukaryotic cells.
6. Before pyruvate enters the citric acid cycle, it is decarboxylated, oxidized, and combined with coenzyme A, forming acetyl CoA, carbon dioxide, and ________ .
7. In the first step of the citric acid cycle, acetyl CoA reacts with oxaloacetate to form ________ .
8. Carbons leave the ________ ________ ________ as molecules of carbon dioxide.
9. Dehydrogenase enzymes remove hydrogens from fuel molecules and transfer them to primary acceptors such as ________ and ________ .
10. ________ are intermediate electron carriers in the mitochondrial electron transport chain.
11. In the process of ________, electron transport and ATP synthesis are coupled by a proton gradient across the inner mitochondrial membrane.
12. The diffusion of protons through the ________ ________ complex results in ATP synthesis.
13. One important part of the feedback inhibition of aerobic respiration is the inhibitory effect of ATP on phosphofructokinase, an enzyme required in ________ .
14. A net profit of only two ATPs can be produced anaerobically from the ________ of one molecule of glucose, compared with a maximum of 38 ATPs produced in ________ ________ .
15. Organisms that can respire aerobically but switch to fermentation when oxygen is in short supply are called ________ ________ .
16. When deprived of oxygen, yeast cells obtain energy by fermentation, producing carbon dioxide, ATP, and ________ .
17. During strenuous muscle activity, the pyruvate in muscle cells may accept hydrogen to become ________ .
18. The role of the reaction described in Question 17 is to regenerate ________ required in glycolysis.

REVIEW QUESTIONS

1. What is the specific role of oxygen in most cells? What happens when cells that can only respire aerobically are deprived of oxygen?
2. Mitochondria are often referred to as the "power plants" of the cell. Justify this with a specific explanation.
3. What is the evolutionary significance of glycolysis?
4. Refer to Figure 7–8, the diagram of the steps in the citric acid cycle. Look at each reaction and, without reading the description, determine if it is a dehydrogenation, decarboxylation, or preparation reaction.
5. Sketch a mitochondrion and indicate the locations of the following: (a) enzymes of the citric acid cycle; (b) the electron transport system; (c) the proton gradient that drives ATP production.
6. Why is each of the following essential to chemiosmotic ATP synthesis? (a) electron transport chain; (b) proton gradient; (c) ATP synthase complex.
7. Explain the roles of the following in aerobic respiration: (a) NAD^+; (b) oxygen; (c) FAD.
8. Sum up how much energy (as ATP) is made available to the cell from a single glucose molecule by the operation of glycolysis, the formation of acetyl CoA, the citric acid cycle, and the electron transport system.
9. Trace the fate of hydrogens removed from glucose during glycolysis when oxygen is present in muscle cells; compare this to the fate of hydrogens removed from glucose when the amount of available oxygen is insufficient to support aerobic respiration.
10. Compare the ATP yields of aerobic respiration and fermentation.

YOU MAKE THE CONNECTION

1. Why is it advantageous that most anabolic reactions are generally not the reverse of catabolic reactions?
2. Why are the vast majority of organisms so similar in their energy metabolism?
3. What is the importance of the mitochondrial inner membrane in the coupling of electron transport and ATP synthesis?

RECOMMENDED READINGS

Alberts, B., D. Bray, J. Lewis, M. Raff, K. Roberts, and J.D. Watson. *Molecular Biology of the Cell,* 3rd ed. Garland Publishing, New York, 1994. An in-depth treatment of catabolic pathways.

Garrett, R.H., and C.M. Grisham. *Biochemistry.* Saunders College Publishing, 1995. A comprehensive biochemistry text with good coverage of cellular respiration and related aspects of metabolism.

Lodish, H., D. Baltimore, A. Berk, S.L. Zipursky, P. Matsudaira, and J. Darnell. *Molecular Cell Biology,* 3rd ed. Scientific American Books, New York, 1995. A readable and detailed account of cellular energy metabolism.

C H A P T E R 8

Photosynthesis: Capturing Energy

This young bean seedling is beginning to function as a photoautotroph. During germination, it lived as a heterotroph, using organic molecules stored in the seed. (Dwight Kuhn)

All living things are either producers or dependent on the activities of producers. Producers are **autotrophs** (from the Greek *auto,* "self," and *trophos,* "nourishing"), organisms that can make organic molecules from inorganic raw materials. Most producers, including plants, algae, and certain bacteria, are **photosynthetic autotrophs (photoautotrophs),** producers uniquely capable of absorbing and converting light energy into stored chemical energy by the process of **photosynthesis** (Fig. 8–1). Photoautotrophs use light energy to make ATP and other molecules that temporarily hold chemical energy. These molecules are unstable and cannot be stockpiled in the cell. Their energy drives the anabolic pathway by which a photosynthetic cell synthesizes stable organic molecules from simple inorganic compounds (carbon dioxide and water). These organic compounds are used not only as starting materials to synthesize all the other organic compounds the photosynthetic organism needs, but also for energy storage. They include relatively reduced compounds, such as glucose and other carbohydrates, that can be subsequently oxidized by cellular respiration or by some other catabolic pathway (see Chapter 7).

Consumers and decomposers are **heterotrophs** (from the Greek *heter,* "other," and *trophos,* "nourishing"), organisms that cannot make their own organic compounds and so must obtain them from other organisms. Heterotrophs live by consuming producers or organisms that have eaten producers. Photosynthesis sustains not only plants and most other producers, but also indirectly supports almost all animals and other organisms in the biosphere. Each year photosynthetic organisms convert carbon dioxide into billions of tons of organic molecules. The chemical energy stored in these molecules fuels the metabolic reactions that sustain almost all life.

LEARNING OBJECTIVES

After you have studied this chapter you should be able to

1. Diagram the internal structure of a chloroplast, and explain how its components interact and facilitate the process of photosynthesis.
2. Write a summary reaction for photosynthesis, showing the origin and fate of each substance involved.
3. Describe photosynthesis as a redox process.
4. Explain the relationship between the wavelength of light and its energy. Describe the physical properties of light, and explain how the absorption of photons can activate a pigment such as chlorophyll.
5. Describe what can happen to an electron in a pigment molecule such as chlorophyll when light energy is absorbed.
6. Distinguish between the light-dependent reactions and carbon fixation reactions of photosynthesis.
7. Contrast cyclic and noncyclic photophosphorylation.
8. Explain how a proton (H^+) gradient is established across the thylakoid membrane, and how this gradient functions in ATP synthesis.
9. Summarize the chemical reactions involved in the conversion of CO_2 to carbohydrate in the Calvin cycle, and indicate the roles of ATP and NADPH in the process.
10. Discuss how the C_4 pathway increases the effectiveness of the Calvin cycle in certain types of plants.

(a)

(b)

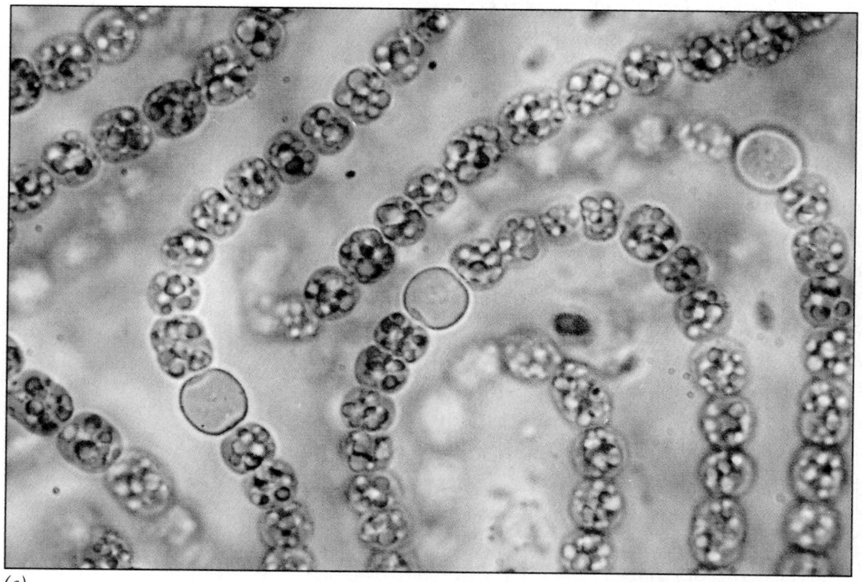

(c)

5 μm

FIGURE 8–1 Photosynthetic organisms include plants, algae, and prokaryotes. (***a***) Plants, such as the prayer plant (*Maranta* sp.), (***b***) algae (the kelp *Macrocystis integrifolia*), and (***c***) cyanobacteria (*Nostoc* sp.) are all organisms that obtain energy from light by the process of photosynthesis. Plants are primarily terrestrial, whereas algae and cyanobacteria are primarily aquatic. Algae represent a large size range, from those that are microscopic to large seaweeds. Cyanobacteria are prokaryotic organisms that photosynthesize like plants and algae. In addition, some photosynthetic bacteria trap the light's energy in a different way than the organisms shown. (*a*, James L. Castner; *b*, Flip Nicklin/Minden Pictures; *c*, Visuals Unlimited/R. Calentine)

Photosynthesis in Eukaryotes Takes Place in Chloroplasts

When a section of leaf tissue is examined under the microscope, we can see that the green pigment, **chlorophyll,** is not uniformly distributed in the cell but is confined to organelles called **chloroplasts.** In plants, chloroplasts are located mainly in the cells of the **mesophyll,** a tissue inside the leaf. Each mesophyll cell has 20 to 100 chloroplasts.

Electron microscopy reveals that the chloroplast, like the mitochondrion, is bounded by an outer and an inner membrane (Fig. 8–2). The inner membrane encloses a fluid-filled region called the **stroma,** which contains most of the enzymes required to produce carbohydrate. The inner chloroplast membrane also encloses a third system of membranes that forms an interconnected set of flat, disclike sacs called **thylakoids.** The thylakoid membranes enclose a fluid-filled interior space, the **thylakoid interior space** or simply the **thylakoid space.** In some regions, thylakoid sacs are arranged in stacks referred to as **grana** (sing., *granum*). Each granum looks something like a stack of coins, with each "coin" being a thylakoid disc (Fig. 8–3*e*). Some thylakoid membranes extend from one granum to another.

Chlorophyll and other photosynthetic pigments are part of the thylakoid membranes. These membranes, like the inner mitochondrial membrane (see Chapter 7), are involved in ATP synthesis.

Figure 8–4 presents a simplified view of chloroplast structure, emphasizing the physical relationships among the parts of the chloroplast involved in photosynthesis: the thylakoid membrane, the thylakoid interior space, and the stroma.

Photosynthetic prokaryotes have no chloroplasts, but thylakoid membranes are often arranged around the periphery of the prokaryotic cell, as infoldings of the plasma membrane.

Photosynthesis Is the Conversion of Light Energy to Chemical Bond Energy

During photosynthesis, a cell uses light energy captured by chlorophyll to power the synthesis of carbohydrates. The overall reactions of photosynthesis can be summarized as follows:

$$\underset{\text{Carbon dioxide}}{6\,CO_2} + \underset{\text{Water}}{12\,H_2O} \xrightarrow[\text{Chlorophyll}]{\text{Light}} \underset{\text{Glucose}}{C_6H_{12}O_6} + \underset{\text{Oxygen}}{6\,O_2} + \underset{\text{Water}}{6\,H_2O}$$

The equation is typically written in the form given above, with water on both sides, because water is a reactant in some reactions and a product in others. However, because there is no net yield of water, we can simplify the summary for purposes of discussion:

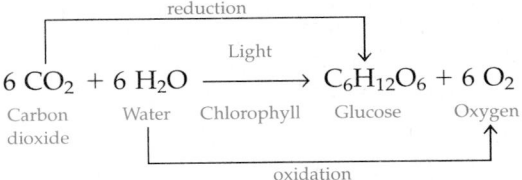

$$\underset{\text{Carbon dioxide}}{6\,CO_2} + \underset{\text{Water}}{6\,H_2O} \xrightarrow[\text{Chlorophyll}]{\text{Light}} \underset{\text{Glucose}}{C_6H_{12}O_6} + \underset{\text{Oxygen}}{6\,O_2}$$

(reduction: CO_2 → $C_6H_{12}O_6$; oxidation: H_2O → O_2)

When we analyze this process, it appears that hydrogen atoms are transferred from water to carbon dioxide to form carbohydrate, and so we recognize it as an oxidation-reduction (redox) reaction (see Chapter 6). In a redox reaction, one or more electrons (usually as part of one or more hydrogen atoms) are transferred from an electron donor (a reducing agent) to an electron acceptor (an oxidizing agent). When the electrons are transferred, some of their energy is transferred as well. However, the summary is somewhat misleading, because no direct

(*Text continues on page 195*)

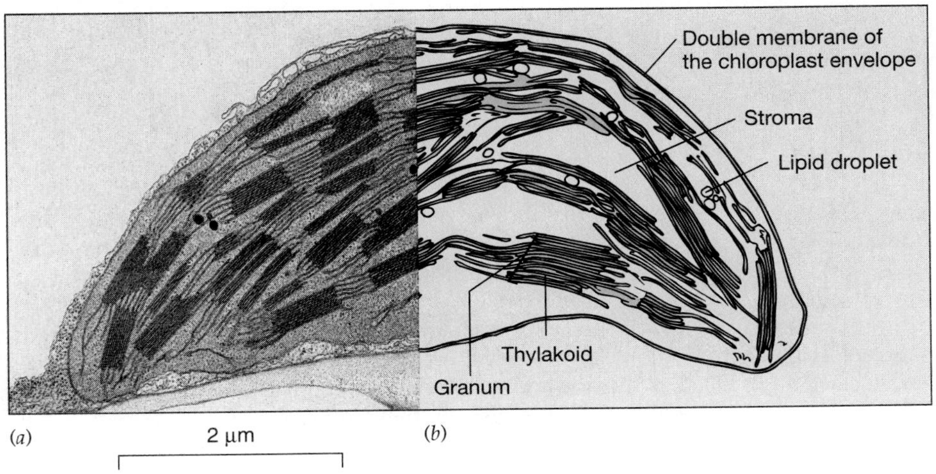

FIGURE 8–2 This transmission electron micrograph (*a*) of a chloroplast from a leaf cell of corn (*Zea mays*) is paired with an interpretive drawing (*b*). Note the grana, which are stacks of thylakoids. The pigments necessary for the light-capturing reactions of photosynthesis are part of the thylakoid membranes, whereas the enzymes necessary for the manufacture of carbohydrate are found in the stroma. (*a*, E. H. Newcomb and W. P. Wergin, University of Wisconsin/Biological Photo Service)

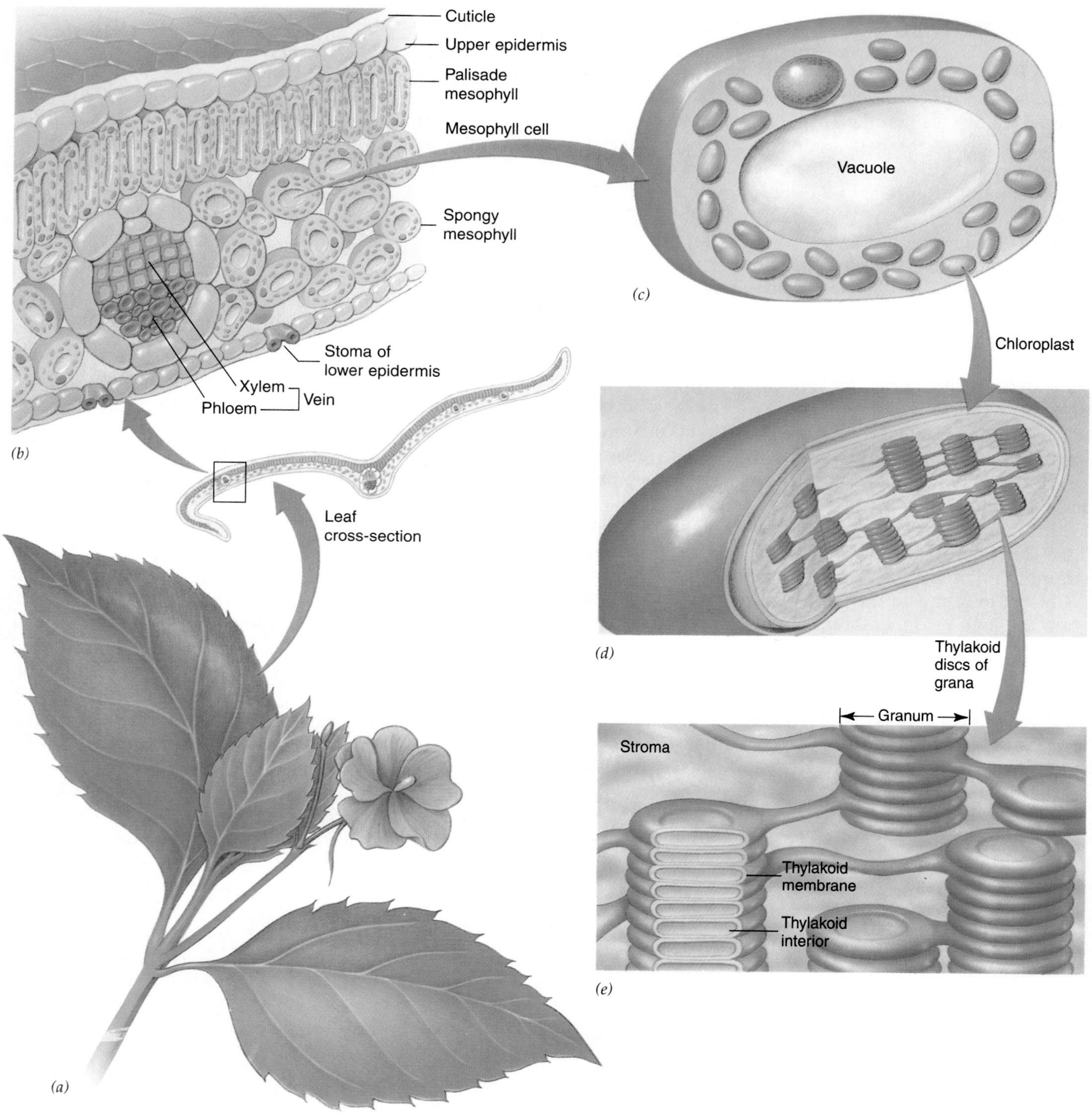

FIGURE 8–3 These drawings show the chloroplast in the context of the plant of which it is a part. (*a*) Photosynthesis occurs in the chlorophyll-containing tissues of the plant. (*b*) A cross section of a leaf reveals a structure marvelously adapted for photosynthesis. The middle portion of the leaf, the mesophyll, is the photosynthetic tissue. Carbon dioxide enters the leaf through tiny pores called stomata, and water is carried to the mesophyll in veins. (*c*) A typical mesophyll cell contains numerous chloroplasts. (*d*) Each chloroplast is surrounded by a double membrane. Within the chloroplast, membranous thylakoids are stacked to form grana. The fluid-filled matrix surrounding the grana is the stroma. (*e*) A close-up of the interior of the chloroplast. Chlorophyll is located in the membranes of the thylakoids, which are involved in the light-dependent reactions of photosynthesis. Synthesis of carbohydrate from carbon dioxide and water (the carbon fixation reactions of photosynthesis) takes place in the stroma.

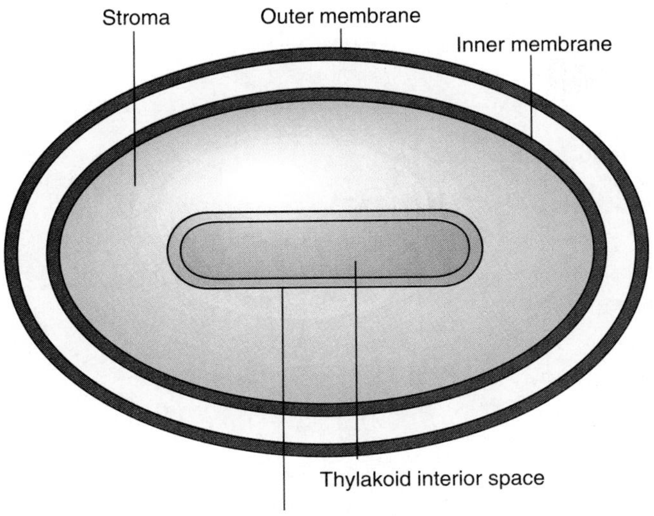

FIGURE 8–4 Although chloroplast structure appears complicated, the spatial relationships among the membranes and spaces that play important roles in photosynthesis are relatively simple. The thylakoid membranes, which contain chlorophyll (*green*) and enclose the thylakoid interior space, are responsible for reactions that require light. Carbohydrate synthesis (carbon fixation) takes place in the stroma, the fluid-filled space outside the thylakoids. The chloroplast's inner and outer membranes are not involved in photosynthesis.

transfer of hydrogen atoms actually occurs. The summary equation for photosynthesis describes what happens but not how it happens. The "how" is much more complex and involves many steps, many of which are redox reactions. The reactions of photosynthesis are divided into two parts: the *light-dependent reactions* and the *carbon fixation reactions* by which carbon dioxide is converted to carbohydrates (Table 8–1).

ATP and NADPH are the products of the light-dependent reactions

Light energy is converted to chemical energy in the **light-dependent reactions,** which are associated with the thylakoids. The role of light in these reactions is the key to resolving an apparent paradox associated with the summary of photosynthesis. During photosynthesis, electrons appear to be transferred from water to carbon dioxide, with the formation of carbohydrates. The electrons that make up the covalent bonds joining each hydrogen and the single oxygen in water are at a very low energy state; they do not have enough energy to overcome the attraction of the electronegative oxygen atom (see Chapter 2). Remember that in a redox reaction, an electron can move only from a higher energy level to a lower energy level. How, then, can electrons be transferred from water to or-

Table 8–1 SUMMARY OF PHOTOSYNTHESIS

Reaction Series	*Summary of Process*	*Needed Materials*	*End Products*
A. Light-dependent reactions (take place in thylakoid membranes)	Energy from sunlight used to split water, manufacture ATP, and reduce $NADP^+$		
1. Photochemical reactions	Chlorophyll activated; reaction center gives up photoexcited electron to electron acceptor	Light energy; pigments (chlorophyll)	Electrons
2. Electron transport	Electrons are transported along chain of electron acceptors in thylakoid membranes; electrons reduce $NADP^+$; splitting of water provides some of H^+ that accumulates inside thylakoid space	Electrons, $NADP^+$, H_2O, electron acceptors	NADPH, O_2
3. Chemiosmosis	H^+ are permitted to move across the thylakoid membrane down their gradient; they cross the membrane through special channels in ATP synthase complex; energy released is used to produce ATP	Proton gradient, ADP + P_i	ATP
B. Carbon fixation reactions (take place in stroma)	Carbon fixation: carbon dioxide is used to make sugar	Ribulose bisphosphate, CO_2, ATP, NADPH, necessary enzymes	Carbohydrates, ADP + P_i, $NADP^+$

Making the Connection

Photosynthesis and Aerobic Respiration

In what ways are photosynthesis and aerobic respiration alike, and how do they differ? Both are redox processes that are intimately connected with the energy requirements of living things. Although the series of steps by which the processes occur are quite different, their overall equations appear to be almost exactly opposite.

In photosynthesis:

$$6\ CO_2 + 12\ H_2O \xrightarrow{\text{Light energy}} C_6H_{12}O_6 + 6\ O_2 + 6\ H_2O$$

In aerobic respiration:

$$C_6H_{12}O_6 + 6\ O_2 + 6\ H_2O \longrightarrow 6\ CO_2 + 12\ H_2O + \text{ATP energy}$$

The following table compares other aspects of these processes. ▲

	Photosynthesis	*Aerobic Respiration*
Raw materials	CO_2, H_2O	$C_6H_{12}O_6$, O_2
End products	$C_6H_{12}O_6$, O_2	CO_2, H_2O
Which cells have these processes	Cells that contain chlorophyll (certain cells of plants, algae, and some bacteria)	Every actively metabolizing cell has aerobic respiration or some other energy-releasing pathway.
Sites involved (in eukaryotic cells)	Chloroplasts	Cytosol (glycolysis); mitochondria
ATP production	By photophosphorylation (a chemiosmotic process)	By substrate-level phosphorylation and by oxidative phosphorylation (a chemiosmotic process)
Principal electron transfer compound	$NADP^+$ is reduced to form NADPH*	NAD^+ is reduced to form NADH*
Location of electron transport chain	Thylakoid membrane	Mitochondrial inner membrane (cristae)
Source of electrons for electron transport chain	In noncyclic phosphorylation: H_2O (undergoes photolysis to yield electrons, protons, and oxygen)	Immediate source: NADH Ultimate source: glucose or other carbohydrate
Terminal electron acceptor for electron transport chain	In noncyclic phosphorylation: $NADP^+$ (becomes reduced to form NADPH)	O_2 (becomes reduced to form H_2O)

*NADPH and NADH are very similar hydrogen (i.e. electron) carriers, differing only in a single phosphate group. However, NADPH generally works with enzymes in anabolic pathways, such as photosynthesis. NADH is associated with catabolic pathways, such as cellular respiration.

ganic molecules? Light energy supplied to the system allows this transfer to occur, although by a rather indirect route.

The light-dependent reaction sequence begins as chlorophyll captures light energy, which causes one of its electrons to move to a higher energy state. The high-energy electron is transferred to an acceptor molecule and is replaced by an electron from water. When this happens, water is split and molecular oxygen is released (Fig. 8–5). The energy of the high-energy electrons is needed in the synthesis of the molecules that transfer the energy required for the endergonic carbon fixation reactions. These are ATP and a reduced coenzyme, NADPH[1] (nicotinamide adenine dinucleotide phosphate; see Chapter 6).

[1]Although the correct way to write the reduced form of $NADP^+$ is $NADPH + H^+$, for simplicity's sake we present the reduced form as NADPH.

FIGURE 8–5 On sunny days the oxygen released by aquatic plants may sometimes be visible as bubbles in the water. This plant *(Anacharis)* is actively carrying on photosynthesis, as evidenced by the oxygen bubbles. (Visuals Unlimited/ Bernard Wittich)

Carbohydrates are produced during the carbon fixation reactions

The ATP and NADPH molecules produced during the light-dependent phase are suited for transferring chemical energy but not for long-term energy storage. For this reason, some of their energy is transferred to chemical bonds in carbohydrates. In these reactions, which take place in the stroma, carbon atoms (from carbon dioxide) become "fixed" as part of the skeletons of organic molecules, a process called **carbon fixation** or **CO_2 fixation.** Because the carbon fixation reactions have no direct requirement for light, they are often referred to as the light-independent reactions or the "dark" reactions. However, they certainly do not require darkness; in fact, many of the enzymes involved are much more active in the light than in the dark. Furthermore, they do depend on the *products* of the light-dependent reactions (Fig. 8–6).

These carbohydrates are a source of energy because of their relatively high-energy electrons. The catabolism of glucose, which occurs most commonly through cellular respiration (see Chapter 7), involves a series of redox reactions in which the energy of these electrons is used to synthesize ATP.

Many of the simple carbohydrates produced in the carbon fixation reactions are used as starting materials in the anabolic reactions by which plant cells synthesize various types of organic compounds, including more complex carbohydrates, amino acids, lipids, etc.

Now that we have presented an overview of photosynthesis, let's examine the process more closely.

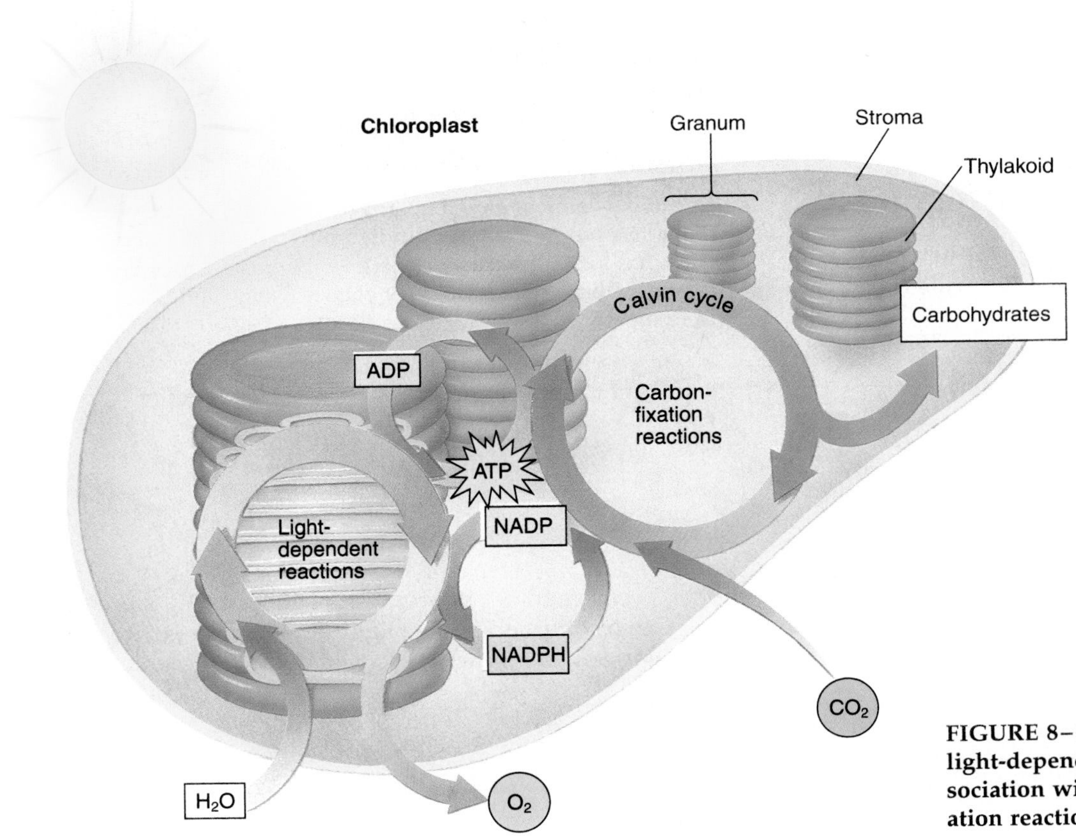

FIGURE 8–6 Photosynthesis consists of light-dependent reactions, which occur in association with the thylakoids, and carbon fixation reactions, which occur in the stroma.

Making the Connection

Nutrition and Metabolic Diversity

How can we categorize organisms on the basis of their nutritional needs? It is helpful to understand that nutrition has two main components: (1) how the organism obtains the carbon atoms required to make up the carbon skeletons of its organic molecules, and (2) how the organism obtains energy. Organisms obtain carbon in two ways. **Autotrophs** are able to carry out carbon fixation; they use carbon dioxide as a carbon source. **Heterotrophs** cannot fix carbon; they use preformed organic molecules (produced by other organisms) as a carbon source.

Energy can come from chemical nutrients or from light. **Phototrophs** use light as their primary energy source. **Chemotrophs** must obtain their energy from chemicals. Most chemotrophs oxidize organic molecules, but some are capable of capturing energy released by the oxidation of inorganic substances.

Animals and most decomposers, such as fungi and many bacteria, are **chemosynthetic heterotrophs (chemoheterotrophs);** they use organic molecules as a source of both energy and carbon. Plants and algae, as well as some bacteria, are **photosynthetic autotrophs (photoautotrophs);** they use light as a source of the energy required to carry out carbon fixation.

Because most commonly encountered organisms (even most bacteria) are either chemoheterotrophs or photoautotrophs, it is easy to be misled into thinking that all chemotrophs are also heterotrophs and that all phototrophs are also autotrophs. However, bacteria exhibit a great deal of metabolic diversity (see Chapter 23). A few bacteria are **chemosynthetic autotrophs (chemoautotrophs).** These producers obtain the energy they require by oxidizing simple inorganic compounds such as sulfur and ammonia; some of this energy is then used to carry out carbon fixation (see *Focus On: Life without the Sun* in Chapter 53).

Some other bacteria are **photosynthetic heterotrophs (photoheterotrophs),** organisms able to use light energy but unable to carry out carbon fixation. Photoheterotrophs must obtain carbon from organic sources (as "food"). ▲

In biology, almost any generalization has exceptions. This plant, commonly known as Indian pipes *(Monotropa uniflora)*, has lost the capacity to perform photosynthesis. It is a chemosynthetic heterotroph that obtains energy and carbon from organic molecules in the soil. (Richard Sheill/Dembinsky Photo Associates)

THE LIGHT-DEPENDENT REACTIONS CONVERT LIGHT ENERGY TO CHEMICAL BOND ENERGY

In the light-dependent reactions, the radiant energy from sunlight is used to make ATP and to reduce the electron acceptor molecule $NADP^+$, forming NADPH. The light energy captured by chlorophyll is temporarily stored in these two compounds. The light-dependent reactions are summarized as follows:

$$12\ H_2O + 12\ NADP^+ + 18\ ADP + 18\ P_i \xrightarrow[\text{Chlorophyll}]{\text{Light}} 6\ O_2 + 12\ NADPH + 18\ ATP$$

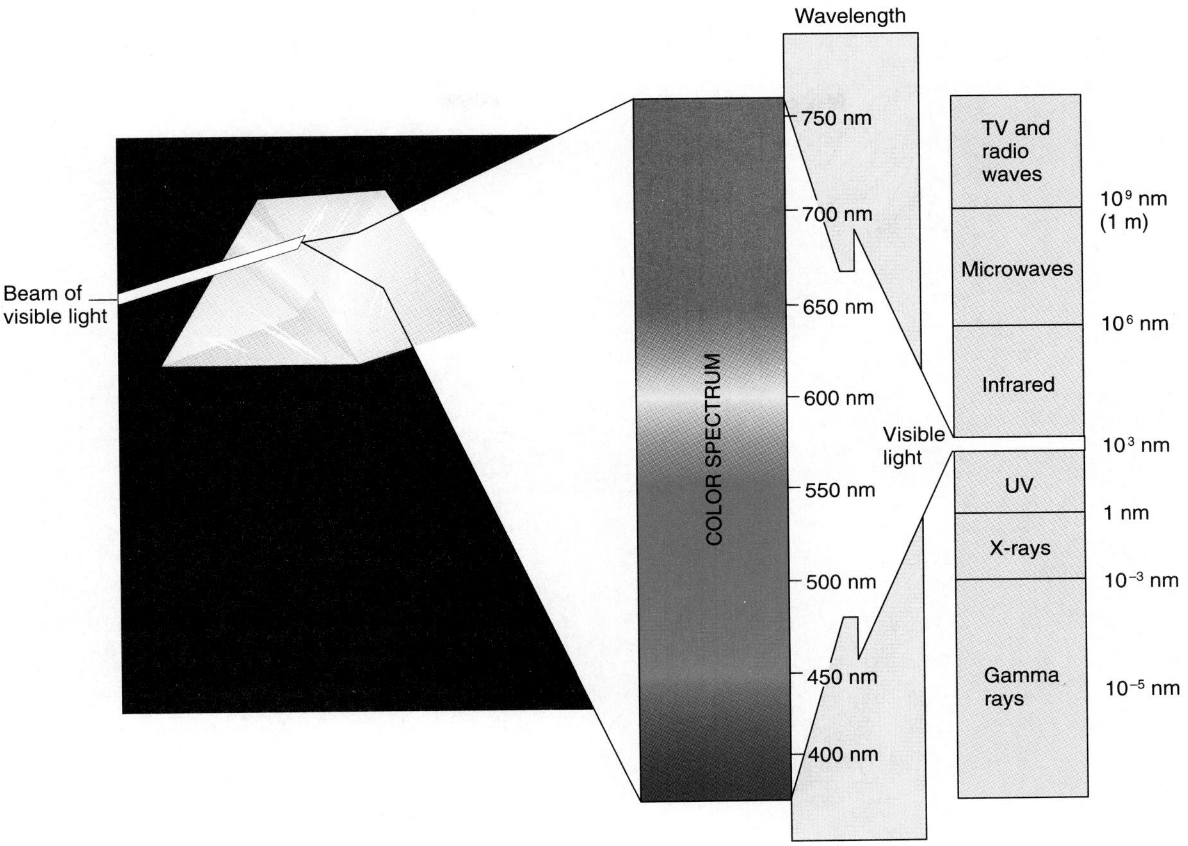

FIGURE 8–7 Visible light represents a small fraction of the wide range of wavelengths that make up the electromagnetic spectrum. White light consists of a mixture of wavelengths ranging from approximately 380 to 760 nm. A prism sorts light into its component colors by bending light of different wavelengths by different degrees. During photosynthesis, energy from visible light is used to synthesize organic compounds.

Light is composed of particles that travel as waves

Because most life on our planet depends on light, at least indirectly, it is important to understand the nature of light and how it permits photosynthesis to occur. Light is a very small portion of a vast, continuous spectrum of radiation called the electromagnetic spectrum (Fig. 8–7). All radiations in this spectrum travel in waves. A **wavelength** is the distance from one wave peak to the next. At one end of the electromagnetic spectrum are gamma rays, which have very short wavelengths (measured in nanometers). At the other end of the spectrum are radio waves with wavelengths so long they can be measured in kilometers. The portion of the electromagnetic spectrum from 380 nanometers to 760 nanometers is called the visible spectrum because humans can see it. The visible spectrum includes all the colors of the rainbow; violet has the shortest wavelength and red has the longest.

Light behaves not only as waves do but also as particles. Light is composed of small particles, or packets, of energy called **photons.** The energy of a photon is *inversely* proportional to its wavelength; shorter wavelength light has more energy per photon than does longer wavelength light.

Why does photosynthesis depend on light detectable by the human eye (visible light) rather than on some other wavelength of radiation? We can only speculate on the answer. One reason may be that much of the radiation reaching our planet from the sun is within this portion of the electromagnetic spectrum; thus, organisms may have evolved the ability to use visible light because it was the most abundant form available. Perhaps more importantly, radiation within the visible light portion of the spectrum excites certain types of biological molecules, moving electrons into higher energy levels. Radiation with wavelengths longer than those of visible light does not possess enough energy to excite these biological mol-

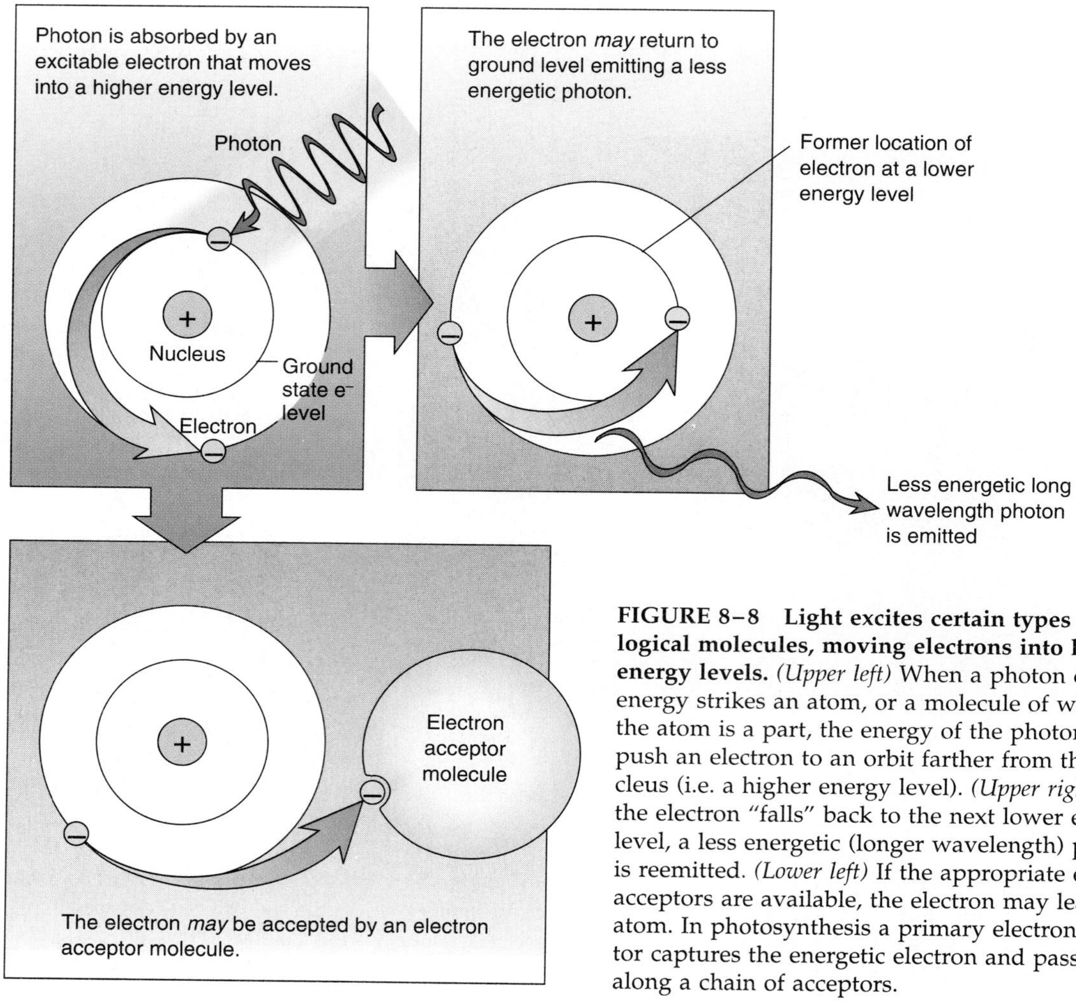
Photon is absorbed by an excitable electron that moves into a higher energy level.
Photon
Nucleus
Ground state e⁻ level
Electron
The electron *may* return to ground level emitting a less energetic photon.
Former location of electron at a lower energy level
Less energetic long wavelength photon is emitted
Electron acceptor molecule
The electron *may* be accepted by an electron acceptor molecule.

FIGURE 8–8 Light excites certain types of biological molecules, moving electrons into higher energy levels. *(Upper left)* When a photon of light energy strikes an atom, or a molecule of which the atom is a part, the energy of the photon may push an electron to an orbit farther from the nucleus (i.e. a higher energy level). *(Upper right)* If the electron "falls" back to the next lower energy level, a less energetic (longer wavelength) photon is reemitted. *(Lower left)* If the appropriate electron acceptors are available, the electron may leave the atom. In photosynthesis a primary electron acceptor captures the energetic electron and passes it along a chain of acceptors.

ecules. Radiation with wavelengths shorter than those of visible light is so energetic that it disrupts the bonds of many biological molecules.

Photons interact with atoms in a variety of ways, all of which depend on the electron configuration of the atom. Recall that an atom consists of an atomic nucleus surrounded by electrons located in one or more energy levels (see Chapter 2). The lowest energy state an atom possesses is called the **ground state,** but energy can be added to an electron so that it attains a higher energy level. When an electron is raised to a higher energy level than its ground state, the atom is said to be *excited,* or energized.

When a molecule absorbs a photon of light energy, one of its electrons is raised to a higher energy state. One of two things then happens, depending on the atom and its surroundings (Fig. 8–8). The electron may return to its ground state. If this happens its energy is dissipated as heat or as light of a longer wavelength than the wavelength of the absorbed light; this emission of light is called **fluorescence.** Alternatively, the excited electron may leave the atom and be accepted by an electron acceptor molecule; this is what occurs in photosynthesis.

Chlorophyll is found in the thylakoid membrane

Thylakoid membranes contain several kinds of pigments, which are substances that absorb visible light. Different pigments absorb light of different wavelengths. Chlorophyll, the main pigment of photosynthesis, absorbs light primarily in the blue and red regions of the visible spectrum. Green light is not appreciably absorbed by chlorophyll; instead, it is reflected. Plants usually appear green because their leaves reflect most of the green light that strikes them.

A chlorophyll molecule (Fig. 8–9) has two main parts: one captures energy and the other holds the molecule in place in the thylakoid membrane. Light energy is absorbed by a complex ring, called a *porphyrin ring,* made up of joined smaller rings composed of carbon and nitrogen atoms. The porphyrin ring of chlorophyll is strikingly

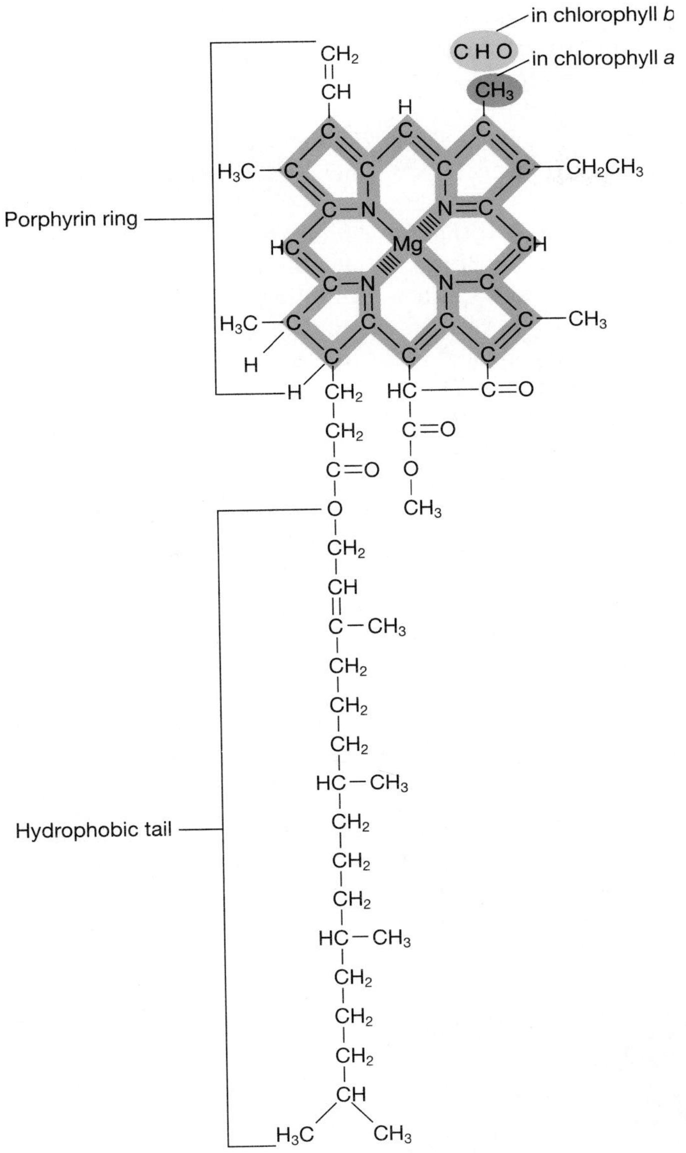

FIGURE 8–9 Chlorophyll consists of a porphyrin ring and a hydrocarbon tail. The porphyrin ring, which contains a magnesium atom in its center, is excited by light. Note that the ring contains a system of alternating double and single bonds; these are commonly found in molecules that strongly absorb visible light. At the top right corner of the diagram, the methyl group ($—CH_3$) distinguishes chlorophyll *a* from chlorophyll *b*, which is an aldehyde because it has a terminal carbonyl group ($—CHO$) in this position. The hydrophobic tail is embedded in the thylakoid membrane.

similar to the heme portion of the red pigment hemoglobin in red blood cells (see Fig. 3–24). However, unlike heme, which contains an atom of iron in the center of the ring, chlorophyll contains an atom of magnesium in that position. The chlorophyll molecule is embedded in the thylakoid membrane by a long hydrophobic hydrocarbon tail. Because of their shape, many chlorophyll molecules can be grouped together like a stack of saucers. Each thylakoid membrane is filled with precisely oriented chlorophyll molecules, an arrangement that permits transfer of energy from one molecule to another.

There are several kinds of chlorophyll. The most important is **chlorophyll *a*,** the pigment that initiates the light-dependent reactions. **Chlorophyll *b*** is an accessory pigment that also participates in photosynthesis. It differs from chlorophyll *a* only in a functional group on the porphyrin ring: the methyl group ($—CH_3$) in chlorophyll *a* is replaced in chlorophyll *b* by a terminal carbonyl group ($—CHO$). This difference shifts the wavelengths of light absorbed and reflected by chlorophyll *b*, making it yellow-green, whereas chlorophyll *a* is bright green.

Plant cells also have other accessory photosynthetic pigments, such as **carotenoids,** which are yellow and orange (see Fig. 3–15). Carotenoids absorb different wavelengths of light than chlorophyll does and so broaden the spectrum of light that provides energy for photosynthesis. Chlorophyll may be excited by light directly or by energy passed to it from accessory pigments that have become excited by light. Thus, when a carotenoid molecule is excited, its energy can be transferred to chlorophyll *a*.

Chlorophyll is the main photosynthetic pigment

As we have seen, the thylakoid membrane contains more than one kind of pigment. It is possible to determine which of these pigments is mainly responsible for photosynthesis by comparing the wavelengths of light absorbed by each pigment with the wavelengths of light that are most effective in promoting the reactions of photosynthesis. An instrument called a spectrophotometer is used to measure the relative abilities of different pigments to absorb different wavelengths of light. The **absorption spectrum** of a pigment is a plot of its absorption of light of different wavelengths (Fig. 8–10).

The relative effectiveness of these different wavelengths of light in photosynthesis is given by an **action spectrum** of photosynthesis. The first action spectrum was obtained in one of the classic experiments in biology. In 1883 the German biologist T. W. Engelmann carried out an experiment that took advantage of the shape of the chloroplast in *Spirogyra*, a green alga that occurs as long, filamentous strands in freshwater habitats, especially slow-moving or still waters (Fig. 8–11*a*). The individual cells of *Spirogyra* are exquisitely beautiful, each containing a long, spiral-shaped, emerald-green chloroplast embedded in cytoplasm. Engelmann exposed these cells to a color spectrum produced by passing light through a prism. He reasoned that if chlorophyll were indeed responsible for photosynthesis, then it would take place most rapidly in the areas where the chloroplast was illuminated by the colors most readily absorbed by chlorophyll.

Yet how could photosynthesis be measured in those technologically unsophisticated days? Engelmann knew that photosynthesis produces oxygen, and that certain motile bacteria are attracted to areas of high oxygen concentration (Fig. 8–11*b*). He determined the action spectrum of photosynthesis by observing that the bacteria swam toward the portions of *Spirogyra* located in the red and blue regions of the spectrum. The fact that the bacteria did not move toward red and blue areas when *Spirogyra* was absent showed that bacteria are not merely attracted to any region where red or blue light is present. Because the action spectrum of photosynthesis closely matched the absorption spectrum of chlorophyll, Engelmann concluded that chlorophyll in the chloroplasts (and not another compound in another organelle) is responsible for photosynthesis. Numerous studies using sophisticated instruments have since confirmed Engelmann's conclusions.

The action spectrum of photosynthesis does not parallel the absorption spectrum of chlorophyll exactly (Fig. 8–10). This is because accessory pigments, such as carotenoids, transfer some of the energy of excitation produced by green light to chlorophyll molecules. The pres-

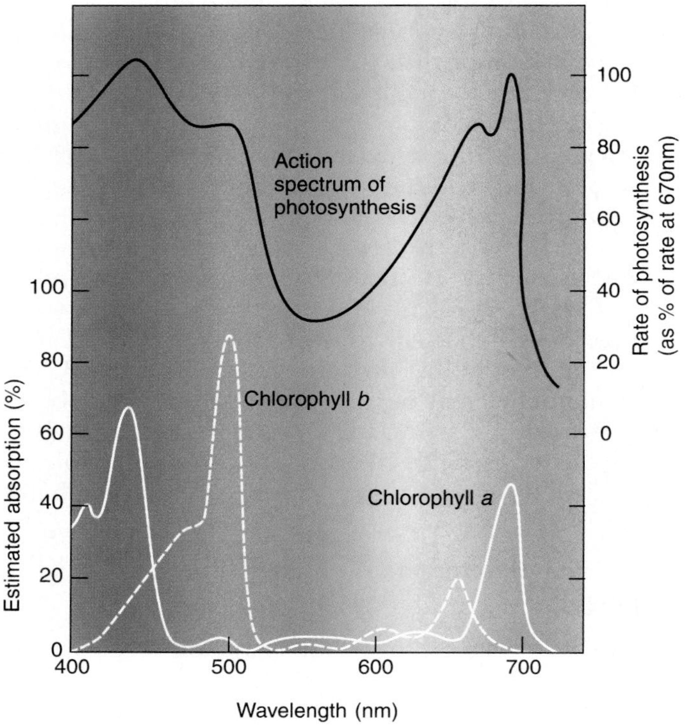

FIGURE 8–10 The combined absorption spectra for chlorophyll *a* and chlorophyll *b* roughly parallel the action spectrum for photosynthesis. The white curves at the bottom illustrate the absorption spectra for chlorophyll. Chlorophyll *a* *(solid curve)* and *b* *(dashed curve)* absorb light mainly in the blue and red regions of the spectrum. The black curve at the top (action spectrum of photosynthesis) illustrates the effectiveness of various wavelengths of light in powering photosynthesis.

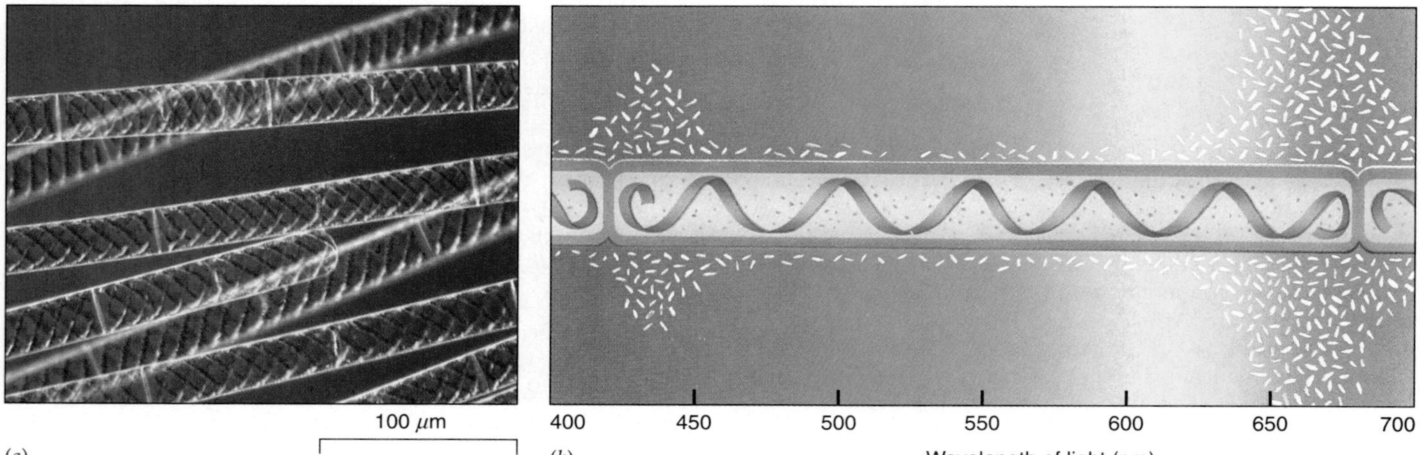

(*a*) (*b*)

FIGURE 8–11 Engelmann demonstrated in a classic experiment that the wavelengths of light most effective for photosynthesis are those most strongly absorbed by chlorophyll. (***a***) Filaments of *Spirogyra*, the green alga that Engelmann used in his experiment. (***b***) Engelmann illuminated a filament of *Spirogyra* with light that had been passed through a prism, producing a spectrum. In this way, different parts of the filament were exposed to different wavelengths of light. He estimated the rate of photosynthesis indirectly, by observing the movement of aerobic bacteria toward the portions of the algal filament emitting the most oxygen. Watching through a microscope, Engelmann observed that the bacteria aggregated most densely along the cells in the blue and red portions of the spectrum. This indicated that blue and red light works most effectively for photosynthesis. (*a*, Visuals Unlimited/T. E. Adams) ence of these accessory photosynthetic pigments can be demonstrated by chemical analysis of almost any leaf, although it is obvious in temperate climates when leaves change color in the fall. Toward the end of the growing season, chlorophyll is broken down (and its magnesium is stored in the permanent tissues of the tree), leaving accessory pigments in the leaves.

Photosystems I and II include light-harvesting antenna complexes of pigment molecules

The light-dependent reactions of photosynthesis begin when chlorophyll *a* and/or accessory pigments absorb light. According to the currently accepted model, chlorophyll molecules, accessory pigments, and associated electron acceptors are organized in the thylakoid membrane into units called **photosystems.** They are referred to as Photosystem I and Photosystem II for reasons that are mainly historical: Photosystem I was described first, probably evolved first, and can sometimes function independently. The pigment molecules are arranged as highly ordered groups of 200 to 300 molecules associated with specific enzymes and other proteins. Each such array of pigments acts as a "light-harvesting antenna complex," each of which includes a **reaction center** made up of a few special chlorophyll *a* molecules. These are distinguishable because they are associated with proteins in a way that causes a slight shift in their absorption spectrum. Ordinary chlorophyll *a* has a strong absorption peak at about 660 nanometers. In contrast, the chlorophyll *a* molecules that make up the reaction center associated with Photosystem I have an absorption peak at 700 nanometers and are referred to as **P700.** The reaction center of Photosystem II (P680) is made up of chlorophyll *a* molecules with an absorption peak at 680 nanometers.

When a pigment molecule absorbs light energy (Fig. 8–12), that energy is passed from one pigment molecule to another until it reaches the reaction center. When the energy reaches a molecule of P700 or P680 at the reaction center, an electron is raised to a high energy level. This excited electron can be donated to a primary electron acceptor, which is reduced in the process.

The photosynthetic electron transport system is coupled to ATP synthesis

The primary electron acceptor is part of an **electron transport chain** embedded in the thylakoid membrane. Each member of the chain can exist in an oxidized (lower energy) form and a reduced (higher energy) form. The electron accepted by the primary acceptor is at a high energy level; it is passed from one carrier to the next in a series of exergonic redox reactions, losing some of its energy at each step. Some of the energy given up by the electron is not lost to the system, however; it is used to drive the synthesis of ATP (an endergonic reaction). Because the synthesis of ATP (that is, the phosphorylation of ADP) is coupled to the transport of electrons that have been excited by photons of light, the process is called **photophosphorylation.**

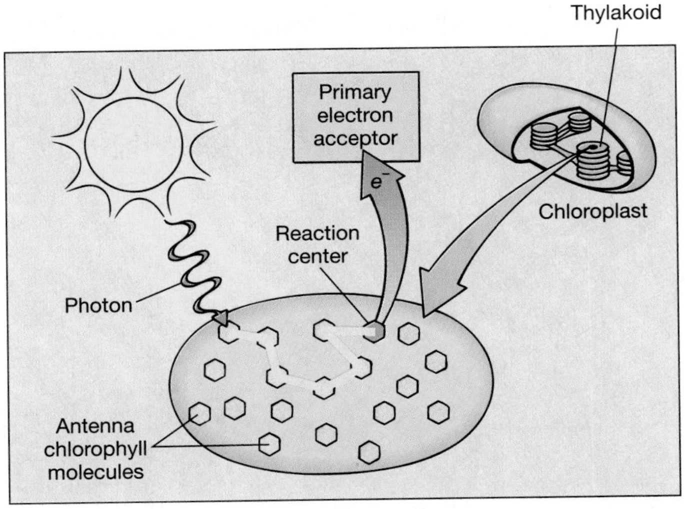

FIGURE 8–12 Chlorophyll molecules are arranged in arrays that act as light harvesting antennae. When a molecule in the complex absorbs a photon, the energy of that photon is funneled into the reaction center of the complex. The very orderly arrangement of the pigment molecules allows for direct transfer of energy within the complex (represented by the jagged yellow line). Light is not reflected from molecule to molecule, nor are electrons transferred. When this energy reaches a chlorophyll in the reaction center, an electron becomes excited (raised to a high energy level) and is accepted by a primary acceptor.

The chemiosmotic model explains the coupling of ATP synthesis and electron transport

The pigments and electron acceptors of the light-dependent reactions are embedded in the thylakoid membrane. Energy released from electrons traveling through the chain of acceptors is used to pump protons from the stroma, across the thylakoid membrane, and into the thylakoid interior space. Because protons are actually hydrogen ions (H^+), the accumulation of protons causes the pH of the thylakoid interior to fall. In fact, the pH approaches 4 in bright light. This produces a difference of about 3.5 pH units across the thylakoid membrane, more than a 1000-fold difference in hydrogen ion concentration.

Thus the pumping of protons in the chloroplast results in the formation a proton gradient across the thylakoid membrane (see Chapter 5). The gradient has a great deal of free energy because of its state of low entropy. How does the chloroplast convert that energy to a more useful form?

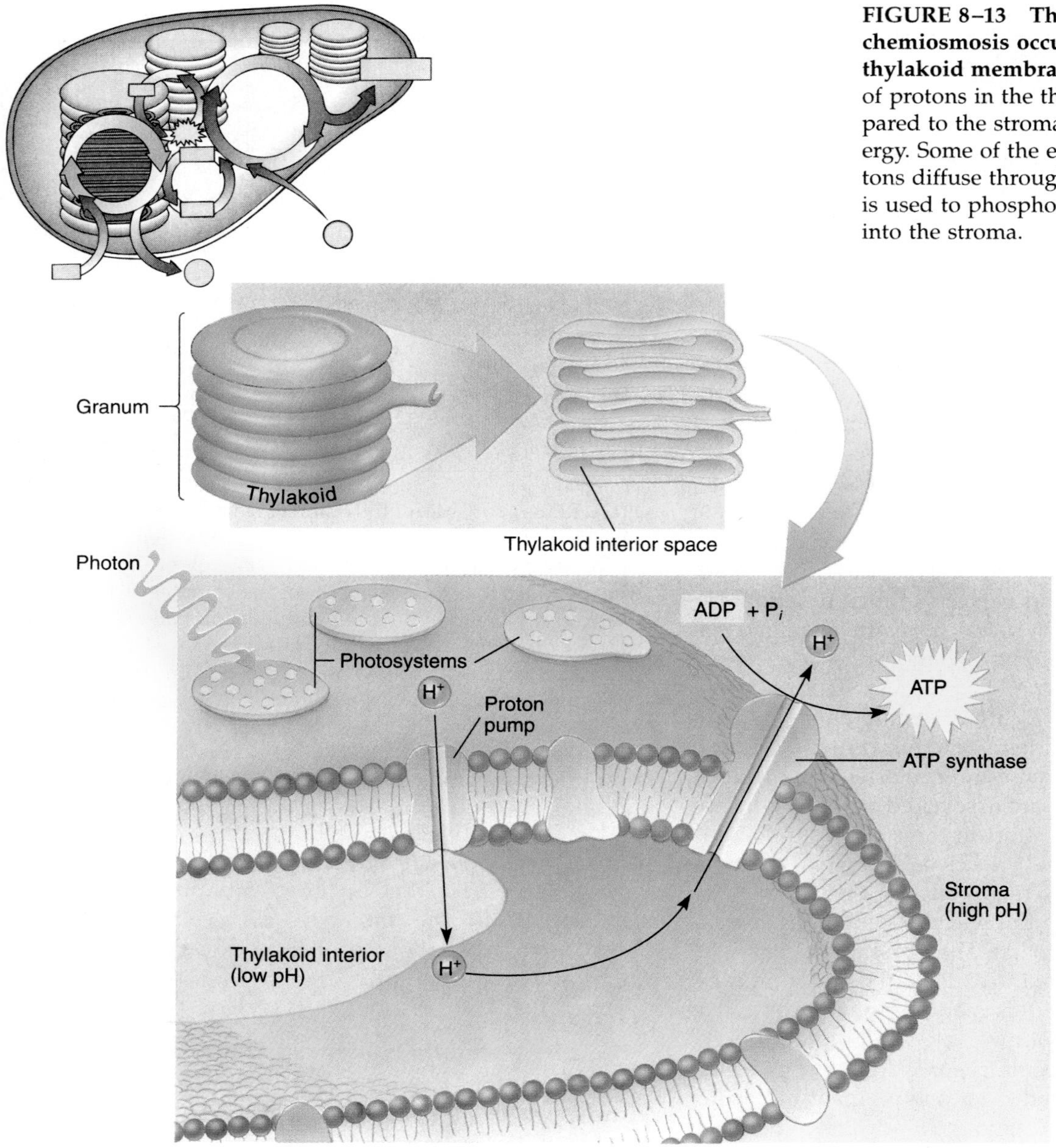

FIGURE 8–13 The production of ATP by chemiosmosis occurs in association with the thylakoid membrane. The high concentration of protons in the thylakoid interior space (compared to the stroma) is a form of potential energy. Some of the energy released as the protons diffuse through the ATP synthase complex is used to phosphorylate ADP. ATP is released into the stroma.

In accordance with the general principles of diffusion (see Chapter 5), the highly concentrated protons inside the thylakoid might be expected to diffuse out readily. However, they are prevented from doing so because the thylakoid membrane is impermeable to H^+ except through certain channels formed by an enzyme called **ATP synthase.** ATP synthase, a transmembrane protein, forms complexes so large they can be seen in electron micrographs; these project into the stroma. As the protons diffuse through an ATP synthase complex, free energy decreases as a consequence of an increase in entropy. Each ATP synthase complex couples this exergonic process of diffusion down a concentration gradient (see Chapter 6) to the endergonic process of phosphorylation of ADP to form ATP, which is released into the stroma (Fig. 8–13).

This mechanism by which phosphorylation of ADP is coupled to diffusion down a proton gradient is called **chemiosmosis.** As the essential connection between electron transport chains and phosphorylation of ADP, chemiosmosis is a basic mechanism of energy coupling in cells. Chemiosmosis also occurs in cellular respiration (see Chapter 7).

Noncyclic photophosphorylation produces ATP and NADPH

Both Photosystems I and II are used in **noncyclic photophosphorylation,** which is the more common light-dependent reaction. Light energizes electrons, which pass down an electron transport chain from the ultimate elec-

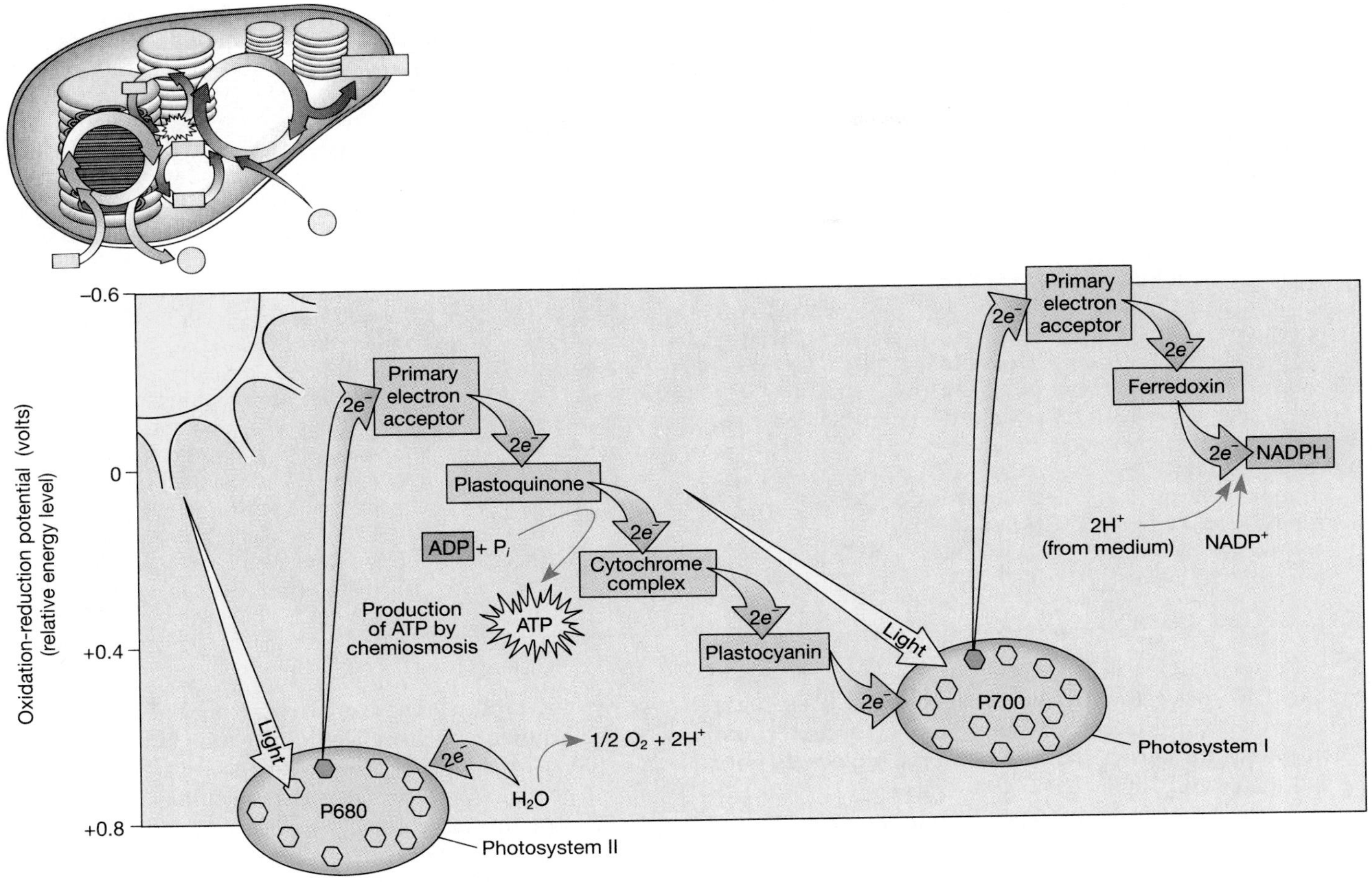

FIGURE 8–14 In noncyclic photophosphorylation, the formation of ATP is coupled to the one-way flow of energized electrons from water to NADP$^+$. Single electrons actually pass down the electron transport chain; two are shown in the figure because two are required to form one NADPH. Electrons are supplied to the system by the splitting of water by Photosystem II, with the release of molecular oxygen as a waste product. When Photosystem II is activated by absorbing photons, electrons are passed along an electron transport chain and are eventually donated to Photosystem I. Electrons in Photosystem I are "re-excited" by the absorption of additional light energy and are finally passed to NADP$^+$. This pathway is sometimes referred to as the Z scheme because of its zigzag route, which is a consequence of the fact that light energy is required to excite electrons in Photosystem II and again in Photosystem I.

tron source, water, to the terminal electron acceptor, NADP$^+$. For every two electrons that enter this pathway, there is an energy yield of two ATP molecules (photophosphorylation) and one NADPH molecule (Fig. 8–14).

NADPH is formed by transfer of high-energy electrons to NADP$^+$

We will begin our discussion with the events associated with Photosystem I. A pigment molecule in an antenna complex in Photosystem I absorbs a photon of light. The absorbed energy is transferred to the reaction center, where it excites an electron in a molecule of P700. This excited electron is transferred to a primary acceptor, which transfers it to an iron-containing intermediate carrier, ferredoxin. Ferredoxin then transfers the electron to NADP$^+$. The electron transport chain must furnish two electrons to reduce NADP$^+$ to NADPH. When NADP$^+$ accepts the two electrons, they unite with a proton; hence the reduced form of NADP$^+$ is NADPH, which is released into the stroma.

P700 ionizes, becoming positively charged when it gives up an electron to the primary acceptor; the missing electron is replaced by one donated by Photosystem II. Photosystem II must donate two electrons to Photosystem I for every NADPH produced.

Water is the electron source for noncyclic photophosphorylation

Where do the electrons donated by Photosystem II originate? Like Photosystem I, Photosystem II is activated when a pigment molecule in an antenna complex absorbs a photon of light energy. The energy is transferred to the reaction center, where it causes an electron in a molecule

Focus On

The Evolution of Photosystems I and II

Photosynthesis is an extremely ancient biological process that has apparently changed a great deal since it first appeared more than 3 billion years ago. The use of light energy to manufacture organic molecules first evolved in ancient bacteria that were similar to the green sulfur bacteria existing today.

Initially there was only one photosystem, Photosystem I, which used a green pigment called **bacteriochlorophyll** to gather light energy. Photosystem I operated alone, generating ATP from light energy by cyclic photophosphorylation. However, cyclic photophosphorylation does not provide the reducing power of NADPH, which is needed to manufacture carbohydrate molecules from CO_2. (Recall that in cyclic photophosphorylation, the electrons from chlorophyll are not passed to $NADP^+$ but are instead returned to chlorophyll.) Ancient photosynthetic bacteria, like some of their modern counterparts, used electron donors such as hydrogen sulfide (H_2S) rather than H_2O to generate the reducing power needed to manufacture carbohydrates in photosynthesis:

$$H_2S \longrightarrow S + 2\,H^+ + 2\,e^-$$

This process is not very efficient, however.

Around 3.1 billion years ago, a new group of prokaryotes called **cyanobacteria** evolved (see Chapter 20). These ancient cyanobacteria were probably very similar to modern cyanobacteria. The light-requiring reactions of living cyanobacteria are similar to those of photosynthetic eukaryotes, including plants. They possess chlorophyll *a* instead of bacteriochlorophyll, and can carry out noncyclic photophosphorylation because they have Photosystem II in addition to Photosystem I. Water provides the electrons to generate NADPH, which in turn provides the reducing power required to manufacture carbohydrate molecules from CO_2. ■

of P680 to move to a higher energy level. This high-energy electron is accepted by a primary acceptor and then passes through a chain of acceptor molecules until it is donated to P700.

A molecule of P680 that has given up an excited electron to the primary acceptor is positively charged. This activated P680 is an oxidizing agent so strong that it is capable of oxidizing an oxygen atom that is part of a water molecule. In a reaction catalyzed by a unique enzyme, water is split by a process called **photolysis** ("breaking by light") into its components: two electrons, two protons (H^+), and oxygen. Each electron is donated to a P680 molecule and the protons are released into the thylakoid interior space. Because oxygen does not exist in atomic form, the oxygen produced by splitting one water molecule is written $\frac{1}{2}\,O_2$. Two water molecules must be split to yield one molecule of oxygen, which is a waste product released into the atmosphere. The photolysis of water is a remarkable reaction, but its name is somewhat misleading because it implies that water is broken by light. Actually, light breaks water indirectly, by activating P680 molecules.

ATP is produced by chemiosmosis

Photophosphorylation occurs by a chemiosmotic mechanism. As electrons are transferred along the electron transport chain that connects Photosystem II with Photosystem I, they lose energy. Some of the energy released is used to pump protons across the thylakoid membrane, from the stroma to the thylakoid interior space, producing a proton gradient. By contrast, the electron transport chain that leads from Photosystem I to $NADP^+$ does not pump any protons. There are two additional contributors to the gradient: each molecule of water that undergoes photolysis yields two protons to the thylakoid interior space, and the formation of one molecule of NADPH removes one proton from the stroma. ATP is released into the stroma as protons diffuse down the gradient through the ATP synthase complex.

Noncyclic photophosphorylation is a continuous linear process

In the presence of light, there is a continuous, one-way flow of electrons from the ultimate electron source, water, to the terminal electron acceptor, $NADP^+$. Water undergoes enzymatically catalyzed photolysis to replace photoexcited electrons donated to the electron transport chain by molecules of P680 in Photosystem II. Photoexcited electrons donated to the electron transport chain by P700 (in Photosystem I) are replaced by electrons traveling down the electron transport chain that connects Photosystem II with Photosystem I. Two electrons (the number released from one water molecule) are required to form one molecule of NADPH. ATP is produced by chemiosmosis and, like NADPH, is released into the stroma, where both are required in the carbon fixation reactions.

Cyclic photophosphorylation produces ATP only

Only Photosystem I is involved in **cyclic photophosphorylation,** the simplest light-dependent reaction. The pathway is cyclic because photoexcited electrons that

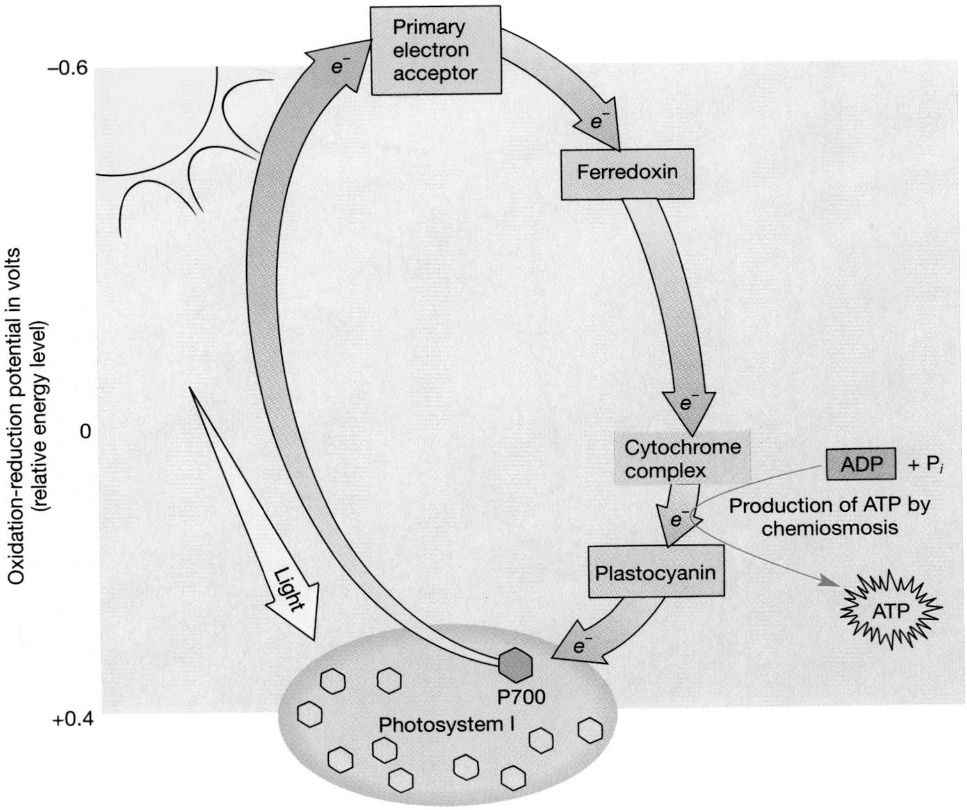

FIGURE 8–15 Photosystem I can work independently to produce ATP by cyclic photophosphorylation. When a pigment molecule in Photosystem I absorbs a photon, energy is transferred to a P700 molecule in the reaction center. P700 then gives up its excited electron to a primary electron acceptor. Electrons are transferred through a sequence of electron acceptors (losing energy with each transfer) and returned to P700. Because the system is cyclic, there is no electron source and no terminal electron acceptor. Photolysis of water does not occur, and neither oxygen nor NADPH is produced. Note that the electron transport chain includes a cytochrome complex. This is the same cytochrome complex that is part of the chain connecting Photosystem II with Photosystem I in noncyclic phosphorylation (Fig. 8–14). Because this cytochrome complex pumps protons across the thylakoid membrane, ATP synthesis can occur by chemiosmosis.

originate from P700 at the reaction center eventually return to P700 (Fig. 8–15). In the presence of light, there is a continuous flow of electrons through an electron transport chain within the thylakoid membrane. As they are passed from one acceptor to another, the electrons lose energy, some of which is used to pump protons across the thylakoid membrane. An ATP synthase enzyme in the thylakoid membrane uses the energy of the proton gradient to manufacture ATP. NADPH is not produced, water is not split, and oxygen is not generated. By itself, cyclic photophosphorylation could not serve as the basis of photosynthesis because, as we explain in the next section, NADPH is required to reduce CO_2 to carbohydrate.

Cyclic photophosphorylation occurs in plant cells when there is too little $NADP^+$ to accept electrons from ferredoxin. Biologists think that this process was used by ancient bacteria to produce ATP from light energy (see *Focus On: The Evolution of Photosystems I and II*). A reaction pathway analogous to cyclic photophosphorylation in plants is present in some modern photosynthetic bacteria. Noncyclic and cyclic photophosphorylation are compared in Table 8–2.

Table 8–2 A COMPARISON OF NONCYCLIC AND CYCLIC PHOTOPHOSPHORYLATION

	Noncyclic photophosphorylation	Cyclic photophosphorylation
Ultimate electron source	H_2O	None
Oxygen released?	Yes (from H_2O)	No
Terminal electron acceptor	$NADP^+$	None
Form in which energy is temporarily captured	ATP (by chemiosmosis); NADPH	ATP (by chemiosmosis)
Photosystem(s) required	I & II	I only

FIGURE 8–16 The classic experiments carried out by Calvin, Benson, and others in the 1950s elucidated the steps in carbon fixation of photosynthesis, also known as the Calvin cycle. Calvin and his colleagues grew algae in the green "lollipop." CO_2 labeled with carbon-14 was bubbled through the algae, and they were periodically killed by dumping the "lollipop" contents into a beaker of boiling alcohol. By identifying which compounds contained the ^{14}C at different times, Calvin was able to determine the steps of carbon fixation in photosynthesis. (Courtesy of Melvin Calvin, University of California, Berkeley)

THE CARBON FIXATION REACTIONS REQUIRE ATP AND NADPH

In carbon fixation, the energy of ATP and NADPH is used in the formation of organic molecules from carbon dioxide. The carbon fixation reactions may be summarized as follows:

$$12\ \text{NADPH} + 18\ \text{ATP} + 6\ \text{CO}_2 \longrightarrow \text{C}_6\text{H}_{12}\text{O}_6 + 12\ \text{NADP}^+ + 18\ \text{ADP} + 18\ \text{P}_i + 6\ \text{H}_2\text{O}$$

Most plants use the Calvin (C_3) cycle to fix carbon

Carbon fixation occurs in the stroma through a sequence of reactions known as the **Calvin cycle.** M. Calvin, A. Benson, and others at the University of California were able to elucidate the details of this cycle; for his work Dr. Calvin was awarded a Nobel Prize in 1961 (Fig. 8–16).

As you read the following description of the Calvin cycle, follow the reactions illustrated in Figure 8–17. The cycle begins when a molecule of carbon dioxide reacts with a highly reactive phosphorylated five-carbon compound, **ribulose bisphosphate (RuBP).** This reaction is catalyzed by the enzyme **ribulose bisphosphate carboxylase,** also known as **Rubisco.** The product of this reaction is an unstable six-carbon intermediate, which immediately breaks down into two molecules of **phosphoglycerate (PGA)** with three carbons each. The carbon that was originally part of a carbon dioxide molecule is now part of a carbon skeleton; the carbon has been "fixed." Because the product of the initial carbon fixation reaction is a three-carbon compound, the Calvin cycle is also known as the **C_3 pathway.** A total of six carbons must be fixed in this way to produce the equivalent of one molecule of glucose (or some other hexose sugar, such as fructose). At the end of each cycle, the starting material, ribulose bisphosphate, is re-formed.

With the energy and reducing power from ATP and NADPH (both produced in the light-dependent reactions), the PGA molecules are converted to **glyceraldehyde-3-phosphate,** known simply as **PGAL.** For every six carbons that enter the cycle as carbon dioxide, six carbons can leave the system as two molecules of PGAL, to be used in carbohydrate synthesis. Each of these three-carbon molecules of PGAL is essentially half a hexose (six-carbon sugar) molecule. (In fact, PGAL is a key intermediate in the splitting of sugar in glycolysis; see Figs. 7–4 and 7–5.) The reaction of two molecules of PGAL is exergonic and can lead to the formation of glucose or fructose. In some plants, glucose and fructose are then joined to produce sucrose (common table sugar). This we harvest from sugar cane, sugar beets, and maple sap. The plant cell also uses glucose to produce starch or cellulose.

Notice that, although two PGAL molecules are removed from the cycle, ten PGAL molecules remain; this represents 30 carbon atoms in all. Through a complex series of reactions, these 30 carbons and their associated atoms are rearranged into six molecules of ribulose phosphate, each of which is phosphorylated to produce RuBP,

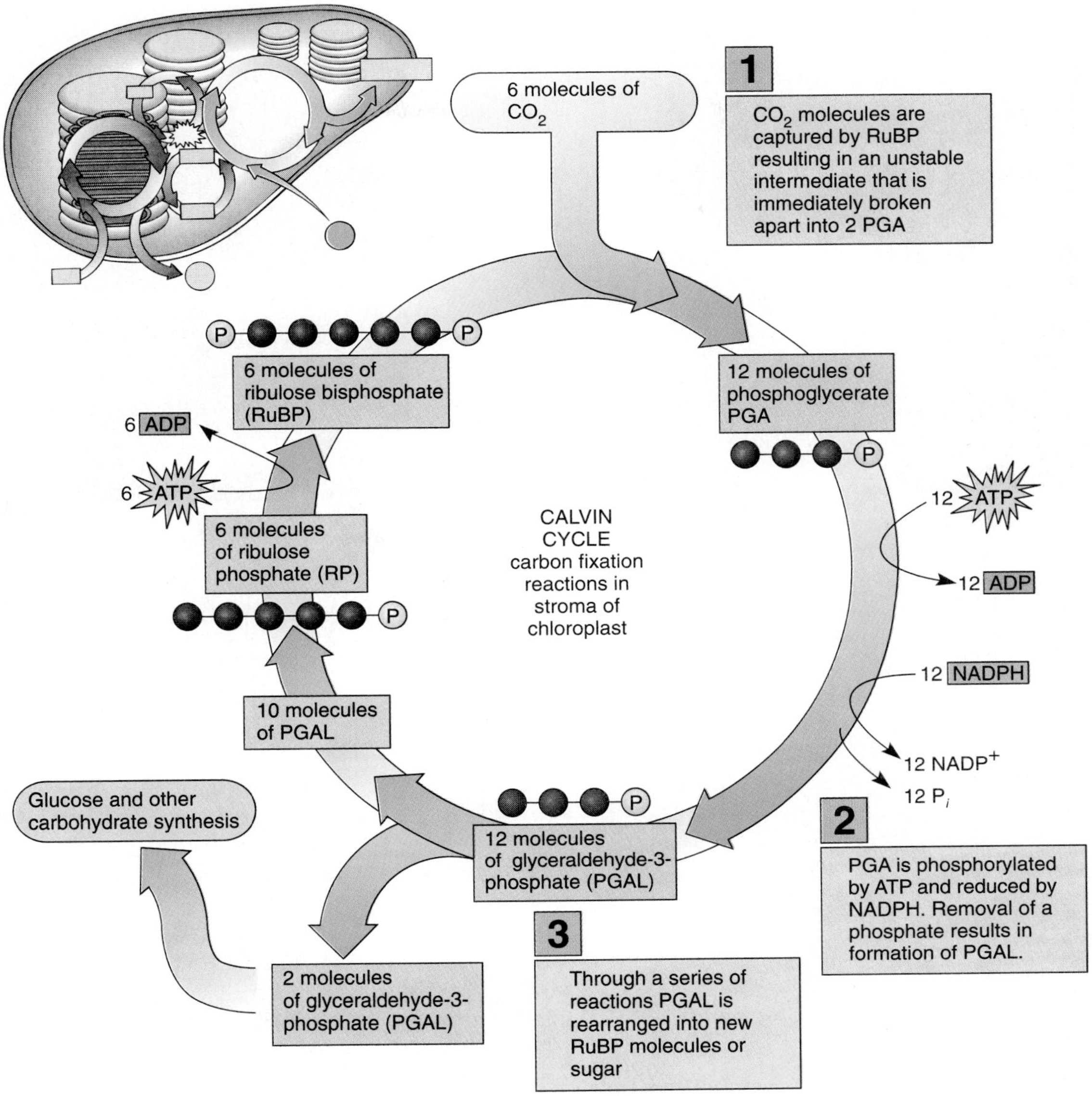

FIGURE 8–17 The Calvin cycle is the most common pathway for carbon fixation in plants. This diagram shows that six molecules of CO_2 must be "fixed" (incorporated into pre-existing carbon skeletons) to produce one molecule of a six-carbon sugar such as glucose. Two glyceraldehyde-3-phosphate (PGAL) molecules "leave the cycle" for every glucose formed. Although these reactions do not require light directly, the energy that drives the Calvin cycle comes from ATP and NADPH, which are the products of the light-dependent reactions.

the very five-carbon compound with which the cycle started. These RuBP molecules can begin the process of CO_2 fixation and eventual PGAL production once again.

In summary, the inputs required for the light-independent reactions are six molecules of CO_2, phosphates transferred from ATP, and electrons (as hydrogen) from NADPH. In the end, the six carbons from the CO_2 can be accounted for by the harvest of a hexose molecule. The remaining PGAL molecules are used to synthesize the RuBP molecules with which more CO_2 molecules may combine.

The initial carbon fixation step differs in C_4 plants and in CAM plants

Because carbon dioxide is not a very abundant gas (comprising only about 0.03% of the atmosphere), it is not easy for plants to obtain the carbon dioxide they need. This problem is complicated by the fact that gas exchange can occur only across a moist surface. The surfaces of leaves and other exposed plant parts are covered with a waterproof layer that helps prevent excess loss of water vapor. Entry and exit of gases is therefore limited to tiny pores,

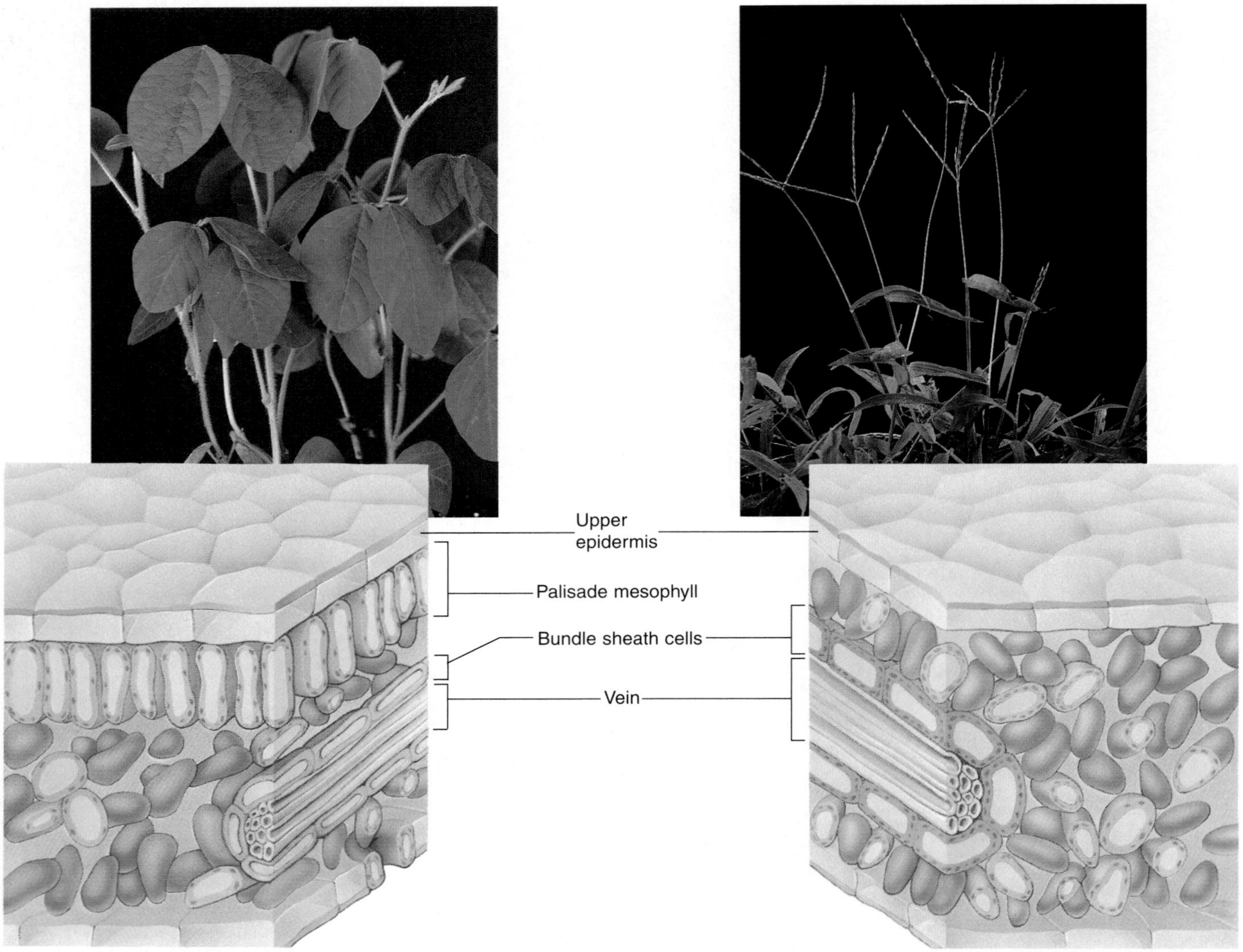

(a) Arrangement of cells in a C_3 leaf

(b) Arrangement of cells in a C_4 leaf

FIGURE 8–18 The leaf anatomy of a C_3 plant differs from that of a C_4 plant. These differences involve the mesophyll cells and the bundle sheath cells, which surround the veins of the leaf. (***a***) In C_3 plants, such as soybeans, the Calvin cycle takes place within the mesophyll cells. (***b***) In C_4 plants, such as crabgrass, reactions that fix CO_2 into four-carbon compounds take place in the mesophyll cells. These are closely associated with the bundle sheath cells, which also contain chloroplasts. The four-carbon compounds are transferred from the mesophyll cells to the bundle sheath cells, where the Calvin cycle takes place. (Dennis Drenner)

called **stomata** (sing., *stoma*) usually concentrated on the undersides of the leaves. These openings lead to the interior of the leaf, which is made up of a layer of cells, known as the **mesophyll** layer, with many air spaces and a very high concentration of water vapor (Figs. 8–3 and 8–18). The stomata open and close in response to such factors as water content and light intensity. When conditions are hot and dry, the stomata close to reduce the loss of water vapor. As a result, the supply of carbon dioxide is greatly diminished.

Ironically, carbon dioxide is potentially less available at the very times when maximum sunlight is available to power the light-dependent reactions. Many plant species living in hot, dry environments have evolved adaptations that allow them to initially fix carbon dioxide through one of two pathways that help minimize water loss. These pathways, known as the C_4 pathway and the CAM pathway, take place in the cytosol. They merely *precede* the Calvin cycle (C_3 pathway); they do not replace it.

The C_4 pathway efficiently fixes CO_2 at low concentrations

Some plants, known as **C_4 plants**, first fix CO_2 into a four-carbon compound, **oxaloacetate,** prior to the Calvin cycle (C_3 pathway). The C_4 pathway not only occurs before the C_3 pathway, it also occurs in different cells.

Leaf anatomy is usually distinctive in C_4 plants. In addition to having mesophyll cells (photosynthetic cells in the middle of the leaf), C_4 plant leaves have prominent photosynthetic **bundle sheath cells** (Fig. 8–18). These cells are tightly packed and form sheaths around the veins of the leaf. The mesophyll cells in C_4 plants are closely associated with the bundle sheath cells. The **C_4 pathway** (also called the **Hatch-Slack pathway,** after M. D. Hatch and C. R. Slack, who worked out many of its steps) occurs in the mesophyll cells, whereas the Calvin cycle takes place within the bundle sheath cells.

The key component of the C_4 pathway is a remarkable enzyme that has an extremely high affinity for carbon dioxide, binding it effectively even at unusually low concentrations. This enzyme, **PEP carboxylase,** catalyzes the reaction by which carbon dioxide reacts with the three-carbon compound **phosphoenolpyruvate (PEP),** forming oxaloacetate (Fig. 8–19).

In a step that requires NADPH, oxaloacetate is converted to some other four-carbon compound, usually malate. The malate then passes to the bundle sheath cells, where a different enzyme catalyzes the decarboxylation of malate to yield pyruvate (which has three carbons) and CO_2. NADPH is formed, replacing the one used earlier.

$$\text{Malate} + \text{NADP}^+ \longrightarrow \text{Pyruvate} + CO_2 + \text{NADPH}$$

The CO_2 released in the bundle sheath cell combines with ribulose bisphosphate and goes through the Calvin cycle in the usual manner. The pyruvate formed in the decarboxylation reaction returns to the mesophyll cell, where it reacts with ATP to regenerate phosphoenolpyruvate.

The role of the C_4 pathway is to efficiently capture CO_2 and ultimately increase its concentration within the bundle sheath cells; this mass action effect essentially drives the Calvin cycle. The operation of the C_4 pathway serves to increase the concentration of CO_2 within the bundle sheath cells to some 10 to 60 times over that in the mesophyll cells of plants having only the C_3 pathway.

The combined C_3–C_4 pathway involves the expenditure of 30 ATPs per hexose, rather than the 18 ATPs used in the absence of the C_4 pathway. The extra energy expense is worthwhile at high light intensity because it ensures a high concentration of CO_2 in the bundle sheath cells and permits them to carry on photosynthesis at a rapid rate. The C_4 pathway is present in addition to the Calvin cycle in a number of plant orders and apparently has evolved independently several times. Because PEP carboxylase fixes carbon dioxide so efficiently, C_4 plants do not need to have their stomata open as much; they therefore tolerate higher temperatures and higher light intensities, lose less water by transpiration (evaporation), and have higher rates of photosynthesis and growth than plants that use only the Calvin cycle. Among the many quick-growing and aggressive plants that use the C_4 pathway are sugar cane, corn, and crabgrass (Fig. 8–18). If sunlight is not limiting, the yields of C_4 crop plants can be two to three times greater than those of C_3 plants. If this pathway could be incorporated into more of our crop plants by genetic manipulation, we might well be able to greatly increase food production in some parts of the world.

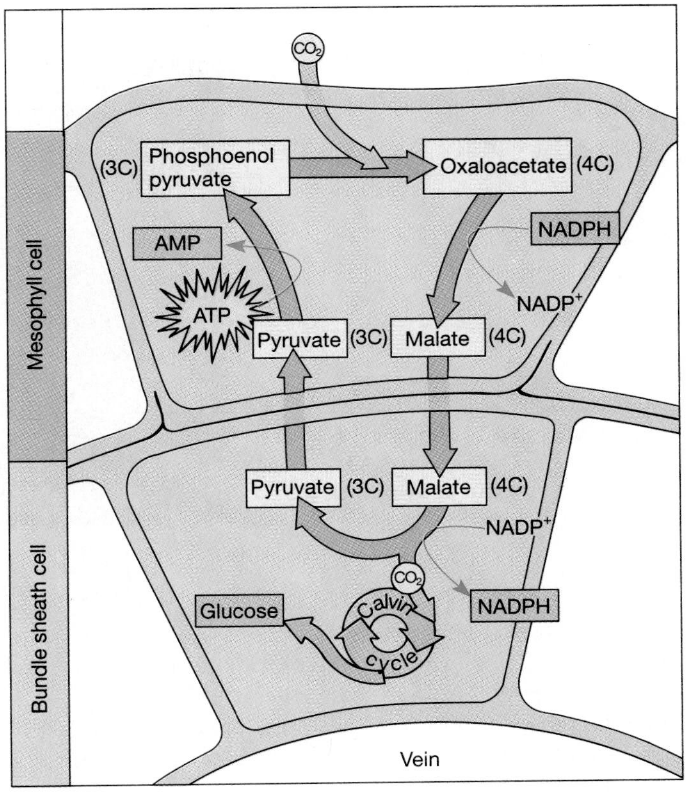

FIGURE 8–19 In the C_4 pathway, CO_2 initially captured by the mesophyll cells is transferred to the bundle sheath cells. CO_2 combines with phosphoenolpyruvate (PEP) in mesophyll cells, forming a four-carbon compound that is converted to malate. Malate goes to the bundle sheath cell, where it is decarboxylated. The CO_2 thus released in the bundle sheath cell is used to make carbohydrate by way of the Calvin cycle. Because the C_4 system consumes some energy, ultimately made available only by photosynthesis, this system is important to the plant only at high light intensities. Under these conditions, the C_4 pathway can fix more carbon than the Calvin cycle can fix by itself.

When light is abundant, the rate of photosynthesis is limited by the concentration of CO_2, so C_4 plants, with their higher levels of CO_2 in bundle sheath cells, have the advantage. At lower light intensities and temperatures, C_3 plants are favored. For example, winter rye, a C_3 plant, grows lavishly in cool weather when crabgrass cannot because it requires more energy to fix CO_2.

CAM plants fix CO_2 at night

Plants living in very dry, or **xeric,** conditions have a number of special anatomical adaptations that enable them to

FIGURE 8–20 ***Echeveria derembergii,*** **a member of the family Crassulaceae, is a typical CAM plant that grows in xeric habitats.** (Michel Viard/Peter Arnold, Inc.)

survive. Many xeric plants have physiological adaptations as well. For example, their stomata may open during the cooler night and close during the hot day to reduce water loss from transpiration. This is in contrast to most plants, which have stomata that are open during the day and closed at night. But xeric plants that have their stomata closed during the day cannot exchange gases for photosynthesis. (Other plants typically fix carbon dioxide during the day, when sunlight is available.)

Many xeric plants evolved a special carbon fixation pathway called **crassulacean acid metabolism,** or **CAM,** that in effect solves this dilemma. The name comes from the stonecrop plant family (the Crassulaceae), which possesses the CAM pathway, although it has evolved independently in some members of more than 25 other plant families as well (Fig. 8–20).

CAM plants fix CO_2 during the night when stomata are open, forming malate (a four-carbon compound), which is stored in cell vacuoles. During the day, when stomata are closed and gas exchange cannot occur between the plant and the atmosphere, CO_2 is removed from malate by a decarboxylation reaction. Now the CO_2 is available *within the leaf tissue* to be fixed into sugar by the Calvin cycle (C_3 pathway).

The CAM pathway is very similar to the C_4 pathway, but with important differences. C_4 plants initially fix carbon dioxide into four-carbon organic acids in mesophyll cells. The acids are later decarboxylated to produce CO_2, which is fixed by the C_3 pathway in the bundle sheath cells. In other words, the C_4 and C_3 pathways occur in *different locations* within the leaf of a C_4 plant. In CAM plants, the initial fixation of CO_2 occurs at night. Decarboxylation of malate and subsequent production of sugar from CO_2 by the normal C_3 photosynthetic pathway occur during the day. In other words, the C_4 and C_3 pathways occur at *different times* within the same cell of a CAM plant.

Although it does not promote rapid growth the way that the C_4 pathway does, the CAM pathway is a very successful adaptation to xeric conditions. CAM plants are able to carry out gas exchange for photosynthesis *and* to reduce water loss significantly. Plants with CAM photosynthesis survive in deserts where neither C_3 nor C_4 plants can.

Photorespiration reduces photosynthetic efficiency

Many C_3 plants, including certain agriculturally important crops such as soybeans, wheat, and potatoes, do not yield as much carbohydrate from photosynthesis as might be expected. This reduction in yield is especially significant during very hot spells in summer. On hot, dry days plants close their stomata to conserve water. Once the stomata close, photosynthesis rapidly uses up the CO_2 remaining in the leaf and produces O_2, which accumulates in the chloroplasts. Recall that RuBP carboxylase (Rubisco) is responsible for CO_2 fixation in the Calvin cycle by attaching CO_2 to RuBP. O_2 competes with CO_2 for binding to the active site of Rubisco. Therefore, when chloroplast oxygen levels are high and CO_2 levels are low, Rubisco is more likely to catalyze the reaction of RuBP with O_2 instead of with CO_2. When this occurs, some of the intermediates involved in the Calvin cycle are degraded to CO_2 and H_2O. This process is called **photorespiration** because (1) it occurs in the presence of light; (2) it requires oxygen, like aerobic respiration; and (3) it produces CO_2 and H_2O, like aerobic respiration. Unlike aerobic respiration, however, ATP is not produced during photorespiration. Photorespiration reduces photosynthetic efficiency because it removes some of the intermediates used in the Calvin cycle.

The reasons for photorespiration are incompletely understood, although it is thought to possibly reflect the origin of Rubisco at an ancient time when carbon dioxide levels were high and molecular oxygen levels were low.

Photorespiration is negligible in C_4 plants because the concentration of CO_2 in bundle sheath cells is always high and because oxygen does not compete for binding to the active site of PEP carboxylase. However, many important crop plants are C_3 plants that carry out photorespiration. This is yet another reason that some scientists are attempting to transfer genes for the C_4 pathway to C_3 crops such as soybeans and wheat. If this is accomplished, these plants should be able to produce much more carbohydrate during hot weather.

SUMMARY

I. Most producers are photosynthetic autotrophs and use light as an energy source for manufacturing organic compounds from carbon dioxide and water.

II. In plants, photosynthesis occurs in chloroplasts, which are located mainly within mesophyll cells inside the leaf.
   A. Chloroplasts are organelles bounded by a double membrane; the inner membrane encloses the stroma and thylakoids.
   B. Chlorophyll and other photosynthetic pigments are components of the thylakoid membranes of chloroplasts.
   C. The thylakoids enclose the thylakoid interior space and are arranged in stacks called grana.

III. During photosynthesis, light energy is captured by chlorophyll and converted to chemical energy in a way that ultimately results in carbohydrate synthesis. Oxygen is released as a byproduct.
   A. During the noncyclic, light-dependent reactions of photosynthesis, chlorophyll electrons become photoenergized. These electrons reduce $NADP^+$, forming NADPH, and some of their energy is used to phosphorylate ADP, forming ATP.
   B. The carbon fixation reactions of photosynthesis use the energy of ATP and NADPH to manufacture carbohydrate molecules from CO_2.

IV. Chlorophyll is a pigment that captures light energy. Its absorption spectrum is very similar to the action spectrum for photosynthesis.

V. Light behaves as both a wave and a particle. Particles of light energy, called photons, can excite pigment molecules such as chlorophyll. The resulting high-energy electrons are accepted by electron acceptor compounds.
   A. Chlorophyll molecules and accessory pigments are organized into photosystems.
   B. Only a special chlorophyll *a* in the reaction center of a photosystem actually gives up its energized electrons to a nearby electron acceptor.

VI. In noncyclic photophosphorylation, the electrons emitted by Photosystem I are passed through a chain of electron acceptors to $NADP^+$.
   A. Electrons given up by P700 in Photosystem I are replaced by electrons from P680 in Photosystem II.
   B. Electrons given up by P680 in Photosystem II are replaced by electrons made available by the photolysis of water. Oxygen is released in the process.
   C. A series of redox reactions takes place as excited electrons are passed from Photosystem II to Photosystem I along a chain of electron acceptors. Some of the energy is used to pump protons across the thylakoid membrane, providing the energy to generate ATP.

VII. In cyclic photophosphorylation, electrons from Photosystem I are eventually returned to Photosystem I; ATP but no NADPH is produced, and no oxygen is generated.

VIII. During the carbon fixation reactions, energy from ATP, and NADPH, is used to chemically combine carbon dioxide with hydrogen.
   A. The carbon fixation reactions proceed by way of the Calvin cycle.
   B. In the Calvin cycle, carbon dioxide is combined with ribulose bisphosphate (RuBP), a five-carbon sugar.
   C. For every six carbon dioxide molecules fixed, two molecules of PGAL can leave the cycle. Two PGAL molecules are required to produce the equivalent of one molecule of glucose; the remaining PGAL molecules are modified to regenerate the CO_2 acceptor molecules (RuBP).
   D. ATP and NADPH molecules are consumed in the conversion of CO_2 into carbohydrate.

IX. In the C_4 pathway, the enzyme PEP carboxylase binds CO_2 effectively, even when CO_2 is at a low concentration.
   A. The initial reactions take place within mesophyll cells. The carbon dioxide is fixed in oxaloacetate, which is then converted to malate.
   B. The malate moves into a bundle sheath cell and CO_2 is removed from it. The released CO_2 then enters the Calvin cycle.

X. The CAM (crassulacean acid metabolism) pathway is similar to the C_4 pathway; PEP carboxylase fixes carbon at night in the mesophyll cells, and the Calvin cycle occurs during the day in the same cells.

XI. In photorespiration, C_3 plants consume oxygen and generate carbon dioxide. This process, which decreases photosynthetic efficiency, occurs on bright, hot, dry days when plant cells close their stomata, conserving water but preventing the passage of CO_2 into the leaf.

Summary Equations for Photosynthesis

The light-dependent reactions:

$$12\ H_2O + 12\ NADP^+ + 18\ ADP + 18\ P_i \dashrightarrow 6\ O_2 + 12\ NADPH + 18\ ATP$$

The carbon fixation reactions:

$$12\ NADPH + 18\ ATP + 6\ CO_2 \dashrightarrow C_6H_{12}O_6 + 12\ NADP^+ + 18\ ADP + 18\ P_i + 6\ H_2O$$

By canceling out the common items on opposite sides of the arrows in these two coupled equations, we obtain the simplified overall equation for photosynthesis:

$$6\ CO_2 + 12\ H_2O \dashrightarrow C_6H_{12}O_6 + 6\ O_2 + 6\ H_2O$$

SELECTED KEY TERMS

absorption spectrum
action spectrum
ATP synthase
autotroph
bundle sheath cells
C_3 pathway
C_4 pathway (Hatch-Slack pathway)
Calvin cycle
CAM (crassulacean acid metabolism)
carbon fixation reactions
carotenoid
chemiosmosis
chemoautotroph
chemoheterotroph
chlorophyll (*a*, *b*)
chloroplast
cyclic photophosphorylation
fluorescence
glyceraldehyde-3-phosphate (PGAL)
granum (grana)
ground state
heterotroph
light-dependent reactions
$NADP^+$/NADPH
noncyclic photophosphorylation
oxaloacetate
P680
P700
phosphoenolpyruvate (PEP)
phosphoglycerate (PGA)
photoautotroph
photoheterotroph
photolysis
photon
photophosphorylation
photorespiration
photosynthesis
photosystem (I and II)
reaction center
ribulose bisphosphate (RuBP)
ribulose bisphosphate carboxylase (Rubisco)
stomata (stoma)
stroma
thylakoid
wavelength

POST-TEST

1. Most producers are photosynthetic ________ .
2. Chlorophyll is located in the ________ membranes of a chloroplast.
3. In ________, some of the energy captured by chlorophyll is used to split water.
4. Light is composed of particles of energy called ________ .
5. The relative effectiveness of different wavelengths of light in photosynthesis is demonstrated by a(an) ________ ________ .
6. In plants, the final electron acceptor in the light-dependent reactions is ________ .
7. In addition to chlorophyll, most plants contain accessory photosynthetic pigments such as ________ .
8. Only the special chlorophyll *a* in the reaction center of a(an) ________ gives up its electron to an electron acceptor.
9. In ________, electrons that have been energized by light indirectly contribute their energy to add phosphate to ADP, producing ATP.
10. In ________ ________, there is a one-way flow of electrons to $NADP^+$, forming NADPH.
11. The mechanism by which electron transport is coupled to ATP production is called ________ .
12. The transfer of electrons through a sequence of electron acceptors provides energy to pump protons across the ________ membrane.
13. The inputs for the ________ ________ reactions are CO_2, NADPH, and ATP.
14. The ________ cycle begins when carbon dioxide reacts with ribulose bisphosphate.
15. When light is abundant and carbon dioxide is the limiting factor, plants with the ________ pathway have an advantage.
16. In C_4 plants, the Calvin cycle takes place in ________ ________ cells, where malate is decarboxylated, yielding CO_2 for photosynthesis.
17. In ________, ribulose bisphosphate reacts with O_2 instead of CO_2.

REVIEW QUESTIONS

1. Why does photosynthesis require light energy?
2. What is the role of chlorophyll in photosynthesis?
3. What is the significance of the fact that the combined absorption spectra of chlorophyll *a* and *b* roughly parallels the action spectrum of photosynthesis? Why do they not coincide exactly?
4. How is oxygen produced during photosynthesis?
5. How are ATP and NADPH produced and used in the process of photosynthesis?
6. How are carbohydrates produced from CO_2 in photosynthesis?

YOU MAKE THE CONNECTION

1. The electrons in glucose have relatively high free energies. How did they become so energetic?
2. Only some plant cells have chloroplasts, but all living plant cells have mitochondria. Why?
3. What strategies might be employed in the future to increase world food supply? Base your answer on your knowledge of photosynthesis and related processes.

RECOMMENDED READINGS

Kemp, P. R., G. L. Cunningham, and H. P. Adams. "Specialization of Mesophyll Structure in C_4 Grasses." *BioScience,* July/August 1983. Discusses the relationship between structure and function as it relates to C_4 photosynthesis.

Mauseth, J. D. *Botany: An Introduction to Plant Biology*, 2nd edition. Saunders College Publishing, Philadelphia, 1995. A readable introduction to photosynthesis, with emphasis on environmental and internal factors that affect photosynthesis.

Nakatani, H. Y. "Photosynthesis." *Carolina Biology Reader.* Carolina Biological Supply Company, Burlington, NC, 1988. A clear description of photosynthesis with emphasis on the light-dependent reactions.

Taiz, L., and E. Zeiger. *Plant Physiology.* Benjamin Cummings, Redwood City, CA, 1991. An in-depth examination of the photochemistry of photosynthesis.

Youvan, D. C., and B. L. Marrs. "Molecular Mechanisms of Photosynthesis." *Scientific American,* June 1987. Explores molecular aspects of photosynthesis.

Tissue Bank Director

MARTHA ANDERSON

Martha Anderson originally planned to attend medical school after earning a biology degree at Colorado State University. But by the time she transferred to complete her biology studies at the University of Colorado, her focus had changed. Upon graduating, she began a career that enabled her to help many more patients and their families than a single doctor ever could. Martha is Vice President of Donor Services for the Musculoskeletal Transplant Foundation, a non-profit organization that "banks" human tissue for transplant use and conducts research to further develop transplantation techniques. The Foundation stores tissues and bone that are used in a variety of procedures, from periodontal reconstructive surgery to replacement of cancerous tissue. The Foundation is a membership organization composed of medical and academic institutions as well as other organ and tissue procurement agencies.

You had planned on going to medical school and decided against that profession. What did you do after college?

I went to work in one of Denver's trauma hospitals where I took an administrative position. Eventually I was promoted into what was probably the best job of my life as a patient representative at St. Anthony's Hospital.

What does a patient representative do?

A patient representative acts as an advocate for patients. You resolve a lot of patient care related issues and are responsible for crisis intervention with many families. Sometimes you serve as a person to talk to for the patient who doesn't have any family or friends around. We had the first helicopter rescue system in the country at St. Anthony's, and as a result, we had a great deal of trauma victims. Often these patients arrived unconscious and without any identification. It was my job to find out who they were and contact their families. It was a very stressful job, but it was rewarding because we could help people negotiate the health care system. Many people don't know anything about health care, and if they are incapacitated by their illness or by grief it's difficult for them to know what questions to ask so they can get the help that they need. Through my work as a patient representative I became involved with the organ and tissue donation process.

And that's how you got into your current position?

Yes, I helped set up the largest donor program in Colorado. We had more donors from St. Anthony's in Denver than from any other hospital in the state.

How did you do that?

We were very pro-active in identifying potential donors and offering their families the opportunity to talk about donation and find out if they wanted to be donors.

I suppose there are other organizations that "bank" bone or tissue. What's different about your organization?

There are probably about 100 or 150 tissue banks in the country. Ours is the largest and we have a unique structure. Many organ and tissue banks are regional or local, but we're a national consortium of medical schools and other recovery agencies. When someone has the potential of being an organ or tissue donor the local organ procurement agency is usually contacted. There are over 25 locations around the country that recover tissue for us — the bones, tendons and ligaments — which we then provide to surgeons and hospitals throughout the U.S.

Then you process them for storage so they can be transplanted?

Yes, we have them processed in our New Jersey facility. The tissue is formed into different sorts of shapes depending on whatever the local need is. For example, there may be a hospital that does a lot of spine surgery so we would cut the bone into sizes for spine surgeons to use. Or there may be a need for an entire femur which would probably be used for someone who has cancer. In the past the only option for removing a cancerous tumor in a femur was to amputate the leg. Now we can take out the portion of bone that has the tumor in it and replace it with another that's the right size. Such a person can now keep their leg, and maintain an active lifestyle. This procedure is called limb salvage.

Do you feel that you use your biology degree in your current job? And, if so, how do you use it?

I used my biology degree when I was a patient representative because my studies focused on human physiology and anatomy. I had a very good basic understanding of how the human body works and that helped me a lot. One of the things that I do use every day are the critical thinking skills I developed as an undergraduate. I have to figure out answers to problems that have a basis in science. One of the things you learn in basic biology is that you have to think and figure out how things work. These are skills I use quite frequently.

Is there a specific kind of science or education background required to work in your field?

A lot of the people who work for me are either nurses or operating room technicians. A few people have bachelors degrees in biology. Having only a bachelor's degree in biology may not be enough anymore, and I think a graduate degree is probably required to advance in the field.

PART 3

The Continuity of Life: Genetics

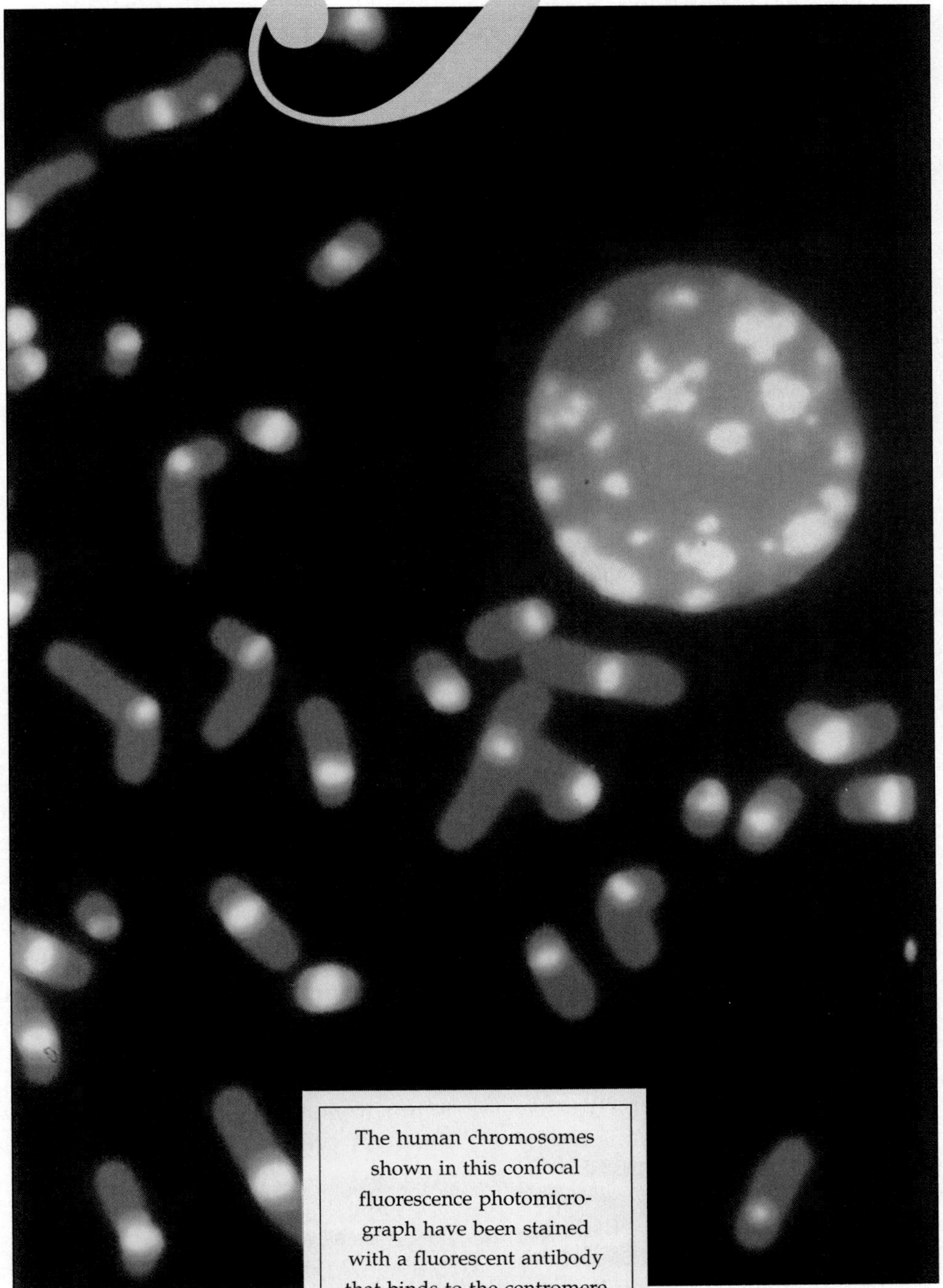

The human chromosomes shown in this confocal fluorescence photomicrograph have been stained with a fluorescent antibody that binds to the centromere regions (yellow). (Courtesy of Oncor, Inc.)

CHAPTER 9

Chromosomes, Mitosis, and Meiosis

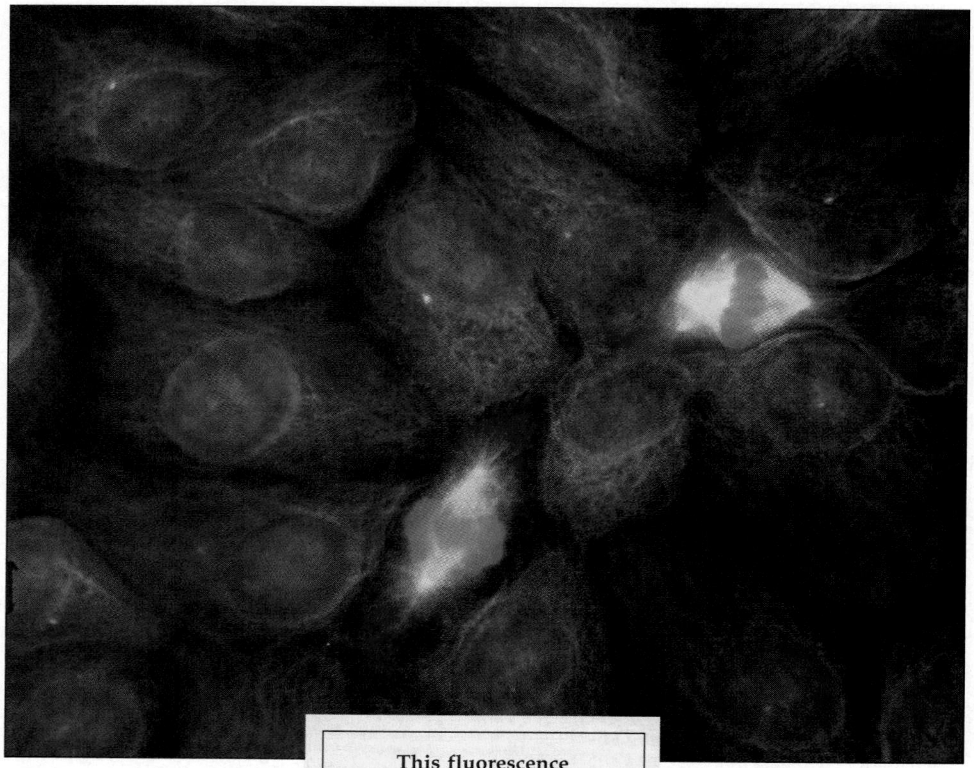

This fluorescence micrograph shows dividing squamous carcinoma (cancer) cells. (Nancy Kedersha, Immunogen, Inc.)

Every organism, even the simplest, contains a massive amount of information in the form of DNA. In every cell, copies of the DNA are organized into informational units called **genes** that ultimately control all aspects of the life of the organism. The transmission of genetic information from parent to offspring is called **heredity,** and the branch of biology concerned with the structure, transmission, and expression of genetic information is called **genetics.**

Cells are responsible for the continuity of life because they provide the essential link in the transmission of genes between generations. All cells are formed by the division of preexisting cells. When a cell divides, copies of its genes must be precisely transmitted to each daughter cell through a complex series of processes. Both prokaryotic and eukaryotic cells must first faithfully copy (replicate) their DNA.

A eukaryotic nucleus contains multiple DNA molecules. These molecules are very long and thin, and to prevent tangling they must be packaged with proteins and organized into structures called **chromosomes.** Prokaryotic cells do not have such chromosomes because they contain much less DNA and its packaging is simpler.

Chromosomes must be distributed to the daughter cells in a highly organized way. Most cell divisions in the body cells of eukaryotes involve a process called **mitosis,** which ensures that each daughter cell receives one copy of every chromosome (and therefore one copy of every gene) in the parent cell. The distribution of genetic material in prokaryotic cells is much simpler, but it too must be very precise if the daughter cells are to be genetically identical to the parent cell.

Sexual reproduction in eukaryotes occurs by fusion of two sex cells, or **gametes,** to form a single cell called a **zygote.** In higher plants and animals, the gametes are the eggs and sperm. To prevent zygotes from having twice as many chromosomes as the parents, each gamete must contain only half the number of parental chromosomes. For this reason, sexual life cycles include a special type of cell division, called **meiosis,** which reduces the chromosome number by half.

LEARNING OBJECTIVES

After you have studied this chapter you should be able to

1. Discuss the significance of chromosomes in terms of their information content.
2. Identify the stages in the eukaryotic cell cycle, describe the principal events characteristic of each, and point out some ways in which the cycle is controlled.
3. Illustrate the structure of a duplicated chromosome, labeling the sister chromatids, sister centromeres, and sister kinetochores.
4. Explain the significance of mitosis, and diagram the process.
5. Discriminate between asexual and sexual reproduction.
6. Distinguish between haploid and diploid cells, and define homologous chromosomes.
7. Contrast the events of mitosis and meiosis.
8. Compare the roles of mitosis and meiosis and of haploidy and diploidy in various generalized life cycles.

EUKARYOTIC CHROMOSOMES CONTAIN DNA, PROTEIN, AND RNA

The carriers of genetic information in eukaryotes are the **chromosomes** (Fig. 9–1) contained within the cell nucleus. Although the term *chromosome* means "colored body," chromosomes are virtually colorless; the name refers to their ability to be stained by certain dyes.

Chromosomes are made up of **chromatin,** a complex material that consists of fibers containing about 60% protein, 35% deoxyribonucleic acid (DNA), and 5% ribonucleic acid (RNA). When a cell is not dividing, the chromosomes are present but in a decondensed form. The chromatin consists of long, thin threads that are somewhat aggregated, which gives them a granular appearance when viewed with the light microscope. At the time of cell division, the chromatin fibers condense and the chromosomes become visible as distinct structures. The structure of chromatin is described in more detail in Chapter 11.

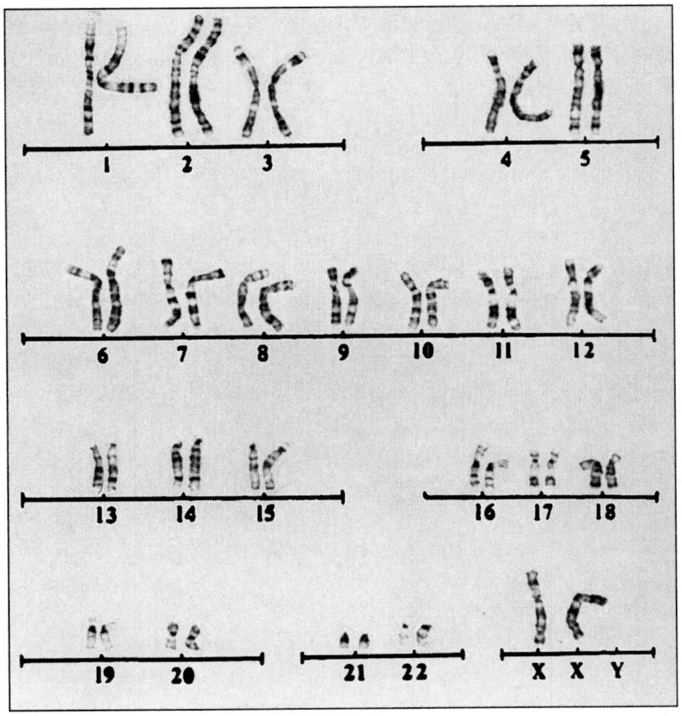

(*a*)

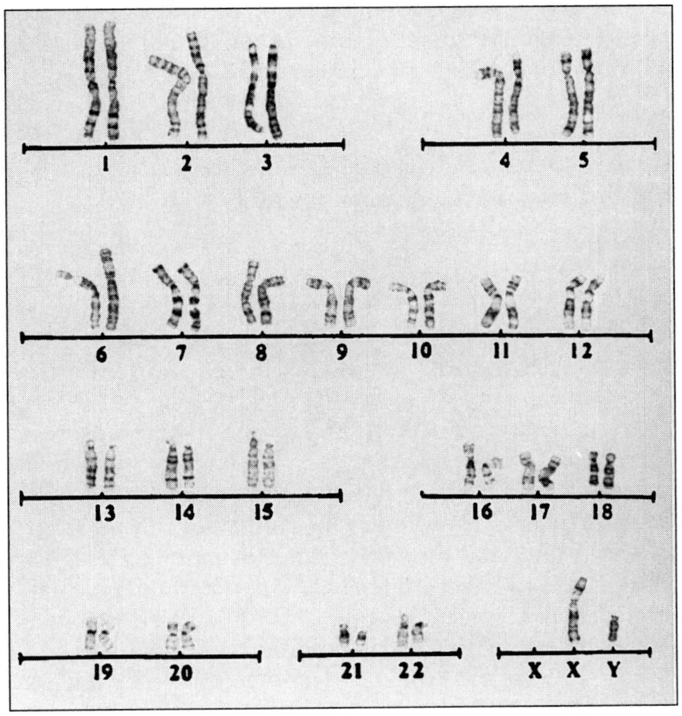

(*b*)

FIGURE 9–1 The chromosome constitution of an individual is its karyotype. Shown here are the karyotypes of a normal human female (*a*) and (*b*) male. The apparent size differences between the female and male chromosomes have no significance; they merely reflect different degrees of chromosomal contraction in the particular cells photographed. The process of karyotyping is discussed in Chapter 15. (Courtesy of Dr. Leonard Sciorra)

A typical prokaryotic cell contains a single circle of DNA with very few associated proteins. Bacterial reproduction is discussed in more detail in Chapter 23.

DNA is organized into informational units called genes

Each chromosome may contain hundreds or even thousands of genes. For example, humans are thought to have 50,000 or more genes, although the actual number is not known.

Our concept of the gene has changed considerably since the beginnings of the science of genetics, but our definitions have always centered on the gene as an informational unit that ultimately affects some characteristic of the organism. For example, we speak of genes controlling eye color in humans, wing length in flies, seed color in peas, and so on.

Today a **gene** can be roughly defined as a part of a DNA molecule that can be copied in the form of an RNA molecule through a process known as *transcription* (see Chapter 12), plus adjacent regions of the DNA that are required to regulate transcription. Different kinds of RNA molecules have different specific functions. Many are responsible for carrying the code that specifies a particular sequence of amino acids in a polypeptide chain. This means that genes control the structure of all of the proteins of the organism, including the enzymes. In their role as catalysts, enzymes regulate the metabolic reactions of the organism. (We discuss some of the limitations of this definition of the gene in Chapter 13.)

Chromosomes of different species differ in number and informational content

Every individual of a given species has a characteristic number of chromosomes in most nuclei of its body cells. Most human body cells have exactly 46 chromosomes. Figure 9–1 compares the normal chromosome constitutions, or **karyotypes,** of human females and males (see Chapter 15).

Humans are not unique in having 46 chromosomes; some other species of animals and plants also have 46, whereas others have different chromosome numbers. A certain species of roundworm has only two chromosomes in each cell, while some crabs have as many as 200, and some ferns have more than 1000. Most animal and plant species have between 10 and 50 chromosomes. Numbers above and below this are uncommon.

Humans are not humans merely because they have 46 chromosomes; in fact, some humans have abnormal karyotypes with more or fewer than 46 (see Chapter 15). The *number* of chromosomes is not what makes each species unique, but rather the *information* specified by the genes on the chromosomes.

THE CELL CYCLE IS A SEQUENCE OF CELL GROWTH AND DIVISION

Usually when cells reach a certain size, they must either stop growing or divide. Some cells, such as nerve, skeletal muscle, and red blood cells, do not normally divide once they are mature. The activities of growing and dividing cells can be described in terms of the life cycle of the cell, or the **cell cycle.**

In cells capable of dividing, the cell cycle is the period from the beginning of one division to the beginning of the next and is represented in diagrams as a circle (Fig. 9–2). The time it takes to complete one cell cycle is the **generation time, T.** The generation time can vary widely, but in actively growing plant and animal cells it is often about 8 to 20 hours.

Cell division involves two main processes, mitosis and cytokinesis. **Mitosis,** a complex process involving the nucleus, ensures that each new nucleus receives the same number and types of chromosomes as were present in the original nucleus. **Cytokinesis,** which generally begins before mitosis is complete, is the division of the cytoplasm of the cell to form two cells. Multinucleate cells are formed if mitosis is not followed by cytokinesis; this is a normal condition for some kinds of cells.

Chromosomes become duplicated during interphase

Most of the life of the cell is spent in **interphase,** the stage between successive cell divisions. The cell is very active

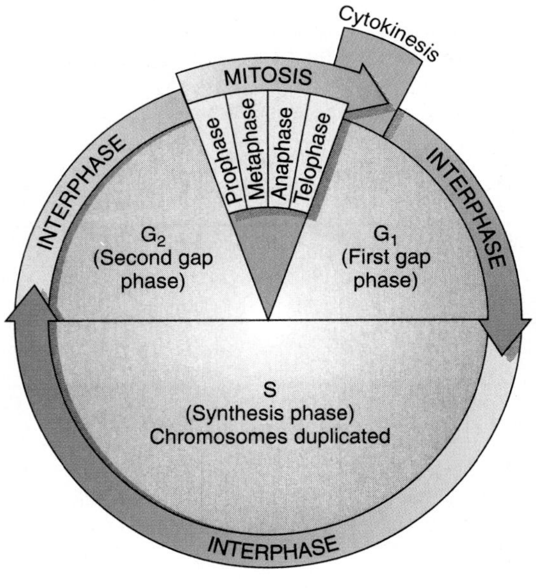

FIGURE 9–2 This circular diagram represents a single cell cycle. The generation time (T), the time required to complete one cycle, includes cell division (mitosis and cytokinesis) and interphase. Actual times vary with the species, cell type, and growth conditions.

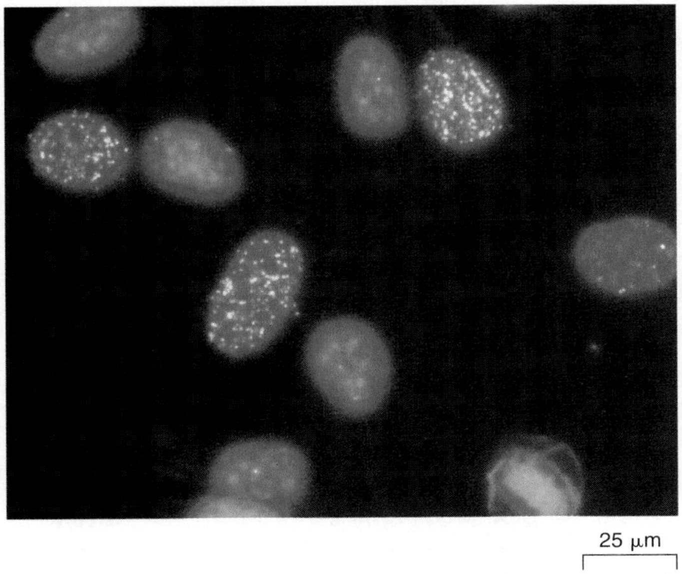

FIGURE 9–3 The orange grains mark interphase nuclei that previously incorporated thymidine (a DNA precursor) into their DNA, during the S phase. Nuclei exhibiting grains contain newly synthesized DNA and are at the G_2 phase of the cell cycle. Nuclei without grains are in the G_1 phase. (Jonathan G. Izant)

during this time, synthesizing needed materials and growing. Most proteins and other materials are synthesized throughout interphase. The major exceptions are molecules required for cell division. The relatively restricted times during which these molecules are synthesized provide a basis for subdividing interphase.

During the early 1950s it was recognized that chromosomes undergo duplication during interphase and are later separated and distributed to the daughter nuclei during mitosis. The period of DNA replication during interphase serves as a major landmark, termed the *synthesis phase,* or **S phase** (Figs. 9–2 and 9–3). Other chromosomal components, such as the chromosomal proteins, are also synthesized at this time. Chromosome duplication is a complex process discussed in Chapter 11.

The time between mitosis and the beginning of the S phase is termed the **G_1 phase,** or first gap phase. Growth takes place during the G_1 phase of an actively cycling cell; toward the end of G_1, there is increased activity of enzymes required for DNA synthesis. These activities make it possible for the cell to enter the S phase. Noncycling cells usually remain in a stage called G_0, roughly equivalent to G_1.

After it completes the S phase, the cell enters a second gap phase, the **G_2 phase.** At this time, increased protein synthesis occurs as the final steps in the cell's preparation for division take place. The completion of the G_2 phase is marked by the beginning of mitosis. The sequence of the substages of interphase is therefore:

G_1 phase $\longrightarrow$ S phase $\longrightarrow$ G_2 phase

Mitosis ensures orderly distribution of chromosomes

Each mitotic division is a continuous process, with each stage merging into the next. However, for descriptive purposes mitosis has been divided into stages. Refer to Figure 9–4 as you read the description of the main phases, which follow an orderly sequence:

prophase $\longrightarrow$ metaphase $\longrightarrow$ anaphase $\longrightarrow$ telophase

Duplicated chromosomes become visible with the microscope during prophase

The first stage of mitosis, **prophase,** begins when the long chromatin threads begin to condense and appear as mitotic chromosomes. This condensation is accomplished mainly by a coiling process in which chromosomes become simultaneously shorter and thicker. The chromatin can thus be distributed to the daughter cells without tangling.

When stained with certain dyes and viewed through the light microscope, chromosomes are visible as darkly staining bodies during prophase. Each chromosome has been duplicated during the preceding S phase and consists of a pair of identical units, termed **sister chromatids.** Each chromatid contains a nonstaining, constricted region called the **centromere.** Sister chromatids are tightly associated in the vicinity of their centromeres (Fig. 9–5). Although the chemical basis for this close association is not completely understood, evidence suggests that special kinds of DNA and special proteins that bind to those DNA sequences are involved. Each centromere contains a structure called the **kinetochore** to which microtubules can bind.

A dividing cell is usually described as a globe, with an equator that determines the midplane, or equatorial plane, and two opposite poles. This terminology is used for all cells regardless of their actual shape.

Fibers form between the poles and begin to organize into the **mitotic spindle,** a complex structure consisting mainly of microtubules. The mitotic spindle is responsible for the separation of the chromosomes during anaphase.

Animal cells differ from the cells of complex plants in the details of mitotic spindle formation. In both types of dividing cell, each pole contains a region, referred to as a **microtubule-organizing center (MTOC),** from which the microtubules radiate outward. When viewed with the electron microscope, microtubule-organizing centers in the cells of higher plants consist of rather dense fibrillar matter with little or no discernible structure.

A pair of centrioles (see Chapter 4) resides in the middle of each microtubule-organizing center of an animal cell. The centrioles are surrounded by a cloud of matter referred to as the **pericentriolar material.** This pericentriolar material is similar in appearance to the material in

(*Text continues on page 223*)

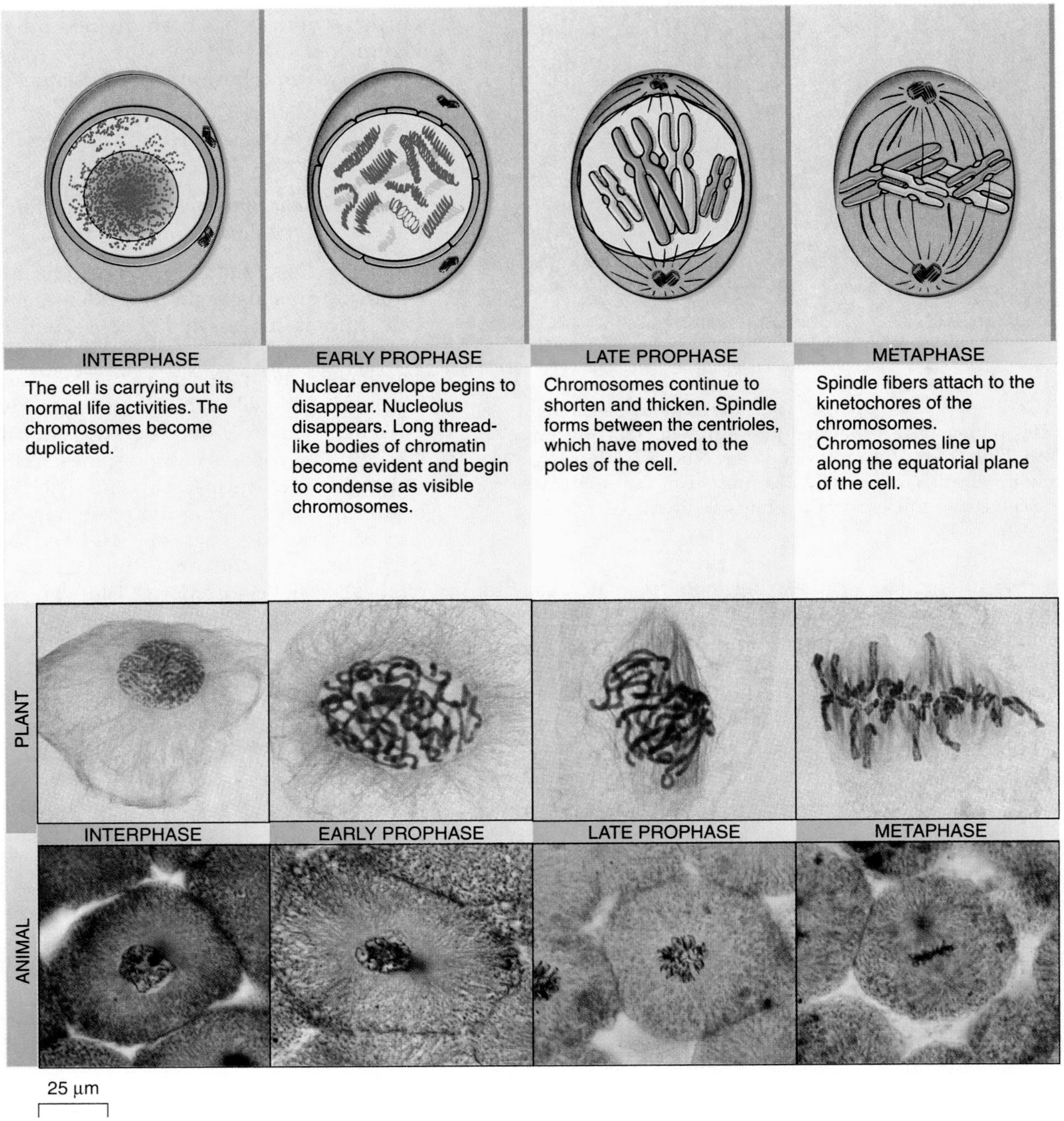
INTERPHASE
The cell is carrying out its normal life activities. The chromosomes become duplicated.
EARLY PROPHASE
Nuclear envelope begins to disappear. Nucleolus disappears. Long thread-like bodies of chromatin become evident and begin to condense as visible chromosomes.
LATE PROPHASE
Chromosomes continue to shorten and thicken. Spindle forms between the centrioles, which have moved to the poles of the cell.
METAPHASE
Spindle fibers attach to the kinetochores of the chromosomes. Chromosomes line up along the equatorial plane of the cell.
PLANT
INTERPHASE
EARLY PROPHASE
LATE PROPHASE
METAPHASE
ANIMAL
25 μm

FIGURE 9–4 Interphase and the stages of mitosis are similar in plant and animal cells. The chief difference between these cells is the lack of centrioles in the plant cells. Individual steps are explained in labels within the figure. The drawings depict generalized animal cells with a diploid chromosome number of four. The sizes of the nuclei and chromosomes are exaggerated to show the structures more clearly. The upper row of photomicrographs depicts cells of a plant, the blood lily, *Haemanthus.* Cells of an animal (the whitefish) are shown in the lower row. The chromosomes have been stained and the cells flattened on microscope slides. (*Plant cells,* Andrew S. Bajer, University of Oregon; *animal cells,* Ed Reschke)

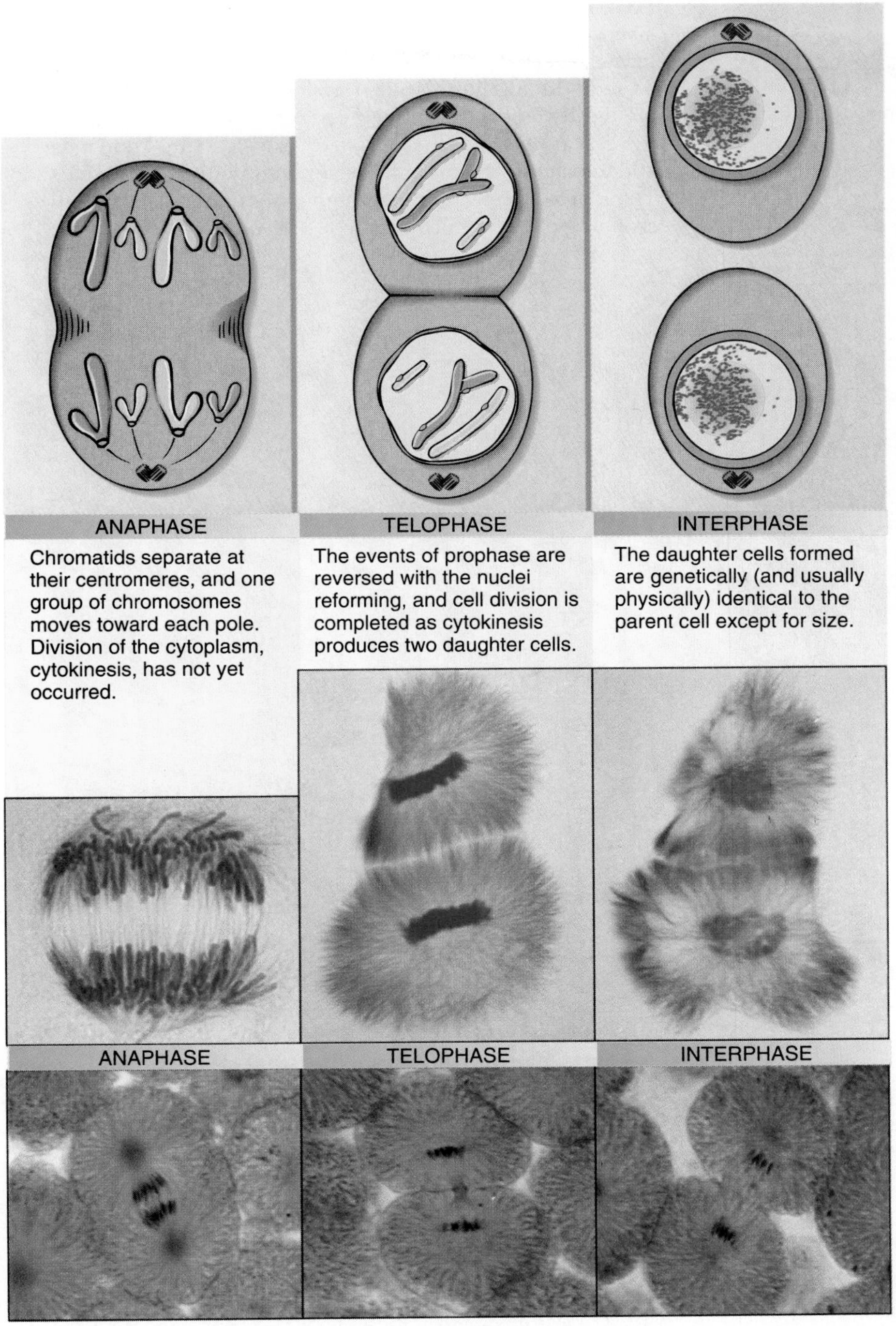
ANAPHASE
Chromatids separate at their centromeres, and one group of chromosomes moves toward each pole. Division of the cytoplasm, cytokinesis, has not yet occurred.
TELOPHASE
The events of prophase are reversed with the nuclei reforming, and cell division is completed as cytokinesis produces two daughter cells.
INTERPHASE
The daughter cells formed are genetically (and usually physically) identical to the parent cell except for size.
ANAPHASE
TELOPHASE
INTERPHASE

the MTOCs of plants, and recently has been shown to be chemically similar as well. The spindle microtubules end in the pericentriolar material, but they do not actually touch the centrioles themselves. Each of the two centrioles becomes duplicated during interphase, yielding two pairs of centrioles. Late in prophase, microtubules radiate from the pericentriolar material surrounding the centrioles, and one pair of centrioles migrates to each pole. The migration of the centrioles to the poles essentially marks the migration of microtubule-organizing centers. Additional microtubules form clusters extending outward in many directions from the MTOCs at the poles; these structures are called **asters** (Fig. 9–6).

(Text continues on page 225)

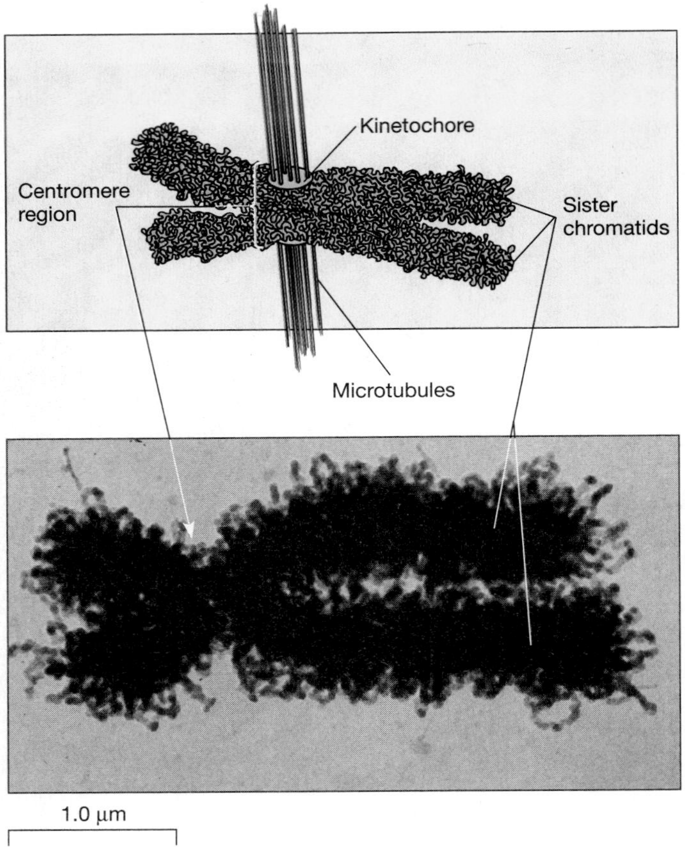

FIGURE 9–5 A metaphase chromosome contains a pair of sister chromatids, each consisting of tightly coiled chromatin fibers. A scanning electron micrograph *(bottom)* is paired with an interpretive drawing. The sister chromatids are tightly associated at their centromere regions, indicated by the brace. Within each centromere is a structure known as a kinetochore, which serves as a microtubule attachment site. The kinetochores and microtubules are not evident in the scanning electron micrograph. (E.J. DuPraw)

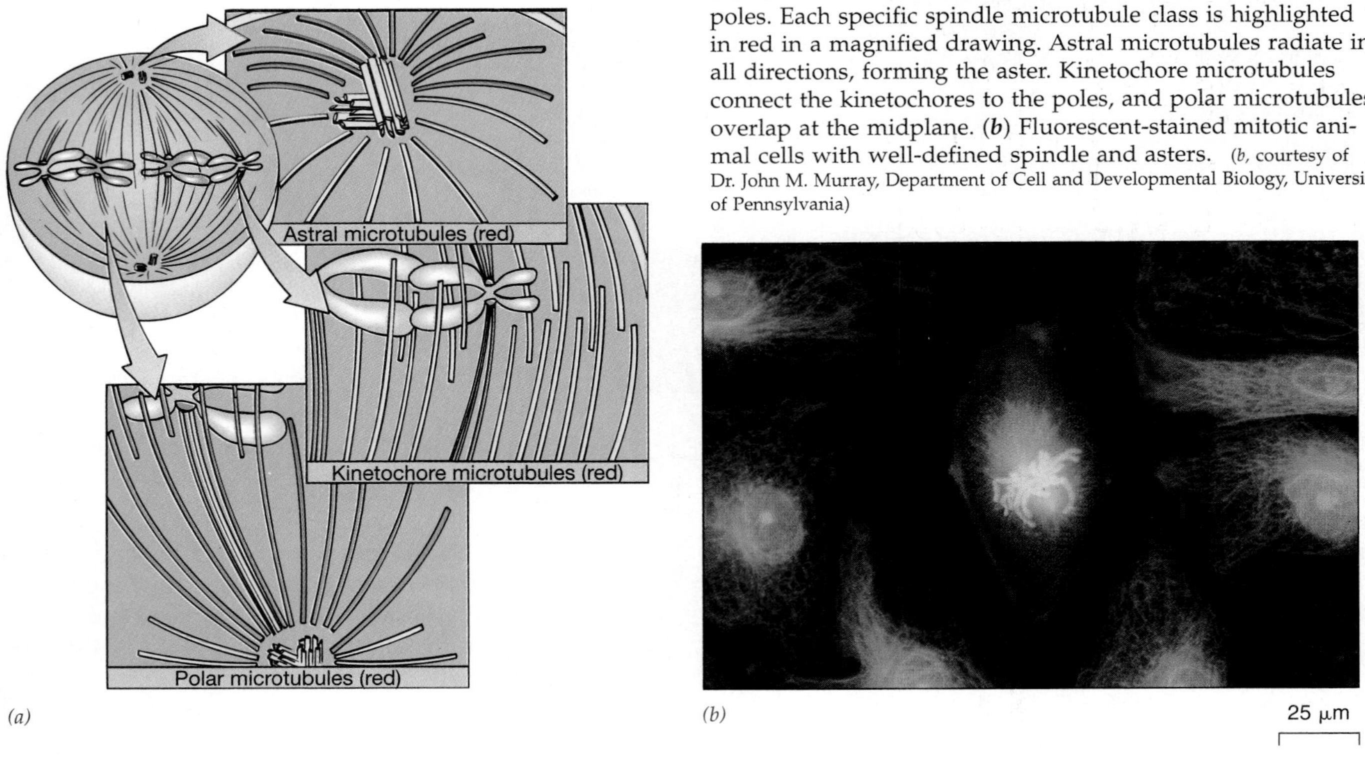

FIGURE 9–6 The mitotic spindle of an animal cell includes kinetochore microtubules, polar microtubules, and astral microtubules. Centrioles and astral microtubules are not found in a plant cell. (***a***) Interpretive drawing showing that one end of each microtubule is associated with one of the poles. Each specific spindle microtubule class is highlighted in red in a magnified drawing. Astral microtubules radiate in all directions, forming the aster. Kinetochore microtubules connect the kinetochores to the poles, and polar microtubules overlap at the midplane. (***b***) Fluorescent-stained mitotic animal cells with well-defined spindle and asters. (*b*, courtesy of Dr. John M. Murray, Department of Cell and Developmental Biology, University of Pennsylvania)

It is likely that both plant and animal spindles are organized by similar MTOCs. Centrioles were long thought to be involved in organizing the spindle in animal cells. However, their apparent involvement is probably coincidental. The association of the centrioles with the MTOC may have evolved to provide for the orderly distribution of these organelles, ensuring that each daughter cell receives a pair of centrioles. Centrioles are important to animal cells because they are thought to be necessary for the formation of the basal bodies of cilia and flagella (see Chapter 4). Centrioles are not found in the cells of flowering plants and more advanced gymnosperms (see Chapter 27). Both of these groups lack flagellated sperm and other flagellated or ciliated cells.

During prophase, the nucleolus (see Chapter 4) diminishes in size and usually disappears. Toward the end of prophase, the nuclear envelope breaks down, and each chromatid becomes attached to some of the spindle microtubules at its kinetochore (Fig. 9–6*a*). The chromosomes then move from pole to pole and finally become aligned along the equatorial plane of the cell, midway between the two poles.

Duplicated chromosomes line up on the midplane at metaphase

The period during which the chromosomes are lined up along the equatorial plane of the cell (the *metaphase plate*) constitutes **metaphase.** The mitotic spindle is complete. It is composed of numerous microtubules extending from each pole to the equatorial region (**polar microtubules**), where they generally overlap, and from the kinetochores to the poles (**kinetochore microtubules**) (Fig. 9–6*a*). At mitotic metaphase the individual sister kinetochores are attached by spindle microtubules to *opposite* poles of the cell.

During metaphase each chromatid is completely condensed and appears quite thick and distinct. Because chromosomes can generally be seen more clearly at metaphase than at any other time, they are usually photographed at this stage to be studied for certain chromosome abnormalities (see Chapter 15).

Chromosomes move toward the poles during anaphase

Anaphase begins as the forces holding the sister chromatids together in the vicinity of their centromeres are released. *Each chromatid is now referred to as an independent chromosome.* The separated chromosomes slowly move to opposite poles. The kinetochores of the chromosomes, still attached to kinetochore microtubules, lead the way, with the chromosome arms trailing behind. Anaphase ends when all the chromosomes have reached the poles.

The overall mechanism of chromosome movement in anaphase is still poorly understood, although significant progress is being made in this area. Microtubules lack elastic or contractile properties. So how do the chromosomes move apart? Are they pushed or pulled, or do other forces operate?

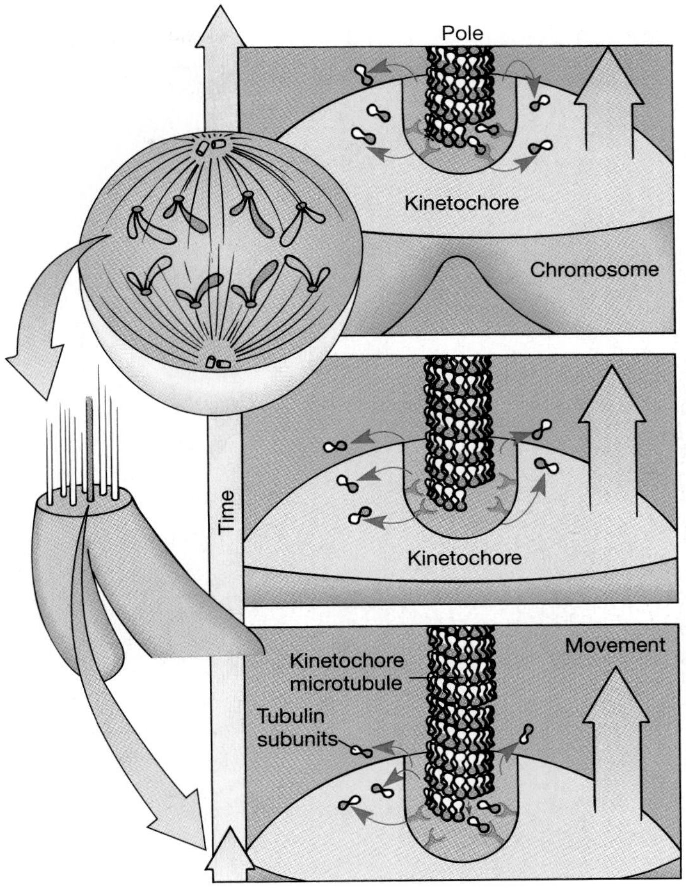

FIGURE 9–7 Available evidence suggests that a chromosome may move toward a pole by disassembling microtubules at its kinetochore. According to the model, each kinetochore is involved in the simultaneous disassembly of many microtubules.

Careful analyses of electron micrographs have yielded two main findings: (1) As anaphase continues, the kinetochore microtubules shorten. One current hypothesis suggests that chromosomes move as tubulin subunits are removed from the ends of the microtubules, particularly at the kinetochore (Fig. 9–7). If the kinetochore remains firmly attached to the ends of the microtubules, even as they are disassembled, the net result is chromosome movement toward the poles. (2) During anaphase the spindle as a whole elongates and may thus help to "push" the chromosomes apart. The spindle may lengthen as microtubules originating at opposite poles slide past one another in the region of overlap at the equator.

Two separate nuclei are formed during telophase

The final stage of mitosis, **telophase,** is characterized by a return to interphase-like conditions. The chromosomes

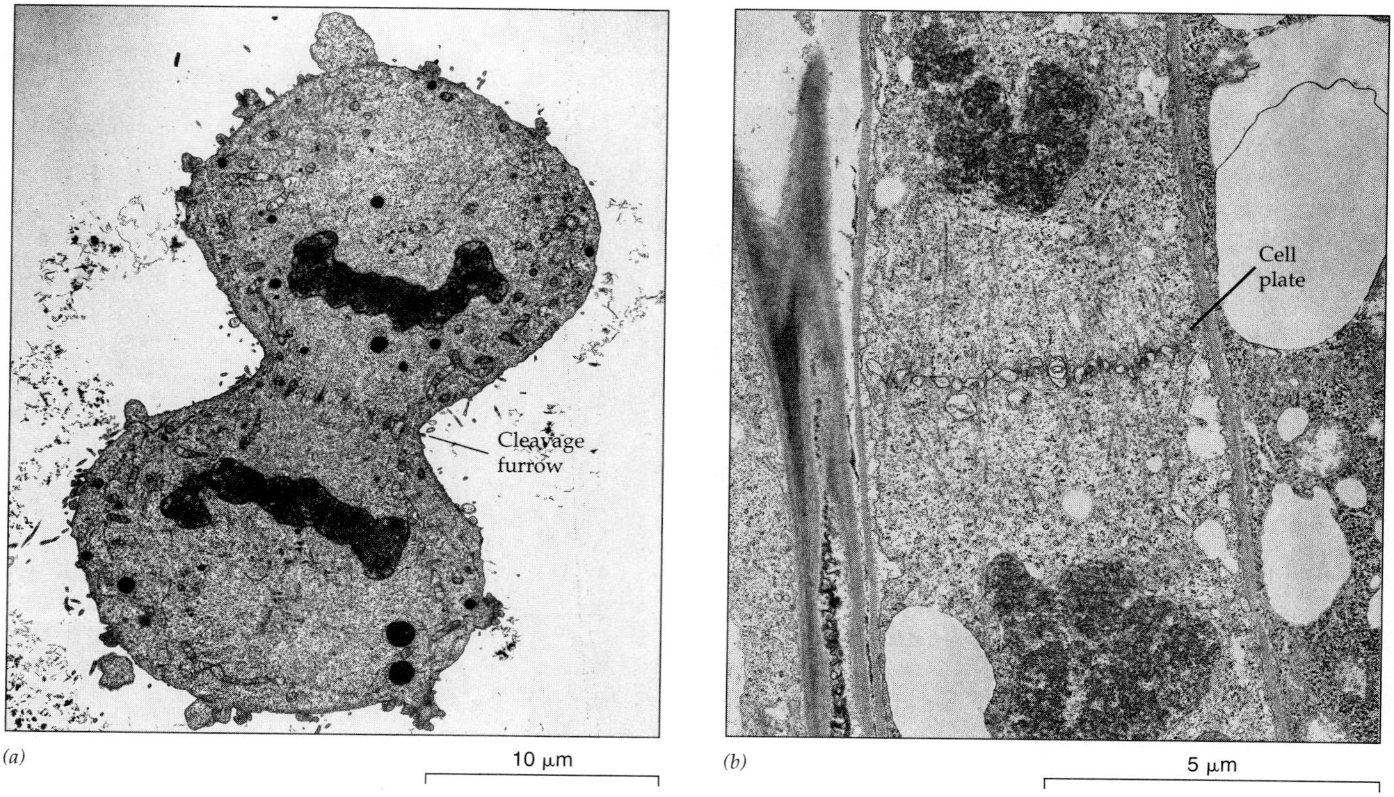

FIGURE 9–8 The mechanism of cytokinesis differs in plant and animal cells. (*a*) This TEM shows a cleavage furrow forming in the equatorial region of a cultured animal cell undergoing cytokinesis. (*b*) Cytokinesis is occurring by cell plate formation in this TEM of a maple leaf cell, *Acer saccharinum.* The nuclei are at the telophase stage. (*a,* T. E. Schroeder, University of Washington/Biological Photo Service; *b,* E. H. Newcomb and B. A. Palevitz, University of Wisconsin/Biological Photo Service)

decondense by uncoiling. A new nuclear envelope forms around each set of chromosomes, made at least in part from small vesicles and other components derived from the old nuclear envelope. The spindle microtubules disappear, and the nucleoli become apparent.

Two separate daughter cells are formed by cytokinesis

Cytokinesis, the division of the cytoplasm to yield two daughter cells, usually overlaps mitosis, generally beginning during telophase (see Fig. 9–4).

Cytokinesis of an animal cell begins with a furrow that encircles the cell in the equatorial region (Fig. 9–8*a*). The furrow, formed by a ring of microfilaments, gradually deepens and separates the cytoplasm into two daughter cells, each with a complete nucleus.

In plant cells, cytokinesis occurs by the formation of a **cell plate** (Fig. 9–8*b*), a partition constructed in the equatorial region of the spindle and growing laterally to the cell wall. The cell plate forms from vesicles that originate in the Golgi complex (see Chapter 4). Each daughter cell forms a plasma membrane and a cellulose cell wall *outside of the plasma membrane* on its side of the cell plate.

Mitosis typically produces two cells genetically identical to the parent cell

The remarkable regularity of the process of cell division ensures that each of the daughter nuclei receives exactly the same number and kind of chromosomes that the parent cell had. Thus, with a few exceptions, every cell of a multicellular organism has exactly the same genetic makeup. If a cell receives more or fewer than the characteristic number of chromosomes through some malfunction of the cell division process, the resulting cell may show marked abnormalities and be unable to survive.

Most cytoplasmic organelles are distributed randomly to the daughter cells

Mitosis provides for the orderly distribution of chromosomes (and of centrioles, if present), but what about the various cytoplasmic organelles? For example, all eukaryotic cells, even plant cells, require mitochondria. Likewise, photosynthetic plant cells cannot carry out photosynthesis without chloroplasts. These organelles contain their own DNA and appear to form by the division of previously existing mitochondria or plastids or their precursors. However, they generally divide during interphase, not when the cell divides. Because many copies of each organelle are present in each cell, organelles are apportioned more or less equally between the daughter cells at cytokinesis.

The cell cycle is controlled by an internal genetic program interacting with external signals

When conditions are optimal, some prokaryotic cells (which divide nonmitotically; see Chapter 23) can divide every 20 minutes. The generation times of eukaryotic cells are generally much longer, although the frequency of cell division varies widely among different species and among different tissues of the same species. Some cells in the central nervous system usually cease dividing after the first few months of life, whereas blood-forming cells, digestive tract cells, and skin cells divide frequently throughout life. Under optimal conditions of nutrition, temperature, and pH, the eukaryotic cell cycle length is constant for any cell type. Under less favorable conditions, however, the generation time may be longer.

The length of a cell cycle is the time required for the cell to carry out, through interaction with certain external signals, a precise program that has been built into it.

Recent studies indicate that certain basic mechanisms of genetic control of the cell cycle are common to all eukaryotes. Among the key components of the regulatory system are **protein kinases,** enzymes that add phosphate groups at specific locations in other proteins. The particular protein kinases involved in controlling the cell cycle are called cyclin-dependent protein kinases because they are only active when complexed with regulatory proteins called **cyclins.** The cyclins are so-named because their levels oscillate during the cell cycle.

When a specific cyclin-dependent protein kinase is complexed with a specific cyclin, it actively phosphorylates certain cellular proteins. Some of these proteins, including certain enzymes, become activated when they are phosphorylated. Conversely, phosphorylation inactivates some other enzymes and proteins. As some enzymes are activated and others are inactivated, the activities of the cell (as they relate to the steps of the cell cycle) change (Fig. 9–9).

For example, an active cyclin-dependent protein kinase called **mitosis-promoting factor (MPF)** is required for the cell to make the transition from G_2 to mitosis. Active MPF is produced when the inactive enzyme becomes complexed with a form of cyclin known as cyclin B. Active MPF then phosphorylates many proteins, activating those needed for entry to mitosis, and inactivating those

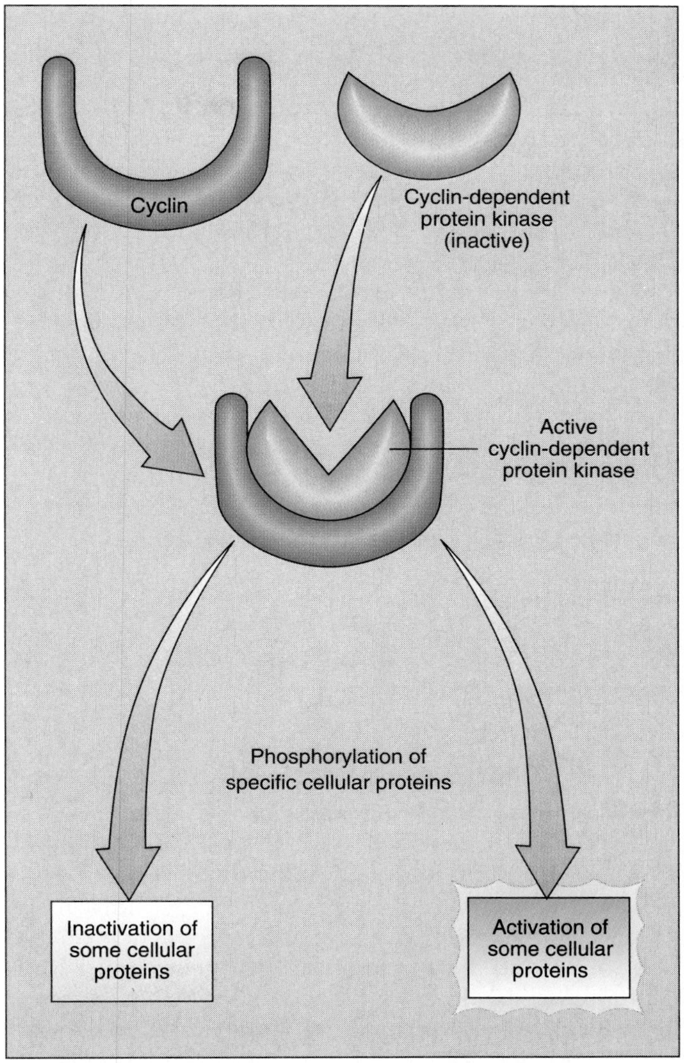

FIGURE 9–9 Protein kinases (enzymes that phosphorylate proteins) are instrumental in the control of the cell cycle. These are known as cyclin-dependent protein kinases because they are active only when they are complexed with one of a class of regulatory proteins called cyclins. The cyclins are so-named because their levels rise and fall during the cell cycle. The active cyclin-dependent protein kinase phosphorylates many cellular proteins, activating some and inactivating others.

that would impede mitosis. Later in mitosis, the cyclin B is degraded, rendering MPF inactive. Although not all of the details are understood, these systems of regulating the cell cycle have been highly conserved during the evolution of eukaryotes; they are found in organisms as diverse as yeast (a simple single-celled fungus), frogs, clams, and plants.

Certain drugs can stop the cell cycle. Some of these prevent DNA synthesis, whereas others inhibit the synthesis of proteins that control the cycle, as well as the synthesis of structural proteins that contribute to the mitotic spindle. Because cancer cells (see Chapter 16) often divide much more rapidly than most normal body cells, they may be most affected by these drugs. Many of the side effects of certain anticancer drugs (e.g., nausea, hair loss) are due to the drugs' effects on rapidly dividing cells in the digestive system, hair follicles, and so forth.

Certain plant hormones are known to stimulate mitosis. Chief among these are the **cytokinins,** which act as promoters of mitosis both in normal growth and in wound healing (see Chapter 36).

The cell cycle can also be affected by cold temperatures and other agents that interfere with the normal function of the mitotic spindle. **Colchicine,** a drug used to block cell division in eukaryotic cells, binds with tubulin, the major microtubule protein. This causes the spindle to break down and prevents the chromosomes from moving to the opposite poles of the cell. As a result, a cell may end up with extra sets of chromosomes (a condition known as *polyploidy,* which is discussed in the next section).

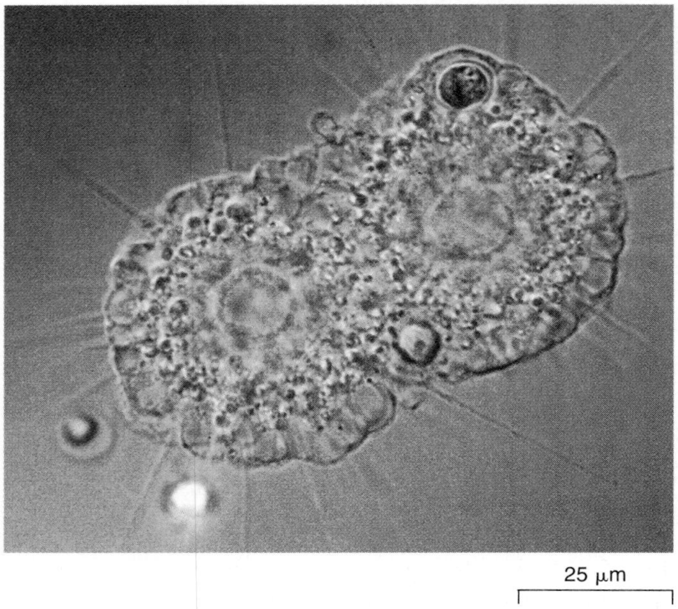

FIGURE 9–10 This amoeboid protozoon (*Actinophrys sol*) is undergoing binary fission, a form of asexual reproduction that involves mitosis. (Phillip A. Harrington/Peter Arnold, Inc.)

Sexual life cycles require a mechanism to reduce the chromosome number

Although the details of the reproductive process vary greatly among different kinds of eukaryotes, we can distinguish two basic types of reproduction: asexual and sexual. In **asexual** reproduction a single parent usually splits, buds, or fragments to produce two or more individuals (Fig. 9–10). In most forms of asexual reproduction, all the cells are produced by mitosis, so their genes and inherited traits are identical to those of the parent. Such a group of genetically identical organisms is termed a **clone.** Asexual reproduction is usually a rapid process; it permits organisms well adapted to their environment to produce new generations of similarly adapted organisms.

In contrast, **sexual** reproduction involves the union of two specialized sex cells, or **gametes,** to form a single cell called a **zygote.** Usually the gametes are contributed by two different parents, but in some cases a single parent furnishes both gametes. In the case of animals and plants, the egg and the sperm are the gametes, and the fertilized egg is the zygote.

Offspring produced sexually are not genetically identical to their parents, so some may be able to survive environmental changes or other stresses better than either parent, whereas others, with a different combination of traits, may be less likely to survive.

Because you now understand the roles of chromosomes in inheritance, you may recognize a problem in eukaryotic sexual reproduction: if each gamete has the same number of chromosomes as did the parental cell that produced it, then the zygote would be expected to have twice as many chromosomes, and this doubling would occur generation after generation. How do organisms avoid producing zygotes with ever-increasing chromosome numbers? To answer this question, we need more information about the types of chromosomes found in cells.

Each chromosome found in a somatic (body) cell of a higher plant or animal normally has a partner chromosome. The two partners, known as **homologous chromosomes,** are similar in size, shape, and the position of their centromeres. When stained by special techniques, the members of a pair generally share a characteristic pattern of bands. In most species, chromosomes vary enough in their morphological features that cytologists can distinguish the different homologous pairs and match up the partners. The 46 chromosomes in human cells constitute 23 different pairs of homologous chromosomes (see Fig. 9–1).

The most important feature of homologous chromosomes is that they carry information for controlling the same kinds of genetic traits, although the information is not necessarily identical. For example, members of a pair of homologous chromosomes might each carry a gene that specifies hemoglobin structure; however, one member might have the information for normal hemoglobin, whereas the other might specify the abnormal form of hemoglobin associated with sickle cell anemia (see Chapter 15).

A *set* of chromosomes has one of each kind of chromosome; in other words, it has one member of each homologous pair. If a cell or nucleus contains two sets of chromosomes, it is said to have a **diploid** chromosome number. If it has only a single set of chromosomes, it has the **haploid** number. In humans the diploid chromosome number is 46 and the haploid number is 23. When the sperm and egg fuse at fertilization, each gamete contributes a haploid set of chromosomes; the diploid number is thereby restored in the fertilized egg (zygote). When the zygote divides by mitosis to form the first two cells of the embryo, each daughter cell receives the diploid number of chromosomes and this is repeated in subsequent mitotic divisions. Thus, most human body cells are diploid.

If a cell or an individual has three or more sets of chromosomes, we say that it is **polyploid.** Polyploidy is relatively rare among animals but quite common among plants (see Chapter 19). In fact, polyploidy has been an important factor in plant evolution. Between 30 and 80% of all flowering plants are polyploid. Polyploid plants are often larger and more hardy than diploid members of the same group and may be important commercially. Modern bread wheat, *Triticum aestivum* (Fig. 9–11), is a hexaploid with 42 chromosomes, derived from three different diploid species with 14 chromosomes each.

FIGURE 9–11 Modern bread wheat is a hexaploid plant.
(Sharon Cummings/Dembinsky Photo Associates)

The abbreviation for the chromosome number found in the gametes of a particular species is ***n*,** and the zygotic chromosome number is given as **2*n*.** For organisms that are not polyploid, the haploid chromosome number is equal to n and the diploid number is equal to $2n$. For simplicity, in the rest of this chapter we assume that the organisms used as examples are not polyploid. We therefore use the designations diploid and $2n$, and haploid and n, interchangeably, although these terms are not strictly synonymous.

Diploid cells undergo meiosis to form haploid cells

We have examined the process of mitosis, which ensures that each daughter cell receives exactly the same number and kinds of chromosomes that the parent cell had. A diploid cell that undergoes mitosis produces two diploid cells; similarly a haploid cell produces two haploid cells. A constant chromosome number in successive generations of sexually reproducing organisms is ensured by a special type of cell division called **meiosis.** The term *meiosis* means "to make smaller," referring to the fact that the process involves two successive cell divisions during which the chromosome number is reduced by one half.

Meiosis is directly involved in gamete (egg and sperm) formation in animals. In plants and some other organisms, meiosis gives rise to haploid *spores.* These then divide mitotically, and some of their descendants become haploid gametes. When two haploid gametes unite, the fusion of their nuclei reconstitutes the diploid chromosome number in the zygote. The role of meiosis in sexual life cycles is discussed in *Making the Connection: Mitosis, Meiosis, and Life Cycles* on page 232.

Meiosis produces haploid cells with unique gene combinations

The events of meiosis are similar to the events of mitosis, with four important differences: (1) Meiosis involves two successive nuclear and cytoplasmic divisions, with the potential to yield a total of four cells. (2) Despite two successive nuclear divisions, the DNA and other chromosomal components are duplicated only once, during the interphase preceding the first meiotic division. (3) Each of the four cells produced by meiosis contains the haploid chromosome number — that is, only one set with only one representative of each homologous pair. (4) During meiosis, the genetic information from both parents is shuffled, so each resulting haploid cell (which either is a gamete itself or ultimately gives rise to one or more gametes) has a virtually unique combination of genes.

(*Text continues on page 231*)

FIGURE 9–12 Meiosis consists of two nuclear divisions, meiosis I and meiosis II. The process begins with a cell that has a diploid chromosome number of four and ends with the formation of four haploid cells, with two chromosomes each. The maternal chromosomes are shown in pink; the paternal chromosomes are purple.

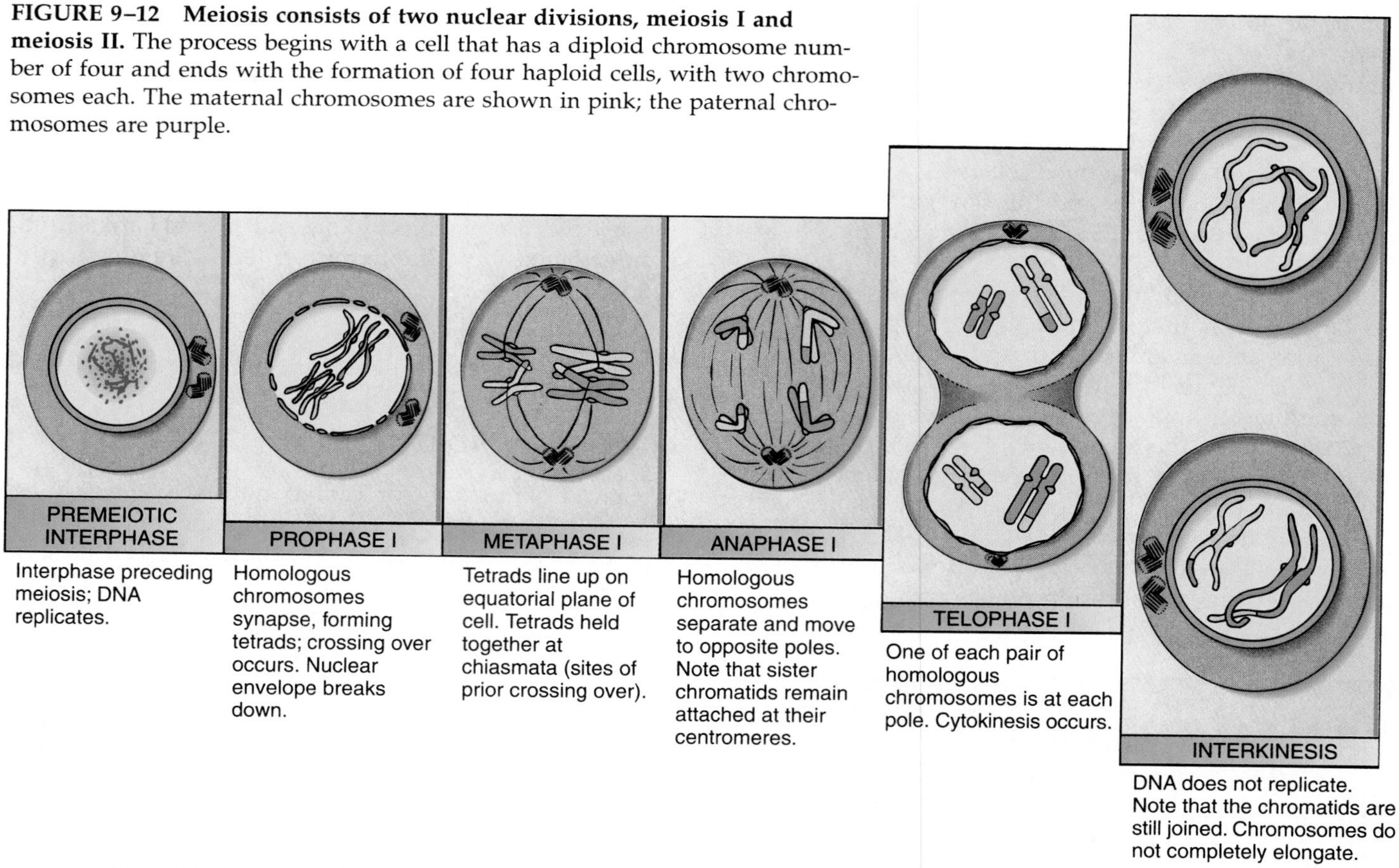
PREMEIOTIC INTERPHASE
Interphase preceding meiosis; DNA replicates.
PROPHASE I
Homologous chromosomes synapse, forming tetrads; crossing over occurs. Nuclear envelope breaks down.
METAPHASE I
Tetrads line up on equatorial plane of cell. Tetrads held together at chiasmata (sites of prior crossing over).
ANAPHASE I
Homologous chromosomes separate and move to opposite poles. Note that sister chromatids remain attached at their centromeres.
TELOPHASE I
One of each pair of homologous chromosomes is at each pole. Cytokinesis occurs.
INTERKINESIS
DNA does not replicate. Note that the chromatids are still joined. Chromosomes do not completely elongate.

FIGURE 9–13 These light photomicrographs illustrate meiosis in the trumpet lily, *Lilium longiflorum*. The chromosomes have been stained and the cells flattened on microscope slides. (***a***) Mid-prophase I. (***b***) Late prophase I. (***c***) Metaphase I. (***d***) Anaphase I. (***e***) Prophase II. (***f***) Metaphase II. (***g***) Anaphase II. (***h***) Four daughter cells. (Clare Hasenkampf, University of Toronto/Biological Photo Service)

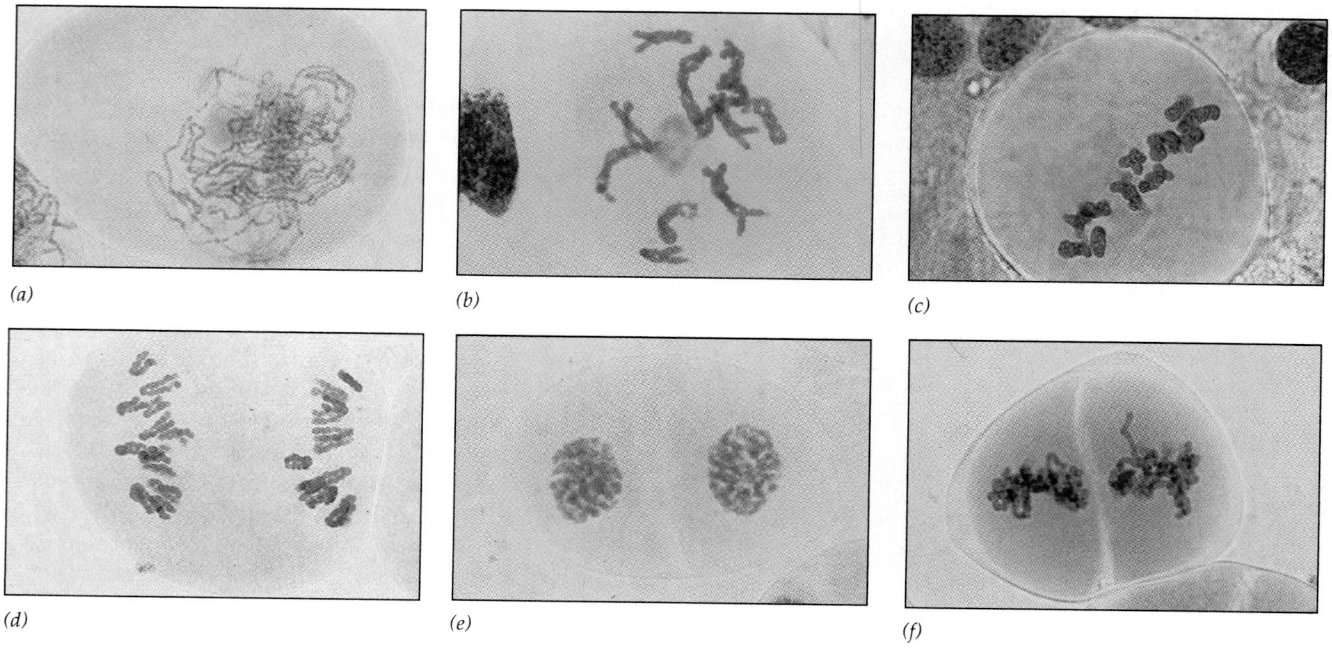
(a)
(b)
(c)
(d)
(e)
(f)

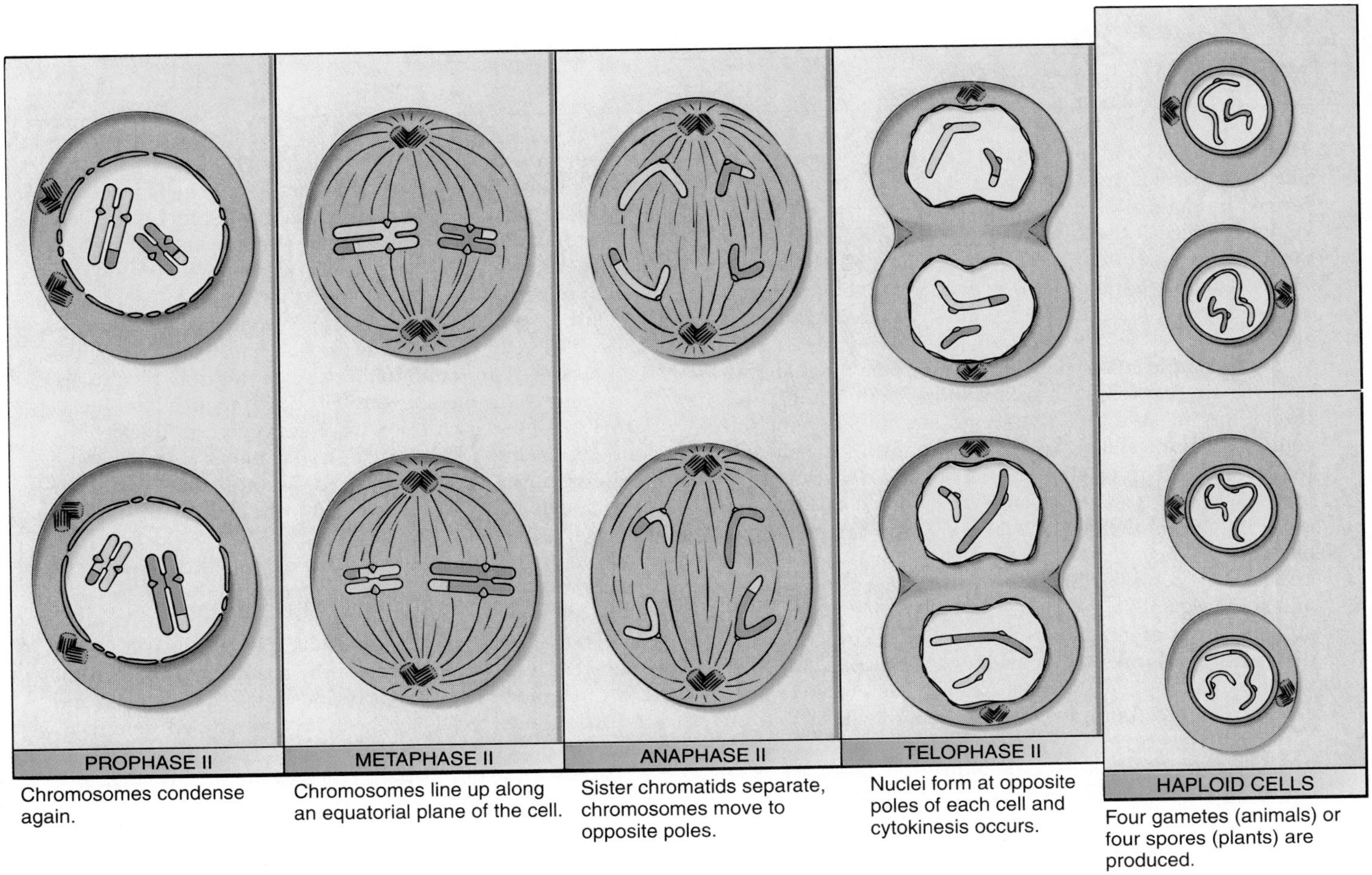
PROPHASE II
Chromosomes condense again.
METAPHASE II
Chromosomes line up along an equatorial plane of the cell.
ANAPHASE II
Sister chromatids separate, chromosomes move to opposite poles.
TELOPHASE II
Nuclei form at opposite poles of each cell and cytokinesis occurs.
HAPLOID CELLS
Four gametes (animals) or four spores (plants) are produced.

In meiosis homologous chromosomes become distributed into different daughter cells

Meiosis typically consists of two nuclear and cytoplasmic divisions, designated the *first* and *second meiotic divisions,* or simply **meiosis I** and **meiosis II.** Each includes prophase, metaphase, anaphase, and telophase stages. During meiosis I, the members of each homologous pair of chromosomes first join together and then separate and are distributed into different nuclei. In meiosis II the chromatids that make up each chromosome separate and are distributed to the nuclei of the daughter cells. The following discussion describes meiosis in an animal with a diploid chromosome number of four. Refer to Figures 9–12 and 9–13 as you read.

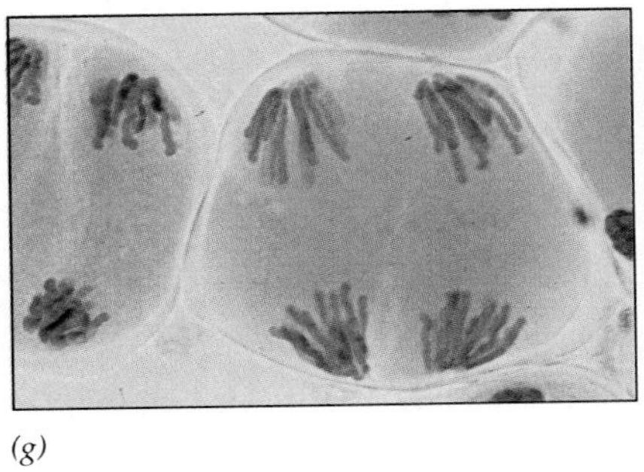

(g)

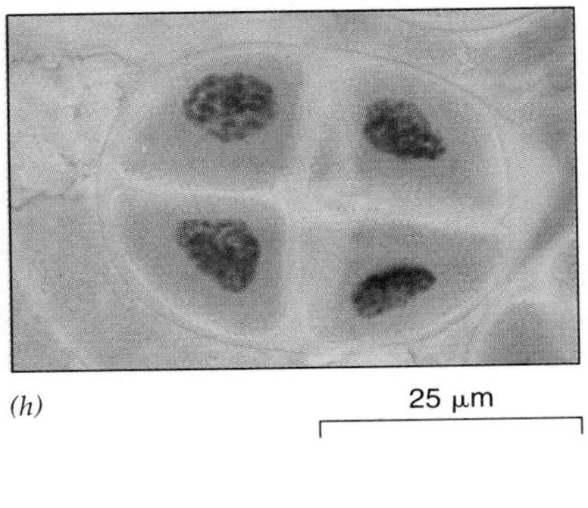
25 µm

(h)

Making the Connection

Mitosis, Meiosis, and Life Cycles

How do mitosis and meiosis fit into the life cycles of sexually reproducing organisms? Sexual reproduction is characterized by the fusion of two haploid sex cells to form a diploid zygote. It therefore follows that, in a sexual life cycle, meiosis must occur before gametes can be produced. However, although meiosis occurs at some point in a sexual life cycle, it does not always *directly* precede gamete formation.

Many simple eukaryotes (including some fungi and algae) remain haploid (their cells dividing mitotically) throughout most of their lives, with individuals being unicellular or multicellular. Two haploid gametes (produced by mitosis) fuse to produce a diploid zygote that undergoes meiosis to restore the haploid state (Fig. *a*). Examples of these types of life cycles can be found in Chapters 24 and 25.

The most complex life cycles are displayed by plants and some algae (Fig. *b*). These life cycles, characterized by an **alternation of generations,** consist of a multicellular diploid stage, termed the **sporophyte generation,** and a multicellular haploid stage, termed the **gametophyte generation**. Diploid sporophyte cells undergo meiosis to form haploid spores, each of which then divides mitotically to produce a multicellular haploid gametophyte. Gametophytes produce gametes by mitosis. The female and male gametes (eggs and sperm) then fuse to form a diploid zygote that divides mitotically to produce a multicellular diploid sporophyte.

In higher plants, including flowering plants, the diploid sporophyte — which includes the roots, stems, and leaves of the plant body — is the dominant form. The gametophytes are small and inconspicuous. For example, a pollen grain contains a haploid male gametophyte that forms haploid sperm by mitosis. More detailed descriptions of alternation of generations can be found in Chapters 26 and 27.

Only in animals and a few other organisms does meiosis lead directly to gamete formation (Fig. *c*). The body (somatic) cells of an individual organism multiply by mitosis and are diploid; the only haploid cells produced are the gametes. These are formed when certain **germ line** cells undergo meiosis. The formation of gametes is known as **gametogenesis.** Male gametogenesis, termed **spermatogenesis,** results in the formation of four haploid sperm cells for each cell that enters meiosis.

In contrast, female gametogenesis, termed **oogenesis,** results in the formation of a single egg cell, or **ovum,** for every cell that enters meiosis. This is accomplished by a process that apportions virtually all of the cytoplasm to only one of the two nuclei at each of the meiotic divisions. At the end of the first meiotic division, one nucleus is retained and the other, called the first **polar body,** is excluded from the cell and ultimately degenerates. Similarly, at the end of the second division, one nucleus becomes the second polar body and the other nucleus survives. In this way, one haploid nucleus becomes the recipient of most of the accumulated cytoplasm and nutrients from the original meiotic cell. (See Chapter 48 for a more detailed description.) ▲

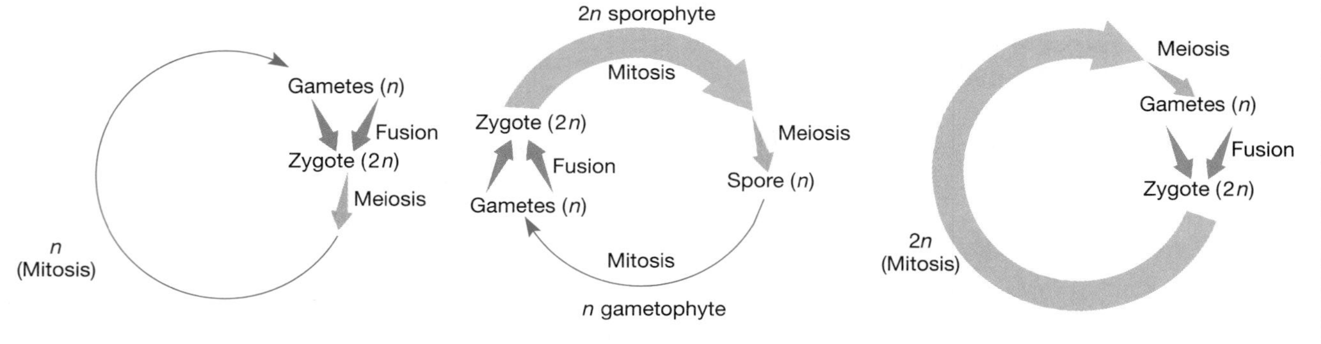

(*a*) Simple organisms (*b*) Alternation of generations (*c*) Animals

Prophase I includes synapsis and crossing over

As in mitosis, the chromosomes are duplicated during the S phase of interphase, before meiosis actually begins. Each duplicated chromosome consists of two chromatids, joined in the vicinity of their centromeres. During *prophase I,* while the chromatids are still elongated and thin, the homologous chromosomes come to lie lengthwise side by side. This process is called **synapsis,** which means "fastening together." In our example, because the diploid number is four, there are two homologous pairs.

It is customary when discussing higher organisms to refer to one member of each homologous pair as a **maternal homologue** because it was originally inherited

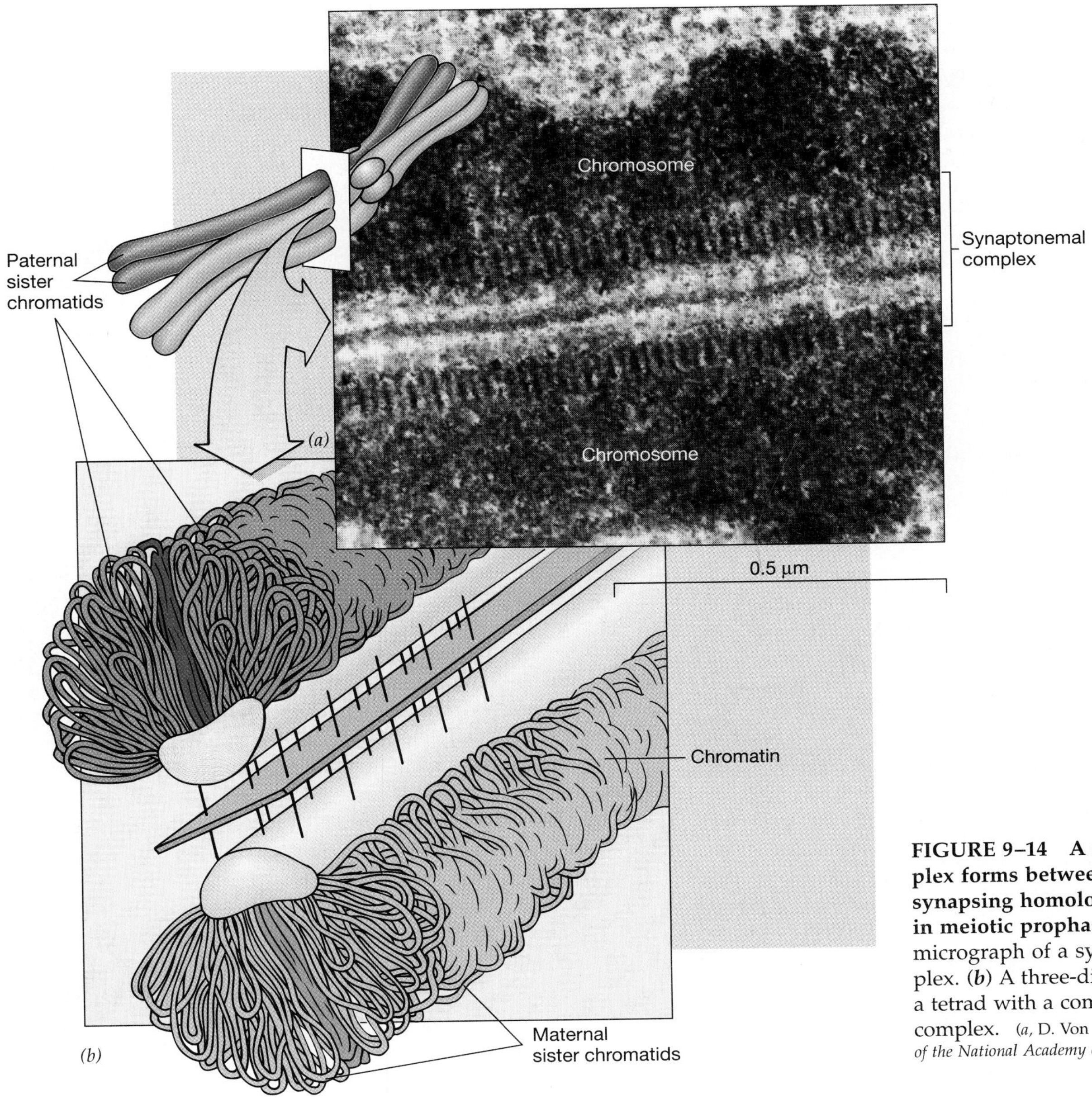

FIGURE 9–14 A synaptonemal complex forms between each pair of synapsing homologous chromosomes in meiotic prophase I. (***a***) An electron micrograph of a synaptonemal complex. (***b***) A three-dimensional model of a tetrad with a complete synaptonemal complex. (*a*, D. Von Wettstein, *1971 Proceedings of the National Academy of Science* 68, pp. 851–855)

from the female parent, and to the other as a **paternal homologue** because it was contributed by the male parent during the formation of the zygote. Because each chromosome was duplicated during the premeiotic interphase and now consists of two chromatids, synapsis results in the association of *four* chromatids. The resulting complex is known as a **bivalent** or a **tetrad.** The term *bivalent,* in which the prefix *bi-* refers to the two homologous chromosomes, is commonly used by cytogeneticists. The term *tetrad* (*tetra* means "four") is preferred by some geneticists interested in following the fates of the four chromatids. We will use *tetrad* in further discussions.

The number of tetrads equals the haploid number of chromosomes. In our example there are two tetrads; in human cells there are 23 tetrads (and a total of 92 chromatids) at this stage.

Homologous chromosomes become closely associated during synapsis. Electron microscopic observations reveal that a characteristic structure, known as the **synaptonemal complex,** forms between the synapsed homologues (Fig. 9–14). Genetic material may be exchanged between homologous (nonsister) chromatids by **crossing over,** a process in which enzymes break the chromatids, exchange parts, and then rejoin the chromatids to produce new combinations of genes. The resulting **genetic recombination** greatly enhances the amount of genetic variation among sexually produced offspring. This process is discussed in more detail in Chapter 10.

In many species, prophase I is a lengthy phase during which the cell grows and synthesizes nutrients. This is especially true during the formation of some egg cells because materials need to be made for the benefit of the future embryo. In many types of meiotic cells, the chromosomes assume unusual shapes during this phase. For example, lampbrush chromosomes, found in the female meiotic cells (oocytes) of some amphibians, are composed of hundreds of pairs of loops projecting from the chromatid axis. They owe their name to their resemblance to the brushes used to clean old-fashioned oil lamps (Fig. 9–15). The loops are sites of intense RNA synthesis. The RNA is used to direct the synthesis of specific proteins.

In addition to the unique processes of synapsis and crossing over, events similar to those seen during mitotic prophase also take place. A spindle composed of microtubules and other components forms. In animal cells a pair of centrioles moves to each pole and astral rays are formed.

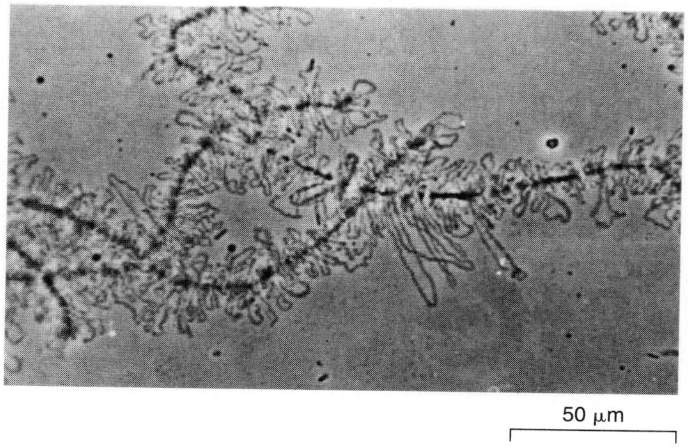

FIGURE 9–15 Lampbrush chromosomes are meiotic prophase I tetrads that show characteristic loops extending from a central axis. The loops are sites of very active RNA synthesis. This photomicrograph shows parts of several tetrads from a female meiotic cell (oocyte) of the newt *Triturus viridescens*. (Dennis Gould)

The nuclear envelope disappears in late prophase I, and in favorable material the structure of the tetrads can be seen clearly with the microscope (Fig. 9–16). The sister chromatids continue to be closely aligned along their lengths, but the homologous chromosomes are no longer closely associated and their centromeres (and kinetochores) are separated from one another. In late prophase I, the homologous chromosomes are held together only at specialized regions, termed **chiasmata** (sing., *chiasma*). Each chiasma is a site of crossing over—that is, a site at which homologous chromatids previously broke, exchanged genetic material, and rejoined, producing an X-shaped configuration.

Homologous chromosomes separate during meiosis I

Prophase I ends when the tetrads become aligned on the equatorial plane; the cell is now said to be at *metaphase I.* The sister kinetochores of one chromosome are attached by spindle fibers to only one of the two poles, and those of the homologous chromosome are attached to the opposite pole. (By contrast, in mitosis the *sister* kinetochores are attached to opposite poles.) During *anaphase I,* the homologous chromosomes of each pair separate, or disjoin, and move toward opposite poles. Each pole receives a random mixture of maternal and paternal chromosomes, but only one member of each pair is present at each pole. The sister chromatids are still united at their centromere regions. Again, this differs from mitotic anaphase, in which the sister chromatids pass to opposite poles.

During *telophase I,* the chromatids generally decondense somewhat, the nuclear envelope may reorganize, and cytokinesis usually takes place. In telophase I in our

(*Text continues on page 236*)

Making the Connection

Mitosis and Meiosis

How are mitosis and meiosis similar and how do they differ? The diagram compares mitosis and meiosis, beginning with a diploid cell with four chromosomes (i.e., two pairs of homologous chromosomes). Some of the stages have been omitted for simplicity. Because the chromosomes were duplicated in the previous interphase, each chromosome consists of two sister chromatids. The chromosomes derived from one parent are shown in blue and those from the other parent are pink. The homologous partner chromosomes (one blue and the other pink) can be recognized by their similarity in size and shape.

Mitosis is a single division in which *sister chromatids* disjoin (separate) from each other, ultimately producing two diploid daughter cells identical to each other and to the original cell. Note that homologous chromosomes do not associate physically at any time in mitosis.

Meiosis consists of two successive divisions, meiosis I and meiosis II, ending with the formation of four haploid daughter cells that differentiate as gametes (in animals) or as spores (in plants). In prophase I of meiosis the homologous chromosomes undergo synapsis to form tetrads. If we ignore crossing over, we can say that *homologous chromosomes* disjoin during meiosis I, and *sister chromatids* disjoin during meiosis II. It is also correct to say that homologous centromeres (or kinetochores) disjoin during meiosis I, and sister centromeres (or kinetochores) disjoin during meiosis II. ▲

Mitosis and Meiosis (*continued*)

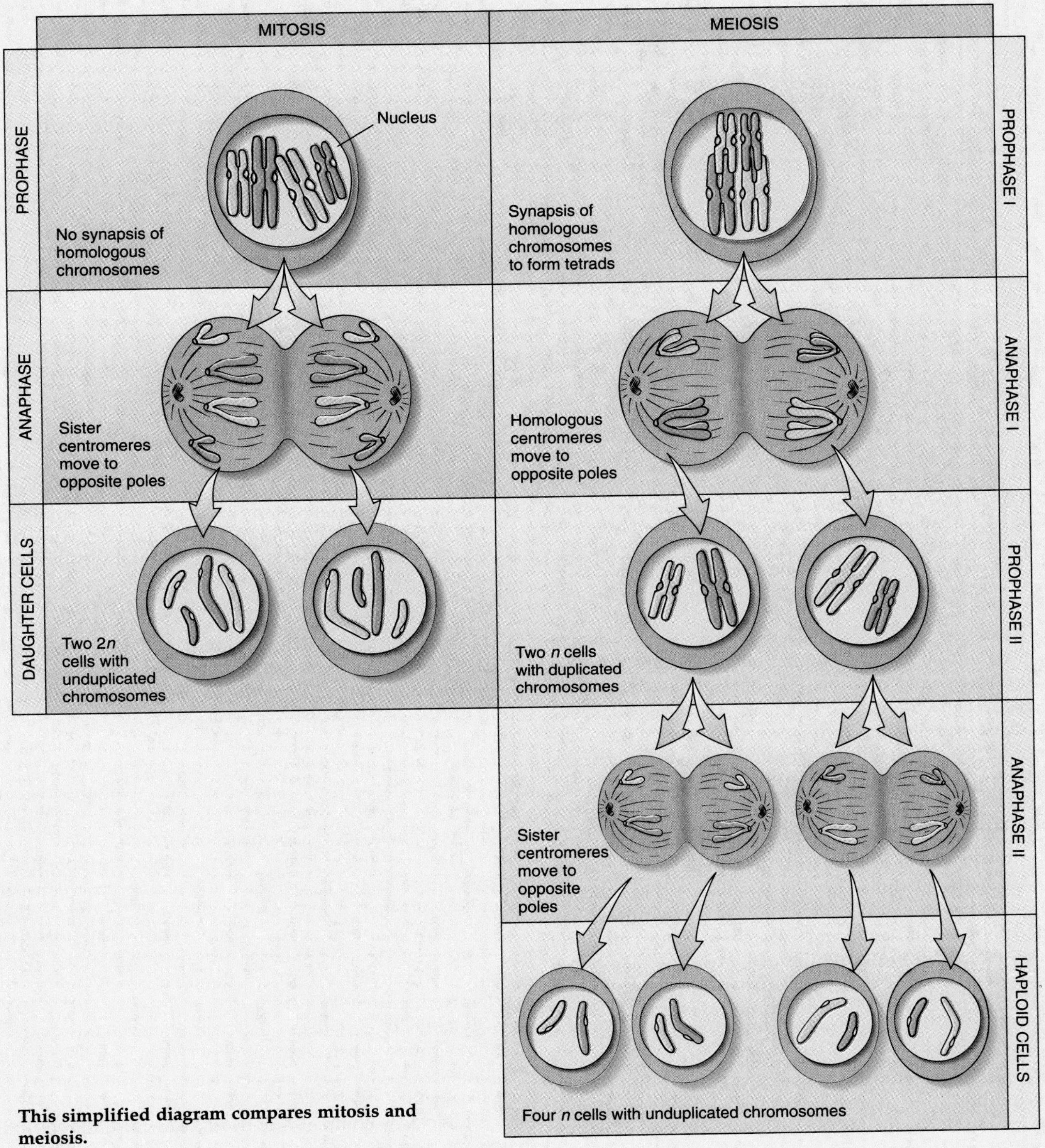

This simplified diagram compares mitosis and meiosis.

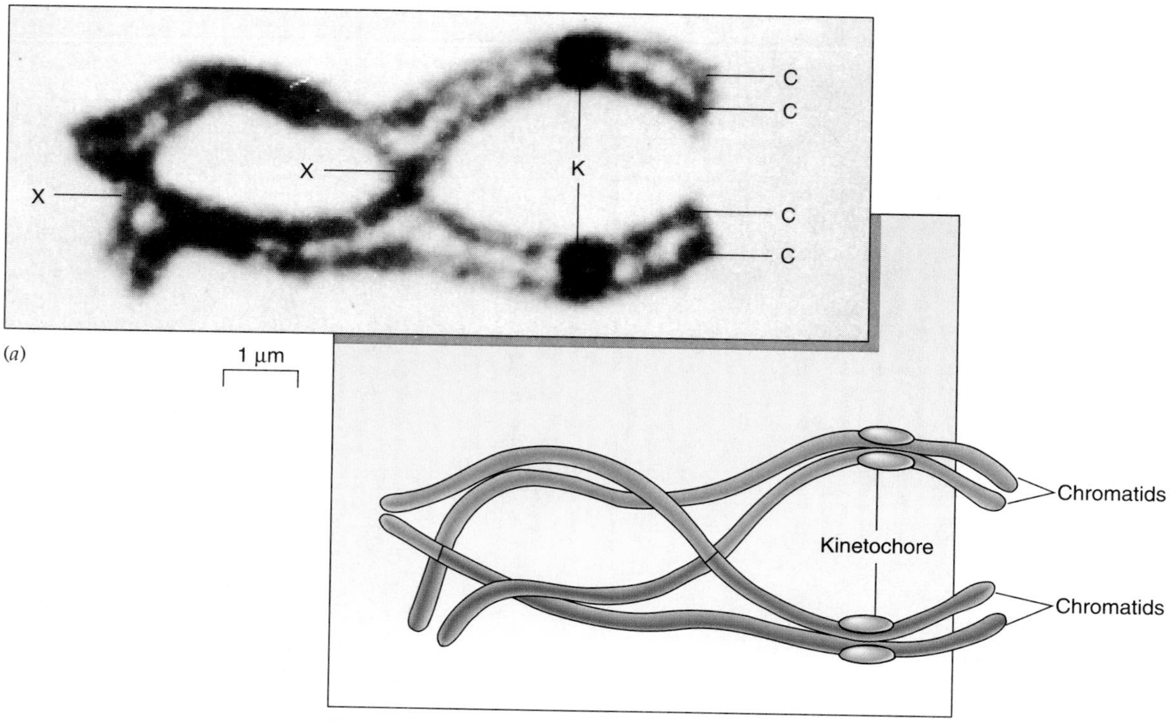

FIGURE 9–16 A tetrad consists of four chromatids. (***a***) A photomicrograph of a tetrad during late prophase I of a male meiotic cell (spermatocyte) from a salamander. The chromatids are labelled *C*, the kinetochores are visible at *K*, and chiasmata (resulting from crossing over) are indicated at each X. (***b***) Interpretive drawing indicating the structure of the tetrad. The paternal chromatids are purple, and the maternal chromatids are pink. (*a*, Courtesy of J. Keezer)

example, there are two duplicated chromosomes at each pole, for a total of four chromatids; in humans there are 23 duplicated chromosomes (46 chromatids) at each pole.

During the interphase-like stage that follows, called **interkinesis,** there is no S phase because no further chromosome replication takes place. Interkinesis is very brief in most organisms and absent in some.

Chromatids separate in meiosis II

Because the chromosomes usually remain partially condensed between divisions, the prophase of the second meiotic division is also brief. **Prophase II** is similar to mitotic prophase in many respects. There is no pairing of homologous chromosomes (indeed, only one member of each pair is present in each nucleus) and no crossing over.

During **metaphase II,** the chromosomes line up on the equatorial planes of their cells. The first and second metaphases can be easily distinguished in diagrams; at metaphase I the chromatids are arranged in bundles of four (tetrads), and at metaphase II they are in groups of two (as in mitotic metaphase). This is not always so obvious in living cells.

During *anaphase II* the chromatids, attached to spindle fibers at their kinetochores, separate and move to opposite poles, just as they would at mitotic anaphase. As in mitosis, each former chromatid is now referred to as a chromosome. Thus, at *telophase II* there is one representative for each homologous pair at each pole. Each is an unduplicated chromosome. Nuclear envelopes then re-form, the chromosomes gradually elongate to form chromatin threads, and cytokinesis occurs.

The two successive divisions have the potential to yield four haploid nuclei, each containing *one* of each kind of chromosome. Each resulting haploid cell has a different combination of genes. This genetic variation has two sources: (1) During meiosis the maternal and paternal chromosomes are "shuffled" and one of each pair is randomly distributed to the poles at anaphase I. (2) DNA segments are exchanged between maternal and paternal homologues during crossing over. The genetic consequences of these events are discussed in more detail in Chapter 10.

Meiosis is compared with mitosis in *Making the Connection: Mitosis and Meiosis* on pages 234 and 235.

SUMMARY

I. In the production of a new generation, cells transfer genetic information from parent to offspring; this process is termed heredity. Genetics is the study of the structure, transmission, and expression of genes.

II. Genes are made of DNA. DNA is complexed with protein and RNA to form the chromatin fibers that make up chromosomes.
   A. A gene is the portion of chromosomal DNA needed to produce a specific RNA molecule, which may in turn code for a specific polypeptide.
   B. A diploid organism of a given species has a characteristic number of chromosome pairs per cell. The two members of each pair, called homologous chromosomes, are similar in length, shape, and other structural features, and carry genes affecting the same kinds of features of the organism.

III. The eukaryotic cell cycle is the period from the beginning of one division to the beginning of the next division.
   A. Interphase can be divided into the first gap phase (G_1), the chromosomal synthesis phase (S), and the second gap phase (G_2).
      1. During the G_1 phase, the cell grows and prepares for the S phase.
      2. DNA and the chromosomal proteins are synthesized during the S phase.
      3. During the G_2 phase, protein synthesis increases in preparation for cell division.
   B. During mitosis, identical chromosomes are distributed to each pole of the cell, and a nuclear envelope forms around each set.
      1. During prophase, the chromosomes become visible with the microscope, the nucleolus disappears, the nuclear envelope breaks down, and the mitotic spindle begins to form.
      2. During metaphase, the duplicated chromosomes, each composed of a pair of sister chromatids, line up along the equatorial plane of the cell; the mitotic spindle is complete.
      3. During anaphase, the sister chromatids separate from one another and move to opposite poles of the cell. Each former chromatid is now referred to as a chromosome.
      4. During telophase, a nuclear envelope re-forms around each set of chromosomes, nucleoli become apparent, the chromosomes uncoil, and the spindle disappears.
   C. During cytokinesis, which generally begins in telophase and therefore overlaps mitosis, the cytoplasm divides to form two individual cells.

IV. There are two major forms of reproduction: asexual and sexual.
   A. Offspring produced by asexual reproduction usually have hereditary traits identical to those of the single parent. These offspring constitute a clone. Usually all the cells involved are produced by mitosis.
   B. In sexual reproduction two haploid sex cells, or gametes, fuse to form a single diploid zygote.
      1. A diploid cell undergoing meiosis completes two successive cell divisions to give rise to four haploid cells. Meiosis must take place at some point in the life cycle of a sexually reproducing diploid organism if gametes are to be haploid.
         a. During meiotic prophase I, the members of a homologous pair of chromosomes undergo synapsis and crossing over, during which segments of DNA strands are exchanged between homologous (nonsister) chromatids.
         b. The members of each pair of homologous chromosomes separate during meiotic anaphase I and are distributed to different daughter cells.
         c. During meiosis II, the two chromatids of each chromosome separate and one is distributed to each daughter cell. Each former chromatid is now referred to as a chromosome.
      2. When a zygote is formed, each parent contributes one member of each homologous pair.

V. Various groups of organisms differ with respect to the roles of mitosis and meiosis in their life cycles.
   A. Simple eukaryotes may be regularly haploid; the only diploid stage is the zygote, which undergoes meiosis to restore the haploid state.
   B. Plants (and some algae) have alternation of generations. A multicellular diploid sporophyte forms haploid spores by meiosis. These divide mitotically to form multicellular haploid gametophytes, which produce gametes mitotically. Two haploid gametes then fuse to form a diploid zygote, which divides mitotically to produce a new diploid sporophyte.
   C. The somatic cells of animals are diploid; the only haploid cells are the gametes (produced by meiosis).

SELECTED KEY TERMS

anaphase
asexual reproduction
aster
bivalent
cell cycle
cell plate
centromere
chiasma (chiasmata)
chromatin
chromosome
clone
colchicine
crossing over
cyclins
cytokinesis
cytokinins
diploid
G_1 phase
G_2 phase
gamete
generation time (T)
genetic recombination
haploid
homologous chromosomes
interkinesis
interphase
karyotype
kinetochore
meiosis
metaphase
microtubule
microtubule-organizing center (MTOC)
mitosis
mitosis-promoting factor (MPF)
mitotic spindle
n, $2n$
pericentriolar material
ploidy

(Key Terms continued on next page)

polyploid
prophase
protein kinase
S phase
sexual reproduction
sister chromatid
synapsis
synaptonemal complex
telophase
tetrad
zygote

POST-TEST

1. Chromosomes are composed of ________ fibers, which are made up of DNA, RNA, and protein.
2. The chromosomal constitution of an individual is its ________.
3. The period from the beginning of one cell division to the beginning of the next is termed the ________ ________.
4. DNA for a new set of chromosomes is synthesized during the ________ phase of the cell cycle.
5. To facilitate description of the process, mitosis has been divided into four stages, which occur in the following order: ________, ________, ________, and ________.
6. A duplicated chromosome consists of a pair of ________ ________.
7. The centromere of each chromatid contains a specialized structure, the ________, to which some of the spindle fibers attach.
8. The period during which the chromosomes are lined up on the equatorial plane of the cell constitutes ________.
9. The division of the cytoplasm to yield two daughter cells is called ________.
10. In cell division in plants, a structure known as the ________ ________ forms between two daughter cells.
11. The drug ________ binds to tubulin, the major microtubule protein, thereby blocking cell division.
12. The splitting, budding, or fragmenting of a single parent to give rise to offspring with hereditary traits identical to those of the parent is termed ________ reproduction.
13. A group of genetically identical individuals is called a(an) ________.
14. Chromosomes that are similar in size, shape, and genetic content are referred to as ________ chromosomes.
15. Cells containing two complete sets of chromosomes are ________, whereas those containing more than two complete sets are ________.
16. The chromosome number of a zygote is designated ________, while the chromosome number of a gamete is designated ________.
17. The pairing of homologous chromosomes during prophase I is known as ________.
18. The chromosome configuration produced by the physical association of homologous chromosomes is termed a(an) ________ or a(an) ________.
19. The exchange of segments of homologous chromatids during meiotic prophase I is known as ________ ________.

REVIEW QUESTIONS

1. Two species may have the same chromosome number and yet be very different. Explain.
2. Sketch a duplicated chromosome and label the sister chromatids, the centromeres, and the kinetochores. What are the functions of centromeres and of kinetochores?
3. How does the DNA content of the cell change from the beginning of interphase to the end of interphase? Does the chromosome number change?
4. Are homologous chromosomes present in a diploid cell? Are they present in a haploid cell?
5. How does meiosis differ from mitosis? Are there any points of similarity between these two processes? Explain.
6. What kinds of life cycles include a multicellular haploid stage? Can haploid cells divide by mitosis? By meiosis?
7. Assume that an animal has a diploid chromosome number of ten. (a) How many chromosomes would it have in a typical body cell, such as a skin cell? (b) How many chromosomes would be present in a cell at mitotic prophase? How many chromatids? (c) How many chromosomes would be present in each daughter cell produced by mitosis? Are these duplicated chromosomes? (d) How many tetrads would form in prophase I of meiosis? (e) How many chromosomes would be present in each gamete? Are these duplicated chromosomes?

YOU MAKE THE CONNECTION

1. Decide whether each of the following is an example of sexual or asexual reproduction, and state why. (a) A diploid queen honeybee produces haploid eggs by meiosis. Some of these eggs are never fertilized and develop into haploid male honeybees (drones). (b) Haploid male honeybees produce haploid sperm by mitosis. These sperm fertilize haploid eggs produced by the queen, resulting in the development of diploid female worker bees. (c) Seeds develop after a flower has been pollinated with pollen from a different plant of the same species. (d) Seeds develop after a flower has been pollinated with pollen from the same plant. (e) A cutting from a plant develops roots after it has been placed in water. The plant survives and grows after it is transplanted to soil.

RECOMMENDED READINGS

Alberts, B., D. Bray, J. Lewis, M. Raff, K. Roberts, and J. D. Watson. *Molecular Biology of the Cell,* 3rd ed. Garland, New York, 1994. An extensive, detailed, and well-written discussion of cell growth and division, covering the control of cell division, the cell cycle, and the events of mitosis and meiosis.

McIntosh, J. R., and K. L. McDonald. "The Mitotic Spindle." *Scientific American,* October 1989. A review of current understanding of the mechanisms involved in mitotic chromosome separation.

Murray, A. W., and M. W. Kirschner. "What Controls the Cell Cycle." *Scientific American,* March 1991. A review of the universal regulators involved in controlling the cell cycle.

CHAPTER 10

The Basic Principles of Heredity

Gregor Mendel (1822–1884).
(The Bettmann Archive)

The basic rules of inheritance in eukaryotes were discovered by Gregor Mendel (1822–1884), an abbot who bred pea plants in his monastery garden at Brünn, Austria (now Brno, Czech Republic). Mendel was the first scientist to effectively apply quantitative methods to the study of inheritance. He did not merely describe his observations; he planned his experiments carefully, recorded the data, and subjected the results to mathematical analysis. Although his work was unappreciated in his lifetime, it was "rediscovered" in 1900. Three of his major findings, now known as *Mendel's principles of dominance, segregation, and independent assortment*, became the foundation of the science of genetics.

Early geneticists initially extended Mendel's principles by correlating the *transmission of genetic information* from generation to generation with the behavior of chromosomes during meiosis. They also refined his methods and, through their studies on a variety of organisms, both verified Mendel's findings and added to a growing list of so-called exceptions to his principles. These include such phenomena as linkage, sex linkage, and polygenic inheritance.

Geneticists study not only the transmission of genes, but also the *expression of genetic information.* As you will see in this and succeeding chapters, our understanding of the relationship between an organism's genes and its characteristics has become increasingly sophisticated as we have learned more about the flow of information in cells.

LEARNING OBJECTIVES

After you have studied this chapter you should be able to

1. Define and use correctly the terms *allele, locus, genotype, phenotype, dominant, recessive, homozygous, heterozygous,* and *test cross.*
2. Apply Mendel's principles to solve genetics problems involving monohybrid and dihybrid crosses.
3. Discriminate between independent and mutually exclusive events; apply the product law and sum law appropriately when predicting the outcomes of genetic crosses.
4. Solve genetics problems involving incomplete dominance, epistasis, polygenes, multiple alleles, and X-linked traits.
5. Explain some of the ways in which genes may interact to affect the appearance of a single trait; discuss how it is possible for a single gene to affect many features of the organism simultaneously.
6. Analyze data from a test cross involving alleles of two loci. Show how such data can be used to distinguish between independent assortment and linkage. Relate independent assortment and linkage to specific events in meiosis.
7. Discuss the genetic determination of sex and the role of the Y chromosome in determining male sex in humans; contrast the mechanism of sex determination in humans and other mammals with that in various other animals and some plants; compare dosage compensation of X-linked genes in mammals and *Drosophila.*
8. Assess the effects of inbreeding versus outbreeding on a population; illustrate the genetic basis of hybrid vigor.

MENDEL FIRST DEMONSTRATED THE PRINCIPLES OF INHERITANCE

Gregor Mendel was not the first plant breeder; at the time he began his work, **hybrid** plants and animals (offspring of two genetically dissimilar parents) had been known for a long time. When Mendel began his breeding experiments in 1856, two main facts about inheritance were widely recognized: (1) All hybrid plants with the same two kinds of parents are similar in appearance. (2) When hybrids themselves are mated they do not breed true; their offspring show a mixture of traits. Some look like their parents, and some have features like their grandparents. Mendel's genius lay in his ability to recognize a pattern in the way the parental traits reappear in the offspring of hybrids. No one before had categorized and counted the offspring and analyzed these regular patterns over several generations.

Just as geneticists do today, Mendel chose the organism for his experiments very carefully. The garden pea, *Pisum sativum,* had a number of advantages. Pea plants are easy to grow, and many varieties were available through commercial sources. It is impossible to study inheritance without such genetic **variation.** (If every person in the population had blue eyes, it would be impossible to study the inheritance of eye color.)

Another advantage of pea plants is that **controlled pollinations** are relatively easy to conduct. Pea flowers (Fig. 10–1) have both male and female parts. The male **anthers** (pollen-producing parts of the flower) can be removed to prevent self-fertilization. Pollen from a different source can then be applied to the **stigma** (receptive surface of the female parts). Pea flowers are easily protected from other sources of pollen because the reproductive structures are completely enclosed by the petals. Covering the pollinated flowers with small bags provides additional protection from pollinating insects.

Although his original pea seeds were obtained from commercial sources, Mendel did some important preliminary work before he started his actual experiments. For several years he worked to develop genetically pure, or **true breeding,** lines for a number of inherited traits. A true breeding line for a given trait produces only organisms expressing that trait, generation after generation.

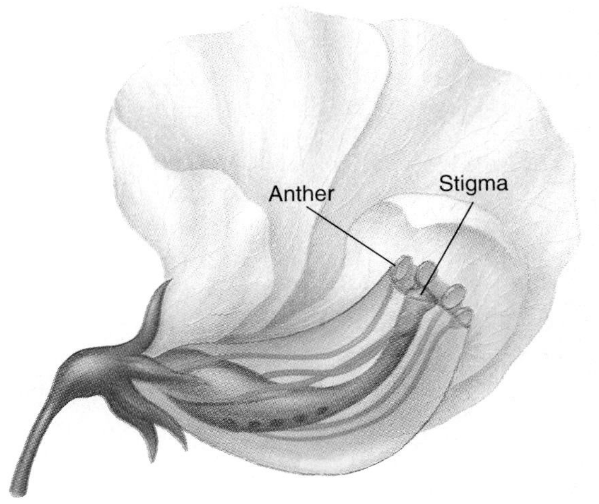

FIGURE 10–1 The reproductive structures of a pea flower are enclosed by the petals. This cut-away drawing illustrates the pollen-producing anthers and the stigma, that portion of the female part of the flower that receives the pollen.

During this time he apparently chose those characteristics of his pea strains that could be studied most easily, and probably discarded or ignored a number of others. He probably also made the initial observations that would later form the basis of his theories.

Mendel eventually chose strains with seven clearly contrasting pairs of traits: yellow versus green seeds; round versus wrinkled seeds; green versus yellow pods; tall versus short plants; inflated versus constricted pods; white seed coats versus gray seed coats; and flowers borne on the ends of the stems versus flowers appearing all along the stems. Because other plant breeders typically studied hybrids between parents that differed in many traits, some of which were not clearly distinguishable, their results were confusing. Mendel's results were much easier to analyze because he limited the number of traits studied in each experiment.

Mendel began his experiments by crossing plants from two true breeding lines with contrasting traits; these genetically pure individuals constituted the **parental,** or **P, generation.** In every case, the members of the first generation of offspring all looked alike and resembled one of the two parents (Fig. 10–2). For example, when he crossed tall plants with short plants, all the progeny were tall. These offspring were the **first filial** (*filial* comes from Latin for "sons and daughters") **generation,** or **F_1 generation.** The second filial generation, or **F_2 generation,** was produced by a cross between F_1 individuals, or by self-pollination of F_1 individuals. Mendel's F_2 generation included 787 tall plants and 277 short plants. Because the hereditary factor for shortness reappeared in the F_2 generation, it clearly had not been "lost" in the F_1 generation.

Mendel's experiments led to his discovery and explanation of three major principles of heredity, now known as the principles of *dominance, segregation,* and *independent assortment.* We consider the first two now and the third later in the chapter.

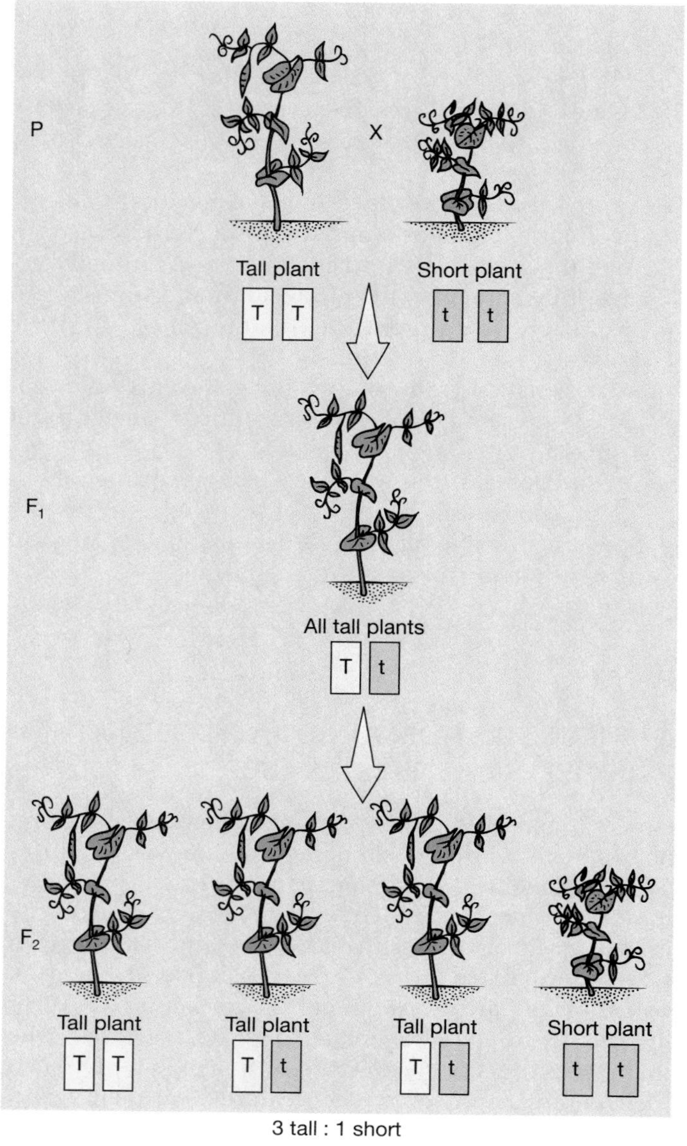

FIGURE 10–2 This diagram illustrates one of Gregor Mendel's many crosses between different strains of peas. Crossing a tall pea plant with a short pea plant yielded only tall offspring in the F_1 generation. However, when these F_1 individuals self-pollinated, or when two F_1 individuals were crossed, the resulting F_2 generation included tall and short plants in a ratio of about 3:1.

The principle of dominance states that one gene can mask the expression of another in a hybrid

Based on results such as those just discussed, Mendel proposed that each kind of inherited feature of an organism is controlled by two "factors" that are present in every individual. Mendel's "hereditary factors" are essentially what we call *genes* today, so we will use that term in our discussion.

On the basis of his findings, Mendel formulated an idea that has become known as his **principle of dominance;** this states that in an F_1 hybrid, the gene from one of the parents masks expression of the gene from the other parent. The gene expressed in the F_1 generation (tallness in our example) is said to be **dominant;** the one hidden (shortness) is said to be **recessive.** This finding was at odds with the prevailing idea of **blending inheritance,** which implied that a hybrid should be intermediate between the two parents. Although we know today that the principle of dominance does not always apply (exceptions are considered later in this chapter), the recognition that one gene can mask the expression of another was an important intellectual leap on Mendel's part.

The principle of segregation states that the genes of a pair separate before gametes are formed

Mendel further proposed that the genes from the parents behave as particles and become separated so that each sex cell (egg or sperm) contains only one member of each pair. The two genes remain intact during this process (one does not "contaminate" or eliminate the other); thus, recessive traits are not lost and so can reappear in the F_2 generation. This idea, referred to as the **principle of segregation,** also ran counter to the notion of blending inheritance, which envisioned hereditary determinants as fluids that became inseparably mixed once they were combined in a hybrid.

In our example, the F_1-generation tall plants had two genes, one for tallness (which we designate *T*) and one for shortness (which we designate *t*), but because the tall gene was dominant, these plants were tall. However, before these F_1 plants formed gametes, the gene for tallness separated (segregated) from the gene for shortness, so that half of the gametes contained a gene for tallness and the other half contained a gene for shortness. The random process of fertilization led to three possible combinations of factors in the F_2 offspring: one fourth with two tallness genes (*TT*), one fourth with two shortness genes (*tt*), and one half with one gene each for tallness and shortness (*Tt*). Because both *TT* and *Tt* plants are tall, on average one could expect approximately three fourths (787/1064) to express the dominant gene (tall) and about one fourth (277/1064) the recessive gene (short).

Today we know that segregation of genes is a direct result of the separation of homologous chromosomes during meiosis. Later, at the time of fertilization, each haploid gamete contributes one chromosome of each homologous pair and therefore one gene for each gene pair (either *T* or *t* in our example). Although gametes and fertilization were known at the time Mendel carried out his research, mitosis and meiosis had not yet been discovered. It is truly remarkable that Mendel was able to formulate his ideas mainly on the basis of mathematical abstractions. Today his principles are much easier to understand because we are able to think about them in concrete terms, by relating the transmission of genes to the behavior of chromosomes.

Mendel reported these and other findings (discussed later in this chapter) at a meeting of the Brünn Society for the Study of Natural Science, and published his results in the transactions of that society in 1866. At that time biology was largely a descriptive science, and biologists had little interest in applying quantitative and experimental methods such as Mendel had used. The importance of his results and his interpretations of those results was not appreciated by other biologists of the time, and his findings were neglected for nearly 35 years.

In 1900 Hugo DeVries in Holland, Karl Correns in Germany, and Erich von Tschermak in Austria each rediscovered Mendel's paper and found that it provided explanations for their own research findings. They gave credit to Mendel by naming the basic laws of inheritance after him. By this time biologists had a much greater appreciation of the value of experimental methods. The details of mitosis, meiosis, and fertilization had been described, and in 1903 W. S. Sutton pointed out the connection between Mendel's segregation of genes and the separation of homologous chromosomes during meiosis. The time was right for wider acceptance and extension of these ideas and their implications.

Alleles Occupy Corresponding Loci on Homologous Chromosomes

Today we know that each chromatid is made up of one long DNA molecule and that each gene is actually a segment of that DNA molecule. We also know that homologous chromosomes usually have similar genes located in corresponding positions. The term **locus**[1] (pl., *loci*) was originally coined to designate the location of a particular gene on the chromosome. Of course we are actually referring to a segment of the DNA that has the information required to control some aspect of the organism. One locus may determine seed color, another seed shape, still another the shape of the pods, and so on. A particular locus can be identified (by traditional genetic methods at least) only if at least two genes produce contrasting traits, such as yellow peas versus green peas. In the simplest cases an individual can express one (yellow) or the other (green) but not both.

Genes that govern variations of the same feature (yellow versus green seed color) and occupy corresponding loci on homologous chromosomes are called **alleles** (Fig. 10–3). Each allele (variant) of a locus is assigned a single letter (or group of letters) as its symbol. Although more complicated forms of notation are often used by geneticists, it is customary when working simple genetics problems to indicate a dominant allele with a capital letter and a recessive allele with the same letter in lowercase. The choice of the letter is generally determined by the first allelic variant found for that locus. For example, the dominant allele that governs the yellow color of the seed might be designated *Y*, and the recessive allele responsible for the green color would then be designated *y*. Because discovery of the yellow allele made identification of this locus possible, we refer to the locus as the *yellow* locus, although pea seeds are most commonly green.

Remember that the term *locus* is used to designate not only a position on a chromosome, but also a type of gene controlling a particular *kind* of characteristic; thus, *Y* (yellow) and *y* (green) represent a specific pair of alleles of a locus involved in determining seed color in peas. Although you may initially be uncomfortable with the fact that geneticists sometimes use the term *gene* to specify a locus and at other times to specify one of the alleles of that locus, the meaning is usually clear from the context.

---

[1]In mathematics a locus is a dimensionless point; a genetic locus, being a segment of DNA, is obviously not dimensionless!

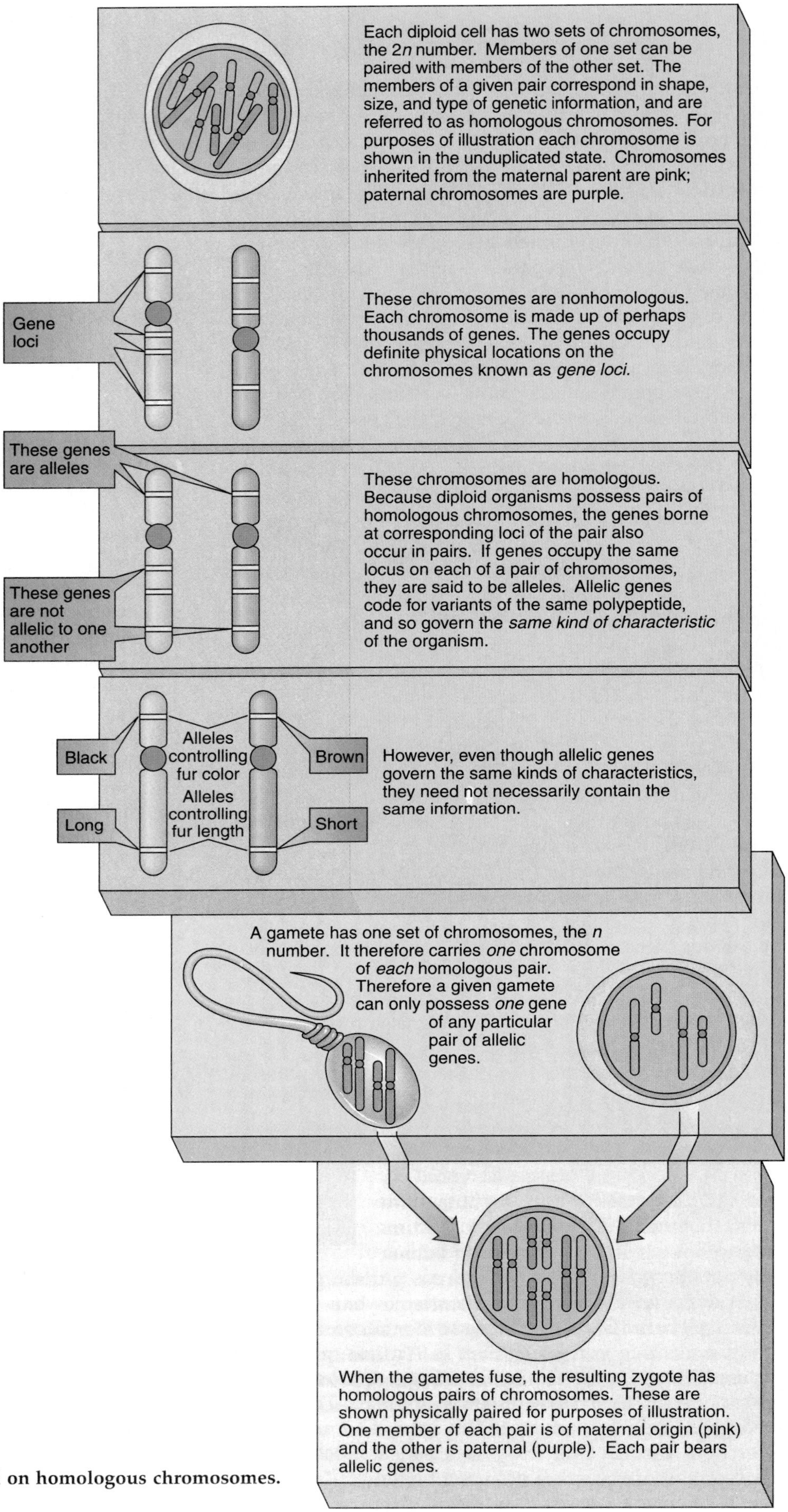
Each diploid cell has two sets of chromosomes, the 2n number. Members of one set can be paired with members of the other set. The members of a given pair correspond in shape, size, and type of genetic information, and are referred to as homologous chromosomes. For purposes of illustration each chromosome is shown in the unduplicated state. Chromosomes inherited from the maternal parent are pink; paternal chromosomes are purple.
Gene loci
These chromosomes are nonhomologous. Each chromosome is made up of perhaps thousands of genes. The genes occupy definite physical locations on the chromosomes known as gene loci.
These genes are alleles
These genes are not allelic to one another
These chromosomes are homologous. Because diploid organisms possess pairs of homologous chromosomes, the genes borne at corresponding loci of the pair also occur in pairs. If genes occupy the same locus on each of a pair of chromosomes, they are said to be alleles. Allelic genes code for variants of the same polypeptide, and so govern the same kind of characteristic of the organism.
Black
Alleles controlling fur color
Brown
Long
Alleles controlling fur length
Short
However, even though allelic genes govern the same kinds of characteristics, they need not necessarily contain the same information.
A gamete has one set of chromosomes, the n number. It therefore carries one chromosome of each homologous pair. Therefore a given gamete can only possess one gene of any particular pair of allelic genes.
When the gametes fuse, the resulting zygote has homologous pairs of chromosomes. These are shown physically paired for purposes of illustration. One member of each pair is of maternal origin (pink) and the other is paternal (purple). Each pair bears allelic genes.

FIGURE 10–3 Allelic genes are carried on homologous chromosomes.

A MONOHYBRID CROSS INVOLVES INDIVIDUALS WITH DIFFERENT ALLELES FOR A GIVEN GENE LOCUS

The basic principles of genetics and the use of genetic terms are best illustrated by examples. The simplest case, a **monohybrid cross,** involves the mating of two individuals that carry different alleles for a single locus. Our first example in this section deals with the expected ratios in the F_2 generation, as did our previous example of Mendel's work on tall and short pea plants.

Heterozygotes carry two different alleles for a locus; homozygotes carry identical alleles

Figure 10–4 illustrates a monohybrid cross featuring a locus that governs coat color in guinea pigs. The male comes from a true breeding line of black guinea pigs. We say that he is **homozygous** for black because the two alleles he carries for this locus are *identical.* The brown female is also from a true breeding line and is homozygous for brown. What color would you expect the F_1 offspring to be? Dark brown? Spotted? It is impossible to make such a prediction without more information.

In this particular case, the F_1 offspring are black, but they are **heterozygous,** meaning that they carry two *different* alleles for this locus. The allele for brown coat color can be expressed only in a homozygous brown individual; it is referred to as a *recessive allele.* The allele for black coat color can be expressed in both homozygous black and heterozygous individuals; it is said to be a *dominant allele.* On the basis of this information, we can use standard notation to designate the dominant black allele as *B* and the recessive brown allele as *b.*

During meiosis in the male parent (*BB*), the two *B* alleles separate according to Mendel's principle of segregation so that each sperm has only one *B* allele. In the formation of eggs in the female (*bb*), the two *b* alleles separate so that each egg has only one *b* allele. The fertilization of each *b* egg by a *B* sperm results in heterozygous F_1 offspring, each with the alleles *Bb*. That is, each individual has one allele for brown coat and one for black coat. Because this is the only possible combination of alleles present in the eggs and sperm, all of the offspring are *Bb*.

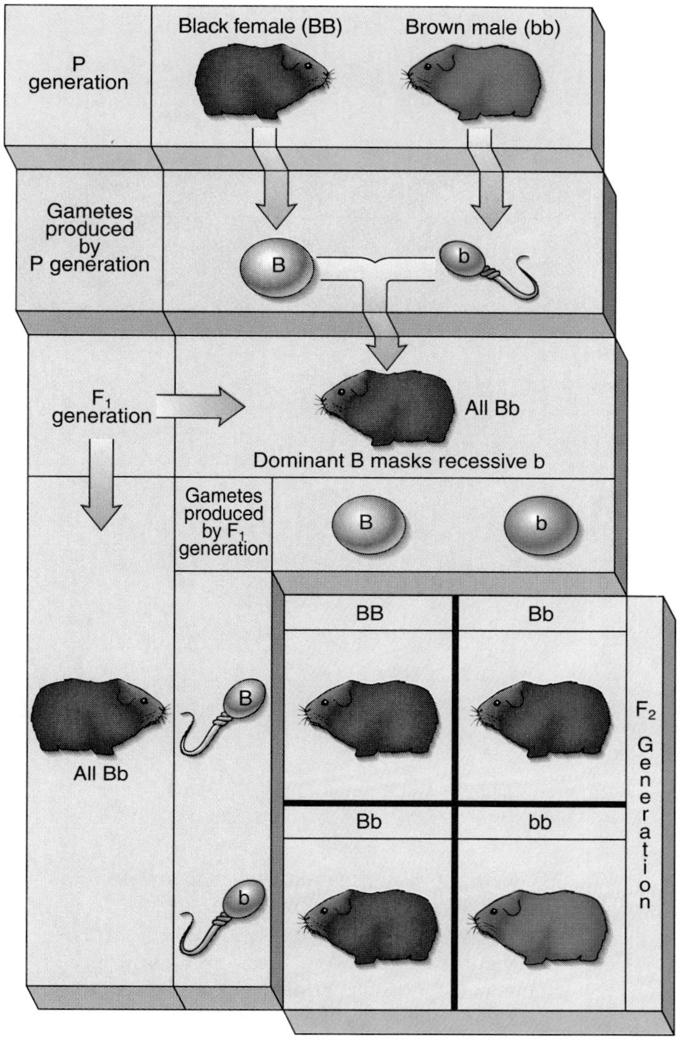

FIGURE 10–4 A monohybrid cross follows the inheritance of alleles of a single locus. In the monohybrid cross illustrated, a homozygous brown guinea pig is mated with a homozygous black guinea pig. The F_1 generation includes only black individuals. However, the mating of two of these offspring yields F_2 generation offspring in the expected black:brown ratio of 3:1, indicating that the F_1 individuals are heterozygous.

The phenotype of an individual does not always reveal its genotype

The fact that some alleles may be dominant and others recessive means that we cannot always determine which alleles are carried by an organism simply by looking at it. The term used to specify the *appearance* of an individual in a given environment with respect to a certain inherited trait is known as its **phenotype.** The *genetic constitution* of that organism, most often expressed in symbols, is its **genotype.** In the cross we have been considering, the genotype of the female parent is homozygous recessive, *bb,* and her phenotype is brown. The genotype of the male parent is homozygous dominant, *BB,* and his phenotype is black. The genotype of all the F_1 offspring is heterozygous, *Bb,* and their phenotype is black. To prevent confusion we always indicate the genotype of a heterozygous individual by writing the symbol for the dominant allele first and the recessive allele second (always *Bb,* never *bB*).

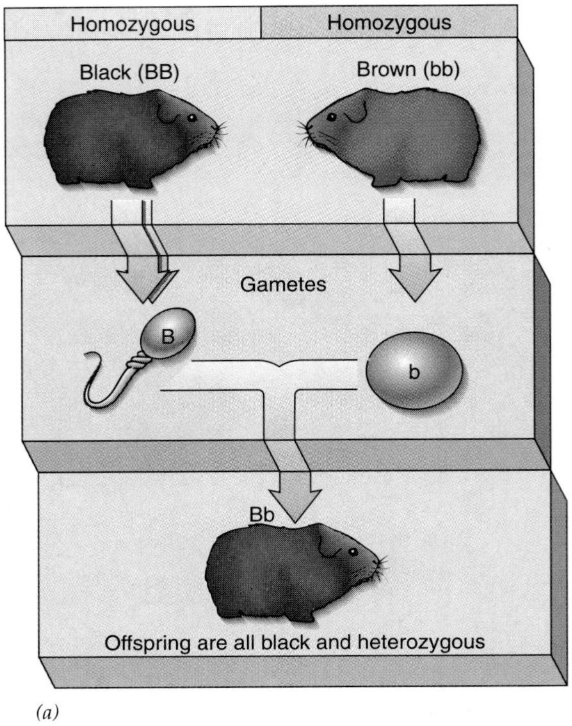

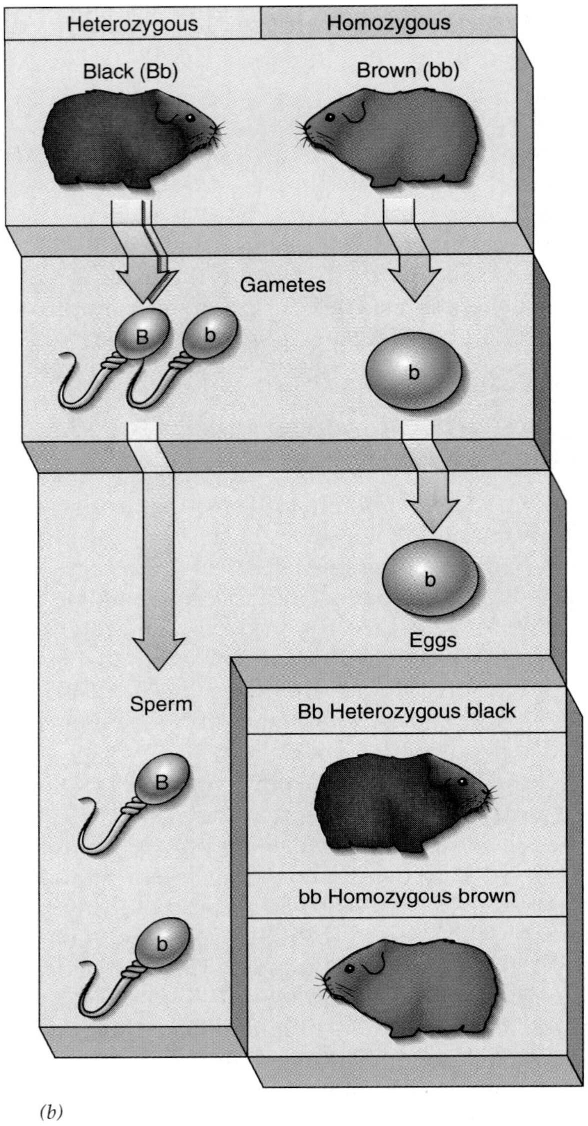

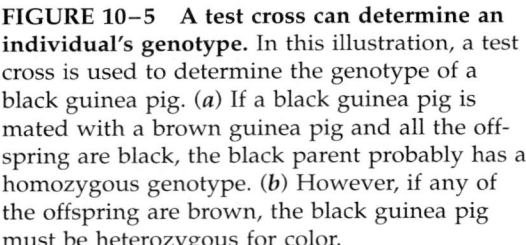

FIGURE 10–5 A test cross can determine an individual's genotype. In this illustration, a test cross is used to determine the genotype of a black guinea pig. (***a***) If a black guinea pig is mated with a brown guinea pig and all the offspring are black, the black parent probably has a homozygous genotype. (***b***) However, if any of the offspring are brown, the black guinea pig must be heterozygous for color.

The phenomenon of dominance partly explains why an individual may resemble one parent more than the other, even if the two parents make equal contributions to their offspring's genetic constitution. Dominance is not completely predictable and can be determined only by experiment. In one species of animal, black coat may be dominant to brown; in another species, brown may be dominant to black.

A Punnett square predicts the ratios of genotypes and phenotypes of the offspring of a cross

During meiosis in heterozygous (*Bb*) black guinea pigs, the chromosome containing the *B* allele becomes separated from its homologue (the chromosome containing the *b* allele), so each sperm or egg contains *B* or *b* but never both. Gametes containing *B* alleles and those containing *b* alleles are formed in equal numbers by heterozygous *Bb* individuals. Because no special attraction or repulsion occurs between an egg and a sperm containing the same allele, fertilization is a random process.

The possible combinations of eggs and sperm at fertilization may be represented in the form of a "checkerboard" devised by an early geneticist, Sir Reginald Punnett, and known as a **Punnett square** (Fig. 10–4). The types of gametes from one parent are represented across the top, and those from the other parent are indicated along the left side; the squares are then filled in with the resulting F_2 zygote combinations. Three fourths of all F_2 offspring are genotypically *BB* or *Bb* and phenotypically black; one fourth are genotypically *bb* and phenotypically brown. The genetic mechanism responsible for the approximate 3:1 F_2 ratios (called *monohybrid F_2 phenotypic ratios*) obtained by Mendel in his pea-breeding experiments is again evident. The corresponding genotypic ratio is 1*BB*: 2*Bb*: 1*bb*.

A test cross can detect heterozygosity

One third of the black guinea pigs in the F_2 generation derived from the mating of F_1 hybrids are themselves homozygous, *BB*; the other two thirds are heterozygous, *Bb*. Guinea pigs with the genotypes *BB* and *Bb* are alike phenotypically; they both have black coats. Geneticists distinguish the homozygous (*BB*) and heterozygous (*Bb*) black-coated guinea pigs by a **test cross** in which each black guinea pig is mated with a homozygous brown (*bb*) guinea pig (Fig. 10–5). In a test cross, the two types of gametes produced by the heterozygous parent are not "hidden" in the offspring by dominant alleles coming from the other parent. Therefore, through a test cross one can deduce the genotypes of all the classes of offspring directly from their phenotypes. If all the offspring were black, what inference would you make about the genotype of the black parent? If any of the offspring were brown, what conclusion would you draw regarding the genotype of the black parent? Would you be more certain about one of these inferences than the other?

Mendel did just these sorts of experiments, breeding heterozygous tall (*Tt*) pea plants with homozygous recessive (*tt*) short ones. He predicted that the heterozygous parent would produce equal numbers of *T* and *t* gametes, whereas the homozygous short parent would produce only *t* gametes, and that this should lead to equal numbers of tall (*Tt*) and short (*tt*) individuals among the progeny. This was essentially a test of the hypothesis that there is 1:1 segregation of the alleles of the heterozygous parent. Thus, Mendel's principles of dominance and segregation not only explained the known facts, such as the monohybrid F_2 3:1 phenotypic ratio, but also enabled him to predict the results of other experiments, in this case the 1:1 test cross phenotypic ratio.

THE LAWS OF PROBABILITY ARE USED TO PREDICT THE LIKELIHOOD OF GENETIC EVENTS

All genetic ratios are properly expressed in terms of probabilities. In the examples just discussed, among the offspring of two individuals heterozygous for the same gene pair, the ratio of the phenotypes of the dominant and recessive alleles is 3:1. A better way to express our expectations is to say that there are 3 chances in 4 (3/4) that any particular individual offspring of two heterozygous individuals will express the phenotype of the dominant allele and 1 chance in 4 (1/4) that it will express the phenotype of the recessive allele. Although we sometimes speak in terms of percentages, probabilities must always be calculated as fractions (e.g., 3/4) or decimal fractions (e.g., 0.75). If an event is certain to occur, its probability is 1; if it is certain not to occur, its probability is 0. A probability can be 0, 1, or some number between 0 and 1.

Often we wish to *combine* two or more probabilities. The Punnett square, which we use to predict the results of genetic crosses, is a device that allows us to combine probabilities. When we use a Punnett square we are intuitively following two important rules, known as the **product law** and the **sum law.**

The product law predicts the combined probabilities of independent events

Events are independent if the occurrence of one does not affect the probability that the other will occur. For example, the probability of obtaining heads on the first toss of a coin is 1/2; the probability of obtaining heads on the second toss (an independent event) is also 1/2. If two or more events are *independent* of each other, the probability of their both occurring is the *product* of their individual probabilities. If this seems strange to you, keep in mind that when we multiply two numbers that are less than 1, the product is a smaller number. The probability of obtaining heads first and also second on successive tosses of the coin is the product of their individual probabilities ($1/2 \times 1/2 = 1/4$, or 1 chance in 4) (Fig. 10–6).

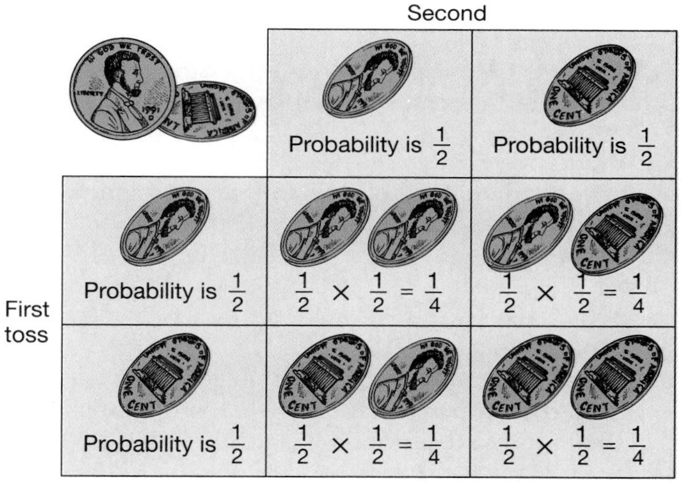

FIGURE 10–6 The laws of probability can be illustrated by two successive coin tosses. When one tosses a coin twice, the probability of getting heads on the first toss is 1/2, and the probability of getting tails on the first toss is also 1/2. Identical predictions apply to the second toss. The outcome of the first toss does not affect the outcome of the second toss, so these events are independent, and we combine them by multiplying the individual probabilities (according to the product law). As shown in the Punnett square, there are four different classes of combined outcomes of two successive tosses. These are mutually exclusive, and we therefore combine them by adding them (according to the sum law). For example, heads/heads and tails/tails are mutually exclusive outcomes. If we wish to calculate the probability that we will get either heads/heads or tails/tails, we add: $1/4 + 1/4 = 1/2$. These same laws of probability are used to predict genetic events.

Similarly, we can apply the product law to genetic events. If both parents are *Bb*, what is the probability that they will produce a child who is *bb*? For the child to be *bb*, he or she must receive a *b* gamete from each parent. The probability of a *b* egg is 1/2 and the probability of a *b* sperm is also 1/2. These probabilities are independent, so we combine them by the product rule ($1/2 \times 1/2 = 1/4$). You may wish to check this result using a Punnett square.

The sum law predicts the combined probabilities of mutually exclusive events

Events are mutually exclusive if the occurrence of one *precludes* the occurrence of the other. Mutually exclusive events can be thought of as different ways of obtaining some specified result. Naturally, if there is more than one way to obtain a result, the chances of its being obtained are improved; we therefore combine the probabilities of mutually exclusive events by summing (adding) their individual probabilities.

For example, if we flip a coin twice, what is the probability that it will come up heads one time and tails the other time if we do not specify the order in which these events are to occur? There are two mutually exclusive ways to obtain this outcome. We could get heads the first time (probability 1/2) and tails the second (probability 1/2); we use the product law to calculate the combined probability of these independent events, which is $1/2 \times 1/2 = 1/4$.

Alternatively, we could also get tails the first time and heads the second; the probability of this occurring is also 1/4. We combine the probabilities of these mutually exclusive outcomes using the sum law: $1/4 + 1/4 = 1/2$. That is, the probability of getting heads once (and only once) and tails once (and only once) on two successive tosses of the coin is 1/2.

We can also apply the sum law to genetic events. For example, if both parents are *Bb*, what is the probability that they will produce a child like themselves (*Bb*)? There are two mutually exclusive ways of obtaining a *Bb* child. A *B* egg can combine with a *b* sperm. The probability of this outcome is 1/4 (calculated by the product rule). A *b* egg can combine with a *B* sperm; this probability is also 1/4. Because these two ways of obtaining a *Bb* child are mutually exclusive, we combine their probabilities using the sum law ($1/4 + 1/4 = 1/2$). Again, a Punnett square serves as a useful check.

The laws of probability can be applied to a variety of calculations

The laws of probability have wide applications. For example, what are the probabilities that a family with two (and only two) children will have two girls, two boys, or one girl and one boy? For purposes of discussion we will assume that male and female births are equally probable. The probability of having a girl first is 1/2, and the probability of having a girl second is also 1/2. These are independent events, so we combine their probabilities by multiplying: $1/2 \times 1/2 = 1/4$. Similarly, the probability of having two boys is also 1/4.

In families with both a girl and a boy, the girl can be born first or the boy can be born first. The probability that a girl will be born first is 1/2, and the probability that a boy will be born second is also 1/2. We use the product law to combine the probabilities of these two independent events: $1/2 \times 1/2 = 1/4$. Similarly, the probability that a boy will be born first and a girl second is also 1/4. These two kinds of families represent mutually exclusive outcomes—that is, two different ways of obtaining a family with one boy and one girl. Having two different ways of obtaining the desired result improves our chances, so we use the sum law to combine the probabilities: $1/4 + 1/4 = 1/2$. Notice that the probabilities of the three types of families (all mutually exclusive outcomes) add up to 1. This serves as a useful check that the calculations have been done correctly. You may also wish to confirm these results by making a Punnett square.

In working with probabilities, it is important to keep in mind a point that many gamblers forget. We can say that "chance has no memory." This means that if events are truly random, past events have no influence on the probability of the occurrence of independent future events. For example, if two brown-eyed people have a child, what is the probability that it will have blue eyes? If their first child has blue eyes, what is the probability that their second child will also have blue eyes? The color of the iris of the human eye is controlled by alleles at several loci, but alleles at one locus are primarily responsible. The allele for brown eye color, *B*, is usually dominant to the allele for blue, *b*. If the two brown-eyed parents are heterozygous, there is 1 chance in 4 that any child of theirs will have blue eyes. Each fertilization is a separate, independent event; its result is not affected by the results of any previous fertilizations. If these two heterozygous brown-eyed parents have had three brown-eyed children and are expecting their fourth child, what is the probability that the child will have blue eyes? The uninformed might guess that this one *must* have blue eyes, but in fact there is still only 1 chance in 4 that the child will have blue eyes and 3 chances in 4 that the child will have brown eyes.

If we phrase the question differently, however, we obtain a very different answer. If two heterozygous people marry and expect to have four children, what is the probability that all four will have brown eyes? The probability of brown eyes for each child is 3/4, so we combine these independent events by the product rule: $3/4 \times 3/4 \times 3/4 \times 3/4 = 81/256$ or 0.32. Why the different answers for the two types of problems? Remember that once a brown-eyed child is born, chance (3/4) is replaced by certainty (1), so the calculation becomes $1 \times 1 \times 1 \times 3/4 = 3/4$. The chance that the fourth (as yet unborn) child will have brown eyes is therefore 3/4 (and the chance of blue eyes is 1/4).

When working probability problems, common sense is more important than blindly memorizing rules. Examine your results to see if they appear reasonable; if they do not you should reevaluate your assumptions.

A DIHYBRID CROSS INVOLVES INDIVIDUALS THAT HAVE DIFFERENT ALLELES AT TWO LOCI

Simple monohybrid crosses each involve a pair of alleles representing a single locus. Mendel also analyzed crosses involving alleles representing two or more loci. A mating between individuals having different alleles at two loci is called a **dihybrid cross.** When two pairs of alleles are located in nonhomologous chromosomes, each pair is inherited independently; that is, each pair segregates during meiosis independently of the other.

An example of a dihybrid cross carried through the F_2 generation is illustrated in Figure 10–7. When a homozygous black, short-haired guinea pig (*BBSS,* because black is dominant to brown and short hair is dominant to long hair) and a homozygous brown, long-haired guinea pig (*bbss*) are mated, the *BBSS* animal produces gametes that are all *BS,* and the *bbss* individual produces gametes that are all *bs.* Each gamete contains one and only one allele for each of the two loci. The union of the *BS*

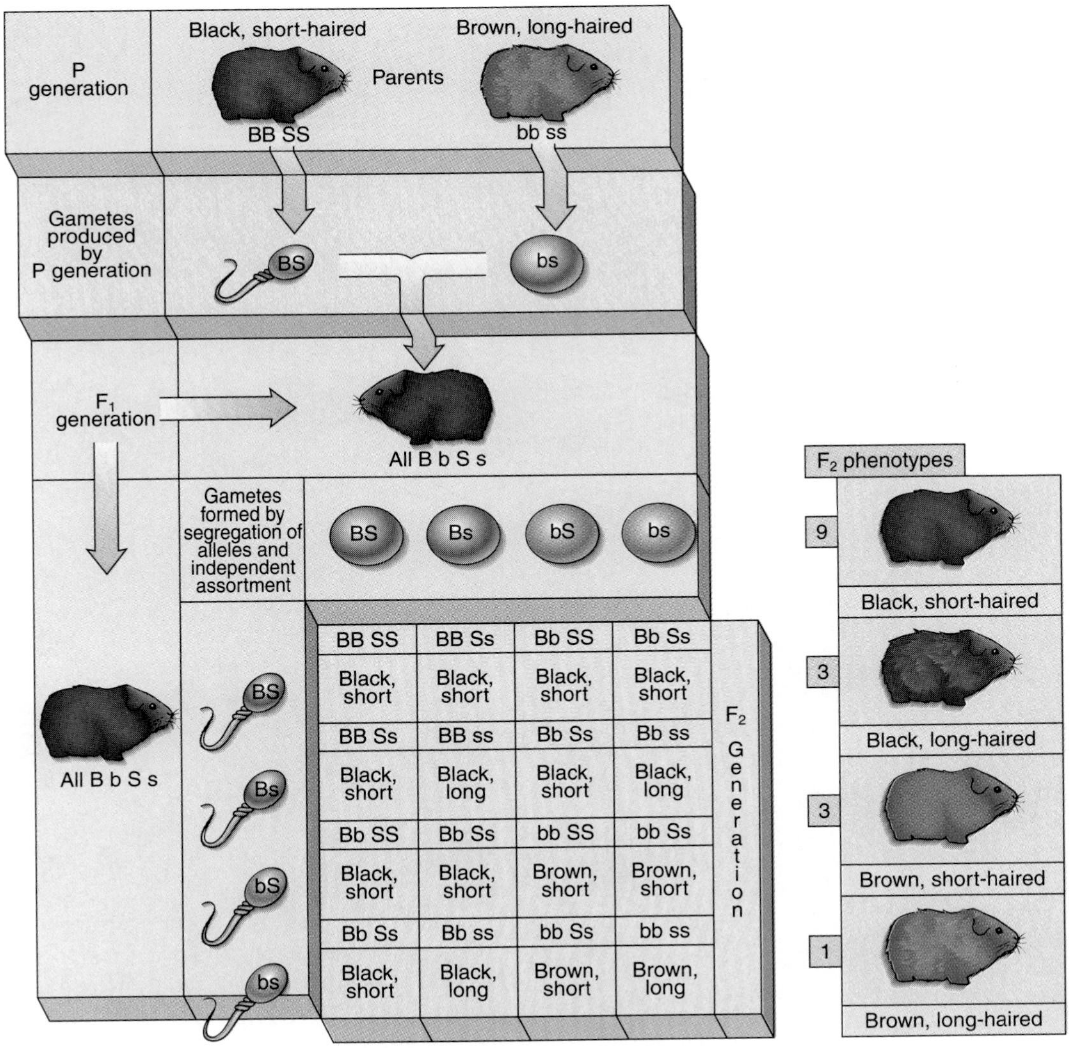

FIGURE 10–7 A dihybrid cross can be used to trace the inheritance of alleles of two different loci. When a black, short-haired guinea pig is crossed with a brown, long-haired one, all the offspring are black and have short hair. However, when two members of the F_1 generation are crossed, the ratio of phenotypes is 9:3:3:1. Note that the two pairs of alleles considered here assort independently.

Making the Connection

Independent Assortment and the Mechanics of Meiosis

How does Mendel's principle of independent assortment relate to the behavior of the chromosomes during meiosis? Independent assortment occurs only if the alleles of one locus are on one pair of homologous chromosomes and the alleles of the other locus are on a different homologous pair. As shown in the figure, there are two different ways for two pairs of homologous chromosomes to be distributed during meiosis. The orientation shown at the left produces *BS* and *bs* gametes in a 1:1 ratio. The orientation shown at the right produces *Bs* and *bS* gametes in a 1:1 ratio. Because these two orientations occur randomly, approximately half the meiotic cells have the first orientation and the other half have the second, resulting in an overall 1:1:1:1 ratio for the four possible types of gametes. ▲

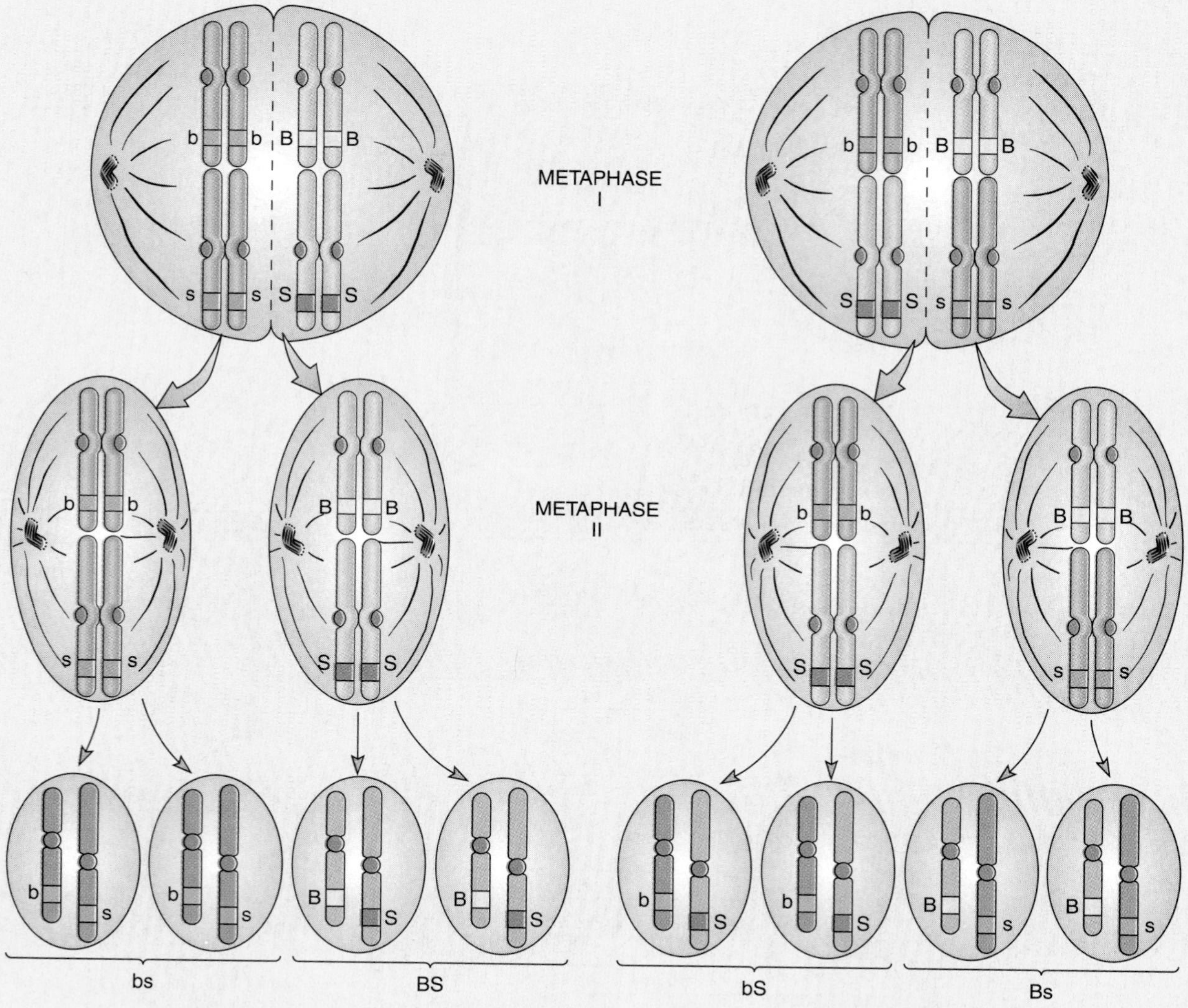

and *bs* gametes yields only individuals with the genotype *BbSs*. All these F_1 offspring are heterozygous for hair color and for hair length, and all are phenotypically black and short-haired.

The principle of independent assortment states that the alleles of different loci are randomly distributed into gametes

Each F_1 individual produces four kinds of gametes with equal probability: *BS, Bs, bS,* and *bs*. Hence, the Punnett square has 16 squares representing the zygotes, some of which are genotypically or phenotypically alike. There are 9 chances in 16 of obtaining a black, short-haired individual; 3 chances in 16 of obtaining a black, long-haired individual; 3 chances in 16 of obtaining a brown, short-haired individual; and 1 chance in 16 of obtaining a brown, long-haired individual. This 9:3:3:1 phenotypic ratio is expected in a dihybrid F_2 if the loci are on nonhomologous chromosomes.

On the basis of similar results, Mendel formulated his third principle of inheritance, now called Mendel's **principle of independent assortment,** which states that members of any gene pair segregate from one another independently of the members of the other gene pairs. Each gamete contains one allele for each locus, but the alleles of different loci are assorted at random with respect to each other in the gametes.

Today we recognize that independent assortment is a consequence of the events of meiosis (see *Making the Connection: Independent Assortment and the Mechanics of Meiosis*). As we shall see in the next section, independent assortment is actually a special case. This principle does not apply if the two gene pairs are **linked**—that is, located on the same pair of homologous chromosomes.

Procedures used in solving genetics problems that illustrate Mendel's principles of dominance, segregation, and independent assortment are summarized in *Focus On: Solving Genetics Problems* and *Focus On: Deducing Genotypes*.

THE LINEAR ORDER OF LINKED GENES ON A CHROMOSOME CAN BE "MAPPED" BY CALCULATING THE FREQUENCY OF CROSSING OVER

Chromosomes are inherited as units and they pair and separate during meiosis as units. For this reason, all the alleles at different loci on a given chromosome tend to be inherited together. If the chromosomal units never changed, the genes on any one chromosome would always be inherited together. However, during meiosis, when the chromosomes pair and undergo synapsis (see Chapter 9), crossing over may occur.

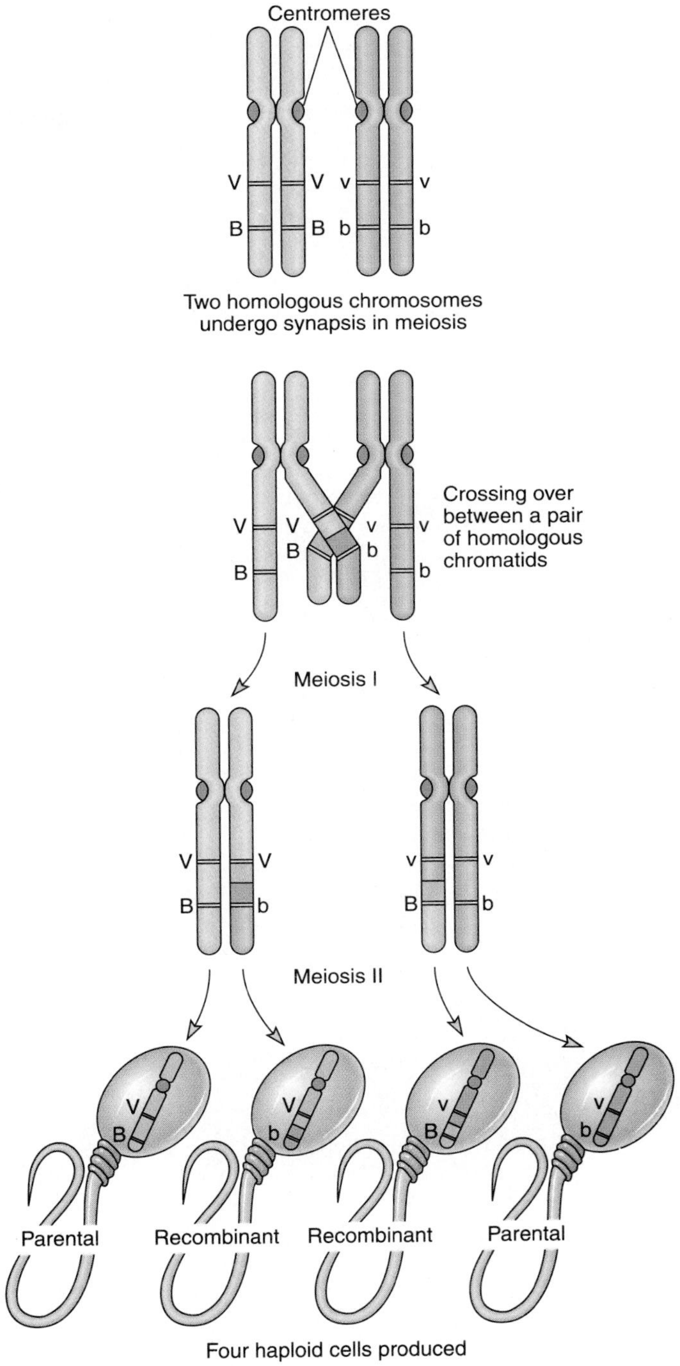

FIGURE 10–8 Crossing over, the term used to describe the exchange of segments between chromatids of homologous chromosomes, permits the recombination of linked genes. Genes located far apart on a chromosome have a greater probability of being separated by an exchange of segments than do genes that are closer together.

During **crossing over,** homologous (nonsister) chromatids exchange segments of chromosomal material by a process of breakage and rejoining (Fig. 10–8). This exchange of chromatid segments occurs at random along

Focus On

Solving Genetics Problems

Simple Mendelian genetics problems are like puzzles. They can be fun and easy to work if you follow certain conventions and are methodical in your approach.

1. Always use standard designations for the generations. The generation with which a particular genetic experiment is begun is called the P, or *parental, generation.* Offspring of this generation (the "children") are called the F_1, or *first filial, generation.* The offspring resulting when two F_1 individuals are bred constitute the F_2, or second filial, generation (the "grandchildren").
2. Write down a key for the symbols you are using for the allelic variants of each locus. Use uppercase to designate a dominant allele and lowercase to designate a recessive allele. Use the same letter of the alphabet to designate both alleles of a particular locus. If you are not told which is dominant and which is recessive, the phenotype of the F_1 generation is a good clue.
3. Determine the genotypes of the parents of each cross by making use of the following types of evidence:
   a. Are they from true breeding lines? If so, they should be homozygous.
   b. Can their genotypes be reliably deduced from their phenotypes? This is usually true if they express the recessive phenotype.
   c. Do the phenotypes of their offspring provide any information? See *Focus On: Deducing Genotypes* for an example of how these determinations can be made.
4. Indicate the possible kinds of gametes formed by each of the parents. It is helpful to draw a circle around the symbols for each kind of gamete.
   a. If it is a monohybrid cross, we must apply the principle of segregation; i.e., a heterozygote *Aa* forms two kinds of gametes: *A* and *a*. Of course a homozygote, such as *aa*, forms only one kind of gamete: *a*.
   b. If it is a dihybrid cross, we must apply both the principle of segregation *and* the principle of independent assortment. For example, an individual heterozygous for two loci would have the genotype *AaBb*. *A* segregates from *a*, and *B* segregates from *b*. The assortment of *A* and *a* into gametes is independent of the assortment of *B* and *b*. Therefore *A* is equally likely to end up in a gamete with *B* or *b*. The same is true for *a*.

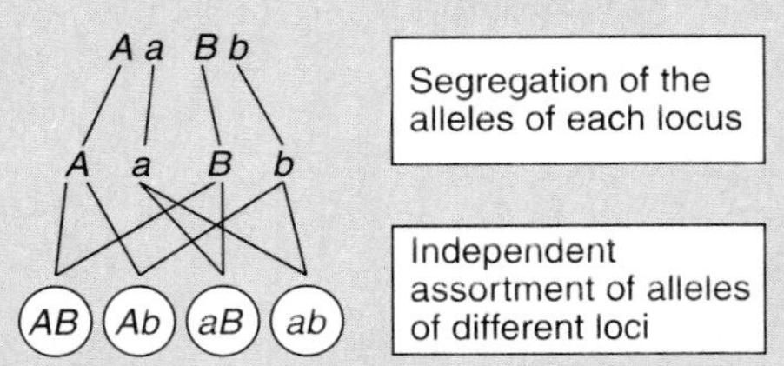

5. Set up a Punnett square, placing the possible types of gametes from one parent down the left side and the possible types from the other parent across the top.
6. Fill in the Punnett square and read off (and sum up) the genotypic and phenotypic ratios of the offspring. Avoid confusion by consistently placing the dominant allele first and the recessive allele second in heterozygotes (*Aa*, never *aA*). If it is a dihybrid cross, it is very important to always write the two alleles of one locus first and the two alleles of the other locus second. It does not matter which locus you choose to write first, but once you have decided on the order, it is critical that you maintain it consistently. This means that if the individual is heterozygous for both loci you will always use the form *AaBb*. Writing this particular genotype as *aBbA*, for example, would cause confusion.
7. If you do not need to know the frequencies of all of the expected genotypes and phenotypes, you may use the rules of probability as a shortcut. For example, if both parents are *AaBb*, what is the probability of an *AABB* offspring? To be *AA*, the offspring must receive an *A* gamete from each parent. The probability that a given gamete is *A* is 1/2 and each gamete represents an independent event, so we combine their probabilities by multiplying ($1/2 \times 1/2 = 1/4$). The probability of *BB* is calculated similarly and is also 1/4. The probability of *AA* is independent of the probability of *BB*, so again we use the product rule to obtain their combined probabilities ($1/4 \times 1/4 = 1/16$). ■

the length of the paired homologous chromosomes. Several exchanges may occur at different points during a single meiotic division. In general, the probability that two genes will be separated by crossing over is proportional to the physical distance between them in the chromosome.

In fruit flies a locus controlling wing shape (the dominant allele *V* for normal wings and the recessive allele *v* for vestigial wings) and a locus controlling body color (the dominant allele *B* for gray and the recessive allele *b* for black) are located in the same pair of homologous chromosomes (Fig. 10–9). They therefore tend to be inherited together and are said to be *linked.* If a homozygous *BBVV* fly is crossed with a homozygous *bbvv* fly, the F_1 flies all have gray bodies and normal wings, and their genotype is *BbVv*. Linkage is most readily observed by analyzing the results of a test cross of heterozygous F_1 flies *(BbVv)* mated with homozygous *bbvv* flies. This test cross is similar to the test cross described previously in that heterozygous individuals are mated to homozygous individuals. However, it is called a **two-point test cross** because alleles of two loci are involved.

If the loci governing these characteristics were *not* linked, their alleles would undergo *independent assortment* during meiosis. The heterozygous parent would produce four kinds of gametes (*BV*, *Bv*, *bV*, and *bv*) in equal numbers. As a result of independent assortment, new gene combinations not present in the parental generation would be produced. Any process that leads to new gene combinations is called **recombination.** In our example, *Bv* and *bV* are both **recombinant** gametes. The other two kinds of gametes, *BV* and *bv*, are called **parental** gametes because they are identical to the gametes produced by the P generation. Of course, the homozygous recessive parent produces only one kind of gamete, *bv*. Thus, if independent assortment were to occur, approximately 1/4 of the offspring would be gray-bodied and normal-winged (*BbVv*), 1/4 black-bodied and normal-winged (*bbVv*), 1/4 gray-bodied and vestigial-winged (*Bbvv*), and 1/4 black-bodied and vestigial-winged (*bbvv*). Notice that the two-

Focus On

Deducing Genotypes

The science of genetics resembles mathematics in that it consists of a few basic principles, which, once grasped, enable the student to solve a wide variety of problems. Very often the genotypes of the parents can be deduced from the phenotypes of their offspring. In chickens, for example, the allele for rose comb (*R*) is dominant to the allele for single comb (*r*). Suppose that a cock is mated to three different hens, as shown in the figure. The cock and hens A and C have rose combs; hen B has a single comb. Breeding the cock with hen A produces a rose-combed chick, with hen B a single-combed chick, and with hen C a single-combed chick. What types of offspring can be expected from further matings of the cock with these hens?

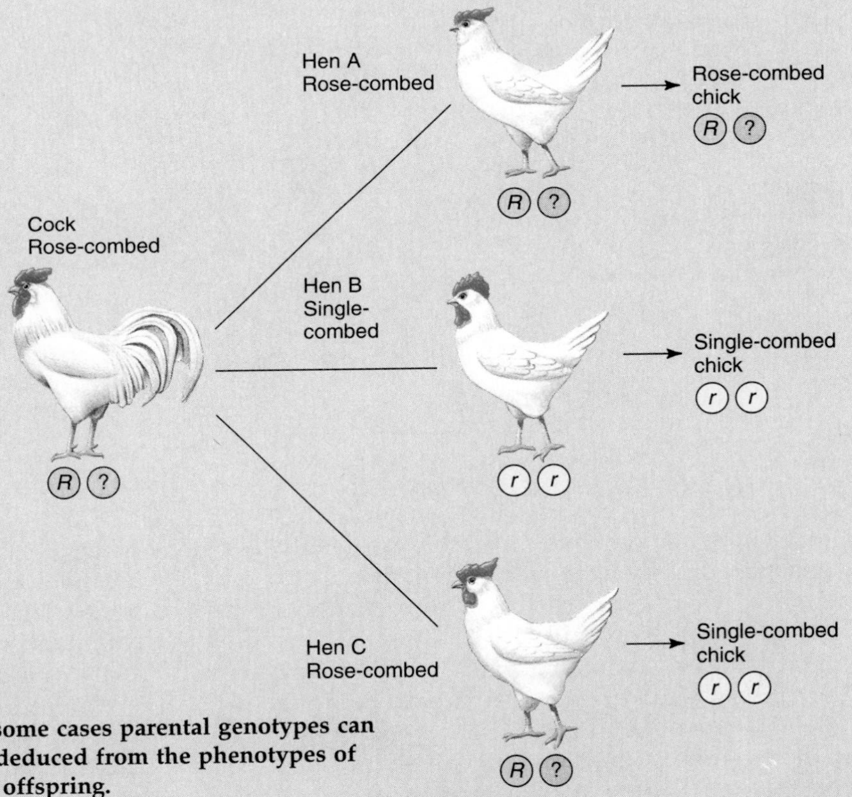

In some cases parental genotypes can be deduced from the phenotypes of the offspring.

Because the allele for single comb, *r*, is recessive, all of the hens and chicks that are phenotypically single-combed must be *rr*. We can deduce that hen B and the offspring of hens B and C are genotypically *rr*.

All individuals that are phenotypically rose-combed must have at least one *R* allele. The fact that the offspring of the cock and hen B was single-combed proves that the cock is heterozygous *Rr*, because, although the single-combed chick received one *r* allele from its mother, it must have received the second one from its father.

The fact that the offspring of the cock and hen C had a single comb proves that hen C is heterozygous, *Rr*. It is impossible to decide from the data given whether hen A is homozygous *RR* or heterozygous *Rr*; further breeding would be necessary to determine this. (Can you suggest an appropriate mating?)

Additional matings of the cock with hen B should result in half rose-combed and half single-combed individuals; additional matings of the cock with hen C should produce three fourths rose-combed and one fourth single-combed chicks. ■

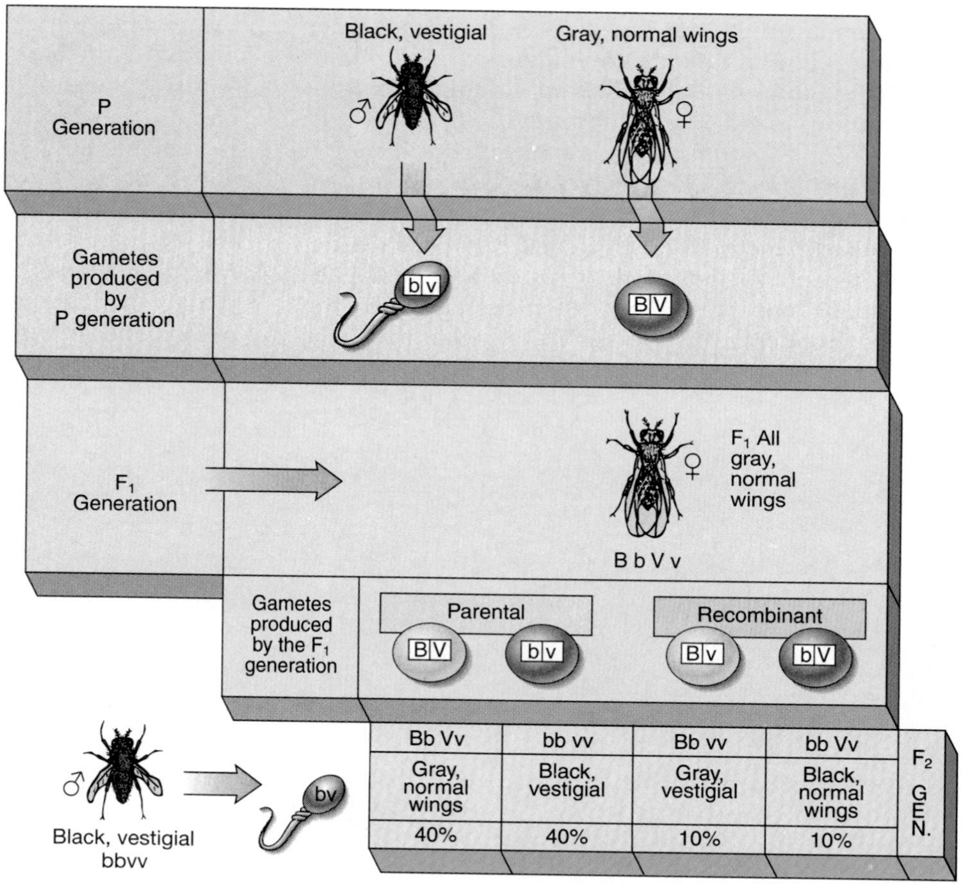

FIGURE 10–9 Linkage can be detected by a two-point test cross. In fruit flies, loci for wing length (vestigial versus normal wings) and for body color (black versus gray body) are linked; they are located on a homologous chromosome pair. Linkage can be recognized when an excess of parental type offspring and a deficiency of recombinant type offspring are produced. (Fruit flies are unusual in that crossing over occurs only in females and not in males. It is far more common for crossing over to occur in both sexes of a species.)

point test cross allows us to determine the genotypes of the offspring directly from their phenotypes.

If the loci were absolutely linked, no exchange of chromosomal segments would occur. In that case, only parental gametes would be formed by the heterozygous F_1 individuals. The two classes of parental type offspring, with phenotypes identical to the phenotypes of the P generation, would be produced in equal numbers. The parental type offspring in the example have gray bodies and normal wings *(BbVv)* or black bodies and vestigial wings *(bbvv)*.

However, in our example there is actually an exchange between these loci in some of the meiotic cells of the heterozygous female flies (Fig. 10–8). Because of this crossing over, some gray-bodied, vestigial-winged flies and some black-bodied, normal-winged flies are seen, in addition to the parental types. These **recombinant types** are the flies that received a **recombinant** gamete from the heterozygous F_1 parent.

Remember that a recombinant gamete is one that contains a *combination* of genes that was not present in the parental (P) generation. If the loci are unlinked, recombination results from independent assortment. However, if the loci are linked, a recombinant gamete must carry a chromatid that has undergone crossing over so it now contains a *new combination* of alleles for these two loci.

In this instance, about 20% of the gametes are recombinant (*bV* or *Bv*). These account for the two classes of recombinant offspring: gray flies with vestigial wings, *Bbvv* (approximately 10% of the total); and black flies with normal wings, *bbVv* (also about 10% of the total). In such crosses, about 40% of the offspring are gray flies with normal wings, *BbVv;* and another 40% are black flies with vestigial wings, *bbvv.* These two make up the parental, or nonrecombinant, class of offspring.

The genetic distance between two loci in a chromosome is measured in **map units,** or recombination units, which are a measure of the percentage of crossing over between them. There is a rough correlation between this genetic distance and the actual physical distance along the chromosome. One can calculate the percentage of recombination by adding the number of individuals in the

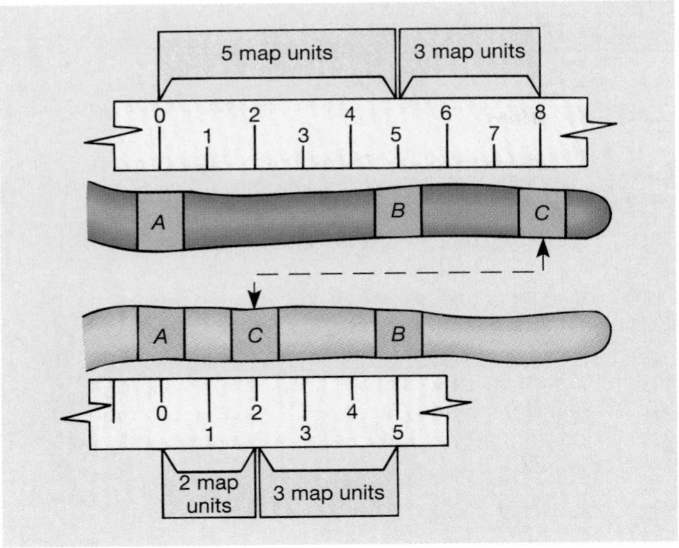

FIGURE 10–10 Gene loci can be mapped to positions on chromosomes. Gene order (i.e., which locus lies between the other two) is determined by the percentage of recombination between each of the possible pairs. In this hypothetical example, the percentage of recombination between locus *A* and locus *B* is 5% (corresponding to 5 map units), and that between *B* and *C* is 3% (3 map units). If the recombination between *A* and *C* is 8% (8 map units), *B* must be in the middle. However, if the recombination between *A* and *C* is 2%, then *C* must be in the middle.

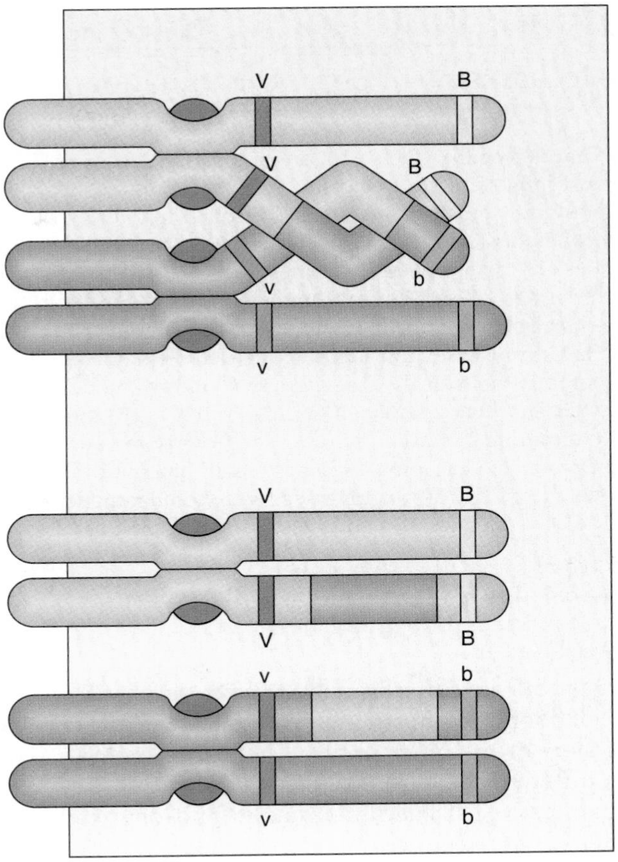

FIGURE 10–11 Double crossing over involving the same homologous chromatids does not result in the formation of recombinant gametes.

two recombinant classes of offspring (10 + 10), dividing by the *total number* of offspring (40 + 40 + 10 + 10), and multiplying by 100. Thus, the *V* locus and the *B* locus have 20% recombination between them. By convention, 1% recombination between two loci equals a distance of 1 map unit, so they are said to be 20 map units apart.

The frequencies of recombination between specific linked loci have been measured in a number of species. All of the experimental results are consistent with the hypothesis that genes are present in a linear order in the chromosomes. Figure 10–10 illustrates the method for determining the order of genes in a chromosome.

Crossing over occurs at random, and more than one crossover between two loci in a single tetrad (bivalent) can occur in a given cell undergoing meiosis. We can observe only the frequency of offspring receiving recombinant gametes from the heterozygous parent, not the actual number of crossovers. In fact, the actual frequency of crossing over is slightly more than the observed frequency of recombinant gametes. This is because the simultaneous occurrence of two crossovers involving the same two homologous chromatids reconstitutes the original combination of genes (Fig. 10–11). When two loci are relatively close together, this effect is minimized.

All the genes in a particular chromosome tend to be inherited together and therefore are said to constitute a **linkage group.** The number of linkage groups determined by genetic tests is equal to the number of pairs of chromosomes. By putting together the results of many crosses, scientists have developed detailed linkage maps for a number of eukaryotes, including the fruit fly (which has four pairs of chromosomes), the mouse, yeast, and *Neurospora* (a fungus). In addition, special genetic methods have made possible the development of a detailed map for *Escherichia coli*, a bacterium with a single circular DNA molecule, and a number of other prokaryotes and viruses. More sophisticated maps of chromosomes are currently made by means of recombinant DNA technology (see Chapter 14). These methods have been particularly useful in producing maps of human chromosomes (see Chapter 15).

SEX IS COMMONLY DETERMINED BY SPECIAL SEX CHROMOSOMES

Mechanisms of sex determination vary considerably (see *Making the Connection: Mechanisms of Sex Determination*). Most animals have special sex chromosomes. Typically, members of one sex (the **homogametic** sex) have a pair

Making the Connection

Mechanisms of Sex Determination

What determines the sex of an organism? Genes are the most important sex determinants in most organisms, although in some species sex is controlled mainly by the environment. The major sex-determining genes of most animals are carried by sex chromosomes.

An XX/XY sex chromosome mechanism, similar to that of humans, operates in many species of animals. However, it is not universal and many of the details may vary. For example, the fruit fly, *Drosophila*, has homogametic (XX) females and heterogametic (XY) males, but the Y is not male-determining; a fruit fly with an X chromosome and no Y chromosome has a male phenotype. In birds and butterflies the mechanism is reversed, with homogametic males (the equivalent of XX) and heterogametic females (the equivalent of XY).

In **hermaphroditic** organisms, organs of both sexes are found in the same individual. Hermaphroditic animals (see Chapters 28, 29, and 48) do not have sex chromosomes. Most flowering plants are hermaphrodites. When the male and female sexual organs are in separate flowers on the same plant, the plants are said to be **monoecious;** corn, walnuts, and oaks are examples. Far fewer flowering plants are **dioecious,** having male and female floral organs on separate plants (see Chapter 27). A few dioecious plants, such as asparagus, apparently have sex chromosomes, although they are not necessarily comparable to those of animals. ▲

of identical sex chromosomes and produce gametes that are all identical in sex chromosome constitution. The members of the other sex (the **heterogametic** sex) have two different sex chromosomes and produce two kinds of gametes, each bearing a single kind of sex chromosome.

The females of many animal species (including humans) are homogametic; their cells contain two identical sex chromosomes, called **X chromosomes.** In contrast, the males are heterogametic, having a single X chromosome and a smaller **Y chromosome.** Human males have 22 pairs of **autosomes,** which are chromosomes other than the sex chromosomes, plus one X chromosome and one Y chromosome; females have 22 pairs of autosomes plus two X chromosomes.

The Y chromosome determines male sex in most species of mammals

Do human males have a male phenotype because they have only one X chromosome or because they have a Y chromosome? Much of the evidence bearing on this question comes from studies of persons with abnormal sex chromosome constitutions (see Chapter 15). A person with an XXY constitution is a nearly normal male in external appearance but has underdeveloped testes (Klinefelter syndrome). A person with one X but no Y chromosome has the appearance of an immature female (Turner syndrome). Hence we think that the Y is the male-determining chromosome, and at least one gene on the Y thought to be involved in this process has been identified.

In humans and other species in which the normal male has one X and one Y chromosome, these chromosomes regularly synapse and disjoin from one another during meiosis. The X and Y chromosomes therefore are an exception to the general rule that the members of a homologous pair are similar in size, shape, and genetic constitution. Half the sperm contain an X chromosome and half contain a Y chromosome. All eggs bear a single X chromosome (Fig. 10–12). Fertilization of an X-bearing egg by an X-bearing sperm results in an XX (female) zygote; fertilization by a Y-bearing sperm results in an XY (male) zygote.

We would expect to have equal numbers of X- and Y-bearing sperm and a 1:1 ratio of females to males. In fact, however, more males are conceived than females and more males die before birth. Even at birth the ratio is not 1:1; about 106 boys are born for every 100 girls. It is not known why this occurs, but the Y-bearing sperm is assumed to have some competitive advantage.

X-linked genes have unusual inheritance patterns

The human X chromosome contains many loci that are required in both sexes, whereas the Y chromosome contains only a few genes, including one or more genes for maleness. Traits controlled by genes located in the X chromosome, such as color blindness and hemophilia, are sometimes called **sex-linked** traits. It is more appropriate, however, to refer to them as **X-linked** traits because they follow the transmission pattern of the X chromosome and strictly speaking are not linked to the sex of the organism per se.

A female receives one X from her mother and one X from her father. A male receives his Y chromosome, which makes him male, from his father. From his mother he inherits a single X chromosome and therefore all of his genes for X-linked traits. In the male, every X chromo-

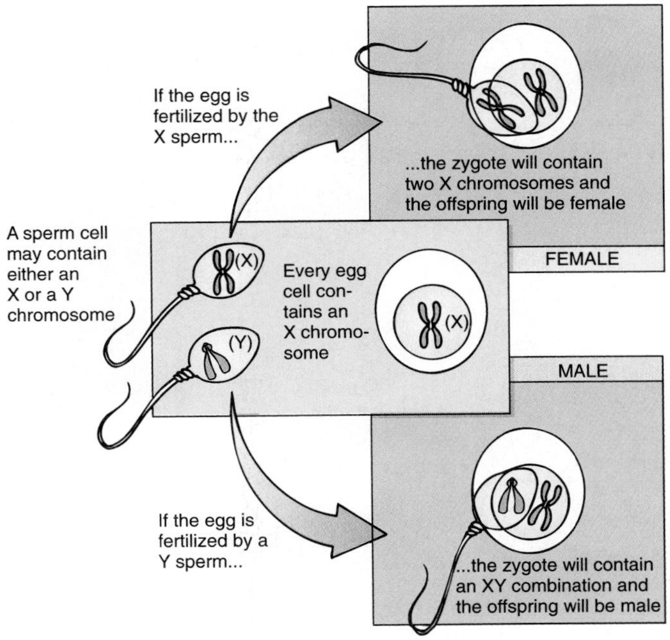

FIGURE 10–12 Sex is determined at the time of fertilization by the sperm. An X-bearing sperm produces a female; a Y-bearing sperm produces a male.

some allele present is expressed, whether that allele was dominant or recessive in the female parent. A male is always **hemizygous** for every X-linked locus. The term *hemi* means "half"; a hemizygous male is neither homozygous nor heterozygous for X-linked traits.

Various forms of notation are used for problems involving X linkage. We will use a simple system, indicating the X and incorporating specific alleles as superscripts. For example, the symbol X^c signifies a recessive X-linked allele for color blindness, and X^C the dominant X-linked allele for normal color vision. The Y chromosome is written without superscripts because it does not carry the locus of interest.

For most X-linked loci, the abnormal or uncommon allele is recessive in the female and the normal or most common allele is dominant. Therefore, two recessive X-linked alleles must be present in a female for the abnormal phenotype to be expressed, whereas in the hemizygous male a single abnormal allele is expressed. As a practical consequence, these abnormal alleles are usually expressed only in male offspring, although they may be carried by a female.

To be expressed in a female, a recessive X-linked allele must be present on both X chromosomes; that is, the alleles must be inherited from both parents. A color-blind female, for example, must have a color-blind father and a mother who is heterozygous or homozygous for color blindness (Fig. 10–13). Such a combination is unusual. In

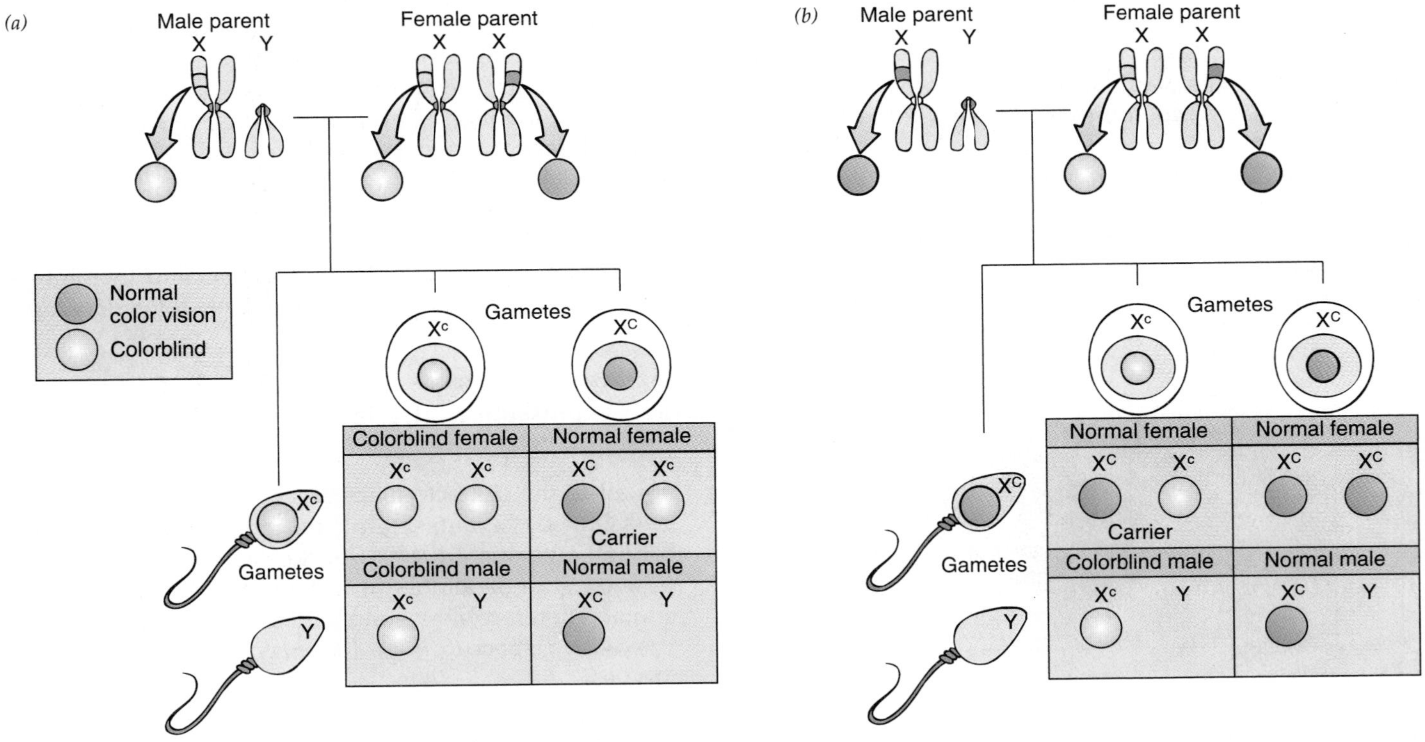

FIGURE 10–13 The most common form of color blindness is an X-linked recessive trait. Note that the Y chromosome does not carry a gene for color vision. (***a***) To be color blind, a female must inherit alleles for color blindness from both parents. (***b***) If a normal male mates with a carrier female, one-half of their sons would be expected to be color blind and one-half of their daughters would be expected to be carriers.

contrast, a color-blind male need only have a mother who is heterozygous for color blindness; his father can be normal. Hence, X-linked recessive traits are generally much more common in males than in females, a fact that may partially explain why human male embryos are more likely to die.

Dosage compensation equalizes the expression of X-linked genes in males and females

The X chromosome contains numerous genes required by both sexes, yet a normal female has two copies ("doses") for each locus, whereas a normal male has only one. Generally, a mechanism of **dosage compensation** is required to make the two doses in the female and the single dose in the male equivalent. Male fruit flies accomplish this by increasing the metabolic activity of their single X chromosome. In most tissues the male X chromosome is just as active as the two X chromosomes present in the female.

Dosage compensation in mammals generally involves inactivation of one of the two X chromosomes in the female. During interphase a dark spot of chromatin, called a **Barr body,** is visible at the edge of the nucleus of each female mammalian cell (Fig. 10–14). The Barr body has been found to represent one of the two X chromosomes, which has become dense and dark-staining. The other X chromosome resembles the autosomes in that during interphase it is a greatly extended thread that is not evident by light microscopy. From this and other evidence, the British geneticist Mary Lyon has suggested that in any one cell of a female mammal, only one of the two X chromosomes is active; the other is inactive and is seen as a Barr body.

Because only one X chromosome is active in any one cell, a female mammal that is heterozygous at an X-linked locus expresses one of the alleles in about half her cells and the other allele in the other half. This is sometimes (but not always) evident in the phenotype. Mice and cats have several X-linked genes for certain coat colors. Females that are heterozygous for such genes may show patches of one coat color in the midst of areas of the other coat color. This phenomenon, termed **variegation,** is evident in calico (Fig. 10–15) and tortoise-shell cats. Early in development, when relatively few cells are present, X chromosome inactivation occurs randomly in each cell. When any one of these cells divides by mitosis, the cells of the resulting clone (group of genetically identical cells) all have the same active X chromosome, and therefore a patch of cells that all express the same color develops.

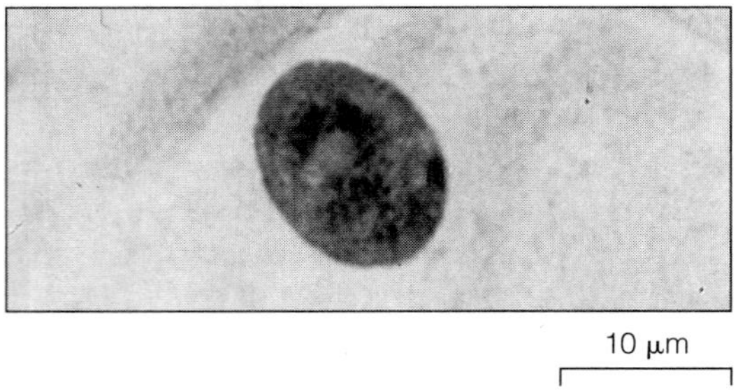

FIGURE 10–14 The darkly-stained Barr body at the edge of the nucleus in this light photomicrograph is an inactivated X chromosome. The entire cell is not shown. (Omikron/Photo Researchers, Inc.)

FIGURE 10–15 A calico cat exhibits the effects of X chromosome inactivation. This cat has X-linked genes for both black and yellow (or orange) pigmentation of the fur, but because of random X chromosome inactivation, black is expressed in some clones of cells and yellow (or orange) is expressed in others. Because other genes affecting fur color are also present, white patches are usually evident as well. (Larime Photographic/Dembinsky Photo Associates)

Sex-influenced genes are autosomal, but their expression is affected by the individual's sex

Not all of the characteristics that differ in the two sexes are X-linked. Certain **sex-influenced** traits are inherited through autosomal genes, but the *expression* of alleles at these loci can be altered or influenced by the sex of the animal. Therefore, males and females with the same genotype with respect to these loci may have different phenotypes.

Pattern baldness in humans, characterized by premature loss of hair on the front and top of the head, but not on the sides, is far more common among males than among females. It has been proposed that a single pair of alleles is involved, with the allele responsible for pattern baldness being dominant in males and recessive in females. Because of this unusual situation we modify our notation, designating the pattern baldness allele as B_1 and the allele for normal hair growth as B_2. Individuals with the genotype B_1B_1 show pattern baldness, regardless of sex. Persons with a B_1B_2 genotype are bald if they are male but not bald if they are female. Individuals with the genotype B_2B_2 are not bald, regardless of sex.

Evidence suggests that the expression of most sex-influenced traits is strongly modified by sex hormones. For example, male hormones (see Chapter 48) are strongly implicated in the expression of pattern baldness.

THE RELATIONSHIP BETWEEN GENOTYPE AND PHENOTYPE IS OFTEN COMPLEX

The relationship between a given locus and the characteristic it controls may be simple: a single pair of alleles of a locus may regulate the appearance of a single characteristic of the organism (e.g., tall versus short). Alternatively, the relationship may be more complex: a pair of alleles may participate in the control of several characteristics, or alleles of many loci may cooperate to regulate the appearance of a single characteristic. Not surprisingly, these more complex relationships are quite common.

As you will learn in Chapters 11 and 12, each locus is a segment of DNA in which biological information is stored as a triplet (three-base) code in the sequence of nucleotides that compose the double helix of the DNA molecule. The information is "read out," and in a great many cases a specific protein is ultimately synthesized. The presence of a specific protein, such as an enzyme, usually provides the chemical basis for the genetic trait. Because most biologically important molecules are synthesized by complex metabolic pathways involving a number of enzymes, it is not difficult to appreciate why relationships between genes and the characteristics of the organism are complex.

We may assess the phenotype on one or many levels. It may be a morphological characteristic such as shape, size, or color. It may be a physiological characteristic or even a biochemical trait, such as the presence or absence of a specific enzyme required for the metabolism of some specific molecule. In addition, the phenotypic expression of genes may be altered by changes in the environmental conditions under which the organism develops.

Dominance is not always complete

Studies of the inheritance of many traits in a wide variety of organisms have clearly shown that one member of a pair of alleles may not be completely dominant to the other. Indeed, it is improper to use the terms *dominant* and *recessive* in such instances.

For example, red and white are common flower colors in Japanese four o'clocks. Each color breeds true when these plants are self-pollinated. What flower color might we expect in the offspring of a cross between a red-flowering plant and a white-flowering one? Without knowing which is dominant, we might predict that all would have red flowers or all would have white flowers. This cross was first made by the German botanist Karl Correns (one of the rediscoverers of Mendel's work), who found that all F_1 offspring have pink flowers! Does this result in any way prove that Mendel's assumptions about inheritance are wrong? Quite the contrary, for when two of these pink-flowered plants are crossed, red-flowered, pink-flowered, and white-flowered offspring appear in a ratio of 1:2:1 (Fig. 10–16).

In this instance, as in all other aspects of science, results that differ from those predicted prompt scientists to reexamine and modify their assumptions to account for the exceptional results. The pink-flowered plants are clearly the heterozygous individuals, and neither the red allele nor the white allele is completely dominant. When the heterozygote has a phenotype that is intermediate between those of its two parents, the genes are said to show **incomplete dominance.** In these crosses the genotypic and phenotypic ratios are identical.

Incomplete dominance is not unique to Japanese four o'clocks. Red- and white-flowered sweet pea plants also produce pink-flowered plants when crossed, and numerous additional examples are known in both plants and animals.

In both cattle and horses, reddish coat color is not completely dominant to white coat color. Heterozygous individuals have reddish coats with spots of white hairs, and are referred to as roan-colored. If you saw a white mare nursing a roan-colored foal, what would you guess was the coat color of the foal's father? Because the reddish and white colors are expressed independently in the roan heterozygote, we sometimes refer to this as a case of **codominance.** Strictly speaking, *incomplete dominance* refers to instances in which the heterozygote is intermediate in phenotype, and *codominance* refers to instances in which two alleles are expressed independently in the heterozygote. The human ABO blood group (see Chapter 15) provides a classic example of codominant alleles.

Multiple alleles for a locus may exist in a population

The examples given so far have dealt with situations in which each locus was represented by a maximum of two allelic variants, and in most of these examples one of the alleles has been dominant and one recessive. It is true that a single diploid individual has a maximum of two different alleles for a particular locus and that a haploid ga-

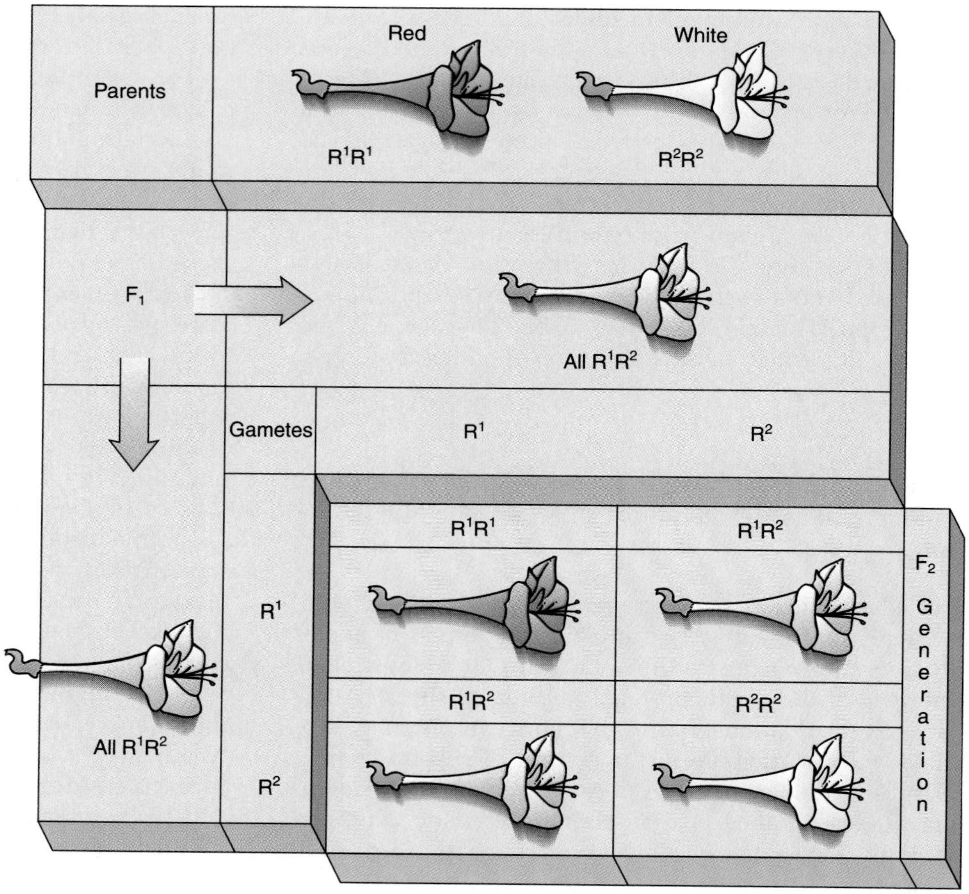

FIGURE 10–16 If a pair of alleles is incompletely dominant to each other, a heterozygote has a phenotype intermediate between its parents. Two incompletely dominant alleles, R^1 and R^2, are responsible for red, white, and pink flower colors in Japanese four o'clocks. Red individuals are R^1R^1; white individuals are R^2R^2, and heterozygotes (R^1R^2) are pink. Note that uppercase notation is used for both alleles, because neither is recessive to the other.

mete has only one allele for each locus. However, if we survey a population, we may find more than two alleles for a particular locus. If three or more alleles for a given locus exist within the population, we say that locus has **multiple alleles.** A great many loci can be shown to have multiple alleles if the population is surveyed carefully. Some alleles can be identified biochemically but do not produce a distinct phenotype. Others are distinguishable by phenotype, and certain patterns of dominance can be discerned when the alleles are combined in various ways.

In rabbits, for example, a C allele causes a fully colored coat. The homozygous recessive genotype, cc, causes albino coat color. There are two additional allelic variants of the same locus, c^h and c^{ch}. In a homozygous rabbit, the allele c^h causes the "Himalayan" pattern, in which the body is white but the tips of the ears, nose, tail, and legs are colored (similar to the color pattern of a Siamese cat). An individual homozygous for the c^{ch} allele has the "chinchilla" pattern, in which the entire body has a light gray color. On the basis of the results of genetic crosses, these alleles can be arranged in the following series: $C > c^h > c^{ch} > c$. Each allele is dominant to those following it and recessive to those preceding it. In some other series of multiple alleles, certain alleles may be codominant and others incompletely dominant; hence the heterozygotes commonly have phenotypes intermediate between those of their parents.

A single gene may affect multiple aspects of the phenotype; alleles of different loci may interact to produce a phenotype

In the examples presented so far, the relationship between a gene and its phenotype has been direct, precise, and exact, and the loci considered have controlled the appearance of single traits. However, the relationship of gene to characteristic may be quite complex.

Most genes probably have many different effects, a quality referred to as **pleiotropy.** This is dramatically evident in many genetic diseases, such as cystic fibrosis and sickle cell anemia (see Chapter 15), in which multiple symptoms can be traced to a single pair of alleles. Albino individuals lack pigment in the skin, hair, and eyes, demonstrating that a single locus can simultaneously affect a number of characteristics. Conversely, virtually every feature of the organism is actually controlled by a large number of loci. We are not always aware of this because not all of these loci have been identified.

Epistasis is a common type of gene interaction in which the presence of a particular allele of one gene pair

FIGURE 10–17 The interaction of two gene pairs governs the inheritance of these types of genetically determined combs in roosters.

determines whether certain alleles of another gene pair are expressed. Several pairs of alleles may interact to affect a single trait, or one pair may inhibit or reverse the effect of another pair. More than 12 pairs of alleles interact in various ways to produce coat color in rabbits, and more than 100 pairs are concerned with eye color and shape in fruit flies.

One of the simplest types of gene interaction is illustrated by the inheritance of combs in poultry (Fig. 10–17). The allele for a rose comb, *R*, is dominant to that for a single comb, *r*. Another gene pair governs the inheritance of a pea comb, *P*, versus a single comb, *p*. A single-combed fowl must therefore have the genotype *pprr*; a pea-combed fowl is either *PPrr* or *Pprr*; and a rose-combed fowl is either *ppRR* or *ppRr*. When a homozygous pea-combed fowl is mated to a homozygous rose-combed one, the offspring have neither a pea nor a rose comb, but a completely different type, called a *walnut comb*. The walnut comb phenotype is produced whenever a fowl has one or two *R* alleles, plus one or two *P* alleles (that is, *PPRR*, *PpRR*, *PPRr*, or *PpRr*). What would you predict about the types of combs among the offspring of two heterozygous walnut-combed fowl, *PpRr*? How does this form of epistasis affect the ratio of phenotypes in the F_2 generation? Is it the typical Mendelian 9:3:3:1 ratio?

We have already seen that coat color in guinea pigs is determined by the *B* and *b* allelic pair, with the *B* allele for black coat dominant to the *b* allele for brown coat. The expression of either phenotype, however, depends on the presence of a dominant allele at yet another locus. This allele, *C*, codes for the enzyme tyrosinase, which converts a colorless precursor to the pigment melanin and hence is required for the production of any kind of pigment. The recessive allele (*c*) codes for an inactive form of the enzyme. Thus, an animal that is homozygous recessive for this allele essentially lacks the enzyme and produces no melanin. It is therefore a white-coated, pink-eyed albino, regardless of the combination of *B* and *b* alleles. Albinism, or lack of melanin pigment, is not restricted to guinea pigs, but is found in humans (see Fig. 15–1) and a variety of other animals (Fig. 10–18).

When an albino guinea pig with the genotype *ccBB* is mated to a brown guinea pig with the genotype *CCbb*, the F_1 generation is black coated, *CcBb*. When two such animals are mated, their offspring appear black-coated, brown-coated, and albino in a ratio of 9:3:4. (Make a Punnett square to verify this.)

FIGURE 10–18 Alleles for albinism exhibit epistasis. When homozygous, they mask the expression of alleles of other loci that govern production of melanin pigment. Albino individuals, such as this albino koala, occur occasionally in nature. (Tom McHugh/Photo Researchers, Inc.)

You might wonder why heterozygous *Cc* individuals do not show at least some lightening of the coat color, given the fact that they produce only about half the nor-

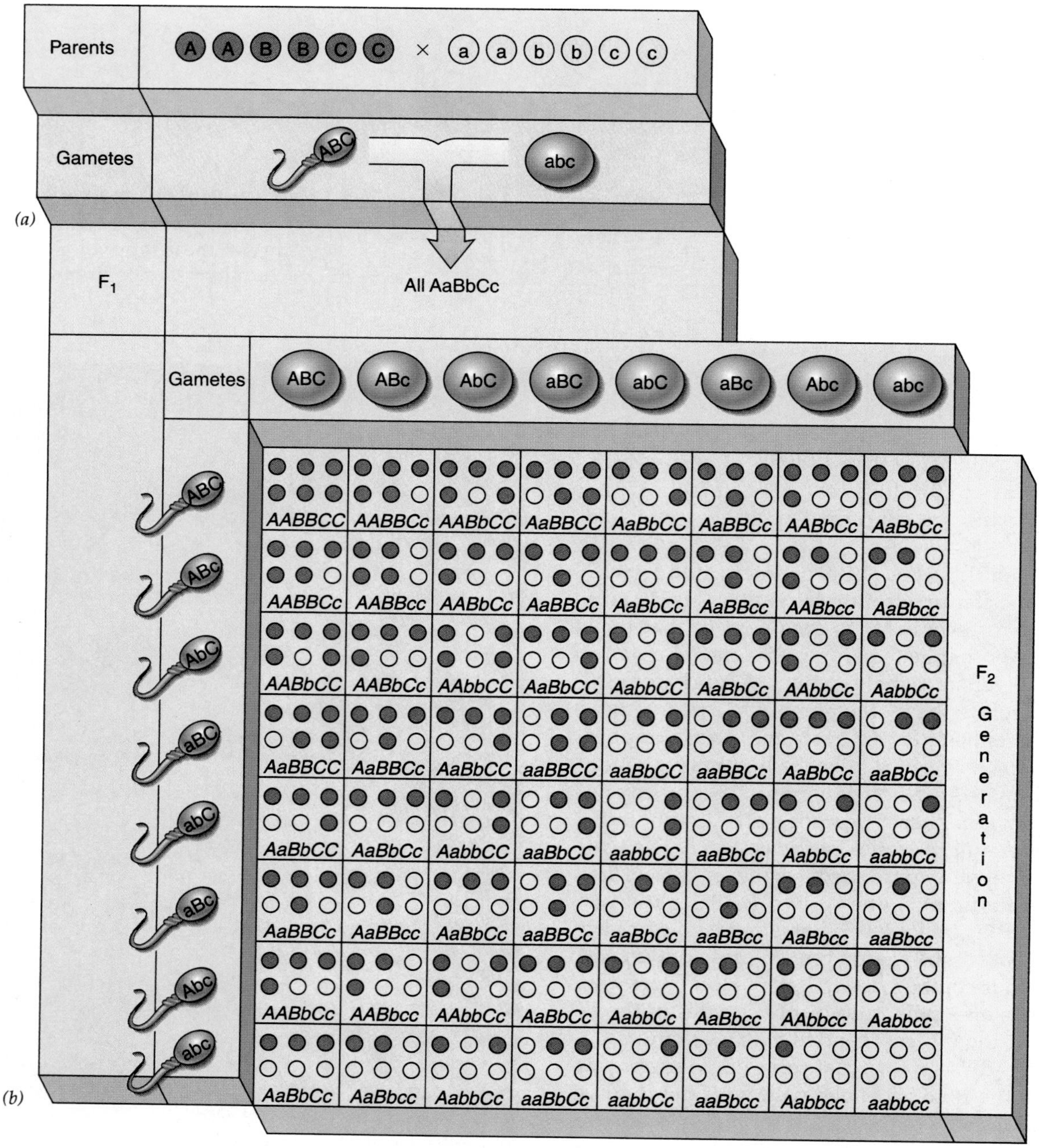

FIGURE 10–19 This simplified model of polygenic inheritance assumes that skin color in humans is governed by alleles of three unlinked loci. The alleles producing dark skin (*A*, *B*, and *C*) are represented by capital letters, but they are not dominant. Instead they have additive effects. If one parent is very dark and the other very light, their children (F_1) are intermediate in skin color. A wide range of skin colors are expected in the F_2. The number of dark dots (each signifying an allele producing dark skin) is counted to determine the phenotype. The results are summarized in Figure 10–20.

mal amount of tyrosinase enzyme. It turns out that half the normal amount of enzyme is usually adequate to produce normal amounts of pigment. This is a very common phenomenon and accounts for many (although certainly not all) cases of dominance.

Polygenes act additively to produce a phenotype

The inherited components of many human characteristics, such as height, body form, and skin color, are not in-

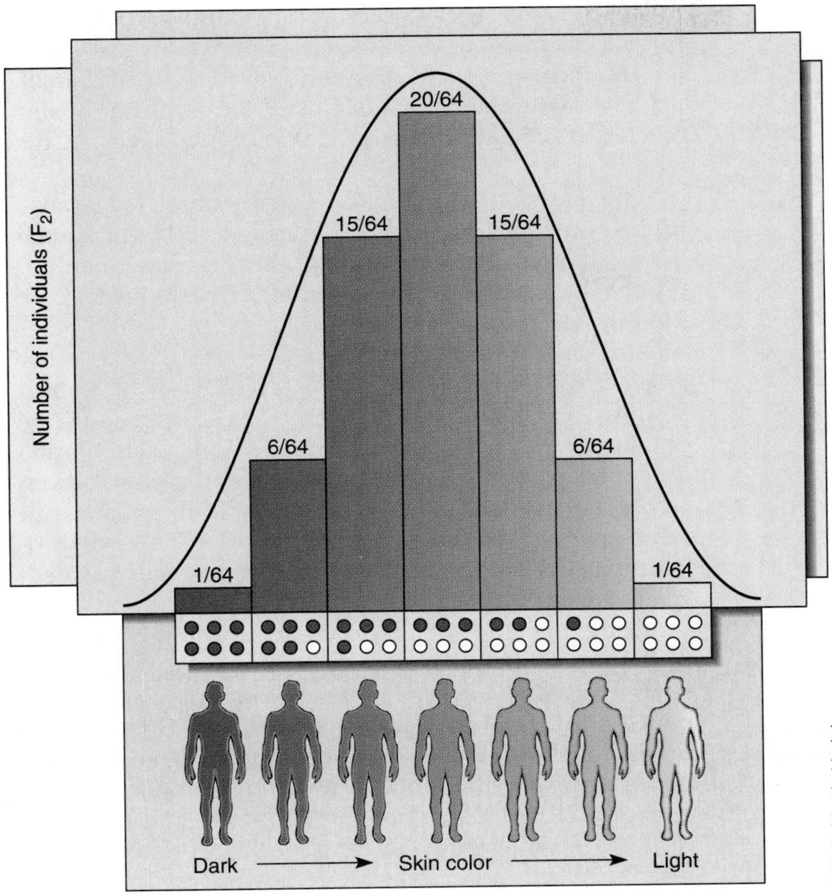

FIGURE 10–20 Polygenic inheritance is characterized by a wide range of phenotypes in the F_2 generation. The bars indicate the expected phenotypic ratios in the F_2 generation shown in Figure 10–19. This expected distribution of phenotypes is consistent with the superimposed normal curve.

herited through alleles at a single locus. The same holds true for many commercially important characteristics in domestic plants and animals, such as milk and egg production. Alleles at several, perhaps many, different loci affect each characteristic. The term **polygenic inheritance** is applied when multiple independent pairs of genes have similar and additive effects on the same characteristic.

Polygenes are responsible for the inheritance of skin color in humans. It is now thought that alleles representing four or more different loci are involved in determining skin color, but the principle of polygenic inheritance can be illustrated with pairs of alleles at only three unlinked loci (Fig. 10–19). These can be designated *A* and *a*, *B* and *b*, and *C* and *c*. The capital letters represent *incompletely dominant* alleles producing dark skin. The more capital letters, the darker the skin, because the alleles affect skin color in an *additive* fashion. A person with the darkest skin would have the genotype *AABBCC*, and a person with the lightest skin would have the genotype *aabbcc*. The F_1 offspring of an *aabbcc* person and an *AABBCC* person are all *AaBbCc* and have an intermediate skin color. The F_2 offspring of two such triple heterozygotes would have skin colors ranging from very dark to very light.

Polygenic inheritance is therefore characterized by an F_1 generation that is intermediate between the two completely homozygous parents and by an F_2 generation that shows wide variation between the two parental types. Most of the F_2-generation individuals have one of the intermediate phenotypes; only a few show the extreme phenotypes of the grandparents (P generation). On average, only 1 of 64 is as dark as the very dark grandparent, and only 1 of 64 is as light as the very light grandparent (Fig. 10–20). The alleles *A*, *B*, and *C* each produce about the same amount of darkening of the skin; hence, the genotypes *AaBbCc*, *AABbcc*, *AAbbCc*, *AaBBcc*, *aaBBCc*, *AabbCC*, and *aaBbCC* all produce similar intermediate phenotypes.

The model used here for the inheritance of skin color in humans is a rather simple example of polygenic inheritance because only three major allelic pairs are used. The inheritance of height in humans involves alleles representing ten or more loci. Because many allelic pairs are involved and because height is modified by a variety of environmental conditions, the heights of adults range from perhaps 125 to 215 cm. If we were to measure the heights of 1000 adult American men selected at random, we would find that only a few are as tall as 215 cm or as short as 125 cm. The heights of most would cluster around the mean, about 170 cm. When the number of men at each height is plotted against height (in centimeters) and the points are connected, the result is a bell-shaped curve, called a **normal distribution curve** (Fig. 10–21).

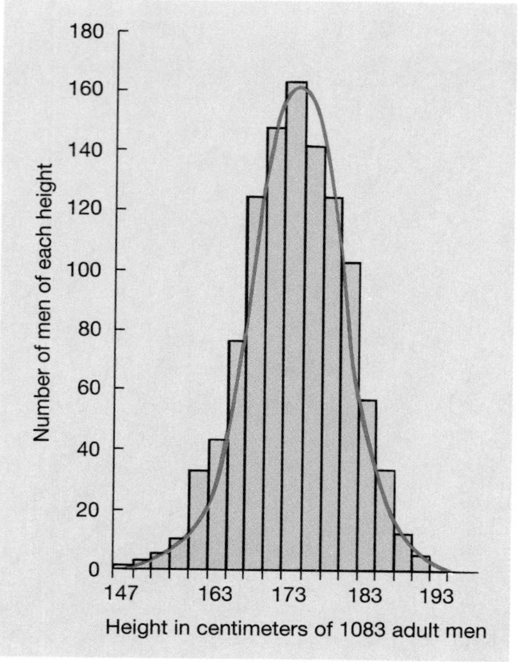

FIGURE 10–21 Polygenic inheritance is associated with continuous phenotypic variation in a population. The distribution of heights of 1083 adult males approximates a normal curve. The bars indicate the actual number of men whose heights were within the unit range. For example, there were 163 men whose heights were between 170 and 173 cm.

SELECTION, INBREEDING, AND OUTBREEDING ARE USED TO DEVELOP IMPROVED STRAINS

How do geneticists go about establishing a breed of cow that will give more milk, a strain of hens that will lay bigger eggs, or a variety of corn with more kernels per ear? By selecting organisms that manifest the desired phenotype, and using these organisms in further matings, a true breeding strain with the commercially advantageous trait is gradually developed. Such a strain should be homozygous for all of the genes involved, whether they be dominant, recessive, or additive in their effects.

There is a limit to the effectiveness of breeding by selection. When a strain becomes homozygous for all of the genes involved, further selective breeding cannot increase the desired quality. Moreover, because of **inbreeding**—the mating of two closely related individuals—the strain may become homozygous for multiple undesirable traits as well. Certain dog breeds, for instance, are known for their susceptibility to congenital dislocation of the hip. Evidence suggests that human inbreeding increases the frequency of genetic disease in the population (see Chapter 15), although the individual risk is relatively small. For this reason marriages of close relatives (first cousins or closer) are forbidden by law in many states in the United States.

The mating of individuals of totally unrelated strains, termed **outbreeding,** frequently leads to offspring much better adapted for survival than either parent. Such improvement reflects a phenomenon called **hybrid vigor.** Mongrel dogs are often hardier than highly inbred purebreds. A large proportion of the corn, wheat, and other crops grown in the United States consists of hybrid strains. Each year the seed to grow these crops must be obtained by mating of the original strains. The hybrids are heterozygous at a great many loci and give rise, even when self-fertilized, to a wide variety of forms, none of which is as good as the original hybrid. (The seeds produced by F_1 hybrid corn plants are not normally planted, but eaten instead!)

One explanation of hybrid vigor is the following: Each of the parental strains is homozygous for certain undesirable recessive genes, but any two strains are homozygous for different undesirable genes. Each strain contains dominant genes to make up for the recessive undesirable genes of the other strain. One strain then might have the genotype *AAbbCCdd,* and another strain the genotype *aaBBccDD.* (The capital letters represent dominant genes for desirable traits, and the lowercase letters represent recessive genes for undesirable traits.) The hybrid offspring, with the genotype *AaBbCcDd,* would express all of the desirable and none of the undesirable traits of the two parental strains.

Sometimes an individual that is heterozygous for a particular locus expresses a more extreme phenotype than either of the parental homozygotes. Geneticists call this phenomenon **overdominance.** If the phenotype of the heterozygote is more desirable, we say that there is a **heterozygote advantage.** The use of this term implies that at least sometimes there is an advantage to heterozygosity for its own sake.

In humans, individuals who are heterozygous for the recessive sickle cell anemia allele (*s*) and the normal dominant allele (*S*) appear to have increased resistance to the parasite that lives inside red blood cells and causes malaria. Such resistance is a significant advantage in areas of the world where malaria is still uncontrolled. Homozygous normal individuals (*SS*) appear to be less resistant to malaria. Homozygous sickle cell individuals (*ss*) are at a distinct disadvantage due to severe anemia and other serious effects of the sickle cell allele (see Chapter 18).

SUMMARY

I. Mendel's inferences from his garden pea breeding experiments have been tested repeatedly in all kinds of diploid organisms and found to be generally true. These principles have been extended and now can be stated in a more modern form.
   A. Today we know that the genes are in chromosomes; the site a gene occupies in the chromosome is its locus.
   B. Different forms of a particular locus are alleles; they occupy corresponding loci on homologous chromosomes. Genes therefore exist as pairs of alleles in diploid individuals.
   C. An individual that carries two identical alleles for a given locus is said to be homozygous for that locus. If the two alleles are different, that individual is said to be heterozygous for that locus.
   D. According to Mendel's principle of dominance, one allele (the dominant allele) may mask the expression of the other allele (the recessive allele) in a heterozygous individual. For this reason two individuals with the same appearance (phenotype) may differ from each other in genetic constitution (genotype). Dominance does not always apply, and alleles can be incompletely dominant or codominant.
   E. According to Mendel's principle of segregation, during meiosis the alleles for each locus separate, or segregate, from each other as the homologous chromosomes separate. When haploid gametes are formed, each contains only one allele for each locus.
   F. According to Mendel's principle of independent assortment, alleles of different loci are distributed randomly into the gametes. This can result in recombination — that is, production of new gene combinations that were not present in the parental (P) generation.
   G. Each chromosome behaves genetically as if it were composed of genes arranged in a linear order. Genes in the same chromosome are linked and do not assort independently. Recombination of linked genes can occur as a result of crossing over (breaking and rejoining of homologous chromatids) in meiotic prophase I. By measuring the frequency of recombination between linked genes, it is possible to construct a genetic map of a chromosome.
   H. A cross between homozygous parents (P generation) that differ from each other with respect to their alleles at one locus is called a monohybrid cross; if they differ at two loci, it is a dihybrid cross. The first generation of offspring is heterozygous and is called the first filial, or F_1, generation; the generation produced by a cross of two F_1 individuals is the second filial, or F_2, generation. A test cross is between F_1 and homozygous recessive individuals.

II. Genetic ratios can be expressed in terms of probabilities.
   A. Any probability is expressed as a fraction or decimal fraction, calculated as the number of favorable events divided by the total number of events. This can range from 0 (an impossible event) to 1 (a certain event).
   B. The probability of two independent events occurring together is the product of the probabilities of each occurring separately.
   C. The probability that one or the other of two mutually exclusive events will occur is the sum of their separate probabilities.

III. The sex of humans and many other animals is determined by the X and Y sex chromosomes or their equivalents. Autosomes are not sex chromosomes.
   A. Normal female mammals have two X chromosomes; normal males have one X and one Y.
   B. The fertilization of an X-bearing egg by an X-bearing sperm results in a female (XX) zygote. The fertilization of an X-bearing egg by a Y-bearing sperm results in a male (XY) zygote.
   C. The Y chromosome is responsible for determining male sex in mammals.
   D. The X chromosome contains many important genes unrelated to sex determination and required by both males and females. A male receives all his X-linked genes from his mother. A female receives X-linked genes from both parents.
   E. A female mammal shows dosage compensation of X-linked genes. Only one of the two X chromosomes is expressed in each cell; the other is inactive and is seen as a dark-staining Barr body at the edge of the interphase nucleus.

IV. Multiple alleles (three or more alleles that can potentially occupy a particular locus) may exist in a population. A diploid individual has any two of the alleles; a haploid individual or gamete has only one.

V. The relationship between a gene and its phenotype may be quite complex.
   A. Most genes have many different effects; they are pleiotropic.
   B. The presence of an epistatic allele at one locus affects the expression of alleles of a different locus.
   C. In polygenic inheritance, multiple independent pairs of genes may have similar and additive effects on the phenotype.
      1. Many human characteristics showing continuous variation, such as height and skin color, as well as many characteristics in other animals and plants, are inherited through polygenes.
      2. In polygenic inheritance, the F_1 generation is intermediate between the two parental types and shows little variation; the F_2 generation shows wide variation.

VI. Inbreeding, the mating of two closely related individuals, greatly increases the probability that an individual offspring will be homozygous for recessive genes. Outbreeding, the mating of totally unrelated individuals, increases the probability that the offspring will be heterozygous at many loci and likely to exhibit hybrid vigor.

SELECTED KEY TERMS

allele
Barr body
dihybrid cross
dominant
dosage compensation
epistasis
F_1 generation
F_2 generation
genotype
hemizygous
heterogametic
heterozygote advantage
heterozygous
homogametic
homozygous
hybrid
hybrid vigor
inbreeding
incomplete dominance
linkage
locus (loci)
monohybrid cross
multiple alleles
overdominance
P generation
phenotype
pleiotropy
polygenic inheritance
principle of dominance
principle of independent assortment
principle of segregation
product law
Punnett square
recessive
sex-influenced
sex-linked
sum law
test cross
true breeding line
variegation
X-linked

POST-TEST

1. The specific site in the chromosome occupied by a given gene is termed its __________ .
2. Genes governing different forms of the same character (e.g., green seeds versus yellow seeds) and occupying corresponding loci in homologous chromosomes are termed __________ .
3. A cross between two organisms differing with respect to alleles of a single locus is a(an) __________ cross.
4. An organism's genetic constitution, expressed in symbols, is called its __________ .
5. The appearance of an individual with respect to a given inherited characteristic is known as its __________ .
6. An allele that is fully expressed in the phenotype of a heterozygous individual is a(an) __________ allele.
7. A(an) __________ allele can only be expressed in the phenotype of a homozygous individual.
8. An organism with two identical alleles for a particular locus is said to be __________ for that locus; an organism with two different alleles at a particular locus is said to be __________ at that locus.
9. The offspring of the parental (P) generation are called the __________ generation.
10. The probabilities of independent events are combined by applying the __________ law.
11. The probabilities of mutually exclusive events are combined by applying the __________ law.
12. A mating of individuals that have different alleles at two loci is called a(an) __________ cross.
13. The genes in a given chromosome tend to be inherited together and are said to be __________ .
14. If a particular characteristic of an organism is governed by several pairs of genes that have similar and additive effects, we say that characteristic is under __________ control.
15. If three or more alleles for a given locus are present in the population, we say that locus has __________ alleles.
16. The mating of two closely related individuals, such as first cousins, is termed __________ .
17. The offspring of totally unrelated parents may be better adapted for survival than either parent. This phenomenon is called __________ __________ .

REVIEW QUESTIONS

1. In peas, yellow seed color is dominant to green. Predict the phenotypes (and their proportions) of the offspring of the following crosses: (a) homozygous yellow × green; (b) heterozygous yellow × green; (c) heterozygous yellow × homozygous yellow; (d) heterozygous yellow × heterozygous yellow.
2. If two animals heterozygous for a single pair of alleles are mated and have 200 offspring, about how many would be expected to have the phenotype of the dominant allele (i.e., to look like the parents)?
3. When two long-winged flies were mated, the offspring included 77 with long wings and 24 with short wings. Is the short-winged condition dominant or recessive? What are the genotypes of the parents?
4. A blue-eyed man, both of whose parents were brown-eyed, married a brown-eyed woman whose father was blue-eyed and whose mother was brown-eyed. If brown is dominant to blue, what are the genotypes of the individuals involved?
5. Outline a breeding procedure whereby a true-breeding strain of red cattle could be established from a roan bull and a white cow.
6. What is the probability of rolling a seven with a pair of dice? Which is a more likely outcome, rolling a six with a pair of dice, or rolling an eight?
7. In rabbits, spotted coat (*S*) is dominant to solid color (*s*), and black (*B*) is dominant to brown (*b*). These loci are unlinked. A brown spotted rabbit from a pure line is mated to a solid black one, also from a pure line. What are the genotypes of the parents? What would be the genotype and phenotype of an F_1 rabbit? What would be the expected genotypes and phenotypes of the F_2 generation?
8. The long hair of Persian cats is recessive to the short hair of Siamese cats, but the black coat color of Persians is dominant to the brown-and-tan coat color of Siamese. Make up appropriate symbols for the alleles of these two unlinked loci. If a pure black, long-haired Persian is mated to a pure brown-and-tan, short-haired Siamese, what will be the appearance of the F_1 offspring? If two of these F_1 cats are mated, what is the chance that a long-haired, brown-and-tan cat will be produced in the F_2 generation? (Use the short-cut probability method to obtain your answer; then check it with a Punnett square.)

9. The expression of an allele called *frizzle* in fowl causes abnormalities of the feathers. As a consequence, the animal's body temperature is lowered, adversely affecting the functions of many internal organs. When one gene affects many characteristics of the organism in this way, we say that gene is __________.
10. A walnut-combed rooster is mated to three hens. Hen A, which is walnut-combed, has offspring in the ratio of 3 walnut:1 rose. Hen B, which is pea-combed, has offspring in the ratio of 3 walnut:3 pea:1 rose:1 single. Hen C, which is walnut-combed, has only walnut-combed offspring. What are the genotypes of the rooster and the three hens?
11. What kinds of matings result in the following phenotypic ratios? (a) 3:1; (b) 1:1; (c) 9:3:3:1; (d) 1:1:1:1.
12. The weight of the fruit in a certain variety of squash is determined by two pairs of genes: *AABB* produces fruits weighing 4 pounds each, and *aabb* produces fruits weighing 2 pounds each. Each allele represented by a capital letter adds 0.5 pound to the weight. When a plant that produces 4-pound fruits is crossed with a plant that produces 2-pound fruits, all of the offspring produce fruits that weigh 3 pounds each. What would be the weights of the fruits produced by the F_2 plants if two of these F_1 plants were crossed?
13. The X-linked *barred* locus in chickens controls the pattern of the feathers, with the alleles *B* for barred pattern and *b* for no bars. If a barred male (X^BY) is mated to a nonbarred female (X^bX^b), what will be the appearance of the male and female progeny? Do you see any commercial usefulness for this? (*Hint*: It is notoriously difficult to determine the sexes of newly hatched chicks.)
14. Individuals of genotype *AaBb* were mated to individuals of genotype *aabb*. One thousand offspring were counted, with the following results: 474 *Aabb*; 480 *aaBb*; 20 *AaBb*; 26 *aabb*. What is this type of cross known as? Are these loci linked? What are the two parental classes and the two recombinant classes of offspring? What is the percentage of recombination between these two loci? How many map units apart are they?
15. Genes *A* and *B* are 6 map units apart, and *A* and *C* are 4 map units apart. Which gene is in the middle if *B* and *C* are 10 map units apart? Which is in the middle if *B* and *C* are 2 map units apart?

YOU MAKE THE CONNECTION

1. Would the development of the science of genetics in the 20th century have been any different if Gregor Mendel had never lived?
2. Sketch a series of diagrams showing each of the following, making sure to end each series with haploid gametes: (a) how a pair of alleles for a single locus segregates in meiosis; (b) how the alleles of two unlinked loci assort independently in meiosis; (c) how the alleles of two linked loci undergo genetic recombination.
3. Can you always ascertain an organism's genotype for a particular locus if you know its phenotype? Conversely, if you are given an organism's genotype for a locus, can you always reliably predict its phenotype? Explain.

RECOMMENDED READINGS

Corcos, A., and F. Monaghan. *Mendel's Experiments on Plant Hybrids: A Guided Study.* Rutgers University Press, New Brunswick, 1993. This interpretive study of Mendel's paper includes information on his life as a monk and a scientist.

Mendel, G. "Experiments in Plant Hybridization." Reprinted in *Genetics: Readings from Scientific American.* W. H. Freeman and Co., San Francisco, 1990. Try reading this translation of Mendel's classic paper from the perspective of other scientists of his time who lacked knowledge of chromosomes, mitosis, and meiosis.

There are a number of well-written college level genetics texts that cover the principles of genetics in eukaryotes. The following are two representative examples:

Griffiths, A.J.F, J.H. Miller, D.T. Suzuki, R.C. Lewontin, and W.M. Gelbart. *An Introduction to Genetic Analysis,* 5th ed. W. H. Freeman, New York, 1993.

Russell, P. *Genetics,* 4th ed. Harper Collins, New York, 1995.

C H A P T E R 11

DNA: The Carrier of Genetic Information

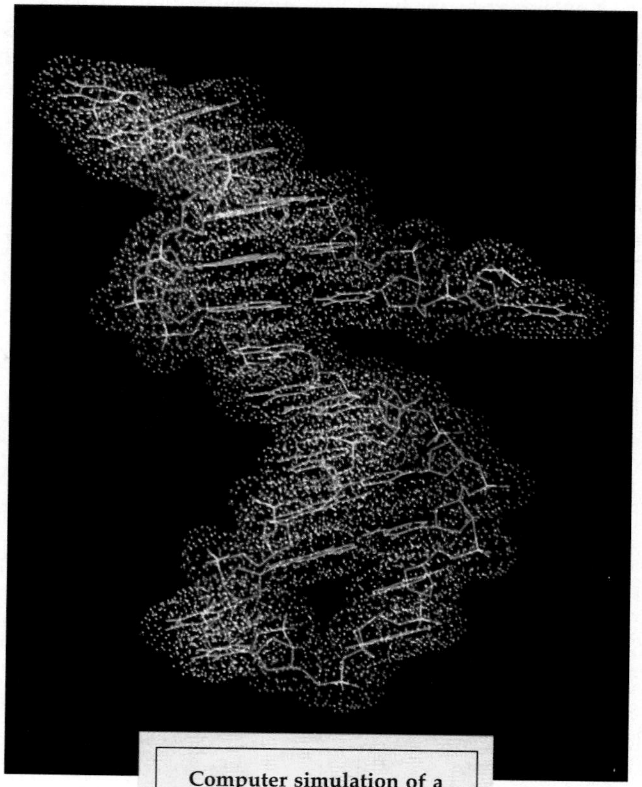

Computer simulation of a piece of synthetic DNA used in biophysical studies of the different structural forms of DNA.
(Dr. Kenneth Breslauer, Rutgers University)

Following the rediscovery of Mendel's principles, geneticists conducted elegant experiments to learn how genes are arranged in chromosomes and how they are transmitted from generation to generation. However, two very basic questions remained unanswered: What are genes made of? How do genes work? Although studies of inheritance patterns did not answer these questions, they contributed to an emerging set of predictions about the molecular characteristics of genes and how they might function.

Among the properties attributed to genes was their ability to store information in a stable form that can be accurately copied and passed from generation to generation. However, it was thought that the genetic material could not always be perfectly constant, because genetic changes, called **mutations,** had been observed to suddenly appear and to subsequently be transmitted to future generations. Therefore it was thought that the molecule responsible for inheritance would have to be relatively stable, but have the capacity to change under some circumstances.

It was also recognized that a mechanism was required to retrieve the information stored in the genetic material and to use it to direct cellular functions. Because proteins are so important to every aspect of cellular structure and metabolism, they were considered the prime candidates for the genetic material. Today we know that **deoxyribonucleic acid (DNA),** not protein, is the molecule responsible for inheritance. In this chapter we examine the unique features of DNA that allow it to carry out this role.

LEARNING OBJECTIVES

After you have studied this chapter you should be able to

1. Summarize the evidence that accumulated during the 1940s and early 1950s demonstrating that DNA is the fundamental genetic material.
2. Relate the chemical and physical features of DNA to the structure proposed by Watson and Crick.
3. Sketch how nucleotide subunits are linked together to form a single DNA strand.
4. Illustrate how the two strands of DNA are oriented with respect to each other.
5. State the base-pairing rules for DNA, and describe how complementary bases bind to each other.
6. Cite experimental evidence that allowed scientists to differentiate between semiconservative replication of DNA and alternative models (conservative and dispersive replication).
7. Summarize how DNA replicates, and identify some of the unique features of the process.
8. Explain why DNA replication is discontinuous in one strand and continuous in the other, and why the process is bidirectional.
9. Compare the organization of DNA in prokaryotic and eukaryotic cells.

MOST GENES CARRY INFORMATION FOR MAKING PROTEINS

The idea that genes and enzymes (which we now know are proteins) are related in some way was first clearly stated in 1908 by an English physician, Archibald Garrod, who proposed that certain inherited human diseases are caused by a block in a sequence of chemical reactions within the body.

In his book, *Inborn Errors of Metabolism,* Garrod discussed a genetic disease called **alkaptonuria,** which is inherited as a simple Mendelian trait. The condition involves a block in the metabolic reactions that break down the amino acids phenylalanine and tyrosine; the urine of affected persons turns black when exposed to air (Fig. 11–1). Homogentisic acid, the substance that turns black, is an intermediate in this breakdown pathway; it is normally oxidized and eventually converted to carbon dioxide and water.

Garrod hypothesized that persons with alkaptonuria lack the oxidation enzyme, causing homogentisic acid to accumulate in their tissues and blood, and to be excreted in their urine. Before the second edition of his book had been published in 1923, it was found that affected persons do indeed lack the enzyme that oxidizes homogentisic acid. Garrod was right: a mutation in a specific gene could be associated with the absence of a specific enzyme. Despite the implications of this finding, little work was done in this area, primarily because errors in metabolism appeared to occur only in patients with rare diseases, which made genetic experiments impossible.

A major advance in understanding the relationship between genes and enzymes came in the early 1940s, when George Beadle and Edward Tatum developed a new approach to the problem. Most efforts until that time had centered on studying known genes and attempting to determine what biochemical reactions they affected. Experimenters examined previously identified genes, such as those controlling eye color in *Drosophila* or pigments in plants. They found that such traits are controlled

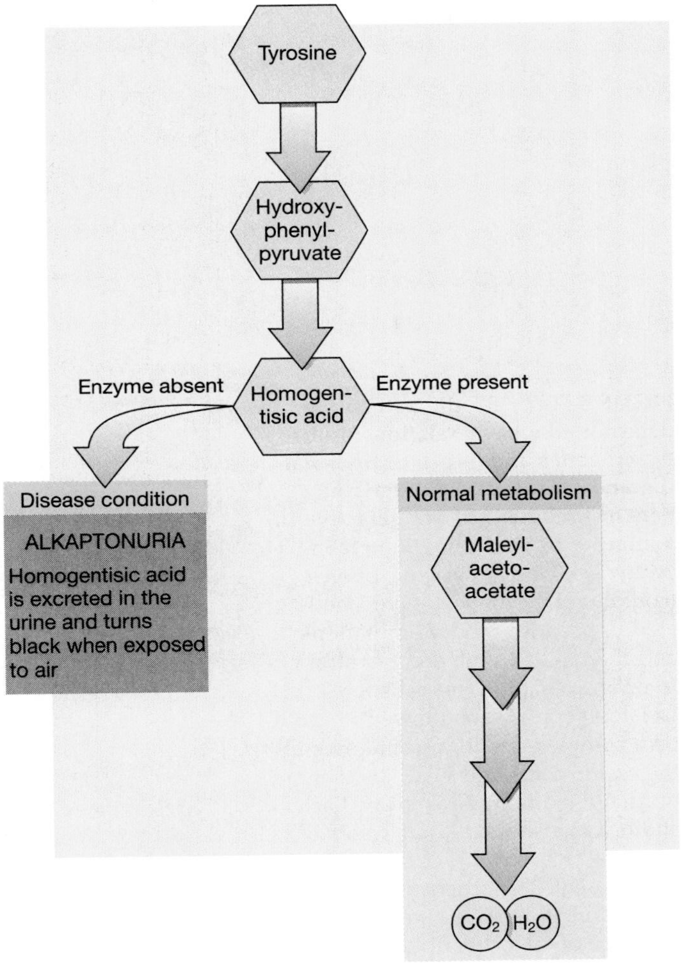

FIGURE 11–1 Garrod proposed that the alkaptonuria gene causes the absence of a specific enzyme, or an "inborn error of metabolism." The mutation that causes alkaptonuria produces a defect in an enzyme that is part of the pathway by which the amino acid tyrosine is catabolized. That enzyme normally converts homogentisic acid to maleylacetoacetate. Homogentisic acid thus accumulates in the blood and is excreted through the urine. When the homogentisic acid in the urine comes in contact with air, it oxidizes and turns black.

by a series of biosynthetic reactions, but it was not clear to the investigators whether the genes themselves were acting as enzymes or if they determined the specificities of enzymes in more complex ways.

Beadle and Tatum decided to take the opposite approach. Rather than try to identify the enzymes affected by single genes, they decided to look for mutations interfering with known metabolic reactions that produce essential molecules such as amino acids and vitamins. They studied the bread mold *Neurospora* for several reasons. First, wild-type[1] *Neurospora* can make all of its essential biological molecules when it is grown on a simple minimal medium containing sugar, salts, and the vitamin biotin. A mutant that cannot make a substance such as an amino acid can still grow if that substance is simply added to the minimal medium.

Second, *Neurospora* grows primarily as a haploid organism. Thus, a mutation that occurs in a gene can be immediately identified because it is not masked by a normal allele on a homologous chromosome. Third, *Neurospora* produces haploid spores, known as *conidia.* These can grow and divide mitotically to produce more haploid cells. Two haploid cells can fuse and undergo a brief sexual phase of the life cycle. (For an illustration of the generalized life cycles of simple organisms, see the figure in

[1]Wild-type is a term commonly applied to nonmutant strains or individuals.

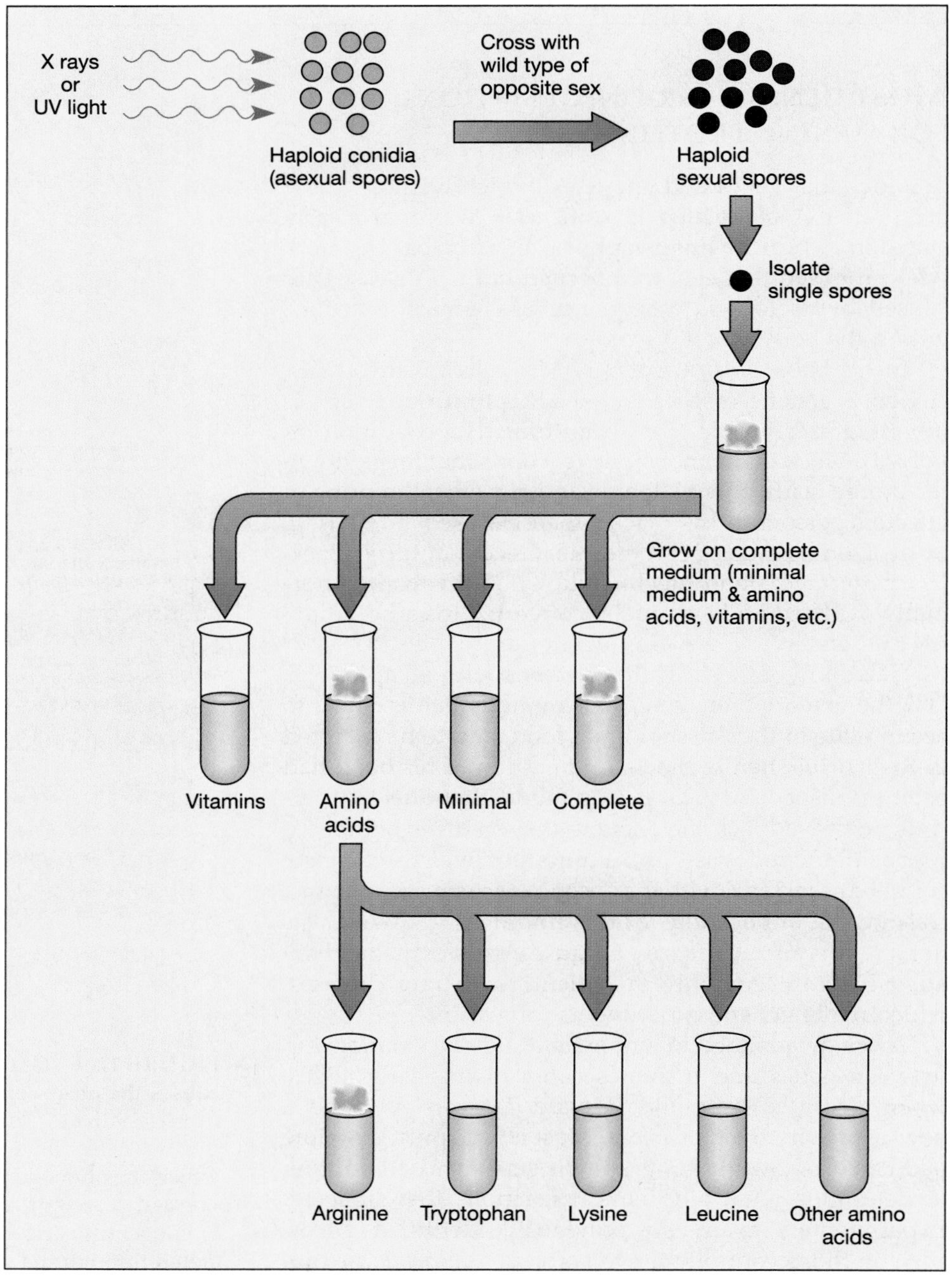

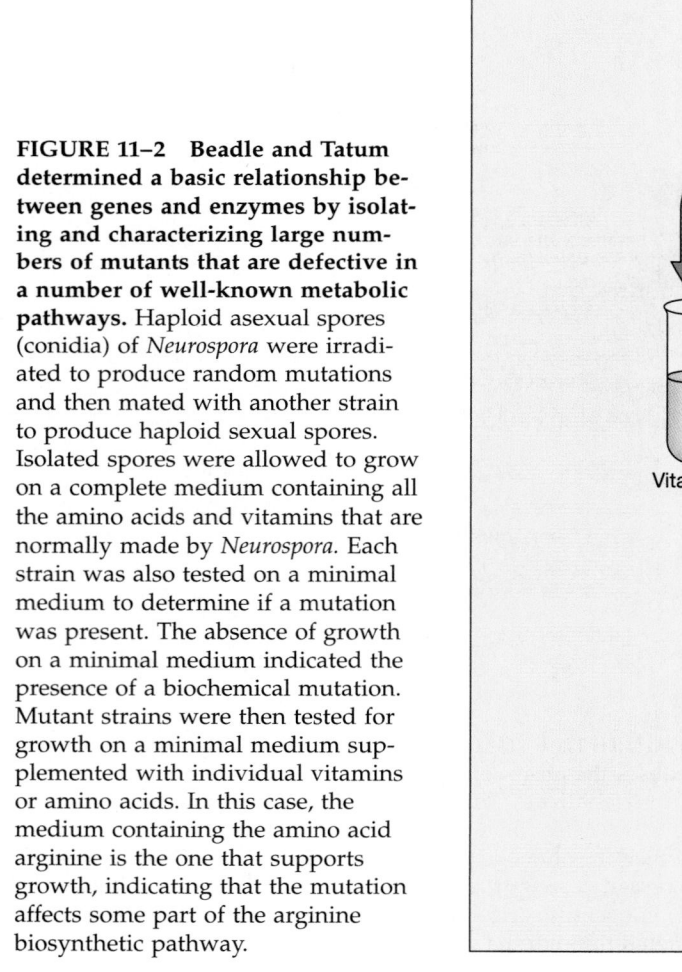

FIGURE 11–2 Beadle and Tatum determined a basic relationship between genes and enzymes by isolating and characterizing large numbers of mutants that are defective in a number of well-known metabolic pathways. Haploid asexual spores (conidia) of *Neurospora* were irradiated to produce random mutations and then mated with another strain to produce haploid sexual spores. Isolated spores were allowed to grow on a complete medium containing all the amino acids and vitamins that are normally made by *Neurospora.* Each strain was also tested on a minimal medium to determine if a mutation was present. The absence of growth on a minimal medium indicated the presence of a biochemical mutation. Mutant strains were then tested for growth on a minimal medium supplemented with individual vitamins or amino acids. In this case, the medium containing the amino acid arginine is the one that supports growth, indicating that the mutation affects some part of the arginine biosynthetic pathway.

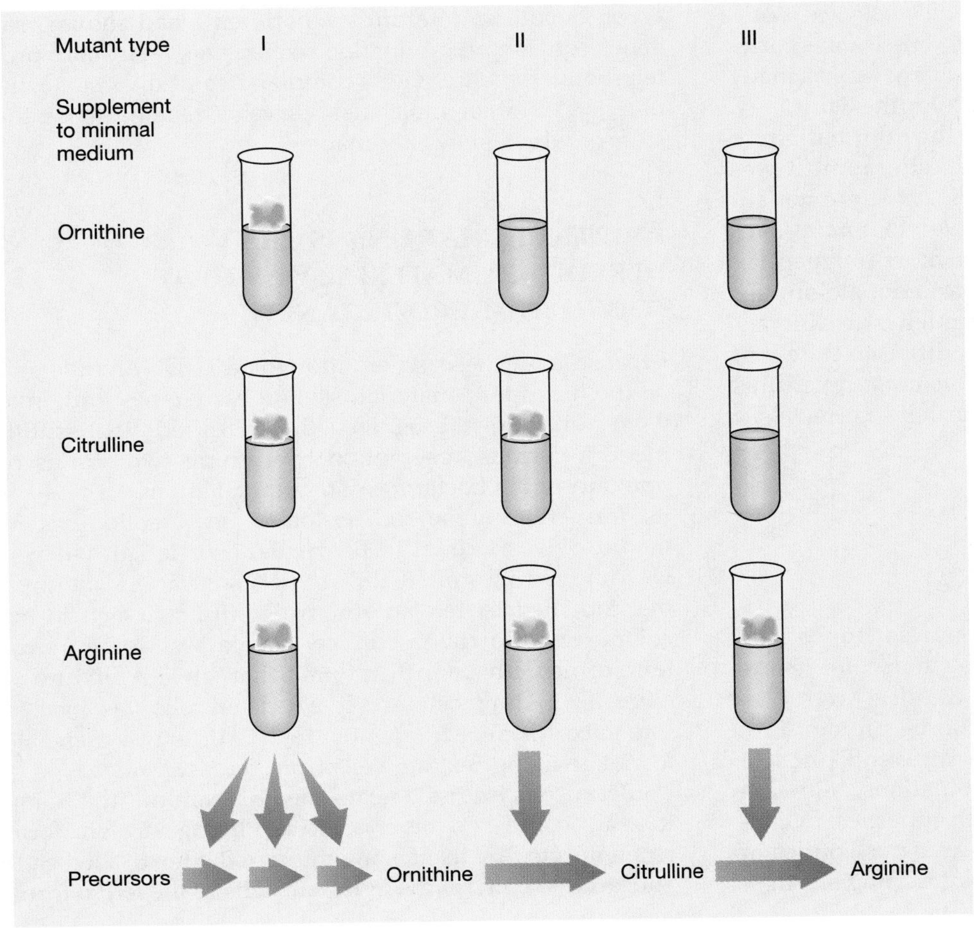

FIGURE 11–3 By analyzing different mutant strains that require the same amino acid or vitamin, Beadle and Tatum were able to verify that each mutant gene affects only a single enzymatic step in the metabolic pathway. In this example, a number of different mutant strains that require the amino acid arginine were tested to determine which step in the pathway was blocked. Arginine-requiring strains I, II, and III were all found by genetic crossing methods to have mutations in different genes. Mutant type I grows on ornithine, citrulline, and arginine, and must therefore be blocked prior to the formation of all three compounds. Mutant type II is unable to grow on ornithine, but allows the conversion of citrulline to arginine. Mutant type III can grow only on arginine, and must therefore be blocked at a point after the synthesis of both ornithine and citrulline.

Making the Connection: Mitosis, Meiosis, and Life Cycles, in Chapter 9.) The resulting zygote undergoes meiosis to form haploid sexual spores. Thus, researchers can use sexual crosses to perform genetic analyses of isolated mutants.

Beadle and Tatum searched for large numbers of mutants that were unable to synthesize biologically important molecules (e.g., amino acids, purines, pyrimidines, vitamins). They first exposed the haploid conidia to x rays or ultraviolet radiation, both of which were known to produce mutations. Following exposure, the conidia were grown on a *complete medium,* one containing all essential molecules, so that many different types of mutants produced by the irradiation could survive and reproduce. If an isolated mutant grew on the complete medium but failed to grow after transfer to the minimal medium, Beadle and Tatum reasoned that it was unable to produce one of the compounds essential for growth. Further testing of the mutant on media containing different combinations of amino acids, purines, vitamins, and so on enabled the investigators to determine the exact essential compound (Fig. 11–2). Each mutant strain isolated in that way was then verified by genetic crossing experiments to have a mutation in only one gene.

Their findings can be illustrated with a class of mutants that require the amino acid arginine. Two compounds, ornithine and citrulline, are known to be precursors in the biosynthetic pathway that leads to arginine. The order of these intermediates in the pathway is shown below. Beadle and Tatum found that some of the arginine-requiring mutants could grow on ornithine, citrulline, or arginine; others could grow only on citrulline or arginine; and still others could grow only on arginine.

$$\text{X} \xrightarrow[\text{Enzyme A}]{} \text{ornithine} \xrightarrow[\text{Enzyme B}]{} \text{citrulline} \xrightarrow[\text{Enzyme C}]{} \text{arginine}$$

A mutation in which enzyme A is inactive would belong to the first group because enzymes B and C would still be able to convert ornithine or citrulline to arginine. Any mutant lacking enzyme B would be placed in the second group, because in these mutants citrulline could be converted to arginine but ornithine could not. Neither ornithine nor citrulline would be able to support growth in a mutant lacking enzyme C, because both of these precursors are produced before the blocked step (Fig. 11–3).

Using this approach, Beadle and Tatum analyzed large numbers of mutants affecting a number of metabolic pathways. They found that *for each individual gene identified, only one enzyme was affected.* This one-to-one correspondence between genes and enzymes was succinctly stated as the **one gene, one enzyme hypothesis.**

The idea that a gene might encode the information for a single enzyme held for almost a decade, until it was found that many genes encode proteins that are not enzymes. In addition, some studies showed that many proteins may be constructed from two or more polypeptide chains (e.g., the α and β subunits of hemoglobin; see Chapter 3), each of which may be encoded by a different gene. The definition was therefore modified to state that *one gene is responsible for one polypeptide chain.* Even this definition has proved to be only partially correct (see Chapter 13).

GENES ARE MADE OF DEOXYRIBONUCLEIC ACID (DNA)

Although we now know that genes are made of deoxyribonucleic acid, or DNA, when the molecular nature of the gene was first studied most scientists were convinced that the genetic material had to be protein. Proteins are made up of more than 20 different kinds of amino acids in many different combinations, allowing each type of molecule to have unique properties; in contrast, nucleic acids are made of only four nucleotides in what appeared to be a regular and uninteresting arrangement. Beadle and Tatum's experiments had shown that genes control the production of enzymes, which are proteins. Given their obvious complexity and diversity compared with other molecules, proteins seemed to be the stuff of which genes are made.

EVIDENCE THAT DNA IS THE HEREDITARY MATERIAL WAS FIRST FOUND IN MICROORGANISMS

Because many researchers thought that DNA could not be the hereditary material, several early clues to its role were not widely recognized. In 1928 Frederick Griffith made a curious observation concerning two strains of pneumococcus bacteria. A smooth (S) strain, named for its formation of smooth colonies on a solid growth medium, was known to be **virulent,** or lethal; when it was injected into mice, the animals contracted pneumonia and died. A related rough (R) strain, which forms colonies with a rough surface, was known to be **avirulent,** or nonlethal. Griffith found that when a mixture of *heat-killed* S-strain cells and live R-strain cells was injected into mice, the mice frequently died. Griffith was then able to isolate living S-strain cells from the dead mice.

Because neither the heat-killed S strain nor the living R strain could be converted to the living virulent form when injected by itself, something in the heat-killed cells appeared to convert the avirulent cells to the lethal form.

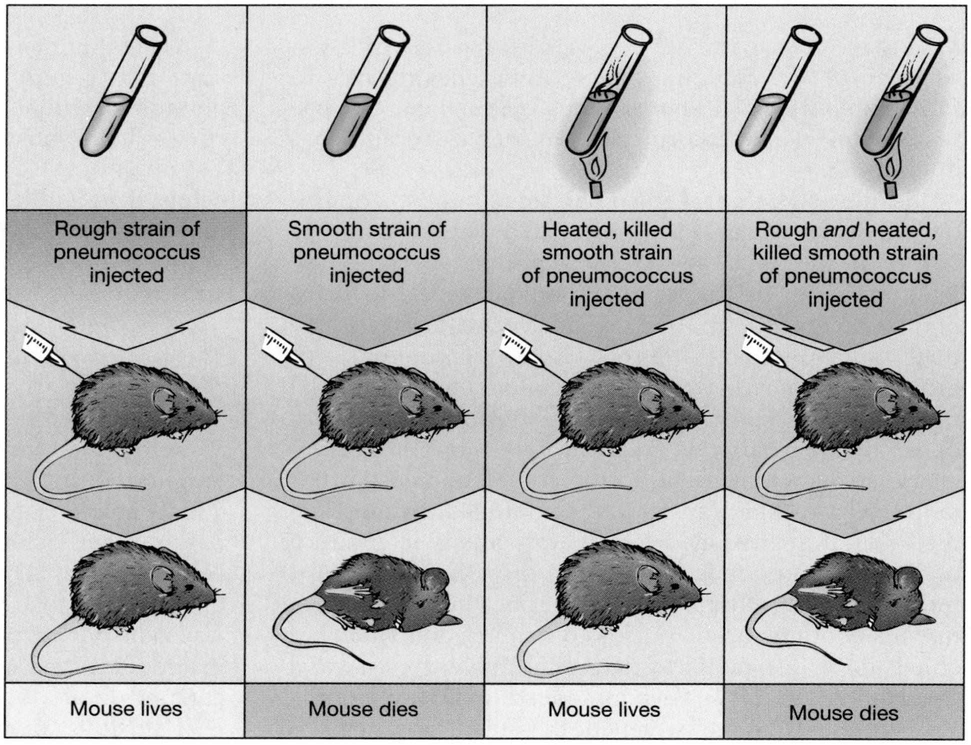

FIGURE 11–4 Frederick Griffith demonstrated the transfer of genetic information from dead, heat-killed pneumococcus bacteria to living pneumococci of a different strain. Although neither the rough (R) strain nor the heat-killed smooth (S) strain could kill a mouse, a combination of the two did. Autopsy of the dead mouse showed the presence of living, S-strain pneumococci. These results indicated that some substance in the heat-killed S-strain was responsible for the transformation of the R-strain into virulence. Avery and co-workers later demonstrated that purified DNA that is isolated from the S-strain confers virulence on the R-strain bacteria, establishing that the DNA carries the necessary information for bacterial transformation.

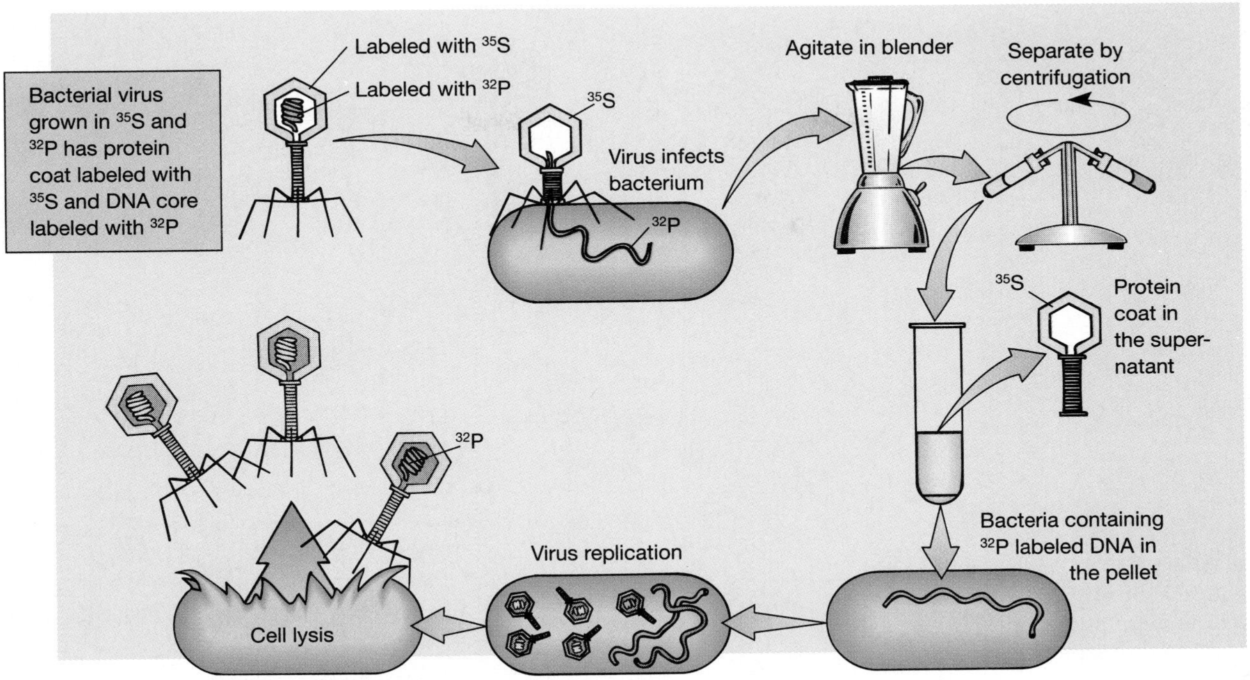

FIGURE 11–5 The Hershey-Chase experiment demonstrated that only the DNA of a bacterial virus is necessary for the reproduction of new viruses. The virus consists of a DNA core, which contains phosphorus but no sulfur atoms, surrounded by a protein coat, which contains sulfur but no phosphorus atoms. By growing the virus on a medium containing the radioactive isotopes ^{32}P and ^{35}S, Hershey and Chase were able to label the protein and DNA specifically with different isotopes. They found that only the ^{32}P-labeled DNA entered the bacterium, and that the ^{35}S-labeled viral protein could be separated from the cells after infection without interfering with the replication of the virus. All the genetic information needed for the synthesis of new protein coats and new viral DNA was provided by the parental viral DNA.

This phenomenon, called **transformation,** was thought to be caused by some chemical substance (called the "transforming principle") in the dead bacteria that "transformed" a related strain to a genetically stable new form (Fig. 11–4). Later, in 1944, Avery, MacLeod, and McCarty of the Rockefeller Institute chemically identified the transforming principle as DNA.

During the next few years, new evidence accumulated that the haploid nuclei of pollen grains and gametes such as sperm contain only half the amount of DNA found in diploid somatic cells of the same species. Because the idea that genes are on chromosomes was generally accepted, these findings correlating DNA content with chromosome number provided strong circumstantial evidence of DNA's importance in inheritance.

In 1952 Alfred Hershey and Martha Chase performed a series of elegant experiments demonstrating that **bacteriophages** (viruses that infect bacteria) do so by injecting their DNA into the bacterial cells and leaving most of their protein on the outside (Fig. 11–5). This finding emphasized the importance of DNA in the reproduction of the virus and was seen by many to be another important indication of the role of DNA as the hereditary material.

THE STRUCTURE OF DNA ALLOWS IT TO CARRY INFORMATION AND TO SELF-REPLICATE

DNA was not widely accepted as the genetic material until James Watson and Francis Crick proposed a model for its structure that had extraordinary explanatory power. The story of how the structure of DNA came to be determined is one of the most remarkable chapters in the history of modern biology.

As we shall see, a great deal was known about the physical and chemical properties of DNA when Watson and Crick became interested in the problem. Their all-important contribution was to integrate all of this information into a model that demonstrated how the molecule can both carry information and serve as its own template (pattern) for self-duplication.

Nucleotides can be covalently linked in any order to form long polymers

As discussed in Chapter 3, each DNA building block is a **nucleotide** consisting of a pentose sugar **(deoxyribose),**

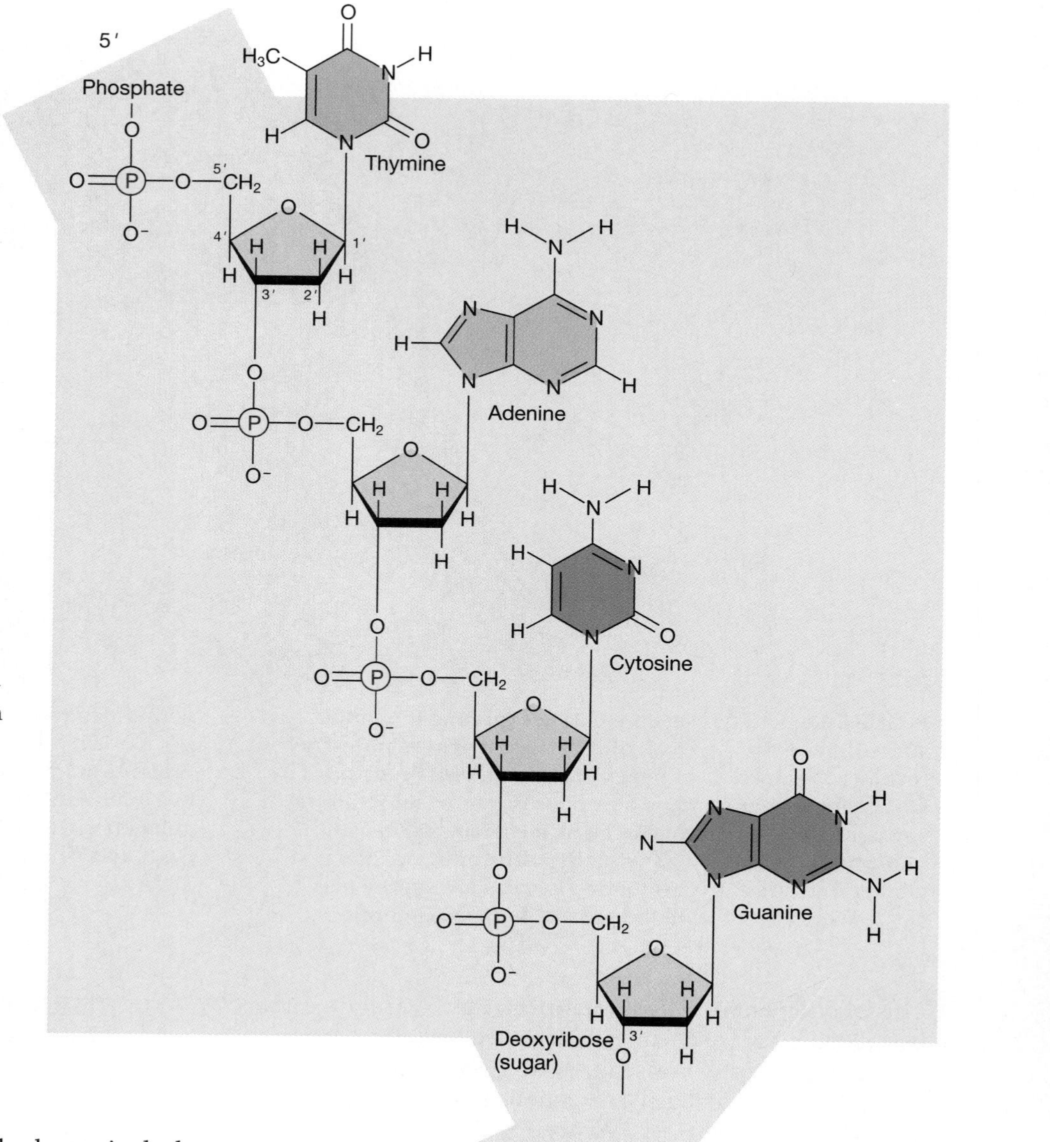

FIGURE 11–6 A DNA molecule is made of nucleotide subunits that are deoxyribonucleoside monophosphates. Each subunit is composed of a phosphate group linked to the sugar deoxyribose at its 5′ carbon atom. Deoxyribose sugars are shown in tan. Linked to the 1′ carbon of the sugar is one of four nitrogenous bases. The purine bases, adenine and guanine, have two-ring structures; the pyrimidine bases, thymine and cytosine, have one-ring structures. Phosphodiester linkages connect the 5′ and 3′ carbon atoms of adjacent deoxyribose sugars. The schematic drawing illustrates the polarity of the polynucleotide chain, with the 5′ end at the top of the figure and the 3′ end at the bottom.

a phosphate, and a nitrogenous base. The bases include the **purines—adenine (A)** and **guanine (G),**—and the **pyrimidines—thymine (T)** and **cytosine (C).** As shown in Figure 11–6, the nucleotides are linked together by covalent bonds to form an alternating sugar-phosphate backbone.

The structure of the nucleotide subunits in DNA is identical to that of AMP (see Chapter 3), except that the sugar has a hydrogen atom instead of a hydroxyl group on the 2′ carbon atom (hence the name *deoxyribose*).[2] The nitrogenous base is attached to the 1′ carbon of the sugar, and the phosphate is attached to the 5′ carbon. As shown in Figure 11–7, the nucleotides are linked by covalently joining the 3′ carbon of one sugar to the 5′ phosphate of the adjacent sugar to form a 3′, 5′ **phosphodiester linkage.** It is therefore possible to form a polymer of indefinite length. We now know that most DNA molecules found in cells are millions of bases long and that the nucleotides can be linked together in any order. Figure 11–6 illustrates that a single polynucleotide chain is directional. No matter how long the chain may be, one end (the **5′ end**) has a 5′ carbon and the other (the **3′ end**) has a 3′ carbon that is not linked to another nucleotide.

[2]It is conventional to number the carbon atoms in a molecule, using a system devised by organic chemists. In nucleic acid chemistry the "prime" designations, such as 2′, designate individual carbon atoms in the sugar, to distinguish them from those in the base.

DNA is made of two polynucleotide chains intertwined to form a double helix

Important information about the structure of DNA came from **x-ray diffraction** studies on crystals of purified

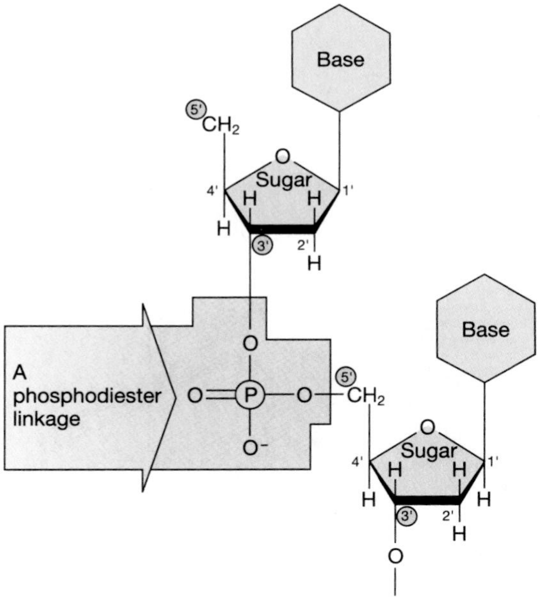

FIGURE 11–7 A phosphodiester linkage joins two deoxyribose sugars in the DNA backbone.

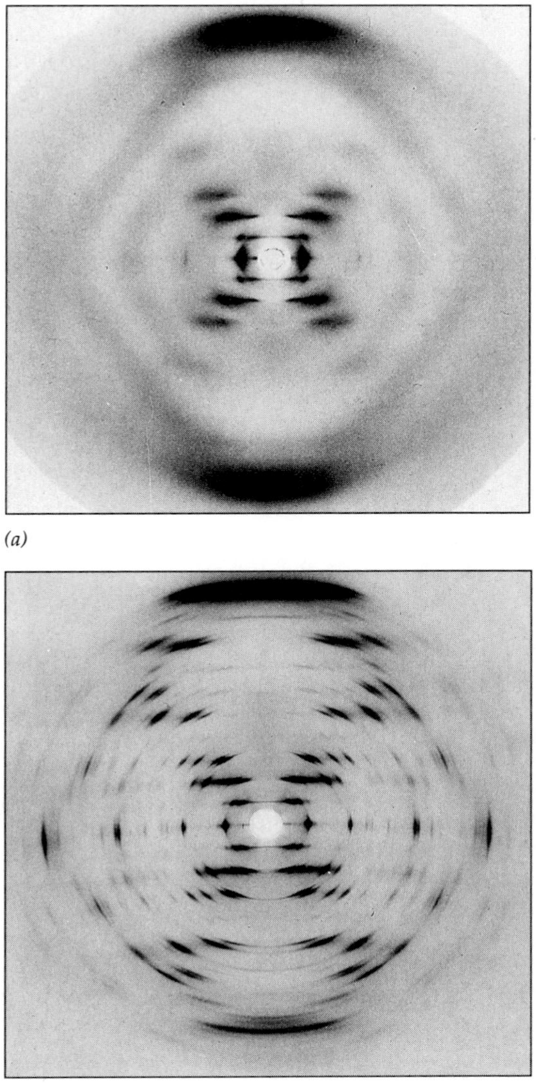

(a)

(b)

FIGURE 11–8 Analysis of x-ray diffraction photographs of suitably hydrated fibers of DNA provided important clues about the structure of DNA. (*a*) Pattern obtained using the sodium salt of DNA. (*b*) Pattern obtained using the lithium salt of DNA. This pattern permits a thorough analysis of DNA. The diagonal pattern of spots (reflections) stretching from 11 o'clock to 5 o'clock and from 1 o'clock to 7 o'clock provides evidence for the helical structure of DNA. The elongated horizontal reflections at the tops and bottoms of the photographs provide evidence that the purine and pyrimidine bases are stacked 0.34 nanometers apart and are perpendicular to the axis of the DNA molecule. (Dr. S. D. Dover, Division of Biomolecular Sciences, Kings College, London)

DNA carried out by Rosalind Franklin in the laboratory of M. H. F. Wilkins. X-ray diffraction is a powerful method for determining distances between atoms of molecules arranged in a regular, repeating crystalline structure (Fig. 11–8). X rays have such extremely short wavelengths that they can be scattered by the electrons surrounding the atoms in a molecule. Atoms with dense electron clouds (e.g., phosphorus, oxygen) tend to deflect electrons more strongly than do atoms with lower atomic numbers.

When a crystal is exposed to an intense beam of x rays, the regular arrangement of the atoms in the crystal causes the x rays to be diffracted, or bent, in specific ways. The pattern of diffracted x rays is seen on film as dark spots. Mathematical analysis of the arrangement and distances between the spots can then be used to determine precise distances between atoms and their orientation within the molecules.

Franklin had already produced clear x-ray crystallographic films of DNA patterns when Watson and Crick began to pursue the problem of DNA structure. The pictures clearly showed that DNA has a type of helical structure and three major types of regular, repeating patterns in the molecule with the dimensions 0.34 nanometer, 3.4 nanometers, and 2.0 nanometers. Franklin and Wilkins had inferred from these patterns that the nucleotide bases (which are flat molecules) are stacked like rungs of a ladder. Using this information, Watson and Crick began to build scale models of the DNA components and then fit them together to agree with the experimental data.

After a number of trials, the two worked out a model that fit the existing data (Fig. 11–9). The nucleotide chains conformed to the dimensions of the x-ray data only if each DNA molecule consisted of *two* polynucleotide chains arranged in a coiled **double helix.** In their model, the sugar-phosphate backbones of the two chains form the outside of the helix. The bases belonging to the two chains

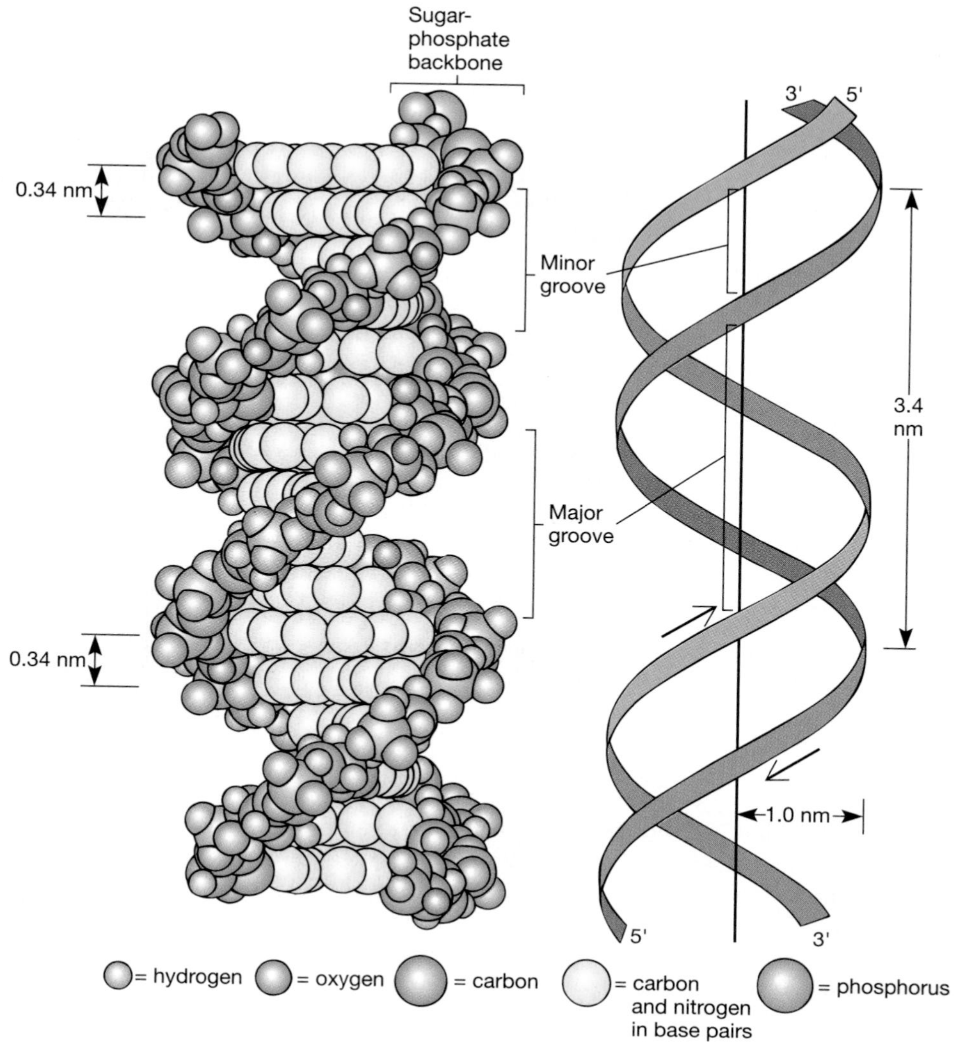

FIGURE 11–9 The structure of DNA is highly regular. On the left is a space-filling model of the DNA double helix. On the right is a diagrammatic model of the DNA double helix with certain dimensions shown in nanometers (nm). The ribbons represent the sugar-phosphate backbone of each strand; the arrows indicate that the two strands extend in opposite directions.

associate as pairs in the center. The reasons for the 0.34-nanometer and 3.4-nanometer periodicities are readily apparent from the model: each pair of bases is exactly 0.34 nanometer from the adjacent pairs above and below. Because exactly ten base pairs are present in each full turn of the helix, each turn is 3.4 nanometers high. To fit the data, the two chains must run in opposite directions; therefore, each end of the double helix must have an exposed 5′ phosphate on one strand and an exposed 3′ hydroxyl group on the other. Because the two strands run in opposite directions, they are said to be **antiparallel** to each other.

In double-stranded DNA, hydrogen bonds form between adenine and thymine and between guanine and cytosine

Other features of the model required additional integration of chemical and x-ray diffraction data. By 1950, the base composition of DNA from a number of organisms and tissues had been determined by Erwin Chargaff and his coworkers at Columbia University. They found a simple relationship among the bases that turned out to be an important clue to the structure of DNA. Regardless of the source of the DNA, in Chargaff's words, the "ratios of purines to pyrimidines and also of adenine to thymine and of guanine to cytosine were not far from 1." In other words, in DNA molecules, A = T and G = C.

The x-ray diffraction studies indicated that the double helix has a precise and constant width, as shown by the 2.0-nanometer reflections. This finding is actually connected to Chargaff's rules. Notice in Figure 11–6 that the pyrimidines (cytosine and thymine) contain only one ring of atoms and are smaller than the purines (guanine and adenine), which contain an additional five-membered ring. Study of the models made it clear to Watson and Crick that if each cross-rung of the ladder were to contain one purine and one pyrimidine, the width of the helix at that point would be exactly 2.0 nanometers; the combination of two purines (each of which is 1.2 nanometers wide) would be wider, and that of two pyrimidines would be narrower.

Further examination of the model showed that adenine can pair with thymine (and guanine with cytosine) in such a way that hydrogen bonds form between them; the opposite combinations, cytosine with adenine and guanine with thymine, do *not* lead to favorable hydrogen-bonding.

The nature of the hydrogen-bonding between adenine and thymine and between guanine and cytosine is shown in Figure 11–10. Two hydrogen bonds can form between adenine and thymine, and three between guanine and cytosine. This concept of *specific base-pairing* neatly explains Chargaff's rules. The amount of cytosine has to equal the amount of guanine, because every cytosine in one chain must have a paired guanine in the other chain. Similarly, every adenine in the first chain must have a thymine in the second chain. Thus, the sequences of bases in the two chains are **complementary,** but not identical, to each other; in other words, *the sequence of nucleotides in one chain*

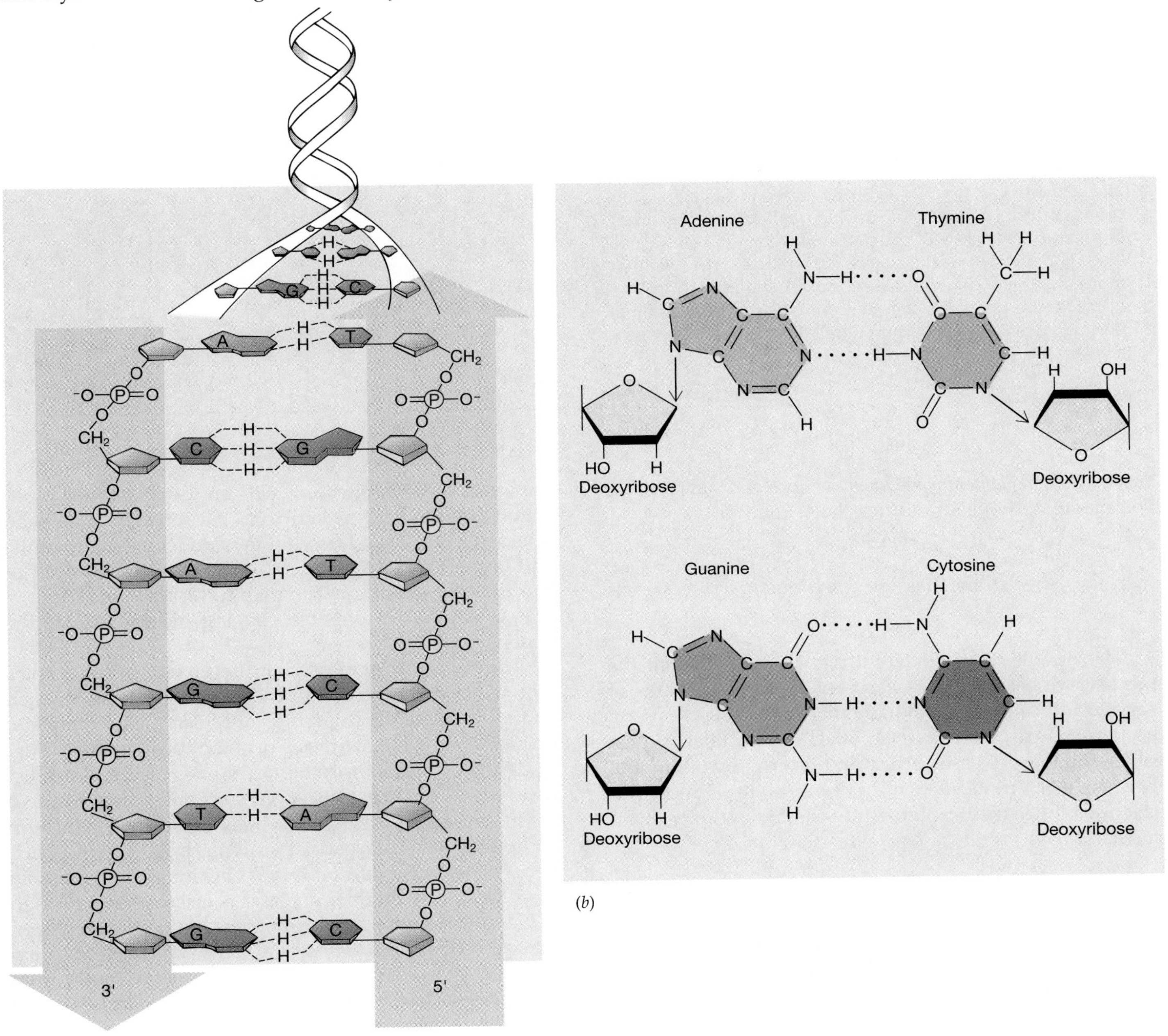

(*a*) (*b*)

FIGURE 11–10 The two strands of a DNA double helix are associated by hydrogen bonding between the bases. (*a*) The two sugar-phosphate chains run in opposite directions. This orientation permits the complementary bases to pair. (*b*) Diagram of the hydrogen bonding between base pairs adenine (A) and thymine (T) (*top*) and guanine (G) and cytosine (C) (*bottom*). The AT pair has two hydrogen bonds; the GC pair has three.

Making the Connection

Mutations and the Structure of DNA

How do new genetic variants arise? This question was of great interest to geneticists, who had long known that mutations, or genetic changes, could arise in genes and then be transmitted faithfully to succeeding generations. According to the double-helix model, mutations could represent a *change in the sequence of bases in the DNA.* If DNA is copied by a mechanism involving complementary base-pairing, any change in the sequence of bases on one strand would result in a new sequence of complementary bases during the next replication cycle. The new base sequence would then be passed on to daughter molecules by the same mechanism used to copy the original genetic material, as if no change had occurred.

In the example shown in the figure, an adenine base in one of the DNA strands has been changed to guanine (this could occur by a rare error in DNA replication or by one of several other known mechanisms). When the DNA molecule is replicated again, one of the strands gives rise to a molecule exactly like the parent strand; the other (mutated) strand gives rise to a molecule with a new combination of bases that will be perpetuated generation after generation. ▲

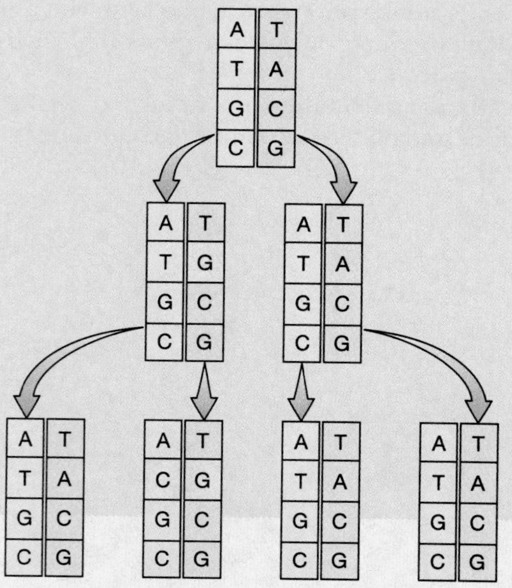

A mutation can be stabilized by DNA replication.

dictates the complementary sequence of nucleotides in the other. For example, if one strand has the sequence:

3′—A G C T A C—5′

then the other strand has the complementary sequence:

5′—T C G A T G—3′

The double-helix model strongly suggested that the sequence of bases in DNA can provide for the storage of genetic information. Although there are restrictions on how the bases pair with each other, the number of possible sequences of bases in a strand is virtually unlimited. Because DNA molecules in a cell can be millions of nucleotides long, they can store enormous amounts of information.

DNA REPLICATION IS SEMICONSERVATIVE: EACH DOUBLE HELIX CONTAINS AN "OLD" STRAND AND A NEWLY SYNTHESIZED STRAND

Two immediately apparent and distinctive features of the Watson-Crick model made it seem more likely that DNA is the genetic material. We have already mentioned that DNA can carry coded information in its sequence of bases. The model also suggested a way in which information in DNA could be precisely copied—a process known as **DNA replication.** The importance of the replication mechanism was known to Watson and Crick, who noted in a classic and now famous understatement at the end of their first brief paper, "It has not escaped our notice that the specific pairing we have postulated immediately suggests a possible copying mechanism for the genetic material."

The model suggested that, because the nucleotides pair with each other in a complementary fashion, each strand of the DNA molecule could serve as a template, or pattern, for the synthesis of the opposite strand (Fig. 11–10). It would simply be necessary for the hydrogen bonds between the two strands to break and the two chains to separate. Each half-helix could then pair with complementary nucleotides to replace its missing partner. The result would be two DNA double helices, each identical to the original one and consisting of one original strand from the parent molecule and one newly synthesized complementary strand. This type of information copying is known as a **semiconservative replication** mechanism. The recognition that DNA could be copied in this way suggested how DNA could provide a third essential characteristic of genetic material—the ability to mutate (see *Making the Connection: Mutations and the Structure of DNA*).

Although the semiconservative replication mechanism suggested by Watson and Crick was (and is) a simple and compelling model, experimental proof was needed to establish that DNA in fact duplicates in that manner. First it was necessary to rule out several other possibilities. For example, with a *conservative* replication mechanism, both parent (or old) strands would remain together, and the two newly synthesized strands would form a second double helix. With a *dispersive* mechanism, after replication the DNA would contain random regions of both parental and newly synthesized strands. To discriminate between the semiconservative replication mechanism and other possibilities, it was necessary to distinguish between old and newly synthesized strands of DNA.

One way to accomplish this is to use a heavy-nitrogen isotope, nitrogen-15 (ordinary nitrogen is nitrogen-14), to label DNA strands. Large molecules such as DNA can be separated on the basis of differences in their density, using a technique known as **density gradient centrifugation.** When DNA is mixed with a solution containing cesium chloride (CsCl) and centrifuged at high speed, the solution forms a density gradient in the centrifuge tube, ranging from a low density at the top to the highest density at the bottom. During the centrifugation the DNA molecules migrate to the region of the gradient identical to their own density.

In 1957, Matthew Meselson and Franklin Stahl grew cells of the bacterium *Escherichia coli* on a medium that contained nitrogen-15 in the form of ammonium chloride (NH_4Cl). The cells used the nitrogen-15 to synthesize bases, which were then incorporated into DNA (Fig. 11–11). The resulting heavy nitrogen-containing DNA

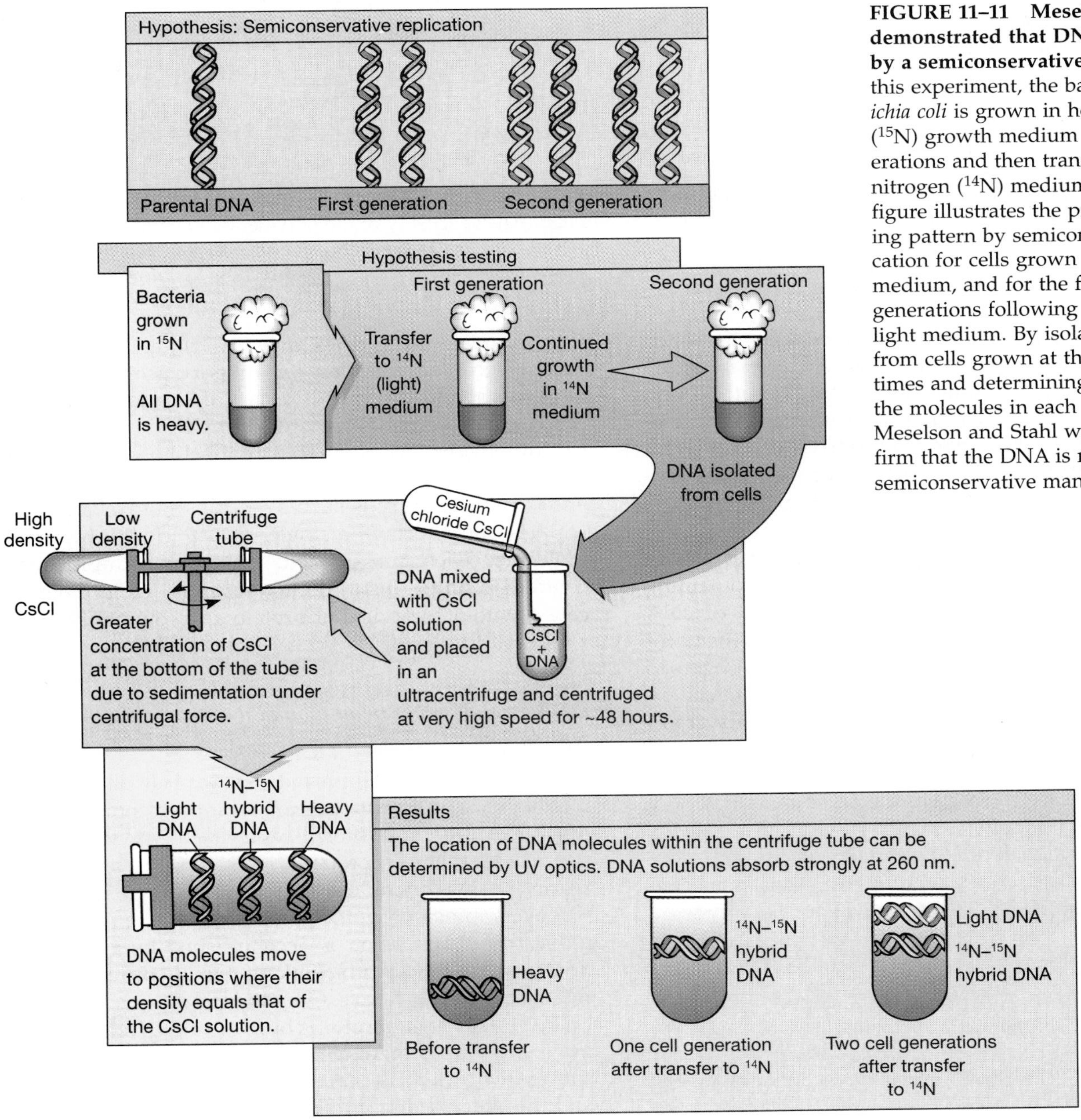

FIGURE 11–11 Meselson and Stahl demonstrated that DNA is replicated by a semiconservative mechanism. In this experiment, the bacterium *Escherichia coli* is grown in heavy-nitrogen (^{15}N) growth medium for many generations and then transferred to light-nitrogen (^{14}N) medium. The top of the figure illustrates the predicted labeling pattern by semiconservative replication for cells grown in the heavy medium, and for the first and second generations following transfer to the light medium. By isolating the DNA from cells grown at these different times and determining the density of the molecules in each generation, Meselson and Stahl were able to confirm that the DNA is replicated in a semiconservative manner.

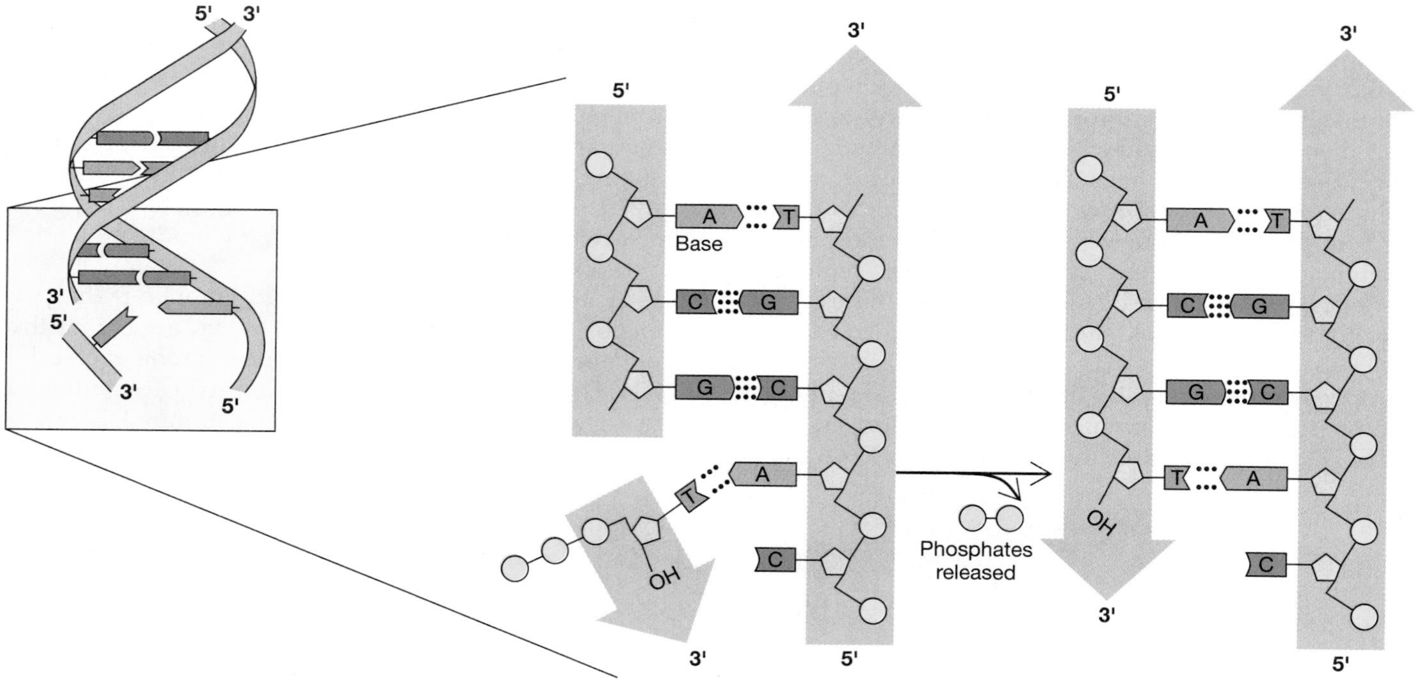

FIGURE 11–12 DNA replication proceeds by the addition of nucleotides. The building blocks for the DNA molecule are nucleotides with three phosphates (nucleoside triphosphates). Two of the phosphates are released when the nucleotides are linked to the 3′ carbon of the sugar at the end of the growing chain. The specificity of the polymerase enzymes that catalyze the polymerization reactions requires that each growing chain always elongate in the 5′ → 3′ direction.

molecules were extracted from some of the cells; when they were subjected to density gradient centrifugation they accumulated in the high-density region of the gradient. The rest of the bacteria (which also contained nitrogen-15-labeled DNA) were transferred to a different growth medium in which the NH_4Cl contained the naturally abundant, lighter nitrogen-14 isotope; they were then allowed to undergo several more cell divisions.

The newly synthesized DNA strands were expected to be less dense because they incorporated bases containing the lighter nitrogen-14 isotope. Indeed, molecules of DNA from cells isolated after one generation had a density indicating that they contained half as many nitrogen-15 atoms as the "parent" DNA. After another cycle of cell division, two kinds of DNA appeared in the density gradient at levels indicating that one consisted of "hybrid" DNA helices (labeled with equal amounts of nitrogen-15 and nitrogen-14 DNA), whereas the other contained only DNA with the naturally occurring light isotope. Each strand of the parental double-helix molecule was thus conserved in a *different* daughter molecule, exactly as predicted by the semiconservative replication model.

DNA replication is complex and has a number of unique features

Although the general principles of DNA replication are simple and straightforward predictions from the Watson-Crick model, the process actually requires a complex structure containing a large number of proteins and enzymes that work together as a "replication machine." Many of the essential features of replication are universal, although some differences exist between prokaryotes and eukaryotes because their DNA is organized differently. In bacterial cells such as *E. coli,* most or all of the DNA is in the form of a single, *circular,* double-stranded molecule. Each unreplicated eukaryotic chromosome contains a single, *linear,* double-stranded molecule associated with a great deal of protein and some RNA.

DNA strands must be unwound during replication

Watson and Crick recognized that in their double-helix model the two DNA strands are wrapped around one another like the strands of a rope. If we try to pull the strands apart, the rope must either rotate or twist into tighter coils. We would expect similar things to happen when complementary DNA strands are separated for replication. Unwinding is accomplished by **DNA helicase enzymes** that travel along the helix, unwinding the strands as they move. Once the strands are separated, **helix-destabilizing proteins** bind to single DNA strands, preventing re-formation of the double helix until the strands are copied. Because DNA molecules are very long and thin, the strain on the molecule must be relieved as the two strands are unwound. Special enzymes, called **topoisomerases,** produce breaks in the DNA molecules and then rejoin the strands, relieving strain by effectively untying knots that develop during replication.

DNA synthesis always proceeds in a 5′ → 3′ direction

The enzymes that catalyze the linking together of the nucleotide subunits are called **DNA polymerases.** They have several limitations that contribute to the complexity of the replication process. They are able to add nucleotides only to the 3′ end of a polynucleotide strand that is *paired* to the strand being copied (Fig. 11–12). Nucleotides known as **nucleoside triphosphates** are used as substrates for the polymerization reaction; these are similar to ATP in that they contain three phosphate groups linked to the 5′ carbon of the sugar group, plus a base. As the nucleotides are linked together, two of the phosphates are removed. Like the hydrolysis of ATP, these reactions are strongly exergonic (see Chapter 6) and do not require additional energy. Because the new polynucleotide chain is elongated by the linkage of the 5′ phosphate group of the next nucleotide subunit to the 3′ hydroxyl group of the sugar at the end of the strand, the new strand of DNA always grows in the 5′ → 3′ direction.

DNA synthesis requires an RNA primer

A second limitation of the DNA polymerases is that they can add nucleotides only to the 3′ end of an *existing* polynucleotide strand. So how can DNA synthesis be initiated once the two strands are separated? The answer is that a short piece (usually about five nucleotides) of an **RNA primer** (Fig. 11–13) is first synthesized by an aggregate of proteins called a **primosome.** RNA, or **ribonucleic acid** (see Chapter 12), is a nucleic acid polymer consisting of nucleotide subunits that associate by complementary base-pairing with the single-stranded DNA template at the point of initiation of replication. DNA polymerase can then add subunits to the 3′ end of

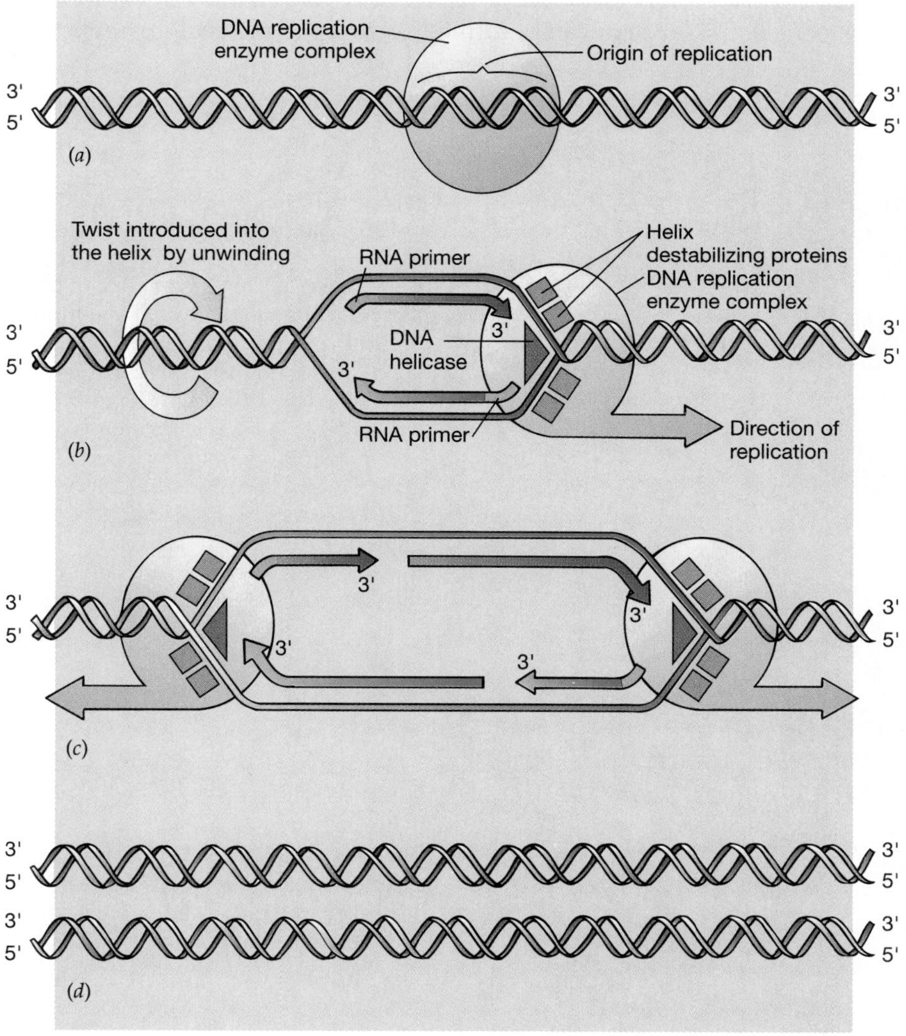

DNA synthesis begins at a specific base sequence, termed the *origin of replication.*

Strands are separated at the origin of replication and unwound by DNA helicase, which "walks" along the DNA molecule preceding the DNA-synthesizing enzymes. Single-stranded regions are prevented from re-forming into double strands by helix-destabilizing proteins, which bind to single-stranded DNA. The region of active DNA synthesis is associated with the "replication fork," formed at the junction of the single strands and the double-stranded region. Both strands are synthesized in the vicinity of the fork (each in a 5'→3' direction).

As the new strands continue to grow in the first direction, unwinding and replication initiate on the other side of the origin of replication, forming a second replication fork. Thus replication proceeds in opposite directions.

Completion of replication results in the formation of two daughter molecules, each containing one old and one newly synthesized strand.

FIGURE 11–13 DNA replication requires a number of steps.

the RNA primer. The primer is later degraded by specific enzymes and the space is filled in with DNA.

DNA replication is discontinuous in one strand and continuous in the other

A major obstacle in understanding DNA replication was the fact that the complementary DNA strands run in opposite directions. Because DNA synthesis proceeds only in the direction of 5′ → 3′ (which means that the strand being copied is being read in a 3′ → 5′ direction), it appears necessary to copy one of the strands starting at one end of the double helix and the other strand starting at the opposite end. We know, however, that this is not the case. DNA replication begins at specific sites on the DNA molecule, termed **origins of replication,** and both strands are replicated at the same time at a Y-shaped structure called the **replication fork** (Fig. 11–14*a*). The 3′ end of one of the new strands is always growing *toward* the replication fork. Because this strand can be formed smoothly and continuously, it is called the **leading strand.** The 3′ end of the other new strand is always growing *away* from the replication fork. Therefore this strand, termed the **lagging strand,** must be synthesized in short (100- to 1000-nucleotide) pieces, called **Okazaki fragments** after their discoverer, Reijii Okazaki.

Each Okazaki fragment is initiated by a separate RNA primer and is then extended toward the 5′ end of the previously synthesized fragment by DNA polymerase. More than one type of DNA polymerase is involved in this process, and each is a complex enzyme with several functions. As the growing fragment approaches the one synthesized previously, one part of DNA polymerase degrades the previous RNA primer, allowing a different polymerase to fill in the gap between the two fragments. The fragments are then joined together by **DNA ligase,** an enzyme that links the 3′ end of one DNA fragment to the 5′ end of another.

It has been suggested (Fig. 11–14*b*) that simultaneous synthesis of both the leading and lagging strands is possible because the lagging strand forms a loop. This allows the DNA polymerase to remain at the fork while adding subunits to the 3′ end of each strand.

Most DNA synthesis is bidirectional

When double-stranded DNA is separated, two forklike structures are formed so that the molecule is replicated

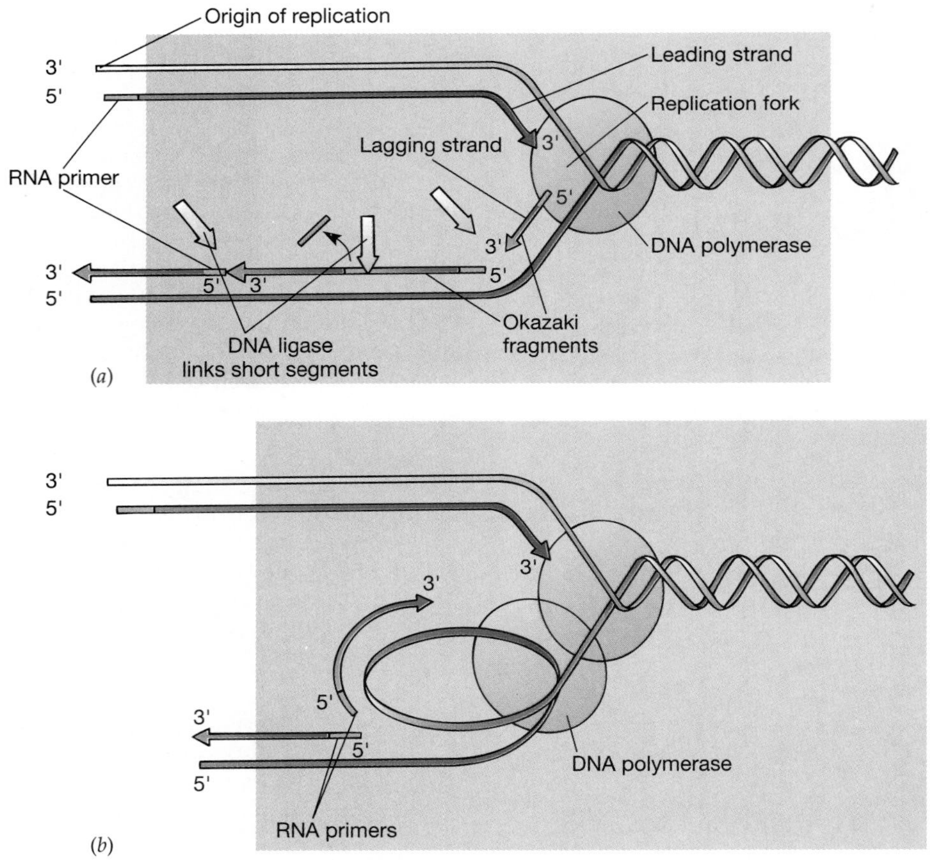

The *leading strand* is synthesized continuously in a direction toward the replication fork, whereas the *lagging strand* is synthesized in short pieces called *Okazaki fragments*, in a direction apparently away from the replication fork. Initiation of synthesis for both strands requires an *RNA primer* because DNA can be elongated only by addition to the 3' end of an existing polynucleotide strand. After elongation has begun, the RNA primer is degraded, the gaps are filled in with DNA, and the adjoining fragments are linked together by DNA ligase.

There is evidence that the lagging strand forms a loop, allowing the leading and lagging strands to be synthesized simultaneously.

FIGURE 11–14 Because elongation can proceed only in a 5′ → 3′ direction, the two strands at the replication fork are copied in different ways.

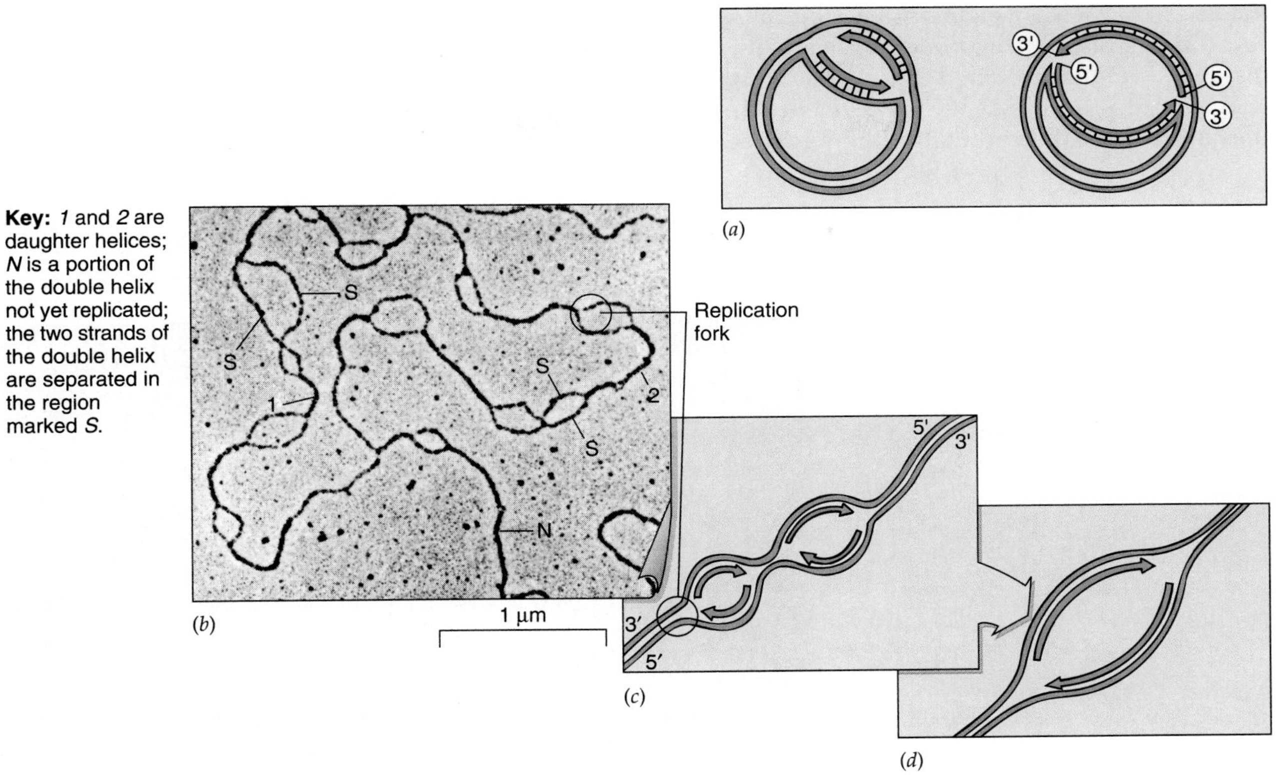
Key: *1* and *2* are daughter helices; *N* is a portion of the double helix not yet replicated; the two strands of the double helix are separated in the region marked *S*.
3'
5'
Replication fork
1 µm
(a)
(b)
(c)
(d)

FIGURE 11–15 DNA replication is bidirectional in bacterial DNA and in eukaryotic chromosomes. The leading strands and lagging strands are not represented in the illustrations. (***a***) The circular DNA in *E. coli* has only one origin of replication. DNA synthesis proceeds from that point in both directions until the two replication forks meet. (***b***) The photograph shows a segment of a eukaryotic chromosome that has been partially replicated. The conditions used to produce this image resulted in the separation of some of the newly formed DNA double helixes. (***c***) Eukaryotic chromosomal DNA contains multiple origins of replication. DNA synthesis proceeds in both directions from each origin. (***d***) Adjacent "replication bubbles" eventually merge. (*b*, Courtesy of H. J. Kriegstein and D. S. Hogness)

in both directions from the origin of replication. Prokaryotic cells usually have only one origin of replication on each circular DNA molecule (Fig. 11–15*a*), so the two replication forks proceed around the circle and eventually meet at the other side to complete the formation of two new DNA molecules.

A eukaryotic chromosome is composed of one extremely long linear DNA molecule, so the process is speeded up by having multiple origins of replication (Fig. 15*b–d*). Synthesis continues at each replication fork until it meets one coming from the opposite direction, resulting in the formation of a chromosome containing two DNA double helices. Each double helix corresponds to a chromatid.

DNA IN CHROMOSOMES IS PACKAGED IN A HIGHLY ORGANIZED WAY

Prokaryotic and eukaryotic cells differ markedly in their DNA content as well as in the organization of DNA molecules. An *E. coli* cell normally contains about 4×10^6 base pairs (almost 1.35 millimeters) of DNA in its single circular DNA molecule. In fact, the total length of the DNA is about 1000 times greater than the length of the cell itself. Therefore the DNA molecule must, with the help of special proteins, be twisted and folded compactly to fit inside the bacterial cell.

A typical eukaryotic cell contains much more DNA than a bacterium does, and it is organized in the nucleus as multiple chromosomes; these vary widely in size and number among different species. Although a human cell nucleus is about the size of a large bacterial cell, it contains almost 1000 times the amount of DNA found in *E. coli*. The haploid DNA content of a human cell is about 3×10^9 base pairs; if stretched end to end, it would be almost 1 meter long.

In eukaryotes, DNA, which is acidic and negatively charged, is associated with basic (positively charged) **histone** proteins to form structures called **nucleosomes.** The fundamental unit of the complex consists of a beadlike structure with about 140 base pairs of DNA wrapped around a disc-shaped core of eight histone molecules (Fig. 11–16). Additional histones are bound to the segment of DNA that links two neighboring beads. A nucleosome is defined as a bead plus a linker DNA segment. The nucleosomes are part of the **chromatin,** the nucleoprotein complex that makes up the chromosomes (Fig. 11–17). They are organized into large coiled loops held together by a set of nonhistone **scaffolding proteins.** Figure 11–18 illustrates the dense packing of DNA fibers on a histone-depleted mouse chromosome.

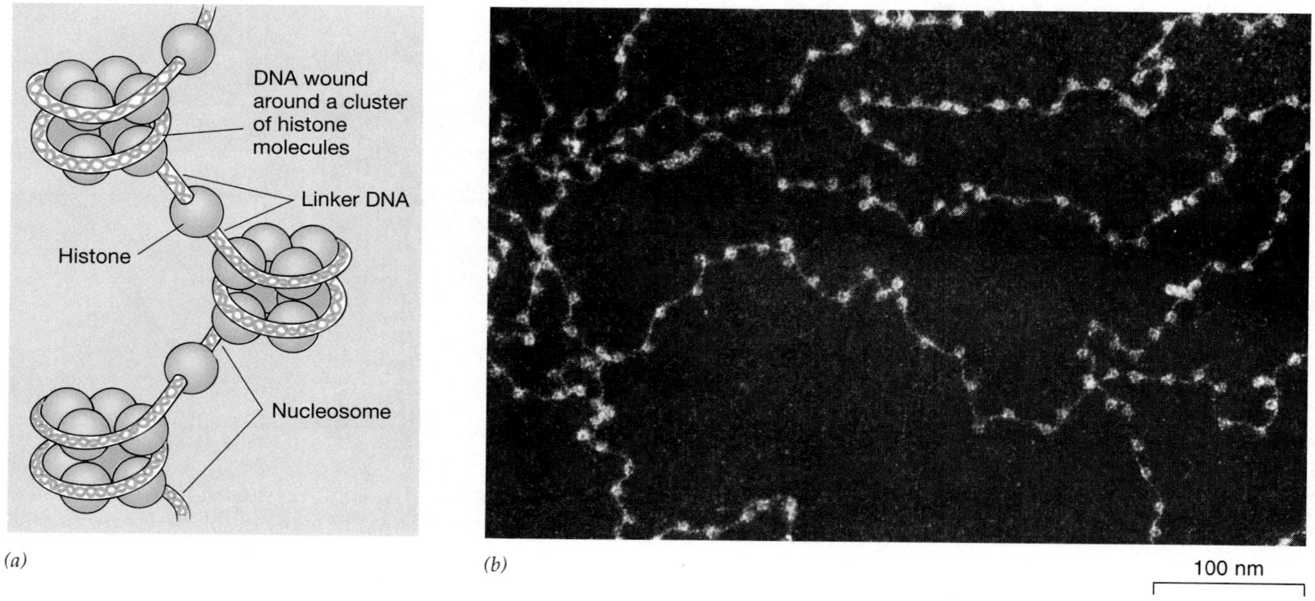

FIGURE 11–16 The units of histone surrounded by DNA in the chromosome are called nucleosomes. (*a*) A model for the structure of a nucleosome. Each nucleosome bead contains a set of eight histone molecules; these form a protein core around which the double-stranded DNA is wound. The DNA surrounding the histone consists of 146 nucleotide pairs; another segment of DNA, about 60 nucleotide pairs long, links nucleosome beads. A linker DNA segment plus one nucleosome bead together constitute a nucleosome. One type of histone covers the linker DNA between adjacent nucleosome beads. This histone appears to be responsible for packing nucleosomes and may help link them to one another. (*b*) Nucleosomes from the nucleus of a chicken red blood cell. Each spherical structure and its adjacent linker are a nucleosome. Normally nucleosomes are packed more closely together, but the preparation procedure has spread them apart, revealing the DNA linkers. (*b*, courtesy of D. E. Olins and A. L. Olins)

FIGURE 11–18 These electron micrographs show the residual structure of a mouse chromosome that has been depleted of histones. (*a*) Note how densely packed the DNA fibrils are, even though they have been released from the proteins that organize them into tightly coiled structures. The dark structure extending from left to right across the bottom of the photograph is composed of scaffolding proteins. (*b*) Higher magnification of a portion of part (*a*). The DNA is still organized in the form of loops that protrude from the scaffolding. (Courtesy of U. Laemmli, from *Cell* 12:817, 1988. Copyright by Cell Press)

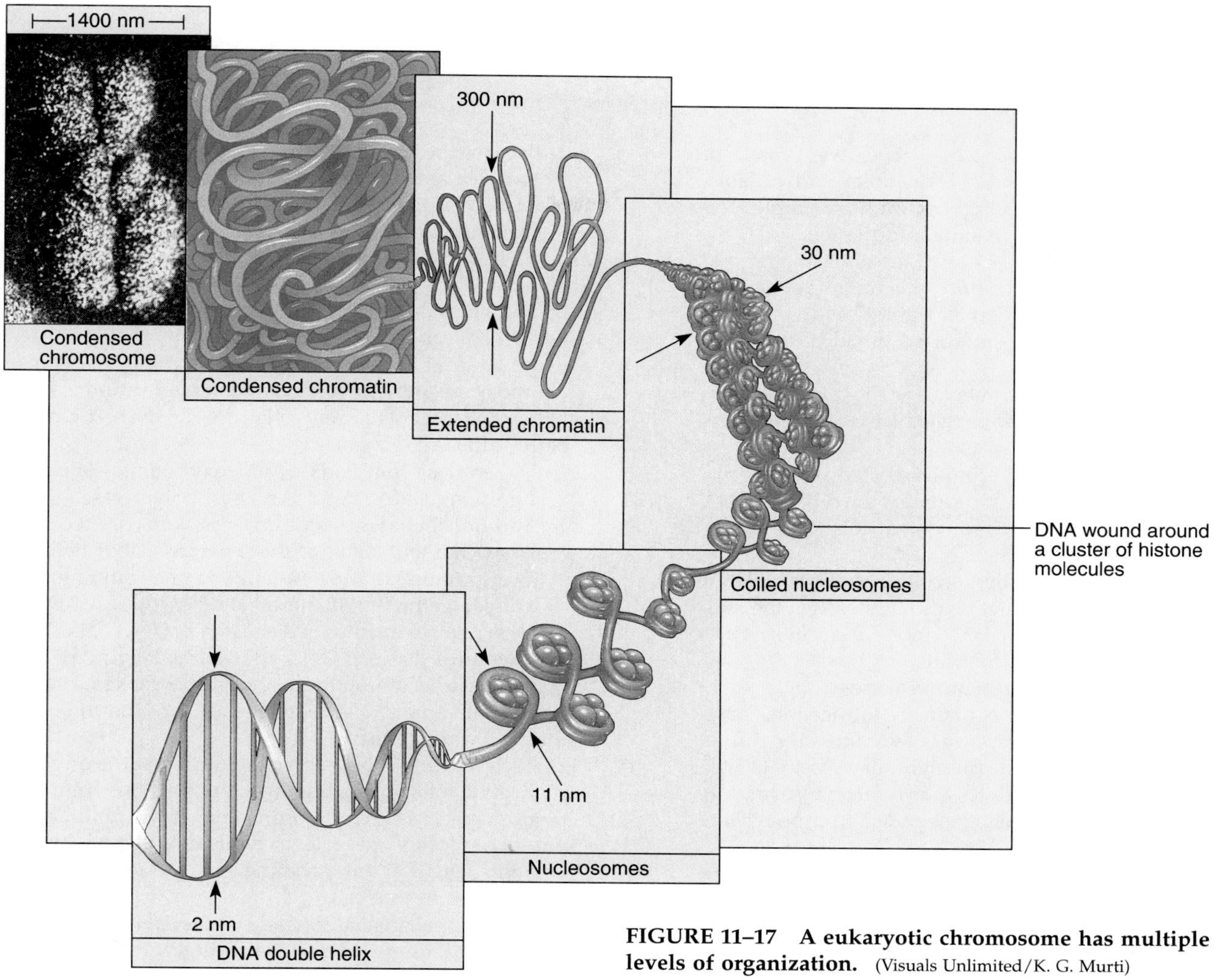

FIGURE 11–17 A eukaryotic chromosome has multiple levels of organization. (Visuals Unlimited/K. G. Murti)

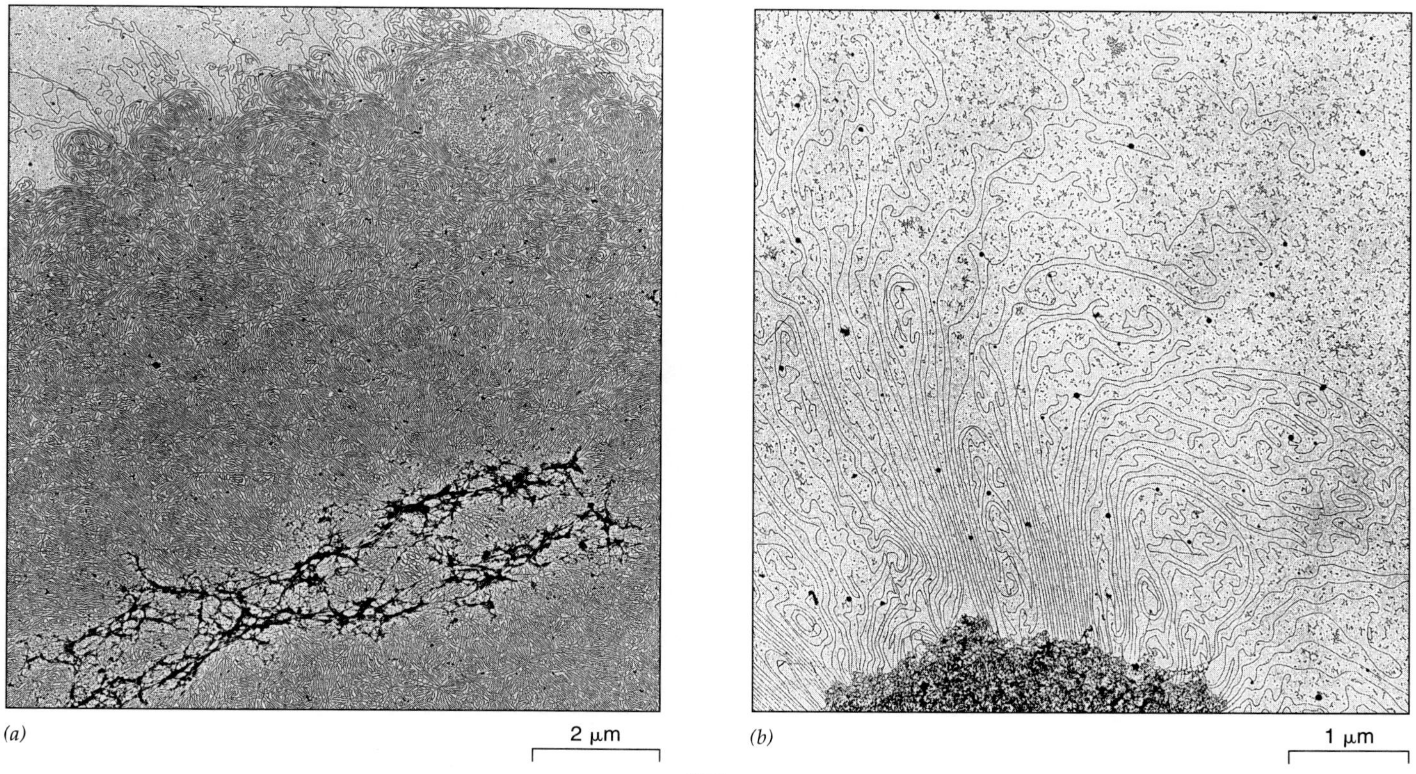

SUMMARY

I. Many early geneticists thought that genes were made of proteins. Proteins were known to be complex and variable, whereas nucleic acids were thought of as rather simple molecules with a limited ability to store information.
   A. Garrod's work on inborn errors of metabolism and that of Beadle and Tatum with *Neurospora* mutants suggested that each protein is specified by a single gene.
   B. Several lines of evidence supported the idea that DNA is the genetic material.
      1. In transformation experiments, the DNA of one strain of bacterium can endow a related bacterium with new genetic characteristics.
      2. When a bacterial cell becomes infected with a virus, only the DNA from the virus enters the cell; this DNA is sufficient for the virus to reproduce and form new virus particles.
   C. Watson and Crick's studies on the structure of DNA demonstrated how information can be stored in the molecule's structure and how DNA molecules can serve as templates for their own duplication.

II. DNA is a very regular polymer of nucleotides.
   A. Each nucleotide subunit contains a nitrogenous base, which may be one of the purines (adenine or guanine) or one of the pyrimidines (thymine or cytosine). Each base is covalently linked to a five-carbon sugar, deoxyribose, which is covalently bonded to a phosphate group.
   B. The backbone of each single DNA chain is formed by alternating sugar and phosphate groups that are linked by covalent bonds. Each phosphate group is attached to the 5′ carbon of one deoxyribose and to the 3′ carbon of the neighboring deoxyribose.
   C. Each DNA molecule is composed of two polynucleotide chains that associate as a double-helix structure. The two chains are antiparallel (meaning they run in opposite directions); at each end of the DNA molecule one chain has an exposed 5′ deoxyribose carbon and the other has an exposed 3′ deoxyribose carbon.
   D. The two chains of the helix are held together by hydrogen-bonding between specific base pairs. Adenine (A) forms two hydrogen bonds with thymine (T); guanine (G) forms three hydrogen bonds with cytosine (C).
      1. Complementary base-pairing between A and T and between G and C is the basis of Chargaff's rules, which state that A = T and G = C.
      2. Because the two strands of DNA are held together by complementary base-pairing, it is possible to predict the base sequence of one strand if one knows the base sequence of the other strand.

III. During DNA replication, the two strands of the double helix unwind. Each strand serves as a template for the formation of a new complementary strand.
   A. DNA replication is semiconservative; that is, each daughter double helix contains one strand from the parent molecule and one newly synthesized strand.
   B. DNA replication is a complex process requiring a number of different enzymes.
      1. The enzyme that adds new deoxyribonucleotides to a growing DNA strand is a DNA polymerase.
      2. Additional enzymes and other proteins are required to unwind and stabilize the separated DNA helix, to form primers, to prevent tangling and knotting, and to link together fragments of newly synthesized DNA.
   C. DNA synthesis always proceeds in a 5′→3′ direction. This requires that one DNA strand (the lagging strand) be synthesized discontinuously, as short Okazaki fragments. The opposite strand (the leading strand) is synthesized continuously.
   D. DNA replication is bidirectional, starting at the origin of replication and proceeding in both directions from that point. A eukaryotic chromosome may have multiple origins of replication and may be replicating at many points along its length at any one time.

IV. DNA is organized in a cell.
   A. Prokaryotic cells usually have circular DNA molecules.
   B. Eukaryotic chromosomes have several levels of organization.
      1. The DNA is associated with histones (basic proteins) to form nucleosomes, each of which consists of a histone bead with DNA wrapped around it, plus an adjacent linker DNA with a histone attached.
      2. The nucleosomes are organized into large coiled loops held together by nonhistone scaffolding proteins.
      3. DNA molecules are much longer than the nuclei or the cells that contain them. The organization of DNA into chromosomes allows the DNA to be accurately replicated and segregated into daughter cells without tangling.

SELECTED KEY TERMS

adenine (A)
antiparallel
avirulent
bacteriophage
bidirectional replication
chromatin
complementary base sequences
cytosine (C)
density gradient centrifugation
deoxyribonucleic acid (DNA)
deoxyribose
DNA helicase enzyme
DNA ligase
DNA polymerase
DNA replication
double helix
3′ end
5′ end
guanine (G)
helix-destabilizing protein
histone
lagging strand
leading strand
mutation
nucleoside triphosphate
nucleosome
nucleotide
Okazaki fragment
one gene, one enzyme hypothesis
origin of replication
phosphodiester linkage
primosome
purine
pyrimidine
replication fork
ribonucleic acid (RNA)
RNA primer
scaffolding protein
semiconservative replication
thymine (T)
topoisomerase
transformation
virulent
x-ray diffraction

POST-TEST

1. Early evidence that DNA is the genetic material came from ____________ experiments showing that purified ____________ is capable of changing a bacterial strain to a genetically stable new form.
2. Nucleotides found in DNA contain the five-carbon sugar ____________ .
3. The bases adenine and guanine are called ____________ ; the bases thymine and cytosine are referred to as ____________ .
4. The backbone of a DNA strand is formed from alternating ____________ and phosphates, joined by ____________ linkages.
5. Chargaff's rules were formulated by analysis of the base composition of DNA from different species. These findings stated that the number of ____________ bases equals the number of ____________ bases, and the number of ____________ bases equals the number of ____________ bases.
6. The basic information that Watson and Crick used to construct their DNA model came from knowledge of the chemical composition of DNA and the ____________ ____________ studies of Franklin and Wilkins.
7. Each DNA molecule consists of two anti____________ chains.
8. The process of DNA synthesis is called DNA ____________ .
9. DNA is synthesized by a mechanism known as ____________ ; that is, each double helix contains one old strand and one newly synthesized strand.
10. For DNA replication to start, the two strands must be separated at a point in the molecule known as the origin of ____________ .
11. The new strand of DNA is always synthesized starting at its ____________ end; additional nucleotides are added at its growing ____________ end.
12. The DNA molecules are formed from precursors known as ____________ ____________ .
13. DNA replication is referred to as a discontinuous process, which means that the lagging strand must be synthesized in short pieces, called ____________ ____________ . They are joined by an enzyme called ____________ ____________ .
14. The ____________ strand of DNA is synthesized as a continuous molecule.
15. DNA is synthesized by an enzyme called ____________ ____________ .
16. Most DNA synthesis in both prokaryotic and eukaryotic cells is ____________ , which means that it proceeds in both directions from the origin of replication.
17. DNA in eukaryotic cells is organized into structures called ____________ , which are composed of about 140 base pairs of DNA wrapped around a core of histone proteins, plus an adjacent region of linker DNA with an additional histone.
18. ____________ , composed of nucleosomes, is held together by nonhistone ____________ ____________ .

REVIEW QUESTIONS

1. How did the experiments of Avery and coworkers point to DNA as the essential genetic material? Did the Hershey-Chase experiment establish that DNA is the genetic material in all organisms? Did either of these experiments demonstrate how DNA could function as the chemical basis of genes?
2. Sketch the structure of a single strand of DNA. What types of subunits make up the chain? How are they linked?
3. Describe the structure of double-stranded DNA as determined by Watson and Crick.
4. Does a single strand of DNA obey Chargaff's rules? How do Chargaff's rules relate to the structure of DNA?
5. What are some of the mechanical problems encountered in DNA replication? How are they dealt with by the cell?
6. Why is DNA replication continuous for one strand but discontinuous for the other?
7. Compare the structures of a bacterial DNA molecule and a eukaryotic chromosome. What effects do these differences have on replication?
8. How do the dimensions of a bacterial cell compare with the length of the DNA contained within it? What is the relationship between the diameter of a human cell nucleus and the average length of DNA per human chromosome?

YOU MAKE THE CONNECTION

1. What characteristics must a molecule have if it is to serve as genetic material? What important features of the structure of DNA are consistent with its role as the chemical basis of heredity?
2. In Chapter 10 we discussed the fact that heritable variation is essential for the study of inheritance. What role did mutant strains play in Beadle and Tatum's development of the one gene, one enzyme hypothesis?

RECOMMENDED READINGS

Felsenfeld, G. "DNA." *Scientific American,* October 1985. An excellent, well-illustrated article on the structure and organization of DNA.

Judson, H. F. *The Eighth Day of Creation: Makers of the Revolution in Biology.* Simon & Schuster, New York, 1979. A beautifully written and fascinating account of the early history of molecular biology.

Rennie, J. "DNA's New Twists." *Scientific American,* March 1993. A fascinating summary of novel studies in genetics that challenge previous ideas and open the way for a new understanding of the role of DNA in evolution and in genetic diseases.

Watson, J. D. *The Double Helix.* Atheneum, New York, 1968. Watson's view of the discovery of the structure of DNA. Somewhat controversial, but entertaining and insightful reading.

Watson, J. D., and F. H. C. Crick. "Molecular Structure of Nucleic Acids: A Structure for Deoxyribose Nucleic Acid." *Nature* Vol. 171, 1953. Watson and Crick's original report—a simple, clearly written two-page paper that shook the scientific world.

CHAPTER 12

RNA and Protein Synthesis: The Expression of Genetic Information

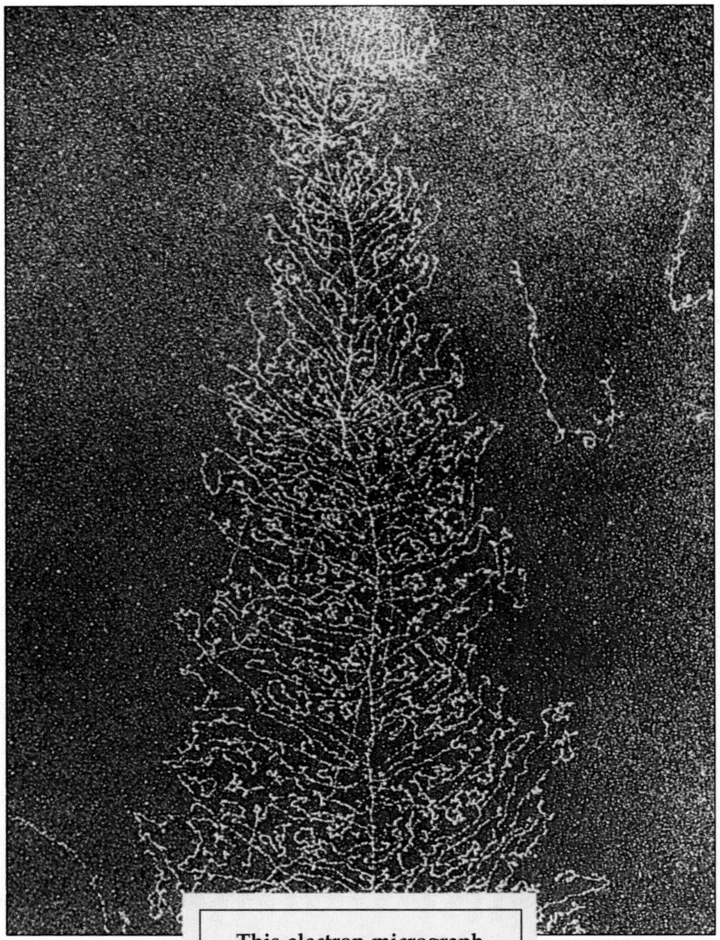

This electron micrograph shows RNA molecules (lateral strands) in the process of being synthesized as complementary copies of a DNA template (central axis). (Professor Oscar Miller/Science Photo Library/Photo Researchers, Inc.)

In Chapter 11 we saw that DNA molecules can store information. We also examined how that information is replicated by a cell so that it can be passed accurately to its descendants. The basic features of DNA originally described by Watson and Crick are now known to be the same in virtually all cells, from bacteria to humans.

By the mid-1950s it became evident that the genetic information in DNA contains the code for all the proteins needed by the cell. However, it was more than a decade after Watson and Crick's famous paper before scientists finally understood how cells are able to convert DNA information into amino acid sequences of proteins. Much of that understanding came from studying the functions of bacterial genes. After the discovery of the structure of DNA, prokaryotic cells quickly became the organisms of choice for these investigations because they could be grown quickly and easily and because they seemed to contain only the minimal amount of DNA needed for growth and reproduction.

In this chapter we examine how genes are expressed—that is, how their genetic information is decoded and used to make proteins. Gene expression includes a complex series of events involving the synthesis of RNA molecules complementary to the DNA (*transcription*), as well as protein synthesis (*translation*). In Chapter 13 we consider some of the ways the entire process is controlled.

We first focus our attention on gene expression in prokaryotic cells, because these cells are best understood. We then extend our discussion to include eukaryotic cells, for our understanding of these cells is increasing rapidly as a result of groundwork laid by study of the simpler bacterial systems.

LEARNING OBJECTIVES

After you have studied this chapter you should be able to

1. Outline the flow of genetic information in cells, from DNA to protein.
2. Compare the structures of DNA and RNA, and explain how the structure of each is related to its role in the cell.
3. Compare the processes of transcription and replication, identifying both similarities and differences.
4. Identify the features of tRNA that are important in decoding genetic information and translating it into "protein language."
5. Explain why the ribosome has a central role in protein synthesis.
6. Diagram the processes of initiation, chain elongation, and chain termination in protein synthesis.
7. Compare eukaryotic and prokaryotic mRNAs, and explain the functional significance of their structural differences.
8. Analyze the differences in translation in prokaryotic and eukaryotic cells.
9. Explain why the genetic code is said to be redundant and virtually universal, and examine some of the evolutionary implications of these aspects of the code.
10. Give examples of the different classes of mutations that affect the base sequence of DNA, and demonstrate the effects that each has on the protein produced.

BASE SEQUENCES IN DNA ARE TRANSCRIBED AS BASE SEQUENCES IN RNA, THEN TRANSLATED INTO AMINO ACID SEQUENCES IN PROTEINS

Although the sequence of bases in DNA determines the sequence of amino acids in proteins, the information in DNA is not used directly. Instead, a related nucleic acid, **RNA,** or **ribonucleic acid,** serves as an intermediary between DNA and protein (Fig. 12–1). Like DNA, RNA is a polymer of nucleotides, but it has some important differences.

RNA is usually single-stranded, although internal regions of some RNAs may have complementary sequences that allow them to fold back and pair to form short, double-stranded segments. As shown in Figure 12–2, the sugar in RNA is **ribose** (rather than deoxyribose), and the base **uracil** substitutes for thymine. Uracil, like thymine, is a pyrimidine and can form two hydrogen bonds with adenine. Hence, uracil and adenine are a complementary pair.

When a protein-coding gene is expressed, an RNA copy is made of the information in the DNA (Fig. 12–3). This process resembles DNA replication in that the sequence of bases in the RNA strand is determined by complementary base-pairing with one of the DNA strands. Because RNA synthesis involves making a copy of information in one kind of nucleic acid (DNA) in the form of another nucleic acid (RNA), we refer to this process as **transcription.** The RNA that carries the specific information for making a protein is called **messenger RNA,** or **mRNA.**

In the second stage of gene expression, the transcribed information in the mRNA is converted into the amino acid sequence of a protein. This process is called **translation** because it involves transformation of the "nucleic acid language" in the mRNA molecule into the "amino acid language" of the protein.

The protein-coding information contained within mRNA is specified by **codons.** Because each codon is a combination of three consecutive bases, the code is referred to as a **triplet code.** Each codon in the mRNA specifies one amino acid; for example, one codon that corresponds to the amino acid threonine is 5′—ACG—3′ (Table 12–1). Translation requires cellular machinery that can recognize and decode the codons in the mRNA. **Transfer RNAs (tRNAs)** are critical parts of the decoding machinery; each tRNA is an "adapter" molecule that can link with a specific amino acid and recognize the appropriate mRNA codon for that particular amino acid. Codon recognition is possible because each tRNA molecule has a sequence of three bases, called the **anticodon,** that associates with the mRNA codon by complementary base-pairing. In our example, the exact anticodon for threonine is 3′—UGC—5′.

Translation also requires the linking of amino acids in the correct order. This is accomplished by **ribosomes** (see Chapter 4), complex organelles composed of two different subunits, each containing a number of proteins and **ribosomal RNA (rRNA).** Ribosomes attach to the end of the mRNA and travel along it, allowing the tRNAs to decode the message so that the amino acids are properly positioned and joined by peptide bonds in the correct sequence to form a polypeptide.

These events could not take place without direction by the genetic code. Before the discoveries of Watson and

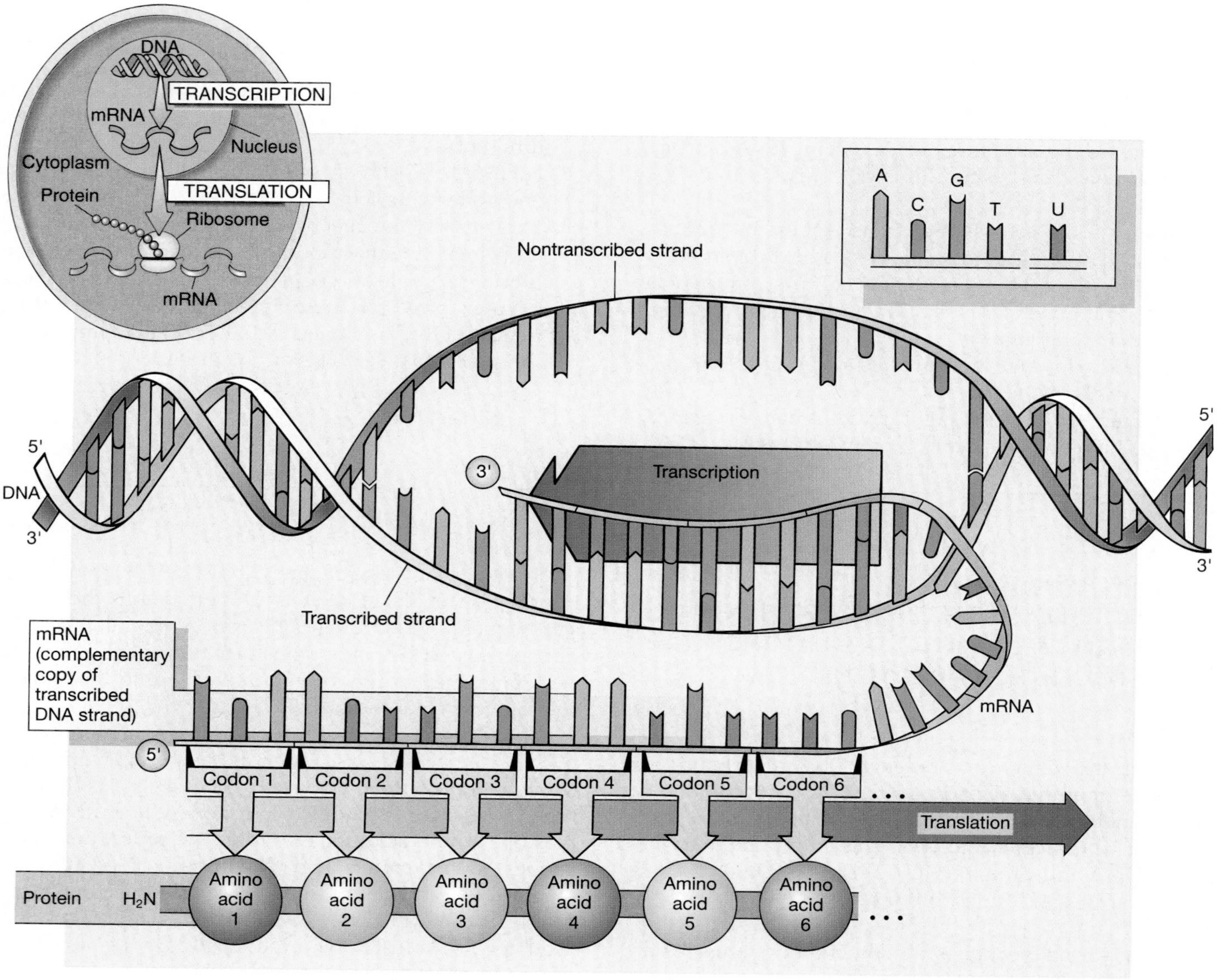

FIGURE 12–1 Messenger RNA carries genetic information in the form of sets of three bases called codons, each of which specifies one amino acid. In the process of transcription, messenger RNA is synthesized as a complementary copy of one of the DNA strands. Messenger RNA codons are translated consecutively, thus specifying the linear sequence of amino acids in the polypeptide chain. In eukaryotes transcription takes place in the nucleus and translation occurs in association with ribosomes in the cytoplasm.

Crick it was not clear how the four bases in DNA could be used to govern the assembly of 20 amino acids into a vast variety of cellular proteins. When scientists examined this problem in light of the new model of DNA, they found that the DNA bases could serve as a four-letter alphabet. Three-letter combinations of the four bases (4^3) made it possible to form a total of 64 "words," more than sufficient to specify all of the naturally occurring amino acids. More than 10 years later the "cracking" of the genetic code was completed, verifying the existence of the three-base triplet code that is common to all organisms.

Transcription Is the Synthesis of RNA from a DNA Template

Three main kinds of RNA are transcribed from DNA, including ribosomal RNA (rRNA) and transfer RNA (tRNA), as well as messenger RNA (mRNA). Most RNA is synthesized by **DNA-dependent RNA polymerases,** enzymes that are present in all cells. These enzymes require DNA as a template and have many similarities to DNA polymerase. They use nucleoside triphosphates (nucleotides with three phosphate groups) as substrates,

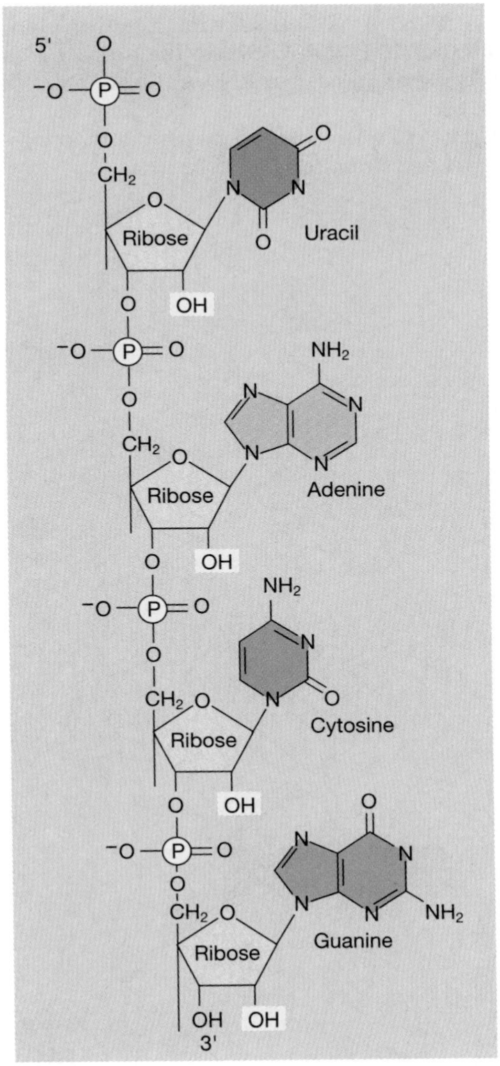

FIGURE 12–2 RNA consists of ribonucleotide subunits joined by 5′ → 3′ phosphodiester linkages, like those found in DNA. Three of the nitrogenous bases—adenine, guanine, and cytosine—are the same as those found in DNA. Instead of having thymine, however, RNA has the base uracil, which pairs with adenine. All four nucleotides contain the five-carbon sugar ribose, which has a hydroxyl group on the 2′ carbon atom.

removing two of the phosphates as the subunits are covalently linked to the 3′ end of the RNA (Fig. 12–4).

Messenger RNA contains base sequences that code for protein

Usually only one of the strands in a protein-coding region of DNA is transcribed (Fig. 12–5*a*). Consider a segment of DNA that contains the following DNA base sequence:

5′—ATTGCCAGA—3′

Its complementary strand would read:

3′—TAACGGTCT—5′

If both were transcribed, complementary base strands would specify entirely different amino acid sequences. Thus, only one of the DNA strands in a gene is complementary to the mRNA and that is the strand that is transcribed. The transcribed strand is also referred to as the template strand. Because a chromosome-sized, double-stranded DNA molecule includes thousands of genes, a particular strand may serve as the transcribed strand for some genes and the nontranscribed strand for others (Fig. 12–5*b*).

Table 12–1 THE GENETIC CODE: CODONS OF mRNA THAT SPECIFY A GIVEN AMINO ACID

First Position (5′ end)	*Second Position*	*Third Position (3′ end)* *U*	*C*	*A*	*G*
U	U	UUU	UUC	UUA	UUG
		Phenylalanine		Leucine	
	C	UCU	UCC	UCA	UCG
		Serine			
	A	UAU	UAC	UAA	UAG
		Tyrosine		Stop	
	G	UGU	UGC	UGA Stop	UGG Tryptophan
		Cysteine			
C	U	CUU	CUC	CUA	CUG
		Leucine			
	C	CCU	CCC	CCA	CCG
		Proline			
	A	CAU	CAC	CAA	CAG
		Histidine		Glutamine	
	G	CGU	CGC	CGA	CGG
		Arginine			
A	U	AUU	AUC	AUA	(start) AUG
		Isoleucine			Methionine
	C	ACU	ACC	ACA	ACG
		Threonine			
	A	AAU	AAC	AAA	AAG
		Asparagine		Lysine	
	G	AGU	AGC	AGA	AGG
		Serine		Arginine	
G	U	GUU	GUC	GUA	GUG
		Valine			
	C	GCU	GCC	GCA	GCG
		Alanine			
	A	GAU	GAC	GAA	GAG
		Aspartic acid		Glutamine	
	G	GGU	GGC	GGA	GGG
		Glycine			

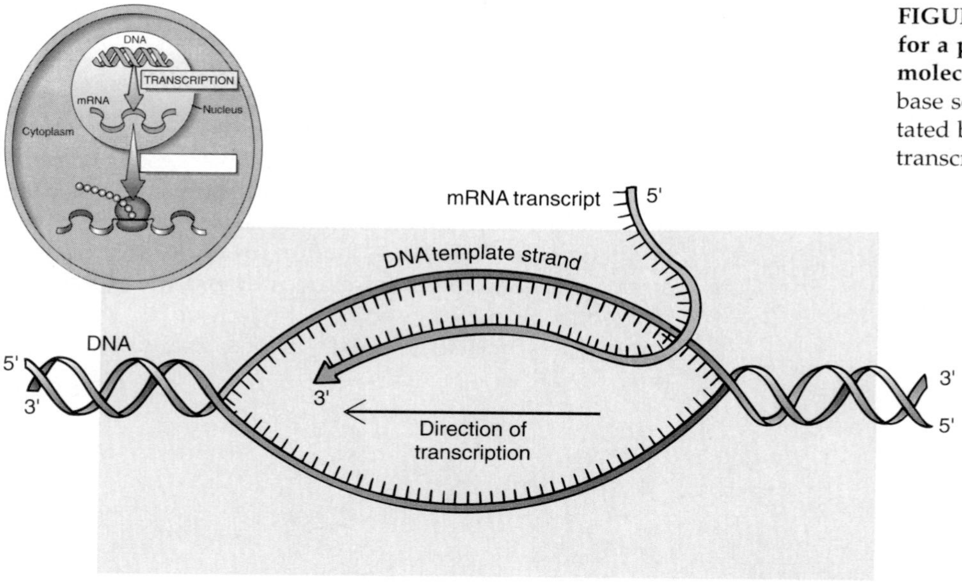

FIGURE 12–3 The DNA coding information for a protein is transcribed in the form of a molecule of messenger RNA (mRNA). The base sequence in the mRNA molecule is dictated by complementary base-pairing with the transcribed strand of the DNA.

Whenever nucleic acid molecules associate by complementary base-pairing, the two strands are antiparallel. Just as the two paired strands of DNA are antiparallel (see Chapter 11), the transcribed strand of the DNA and the complementary RNA strand are also antiparallel.

RNA polymerase begins transcription by recognizing a specific **promoter** base sequence at the beginning of a gene. Unlike DNA synthesis, RNA synthesis does not require a primer. The first nucleotide at the 5′ end of a new mRNA chain retains its triphosphate group, but as each additional nucleotide is incorporated at the 3′ end of the growing molecule, two of its phosphates are removed, leaving the remaining phosphate to become part of the sugar/phosphate backbone (as in DNA). The last nucleotide to be incorporated has an exposed 3′ hydroxyl group (Fig. 12–6).

(*Text continues on page 294*)

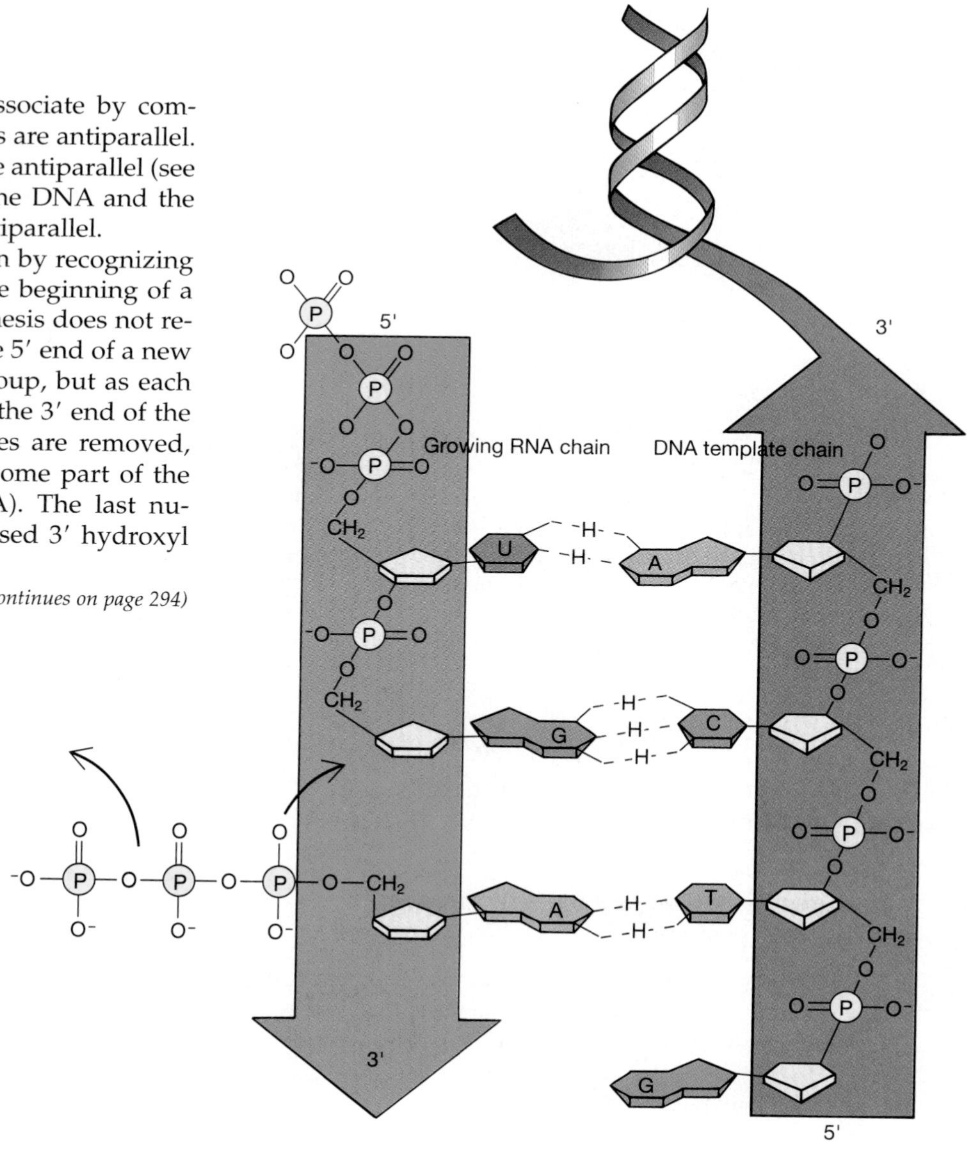

FIGURE 12–4 During transcription, the DNA helix is unwound and the bases of incoming nucleoside triphosphates pair with complementary bases on the DNA template strand (*right*). RNA polymerase cleaves two phosphates from each nucleoside triphosphate and covalently links the remaining phosphate to the 3′ end of the growing RNA chain. Thus, RNA, like DNA, is synthesized in a 5′ → 3′ direction.

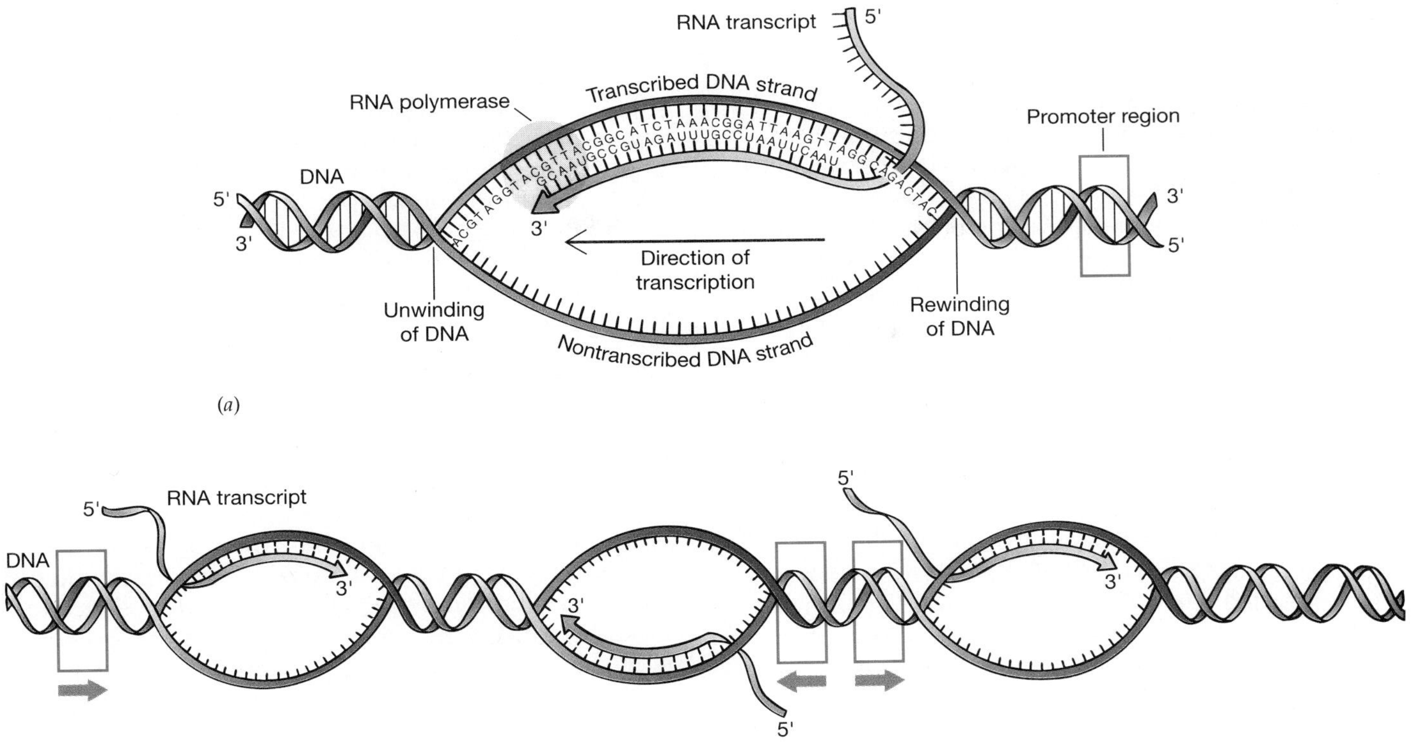

FIGURE 12–5 Transcription was first studied using bacteria as a model system. (*a*) The RNA is synthesized in a $5' \rightarrow 3'$ direction from the transcribed, or template, strand of the DNA molecule. Transcription starts downstream from DNA promoter sequences, which serve as the RNA polymerase recognition site. The promoter sequences are not transcribed. Termination sequences downstream from the coding sequences in the gene signal the RNA polymerase to stop transcription and be released from the DNA. (***b***) Only one of the two strands is transcribed for a given gene, but the opposite strand may be transcribed for a neighboring gene. Each transcript is started at its own promoter.

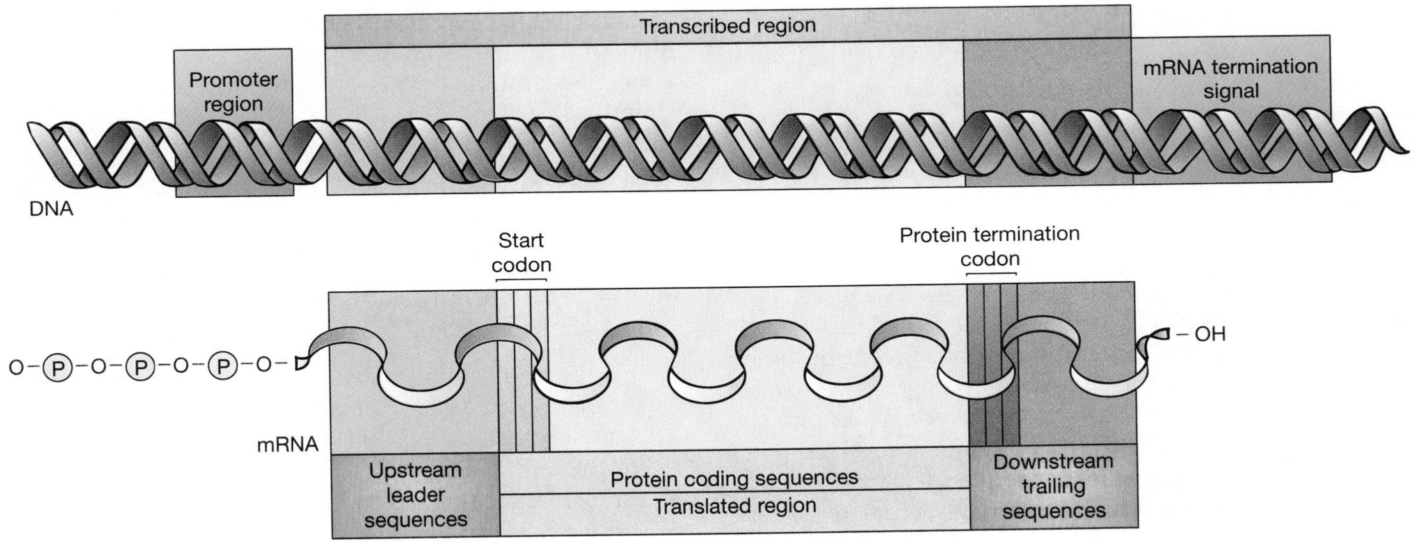

FIGURE 12–6 In addition to its protein-coding sequence, a bacterial mRNA molecule contains leader and trailing sequences. This figure compares the structures of a bacterial mRNA and the region of DNA from which it was transcribed. RNA polymerase recognizes promoter sequences in the DNA located five to eight bases upstream from the base where RNA synthesis is initiated. RNA synthesis stops when the polymerase encounters termination signals downstream from the protein-coding sequences. The ribose at the 5′ end of the mRNA molecule has three phosphate groups attached to its 5′ carbon. The ribose at the 3′ end of the molecule has an exposed hydroxyl group attached to its 3′ carbon. Ribosome-recognition sites are located in 5′ leader sequences of the mRNA, which are upstream from the protein-coding sequences. Protein-coding sequences begin at a start codon, which follows the leader sequences, and end at a termination codon near the 3′ end of the molecule. Noncoding trailing sequences, which can vary in length, follow the protein-coding sequences.

It is conventional to refer to a sequence of bases in a gene or the mRNA sequence transcribed from it as *upstream* or *downstream* of some reference point. **Upstream** means toward the 5′ end of the mRNA sequence or the 3′ end of the transcribed DNA strand. **Downstream** means toward the 3′ end of the RNA or the 5′ end of the transcribed DNA strand.

Upstream — Downstream

5′ — A — T — G — A — C — T — 3′ (nontranscribed DNA strand)

3′ — T — A — C — T — G — A — 5′ (transcribed DNA strand)

Direction of transcription →

Triphosphate 5′ — A — U — G — A — C — U — 3′ OH (RNA)

In the bacterium *Escherichia coli,* transcription of a gene is initiated when RNA polymerase (with the help of another protein) recognizes a specific promoter sequence of bases upstream from the protein-coding sequence. Different genes may have slightly different promoter sequences, so the cell can direct which genes are transcribed at any one time. Bacterial promoters are usually about 40 bases long and are positioned in the DNA just upstream of the point at which transcription will begin. Once the polymerase has recognized the correct promoter, it unwinds the helix and transcribes only the template strand of the DNA molecule.

The termination of transcription, like its initiation, is controlled by a set of specific base sequences. These sequences at the end of the gene act as "stop" signals for the RNA polymerase.

Messenger RNA contains additional base sequences that do not directly code for protein

The completed bacterial RNA contains more than the nucleotide sequence that codes for the protein (see Fig. 12–6). RNA polymerase starts transcription of a gene well upstream of the protein-coding sequences. As a result, the mRNA has a noncoding **leader sequence** at its 5′ end. The leader contains recognition signals for ribosome binding, which allow the ribosomes to be properly positioned to translate the message. In bacterial cells, one or more proteins may be encoded by a single mRNA molecule (see Chapter 13). The leader sequence is followed by the **coding sequences,** which contain the actual messages for the proteins. At the end of the coding sequences are special termination signals that specify the end of the protein. These are followed by noncoding 3′ trailing sequences, which can vary in length.

THE NUCLEIC ACID MESSAGE IS DECODED DURING TRANSLATION

Translation, or protein synthesis, adds another level of complexity to the process of information transfer because it involves the conversion of the four-base nucleic acid code to the 20-amino acid alphabet of proteins. Translation requires the coordinated functioning of more than 100 kinds of macromolecules, including the protein and RNA components of the ribosomes, mRNA, and amino acids linked to tRNAs.

An amino acid must be attached to its specific transfer RNA prior to becoming incorporated into a polypeptide

Amino acids are joined together by peptide bonds to form proteins (see Chapter 3). This joining involves linking the amino and carboxyl groups of adjacent amino acids. Peptide bond formation is only one aspect of the translation process, however, because the amino acids must be joined together in the correct sequence specified by the codons in the mRNA. The structural differences between a polynucleotide chain and a polypeptide chain are so great that no simple way exists for amino acids to interact directly with an mRNA molecule to make a protein.

Francis Crick recognized this problem and proposed that a molecule was needed to serve as an "adapter" in protein synthesis and bridge the gap between mRNA and proteins. Crick's adapters turned out to be transfer RNA (tRNA) molecules. DNA contains special tRNA genes that are transcribed to form the tRNA. Amino acids are covalently linked to their respective tRNA molecules by specific enzymes called **aminoacyl-tRNA synthetases,** which use ATP as an energy source (Fig. 12–7). The resulting complexes, called **aminoacyl-tRNAs,** are able to bind to the mRNA coding sequence so as to align the amino acids in the correct order to form the polypeptide chain.

Transfer RNA molecules have specialized regions with specific functions

Although tRNA molecules are considerably smaller than mRNA or rRNA molecules, they have a complex structure. A tRNA molecule must have several properties:

1. It must be recognized by a specific aminoacyl-tRNA synthetase that adds the correct amino acid.
2. It must have a region that serves as the attachment site for the amino acid.
3. It must be recognized by ribosomes.
4. It must have an **anticodon,** a specific complementary binding sequence for the correct mRNA codon.

The tRNAs are polynucleotide chains about 70 nucleotides long (Fig. 12–8), each with a number of unique base sequences as well as some that are common to all. Complementary base-pairing within each tRNA molecule causes it to be doubled back and folded to form three or more loops of unpaired nucleotides (Fig. 12–8*b*). The amino acid–binding site is at the end of a "stem" that forms at the 3′ end of the molecule (Fig. 12–8*c*). The *car-*

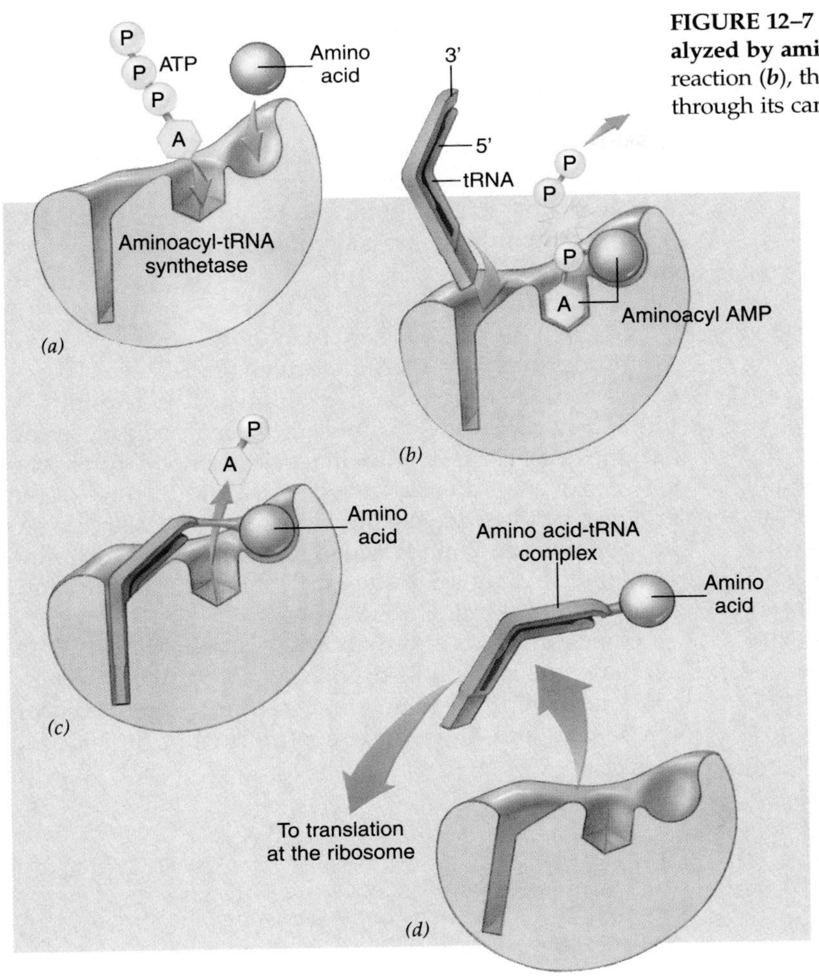

FIGURE 12–7 Formation of an amino acid-tRNA complex is catalyzed by aminoacyl-tRNA synthetase (*a*) Through an ATP-requiring reaction (*b*), the amino acid is coupled to the 3′ end of the tRNA through its carboxyl group (*c*), and released (*d*).

FIGURE 12–8 The genetic code is "read" by tRNA molecules, which have characteristic structures. (*a*) Diagram of the actual shape of a tRNA molecule. Its three-dimensional shape is determined by hydrogen bonds that form between complementary bases, which are most clearly observed in the two-dimensional cloverleaf form. (*b*) One loop contains the triplet anticodon that forms specific base pairs with the mRNA codon. The amino acid is attached to the terminal ribose at the 3′ OH end, which has the nucleotide sequence CCA. Each tRNA has G at its 5′ end and also contains several modified nucleotides. The pattern of folding results in a constant distance between the anticodon and amino acid in all tRNAs examined. (*c*) Schematic diagram of how the amino acid is attached to its tRNA by its carboxyl group, leaving its amino group exposed for peptide bond formation. ▼

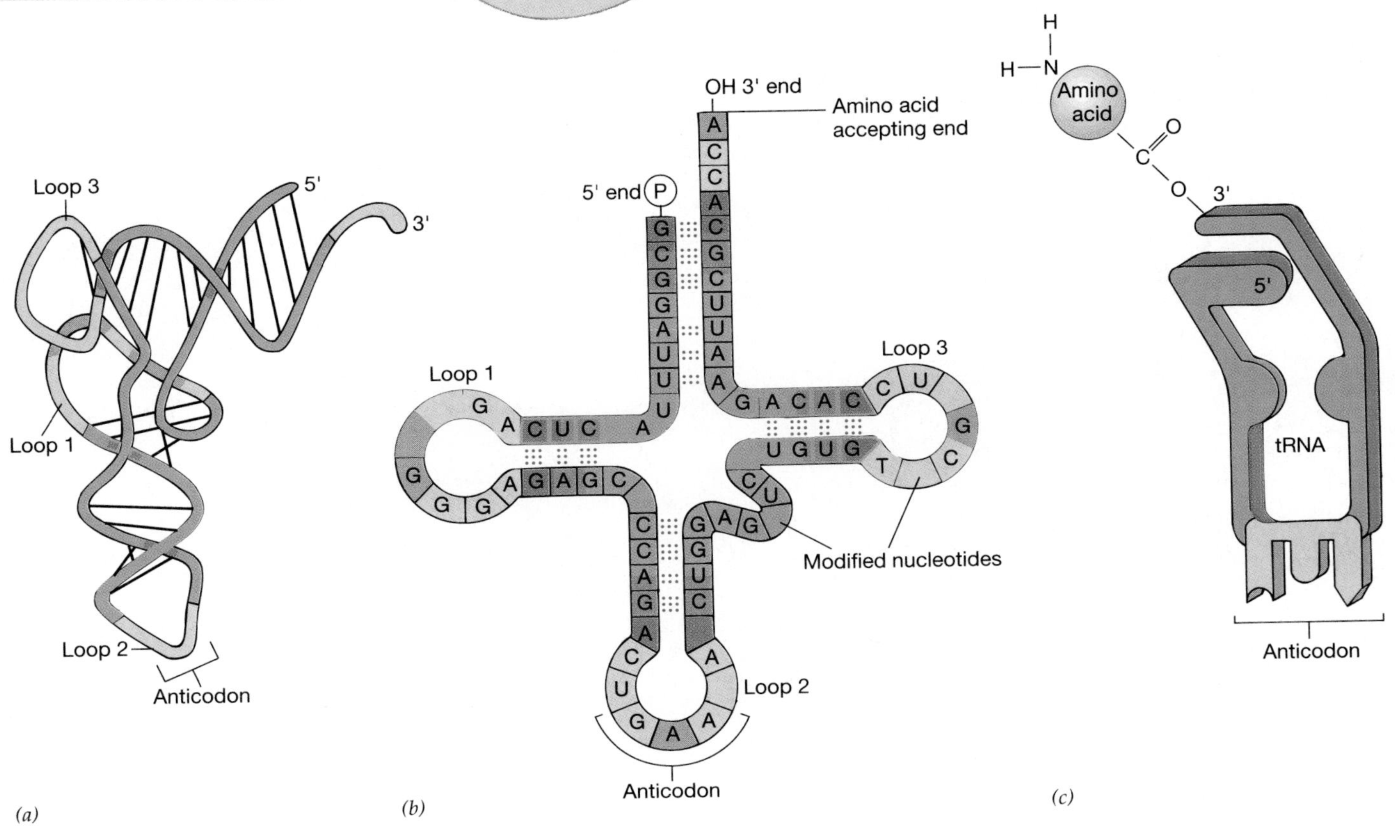

boxyl group of the amino acid is bound to the exposed 3′ hydroxyl group of the terminal adenine nucleotide, leaving the *amino group* free to participate in peptide bond formation. The three-base sequence that serves as the anticodon site is in the middle of the second loop.

Ribosomes bring together all the components of the translational machinery

The importance of ribosomes and of protein synthesis in cellular metabolism is exemplified by a rapidly growing *E. coli* cell, which contains some 15,000 ribosomes, comprising nearly one third of the total mass of the cell. Although prokaryotic and eukaryotic ribosomes are not identical, ribosomes from all organisms are composed of two subunits. In bacteria the smaller of these subunits contains 21 proteins and one RNA molecule, and the larger contains 35 proteins and two RNA molecules. Ribosomal RNA is transcribed from DNA. It does not transfer specific information, but instead has a structural and catalytic function.

Each ribosomal subunit can be isolated intact in the laboratory and then separated into each of its RNA and protein constituents. Under certain conditions it is then possible to reassemble each subunit in a functional form by adding each component in its correct order. Through this approach, together with sophisticated electron microscopic studies, it has been possible to determine the three-dimensional structure of the ribosome (Fig. 12–9*a*) as well as how it is assembled in the living cell. The large subunit contains a depression on one surface into which the small subunit fits. The mRNA fits in a groove formed between the contact surfaces of the two subunits.

Within each ribosome are two depressions, the **A** and **P** binding sites for tRNA molecules (Fig. 12–9*b*). The A site is so named because the *a*minoacyl-tRNA binds at this location. The tRNA holding the *p*olypeptide chain occupies the P site. Following peptide bond formation between the amino acid at the A site and the end of the growing polypeptide chain, the tRNA (now with the entire polypeptide chain attached) moves to the P site of the ribosome, leaving the A site available for the next aminoacyl-tRNA molecule.

One of the roles of the ribosome is to hold the mRNA template, the aminoacyl-tRNA, and the growing peptide chain in the correct orientation so that the genetic code can be read and the peptide bond formed.

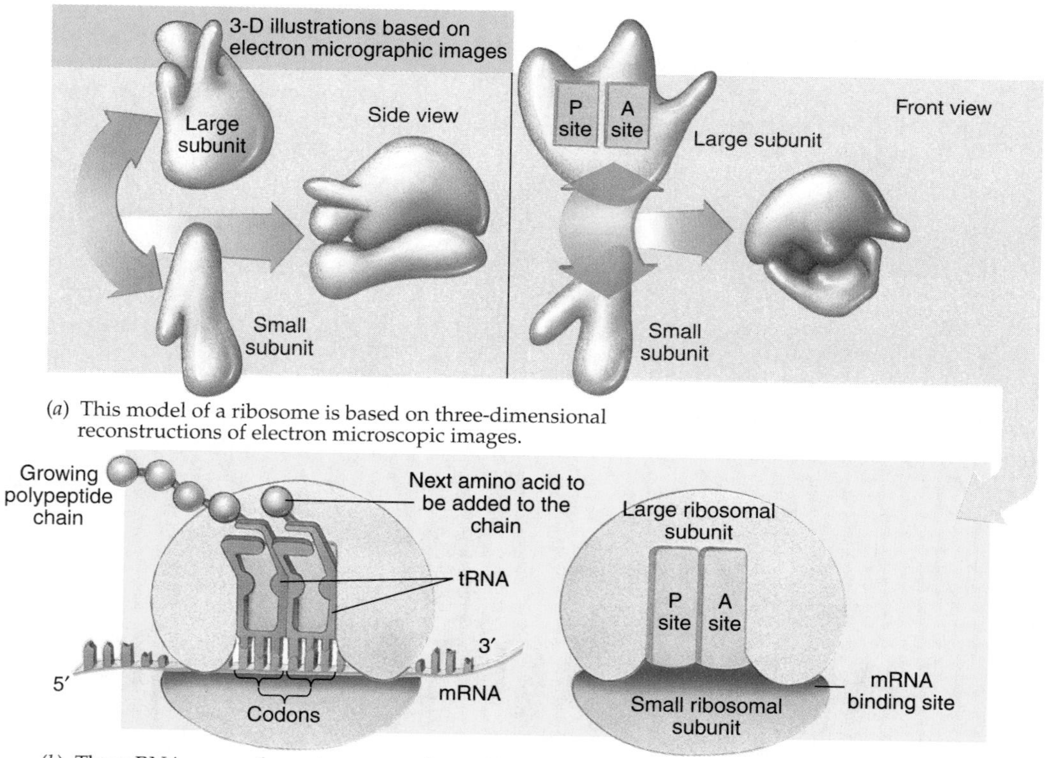

(*a*) This model of a ribosome is based on three-dimensional reconstructions of electron microscopic images.

(*b*) The mRNA passes through a groove formed between the two ribosomal subunits. A ribosome contains two binding sites for tRNAs that recognize adjacent codons. The A site (aminoacyl-tRNA site) binds an aminoacyl tRNA that will be used to add an amino acid to the growing chain. The P site (peptidyl-tRNA site) binds the tRNA that is linked to the growing polypeptide chain.

FIGURE 12–9 A ribosome consists of two subunits, one larger and one smaller, and contains two binding sites for aminoacyl tRNA molecules.

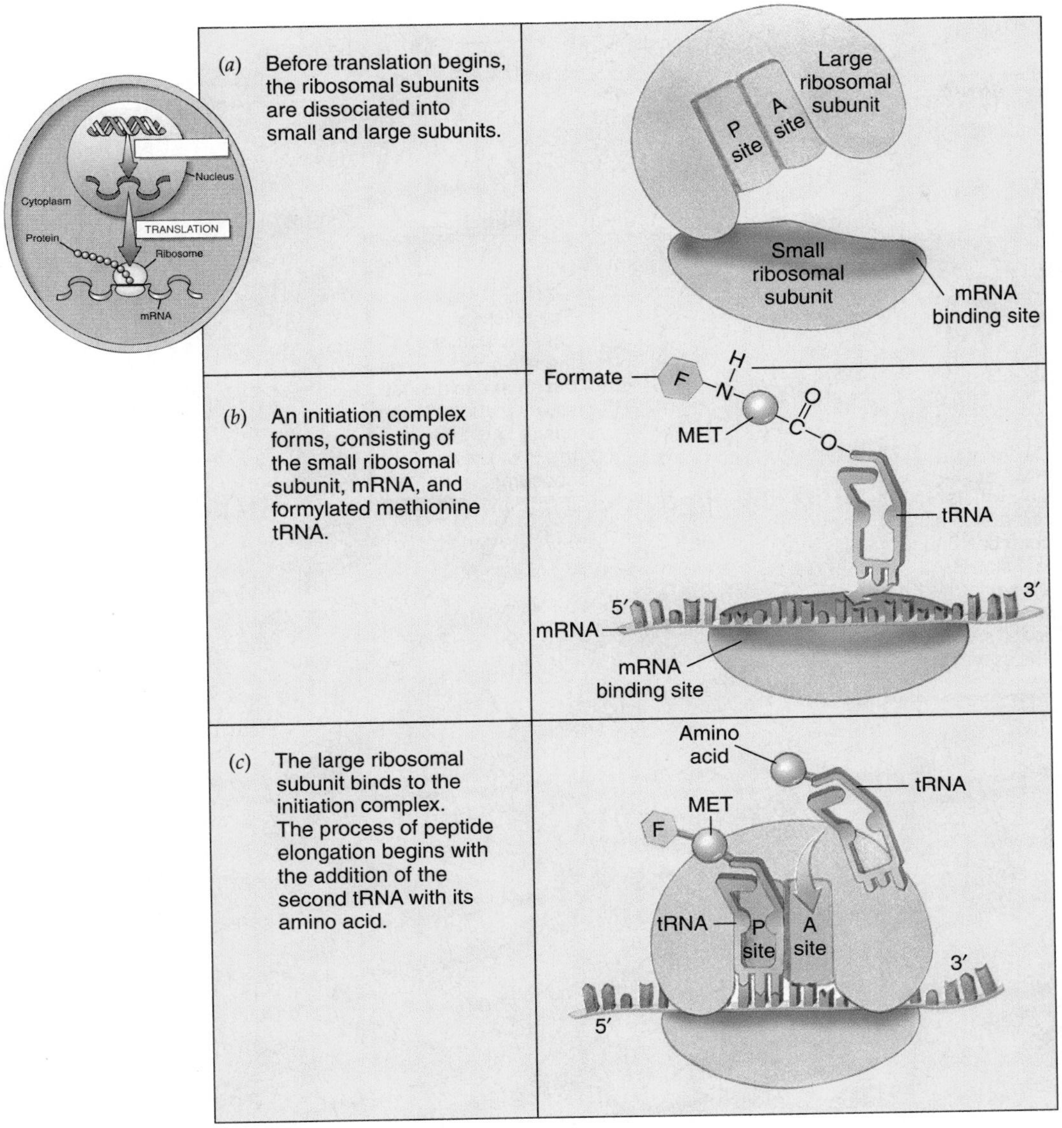

Cytoplasm
Nucleus
Protein
TRANSLATION
Ribosome
mRNA
(a) Before translation begins, the ribosomal subunits are dissociated into small and large subunits.
Large ribosomal subunit
P site
A site
Small ribosomal subunit
mRNA binding site
(b) An initiation complex forms, consisting of the small ribosomal subunit, mRNA, and formylated methionine tRNA.
Formate
F
H
N
MET
O
C
O
tRNA
5′
3′
mRNA
mRNA binding site
(c) The large ribosomal subunit binds to the initiation complex. The process of peptide elongation begins with the addition of the second tRNA with its amino acid.
Amino acid
tRNA
MET
F
tRNA
P site
A site
3′
5′

FIGURE 12–10 During the initiation phase of protein synthesis the small ribosomal subunit participates in the formation of an initiation complex, which then associates with the large subunit.

Translation includes initiation, elongation, and termination

For purposes of discussion the process of protein synthesis is generally divided into three distinct stages: **initiation,** repeating cycles of **elongation,** and **termination.**

The initiation process consists of a number of steps and requires a number of proteins, called **initiation factors.** Initiation begins with the loading of a special **initiation tRNA** onto the small ribosomal subunit. In all organisms the codon for the initiation of protein synthesis is AUG, which codes for the amino acid methionine (Table 12–1).

E. coli contains two types of methionine tRNA that recognize the codon AUG: an initiator tRNA and a regular tRNA. After methionine has been attached to the initiator tRNA, it is modified by the addition of a one-carbon group derived from formic acid to its amino group. Every protein in *E. coli* is synthesized with the modified amino acid **N-formyl-methionine** (Fig. 12–10) at its amino terminal end. Formylated methionine is used only for the first amino acid in a polypeptide chain; if an AUG codon appears in the middle of a protein-coding sequence, a regular methionine tRNA is used.

Once the initiator tRNA is loaded on the small subunit, the initiation complex binds to the special **ribosome-recognition sequences** near the 5′ end of the mRNA; these are upstream of the coding sequences. Binding results in alignment of the anticodon of the initiator tRNA with the AUG initiation codon of the mRNA. The large ribosomal subunit then binds to the complex, forming the completed ribosome.

The addition of other amino acids to the growing polypeptide is called **elongation.** The initiator tRNA is

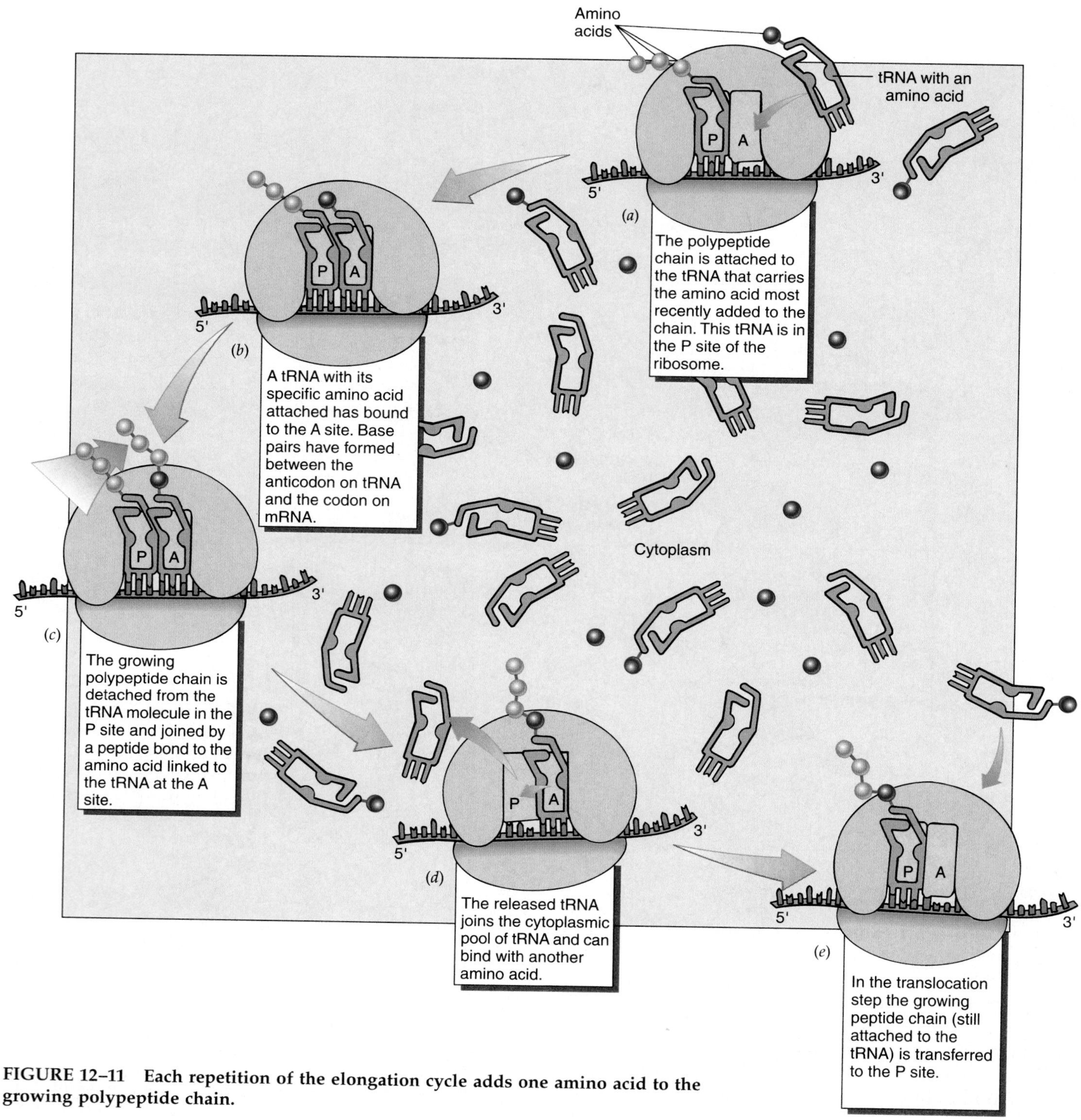

FIGURE 12–11 Each repetition of the elongation cycle adds one amino acid to the growing polypeptide chain.

bound to the P site of the ribosome, leaving the A site unoccupied so that it can be filled by the aminoacyl-tRNA specified by the next codon. Figure 12–11 outlines the events involved in elongation. The appropriate aminoacyl-tRNA binds to the A site by specific base-pairing of its anticodon with the complementary mRNA codon. This binding step requires energy, in this case supplied by guanosine triphosphate (GTP). (GTP, like ATP, transfers energy by donating a phosphate group. You may recall from Chapter 7 that GTP is produced during one of the reactions of the citric acid cycle.)

The amino group of the amino acid at the A site is now aligned with the carboxyl group of the preceding amino acid at the P site. Peptide bond formation then takes place between the amino group of the new amino acid and the carboxyl group of the preceding amino acid. This reaction is spontaneous and does not require additional energy. In this process, the amino acid attached at the P site is released from its tRNA and becomes attached to the aminoacyl-tRNA at the A site.

Recall from Chapter 3 that polypeptide chains have direction, or polarity. The amino acid on one end has a free amino group (the amino end), and the amino acid at the other end has a free carboxyl group (the carboxyl end). *Protein synthesis always proceeds from the amino end to the carboxyl end of the growing peptide chain.*

After the peptide bond is formed, the tRNA molecule is removed from the P site and released so that a new amino acid can be added to it. The growing peptide chain, *which is now attached to the tRNA in the A site,* is then translocated to the P site, leaving the A site open for the next tRNA–amino acid complex. This **translocation** process requires energy, which is again supplied by GTP.

The ribosome and the message move in relation to each other so that the codon specifying the next amino acid in the polypeptide chain becomes positioned in the unoccupied A site. This process involves movement of the ribosome in the 3′ direction along the mRNA molecule; thus, *translation of the mRNA always proceeds in a 5′ to 3′ direction.* The end of the mRNA molecule that is synthesized first during transcription is also the first to be translated into protein.

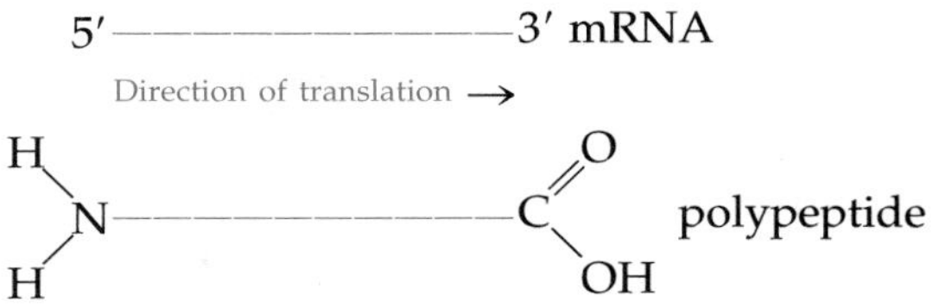

Formation of each peptide bond requires only about 1/20 of a second, so an average-sized protein of about 360 amino acids is completed in about 18 seconds.

The synthesis of the peptide chain is terminated by "release factors" that recognize **termination,** or **stop, codons** at the end of the coding sequence. The codons UAA, UGA, and UAG are special stop signals that do not code for any amino acid. Recognition of a termination codon by the release factors causes the ribosome to dissociate into its two subunits, which can then be used to form a new initiation complex with another mRNA molecule.

A polyribosome is a complex of one mRNA and many ribosomes

In *E. coli* and other prokaryotes, transcription and translation are *coupled* (Fig. 12–12). Ribosomes can bind to the

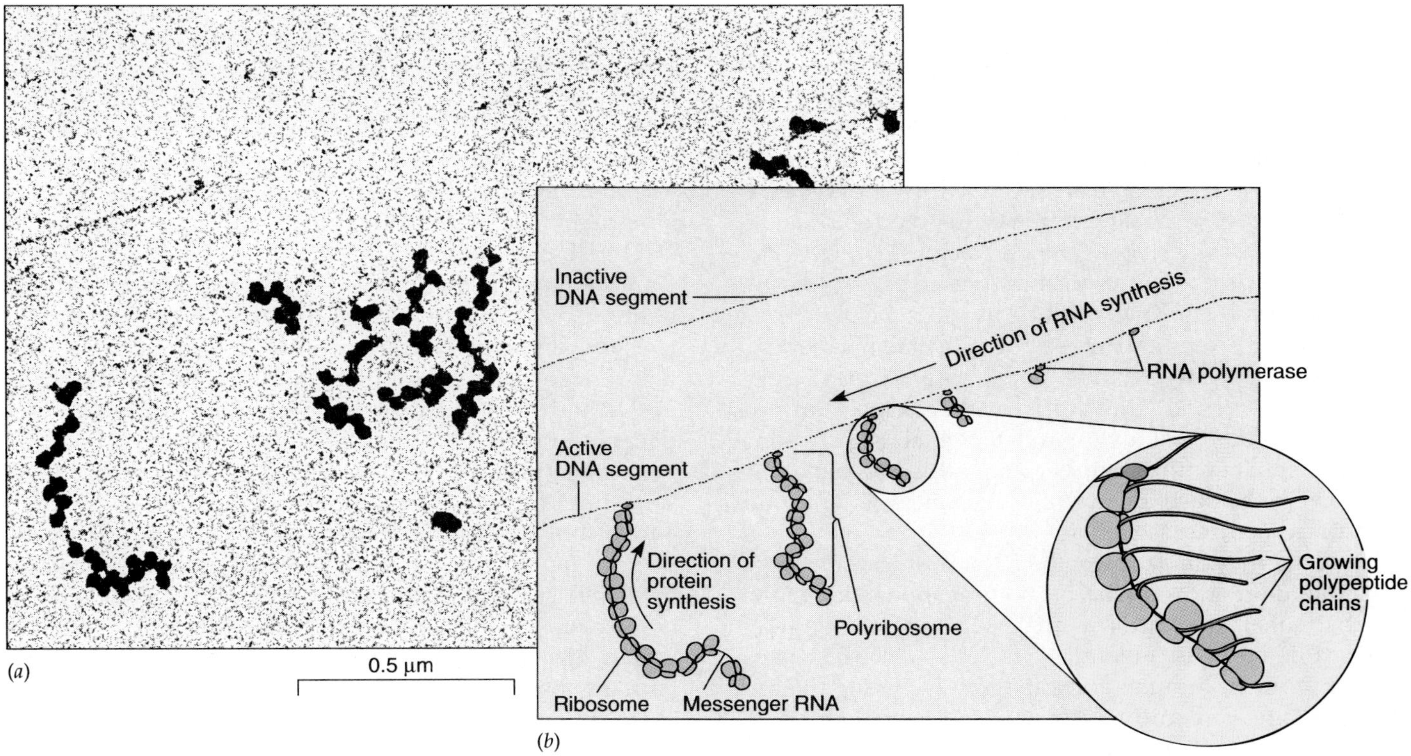

FIGURE 12–12 Transcription and translation are coupled in bacteria, such as *E. coli.* (***a***) Electron micrograph of two strands of DNA, one inactive and the other actively producing mRNA. Protein synthesis begins while the mRNA is being completed, as multiple ribosomes attach to the mRNA to form a polyribosome. (***b***) Diagrammatic representation of the coupled transcription and translation processes. (Courtesy of Dr. Barbara Hankalo, University of California, Irvine)

5′ end of the growing mRNA and initiate translation long before the message is completed. As many as 15 ribosomes may be bound to a single mRNA molecule, spaced as close together as 80 nucleotides. Messenger RNA molecules bound to clusters of ribosomes are referred to as **polyribosomes,** or sometimes **polysomes.**

Although a number of polypeptide chains can be actively synthesized on a single messenger RNA at any one time, the half-life (the time it takes for half of the molecules to be degraded) of mRNA molecules in bacterial cells is only about 2 minutes. Usually, degradation of the 5′ end of the mRNA begins even before synthesis is complete, and once the ribosome recognition sequences are degraded, no more ribosomes can attach to the mRNA and initiate protein synthesis.

TRANSCRIPTION AND TRANSLATION ARE MORE COMPLEX IN EUKARYOTES THAN IN PROKARYOTES

Although the basic mechanisms of transcription and translation are quite similar in all organisms, some significant differences exist between eukaryotes and prokaryotes, particularly with regard to the characteristics of their mRNAs. Whereas bacterial mRNAs are used immediately after transcription without further processing, eukaryotic mRNA molecules undergo specific **posttranscriptional modification and processing.**

Although bacterial mRNA is translated as it is being transcribed from the DNA, eukaryotic RNA is not. Eukaryotic chromosomes are confined to the nucleus of the cell, and protein synthesis takes place in the cytoplasm. The mRNA must be transported through the nuclear envelope and into the cytoplasm before it can be translated. In addition, the original transcript must be modified in several ways (while it is still in the nucleus) before it becomes competent for transport and translation (Fig. 12–13).

Modification of the eukaryotic message begins when the growing RNA transcript is about 20 to 30 nucleotides long. At that point enzymes add a **cap** to the 5′ end of the mRNA chain. The cap is in the form of an unusual nucleotide, 7-methylguanylate, which is guanosine monophosphate with a methyl group added to one of the nitrogens in the base. Eukaryotic ribosomes cannot bind to an uncapped message.

Capping may also protect the RNA from certain types of degradation and may be partially responsible for the fact that eukaryotic mRNAs are much more stable than prokaryotic mRNAs. Eukaryotic mRNAs have half-lives ranging from 30 minutes to as long as 24 hours; the average half-life of an mRNA molecule in a mammalian cell is about 10 hours (compared with 2 minutes in a bacterial cell).

A second modification of eukaryotic mRNA occurs at the 3′ end of the molecule. Near the 3′ end of each completed message there is a sequence of bases that serves as a signal for the addition of a "tail" with many adenines, known as a **polyadenylated** (or **poly-A**) **tail.** Within about 1 minute of completion of the transcript, enzymes in the nucleus recognize this **polyadenylation signal** and cut the mRNA molecule at that site. This is followed by the addition of a string of 100 to 250 adenine nucleotides to the 3′ end. The function of polyadenylation is not clear; perhaps it helps stabilize the mRNA against degradation.

Both noncoding nucleotide sequences (introns) and coding sequences (exons) are transcribed from eukaryotic genes

The final step in mRNA modification is one of the most surprising findings in molecular biology. Most eukaryotic genes have **interrupted coding sequences;** that is, there are long sequences of bases within the protein-coding sequences of the gene that do not code for amino acids in the final protein product! The noncoding regions within the gene are called **introns** (*in*tervening sequences), as opposed to **exons** (*ex*pressed sequences), which are parts of the protein-coding sequence. The reason for this complex structure of eukaryotic genes is a matter of ongoing debate among molecular biologists (see *Making the Connection: "Split Genes" and Evolution*).

The number of introns found in genes is quite variable. For example, the β-globin gene, which produces one component of hemoglobin, contains two introns; the ovalbumin gene of egg white contains seven; and the gene specifying another egg-white protein, conalbumin, contains 16. In many cases the combined lengths of the introns are much greater than those of the exon sequences. For instance, the ovalbumin gene contains about 7700 base pairs, whereas the exon sequences together are only 1859 base pairs long.

When a gene that contains introns is transcribed, the entire gene is copied as a large RNA transcript referred to as **precursor mRNA,** or **pre-mRNA.** This molecule contains both exon and intron sequences. (Note that the terms *intron* and *exon* refer to corresponding nucleotide sequences in both DNA and RNA.)

For the pre-mRNA to be made into a functional message, not only must it be capped and have a poly-A tail added, but the introns must be removed and the exons spliced together to form a continuous protein-coding message. The splicing reactions are mediated by special base sequences within and to either side of the introns. Splicing itself can occur by several different mechanisms. In many instances it involves the association of **small nuclear ribonucleoprotein complexes (snRNPs),** which bind to the introns and catalyze the excision and splicing reactions.

Surprisingly, in some cases the RNA within the intron has the ability to splice itself without the use of protein

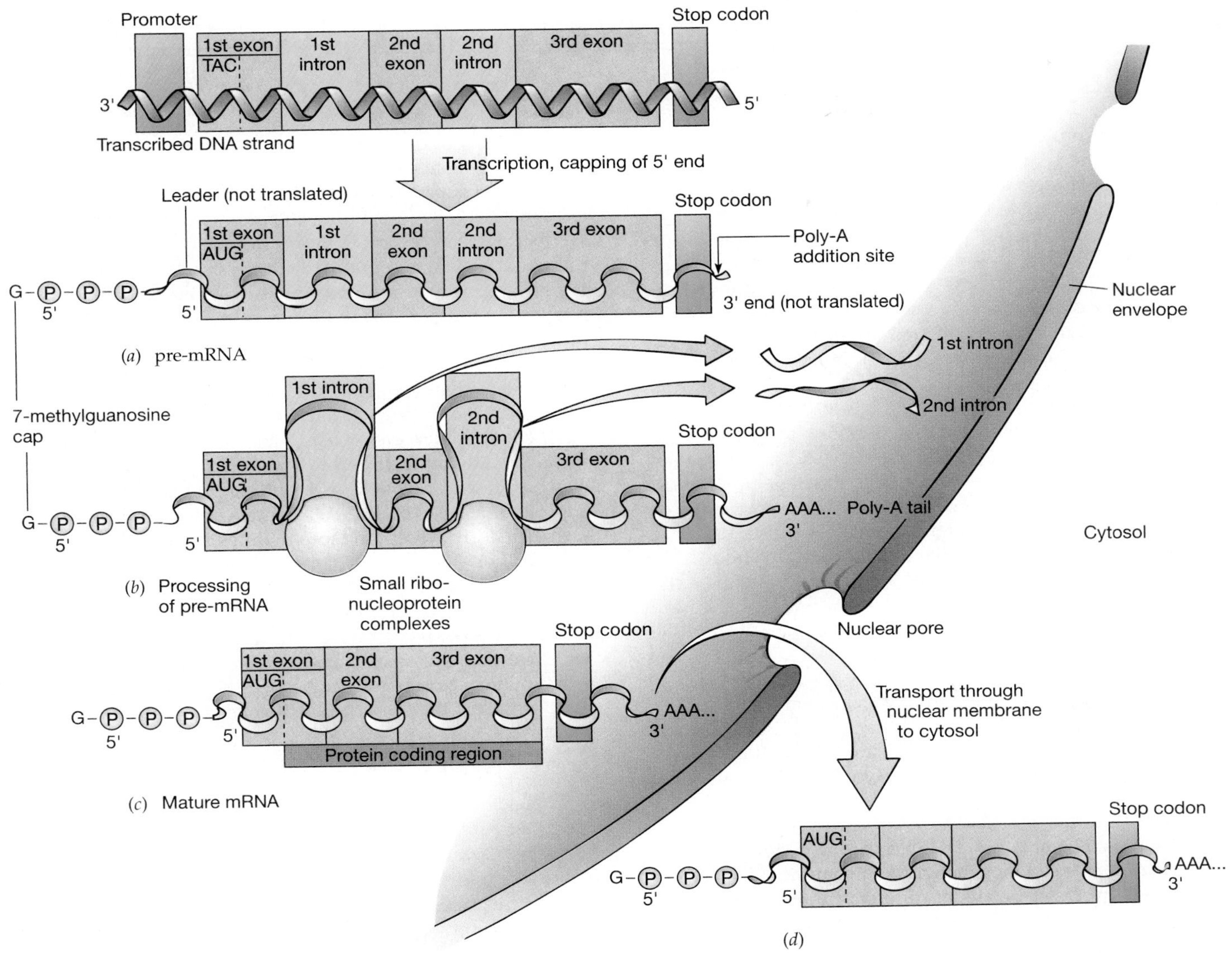

FIGURE 12–13 Eukaryotic RNA undergoes extensive post-transcriptional modification and processing in the nucleus. A typical eukaryotic gene may have multiple exons (coding sequences) and introns (noncoding sequences). (***a***) A DNA sequence containing both exons and introns is transcribed by RNA polymerase to make the primary transcript, or mRNA precursor. As the pre-mRNA is synthesized, the molecule is "capped" by the addition of a modified base to its 5′ end. (***b***) The 3′ end of the synthesized RNA molecule is cleaved at a sequence that designates the poly-A addition site. A poly-A tail (50 to 200 nucleotides long) is then added to the exposed 3′ OH group. (***c***) Introns are removed from the molecule and the exons are spliced together. (***d***) The mature mRNA is transported through the nuclear envelope and into the cytoplasm to be used for protein synthesis.

enzymes. This ability may be related to the fact that some RNAs are capable of acting as enzymes. The discovery of this property led to the recognition of a new class of biological catalysts, termed **ribozymes,** which are formed from RNA molecules rather than proteins.

Although most eukaryotic mRNAs require capping, tailing, and splicing reactions, not all mRNA molecules are modified by all three mechanisms. Some messages, for example, do not contain introns and so do not require splicing. Other eukaryotic mRNAs are capped but do not contain introns or poly-A tails.

THE GENETIC CODE IS READ AS A SERIES OF CODONS

Before the genetic code was deciphered, a number of scientists had become interested in how a genetic code might work. In 1961, Francis Crick and his coworkers concluded from a mathematical analysis that the code was based on nonoverlapping triplets of bases. They predicted that the code is read, one triplet at a time, from a fixed starting point that establishes the **reading frame.** Because there are no "commas" separating the triplets, an alteration in

Making the Connection

"Split Genes" and Evolution

Why do introns occur in most eukaryotic nuclear genes but not in the genes of prokaryotes (or of mitochondria and chloroplasts)? How did this remarkable genetic system involving interrupted coding sequences ("split genes") evolve and why has it survived? It seems incredible that as much as 75% of the original transcript of a eukaryotic nuclear gene has to be removed to make a working message.

Scientific debate has centered around two major questions: How did "split genes" first originate? What role have they played in the evolution of organisms that have them?

One idea that is central to both questions has been advanced. This is the proposal that, although proteins are synthesized as continuous linear amino acid sequences, they actually are *modular* in that they are made up of various functional regions called **domains.** For example, the active site of an enzyme might comprise one domain. A different domain might enable that enzyme to bind to a particular cellular structure and yet another might be a site involved in allosteric regulation (see Chapter 6).

In the early 1980s, Walter Gilbert of Harvard University proposed that exons are nucleotide sequences that code for different structural and functional protein domains. This has turned out to be only partially true. Analyses of the DNA and amino acid sequences of a number of eukaryotic genes have shown that most exons are too small to code for an entire protein domain. However, a block of several exons can code for a domain.

Gilbert further postulated that new proteins with new functions can emerge rapidly when novel combinations of exons are produced by genetic recombination within intron regions of genes that code for different proteins. This hypothesis has become known as *evolution by "exon shuffling."* For example, the low-density lipoprotein (LDL) receptor protein (a protein found on the surface of human cells that binds to cholesterol transport molecules; see Chapter 5) has a number of domains that are related to parts of several other proteins with totally different functions.

Evidence supporting evolution by exon shuffling has led some scientists, including Gilbert, to speculate that the number of basic "exon families" (and their corresponding protein domains) is actually relatively small, and that all of the diverse array of proteins found today evolved from just a few thousand domain prototypes. They argue that these were coded for by "mini-genes" (corresponding to exons) separated by "spacers" (corresponding to introns) in organisms that were ancestral to both prokaryotes and eukaryotes. According to their view, known as the *exon theory of genes*, prokaryotes subsequently lost their introns and retained only their exons. Many scientists disagree; they contend that while exon shuffling probably has been important in the evolution of recently evolved proteins, such as the LDL receptor, these findings do not necessarily support the exon theory of genes. They consider studies of proteins thought to be of "ancient" origin (that is, present in the common ancestor of modern prokaryotes and eukaryotes) to be a better test. Sophisticated biochemical and statistical analyses of such ancient genes have so far failed to reveal any correspondence between exons and protein structure. Thus the exon theory of genes has become less attractive.

Another view is that introns first evolved in the nucleus of an early eukaryote and were propagated as mobile genetic elements, known as transposons (see *Focus On: Reverse Transcriptase, Jumping Genes, and Pseudogenes*). Regardless of how split genes orginated, intron excision provides one of the many ways in which present-day eukaryotes regulate the expression of their genes (see Chapter 13). This opportunity for control, together with the fact that eukaryotic RNAs are far more stable than those of prokaryotes, may balance the energy cost of maintaining a large load of noncoding DNA. ▲

the reading frame would result in the incorporation of incorrect amino acids.

Experimental evidence allowing the assignment of specific triplets to specific amino acids was first obtained by Marshall Nirenberg and Heinrich Matthaei. By constructing artificial mRNA molecules with known base sequences, they were able to determine which amino acids would be incorporated into protein in purified protein synthetic systems. For example, when the synthetic mRNA polyuridylic acid (UUUUUUUUU . . .) was added to a mixture of purified ribosomes, aminoacyl tRNAs, and essential cofactors needed to synthesize protein, only phenylalanine was incorporated into the resulting polypeptide chain. The inference that UUU is the triplet that codes for phenylalanine was inescapable. Similar experiments showed that polyadenylic acid (AAAAAAAAA . . .) codes for a polypeptide of lysine, and polycytidylic acid (CCCCCCCCC . . .) codes for a polypeptide of proline.

By using mixed nucleotide polymers (such as a random polymer of A and C) as artificial messengers, it became possible to assign the other nucleotide triplet codons to specific amino acids. However, three of the codons—UAA, UGA, and UAG—were not found to specify any amino acid. These codons (the stop, or termination, codons mentioned earlier) are now known to be the signals that specify the end of the coding sequence for a polypeptide chain.

Taken together, these experiments led to the coding assignments of all 64 possible codons, listed in Table 12–1, which are essentially universal in living organisms (see

Making the Connection: The Genetic Code and Evolution). Investigators were also able to demonstrate conclusively that the code is a nonoverlapping triplet code.

Remember that *the genetic code we define and use is an mRNA code.* The tRNA anticodon sequences as well as the DNA sequence from which the message is transcribed are complementary to the sequences shown in Table 12–1. For example, the mRNA codon for the amino acid methionine is 5′—AUG—3′. It is transcribed from the DNA base sequence 3′—TAC—5′, and the corresponding tRNA anticodon is 3′—UAC—5′.

The genetic code is redundant

Some of the amino acids in the codon assignments in Table 12–1 are specified by more than one codon. This **redundancy** in the code has certain characteristic patterns. The codons CCU, CCC, CCA, and CCG are synonymous in that they all code for the amino acid proline. The only difference among the four codons involves the nucleotide at the 3′ end of the triplet. Although the code may be read three nucleotides at a time, only the first two nucleotides appear to contain specific information for proline. A similar pattern can be seen for many other amino acids. Only methionine and tryptophan have single-triplet codes. All other amino acids are specified by two to six different codons.

There are 61 codons that specify amino acids. Although most cells contain only about 40 different tRNA molecules, some of these tRNAs can pair with more than one codon, so all of the codons can still be used. This apparent breach of the base-pairing rules was first proposed by Francis Crick as the **wobble hypothesis.** Crick reasoned that the third nucleotide of a tRNA anticodon (which is the 5′ base of that sequence) may sometimes be capable of forming hydrogen bonds with more than one kind of third nucleotide (the 3′ base) of an mRNA codon. Investigators later established this experimentally by determining the anticodon sequences of tRNA molecules and testing their specificities in artificial systems. Some tRNA molecules can recognize as many as three separate codons specifying the same amino acid.

A GENE IS DEFINED AS A FUNCTIONAL UNIT

In Chapter 11 we traced the development of ideas regarding the nature of the gene. For a time it was useful to define a gene as a sequence of nucleotides that codes for one polypeptide chain. As we have continued to learn more about how genes work, we have revised our definition. We now know that some genes produce RNA molecules such as rRNA and tRNA, whereas others code for the RNA component of the small nuclear ribonucleoprotein complexes used to modify complex mRNA molecules. Studies have also shown that in eukaryotic cells a single gene may be capable of producing more than one polypeptide chain by modifications in the way the mRNA is processed (see Chapter 13).

A gene may be defined in terms of its product. One useful definition is that a gene includes a *transcribed nucleotide sequence (plus associated sequences regulating its transcription) that yields a product with a specific cellular function.*

MUTATIONS ARE CHANGES IN DNA

One of the first major discoveries about genes was that they can undergo changes, called **mutations.** We now know that mutations are caused by changes in the nucleotide sequence of the DNA. As explained in Chapter

Making the Connection

The Genetic Code and Evolution

Why does this chapter have only one table to represent the genetic code? This is because of the single most remarkable feature of the code: *it is essentially universal!* Over the years the genetic code has been examined in a diverse array of species and found to be the same in organisms as different as *E. coli*, rose bushes, and humans. These findings strongly suggest that the code is an ancient legacy, derived by the evolution of all living organisms from a common ancestor.

It is thought that the code evolved very early in the history of life (see Chapter 20), and that it has been retained as a kind of "frozen accident" because all but the most minimal changes would be lethal. Recently some very minor exceptions to the universality of the genetic code have been discovered. In several single-celled protozoa, two of the stop codons, UAA and UGA, code for the amino acid glutamine. The other exceptions are found in mitochondria, which contain their own DNA and protein-synthesis machinery for a small number of genes (see Chapters 4 and 20). These slight coding differences vary with the organism, but it is important to keep in mind that in each case all of the other coding assignments are identical to the standard genetic code. ▲

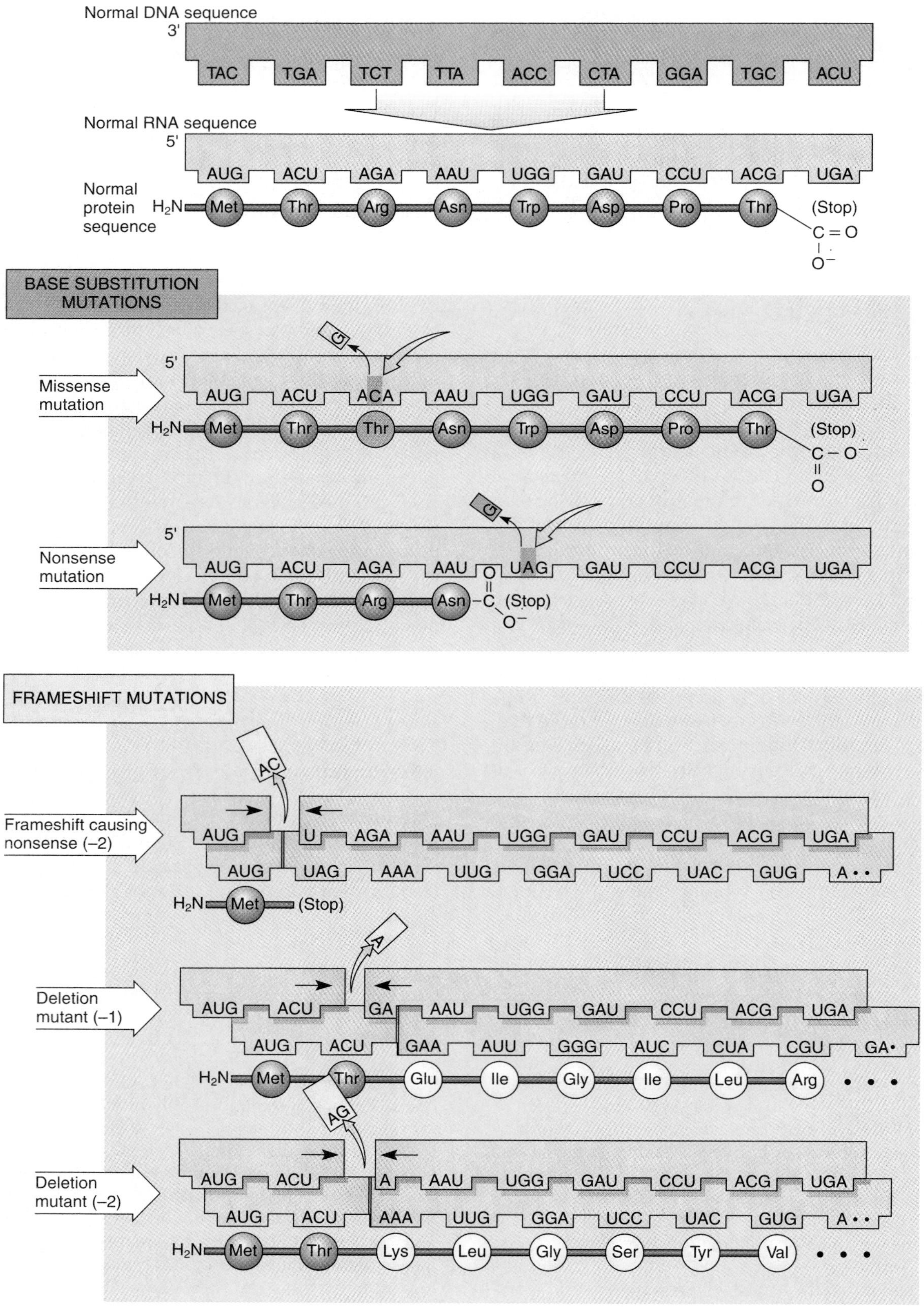
Normal DNA sequence
3'
TAC TGA TCT TTA ACC CTA GGA TGC ACU
Normal RNA sequence
5'
AUG ACU AGA AAU UGG GAU CCU ACG UGA
Normal protein sequence
H2N Met Thr Arg Asn Trp Asp Pro Thr (Stop)
C=O
O−
BASE SUBSTITUTION MUTATIONS
Missense mutation
G
5'
AUG ACU ACA AAU UGG GAU CCU ACG UGA
H2N Met Thr Thr Asn Trp Asp Pro Thr (Stop)
C−O−
O
Nonsense mutation
G
5'
AUG ACU AGA AAU UAG GAU CCU ACG UGA
H2N Met Thr Arg Asn −C (Stop)
O
O−
FRAMESHIFT MUTATIONS
Frameshift causing nonsense (−2)
AC
AUG U AGA AAU UGG GAU CCU ACG UGA
AUG UAG AAA UUG GGA UCC UAC GUG A• •
H2N Met (Stop)
Deletion mutant (−1)
A
AUG ACU GA AAU UGG GAU CCU ACG UGA
AUG ACU GAA AUU GGG AUC CUA CGU GA•
H2N Met Thr Glu Ile Gly Ile Leu Arg • • •
Deletion mutant (−2)
AG
AUG ACU A AAU UGG GAU CCU ACG UGA
AUG ACU AAA UUG GGA UCC UAC GUG A• •
H2N Met Thr Lys Leu Gly Ser Tyr Val • • •

11, once the DNA sequence is changed, DNA replication copies the altered sequence just as it would copy a normal sequence, making the mutation stable over an indefinite number of generations. In most cases the mutant gene has no greater tendency than the original gene to mutate again. Mutations provide the diversity of genetic material that makes it possible to study inheritance and the molecular nature of genes. As we shall see in later chapters, mutations also provide the variation necessary for evolution to occur within a given species.

Genes can be altered by mutation in a number of ways (Fig. 12–14). The simplest type of mutation, called a **point mutation,** or **base-substitution mutation,** involves a change in only one pair of nucleotides. It is now possible to determine where a specific point mutation occurs in a gene by using recombinant DNA methods to isolate the gene and determine its sequence of bases (see Chapter 14). Often these mutations result from errors in base-pairing that occurred during the replication process. For example, an AT base pair might be replaced by a GC, CG, or TA pair. Such a mutation may cause the altered DNA to be transcribed as an altered mRNA. The altered mRNA may then be translated into a peptide chain with only one amino acid different from the normal sequence.

Mutations that result in the substitution of one amino acid for another are sometimes referred to as **missense mutations.** Substitution of a different amino acid into a protein in this way can have a wide range of effects. If the amino acid substitution occurs at or near the active site of an enzyme, the activity of the altered protein may be decreased or even destroyed. Some missense mutations involve a change in an amino acid that is not part of the active site. Others may result in the substitution of a closely related amino acid (one with very similar chemical characteristics). Such mutations may be *silent* (undetectable), at least if one simply examines its effects on the whole organism. Because silent mutations occur relatively frequently, the true number of mutations in an organism or a species is much greater than what is actually observed.

Nonsense mutations are point mutations that can change an amino acid-specifying codon into a termination codon. A nonsense mutation in a gene usually destroys the function of the gene product; in the case of a protein-specifying gene, the part of the polypeptide chain that follows the termination codon is missing.

In **frameshift mutations,** nucleotide pairs are *inserted into* or *deleted from* the molecule. Insertion or deletion of a base in a DNA sequence causes an alteration of the *reading frame.* As a result of this shift, codons downstream of the insertion site specify an *entirely new sequence of amino acids.* Depending on where the insertion or deletion occurs in the gene, a number of different effects can be generated. In addition to producing an entirely new polypeptide sequence immediately after the change, frameshift mutations usually produce a stop or termination codon within a short distance of the mutation. This codon terminates the already altered polypeptide chain. A frame shift in a gene specifying an enzyme usually results in a loss of enzyme activity. If the enzyme is an essential one, the effect on the organism can be disastrous.

Other types of mutations may be due to a change in chromosome structure (see Chapters 13, 15, and 16). These changes usually have a wide range of effects because they involve large numbers of genes.

One type of mutation whose mechanism of action has only recently been understood is caused by DNA sequences that "jump" into the middle of a gene. These movable sequences of DNA, called **transposons,** not only disrupt the functions of some genes but under some conditions also activate previously inactive genes (see *Focus On: Reverse Transcription, Jumping Genes, and Pseudogenes*).

All of the mutations discussed so far can occur infrequently but spontaneously, as a consequence of either mistakes in DNA replication or defects in the mitotic or meiotic separation of chromosomes. Some regions of DNA are much more likely than others to undergo mutation. Such **hot spots** are often single nucleotides or short stretches of repeated nucleotides. They may consist of unusual bases that spontaneously change their structure, or short stretches of repeated nucleotides, which can cause DNA polymerase to "slip." Mutations in certain genes can increase the overall mutation rate, probably by making DNA replication less precise.

Not all mutations occur spontaneously; many of the types of mutations discussed above can also be caused by agents known as **mutagens.** Among these are various types of ionizing radiation, including x rays, gamma rays,

(Text continues on page 308)

◀ **FIGURE 12–14 Some mutations result from changes in only one or a few base pairs.** In each example the normal DNA sequence has been specifically mutated to produce the mRNA molecule shown. Base-pair substitutions can produce several different types of mutations. Silent mutations have no visible effect on the protein because, although a codon has been changed, it still specifies the same or a related amino acid. Missense mutations produce proteins of the same length as the normal protein. These mutant proteins range from being completely functional to having no activity, depending on the type of amino acid that was changed and its location in the polypeptide chain. Nonsense mutations, caused by a change of an amino acid–specifying codon to a termination codon, result in the production of a truncated protein, which is usually not functional. Frameshift mutations, which result from the insertion or deletion of one or two bases, usually have more drastic effects; they cause the base sequence following the mutation to shift to a new reading frame, altering the structure and function of the protein. A frame shift may also produce a termination codon downstream from the mutation, which would have the same effect as a nonsense mutation caused by base substitution.

Focus On

Reverse Transcription, Jumping Genes, and Pseudogenes

For several decades, one of the central premises of molecular biology was that genetic information always flows from DNA to RNA to protein. An important exception to this rule was discovered by Howard Temin in 1964 through his studies on certain viruses. Although viruses are noncellular, they contain a single type of nucleic acid and are capable of reproducing in a host cell. Temin was studying certain unusual, cancer-causing tumor viruses that have RNA, rather than DNA, as their genetic material. He found that infection of a host cell by one of these particular viruses is blocked by inhibitors of DNA synthesis and also by inhibitors of transcription. These findings suggested that DNA synthesis and transcription are required for the multiplication of RNA tumor viruses and that there must be a way for information to flow in the "reverse" direction (that is, from RNA to DNA).

Temin proposed that a **DNA provirus** is formed as an intermediary in the replication of RNA tumor viruses. This hypothesis required a new kind of enzyme—one that would synthesize DNA using RNA as a template. In 1970 Temin and David Baltimore discovered just such an enzyme, and in 1975 they shared the Nobel Prize for their discovery. This RNA-directed DNA polymerase, also known as **reverse transcriptase,** was found in all RNA tumor viruses. (Some RNA viruses that do not produce tumors, however, are replicated directly without using a DNA intermediate.)

After an RNA tumor virus enters the host cell, the viral reverse transcriptase synthesizes a DNA strand that is complementary to the viral RNA. Next, a complementary DNA strand is synthesized, thus completing the double-stranded DNA provirus, which is then integrated into the host cell's DNA. The provirus DNA is transcribed, and the resulting viral mRNA is translated to form specific viral proteins. Additional viral RNA molecules are produced and then incorporated into mature virus particles enclosed by protein coats. Because of their reversal of the usual direction of information flow, such viruses have become known as **retroviruses** (Fig. *a*). The AIDS virus (HIV-1) is the most widely known retrovirus.

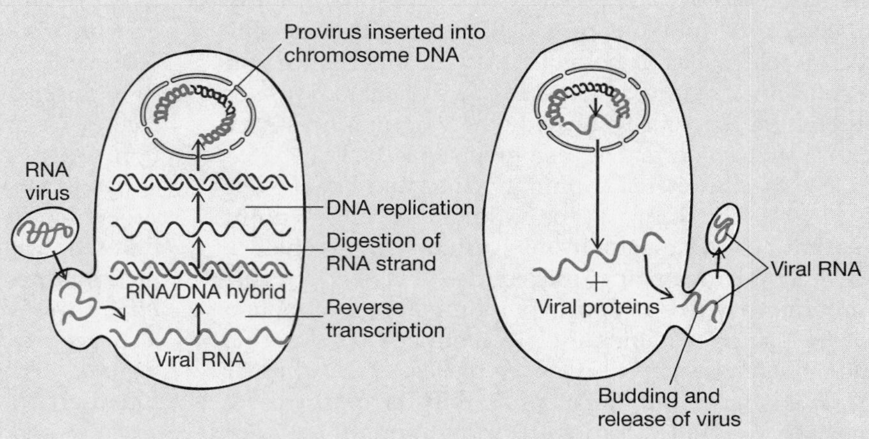

(*a*)

Until a few years ago, reverse transcription was thought to be associated only with retroviruses. Evidence now suggests that reverse transcription may be quite common, which may partially explain such curious phenomena as "jumping" genes and pseudogenes.

Jumping genes, or **mobile genetic elements,** were discovered in maize (corn) by Barbara McClintock in the 1950s. She observed that certain genes appeared to be "turned off" and "turned on" spontaneously. She deduced that the mechanism involved a gene that moved from one region of a chromosome to another, where it would either activate or inactivate genes in that vicinity. It was not until the development of recombinant DNA methods (see Chapter 14) and the discovery of jumping genes in a wide variety of organisms that this phenomenon began to be understood. In recognition of her insightful findings, McClintock was awarded the Nobel Prize in 1983.

Jumping genes, also called **transposable elements** or **transposons,** are segments of DNA that range from a few hundred to several thousand bases. The elements themselves seem to require a special **transposase** enzyme in order to be incorporated into a new location within the chromosome. The longer elements may contain other genes that "go along for the ride."

Many transposable elements have been found to have similarities to retroviruses. Their DNA has unusual base sequences at each end, and their genes are remarkably similar, especially those that code for the proteins required for reverse transcription and integration into the chromosome.

Experiments in Gerald Fink's laboratory have provided evidence that reverse transcriptase is involved in the mechanism by which some transposons move. In these cases the DNA sequence itself does not jump from one location to another; instead, the *information moves through an RNA intermediate* (Fig. *b*). Fink and his colleagues used recombinant DNA methods (see Chapter 14) to insert an intron into the DNA sequence of a yeast transposon as a way of identifying it. They then set up conditions that allowed them to recover and analyze the transposed sequence once it had "jumped." When the transposed sequence appeared at a new location, the intron had been removed, just as introns are removed during the processing of normal mRNA molecules.

Because the enzymes are known only to splice RNA, it appeared that the transposed DNA sequence had been produced from a processed RNA copy of the original DNA rather than from the DNA itself. This would have required the RNA sequence to be converted back to DNA by reverse transcriptase activity within the yeast cells.

Other evidence of nonviral reverse transcriptase activities in cells comes from analyses of **pseudogenes,** which closely resemble certain types of normal genes in mammalian cells. Pseudogenes are DNA sequences that are almost identical to those of normal genes, except that they are riddled with mutations that prevent them from functioning in normal protein synthesis. Many pseudogenes resemble DNA copies of mRNA, for where a normal gene would contain one or more introns, these pseudogenes do not. In addition, many pseudogene DNA sequences end with long poly-A tails. One hypothesis concerning the origin of pseudogenes is that some of them may be derived from the processed mRNAs of normal genes.

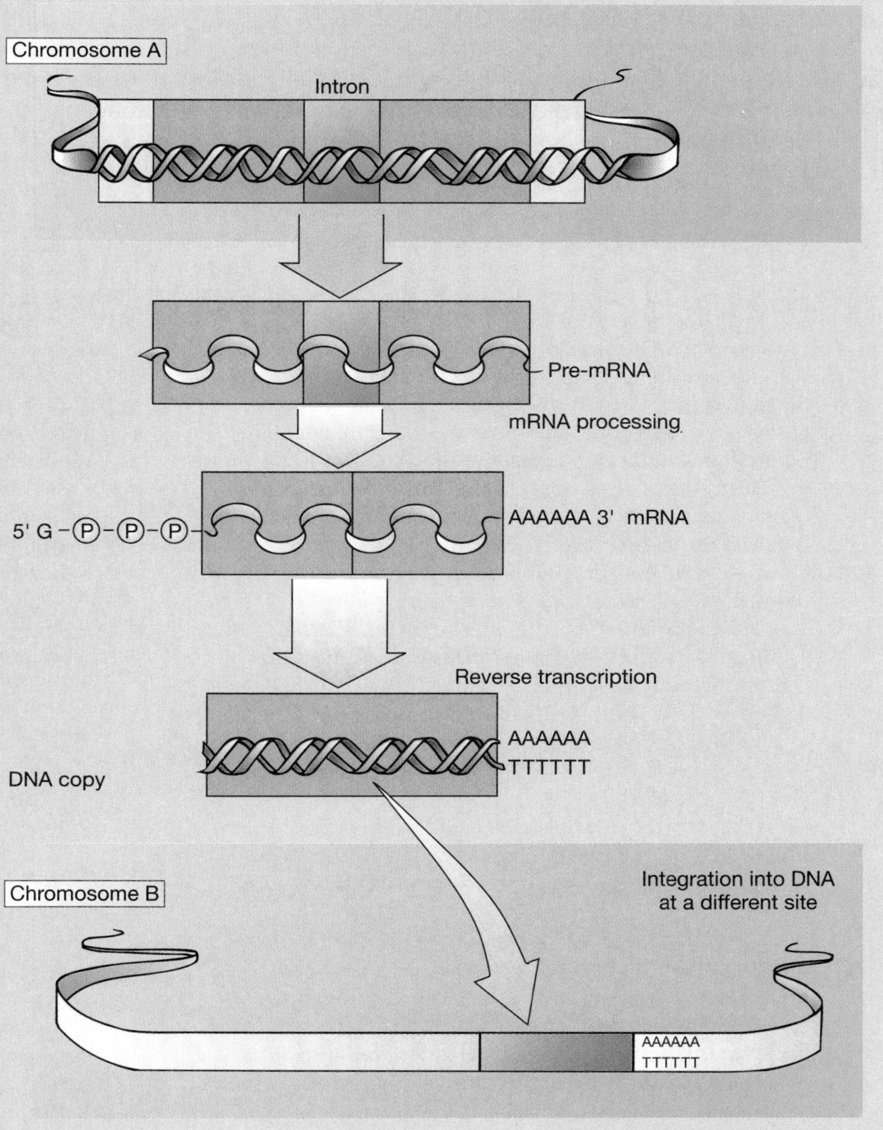

(*b*)

Reverse transcriptase may have synthesized DNA copies, which were reinserted into the chromosome. Because they lack promoter sequences, they are not expressed and simply act as excess baggage, silently accumulating mutations. It is estimated that there may be hundreds or thousands of such sequences in normal human DNA. ■

cosmic rays, and ultraviolet rays. Some chemical mutagens react with and modify specific bases in the DNA, leading to mistakes in complementary base-pairing when the DNA molecule is replicated. Other mutagens are inserted into the DNA molecule and change the normal reading frame during replication.

The overall observed mutation rate is much lower than the frequency of damage to DNA, because all organisms have special systems of enzymes that can repair certain kinds of alterations in the DNA. Nevertheless, some new mutations do persist. In fact, each of us is very likely to have some mutant gene that was not present in either of our parents. Although some of these mutations can produce an altered phenotype, most are not noticeable because they are recessive.

Mutations that occur in the cells of the body (somatic cells) are not passed on to the offspring. However, these mutations are of concern because there is a close relationship between somatic mutations and cancer. Many mutagens are also **carcinogens,** agents that produce cancer.

SUMMARY

I. The mechanism by which information encoded in DNA is used to specify the sequences of amino acids in proteins involves two processes: transcription and translation.
   A. During transcription, an RNA molecule that is complementary to the transcribed or template DNA strand is synthesized. Messenger RNA (mRNA) molecules contain information that specifies the amino acid sequences of polypeptide chains.
   B. During translation, a polypeptide chain specified by the mRNA is synthesized.
      1. Each triplet (three-base sequence) in the mRNA constitutes a codon, which specifies one amino acid in the polypeptide chain.
      2. Translation requires tRNAs and complex machinery, including ribosomes.

II. Messenger RNA is synthesized by DNA-dependent RNA polymerase enzymes.
   A. RNA is formed from ribonucleoside triphosphate subunits, each of which contains the sugar ribose, a base (uracil, adenine, guanine, or cytosine), and three phosphates.
   B. RNA polymerase initially binds to a special DNA sequence called the promoter region.
   C. Like DNA, RNA subunits are covalently joined by a 5′—3′ linkage to form an alternating sugar/phosphate backbone. The same base-pairing rules are followed as in DNA replication, except that uracil is substituted for thymine.
   D. RNA synthesis proceeds in a 5′ → 3′ direction, which means that the template DNA strand is "read" in a 3′ → 5′ direction.

III. Transfer RNAs (tRNAs) are the "decoding" molecules in the translation process.
   A. Each tRNA molecule is specific for only one amino acid. One part of the molecule contains a three-base anticodon, which is complementary to a codon on the mRNA. Attached to one end of the tRNA molecule is the amino acid specified by the complementary mRNA codon.
   B. Amino acids are covalently bound to tRNA by aminoacyl-tRNA synthetase enzymes.

IV. Ribosomes bring together all of the mechanical machinery necessary for translation. They couple the tRNAs to their proper codons on the mRNA, facilitate the formation of peptide bonds between amino acids, and translocate the mRNA so that the next codon can be read.
   A. Each ribosome is made of a large and a small subunit; each subunit contains rRNA and a large number of proteins.
   B. Initiation, the first stage of translation, includes the binding of the small ribosomal subunit protein, plus initiation factors and the initiation tRNA, to the 5′ region of the mRNA, followed by binding of the large ribosomal subunit.
   C. During the elongation cycle, amino acids are added one by one to the growing polypeptide chain.
      1. Elongation proceeds in a 5′ → 3′ direction along the mRNA.
      2. The polypeptide chain grows from its amino end to its carboxyl end.
   D. Termination, the final stage of translation, occurs when the ribosome reaches one of three special termination, or stop, codons, which triggers release of the completed polypeptide chain.

V. In bacterial cells, transcription and translation are coupled. Translation of the mRNA molecule usually begins before the 3′ end of the transcript is completed. A single mRNA molecule can be translated by groups of ribosomes called polyribosomes.

VI. The basic features of transcription and translation are the same in prokaryotic and eukaryotic cells, but eukaryotic genes and their mRNA molecules are more complex than those of bacteria.
   A. After transcription, eukaryotic mRNA molecules are capped at the 5′ end with a modified guanosine triphosphate. Many also have a tail of poly-A nucleotides added at the 3′ end. These modifications appear to protect the molecules from degradation, giving them longer lifetimes than bacterial mRNA.
   B. In many eukaryotic genes the coding regions, called exons, are interrupted by noncoding regions, called introns. Both introns and exons are transcribed, but the introns are later removed from the mRNA precursor and the exons are spliced together to produce a continuous protein-coding sequence.

VII. The genetic code is defined at the mRNA level. There are 61 codons that code for amino acids, plus three codons that serve as stop signals.

A. The start signal for all proteins is the codon AUG, which also specifies the amino acid methionine.
B. The genetic code is virtually universal, strongly suggesting that all organisms are descended from a common ancestor. The only exceptions to the standard code are minor variations.
C. The code is said to be redundant because some amino acids are specified by more than one codon.
D. The genetic code is read from mRNA as a series of nonoverlapping triplets that specify a single sequence of amino acids.

VIII. A gene can be defined as a sequence of nucleotides (plus closely associated regulatory sequences) that can be transcribed to yield a product with a specific cellular function.

IX. Types of mutations range from disruption of the structure of a chromosome to a change in only a single pair of nucleotide bases.
A. A point mutation can destroy the function of a protein if it alters a codon so that it specifies a different amino acid (missense mutation) or becomes a termination codon (nonsense mutation). A point mutation has minimal effects if the amino acid is not altered or if the codon is changed to specify a similar amino acid.
B. Insertion or deletion of one or two base pairs in a gene invariably destroys the function of that protein because it results in a frameshift mutation that changes the codon sequences downstream from the mutation.
C. Mutations can be produced by errors in DNA replication, by physical agents such as x rays or ultraviolet rays, or by chemical mutagens. Mutations can also occur through transposable genetic elements, or "jumping genes," which move from one part of a chromosome to another, disrupting the function of a part of the DNA. Some damage to DNA can be repaired by special systems of enzymes.

SELECTED KEY TERMS

aminoacyl-tRNA
aminoacyl-tRNA synthetase
anticodon
base-substitution mutation
cap
carcinogen
coding sequence
codon
DNA-dependent RNA polymerase
DNA provirus
elongation
exon
frameshift mutation
initiation
interrupted coding sequence
intron
leader sequence
messenger RNA (mRNA)
missense mutation
mobile genetic element
mutagen
N-formyl-methionine
nonsense mutation
point mutation
polyadenylated (poly-A) tail
polyribosome (polysome)
posttranscriptional modification and processing
precursor mRNA (pre-mRNA)
promoter
protein domain
pseudogene
reading frame
redundancy
retrovirus
reverse transcriptase
ribonucleic acid (RNA)
ribosomal RNA (rRNA)
ribosome
ribosome recognition sequence
small nuclear ribonucleoprotein complex (snRNP)
termination
transcription
transfer RNA (tRNA)
translation
translocation
transposon
triplet code
uracil
wobble hypothesis

POST-TEST

1. The process by which information is copied from DNA to mRNA is called __________ .
2. The process by which genetic information in mRNA is decoded to specify the amino acid sequence of a protein is called __________ .
3. An amino acid is specified in the genetic code as a sequence of nucleotides, called a(an) __________ .
4. The type of RNA molecule that "decodes" the information in a codon and translates it into an amino acid is __________ RNA.
5. The "machine" that facilitates formation of peptide bonds between amino acids during translation is a(an) __________ .
6. Messenger RNA is synthesized by DNA-dependent __________ __________ .
7. Ribonucleotides differ from the deoxyribonucleotide subunits found in DNA in that the sugar is ribose and the base __________ is substituted for thymine.
8. The "start" signals for transcription on DNA are __________ regions, which occur just before the point at which synthesis of the 5′ end of the mRNA begins.
9. The nucleotide sequence of the tRNA molecule that is complementary to the appropriate codon on the mRNA is called the __________ .
10. The energy for forming a peptide bond is derived from the process of amino acid activation, an ATP-requiring process that results in the attachment of the amino acid to its __________ __________ .
11. A ribosome is made of two subunits, each of which contains protein and __________ __________ molecules.
12. The first stage of protein synthesis is called __________ . The first step involves the binding of the small ribosomal subunit, protein __________ factors, and aminoacyl tRNA to the 5′ end of the mRNA.
13. The second stage of protein synthesis is the __________ cycle, which involves the sequential binding of the tRNA

specified by each codon, peptide bond formation, and __________ of the ribosome to the next codon on the mRNA.

14. The final stage of protein synthesis, called __________, occurs when the ribosome reaches a stop codon.
15. A complex consisting of a group of ribosomes bound to a single mRNA molecule is a(an) __________.
16. Many eukaryotic genes contain noncoding sequences, called __________, which interrupt protein-coding __________ sequences.
17. After it has been transcribed, the 5′ end of a eukaryotic mRNA is __________ by the addition of a modified nucleotide.
18. Many eukaryotic mRNAs have a modified 3′ end consisting of a(an) __________ tail.
19. Mutations that change a single base pair in a gene, thereby converting an amino acid–specifying codon to a termination codon, are called __________ mutations. Mutations caused by the insertion or deletion of one or two bases in a gene are called __________ mutations.

REVIEW QUESTIONS

1. A transcribed DNA strand has the following nucleotide sequence:

   3′ — TACTGCATAATGATT — 5′

   What would be the sequence of codons in the mRNA transcribed from this strand and also the nucleotide sequence of the complementary nontranscribed DNA strand? What would be the exact anticodon for each codon? Use Table 12–1 to determine the amino acid sequence of the polypeptide. Be sure to label the 5′ and 3′ ends of the nucleic acids, and the carboxyl and amino ends of the polypeptide.
2. What are ribosomes made of? Do ribosomes themselves carry information to specify the amino acid sequence of proteins?
3. In what ways are DNA polymerase and RNA polymerase similar? How do they differ?
4. Outline the steps involved in protein synthesis. Describe the steps of initiation, elongation, and termination.
5. Explain how the genetic code was deciphered. What experimental procedures needed to be developed before this could be accomplished?
6. What are the main types of mutations? What effects does each have on the protein product produced?
7. Why can't amino acids become incorporated into polypeptides without the aid of tRNA?

YOU MAKE THE CONNECTION

1. How many amino acids could be specified if two bases coded for one amino acid? Why is redundancy important to the idea of wobble? If you could "reinvent" the genetic code, would you make any changes? Why or why not?
2. Compare and contrast the formation of mRNA in prokaryotic and eukaryotic cells. How do the differences affect the way in which each type of mRNA is translated? Does one system have any obvious advantage in terms of energy cost? Which system offers greater opportunities for control of gene expression?

RECOMMENDED READINGS

Craig, P.P. "Jumping Genes: Barbara McClintock's Scientific Legacy." *Carnegie Institution of Washington Perspectives in Science,* No. 6, 1994. Copies may be obtained for $1.00 each by calling 202-387-6411 or by writing The Carnegie Institution of Washington, 1530 P Street N.W., Washington, DC, 20005. A fascinating illustrated essay on the life and work of Barbara McClintock and the far-reaching implications of her work.

Darnell, J.E., Jr. "RNA." *Scientific American,* October 1985. An excellent discussion of the role of RNA in the translation of nucleic acid information into protein, and interesting speculations on how RNA itself may have been the first genetic material.

Doolittle, R.F., and P. Bork. "Evolutionarily Mobile Modules in Proteins." *Scientific American,* Vol. 269, No. 4 October 1993. A discussion of the possible roles of exons and introns in protein evolution.

CHAPTER 13

Gene Regulation: The Control of Gene Expression

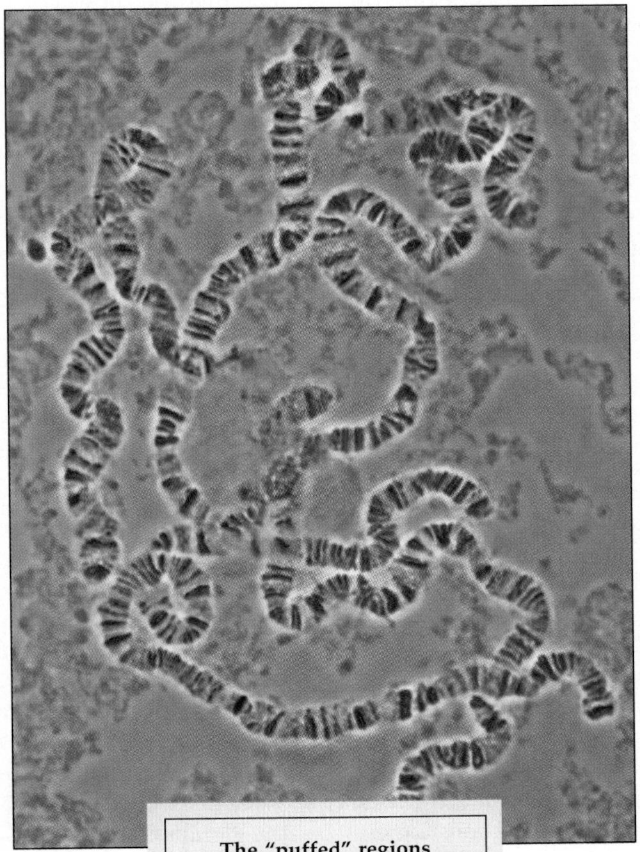

The "puffed" regions of these chromosomes from the salivary gland of a *Drosophila* larva are sites of actively transcribed genes.

(Peter J. Bryant/Biological Photo Service)

Each type of cell in a multicellular organism has a characteristic shape, carries out very specific activities, and makes a distinct set of proteins. Yet, with few exceptions, they all contain the same genetic information. Why, then, are they not identical in structure and molecular composition? Genes are regulated, and only certain subsets of the total genetic information are expressed in any given cell.

What are the mechanisms controlling the expression of a gene? Let us consider a gene that codes for a protein that is an enzyme. It is fully expressed only when it is transcribed into mRNA, the mRNA is translated into protein, and the protein actively catalyzes a specific reaction. Gene expression, then, is the result of a series of processes, each of which can be controlled in many different ways. The control mechanisms require information in the form of various signals (some originating within the cell and others coming from the environment) that interact with DNA, RNA, or protein.

Some of the main strategies used to regulate gene expression include controlling (1) the amount of mRNA that is available, (2) the rate of translation of the mRNA, and (3) the activity of the protein product.

In bacterial cells, energy efficiency and economical use of resources are usually the primary considerations. As a result, most gene regulation in prokaryotes is at the transcriptional level. In eukaryotes there is a much greater emphasis on fine-tuning the control systems, which is consistent with the greater complexity of these cells and the need to control development in multicellular organisms (see Chapter 16). Consequently, eukaryotic gene regulation occurs at many levels.

LEARNING OBJECTIVES

After you have studied this chapter you should be able to

1. Explain why the organization of genes into operons is advantageous to bacteria. Explain why some genes, such as those of the lactose operon, are inducible, while others, such as those of the tryptophan operon, are repressible.
2. Diagram the main components of an inducible operon, such as the lactose operon, and explain the functions of the operator and promoter regions.
3. Sketch the structure of an mRNA molecule produced by the lactose operon and relate that structure to the organization of the DNA in the operon.
4. Differentiate between positive and negative control, and show how both types of control operate in the regulation of the lactose operon.
5. Sketch the structure of a typical eukaryotic gene and the DNA sequences involved in the regulation of that gene.
6. Describe the different functional domains that might be found in a eukaryotic DNA-binding protein.
7. Illustrate how a change in chromosome structure might affect the activity of a gene.
8. List two ways that a gene in a multicellular organism might be able to produce different products in different types of cells.
9. Identify some of the types of regulatory controls that can be exerted in eukaryotes after mature mRNA is formed.

GENE REGULATION IN PROKARYOTES EMPHASIZES ECONOMY

An *Escherichia coli* cell has between 2000 and 4000 genes. Some encode proteins that are always needed (e.g., enzymes involved in glycolysis). These genes, which are constantly transcribed, are called **constitutive genes.** Other gene products are needed only when the bacterium is growing under special conditions.

For instance, the bacteria living in the colon of an adult cow are not normally exposed to the milk sugar lactose. If those cells were to end up in the colon of a calf, however, they would have lactose available as a source of energy. This poses a dilemma. Should a bacterial cell invest energy and materials to produce lactose-metabolizing enzymes just in case it ends up in the digestive system of a calf? Given that the average lifetime of an actively growing *E. coli* cell is about 30 minutes, such a strategy appears wasteful. Yet if *E. coli* cells do not have the capacity to produce those enzymes, they might starve in the midst of an abundant food supply. *E. coli* handles this problem by regulating the production of many of its enzymes in order to efficiently use available organic molecules.

Cells have two basic ways of controlling their metabolic activity. They can regulate the *activity* of certain enzymes (how effectively an enzyme molecule works), and/or they can control the *number* of enzyme molecules present in each cell. Some enzymes may be regulated in both ways in the same cell. An *E. coli* cell growing on glucose is estimated to need about 800 different enzymes. Some of these must be present in large amounts, whereas others are required only in small quantities. In order for the cell to function properly, each enzyme must be efficiently controlled.

Operons in prokaryotes permit coordinated control of functionally related genes

The French researchers François Jacob and Jacques Monod are credited with the first demonstration, in 1961, of how some genes are regulated at the biochemical level. They studied the genes that code for the enzymes that metabolize the disaccharide lactose (see Chapter 3). For *E. coli* to use lactose as an energy source, the sugar must be cleaved into the monosaccharides glucose and galactose by the enzyme **β-galactosidase.** Galactose is then converted to glucose by another enzyme, and the resulting two glucose molecules are further broken down by the glycolysis pathway (see Chapter 7).

E. coli cells growing on glucose contain very little of the β-galactosidase enzyme, perhaps no more than one to three molecules per cell. However, cells grown on lactose have as many as 3000 β-galactosidase molecules per cell, accounting for about 3% of the total cellular protein. Levels of two other enzymes, galactose permease and galactoside transacetylase, also increase when the cells are grown on lactose. The permease is needed to transport lactose efficiently across the bacterial plasma membrane; without it, the lactose cannot enter the cell and be cleaved by β-galactosidase. The transacetylase is also involved in lactose metabolism, although its function is less clear.

Jacob and Monod were able to identify mutant strains of *E. coli* in which a single genetic defect resulted in the loss of all three enzymes. This finding, along with other information, led them to the conclusion that the coding DNA sequences for all three enzymes are linked together on the bacterial DNA as a unit, the **lactose operon** (Fig. 13–1), and are subject to a common control mechanism.

RNA polymerase binds to a single **promoter** site upstream from the coding sequences and proceeds to tran-

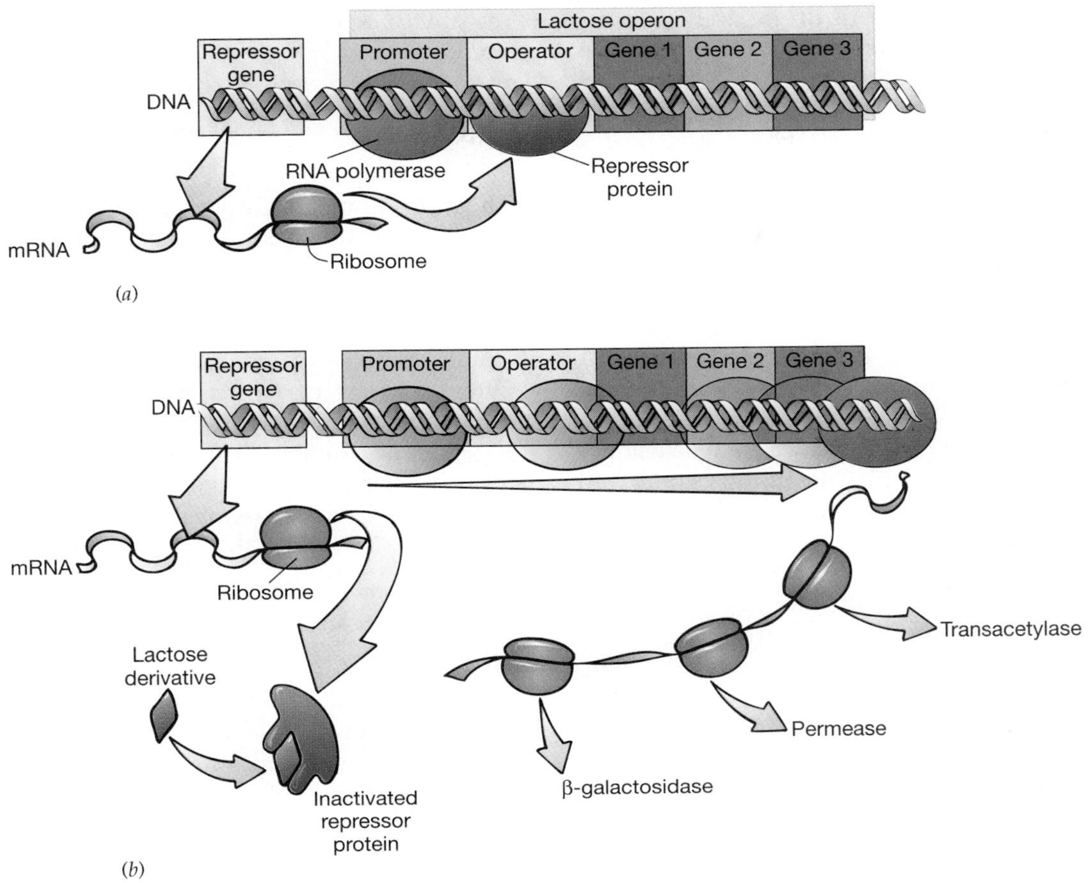

FIGURE 13–1 The genes for three enzymes used by *E. coli* to metabolize the disaccharide lactose are organized as a single unit, called an operon. All three genes are transcribed as part of a single mRNA molecule. A sequence of bases called the operator is next to the promoter region. (*a*) In the absence of lactose, a repressor protein, encoded by a gene separate from the operon, binds to the operator region. By preventing RNA polymerase from binding to the promoter, the bound repressor protein blocks transcription of the structural genes. (*b*) When lactose is present, a metabolic derivative of the sugar binds to the repressor at an allosteric site, altering the structure of the protein so that it can no longer bind to the operator. This allows RNA polymerase to bind to the promoter and synthesize the mRNA. The lactose operon is referred to as an inducible operon because it is normally inactive and requires an inducer (in the form of the metabolic derivative of lactose) to turn on transcription.

scribe the DNA, forming a single mRNA molecule that contains the coding information for all three enzymes. This mRNA contains translation termination and initiation codons between the enzyme-coding sequences; hence it is translated as three separate protein molecules. Because all three enzymes are translated from the same mRNA molecule, their synthesis can be coordinated by turning a single molecular "switch" off or on.

The switch that controls mRNA synthesis is called the **operator;** it is a sequence of bases that overlaps part of the promoter region and is upstream from the first structural (protein-coding) gene in the lactose operon. When lactose is absent, a **repressor protein** called the *lactose repressor* binds tightly to the operator region. Because the repressor protein is large enough to cover part of the promoter sequence, RNA polymerase is unable to bind to the lactose promoter site, and transcription of the lactose operon is effectively blocked.

The lactose repressor protein is encoded by a regulatory gene located upstream from the operator site. Unlike the lactose operon genes, the repressor gene is constitutive; that is, it is always "on," so small amounts of the repressor protein are produced continuously. This protein is able to diffuse throughout the cell and bind specifically to the lactose operator sequence. When cells are grown in the absence of lactose, the operator site is nearly always occupied by a repressor molecule. Only on rare occasions, when the operator site is briefly unoccupied, can RNA polymerase bind and initiate transcription of the structural genes. Because *E. coli* mRNA is rapidly degraded (having a half-life of about 2 to 4 minutes), very few proteins are translated from that mRNA.

Lactose is able to "turn on," or *induce*, the transcription of the lactose operon because the lactose repressor protein contains a second functional region separate from its DNA binding site. This is an **allosteric binding site** (see Chapter 6) for allolactose, a metabolite of lactose. If lactose is present in the growth medium, a few molecules are able to enter the cell and are converted to allolactose. When a molecule of allolactose binds to the repressor at the allosteric site, it alters the conformation of the protein so that its DNA binding site can no longer recognize the operator. When all the repressor molecules have allolactose bound to them and are therefore inactivated, RNA polymerase binds to the unblocked promoter, and the operon is actively transcribed.

The *E. coli* cell continues to produce β-galactosidase and the other lactose operon proteins until virtually all of the lactose is used up. When intracellular levels of lactose drop, the repressor proteins no longer have the allolactose sugar bound to them. They then assume a conformation that allows them to bind to the operator region and shut down transcription of the operon.

Mutants play an essential role in allowing researchers to dissect a regulatory system such as the lactose operon. They allow investigators to map the positions of the genes on the DNA and to infer normal gene functions by studying what happens when they are missing or altered. This information is usually combined with results of direct biochemical studies.

For example, Jacob and Monod studied various constitutive lactose operon mutants that transcribed the structural genes of the lactose operon at a significant rate even in the absence of lactose. One group of constitutive mutants had abnormal genes with map positions in a region that is not close to the lactose operon itself. The genes responsible for the behavior of a second group of constitutive mutants had map positions in the region of the lactose operon but did not directly involve any of the three structural genes.

On the basis of these findings, Jacob and Monod hypothesized the existence of a regulatory gene that is separate from the genes of the lactose operon and responsible for producing a repressor protein. Although the specific defect may vary, the mutants of the first group do not produce active repressor proteins; hence no binding to the lactose operator and promoter takes place, and the lactose operon is transcribed constitutively. The members of the second group of constitutive mutants produce normal repressor molecules but have abnormal operator sequences incapable of binding the repressor.

In contrast to the constitutive mutants, other mutants failed to transcribe the lactose operon even when lactose was present. Some of the abnormal genes had the same map position as the regulatory gene. They were found to have an altered allosteric binding site on the repressor protein that prevented allolactose from binding, although the ability of the repressor to bind to the operator was unaffected

An inducible gene is not transcribed unless a specific inducer inactivates its repressor

The lactose operon is called an **inducible system.** An inducible gene or operon is usually controlled by a repressor that keeps it in the "off" state. The presence of an **inducer molecule** (in this case allolactose) renders the repressor inactive and so the gene or operon is turned "on." Inducible genes or operons usually code for enzymes that are part of catabolic pathways; these break down molecules to provide both energy and components for anabolic reactions. This type of regulatory system enables the cell to save the energy costs of making enzymes when no substrates are available for them to act on.

A repressible gene is transcribed unless a specific repressor/corepressor complex is bound to the DNA

Another type of gene regulation system in bacteria is associated mainly with anabolic pathways in which amino acids, nucleotides, and other essential molecules are synthesized from simpler precursors. Regulation of these pathways normally involves **repressible** enzymes, which are coded for by repressible genes.

Repressible genes and operons are usually "on"; they are turned off only under special conditions. In most cases the molecular signal used to regulate these genes is the end product of the metabolic pathway. When the supply of the end product (e.g., an amino acid) is low, all enzymes in the pathway are actively synthesized. When intracellular levels of the end product are high, enzyme synthesis is repressed. Because compounds such as amino acids are continuously needed by the growing cell, the most effective strategy is to keep the genes that control their production "on" except when a large supply of the amino acid is available. The ability to turn the genes off allows cells to avoid overproduction of amino acids and other molecules that are essential but energetically expensive to make.

The tryptophan operon is an example of a repressible system. In both *E. coli* and a related bacterium, *Salmonella*, the operon consists of five structural genes that code for the enzymes required for synthesis of the amino acid tryptophan; these are clustered together as a transcriptional unit with a single promoter and a single operator (Fig. 13–2). A distant regulator gene codes for a diffusible repressor protein, which differs from the lactose repressor in that it is synthesized in an inactive form and is unable to bind to the operator region.

The DNA-binding site of the repressor becomes effective only when tryptophan, its **corepressor,** binds to an

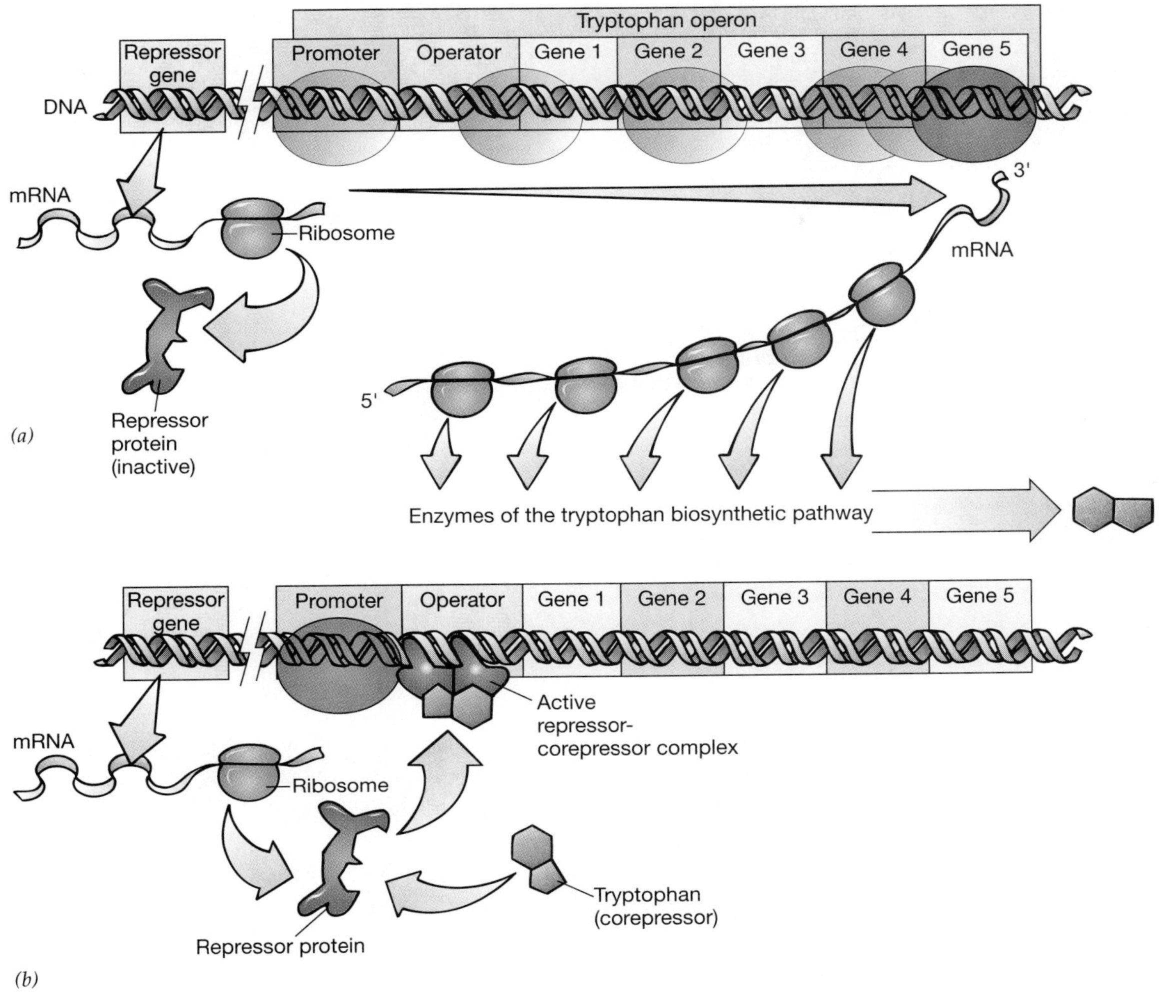

FIGURE 13–2 Genes coding for enzymes that synthesize the amino acid tryptophan are organized in a repressible operon (one that is normally actively transcribed). (*a*) A regulatory gene encodes a repressor protein that is initially inactive; it is unable to prevent transcription because it cannot bind to the operator. (*b*) When intracellular tryptophan levels are high, the amino acid binds to an allosteric site on the repressor protein, changing its conformation. The resulting active form of the repressor binds to the operator region, blocking transcription of the operon until tryptophan is again required by the cell.

allosteric site on the repressor. When intracellular tryptophan levels are low, the repressor protein is inactive and unable to bind to the operator region of the DNA. As the concentration of intracellular tryptophan rises, some tryptophan binds to the allosteric site of the repressor, altering its conformation so that it binds tightly to the operator. This has the effect of switching the operon off, thereby blocking transcription.

Negative regulators repress transcription; positive regulators activate transcription

The features of the lactose and tryptophan operons described so far are examples of **negative control.** Systems under negative control are those in which the DNA-binding regulatory protein is a *repressor* that turns *off* transcription of the gene. Some regulatory systems involve **positive control**—that is, regulation by **activator proteins** that bind to DNA and thereby stimulate transcription of a gene. The lactose operon contains both negative and positive controlling elements (Fig. 13–3).

Positive control of the lactose operon requires that the cell be able to sense the absence of the sugar glucose, which is the initial substrate in the glycolysis pathway (see Chapter 7). Lactose, like glucose, is a catabolite and can undergo stepwise breakdown to yield energy. However, because lactose must first be converted to glucose, it is most efficient for *E. coli* cells to use the available sup-

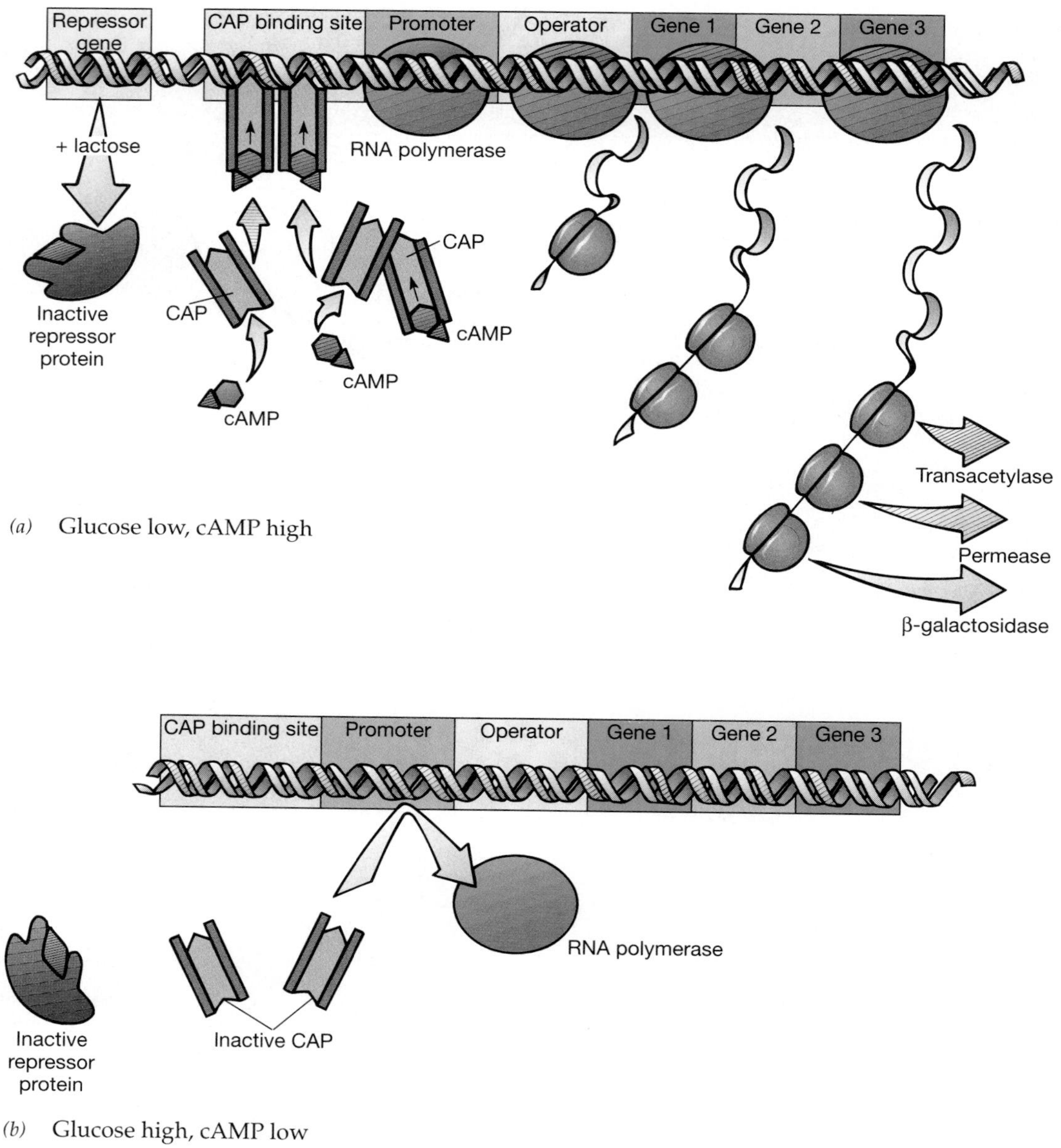

FIGURE 13–3 The lactose operon is controlled positively as well as negatively. The lactose promoter by itself is weak and binds RNA polymerase inefficiently even when the lactose repressor is inactive because lactose is present. The catabolite activator protein is an allosteric regulator that can bind to a sequence of bases adjacent to the promoter, allowing RNA polymerase to bind efficiently, thereby stimulating transcription of the operon. The CAP molecule contains two polypeptides. When each has cyclic AMP (cAMP) bound to its allosteric site, the protein can bind to the DNA sequence. The cell's cAMP concentration is inversely proportional to the glucose concentration. (***a***) When glucose levels are low, cAMP binds to CAP, which then activates transcription of the operon by binding to the DNA. (***b***) When glucose levels are high, cAMP is low. CAP is in an inactive form and cannot activate transcription.

ply of glucose first. By using glucose as the preferred substrate, the cell is spared the considerable energy cost of making additional enzymes such as β-galactosidase.

The lactose operon actually has a very inefficient promoter sequence; that is, it has a low affinity for RNA polymerase even when the repressor protein is inactivated. However, a DNA sequence adjacent to the promoter site is a binding site for another protein, called the **catabolite activator protein (CAP)** (Fig. 13–4).

CAP increases the affinity of the promoter region for RNA polymerase, allowing the enzyme to recognize the promoter efficiently and to bind tightly to the DNA. In its

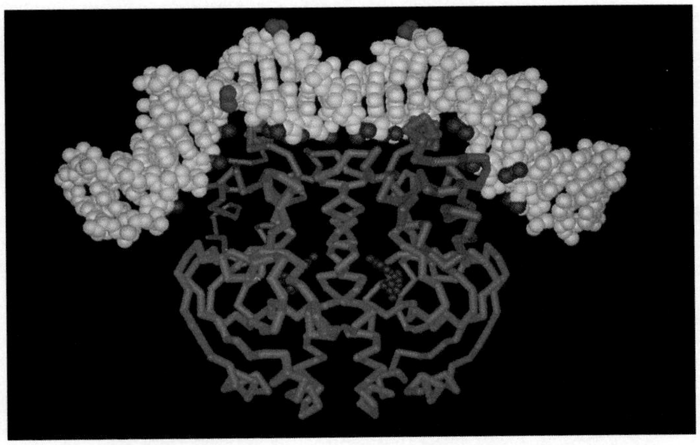

FIGURE 13–4 This computer-generated picture depicts the three-dimensional structure of active CAP binding to DNA. Notice that the binding of CAP causes bending of the DNA double helix. CAP is a dimer consisting of two identical polypeptide chains, each of which binds one molecule of cAMP. (Courtesy of S. C. Schultz, G. C. Shields, and T. A. Steitz, Yale University)

active form CAP has **cyclic AMP,** or **cAMP** (see Fig. 3–29), an altered form of adenosine monophosphate, bound to an allosteric site. As the cells become depleted of glucose, cAMP levels increase. The cAMP molecules bind to CAP, and the resulting complex then binds to the CAP-binding site near the lactose operon promoter and stimulates the transcription of the operon. Thus, the operon is fully active only if lactose is present and intracellular glucose levels are low. The properties of negative and positive control systems are summarized in Table 13–1.

A regulon is a group of functionally related operons controlled by a common regulator

CAP differs from the lactose and tryptophan repressors in that it can control transcription of operons involved in the metabolism of a number of other catabolites—such as the sugars galactose, arabinose, and maltose—as well as of lactose. A group of operons controlled by one regulator of this type is generally referred to as a **regulon** (Fig. 13–5).

Other multigene systems in bacteria are also controlled in this manner. For example, genes involved in nitrogen and phosphate metabolism are organized into regulons that consist of multiple sets of operons controlled by one or more combinations of regulatory genes. Other complex multigene systems respond to changes in environmental conditions, such as rapid shifts in temperature, exposure to radiation, changes in osmotic pressure, and changes in oxygen levels. Specific mutants often provide clues to the existence of a regulon system. A single mutation that destroys the activity of CAP, for example, prevents the cell from efficiently metabolizing not only lactose but many other sugars also regulated by CAP.

Table 13–1 TYPES OF TRANSCRIPTIONAL CONTROL IN PROKARYOTES*

NEGATIVE CONTROL		
Inducible genes		
Repressor protein alone	→	**Active repressor "turns off" regulated gene(s)**
Lactose repressor alone	→	Lactose operon not transcribed
Repressor protein + inducer	→	**Inactive repressor/inducer complex fails to "turn off" regulated gene(s)**
Lactose repressor + allolactose	→	Lactose operon transcribed
Repressible genes		
Repressor protein alone	→	**Inactive repressor fails to "turn off" regulated gene(s)**
Tryptophan repressor alone	→	Tryptophan operon transcribed
Repressor protein + corepressor	→	**Active repressor/corepressor complex "turns off" regulated gene(s)**
Tryptophan repressor + tryptophan	→	Tryptophan operon not transcribed
POSITIVE CONTROL		
Activator protein alone	→	**Activator alone cannot "turn on" regulated gene(s)**
CAP alone	→	Transcription of lactose operon not stimulated
Activator protein + coactivator	→	**Functional activator/coactivator complex "turns on" regulated gene(s)**
CAP + cAMP	→	Transcription of lactose operon stimulated

*A general description of each type is followed by a specific example. A negative regulator is a repressor that "turns off" transcription of the regulated gene(s). Conversely, a positive regulator is an activator that "turns on" transcription.

(a) Low glucose

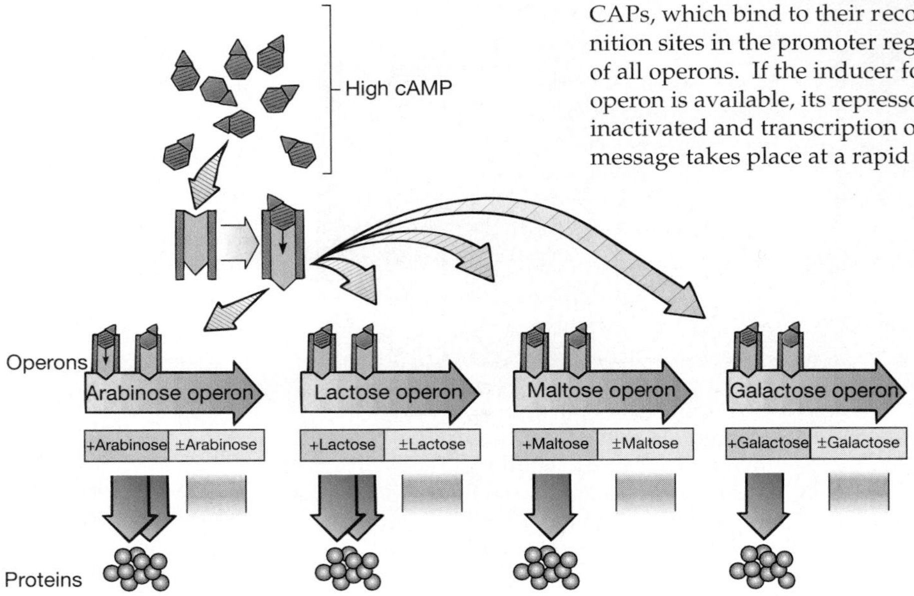

(a) When glucose levels are low, cAMP levels increase, activating CAPs, which bind to their recognition sites in the promoter regions of all operons. If the inducer for an operon is available, its repressor is inactivated and transcription of the message takes place at a rapid rate.

(b) High glucose

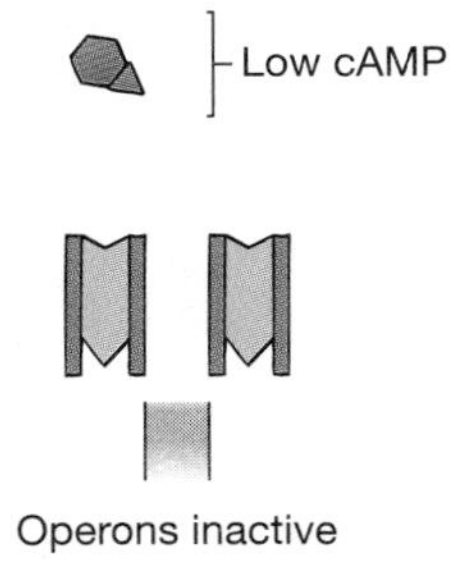

(b) When glucose levels are high, cAMP levels are low, rendering CAP inactive. Under these conditions, none of the operons is active, even if the appropriate carbon source is available.

FIGURE 13–5 Operons that convert a number of different sugars to glucose in *E. coli* make up a regulon that is under positive control by CAP.

Not all constitutive genes are transcribed at the same rate

Many of the gene products encoded by the *E. coli* DNA are needed only under certain environmental or nutritional conditions. As we have seen, these genes are generally regulated at the level of transcription. They can be turned on and off as metabolic and environmental conditions change. By contrast, constitutive genes are continuously transcribed, but they are not necessarily transcribed (or their mRNAs translated) at the same rate. Some enzymes work more effectively or are more stable than others and consequently need to be present in smaller amounts. Constitutive genes that encode proteins required in large amounts are generally transcribed more rapidly than genes for proteins required at lower levels. The transcription rate of these genes is controlled by their promoter sequences. Genes with efficient ("strong") promoters bind RNA polymerase more frequently and consequently transcribe more mRNA molecules than those with inefficient ("weak") promoters.

Genes coding for repressor or activator proteins that regulate metabolic enzymes are usually constitutive and produce their protein products constantly. Because each cell usually needs relatively few molecules of any specific repressor or activator protein, promoters for those genes tend to be relatively weak.

Some posttranscriptional regulation occurs in prokaryotes

Although much of the variability in protein levels in *E. coli* is determined by **transcriptional level control,** for some genes other regulatory mechanisms operate after transcription. These **posttranscriptional controls** may work at various levels of gene expression.

Translational controls are posttranscriptional controls that regulate the rate at which a particular mRNA molecule is translated. Because the lifetime of an mRNA molecule in a bacterial cell is very short, a molecule that is translated rapidly can produce more proteins than one that is translated slowly. Some mRNA molecules in *E. coli* are translated as much as 1000 times faster than others. Most of the differences appear to be due to the rate at which ribosomes can attach to the mRNA and begin translation.

Posttranslational controls generally act as switches that activate or inactivate one or more existing enzymes. These systems allow the cell to respond to changes in the intracellular concentrations of essential molecules, such as amino acids, by rapidly adjusting the activities of its enzymes. A common posttranslational control adjusts the rate of synthesis in a metabolic pathway through **feedback inhibition** (see Chapter 6). The end product binds to the first enzyme in the pathway at an allosteric site, temporarily inactivating the enzyme. When the first enzyme in the pathway does not function, all of the succeeding enzymes are deprived of substrates. Notice that this differs from the end-product repression of the tryptophan operon discussed previously. In that case, the end product of the pathway prevented the formation of new enzymes. Feedback inhibition acts as a fine-tuning mechanism that regulates the activity of the existing enzymes in a metabolic pathway.

GENE REGULATION IN EUKARYOTES IS MULTIFACETED

Like bacteria, eukaryotic cells must respond to changes in their environment by turning appropriate sets of genes on and off. Multicellular eukaryotes require additional modes of regulation that permit individual cells to become committed to specialized roles, and groups of cells to organize into tissues and organs. In previous chapters we discussed the fact that all aspects of information transfer—including replication, transcription, and translation—are far more complicated in eukaryotes. Not surprisingly, this complexity provides additional opportunities for control of gene expression.

Unlike many of the prokaryotic genes, most eukaryotic genes are not found in operon-like clusters. However, each eukaryotic gene has specific regulatory sequences that are essential in the control of transcription.

Many of the "housekeeping" enzymes (those needed by all cells) appear to be encoded by constitutive genes, which are expressed in all cells at all times. Some inducible genes have also been found; these respond to environmental threats or stimuli such as heavy metal ingestion, viral infection, and heat shock.

Some genes appear to be inducible only at certain periods in the life of the organism; they are thought to be controlled by **temporal regulation** mechanisms. Finally, a number of genes are under the control of **tissue-specific regulation.** For example, a gene involved in the production of a particular enzyme may be regulated by one stimulus (e.g., a hormone) in muscle tissue, by an entirely different stimulus in pancreatic cells, and by a third stimulus in liver cells.

Eukaryotic transcription is controlled at many sites and by many different regulatory molecules

Various DNA segments, including upstream promoter elements, enhancers, and transcription initiation sites, are important in transcriptional control. In addition, the rate of transcription is affected by regulatory proteins known as transcription factors and by the way the DNA is organized in the chromosome.

Eukaryotic promoters vary in efficiency, depending on their upstream promotor elements

In eukaryotic as well as prokaryotic cells, the transcription of any gene requires a promoter to which RNA polymerase binds. In multicellular eukaryotes the RNA polymerase binds to a sequence of bases, known as a **TATA box,** about 30 base pairs upstream from the transcription initiation site (Fig. 13–6). The promoter region also contains one or more sequences of 8 to 12 bases known as **upstream promoter elements (UPEs)** within a short distance of the RNA polymerase-binding site. The efficiency of the promoter seems to depend on the number and type of UPEs. Thus, a constitutive gene containing only one UPE would be weakly expressed, whereas one containing five or six UPEs would be actively transcribed.

Enhancers are DNA sequences that increase the rate of transcription

Regulated eukaryotic genes require not only the promoter elements but also DNA sequences called **enhancers.** Whereas the promoter elements are required for accurate and efficient initiation of mRNA synthesis, enhancers increase the *rate* of RNA synthesis after initiation.

Enhancer sequences are remarkable in many ways. Although present in all cells, a particular enhancer is functional only in certain types of cells. An enhancer can regulate a gene on the same DNA molecule from very long distances (up to thousands of bases away from the promoter) and can be either upstream or downstream of the promoters it controls (Fig. 13–7). Furthermore, if an enhancer sequence is experimentally cut out of the DNA and inverted, it still regulates the gene it normally controls. Evidence suggests that at least some enhancers interact with proteins that regulate transcription.

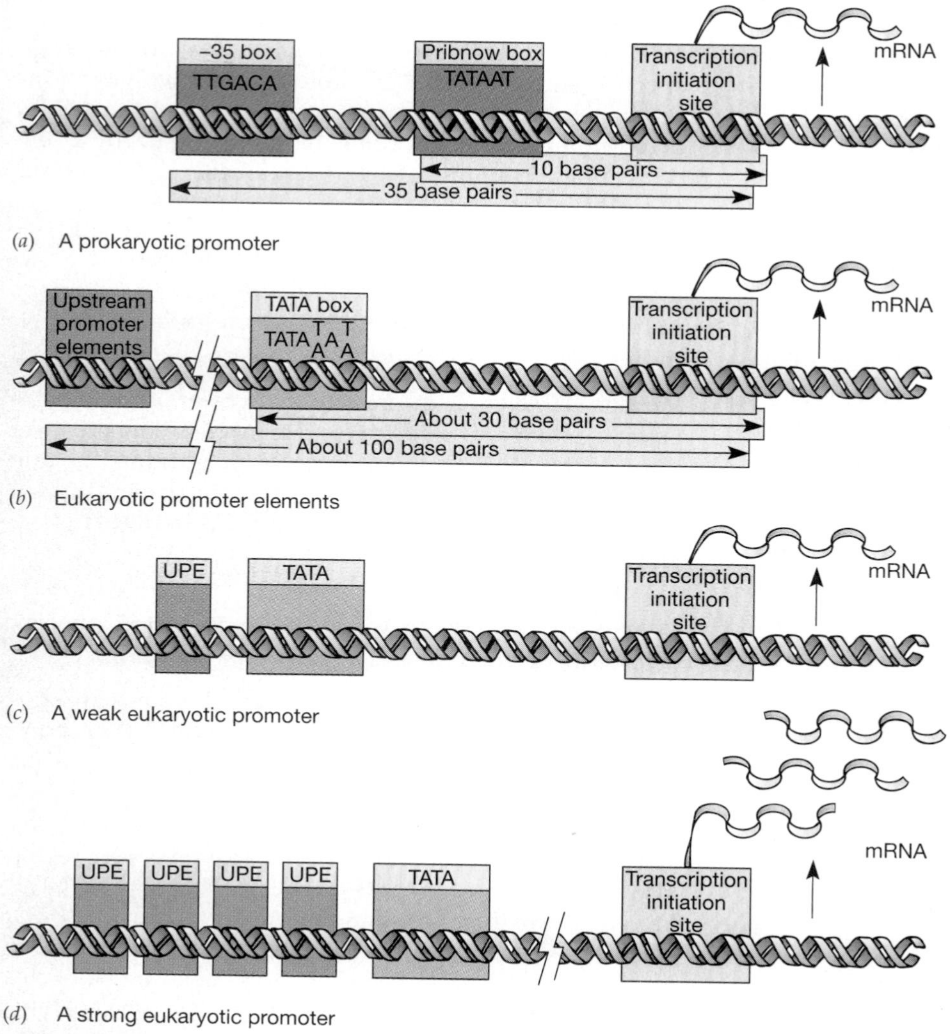

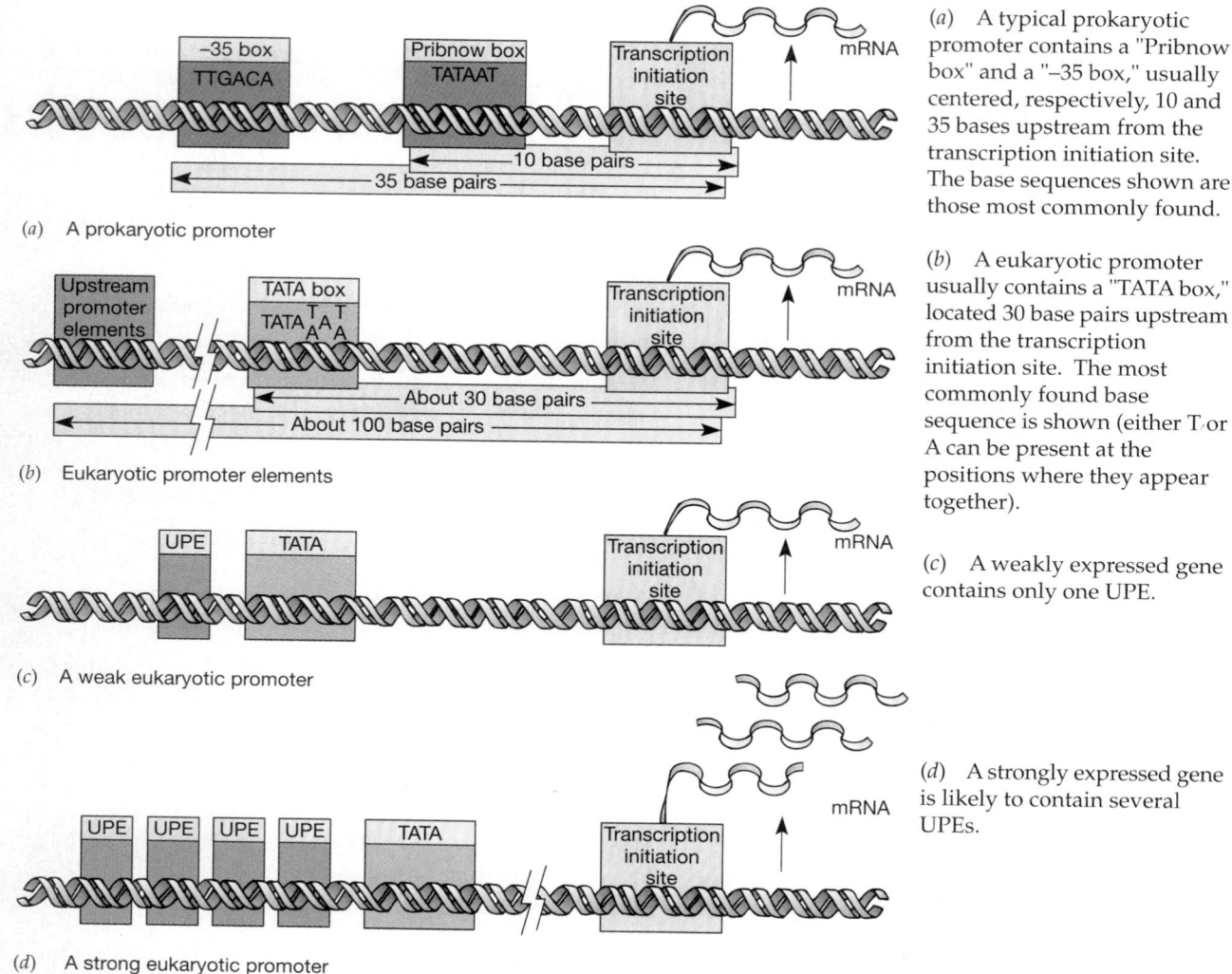

(*a*) A typical prokaryotic promoter contains a "Pribnow box" and a "–35 box," usually centered, respectively, 10 and 35 bases upstream from the transcription initiation site. The base sequences shown are those most commonly found.

(*b*) A eukaryotic promoter usually contains a "TATA box," located 30 base pairs upstream from the transcription initiation site. The most commonly found base sequence is shown (either T or A can be present at the positions where they appear together).

(*c*) A weakly expressed gene contains only one UPE.

(*d*) A strongly expressed gene is likely to contain several UPEs.

FIGURE 13–6 Prokaryotic promoters are simpler than those of eukaryotes.

Transcription factors are regulatory proteins that have several functional domains and may work in various combinations

We previously discussed some DNA-binding proteins that regulate transcription in prokaryotes. These **transcription factors** include the lactose repressor, the tryptophan repressor, and the catabolite activator protein (CAP). Although transcription factors were first studied in prokaryotes, many have been identified in eukaryotes.

It is useful to compare prokaryotic and eukaryotic transcription factors. Many transcription factors are modular molecules; that is, they have more than one structural region (domain) and each region has a different function. For example, every prokaryotic regulator has a DNA-binding domain plus other domains that can activate RNA polymerase, or bind inducers such as allolactose or corepressors such as tryptophan. The DNA-binding regions of the lactose repressor and CAP contain two α-helical segments that are inserted into the grooves of the DNA without unwinding the double helix. Recall from Chapter 3 that an α-helix of a protein is arranged such that the peptide bonds are in the interior of the helix and the functional groups of the amino acids are on the surface of the helix. This allows certain functional groups of amino acids to form hydrogen bonds with specific base pairs in the DNA. The hydrogen bonds that form between the regulatory protein and the DNA differ from those involved in complementary base-pairing (Fig. 13–8).

Transcription in eukaryotes requires multiple regulatory proteins that are bound to different parts of the promoter. The "general transcription machinery" is a protein complex that binds to the TATA region of the promoter

Making the Connection

Regulation in Prokaryotes and Eukaryotes

Why do prokaryotic and eukaryotic cells have distinctly different strategies for regulating the activity of their genes? In large part these differences reflect the ways in which the organisms make their living. Because bacterial cells exist independently, each cell must be able to perform all of its own essential functions. And because they grow rapidly and have relatively short lifetimes, they carry little excess baggage.

The dominant theme of prokaryotic gene regulation is *economy,* and controlling transcription is usually the most cost-effective way to regulate gene expression. The organization of related genes into operons and regulons that can be rapidly turned on and off as units allows these cells to synthesize only the gene products needed at any particular time. This type of regulation requires rapid turnover of mRNA molecules to prevent messages from accumulating and continuing to be translated when they are not needed.

Bacteria rarely regulate enzyme levels by degrading proteins. Once the synthesis of a protein ends, the previously synthesized protein molecules are diluted out so rapidly in subsequent cell divisions that breaking them down is usually not necessary. Only when cells are starved or deprived of essential amino acids are protein-digesting enzymes used to recycle their amino acids by breaking down proteins no longer needed for survival.

Eukaryotic cells have different regulatory requirements. In multicellular organisms, groups of cells cooperate with each other in a division of labor. Because a single gene may need to be regulated in different ways in different types of cells, eukaryotic gene regulation is complex, occurring not only at the level of transcription but also at other levels of gene expression. Eukaryotic cells also usually have long lifetimes, during which they may need to respond repeatedly to many different stimuli. Rather than synthesize new enzymes each time they respond to a stimulus, these cells make extensive use of preformed enzymes and other proteins that can be rapidly converted from an inactive to an active state.

Much of the emphasis of gene regulation in multicellular organisms is on *specificity* in the form and function of the cells in each tissue. Each type of cell has certain genes that are active and others that may never be used. Apparently the adaptive advantages of cellular cooperation in eukaryotes far outweigh the detrimental effects of carrying a load of inactive genes through many cell divisions. ▲

near the transcription initiation site. That complex is required for RNA polymerase to bind and initiate transcription. Other combinations of regulatory proteins are bound to the more distant enhancer or UPE regions of the promoter. Those proteins then make contact with the general machinery and control the activity of the RNA polymerase.

Eukaryotic regulators, like those of prokaryotes, may be either activators or repressors; we discuss only eukaryotic activators here because they seem to be more common and have been studied more intensively.

Each activator has several functional domains, including a DNA-binding region. In some cases, the binding region consists of one or more recognition α-helices that can be inserted into specific regions of the DNA in a manner similar to that described previously for some prokaryotic regulators. Some other activators have multiple "zinc fingers," loops of amino acids held together by zinc ions (Fig. 13–9). Certain amino acid functional groups exposed in each finger have been shown to recognize specific DNA sequences.

Both enhancers and UPEs apparently become functional when specific regulatory proteins are bound to them. The DNA between the enhancer and promoter se-

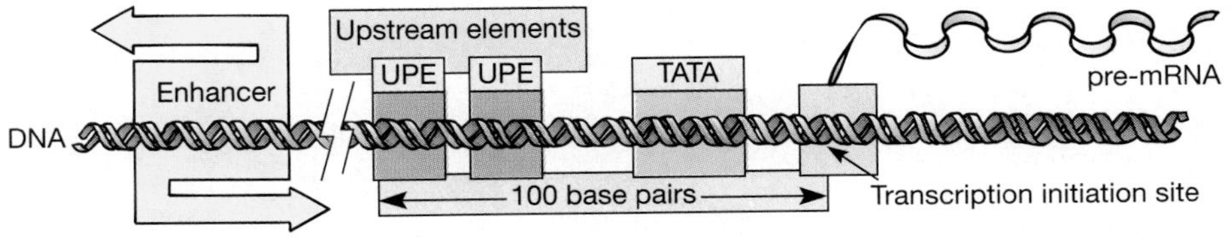

FIGURE 13–7 Enhancers are eukaryotic DNA sequences that can stimulate transcription of a gene by several orders of magnitude, at distances thousands of bases from the promoter. An enhancer can work in either direction, upstream or downstream, from the promoter. The figure shows a eukaryotic gene controlled by an upstream enhancer. The promoter, consisting of the TATA box and UPEs, is within about 100 bases of the transcription initiation site; the enhancer sequence is several thousand bases from the promoter.

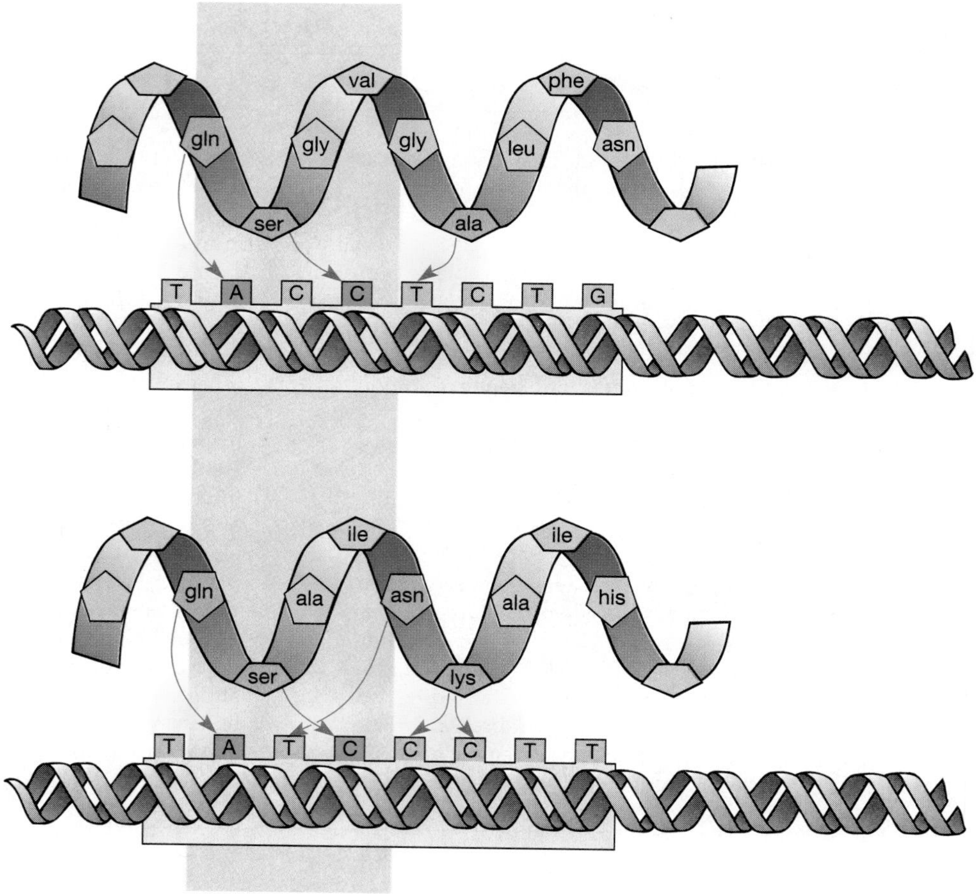

FIGURE 13–8 The DNA binding region of a transcription factor has a specific amino acid sequence that recognizes a specific base sequence in the DNA. This figure compares the amino acid base pair interactions of two related DNA-binding proteins. Only certain amino acids in the "recognition helix" regions bond with the bases of DNA. Two of the amino acids are the same in the two proteins and recognize the same bases; other amino acids differ, allowing the two proteins to recognize specific combinations of bases.

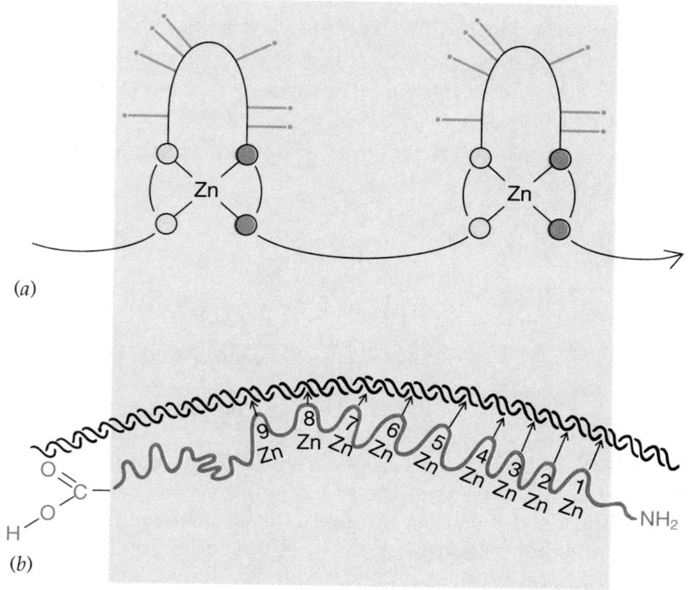

FIGURE 13–9 "Zinc-finger" DNA-binding proteins have projections that insert into DNA. (***a***) Zinc atoms cause regions of the polypeptides to form finger-like loops, which can insert into the grooves of the DNA and bind to specific base sequences. The colored circles represent amino acids that bind to the zinc atoms and form the loop. Each loop consists of about 13 amino acids. The lines projecting from the loop represent amino acids that are thought to recognize specific base sequences in the DNA. (***b***) Zinc-finger proteins that have been described have between two and nine fingers. Each finger is thought to fit into a separate groove in the DNA of a control region for a specific gene.

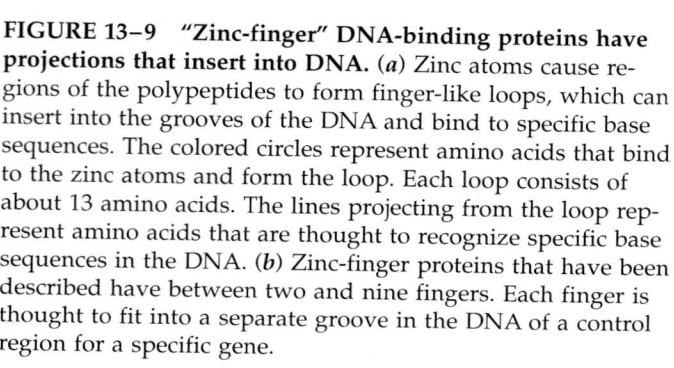

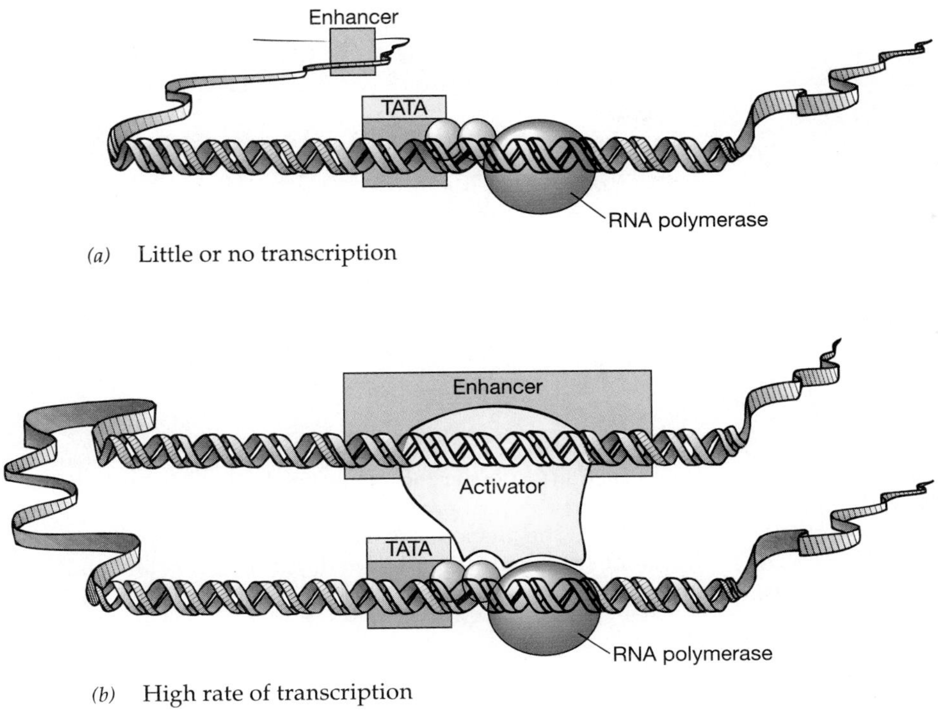

FIGURE 13–10 An enhancer can be located a considerable distance from the promoter it controls. (*a*) The general transcriptional machinery, including RNA polymerase, is bound to the promoter, but the gene is transcribed at a very low rate or not at all. (*b*) A regulatory protein that functions as a transcriptional activator has become bound to the enhancer. The intervening DNA forms a loop, allowing the activator to contact one or more target proteins in the general transcriptional machinery, thus stimulating transcription.

quences is thought to form a loop that allows an activator bound to an enhancer to come in contact with one or more target proteins associated with the general transcriptional complex. When this occurs, transcription is stimulated (Fig. 13–10).

Notice that each activator must have at least two functional domains: a DNA recognition site that usually binds to an enhancer or UPE, and a "gene activation site" that

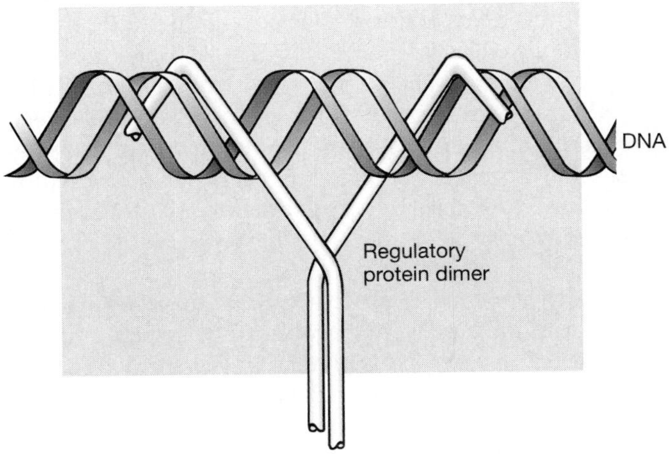

FIGURE 13–11 Some regulatory proteins associate as dimers before binding to DNA.

contacts the target in the general transcriptional complex. In addition, many activators are functional only as pairs or *dimers*, and these have special domains required for dimer formation. Figure 13–11 shows how one type of transcription factor is thought to form dimers and bind to specific base sequences in the DNA.

In some cases the two polypeptides that make up the dimer may be identical and form a *homodimer.* In other instances they are different, and the resulting *heterodimer* may have very different regulatory properties.

For a very simple example, let us assume that three regulatory proteins—A, B, and C—are involved in controlling a particular gene. These three proteins can associate as dimers in six different ways: three kinds of homodimers (AA, BB, and CC) and three kinds of heterodimers (AB, AC, and BC). The fact that some regulatory proteins can join together in different combinations that have different properties greatly increases the number of ways that transcription can be controlled.

Transcriptional units may overlap in eukaryotes

A given gene may sometimes be transcribed in more than one way, and somewhat different forms of its protein product may be produced as a result. For example, the enzyme invertase, which cleaves the disaccharide sucrose, exists in two forms in yeast: an intracellular form and a form secreted from the cell into the growth medium. Both forms are encoded by the same gene (Fig.

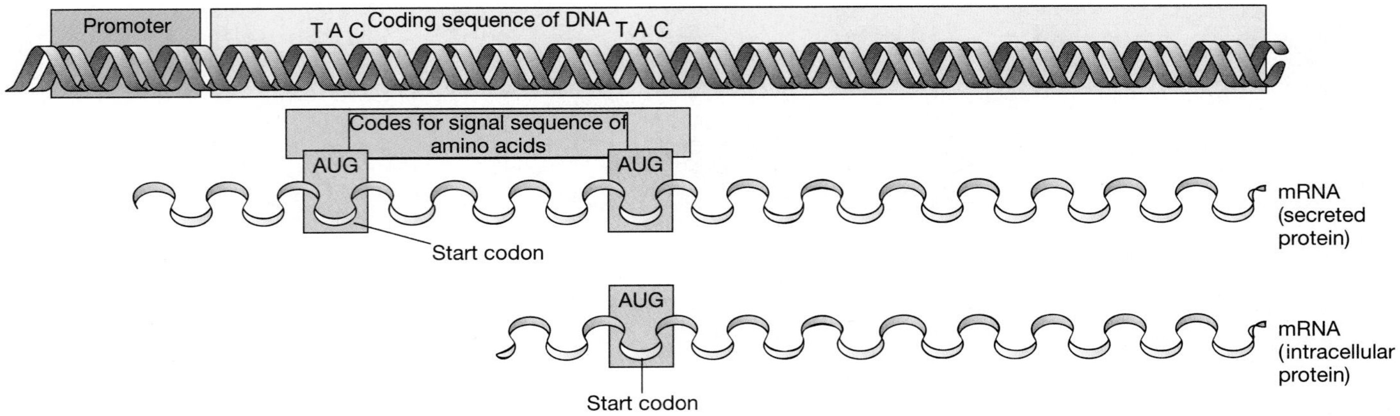

FIGURE 13–12 In some cases two forms of the same protein can be made by overlapping transcription units. This is possible if transcription can be initiated at two different sites in a single gene. The yeast invertase gene encodes an mRNA with two start codons (AUG) in its coding region. If transcription is initiated at a downstream site, a short mRNA is made that yields, when translated, an intracellular form of the enzyme. If transcription starts at a point upstream from both AUG codons, a longer message is made. Translation begins at the first start codon, producing an enzyme with a "signal sequence" at the amino-terminal end of the polypeptide chain that targets the protein to the endoplasmic reticulum and then to the Golgi complex for secretion from the cell.

13–12), but their mRNAs are transcribed from two different **transcription initiation sites.** The longer mRNA encodes the extracellular form of the enzyme, which is a longer polypeptide with extra amino acids at its amino-terminal end. Those amino acids serve as a "signal sequence" indicating that the protein is to be processed through the Golgi complex and secreted from the cell. The smaller mRNA encodes a protein that lacks the signal sequence and remains in the cytosol.

Other genes have tissue-specific overlapping transcriptional units. The primary pre-mRNA transcript of the gene for amylase (an enzyme that breaks down starch) in the mouse salivary gland is several thousand bases longer than its counterpart in the mouse liver. This is because in the salivary gland transcription starts at a site on the DNA that is farther upstream. Although the coding portions of the mRNAs in both cell types are identical after splicing and processing, transcription occurs about 100 times more frequently in the salivary gland than in the liver, resulting in the production of higher levels of the salivary amylase enzyme.

The organization of the chromosome may affect the expression of some genes

A chromosome is not simply a bearer of genes. Various arrangements of its ordered components can result in increased or decreased expression of the genes it contains.

Multiple copies of some genes are required A single gene cannot always provide enough copies of its mRNA to meet the cell's needs. The requirement for high levels of certain products may be met if multiple copies of the genes that encode them are present in the chromosome. Genes of this type, whose products are essential for all cells, may occur as tandemly repeated gene sequences in all cells. Other genes, which may be required by only a small group of cells, may be selectively replicated in those cells in a process called **gene amplification** (see Chapter 16).

Within an array of repeated genes, each copy is almost identical to the others. Histone genes, which code for the proteins that associate with DNA to form nucleosomes (see Chapter 11), are usually found as multiple copies of 50 to 500 genes in multicellular organisms.

Genes for rRNA and tRNA also occur in multiple copies in all cells. To ensure that the rRNA molecules are made in equal amounts, the RNA genes are arranged as multiple transcription units, each containing one copy of each of the three rRNA genes (Fig. 13–13). Most eukaryotic species contain 150 to 450 such transcription units per cell. The demand for rRNA is so great in actively growing mammalian cells that, although hundreds of copies of the genes are present, each gene is usually copied simultaneously by approximately 100 RNA polymerase enzymes.

Changes in chromatin structure may inactivate some genes In multicellular eukaryotes, only a subset of the genes present in a cell are active at any one time. The inactivated genes differ among cell types and in many cases seem to be irreversibly quiescent.

Some of the inactive genes appear to be associated with highly compacted chromatin, which can be seen as densely staining regions of chromosomes during cell division. These regions of chromatin remain tightly coiled

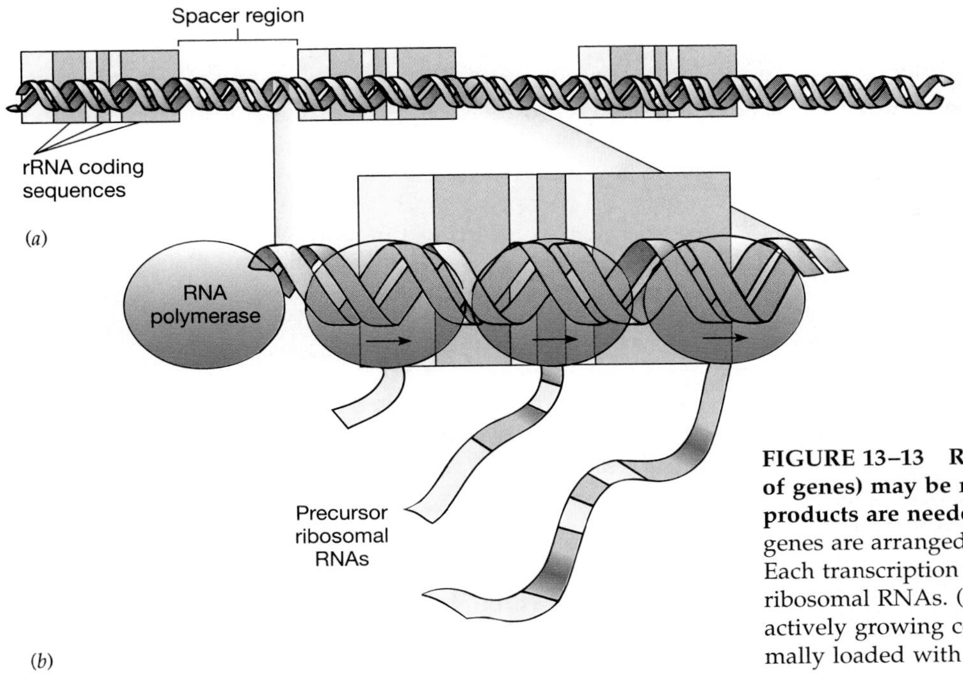

FIGURE 13–13 Repeated gene sequences (multiple copies of genes) may be required when large amounts of their products are needed by the cell. (***a***) Human ribosomal RNA genes are arranged as 200 to 300 randomly repeated copies. Each transcription unit encodes a single copy of three of the ribosomal RNAs. (***b***) The requirement for rRNA is so great in actively growing cells that each of the units must be maximally loaded with RNA polymerases.

throughout the cell cycle, and even during interphase are visible as darkly staining fibers called **heterochromatin.** Evidence suggests that the DNA of heterochromatin is not transcribed. When one of the two X chromosomes is inactivated in female mammals, most of the inactive X chromosome becomes heterochromatic and is seen as the Barr body (see Chapter 10). Active genes are associated with a more loosely packed chromatin structure called **euchromatin** (Fig. 13–14).

The long-lived, highly processed mRNAs of eukaryotes provide many opportunities for posttranscriptional control

The half-life of prokaryotic mRNA is usually measured in minutes; eukaryotic mRNA, even when it turns over rapidly, is far more stable. Prokaryotic mRNA is transcribed in a form that can be translated immediately. In contrast, eukaryotic mRNA molecules require further

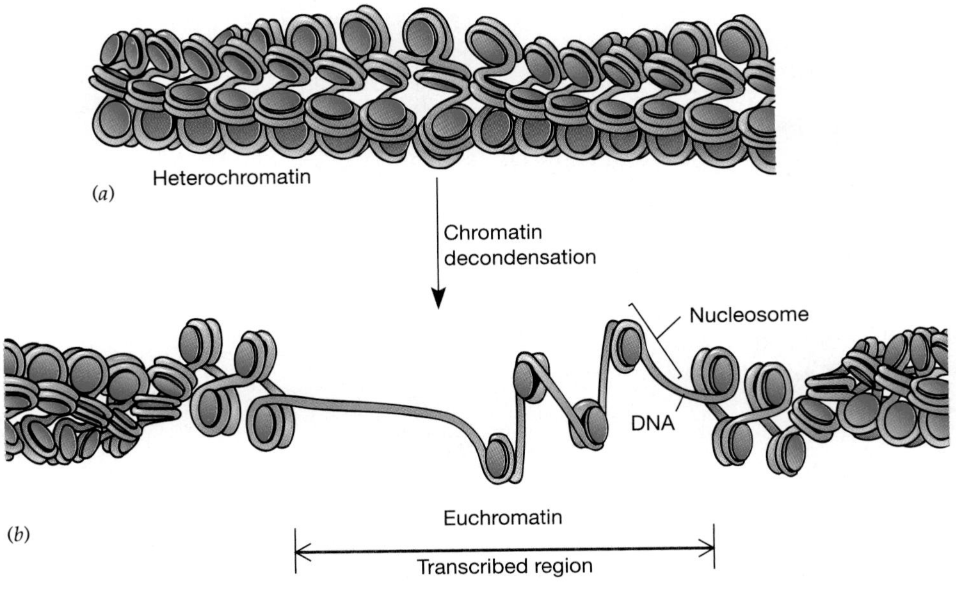

FIGURE 13–14 The structure of a chromosome affects transcription. (***a***) An inactive region of DNA (heterochromatin) is organized into tightly associated nucleosomes. (***b***) Active genes are associated with decondensed chromatin (euchromatin). Chromatin decondensation is often a response to specific inducing signals. The loosely packed chromatin increases the accessibility to RNA polymerase required for transcription of the region.

modification and processing before they can be used in protein synthesis (see Chapter 12). The message is capped, spliced, polyadenylated, and then transported from the nucleus to the cytoplasm to initiate translation. These events represent potential control points at which translation of the message and production of its encoded protein can be regulated.

Some pre-mRNAs can be processed in more than one way

Several forms of regulation involving mRNA processing have been discovered. In some instances, the same gene is used to produce one type of protein in one tissue and a related but somewhat different type of protein in another tissue (Fig. 13–15). This is possible because some genes produce pre-mRNA molecules that have multiple splicing patterns; that is, they can be spliced in more than one way depending on the tissue. Typically, such a gene includes at least one segment that can be either an intron or an exon. Through **differential nuclear RNA processing** the cells in each tissue produce their own version of mRNA corresponding to the particular gene.

The stability of mRNA molecules can vary

Controlling the lifetime of a particular kind of mRNA molecule makes it possible to control the number of protein molecules translated from it. In some cases messenger RNA stability is under hormonal control. This is true for mRNA that codes for vitellogenin, a protein made in the livers of certain female animals such as frogs and chickens.

After it is synthesized, vitellogenin is transported to the oviduct, where it is used in the formation of yolk proteins in the egg. Vitellogenin synthesis is regulated by the hormone estradiol. When estradiol levels are high, the half-life of vitellogenin mRNA in frog liver is about 500 hours. When cells are deprived of estradiol, the half-life of the mRNA drops rapidly to less than 165 hours. This leads to a rapid decrease in cellular vitellogenin mRNA levels and decreased synthesis of the vitellogenin protein. In addition to affecting the stability of the mRNA, the hormone seems to control the rate at which the mRNA is synthesized.

The activity of eukaryotic proteins may be altered by posttranslational chemical modifications

Eukaryotic enzyme activity can also be regulated after the protein is synthesized. As in bacteria, many metabolic pathways in eukaryotes contain allosteric enzymes that are regulated through feedback inhibition. In addition, many eukaryotic proteins are extensively modified after they are synthesized.

In *proteolytic processing* the proteins are synthesized as inactive precursors, which are converted to an active form by removal of a portion of the polypeptide chain. Other proteins may be regulated in part by a process of *selective degradation,* which keeps their numbers constant within the cell. *Chemical modification,* through the addition or removal of functional groups, can reversibly alter the activity of an enzyme. One very common way of modifying the activity of an enzyme or other protein is the addition or removal of phosphate groups. These alterations allow the cell to respond rapidly to fast-changing environmental or nutritional conditions.

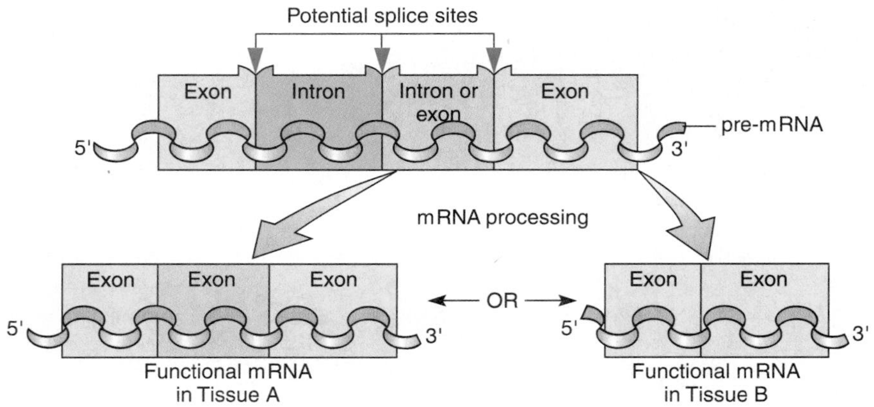

FIGURE 13–15 In differential mRNA processing, a complex transcription unit can be processed in more than one way to yield two or more mRNAs, each of which encodes a related, but different protein. In this example the gene contains a segment that can be an exon in Tissue A *(left),* but an intron in Tissue B *(right).*

SUMMARY

I. Most regulated genes in bacteria are organized into operons, which may encode several proteins.
   A. Each operon is controlled by a single promoter region upstream from the protein-coding regions.
   B. The operator is a sequence of bases that overlaps the promoter and serves as the regulatory switch controlling the operon.
      1. A repressor protein binds specifically to the operator sequence and blocks transcription by preventing RNA polymerase from binding to the promoter.
      2. When the repressor is not bound to the operator, RNA polymerase can bind to the promoter and transcription can proceed.
   C. An inducible operon such as the lactose operon is normally turned off. The repressor protein is synthesized in an active form that binds to the operator. If cells are exposed to lactose, a metabolite of that sugar binds to an allosteric site on the repressor protein, causing it to change shape. The altered repressor cannot bind to the operator, and the operon is transcribed.
   D. A repressible operon such as the tryptophan operon is normally turned on. The repressor protein is synthesized in an inactive form that cannot bind to the operator. A metabolite (usually the end product of a metabolic pathway) acts as a corepressor. When intracellular corepressor levels are high, one of the molecules binds to an allosteric site on the repressor, changing its shape so that it can bind to the operator and thereby turn off transcription of the operon.
   E. Repressible and inducible operons are under negative control. When the repressor protein binds to the operator, transcription of the operon is turned off.
   F. Some inducible operons are also under positive control. A separate protein can bind to the DNA and activate transcription of the gene.
      1. The lactose operon is activated by CAP (catabolite activator protein), which binds to the promoter region, stimulating transcription by binding RNA polymerase tightly.
      2. To bind to the lactose operon, CAP requires cAMP (cyclic adenosine monophosphate). Levels of cAMP increase as levels of glucose decrease.
   G. A group of operons can be organized into a multigene system, known as a regulon, which is controlled by a single regulatory protein. CAP activates a number of operons associated with the metabolism of carbohydrates.

II. Constitutive genes are neither inducible nor repressible; they are active at all times. Regulatory proteins such as CAP and the repressor proteins are produced constitutively. The activity of these genes is controlled by how efficiently RNA polymerase binds to their promoter regions.

III. DNA-binding regulatory proteins work by recognizing and binding to specific base sequences in the DNA.

IV. The expression of some prokaryotic genes is modified after the mRNA is translated.
   A. The rate of translation of a particular mRNA may be regulated.
   B. The activity of key enzymes in some metabolic pathways can be controlled by feedback inhibition.

V. Eukaryotic genes are generally not organized into operons. Regulation of eukaryotic genes can occur at the levels of transcription, mRNA processing, translation, and the protein product.
   A. The promoter of a regulated eukaryotic gene consists of an RNA polymerase-binding site and short DNA sequences known as upstream promoter elements (UPEs). The efficiency of the promoter is determined by the number and types of UPEs within the promoter region.
   B. Inducible eukaryotic genes are controlled by enhancer elements, which can operate thousands of bases away from the promoter. Proteins that bind to enhancers appear to facilitate the binding of RNA polymerase to the promoter.
   C. Transcription factors are regulatory proteins that bind to DNA and either activate or repress transcription.
   D. The activity of eukaryotic genes is affected by chromosome structure.
      1. Some genes whose products are required in large amounts exist as multiple copies in the chromosome. Other genes may be selectively amplified in only some cells by DNA replication.
      2. Genes can be inactivated by changes in chromosome structure. Densely packed regions of chromosomes called heterochromatin contain inactive genes. Active genes are associated with a loosely packed chromatin structure called euchromatin.
   E. Many eukaryotic genes are regulated after the RNA transcript is made.
      1. Gene regulation can occur as a consequence of mRNA processing. In some cases a single gene can produce different forms of a protein, depending on how the pre-mRNA is polyadenylated or spliced.
      2. Certain regulatory mechanisms increase the stability of mRNA, allowing more proteins to be formed per mRNA molecule prior to degradation.
      3. Posttranslational control of eukaryotic genes can occur by feedback inhibition or by modification of the protein structure.

SELECTED KEY TERMS

activator protein
allosteric binding site
β-galactosidase
catabolite activator protein (CAP)
constitutive gene
corepressor
enhancer
euchromatin
feedback inhibition
gene amplification
heterochromatin
inducer molecule
inducible system
negative control
operator
operon
postive control
posttranscriptional control
posttranslational control
promotor
regulon
repressible system
repressor protein

(Selected Key Terms continued on next page)

TATA box
temporal regulation
tissue-specific regulation
transcription factor
transcription initiation site
transcriptional control
translational control
upstream promoter element (UPE)

POST-TEST

1. Regulated prokaryotic genes are organized into clusters called ____________ .
2. A regulatory DNA sequence that overlaps the promoter sequences and is associated with inducible and repressible operons is called the ____________ .
3. A ____________ protein that binds to the operator region of an operon blocks transcription of the structural genes by preventing RNA polymerase from binding to the promoter. If a regulator blocks transcription when it binds to the DNA, we refer to this as an example of ____________ control.
4. ____________ operons are controlled by regulatory proteins that are normally in an active state. When a metabolite binds to a(an) ____________ binding site on the regulatory protein, it changes its conformation and renders it inactive.
5. ____________ operons are controlled by regulatory proteins that are normally inactive. When the ____________ binds to the allosteric site of the repressor protein, it is converted to an active repressor.
6. Positive gene control involves ____________ proteins that bind to the DNA.
7. ____________ ____________ protein is a positive regulator that requires cyclic AMP to bind to a specific site on an operon.
8. CAP is a controlling element of a system of multiple operons involved in carbohydrate metabolism called a(an) ____________ .
9. Genes not under regulatory control are termed ____________ genes. They encode enzymes needed by cells at all times and are transcribed continuously.
10. An example of a constitutive gene is the lactose ____________ gene.
11. ____________ ____________ are DNA-binding regulatory proteins.
12. Some metabolic pathways are regulated by ____________ ____________ , in which the end product blocks the activity of the first enzyme in the pathway.
13. In a eukaryotic cell, the promoter region of a gene consists of an RNA polymerase–binding site and adjacent ____________ ____________ elements.
14. Inducible eukaryotic genes may be controlled by ____________ , which can be positioned thousands of bases away from the promoter.
15. Some genes that are required in large numbers in some cells but not in others undergo gene ____________ , which is selective replication of a small part of the chromosome.
16. Genes present in tightly coiled regions of chromosomes called ____________ are generally inactive.
17. Active genes are found in loosely packed chromatin called ____________ .

REVIEW QUESTIONS

1. Make a sketch of the lactose operon and briefly describe its function. Be sure to include the following elements: (a) structural genes; (b) promoter; (c) operator; (d) CAP-binding site.
2. What structural features does the tryptophan operon have in common with the lactose operon? What features are different?
3. Why do we define the tryptophan operon as repressible and the lactose operon as inducible?
4. The genes that code for the lactose repressor and the tryptophan repressor are not tightly linked to the operons they regulate. Would it be advantageous if they were? Explain your answer.
5. How is glucose involved in the positive control of the lactose operon? How is CAP similar to the lactose repressor protein? How is it different?
6. Compare the structure of a prokaryotic promoter region with known eukaryotic promoter regions. How does the regulation of inducible eukaryotic genes differ from the regulation of inducible prokaryotic genes?
7. Explain why it is necessary for certain genes in eukaryotic cells to be present in multiple copies.
8. How can the activity of some eukaryotic genes be affected by the structure of the chromosome?
9. Make a sketch showing how differential mRNA processing can give rise to different forms of a eukaryotic protein.

YOU MAKE THE CONNECTION

1. Develop a simple hypothesis that would explain the behavior of each of the following types of mutants in *E. coli*:
   (a) *Mutant a:* The map position of this mutation is in the tryptophan operon. The mutant cells are constitutive; that is, they produce all of the enzymes coded for by the tryptophan operon, even if large amounts of tryptophan are present in the growth medium.
   (b) *Mutant b:* The map position of this mutation is in the tryptophan operon. The mutant cells do not produce any of the enzymes coded for by the tryptophan operon under any conditions.
   (c) *Mutant c:* The map position of this mutation is some distance from the tryptophan operon. The mutant cells are constitutive; that is, they produce all of the enzymes coded for by the tryptophan operon, even if the growth medium contains large amounts of tryptophan.
   (d) *Mutant d:* The map position of this mutation is some distance from the tryptophan operon. The mutant cells do

not produce any of the enzymes coded for by the tryptophan operon under any conditions.

2. Compare the types of bacterial genes associated with inducible operons, those associated with repressible operons, and those that are constitutive. Predict the category into which each of the following would most likely fit: (a) a gene that codes for RNA polymerase; (b) a gene that codes for an enzyme required to break down maltose; (c) a gene that codes for an enzyme used in the synthesis of adenine.

RECOMMENDED READINGS

Grunstein, M. "Histones as Regulators of Genes." *Scientific American,* Vol. 67, No. 4, October 1992. A discussion of how DNA transcription can be regulated through the way it is packaged with histones in chromatin.

McKnight, S. L. "Molecular Zippers in Gene Regulation." *Scientific American,* Vol. 264, No. 4, April 1991. Considers how some transcription factors associate in various combinations to form dimers that affect transcription in specific ways.

Ptashne, M. "How Gene Activators Work." *Scientific American,* Vol. 260, No. 1, January 1989. Discusses the parallels between prokaryotic and eukaryotic transcription factors.

Rhodes, D., and A. Klug. "Zinc Fingers." *Scientific American,* Vol. 268, No. 2, February 1993. Considers the role of regulatory proteins possessing the zinc fingers motif in controlling gene transcription in a wide range of organisms.

Tjian, R. "Molecular Machines That Control Genes." *Scientific American,* Vol. 272, No. 2, February 1995. A discussion of the current state of knowledge of the transcriptional machinery, with a perspective on how this information might be used in the future to treat certain diseases

CHAPTER 14

Genetic Engineering

Genetic engineering delays softening in FLAVR SAVR™ tomatoes, developed by Calgene, Inc.
(Courtesy of Calgene, Inc.)

Beginning in the mid-1970s, a revolution in the field of biology occurred as the development of **recombinant DNA technology** led to radically new research approaches. This technology has not only been applied to genetic studies but has also had a major impact in areas ranging from cell biology to evolution.

Recombinant DNA techniques were initially developed so scientists could obtain a great many copies of any specific DNA segment in order to study it biochemically. This can now be done in a number of ways, but most methods involve introducing foreign DNA into the cells of microorganisms. Under the right conditions, this DNA is replicated and transmitted to the daughter cells when a cell divides. In this way a particular DNA sequence can be amplified, or **cloned,** to provide millions of identical copies that can be isolated in pure form. Methods for cloning DNA *in vitro* (outside of a living organism) have also been developed.

Studies of cloned DNA sequences have been of immense value in allowing scientists to understand the organization of genes and the relationship between genes and their products. In fact, most of our knowledge of the complex structure and control of eukaryotic genes (see Chapters 12 and 13) is derived from the application of these methods.

Recombinant DNA technology also has many practical applications. One of the rapidly advancing areas of study today is **genetic engineering**—the modification of the DNA of an organism to produce new genes with new characteristics. Genetic engineering can take many forms, ranging from the production of strains of bacteria that manufacture useful protein products to the development of plants and animals that express foreign genes. Unprecedented advances in fields such as pharmacology, medicine and human genetics, and agriculture have resulted.

LEARNING OBJECTIVES

After you have studied this chapter you should be able to

1. Draw a sketch that demonstrates how a typical restriction enzyme cuts DNA molecules, and give examples of the ways in which these enzymes are used in recombinant DNA technology.
2. Summarize the properties of plasmids that allow them to be used as DNA cloning vectors.
3. Differentiate between a genomic DNA library and a cDNA library.
4. Explain why one would clone the same eukaryotic gene from both a genomic library and a cDNA library.
5. Identify some of the uses of DNA hybridization probes.
6. Describe how a gene is restriction mapped, and explain why restriction mapping is useful.
7. Draw a diagram that illustrates the most widely used DNA sequencing technique.
8. List some important proteins and other products that can be produced by genetic engineering techniques.
9. List some of the difficulties encountered in using *Escherichia coli* to produce proteins coded by eukaryotic genes, and explain the rationale behind using transgenic plants and animals to solve some of those problems.

RECOMBINANT DNA METHODS GREW OUT OF RESEARCH IN MICROBIAL GENETICS

Recombinant DNA technology was not developed quickly. It actually began with the first genetic studies of bacteria and the viruses that infect them, the **bacteriophages** (literally "bacteria eaters"; see Chapter 23). Only after decades of basic research and the accumulation of extensive knowledge did the current technology become feasible and available to the many scientists who now use these methods (Fig. 14–1).

Among other things, bacteria have provided researchers with special enzymes, known as restriction en-

(a)

(b)

FIGURE 14–1 Sophisticated applications of recombinant DNA technology generally require the cooperative efforts of teams of scientists. (*a*) Researchers study proteins produced by recombinant DNA technology. (*b*) Scientists use gene splicing to introduce bacterial genes into plants. (*a*, Will & Deni McIntyre/Photo Researchers, Inc.; *b*, Matt Meadows/Peter Arnold, Inc.)

zymes, that cut DNA molecules only in specific places. In addition, recombinant DNA molecules are most often introduced into bacterial cells or bacteriophages so that many identical copies can be made. Certain aspects of their genetic systems facilitate this process.

Restriction enzymes are "molecular scissors" that cleave DNA reproducibly

A major breakthrough in the development of recombinant DNA technology was the discovery of bacterial enzymes called **restriction enzymes,** which are able to cut DNA molecules only at specific base sequences. One restriction enzyme may recognize and cut a DNA molecule at the base sequence 5′—AAGCTT—3′, whereas another cuts only the sequence 5′—GATC—3′. Bacteria normally use these enzymes as a defense mechanism to attack bacteriophage DNA that enters the cell. The bacteria protect their own DNA from attack by altering it in some way after it is synthesized. Purification of these enzymes enabled scientists to cut DNA from chromosomes into shorter fragments in a controlled way (Fig. 14–2).

Many of the restriction enzymes used for recombinant DNA studies cut **palindromic** sequences, which means that the base sequence of one strand reads the same as its complement, but in the opposite direction. (Thus, the complement of our example, 5′—AAGCTT—3′, reads 3′—TTCGAA—5′.) By cutting both strands of the DNA in an asymmetric manner, these enzymes leave fragments with complementary, single-stranded ends. These ends are called "sticky ends" because they can pair (by hydrogen-bonding) with the complementary single-stranded ends of other DNA molecules that have been cut with the same enzyme. Once two molecules have been joined together in this way, they can be treated with **DNA ligase** (see Chapter 11), an enzyme that covalently links the two fragments to form a stable recombinant DNA molecule.

Restriction enzymes vary widely in the number of DNA bases that they recognize, ranging from as few as 4 to as many as 23 bases. Based on probability alone, we expect the restriction sequence of a "four-base cutter" to occur in a DNA molecule once on the average of every 4^4, or 256, bases, whereas one that recognizes six bases would cut fragments that average 4^6, or 4096, bases in size. Restriction enzymes that recognize sequences with large numbers of bases are particularly suited for studying very large DNA molecules such as those that make up entire chromosomes.

Recombinant DNA is formed when DNA is spliced into a vector (DNA carrier)

Most recombinant DNA molecules are isolated and amplified by introducing them into cells of the bacterium *E. coli.* To isolate a specific piece of DNA (after it has been cut by a restriction enzyme), that fragment must first be incorporated into a suitable carrier, or **vector molecule** (Fig. 14–3). Bacteriophages or special DNA molecules called **plasmids** are commonly used as vectors. A plasmid is a small circular DNA molecule that can replicate inside a bacterial cell. These plasmids can be isolated from bacterial cells in pure form and then be introduced into other cells by a method called **transformation** (see Chapter 11), which involves altering the bacterial cell wall to make it permeable to the plasmid DNA molecules. Once a plasmid enters a cell, it is replicated and distributed to the daughter cells during cell division. Plasmids do not carry genes that are essential to the *E. coli* cells, but they often carry genes that are useful under some environ-

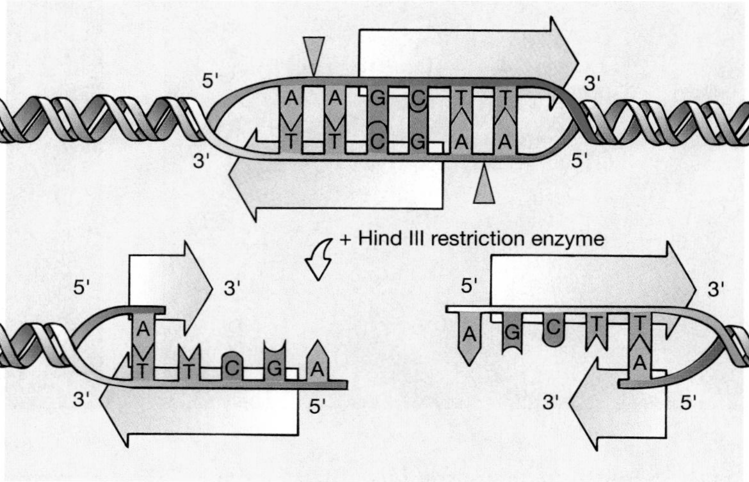

FIGURE 14–2 A restriction enzyme acts at a specific base sequence. Many restriction enzymes cut DNA at sequences of bases that are palindromic (each strand has the same base sequence, but in the opposite direction). Cutting the sequence produces complementary sticky ends.

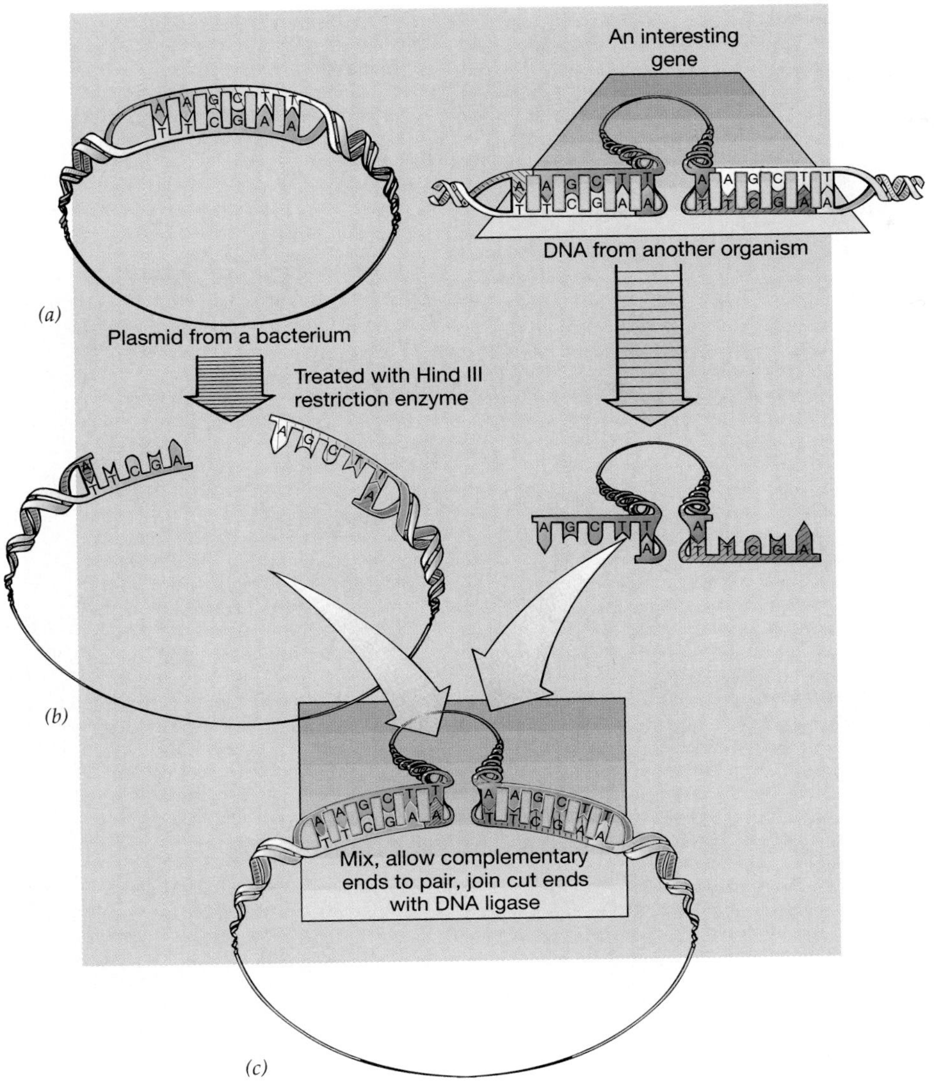

FIGURE 14–3 A recombinant DNA molecule can be produced by joining DNA molecules from different sources. (***a***) First the molecules are cut, each by the same restriction enzyme, to produce (***b***) smaller molecules with complementary single-stranded ends. In this example one molecule is a circular plasmid from a bacterium. (***c***) The recombinant DNA is constructed by mixing the two types of molecules so that their cohesive ends pair. DNA ligase then forms covalent 5′—3′ phosphodiester linkages between the junctions of the two molecules.

mental conditions, such as those that confer resistance to particular antibiotics.

The plasmids now used in recombinant DNA work have been extensively "engineered" to include a number of features helpful in the isolation and analysis of cloned DNA (Fig. 14–4). A limiting property of any plasmid, however, is the size of the DNA fragment that it can effectively carry. The size of a DNA segment is often given in kilobases, with 1 **kilobase (kb)** being equal to 1000 bases. Fragments smaller than 10 kb can usually be inserted into plasmids for use in *E. coli.* However, larger fragments require the use of bacteriophage vectors, which can handle up to 15 kb of DNA.

Recombinant DNA can also be introduced into cells of complex organisms. For example, engineered viruses are used as vectors in mammalian cells. These viruses have been disabled in such a way that they do not kill the cells they infect; instead their DNA, and any foreign DNA they carry, becomes incorporated into the chromosomes of the cell following infection. As discussed later, other methods have been developed that do not require a biological vector. These involve injecting the DNA directly into the cell nucleus or allowing cells to incorporate DNA that has been adsorbed onto the surface of calcium phosphate crystals.

Cloning techniques provide the means for replicating and isolating many copies of a specific recombinant DNA molecule

Because a single gene is only a small part of the total DNA in an organism, isolating the piece of DNA containing that particular gene is like finding a needle in a haystack. A powerful detector is needed.

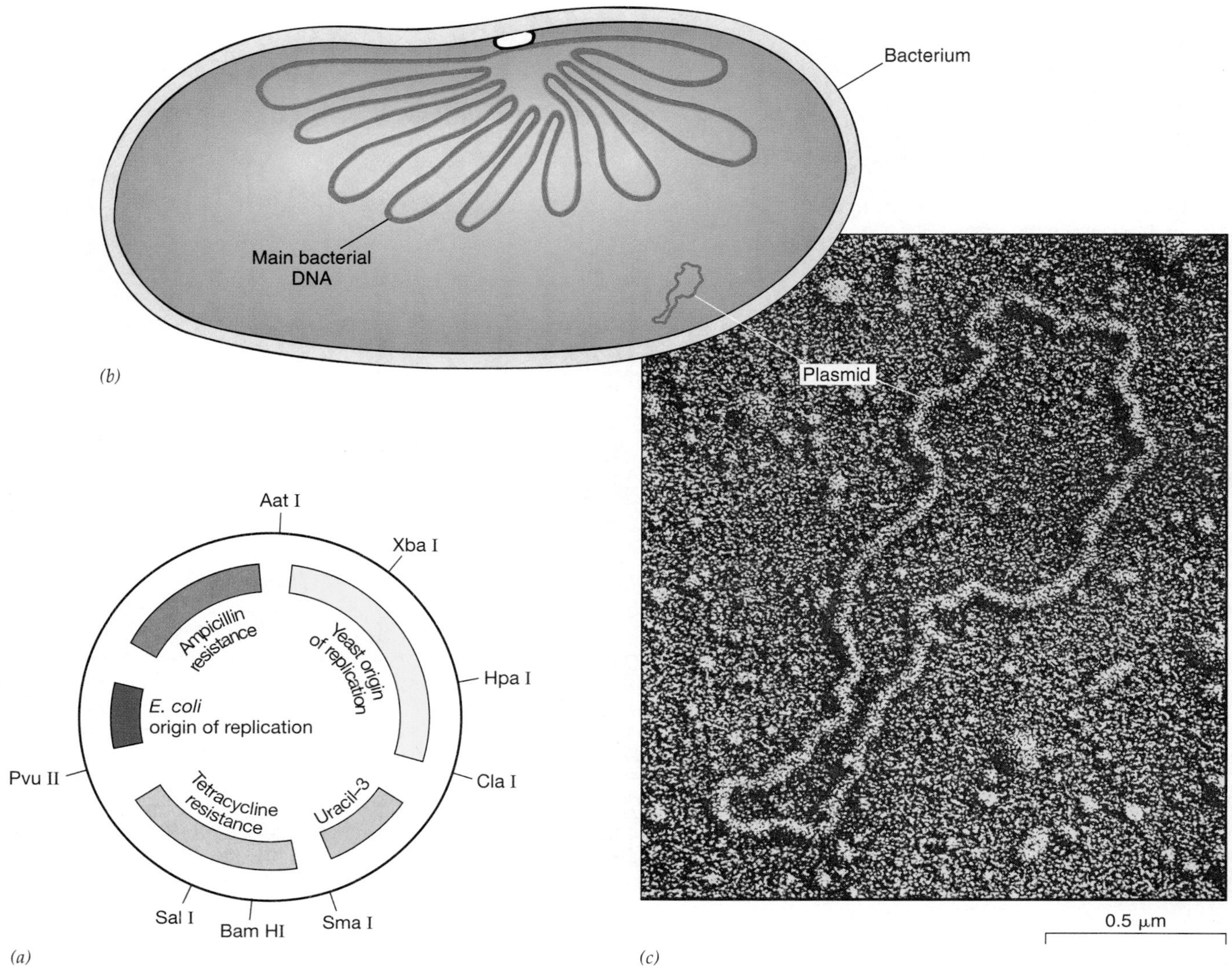

FIGURE 14–4 Plasmids used in recombinant DNA technology have been designed to have certain desirable characteristics. (*a*) The genetically engineered plasmid vector shown in this map can replicate in cells of both *E. coli* and the yeast *Saccharomyces cerevisiae*. This vector has been constructed from DNA fragments isolated from plasmids, *E. coli* genes, and yeast genes in order to have useful features. Letters on the outer circle designate sites for restriction enzymes that cut the plasmid only at that one position. The plasmid has two origins of replication, one for *E. coli* and one for yeast, allowing it to replicate independently in either type of cell. Resistance genes for the antibiotics ampicillin and tetracycline, and the yeast URA-3 gene (for an enzyme involved in uracil biosynthesis) are also shown. The URA-3 gene is used to transform yeast cells lacking that particular enzyme; cells that take up the plasmid are able to grow on a uracil-deficient medium. (*b*) The relative sizes of a plasmid and the main DNA of a bacterium. (*c*) Electron micrograph of a plasmid from *E. coli*. (c, Dr. Stanley Cohen/Science Photo Library/Photo Researchers, Inc.)

Isolating a gene from an organism such as a human first requires the construction of a **library,** or gene bank, from the human DNA (Fig. 14–5). The first step is to cut the DNA with a restriction enzyme, generating a population of DNA fragments. These fragments vary in size and in the genetic information they carry, but they all have identical sticky ends. The plasmid vector DNA is treated with the same restriction enzyme, which converts the circular plasmids into linear molecules with sticky ends complementary to those of the human DNA fragments. The two kinds of DNA (human and plasmid) are mixed together under conditions that promote hydrogen-bonding of complementary bases, and the paired ends of the plasmid and human DNA are then joined by DNA ligase.

The result is a mixture of recombinant plasmids, each containing a different human DNA sequence. To allow identification of the plasmid containing the sequence of interest, each plasmid then has to be amplified, or cloned, until there are millions of copies to work with. This process occurs inside the cells of *E. coli.* First the recombinant plasmids are inserted into antibiotic-sensitive *E. coli* cells by transformation. This is done under conditions with a low ratio of plasmids to cells, so it should be rare for a cell to receive more than one plasmid molecule, and not all cells receive a plasmid. The cells are incubated on a nutrient medium that also contains antibiotics, so only cells that have incorporated a plasmid (which contains a gene for antibiotic resistance) grow.

The researcher may then spread a sample of the bacterial culture on solid growth medium. If the suspension of cells is dilute enough, the cells will be widely separated. When each cell reproduces it gives rise to a **colony,** which is a clone of genetically identical cells. All of the cells of a particular colony contain the same recombinant plasmid, so during this process a specific sequence of human DNA has also been cloned. The major task is to determine which colony (out of thousands) contains the cloned fragment of interest. There are a number of ways in which this can be done.

A specific DNA sequence can be detected by a complementary genetic probe

A common approach to the problem of detecting the DNA of interest involves the use of a **genetic probe,** which is usually a radioactively labeled segment of RNA or single-stranded DNA that is complementary to the target gene. Suppose that a researcher wishes to identify the gene that codes for insulin. Because the amino acid sequence of insulin is known, it is possible to synthesize a radioactive single-stranded DNA molecule complementary to the DNA sequence that codes for insulin. This is not as simple as it may sound, because even if the amino acid sequence of all or part of a protein is known, those amino acids could be coded for by a number of different base sequences (see Chapter 12). The single-stranded DNA probe **hybridizes** (becomes attached by comple-

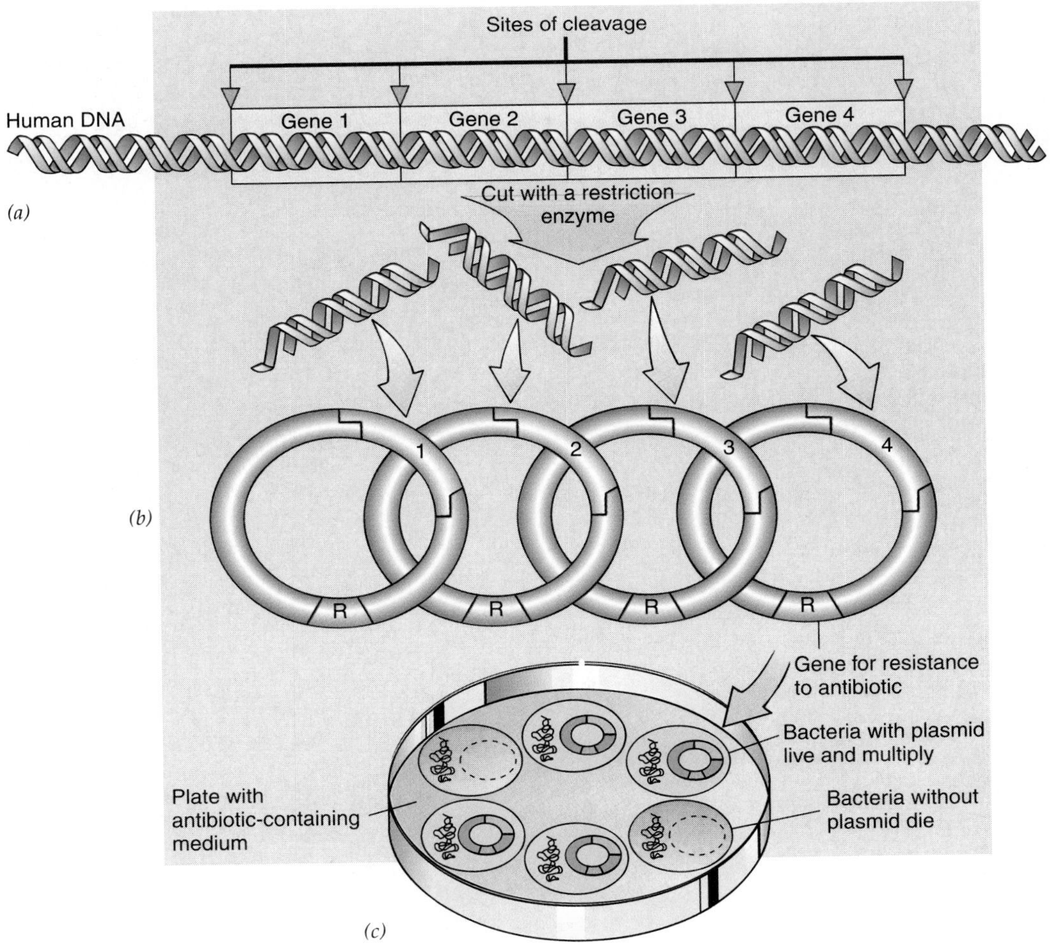

(a) A specific restriction enzyme is used to cut DNA from a human into fragments; these are then spliced into a complementary restriction site in a plasmid that contains an antibiotic resistance gene (R).

(b) The recombinant plasmids are used to transform an antibiotic-sensitive *E. coli* strain under conditions that ensure that each bacterial cell receives no more than one plasmid molecule.

(c) The cells are then grown on antibiotic-containing solid nutrient medium. Only those that receive the plasmid survive.

FIGURE 14–5 The construction of a genomic library requires a series of steps.

mentary base-pairing) with the DNA sequence that codes for insulin (Fig. 14–6). If the DNA from a small group of cells from a particular colony binds to the probe (Fig. 14–7), that DNA becomes radioactive and can be detected by x-ray film. The rest of the cells in the colony can then be grown in quantity for further testing.

A genomic library contains fragments of all the DNA in the genome

The total DNA per cell of an organism is referred to as that organism's **genome.** If the DNA is extracted from human cells, as in our example, we refer to it as human genomic DNA. A very large population of recombinant plasmids, each containing a fragment of the genome, is referred to as a **genomic library.** A library (plasmid population) will contain redundancies; that is, many human DNA sequences will be duplicated many times, purely by chance. However, each individual recombinant plasmid (analogous to a book in the library) contains only a single fragment of the total human genome. Each of these fragments is usually smaller than a gene; therefore several clones must usually be isolated to study the complete gene.

A cDNA library is complementary to mRNA and does not contain introns

Many eukaryotic genes contain introns, and bacterial cells cannot remove introns from RNA. To avoid cloning those parts of a gene that do not code for proteins, libraries can also be constructed from DNA copies of the eukaryotic mRNA. Those copies are made by isolating mRNA and making DNA copies of the message using **reverse transcriptase** (see Chapter 12). The **complementary DNA (cDNA)** copies of the message can then be inserted into the DNA of a plasmid or virus vector to form a **cDNA library** (Fig. 14–8).

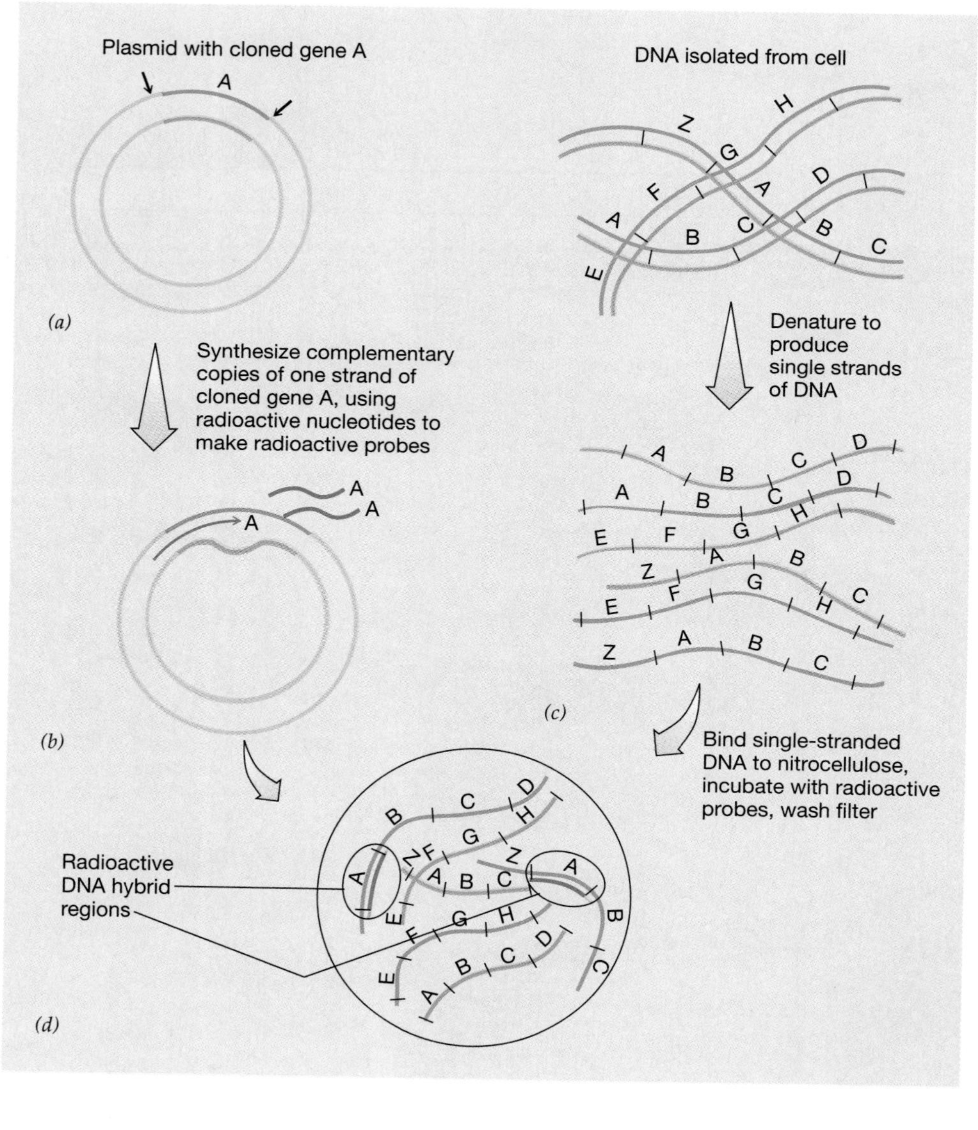

FIGURE 14–6 A single-stranded radioactive DNA molecule can hybridize with single-stranded DNA that has a complementary sequence. Total cellular DNA contains a large number of different genes. A single gene (or a sequence of bases) in that DNA can be detected by using a cloned copy of the gene carried by a vector (*a*) to make complementary radioactive probe molecules (*b*), shown in red. DNA isolated from the cell is denatured to produce single strands of DNA (*c*), which are then bound to the surface of a nylon membrane (*d*). The membrane is incubated with the radioactive DNA probe, which specifically hybridizes (pairs) with complementary regions of the bound DNA.

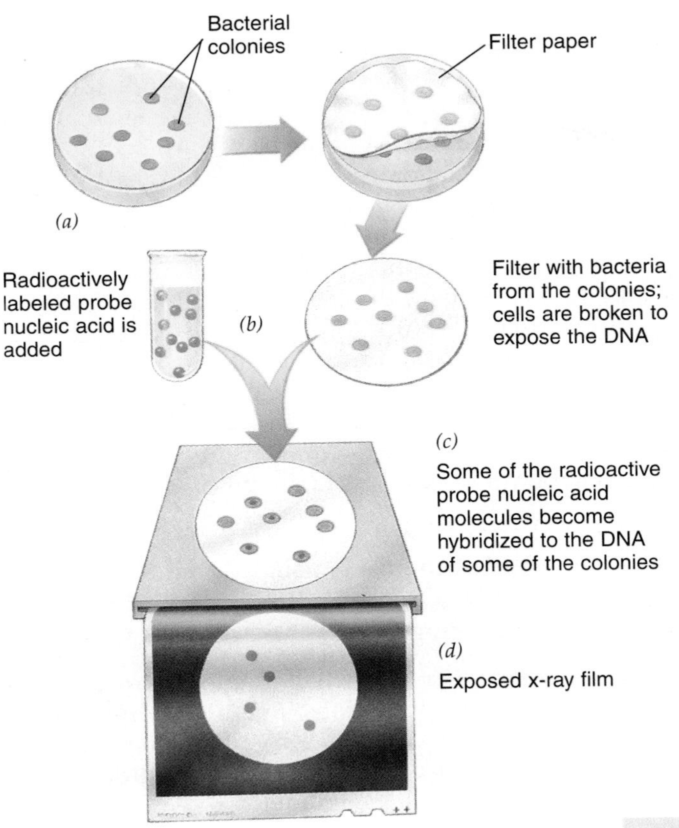

FIGURE 14–7 A genetic probe can be used to identify DNA corresponding to a specific gene. (*a*) *E. coli* cells containing a DNA library are spread on solid nutrient medium in dishes so that only one cell is found in each location. Each cell gives rise to genetically identical descendants to form a colony on the medium. Each colony contains plasmids with only a single DNA fragment from the library. (***b***) To identify which colonies contain at least part of the required gene, a few cells from each colony are transferred to nitrocellulose filters, which bind the DNA from the cells. (***c***) The filter is incubated with a radioactive DNA probe that is complementary to the desired gene. (***d***) DNA from cells that contain the sequence complementary to the probe becomes radioactive and can be detected by x-ray film. The pattern of spots on the film allows one to identify and isolate colonies containing the correct plasmid.

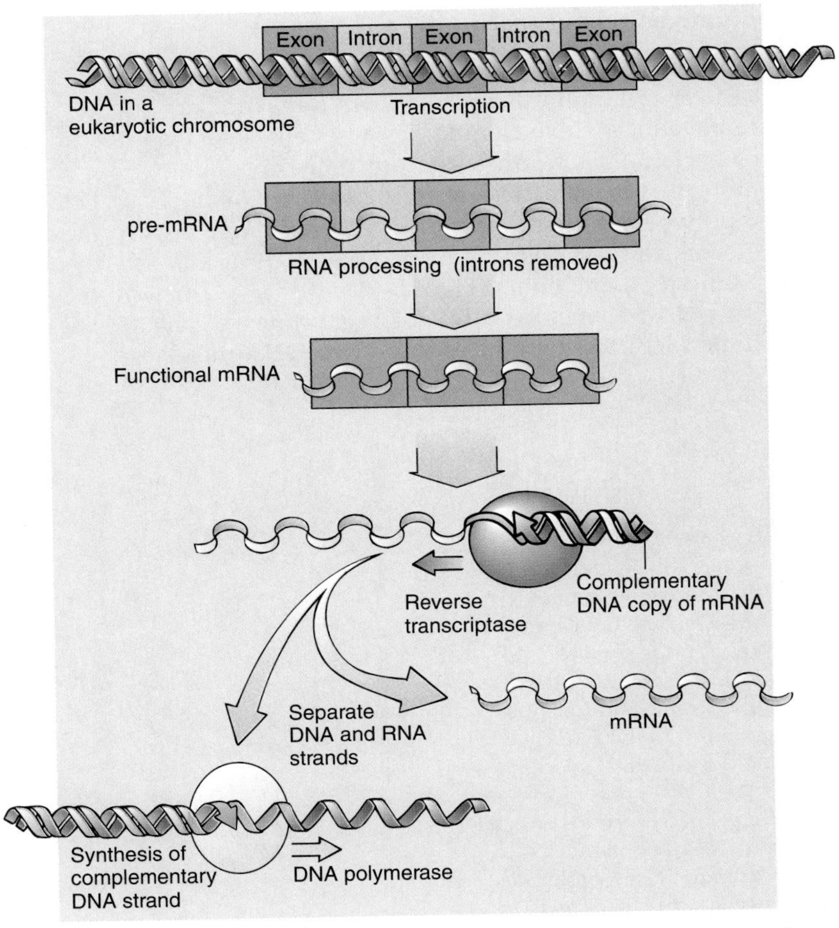

FIGURE 14–8 A cDNA library contains base sequences complementary to mRNA. Reverse transcriptase is used to make a complementary DNA (cDNA) copy of mRNA that has been isolated from cells. DNA polymerase is used to make a complementary DNA copy of the cDNA strand.

Cloning a gene from both a cDNA library and a genomic library has several advantages. Analysis of the genomic DNA clones gives useful information about the structure of the gene in the chromosome and the structure of the primary pre-mRNA transcript, as well as nontranscribed regulatory regions.

Analysis of cDNA clones allows investigators to determine certain characteristics of the protein encoded by the gene, including its exact amino acid sequence. The structure of the processed mRNA can also be studied. Furthermore, because the cDNA copy of the mRNA does not contain intron sequences, comparison of the cDNA and genomic DNA base sequences reveals the locations of intron and exon coding sequences on the chromosome.

Cloned cDNA sequences are also useful when it is desirable to produce a eukaryotic protein in *E. coli.* When an intron-containing human gene such as the gene for human growth hormone is introduced into *E. coli,* the bacterium is unable to remove the introns from the transcribed RNA to make a functional mRNA for the production of its protein product. If a cDNA clone of the gene is inserted into the bacterium, however, its transcript contains an uninterrupted coding region. A functional protein can be synthesized if the gene is inserted downstream of an appropriate bacterial promoter.

The polymerase chain reaction is a technique for amplifying DNA in vitro

The methods to amplify a specific DNA sequence described above all involve cloning DNA in cells, usually those of bacteria. These processes are time-consuming and require an adequate DNA sample as starting material. The **polymerase chain reaction (PCR)** technique allows researchers to amplify a tiny sample of DNA millions of times in a few hours (Fig. 14–9).

Using DNA polymerase, a DNA target sequence can be replicated in a test tube to produce two DNA molecules. The double strands of each molecule are separated by heating and replicated again, so then there are four molecules. After the next cycle of heating and replication there are eight molecules, and so on, with the number of DNA molecules doubling in each cycle. After 20 such cycles this exponential process yields 2^{20}, or over 1 million, copies of the target sequence!

A special heat-resistant DNA polymerase (which originated in a bacterium that lives in hot springs) is used because it remains stable through many heating cycles. Despite occasional technical problems, the PCR technique has been invaluable to researchers. It allows the amplification and analysis of tiny DNA samples from seemingly unlikely sources, ranging from fossil leaves and archeological remains to evidence from crime scenes. The potential applications seem virtually limitless.

A cloned gene sequence is usually first analyzed by restriction mapping, followed by DNA sequencing

A cloned piece of DNA can be used as a research tool for a wide variety of applications. Even if the purpose of cloning the gene is to obtain the encoded protein for some industrial or pharmaceutical process, a great deal of information must be obtained about the gene and how it functions before it can be "engineered" for a particular application.

Restriction mapping establishes landmarks in a cloned DNA fragment

One of the first things done with a gene after cloning is the construction of a **restriction map** of the DNA fragment. Restriction mapping of DNA involves identifying sites that are attacked by specific restriction enzymes. These "restriction sites" serve as landmarks for further studies. That information can be used to isolate (subclone)

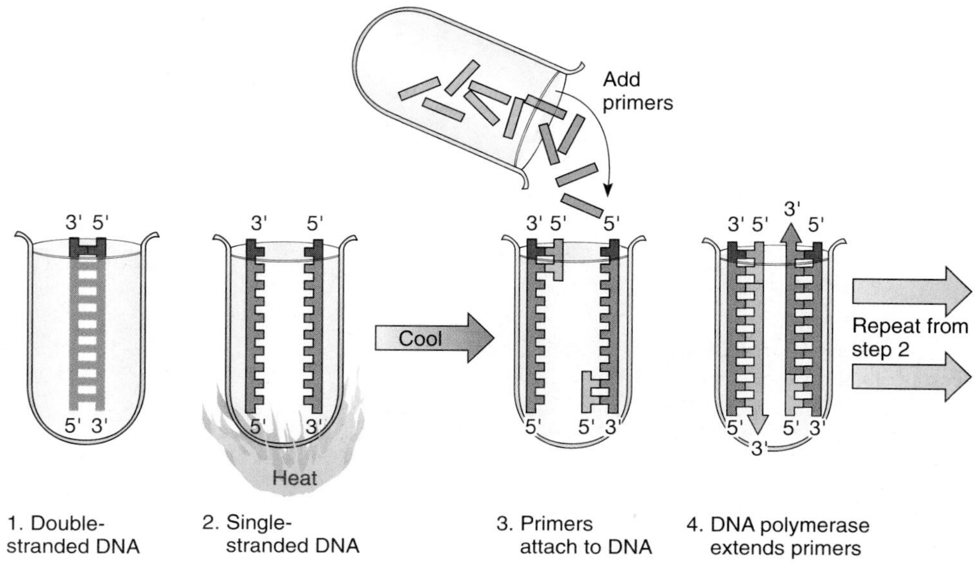

FIGURE 14–9 The polymerase chain reaction (PCR) permits scientists to produce millions of copies of a DNA sequence in vitro. In each cycle double-stranded DNA is heated to produce single strands, and primers complementary to the sequence to be copied are added. A special heat-resistant DNA polymerase extends the primers, forming double-stranded DNA. The number of double-stranded DNA molecules doubles each time the cycle is repeated.

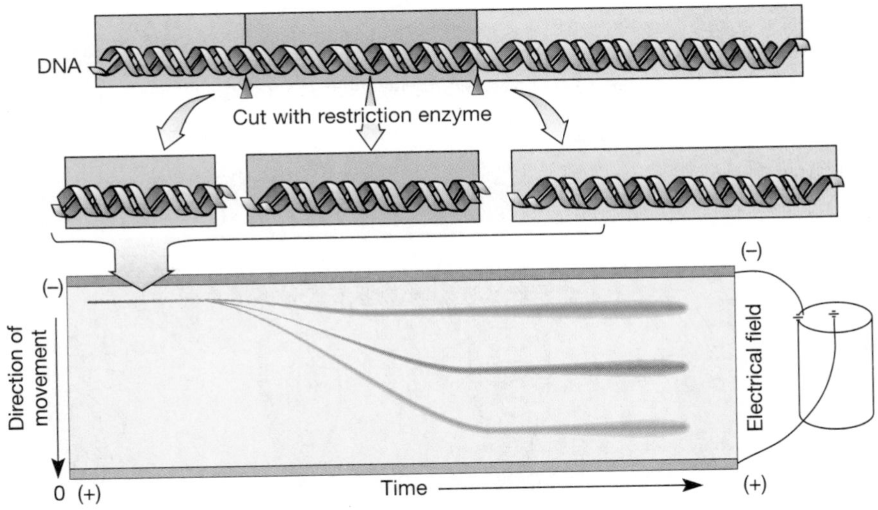

(a)

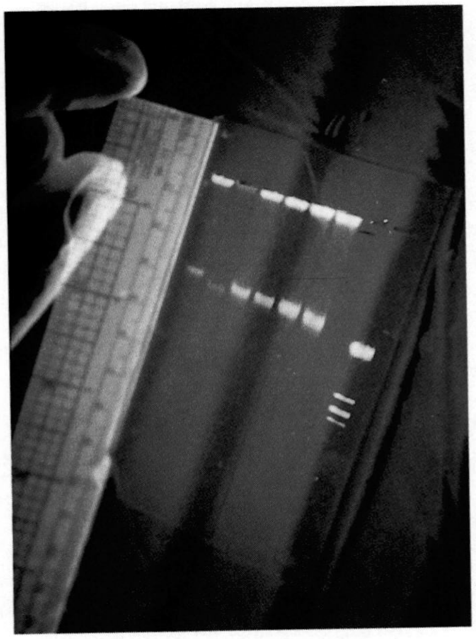
(b)

FIGURE 14–10 Gel electrophoresis is a technique that can be used to separate DNA fragments on the basis of size. (*a*) The gel material, which consists of agarose or polyacrylamide, is poured as a thin slab on a glass or Plexiglas holder. After the gel has solidified, samples containing DNA fragments of different sizes are loaded in wells formed at one end of the gel. DNA molecules are negatively charged, so the molecules migrate through the gel toward the positive pole of an electrical field. The rate at which the molecules travel through the gel is inversely proportional to their molecular weight. Therefore, the smallest DNA fragments (*green*) travel the longest distance. Including DNA fragments of a known size in some of the wells allows accurate measurement of the molecular weights of the unknown fragments. (*b*) A gel containing separated DNA fragments. The gel is stained with ethidium bromide, a dye that binds to DNA and is fluorescent under UV light. (*b*, Visuals Unlimited/Michael Gabridge)

smaller DNA fragments for a variety of purposes. The mapping procedure involves cutting the DNA fragment with various combinations of restriction enzymes and then separating the DNA fragments by **gel electrophoresis** (Fig. 14–10) to determine the molecular weight of each cut fragment. After the size of each fragment is known, the positions of the restriction sites and the distances between them can be determined (Fig. 14–11).

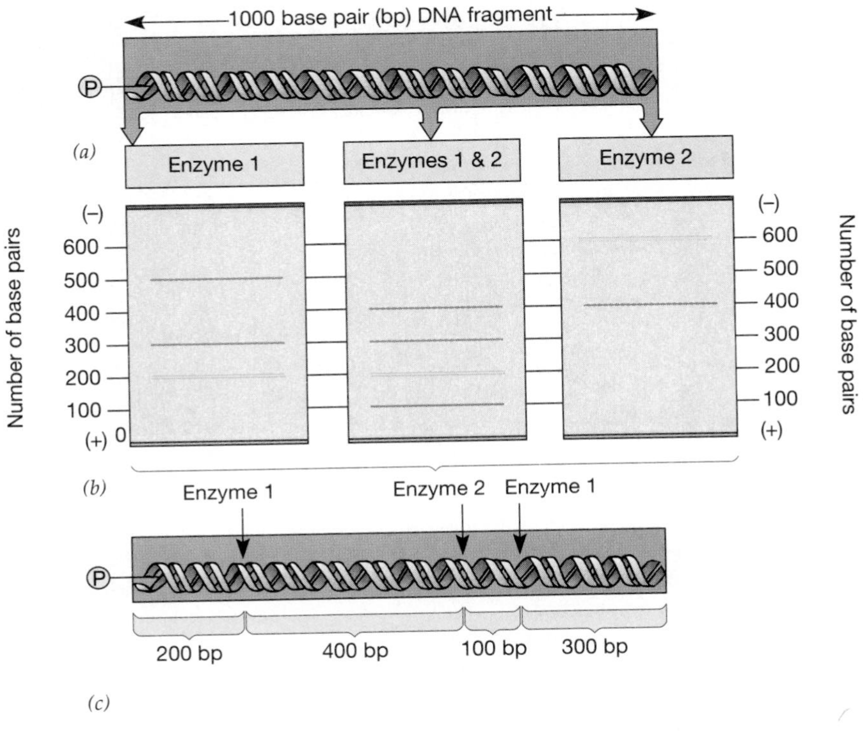

FIGURE 14–11 A restriction map is constructed by determining the sizes of fragments produced when purified DNA is digested by a restriction enzyme. (*a*) One end of a 1000-base-pair DNA fragment is labeled with radioactive phosphorus. Samples of the DNA are cut by either or both of two different restriction enzymes. (*b*) The sizes of DNA fragments and the location of the radioactive fragment (*yellow*) are determined by gel electrophoresis. (*c*) The positions of the restriction sites in the original fragment with respect to the radioactive end of the molecule are then deduced.

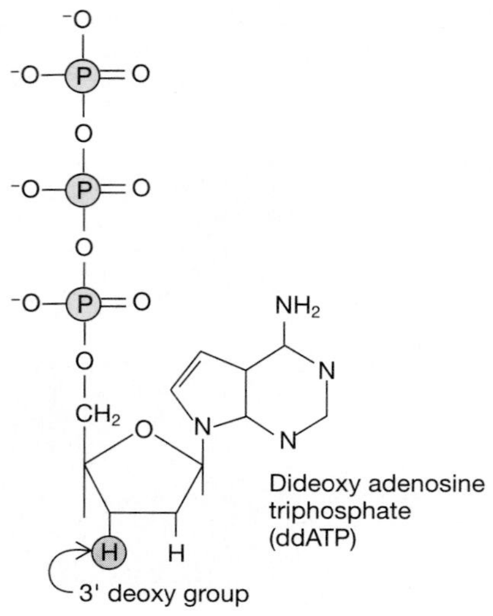

(a) Dideoxynucleotides are modified nucleotides that lack a 3' hydroxyl group and thus block further elongation of a new DNA chain.

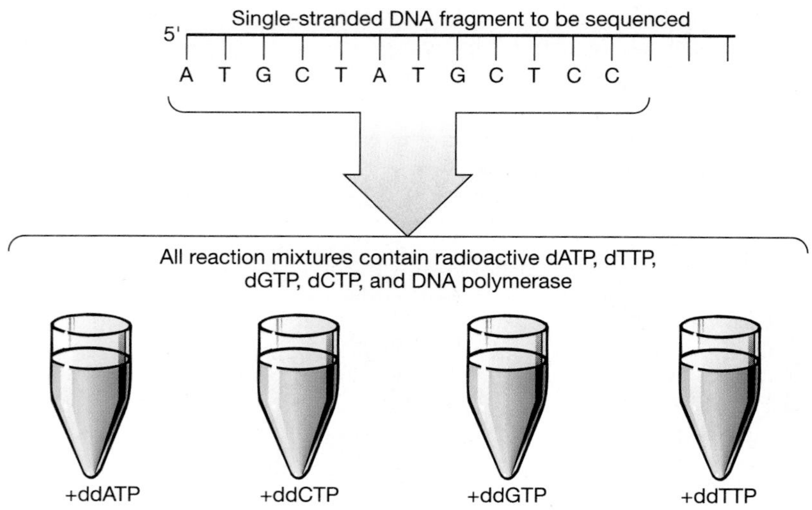

(b) Four different reaction mixtures are used to sequence a DNA fragment; each contains a small amount of a single dideoxynucleotide, such as dideoxy ATP (adenine), and larger amounts of the four normal deoxynucleotides.

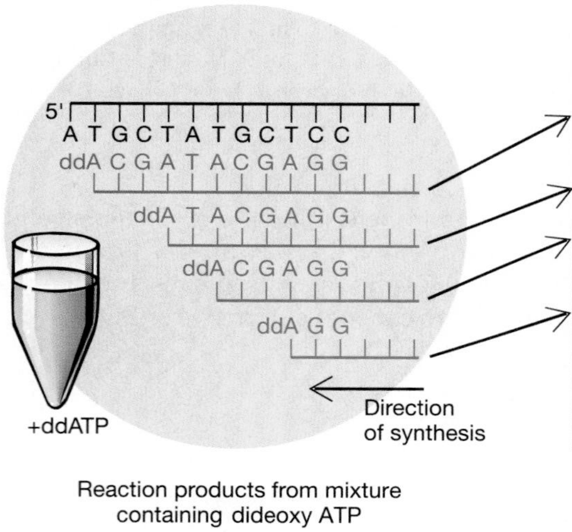

(c) The random incorporation of dideoxy ATP into the growing chain generates a series of smaller DNA fragments ending at all the possible positions where adenine is found in the original fragment.

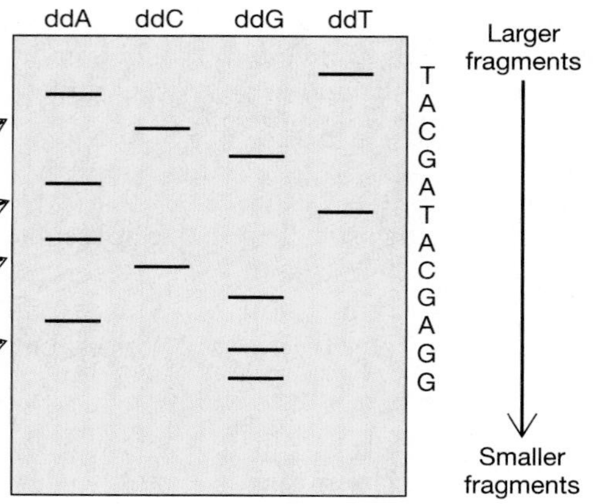

(d) The radioactive products of each reaction mixture are separated by gel electrophoresis and located by exposing the gel to x-ray film. The nucleotide sequence of the newly synthesized DNA is read directly from the film.

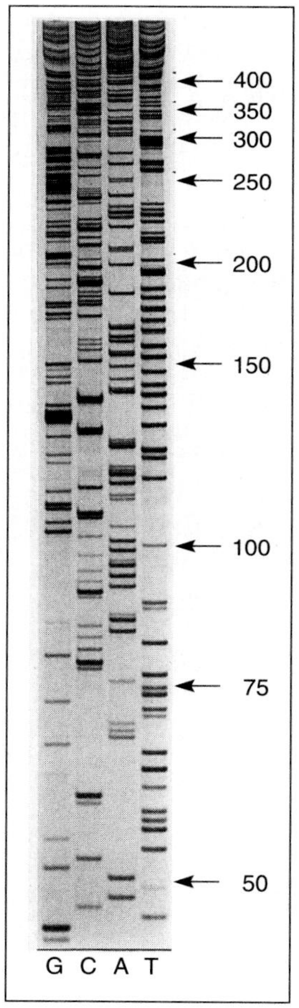

(e) An exposed x-ray film of a DNA sequencing gel. The four lanes represent G, C, A, and T dideoxy reaction mixes, respectively.

FIGURE 14–12 DNA sequencing is a relatively straight forward procedure. The most commonly used method involves synthesis of complementary copies of a single DNA strand by DNA polymerase. The newly synthesized DNA strands are radioactively labeled at one end so that they can be identified. (*e*, Courtesy of B. Slatko, New England Biolabs)

Once the restriction map has been established, subcloned regions of the DNA fragment can be sequenced or used as DNA probes for analytical purposes.

A great deal of information can be inferred from a DNA nucleotide sequence

DNA sequencing (Fig. 14–12) is usually done by copying a strand of the cloned DNA in four different reaction mixtures, using a modified form of DNA polymerase. In each reaction mixture the copies are made so that the newly synthesized DNA chains are terminated randomly at positions corresponding to one of the four bases. The lengths of the fragments from each reaction are then determined by gel electrophoresis, allowing the sequence of bases in the cloned DNA fragment to be read off from one end to the other.

Knowing the DNA sequence in the cloned gene allows investigators to identify which parts of the DNA molecule contain the actual protein-coding sequences, as well as which parts may be regulatory regions involved in gene expression (see Chapter 13). The amino acid sequence of the protein can then be read directly from the DNA, along with other signals involved in mRNA processing and modification. This in itself represents a tremendous advance for research in molecular biology. Prior to the development of DNA sequencing methods, protein sequences were determined by laborious methods from highly purified protein samples. Although protein microsequencing technology has also advanced rapidly, in most cases cloning and sequencing the gene is easier than purifying and sequencing a particular protein.

DNA sequence information is now kept in large computer databases that are available to investigators for comparing newly discovered protein sequences with those already known. By searching for DNA (and amino acid) sequences in the database, researchers can gain a great deal of information about the function and structure of the gene product as well as the evolutionary relationships among genes.

A restriction fragment can be used as a genetic probe

Radioactive genetic probes made from restriction fragments of cloned genes can also be used to detect related sequences in DNA or RNA from other cells. Because the DNA probe binds only to complementary base sequences, it is not necessary to purify those fragments from the rest of the cellular DNA or RNA. The DNA to be probed is simply cut with restriction enzymes, and the entire collection of fragments is separated by electrophoresis. The gel has a banded appearance, with each band containing fragments of a particular size. The separated fragments are then bound to nitrocellulose membranes that pick up the DNA much as a blotter picks up ink. When the DNA on the membrane is incubated with the radioactive probe, the probe binds to any complementary fragments, thus marking their locations. This type of **blot hybridization** (called a **"Southern blot"** after its inventor, E. M. Southern) has widespread applications.[1] It is often used to diagnose certain types of genetic disorders because the radioactive probes can sometimes be used to identify DNA sequences associated with certain genetic defects (see *Making the Connection: Molecular Biology and Diagnosing Genetic Diseases,* Chapter 15).

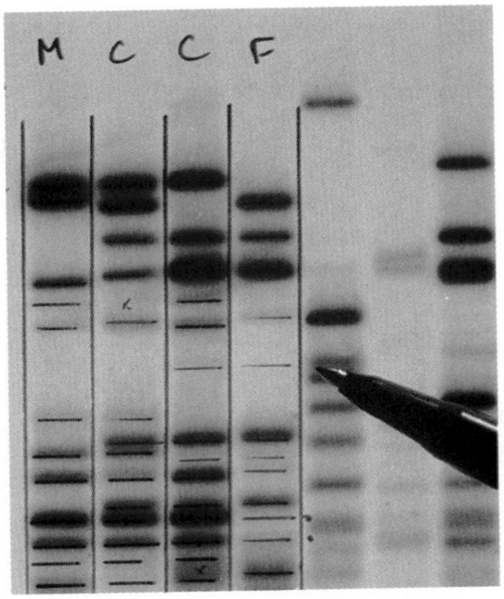

FIGURE 14–13 A restriction fragment length polymorphism (RFLP) is an indication of genetic variability from individual to individual. Restriction enzymes are used to cut the DNA from two or more individuals and the fragments are separated by gel electrophoresis. The DNA on the gel is then allowed to hybridize with a genetic probe representing a sequence that is repeated and interspersed throughout the genome. The resulting patterns can then be compared. (David Parker/Science Photo Library/Photo Researchers, Inc.)

Restriction fragment length polymorphisms (RFLPs) allow detection of genetic relationships

Restriction enzymes can also be used to examine the variability of genes within a population of organisms. Random DNA mutations and recombination may result in individual differences in the number and location of sites where a particular restriction enzyme acts, and therefore differences in the lengths of the fragments produced. Such **restriction fragment length polymorphisms**[2] (commonly known as **RFLPs,** or "Riflips") can be used to determine how closely related different members of the population are (Fig. 14–13).

---

[1] Similar techniques are used to study RNA and proteins. When RNA molecules separated by electrophoresis are transferred to a membrane, the result is a **"Northern blot"**; while a **"Western blot"** consists of separated protein molecules.

[2] The term *polymorphism* literally means "many forms." A genetic polymorphism is said to exist if individuals of two or more discrete genetic types ("morphs") are found in a population or species (see Chapter 18).

RFLP analysis has been found to be an especially powerful tool in the fields of population and evolutionary biology because it can measure the degree of genetic relatedness between individuals. It is also very useful in settling cases of disputed parentage.

The most controversial use of this technology is in the field of forensics. If even small amounts of blood, semen, or other DNA-containing tissue are left at the scene of a crime, one or more target DNA sequences can be amplified by the PCR technique, cut with appropriate restriction enzymes, and subjected to electrophoresis. The resulting pattern of bands is commonly referred to as a "DNA fingerprint."

If applied properly, DNA fingerprinting has the power to exonerate the innocent and may identify the guilty with a high degree of certainty. Such evidence has been ruled admissible in many court cases. One limitation arises from the fact that the DNA samples are usually small and may have been degraded. Secondly, concerns have been raised over just how "unique" each individual pattern might be. For example, a pattern that is quite rare in the general population might be more common in a particular ethnic group. As data have accumulated it has been learned that these concerns are of less practical importance than once thought. For example, the odds that two persons taken at random from the general population would have identical DNA fingerprints may be as low as one in several billion. If two persons are members of the same ethnic group the odds of a match may increase, but are usually still extremely low (perhaps one in several million).

GENETIC ENGINEERING HAS MANY APPLICATIONS

Recombinant DNA technology has provided not only a new and unique set of tools for examining fundamental questions about how living cells work, but also new approaches to problems of applied technology in many other fields. In some cases the production of genetically engineered proteins and organisms has begun to have considerable impact on our lives. The most striking of these have been in the fields of pharmacology and medicine.

Human insulin produced by *E. coli* was one of the first genetically engineered proteins to be commercially produced. Prior to the development of the altered bacterium to produce the human hormone, insulin was derived exclusively from other animals. Many diabetic persons become allergic to the insulin from animal sources because its amino acid sequence differs slightly from human insulin. The ability to produce the human protein by recombinant DNA methods has resulted in significant medical benefits to diabetics.

Genetically engineered human growth hormone (see Chapter 47) is required by some children to overcome growth deficiencies. Human growth hormone could previously be obtained only from cadavers. Only small amounts were available, and evidence suggested that some of the preparations were contaminated with viruses. The list of products that can be produced by genetic engineering is ever growing.

Additional engineering is required for a recombinant eukaryotic gene to be expressed in bacteria

Even if a gene has been isolated and successfully introduced into *E. coli,* the bacterium does not necessarily make the encoded protein in large quantities. Several obstacles stand in the way of producing gene products of eukaryotes in bacteria. One is that the gene has to be correctly associated with an appropriate set of regulatory and promoter sequences that the bacterial RNA polymerase can recognize. Recall from Chapters 12 and 13 that the regulatory regions of prokaryotic and eukaryotic genes are quite different. A usual approach to this problem is to combine the amino acid coding portion of a eukaryotic gene with a bacterial promoter sequence that can be strongly expressed. Some eukaryotic genes, for example, are fused to the lactose operon regulatory region (see Chapter 13); the protein product of the eukaryotic gene is synthesized when the bacterium is fed lactose in the growth medium.

We have already discussed the fact that bacterial cells cannot process RNA molecules containing eukaryotic intron sequences, and that one solution to this problem is to introduce a cDNA copy of the gene. Other problems may arise in the expression of a recombinant protein in *E. coli* because of differences in the ways the proteins are expressed in prokaryotic and eukaryotic cells.

Insulin, for example, is made in human cells from a large polypeptide that is folded in a specific way by the formation of three disulfide bonds (see Fig. 3–21) between six sulfur-containing amino acids. After the polypeptide is folded, parts of the protein are removed by proteolytic (protein-digesting) enzymes, leaving the insulin as two separate polypeptide chains held together by the disulfide bonds. *E. coli* lacks the specific enzymes necessary to cut the larger protein and is not able to fold the molecule properly. To overcome these problems, the gene was engineered to produce the two polypeptides separately. The recombinant proteins are then purified from the cells and allowed to associate in vitro. This procedure results in a relatively low yield of the active hormone, because the insulin can fold in several ways, only one of which results in a functional hormone. It has been possible to circumvent some of these problems by introducing the gene into eukaryotic cells, such as yeast or cultured mammalian

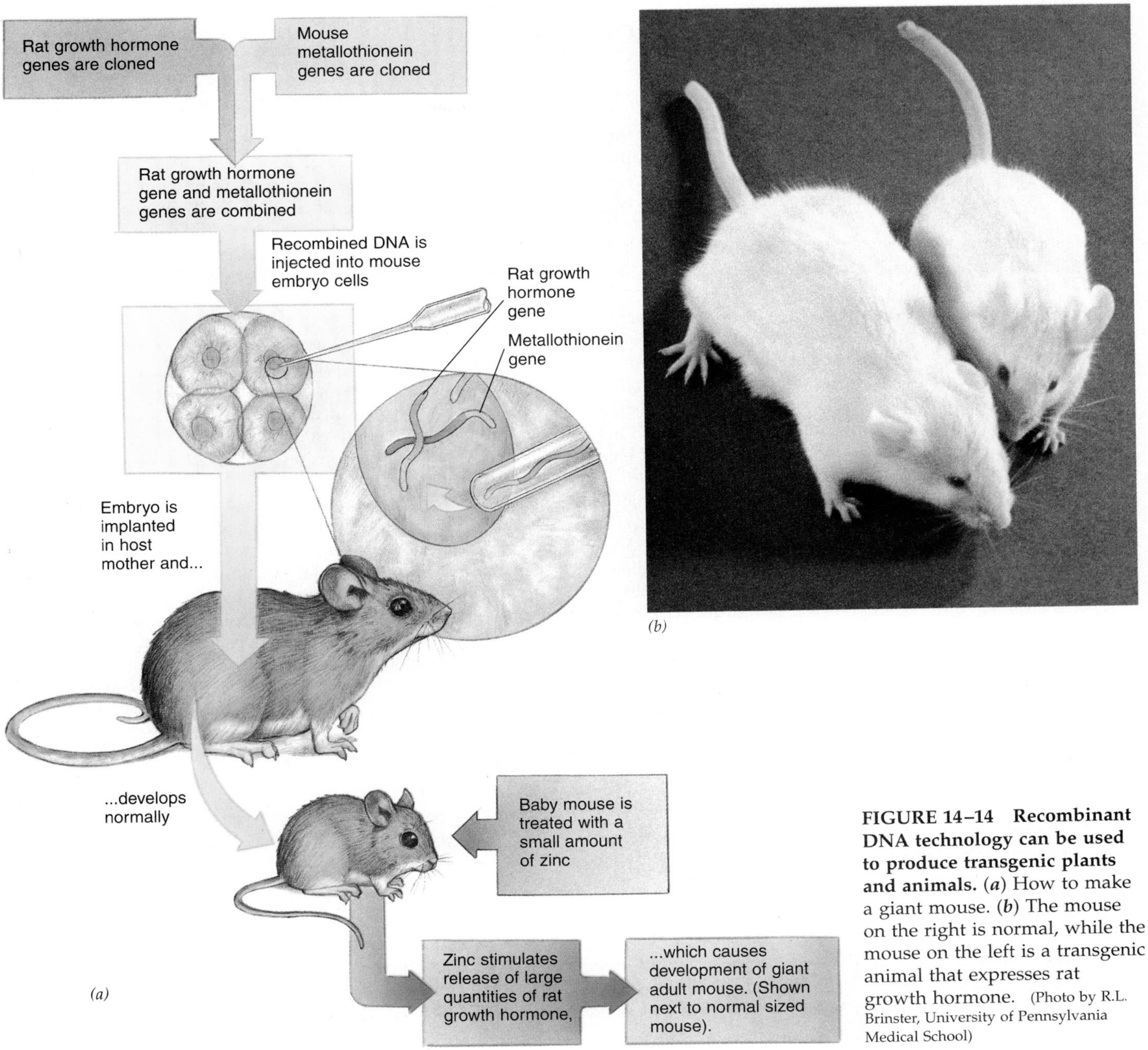

FIGURE 14–14 Recombinant DNA technology can be used to produce transgenic plants and animals. (*a*) How to make a giant mouse. (*b*) The mouse on the right is normal, while the mouse on the left is a transgenic animal that expresses rat growth hormone. (Photo by R.L. Brinster, University of Pennsylvania Medical School)

cells, that contain the protein-processing machinery required to produce fully functional proteins.

Transgenic organisms have incorporated foreign DNA into their cells

Plants and animals that have incorporated foreign genes are referred to as **transgenic** organisms. A number of approaches are being used to insert foreign genes into plant or animal cells. Viruses are often used as vectors, although other methods, such as direct injection of DNA into cells, have also been used.

Transgenic animals are valuable in research and may have commercial uses

One approach to genetic engineering of animal proteins is to use live animals that have incorporated a foreign gene to make the recombinant protein. These transgenic animals are usually produced by microinjecting the DNA of a particular gene into the nucleus of a recipient fertilized egg cell (Fig. 14–14; also see Fig. 16–19). The eggs are then implanted into the uterus of a female and allowed to develop.

Making the Connection

The Genetics of Mice and Humans

How can the study of the genetics of other organisms further our understanding of human genetics? In this chapter we have seen how recombinant DNA technology was developed using knowledge of the genetics of bacteria, yeast, and other microorganisms. As the available techniques have become more sophisticated, it has become possible to genetically engineer complex organisms, producing transgenic plants and animals.

The laboratory mouse, *Mus,* has become a particularly useful model for studying human genes. One extremely powerful research tool is **gene targeting,** a procedure in which a single gene is chosen and "knocked out" (inactivated) in a mouse. The roles of the inactivated gene can be determined by observing the phenotype of the mouse bearing the knocked out gene. Because at least 99% of the genes of humans and mice are essentially the same (although not necessarily identical), information about knockout genes in mice provides details about human genes as well.

Gene targeting, pioneered by Mario Capecchi, a molecular geneticist at the University of Utah School of Medicine, is a rather complex and lengthy procedure; it takes about a year to develop a new strain of knockout mice. The process makes use of results that geneticists have found to be fairly typical whenever a foreign gene is inserted into a cell. That is, once inside a cell, the introduced gene may become physically associated with a corresponding gene in a chromosome. In a poorly understood process known as *homologous recombination,* the two tend to exchange DNA segments. Thus, the introduced modified gene replaces the normal gene in the mouse chromosome.

Researchers place knock-out genes into embryonic mouse cells, inject the genetically modified cells into early mouse embryos, and allow the mice to develop to maturity. The mice are then bred for several generations to develop homozygous offspring that carry the modified gene in every cell.

Many genes are essential to life (and therefore lethal when inactivated). For this reason researchers have modified the knockout technique to develop strains in which a specific gene is selectively inactivated in only one cell type. Today more than 250 different strains of knockout mice, each displaying its own characteristic phenotype, have been developed in various research labs, and the number continues to grow.

Gene targeting is providing answers to basic biological questions, including the development of embryos (see Chapter 16), the development of the nervous system (see Chapter 39), and the normal functioning of the immune system (see Chapter 43).

Gene targeting has great potential for learning more about various human diseases, more than 5000 of which have a genetic component. Gene targeting is being used to study cancer, heart disease, respiratory diseases such as cystic fibrosis (see *On the Cutting Edge: Using a Mouse Model to Study a Human Genetic Disease,* Chapter 15), and other health problems. ▲

In a pioneering study of this type, the gene for growth hormone was isolated from a library of genomic rat DNA. It was combined with the promoter region of a mouse gene that normally produces metallothionein, a protein whose synthesis is stimulated by the presence of heavy metals such as zinc. The metallothionein regulatory sequences were used as a switch to turn the production of rat growth hormone on and off at will. After the engineered gene was injected into fertilized mouse egg cells, the eggs were implanted into the uterus of a mouse and allowed to develop. Embryos in which the gene transplant had been successful grew rapidly when exposed to small amounts of zinc. One mouse, which developed from an egg that had received two copies of the growth hormone gene, grew to more than double the normal size. As might be expected, such mice are often able to transmit their increased growth capability to their offspring.

Transgenic offspring have already been shown to have valuable research applications in a wide range of studies. These include regulation of gene expression (see Chapter 13), immune system function, genetic diseases, viral diseases, and genes responsible for the development of cancer (see *Making the Connection: The Genetics of Mice and Humans*).

Transgenic animals have been used to develop strains that secrete important proteins in milk. For example, the gene for tissue plasminogen activator (TPA), a protein that dissolves blood clots that cause heart attacks, has been introduced into transgenic mice. The gene for human blood clotting factor has been similarly introduced into sheep. These recombinant genes have been fused to the regulatory sequences of the milk protein genes and are therefore activated only in mammary tissues involved in milk production.

The advantage of obtaining the protein from milk is that it is produced in large quantities and can be harvested simply by milking the animal. The protein is then purified from the milk. The animals are not harmed by the introduction of the gene and, because the progeny of the transgenic animal usually also produce the recombinant protein, transgenic strains can be established simply by breeding the animals.

Sometimes viruses are used as recombinant DNA vectors. RNA viruses called **retroviruses** make DNA copies of themselves by reverse transcription (see *Focus On: Reverse Transcription, Jumping Genes, and Pseudogenes* in Chapter 12). Sometimes the DNA copies become integrated into the host chromosomes, where they are replicated along with host DNA. For example, genetically altered mouse leukemia viruses are retroviruses that can be used as vectors to incorporate recombinant genes into cultured cells. Under certain conditions genes carried by the engineered virus can be expressed in the animal cells to produce genetically engineered proteins.

A major disadvantage of introducing genes into cultured animal cells is that the yields of the proteins encoded by the foreign DNA carried by the viruses are generally low. However, these types of vectors show some promise as a means of **gene therapy** for human genetic disorders (see Chapter 15).

Transgenic plants are increasingly important in agriculture

Plants have been selectively bred for thousands of years. The success of such efforts depends on the presence of desirable traits in the variety of plant being selected, or in closely related wild or domesticated plants whose traits can be transferred by cross-breeding. Primitive varieties or closely related species of cultivated plants often have traits, such as disease resistance, that could be advantageously introduced into varieties more suited to modern needs.

If genes are introduced into plants from strains or species with which they do not ordinarily interbreed, the possibilities for improvement are greatly increased. Unfortunately, a suitable vector for the introduction of recombinant genes into many types of plant cells has proved very difficult to find. The most widely used vector system employs the crown gall bacterium, *Agrobacterium tumefaciens.* This bacterium normally produces plant tumors by introducing a special plasmid, called the *Ti* (for *tumor-inducing*) *plasmid,* into the cells of its host (Fig. 14–15). The Ti plasmid induces abnormal growth by forcing the plant cells to produce elevated levels of a plant growth hormone called cytokinin (see Chapter 36).

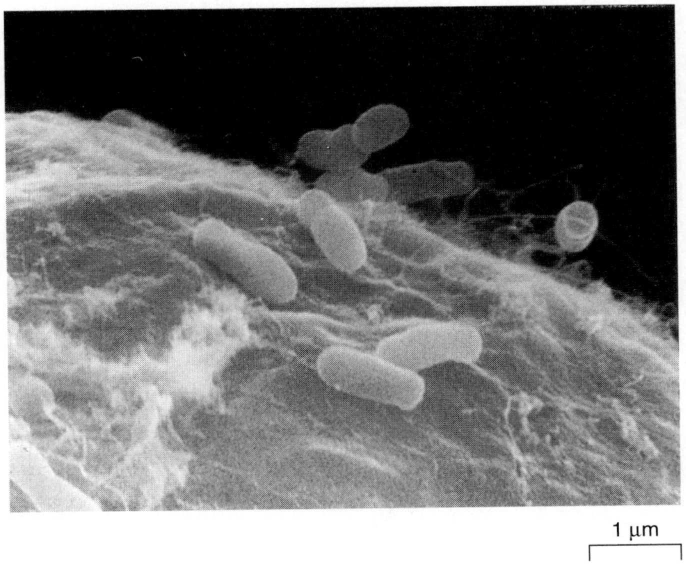

FIGURE 14–15 This scanning electron micrograph shows *Agrobacterium tumefaciens* infecting cultured plant cells. The close contact permits the transfer of plasmid DNA from the bacteria to the larger plant cells. (Courtesy Ann G. Matthysse)

It is possible to "disarm" the Ti plasmid so that it does not induce tumor formation, and then to use it as a vector to insert genes into plant cells. The cells into which the altered plasmid is introduced are essentially normal except for the genes that have been inserted. Genes placed in the plant genome in this fashion may be transmitted sexually, via seeds, to the next generation, but they can also be propagated asexually if desired.

A major problem with the Ti vector is that, with a few exceptions, only dicotyledonous plants (see Chapter 27) can be infected by *A. tumefaciens.* Unfortunately, the grain plants that are the main food source for humans are monocotyledonous plants and therefore outside the host range of the bacterium. Intensive research is under way to develop vector systems for monocotyledonous plants. One approach has been the development of a genetic "shotgun." Microscopic metal fragments are coated with DNA and then shot into plant cells, penetrating the cell walls. Some of the cells retain the DNA and are transformed by it. Those cells can then be cultured and used to regenerate an entire plant (see Chapter 16).

An additional complication of plant genetic engineering is that a number of important plant genes are located in the DNA of the chloroplasts (see Chapter 4). Chloroplasts are essential in photosynthesis, which is the basis for plant productivity. Obviously, it would be useful to develop methods for changing the portion of the plant's DNA that resides within the chloroplast. Methods of chloroplast engineering are currently the focus of intense research.

SAFETY GUIDELINES HAVE BEEN DEVELOPED FOR RECOMBINANT DNA TECHNOLOGY

People who have experienced the direct applications of recombinant DNA technology today would undoubtedly agree that those developments have been important and beneficial. When the new technology was introduced in the 1970s, however, many scientists considered the potential misuses to be at least equally significant. The possibility that an organism with undesirable environmental effects might be accidentally produced was a concern. Totally new strains of bacteria or other organisms, with which the world has no previous experience, might be difficult to control. This possibility was recognized by those who developed the recombinant DNA methods and led them to insist on stringent guidelines for making the new technology safe.

Recent history has failed to bear out these worries. Experiments over the past years in thousands of university and industrial laboratories have seen recombinant DNA manipulations carried out safely. One of the main concerns — the accidental release of laboratory bacterial strains containing dangerous genes into the environment — has turned out to be groundless. Laboratory strains of *E. coli* are poor competition for the wild strains in the outside world and quickly perish. Experiments thought to entail unusual risks are carried out in special facilities designed to contain dangerous disease-causing organisms and allow researchers to work with them safely. The fears of accidentally cloning a dangerous gene or releasing a dangerous organism into the environment seem to be laid to rest. This does not mean, however, that *intentional* manipulations of dangerous genes are not a possibility.

Most scientists today recognize the importance of recombinant DNA technology and agree that the perceived threat to humans and the environment was overestimated. Many of the restrictive guidelines for using recombinant DNA have been relaxed as the safety of the experiments has been established. Stringent restrictions still exist, however, in certain areas of recombinant DNA research where there are known dangers, or where questions about possible effects on the environment are still unanswered.

These restrictions are most evident in research that proposes to introduce recombinant organisms into the wild, such as agricultural strains of plants whose seeds or pollen might spread in an uncontrolled manner. A great deal of research activity is now concentrated on determining the effects of introducing recombinant organisms into a natural environment. Carefully conducted tests have shown that recombinant organisms are not dangerous to the environment simply because they are recombinant. However, it is important to assess the biology of each new recombinant organism. In this way scientists will be able to determine if it has characteristics that might cause it to present an environmental hazard under certain conditions.

SUMMARY

I. Recombinant DNA technology is concerned with isolating and amplifying specific sequences of DNA by incorporating them into vector DNA molecules. The resulting recombinant DNA can then be propagated and amplified in organisms such as *E. coli.*
   A. Restriction enzymes are used to cut DNA into specific fragments.
      1. Each type of restriction enzyme recognizes and cuts DNA at a highly specific base sequence.
      2. Many restriction enzymes cleave DNA sequences to produce complementary single-stranded cut ends (sticky ends).
   B. The most common recombinant DNA vectors are constructed from naturally occurring circular DNA molecules called plasmids, or from bacterial viruses called bacteriophages; both of these are found in some bacteria.
   C. Recombinant DNA molecules are often constructed by allowing the ends of a DNA fragment and a plasmid (which have both been cut with the same restriction enzyme) to associate by complementary base-pairing. The DNA strands are then covalently linked by DNA ligase to form the recombinant molecule.
   D. Single genes are isolated from recombinant DNA libraries, which are mixtures of DNA fragments inserted into appropriate vectors.
      1. Genomic libraries are formed from the total DNA of an organism. Genes present in recombinant DNA genomic libraries from eukaryotes contain introns. Those genes can be amplified in *E. coli,* but the protein is not properly expressed.
      2. When a cDNA library is produced, DNA copies are made from mRNA isolated from eukaryotic cells; these are then incorporated into recombinant DNA vectors. Because the introns have been removed from mRNA molecules, eukaryotic genes in cDNA libraries can sometimes be expressed in *E. coli* to make their protein products.
   E. Analysis of a cloned sequence can yield useful information about the gene and its protein, and can enable investigators to identify and subclone DNA fragments for use as molecular probes.
      1. The first step in analyzing a cloned DNA sequence is to construct a restriction map, thereby identifying sites that are cut by specific restriction enzymes.
      2. Determining the nucleotide sequence of a cloned DNA fragment gives information about the structure of the gene and the probable amino acid sequence of the encoded proteins.
      3. Subcloned restriction fragments of a cloned gene can be made radioactive and used as DNA probes to identify related complementary DNA and RNA sequences. The DNA or RNA to be identified is separated from other nucleic acids by gel electrophoresis and then blotted onto special paper. The radioactive DNA probe is then hybridized by complementary base-pairing to the DNA bound to the paper, and the radioactive band or bands of DNA can be identified.

a. DNA blotting methods are used in diagnosis and in identifying carriers of some genetic diseases.
b. The degree of genetic relationship among the individuals in a population can be estimated by studying restriction fragment length polymorphisms (RFLPs).

II. Genetic engineering is a technology that uses genetic and recombinant DNA methods to devise new combinations of genes to produce improved pharmaceutical and agricultural products.
A. Genes isolated from one organism can be modified and expressed in other organisms ranging from *E. coli* to transgenic plants and animals.
1. Expression of eukaryotic proteins in bacteria such as *E. coli* requires that the gene be linked to regulatory elements that the bacterium can recognize. In addition, bacterial cells do not contain many of the enzymes needed for the posttranslational processing of eukaryotic proteins.
2. Expression of eukaryotic genes in eukaryotic host organisms shows great promise, because the processing and modification machinery for eukaryotic proteins is already present in these cells.
a. Production of important pharmaceutical products can be engineered in transgenic animals so that the products are secreted in milk.
b. Genetic engineering of plants and domestic animals holds the promise of increasing the availability of food.

B. Recombinant DNA technology is carried out under certain safety guidelines.

SELECTED KEY TERMS

bacteriophage
blot hybridization
cDNA library
clone
colony
complementary DNA (cDNA)
DNA ligase
DNA sequencing
gel electrophoresis
gene therapy
genetic engineering
genetic probe
genome
genomic DNA library
kilobase
palindrome
plasmid
polymerase chain reaction (PCR)
recombinant DNA
restriction enzyme
restriction fragment length polymorphism (RFLP)
restriction map
retrovirus
reverse transcriptase
Southern blot
transformation
transgenic organism
vector molecule

POST-TEST

1. Transferring genes from one organism to another requires a DNA carrier, or __________ molecule.
2. Specific DNA sequences can be isolated by combining a DNA fragment from one organism with a vector DNA molecule so that the resulting __________ DNA molecule can be amplified in an organism such as *E. coli.*
3. Many restriction enzymes cut DNA at base sequences that are __________, yielding DNA fragments with single-stranded complementary, or sticky, ends.
4. Recombinant DNA vectors used in *E. coli* are usually small, circular DNA molecules called __________, or bacterial viruses called __________.
5. A recombinant DNA plasmid is usually constructed by allowing the complementary ends of a DNA fragment and a plasmid that have been cut with the same __________ __________ to associate by hydrogen-bonding. The single-stranded ends are then covalently linked together by __________ __________.
6. A colony of genetically identical bacteria grown from a single cell is a __________.
7. A __________ DNA library is composed of the total DNA of an organism.
8. A __________ library is composed of DNA fragments made from mRNA molecules that have been copied by __________ __________.
9. Intron sequences are expected in a eukaryotic __________ DNA library, but not in a __________ library.
10. The __________ __________ __________ technique allows scientists to produce millions of copies of DNA from a tiny sample in just a few hours.
11. A first step in analyzing a DNA sequence once it has been cloned is to construct a __________ __________ of the DNA in order to locate landmarks that can be used for further studies.
12. DNA __________ usually involves making copies of a cloned DNA sequence under conditions that block elongation of the DNA chain at points where specific bases are located.
13. DNA molecules can be separated on the basis of size by __________ __________.
14. Short fragments of a cloned DNA sequence can be used as radioactive DNA genetic __________, which can be used to locate complementary nucleic acid fragments by blot hybridization methods.
15. Plants and animals that have been modified by the introduction of recombinant DNA are referred to as __________ organisms.

REVIEW QUESTIONS

1. What is meant by the term *genetic engineering*?
2. What are restriction enzymes? How are they used in recombinant DNA research?
3. What characteristics should be engineered into a plasmid to make it a useful cloning vector?
4. Diagram the process by which recombinant DNA molecules are usually constructed.
5. How is a gene library constructed? What are the relative merits of genomic libraries and cDNA libraries?
6. Sketch an example illustrating how a restriction map of a gene is made.
7. Why is the PCR technique valuable?

YOU MAKE THE CONNECTION

1. What are some of the problems that might arise if you were trying to produce a eukaryotic protein in a bacterium? How might some of these problems be solved by using transgenic plants or animals?
2. Would genetic engineering be possible if we did not know a great deal about the genetics of bacteria? Explain.
3. What are some of the ecological concerns regarding transgenic organisms? What kinds of information are needed to allay these fears?

RECOMMENDED READINGS

Capecchi, M.R. "Targeted Gene Replacement." *Scientific American,* Vol. 270, No. 3, March 1994. Biologists can now produce mice that contain a mutation in any gene desired, thereby enabling them to determine the function of that gene.

Franklin-Barbajosa, C. "DNA Profiling: The New Science of Identity." *National Geographic,* May 1992. Examines some of the varied uses of DNA fingerprinting.

Gasser, C.S., and R.T. Fraley. "Transgenic Crops." *Scientific American,* Vol. 266, No. 6, June 1992. An overview of the use of biotechnology to produce plants with commercially desirable characteristics.

Mullis, K.B. "The Unusual Origin of the Polymerase Chain Reaction." *Scientific American,* Vol. 262, No. 4, April 1990. A highly personal first-hand account of the development of the PCR technique and an excellent illustration of the nature of scientific insight.

Neufield, P.J., and N. Colman. "When Science Takes the Witness Stand." *Scientific American,* Vol. 262, No. 5, May 1990. A discussion of the use of DNA fingerprinting in forensics, with emphasis on the need for appropriate standards.

Pääbo, S. "Ancient DNA." *Scientific American,* Vol. 269, No. 5, November 1993. A fascinating account of how the PCR technique is enabling researchers to study DNA fragments obtained from fossils.

Watson, J.D., M. Gilman, J. Witkowski, and M. Zoller. *Recombinant DNA,* 2nd ed. W.H. Freeman, New York, 1992. A comprehensive source of information about genetic engineering.

CHAPTER 15

Human Genetics

Studies of large extended families have contributed to our understanding of human genetics. (Rafael Macia/Photo Researchers, Inc.)

Geneticists quite naturally have great interest in the study of human genetics. Humans do not serve well as the subjects of most types of genetic research, however. To study the mode of inheritance in any species, geneticists should ideally (1) have standard stocks of genetically identical individuals—that is, **isogenic strains**—that are homozygous at virtually all of their loci; (2) conduct **controlled matings** between members of different isogenic strains; and (3) raise the offspring under carefully controlled conditions. Of course, the human population is very diverse and individuals are heterozygous for many genes. In addition, human families are small, and 20 to 30 years or more elapse between generations. It is therefore virtually impossible, as well as unethical, to conduct such research on humans.

Despite the inherent difficulties, study of human genetics is progressing very rapidly. The field has been revolutionized by combining genetic engineering technology (see Chapter 14) with a variety of more traditional approaches, including pedigree analysis and population studies. This work has been greatly facilitated by the medical attention given to genetic diseases in humans. The extensive medical records of diseases serve as a very useful data pool upon which hypotheses may be based and against which they may be tested. Furthermore, genetic studies of other organisms have provided invaluable insights. Indeed, many phenomena in human inheritance that were initially quite puzzling have been explained by solving analogous problems in the inheritance of bacteria, yeasts, fruit flies, and mice.

LEARNING OBJECTIVES

After you have studied this chapter you should be able to

1. Explain why humans have traditionally been considered poor subjects for the study of inheritance. Point out some of the ways in which these limitations have been overcome.
2. Distinguish between environmentally induced and inherited abnormalities, and between chromosome abnormalities and single gene defects.
3. Make a sketch illustrating how nondisjunction can occur in meiosis. Show how nondisjunction can be responsible for specific chromosome abnormalities such as Down syndrome, Klinefelter syndrome, and Turner syndrome.
4. Describe how amniocentesis is used in the prenatal diagnosis of human genetic abnormalities; state the relative advantages and disadvantages of amniocentesis and chorionic villus sampling.
5. Discuss the scope and implications of genetic counseling.
6. Relate each ABO type to the appropriate genotype(s). Explain the genetic and physiological basis for Rh incompatibility between a mother and a fetus.
7. Explain why quantitative traits in humans are thought to be under the control of polygenes.
8. Discuss the implications of the Human Genome Initiative, including the costs and possible benefits.
9. List some of the ways that genetics affects human society.

ANALYSIS OF INHERITANCE PATTERNS IN HUMANS REQUIRES ALTERNATIVE METHODS

Early studies of human heredity usually dealt with readily identified pairs of contrasting traits and their distribution among members of a family, as illustrated by the **pedigree** in Figure 15–1. This method is still useful, but because human families tend to be small and information on certain family members may be lacking, it has serious limitations.

Human geneticists therefore also use methods that allow them to make inferences about a trait's mode of inheritance based on studies of its distribution in an entire

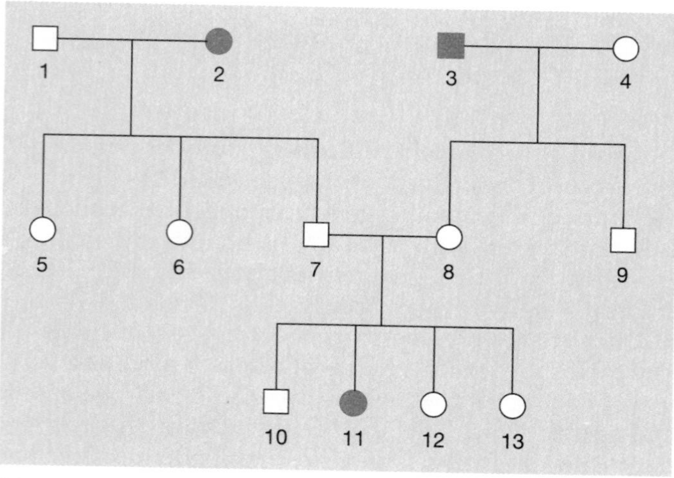

(a)

(b)

FIGURE 15–1 Albinism (lack of pigment in the hair, skin, and eyes) is inherited as an autosomal recessive trait, which means it is not carried on a sex chromosome. (*a*) In this pedigree, males are indicated by squares and females by circles. Individuals in the study showing the albino trait are indicated by red symbols; those not showing the trait are indicated by white symbols. Relationships are indicated by connecting lines, and all members of the same generation are placed on the same row. Thus, 11 is an albino girl whose paternal grandmother (2) and maternal grandfather (3) are also albinos. All of her other relatives depicted here are phenotypically normal. Notice that the inheritance pattern could not be autosomal dominant or X-linked dominant because neither of 11's parents is albino. If the trait were X-linked recessive, her mother would have to be a heterozygous carrier, and her father would have to be an albino (which he is not). (***b***) Albinism is very common among the Cuna people living on the San Blas Islands of Panama. (*b*, Anna Zuckerman/Photo Edit) population. By applying the laws of probability to data obtained from a relatively large sample of individuals who are representative of the population, it is often possible to determine if the mode of inheritance is simple or complex, and if more than one locus is involved. Some of the methods used by population geneticists are introduced in Chapter 18. As we shall see later in this chapter, human inheritance can be studied most effectively by combining these and other approaches with the methods of molecular biology and recombinant DNA technology.

MOST HUMAN TRAITS RESULT FROM COMPLEX GENETIC AND ENVIRONMENTAL INTERACTIONS

The development of each organ of the body is regulated by a large number of genes that interact in complex ways. The mechanisms of inheritance of many physical traits and hundreds of specific enzymes are now known. In fact, the loci of many genes have been identified, and chromosome maps, although incomplete, have been worked out for each human chromosome.

The age at which a particular gene expresses itself phenotypically may vary widely. Most characteristics develop before birth, but some, such as hair and eye color, are not fully expressed until shortly after birth. Others, such as muscular dystrophy, become evident in early childhood, while still others, such as glaucoma and Huntington's disease, develop only in adulthood.

SOME BIRTH DEFECTS ARE INHERITED

A **birth defect,** or congenital defect, is one that is present at birth; it may or may not be inherited. Some birth defects are inherited. Others are produced by environmental factors that affect the developmental process. For example, if a woman contracts the viral disease rubella (commonly known as German measles) during the first three months of pregnancy, a substantial risk exists that her offspring will develop malformations. Environmental factors that have been linked to birth defects are discussed in Chapter 49.

Certain inherited abnormalities are the result of mutations involving a single locus. Sickle cell anemia and albinism are examples of this type of defect. Other abnormalities, such as Down syndrome, result from an abnormal number of chromosomes.

CHROMOSOMAL ABNORMALITIES ARE RESPONSIBLE FOR SOME BIRTH DEFECTS

Cytogenetics is the study of chromosomes and their role in inheritance. Many of the basic principles of genetics were discovered by experiments with simpler organisms, in which it was possible to relate genetic data with the number and structure of the chromosomes. Some of the organisms used in genetics, such as the fruit fly *Drosophila,* have very few chromosomes (only four pairs in *Drosophila*). In *Drosophila* salivary glands and certain other tissues, the chromosomes are large enough that their structural details are readily evident (see Chapter 16). Therefore, this organism has provided unique opportunities for correlating certain kinds of genetic changes with certain kinds of alterations in chromosome structure. Although the science of human cytogenetics is not nearly as refined, many useful determinations are still possible.

Karyotyping is the analysis of chromosomes

Representative normal human karyotypes for males and females are shown in Figure 9–1. The term **karyotype** refers both to the chromosome composition of an individual and to a photomicrograph showing that composition. In karyotyping, cells from the bone marrow, blood, or skin are incubated with chemicals that stimulate mitosis. These chemicals, derived from certain plants, are called **lectins.** The cells are then treated with the drug colchicine (see Chapter 9), which arrests them at mitotic metaphase. Next they are placed into a hypotonic solution (see Chapter 5) that causes them to swell; this spreads out the chromosomes so they can be readily observed. The cells are then flattened on microscope slides, and the chromosomes are stained to reveal the patterns of bands, which are unique for each homologous pair.

After the chromosomes have been photographed, each one is cut out of the photographic print, and the homologous pairs are matched and placed together. Chromosomes can then be identified by length, position of the centromere, banding patterns, and other morphological features such as knobs. The largest chromosome is about five times as long as the smallest one, but there are only slight size differences among some of the intermediate ones.

Most chromosome abnormalities are lethal or cause serious defects

Polyploidy, the presence of multiple chromosome sets, is common in plants but rare in animals. It may arise as a result of the chromosomes' failure to separate during meiosis, or of fertilization of an egg by more than one sperm. Polyploidy is lethal in humans and many other animals when it occurs in all the cells of the body. Triploidy is sometimes found in embryos that have been spontaneously aborted in early pregnancy. The few triploid or tetraploid individuals who have been born alive and survived for a few days have been found to contain a mixture of diploid and polyploid cells.

Abnormalities involving the presence of an extra chromosome or the absence of a chromosome are much more common in humans. These conditions are called **aneu-**

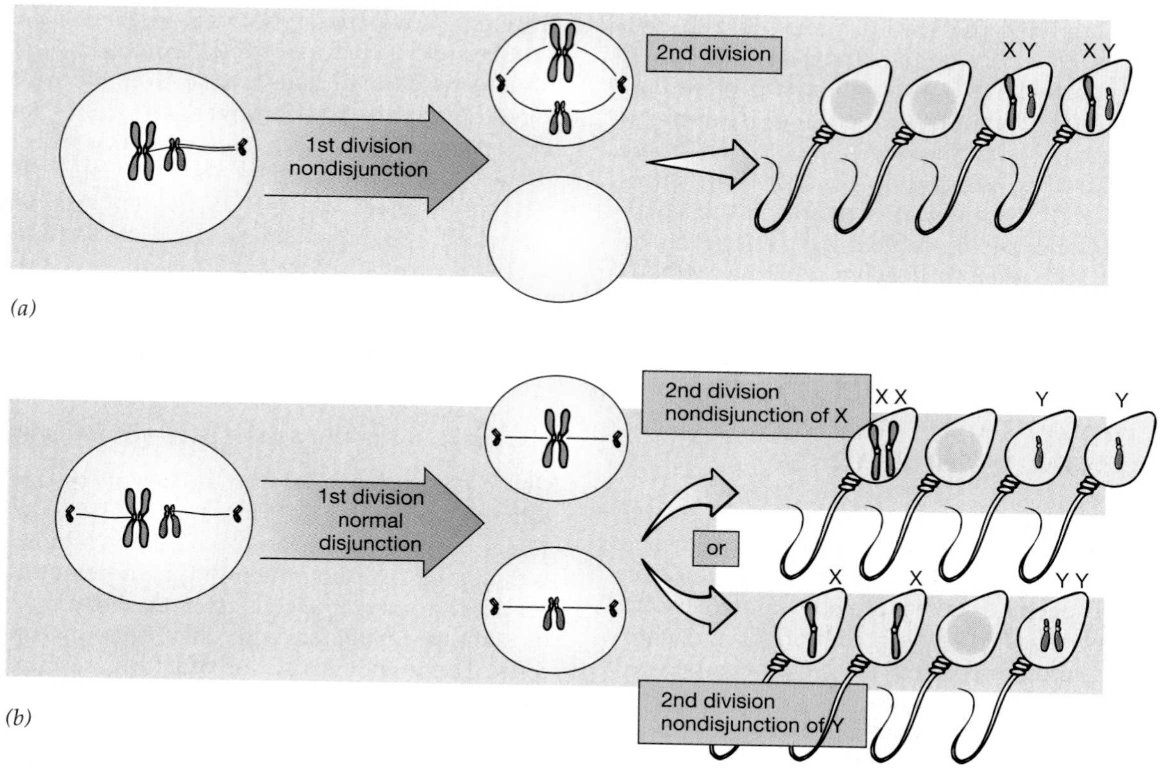

FIGURE 15–2 Nondisjunction may occur during the first or second meiotic division, as shown in these examples of meiotic nondisjunction of the sex chromosomes in the human male. Only the X *(red)* and Y *(brown)* chromosomes are shown. (***a***) Nondisjunction in the first meiotic division results in two XY sperm and two sperm with neither an X nor a Y. (***b***) Second-division nondisjunction of the X chromosome results in one sperm with two X chromosomes, two with one Y each, and one with no sex chromosomes. Similarly, nondisjunction of the Y results in one sperm with two Y chromosomes, two with one X each, and one with no sex chromosome. By contrast, nondisjunction in the female (not shown) results in the formation of eggs with two X chromosomes or no sex chromosomes, regardless of whether it occurs in the first or second meiotic division.

ploidies. Recall that ordinarily there are two of each kind of chromosome; this is the normal **disomic** condition. An individual with an extra chromosome—that is, with three of one kind—is said to be **trisomic** for that kind of chromosome. An individual lacking one member of a pair of chromosomes is said to be **monosomic.**

Aneuploidies generally arise as a result of an abnormal meiotic (or mitotic) division in which chromosomes fail to separate at anaphase. This phenomenon is called **nondisjunction.** In meiosis, chromosomal nondisjunction may occur during the first or second meiotic division (or both). For example, two X chromosomes that fail to separate at either the first or the second meiotic division might both enter the egg nucleus. Alternatively, the two joined X chromosomes might go into a polar body, leaving the egg with no X chromosome.

Nondisjunction of the XY pair in the male might lead to the formation of sperm with both an X and a Y chromosome, or sperm with neither an X nor a Y chromosome. Similarly, nondisjunction at the second meiotic division can produce sperm with two X's or two Y's. Some of these examples of meiotic nondisjunction are illustrated in Figure 15–2. When an abnormal gamete unites with a normal one, the resulting zygote has a chromosome abnormality that is present in every cell of the body.

Nondisjunction during a mitotic division leads to the establishment of a clone of abnormal cells in an otherwise normal individual. Such an individual therefore contains a mixture of normal and abnormal cells.

In some cases part of one chromosome may break off and attach to a nonhomologous chromosome, or two nonhomologous chromosomes may exchange parts. Such an event is called a **translocation.** The consequences of translocations vary considerably, but include situations in which some genes are missing (*deletions*) and extra copies of other genes are present (*duplications*). Table 15–1 summarizes some disorders produced by aneuploidies.

Persons with Down syndrome usually have three copies of chromosome 21

Down syndrome is one of the most common chromosomal abnormalities in humans. Affected individuals have abnormalities of the face, eyelids, tongue, hands, and other parts of the body, and are mentally and physically retarded (Fig. 15–3*a*). They are also unusually suscepti-

Table 15–1 SOME CHROMOSOME ABNORMALITIES

Karyotype	*Common Name*	*Clinical Description*
Trisomy 13	Patau syndrome	Multiple defects, with death by age 1 to 3 months
Trisomy 18	Edwards syndrome	Ear deformities, heart defects, spasticity, and other damage; death by age 1 year
Trisomy 21	Down syndrome	Overall frequency is about 1 in 700 live births. True trisomy is most often found among children of older (age 40+) mothers, but translocation resulting in the equivalent of trisomy may occur in children of younger women. A similar, though less marked, influence is exerted by the age of the father. Trisomy 21 is characterized by a fold of skin above the eye, varying degrees of mental retardation, short stature, protruding furrowed tongue, transverse palmar crease, and cardiac deformities.
Trisomy 22	—	Similar to Down syndrome, but with more skeletal deformities
XO	Turner syndrome	Short stature, webbed neck, sometimes slight mental retardation; ovaries degenerate in late embryonic life, leading to rudimentary sexual characteristics; gender is female; no Barr bodies
XXY	Klinefelter syndrome	Male with slowly degenerating testes, enlarged breasts; one Barr body per cell
XYY	XYY karyotype	Unusually tall male with heavy acne; some tendency to mild mental retardation
XXX	Triplo-X	Despite three X chromosomes, usually fertile, fairly normal females

(a)

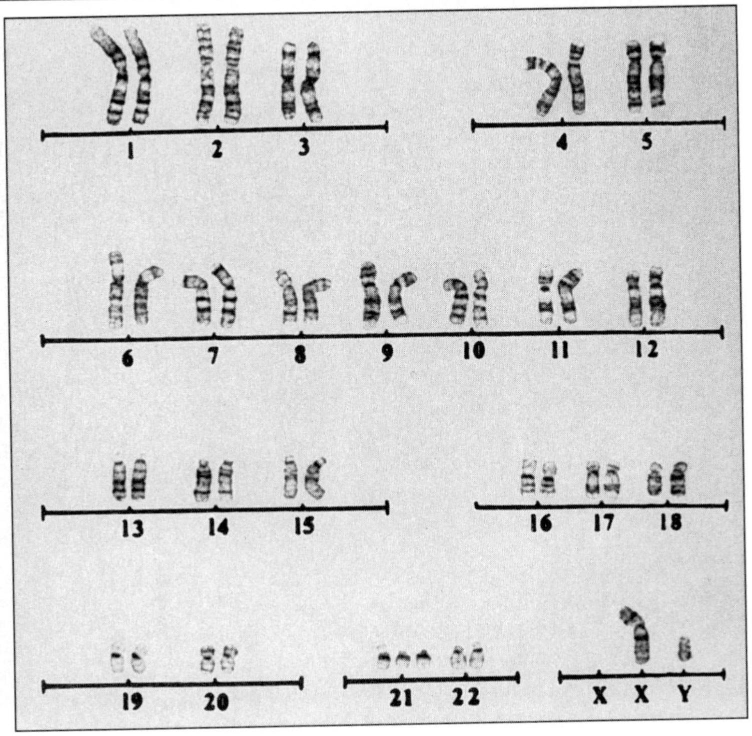

(b)

FIGURE 15–3 Most individuals with Down syndrome have trisomy 21. (*a*) This child with Down syndrome is participating in the Special Olympics. (*b*) Note the presence of an extra chromosome 21 in this karyotype. (*a*, Jose Carrillo/Photo Edit; *b*, Courtesy of Dr. Leonard Sciorra) ble to certain diseases, such as leukemia and Alzheimer's disease. The term *mongolism,* which is no longer used because of its racist connotations, was originally applied to this condition because of a characteristic eyelid fold in affected persons that superficially resembles that typically found in oriental populations.

Cytogenetic studies have revealed that most people with Down syndrome have 47 chromosomes because they are trisomic for chromosome 21, one of the smaller chromosomes (Fig. 15–3*b*). Nondisjunction during meiosis is thought to be responsible for the presence of the extra chromosome.

Down syndrome occurs in only about 0.15% of all births, but its incidence increases markedly with increasing maternal age (Fig. 15–4). It is 100 times more likely in the offspring of mothers who are 45 years of age or older than it is in the offspring of mothers who are under 19 years of age. The occurrence of Down syndrome is affected much less by the age of the father.

Because of this striking correlation with increased maternal age, Down syndrome is thought to be due in most (but certainly not all) cases to meiotic nondisjunction in the mother. The reason for this is not fully understood. One proposed explanation relates to differences in meiosis in human males and females (see Chapter 48). All of the cells in a human female that will ever enter meiosis do so before she is even born. They become arrested in meiotic prophase and remain in that state until she reaches puberty, after which one cell per month resumes meiosis. Therefore, when a woman produces an egg to be fertilized, that egg is essentially as old as she is. In contrast, in human males new cells are continually entering meiosis, and the entire process of sperm production takes only about 50 days.

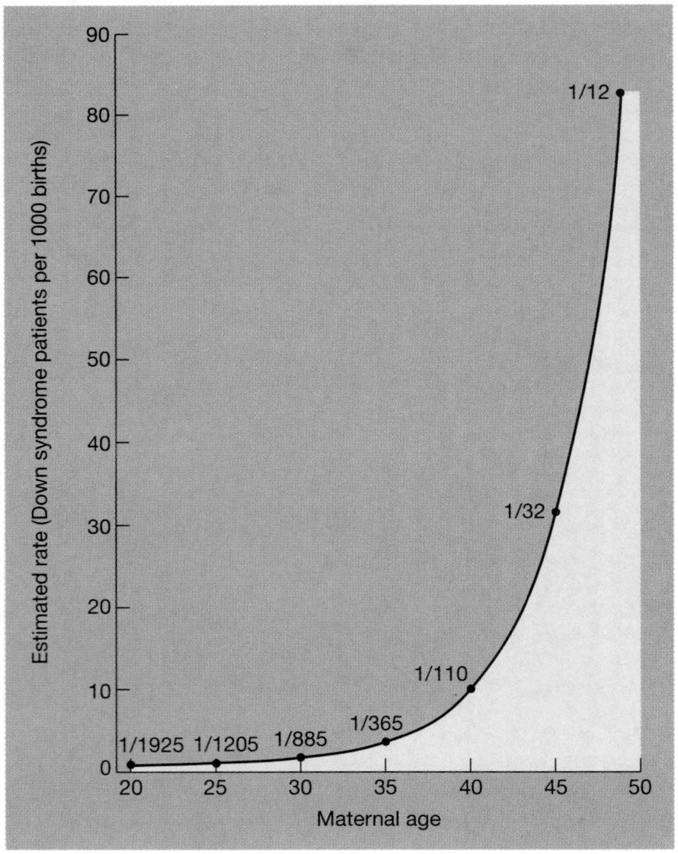

FIGURE 15–4 The probability of the birth of a child with Down syndrome rises as maternal age increases.

In about 4% of patients with Down syndrome, only 46 chromosomes are present, but one is abnormal. Extra genetic material from chromosome 21 has been translocated onto one of the larger chromosomes, such as chromosome 14. We refer to the abnormal translocation chromosome as a *14/21 chromosome.* Affected persons have one chromosome 14, one 14/21 chromosome, and two copies of chromosome 21. The genetic material from chromosome 21 is thus present in triplicate. When the karyotypes of such an individual's parents are studied, either the mother or the father is usually found to have only 45 chromosomes, although he or she is generally phenotypically normal. Such a person has one chromosome 14, one 14/21 chromosome, and one chromosome 21. Although the karyotype is abnormal, there is no extra genetic material. In contrast to true, or free, **trisomy 21,** this translocation form of Down syndrome can run in families (Fig. 15–5) and its incidence does not increase with maternal age.

The presence of this extra chromosomal material leads to the complex physical and mental abnormalities that characterize Down syndrome. Paradoxically, although no genetic information is missing in these individuals, the extra "doses" of chromosome 21 genes bring about some type of genetic imbalance that is responsible for abnormal physical and mental development. Down syndrome is quite variable in expression, with some individuals far more severely affected than others. Genetic imbalances produced by the addition or deletion of all or part of a chromosome typically result in multiple defects.

When a disease causes multiple symptoms, we refer to it as a **syndrome.** Virtually all chromosomal abnormalities fall into this category. Because the nervous system is so complicated in its development, it appears to be quite sensitive to altered gene dosages, and some form of mental retardation commonly accompanies most chromosome abnormalities. Researchers are using genetic engineering methods to attempt to pinpoint genes on chromosome 21 that affect mental development, as well as possible oncogenes (see *Focus On: Oncogenes and Cancer* in Chapter 16) and genes that may be involved in Alzheimer's disease.

In general, chromosome abnormalities involving the autosomes are devastating in their consequences. Other than Down syndrome, very few autosomal trisomies are known (Table 15–1). The condition known as autosomal **monosomy,** in which only one member of a pair is pres-

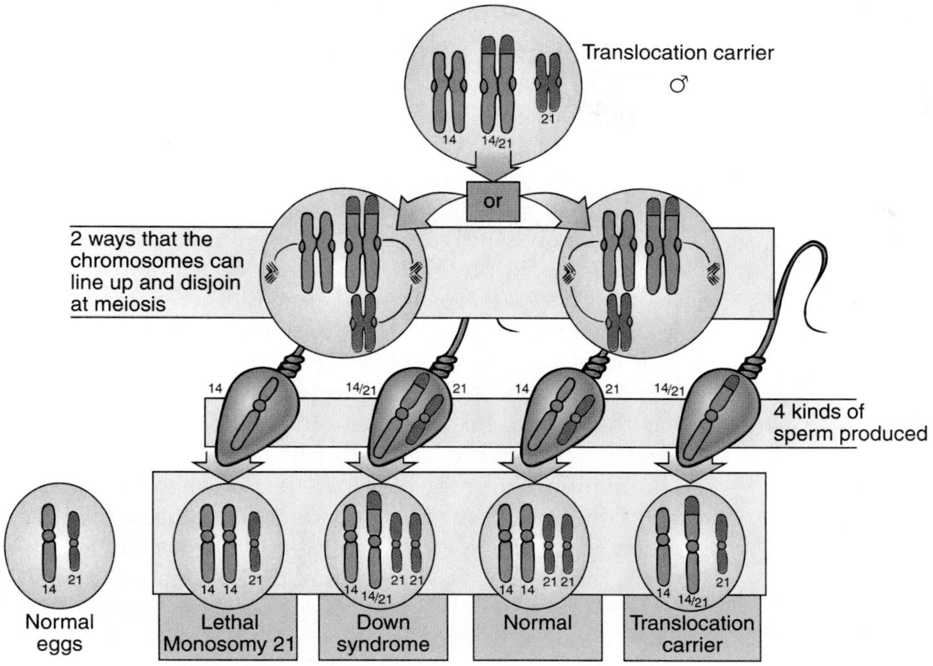

FIGURE 15–5 A person can be a phenotypically normal carrier of the translocation form of Down syndrome. In a carrier individual (could be the mother or the father), most of the genetic material from a chromosome 21 has become fused to a chromosome 14, forming a 14/21 translocation chromosome. In this example the father is the carrier; he has 45 chromosomes, with one 14, one 21, and one 14/21. At anaphase I of meiosis, the chromosome 14 and the 14/21 translocation chromosome usually disjoin from one another. The chromosome 21 can go either to the pole with the chromosome 14 or to the pole with the 14/21 translocation chromosome. On average, such a carrier produces four kinds of sperm; when they fertilize normal eggs, these sperm produce four kinds of zygotes. One quarter of the zygotes have only one chromosome 21 (monosomy 21), a lethal condition; one quarter have the translocation form of Down syndrome; one quarter are genotypically and phenotypically normal; and one quarter are translocation carriers like the father.

ent, is apparently incompatible with life because it is not seen in live births.

Sex chromosome abnormalities are usually less severe than autosomal abnormalities

Sex chromosome abnormalities appear to be relatively well tolerated (see Table 15–1), apparently at least in part because of the phenomenon of dosage compensation discussed in Chapter 10. According to the *single active X hypothesis,* mammals compensate for extra X chromosomal material by rendering all but one X chromosome inactive. The inactive X is seen as a Barr body, a region of darkly staining, condensed chromatin next to the nuclear envelope of an interphase nucleus (see Fig. 10–14).

The presence of the Barr body in the cells of normal females (but not of normal males) has been used as an initial screen to determine whether an individual is genetically female or male. As we shall see in our discussion of abnormal sex chromosome constitutions, the Barr body test, also known as **nuclear sexing,** has serious limitations. Unfortunately the test has sometimes been misused, particularly in judging the eligibility of individuals to participate as females in athletic contests. If it is used at all, any unusual findings should be followed by karyotype analysis and other tests.

Individuals with Klinefelter syndrome have an XXY karyotype

Persons with **Klinefelter syndrome** have 47 chromosomes, including two X's and one Y. They are nearly normal males, but they have small testes and produce few or no sperm. Evidence that the Y chromosome is the major determinant of the male phenotype has been substantiated by the fact that there is at least one gene on the Y chromosome that appears to act as a "genetic switch," directing male development.

People with Klinefelter syndrome tend to be unusually tall and to have female-like breasts. About half show some degree of mental retardation but most live relatively normal lives. However, when their cells are examined they are found to have one Barr body per cell. On the basis of such a test, they would be erroneously classified as females.

Persons with Turner syndrome have only one X chromosome and no Y

We designate the sex chromosome constitution for **Turner syndrome** as *XO,* the *O* referring to the absence of a second sex chromosome. Because of the absence of the strong male-determining effect of the Y chromosome, these persons develop essentially as females. However, both their internal and external genital structures are undeveloped, and they are sterile. Apparently a second X chromosome is necessary for the normal development of the ovaries in a female embryo. Examination of their cells reveals no Barr bodies, because there is no "extra" X chromosome to be inactivated.

Some essentially normal males have an XYY karyotype

People with an X chromosome plus two Y chromosomes are phenotypically fertile males. Other characteristics of these individuals (unusually tall, with severe acne) hardly merit the term *syndrome;* hence the designation **XYY karyotype.** Some years ago there was a widely publicized suggestion that persons with this condition are more likely to display criminal tendencies and be imprisoned, but further studies have failed to substantiate this.

Chromosome abnormalities are relatively common at conception but usually result in prenatal death

Recognizable chromosome abnormalities are seen in less than 1% of all live births, but substantial evidence suggests that the rate at conception is much higher. About 17% to 20% or more of all recognized pregnancies end in a spontaneous abortion ("miscarriage"). Approximately half of these spontaneously aborted embryos have major chromosome abnormalities, including autosomal trisomies (e.g., trisomy 21), triploidy and tetraploidy, and Turner syndrome (XO), the last of which is the most common.

Autosomal monosomies are exceedingly rare. It is *unlikely* that they never occur; it is far more probable that autosomal monosomy is so incompatible with life that a spontaneous abortion occurs very early, before the woman is even aware that she is pregnant. Some investigators place surprisingly high estimates (50% or more) on the rate of loss of very early embryos. It is widely assumed that chromosome abnormalities are responsible for a substantial fraction of these.

MOST GENETIC DISEASES ARE INHERITED AS AUTOSOMAL RECESSIVE TRAITS

Hundreds of human disorders involving enzyme defects have been found to be caused by genetic mutations. These disorders, sometimes referred to as **inborn errors of metabolism,** include phenylketonuria (PKU) and alkaptonuria (see Chapter 11). Two other such genetic disorders associated with single gene defects (although not necessarily in genes coding for enzymes) are cystic fibrosis and sickle cell anemia. Not all human genetic diseases have a simple inheritance pattern, but most of those that do are transmitted as autosomal recessive traits and so are expressed only in the homozygous state.

Phenylketonuria (PKU) is due to an enzyme deficiency and can be treated with a special diet

It is not possible today to cure any genetic disease. The best one can hope for is successful treatment of the symptoms. Perhaps the most dramatic success to date has been in the treatment of **phenylketonuria (PKU).** Homozygous recessive individuals lack an enzyme that converts the amino acid phenylalanine to another amino acid, tyrosine. These persons instead convert phenylalanine into toxic phenylketones that accumulate and damage the central nervous system. The ultimate result is severe mental retardation. A homozygous PKU infant is usually healthy at birth because its mother, who is heterozygous, produces enough enzyme to prevent phenylalanine accumulation before birth. However, during infancy and early childhood, the toxic products eventually cause irreversible damage to the central nervous system.

In the 1950s it was found that if PKU infants are identified and placed on a special low-phenylalanine diet early enough, the symptoms can be dramatically alleviated. Biochemical tests for PKU have been developed, and screening of newborns through a simple blood test is widespread in the United States, with more than 90% of all infants being tested. Because of these screening programs and the availability of effective treatment, thousands of children have been saved from severe mental retardation. Most such children are able to discontinue the diet by adolescence. Although they still produce phenylketones, the sensitive period is past.

Ironically, the success of PKU treatment in childhood presents a new problem today. If a homozygous female who was saved from mental retardation becomes pregnant, the high phenylalanine levels in her blood can damage the brain of the fetus she is carrying, even though that fetus is only heterozygous. Therefore, she must resume the diet, preferably before becoming pregnant. This procedure is usually (although not always) successful in preventing the effects of **maternal PKU.** Therefore it is especially important that females with PKU be aware of their condition so that they may obtain appropriate counseling and medical treatment during pregnancy.

Sickle cell anemia results from a hemoglobin defect

Sickle cell anemia is inherited as an autosomal recessive trait. The disease is most common in persons of African descent, and about 1 in 12 African-Americans is heterozygous for it. The blood cells of a person with sickle cell anemia are shaped like sickles, or half-moons, whereas normal red blood cells are biconcave discs.

The sickle cell contains abnormal hemoglobin molecules, which have the amino acid valine instead of glutamic acid at position 6 (the sixth amino acid from the amino terminal end) in the beta chain (see Chapter 3).

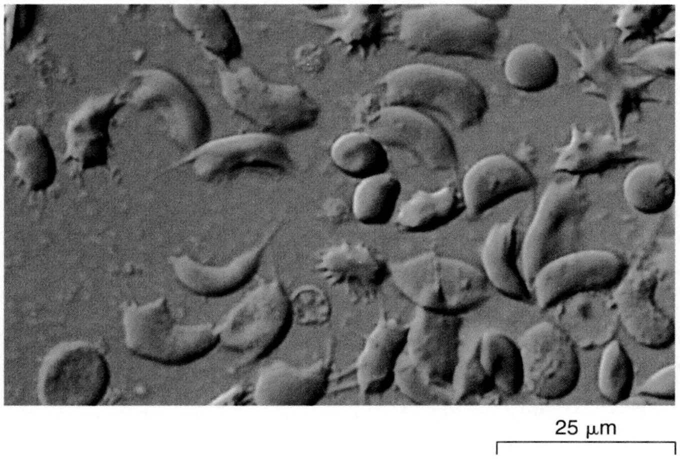

FIGURE 15–6 Some of the red blood cells from this patient with sickle cell anemia show an abnormal "sickled" shape. (G. W. Willis, Ochsner Medical Institution/Biological Photo Service)

The substitution of an amino acid with a hydrophobic, uncharged side chain (valine) for one with a hydrophilic, charged side chain (glutamic acid) makes the hemoglobin less soluble. As a result, it tends to form crystal-like structures that change the shape of the red blood cells (Fig. 15–6). This sickling occurs in the veins after the oxygen has been released from the hemoglobin. The blood cells' abnormal sickle shape slows blood flow and blocks small blood vessels, with resulting tissue damage and painful episodes. Sickle cells also have short life spans, leading to anemia in affected persons.

Available treatments for sickle cell anemia include pain-relief measures, transfusions, and some forms of drug therapy, but these are of limited effectiveness. Children with sickle cell anemia generally lead short, painful lives. Ongoing research is directed toward eventually providing gene therapy for these patients.

Cystic fibrosis results from defective ion transport

Cystic fibrosis is the most common autosomal recessive disorder in Caucasian children. About 1 of every 20 persons in the United States is a heterozygous carrier of the cystic fibrosis gene. This disorder is characterized by abnormal secretions in the body; its most severe effect is on the respiratory system, which produces abnormally viscous mucus. The cilia that line the bronchi (see Chapter 44) cannot easily remove the mucus, and it thus becomes a culture medium for dangerous bacteria. These bacteria or their toxins attack the surrounding tissues, leading to recurring pneumonia and other complications. The heavy mucus also occurs elsewhere in the body (e.g., in the ducts of the pancreas and liver and in the intestines), causing digestive difficulties and other effects.

The cystic fibrosis gene has been cloned. It codes for a protein that controls the transport of chloride ions across cell membranes. The malfunction caused by the mutant protein is responsible for the production of the unusually thick mucus that eventually leads to tissue damage in the respiratory and digestive systems. Although many mutant forms exist and these vary somewhat in severity of symptoms, the disease is usually serious.

Antibiotics are used to control bacterial infections and daily physical therapy is required to clear mucus from the respiratory system. Treatment with an enzyme (produced by recombinant DNA technology) that breaks down the mucus is also helpful. Without such treatment, death would occur in infancy. With treatment, about 50% of affected persons live into their 20s, only to die in what should be the prime of life after having spent the equivalent of about four years in the hospital. Because of the serious limitations of available treatments, gene therapy for cystic fibrosis is under development (Fig. 15–7). This and other aspects of the disease are discussed later in this chapter.

Tay-Sachs disease is a result of abnormal lipid metabolism in the brain

Tay-Sachs disease is an autosomal recessive disease of the central nervous system that results in blindness and severe mental retardation. The symptoms begin within the first year of life and result in death before the age of five. Because of the absence of an enzyme, a normal membrane lipid in the brain fails to break down properly and accumulates in the lysosomes (see Chapter 4). Unfortunately, although research is ongoing, no effective treatment for Tay-Sachs disease is available at this time.

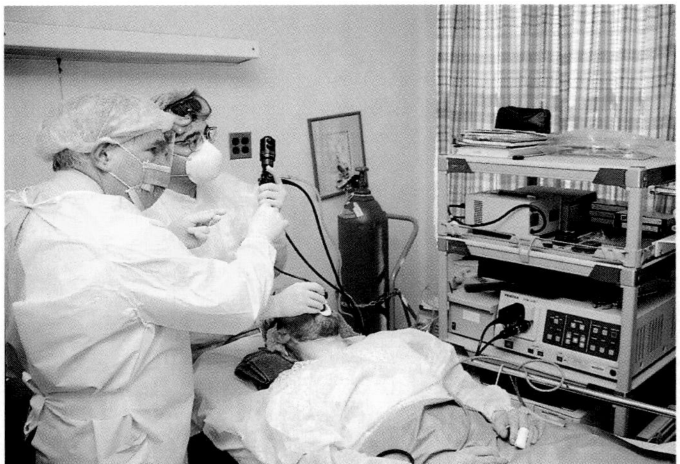

FIGURE 15–7 This patient is receiving gene therapy for cystic fibrosis. A vector carrying the normal allele is introduced into the patient's respiratory system through a fiberoptic bronchoscope. The physician (left foreground) is positioning a catheter through the bronchoscope, enabling the vector to be administered. (Courtesy of Dr. Ronald Crystal, The New York Hospital–Cornell Medical Center)

The abnormal allele is especially common in the United States among Jews whose ancestors came from Eastern Europe (Ashkenazi Jews). By contrast, Jews whose ancestors came from the Mediterranean region (Sephardic Jews) have a very low frequency of the allele.

HUNTINGTON'S DISEASE IS AN AUTOSOMAL DOMINANT DISORDER THAT AFFECTS THE NERVOUS SYSTEM

Huntington's disease (formerly called *Huntington's chorea*) is due to a rare autosomal dominant allele that causes severe mental and physical deterioration, uncontrollable muscle spasms, personality changes, and ultimately insanity. No effective treatment has been found. Every child of an affected individual has a 50% chance of also being affected (and of course of passing the abnormal allele to his or her offspring).

Ordinarily we would expect a dominant allele with such devastating effects to occur only as a new mutation and not to be transmitted to future generations. However, this disease is characterized by onset relatively late in life (usually between the ages of 35 and 50), so an individual may have children before knowing that the allele is present.

The gene responsible for Huntington's disease was cloned in 1993. More than a decade of basic research was required to pin down the gene after its chromosomal location was first identified by recombinant DNA technology. These accomplishments are an excellent illustration of the fact that much of the best biological research today involves a combination of approaches. The complexity of the task requires extensive collaboration among researchers in different disciplines.

One gene isolation strategy is to isolate DNA from persons who carry the abnormal allele and compare it to DNA from close relatives who do not have the allele. This is generally done by restriction fragment mapping (see Chapter 14). One advantage of studying close relatives is that one can minimize other genetic differences and concentrate on the gene of interest.

The optimal approach is to study large families, carefully construct pedigrees, and obtain DNA samples from affected and unaffected individuals across generations. A large extended family with a very high incidence of Huntington's disease was discovered in Venezuela and has been the subject of exhaustive pedigree and DNA analysis. This made possible the identification of a **DNA marker** for Huntington's disease. A DNA marker is not the DNA of the gene of interest, but it is the next best thing: a piece of DNA that is closely associated with the abnormal allele on the chromosome and inherited along with it. The Huntington's disease marker became the basis for tests that allowed those at risk to learn if they carried the allele. (See *Making the Connection: Molecular Biology and Diagnosing Genetic Diseases.*)

The decision to be tested for any genetic disease is understandably a highly personal one. Certainly, the information can be very useful for those who must make decisions such as whether or not to have children. However, someone who tests positive for the Huntington's disease allele must then live with the virtual certainty of eventually developing this devastating and incurable disease. It is hoped that identifying affected persons before the onset of symptoms may ultimately contribute to the development of effective treatments. If those with the Huntington's allele choose not to reproduce, the frequency of this allele in the population will decrease.

HEMOPHILIA A IS AN X-LINKED RECESSIVE DISORDER THAT AFFECTS BLOOD CLOTTING

Hemophilia A is sometimes referred to as a disease of royalty because of its incidence among male descendants of Queen Victoria, but it is also found in many nonroyal pedigrees. Characterized by the lack of a blood-clotting factor, Factor VIII, it causes severe bleeding from even a slight wound. Because the mode of inheritance is X-linked recessive, affected persons are almost exclusively male, having inherited the abnormal allele on the X chromosome from their heterozygous carrier mothers.

Today, treatments consist of blood transfusions and administration of Factor VIII by injection. Unfortunately, these treatments are costly and have been associated with infection with HIV-1, the virus that causes AIDS. Factor VIII produced using recombinant DNA technology (see Chapter 14) will provide a safer source of the clotting factor.

SOME GENETIC ABNORMALITIES AND OTHER BIRTH DEFECTS CAN BE DETECTED BEFORE BIRTH

Genetic abnormalities may become apparent during early intrauterine life or not until late in adult life. Given that early detection increases the possibilities for prevention or alleviation of the effects of genetic abnormalities, efforts have been made over the years to detect such abnormalities before birth. In the past 20 years physicians have become increasingly successful at prenatal diagnosis and treatment, including transfusion and surgical correction of malformations. Intrauterine diagnosis of a number of genetic abnormalities is now possible.

In one diagnostic technique known as **amniocentesis,** a sample of the fluid surrounding the fetus (the *amniotic fluid*; see Chapter 49) is obtained (Fig. 15–8). A needle is inserted through the walls of the pregnant woman's abdomen and uterus, and into the uterine cavity, where some of the fluid is drawn into a syringe. This procedure carries some risk but is relatively safe, partly because the

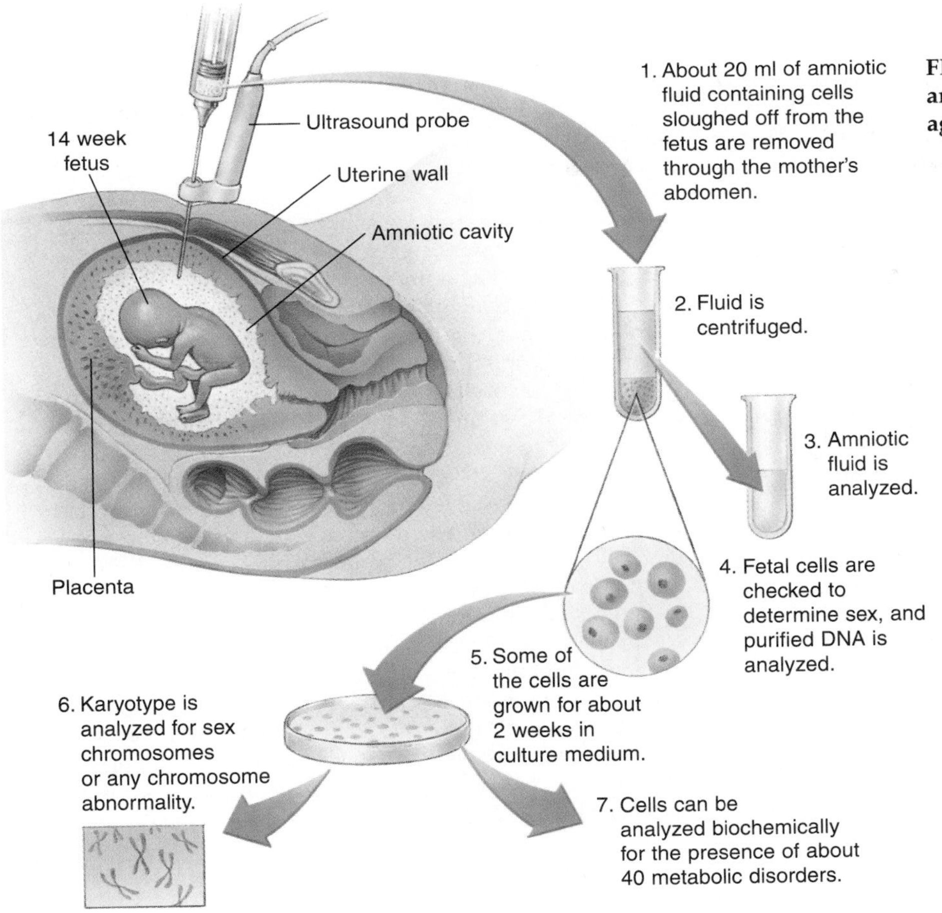
14 week fetus
Ultrasound probe
Uterine wall
Amniotic cavity
Placenta
1. About 20 ml of amniotic fluid containing cells sloughed off from the fetus are removed through the mother's abdomen.
2. Fluid is centrifuged.
3. Amniotic fluid is analyzed.
4. Fetal cells are checked to determine sex, and purified DNA is analyzed.
5. Some of the cells are grown for about 2 weeks in culture medium.
6. Karyotype is analyzed for sex chromosomes or any chromosome abnormality.
7. Cells can be analyzed biochemically for the presence of about 40 metabolic disorders.

FIGURE 15–8 Certain genetic diseases and other abnormal conditions can be diagnosed prenatally by amniocentesis.

positions of the fetus and the needle can be determined through **ultrasound imaging** (Fig. 15–9).

The amniotic fluid contains living cells sloughed off the body of the fetus and hence genetically identical to the cells of the fetus. These amniotic fluid cells can be cultured in the laboratory. After two to three weeks, dividing cells from the culture can be studied for evidence of chromosomal abnormalities and other genetic defects.

Amniocentesis is performed mostly on pregnant women over 35 years of age, whose offspring have a higher than normal risk of Down syndrome. Many other tests have been developed to detect a number of simply

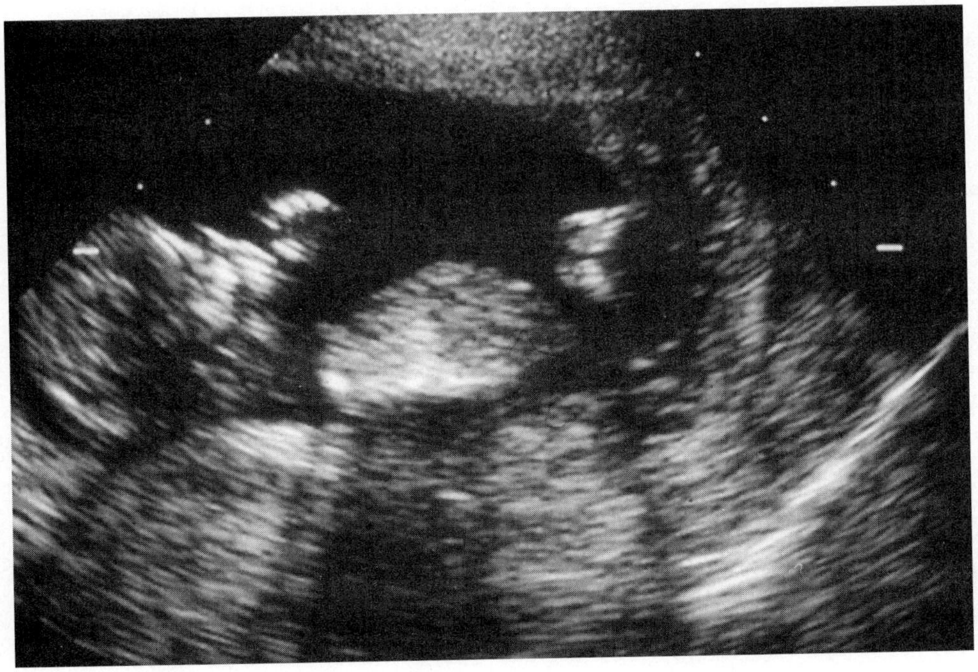

FIGURE 15–9 Ultrasound image of a 12 1/2-week-old fetus. A fetus of this age is typically about 8.5 centimeters long; the head is to the left. (Courtesy of F. R. Batzer, M.D., Philadelphia Fertility Institute)

Making the Connection

Molecular Biology and Diagnosing Genetic Diseases

How can genetic engineering techniques be used in characterizing and detecting genetic disease? The earliest studies of this sort were carried out on sickle cell anemia. Carriers and affected individuals can usually be detected by a simple blood test. Because of the great difficulty in obtaining a reliable blood sample from an embryo, prenatal diagnosis is much more difficult. However, through the use of molecular genetic techniques it is possible to examine fetal DNA (obtained from amniocentesis or CVS) and test for the presence of sequences that code for sickle cell hemoglobin.

Hemoglobin (see Chapter 3) contains four polypeptide chains, two identical alpha chains and two identical beta chains. The beta chains of sickle cell hemoglobin differ from those of normal hemoglobin by only a single amino acid at position 6. The base sequence of the normal hemoglobin allele that codes for that amino acid and its two neighbors includes a recognition site for the restriction enzyme MstII. The base sequence of this recognition site is:

CCTGAGG

The sickle cell hemoglobin allele differs by only one base, altering the sequence to read:

CCTGTGG

This change is sufficient to prevent the restriction enzyme from cutting the DNA at that point. This single base change therefore results in a restriction fragment length polymorphism, which can be detected by blot hybridization methods. By synthesizing a radioactive probe complementary to the DNA sequence on one side of the restriction site, it is possible to differentiate that specific fragment of DNA from all of the other DNA fragments in the genome. When normal human DNA is cut with the restriction enzyme MstII, a single radioactive band can be detected on the blotting filter. That band corresponds to a DNA molecule that is about 1.15 kilobases (1.15 kb, or 1150 bases) long. Because the restriction site is abolished in individuals who have the sickle cell allele, a longer DNA fragment, 1.35 kb in length, is detected. Heterozygous individuals have one copy of the normal allele and one copy of the sickle cell allele. When their DNA is analyzed, two bands can be detected by the probe, corresponding to the 1.15- and 1.35-kb fragments. It is thus possible to distinguish fetuses who are homozygous normal, homozygous recessive, and heterozygous carriers of the trait.

Few genetic diseases are as well understood as sickle cell anemia (see figure). As more human genes are identified and sequenced, an increasing number of abnormal alleles will be detectable in healthy carriers and in fetuses. Even when a gene has been cloned, however, problems can occur in developing screening methods. For example, when the cystic fibrosis gene was cloned, it was found that the mutant alleles in the population are not all identical. The first test to be developed could identify only the most common mutant allele, present in about 70 percent of carriers. It is now possible to identify carriers of many of the less common alleles. The ultimate goal is to develop a test that will detect 95 percent or more of the carriers. With hundreds of different mutant cystic fibrosis alleles detected so far, this is a formidable task.

For many genetic diseases it is not yet possible to detect the actual mutant allele. However, for some of these, including Huntington's disease, it has been possible to identify a particular DNA sequence in some families that is closely linked to the mutant allele and inherited along with it. Such a sequence, called a **DNA marker,** can be detected by blot hybridization methods. ▲

Detecting the sickle cell allele with a genetic probe. (*a*) The normal hemoglobin gene has two sites recognized by the restriction enzyme MstII, which cuts the DNA to yield a 1.15-kb DNA fragment. The sickle cell allele has an altered base sequence, so one of these sites is not recognized by MstII. When that DNA is cut by the enzyme, a 1.35-kb fragment is produced. (*b*) The gene is detected by isolating total DNA from amniotic fluid cells and cutting it with MstII. Fragments from the total DNA are separated by gel electrophoresis (*c*). DNA from the gel is transferred (blotted) to nitrocellulose filters and hybridized with a radioactive probe (the cloned 1.15-kb MstII fragment derived from the normal hemoglobin gene). (*d*) DNA sequences complementary to the probe are detected by exposing the filter to x-ray film. Homozygous normal individuals show a single exposed band on the film at a position corresponding to a molecular weight of 1.15 kb. DNA from individuals homozygous for the sickle cell gene displays a single band at 1.35 kb. Heterozygotes have a single copy of each gene; their DNA produces two bands, at the 1.35- and 1.15-kb positions. ▶

inherited genetic disorders, but these disorders are rare enough that the tests are usually done only if a particular problem is suspected. Enzyme deficiencies can often be detected through incubation of cells recovered from amniotic fluid with the appropriate substrate and measurement of the product; this technique has been useful in the prenatal diagnosis of disorders such as Tay-Sachs disease. The tests for a number of other diseases, including sickle cell anemia, Huntington's disease, and cystic fibrosis, are less direct, requiring the use of genetic engineering methods. Methods for detecting many more genetic diseases are now being actively sought by researchers.

Amniocentesis is also useful in detecting a condition known as *spina bifida,* in which the spinal cord does not

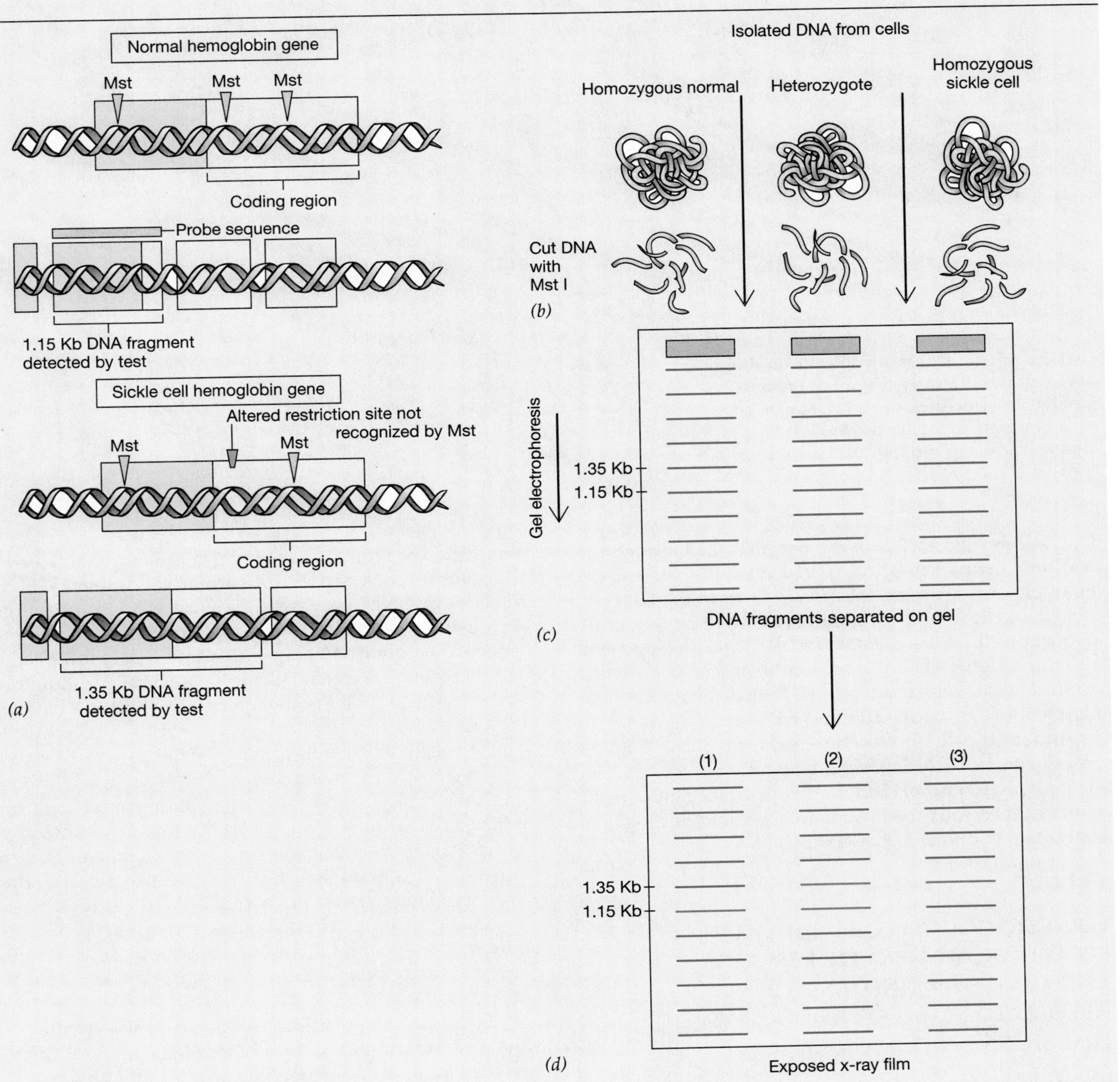

close properly during development. A relatively common (about 1 in 300 births) nongenetic malformation, this birth defect is associated with abnormally high levels of a normally occurring protein, alpha-fetoprotein, in the amniotic fluid. Abnormally low levels are associated with Down syndrome. Alpha-fetoprotein levels can also be measured by testing the mother's blood, but the results are less reliable and any unusual findings should be confirmed by amniocentesis or other tests.

One problem with amniocentesis is that most of the conditions it detects are incurable and the results are generally not obtained until well into the second trimester when abortion is both psychologically and medically difficult. Therefore, efforts have been made to develop tests

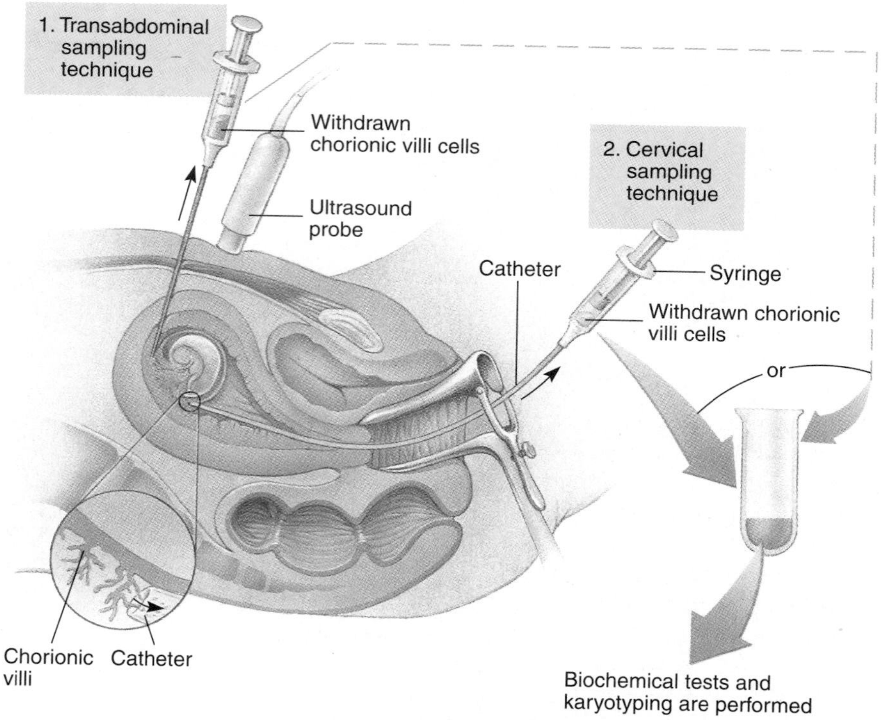

FIGURE 15–10 Chorionic villus sampling (CVS) allows the early diagnosis of some genetic abnormalities. Samples may be obtained by inserting a needle through the uterine wall (1) or through the cervical opening (2).

that yield results earlier in the pregnancy. One such test, **chorionic villus sampling (CVS)** (Fig. 15–10), involves removing and studying cells that will form the fetal contribution to the placenta (and hence should be genetically identical to the fetal cells). CVS may be associated with a slightly higher risk of infection or miscarriage than is amniocentesis, but its advantage is that results can usually be obtained within the first trimester.

Although both amniocentesis and CVS can diagnose certain genetic disorders with a high degree of accuracy, they are not foolproof, and many disorders cannot be diagnosed; therefore, the lack of an abnormal finding is no guarantee of a normal pregnancy.

GENETIC COUNSELORS EDUCATE PEOPLE ABOUT GENETIC DISEASES AND GIVE THEM INFORMATION NEEDED TO MAKE REPRODUCTIVE DECISIONS

Couples who are concerned about the risk of abnormality in their children, either because they have had an abnormal child or a relative affected by a hereditary disease, may seek **genetic counseling.** Genetics clinics are available in most metropolitan centers.

Advice can be given only in terms of the *probability* that any given offspring will have a particular condition. The geneticist needs complete family histories of both the man and the woman and may use tests for the detection of heterozygous carriers of certain conditions. When a disease involves only a single gene locus, probabilities can usually be easily calculated. For example, if one prospective parent is affected with a trait that is inherited as an autosomal dominant disorder, such as Huntington's disease, the probability that any given child will have the disease is 0.5.

The birth to phenotypically normal parents of a child affected with an autosomal recessive trait, such as albinism or PKU, establishes that both parents are heterozygous carriers. The probability that any subsequent child will be affected is therefore 0.25. (In this context, the term *carrier* is used specifically to refer to an individual who is heterozygous for a recessive allele that causes a genetic disease. Homozygous recessive persons are referred to as *affected* individuals; they are not called *carriers* even though they also "carry" alleles for the disease.)

For a disease inherited through a recessive allele on the X chromosome, such as hemophilia A, the probability depends on the genotypes of the parents. A normal woman and an affected man will have daughters who are carriers and sons who are normal. The probability that the son of a carrier mother and a normal father will be affected is 0.5; the probability that their daughter will be a carrier is also 0.5.

It is now possible to detect carriers of several genetic diseases. In these cases counseling can be provided when both husband and wife are heterozygous. For diseases that involve an enzyme defect, carriers often show only half the level of enzyme activity characteristic of normal homozygotes. Voluntary screening programs have been set up by synagogues and other organizations to detect couples who are carriers of Tay-Sachs disease among Jews of Eastern European descent.

Persons heterozygous for sickle cell anemia can be readily identified with a simple blood test. They have a mixture of normal and abnormal hemoglobins in their red blood cells, with about 45% of their total hemoglobin being abnormal. Such persons, said to have **sickle cell trait,** are not ill, and their blood cells do not usually undergo sickling (although they can be made to do so by reducing the amount of oxygen). Carrier testing for some other diseases, such as cystic fibrosis and hemophilia A, is much more complicated and is usually done only if family history suggests that a person may be a carrier.

It is very important that identified carriers receive appropriate genetic counseling. A genetic counselor is trained not only to give information needed for reproductive decisions, but also to help individuals understand their situation and avoid feeling stigmatized.

Inquiries are commonly made about mental retardation, epilepsy, deafness, congenital heart disease, and other conditions. It is possible that some environmental factor may have played a role in producing the abnormality in the affected child. Did the mother have an infectious disease during pregnancy (e.g., rubella)? Was she receiving some kind of drug therapy, or was she subjected to ionizing radiation? Had the father been exposed to any potentially hazardous agents? By dissecting the environmental contributions, the geneticist can more accurately estimate the probability of the trait's recurrence in subsequent offspring.

Gene Replacement Therapy Is Being Developed for Several Genetic Diseases

Because many difficulties are inherent in treating most serious genetic diseases, scientists have dreamed of developing actual cures. Today, genetic engineering is bringing these dreams closer to reality. Such therapy could take two main forms.

One approach would be to introduce copies of a normal gene into a fertilized egg, using modifications of the technology already used to produce transgenic animals (see Chapter 14). In some transgenic animals the introduced gene can remain stable from generation to generation, constituting a true "genetic cure." However, this approach raises such complex ethical problems that it is not being actively pursued at this time.

A second strategy — to introduce the normal gene into only some body cells (somatic cell gene therapy) — is receiving increased attention today. The rationale is that, although a particular gene may be present in all cells, it is expressed only in some (see Chapter 16). Expression of the normal allele in only the cells that require it may be sufficient to give a normal phenotype. Needless to say, this approach presents a number of technical obstacles.

The solutions to these problems must be tailored to the nature of the gene itself, as well as to its product and the types of cells in which it must be expressed. First the gene must be cloned and the DNA introduced into the appropriate cells. One of the most successful techniques is to use a virus as the vector. Ideally the virus should infect a high percentage of the cells and facilitate the integration of the introduced gene into a chromosome. Most importantly, the virus should do no harm, especially over the long term. This is a large order, and a great deal of attention is being paid to the development of viral strains with just such desirable characteristics.

The overall process can be illustrated by the gene therapy first approved for clinical trials on patients. A genetic disorder called *severe combined immune deficiency* (SCID) renders the immune system (see Chapter 43) essentially nonfunctional. Because an affected baby has no defenses, it will die unless isolated from possible sources of infection. Because the cells that constitute the immune system are certain white blood cells that originate in the bone marrow, SCID has been treated successfully in some cases by transplanting healthy bone marrow cells from a normal donor.

About one-quarter of all SCID patients are known to lack a specific enzyme, adenosine deaminase (ADA); it is the absence of ADA in the immune system cells that causes them to die. These patients have been treated successfully by providing them with ADA, linked to a molecule that makes the enzyme more stable in the bloodstream. This enzyme-replacement therapy is not a cure, for the ADA treatments must continue throughout life.

Gene therapy for this condition was thought to be feasible because the normal ADA gene had been cloned and successfully introduced into cells through a viral vector. In the first clinical trials a virus was used to introduce the ADA gene into white blood cells from SCID patients; these cells were then transfused back into the patients. The results have been encouraging but, because white blood cells have limited lifetimes, such treatments must be repeated.

Although many obstacles must be overcome, gene therapies for a number of other genetic diseases (including cystic fibrosis) are undergoing development or are being tested on patients in clinical trials. Scientists are currently addressing some of the unique problems presented by each disease. Not all types of cells can be removed from the body, infected with a virus, and then replaced. For example, successful gene therapy for cystic fibrosis will require introduction of normal genes directly into

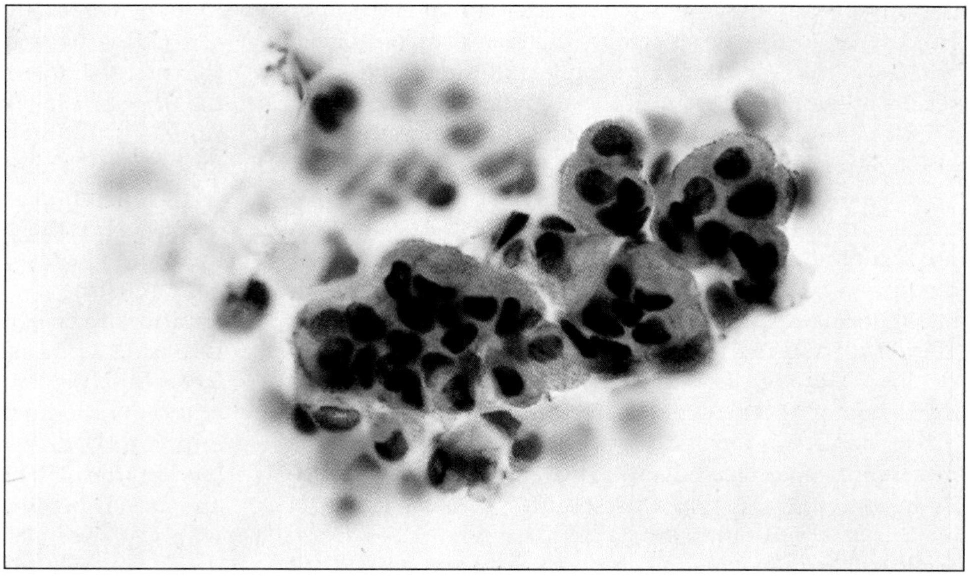

FIGURE 15–11 These epithelial cells from the respiratory system of a patient with cystic fibrosis were photographed four days after administration of an adenovirus vector containing the normal human cystic fibrosis transmembrane conductance regulator (CFTR) DNA. The adenovirus vector is a weakened form of a virus that can cause the common cold. The red-pink color indicates that successful *in vivo* transfer and expression of the normal CFTR DNA has occurred. (Courtesy of Dr. Ronald Crystal, The New York Hospital–Cornell Medical Center)

lung cells. A method that has shown some promise is to use as the vector a virus that causes the common cold (Fig. 15–11). The field of gene therapy is expected to develop considerably in the future, although such treatments may not become routinely available for some years.

A GREAT DEAL OF NATURAL VARIATION EXISTS IN THE HUMAN POPULATION

One does not need to be a geneticist to recognize that humans are very diverse, and it is widely acknowledged that much human variation has a genetic basis. However, it is difficult to determine the genetic contribution to characteristics that are hard to assess, such as intelligence and behavior.

Studies on blood contribute to our understanding of genetic diversity in humans

It is not difficult to understand why so much knowledge of human variation is based on studies of blood. Blood samples are relatively easy to obtain, and blood is a complex tissue consisting of a number of cell types and extracellular molecules that can be studied.

The ABO blood group alleles control the expression of certain red blood cell antigens

The human blood cell types O, A, B, and AB are inherited through multiple alleles representing a single locus. Allele I^A provides the code for the synthesis of a specific glycoprotein, antigen A, which is expressed on the surface of the red blood cells. (Immunity is discussed in Chapter 43; for now we define antigens simply as substances capable of stimulating an immune response.) Allele I^B leads to the production of a different (but related) glycoprotein, antigen B. The allele i^O does not code for an antigen, although it is allelic to I^A and I^B. Allele i^O is recessive to the other two. Neither allele I^A nor allele I^B is dominant to the other; they are both expressed phenotypically and are therefore **codominant.**

Persons with the genotype I^AI^A or I^Ai^O have **blood type A** (Table 15–2); those with genotype I^BI^B or I^Bi^O have **blood type B;** and those with genotype i^Oi^O have **blood type O.** When both the I^A and I^B alleles are present, both antigen A and antigen B are produced in the red blood cells; persons with this I^AI^B genotype have **blood type AB.**

Antibodies anti-A and anti-B are proteins that appear in the plasma (the fluid component of blood) of persons lacking the corresponding antigens on their red blood cells. (Antibodies are proteins produced by the immune system that combine with specific antigens; hence, anti-A combines with antigen A.) Because of their specificity for the corresponding antigens, these antibodies are used in standard tests to determine blood type.

Determining blood types was one of the traditional ways of settling cases of disputed parentage. However, blood type tests can never prove that a certain person *is* the parent of a particular child; they can determine only whether he or she *could* be. Could a man with blood type AB be the father of a child with blood type O? Could a woman with blood type O be the mother of a child with blood type AB? Could a type-B child with a type-A mother have a type-A father or a type-O father?[1]

More than a dozen other sets of blood types, including the Rh group (discussed in the next section), are in-

[1]The answer to all these questions is *no*.

Table 15–2 ABO BLOOD TYPES*

Phenotype (blood type)	Genotypes	Antigen on RBC	Antibodies in Plasma	Frequency in U.S. Population (%)	
				Western European Descent	African Descent
A	I^AI^A, I^Ai^O	A	Anti-B	45	29
B	I^BI^B, I^Bi^O	B	Anti-A	8	17
AB	I^AI^B	A, B	None	4	4
O	i^Oi^O	None	Anti-A, anti-B	43	50

*This table and the discussion of the ABO system have been simplified somewhat. Note that persons produce antibodies against the antigens *lacking* on their own red blood cells (RBCs).

herited through other loci, independently of the ABO blood types. Determining some of these types in a given person may be useful in establishing relationships that could not be made with certainty by ABO blood typing alone.

Today more sophisticated types of genetic tests are used to determine parentage. These include **DNA fingerprinting** (see Chapter 14) and **tissue typing,** which is an examination of inherited antigens found on the surfaces of the body's cells (see Chapter 43). Theoretically, only identical twins would be expected to have the same DNA fingerprint and the same tissue type. If properly performed, these tests have greater than 99% certainty and can come close to proving parentage.

The Rh alleles are involved in the expression of other red blood cell antigens

Named for the rhesus monkeys in whose blood it was first found, the Rh system consists of at least eight different kinds of Rh antigens, each referred to as an **Rh factor.** By far the most important of these factors is **antigen D.** About 85% of those United States residents who are of Western European descent are Rh-positive. This means that they have antigen D on the surfaces of their red blood cells (in addition to the antigens of the ABO system and other blood groups). The 15% or so of this population who are Rh-negative have no antigen D. Unlike the situation discussed for the ABO blood group, Rh-negative persons do not naturally produce antibodies against antigen D (anti-D). However, they produce anti-D antibodies if they are exposed to Rh-positive blood. The allele coding for antigen D is dominant to the allele for the absence of antigen D. Hence, Rh-negative persons are homozygous recessive, and Rh-positive persons are heterozygous or homozygous dominant.

Although several kinds of maternal-fetal blood type incompatibilities are known, **Rh incompatibility** is probably the most important (Fig. 15–12). If a woman is Rh-negative and the father of the fetus she is carrying is Rh-

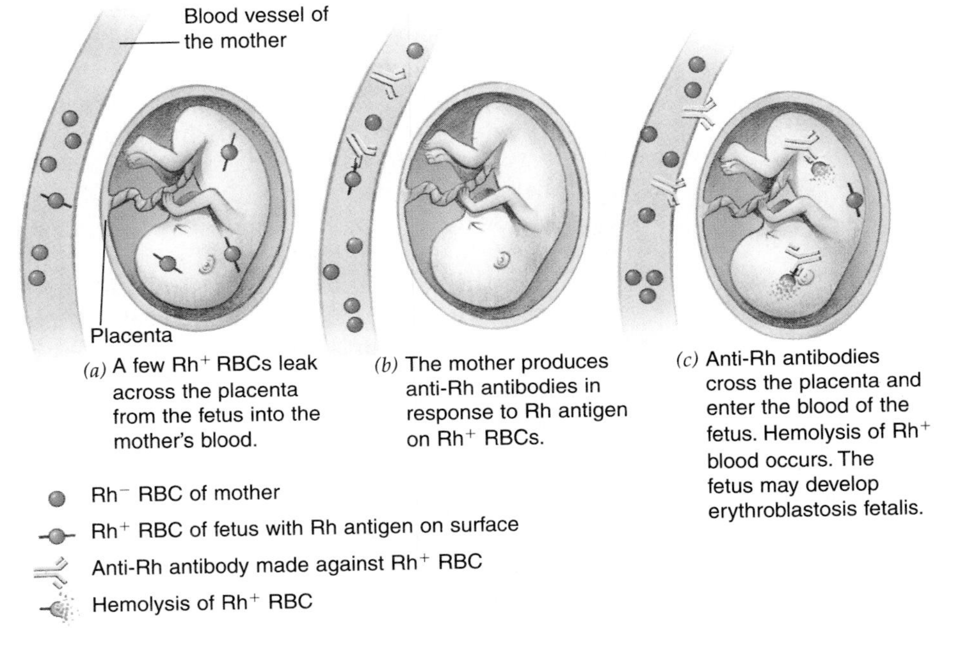

Blood vessel of the mother
Placenta
(a) A few Rh+ RBCs leak across the placenta from the fetus into the mother's blood.
(b) The mother produces anti-Rh antibodies in response to Rh antigen on Rh+ RBCs.
(c) Anti-Rh antibodies cross the placenta and enter the blood of the fetus. Hemolysis of Rh+ blood occurs. The fetus may develop erythroblastosis fetalis.
Rh− RBC of mother
Rh+ RBC of fetus with Rh antigen on surface
Anti-Rh antibody made against Rh+ RBC
Hemolysis of Rh+ RBC

FIGURE 15–12 Rh incompatibility can cause serious problems when an Rh-negative woman and an Rh-positive man produce Rh-positive offspring. (*a*) Some antigen D-bearing red blood cells leak across the placenta from the fetus into the mother's blood. (*b*) The mother produces anti-D antibodies in response to the D antigens on the fetal red blood cells. (*c*) In her next pregnancy, some of the mother's anti-D antibodies cross the placenta and enter the blood of her fetus, causing red blood cells to rupture (undergo hemolysis) and release hemoglobin into the circulation. As a result the fetus may develop erythroblastosis fetalis.

positive, the fetus may also be Rh-positive, having inherited the D allele from the father. Ordinarily no mixing of maternal and fetal blood occurs; nutrients, oxygen, and other substances are exchanged between these two circulatory systems across the placenta. However, late in pregnancy or during the birth process, a small quantity of blood from the fetus may pass through some defect in the placenta.

The fetus's red blood cells, which bear antigen D, sensitize the mother's white blood cells, inducing them to form antibodies to antigen D. When the woman becomes pregnant again, her sensitized white blood cells produce anti-D antibodies that can cross an intact placenta and enter the fetal blood. There they combine with the antigen D molecules on the surface of the fetal red blood cells, causing the cells to rupture. Breakdown products of the hemoglobin released into the circulation damage many organs, including the brain. In extreme cases of this disease, known as **erythroblastosis fetalis,** so many fetal red blood cells are destroyed that the fetus may die.

When Rh-incompatibility problems are suspected, fetal blood can be exchanged by transfusion before birth, but this is a risky procedure. Rh-negative women are now treated just after childbirth (or at termination of pregnancy by miscarriage or abortion) with a preparation of anti-D antibodies known as RhoGAM. These antibodies apparently clear the Rh-positive fetal red blood cells from the mother's blood very quickly, minimizing the chance for her own white blood cells to be sensitized. The antibodies are also soon eliminated from her body. As a result, when she becomes pregnant again her blood does not contain the anti-D that could harm her baby.

Quantitative traits are controlled by polygenes

Many human characteristics are **quantitative traits;** that is, they represent some measurable quantity such as height. Such characteristics show continuous variation in the population because of the number of loci involved, which can range from a few to a great many (polygenic inheritance), and also because of environmental factors, whose role is both significant and difficult to quantify. In addition to height and skin color (see Chapter 10), the inherited components of human intelligence are apparently under polygenic control. It is impossible to specify what these are, given our current inability even to define intelligence in a way that is universally accepted. However, many researchers think that polygenic factors determine the upper limit of mental ability in persons within the normal population; how close each individual comes to that limit depends on a variety of environmental factors, including nutrition and experience.

Many common physical characteristics are inherited

We are often curious about the inheritance of certain physical characteristics. Some of these have relatively simple modes of inheritance. For example, dark hair color is due to heavy deposits of the pigment melanin in the hair shaft. There are probably multiple alleles of the locus governing the deposition of melanin, the "darker" alleles being somewhat, but incompletely, dominant to the "lighter" alleles. Red hair color is governed by another unlinked locus with at least two incompletely dominant alleles, one coding for reddish pigment, the other for lack of such pigment. The appearance of red hair is determined by the presence of one or two "red" alleles, but the expression of these alleles can be partially or completely blocked by heavy melanin deposits.

Eye color is determined by the pattern of melanin distribution in the iris. Blue pigment does not exist, only brown or yellowish melanin. If the melanin is deposited in such a way that much of the light is reflected back from the eye, the iris appears blue. Although inheritance of eye color is not simple, the inheritance patterns seen in most families can be explained if we assume that one locus is involved and that the alleles that govern the darker colors are dominant to the alleles that govern the lighter colors (e.g., blue, gray, hazel). It is not uncommon for two dark-eyed parents to have a light-eyed child; it is relatively rare for two light-eyed parents to have a dark-eyed child, but this does occasionally occur.

A number of other human characteristics show relatively simple inheritance patterns. For example, the allele for dimples is generally dominant to the allele for no dimples, and the allele for freckling is usually dominant to that for no freckles.

THE HUMAN GENOME INITIATIVE IS A SYSTEMATIC STUDY OF ALL HUMAN GENES

Along with advances in technology to determine the base sequences of DNA came the realization that it is possible to study the human genome in great depth. Some scientists began to envision a coordinated effort to determine the total informational content of the human genome. The project, initiated in the late 1980s, is an international undertaking that will take place over a 15-year period. In its most direct sense the information content is the sequence of 3 billion base pairs in a human haploid genome. However, only a tiny fraction of human DNA is known to code for protein or RNA; the rest (95% or more) is either nonfunctional or has some function that has not yet been identified.

The project includes not only sequencing but also a broad-based, multifaceted strategy that involves scientists from many disciplines. This approach has concentrated on various types of mapping studies that will allow us to understand the physical and functional relationships among genes and groups of genes as revealed by their order on the chromosomes.

Comparative mapping studies are being carried out simultaneously in a number of other organisms, especially the laboratory mouse, which has long been the favorite model organism for the study of mammalian genetics (see *Making the Connection: The Genetics of Mice and Humans,* Chapter 14). Comparisons of the DNA sequences and chromosomal organization of related genes and clusters of genes from different organisms are powerful tools for identifying the elements essential for their functions. Large scale analyses of base sequences are also underway. Automation is making the task of sequencing less laborious, and powerful computer programs manage and analyze the data.

The project has been controversial for many reasons. Some scientists have argued that the conventional approach of first identifying an important gene, then cloning and studying it, is scientifically more interesting and more cost-effective over the long run. Supporters of the project argue that many important genes that might be very difficult to identify will be uncovered in the course of the investigation. Even apparently nonfunctional DNA (sometimes called "junk DNA") may turn out to be important.

Our knowledge of human genetics clearly will be expanding at a great rate over the coming years. The information gained from the comparative mapping studies of other organisms will add greatly to our understanding of evolutionary relationships. Many genes known to be responsible for genetic diseases are being studied, and many more already being uncovered appear to be associated with predispositions to diseases such as heart disease and cancer.

BOTH HUMAN GENETICS AND OUR BELIEFS ABOUT GENETICS HAVE AN IMPACT ON SOCIETY

Some people harbor misconceptions about genetic diseases and their effect on society. There is a widespread tendency to think of certain individuals or populations as "genetically unfit" and responsible for many of society's ills. Actually, abnormal alleles are present in all individuals and all groups; no one is exempt.

It is not always easy to explain why a certain allele is present at a particularly high frequency in a certain population. To find such answers, a great deal of information must be gathered from many sources. Much must be known about the mutant allele and its effects in both homozygous and heterozygous individuals. Knowledge of the social structure, demographics, and history of a population often provides important clues.

"Heterozygote advantage" in the form of resistance to malaria is the widely accepted explanation for the high frequency of the sickle cell allele among persons of African descent (see Chapter 10). Various hypotheses have been advanced to explain the high frequencies of other genetic diseases in certain populations or ethnic groups.

Sometimes history provides possible explanations. For example, in the Middle Ages Eastern European Jews were subjected to widespread persecution. People were forced into ghettos and the population decreased (a phenomenon known to population geneticists as a **population bottleneck;** see Chapter 18). A coincidental high frequency of the Tay-Sachs allele in the small surviving population might explain the high frequency (1/28) of this allele in the Ashkenazi Jewish population today. However, molecular analysis has revealed that there are at least two mutant alleles for Tay-Sachs disease in that population. The simplest hypothesis is that these arose independently and were maintained because they conferred some kind of heterozygote advantage. People living in ghettos were vulnerable to a variety of infections, such as tuberculosis; resistance to one of these diseases could have been the basis of a heterozygote advantage.

Molecular studies have shown that a number of mutant alleles can be responsible for cystic fibrosis, with the most severe form predominating in northern Europe and another, somewhat less serious, form being more prevalent in southern Europe. Presumably these mutant alleles are all independent mutations, maintained by natural selection. Experimental evidence has been obtained that supports the hypothesis that heterozygous individuals are less likely to die from infectious diseases that produce severe diarrhea. (See *On the Cutting Edge: Using a Mouse Model to Study a Human Genetic Disease.*)

It is often argued that medical treatment of persons affected with genetic diseases, especially those who are able to reproduce, greatly increases the frequency of abnormal alleles in the population. This is true for autosomal dominant and X-linked diseases, but most genetic diseases that are simply inherited show an autosomal recessive inheritance pattern. Only homozygous persons actually have the disease; heterozygous carriers, who are far greater in number, are phenotypically normal. For example, if 1 in 20 persons in the United States is heterozygous for cystic fibrosis, the chance that a husband and wife will both be heterozygous is $1/20 \times 1/20 = 1/400$. On average, one-fourth of the children of such a couple would have cystic fibrosis, so the frequency of affected individuals in the population would be about $1/400 \times 1/4 = 1/1600$. Because their numbers are usually very small compared with heterozygotes, reproduction by homozygotes contributes very little to overall frequencies of abnormal alleles.

It has been estimated that each of us is heterozygous for several (3–15) very harmful alleles, any of which could cause debilitating illness or death in the homozygous state. Why aren't genetic diseases more common? Each of us has many thousands of essential genes, any of which can be mutated. It is very unlikely that the abnormal alleles carried by one person are also carried by that per-

ON THE CUTTING EDGE

Using a Mouse Model to Study a Human Genetic Disease

Objective: To test the hypothesis that the alleles responsible for cystic fibrosis have a heterozygote advantage because heterozygous individuals are less likely to die from certain types of life-threatening diarrhea.

Method: The effects of cholera toxin (which is produced by a bacterial infection and causes severe diarrhea) were studied in mice. The responses of (1) mice homozygous for an allele that causes cystic fibrosis, (2) heterozygous mice, and (3) normal mice were evaluated.

Results: Mice homozygous for a cystic fibrosis allele did not respond to cholera toxin. Cholera toxin caused heterozygotes to lose only half as much fluid from the cells of the intestinal lining as did homozygous "normal" animals.

Conclusion: These results support the hypothesis that individuals heterozygous for cystic fibrosis are less likely to die from certain kinds of diarrhea.

Human geneticists have long been puzzled by the fact that heterozygosity for cystic fibrosis alleles is so common among Caucasians (1/20), particularly among Northern Europeans and those of Northern European descent. Recent advances in our understanding of the molecular biology of this genetic disease have led to a new hypothesis to explain why these alleles are maintained in the population: Heterozygous individuals are less likely to die from certain types of potentially fatal diarrhea.

In severe diarrhea, large amounts of water and electrolytes are lost from the intestine. If unchecked (particularly in infants and young children) this condition can result in death. It has been found that the allele that causes cystic fibrosis is a mutant form of a gene involved in controlling the body's water and electrolyte balance. This gene has been cloned and found to code for a plasma membrane chloride ion channel protein known as the *CFTR protein.* (CFTR stands for *cystic fibrosis transmembrane conductance regulator.*) This ion channel is responsible for transporting chloride ions out of the cells lining the digestive tract and the respiratory system. When the chloride ions leave the cells, water follows by osmosis (see Chapter 5). Thus the normal secretions of these cells are relatively watery.

Because the cells of individuals with cystic fibrosis lack normal ion channels, their secretions have a very low water content. Cells of heterozygous individuals have only half the usual number of functional CFTR ion channels, but these are sufficient to maintain adequate fluidity of their secretions. However, this ion channel deficiency might be an advantage if an individual is infected by a pathogen that produces a toxin that causes the ion channels to remain constantly open (precipitating diarrhea). With only half as many ion channels, a heterozygote might lose only half as much chloride and water.

Research on any disease is greatly facilitated if an animal model can be used for experimentation. Such a model became available when strains of mice homozygous for cystic fibrosis, as well as those that are heterozygous, were produced by *targeted gene replacement* (see *Making the Connection: The Genetics of Mice and Humans,* Chapter 14).

Sherif E. Gabriel and his colleagues at the University of North Carolina used the cystic fibrosis mouse model to test the heterozygote advantage hypothesis.* They treated mice with cholera toxin produced by *Vibrio cholerae* (the bacterium that causes cholera). Cholera toxin is known to affect the functioning of the CFTR ion channels, causing the uncontrolled loss of chloride ions and water.

Their results neatly fit the predictions of the heterozygote advantage hypothesis. Animals homozygous for cystic fibrosis (those with no CFTR channels) did not lose any fluid through their intestinal cells when exposed to cholera toxin. The toxin caused heterozygotes (with only half the normal number of CFTR channels) to lose only half as much fluid as mice with the normal number of channels.

Despite the success of this demonstration, it is not thought that cholera itself is the selective force responsible for the high incidence of cystic fibrosis alleles today. Cystic fibrosis is thought to have arisen over 50,000 years ago, whereas European cholera epidemics were first recognized in the early 1800s. Rather, other diarrhea-causing infections are the probable culprits. Researchers are continuing to test likely candidates.

*Gabriel, S.E., K.N. Brigman, B.H. Koller, R.C. Boucher, and M.J. Stutts. "Cystic fibrosis heterozygote resistance to cholera toxin in the cystic fibrosis mouse model." *Science,* Vol. 266, October 7, 1994.

son's mate. Of course, this possibility is more likely if the harmful allele is a relatively common one, such as the one responsible for cystic fibrosis.

Relatives are more likely than nonrelatives to carry the same harmful alleles, having inherited them from a common ancestor. In fact, a greater than normal frequency of a particular genetic disease among offspring of **consanguineous matings** (matings of close relatives) is often the first clue that the mode of inheritance is autosomal recessive. The offspring of consanguineous matings have a small but significantly increased risk of genetic disease. In fact, they can account for a disproportionately high percentage of those individuals in the population with autosomal recessive disorders. Because of this perceived social cost, first-cousin marriages are prohibited in most states in the United States. However, consanguineous marriages are still relatively common in many developing countries, where other factors may outweigh the genetic costs.

As scientists learn more about human genetics, more and more alleles associated with human disease are being identified. It is important that we prepare ourselves to use this knowledge wisely, not to stigmatize or discriminate, but to improve human health.

SUMMARY

I. Geneticists investigating human inheritance cannot make specific crosses of pure genetic strains; instead, they must rely on studies of populations, analyses of family pedigrees, and molecular studies.

II. Studies of the karyotype (the number and kinds of chromosomes present in the nucleus) permit detection of individuals with various chromosome abnormalities.
   - A. Such studies can detect trisomy, in which an individual possesses an extra chromosome, and monosomy, in which one member of a pair of chromosomes is lacking.
   - B. The most common form of Down syndrome (trisomy 21) and Turner syndrome (XO) are examples of trisomy and monosomy, respectively.

III. Abnormal alleles of a number of loci are responsible for many inherited diseases. Most human genetic diseases that show a simple inheritance pattern are transmitted as autosomal recessive traits.
   - A. Phenylketonuria is an autosomal recessive disorder in which toxic phenylketones damage the developing nervous system.
   - B. Sickle cell anemia is an autosomal recessive disorder in which abnormal hemoglobin (the protein needed to carry oxygen in the blood) is produced.
   - C. Cystic fibrosis is an autosomal recessive disorder in which abnormal secretions are produced in the respiratory and digestive systems.
   - D. Huntington's disease has an autosomal dominant inheritance pattern. It results in mental and physical deterioration, usually beginning in middle age.
   - E. Hemophilia A is an X-linked recessive disorder. It results in a defect in one of the components required for blood clotting.

IV. Some genetic diseases and chromosome abnormalities can be diagnosed long before birth.
   - A. In amniocentesis, the amniotic fluid surrounding the fetus is removed and the fetal cells suspended in the fluid are cultured and screened for genetic defects. Amniocentesis provides results in the second trimester of pregnancy.
   - B. In chorionic villus sampling (CVS), cells from the chorion are removed and studied. CVS provides results in the first trimester of pregnancy.

V. Genetic counselors can advise prospective parents with a family history of genetic disease about the probabilities of giving birth to affected offspring. It is possible to detect the presence of a number of harmful alleles.

VI. Variation in the human population is exemplified by human blood types, among which are the ABO blood group and the Rh system. The Rh-positive offspring of an Rh-negative mother may develop a very serious disease known as erythroblastosis fetalis.

VII. The effect of genetics on society is complex.
   - A. The fact that a particular abnormal allele is especially common in a certain population does not mean that group has a higher frequency of abnormal alleles in general.
   - B. Some alleles that cause a genetic disease when they are homozygous may be advantageous in heterozygous individuals, at least in a particular environment. For example, the allele responsible for sickle cell anemia in homozygous individuals appears to confer resistance to malaria in heterozygotes.
   - C. Because most abnormal alleles are recessive, they are manifested phenotypically only in homozygotes, who constitute a tiny fraction of the individuals with the allele. Virtually every individual in the population is a heterozygous carrier of several abnormal alleles.

SELECTED KEY TERMS

ABO blood group
amniocentesis
aneuploidy
birth defect
chorionic villus sampling (CVS)
consanguineous mating
controlled mating
cystic fibrosis
cytogenetics
disomic
DNA marker
Down syndrome
erythroblastosis fetalis
genetic counseling
hemophilia A
Huntington's disease
inborn error of metabolism
isogenic strain
karyotype
Klinefelter syndrome
lectin
monosomic
nondisjunction

(Selected Key Terms continued on next page)

nuclear sexing
pedigree
phenylketonuria (PKU)
polyploidy
population bottleneck
quantitative trait
Rh incompatibility
sickle cell anemia
Tay-Sachs disease
translocation
trisomic
Turner syndrome
ultrasound imaging
XYY karyotype

POST-TEST

1. Standard stocks of genetically identical individuals are called __________ strains.
2. The array of chromosomes present in a given cell is called the __________ .
3. An abnormality that is present and evident at birth is called a(an) __________ defect.
4. An abnormality in which there is one more or one fewer than the normal number of chromosomes is called a(an) __________ .
5. A person with an extra chromosome (three of one kind) is said to be __________ .
6. A person who is missing a chromosome, having only one member of a pair, is termed __________ .
7. The failure of chromosomes to separate normally during cell division is called __________ .
8. The transfer of a part of one chromosome to a nonhomologous chromosome is called a(an) __________ .
9. Individuals with trisomy 21, or __________ syndrome, are mentally and physically retarded and have abnormalities of the face, tongue, and eyelids.
10. An XXY individual has the disorder known as __________ syndrome; an XO individual has __________ syndrome.
11. An inherited disorder caused by a defective or absent enzyme is called an inborn error of __________ .
12. The __________ __________ allele codes for an altered hemoglobin molecule, which is less soluble than usual and more likely than normal to crystallize and deform the shape of the red blood cell.
13. In a person with __________ __________, the mucus is abnormally viscous and tends to plug the ducts of the pancreas and liver, and to accumulate in the lungs.
14. In the process of __________, a sample of amniotic fluid is obtained by insertion of a needle through the walls of the abdomen and uterus and into the uterine cavity.
15. __________ __________ sampling involves removal and study of cells that will make up the fetal contribution to the placenta.
16. Couples who have some reason to be concerned about the possibility of genetic abnormalities in their offspring may seek __________ __________ .

REVIEW QUESTIONS

1. What means have been devised for overcoming some of the difficulties in studying human inheritance?
2. What is meant by nondisjunction? What are some human abnormalities that appear to be the result of nondisjunction?
3. Are all birth defects hereditary?
4. What are the relative advantages and disadvantages of amniocentesis and chorionic villus sampling?
5. What are some of the ways that carriers of certain genetic diseases can be identified?
6. What is meant by inborn errors of metabolism? Give an example.
7. To be expressed, an autosomal recessive genetic disease must be homozygous. What relationship does this fact have to consanguineous matings?

YOU MAKE THE CONNECTION

1. Mrs. Doe and Mrs. Smith had babies in the same hospital at the same time. Mrs. Doe took home a girl and named her Nancy. Mrs. Smith took home a boy and named him Richard. However, she was sure that she had given birth to a girl and brought suit against the hospital. Blood tests showed that Mr. Smith was blood type A, Mrs. Smith was type B, and Mr. and Mrs. Doe were both type AB. Nancy was type A, and Richard was type B. Had an exchange occurred? What other kinds of genetic information would help resolve this question?
2. Imagine that you are a genetic counselor. What advice or suggestions might you give in the following situations?
   (a) A couple has come for advice because the woman had a sister who died of Tay-Sachs disease.
   (b) A pregnant woman has learned that as a newborn she suffered a mild case of erythroblastosis fetalis; she is concerned that she might have a similarly affected child.
   (c) A young man and woman who are not related are engaged to be married. However, they have learned that the man's parents are first cousins. They are worried about the possibility of increased risk of genetic defects in their own children.
   (d) A young woman's paternal uncle (her father's brother) has hemophilia A. Her father is free of the disease and there has never been a case of hemophilia A in her mother's family. Should she be concerned about the possibility of hemophilia A in her own children?
   (e) A 20-year-old man is seeking counseling because his father has just been diagnosed with Huntington's disease.
3. Why is the perpetuation of various myths about human genetics harmful to society?

RECOMMENDED READINGS

Anderson, W.F. "Gene Therapy," *Scientific American*, Vol. 273, No. 3, September, 1995. A pioneer in gene therapy discusses progress in the field, and prospects for the future.

Gibbons, A. "The Risks of Inbreeding." *Science*, Vol. 259, February 26, 1993. Reviews current world-wide data on the increased risk of genetic disease from consanguineous matings.

Glausiusz, J. "Hidden Benefits." *Discover*, March 1995. Discusses the hypothesis that heterozygosity for cystic fibrosis may confer resistance to certain types of diarrhea.

Horgan, J. "Eugenics Revisited." *Scientific American*, Vol. 268, No. 6, June 1993. An article warning of potential misuses of findings in human genetics. Points out that many studies claiming to demonstrate a relationship between specific genes and certain types of human behavior, such as crime, mental illness, and alcoholism, have not been substantiated.

McKusick, V.A. *Mendelian Inheritance in Man*, 11th ed. Johns Hopkins University Press, Baltimore, 1994. An up-to-date compilation of human genetic data.

Patterson, D. "The Causes of Down Syndrome." *Scientific American*, Vol. 257, No. 2, August 1987. A discussion of efforts to identify sites on chromosome 21 responsible for the abnormalities associated with Down syndrome.

Rennie, J. "Grading the Gene Tests." *Scientific American*, Vol. 270, No. 6, June 1994. A discussion of the ethical and practical implications of genetic testing in humans.

Verma, I.M. "Gene Therapy." *Scientific American*, Vol. 263, No. 5, November 1990. A consideration of future prospects for gene therapy and some of the problems to be solved.

Watson, J.D., M. Gilman, J. Witkowski, and M. Zoller. *Recombinant DNA*, 2nd ed. W. H. Freeman, New York, 1992. Chapters 26 through 30 deal with the uses of recombinant DNA technology in human genetics, including diagnosis, prospects for gene therapy, and the Human Genome Initiative.

CHAPTER 16

Genes and Development

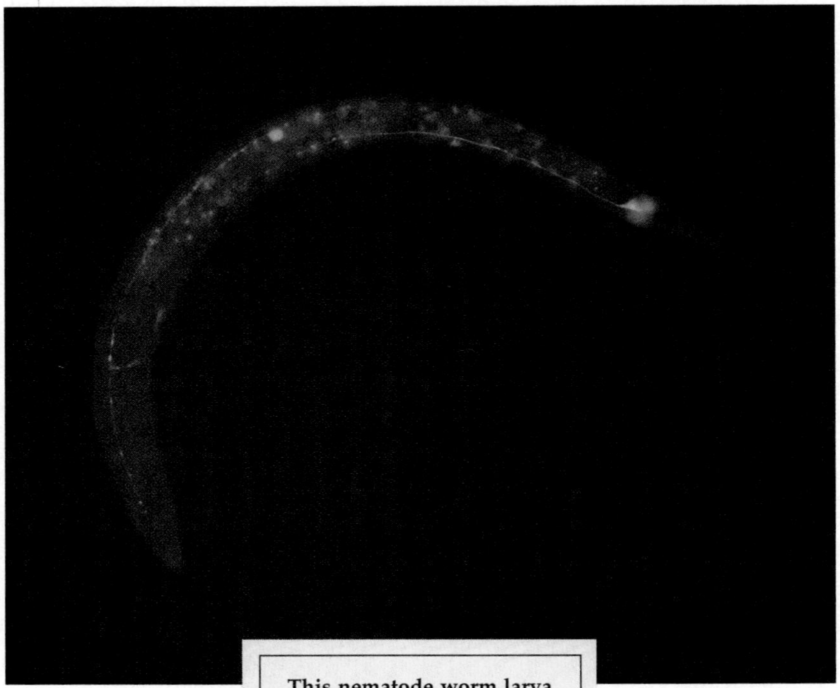

This nematode worm larva (*Caenorhabditis*) demonstrates one of the techniques used to mark cells in which a developmentally important gene is expressed. The worm has been genetically engineered to produce a green fluorescent protein, known as GFP (*green spots*), in cells in which the gene of interest is active. (Courtesy of Dr. Martin Chalfie, Columbia University)

The study of **development,** the process by which cells specialize and organize into a complex organism, encompasses some of the most fascinating and difficult problems in biology today. During the many cell divisions required for a single cell to develop into a multicellular organism, groups of cells become gradually committed to specific patterns of gene activity through a process called **determination.** The final step leading to cell specialization is **cellular differentiation.** A differentiated cell can be recognized by its characteristic appearance and activities.

Morphogenesis, the development of form, is an even more intriguing piece of the developmental puzzle. It involves a multistep process known as **pattern formation,** by which cells in specific locations become progressively organized into recognizable structures.

Until recently, little was known about how certain genes interact with various signals from other genes and from the environment to control development. These networks are too complex to unravel using only traditional methods.

Today scientists interested in development study a variety of carefully chosen mutant organisms with altered developmental patterns. They use the tools of genetic engineering combined with more conventional descriptive and experimental approaches to derive fresh insights into the role of genetic information in the control of development. They are also finding new ways to unravel evolutionary relationships through the study of developmentally important genetic mechanisms that appear to be deeply rooted in the evolutionary history of multicellular organisms.

LEARNING OBJECTIVES

After you have studied this chapter you should be able to

1. Distinguish between cellular determination and cellular differentiation.
2. Relate the process of pattern formation to morphogenesis.
3. Describe the kinds of experiments that indicate the totipotency of at least some differentiated plant cells and animal nuclei. Discuss how these findings support the idea of nuclear equivalence.
4. Identify the attributes of an organism that would make it especially useful in studies on the genetic control of development. Discuss the value of transgenic organisms in research on the genetic control of development.
5. Indicate the features of the development and genetics of *Drosophila, Caenorhabditis,* and the mouse (*Mus*) that have made these organisms so valuable to researchers.
6. Distinguish among maternal effect genes, zygotic genes, and homeotic genes in *Drosophila.*
7. Explain the relationship between transcription factors and genes that control development. Provide some examples of genes that are known to function as genetic switches in development.
8. Define the phenomena of induction and programmed cell death, and give examples of the roles they play in development.
9. Describe the functions of some homeotic-like genes in plants.
10. Point out some of the known exceptions to the general phenomenon of nuclear equivalence.

CELLULAR DIFFERENTIATION USUALLY DOES NOT INVOLVE CHANGES IN DNA

The human body contains more than 200 recognizably different types of cells (Fig. 16–1). Combinations of those cells are organized into remarkably diverse and complex structures such as the eye, the hand, and the brain, each one capable of carrying out many sophisticated activities. Most remarkable of all, however, is the fact that all of the structures of the body and the different cells within them are descended from a single fertilized egg.

All multicellular plants and animals undergo complex patterns of development. The root cells of plants, for example, have structures and functions very different from those of the various types of cells located in plant leaves. Remarkable diversity can also be found at the molecular level; most strikingly, each type of plant or animal cell makes a highly specific set of proteins (Fig. 16–2). In some cases, such as the protein hemoglobin in red blood cells, one cell-specific protein may make up more than 90% of the total mass of protein in the cell. Other cells may have a complement of cell-specific proteins that are present in small amounts but still play an essential role. However, because certain proteins are required in every type of cell (all cells, for example, require certain enzymes for glycolysis), cell-specific proteins usually make up only a fraction of the total number of different kinds of proteins.

One explanation for the fact that each type of differentiated cell makes a unique set of proteins might be that during development each group of cells loses the genes it does not need and retains only those that are required. With just a few exceptions, however, this does not seem to be true. According to the concept of **nuclear equivalence,** the nuclei of essentially all differentiated adult cells of an individual are genetically (though not necessarily metabolically) identical to each other, and to the nucleus of the fertilized egg cell from which they descended. This means that virtually all the **somatic**[1] cells in an adult have the same genes; they are simply expressed in different tissues in different ways.

The evidence for nuclear equivalence comes from cases in which differentiated cells or their nuclei have been found to be capable of supporting normal development. Such cells or nuclei are said to be **totipotent.**

A totipotent nucleus contains all the information required to direct normal development

In plants it is possible to demonstrate that at least some differentiated cells can be induced to become the equivalent of embryonic cells (Fig. 16–3). **Tissue culture** techniques are used to isolate cells from certain plants and allow them to grow in a nutrient medium.

In some of the first experiments, single root cells from a carrot were induced to divide in a liquid nutrient medium and to form groups of cells called "embryoid" (embryo-like) bodies. These clumps of dividing cells could then be transferred to an agar medium, which provides nutrients plus a solid supporting structure for the developing plant cells. After transfer to the agar, some of

[1]Somatic cells are cells of the body and are distinguished from **germ-line** cells, which ultimately give rise to a new generation. The distinction between somatic cells and germ-line cells is not clear-cut in plants. In animals, however, germ-line cells, whose descendants ultimately undergo meiosis and differentiate as gametes, are generally set aside early in development.

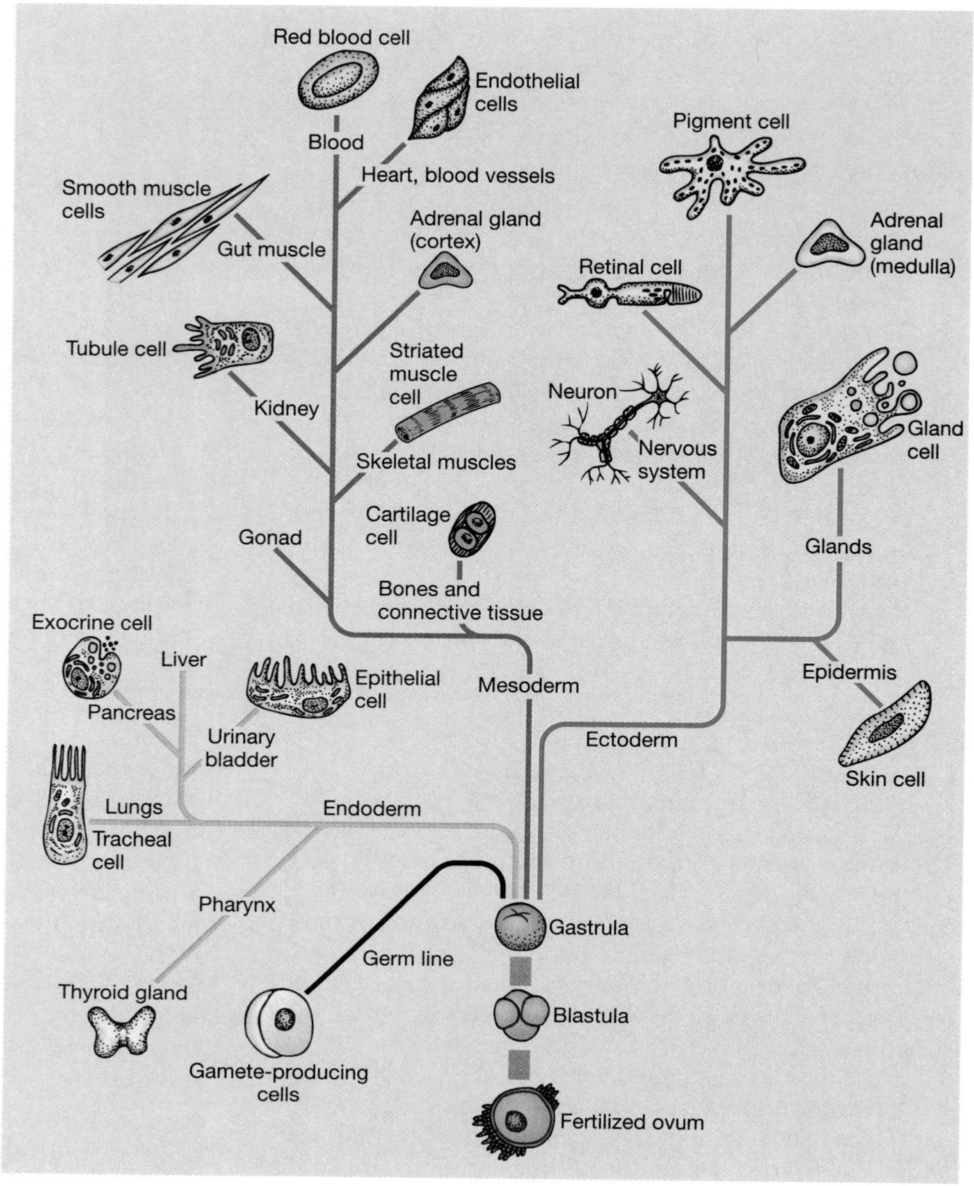

FIGURE 16–1 Animal cell lineages lead to differentiated cell types. Repeated divisions of the fertilized egg result in the establishment of tissues from which groups of specialized cells are produced (see Chapters 37 and 49 for a discussion of how these are formed). Germ-line cells (cells that produce the gametes) are set aside early in development. Somatic cells progress along the developmental pathways, undergoing a series of developmental commitments that progressively determine the fates of different lines of cells.

the embryoid cells gave rise to roots, stems, and leaves. The resulting "plantlets" were then transplanted to soil, where they ultimately developed into adult plants capable of producing flowers and viable seeds. The methods of plant tissue culture are now extensively used to produce genetically engineered plants, for they allow the regeneration of whole plants from cells that have incorporated recombinant DNA molecules.

Similar experiments have been attempted with animal cells, but so far it has not been possible to induce a fully differentiated somatic cell to behave like a zygote. Instead, it has been possible to test whether steps in the process of determination are reversible by transplanting the *nucleus* of a cell in a relatively late stage of development into an egg cell whose own nucleus has been destroyed (Fig. 16–4).

Nuclei from amphibian cells at different stages of development were transplanted into egg cells. Some of the transplants proceeded normally through a number of developmental stages, and a few even developed into normal tadpoles. As a rule, the nuclei transplanted from cells at earlier stages were most likely to support development to the tadpole stage. As the fate of the cells became more and more determined, the probability that a transplanted nucleus could control normal development diminished rapidly.

In a few cases, nuclei isolated from the specialized intestinal cells of a tadpole were able to direct development up to the tadpole stage. This occurred infrequently, but in such experiments success counts more than failure, and we can safely conclude that at least some nuclei of differentiated cells are in fact totipotent and have not lost any genetic material.

These results lead to several important interpretations. When specialized plant cells and animal nuclei are totipotent, it is clear that genes have not been lost as a

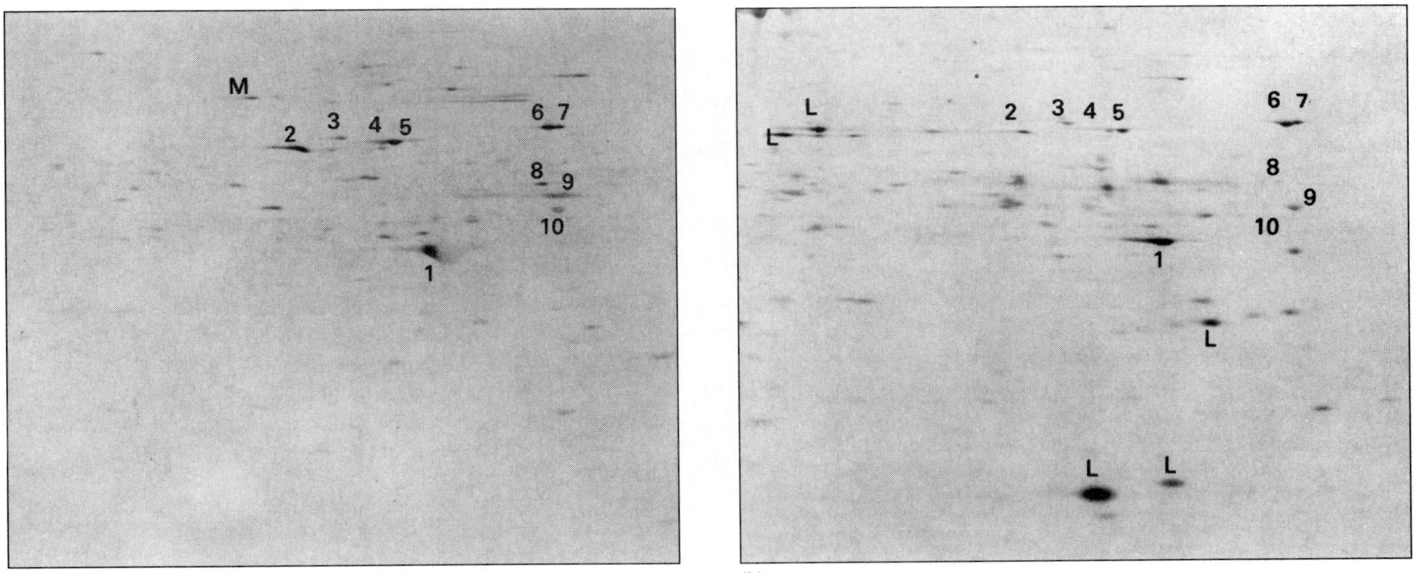

(a) (b)

FIGURE 16–2 Each cell type makes a highly specific set of proteins. The spots in the photographs are proteins from (*a*) muscle and (*b*) liver cells of a mouse. The proteins were separated by two-dimensional gel electrophoresis, a method that separates the proteins in the horizontal direction by their electric charge, followed by a second separation in the vertical direction by molecular weight. Several hundred proteins can be distinguished in each panel. The spots that are labeled with numbers are present in all tissues, but notice that many of them are present in different amounts from one tissue to another. The proteins that are labeled with letters are found only in that specific tissue. (Patrick O'Farrell, from Darnell, Lodish, and Bathmore, *Molecular Cell Biology,* Fig. 12–1, Scientific American Books)

consequence of development. That is, genes that were apparently inactive were, in fact, capable of being reactivated when the cells or nuclei were placed in a new environment. Second, even if no genetic material is lost during development, the nuclei of cells undergo metabolic changes that make it progressively more difficult to remain in a totipotent state. This is especially true of animal nuclei, although various kinds of animals differ considerably in this regard.

Most differences among cells are due to differential gene expression

Because genes do not appear to be lost regularly during development, the differences in the molecular composition of cells must occur by *regulating the activities of different genes.* This process of developmental gene regulation is often referred to as **differential gene expression.** As discussed in Chapter 13, the expression of eukaryotic

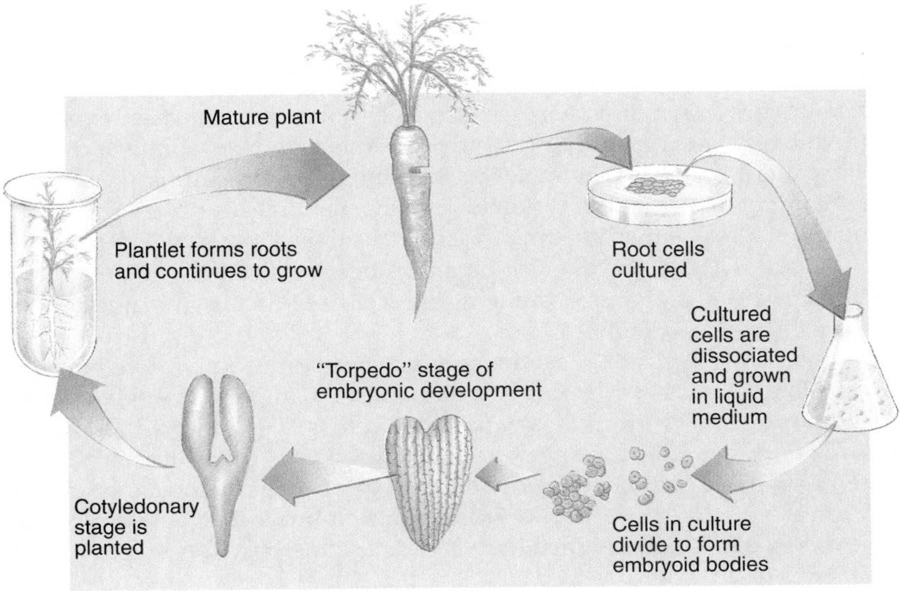

FIGURE 16–3 A complete carrot plant can develop from differentiated somatic cells. Discs of phloem cells, which are specialized for nutrient transport, were isolated from carrot root tissues. When the cells were cultured in a liquid nutrient medium, clumps of dedifferentiated cells developed from individual phloem cells. These clumps (embryoids) closely resembled plant embryos in their early stages of development and then progressed to form embryonic shoots and roots. Transferring the embryonic tissue to a solid nutrient medium stimulated the tissues to form small plants, called plantlets, which then developed into mature plants.

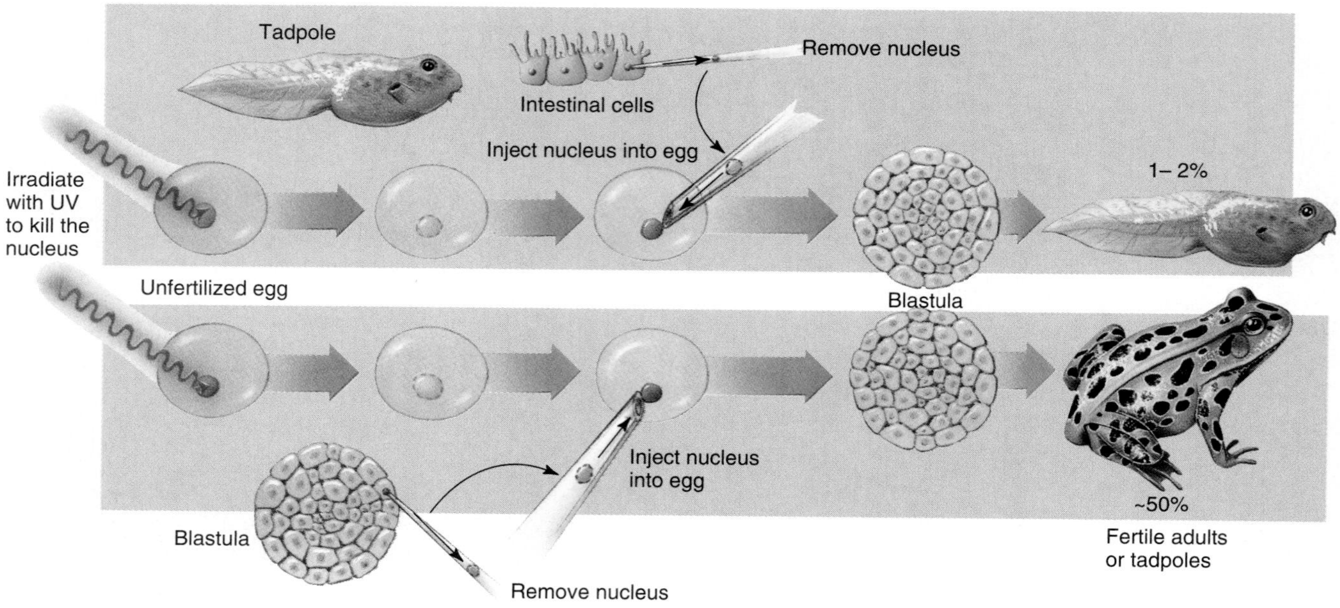

FIGURE 16–4 Nuclear transplantation experiments demonstrate that a nucleus from a differentiated amphibian cell can program development. In the experiments of R. Briggs and T. J. King, and later J. Gurdon, a nucleus from a differentiated cell was injected into an egg whose own nucleus had been destroyed by ultraviolet radiation. The probability of success of the procedure depended on the developmental stage of the transplanted nucleus. As shown in the lower panel, if a nucleus was taken from a cell at the blastula stage of development (when cell division has produced about 1000 cells formed in the shape of a ball), there was a high probability that it would program normal development, resulting in a fertile adult. However, as seen in the upper panel, most trials using nuclei from tadpole intestinal cells (a much later developmental stage) resulted in no growth, probably as a result of damage to the egg or the nucleus by the procedure. In a small number of trials, however, normal development proceeded until the tadpole stage, indicating that the genes necessary to program development to that point were still present.

genes can be regulated in many different ways and at many levels. For example, a particular enzyme may be produced in an inactive form and then be activated at a later time. However, much of the regulation that is important in development occurs at the transcriptional level. The transcription of certain sets of genes is repressed, whereas others are activated. Even expression of genes that are constitutive and active in all cells can be regulated during development so that the *quantity* of each product varies from one tissue type to another.

We can think of differentiation as a series of pathways leading from a single cell to cells in each of the different specialized tissues, arranged in an appropriate pattern. There are times when a cell makes genetic "commitments" to the developmental path its descendants will follow. These commitments gradually *restrict* the development of the descendants to a limited set of final tissue types. This progressive fixation of the fate of a cell's descendants is the process of **determination.**

As the development of a cell is determined along a particular differentiation pathway, those changes may not be obvious. Nevertheless, when a particular stage of determination is complete, the changes in the cell usually become self-perpetuating and are not easily reversed. **Differentiation** is usually the last stage in the developmental process. At this stage, a precursor cell becomes structurally and functionally recognizable as a bone cell, for example, and its pattern of gene activity is different from that in a nerve cell.

MOLECULAR GENETICS IS REVOLUTIONIZING THE STUDY OF DEVELOPMENT

Development has been an important area of research for many years, and considerable effort has been expended on studying the development of invertebrate and vertebrate animals. By identifying patterns of tissue development in different animals, researchers have been able to identify similarities, as well as differences, in the basic plan of development from a fertilized egg to an adult in organisms ranging from the sea urchin to mammals (see Chapter 49).

In addition to descriptive studies, a number of classic experiments have established important evidence concerning how groups of cells are determined along particular developmental pathways. Today researchers are combining a wide variety of methodologies to identify genes in both plants and animals that actually control development, and to determine how those genes work.

Certain organisms are particularly well suited for studies on the genetic control of development

In studies of the genetic control of development, the choice of an organism to use as an experimental system has become increasingly important. One of the most powerful approaches involves the isolation of mutants with arrested or abnormal development at a particular stage. Not all organisms have useful characteristics that allow developmental mutants to be isolated and maintained for future study. The fruit fly, *Drosophila melanogaster,* has such thoroughly understood genetics that it has become one of the most important systems for such studies. Other organisms such as the nematode worm, *Caenorhabditis elegans,* and the laboratory mouse, *Mus musculus,* as well as various plants and some simple eukaryotes, have also become important in developmental genetics. Each of these organisms has attributes that make it particularly useful for examining certain aspects of development.

DROSOPHILA MELANOGASTER PROVIDES RESEARCHERS WITH A WEALTH OF DEVELOPMENTAL MUTANTS

Undoubtedly the most extensive (and spectacular) examples of genes that control development have been identified in the fruit fly, *Drosophila melanogaster.* One of the main advantages of using *Drosophila* as a research organism is the abundance of mutants (including developmental mutants) available for study, and the relative ease with which a new mutation can be directly mapped on the chromosomes. The genetic analysis is greatly facilitated by special chromosomes found in certain tissues with large, metabolically active cells, including the salivary glands of the larvae. These **polytene** ("many-stranded") chromosomes (Fig. 16–5) are unusual interphase chromosomes formed when the DNA replicates many times but without mitosis and cytokinesis.

A typical polytene chromosome may consist of more than 1000 DNA double helices (along with associated histones and other proteins) aligned side by side. Polytene chromosomes are therefore quite large and show a pattern of bands that is very useful in assigning a particular gene to a particular location on the chromosome. When a gene is active, the chromosome band in which it resides decondenses and forms a **puff,** which is a site of intense RNA synthesis. This evidence of gene activity is similar to that observed in lampbrush chromosomes of certain female meiotic cells (see Chapter 9).

Once the chromosomal position of the mutant gene is determined, the gene can be cloned—using a technique called **chromosome walking**—from a nearby gene that has been previously cloned. Studies of *Drosophila* are also facilitated by the fact that foreign DNA can be injected into eggs and become incorporated into the fly's DNA in a process called **transformation** (by analogy with transformation in prokaryotes).

The *Drosophila* life cycle includes egg, larval, pupal, and adult stages

The life cycle of *Drosophila* consists of several distinct stages (Fig. 16–6). After the egg is fertilized, a period of

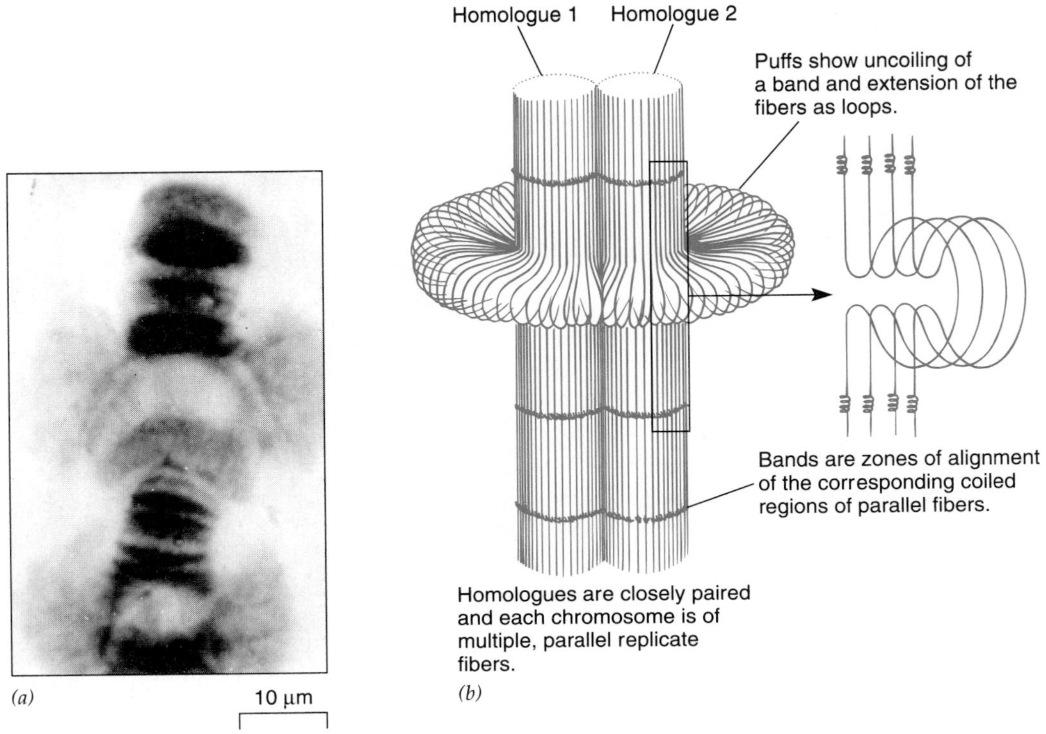
(a)
10 μm
Homologue 1
Homologue 2
Puffs show uncoiling of a band and extension of the fibers as loops.
Bands are zones of alignment of the corresponding coiled regions of parallel fibers.
Homologues are closely paired and each chromosome is of multiple, parallel replicate fibers.
(b)

FIGURE 16–5 Polytene chromosomes, such as those in the salivary gland cells of *Drosophila,* aid in locating genes. (***a***) A region of a polytene chromosome showing the pattern of stained bands of condensed chromatin (B) and decondensed puffed bands (P), which are the sites of intense gene activity. The chromosome banding patterns in a particular tissue are constant and can be associated with the locations of mutant genes by genetic mapping and DNA hybridization methods. (***b***) In contrast to the chromosomes of most somatic cells, homologous polytene chromosomes are paired and each consists of more than 1000 parallel longitudinal DNA fibers. (*a,* Courtesy of U. Clever)

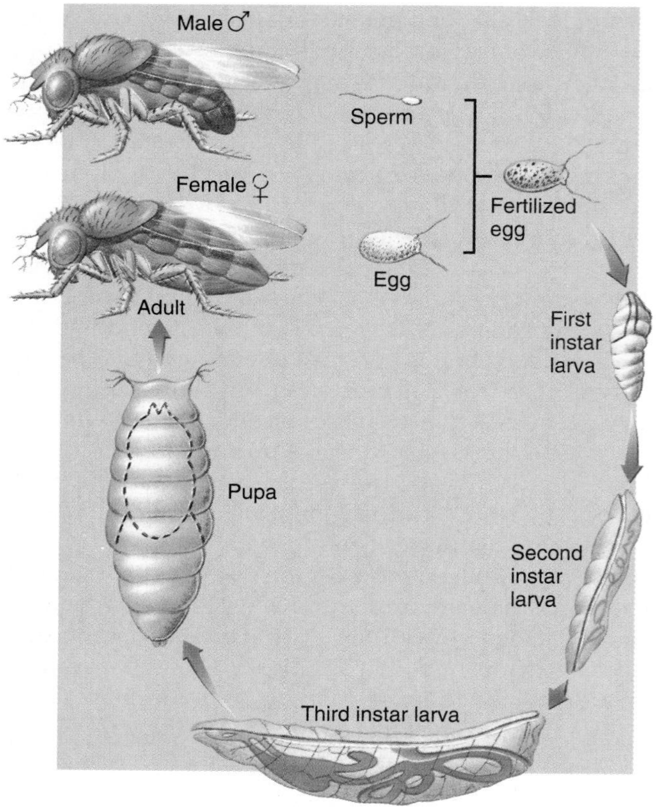

FIGURE 16–6 A *Drosophila* passes through a number of stages as it develops from the egg to the adult fly.

embryogenesis occurs during which the zygote develops into a sexually immature form known as a **larva** (pl., *larvae*). After hatching from the egg, each larva undergoes several molts (shedding of the external covering or cuticle). Each molt allows an increase in size until the larva is ready to pupate. **Pupation** involves a molt and the hardening of the new external cuticle, so that the pupa is completely encased. The insect then undergoes a complete metamorphosis (change in form). During that time, most of the larval tissues degenerate and other tissues differentiate to form the body parts of the sexually mature adult fly.

The larvae are wormlike in appearance and look nothing like the adult flies. However, precursor cells of many of the adult structures are organized—as relatively undifferentiated paired structures called **imaginal discs**—very early in embryogenesis of the developing larvae. This term comes from **imago,** the name given to the adult form of the insect. Each imaginal disc occupies a definite position in the larva and will form a specific structure, such as a wing or a leg, in the adult body (Fig. 16–7). The discs are formed by the time embryogenesis is complete and the larva is ready to begin feeding.

In some respects the larva can be thought of as a complex developmental stage that is simply used to feed and nurture the precursor cells for the adult fly (which is the only form that can reproduce).

The organization of the precursors of the adult structures, including the imaginal discs, is under complex genetic control. So far more than 50 different genes have been identified that specify the formation of the discs, their positions within the larva, and their ultimate functions within the adult fly. Those genes have been identified through mutations that either prevent certain discs from forming, or alter their structure or ultimate fate.

Many *Drosophila* developmental mutants affect the body plan

Many types of developmental mutants of *Drosophila* have been identified. Their effects on development in various combinations have been examined and studied extensively at the molecular level. In our discussion we pay particular attention to those that affect the segmented body plan of the organism, both in the larva and in the adult.

Maternal effect segmentation genes organize the egg cytoplasm

The earliest stages of *Drosophila* development are controlled by maternal genes that act to organize the structure of the egg cell. As the egg develops in the ovary of the female, stores of messenger RNA (mRNA), along with yolk proteins and other cytoplasmic molecules, are passed into it from the surrounding maternal cells. Therefore, all these mRNA molecules are transcribed exclusively from genes found in the mother. The genes that code for these mRNA molecules are referred to as **maternal effect** genes. Analysis of mutants defective in these genes has revealed that many are involved in establishing the polarity of the embryo by designating which parts of the egg are dorsal or ventral, and which are anterior or posterior (see Chapter 28).

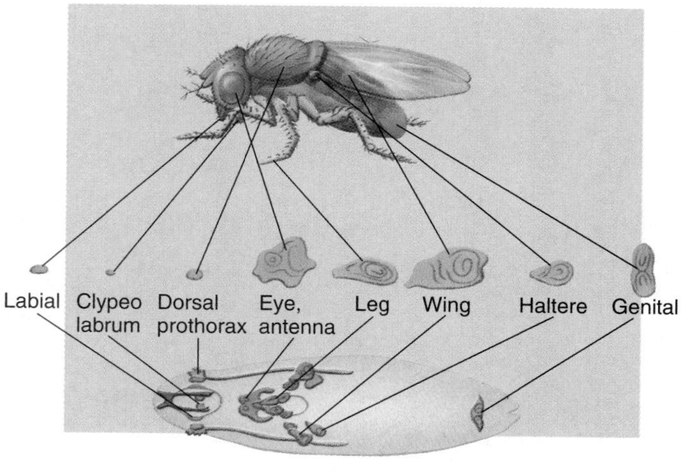

FIGURE 16–7 Each pair of imaginal discs in a *Drosophila* larva develops into a specific pair of structures in the adult fly.

ON THE CUTTING EDGE

Is There a Master Control Gene for Eye Development?

Objective: To test the hypothesis that the *eyeless* locus of *Drosophila* and homologous loci found in other animals are very similar versions of a "master control gene" for eye development.

Method: The wild type allele of the *eyeless* locus of *Drosophila* was activated in imaginal discs other than the eye discs (in which it is normally expressed). The normal allele of a related locus in mice (*Pax-6*, also known as *Small eye*) was expressed in *Drosophila* in similar experiments.

Results: Normal compound eyes were produced ectopically (that is, in unusual locations) in the structures that developed from *Drosophila* imaginal discs in which either the normal fly allele or the normal mouse allele was activated. Eyes were formed on wings, legs, and antennae.

Conclusion: These results support the hypothesis that the *eyeless* locus of *Drosophila* and the *Pax-6* locus of the mouse are homologous genes that function as developmental switches in eye formation.

Is it possible that there is a master control gene for the development of eyes? If we survey the animal kingdom, we find several types of apparently very dissimilar types of eyes. For example, if we compare the eyes of vertebrates with the compound eyes of insects (see Chapter 41), they appear so different morphologically that it would seem unlikely that they could develop under the control of similar genes.

The first clues that there could be a link between the genes responsible for eye formation in *Drosophila* and in vertebrates came from the work of R. Quiring and other researchers working in the laboratory of Walter Gehring in Basel, Switzerland.* They discovered a *Drosophila* transcription factor that was coded by *eyeless*, a well-known *Drosophila* locus. Mutant alleles of the *eyeless* locus cause the compound eyes to be small or absent. Subsequently, these researchers found that the base sequence of the *eyeless* locus is very similar to that of the *Pax-6* locus of mice (also known as *Small eye*). Mutant *Pax-6* alleles cause the mice to develop small eyes when heterozygous, and no eyes when homozygous. There is even a human counterpart to these genes; the eyes of a person heterozygous for a mutant allele of the locus known as *Aniridia* have defects of the iris, cornea, lens, and retina.

G. Halder and P. Callaerts, two other scientists in Gehring's laboratory, decided to test the hypothesis that these genes are master switches in eye development.** They knew that the *eyeless* locus is normally active in the paired eye imaginal discs. They reasoned that if the *eyeless* locus were really a master switch, it would be capable of causing eye development wherever it was active. Using a technique known as *targeted gene expression*, they activated the normal *eyeless* allele in various imaginal discs. The experiment worked; ectopic compound eyes formed on legs, wings, and antennae! Even more surprisingly, similar results were obtained when normal *Pax-6* alleles (from mice) were expressed in *Drosophila*.

These findings are consistent with the hypothesis that the transcription factors encoded by *eyeless* and by *Pax-6* act as developmental switches, turning on the genes necessary for the formation of the compound eyes of *Drosophila* and the vertebrate eye, respectively.

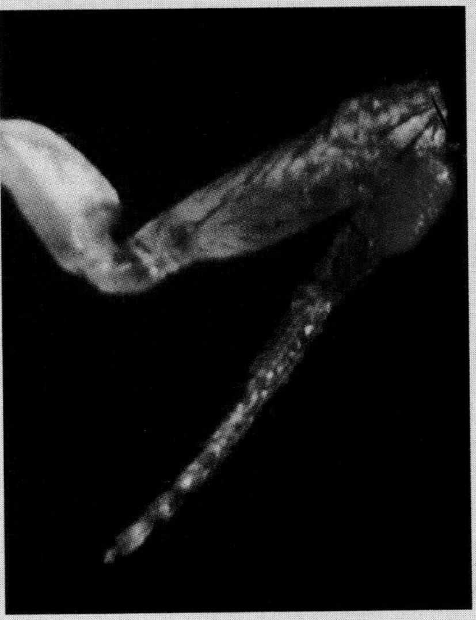

The activation of a normal allele of the *eyeless* locus resulted in the formation of a compound eye on this *Drosophila* leg. (BIOZENTRUM/University of Basel, Switzerland/Courtesy of Professor W.H. Gehring)

*Quiring, R., Walldorf, U., Klotter, U., and W.J. Gehring. "Homology of the *eyeless* Gene of *Drosophila* to the *Small eye* Gene in Mice and Aniridia in Humans." *Science*, Vol. 165, August 5, 1994.

**Halder, G., P. Callaerts, and W.J. Gehring. "Introduction of Ectopic Eyes by Targeted Expression of the *eyeless* Gene in *Drosophila*." *Science*, Vol. 267, March 24, 1995.

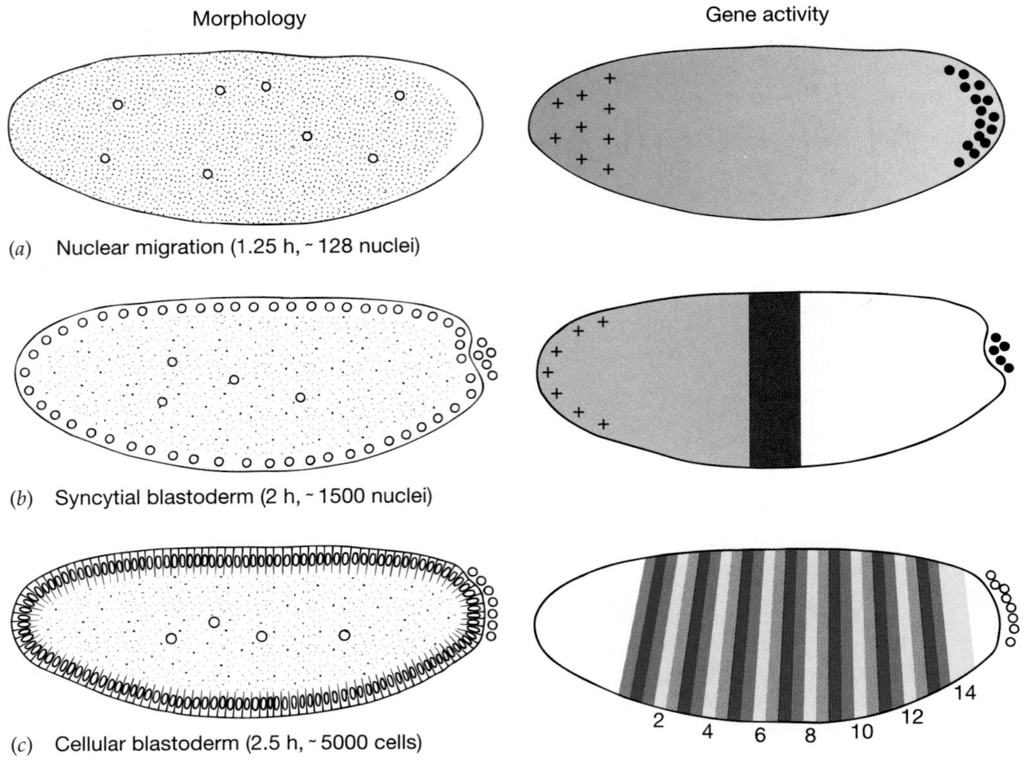

FIGURE 16–8 The early development of a *Drosophila* embryo includes an orderly sequence of events. The diagrams on the left show the structure of the embryo at different times after fertilization. The panels on the right show the patterns of activity of particular genes at each of those stages. (***a***) At 1.25 hours (about 128 nuclei). Between the seventh and eighth nuclear divisions, the nuclei start to migrate to the periphery of the egg. The products of several maternal genes can be located in different regions of the egg. The crosses mark the location of maternal mRNA transcribed from a gene that defines the anterior (head) end of the egg. The dots represent the location of mRNA transcribed from a gene that specifies how cells located in the posterior of the embryo develop. The pink region represents a concentration gradient of a maternal mRNA extending from the anterior to the posterior end. The protein produced by translation of the mRNA appears to be part of a system of determinants that organize the early pattern of development in the embryo. (***b***) The pattern at 2 hours (about 1500 nuclei). Most of the nuclei have reached the perimeter of the egg and have started to make their own mRNA. The maternal mRNA shown in pink in the previous panel is now being transcribed from the corresponding zygotic gene by the nuclei in the anterior part of the embryo. The mRNA from a zygotic gap gene is transcribed from cells in only one segment in the middle of the embryo. (***c***) The pattern at 2.5 hours (about 5000 cells). Membranes start to form around nuclei located at the perimeter of the egg. Messenger RNAs from two pair-rule genes can be detected as a series of stripes around the embryo. These stripes mark the prepattern used to form the segments found in the mature larva. The boundaries of each stripe in the prepattern are defined by two or more different genes. The genes that form the prepattern affect another set of genes that defines the actual pattern of segments in the embryo. (After Akam, *Development* 101:1–22.)

For example, due to the absence of specific signals in the egg, maternal effect mutations can produce an embryo with two heads or two posterior ends. The mRNA transcripts from some of the maternal effect genes can be identified by their ability to hybridize with radioactive DNA probes from the cloned genes. Alternatively, their protein products can be identified by antibodies that specifically bind to them. In some cases, the mRNA or its protein product can be seen to form a concentration gradient in the embryo (Fig. 16–8). These gradients may provide positional information that specifies the fate of each nucleus or cell within the embryo. That information may then be interpreted by a cell as signals specifying the developmental path it should follow.

In many cases, the phenotype associated with a maternal effect mutation can be reversed by injecting normal maternal mRNA into the mutant embryo. When this is done, the fly develops normally, indicating that the gene product is needed only for a short time at the earliest stages of development.

Zygotic segmentation genes continue and extend the developmental program

Immediately after fertilization, the zygote nucleus in the *Drosophila* egg divides, beginning a remarkable series of 13 mitotic divisions (Fig. 16–8*b*). Each of these divisions takes only 5 or 10 minutes, which means that the DNA

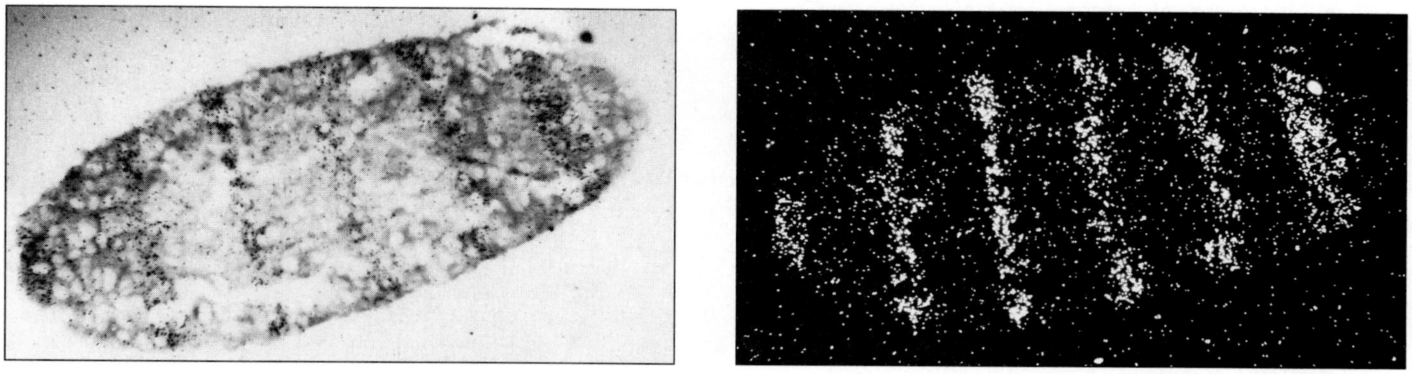

FIGURE 16–9 mRNA molecules transcribed from a specific gene may be present in a regular pattern in an embryo. The illustrations indicate the locations of the mRNA molecules transcribed from the *engrailed* gene, a gene that specifies the boundaries of each segment in a *Drosophila* embryo. The dots in the panels are silver grains produced on x-ray film by radioactive DNA from the cloned *engrailed* gene sequence that hybridized to its complementary mRNA in the embryo. The autoradiographs are photographed through a microscope in two ways: (***a***) through normal optics, which cause the silver grains to appear as black dots, or (***b***) through darkfield optics, which reverse the image. (Courtesy of W. J. Gehring, from A. Fjose, W. J. McGinnis, and W. J. Gehring, 1985, Reprinted by permission from *Nature* Vol. 313, p. 284; Copyright © 1996 Macmillan Magazines Limited)

in the nuclei is replicated constantly at a very rapid rate. During that time the nuclei do not synthesize RNA. Cytokinesis does not take place, and the several thousand nuclei produced by the first seven divisions remain at the center of the egg until the eighth division occurs.

At that time, most of the nuclei start to migrate to the periphery of the egg. Membranes begin to form around the nuclei in the periphery. Embryonic mRNA production begins, and some of the **zygotic genes** begin to be expressed. (It is customary to refer to the genes of the embryo itself as zygotic genes, even though the embryo is no longer a zygote.) Certain zygotic genes begin to extend the developmental program beyond the pattern established by the maternal genome.

So far geneticists have identified at least 24 **zygotic segmentation genes** that are responsible for generating a repeating pattern of segments within the embryo (Figs. 16–8 to 16–10). The segmentation genes fall into three classes — gap genes, pair-rule genes, and segment polarity genes — representing a rough hierarchy of gene action (Table 16–1).

The **gap genes** are apparently the first sets of zygotic genes to act. These genes seem to interpret the maternal anterior-posterior information in the egg and begin organization of the segments. A mutation in one of the gap genes usually causes one or more missing segments in an embryo.

The other two classes of segmentation genes do not act on small groups of segments but rather affect all of the segments. For example, mutations in the **pair-rule** genes delete every other segment, whereas mutations in the **segment polarity** class of genes produce segments in which one part is missing and the remaining part is duplicated as a mirror image. The effects of the different classes of mutants are summarized in Figure 16–11.

Each gene can be shown to have distinctive times and places in the embryo in which it is active (Figs. 16–9 and 16–10). The observed pattern of expression of the maternal and zygotic genes controlling segmentation indicates that cells destined to form adult structures are determined by a progressive series of developmental decisions. First, the anterior-posterior (head-to-tail) axis and the dorsal and ventral regions of the embryo are determined by maternal segmentation genes thought to form

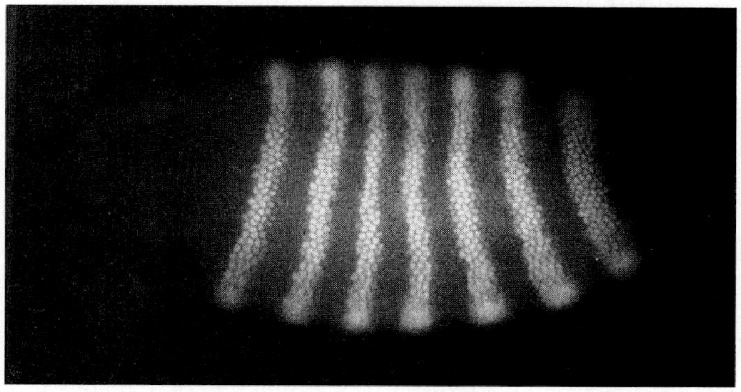

FIGURE 16–10 The activity of a gene that regulates development can be detected by locating its protein product using fluorescent antibody molecules. This example shows the normal pattern of expression of the protein product of one of the zygotic pair-rule loci known as *fushi tarazu* (Japanese for "not enough segments"). The protein molecules are localized in the nuclei of alternate segments. These segments are missing in mutants that lack the protein. (Courtesy of W. J. Gehring, from *Science* 236 (1987):1245–1252; Copyright 1987 by the American Association for the Advancement of Science)

Table 16–1 CLASSES OF GENES INVOLVED IN PATTERN FORMATION OF EMBRYONIC SEGMENTS IN *DROSOPHILIA*

Type of Gene	*Site of Gene Activity*	*Effects of Mutant Alleles and Proposed Function(s) of Genes*
Maternal effect genes	Maternal tissues (ovary)	Many maternal effect mutations alter the polarity of the embryo; initiate pattern formation by activating regulatory genes in nuclei in certain locations in embryo
Zygotic segmentation genes		
Gap genes	Embryo	Mutant alleles cause alternate segments to be missing; some may influence activity of pair-rule genes, segment polarity genes, and homeotic genes
Pair-rule genes	Embryo	When mutated, cause parts of segments to be missing; some may influence activity of segment polarity genes and homeotic genes
Segment polarity genes	Embryo	Mutant alleles delete part of every segment; replace with mirror image of remaining structure; may influence activity of homeotic genes
Homeotic genes	Embryo	Homeotic mutations cause parts of fly to form structures normally formed in other segments; control the identities of the segments

gradients of **morphogens** in the egg. (A morphogen is a chemical agent that affects the differentiation of cells and the development of form.)

Zygotic segmentation genes then respond to the amounts of various morphogens at each location to control the production of a series of segments from the head to the posterior region. Then, within each segment, other genes are activated that read the position of the segment and interpret that information to specify which body part that segment should become. Within every segment, each cell's position is further specified so that it now has a specific "address," that is designated by combinations of the regulatory genes' activities.

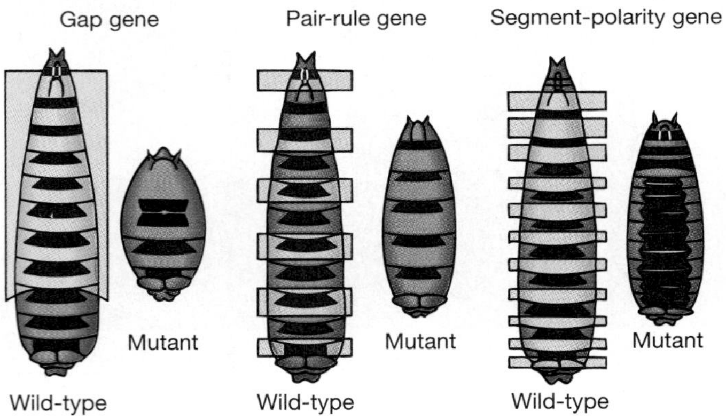

FIGURE 16–11 Three main classes of zygotic segmentation genes — gap genes, pair-rule genes, and segment-polarity genes — control the pattern of segments in a *Drosophila* embryo. The blue bands mark the regions in which the protein products of these genes are normally expressed in wild type embryos. When mutated, each gene produces a phenotype that is characteristic of the class to which it belongs. (After Nüsslein-Volhard and Wieschaus, 1980)

It is thought that the zygotic segmentation genes act in sequence, with the gap genes acting first, then the pair-rule genes, and finally the segment polarity genes. In addition, members of each group can interact with each other. Each time a new group of genes acts, cells of a particular group become more finely restricted in the way that they will develop. As the embryo develops, it is progressively subdivided into smaller specified regions.

Most, if not all, of the segmentation genes (maternal and zygotic) code for **transcription factors** (see Chapter 13). For example, some of the segmentation genes code for a "zinc-finger" type of DNA-binding regulatory protein (see Fig. 13–9). Others code for other types of transcription factors; these are discussed in the next section.

Evidence exists that many of the genes involved in the control of development code for transcription factors. This indicates that those proteins indeed act as genetic "switches" regulating the expression of other genes. Once proteins that function as transcription factors have been identified, it is possible to use the purified proteins to identify the DNA target sequences to which they bind. This approach has been increasingly useful in identifying additional parts of the regulatory pathway involved in different stages of development.

Homeotic selector genes specify the identity of each segment

One function of the zygotic segmentation genes is to regulate the expression of a separate set of genes that actually designates the final adult structure formed by each of the imaginal discs. These genes are called **homeotic** genes. Because of their involvement in segment identity, mutations in homeotic genes cause one body part to be substituted for another and therefore produce some very peculiar changes in the adult. Among the most striking examples are the *Antennapedia* mutants, which have legs that grow from the head at a position where the antennae would normally be found (Fig. 16–12).

Homeotic genes in *Drosophila* were originally identified by the altered phenotypes produced by mutant alleles. When geneticists analyzed the DNA sequences of a number of homeotic genes, they discovered a short DNA sequence of approximately 180 base pairs that is characteristic of many homeotic genes as well as some other genes that play a role in development. This sequence has been termed the **homeobox.** Using the homeobox sequence of bases as a molecular probe made it possible to clone new homeotic genes in *Drosophila* that had not been previously identified. Surprisingly, the homeobox probe has detected homologous DNA sequences in a wide range of other organisms, including humans. This finding generated considerable excitement because developmental mutants can be difficult to obtain in many organisms, especially vertebrates. The homeobox has allowed researchers to identify and clone a number of genes thought to control development in complex organisms.

The homeobox sequences of a large number of genes have been determined. Comparisons have shown that the DNA sequence itself has been highly conserved during evolution and shows remarkable similarities among organisms as diverse as sea urchins, yeasts, and humans. Each homeobox codes for a protein functional region called a **homeodomain,** consisting of 60 amino acids that form four alpha helices. One of these serves as a recognition helix, which can bind to specific DNA sequences and affect transcription. Thus the products of the homeotic selector genes, like those of the earlier-acting segmentation genes, are transcription factors. In fact, some of the segmentation genes also contain homeoboxes.

CAENORHABDITIS ELEGANS IS A ROUNDWORM WITH A VERY RIGID EARLY DEVELOPMENTAL PATTERN

Caenorhabditis elegans, a roundworm or nematode (see Chapter 28), has one of the simplest systems of genetic control of development. The study of this animal was begun in the 1960s by Sydney Brenner, a molecular biologist. Today it is an important tool for answering basic questions about the development of individual cells within a multicellular organism.

Even as an adult, *Caenorhabditis* is only 1.5 millimeters long and contains only about 1000 somatic cells (the exact number depends on the sex) and about 2000 germ-line cells. Individuals can be either **hermaphrodites** (organisms with both sexes in the same individual) or males. Hermaphroditic individuals are self-fertilizing, which makes it easy to obtain offspring homozygous for newly induced recessive mutations. The availability of males

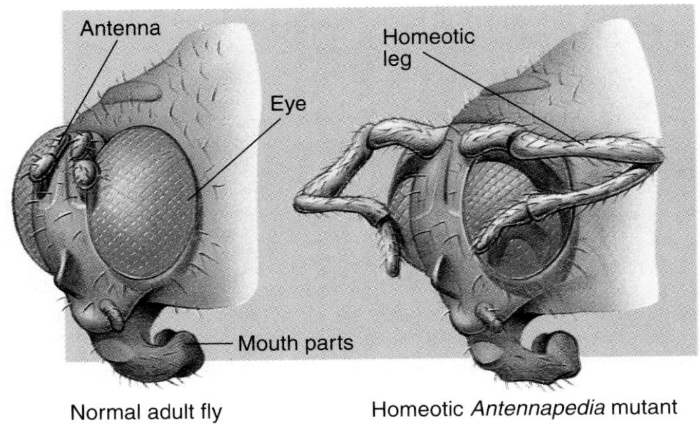

(*a*)

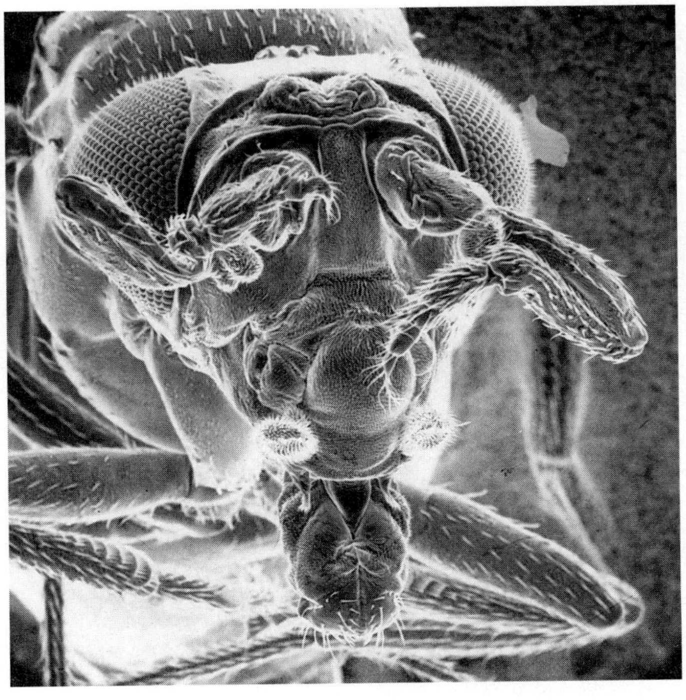

(*b*)

FIGURE 16–12 Mutant alleles of the *Antennapedia* locus of *Drosophila* cause homeotic transformations in which the antennae are replaced by legs or parts of legs. (***a***) Head of a normal fly and a fly with the *Antennapedia* mutation. (***b***) Head of a fly with a mutant allele of the *Antennapedia* locus that produces an extreme phenotype. Most of the mutant alleles produce only incomplete legs in place of the structures of the antennae. (*b*, Dr. Thomas Kaufman) that can mate with the hermaphrodites makes it possible to do genetic crosses as well.

Because the worm's body is transparent, researchers can follow the development of literally every one of its somatic cells (Fig. 16–13) using a Nomarski differential interference microscope (see Chapter 4). As a result of efforts by several laboratories, the lineage of each somatic cell in the adult has now been determined. Those studies have shown that the nematode has a very rigid developmental pattern. After fertilization, the egg undergoes repeated divisions to produce about 550 cells that make up the small, sexually immature larva. After the larva hatches from the egg case, further cell divisions give rise to the adult worm.

The lineage of each somatic cell in the adult can be traced to a single cell in a small group of **stem cells,** or **founder cells,** that are formed early in development (Figs. 16–14 and 16–15). If a particular founder cell is destroyed or removed, the structures that would normally develop from that cell are missing. An embryo with such an invariant developmental pattern is said to be highly **mosaic,** meaning that the fates of cells are largely predetermined.

It was originally thought that each founder cell gives rise to only one organ. Detailed analysis of cell lineages, however, reveals that many of the structures found in the adult, such as the nervous system and the musculature, are in fact derived from more than one founder cell. Conversely, a few lineages have been identified in which a nerve cell and a muscle cell are derived from the division of a single cell. A number of mutations affecting cell lineages have been isolated, and many of these appear to have properties that would be expected of genes involved in control of developmental decisions.

By using microscopic laser beams small enough to destroy individual cells, it is possible to determine what influence one cell may have on the development of a neighbor. Consistent with the rigid pattern of cell lineages, destruction of an individual cell in *Caenorhabditis* results, in most cases, in the absence of all of the structures derived from that cell, but with the normal differentiation of all of the neighboring somatic cells. This suggests that development in each cell is regulated through its own internal program.

However, there are cases in which differentiation of a cell can be influenced by interactions with particular neighboring cells, a phenomenon known as **induction.** One example is the formation of the vulva (pl., *vulvae*), the structure through which the eggs are laid. A single nondividing cell, called the *anchor cell,* is a part of the ovary (the structure in which the germ-line cells undergo meiosis to produce the eggs). The anchor cell attaches to the ovary and to a point on the outer surface of the animal, triggering the formation of a passage through which the eggs pass to the outside. When the anchor cell is present, cells on the surface organize to form the vulva and its opening. If the anchor cell is destroyed by a laser beam, however, the vulva does not form and the cells that would normally form the vulva remain as surface cells (Fig. 16–16). The anchor cell therefore induces the surface cells to form a vulva.

Analysis of certain cell lineage mutations has been

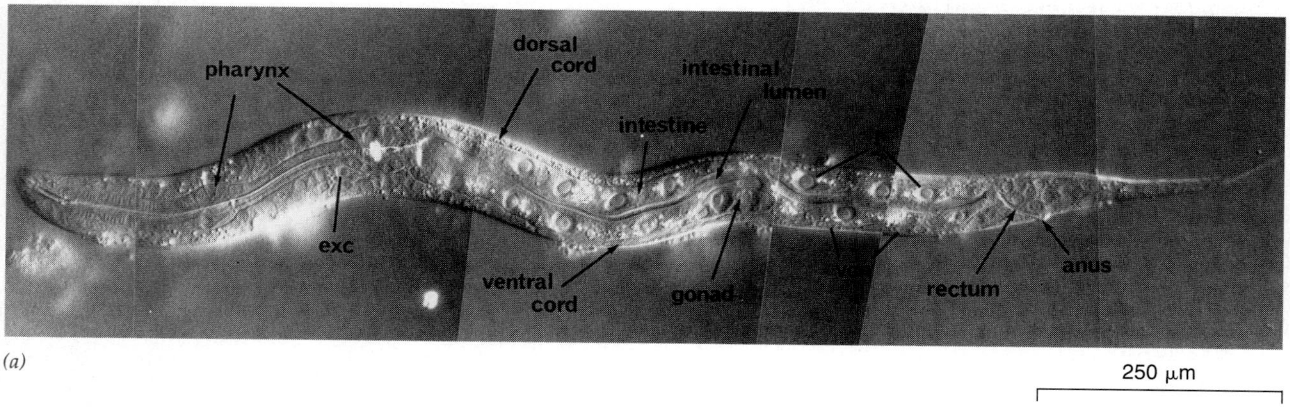

(a)

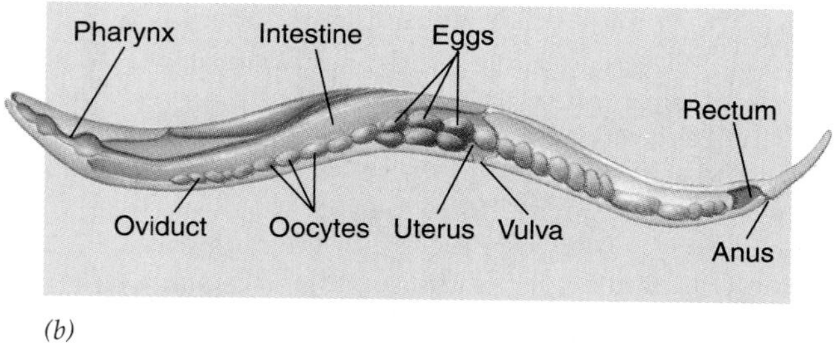

(b)

FIGURE 16–13 ***Caenorhabditis elegans*** **is a transparent organism with a fixed number of somatic cells.** (***a***) Nomarski interference micrograph of the adult hermaphrodite nematode. (***b***) Diagram illustrating structures in the adult hermaphrodite. The sperm-producing structures are not shown. (*a,* Courtesy of Dr. John Sulston, Medical Research Council, from Walbot and Holder, *Developmental Biology,* p. 607, Figure 22.6a, Random House)

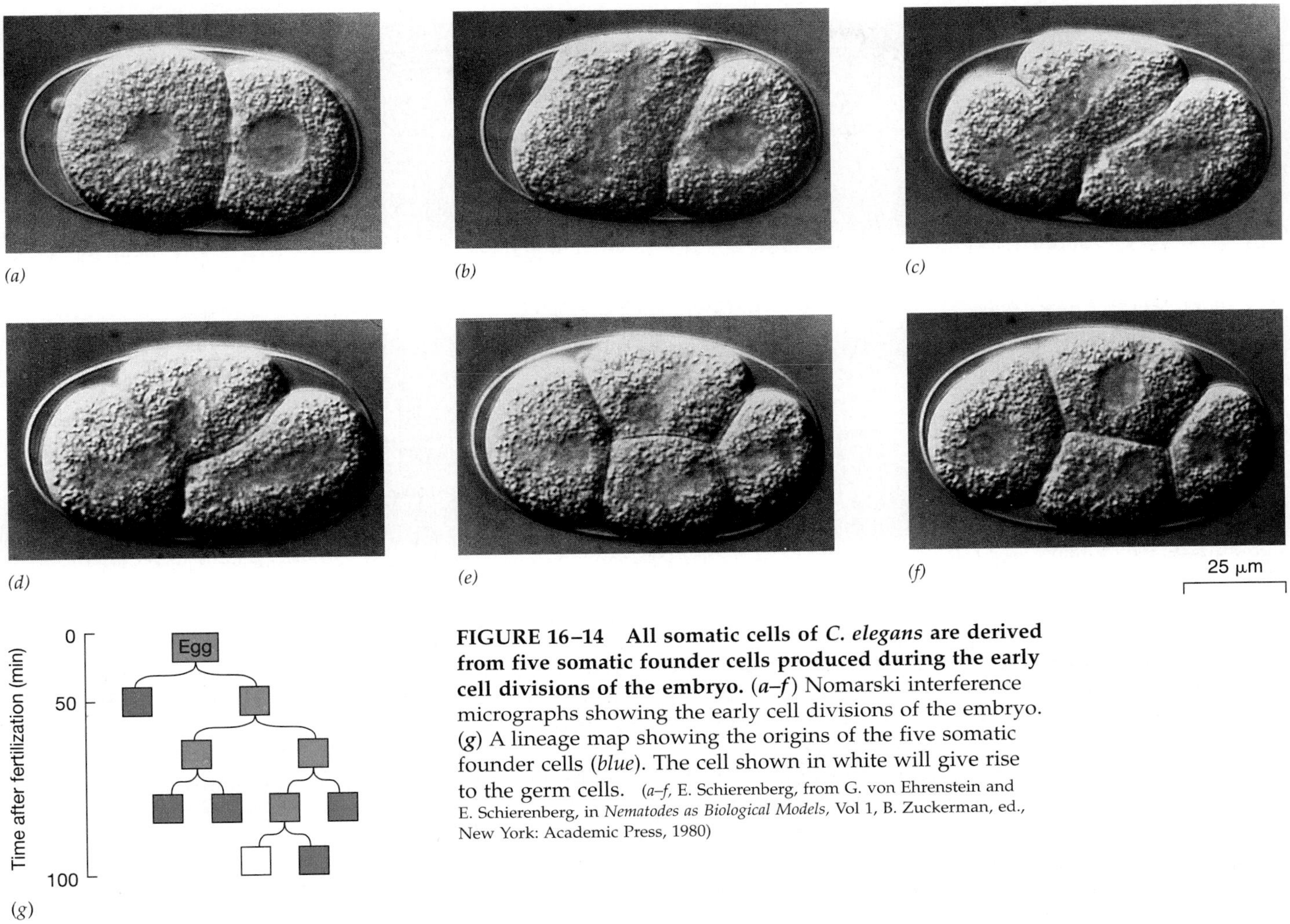

FIGURE 16–14 All somatic cells of *C. elegans* are derived from five somatic founder cells produced during the early cell divisions of the embryo. (*a–f*) Nomarski interference micrographs showing the early cell divisions of the embryo. (*g*) A lineage map showing the origins of the five somatic founder cells (*blue*). The cell shown in white will give rise to the germ cells. (*a–f,* E. Schierenberg, from G. von Ehrenstein and E. Schierenberg, in *Nematodes as Biological Models,* Vol 1, B. Zuckerman, ed., New York: Academic Press, 1980)

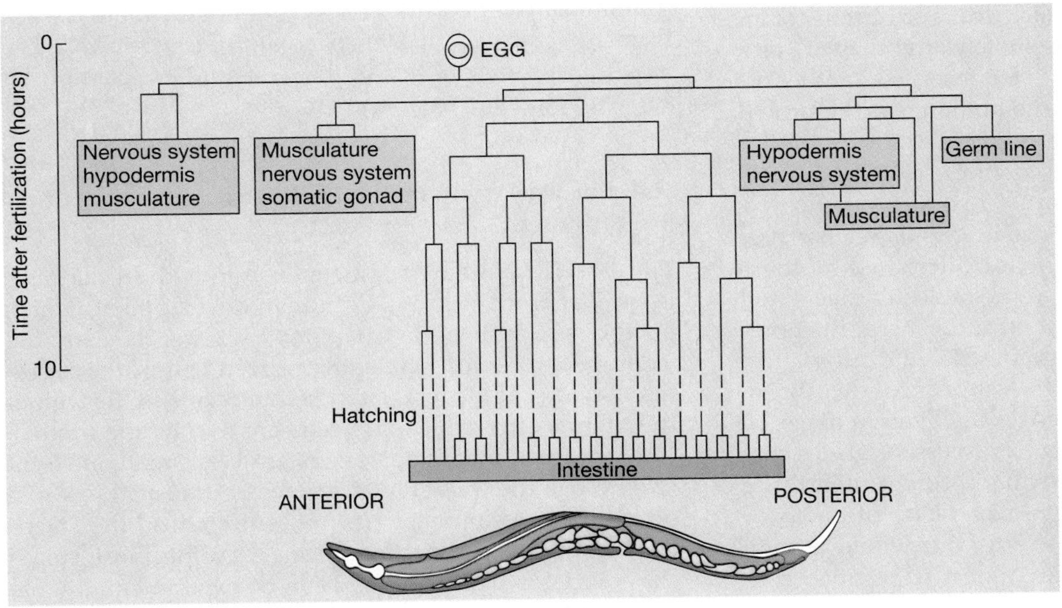

FIGURE 16–15 This lineage map traces the development of the cells that form the intestine in *C. elegans*.

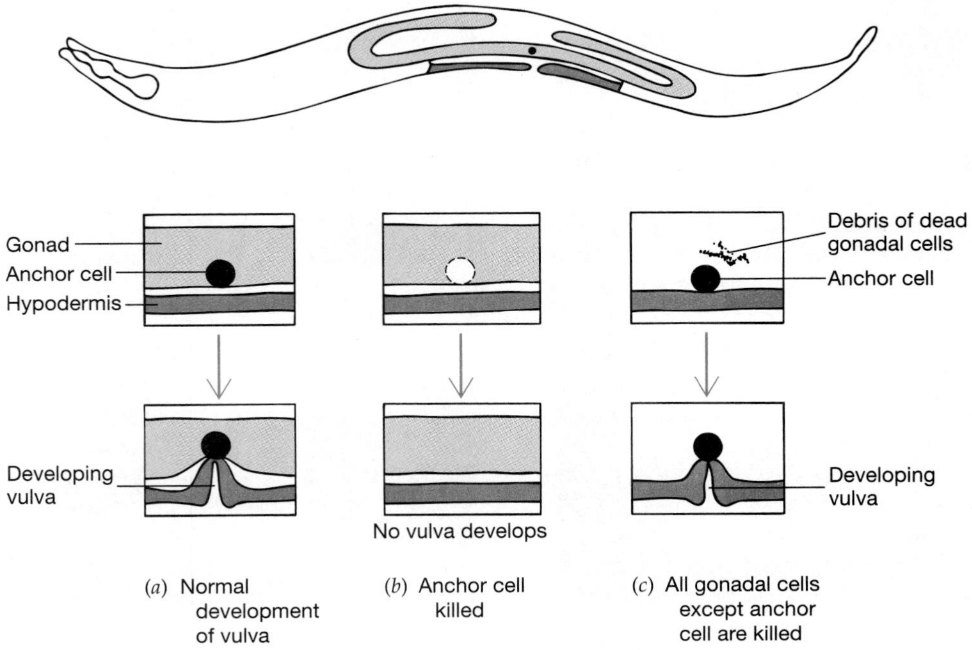

FIGURE 16–16 A single anchor cell induces neighboring cells to form the vulva in *C. elegans*. This schematic diagram shows how laser destruction of single cells or a group of cells can be used to demonstrate the influence of a cell on its neighbors.

useful in understanding such inductive interactions. For example, several types of mutations cause more than one vulva to form. In such mutant animals, multiple vulvae form even if the anchor cell is destroyed. Thus, the mutant cells do not require an inductive signal from an anchor cell to form a vulva. Evidently the gene or genes responsible for vulva formation are constitutively expressed in these mutants. Conversely, mutants lacking a vulva are also known. In some of these, the cells that would normally form the vulva appear unable to respond to the inducing signal from the anchor cell.

During development in *Caenorhabditis,* a number of instances occur in which cells die shortly after they are produced. Such phenomena have been observed in other organisms as well. For example, the human hand is formed as a webbed structure, but the fingers become individualized when the cells between them die. In *Caenorhabditis,* these **programmed cell deaths** are under genetic control and a number of mutants have been isolated that alter the pattern of these deaths. The loci identified by these mutations are being analyzed at the molecular level and should shed considerable light on the general phenomena of cellular aging and programmed cell death.

Mutations are also known that appear to identify so-called **chronogenes**—that is, genes involved in developmental timing. One such locus has recessive alleles that cause certain cells to adopt fates that would ordinarily be seen later in development. Dominant alleles of the same locus cause certain cells to adopt fates that would usually be expressed earlier. Such genes appear to be good candidates for "switches" that control developmental timing.

Genes that contain homeobox-like sequences have been discovered in *Caenorhabditis.* They are sufficiently different from the *Drosophila* homeobox genes that they were not identified by molecular probes from *Drosophila.* Now that the *Caenorhabditis* sequences can be used as probes, they are allowing the identification of additional homeobox genes in it and other organisms.

THE MOUSE IS A MODEL FOR MAMMALIAN DEVELOPMENT

Mammalian embryos develop in markedly different ways from the embryos of *Drosophila* and *Caenorhabditis.* The laboratory mouse, *Mus,* is the best-studied example of early mammalian development.

Cells of very early mouse embryos are totipotent

The early development of the mouse and other mammals is similar in many ways to human development, which is described in detail in Chapter 49. During the early developmental period, the embryo lives free in the reproductive tract of the female. It then implants in the wall of the uterus, after which its needs are met by the mother. Consequently, mammalian eggs are very small and contain little in the way of food reserves. Almost all research on mouse development has concentrated on the stages leading up to implantation because during those stages the embryo is free-living and can be experimentally manipulated. During that period, a number of critical developmental commitments take place that have a significant effect on the future organization of the embryo.

Following fertilization, a series of cell divisions gives rise to a loosely packed group of cells. It has been possible to show that all the cells in the very early mouse embryo are equivalent. For example, at the two-cell stage of mouse embryogenesis, one of the two cells can be destroyed by pricking it with a fine needle. Implanting the remaining cell into the uterus of a surrogate mother in most cases leads to the development of a normal mouse.

Conversely, two embryos at the eight-cell stage of development can be fused together and implanted into a surrogate mother, resulting in the development of a normal-sized mouse (Fig. 16–17). By using two embryos with different genetic markers (such as coat color), it can be demonstrated that the resulting mouse has four genetic parents. These mice have fur with patches of different colors derived from clusters of genetically different cells. Animals formed in this way are called **chimeras.** (The term *chimera,* derived from the name of a mythical beast that had the head of a lion, the body of a goat, and the tail of a snake, is used today to refer to any organism that contains two or more kinds of genetically dissimilar cells arising from different zygotes.) Chimeras have been important in allowing the use of genetically marked cells to trace the fates of certain cells during development.

The responses of mouse embryos to these kinds of manipulations are in marked contrast to the mosaic or predetermined nature of early *Caenorhabditis* development, in which the destruction of one of the founder cells results in loss of a significant portion of the embryo. For this reason, we say that very early development of the mouse (and presumably of other mammals) is highly **regulative.** This means that the early embryo acts as a self-regulating whole that can accommodate missing or extra parts. On the other hand, it has not been possible so far to demonstrate totipotency of either cells or nuclei from slightly later stages of mouse development.

Transgenic mice are used in studies on developmental regulation

In transformation experiments similar to those done with *Drosophila,* foreign DNA injected into fertilized mouse eggs can be incorporated into the chromosomes and expressed (Figs. 16–18 and 16–19). The resulting **transgenic** (see Chapter 14) mice have given researchers insights into how genes are activated during development. In addition, mouse genes can be inactivated ("knocked out") by the technique of targeted gene replacement (see *Making the Connection: The Genetics of Mice and Humans,* Chapter 14).

Scientists can identify a transgene (foreign gene) that has been introduced into a mouse, and determine whether it is active, by marking the gene in several ways. Sometimes a similar gene from a different species is used; its protein can be distinguished from the mouse protein by specific antibodies. It is also possible to construct a "hybrid gene" that contains the regulatory elements of a mouse gene of interest together with part of another gene

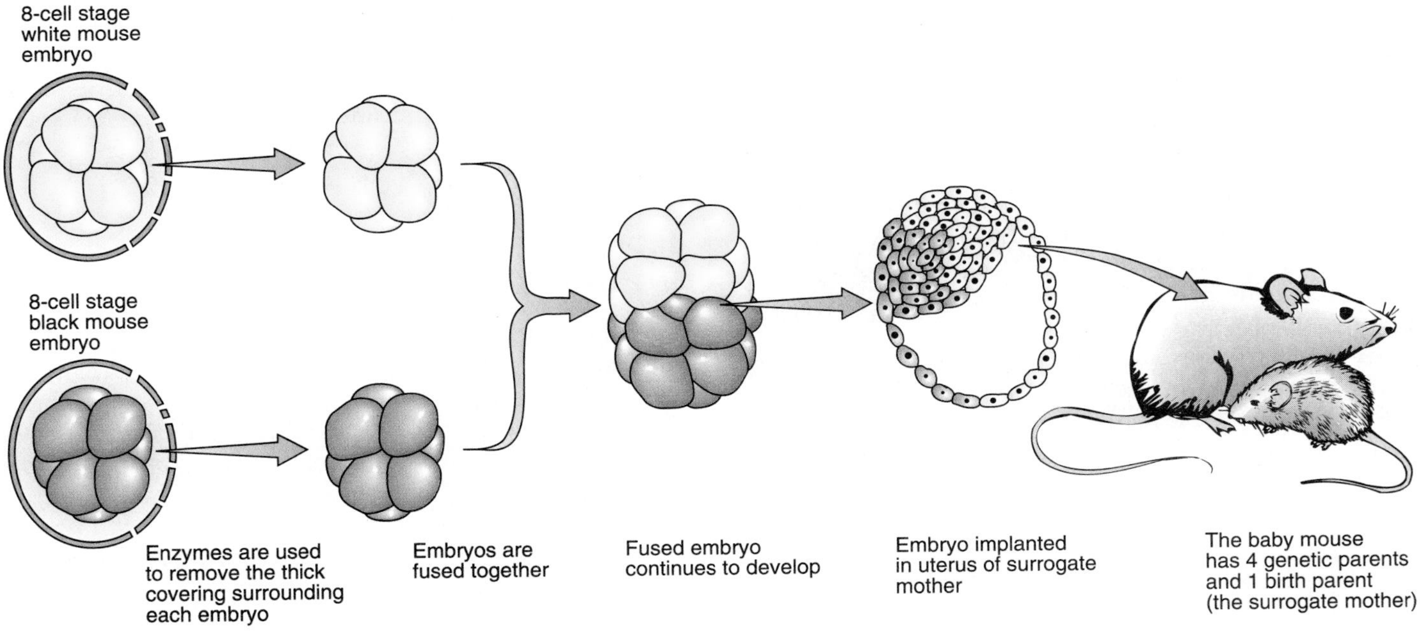
8-cell stage white mouse embryo
8-cell stage black mouse embryo
Enzymes are used to remove the thick covering surrounding each embryo
Embryos are fused together
Fused embryo continues to develop
Embryo implanted in uterus of surrogate mother
The baby mouse has 4 genetic parents and 1 birth parent (the surrogate mother)

FIGURE 16–17 Chimeric mice can be produced by removing embryos from females of two different strains and combining the cells in vitro. The resulting aggregate embryo continues to develop and is implanted in the uterus of a surrogate mother. The offspring has four different genetic parents. Although the surrogate mother is the birth mother, she is not genetically related.

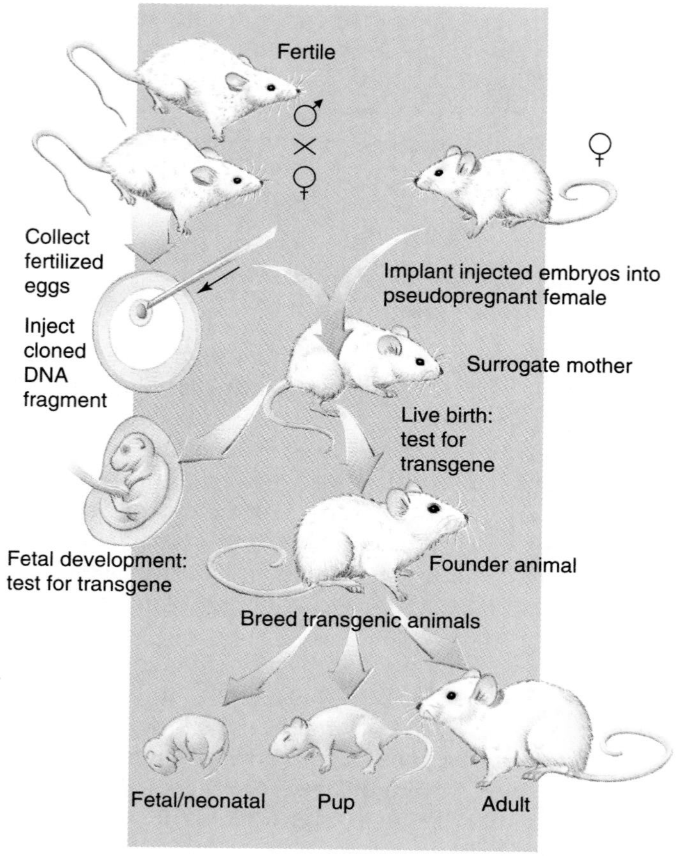

FIGURE 16–18 The production of transgenic mice is a multi-step process. Cloned DNA fragments are injected into the nucleus of a fertilized egg. The eggs are then surgically transferred to a surrogate mother. The presence of the foreign gene can be examined in the transgenic animal, or the animal can be bred to establish a transgenic line of mice.

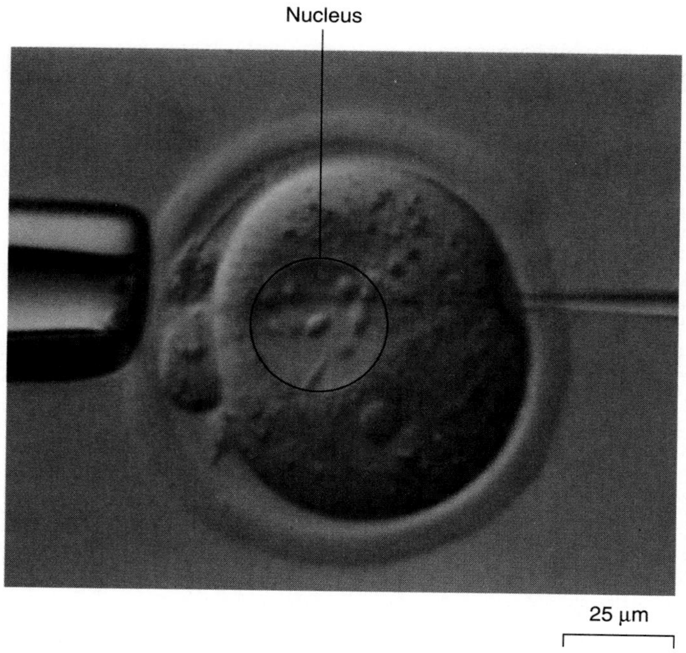

FIGURE 16–19 The first step in producing transgenic mice is the injection of fertilized mouse eggs with foreign DNA. The egg is held by suction on a holding pipet (*left*). The DNA is injected into the nucleus by the glass needle, shown entering from the right, which is about 1 μm in diameter at the tip. (R. L. Brinster, University of Pennsylvania School of Veterinary Medicine)

that codes for a "reporter" protein, such as an enzyme not normally found in the mouse. Such studies have been important in showing which DNA sequences of a mouse homeobox gene determine where the gene is expressed in the embryo.

A number of developmentally controlled genes have been introduced into mice and have yielded important information about gene regulation. Most importantly, when developmentally controlled genes from other species such as humans or rats have been introduced into mice, they are regulated in the same way that they normally are in the donor animal.

For example, when introduced into the mouse, human genes encoding insulin, globin, and crystallin—which are normally expressed in cells of the pancreas, blood, and eye lens, respectively—are expressed only in those same tissues in the mouse. The fact that these genes are correctly expressed in their appropriate tissues indicates that the signals for tissue-specific gene expression are highly conserved through evolution. This is an exciting finding because it means that information on the regulation of genes controlling development in one organism can have valuable applications to other organisms such as humans.

HOMEOTIC-LIKE MUTATIONS OCCUR IN PLANTS

Certain well-characterized plants are also being used in the study of the genetic control of development. Many of these are economically important crop plants such as corn, *Zea mays.* A number of genes with developmental effects are known in corn, including some that can be thought of as analogous to the homeotic genes of *Drosophila.*

Another plant being used increasingly to study genetics and development in plants is a member of the mustard family, *Arabidopsis.* Although *Arabidopsis* itself is of no economic importance, it has a number of advantages for research. The plant is quite small, so thousands of individuals can be grown in limited space. Chemical mutagens can be used to produce mutant strains, and a number of developmental mutants, including some that have

(Text continues on page 390)

Evolution of Gene Complexes That Control the Body Plan

Why do scientists use a wide variety of organisms to study genes that control development? Not only do certain organisms have particular advantages for certain types of investigations, but comparisons among organisms are providing important (and often surprising) insights into evolutionary relationships. This approach is exemplified by studies on clusters of homeobox-containing genes, which were initially discovered in *Drosophila*. These genes of *Drosophila* and other invertebrates are called **HOM genes;** related genes of vertebrates are referred to as **Hox genes.**

The HOM genes of *Drosophila* form two adjacent groups on the chromosome, the *Antennapedia* complex and the *bithorax* complex. As homeobox-containing genes have been identified in other animals, it has been found that these genes are also clustered and their organization is remarkably similar to that seen in *Drosophila*. The figure compares the organization of the HOM gene clusters of *Drosophila* and *Caenorhabditis* with the Hox genes of the mouse. These images are matched with the regions where they are expressed in the animals. Remarkably, the *Drosophila* HOM genes and the equivalent mouse Hox genes are located in the same order along the chromosome, although the correlation is less clear for *Caenorhabditis*. Furthermore, the order of the genes on the chromosome reflects the order of the corresponding segments they control (from anterior to posterior) in the animal. This organization apparently reflects the need for these genes to be transcribed in a specific sequence.

Drosophila has only the *Antennapedia/bithorax* HOM complex. However, vertebrates have four similar Hox complexes, each located in a different chromosome. These complexes probably arose through gene duplication. The fact that extra copies of these genes are present may explain why mutations causing homeotic-like transformations (e.g., the substitution of a leg for an arm) have not been found in vertebrate animals. It is extremely unlikely that corresponding loci in all four complexes would be mutated simultaneously.

It is now thought that the homeobox genes are generally responsible for specifying position in the developing animal embryo, particularly the position along the anterior-posterior axis. The fact that very similar developmental controls are seen in organisms as diverse as insects, unsegmented roundworms, and vertebrates (including humans) indicates that the basic mechanism evolved early and has been highly conserved in all animals that have an anterior-posterior axis, even those that are not segmented (see Chapters 28, 29, and 30). The system has apparently been modified in segmented animals such as insects and vertebrates to provide for control of segmentation and specification of segment identity.

The idea that homeobox genes are involved in specifying position in the embryo has been strengthened by findings that they not only control the formation of the body axis, but also have a role in vertebrate limb development. It is becoming clear that once a successful way of controlling groups of genes and integrating their activities evolved, it was retained; although it has apparently been modified in various ways to provide for alterations of the body plan.

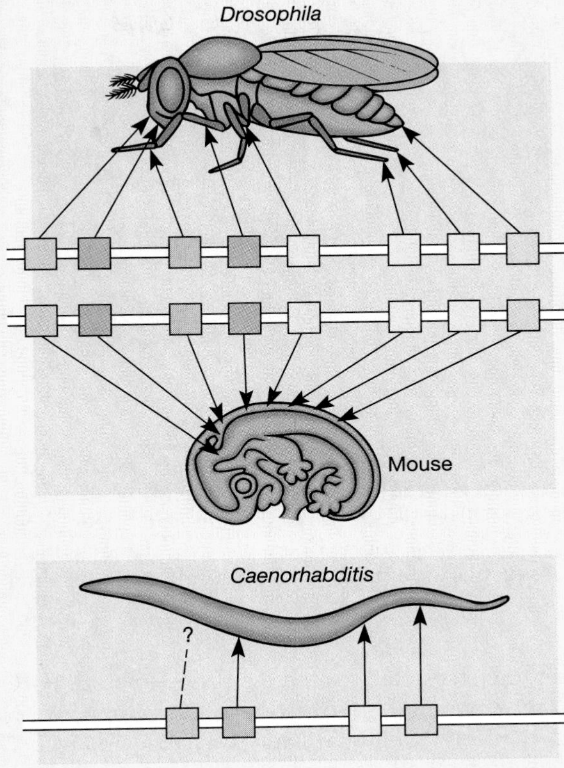

A comparison of the organization of the HOM gene cluster in *Drosophila* and one of the four Hox gene clusters of the laboratory mouse, correlated with the parts of the body in which each gene is expressed. (Although not evident in the figure, some of these regions of expression overlap.) Note that in each organism the order of the genes on the chromosome reflects their spatial order of expression in the embryo. The most anteriorly expressed genes are shown to the left, while those expressed most posteriorly are at the right. *Caenorhabditis* also has clustered HOM genes, although their relationships are less well understood. (After Lenyon et al., *Science* 253: 516)

The finding of homeobox-like genes in plants suggests that these genes may have had a very ancient origin and may in fact be the genes that made multicellularity possible. Further investigations may allow researchers to develop an overall model of how the rudiments of morphogenesis are controlled in both plants and animals. These systems of master genes that control development are proving to be a rich source of "molecular fossils" that are illuminating evolutionary history in new and exciting ways. ▲

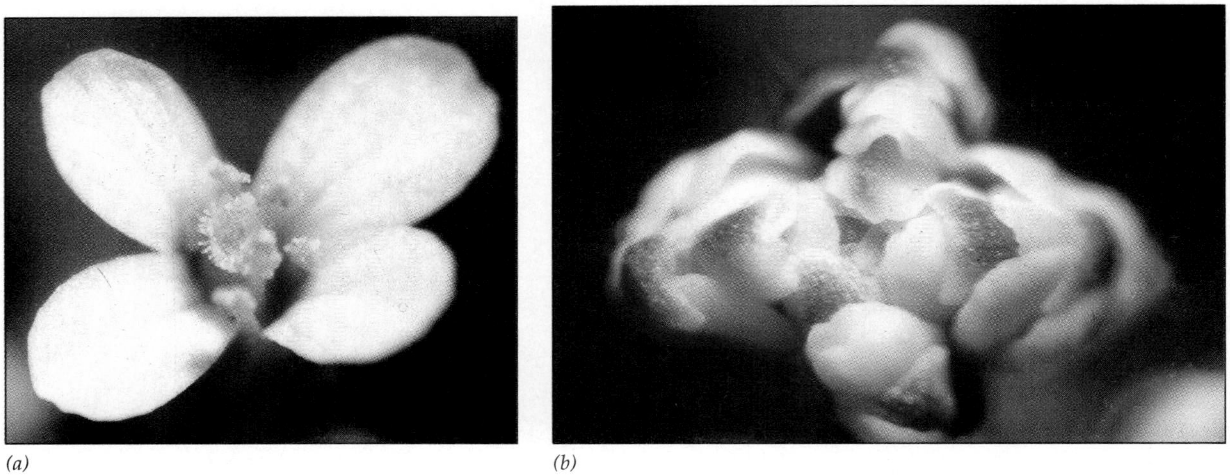

FIGURE 16–20 **Numerous homeotic transformations of flowers have been studied in *Arabidopsis.*** (*a*) A normal flower of *A. thaliana,* which has four outer leafy green sepals (hidden by the petals), four white petals, six stamens (the male reproductive structures), and a central pistil (the female reproductive structure). (*b*) This homeotic mutant has only sepals and petals because the gene normally responsible for the development of the stamens and pistil is missing. (Dr. Elliot Meyerowitz, California Institute of Technology)

homeotic-like characteristics, have been isolated. For example, the genes that control flower development have been particularly well-characterized (Fig. 16–20). The plant has a very small and simple genome, which greatly facilitates cloning of genes. In addition, cloned foreign genes can be inserted into *Arabidopsis* cells, and these can be integrated into the chromosomes and expressed. These transformed cells can be induced to differentiate into transgenic plants.

Several homeotic-like genes in plants have been shown to code for transcription factors. For example, homeobox-containing genes have been shown to be involved in the development of the shoot (above-ground portion) of the corn plant. Similarly, the genes that specify the identities of the parts of the *Arabidopsis* flower code for transcription factors, although they do not contain homeoboxes. Now that suitable molecular probes are available from plants, many more such genes should be identified in a wide range of organisms. This information will lead to a deeper understanding of the functions and evolutionary history of these genes (see *Making the Connection: Evolution of Gene Complexes That Control the Body Plan*).

SOME EXCEPTIONS TO THE PRINCIPLE OF NUCLEAR EQUIVALENCE HAVE BEEN FOUND

Although the concept of nuclear equivalence appears to apply to most cells in higher organisms, certain types of developmental regulation can involve physical changes in the DNA. Such changes in the structure of the genome are not common.

Genomic rearrangements involve structural changes in the DNA

The activity of some genes may be modified during development by different types of **genomic rearrangements** that lead to actual physical changes in the structure of the gene. In some cases, parts of genes are rearranged to make new coding sequences. This is an im-

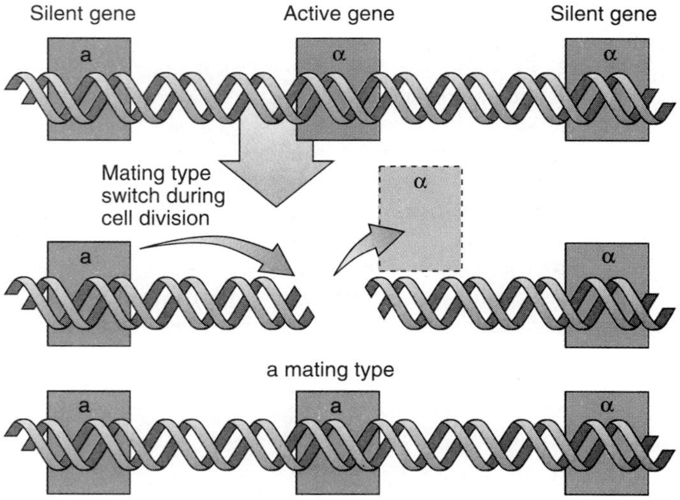

FIGURE 16–21 **Mating type switching in yeast involves rearrangements of the genes in the chromosome.** The active form of the mating type gene resides at the *MAT* locus near the center of the chromosome. Silent copies of the ***a*** and *α* genes are located near either end of the chromosome. During cell division, a copy of the opposite mating type is transferred to the MAT locus, resulting in the reversal of the mating type of the cell by the new resident gene.

portant mechanism for the development of the immune system (see Chapter 43).

Another type of rearrangement involves the replacement of an active gene with a copy of a "silent" gene located on a different part of the same chromosome. The baker's yeast *Saccharomyces cerevisiae* is a simple eukaryote that has two sexes or mating types called ***a*** and **α.** The mating type of a cell is determined by an active gene located close to the middle of one of the yeast chromosomes; this is called the mating type locus. At some distance on either side of the active gene are two silent genes called *MAT a* and *MAT* α. If a copy of the *MAT a* gene occupies the mating type locus, the mating type is *a*; if a copy of the *MAT* α gene occupies that site, the mating type of the cell is α. These yeast strains can switch their mating type from one form to the other as frequently as every generation (Fig. 16–21). The gene located at the mating type locus is removed and replaced by a DNA sequence copied from the silent gene that corresponds to the opposite mating type.

A somewhat similar system of gene replacement takes place in the unicellular parasite *Trypanosoma brucei,* which causes sleeping sickness in humans and related diseases in other animals (Fig. 16–22). When the parasite infects

(Text continues on page 394)

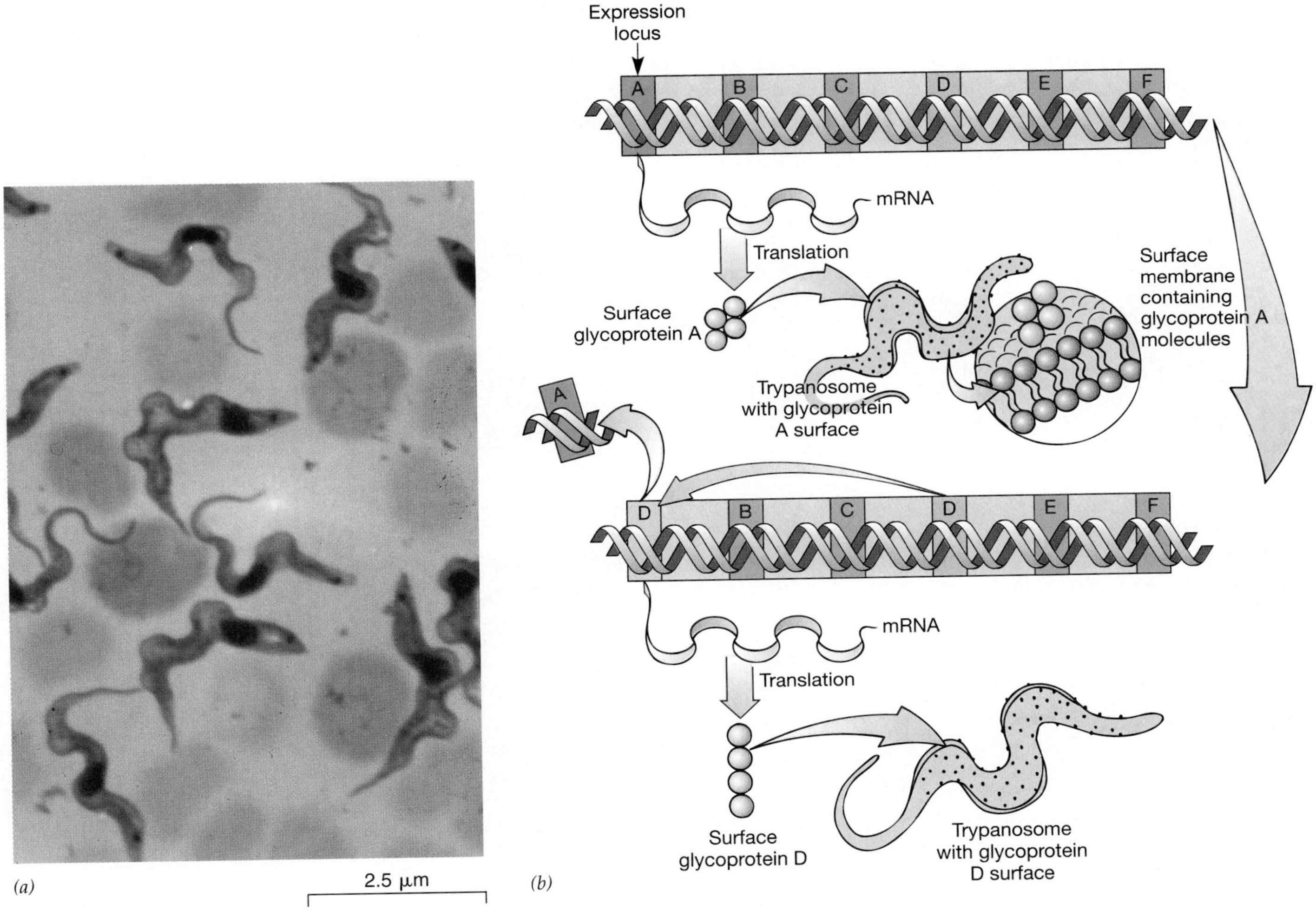

FIGURE 16–22 Trypanosomes rearrange the genes coding for the molecules on their cell surfaces. (*a*) Blood infected with *Trypanosoma brucei,* which is a single-celled protozoon (see Chapter 24) carried by the tsetse fly in Africa. This organism causes sleeping sickness in humans and nagana in cattle. (*b*) Trypanosomes change the molecules coating their surfaces frequently and are thus able to "outrun" the immune system of their human host. Each cell contains as many as a thousand silent genes, each coding for a different surface coat glycoprotein. Only one of those genes, which is located at a position near the end of a chromosome, called the expression site, is active at any one time. As the trypanosomes multiply in the blood, occasionally a copy of one of the silent genes replaces the gene currently in the expression site, leading to the production of a new group of organisms with a new surface glycoprotein. These new trypanosomes appear every 7 to 10 days, preventing the immune system of the body from developing antibodies against the surface protein in time to defeat the infection. (*a,* Ed Reschke)

Focus On

Oncogenes and Cancer

A cancer cell lacks normal biological inhibitions. Normal cells are tightly regulated by control mechanisms that cause them to divide when necessary and prevent them from growing and dividing at inappropriate times. Cells of many tissues in the adult are normally prevented from dividing; they reproduce only to replace a neighboring cell that has died or become damaged. Cancer cells have escaped such controls and can divide continuously.

As a consequence of their abnormal growth pattern, some cancer cells eventually form a mass of tissue called a **tumor.** If the tumor remains at the spot where it originated, it can usually be removed by surgery. One of the major problems with certain forms of cancer is that the cells can escape from the controls that maintain them in their proper location. These cells can **metastasize,** or spread, to different parts of the body, invading other tissues and forming multiple tumors. Lung cancer, for example, is particularly deadly because its cells are highly metastatic and can enter the blood and spread to form tumors in other parts of the lungs, or in other organs such as the liver and the brain. Tumors with cells that can metastasize are referred to as **malignant tumors.**

We now know that cancer is a disease caused by altered gene expression. Using recombinant DNA methods, researchers identified many of the genes that transform normal cells into cancer cells when they function abnormally. Each kind of cancer cell apparently owes its traits to at least one, and possibly several, of a relatively small set of genes known as **oncogenes** (cancer-causing genes). Oncogenes arise from changes in the expression of certain genes called **proto-oncogenes,** which are *normal* genes found in all cells and involved in the control of growth and development.

Oncogenes first were discovered in viruses that can infect mammalian cells and transform them into cancer cells (malignant transformation). Such viruses can incorporate DNA sequences of the cellular proto-oncogenes into their own nucleic acid. In some cases, the viruses alter the expression of the proto-oncogenes, converting them into oncogenes. This may happen if the DNA sequences come under the control of viral regulatory elements, which cause the gene to be transcribed at much higher than normal levels, or if the captured gene mutates so that its protein product is more active than the product of the normal proto-oncogene.

A proto-oncogene in a cell that has not been infected by a virus can also mutate and become an oncogene. One of the first oncogenes identified was isolated from a bladder tumor. In the cell that gave rise to the tumor, a proto-oncogene had undergone a single base-pair mutation; the result was that the amino acid glycine was replaced by a valine in the protein product of the gene. This subtle change was apparently a critical factor in the conversion of the normal cell into a cancer cell.

Not all oncogenes, however, code for proteins with amino acid substitutions. The position of the oncogene on the chromosome may also be an important factor in malignant transformation. Some cancers occur more frequently in individuals with certain chromosomal abnormalities. For example, persons with Down syndrome (see Chapter 15) are unusually susceptible to leukemia. It is possible that cellular proto-oncogenes are activated into oncogenes when they become associated with new regulatory regions by chromosome breakage or rearrangements, or when they become duplicated, which results in overexpression of the gene.

By means of recombinant DNA technology and other techniques of molecular biology, it has been possible for researchers to identify more than 60 oncogenes and their corresponding proto-oncogenes. Because the fundamental controls of normal cell division and differentiation probably evolved very early in the evolutionary history of eukaryotes, it is not surprising that very similar proto-oncogenes have been found in a diverse array of organisms, ranging from yeasts to humans. For example, the proto-oncogene counterpart of the oncogene found in some bladder tumors (mentioned previously) has also been found in yeast cells.

Some of these controls are illustrated in greatly simplified form in the accompanying figure. The growth and division of cells can be triggered by one or more external signal molecules (see *Making the Connection: Information Transfer Across the Plasma Membrane,* Chapter 5). These substances, known as **growth factors,** bind to specific **growth factor receptors** associated with the cell surface, initiating a cascade of events inside the cell. Often the growth factor-receptor complex acts as a **protein kinase** (an enzyme that phosphorylates proteins), which then phosphorylates specific amino acids of a number of cytoplasmic proteins. This posttranslational modification usually results in the activation of previously inactive enzymes. These activated enzymes are then able to catalyze the activation of certain nuclear proteins, many of which are **transcription factors.**

Activated transcription factors bind to their DNA targets and stimulate transcription of specific sets of genes that initiate growth and cell division. Even in the simplified scenario presented in the figure, it is evident that multiple steps are required to control cell proliferation. Remarkably, the

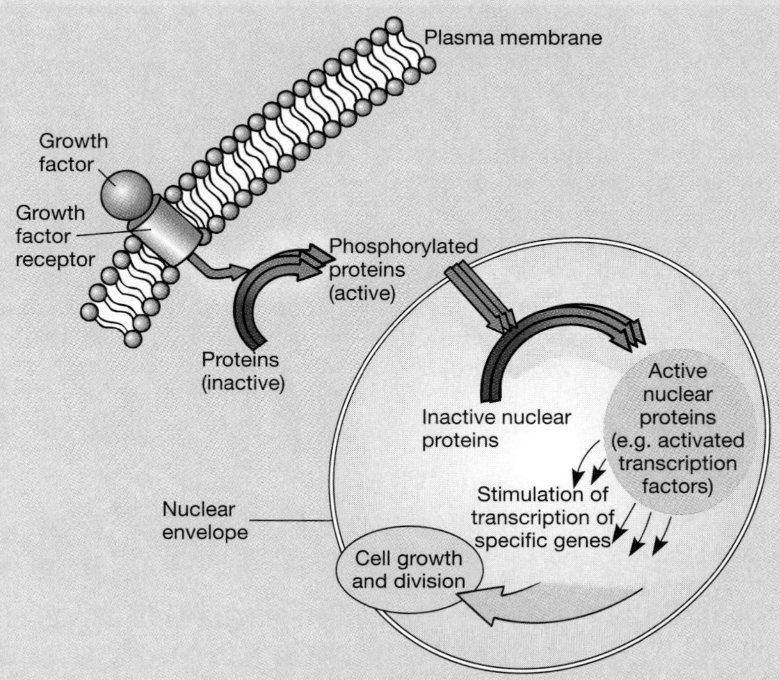

This is a simplified view of part of a growth control cascade in which a growth factor stimulates cell growth. The growth factor receptor, as well as some of the other components of the system, are coded for by proto-oncogenes. When a proto-oncogene mutates, becoming an oncogene, the cell grows and divides even in the absence of the growth factor. Conversely, other growth control cascades are regulated by growth inhibiting factors. The receptor for the growth inhibiting factor, as well as other parts of the system, are coded for by tumor suppressor genes. When a tumor suppressor gene mutates, the cell grows and divides even if the growth inhibiting factor is present.

proto-oncogenes that encode the products responsible for a great many of these steps have been identified. The current list of known proto-oncogenes includes genes that code for various growth factors or growth factor receptors, and genes that respond to stimulation by growth factors (including a number of transcription factors). When one of these proto-oncogenes is expressed inappropriately, the cell may misinterpret the signal and respond by growing and dividing.

Not all genes that cause cancer when mutated are proto-oncogenes. Some apparently interact with growth inhibiting factors to block cell division; these are sometimes referred to as **tumor suppressor genes (anti-oncogenes).**

Certain oncogenes appear to be particularly common and are found in a variety of tumors. However, a change in a single proto-oncogene is usually insufficient to cause a cell to become malignant. The development of cancer is usually a multistep process involving both oncogenes and mutated tumor suppressor genes. As more oncogenes and anti-oncogenes are discovered and their complex interactions are unraveled, we will gain a fuller understanding of the control of growth and development. This understanding is leading to improved diagnosis and treatment of various cancers. ■

humans, it is able to defeat the immune system by constantly changing the glycoprotein molecules that are exposed on the surface of its cell.

Unlike yeast, which has only two basic copies of the mating type gene per cell, the trypanosome cell contains as many as 1000 different genes for cell surface molecules. These have amino acid sequences that are so different that an antibody that recognizes one of them would not recognize another. Only one or a few of those copies are expressed at any one time, depending on which copy is present at an **expression site,** which is usually located near the end of a chromosome. The genes in the expression site are exchanged in about one out of every 10^4 to 10^6 cells, providing a constant supply of new cells that cannot be recognized by the immune system and thus maintaining the infection. Although gene replacement clearly offers a mechanism that could serve as a regulatory "genetic switch," it is not known at present whether these mechanisms are relevant to development in multicellular eukaryotes.

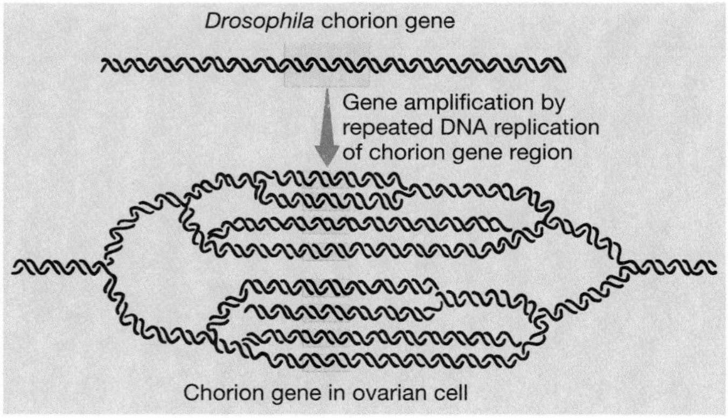

FIGURE 16–23 *Drosophila* chorion genes are amplified during development. This is accomplished by multiple replications of a small region of the chromosome containing the chorion protein genes. Replication is initiated at a discrete chromosome origin of replication for each copy of the gene that is produced and is randomly terminated, resulting in a series of forked structures in the chromosome.

Gene amplification increases the number of copies of specific genes

Some gene products are required in such large quantities during certain stages of development that a single copy of a gene cannot be transcribed, nor can its mRNA be translated, rapidly enough to fill the needs of the developing cells. In certain cases, the number of gene copies may be increased, through a process known as **gene amplification,** to meet the demand. For example, the *Drosophila* chorion (eggshell) gene product is a protein made specifically in cells of the insect oviduct. These cells make massive amounts of the particular protein that envelops and protects the fertilized egg. The demand for chorion mRNA in those cells is met by specifically amplifying the gene by DNA replication so that the DNA in that small region of the chromosome is copied many times (Fig. 16–23). In other cells of the insect body, however, the gene appears to exist as a single copy in the chromosome.

THE STUDY OF DEVELOPMENTAL BIOLOGY PRESENTS MANY FUTURE CHALLENGES

Scientists are now beginning to learn how genes are activated, inactivated, and modified, and how batteries of master regulatory genes interact to control development in a wide variety of organisms. Eventually we hope to understand not only how differentiation and morphogenesis are controlled, but also how the basic control systems have evolved. The identification of certain features common to many organisms, such as homeobox genes, will make the task easier, but we have only scratched the surface. Many complex interactions remain to be explored, and many revelations await us.

SUMMARY

I. Development is the process by which the descendants of a single cell specialize and organize into a complex organism.
   A. An organism contains many types of cells that are specialized both structurally and chemically to carry out specific functions. These cells are the product of a process of gradual commitment, called cellular determination, which ultimately leads to the final step in cell specialization, called differentiation.
   B. Morphogenesis, the development of form, occurs through stages, referred to as pattern formation, in which the various specialized cells become organized into structures.
   C. There is no evidence that genes are normally lost during most developmental processes.
      1. At least some nuclei from differentiated plant and animal cells contain all the genetic material that would be present in the nucleus of a zygote and are therefore said to be totipotent.

2. Nuclear equivalence is the concept that all (with a few exceptions) of the nuclei of the differentiated somatic cells of an organism are identical to each other, and to the nucleus of the single cell from which they descended.
3. Differences among various cell types are apparently due to differential gene activity.

II. Several organisms have characteristics that make them especially useful in studies of the genetic control of development.

A. Many types of developmental mutants have been identified in the fruit fly, *Drosophila melanogaster.* Many of these affect the segmented body plan of the organism.

1. The earliest developmental program to operate in the egg is established by maternal genes; these are active prior to fertilization, and some affect the segmentation pattern that is progressively established in the embryo.
2. Zygotic segmentation genes do not become active until much later, at a time when the embryo is no longer a zygote. They continue and extend the developmental program initiated by the maternal effect genes.
3. The zygotic genes and their products interact with each other and with the products of the maternal genes according to a hierarchical pattern, with certain earlier-acting genes controlling particular later-acting genes.
4. The later-acting homeotic selector genes are responsible for specifying the identity of each segment.
5. Many of the segmentation genes are known to code for transcription factors. Some of these contain a DNA sequence called a homeobox, which codes for a protein with a DNA-binding region called a homeodomain.

B. *Caenorhabditis elegans* is a roundworm with an extremely rigid developmental pattern in which the fates of cells are largely predetermined.

1. The lineage of every somatic cell in the adult is known, and each can be traced to a single founder cell in the early embryo.
2. A number of mutations affecting cell lineages have been identified, and many of these appear to identify genes that control developmental processes such as induction (developmental interactions with neighboring cells), programmed cell death, and developmental timing.

C. The laboratory mouse, *Mus,* is extensively used in studies of mammalian development.

1. In contrast to *Caenorhabditis,* mouse development is highly regulative, which means that the very early embryo is a self-regulating whole, and hence that an embryo containing extra or missing cells can still develop normally.
2. Transgenic mice have been extremely useful in determining how genes are activated and regulated during development.

D. Genes affecting the developmental pattern have also been identified in certain plants, including *Zea mays* (corn) and *Arabidopsis.* Some of these have been shown to contain homeobox-like sequences or to have other characteristics that make them good candidates for genetic switch genes.

E. Some homeobox genes are organized into complexes that appear to be systems of master genes specifying an organism's body plan. Remarkable parallels exist between the homeobox complex of *Drosophila* and those of other animals, including the laboratory mouse and *Caenorhabditis.*

III. A few exceptions to the general rule of nuclear equivalence are known. Among these are physical rearrangements of the DNA of the genome, and amplification of certain genes to provide more copies for transcription.

SELECTED KEY TERMS

anti-oncogene
cellular differentiation
chimera
chromosome walking
determination
development
differential gene expression
expression site
founder cell
gap gene
gene amplification
genomic rearrangement
growth factor
hermaphrodite
homeobox
homeodomain
homeotic gene
imaginal disc
imago
induction
larva (larvae)
malignant
maternal effect gene
metastasis
morphogen
morphogenesis
mosaic development
nuclear equivalence
oncogene
pair-rule gene
pattern formation
polytene chromosome
programmed cell death
protein kinase
proto-oncogene
puff
pupa (pupae)
regulative development
segment polarity gene
segmentation gene
somatic cell
stem cell
totipotent
transcription factor
transformation
transgenic organism
tumor
tumor suppressor gene
zygotic gene

POST-TEST

1. Cells become specialized or ________ by a gradual process called ________ .
2. The development of form is called ________ . This occurs through a series of stages known as ________ .
3. The idea that the nuclei of essentially all cells of an organism contain the same genetic information is referred to as the concept of ________ ________ .
4. If a nucleus is found to contain all the information required to support normal development from the embryonic stages to the adult, it is said to be ________ .
5. Differences among differentiated cells with regard to their structures, functions, and molecular compositions are generally attributed to differential ________ ________ .
6. The ________ chromosomes of *Drosophila* make it relatively easy to establish the physical position of a gene.
7. Prior to fertilization, the *Drosophila* egg is organized by the activity of certain ________ ________ genes.
8. ________ genes act after the embryo has formed.
9. Chemical agents that affect cellular differentiation and development of form are generally referred to as ________ .
10. ________ genes are involved in specifying the identities of body parts; when mutated, they generally cause one body part to be substituted for another.
11. ________ are found in certain proteins that bind to DNA.
12. Cell lineage studies in *Caenorhabditis* are meaningful because the developmental pattern in this organism is highly ________ .
13. When one cell or group of cells influences the differentiation of another, we refer to this phenomenon as ________ .
14. Early embryological development in the mouse is highly ________ , meaning that if cells are lost or added during that time, the embryo still develops normally.
15. ________ mice are important in the study of developmental regulation of certain genes.
16. Mutations that transform one plant part into another are thought to identify genes that are at least somewhat analogous to the ________ genes of *Drosophila.*
17. Physical changes in the structure of a gene (at the DNA level) that may be occasionally important in development are referred to as ________ ________ .
18. ________ ________ sometimes occurs when extra copies of a gene are required to meet a great demand for its product in certain tissues.

REVIEW QUESTIONS

1. Development consists of four main processes: cell determination, differentiation of cells, pattern formation, and morphogenesis. Define each process and describe how they relate to one another.
2. What lines of evidence support the concept of nuclear equivalence?
3. What are the relative merits of *Drosophila, Caenorhabditis, Mus,* and *Arabidopsis* as model organisms for the study of development?
4. What is the value of homeotic genes in developmental studies?
5. Describe how transgenic organisms are useful in the study of gene regulation in development.
6. Give some examples of genomic rearrangements that are known to occur as a part of some developmental processes.
7. Under what conditions are examples of gene amplification seen?
8. What are oncogenes and what is their relationship to cellular genes involved in the control of normal growth and development?

YOU MAKE THE CONNECTION

1. Why is an understanding of gene regulation in eukaryotes crucial to an understanding of developmental processes?
2. Why do developmental biologists need to understand biological diversity? Why is it necessary for scientists to study development in more than one type of organism?

RECOMMENDED READINGS

Barinaga, M. "Looking to Development's Future." *Science,* October 28, 1994. This article introduces a special section entitled "Frontiers in Biology: Development," which includes a collection of articles and perspectives by well-known developmental biologists. A wide range of organisms, including *Drosophila, C. elegans,* the mouse, and plants, are discussed.

Browder, L., C. Erickson, and W. Jeffery. *Developmental Biology,* 3rd ed. Saunders College Publishing, Philadelphia, 1991. A thorough coverage of gene expression at the molecular level, as well as experimental and descriptive aspects of development.

Cavenee W.K. and R.L. White. "The Genetic Basis of Cancer." *Scientific American,* Vol. 272, No. 3 March, 1995. Discusses the roles of both oncogenes and mutated tumor suppressor genes in malignancy.

Fletcher, C. "A Garden of Mutants." *Discover,* August 1995. A simplified discussion of the genetic control of flower development in *Arabidopsis.*

McGinnis, W. and M. Kuziora. "Molecular Architects of Body Design." *Scientific American,* Vol. 270, No. 2 February 1994. A comparison of the HOM genes of *Drosophila* and the Hox genes of vertebrates.

Meyerowitz, E.M. "The Genetics of Flower Development." *Scientific American,* Vol. 271, No. 5 November, 1994. Presents a model to explain the genetic control of flower development in *Arabidopsis.*

Watson, J. D., M. Gilman, J. Witkowski, and M. Zoller. *Recombinant DNA,* 2nd ed. W. H. Freeman and Company, New York, 1992. A discussion of the uses of recombinant DNA technology in studies on oncogenes and cancer (Chapter 18) and genes controlling *Drosophila* development (Chapter 20).

Career Visions

Director of Strategic Development

ANDREA WEISMAN

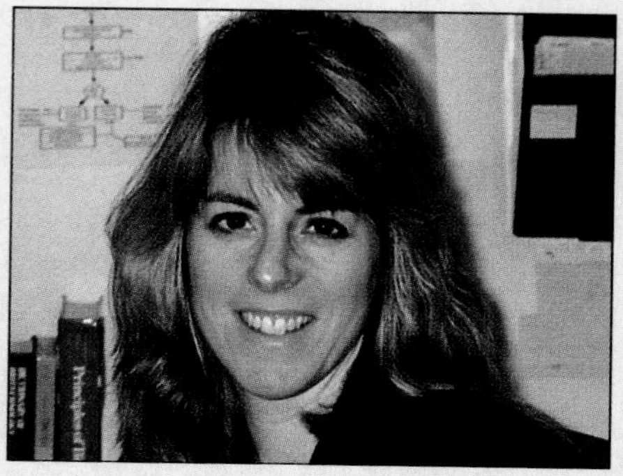

Andrea Weisman's father—a chemical engineer and inventor—sparked her interest in science with gifts of scientific texts and a microscope when she was ten years old. As director of Strategic Development at Chiron Corporation, it's Andrea's job to discover and help develop the work of scientists worldwide. Andrea has followed a circuitous path to her current position. Beginning her undergraduate educational career at the University of California at Berkeley, and continuing in the Ph.D. program at the University of California at San Francisco—followed by postdoctoral work at McGill University in Montreal, and a drug company in Paris—she fulfilled her dream of becoming a research scientist. However, after several years as a cancer researcher, she began a second career helping to organize corporate collaborations and develop scientific programs with the potential to produce new technologies and therapeutics to treat human disease.

You began your college career with the goal of becoming a research scientist. What did you do after school?

One of my dreams was to live and work in Paris, France. I accepted a postdoctoral fellowship to perform cancer research at McGill University, with the promise of spending my third year in Paris. I spent my third year as a postdoctoral fellow at Roussel Uclaf (now Roussel Hoescht), a French drug company just outside of Paris. I taught their scientists brain tumor tissue culture techniques, and, in return, they taught me molecular biology. After completing my fellowship there I was able to obtain a research position in an academic institution in Paris.

How did you wind up at Genentech?

I came back to the Bay area and tried to get a job "off the bench," (out of the research lab) but I wasn't successful. It's a catch 22; it's almost impossible to get a job off the bench if you have no experience off the bench! So I accepted a position with a small biotechnology company called Triton Biosciences (now Berlex) and performed cancer research. However, I did everything I possibly could to learn other aspects of the biotechnology industry. I organized research collaborations and worked with the intellectual property lawyers in an effort to gain experience off the bench. After a year and a half at Triton, I called up Genentech—it was literally a cold call—and asked if they had a Collaborations Program and required the services of someone like myself. Timing is everything because they were just forming a new program, and after a grueling interview process, they hired me!

Is "collaborations" just what it sounds like?

Yes. At Genentech I was actually a "New Ideas" Manager. I established academic research collaborations and helped to identify new areas of research. Every year, I attended between ten and fifteen scientific conferences worldwide. I was also the scientific liaison to Genentech's Canadian and Japanese offices. In addition, I wrote in-depth reports on new ideas in the areas of cancer and cardiovascular research. Unfortunately, after almost four years, Genentech decided to eliminate the program, but my experience there was invaluable.

What do you do at Chiron?

I can divide my job into three areas of responsibility. One third of my job involves business development. I keep my ears and eyes open for any new technologies that can be used to develop or support R&D programs. Specifically, I invite prospective corporate collaborators to present their technology to our research scientists, and I am responsible for the scientific and business reviews of these technologies. I meet with approximately 30–40 companies per year.

Another part of my job involves the identification of new technologies. I am currently focused on the discovery and development of novel delivery systems for gene therapy, which is very important because delivery is a major hurdle for potential gene therapy products. I'm currently participating in the organization of a delivery systems program at Chiron.

The final part of my job involves R&D operations. I am helping to formulate our current strategic plan. I also organize the review of our scientific programs and often arrange scientific symposia or roundtables to help generate new ideas and foster research collaborations.

What kind of background should a student have to do what you do?

In this highly specialized world, credentials are essential, so I strongly recommend that they obtain a Ph.D. or an M.D. My job also requires a broad scientific background and a lot of native curiosity. Essentially, one needs to be a scientific Sherlock Holmes. It also helps to have an intuitive sense of what is a good idea or a truly novel technology, or what we call in the industry a "nose" for discovery.

PART 4

The Continuity of Life: Evolution

Cypress (*Taxodium distichum*) trees in Lake Matamuskee Swamp, North Carolina. Cypresses, like all organisms, possess adaptations that make them well-suited for the environment in which they live. (Ian Adams/Dembinsky Photo Associates)

CHAPTER 17

Evolution: Mechanisms and Evidence

A twobar anemone fish hovers over a carpet anemone in the Red Sea, while a school of domino damsel fish swims by.
(Sharksong/M. Kazmers/Dembinsky Photo Associates)

The life forms present in vast diversity on our planet are thought to have evolved during Earth's long history from one simple kind of organism. This means that seemingly unrelated organisms such as slime molds and blue whales are in fact distantly related to one another and share a common ancestor. All of the organisms that exist today arose from earlier ones by a process of gradual divergence (separation) that Darwin originally described as "descent with modification," or **evolution.**

Evolution can be defined as genetic changes occurring over time in a population of organisms. It does not refer to changes that occur in an individual organism within its lifetime, but to changes in the characteristics of populations over many generations. Evolution involves shifts in the frequencies of alleles (see Chapter 10) in a population. All the alleles of all the genes present in a population at a given time comprise its **gene pool.** Evolution, then, is a change in the allele frequencies within a gene pool over time (see Chapter 18).

Consider the evolution of bacterial resistance to antibiotics evident in Figure 17–1. When antibiotics were first used to treat human and animal infections, it was thought that these drugs would eliminate bacterial diseases. This has not been the case, however. Bacteria are continually evolving, even inside the bodies of human and animal hosts. Each time an antibiotic is used to treat a bacterial infection, a few bacteria survive because they are genetically resistant to the antibiotic, and they pass this trait on to future generations. As a result, the bacterial population contains a larger percentage of antibiotic-resistant bacteria than before. The increase in antibiotic resistance in bacteria is an example of evolution, any cumulative genetic change in a population from one generation to another.

The concept of evolution is the cornerstone of biology because it links all fields of the life sciences. Biologists attempt to understand both remarkable variety as well as the fundamental similarities of living organisms within the context of evolution. While biologists agree that evolution is the best scientific explanation of the origin of Earth's organisms, they currently study and actively debate the actual mechanisms that cause evolution.

LEARNING OBJECTIVES

After you have studied this chapter you should be able to

1. Discuss the historical development of the theory of evolution.
2. Explain the four premises of evolution by natural selection as proposed by Darwin.
3. Explain how the synthetic theory of evolution differs from Darwin's original theory of evolution.
4. Summarize the evidence for evolution obtained from the fossil record.
5. Summarize the evidence for evolution derived from comparative anatomy.
6. Define and give examples of homologous, analogous, and vestigial structures.
7. Explain how developmental biology provides insights into the evolutionary process.
8. Define biogeography and summarize the types of evidence it provides for evolution.
9. Relate how scientists make inferences about evolutionary relationships from the sequence of amino acids in proteins and the sequence of nucleotides in genes.

IDEAS ABOUT EVOLUTION ORIGINATED BEFORE DARWIN

Although Charles Darwin (1809–1882) is universally associated with evolution, ideas of evolution predate Darwin by centuries. Aristotle (384–322 B.C.) saw much evidence of natural affinities among organisms. This led him to arrange all of the organisms he knew in a "Scale of Nature" that extended from the exceedingly simple to the most complex. He visualized living organisms as being imperfect but "moving toward a more perfect state." This idea has been interpreted by some scientific historians as the forerunner of evolutionary theory, but Aristotle was vague on the nature of this "movement toward perfection" and certainly did not propose that the process of evolution was driven by natural selection. Furthermore, modern evolutionary theory now recognizes that evolution does *not* move toward more "perfect" states, nor even necessarily toward greater complexity.

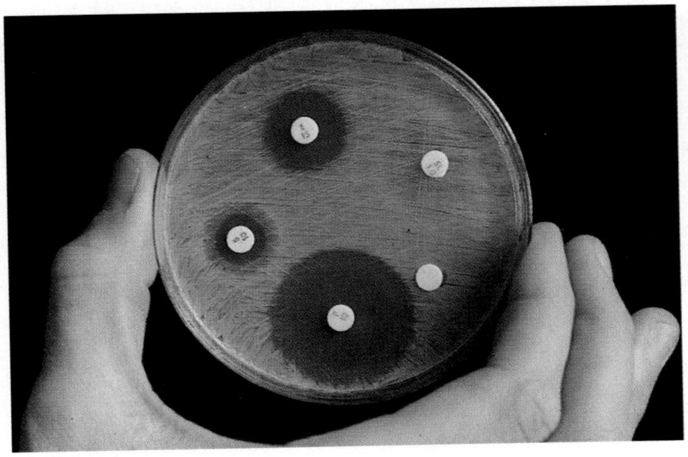

FIGURE 17–1 An even coating of bacteria covers the surface of this culture dish, except where different antibiotics (disks) to which the bacteria are sensitive prevent this growth. Note that certain antibiotics are more effective than others at preventing bacterial growth. The bacteria on this plate are completely resistant to two of the antibiotics (two disks on right). Varieties of bacteria that have developed resistance to certain antibiotics are common today, especially in hospitals. (Dennis Drenner)

Long before Darwin, fossils had been discovered embedded in rocks. Some of these corresponded to parts of familiar living organisms, but others were strangely unlike any known form. Fossils were also often found in unexpected contexts; marine invertebrates (sea animals without backbones), for instance, were sometimes discovered in rocks high on mountains. Leonardo da Vinci was among the first to correctly interpret these unusual finds as the remains of animals that had existed in previous ages but had become extinct.

During the Renaissance there was an increased interest in the study of nature and a movement away from the simple acceptance of medieval ideas. Modern scientific thought, intent on elucidating natural laws and based on the scientific method of observation, experimentation, and rigorous inductive and deductive logic, emerged in the 17th century (see Chapter 1). In the 18th century this new scientific thought began to have a significant effect on interpretations of the biological world. As new continents were explored, the discovery of previously unknown species[1] and more fossils led many to think that, like the physical world, the world of living organisms must be guided by natural laws.

Darwin's observations were substantially at odds with the most accepted view of evolution as expressed in 1809 by the French naturalist Jean Baptiste de Lamarck in his *Philosophie Zoologique.* Like most biologists of that era, Lamarck thought that all living organisms were endowed with a vital force that drove them to change toward greater complexity over time. He also thought that organisms could pass traits acquired during their lifetimes on to their offspring. As an example of this line of

[1]The concept of a species is developed extensively in Chapter 19. A simple working definition is that a species comprises a group of similar organisms capable of interbreeding.

ON THE CUTTING EDGE

Test-Tube Evolution

Objective: To study bacterial evolution in the laboratory.

Method: Expose starving bacteria to a new food that they cannot metabolize.

Results: Mutations appear more frequently than would be expected in the starving cells, and some of these mutations are adaptive, allowing the bacteria to metabolize the new food. As a result, the bacteria survive and pass the ability to metabolize the new food on to their offspring.

Conclusion: Under conditions of starvation, adaptive mutation may occur in bacteria.

In 1988 a well-known British biologist, John Cairns, published the results of a study that indicated a different aspect of evolution known as **adaptive mutation.** By starving bacteria, Cairns found that they mutate in a nonrandom way and thus obtain the mutation they need to survive. His experiment startled many biologists because it seemed to support Lamarck's discredited idea that evolution is directed and purposeful. Like Lamarck's classical example of giraffes needing—and therefore evolving—longer necks to stretch into treetops, Cairn's starving bacteria needed—and seemed to evolve—a certain mutation that allowed them to use lactose for food.

The mutations produced by these bacteria did not appear to be the random mutations (that is, produced without regard for usefulness) characteristic of neo-Darwinian evolution. Adaptive mutation allowed the bacteria to obtain the needed mutation in a fraction of the number of cell divisions that "normal" random mutations would require.

Several other biologists repeated Cairn's experiment and reproduced his results, but the mechanism or mechanisms responsible for such apparently directed evolution remained unknown. Then in 1994 a group of researchers headed by Susan Rosenberg from the University of Alberta published two papers in the journal *Science** that reported a probable molecular mechanism for the variation produced by adaptive mutation. Certain proteins used during the recombination of bacterial DNA are necessary for adaptive mutation. The absence or inactivation of the genes encoding these proteins results in a strain of bacteria that cannot undergo rapid mutation when they are starving.

Many questions remain unanswered. Are these mutations occurring *because* they are needed? Why do the recombination genes cause adaptive mutation in starving bacteria but not in well-fed, rapidly growing cells? How important is adaptive mutation in bacterial evolution? Certain eukaryotic cells (yeasts) appear to have a process like adaptive mutation. Does it use a similar mechanism? Biologists disagree about the evolutionary implications of adaptive mutation, which will likely make this controversial phenomenon the target of investigation for many years.

*1. Harris, R.S., S. Longerich, and S.M. Rosenberg. "Recombination in adaptive mutation." *Science,* Vol. 264, 8 April 1994. 2. Rosenberg, S.M., S. Longerich, P. Gee, and R.S. Harris. "Adaptive mutation by deletions in small mononucleotide repeats." *Science,* Vol. 265, 15 July 1994.

reasoning, Lamarck suggested that the long neck of the giraffe developed when a short-necked ancestor began browsing on the leaves of trees instead of on grass. Lamarck speculated that the ancestral giraffe, by reaching up, stretched and elongated its neck. Its offspring inherited the longer neck, which stretched still further. This process, repeated over many generations, supposedly resulted in the long necks of modern giraffes.

The proposed mechanism for Lamarckian evolution was an "inner drive" for self-improvement, a notion that was discredited when the actual basis of heredity was later discovered. (However, see *On the Cutting Edge: Test-Tube Evolution* for a possible example of "Lamarckian" evolution in bacteria.) Lamarck's contribution to science is important because he was the first to propose that organisms undergo change over time as a result of some natural phenomenon rather than divine intervention. It remained for Charles Darwin to discover the mechanism of evolution by natural selection.

DARWIN WAS INFLUENCED BY HIS CONTEMPORARIES

Charles Darwin, the son of a prominent physician, was sent at the age of 15 to study medicine at the University of Edinburgh. Finding himself unsuited for medicine, he transferred to Cambridge University to study theology. Shortly after receiving his degree, Darwin embarked as a naturalist on the *H.M.S. Beagle,* which was taking a five-year exploratory cruise around the world to prepare navigation charts for the British Navy.

FIGURE 17–2 This portrait was made shortly after Darwin returned to England from his voyage around the world (see map) on the *H.M.S. Beagle.* Observations made during this voyage helped him formulate the concept of evolution by natural selection. (William E. Ferguson)

The *Beagle* left Plymouth, England, in 1831 and cruised slowly along the east and west coasts of South America (Fig. 17–2). While other members of the company mapped the coasts and harbors, Darwin spent many weeks ashore studying the animals, plants, fossils, and geological formations of both coastal and inland regions, areas that had not been extensively explored. He collected and catalogued thousands of plant and animal specimens and kept notes of his observations.

The *Beagle* spent almost two months at the Galapagos Islands, 965 kilometers (600 miles) west of Ecuador, where Darwin continued his observations and collections. He compared the animals and plants of the Galapagos with those of the South American mainland. He was particularly impressed by their similarities and wondered why the organisms of the Galapagos should resemble those from South America more than those from other islands in different parts of the world. Moreover, although there were similarities between Galapagos and South American species, there were also distinct differences. There were even recognizable differences in the birds (Fig. 17–3) and reptiles from one island to the next! Darwin pondered these observations and tried to develop a satisfactory explanation for the distribution of species among the islands.

Despite the work of Lamarck, the general notion in the mid-1800s was that the Earth was too young for organisms to have changed significantly since they first appeared. There were some obvious exceptions to this idea, however. For example, breeders could develop many varieties of domesticated animals and plants in just a few generations (Figs. 17–4 and 17–5). This was accomplished by choosing certain traits and by breeding only individuals that possessed the desired traits, a procedure known as **artificial selection.**

Evidence found in rocks was also beginning to contradict the accepted views, as an increasing number of fossil specimens were discovered that did not have living counterparts. Even more troublesome to popular opinion was geological evidence suggesting that the Earth was far older than had been thought. During the early 19th century, geologist Charles Lyell advanced the idea that mountains, valleys, and other physical features of Earth's surface were not created in their present forms. Instead, they were formed slowly over long periods of time by the geological processes of volcanic activity, uplift, erosion, and glaciation. The slow pace of these geological processes, which still occurs today, indicated that the Earth was *extremely old.*

The ideas of Thomas Malthus (1766–1834), a clergyman and economist, were another important influence on Darwin. Malthus noted that populations have the capacity to increase geometrically ($1 \rightarrow 2 \rightarrow 4 \rightarrow 8 \rightarrow 16$) and thus outstrip the food supply, which has the capacity to increase only arithmetically ($1 \rightarrow 2 \rightarrow 3 \rightarrow 4 \rightarrow 5$). In the case of humans, Malthus suggested that wars, famine, and disease serve as inevitable brakes on population growth.

(a) (b) (c)

FIGURE 17–3 The various species of Galapagos Island finches have beaks specialized for different kinds of food. Darwin inferred that these drab, unremarkable-appearing birds are derived from a common ancestral population of seed-eating birds from South America. The likely derivation of such different birds from a common ancestor suggested to Darwin that species originate by natural selection. (***a***) Cactus finch, *Geospiza scandens.* The cactus finch feeds on the fleshy parts of cacti. (***b***) A large ground finch, *Geospiza magnirostris.* This bird has an extremely heavy, nutcracker-type bill adapted for eating heavy-walled seeds. (***c***) Woodpecker finch, *Camarhyncus pallidus.* This remarkable bird has insectivorous habits similar to those of woodpeckers but lacks the complex beak and tongue adaptations that permit woodpeckers to reach their prey. The adaptations of the woodpecker finch to this lifestyle are almost entirely behavioral. In one of the few known instances of animal tool use, this bird digs insects out of bark and crevices using cactus spines, twigs, or even dead leaves. (*a* and *b*, Frans Lanting/Minden pictures; *c*, Miguel Castro/Photo Researchers, Inc.)

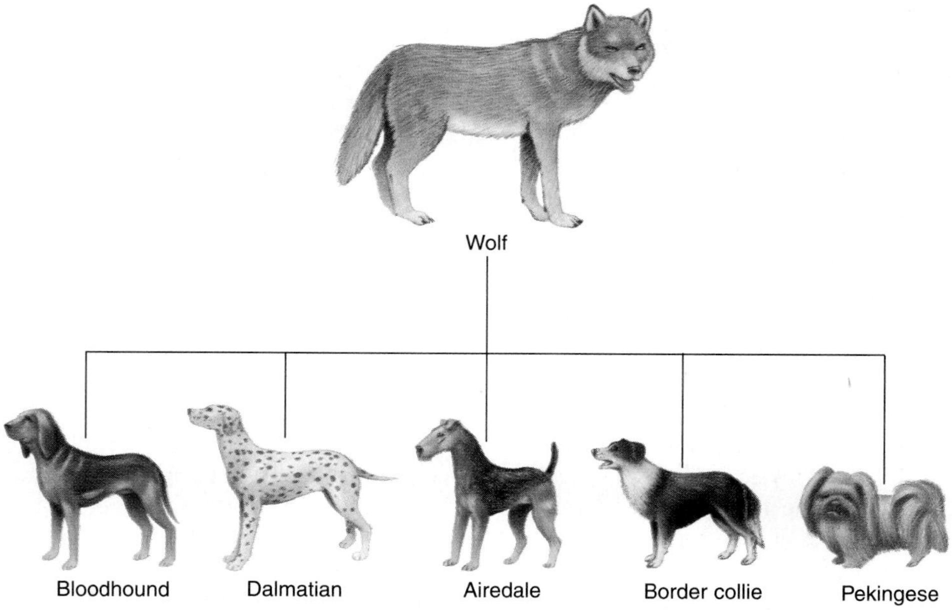

FIGURE 17–4 Numerous dog varieties have been produced by artificial selection. Bloodhounds, dalmatians, airedales, border collies, and Pekingese are some of the 136 breeds currently recognized by the American Kennel Club. The ancestor of dogs was most likely the wolf.

FIGURE 17–5 A number of common vegetables are members of the species *Brassica oleracea*, including cauliflower, broccoli, cabbage, brussels sprouts, and kale. Artificial selection is responsible for the variation shown within this species. (Raymond Tschoepe)

DARWIN PROPOSED THAT EVOLUTION OCCURS BY NATURAL SELECTION

Darwin's years of observing the habits of animals and plants had introduced him to the struggle for existence described by Malthus. It occurred to Darwin that in this struggle, inherited variations favorable to survival would tend to be preserved, while unfavorable ones would be eliminated. The result would be **adaptation** (evolutionary modification that improves the chances of survival and reproductive success) of the population to the environment. Eventually, the accumulation of modifications might result in a new species. Time was the only thing required for new species to originate, and the geologists of the era, including Darwin's friend Lyell, had supplied evidence that Earth was indeed old enough to provide an adequate period of time.

Darwin had at last obtained a workable mechanism of evolution—that of **natural selection,** in which better-adapted organisms are more likely to survive and reproduce, thereby increasing their proportion in the population. As a result of natural selection, the population changes over time; the frequency of favorable traits increases in successive generations while less favorable traits decrease or disappear. Darwin spent the next 20 years accumulating an immense body of evidence to support his theory and formulating his arguments for natural selection.

As Darwin was pondering his ideas, Alfred Russel Wallace, a biologist who was studying the plants and animals of Malaysia and Indonesia, was similarly struck by the diversity of species and the peculiarities of their distribution. Wallace independently arrived at the conclusion that evolution occurs by natural selection. In 1858, he sent a brief essay to Darwin, by then a world-renowned biologist, asking his opinion. Darwin's friends persuaded him to present Wallace's findings along with an abstract of his own work, which he had prepared and circulated to a few friends several years earlier. Both papers were presented in July 1858, at a London meeting of the Linnaean Society. Darwin's monumental book, *The Origin of Species by Means of Natural Selection,* was published in 1859.

Darwin's mechanism of evolution by natural selection consists of four observations about the natural world: overproduction, variation, limits on population growth, and differential reproductive success.

1. **Overproduction.** Each species has the capacity to produce more offspring than will survive to maturity. Through reproduction, natural populations may geometrically increase in number over time. For example, if each breeding pair of elephants produces six offspring during its 90-year lifespan, in 750 years a single pair of elephants will give rise to a population of 19 million! Yet elephants have not overrun the planet.
2. **Variation.** The individuals in a population exhibit variation in their traits. Some of these traits improve the chances of an individual's survival and reproductive success, whereas other traits do not. It is important to remember that the variation necessary for evolution by natural selection must be heritable (able to be passed on to offspring; Fig. 17–6). (Although Darwin recognized the importance of inherited variation to evolution, he did not know about the mechanism of inheritance.)
3. **Limits on population growth,** or **a struggle for existence.** There is only so much food, water, light, growing space, and other resources available to a population, and organisms compete with one another for

FIGURE 17–6 Individuals within a species exhibit variation. A mother rabbit with her young, which differ from one another in several ways, including coat color. One premise of natural selection is that sexual reproduction results in offspring that are not identical to one another. (BIOS/M. Gunther/Peter Arnold, Inc.)

these limited resources. Because there are more individuals than the environment can support, not all will survive to reproductive age. Other limits on population growth include predators and disease organisms.

4. **Differential reproductive success,** or **"survival of the fittest."** Those individuals that possess the most favorable combination of characteristics are most likely to survive and reproduce, passing their heritable traits on to the next generation.

The process of natural selection thus causes an increase of favorable alleles and a decrease of unfavorable alleles within a population. Over succeeding generations, individual members become better adapted to local conditions. Successful reproduction is the key to natural selection: the "fittest" individuals are those that reproduce most successfully.

Over time, enough changes may accumulate in geographically separated populations (often with slightly different environments) to form new species. Darwin noted that the Galapagos finches appear to have evolved in this way. The different islands in the Galapagos kept the finches isolated from one another, thereby allowing them to diverge into separate species. The evolution of new species is considered in greater detail in Chapter 19.

THE SYNTHETIC THEORY OF EVOLUTION COMBINES DARWIN'S THEORY WITH GENETICS

One of the premises on which Darwin based his theory of evolution by natural selection is that individuals pass traits on to the next generation. However, Darwin was unable to explain *how* this occurs. He was also unable to explain *why* individuals vary within a population. Darwin was a contemporary of Gregor Mendel (see Chapter 10), who worked out the basic patterns of inheritance. However, Darwin was apparently not acquainted with Mendel's work, which was not recognized by the scientific community until the early part of the 20th century.

During the 1920s to 1940s, biologists combined Mendelian genetics with Darwin's theory to form a unified explanation of evolution known as **neo-Darwinism** or, more commonly, the **synthetic theory of evolution.** (*Synthesis* in this context refers to putting together parts of several previous theories to form a whole.) The synthetic theory of evolution explains Darwin's observation of variation among offspring in terms of mutation; that is, mutation provides the genetic variability that natural selection acts on during evolution. The synthetic theory of evolution, which emphasizes the genetics of *populations* (rather than individuals) as the central focus of evolution, has held up well since it was formulated (see Chapter 18). It has dominated the thinking and research of biologists working in many areas, and has resulted in an enormous accumulation of scientific evidence to support evolution by natural selection.

Most biologists accept the basic principles of the synthetic theory of evolution but have recently scrutinized some of its aspects. For example, what is the role of chance in evolution? How rapidly do new species evolve? These questions have arisen in part from a reevaluation of the fossil record and in part from discoveries in molecular aspects of inheritance. Such critical analyses are an integral part of the scientific process because they stimulate additional observation and experimentation, along with reexamination of previous evidence. Science is an ongoing process, and information obtained in the future may require modifications to certain parts of the synthetic theory of evolution.

MANY TYPES OF SCIENTIFIC EVIDENCE SUPPORT EVOLUTION

A vast body of scientific evidence supports evolution, including observations from the fossil record, comparative anatomy, developmental biology, biogeography, and molecular biology.

The fossil record indicates that evolution occurred in the past

Perhaps the most direct evidence for evolution comes from the discovery, identification, and interpretation of **fossils,** which are the remains or traces left in rock strata (layers) by previous organisms. (The term *fossil* comes from the Latin word *fossilis,* meaning "something dug up.") Although most fossils are preserved in rock strata, some more recent remains have been exceptionally well preserved in bogs, tar, amber (ancient tree resin), or ice (Fig. 17–7). For example, the remains of a woolly mammoth deep-frozen in Siberian ice for more than 25,000 years were so well preserved that part of its DNA could be analyzed.

The most common vertebrate (animal with a backbone) fossils are teeth and bones. From the shapes of the bones and the positions of the bone scars indicating points of muscle attachment, biologists can infer an animal's posture and style of walking, the position and size of its muscles, and the contours of its body. By a careful study of fossil remains, biologists can often reconstruct what an animal probably looked like in life.

Fossils provide a record of animals and plants that lived earlier and some understanding of where and when they lived. Using fossils of organisms from different geological ages, the lines of descent (evolutionary relationships) that gave rise to modern-day organisms can often be inferred. Sometimes fossils provide direct evidence of the origin of new species from preexisting species, including intermediate forms.

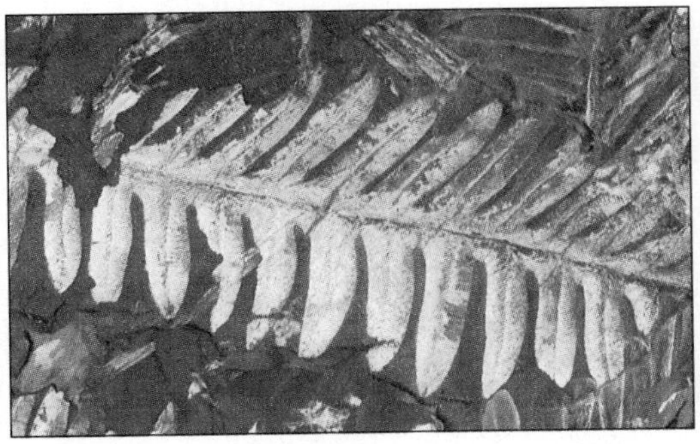
(a)

(b)

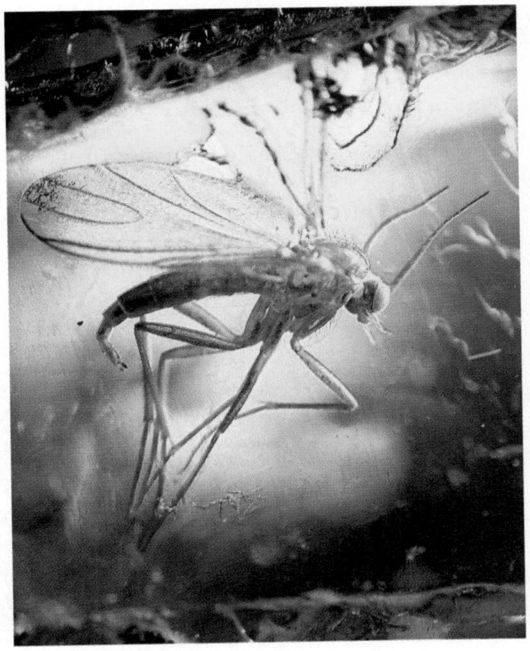
(c)

(d)

(e)

FIGURE 17–7 Fossils form in different ways. (*a*) Fossil of a seed fern leaf. Although some fossils contain traces of organic matter, all that remains in this particular fossil is an impression, or imprint, in the rock. (*b*) Petrified wood from the Petrified Forest National Park in Arizona consists of trees that were buried and infiltrated with minerals. (*c*) An insect (a midge) embedded in amber. Approximately two million years old, it has been preserved almost perfectly. (*d*) Cast fossil of ancient echinoderms called crinoids. A cast forms when an organism decomposes, leaving a mold that later fills with dissolved minerals that then harden. (*e*) Dinosaur footprints, each 75 to 90 centimeters (2.5 to 3 feet) in length, occur in sedimentary rock in Texas. Dinosaur footprints provide clues about the locomotion, behavior, and ecology of these extinct animals. (*a*, Carolina Biological Supply Company/Phototake–NYC; *b*, B. Gruliow; *c*, Alfred Pasieka/Science Photo Library/Photo Researchers, Inc.; *d*, Visuals Unlimited/A.J. Copley; *e*, Visuals Unlimited/Scott Berner)

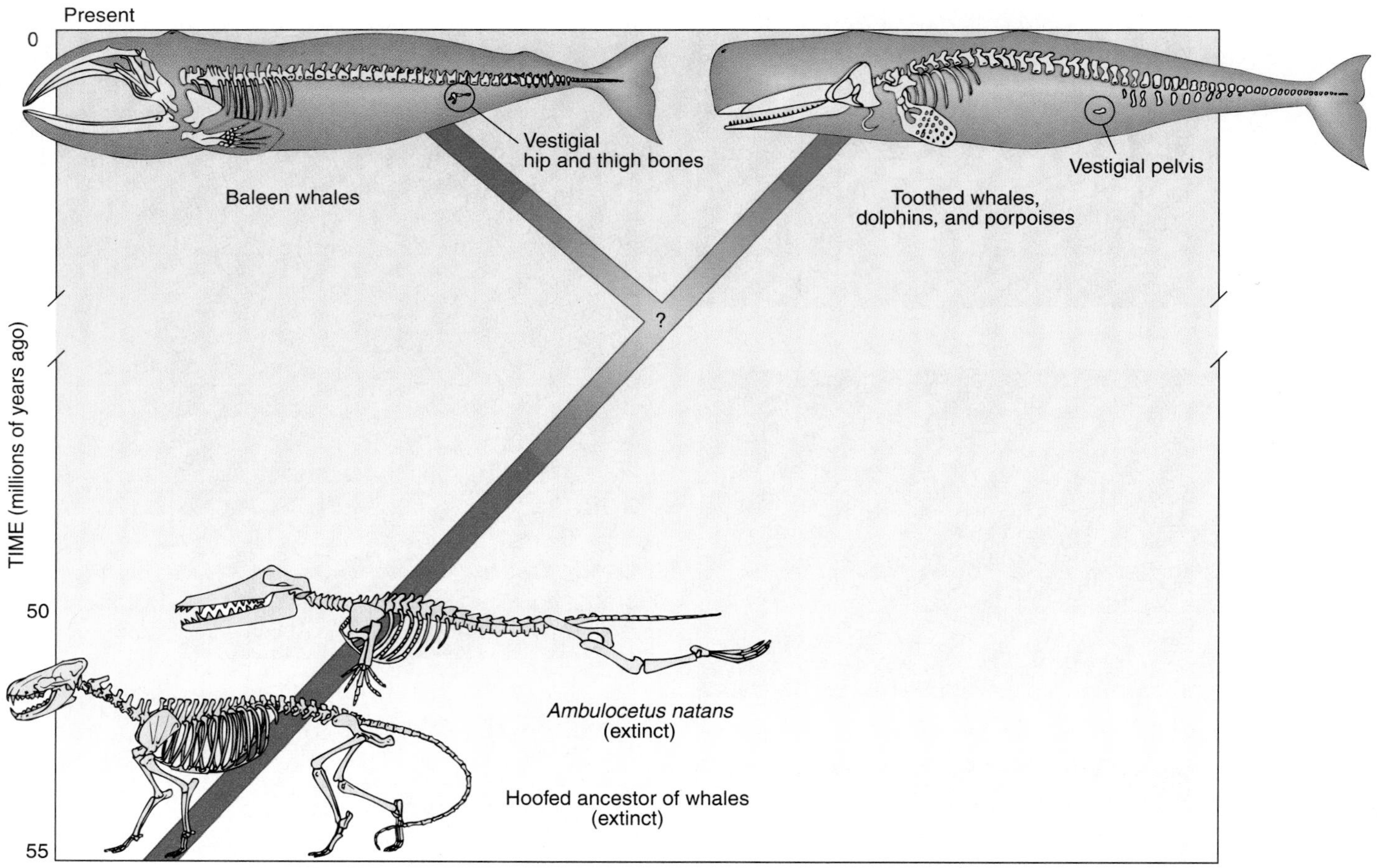

FIGURE 17–8 The whale family tree has a new addition: an intermediate in whale evolution. The now-extinct *Ambulocetus* was a transitional form between modern whale descendants and their terrestrial ancestors. Although its exact relationship to other whales is uncertain, *Ambulocetus* possessed a number of recognizable whale features *and* retained the hindlimbs of its four-legged ancestors. This ancient whale may have moved about on land (perhaps as sea lions do today) in addition to swimming in shallow seas.

For example, an exciting fossil intermediate in whale evolution was reported in *Science* in 1994 (Fig. 17–8). Scientific evidence indicates that whales descended from a now-extinct group of four-legged terrestrial mammals. (The closest living relatives of whales and porpoises are hoofed mammals such as antelopes, deer, and giraffes.) Fossils of *Ambulocetus,* which was discovered in Pakistan, had many features of modern whales but also possessed partial hindlimbs and feet. (Modern whales do not have hindlimbs, although vestigial pelvic and hindlimb bones persist. Vestigial organs will be discussed later in this chapter.)

The formation and preservation of a fossil require that an organism be buried under conditions that slow or prevent the decay process. This is most likely to occur if an organism's remains are covered quickly by a sediment of fine soil particles suspended in water. In this way remains of aquatic organisms may be trapped in bogs, mud flats, sand bars, or deltas. Remains of terrestrial organisms that lived on a flood plain may also be covered by water-borne sediments or, if the organism lived in an arid region, by wind-blown sand. Over time, the sediments harden to form sedimentary rock, and the organism's remains are usually replaced by minerals so that many details of its structure—even cellular details—remain.

Layers of sedimentary rock occur naturally in the sequence of their deposition, with the more recent layers on top of the older, earlier ones (Fig. 17–9). However, geological events that occurred after the rocks were initially formed occasionally change the relationships of some rock layers. Geologists identify specific sedimentary rocks not only by their positions in layers but also by features such as mineral content, and by the fossilized remains of certain organisms, known as **index fossils,** that characterize a specific layer over large geographical areas. Index fossils are fossils of organisms that existed for a relatively short geological time but were preserved as fossils in large numbers. With this information, geologists

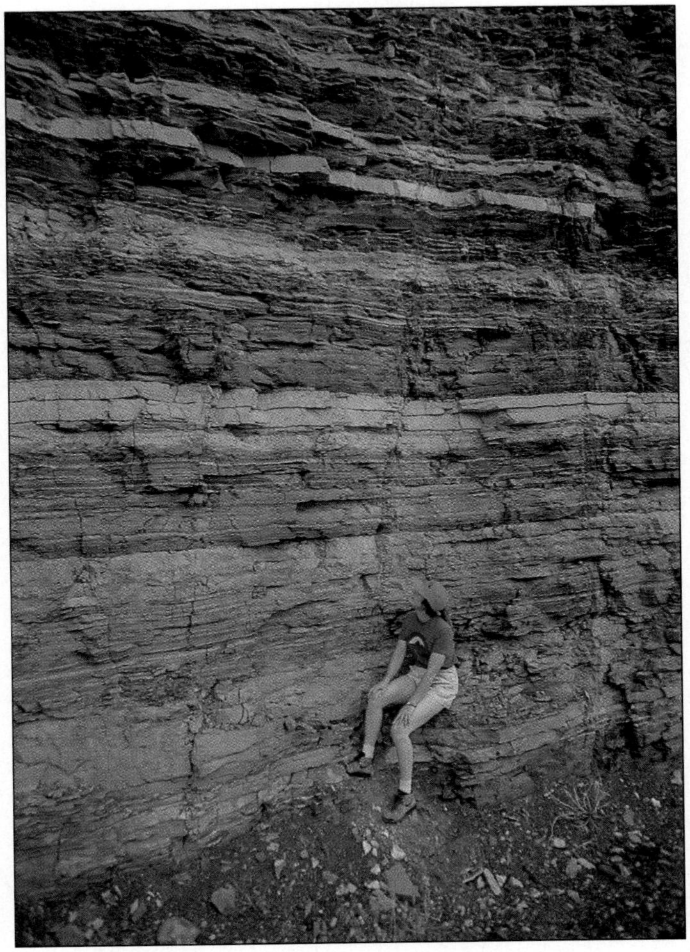

FIGURE 17–9 Layers characterize sedimentary rock. The strata shown here were exposed when a road cut was made for a highway in Utah. (Tom Bean)

arrange strata and the fossils they contain in chronological order, and identify comparable layers in widely separated locations.

Because of the conditions required for preservation, the fossil record is not a random sample of past life, but instead is biased toward aquatic organisms and those living in the few terrestrial habitats conducive to fossil formation. For example, relatively few fossils of tropical forest organisms have been found because plant and animal remains decay extremely rapidly on the forest floor, before fossils can form. Another reason for bias is that organisms with hard body parts such as bones and shells are more likely to form fossils than those with soft body parts.

Radioactive dating can be used to determine the age of fossils

The most useful fossil evidence comes from rocks that have been accurately dated. Radioactive isotopes, also called **radioisotopes,** present in a rock give us an accurate measure of its age (see Chapter 2). Radioisotopes emit powerful, invisible radiations. As a radioisotope emits radiation, its nucleus changes into the nucleus of a different element in a process known as **radioactive decay.** For example, the radioactive nucleus of uranium-235 decays over time into lead-207.

Each radioisotope has its own characteristic rate of decay. The period of time required for one half of the atoms of a radioisotope to change into a different atom is known as its **half-life** (Fig. 17–10). Different radioisotopes have enormous variations in their half-lives. For example, the half-life of iodine-132 is only 2.4 hours, whereas the half-life of uranium-235 is 704 million years. The half-life of a

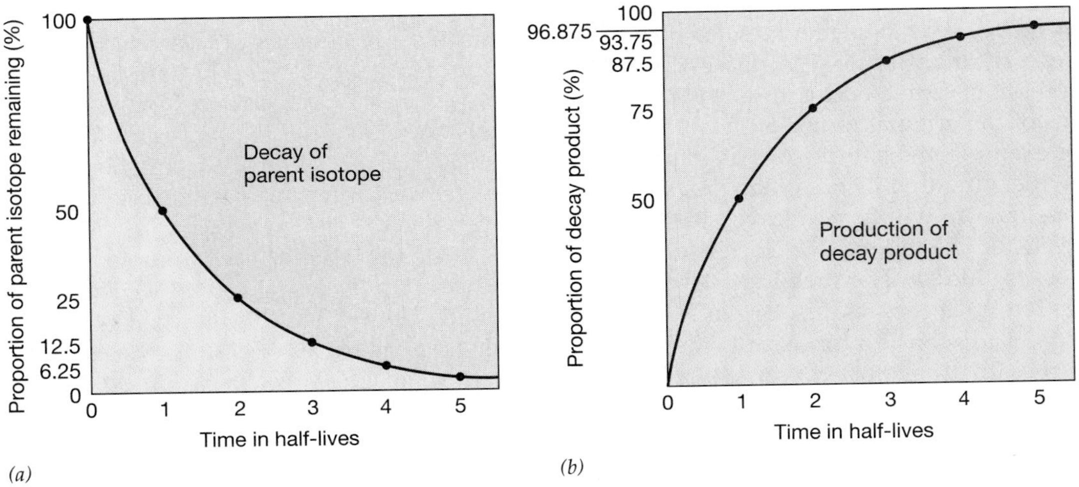

FIGURE 17–10 As radioisotopes decay, their decay products accumulate. (*a*) At time zero, the radioactive clock begins ticking. At this point a sample is composed entirely of the radioisotope. After one half-life, only 50% of the original radioisotope remains. (**b**) At time zero, the same sample contains no decay product(s). After one half-life, 50% of the original radioisotope has decayed into the decay product(s). During each succeeding half-life, one half of the remaining radioisotope is converted to decay product(s).

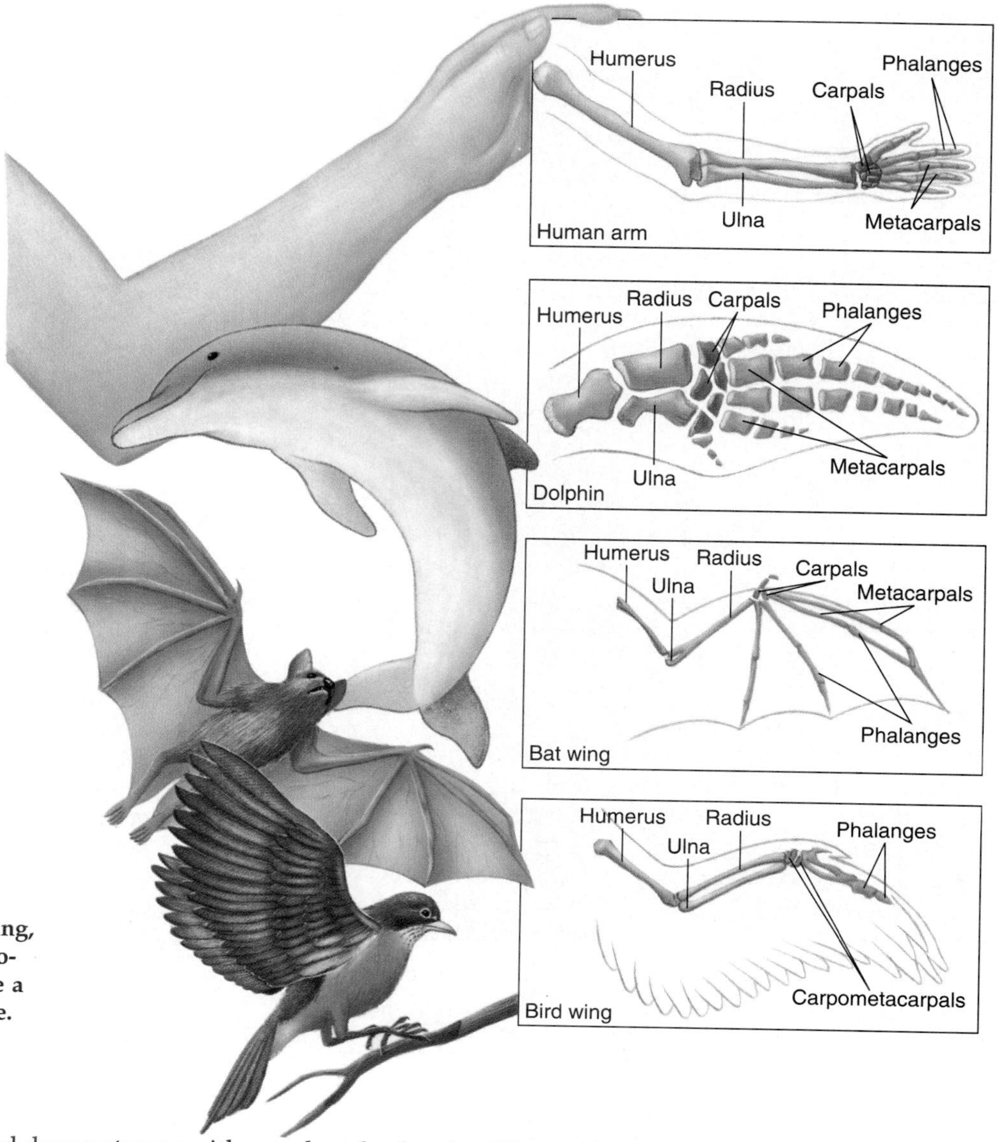

FIGURE 17–11 The bird wing, bat wing, dolphin flipper, and human arm are homologous structures because they have a basic underlying similarity of structure.

particular radioisotope is constant and does not vary with temperature, pressure, or any other environmental factor.

The age of a fossil is estimated by measuring the relative proportions of the original radioisotope and its decay product. For example, the half-life of potassium-40 is 1.3 billion years, meaning that in 1.3 billion years half of the radioactive potassium will have decayed into its decay product, argon-40. The radioactive clock begins ticking when the rock solidifies. The rock initially contains some potassium, but no argon. Because argon is a gas, it escapes from hot rock as soon as it forms, but when potassium decays in rock that has cooled and solidified, the argon accumulates in the crystalline structure of the rock. If the ratio of potassium-40 to argon-40 in the rock being tested is 1:1, the rock is 1.3 billion years old.

Several different radioisotopes are commonly used to date fossils. These include potassium-40 (half-life 1.3 billion years), uranium-235 (half-life 704 million years), and carbon-14 (half-life 5730 years). Potassium-40, with its long half-life, can be used to date fossils that are many hundreds of millions of years old. Radioisotopes other than carbon-14 are used to date the *rock* in which fossils are found, whereas carbon-14 is used to date the *carbon remains* of anything that was once living, such as wood, bones, or shells. Whenever possible, the age of a fossil is independently verified using two or more different radioisotopes.

Carbon-14, which is continuously produced in the atmosphere from nitrogen-14 (by cosmic radiation), subsequently decays back to nitrogen-14. Because the formation and the decay of carbon-14 occur at constant rates, the ratio of carbon-14 to carbon-12 (the more abundant, stable isotope of carbon) is constant in the atmosphere. Since each living organism absorbs carbon from the atmosphere,[2] its ratio of carbon-14 to carbon-12 is the same as the atmosphere. When an organism dies, however, it

[2]Living organisms absorb carbon from the atmosphere either directly (by photosynthesis) or indirectly (by consuming photosynthetic organisms).

no longer absorbs carbon, and the proportion of carbon-14 in its remains declines as carbon-14 decays to nitrogen-14. Because of its relatively short half-life, carbon-14 is useful for dating fossils that are 50,000 years old or less.

Comparative anatomy of related species demonstrates similarities in their structures

Comparing the structural details of any particular organ found in different but related organisms reveals a basic similarity of form. For example, a bird's wing, a dolphin's front flipper, a bat's wing, and a human arm and hand, although quite different in appearance, have strikingly similar arrangements of bones, muscles, and nerves. Figure 17–11 shows a comparison of their skeletal structures. Each has a single bone (the humerus) in the part of the limb nearest the trunk of the body, followed by the two bones (radius and ulna) of the forearm, a group of bones (carpals) in the wrist, and a variable number of digits (metacarpals and phalanges). This similarity is particularly striking because wings, flippers, and the human arm are used in different ways for different functions, and there is no overriding mechanical reason for them to be so similar structurally.

Similar arrangements of parts of the forelimb are evident in ancestral reptiles and amphibians and even in the first fishes that came out of water onto land hundreds of millions of years ago. Darwin pointed out that such basic structural similarities in organs used in different ways are precisely the expected outcome of a common evolutionary origin. The basic structure present in a common ancestor was modified in different ways as various descendants subsequently evolved. Features in different species that are similar in underlying structure due to a common evolutionary origin are termed **homologous structures.**

Not all species with "similar" structures have descended from a common ancestor, however. Organs that are not homologous but simply have similar functions in unrelated organisms are termed **analogous structures.** For example, the lungs of mammals and the tracheae (air tubes) of insects are analogous structures that have evolved over time to meet, in quite different ways, the common problem of exchanging gases. Also, the wings of various unrelated flying animals, such as insects and birds, resemble one another superficially (Fig. 17–12) but are different in more fundamental aspects. Bird wings are modified forelimbs supported by bones, whereas insect wings are outgrowths of the upper wall of the thorax and are supported by chitinous veins.

Like homologous structures, analogous structures offer crucial evidence of evolution. Analogous structures are of evolutionary interest because they demonstrate that populations with separate ancestries may adapt in similar ways to similar environmental demands. This is known as **convergent evolution** (Fig. 17–13).

Comparative anatomy reveals the existence of **vestigial organs.** Many organisms contain organs or parts of organs that are seemingly nonfunctional and degenerate, often undersized or lacking some essential part. Vestigial organs are remnants of more developed organs that were present in ancestral organisms. In the human body, more than 100 structures are considered vestigial, including the appendix, the coccyx (fused tail bones), the wisdom teeth, and the muscles that move our ears. Whales (see Fig. 17–8) and pythons have vestigial hindlimb bones; wingless birds have vestigial wing bones; and many blind, burrowing, or cave-dwelling animals have vestigial eyes.

Darwin was interested in vestigial organs because they conflicted with the theologically based view of creation. He wondered why organisms that were the product of a divine creation would have useless parts. The scientific model of evolution, however, easily explained the existence of vestigial structures. The occasional presence of a vestigial organ is to be expected as a species adapts to a changing mode of life. Some organs become much

(a) *(b)*

FIGURE 17–12 The wings of birds and insects are analogous structures. Although used for similar functions, bird wings (***a***) and insect wings (***b***) have no underlying structural similarity.
(*a*, Dennis Drenner; *b*, Skip Moody/Dembinsky Photo Associates)

(a) (b)

FIGURE 17–13 Two unrelated plant families exhibit structural similarities, such as thick, succulent stems and leaves modified into protective spines, as a result of convergent evolution. These plants evolved in similar desert environments in different parts of the world. (***a***) *Euphorbia ingens,* a member of the spurge family native to Africa. (***b***) A member of the cactus family native to North America. (Dennis Drenner)

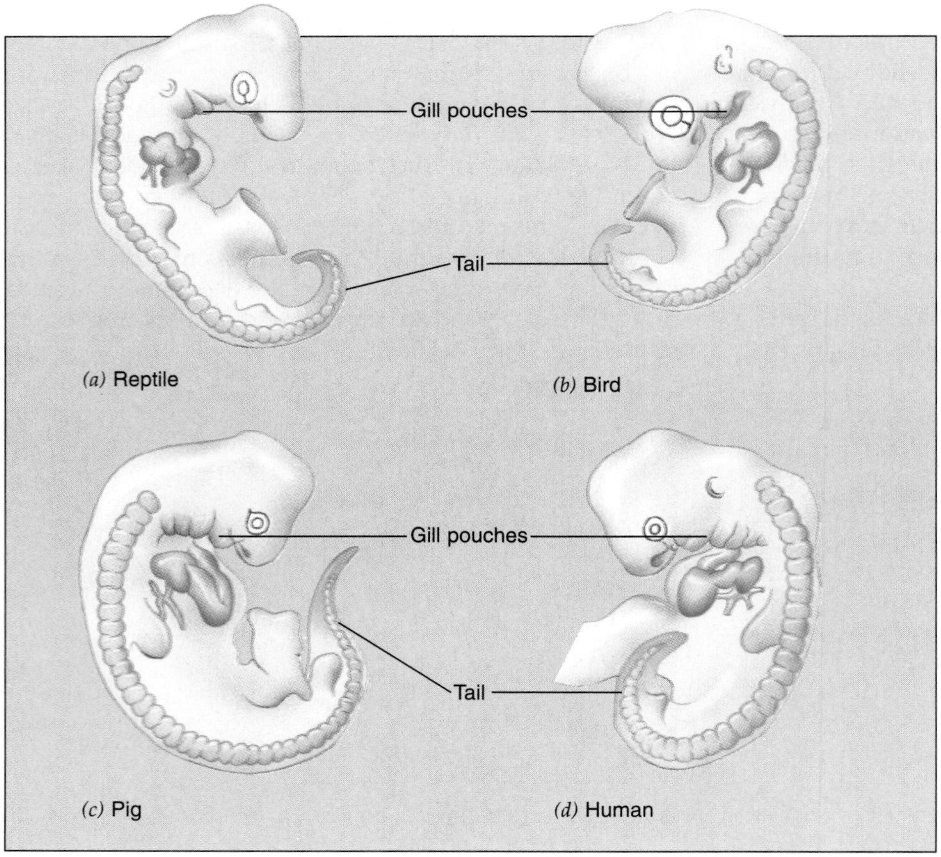

FIGURE 17–14 The early stages of embryonic development are almost identical in different vertebrate species. Numerous structural similarities are shared by the early stages, including the presence of gill pouches and a tail.

less important for survival and may end up as vestiges. When an organ no longer confers a selective advantage, it usually becomes smaller and loses much or all of its function with the passage of time. Since the presence of the vestigial organ is usually not harmful to the organism, however, selective pressure for completely eliminating it is weak, and the vestigial organ can be found in many subsequent generations.

Related species have similar patterns of embryological development

The resemblance among embryos of different vertebrates is closer than the resemblance among adults. In fact, it is difficult to distinguish among the early embryos of a reptile, a bird, a pig, or a human (Fig. 17–14). Segmented muscles, gill pouches, a tubular heart without left and right sides, a system of arteries known as aortic arches in the gill region, and many other features are found in all vertebrate embryos. All of these structures are necessary and functional in the developing fish. The small segmented muscles of the fish embryo give rise to the segmented muscles used by the adult fish in swimming. The gill pouches break through to the surface as gill slits. The adult fish heart remains undivided because it pumps blood forward to the gills that develop in association with the aortic arches.

None of these embryonic features persists in the adults of reptiles, birds, or mammals, so why are these fishlike structures present in their embryos? Evolution is a conservative process, and natural selection builds on what has come before rather than starting from scratch. Because terrestrial vertebrates are thought to have evolved from fishlike ancestors, they share some of the early stages of development still found in fish today. The accumulation of genetic changes over time in these vertebrates has modified the basic body plan laid out in fish development.

The distribution of plants and animals supports evolution

The study of the distribution of plants and animals is called **biogeography.** One of its basic tenets is that each species originated, or evolved, only once. The particular place where this occurred is known as the species' **center of origin.** The center of origin is not a single point but the distribution of the population when the new species originated. From its center of origin, each species spreads out until halted by a barrier of some kind—physical, such as an ocean, desert, or mountain; environmental, such as an unfavorable climate; or ecological, such as the presence of organisms that compete with it for food or shelter (Fig. 17–15).

Most plant and animal species have their own characteristic geographical distributions. The **range** of a particular species—that is, the portion of Earth over which it is found—may be only a few square kilometers or, as with humans, almost the entire world. In general, closely related species do not have identical ranges, nor are their ranges far apart. They are usually adjacent but separated by a barrier of some sort, such as a mountain or desert.

If evolution were not a factor in distribution, we would expect to find a given species everywhere that it could survive. However, Central Africa has elephants, gorillas, chimpanzees, lions, and antelopes, whereas Brazil, with similar climate and environmental conditions, has none of these. They originated in Africa and could not expand their range into South America because the Atlantic Ocean was an impassable barrier to them. Likewise, South America has prehensile-tailed monkeys, sloths, and tapirs, none of which is found in Africa. Thus, the natural distribution of organisms seems understandable only in the context of evolution.

Regions such as Australia and New Zealand, which have been separated from the rest of the world for a long time, have distinctive sets of organisms. Australia has populations of egg-laying mammals (monotremes) and pouched mammals (marsupials) not found anywhere else. Two hundred million years ago, Australia and the other continents were joined together in a major land mass (see Chapter 20). Over the course of millions of years, the Australian continent gradually separated from the others. The original monotremes and marsupials that were present when Australia broke its connection gave rise to a variety of species able to take advantage of the different environments available to them. The isolation of Australia also prevented placental mammals, which arose elsewhere at a later time, from competing with its more primitive monotremes and marsupials. (In those areas where placental mammals existed, most monotremes and marsupials became extinct.)

The kinds of animals and plants found on oceanic islands in general resemble those of the nearest mainland, yet they include some species found nowhere else. Darwin studied the plants and animals of the Cape Verde Islands, some 650 kilometers (about 400 miles) west of Dakar, Africa, and of the Galapagos Islands, a comparable distance west of Ecuador, South America. On each group of islands, the plants and terrestrial animals were indigenous (native), but those of the Cape Verdes resembled African species and those of the Galapagos resembled South American species. Darwin concluded that species from the neighboring continent migrated or were carried to the islands, where they subsequently evolved into new species. His insight proved to be critical to the development of his theory of natural selection.

The only animals and plants indigenous to oceanic is-

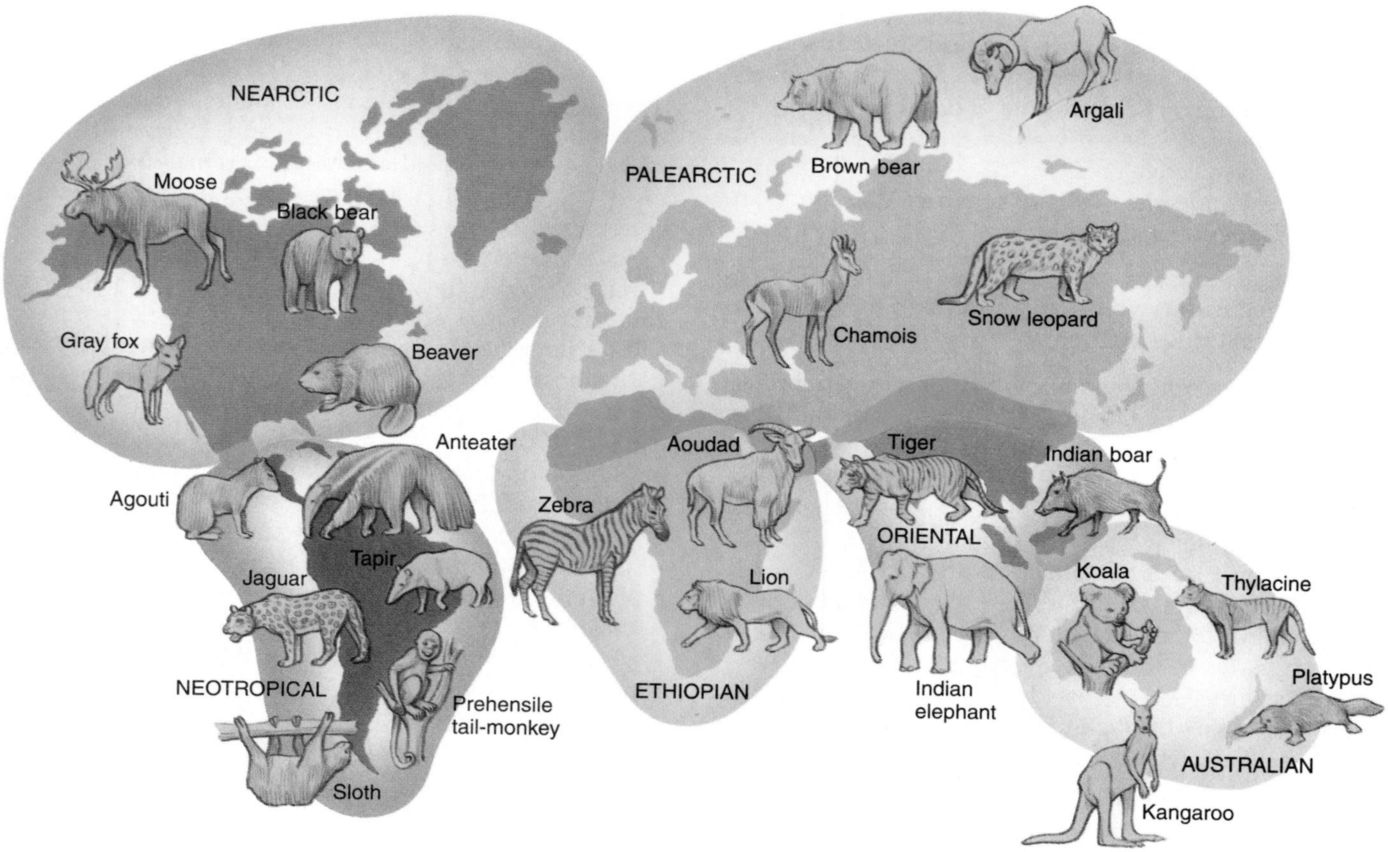

FIGURE 17–15 Animals and plants are distributed around the world in a distinctive pattern that reveals the existence of six major biogeographical realms. Each of these is characterized by the presence of certain unique species. These biogeographical realms are the direct outcome of the centers of origin of certain species, of their past migrations, and of the barriers they encountered.

lands are those that could survive the trip there. No frogs or toads live on the Galapagos, for example, even though there are woodland spots ideally suited to them, because neither these animals nor their eggs can survive exposure to sea water. No native terrestrial mammals (other than bats) live there either, although many land and sea birds are present. The occurrence in the Galapagos of these particular forms—closely related, yet not identical, to those of the Ecuador coast—suggests strongly that evolution has occurred and that the descendants of the first animals and plants to reach the islands are different from their ancestors.

Molecular comparisons among organisms provide evidence for evolution

Evidence for evolutionary relationships is provided by similarities and differences in the biochemistry and molecular biology of various organisms. Indeed, lines of descent based solely on biochemical and molecular characters closely resemble lines of descent based on structural and fossil evidence. Molecular evidence for evolution includes the genetic code and the conserved sequences of amino acids in proteins and of nucleotides in DNA.

The genetic code is universal

Organisms owe their characteristics to the types of proteins that they possess, which in turn are determined by the sequence of nucleotides in their mRNA (as specified by the order of nucleotides in their DNA). Evidence that all life is related comes from the fact that all organisms use a genetic code that is virtually identical. Recall from Chapter 12 that the genetic code specifies a triplet (a sequence of three nucleotides in DNA) that codes for a particular codon (a sequence of three nucleotides in mRNA) that codes for a particular amino acid in a polypeptide chain. For example, "AAA" in DNA codes for "UUU" in mRNA, which codes for the amino acid phenylalanine in organisms as diverse as shrimp, humans, bacteria, and

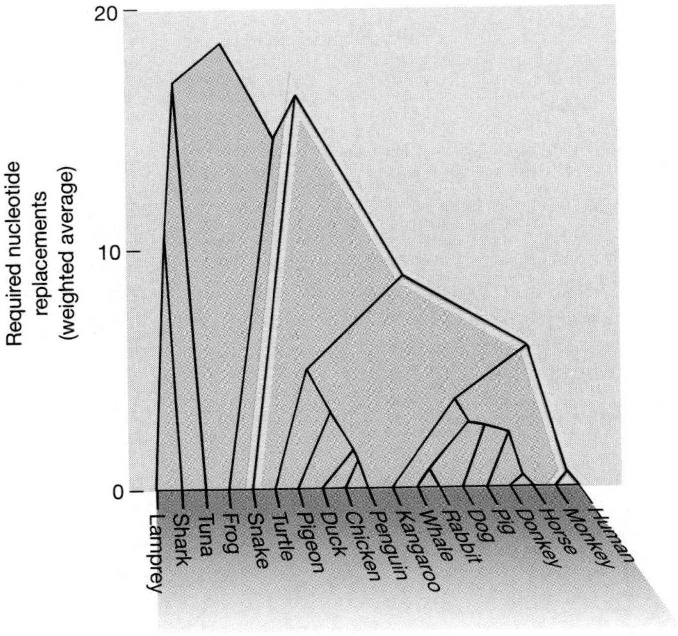

FIGURE 17–16 Lines of descent of selected vertebrates that are based on differences in the amino acid sequence of cytochrome *c* are compatible with fossil and structural evidence. Closely related organisms, such as monkeys and humans, show fewer differences in their nucleotide sequences (a relative distance of 1 as shown by the point at which monkeys and humans diverge). More distantly related organisms, such as snakes and humans, show greater differences in their nucleotide sequences (a relative difference of 17). (Adapted from Fitch and Margoliash, *Evolutionary Biology,* Plenum Publishing, 1970.)

tulips. In fact, "AAA" codes for phenylalanine in *all* organisms examined to date.

The universality of the genetic code—no other code has been found in any living organism—is compelling evidence that organisms arose from a common ancestor. The genetic code has been passed along through all branches of the evolutionary tree (another example of the conservative nature of evolution) since its origin in some extremely early (and successful) form of life.

Proteins contain a record of evolutionary change

Darwin's theory that all forms of life are related through descent with modification from the earliest organisms has been further verified as we have learned more about molecular biology. Investigations of the sequence of amino acids in proteins playing the same roles in different species have revealed both great similarities and certain specific differences.

Even organisms that are remotely related, such as humans, oaks, and the intestinal bacterium *Escherichia coli,* have some proteins such as cytochrome *c* in common. In order to survive, all aerobic organisms need a respiratory protein with the same basic structure and function as the cytochrome *c* of their common ancestor. Consequently, not all of the amino acids that give the protein those features are free to change. However, in the course of the long, independent evolution of different organisms, mutations have resulted in the substitution of many amino acids at less important locations in the cytochrome *c* protein.

The longer it has been since two organisms diverged, or took separate evolutionary pathways, the greater the differences in the amino acid sequences of their cytochrome *c* molecules. A **phylogenetic tree,** a diagram showing lines of descent, can be derived from differences in the amino acid sequence of a common protein like cytochrome *c*. Such a phylogenetic tree for vertebrates is depicted in Figure 17–16.

DNA contains a record of evolutionary change

Because DNA codes for proteins,[3] the differences in amino acid sequences indirectly demonstrate the nature and number of underlying DNA base-pair changes that must have occurred during evolution. Such molecular information is determined directly by **DNA sequencing,** in which the order of nucleotide bases in DNA that codes for a gene shared by several organisms is determined. Generally, the more closely species are thought to be related on the basis of other scientific evidence, the greater the percentage of nucleotide sequences that their DNA molecules have in common (Table 17–1).

[3]Of course, not all DNA codes for proteins (witness tRNA genes). However, DNA sequencing of non-protein-coding DNA is also useful in determining evolutionary relationships.

Table 17–1 DIFFERENCES IN NUCLEOTIDE SEQUENCES IN DNA AS EVIDENCE OF PHYLOGENETIC RELATIONSHIPS

Species Pairs	*Percentage Differences in Nucleotide Sequences between Pairs of Species*
Human–human	0.5–1.0
Human–chimpanzee	2.5
Human–gibbon	5.1
Human–Old World monkey	9.0
Human–New World monkey	15.8
Human–lemur	42.0

From Stebbins, G.L. *Darwin to DNA, Molecules to Humanity.* W.H. Freeman, San Francisco, 1982.

Making the Connection

Evolution and Genetics

Can biologists make use of the fact that mutations in DNA sequences have been occurring throughout the course of evolution? Within a given taxonomic group, mutations appear to have occurred at a uniform rate over millions of years. From the number of alterations in the DNA nucleotide sequence of one organism compared with another, we can develop a **molecular clock** to estimate the time of divergence between two closely related species or higher taxonomic groups. Such molecular clocks can be used to complement geological estimates of the divergence of species or to assign tentative dates to evolutionary events that lack fossil evidence.

Molecular clocks must be developed and interpreted with care. The rates of mutation appear to vary among different genes and among different taxonomic groups. Therefore, although some mutations occur at a fairly uniform rate, a single molecular clock for all genes in all organisms cannot be elucidated. ▲

SUMMARY

I. Evolution, the genetic change in a population of organisms over time, is the unifying concept of biology.

II. Charles Darwin and Alfred Wallace independently proposed the theory of evolution by natural selection, which is based on four observations:
   A. Overproduction: Each species produces more offspring than will survive to maturity.
   B. Variation: Genetic variation exists among the individuals in a population.
   C. Limits on population growth: Organisms compete with one another for the resources needed for life, such as food, living space, water, light, and so on.
   D. Differential reproductive success: The offspring with the most favorable combination of characteristics are most likely to survive and reproduce, passing those genetic characteristics on to the next generation.

III. The synthetic theory of evolution combines Darwin's theory of evolution by natural selection with the genetic mechanisms to explain evolution.
   A. Mutation provides the genetic variability that natural selection acts on during evolution.
   B. The synthetic theory of evolution emphasizes genetics of populations rather than of individuals.

IV. The concept that evolution has occurred and is occurring is now well documented.
   A. Direct evidence of evolution comes from fossils, the remains or traces of ancient organisms.
   B. Evidence supporting evolution is derived from comparative anatomy.
      1. Homologous structures (those related by descent) have basic structural similarities, even though the structures may be used in different ways. Homologous structures indicate evolutionary ties among the organisms possessing them.
      2. Analogous organs (those not related by descent) have similar functions in quite different, genetically unrelated organisms. Analogous structures demonstrate that organisms with separate ancestries can adapt in similar ways to comparable environmental demands (convergent evolution).
      3. The occasional presence of a vestigial organ is to be expected as a species adapts to different modes of life.
   C. Developmental biology provides evidence of evolution.
      1. The embryos of related animals are more similar than the adults.
      2. The accumulation of genetic changes since organisms diverged in evolution modifies the pattern of development in more complex vertebrate embryos.
   D. Biogeography, the distribution of plants and animals, supports evolution.
      1. Each species originated only once (at its center of origin).
      2. From its center of origin, each species spread out until halted by a barrier of some kind.
      3. Areas that have been separated from the rest of the world for a long time have organisms that are unique to those areas.
   E. Molecular biology provides evidence of evolution.
      1. The universality of the genetic code is compelling evidence that all life is related.
      2. The sequence of amino acids in common proteins reveals greater similarities in closely related species.
      3. A greater proportion of the sequence of nucleotides in DNA is identical in closely related organisms.

SELECTED KEY TERMS

adaptation
analogous structure
artificial selection
biogeography
center of origin
convergent evolution
DNA sequencing
evolution
fossil
gene pool
half-life
homologous structure
index fossil
molecular clock
natural selection
neo-Darwinism
phylogenetic tree
radioactive decay
radioisotope
range
synthetic theory of evolution
vestigial organ

POST-TEST

1. The fact that all species developed from earlier forms by the accumulation of genetic changes over many successive generations is known as the theory of __________ .
2. The genetic constitution of an entire population of a given species is termed its __________ __________ .
3. Darwin proposed __________ __________ as the mechanism by which evolutionary change takes place.
4. The four premises of __________ __________ are overproduction, variation, limits on population growth, and differential reproductive success.
5. In natural selection, the selecting agent is the environment, whereas in __________ __________ , the selecting agent is humans.
6. The modern theory of evolution in which Darwin's observation of variation is explained by mutation is known as the __________ __________ of evolution.
7. The synthetic theory of evolution is also called __________ .
8. Geologists can identify specific layers of rock by the presence of certain key fossils, known as __________ __________ .
9. Unstable isotopes that spontaneously emit radiation, known as __________ , are used to date the rocks in which fossils appear.
10. An organ that appears to have little or no function, and is smaller than a similar, fully functional equivalent in the organism's ancestor or relatives, is known as a (an) __________ organ.
11. The wings of butterflies and bats have similar functions but are quite different in structure. This is an example of __________ structures.
12. The independent evolution of similar structures in two unrelated organisms is known as __________ __________ .
13. The study of the distribution of plants and animals is called __________ .
14. The portion of Earth over which a given species is found is its __________ .
15. One type of molecular evidence for evolution is __________ __________ , in which the order of nucleotide bases in a strand of DNA that codes for a gene shared by several organisms is determined.

REVIEW QUESTIONS

1. Explain briefly the concept of evolution by natural selection.
2. Why are only inherited variations important in the evolutionary process?
3. What part of Darwin's theory was he unable to explain? How does the synthetic theory of evolution explain this?
4. How do scientists date fossils? How do fossils provide evidence of evolution?
5. Distinguish among homologous, analogous, and vestigial structures. How does each provide evidence of evolution?
6. How does developmental biology provide evidence of a common ancestry for vertebrates as diverse as reptiles, birds, pigs, and humans?
7. Explain why marsupials are widespread in Australia and almost nonexistent elsewhere.
8. What is indicated if the DNA from two species is found to be almost identical?

YOU MAKE THE CONNECTION

1. Although most salamanders have four legs, a few species that live in shallow water lack hindlimbs and have extremely tiny forelimbs. Explain how limbless salamanders came about according to Lamarck's concept of evolution. Then explain these amphibians using Darwin's mechanism of evolution by natural selection.
2. Based on what you have learned in this chapter, explain why such genetically different organisms as porpoises, which are mammals, and sharks, which are fish, are so similar in body form.
3. Discuss these statements:
   a. Natural selection chooses from among the individuals in a population those most suited to *current* environmental conditions. It does not guarantee survival under future conditions.
   b. Evolution is not hierarchical. For example, humans are not "better" or "more highly evolved" than the bacteria living in our intestines.
   c. Individuals do not evolve, but populations do.
   d. Evolution is not purposeful, but is based on chance.

The narrow-striped dwarf siren (*Pseudobranchus striatus axanthus*) is an aquatic salamander that resembles an eel. It is native to Florida. (Suzanne L. Collins and Joseph T. Collins/Photo Researchers, Inc.)

RECOMMENDED READINGS

Gould, S.J. "Hooking Leviathan by Its Past." *Natural History,* May 1994. Discusses the latest evidence used to reconstruct the evolution of whales from land-dwelling mammals.

Grant, P.R. "Natural Selection and Darwin's Finches." *Scientific American,* Vol. 265, No. 4 October 1991. A study of the finches of the Galapagos Islands reveals natural selection in action during a recent drought.

Mestel, R. "Ascent of the dog." *Discover,* October 1994. All that is known and not known about the evolution of the dog.

Moore, John A. *Science as a Way of Knowing: The Foundations of Modern Biology,* Harvard University Press, Cambridge, Mass., 1993. Contains an excellent introduction to the history of evolutionary thought.

Pääbo, S. "Ancient DNA." *Scientific American,* Vol. 269, No. 5 November 1993. A review of how DNA is recovered from the tissue remains of long-dead animals and plants.

Porter, D.M. and P.W. Graham. *The Portable Darwin.* Penguin Books, New York, 1993. This collection of Darwin's writings reveals the diverse interests of the man who profoundly changed the intellectual climate of the 19th and 20th centuries.

Smith, J.M. "Bacteria Break the Antibiotic Bank." *Natural History,* June 1994. Bacterial evolution is causing the spread of antibiotic resistance among many disease-causing bacteria.

Thewissen, J.G.M., S.T. Hussain, and M. Arif. "Fossil evidence for the Origin of Aquatic Locomotion in Archaeocete Whales." *Science,* Vol. 263, 14 January 1994. This paper and the accompanying perspective describe the recently discovered fossil intermediate in whale evolution.

C H A P T E R 18

Population Genetics

Genetic variation is evident in a population of emerald tree boas in French Guyana. Many species of snakes have considerable variation in their coloration and patterns.

(J. Sauvanet, Peter Arnold Inc.)

A **population** consists of all the individuals of the same species that live in a particular place at a specific time. Individuals in a population exhibit genetic variation—as represented by the number and kinds of alleles in a population—for the traits characteristic of that population. As shown in the photograph, a population of emerald tree boas may all possess the same trait (a scaly skin, for example) but they vary from one individual to another in coloration and pattern.

Each population possesses an isolated **gene pool,** which is the total genetic material of all the individuals within the population (see Chapter 17). The gene pool includes all possible alleles of each gene present in the population. A single individual has only a small fraction of the alleles found in a population's gene pool. Because most species are diploid, each individual member of a population contains only two alleles for each locus, one on each of the homologous chromosomes (members of a chromosome pair).

Evolution occurs in populations, not individuals. Although natural selection acts on individuals by determining which of them will survive to reproduce, individuals do not evolve during their lifetimes. Populations, however, undergo evolutionary change over many generations. Evolutionary change, which includes modifications in structure, physiology, ecology, and behavior, is inherited from one generation to the next. Although Darwin recognized that evolution occurs in populations, he did not understand how traits are passed on to successive generations. One of the most significant advances in biology since Darwin's time has been the demonstration of the genetic basis of evolution.

If a population is not undergoing evolutionary change, the frequencies of each allele in the gene pool remain constant from generation to generation. During the course of evolution, however, some alleles increase in frequency within a population, whereas other alleles decrease or even disappear. Thus, populations evolve by changes in allele frequencies—that is, by changes in the relative proportions of their various alleles.

Evolution that involves changes in allele frequencies over successive generations is called **microevolution** (micro- comes from the Greek *mikros,* meaning "small") because it involves small, gradual changes within a population. In this chapter, we examine the factors responsible for microevolution: how the relative abundance of different alleles changes in a population during the course of evolution.

LEARNING OBJECTIVES

After you have studied this chapter you should be able to

1. Define population, gene pool, allele frequency, and microevolution.
2. Distinguish between the gene pool of a population and the genotype of an individual.
3. Discuss the significance of the Hardy-Weinberg principle as it relates to evolution, and list the five conditions required for genetic equilibrium.
4. Discuss how each of the following alters allele frequencies in populations: genetic drift, gene flow, mutation, and natural selection.
5. Distinguish among stabilizing selection, directional selection, and disruptive selection, and give an example of each.
6. Describe the nature and extent of genetic variation, including genetic polymorphism.
7. Explain how the sickle cell allele illustrates heterozygote advantage.
8. Relate how frequency-dependent selection affects genetic variation.

THE HARDY-WEINBERG PRINCIPLE DEMONSTRATES GENETIC EQUILIBRIUM

If we set a net over a bunch of ripe bananas, we can sample a population of fruit flies, *Drosophila melanogaster*. After anesthetizing and counting them, we might find that we have 1000 fruit flies, 910 with gray bodies and 90 with black bodies. After the fruit flies are released, they will mate. If we trap and count the next generation of fruit flies, we might find a population that is essentially the same as the previous one, with roughly nine gray flies to every one black fly. If we do this for a succession of generations, we might always get the same result.

The explanation for this stability of successive generations of populations in genetic equilibrium was provided independently by Godfrey Hardy, an English mathematician, and Wilhelm Weinberg, a German physician, in 1908. They pointed out that the frequencies of various genotypes (genetic makeups) in a population can be described mathematically. The resulting **Hardy-Weinberg principle** represents an ideal situation that probably never occurs in the natural world. However, it is useful because it provides a model to help us understand the real world.

The Hardy-Weinberg principle shows that in large populations, the process of inheritance does not by itself cause changes in allele frequencies. It also explains why dominant alleles are not necessarily more common than recessive ones. Thus, knowledge of the Hardy-Weinberg principle is essential to understanding the mechanisms of evolutionary change in sexually reproducing populations.

We now expand the fruit fly example to explain the Hardy-Weinberg principle. A few simple crosses (controlled mating experiments) of black and gray fruit flies reveal that the allele for gray body, *B*, is dominant over the allele for black body, *b*. Gray-bodied flies include some that are homozygous, *BB*, and some that are heterozygous, *Bb*. Obviously, all the black flies are homozygous, *bb*. The frequency of either allele, *B* or *b*, is described by a number from zero to one. An allele that is totally absent from the population has a frequency of zero. If all of the alleles of a given locus are the same in the population, then the frequency of that allele is one.

Because only two alleles, *B* and *b*, exist for the locus, the sum of their frequencies must equal one. If we let p represent the frequency of the dominant (*B*) allele, and q the frequency of the recessive (*b*) allele in the population, then we can summarize their relationship with a simple binomial equation, $p + q = 1$. (A binomial equation is an algebraic expression that consists of two quantities connected by a plus or minus sign.) When we know the value of either p or q, we can calculate the value of the other: $p = 1 - q$ and $q = 1 - p$.

Because $p + q = 1$, then $(p + q)^2 = 1$. This binomial equation can be expanded to describe the relationship of the allele frequencies to the genotypes in the population. When it is expanded, we obtain the frequency of the offspring genotypes:

p^2	+	$2pq$	+	q^2	=	1
Frequency of *BB*		Frequency of *Bb*		Frequency of *bb*		All the individuals in a population

From the fact that we had 90 black flies in our population of 1000, we know that the frequency of the *bb* genotype, q^2, is 90/1000, or 0.09. Since q^2 equals 0.09, q is equal to the square root of 0.09, or 0.3. From the relationship between p and q, we know that $p = 1 - q = 1 - 0.3 = 0.7$.

Based on this information, we can calculate the frequency of homozygous gray flies, *BB*: $p^2 = 0.7 \times 0.7 = 0.49$. The frequency of heterozygous gray flies, *Bb*, would be: $2pq = 2 \times 0.7 \times 0.3 = 0.42$. Thus, approximately 490 of the gray flies are homozygous and 420 are heterozygous. Note that the sum of homozygous and heterozygous gray flies equals 910, the number with which we began.

Any population in which the distribution of genotypes conforms to the relation $p^2 + 2pq + q^2$, whatever

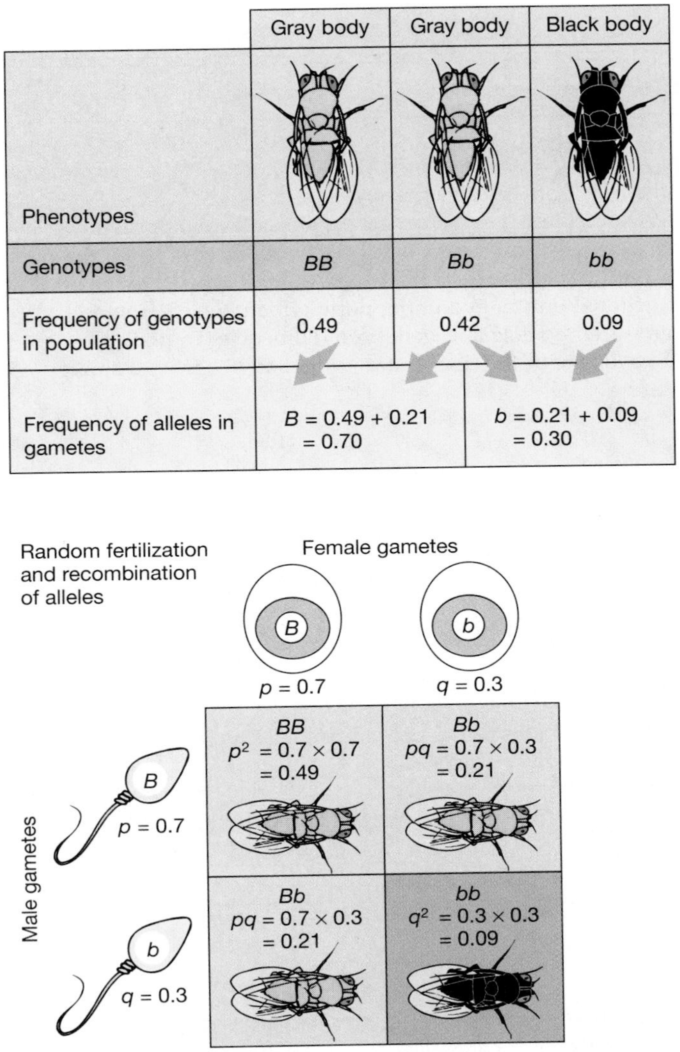

FIGURE 18–1 How to use the Hardy-Weinberg principle of genetic equilibrium. When eggs and sperm containing *B* or *b* alleles unite randomly, the frequency of appearance of each of the possible genotypes *(BB, Bb, bb)* in the offspring is calculated by multiplying the frequencies of the alleles *B* and *b* in eggs and sperm.

the absolute values for p and q may be, is at genetic equilibrium (Fig. 18–1). *The Hardy-Weinberg principle of genetic equilibrium tells us what to expect when a sexually reproducing population is not evolving.* The proportion of alleles in successive generations will always be the same, provided the following criteria are met:

1. **Random mating.** In unrestricted random mating, each individual in a population has an equal chance of mating with any individual of the opposite sex. In our example, the individuals represented by *BB* (homozygous dominant), *Bb* (heterozygous), and *bb* (homozygous recessive) must mate with one another at random and must not select their mates on the basis of genotype. There must be equal probabilities of matings among genotypes; that is, matings among genotypes must occur in proportion to the frequencies of the genotypes. (For example, if *BB* and *Bb* flies occur in the proportions of 0.49 and 0.42, respectively, $BB \times Bb$ matings should occur at a frequency of $0.49 \times 0.42 = 0.21$.)
2. **No mutation.** There must be no mutations of *B* or *b*. That is, the relative numbers of *B* and *b* must not change due to mutations.
3. **Large population size.** Because the Hardy-Weinberg principle is a statistical tool, the population must be large enough that the effect of chance is small. Allele frequencies in a small population are more likely to be affected by chance events (genetic drift, discussed shortly) than allele frequencies in a large population.
4. **No migration.** There can be no exchange of genes with other populations that might have different allele frequencies. In other words, there can be no migration of individuals into or out of a population.
5. **No natural selection.** If natural selection is occurring, certain genotypes are favored over others. Consequently, the allele frequencies will change and the population will evolve.

Evolution occurs when allele frequencies undergo change in a gene pool

Allele frequencies in nature are often significantly different from those that the Hardy-Weinberg principle would predict. Evolution represents a departure from the Hardy-Weinberg principle of genetic equilibrium. Changes in the gene pool of a population result from four microevolutionary processes: mutation, genetic drift, gene flow, and natural selection. When one or more of these processes is acting on a population, allele frequencies will change from one generation to the next.

Mutation increases variation within a gene pool

Variation is introduced into a gene pool through **mutation,** which is an unpredictable, permanent change in DNA (Fig. 18–2). Mutations are the ultimate source of all new alleles. Although allele frequencies may be changed by mutation, these changes are typically several orders of magnitude smaller than changes caused by other evolutionary forces. As an evolutionary force, mutation is often negligible, but it is important as the ultimate source of variation.

Mutations result from a change in the nucleotide base pairs of a gene, from a rearrangement of genes within chromosomes so that their interactions produce different

Making the Connection

Relating Mendelian Genetics to Population Genetics

How does the Hardy-Weinberg principle of genetic equilibrium relate to Mendel's principles of inheritance (discussed in Chapter 10)? The Hardy-Weinberg principle describes the frequencies of alleles in the genotypes of an entire breeding population. In contrast, the principles of inheritance worked out by Gregor Mendel describe the frequencies of genotypes among offspring of a single mating. Thus, the emphasis in population genetics is on populations, whereas the emphasis in Mendelian genetics is on individuals. Population genetics studies populations over many generations, whereas Mendelian genetics focuses on offspring from a single mating.

The field of population genetics enables biologists to study the role of genetics in the evolutionary process. Because the population (rather than the individual) is the functional unit in evolution, population genetics is concerned with allele and genotype frequencies in populations, not with the distribution of genotypes from a single mating. ▲

A COMPARISON OF INDIVIDUALS AND POPULATIONS

	Individual	*Population*
Period of existence	One generation	Many generations
Genetic description	Genotype	Gene pool
Genetic variation	No more than two different alleles for each locus	Substantial polymorphism involving many loci
Ability to evolve	No	Yes (change in allele frequencies)

(a)

(b)

FIGURE 18–2 Hundreds of mutations are known in fruit flies *(Drosophila melanogaster)*. (*a*) A normal fly. (*b*) A mutant with vestigial wings. Because mutations are random changes in genetic material, most mutations are neutral or harmful to the organism (as is obviously the case in the vestigial wing mutation shown). Yet for island-dwelling insects, fully developed wings might be more of a disadvantage than an advantage, permitting the insect to be too easily blown away from land. Perhaps for this reason, flies and other insects that dwell on small islands frequently have reduced wings or are entirely wingless. (*a, b*, Peter J. Bryant/Biological Photo Service) effects, or from a change in the chromosomes. Mutations occur unpredictably and spontaneously. The rates of mutation appear to be relatively constant for a particular gene but may vary by several orders of magnitude among genes within a single species and among different species.

Not all mutations are passed from one generation to the next. Those occurring in somatic (body) cells are not heritable. When an individual with such a mutation dies, the mutation dies with it. Some mutations, however, alter the DNA in reproductive cells. These mutations may or may not affect the offspring, because most of the DNA in a cell is "silent" and does not code for specific polypeptides or proteins that are responsible for physical characteristics. Even if a mutation occurs in the DNA that codes for a polypeptide, it may still have little effect on the structure or function of that polypeptide. However, when the polypeptide is altered enough to change how it functions, the mutation is usually harmful, just as tinkering with the inner workings of a watch sometimes results in improved performance, but is much more often disastrous.

Mutations do not determine the *direction* of evolutionary change. Consider a population living in an increasingly dry environment. A mutation producing a new allele that helps an individual adapt to dry conditions is no more likely to occur than one for adapting to wet conditions or one with no relationship to the changing environment. The production of new mutations simply increases the genetic variability within a population and, therefore, the potential for new adaptations.

Most mutations produce small changes in the phenotype that are often detectable only by sophisticated biochemical techniques. By acting against seriously abnormal phenotypes, natural selection eliminates or reduces to low frequencies the most harmful mutations. Small mutations, even those with slightly harmful phenotypic effects, have a better chance of being incorporated into the gene pool, where at some later time they may produce traits that are helpful or adaptive for the population.

Genetic drift causes changes in allele frequencies by random, or chance, events

The size of a population has important effects on allele frequencies because random events, or chance, will tend to cause changes in a small population. If a population consists of only a few individuals, an allele present at a low frequency in the population could be lost purely by chance. Such an event would be most unlikely in a large population.

For example, consider two populations, one with 10,000 individuals and one with 10 individuals. If an uncommon allele occurs at a frequency of 10%, or 0.1, then 1900 individuals in the large population possess the allele [$2pq + q^2 = 2 \quad (.9)(.1) + (.1)^2 = 0.18 + 0.01 = 0.19$; $0.19 \times 10{,}000 = 1900$]. That same frequency, 0.1, in the smaller population means that no more than 2 individuals possess the allele ($0.19 \times 10 = 1.9$). From this exercise, it is easy to see that there is a greater likelihood of losing the uncommon allele from the smaller population than from the larger one. Predators, for example, might happen to kill one or two individuals possessing the uncommon allele in the smaller population purely by chance.

The production of random evolutionary changes in small breeding populations is known as **genetic drift.** Genetic drift results in allele frequency changes in the gene pool of a population from one generation to another. One allele may be eliminated from the population by chance, regardless of whether that allele is beneficial, harmful, or of no particular advantage or disadvantage. Thus, genetic drift can decrease genetic variation *within* a population, although it tends to increase the genetic differences *among* different populations.

Genetic bottlenecks cause genetic drift

Because of fluctuations in the environment, such as depletion in food supply or an outbreak of disease, a population may periodically experience a rapid and marked decrease in the number of individuals. The population is said to go through a **genetic bottleneck** in which genetic drift can occur in the few remaining survivors. As the population again increases in size, many allele frequencies may be quite different from those in the population preceding the decline (Fig. 18–3).

Genetic variation in the cheetah was considerably reduced by a genetic bottleneck at the end of the last Ice Age, some 10,000 years ago. At that time, cheetahs almost became extinct. The few surviving cheetahs had a greatly reduced genetic variability, and as a result the cheetah population today is genetically uniform or homogeneous.

The founder effect occurs when a few "founders" establish a new colony

When one or a few individuals from a large population establish, or found, a colony (as when a few birds separate from the rest of the flock and fly to a new area), they bring with them only a small fraction of the genetic variation present in the original population. As a result, the only alleles represented among their descendants will be those few that the colonizers happened to possess. This often results in allele frequencies in the newly founded population that are very different from those possessed by other populations of the same species. The genetic drift that results when a small number of individuals from a large population colonize a new area is called the **founder effect** (Fig. 18–4; see *Focus On: Evolution of the Africanized Honeybee*).

The ABO blood groups in Australian Aborigines illustrate the founder effect. Recall that there are three different alleles (I^A, I^B, and i^O) that result in four different phenotypic blood types (A, B, AB, and O) (see Chapter

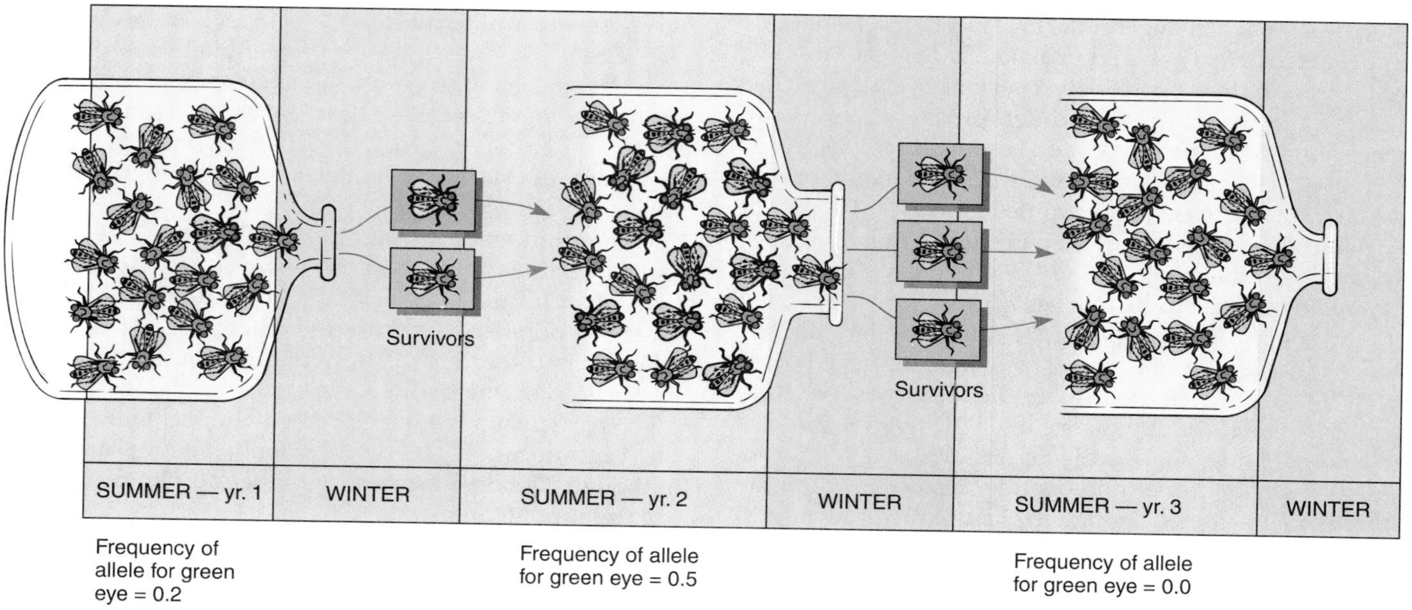

FIGURE 18–3 The result of a genetic bottleneck is genetic drift. Because only a small population of flies survives the winter, its genotypes, not necessarily resulting from natural selection, determine the allele frequencies of the entire succeeding summer population. (The allele frequencies do not match the phenotypic ratios depicted because heterozygous individuals exhibit the dominant phenotype but contribute recessive alleles to the allele frequencies.)

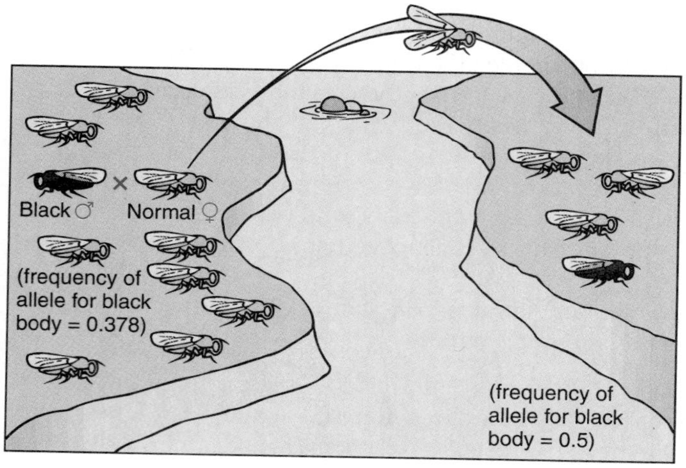

FIGURE 18–4 The founder effect causes genetic drift. In this hypothetical example, a single inseminated female fruit fly is blown to an island. The alleles of her offspring, which occur in frequencies different from those in the original population, will serve as the foundation for the gene pool of all future fruit fly populations on that island. (The allele frequencies do not match the phenotypic ratios depicted because heterozygous individuals exhibit the dominant phenotype but contribute recessive alleles to the allele frequencies.)

15). Apparently the small founder population of humans who colonized Australia, eventually becoming Aborigines, lacked the I^B allele. Consequently, Australian Aborigines do not have either type B or type AB blood.

Gene flow, caused by migration, generally increases variation in the gene pool

Members of a species tend to be distributed in local populations that are more or less isolated genetically from other populations. For example, the bullfrogs of one pond form a population separated from those in an adjacent pond (Fig. 18–5). Some exchanges occur by migration between ponds, but the frogs in one pond are much more likely to mate with those in the same pond. Members of most species tend to be distributed in such local populations. Because each population is more or less isolated from other populations of the species, they can have distinct genetic traits.

The migration of breeding individuals between populations causes a corresponding movement of alleles, or **gene flow,** that can have significant evolutionary consequences. As alleles flow from one population to another, they usually increase the amount of genetic variability within the population that receives them. If the gene flow between two populations is great enough, they become more similar genetically. Because gene flow has a tendency to reduce the amount of variation between two populations, it tends to counteract the effects of natural selection and genetic drift, both of which cause populations to become increasingly distinct.

If migration by members of a population is considerable, and if populations differ in their allele frequencies, then significant genetic changes can result. For example, by 10,000 years ago modern humans occupied all of Earth's major land areas except a few islands. Because the

Focus On

Evolution of the Africanized Honeybee

Periodically newspapers report on the migration of "killer bees," or Africanized honeybees, into North America (see figure). The movement of these bees is of concern for several reasons. First, the Africanized honeybee, which looks just like any other honeybee, is considered dangerous because it is unpredictable and aggressive. Groups of them have been reported to attack at the slightest provocation. Also, the Africanized honeybee interbreeds or competes with the European honeybee, the type of bee raised commercially in this country. It is feared that interbreeding will adversely affect the European honeybees' important roles in pollination and honey production.

Africanized honeybees are descended from a small number of African honeybees introduced into Brazil in 1956. (The European honeybee does not fare well in the tropical climate of Brazil.) The number of introduced African honeybees was small, and as a result, they contained only a fraction of the genes present in the gene pool of African honeybees. Because of the founder effect, the few alleles present in the introduced population formed the gene pool on which natural selection acted in the new South American environment. Natural selection of this limited gene pool resulted in the Africanized honeybee.

Africanized honeybees, which are better adapted than European honeybees to tropical areas, have spread beyond their original point of origin in Brazil to occupy large areas of Latin America. Wherever they have migrated, decreases in honey production have initially occurred. Beekeeping in countries occupied by the Africanized bees has also changed, because the beekeepers must now wear more protective equipment. However, as beekeepers develop new methods to handle the Africanized bees, honey production often increases. For example, Brazil was 47th in world honey production before the African bees were introduced; today it ranks 7th.

A swarm of Africanized honeybees covers a tree trunk in Mexico. (Scott Camazine/Photo Researchers, Inc.)

Africanized honeybees currently range from Argentina to the southern United States. How far will they migrate? Will they stop in the southern United States because of a climate barrier? Or is it possible that they can expand into the northerly areas of North America? Because Africanized honeybees survive on snow-clad mountains in Latin America, biologists predict that climate will pose no barrier.

Some biologists advocate using artificial selection (see Chapter 17) to improve Africanized honeybees. The bees have a number of desirable traits, including resistance to disease and mites, a particularly serious problem for commercial beekeepers in the United States. Artificial selection might even give rise to a honeybee variety that is as important commercially as the European honeybee. ■

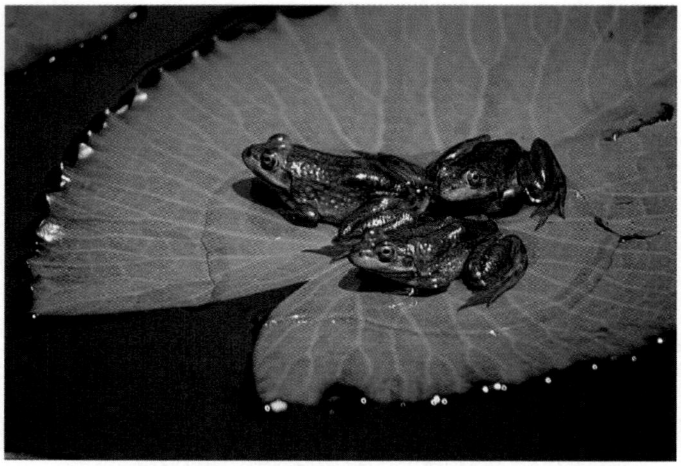

FIGURE 18–5 Three members of a population of bullfrogs *(Rana catesbiana)* rest on a lily pad. The frogs in one pond are somewhat isolated from frogs in other ponds and therefore tend to mate with frogs in their home pond. (William E. Ferguson)

population density was low, the small, isolated human populations underwent random genetic drift and natural selection. More recently (during the past 300 years or so), major migrations have caused an increase in gene flow, significantly altering allele frequencies within various human populations.

Natural selection changes allele frequencies in a way that increases adaptation

Natural selection is the mechanism of evolution first proposed by Darwin in which members of a population that possess more successful adaptations to the environment are more likely to survive and reproduce (see Chapter 17). Over successive generations, the proportion of more favorable alleles increases in the population. Natural selection counteracts the random effects of the other microevolutionary processes (mutation, genetic drift, and gene flow) and leads to adaptive evolutionary change. As a result, a population becomes better adapted to the environment in which it lives.

Natural selection not only explains why organisms are well-adapted to the environments in which they live, but also helps account for the remarkable diversity of life. Natural selection enables populations to change, thereby adapting to different environments and different ways of life.

Natural selection results in the differential reproduction of individuals with different observable traits, or phenotypes (and therefore different genotypes) in response to the environment. Natural selection functions to preserve individuals with favorable genotypes and eliminate those with unfavorable genotypes. Individuals have a selective advantage if they are able to survive and produce fertile offspring. Natural environmental pressures, such as competition for food or water or living space, "select" the individuals that survive to reproduce.

Fitness is a measure of the ability of an organism, owing to its genotype, to compete successfully and make a genetic contribution to subsequent generations. Organisms favored by natural selection exhibit high fitness, whereas organisms exposed to adverse selection pressures exhibit low fitness.

The mechanism of natural selection does not cause the development of a "perfect" organism. Rather, it "weeds out" those individuals whose phenotypes are less adapted to environmental challenges, while allowing better adapted individuals to survive and pass their alleles on to their offspring. Natural selection is the only process known that brings genetic variation into harmony with the environment and leads to adaptation. By reducing or eliminating alleles that result in the expression of less favorable traits, the probability that favorable alleles responsible for an adaptation will come together in the offspring is increased. Natural selection changes the composition of a gene pool in a direction that causes the population as a whole to be better suited to its environment.

To summarize, using the Hardy-Weinberg principle of genetic equilibrium as our base of comparison, we can see that evolution is occurring in most populations because allele frequencies are always changing. In most instances, microevolutionary changes in allele frequencies are inevitable in natural populations of organisms.

SELECTION OPERATES ON AN ORGANISM'S PHENOTYPE

Natural selection does not act directly on an organism's genotype. Rather, it acts on the phenotype, which is at least in part, an expression of the genotype. Natural selection acts on the heritable component of the phenotype, which represents an interaction of all the alleles in the organism's genotype. It is rare that a single gene pair has complete control over a single phenotypic trait, such as Mendel originally observed in garden peas. Much more common is the interaction of several genes for the expression of a single trait (see Chapter 10). Many plant and animal characteristics are under this type of polygenic control.

When traits (human height, for example) are under polygenic control, a range of phenotypes occurs, with most of the population located in the median range and fewer at either extreme. This is a normal distribution or standard bell curve (Fig. 18–6*a*; see also Fig. 10–21).

Three kinds of selection occur that cause changes in the normal distribution of phenotypes in a population: stabilizing, directional, and disruptive selection. Although we consider each process separately, their influences generally overlap in nature.

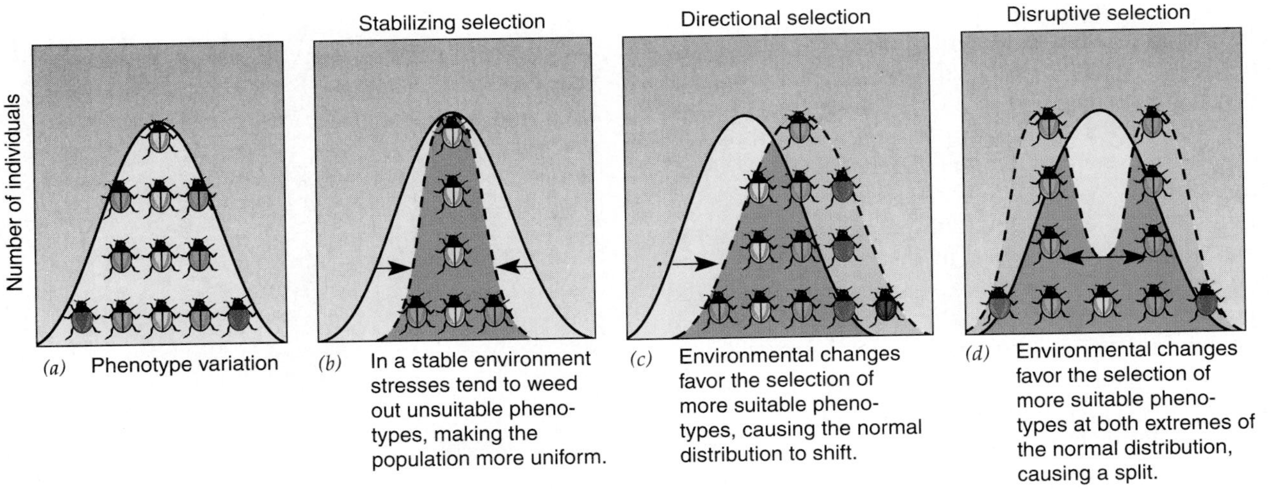

Stabilizing selection
Directional selection
Disruptive selection
Number of individuals
(a) Phenotype variation
(b) In a stable environment stresses tend to weed out unsuitable phenotypes, making the population more uniform.
(c) Environmental changes favor the selection of more suitable phenotypes, causing the normal distribution to shift.
(d) Environmental changes favor the selection of more suitable phenotypes at both extremes of the normal distribution, causing a split.

FIGURE 18–6 The distribution of wing colors in a hypothetical population of beetles demonstrates different types of natural selection. (*a*) A trait that is under polygenic control exhibits a normal distribution of phenotypes. (*b*) As a result of stabilizing selection, the curve is narrower. (*c*) Directional selection moves the curve in one direction. (*d*) Disruptive selection results in two or more peaks.

Stabilizing selection favors intermediate phenotypes

The process of natural selection associated with a population that is well-adapted to its environment is known as **stabilizing selection.** Most populations are probably under the influence of stabilizing forces most of the time. In stabilizing selection, phenotypic extremes are selected against. In other words, individuals with a phenotype near the mean are favored.

One of the most widely studied cases of stabilizing selection involves human birth weight, which is under polygenic control and is influenced by environmental factors. Extensive data from hospitals have shown that infants born with intermediate weights are more likely to survive (Fig. 18–7). Infants at either extreme (too small or too large) have higher rates of mortality. When newborn infants are too small, their body systems are immature, and when they are too large, they have difficult deliveries because they cannot pass as easily through the birth canal. Stabilizing selection operates to reduce the variability in birth weight so that it is close to the weight with the minimum mortality rate.

Because stabilizing selection tends to decrease variation by favoring individuals near the mean of the normal distribution at the expense of those at either extreme, the bell curve narrows (Fig. 18–6*b*). Although stabilizing selection decreases the amount of variation in a population, variation is rarely eliminated by this process because other microevolutionary processes act against a decrease in variation. For example, mutation is slowly but continually adding to the genetic variation within a population.

Directional selection favors one phenotype over another

If an environment changes over time, **directional selection** may favor phenotypes at one of the extremes of the normal distribution (Fig. 18–6*c*). Over successive gener-

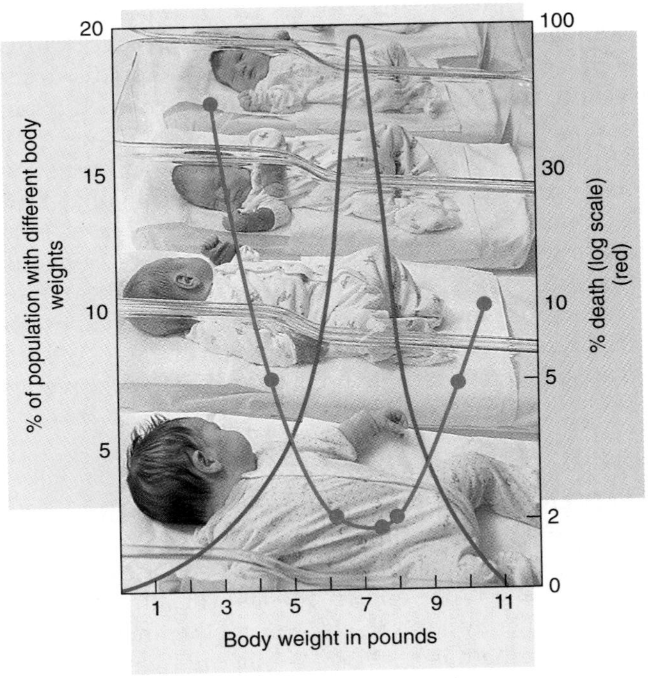

FIGURE 18–7 Human birth weight is an example of stabilizing selection. Infants with very low or very high birth weights have higher death rates. The blue line indicates percentage of infants born at each weight; the red line indicates percentage of deaths at each birth weight. (Terry Wild Studio)

ations, one phenotype gradually replaces another. So, for example, if greater size is advantageous in a new environment, larger individuals will become increasingly more common in the population. Directional selection can only occur, however, if the appropriate alleles (those favored under the new circumstances) are already present in the population.

A classic example of directional selection is the peppered moth population studied in England (see *Focus on Evolution in Action* in Chapter 1). Recall that most of the peppered moths *(Biston betularia)* in rural England have a light (black and white peppered) wing color and that only a few are melanic, or all black. In industrial regions, the situation is reversed: most of the moths are black and only a few are light. In other words, natural selection is directional; in some places, natural selection operates in favor of the light phenotype, whereas in other places, selection occurs in the opposite direction toward the black form. Seldom, however, does a population become entirely one type. Male peppered moths fly considerable distances, so gene flow between populations tends to maintain both forms.

Disruptive selection selects for phenotypic extremes

Sometimes extreme changes in the environment may favor two or more different phenotypes at the expense of the mean. That is, more than one phenotype may be favored in the new environment, whereas the average, or intermediate, phenotype is selected against. **Disruptive selection** is a special type of directional selection in which there is a trend in several directions rather than just one (Fig. 18–6*d*). It results in a divergence, or splitting apart, of distinct groups of individuals within a population.

Limited food supply during a severe drought caused a population of Galapagos finches to undergo disruptive selection. The finch population initially exhibited a variety of beak sizes and shapes. Because the only food available during the drought was wood-boring insects and seed from cactus fruits, natural selection favored birds with beaks suitable for obtaining these types of food. Finches with longer beaks survived because they could open cactus fruits, whereas those with wider beaks survived because they could strip off tree bark to expose insects. Thus, finches with beaks at two extremes of the normal distribution were favored over birds with average beaks.

GENETIC VARIATION IS NECESSARY IF EVOLUTION IS TO OCCUR

Change in the types and frequencies of alleles in gene pools, whether by natural selection or other means, is possible only if there is a source of inherited variation. Genetic variation is the raw material for evolutionary change. It provides the diversity on which natural selection can act. Without genetic variation there can be no heritable differences in the ability to reproduce, and therefore no evolution driven by natural selection (Fig. 18–8).

FIGURE 18–8 Genetic variation is evident in the shell patterns and colors that occur in this tree snail species *(Polymita picta).* Variation in shell color has been demonstrated to have adaptive value in some snails. Some colors predominate in cooler environments, whereas other colors are adaptively superior in warmer habitats. (Chip Clark)

The gene pools of populations often contain large reservoirs of genetic variation that have been introduced by mutation. Sexual reproduction, with its associated crossing over, independent assortment of chromosomes during meiosis, and union of gametes, also contributes to genetic variation. The sexual process allows the variability introduced by mutation to be combined in new ways, which may be expressed as new phenotypes.

The effect of recombination can be surprisingly great. Nine different genotypes are generated in a dihybrid cross (*AaBb* × *AaBb*) involving only two loci, each with only two alleles.[1] If we were dealing with five different unlinked loci, each with six alleles, the number of different genotypes possible would be 4,084,101![2] Since most organisms possess thousands of different loci, the num-

[1]The nine genotypes that result from *AaBb* × *AaBb* are *AABB, AABb, AAbb, AaBB, AaBb, Aabb, aaBB, aaBb,* and *aabb.*

[2]There are 21 different ways that six alleles can combine in diploid individuals. Because there are five different genes, each with six alleles, the total number of different genotypes possible is 21^5, or 4,084,101.

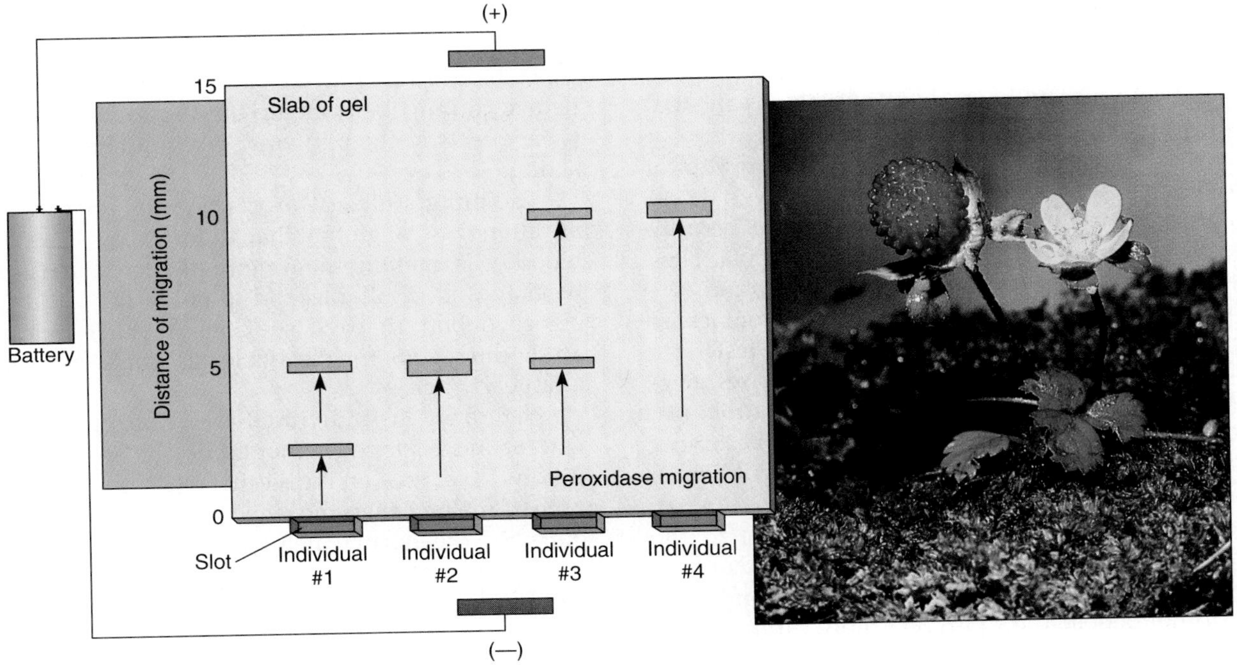

FIGURE 18–9 Gel electrophoresis demonstrates genetic polymorphism in a wild strawberry population. Tissue extracts containing the enzyme peroxidase from different individuals were placed in slots in a slab of gel *(bottom of drawing)*. An electric current was applied to the gel, and because peroxidase has a net negative charge, it migrated toward the positive side. Slight variations in amino acid sequence in the peroxidase molecules caused them to have slightly different negative charges and therefore to migrate at different rates. Three different forms of peroxidase were found among the individuals studied. Of course, each individual can only possess a maximum of two different forms because strawberries are diploid and therefore carry two alleles for this protein. Individuals with two different forms of peroxidase are heterozygous, whereas those with only one form are homozygous. (Richard H. Gross)

ber of possible allele combinations and resulting phenotypes is staggering! Some of the combinations may be adaptively superior and favored by natural selection.

Genetic polymorphism is an example of variation

One way of evaluating genetic variation in a population is to examine **genetic polymorphism,** which is the presence in a population of two or more alleles for a given locus. Gene pools contain a tremendous reservoir of genetic polymorphism, much of which is present at low frequencies and much of which is hidden because it does not produce distinct phenotypes.

Biologists estimate the total amount of genetic polymorphism in populations by comparing their proteins (each variety of a particular protein is coded by a different allele; Fig. 18–9) or DNA. Determining the sequence of nucleotides in DNA (see Chapter 17) from individuals in a population provides a *direct* estimate of genetic polymorphism. One method of DNA sequencing is described in Figure 14–12. DNA sequencing of specific genes in an increasing number of organisms, including humans, indicates that the reservoir of genetic polymorphism in most populations is immense.

The genes that code for human HLA proteins are polymorphic

Although their exact function is unknown, HLA proteins appear to be important in the recognition of infectious organisms by the body (HLA stands for human leukocyte antigen; see Chapter 43). Once a cell or molecule is recognized as foreign, the body mounts an immune response against the invader.

HLA proteins are highly variable in the human population. (The rejection of heart, kidney, or other organ transplants is caused by this variability. The body's immune system recognizes that the HLA proteins on the transplanted tissues are foreign.) Humans possess at least seven different loci that code for HLA proteins, and most of these have a remarkable number of different possible alleles. Because humans are diploid organisms, each individual contains only two alleles for each locus. However, the extremely polymorphic nature of the different HLA loci means that the human population contains thousands of individual variants.

HLA proteins are linked to susceptibility to viral infections; that is, certain combinations of HLA proteins cause an increased resistance to specific viral infections. Acquired immune deficiency syndrome (AIDS), for example, is a deadly immune disease caused by a virus. The virus, called human immunodeficiency virus (HIV-1), infects and destroys cells of the immune system, thereby making the body susceptible to deadly infections and diseases (see Chapter 43).

Medical researchers report that a small group of women in Africa appear to be resistant to HIV-1. These women are regularly exposed to the AIDS virus because they are prostitutes in the slums of Nairobi, Kenya. They have as many as 2000 sexual partners each year, and 10% to 15% of the men with which they come into contact are HIV-positive. The AIDS-resistant prostitutes are tested for HIV infection on a regular basis and repeatedly test negative. (In contrast, most of the prostitutes in Nairobi test positive for HIV infection a few months after becoming prostitutes. These women subsequently develop AIDS an average of four years after first being exposed.)

Why are some women apparently immune to the AIDS virus? Early in their study, researchers ruled out behavior as a possible cause. That is, the AIDS-immune women do not choose their partners more carefully, nor do they practice safe sex more regularly than other prostitutes. One thing that the AIDS-immune women appear to have in common is the presence of up to three specific HLA protein molecules on the surface of their blood cells. If the presence of AIDS immunity in a human population is substantiated, it may ultimately lead to the development of a vaccine to prevent AIDS. Researchers caution that it will take years to duplicate that immunity in the lab, and that is a prerequisite for producing a vaccine.

Balanced pomorphism can exist for long periods of time

Balanced polymorphism is a special type of genetic polymorphism in which two or more alleles persist in a population over many generations as a result of selection. Heterozygote advantage and frequency-dependent selection are mechanisms that preserve balanced polymorphism.

Genetic variation can be maintained by heterozygote advantage

We have seen that natural selection often proceeds in such a way that unfavorable alleles are eliminated from a population while favorable alleles are retained. However, natural selection sometimes helps to maintain genetic diversity, including alleles that by themselves are unfavorable, in a population. This happens, for example, when the heterozygote, *Aa*, has a higher degree of fitness than either homozygote, *AA* or *aa*. This phenomenon, known as **heterozygote advantage,** is demonstrated in humans by the selective advantage of heterozygous carriers of the sickle cell allele.

Heterozygote Advantage of the Sickle Cell Allele. Many genes are pleiotropic and affect more than one trait (see Chapter 10). Whether a mutant allele is harmful or beneficial must be evaluated in the context of its phenotypic effects in the particular environment in which it is acting.

The mutant allele *(s)* for sickle cell anemia produces an altered hemoglobin that deforms or sickles the red blood cells, making them more likely to form dangerous blockages in capillaries and to be destroyed in the liver, spleen, or bone marrow (see Chapter 15). Individuals who are homozygous for the sickle cell allele *(ss)* usually die at an early age.

Heterozygous individuals carry alleles for both normal *(S)* and sickle cell hemoglobin. The heterozygous condition *(Ss)* causes an individual to be more resistant to a type of malaria than those individuals who are homozygous for the normal hemoglobin allele *(SS)*. In a heterozygous individual, each allele produces its own specific kind of hemoglobin, and the red blood cells contain the two kinds in roughly equivalent amounts. Such cells do not ordinarily sickle as readily as cells containing only the *s* allele, and the red blood cells containing the abnormal hemoglobin are more resistant to infection from the malarial parasite (which lives in red blood cells) than are the red blood cells containing only normal hemoglobin.

Each of the two types of homozygous individuals is at a disadvantage. Those homozygous for the sickle cell allele are likely to die of sickle cell anemia, whereas those homozygous for the normal allele may suffer or die of malaria. The heterozygote is therefore more fit than either homozygote. In certain parts of Africa, India, and southern Asia, where malaria is prevalent, heterozygous individuals survive in greater numbers than either homozygote (Fig. 18–10). Both alleles are maintained in the population even though the homozygous recessive condition is lethal.

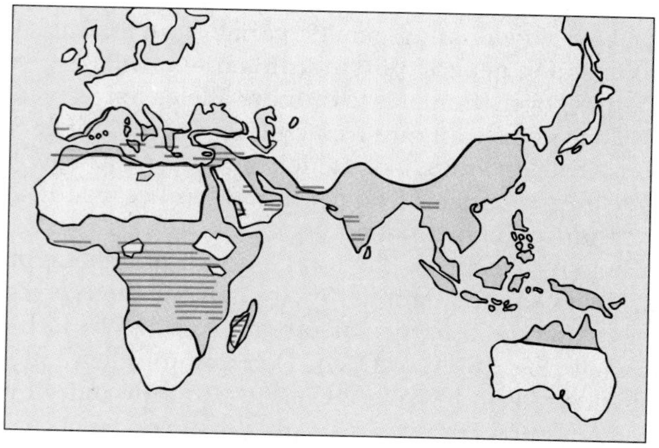

FIGURE 18–10 The distribution of sickle cell anemia *(bars)* is compared with the distribution of falciparum malaria *(dark tan region)*. The correlation strongly suggests that the resistance of heterozygous individuals to malaria has served to balance the harmful effects of sickle cell anemia.

Genetic variation may be maintained by frequency-dependent selection

Thus far in our discussion of natural selection, we have assumed that the fitness of particular phenotypes (and their corresponding genotypes) is independent of their frequency in the population. There are, however, cases of **frequency-dependent selection,** in which the fitness of a particular phenotype depends on how frequently it appears in the population. Often, a phenotype has a greater adaptive value when it is rare than when it is common in the population. Such phenotypes lose their advantage as they become more common.

Frequency-dependent selection often acts to maintain genetic variation in populations of prey. In this case, the predator catches and consumes the more common phenotype, but may ignore the rarer phenotypes. Frequency-dependent selection has been demonstrated with aquatic insects called water boatmen (Fig. 18–11), which have three distinct color phenotypes. When all three phenotypes are present in equal frequencies, fish are more likely to consume the most obvious (least camouflaged) form. However, in populations where one phenotype is present in greater numbers than the other two, the most abundant form is preferentially eaten by fish, regardless of its coloration. (The reason for this is unknown.) Thus, frequency-dependent selection acts to decrease the frequency of the more common phenotypes (and their genotypes) and increase the frequency of the less common types.

FIGURE 18–11 Frequency-dependent selection has been demonstrated in water boatmen, aquatic insects that swim in ponds and streams by using their middle and hind legs as oars. These insects occur in three color forms (only one is shown). Frequency-dependent selection operates to maintain all three phenotypes within a given population. (Stephen Dalton/ Photo Researchers, Inc.)

Neutral mutations give no selective advantage or disadvantage

Some of the genetic variations observed in a population are caused by mutations that may confer no apparent selective advantage or disadvantage in a particular environment. For example, changes in DNA that do not alter protein structure usually do not affect phenotype. Such **neutral mutations** do not alter the fitness of an individual to survive and reproduce and are, therefore, not adaptive.

The extent of neutral mutations in organisms is difficult to determine. It is relatively easy to demonstrate that an allele is beneficial or harmful, provided that its effect is observable. But the variation in alleles that involves only slight differences in the proteins they code for may or may not be neutral. These alleles may be influencing the organism in subtle ways that are difficult to measure or assess. Also, an allele that is neutral in one environment may be beneficial or harmful in another.

SUMMARY

I. Evolution is a change in the allele frequencies in a population's gene pool. Each individual within a population contains only a portion of the alleles in the gene pool.

II. The Hardy-Weinberg principle states that allele frequencies in a population tend to remain constant in successive generations unless evolutionary forces are operating.

III. Allele frequencies may be changed by mutation, genetic drift, gene flow, and natural selection.
   A. The source of new alleles in a gene pool is mutation.
   B. Genetic drift is a random change in the allele frequencies of a small population. The changes caused by genetic drift are usually not adaptive.

C. The migration of individuals between populations causes a corresponding movement of alleles, or gene flow, that can cause changes in allele frequencies.
D. Changes in allele frequencies that lead to adaptation are caused by natural selection.

IV. Selection can change the composition of a gene pool in a favorable direction for a particular environment.
A. Stabilizing selection favors the mean at the expense of phenotypic extremes.
B. Directional selection favors one phenotype over another, causing a shift in the phenotypic mean.
C. Disruptive selection favors two or more phenotypic extremes.

V. Most populations have a large reservoir of genetic variability. Genetic polymorphism is the presence in a population of two or more alleles for a given locus.
A. Heterozygote advantage occurs when the heterozygote has a higher degree of fitness than either homozygote. Both alleles are maintained in the population.
B. In frequency-dependent selection, a genotype's selective value varies with its frequency of occurrence.
C. Genetic variation that confers no detectable selective advantage is caused by neutral mutation.

SELECTED KEY TERMS

balanced polymorphism
directional selection
disruptive selection
fitness
founder effect
frequency-dependent selection
gene flow
gene pool
genetic bottleneck
genetic drift
genetic polymorphism
Hardy-Weinberg principle
heterozygote advantage
microevolution
mutation
natural selection
neutral mutations
population
stabilizing selection

POST-TEST

1. A(an) __________ consists of all of the individuals of the same species that live in a particular place at a particular time.
2. Evolution that involves small, gradual changes within a population is known as __________ .
3. The __________ - __________ principle demonstrates that the process of inheritance does not, by itself, cause changes in allele frequencies.
4. Random genetic events, called __________ __________ , may have a major effect on allele frequencies, particularly in small populations.
5. The movement of alleles between populations, called __________ __________ , is caused by the migration of breeding individuals.
6. The source of the genetic variability that is the raw material of evolution is __________ .
7. __________ __________ leads to adaptive evolutionary change.
8. In __________ selection, individuals with a phenotype near the mean are favored over those with phenotypic extremes.
9. The __________ selection of peppered moths is an indirect consequence of air pollution.
10. __________ selection is a special type of directional selection in which there is a trend in several directions at once, resulting in a divergence within a population.
11. The presence in a population of two or more alleles for a given locus is known as __________ __________ .
12. A human with alleles for both normal hemoglobin and sickle cell hemoglobin demonstrates __________ __________ in an area where malaria is prevalent.
13. In __________-__________ selection, the fitness of a particular genotype varies with its frequency of occurrence in the population.
14. Genetic variation that does not confer a selective advantage on the individual possessing it is caused by __________ __________ .

REVIEW QUESTIONS

1. Define population, gene pool, and allele frequency. How does each relate to microevolution?
2. Given that genetic equilibrium demonstrated by the Hardy-Weinberg principle only occurs under conditions that populations in nature seldom, if ever, experience, why is it important?
3. Explain the effect of each of the following on allele frequencies: (a) natural selection; (b) mutation; (c) gene flow; (d) genetic drift.
4. Draw three graphs to represent stabilizing, directional, and disruptive selection.
5. Explain the evolution of the giraffe's long neck by directional selection. Explain how stabilizing selection accounts for the giraffe's long neck today.
6. Insect populations that have been exposed to an insecticide such as DDT develop resistance to the insecticide over time. Would this be an example of stabilizing selection, directional selection, or disruptive selection? Explain.
7. What is genetic polymorphism? Balanced polymorphism?
8. Distinguish between heterozygote advantage and frequency-dependent selection.

YOU MAKE THE CONNECTION

1. Why are mutations almost always neutral or harmful?
2. Explain this apparent anomaly: We discuss evolution in terms of *genotype* fitness (the selective advantage that a particular genotype confers on an individual), yet natural selection acts on an organism's *phenotype*.

RECOMMENDED READINGS

Cohn, J.P. "Genetics for Wildlife Conservation." *BioScience*, Vol. 40, No. 3, March 1990. Genetic analysis of endangered species offers hope for their survival.

Mayr, E. *Population, Species, and Evolution.* Harvard University Press, Cambridge, MA, 1970. A classic discussion of evolution.

Rinderer, T.E., B.P. Oldroyd, and W.S. Sheppard. "Africanized Bees in the U.S." *Scientific American*, Vol. 269, No. 6, December 1993. An update on the characteristics and spread of Africanized bees.

Weiner, Jonathan. *The Beak of the Finch: A Story of Evolution in our Time.* New York, Alfred A. Knopf, 1994. An award-winning account of various aspects of evolutionary biology.

C H A P T E R 19

Speciation and Macroevolution

Members of different species, such as lions and tigers, sometimes interbreed in artificial surroundings. Shown on the right is a liger, the sterile offspring of such a mating. (Volker Bartholdt)

The concept of distinct groups of living organisms, known as **species** (from Latin, meaning "kind") is not new. However, every definition of species has some sort of limitation. Linnaeus, the 18th century biologist who is considered the founder of modern taxonomy, classified plants into separate species based on structural differences (see Chapter 22). This method is still used to help characterize species, but structure alone is not adequate to explain what constitutes a species. For example, dogs come in a wide variety of sizes and shapes, yet all dogs are clearly the same *kind* of organism and are classified as members of the same species.

Population genetics did much to clarify the concept of species. A species is a group of organisms with a common gene pool. Often referred to as the **biological species concept,** this definition is based on reproductive isolation. Members of a species freely interbreed with other members of the same species to produce fertile offspring, and do not interbreed with (that is, are reproductively isolated from) members of different species. In other words, each species has a gene pool that is isolated from that of other species, and each is restricted by reproductive barriers from genetic mixing with other species.

One of the problems with the biological species concept is that it applies only to sexually reproducing organisms. Organisms that reproduce asexually do not interbreed, so we cannot think of them in terms of reproductive isolation. For these organisms and for extinct organisms, the species concept is still valid; they are classified on the basis of structural and biochemical characteristics.

Another potential problem with the biological species concept is that organisms assigned to different species in the wild may interbreed if brought into a circus, zoo, greenhouse, aquarium, or laboratory. On the right side of the figure is shown a liger, a hybrid between a lion and tiger, in a state circus in the former East Germany. Compare the liger's features with those of the tiger to its left. Although the geographical ranges of lions and tigers overlap in parts of Asia, a liger has never been found in nature. Therefore, we usually include in our definition of species that members of different species do not normally interbreed *in nature.*

To summarize, a working definition of the term *species* is a group of more or less distinct organisms capable of interbreeding with one another in nature but reproductively isolated from other species. This definition is far from perfect, however, and the biological species concept has limitations.

Biologists have identified and named only about 2 million species of organisms; millions more remain to be identified. This chapter introduces the mechanisms by which they are thought to have originated.

LEARNING OBJECTIVES

After you have studied this chapter you should be able to

1. Define a species and explain the limitations of the biological species concept.
2. Explain the significance of reproductive isolating mechanisms and distinguish among the different prezygotic and postzygotic isolating mechanisms.
3. Explain the mechanism of allopatric speciation and give an example.
4. Explain the mechanism of sympatric speciation and give an example.
5. Take either side in a debate on the pace of evolution, by representing the opposing views of gradualism and punctuated equilibrium.
6. Define macroevolution and distinguish among microevolution, speciation, and macroevolution.
7. Discuss macroevolution in the context of novel features, including preadaptations, allometric growth, and paedomorphosis.
8. Discuss the significance of evolutionary novelties, adaptive radiation, and extinction.

SPECIES ACHIEVE REPRODUCTIVE ISOLATION IN VARIOUS WAYS

A number of **reproductive isolating mechanisms** prevent interbreeding between two different species whose ranges overlap. These mechanisms preserve the genetic integrity of each species because gene flow between species is prevented. To block a chance occurrence of individuals from two different species overcoming a single reproductive isolating mechanism, most species have two or more mechanisms. Many work before fertilization occurs (prezygotic), whereas others work after fertilization has taken place (postzygotic) (Table 19–1).

Prezygotic barriers interfere with mating

Prezygotic barriers are reproductive isolating mechanisms that prevent fertilization from taking place. Because male and female gametes never come into contact, an interspecific zygote (fertilized egg formed by the union of an egg from one species and a sperm from another species) never forms. Prezygotic barriers include temporal isolation, behavioral isolation, mechanical isolation, and gametic isolation.

Sometimes genetic exchange between two groups is prevented because they reproduce at different times of the day, season, or year. Such examples demonstrate **tem-**

Table 19–1 REPRODUCTIVE ISOLATING MECHANISMS

Mechanism	*Type*	*How It Works*
Temporal isolation	Prezygotic	Similar species reproduce at different times
Behavioral isolation	Prezygotic	Similar species have distinctive courtship behaviors
Mechanical isolation	Prezygotic	Similar species have structural differences in their reproductive organs
Gametic isolation	Prezygotic	Gametes of similar species are chemically incompatible
Hybrid inviability	Postzygotic	Interspecific hybrid dies at early stage of embryonic development
Hybrid sterility	Postzygotic	Interspecific hybrid survives to adulthood but is unable to reproduce successfully
Hybrid breakdown	Postzygotic	Offspring of interspecific hybrid are unable to reproduce successfully

poral isolation. For example, two very similar species of fruit flies, *Drosophila pseudoobscura* and *D. persimilis*, have ranges that overlap to a great extent, but they do not interbreed. *Drosophila pseudoobscura* is sexually active only in the afternoon and *D. persimilis* only in the morning. Similarly, two species of sage (*Salvia*) have overlapping ranges in southern California. Black sage flowers in early spring, whereas white sage flowers in late spring and early summer (Fig. 19–1).

Many animal species exchange a distinctive series of signals before mating. Such courtship behaviors illustrate **behavioral isolation** (also known as **sexual isolation**). Bowerbirds, for example, exhibit species-specific courtship patterns. The male satin bowerbird of Australia constructs an elaborate bower of twigs, adding decorative blue parrot feathers and white flowers at the entrance (Fig. 19–2). When a female approaches the bower, the male dances about her, holding a particularly eye-catching decoration in his beak. While dancing, he sings a courtship song that consists of a variety of sounds, including buzzes and laughlike hoots. These specific courtship behaviors keep similar bird species reproductively isolated from the satin bowerbird. If a male and female of two different species begin courtship, it stops when one member does not recognize the signals of the other.

Sometimes members of different species court and even attempt copulation, but the incompatible structures of their genital organs prevent successful mating. Structural differences that inhibit mating between species are known as **mechanical isolation.** For example, many flowering plants have physical differences in their flower parts that help them maintain their reproductive isolation from one another. The sage plants mentioned earlier as an example of temporal isolation also have mechanical isolation. Black sage, which is pollinated by small bees, has a floral structure different from that of white sage, which is pollinated by large carpenter bees (Fig. 19–3). These differences prevent the insects from cross-pollinating the two species should they happen to flower at the same time.

FIGURE 19–2 Different bowerbird species exhibit behavioral isolation by having a variety of highly specialized courtship patterns that prevent mating between species. The male satin bowerbird (shown) constructs a bower to attract a female. Note the flowers and blue decorations that he has arranged at the entrance to his bower. (Patti Murray)

(a) (b)

FIGURE 19–1 Black sage and white sage, photographed at the same time of year (early spring), are an example of temporal isolation. Note that the flowers of black sage (*a*) have bloomed, whereas the flower buds of white sage (*b*) are still unopened. (Courtesy of Robert Thorne, Rancho Santa Ana Botanic Garden)

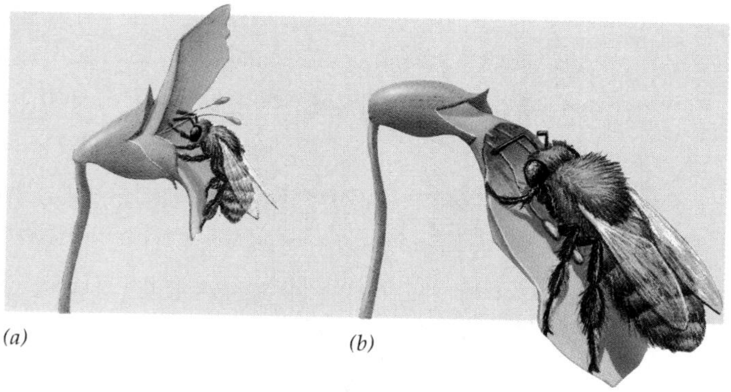

(a) (b)

FIGURE 19–3 Differences in floral structures between black and white sage—an example of mechanical isolation—allow them to be pollinated by different insects. Because the two species use different pollinators, they cannot interbreed. (*a*) The petal of the black sage functions as a landing platform for small bees. Larger bees cannot fit on this platform. (*b*) The larger landing platform and longer stamens of white sage allow pollination by larger carpenter bees. If smaller bees land on white sage, their bodies cannot brush against the pollen-bearing stamens.

FIGURE 19–4 **A basket sponge releases a cloud of gametes into the water, but if one of these gametes encounters a gamete of a different species, molecular and chemical differences, known as gametic isolation, will prevent their union.** (Visuals Unlimited/Marty Snyderman)

If mating has taken place between two species, their gametes may still not combine. Molecular and chemical differences between species cause **gametic isolation,** in which the egg and sperm of different species are incompatible. In aquatic animals that release their eggs and sperm into the surrounding water simultaneously, interspecific fertilization is extremely rare (Fig. 19–4). This is because the surface of the egg contains specific proteins that bind only to complementary molecules on the surface of sperm cells of the same species (see Chapter 49). A similar type of recognition often occurs between pollen grains and the stigma (receptive surface of the female part of the flower) so that pollen does not germinate on the stigma of a different plant species.

Postzygotic barriers prevent successful reproduction if mating occurs

Fertilization sometimes occurs between gametes of two different species despite the existence of prezygotic barriers. When this happens, postzygotic barriers that increase the likelihood of reproductive failure come into play.

Generally, the embryo of an interspecific hybrid spontaneously aborts. Embryonic development is a complex process requiring the precise interaction and coordination of many genes. Apparently the genes from parents belonging to different species do not interact properly in regulating the mechanisms for normal development. In this case, reproductive isolation is achieved by **hybrid inviability.** For example, nearly all of the hybrids die in the embryonic stage when the eggs of a bullfrog are fertilized artificially with sperm from a leopard frog. Similarly, in crosses between different species of irises, the embryos die before reaching maturity.

If an interspecific hybrid does live, it may not be able to reproduce. There are several reasons for this. Hybrid animals may exhibit courtship behaviors incompatible with those of either parental species and, as a result, they will not mate. More often, **hybrid sterility** occurs in which the gametes of an interspecific hybrid are abnormal because of problems during meiosis (see Chapter 9). Hybrid sterility is particularly common if the two parental species have different chromosome numbers. For example, a mule is the offspring of a female horse ($2n = 64$) and a male donkey ($2n = 62$). This type of union almost always results in sterile offspring ($2n = 63$) because synapsis, the pairing of homologous chromosomes during meiosis, cannot occur properly.

Occasionally, a fertile interspecific hybrid develops that produces a second generation from a cross between two hybrids or between a hybrid and one of the parental species. The F_2 hybrid generally exhibits **hybrid breakdown,** the inability of a hybrid to reproduce due to some defect. For example, hybrid breakdown in the F_2 generation of a cross between two sunflower species was 80%. In other words, 80% of the F_2 generation were defective in some way and could not reproduce successfully. Hybrid breakdown can also occur in the F_3 and later generations.

THE KEY TO SPECIATION IS REPRODUCTIVE ISOLATION

We are now ready to consider how entirely new species may arise from previously existing ones. The evolution of a new species, known as **speciation,** occurs when a population becomes reproductively isolated from other members of the species. Over time the gene pools of the two separated populations begin to diverge in genetic composition. When a population is sufficiently different from its ancestral species that no genetic exchange can occur between them, we say that speciation has occurred. Such a situation is thought to arise in two ways, through allopatric and sympatric speciation.

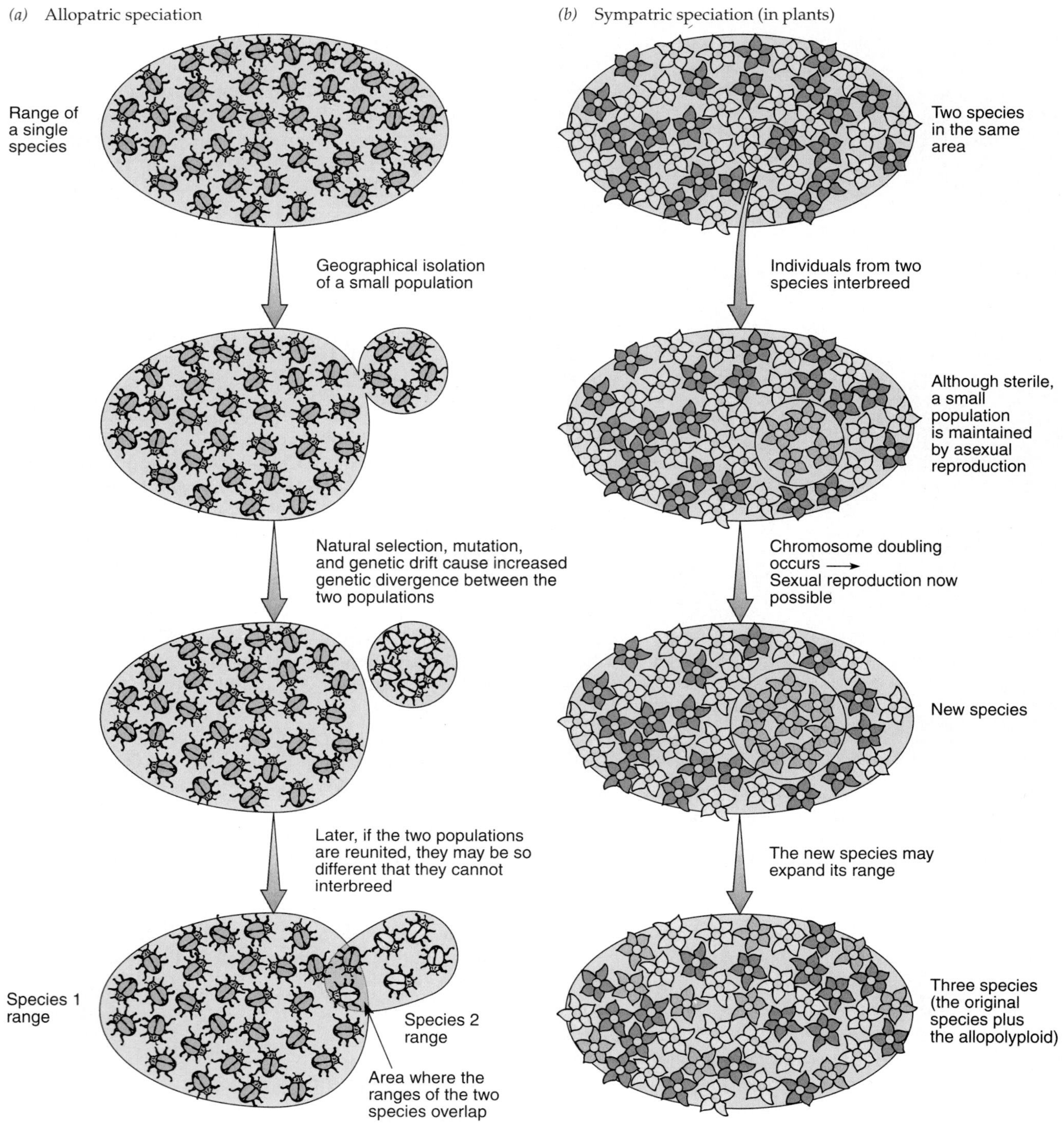
(a) Allopatric speciation
Range of a single species
Geographical isolation of a small population
Natural selection, mutation, and genetic drift cause increased genetic divergence between the two populations
Later, if the two populations are reunited, they may be so different that they cannot interbreed
Species 1 range
Species 2 range
Area where the ranges of the two species overlap
(b) Sympatric speciation (in plants)
Two species in the same area
Individuals from two species interbreed
Although sterile, a small population is maintained by asexual reproduction
Chromosome doubling occurs ⟶ Sexual reproduction now possible
New species
The new species may expand its range
Three species (the original species plus the allopolyploid)

FIGURE 19–5 The mechanism of allopatric speciation is compared to that of sympatric speciation. (*a*) Allopatric speciation. (***b***) Sympatric speciation in plants. (The role of sympatric speciation in animal evolution is unclear.)

Long physical isolation and different selective pressures result in allopatric speciation

Speciation that occurs when one population becomes geographically separated from the rest of the species and subsequently evolves by natural selection and/or genetic drift is known as geographical speciation, or **allopatric speciation** (from the Greek *allo,* "different," and *patri,* "fatherland"). Allopatric speciation is thought to be the most common method of speciation, and the evolution of new animal species has been almost exclusively by allopatric speciation (Fig. 19–5*a*). (See *Focus On: The Kaibab Squirrel: Evolution in Action* for an example of allopatric speciation in progress.)

Focus On

The Kaibab Squirrel: Evolution in Action

About ten thousand years ago, when the American Southwest was less arid, the forests in the area supported a tree squirrel with conspicuous tufts of hair sprouting from its ears. A small tree squirrel population living on the Kaibab Plateau of the Grand Canyon became geographically isolated when the climate changed, causing areas to the north, west, and east to become desert. Just a few miles to the south lived the rest of the squirrels, known as Abert squirrels, but the two groups were separated by the Grand Canyon. With changes over time in both its appearance and its ecology, the Kaibab squirrel is on its way to becoming a new species.

The Kaibab squirrel. (Tom and Pat Leeson)

The Abert squirrel. (Kent and Donna Dannen)

During its many years of geographical isolation, the small population of Kaibab squirrels has diverged from the widely distributed Abert squirrels in a number of ways. Perhaps most evident are changes in fur color. The Kaibab squirrel now has a white tail and a black belly, in contrast to the gray tail and white belly of the Abert squirrel (see figures). It is not clear why these striking changes arose in Kaibab squirrels. ■

The geographical isolation required for allopatric speciation may occur in several ways. Earth's surface is in a constant state of change. Such change includes rivers shifting their courses; glaciers migrating; mountain ranges forming; land bridges developing that separate previously united aquatic populations; and large lakes diminishing into several smaller, geographically separated pools.

What might be an imposing geographical barrier to one species may be of no consequence to another. Birds and cattails, for example, do not become isolated when a lake subsides into smaller pools; birds can easily fly from one pool to another, and cattails disperse their fruits by air currents. Fish, on the other hand, are usually unable to cross the land barriers between the pools and so become reproductively isolated. In the Death Valley region of California and Nevada, large interconnected lakes formed as glaciers melted at the end of the last ice age. These lakes were populated by one or several species of pupfish. Over time, as the glaciers retreated and the climate became drier, the large glacial lakes dried up, leaving isolated pools. Today, there are numerous species of pupfish, but each is restricted to a single water hole (Fig. 19–6). It is thought that each pool contained a small population of pupfish that gradually diverged from the common ancestral species by genetic drift and natural selection.

FIGURE 19–6 **Allopatric speciation of pupfish is thought to have occurred in Death Valley.** Shown is one of the various pupfish species that apparently evolved when large glacial lakes dried up about 10,000 years ago, leaving behind small, isolated pools fed by springs. Each pool contained a small population of pupfish that gradually diverged from the common ancestral species by genetic drift and natural selection. (Steinhart Aquarium, Tom McHugh/Photo Researchers, Inc.)

Allopatric speciation also occurs when a small population migrates and colonizes a new area away from the range of the original species. This colony is geographically isolated from its parental species, and the small changes that accumulate over many generations as a result of microevolution (see Chapter 18) may eventually be enough to form a new species. Because islands pro-

(a) (b)

FIGURE 19–7 Allopatric speciation can occur when a small population colonizes a geographically isolated area such as an island. (***a***) The nene (pronounced "nay-nay"), *Branta sandvicensis,* is a goose found in the Hawaiian Islands. It is thought to have evolved from a small population of geese that originated in North America. Although the nene is an endangered species, strict conservation measures have brought it back from the brink of extinction. (***b***) The Canada goose, *B. canadensis,* is a close relative of the Hawaiian goose. (*a,* M. J. Rauzon/VIREO; *b,* Mary Clay/Tom Stack & Associates)

vide the geographical isolation required for allopatric speciation, they offer excellent opportunities to study this mechanism. The Galapagos Islands and the Hawaiian Islands, for example, were probably originally colonized by a few individuals of a few species. The hundreds of unique species found on each island today are thought to have descended from these original colonizers (Fig. 19–7).

Speciation is more likely to occur if the original isolated population is small. Recall that genetic drift, including the founder effect, is more influential in small populations (see Chapter 18). Genetic drift tends to result in rapid changes in allele frequencies in the isolated population. The divergence caused by genetic drift is further accentuated by the different selective pressures of the environment to which the population is exposed.

Sometimes allopatric speciation occurs quite rapidly. Early in the 15th century, a small population of rabbits was released on Porto Santo, a small island off the coast of Portugal. Because there were no other rabbits or competitors and no carnivorous enemies on the island, the rabbits thrived. By the 19th century, these rabbits were markedly different from their European ancestors. They were only half as large, with a different color pattern and a more nocturnal lifestyle. Most significantly, they could not produce offspring when bred with members of the ancestral European species. Within 400 years, an extremely short period of time in evolutionary history, a new species of rabbit had evolved.

Two populations diverge in the same physical location by sympatric speciation

Although geographical isolation is an important factor in many cases of evolution, it may not be an absolute requirement. When a population forms a new species within the same geographical region as the parental species, **sympatric speciation** (from the Greek *sym,* "together," and *patri,* "fatherland") has occurred (Fig. 19–5*b*). The divergence of two gene pools in the same geographical range is especially common in plants. The role of sympatric speciation in animal evolution, however, is unclear and has been difficult to demonstrate in nature.

How does sympatric speciation occur in plants? As discussed earlier, the union of two gametes from different species rarely forms viable offspring; if offspring *are* produced, they are usually sterile. Before gametes form, meiosis occurs to reduce the chromosome number. In order for the chromosomes to be parcelled correctly into the gametes, homologous chromosome pairs must come together (a process called *synapsis*) during prophase I. This cannot usually occur in interspecific hybrid offspring because the chromosomes are not homologous. However, *if* the $2n$ chromosome number doubles *before* meiosis, then pairing of homologous chromosomes can take place. This spontaneous doubling of chromosomes has been documented in both plants and animals. It is not a common occurrence, but neither is it rare. It produces nuclei with multiple sets of chromosomes.

Polyploidy, the possession of more than two sets of chromosomes (see Chapter 9), is a major factor in plant evolution. When polyploidy occurs in conjunction with **hybridization** (sexual reproduction between individuals from different species), it is known as **allopolyploidy.** Allopolyploidy can produce a fertile interspecific hybrid because the polyploid condition provides the homologous chromosome pairs necessary for synapsis during meiosis. As a result, gametes may be viable (Fig. 19–8). An allopolyploid — that is, an interspecific hybrid produced by allopolyploidy — can reproduce with itself (self-fertilize) or with a similar individual. However, allopolyploids are reproductively isolated from both parents because the gametes of the allopolyploid have a different number of chromosomes than do those of either parent.

If a population of allopolyploids (that is, a new species) becomes established, selective pressures cause one of three outcomes. One, the new species may be unable to compete successfully against species that are already established, and so it becomes extinct. Two, the allopolyploid individuals may fill a new role in the environment and so coexist with both parental species. Three, the new species may compete with either or both of its parental species. If it has a combination of traits that confers greater fitness than one or both parental species for all or part of the original range of the parent(s), the hybrid species may replace the parent(s).

The significance of sympatric speciation in animal evolution is highly disputed among biologists. The few examples almost exclusively involve parasitic insects and rely on mechanisms other than polyploidy.[1] It is thought that a mutation arises in an individual and spreads through a small group of insects as a result of sexual reproduction. The mutation isolates the insects reproductively from the rest of the population, perhaps by allowing them to parasitize a different host species. Further mutations may occur that cause the mutant population to diverge even further from the original population. The lineage is maintained because these insects either self-fertilize or breed with brothers or sisters (copies of the same mutant alleles are more likely in close relatives). Although the original species and the mutant population continue to occupy the same geographical area, no gene flow occurs between them because the mutant population now lives on a different host. In other words, the two groups have separate mating locations in the same geographical area.

Although allopolyploidy is extremely rare in animals, it is considered a significant factor in the evolution of flowering plants. Between 30 and 80% of all flowering plants are thought to be polyploids, and most of these are allopolyploids (Fig. 19–9; see *Making the Connection* in

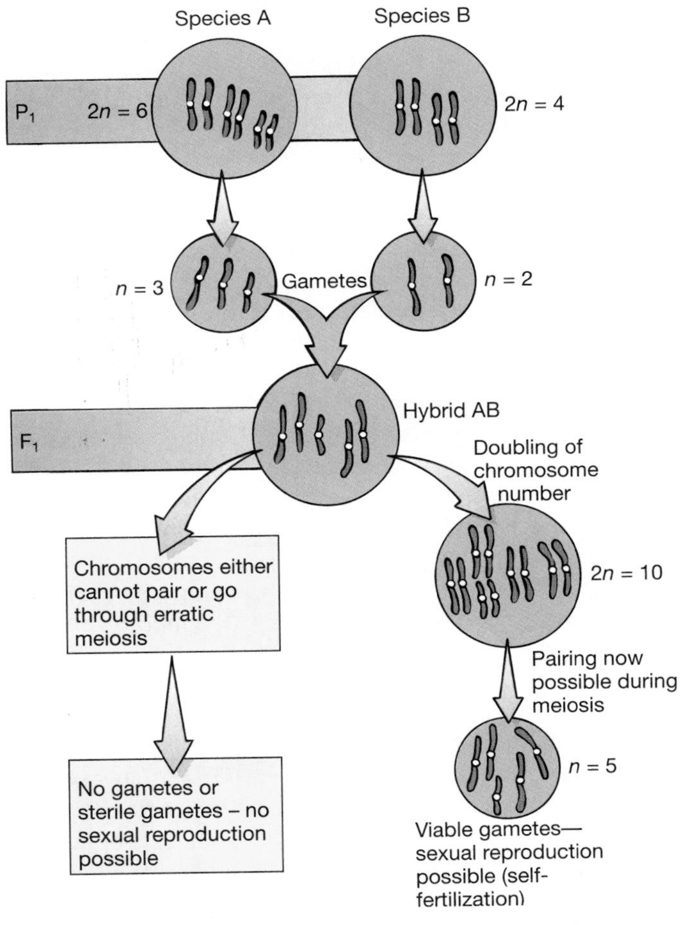

FIGURE 19–8 How a new plant species arises as a fertile allopolyploid. Interspecific hybridization occurs between two species, yielding a hybrid F_1 generation. If the chromosome number does not double, they are unable to undergo normal meiosis, and the hybrid is sterile (*left*). If the chromosome number doubles, the hybrid is able to undergo meiosis and is fertile (*right*).

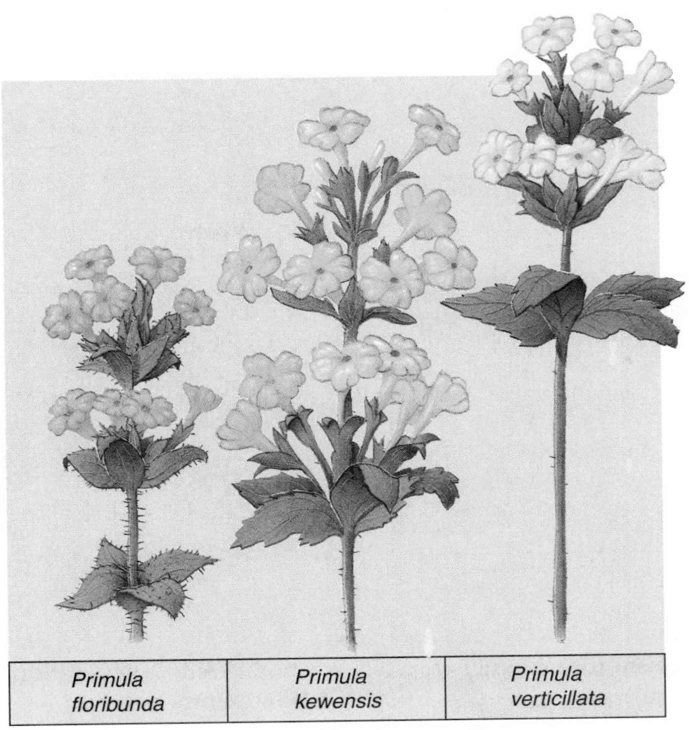

FIGURE 19–9 An allopolyploid primrose, *Primula kewensis*, arose during the early part of the 20th century. The F_1 hybrid of *P. floribunda* ($2n = 18$) and *P. verticillata* ($2n = 18$) was a sterile diploid perennial ($2n = 18$). On three different occasions it spontaneously formed a fertile polyploid branch ($2n = 36$) that produced seeds. The plants that grew from these seeds were a new species, *P. kewensis*. (The specific epithet *kewensis* was chosen because this species arose accidentally at the Royal Botanic Gardens at Kew, England.)

[1]In the April 14, 1994 issue of *Nature*, a group of German biologists reported that their study of the mitochondrial DNA of cichlids in East African lakes suggested that these fish species evolved sympatrically.

Making the Connection

Evolution and the Diversity of Life

Are all organisms related to one another? In Chapter 18 we discussed how the processes of natural selection, mutation, genetic drift, and gene flow explain microevolution (the evolution within a population). These ideas were extended in this chapter with evidence that, if a population remains reproductively isolated from its parental species, it diverges even more. Given enough changes, the population may become a new species.

Thus, new species arise from preexisting species, which had evolved from other species that came before them. The family tree of living organisms can be traced back to the origin of life. All life is thought to be connected; that is, all organisms existing today are related to one another because if we go back far enough in geological time, all share a common ancestor. In Chapter 20 we will reconstruct the origin and history of life. We will continue this process by examining the evolution of primates, including humans, in Chapter 21. Chapters 22 through 30 will explore life's diversity and continue our examination of how species that exist today are interrelated. ▲

Chapter 27). Moreover, allopolyploidy provides a mechanism for extremely rapid speciation. A single generation is all that is needed to form a new, reproductively isolated species. Allopolyploidy may explain the rapid appearance of flowering plants in the fossil record and their remarkable diversity (about 235,000 species) today.

The mechanism of sympatric speciation has been experimentally verified for many plant species. One example is hemp nettle, several species of which occur in temperate parts of Europe and Asia. One hemp nettle species, *Galeopsis tetrahit* ($2n = 32$), is a naturally occurring allopolyploid thought to have formed by the hybridization of two species, *G. pubescens* ($2n = 16$) and *G. speciosa* ($2n = 16$). This process occurred in nature but was experimentally reproduced in the laboratory. *G. pubescens* and *G. speciosa* were crossed to produce F_1 hybrids, most of which were sterile. Nevertheless, both F_2 and F_3 generations were produced. The F_3 generation included a polyploid plant with $2n = 32$; it then yielded fertile F_4 offspring that could not mate with either of the parental species. These allopolyploid plants had the same appearance and chromosome number as the naturally occurring *G. tetrahit.* When the experimentally produced plants were crossed with the naturally occurring *G. tetrahit,* a fertile F_1 generation was formed. Thus, the experiment duplicated the speciation process that occurred in nature.

EVOLUTIONARY CHANGE CAN OCCUR RAPIDLY OR GRADUALLY

Does the fossil record provide clues about how rapidly new species arise? Biologists have long recognized that the fossil record lacks many transitional forms; the starting points (ancestral species) and the end points (new species) are present, but the intermediate stages in the evolution from one species to another are "missing." This observation has traditionally been explained by the incompleteness of the fossil record. Biologists have attempted to fill in the missing parts, much as a writer might fill in the middle of a novel when the beginning and end are already there.

Two different models, punctuated equilibrium and gradualism, have been developed to explain evolution as observed in the fossil record (Fig. 19–10). The **punctuated equilibrium** model was proposed by biologists who question whether the fossil record really is as incomplete as it initially appeared. First advanced by Stephen Jay Gould and Niles Eldredge in 1972, the punctuated equilibrium model suggests that the fossil record accurately reflects evolution as it actually occurs, with long periods of stasis (no evolution) punctuated, or interrupted, by short periods of rapid speciation that are perhaps triggered by changes in the environment. Thus, speciation normally proceeds in "spurts." These relatively short periods of active evolution are followed by long periods (often millions of years) of stability. Later, when evolution resumes, new species form, and many old ones are successfully competed against and become extinct.

With punctuated equilibrium, speciation can occur in a relatively short period of time. Keep in mind, however, that a "short" amount of time for speciation may be thousands of years. Such a span is short only when compared with the several million years of a species' existence. Biologists who support punctuated equilibrium point out that sympatric speciation and even allopatric speciation can occur in such relatively short periods.

Punctuated equilibrium accounts for the abrupt appearance of a new species in the fossil record, with little or no evidence of intermediate forms. That is, proponents think that few transitional forms appear in the fossil record because few transitional forms occurred during speciation.

In contrast, the traditional Darwinian view of evolution espouses the **gradualism** model. According to this model, evolution proceeds at a more or less constant rate but is not observed in the fossil record because the record is incomplete. (Recall from Chapter 17 that the conditions

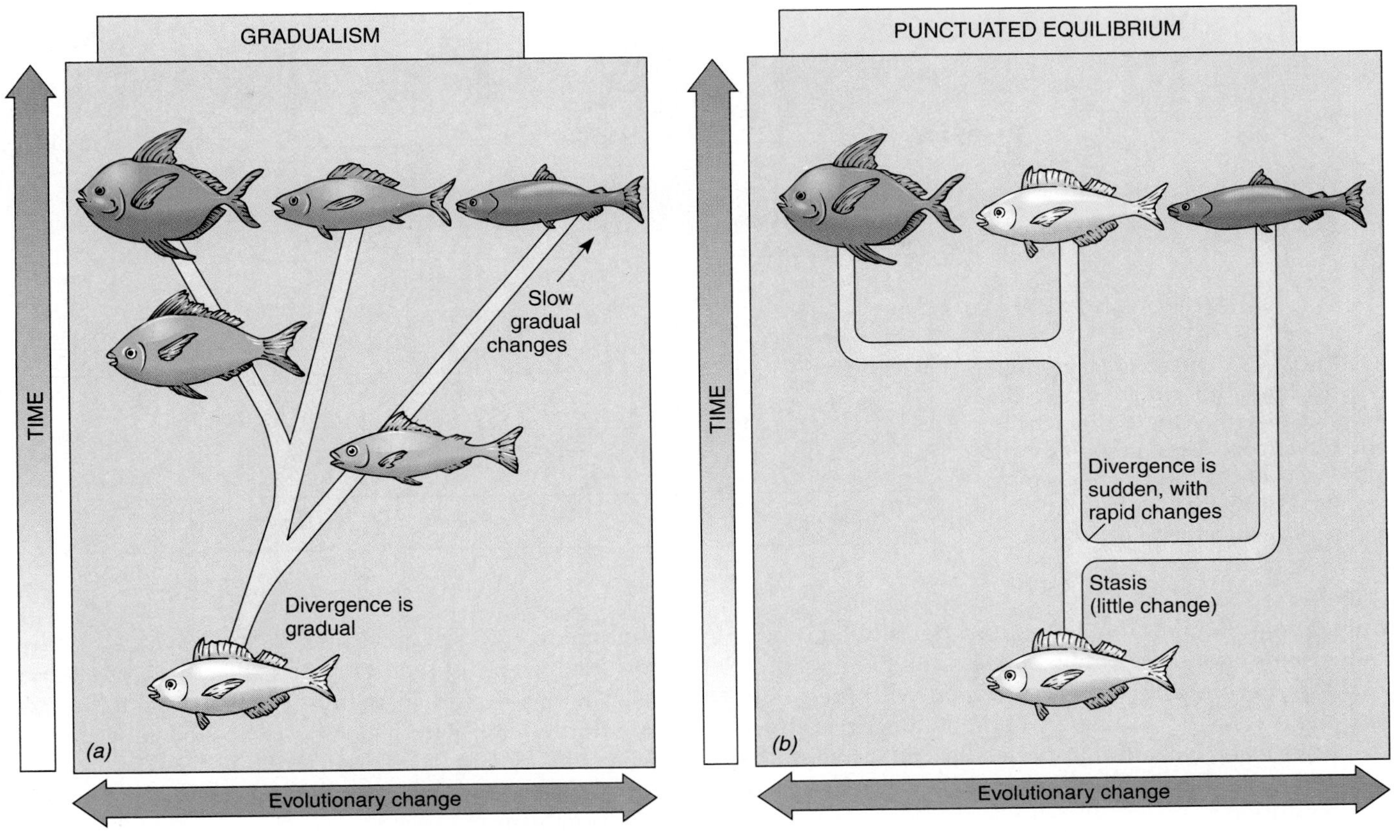

FIGURE 19–10 Although biologists agree that new species arise from preexisting species, they present two different models, gradualism and punctuated equilibrium, for the pace of evolution. (*a*) In gradualism a slow, steady change in species occurs over time. (*b*) In punctuated equilibrium long periods of little evolution (stasis) are followed by short periods of rapid speciation.

required for fossil formation are quite precise. The vast majority of organisms decompose when they die, leaving no trace of their existence.) Occasionally, a complete fossil record of transitional forms is discovered and cited as a strong case for gradualism. The gradualism model maintains that populations slowly diverge from one another by the gradual accumulation of adaptive characteristics within each population. These adaptive characteristics accumulate as a result of different selective pressures exerted by different environments.

The abundant evidence in the fossil record of long periods with no change in a species has been used to argue against gradualism. Gradualists, however, think that any periods of stasis evident in the fossil record are the result of stabilizing selection (see Chapter 18). They also point out that stasis in fossils is deceptive because fossils do not reveal all aspects of evolution. Fossils show changes in external structure and skeletal structure, but inherited characteristics such as changes in physiology, internal structure, and behavior (all of which also represent evolution) are not revealed by fossils. Gradualists recognize rapid evolution only when strong directional selection occurs.

Thus, while biologists generally agree that natural selection is the main mechanism responsible for evolution, they disagree as to the timing, or pace, of evolution during a given species' existence. Most biologists think that both punctuated equilibrium and gradualism can occur —that the pace of evolution may be abrupt in certain instances and gradual in others.

Macroevolution Involves Major Evolutionary Events

Macroevolution refers to dramatic changes that sometimes occur in evolution. These major evolutionary events are large phenotypic changes such as the appearance of wings with feathers during the evolution of birds from reptiles. The phenotypic changes are so great that the new species possessing them are assigned to different genera or higher taxonomic categories. (The taxonomic categories above the level of species are artificial constructs used by humans to indicate degrees of relatedness among organisms. Thus, closely related species are grouped into the same genus (pl., *genera*), similar genera into the same

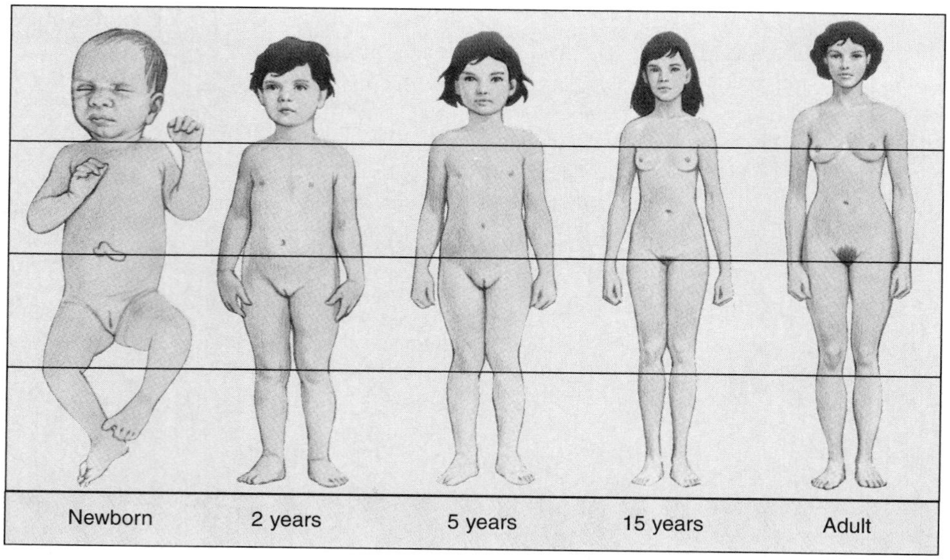

FIGURE 19–11 Different stages in the growth of a human are drawn the same size to demonstrate allometric growth (varied rates of growth for different parts of the body). For example, as humans develop, their legs grow more rapidly than their heads.

family, similar families into the same order, similar orders into the same class, and similar classes into the same phylum.) Macroevolution is thus concerned with the origin of species classified into new taxonomic categories above the species level. Evolutionary novelties, adaptive radiation, and mass extinction are important aspects of macroevolution.

Evolutionary novelties originate from mutations that alter development

New designs sometimes arise from structures already in existence. A change in the basic pattern of an organism can produce something unique. Examples of "new" and unusual features include wings on insects, flowers on plants, and feathers on birds. Usually these evolutionary novelties, called **preadaptations,** are variations of some structure that originally fulfilled one role, but changed in a way that was adaptive for a different role. Bird feathers are a good example of a preadaptation because they evolved from reptilian scales. Feathers, which may have originally provided thermal insulation, preadapted primitive birds for flight, as feathers turned out to be useful for flying. In like manner, the mammalian middle ear bones originated from modified jaw bones of reptiles.

How do such evolutionary novelties occur? Many are probably due to changes in development, which is the orderly sequence of changes that take place as an organism grows and matures. Regulatory genes may exert control over hundreds of other genes, and very slight genetic changes in regulatory genes could cause major structural changes in the organism.

For example, during development, most organisms have **allometric growth,** varied rates of growth for different parts of the body. The size of the head in human newborns is large in proportion to the rest of the body. As a human grows and matures, its torso, hands, and legs grow more rapidly than the head (Fig. 19–11). Allometric growth is found in many organisms, including the male fiddler crab with its single, oversized claw, and the ocean sunfish with its enlarged tail (Fig. 19–12). If growth rates are altered even slightly, drastic changes in the shape of an organism may result.

Sometimes novel changes are the result of changes in the *timing* of development. Consider, for example, the changes that would occur if a juvenile characteristic were retained in an adult, a phenomenon known as **paedomorphosis.** Adults of some salamander species have gills, a feature found only in the larval (immature) stages of other salamanders. Retention of gills throughout life obviously alters the salamander's behavioral and ecological characteristics (Fig. 19–13). Perhaps such salamanders succeeded because they had a selective advantage over "normal" adult salamanders. The gilled forms remained aquatic and did not have to compete for food with the adult forms, most of which lived on land. The gilled forms also escaped the typical predators of terrestrial salamanders (although they had other predators in their watery environment).

Adaptive radiation is the diversification of an ancestral species into many species

Once a novel feature arises, **adaptive radiation** may occur, in which an ancestral species diversifies over time to fill a variety of different ecological roles. Adaptive radiation is the evolution of many related species from one or a few ancestral species in a relatively short period of time.

The concept of **adaptive zones** was developed to help explain why adaptive radiations take place. Adaptive zones are new ecological roles or ways of living that were

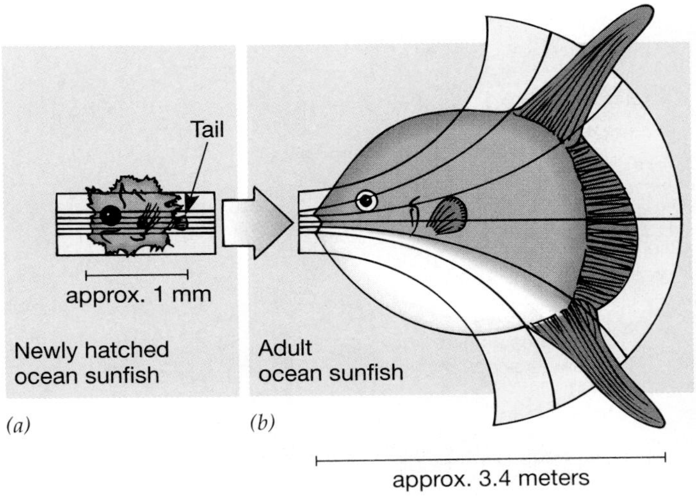

(c)

FIGURE 19–12 Allometric growth occurs in the ocean sunfish. The tail end of an ocean sunfish grows faster than the head end, resulting in the unique shape of the adult ocean sunfish. (***a***) A newly hatched ocean sunfish, only 1 mm long, has an extremely small tail. (***b***) This allometric transformation can be visualized by drawing rectangular coordinate lines through a picture of the juvenile fish and then changing the coordinate lines mathematically. (***c***) An ocean sunfish, also known as a mola, swims off the coast of southern California. The adult ocean sunfish is 3.4 meters (11 feet) long and weighs 1 metric ton (2206 pounds). (Richard Herrmann)

not used by an ancestral organism. At the species level, an adaptive zone is essentially identical to an *ecological niche* (an organism's role within a community; see Chapter 53). Examples of adaptive zones include nocturnal flying to catch insects, grazing on grass while migrating across a savanna, and swimming at the ocean's surface to filter out plankton.

Each adaptive zone can be occupied by only one group of organisms. A newly evolved species can take over an adaptive zone that is already occupied if the new species has features that make it competitively superior to the original occupant. If a number of adaptive zones are empty, they may be exploited by adaptive radiation. Consider the honeycreepers, a group of related birds found on the Hawaiian Islands. When the honeycreeper ancestors reached Hawaii, few other birds were present. The succeeding generations of honeycreepers quickly diversified and, in the process, occupied the many available adaptive zones. The diversity of their bills, like those of Galapagos finches (see Chapter 17), is a particularly good illustration of adaptive radiation (Fig. 19–14). Some honeycreeper bills are curved to extract nectar out of tubular flowers, for example, whereas others are short and thickened for ripping away bark in search of insects.

Adaptive radiation appears to be more common during periods of major environmental change, but it is difficult to determine if these changes actually trigger adaptive radiation. It is possible that major environmental change indirectly affects adaptive radiation by increasing the rate of extinction. Extinction produces empty adaptive zones, which are then available for adaptive radiation. Mammals, for example, had existed for millions of

FIGURE 19–13 An adult axolotl salamander (*Ambystoma mexicanum*), which has retained the juvenile characteristic of external gills, exhibits paedomorphosis. The axolotl remains permanently aquatic and is able to reproduce without developing other adult characteristics. (In most salamander species, the external gills found in the larval stage are not present in the adult.) (Jane Burton/Bruce Coleman, Inc.)

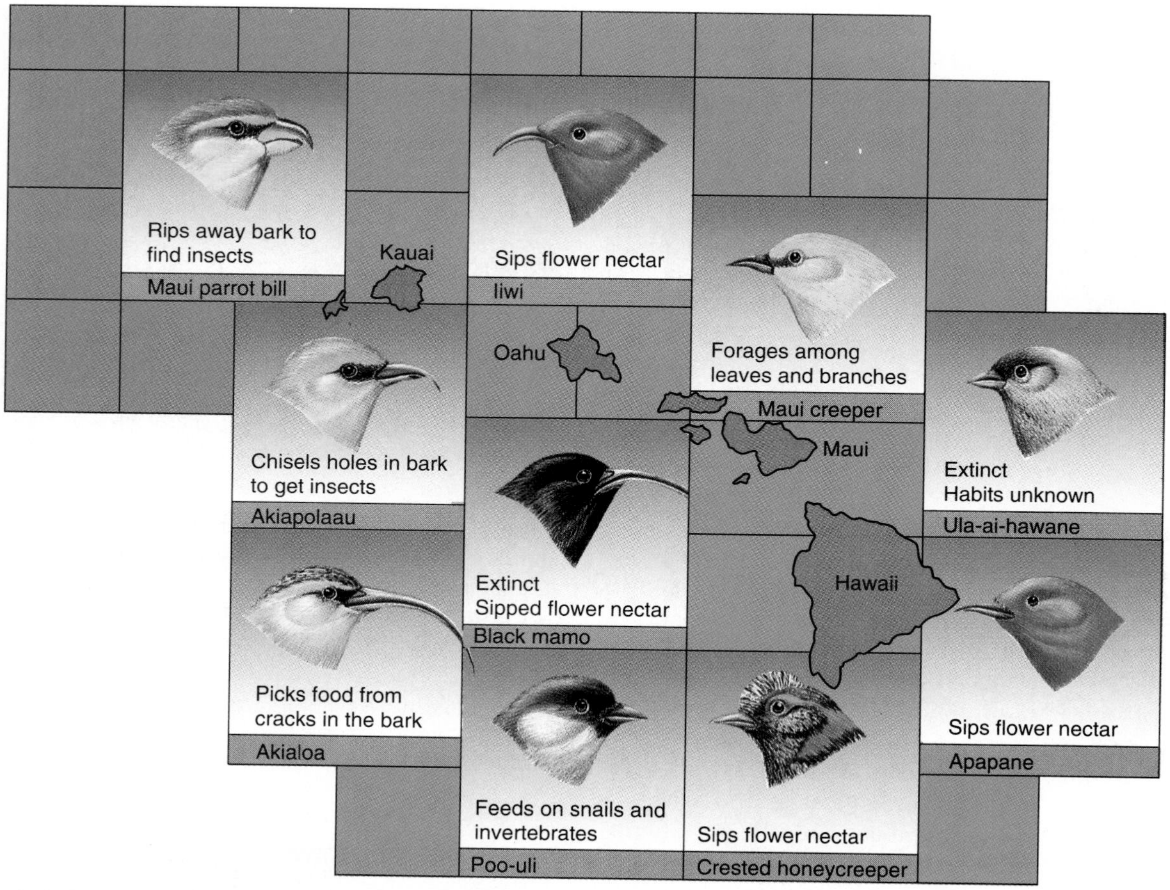

FIGURE 19–14 The many species of Hawaiian honeycreepers are a dramatic example of adaptive radiation. Compare the various beak shapes and methods of obtaining food. Many honeycreepers are now extinct as a result of human activities, including the destruction of habitat and the introduction of predators such as rats, dogs, and pigs.

years before undergoing adaptive radiation, which is thought to have been triggered by the extinction of the dinosaurs. The original mammals were small animals that ate insects. Relatively soon after the dinosaurs' demise, mammals diversified, exploiting a variety of adaptive zones. Flying bats, running gazelles, burrowing moles, and swimming whales all originated from the small, insect-eating ancestral mammals.

The appearance of novel features is associated with each major period of adaptive radiation. For example, shells and skeletons may have been the evolutionary novelties responsible for a period of adaptive radiation at the beginning of the Paleozoic era (see Chapter 20), in which most animal phyla, living and extinct, appeared.

Care must be taken in interpreting a cause-and-effect relationship between the appearance of a novel feature and adaptive radiation, however. It is tempting to take a simplistic approach and state, for example, that the evolution of the flower triggered the adaptive radiation of thousands of species of flowering plants. Perhaps it is true that the flowering plants diversified after the evolution of the flower, which presented a more competitive method of sexual reproduction because it permitted pollination by insects and other animals. However, adaptive radiation in the flowering plants may instead be a consequence of other adaptations that evolved. (Chapter 27 discusses other flowering plant adaptations besides their highly successful mode of reproduction.)

Extinction is an important aspect of evolution

Extinction, the end of a lineage, occurs when the last individual of a species dies. The loss is permanent, for once a species is extinct it can never reappear. Extinctions have occurred continually since the origin of life. By one estimate, only one species is alive today for every 2000 that have become extinct. Extinction is the eventual fate of all species, in the same way that death is the eventual fate of all individual organisms.

Although extinction has a negative short-term impact on biological diversity, it can have a positive evolutionary aspect over a period of thousands to millions of years. As mentioned previously, when species become extinct, their adaptive zones become vacant. Consequently, those organisms still living are presented with new opportunities for adaptive radiation and can diverge to fill the unoccupied adaptive zones. In other words, the extinct species are eventually replaced by new species.

During the long history of life, extinction appears to have occurred at two different rates. The continuous, low-level extinction of species is sometimes called *background extinction.* In contrast, five or possibly six times during Earth's history, *mass extinctions* of numerous species and higher taxa have taken place in both terrestrial and marine environments. The most recent mass extinction, which occurred about 65 million years ago, killed off the dinosaurs (Fig. 19–15). The time span over which a mass extinction occurred may have lasted several million years, but that is relatively short compared with the history of life. Each period of mass extinction has been followed by a period of adaptive radiation.

The causes of past episodes of mass extinction are not well understood. Both environmental and biological factors seem to have been involved. Major changes in the climate could have adversely affected those plants and animals that lacked the genetic flexibility to adapt. Marine organisms, in particular, are adapted to a steady, unchanging climate. If Earth's temperature were to increase or decrease by just a few degrees overall, many marine species would probably perish.

It is also possible that past mass extinctions were due to changes in the environment triggered by catastrophes. If a large comet or small asteroid collided with Earth, for example, the dust ejected into the atmosphere on impact could have blocked much of the sunlight. In addition to killing many plants (and therefore terrestrial animals), this event would have lowered Earth's temperature, leading to the death of many marine organisms. Evidence that the extinction of dinosaurs was caused by an extraterrestrial object's collision with Earth continues to mount.

Biological factors also trigger extinction. Competition among species may lead to the extinction of those species that cannot compete effectively. The human species, in particular, has had a profound impact on the rate of extinction (see Chapter 55). The habitats of many animal and plant species have been altered or destroyed by humans, and habitat destruction can result in a species' extinction (Fig. 19–16). Indeed, some biologists think that Earth has entered the greatest mass extinction episode in its entire history, and that this has been triggered by human activities.

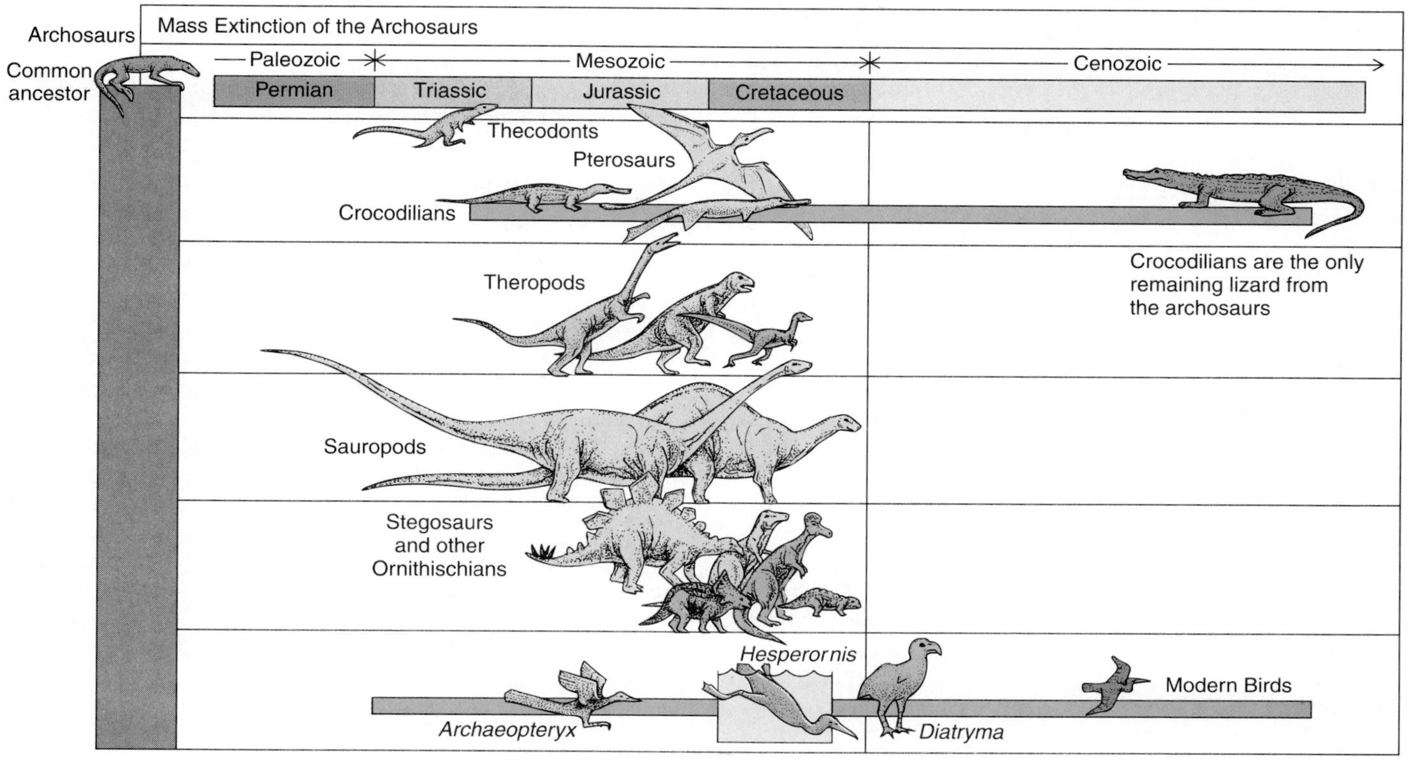

FIGURE 19–15 At the end of the Cretaceous period, approximately 65 million years ago, a mass extinction of many organisms, including the dinosaurs, occurred. At that time most of the archosaurs (one of five main groups of reptiles) became extinct. The only lines to survive were crocodiles and birds, both of which are archosauran descendants.

FIGURE 19–16 The dusky seaside sparrow became extinct in 1987, largely owing to human destruction of its habitat in Florida. (U.S. Fish and Wildlife Service)

IS MICROEVOLUTION RELATED TO SPECIATION AND MACROEVOLUTION?

The concepts presented in Chapters 17 and 18 represent the synthetic theory of evolution, in which mutation provides the genetic variation on which natural selection acts. The synthetic theory of evolution combines Darwin's theory with important aspects of genetics. All aspects of the synthetic theory of evolution have been tested and verified at the population and subspecies levels.

Many biologists think that microevolutionary processes (natural selection, mutation, genetic drift, and gene flow) account for the genetic variation within species *and* for the origin of new species. These biologists also think that macroevolution can be explained by microevolutionary processes. The synthetic theory of evolution *as it relates to speciation and macroevolution* is supported by a considerable body of data from many fields.

Some biologists who support the punctuated equilibrium model, however, question whether natural selection is as significant in the evolution of species as it is in microevolution. They think that evolutionary mechanisms different from those of microevolution have been responsible for large-scale change. To these biologists, microevolution is associated with the long periods of stasis in the fossil record, while speciation and macroevolution are associated with the rapid[2] bursts of evolution that punctuate stasis. Additional research may clarify the significance of natural selection in speciation and macroevolution.

[2]"Rapid" bursts of evolution are indeed faster than the more static periods, but even so they may last several million years.

SUMMARY

I. According to the biological species concept, a species consists of a group of similar organisms with the potential to interbreed with one another but not with members of different species.

II. Reproductive isolating mechanisms restrict the gene flow between species.
   A. Prezygotic barriers are reproductive isolating mechanisms that prevent fertilization from taking place.
      1. Temporal isolation occurs when two species reproduce at different times of the day, season, or year.
      2. In behavioral isolation, distinctive courtship behaviors prevent mating between species.
      3. Mechanical isolation is due to structural differences in the reproductive organs of similar species.
      4. In gametic isolation, gametes from different species are incompatible due to molecular and chemical differences.
   B. Postzygotic barriers are reproductive isolating mechanisms that ensure reproductive failure when fertilization has taken place.
      1. Hybrid inviability is the death of an interspecific embryo at an early stage of development.
      2. Hybrid sterility prevents interspecific hybrids that survive to adulthood from reproducing successfully.
      3. Hybrid breakdown prevents the offspring of hybrids that survive to adulthood and successfully reproduce from reproducing beyond one or a few generations.

III. Speciation is the evolution of a new species from an ancestral population.
   A. Allopatric speciation occurs when one population becomes geographically isolated from the rest of the species and subsequently evolves.
   B. Sympatric speciation does not require geographical isolation. It is extremely rare in animals; in plants it results almost exclusively from allopolyploidy.

IV. The pace of evolution is currently being debated.

A. According to the punctuated equilibrium model, evolution proceeds in spurts. Short periods of active evolution are followed by long periods of stasis.
B. According to the gradualism model, populations slowly diverge from one another by the accumulation of adaptive characteristics within a population.

V. Macroevolution refers to major evolutionary events—large phenotypic changes—that occur in groups of species over geological time. The phenotypic changes are so great that the organisms possessing them are assigned to a new genus or higher taxonomic category.
A. Macroevolution is concerned with the origin of taxonomic categories above the species level.
B. Evolutionary novelties, adaptive radiation, and extinction are aspects of macroevolution.
1. Evolutionary novelties may originate from mutations that alter developmental pathways.
2. Adaptive radiation is the process of diversification of an ancestral species into many new species.
3. Extinction is the death of a species. When species become extinct, the adaptive zones that they occupied become vacant, allowing new species to fill them.

SELECTED KEY TERMS

adaptive radiation
adaptive zone
allometric growth
allopatric speciation
allopolyploidy
behavioral isolation
biological species concept
extinction
gametic isolation
gradualism
hybrid breakdown
hybrid inviability
hybrid sterility
hybridization
macroevolution
mechanical isolation
paedomorphosis
polyploidy
preadaptation
punctuated equilibrium
reproductive isolating mechanism
sexual isolation
speciation
species
sympatric speciation
temporal isolation

POST-TEST

1. According to the ________ ________ ________, a species is a group of organisms with a common gene pool.
2. The reproductive isolating mechanism in which two closely related species reproduce at different times of the year is known as ________ ________.
3. The reproductive isolating mechanism in which two similar species have differences in their reproductive structures that prevent mating is known as ________ ________.
4. The reproductive isolating mechanism in which an interspecific hybrid embryo dies at an early stage of development is known as ________ ________.
5. The most important method of speciation in animal evolution is ________ speciation.
6. In ________ speciation, a population diverges to form a new species within the same geographical region that its parental species lives.
7. Possession of multiple sets of chromosomes, one or more of which came from a different species, is known as ________.
8. Evidence in the fossil record of long periods of stasis and few transitional forms during speciation is used to support the ________ ________ model.
9. According to the ________ model, evolution proceeds at a more or less constant rate but is not observed in the fossil record because the record is incomplete.
10. Evolution that involves taxa above the level of species is known as ________.
11. A "new" structure that is a variation of some structure already in existence is called a(an) ________.
12. The unusual shape of the ocean sunfish is the result of ________ ________.
13. The permanent loss of a species that occurs when the last member of a species dies is known as ________.
14. The evolution of several or many species from a single ancestral species is known as ________ ________.

REVIEW QUESTIONS

1. Compare the biological species concept with Linnaeus' concept of species.
2. Give an example of each of the following: (a) temporal isolation; (b) behavioral isolation; (c) mechanical isolation; (d) gametic isolation.
3. Describe the three types of postzygotic barriers and give an example of each.
4. Identify at least five geographical barriers that might lead to allopatric speciation.
5. Explain how hybridization and polyploidy can cause a new plant species to form in as little as one generation.
6. If you were in a debate and had to support the gradualism model, what would you say? How would you support the punctuated equilibrium model? Are these two ideas mutually exclusive?
7. In macroevolution, how are novel changes in structure related to development?
8. Give an example of each of the following: (a) preadaptation; (b) allometric growth; (c) paedomorphosis.
9. What role does extinction play in evolution?

YOU MAKE THE CONNECTION

1. Why is the biological species concept not valid for certain plants discussed in this chapter?
2. Why is allopatric speciation more likely to occur if the original isolated population is small?
3. Since extinction is a natural process, should humans be concerned about the current mass extinctions that we are causing? Why or why not?

RECOMMENDED READINGS

Arnold, M.L. "Natural Hybridization and Louisiana Irises: defining a major factor in plant evolution." *BioScience,* Vol. 43, No. 3, March 1994. Interspecific hybridization appears to be a significant factor in the evolution of the Louisiana irises.

Bean, M.J. "Where Late the Sweet Birds Sang." *Natural History,* February 1993. The tale of the dusky seaside sparrow's extinction.

Jackson, J. and A. Cheetham. "On the Importance of Doing Nothing." *Natural History,* June 1994. A study of fossil bryozoa supports the punctuated equilibrium model.

Knowlton, N. "A Tale of Two Seas." *Natural History,* June 1994. The author describes her work on look-alike species of snapping shrimp.

Mayr, E. *Animal Species and Evolution.* Harvard University Press, Cambridge, MA, 1963. A classic on animal evolution by the biologist who introduced the biological species concept and is a strong proponent of allopatric speciation.

Rennie, J. "Profile: Ernst Mayr, Darwin's Current Bulldog." *Scientific American,* Vol. 271, No. 2, August 1994. Portrays one of the 20th century's leading biologists.

Thompson, J.D. "The Biology of an Invasive Plant." *BioScience,* Vol. 41, No. 6, June 1991. A species of cordgrass that arose as an allopolyploid approximately 100 years ago has become a common plant in Great Britain's salt marshes.

C H A P T E R 20

The Origin and Evolutionary History of Life

Allosaurus was a formidable predatory dinosaur. One of the largest carnivorous dinosaurs of the Jurassic period, *Allosaurus* was up to 11 meters long. (Chris Butler/Science Photo Library/Photo Researchers, Inc.)

The preceding three chapters were concerned with the evolution of organisms, but we have not yet dealt with a fundamental question involving biological evolution: How did life begin? Biologists generally accept the hypothesis that life developed from nonliving matter. This process, called **chemical evolution,** probably involved several stages. Current evidence indicates that small organic molecules first formed spontaneously and accumulated over time. They were able to accumulate rather than being broken down (as occurs today) because the two factors responsible for breaking down organic molecules—free oxygen and living organisms—were absent from early Earth.

Then, large organic macromolecules such as proteins and nucleic acids assembled from the smaller molecules. The macromolecules interacted with one another, combining into more complicated structures that could eventually metabolize and replicate. Natural selection favored macromolecular assemblages with cell-like structures. Their descendants eventually became the first true cells.

After the first cells originated, they diverged over several billion years into the rich biological diversity that characterizes our planet today. Photosynthesis, aerobic respiration, and eukaryotic cell structure represent several major advances that developed during the history of life.

Geological evidence, in particular the fossil record, provides us with much of what we know about the history of life: what kinds of organisms existed, and where and when they lived. Certain organisms appear in the fossil record, then disappear, replaced by others. Initially, unicellular prokaryotes predominated, followed by unicellular eukaryotes. The first multicellular eukaryotes—soft-bodied animals that did not leave many fossils—appeared about 630 million years ago in the sea. Shelled animals and many other marine invertebrates (animals without backbones) appeared next, followed by the first vertebrates. The first fishes with jaws appeared and diversified; some of these gave rise to amphibians, which also spread and diversified. About 300 million years ago, amphibians gave rise to reptiles, which diversified and populated the land. Reptiles in turn gave rise to birds and mammals. Plants underwent a comparable evolutionary history and diversification.

In this chapter, we survey life over a vast span of time, starting some 3.5 billion years ago when our planet was relatively young. We examine ideas about how life began and trace life's long evolutionary history from its beginnings to the present.

LEARNING OBJECTIVES

After you have studied this chapter you should be able to

1. Describe the conditions that are thought to have existed on early Earth.
2. Outline the major steps that are thought to have occurred in the origin of cells.
3. Explain how the evolution of photosynthetic autotrophs affected the atmosphere and other organisms.
4. Describe the endosymbiont theory and summarize the supporting evidence.
5. List the four geological eras in chronological order and give approximate dates for each.
6. Briefly describe the climate, geological features, and distinguishing organisms for the Precambrian, Paleozoic, Mesozoic, and Cenozoic eras.
7. Explain how the course of evolution was affected by continental drift.

EARLY EARTH PROVIDED THE CONDITIONS FOR CHEMICAL EVOLUTION

It is thought that life originated only once and that life's beginnings occurred under environmental conditions quite different from those of today. We must therefore examine the conditions of early Earth to understand the origin of life. Although we will never be certain about the exact conditions when life arose, scientific evidence from a number of sources provides us with valuable clues.

Astrophysicists and geologists have determined that Earth is approximately 4.6 billion years old. The atmosphere of early Earth included carbon dioxide (CO_2), water vapor (H_2O), carbon monoxide (CO), hydrogen (H_2), and nitrogen (N_2). It contained little or no free oxygen (O_2). The early atmosphere may also have contained some ammonia (NH_3), hydrogen sulfide (H_2S), and methane (CH_4), although these reduced molecules may have been rapidly broken down by ultraviolet radiation from the sun.

Four requirements existed for the chemical evolution of life: no free oxygen, a source of energy, the availability of chemical building blocks, and time. First, life could have begun only in the absence of free oxygen. Oxygen is quite reactive and would have broken down the organic molecules that are a necessary step in the origin of life. Earth's early atmosphere was probably strongly reducing, which means that any free oxygen would have reacted with other elements to form oxides. Thus, oxygen would have been tied up in compounds.

A second requirement for the origin of life was energy. Early Earth was a place of high energy, with violent thunderstorms, widespread volcanic activity, bombardment from meteorites and other extraterrestrial objects, and intense radiation, including ultraviolet radiation from the sun (Fig. 20–1). The young sun probably produced more ultraviolet radiation than it does today, and Earth had no protective ozone layer to filter it.

Third, the chemical building blocks needed for chemical evolution must have been present. These included water, dissolved inorganic minerals (present as ions), and the gases present in the early atmosphere. A final requirement for the origin of life was time—time for molecules to accumulate and react with one another. The approximately 4.6 billion years that Earth has existed has provided adequate time for chemical evolution.

Organic molecules formed on primitive earth

Because organic molecules are the building materials for living organisms, it is reasonable to consider how they might have originated. The concept that simple organic molecules such as sugars, nucleotide bases, and amino acids could form spontaneously from simpler raw materials was first hypothesized in the 1920s by two scientists working independently: A.I. Oparin, a Russian biochemist, and J.B.S. Haldane, a Scottish physiologist and geneticist.

Their hypothesis was tested in the 1950s by Stanley Miller and Harold Urey, who designed an apparatus that simulated conditions then thought to be prevalent on early Earth (Fig. 20–2). They exposed an atmosphere rich in hydrogen (H_2), methane (CH_4), water (H_2O), and ammonia (NH_3) to an electrical discharge that simulated lightning. Their analysis of the chemicals produced in a week revealed that amino acids and other organic molecules had formed. Although more recent data suggest that Earth's early atmosphere was not rich in methane or ammonia, similar experiments using different combinations of gases have produced a wide variety of organic molecules, including the nucleotide bases of RNA and DNA.

Oparin envisioned that the organic molecules would, over vast spans of time, accumulate in the shallow seas to form a "sea of organic soup." Under such conditions, he thought that smaller organic molecules (monomers) would combine to form larger ones (polymers). Based on evidence gathered since Oparin's time, most scientists

FIGURE 20–1 Conditions on early Earth would have been inhospitable for most of today's life forms. The strongly reducing atmosphere lacked oxygen; volcanoes erupted, spewing gases that contributed to the atmosphere; and violent thunderstorms produced torrential rainfall that eroded the land. Meteorites and other extraterrestrial objects continually bombarded Earth, causing cataclysmic changes in the crust, oceans, and atmosphere. (Tsuyoshi Nishiinoue and Orion Press)

think it more likely that organic polymers formed and accumulated on rock or clay surfaces rather than in the primordial seas. Clay, which consists of microscopic particles of weathered rock, is particularly intriguing as a possible site for early polymerizations because it contains zinc and iron ions that might have served as catalysts. Laboratory experiments have confirmed that organic polymers form spontaneously from monomers on rock or clay surfaces.

Another possible site of early polymerizations that led to the origin of life was **hydrothermal vents,** cracks in the deep ocean floor where hot water and minerals such as sulfur spew forth. Such a location would have been better protected than Earth's surface from the catastrophic effects of meteorite bombardment. These hot springs produce precursors of biological molecules and of energy-rich "food," including hydrogen sulfide and methane.

After the first polymers formed, could they have assembled spontaneously into more complex structures? Scientists have synthesized several different **protobionts,** which are assemblages of abiotically produced organic polymers. They have been able to make protobionts that resemble living cells in several ways, helping us to envision how aggregations of complex nonliving molecules took that "giant leap" and became living cells. Protobionts exhibit some of the properties of life. They often

(*a*)

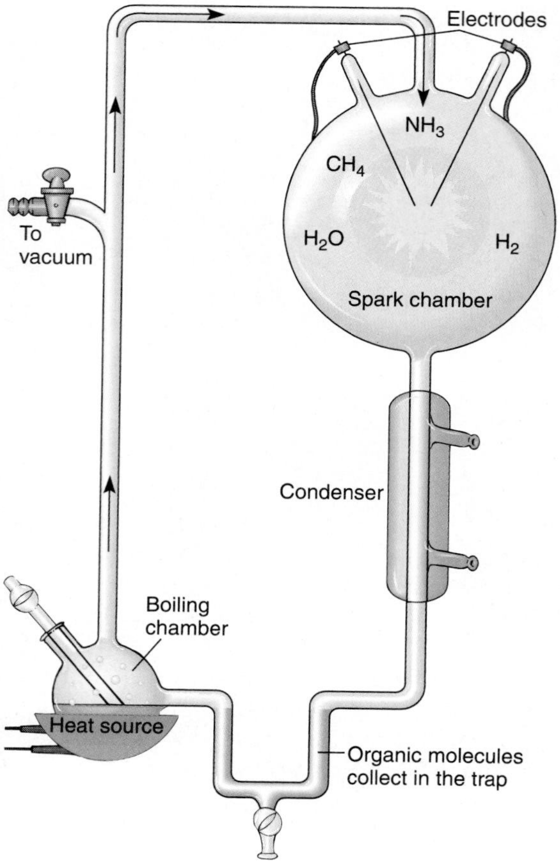

(*b*)

FIGURE 20–2 American biochemists Stanley Miller and Harold Urey demonstrated the synthesis of organic molecules from inorganic substances under conditions mimicking those of early Earth. (*a*) Stanley Miller operates the apparatus that he and Urey used to simulate the reducing atmosphere of early Earth. (*b*) Diagram of the apparatus. An electrical spark was produced in the upper right flask to simulate lightning. The gases present in the flask reacted together, forming a number of simple organic compounds that accumulated in the trap at the bottom. (© 1988 Roger Ressmeyer/Starlight)

divide in half after they have "grown." Protobionts maintain an internal chemical environment that is different from the external environment, and some of them show the beginnings of metabolism. They are highly organized, considering their relatively simple composition.

Microspheres are a type of protobiont formed by adding water to abiotically formed polypeptides (Fig. 20–3). Some microspheres demonstrate excitability: they produce an electrical potential across their surfaces, reminiscent of electrochemical gradients in cells. Microspheres can also absorb materials from their surroundings and respond to changes in osmotic concentration as though they were enveloped by membranes, even though they contain no lipid.

The first cells probably assembled from organic molecules

Studying protobionts helps us appreciate that relatively simple "pre-cells" can exhibit some of the properties of life. However, it is a major step (or several steps) to go from simple molecular aggregates such as protobionts, despite their lifelike properties, to living cells. Although much has been learned about the synthesis of organic molecules on primitive Earth, the problem of how pre-cells evolved into living cells remains to be solved. Nonetheless, fossil evidence indicates that cells were thriving 3.5 billion years ago, soon after Earth's crust cooled.

Those earliest cells were prokaryotic. Australian and South African rocks have yielded microscopic fossils of prokaryotic cells 3.1 to almost 3.5 billion years old. **Stromatolites** are another type of fossil evidence of the earliest cells (Fig. 20–4). These rocklike columns are composed of many minute layers of prokaryotic cells, usually cyanobacteria. Over time, sediment collects around the cells and gradually becomes mineralized. Meanwhile, a new layer of living cells grows over the older, dead cells. Fossil stromatolite reefs are found in a number of places in the world, including Great Slave Lake in Canada and the Gunflint Iron Formations along Lake Superior in the United States. Some fossil stromatolites are extremely ancient. One group in Western Australia, for example, is several billion years old. Living stromatolite reefs are still found in hot springs and in warm, shallow pools of fresh and salt water.

We have said that the origin of cells from macromolecular assemblages was a major step in the origin of life. Actually, the evolution of cells probably occurred in a series of small steps. One of the most significant parts of that process would have been the evolution of molecular reproduction.

A crucial step in the origin of cells was molecular reproduction

In living cells, genetic information is stored in the nucleic acid DNA, which is transcribed into the message in RNA, which in turn is translated into the proper amino acid sequence in proteins. All three macromolecules in the DNA → RNA → protein sequence contain precise information, but only DNA and RNA are capable of self-replication (copying themselves). Because both RNA and DNA can form spontaneously on clay in much the same way that polypeptides do, biologists have wondered which molecule—DNA or RNA—first appeared in the prebiotic world.

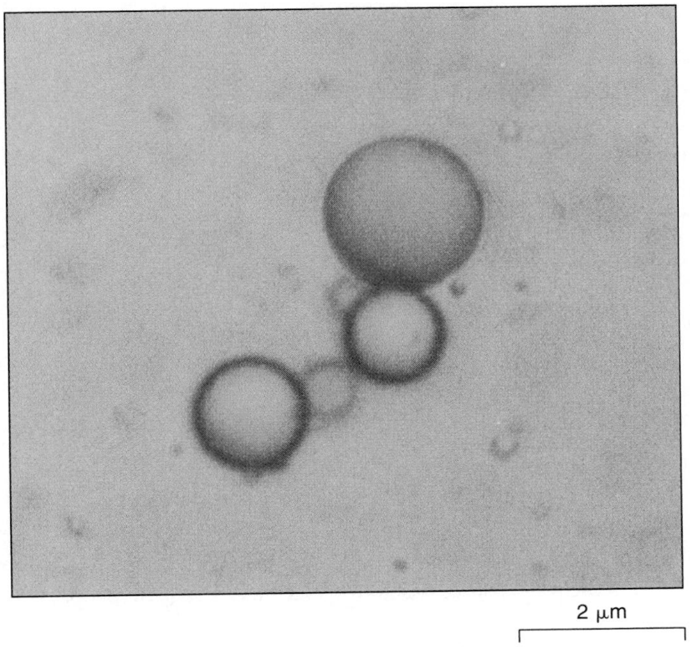

FIGURE 20–3 Microspheres are tiny protobionts that exhibit some of the properties of life. (Steven Brooke and Richard LeDuc)

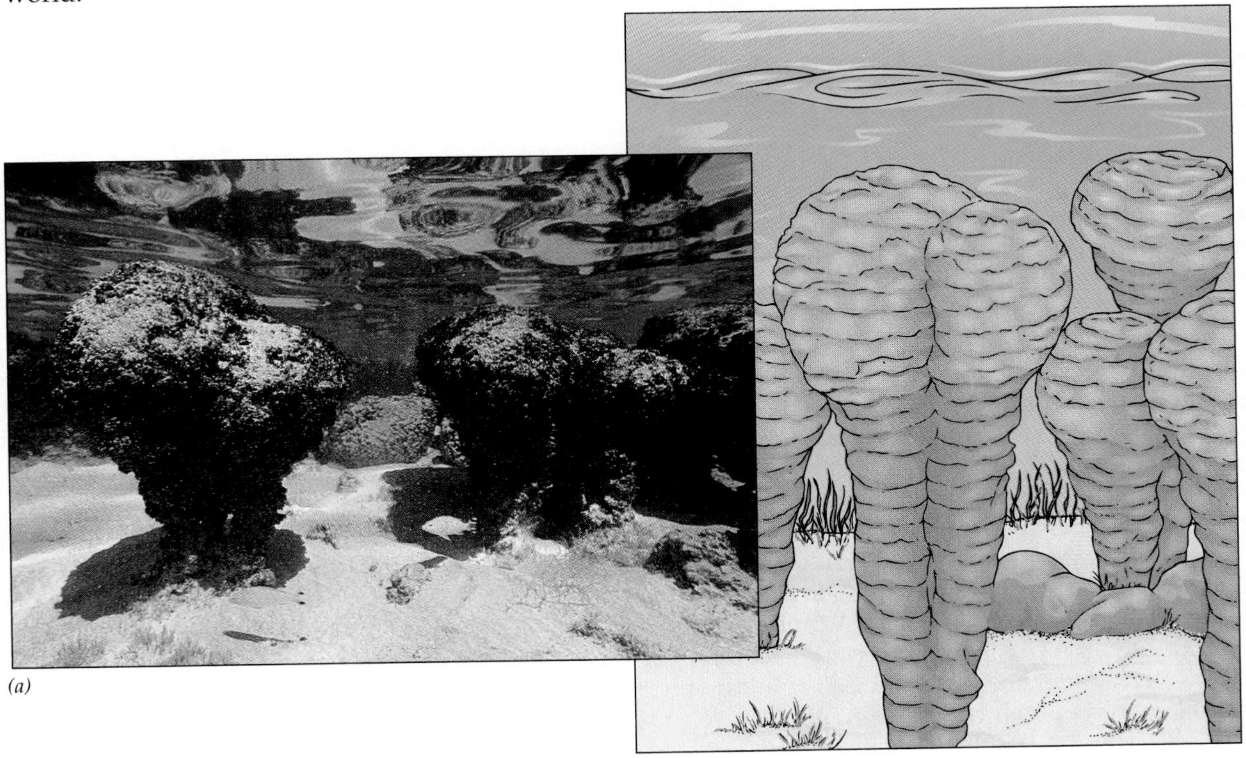

(a)

(b)

FIGURE 20–4 Stromatolites are rocklike structures found in warm, shallow marine environments. (*a*) These stromatolites at Hamlin Pool, West Australia, are composed of mats of cyanobacteria and minerals like calcium carbonate, and are several thousand years old. Some fossil stromatolites are 3.1 to 3.4 billion years old. (*b*) Diagram of stromatolites, showing the layers of cyanobacteria and sediments that accumulated over time. (Fred Bavendam/Peter Arnold, Inc.)

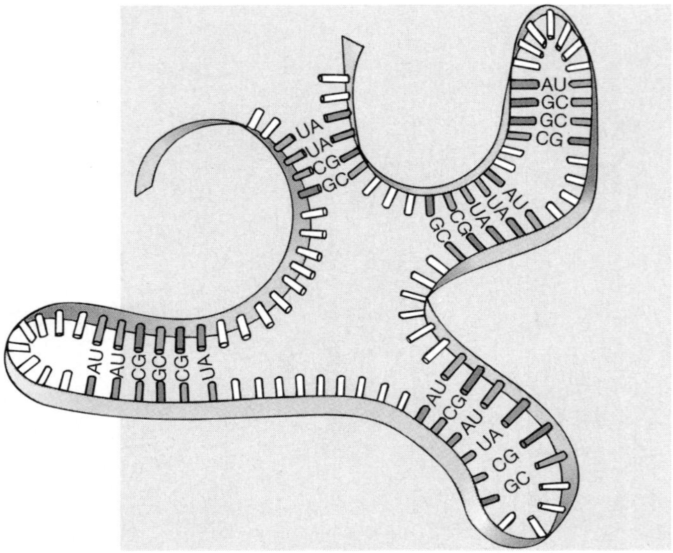

FIGURE 20–5 Some types of RNA have catalytic properties. Single-stranded RNA can form base pairs with itself, producing a precise conformation that may have catalytic properties. The order of the nucleotides determines the ultimate shape of the molecule.

Some scientists have suggested that RNA was the first information molecule to evolve in the progression toward the first cell, and that proteins and DNA came along later. One of the surprising features of RNA is that it often has catalytic properties (Fig. 20–5). Such enzymatic RNAs are called **ribozymes.** In present-day cells ribozymes help catalyze the synthesis of RNA, and process RNA into rRNA, tRNA, and mRNA (see Chapter 12). Before the evolution of true cells, ribozymes may have catalyzed the formation of more RNA in the clays, shallow rock pools, or hydrothermal vents where life originated. When RNA strands are added to a test tube containing RNA nucleotides, replication occurs without enzymes. The rate of this reaction is increased if zinc is added as a catalyst. (Recall that zinc is bound to clay.)

RNA can also direct protein synthesis by catalyzing peptide bond formation. Some single-stranded RNA molecules fold back on themselves as a result of interactions among the nucleotides composing the RNA strand. Sometimes the conformation (shape) of the folded RNA molecule is such that it weakly binds to an amino acid. If amino acids are held closely together by RNA molecules, they may bond together, forming a polypeptide.

We have considered how the evolution of information molecules may have given rise to RNA and later to proteins. If a self-replicating RNA capable of coding for proteins appeared before DNA, how did DNA, the universal molecule of heredity, get involved? In other words, how was DNA incorporated into the information transfer system? DNA is a double helix and is therefore more stable (that is, less reactive) than RNA. Such stability in a molecule that stores genetic information would have provided a decided advantage in the prebiotic world (as it does today). RNA would still have been needed, however, because DNA is not catalytic. Thus, natural selection at the molecular level resulted in the DNA $\rightarrow$ RNA $\rightarrow$ protein information sequence. Once DNA was incorporated into the molecular information sequence, RNA molecules assumed their present role as an intermediary in the transfer of genetic information.

Several additional steps had to occur before a true living cell could develop from macromolecular aggregations. For example, how did the genetic code, which must have arisen extremely early in the prebiotic world because all living organisms possess it, originate? Also, how did a plasma membrane of lipid and protein come to envelop the pre-cell assemblages, thereby permitting the accumulation of some molecules and the exclusion of others?

The first cells were probably heterotrophs, not autotrophs

The earliest cells may have been **heterotrophic** cells that obtained the organic molecules they needed from the environment, rather than synthesizing them. These primitive organisms probably consumed many types of organic molecules that had spontaneously formed—sugars, nucleotides, and amino acids, to name a few. By fermenting these organic compounds, they obtained the energy needed to support life. Fermentation is, of course, an anaerobic process (that is, performed in the absence of oxygen), and the first cells were almost certainly **anaerobes.**

When the supply of spontaneously generated organic molecules decreased, only certain organisms could survive. Mutations had probably already occurred that permitted some cells to obtain energy directly from sunlight (perhaps by using sunlight to make ATP). These cells, which did not require the energy-rich organic compounds that were now in short supply in the environment, had a distinct selective advantage.

Photosynthesis requires not only light energy but also a source of hydrogen, which is used to reduce CO_2 when organic molecules such as glucose are synthesized (see Chapter 8). Most likely, the first photosynthetic **autotrophs** (organisms that produce their own food from simple raw materials) used the energy of sunlight to split hydrogen-rich molecules such as H_2S, releasing elemental sulfur (not oxygen) in the process. Indeed, the green sulfur bacteria and the purple sulfur bacteria still use H_2S as a hydrogen source for photosynthesis.[1]

The first photosynthetic autotrophs to obtain hydrogen by splitting water were the cyanobacteria. Water was

[1]A third group of bacteria, the purple nonsulfur bacteria, uses other organic molecules or hydrogen gas as a hydrogen source.

quite abundant on early Earth (as it is today), and the selective advantage that splitting water bestowed on them allowed the cyanobacteria to thrive. In the process of splitting water, oxygen was released as a gas (O_2). Initially, the oxygen released from photosynthesis oxidized minerals in the ocean and in Earth's crust. Over time, oxygen began to accumulate in the oceans and the atmosphere.

The timing of the events just described has been estimated on the basis of geological and fossil evidence. The first autotrophs (which were prokaryotes and included the cyanobacteria) appeared approximately 3.1 to 3.4 billion years ago. Fossil evidence includes rocks from that period that contain traces of chlorophyll, as well as the fossil stromatolites discussed previously. By two billion years ago, the cyanobacteria had produced enough oxygen to begin significantly changing the composition of the atmosphere.

Aerobes appeared after oxygen increased in the atmosphere

The increase in atmospheric oxygen had a profound effect on life. Obligate anaerobes (those organisms that cannot use oxygen for cellular respiration) were poisoned by the oxygen, and many species undoubtedly perished. Some anaerobes, however, survived in environments where oxygen does not penetrate; others developed ways to neutralize the oxygen so that it could not harm them. Some organisms, called **aerobes,** developed a respiratory pathway that *used* the oxygen to extract more energy from food and convert it to ATP energy. Aerobic respiration was tacked onto the existing anaerobic process of glycolysis.

The evolution of organisms that could use oxygen had several consequences. Organisms that respire aerobically gain much more energy from a single molecule of glucose than anaerobes gain by fermentation. As a result, aerobic organisms were more efficient and more competitive than anaerobes. Coupled with the poisonous nature of oxygen to anaerobes, the efficiency of aerobes forced anaerobes into relatively minor roles. Today the vast majority of organisms, including plants, animals, and most fungi, protists, and prokaryotes, use aerobic respiration.

The evolution of aerobic respiration had a stabilizing effect on both oxygen and carbon dioxide in the ecosphere. Photosynthetic organisms used carbon dioxide as their carbon source. This raw material would have been depleted from the atmosphere in a relatively short time without the advent of aerobic respiration. Aerobic respiration released carbon dioxide as a waste product from the complete breakdown of organic molecules. Carbon thus started cycling in the ecosphere, moving from the nonliving physical environment to photosynthetic organisms to heterotrophs that ate the plants (see Chapter 54). Carbon was released back into the physical environment as carbon dioxide by aerobic respiration, and the cycle continued. In a similar manner, oxygen was produced by photosynthesis and used during aerobic respiration.

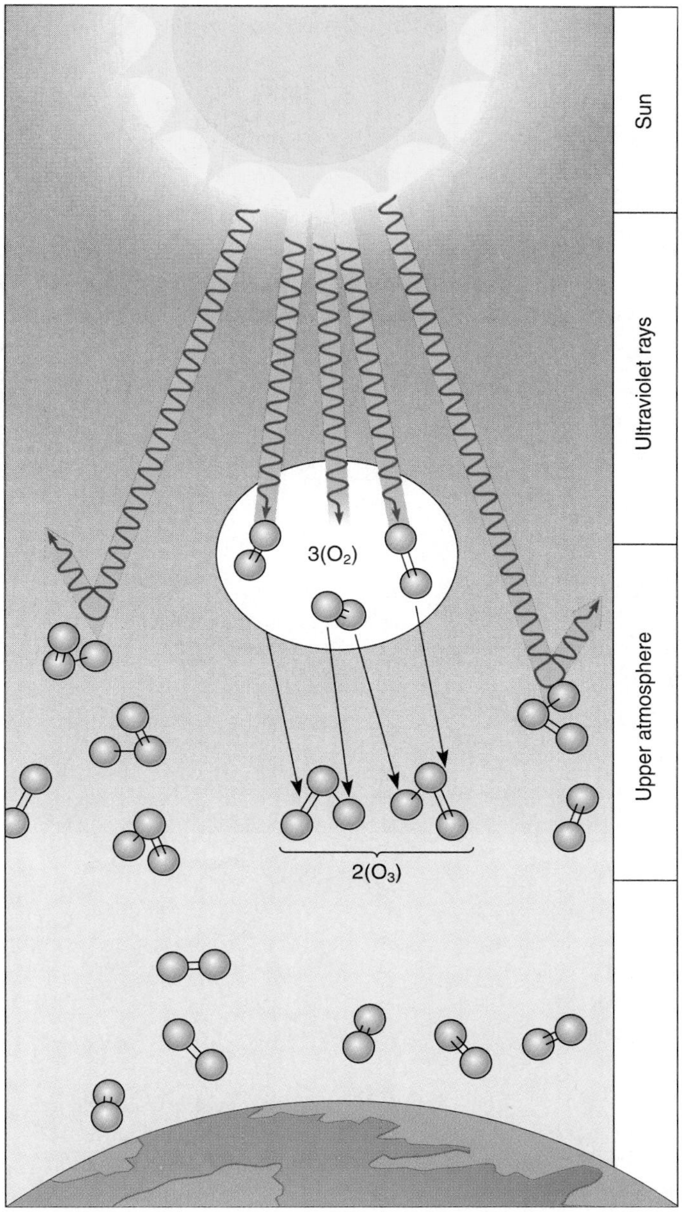

FIGURE 20–6 Ozone (O_3) forms in the upper atmosphere when ultraviolet radiation from the sun breaks the double bonds of oxygen molecules.

Another significant consequence of photosynthesis occurred in the upper atmosphere, where molecular oxygen reacted to form **ozone,** O_3 (Fig. 20–6). A layer of ozone eventually blanketed Earth, preventing much of the sun's ultraviolet radiation from penetrating to the surface. With the ozone layer's protection from the disruptive effect of ultraviolet radiation, organisms were able to live closer to the surface in aquatic environments and eventually to move onto land. Because the energy in ul-

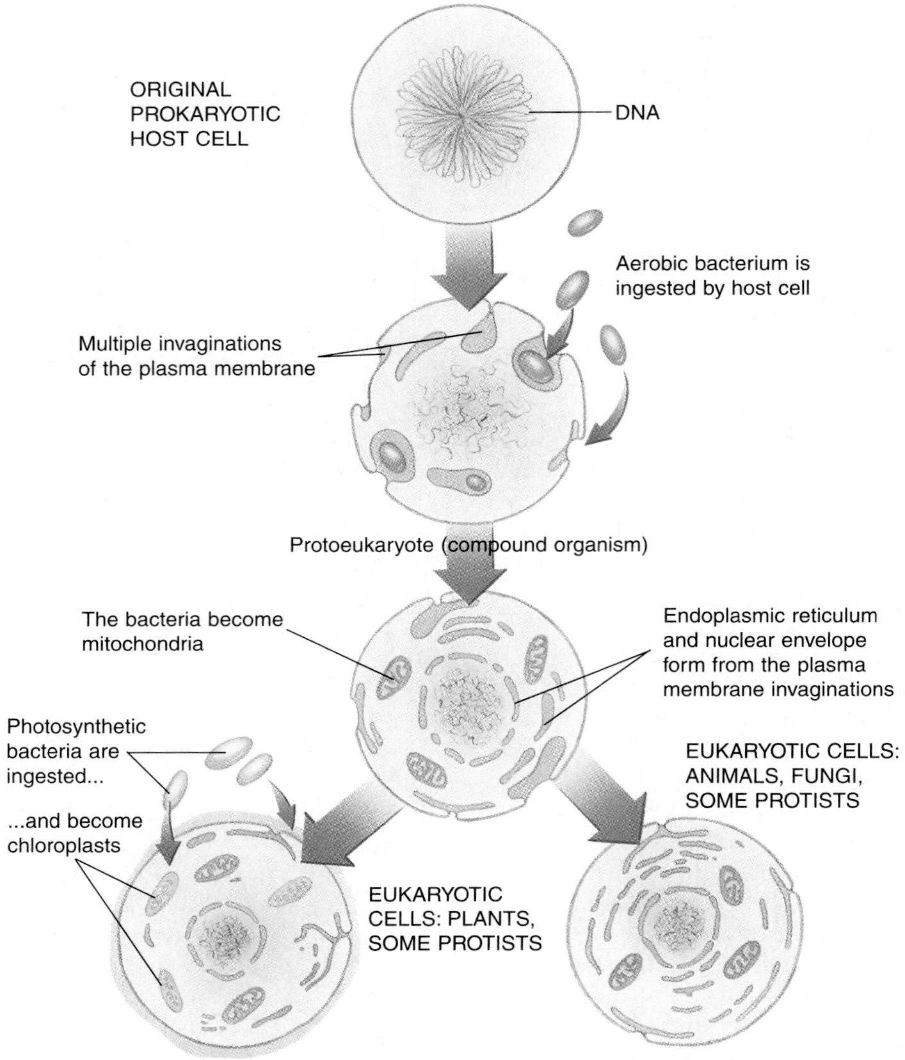
ORIGINAL PROKARYOTIC HOST CELL
DNA
Aerobic bacterium is ingested by host cell
Multiple invaginations of the plasma membrane
Protoeukaryote (compound organism)
The bacteria become mitochondria
Endoplasmic reticulum and nuclear envelope form from the plasma membrane invaginations
Photosynthetic bacteria are ingested...
...and become chloroplasts
EUKARYOTIC CELLS: ANIMALS, FUNGI, SOME PROTISTS
EUKARYOTIC CELLS: PLANTS, SOME PROTISTS

FIGURE 20–7 The endosymbiont theory explains the origin of certain eukaryotic organelles. Chloroplasts and mitochondria of eukaryotic cells are thought to have originated from various bacteria that lived as endosymbionts inside other cells.

traviolet radiation was necessary to form organic molecules, however, their spontaneous synthesis decreased.

Eukaryotic cells descended from prokaryotic cells

Eukaryotes appeared in the fossil record 1.9 to 2.1 billion years ago. They arose from prokaryotes. Recall that prokaryotic cells lack nuclear envelopes as well as other membranous organelles such as mitochondria and chloroplasts. How did eukaryotic cells arise from prokaryotes?

The **endosymbiont theory,** advanced by Lynn Margulis, suggests that organelles such as mitochondria and chloroplasts may have originated from mutually advantageous symbiotic relationships between two prokaryotic organisms (Fig. 20–7). Chloroplasts are thought to have evolved from photosynthetic bacteria that lived inside other cells, while mitochondria evolved from aerobic bacteria that lived inside other cells. Thus, early eukaryotic cells contained a group of formerly free-living prokaryotes.

How did these bacteria come to be **endosymbionts,** which are organisms that live symbiotically inside a host cell? They may have originally been ingested, but not digested, by a host cell. They could have survived and reproduced along with the host cell so that future generations of the host also contained endosymbionts. The two organisms developed a mutualistic relationship, in which each contributed something to the other. Eventually the endosymbiont lost the ability to exist outside its host, and the host cell lost the ability to survive without the endosymbionts.

This theory stipulates that each of these partners brought to the relationship something the other lacked.

For example, mitochondria provided the ability to carry out the aerobic respiration lacking in the original host cell; chloroplasts provided the ability to use a simple carbon source (carbon dioxide) to produce needed organic molecules. The host cell provided a safe habitat and raw materials or nutrients.

The principal evidence in favor of the endosymbiont theory is that mitochondria and chloroplasts possess some (although not all) of their own genetic material. They have their own DNA (as a circular molecule similar to that of prokaryotes; see Chapter 23) and their own ribosomes (which resemble prokaryotic rather than eukaryotic ribosomes). Mitochondria and chloroplasts have some of the machinery for protein synthesis, including tRNA molecules, and are able to conduct protein synthesis on a limited scale independent of the nucleus. Further, it is possible to poison mitochondria and chloroplasts with an antibiotic that affects bacteria but not eukaryotic cells. Mitochondria and chloroplasts are enveloped by double membranes. Biologists hypothesize that the outer membrane developed from the invagination of the host cell's plasma membrane, while the inner membrane developed from the endosymbiont's plasma membrane.

A number of endosymbiotic relationships exist today. Many corals have algae living as endosymbionts within their cells. In the gut of the termite lives a protozoon (*Myxotricha paradoxa*) that in turn has several different endosymbionts, including spirochete bacteria that are attached to the protozoon and function as whiplike flagella, allowing it to move.

The endosymbiont theory does not completely explain the evolution of eukaryotic cells from prokaryotes. It does not explain how the genetic material in the nucleus came to be surrounded by a membranous envelope, for example. However, the advent of eukaryotic cells set the stage for further evolutionary developments.

THE FOSSIL RECORD PROVIDES US WITH CLUES TO THE HISTORY OF LIFE

The sediments of Earth's crust consist of five major rock strata (layers), each subdivided into minor strata, lying one on top of the other. These sheets of rock were formed by the accumulation of mud and sand at the bottoms of oceans, seas, and lakes. Each layer contains certain characteristic fossils that serve to identify deposits made at approximately the same time in different parts of the world.

Geologists divide the Earth's 4.6-billion-year history into units of time based on major geological, climatic, and biological events. Relatively little is known about Earth from its beginnings approximately 4.6 billion years ago up to 570 million years ago, a period known informally as **Precambrian time.** The fossil record of ancient organisms is abundant beginning about 570 million years ago. This most recent time is divided into **eras** based primarily on organisms that were characteristic of each era (Table 20–1). Eras are subdivided into **periods,** which in turn are composed of **epochs.**

Between the major eras, and serving to distinguish them, widespread geological disturbances occurred, which raised or lowered vast regions of Earth's surface and produced or eliminated shallow inland seas. These disturbances altered the distribution of sea and land organisms and may have triggered the mass extinction of many organisms. The raising and lowering of portions of Earth's crust result from the slow movements of the enormous plates that compose its crust (see *Focus On Continental Drift*).

Evidence of living cells is found in Precambrian times

Signs of Precambrian life date back to about 3.5 billion years ago. Not much physical evidence is available to us because the rocks of Precambrian time, being extremely ancient, are deeply buried in most parts of the world. Precambrian rocks are exposed in a few places, including the bottom of the Grand Canyon and along the shores of Lake Superior. More than 400 Precambrian rock formations have revealed **microfossils,** remains of microscopic organisms.

Precambrian time was characterized by widespread volcanic activity and giant upheavals that raised mountains. The heat, pressure, and churning associated with these movements probably destroyed most of whatever fossils may have been formed, but some evidence of life still remains. This evidence consists of traces of graphite or pure carbon, which may be the transformed remains of primitive life. These remains are especially abundant in what were the oceans and seas of that time. Fossils of what appear to be cyanobacteria have been recovered from several Precambrian formations.

The fossils found in later (more recent) Precambrian rocks show clear-cut examples of some major groups of bacteria, fungi, protists (including multicellular algae), and animals.

One rich source of Precambrian fossil deposits is the Ediacaran Hills in South Australia. The animals found there—all invertebrates—include jellyfish, segmented worms, soft-bodied arthropods, and several animals with no resemblance to any other known fossil or living form (Fig. 20–8). Ediacaran fossils, the oldest known fossils of complex, multicellular animals, are from very late in Precambrian time.

A considerable diversity of organisms evolved during the Paleozoic era

The **Paleozoic era** began approximately 570 million years ago and lasted approximately 322 million years. It is divided into six periods: Cambrian, Ordovician, Silurian, Devonian, Carboniferous, and Permian.

Table 20–1 SOME IMPORTANT BIOLOGICAL EVENTS IN GEOLOGICAL TIME

Million Years Before Present

Time	*Era*	*Period*	*Epoch*	*Geological/Climatic Conditions*
10,000 years ago to present	Cenozoic	Quaternary	Holocene	End of last Ice Age; warmer climate; higher sea levels as glaciers melt
2			Pleistocene	Four Ice Ages; glaciers in Northern Hemisphere
5		Tertiary	Pliocene	Uplift and mountain-building; volcanoes; climate much cooler; North and South America join at Isthmus of Panama
25			Miocene	Mountains form; climate drier and cooler
38			Oligocene	Rise of Alps and Himalayas; most land low; volcanic activity in Rockies; climate cool and dry
55			Eocene	Climate warmer
65			Paleocene	Continental seas disappear; climate mild to cool and wet
144	Mesozoic	Cretaceous		Continents separate; Rockies form; other continents low; large inland seas and swamps; climate warm
213		Jurassic		Continents low; inland seas; mountains form; continental drift begins; climate mild
248		Triassic		Many mountains form; widespread deserts; climate warm and dry
286	Paleozoic	Permian		Glaciers; Appalachians form; continents rise and merge as Pangaea; climate variable
360		Carboniferous		Lands low and swampy; climate warm and humid, becoming cooler later
408		Devonian		Glaciers; inland seas
438		Silurian		Continents mainly flat; flooding; climate warm
505		Ordovician		Sea covers continents; climate warm
570		Cambrian		Oldest rocks with abundant fossils; lands low; climate mild and wet

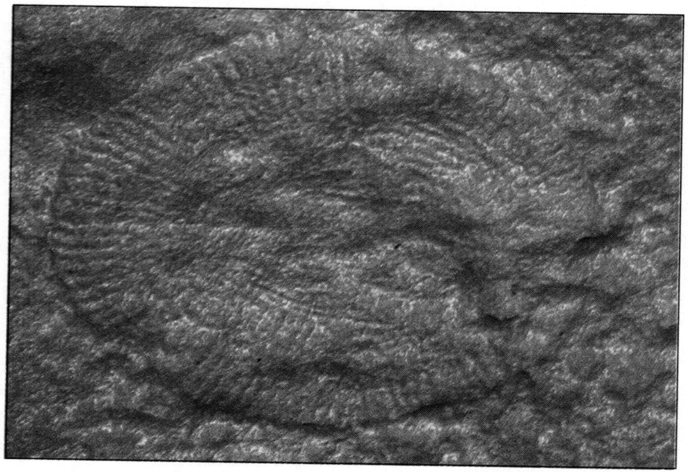

The oldest subdivision of the Paleozoic era, the Cambrian period, is represented by rocks rich in fossils. Evolution was in such high gear that this period has been nicknamed the "Cambrian explosion." Fossils of all of the present-day animal phyla except the chordates (the animal phylum that includes the vertebrates) are present, at least in marine sediments. The sea floor was covered with sponges, corals, snails, feather stars, clamlike bivalves, primitive squidlike cephalopods, lamp shells (brachio-

FIGURE 20–8 This Precambrian fossil was found in the Ediacaran Hills of South Australia. The organism, which is unlike any known modern organism, lived in shallow marine waters. (William E. Ferguson)

Table 20–1 continued

Plants and Microorganisms	*Animals*
Decline of some woody plants; rise of herbaceous plants	Age of *Homo sapiens*
Extinction of many plant species	Extinction of many large mammals; humans evolve
Development of grasslands; decline of forests	Many grazing mammals; large carnivorous mammals; first known human-like primates
Flowering plants continue to diversify	Many new mammal species appear
Spread of forests; flowering plants	Apes appear; all present mammalian families are represented
Gymnosperms and flowering plants dominant	Mammals diversify; modern birds diverge
Many now-extinct woody flowering plants	Primitive mammals diversify
Rise of flowering plants	Dinosaurs reach peak, then become extinct; toothed birds become extinct; first modern birds; primitive mammals
Gymnosperms common	Large, specialized dinosaurs; first toothed birds; insectivorous marsupials
Gymnosperms dominant; ferns common	First dinosaurs; egg-laying mammals
Conifers diversify; cycads appear	Modern insects appear; mammal-like reptiles; extinction of many Paleozoic invertebrates
Forests of ferns, club mosses, horsetails, and gymnosperms; mosses and liverworts	First reptiles; spread of ancient amphibians; many insect forms; ancient sharks abundant
Plants diversify and become well-established; first forests; gymnosperms appear; bryophytes appear	Fishes diversify; amphibians appear; wingless insects appear; many trilobites
Vascular plants appear; algae dominant in aquatic environments	Fishes diversify; terrestrial arthropods; coral reefs common
Marine algae dominant	Invertebrates dominant; first fishes appear
Algae; bacteria and cyanobacteria; fungi	Age of marine invertebrates; most modern animal phyla represented

pods), primitive aquatic arthropods known as trilobites, and other marine animals (Fig. 20–9).

Except for the chordates, the major animal body plans were established so early in the history of the eukaryotes that little change of a basic nature has occurred since. This probably indicates that by the early Cambrian period, each major animal form had reached a degree of adaptation that allowed it to exploit its environment and accommodate changes in its surroundings with only limited modifications in its body plan.

According to geologists, the continents were gradually flooded during the Cambrian period. In the Ordovician period, much of what is now land was covered by shallow seas. Inhabiting the seas were giant cephalopods, squidlike animals with straight shells 5 to 7 meters (16 to 23 feet) long and 30 centimeters (12 inches) in diameter. The first traces of the earliest vertebrates, the jawless, bony-armored fish called *ostracoderms,* are also found in Ordovician rocks.

Two life forms of great biological significance appeared in the Silurian period: terrestrial plants and air-breathing animals. The first known plants resembled ferns in that they possessed vascular (conducting) tissue and reproduced by spores. The evolution of plants allowed terrestrial animals to colonize the land because plants provided the first land animals with food and shelter. The only air-breathing land animals that have been discovered in Silurian rocks were arachnids resembling scorpions.

(Text continues on page 464)

Focus On

Continental Drift

In 1915 the German scientist Alfred Wegener, who had noted a similarity between the geographical shapes of South America and Africa, proposed that all the land masses had at one time been joined into one huge supercontinent, which he called Pangaea (see figure). He further suggested that Pangaea had subsequently broken apart and that the various land masses had separated in a process known as **continental drift.** Wegener did not know of any mechanism that could have caused continental drift, and so his idea, although debated initially, was largely ignored.

In the 1960s, scientific evidence accumulated that provided the explanation for continental drift. Earth's crust is composed of seven large plates (plus a few smaller ones) that float on the mantle (the mostly solid layer of Earth lying beneath the crust and above the core). The land masses are

Continents have slowly shifted their positions relative to one another as a result of continential drift. (***a***) The supercontinent Pangaea, about 240 million years BP (before present). (***b***) Breakup of Pangaea into Laurasia (northern hemisphere) and Gondwana (southern hemisphere), 120 million years BP. (***c***) Further separation of land masses, 60 million years BP. Note that Europe and North America were still joined and that India was a separate land mass. (***d***) The continents today. (***e***) Projected positions of the continents in 50 million years.

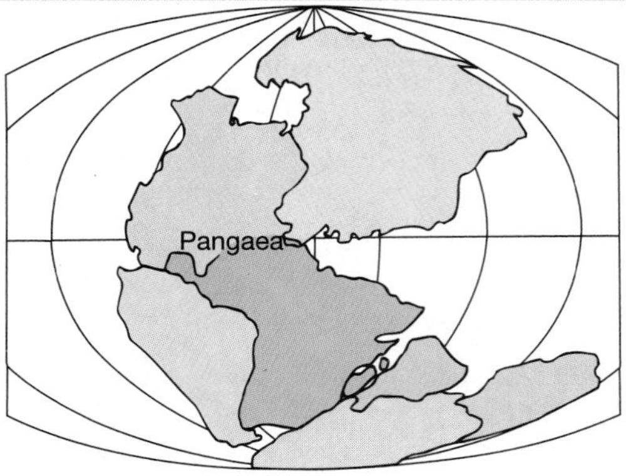

(a) 240 million years ago (Triassic period)

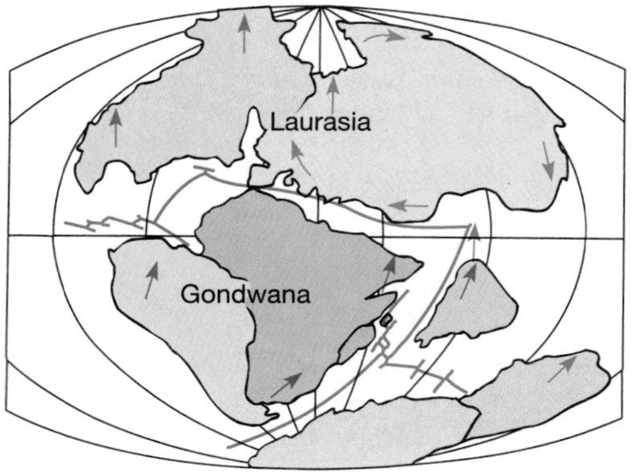

(b) 120 million years ago (Cretaceous period)

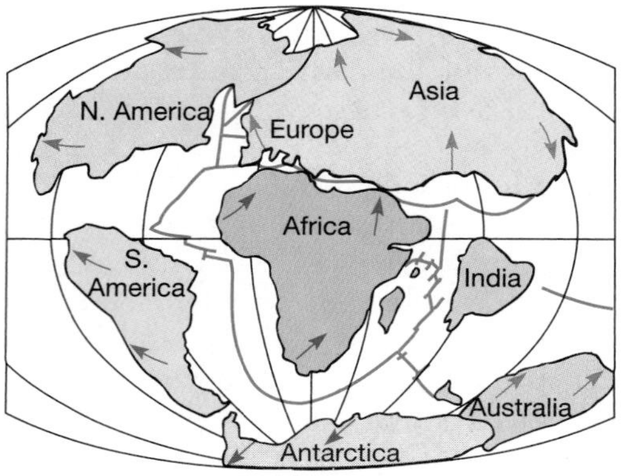

(c) 60 million years ago (early Tertiary period)

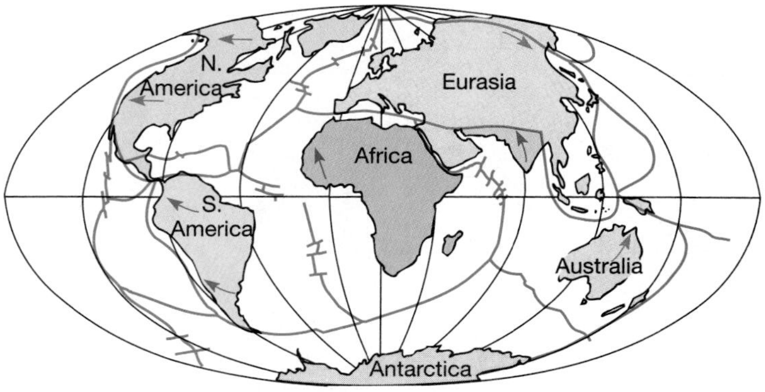

(d) Today

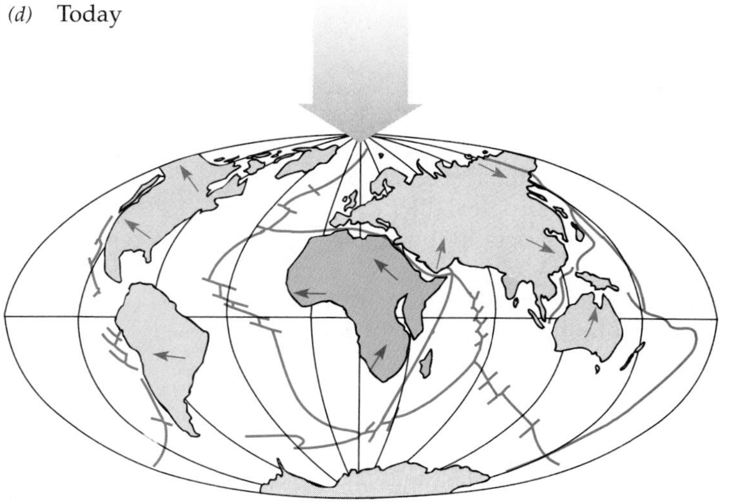

(e) 50 million years from now

situated on some of these plates. As the plates move, the continents change their relative positions (see figure). The movement of the crustal plates is termed **plate tectonics.**

Any area where two plates meet is a site of intense geological activity. Earthquakes and volcanoes are common in such a region. Both San Francisco, noted for its earthquakes, and the Mount Saint Helens volcano are situated where two plates meet. If land masses lie on the edges of two adjacent plates, mountains may be formed. The Himalayas formed when the plate carrying India rammed into the plate carrying Asia. When two plates grind together, one of them is sometimes buried under the other in a process known as subduction. When two plates move apart, a ridge of lava forms between them. The Atlantic Ocean is getting larger because of the expanding zone of lava along the mid-Atlantic ridge, where two plates are separating.

Knowledge that the continents were at one time connected and have since drifted apart is useful in explaining the geographical distribution of plants and animals, or biogeography (see figure; also see Chapter 17). Likewise, continental drift has played a major role in the evolution of different organisms. When Pangaea originally formed during the late Permian period, it brought together terrestrial species that had evolved separately from one another, leading to competition and some extinctions. Marine life was adversely affected, largely because, with the continents joined as one large mass, less coastline existed. (Because coastal areas are shallower, they contain high concentrations of marine organisms.)

Pangaea separated into several land masses approximately 180 million years ago. As the continents began to drift apart, populations became geographically isolated in different environmental conditions and began to diverge along separate evolutionary pathways. As a result, the plants, animals, and other organisms of previously connected continents, South America and Africa for example, differ. ■

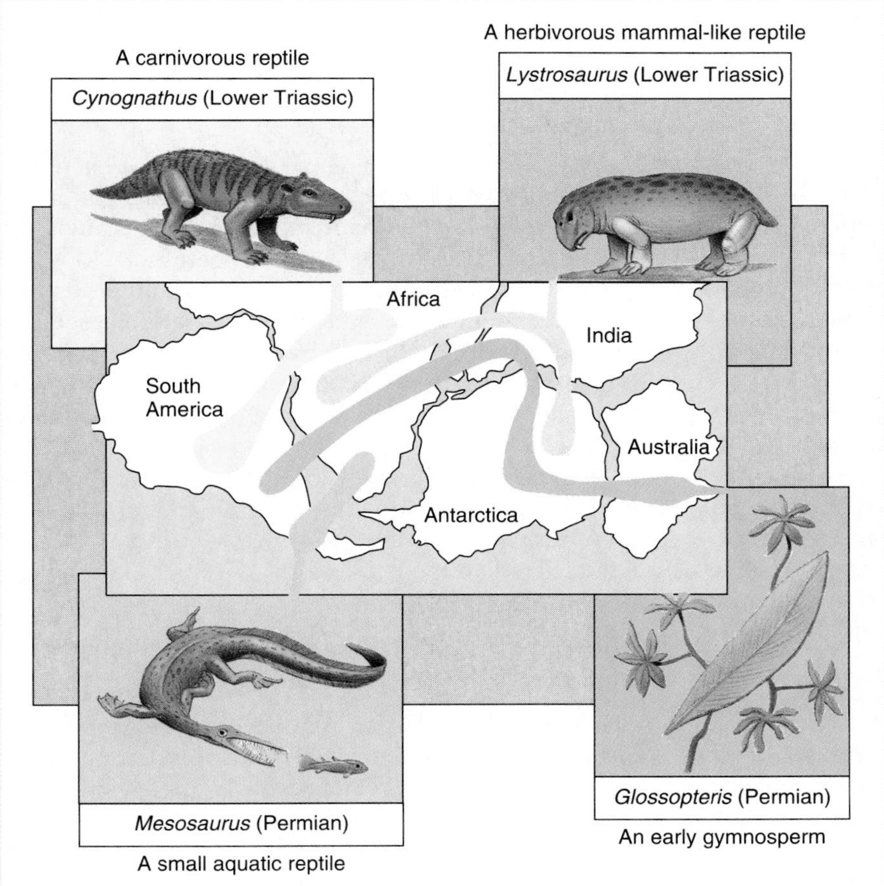

The distribution of fossils of the same animal and plant species on four continents suggests that the continents were once joined. (Adapted from Colbert)

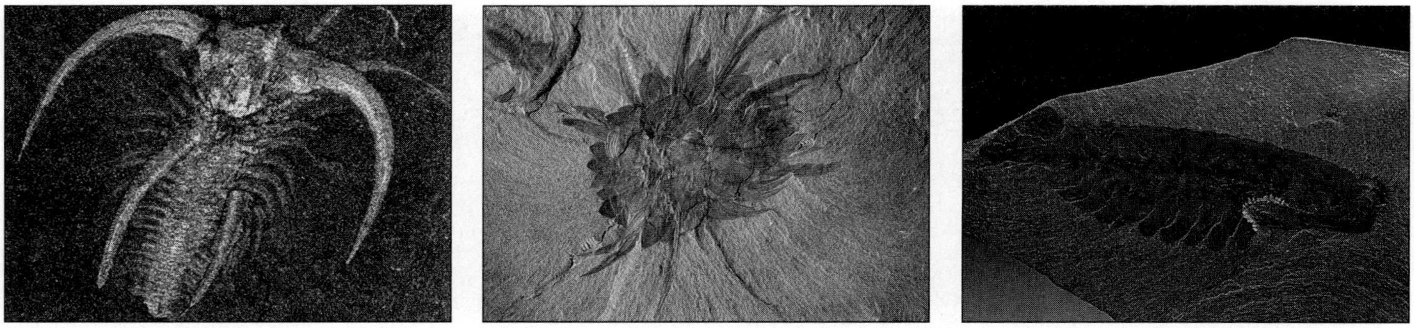

(a) (b) (c)

FIGURE 20–9 Many unusual animals evolved during the "Cambrian explosion" early in the Paleozoic era. (*a*) *Marrella splendens* was a small arthropod. (*b*) In spite of its unusual appearance, *Wiwaxia* was a worm that was distantly related to earthworms. (*c*) *Opabinia regalis* had a segmented body, five eyes, and many "legs." All three of these fossils were obtained from the Burgess Shale in the Canadian Rockies. (Chip Clark)

A great variety of fishes appeared in the Devonian period. In fact, the Devonian period is frequently called the "Age of Fishes." Unlike the jawless ostracoderms, the Devonian fishes typically had jaws, an adaptation that enables a vertebrate to chew and bite. Appearing in Devonian deposits are sharks and the three main types of bony fish: lungfishes, lobe-finned fishes, and ray-finned fishes. A few lungfish species persist even today. The ray-finned fishes later gave rise to the major modern orders of fishes. The lobe-finned fishes, some of which are considered ancestral to the land vertebrates, were thought to have been extinct. However, in 1939 the first living *coelacanth,* a primitive bony fish with lobed fins, was discovered off the coast of Madagascar (see Fig. 30–15).

Upper (more recent) Devonian sediments contain fossil remains of salamander-like amphibians that were often quite large, with short necks and heavy, muscular tails. These animals, whose skulls were encased in bony armor, were quite similar in many respects to the lobe-finned fishes.

The early vascular plants diversified during the Devonian period in a burst of evolution that rivaled that of animal evolution during the "Cambrian explosion." All major plant groups appeared during the Devonian except for flowering plants. Forests of ferns, club mosses, horsetails, and seed ferns (an extinct group of ancient plants that had fernlike foliage but reproduced by forming seeds) flourished. Wingless insects and millipedes also originated in the late Devonian period.

The Carboniferous period is named for the great swamp forests whose remains persist today as major coal deposits. Much of the land during this time was covered with low swamps filled with horsetails, club mosses, ferns, seed ferns, and gymnosperms (seed-bearing plants such as conifers) (Fig. 20–10).

An early group of reptiles, the *cotylosaurs,* appeared in the Carboniferous period, flourished in the Permian period, and became extinct early in the Mesozoic era. Two groups of winged insects, cockroaches and dragonflies, appeared in the Carboniferous period. The dragonflies ranged in size from those smaller than today's dragonflies to some with wingspans of 75 centimeters (2.5 feet).

The final period of the Paleozoic era, the Permian period, was characterized by great changes in climate and topography. At the end of the Permian period, mountain ranges formed in North America and Europe. An ice sheet, spreading northward from the Antarctic, covered most of the Southern Hemisphere, extending almost to the equator in Brazil and Africa.

Many Paleozoic forms of life may have been unable to adjust to the climatic and geological changes and thus became extinct. Even many marine forms became extinct, perhaps owing to cooler water temperatures. Most of the plants that were dominant in the Carboniferous period became extinct during the Permian period. Seed plants grew dominant, with the diversification of conifers and the appearance of cycads (plants resembling palms with crowns of fernlike leaves and large, seed-containing cones).

During the Permian period, a group of reptiles evolved that was ancestral to the mammals. These carnivorous reptiles, called *therapsids,* were more slender and lizard-like than the cotylosaurs.

The dinosaurs and other reptiles dominated the Mesozoic era

The **Mesozoic era** began about 248 million years ago and lasted some 183 million years. It is divided into the Triassic, Jurassic, and Cretaceous periods. The outstanding feature of the Mesozoic era was the origin, differentiation, and finally the extinction of a large variety of reptiles. For this reason, the Mesozoic era is commonly called the "Age of Reptiles." From a botanical viewpoint, the

FIGURE 20–10 The plants of the Carboniferous period included giant ferns, horsetails, and club mosses. (No. GEO85638c, Field Museum of Natural History, Chicago)

Mesozoic era was dominated by gymnosperms until the mid-Cretaceous period, when they were largely replaced by the flowering plants.

Turtles were one of the most ancient reptilian lines present in the Mesozoic era. Fossils of the earliest turtles are some 200 million years old, and ancient reptiles interpreted as turtle ancestors are 250 million years old. Both marine and land turtles have survived to the present with few skeletal changes since before the time of the dinosaurs. Most of the snakes and lizards found in Mesozoic formations are also similar to their present-day descendants. The marine lizards of the Cretaceous period, which attained lengths of 10 meters (33 feet) or more, did not survive to the present (Fig. 20–11*a*).

FIGURE 20–11 Reptiles (not drawn to scale) dominated Earth during the Mesozoic era. (***a***) *Tylosaurus,* a large marine reptile, belongs to a group ancestral to modern lizards. (***b***) *Tyrannosaurus,* one of the largest of the carnivorous dinosaurs. (***c***) *Ankylosaurus,* a heavily armored ornithischian. Two marine reptiles, (***d***) *Plesiosaurus* and (***e***) *Ichthyosaurus.* Note the superficial similarity of *Plesiosaurus* to animals such as seals and the similarity of *Ichthyosaurus* to porposies.

Of all the reptilian branches, dinosaurs are the most famous. There were two main groups of dinosaurs, the *saurischians,* with pelvic bones similar to those of modern-day lizards, and the *ornithischians,* with pelvic bones similar to those of birds (Fig. 20–12). Some saurischians were fast, two-legged forms ranging from those the size of a dog to the ultimate representative of this group, the gigantic carnivore of the Cretaceous period, *Tyrannosaurus* (Fig. 20–11*b*). *Tyrannosaurus,* one of the largest known predators to ever live on land, stood 6 meters (20 feet) tall. Other saurischians were huge, four-legged dinosaurs that ate plants. Some of these were the largest terrestrial animals that have ever lived, including *Argentinosaurus,* discovered in the 1980s in Argentina, with an estimated length of 30 meters (98 feet) and an estimated weight of 72 to 90 metric tons (80–100 tons). It is thought that *Argentinosaurus* and other plant-eating saurischians ate huge quantities of vegetation such as needles (leaves) from tall conifers.

The other group of dinosaurs, the ornithischians, were entirely herbivorous. Although some of them walked upright, the majority walked on four legs. Some had no front

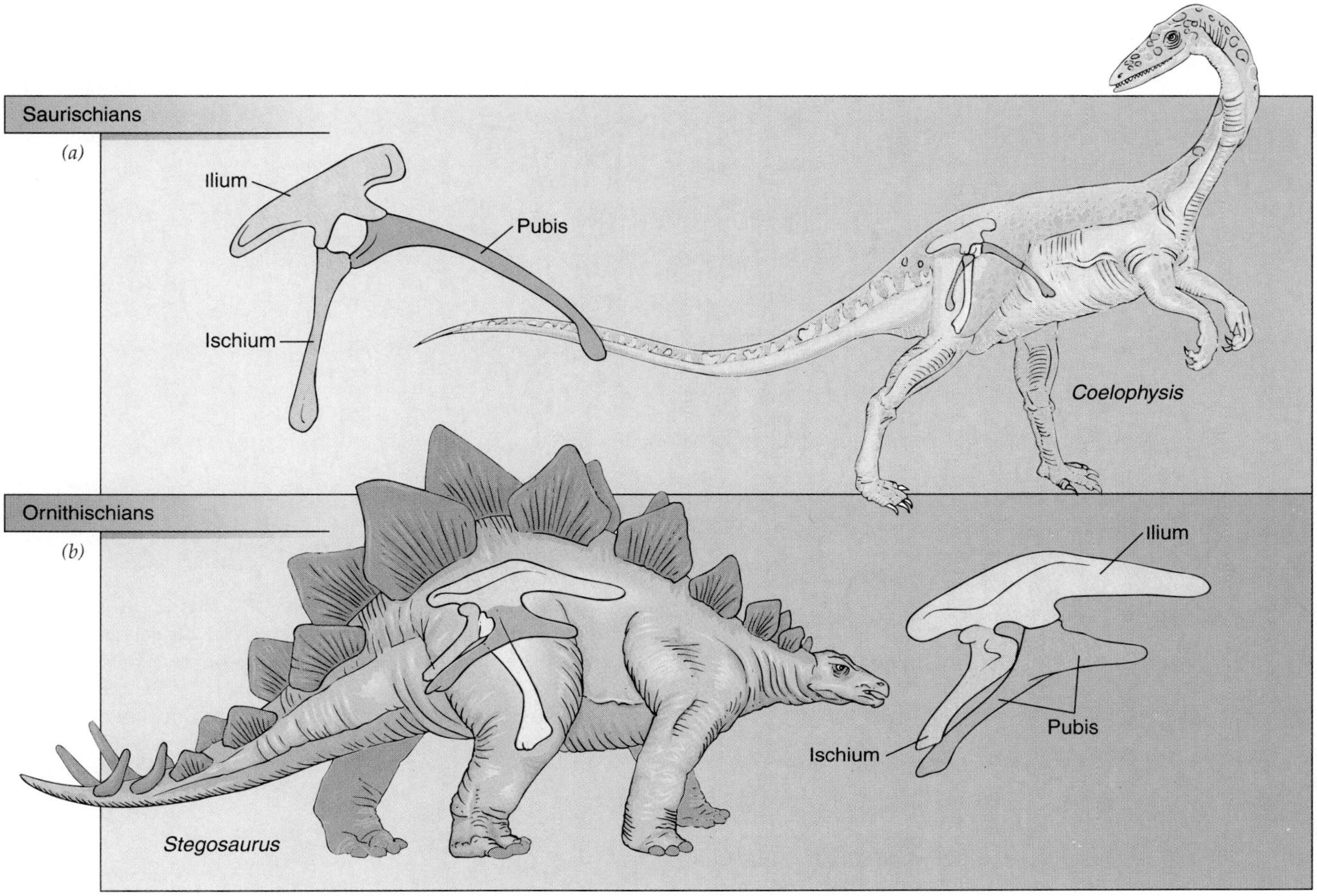

FIGURE 20–12 Although dinosaurs are considered a single group of reptiles, they are classified in two orders based primarily on differences in their pelvic bones. It is not certain whether the two dinosaur groups share a common reptilian ancestor. (***a***) Saurischian pelvis. (***b***) Ornithischian pelvis.

teeth and possessed stout, horny, birdlike beaks. In some species these beaks were broad and ducklike, hence the common name, duck-billed dinosaurs. Webbed feet were also characteristic of duck-billed dinosaurs. Other ornithischians had great armor plates, possibly as protection against the carnivorous saurischians. *Ankylosaurus,* for example, had a broad, flat body covered with armor plates (actually bony scales embedded in the skin) and large, laterally projecting spikes (Fig. 20–11*c*).

Many early ideas about dinosaurs—that they were cold-blooded, slow-moving monsters living in swamps, for example—have been reconsidered. Recent evidence suggests that dinosaurs may have been warm-blooded and capable of moving extremely fast. Many appear to have had complex social behaviors, including courtship rituals and nurturing of their young. Some species lived in groups and hunted in packs.

Two other groups of Mesozoic reptiles, the plesiosaurs and ichthyosaurs, were not dinosaurs. *Plesiosaurs* were aquatic reptiles with bodies up to 15 meters (50 feet) long and paddlelike fins (Fig. 20–11*d*). *Ichthyosaurs,* also aquatic reptiles, had body forms superficially resembling those of sharks or porpoises, with short necks, large dorsal fins, and shark-type tails (Fig. 20–11*e*).

Although reptiles were the dominant animals of the Mesozoic era, many other animals lived at that time. Most of the modern orders of insects appeared during the Mesozoic era. Snails and bivalves (clams and their relatives) increased in number and diversity, and sea urchins reached their peak diversity. Mammals first appeared in the Triassic period, and birds in the Jurassic. Excellent bird fossils, some even showing the outlines of feathers, have been preserved from the Jurassic period. *Archaeopteryx,* a primitive bird that lived about 150 million years ago, was about the size of a crow and had rather feeble wings (Fig. 20–13; see Fig. 30–22). Although *Archaeopteryx* is generally considered a bird (witness the feathers), it had many reptilian features, including teeth and a long bony tail.

At the end of the Cretaceous period, 65 million years ago, a great many animals abruptly became extinct. Most gymnosperms, with the exception of conifers, also per-

FIGURE 20–13 ***Archaeopteryx,* a tailed, toothed, primitive bird from the Jurassic period, had both reptilian and avian (birdlike) characteristics.** (Dennis Drenner)

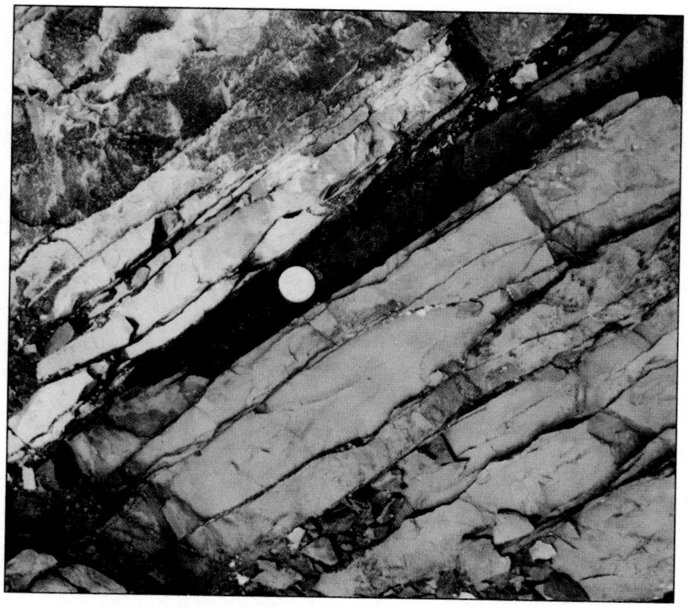

FIGURE 20–14 Evidence exists of a major collision between Earth and a large extraterrestrial object around the time of the dinosaurs' extinction. The dark band of iridium-enriched clay located between Mesozoic and Cenozoic sediments suggests that a meteorite slammed into Earth at that time. The coin on the iridium-rich layer demonstrates the relative size of the layer. (Lawrence Berkeley Laboratory, University of California)

ished. Many explanations for the mass extinction at the end of the Cretaceous period have been proposed. Interestingly, an increasing amount of scientific evidence suggests that a catastrophic collision of Earth with a large extraterrestrial body resulted in dramatic climatic changes that played a factor in their demise. Part of the evidence is a small band of dark clay with a high concentration of iridium located between Mesozoic and Cenozoic sediments (Fig. 20–14). Iridium is rare on Earth but abundant in meteorites, leading many to conclude that Earth was hit by a large extraterrestrial object at that time. (The force of the impact drove the iridium into the atmosphere; it was deposited on the land by precipitation.)

The Chicxulub crater in the Yucatan Peninsula in Mexico is thought to be the site of the collision (Fig. 20–15). The impact produced a giant tidal wave that deposited materials from the extraterrestrial body around the perimeter of the Gulf of Mexico, from Alabama to Guatemala.

Although most scientists accept that a collision with an extraterrestrial body occurred 65 million years ago, there is no consensus about the effects of such a collision on living organisms. Many marine microorganisms became extinct at or immediately after the time of the impact, likely as a result of the environmental upheaval produced by the collision. However, a number of clam species associated with the mass extinction at the end of the Cretaceous period appear to have become extinct *before* the impact, suggesting that some of the massive extinctions occurring then were caused by other factors.

The Cenozoic era is known as the "Age of Mammals"

With equal justice the **Cenozoic era** could be called the "Age of Mammals," the "Age of Birds," the "Age of Insects," or the "Age of Flowering Plants." It is marked by the appearance of all these forms in great variety and numbers of species. The Cenozoic era extends from 65 million years ago to the present. It is subdivided into two periods: the Tertiary period, encompassing some 63 million years, and the Quaternary period, which covers the last 2 million years.

The Tertiary period is subdivided into five epochs, named from earliest to latest: Paleocene, Eocene, Oligocene, Miocene, and Pliocene. The Quaternary period is subdivided into the Pleistocene and Holocene

(Text continues on page 470)

Making the Connection

Mammalian Diversity and the Carrying Capacity of the Environment

Has mammalian diversity peaked or is it declining? The fossil record indicates that during the 10 million years or so following the extinction of the dinosaurs, mammals underwent adaptive radiation. According to conventional wisdom, mammals have continued to diversify so that the present is the time of maximum mammalian diversity.

John Alroy, a paleontologist at the University of Arizona, disagrees. He assembled an extensive database of all mammalian genera* in North America (excluding bats and marine mammals) from the demise of the dinosaurs about 65 million years ago to the present. His data, which were reported at several scientific meetings in 1994, indicate that for the past 55 million years, North America has been home to about 90 genera of mammals. (The actual number of genera has fluctuated over time, but appears to converge at an equilibrium number of 90.) What is remarkable about Alroy's data is that the players during the past 55 million years have changed many times as existing mammalian species became extinct and others evolved, yet the number of mammalian genera always seems to level out at 90.

Alroy's data also indicate that mammals as a group are robust in an evolutionary sense. Mammals have not experienced a large mass extinction event (of the magnitude of the one at the end of the Cretaceous period) during the past 65 million years. Instead, the mammalian line has undergone only minor increases and decreases in rates of extinction and speciation.

Why has North America been home to about 90 mammalian genera for the past 55 million years? Why not 50 genera, or 150? Alroy's research has sparked a connection between paleontology and ecology. Some scientists have hypothesized that the North American continent has a finite amount of food and other resources to support mammals. In ecological terms, an environment's ability to support a group of organisms is known as its *carrying capacity* (see Chapter 52). These scientists suggest that North America's carrying capacity for mammals is 90 genera. Perhaps when the number of genera rises above 90, greater competition among species decreases the rate of speciation (or increases the extinction rate) so that mammalian diversity declines. Using the same line of reasoning, perhaps when the number of genera falls below 90, mammalian diversification increases because competition among species is lower and, as a result, the rate of speciation increases (or the rate of extinction decreases). The actual relationships among mammalian diversity, speciation rates, and extinction rates remain to be determined by further research. ▲

*Recall from Chapter 1 that a genus (pl, genera) is a group of closely related species.

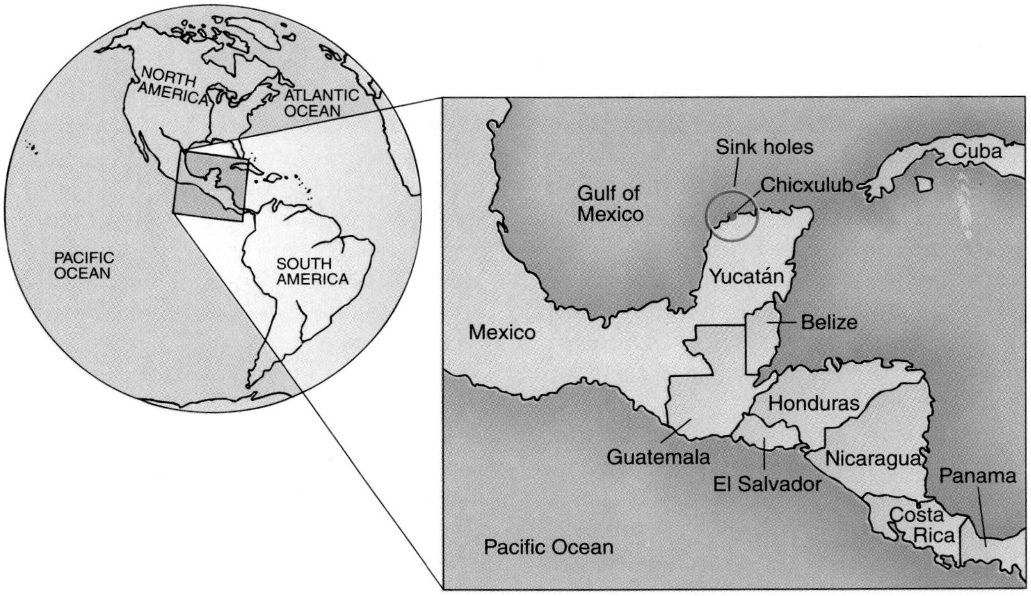

FIGURE 20–15 The Yucatán was the site of a collision with an extraterrestrial object some 65 million years ago. The Chicxulub crater on the Yucatán peninsula is the largest known crater on our planet. The site, entirely under water at the time of impact, is buried underground today.

FIGURE 20–16 Mammals replaced reptiles as Earth's dominant terrestrial animals during the Cenozoic era. (*a*) The woolly mammoth existed during the Pleistocene epoch but disappeared at the end of the Ice Age. (*b*) *Megatherium* was a giant ground sloth nearly the size of an elephant. (*c*) A *glyptodont,* found in what is now the southern United States, weighed more than 1 ton and resembled a cross between a turtle and an armadillo. (*d*) *Smilodon,* the saber-toothed cat, was found in North and South America.

epochs. The Rocky Mountains, formed at the beginning of the Tertiary period, were considerably eroded by the time of the Oligocene epoch, giving the North American continent a gently rolling topography. During the Miocene epoch, another series of uplifts raised the Sierra Nevada and a new set of Rockies, and resulted in the formation of the western deserts.

The uplift begun in the Miocene epoch continued in the Pliocene and, coupled with the ice ages of the Pleistocene epoch, may have killed many of the contemporary mammals and other organisms. The final elevation of the Colorado Plateau, which also caused the cutting of the Grand Canyon, occurred almost entirely in the short Pleistocene and Holocene epochs.

During the Tertiary period, grasses, which served as food, and dense forests, which afforded protection from predators, may have contributed to changes in the mammalian body form. Along with the tendency toward increased size, mammals displayed an increase in the relative size of the brain and changes in the teeth and feet.

Evidence of the first known carnivorous mammals, the *creodonts,* appears in Paleocene and Eocene formations. They were replaced in the Eocene and Oligocene epochs by more modern forms ancestral to present-day carnivores such as cats, dogs, bears, and weasels, as well as by the web-footed marine carnivores, the seals and walruses.

Remains of the earliest ungulates also appear in Paleocene formations. *Ungulates* are large, hoof-bearing, grazing mammals. They do not form a single, natural group, but consist of several independent lines. The molar teeth of ungulates are flattened and enlarged to facilitate chewing leaves and grass. Their legs are elongated and adapted for the rapid movement necessary to escape predators.

The Pleistocene epoch of the Quaternary period was marked by four glacial episodes. In North America, these ice sheets covered nearly 4 million square miles at the time of their greatest extent, spreading south as far as the Ohio and Missouri Rivers. During the Pleistocene glaciation, enough water was removed from the oceans and locked in the ice to lower the sea level by 65 to 100 meters (213 to 328 ft). This formed land bridges, highways for the dispersal of many life forms. Examples include a land bridge that connected Siberia to Alaska at the Bering Strait and one that connected England to the European continent.

The plants and animals of the Pleistocene epoch were similar to those alive today. For this reason, it is sometimes difficult to distinguish between Pleistocene and Holocene deposits. A considerable number of mammals, including the saber-toothed cat, the mammoth, and the giant ground sloth, became extinct during the Pleistocene epoch, possibly as a result of early human hunting (Fig. 20–16). The Pleistocene epoch was marked by the extinction of many plant species, especially woody ones, and the appearance of numerous herbaceous plants.

SUMMARY

I. It is thought that life originated from nonliving matter by chemical evolution. While we may never know exactly how life began, a number of hypotheses about the origin of life are testable.
   A. Four requirements for chemical evolution are:
      1. The absence of oxygen (because free oxygen would have reacted with and broken down organic molecules).
      2. Energy (to form organic molecules).
      3. Chemical building blocks (including water, minerals, and gases present in the atmosphere) to form organic molecules.
      4. Sufficient time (for molecules to accumulate and react).
   B. Four steps are hypothesized in chemical evolution.
      1. Small organic molecules formed and accumulated.
      2. Macromolecules assembled from the small organic molecules.
      3. Macromolecular assemblages (pre-cells) formed from macromolecules.
      4. Cells arose from the assemblages of organic polymers.

II. The first cells were prokaryotic anaerobes.
   A. The oldest cells in the fossil record are 3.1 to almost 3.5 billion years old.
   B. The evolution of photosynthesis ultimately changed early life because it generated oxygen, which accumulated in the atmosphere.
   C. Aerobic organisms, which could use oxygen for a more efficient type of cellular respiration, evolved.
   D. Certain eukaryotic organelles (mitochondria and chloroplasts) probably evolved from prokaryotic endosymbionts.

III. Earth's history is divided into eras, periods, and epochs.
   A. Life began and diverged into different groups of bacteria, protists (including algae), fungi, and animals during Precambrian time.
   B. During the Paleozoic era, all major groups of plants except for flowering plants appeared, and fish and amphibians flourished.
   C. The Mesozoic era was characterized by the evolution of flowering plants and reptiles. Insects flourished, and birds and early mammals evolved.
   D. In the Cenozoic era, which extends to the present time, flowering plants, birds, insects, and mammals diversified greatly.

SELECTED KEY TERMS

aerobes
anaerobes
autotroph
Cenozoic era
chemical evolution
continental drift
endosymbiont
endosymbiont theory
epoch
era
heterotroph
hydrothermal vent
Mesozoic era
microfossil
microsphere
ozone
Paleozoic era
period
plate tectonics
Precambrian time
protobionts
ribozymes
stromatolites

POST-TEST

1. Energy, the absence of oxygen, chemical building blocks, and time were the requirements for __________ __________ .
2. Some biologists think that life began in __________ __________ in the ocean floor.
3. __________ are assemblages of abiotically produced organic polymers that resemble living cells in several ways.
4. __________ obtain the organic molecules they need from the environment, whereas __________ synthesize them.
5. The first __________ probably used sunlight to split hydrogen sulfide.
6. Fossilized mats of cyanobacteria are known as __________ .
7. Organisms that metabolize in the absence of oxygen are known as __________ .
8. According to the __________ __________ , chloroplasts, mitochondria, and possibly other organelles originated from symbiotic relationships among prokaryotic organisms.
9. __________ __________ encompasses all geological time prior to the beginning of the Paleozoic era some 570 million years ago.
10. The correct chronological order of geological eras, starting with the oldest, is __________ , __________ , and __________ .
11. Eras are divided into __________ , which may be divided into __________ .
12. Terrestrial plants and air-breathing animals appeared during the Silurian period of the __________ era.
13. The __________ era is known as the "Age of Reptiles."
14. Flowering plants and mammals diversified and became dominant during the __________ era.
15. The idea originally suggested by Alfred Wegener that continents slowly shift their positions is known as __________ __________ .
16. Continental drift is explained by __________ __________ .

REVIEW QUESTIONS

1. What are the four requirements for chemical evolution, and why is each essential?
2. Discuss two ways that the presence of molecular oxygen in the atmosphere affected early life.
3. Give at least two types of evidence that support the endosymbiont theory.
4. Put the following organisms in order of appearance in the fossil record, starting with the earliest: (a) reptiles, mammals, amphibians, fish; (b) flowering plants, ferns, gymnosperms.
5. How does continental drift occur?
6. Explain why fossils of *Mesosaurus,* an extinct reptile that could not swim across open water, are found in the southern parts of both Africa and South America.

YOU MAKE THE CONNECTION

1. If you were experimenting on how protobionts evolved into cells, and you developed a protobiont that was capable of self-replication, would you consider it a living cell? Why or why not?
2. Why did the evolution of complex multicellular organisms such as plants and animals have to be preceded by the evolution of oxygen-producing photosynthesis?
3. How might studying outer space help us reconstruct the evolutionary history of life on Earth?

RECOMMENDED READINGS

Dalziel, I.W.D. "Earth before Pangaea." *Scientific American,* January 1995. This article discusses continental drift that occurred *before* Pangaea formed some 260 million years ago.

Gore, R. "The Cambrian Period: Explosion of Life." *National Geographic,* Vol. 184, No. 4, October 1993. Examines the wealth of Cambrian fossils from such famous locations as the Burgess Shale in the Canadian Rockies.

Gould, S.J. "The Evolution of Life on the Earth." *Scientific American,* October 1994. An eminent paleontologist discusses some common misconceptions associated with the evolutionary history of life.

Knoll, A.H. "Life's Expanding Realm." *Natural History,* June 1994. How the production of oxygen as a byproduct of photosynthesis affected the evolution of new species.

Orgel, L.E. "The Origin of Life on the Earth." *Scientific American,* October 1994. Examines the evidence that RNA was the first information molecule to emerge during the origin of life.

Trefil, J. "However It began on Earth, Life May Have Been Inevitable." *Smithsonian,* February 1995. An interesting summary of scientific research on the origin of life.

Zimmer, C. "Masters of an Ancient Sky." *Discover,* February 1994. Recent discoveries of well-preserved *Pterosaur* fossils in Brazil are helping scientists understand Earth's first flying vertebrates.

C H A P T E R 21

The Evolution of Primates

The common tree shrew resembles the ancient mammals that gave rise to the primates. (Warren & Genny Garst/Tom Stack & Associates)

Most people have an interest in their roots. To many of us, this means trying to discover the immediate ancestors of our great grandparents. In this chapter we examine what we might call our "deep roots" as we trace the origin of humans, *Homo sapiens,* back some 50 million years to the earliest primates.

Twelve years after Darwin wrote *The Origin of Species,* he published another controversial book, *The Descent of Man,* which addressed human evolution. In it, Darwin hypothesized that humans and apes shared a common ancestry. For nearly a century after Darwin's *The Descent of Man,* fossil evidence of human ancestry was extremely fragmentary. However, research over the last few decades, especially in Africa, has yielded fossils that provide some reasonable answers to the question, "Where did we come from?" Fossil evidence has allowed **paleoanthropologists,** scientists who study human evolution, to infer not only the structure but also the habits of early humans.

Humans and other primates are mammals, members of the class Mammalia. Mammals are **endothermic** (they use metabolic energy to maintain a constant body temperature), produce body hair, and feed their young with milk from mammary glands. Most mammals are **viviparous,** which means that their eggs develop into young offspring within the female body.

Mammals arose from reptiles approximately 240 million years ago, during the Mesozoic era (see Chapter 20). It was, however, the "Age of Reptiles," and reptiles (including dinosaurs) were the dominant animals, occupying almost every habitat that supported life. Three main lines of mammals existed during the Mesozoic era: (1) the *multituberculates,* which may have given rise to monotremes (mammals that lay eggs) such as the duck-billed platypus; (2) the *marsupials,* which were the ancestors of modern-day kangaroos and opossums; and (3) small shrew-like *placental* mammals that ate insects and lived a nocturnal existence in the trees. (Placental mammals have an organ of exchange—a placenta—between the mother and the developing fetus.) These early mammals remained a minor component of life on Earth for almost 150 million years.

Approximately 65 million years ago the dinosaurs and many other species became extinct. This provided numerous opportunities for mammals to diverge and fill the ecological niches (roles, or lifestyles, of species; see Chapter 53) vacated by the dinosaurs. In addition, various species of flowering plants, including many trees, evolved, providing mammals with new habitats, sources of food, and protection from predators.

During the early Cenozoic era, mammals flourished and underwent adaptive radiation. The first primates appeared at this time, apparently descendants of the small shrewlike placental mammals that lived in trees and ate insects, much like the tree shrew of today (see figure). Many traits of humans and other primates are related to their **arboreal** (tree-dwelling) past.

LEARNING OBJECTIVES

After you have studied this chapter you should be able to

1. Describe the structural adaptations that primates possess for life in trees.
2. Explain why primates have adaptations for an arboreal existence, even though many primates live on the ground.
3. List the two groups of primates and give several distinguishing features and representative examples of each.
4. Distinguish among mammals, primates, prosimians, anthropoids, hominoids, and hominids.
5. Describe skeletal and skull differences between apes and hominids.
6. Compare the following hominids: *Australopithecus* spp., *Homo habilis, Homo erectus,* and *Homo sapiens.*
7. Discuss the current debate over the origin of modern humans and briefly describe the opposing "out of Africa" and "multiregional" hypotheses.
8. Describe cultural evolution and its impact on the ecosphere.

PRIMATES ARE ADAPTED FOR AN ARBOREAL EXISTENCE

One of the most significant features of primates is that they have five grasping digits: four fingers plus an opposable thumb. The opposable thumb enables primates to grasp objects such as tree branches. Nails (instead of claws) provide a protective covering for the tips of the digits, and the fleshy pads at the ends of the digits are sensitive to touch. Another arboreal feature is long, slender limbs that rotate freely at hips and shoulders, giving primates full mobility to climb and search for food in the tree tops. The location of the eyes in front of the head provides stereoscopic, or three-dimensional, vision. Stereoscopic vision is essential for arboreal animals, as an error in depth perception might cause a fatal fall. In addition to sharp sight, hearing is acute in primates, although their sense of smell is poor.

Primates share several other characteristics, including large brain size. It has been suggested that increased sensory input associated with their sharp vision and greater agility favored the evolution of larger brains. Primates also have complex social behaviors. Females usually bear one offspring, which is helpless and requires a long period of nurturing and protection, at a time.

The order Primates consists of two groups, prosimians and anthropoids (Table 21–1). The **prosimians** (which means "before apes") include lemurs, lorises, and tarsiers. The **anthropoids,** which comprise monkeys, apes, and humans, are primates with larger brains.

Table 21–1 CLASSIFICATION OF THE PRIMATES

Order Primates
Suborder Prosimii (lower primates)
Family Cheirogallidae (dwarf lemurs)
Family Lemuridae (lemurs)
Family Indriidae (indris)
Family Daubentoniidae (aye-ayes)
Family Lorisidae (lorises, bush babies)
Family Tarsiidae (tarsiers)
Suborder Anthropoidea (higher primates)
Family Callitrichidae (marmosets)
Family Cebidae (New World monkeys)
Family Cercopithecidae (Old World monkeys)
Family Hylobatidae (gibbons)
Family Pongidae (great apes: orangutan, gorilla, chimpanzee)
Family Hominidae (ancient and modern humans)

Prosimians are primitive, arboreal primates

The first primates to appear are classified as prosimians. They flourished during the Eocene epoch just over 50 million years ago (Fig. 21–1). Fossils indicate that primitive prosimians had opposable thumbs, digits with nails, and eyes directed somewhat forward. The climate was milder then, and prosimians were widely distributed over much of North America, Europe, and Asia. (Recall from Chapter 20 that North America was still attached to Europe at that time.) As the climate became cooler and drier toward the end of the Eocene epoch, many of these early prosimians became extinct. Modern prosimians have changed relatively little from their ancestors. Several species are now endangered, especially in Madagascar, where they occur in small, isolated populations.

All lemurs are restricted to the island of Madagascar off the coast of Africa (Fig. 21–2*a*). Because of extensive habitat destruction, they are endangered. Lorises, which are found in tropical areas of Southeast Asia and Africa, resemble lemurs in many respects. Both lemurs and lorises have retained some early mammalian features, such as elongated, pointed faces and somewhat lateral-facing eyes.

(Text continues on page 476)

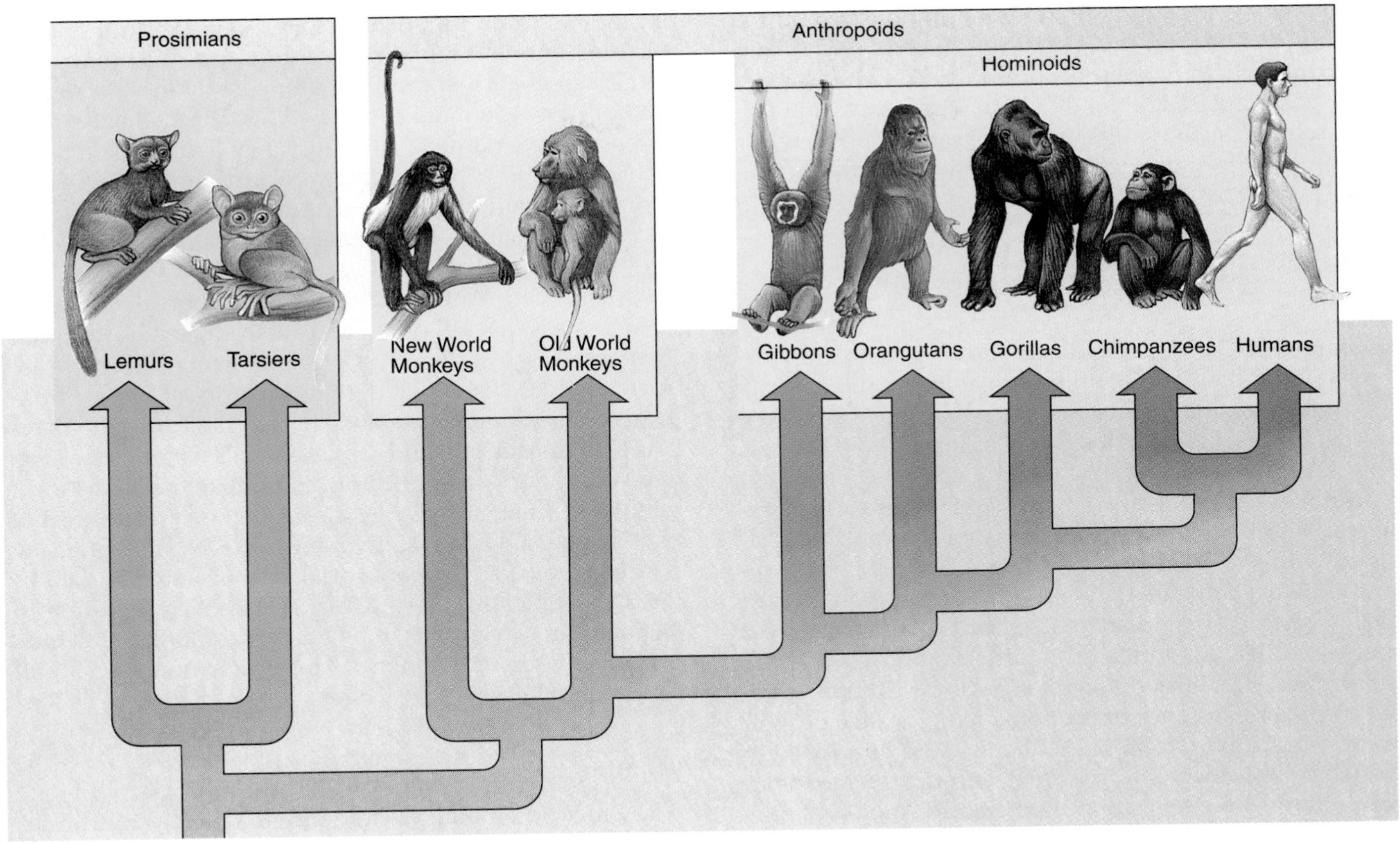

FIGURE 21–1 Relationships among modern primates reflect primate evolution. The first primates were the prosimians, which evolved from a small, insect-eating mammal. The other major group of primates, the anthropoids (which include monkeys, apes, and humans), arose from a line of prosimians. (Figures are not drawn to scale.)

(*a*)

(*b*)

FIGURE 21–2 Lemurs and tarsiers are prosimians. (***a***) A mother lemur and her baby share a piece of fruit. Lemurs are native to Madagascar. (***b***) The huge eyes of the tarsier help it find insects, lizards, and other prey when it hunts at night. Tarsiers live in the rain forests of Indonesia and the Philippines. (*a,* Frans Lanting/Minden Pictures; *b,* Doug Wechsler)

Tarsiers are found in rain forests of Indonesia and the Philippines. They are small primates the size of squirrels and are adept climbers and leapers (Fig. 21–2*b*). Although tarsiers are prosimian tree dwellers, these nocturnal primates resemble the anthropoids in a number of ways, including their shortened snouts and forward-pointing eyes.

Anthropoids include monkeys, apes, and humans

The anthropoids arose from a group of prosimians during the Oligocene epoch, approximately 38 million years ago. While fossil evidence indicates that they originated in Africa or Asia, they quickly spread throughout Europe, Asia, and Africa.

Monkeys are generally larger than their prosimian relatives. Most are diurnal (active during the day) as compared with the nocturnal prosimians. Like the prosimians, monkeys are generally tree dwellers, although their diet is more varied than that of prosimians. Different species eat leaves, fruits, buds, insects, and even small vertebrates. Probably the most significant difference between prosimians and anthropoids is in the size of their brains. The cerebrum, in particular, is more developed in monkeys and apes. The two main groups of monkeys, New World monkeys and Old World monkeys, are named for the hemispheres where they originated.

New World monkeys are arboreal and some possess long, slender limbs that permit easy movement in the trees (Fig. 21–3*a*). Many have **prehensile** tails capable of wrapping around branches. Some New World monkeys have smaller thumbs, and in certain cases the thumbs are totally absent. Their facial anatomy is different from that of the Old World monkeys; they have flattened noses with the nostrils opening to the side. They live in groups and exhibit social behavior. New World monkeys are restricted to Central and South America and include marmosets, capuchins, howler monkeys, squirrel monkeys, and spider monkeys.

Many Old World monkeys are arboreal, although some, such as baboons and macaques, are ground dwellers (Fig. 21–3*b*). The ground dwellers, which are **quadrupedal** (four-footed; they walk on all fours), arose from arboreal monkeys. None of the Old World monkeys has a prehensile tail, and some lack tails completely. They have a fully opposable thumb, and unlike the New World monkeys, their nostrils are closer together and directed downward. Old World monkeys are larger than New World monkeys. They are social animals and are distributed in tropical parts of Africa and Asia. In addition to baboons and macaques, the Old World monkeys include langurs, colobus monkeys, proboscis monkeys, and vervet monkeys.

Hominoids include apes and humans

The Old World monkeys were ancestral to the **hominoids,** a group composed of apes and **hominids** (humans and their ancestors). One of the earliest anthropoids was discovered in Egypt and named *Aegyptopithecus* (Fig. 21–4*a*). *Aegyptopithecus,* a cat-sized, forest-dwelling arboreal

(*a*)

(*b*)

FIGURE 21–3 New World and Old World monkeys are anthropoids. (***a***) Geoffrey's spider monkey, like most New World monkeys, has a prehensile tail that functions almost as effectively as another limb. (***b***) The lion-tailed monkey *(Macaca silenus)* is an Old World monkey native to India.
(*a,* Frans Lanting/Minden Pictures; *b,* Dennis Drenner)

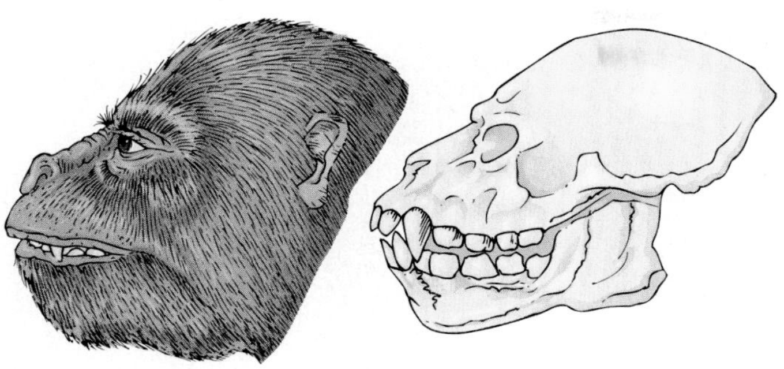

(a) Early anthropoid, *Aegyptopithecus*

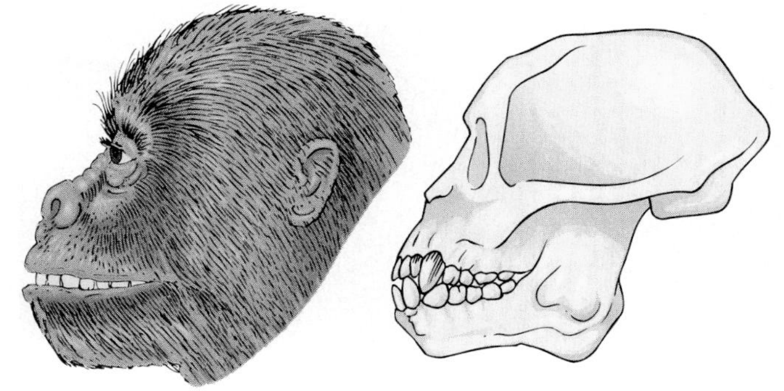

(b) Early ape, *Dryopithecus*

FIGURE 21–4 ***Aegyptopithecus*** **was an early anthropoid;** ***Dryopithecus*** **was an early ape.** (***a***) Fossils of *Aegyptopithecus*, an early apelike anthropoid, were discovered in Egypt. (***b***) *Dryopithecus*, a primitive ape, may have given rise to modern hominoids.

monkey with a few apelike characteristics, lived approximately 35 million years ago. During the Miocene epoch, which began approximately 25 million years ago, the apes and Old World monkeys diversified. Fossils of an early forest-dwelling ape, *Dryopithecus*, are of special interest because this hominoid may have given rise to modern apes as well as to the human line (Fig. 21–4*b*). Dryopithecines were arboreal but may have also spent a significant amount of time on the ground. Although quadrupedal, they lacked the long forearms characteristic of apes today and had sloping craniums with bony ridges above the eyes. The dryopithecines were distributed widely across Europe, Africa, and Asia. As the climate gradually cooled and became drier, their range became more limited.

By the beginning of the Pliocene epoch, approximately 5 million years ago, the apes were restricted primarily to tropical rain forests. Unfortunately, moist conditions of the tropics preclude the formation of many fossils, so our knowledge of ape evolution is quite sketchy.

The four genera of apes alive today are usually classified into two families (Fig. 21–5): Gibbons *(Hylobates)* are known as lesser apes and are placed in a separate family, Hylobatidae, whereas the other family, Pongidae, includes orangutans *(Pongo)*, gorillas *(Gorilla)*, and chimpanzees *(Pan)*. Gibbons are well-adapted for an arboreal existence. They are natural acrobats and can **brachiate,** or swing, with their weight supported by one arm at a time. Orangutans are also tree dwellers, but chimpanzees and especially gorillas have adapted to life on the ground. They have retained elongated forearms typical of tree-dwelling primates but use these to assist in quadrupedal walking, sometimes known as **knuckle-walking** because of the way they fold their digits when moving. Apes, like humans, lack tails. They are generally larger than monkeys; gibbons are a notable exception.

Evidence of the close relatedness of gorillas, chimps, and humans is abundant at the molecular level. The amino acid sequence of the chimpanzee's hemoglobin is identical to that of the human; those of the gorilla and rhesus monkey differ from the human's in 2 and 15 amino acids, respectively. DNA sequence analyses indicate that chimpanzees are likely to be our nearest living relatives among the apes. Molecular and fossil evidence demonstrates that gorillas may have diverged from the chimpanzee and hominid lines some 8 to 10 million years ago, whereas chimpanzees and hominids probably separated about 6 million years ago (Fig. 21–6).

THE FOSSIL RECORD PROVIDES CLUES TO HOMINID EVOLUTION

General trends in hominid body design are evident from the fossil record, but we do not have enough evidence to draw specific conclusions. There are simply too few early hominid fossils, and the ones we do have are represented by only a few bones. Moreover, it is impossible to determine many aspects of early hominid biology, appearance, and behavior from fossilized bones. Nevertheless, it is evident that early hominids adopted a **bipedal** (two-footed) posture before their brains enlarged.

Evolutionary changes from the earliest hominids to modern humans are evident in some of the characteristics of the skeleton and skull. Compared with the ape skeleton, the human skeleton possesses distinct differences that reflect our ability to stand erect and walk on two feet (Fig. 21–7). These differences also reflect the habitat change for early hominids, from an arboreal existence in the forest to a life spent at least partly on the ground. The curvature of the human spine provides better balance and weight distribution for bipedal locomotion. The human pelvis is shorter and more rounded than the ape pelvis, providing a better attachment of muscles used for upright walking. The hole in the base of the skull for the spinal cord, called the **foramen magnum,** is located in the middle of the rear of the skull in apes. In contrast, the human foramen magnum is centered at the base of the skull, positioning the head for erect walking. An increase in the

(Text continues on page 479)

(a)

(b)

(c)

(d)

FIGURE 21–5 Gibbons, orangutans, gorillas, and chimpanzees, all apes, have no tails. (***a***) White-handed gibbons are extremely acrobatic and often move through the trees by brachiation. (***b***) An orangutan mother and baby. (***c***) A young lowland gorilla in knuckle-walking stance. (***d***) A chimpanzee family in West Africa. Chimpanzees live in groups and interact in complex ways. (*a,* Visuals Unlimited/Joe McDonald; *b,* A. Compost/Peter Arnold, Inc.; *c,* David J. Cross/Peter Arnold, Inc.; *d,* Animals Animals © 1996 Mike Birkhead, Oxford Scientific Films)

FIGURE 21–6 Chimpanzees and humans are closely related. The similarities between chimpanzees and humans are particularly striking in this photo of the stars of the 1951 movie, "Bedtime for Bonzo." (The Kobal Collection)

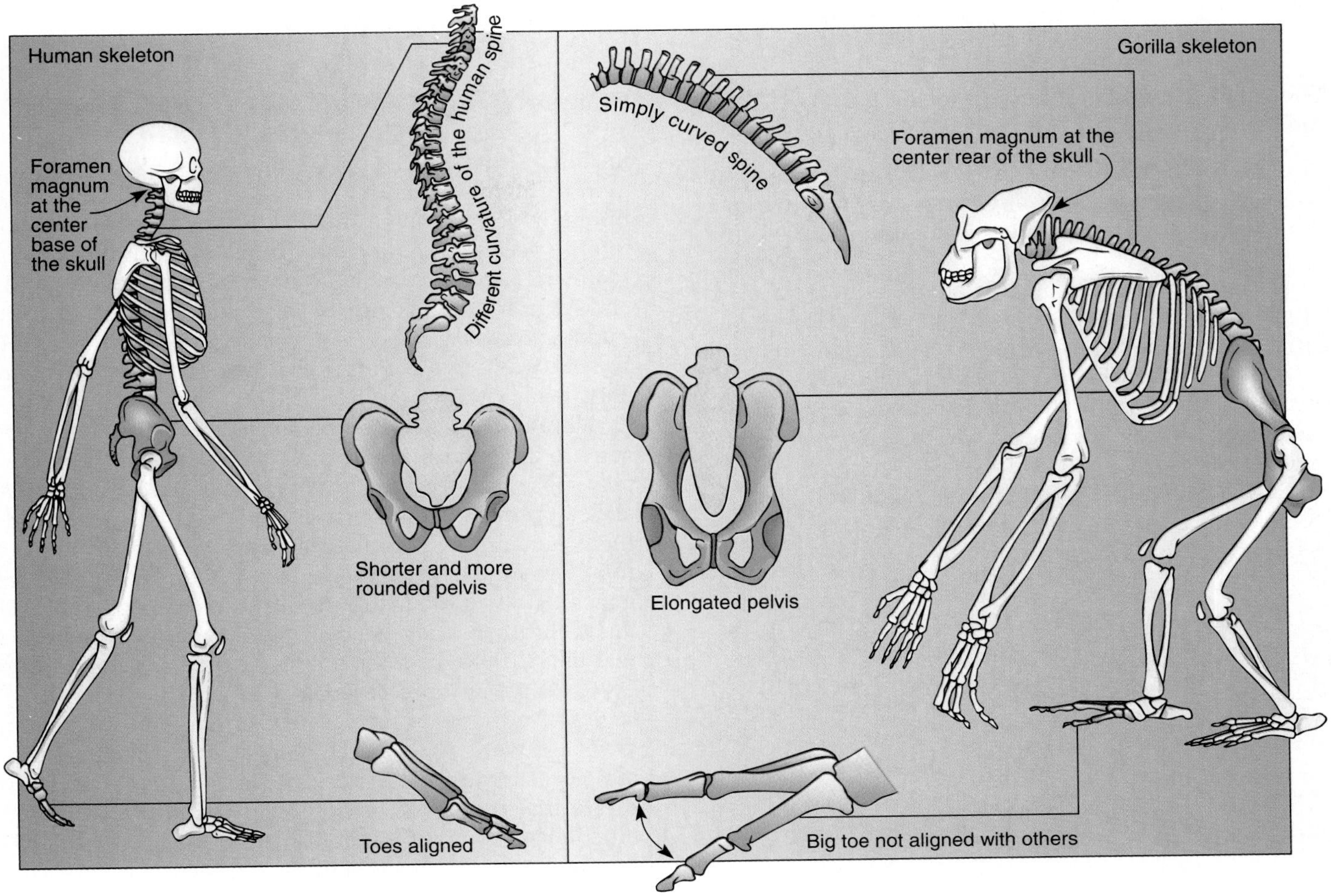

FIGURE 21–7 When gorilla and human skeletons are compared, the skeletal adaptations for bipedalism in humans become apparent.

length of the legs relative to the arms, and alignment of the big toe with the rest of the toes, further adapted the early hominids for bipedalism.

Another major trend in human evolution was an increase in the size of the brain relative to the size of the body (Fig. 21–8). In addition, the ape skull possesses prominent bony ridges above the eye sockets, whereas these **supraorbital ridges** are lacking in modern human skulls. Human faces are flatter than those of apes, and the jaws are different. The arrangement of teeth in the ape jaw is somewhat rectangular, compared with a rounded, or U-shaped, arrangement in humans. Apes have larger teeth than humans do, and their canines are especially large.

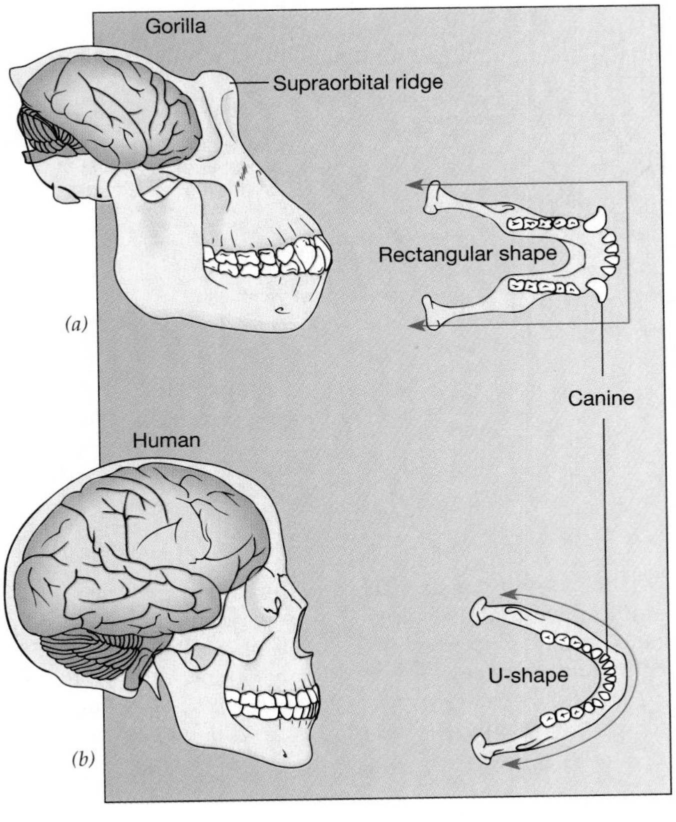

FIGURE 21–8 A comparison of gorilla and human heads reveals certain basic differences. (*a*) The ape skull has a pronounced supraorbital ridge. (*b*) The human skull is flatter in the front and has a more pronounced chin. The human brain, particularly the cerebrum *(purple)*, is larger than that of an ape, and the human jaw is structured so that the teeth are arranged in a U-shape. Human canines are also smaller than ape canines.

The earliest hominids belong to the genus *Australopithecus*

Hominid evolution began in Africa. The earliest hominids belong to the genus ***Australopithecus,*** or "southern man ape," which appeared about 4.4 million years ago (Fig. 21–9).

The actual number of australopithecine species for which fossil evidence has been found is under debate. Differences in the relatively few skeletal fragments could indicate either variation among individuals within a species or evidence of separate species. Most paleoanthropologists recognize at least two to four species of australopithecines.

Fossils of the earliest hominids, tentatively assigned to the species *A. ramidus,* were discovered in Africa in 1992 and reported in the journal *Nature* by Tim White (University of California, Berkeley) and co-workers in 1994. The specific epithet *ramidus* is derived from a word meaning "root" in the Afar language, spoken in the region of Ethiopia where the fossils were found. This hominid, which is more primitive than any other known hominid species, is quite close to the "root" of the human family tree—that is, to the common ancestor of hominids and African apes. Because no leg bones were found in the initial discovery, it has not yet been determined if *A. ramidus* was bipedal. Future discoveries are likely to clarify this important point.

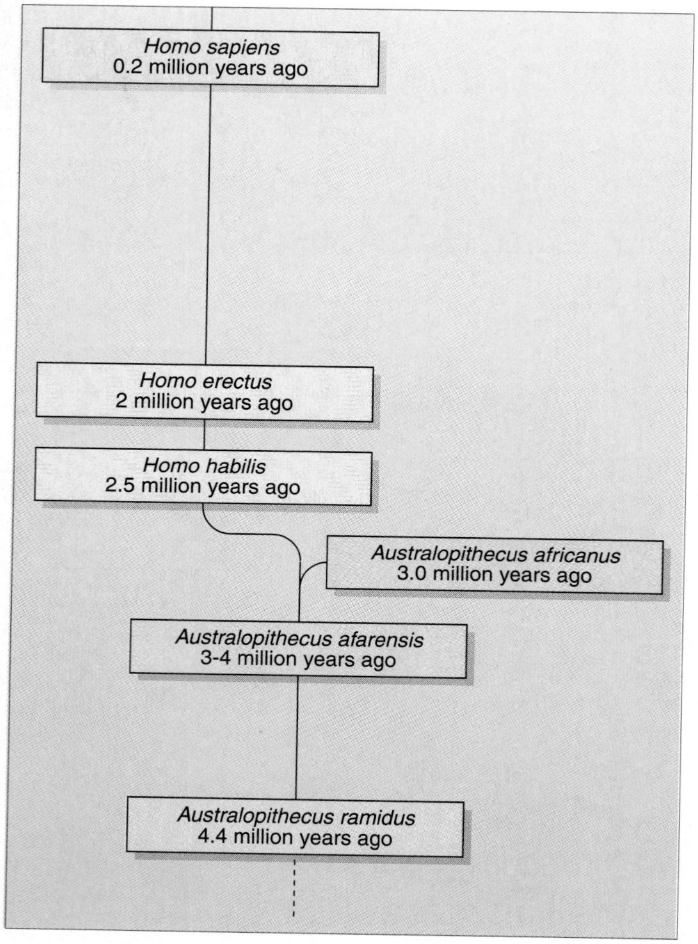

FIGURE 21–9 One of many possible interpretations of hominid evolution is shown. Paleoanthropologists are not in complete agreement about the details of our lineage, but many think that the current evidence supports the evolution of *Homo habilis* from *Australopithecus afarensis.* It is generally thought that *A. africanus* is not in a direct line to *H. sapiens.*

Hominids that existed between 3 and 4 million years ago are assigned to the species *A. afarensis* and presumably arose from *A. ramidus.* Several fossils of *A. afarensis* skeletal remains have been discovered in Africa, including a remarkably complete skeleton nicknamed "Lucy" found in Ethiopia in 1974 by a team led by Donald Johanson, now of the Institute of Human Origins in Berkeley, California. Lucy, a small hominid approximately 3 feet tall, is thought to be about 3.2 million years old. In 1976, beautifully preserved fossil footprints of three *A. afarensis* individuals who walked more than 3.6 million years ago were discovered by Mary Leakey and co-workers. These footprints, plus pelvis, leg, and foot bones, indicate that the development of an upright posture and bipedalism occurred early in human evolution. In 1994 the first complete adult skull of *A. afarensis* was discovered by William Kimbel and other paleoanthropologists from the Institute of Human Origins. The skull, characterized by a relatively small brain, pronounced supraorbital ridges, a jutting jaw, and large canine teeth, is an estimated 3 million years old. It is probable that *A. afarensis* and other australopithecines did not talk (their bone structure was inappropriate) or construct tools or make fires (no evidence of tools or fire has been found at fossil sites).

Many paleoanthropologists think *A. afarensis* gave rise to a whole family of australopithecines, including *A. africanus,* which appeared approximately 3 million years ago. The first *A. africanus* fossil was discovered in South Africa in 1924, and since then a number of others have been found. This rather small hominid walked erect and possessed hands and teeth that were distinctly humanlike. Based on characteristics of the teeth, it is thought that *A. africanus* ate both plants and animals. Like *A. afarensis,* it had a small brain, more like that of its primate ancestors than present-day humans.

Homo habilis is the oldest member of the genus *Homo*

The first hominid to have enough human features to be placed in the same genus as modern humans is ***Homo habilis.*** *Homo habilis* had a larger brain than the australopithecines. This early human appeared approxi-

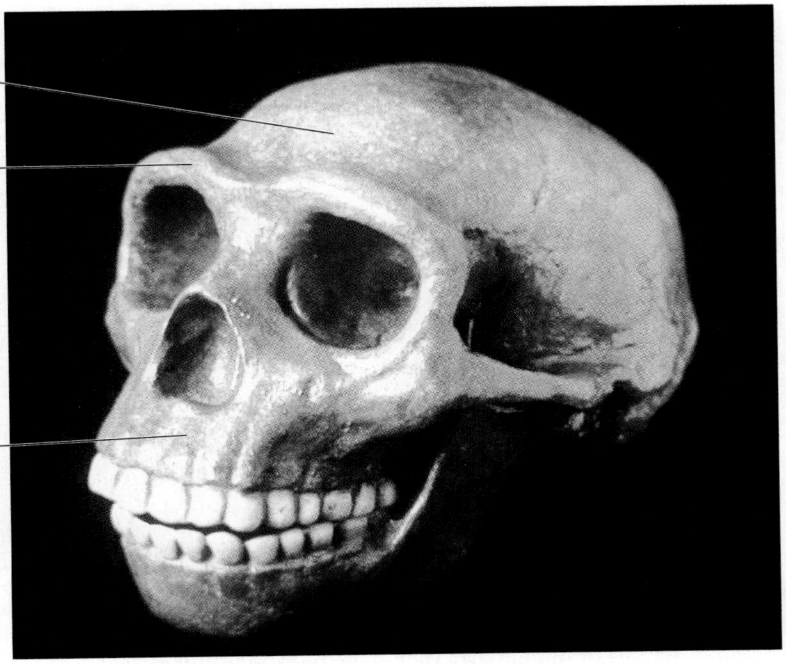

FIGURE 21–10 A replica of a *Homo erectus* skull reveals a receding forehead, pronounced supraorbital ridge, and projecting face and jaws. (Dennis Drenner)

mately 2.5 million years ago and persisted for more than half a million years. Fossils of *H. habilis* have been found in numerous areas in Africa. These sites contain the first primitive tools, stones that had been chipped, cracked, or hammered to make sharp edges for cutting or scraping. Although primates other than humans occasionally use tools, *H. habilis* represents the first primate to consciously design them.

The relationship between the australopithecines and *H. habilis* is not clear. Using physical characteristics of their fossilized skeletons as evidence, many paleoanthropologists have inferred that the australopithecines were ancestors of *H. habilis*. Others think that *H. habilis* and *A. africanus* were contemporaries for much of their existence and that *H. habilis* was in a direct line to humans while *A. africanus* was not. Discoveries of additional fossils may help clarify these relationships.

Homo erectus apparently evolved from *Homo habilis*

Numerous fossils of ***Homo erectus*** have been found throughout Africa and Asia. *Homo erectus* is thought to have originated in Africa about 2 million years ago and then to have spread quickly via Europe into Asia. The oldest fossils of *H. erectus* that have been found in Southeast Asia, for example, were recently dated by some researchers at almost 2 million years old, although the most widely accepted date is about 1 million years. Peking man and Java man, discovered in Asia, were later examples of *H. erectus*, which existed until approximately 200,000 years ago.

Homo erectus was taller than *H. habilis*. Its brain, which was larger than that of *H. habilis*, got progressively larger during the course of its existence. Its skull, although larger, did not possess totally modern features, retaining the heavy supraorbital ridge and projecting face that are more characteristic of its ape ancestors (Fig. 21–10).

The increased mental faculties associated with an increased brain size enabled these early humans to make more advanced stone tools, known as Acheulian tools, including hand axes and other implements that have been interpreted as choppers, borers, and scrapers. Their intelligence also allowed them to survive in cold areas. *Homo erectus* wore clothing, built fires, lived in caves or shelters, and obtained food by hunting or scavenging. To date, no evidence of weapons has been unearthed at *Homo erectus* sites.

Homo sapiens appeared approximately 200,000 years ago

Humans having features modern enough to classify them within our species appeared approximately 200,000 years ago. One of the earliest groups of ***Homo sapiens*** was the Neandertals.

Neandertals may have arisen in Eurasia

Neandertals were first discovered in the Neander Valley in Germany, but they lived throughout Europe and Asia from about 200,000 to 27,000 years ago. These early humans had short, sturdy builds. Their faces projected slightly, their chins receded, and they had heavy brow ridges.

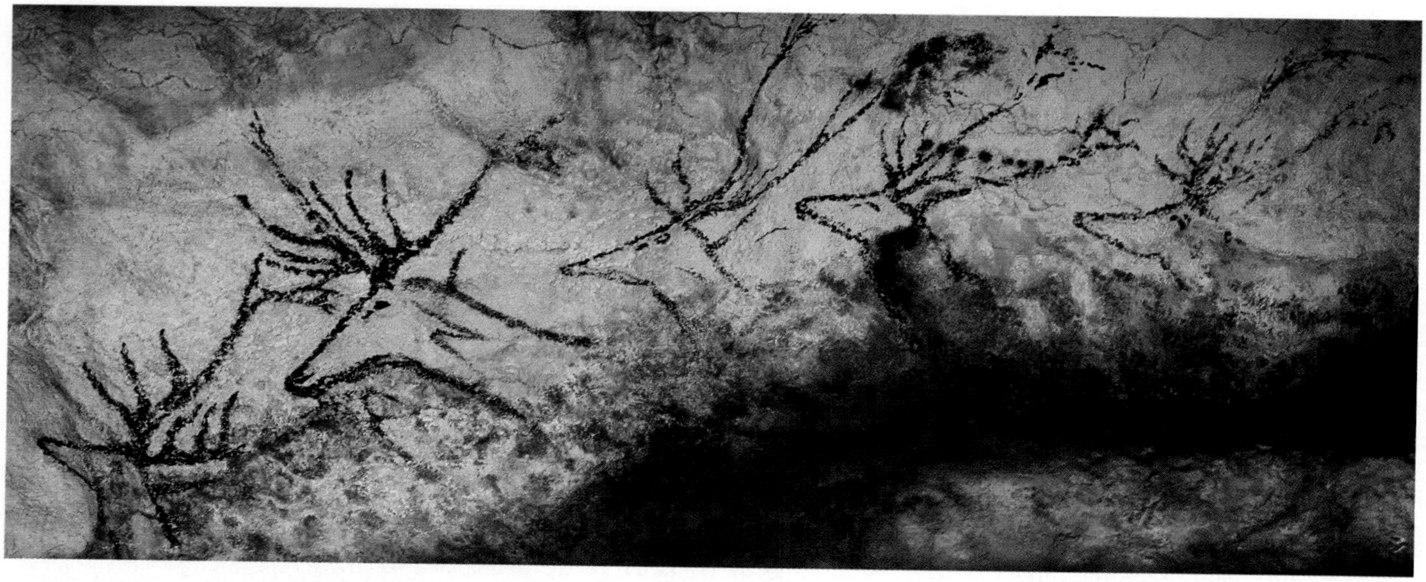

FIGURE 21–11 The Cro-Magnon people painted animals on cave walls in Europe. These are some of the earliest known examples of human art. They have been interpreted as having religious significance, possibly to guarantee a successful hunt. (Photo by J. Beckett/D. Stipkovich, courtesy Department of Library Services, American Museum of Natural History)

Neandertal tools, including the oldest known spear points, were more sophisticated than those of *H. erectus.* Studies of Neandertal sites indicate that they hunted large animals. The existence of skeletons of elderly Neandertals and of Neandertals with healed fractures demonstrates that they cared for the aged and the sick, an indication of advanced social cooperation. They apparently had rituals, possibly of religious significance, and buried their dead. The presence of food, weapons, and flowers in their graves suggests that they possessed the abstract concept of an afterlife.

The disappearance of the Neandertals some 27,000 years ago is a mystery that has sparked debate among paleoanthropologists. Other groups of *H. sapiens* with more modern features coexisted for thousands of years with the Neandertals. It is possible that the Neandertals interbred with these humans, diluting their features beyond recognition. Alternatively, perhaps the other humans out-competed or exterminated them. It is also possible that the Neandertals could not adapt to the climate changes of the Pleistocene epoch and that their disappearance was unrelated to the presence of other humans.

The origin of modern **Homo sapiens** *is hotly debated*

Homo sapiens with thoroughly modern features existed 40,000 years ago and possibly earlier (some evidence from South Africa indicates that modern *H. sapiens* may have existed 100,000 years ago). The early *H. sapiens* skull lacked a heavy brow ridge and possessed a distinct chin. The **Cro-Magnon** culture in France and Spain exemplifies these humans. Their weapons and tools were complex and often made of materials other than stone, including bone, ivory, and wood. They made stone blades that were extremely sharp. Cro-Magnons developed art, including cave paintings, engravings, and sculpture, possibly for ritualistic purposes (Fig. 21–11). Their sophisticated tools and art indicate that they may have possessed language, which would have been used to transmit their culture to younger generations.

Two opposing hypotheses currently exist about the origin of these modern humans: the "out of Africa" hypothesis and the "multiregional" hypothesis. The "out of Africa" hypothesis holds that modern *H. sapiens* arose in Africa and then migrated to Europe and Asia, displacing the more primitive humans living there. According to the "multiregional" hypothesis, modern humans originated at about the same time from *H. erectus* populations living in various parts of Africa, Asia, and Europe. Data from *Homo* fossils, as well as molecular biology and population genetics studies of modern humans, have been cited in support of both hypotheses, and both have vigorous defenders and strong detractors. Such disagreement is an important part of the scientific process because it stimulates research that may ultimately resolve the issue.

Making the Connection

DNA and Human Evolution

Can molecular biology provide clues about the origin of modern humans? The "out of Africa" hypothesis was originally supported by studies in the late 1980s of mitochondrial DNA from various human populations. In 1992 the statistical assumptions used in one analysis of mitochondrial DNA were found to be erroneous, leading to questions about the validity of this one purported demonstration of the "out of Africa" hypothesis. Several other molecular studies, however, have all produced essentially the same answer.

One of the most recent studies, reported in 1994 by Sarah Tishkoff, a graduate student at Yale University,[1] indicates an African origin for humans. She examined the genetic variation in two stretches of noncoding DNA on human chromosome 12 from 1000 people living in 31 different populations around the world. One of the segments of noncoding DNA varied in humans depending on where they lived. Based on her research, Tishkoff divided the present-day human population into three groups: Sub-Saharan Africans, Northeastern Africans, and non-Africans. The Sub-Saharan Africans exhibited the greatest genetic diversity, while the non-African populations were the least diverse.

These data and similar studies support an African origin for modern humans because there are much greater differences in the DNA of Sub-Saharan Africans than in other groups. Presumably, the Sub-Saharan populations are older than other human populations and hence have had longer to accumulate that diversity.

Tishkoff's work is significant because it provides scientists with another way to evaluate human origins. Her research has not eliminated the "multiregional" hypothesis, but points the direction for additional research on other segments of human DNA. ▲

[1]Tishkoff's paper was delivered at the 63rd Annual Meeting of the American Association of Physical Anthropologists.

HUMANS UNDERGO CULTURAL EVOLUTION

Genetically speaking, humans are not very different from other primates. At the level of our DNA sequences, we are roughly 98% identical to gorillas and chimpanzees. Our relatively few genetic differences, however, give rise to several important distinguishing features, such as greater intelligence and the ability to capitalize on it through **cultural evolution,** which is the transmission of knowledge from one generation to the next. Human culture is dynamic; it is modified as we obtain new knowledge (Fig. 21–12). Human cultural evolution is generally divided into three stages: (1) the development of hunter-gatherer societies; (2) the development of agriculture; and (3) the Industrial Revolution.

Early humans were hunters and gatherers who relied on what was available in their immediate environment. They were nomadic, and as the resources in a given area were exhausted or as the population increased, they migrated to a different area. These societies required a division of labor and the ability to make tools and weapons, which were needed not only to kill game but also to scrape hides, dig up roots and tubers, and cook food. Although we are not certain when hunting was incorporated into human society, we do know that it declined in importance approximately 15,000 years ago. This may have been due to a decrease in the abundance of large animals, triggered in part by a change in climate. A few isolated groups of hunter-gatherer societies, including the Inuit of northern polar regions and the Aborigines of Australia, have survived into the 20th century.

Development of agriculture resulted in a more dependable food supply

Evidence that humans had begun to cultivate crops approximately 10,000 years ago includes the presence of agricultural tools and plant material at archaeological sites. Agriculture, which involves keeping animals as well as cultivating plants, resulted in a more dependable food supply. Recent archaeological evidence suggests that agriculture arose in several steps. Although there is much variation from one site to another, plant cultivation, in combination with hunting, usually occurred first. Animal domestication followed later. Agriculture, in turn, often led to more permanent dwellings because considerable time was invested in growing crops in one area. Villages and cities often grew up around the farmlands,. but tying the advent of agriculture to the establishment of villages and towns is complicated by recent discoveries. For example, Abu Hureyra in Syria was a village founded *before* agriculture arose. The villagers subsisted on the rich plant life of the area and the migrating herds of gazelle. Once people turned to agriculture, however, they seldom went back to hunting and gathering to obtain food.

(a)

(b)

(c)

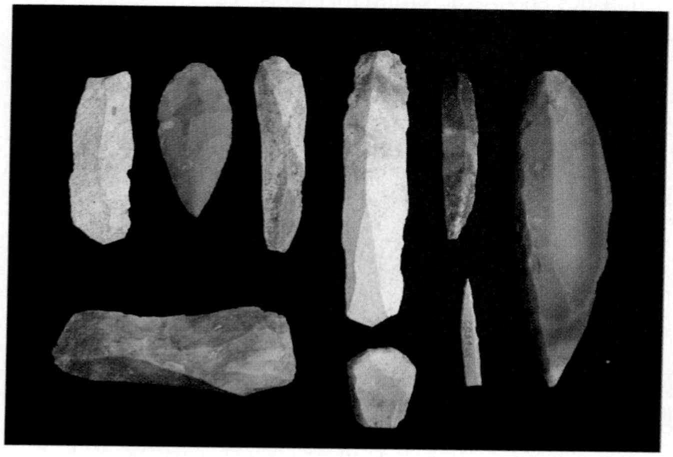

(d)

FIGURE 21–12 The progressive improvement of stone tools is evidence of cultural evolution. (*a*) Hand axes appeared approximately 1.5 to 2.0 million years ago. Shown are Oldowan pebble choppers that were used to cut through animal hides to obtain meat. The users of these tools, members of *H. habilis,* probably did not kill the animals; instead, they are thought to have scavenged the remains of animals killed by large predators. (*b*) The Acheulean hand axes from Europe are more advanced than the earliest axes discovered in Africa. They were produced by *H. erectus.* (*c*) These Neandertal Mousterian tools represent more advanced examples of stone tools. Each is specialized for a particular task. (*d*) Stone blades were fashioned by Cro-Magnon humans. Note that the length of the blade is greater than its width. These are examples of the most advanced tools made from stone. (*a–d,* Dennis Drenner)

Archaeological evidence indicates that agriculture developed independently in several different regions. There were three main centers of agriculture and several minor ones. Each of the main centers was associated with cultivation of a cereal crop, although other foods were grown as well. Cereals are grasses, which are members of the monocot group of flowering plants (see Chapter 27). The cereals associated with the three main centers of agriculture are wheat, corn, and rice.

Wheat was cultivated in the semiarid regions along the eastern edge of the Mediterranean. Other crops that originated there include peas, lentils, grapes, and olives. Central and South America were the sites of the maize, or corn, culture. Squash, chili peppers, beans, and potatoes were also cultivated there. In the Far East, in southern China, evidence exists of the early cultivation of rice and other crops such as soybeans. The actual date for the domestication of rice is unknown because rice is cultivated in a wet environment, a condition that prevents preservation of archaeological evidence.

Corn, wheat, and rice are all propagated by seed, which requires fairly sophisticated agricultural practices. Cultivation of plants that could be propagated vegetatively may have occurred earlier. Plants cultivated in this manner, such as bananas, yams, potatoes, and manioc, do not preserve as well as grains because of their high water content. For that reason, we may have no archaeological evidence of their cultivation.

Other advances in agriculture include the domestication of animals, which were kept to supply food, milk, and hides. In the Old World, animals were also used to prepare fields for planting. Another major advance in agriculture was irrigation, which dates to 7000 years ago in the Near East.

Producing food agriculturally was more time-consuming than hunting and gathering, but it was also more productive. In hunter-gatherer societies, everyone shares the responsibility for obtaining food. In agricultural societies, fewer people are needed to provide food for everyone. Thus agriculture freed some people to pursue other endeavors, including religion, art, and various crafts.

Cultural evolution has had a profound impact on the ecosphere

Cultural evolution has had far-reaching effects on both human society and on other organisms. The Industrial Revolution, which began in the 18th century, caused populations to concentrate in urban areas near centers of manufacturing. Advances in agriculture encouraged urbanization, as fewer and fewer people were needed in rural areas to provide food for everyone. The spread of industrialization has increased the demand for natural resources to supply the raw materials for industry. The human population has expanded so dramatically that there are serious questions about Earth's ability to support our members (see Chapter 52). As it is, millions of people are malnourished or undernourished. Almost all of the arable land on Earth is under cultivation.

Cultural evolution has resulted in large-scale disruption and degradation of the environment. Tropical rain forests and other natural environments are rapidly being eliminated. Soil, water, and air pollution occur in many places. Desertification (the spread of deserts) is increasing, largely because of human activities such as overgrazing, removal of forests, soil erosion, and improper irrigation leading to salty soil. Many plant and animal species cannot adapt to the rapid environmental changes caused by humans and thus are perishing. The decrease in biological diversity due to extinction is alarming.

On a positive note, we are aware of the damage we are causing, and we have the intelligence to further modify our behavior to improve these conditions. Education, including the study of biology, may help future generations develop environmental sensitivity, making cultural evolution our salvation rather than our destruction.

SUMMARY

I. Primates arose from small, arboreal, shrewlike mammals.
   A. Primates are adapted for an arboreal (tree-dwelling) existence by: the presence of five grasping digits, including an opposable thumb; long, slender limbs that move freely at the hips and shoulders; and eyes located in front of the head.
   B. Primates are divided into two groups, the prosimians and the anthropoids.
      1. Prosimians include lemurs, tarsiers, and lorises.
      2. Anthropoids include monkeys, apes, and humans.

II. Anthropoids arose from prosimian ancestors.
   A. The early anthropoids branched into two groups, the New World monkeys and the Old World monkeys.
   B. Apes arose from the Old World monkey lineage.
   C. There are four modern genera of apes: gibbons, orangutans, gorillas, and chimpanzees.

III. The hominid line separated from the ape line.
   A. The earliest hominids belong to the genus *Australopithecus*. At least some australopithecines walked on two feet, a hominid feature.
   B. *Homo habilis* was the earliest known hominid with some of the human features lacking in the australopithecines, including a slightly larger brain. *H. habilis* fashioned crude tools from stone.
   C. *Homo erectus* had a larger brain than *H. habilis*, made more sophisticated tools, and used fire.
   D. *Homo sapiens* appeared approximately 200,000 years ago.
      1. Neandertals are the earliest hominids classified as *Homo sapiens*. Their disappearance is a mystery.
      2. The origin of modern humans is controversial. Two different hypotheses, the "out of Africa" and the "multiregional" hypotheses, purport to explain the origin of modern humans.

IV. Cultural evolution is the transmission of knowledge from one generation to the next.
   A. An evolutionary increase in human brain size makes cultural evolution possible.
   B. Two of the most significant advances in cultural evolution were the development of agriculture and the Industrial Revolution.

SELECTED KEY TERMS

anthropoid
arboreal
Australopithecus
bipedal
brachiate
Cro-Magnon
cultural evolution
endothermic
foramen magnum
hominid
hominoid
Homo erectus
Homo habilis
Homo sapiens
knuckle-walking
Neandertal
paleoanthropologist
prehensile
prosimian
quadrupedal
supraorbital ridge
viviparous

POST-TEST

1. The two groups of the order Primates are the prosimians and the ________ .
2. Tarsiers and lemurs are examples of ________ .
3. Unlike the Old World monkeys, some New World monkeys possess a(an) ________ tail.
4. Apes and humans are collectively called ________ .
5. A gibbon ________ by swinging through the trees with its weight supported by one arm at a time.
6. The presence of large ________ ridges is more characteristic of apes than of humans.
7. The ________ ________ in humans is centered at the base of the skull, positioning the head for erect walking.
8. Humans and their ancestors are collectively called ________ .
9. The earliest hominids belong to the genus ________ .
10. The earliest hominid to be placed in the genus *Homo* was *Homo* ________ .
11. *Homo* ________ made sophisticated tools and discovered how to use fire.
12. *Homo* ________ appeared approximately 200,000 years ago.
13. ________ were an early group of *Homo sapiens* with a short, sturdy build and heavy brow ridges.
14. The progressive addition of knowledge to the human experience is known as ________ ________ .

REVIEW QUESTIONS

1. Distinguish between each of the following pairs: (a) mammals and primates; (b) prosimians and anthropoids; (c) anthropoids and hominoids; (d) hominoids and hominids.
2. Describe three different ways primates are adapted to an arboreal existence.
3. Identify at least three differences between the skulls of apes and humans.
4. List at least three ways an ape skeleton differs from a human skeleton.
5. How do *Australopithecus afarensis, Homo habilis, Homo erectus, Homo sapiens* (Neandertal), and *Homo sapiens* (modern) differ from one another?
6. Describe the two currently proposed hypotheses that explain where modern humans originated.
7. What is cultural evolution and how has it affected Earth?

YOU MAKE THE CONNECTION

1. If you were evaluating whether other early humans exterminated the Neandertals, what kinds of archaeological evidence might you look for?
2. The remains of Cro-Magnons have been found in *southern* Europe alongside reindeer bones, but reindeer currently exist only in *northern* Europe and Asia. Present two different hypotheses to account for this fact. How could you test these hypotheses?
3. How has cultural evolution helped humans?

RECOMMENDED READINGS

Coppens, Y. "East Side Story: The Origin of Humankind." *Scientific American,* Vol. 270, No. 5 May 1994. An overview of human evolution in East Africa along with a review of the major discoveries there.

Kates, R.W. "Sustaining Life on The Earth." *Scientific American,* Vol. 271, No. 4 October 1994. Examines whether humans have a long-term future on Earth and provides an overview of cultural evolution.

Savage-Rumbaugh, S. and R. Lewin. "Ape at the Brink." *Discover,* September 1994. A chimp named Kanzi has started chipping and flaking simple tools from rocks, providing insights into how our early human ancestors may have fashioned tools.

Shreeve, J. "*Erectus* Rising." *Discover,* September 1994. Highlights one of the most heated debates in the field of paleoanthropology: where did *Homo sapiens* arise?

Wilson, A.C. and R.L. Cann. "The Recent African Genesis of Humans." *Scientific American,* Vol. 266, No. 4 April 1992. One of two articles in this issue that debate the origin of humans. Wilson and Cann present molecular evidence in support of an African genesis of modern humans.

Career Visions

Biotechnology Patent Lawyer

KIT CHAN

Kit Chan came to the United States with a degree in biology and biochemistry from The Chinese University of Hong Kong. He completed his undergraduate studies during a time when there was much excitement about the prospects of genetic engineering, and Kit was particularly fascinated by molecular biology. He went on to obtain a Ph.D. in molecular virology at Baylor College of Medicine and proceeded to become a laboratory scientist. After working in research science for several years he attended Columbia School of Law and now applies both his scientific and legal knowledge to patent law.

How did you decide to become a lawyer?

After I finished my graduate studies I entered a post-doctoral program at Cold Spring Harbor Laboratory on Long Island. I was working in molecular oncology—how an oncogene is turned on and how that affects the cell, particularly the cell cycle. The oncogene I was working on, called *myc*, encodes a nuclear protein that regulates transcription. My project was to try to see how the protein is made and how it interacts with different cellular machinery. It was a very challenging project and I stayed at Cold Spring Harbor for more than two years. However, family obligations required me to improve my finances, and I looked for ways to utilize my science background at an increased compensation. At that time there were constant discussions at Cold Spring Harbor Laboratory about the Human Genome Project and the legal ramifications of biotechnology. I did some research and discovered that I could apply my science background to a career in law as a biotechnology patent lawyer.

You decided to attend Columbia School of Law.

Yes. I researched various law schools and I was surprised to find quite a few actually had programs in science technology and law. Columbia had such a program, so I contacted them and spoke with a professor there. He introduced me to several practitioners of this kind of law and I decided that it was what I wanted to do.

How did your biology background help you?

A biology background is useful in evaluating the prospects for obtaining a biotechnology patent. For example, if someone discovers a gene that is responsible for Alzheimer's disease, the patent application could include the DNA and RNA probes for this gene. The application might also include the protein that is encoded by the Alzheimer's disease gene and antibodies directed against the protein. The antibody might be useful for diagnosis and treatment as well.

My biology background also enables me to determine the further ramifications of an invention. For instance, if someone wishes to patent a drug for the herpes virus, I know that the same drug might block the replication of other viruses as well. This helps me to devise a patent application that covers other potential uses of the drug.

Do many of your colleagues have the kind of background that you do?

Yes they do. A doctorate degree is still a rarity, but more of my colleagues now have master's degrees in biology, chemistry, or even chemical engineering.

What does an inventor obtain by filing a patent?

The patent is a contract between the government and the applicant. For a definite period of time an applicant obtains exclusive rights to his or her invention, and in exchange, the inventor provides full disclosure of the invention in the public domain. Currently, the patent expiration period is twenty years. This means that while the patent holder retains rights to the invention, other scientists can have access to the invention and may be able to build upon the original work.

What else does your job entail beyond preparing patent applications?

I often see clients to evaluate their inventions and determine whether a patent is warranted in their case. Other scientists or venture capitalists may come to see me because they're interested in a technology transfer—they want to use a patented invention—I evaluate the prospects for their project. I may also handle patent litigation if a client's patent has been infringed upon.

PART 5

The Diversity of Life

A clouded yellow butterfly (*Colias croceus*) hovers over a bull thistle (Cirsium vulgare). Both organisms are common throughout many parts of North Amercia, particularly open woods, pastures, and roadsides.

(Animals Animals © 1996 Stephen Dalton)

CHAPTER 22

The Classification of Organisms

The curved beak of the 'i'iwi bird, *Vestiaria coccinea,* is beautifully adapted to reach the nectar in the tubular flowers of the Lobelia plant.

(Frans Lanting/Minden Pictures)

About two million species of living organisms have been identified on our planet, and biologists speculate that several million additional species remain to be identified. In order to organize these life forms so that we can study them and effectively communicate knowledge about them, we need a system of **classification.**

Imagine that you were going to develop a classification system. How would you use what you already know about living things to assign them to categories? Would you place insects, bats, and birds in one category because they all have wings and fly? And would you, perhaps, place squid, whales, fish, penguins, and Olympic backstroke champions in another category because they all swim? Or would you classify organisms according to a culinary scheme, placing lobsters and tuna in the same part of the menu, perhaps identifying them as "seafood"?

All of these schemes might be valid, depending on your purpose. Similar methods have been used throughout history. Animals, for example, were classified by St. Augustine in the fourth century as useful, harmful, or superfluous—to humans. Anthropologists have discovered that some cultures still use a similar system.

During the Renaissance scholars began to develop categories based more on the characteristics of the organisms themselves. These categories were originally arranged roughly in order from the simple to the complex. Out of the many classification systems that were developed, the one designed by Carolus Linnaeus in the mid-18th century (described briefly in Chapter 1) has survived with some modification to the present day.

Linnaeus, a Swedish botanist and natural historian, simplified the scientific classification of organisms. In classification schemes before Linnaeus, each organism had a lengthy descriptive name, sometimes composed of ten or more Latinized words! Linnaeus developed a binomial system of nomenclature in which each species is assigned a unique two-part name (see Chapter 1). He also devised a system for assigning species to a hierarchy of increasingly general groups. Linnaeus probably had no theory of evolution in mind when he set up his system. Neither did he have any concept of the vast number of living and extinct organisms that would later be discovered. Yet his system has proved to be remarkably flexible and adaptable to new biological knowledge and theory. Very few other 18th century inventions survive today in a form that would still be recognizable to their originators. The hierarchical system presently used includes kingdom, phylum, class, order, family, genus, and species.

LEARNING OBJECTIVES

After you have studied this chapter, you should be able to

1. Offer at least two justifications for the use of scientific names and classifications of organisms.
2. Arrange the Linnaean categories in hierarchical fashion, from most inclusive to least inclusive.
3. List the five kingdoms of organisms recognized by modern biologists, give the rationale for this system of classification, and describe the distinguishing characters of the organisms assigned to each.
4. Critically summarize the difficulties encountered in choosing taxonomic criteria.
5. Apply the concept of shared derived characteristics to the classification of organisms.
6. Describe the methods of molecular biology now used by taxonomists, and summarize their advantages.
7. Contrast the three major approaches to classification: phenetics, cladistics, and classical evolutionary taxonomy.

ORGANISMS ARE NAMED USING THE BINOMIAL SYSTEM OF NOMENCLATURE

The science of naming and classifying organisms is known as **taxonomy.** In the **binomial system of nomenclature,** each species is assigned a two-part name. The first part designates the **genus** (pl., *genera*), and the second part is called the **specific epithet.** The same specific epithet can be used as the second name of species of different genera. For example, *Quercus alba* is the scientific name for the white oak, and *Salix alba* is the name for the white willow. (Alba comes from a Latin word meaning "white.") Thus, both parts of the name must be used to ensure that the species is being accurately identified.

The genus name is always capitalized, whereas the specific epithet is not. Both names must be underlined or italicized. The genus name can be used alone to designate all species in the genus, but the specific epithet is never used alone; it must always be preceded by the full or abbreviated genus name (e.g., *Quercus alba* or *Q. alba*).

Scientific names are generally derived from Greek or Latin roots or from Latinized versions of the names of persons, places, or characteristics. For example, the genus name for the bacterium *Escherichia coli* is based on the name of the scientist Theodor Escherich, who first described it. The specific epithet *coli* reminds us that *E. coli* lives in the colon (large intestine).

Scientific names permit taxonomy to be a truly international study. Even though many scientifically important organisms do not have common names, and even though the common names often vary in different locations and languages, organisms can be universally identified by their scientific names. A researcher in Puerto Rico can know exactly which organisms were used in a study published by a Russian scientist, and therefore can repeat or extend the Russian's experiments using the same species.

Each taxonomic level is more general than the one below it

Recall from Chapter 1 that the narrowest category in the Linnaean system is the species, and that the broadest is the kingdom. The range of categories in between form a hierarchy (Fig. 22–1; Table 22–1). A **taxon** (pl., *taxa*) is a taxonomic grouping at any level, such as species, genus, or phylum. For example, the phylum Chordata is a taxon that contains several classes, including Mammalia and Amphibia. Similarly, the class Mammalia is a taxon that includes many different orders. (See *Focus On: Classifying Yourself*.)

Recall that a species is a group of similar organisms that can interbreed in their natural environment and are reproductively isolated from other organisms (see Chapter 19). Closely related species are assigned to the same genus, and closely related genera may be grouped together in a single **family.** Families are grouped into **orders,** orders into **classes,** classes into **phyla,**[1] and phyla into **kingdoms.** These groupings can also be separated into subgroupings—for example, subphyla or subclasses.

The **species** is the only taxon that actually exists in nature. Members of a species are defined by their common gene pool and ability to interbreed. All higher taxa are artificial constructs designed by taxonomists. No operational definition exists for a genus, family, order, class, phylum, or kingdom.

Just how organisms are grouped can vary according to the basis used for classification and the judgment of the taxonomists. Some taxonomists ignore minor varia-

[1]Plants and fungi have been traditionally classified into divisions rather than phyla. However, the International Botanical Congress recently approved the phylum designation for plants and fungi.

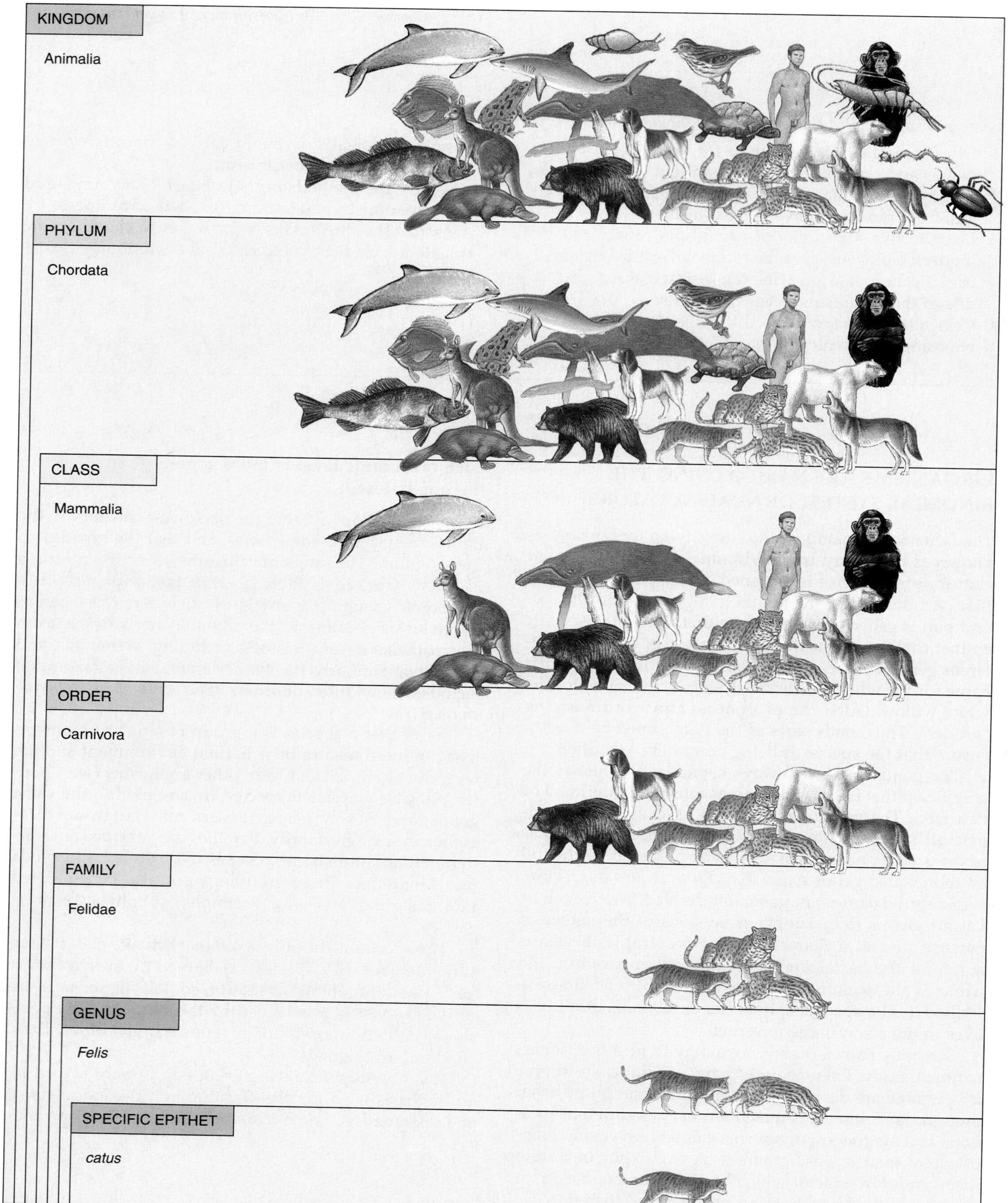

FIGURE 22–1 The principal categories used in classifying an organism. The domestic cat (*Felis catus*) illustrates the hierarchical organization of our taxonomic system.

Table 22–1 CLASSIFICATION OF CORN

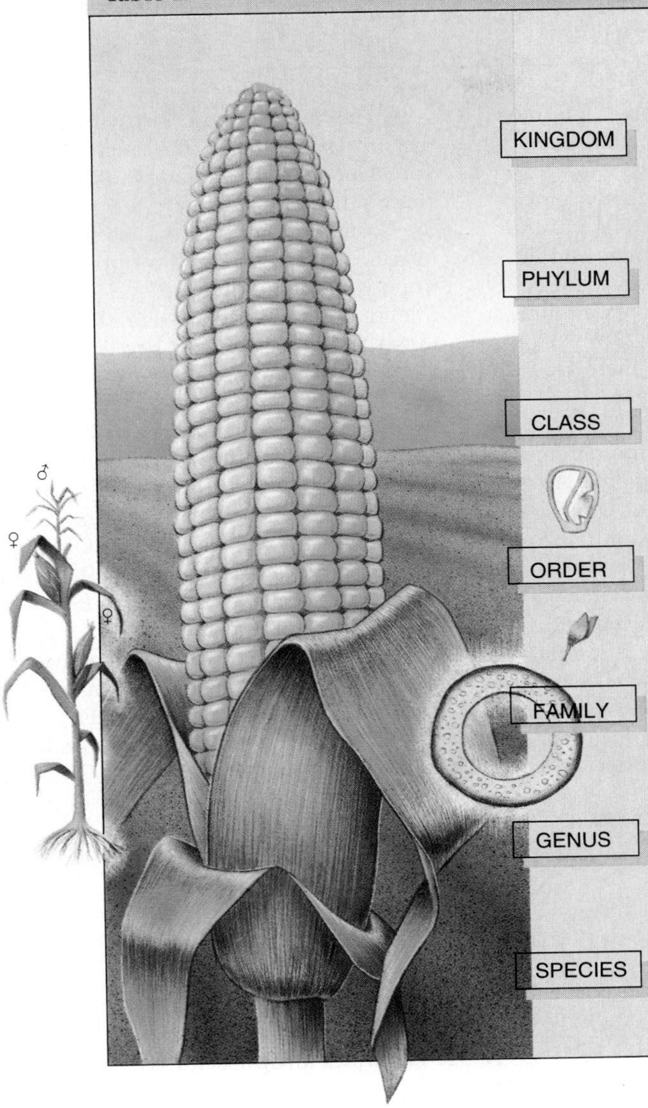

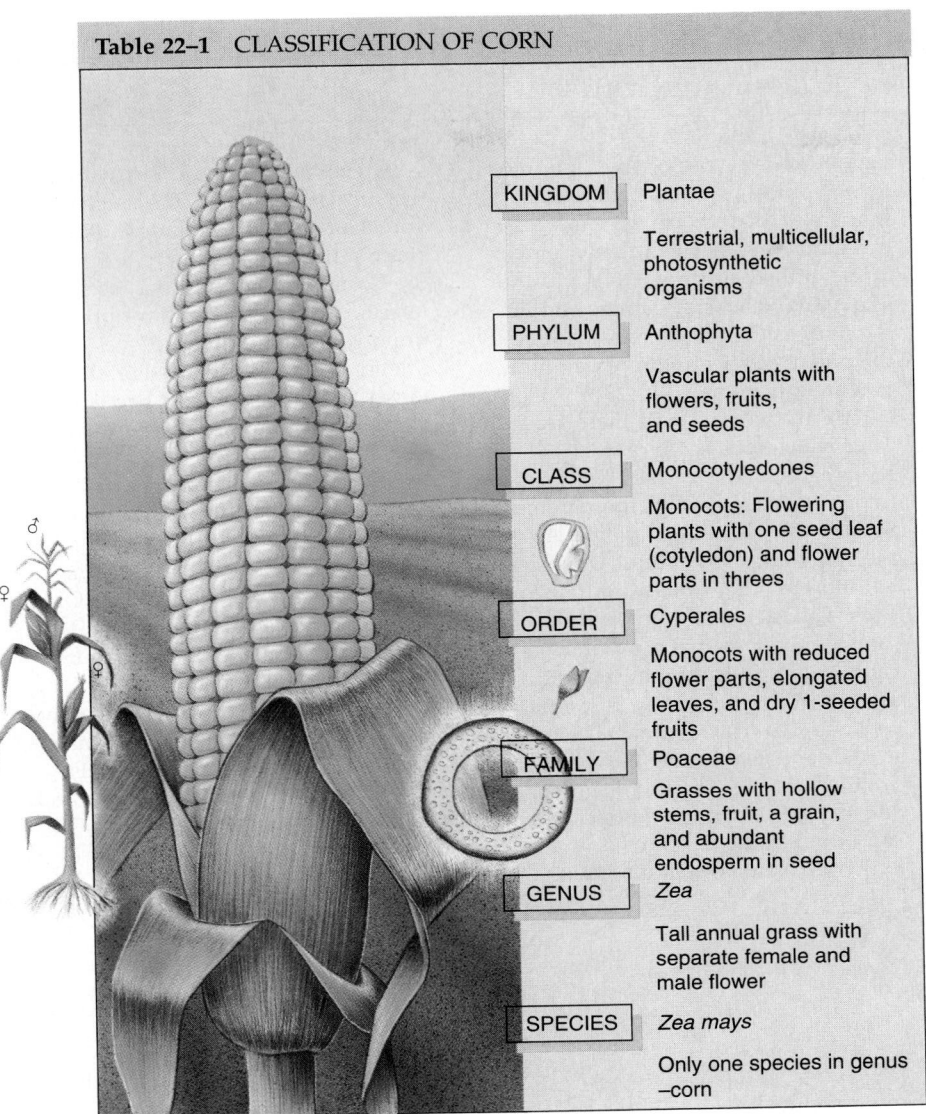

KINGDOM	Plantae	Terrestrial, multicellular, photosynthetic organisms
PHYLUM	Anthophyta	Vascular plants with flowers, fruits, and seeds
CLASS	Monocotyledones	Monocots: Flowering plants with one seed leaf (cotyledon) and flower parts in threes
ORDER	Cyperales	Monocots with reduced flower parts, elongated leaves, and dry 1-seeded fruits
FAMILY	Poaceae	Grasses with hollow stems, fruit, a grain, and abundant endosperm in seed
GENUS	*Zea*	Tall annual grass with separate female and male flower
SPECIES	*Zea mays*	Only one species in genus –corn

tions and group organisms into already existing taxa. This practice is referred to as "lumping." Other taxonomists subdivide taxa on the basis of minor differences, establishing separate categories for forms that do not fall naturally into one of the existing classifications. This practice is called "splitting." Lumpers acknowledge as few as 10 animal and 4 plant phyla, whereas "splitters" may recognize more than 30 animal and up to 12 plant phyla.

Some biologists use a level of classification above the kingdom, called a **domain,** to divide the prokaryotes and eukaryotes. Carl Woese of the University of Illinois has proposed that organisms be classified in three domains, consisting of the archaebacteria, the eubacteria, and the eukaryotes.

Subspecies may become species

The species is the basic unit of classification, but not the smallest taxon in use. Geographically different populations within a species often display certain consistent characteristics that serve to distinguish them from other populations of the same species. If they can interbreed, however, they are not truly separate species, but **subspecies.** For some kinds of microorganisms, such as bacteria, the term **strain** is used.

Experts are usually able to distinguish subspecies. However, subspecies may grade imperceptibly into one another at the borders of their geographical ranges, where there is opportunity to interbreed. Some of these subspecies may be in the process of becoming reproductively isolated and may, in the course of time, become separate species. Thus, they provide an opportunity for field studies of gene pools and of the speciation process.

Many biologists recognize five kingdoms

From the time of Aristotle to the mid-20th century, biologists divided living things into two kingdoms, **Plantae**

Focus On

Classifying Yourself

Because your cells have discrete nuclei surrounded by nuclear envelopes, you are a eukaryote. Your cells lack chloroplasts and cell walls, and you are a multicellular heterotroph with highly differentiated tissues and organ systems. That makes you a member of the kingdom Animalia.

What kind of an animal are you? You possess a spinal column, composed of bony vertebrae, that has largely replaced the cartilaginous rod you had as an embryo, the notochord. At that time you also had pharyngeal grooves that, had you been a fish, would have developed into gill slits. You have a hollow dorsal nerve cord and a brain. These traits mark you as a chordate and a vertebrate; that is, you belong to the phylum Chordata (because you had a notochord), and to the subphylum Vertebrata (because you have vertebrae that replaced the notochord).

Among the vertebrates there are several classes: jawless fishes, cartilaginous fishes, bony fishes, amphibians, reptiles, mammals, and birds. You are an endotherm (that is, you can maintain a constant body temperature) and so must be either a bird or a mammal. Lacking feathers and having hair, teeth, and (if you are female) the potential for nursing your young, you are a mammal.

Within the mammals there are three subclasses: the Prototheria, Metatheria, and Eutheria. The Prototheria (also known as monotremes) include the duck-billed platypus and the spiny anteater, both of which, in addition to other unusual traits, lay eggs. The Metatheria (more commonly known as marsupials) usually carry their still-embryonic young around in a pouch. They lack a placenta, an organ of exchange between mother and developing embryo. If you did not hatch from an egg or spend your infancy in a pouch, you can be confident of your status as a eutherian, or placental mammal.

A number of orders exist within the subclass Eutheria. The insectivores, for instance, include moles and shrews; the Chiroptera are bats; and the Carnivora include dogs, cats, and ferrets, among others. Your opposable thumbs, frontally directed eyes, flat fingernails, and several other characteristics identify you as a primate, along with monkeys, apes, and tarsiers.

Primates include a number of families. You and the New World monkeys are obviously very different: they generally have prehensile tails, for instance, which you, apes, and all Old World monkeys lack. Indeed, you and the apes lack tails altogether. Your posture is upright and you have long legs, short arms, and not much body hair—all of which make you a member of the family Hominidae. This family has only one living genus and one living species: *Homo sapiens*. ■

---

and **Animalia.** After the development of microscopes, it became increasingly obvious that many organisms could not be easily assigned to either the plant or the animal kingdom. More than a century ago a German biologist, Ernst Haeckel, suggested that a third kingdom be established, the kingdom **Protista.** Although Haeckel changed the defining characters of this new kingdom during the course of his work, his goal was to include simple organisms, such as bacteria and most other microorganisms, that did not appear to fit into the plant or animal kingdoms.

In 1937 the French marine biologist Edouard Chatton suggested the term *procariotique* ("before nucleus") to describe bacteria, and the term *eucariotique* ("true nucleus") to describe all other cells. This dichotomy is now universally accepted by biologists as a fundamental evolutionary divergence.

In the 1960s advances in electron microscopy and biochemical techniques revealed basic cellular differences that inspired many new proposals for classifying organisms. In 1969, R. H. Whitaker proposed a five-kingdom classification that has been accepted by the majority of biologists. Whitaker suggested that the fungi be assigned to kingdom **Fungi** rather than included as part of the plant kingdom. After all, fungi are not photosynthetic and must absorb nutrients produced by other organisms. Fungi also differ from plants in the composition of their cell walls, in their body structures, and in their modes of reproduction.

Kingdom **Prokaryotae** (formerly called *Monera*) was established to accommodate the bacteria, which are fundamentally different from all other organisms in that they lack distinct nuclei and other membranous organelles. Some biologists have recently added a sixth kingdom by dividing kingdom Prokaryotae into kingdom Archaebacteria and kingdom Eubacteria. The bacteria in these two groups have important differences justifying such separation (see Chapter 23). However, in this edition of *Biology,* we use the still more widely accepted five-kingdom approach: Prokaryotae (bacteria), Protista (algae, includ-

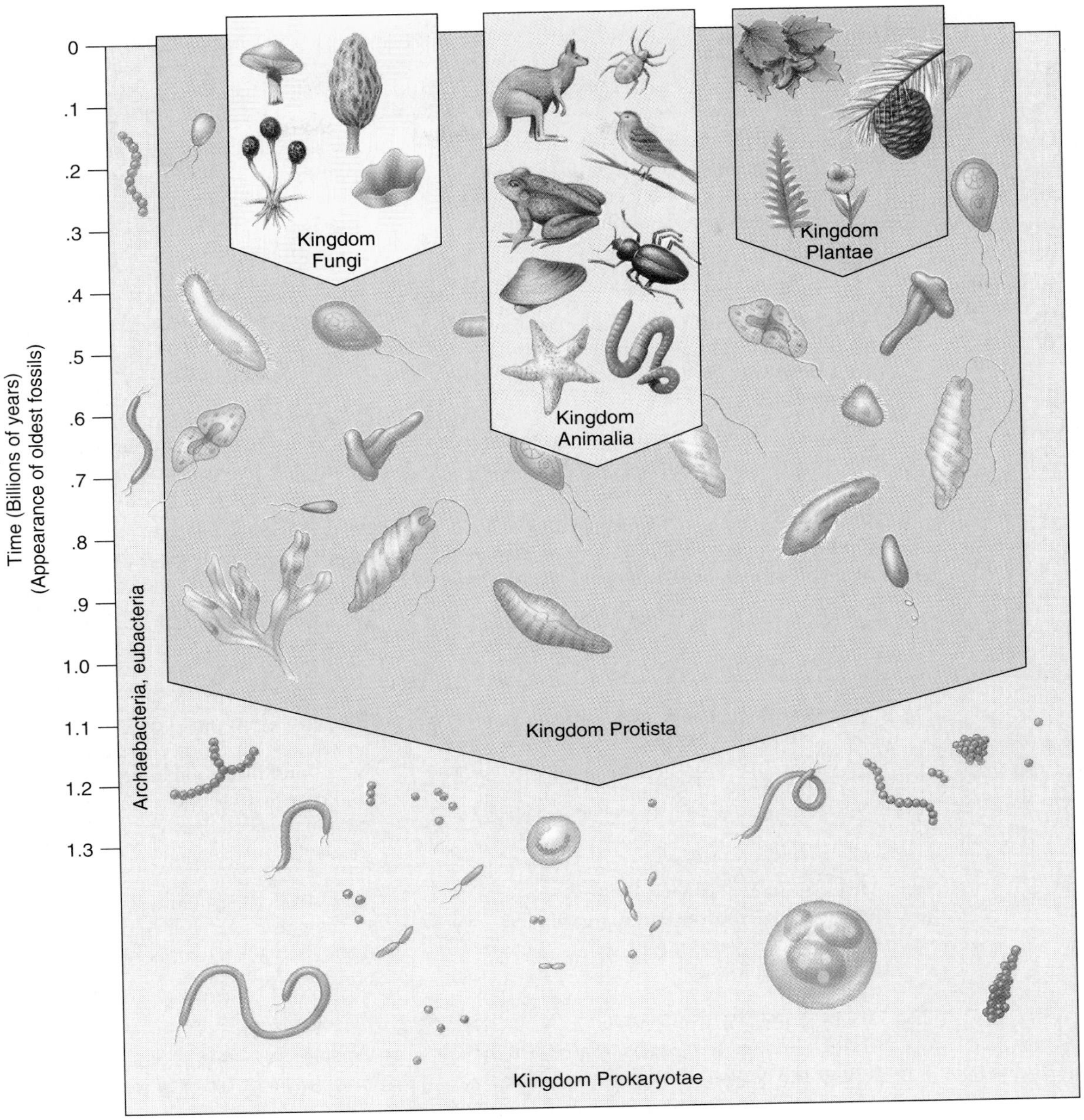

FIGURE 22–2 Most biologists currently use a five-kingdom system of classification.

ing multicellular forms; protozoa; water molds; and slime molds), Fungi (mushrooms and molds), Plantae, and Animalia (Fig. 22–2 and Table 22–2).

Systematics is concerned with reconstructing phylogeny

Modern classification is often referred to as **systematics** because it is based on evolutionary relationships. Many taxonomists are systematists, who seek to reconstruct the evolutionary history, or **phylogeny** (literally, "production of phyla"), of organisms. Once these relationships are defined, the classification of organisms can be based on common ancestry.

An ideal taxon is monophyletic

A population of organisms has a dimension in space—its geographical range—and also a dimension in time. Each population extends backward in time. Somewhat like branches of a tree, a population diverges from populations of other species (Fig. 22–3). Species have various degrees of evolutionary relationship with one another,

Table 22–2 FIVE KINGDOMS: PROKARYOTAE, PROTISTA, FUNGI, PLANTAE, AND ANIMALIA

Kingdom	*Characteristics*	*Ecological Role and Comments*
Prokaryotae	Prokaryotes (lack distinct nuclei and other membranous organelles); single-celled; microscopic; cell walls generally composed of peptidoglycan; metabolically varied.	Most are decomposers; some parasitic (and pathogenic); some chemosynthetic autotrophs; some photosynthetic; important in recycling nitrogen and other elements; some used in industrial processes.
Protista	Eukaryotes; mainly unicellular or simple multicellular. Three informal groups (not taxa) include protozoa; algae; and slime molds and water molds.	
Protozoa	Microscopic; heterotrophic; most move by means of flagella, cilia, or pseudopodia.	Important part of zooplankton; near base of many food webs; some are parasitic (and pathogenic).
Algae	Photosynthetic; sometimes hard to differentiate from protozoa; some have brown or red pigments in addition to chlorophyll.	Very important producers, especially in marine and freshwater ecosystems.
Slime molds and water molds	Heterotrophic; reproduce by forming spores.	Aquatic or terrestrial; varied modes of nutrition.
Fungi	Heterotrophic; absorb nutrients; do not photosynthesize; body composed of threadlike hyphae that form tangled masses that infiltrate food or habitat.	Decomposers; some parasitic (and pathogenic); some used as food; yeast used in making bread and alcoholic beverages; some used to make industrial chemicals or antibiotics; responsible for much spoilage and crop loss.
Plantae	Multicellular; photosynthetic; possess multicellular reproductive organs; alternation of generations; cell walls of cellulose.	Terrestrial biosphere depends on plants in their role as primary producers; important source of oxygen in Earth's atmosphere.
Animalia	Multicellular heterotrophs, many of which exhibit advanced tissue differentiation and complex organ systems; most able to move about by muscular contraction; specialized nervous tissue coordinates responses to stimuli.	Consumers; some specialized as herbivores, carnivores, or detritus feeders.

depending on the length of time that has elapsed since their populations diverged from a common ancestor.

If all of the subgroups within any taxon share the same common ancestor, the grouping is referred to as **monophyletic** (one branch). Monophyletic taxa are, therefore, natural groupings because they represent true evolutionary relationships and include all close relatives. A taxon containing a common ancestor and all the taxa descended from it is called a **clade.**

Many currently recognized taxa are actually **polyphyletic,** consisting of several evolutionary lines and not including a common ancestor. Polyphyletic taxa, therefore, may misrepresent evolutionary relationships. For this reason, taxonomists try to avoid constructing polyphyletic taxa.

Homologous structures are important criteria for classification

Just how to group species into higher taxonomic groups is sometimes a difficult decision. For example, in Figure 22–3, should species A and B be placed within a single genus or do they represent two distinct genera? If species C and E are distinct genera, should D be part of either of

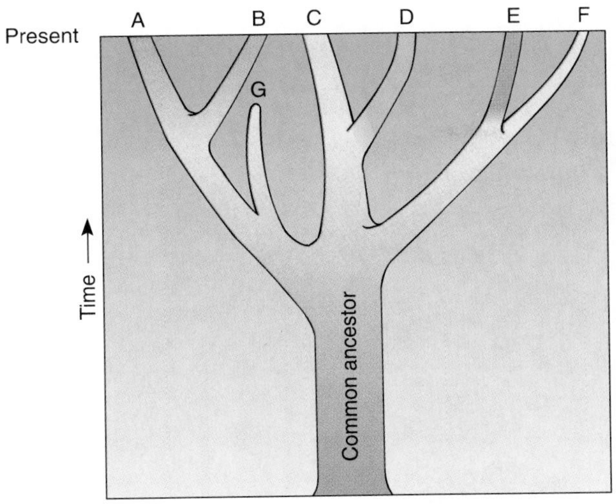

FIGURE 22–3 This tree depicts the evolutionary relationships of several hypothetical monophyletic species. The upper ends of the branches represent the species at the present time. Junctions of the branches represent points of common ancestry. Species G is extinct. If you go far enough back in time, all taxa share a common ancestor. The base of the tree represents the common ancestor of species A, B, C, D, E, F, and G. Which groupings of these species might be considered a genus? A family?

these genera? Similar difficulties exist in determining the assignments to families, orders, classes, and phyla. Most biologists base their judgments about the degree of relationship on the extent of similarity between living species, and, when available, on the fossil record.

In evaluating similarities, biologists consider structural, physiological, behavioral, and molecular traits. When comparing structural similarities, we look for **homologous structures** in different organisms (see Chapter 17). The presence of homologous structures suggests that two species of organisms evolved from a common ancestor. In contrast, similarities among analogous structures result not from shared ancestry but from convergent evolution (Fig. 22–4). This sometimes occurs when unrelated or distantly related organisms become adapted to similar environmental conditions. For example, the shark and the dolphin have similar body forms because they have become adapted to similar environments (aquatic) and life styles (predatory).

Derived characters have evolved more recently than ancestral characters

Organisms sharing many homologous structures are thought to be closely related, and less similar organisms sharing few homologous characters are less closely related. However, distinguishing between homologous and analogous structures is not always easy. Therefore, the choice of which similarities to use to show evolutionary relationships is extremely important.

How does a systematist interpret the significance of these similarities? In making decisions about taxonomic relationships, the biologist first examines the characteristics common to the largest category of organisms and interprets them as indicating the most remote common ancestry. These **ancestral characters** are traits that were

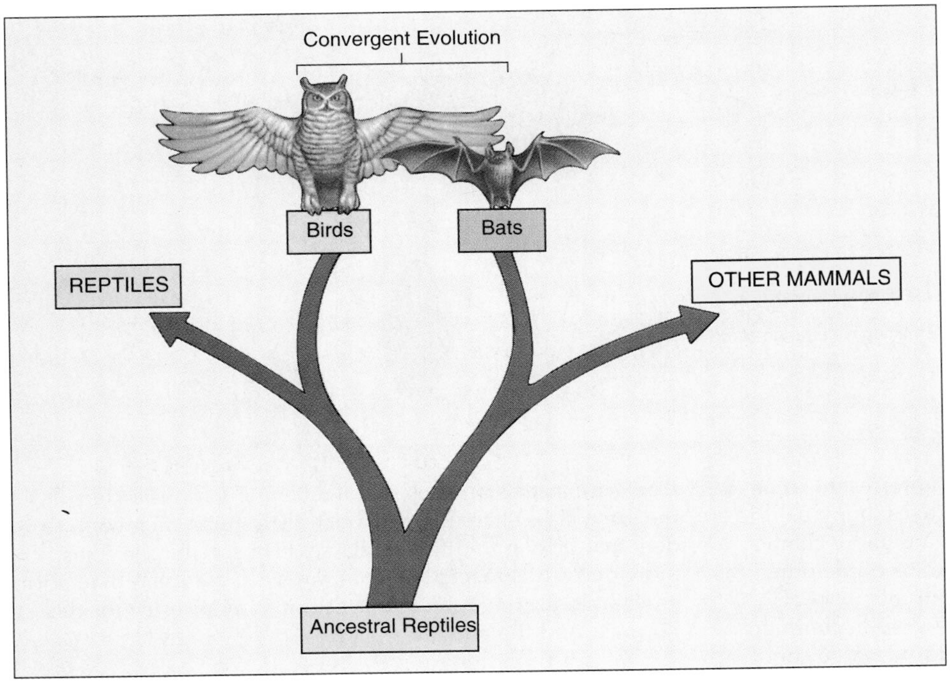

FIGURE 22–4 In convergent evolution, distantly related groups may come to resemble one another in structure and function as they become adapted to similar modes of life. Their analogous structures, such as wings of birds and bats, do not indicate an immediate common ancestor. In contrast, the presence of homologous structures suggests evolution from a common ancestor.

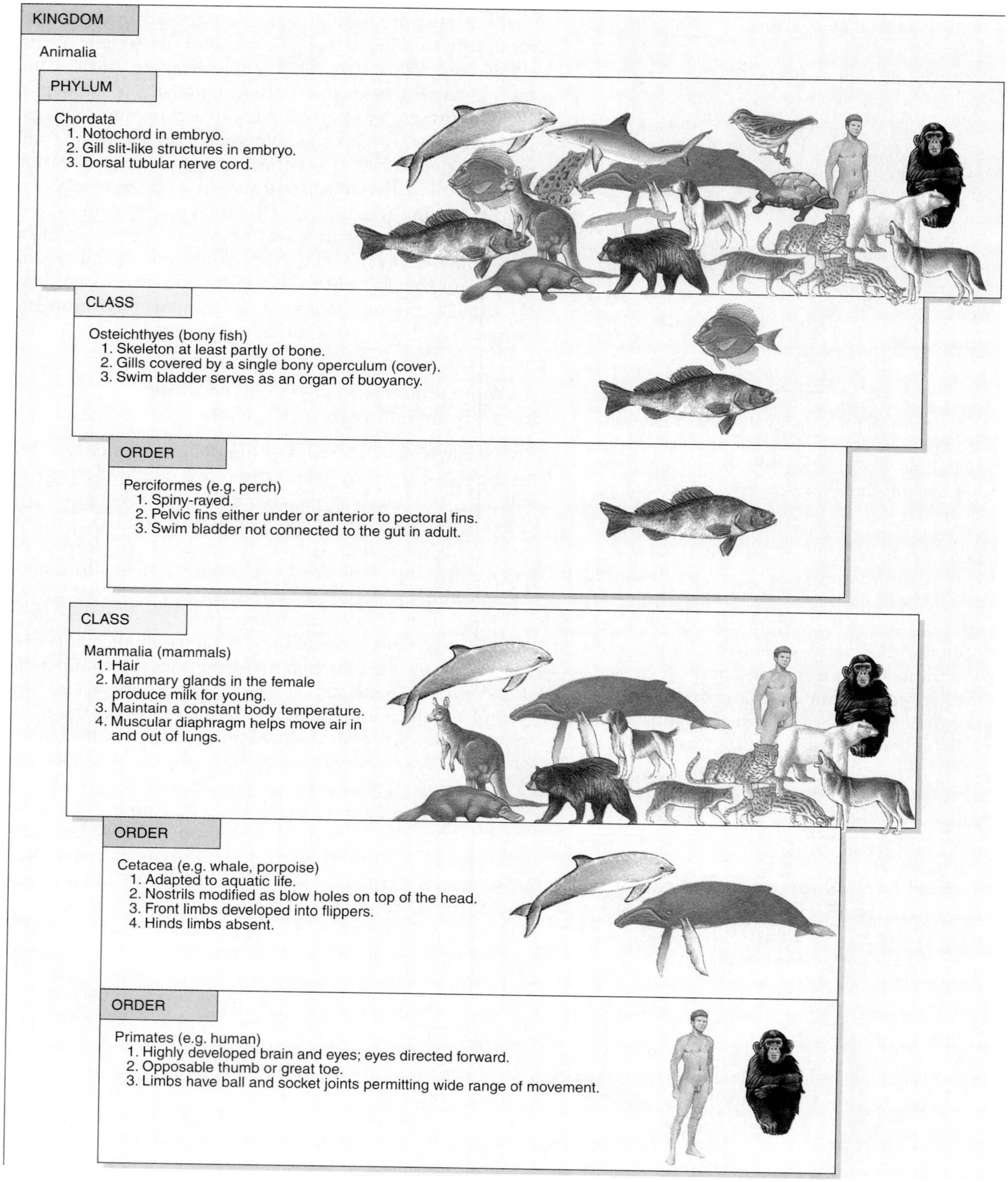

KINGDOM
Animalia
PHYLUM
Chordata
1. Notochord in embryo.
2. Gill slit-like structures in embryo.
3. Dorsal tubular nerve cord.
CLASS
Osteichthyes (bony fish)
1. Skeleton at least partly of bone.
2. Gills covered by a single bony operculum (cover).
3. Swim bladder serves as an organ of buoyancy.
ORDER
Perciformes (e.g. perch)
1. Spiny-rayed.
2. Pelvic fins either under or anterior to pectoral fins.
3. Swim bladder not connected to the gut in adult.
CLASS
Mammalia (mammals)
1. Hair
2. Mammary glands in the female produce milk for young.
3. Maintain a constant body temperature.
4. Muscular diaphragm helps move air in and out of lungs.
ORDER
Cetacea (e.g. whale, porpoise)
1. Adapted to aquatic life.
2. Nostrils modified as blow holes on top of the head.
3. Front limbs developed into flippers.
4. Hinds limbs absent.
ORDER
Primates (e.g. human)
1. Highly developed brain and eyes; eyes directed forward.
2. Opposable thumb or great toe.
3. Limbs have ball and socket joints permitting wide range of movement.

FIGURE 22–5 Derived characteristics are not present in an ancestral species. Members of class Osteichthyes (bony fish) and class Mammalia (the mammals) share many more characteristics with one another and with the members of the other classes of phylum Chordata than they do with members of any other phylum. For example, a perch has more in common with a monkey (a notochord, gill slits, and dorsal nerve cord) than with a sea star or clam. Members of various orders of the same class share more characteristics than members of orders that belong to different classes. Thus, a porpoise has more derived characters in common with a human than with a perch, indicating a more recent common ancestry for the porpoise and the human.

present in an ancestral species and remained essentially unchanged.

Derived characters are those traits not present in ancestral species because they evolved more recently. A feature viewed as a derived character in a more inclusive (broader) taxon may also be considered an ancestral character in a less inclusive (narrower) taxon. More recent common ancestry is indicated by classification into less and less inclusive taxonomic groups with more and more specific shared derived characters.

For example, the three small bones in the middle ear are useful in identifying a branching point between mammals and reptiles. The evolution of this derived character was a unique event, and only mammals have these bones. However, if we are comparing mammals with one another, the three ear bones are an ancestral character because all mammals have them. They have no value in distinguishing among mammalian groups. Different characters must be used to establish branching points *among* the mammals.

FIGURE 22–6 A few mammals share important characteristics with birds. The duck-billed platypus, a monotreme, lays eggs, has a beak, and lacks teeth. Should we classify it as a bird? (Jean Philippe Varin/Jacana/Photo Researchers, Inc.)

Biologists carefully choose taxonomic criteria

Although both fish and porpoises have streamlined body forms, this characteristic is an analogous adaptation and does not indicate evolutionary relationships. The porpoise shares important homologous derived characters with mammals such as humans: the ability to breathe air, nurse young, maintain a constant body temperature, and grow hair. Thus, the porpoise is classified as a mammal and thought to have descended from a terrestrial mammalian ancestor.

Although porpoises have more features in common with humans than with fishes, some characteristics are shared by all three of these animals. Among these are a dorsal tubular nerve cord, and, during embryonic development, a notochord (skeletal rod) and rudimentary gill slits. These shared ancestral characteristics indicate a common ancestry and serve as a basis for classification. This ancestry is more remote between the porpoise and the fish than between the porpoise and the human. Therefore, fishes, humans, and porpoises are grouped together in a more inclusive taxon, the phylum Chordata, and humans and porpoises are also classified together in class Mammalia, a less inclusive taxon within phylum Chordata, indicating their closer relationship (Fig. 22–5).

Deciding which traits best illustrate evolutionary relationships is not always simple. What, for example, are the most important taxonomic characteristics of a bird? We might list feathers, beak, wings, absence of teeth, egg-laying, and the fact that they are **endotherms** (animals able to maintain a constant body temperature). Some mammals (monotremes such as the duck-billed platypus) have many of these same characteristics: they have beaks, lack teeth, lay eggs, and are endotherms. Yet we do not classify them as birds (Fig. 22–6).

No mammal, however, has feathers. Is this trait absolutely diagnostic of birds? According to the conventional taxonomic wisdom, the presence or absence of feathers determines what is and is not a bird. This applies only to modern birds, however. Some extinct reptiles may have been covered with feathers.

Usually, organisms are classified on the basis of a combination of traits rather than on any single trait such as the ability to live in water. The significance of these combinations is determined inductively—that is, by an integration and interpretation of data. Such induction is a necessary part of science.

Taxonomists hypothesize, for example, that birds should all have beaks, feathers, no teeth, and so on. Then they reexamine the living world and observe whether there are organisms that might reasonably be called birds that do not fit the current definition of "birdness." If not, the definition is permitted to stand. If too many exceptions emerge, the definition may be modified or abandoned. Sometimes, the taxonomist persuades the world that an apparent exception—the bat, for instance—resembles a bird only superficially and should not be considered one. The bat has all of the basic characteristics of a mammal, such as hair and mammary glands that produce milk for the young.

Taxonomy is a dynamic science that proceeds by the constant reevaluation of data, hypotheses, and theoretical constructs. As new data are discovered and old data are subjected to reinterpretation, the ideas of taxonomists change. During the 1980s, for example, a new group of

Making the Connection

Molecular Biology, Evolution, and Taxonomy

How do molecular biology and evolution come together in the laboratory to provide important data that can be used by taxonomists? Advances in molecular biology have provided the tools for biologists to compare the macromolecules of various organisms. Amino acid sequencing techniques, immunological methods, and DNA and RNA sequencing are among the procedures now used to compare macromolecules. Such comparisons of molecular structure provide objective, quantifiable measures of evolutionary relationships.

Cytochrome *c*, the respiratory protein mentioned in Chapter 7, provides a good example of how data gained through amino acid sequencing contribute to taxonomic decisions. In Chapter 17 we pointed out that although the structure and function of cytochrome *c* are similar in all aerobic organisms, some differences in amino acid sequences exist among species. In fact, the extent of the differences in amino acid sequences reflects the time since two species diverged in their evolutionary history. Human and chimpanzee cytochrome *c* molecules have identical amino acid sequences. In a more distantly related primate, the rhesus monkey, one of the 104 amino acids in the sequence is different from that in human cytochrome *c*. In the dog, a nonprimate, 13 amino acids are different. Taxonomists use this type of information to help make decisions about classifying organisms. Such decisions are generally based on data derived from many studies of different proteins.

We have learned that among related species the DNA sequences for the same structural genes are very similar. Detailed restriction maps within large homologous regions of chromosomes of related organisms are also very similar (see Chapter 14). For example, the DNA region that codes for the equivalent of the human hemoglobin beta chain has been mapped in several primates. Even though the gorilla diverged from humans 4 to 5 million years ago, 65 of the 70 restriction sites are identical.

Researchers have determined the nucleotide sequence of a portion of DNA from each of three species of primates (humans, gorillas, and chimpanzees). From their analysis of this 7000 nucleotide sequence, the investigators inferred a common ancestral gene. The simplest branching pattern that would account for the results suggests that the gorilla first split off from the common ancestor it shared with the chimpanzee and human. Later the chimpanzee and human lines diverged. ▲

organisms, the Loricifera, was discovered whose combination of traits did not fit those of any existing phylum. A new phylum—Phylum Loricifera—was established to accommodate this animal.

Molecular biology provides taxonomic tools

When a new species evolves, it does not always have obvious phenotypic differences. For example, two distinct species of fruit flies may appear identical. Some of their macromolecules, however, are different. Variations in the structure of specific macromolecules among species, just like differences in anatomical structure, result from mutations. Macromolecules that are functionally similar in two different types of organisms are considered homologous if their subunit structure is similar.

Methods that enable biologists to compare the nucleotide sequences of nucleic acids and the amino acid sequences of proteins have become extremely important taxonomic tools. These comparisons can give us some idea of the degree of relatedness of two organisms. The greater the correspondence in amino acid sequences, the more closely organisms are thought to be related. The number of differences in DNA or RNA nucleotide sequences or in certain protein amino acid sequences in two groups of organisms may reflect how much time has passed since the groups branched off from a common ancestor. Thus, specific genes and specific proteins can be used as **molecular clocks** (see Chapter 17). Biologists can use such clocks to help date the divergence of two groups from a common ancestor.

Comparison of ribosomal RNA sequences has recently been used to challenge the widely accepted idea that fungi are more closely related to plants than to animals. In a study reported in *Science* in 1993, investigators suggested that, based on ribosomal RNA analysis, fungi are more closely related to animals than to plants. These biologists hypothesize that animals and fungi share a more recent common ancestor, perhaps a flagellated protist.

Biologists are using polymerase chain reaction (PCR), a method described in Chapter 14, to amplify small amounts of DNA extracted from fossils. With this method researchers can obtain sufficient DNA to compare with that of modern organisms, a strategy that may contribute to some important taxonomic decisions.

TAXONOMISTS USE THREE MAIN APPROACHES

In determining the relationships among organisms, taxonomists work to identify monophyletic groups. They consider branch points that indicate the time at which a particular group of organisms evolved. They also consider the degree of divergence between branches, or how different two groups have become since they originated from a common ancestor and evolved along different pathways. Which of these types of evolutionary data is used more in classifying a group of organisms depends on the taxonomist's approach. Three major approaches to the classification of organisms are phenetics, cladistics, and classical evolutionary taxonomy. Many taxonomists now use a combination of these approaches.

Phenetics is based on phenotypic similarities

Pheneticists argue that we cannot be sure that our view of evolutionary history is correct. These taxonomists base their classification on traits they can measure. The **phenetic (phenotypic) system,** sometimes called numerical taxonomy, is based on similarities of many characteristics. In this system organisms are grouped according to the number of characteristics they share, with no attempt to determine whether their similarities arose from a common ancestor or from convergent evolution. Pheneticists suggest that it is not important to try to sort homologous and analogous characteristics, because many more similarities are due to homology than to analogy. Thus, the number of similarities that two organisms have in common reflects the degree of homology.

A taxonomist who follows the phenetic system would explain that porpoises are classified along with humans as mammals rather than as fish because they share more similarities with mammals. Pheneticists assign numbers to many arbitrarily chosen traits that are given equal weight. They designate these traits as present (+) or absent (−) in the organisms of a particular taxon. This information is fed into a computer, which indicates which groups have the most traits in common.

Phenetics is not widely used by taxonomists today because the use of analogous similarities can lead to inaccurate conclusions about evolutionary relationships. However, the phenetic emphasis on quantitative comparisons of characters has been an important contribution to taxonomy.

Cladistics emphasizes phylogeny

The **cladistic** approach to taxonomy emphasizes phylogeny, focusing on *when* evolutionary lineages divided into two different branches. Cladists determine branch points using carefully defined objective criteria, and insist that taxa be monophyletic. Thus, common ancestry, rather than data on phenetic similarity, is the basis for classification. A cladist would say that porpoises are classified with mammals rather than with fish because porpoises and mammals share a more recent common ancestor than porpoises and fish.

Cladists develop branching diagrams called **cladograms.** Each branch on a cladogram represents the divergence, or splitting, of two new groups from a common ancestor. Consider the evolutionary grouping of turtles, lizards, snakes, dinosaurs, crocodiles, and birds. The birds, along with dinosaurs, are thought to share a common ancestor with the modern crocodiles and alligators. Birds evolved from reptiles that also gave rise to the crocodiles. Crocodiles, dinosaurs, and birds, then, comprise a monophyletic group, and cladists would classify them in the same taxon (Fig. 22–7). Similarly, snakes and lizards form a monophyletic group that is a sister group to birds, dinosaurs, and crocodiles. Finally, mammals and turtles form two additional groups.

Notice that in this interpretation, reptiles (snakes, lizards, crocodiles, dinosaurs, and turtles) are not recog-

(Text continues on page 504)

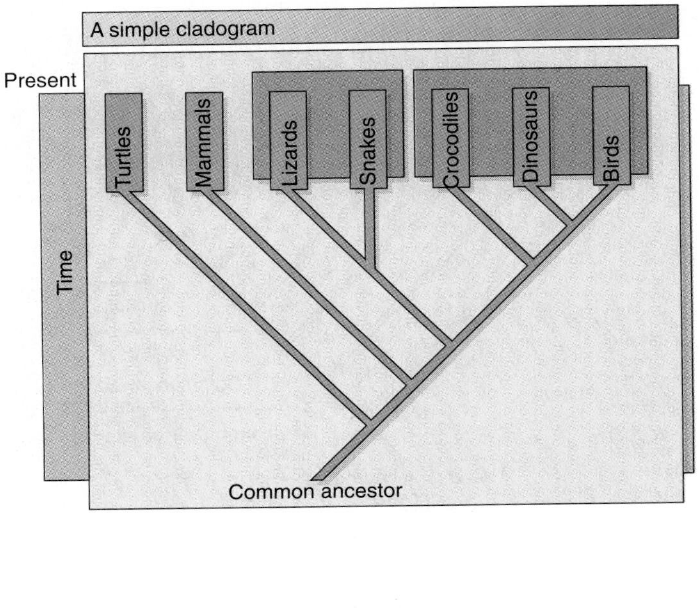

FIGURE 22–7 According to the cladistic approach, birds and some reptiles are classified together because they have a common ancestor.

Focus On

Building and Interpreting Cladograms

The general goal of cladistics is to reconstruct phylogenies using an analysis of the evolutionary changes in specific traits or characters. The kinds of characters used for the analysis can be structural, behavioral, physiological, or molecular. The only requirements are that the characters be homologous (identical by common descent) and that they have evolved independently of each other.

Collecting and preparing the data

The first step in constructing a cladogram is to select the taxa, which may consist of individuals, species, genera, or other taxonomic levels. Here, we use a representative group of eight chordates (Table 22–A).

Next, the homologous characters to be analyzed must be selected. In our example we use seven characters. For each character, we must define all of the different conditions, or states, as they exist in our taxa. For simplicity, we will consider our characters to have only two different states: present or absent. Keep in mind that many characters used in cladistics have more than two states. For example, black, brown, yellow, and red may be only a few of the many possible states for the character of hair color.

The last, and often the most difficult, step in preparing the data is to organize the character states into their correct evolutionary order. The most common method of accomplishing this task is by outgroup analysis. An **outgroup** is a taxon that is considered to have diverged earlier than any of the other taxa under investigation and thus to represent an approximation of the ancestral condition. In our example, amphioxus is the chosen out-

Table 22–A
A COMPARISON OF EIGHT CHORDATES.

A = Absent
P = Present

TAXA	CHARACTERS: Vertebrae (backbones)	Jaws	Tetrapod (4 limbs)	Amniotic egg	Mammary glands	Opposable thumb	Upright posture
Amphioxus (outgroup)	A	A	A	A	A	A	A
Hagfish	P	A	A	A	A	A	A
Sunfish	P	P	A	A	A	A	A
Newt	P	P	P	A	A	A	A
Lizard	P	P	P	P	A	A	A
Bear	P	P	P	P	P	A	A
Chimpanzee	P	P	P	P	P	P	A
Human	P	P	P	P	P	P	P

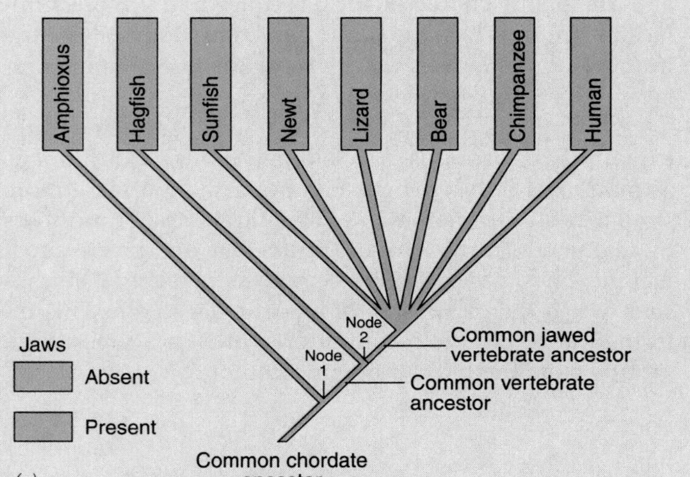

(a)

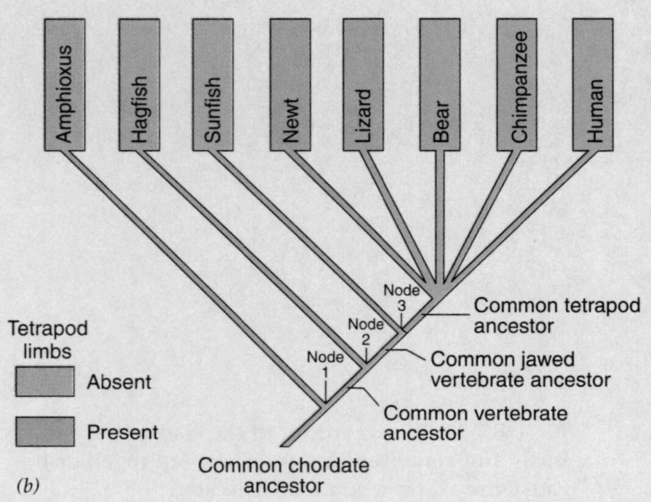

(b)

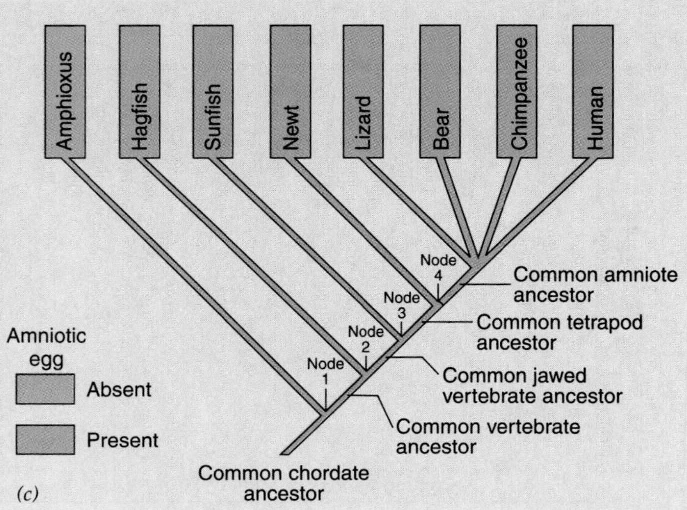

(c)

group. Therefore, the character state "absent" is the ancestral condition and the character state "present" is the derived condition for all seven characters (Table 22–A).

Constructing the cladogram from the data

Our objective is to find the cladogram that requires the fewest number of evolutionary changes in the characters. The reason for this criterion is that the cladogram requiring the least evolutionary change is also the most likely cladogram. In cladistics, taxa are grouped by the presence of shared derived character states. In order to form a valid monophyletic group, all members must share at least one derived character state. Membership in a group cannot be established by shared ancestral character states.

In our example, notice that all taxa except the outgroup possess vertebrae. We may therefore conclude that these seven vertebrate taxa form a valid monophyletic group. Next, among the seven vertebrate taxa, notice that jaws are present in all groups except for hagfish. Using these data, we may construct a preliminary cladogram (Figure *a*). The base of the tree represents the common ancestor for all taxa being analyzed and the branch points (referred to as nodes) represent the divergence of the ancestral lineage into two derived lineages, or clades. In Figure *a*, node 1 represents the divergence of the outgroup (amphioxus) and the common ancestor of the seven vertebrate taxa from the common chordate ancestor. Similarly, node 2 represents a subsequent divergence of hagfish and the ancestor of the six vertebrates with jaws. Continuing with this procedure, notice that among the six jawed taxa, all but sunfish are tetrapods (Figure *b*). Among the five tetrapods, all but newts have amniotic eggs (Figure *c*). The branching process is continued using Table 22–A data until all clades are established (Figure *d*).

How to interpret the cladogram

In Figure *d*, notice that humans and chimpanzees are more similar to each other than to any other clade. This relationship is indicated by the presence of a common ancestor at node 7. In the same way, bears are more similar to the human-chimpanzee clade than to any other clade as indicated by the common ancestor at node 6. In comparing the nodes, time of divergence is indicated by distance from the base of the tree. The further a node is located up the tree, the more recent the time of divergence. In our example, node 7 represents the most recent divergence, and node 1 represents the least recent (most ancient) divergence. Thus, in our example, humans are closely related to chimpanzees (through node 7) but more distantly related to bears (through node 6). Therefore, humans and chimpanzees would be assigned to a common taxon (order Primates) whereas humans, chimpanzees, and bears would be assigned to a broader, less exclusive taxon (class Mammalia). In addition, the cladogram reveals that lizards are more closely related to the clade Mammalia than to newts, sunfish or any other clade. Can you explain why?

In interpreting cladograms, two points must be kept in mind. First, the relationships between taxa can only be determined by tracing along the branches back to the most recent common ancestor (i.e., node) and not by the relative placement of the branches along the horizontal axis. It is possible to represent the same relationships with many different branching diagrams. For example, the cladogram in Figure *e* is equivalent to the one in Figure *d* (you should verify this by comparing the numbered nodes and by checking the relationships described earlier). Second, the cladogram does not establish ancestor-descendant relationships between taxa. In other words, a cladogram does not suggest that a taxon gave rise to any other taxon; rather it tells us which taxa shared a common ancestor and how recently they shared a common ancestor. The ancestor itself remains unspecified. ■

Essay contributed by Dr. John Beneski, Department of Biology, West Chester University, West Chester, PA

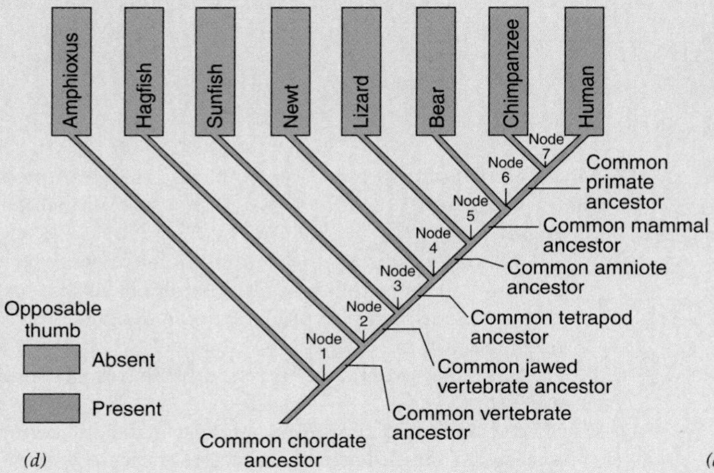

(*d*)

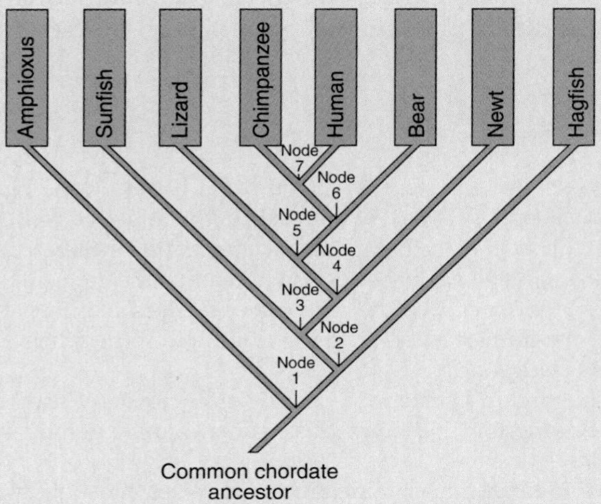

(*e*)

nized as a natural grouping because they do not form a monophyletic clade. One explanation for this apparent dilemma is that taxonomists who consider reptiles to be a valid group base their conclusion on different data. These taxonomists consider shared *ancestral* characters such as epidermal scales and **ectothermy** (fluctuation of the body temperature with the temperature of the surrounding environment). In contrast, cladists base their assessment on shared *derived* characters, such as heart structure, skeletal modifications, and reproductive behavior. The cladistic approach challenges us to reconsider and reevaluate our view of evolutionary relationships (see *Focus on: Building and Interpreting Cladograms*).

Classical evolutionary taxonomy uses phylogenetic trees

Classical evolutionary taxonomy (the approach used in this book) uses a system of phylogenetic classification and presents evolutionary relationships in phylogenetic trees. Classical taxonomists consider both evolutionary branching (as do cladists) and the extent of divergence that has occurred in a lineage since it branched from a stem group (Fig. 22–8). They base their classification decisions on shared ancestral characters.

A taxonomist using the classical approach might explain that porpoises are mammals rather than fish because they share many characteristics with other mammals and because these characteristics can be traced to a common ancestor. Organisms are classified in the same taxon according to their shared characteristics only if those traits are derived from a demonstrable common ancestor. The significance of the adaptations possessed by related organisms is also considered. If, for example, egg-laying mammals could be shown to have a very different ancestry from the other mammals, the classical taxonomist might erect a separate class to accommodate them. On the other hand, common ancestry, although necessary for inclusion in the same category, would not by itself be sufficient grounds for inclusion.

A classical taxonomist would classify birds and crocodiles separately, for example, even though they share a common ancestor. They would place birds in class Aves because they are endotherms and have feathers, wings, and other features that indicate extensive divergence since branching from the early reptiles. These taxonomists consider the shared ancestral characters, such as horny scales and ectothermy, of turtles, lizards, snakes, crocodiles, and dinosaurs,[2] and assign these animals to a separate class, Reptilia.

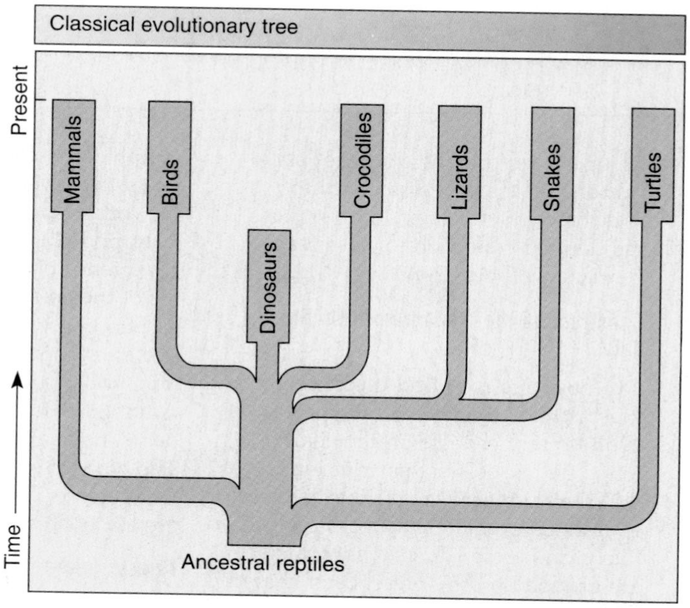

FIGURE 22–8 Classical evolutionary taxonomists consider both common ancestry and extent of divergence that has occurred since two taxa split. The branching points and degrees of difference in the evolution of the major groups of reptiles are depicted. Turtles, snakes, lizards, and crocodiles are most similar, but birds, dinosaurs, and crocodiles are most closely related because they branched most recently from a common ancestor.

[2]Recent evidence suggsts that many dinosaurs may have been endothermic.

SUMMARY

I. Biologists use a system of classification based on the binomial system developed by Linnaeus in the mid-18th century.
   A. In this system the basic unit of classification is the species.
   B. The name of each species has two parts: the genus name and the specific epithet. For example, the scientific name of the human is *Homo sapiens*, and that of the white oak is *Quercus alba*.

II. The hierarchical system of classification currently used includes kingdom, phylum, class, order, family, genus, and species.

III. The five-kingdom classification recognizes the kingdoms Prokaryotae, Protista, Fungi, Plantae, and Animalia.

IV. Modern taxonomy is based on similarity, as determined by shared characteristics, and on evolutionary relationships, or phylogeny.
   A. All of the organisms in a monophyletic taxon have a common ancestor that was also a member of that taxon; the organisms in a polyphyletic taxon evolved from different ancestors.
   B. Homologous structures imply evolution from a common ancestor.
   C. Shared ancestral characters suggest a distant common ancestor; shared derived characters indicate a more recent common ancestor.

D. Comparison of nucleic acid and protein structure provides a powerful tool for confirming evolutionary relationships.

V. Three main approaches to taxonomy are phenetics, cladistics, and classical evolutionary taxonomy.
A. The phenetic system is a numerical taxonomy based on similarities of many characters. Organisms are classified according to the number of characteristics they share without trying to determine whether their similarities are homologous or analogous.
B. The cladistic approach insists that taxa be monophyletic. Each taxon consists of a common ancestor and all of its descendants, and is based on shared derived characters.
C. Classical evolutionary taxonomy considers both evolutionary branching and the extent of divergence. Classical taxonomy is based on shared ancestral characters.

SELECTED KEY TERMS

ancestral character
binomial system
clade
cladistics
cladogram
class
classical evolutionary taxonomy
classification
derived character
family
genus (genera)
homologous structure
kingdom
kingdom Animalia
kingdom Fungi
kingdom Plantae
kingdom Prokaryotae
kingdom Protista
molecular clock
monophyletic taxon
order
phenetics
phylogeny
phylum (phyla)
polyphyletic taxon
species
specific epithet
subspecies
systematics
taxon
taxonomy

POST-TEST

1. The science of describing, naming, and classifying organisms is ___________.
2. Using the binomial system of nomenclature, the scientific name of each species consists of two parts, the ___________ and the ___________ epithet.
3. The mold that produces penicillin is *Penicillium notatum*. *Penicillium* is the name of its ___________.
4. Closely related genera may be grouped together in a single ___________.
5. Related classes are grouped together in the same ___________.
6. The kingdom that includes the algae is ___________.
7. Kingdom ___________ consists of decomposers such as molds and mushrooms.
8. The members of a(an) ___________ taxon have a common ancestor that is classified as a member of that group.
9. The presence of ___________ structures in different organisms suggests that they evolved from a common ancestor.
10. The porpoise and the human both have the ability to nurse their young, whereas the less closely related fish does not. The ability to nurse their young is a shared ___________ character of mammals.
11. The constancy in DNA and protein evolution permits biologists to use these macromolecules as molecular ___________.
12. ___________ is a numerical taxonomy based on phenotypic similarities.
13. Taxonomists who follow the ___________ school of taxonomy might classify crocodiles and birds in the same group based on recent common ancestry.
14. A system of classification in which both evolutionary branching and extent of divergence are considered is used by ___________ ___________ taxonomists.
15. The only taxon that exists in nature is the ___________.
16. Taxa containing organisms that do not share a common ancestor are described as ___________.

REVIEW QUESTIONS

1. Briefly describe the binomial system of nomenclature.
2. Describe in modern terms: (a) species; (b) class; (c) phylum.
3. What are the advantages of a "five-kingdom" system over a "two-kingdom" one? What types of organisms are especially difficult to assign a place in the taxonomic hierarchy?
4. In which kingdom would you classify each of the following? (a) an oak tree; (b) an amoeba; (c) *Escherichia coli* (a bacterium); (d) a tapeworm.
5. Compare the phenetic, cladistic, and classical evolutionary approaches to taxonomy.
6. Of what use to a taxonomist is knowledge of the amino acid sequences of various organisms?

YOU MAKE THE CONNECTION

1. What difficulties do we encounter when we attempt to use the concept of a species? Are members of a genus similar because they share a common ancestor, or do they belong to the same genus because they are similar? How might your answer vary depending on which approach to taxonomy you are following?
2. After many years of being considered "old-fashioned," the science of systematics has reemerged on the cutting edge of biological research. Why do you think this shift has occurred?

RECOMMENDED READINGS

Gaffney, E.S., L. Dingus, and M.K. Smith. "Why Cladistics?" *Natural History,* June, 1995. An interesting introduction to building cladograms using the dinosaurs as an example.

Margulis, L., and K.V. Schwartz. *Five Kingdoms: An Illustrated Guide to the Phyla of Life on Earth,* 2nd ed. W.H. Freeman, San Francisco, 1988. The great diversity of living things, beautifully illustrated.

May, Robert M. "How Many Species Inhabit the Earth?" *Scientific American,* Vol. 267, No. 4, October 1992. An argument for the importance of identifying and classifying the organisms that inhabit our planet; this information impacts on environmental issues.

Sibley, C.G., and J.F. Ahlquist. "Reconstructing Bird Phylogeny by Comparing DNAs." *Scientific American,* Vol. 254, No. 2, February 1986. An interesting account of a modern taxonomic method.

Wainright, P.O., G. Hinkle, M.L. Slogin, and S.K. Stickel. "Monophyletic Origins of the Metazoa: An Evolutionary Link with Fungi." *Science,* Vol. 260, April 16, 1993. The authors use comparisons of ribosomal RNA sequences to hypothesize evolutionary relationships of animals and fungi.

Whittaker, R.H. "New Concepts of Kingdoms of Organisms." *Science,* Vol. 163, 1969. A proposal for classifying organisms according to a five-kingdom system.

CHAPTER 23

Viruses and Bacteria

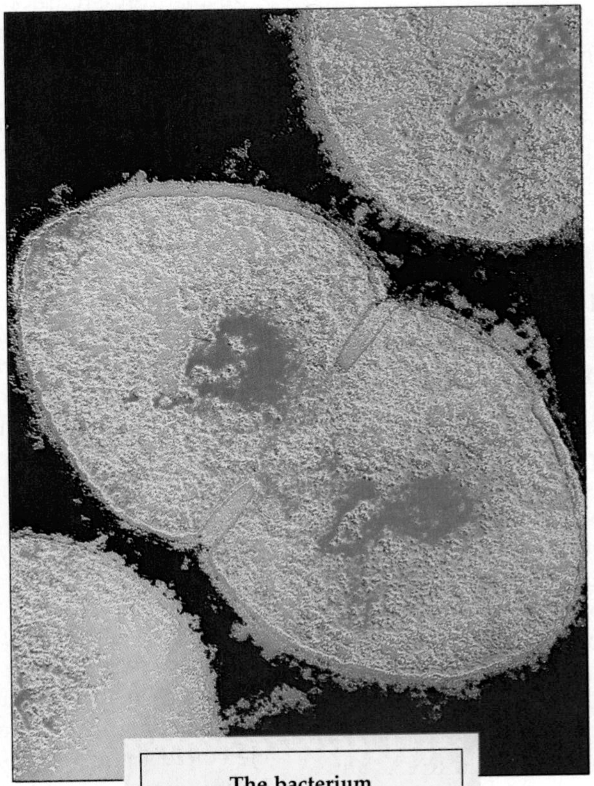

The bacterium *Streptococcus pyogenes* normally inhabits the human nose and throat. Sometimes referred to as killer bacteria, members of the strain shown dividing in this EM are resistant to antibiotics and infection can be fatal. (Dr. Kari Lounatmaa/Science Photo Library/Photo Researchers, Inc.)

Anton von Leeuwenhoek discovered bacteria and other microorganisms in 1674 when he looked at a drop of lake water through a glass lens. During the late 1800s, many microorganisms—bacteria, protozoa, and fungi—were identified as **pathogens,** agents that cause disease. About the same time, botanists were looking for the cause of tobacco mosaic disease, which stunts the growth of tobacco plants and gives the infected tobacco leaves a spotted, mosaic appearance. They found that the disease could be transmitted to healthy plants by daubing their leaves with the sap of diseased plants. In 1892 Ivanowsky, a Russian botanist, showed that the sap was infective even after it had been passed through filters fine enough to remove all known bacteria.

Early in the 20th century, scientists discovered infectious agents that could kill bacteria. Like the agents that caused tobacco mosaic disease, these pathogens passed through filters that removed known bacteria and were so small that they could not be seen with the light microscope. Curiously, they could not be grown in laboratory cultures unless living cells were present. These pathogens came to be known as **viruses,** infectious agents that lie on the threshold between life and nonlife. In addition to causing many plant diseases, viruses are responsible for a wide variety of diseases in humans and other animals; these include chickenpox, rabies, and AIDS.

In contrast to viruses, which consist only of nucleic acid and protein, **bacteria** are cellular organisms. Their prokaryotic cell structure is fundamentally different from the cells of other living organisms. For this reason they have been assigned to their own kingdom, Prokaryotae (also called kingdom Monera). Recently, some biologists have divided the bacteria into two kingdoms: Archaebacteria and Eubacteria (discussed later in this chapter). Although bacteria cause many diseases, including respiratory infections and food poisoning in humans, only a small minority of bacteria are pathogens.

Bacteria play an essential role in the biosphere as decomposers, breaking down organic molecules into their components. Along with fungi, bacteria are nature's recyclers. Without these microorganisms, the available carbon, nitrogen, phosphorus, and sulfur would eventually be tied up in the wastes and dead bodies of plants and animals. Life would soon cease to exist because of the lack of raw materials for the synthesis of new cellular components.

Bacteria also perform a key role in agriculture by fixing nitrogen. In this process, they change atmospheric nitrogen to a form that can be used by plants. Nitrogen fixation enables plants and animals (because they eat plants) to manufacture essential compounds such as proteins and nucleic acids.

This chapter examines the diversity and characteristics of viruses and bacteria. They are *not* a natural group of closely related organisms and we discuss them in a single chapter only for convenience.

LEARNING OBJECTIVES

After you have studied this chapter you should be able to

1. Describe the structure of a virus, and compare a virus with a free-living cell.
2. Characterize bacteriophages and contrast a lytic infection with a lysogenic infection.
3. Explain how a virus infects an animal cell.
4. Identify two viral infections of plants.
5. Speculate about the evolutionary origin of viruses.
6. Describe the distinguishing characteristics of members of the kingdom Prokaryotae, and contrast arachaebacteria with eubacteria.
7. Characterize the metabolic diversity of both autotrophic and heterotrophic bacteria, including aerobes, facultative anaerobes, and obligate anaerobes.
8. Summarize the three mechanisms (transformation, conjugation, and transduction) that may lead to genetic recombination in bacteria.
9. Distinguish among the following groups of bacteria: archaebacteria, wall-less bacteria, gram-negative bacteria, and gram-positive bacteria. Give examples of each.
10. Discuss the important ecological roles of bacteria, and their importance as pathogens.

VIRUSES ARE TINY, INFECTIOUS AGENTS THAT ARE NOT ASSIGNED TO ANY OF THE FIVE KINGDOMS

A **virus** is a tiny, infectious particle consisting of a nucleic acid core (its genetic material) surrounded by a protein coat called a **capsid.** Some viruses are also surrounded by an outer membranous envelope containing proteins, lipids, carbohydrates, and traces of metals.

Viruses lie on the threshold between life and nonlife, and, as such, are not true living organisms. Viruses are not cellular and cannot independently perform metabolic activities. They do not have the components necessary to carry on cellular respiration or to synthesize proteins and other molecules.

All living organisms contain both DNA and RNA, but a virus contains *either* DNA *or* RNA, not both. Viruses can reproduce, but only within the complex environment of the living cells that they infect. In a sense, viruses come "alive" only when they infect a cell. Viruses have sufficient genetic information to replicate their own nucleic acids, to make the protein coat, and to move in and out of the host cell.

Where did viruses come from? The hypothesis currently considered most likely is that viruses are bits of nucleic acid that "escaped" from cellular organisms. According to this view, some viruses may trace their origin to animal cells, others to plant cells, and still others to bacterial cells. Their multiple origins might explain why viruses are species-specific; perhaps they infect only those species that are the same as or closely related to the organisms from which they originated. This hypothesis is supported by the genetic similarity between a virus and its host cell—a closer similarity than exists between one virus and another.

The shape of a virus is determined by the organization of protein subunits that make up the capsid. Viral capsids are generally either helical, polyhedral, or a complex combination of both shapes (Fig. 23–1). Helical viruses, such as the tobacco mosaic virus, appear as long rods or threads. The capsid is a hollow cylinder that encloses the nucleic acid. Polyhedral viruses, such as the adenovirus (which causes human respiratory infection), are somewhat spherical. Viruses that infect bacteria may consist of a polyhedral "head" attached to a helical "tail."

Bacteriophages are viruses that attack bacteria

Among the most complex viruses are some of those that infect bacteria. These viruses are known as **bacteriophages** ("bacteria eaters"), or simply, **phages** (Fig. 23–1*d*). The most common phage structure consists of a long nucleic acid molecule (usually DNA) coiled within a polyhedral head. Many phages have a tail attached to the head. Fibers extending from the tail may be used to attach the virus to a bacterium.

Much of our knowledge of viruses has come from studying phages because they can be cultured easily within living bacteria in the laboratory. Phages were also used to determine that DNA, not protein, is the genetic material (see Chapter 11). Currently phages are used in genetic engineering research.

Lytic viruses destroy the host cell

Bacteriophages are either lytic or temperate. **Lytic** phages lyse (destroy) the host cell. When a lytic virus infects a susceptible host cell, it uses the host's metabolic machinery to replicate viral nucleic acid and produce viral proteins. Several steps in the process of viral infection are common to almost all phages (see Fig. 23–2):

1. **Attachment.** The phage attaches to the host cell wall.
2. **Penetration.** After the phage has attached to the cell surface, its tail contracts and pushes a hole through the cell wall; nucleic acid is then injected through the

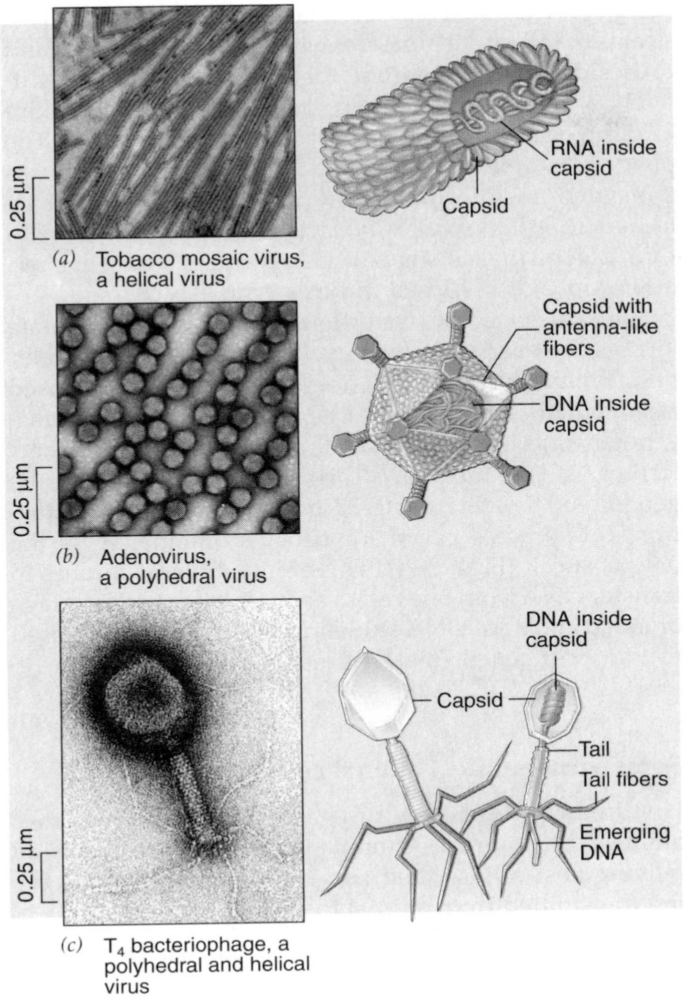

FIGURE 23–1 Viruses are generally either helical or polyhedral in shape, or a combination of both. (*a*) Tobacco mosaic virus has a helical capsid and appears rod-shaped. (*b*) Adenovirus has a polyhedral capsid with projecting protein spikes. (*c*) The bacteriophage known as T_4 has a polyhedral "head" and a helical "tail." (*a,* Visuals Unlimited/K.G. Murti; *b,* Visuals Unlimited/Hans G. Elderblom; *c,* Lee D. Simon/Science Source/Photo Researchers, Inc.)

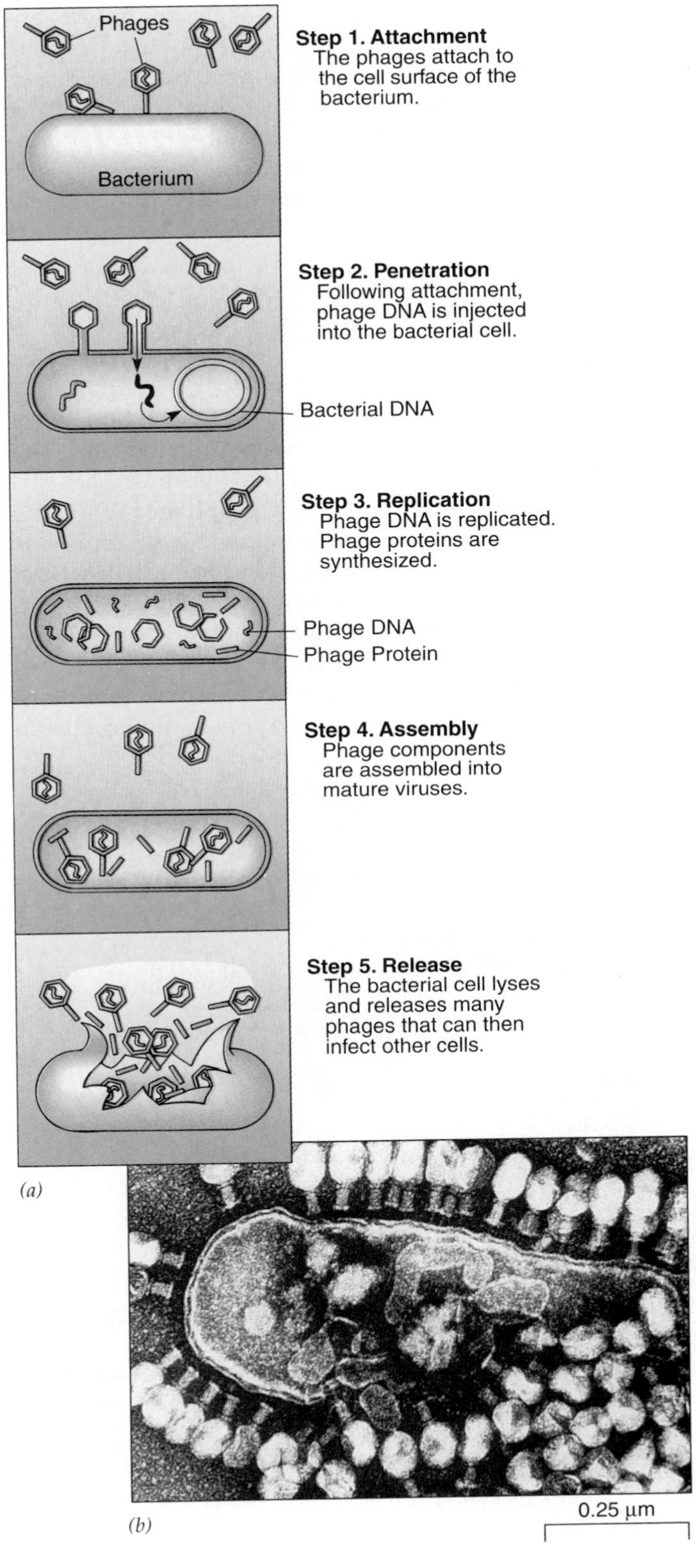

Phages
Bacterium
Step 1. Attachment
The phages attach to the cell surface of the bacterium.
Step 2. Penetration
Following attachment, phage DNA is injected into the bacterial cell.
Bacterial DNA
Step 3. Replication
Phage DNA is replicated. Phage proteins are synthesized.
Phage DNA
Phage Protein
Step 4. Assembly
Phage components are assembled into mature viruses.
Step 5. Release
The bacterial cell lyses and releases many phages that can then infect other cells.
(a)
(b)
0.25 µm

FIGURE 23–2 The host cell is destroyed in a lytic infection. (*a*) The sequence of events in a lytic infection is (1) attachment, (2) penetration, (3) replication, (4) assembly, and (5) release. (*b*) Phages infecting a bacterium, *Escherichia coli.* (*b,* Lee D. Simon/Science Source/Photo Researchers, Inc.)

plasma membrane and into the cytoplasm of the host cell. The capsid of a phage remains on the outside. (In contrast, most viruses that infect animal cells enter the host cell intact.)

3. **Replication.** Bacteriophage genes contain all the information necessary to produce new phages. Once inside, the phage DNA takes over the metabolic machinery of the cell. Using the host cell's ribosomes, its energy, and many of its enzymes, the phage synthesizes its own molecules.
4. **Assembly.** The newly synthesized viral components are assembled into new phages.

5. **Release.** In a lytic infection, the phage produces an enzyme that degrades the cell wall of the host cell. The host then lyses (ruptures), releasing about 100 phages. These new viruses infect other bacteria, and the process begins anew. The time required for viral multiplication, from the attachment to the bacterium to the release of phages, has been measured at approximately 30 to 35 minutes.

Temperate viruses can integrate their DNA into the host DNA

Unlike lytic viruses that lyse their host cells, **temperate** bacteriophages do not always destroy their hosts. The phage DNA becomes integrated into the host bacterial DNA, and is then referred to as a **prophage.** When the bacterial DNA replicates, the prophage also replicates (Fig. 23–3). The viral genes that code for viral structural proteins may be repressed indefinitely. Bacterial cells carrying prophages are referred to as **lysogenic cells.**

Bacterial cells containing certain temperate viruses may exhibit new properties. This is called **lysogenic conversion.** An interesting example involves the bacterium (*Corynebacterium diphtheriae*) that causes diphtheria. Two strains of this species exist, one that produces a toxin (and causes diphtheria) and one that does not. The only difference between these two strains is that the toxin-forming bacteria contain a specific temperate phage. The phage DNA encodes for the powerful toxin that causes the symptoms of diphtheria. In the same way, the bacterium (*Clostridium botulinum*) that causes botulism (a form of food poisoning) is harmless unless it contains certain prophage DNA that induces synthesis of toxin.

Certain external conditions can cause temperate viruses to revert to a lytic cycle and then destroy their host. When a lysogenic cell does lyse, the phages released may contain some bacterial DNA in place of their own genetic material. When such a phage infects a new bacterium, it can introduce this bacterial DNA into the genome of the new bacterial host. Known as **transduction,** this process permits genetic recombination in the new host cell (Fig. 23–4). This ability of some viruses to transfer DNA from one cell to another is taken advantage of in recombinant DNA studies in which viruses are used to transport genetic material inside a cell.

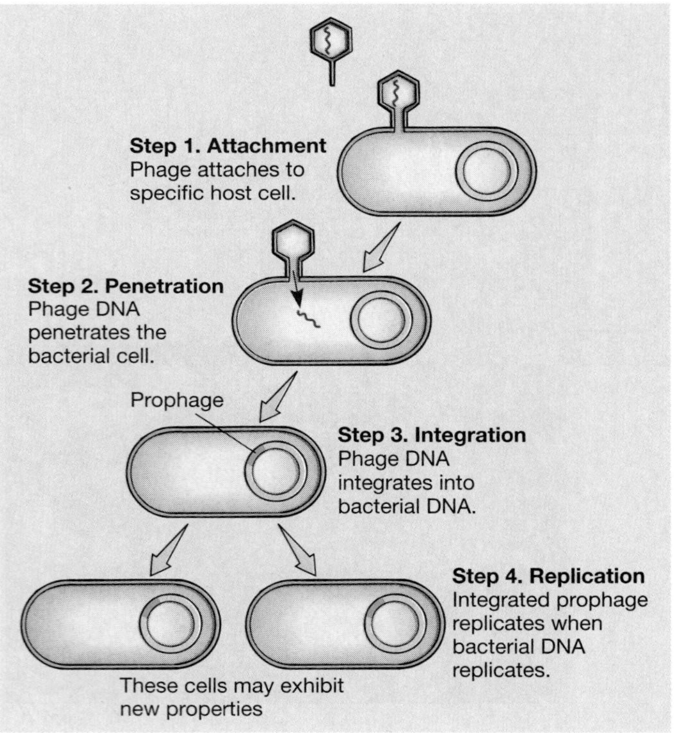
Step 1. Attachment
Phage attaches to specific host cell.
Step 2. Penetration
Phage DNA penetrates the bacterial cell.
Prophage
Step 3. Integration
Phage DNA integrates into bacterial DNA.
Step 4. Replication
Integrated prophage replicates when bacterial DNA replicates.
These cells may exhibit new properties

FIGURE 23–3 Temperate phages can integrate their nucleic acid into the host cell DNA, making it a lysogenic cell. The sequence of events in a lysogenic infection includes (1) attachment, (2) penetration, (3) integration into bacterial DNA, and (4) replication when the bacterial DNA replicates.

Some viruses infect animal cells

Hundreds of different viruses infect humans and other animals. Most viruses cannot survive very long outside a living host cell, so their survival depends on their being transmitted from animal to animal.

The type of attachment proteins on the surface of a virus determines what type of cell it can infect. Some viruses, such as the adenoviruses, have fibers that project from the capsid and are thought to help the virus adhere to complementary receptor sites on the host cell. Other viruses, such as those that cause herpes, influenza, and rabies, are surrounded by a lipoprotein envelope with projecting glycoprotein spikes that aid in attachment to a host cell.

Receptor sites vary with each species and sometimes with each type of tissue. Thus, some viruses can infect only humans, because their attachment proteins combine only with receptor sites found on human cell surfaces. The measles virus and pox viruses can infect many types of tissue because their attachment proteins combine with receptor sites on a variety of cells. In contrast, polioviruses can attach only to certain types of cells such as motor neurons in the brain and spinal cord.

Viruses have several ways to penetrate animal cells. After attachment to a host-cell receptor, some enveloped viruses fuse with the animal cell's plasma membrane. The viral capsid and nucleic acid are both released into the animal cell. Other enveloped viruses and naked viruses enter the host cell by endocytosis. In this process, the animal cell plasma membrane invaginates to form a membrane-bound vesicle that contains the virus (Fig. 23–5).

Like other viruses, those that infect animal cells multiply and produce new virus particles. While viral nucleic acid is replicated and viral proteins are synthesized, the synthesis of host DNA, RNA, and protein is inhibited.

Animal viruses can contain either DNA or RNA. In DNA viruses, the synthesis of viral DNA and protein is similar to the processes by which the host cell would normally carry out its own DNA and protein synthesis. In most RNA viruses, transcription takes place with the help of an RNA polymerase. However, **retroviruses** are RNA viruses that use a DNA polymerase called **reverse transcriptase** to transcribe the RNA genome into a DNA intermediate (see Chapter 12). This DNA is then used to synthesize copies of the viral RNA. The human immunodeficiency virus (HIV) that causes AIDS is a retrovirus. Certain cancer-causing viruses are also retroviruses.

After the viral genes are transcribed, the viral structural proteins are synthesized. The capsid is produced and new virus particles are assembled. Viruses that do not have an outer envelope exit by cell lysis. The plasma membrane ruptures, releasing the viral particles. Enveloped viruses receive their lipoprotein envelopes as they pass through the plasma membrane (or, in some types, the nuclear envelope). Because they are released slowly (by a process called budding), these viruses do not usually destroy the host cell as they exit.

Viral proteins damage the host cell in a variety of ways. These proteins may alter the permeability of the plasma membrane or may inhibit synthesis of host nucleic acids or proteins. Viruses sometimes damage or kill their host cells by their sheer numbers. A poliovirus may produce 100,000 new viruses within a single host cell!

Most of us suffer from two to six viral infections each year. Fortunately, many of these are relatively benign forms such as the common cold. More serious human diseases caused by viruses include chickenpox, herpes simplex (one type causes genital herpes), mumps, rubella (German measles; Fig. 23–5), rubeola (measles), rabies, warts, infectious mononucleosis, influenza, hepatitis, AIDS, and Ebola (Table 23–1). Other animal viruses are known to cause certain types of cancer (for example, feline leukemia), hog cholera, foot-and-mouth disease, canine distemper, and swine influenza.

Some viruses infect plants

Viral diseases can be spread among plants by insects such as aphids and leafhoppers as they feed on plant tissues. Plant viruses can be inherited by way of infected seeds or by asexual propagation.

Once a plant is infected, the virus spreads through the plant body by passing through the cytoplasmic connec-

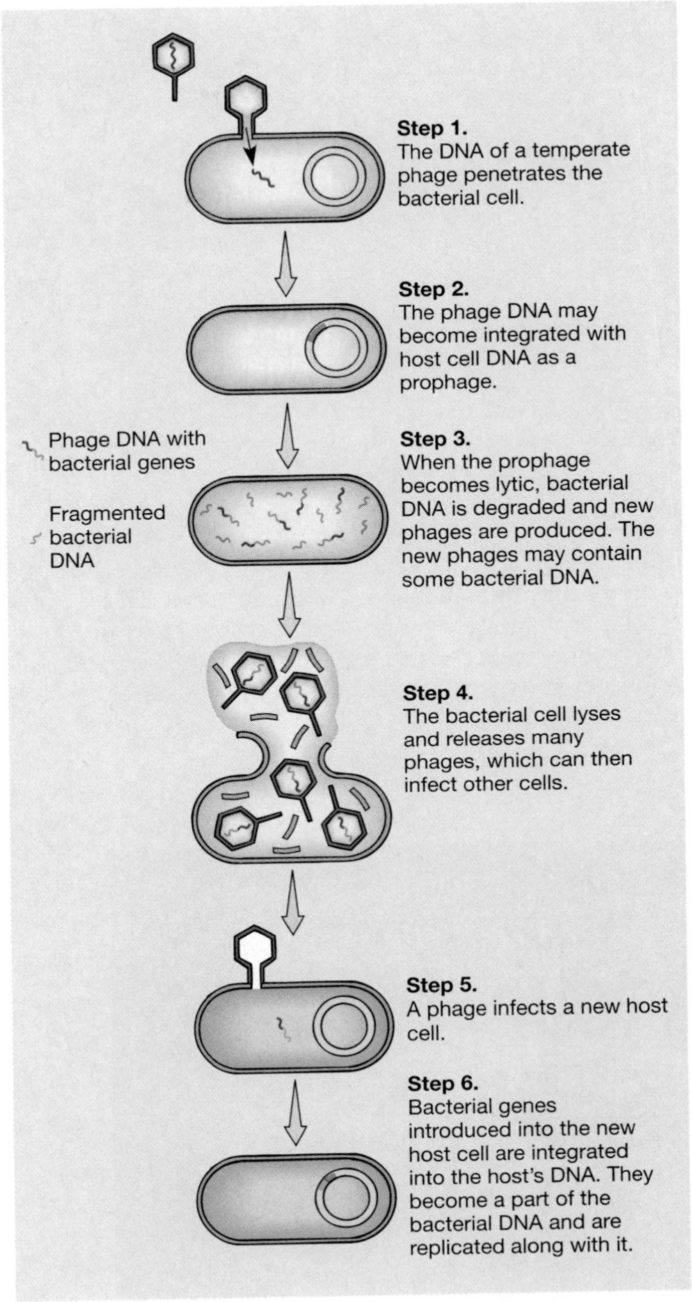

Step 1.
The DNA of a temperate phage penetrates the bacterial cell.
Step 2.
The phage DNA may become integrated with host cell DNA as a prophage.
Phage DNA with bacterial genes
Fragmented bacterial DNA
Step 3.
When the prophage becomes lytic, bacterial DNA is degraded and new phages are produced. The new phages may contain some bacterial DNA.
Step 4.
The bacterial cell lyses and releases many phages, which can then infect other cells.
Step 5.
A phage infects a new host cell.
Step 6.
Bacterial genes introduced into the new host cell are integrated into the host's DNA. They become a part of the bacterial DNA and are replicated along with it.

FIGURE 23–4 A phage can transfer bacterial DNA from one bacterium to another in the process called transduction. (1) The DNA of the temperate phage penetrates the bacterium. (2) The phage DNA becomes a prophage. (3) Bacterial DNA is degraded and new phages are produced. (4) The bacterial cell lyses, releasing many phages. (5) A phage infects a new bacterium. (6) The phage introduces bacterial DNA into the new host cell.

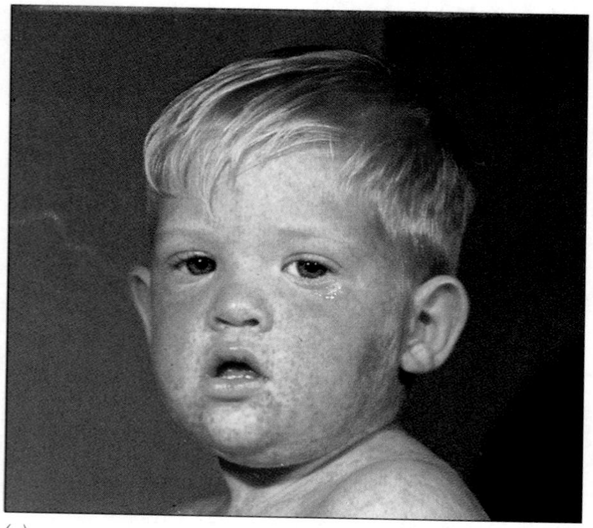

(a)

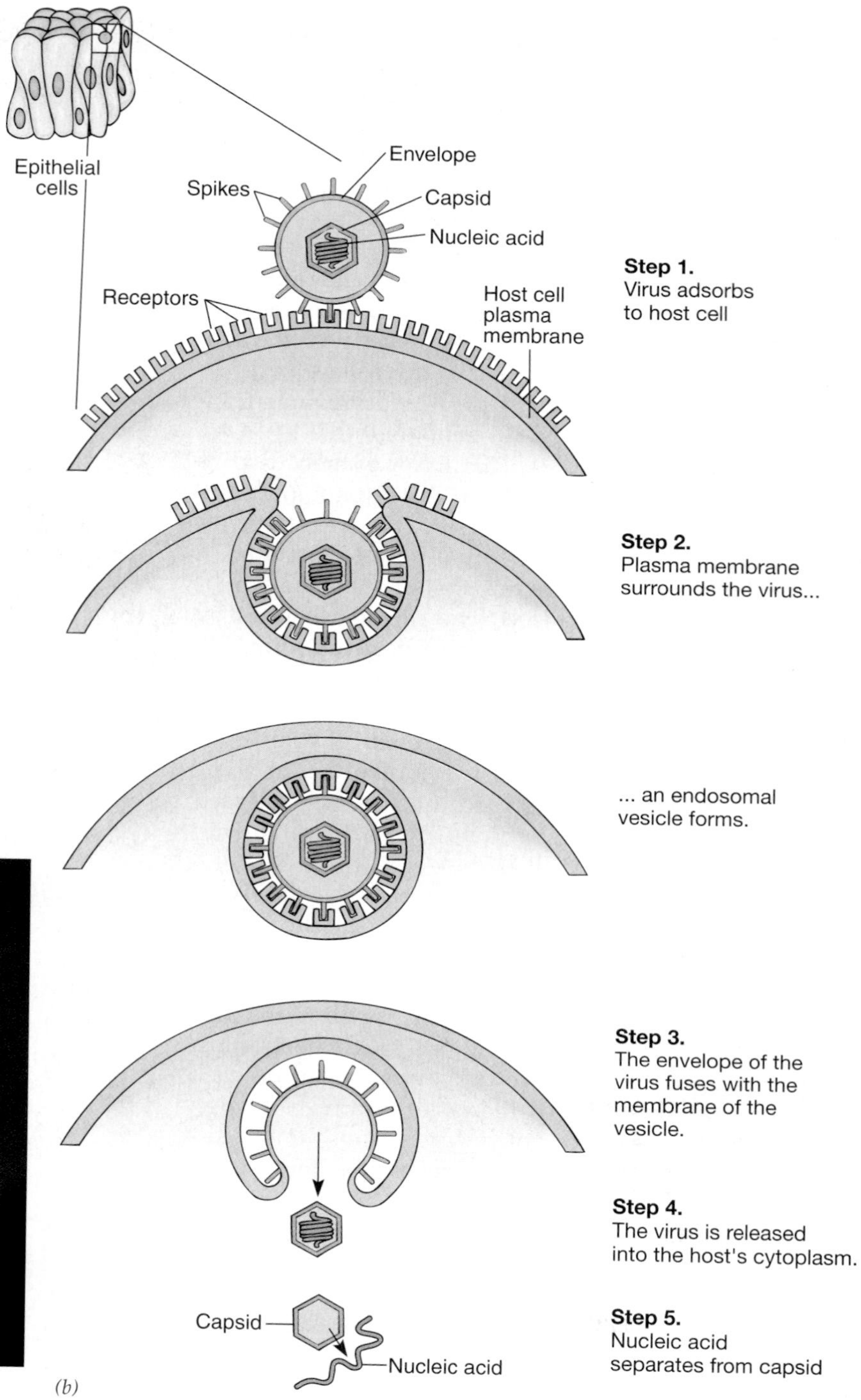
Epithelial cells
Spikes
Envelope
Capsid
Nucleic acid
Receptors
Host cell plasma membrane
Step 1.
Virus adsorbs to host cell
Step 2.
Plasma membrane surrounds the virus...
... an endosomal vesicle forms.
Step 3.
The envelope of the virus fuses with the membrane of the vesicle.
Step 4.
The virus is released into the host's cytoplasm.
Capsid
Nucleic acid
Step 5.
Nucleic acid separates from capsid

(b)

FIGURE 23–5 Many diseases in humans and other animals are caused by viruses. (*a*) Rubella (German measles) is caused by an RNA virus spread by close contact. When contracted during pregnancy, it can cause birth defects. Vaccination has greatly decreased the incidence of this disease. (*b*) Some viruses enter animal cells by endocytosis. (1) A virus combines with receptors on host cell. (2) The plasma membrane surrounds the virus, forming an endosome. (3) The virus envelope fuses with the endosome membrane. (4) The virus is released into the host cytoplasm. (5) The nucleic acid separates from the capsid. (*a*, Centers for Disease Control and Prevention, U.S. Department of Health and Human Services)

Focus On

Viroids and Prions

The discovery of infective agents even smaller than viruses has challenged old ideas. **Viroids** cause a variety of plant diseases and may also infect animals. Each viroid consists of a very short strand of RNA (only 250–400 nucleotides) with no protective coat. Until viroids were studied, most biologists assumed that protein was necessary for an infectious agent to duplicate itself. However, no proteins are associated with viroids, and evidence suggests that the viroid genome does not code for any proteins. The viroid uses its host's enzymes to replicate its RNA. Viroids are generally found within the host cell nucleus, but how they cause disease is not clear at this time.

Even more heretical is the idea that a pathogen could exist without nucleic acids. Certain degenerative brain diseases in sheep, cattle, and humans are apparently caused by a structure called a **prion.** These tiny agents may be responsible for certain inherited, as well as infectious, diseases.

The prion, a protein-like infectious particle, appears to consist only of a glycoprotein. The glycoprotein contains at least one polypeptide about 250 amino acids long. Despite extensive searching, no nucleic acid component has been found.

Because nucleic acids are the molecules replicated during cell division and reproduction, exactly how prions multiply is of great biological interest. Stanley Prusiner, professor of neurology and biochemistry at the University of California School of Medicine, San Francisco, and other researchers suggest that prions multiply by converting normal protein molecules in brain neurons to disease-causing prion proteins by changing their shape.

The diseases caused by prions are all fatal. These diseases are sometimes called spongiform encephalopathies because the infected brain appears to develop holes, becoming somewhat spongelike. The best studied prion disease is scrapie in sheep and goats. When infected, animals lose coordination, become irritable, and itch so severely that they scrape off their wool or hair. Researchers suggest that prions may be responsible for some human neuromuscular diseases as well. ■

tions (plasmodesmata; see Fig. 5–26) that penetrate the walls between adjacent cells. Most plant viruses contain RNA. After infecting a host cell, the viral RNA attaches to the host's ribosomes and is translated as though it were mRNA.

Symptoms of viral infection include reduced plant size, and spots, streaks, or mottled patterns on leaves, flowers, or fruits (Fig. 23–6). Infected crops almost always produce lower yields. Cures are not known for most viral diseases of plants, and so it is common to burn plants

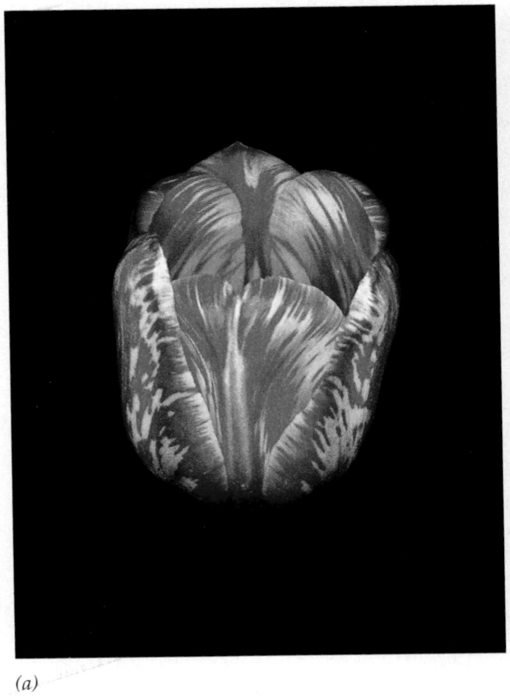

(a)

(b)

FIGURE 23–6 Many types of viruses infect plants. (*a*) Virus-streaked tulips. The virus that causes this relatively harmless disease affects pigment formation in the petals. (*b*) Tobacco leaf infected with the tobacco mosaic virus. The leaf is characteristically mottled with light green areas. (Kenneth M. Corbett)

Table 23–1 ANIMAL VIRUSES

Group	Diseases Caused
DNA Viruses	
Poxviruses	Smallpox, cowpox, and economically important diseases of domestic fowl
Herpesviruses	Herpes simplex type 1 (cold sores); herpes simplex type 2 (genital herpes, a sexually transmitted disease); varicella-zoster (chickenpox and shingles). The Epstein–Barr virus causes infectious mononucleosis and Burkitt's lymphoma
Adenoviruses	About 40 types known to infect human respiratory and intestinal tracts; common cause of sore throat, tonsillitis, and conjunctivitis; other varieties infect other animals
Papovaviruses	Human warts and some degenerative brain diseases; some cancers
Parvoviruses	Infections in dogs, swine, arthropods, rodents; cause gastroenteritis in humans after eating infected shellfish
RNA Viruses	
Picornaviruses	About 70 types infect humans, including polioviruses; hepatitis A virus; enteroviruses infect intestine; rhinoviruses infect respiratory tract and are main cause of human colds; coxsackievirus and echovirus cause aseptic meningitis
Togaviruses	Rubella, yellow fever, encephalitis
Orthomyxoviruses	Influenza in humans and other animals
Paramyxoviruses	Rubeola (measles), mumps, distemper in dogs
Rhabdoviruses	Rabies
Reoviruses	Vomiting and diarrhea
Retroviruses	AIDS, some types of cancer

that have been infected. Some agricultural scientists are focusing their efforts on prevention of viral disease by developing virus-resistant strains of important crop plants.

BACTERIA MAKE UP KINGDOM PROKARYOTAE

Bacterial cells are typically very small (See *Focus on Giant Bacteria* for an exception.). Their cell volume is only about one thousandth that of small eukaryotic cells, and their length is only about one tenth. Most prokaryotes are unicellular organisms, but some form colonies or filaments containing specialized cells.

Bacteria have three main shapes

Although thousands of kinds of bacteria are known, they have three main shapes: spherical, rod-shaped, and spiral (Fig. 23–7). Spherical bacteria, known as **cocci** (sing., *coccus*), occur singly in some species and in groups of independent cells in others. Cells may be grouped in twos (diplococci), in long chains (streptococci), or in irregular clumps that look like bunches of grapes (staphylococci). Rod-shaped bacteria, called **bacilli** (sing., *bacillus*), may occur as single rods or as long chains of rods. Spiral, or helical, bacteria are known as **spirilla** (sing., *spirillum*).

Prokaryotic cells lack membrane-bounded organelles

Recall that prokaryotic cells contain ribosomes but lack membrane-bounded organelles typical of eukaryotic cells. Thus, bacterial cells have no nuclei, no mitochondria, no chloroplasts, no endoplasmic reticulum, no Golgi complex, and no lysosomes (Fig. 23–8).

The dense cytoplasm of the bacterial cell contains ribosomes and storage granules that hold glycogen, lipid, or phosphate compounds. Most enzymes needed for metabolic activities are located in the cytoplasm.

Characteristics
Large, complex double-stranded DNA; replicate in the cytoplasm of the host cell. The vaccinia virus is used to produce genetically engineered vaccines
Medium to large, enveloped viruses; double-stranded DNA; replicate in host nucleus; frequently cause latent infections; some cause tumors
Double-stranded DNA; replicates in host nucleus
Double-stranded DNA; the virus SV40 has been used as a vector to transport genes into cells
Single-stranded DNA; some require a helper virus in order to multiply
Diverse group of small viruses; single-stranded RNA that can serve as mRNA
Large, diverse group of medium-sized, enveloped viruses; single-stranded RNA that can serve as mRNA; many transmitted by arthropods
Medium-sized viruses that often exhibit projecting spikes; single-stranded RNA that serves as template for mRNA
Resemble orthomyxoviruses but somewhat larger; related to the Ebola virus
Single-stranded RNA
Double-stranded RNA; no envelope
RNA viruses that contain reverse transcriptase for transcribing the RNA genome into DNA; two identical molecules of single-stranded RNA

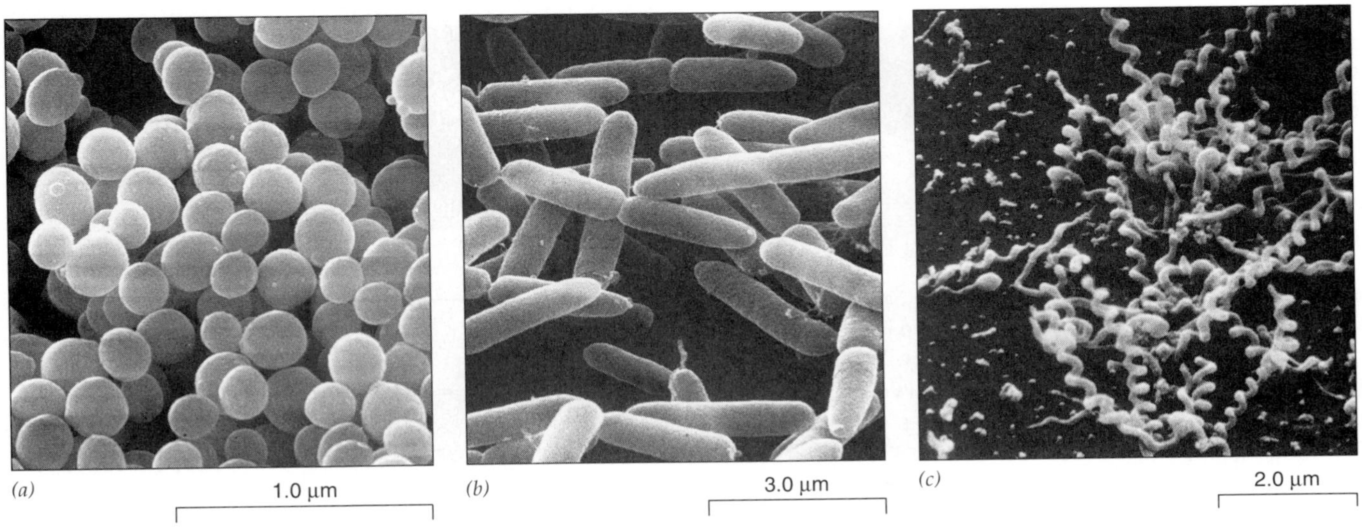

FIGURE 23–7 Bacteria have three main shapes. (*a*) Cocci, like these *Micrococcus* bacteria, are spherical. (*b*) *Salmonella* are bacilli, rod-shaped bacteria. (*c*) Spiral, or helical, bacteria, like the *Spiroplasma* shown here, are called spirilla.
(Visuals Unlimited/David M. Phillips)

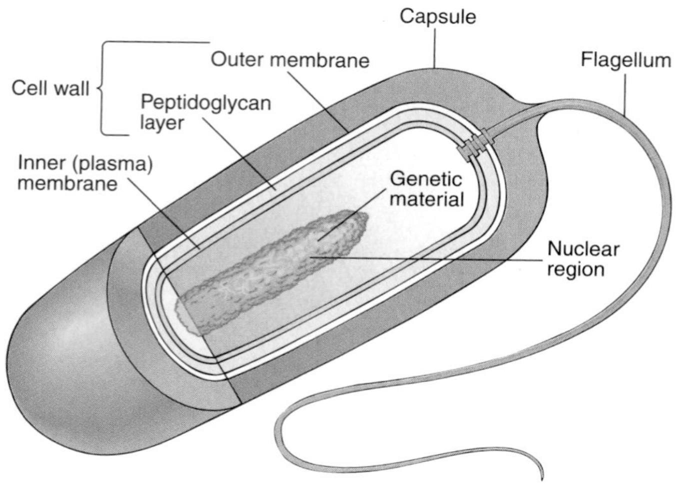

FIGURE 23–8 Bacteria are prokaryotic cells. This bacillus is a gram-negative bacterium (discussed in text). Note the absence of a nuclear envelope surrounding the bacterial DNA.

Although the membranous organelles of eukaryotic cells are absent, in some bacterial cells the plasma membrane is elaborately folded inward. Enzymes needed for cellular respiration, photosynthesis, or nitrogen fixation may be attached to the plasma membrane or its folds.

The cell surface is generally covered by a cell wall

Most prokaryotic cells have a cell wall surrounding the plasma membrane, but its structure and composition differ from those of eukaryotic cell walls. The cell wall provides a rigid framework that supports the cell, maintains its shape, and keeps it from bursting because of osmotic pressure (see Chapter 5). (Most bacteria seem to be adapted to hypotonic surroundings.) Normally, bacteria cannot survive without their cell walls. When wall-less forms are produced experimentally, they must be maintained in isotonic solutions to keep them from bursting. However, cell walls are of little help when the bacterium is in a hypertonic environment, as found in food preserved by means of a high sugar or salt content. That is why most bacteria grow poorly in jellies, jams, salted fish, and other foods preserved in these ways.

The bacterial cell wall is composed of **peptidoglycan,** a complex organic molecule that consists of two unusual types of sugars linked with short peptides. The sugars and peptides are linked to form a single macromolecule that surrounds the entire plasma membrane.

Differences in bacterial cell wall composition are of great interest to biologists and are important clinically. In 1884 the Danish physician Christian Gram developed the Gram staining procedure. Bacteria that absorb and retain crystal violet stain in the laboratory are referred to as **gram-positive,** whereas those that do not retain the stain are **gram-negative.** The cell walls of gram-positive bacteria are very thick and consist primarily of peptidoglycan. The cell walls of gram-negative bacteria consist of two layers, a thin peptidoglycan wall and a thick outer membrane containing carbohydrates bonded to lipids (Fig. 23–9).

Distinguishing between gram-positive and gram-negative bacteria is important in treating certain diseases.

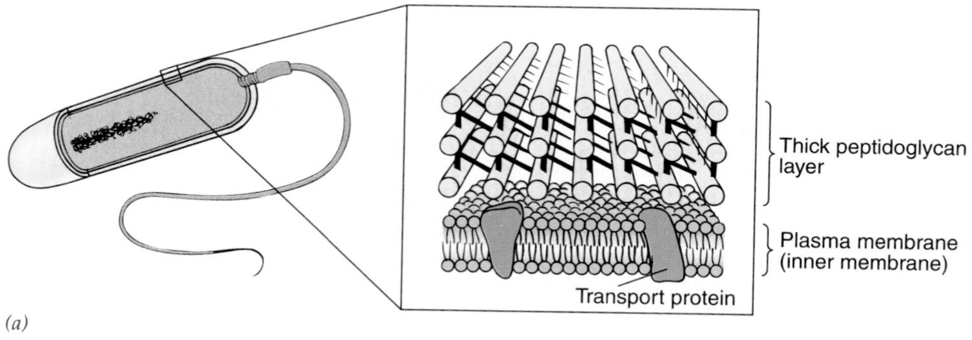

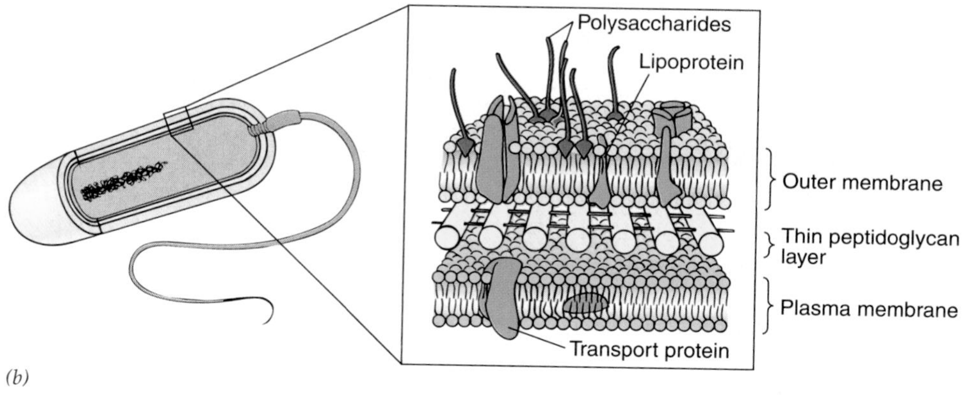

FIGURE 23–9 The bacterial cell wall has a unique structure. (***a***) In the gram-positive cell wall, many layers of peptidoglycan are held together by amino acids. (***b***) In the gram-negative cell wall, a thin peptidoglycan layer is covered by a thick outer membrane.

Focus On

Giant Bacteria

The generalization that all bacteria are microscopic—far too small to be viewed with the naked eye—was long accepted by biologists. When investigators discovered a new organism (*Epulopiscium fishelsoni*) in the intestine of surgeonfish caught in the Red Sea and off the Great Barrier Reef in Australia, they thought it was too large to be a bacterium. After all, it was about 600 μm long and 80 μm wide, about a million times larger than a typical bacterium. They guessed that it must be a protozoon—a single-celled organism (such as *Paramecium*) belonging to kingdom Protista. Investigators then studied the structure of this new organism with the electron microscope. They found that it lacked a distinct nucleus and had other structural features more typical of bacteria than protists. This new information challenged old beliefs. Could this organism be a giant bacterium?

Esther Angert, a graduate student at Indiana University, and her colleagues carried out a molecular study in which they compared gene sequences in *E. fishelsoni* to those in other microorganisms. Results of these molecular studies by Angert and her coworkers (reported in the journal *Nature* in 1993) confirmed that these organisms are indeed bacteria—the largest bacteria ever imagined by biologists. The newly discovered bacteria live within the intestine of surgeonfish. The fish graze on algae, and the bacteria are thought to digest the algae.

The discovery of the giant bacteria illustrates some important features of the scientific method. Scientists must be willing to have their most strongly held beliefs challenged. When new evidence casts doubt on an accepted theory, they are willing to reexamine their assumptions and modify the theory. Thus, science is a self-correcting process. Its method encourages continuous observing, questioning, recognizing problems, developing hypotheses, making predictions, testing, and, when indicated, modifying accepted theory. ■

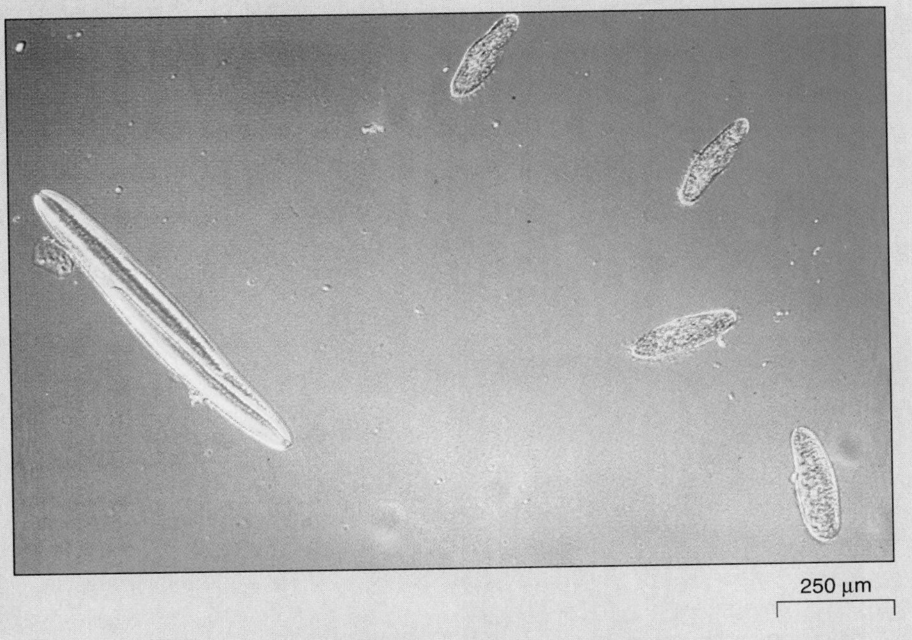

This giant bacterium (*Epulopiscium*) was photographed with four paramecia. The bacterium is about a million times larger than a typical bacterium. (Esther R. Angert and Norman R. Pace, Indiana University)

For example, the antibiotic penicillin interferes with peptidoglycan synthesis, ultimately resulting in a fragile cell wall that cannot protect the cell (see Chapter 6, *Focus On: Enzyme Inhibition and Antibacterial Drugs*). Predictably, penicillin works most effectively against gram-positive bacteria.

Some species of bacteria produce a **capsule** or slime layer that surrounds the cell wall. In free-living species, the capsule may provide the cell with added protection against phagocytosis by other microorganisms. In disease-causing bacteria, the capsule may protect against phagocytosis by the host's white blood cells. For example, the ability of *Streptococcus pneumoniae* to cause bacterial pneumonia depends on its capsule. A strain of *S. pneumoniae* that lacks a capsule does not cause the disease. Bacteria also use their capsules to attach to surfaces such as rocks, plant roots, or human teeth (where they cause dental plaque).

Some bacteria have hundreds of hairlike appendages known as **pili** (sing., *pilus*). These structures are organelles of attachment that help the bacteria adhere to certain surfaces, such as the cells they infect. Some pili are involved in the transmission of DNA between bacteria.

Many types of bacteria are motile

Can you imagine trying to swim through molasses? Water has the same relative viscosity to bacteria that molasses has to humans. Some species of bacteria secrete slippery compounds and move through water by gliding. Others move by means of whip-like **flagella.** However, the structure of prokaryotic flagella is quite different from that of

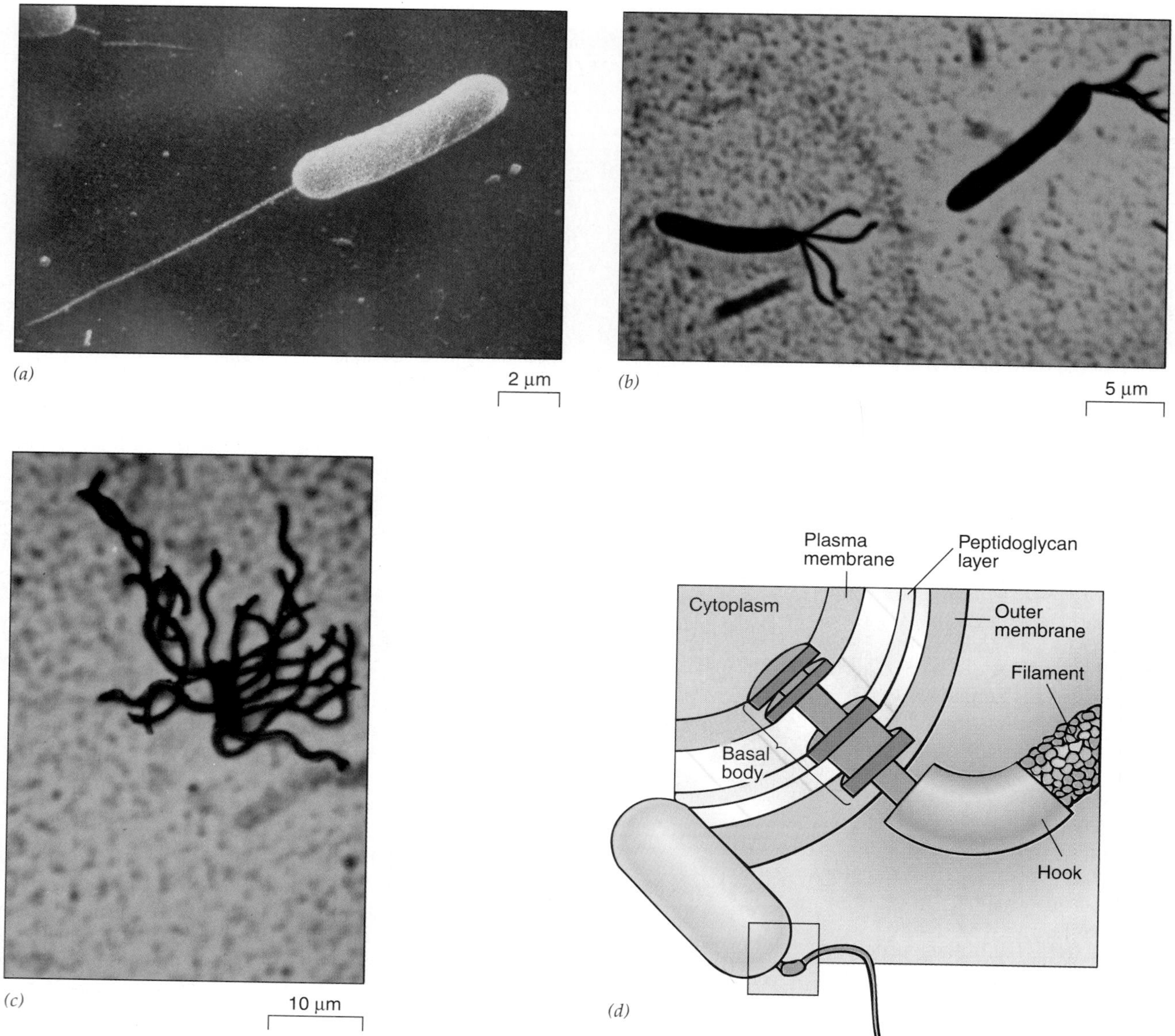

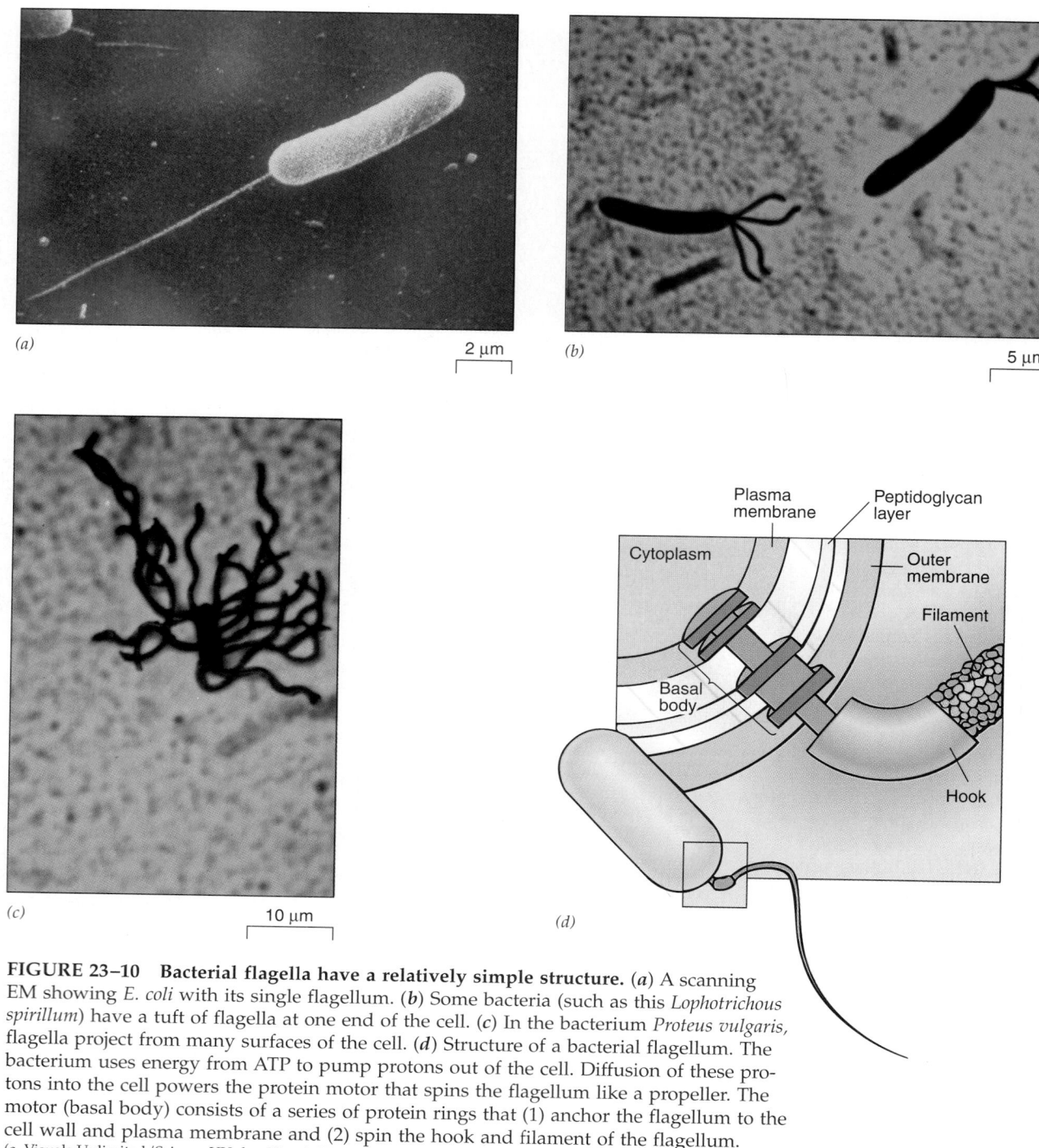

FIGURE 23–10 Bacterial flagella have a relatively simple structure. (***a***) A scanning EM showing *E. coli* with its single flagellum. (***b***) Some bacteria (such as this *Lophotrichous spirillum*) have a tuft of flagella at one end of the cell. (***c***) In the bacterium *Proteus vulgaris,* flagella project from many surfaces of the cell. (***d***) Structure of a bacterial flagellum. The bacterium uses energy from ATP to pump protons out of the cell. Diffusion of these protons into the cell powers the protein motor that spins the flagellum like a propeller. The motor (basal body) consists of a series of protein rings that (1) anchor the flagellum to the cell wall and plasma membrane and (2) spin the hook and filament of the flagellum. (*a,* Visuals Unlimited/Science VU; *b,c,* Visuals Unlimited/E. C. S. Chan)

eukaryotic flagella (see Chapter 4). Bacterial flagella consist of three parts: a basal body, a hook, and a filament (Fig. 23–10). The basal body is a complex structure that produces a rotary motion, pushing the cell much as a propeller pushes a ship through water. The hook connects the basal body to the hollow filament, which consists of several protein chains twisted into a helical structure.

Bacteria have a single DNA molecule

The genetic material of a bacterium lies in the cytoplasm and is not surrounded by a nuclear envelope. It is contained in a single circular DNA molecule. If stretched out to its full length, this molecule would be about 1000 times longer than the cell itself. Unlike eukaryotic chromosomes, the bacterial DNA has little protein associated with it.

In addition to the main bacterial DNA, a small amount of genetic information may be present as smaller DNA circles, called plasmids, which replicate independently of the main DNA (see Chapter 14). Bacterial plasmids often bear genes that confer resistance to antibiotics.

Bacteria reproduce by binary fission

Bacteria generally reproduce asexually by **binary fission,** a process in which one cell divides into two similar cells. First the circular bacterial DNA replicates, and then a transverse wall is formed by an ingrowth of both the plasma membrane and the cell wall.

Binary fission occurs with remarkable speed; under ideal conditions some bacteria divide every 20 minutes. At this rate, if nothing interfered, one bacterium would give rise to more than *one billion* bacteria within 10 hours! However, bacteria cannot reproduce at this rate for very long, because their growth is eventually affected by lack of food or by the accumulation of waste products.

Although sexual reproduction involving the fusion of gametes does not occur in bacteria, genetic material is sometimes exchanged between individuals. This exchange takes place by three different mechanisms: transformation, transduction, and conjugation.

(1) In **transformation,** fragments of DNA released by a broken cell are taken in by another bacterial cell. This mechanism was used experimentally to demonstrate that genes can be transferred from one bacterium to another and that DNA is the chemical basis of heredity (see Chapter 11).
(2) In the second process of gene transfer, **transduction,** a phage carries bacterial genes from one bacterial cell into another (see earlier section on lysogenic infections).
(3) In **conjugation,** two cells of different mating types come together, and genetic material is transferred from one to the other (Fig. 23–11). In contrast to transformation and transduction, conjugation involves contact between two cells.

Conjugation has been most extensively studied in the bacterium *Escherichia coli.* In the *E. coli* population there are male, or donor cells, and female, or recipient cells. DNA is transferred from donor to recipient cell. Often, a plasmid rather than the main bacterial DNA, is transferred. In one strain of *E. coli,* a plasmid, known as the F (fertility) plasmid, codes for the synthesis of protein used to form hollow pili called *F pili.* These serve as conjugation bridges that pass from the donor to the recipient cell. The F pili are long and narrow and have a hole through which fragments of DNA pass from one bacterium to the other.

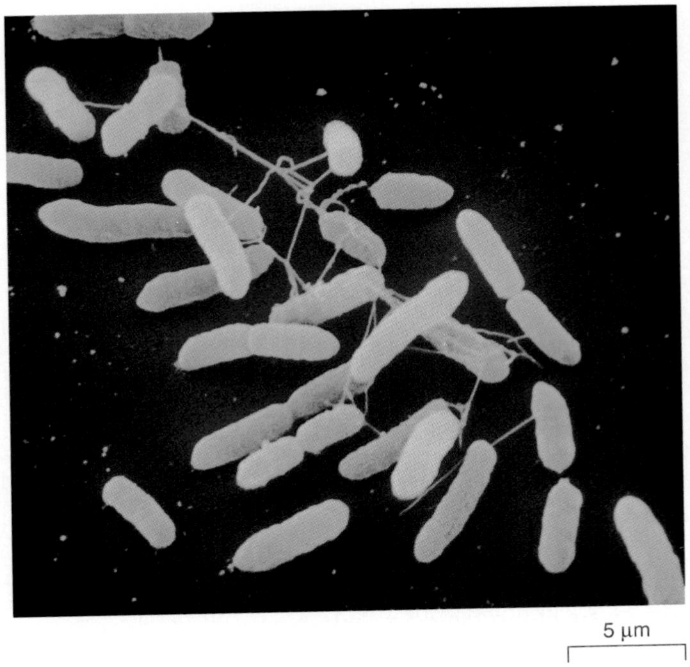

FIGURE 23–11 F pili connects these *Escherichia coli* bacteria. Plasmid DNA is transferred during conjugation. (Manfred Kage/Peter Arnold, Inc.)

Some bacteria form endospores

When the environment of a bacterium becomes unfavorable, such as when it gets very dry, many species become dormant. The cell loses water, shrinks slightly, and remains inactive until water is again available. Other species form dormant, extremely durable cells called **endospores.** After the endospore forms, the cell wall of the original cell lyses, releasing the endospore. Endospores can survive in very dry, hot, or frozen environments, or when food is scarce (Fig. 23–12). Some endospores are so resistant that they can survive an hour or more of boiling, or centuries of freezing. When environmental conditions are again suitable for growth, the endospore once more becomes an active, growing bacterial cell. In May 1995, two researchers reported in *Science* that they had revived bacterial endospores from a bee that had been fossilized in amber more than 25 million years ago. Using polymerase chain reaction (PCR) to copy and sequence DNA from the bacteria, they are studying biochemical differences between these ancient bacteria and contemporary forms of the same species (*Bacillus sphaericus*).

Endospores are not comparable to the reproductive spores of fungi and plants, and endospore formation is not really a kind of reproduction in bacteria. Only one endospore is formed per original cell, so the total number of individuals does not increase.

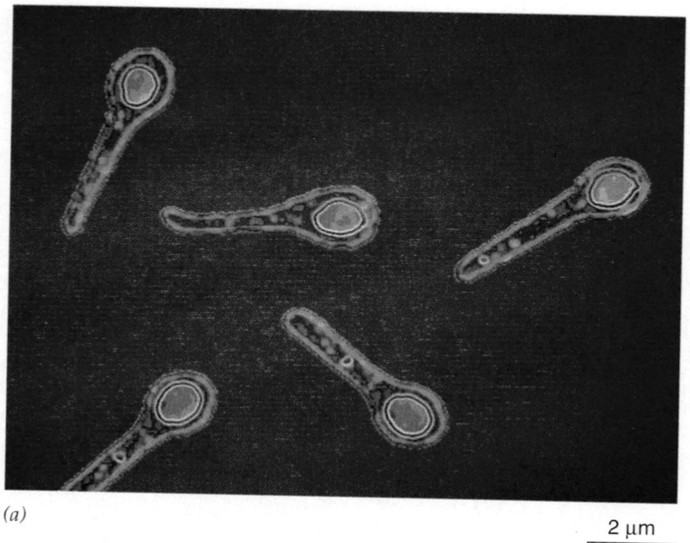

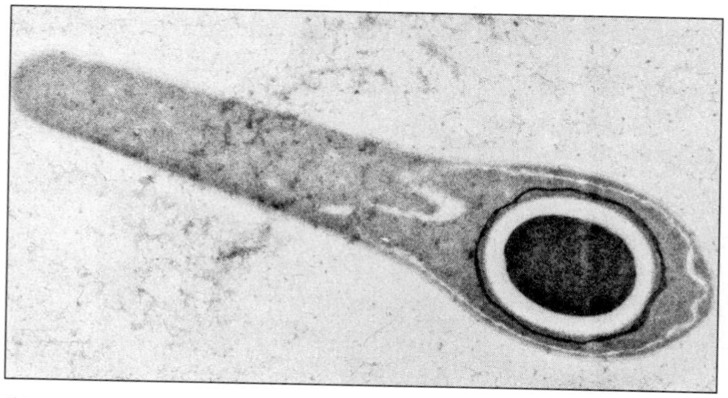

FIGURE 23–12 Some bacteria form endospores when conditions are unfavorable. (*a*) Endospores within cells of *Clostridium tetani*, the bacterium that causes tetanus. Each bacterial cell contains one endospore, a resistant, dehydrated cell that develops within the original cell. (*b*) Magnified view of endospore within a cell of *Clostridium tetani*. (*a*, Alfred Pasieka/Peter Arnold Inc.; *b*, T.J. Beveridge/Biological Photo Service)

Metabolic diversity is evident among bacteria

A bacterial cell contains about 5000 different chemical compounds. The functions of these compounds, how they interact, and how the bacterium synthesizes them from the nutrients it takes in are complex biochemical problems that have interested researchers for years. Much of the knowledge gained from studying these mechanisms in bacterial cells has been successfully applied to cells of humans and other organisms. The basic biochemical processes of all living organisms are surprisingly similar.

Bacteria are either heterotrophic or autotrophic. Most bacteria are **heterotrophs** and must obtain organic compounds from other organisms. The majority of these heterotrophic bacteria are free-living **saprobes** (also called saprotrophs), organisms that get their nourishment from dead organic matter. Other heterotrophic bacteria obtain their nourishment from living organisms, in some cases harming them by causing diseases, and in other cases actually providing a beneficial service for their host.

Autotrophic bacteria are either photosynthetic or chemosynthetic and are able to manufacture their own organic molecules. **Photosynthetic autotrophs,** or simply **photoautotrophs,** obtain their energy from light. Chemosynthetic autotrophs, or **chemoautotrophs,** obtain energy by oxidizing inorganic chemicals.

Whether they are heterotrophs or autotrophs, most bacterial cells are **aerobic** (like animal and plant cells), requiring atmospheric oxygen for cellular respiration. Some bacteria are **facultative anaerobes,** meaning that they can use oxygen for cellular respiration if it is available, but can also carry on metabolism anaerobically when necessary. Other bacteria are **obligate anaerobes** that can carry on energy-yielding metabolism only anaerobically. Some obligate anaerobes are actually killed by even low concentrations of oxygen.

TWO FUNDAMENTALLY DIFFERENT GROUPS OF PROKARYOTES ARE THE ARCHAEBACTERIA AND THE EUBACTERIA

Under a microscope, most bacteria appear similar in size and form. However, evidence from molecular biology has led biologists to conclude that ancient prokaryotes split into two lineages early in the history of life. The modern descendants of these two ancient lines are the **archaebacteria,** which include several groups of prokaryotes able to live in extreme environments, and the **eubacteria,** which comprise all other prokaryotes.

In recognition of the significant differences between archaebacteria and eubacteria, some biologists have proposed that the prokaryotes be divided into two kingdoms: kingdom Eubacteria and kingdom Archaebacteria. The eubacteria include most of the common bacteria familiar to biologists, whereas the archaebacteria are less familiar. They inhabit certain very harsh environments such as acidic hot springs. In this book we use the more traditional, five-kingdom classification scheme that places both archaebacteria and their distant relatives, the eubacteria, within the kingdom Prokaryotae (or Monera; see Chapter 22).

More than 400 genera of prokaryotes have been classified, and hundreds more species have been identified but not yet classified. Prokaryote classification has been problematic and is a matter of controversy. Although an important goal is to classify bacteria according to their evolutionary relationships, insufficient molecular information currently prevents taxonomists from accomplishing this task.

The editors of *Bergey's Manual of Systematic Bacteriology*[1], considered the definitive reference text by many microbiologists, have divided prokaryotes into four groups based on their cell wall type. They further group them according to their shape, mode of nutrition, and metabolism. The four groups are: (1) bacteria with a gram-negative type of cell wall; (2) bacteria with a gram-positive type of cell wall; (3) bacteria that lack cell walls (mycoplasmas); and (4) bacteria with cell walls that lack peptidoglycan (archaebacteria). Representative groups are described in Table 23–2.

The archaebacteria include methanogens, halobacteria, and thermoacidophiles

Biochemically, archaebacteria are very different from other prokaryotes. One of their most distinguishing features is the absence of peptidoglycan in their cell walls. They also have unusual lipids in their plasma membranes, and distinctive RNA molecules and enzymes. In some ways, the archaebacteria resemble the eukaryotes. For example, their RNA polymerase is more like that of eukaryotes than that of eubacteria.

Many of the extreme environments to which the modern archaebacteria are adapted resemble conditions that were common on primitive Earth but are somewhat rare today. These include hot springs whose temperatures may exceed 100° C and deep-sea vents that spew sulfide gases.

The **methanogens,** probably the most common of the archaebacteria, are strict anaerobes. In fact, they are quickly killed by oxygen. Methanogens inhabit oxygen-free environments in sewage and swamps, and are common in the digestive tracts of humans and other animals. These bacteria produce methane gas from carbon dioxide and hydrogen.

The **halobacteria** live only in extremely salty environments such as salt ponds (Fig. 23–13). Some are capable of photosynthesis in which they capture the energy of sunlight with a purple pigment.

The **thermoacidophiles** normally grow in hot (45–110° C), sometimes acidic, environments. One species is found in the hot sulfur springs of Yellowstone Park at temperatures near 60° C and pH values of 1 to 2 (the pH of concentrated sulfuric acid). Another species, isolated from hot deep-sea vents on the sea floor, grows at temperatures from 80 to 110° C.

FIGURE 23–13 The halobacteria are archaebacteria that thrive in salty environments. Seawater evaporating ponds near San Francisco Bay are colored pink, orange, and yellow from the large number of extreme halophiles (salt-loving archaeobacteria) growing in them. The salt that remains after the water has evaporated has commercial value. (Helen E. Carr, Biological Photo Service)

Eubacteria are the most familiar prokaryotes

The eubacteria are widely distributed in the environment and are better known to microbiologists than the archaebacteria. These prokaryotes exhibit great diversity. Several groups of eubacteria are described in Table 23–2.

Bacteria are ecologically important

Bacteria are vital members of the ecosphere. As described in the chapter introduction, they play essential roles as decomposers and as organisms that fix nitrogen. Bacte-

[1]*Bergey's Manual of Systematic Bacteriology, 9th ed.,* Williams & Wilkins, Baltimore, Philadelphia, 1994.

(Text continues on page 524)

Table 23–2 SOME MAJOR GROUPS OF BACTERIA

Bacteria with Cell Walls Lacking Peptidoglycan	
Archaebacteria  5 µm	Bacteria that grow in extreme environments. Have unusual lipids in their cell membranes and have distinctive RNA and enzymes. Three main groups are methanogens, halobacteria, and thermoacidophiles. ***Methanosarcina* colony.** (Color SEM; Visuals Unlimited/Ralph Robinson)
Eubacteria with Gram-Negative Cell Walls	
Spirochetes 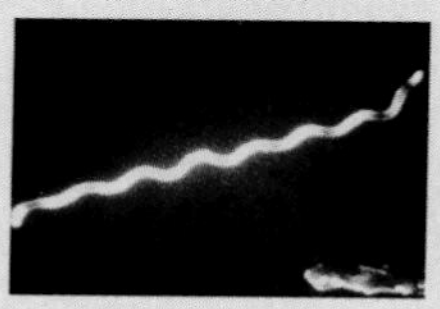 5 µm	Gram-negative, spiral-shaped bacteria with flexible cell walls. Move by means of unique internal flagella called **axial filaments.** Some species are free-living and inhabit freshwater and marine habitats, whereas other species form symbiotic associations; a few are parasitic. The spirochete of greatest medical importance is *Treponema pallidum,* the pathogen that causes syphilis. Lyme disease, a tick-borne disease of humans and some other animals, is caused by a spirochete belonging to the genus *Borrelia.* ***Treponema pallidum,* the spirochete that causes syphilis.** (Visuals Unlimited/Science VU—Charles W. Stratton)
Gram-negative aerobic rods and cocci 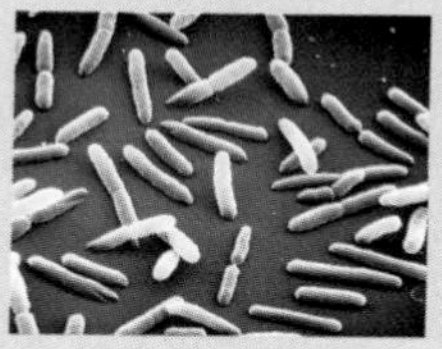 1 µm	***Rhizobium*** species convert atmospheric nitrogen to a form usable by plants (nitrogen fixation). **Pseudomonads** are heterotrophs capable of metabolizing unusual sugars, amino acids, and other nutrients. Can break down many natural and synthetic compounds that resist decomposition by other bacteria. Produce non-photosynthetic pigment. Can cause disease in plants and animals, including humans. **Azotobacteria** inhabit the soil. Fix nitrogen under aerobic conditions. Can form a resting cell termed a cyst that is resistant to drying. Medically important gram-negative aerobes include *Neisseria gonorrhoeae* which causes gonorrhea and *Legionella pneumophila* which causes Legionnaires' disease. ***Pseudomonas aeruginosa* plays an important role in the nitrogen cycle and can cause infections in immunologically compromised humans.** (Color SEM; Manfred Kage/Peter Arnold, Inc.)
Gram-negative rods that are facultative anaerobes 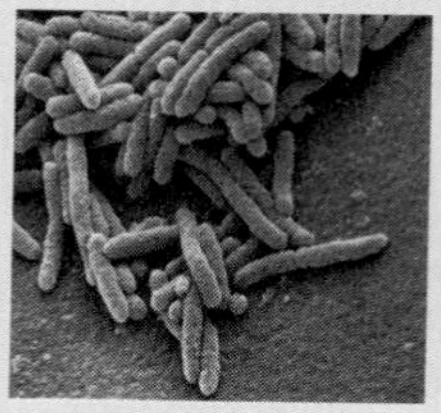 10.0 µm	**Enterobacteria.** Group includes decomposers that live on decaying plant matter, pathogens, and a variety of bacteria that inhabit humans. *Escherichia coli* inhabits the intestinal tracts of humans and other animals as part of the normal microorganism population. Certain strains of *E. coli* can cause moderate to severe diarrhea. For example, in 1993 almost 500 people developed bloody diarrhea and three people died in the Pacific Northwest after ingesting hamburger meat contaminated with a new and deadly strain of *E. coli.* One species of *Salmonella* causes a form of food poisoning; another species causes typhoid fever. **Vibrios.** Mainly marine; some luminescent. **SEM of *Escherichia coli* colony.** (David Scharf/Peter Arnold, Inc.)
Rickettsias and Chlamydias 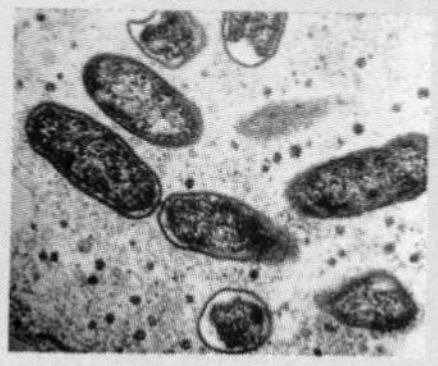 2 µm	Small, intracellular parasites. **Rickettsias** are rod-shaped. A few species are pathogenic to humans (and other animals); transmitted by arthropods through bites or contact with their excretions. Among these are typhus (transmitted by fleas and lice) and Rocky Mountain spotted fever (transmitted by ticks). ***Rickettsia rickettsia,* the causative agent of spotted fever.** (Visuals Unlimited/Science VU-ASM)

Table 23–2 (Continued)

Rickettsias and Chlamydias

Chlamydias are more spherical than rod-shaped, lack peptidoglycan in their cell walls, and do not depend on arthropod vectors for transmission. Although they do contain many enzymes and can carry on some metabolic processes, chlamydias are energy parasites—that is, completely dependent on their host for ATP. Chlamydias infect almost every species of bird and mammal. Perhaps 10% to 20% of the human population of the world is infected. Trachoma, the leading cause of blindness in the world, is caused by a strain of *Chlamydia*; sexually transmitted chlamydias are a major cause of pelvic inflammatory disease (PID).

Myxobacteria

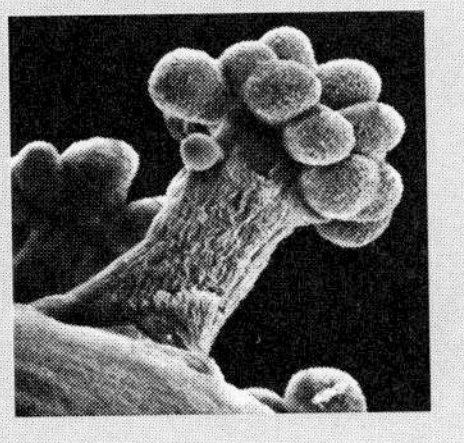

10 μm

Secrete slime and glide or creep along. Most **Myxobacteria** are saprobes that break down organic matter in the soil, manure, or rotting wood. Some species break down complex substrates such as cellulose and peptidoglycan. When nutrients are exhausted, these bacteria form masses, which develop into stalked, multicellular reproductive structures called **fruiting bodies.** During this process, bacterial cells within the fruiting body enter a resting stage equivalent to spores. When conditions are favorable, the spores break open and the resting cells become active.

The fruiting body of the myxobacterium *Stigmatella aurantiaca.* Protective resting cells form within the cyst that are very resistant to heat and drying. (From Grilicone, P. L., and Pangborn, J., *Journal of Bacteriology* 124:1558, 1975)

Cyanobacteria

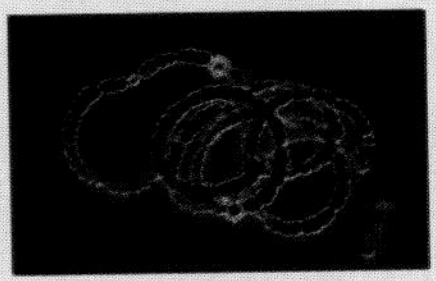

50 μm

(Formerly known as blue-green algae.) Gram-negative, photosynthetic bacteria. Inhabit ponds, lakes, swimming pools, moist soil, dead logs, and the bark of trees. Contain chlorophyll *a* and use a photosynthetic process similar to that of plants and algae. Many species also fix nitrogen.

***Anabaena*, a filamentous cyanobacterium that fixes nitrogen.** Nitrogen fixation occurs in the rounded cells, called heterocysts. (Dwight Kuhn)

Eubacteria with Gram-Positive Cell Walls

Gram-positive cocci

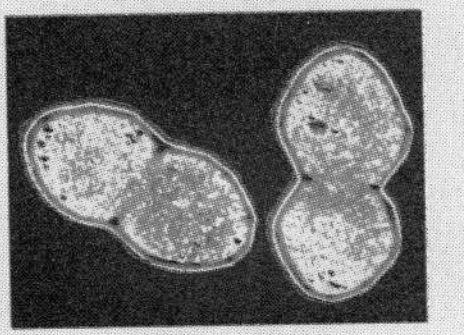

1 μm

Streptococci are found in the mouth as well as in the digestive tract of humans and some other animals. Among the harmful species are those that cause "strep throat," dental caries, a form of pneumonia, scarlet fever, rheumatic fever. One particularly virulent strain of *S. pyogenes* is resistent to antibiotics and can cause death within a few hours (see Chapter Opening figure).

Staphylococci normally live in the nose and on the skin. They are opportunistic pathogens, which means that they cause disease when the immunity of the host is lowered. *Staphylococcus aureus* causes boils and skin infections and may infect wounds. Certain strains of *S. aureus* cause a form of food poisoning, and some cause toxic shock syndrome.

***Streptococcus pneumoniae* can cause sinusitis, middle ear infections, and meningitis, as well as pneumonia.** (Alfred Paseika/Peter Arnold, Inc.)

Gram-positive rods

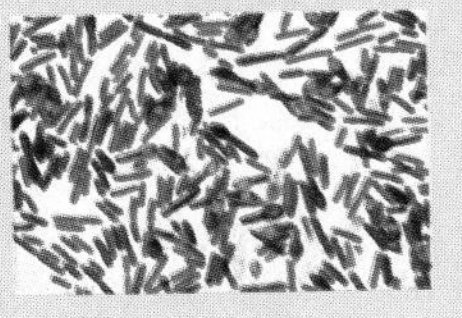

20 μm

Clostridia are anaerobic. One species causes tetanus, another causes gas gangrene, and *Clostridium botulinum* can cause botulism, a highly fatal type of food poisoning. *C. botulinum* produces endospores that are resistant to heat. Botulism results from consuming foods, such as canned vegetables and smoked meats and fish, that have been inadequately sterilized; the endospores grow and produce poisons that are among the most potent toxins known. Approximately one microgram of toxin is enough to kill a human.

***Clostridium botulinum* is responsible for serious food poisoning in humans.** Its toxin affects the central nervous system. (Michael Abbey/Photo Researchers, Inc.)

(Table continues on next page)

Table 23–2 (Continued)

Lactic acid bacteria 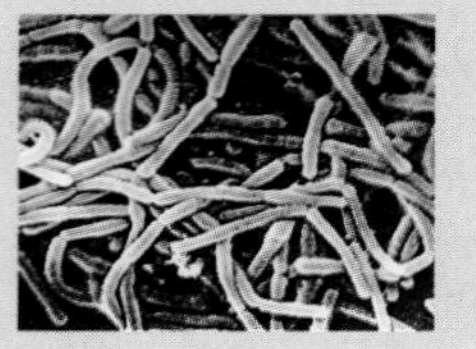 10 µm	Gram-positive bacteria that ferment sugar, producing lactic acid as the main end-product. Inhabit decomposing plant material, milk, and other dairy products. The characteristic taste of yogurt, acidophilus milk, pickles, sauerkraut, and green olives is due to the action of lactic acid bacteria. Also among the normal inhabitants of the human mouth and vagina. **False-color SEM of *Lactobacillus bulgaricus*, taken from live yogurt.** (Moredun Animal Health Ltd./Science Photo Library/Photo Researchers, Inc.)
Mycobacteria 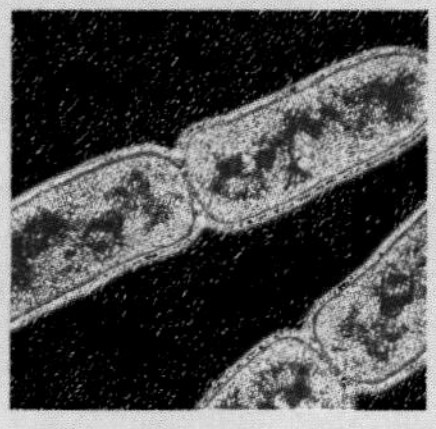 0.5 µm	Slender, irregular rods. Contain a waxy substance in their cell walls. One species causes tuberculosis; another causes leprosy. **TEM of *Mycobacterium tuberculosis*, the causative agent of human tuberculosis.** (Alfred Paseika/Science Photo Library/Photo Researchers, Inc.)
Actinomycetes 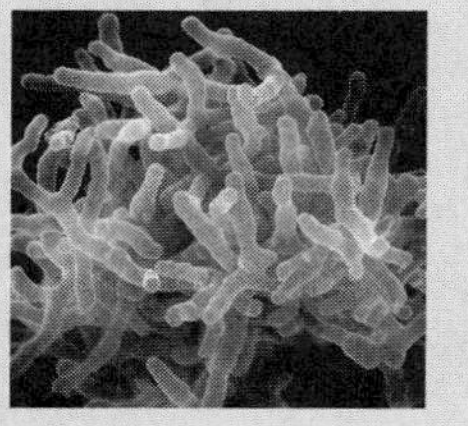 5 µm	Resemble fungi in that their cells remain together, forming branching filaments, and many produce moldlike spores. However, they have peptidoglycan in their cell walls, lack nuclear envelopes, and have other prokaryotic characteristics. Decompose organic materials in soil. Most members of this group are saprobes, and some are anaerobic. Several species of the genus *Streptomyces* produce antibiotics such as streptomycin, erythromycin, chloramphenicol, and the tetracyclines. In fact, most known antibiotics are derived from actinomycetes. Some actinomycetes cause serious lung disease or generalized infections in humans and other animals. ***Actinomyces naeslundi* form filamentous colonies.** (Visuals Unlimited/David M. Phillips)
Eubacteria That Lack Cell Walls	
Mycoplasmas 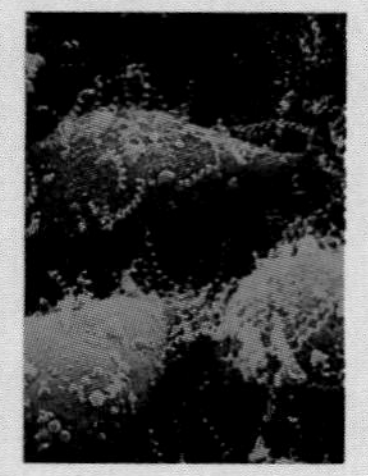 5 µm	Lack cell wall. Inhabit soil, sewage; some parasitic on plants or animals. Some inhabit human mucous membranes but do not generally cause disease; one species causes a mild type of bacterial pneumonia in humans. ***Spiroplasma* sp., shown on plant cells, can cause certain plant and insect diseases.** (Visuals Unlimited/David M. Phillips)

ria, especially actinomycetes and the myxobacteria, are the most numerous inhabitants of soil. Soil bacteria are important in cycling nutrients, including nitrogen, oxygen, carbon, phosphorus, sulfur, and many trace elements. A discussion of biogeochemical cycles is found in Chapter 54.

Nitrogen is constantly removed from the soil by plants and other natural processes, as well as by human activity. Nitrogen must be continually added to the soil because plant growth depends on the availability of usable nitrogen. Several types of bacteria transform atmospheric nitrogen to forms that can be used by plants (Fig. 23–14).

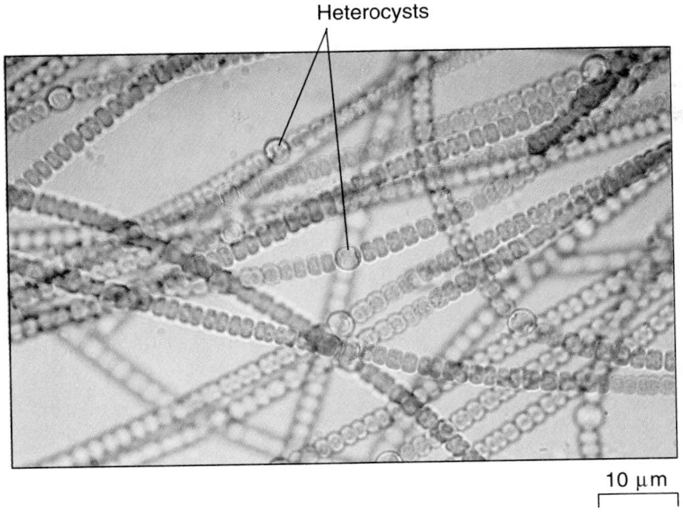

FIGURE 23–14 ***Anabaena*** **is a cyanobacterium that fixes nitrogen.** Nitrogen fixation takes place in the rounded cells, called heterocysts. (Dennis Drenner)

Some bacteria cause disease

Bacteria are important pathogens of humans and many other types of organisms (Fig. 23–15). A wide variety of bacteria normally inhabit various parts of the human body—including the skin, mouth, nose, throat, intestine, vagina, and urethra. Most of these residents are harmless symbiotic bacteria that prevent harmful microorganisms (including other bacteria) from flourishing. Some of the normal bacterial (and viral) inhabitants are opportunistic pathogens that can cause disease only under certain conditions. For example, if one's immune system is compromised, opportunistic bacteria may increase in numbers and cause disease.

Pathogens can enter the body in food or air, through damaged skin, or from contact with infected organisms. In order to cause disease, pathogens must multiply, increasing their numbers. They must successfully compete with normal bacterial residents and also overcome the host's defense mechanisms.

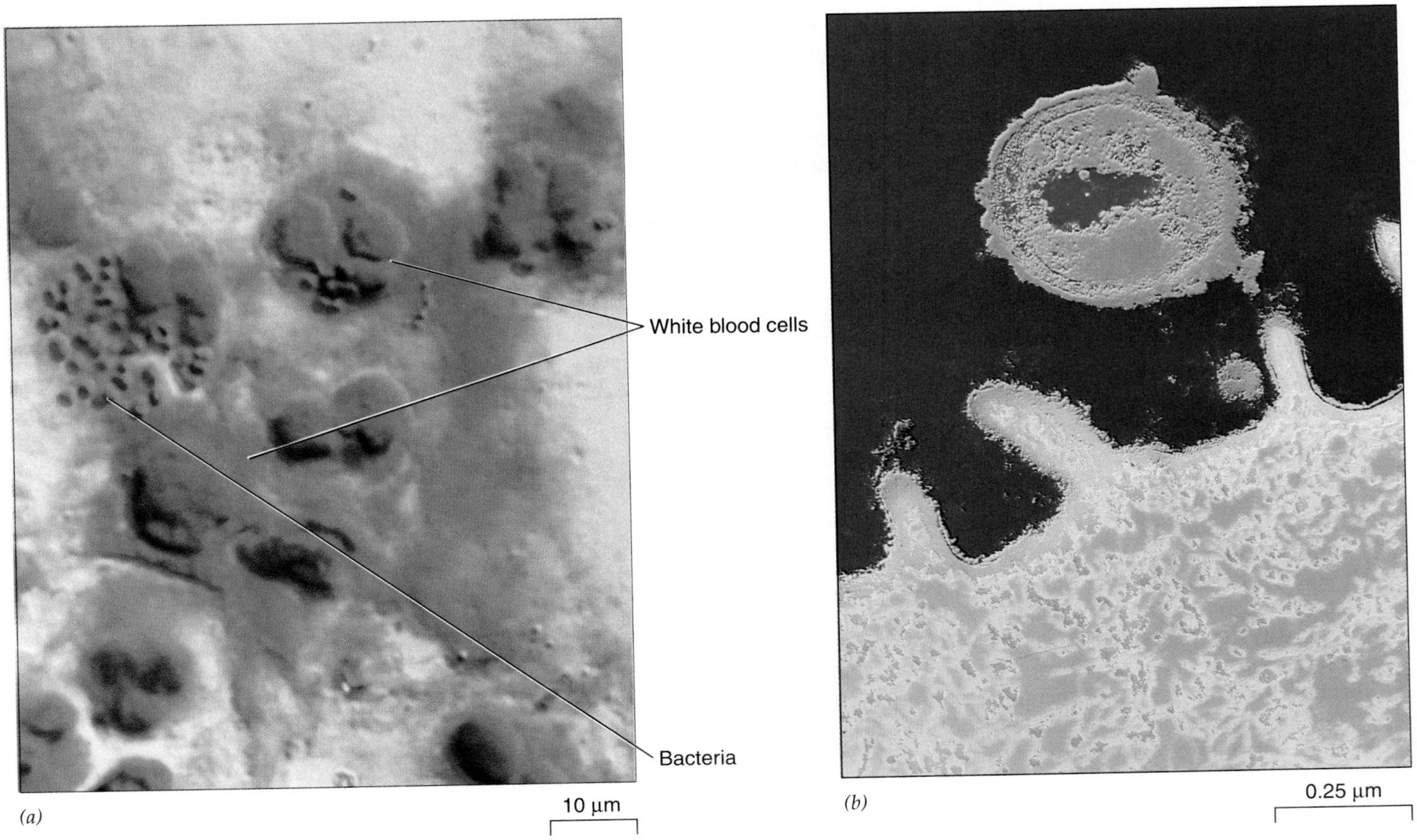

FIGURE 23–15 Some types of bacteria are pathogens. (***a***) Pus containing *Neisseria gonorrhea* (small pink spheres), the bacteria that cause gonorrhea, a common sexually transmitted disease. The larger cells are white blood cells. (***b***) Colored transmission EM of the bacterium (*Corynebacterium diphtheriae*) that causes diphtheria. This gram-positive bacterium is shown adhering to the surface of an epithelial cell of the throat. The multiplying bacteria release toxins that damage the heart and nerves. (*a*, M.I. Walker, Science Source/Photo Researchers, Inc.; *b*, Dr. Kari Lounatmaa/Science Photo Library/Photo Researchers, Inc.)

ON THE CUTTING EDGE

Containing Viral Outbreaks

Objective: To contain viruses such as *Ebola* in order to prevent the spread of infectious disease.

Method: Identify the population and the infectious agent; identify mode of transmission and implement strategies to reduce transmission.

Results: Global cooperation and rapid response by public health agencies have successfully contained outbreaks of *Ebola* and some other deadly viruses.

Conclusions: Lessons learned from dealing with emerging viruses and the resurgence of old ones will help us contain future epidemics, but much more research is needed.

In 1995 an outbreak of the deadly *Ebola* virus occurred in Zaire, a country in Central Africa. Named after a river in Zaire, *Ebola* is an elongated, single-stranded RNA virus that causes a fatal disease. Within about three days of infection, victims develop a fever and weakness, followed by a rash and vomiting. The victim hemorrhages internally, and bleeds from the mouth, eyes, ears, and other body orifices. Internal organs shut down, and 50 to 90% of victims die within about two weeks after infection.

Early symptoms of *Ebola* resemble those of influenza or dysentery, and in the 1995 outbreak, health care workers did not immediately recognize the danger. When (in late April) they became suspicious, blood samples from 16 patients were flown to the U.S. Centers for Disease Control and Prevention in Atlanta. Within 36 hours the *Ebola* virus was identified in blood from 14 of the 16 patients.

As soon as *Ebola* was identified in the 1995 outbreak, global health care teams, including epidemiologists, were sent into Central Africa to prevent its spread. Health officials confirmed that, like HIV (the virus that causes AIDS), *Ebola* is not spread by casual contact, but by direct contact with infected body fluids. Because victims of *Ebola* hemorrhage, infected blood appears to be the principal method of transmission. Infection-control methods, such as using gloves, gowns, and masks, were implemented. Patients were quarantined, and travel out of the infected area was monitored. Global cooperation and rapid response successfully contained the 1995 *Ebola* outbreak.

An important aspect of understanding an infectious agent is identifying patient zero, the first patient who contracted the virus. If the origin of the epidemic can be found, investigators might be able to trace where the virus came from and how it infects humans. *Ebola* is known to infect chimpanzees, but because they die quickly they are not thought to be the reservoir species. However, some other primate host might be the reservoir that would allow the

Making the Connection

Bacteria, Nitrogen Fixation, and Agriculture

How does the mutualistic relationship between certain bacteria and plants benefit agriculture? Rhizobial bacteria (bacteria in the genus *Rhizobium*) form symbiotic associations with the roots of **legumes,** a large family of herbs, shrubs, and trees. Legumes include such important crops as peas, beans, lentils, soybeans, and peanuts. Clover and alfalfa, which are grown for livestock feed and to fertilize the soil, are also important legumes.

Rhizobial bacteria are motile rod-shaped bacteria that inhabit the soil. After they infect the roots of a legume, nodules are produced on the roots. The nodules consist of plant tissue in which the bacteria reside and fix nitrogen.

The relationship between rhizobial bacteria and the roots of legumes is a type of symbiotic relationship referred to as **mutualistic.** In a mutualistic relationship both partners benefit (see Chapter 53). The bacteria living in nodules supply the plant with all the nitrogen that it requires, and the plant provides the bacteria with sugar needed for cellular respiration.

Because legumes, like other plants, produce sugar by photosynthesis, a correlation exists between photosynthesis and nitrogen fixation. When a legume is photosynthesizing at a higher rate, it provides more sugar for its bacterial partners. The bacteria are then able to fix larger amounts of nitrogen.

Plants without nodules must obtain nitrogen from the soil (see Chapter 34), and many soils are deficient in nitrogen. Therefore, legumes that have formed a mutualistic association with rhizobial bacteria are able to thrive in nitrogen-deficient soils, which gives them a decided advantage over other plants. ▲

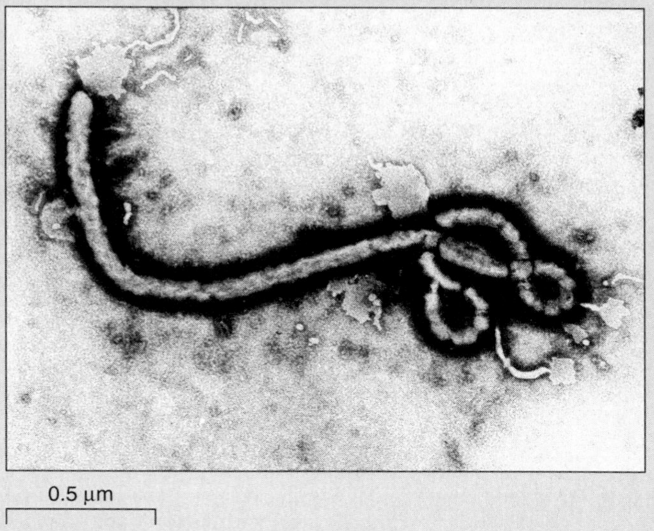

Transmission electron micrograph of Ebola virus.
(Courtesy of Frederick A. Murphy)

virus to maintain itself indefinitely. Researchers continue to search for clues to the origin of *Ebola* and are working on the development of effective antiviral drugs and vaccines.

Ebola outbreaks have occurred several times during the past few decades. The virus was first identified in 1976 in Zaire and Sudan (outbreaks that killed more than 400 people). Between outbreaks the virus remains hidden. Like other RNA viruses, *Ebola* makes frequent mistakes when it replicates its RNA. This leads to a high mutation rate, and thus to the rapid evolution of new strains.

The 1995 *Ebola* outbreak serves as a grim reminder that pathogens can strike quickly and fatally. According to the U.S. Centers for Disease Control and Prevention, more than 200 new, continual, or re-emerging pathogens have the potential to strike globally. Historically, new viral strains have claimed many human lives. For example, in 1918 an influenza epidemic killed more than 20 million people. Even at the level of our current knowledge about viruses and epidemiology, just how well could we contain a particularly virulent virus even now? How well are we containing HIV (the virus that causes AIDS), which has already claimed more than one million lives?

Human activity, including social factors, such as urbanization and global travel, contribute to epidemics of infectious disease. For example, as human populations concentrate in cities, large numbers of people are in close contact, permitting rapid spread of viruses. Standards of living, including sanitation, nutrition, physical stress, level of health care, and sexual practices, are important factors in the spread of disease. In the U.S. and other developed countries, infectious disease accounts for about 4 to 8% of deaths compared to death rates of 30 to 50% in developing regions.

Changing natural habitats can create the conditions necessary for new pathogens to emerge. For example, cutting down forests can bring disease-carrying insects into contact with humans. Sometimes, even naturally occurring ecological changes can spawn outbreaks of disease. The 1993 outbreak of hantavirus in the American Southwest killed more than 50 persons. A mild winter coupled with heavy rainfall resulted in a large crop of seeds, which supported a population explosion of field mice. The mice carry the virus.

Pathogens produce a variety of substances that increase their success. Some bacteria produce **exotoxins,** strong poisons that either are secreted from the cell or leak out when the bacterial cell is destroyed. The toxin, not the presence of the bacteria themselves, is responsible for the disease. Botulism, a type of food poisoning that can lead to paralysis and sometimes death, results from ingestion of improperly canned food. Botulism is caused by an exotoxin released by the gram-positive bacterium, *Clostridium botulinum.* This exotoxin is so powerful that one gram could kill a million humans! Like many exotoxins, the one that causes botulism can be inactivated by heating. (Food must be heated to 80° C for 10 minutes, or boiled for 3 to 4 minutes.)

Endotoxins are not secreted by pathogens, but are components of the cell wall of most gram-negative bacteria. Endotoxins are not destroyed by heating. These compounds affect the host only when the bacteria die and release them. Endotoxins bind to macrophages and stimulate them to release substances that cause fever and other symptoms of infection. Unlike exotoxins, which cause specific symptoms, all endotoxins appear to cause the same general symptoms such as fever.

Humans harness bacteria for medical and industrial purposes

Bacteria have been producing antibiotics for millennia. These compounds evolved as part of the bacterial (and fungal) arsenal of weapons that inhibit or destroy other microorganisms. Pharmaceutical companies obtain most antibiotics from three groups of microorganisms: actinomycetes, gram-positive bacteria of the genus *Bacillus,* and molds (eukaryotes belonging to kingdom Fungi). By the 1950s antibiotics had become important clinical tools that transformed the treatment of infectious disease. Today, literally tons of antibiotics are produced annually.

Many foods and beverages are produced with the help of microorganisms. Lactic acid bacteria are used in the production of acidophilus milk, yogurt, pickles, olives, and sauerkraut, and several types of bacteria are used to produce cheese.

Bacteria are also used to make many industrial compounds such as acetone. Methanogens are used to decompose wastes in sewage treatment. In the process known as *bioremediation,* a contaminated site is exposed to microorganisms that produce enzymes that break down the toxins, leaving behind harmless metabolic byproducts such as carbon dioxide and chlorides. To date, more than 1,000 different species of bacteria and fungi have been used to clean up various forms of pollution. For example, bioremediation was used in the early 1990s to clean up an Iowa site contaminated by a poisonous wood preservative called pentachlorophenol (or more simply, penta). A bacterium called *Flavobacterium* was introduced to clean up the site because it breaks down the penta molecule.

During bioremediation, conditions at the hazardous waste site are modified so that the bacterium will thrive in large enough numbers to be effective. The site in Iowa was small, so engineers mixed the contaminated soil with sand (to make it more porous) and then injected the soil with *Flavobacterium.* They pumped air through the soil (to increase its oxygen level) and added a few soil nutrients like phosphorus; *Flavobacterium* requires both oxygen and phosphorus in order to flourish. Engineers also built a drainage system at the bottom of the pit to pipe any penta-laden water that leached through the soil back to the surface for another encounter with the bacteria.

SUMMARY

I. A virus is a tiny particle consisting of a DNA or RNA core surrounded by a capsid (protein coat).
   A. Viruses are not cellular, cannot metabolize on their own, and are not considered to be truly living organisms.
   B. Viruses reproduce inside living cells.
      1. Bacteriophages (phages) are viruses that infect bacteria.
      2. Many different viruses infect humans and other animals. Examples of viral diseases in humans include chickenpox, herpes simplex, infectious mononucleosis, mumps, warts, influenza, hepatitis, AIDS, and Ebola. Viruses also cause certain types of cancer.
      3. Plant viruses cause serious agricultural losses. Viral diseases can be spread among plants by insect vectors.

II. Kingdom Prokaryotae contains the bacteria.
   A. Bacteria have a prokaryotic cell structure and lack membrane-bounded organelles such as nuclei and mitochondria. The three main shapes of bacterial cells are spherical (cocci), rod-shaped (bacilli), and spiral (spirilla).
      1. The genetic material of a prokaryote is a single circular DNA molecule. Plasmids may also be present.
      2. Most bacteria have cell walls composed of peptidoglycan. The walls of gram-positive bacteria are very thick and consist mainly of peptidoglycan. The cell walls of gram-negative bacteria consist of a thin peptidoglycan layer and a thick outer membrane.
      3. Bacterial flagella are structurally different from eukaryotic flagella.
      4. Bacteria are metabolically diverse; some are heterotrophic, whereas others are autotrophic. The majority of heterotrophic bacteria are free-living saprobes. Autotrophs may be photosynthetic or chemosynthetic. Most bacteria are aerobic, but some are facultative anaerobes; others are obligate anaerobes.
      5. Bacteria reproduce asexually by binary fission.
      6. Genetic material may be exchanged by transformation, transduction, or conjugation.
   B. Prokaryotes are divided into two groups, archaebacteria and eubacteria.
      1. The archaebacteria have cell walls with an unusual chemical composition, can live in oxygen-deficient environments, and are often adapted to harsh conditions. The three main groups of archaebacteria are the methanogens, the halobacteria, and the thermoacidophiles.
      2. The remaining bacteria are collectively known as the eubacteria. Eubacteria can be assigned to three groups based on their cell wall composition.
         a. Mycoplasmas are bacteria that lack cell walls.
         b. Gram-negative bacteria have cell walls composed of a thin peptidoglycan layer, but surrounded by a thick membrane.
         c. Gram-positive bacteria have thick-layered cell walls of peptidoglycan.

SELECTED KEY TERMS

archaebacteria
autotroph
bacillus (bacilli)
bacteria
bacteriophage (or phage)
binary fission
capsid
capsule
chemoautotroph
coccus (cocci)
conjugation
endospore
endotoxin
eubacteria
exotoxin
facultative anaerobe
flagellum
gram-negative
gram-positive
heterotroph
lysogenic conversion
lytic virus
obligate anaerobe
pathogen
peptidoglycan
pilus (pili)
retrovirus
saprobe
spirillum (spirilla)
temperate virus
transduction
transformation
virus

POST-TEST

1. Microorganisms that cause disease are referred to as ____________ .
2. The core of a(an) ____________ consists of DNA or RNA, but not both.
3. The protein coat surrounding the nucleic acid core of a virus is called a(an) ____________ .
4. Bacteriophages are viruses that infect ____________ .
5. ____________ viruses kill the host cell.
6. Lysogenic viruses are also known as ____________ viruses.
7. Lysogenic phages can transfer nucleic acid from one bacterium to another, resulting in genetic recombination; this process is known as ____________ .
8. ____________ is a chemical compound found in the cell walls of most eubacteria.
9. The majority of heterotrophic bacteria are free-living ____________ that get their nourishment from dead organic matter.
10. Spherical bacteria are referred to as ____________ , rod-shaped bacteria as ____________ , and helical bacteria as ____________ .
11. Bacteria that absorb and retain crystal violet stain (the Gram stain) are known as ____________ - ____________ bacteria.
12. Bacteria that produce organic molecules from simple organic ingredients, using energy obtained from oxidizing inorganic compounds, are called ____________ .
13. In ____________ , genetic material is transferred from one bacterium to another of a different mating type.
14. The methanogens are ____________ that produce methane from carbon dioxide and hydrogen.
15. Some bacteria produce and secrete disease-causing poisons known as ____________ .

REVIEW QUESTIONS

1. What characteristics does a virus share with a living cell? What characteristics of life are lacking in a virus?
2. List the steps in bacteriophage infection. Draw diagrams to illustrate your answer.
3. Contrast the sequence of events in a lysogenic infection with that of a lytic infection.
4. What are the distinguishing characteristics of bacteria?
5. What are the differences between the archaebacteria and the eubacteria?
6. Contrast the cell wall of a gram-positive bacterium with that of a gram-negative bacterium.
7. Using Table 23–2 as a guide, give the distinguishing characteristics of each of the following groups: wall-less bacteria, gram-negative bacteria, and gram-positive bacteria. Give examples of each group.

YOU MAKE THE CONNECTION

1. Historically, biologists thought that viruses, because of their simple structure, evolved before cellular organisms. Based on what you have learned about viruses, present an argument against this hypothesis.
2. Imagine that you discover a new microorganism. After careful study you determine that it should be classified in the kingdom Prokaryotae, with the archaebacteria. What characteristics might lead you to this classification?
3. How might life on planet Earth be different if bacteria had never evolved?

RECOMMENDED READINGS

Angert, E.R., K.D. Clements, and N.R. Pace. "The Largest Bacterium." *Nature,* March 18, 1993. Discovery of a novel bacterium so large that it can be seen *without* a microscope.

Caldwell, M. "Prokaryotes at the Gate." *Discover,* August 1994. An interesting account of bacterial resistance to antibiotics.

Canby, T.Y. "Bacteria: Teaching Old Bugs New Tricks." *National Geographic,* August 1993. Considers the industrial potential of bacteria, from cleaning up toxic waste to degrading oil spills.

Carmichael, W.W. "The Toxins of Cyanobacteria." *Scientific American,* January 1994. Many cyanobacteria produce unusual toxins that are deadly but potentially valuable.

Finlay, B.B. "Bacterial Virulence Factors," *Scientific American,* May/June 1995. A facinating discussion of the mechanisms that make pathogenic bacteria successful.

Hively, W. "Life Beyond Boiling." *Discover,* May 1993. Research discoveries about bacteria that thrive at temperatures well above 212° F. Most of these bacteria live in hot springs at the bottom of the ocean.

Monastersky, R. "Ancient Bacteria Brought Back To Life," *Science News,* May 20, 1995. A brief review of a study reported in *Science,* May 9, 1995 on reviving endospores preserved in amber.

Nester, E.W., C.E. Roberts, and M.T. Nester. *Microbiology: A Human Perspective.* Wm. C. Brown, Dubuque, Iowa, 1995. A comprehensive introduction to microbiology.

Prusiner, S.B. "The Prion Diseases." *Scientific American,* January 1995. Infectious agents, smaller and simpler than viruses, may be responsible for a number of disorders.

Zimmer, C. "Triumph of the Archaea," *Discover,* February 1995. Discussion of a molecular approach to the study of archaebacteria.

CHAPTER 24

The Protist Kingdom

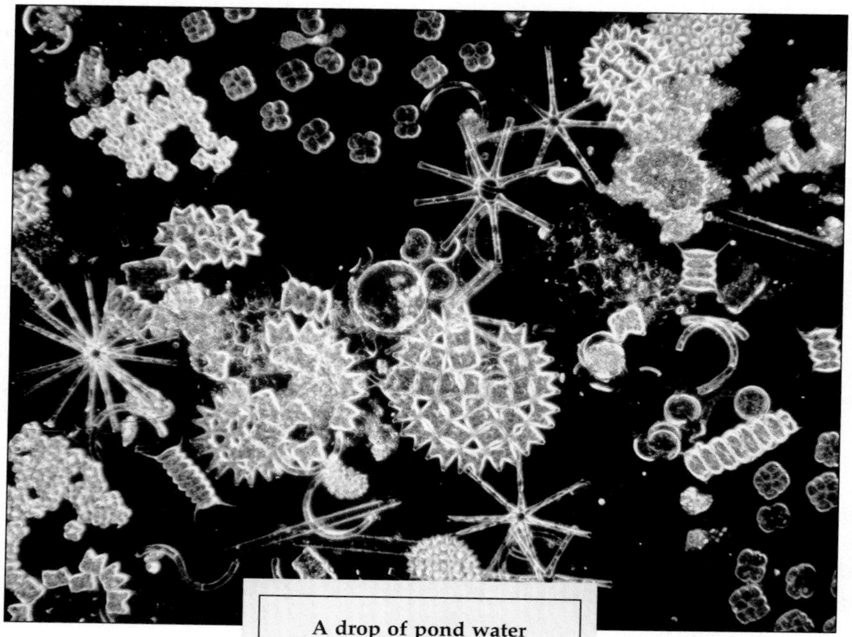

A drop of pond water reveals various photosynthetic protists. The protists are an extremely diverse kingdom of organisms.

(Roland Birke/Peter Arnold, Inc.)

Kingdom Protista consists of a vast assortment of primarily aquatic organisms whose diverse body forms, types of reproduction, modes of nutrition, and life styles make them difficult to characterize. Biologists estimate that there are as many as 200,000 extant (living) species of **protists**—unicellular or simple multicellular organisms that possess a eukaryotic cellular organization. The word *protist,* from the Greek, meaning "the very first," reflects the idea that protists were the first eukaryotes to evolve (as discussed later in the chapter).

Eukaryotic cells, the unifying feature of protists, are common to complex multicellular organisms from three other kingdoms (fungi, animals, and plants) but clearly separate protists from members of the kingdom Prokaryotae (bacteria). Recall from Chapter 4 that eukaryotic cells have nuclei and other membrane-bounded organelles such as mitochondria and plastids.

Size varies considerably within the protist kingdom, from microscopic protozoa to giant kelps, which are brown algae that can reach 60 meters (almost 200 feet) in length. Although most protists are unicellular, some have a colonial organization (a colony is a loose aggregation of cells), some are **coenocytic** (multinucleate but not multicellular), and some are multicellular. Unlike animals, fungi, and plants, multicellular protists have relatively simple body forms without specialized tissues.

There is no universal acceptance among biologists about what comprises a "protist." As do many biologists, we interpret the protist kingdom broadly to include heterotrophic protists (the protozoa, slime molds, and water molds) and autotrophic protists (the algae).

The protist kingdom is a polyphyletic group of organisms; that is, protists do not share a single common ancestor. Any eukaryotic organism not considered a fungus, animal, or plant is classified in the protist kingdom solely for convenience. If a cladist were classifying these organisms into monophyletic kingdoms, the kingdom Protista would be split into numerous kingdoms—perhaps as many as twenty!

Because of their huge numbers, members of kingdom Protista are important to the natural balance of the living world. Protists are an important source of food for other organisms, and photosynthetic protists also supply oxygen to aquatic and terrestrial ecosystems. Certain protists are economically important, while others cause diseases. Consideration of all protist phyla is beyond the scope of this text, but we will discuss fifteen representative groups to demonstrate the mind-boggling diversity in kingdom Protista.

LEARNING OBJECTIVES

After you have studied this chapter you should be able to

1. Characterize the common features of the members of kingdom Protista.
2. Discuss in general terms the diversity inherent in the protist kingdom, including modes of nutrition, body forms, and methods of reproduction.
3. Briefly describe these representative protozoan groups: amoebas, foraminiferans, actinopods, flagellates, ciliates, and sporozoans.
4. Briefly characterize the representative groups of algae: dinoflagellates, diatoms, euglenoids, green algae, red algae, and brown algae.
5. Briefly discuss the representative fungus-like protists: plasmodial slime molds, cellular slime molds, and water molds.
6. Describe how multicellularity may have arisen within the protists, using *Chlamydomonas* as an example.
7. Discuss the evolutionary relationships among certain protists and the other eukaryotic kingdoms.

PROTISTS EXHIBIT REMARKABLE VARIATION

Size and structural complexity are not the only variable features of protists. Organisms in the kingdom Protista are also diverse in how they obtain nutrients, in their relationships with other organisms, in where they live, in how they reproduce, and in their means of locomotion.

Methods of obtaining nutrients differ widely in the kingdom Protista. Autotrophic protists (the algae) have chlorophyll and photosynthesize as plants do. Some of the heterotrophic protists (the water molds) obtain their food by absorption, as fungi do. Other heterotrophs (protozoa and slime molds) resemble animals in that they ingest food derived from the bodies of other organisms. Some protists switch their modes of nutrition and are autotrophic at certain times and heterotrophic at others.

Many protists are free-living, while others form symbiotic associations with different organisms. These intimate associations range from *mutualism,* a more or less equal partnership in which both organisms benefit, to *parasitism,* in which one organism lives on or in another and is metabolically dependent on it (see Chapter 53). Some parasitic protists are important pathogens (disease-causing agents) of plants or animals. Specific examples of symbiotic associations involving protists are given throughout this chapter.

Most protists are aquatic and live in oceans or freshwater ponds, lakes, and streams. They make up part of the **plankton,** the floating, often microscopic organisms that inhabit surface waters and are the base of the food web in aquatic ecosystems. Other aquatic protists attach to rocks or other surfaces in the water. Terrestrial (land-dwelling) protists are restricted to damp places like soil and leaf litter. Even the parasitic protists are aquatic because they live in the watery environments of other organisms' body fluids.

Reproduction is quite varied in the kingdom Protista. All protists reproduce asexually, and many also reproduce sexually, with both meiosis and **syngamy,** the union of gametes. However, most protists do not develop multicellular reproductive organs, nor do they form embryos the way more complex organisms do.

Protists, most of which are motile at some point in their life cycle, have various means of locomotion. Movement may be accomplished by amoeboid motion (extending cell protrusions), by flexing individual cells, by waving cilia, or by lashing flagella. Many protists use a combination of two or more means of locomotion—for example, both flagellar and amoeboid motion. Their cilia and flagella, unlike those of prokaryotes but like those of all eukaryotes, possess a 9 + 2 arrangement of microtubules—that is, nine outer doublet microtubules encircling two single microtubules (see Chapter 4).

PROTOZOA ARE ANIMAL-LIKE PROTISTS

The name **protozoa** (from the Latin, meaning "first animals"; sing., *protozoon*) was originally given to animal-like organisms that are not multicellular. The term *protozoa* is used today to designate an informal group of protists that ingest food (as animals do). Protozoa are a polyphyletic group, and their relationships are continually evaluated as additional evidence becomes available. In this chapter we will consider six groups of protozoa: amoebas, foraminiferans, actinopods, flagellates, ciliates, and sporozoans (Table 24–1).

Amoebas move by forming pseudopodia

Amoebas (phylum Rhizopoda) are unicellular organisms found in soil, fresh water, and oceans. Many members of this group have no definite body form and continually change shape as they move. (The word *amoeba* is derived from a Greek word meaning "change.") An amoeba moves by pushing out temporary cytoplasmic projections called **pseudopodia** (sing., *pseudopodium,* meaning "false foot") from the surface of the cell. More cytoplasm flows into the pseudopodia, enlarging them until all the cyto-

(Text continues on page 533)

Table 24–1 A COMPARISON OF REPRESENTATIVE PHYLA IN THE PROTIST KINGDOM

Common Name	*Phylum*	*Morphology*	*Locomotion*	*Photosynthetic Pigments*	*Special Features*
Amoebas	Rhizopoda	Single cell, no definite shape	Pseudopodia	—	Some have shells (tests)
Foraminiferans	Foraminifera	Single cell	Cytoplasmic projections	—	Pore-studded shells (tests)
Actinopods	Actinopoda	Single cell	Some produce flagellated reproductive cells	—	Axopods protrude through pores in skeleton
Flagellates	Zoomastigina	Single cell	One to many flagella; some amoeboid	—	Symbiotic forms often highly specialized
Ciliates	Ciliophora	Single cell	Cilia	—	Macronuclei; micronuclei
Sporozoans	Apicomplexa	Single cell	Nonmotile	—	All parasitic; develop resistant spores
Dinoflagellates	Dinoflagellata	Single cell, some colonial	Two flagella	Chlorophylls *a* and *c*; carotenoids, including fucoxanthin	Many covered with cellulose plates
Diatoms	Bacillariophyta	Single cell, some colonial	Most nonmotile; some move by gliding over secreted slime	Chlorophylls *a* and *c*; carotenoids, including fucoxanthin	Silica in shell
Euglenoids	Euglenophyta	Single cell	Two flagella (one very short)	Chlorophylls *a* and *b*; carotenoids	Flexible outer covering
Green algae	Chlorophyta	Single cell, colonial, siphonous, multicellular	Most flagellated at some stage in life; some entirely nonmotile	Chlorophylls *a* and *b*; carotenoids	Reproduction highly variable
Red algae	Rhodophyta	Most multicellular, some single cell	Nonmotile	Chlorophyll *a*; carotenoids; phycocyanin; phycoerythrin	Some reef builders
Brown algae	Phaeophyta	Multicellular	Two flagella on reproductive cells	Chlorophylls *a* and *c*; carotenoids, including fucoxanthin	Differentiation of body into blade, stipe, and holdfast
Plasmodial slime molds	Myxomycota	Multinucleate plasmodium	Streaming cytoplasm, flagellated or amoeboid reproductive cells	—	Reproduce by spores formed in sporangia

Table 24–1 *(Continued)*

Common Name	*Phylum*	*Morphology*	*Locomotion*	*Photosynthetic Pigments*	*Special Features*
Cellular slime molds	Acrasiomycota	Vegetative form—single cell; reproductive form—multicellular (slug and fruiting body)	Pseudopods (for single cells); cytoplasmic streaming (for multicellular)	—	Aggregation of cells signaled by cyclic AMP
Water molds	Oomycota	Coenocytic mycelium	Biflagellate zoospores	—	Cellulose and/or chitin in cell walls

plasm has entered and the organism as a whole has moved. Pseudopodia are also used to engulf and capture food by forming a vacuole around it (Fig. 24–1). A food vacuole encompasses and digests the food particles using digestive enzymes released by lysosomes (see Chapter 4). Digested materials are absorbed from the food vacuole, which gradually shrinks as it empties. Amoebas reproduce asexually by cell division; sexual reproduction has not been observed.

Parasitic amoebas include *Entamoeba histolytica,* which causes amoebic dysentery, a serious human intestinal disease that causes severe diarrhea. Some amoebas, like *Acanthamoeba,* are usually free-living but can produce opportunistic infections such as eye infections in contact lens users.

Foraminiferans extend cytoplasmic projections through tests

Foraminiferans (phylum Foraminifera) are almost all marine organisms that produce shells, or **tests** (Fig. 24–2).

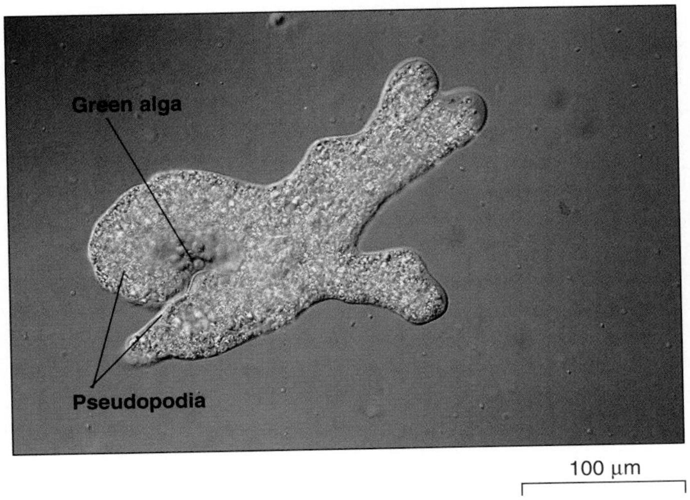

FIGURE 24–1 Amoebas are single-celled protists that move and feed by means of pseudopodia. A giant amoeba's pseudopodia surround a colonial green alga, ingesting it. (Michael Abbey/Photo Researchers, Inc.)

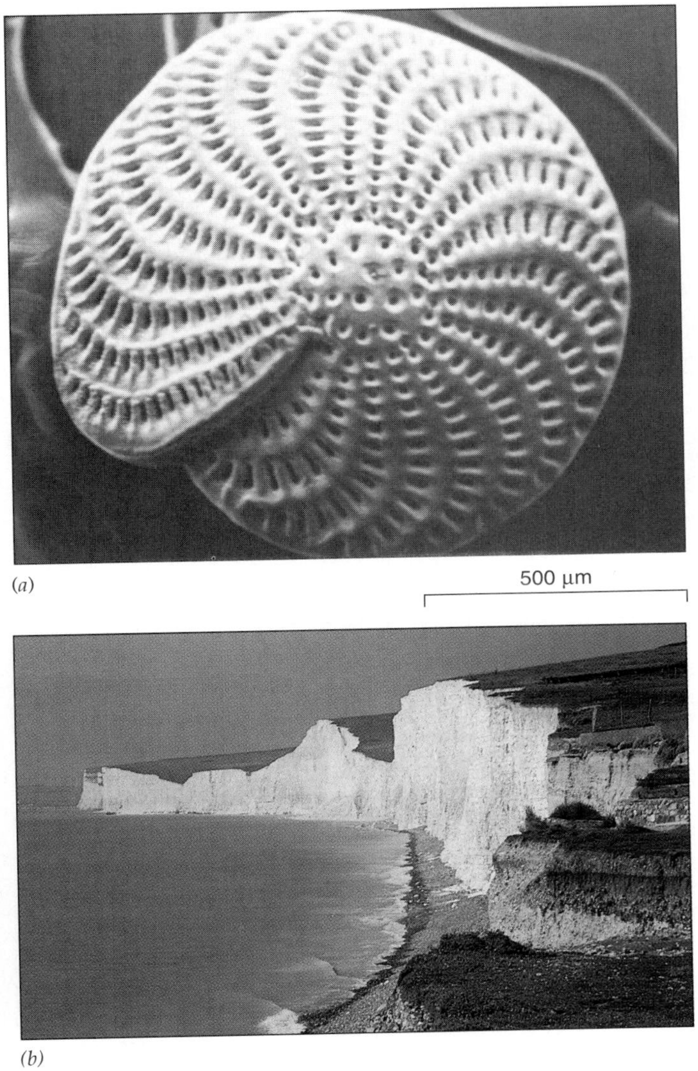

FIGURE 24–2 Foraminiferans are mainly marine protists with shells, or tests. (*a*) Scanning electron micrograph of a test. Note the pores through which cytoplasm extrudes. (*b*) The white cliffs of Dover, England, consist largely of the tests of foraminiferans. (*a*, Biophoto Associates; *b*, Lynn McLaren/Photo Researchers, Inc.)

The oceans contain enormous numbers of foraminiferans, which secrete chalky, many-chambered tests with pores through which cytoplasmic projections can be extended. The group derives its phylum name from this characteristic, as *Foraminifera* is derived from Latin words that mean "bearing openings." The cytoplasmic projections form a sticky, interconnected net that entangles its prey.

Dead foraminiferans sink to the bottom of the ocean, where their shells form a gray mud that is gradually transformed into chalk. With geological uplifting, these chalk formations can become part of the land, like the white cliffs of Dover in England. (The white cliffs of Dover are composed of the remains of a variety of calcareous organisms, not just foraminiferans.) Because foraminiferan tests are often found in rock layers covering oil deposits, geologists involved in oil exploration look for foraminiferan tests in rock strata.

Actinopods project slender axopods

Actinopods (phylum Actinopoda) have long, filamentous cytoplasmic projections called **axopods** that protrude through pores in their skeletons (Fig. 24–3). Each axopod is strengthened by a cluster of microtubules. Unicellular algae and other prey become entangled in these axopods and are engulfed outside the main body of the actinopod; cytoplasmic streaming carries the prey back within the shell. Many actinopods contain symbiotic algae that provide them with the products of photosynthesis.

Some actinopods secrete elaborate and beautiful skeletons made of silica. When actinopods die, their skeletons settle and become mud on the ocean floor; eventually they are compressed into sedimentary rock.

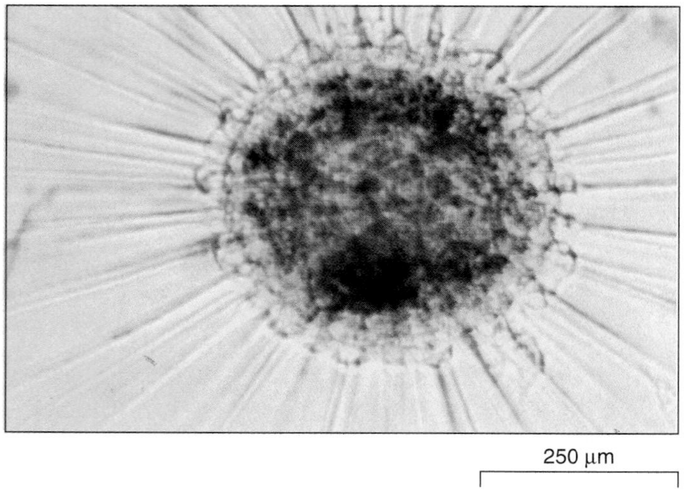

FIGURE 24–3 Actinopods are mainly marine protists that trap prey with axopods. Note the many slender axopods that project from the cell of *Actinosphaerium eichornia.* (Visuals Unlimited/Arthur M. Siegelman)

Flagellates move by means of flagella

Flagellates (phylum Zoomastigina) are unicellular organisms with spherical or elongated bodies, a single central nucleus, and from one to many long, whiplike flagella that enable them to move. Flagellates move rapidly, pulling themselves forward by lashing flexible flagella that are usually located at the anterior (front) end. Some flagellates are also amoeboid and engulf food by forming pseudopodia. Others have a definite "mouth" or *oral groove,* a "throat" or *cytopharynx,* and specialized organelles for processing food.

Flagellates are heterotrophic and obtain their food either by ingesting living or dead organisms or by absorbing nutrients from dead or decomposing organic matter. They may be free-living or symbionts. For example, flagellates with a large number of flagella and extremely specialized cells live in the guts of termites (Fig. 24–4*a*). These flagellates apparently possess the enzymes to digest cellulose in the wood that termites eat, and both the termite and the flagellates obtain their nutrients from this source.[1] Termites would starve to death without their **endosymbionts** (an endosymbiont lives *inside* the body of the organism with which it has formed a close relationship). Some parasitic flagellates cause disease. For example, the flagellate *Trypanosoma* is a human parasite that causes African sleeping sickness (Fig. 24–4*b*).

Choanoflagellates (collared flagellates) are one of the classes of flagellates in the phylum Zoomastigina. They are of special interest because of their striking resemblance to certain cells in sponges. Most biologists think choanoflagellates are related to sponges but probably not to other animals. These sedentary flagellates are attached to a substrate by a stalk, and their single flagellum is surrounded by a delicate collar of cytoplasm (Fig. 24–5).

Ciliates use cilia for locomotion

Ciliates (phylum Ciliophora) are unicellular organisms with a flexible outer covering that gives them a definite but somewhat changeable shape. In *Paramecium,* the surface of the cell is covered with several thousand fine short hairs, called cilia, that extend through pores in the outer covering and permit movement (Fig. 24–6). The cilia beat in such a precisely coordinated fashion that the organism not only can go forward but also can back up and turn around. Near their surface, many ciliates possess numerous small **trichocysts,** organelles that discharge filaments thought to aid in trapping and holding prey.

(Text continues on page 536)

[1]The flagellate endosymbionts of termites in turn possess endosymbionts — bacteria that reside within the flagellates. These bacteria, rather than the flagellates, may produce the enzymes that digest cellulose.

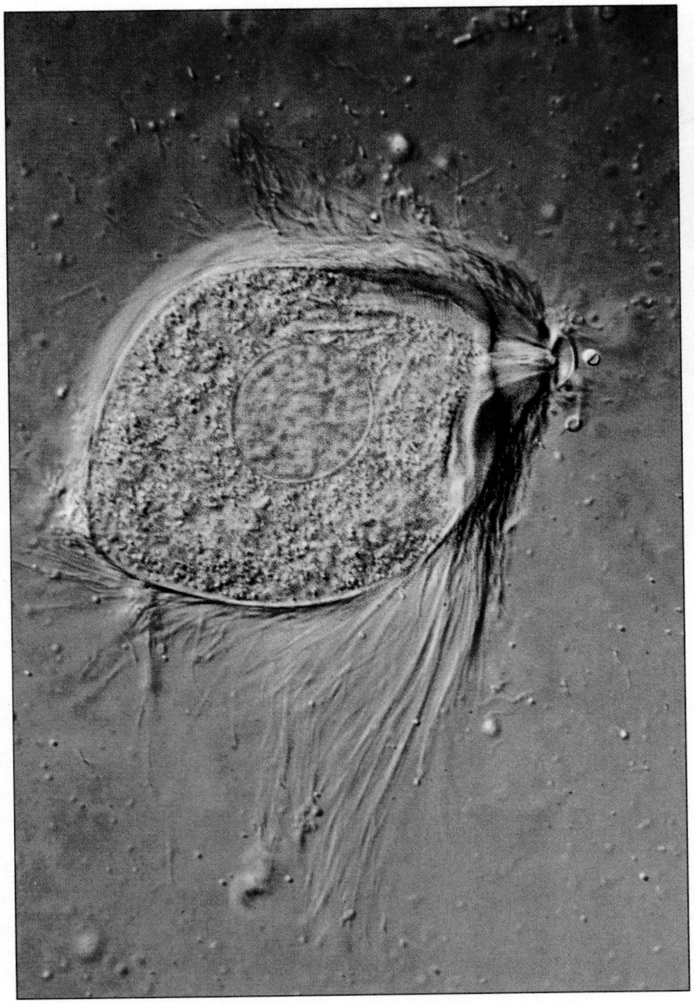

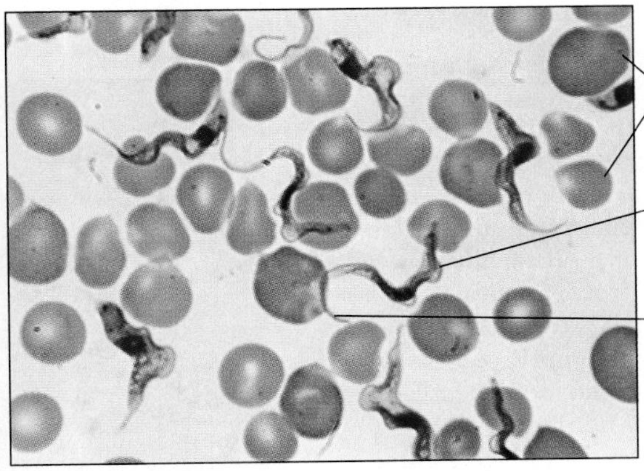

FIGURE 24–4 Flagellates are protists that move by means of flagella. (***a***) A flagellate (*Trichonympha* sp.) that lives inside the guts of wood-eating termites and digests the cellulose of the wood particles, from which it obtains sugar for itself and its host. (***b***) *Trypanosoma gambiense,* the flagellates that cause sleeping sickness, swarm among red blood cells in a blood smear. (*a,* Visuals Unlimited/M.A. Abbey; *b,* Biophoto Associates/Photo Researchers, Inc.)

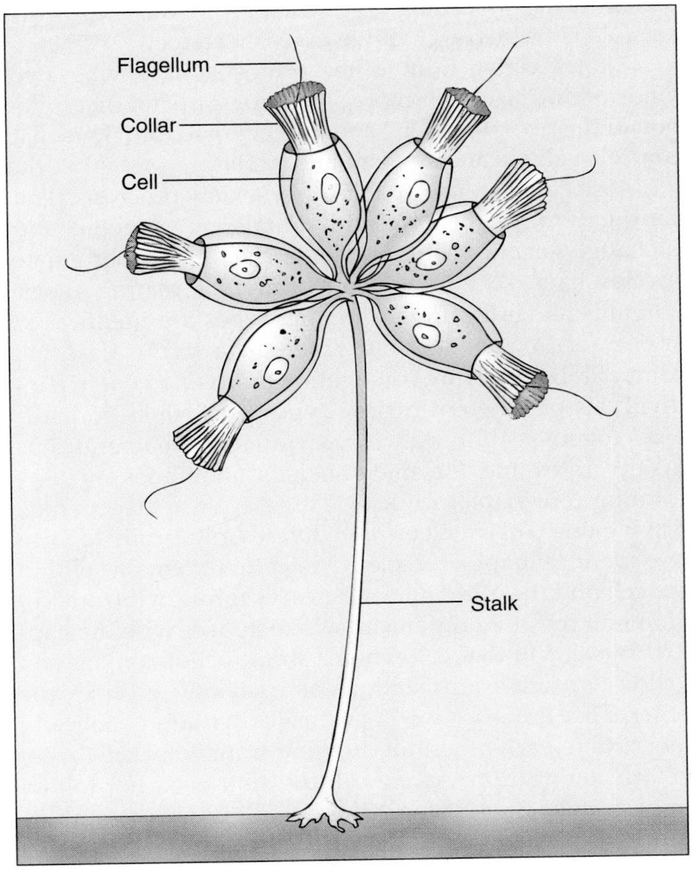

FIGURE 24–5 Choanoflagellates are sedentary flagellates that are probably related to sponges. These flagellates obtain food by waving their flagella, causing water currents to carry small particles of food into the collar. Shown is a colonial form.

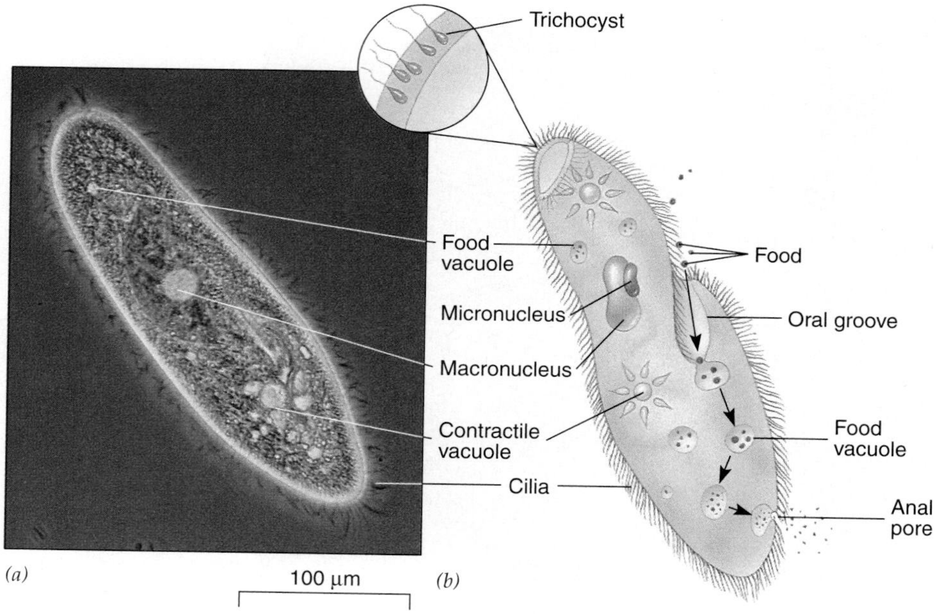

FIGURE 24–6 Ciliates are unicellular protists that move by means of cilia. (*a*) Note the complex cellular structure of *Paramecium,* a freshwater protozoon. Like many ciliates, *Paramecium* has multiple nuclei (a macronucleus and one or more smaller micronuclei). (*b*) Food particles are swept into its ciliated oral groove and incorporated into food vacuoles. Lysosomes fuse with the food vacuoles, and the food is digested and absorbed; undigested wastes are eliminated through the anal pore. *Paramecium* absorbs water by osmosis from its freshwater surroundings, but it does not swell up because contractile vacuoles fill up with excess water and then contract to void the contents.
(M. Abbey/Photo Researchers, Inc.)

Most ciliates ingest bacteria or other tiny organisms; their cilia draw the food into a funnel-like oral opening. A vacuole forms around the food at the end of the opening, and the food is digested. Water regulation in freshwater ciliates is controlled by special organelles called *contractile vacuoles.* Because freshwater ciliates continually take in water by osmosis (see Chapter 5), the contractile vacuole is needed to remove excess water.

Ciliates differ from other protozoa in having two kinds of nuclei: one or more small **micronuclei** that function in the sexual process, and a larger **macronucleus** that controls cellular metabolism and growth.

Most ciliates are capable of a sexual process called **conjugation,** in which two individuals come together and exchange genetic material. *Paramecium* and other ciliate species have several to many different mating types. (Organisms with different mating types are identical in appearance but genetically different in terms of sexual compatibility.) During conjugation in *Paramecium,* two individuals of different mating types press their oral surfaces together. Within each individual the macronucleus disintegrates and the micronucleus undergoes meiosis, forming four haploid nuclei. Three of these degenerate, leaving one. This nucleus then divides mitotically to form two identical haploid nuclei. One of these remains within the cell and the other nucleus crosses through the oral region into the other organism, where it fuses with the haploid nucleus in that cell. Thus a single act of conjugation yields two cross fertilizations as each cell fertilizes the other. This leads to two "new" cells that are genetically identical to each other but different from what they were before conjugation. Actual cell division need not follow immediately after conjugation. Cell division is a complex process involving more than simply splitting in half because of the presence of complex organelles that must be duplicated. In addition, after the new micronucleus divides, a new macronucleus must be formed from one of the micronuclei.

Not all ciliates are motile. Some sessile forms are stalked, and others, while capable of some swimming, are more likely to remain attached to a rock or other surface at one spot. Their cilia set up water currents that draw food toward them.

Sporozoans are spore-forming parasites of animals

Sporozoans (phylum Apicomplexa) are a large group of parasitic protozoa, some of which cause serious diseases such as malaria in humans. Sporozoans lack structures for locomotion. They do move, however, by flexing. At some stage in their lives many develop a **spore,** a small infective agent transmitted to the next host. Sporozoans often spend part of their lives in one host species and part in a different host species.

Malaria, which is caused by a sporozoan, is one of the world's most common serious infectious diseases. According to the World Health Organization, approximately 100 million people currently have malaria, and one to two million people die each year from it.[2] *Plasmodium,* the sporozoan that causes malaria, enters human blood

[2]Despite a discouraging history, slow progress is being made against malaria. A vaccine developed in the early 1990s by Manuel Patarroyo, a biochemist at the Immunology Institute of the National University in Bogota, Colombia, is being tested in several countries with some success.

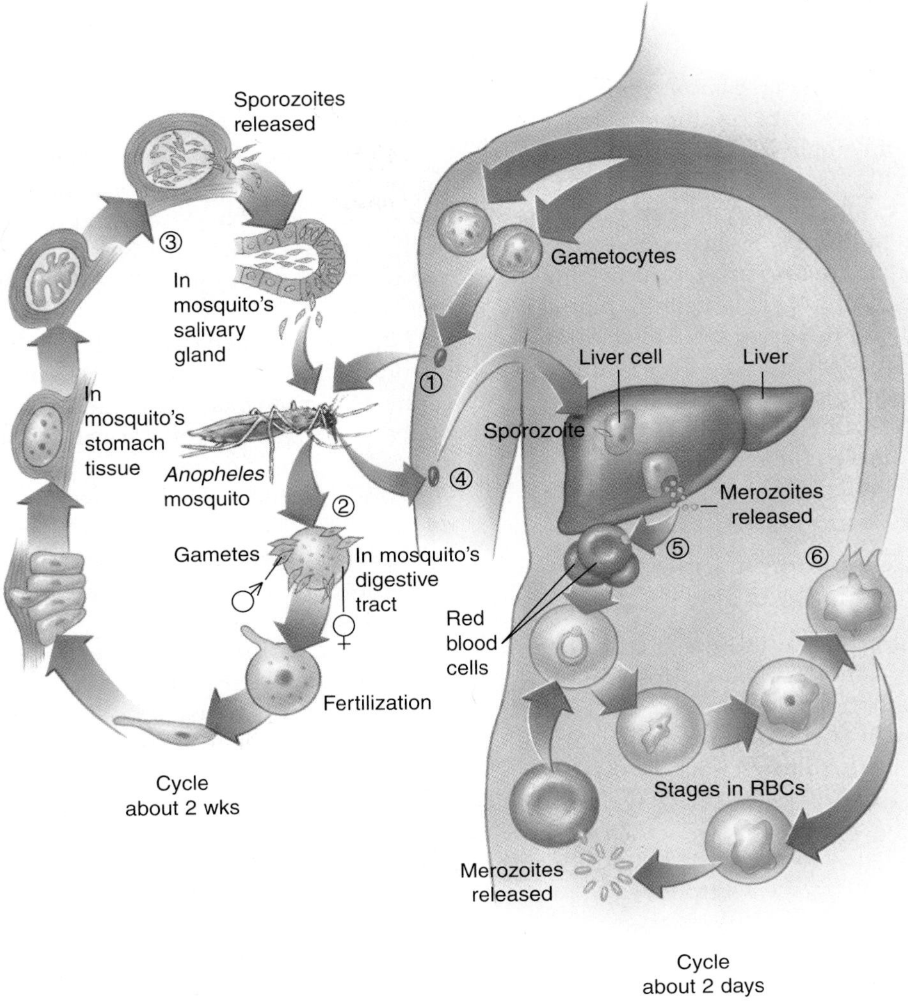

FIGURE 24–7 ***Plasmodium*, the sporozoan that causes malaria, has a complex life cycle that involves two different hosts.** **(1)** A female *Anopheles* mosquito bites an infected person and obtains gametocytes. **(2)** In the mosquito's digestive tract, the gametocytes develop into gametes and fertilization occurs. **(3)** The zygote embeds in the mosquito's stomach lining and produces sporozoites, which are released and migrate to the salivary gland. **(4)** The mosquito bites an uninfected human and transmits sporozoites to the human's blood. **(5)** The sporozoites enter liver cells and divide to produce merozoites that infect red blood cells. **(6)** In the blood cells, merozoites divide to form more merozoites, which infect more red blood cells. Some merozoites form gametocytes, which can be transmitted to the next mosquito that bites that human, and the process is repeated.

through the bite of an infected female *Anopheles* mosquito (Fig. 24–7). *Plasmodium* first enters liver cells and then red blood cells, where it multiplies. When each infected red blood cell bursts, many new parasites are released. The released parasites infect new red blood cells and the process is repeated. The simultaneous bursting of millions of red cells causes the symptoms of malaria: a chill followed by high fever (as toxic substances are released and affect other organs of the body).

When a mosquito bites an infected human, it sucks up some malarial parasites along with human blood. A complicated process of sexual reproduction then occurs within the mosquito's stomach, and new malarial parasites develop, some of which migrate into the mosquito's salivary glands to infect the next person bitten.

ALGAE ARE PLANTLIKE PROTISTS

Algae (sing., *alga*) are an informal group of mostly photosynthetic protists that range in size from unicellular, microscopic forms to large, multicellular seaweeds. (The word *alga* is derived from a Latin word for seaweed.)

Although most algae are photosynthetic like plants, they are not considered plants because they lack many plant structures. For example, algae lack a cuticle, which is a waxy covering over plants that reduces water loss. When actively growing, algae are restricted to damp or wet environments, such as oceans; freshwater ponds, lakes, and streams; hot springs; polar ice; moist soil, trees, and rocks; and the bodies of certain animals including sloths, sea anemones, corals, and worms. Also, most algae do not have multicellular **gametangia** (sing. *gametangium;* reproductive organs in which gametes are produced); algal gamentangia are formed from single cells.

In addition to green chlorophyll *a* and yellow and orange **carotenoids,** which are photosynthetic pigments found in all algae, different groups of algae possess a variety of other pigments that are also important in photosynthesis. Classification into phyla is largely by pigment composition and type of storage products. Other characteristics used to classify algae include their cell wall composition, the number and placement of flagella, and chloroplast structures. We will consider six groups of algae: dinoflagellates, diatoms, euglenoids, green algae, red algae, and brown algae (Table 24–1).

Most dinoflagellates are a part of marine plankton

One of the most unusual protistan groups is that of the **dinoflagellates** (phylum Dinoflagellata). Most dinoflagellates are unicellular, although a few are colonial. Their cells are often covered with shells of interlocking cellulose plates impregnated with silicates (Fig. 24–8*a*). The typical dinoflagellate has two flagella: one flagellum is wrapped around a transverse groove in the center of the cell like a belt, and the other is located in a longitudinal groove (perpendicular to the transverse groove) and projects behind the cell. The undulation of these flagella propels the dinoflagellate through the water like a spinning top. Indeed, the dinoflagellates' name is derived from the greek *dinos,* meaning "whirling."

Most dinoflagellates are photosynthetic and possess the photosynthetic pigments chlorophyll *a*, chlorophyll *c*, and carotenoids. However, a number are colorless; some of these ingest other microorganisms for food. The storage products of dinoflagellates are usually oils or polysaccharides.

Many dinoflagellates are endosymbionts that reside in the bodies of marine invertebrates such as jellyfish, corals, and mollusks. These dinoflagellates lack cellulose plates and flagella and are called **zooxanthellae.** Zooxanthellae photosynthesize and provide food for their invertebrate partners. Their contribution to the productivity of coral reefs is substantial. Other dinoflagellates that are endosymbionts lack pigmentation and do not photosynthesize; these heterotrophs are parasitic on their hosts.

Reproduction in the dinoflagellates is primarily asexual, by longitudinal cell division, although a few species have been reported to reproduce sexually. The dinoflagellate nucleus is unusual because the chromosomes are permanently condensed and always evident. Meiosis and mitosis are unique because the nuclear envelope remains intact throughout cell division, and the spindle is located *outside* the nucleus. (The chromosomes do not make direct contact with the spindle microtubules; instead, the chromosomes appear to be attached to the nuclear envelope and the spindle separates the new nuclei from each other.) Based on the uniqueness of their chromosome

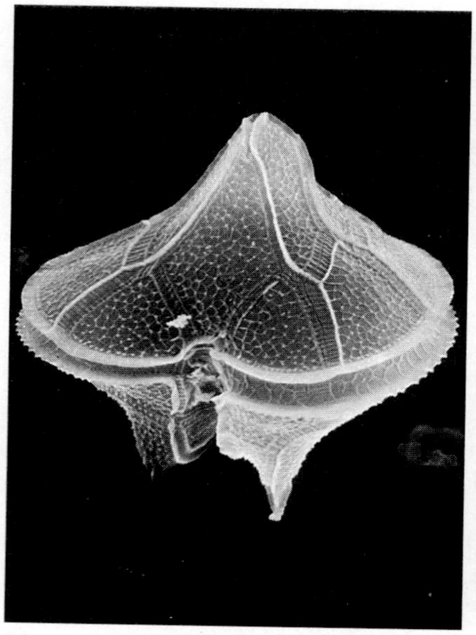

(*a*) 25 μm

(*b*)

FIGURE 24–8 Most dinoflagellates are unicellular, photosynthetic, and biflagellate. (*a*) Scanning electron micrograph of *Protoperidinium,* a dinoflagellate. Note the cellulose plates that encase the unicellular body. The two flagella (not visible) are located in grooves. (***b***) A red tide in Baja California, Mexico. The water is colored due to the presence of billions of dinoflagellates. (*a*, Courtesy of T.K. Maugel, University of Maryland; *b*, Kevin Schafer) structure and mitosis, the dinoflagellates are thought to have no close extant relatives.

Ecologically, dinoflagellates are one of the most important groups of producers in marine ecosystems. A few dinoflagellates are known to have occasional population explosions, or blooms. These blooms frequently color the water orange, red, or brown and are known as **red tides** (Fig. 24–8*b*). It is not known what environmental conditions initiate dinoflagellate blooms, but they are more common in the warm waters of late summer. Many experts think that human-produced coastal pollution triggers red tides, presumably by providing nutrients to the dinoflagellates. Some of the dinoflagellate species that form red tides produce a toxin that attacks the nervous systems of fishes, leading to massive fish kills. Humans sometimes get paralytic shellfish poisoning by eating filter-feeding mollusks, such as oysters, mussels, or clams, that fed on certain dinoflagellates. Paralytic shellfish poisoning causes respiratory problems in humans who eat the contaminated shellfish; death from respiratory failure occasionally occurs. (The dinoflagellates do not appear to hurt the shellfish.)

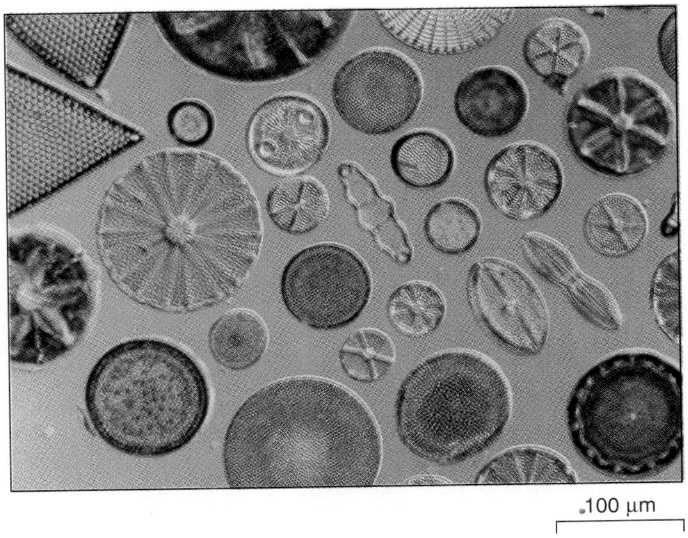

FIGURE 24–9 Most diatoms are unicellular algae with shells that contain silica. These algae have strikingly beautiful patterns on their symmetrical shells. (The Stock Market/Phillip Harrington)

Diatoms have shells composed of two parts

Most **diatoms** (phylum Bacillariophyta) are unicellular, although a few exist as colonies. The cell wall of each diatom consists of two shells that overlap where they fit together, much like a petri dish. Silica is deposited in the shell, and this glasslike material is laid down in striking, intricate patterns (Fig. 24–9). There are two basic groups of diatoms, those with radial symmetry (wheel-shaped) and those with bilateral symmetry (boat-shaped or needle-shaped). Although most diatoms are part of the floating plankton, some live on rocks and other surfaces, where they move by gliding. This gliding movement is facilitated by the secretion of a slimy material from a small groove along the shell.

Most diatoms are photosynthetic and contain the photosynthetic pigments chlorophyll *a*, chlorophyll *c*, and carotenoids; their pigment composition gives them a yellow or brown color. Food reserves are stored as oils or carbohydrates.

Diatoms most often reproduce asexually by cell division. When a diatom divides, the two halves of its shell separate and each becomes the larger half for a new diatom cell. Therefore, some diatom cells get progressively smaller with each succeeding generation (the glass shell does not grow). When diatoms are a fraction of their original size, sexual reproduction is triggered, with the production of shell-less gametes. Sexual reproduction restores the diatom to its original size because the resulting zygote (fertilized egg) grows substantially before producing a new shell.

Diatoms are common in fresh water, but they are especially abundant in relatively cool marine water. They are major producers in aquatic ecosystems because of their extremely large numbers. When diatoms die, their shells trickle down to the ocean floor and accumulate in layers of what eventually becomes sedimentary rock. After millions of years, some of these deposits have been exposed on land by geological upheaval. Called diatomaceous earth, these deposits are mined and used as a filtering, insulating, and soundproofing material.

Euglenoids are unique freshwater unicellular flagellates

Most **euglenoids** (phylum Euglenophyta) are unicellular flagellates, and about one-third of them are photosynthetic (Fig. 24–10). They generally possess two flagella, one long and whiplike and one so short that it does not protrude outside the cell. Euglenoids change shape continually as they move through the water because their outer covering is flexible rather than rigid. Euglenoids reproduce asexually by longitudinal cell division; none has ever been observed to reproduce sexually.

Euglenoids are included in our discussion of algal protists rather than being classified as flagellates because so many of them contain chloroplasts and photosynthesize. They have chlorophyll *a*, chlorophyll *b*, and carotenoids, which are the same pigments found in green algae and plants.[3] Their food is stored as paramylon, a polysaccha-

[3]Although the euglenoids have the same pigments as the green algae and plants, they are not thought to be closely related to either group.

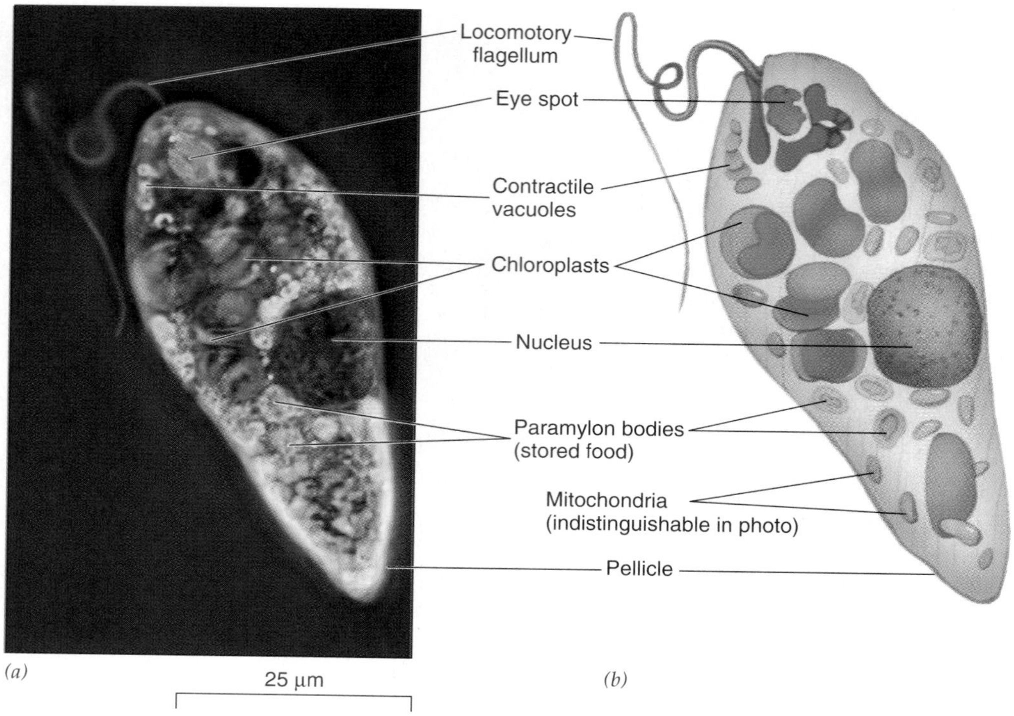

FIGURE 24–10 Euglenoids are unicellular, flagellated algae. Euglenoids have at various times been classified in the plant kingdom (with the algae) and in the animal kingdom (when protozoa were considered animals). (***a***) A living *Euglena.* (***b***) *Euglena* has a complex cellular structure. The eyespot is a light-sensitive organelle that helps it react to light. Its outer covering, called a pellicle, is flexible and enables *Euglena* to change shape easily. (Biophoto Associates/Photo Researchers, Inc.)

ride. Some photosynthetic euglenoids lose their chlorophyll when grown in the dark and obtain their nutrients heterotrophically by ingesting organic matter. Other species of euglenoids are always colorless and heterotrophic.

Euglenoids inhabit freshwater ponds and puddles, particularly those with large amounts of organic material. For that reason they are used as indicator species of organic pollution. If a body of water has unusually large numbers of euglenoids, it is probably polluted. Marine waters and mud flats are also inhabited by some euglenoid species.

Green algae share many similarities with plants

Green algae (phylum Chlorophyta) have pigments, storage products, and cell walls that are chemically identical to those of plants. Green algae are photosynthetic, with chlorophyll *a,* chlorophyll *b,* and carotenoids present in chloroplasts of a wide variety of shapes. Starch is the main food reserve. Most green algae possess cell walls with cellulose, although some lack walls. Because of these and other similarities, it is generally accepted that plants arose from ancestral green algae. Taxonomy of the green algae is currently under study, and research in ultrastructure (cell structure studied with the aid of electron microscopy) and biochemistry is providing insights into this extremely diverse group.

Green algae exhibit a variety of body forms, from single cells to colonial forms to coenocytic, siphonous (tubular) algae to multicellular filaments and sheets (Fig. 24–11). The multicellular forms do not have cells differentiated into tissues, however. Most green algae are flagellated during at least part of their life history, although a few are totally nonmotile.

Reproduction in the green algae is as varied as their body forms. Both sexual and asexual reproduction occur. Asexual reproduction may be by cell division in single cells or by fragmentation in multicellular forms. Many green algae produce spores asexually by mitosis; if these spores are flagellated and motile, they are called **zoospores** (Fig. 24–12). Sexual reproduction in the green algae involves the formation of gametes in unicellular gametangia. Three types of sexual reproduction—isogamous, anisogamous, and oogamous—are recognized in green algae. If the two flagellated gametes that fuse are identical in size and appearance, sexual reproduction is said to be **isogamous** (Fig. 24–12). **Anisogamous** sexual reproduction involves the fusion of two flagellated gametes of different sizes. Some green algae are **oogamous** and produce a nonmotile egg and a flagellated male ga-

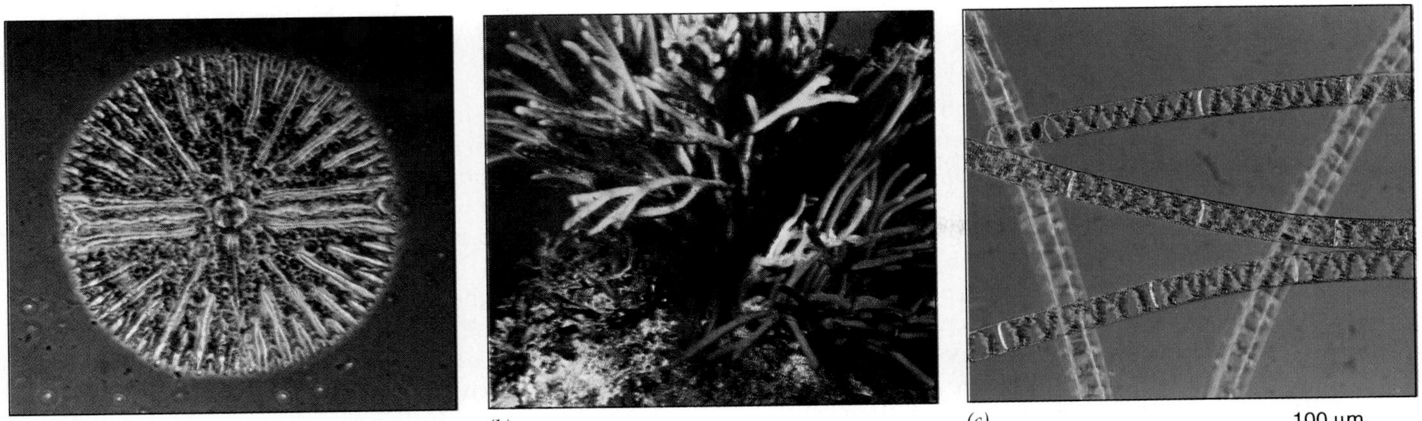

FIGURE 24–11 Green algae are a large, extremely diverse group. (***a***) A desmid *(Micrasterias)* is a unicellular green alga with mirror-image halves. (***b***) Underwater view of dead man's fingers *(Codium fragile)*, common off the northeastern coast of the United States. A siphonous green alga like *Codium* is coenocytic, which means that its body is composed of one giant cell with multiple nuclei. (***c***) *Spirogyra* is a multicellular green alga with a filamentous body form. Note the spiral-shaped chloroplasts. (*a,* Ronald W. Hoham; *b,* Visuals Unlimited/William C. Jorgensen; *c,* Brian Parker/Tom Stack & Associates)

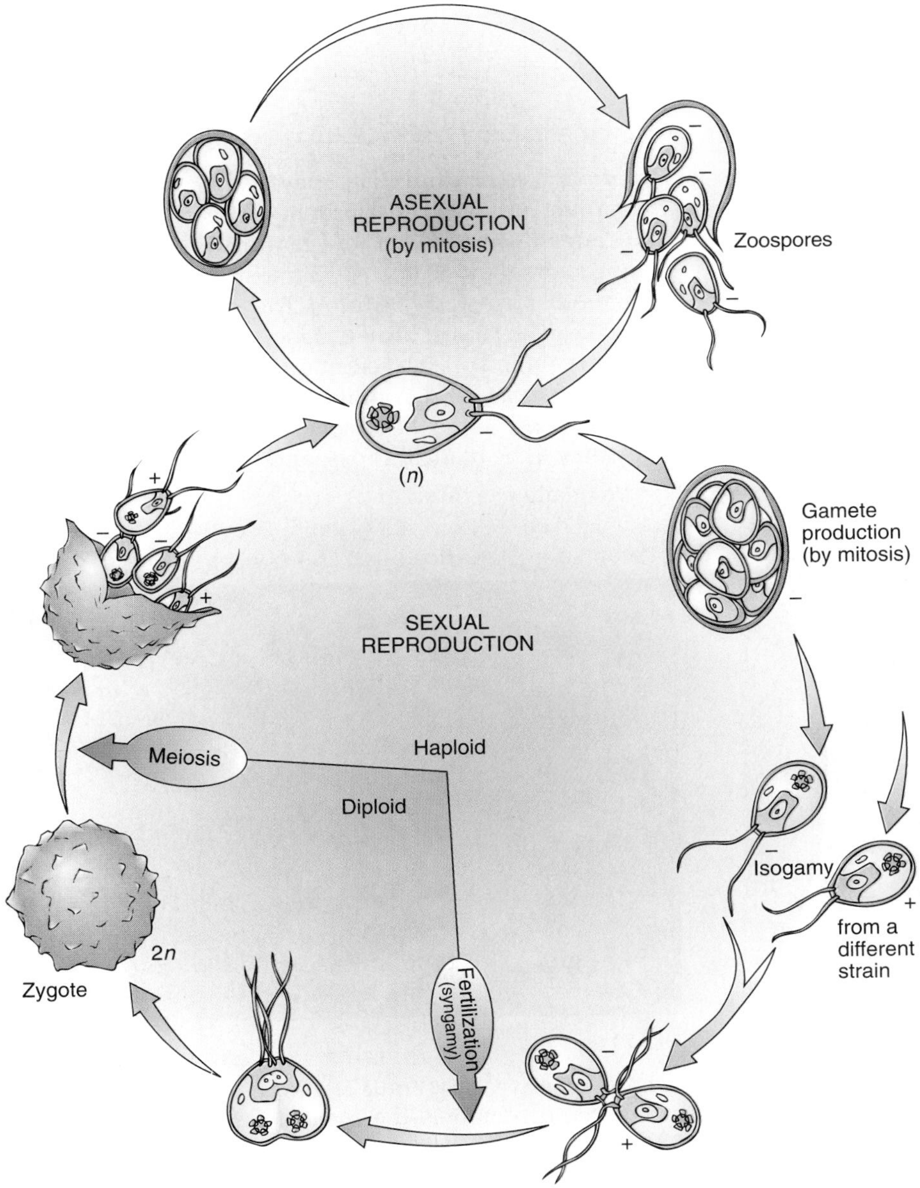

FIGURE 24–12 The life cycle of *Chlamydomonas* includes both asexual and sexual reproduction. *Chlamydomonas* is a haploid green alga with two strains, (+) and (−), that are visually indistinguishable. Both strains reproduce asexually by mitosis. During sexual reproduction, a (+) gamete fuses with a (−) gamete, forming a diploid zygote. (This is an example of isogamous sexual reproduction.) Meiosis occurs, and four haploid cells emerge, two (+) and two (−).

mete. In addition to sexual reproduction by the fusion of gametes, some green algae exchange genetic information by a form of conjugation, in which the genetic material of one cell passes into a recipient cell.

Both aquatic and terrestrial forms of green algae occur. Aquatic green algae primarily inhabit fresh water, although there are a number of marine species. Terrestrial green algae are restricted to damp soil, cracks in tree bark, and other moist places. Many of the green algae are symbionts with other organisms; some live in body cells of invertebrates, and others grow together with fungi as "dual organisms" called lichens (see Chapter 25). Regardless of where they live, green algae are ecologically important as the base of the food web.

Red algae do not produce motile cells

The vast majority of **red algae** (phylum Rhodophyta) are multicellular organisms, although there are a few unicellular species. The multicellular body form of red algae is commonly composed of complex, interwoven filaments that are delicate and feathery (Fig. 24–13*a*), although a few red algae are flattened sheets of cells. Most multicellular red algae attach to rocks or other substrates by a rootlike **holdfast.** The chloroplasts of red algae contain **phycoerythrin,** a red pigment, and **phycocyanin,** a blue pigment, in addition to chlorophyll *a* and carotenoids. Their storage product is floridean starch, a polysaccharide similar to glycogen. The red algae have the same pigment composition as the cyanobacteria (a group of photosynthetic bacteria; see Chapter 23), supporting the hypothesis that red algal chloroplasts descended from cyanobacterial endosymbionts (discussed later in this chapter).

Reproduction in the red algae has been studied in detail for only a few species, but it is amazingly complex, with an alternation of sexual and asexual stages. Although sexual reproduction is common, at no stage in the life history of red algae do any flagellated cells develop.

The cell walls of red algae often contain mucilaginous polysaccharides that are of commercial value. For example, gelatin-like agar is extracted from certain red algae and used to make a culture medium for growing microorganisms such as bacteria. A second polysaccharide extracted from red algae is carrageenan, used to stabilize puddings, laxatives, ice creams, and toothpastes. People consume red algae as vegetables, particularly in Japan, Korea, and China.

The red algae are primarily found in warm tropical oceans, although many species occur in cooler ocean waters and a few in fresh water or soil. Some red algae incorporate calcium carbonate into their cell walls (Fig. 24–13*b*). These coralline red algae are extremely important in building coral reefs, along with the coral animals themselves.

Brown algae are multicellular seaweeds

Brown algae (phylum Phaeophyta) include the giants of the protist kingdom. All brown algae are multicellular and range in size from a few centimeters (an inch or so) to approximately 60 meters (almost 200 feet) in length. Their body forms may be tufts, "ropes," or thick, flattened branches. The largest brown algae, called kelps, are tough and leathery in appearance; many possess leaflike **blades,** stemlike **stipes,** and rootlike anchoring holdfasts (Fig. 24–14*a*). They often have gas-filled floats that provide buoyancy. (The blades, stipes, and holdfasts of brown al-

(*a*)

(*b*)

FIGURE 24–13 Red algae are a diverse group characterized by a complete lack of flagellated cells. (***a***) Most red algae are multicellular, many with complex filamentous bodies, such as *Bonnemaisonia hamifera.* (***b***) Coralline red algae, such as *Bossiella* sp., have hard, brittle bodies encrusted with calcium carbonate. (*a,* D.P. Wilson and David Hosking/Photo Researchers, Inc.; *b,* Visuals Unlimited/D. Gotshall)

(a)

(b)

FIGURE 24–14 Brown algae are structurally diverse, and all are multicellular. (**a**) *Laminaria* is a typical brown alga. Note its blade, stipe, and holdfast (visible in this photograph because *Laminaria* was removed from the water). (**b**) A kelp bed off the coast of California. These underwater forests are ecologically important, supporting large numbers of aquatic organisms. (*a,* J. Robert Waaland, University of Washington/Biological Photo Service; *b,* Richard Herrmann)

gae are not homologous to the leaves, stems, and roots of plants. Brown algae and plants arose from different unicellular ancestors.)

Brown algae are photosynthetic and possess chlorophyll *a*, chlorophyll *c*, and carotenoids in their chloroplasts. A special yellow-brown carotenoid, **fucoxanthin,** is found only in brown algae, dinoflagellates, and diatoms. The main food storage reserve in brown algae is a carbohydrate called laminarin.

Reproduction is varied and complex in the brown algae. They reproduce sexually, and most spend a portion of their life cycles as haploid organisms and a portion as diploid organisms. Their reproductive cells, both asexual zoospores and sexual gametes, are flagellated.

Brown algae are commercially important for several reasons. They have a polysaccharide called algin in their cell walls, possibly to help cement the cell walls together. Algin is used as a thickening agent in ice cream, marshmallows, and cosmetics. Brown algae are an important human food, particularly in East Asian countries, and they are rich sources of minerals such as iodine.

Brown algae are common in cooler marine waters, especially along rocky coastlines, where they can be found mainly in the intertidal zone or relatively shallow offshore waters. Kelps form extensive underwater "forests" called kelp beds (Fig. 24–14*b*). They are essential in that ecosystem for two reasons: they are the primary food producers and they provide habitats for many marine invertebrates, fish, and mammals. The diversity of life supported by kelp beds rivals that found in coral reefs. There is also an extensive population of floating brown algae in a central area of the North Atlantic Ocean called the Sargasso Sea, named for the brown alga *Sargassum*. (The Sargasso Sea is not greatly affected by the surface ocean currents rotating around the margins of the North Atlantic, and so the floating *Sargassum* remain there.)

SLIME MOLDS AND WATER MOLDS ARE FUNGUS-LIKE PROTISTS

Some protists superficially resemble fungi in that they are not photosynthetic and their bodies are often formed of threadlike structures called *hyphae*. However, fungus-like protists are not fungi for several reasons. Many produce flagellated cells, which fungi lack. Many of these protists

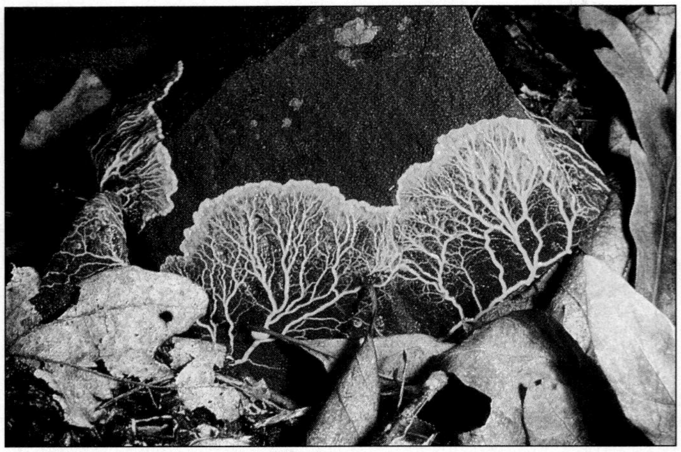

(*a*)

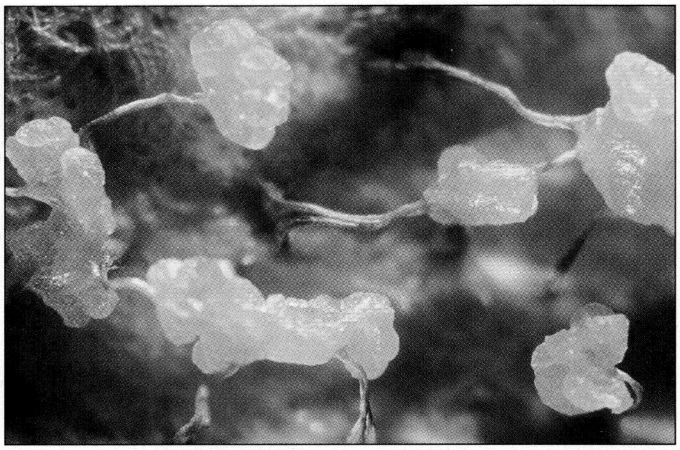

(*b*)

FIGURE 24–15 Plasmodial slime molds are characterized by a creeping plasmodium and the production of spores. (***a***) The plasmodium of *Physarum* is colored bright yellow. This naked mass of protoplasm is multinucleate and feeds on bacteria and other microorganisms. (***b***) The reproductive structures of plasmodial slime molds such as *Physarum* are often stalked sporangia. (*a,* P. W. Grace/Photo Researchers, Inc.; *b,* Carolina Biological Supply Company/Phototake–NYC)

also have centrioles and produce cellulose as a major component of their cell walls, whereas fungi lack centrioles and have cell walls of chitin. We will consider three groups of fungus-like protists: plasmodial slime molds, cellular slime molds, and water molds (see Table 24–1).

Plasmodial slime molds feed as multinucleate plasmodia

The feeding stage of a **plasmodial slime mold** (phylum Myxomycota) is a **plasmodium,** a multinucleate mass of cytoplasm (Fig. 24–15*a*). The plasmodium, which is slimy in appearance, streams over damp, decaying logs and leaf litter, often forming a network of channels to cover a larger surface area. As it creeps along, it ingests bacteria, yeasts, spores, and decaying organic matter.

When the food supply dwindles or there is insufficient moisture, the plasmodium crawls to an exposed surface and initiates reproduction. Stalked structures of intricate complexity and beauty usually form from the drying plasmodium (Fig. 24–15*b*). Within these structures, called **sporangia,** meiosis produces haploid **spores** that are extremely resistant to adverse environmental conditions.

When conditions become favorable again, the spores germinate and a haploid reproductive cell emerges from each. Each cell is either biflagellate (called a *swarm cell*) or amoeboid (called a *myxamoeba*), depending on the moisture available, and can readily convert from one form to the other as conditions change. Swarm cells and myxamoebas act as gametes. Eventually two fuse, and the resultant diploid zygote divides by mitosis without cytoplasmic division to form a multinucleate plasmodium.

Cellular slime molds feed as single amoeboid cells

Although organisms in the phylum Acrasiomycota are called **cellular slime molds,** their resemblance to the plasmodial slime molds is superficial. Indeed, they have much closer affinities with the amoebas. During its feeding stage, each cellular slime mold is an individual amoeboid cell that behaves as a separate, solitary organism. Each cell creeps over rotting logs and soil or swims in fresh water, ingesting bacteria and other particles of food as it goes. Each cell has a haploid nucleus.

When moisture or food becomes inadequate, the individual cells send out a chemical signal, cyclic AMP (cyclic 3′, 5′–*a*denosine *m*ono*p*hosphate) (see Fig. 3–29), that causes them to aggregate by the hundreds or thousands for reproduction. During this stage the cells creep about for short distances as a single multicellular unit, called a **pseudoplasmodium,** or "slug." Each cell of the slug retains its plasma membrane and individual identity. Eventually the slug settles and constructs a stalked fruiting body, which is a reproductive structure that bears spores. After being released, each spore opens, a single amoeboid cell emerges, and the cycle repeats itself (Fig. 24–16). This reproductive cycle is asexual, although sexual reproduction has been observed occasionally. There are no flagellated stages for most cellular slime molds.

Water molds produce flagellated reproductive cells

Water molds (phylum Oomycota) were once classified as fungi because of their superficial resemblance. Both wa-

(Text continues on page 547)

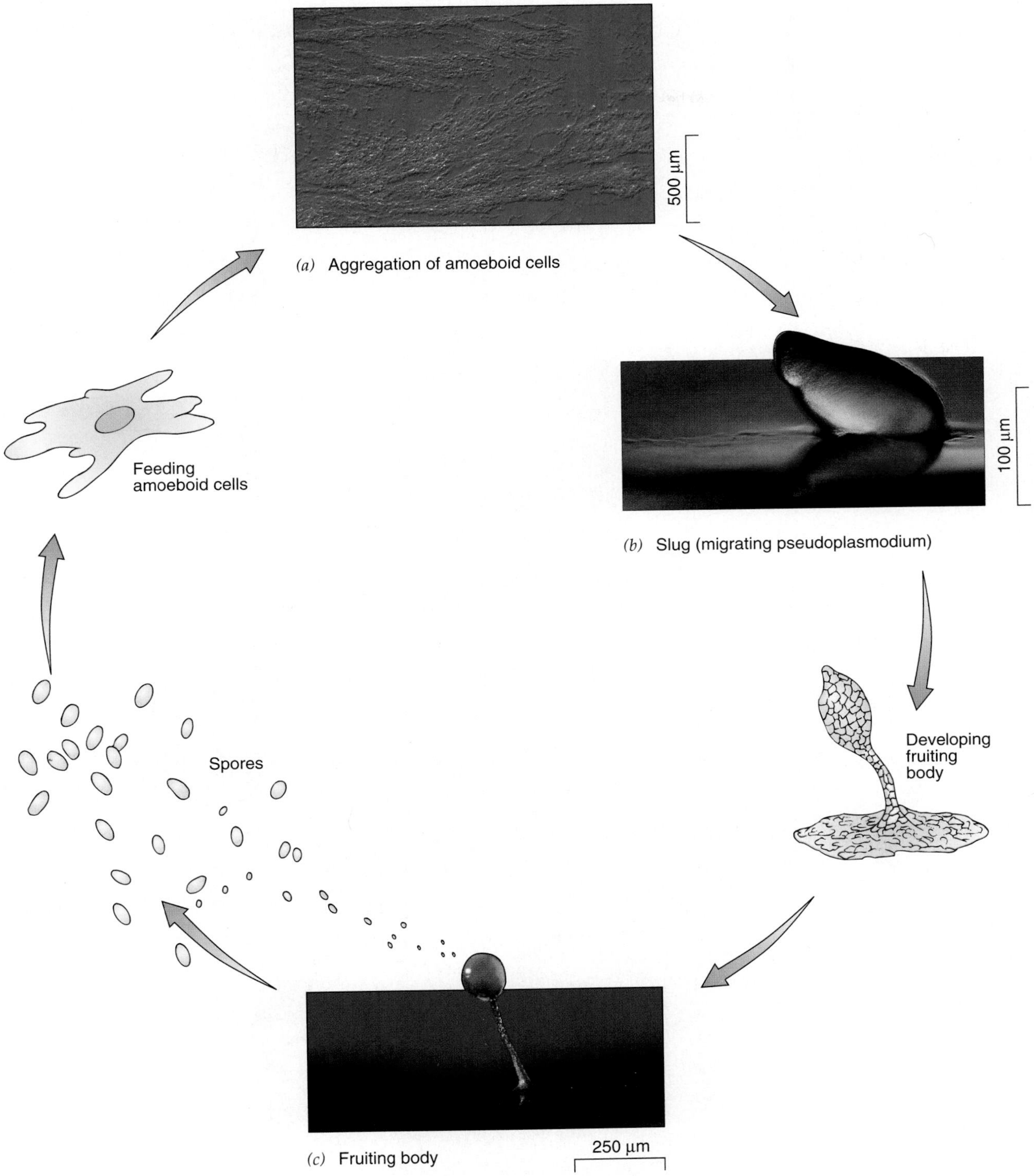

FIGURE 24–16 The life cycle of the cellular slime mold *Dictyostelium discoideum* includes a unicellular stage and a multicellular pseudoplasmodium. Unicellular amoeboid cells (*upper left*) ingest food, grow, and reproduce by cell division. (***a***) After their food supply is depleted, the cells stream together. (***b***) The aggregation organizes into a slug-shaped, multicellular pseudoplasmodium that migrates for a period of time before forming a stalked fruiting body (***c***). The mature fruiting body releases spores, each of which opens to liberate an amoeba-like unicellular organism, and the cycle continues. (*a, b, c,* Carolina Biological Supply Company/Phototake–NYC)

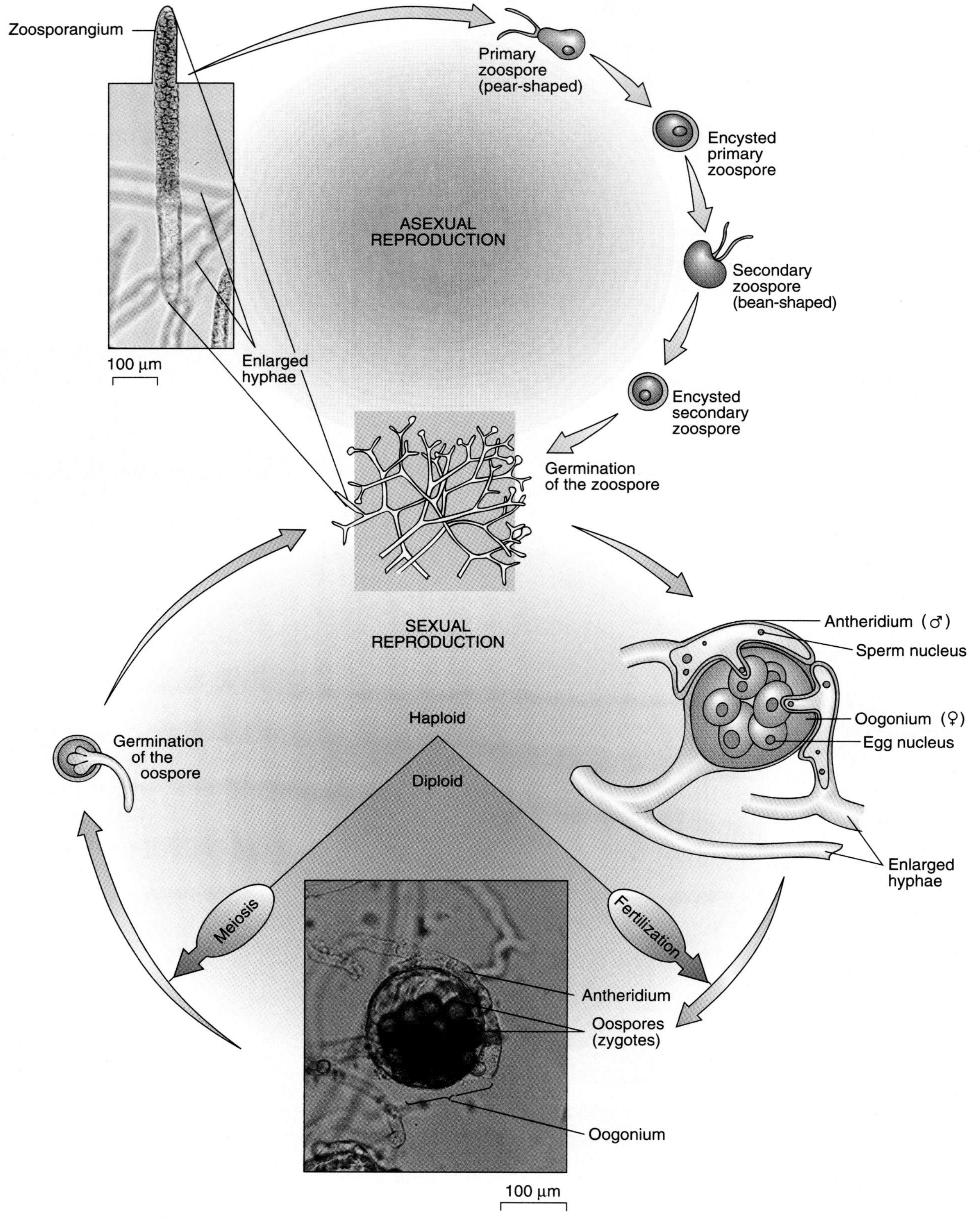

FIGURE 24–17 The life cycle of a typical water mold includes both sexual and asexual reproduction. *Saprolegnia*, a common water mold, reproduces both sexually (by antheridia and oogonia) and asexually (by zoosporangia). (*Top left,* Carolina Biological Supply Company/Phototake–NYC; *Bottom,* J.W. Richardson/CBR Images)

Making the Connection

The Evolutionary Relationships Between Protists and Other Eukaryotes

How are protists related to the other eukaryotic kingdoms? Plants, animals, and fungi are thought to have their ancestry in the protist kingdom. The protistan ancestor of plants is generally accepted, whereas both animal and fungal origins are unclear.

Green algae are regarded as the ancestors of plants, in part because of identical pigments and storage products. In addition, some green algae share other traits with plants. Both groups have cellulose in their cell walls and share details of mitosis, including the production of a cell plate during cytokinesis.

Choanoflagellates bear a striking resemblance to the sponges, and some biologists have suggested that an ancient choanoflagellate may be the ancestor of sponges. One or several other flagellates are probably the ancestors of all other animals, which differ from the sponges because they have tissues organized into organs and organs organized into organ systems. However, at this time biologists are not certain whether there was a single protistan ancestor for animals other than sponges, or several different protistan ancestors for different animal groups.

Fungi are thought to have had protistan ancestors, but their lineage is uncertain. Both fungi and red algae lack motile cells and share similarities in aspects of their sexual reproduction. The resemblance is strongest between the red algae and more complex fungi,* suggesting that these fungi descended from an ancient red alga. Thus, the evolution of the fungi is not clear. Did the more complex fungi evolve from red algae or from other, less complex fungi? If an ancient red alga is the ancestor of more complex fungi, how did other fungi originate? To further complicate the picture, recent molecular evidence suggests that fungi are more closely related to animals than to plants (see Chapter 25). ▲

*The more complex fungi include the ascomycetes and basidiomycetes; the zygomycetes are less complex and considered less advanced (see Chapter 25).

ter molds and fungi have a body, termed a **mycelium,** that grows over organic material, digesting it and then absorbing the predigested nutrients. The thread-like **hyphae** that make up the mycelium in water molds are coenocytic; there are no cross walls, and the body is like one giant multinucleate cell. The cell walls of water molds may be composed of cellulose (as in plants), chitin (as in fungi), or both. Because water molds produce flagellated reproductive cells at some point in their life cycles, whereas fungi *never* produce motile cells, most biologists classify the water molds as protists rather than as fungi.

When food is plentiful and environmental conditions are favorable, water molds reproduce asexually (Fig. 24–17). A hyphal tip swells and a cross wall is formed, separating the hyphal tip from the rest of the mycelium. Within this structure, tiny biflagellate zoospores form, each of which can develop into a new mycelium. When environmental conditions worsen, water molds initiate sexual reproduction. After fusion of male and female nuclei, a thick-walled **oospore** develops from the zygote. Water molds often overwinter as oospores.

Some water molds have played infamous roles in human history. For example, the Irish potato famine of the 19th century was precipitated by the water mold that causes late blight of potatoes. Due to several rainy, cool summers in Ireland in the 1840s, the water mold multiplied unchecked, causing potato tubers to rot in the fields. Since potatoes were the staple of the Irish peasants' diet, many people starved. Estimates of the number of deaths that resulted from the outbreak of this plant disease range from one-quarter million to more than one million. A mass migration out of Ireland to such countries as the United States also ensued. Late blight is still a problem today. In 1992, new strains of the water mold that causes it were reported in New York and Washington states; one year later, late blight was detected in 11 states. These water mold strains are resistant to the fungicide usually used to control the disease.

THE EARLIEST EUKARYOTES WERE PROTISTS

Protists are thought to have been the first eukaryotic cells. They may have originated from prokaryotes as early as 2.1 billion years ago. There is compelling evidence that two eukaryotic organelles, mitochondria and chloroplasts, arose from endosymbiotic relationships between larger prokaryotes and the smaller prokaryotes that lived within them (see Chapter 20). Chloroplasts are thought to have originated from photosynthetic prokaryotes, and mitochondria from aerobic bacteria.

A few protists with hardened shells — for example, diatoms and foraminiferans — produced abundant fossils. However, most ancient protists did not leave extensive fossil records because their bodies were too soft to leave permanent traces. Therefore, evolutionary studies of protists are based primarily on comparisons of present-day organisms, which contain many clues about their evolutionary history. Some of the most useful data for evolu-

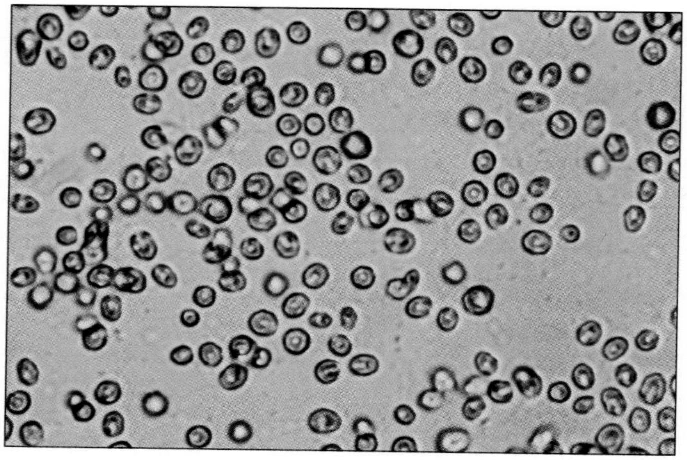

(a)

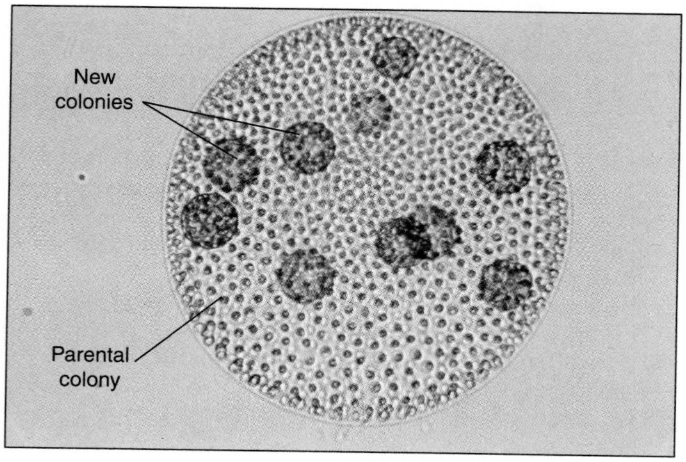

(b)

FIGURE 24–18 Multicellularity may have evolved in this sequence: unicellular organisms → colonial organisms → multicellular organisms. (*a*) *Chlamydomonas* is a biflagellate unicellular green alga. (*b*) *Volvox* is a colonial green alga composed of up to 50,000 *Chlamydomonas*-like cells. New colonies can be observed inside the parent colony, which eventually breaks apart to release them. (*a, b,* Visuals Unlimited/James W. Richardson)

tionary interpretations are ultrastructural studies of cellular organelles. In addition to electron microscopy, biochemistry and molecular biology also provide important insights (see section on molecular comparisons among organisms in Chapter 17).

Multicellularity evolved in the protist kingdom

Green algae, red algae, and brown algae are examples of protist groups with multicellular species. However, multicellular green algae have more in common with unicellular green algae than with other multicellular protists. Similarly, multicellular forms of both red and brown algae have little in common with each other or with the multicellular green algae. Because these groups are so different, it is likely that multicellularity arose in the protist kingdom several different times. That is, the multicellular green algae, red algae, and brown algae probably had different unicellular protistan ancestors.

Studying the protists living today provides clues about how multicellularity may have arisen. For example, *Chlamydomonas* (Fig. 24–18*a*) is a unicellular green alga that uses two flagella for motility. Green algae include a number of colonial species composed of attached *Chlamydomonas*-like cells. For example, *Gonium* is a colonial green alga that consists of four *Chlamydomonas*-like cells, and *Pandorina* is a colonial organism composed of 16 to 32 *Chlamydomonas*-like cells. The largest colonies, containing from 1000 to 50,000 *Chlamydomonas*-like cells, are in the genus *Volvox* (Fig. 24–18*b*). As colonies increase in size and number of cells, specialization in cell structure and function occurs, with *Volvox* demonstrating an obvious division of labor among the cells.

The *Chlamydomonas* line is an evolutionary dead end that did not give rise to organisms with greater complexity. However, the trend in increasing colony size and cell differentiation within the *Chlamydomonas* line indicates one possible way that multicellularity may have originated:

single cells → colonies → multicellular organisms

SUMMARY

I. Kingdom Protista is composed of "simple" eukaryotic organisms, most of which are unicellular and live in aquatic environments.
   A. Protists range in size from microscopic single cells to multicellular organisms 60 meters (nearly 200 feet) long.
   B. Protists obtain their nutrients autotrophically or heterotrophically.
   C. Protists may be free-living or endosymbiotic, with symbiotic relationships ranging from mutualism to parasitism.
   D. Many protists reproduce both sexually and asexually; others reproduce only asexually.
   E. Protists have various means of locomotion, including pseudopodia, flagella, and cilia. Some are nonmotile.

II. Protozoa are the heterotrophic, animal-like protists.

A. Amoebas move and obtain food using cytoplasmic extensions called pseudopodia.
B. Foraminiferans secrete many-chambered shells with pores through which cytoplasmic projections extend to move and obtain food.
C. Actinopods move and obtain food by means of axopods, slender cytoplasmic projections that extend through pores in their skeletons.
D. Flagellates are heterotrophic protozoa that move by means of flagella.
E. Ciliates move by cilia, have micronuclei and macronuclei, and undergo complex reproduction.
F. Sporozoans are parasites that produce spores and are nonmotile. A sporozoan causes malaria.

III. Algae are autotrophic, plantlike protists.
A. Dinoflagellates are mostly unicellular, biflagellate, photosynthetic organisms of great ecological importance.
B. Diatoms, which are major producers in aquatic ecosystems, are mostly unicellular, with shells containing silica.
C. Euglenoids are unicellular, flagellated algae. Many are not photosynthetic.
D. Green algae exhibit a wide diversity in size, structural complexity, and reproduction.
E. Red algae, which are mostly multicellular seaweeds, are ecologically important in warm tropical oceans. They lack motile cells and have complex reproductive cycles.
F. Brown algae are multicellular seaweeds that are ecologically important in cooler ocean waters. They produce flagellated cells during their complex reproductive cycles.

IV. Fungus-like protists were originally classified with the fungi, but have features that are clearly protistan.
A. The feeding stage of plasmodial slime molds is a multinucleate plasmodium. Reproduction is by spores.
B. Cellular slime molds feed as individual amoeboid cells. They reproduce by aggregating into a pseudoplasmodium (slug), then forming asexual spores.
C. Water molds have a coenocytic mycelium. They reproduce asexually by forming biflagellate zoospores and sexually by forming oospores.

V. Protists originated about 2.1 billion years ago and were the first eukaryotes.
A. Multicellularity probably evolved several times within the kingdom Protista, possibly by following a trend in greater complexity from single cells to colonies to multicellular organisms.
B. Plants, animals, and fungi originated from protistan ancestors. Green algae are the probable ancestors of plants, red algae may have given rise to fungi, and choanoflagellates may have been the ancestors of sponges. The ancestor of other animals is probably one or more other flagellates.

SELECTED KEY TERMS

actinopod
alga
amoeba
anisogamy
axopod
blade
brown alga
carotenoid
cellular slime mold
ciliate
coenocytic
conjugation
diatom
dinoflagellate
endosymbiont
euglenoid
flagellate
foraminiferan
fucoxanthin
gametangium
green alga
holdfast
hyphae
isogamy
macronucleus
micronucleus
mycelium
oogamy
oospore
phycocyanin
phycoerythrin
plankton
plasmodial slime mold
plasmodium
protist
protozoa
pseudoplasmodium
pseudopodium
red alga
red tide
sporangium
spore
sporozoan
stipe
syngamy
test
trichocyst
water mold
zoospore
zooxanthellae

POST-TEST

1. Unicellular or simple multicellular organisms that possess a eukaryotic cellular organization are called __________ .
2. The floating, often microscopic organisms that are the base of food webs in aquatic ecosystems are collectively called __________ .
3. Protists that ingest their food as animals do are informally called __________ .
4. Amoebas move and obtain food by means of __________ .
5. Foraminiferans secrete many-chambered shells called __________ .
6. Some actinopods have long __________ that protrude through pores in their skeletons.
7. Unicellular protozoa that are free-living or parasitic, move by means of flagella, and do not photosynthesize are called __________ .
8. *Paramecium* and other __________ move by means of cilia.
9. The ciliates often display a sexual phenomenon called __________ .
10. The __________ are a group of parasitic protozoa that form spores at some stage in their life.

11. __________ are algae characterized by two flagella, one wrapped around the center of the cell like a belt and the other projecting behind the cell.
12. A dinoflagellate bloom is known as a __________ __________.
13. The __________ are photosynthetic protists with shells composed of two halves that fit together like a petri dish.
14. Chlorophyll *a*, chlorophyll *b*, and carotenoids are found in green algae, __________, and plants.
15. The __________ __________ have pigmentation similar to the cyanobacteria.
16. Agar and carrageenan are economically important products derived from __________ __________.
17. The multicellular bodies of __________ __________ are differentiated into blades, stipes, holdfasts, and gas-filled floats.
18. The feeding stage of plasmodial slime molds is a multinucleate __________.
19. The __________ __________ __________ behave as unicellular organisms until reproduction, when they aggregate to form a slug.
20. Water molds reproduce asexually by forming biflagellate __________, and sexually by forming __________.

REVIEW QUESTIONS

1. What are the features of a typical protist? Why are protists so difficult to characterize?
2. How are the protists important to humans? How are they important ecologically?
3. What are the three main informal groups of protists? Describe and give at least three examples of each group.
4. Distinguish among each of the following protozoan groups: (a) amoebas, foraminiferans, and actinopods; (b) flagellates, ciliates, and sporozoans.
5. Distinguish among each of the following algal groups: (a) Dinoflagellates, diatoms, and euglenoids; (b) Green algae, red algae, and brown algae.
6. Distinguish among plasmodial slime molds, cellular slime molds, and water molds. What features do they share?
7. How is the structure of each of the following types of protists related to its way of life? (a) flagellates; (b) green algae; (c) water molds.

YOU MAKE THE CONNECTION

1. A few biologists still classify the protozoa as animals, the algae as plants, and the water molds as fungi. Explain their rationale. Why do most biologists classify these organisms as protists?
2. Plasmodial slime molds reproduce sexually, but cellular slime molds do not. What advantages might there be for each organism to have the type of reproduction it does?
3. How does the study of the evolutionary history of protists pertain to the Cheshire cat? (Recall that in Lewis Carroll's *Alice's Adventures in Wonderland,* the Cheshire cat would vanish, leaving only its smile.)

RECOMMENDED READINGS

Anderson, D.M. "Red Tides." *Scientific American,* August 1994. Red tides have proliferated in recent years, possibly the result of increased pollution.

Kunzig, R. "Invisible Garden." *Discover,* April 1990. Although they are invisible, microscopic algae are ecologically important to the biosphere.

Margulis, L. and R. Gurrero. "Kingdoms in Turmoil." *New Scientist,* March 23, 1991. Highlights the widely differing interpretations of protist taxonomy.

Sharnoff, S.D. "Beauties from a Beast: Woodland Jekyll and Hydes." *Smithsonian,* July 1991. Beautiful photographs of the sporangia of slime molds.

C H A P T E R 25

Kingdom Fungi

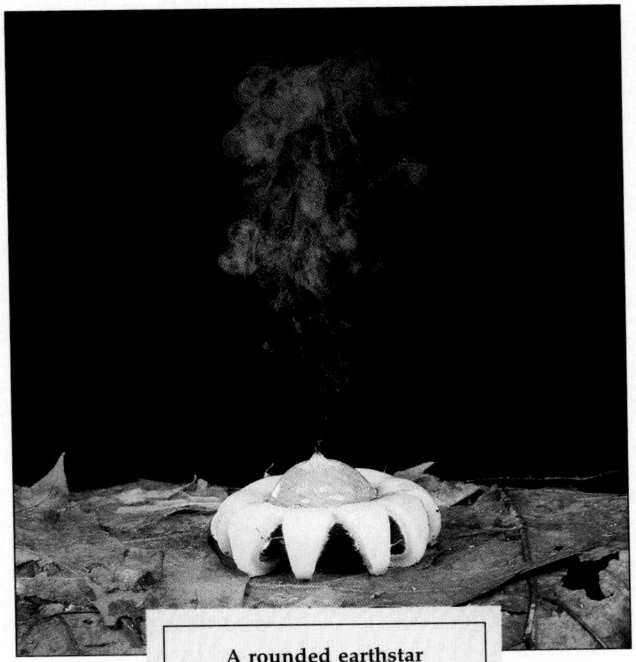

A rounded earthstar (*Geastrum saccatum*) releases a puff of microscopic spores after the sac wall is hit by a raindrop. This fungus is common in leaf litter under trees. (Jeff Lepore/Photo Researchers, Inc.)

Mushrooms, morels, and truffles—delights of the gourmet—have much in common with the black mold that forms on stale bread and the mildew that collects on damp shower curtains. These life forms belong to the kingdom Fungi, a diverse group of more than 80,000 known species, most of which are terrestrial.

Although they vary strikingly in size and shape, all **fungi** (sing. *fungus*) are eukaryotes; their cells contain membrane-bounded nuclei, mitochondria, and other organelles.

Like plant cells, fungal cells are enclosed by cell walls during at least some stage in their life cycle. Fungal cell walls, however, have a different chemical composition than that of plant cell walls. In most fungi, the cell wall is composed of **chitin,** a polymer that consists of subunits of a nitrogen-containing sugar (see Fig. 3–11). Chitin, which is also a component of the external skeletons of insects and other arthropods, is far more resistant to breakdown by microorganisms than is cellulose, which makes up plant cell walls.

Fungi lack chlorophyll and chloroplasts and are not photosynthetic. As heterotrophs they are unable to synthesize their own organic material, but they do not ingest food as animals do. Instead, fungi secrete digestive enzymes and then *absorb* the predigested food (as small organic molecules) through their cell walls and plasma membranes. They obtain their nutrients from dead organic matter (as decomposers) or from other living organisms (as parasites). As decomposers, fungi (along with the bacteria; see Chapter 23) play an important ecological role.

Fungi grow best in dark, moist habitats, but they are found universally wherever organic material is available. They require moisture to grow, and they can obtain water from a humid atmosphere as well as from the medium on which they live. When the environment becomes extremely dry, fungi survive by going into a resting stage or by producing spores that are resistant to desiccation (drying out). Although the optimum pH for most species is about 5.6, different fungi can tolerate and grow in environments where the pH ranges from 2 to 9. Many fungi are less sensitive to high osmotic pressures than are bacteria. As a result, they can grow in concentrated salt solutions, or in sugar solutions such as jelly, that discourage or prevent bacterial growth. Fungi may also thrive over a wide temperature range. Even refrigerated food is not immune to fungal invasion.

Fungi were originally classified in the plant kingdom, but biologists today recognize that they are not plants. Interestingly, recent studies suggest that fungi are more closely related to animals than to plants. Because fungi are distinct from plants, animals, and other eukaryotes in many ways, they are assigned to a separate kingdom.

LEARNING OBJECTIVES

After you have studied this chapter you should be able to

1. Describe the distinguishing characteristics of the kingdom Fungi.
2. Contrast the body plan of a yeast with that of a mold.
3. Trace the fate of a fungal spore that lands in an appropriate location, such as an overripe peach, and describe conditions that permit fungal growth.
4. List distinguishing characteristics, describe a typical life cycle, and give examples of each of the following fungal groups: zygomycetes, ascomycetes, basidiomycetes, and imperfect fungi.
5. Characterize the unique nature of a lichen.
6. Explain the ecological significance of fungi as decomposers, and describe the special ecological role of mycorrhizae.
7. Summarize the various uses of fungi.
8. Identify at least three fungal diseases of plants and three of humans.

MOST FUNGI HAVE A FILAMENTOUS BODY PLAN

The body structures of fungi vary in complexity, ranging from the single-celled yeasts to multicellular, filamentous molds (a term used loosely to include the mildews, rusts and smuts, mushrooms, and many other fungi). Most fungi are filamentous molds (Fig. 25–1*a*). A mold consists of long, branched threads (or filaments) of cells called **hyphae** (sing., *hypha*) (Fig. 25–1*b*). Hyphae form a tangled mass or tissue-like aggregation known as a **mycelium** (pl., *mycelia*). The cobweb-like mold sometimes seen on bread is the mycelium of a fungus. What is not seen is the extensive mycelium that grows down into the bread.

Some **coenocytic** hyphae are not divided into individual cells, and are something like an elongated, multinucleated giant cell (Fig. 25–1*c*). Other hyphae are divided by cross walls, called **septa** (sing., *septum*) into individual cells containing one or more nuclei (Fig. 25–1*d,e*). The septa of many septate fungi are perforated, containing large pores that permit organelles to flow from cell to cell. Cytoplasm flows within the hypha, providing a system of internal transport.

MOST FUNGI REPRODUCE BY SPORES

Fungi reproduce by means of microscopic **spores,** which have no flagella and are nonmotile reproductive cells dispersed by wind or by animals. Spores are usually produced on hyphae that project up into the air. This arrangement permits the spores to be blown to new areas by air

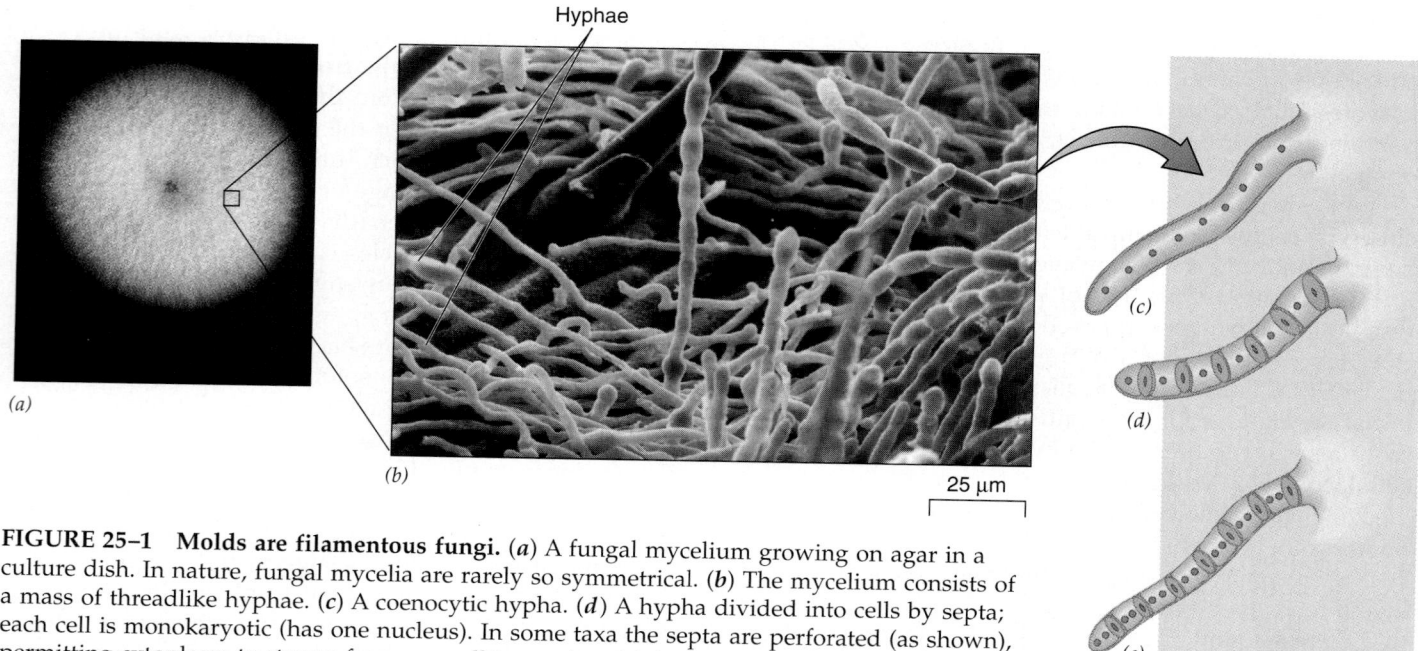

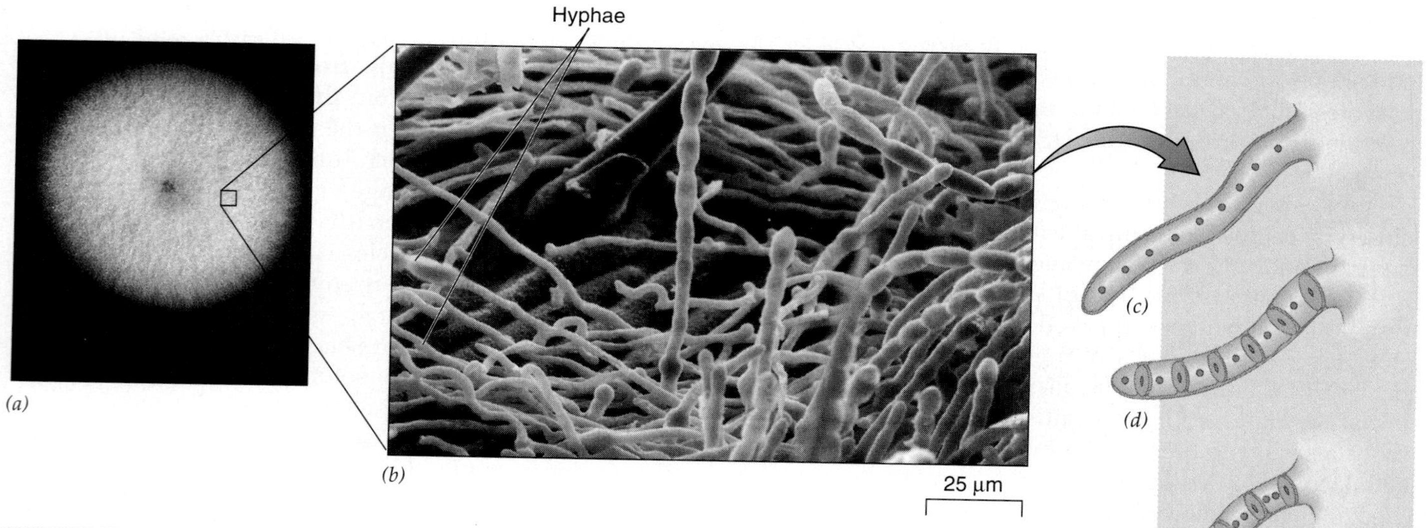

FIGURE 25–1 Molds are filamentous fungi. (***a***) A fungal mycelium growing on agar in a culture dish. In nature, fungal mycelia are rarely so symmetrical. (***b***) The mycelium consists of a mass of threadlike hyphae. (***c***) A coenocytic hypha. (***d***) A hypha divided into cells by septa; each cell is monokaryotic (has one nucleus). In some taxa the septa are perforated (as shown), permitting cytoplasm to stream from one cell to another. (***e***) A septate hypha in which each cell is dikaryotic (has two nuclei). (*a*, Dennis Drenner; *b*, G.T. Cole, University of Texas/BPS)

Table 25–1 PHYLA OF KINGDOM FUNGI

Phylum	*Common Types*	*Asexual Reproduction*	*Sexual Reproduction*
Zygomycota	Black bread mold	Nonmotile spores form in a sporangium	Zygospores
Ascomycota (sac fungi)	Yeasts, powdery mildews, molds, morels, truffles	Conidia pinch off from conidiophores	Ascospores
Basidiomycota (club fungi)	Mushrooms, bracket fungi, puffballs, rusts, smuts	Uncommon	Basidiospores
Deuteromycota (imperfect fungi)	Molds	Conidia	Sexual stage not observed

currents. In some fungi the aerial hyphae form large, complex reproductive structures in which spores are produced. These structures are called **fruiting bodies.** The familiar part of a mushroom or toadstool is a large fruiting body. We do not normally see the bulk of the fungus, a nearly invisible network of hyphae buried out of sight in the rotting material or soil on which it grows.

Fungi may produce spores either sexually or asexually. Unlike most animal and plant cells, fungal cells usually contain haploid nuclei. In sexual reproduction, the hyphae of two genetically different mating types come together and their nuclei fuse, forming a diploid zygote. In two fungal groups, the ascomycetes and basidiomycetes, the hyphae fuse but the two different nuclei do not fuse immediately; rather, they remain separate within the fungal cytoplasm. Hyphae that contain two genetically distinct nuclei within each cell are **dikaryotic,** which is described as $n + n$ (because there are two separate haploid nuclei) rather than $2n$ (which implies a single diploid nucleus). Hyphae that contain only one nucleus per cell are said to be **monokaryotic.**

When a fungal spore comes into contact with an appropriate food source, perhaps an overripe peach that has fallen to the ground, it germinates and begins to grow. A threadlike hypha emerges from the tiny spore and grows, branching frequently. Soon a tangled mat of hyphae infiltrates the peach, degrading its organic compounds to small molecules that the fungus can absorb. Fungi are extremely efficient at converting nutrients into new cell material. If excessive amounts of nutrients are available, fungi are able to store them in the mycelium.

Fungi are classified into three phyla

Classification of fungi is based mainly on the characteristics of their sexual spores and fruiting bodies. Biologists do not agree on how to classify these diverse organisms, but most assign them to three phyla: Zygomycota, Ascomycota, and Basidiomycota. In addition, fungi with no known sexual stage are assigned to a *form phylum,* Deuteromycota. Members of a form phylum are similar to one another in certain respects but probably do not share a common ancestry; that is, the group is polyphyletic (see Chapter 22). They are classified together simply as a matter of convenience. Table 25–1 summarizes fungal classification. Slime molds and water molds were originally classified as fungi, but are now generally considered protists (see Chapter 24).

Zygomycetes reproduce sexually by forming zygospores

About 800 species of **zygomycetes** (phylum Zygomycota) produce sexual spores, called **zygospores.** Their hyphae are coenocytic; that is, they lack septa. (Septa do form, however, to separate the hyphae from reproductive structures.) Most zygomycetes are decomposers that live in the soil on decaying plant or animal matter, although some are parasites of plants and animals.

One common zygomycete is the black bread mold, *Rhizopus stolonifer,* a decomposer that breaks down bread and other foods (Fig. 25–2). If preservatives are not added, bread left at room temperature often becomes covered with a black, fuzzy growth in a few days. Bread becomes moldy when a spore falls on it and then germinates and grows into a tangled mass of threads, the mycelium. Hyphae penetrate the bread and absorb nutrients. Eventually, certain hyphae grow upward and develop spore sacs called **sporangia** (sing., *sporangium*) at their tips. Clusters of black asexual spores develop within each sporangium and are released when the delicate sporangium ruptures. The spores give the black bread mold its characteristic color.

Sexual reproduction in the black bread mold occurs when the hyphae of two different mating types, designated as plus (+) and minus (−), grow into contact with one another. The bread mold is **heterothallic,** meaning that an individual fungal hypha is self-sterile and mates

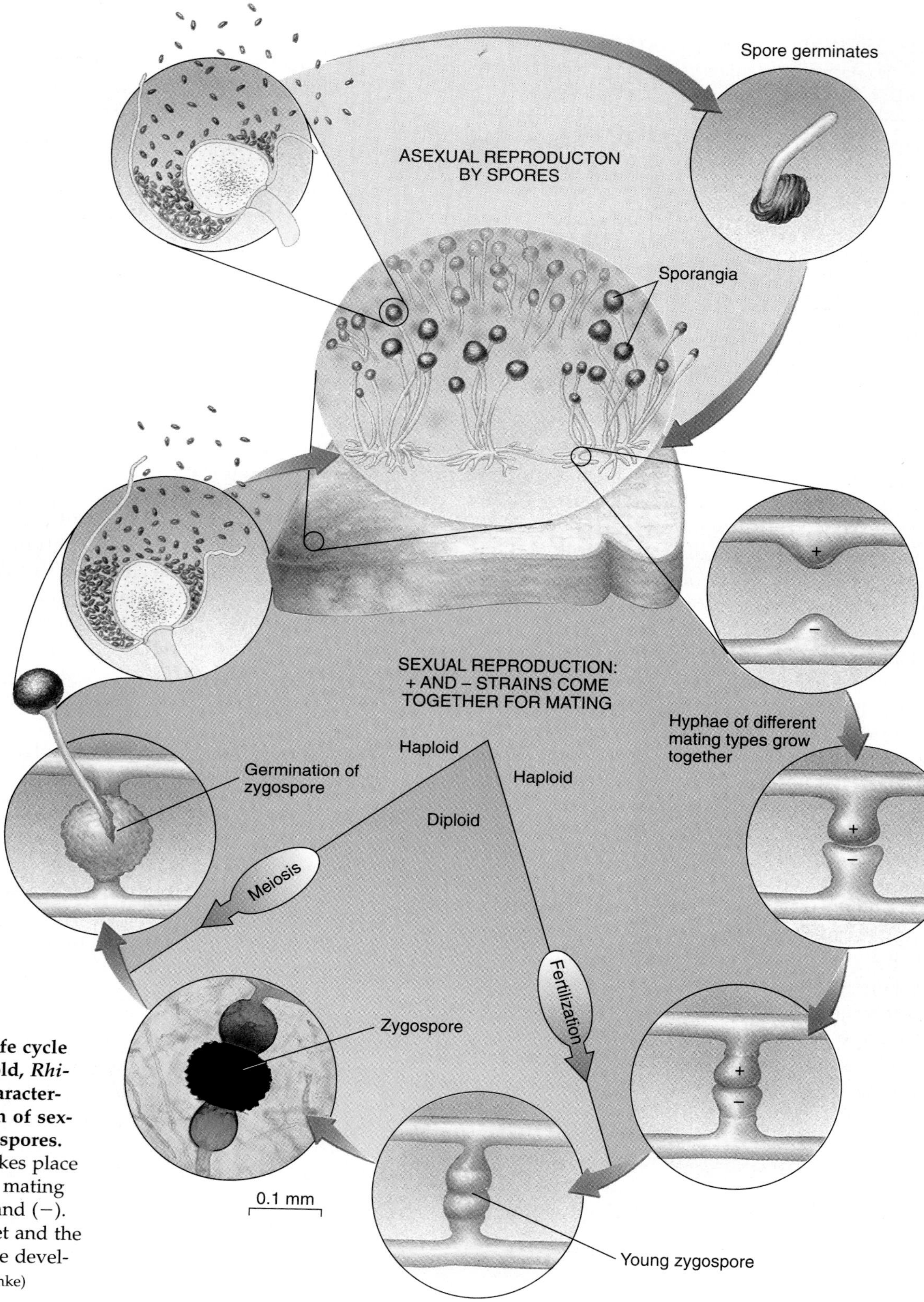

FIGURE 25–2 The life cycle of the black bread mold, *Rhizopus stolonifer*, is characterized by the production of sexual spores called zygospores. Sexual reproduction takes place only between different mating types, designated (+) and (−). After their hyphae meet and the nuclei fuse, a zygospore develops *(inset)*. (Inset, Ed Reschke)

only with a hypha of a different mating type. That is, sexual reproduction occurs only between a member of a (+) strain and one of a (−) strain. Because there are no physical differences between the two mating types, it is not appropriate to refer to them as "male" and "female."

When hyphae of opposite mating types meet, hormones are produced that cause the tips of the hyphae to come together. (+) and (−) nuclei then fuse to form a diploid nucleus, the zygote. A zygospore develops, providing a thick protective covering around the zygote. The zygospore may lie dormant for several months and can survive desiccation and extreme temperatures. Meiosis probably occurs at or just before germination of the zygospore. When the zygospore germinates, an aerial hypha develops (by mitosis) with a sporangium at the tip. Mitosis within the sporangium forms haploid spores, which are released and may germinate to form new hyphae. Only the zygote of a black bread mold is diploid; all of the hyphae and the asexual spores are haploid.

Ascomycetes (sac fungi) reproduce sexually by forming ascospores

Ascomycetes (phylum Ascomycota) comprise a large group of fungi consisting of about 30,000 described species. Ascomycetes are sometimes referred to as *sac fungi* because their sexual spores are produced in little sacs called **asci** (sing., *ascus*). Their hyphae usually have septa, but these cross walls are perforated so that cytoplasm can move from one compartment to another.

The diverse ascomycetes include most yeasts; the powdery mildews; most of the blue-green, pink, and brown molds that cause food to spoil; decomposer cup fungi; and the edible morels and truffles. Some ascomycetes cause serious plant diseases such as Dutch elm disease, chestnut blight, ergot disease (on rye), and powdery mildew (on fruits and ornamental plants).

In most ascomycetes, asexual reproduction involves production of spores called **conidia** (sing., *conidium*), which are pinched off at the tips of certain specialized hy-

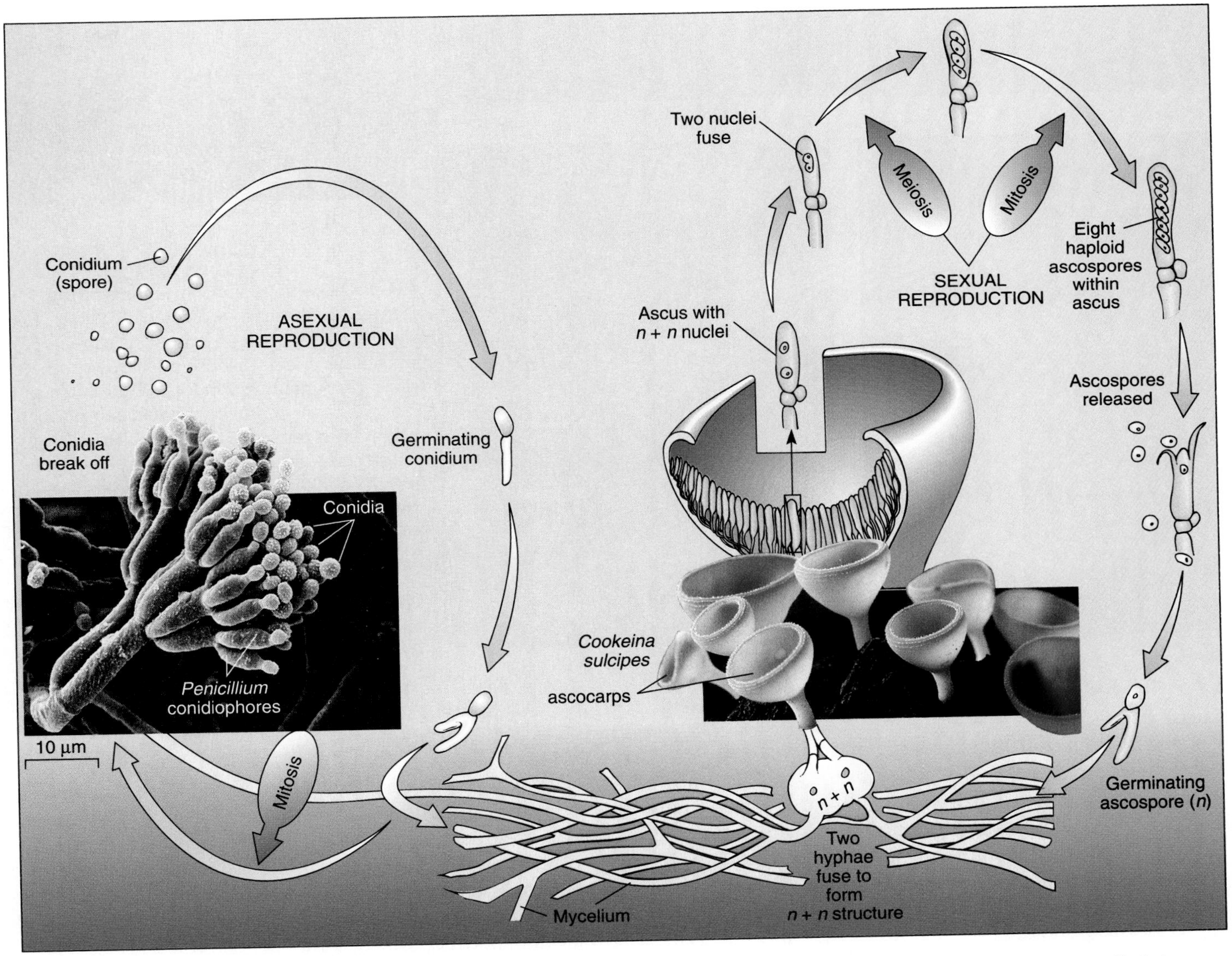

FIGURE 25–3 The life cycle of ascomycetes is characterized by the production of sexual spores called ascospores. *(Right)* Sexual reproduction involves the fusion of two haploid hyphae to form a dikaryotic ($n + n$) structure. Asci form from this structure; in each ascus the $n + n$ nuclei fuse, followed by meiosis and mitosis to produce eight ascospores. The asci form the inner layer of a fruiting body called the ascocarp. Shown are ascocarps of the rainforest cup fungus *Cookeina sulcipes. (Left)* Asexual reproduction involves the formation of haploid conidia. Shown are conidia-bearing conidiophores of *Penicillium.* (*Right,* Animals Animals © 1996 Michael Fogden/Oxford Scientific Films; *Left,* Biophoto Associates/Photo Researchers, Inc.)

(a)

(b)

(c)

(d)

FIGURE 25–4 Basidiomycete fruiting bodies are diverse in form. (***a***) Basidia line the gills of the jack-o'-lantern mushroom, a poisonous species whose gills glow in the dark. (***b***) A giant puffball in White Plains, New York. At maturity, a dried-out puffball often has a pore through which the basidiospores are discharged as a puff of dust. (***c***) The stinkhorn has a foul smell that attracts flies, which help disperse the slimy mass of basidiospores. (***d***) Bracket fungi grow on both dead and living trees, producing shelflike fruiting bodies. Basidiospores are produced in pores located underneath each shelf. (*a,* Dennis Drenner; *b,* Ed Kanze/Dembinsky Photo Associates; *c,* Visuals Unlimited/Richard D. Poe; *d,* Richard H. Gross)

phae known as **conidiophores** (from the Greek, meaning "conidia-bearers") (Fig. 25–3). Sometimes called summer spores, conidia are a means of rapidly propagating new mycelia when environmental conditions are favorable. Conidia occur in various shapes, sizes, and colors in different species. The color of the conidia is what gives the characteristic brown, blue, green, pink, or other tints to many of these molds.

Some species of ascomycetes are heterothallic and have different mating strains; others are **homothallic,** which means that they are self-fertile and have the ability to mate with themselves. In both heterothallic and homothallic ascomycetes, sexual reproduction takes place after two hyphae grow together and their cytoplasm mingles (Fig. 25–3). Within this fused structure, nuclei from the parent hyphae pair but do not fuse. New hyphae develop from the fused structure, and the cells of these hyphae are dikaryotic. These $n + n$ hyphae form a fruiting body, known as an **ascocarp,** that is characteristic of the species. The asci develop in the ascocarp.

Within a cell that develops into an ascus, the two nuclei fuse and form a diploid nucleus, the zygote, which then undergoes meiosis to form four haploid nuclei. This process is usually followed by one mitotic division of each of the four nuclei, resulting in eight haploid nuclei. Each haploid nucleus develops into an **ascospore,** so there are usually eight haploid ascospores within the ascus. The ascospores are released when the tip of the ascus breaks open. Individual ascospores are carried by air currents, often for long distances. If one lands in a suitable location, it germinates and forms a new mycelium.

Phylum Ascomycota includes more than 300 species of unicellular **yeasts.** Asexual reproduction of yeasts is mainly by **budding;** in this processs a small protuberance (bud) grows and eventually separates from the parent cell. Each bud can grow into a new yeast cell. Yeasts also reproduce asexually by fission and sexually by forming ascospores. During sexual reproduction, two haploid yeasts fuse, forming a diploid zygote. The zygote undergoes meiosis, and the resulting haploid spores remain enclosed for a time within the original diploid cell wall. This sac of spores corresponds to an ascus and ascospores. Yeasts are essential in making bread and fermenting alcoholic beverages (discussed later in this chapter).

Basidiomycetes (club fungi) reproduce sexually by forming basidiospores

The 25,000 or more species of **basidiomycetes** (phylum Basidiomycota) include the most familiar of the fungi: the mushrooms, bracket fungi, and puff balls (Fig. 25–4). Some destructive plant parasites of important crops, such as wheat rust and corn smut, are also basidiomycetes.

Basidiomycetes, also called *club fungi,* derive their name from the fact that they develop a club-shaped **basidium** (pl., *basidia*), a structure comparable in function to the ascus of ascomycetes. Each basidium is an enlarged hyphal cell, on the tip of which develop four **basidiospores.** Note that basidiospores develop on the *outside* of a basidium, whereas ascospores develop *within* an ascus.

Each individual fungus produces millions of basidiospores, and each basidiospore has the potential to give rise to a new **primary mycelium** (Fig. 25–5). Hyphae of a primary mycelium are composed of monokaryotic cells. The mycelium of a basidiomycete such as the cultivated mushroom, *Agaricus campestris,* consists of a mass of white, branching, threadlike hyphae that occur mostly below ground (see *Focus On: The Giant Fungus*). The hyphae are divided into cells by septa, but as in ascomycetes, the septa are perforated and allow cytoplasmic streaming between cells.

When, in the course of its growth, a hypha of a primary mycelium encounters another monokaryotic hypha of a different mating type, the two hyphae fuse. As in the ascomycetes, however, the two haploid nuclei remain separate within each cell. In this way a **secondary mycelium** with dikaryotic hyphae is produced, in which each cell contains two haploid nuclei.

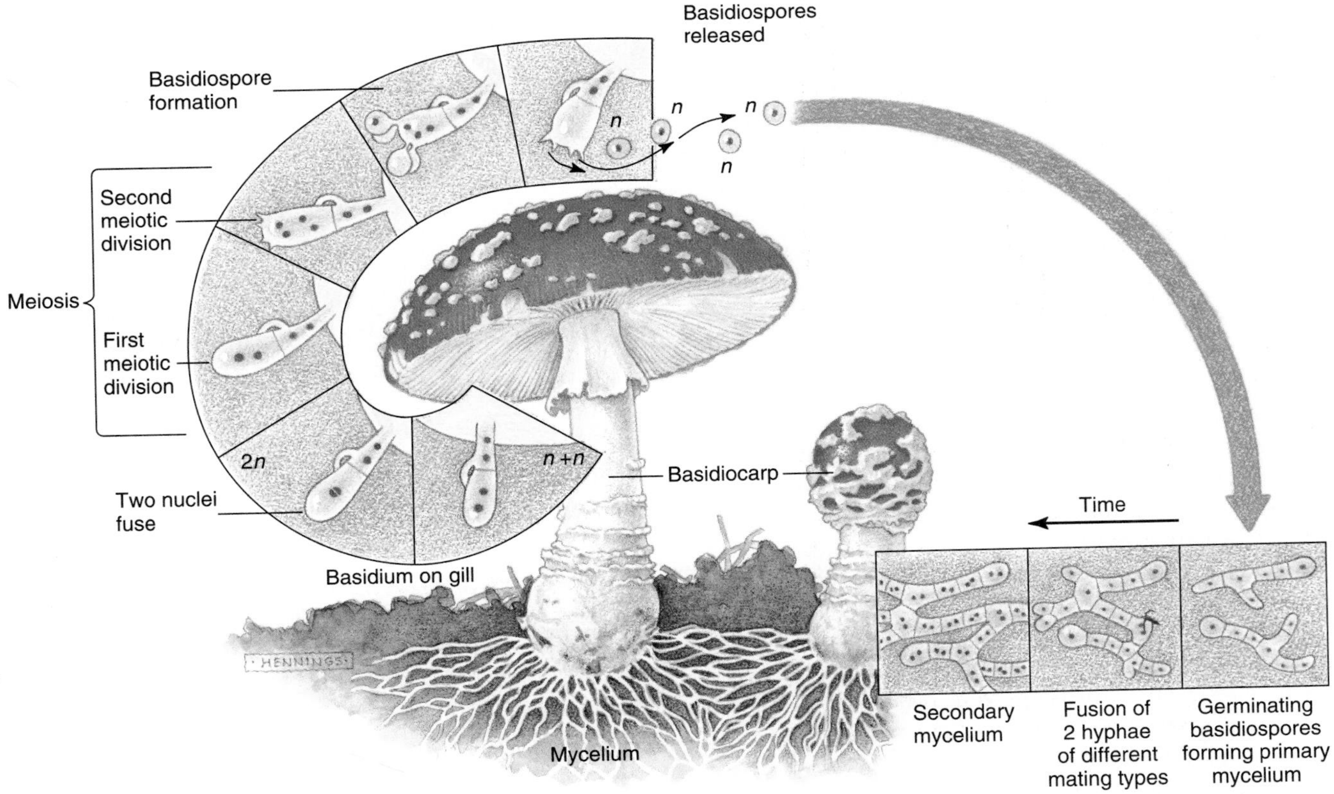

FIGURE 25–5 The life cycle of a typical mushroom-forming basidiomycete is characterized by the production of sexual spores called basidiospores. Details are given in the text.

Focus On

The Giant Fungus

Giants are known in the plant and animal kingdoms (for example, the giant sequoia and the blue whale), but until recently, the fungi were all thought to be relatively small organisms. In 1992 several biologists discovered, living in a forest in Michigan, a saprotrophic fungus *(Armillaria bulbosa)* that is one of Earth's largest and oldest organisms (see figure). Commonly called the honey mushroom, this species of fungus digests and absorbs dead or dying tree roots. The honey mushroom has a vast underground network of hyphae and branching cord-like structures called rhizomorphs. The rhizomorphs, which grow slowly through the upper layers of the soil, are composed of interwoven strands of hyphae and often resemble shoestrings in thickness.

The biologists used a type of genetic fingerprinting to confirm that a single fungus individual extends through at least 15 hectares (37 acres) of the forest floor — an area equivalent in size to more than 33 football fields! Moreover, they found that the fungus is territorial; no other honey mushroom individual was found in that area. Near the end of each summer, the giant fungus produces thousands of honey-colored mushrooms. But, according to the DNA data, none of the millions of spores produced by these mushrooms appears to have grown into a new individual within the 15 hectares. (Of course, many other species of fungi and other organisms cohabit the 15 hectares occupied by the giant honey mushroom.)

The honey mushroom (*Armillaria bulbosa*) is among the Earth's largest and longest-lived organisms. (Courtesy of Johann N. Bruhn)

Because rhizomorphs grow at a rate of about 20 centimeters (8 inches) per year in Northern Michigan, the fungus was calculated to be over 1,500 years old. At least one forest fire (in 1928) is known to have destroyed the trees in the area inhabited by this fungal giant, but it survived underground. It probably thrived immediately after the fire because of the large number of dead tree roots that became available for food.

The discovery of this giant fungus has prompted biologists to speculate that even larger fungi must exist elsewhere in soil. For example, a single fungus of a related species of *Armillaria* in the state of Washington is claimed to occupy over 600 hectares (1,500 acres), but so far this claim has not been substantiated with DNA evidence. Clearly, fungi are the giants of the forest, a fact that helps confirm their pivotal role in the forest ecosystem. ■

The $n + n$ hyphae of the secondary mycelium grow extensively and eventually form compact masses, called buttons, along the mycelium. Each button grows into a fruiting body that we call a mushroom. A mushroom, which consists of a stalk and a cap, is more formally referred to as a **basidiocarp.** Each basidiocarp actually consists of intertwined hyphae that are matted together.

The lower surface of the cap usually consists of many thin perpendicular plates called **gills** that radiate from the stalk to the edge of the cap (Fig. 25–6*a*). On the gills of the mushroom, haploid nuclei of the dikaryotic cells fuse to form diploid zygotes. These are the only diploid cells that form during a basidiomycete's life history. Meiosis then takes place, forming four haploid nuclei that move to the outer edge of the basidium. Finger-like extensions of the basidium develop, into which the nuclei and some cytoplasm move; each of these extensions becomes a basidiospore. A wall forms that separates the basidiospore from the rest of the basidium by a delicate stalk (Fig. 25–6*b*). When the stalk breaks, the basidiospore is released.

Imperfect fungi are fungi with no known sexual stage

About 25,000 species of fungi have been assigned to a group referred to as the **deuteromycetes.** They are also known as **imperfect fungi** because in many of them no sexual stage has been observed at any point during their life cycle. Should further study reveal a sexual stage, these species will be reassigned to a different phylum. Most imperfect fungi reproduce only by means of conidia and so are closely related to the ascomycetes. A few appear to be more closely related to the basidiomycetes.

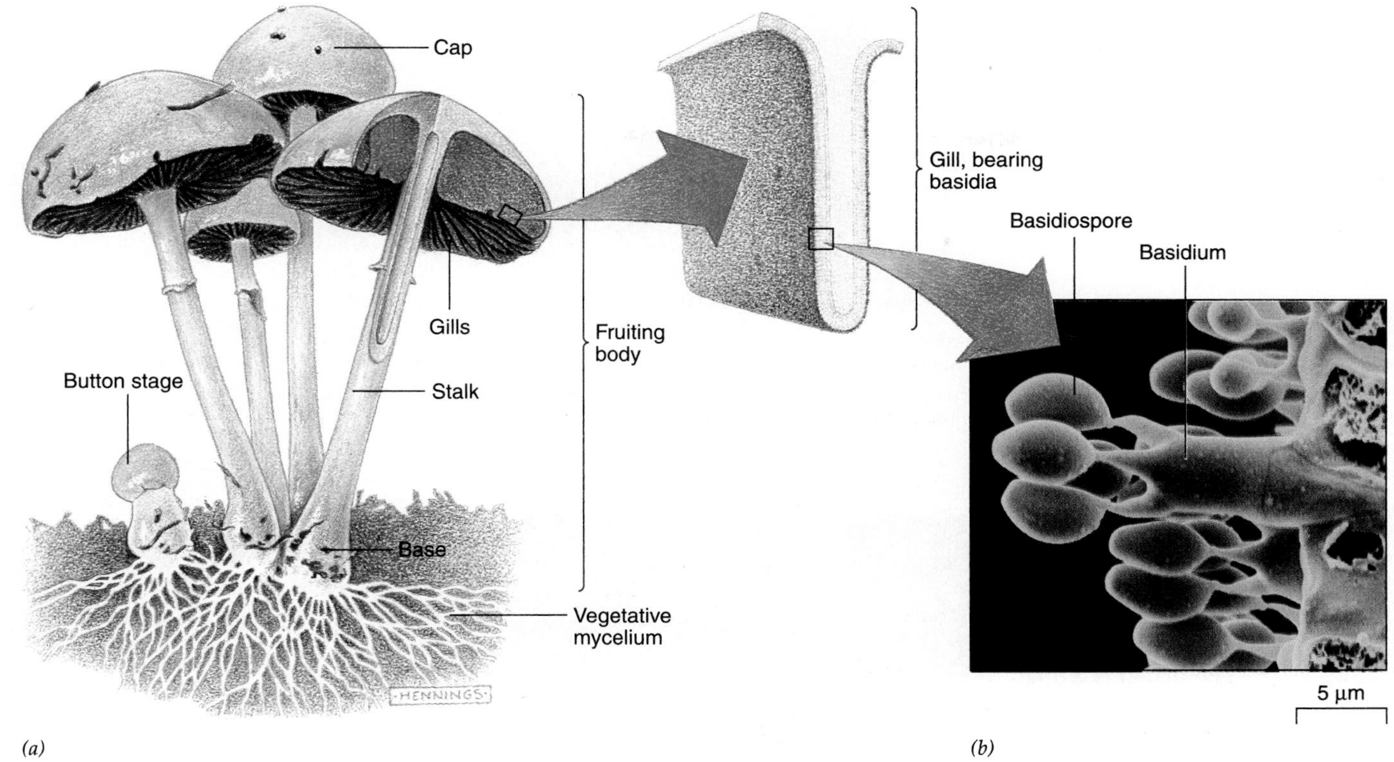

FIGURE 25–6 Mushrooms are umbrella-like basidiocarps. (***a***) Interwoven hyphae from the vegetative mycelium form the basidiocarp commonly called a mushroom. Numerous basidia are borne along the gills. (***b***) Scanning electron micrograph of a basidium. Each basidium produces four basidiospores. (Biophoto Associates)

LICHENS ARE DUAL "ORGANISMS"

Although a **lichen** looks like a single organism, it is actually a symbiotic association between two organisms: a *phototroph* (a photosynthetic organism) and a fungus (Fig. 25–7). (A symbiotic association is an intimate relationship between organisms of different species.) About 20,000 kinds of lichens have been described.

The phototrophic component is either a green alga or a cyanobacterium. The fungus is most often an ascomycete, although in some lichens from tropical regions, the fungal partner is a basidiomycete. Most of the phototrophic organisms found in lichens are also found as free-living species in nature, but the fungal components are generally found only as a part of the lichen.

In the laboratory the fungal and phototrophic components can be isolated and grown separately in appropriate culture media. The phototroph grows more rapidly when separated, whereas the fungus grows more slowly and requires many complex carbohydrates. Neither organism resembles a lichen in appearance when grown separately. The phototroph and fungus can be reassembled as a lichen, but only if they are placed in a culture medium under conditions that cannot support either of them independently.

What is the nature of this partnership? The lichen was originally considered a definitive example of mutualism, a symbiotic relationship that is beneficial to both species. The phototroph carries on photosynthesis, producing food for both members of the lichen, but it is unclear how the phototroph benefits from the relationship. It has been suggested that the phototroph obtains water and minerals from the fungus as well as protection against desiccation. More recently some biologists have suggested that the lichen partnership is not really a case of mutualism but one of controlled parasitism of the phototroph by the fungus.

Lichens typically possess one of three different growth forms. *Crustose lichens* are flat and grow tightly against their substrate (the surface they are growing on); *foliose lichens* are also flat, but they have leaflike lobes and are not so tightly pressed to the substrate; *fruticose lichens* grow erect and are branched and shrublike.

Able to tolerate extremes of temperature and moisture, lichens grow in almost all terrestrial environments except polluted cities. They exist farther north than any plants of the arctic region and are equally at home in the steaming equatorial rain forest. They grow on tree trunks, mountain peaks, and bare rock. In fact, lichens are often the first organisms to inhabit rocky areas. Lichen growth

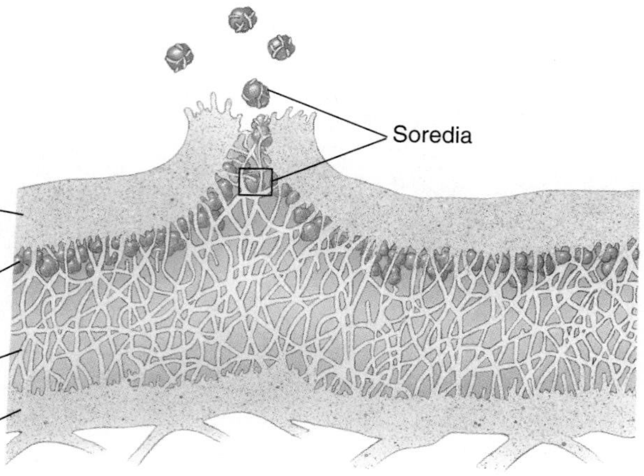

(a)

(b)

FIGURE 25–7 The lichen is a dual "organism." (*a*) This cross section of a typical lichen shows distinct layers. The soredium (pl., *soredia*), an asexual reproductive structure, is composed of clusters of algal or cyanobacterial cells enclosed by fungal hyphae. (*b*) Lichens vary in color, shape, and overall appearance. (*b*, Fred M. Rhoades)

in these areas plays an important role in the formation of soil from rock because they gradually etch tiny cracks in rock (see Chapter 53). This process sets the stage for further disintegration of the rock by wind and rain.

Reindeer mosses of the arctic region are not mosses but lichens that serve as the main source of food for migrating herds of caribou. Some lichens produce colored pigments. One of them, orchil, is used to dye woolens, and another, litmus, is widely used in chemistry laboratories as an acid-base (pH) indicator.

Lichens vary greatly in size. Some are almost invisible, whereas others, like the reindeer mosses, may cover kilometers of land with an ankle-deep growth. Growth proceeds slowly; the radius of a lichen may increase by less than one millimeter each year. Some mature lichens are thought to be thousands of years old.

Lichens absorb minerals mainly from the air, rainwater, and the surface on which they grow. They cannot excrete the elements they absorb, and perhaps for this reason they are extremely sensitive to toxic compounds. A reduction in lichen growth has been used as an indicator of air pollution, particularly sulfur dioxide. Absorption of such toxic compounds results in damage to the chlorophyll of the phototroph. The return of lichens to an area indicates an improvement in air quality. (Recall from Chapter 1 that air pollution killed lichens on tree trunks in England, thereby affecting the evolution of peppered moths.)

Lichens reproduce mainly by asexual means, usually by fragmentation, a process in which special dispersal units of the lichen, called **soredia,** break off and, if they land on a suitable surface, establish themselves as new lichens. Soredia contain cells of both partners. In some lichens, the fungus produces ascospores, which may be dispersed by wind and find an appropriate algal partner only by chance.

FUNGI ARE ECOLOGICALLY IMPORTANT

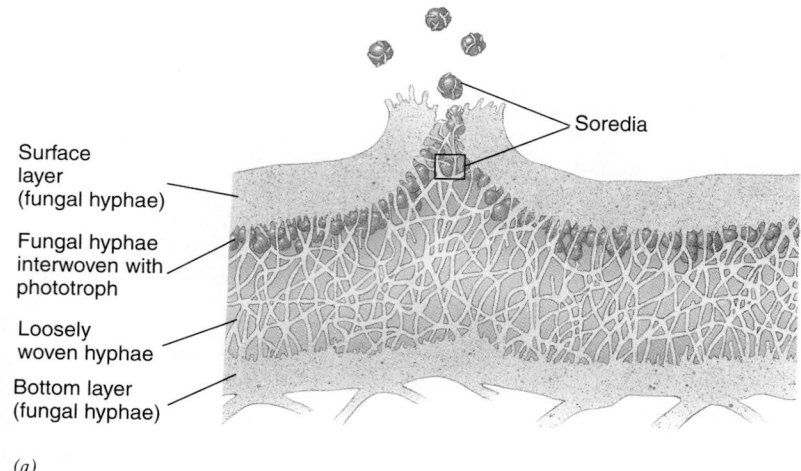

Fungi make important contributions to the ecological balance of our world. Like bacteria, most fungi are **saprotrophs** (also called *saprobes*), decomposers that absorb nutrients from organic wastes and dead organisms. When fungi degrade wastes and dead organisms, water, carbon (as CO_2), and mineral components of organic compounds are released, and these elements are recycled (see Chap-

Making the Connection

Fungi, Flowering Plants, and Mimicry

Do fungi ever resemble flowers? One of the most fascinating phenomena relating to living organisms is mimicry, in which, during the course of evolution, one organism comes to resemble another organism or an inanimate object (see Chapter 53). In 1993 the journal *Nature* reported an unusual example of mimicry that actually requires *two* different organisms, a fungus and a plant, to produce the deception.* The fungus *(Puccinia monoica)* is a plant parasite that causes a rust disease in rock cress *(Arabis holboellii),* a plant widely distributed across the northern part of North America.

When *Puccinia* infects rock cress, it doesn't kill the plant, but it does change the plant's growth pattern. Infected plants grow much taller than normal and produce a cluster of leaves at the top of the plant. The fungal mycelium, which is bright yellow in color, grows over and covers these leaves, giving them the appearance of buttercup flowers. Even botany students have been fooled by the remarkable resemblance. The fungal mimicry is not only visual. The fungus also secretes a sugary solution and produces a strong scent, imitating the nectar and aroma of flowers (see Chapter 35).

What is the advantage of this elaborate mimicry? As you might guess, the answer involves the fungus' reproductive cycle. In the life cycles of flowers and *Puccinia monoica,* insects play the key role of increasing the chances of successful reproduction. The same characteristics that attract insects to real flowers also attract them to the fungal imitation.

In the case of flowers, the insects transfer pollen (the male reproductive part) to the female part of another flower. In the case of *Puccinia,* a basidiomycete, sexual reproduction requires the union of nuclei from two different mating types. Insects, particularly flies, are attracted to the fungal "flowers," where they eat the sugary syrup. As the insects feed, pieces of the fungus cling to their bodies. Flying from one *Puccinia* "flower" to another, the insects distribute complementary mating types from fungus to fungus over a broad range. Thus, for the fungus, flower mimicry leads to better chances for successful reproduction. ▲

*Roy, B.A. "Floral Mimicry Induced by a Plant Pathogen." *Nature,* Vol. 362, March 4, 1993.

ter 54). Without this continuous decomposition, essential nutrients would soon become locked up in huge mounds of dead animals, feces, branches, logs, and leaves. The nutrients would be unavailable for use by new generations of organisms, and life would cease.

Although most fungi are saprotrophs, others form symbiotic relationships of various kinds. Some fungi are **parasites,** organisms that live in or on other organisms and are harmful to their hosts. Parasitic fungi absorb food from the living bodies of their hosts.

Some types of fungi form mutualistic relationships with other organisms. **Mycorrhizae** (from Greek words meaning "fungus-roots") are mutualistic relationships between fungi and the roots of plants (Fig. 25–8). Such

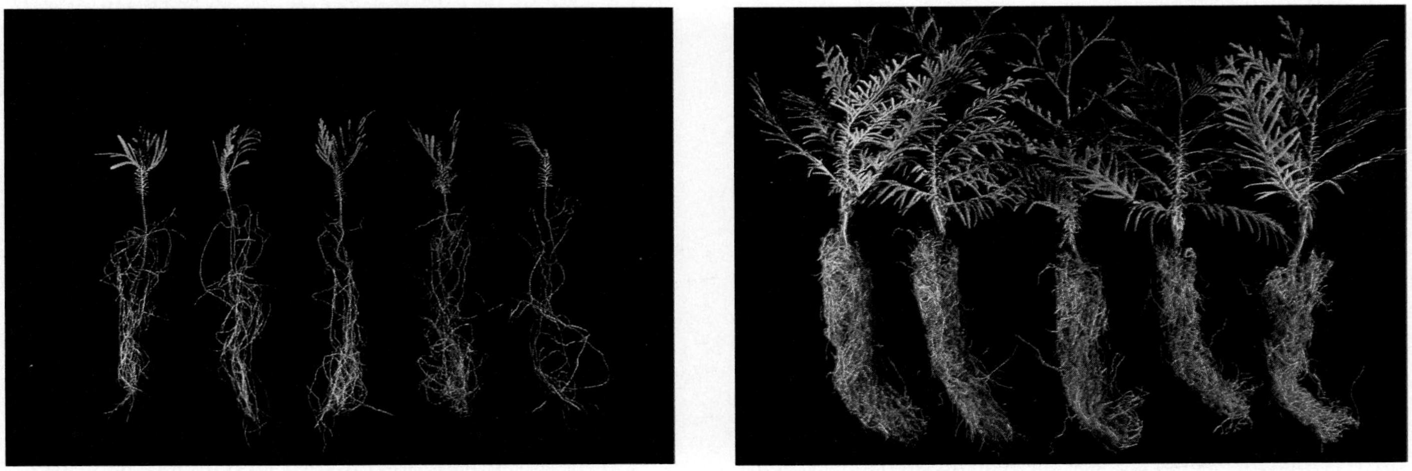

(a) (b)

FIGURE 25–8 Western red cedar *(Thuja plicata)* seedlings respond to mycorrhizae. (*a*) Control plants grown in low phosphorus in the absence of the fungus. (*b*) These seedlings were grown under conditions identical to the control, except that their roots have formed mycorrhizal associations. (*a, b,* Courtesy of Randy Molina, U.S. Forest Service) relationships occur in more than 90% of all plant families. The mycorrhizal fungus benefits the plant by decomposing organic material in the soil and providing water and minerals such as phosphorus to the plant. At the same time, the roots supply sugars, amino acids, and other organic substances to the fungus.

The importance of mycorrhizae first became evident when horticulturalists observed that orchids do not grow unless an appropriate fungus lives with them. Similarly, it has been shown that many forest trees such as pines die from mineral deficiencies when transplanted to mineral-rich grassland soils that lack the appropriate mycorrhizal fungi. When forest soil containing the appropriate fungi or their spores is added to the soil around these trees, they quickly resume normal growth.

FUNGI ARE ECONOMICALLY IMPORTANT

The same powerful digestive enzymes that enable fungi to decompose wastes and dead organisms also permit them to reduce wood, fiber, and food to their basic components with great efficiency. Thus, various fungi cause incalculable damage to stored goods and building materials each year. Bracket fungi, for example, cause enormous losses by decaying wood, both in living trees and in stored lumber.

Fungi cause economic gains as well as losses. People eat them, and grow them to make various chemicals. At the other extreme, some fungi are harmful from a human perspective because they cause diseases in humans and other animals and are the most destructive disease-causing organisms of plants. Their activities cost billions of dollars in agricultural damage yearly.

FIGURE 25–9 Wine, beer, bread, and distinctive cheeses are produced in part by fungi. Yeasts ferment sugars from fruits to make wine or from grains to make beer, producing ethyl alcohol. That same process produces the carbon dioxide bubbles responsible for making bread rise. The bluish areas in the cheese are patches of conidia. (Raymond Tschoepe)

Fungi provide beverages and food

The ability of yeasts to produce ethyl alcohol and carbon dioxide from sugars such as glucose by fermentation is utilized to make wine, beer, and other fermented beverages, and also to make bread. Wine is produced when yeasts ferment fruit sugars, and beer results when yeasts ferment sugars derived from starch in grains (usually barley). During the process of making bread, carbon dioxide produced by yeast becomes trapped in dough as bubbles, causing the dough to rise; this is what gives leavened bread its light texture. Both the carbon dioxide and the alcohol produced by the yeast evaporate during baking (Fig. 25–9).

The unique flavor of cheeses such as Roquefort, Brie, and Camembert is produced by the action of the ascomycete *Penicillium* (Fig. 25–9). *Penicillium roquefortii*, for example, is found in caves near the French village of Roquefort; only cheeses produced in this area can be called Roquefort cheese.

Aspergillus tamarii and other imperfect fungi are used in the Orient to produce soy sauce by fermenting soybeans. Soy sauce enriches other foods with more than just its special flavor. It also adds vital amino acids from both the soybeans and the fungi themselves to supplement the low-protein rice diet.

Among the basidiomycetes, there are some 200 kinds of edible mushrooms and about 70 species of poisonous ones, sometimes called toadstools. Some edible mushrooms are cultivated commercially. In fact, more than 350,000 metric tons (about 780 million pounds) are produced each year in the United States alone. Morels, which superficially resemble mushrooms (Fig. 25–10), and truffles, which produce underground fruiting bodies, are ascomycetes. These delights of the gourmet are now being cultivated as mycorrhizae on the roots of tree seedlings.

Edible and poisonous mushrooms can look very much alike and may even belong to the same genus. There is no simple way to tell them apart; they must be identified by an expert. Some of the most poisonous mushrooms belong to the genus *Amanita.* Toxic species of this genus have been appropriately called such names as "destroying angel" (*Amanita virosa*) and "death cap" (*Amanita phalloides*). Eating a single mushroom of either species can be fatal.

Ingestion of certain species of mushrooms causes intoxication and hallucinations. The sacred mushrooms of the Aztecs, *Conocybe* and *Psilocybe,* are still used in religious ceremonies by Central American Indians and others for their hallucinogenic properties. The chemical ingredient psilocybin, related to lysergic acid diethylamide (LSD), is responsible for the trances and colorful visions experienced by those who eat these mushrooms. Although they are not usually deadly, some people who eat them also experience stomach upsets, sweating, palpitations, and other symptoms of poisoning.

FIGURE 25–10 Morels are an expensive gourmet treat. The edible part of these yellow morels (*Morchella esculenta*) is an ascocarp that produces ascospores. (David M. Dennis/Tom Stack & Associates)

Fungi produce useful drugs and chemicals

In 1928 Alexander Fleming noticed that one of his petri dishes containing bacteria was contaminated by mold. The bacteria were not growing in the vicinity of the mold, leading Fleming to the conclusion that the mold was releasing some substance harmful to them. Within a decade or so of Fleming's discovery, penicillin produced by the ascomycete *Penicillium notatum* was purified and used in treating bacterial infections. Penicillin is still among the most widely used and effective antibiotics. Other drugs derived from fungi include the antibiotic griseofulvin, which is used clinically to inhibit the growth of fungi, and cyclosporine, the drug used to suppress immune responses in patients receiving organ transplants.

The ascomycete *Claviceps purpurea* infects the flowers of rye plants and other cereals. It produces a structure called an *ergot* where a seed would normally form in the grain head. When livestock eat this grain or when humans eat bread made from ergot-contaminated rye flour, they may be poisoned by the extremely toxic substances in the ergot. These substances may cause nervous spasms, convulsions, psychotic delusions, and even gangrene. This condition, called ergotism, was known as St. Anthony's fire during the Middle Ages, when it occurred often. In the year 994, for example, an epidemic of St. Anthony's fire caused more than 40,000 deaths. In 1722, the cavalry of Czar Peter the Great was felled by ergotism on the eve of the battle for the conquest of Turkey. This was one of several recorded times that a fungus changed the course of human history. Lysergic acid, one of the constituents of ergot, is an intermediate in the synthesis of LSD. Some of the compounds produced by ergot are now used clinically in small quantities as drugs to induce labor, to stop uterine bleeding, to treat high blood pressure, and to relieve one type of migraine headache.

Some fungi are grown commercially to produce citric acid and other industrial chemicals. Also, biologists are using recombinant DNA techniques to manipulate yeasts and certain filamentous fungi in order to produce important biological molecules such as hormones.

Fungi cause many important plant diseases

Fungi are responsible for many serious plant diseases, including epidemic diseases that spread rapidly and often result in complete crop failure. All plants are apparently susceptible to some fungal infection. Damage may be localized in certain tissues or structures of the plant, or the disease may be systemic and spread throughout the entire plant. Fungal infections may cause stunting of plant parts or of the entire plant; they may cause growths like warts; or they may kill the plant.

A plant often becomes infected after hyphae enter through stomata (pores) in the leaf or stem or through wounds in the plant body. Alternatively, the fungus may

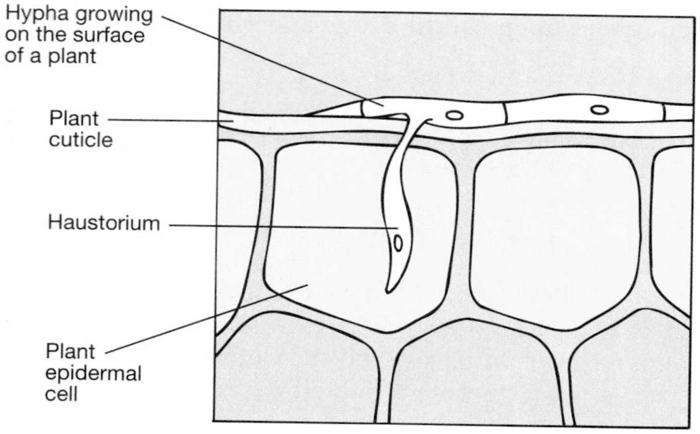

FIGURE 25–11 A haustorium of powdery mildew penetrates a plant epidermal cell.

produce an enzyme called cutinase that dissolves the plant's cuticle (a waxy covering over the surface of leaves and stems); after dissolving the cuticle, the fungus easily invades the plant tissues. As the mycelium grows, it may remain mainly between the plant cells or it may penetrate the cells. Parasitic fungi often produce special hyphal branches called **haustoria** (sing., *haustorium*) that penetrate the host cells and obtain nourishment from the cytoplasm (Fig. 25–11).

Some important plant diseases caused by ascomycetes are powdery mildews, chestnut blight, Dutch elm disease, apple scab, and brown rot, which attacks cherries, peaches, plums, and apricots (Fig. 25–12*a*). Diseases caused by basidiomycetes include smuts and rusts that attack various plants—for example, the cereal crops of corn, wheat, oats, and other grains (Fig. 25–12*b*). Some of these parasites, such as the stem rust of wheat, have complex life cycles that involve two or more different plants; during their life cycles several kinds of spores are produced. For example, wheat rust must infect an American barberry plant at one stage in its life cycle. Since this fact was discovered, the eradication of American barberry plants in wheat-growing regions has reduced infection with wheat rust. Wheat rust has not been eliminated by eradication of the barberry, however, because the fungus overwinters on wheat at the southern end of the Grain Belt and forms asexual spores. During the spring, wind blows these spores for hundreds of miles, reinfecting northern areas of the United States and Canada. Certain imperfect fungi also cause plant diseases.

Fungi cause certain animal diseases

Some fungi cause superficial infections in which only the skin, hair, or nails are infected. Ringworm and athlete's foot are examples of superficial fungal infections; both are caused by imperfect fungi. Candidiasis, a yeast infection of mucous membranes of the mouth or vagina, is among the most common fungal infections; it is also caused by an imperfect fungus.

Other fungi cause systemic infections that infect internal tissues and organs and may spread through many

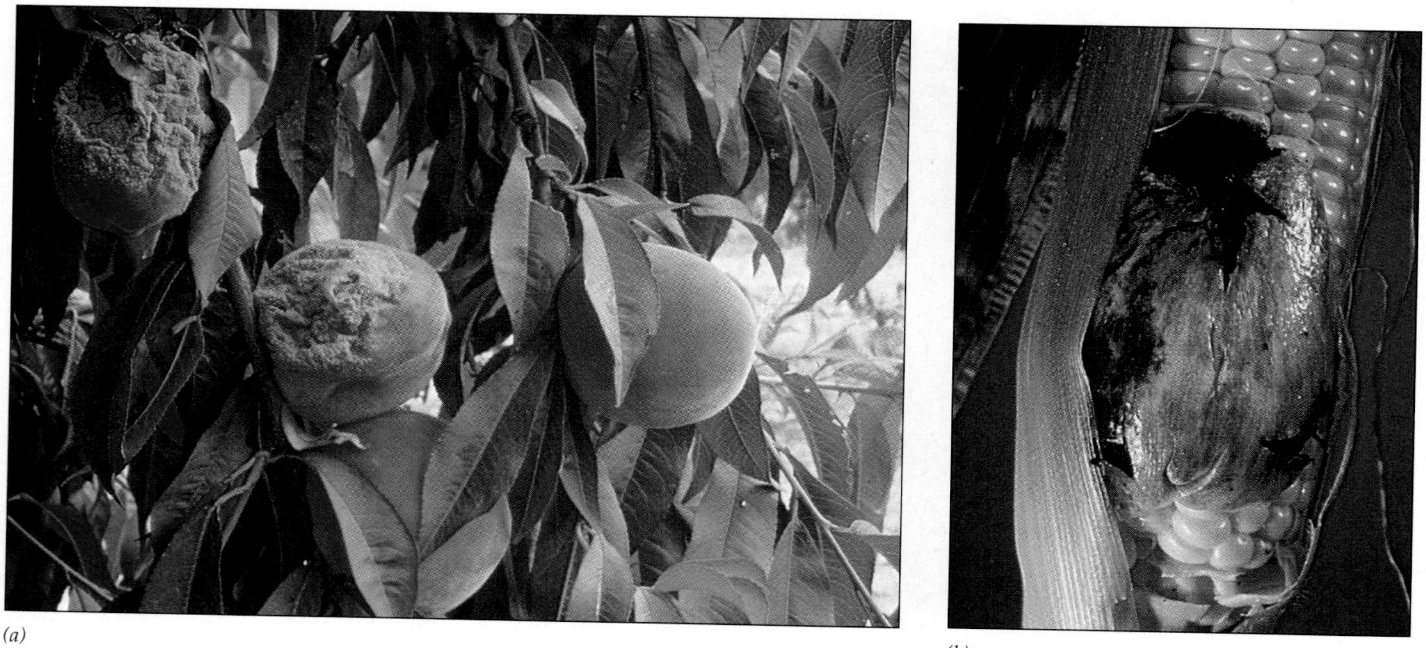

(*a*) (*b*)

FIGURE 25–12 Fungi are important causes of plant disease. (***a***) Brown rot of peaches is caused by *Monilinia fruticola*, an ascomycete. (***b***) Corn smut on an ear of sweet corn is caused by *Ustilago maydis*, a basidiomycete. (*a*, Kathy Merrifield/Photo Researchers, Inc.; *b*, Runk/Schoenberger, from Grant Heilman) regions of the body. Histoplasmosis, for example, is a serious infection of the lungs caused by inhaling spores of a soil fungus. Most people in the eastern and midwestern parts of the United States have been exposed to this fungus at one time or another. Fortunately, the infection usually stays in the lungs and is of short duration, but if the infection spreads into the bloodstream, and from there into the heart, brain, eyes, kidneys, or other parts of the body, it can be quite serious.

Most pathogenic (disease-causing) fungi are opportunists that cause infections only when the body's immunity is lowered. Because the HIV virus disarms the immune system, many AIDS patients have fungal infections that spread throughout their bodies. Other patients at high risk of acquiring life-threatening fungal infections include those with cancer and those with organ transplants.

SUMMARY

I. Fungi are eukaryotes with cell walls composed of chitin.
   A. Fungi lack chlorophyll and are heterotrophic; they absorb predigested food.
   B. A fungus may be unicellular (yeast) or multicellular (mold).
      1. The body of a multicellular fungus consists of long, branched hyphae, which form a mycelium.
      2. In the zygomycetes, the hyphae are coenocytic (undivided by septa).
      3. In other fungi, perforated septa are present that divide the hyphae into individual cells.
   C. Fungi reproduce both sexually and asexually by means of spores. When a fungal spore lands in a suitable spot, it germinates and begins to grow.
      1. Some hyphae infiltrate the food they are growing on and digest its organic compounds.
      2. Spores are produced on aerial hyphae.

II. Fungi are classified into phyla based on their modes of sexual reproduction.
   A. Zygomycetes produce sexual spores called zygospores. The black bread mold is a representative of this group.
   B. Ascomycetes produce asexual spores called conidia; sexual spores called ascospores are produced in asci. Ascomycetes include yeasts, cup fungi, morels, truffles, and pink and green molds.
   C. Basidiomycetes produce sexual spores called basidiospores on the outside of a basidium; basidia develop on the surface of gills in mushrooms. Basidiomycetes include mushrooms, puff balls, rusts, and smuts.
   D. A sexual stage has not been observed in imperfect fungi (deuteromycetes). Most reproduce asexually by forming conidia. Members of this group include *Aspergillus tamarii,* used to produce soy sauce, and fungi that cause certain human fungal infections.
   E. A lichen is a symbiotic combination of a fungus and a phototroph (an alga or cyanobacterium); the association is one of controlled parasitism. Lichens have three main growth forms: crustose, foliose, and fruticose.

III. Fungi are ecologically significant.
   A. Many fungi are saprotrophs that break down organic compounds.
   B. Mycorrhizae are mutualistic relationships between fungi and the roots of plants. The fungus supplies minerals to the plant, and the plant secretes organic compounds needed by the fungus.
   C. Lichens play an important role in soil formation.

IV. Fungi have both positive and negative economic importance.
   A. Mushrooms, morels, and truffles are used as food; yeasts are vital in the production of alcoholic beverages and bread; certain fungi are used to produce cheeses and soy sauce.
   B. Fungi are used to make penicillin and other antibiotics; ergot is used to produce certain drugs; other fungi make citric acid and many other industrial chemicals.
   C. Fungi cause many diseases.
      1. Plant diseases caused by fungi include wheat rust, Dutch elm disease, and chestnut blight.
      2. Human diseases such as ringworm, athlete's foot, candidiasis, and histoplasmosis are caused by fungi.

SELECTED KEY TERMS

ascocarp
ascomycete
ascospore
ascus
basidiocarp
basidiomycete
basidiospore
basidium
budding
chitin
coenocytic
conidiophore
conidium
deuteromycete
dikaryotic
fruiting body
fungus
gill
haustorium
heterothallic
homothallic
hypha
imperfect fungus
lichen
monokaryotic
mycelium
mycorrhizae
parasites
primary mycelium
saprotroph
secondary mycelium
septum
soredium
sporangium
spore
yeast
zygomycete
zygospore

POST-TEST

1. A mold consists of threads of cells called __________, which form a tangled mass called a(an) __________.
2. Some hyphae are divided by walls, called __________, whereas other hyphae lack these walls and are __________.
3. Fungi reproduce both sexually and asexually by forming __________.
4. *Rhizopus* and other zygomycetes form sexual spores called __________.
5. A(an) __________ mycelium is self-sterile and mates only with an individual of a different mating type.
6. In ascomycetes, asexual reproduction involves formation of spores called __________.
7. Yeasts reproduce asexually, mainly by __________.
8. Sexual reproduction in ascomycetes involves production of spores known as __________ within saclike structures called __________.
9. The familiar portion of a mushroom is actually a large fruiting body known as a(an) __________.
10. The type of sexual spore produced by a mushroom is a(an) __________.
11. In mushrooms, basidia develop on the surface of vertical plates called __________.
12. Deuteromycetes are also known as __________ __________.
13. A(an) __________ consists of a phototroph and a fungus that form a symbiotic relationship.
14. Ecologically, fungi serve as decomposers, or __________.
15. Mutualistic relationships between fungi and the roots of plants are known as __________.
16. Special hyphae produced by parasitic fungi that can penetrate host cells are known as __________.

REVIEW QUESTIONS

1. What characteristics distinguish fungi from other organisms?
2. How does the body of a typical yeast differ from that of a mold?
3. What is the ecological importance of saprotrophic fungi? Of lichens? Of mycorrhizae?
4. Draw the life cycle of the black bread mold.
5. Distinguish among each of the following: (a) ascocarp, ascus, and ascospore; (b) basidiocarp, basidium, and basidiospore; (c) conidium and ascospore; (d) ascus and basidium; (e) sporangium and conidium.
6. Some dictionaries erroneously define a morel as a type of mushroom. Why isn't a morel a mushroom?
7. Briefly describe three important fungal diseases of plants and three fungal diseases of humans.
8. Label the following:

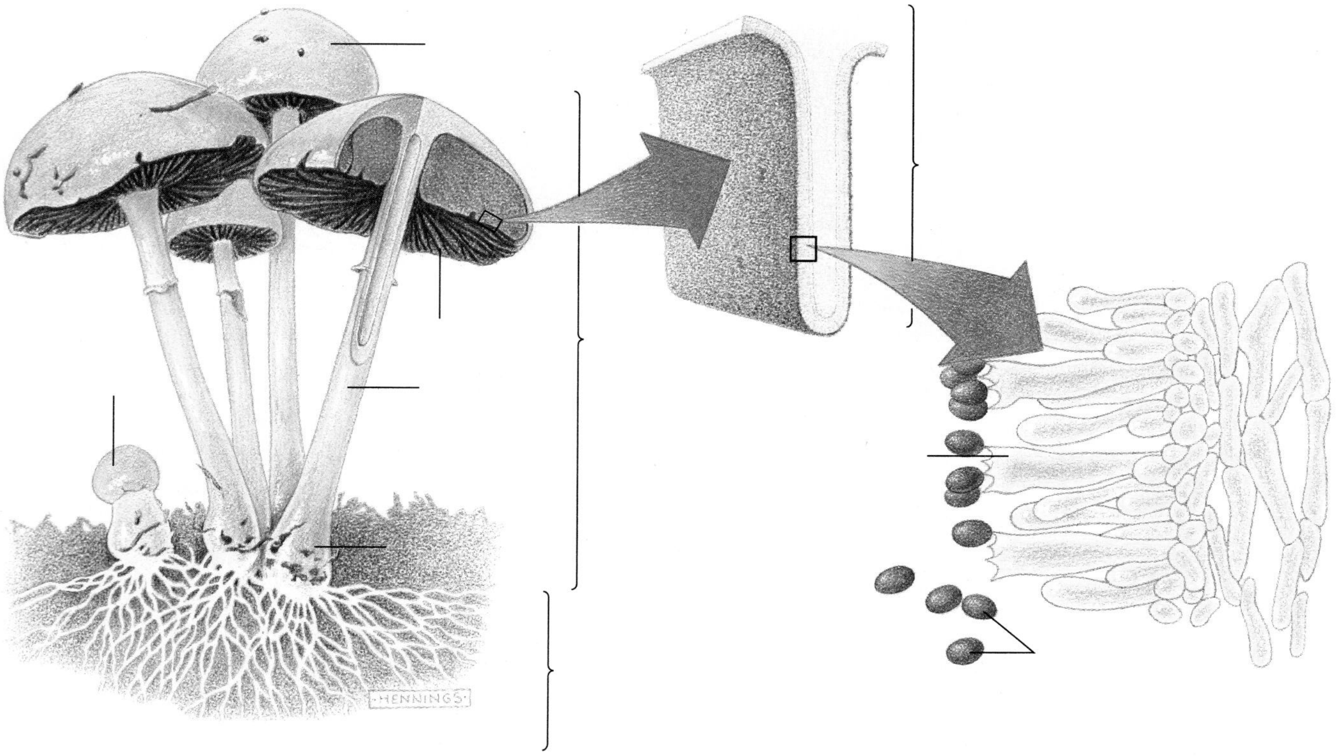

YOU MAKE THE CONNECTION

1. What measures can you suggest to prevent bread from becoming moldy?
2. If you do not see any mushrooms in your lawn, can you conclude that no fungi live there? Why or why not?
3. Under what kinds of environmental conditions might it be more advantageous for a fungus to reproduce asexually? Sexually?

RECOMMENDED READINGS

The Audubon Society Field Guide to North American Mushrooms, Alfred A. Knopf, New York, 8th Printing, 1992. A beautifully illustrated guide to common fungi.

Gould, S.J. "Fungal Forgery." *Natural History,* September 1993. A fascinating account of a fungus that mimics flowers.

Lewis, R. "A New Place for Fungi?" *BioScience* 44:6, June 1994. Several types of molecular evidence suggest that fungi are more closely related to animals than to plants.

Lipske, M. "A New Gold Rush Packs the Woods in Central Oregon." *Smithsonian,* January 1994. A mushroom war is occurring in forests in the Northwest.

Newhouse, J.R. "Chestnut Blight." *Scientific American,* July 1990. An account of the chestnut blight fungus that has almost completely eradicated the American chestnut in the United States.

Radetsky, P. "The Yeast Within." *Discover,* March 1994. The discovery that baker's yeast can be induced to grow like a filamentous mold has far-reaching medical implications.

Scitil, K. "Fungus Among Us," in "Biology 1992." *Discover,* January 1993. Discusses the "monster mushroom" discovered in Michigan in 1992.

CHAPTER 26

The Plant Kingdom: Seedless Plants

Mosses cover several rocks in Marron Creek, Snowmass Wilderness, Colorado. (Sydney Karp: Photo/Nats, Inc.)

The plant kingdom comprises hundreds of thousands of different species that live in every conceivable habitat, from frozen Arctic tundra to lush tropical rain forests to harsh deserts. Plants are complex multicellular organisms that range in size from minute, almost microscopic duckweeds to massive giant sequoias, some of the largest organisms that have ever lived.

One of the unifying characteristics of almost all plants is their mode of nutrition—photosynthesis. Most plants are photosynthetic autotrophs (see Chapter 8). They absorb radiant energy, which is then converted to the chemical energy found in organic molecules such as carbohydrates. In addition to the photosynthetic pigments chlorophyll *a* and chlorophyll *b*, all plants have accessory pigments, the yellow and orange carotenoids: xanthophylls (yellow pigments) and carotenes (orange pigments). Although the ability to photosynthesize is a key feature of plants, it is not a distinguishing trait because other kinds of organisms—the algae and certain bacteria— also photosynthesize, and some plants are not photosynthetic.

Plants demonstrate a remarkable diversity in size, habit, and form, but they are all thought to have descended from a common protistan ancestor, an ancient, freshwater green alga. Because of their common ancestry, the green algae living today share a number of biochemical and metabolic traits with plants. Both contain the same photosynthetic pigments: chlorophylls *a* and *b*, carotenes, and xanthophylls. Also, both green algae and plants store carbohydrates as starch. Cellulose is a major component of the cell walls of both, and certain details of cell division, including the formation of a cell plate during cytokinesis, are shared by plants and many green algae but are found in no other photosynthetic organism. All of these shared characteristics point to the likelihood that modern plants descended from ancient green algae.

Four major groups of plants are living today: bryophytes, seedless vascular plants, gymnosperms, and flowering plants (Table 26–1 and Fig. 26–1). The mosses and other bryophytes are small plants that lack a specialized vascular, or conducting, system. The other three groups of plants—those most familiar to us—possess vascular tissues: **xylem** for water and mineral conduction, and **phloem** for conduction of dissolved sugar. Bryophytes reproduce by spores, as do ferns and their allies, which are seedless vascular plants. The gymnosperms and flowering plants are vascular plants that reproduce by forming seeds (see Chapter 27). Gymnosperms produce seeds borne exposed on a stem or in a cone, whereas flowering plants produce seeds enclosed within a fruit.

LEARNING OBJECTIVES

After you have studied this chapter you should be able to

1. Discuss some of the environmental challenges of living on land and describe several adaptations that plants possess to meet these challenges.
2. Name the protistan group from which plants are thought to have descended and describe supporting evidence.
3. Diagram a generalized plant life cycle, clearly showing alternation of generations.
4. Summarize the features that distinguish bryophytes from green algae and from other plants.
5. Discuss the features that distinguish ferns and fern allies from algae and from other plants.
6. Compare the generalized life cycles of homosporous and heterosporous plants.

PLANTS HAVE ADAPTED TO LIFE ON LAND

What are some of the features of plants that have permitted them to colonize so many different environments? One of the most important adaptations that enables plants to survive on land is a **cuticle,** a waxy covering, over their aerial (above-ground) parts. A cuticle is essential for a terrestrial (land) existence because it helps prevent the desiccation, or drying out, of plant tissues by evaporation. Plants are rooted in the ground and, unlike animals, cannot move to wetter areas during dry spells. In contrast, algae are adapted to an aquatic existence where water conservation is not important, and do not possess a cuticle.

Algae and the few plants that live in aquatic environments are bathed by water containing dissolved materials, including dissolved carbon dioxide (CO_2) and carbonate ions (CO_3^{2-}). Carbonate moves by diffusion into algal cells and is used as the raw material for photosynthesis. Terrestrial plants obtain their carbon from the atmosphere as CO_2, which must be accessible to the chloroplasts inside green plant cells. However, since gas

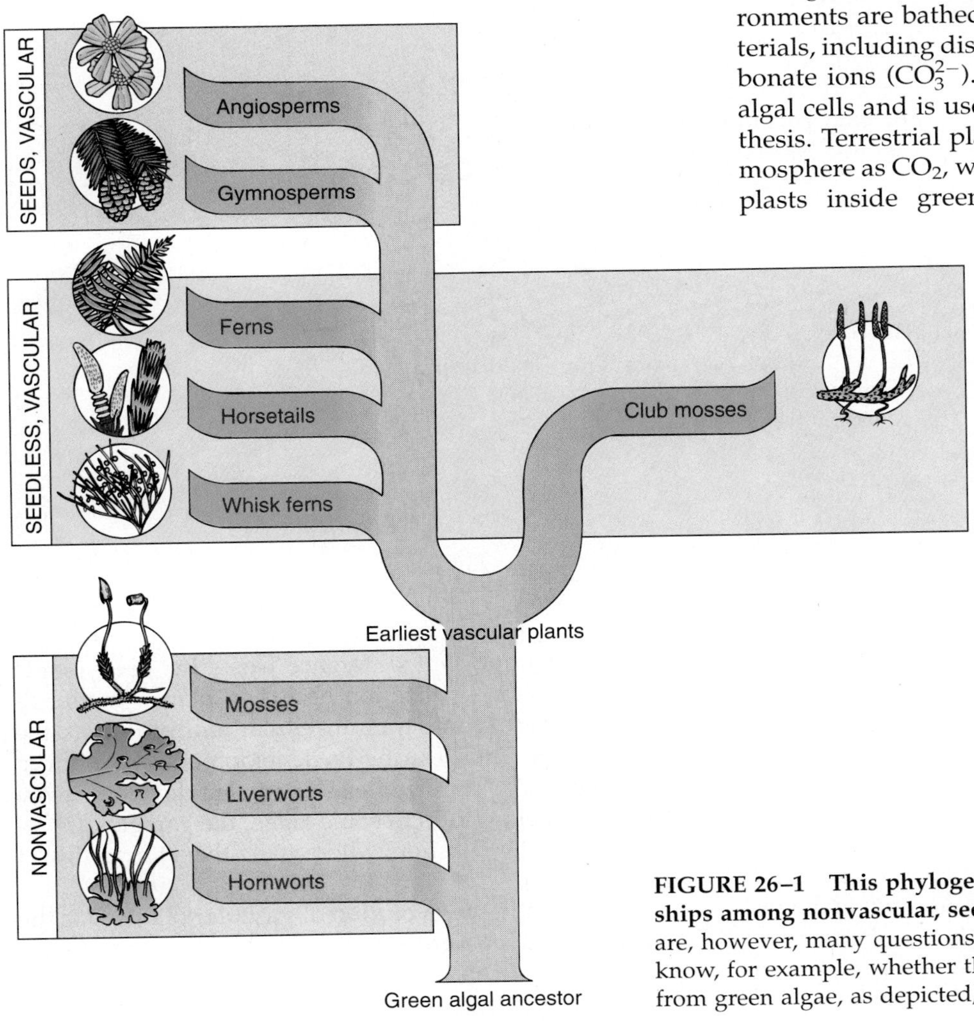

FIGURE 26–1 This phylogenetic tree illustrates possible relationships among nonvascular, seedless vascular, and seed plants. There are, however, many questions that remain to be resolved. We do not know, for example, whether the nonvascular plants descended directly from green algae, as depicted, or from early vascular plants.

Table 26–1 THE PLANT KINGDOM

Nonvascular Plants

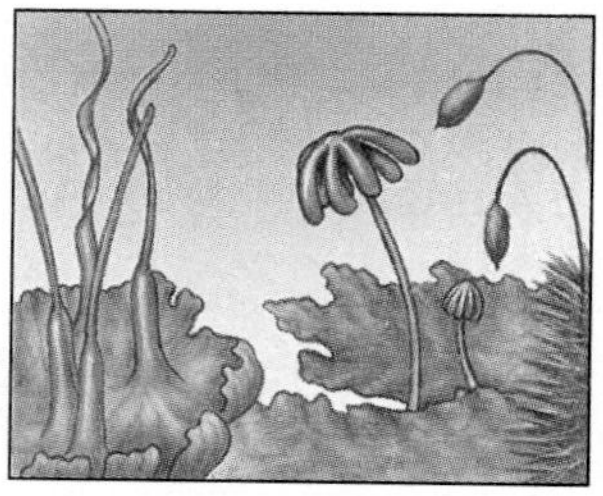

I. Nonvascular plants with a dominant gametophyte generation
- Phylum Bryophyta (mosses)
- Phylum Hepatophyta (liverworts)
- Phylum Anthocerophyta (hornworts)

Vascular Plants

II. Vascular plants with a dominant sporophyte generation

A. Seedless plants
- Phylum Pterophyta (ferns)
- Phylum Psilotophyta (whisk ferns)
- Phylum Sphenophyta (horsetails)
- Phylum Lycophyta (club mosses)

B. Seed plants

1. Plants with naked seeds (gymnosperms)
   - Phylum Coniferophyta (conifers)
   - Phylum Cycadophyta (cycads)
   - Phylum Ginkgophyta (ginkgo)
   - Phylum Gnetophyta (gnetophytes)

2. Seeds enclosed within a fruit
   - Phylum Anthophyta (angiosperms or flowering plants)
     - Class Dicotyledones (dicot)
     - Class Monocotyledones (monocots)

exchange through the waxy cuticle is negligible, plants possess tiny openings, or **stomata** (sing., *stoma*), in the surface tissues of stems and leaves. Stomata permit the gas exchange that is essential for photosynthesis.

Plants have multicellular gametangia (sing., *gametangium*), or sex organs, each of which possesses a sterile (nonreproductive) layer of cells that surrounds and protects the delicate gametes (reproductive cells: eggs and sperm). Each female gametangium, called an **archegonium** (pl., *archegonia*), produces a single egg. Numerous sperm are produced in the male gametangium, called an **antheridium** (pl., *antheridia*). In contrast, algae lack multicellular gametangia.

Another important difference exists between plants and algae. After fertilization occurs in plants, the fertilized egg develops into a multicellular **embryo** *within* the archegonium. Thus, during its development, the embryo is protected. The fertilized egg of an alga develops away from its gametangium. In some algae, the gametes are released before fertilization, whereas in others the fertilized egg is released.

A key step in the evolution of vascular plants was the ability to produce **lignin,** a strengthening polymer found in the walls of cells that function for support and conduction. The stiffening property of lignin enabled plants to grow tall and dominate the landscape. The successful

Making the Connection

Meeting the Environmental Challenges of Living on Land

How have terrestrial organisms met the environmental challenges of living on land? Life began in the oceans, but many life forms have since adapted to terrestrial life in a sea of air. Every single organism living on land has to meet the same environmental challenges: obtaining enough water; preventing excessive water loss; getting enough energy; and, in temperate and polar regions, tolerating widely varying temperature extremes. How those challenges are met varies from one organism to another and in large part explains the diversity of life encountered on land today. Let us compare how vertebrates (animals with backbones) and plants meet several terrestrial challenges.

1. *Obtaining enough water.* Animals are motile; they walk, slither, fly, run, or crawl to water sources. This requires not only the ability to move (skeletal and muscular systems) but also the ability to sense the presence of water (a nervous system). Plants have adapted in a much different way to this challenge. They have roots that not only anchor the plant in the soil but also absorb water and essential dissolved minerals.
2. *Preventing excessive water loss.* The outer layers of terrestrial vertebrates and plants protect the moist inner tissues from drying out. Vertebrates that are adapted to living on land have an accumulation of a water-insoluble protein called keratin in their epithelial cells. Keratin is particularly thick in reptiles, where it forms scales that greatly reduce water loss by evaporation. Plants possess a water-insoluble, waxy coating called a cuticle over their epidermal cells. Plants that are adapted to moister habitats may have a very thin layer of wax, whereas those adapted to drier environments often possess a thick, crusty cuticle.
3. *Obtaining sufficient energy.* Animals are heterotrophs that eat plants or other animals that eat plants. Plants are autotrophs and must absorb enough sunlight for effective photosynthesis. Some plants obtain adequate sunlight by growing tall (to shade out other plants); this adaptive approach required the evolution of strong supporting cells such as fibers because plants lack skeletal systems for support. Other plants have adapted to lower light intensities and so are able to grow in the shade of larger plants, albeit more slowly.
4. *Tolerating widely varying temperature extremes.* Air temperature varies greatly, particularly in temperate and polar regions. Many animals avoid hot temperatures by resting in the shade or by burrowing in the ground during the day when the temperatures are high; these animals become active at night when it is cooler. Sweat glands in mammalian skin produce sweat that cools the body by evaporation. Plants also rely on evaporative cooling; although they don't produce sweat, plants lose large quantities of water through tiny surface pores called stomata.

   Vertebrates deal with the cold temperatures of winter in several ways. Mammalian hair and bird feathers trap air next to the skin's surface, thereby providing insulation and allowing the body to conserve heat. Some animals avoid colder temperatures by migrating to warmer climates for the winter, whereas others avoid the cold by passing the winter in a dormant state called hibernation. Many plants also spend winter in a dormant state. The aerial parts of some plants die during the winter, but the underground parts remain alive; the following spring, they resume metabolic activity and develop new aerial shoots. Many trees are deciduous, that is, they shed their leaves for the duration of their dormancy. Shedding leaves is actually an adaptation to the "dryness" of winter. Roots cannot absorb water from ground that is cold or frozen; by shedding its leaves, the plant reduces water loss during the cold winter months when obtaining water from the soil is impossible. ▲

occupation of the land by plants in turn made possible the colonization of large animals on land by providing them with both habitat and food (see *Making the Connection: Meeting the Environmental Challenges of Living on Land*).

THE PLANT LIFE CYCLE ALTERNATES BETWEEN TWO DIFFERENT GENERATIONS

Plants have a clearly defined **alternation of generations** in which they spend part of their lives in a multicellular haploid stage and part in a multicellular diploid stage (Fig. 26–2).[1] The haploid portion of the life cycle is called the **gametophyte generation** because it gives rise to haploid gametes by mitosis. When two gametes fuse, the diploid portion of the life cycle, called the **sporophyte generation,** begins. The sporophyte generation produces haploid spores immediately following meiosis; these spores represent the first stage in the gametophyte generation.

[1]For convenience we limit our discussion to plants that are not polyploid, although polyploidy is very common in the plant kingdom. We therefore use the terms diploid and $2n$ (and haploid and n) interchangeably, although these terms are not actually synonymous.

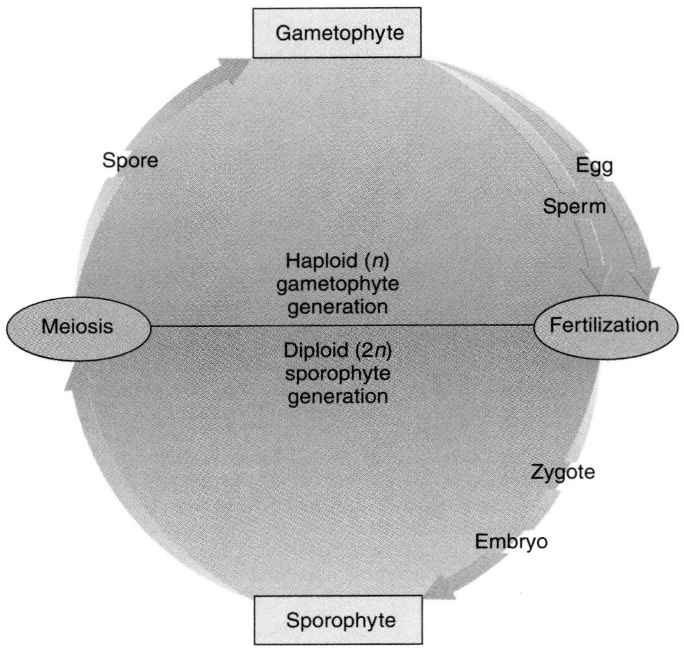

FIGURE 26–2 The basic plant life cycle shows that plants alternate generations, spending part of the cycle in a haploid gametophyte stage and part in a diploid sporophyte stage. All plants have modifications of this cycle.

Let us examine alternation of generations more closely. The haploid gametophyte plants produce antheridia (male gametangia), in which sperm form, and/or archegonia (female gametangia), each bearing a single egg. Sperm reach the female gametangium in a variety of ways, and one sperm fertilizes the egg to form a zygote, or fertilized egg.

The diploid zygote is the first stage in the sporophyte generation. The zygote divides by mitosis and develops into a multicellular embryo. Embryo development takes place *within* the archegonium; thus, during its development, the embryo is protected. Eventually, the embryo matures into a sporophyte plant. The sporophyte plant has special cells called *sporogenous cells* (spore-producing cells, also called *spore mother cells*) that divide by meiosis to form haploid spores. (In contrast to algae and fungi, which may produce spores by meiosis or mitosis, all plant spores are produced by meiosis.)

The spores represent the first stage in the gametophyte generation. Each spore divides by mitosis to produce a multicellular gametophyte plant, and the cycle continues. Plants therefore have an alternation of generations, alternating between a haploid gametophyte generation and a diploid sporophyte generation.

Table 26–2 A COMPARISON OF MAJOR GROUPS OF SEEDLESS PLANTS

Plant Group	*Major Characteristics*	*Dominant Stage of Life Cycle*	*Representative Genera*
Mosses	Nonvascular; reproduce by spores	Gametophyte: leafy plant	*Polytrichum, Sphagnum*
Liverworts	Nonvascular; reproduce by spores	Gametophyte: thalloid or leafy plant	*Marchantia, Porella*
Hornworts	Nonvascular; reproduce by spores	Gametophyte: thalloid plant	*Anthoceros*
Ferns	Vascular; reproduce by spores	Sporophyte: roots, rhizomes, and leaves (megaphylls)	*Pteridium, Polypodium*
Whisk ferns	Vascular; reproduce by spores	Sporophyte: rhizomes and erect stems; no true roots or leaves	*Psilotum, Tmesipteris*
Horsetails	Vascular; reproduce by spores	Sporophyte: roots, rhizomes, erect stems, and leaves (reduced megaphylls)	*Equisetum*
Club mosses	Vascular; reproduce by spores	Sporophyte: roots, rhizomes, erect stems, and leaves (microphylls)	*Lycopodium, Selaginella*

MOSSES AND OTHER BRYOPHYTES ARE NONVASCULAR PLANTS

The **bryophytes,** which comprise over 15,000 species of mosses, liverworts, and hornworts, are the only nonvascular plants (Table 26–2 and Fig. 26–3). Because they are nonvascular and have no means for extensive internal transport of water, dissolved sugar, and essential minerals, bryophytes are typically quite small. They generally require a moist environment for active growth and reproduction, but some bryophytes tolerate dry areas. Although the three groups of bryophytes differ in many ways and may or may not be closely related, their life cycles are similar.

Mosses have a dominant gametophyte generation

Mosses (phylum Bryophyta) usually live in dense colonies or beds. Each individual plant has tiny rootlike structures, called *rhizoids,* that anchor it to the soil. Each plant also has an upright, stemlike structure that bears leaflike blades. Because mosses lack specialized vascular tissues, they do not possess true roots, stems, or leaves. Some moss species possess water-conducting cells and sugar-conducting cells, although these cells are not as specialized or as effective as the conducting cells of vascular plants.

An alternation of generations is clear in the life cycle of mosses (Fig. 26–4). The leafy green moss gametophyte bears its gametangia at the top of the plant. Many moss species have separate sexes: male plants that bear antheridia and female plants that bear archegonia. Other moss species produce antheridia and archegonia on the same plant.

In order for fertilization to occur, one of the sperm must fertilize the egg within the archegonium. Sperm, which are flagellated, are transported from antheridium to archegonium by flowing water—splashing rain droplets, for example. A raindrop lands on the top of a male

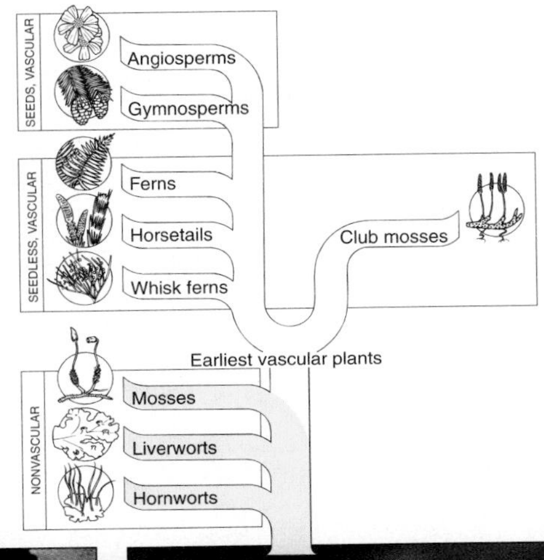

(a) (b) (c)

FIGURE 26–3 Mosses, liverworts, and hornworts are bryophytes. (*a*) A closeup of moss gametophytes, which grow in dense clusters. (*b*) The gametophyte of many liverworts is characterized by flattened, ribbon-like lobes. (*c*) A typical hornwort gametophyte. The "horns" projecting out of the gametophyte body are young sporophytes. (*a*, James Mauseth, University of Texas; *b*, Dennis Drenner; *c*, Robert and Linda Mitchell)

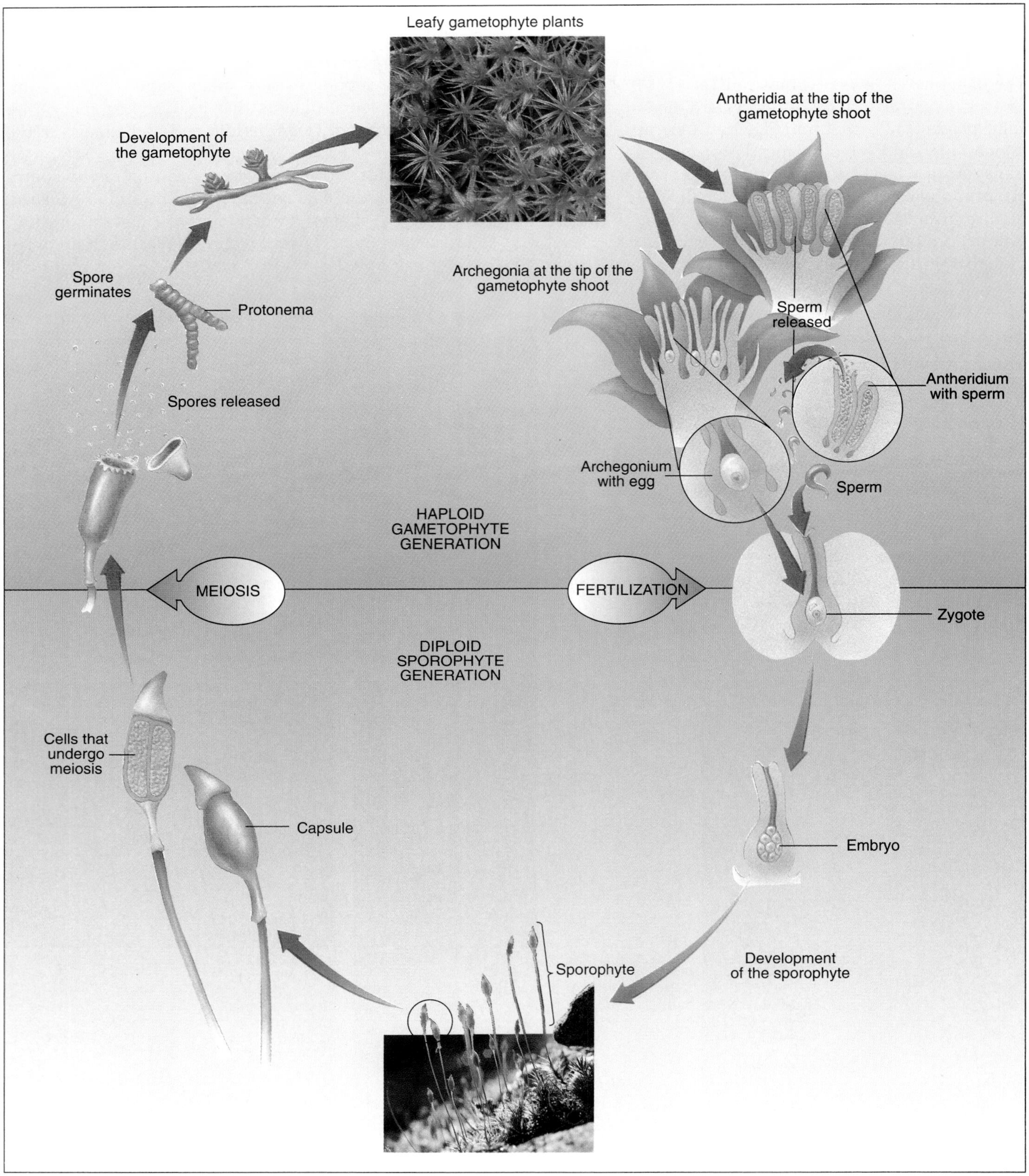

FIGURE 26–4 The gametophyte generation, represented by leafy green plants, is dominant in the moss life cycle. The sporophyte generation grows out of the top of the gametophyte. Mosses require water as a transport medium during fertilization. (*Top,* Rod Planck/Dembinsky Photo Associates; *Bottom,* David Cavagnaro)

gametophyte plant, and sperm are released into it from the antheridia. When another raindrop lands on the male plant, it may splash the sperm-laden droplet into the air and onto the top of a nearby female plant. Or insects may touch the sperm-laden fluid and inadvertently carry it for considerable distances. Once in a film of water on the female moss, a sperm swims down into the archegonium, which secretes sucrose to attract and guide the sperm, and fuses with the egg.

The diploid zygote formed as a result of fertilization grows into a multicellular embryo by mitosis, and matures into a moss sporophyte. This sporophyte plant grows out of the top of the female gametophyte. The sporophyte remains attached and nutritionally dependent on the gametophyte throughout its existence (Fig. 26–5). The sporophyte is initially green in color (and photosynthetic), but becomes a golden brown at maturity. It is composed of three main parts: a *foot,* which anchors the sporophyte to the gametophyte and absorbs minerals and nutrients from it; a *seta,* or stalk; and a *capsule,* which contains sporogenous cells (spore mother cells).

The sporogenous cells undergo meiosis to form haploid spores. When the spores are mature, the capsule opens by various mechanisms to release the spores. These microscopic cells are transported by wind or rain. If a moss spore lands in a suitable spot, it germinates and grows into a filamentous thread of cells called a **protonema.** The protonema, which looks like a filamentous green alga, forms buds, each of which grows into a leafy green gametophyte plant, and the life cycle continues.

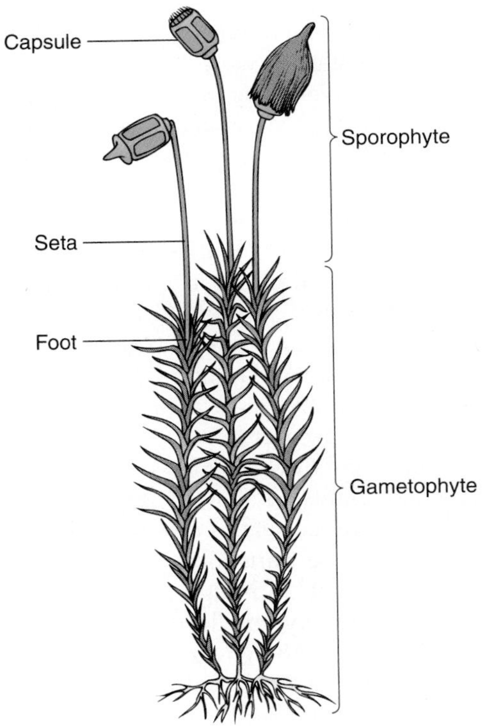

FIGURE 26–5 After sexual reproduction takes place in mosses, the sporophyte grows out of the gametophyte. The darker green represents haploid gametophyte tissue; lighter green is diploid sporophyte tissue.

The haploid gametophyte generation is considered the dominant generation in mosses because it is capable of living independently of the diploid sporophyte. In contrast, the sporophyte generation in mosses is attached to and nutritionally dependent on the gametophyte plant.

Mosses make up an inconspicuous but significant part of their environment. They play an important role in forming soil (see Chapter 53). Because they grow tightly packed together in dense colonies, mosses hold the soil in place and help prevent erosion. They provide food for animals, especially birds and small mammals.

Commercially, the most important mosses are the peat mosses in the genus *Sphagnum.* One of the distinctive features of *Sphagnum* is the presence of large "empty" cells in the "leaves," which apparently function to hold water. This feature makes peat moss particularly beneficial as a soil conditioner. When added to sandy soils, for example, peat moss helps to hold and retain moisture. In some countries such as Ireland and Scotland, layers of dead peat moss that have accumulated for hundreds of years are extracted from peat bogs, dried, and burned for fuel.

The name "moss" is often commonly used for plants that are not truly mosses. For example, reindeer moss is a lichen that is a dominant form of vegetation in the Arctic tundra, Spanish moss is a flowering plant, and club moss (discussed later in this chapter) is a relative of ferns.

Liverworts are either thalloid or leafy

Liverworts (phylum Hepatophyta) are a group of nonvascular plants with a dominant gametophyte generation, but the gametophytes of some liverworts are quite different from those of mosses. Their body form is often a flattened, lobed, leaflike **thallus** (Fig. 26–3*b*). Liverworts are so named because the lobes of their thalli superficially resemble the lobes of the human liver. On the underside of the liverwort thallus are rootlike rhizoids that anchor the plant to the soil. Other liverworts have a leafy appearance rather than a lobed thallus and superficially resemble mosses, with "leaves," "stems," and rhizoids. As with other bryophytes, liverworts are small, generally inconspicuous plants that are largely restricted to damp environments.

Liverworts reproduce both sexually and asexually. Their sexual reproduction involves the production of archegonia and antheridia on the haploid gametophyte. Their life cycle is basically the same as that of mosses, although some of the structures look quite different. The liverwort sporophyte, which is usually somewhat spherical, is attached to the gametophyte plant.

One of the ways that liverworts reproduce asexually is by forming tiny balls of tissue called *gemmae* (sing., *gemma*), which are borne in a saucer-shaped structure, the gemmae cup, directly on the liverwort thallus. Splashing raindrops and small animals aid in the dispersal of gemmae. When a gemma lands in a suitable place, it grows into a new liverwort thallus. Liverworts may also reproduce asexually by thallus branching and growth. The individual thallus lobes elongate and each becomes a separate plant when the older part of the thallus that originally connected the individual lobes dies. Both of these mechanisms of reproduction—gemmae and thallus branching— are asexual because they do not involve fusion of gametes.

Hornworts are inconspicuous thalloid plants

Hornworts (phylum Anthocerophyta) are a small group of bryophytes whose gametophytes superficially resemble those of the thalloid liverworts. Hornworts may or may not be closely related to other bryophytes. For example, their cell structure, particularly the presence of a single large chloroplast in each cell, resembles that of certain algae more than that of plants. Mosses and liverworts, on the other hand, are like all other plants in that they have many disk-shaped chloroplasts per cell.

In hornworts, archegonia and antheridia are embedded in the gametophyte thallus. After fertilization, the sporophyte projects out of the gametophyte tissue, forming a spike or "horn" (Fig. 26–3*c*). A single gametophyte plant often produces a number of sporophytes. Meiosis occurs within each sporangium (pl., *sporangia*), or spore case, and spores are formed.

Seedless vascular plants did not descend from the bryophytes

Although all plants apparently descended from green algal ancestors, mosses and other bryophytes are not in a direct evolutionary path to the vascular plants. That is, the general consensus is that vascular plants did not have mosslike ancestors. Mosses and other bryophytes may represent an evolutionary sideline that arose from ancestral green algae at a different time than did early vascular plants. Alternatively, mosses may have descended from early vascular plants by becoming simpler and losing their vascular tissues. The fossil record of ancient bryophytes does not provide a definitive answer on bryophyte evolution because it can be interpreted in different ways.

SEEDLESS VASCULAR PLANTS INCLUDE FERNS AND THEIR ALLIES

According to the fossil record, the vascular plants arose some 420 million years ago (Fig. 26–6). Ferns and fern allies, an ancient group of plants that extends back nearly 400 million years, were of considerable importance as Earth's dominant plants in past ages. Fossil evidence indicates that many species were immense trees. The few fern allies that survive today are smaller representatives of the ancient groups. Many ferns are also extinct, although 11,000 species of ferns exist today. Ferns are especially common in temperate woodlands and tropical rain forests, where they are found in the greatest variety. Three groups of vascular plants—whisk ferns (several species), club mosses (about 1000 species), and horsetails (15 species)—are considered fern allies because their life cycles are similar to those of ferns (Table 26–2).

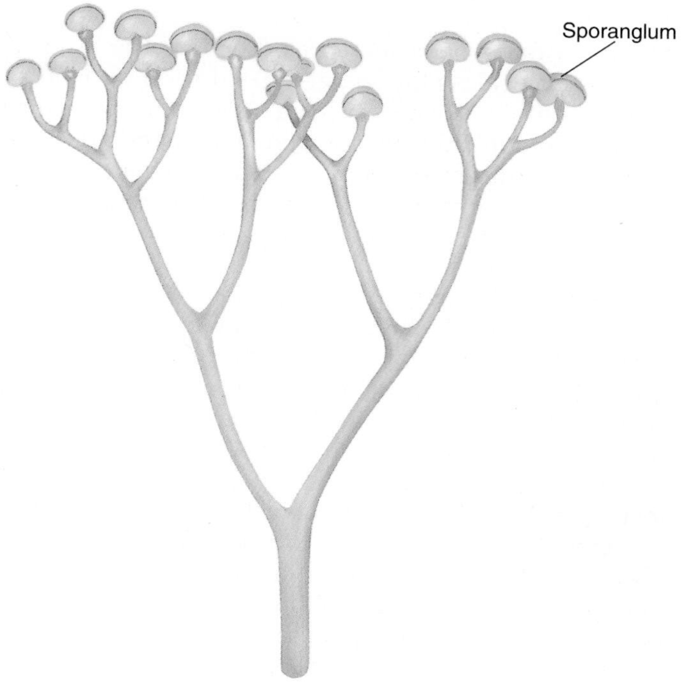

FIGURE 26–6 *Cooksonia caledonica,* **the earliest known vascular plant, is extinct, and its reconstruction is based on fossil evidence.**

The most important adaptation found in ferns and their allies but absent in algae and bryophytes is the presence of specialized vascular tissues—xylem and phloem—for support and conduction. This system of conduction enables vascular plants to achieve larger sizes than the bryophytes do, because water, dissolved minerals, and dissolved sugar can be transported over great distances to all parts of the plant. Although ferns in temperate environments are relatively small, tree ferns in the tropics may grow to heights of 18 meters (60 feet). The ferns and fern allies all have true stems with vascular tissues, and most also have true roots and leaves.

The evolution of the leaf as the main organ of photosynthesis has been studied extensively. There are two basic types of leaves, microphylls and megaphylls (Fig. 26–7). The **microphyll,** which is usually small and pos-

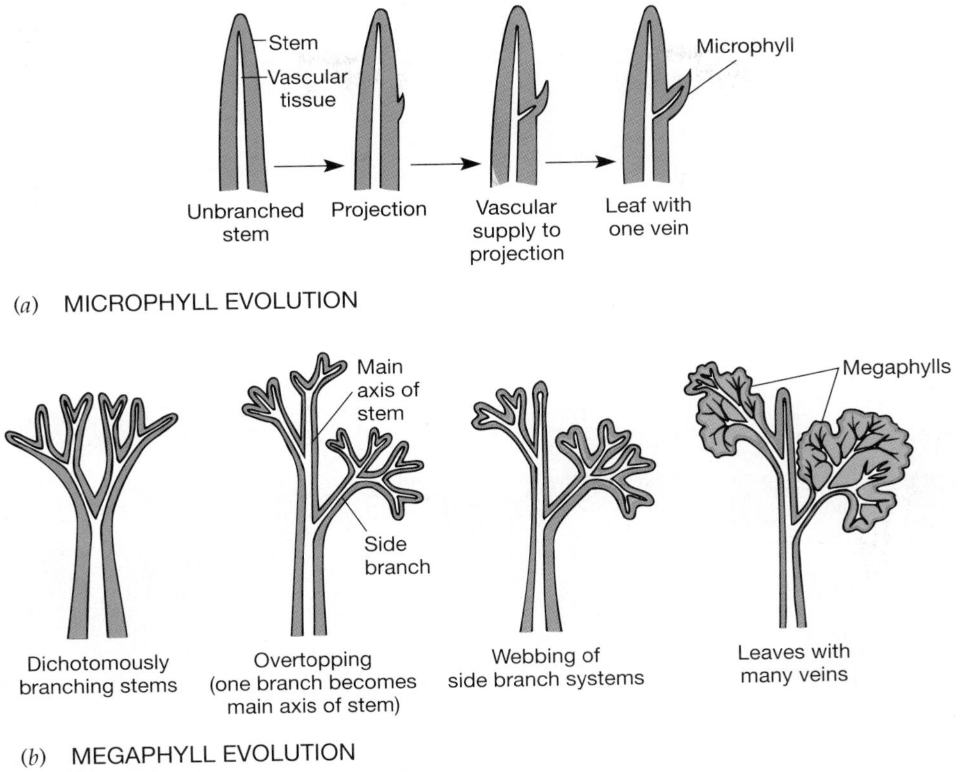

FIGURE 26–7 There are two kinds of leaves, microphylls and megaphylls. (***a***) Microphylls are thought to have originated as outgrowths of stem tissue that developed a single vascular strand later. (***b***) Megaphylls, which are more complex and have multiple veins, probably evolved from the evolutionary modification of side branches. Webbing is the evolutionary process in which the spaces between close branches are filled with chlorophyll-containing cells.

sesses a single vascular strand, is thought to have evolved from small, projecting extensions of stem tissue. Only one group of living plants—the club mosses—possesses microphylls. In contrast, **megaphylls** are thought to have evolved from stem branches, which gradually filled in with additional tissue to form most leaves as we know them today. Megaphylls possess more than one vascular strand, as would be expected if they evolved from branch systems. Ferns, horsetails, gymnosperms, and flowering plants have megaphylls.

Ferns have a dominant sporophyte generation

Although **ferns** (phylum Pterophyta) range from the tropics to the Arctic Circle, most fern species are found in the moist tropics (Fig. 26–8). In temperate areas, ferns commonly inhabit moist woodlands and stream banks. Most ferns are terrestrial, although a few have adapted to aquatic habitats.

The life cycle of ferns involves a clearly defined alternation of generations (Fig. 26–9). The ferns grown as houseplants (such as Boston fern, maidenhair fern, and staghorn fern) represent the sporophyte generation. The fern sporophyte is composed of a horizontal underground stem, or *rhizome,* that bears leaves, called *fronds,* and roots. As each young frond first emerges from the ground, it is tightly coiled and resembles the top of a violin, resulting in the name *fiddlehead.* As fiddleheads grow, they unroll and expand to form fronds. Fern fronds are usually compound (that is, the blade is divided into several leaflets), with the leaflets forming beautifully complex leaves. Fronds, roots, and rhizomes all contain vascular tissues.

Spore production usually occurs in certain areas on the fronds, which develop sporangia in which sporogenous cells (spore mother cells) undergo meiosis to form haploid spores. The sporangia are frequently borne in clusters, called **sori** (sing., *sorus*), on the fronds. When the spores are shed and land in suitable places, they may germinate and grow by mitosis into gametophytes.

The mature fern gametophyte, which bears no resemblance to the sporophyte, is a tiny (about the size of half of one of your fingernails), green, often heart-shaped structure that grows flat against the ground. Called a **prothallus** (pl., *prothalli*), the fern gametophyte lacks vascular tissues and has tiny rootlike rhizoids to anchor it. The prothallus usually produces both archegonia and an-

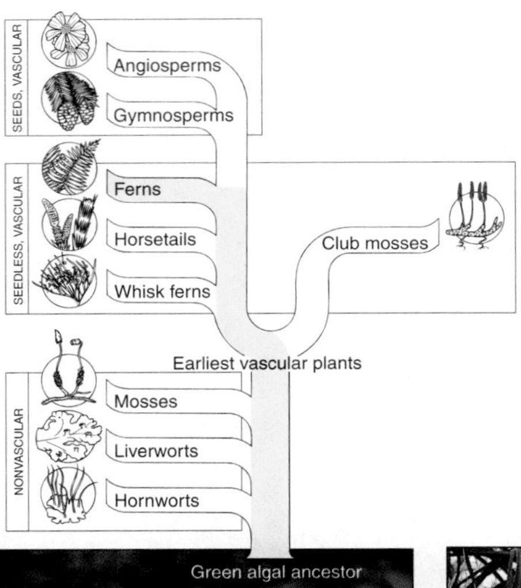

(a) (b) (c)

FIGURE 26–8 Ferns exhibit great diversity in form and growth habit. (***a***) The sword fern *(Polystichum munitum)* is a common fern in Pacific coastal redwood forests. (***b***) The staghorn fern *(Platycerium)* is epiphytic and grows attached to the bark of rainforest trees rather than rooted in the soil. (***c***) The Tasmanian tree fern, native to tropical rain forests, is found in New Zealand, South Africa, and South America. (*a*, Ed Reschke/Peter Arnold, Inc.; *b*, Richard H. Gross; *c*, Dennis Drenner)

theridia on its underside. Each archegonium contains a single egg, whereas numerous sperm are produced in each antheridium.

Although ferns are considered more advanced than mosses because they possess vascular tissues, they have retained a primitive fertilization technique: the use of water as a transport medium. The flagellated sperm swim to the neck of an archegonium through a thin film of water on the ground underneath the prothallus. After one of the sperm fertilizes the egg, a diploid zygote grows by mitosis into a multicellular embryo. At this stage in its life, the sporophyte embryo is attached to and dependent on the gametophyte, but as the embryo matures into a sporophyte plant, the prothallus withers and dies.

The fern life cycle has a clearly defined alternation of generations between the diploid sporophyte with its rhizome, roots, and fronds, and the haploid gametophyte (prothallus). The sporophyte generation is dominant not only because it is larger than the gametophyte, but also because it persists for an extended period of time (many ferns are perennials), whereas the gametophyte dies soon after reproducing.

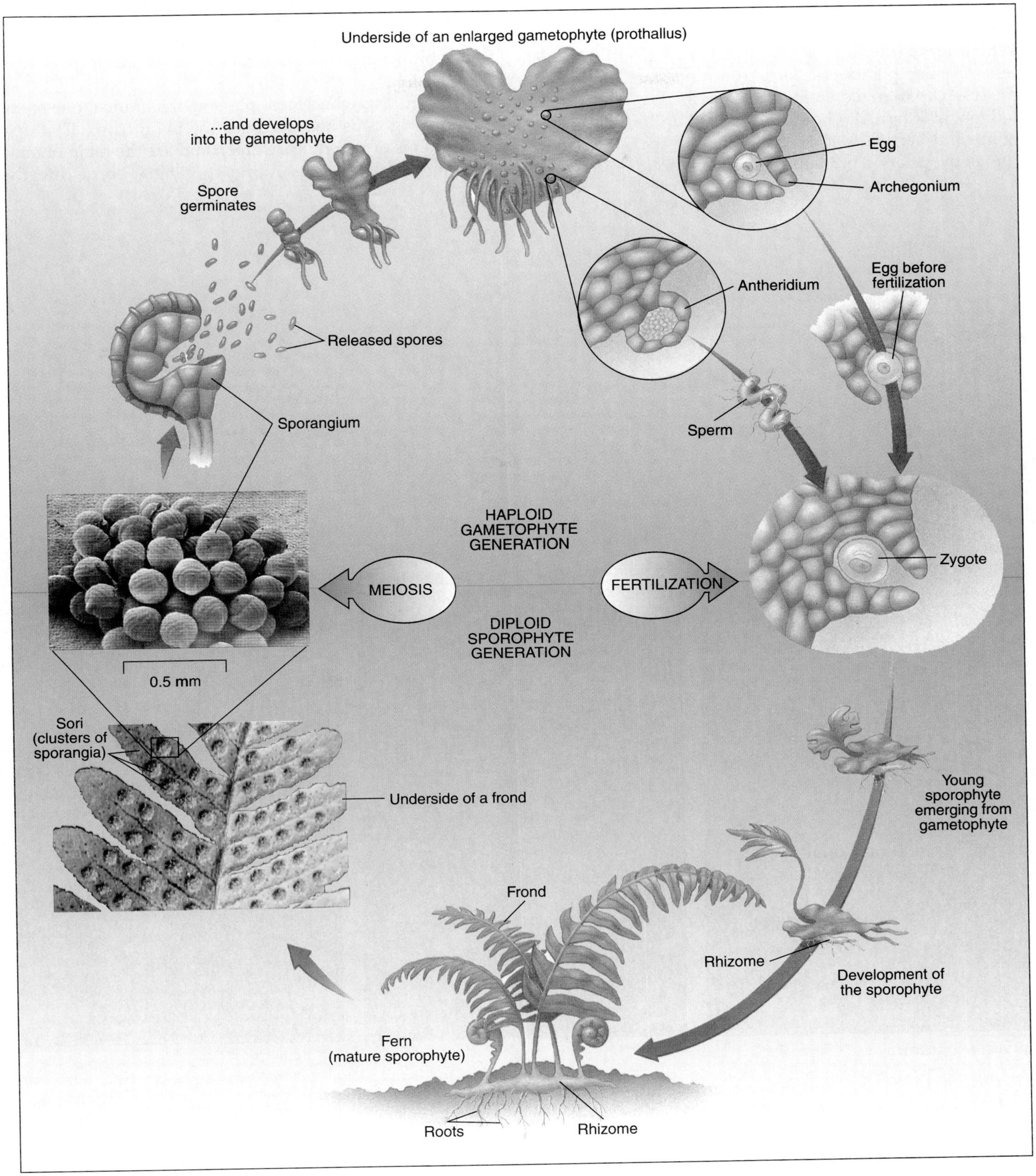

FIGURE 26–9 The sporophyte generation is dominant in the fern life cycle. Note the clearly defined alternation of generations between the gametophyte (prothallus) and sporophyte (leafy plant) generations. (*Bottom left,* David M. Dennis/Tom Stack & Associates; *Top left,* Biophoto Associates/Photo Researchers, Inc.)

Whisk ferns are the simplest vascular plants

Whisk ferns (phylum Psilotophyta) are a group of seedless vascular plants that lack true roots and leaves but possess vascularized stems. *Psilotum nudum,* a representative whisk fern, has both a horizontal underground rhizome and vertical aerial stems (Fig. 26–10*a*). Whenever the stem forks or branches, it always divides into two equal halves. This **dichotomous** branching is considered a primitive characteristic. (For an illustration of dichotomous branching in the earliest known vascular plant, see Fig. 26–6.)

In contrast, when most plant stems branch, one stem is more vigorous and becomes the main trunk. The upright stems of *Psilotum* are green and are the main organs of photosynthesis. Tiny round sporangia, borne directly

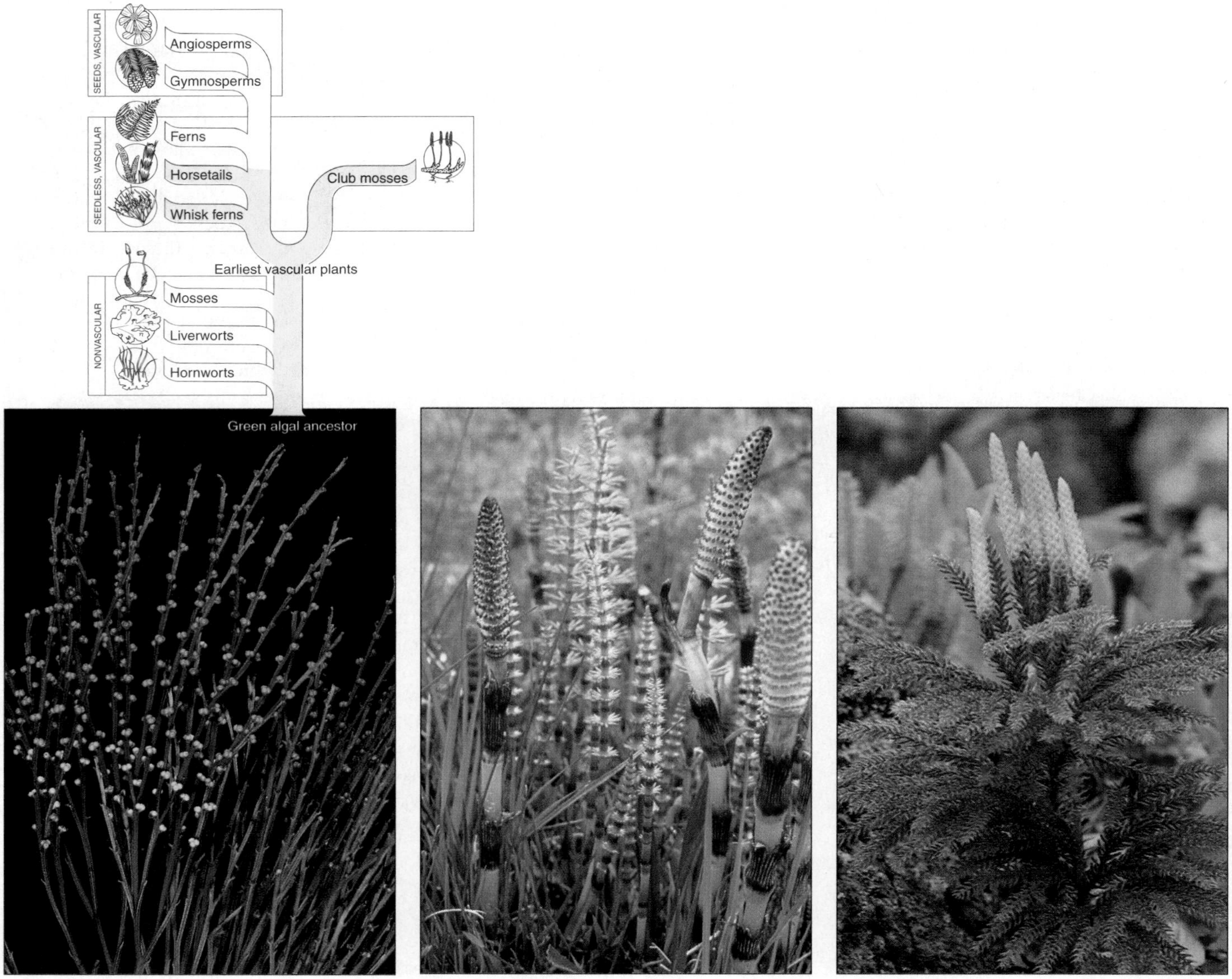

(*a*) (*b*) (*c*)

FIGURE 26–10 Whisk ferns, horsetails, and club mosses are fern allies. (***a***) The growth habit of *Psilotum nudum,* a whisk fern. Lacking leaves, the stem is the main organ of photosynthesis in whisk ferns. (***b***) *Equisetum telematia,* a horsetail with a wide distribution in Eurasia, Africa, and North America, has unbranched, nongreen fertile shoots bearing conelike strobili and separate, highly branched, photosynthetic sterile (nonreproductive) shoots. Both shoots arise from an underground rhizome. (***c***) *Lycopodium,* a club moss. Although club mosses superficially resemble mosses, they are fern allies. The sporophyte plant has reduced, scalelike leaves that are evergreen. Spores are produced in sporangia on fertile leaves clustered in a conelike strobilus *(as shown)* or, in other species, scattered along the stem *(not shown).* (*a,* James Mauseth, University of Texas; *b,* J. Robert Waaland, University of Washington/Biological Photo Service; *c,* Ed Reschke)

Focus On

Ancient Plants and Coal Formation

Our industrial society depends on energy from fossil fuels that formed from the remains of ancient organisms. One of our most important fossil fuels is coal, which is burned to produce electricity and to manufacture items made of steel and iron. Although coal is mined as a mineral, it is not a mineral like gold or aluminum, but an organic material formed from the remains of ancient vascular plants, particularly those of the Carboniferous period (approximately 300 million years ago). Five main groups of plants contributed to coal formation. Three were seedless vascular plants — the club mosses, horsetails, and ferns. The other two important groups were seed plants — seed ferns (now extinct) and primitive gymnosperms.

It is hard to imagine that relatives of the small, relatively inconspicuous club mosses, ferns, and horsetails of today could have been so significant in forming vast beds of coal. However, many of the members of these groups that existed during the Carboniferous period were giants compared with their present-day counterparts, and formed immense forests (see Fig. 20–10).

The climate during the Carboniferous period was warm, moist, and mild. Plants in most locations could grow year-round because of the favorable conditions. Forests of these plants often occurred in low-lying, swampy areas that were periodically flooded when the sea level rose. When the sea level receded, these plants would become re-established.

When these large plants died or were blown over during storms, they decomposed incompletely because they were covered by swamp water. (The anaerobic conditions of the water prevented wood-rotting fungi from decomposing the plants, and anaerobic bacteria do not decompose wood rapidly.) Thus, over time the partially decomposed plant material accumulated and consolidated.

Layers of sediment formed over the plant material each time the water level rose and flooded the low-lying swamps. With time, heat and pressure built up in these accumulated layers and converted the plant material to coal and the sediment layers to sedimentary rock. Much later, geological upheavals raised the layers of coal and sedimentary rock. For example, coal is found in seams (layers) in the Appalachian Mountains. The various grades of coal (lignite, bituminous, and anthracite) were formed as a result of the different temperatures and pressures to which the layers were exposed. ■

on the erect, aerial stems, contain sporogenous cells that undergo meiosis to form haploid spores; after being dispersed, the spores germinate to form haploid prothalli. The prothalli of whisk ferns are difficult to study because they grow underground. They are not photosynthetic, owing to their subterranean location, and they apparently have a symbiotic relationship with fungi, which provide them with sugar and minerals.

Most species of whisk ferns are extinct, and the few surviving species are found mainly in the tropics and subtropics. Although whisk ferns do not closely resemble ferns in appearance, they are considered fern allies because of similarities in their life cycles. Although whisk ferns have been carefully studied in recent years, botanists disagree about how to interpret their structures. Most botanists consider whisk ferns to be surviving representatives of very primitive vascular plants, but others think they are highly modified relatives of ferns.

Horsetails have hollow, jointed stems

About 300 million years ago the **horsetails** (phylum Sphenophyta) were among the dominant plants and grew as large as modern trees. These ancient horsetails are still significant today because they contributed to Earth's vast coal deposits (see *Focus On: Ancient Plants and Coal Formation*). The few surviving horsetails, all in the genus *Equisetum,* grow mostly in wet, marshy habitats and are small but extremely distinctive (Fig. 26–10*b*).

Horsetails have true roots, stems (both rhizomes and erect aerial stems), and small leaves. The hollow, jointed stems are impregnated with silica, which gives them a gritty texture. Small leaves, interpreted as reduced megaphylls, are fused in whorls at each node (the area on the stem where leaves attach). The green stem is the main organ of photosynthesis. Horsetails are so-named because certain vegetative (nonreproductive) stems have whorls of branches that give the appearance of a bushy horse's tail. In pioneer days horsetails were called "scouring rushes" and were used to scrub out pots and pans along stream banks.

Each reproductive branch of a horsetail bears a terminal conelike **strobilus** (pl., *strobili*). The strobilus is composed of a number of umbrella-like structures, each bearing five to ten sporangia in which spores form. The horsetail life cycle is similar in many respects to the fern life cycle.

Club mosses are small plants with rhizomes and short, erect branches

Like horsetails, **club mosses** (phylum Lycophyta) were important plants millions of years ago when species that are now extinct often attained great size. Like the ancient horsetails, ancient club mosses were major contributors to our present-day coal deposits (see *Focus On: Ancient Plants and Coal Formation*). The club mosses today, such as *Lycopodium* (Fig. 26–10*c*), are small, attractive plants commonly found in woodlands. They possess true roots; both rhizomes and erect aerial stems; and small, scale-like leaves (microphylls). Sporangia are borne on fertile leaves in conelike strobili at the tips of, or scattered along, the stems. Club mosses are evergreen and often fashioned into Christmas wreaths and other decorations. In some areas they are endangered by overharvesting.

The fact that common names can be misleading in biology is vividly evident in this group of plants. The most common names for the phylum Lycophyta are "club mosses" and "ground pines," yet these plants are neither mosses nor pines and are most closely allied to the ferns.

More advanced plants are less dependent on water as a transport medium for reproductive cells

Many algae produce flagellated reproductive cells, both spores and sperm, that can swim through the water. Although reproduction by flagellated spores and sperm is an advantage in aquatic environments, it may be detrimental on land, particularly in locations where extended dry periods occur. In such terrestrial sites, the production of nonmotile, airborne spores and sperm may be more advantageous. Thus, a general survey of algae and plants shows that algae have *motile spores and sperm;* the relatively primitive mosses and ferns have *nonmotile spores but have retained motile sperm;* and the more advanced gymnosperms and flowering plants (see Chapter 27) have *nonmotile spores and sperm.*

Some ferns and club mosses are heterosporous

In the life cycles examined thus far, plants produce only one type of spore as a result of meiosis. This condition, known as **homospory,** is characteristic of bryophytes, horsetails, whisk ferns, and most ferns and club mosses. However, certain ferns and club mosses exhibit **heterospory,** in which they produce two different types of spores—microspores and megaspores.

Selaginella, a club moss, is an example of a heterosporous plant; its strobilus bears two kinds of sporangia—microsporangia and megasporangia. *Microsporangia* are sporangia that produce *microsporocytes* (also called *microspore mother cells*), which undergo meiosis to form tiny, haploid **microspores.** Each microspore can develop into a male gametophyte that produces sperm. *Megasporangia* in the *Selaginella* strobilus produce *megasporocytes* (also called *megaspore mother cells*). When megasporocytes undergo meiosis, they form haploid **megaspores,** each of which can develop into a female gametophyte that produces eggs in archegonia. Refer to Figure 26–11 to help visualize this type of life cycle. The development of male gametophytes from microspores and female gametophytes from megaspores occurs within their respective sporangia. The male and female gametophytes are not truly free-living, unlike the gametophytes of other seedless vascular plants.

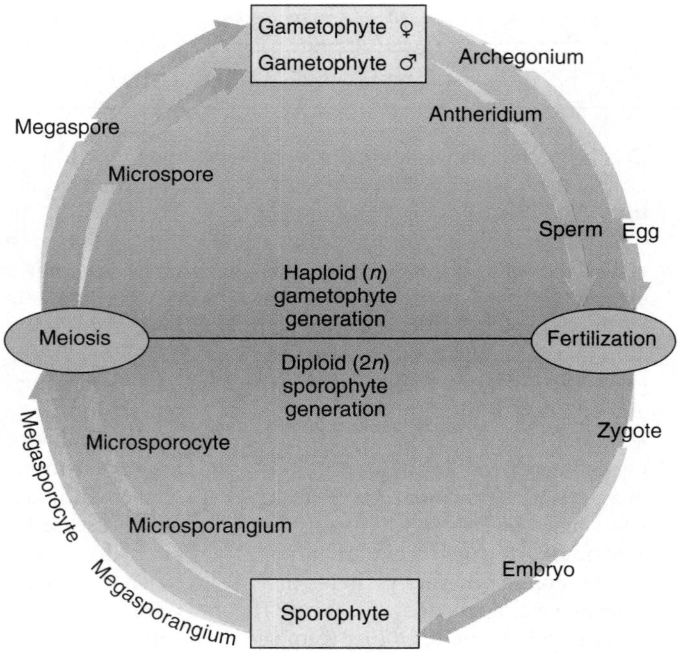

FIGURE 26–11 Two types of spores, microspores and megaspores, are produced during the life cycle of heterosporous plants.

Heterospory was a significant development in plant evolution because it was the forerunner of the evolution of seeds. Heterospory is found in the two most successful groups of living plants—the gymnosperms and the flowering plants—both of which produce seeds (see Chapter 27).

SUMMARY

I. Plants are complex multicellular organisms that obtain energy by photosynthesis. Plants probably arose from green algal ancestors.
   A. Plants and green algae have similar biochemical characteristics: the same photosynthetic pigments, cell wall components, and carbohydrate storage material.
   B. Plants and green algae share similarities in certain fundamental processes like cell division.

II. The colonization of land by plants required the evolution of a number of anatomical, physiological, and reproductive adaptations.
   A. Plants possess (1) a waxy cuticle to protect against water loss and (2) stomata for gas exchange needed for photosynthesis.
   B. An evolutionary trend in plants has been toward a larger, more dominant sporophyte generation and a smaller, less dominant gametophyte generation.
   C. Plants produce multicellular gametangia with a protective jacket of cells surrounding the gametes. Antheridia produce sperm and archegonia produce eggs.
   D. Mosses and ferns, although adapted to life on land, have motile sperm and require water as a transport medium for fertilization.
   E. Vascular plants possess xylem to conduct water and dissolved minerals and phloem to conduct dissolved sugar.

III. Plant life cycles have an alternation of generations in which they spend part of their life cycle in the multicellular haploid gametophyte stage and part in the multicellular diploid sporophyte stage.
   A. The gametophyte produces haploid gametes by mitosis.
   B. These gametes fuse to form a zygote during fertilization.
   C. The first stage in the sporophyte generation is the zygote, which develops into a multicellular embryo that is protected and nourished by the gametophyte.
   D. The mature sporophyte produces sporogenous cells (spore mother cells) that undergo meiosis to form haploid spores, which are the first stage in the gametophyte generation.

IV. Mosses and other bryophytes have several adaptations that green algae lack, including the possession of a cuticle, stomata, and multicellular gametangia. They are nonvascular (lacking xylem and phloem).
   A. Bryophytes are the only plants with a dominant gametophyte generation.
   B. Moss gametophytes are leafy plants that grow from a filamentous protonema.
   C. Liverwort gametophytes are either leafy or thalloid.
   D. Hornworts have thalloid gametophytes.

V. Ferns and fern allies have several adaptations that algae and bryophytes lack, including the possession of vascular tissues and a dominant sporophyte generation.
   A. Ferns and fern allies are seedless vascular plants.
   B. Fern sporophytes have roots, rhizomes, and leaves (megaphylls).
   C. Sporophytes of whisk ferns consist of rhizomes and erect branches. They lack true roots and leaves.
   D. Horsetail sporophytes have roots, rhizomes, aerial stems that are hollow and jointed, and leaves that are reduced megaphylls.
   E. Sporophytes of club mosses consist of roots, rhizomes, erect branches, and leaves that are microphylls.

VI. Homospory is the production of one kind of spore, whereas heterospory is the production of two kinds of spores—microspores and megaspores. The evolution of heterospory was an essential step in the evolution of seeds.

SELECTED KEY TERMS

alternation of generations
antheridium
archegonium
bryophyte
club moss
cuticle
dichotomous
emryo
fern
gametophyte generation
hetrospory
homospory
hornwort
horsetail
lignin
liverwort
megaphyll
megaspore
microphyll
microspore
moss
phloem
prothallus
protonema
sorus
sporophyte generation
stoma (stomata)
strobilus
thallus
whisk fern
xylem

POST-TEST

1. The __________, which include mosses, liverworts, and hornworts, are small plants that lack a vascular system.
2. The waxy layer that covers aerial parts of plants is the __________.
3. A strengthening compound found in cell walls of vascular plants is __________.
4. The openings in plants that allow gas exchange for photosynthesis are called __________.
5. The female gametangium, or __________, produces an egg; the male gametangium, or __________, produces sperm.
6. Plants have an __________ of __________ in which they spend part of their life cycle in the gametophyte stage and part in the sporophyte stage.
7. Liverworts and hornworts share life cycle similarities with __________.
8. The leafy green moss plant is the __________ generation.
9. The flattened, leaflike body form of many liverworts is called a(an) __________.
10. __________ are bryophytes with chloroplasts that are similar to those of certain algae.
11. Seedless vascular plants possess __________ to conduct water and dissolved minerals, and __________ to conduct dissolved sugar.
12. Whisk ferns, horsetails, and club mosses share life cycle similarities with __________.
13. Clusters of sporangia, termed __________, are often found on fern fronds.
14. A(an) __________ is a leaf that arose from a branch system.
15. __________ __________ have vascularized stems but lack true roots and leaves.
16. __________ have hollow, jointed stems that are impregnated with silica.
17. The type of leaf found in club mosses is a(an) __________.
18. Certain seedless vascular plants exhibit __________; that is, they produce two kinds of spores.

REVIEW QUESTIONS

1. What are the most important environmental challenges that plants face living on land, and what adaptations do they possess to meet these challenges?
2. Plants are thought to have descended from which group of protists? What kinds of evidence support this idea?
3. How are mosses, liverworts, and hornworts similar? How is each group distinctive?
4. Name the three groups of plants known as fern allies. What features do these plants share with ferns? How is each group distinctive?
5. State the adaptations of bryophytes that algae lack. What adaptations do ferns have that both algae and bryophytes lack?
6. Compare alternation of generations in the mosses and the ferns. Which stage is dominant in each?
7. How does heterospory modify the life cycle?
8. Label the diagram to the right.

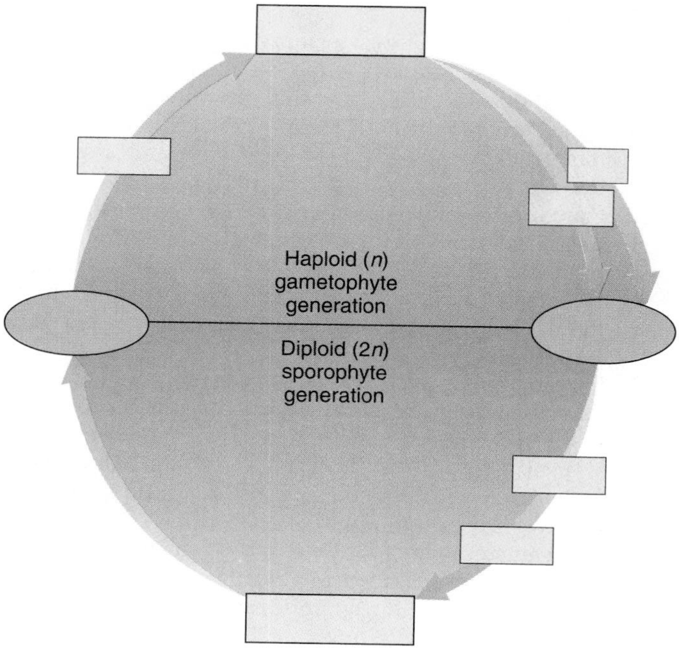

YOU MAKE THE CONNECTION

1. Which group probably colonized the land first, plants or animals? Explain.
2. How might the following trends in plant evolution be adaptive to living on land?
   a. Dependence on water for fertilization → no need for water as a transport medium
   b. Dominant gametophyte generation → dominant sporophyte generation
   c. Homospory → heterospory

RECOMMENDED READINGS

Mauseth, J.D. *Botany: An Introduction to Plant Biology,* 2nd ed. Saunders College Publishing, Philadelphia, 1995. A comprehensive introduction to general botany.

Niklas, K.J. "One Giant Step for Life." *Natural History,* June 1994. An overview of the transition from water to land during the evolution of plants.

Pearce, F. "Peat Bogs Hold Bulk of Britain's Carbon." *New Scientist,* Vol. 144:1952, November 19, 1994. A strong case for the preservation of Scottish peat bogs, which contain far more carbon than would be stored in equivalent areas of forest. Because stored carbon is not in the atmosphere as CO_2, it helps prevent global warming.

Raven, P.H., R.F. Evert, and S.E. Eichhorn. *Biology of Plants,* 5th ed. Worth Publishers, New York, 1992. A general botany text with an evolutionary emphasis.

CHAPTER 27

The Plant Kingdom: Seed Plants

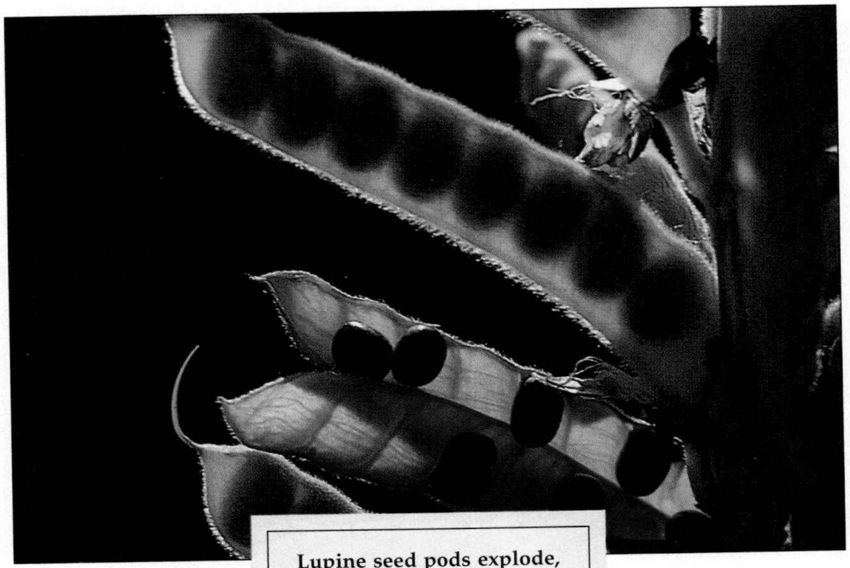

Lupine seed pods explode, dispersing seeds.
(David Cavagnaro)

The primary means of reproduction and dispersal for Earth's most successful plants is seeds, which develop from the female gametophyte and its associated tissues. Seed plants show the greatest evolutionary complexity in the plant kingdom and are the dominant plants in most terrestrial environments.

Seeds are reproductively superior to spores for three main reasons. First, a seed contains a multicellular, well-developed young plant with embryonic root, stem, and leaves already formed, whereas a spore is a single cell. Second, a seed contains a food supply. After germination, the plant embryo is nourished by food stored in the seed until it becomes self-sufficient. Because a spore is a single cell, few food reserves exist for the plant that develops from a spore. Third, a seed is protected by a resistant seed coat. Like spores, seeds can live for extended periods of time at reduced rates of metabolism, germinating when conditions become favorable.

Seeds and seed plants have been intimately connected with the development of human civilization. From prehistoric times, early humans collected and used seeds for food. The food stored in seeds is a concentrated source of proteins, oils, carbohydrates, and vitamins, which are nourishing for humans as well as for germinating plants. Also, seeds are easy to store (provided they are kept dry); this has allowed humans to collect them during times of plenty and save them for times of need. Few other foods can be stored as conveniently or for as long.

In seed plants, each seed develops from an **ovule,** which is a megasporangium and its enclosed structures, following fertilization of an egg cell within. Seed plants are divided into two groups based on whether or not their ovules are protected.

The two groups of seed plants are the **gymnosperms** and the **angiosperms** (see Table 26–1). The word *gymnosperm* is adapted from Greek words meaning "naked seed." These plants produce seeds that are totally exposed or borne on the scales of cones. (In other words, the ovules of gymnosperms are unprotected.) Pine, spruce, fir, and *Ginkgo* are examples of gymnosperms.

The Greek words from which the term *angiosperm* is derived translate as "seed enclosed in a vessel or case." Angiosperms are flowering plants that produce their seeds within a fruit. (Thus, the ovules of angiosperms are protected.) Flowering plants are extremely diverse, and include corn, oaks, water lilies, cacti, apples, palms, and buttercups.

Both gymnosperms and flowering plants possess vascular tissues, xylem for the conduction of water and dissolved minerals, and phloem for the conduction of dissolved sugar. Both have life cycles with an alternation of generations, but the gametophyte generation is significantly reduced in size and is entirely dependent on the sporophyte generation. All gymnosperms and flowering plants are heterosporous and produce two types of spores: microspores and megaspores (see Chapter 26).

LEARNING OBJECTIVES

After you have studied this chapter you should be able to

1. Compare the features of seeds with those of spores and discuss the advantages of plants that reproduce primarily by seeds rather than by spores.
2. Trace the steps in the life cycle of a pine.
3. Summarize the features that distinguish gymnosperms from other plants.
4. Diagram a generalized life cycle of a flowering plant.
5. Contrast dicots and monocots, the two classes of flowering plants.
6. Discuss the evolutionary adaptations of flowering plants.
7. Trace the evolution of gymnosperms from seedless vascular plants, and of flowering plants from gymnosperms.

GYMNOSPERMS ARE THE "NAKED SEED" PLANTS

The gymnosperms include some of the most interesting members of the plant kingdom. For example, a giant sequoia known as the General Sherman tree, located in Sequoia National Park in California is, in terms of sheer bulk, one of the world's most massive organisms. It is only 82 meters (272 feet) tall, but has a diameter of 11 meters (36 feet) and an estimated weight of 6167 tons! Another gymnosperm, a coastal redwood, is probably the world's tallest tree, measuring almost 114 meters (380 feet) in height. One of the oldest living trees, a bristlecone pine in the White Mountains of California, has been determined by tree ring analysis to be 4900 years old!

Gymnosperms are usually classified into four phyla (Fig. 27–1). Numbering 550 species, the largest phylum of gymnosperms is the conifers, woody plants that bear their seeds in cones. Two other phyla of gymnosperms—the ginkgoes and the cycads—represent evolutionary remnants of groups that were more significant in the past. The fourth phylum of gymnosperms, the gnetophytes, is a collection of some extremely unusual plants that share certain traits not found in other gymnosperms.

Conifers are woody plants that bear their seeds in cones

The **conifers** (phylum Coniferophyta), which include pines, spruces, hemlocks, and firs, are the most familiar group of gymnosperms (Table 27–1). They are woody trees or shrubs, most of which are evergreen (Fig. 27–2). Only a few, such as larch and bald cypress, are deciduous and shed their leaves, called **needles,** at the end of the growing season. Most conifers are **monoecious,** which means that they have separate male and female reproductive parts on the same plant. These are generally borne in cones (hence their name, *conifer,* which means "bears cones").

Conifers occupy vast areas, ranging from the Arctic to the tropics, and are the dominant vegetation in the

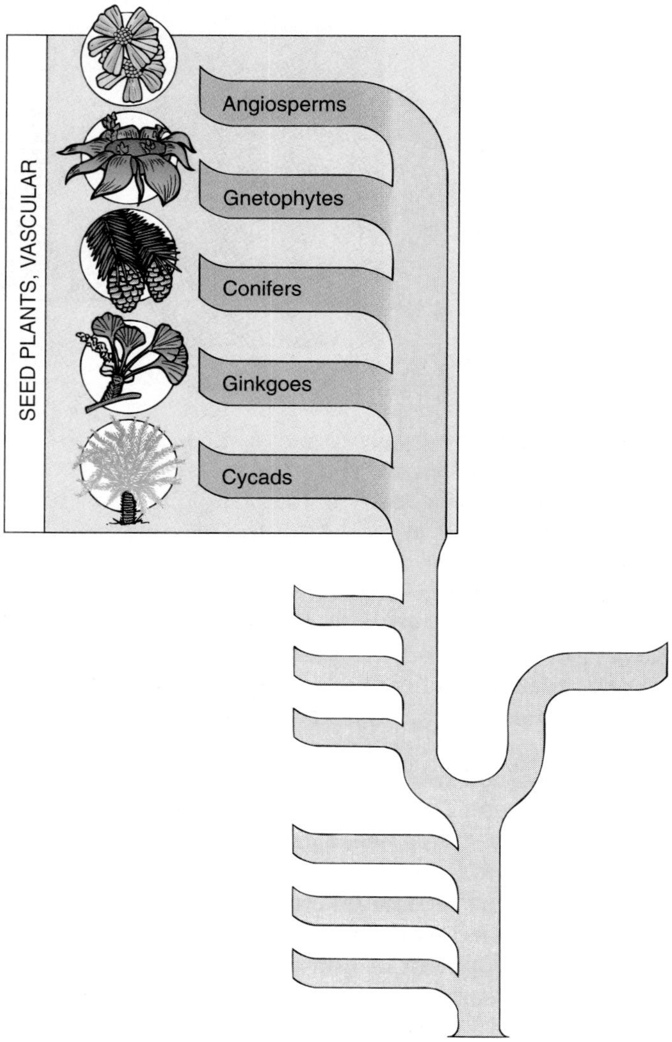

FIGURE 27–1 This phylogenetic tree illustrates possible relationships among plants. The seed plants, which are the focus of this chapter, are the five groups at the top of the tree. Four of these groups (cycads, ginkgoes, conifers, and gnetophytes) are gymnosperms. The angiosperms, or flowering plants, have more living species than all other plant groups combined.

Table 27–1 A COMPARISON OF GYMNOSPERMS AND FLOWERING PLANTS

Characteristic	*Gymnosperms*	*Angiosperms*
Growth habit	Woody trees and shrubs	Woody or herbaceous
Conducting cells in xylem	Tracheids*	Vessel elements and tracheids*
Reproductive structures	Cones (usually)	Flowers
Pollen grain transfer	Wind	Animals or wind
Fertilization	Egg and sperm → zygote	Double fertilization: Egg and sperm → zygote; 2 Polar nuclei and sperm → endosperm
Seeds	Exposed or borne on scales of cones	Enclosed within fruit

*See Chapter 31.

forested regions of Canada, northern Europe, and Siberia. In addition, they are important in the Southern Hemisphere, particularly in areas of South America, Australia, and Malaysia. Ecologically, conifers contribute food and shelter to animals and other organisms, and their roots hold the soil in place and help prevent soil erosion. Humans use conifers for their wood (for building materials as well as paper products), turpentine, and resins. Because of their attractive appearance, conifers are grown for landscape design and Christmas trees.

FIGURE 27–2 Conifers are dominant plants in northern latitudes. A blue spruce is well adapted for surviving in cold environments. (Dennis Drenner)

Pines represent a typical conifer life cycle

A pine tree is the sporophyte generation and therefore forms spores (Fig. 27–3).[1] Pines are heterosporous and produce microspores and megaspores in separate cones. Male cones are smaller than female cones and are generally produced on the lower branches in the spring (Fig. 27–4*a*). Each male cone is composed of **sporophylls,** leaflike structures that bear sporangia. At the base of each sporophyll are two microsporangia, which contain numerous microsporocytes (also called microspore mother cells). Each microsporocyte undergoes meiosis to form haploid microspores. Microspores then develop into extremely reduced male gametophytes; an immature male gametophyte is also called a **pollen grain.** Pollen grains are shed from male cones in great numbers, and some are carried by wind currents to the immature female cones.

The familiar woody pine cones—the female cones—are usually found on the upper branches of the tree. The woody scales of the female cones have megasporangia at their bases. Within each megasporangium, meiosis of a megasporocyte (megaspore mother cell) produces four

[1]It might be helpful to review alternation of generations in Chapter 26 before studying the pine life cycle.

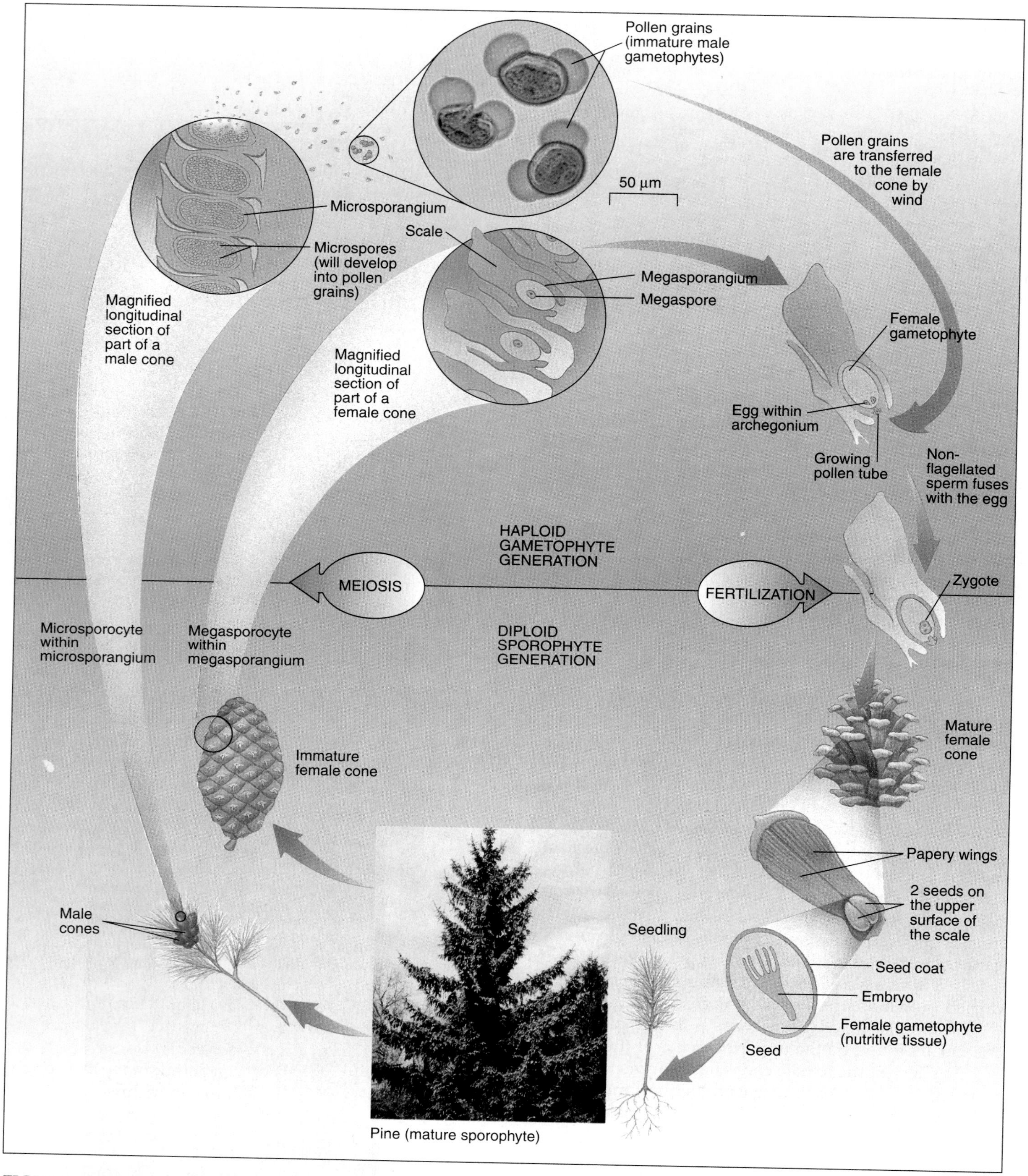

FIGURE 27–3 The sporophyte generation is dominant in the pine life cycle. One major advantage of gymnosperms over the seedless vascular plants is the production of wind-borne pollen grains *(top and top right).* Pines and other gymnosperms do not depend on water as a transport medium for sperm. (*Top,* Manfred Kage/Peter Arnold, Inc.; *Bottom,* Dennis Drenner)

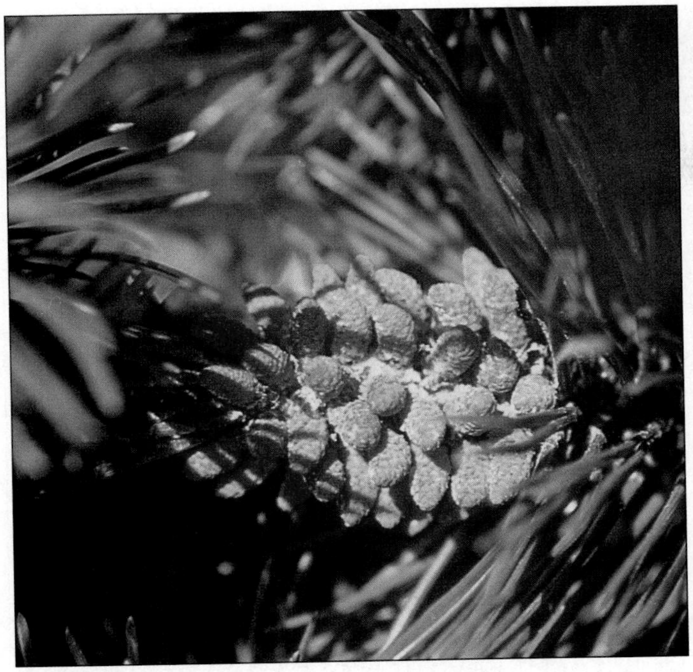

(a)

(b)

FIGURE 27–4 Reproduction in pines takes place in male and female cones. (*a*) Male pine cones, which occur in clusters near the ends of branches, produce copious amounts of pollen grains in the spring. Pine pollen grains are carried by wind to the female cones. (*b*) A mature female pine cone with its many shed seeds. Note the papery wing attached to each seed. (*a*, James Mauseth, University of Texas; *b*, William E. Ferguson)

haploid megaspores. One of these develops into the female gametophyte, which produces an egg within each of several archegonia. The other three megaspores are nonfunctional and soon degenerate.

The scales of the developing female cone open at the time pollen is shed, and pollen grains, carried to the female cones by air currents, drift between the scales and adhere to a drop of sticky fluid. The pollen grains are drawn to the megasporangium as the sticky fluid evaporates, and the scales of the female cone grow together.

Each pollen grain grows a pollen tube that slowly digests its way through the female gametophyte tissue to the egg. Then a cell within the pollen grain divides to form two nonflagellated sperm, which travel down the pollen tube. One sperm fuses with the egg to form a zygote, or fertilized egg, which subsequently grows into a young pine embryo in the seed; the other sperm degenerates. The female gametophyte tissue surrounding the developing embryo becomes the nutritive tissue in the mature seed. In addition to a tough protective seed coat that surrounds the embryo and nutritive tissue, the mature pine seed has a papery wing that allows it to be dispersed by wind currents (Fig. 27–4*b*).

A long time elapses between the appearance of pine cones on a tree and the maturation of seeds. When **pollination,** the transfer of pollen grains to the female cone, occurs in the spring, the female cone is immature and meiosis of the megasporocytes (megaspore mother cells) has not occurred. During the ensuing months, the female tissue gradually matures and eggs are formed within archegonia. Meanwhile, the pollen grain slowly grows a pollen tube through the female tissues to the archegonia. **Fertilization,** the union of egg and sperm, occurs during the spring of the year following pollination. Seed maturation takes several additional months, although some seeds remain within the female cones for a number of years before being shed.

In the pine life cycle, the sporophyte generation is dominant, and the gametophyte generation consists of microscopic structures in the male and female cones. Although the female gametophyte produces archegonia, the male gametophyte is so reduced that it does not produce antheridia. The gametophyte generation depends on the parent sporophyte generation for nourishment.

A major adaptation in the pine life cycle is elimination of the need for external water as a sperm transport medium. Instead, pine pollen grains are carried to female cones by air currents. Gymnosperms apparently were the first plants to have a mode of reproduction totally adapted for life on land.

Cycads are gymnosperms with compound leaves and simple seed cones

Cycads (phylum Cycadophyta) were extremely important during the Triassic period, which began approximately 248 million years ago and is sometimes referred to as the "Age of Cycads." Most species are now extinct, and the few surviving cycads—only 100 species remain—are slow-growing evergreens that are considered very primitive seed plants. (The term *primitive* means that cy-

(a) (b) (c)

FIGURE 27–5 Cycads, ginkgoes, and gnetophytes are gymnosperm groups other than conifers. (*a*) Cycads growing in South Africa. Cycads are tropical gymnosperms with a palmlike appearance. Note the immense seed cones on this plant. (*b*) *Ginkgo biloba*, the maidenhair tree. The unusual leaves of *Ginkgo* resemble those of the maidenhair fern, hence its common name. (*c*) *Welwitschia*, a gnetophyte native to arid deserts in southwestern Africa, is an extremely unusual plant. Its two leaves grow from the edges of a short, wide stem. (*a*, W.H. Hodge/Peter Arnold, Inc.; *b*, William E. Ferguson; *c*, Patti Murray) cads have retained many features of their ancestors.) Many cycads are endangered species, primarily because they are popular as ornamentals and are gathered from the wild and sold to collectors.

Cycads grow in the tropics and subtropics. They have compound leaves that give them a palmlike or fernlike appearance (Fig. 27–5*a*). Cycad reproduction is similar to that in pines except that cycads are **dioecious** and therefore have seed cones and pollen cones on separate plants. Their seed structure is most like that of the earliest seeds found in the fossil record. Cycads have also retained the primitive feature of motile sperm. These flagellated sperm are a vestige, however, because cycad pollen grains are carried by air or possibly insects to the female plants; there the pollen grain germinates and grows a pollen tube down which the sperm pass to get to the egg. In other words, despite having flagellated sperm, cycads do not need external water as a transport medium for fertilization.

Ginkgo is the only living species in its phylum

Ginkgoes (phylum Ginkgophyta) are represented by a single species, the maidenhair tree *(Ginkgo biloba)* (Fig. 27–5*b*). It is a native of southeastern China, where it has been under cultivation for centuries, but is apparently extinct in the wild. *Ginkgo* is the oldest genus of living trees. Fossil ginkgoes 200 million years old have been discovered that are nearly identical to the modern-day *Ginkgo*.

Ginkgo is commonly planted in North America today, particularly in cities, because it is hardy and somewhat resistant to air pollution. Its leaves are deciduous and turn a beautiful golden color in the fall. Like cycads, ginkgoes are dioecious, with separate male and female trees. They have flagellated sperm, a vestige that is not required because ginkgoes produce air-borne pollen grains. Its seeds are completely exposed rather than developing within cones. Male trees are typically planted because the females bear seeds that smell like rancid butter.

Gnetophytes include three unusual gymnosperms

The **gnetophytes** (phylum Gnetophyta), with about 70 species in three genera *(Gnetum, Ephedra, Welwitschia)*, are a remarkably diverse group of gymnosperms. Gnetophytes share a number of angiosperm-like features. In their xylem, they have more efficient water-conducing cells that closely resemble vessel elements (see Chapter 31). (Flowering plants have vessel elements in their xylem, but gymnosperms do not.) Also, the cone clusters produced by some of the gnetophytes resemble some flower clusters.

The genus *Gnetum* contains tropical vines and trees with leaves that resemble those of flowering plants. *Ephedra* species, shrubs found in deserts and other dry regions, resemble horsetails in appearance and are commonly called joint firs. An Asiatic *Ephedra* is the source of the asthma medicine ephedrine. The third gnetophyte genus, *Welwitschia,* contains a single species found in deserts of southwestern Africa (Fig. 27–5*c*). Most of *Welwitschia*'s body grows underground. Its short, wide stem forms a shallow disk, up to 0.9 meter (3 feet) in diameter, from which two ribbon-like leaves extend. These leaves continue to grow from the stem throughout the plant's life, but the ends of the leaves are usually broken and torn by the wind, giving the appearance of numerous leaves. When *Welwitschia* reproduces, it forms cones around the edge of its disklike stem.

FLOWERING PLANTS PRODUCE FLOWERS, FRUITS, AND SEEDS

Flowering plants (phylum Anthophyta), or angiosperms, are the most successful plants today, surpassing even the gymnosperms in importance (Table 27–1). They have adapted to almost every habitat and, with at least 235,000 species, are Earth's dominant plants. Flowering plants reproduce sexually by forming flowers, and the seeds that develop following reproduction are enclosed within a fruit. The fruit protects the developing seeds and often aids in their dispersal. Flowering plants possess efficient water-conducting cells called vessel elements in their xylem, and efficient sugar-conducting cells called sieve tube members in their phloem (see Chapter 31). Their life cycles have a unique double-fertilization process, which we will discuss shortly.

Flowering plants are extremely important to humans because our survival as a species literally depends on them. All of our major food crops are flowering plants, including cereal crops such as rice, wheat, corn, and barley. Woody flowering plants such as oak, cherry, and walnut provide valuable lumber. Flowering plants give us fibers like cotton and linen, and medicines like digitalis and codeine. Products as diverse as rubber, tobacco, coffee, chocolate, and aromatic oils for perfumes come from flowering plants. Economic botany is the subdiscipline of botany that deals with plants of economic importance, most of which are flowering plants.

Monocots and dicots are the two classes of flowering plants

The phylum Anthophyta is divided into two classes, the monocots (class Monocotyledones) and dicots (class Dicotyledones) (Fig. 27–6). Monocots include palms, grasses, orchids, and lilies; and dicots include oaks, roses, cacti, blueberries, and sunflowers. The dicots are much more diverse and include many more species than the monocots. Figure 27–7 provides a comparison of some of the features of the two groups.

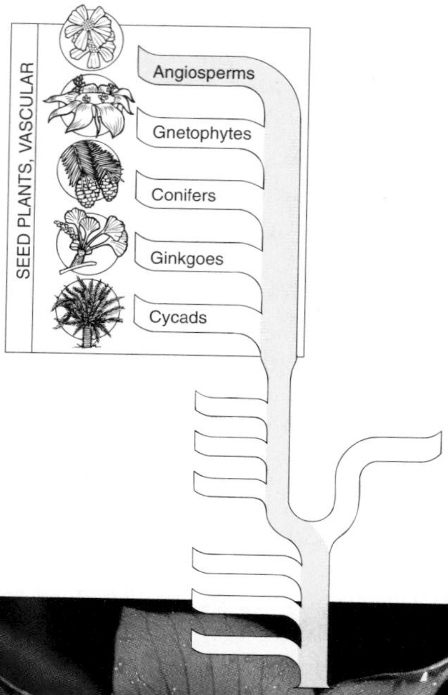

(a)

(b)

FIGURE 27–6 Monocots and dicots are the two classes of flowering plants. (*a*) Most monocots such as this *Trillium* have their floral parts in threes. Note the three green sepals, three rose-colored petals, six stamens, and three stigmas (the compound pistil is composed of three fused carpels). (*b*) Most dicots such as this *Tacitus* have their floral parts in fours or fives. Note the five petals, ten stamens, and five separate pistils. (*a,* Don and Esther Phillips/Tom Stack & Associates; *b,* Richard H. Gross)

Monocots are mostly herbaceous plants with leaves that are usually long and narrow, and have parallel veins (the main leaf veins run parallel to one another). The flower parts of monocot flowers usually occur in three or multiples of three—for example, three petals (flower parts that are usually conspicuously colored) and six stamens (flower parts that produce pollen grains). Monocot seeds have a single **cotyledon** (embryonic seed leaf), and **endosperm** (nutritive tissue) is usually present in the mature seed.

Dicots may be herbaceous (for example, the tomato) or woody (for example, the hickory tree). Their leaves vary in shape, but usually are broader than monocot leaves, with netted veins (branched veins resembling a net). Flower parts usually occur in fours or fives or multiples thereof. Two cotyledons are present in dicot seeds, and endosperm is usually absent in the mature seed, having been absorbed by the two cotyledons prior to germination.

Flowers are involved in sexual reproduction

Flowers are reproductive shoots usually composed of four kinds of organs—sepals, petals, stamens, and carpels—arranged in whorls (circles) on a shortened stem (Fig. 27–8). Sepals and petals are sterile modified leaves, while stamens and carpels are fertile modified leaves. A flower that has all four parts is said to be **complete,** whereas an **incomplete** flower lacks one or more of these parts. All four floral parts are important in the reproductive process, but only the stamens (the "male" organs) and carpels (the "female" organs) participate directly in reproduction. A flower with both stamens and carpels is said to be **perfect,** whereas an **imperfect** flower has stamens *or* carpels, but not both.

Sepals, which make up the lowermost and outermost whorl on a floral shoot, are leaflike in appearance and often green. Sepals cover and protect the other flower parts when the flower is a bud. As the blossom opens from a bud, the sepals fold back to reveal the more conspicuous petals. The collective term for all the sepals of a flower is the **calyx.**

Moving inward, **petals** comprise the next whorl of flower parts. Petals, like sepals, are leaflike in appearance, although they are frequently brightly colored. They play an important role in attracting animal pollinators to the flower (see Chapter 35). Sometimes petals are fused to form a tube or other floral shape. The petals of a flower are referred to collectively as the **corolla.**

Just inside the petals is a whorl of **stamens.** Each stamen is composed of a thin stalk, called a **filament,** and a saclike **anther,** where meiosis occurs to form microspores that develop into pollen grains. Each pollen grain produces two male gametes, or sperm.

In the center of most flowers is one or more **carpels,** the "female" reproductive organs. Carpels bear ovules,

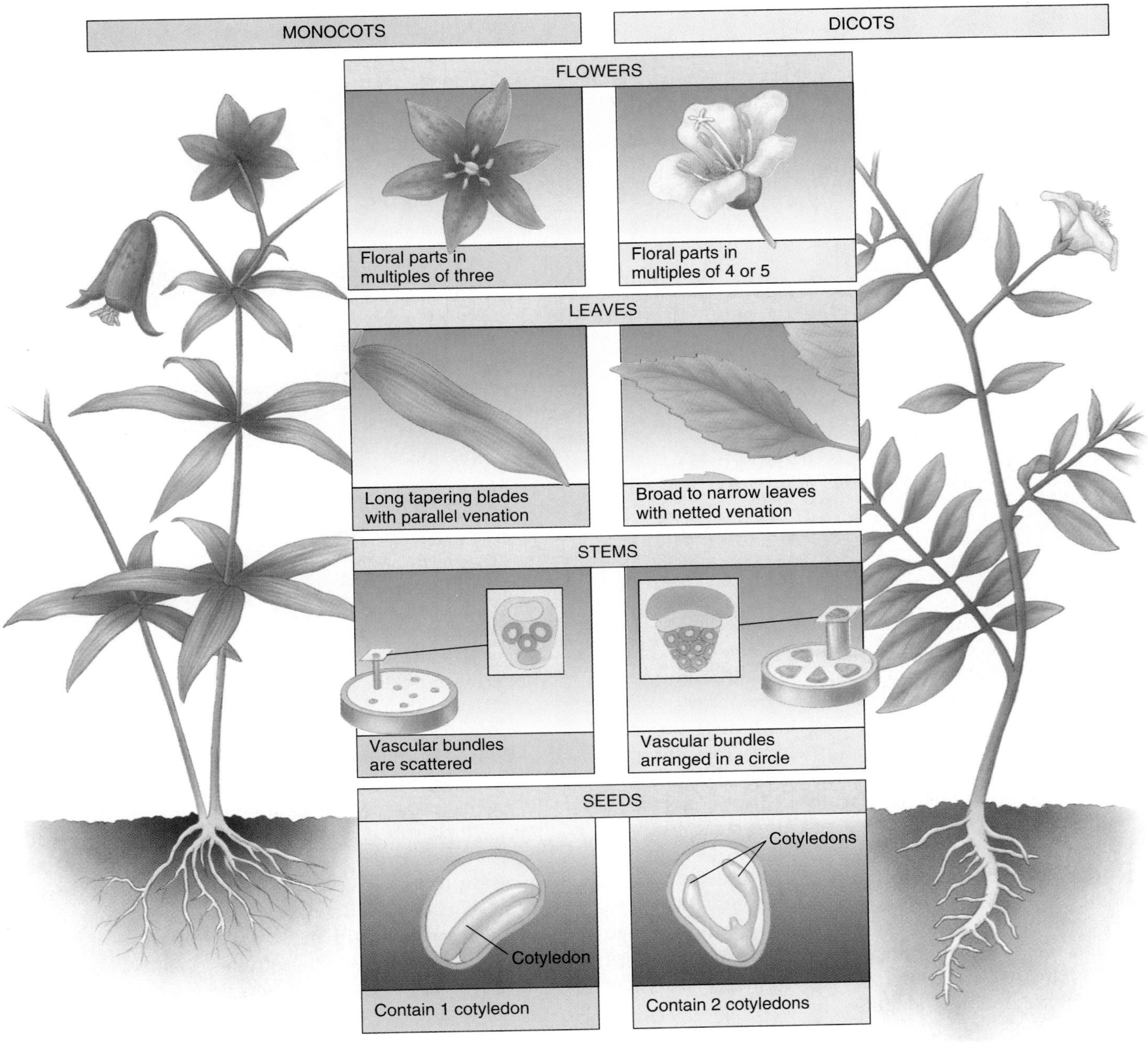

FIGURE 27–7 Monocots and dicots can be distinguished from one another in several ways.

which, as you may recall, are structures with the potential to develop into seeds. The carpels of a flower may be separate or fused together into a single structure. The female part of the flower is sometimes referred to as a **pistil.** A pistil may consist of a single carpel (a *simple* pistil) or a group of fused carpels (a *compound* pistil). Each pistil has three sections: a **stigma,** on which the pollen grain lands; a **style,** a necklike structure through which the pollen tube grows; and an **ovary,** a juglike structure that contains one or more ovules. Each ovule holds a female gametophyte that forms one female gamete (an egg) and two **polar nuclei.** The egg and polar nuclei participate directly in fertilization. After fertilization, the ovule develops into a seed.

The life cycle of flowering plants includes double fertilization

Like all plants, flowering plants have a life cycle that consists of an alternation of generations, but their sporophyte generation is clearly dominant and their gametophyte generation is extremely reduced (to only a few cells) (Fig.

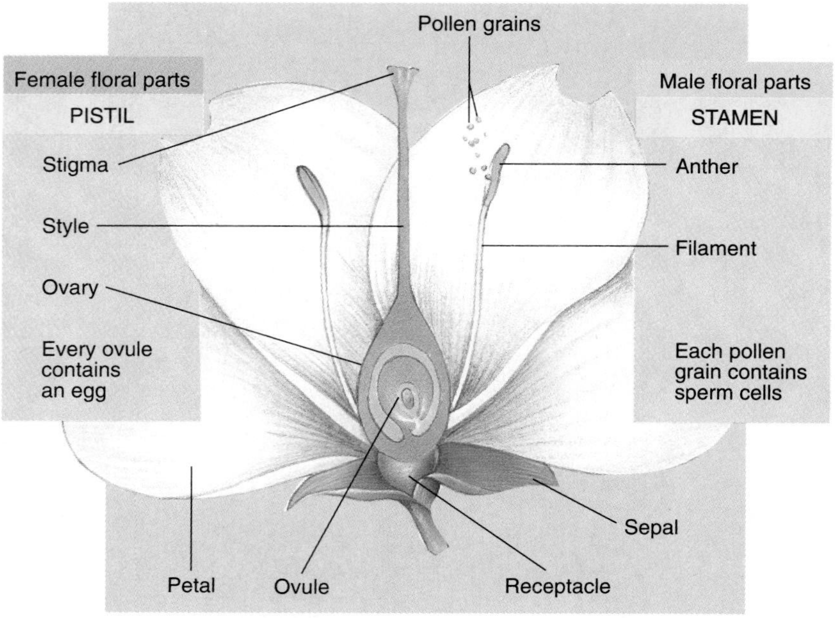

FIGURE 27–8 Flowers are the reproductive structures of flowering plants. This cutaway view of a "typical" flower shows the details of basic floral structure.

27–9). Flowering plants, like gymnosperms and certain other vascular plants, are heterosporous and produce two kinds of spores—microspores and megaspores.

Each ovule within an ovary contains a megasporocyte (megaspore mother cell) that undergoes meiosis, producing four haploid megaspores. Three of these usually disintegrate; one divides mitotically and develops into a female gametophyte, also called an **embryo sac.** The embryo sac contains eight haploid nuclei, three of which participate directly in fertilization (one egg and two polar nuclei).

The anther contains microsporocytes (microspore mother cells) that each undergo meiosis to form four haploid microspores. Each microspore develops into a male gametophyte, also called a pollen grain. Pollen grains are transferred to the stigma of the pistil by wind or animals (see *Focus On: Pollen and Hay Fever*). If compatible with the stigma, the pollen grain germinates and grows a thin pollen tube down the style and into the ovary. A cell within the pollen grain divides to form two nonflagellated sperm. Both sperm are involved in the fertilization process.

Something happens during sexual reproduction in flowering plants that does not occur anywhere else in the living world. When the two sperm enter the embryo sac, *both* of them participate in fertilization. One sperm fertilizes the egg, forming a zygote that develops into the embryo in the seed. The other sperm fuses with the two haploid polar nuclei to form a triploid ($3n$) cell that develops into endosperm in the seed. This fertilization process, which involves two separate cell fusions, is called **double fertilization** and is, with one exception, unique to flowering plants.[2]

As a result of double fertilization, each seed contains a young plant and food or nutritive tissue (the endosperm). In monocots the endosperm persists and is the main source of food in the mature seed. In most dicots the endosperm supplies nutrients to the developing embryo, which subsequently stores food in its cotyledons.

As a seed develops from an ovule, the ovary wall surrounding it enlarges and develops into a fruit. In some instances, other tissues associated with the ovary also enlarge to form the fruit. Fruits serve two purposes: to protect the developing seeds from desiccation during maturation and to aid in the dispersal of seeds (see Chapter 35). For example, dandelion fruits have feathery plumes that enable the entire fruit to be carried by air currents. Once a seed lands in a suitable place, it germinates and develops into a mature sporophyte plant, and the life cycle continues.

Flowering plants are the most successful plant group

The evolutionary adaptations of flowering plants account for their success in terms of ecological dominance and

[2]In 1990, double fertilization was reported in a gymnosperm, *Ephedra nevadensis.* The first fertilization process, between egg and sperm, results in a zygote as in flowering plants. The second produces an additional zygote.

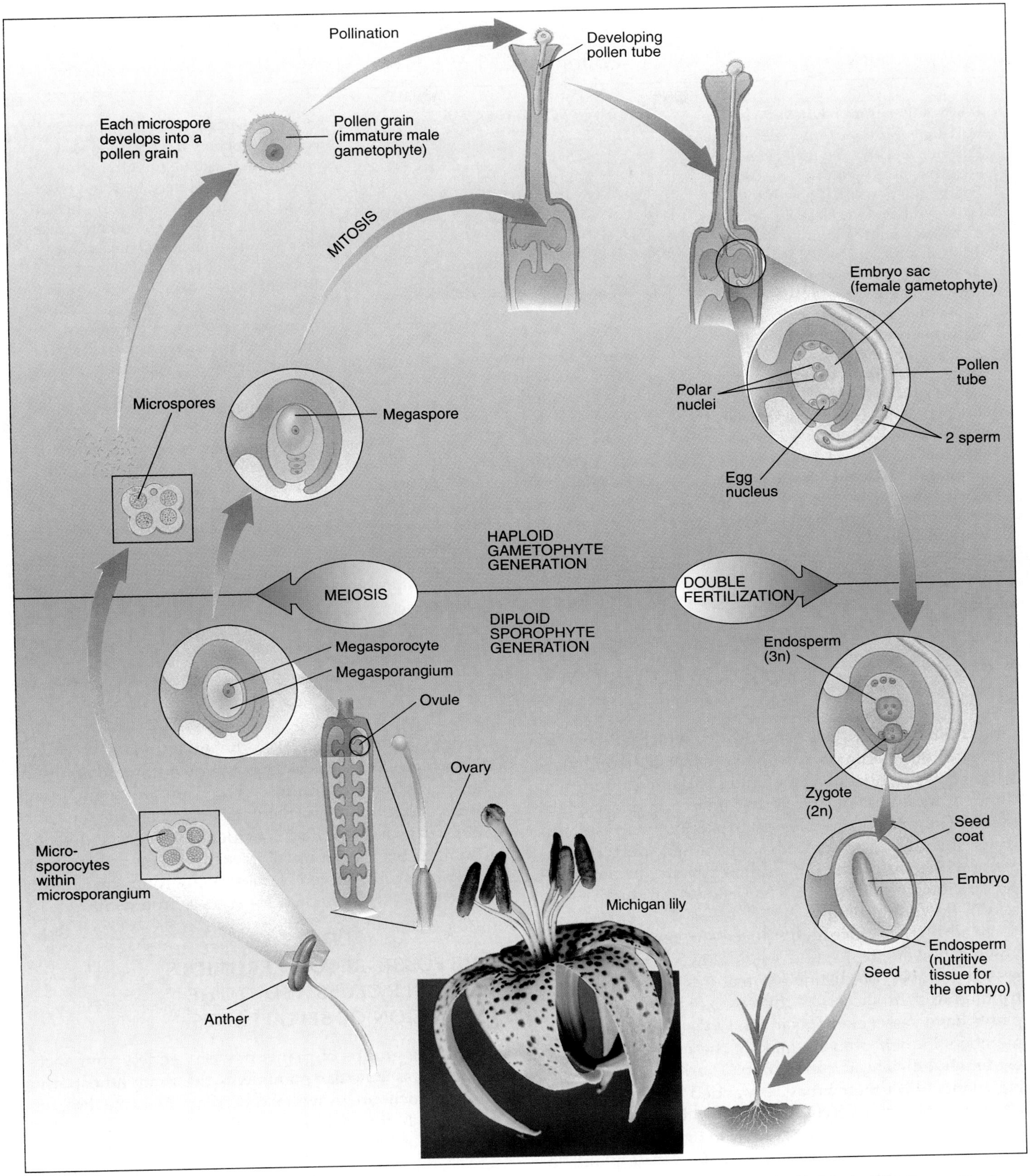

FIGURE 27–9 The sporophyte generation is dominant in the flowering plant life cycle. A significant feature of the flowering plant life cycle is double fertilization. (Visuals Unlimited/John Gerlach)

Focus On
Pollen and Hay Fever

If you suffer from hay fever, you are not alone. Millions of people endure the sneezing and the itchy, watery eyes associated with this condition. Everyone knows that one of the causes of hay fever is pollen, but many blame any plant in bloom when they are suffering. For this reason, roses and goldenrod are often unjustly accused.

Hay fever is caused by certain wind-pollinated plants. Plants that produce large, colorful petals are pollinated by animals and do not cause hay fever because their pollen grains do not get into the air in appreciable quantities. Wind-pollinated plants, on the other hand, must produce large amounts of pollen to ensure that at least some of it lands on the stigmas for successful reproduction. Not all wind-pollinated plants cause an allergic reaction. For example, the conifers are wind-pollinated, yet allergies to conifers are rare.

People with allergies can suffer at different times during the growing season, depending on which plants are pollinating and whether they are sensitized to those plants. In early spring, many trees pollinate before their leaves are fully developed. (Can you explain why pollination at this time would be advantageous to the tree?*) Trees that cause allergic reactions in humans include oaks, ashes, walnuts, maples, and elms. If you suffer in late spring and early summer, you are probably allergic to grass pollen, such as bluegrass, timothy, and redtop. It is interesting that most of our major grass crops (corn, rice, and wheat, for example) do not cause allergies in humans. In late summer and early fall, people are allergic to different plants, depending on their geographical location. Ragweed (see figure) is the culprit in the eastern United States, whereas saltbush and Russian thistle are problems in the West.

Many people appear to be genetically predisposed to pollen allergies, and become sensitized by repeated contact. For that reason, a move to a different geographical location often temporarily halts the suffering. The biology of the allergic reaction is explained in Chapter 43. ■

Scanning electron micrograph of ragweed pollen, which is the most common pollen allergen. (D. Scharf/Peter Arnold, Inc.)

*If pollination occurred when the leaves were fully formed, the leaves would block the effective dispersal of pollen by the wind.

large number of species. The flower, which attracts insects and other animals for pollen dispersal, thereby assuring cross-fertilization (the union of gametes from two different individuals), is a significant evolutionary adaptation. The presence of closed carpels (which results in fruits surrounding the seeds) and double fertilization, along with seed production as their primary means of reproduction and dispersal, increases the reproductive success of flowering plants.

In addition to their highly successful reproduction involving flowers, fruits, and seeds, flowering plants possess a number of distinctive features that have contributed to their success. With few exceptions, flowering plants have vessel elements in their xylem and sieve tube members in their phloem, making these vascular tissues extremely efficient at conduction. The leaves of flowering plants, with their broad, expanded blades, are very efficient at absorbing light for photosynthesis. Abscission (shedding) of these leaves during cold or dry spells is also an advantage that has enabled some flowering plants to expand into habitats that would otherwise be too harsh for survival. Their roots are often modified for food or water storage.

Probably most crucial to the evolutionary success of flowering plants, however, is the overall adaptability of the sporophyte generation. As a group, flowering plants readily adapt to new habitats and changing environments. This adaptability is evident in the large diversity exhibited by the group. For example, the cactus is remarkably well-adapted for desert environments, and the water lily is well adapted for wet environments.

THE FOSSIL RECORD PROVIDES VALUABLE CLUES ABOUT THE EVOLUTION OF SEED PLANTS

One of the groups of plants that descended from ancestral seedless vascular plants was the **progymnosperms,** all of which are now extinct (Fig. 27–10*a*). Progymnosperms had two derived features that their immediate ancestors lacked: leaves that were megaphylls (see Chapter 26) and woody tissue similar to that of modern gymnosperms. Progymnosperms retained a primitive feature of their ancestors, however: they reproduced by spores rather than seeds.

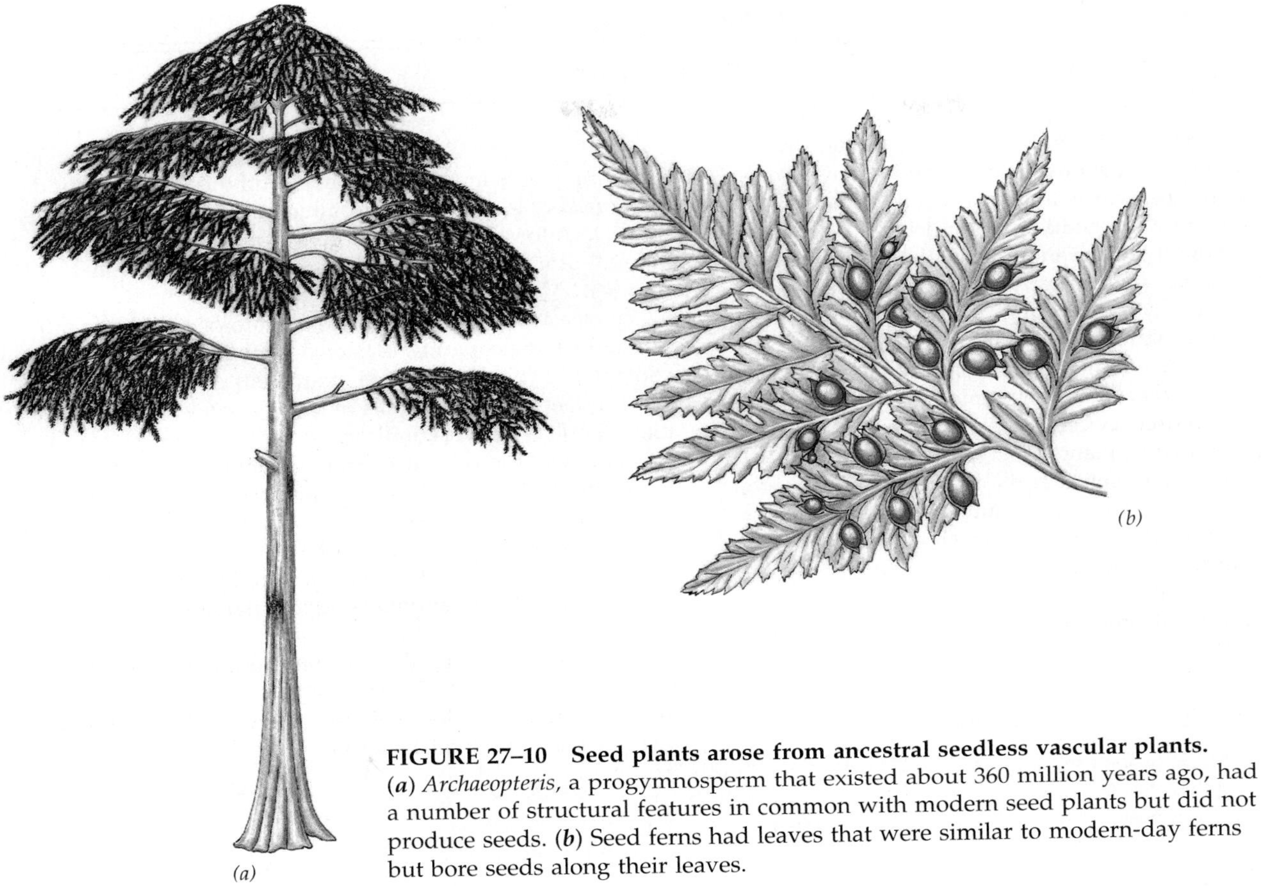

FIGURE 27–10 Seed plants arose from ancestral seedless vascular plants. (***a***) *Archaeopteris,* a progymnosperm that existed about 360 million years ago, had a number of structural features in common with modern seed plants but did not produce seeds. (***b***) Seed ferns had leaves that were similar to modern-day ferns but bore seeds along their leaves.

Fossils of several progymnosperms with reproductive structures intermediate between those of seedless plants and seed plants have been discovered. For example, the evolution of microspores into pollen grains and of megasporangia into ovules can be traced in fossil progymnosperms. Plants producing seeds appeared during the late Devonian period over 360 million years ago, and the fossil record indicates that different groups of seed plants apparently arose independently several times.

Some questions persist about the exact pathways of gymnosperm evolution. The fossil record indicates that progymnosperms probably gave rise to conifers and to another group of extinct plants called **seed ferns,** which were seed-bearing woody plants with fernlike leaves (Fig. 27–10*b*). The seed ferns in turn probably gave rise to cycads and possibly ginkgoes, as well as to several gymnosperm groups that are now extinct. The origin of gnetophytes remains unclear.

Flowering plants are the most recent group of plants to evolve. The fossil record, although incomplete, indicates that flowering plants probably descended from gymnosperms. By the middle of the Jurassic period, approximately 180 million years ago, a number of gymnosperm lines existed with features reminiscent of flowering plants. Among other traits, these derived gymnosperms possessed leaves with broad, expanded blades and the first closed carpels. It is also evident that beetles were visiting these plants, and biologists have suggested that perhaps this was the beginning of coevolution (mutual adaptation) between plants and their animal pollinators.

One important task facing paleobotanists (biologists who study fossil plants) is determining which of the ancient gymnosperms with derived features were in the direct line of evolution leading to the flowering plants. Based on structural data, most botanists think that flowering plants arose only once—that is, that there is only one line of evolution from the gymnosperms to the flowering plants. The gnetophytes are the gymnosperm group that many botanists consider the closest living relatives of flowering plants; this conclusion is supported by both structural and molecular data.

The oldest trace of flowering plants to appear in the fossil record is of pollen grains in Cretaceous rocks some 127 million years old, whereas the oldest fossilized flowers are about 120 million years old. By 90 million years ago, during the late Cretaceous period, flowering plants had diversified and had begun to replace gymnosperms as Earth's dominant plants. Fossils of flowering plants outnumber those of gymnosperms and ferns in late Cretaceous deposits, indicating the rapid success of flowering plants once they appeared.

Making the Connection

Polyploidy, Guard Cells, and Angiosperm Evolution

How are guard cells related to polyploidy and angiosperm evolution? Polyploidy, the possession of more than two sets of chromosomes, is much more common in plants than in animals. However, it has been difficult to assess how important polyploidy has been in the evolutionary history of flowering plants, partly because estimates of the percentage of polyploidy in flowering plants have varied from 30% to 80%. A more precise estimate would be possible if scientists could conclude that certain ancient angiosperms had fewer chromosomes than living members of their families. But how does one determine the number of chromosomes in cells of extinct plants?

Jane Masterson, a biologist at the University of Chicago, was able to estimate chromosome number in ancient angiosperms by measuring the size of fossil guard cells.* (Guard cells surround tiny openings, or stomata, in the surface tissues of leaves and stems.) Masterson first determined that a direct relationship exists between guard cell size and ploidy level (that is, number of chromosome sets) for living plants in three families. To do this, she measured leaf guard cells and then determined the amount of nuclear DNA for each sample of guard cell; the larger the amount of nuclear DNA, the greater the chromosome number.

By measuring guard cell sizes from fossil leaves of various extinct angiosperms, Masterson inferred their DNA content. She found that fossil guard cells of extinct species were *smaller* than guard cells of living species in the same plant family. Her work indicates that these early plants probably had diploid chromosome numbers of 12 to 18. For example, the ancestors of sycamores probably had 14 chromosomes; in comparison, modern-day sycamores have 42 chromosomes — three times the number in their ancestors. This conclusion, which is supported by molecular work, suggests the percentage of polyploidy in angiosperms is high — 70%.

Future studies using this elegant procedure to determine the amount of DNA in extinct plants should provide a greater understanding of how changes in ploidy level affected the evolution of flowering plants. ▲

*Masterson, J. "Stomatal Size in Fossil Plants: Evidence for Polyploidy in the Majority of Angiosperms," *Science*, Vol. 264, April 15, 1994.

Based on the fossil record as well as both structural and molecular data of living angiosperms, the first flowering plants are thought to have been dicots that were weedy shrubs or small herbaceous plants rather than trees. Monocots, which are not as well represented in the Cretaceous fossil record as dicots, apparently originated from dicots.

SUMMARY

I. Seeds have advantages over spores.
   A. Each seed contains a well-developed plant embryo and a food supply surrounded by a protective seed coat.
   B. Gymnosperms and flowering plants reproduce by seeds.

II. Gymnosperms are vascular plants with seeds that are totally exposed or borne on the scales of cones.
   A. Gymnosperms produce wind-borne pollen grains, a feature that seedless vascular plants lack.
   B. There are four phyla of gymnosperms.
      1. Conifers, which are the largest group of gymnosperms, are woody plants that bear needle leaves (usually evergreen) and produce seeds in cones.
      2. Cycads are palmlike or fernlike in appearance but reproduce in a manner similar to pines. There are relatively few living members of this once-large phylum.
      3. *Ginkgo*, the only surviving species in its phylum, is a deciduous, dioecious tree. Female ginkgoes produce fleshy seeds directly on branches.
      4. Gnetophytes share a number of traits with angiosperms, such as efficient water-conducting cells in their xylem.

III. Flowering plants (angiosperms) comprise the phylum of vascular plants that produce seeds enclosed within a fruit. They are the most diverse and most successful group of plants.
   A. Flowering plants have several specialized features that help account for their success.
      1. The flower, which may contain sepals, petals, stamens, and carpels, functions in sexual reproduction.
      2. Double fertilization, which results in the formation of a zygote and endosperm, is characteristic of flowering plants.
      3. Flowering plants possess vessel elements in their xylem.
      4. Various flowering plants use wind or animals to transfer pollen grains.
   B. There are two classes of flowering plants.
      1. Most monocots have floral parts in multiples of three, and their seeds contain one cotyledon. The nutritive tissue in their mature seeds is endosperm.
      2. Dicots usually have floral parts in multiples of four or five, and their seeds contain two cotyledons. The

nutritive organs in their mature seeds are usually the cotyledons, which have absorbed the nutrients in the endosperm.

IV. Seed plants arose from seedless vascular plants.
   A. Progymnosperms were seedless vascular plants that had megaphylls and "modern" woody tissue.
      1. Progymnosperms probably gave rise to conifers.
      2. Progymnosperms probably also gave rise to seed ferns, which in turn probably gave rise to cycads and possibly ginkgoes.
   B. The evolution of gnetophytes is unclear, but they are thought to be the closest living relatives of flowering plants.
   C. Flowering plants probably descended from ancient gymnosperms that had specialized features, such as leaves with broad, expanded blades and closed carpels.
      1. Flowering plants probably arose only once.
      2. The first flowering plants were probably dicots that were weedy shrubs or small herbaceous plants.
      3. Monocots probably arose from dicots.

SELECTED KEY TERMS

angiosperm
anther
calyx
carpel
complete flower
conifer
corolla
cotyledon
cycad
dicot
dioecious
double fertilization
embryo sac
endosperm
fertilization
filament
flowering plant
Ginkgo
gnetophyte
gymnosperm
imperfect flower
incomplete flower
monocot
monoecious
needle
ovary
ovule
perfect flower
petal
pistil
polar nuclei
pollen grain
pollination
progymnosperm
seed fern
sepal
sporophyll
stamen
stigma
style

POST-TEST

1. Conifers, cycads, ginkgoes, and gnetophytes are collectively called ______.
2. Most conifers are ______ and have male and female reproductive parts at different locations on the same plant.
3. The immature male gametophyte of pines is called a(an) ______ ______.
4. The transfer of pollen grains from the male to the female reproductive structure is known as ______.
5. Flagellated sperm are found as vestiges in two gymnosperm groups, the ______ and the ______.
6. The most diverse and most successful group of plants are the ______, also called ______ ______.
7. This class of flowering plants, the ______, includes the palms, grasses, and orchids.
8. The nutritive tissue in flowering plant seeds that is formed as a result of double fertilization is called ______.
9. The pistil has three sections: a(an) ______, ______, and ______.
10. A simple pistil consists of a single ______.
11. A flower that lacks stamens is said to be both a(an) ______ and a(an) ______ flower.
12. After fertilization, the ______ develops into a fruit and the ______ develops into a seed.
13. The female gametophyte in flowering plants is also called a(an) ______ ______.
14. The ______ are extinct seedless vascular plants that had megaphylls and "modern" woody tissue.

REVIEW QUESTIONS

1. Why are seeds such a significant evolutionary development?
2. What distinguishing features do cycads, ginkgoes, and gnetophytes share with conifers?
3. How are flowering plants different from gymnosperms?
4. How does the gymnosperm life cycle differ from that of flowering plants?
5. What are the two classes of flowering plants, and how can one distinguish between them?
6. How does pollination occur in the gymnosperms? In flowering plants?
7. How does fertilization differ in gymnosperms and flowering plants?
8. Describe the evolutionary changes that had to take place as flowering plants arose from ancient gymnosperms.
9. Label the diagram to the right.

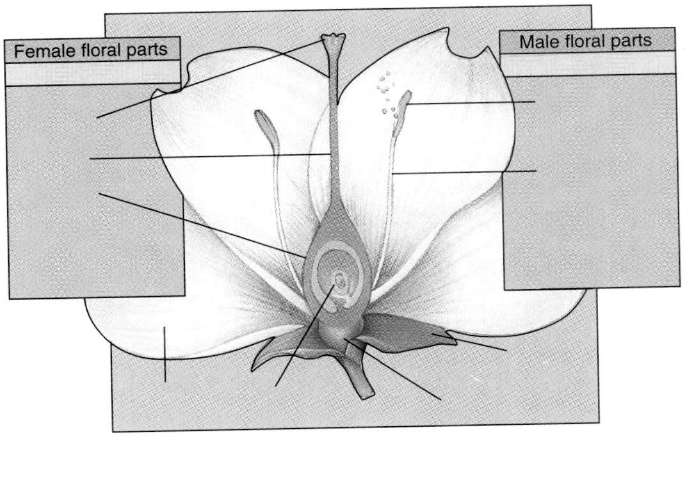

YOU MAKE THE CONNECTION

1. How are cones and flowers alike? How are they different? (Hint: Your answer should consider microspores/megaspores and seeds.)
2. How do the life cycles of seedless plants (see Chapter 26) and seed plants differ? In what fundamental way are they alike?

RECOMMENDED READINGS

Heywood, V.H. *Flowering Plants of the World.* New York, Oxford University Press, 1993. This beautifully illustrated guide describes more than 300 families of angiosperms, including their economic uses.

Mauseth, J.D. *Botany: An Introduction to Plant Biology,* 2nd ed. Saunders College Publishing, Philadelphia, 1995. A comprehensive introduction to general botany.

Raven, P.H., R.F. Evert, and S.E. Eichhorn. *Biology of Plants,* 5th ed. Worth Publishers, New York, 1992. A general botany text with an evolutionary emphasis.

Wolf, T.H. "The Object at Hand." *Smithsonian,* September 1990. This regular feature highlights the dawn redwood, a gymnosperm that was thought to have been extinct for millions of years but was found living in China.

CHAPTER 28

The Animal Kingdom: Animals Without a Coelom

Coral reefs provide shelter for many marine animals. These four-eye butterfly fish *(Chaetodon capistratus)* were photographed along a coral reef in South Caicos, British West Indies. (Fred Bavendam/Peter Arnold, Inc.)

We have no difficulty identifying a horse as an animal and an oak tree as a plant, but many marine animals that live attached to rocks or docks are often mistaken for plants. For example, early naturalists thought that sponges were plants because they did not move from place to place. So many diverse animal forms exist that exceptions can be found to almost any definition of an animal. Still, there are some characteristics that describe at least most animals:

1. All animals are multicellular eukaryotes.
2. The cells that make up the animal body are specialized to perform specific functions. In all but the simplest animals, cells are organized to form tissues, and tissues are organized to form organs. In most animal phyla, specialized organ systems carry on specific functions.
3. Animals are heterotrophs. Most are consumers that ingest their food first and then digest it inside the body, usually within a digestive system.
4. Most animals are capable of locomotion at some time during their life cycle. Some animals (the sponges, for example) move about as **larvae** (immature forms) but are **sessile** (firmly attached to the ground or some other surface) as adults.
5. Most animals have well-developed sense organs and nervous systems and can respond rapidly to stimuli in their environment.
6. Most animals reproduce sexually, with large, nonmotile eggs and small, flagellated sperm. Sperm and egg unite to form a fertilized egg, or zygote, which goes through a series of embryonic stages before developing into a larva or immature form.

In very simple, small animals, life processes such as gas exchange, circulation of materials, and waste disposal take place by diffusion. In large, complex animals, specialized structures and mechanisms have evolved to carry on life processes.

Biologists have identified more than a million species of animals. Several million more remain to be discovered and classified. Most biologists divide **kingdom Animalia** into about 35 different phyla. The animals most familiar to us—dogs, birds, fishes, frogs, snakes—are **vertebrates** (a subphylum of phylum Chordata). A vertebrate is an animal with a backbone (vertebral column). You may be surprised to learn that vertebrates account for fewer than 5% of the species in the animal kingdom. The majority of animals are the less familiar **invertebrates,** animals without backbones. The invertebrates include such diverse forms as sponges, jellyfish, worms, mollusks, insects, crustaceans, and sea stars.

LEARNING OBJECTIVES

After you have studied this chapter you should be able to

1. List the characteristics common to most animals. Using these characteristics, develop a brief definition of an animal.
2. Identify the ecological role of animals and discuss their distribution, comparing the advantages and disadvantages of life in the sea, in fresh water, and on land.
3. Justify classification and proposed relationships of the animal phyla on the basis of symmetry, type of body cavity, and pattern of development (that is, protostomes and deuterostomes).
4. Identify distinguishing characteristics of phyla Porifera, Cnidaria, Ctenophora, Platyhelminthes, Nemertea, Nematoda, and Rotifera.
5. Classify a given animal in the appropriate phylum (from among those listed in Objective 4.)
6. Trace the life cycles of the following parasites: *Ascaris,* tapeworm, hookworm, and *Trichina* worm. Identify several adaptations that these animals possess for their parasitic lifestyles.
7. Describe the adaptive advantages of each of the following characteristics: bilateral symmetry, cephalization, a motile larva, a digestive cavity with two openings, and hermaphroditism.

ANIMALS INHABIT MOST ENVIRONMENTS OF THE ECOSPHERE

As consumers, animals depend on producers for their raw materials, energy, and oxygen. They also depend on decomposers for recycling nutrients.

Animals are distributed in virtually every environment of our planet. They probably evolved in shallow, Precambrian seas, and many animal phyla still inhabit the sea. Of the three environments — salt water, fresh water, and land — the sea is the most hospitable. Sea water is isotonic to the tissue fluids of most marine animals, so they have little problem maintaining fluid and salt balance. The buoyancy of sea water supports its inhabitants, and the temperature is relatively constant due to the large volume of water. **Plankton,** the animals and protists that are suspended in the water and float with its movement, provide a ready source of food.

Life in the sea has certain disadvantages. Although the continuous motion of the water brings nutrients to animals and washes their wastes away, they must be able to cope with the water's constant churning and the currents that could sweep them away. Squids, fishes, and marine mammals are strong enough swimmers that they can direct their movements and maintain their location effectively. However, most invertebrates are unable to swim strongly and so have other adaptations. Some are sessile, attaching to some stable structure like a rock, so that they are not wafted about with the tides and currents. Others cling to the substratum or burrow in the sand and silt that cover the sea bottom. Many invertebrates have adapted by maintaining a small body size and becoming part of the plankton. They survive successfully because while they are tossed about, their food supply continues to surround them.

Fresh water offers a much less constant environment than sea water and generally contains less food. Oxygen content and temperature vary, and turbidity (due to sediment suspended in the water) and even water volume fluctuate. Fresh water is hypotonic to the tissue fluids of animals, so water tends to diffuse into the animal; therefore, freshwater species must have mechanisms for removing excess water while retaining salts. This osmoregulation requires an expenditure of energy. For these reasons, far fewer kinds of animals make their homes in fresh water than in the sea.

Terrestrial life is even more difficult. Dehydration is a serious threat because water is constantly lost by evaporation and is often difficult to replace. Only a few animal groups, most notably representatives of the arthropods (insects, spiders, and related forms) and some vertebrate groups, have successfully made their homes on land.

ANIMALS CAN BE CLASSIFIED ACCORDING TO BODY STRUCTURE OR PATTERN OF DEVELOPMENT

Although the evolutionary origin of animals is unclear, biologists have evidence to support the hypothesis that they evolved from protists, probably from colonial flagellates. Although the relationships among the various animal phyla remain a matter of debate, a few of the more widely held hypotheses are presented in this section.

In comparing groups of animals, it is convenient to use terms like *lower* and *higher, simple* and *complex,* and *primitive* and *advanced.* Such terms as higher, complex, or advanced do not imply that these animals are better or more nearly perfect than others. Rather, they are used in a comparative sense to describe hypothesized evolutionary relationships. For example, the terms *higher* and *lower* usually refer to the level at which a particular group has diverged from a main line of evolution. It is customary, for instance, to refer to sponges and cnidarians as lower invertebrates because they are thought to have originated near the base of the phylogenetic tree of the animal kingdom. However, neither sponges nor cnidarians are primitive in all structural or physiological characteristics. Each has become highly specialized to its own particular lifestyle.

Animals can be classified according to body symmetry

Symmetry refers to the arrangement of body structures in relation to some axis of the body. Sponges have an interesting variety of shapes. Most are not symmetrical, so that when cut in half, the two halves are not similar to one another. Most other animals exhibit either radial or bilateral body symmetry. Members of phylum Cnidaria (jellyfish, sea anemones, and their relatives) and adult echinoderms (sea stars and their relatives) have **radial symmetry.** In radial symmetry, the body has the general form of a wheel or cylinder. Similar structures are regularly arranged as spokes from a central axis. Multiple planes can be drawn through the central axis, each dividing the organism into two mirror images. An animal with radial symmetry receives stimuli equally from all directions in the environment. Many radially symmetrical animals actually have modified radial symmetry. For example, sea anemones and ctenophores (comb jellies) have biradial symmetry (in which parts of the body have become specialized so that only two planes can divide the body into similar halves).

Most animals are **bilaterally symmetrical,** at least in their larval stages. A bilaterally symmetrical animal can be divided through only one plane (which passes through the midline of the body) to produce roughly equivalent right and left halves that are mirror images (Fig. 28–1). Bilateral symmetry is considered an adaptation to locomotion. The front end of the animal generally has a head, where sense organs are concentrated. This end receives most environmental stimuli and generally moves into the environment first. The rear end of the animal may be equipped with a tail for swimming, or it may just follow along.

To locate body structures in bilaterally symmetrical animals, it is helpful to define some basic terms and directions. The back surface of an animal is its **dorsal** surface; the under side (belly) is its **ventral** surface. **Anterior** means toward the front (head-end) of the animal; **posterior** (or caudal) means toward the back (tail-end). A structure is said to be **medial** if it is located toward the midline of the body and **lateral** if it is toward one side of the body; for example, the human ear is lateral to the nose. The terms **cephalic** (and **superior,** in human anatomy) refer to the head end of the body; the term **caudal** refers to structures closer to the tail. (The term **inferior** is used in human anatomy to mean located below some point of reference, or toward the feet.)

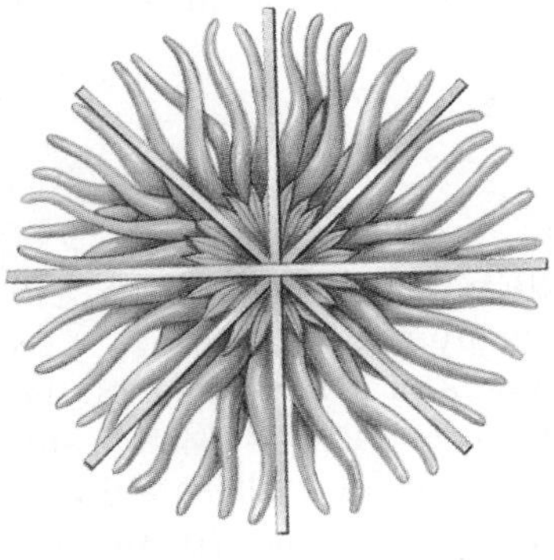

(a)

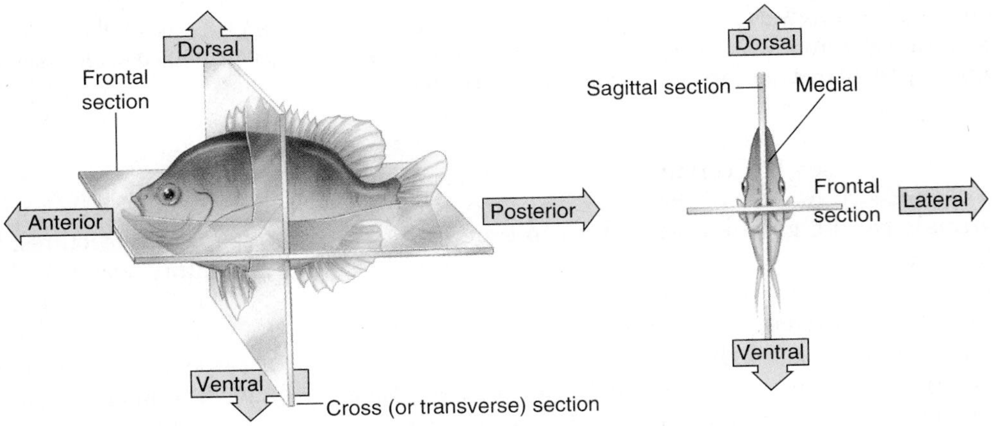

(b)

FIGURE 28–1 Most animals have radial or bilateral symmetry. (*a*) In radial symmetry, multiple planes can be drawn through the central axis; each divides the organism into two mirror images. (*b*) Most animals are bilaterally symmetrical. A sagittal cut, which extends the vertical length, divides the animal into right and left parts (see figure to right). The head-end of the animal is generally referred to as the anterior end, and the opposite end as the posterior end. The back of the animal is its dorsal surface; the belly is its ventral surface. The diagram also illustrates various ways in which the body can be sectioned (cut) in order to study its internal structure. Cross sections and sagittal sections are used in illustrations throughout this book to show relationships among tissues and organs.

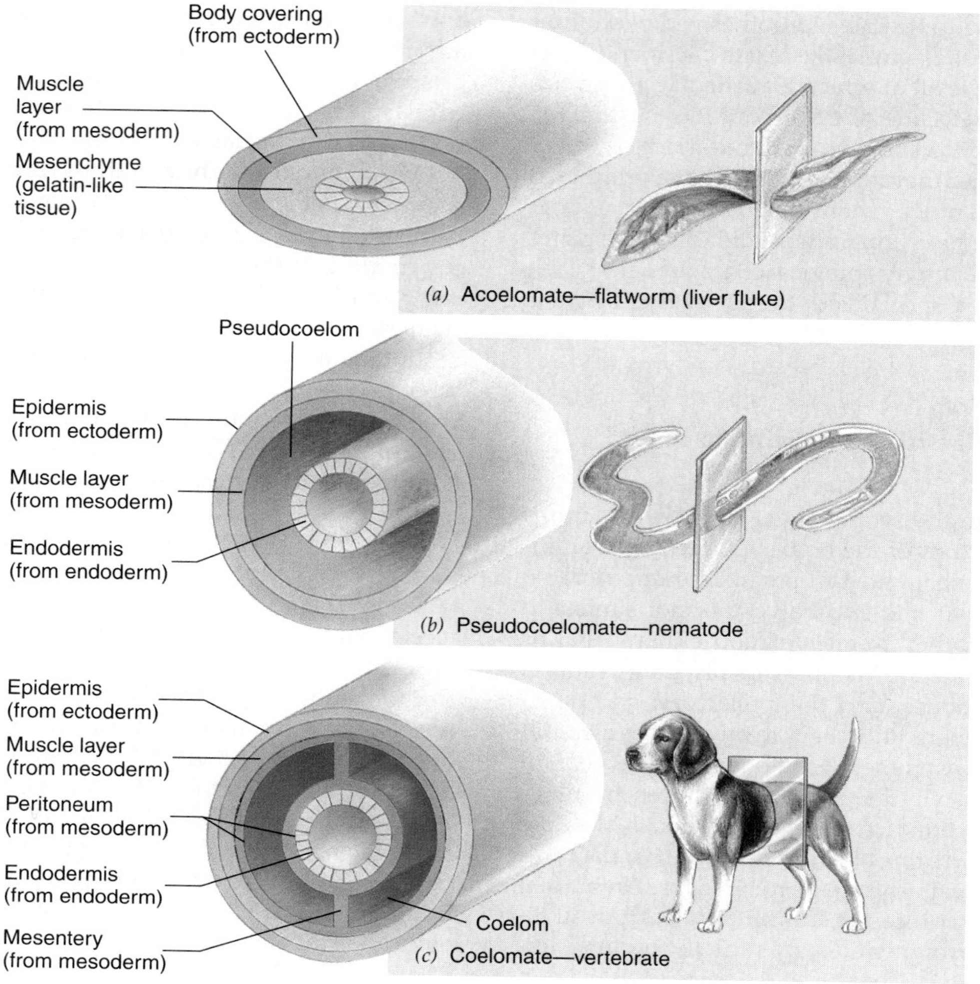

FIGURE 28–2 Three basic animal body plans are illustrated by these cross sections. The term *body cavity* refers to the space between the body wall and the digestive tube. The germ layer from which each tissue was derived is indicated in parentheses. (***a***) An acoelomate animal has no body cavity. (***b***) A pseudocoelomate animal has a "false" body cavity, one that is not completely lined with mesoderm. (***c***) In a coelomate animal the body cavity, or coelom, is completely lined with tissue that develops from mesoderm.

Animals can be grouped according to type of body cavity

A widely held system for grouping animal phyla is based on the presence and type of **body cavity,** a fluid-filled space between the outer body wall and the digestive tube. In order to understand the types of body cavities, we must look briefly at the animal's embryonic development.

The structures of most animals develop from three embryonic tissue layers, called **germ layers,** that are present in the embryo. The outer germ layer, called the **ectoderm,** gives rise to the outer covering of the body and to the nervous system (if the animal has one). The inner layer, or **endoderm,** forms the lining of the digestive tract. **Mesoderm,** the middle layer, gives rise to most of the other body structures, including the muscles, bones, and circulatory system (when they are present).

The flatworms (and members of a few other phyla) form three germ layers and have a solid body with a single opening to the outside—the mouth. These animals have no body cavity, and so are referred to as **acoelomates** (*a,* meaning "without" plus *coelom,* meaning "cavity") (Fig. 28–2).

Other, more complex animals generally have a **tube-within-a-tube body plan.** The outer tube is the body wall. It is covered with tissue that develops from ectoderm. Tissue derived from endoderm lines the inner tube—the digestive tract, or gut—which has an opening at each end: the mouth and the anus. Beneath the ectoderm, the outer tube often consists of tissue derived from mesoderm. The space between the two tubes is the body cavity. If the body cavity is not completely lined with mesoderm it is called a **pseudocoelom** ("false coelom"). Animals with a pseudocoelom are referred to as **pseudocoelomates.** This group includes roundworms and rotifers.

In still more complex animals, the body cavity is completely lined with mesoderm. Such a body cavity is a true **coelom.** Only animals with true coeloms are referred to as **coelomates.** The tree shown in Figure 28–3 indicates the relationships of the major phyla of animals based on their body cavity types.

Animals can be classified as protostomes or deuterostomes

Animals with a true coelom can be divided into two groups: *protostomes* and *deuterostomes.* These groups reflect two main lines of evolution based on their pattern of early development.

Early during development, the embryo consists of a little ball of cells known as a blastula. A group of cells move inward to form an opening called the *blastopore.* In most of the mollusks, annelids, and arthropods, this opening develops into the mouth. These animals are **protostomes** (from Greek words meaning "first, the mouth").

In echinoderms (for example, sea stars and sea urchins) and chordates (the phylum that includes the vertebrates), the blastopore does not give rise to the mouth. Instead it generally develops into the anus. The opening that develops into the mouth forms later in development. These animals are the **deuterostomes** ("second, the mouth").

Another difference in the development of protostomes and deuterostomes is the pattern of **cleavage,** the first

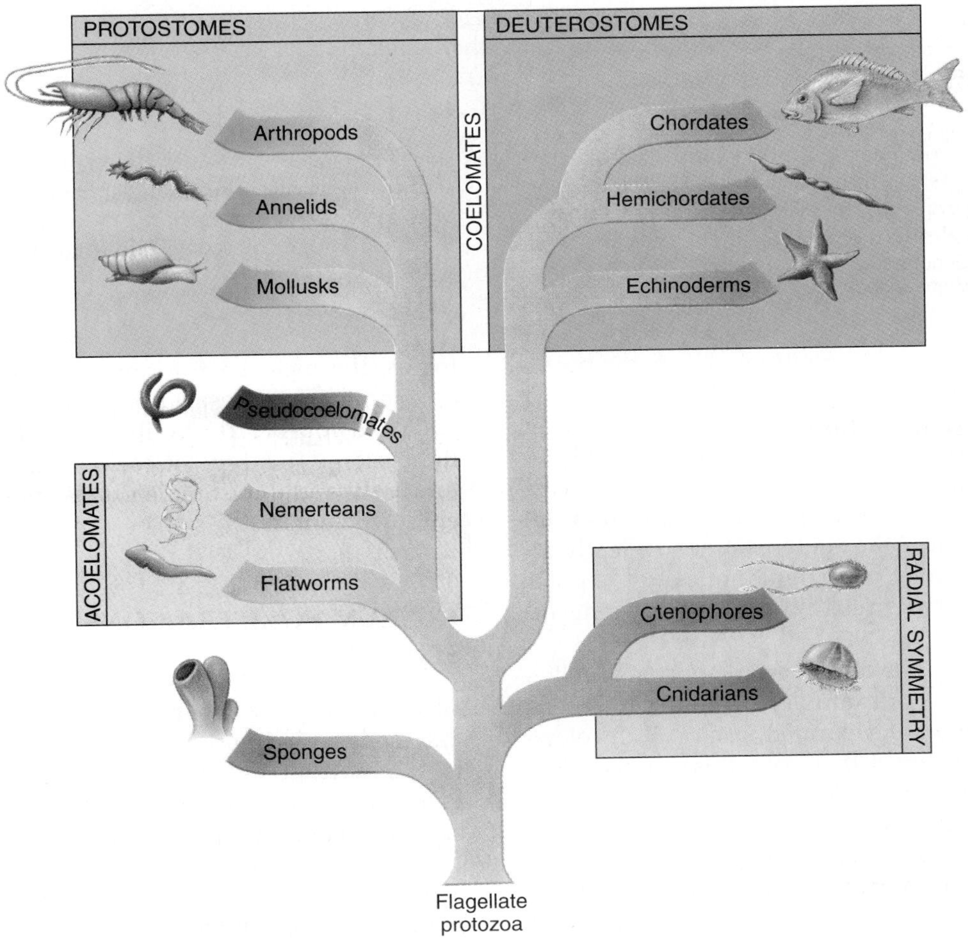

FIGURE 28–3 This phylogenetic tree illustrates some hypothetical evolutionary relationships among acoelomate, pseudocoelomate, and coelomate phyla. The flatworms and nemerteans have a solid body, and therefore are referred to as acoelomate. Nematodes (roundworms) and rotifers have a body cavity called a pseudocoelom. Most other bilateral animals and the echinoderms, however, have a true coelom, a body cavity completely lined with mesoderm. This phylogenetic tree also distinguishes between the protostomes (mollusks, annelids, and arthropods) and the deuterostomes (echinoderms and chordates).

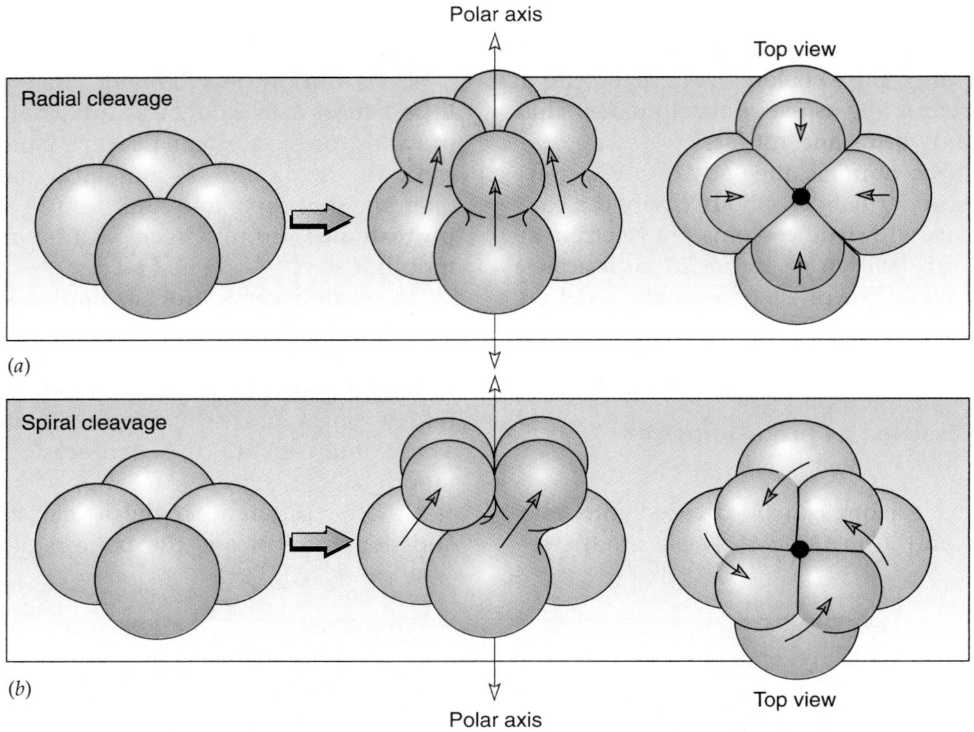

FIGURE 28–4 Deuterostomes and protostomes have different patterns of cleavage. (*a*) In the radial cleavage characteristic of deuterostome embryos, the early divisions are either parallel or at right angles to the polar axis. The cells are stacked in layers. (*b*) Spiral cleavage is characteristic of protostomes. Note the spiral arrangement of the cells. The pattern of cleavage can be appreciated by comparing the position of the dark orange-colored cells in (*a*) and (*b*).

several cell divisions of the embryo. In many protostomes, the early cell divisions are diagonal to the polar axis (the long axis of the egg), resulting in a somewhat spiral arrangement of cells; any one cell is located between the two cells above or below it (Fig. 28–4). This pattern of division is known as **spiral cleavage.** In **radial cleavage,** characteristic of the deuterostomes, the early divisions are either parallel or at right angles to the polar axis; the cells are located directly above or below one another.

In the protostomes, the fate of each embryonic cell is often fixed very early. For example, if the first four cells of an annelid embryo are separated, each cell develops into only a fixed quarter of the larva; this is referred to as **determinate cleavage.** In deuterostomes, cleavage is usually **indeterminate.** If the first four cells of a sea star embryo, for instance, are separated, each cell is capable of forming a complete, though small, larva.

Still another difference between protostome and deuterostome development is the manner in which the coelom is formed. In protostomes, the mesoderm splits, and the split widens into a cavity that becomes the coelom (Fig. 28–5). This method of coelom formation is known as **schizocoely,** and for this reason the protostomes are sometimes called **schizocoelomates.** In many deuterostomes, the mesoderm forms as "outpocketings" of the developing gut. These outpocketings eventually separate and form pouches; the cavity within the pouches becomes the coelom. This type of coelom formation is called **enterocoely,** and these animals are sometimes referred to as **enterocoelomates.**

PHYLUM PORIFERA CONSISTS OF THE SPONGES

About 9000 species of **sponges** have been identified and assigned to phylum **Porifera.** The name Porifera, meaning "to have pores," aptly describes the sponges, which look like sacs perforated by tiny holes. Sponges are aquatic, mainly marine, animals that range in size from 1 to 200 centimeters (0.4–79 inches) in height. Many are asymmetrical, but they vary in shape from flat, encrusting growths to balls, cups, fans, or vases. Living sponges may be brightly colored—green, orange, red, yellow, blue, or purple—or they may be white or drab (Fig. 28–6). Some species are inhabited by symbiotic bacteria or algae that give them color.

Sponges are thought to have evolved from choanoflagellates. Recall from Chapter 24 that these protozoa

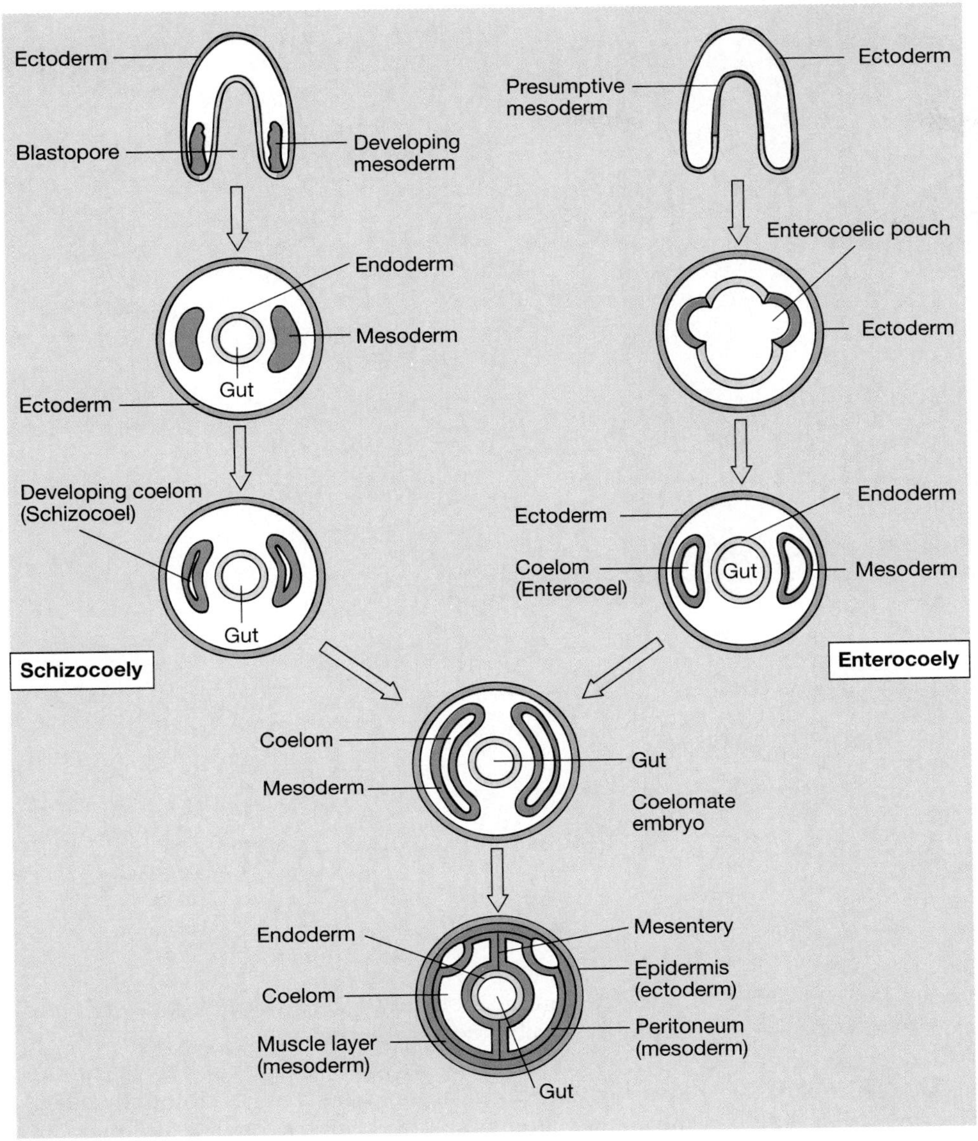

FIGURE 28–5 The coelom forms differently in protostome and deuterostome embryos. The coelom originates in the embryo from blocks of mesoderm that split off from each side of the embryonic gut. In protostomes the coelom forms by a process called schizocoely in which the mesoderm *(red)* splits. The split widens into a cavity that becomes the coelom. In enterocoely, which is characteristic of deuterostomes, the mesoderm outpockets from the gut, forming pouches. The cavity within these pouches becomes the coelom. Ectoderm is shown in blue, endoderm in yellow.

have a single flagellum surrounded by a collar of microvilli. Sponges are the only animals with **collar cells** (choanocytes)—cells that are strikingly similar to the choanoflagellates. In the sense that they apparently did not give rise to any other animal group, sponges seem to represent a dead end in evolution. Of course, sponges themselves continue to change as they are subjected to continual selective pressures from the environment.

Sponges are divided into three classes on the basis of the type of skeleton they secrete. Members of class Calcarea secrete a chalky skeleton composed of small calcium carbonate spikes, or **spicules.** Members of class Hexactinellida, the glass sponges, have a skeleton made of six-rayed spicules containing silicon. Most sponges belong to class Demospongiae, characterized by variable skeletons: some are made of a protein material known as *spongin*, others contain spicules of silicon, and some have a combination of both. What we recognize as a bath sponge is actually a dried spongin skeleton.

In a simple sponge, water enters through hundreds of tiny pores (*ostia*), passes into the central cavity, or **spongocoel** (not a digestive cavity), and flows out through the sponge's open end, the **osculum.** In some types of sponges, the body wall is extensively folded, and there are complicated systems of canals.

Although the sponge is multicellular, its cells are loosely associated and do not form definite tissues. However, a division of labor exists among the several types of cells that make up the sponge, with certain cells specializing in nutrition, support, contraction, or reproduction. Epidermal cells form the outer layer of the sponge and line the canals. Specialized tubelike cells, called *porocytes*, form the pores of a simple sponge. These cells regulate the diameter of the pores by contracting.

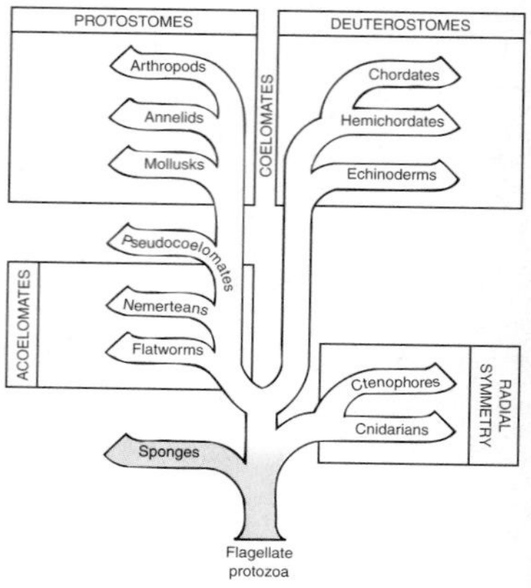

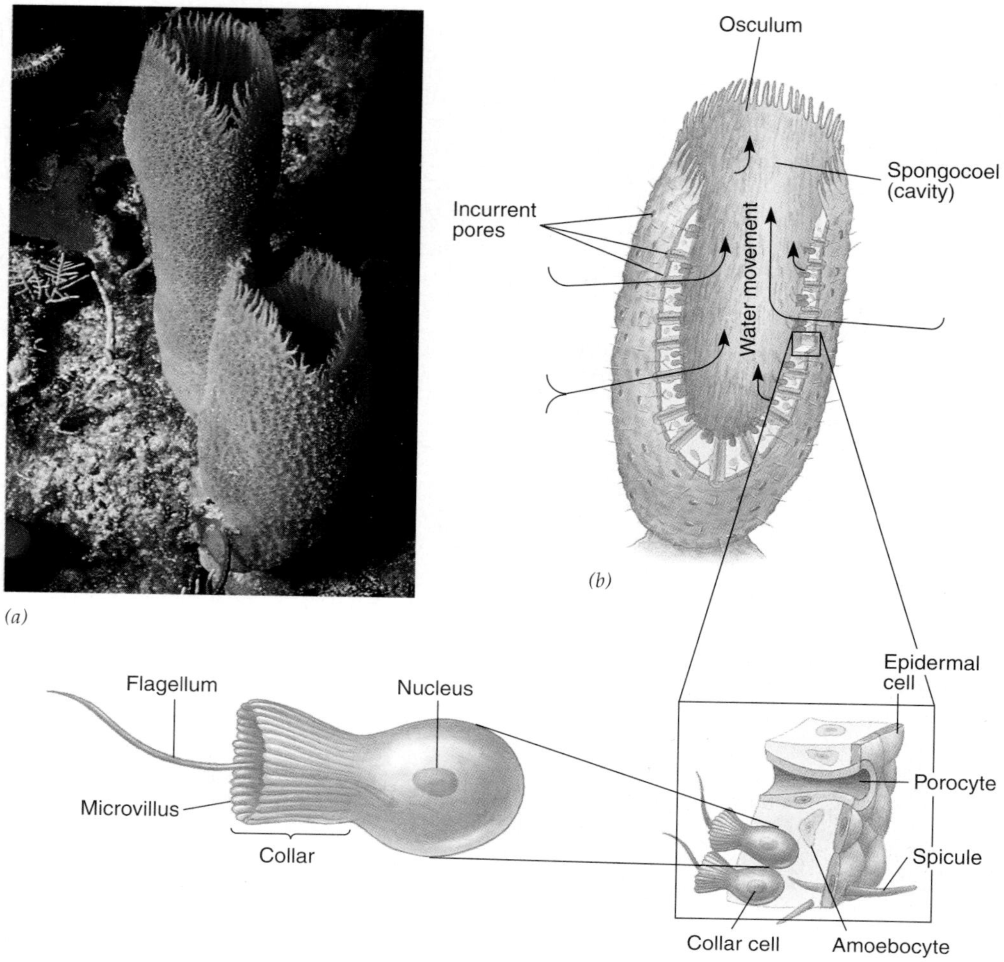

FIGURE 28–6 Sponges vary widely in size, shape, and color. (*a*) Tube sponge *(Spinosella plicifera)* from the Caribbean. (*b*) A simple sponge cut open to expose its cellular organization. (Charles Seaborn/ Odyssey Productions, Chicago)

The collar cells, which make up the inner layer of the sponge, create the water current that brings food and oxygen to the cells and carries away carbon dioxide and other wastes. These cells also trap and phagocytize food particles. Each of these cells is equipped with a tiny collar that surrounds the base of the flagellum. The collar is an extension of the plasma membrane and consists of microvilli. The collar cells of some complex sponges can pump a volume of water equal to the volume of the sponge each minute!

Between the outer and inner cellular layers of the sponge body is a gelatin-like layer, the *mesohyl,* supported by skeletal spicules. Amoeba-like cells, which wander about in this layer, are important in digestion and food transport. Other amoeba-like cells in the mesohyl secrete the spicules.

Sponge larvae are flagellated and able to swim about. However, the adult sponge remains attached to some solid object on the sea bottom and is incapable of locomotion. Sponges are suspension feeders, adapted for trapping and eating whatever food the sea water brings to them. As water circulates through the body, food is trapped along the sticky collars of the choanocytes. Food particles are either digested within the collar cell or transferred to an amoebocyte for digestion. Undigested food is simply eliminated into the water.

Gas exchange and excretion of wastes depend on diffusion into and out of individual cells. Although cells of the sponge are irritable and can react to stimuli, there are no nerve cells that would enable the animal to react as a whole. Behavior appears limited to the basic metabolic necessities such as capturing food and regulating the flow of water through the body.

Sponges can reproduce asexually. A small fragment or bud may break free from the parent sponge and give rise to a new sponge. Such fragments may attach to the parent sponge, forming a colony. Sponges also reproduce sexually. Most sponges are **hermaphroditic,** meaning that the same individual can produce both egg and sperm. Some of the amoeba-like cells develop into sperm cells, others into egg cells. However, hermaphroditic sponges usually produce eggs and sperm at different times and they cross-fertilize with other sponges. Mature sperm are released into the water and are taken in by other sponges. Fertilization and early development take place within the jelly-like mesohyl. Embryos eventually move into the spongocoel and leave the parent along with the stream of outflowing water. After swimming about for a while, the larva finds a solid object, attaches to it, and settles down to a sessile life.

Sponges possess a remarkable ability to repair themselves when injured and to regenerate lost parts. When the cells of a sponge are separated from one another in the laboratory, they reaggregate, forming a complete sponge again.

HYDROZOANS, JELLYFISH, AND CORALS BELONG TO PHYLUM CNIDARIA

Most of the 10,000 or so species of phylum **Cnidaria** (pronounced "nie-dare'-ee-ah") are marine. These animals get their name from their stinging cells, called **cnidocytes** (from a Greek word meaning "sea nettles"). Cnidarians are grouped in three main classes (Table 28–1). Class **Hydrozoa** includes the hydras, the hydroids, such as *Obelia*, and the Portuguese man-of-war; class **Scyphozoa** includes the jellyfish; and class **Anthozoa** includes the sea anemones, true corals, and alcyonarians (sea fans, sea whips, and precious corals) (Fig. 28–7).

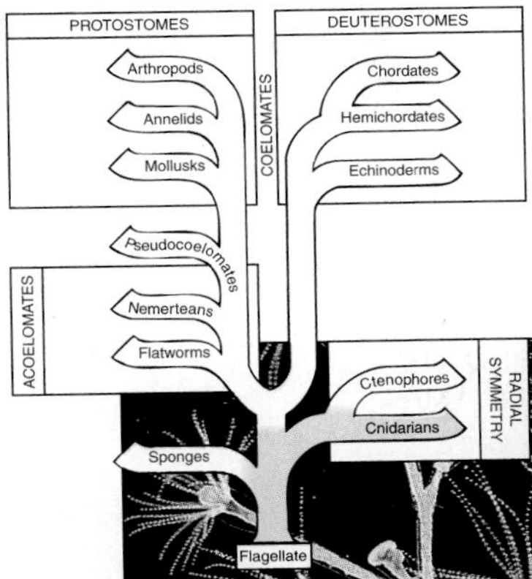

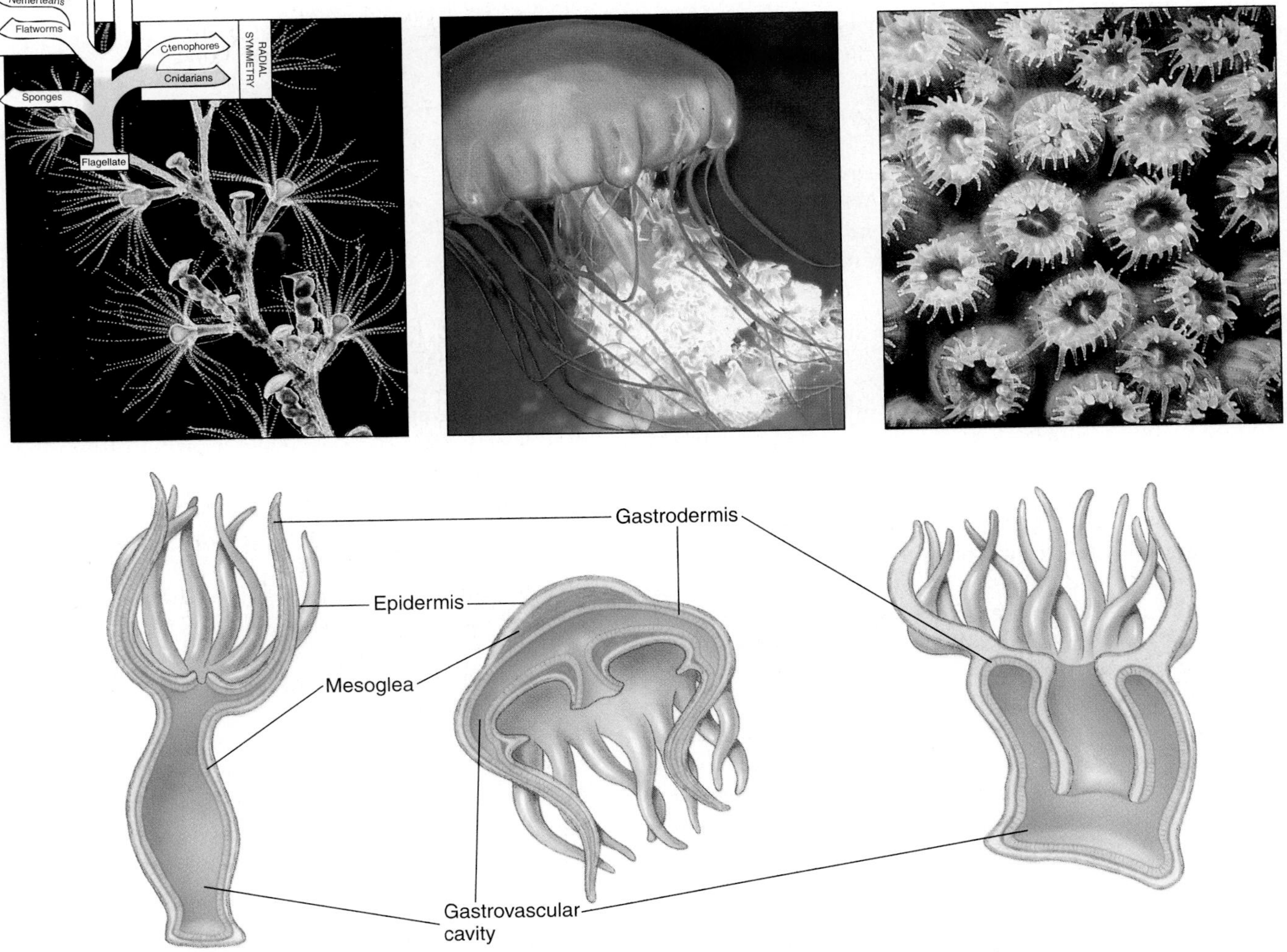

(*a*) Hydrozoa (polyp form) (*b*) Scyphozoa (jellyfish) (*c*) Anthozoa (sea anemone and coral)

FIGURE 28–7 Polyp and medusa body forms are characteristic of phylum Cnidaria. (***a***) This hydrozoan *(Gonothyraea loveni)* forms a colony of polyps. (***b***) This sea nettle *(Chrysaora fuscescens)*, like other jellyfish, is a suspension feeder. It traps plankton in mucus on its ciliated lower surface. Cilia sweep the food to the margin of the bell, and the jellyfish carries it to the mouth using its oral arms, tubelike extensions that surround the mouth. (***c***) Coral polyps *(Montastrea cavernosa)* extended for feeding. (*a*, Robert Brons/Biological Photo Service; *b*, Brian Parker/Tom Stack & Associates; *c*, Mike Bacon/Tom Stack & Associates)

Cnidarians have two body shapes, the polyp and the medusa. The **polyp** form, represented by *Hydra,* resembles an upside-down, slightly elongated jellyfish. Some cnidarians have the polyp shape during one stage and later develop into the **medusa** (pl., *medusae*) (jellyfish) form. In the medusa, the mouth is located in the lower concave, or *oral,* surface; the convex upper surface is the *aboral* surface. Other cnidarians begin as medusae, then develop into polyps.

Although many cnidarians live a solitary existence, others form colonies. Some colonies—for example, the Portuguese man-of-war—consist of many individuals, some of which are polyps and others medusae.

The cnidarian body is radially symmetrical and is organized as a hollow sac with the mouth and surrounding tentacles located at one end. The mouth leads into the digestive cavity, called the **gastrovascular cavity.** The mouth is the only opening into the gastrovascular cavity and so must serve for both ingestion of food and expulsion of wastes.

Much more highly organized than the sponge, the cnidarian is *diploblastic*—has two definite tissue layers. The outer **epidermis** is a protective layer. The inner **gastrodermis** functions in digestion. These layers are separated by a gelatin-like **mesoglea,** which is not itself cellular but which usually contains a few cells.

Table 28–1 MAJOR CLASSES OF PHYLUM CNIDARIA

Class and Representative Animals	*Characteristics*
Hydrozoa *Hydra* *Obelia* Portuguese man-of-war	Mainly marine, but some freshwater species; both polyp and medusa stages in many species (polyp form only in *Hydra*); formation of colonies in some cases.
Scyphozoa Jellyfish	Marine; inhabit mainly coastal water, free-swimming medusae most prominent forms; polyp stage often reduced.
Anthozoa Sea anemones Corals Sea fans	Marine; solitary or colonial polyps; no medusa stage; gastrovascular cavity divided by partitions into chambers, increasing area for digestion; sessile.

Class Hydrozoa includes solitary and colonial forms

Although not really typical, the cnidarian most often studied by beginning biology students is the tiny, solitary *Hydra,* found in freshwater ponds. To the naked eye *Hydra* looks like a bit of frayed string (Fig. 28–8). This animal is named after the multiheaded monster of Greek mythology with the remarkable ability to grow two new heads for each head cut off. *Hydra* also has an impressive ability to regenerate. When cut into several pieces, each piece may grow all the missing parts and become a whole animal.

The hydra's body consists of an outer protective epidermis and an inner gastrodermis that functions in digestion. Both layers have cells specialized to contract. Contractile cells in the epidermis run lengthwise, and those in the gastrodermis run circularly. These two sets of contractile cells, together with the water-filled gastrovascular cavity, form a **hydrostatic skeleton** that supports the body and allows movement. By the contraction of one set of contractile cells or the other, the hydra can shorten, lengthen, or bend its body.

Hydra typically lives attached to a rock, twig, or leaf by a disk of cells at its base. At the other end is the mouth, connecting the gastrovascular cavity with the outside. The mouth is surrounded by tentacles.

Cnidocytes (stinging cells) are located in the epidermis, especially on the tentacles. The cnidocytes contain stinging "thread capsules," or **nematocysts** (Fig. 28–9). When stimulated, the nematocysts release a coiled, hollow thread. Some types of nematocyst threads are sticky. Others are long and coil around prey. A third type bears barbs or spines and can inject a protein toxin that paralyzes the prey. Each cnidocyte has a small projecting trigger (cnidocil) on its outer surface. Stimuli such as touch or chemicals dissolved in the water ("taste") cause the nematocyst to fire its thread.

Captured prey is pushed into the mouth by the tentacles. Digestion begins in the gastrovascular cavity. Partially digested fragments are taken up by pseudopodia of the gastrodermal cells, and digestion is completed within food vacuoles in these cells.

Gas exchange and excretion occur by diffusion. The body of a hydra is small enough so that no cell is far from the surface. The motion of the body as it stretches and shortens circulates the contents of the gastrovascular cavity.

The first true nerve cells in the animal kingdom are found in the cnidarians, but nerve cells are not organized to form a brain or nerve cord. Rather, the nerve cells form irregular **nerve nets** connecting sensory cells in the body wall to contractile and gland cells. An impulse set up in one part of the body passes in all directions more or less equally.

Hydras reproduce asexually by budding during periods when environmental conditions are optimal. However, they differentiate as males and females in the fall or when pond water becomes stagnant. Females develop an *ovary* that produces a single egg, and males form a *testis*

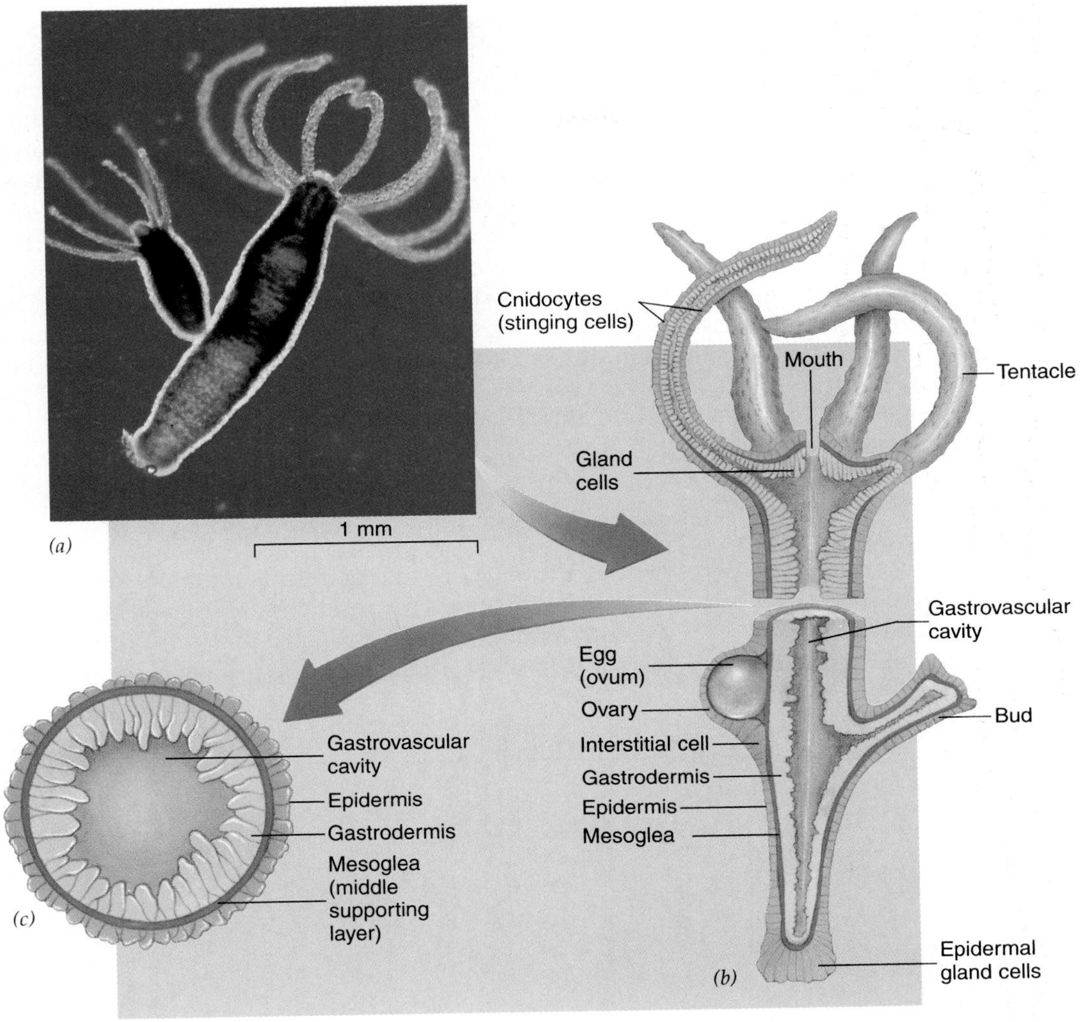

FIGURE 28–8 *Hydra* illustrates the basic cnidarian body plan. (***a***) A brown hydra *(Hydra viridis)* with a large bud that is partially contracted. When the bud separates from its parent, it becomes an independent hydra. (***b***) Hydra cut longitudinally to show its internal structure. Asexual reproduction by budding is represented on the right of the figure; sexual reproduction is represented by the ovary on the left. Male hydras develop testes that produce sperm. (***c***) Cross section through the body of a hydra. (Biophoto Associates/Photo Researchers, Inc.)

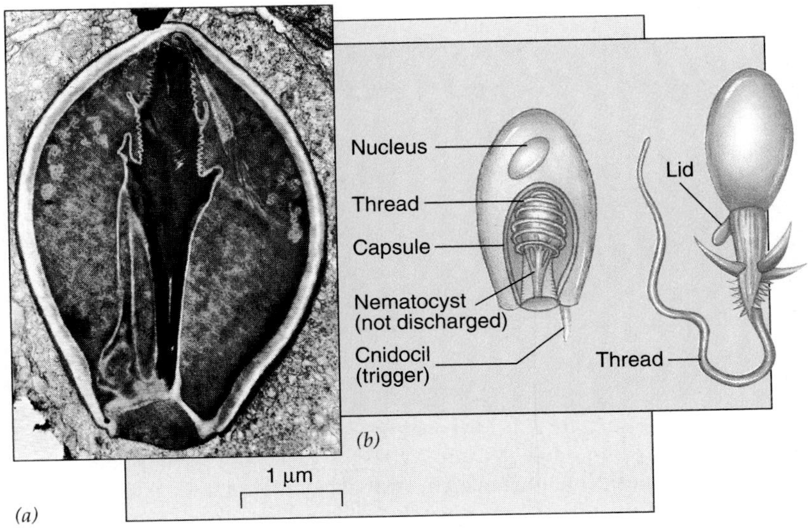

FIGURE 28–9 Nematocysts are the thread capsules within cnidarian stinging cells. (***a***) Electron micrograph of an undischarged nematocyst of *Hydra* (sagittal section). (***b***) Discharge of a nematocyst. When an object comes in contact with the cnidocil, the nematocyst discharges, ejecting a thread that may entangle or penetrate the prey. Some nematocysts secrete a toxic substance that immobilizes the prey. (*a*, G. B. Chapman, Cornell University Medical College)

that produces sperm. After fertilization, the zygote (fertilized egg) becomes covered with a shell. It leaves the parent and remains within the protective shell throughout the winter.

Many marine cnidarians form colonies consisting of hundreds or thousands of individuals. A colony begins with a single individual that reproduces asexually by budding. However, instead of separating from the parent, the bud remains attached and continues to form additional buds. Several types of individuals may arise in the same colony, some specialized for feeding, some for reproduction, and others for defense.

The Portuguese man-of-war, *Physalia,* superficially resembles a jellyfish but is actually a hydrozoan colony of polyps and medusae. A modified colony of medusae serves as a float for the colony in the form of a gas-filled sac colored a vivid, iridescent purple. The long tentacles of this animal may hang down for several meters below the float. Its cnidocytes are capable of paralyzing a large fish and can severely wound a human swimmer.

Some of the marine cnidarians are remarkable for an alternation of sexual and asexual stages. This alternation of stages differs from the alternation of generations in plants in that both sexual and asexual forms are diploid. Only sperm and eggs are haploid. The life cycle of the colonial marine hydrozoan *Obelia* illustrates alternation of sexual and asexual stages (Fig. 28–10). In this polyp colony, the asexual stage consists of two types of polyps: those specialized for feeding and those for reproduction. Free-swimming male and female medusae bud off from the reproductive polyps. These medusae eventually produce sperm and eggs, and fertilization takes place. The zygote develops into a ciliated swimming larva called a **planula.** The larva attaches to some solid object and begins to form a new generation of polyps by asexual reproduction.

The jellyfish belong to class Scyphozoa

Among the jellyfish, the medusa is the more prominent body form. It somewhat resembles an upside-down *Hydra* with a thick, viscous mesoglea that gives firmness to the body. In scyphozoans, the polyp stage is restricted to

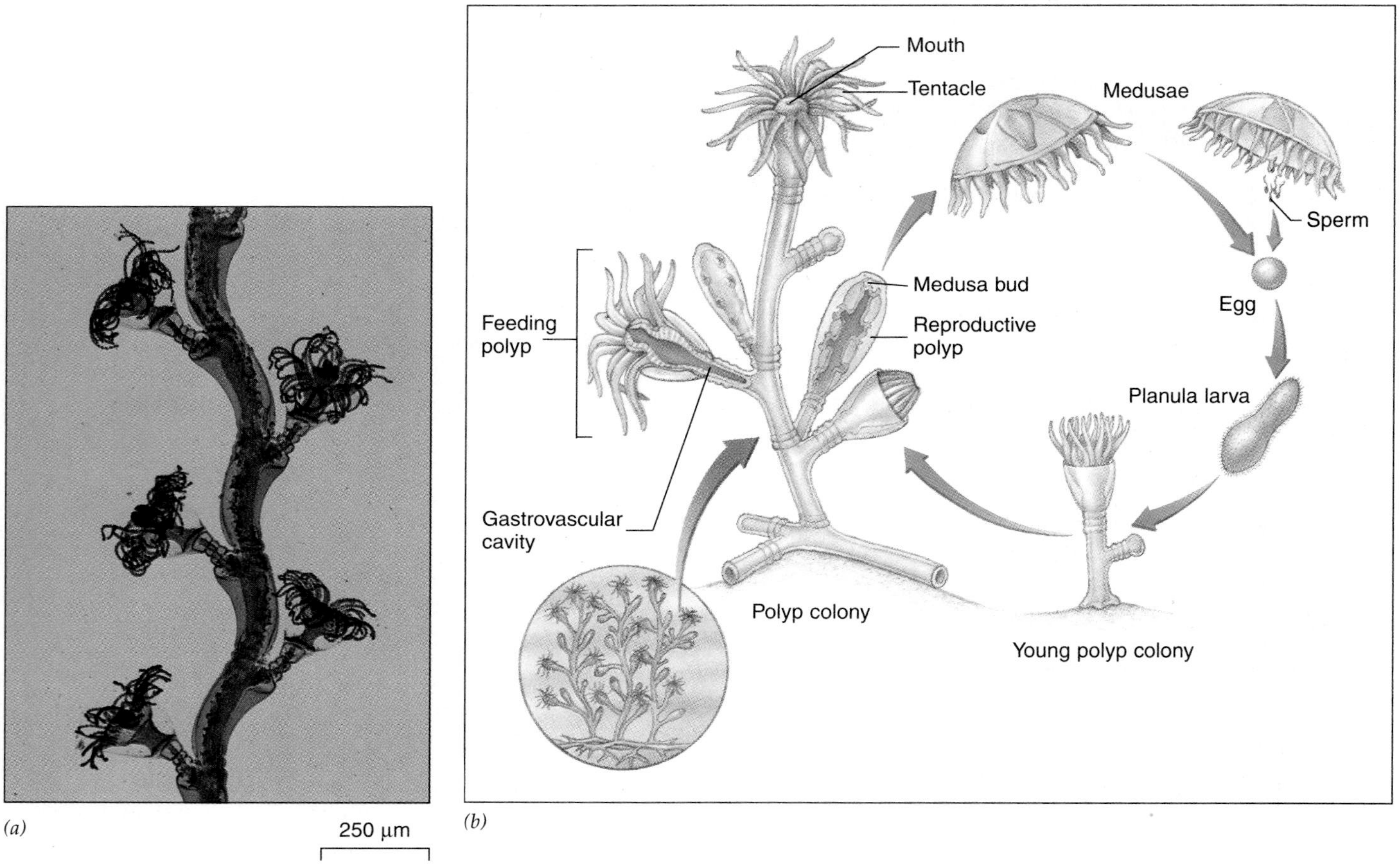

(*a*) (*b*)

FIGURE 28–10 Some hydrozoans form colonies. (***a***) A colonial hydrozoan, *(Obelia,* sp.*)*. (***b***) Life cycle of *Obelia*, a marine hydrozoan that forms colonies. Specialized polyps give rise asexually to medusae. The free-swimming medusae reproduce sexually, and the zygote develops into a planula larva. The larva develops into a polyp which may form a new colony. Note the specialization of members of the polyp colony. (Visuals Unlimited/John D. Cunningham)

Making the Connection

Coral Reefs, Interdependence of Organisms, and Environmental Issues

What are some of the ways that coral reefs are interdependent with other organisms? Coral reefs rival the tropical rain forests in species diversity. A single reef can serve as home for more than 3000 species of fish and other marine organisms, and an estimated one-fourth of all marine species depend on them. Reef organisms form complex food webs. Some organisms attach to the reef, while others find shelter within its crevices. Many terrestrial organisms also benefit from coral reefs, which form and maintain the foundation of thousands of islands. By providing a barrier against waves, reefs also protect shorelines against storms and erosion (see Chapter 51).

Coral reefs are made up of colonies of millions of individual corals and by certain algae (mainly coralline red algae). Coral animals also live in association with photosynthetic algae (called zooxanthellae). Both types of algae contribute to the reef's brilliant colors. The relationship with the zooxanthellae is symbiotic and mutually beneficial. The algae, which live within cells lining the coral's digestive cavity, provide the coral with oxygen and with carbon and nitrogen compounds. In exchange, the coral supplies the algae with waste products such as ammonia, from which the algae make nitrogen compounds for both partners (see Chapter 53).

Human activity, including mining reefs for building materials, polluting reef waters with industrial chemicals, and smothering coral with the silt that washes downstream from clearcut forests, threatens these important ecosystems. Studies indicate that of the 109 countries whose shores are lined with large reef formations, more than 90 have suffered serious reef damage.

The phenomenon of coral bleaching, the whitening of coral, may also be related to human activity. Coral bleaching occurs when the coral lose their colorful symbiotic algae. Without their algae, coral become malnourished and die. Although biologists have known about coral bleaching for more than 75 years, bleaching has recently become more widespread.

The causes of coral bleaching are not well understood; however, suspected environmental factors include pollution, changes in salinity, disease, increased ultraviolet radiation (associated with the destruction of the ozone layer), and unusually high or low temperatures. Many scientists think that one of the most important factors in recent years has been abnormally high water temperature, possibly caused by global warming. Healthy coral thrives in a narrow temperature range. An increase of only one or two degrees above the normal summer maximum temperature can cause widespread coral mortality.

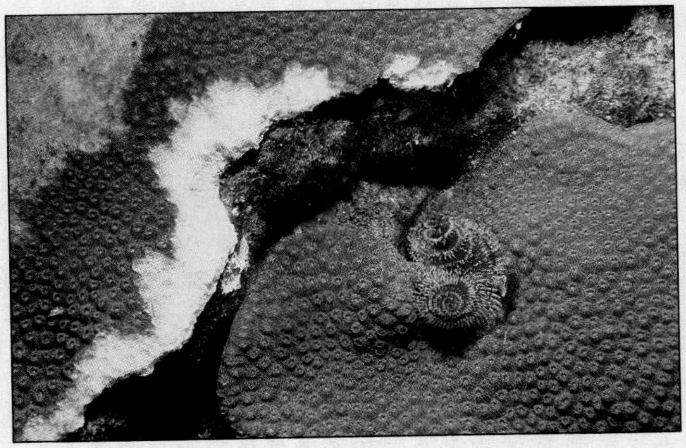

Bleached coral (*Montastrea* sp.). (Marilyn Kazmers, SharkSong/Dembinsky Photo Associates)

Coral bleaching appears to be a response to environmental stress, but some biologists suggest that bleaching may have a positive outcome. Bleaching may be a mechanism that allows a different algal partner to establish residence within the coral. The new partnership may be more resistant to environmental stress. Thus, low-level stress may have adaptive value for the coral and enhance its ability to survive.

Several international monitoring projects are gathering data that are needed to help us understand coral destruction and take action to protect these beautiful ecosystems. Once destroyed, it is difficult to reclaim them due to the long amount of time needed to form new reefs. For example, the major reef builder in Florida and Caribbean waters, a species known as star coral (*Montastrea annularis*), requires about 100 years to form a reef just one meter in height. ▲

a small larval stage. The largest jellyfish, *Cyanea,* may be more than 2 meters (6.5 feet) in diameter and have tentacles 30 meters (98 feet) long. These orange and blue "monsters," among the largest of the invertebrate animals, are a real danger to swimmers in the North Atlantic Ocean.

The corals belong to class Anthozoa

The anthozoans—sea anemones and corals—have no free-swimming medusa stage. The polyps may be either individual or colonial forms. These animals have a small ciliated planula larva, which may swim to a new location before attaching to develop into a polyp.

Anthozoans differ from hydrozoans in that the gastrovascular cavity is partially divided into a number of connected chambers by a series of vertical partitions. The partitions increase the surface area for digestion, so that an anemone can digest an animal as large as a crab or fish. Although corals can capture prey, many tropical species depend for nutrition on photosynthetic algae (zooxanthellae) that live within their cells (Chapter 24).

In warm shallow seas, much of the bottom is covered with coral or anemones, most of them brightly colored. Coral reefs are among the most productive of all ecosystems. The reefs and atolls of the South Pacific are the remains of billions of microscopic, cup-shaped calcareous skeletons, secreted during past ages by coral colonies and by coralline algae. Living colonies occur only in the uppermost regions of such reefs, adding their own skeletons to the forming rock.

PHYLUM CTENOPHORA INCLUDES THE COMB JELLIES

The **ctenophores,** or comb jellies, are a phylum of about 100 marine species. They are fragile, luminescent animals that may be as small as a pea or larger than a tomato. Ctenophores are biradially symmetrical. You could obtain equal halves by cutting through the body (oral-aboral) axis in two ways. Their body plan is somewhat similar to that of a medusa. The body consists of two cell layers separated by a thick jelly-like mesoglea.

The outer surface of a ctenophore bears eight rows of cilia, resembling combs (Fig. 28–11). The coordinated beating of the cilia in these combs moves the animal through the water. A sense organ functions in balance and helps the animal orient itself. Some ctenophores have two tentacles, and they lack the stinging cells characteristic of the cnidarians. However, their tentacles are equipped with adhesive glue cells that trap prey.

FLATWORMS BELONG TO PHYLUM PLATYHELMINTHES

Members of phylum **Platyhelminthes,** the **flatworms,** are flat, elongated animals that exhibit bilateral symmetry. They are acoelomate. The classification of flatworms is controversial. At present the 20,000 species are assigned to four classes. The **Turbellaria** are the free-living flatworms, including planarians and their relatives. Classes **Trematoda** and **Monogenea** include the flukes, which are either internal or external parasites. Class **Cestoda** includes the tapeworms, which as adults are intestinal parasites of vertebrates (Table 28–2).

Some important characteristics of this phylum follow:

1. **Bilateral symmetry and cephalization.** Along with their bilateral symmetry, flatworms have definite anterior and posterior ends. An animal with a front end ("head") generally moves in a forward direction. With a concentration of sense organs in the part of the body that first meets the environment, the animal can detect an enemy quickly. Such an animal is also more likely to see or smell prey quickly enough to capture it. A rudimentary head, the beginnings of **cephalization,** is evident in flatworms.

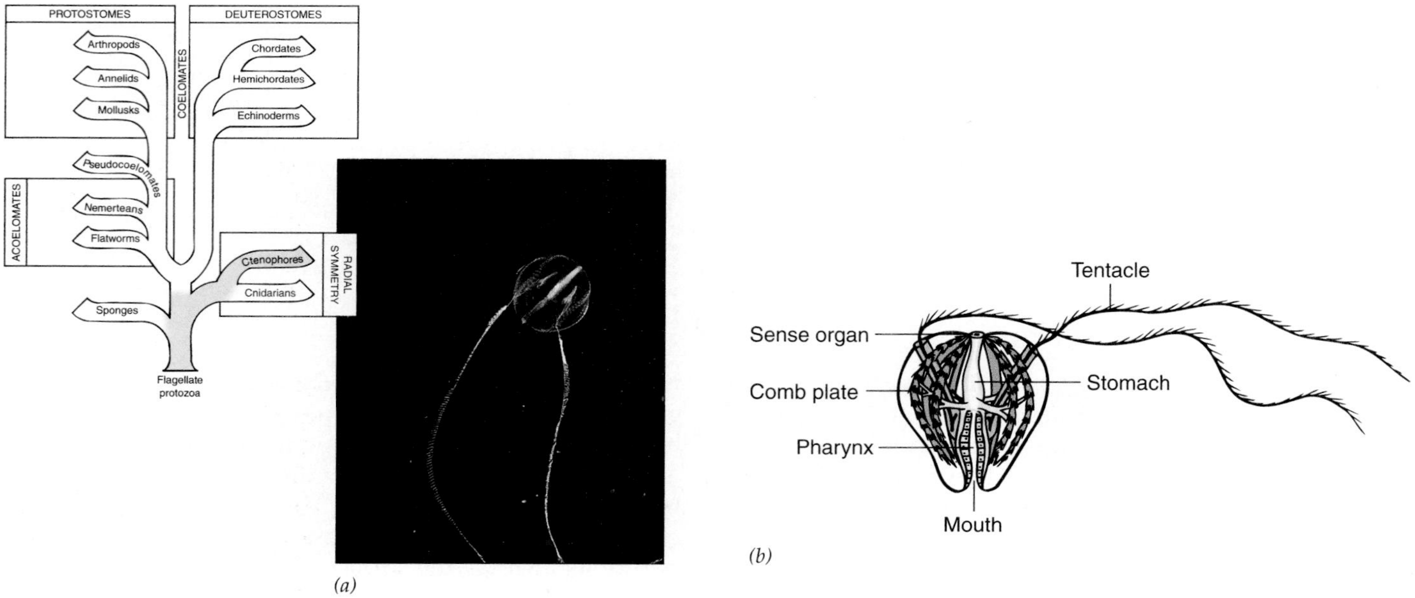

(*a*) (*b*)

FIGURE 28–11 Ctenophores are known as comb jellies. (***a***) The sea gooseberry *(Pleurobranchia pileus).* (***b***) Structure of a ctenophore. (Fred Bavendam/Peter Arnold, Inc.)

2. **Three definite tissue layers** (triploblastic). In addition to an outer epidermis, derived from ectoderm, and an inner endodermis, derived from endoderm, the flatworm has a middle tissue layer that develops from mesoderm.
3. **Well-developed organs.** The flatworms are the simplest animals that have well-developed organs, functional structures made of two or more kinds of tissue. They have a muscular pharynx for taking in food, eyespots and other sensory organs in the head, a simple brain, and complex reproductive organs.
4. **A simple nervous system.** The simple brain consists of two masses of nervous tissue, called **ganglia,** in the head region. In many species, the ganglia are connected to two nerve cords that extend the length of the body. A series of nerves connects the cords like the rungs of a ladder.
5. **Protonephridia,** structures that function in osmoregulation and waste disposal (excretion).
6. **A gastrovascular cavity** in most species. It is often extensively branched and has only one opening, the mouth, usually located on the middle of the ventral surface.

Both groups of parasitic flatworms—the flukes and the tapeworms—are highly adapted to and modified for their parasitic life style. They have suckers or hooks for holding onto the host. The bodies of those who live in digestive tracts are resistant to the digestive enzymes secreted by their hosts. Many have complicated life cycles and they produce large numbers of eggs. Other adaptations include the loss of structures such as sense organs or a digestive system.

Table 28–2 CLASSES OF PHYLUM PLATYHELMINTHES

Class and Representative Animals	*Characteristics*
Turbellaria Planarians	Free-living flatworms; mainly marine; body covered by ciliated epidermis; usually carnivorous forms that prey on tiny invertebrates or on dead organisms.
Trematoda and Monogenea Flukes	All parasites with a wide range of vertebrate and invertebrate hosts; may require intermediate hosts; possess suckers for attachment to host.
Cestoda Tapeworms	Parasites of vertebrates; complex life cycle with one or two intermediate hosts; larval host may be invertebrate; tapeworms have suckers and sometimes hooks for attachment to host; eggs produced within proglottids, which are shed; no digestive system.

Class Turbellaria includes planarians

Members of the class Turbellaria are free-living, mainly marine, flatworms. **Planarians** are turbellarian flatworms found in ponds and quiet streams throughout the world. The common American planarian *Dugesia* is about 15 millimeters (0.6 inch) long, with what appear to be crossed eyes and flapping ears called **auricles** (Fig. 28–12). The auricles actually serve as organs of smell (chemoreceptors).

Planarians are carnivorous, trapping small animals in a mucous secretion. The digestive system consists of a single opening (the mouth), a **pharynx** (the first portion of the digestive tube), and a branched gastrovascular cavity. A planarian can project its pharynx outward through its mouth, using it to ingest its prey. Extracellular digestion takes place in the gastrovascular cavity by enzymes secreted by gland cells. Digestion is completed after the nutrients have been absorbed into individual cells. Undigested food is eliminated through the mouth. The long, highly branched gastrovascular cavity helps to distribute food to all parts of the body, so that each cell is within range of diffusion.

A planarian's flattened body ensures that gases can reach all of its cells by diffusion. It has no specialized respiratory or circulatory structures. Although most excretion takes place by diffusion, some wastes are excreted by the protonephridia. These structures function mainly in fluid balance, or osmoregulation. Protonephridia are tubules that end in **flame cells,** collecting cells equipped with cilia. The beating of the cilia channels waste into the system of tubules.

Planarians are capable of learning. Memory is not localized within the ganglia but appears to be retained throughout the nervous system.

Planarians can reproduce either asexually or sexually. In asexual reproduction, an individual constricts in the middle and divides into two planarians. Each regenerates its missing parts. Sexually, these animals are hermaphroditic. During the warm months of the year, each is equipped with a complete set of male and female organs. Two planarians come together in copulation and exchange sperm cells so that their eggs are cross-fertilized.

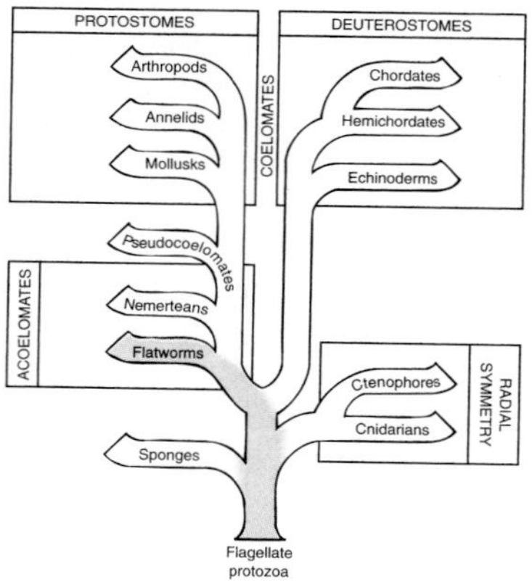

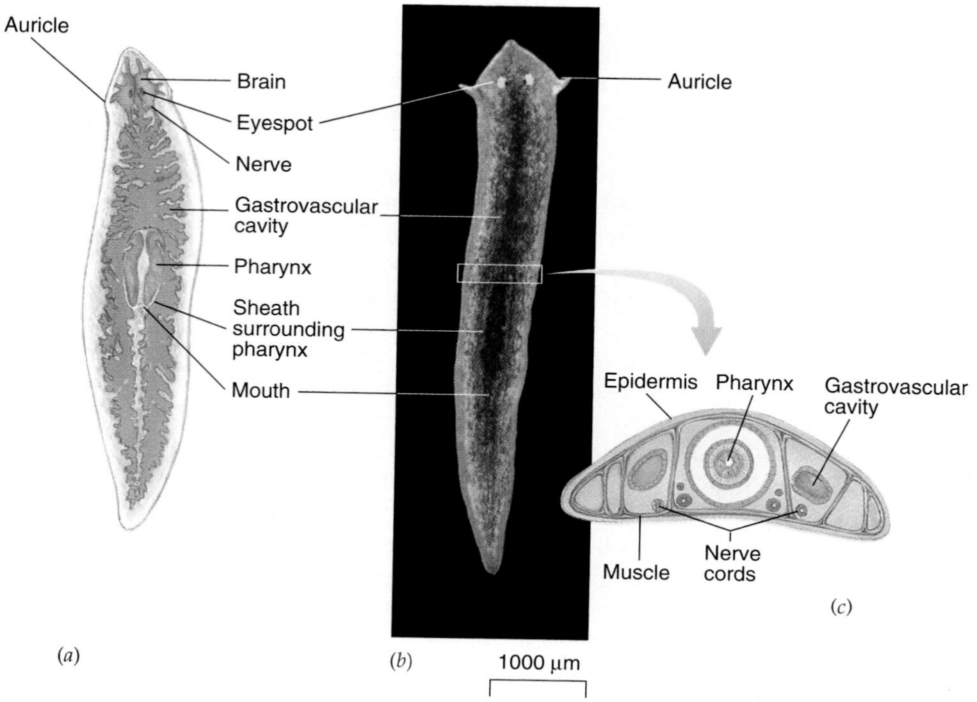

FIGURE 28–12 The common planarian, *Dugesia*, illustrates the platyhelminth life style. (***a***) Internal structure. (***b***) Structure of a living *Dugesia (Dugesia dorotocephala)* and cross section through a planarian. (*b*, T. E. Adams/Peter Arnold, Inc.)

The flukes belong to class Trematoda and Class Monogenea

Although their body plan resembles that of the free-living flatworms, specialized adaptations have evolved in **flukes** that contribute to their success as parasites. Flukes have extremely complex and efficient reproductive organs (Fig. 28–13). Both blood flukes and liver flukes have one or more suckers for clinging to their host.

Among the flukes that are parasitic in humans are the blood flukes, widespread in tropical areas of the world, and the liver flukes, common in Asia, particularly in areas where human feces are used for fertilizing crops. Blood flukes of the genus *Schistosoma* infect about 200 million people who live in tropical areas. Both blood flukes and liver flukes go through complicated life cycles involving a number of different forms, alternation of sexual and asexual stages, and parasitism on one or more intermediate hosts (snails and fishes, for example) (Fig. 28–14). The aquatic snails that serve as intermediate hosts thrive in ponds, rice paddies, and the marshy areas that form when dams are built.

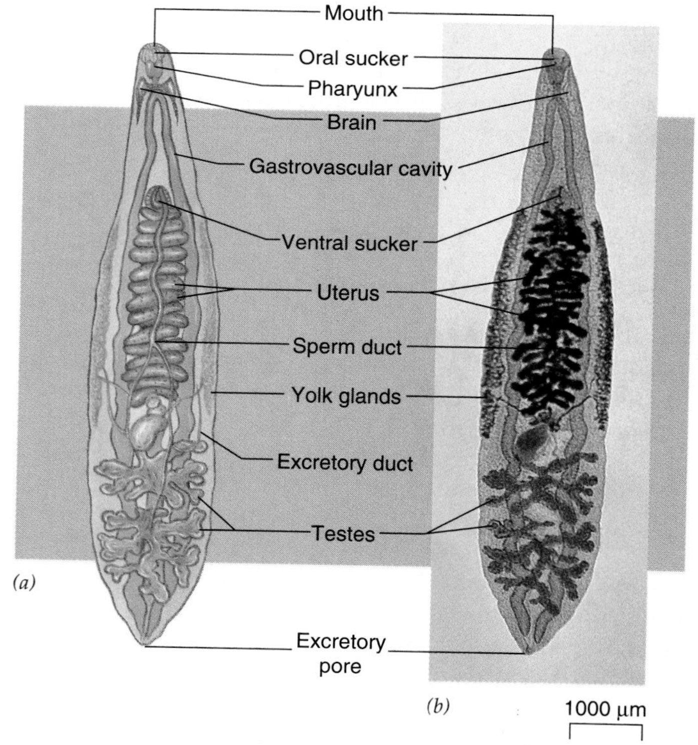

FIGURE 28–13 Flukes are well adapted to their parasitic mode of life. (***a***) Internal structure of a fluke. (***b***) The human liver fluke *(Clonorchis sinensis)*. (*b*, Carolina Biological Supply Company Phototake-NYC)

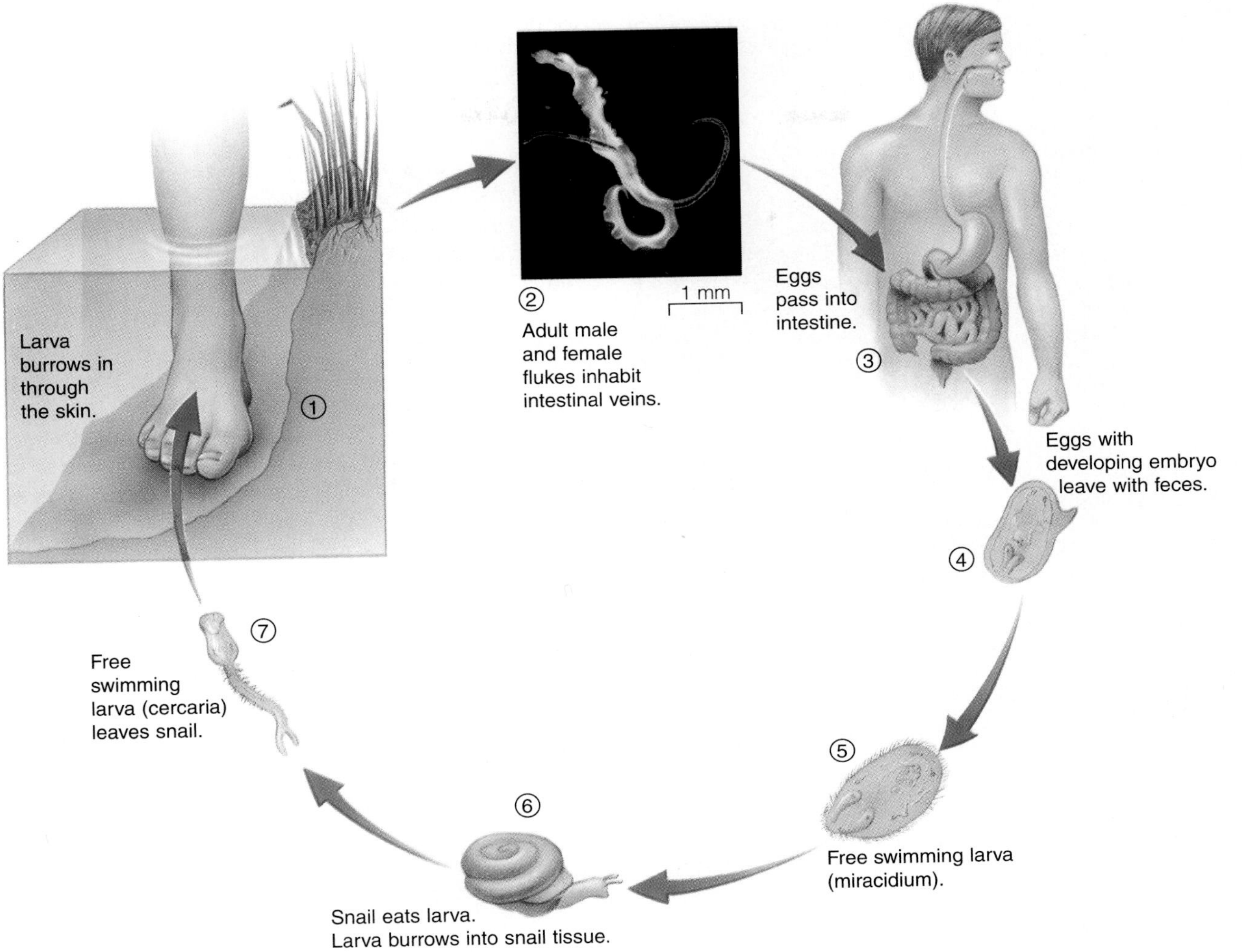

FIGURE 28–14 The blood fluke has a complex life cycle. (*1*) The human is this fluke's final host. Larvae burrow through human skin and make their way to the circulatory system. (*2*) The photograph shows an adult male and female. They remain permanently paired for life. The adult male has a long groove that holds the female during fertilization, which takes place within veins in the intestine. (*3*) Eggs pass into the intestine. (*4*) Eggs containing developing embryos leave the body with the feces. (*5*) If they find their way to fresh water, the eggs hatch, releasing free-swimming larvae (miracidia). (*6*) To survive, the larvae must be ingested by an intermediate host, an aquatic snail. If successful, they burrow into the tissues of the snail and develop into a form that can reproduce asexually. (*7*) Finally, fork-tailed larvae (cercariae) develop and leave the snail. When they contact a human host they burrow into the skin, completing the cycle. (Photograph of adult male and female, Center for Disease Control, Atlanta, Georgia/Biological Photo Service)

The tapeworms belong to class Cestoda

Adult members of the more than 1000 different species of the class Cestoda live as parasites in the intestines of probably every kind of vertebrate, including humans. Tapeworms are long, flat, ribbon-like animals strikingly specialized for their parasitic mode of life. Among their many adaptations are suckers and sometimes hooks on the "head," or **scolex,** which enable the parasite to maintain its attachment to the host's intestine (Fig. 28–15).

The reproductive adaptations and abilities of tapeworms are extraordinary. The body of the tapeworm consists of a long chain of segments called **proglottids.** Each segment is an entire reproductive machine equipped with both male and female reproductive organs and containing up to 100,000 eggs. Because an adult tapeworm may possess as many as 2000 segments, its reproductive potential is staggering. A single tapeworm can produce 600 million eggs in a year. Proglottids farthest from the tapeworm's head contain the ripest eggs; these segments are shed from the host's body daily along with the feces.

FIGURE 28–15 Some species of tapeworms are armed with powerful hooks. These parasites use their hooks to attach to the host. False-color scanning electron micrograph of the head of the small tapeworm *Acanthrocirrus retrisrostris.* This tapeworm reaches maturity in the intestines of wading birds that eat barnacles. The photograph shows the piston-like rostellum, which can be withdrawn into the head or thrust out and buried in the host's tissue. Beneath the rostellum, two of the four powerful suckers are visible. (Cath Ellis, Dept. of Zoology, University of Hull/Science Photo Library/Photo Researchers, Inc.)

The tapeworm lacks certain organs. It absorbs food directly through its body wall from the host's intestine, and it has no mouth and no digestive system of its own. Nor does the tapeworm have well-developed sense organs. Some tapeworms have rather complex life cycles, spending their larval stage within the body of an intermediate host and their adult life within the body of a different, final host. For example, let us consider the life cycle of the beef tapeworm, so named because humans become infected when they eat poorly cooked beef containing the larvae (Fig. 28–16).

The microscopic tapeworm larva spends part of its life cycle encysted within the muscle tissue of a cow. When a human ingests infected meat, digestive juices break down the cyst, releasing the larva. Soon the larva attaches itself to the intestinal lining and within a few weeks matures into an adult tapeworm, which may grow to a length of about 15 meters (50 feet). The parasite reproduces sexually within the human intestine and sheds proglottids filled with ripe eggs. Once established within a human host, the tapeworm makes itself very much at home and may remain for the rest of its life, as long as 10 years. A person infected with a tapeworm may suffer pain or discomfort, increased appetite, weight loss, and other symptoms, or may be totally unaware of its presence.

In order for the life cycle of the tapeworm to continue, its eggs must be ingested by an **intermediate host,** in this case, a cow. This requirement explains why we are not completely overrun by tapeworms, and why a tapeworm must produce millions of eggs to ensure that at least a few survive. When a cow eats grass or other foods contaminated with human feces, eggs may be ingested. The eggs hatch in the cow's intestine, and the larvae make their way into muscle. There they encyst and remain until released by a **final host,** perhaps a human eating rare steak.

Two other tapeworms that infect humans are the pork tapeworm, found in poorly cooked, infected pork, and

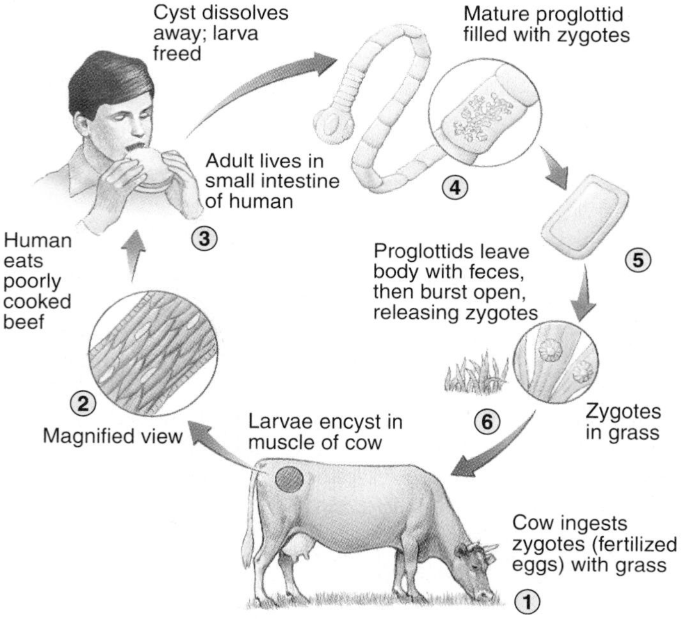

FIGURE 28–16 The beef tapeworm is a parasitic flatworm. (*1*) Cattle, the intermediate hosts for the beef tapeworm, ingest zygotes (fertilized eggs) with grass. Larvae hatch and encyst in the muscle of a cow. (*2*) Magnified view of muscle containing encysted larvae. (*3*) When a human eats poorly cooked, infected beef, digestive juices dissolve the cyst, freeing the larva. (*4*) Humans are the final host for this parasite. The adult tapeworm inhabits the small intestine of a human. Fertilized eggs (zygotes) are produced in body sections, called proglottids. (*5*) Mature proglottids leave the body with the feces and burst open, releasing zygotes. (*6*) Magnified view of zygotes on grass.

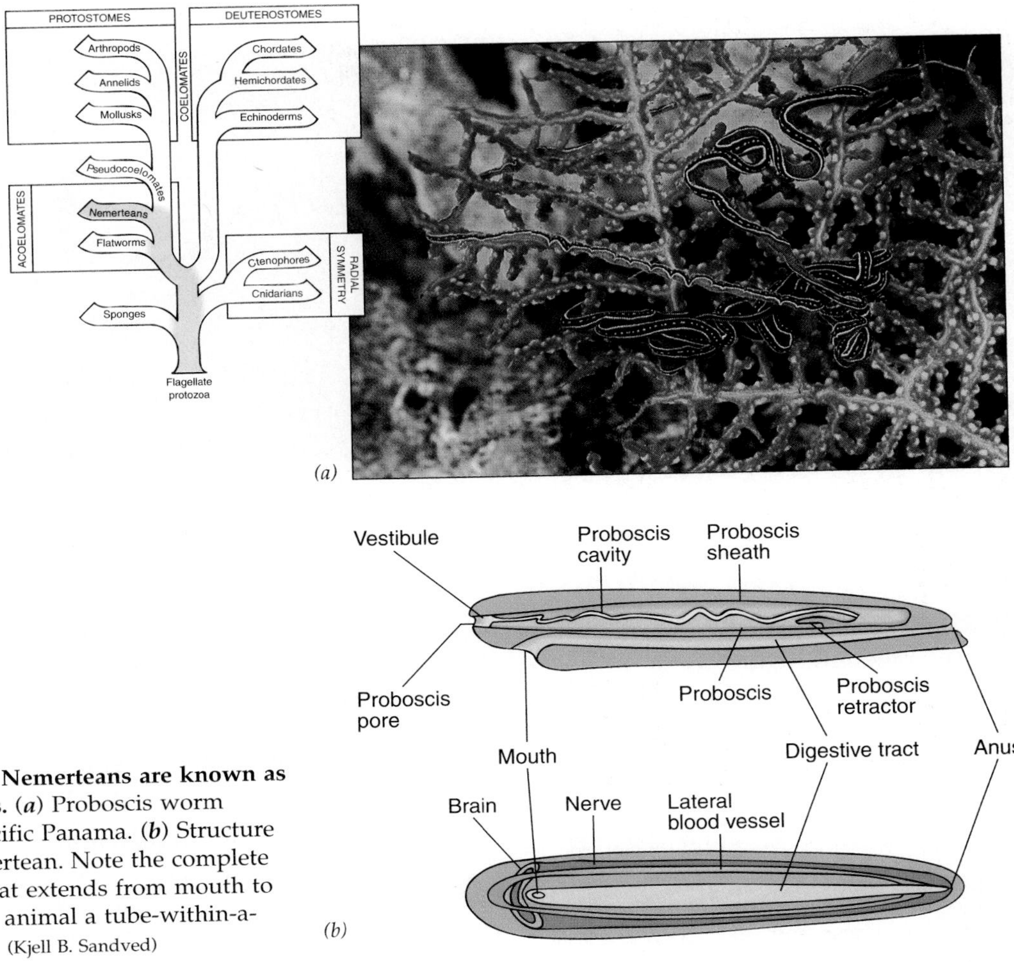

FIGURE 28–17 Nemerteans are known as proboscis worms. (***a***) Proboscis worm (*Lineus*) from Pacific Panama. (***b***) Structure of a typical nemertean. Note the complete digestive tract that extends from mouth to anus, giving this animal a tube-within-a-tube body plan. (Kjell B. Sandved)

the fish tapeworm, found in raw, or poorly cooked, infected fish. Like most parasites, tapeworms tend to be species-specific; that is, each can infect only certain specific species. For example, the beef tapeworm can only spend its adult life in a human host.

PHYLUM NEMERTEA COMPRISES THE PROBOSCIS WORMS

The phylum **Nemertea** is a relatively small group (about 900 species) of free-living animals (Fig. 28–17). Almost all are marine, although a few inhabit fresh water or damp soil. Nemerteans (sometimes called ribbon worms) have long narrow bodies, either cylindrical or flattened, generally varying in length from 5 centimeters (2 inches) to about 2 meters (6.5 feet), although some are much longer. Some are a vivid orange, red, or green, with black or colored stripes.

Their most remarkable organ—the **proboscis,** from which they get one of their common names, *proboscis worm*—is a long, hollow, muscular tube that can be everted from the anterior end of the body for use in seizing food or in defense. The proboscis secretes mucus, which is helpful in catching and trapping prey.

Phylum Nemertea is considered an evolutionary landmark because its members have evolved a circulatory system and a **tube-within-a-tube body plan.** The body wall forms the outer tube while the digestive tract forms the inner tube. The digestive tract is a complete tube, with a mouth at one end for taking in food and an anus at the other for eliminating undigested food. Recall that in the cnidarians and flatworms, food enters and wastes leave through the same opening.

In nemerteans the digestive and circulatory functions are separated. The primitive circulatory system consists simply of muscular tubes—the blood vessels—extending the length of the body and connected by transverse vessels. Nemerteans have no heart to pump the blood. The blood is circulated through the vessels by movements of the body and contractions of the muscular blood vessels. The development of a circulatory system for inter-

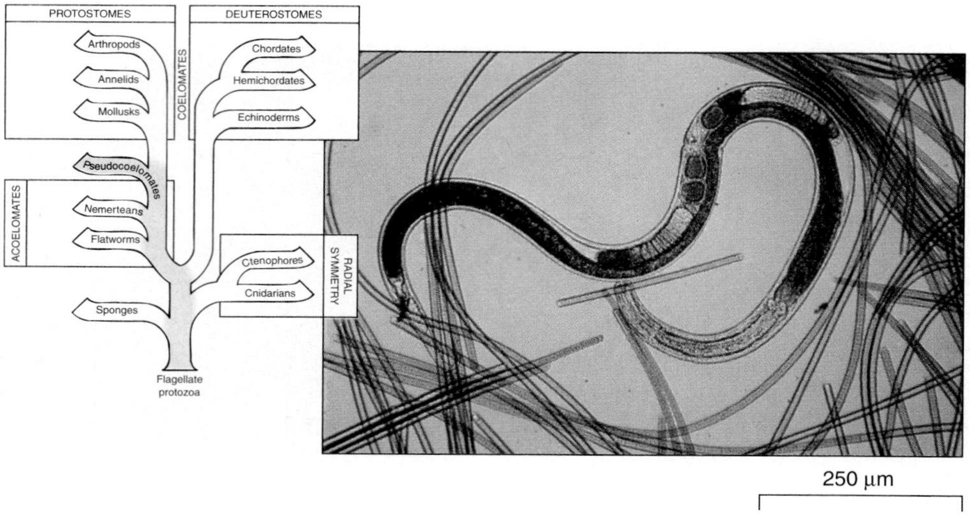

FIGURE 28–18 A free-living nematode is shown among the cyanobacteria *Oscillatoria*, which it eats. (Visuals Unlimited/T. E. Adams)

nal transport of materials provides an alternative to diffusion. The circulatory system and other advances that have evolved in the nemerteans have permitted these animals to be larger and more active than the flatworms.

ROUNDWORMS BELONG TO PHYLUM NEMATODA

Members of phylum **Nematoda,** the **roundworms,** are of great ecological importance. These animals play a key role in decomposition and nutrient recycling. Nematodes are numerous and widely distributed in both the soil and in marine and freshwater sediments. More than 12,000 species have been identified. A spadeful of soil may contain more than a million of these mainly microscopic white worms, which thrash around coiling and uncoiling.

Although many nematodes are free living (Fig. 28–18), others are important parasites in plants and animals. More than 30 roundworms, including hookworms, pinworms, trichina worms, filarial worms, and the intestinal roundworm *Ascaris,* are human parasites. *Caenorhabditis elegans,* a free-living nematode, is an important research organism for biologists studying the genetic control of development (see Chapter 16).

The elongate, cylindrical, threadlike nematode body is pointed at both ends and covered with a tough **cuticle** (Fig. 28–19). Secreted by the underlying epidermis, the cuticle enables nematodes to resist desiccation, permitting them to inhabit dry soils and even deserts. Beneath the epidermis is a layer of longitudinal muscles. No circular muscles are present in the body wall.

Nematodes are the simplest animals known with a body cavity—a pseudocoelom. The fluid-filled pseudocoelom serves as a hydrostatic skeleton that transmits the force of muscle contraction to the enclosed fluid. Like the ribbon worms, the nematodes exhibit bilateral symmetry, a complete digestive tract, three definite tissue layers, and definite organ systems; however, they lack circulatory structures. The sexes are usually separate, and the male is generally smaller than the female.

Ascaris is a parasitic roundworm

A common intestinal parasite of humans, *Ascaris* is a white-colored worm about 25 centimeters (10 inches) long. *Ascaris* spends its adult life in the human intestine, where it makes its living by ingesting partly digested food. Like the tapeworm, it must devote a great deal of effort to reproduction in order to ensure survival of its species. The sexes are separate, and copulation takes place within the host. A mature female may lay as many as 200,000 eggs a day.

Ascaris eggs leave the human body with the feces and, where sanitation is poor (throughout most of the world), they find their way onto the soil. In many parts of the world, human wastes are used as fertilizer, a practice that encourages the survival of *Ascaris* and many other human parasites. People become infected when they ingest *Ascaris* eggs on unwashed vegetables or fruit, or from hands dirty with contaminated soil.

The eggs hatch in the intestine, and the larvae then take a remarkable journey through the body before settling in the small intestine. The larvae burrow through the intestinal wall into blood vessels or lymph vessels. Then they are carried in blood vessels through the heart to the lungs. There, they break through into the air sacs and move up the air passageways to the throat, where they are swallowed. Finally, the larvae pass through the stomach and into the intestine, where they settle and feed on partly digested food. During their migration, the larvae can damage the lungs and other tissues. When they settle in the small intestine, they may cause abdominal pain, allergic reactions, or malnutrition. Sometimes a tangled mass of these worms blocks the intestine.

Several other parasitic roundworms infect humans

The life cycle of a human **hookworm** is somewhat similar to that of *Ascaris*. Only one host is required. Adult worms, which are less than 1.5 centimeter (0.6 inch) long, live in the human intestine and lay eggs, which pass out of the body with the feces. Larvae hatch and feed on bacteria in the soil. After a period of maturation, they become infective. When a potential host walks barefoot, the microscopic larvae bore through the skin and enter the blood. They migrate through the body and find their way to the intestine, where they mature. Hookworms can cause serious tissue damage and loss of blood.

Humans become infected with **trichina worms** by eating poorly cooked, infected pork or bear. The trichina parasite is adapted to live inside many animals (pigs, rats, bears, and others), and the human is an accidental host. Adult trichina worms live in the small intestine of the host. The females produce larvae, which migrate through the body to skeletal muscle, where they encyst. Continuation of the life cycle depends on ingestion by another animal. Because humans are not normally eaten, trichina larvae are not liberated from their cysts and eventually die. The cysts become calcified and remain in the muscles, causing stiffness and discomfort. Other symptoms are caused by the presence of the adults and by the migrating larvae. No cure has been found for this infestation.

Pinworms are the most common worms found in children. The eggs are often ingested by eating with dirty hands. Adult worms, less than 1.3 centimeter (0.5 inch) long, live in the large intestine. Female pinworms often migrate to the anal region at night to deposit their eggs. Irritation and itching caused by this practice induce scratching, which spreads the tiny eggs. Eggs may also be distributed in the air and are in this way scattered throughout the house. Mild infestations may go unnoticed, but those more serious may result in discomfort, irritation, and injury to the intestinal wall.

PHYLUM ROTIFERA ARE WHEEL ANIMALS

Among the more obscure invertebrates are the "wheel animals" of phylum **Rotifera.** Although no larger than many protozoa, these aquatic, microscopic animals are multicellular. Rotifers have a characteristic crown of cilia on the anterior end; these beat rapidly during feeding, giving the appearance of a spinning wheel (Fig. 28–20).

Rotifers are pseudocoelomate animals with a complete digestive tract, including a muscular organ for grinding food. They have a nervous system with a "brain" and sense organs, including eyespots. Protonephridia with flame cells remove excess water from the body and may also excrete wastes.

Like some other pseudocoelomate animals, rotifers are "cell constant": each member of a given species is composed of exactly the same number of cells. Indeed,

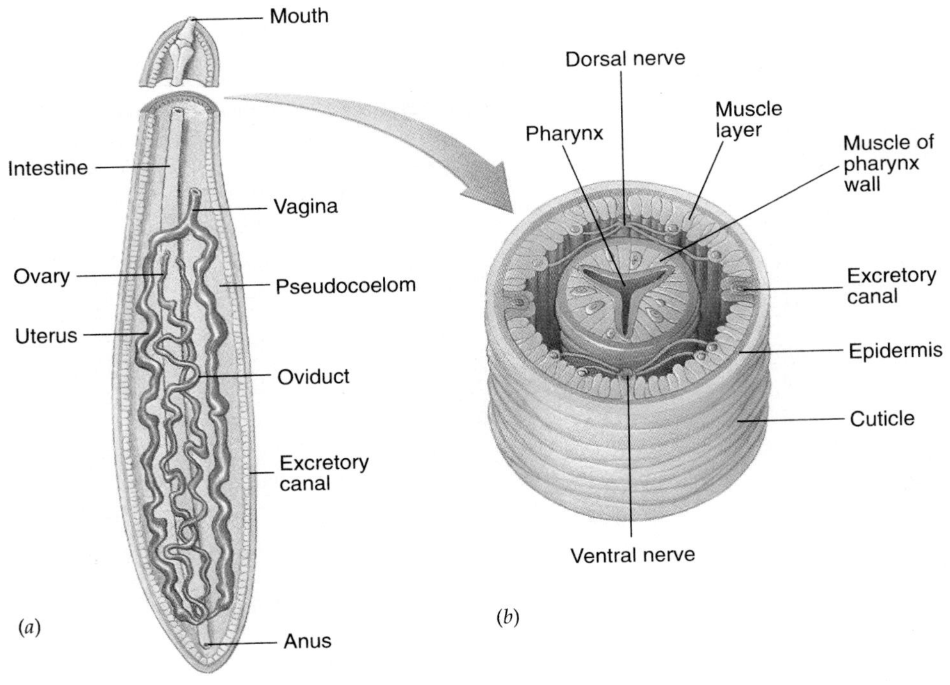

FIGURE 28–19 The roundworm *Ascaris* illustrates the nematode body plan. (*a*) Longitudinal section to show internal anatomy. Note the complete digestive tract that extends from mouth to anus. (*b*) Cross section through the body of *Ascaris*.

FIGURE 28–20 Rotifers are known as wheel animals. (*a*) This Antarctic species of rotifer *(Philodina gregaria)* survives the winter by forming a cyst. Some species reproduce in great numbers, sometimes coloring lake water red. (*b*) Structure of a rotifer. The motion of its cilia draws particles of food such as algae into the mouth.
(*a*, John Walsh/Science Photo Library/Photo Researchers, Inc.)

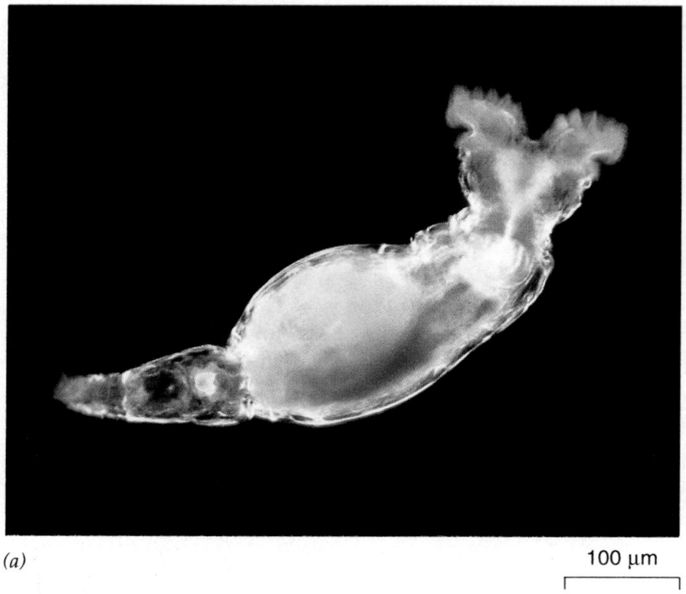

(*a*)

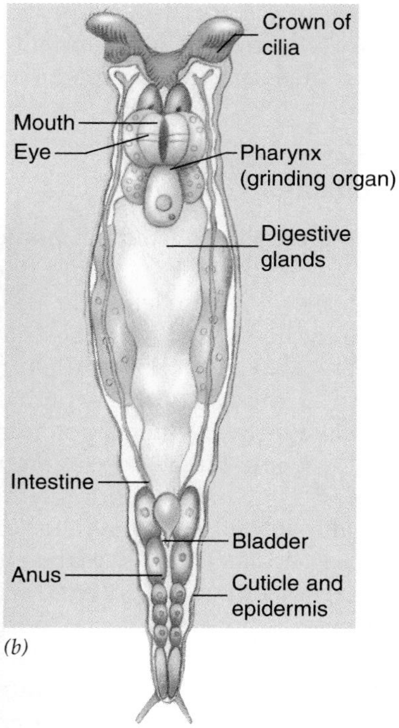

(*b*)

each part of the body is made of a precisely fixed number of cells arranged in a characteristic pattern. Cell division does not take place after embryonic development, and mitosis cannot be induced; growth and repair are not possible. One of the challenging problems of biological research is discovering the difference between such nondividing cells and the dividing cells of other animals. Do rotifers ever develop cancer?

The phyla discussed in this chapter are compared in Table 28–3.

Table 28–3 COMPARISON OF SOME LOWER INVERTEBRATE PHYLA

	Porifera (pore bearers)	*Cnidaria*	*Platyhelminthes (flatworms)*	*Nemertea*	*Nematoda (roundworms)*
Representative Animals	Sponges	Hydras Jellyfish Corals	Planarians Flukes Tapeworms	Proboscis worms	Ascarids Hookworms Nematodes
Level of Organization	Multicellular; cells loosely arranged	Tissues	Organs	Organ systems	Organ systems
Symmetry	Radial or none	Radial	Bilateral; rudimentary head	Bilateral	Bilateral
Digestion	Intracellular	Gastrovascular cavity with only one opening; intra- and extracellular digestion	Gastrovascular cavity with only one opening	Complete digestive tract with mouth and anus	Complete digestive tract with mouth and anus

Table 28–3 Continued

	Porifera (pore bearers)	*Cnidaria*	*Platyhelminthes (flatworms)*	*Nemertea*	*Nematoda (roundworms)*
Circulation	Diffusion	Diffusion	Diffusion	At least two pulsating longitudinal blood vessels; no heart; blood cells with hemoglobin	Diffusion
Gas Exchange	Diffusion	Diffusion	Diffusion	Diffusion	Diffusion
Waste Disposal	Diffusion	Diffusion	Protonephridia; flame cells and ducts	Two lateral excretory canals with flame cells	Excretory canals
Nervous System	Irritability of cytoplasm	Nerve net; no centralization of nerve tissue	Simple brain; two nerve cords; ladder type system; simple sense organs	Simple brain; two nerve cords; cross nerves; simple sense organs	Simple brain; dorsal and ventral nerve cords; simple sense organs
Reproduction	Asexual, by budding; sexual, most are hermaphroditic	Asexual by budding; sexual, sexes separate	Asexual, by fission; sexual, hermaphroditic, but cross-fertilization in some species	Asexual, by fragmentation; sexual, sexes separate	Sexual, sexes separate
Support and Movement	Support by spicules of calcium carbonate, silica, or spongin; contractile cells can change the size of openings	Support by mesoglea, calcareous skeletons (coral), or by fluid in gastrovascular cavity (hydrostatic skeleton); contractile cells	Support by its tissues; well developed muscle tissue	Support by its tissues; locomotion by muscle or cilia	Support by tough cuticle; fluid in pseudocoelom serves as hydrostatic skeleton; longitudinal muscle in body wall
Environment and Lifestyle	Aquatic, mainly marine; ciliated, swimming larvae; adults attach; suspension feeders	Aquatic, mainly marine; some float or swim, others sessile; polyp and medusa forms; some form colonies; capture food with cnidocytes, tentacles	Aquatic, some terrestrial in damp areas; many are carnivores; some parasites	Mainly marine; mainly carnivores; use proboscis for capturing food and defense	Widely distributed in the soil, sea, and fresh water; carnivores, scavengers, parasites
Other Characteristics	Skeleton of chalk, glass, or spongin (a protein material)	Cnidocytes (stinging cells) along their tentacles	Three definite tissue layers; no body cavity	No body cavity	Have pseudocoelom (space between internal organs and body wall)

SUMMARY

I. Animals are eukaryotic, multicellular, heterotrophic organisms with cells specialized to perform specific functions. They generally are capable of locomotion at some time during their life cycle, can reproduce sexually, and can respond adaptively to external stimuli.

II. Animals are consumers that inhabit the sea, fresh water, and land.

III. Animals may be classified in several different ways.
   A. Animals can be classified as acoelomates, pseudocoelomates, or coelomates based on type of body cavity.
   B. In protostomes, the blastopore develops into the mouth; in deuterostomes, it does not, but it often becomes the anus.

IV. Phylum Porifera consists of the sponges.
   A. Sponges are divided into three main classes on the basis of the type of skeleton they secrete.
   B. The sponge body is a sac with tiny openings through which water enters; a central cavity (spongocoel); and an open end, or osculum.

V. Phylum Cnidaria includes the hydras, jellyfish, and corals.
   A. Cnidarians have radial symmetry, cnidocytes, two definite tissue layers, and a nerve net.
   B. In many types of cnidarians a sessile polyp gives rise to free-swimming medusae.

VI. Phylum Ctenophora consists of the comb jellies, which are fragile, luminescent, biradially symmetrical marine animals.

VII. Phylum Platyhelminthes includes the planarians, the flukes, and the tapeworms.
   A. Flatworms are acoelomate animals characterized by bilateral symmetry, cephalization, three definite tissue layers, well-developed organs, a simple brain and nervous system, and protonephridia.
   B. The flukes and tapeworms are specialized for a parasitic life style.

VIII. Members of phylum Nemertea (proboscis worms) have a tube-within-a-tube body plan, a complete digestive tract with mouth and anus, and a separate circulatory system.

IX. Phylum Nematoda, composed of the roundworms, includes species of great ecological importance and species that are parasitic in plants and animals.
   A. Nematodes are characterized by three definite tissue layers, a pseudocoelom, bilateral symmetry, and a complete digestive tract.
   B. Nematodes parasitic in humans include *Ascaris,* hookworms, trichina worms, and pinworms.

X. Members of phylum Rotifera are aquatic, pseudocoelomate, microscopic animals that exhibit cell constancy.

SELECTED KEY TERMS

acoelomate
anterior
auricle
bilateral symmetry
cephalization
Cnidaria
cnidocyte
coelom
collar cells
Ctenophora
cuticle
deuterostome
dorsal
ectoderm
endoderm
flame cells
ganglion
gastrovascular cavity
germ layers
hermaphroditic
invertebrate
medusa
mesoderm
nematocyst
Nematoda
Nemertea
nerve net
Platyhelminthes
polyp
Porifera
posterior
protonephridia
protostome
pseudocoelom
radial symmetry
Rotifera
sessile
spicules
tube-within-a-tube body plan
ventral

POST-TEST

1. Animals without backbones are properly referred to as __________ .
2. Animals that remain stationary and attached to a substrate are described as __________ .
3. __________ animals lack a body cavity.
4. The germ layer that gives rise to the outer covering of the body and the nervous system is __________ .
5. A true coelom is completely lined with __________ .
6. In __________ animals, the blastopore develops into the mouth.
7. __________ symmetry is characteristic of cnidarians.
8. If the first four cells of a(an) __________ embryo are separated, each cell will develop into one-quarter of a larva.
9. The "belly" side of an animal is its __________ surface.
10. Sponges have skeletal elements called __________ .
11. In sponges, a water current is created by the __________ cells.
12. Two body forms found among cnidarians are the __________ and the __________ .
13. The sea anemones and corals belong to phylum __________ .
14. The simplest animals to exhibit cephalization are in phylum __________ .
15. The __________ of a planarian flatworm serve as organs of smell.
16. Phylum __________ contains the simplest animals to have a circulatory system.
17. Members of Phylum __________ are pseudocoelomate and cell constant.
18. Comb jellies belong to Phylum __________ .
19. Tapeworms belong to Phylum __________ .
20. Hookworms belong to Phylum __________ .

REVIEW QUESTIONS

1. For centuries sponges were classified as plants. Justify their current classification as animals.
2. Why is the sea a more hospitable environment for many animals than the land or fresh water?
3. As compared to phylum Cnidaria, what advances do members of phylum Platyhelminthes exhibit? In what ways are these animals alike?
4. What are the advantages of bilateral symmetry and cephalization?
5. Identify the phyla (from among those studied in this chapter) that have the following characteristics: (a) radial symmetry; (b) protonephridia; (c) acoelomate; (d) pseudocoelomate; (e) alternation of sexual and asexual stages; (f) cnidocytes; (g) nerve net; (h) digestive tract with an opening at each end; (i) circulatory system.
6. Describe the alternation of stages exhibited by *Obelia.*
7. How do flatworms survive without specialized structures for gas exchange and internal transport of materials?
8. What special adaptations do tapeworms have for their parasitic mode of life?
9. Describe the life cycles of the following animals: (a) beef tapeworm; (b) *Ascaris*; (c) hookworm.
10. Label the following diagram.

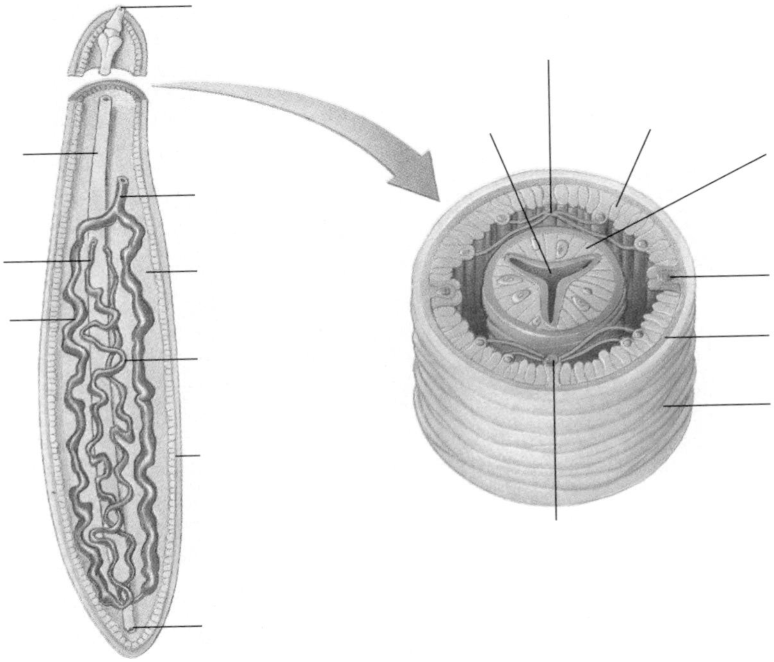

YOU MAKE THE CONNECTION

1. Every evolutionary adaptation has both costs and benefits. Explain this concept in terms of each of the following: (a) cephalization; (b) circulatory system; (c) hermaphroditism in tapeworms.
2. Several international monitoring projects are gathering data needed to help us understand coral reef destruction. What are some reasons that it is important to take action to protect coral reefs?

RECOMMENDED READINGS

Brown, B.E. and J.C. Ogden. "Coral Bleaching." *Scientific American,* Vol. 268, No. 1, January 1993. Environmental stressors can cause permanent damage to coral reefs.

Brusca, R.C. and G.J. Brusca. *Invertebrates.* Sinauer Associates, Inc., Sunderland, Massachusetts, 1990. The authors have combined taxonomic and systems approaches in this comprehensive survey of the invertebrate phyla.

Dorit, R.L., W.F. Walker, and R. Barnes. *Zoology.* Saunders College Publishing, Philadelphia, 1991. A readable, interesting account of animal diversity.

Fabricius, K.E., Y. Benayahu, and A. Genin. "Herbivory in Asymbiotic soft corals," *Science,* Vol. 268, April 7, 1995. A group of soft corals feed on phytoplankton.

Knoll, A.H. "End of the Proterozoic Eon." *Scientific American,* October 1991. A discussion of environmental change that may have led to the evolution of large animals.

Ruppert, E.E. and R.D. Barnes. *Invertebrate Zoology,* 6th ed. Saunders College Publishing, Philadelphia, 1994. This comprehensive textbook discusses the life processes of each invertebrate phylum.

C H A P T E R 29

The Animal Kingdom: The Coelomate Protostomes

The green darner is a large, powerful dragonfly that eats a variety of smaller insects. This dragonfly is covered with dew. (Rod Planck/Dembinsky Photo Associates)

What does an earthworm have in common with a clam, a spider, and a butterfly? All of these animals have a coelom, and all are protostomes. The coelomate protostomes include the annelids, mollusks, and arthropods, as well as several smaller, related phyla. As explained in Chapter 28, the **coelom** is a space completely lined by mesoderm that lies between the digestive tube and the outer body wall. Recall also that protostomes are animals in which the mouth develops from the blastopore, the first opening that forms in the embryonic gut. Coelomate protostomes have a complete digestive tract with separate mouth and anus, and most have well-developed circulatory, excretory, and nervous systems.

Animals with a coelom, and to a lesser extent those with a pseudocoelom, have certain advantages over those lacking a body cavity. The coelom permits a clear separation between the muscles of the body wall and those in the wall of the digestive tract. This allows the digestive tube to move food along independently of body movements. Another benefit is that the coelom can be used as a **hydrostatic skeleton,** an enclosed compartment (or series of compartments) of fluid that can be manipulated by surrounding muscles.

In some animals, fluid within the coelom helps transport materials such as food, oxygen, and wastes. Cells bathed by the coelomic fluid can exchange materials with it. The cells receive nutrients and oxygen from the coelomic fluid and excrete wastes into it. Some coelomates have excretory structures that remove wastes directly from the coelomic fluid.

The coelom also serves as a space in which many organs develop and function. For example, the heart and blood vessels develop within the coelom and can function freely there without being squeezed by other organs. The pumping action of the heart would not be possible without the surrounding space provided by the coelom. The coelom also provides space for the gonads to develop; in animals with breeding seasons, these reproductive structures enlarge periodically as they fill with ripe gametes.

LEARNING OBJECTIVES

After you have studied this chapter you should be able to

1. Identify several advantages of having a coelom.
2. Identify challenges associated with terrestrial living and describe adaptations that enable terrestrial animals to meet those challenges.
3. Describe the distinguishing characteristics of mollusks, annelids, and arthropods, and properly classify an animal that belongs to any of these phyla.
4. Describe the classes of mollusks discussed and give examples of animals that belong to each.
5. Describe and give examples of each of the three classes of annelids discussed.
6. Distinguish among the subphyla and classes of arthropods and give examples of animals that belong to each group.
7. Discuss factors that have contributed to the great biological success of insects.

LIFE ON LAND REQUIRES MANY ADAPTATIONS

Based on the fossil record, most biologists agree that the first air-breathing land animals were scorpion-like arthropods that came ashore in the Silurian period, which began about 438 million years ago. The first land vertebrates, the amphibians, did not appear until the latter part of the Devonian period, about 30 million years later.

Many modern invertebrate coelomates still inhabit the sea. The earthworm is a terrestrial animal, but most annelids are marine. A few snails inhabit the land, but most mollusks also live in the sea. Among the arthropods, the crustaceans (crabs, lobsters, and their relatives) and the merostomes (horseshoe crabs) are also largely marine forms. However, certain modern arthropods—including the insects and spiders—are very successful terrestrial animals.

The chief problem facing all terrestrial organisms is that of drying out in the absence of a surrounding watery medium. A body covering adapted to minimize fluid loss helps solve this problem in many land animals. Location of the respiratory surface deep within the animal also helps prevent fluid loss. Thus, although gills are located externally, lungs and tracheal tubes (found in insects) are internal.

Supporting the body against the pull of gravity in the absence of the buoyant effect of water is another problem associated with life on land. Some animals, such as earthworms, do not share this challenge because they have small bodies and live in the ground. Larger burrowing animals and those living on Earth's surface generally need some sort of supporting skeleton. Arthropods and most mollusks have a tough **exoskeleton,** a supporting armor that covers the body. The vertebrates have an **endoskeleton,** a supporting framework within the body.

Reproduction on land poses still another problem. Aquatic forms generally shed their gametes in the water, where fertilization occurs. The surrounding water serves as an effective shock absorber, protecting the delicate embryos as they develop. Some land animals, including most amphibians, return to the water for reproduction; their larval forms develop in the water. Earthworms, snails, insects, reptiles, birds, and mammals engage in internal fertilization. They transfer sperm from the body of the male directly into the body of the female by copulation. The sperm are surrounded by a watery medium. An important adaptation to reproduction on land is the tough, protective shell secreted by the females of many species around the eggs, preventing the developing embryo from drying out. An alternative adaptation to terrestrial reproduction is the development of the embryo within the moist body of the mother.

MOLLUSKS HAVE A FOOT, VISCERAL MASS, AND MANTLE

Mollusks are among the best known of the invertebrates; most of us have walked along the shore collecting their shells. With more than 50,000 living species and 35,000 fossil species (second only to the arthropods in number), phylum **Mollusca** includes clams, oysters, octopods, snails, slugs, and the largest of all the invertebrates, the giant squid, which may weigh several tons. Although most mollusks are marine, some snails and clams live in fresh water and many species of snails and slugs inhabit the land. Representative mollusks are illustrated in Figure 29–1. Four of the eight recognized classes are discussed in this chapter and are listed in Table 29–1.

Although mollusks vary widely in outward appearance, most share certain basic characteristics:

1. A soft body, usually covered by a dorsal shell.
2. A broad, flat, muscular **foot,** located ventrally, which can be used for locomotion.
3. The body organs (viscera) concentrated as a **visceral mass** located above the foot.

(Text continues on page 629)

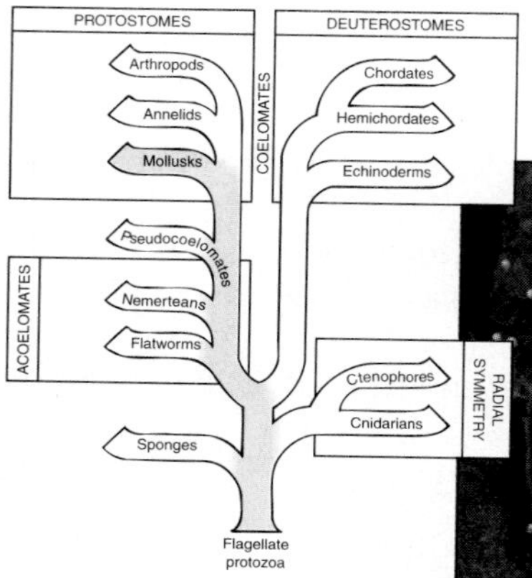
PROTOSTOMES
DEUTEROSTOMES
COELOMATES
Arthropods
Annelids
Mollusks
Chordates
Hemichordates
Echinoderms
Pseudocoelomates
ACOELOMATES
Nemerteans
Flatworms
Ctenophores
Cnidarians
RADIAL SYMMETRY
Sponges
Flagellate protozoa

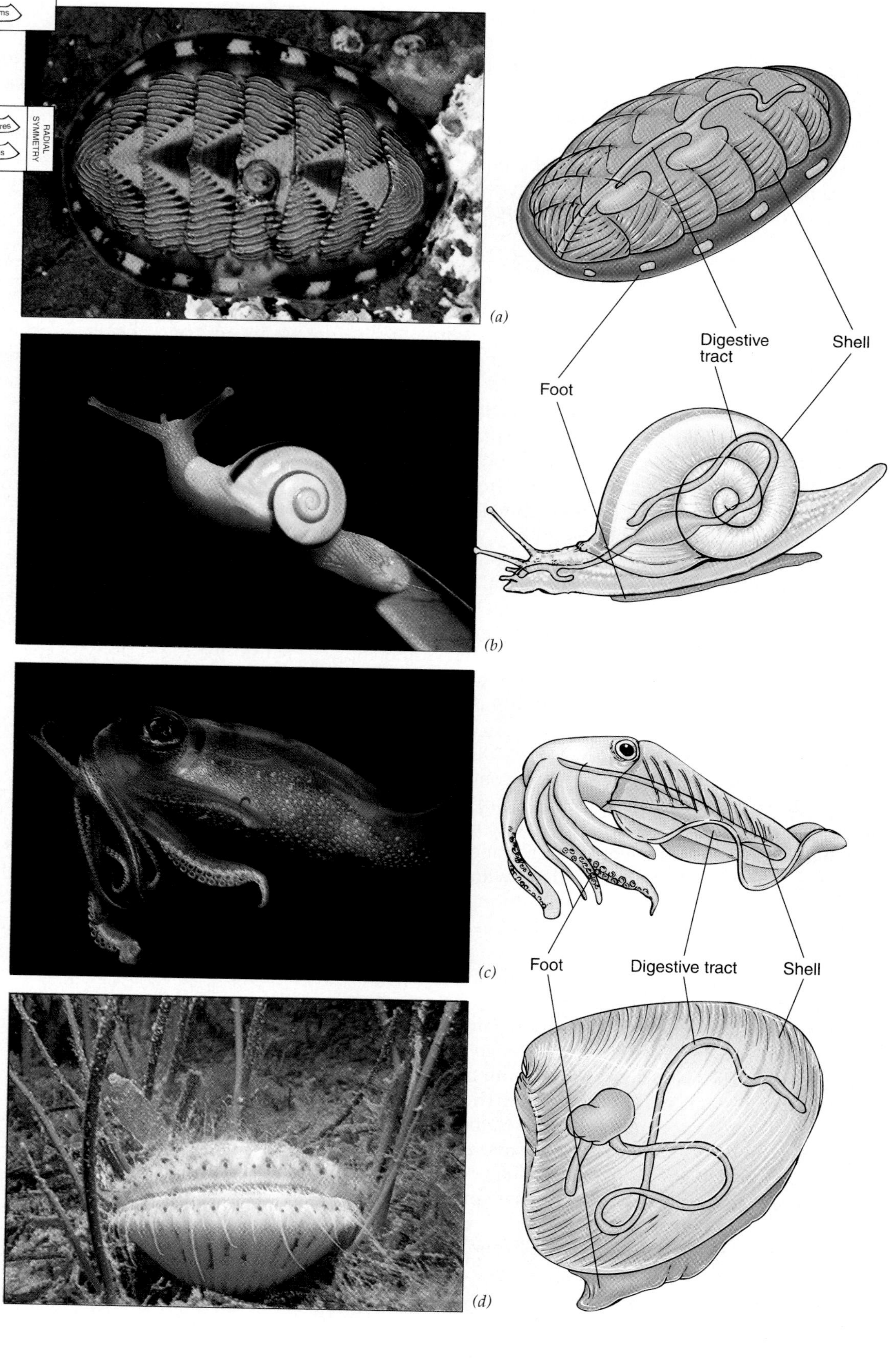
(a)
Foot
Digestive tract
Shell
(b)
(c)
Foot
Digestive tract
Shell
(d)

◀ **FIGURE 29–1 The basic molluskan body plan is varied in chitons, gastropods (snails and their relatives), bivalves (clams, scallops, and their relatives), and cephalopods (squids, octopods, and their relatives).** Note how the foot, shell, and digestive tract have changed their positions in the evolution of the several classes. (*a*) Chitons are sluggish marine animals with shells composed of eight overlapping plates. This sea cradle chiton *(Tonicella lineata)* inhabits coastal waters off the Pacific Northwest. (*b*) The banded garden snail *(Cepaea nemoralis)* photographed in the lowland tropical rain forest. This terrestrial snail inhabits many parts of the world. (*c*) The squid, an example of a cephalopod, can change color to blend with its background. This specimen *(Sepioteuthis lessoniana)* is native to Hawaii. (*d*) A bay scallop *(Argopecten irradians)* photographed in a sea grass bed in Tampa Bay. (*a*, Visuals Unlimited/Kjell B. Sandved; *b*, William E. Ferguson; *c*, Mike Severns/Tom Stack & Associates; *d*, Robin Lewis/Coastal Creations)

4. A **mantle,** a heavy sheet of tissue that covers the visceral mass and usually contains glands that secrete a shell. The mantle generally overhangs the visceral mass, forming a mantle cavity, which may contain gills and other structures.
5. A rasplike structure called the **radula,** which is a belt of teeth in the mouth region. (The radula is not present in clams and their relatives.)
6. The coelom is generally reduced to small compartments around certain organs including the heart and nephridia, and the main body cavity is typically a hemocoel (see discussion of open circulatory system that follows).

All of the organ systems typical of complex animals are present in the mollusks. The digestive system is a tube, sometimes coiled, consisting of a mouth, buccal cavity (mouth cavity), esophagus, stomach, intestine, and anus. The radula, located within the buccal cavity, can be projected out of the mouth and used to scrape particles of food from the surface of rocks or the ocean floor. Sometimes the radula is used to drill a hole in another animal's shell or to tear off pieces of a plant.

Most mollusks have an **open circulatory system** in which the blood bathes the tissues directly. (In a closed circulatory system the blood flows within a complete circuit of blood vessels.) In mollusks, the heart pumps blood into a single blood vessel, the **aorta,** which may branch into other vessels. Eventually blood flows into a network of large spaces called sinuses, where the tissues are bathed directly; this network makes up the **hemocoel,** or blood cavity. From the sinuses, blood drains into vessels that conduct it to the gills, where it is recharged with oxygen. From the gills, the blood returns to the heart. Thus, blood flow in a mollusk follows the pattern

heart → aorta → smaller blood vessels
→ blood sinuses → gills → heart

Table 29–1 MAJOR CLASSES OF PHYLUM MOLLUSCA

Class and Representative Animals	*Characteristics*
Polyplacophora Chitons	Primitive marine animals with segmented shells; shell consists of eight separate transverse plates; head reduced; broad foot used for locomotion.
Gastropoda Snails Slugs Nudibranchs	Marine, freshwater, or terrestrial; body and shell coiled in some species; well-developed head with tentacles and eyes.
Bivalvia Clams Oysters Mussels	Marine and freshwater; body laterally compressed; two-part shell hinged dorsally; hatchet-shaped foot; suspension-feeders.
Cephalopoda Squids Octopods	Marine; predatory; foot divided into tentacles; usually bearing suckers; well-developed eyes.

In open circulatory systems, blood pressure tends to be low, and tissues are not very efficiently oxygenated. However, most mollusks are sluggish animals with low metabolic rates, and so this type of circulatory system is adequate. In the active cephalopods (the class that includes the squids and octopods), the circulatory system is closed.

In mollusks the sexes are usually separate, with fertilization often taking place in the surrounding water. Most marine mollusks pass through one or more larval stages. The first larval stage is typically a **trochophore larva,** a free-swimming, ciliated, top-shaped larva characteristic of mollusks and annelids (Fig. 29–2). In most of the mollusk classes, the trochophore larva develops into a **veliger larva,** which has a shell and foot. The veliger larva is unique to the mollusks.

Similarities in the development of mollusks and annelids — the process of spiral cleavage (Chapter 28) and the trochophore larva — suggest that these two phyla may have had a common protostome ancestor. Mollusks probably evolved early in the protostome clade before the origin of **metamerism,** the repetition of internal and external structures in a series of body segments. Metamerism, which may include the repetition of muscles, nerves, blood vessels, and other organs, is characteristic of the annelids and arthropods.

Class Polyplacophora includes the chitons

Class **Polyplacophora** (meaning "many plates") comprises the **chitons,** sluggish marine animals with flattened

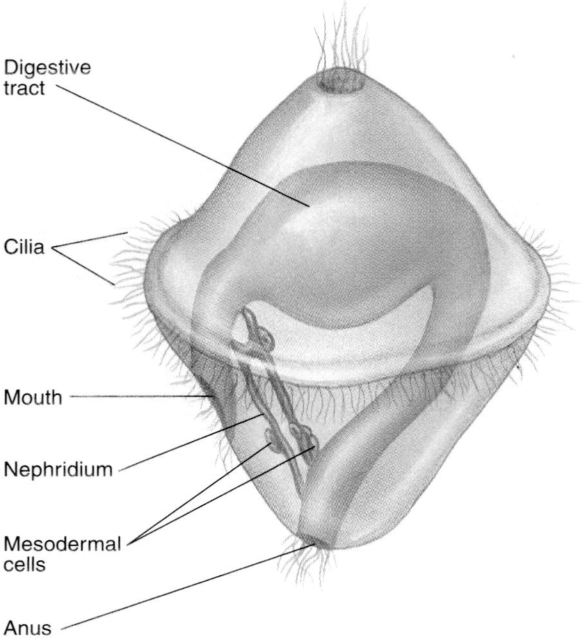

FIGURE 29–2 The trochophore larva is the first larval stage of a marine mollusk. This type of larva is also characteristic of annelids. Note the characteristic ring of ciliated cells just above the mouth.

bodies (Fig. 29–1). Their most distinctive feature is a shell composed of eight separate but overlapping dorsal plates. The head is reduced in this class, and there are no eyes or tentacles.

The chiton inhabits rocky intertidal zones, using its broad, flat foot for locomotion and also to hold firmly onto rocks. In addition, the chiton can press its mantle against the substratum and lift the inner edge of the mantle to produce a partial vacuum. The resulting suction enables the animal to adhere powerfully to its perch. Using its radula for grazing, the chiton scrapes algae and other small organisms off rocks and shells.

Gastropods are the largest group of mollusks

Class **Gastropoda,** which includes the snails and slugs and their relatives, is the largest and most diverse group of mollusks (Fig. 29–3). In fact, gastropods comprise the second largest class in the animal kingdom—second only to the insects. Most gastropods inhabit marine waters, but others make their homes in brackish water, fresh water, or terrestrial areas. Many gastropods have a well-developed head with tentacles. Two simple eyes may be located on stalks that extend from the head. The broad, flat foot is used for creeping.

Most land snails do not have gills; instead, the mantle is highly vascularized and functions as a lung. These garden snails and slugs are described as **pulmonate** (meaning "having a lung").

We think of snails as having a single, spirally coiled shell into which they can withdraw the body, and many do. However, other gastropods, such as limpets, have shells like flattened dunce caps. Still others, such as garden slugs and the beautiful marine snails known as **nudibranchs,** have no shell at all (Fig. 29–3).

A unique feature of gastropods is **torsion,** a twisting of the visceral mass. (This twisting is unrelated to the coiling of the shell.) As the bilateral larva develops, one side of the visceral mass grows more rapidly than the other side. This uneven growth results in rotation of the visceral mass. The visceral mass and mantle twist permanently 90 to 180 degrees, relative to the head. As a result, the digestive tract becomes somewhat U-shaped, and the

(a) *(b)*

FIGURE 29–3 Gastropods are the second largest class in the animal kingdom. (***a***) The garden snail *(Helix aspersa)* feeding on a composite flower. (***b***) Porter's nudibranch *(Chromodoris porterae)* feeding on a lightbulb tunicate *(Clavelina huntsmani).* (*a,* Visuals Unlimited/Kjell B. Sandved; *b,* Animals Animals © 1996 Bruce Watkins) anus comes to lie above the head and gill (Fig. 29–4). Subsequent growth is dorsal and usually in a spiral coil. Torsion limits space in the body, and typically the gill, metanephridium (kidney), and gonad are absent on one side.

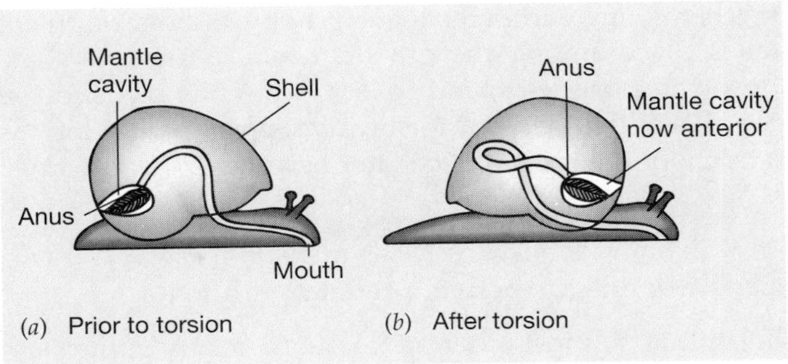

FIGURE 29–4 Embryonic torsion is shown in the gastropod *Acmaea* (a limpet). As the bilateral larva develops, the visceral mass twists 180 degrees relative to the head.

Bivalves typically burrow in the mud

Class Bivalvia includes the clams and oysters and their relatives. The soft body of members of class **Bivalvia** is laterally compressed and completely enclosed by a two-part shell that hinges dorsally and opens ventrally (Fig. 29–5). This arrangement allows the hatchet-shaped foot to protrude ventrally for locomotion and for burrowing in the mud. Large, strong adductor muscles attached to the shell permit the animal to close its shell.

The inner pearly layer of the bivalve shell is made of calcium carbonate secreted in thin sheets by the epithelial cells of the mantle. Known as *mother-of-pearl,* this material is valued for making jewelry and buttons. Should a bit of foreign matter lodge between the shell and the mantle, the epithelial cells covering the mantle may be stimulated to secrete concentric layers of calcium carbonate around the intruding particle. Many species of oysters and clams form pearls in this way.

Some bivalves, such as oysters, attach permanently to the substratum. Others, like clams, burrow slowly through rock or wood, seeking protected dwellings. The shipworm, *Teredo,* which damages dock pilings and other marine installations, is just looking for a home. Finally, some bivalves, such as scallops, swim rapidly by clapping their two shells together with the contraction of a large adductor muscle (the part of the scallop that is eaten by humans).

Clams and oysters are suspension feeders that trap food particles in sea water. They take sea water in through an extension of the mantle called the incurrent siphon. Water leaves by way of an excurrent siphon. As the water passes over the gills, food particles in the water are trapped in mucus secreted by the gills. Cilia move the food to the mouth. An oyster can filter about 3 liters of sea water per hour. Because bivalves are suspension feeders, they have no need for a radula, and indeed they are the only group of mollusks that lack this structure.

Most bivalves have two distinct sexes. In some marine and nearly all freshwater bivalves, sperm are shed into the water and fertilize the eggs within the mantle cavity of the female. In these species, the female also car-

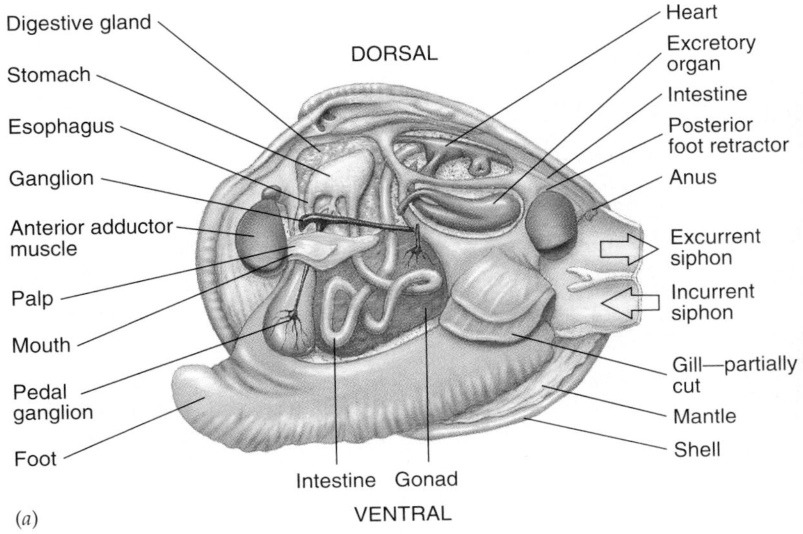

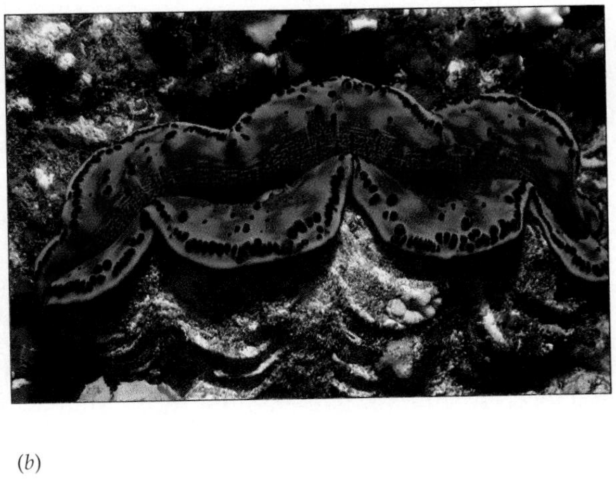

FIGURE 29–5 Bivalves are enclosed by two shells that hinge dorsally and open ventrally. (*a*) Internal anatomy of a clam. (*b*) The giant clam (*Tridacna* sp.) photographed in the Red Sea. The mantle extends over the shell fluting. (*b*, Jeff Rotman/Peter Arnold, Inc.)

ries her young within the mantle cavity and development takes place among the gill filaments. Larvae of some freshwater species spend several weeks as parasites on the gills of fishes. A marine bivalve typically develops as a trochophore larva, which later becomes a veliger larva.

Cephalopods are active, predatory animals

In contrast to most other mollusks, members of the class **Cephalopoda** (meaning "head-feet") are fast-swimming predatory animals. The cephalopod foot is divided into tentacles: ten in squids, eight in octopods, and as many as 90 in the nautilus. The tentacles, or arms, surround the central mouth of the large head (Fig. 29–1*c*). Cephalopods have large, well-developed eyes that form images. Although they develop differently, the eyes are structurally somewhat like vertebrate eyes and function in much the same way. The octopus has no shell, and the shell of the squid is greatly reduced.

Nautilus has a flat, coiled shell consisting of many chambers built up over time. Each year the animal lives in the newest and largest chamber of the series. By secreting a gas resembling air into the other chambers, the *Nautilus* is able to regulate its depth in the water.

The tentacles of squids and octopods are covered with suckers for seizing and holding prey. In addition to a radula, the mouth is equipped with two strong, horny beaks used to kill prey and tear it to bits. The mantle is thick and muscular and fitted with a funnel-like structure. By filling the cavity with water and ejecting it through the funnel, the animal achieves rapid jet propulsion.

Besides its speed, the cephalopod has developed two other important mechanisms that enable it to escape from its predators, which include certain whales and moray eels. One is its ability to confuse the enemy by rapidly changing colors. By expanding and contracting pigment cells in its skin, the cephalopod can display an impressive variety of mottled colors. Another defense mechanism is its ink sac, which produces a thick black liquid that is released in a dark cloud when the animal is alarmed. While its enemy pauses, temporarily blinded and confused, the cephalopod easily escapes. The ink has been shown to inactivate the chemical receptors of some predators, rendering them incapable of detecting their prey.

The octopus feeds on crabs and other arthropods, catching and killing them with a poisonous secretion of its salivary glands. During the day, the octopus usually hides among the rocks; in the evening, it emerges to hunt for food. Its motion is incredibly fluid, giving little hint of the considerable strength in its eight arms.

Small octopods survive well in aquaria and have been studied extensively. They have a relatively high degree of intelligence and can make associations among stimuli. Their very adaptable behavior more closely resembles that of the vertebrates than the more stereotypical patterns of behavior seen in other invertebrates.

ANNELIDS HAVE SEGMENTED BODIES

Members of **phylum Annelida,** the segmented worms, have bilateral symmetry and a tubular body that may be partitioned into more than 100 ringlike segments. This phylum, composed of about 15,000 species, includes three main classes: the polychaetes, a group of marine and freshwater worms; the earthworms; and the leeches (Table 29–2 and Fig. 29–6).

The term *Annelida,* meaning ringed, refers to the series of rings, or segments, that make up the annelid body. Both the body wall and many of the internal organs are segmented. The segments are separated from one another internally by transverse partitions called **septa.** The bilaterally symmetrical, tubular body may consist of about 100 segments. Some structures, such as the digestive tract and certain nerves, extend the length of the body, passing through successive segments. Other structures, such as nephridia, are repeated in each segment.

One advantage of segmentation is that it facilitates locomotion. The coelom is divided into segments, and each segment has its own muscles. This arrangement allows the animal to elongate one part of its body while shortening another part. (The annelid's hydrostatic skeleton, which enables it to do this, is discussed in Chapter 38.) Bristle-like structures called **setae** (sing., *seta*), located on each segment, anchor the worm to the ground while the animal contracts its muscles (first longitudinal muscles, then circular ones) to move along.

Segmentation is very important from an evolutionary perspective because it provides the opportunity for specialization of body regions. In many annelids the indi-

Table 29–2 MAIN CLASSES OF PHYLUM ANNELIDA

Class and Representative Animals	*Characteristics*
Polychaeta Sandworms Tubeworms	Mainly marine; each segment bears a pair of parapodia with many setae; well-developed head; separate sexes; trochophore larva.
Oligochaeta Earthworms	Terrestrial and freshwater worms; few setae per segment; lack well-developed head; hermaphroditic.
Hirudinea Leeches	Most are blood-sucking parasites that inhabit fresh water; lack appendages and setae; prominent muscular suckers.

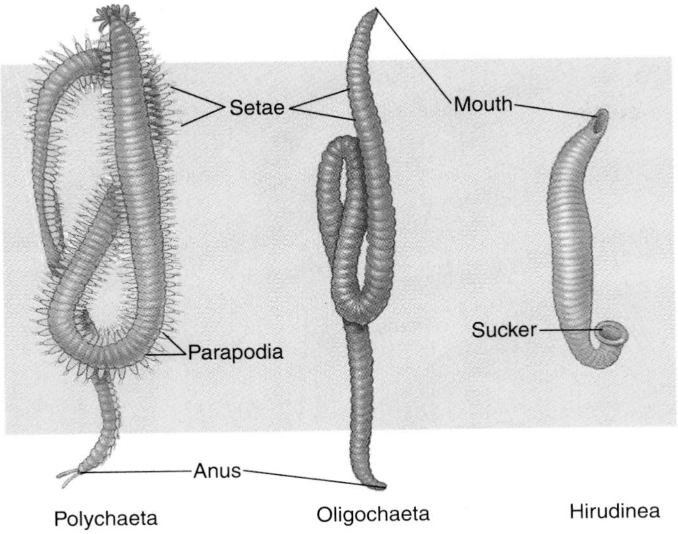

FIGURE 29–6 There are three major classes of annelids. Polychaetes are marine worms with paddle-shaped appendages called parapodia. Oligochaetes, a class that includes the earthworms, inhabit fresh water and moist terrestrial areas. Many leeches, members of class Hirudinea, are blood-sucking parasites equipped with suckers.

vidual segments are almost all alike, but in many segmented animals (arthropods and chordates), different segments and groups of segments are specialized to perform different functions. In some groups, the specialization is so pronounced that the basic segmentation of the body plan may not be apparent. (For example, in humans, although segmentation of the body is not obvious, muscles and nerves are segmentally organized.)

Annelids have a well-developed coelom, a closed circulatory system, and a complete digestive tract extending from mouth to anus. Respiration takes place through the skin or by gills. Typically, a pair of excretory structures called **metanephridia** is found in each segment. The nervous system generally consists of a simple brain composed of a pair of ganglia and a double ventral nerve cord. A pair of ganglia and lateral nerves are repeated in each segment.

Polychaetes have parapodia

Class **Polychaeta** includes marine worms that swim freely in the sea, burrow in the mud near the shore, or live in tubes secreted by the animal or made by cementing bits of shell and sand together with mucus (Fig. 29–7). Each body segment typically bears a pair of paddle-shaped appendages called **parapodia** (sing., *parapodium*) that function in locomotion and in gas exchange. These fleshy structures bear many stiff setae (the name *Polychaeta* means "many bristles"). Most polychaetes have a well-developed head bearing eyes and antennae. The head may also be equipped with tentacles and palps (feelers). Polychaetes develop from free-swimming trochophore larvae similar to those of mollusks.

Many polychaete species have evolved behavioral patterns that ensure fertilization. By responding to cer-

(*a*)

(*b*)

FIGURE 29–7 Polychaete annelids are marine worms. (***a***) The West Indies fireworm *(Hermodice carunculata)* is a crawling polychaete with poisonous setae. (***b***) The Christmas tree worm *(Spirobranchus giganteus)* photographed in a Florida coral reef. (*a,* Charles Seaborn/Odyssey Productions, Chicago; *b,* James H. Carmichael, Coastal Creations)

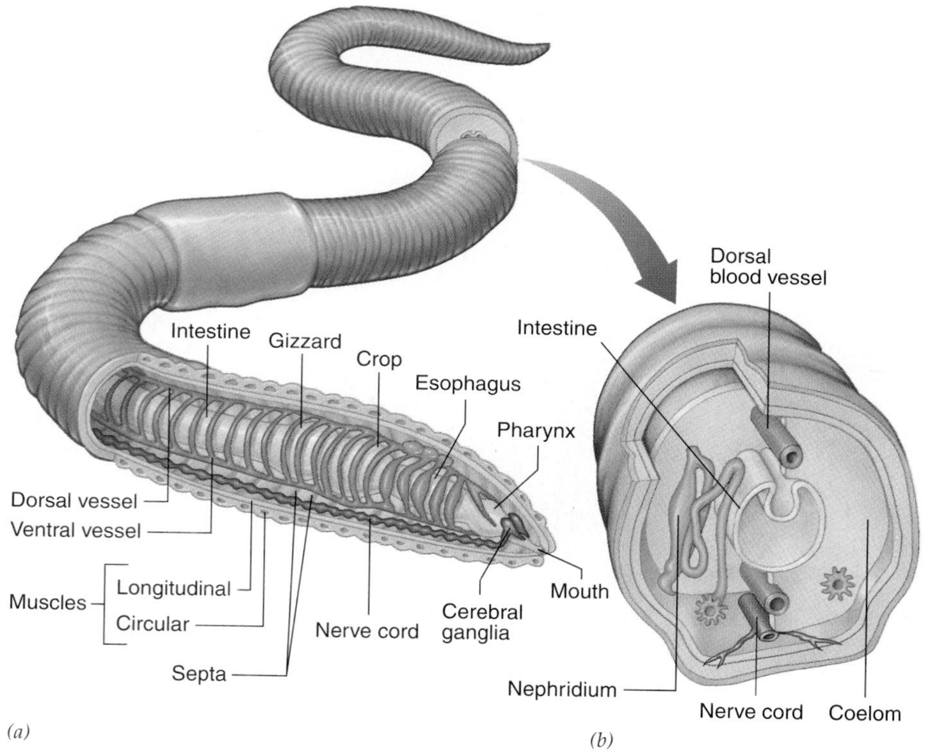

FIGURE 29–8 The earthworm exhibits the characteristic annelid segmented body plan. (*a*) The internal structure has been exposed at the anterior end of an earthworm. (*b*) Cross section of the earthworm.

tain rhythmic variations, or cycles, in the environment, nearly all of the females and males of a given species release their gametes into the water at the same time. For example, more than 90% of reef-dwelling *Palolo* worms of the South Pacific shed their eggs and sperm within a single 2-hour period on one night of the year. In this animal the seasonal rhythm limits the reproductive period to November; the lunar rhythm limits it to a day during the last quarter of the moon when the tide is unusually low; and the time of day limits it to a few hours just after complete darkness. The posterior half of the *Palolo* worm, loaded with gametes, actually breaks off from the rest. It swims backward to the surface, and eventually bursts, releasing the eggs or sperm so that fertilization can occur. Local islanders eagerly await this annual event when they can gather up great numbers of the swarming polychaetes and broil them for dinner.

Earthworms are familiar annelids

The 3000 or so species of the class **Oligochaeta** are found almost exclusively in fresh water and in moist terrestrial habitats. These worms lack parapodia, have few bristles per segment (the name *Oligochaeta* means "few bristles"), and lack a well-developed head. All oligochaetes are hermaphroditic.

Lumbricus terrestris, the common earthworm, is about 20 centimeters (8 inches) long. Its body is divided into more than 100 segments separated externally by grooves and internally by septa (Fig. 29–8). The mouth is located in the first segment, the anus in the last. The earthworm's body is protected from drying by a thin, transparent cuticle, secreted by the cells of the epidermis. Mucus secreted by glandular cells of the epidermis forms an additional protective layer over the body surface. The body wall has an outer layer of circular muscles and an inner layer of longitudinal muscles.

An earthworm literally eats its way through the soil, ingesting its own weight in soil and decaying vegetation every 24 hours. During this process the soil is turned, aerated, and enriched by nitrogenous wastes from the earthworm. This is how earthworms enhance the formation and maintenance of fertile soil. The earthworm's soil meal, containing nutritious decaying vegetation, is processed in its complex digestive system.

Food is swallowed through the muscular **pharynx** and passes through the **esophagus** to the **stomach,** which consists of two parts: a thin-walled **crop** where food is stored, and a thick-walled, muscular **gizzard** where food is ground to bits. The rest of the digestive system is a long, straight **intestine** where food is digested and absorbed. Wastes pass out of the intestine to the exterior through the **anus.**

The efficient, closed circulatory system consists of two main blood vessels that extend longitudinally. The dorsal blood vessel, just above the digestive tract, collects blood from vessels in the segments, then contracts to pump the blood anteriorly. In the region of the esophagus, five pairs of blood vessels propel blood from the dorsal to the ventral blood vessel. Located just below the digestive tract, the ventral blood vessel carries blood posteriorly. Small vessels branch from it and deliver blood to the various structures in each segment as well as to the body wall. Within these structures blood flows through very tiny vessels (capillaries) before returning to the dorsal blood vessel.

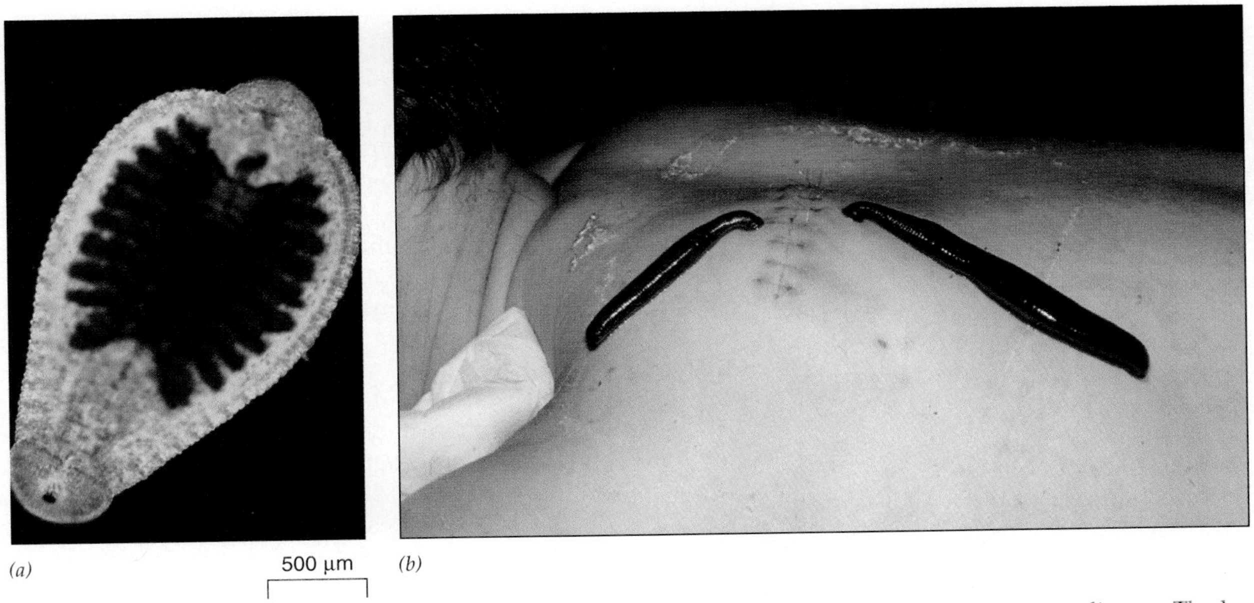

FIGURE 29–9 Most leeches are blood-sucking parasites. (***a***) A blood-sucking leech *(Helobdella stagnalis)* that feeds on mammals. The dark area within its swollen body is recently ingested blood. (***b***) The medicinal leech *(Hirudo medicinalis)* is used to treat hematoma, an accumulation of blood within body tissues that results from injury or disease. The leech releases an anticoagulant, called hirudin, that prevents the blood from clotting and dissolves existing clots. (*a*, Visuals Unlimited/T. E. Adams; *b*, St. Bartholomew's Hospital/Science Photo Library/Researchers, Inc.)

Gas exchange takes place through the moist skin. Oxygen is transported by the respiratory pigment hemoglobin in the blood plasma. The excretory system consists of paired metanephridia, repeated in almost every segment of the body. Each metanephridium consists of a ciliated funnel opening into the next anterior coelomic cavity and connected by a tube to the outside of the body (Fig. 29–8). Wastes are removed from the coelomic cavity partly by the beating of the cilia and partly by currents set up by the contraction of muscles in the body wall. A capillary network surrounding the metanephridium reabsorbs usable materials from the coelomic fluid in the tube. The metanephridia, open at both ends, are quite different from the protonephridia of the flatworms, which are blind tubules that open only to the exterior.

The nervous system consists of a pair of **cerebral ganglia** that serve as a brain, just above the pharynx, and a subpharyngeal ganglion, just below the pharynx. A ring of nerve fibers connects these ganglia. From the lower ganglion a double ventral nerve cord extends beneath the digestive tract to the posterior end of the body. In each segment along the nerve cord, there is a pair of fused segmental ganglia. Nerves extend laterally from the segmental ganglia to the muscles and other structures of that segment. The segmental ganglia coordinate the contraction of the muscles of the body wall, so that the worm can creep along.

Like other oligochaetes, earthworms are hermaphroditic. During copulation, the worms exchange sperm. Two worms, headed in opposite directions, press their ventral surfaces together. These surfaces become glued together by the thick mucous secretions of each worm's clitellum, a thickened ring of epidermis. Sperm from both worms pass posteriorly and are then transferred and stored in the seminal receptacles of the other worm. The worms then separate. A few days later each clitellum secretes a membranous cocoon containing a sticky fluid. As the cocoon is slipped over each worm's head, eggs are laid into it. Sperm are added as the cocoon passes over the seminal receptacles. When the cocoon is free, its openings constrict so that a spindle-shaped capsule is formed, and the fertilized eggs develop into tiny worms within the cocoon without passing through a larval stage. This complex reproductive pattern is an adaptation to terrestrial life.

The leeches belong to class Hirudinea

Most leeches, members of the class **Hirudinea,** inhabit fresh water, but some live in the sea or in moist areas on land. About 75% of the known species of leeches are blood-sucking parasites. Some leeches are nonparasitic predators that capture small invertebrates such as earthworms or snails. Leeches differ from other annelids in having neither setae nor appendages.

Leeches have muscular suckers at both anterior and posterior ends. Most parasitic leeches attach themselves to a vertebrate host, bite through the skin, and suck out a quantity of blood, which is stored in pouches in the digestive tract. Hirudin, an anticoagulant secreted by glands in its crop, ensures leeches a full meal of blood. Their meals may be infrequent, but they can store enough food from one meal to last a long time.

Leeches have been used since ancient times for drawing blood from areas swollen by poisonous stings and bites. During the 19th century they were widely used to remove "bad blood," thought to be the cause of many diseases. Recently, leeches have been employed by modern medicine to remove blood that accumulates within body tissues as a result of injury, disease, or surgery (Fig. 29–9).

The leech attaches its sucker near the site of injury, makes an incision, and deposits hirudin. The hirudin prevents the blood from clotting and dissolves already-existing clots. In 30 minutes, a leech can suck out as much as ten times its own weight in blood.

ARTHROPODS HAVE JOINTED APPENDAGES AND AN EXOSKELETON OF CHITIN

Biologists consider **arthropods** the most diverse and biologically successful group of animals (Fig. 29–10). **Phylum Arthropoda** includes about one million species, and arthropods live in a greater range of habitats than the members of any other phylum. The following adaptations have contributed importantly to their success:

1. **Paired, jointed appendages,** from which they get their name (arthropod means "jointed foot"). These appendages function as swimming paddles, walking legs, mouth parts, or accessory reproductive organs for transferring sperm.
2. A hard, armor-like **exoskeleton,** composed of chitin and protein, that covers the entire body and appendages. The exoskeleton serves as a coat of armor protecting against predators and also against excessive loss of moisture. It also gives support to the underlying soft tissues. Distinct muscle bundles somewhat comparable to individual vertebrate muscles attach to the inner surface of the exoskeleton. The exoskeleton has an important disadvantage, however. As the arthropod grows, it periodically outgrows this nonliving shell. The process of shedding an old shell and growing another larger one is known as **molting.** The shed exoskeleton represents a net metabolic loss, and molting also leaves the arthropod temporarily vulnerable to predators.
3. The arthropod body, like that of the annelid, is **segmented.** In contrast to most annelids, each arthropod

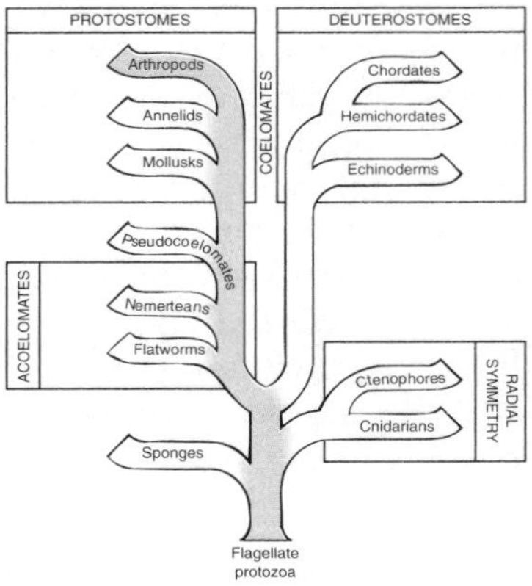

(a)

(b)

FIGURE 29–10 The arthropods are a large, diverse group. (**a**) This sponge crab *(Criptodromia octodenta),* photographed in Edithburg, SA, Australia, wears a living sponge on its back as camouflage. (**b**) Green lacewings *(Chrysopoda* sp.*)* are seldom attacked by predators because they have an unpleasant odor. (*a,* Fred Bavendam/Peter Arnold, Inc.; *b,* Stephen Dalton/Photo Researchers, Inc.)

has a fixed number of segments, which remains the same throughout life. In most arthropods, segments become fused together into groups specialized to perform certain functions. In many arthropods, segments have fused to form three main body regions: the *head*, *thorax*, and *abdomen*.

4. Arthropods have specialized respiratory systems. Most of the aquatic arthropods have a system of gills that function in gas exchange. In contrast, terrestrial forms have a system of internal branching air tubes, called **tracheae,** or plate-like structures called **book lungs.**
5. The nervous system, which resembles that of the annelids, consists of a brain (cerebral ganglia) and paired ventral nerve cords with ganglia. In some arthropods, successive ganglia may fuse together. Arthropods have a variety of very effective sense organs. Many have organs of hearing, and antennae sensitive to touch and chemicals. Many crustaceans and most insects have compound eyes (see Chapter 41).

Arthropods have an open circulatory system. A dorsal, tubular heart pumps blood into a dorsal artery, which may branch into smaller arteries. From the arteries, blood flows into large spaces that collectively make up the hemocoel. Blood in the hemocoel bathes the tissues directly. Eventually blood finds its way back into the heart through openings, called **ostia,** in its walls. The coelom is small and filled chiefly by the organs of the reproductive system.

The arthropod digestive system is a tube similar to that of the earthworm. Excretory structures vary somewhat from class to class.

Living arthropods can be assigned to three subphyla

Arthropod evolution and classification is a controversial issue among invertebrate zoologists. Some investigators hotly debate the majority view that the arthropods are a monophyletic group. In this book we have adopted the classification based on a monophyletic view, and divide the living arthropods into three subphyla: Chelicerata, Crustacea, and Uniramia (Table 29–3; also see Table 29–4).

Subphylum **Chelicerata** includes the horseshoe crabs and arachnids (spiders, scorpions, ticks, mites). Chelicerates, the only arthropods that lack antennae, have claw-like feeding appendages called **chelicerae** (singular, *chelicera*). Members of subphylum **Crustacea** (lobsters, crabs, shrimp, and barnacles) and members of subphylum **Uniramia** (insects, centipedes, and millipedes) all have jaw-like **mandibles** instead of chelicerae. Other than the extinct trilobites (discussed next), the crustaceans are the only group to have **biramous appendages**—that is, appendages with two jointed branches arising from their base. They are also the only group to have two pairs of antennae. Members of subphylum **Uniramia** have **uniramous** (unbranched) **appendages** and a single pair of antennae.

Arthropods are closely related to annelids

Arthropods represent the pinnacle of evolutionary development among the protostomes. Most zoologists think that arthropods evolved either from an ancestor of the annelids or from the annelids directly. The relationship between arthropods and annelids is evident in the basic body plan. Like annelids, arthropods exhibit **metamerism,** meaning that the body is divided into a series of similar segments. The segments develop in the same way in both groups. All arthropods are segmented, at least as embryos, and primitive arthropods exhibit metamerism even as adults. The basic plan of the nervous system is similar in both annelids and arthropods. A double ventral nerve cord extends from a dorsal, anterior brain, and ganglia are present in each segment.

The *trilobites* were among the earliest arthropods to evolve. They inhabited shallow Paleozoic seas more than 600 million years ago. These arthropods, which have been extinct for 250 million years, lived on the sea bottom and dug into the mud. Covered by a hard, segmented exoskeleton, the trilobite body was a flattened oval divided into three parts: an anterior head bearing a pair of antennae and a pair of compound eyes; a thorax; and a posterior abdomen (Fig. 29–11). At right angles to these divisions, two dorsal grooves extended the length of the animal, dividing the body into a median lobe and two lateral lobes. (The name trilobite derives from this division of the body into three longitudinal parts.) Each seg-

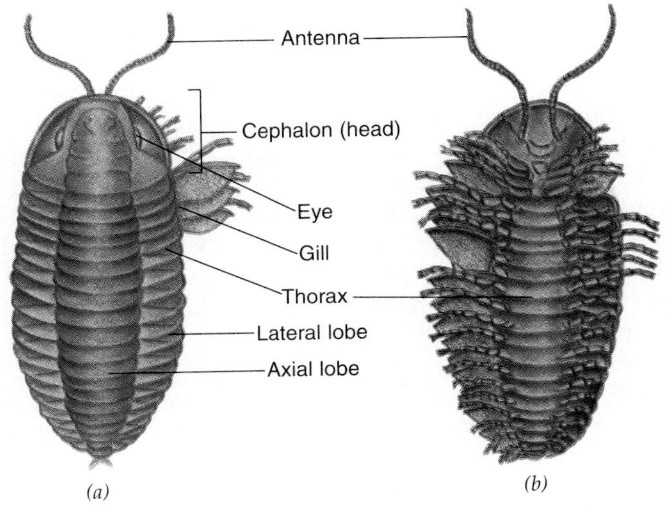

FIGURE 29–11 The trilobites are extinct marine arthropods that are considered the most primitive members of the phylum. (*a*) Dorsal view of a trilobite from the Ordovician period. (*b*) Ventral view of the same trilobite.

Making the Connection

Animal Evolution and Molecular Biology

How can taxonomists use molecular methods to help solve the controversy regarding arthropod evolution? Some zoologists have viewed the onychophorans (velvet worms) as the "missing link" between the annelids and some groups of arthropods. As described in this chapter, these animals have some annelid and some arthropod characteristics. They are internally segmented like annelids, but with an open circulatory system and tracheal tubes like arthropods. Also like arthropods, they have antennae and grow by molting.

Taxonomists have used molecular methods to study the evolutionary position of the onychophorans. Investigators compared the molecular structure of a mitochondrial gene from a variety of arthropods and other groups with those of six onychophoran species. Their results suggested that these animals are not a link between the annelids and arthropods. Rather, the onychophorans may be arthropods that became adapted to a very specialized lifestyle. Or they may be a separate group that branched from the annelid line, perhaps sharing a common ancestor with the arthropods. Are similarities at the molecular level better indicators of evolutionary relationships than anatomical characters? A decision on the evolutionary place of the onychophorans must await additional studies. ▲

ment had a pair of segmented biramous appendages, consisting of an inner walking leg and an outer branch bearing gills.

The trilobites may have given rise to the chelicerates, which also lived in Paleozoic waters. Only a few species of marine chelicerates, most notably the horseshoe crabs, exist today. Fossil evidence suggests that some extinct marine chelicerates invaded fresh water and may have given rise to the arachnids. The early arachnids were aquatic. The first terrestrial arachnids appeared in the Devonian period, and most living arachnids are terrestrial.

Crustacean fossils have been dated back to the Cambrian period, but neither their origin nor their relationship to other arthropod subphyla is clear. A distinctive feature of crustaceans is the nauplius larva, which is often the first stage after hatching. This larva has only the most anterior three pairs of appendages.

Some zoologists propose that the chelicerates and crustaceans evolved from a marine ancestor and that the uniramians (the subphylum that includes the insects) evolved from a terrestrial ancestor. According to this polyphyletic view, basic arthropod features such as jointed appendages and an exoskeleton of chitin evolved independently at least twice. Some zoologists who hold this view suggest that uniramians evolved from members of phylum **Onychophora,** wormlike animals that inhabit humid tropical rain forests. Onychophorans have some annelid and some arthropod features (Fig. 29–12; also see *Making the Connection: Animal Evolution and Molecular Biology*). Like annelids, they are internally segmented, and many organs are duplicated serially. However, the jaws are derived from appendages, as in arthropods. Also in common with arthropods, onychophorans have an open circulatory system, and a respiratory system consisting of tracheal tubes.

The earliest fossil insects—primitive, wingless species—date back to the Devonian period more than 360 million years ago. Insect fossils from the Carboniferous period include both wingless and primitive winged species. Cockroaches, mayflies, and cicadas are among the insects that have survived relatively unchanged from the Carboniferous period to the present day.

Subphylum Chelicerata includes the horseshoe crabs and arachnids

The chelicerate body consists of a cephalothorax and an abdomen, and chelicerates have no antennae and no chewing mandibles (Table 29–3). Instead, the first pair of appendages, located immediately anterior to the mouth, are the chelicerae. The second pair of appendages are the

FIGURE 29–12 Onychophorans have features of both arthropods and annelids. *Peripatus* with young. Note the soft, sluglike body and the series of nonjointed legs. (Visuals Unlimited/Thomas C. Boydean)

Table 29–3 ARTHROPOD SUBPHYLA

Subphylum and Selected Classes	*Subphylum Characteristics*
Subphylum Chelicerata Class Merostomata (horseshoe crab) Class Arachnida (spiders, scorpions, ticks, mites)	First pair of appendages are the chelicerae used to manipulate food; body consists of cephalothorax and abdomen; no mandibles; no antennae.
Subphylum Crustacea Class Malacostraca (lobsters, crabs, shrimp)	Mandibles; two pairs of antennae; biramous appendages.
Subphylum Uniramia Class Insecta (grasshopper, roach) Class Chilopoda (centipedes) Class Diplopoda (millipedes)	Mandibles; single pair of antennae; uniramous appendages.

pedipalps. The chelicerae and pedipalps are modified to perform different functions in various groups, including manipulation of food, locomotion, defense, and copulation. Posterior to the pedipalps are four pairs of legs specialized for walking.

Horseshoe crabs are the only living merostomes

Almost all of the merostomes are extinct. Only the **horseshoe crabs** have survived—essentially unchanged for 350 million years or more. *Limulus polyphemus,* the species common in North America, is horseshoe-shaped (Fig. 29–13). Its long, spikelike tail is used in locomotion, not for defense or offense. Horseshoe crabs feed on mollusks, worms, and other invertebrates that they find on the sandy ocean floor.

The arachnids include the spiders, scorpions, ticks, harvestmen (daddy long-legs), and mites

Most of the 60,000 or so species of arachnids are carnivorous and prey on insects and other small arthropods (Fig. 29–13). The arachnid body consists of a cephalothorax (composed of fused head and thorax) and abdomen.

Table 29–4 GENERAL CHARACTERISTICS OF THE PRINCIPAL ARTHROPOD GROUPS

	Arachnida (About 65,000 Species)	*Crustacea (About 32,000 Species)*	*Insecta (About 800,000 Species)*	*Chilopoda (About 3000 Species*	*Diplopoda (About 7500 Species)*
Main Habitat	Mainly terrestrial	Marine or fresh water; few on land	Mainly terrestrial	Terrestrial	Terrestrial
Body Divisions	Cephalothorax and abdomen	Cephalothorax and abdomen	Head, thorax, and abdomen	Head with segmented body	Head with segmented body
Gas Exchange	Book lungs or tracheae	Gills	Tracheae	Tracheae	Tracheae
Appendages	Uniramous	Biramous	Uniramous	Uniramous	Uniramous
Antennae	None	2 pairs	1 pair	1 pair	1 pair
Mouth parts	Chelicerae, pedipalps	Mandibles, 2 pairs of maxillae (for food handling)	Mandibles, maxillae	Mandibles, maxillae	Mandibles, maxillae
Legs	4 pairs on cephalothorax	1 pair per segment or less	3 pairs on thorax	1 pair per segment	Usually 2 pairs per segment
Development	Direct, except mites and ticks	Usually larval stages (nauplius larva)	Usually larval stages; most with complete metamorphosis	Direct	Direct

(a)

(b)

(c)

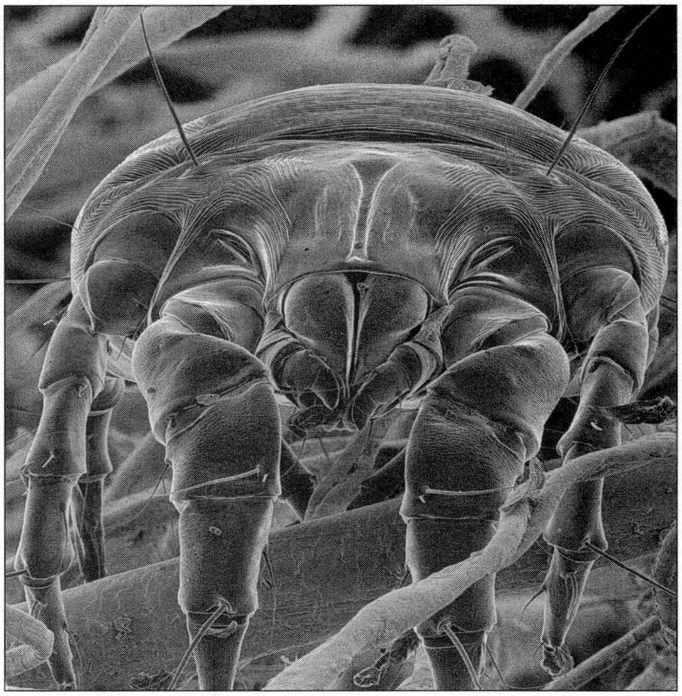

(d)

FIGURE 29–13 Subphylum Chelicerata includes the merostomes (horseshoe crabs) and arachnids. (***a***) The only living merostomes are a few closely related species of horseshoe crabs. Seasonally, horseshoe crabs *(Limulus polyphemus)* return to beaches for mating. In this photograph, males are competing for a female. (***b***) This orb spider *(Araneus quadratus)* is shown building a cocoon. (***c***) This emperor scorpion *(Pandinis imperator)* was photographed in Africa. (***d***) House dust mite *(Dermatophagoides* sp.) on dustball. (*a,* Visuals Unlimited/Milton H. Tierney, Jr.; *b,* Hans Pfletschinger/Peter Arnold, Inc.; *c,* Joe McDonald/Tom Stack & Associates; *d,* David Scharf/Peter Arnold, Inc.)

Arachnids have six pairs of jointed appendages. In spiders, the first pair, the chelicerae, are fanglike structures used to penetrate prey. In some spider species, the chelicerae are used to inject poison into the prey. The second pair of appendages, the pedipalps, are used by spiders to hold and chew food and in some species are modified as sense organs for tasting the food or for reproduction (for sperm transfer and courtship displays). The other four pairs of appendages are used for walking.

Gas exchange in arachnids takes place either by tracheal tubes, by book lungs, or by both. A book lung consists of 15 to 20 plates, like pages of a book, that contain tiny blood vessels. Air enters the body through abdominal slits and circulates between the plates. As air passes over the blood vessels in the plates, oxygen diffuses into the blood and carbon dioxide diffuses out of the blood into the air. An arachnid may have as many as four pairs of book lungs.

(a)

(b)

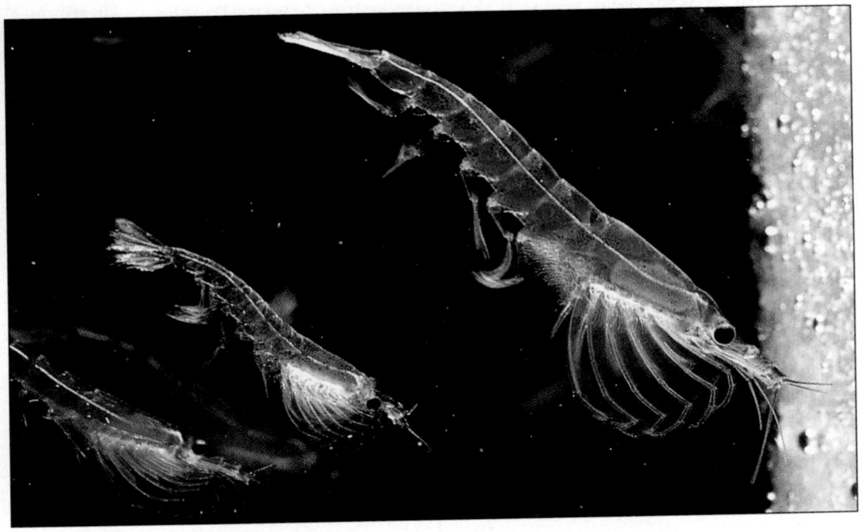

(c)

FIGURE 29–14 Most members of the class Crustacea are aquatic. (***a***) Gooseneck barnacles, *Pollicipes polymerus,* are stalked barnacles that occur in large numbers on intertidal rocks along the west coast of the United States. (***b***) Brokenback shrimp from Monterey Bay. (***c***) Antartic krill *(Euphausia superba).* (*a,* Fred Bavendam/Peter Arnold, Inc.; *b,* Frans Lanting/Minden Pictures; *c,* Flip Nicklin/Minden Pictures)

The spider has glands in its abdomen that secrete silk, an elastic protein that is spun into fibers by organs called spinnerets. The silk is liquid as it emerges from the spinnerets but hardens after it leaves the body. Spiders use silk to build nests, to encase their eggs in a cocoon, and, in some species, to trap prey. Many spiders lay down a silken dragline as they venture forth. This serves as a safety line and also as a means of communication between members of a species. From a dragline a spider can determine the sex and maturity level of the spinner.

Although spiders do have poison glands useful in capturing prey, only a few have poison that is toxic to humans. The most widely distributed poisonous spider in the United States is the black widow (*Latrodectus mactans*). Its poison is a neurotoxin that interferes with transmission of messages from nerves to muscles. The brown recluse (*Loxosceles reclusa*) is smaller than the black widow and has a violin-shaped dorsal stripe on its back. Its venom, which destroys the tissues surrounding the bite, and can occasionally be fatal. Although painful, spider bites cause fewer than five fatalities per year in the U.S.

Mites and ticks are among the most serious arthropod nuisances. They eat crops, infest livestock and pets, and inhabit our own bodies. Many live unnoticed, due to their small size, but others cause disease. Certain mites cause mange in dogs and other domestic animals. Chiggers (red bugs), the larval form of red mites, attach themselves to the skin and secrete an irritating digestive fluid that may cause itchy red welts. Larger than mites, ticks are ectoparasites on dogs and many other animals. They can transmit diseases such as Rocky Mountain spotted fever, Texas cattle fever, relapsing fever, and Lyme disease.

Subphylum Crustacea includes the lobsters, crabs, shrimp, and their relatives

Crustaceans are vital members of marine food webs. Some serve as primary consumers of algae and detritus, and as food for the many carnivores that inhabit the oceans. Countless billions of microscopic crustaceans swarm in the ocean and become food for many fishes and other marine animals such as certain baleen whales.

As discussed earlier, crustaceans are characterized by mandibles, biramous appendages, and two pairs of antennae (Fig. 29–14). Their antennae serve as sensory organs for touch and taste. The hard mandibles, which are the third pair of appendages, are located on each side of the ventral mouth and are used for biting and grinding food. Posterior to the mandibles are two pairs of appendages, the first and second **maxillae,** used for manipulating and holding food. Several other pairs of appendages are present. Usually five pairs are modified for walking. Others may be specialized for swimming, sperm transmission, carrying eggs and young, or sensation.

As the only group of arthropods that are primarily aquatic, crustaceans generally have gills for gas exchange. Two large **antennal glands** located in the head remove metabolic wastes from the blood and body fluids. Wastes are excreted through ducts opening at the base of each antenna. Most adult crustaceans have compound eyes, and many crustaceans have **statocysts,** sense organs that detect the pull of gravity.

Crustaceans characteristically have separate sexes. During copulation, the male uses specialized appendages to transfer sperm into the female. The fertilized eggs are usually carried on some part of the body. The newly hatched animals may resemble the adults, or they may pass by successive molts through a series of larval stages before they develop the adult body. The lobster, for example, molts seven times during its first summer; after each molt it gets larger and more closely resembles the adult. After it becomes a small adult, additional molts allow for growth.

Barnacles, the only sessile crustaceans, differ markedly in their external anatomy from other members of the subphylum. They are exclusively marine and secrete complex limestone cups within which they live. The larvae of barnacles are free-swimming forms that go through several molts. They eventually become sessile and develop into the adult form. The barnacle was described by the 19th century naturalist Louis Agassiz as "nothing more than a little shrimplike animal standing on its head in a limestone house and kicking food into its mouth." The bane of marine boaters, barnacles can proliferate on ship bottoms in great numbers. Their presence may reduce the speed of a ship by more than 30%.

The largest order of crustaceans, **Decapoda,** contains more than 10,000 species of lobsters, crayfish, crabs, and shrimp. Most decapods are marine, but a few, such as the crayfish, certain shrimp, and a few crabs, live in fresh water. The crustaceans in general and the decapods in particular show striking specialization and differentiation of parts in the various regions of the animal. In the lobster, no two of the 19 pairs of appendages are identical, and the appendages in the different parts of the body differ markedly in form and function (Fig. 29–15).

The five segments of the lobster's head and the eight segments of its thorax are fused into a cephalothorax, which is covered on the top and sides by a shield, the **carapace,** composed of chitin impregnated with calcium salts. The two pairs of antennae serve as chemoreceptors and tactile sense organs. The mandibles are short and heavy, with opposing surfaces used in grinding and biting food. Behind the mandibles are two pairs of accessory feeding appendages, the first and second maxillae. The appendages of the first three segments of the thorax are the maxillipeds, which aid in chopping up food and passing it to the mouth. The fourth segment of the thorax has a pair of large **chelipeds,** or pinching claws. The last four thoracic segments bear **walking legs.**

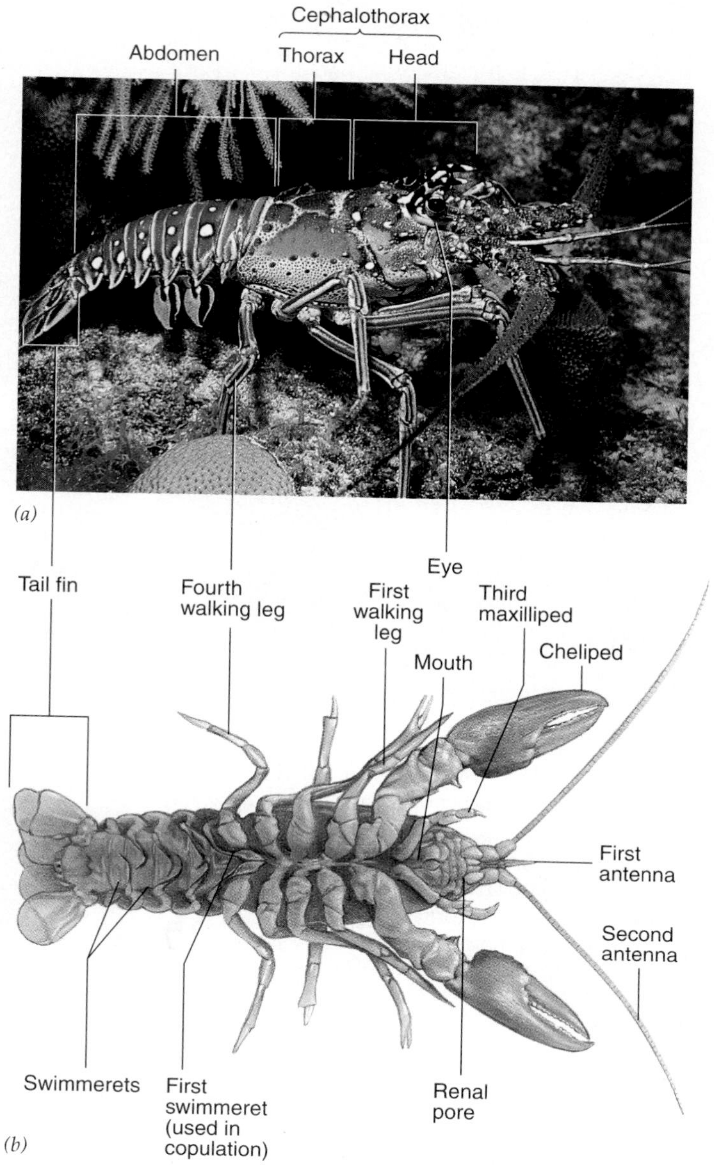

FIGURE 29–15 The structure of the lobster is typical of crustacean anatomy. (***a***) Dorsal view of a spiny lobster *(Panulirus argus)*; (***b***) ventral view. (*a,* Larry Lipsky/Tom Stack & Associates)

(Text continues on page 645)

Table 29–5 SOME ORDERS OF INSECTS

Order, Number of Species, and Examples	*Representative Member*	*Some Characteristics*
Insects with No Metamorphosis (Egg → immature form → adult)		
Thysanura (320) Silverfish Bristletails	 Silverfish (*Lepisma saccharina*) adult and young. (Harry Rogers/Photo Researchers, Inc.)	No wings; biting-chewing mouth parts; 2–3 "tails" extend from posterior tip of abdomen; inhabit dead leaves; eat starch in books
Insects with Incomplete Metamorphosis (Egg → nymph → adult)		
Odonata (5,000) Dragonflies Damselflies	 Black-winged damselfly (*Calopteryx maculata*). (Rod Planck/Dembinsky Photo Associates)	Two pairs of long, membranous wings; chewing mouth parts; large, compound eyes; active predators
Orthoptera (20,000) Grasshoppers Crickets	 Forktailed bush katydid (*Scudderia furcata*) laying eggs between surfaces of a leaf. (William E. Ferguson)	Forewings leathery, hindwings membranous; chewing mouth parts; most herbivorous, some cause crop damage; some predatory
Blattodea (3,700) Cockroaches	 American cockroach (*Periplaneta americana*). (Animals Animals © 1996 George Bryce)	When wings present, forewings leathery, hindwings membranous; chewing mouth parts; legs adapted for running
Isoptera (2,000) Termites	 Dampwood termite (*Zootermopsis* sp.) in wood; found along Pacific coast. (Visuals Unlimited/John D. Cunningham)	2 pairs of wings, or none; wings shed by sexual forms after mating; chewing mouth parts; social insects, form large colonies; eat wood

Table 29–5 Continued

Order, Number of Species, and Examples	*Representative Member*	*Some Characteristics*
Insects with Incomplete Metamorphosis (*continued*)		
Anoplura (500) Sucking lice	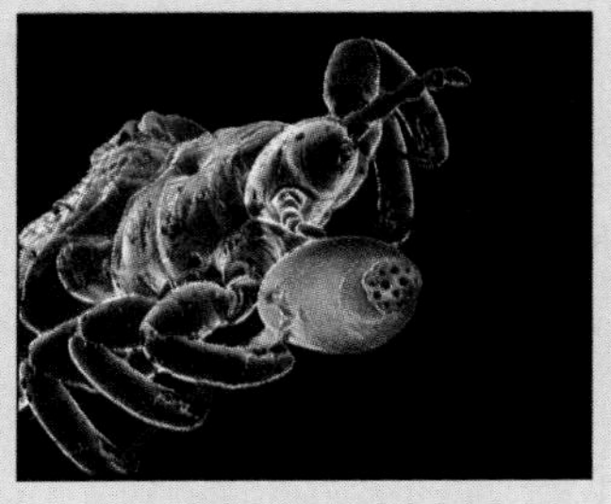 Head louse (*Pediculus humanus capitis*) with hair and egg. (David Scharf/Peter Arnold, Inc.)	No wings; piercing-sucking mouth parts; ectoparasites of mammals; head louse and crab louse are human parasites; vectors of typhus fever
Hemiptera (35,000) (true bugs) Chinch bugs Bed bugs Water striders	 Common bed bug (*Cimex lectularius*). (Manfred Kage/Peter Arnold, Inc.)	Hindwings membranous; forewings smaller; piercing-sucking mouth parts form beak; most herbivorous; some parasitic
Homoptera (33,000) Aphids Leaf hoppers Cicadas Scale insects	 An aphid (*Aphis nerii*) giving birth to a live young. (Peter J. Bryant/Biological Photo Service)	2 pairs membranous wings; piercing-sucking mouth parts form beak; some infect plants; others vectors of plant diseases
Insects with Complete Metamorphosis (Egg → larva → pupa → adult)		
Lepidoptera (120,000) Moths Butterflies	 Bee butterfly (*Zeonia faunus*). (Animals Animals © 1996 Raymond A. Mendez)	Usually 2 pairs of membranous, colorful, scaled wings; sucking mouth parts; larvae are wormlike caterpillars that eat plants; adults suck flower nectar; important pollinators
Diptera (150,000) Houseflies Mosquitos	 Mosquito (*Anophales* sp.) feeding on human. (Rod Planck/Dembinsky Photo Associates)	Only forewings functional in flying; hindwings small, knoblike halteres; mouth parts usually adapted for sucking (and piercing in some); larvae are maggots or wigglers and may damage domestic animals or food; adults may transmit disease such as sleeping sickness or yellow fever

Table 29–5 Continued

Order, Number of Species, and Examples	*Representative Member*		*Some Characteristics*
Siphonaptera (1,750) Fleas		SEM of rat flea (*Xenopsylla cheopis*) clinging to fur of a rat. This flea transmits the bacteria that cause bubonic plague. (Dr. Tony Brain/Science Photo Library/Photo Researchers, Inc.)	No wings; piercing-sucking mouth parts; legs adapted for clinging and jumping; ectoparasites on birds and mammals; vectors of bubonic plague and typhus
Coleoptera (300,000) Beetles Weevils		Milkweed beetle (*Tetraopes tetraophthalmus*) feeding on milkweed. (Skip Moody/Dembinsky Photo Associates)	Forewings modified as protective coverings for membranous hindwings (which are sometimes absent); chewing mouth parts; largest order of insects; most herbivorous; some aquatic
Hymenoptera (130,000) Ants Bees Wasps	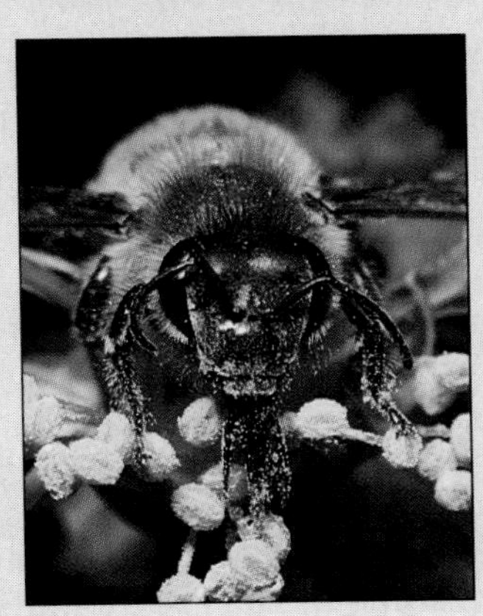	Honeybee worker (*Apis mellifera*) searching for nectar. (J.H. Robinson/Photo Researchers, Inc.)	Usually 2 pairs of membranous wings; mouth parts may be modified for sucking or lapping nectar; many are social insects; some sting

The appendages of the first abdominal segment are part of the reproductive system and function in the male as sperm-transferring structures. The following four abdominal segments bear paired **swimmerets,** small paddle-like structures used by some decapods for swimming and by the females of all species for holding eggs. Each branch of the sixth abdominal appendages (called *uropods*) consists of a large flattened structure. Together with the flattened posterior end of the abdomen (the *telson*), they form a fan-shaped tail fin used for swimming backwards.

Subphylum Uniramia includes the insects, centipedes, and millipedes

The insects, centipedes, and millipedes are grouped together in subphylum Uniramia because they all have uniramous (unbranched) appendages. They also have only a single pair of antennae rather than two pairs as in crustaceans.

Insects are the most successful group of animals

With more than 750,000 described species, the class **Insecta** is the most successful group of animals on our planet in terms of diversity, number of species, and number of individuals (Table 29–5). More species of insects have been identified than of all other classes of animals combined. What they lack in size, insects make up in sheer numbers. If we could weigh all the insects in the world, their weight would exceed that of all of the remaining terrestrial animals. Insects have an extraordinary ability to adapt to changes in the environment. Insects are primarily terrestrial animals, but some species live in

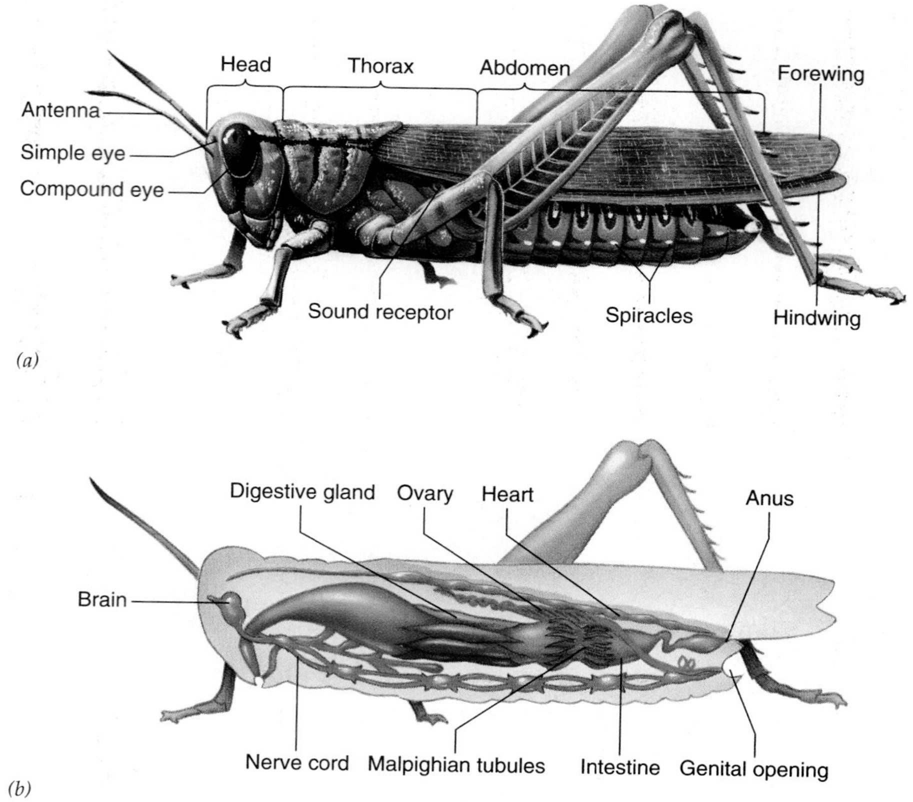

FIGURE 29–16 The grasshopper typifies insect structure. (*a*) External anatomy. Note the three pairs of segmented legs. (*b*) Internal anatomy.

fresh water, a few are truly marine, and others inhabit the shore between the tides.

We can describe an insect as an **articulated** (jointed), **tracheated** (having tracheal tubes for gas exchange) **hexapod** (having six feet). The insect body consists of three distinct parts: head, thorax, and abdomen (Fig. 29–16). Three pairs of legs emerge from the adult thorax, and there may be one or two pairs of wings. One pair of antennae protrudes from the head, and the sense organs include both simple and compound eyes. Their complex mouth parts may be adapted for piercing, chewing, sucking, or lapping.

The tracheal system of insects has made an important contribution to their diversity. Air enters the tracheae through tiny openings, called **spiracles,** in the body wall. The tracheae permit oxygen to pass directly to the internal organs. This effective oxygen delivery system permits the insects to have the high metabolic rate necessary for activities such as flight.

Excretion is accomplished by two to many slender **Malpighian tubules,** which receive metabolic wastes from the blood, concentrate the wastes, and discharge them into the intestine.

The sexes are separate, and fertilization takes place internally. Several molts occur during development. Primitive insects with no wings have direct, or simple, development: the young hatch as juveniles that look like the adult form. Other insects undergo **incomplete metamorphosis** in which the egg gives rise to a nymph that resembles the adult in many ways but lacks functional wings and reproductive structures. There may be several nymphal stages and gradual metamorphosis (change in body form) to the adult form.

Most insects undergo **complete metamorphosis** with four distinct stages in the life cycle: egg, larva, pupa, and adult (Fig. 29–17). The wormlike larva does not look at all like the adult. For example, caterpillars have an entirely different body form than butterflies. Typically, an insect spends most of its life as a larva. Eventually the larva stops feeding, molts, and enters a pupal stage, usually within a protective cocoon or underground burrow. The pupa does not feed and cannot defend itself. Its energy is spent remodeling its body. When it emerges, it is equipped with functional wings and reproductive organs.

Certain species of bees, ants, and termites exist as colonies or societies made up of several different types of individuals, each adapted for some particular function (Fig. 29–18). The members of some insect societies communicate with each other by "dances" and by chemical messengers called pheromones. Social insects and their communication are discussed in Chapter 50.

Many adaptations contribute to the biological success of the insects

What are the secrets of insect success? One important feature is their body plan, which can be modified and specialized in so many ways that insects have been able to

(a) Egg

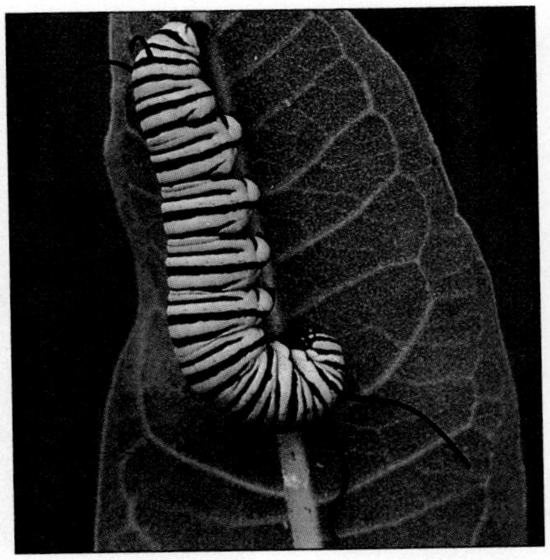

(b) Larva

(c) Early pupa

(e) Adult

(d) Late pupa

FIGURE 29–17 The monarch butterfly undergoes complete metamorphosis. (***a***) Egg. (***b***) The caterpillar (larva) feeds and grows, undergoing several molts. The larva eventually forms a cocoon and (***c***) becomes a pupa. (***d***) The pupa develops into (***e***) an adult—a butterfly that emerges from the cocoon. (*a, c, e,* Sharon Cummings/Dembinsky Photo Associates; *b, d,* Skip Moody/Dembinsky Photo Associates)

adapt to a remarkable number of lifestyles. Another reason for their success is the ability to fly. Unlike other invertebrates, which creep slowly along on (or under) the ground, most insects fly rapidly through the air. Their wings and small size facilitate their wide distribution.

The insect body is well protected by a tough exoskeleton, which also helps to prevent water loss by evaporation. Other protective mechanisms include mimicry, protective coloration, and aggressive behavior. Metamorphosis divides the insect life cycle into different stages. This strategy places larval forms in different lifestyles so that they do not have to compete with adults for food or habitats.

Insects have a great impact on humans

Not all insects compete with us for food or merely cause

FIGURE 29–18 Queen and worker termites (*Nasutitermes* sp.) from Peru interact in a royal cell. The queen, with an enlarged abdomen, occupies the center of the chamber. Most of the individuals are workers. A few soldiers with "squirt gun" heads and reduced mandibles can be seen. (James L. Castner)

(a) (b)

FIGURE 29–19 Chilopods and diplopods have uniramous appendage. (*a*) Centipede *(Lithobius* sp.*)*, a member of the class Chilopoda. Centipedes have one pair of appendages per segment. (*b*) Millipede *(Diplopoda pachydesmus)*, a member of the class Diplopoda. Millipedes have two pairs of appendages per segment. (*a,* Dwight Kuhn; *b,* John R. MacGregor/Peter Arnold, Inc.)

us to scratch, swell up, or recoil from their presence. Bees, wasps, beetles, and many other insects pollinate flowers of crops and fruit trees. Some insects destroy other insects that are harmful. For example, dragonflies eat mosquitos. And some organic farmers and home gardeners purchase ladybird beetles, so adept are they at ridding plants of aphids and other insect pests. Insects are important members of many food webs. Many birds, mammals, amphibians, reptiles, and some fishes depend on insects for food. Many beetles and the larvae (maggots) of flies are detritus feeders; they break down dead plants and animals and their wastes, permitting nutrients to be recycled.

Many insect products are useful to us. Bees produce honey as well as beeswax, which is used in making candles, lubricants, chewing gum, and other products. Shellac is made from lac, a substance given off by certain scale insects that feed on the sap of trees. And the labor of silkworms provides us with silk.

On the negative side, billions of dollars worth of crops are destroyed each year by insect pests. Whole buildings may be destroyed by termites, and clothing can be damaged by moths. Fire ants not only inflict painful stings but cause farmers serious economic loss because of their large mounds, which damage mowers and other farm equipment. Mounds also reduce grazing land because livestock quickly learn to avoid them.

Blood-sucking flies, screwworms, lice, fleas, and other insects annoy and transmit disease in both humans and domestic animals. Mosquitos are vectors of malaria and yellow fever. Body lice may carry the typhus rickettsia, and houseflies sometimes transmit typhoid fever and dysentery. Tsetse flies transmit African sleeping sickness, and fleas may be vectors of bubonic plague.

Classes Chilopoda and Diplopoda include centipedes and millipedes

Members of class Chilopoda are called centipedes ("hundred-legged"), and members of class Diplopoda are known as millipedes ("thousand-legged"). These animals are all terrestrial and are typically found beneath stones or wood in the soil in both temperate and tropical regions.

While they are not clearly related, centipedes and millipedes are similar in having a head and an elongated trunk with many segments, each bearing legs (Fig. 29–19). The centipedes have one pair of legs on each segment behind the head. Most centipedes do not have enough legs to merit their name—the most common number is 30 or so—although a few species have 100 or more. The legs of centipedes are long, enabling them to run rapidly. Centipedes are carnivorous and feed on other animals, mostly insects, but the larger centipedes have been known to eat snakes, mice, and frogs. The prey is captured and killed with poison claws located just behind the head on the first trunk segment.

Millipedes have two pairs of legs on most body segments. Diplopods are not as agile as chilopods, and most species can crawl only slowly over the ground, although they can powerfully force their way through earth and rotting wood. Millipedes are generally herbivorous and feed on both living and nonliving vegetation. The three main phyla discussed in this chapter are compared in Table 29–6.

Table 29–6 COMPARISON OF CHARACTERISTICS OF SOME COELOMATE PROTOSTOME PHYLA*

	Mollusca (100,000 species)	*Annelida (segmented worms) (15,000 species)*	*Arthropoda (joint-footed animals) (over 1 million species)*
Representative Animals	Clams Snails Squids	Earthworms Leeches Marine worms	Crustaceans Insects Spiders
Circulation	Open system (closed in cephalopods)	Closed system	Open system
Gas Exchange	Gills and mantle	Diffusion through moist skin; oxygen circulated by blood	Tracheae in insects; gills in crustaceans; book lungs or tracheae in spider group
Waste Disposal	Metanephridia	Pair of metanephridia in each segment	Malpighian tubules in insects; antennal (green) glands in crustaceans
Nervous System	Three pairs of ganglia; simple sense organs	Simple brain; double ventral nerve cord; simple sense organs	Brain; double ventral nerve cord; well-developed sense organs
Reproduction	Sexual; sexes separate; fertilization in water	Sexual; hermaphroditic but cross-fertilize	Sexual; sexes separate
Support and Movement	Most have hydrostatic skeleton; most have ventral foot for locomotion	Fluid-filled coelom serves as hydrostatic skeleton; well-developed muscles in body wall	Tough exoskeleton; jointed appendages (some have wings); well-developed muscles
Environment and Lifestyle	Mainly marine, some inhabit fresh water or are terrestrial; herbivores, carnivores, scavengers, or suspension feeders	Marine, freshwater, terrestrial; herbivores, carnivores, scavengers, suspension feeders	Most diverse group in habitat and lifestyle; marine, freshwater, and terrestrial; herbivores, carnivores, scavengers
Other Characteristics	Soft-bodied; usually have shell	Earthworms till soil	Hard exoskeleton

*Members of these phyla are at the organ system level of organization, have bilateral symmetry, and have a complete digestive tract.

SUMMARY

I. The coelomate protostomes include the mollusks, annelids, and arthropods, as well as some minor phyla.

II. The coelom provides space for many organs and permits the digestive tract to move independently of body movements. Coelomic fluid helps transport materials and bathes cells that line the coelom.

III. In terrestrial animals, the body covering must prevent fluid loss; some sort of skeleton must be present to withstand the pull of gravity; and there must be reproductive adaptations, such as internal fertilization, development within the mother's body, or shells that prevent drying out of the developing embryo.

IV. Mollusks are soft-bodied animals usually covered by a shell. They possess a ventral foot for locomotion and a mantle that covers the visceral mass.
   A. Class Polyplacophora includes the sluggish marine chitons, which have segmented shells.
   B. Class Gastropoda, the largest and most successful group of mollusks, includes the snails and slugs. In gastropods, the body is twisted, and the shell (when present) is coiled.
   C. Class Bivalvia includes the clams and oysters, animals enclosed by a two-part shell that is hinged dorsally.
   D. Class Cephalopoda includes the squids and octopods,

which are active predatory animals. The foot is divided into tentacles that surround the mouth located in the large head.

V. Phylum Annelida, the segmented worms, includes many aquatic worms, earthworms, and leeches.
   A. Annelids have conspicuously long bodies that are segmented both internally and externally; their large compartmentalized coelom serves as a hydrostatic skeleton.
   B. Class Polychaeta consists of marine worms characterized by bristled parapodia, used for locomotion.
   C. Class Oligochaeta, the earthworms, contains segmented worms characterized by a few short setae per segment. The body is divided into more than 100 segments separated internally by septa.
   D. Class Hirudinea, the leeches, is composed of animals that lack setae and appendages. They are equipped with suckers for sucking blood.

VI. Phylum Onychophora includes animals with both annelid and arthropod characteristics.

VII. Phylum Arthropoda is composed of animals with jointed appendages and an armor-like exoskeleton of chitin.
   A. The trilobites are extinct marine arthropods covered by a hard, segmented shell.
   B. Subphylum Chelicerata includes the merostomes (horseshoe crabs) and the arachnids (spiders, mites, and their relatives).
      1. The first pair of appendages are chelicerae, used to manipulate food. Chelicerates have no antennae and no mandibles.
      2. The arachnid body consists of a cephalothorax and abdomen; there are six pairs of jointed appendages, of which four pairs serve as legs.
   C. Subphylum Crustacea includes the lobsters, crabs, and barnacles. The body consists of a cephalothorax and abdomen; typically, five pairs of walking legs are present. Crustaceans have two pairs of antennae and mandibles for chewing.
   D. Subphylum Uniramia includes class Insecta, class Chilopoda, and class Diplopoda; members of this subphylum have unbranched appendages and a single pair of antennae.
      1. An insect is an articulated, tracheated hexapod; its body consists of head, thorax, and abdomen. Insects are the most biologically successful group of animals on Earth.
      2. The centipedes have one pair of legs per body segment, whereas the millipedes have two pairs of legs per body segment.

SELECTED KEY TERMS

Annelida
antenna
Arthropoda
biramous appendage
chelicerae
Chelicerata
cheliped
coelom
Crustacea
exoskeleton
hemocoel
Malpighian tubule
mandible
mantle
maxilla
metamerism
metamorphosis
Mollusca
molting
nudibranch
open circulatory system
pedipalp
radula
seta (setae)
torsion
tracheae
trochophore larva
Uniramia
visceral mass

POST-TEST

1. The ________ of arthropods and mollusks provides protection and serves as a point of attachment for muscles.
2. The ________ is a belt of teeth within the digestive system of mollusks.
3. A hemocoel is characteristic of animals with a(an) ________ circulatory system.
4. The first larval stage of a mollusk is a free-swimming ________ larva.
5. Segmental, serial repetition of structures within an animal is known as ________ .
6. In pulmonate snails, the highly vascularized ________ functions as a lung.
7. ________ are marine gastropods that lack a shell.
8. ________ are bristle-like structures characteristic of annelids.
9. Animals with an exoskeleton and paired, jointed appendages are ________ .
10. The ________ of a crustacean are used for biting and grinding food.
11. The large, pinching claws of a lobster are called ________ .
12. ________ tubules are excretory organs in insects.
13. A definition of an insect is: an articulated, ________ (referring to respiratory structures) hexapod.
14. The arthropod group whose members have unbranched appendages is subphylum ________ .
15. Arthropods with no antennae or mandibles belong to subphylum ________ .

REVIEW QUESTIONS

1. Identify characteristics that mollusks and annelids share. In what ways are these animals different? What do their characteristics suggest about their evolutionary relationship?
2. Give two distinguishing characteristics for each of the following: (a) mollusks; (b) annelids; (c) arthropods.
3. What are the advantages of each of the following? (a) presence of a coelom; (b) the arthropod exoskeleton; (c) segmentation (metamerism).
4. Contrast the lifestyles of a gastropod and a cephalopod. Identify adaptations that have evolved in each for its particular lifestyle.
5. What is a trochophore larva? A veliger larva?
6. Describe some of the adaptations that have contributed to insect success.
7. Distinguish between insects and spiders.
8. What are the distinguishing features of each of the arthropod subphyla? Identify animals that belong to each group.

YOU MAKE THE CONNECTION

1. Discuss the idea that every evolutionary adaptation has both advantages and disadvantages, using each of the following characteristics as an example. (a) presence of a coelom; (b) the arthropod exoskeleton; (c) segmentation; (d) complete metamorphosis; (e) a complete digestive tract.
2. Hypothesize a benefit that might explain why oysters secrete calcium carbonate layers around foreign particles.
3. Insects that undergo complete metamorphosis outnumber those that do not by more than ten to one. Hypothesize an explanation based on your knowledge of evolution and ecology.

RECOMMENDED READINGS

Berenbaum, M.R., *Bugs in the System: Insects and Their Impact on Human Affairs.* Addison-Wesley. 1995. A readable book on insects describing their lives from physiology to behavior, and including relationships with humans.

Holldobler, B. & Wilson, E.O. *Journey to the Ants: A Story of Scientific Exploration,* 1994, Cambridge, The Belknap Press of Harvard University Press. Everything you have ever wanted to know about ants with emphasis on their biological success.

Knowlten, N. "A Tale of Two Seas," *Natural History,* June 1994. A discussion of reproductive isolation and shrimp diversity.

Morell, V. "Life on a Grain of Sand," *Discover,* April 1995. An interesting look at some unusual marine animals.

Natural History, Vol. 104, No. 3, March 1995. This issue features an interesting collection of articles devoted to spiders.

Zimmer, C. "See How They Run." *Discover,* Vol. 15, No. 9, September 1994. A look at how roaches, ants, centipedes, and other arthropods move.

CHAPTER 30

The Animal Kingdom: The Deuterostomes

The southern biscuit star is an echinoderm, while the Tasmanian blenny, hiding among doughboy scallops, is a chordate. Echinoderms and chordates are deuterostomes. (Photographed in Edithburg, Australia; Fred Bavendam/Peter Arnold, Inc.)

It may seem strange to group the **echinoderms**—the sea stars, sea urchins, and sand dollars—with the **chordates,** the phylum to which we belong. However, there is strong evidence that echinoderms and chordates evolved from a common ancestor, and they are both classified as deuterostomes, the second main branch of the animal kingdom. Several less familiar phyla are also deuterostomes. These include the **lophophorate phyla** (Phoronida, Ectoprocta, and Brachiopoda), the **hemichordates** (a small group of wormlike marine animals), and **chaetognaths,** or arrowworms (a group of about 50 species found in marine plankton).

Recall from Chapter 28 that deuterostomes share similarities in their patterns of development. In contrast to the protostomes, the first opening in the embryo often gives rise to the anus, and the mouth develops from a second opening. Deuterostomes are characterized by radial, rather than spiral, cleavage. And their cleavage is generally indeterminate, which means that the fate of their cells is fixed later in development than is the case in protostomes. In deuterostomes, the mesoderm develops from paired pouches that pocket out from the primitive gut. The coelom forms from cavities within the mesodermal out-pocketings. Some characteristics of echinoderms, chordates, and the less familiar deuterostome phyla are compared in Table 30–1. In this chapter we focus on the echinoderms and chordates (Table 30–2).

The echinoderm larva is bilaterally symmetrical. However, the adult typically exhibits a form of radial symmetry, known as pentaradial symmetry, in which the body is arranged in five parts around a central axis. Echinoderms have an endoskeleton (internal skeleton) consisting of calcium carbonate plates. The endoskeleton typically bears external spines. Echinoderms have a unique water vascular system made up of internal canals and external tube feet. This system functions in locomotion and obtaining food; in some forms it functions in gas exchange as well.

During some time in its life cycle, a chordate has a notochord; a dorsal, tubular nerve cord; and pharyngeal gill slits. The largest chordate subphylum is Vertebrata, animals with backbones (vertebral columns). A vertebrate has a well-developed head with sense organs and a brain encased in a cartilaginous or bony cranium. The cranium and vertebral column are part of the living endoskeleton.

LEARNING OBJECTIVES

After you have studied this chapter you should be able to

1. Discuss the relationship of the echinoderms and chordates, giving specific reasons for grouping them together.
2. Describe the distinguishing characteristics of the echinoderms.
3. Describe and give examples of each of the five main classes of echinoderms.
4. Distinguish among the subphyla of phylum Chordata and describe the characteristics that they have in common; describe the characteristics of subphylum Vertebrata.
5. Distinguish among the classes of vertebrates and assign a given vertebrate to the correct class.
6. Trace the evolution of vertebrates according to current theory.
7. Identify adaptations that reptiles and other terrestrial vertebrates have made to life on land.
8. Contrast monotremes, marsupials, and placental mammals, and give examples of members that belong to each group.
9. Identify the major orders of placental mammals, and give examples of animals that belong to each order.

ECHINODERMS ARE "SPINY-SKINNED" MARINE ANIMALS

All of the members of phylum **Echinodermata** inhabit the sea. They are found in all oceans and at all depths. About 7000 living and more than 13,000 extinct species have been identified. Traditionally, the living species have been divided into five principal classes (Fig. 30–1): class **Crinoidea,** the sea lilies and feather stars; class **Asteroidea,** the sea stars; class **Ophiuroidea,** the basket stars and brittle stars; class **Echinoidea,** the sea urchins and sand dollars; and class **Holothuroidea,** the sea cucumbers. Recently, two new species called *sea daisies* have been discovered on bacteria-rich wood sunk in deep water. These small (less than 1 centimeter in diameter) echinoderms have been assigned to a new class (called Concentricycloidea).

The echinoderms are in many ways unique in the animal kingdom. Echinoderm larvae are bilaterally symmetrical, ciliated, and free-swimming. During development the body form reorganizes and its symmetry changes to pentaradial. Echinoderms have an endoskeleton consisting of small calcareous (composed of $CaCO_3$) plates and often bear spines that project outward; the name *Echinodermata,* meaning "spiny-skinned," reflects this trait. The endoskeleton is covered by a thin, ciliated epidermis.

Also unique to echinoderms is the **water vascular system,** a network of canals filled with sea water. Branches of this system lead to numerous tiny **tube feet** that extend when filled with fluid. The tube feet serve in locomotion and obtaining food and, in some forms, in gas exchange. The water vascular system is a hydraulic system. When the echinoderm begins to extend a foot, a rounded muscular sac, or **ampulla,** at the upper end of the foot contracts, forcing water through a valve into the tube of the foot. At the bottom of the foot is a suction-type structure that adheres to whatever surface the animal has perched upon.

Echinoderms have a well-developed coelom. The complete digestive system is the most prominent body system, although its structure varies in different groups. The coelomic fluid transports materials. A variety of respiratory structures are found in the various classes, including dermal gills in the sea stars and respiratory trees in sea cucumbers. No excretory organs are present. The nervous system is simple, generally consisting of nerve rings with radiating nerves about the mouth. Echinoderms have no brain. The sexes are usually separate, and eggs and sperm are generally released into the water, in which case fertilization takes place externally.

Class Crinoidea includes the feather stars and sea lilies

Class Crinoidea, the oldest class of living echinoderms, includes the feather stars and the sea lilies (Fig. 30–1*a*). Although a great many extinct crinoids are known, there are relatively few living species. The feather stars are motile crinoids, although they often remain in the same location for long periods of time. Sea lilies are sessile and remain attached to the ocean floor by a stalk.

In all other echinoderms, the mouth is directed downward toward the substratum, but in crinoids the oral (mouth) surface is turned upward. A number of branched, feathery arms also extend upward. The crinoids are suspension feeders. Their tube feet, abundant on the feathery arms, are coated with mucus that traps microscopic organisms.

(Text continues on page 656)

Table 30–1 COMPARISON OF DEUTEROSTOME PHYLA

Phylum		*Characteristics*
Lophophorate phyla		Have a lophophore, a ciliated ring of tentacles that surrounds the mouth. Bottom dwellers. Suspension feeders.
Phoronids	 Lophophores project from the tubes of these phoronids (*Phoronis vancouverensis*), inhabitants of the Pacific Coast of the U.S. (Visuals Unlimited/Gary R. Robinson)	Tube-dwelling marine worms.
Ectoprocts (Bryozoans)	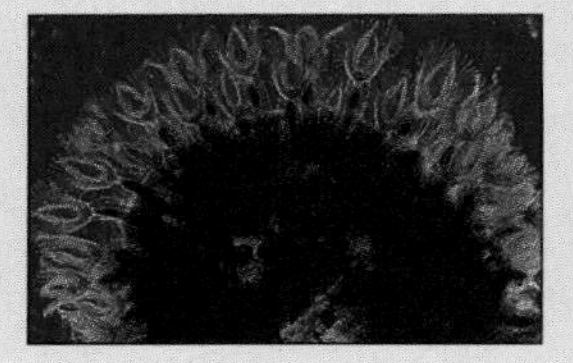 Ectoprocts (*Cristatella* sp.). (M. I. Walker/Photo Researchers, Inc.)	Sessile colonies produced by asexual budding. Moss-like appearance. Mainly marine. Lophophore can be retracted.
Brachiopods	 Brachiopods, Northern lamp shells (*Terebratulina septentrionalis*), photographed in York, Maine. (Fred Bavendam/Peter Arnold, Inc.)	Known as lamp shells. Solitary, marine; live on sea bottom. Body enclosed between two shells (superficially resemble bivalve mollusks like clams).
Echinodermata Sea stars, sand dollars, sea urchins	 Blue sea star (*Linckia lavigata*), photographed on the Great Barrier Reef, Australia. (Fred Bavendam/Peter Arnold, Inc.)	Spiny-skinned marine animals. Endoskeleton bears spines. Water vascular system; tube feet. Adults generally have pentaradial symmetry.
Chaetognatha (Arrow worms)	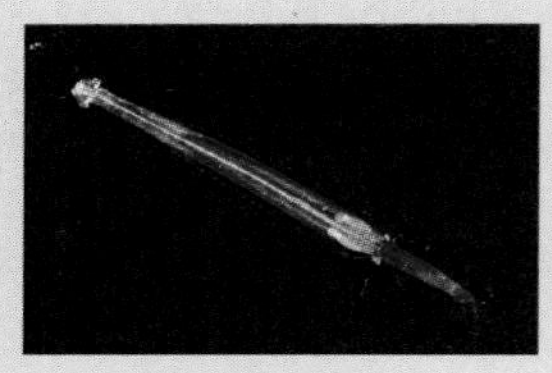 Arrow worm (Animals Animals, © 1996, Peter Parks, Oxford Scientific Films)	Marine; make up part of plankton. Ciliated, pharyngeal gill slits. Bilateral symmetry.
Hemichordata (Acorn worms)	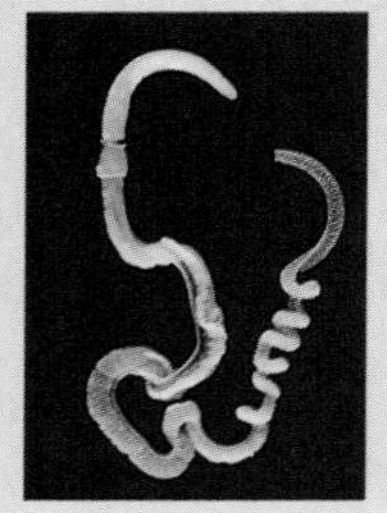 Acorn worm. (*Saccoglossus kowalewski*). (C. R. Wyttenbach, University of Kansas/Biological Photo Service)	Sedentary, wormlike, marine bottom dwellers. Some species construct mucus-lined burrows in the mud.
Chordata Urochordates, cephalochordates, vertebrates	 Four-eye butterfly fish (*Chaetodon capistratus*), photographed in South Caicos, BWI. (Fred Bavendam/Peter Arnold, Inc.)	Diverse organisms that inhabit marine, freshwater, and terrestrial habitats. Have notochord; dorsal, tubular nerve cord; pharyngeal gill grooves; postanal tail.

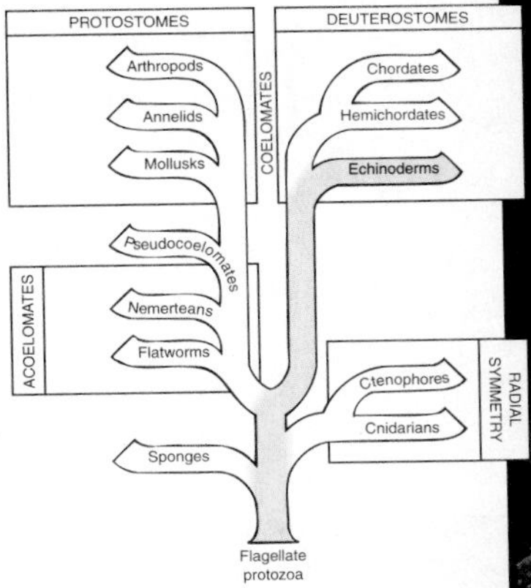

(a)

FIGURE 30–1 Echinoderms are a diverse group. (***a***) Feather stars such as this deep sea crinoid use their slender, jointed appendages to cling to the surface of a rock or coral reef. They are able to creep and often swim away to escape predators. (***b***) The red sea star (*Fromia ghardagana*) on a sponge in the Red Sea. (***c***) This daisy brittle star (*Ophiopholis aculeata*) was photographed in Muscongus Bay, Maine. (***d***) With their flattened, circular bodies, sand dollars (such as *Derdraster excentricus* shown here) are adapted for burrowing in the sea bottom. (***e***) A sea cucumber (*Thelonota* sp.), raising its body to spawn. (*a*, D. J. Wrobel, Monterey Bay Aquarium/Biological Photo Service; *b*, SharkSong/M. Kazmers/Dembinsky Photo Associates; *c*, Robert Dunne/Photo Researchers, Inc.; *d*, D. J. Wrobel, Monterey Bay Aquarium/Biological Photo Service; *e*, Peter Scoones/Seaphot, Ltd.)

(b)

(c)

(d)

(e)

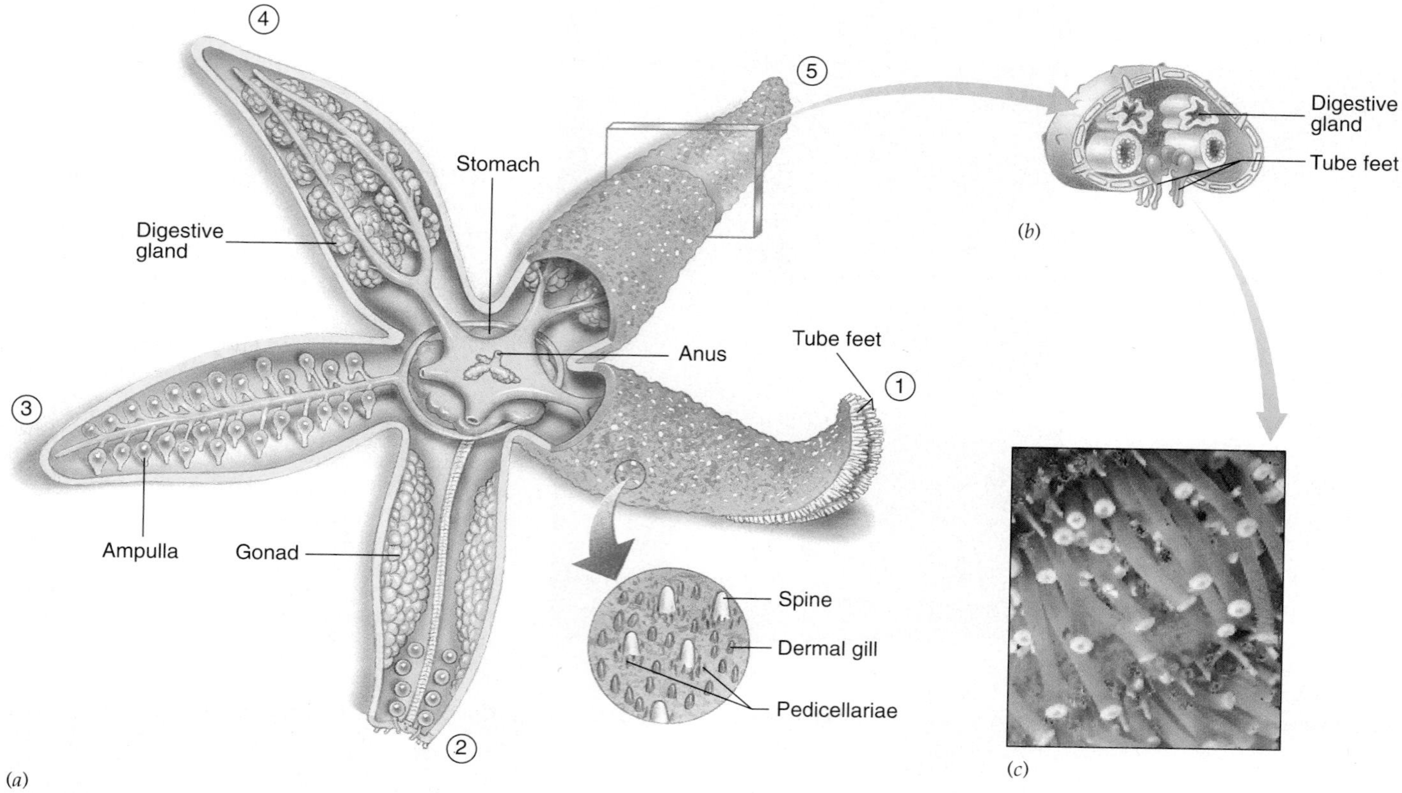

FIGURE 30–2 The sea star has a complex body plan. (*a*) The sea star *Asterias*, viewed from above, with the arms in various stages of dissection. (1) Upper surface with magnified detail showing the features of the surface. The end is turned up to show the tube feet on the lower surface. (2) The arm is dissected to show well-developed gonads. Gonads are present in each arm. (3) Upper body and digestive glands have been removed, exposing the ampullae of some of the hundreds of tube feet (magnified view). (4) Other organs have been removed to show the digestive glands (actually present in each arm). (5) Upper surface. The two-part stomach is in the central disc with the anus dorsal (uppermost) and the mouth beneath on the ventral surface. (*b*) Cross section through arm and tube feet. (*c*) Tube feet of a sea star. (*c*, Charles Seaborn/ Odyssey Productions, Chicago)

Class Asteroidea includes the sea stars

Sea stars, or starfish, are members of class Asteroidea. Their bodies consist of a central disk from which radiate 5 to 20 or more arms, or rays (Fig. 30–2). The undersurface of each arm is equipped with hundreds of pairs of tube feet. The mouth lies in the center of the underside of the disk. The endoskeleton consists of a series of calcareous plates that permit some movement in the arms. Delicate skin gills carry on gas exchange. Around the base of the skin gills, tiny pincer-like structures (pedicellaria) keep the surface of the animal free of debris.

Most sea stars are carnivorous predators and scavengers that feed on crustaceans, mollusks, annelids, and even other echinoderms. Occasionally they catch small fish. The sea star's water vascular system does not permit rapid movement, so its prey are usually slow-moving or stationary animals such as clams.

To attack a clam or other bivalve mollusk, the sea star mounts it and assumes a humped position as it straddles the edge opposite the hinge (Fig. 30–3). Then, holding itself in position with its tube feet, the sea star projects its stomach out through its mouth and onto the soft body of its prey. It can slide its thin, flexible stomach between the closed valves (shell-parts) of the clam. The sea star secretes enzymes that digest the soft parts of the clam to the consistency of a thick soup while the clam is still in its own shell. The partly digested meal passes into the sea star body where it is further digested by enzymes secreted from glands located in each arm.

The blood circulatory system in sea stars is poorly developed and probably of little help in circulating materials. Instead, this function is assumed by the coelomic fluid, which fills the large coelom and bathes the internal tissues. Metabolic wastes pass to the outside by diffusion. The nervous system consists of a ring of nervous tissue encircling the mouth and a nerve cord extending from this ring into each arm.

FIGURE 30–3 A painted sea star (*Orthiasterias koehleri*) attacks a clam. (Richard Chesher/Seaphot, Ltd.)

Class Ophiuroidea includes basket stars and brittle stars

Basket stars and brittle stars (serpent stars) are members of class Ophiuroidea (Fig. 30–1*c*). They make up the largest group of echinoderms, both in number of species and in number of individuals. These animals resemble asteroids in that their bodies also consist of a central disk with arms, but the arms are long and slender and more sharply set off from the central disk (Fig. 30–1*c*). Ophiuroids can move more rapidly than asteroids, using their arms to perform rowing or even swimming movements. The tube feet lack suckers and are not used in locomotion. They are used to collect and handle food and may also serve a sensory function, perhaps that of smell or taste.

Class Echinoidea includes sea urchins and sand dollars

Members of class Echinoidea lack arms, and their skeletal plates are flattened and fused to form a solid shell called a test (Fig. 30–1*d*). The sea urchin body is covered with spines which in some species can penetrate flesh and are difficult to remove. So threatening are these spines that swimmers on tropical beaches are often cautioned to wear shoes when venturing off-shore, where these animated pincushions lurk in abundance.

Sea urchins use their tube feet for moving and their movable spines for pushing themselves along. Most sea urchins graze, scraping the sea floor with their calcareous teeth. They eat algae and tiny protists as well as very small animals. Sand dollars have smaller spines than sea urchins. Their flattened bodies are adapted for burrowing in the sand, where they feed on tiny organic particles.

Class Holothuroidea includes sea cucumbers

Sea cucumbers are appropriately named, for some species are green and about the size and shape of a cucumber. The elongated body is a flexible, muscular sac (Fig. 30–1*e*). The mouth is usually surrounded by a circle of tentacles that are modified tube feet. Another characteristic of holothuroids is the reduction of the endoskeleton to microscopic plates. Like other echinoderms, sea cucumbers have a water vascular system for movement. Their blood circulatory system is more highly developed than that of other echinoderms, and functions to transport oxygen and perhaps nutrients as well.

Sea cucumbers are sluggish animals that usually live on the bottom of the sea, sometimes burrowing in the mud. Some graze with their tentacles, while others stretch their branched tentacles out in the water and wait for dinner to float by. Algae and other morsels are trapped in mucus along the tentacles.

An odd habit of some sea cucumbers is evisceration, in which the digestive tract, respiratory tree, and gonads are expelled from the body, usually when environmental conditions are unfavorable. When conditions improve, the lost parts are regenerated. Even more curious is the fact that when certain sea cucumbers are irritated or attacked, they direct their rear end toward the enemy and shoot red tubules out of their anus! These unusual weapons are sticky, and the attacking animal may become hopelessly entangled. Some of these tubules release a toxic substance.

CHORDATES HAVE A NOTOCHORD, A DORSAL TUBULAR NERVE CORD, AND PHARYNGEAL GILL SLITS DURING SOME TIME IN THEIR LIFE CYCLE

The phylum of animals to which humans belong, **phylum Chordata,** is divided into three subphyla: subphylum **Urochordata,** which consists of marine animals called tunicates; subphylum **Cephalochordata,** composed of marine animals called lancelets; and subphylum **Vertebrata,** the animals with backbones (Fig. 30–4).

Chordates are coelomate animals with bilateral symmetry (Fig. 30–5). They have a segmented body with a tube-within-a-tube body plan and three well-developed germ layers. Chordates have an endoskeleton and a closed circulatory system with a ventral heart. Four characteristics of chordates are:

1. All chordates have a **notochord** during some time in their life cycle. The notochord is a dorsal longitudinal rod that is firm but flexible and supports the body.
2. All chordates have a **dorsal tubular nerve cord** that is different from the nerve cord of most other animals. It

Table 30–2 COMPARISON OF ECHINODERMS AND CHORDATES*

	Echinodermata (Spiny-skinned animals) (6000 Species)	*Chordata (47,000 Species)*
Representative animals	Sea stars Sea urchins Sand dollars	Tunicates Lancelets Vertebrates
Body symmetry	Embryo: bilateral; adult: pentaradial	Bilateral
Gas exchange	Skin; gills	Gills or lungs
Digestion	Complete digestive tract	Complete digestive tract
Circulation	Open system; reduced	Closed system; ventral heart
Fluid balance/Waste disposal	Diffusion	Kidneys and other organs
Nervous system	Nerve rings; no brain	Dorsal nerve cord with brain at anterior end
Reproduction	Sexual; sexes almost always separate	Sexual; sexes separate
Support and movement	Endoskeleton bearing spines; muscles; tube feet	Notochord; endoskeleton of cartilage and/or bone; well-developed muscles
Environment and lifestyle	Marine; mainly carnivores	Diverse habitats and lifestyles; herbivores, carnivores, omnivores, scavengers, suspension feeders
Other characteristics	Water vascular system	Pharyngeal gill slits; post-anal tail

*Members of these phyla are at the organ system level of organization and have a complete digestive tract.

is located dorsally rather than ventrally, is hollow rather than solid, and is single rather than double.

3. All chordates have **pharyngeal gill slits** (also called pharyngeal slits) during some time in their life cycle. In the embryo, a series of alternating branchial arches and grooves develop in the body wall in the pharyngeal (throat) region. Pharyngeal pouches extend laterally from the anterior portion of the digestive tract toward the grooves (described in more detail in Chapter 49).

Biologists think that the earliest chordates (like many living chordates) were suspension feeders. The arrangement of pharyngeal pouches and gill slits permitted them to take water in through the mouth, concentrate small particles of food in the gut, and let the water escape from the body through the slits. The pharyngeal gill slits evolved into functional gill slits in fish. In terrestrial animals they become modified to form entirely different structures more suitable for life on land.

4. Most chordates have a **muscular postanal tail,** an appendage that extends posterior to the anus.

No clear fossil record of the chordate ancestors exists, but they were probably small, soft-bodied animals. A lancelet-like animal, *Pikaia,* has been found in the Burgess Shale of British Columbia. These rocks, which date back to the mid-Cambrian period, are more than 500 million years old.

Subphylum Urochordata includes the tunicates

The **tunicates,** which comprise the subphylum Urochordata, include the sea squirts, or ascidians, and their relatives. Adult sea squirts are barrel-shaped, sessile marine animals unlike other chordates. Indeed, they are often mistaken for sponges or cnidarians (Fig. 30–6).

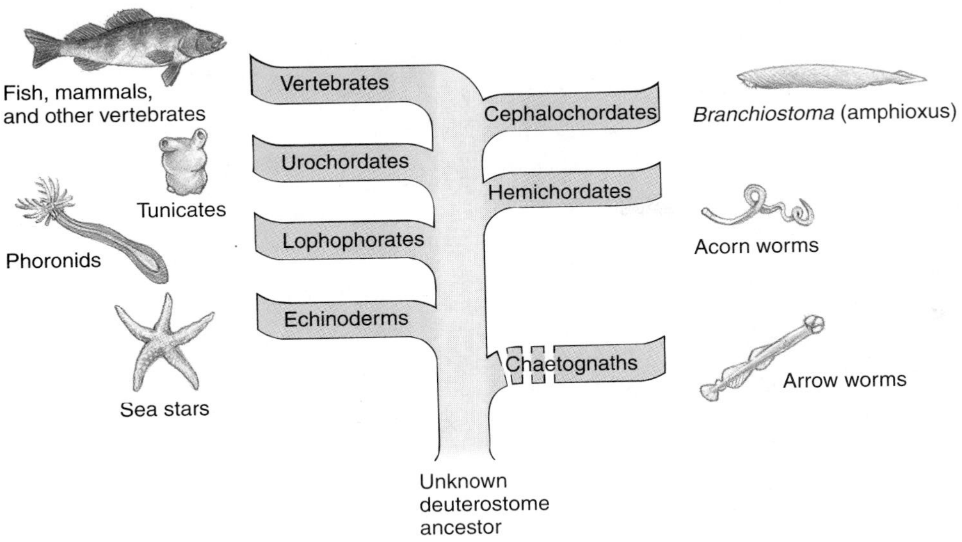

FIGURE 30–4 This family tree suggests the possible course of evolution of the chordates.

Adult tunicates develop a protective covering, or tunic, that may be soft and transparent or quite leathery. Curiously, the tunic is composed of a carbohydrate much like cellulose. The tunic has two openings: the incurrent siphon, through which water and food enter, and the excurrent siphon, through which water, waste products, and gametes pass to the outside.

Tunicates are suspension feeders, removing plankton from the stream of water passing through the pharynx. Food particles are trapped in mucus secreted by cells of the endostyle, a groove that extends the length of the pharynx. Ciliated cells within the pharynx produce a steady current of water that carries food particles down into the esophagus. Much of the water entering the pharynx passes out through gill slits into an **atrium** (chamber). Water is discharged through the excurrent siphon.

Some species of tunicates form large colonies in which members may share a common tunic and excurrent siphon. Colonial forms often reproduce asexually by budding. Sexual forms are usually hermaphroditic.

Larval tunicates have typical chordate characteristics and superficially resemble tadpoles. The expanded body has a pharynx with gill slits, and the long muscular tail contains a notochord and a dorsal, hollow nerve cord. In sea squirts, the larva eventually attaches itself to the sea bottom and loses its tail, notochord, and much of its nervous system. In the adult, only the gill slits suggest that the sea squirt is a chordate. Appendicularians, another class of tunicates, are common members of the zooplankton that retain their chordate features and ability to swim about.

Subphylum Cephalochordata includes the lancelets

Subphylum Cephalochordata consists of the lancelets, which are small, translucent, fish-shaped animals. These segmented animals are 5 to 10 centimeters (approximately 2–4 inches) long and pointed at both ends. They are widely distributed in shallow seas, either swimming freely or burrowing in the sand near the low-tide line. In some parts of the world, lancelets are an important source of food. One Chinese fishery reported an annual catch of 35 tons (about 1 billion lancelets).

Chordate characteristics are highly developed in lancelets. The notochord extends from the tip of the head (hence the name Cephalochordata) to the tip of the tail; many pairs of gill slits are evident in the large pharyngeal region; and a hollow, dorsal nerve cord extends the entire length of the animal (Fig. 30–7). The most common genus, *Branchiostoma* (commonly known as amphioxus), exhibits the basic chordate characteristics so well that it is often the first animal studied in comparative chordate anatomy courses. Although superficially similar to fishes, lancelets have a far simpler body plan.

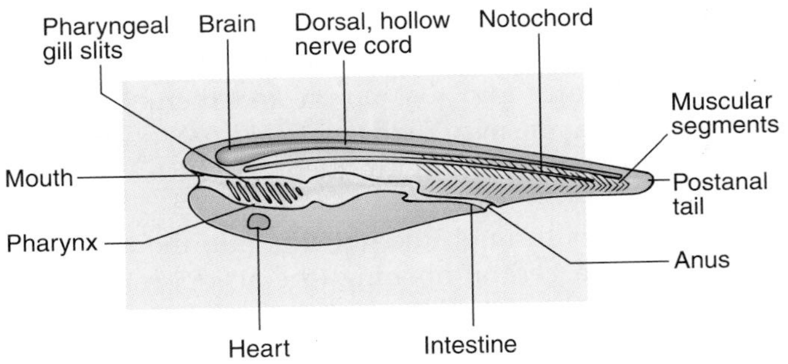

FIGURE 30–5 This generalized chordate illustrates several important characteristics of the phylum. Note the notochord; dorsal, hollow nerve cord; pharyngeal gill slits; and postanal tail.

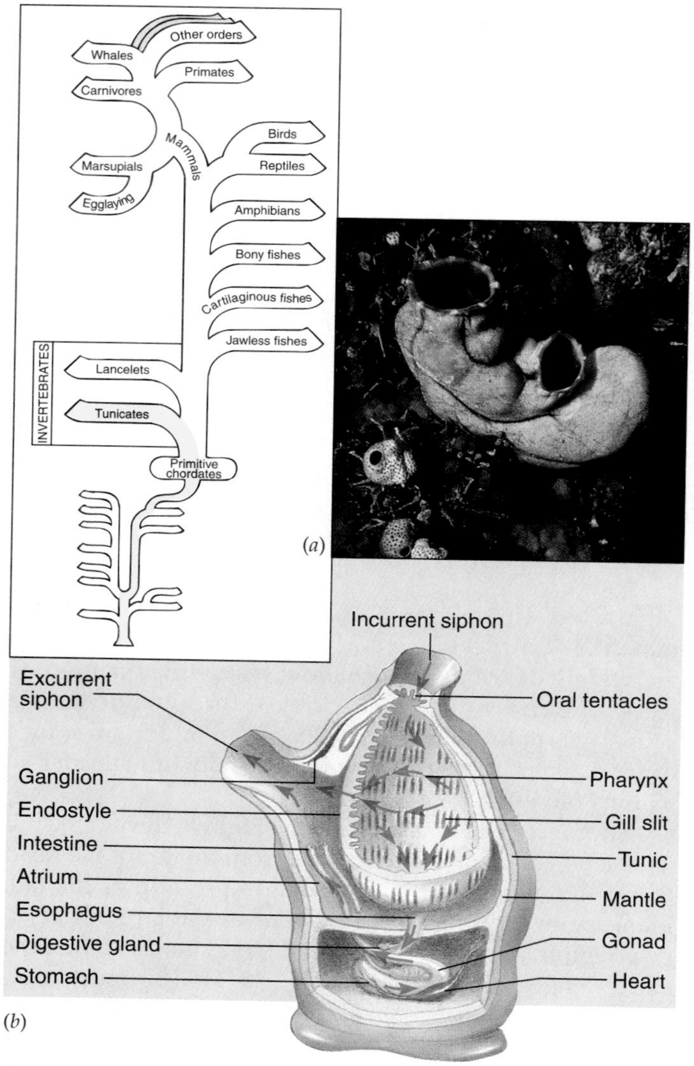

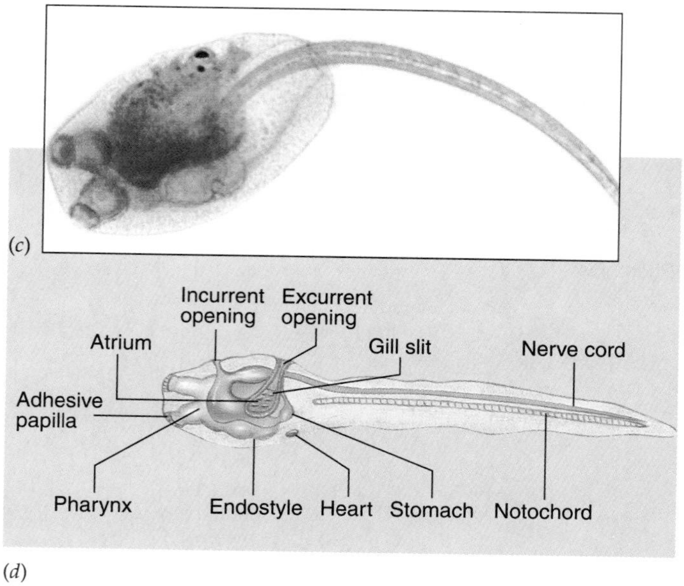

FIGURE 30–6 Tunicates are invertebrate chordates. (***a***) Incurrent (*top*) and excurrent (*side*) siphons of a sea peach tunicate (*Halocynthia auratium*). (***b***) Lateral view of an adult tunicate. The blue arrows represent the flow of water, and the red arrows represent the path of food. The stomach, intestine, and other visceral organs are embedded in the mantle. (***c***) Swimming larval stage of a tunicate, *Distaplia occidentalis*. (***d***) Internal structure of a larval tunicate (lateral view). (*a*, Robert Shupak; *c*, Richard A. Cloney/University of Washington)

They lack paired fins, jaws, sense organs, a heart, and a well-defined brain.

Like the tunicates, lancelets use their cilia to draw a current of water into the mouth and then strain out microscopic organisms. Food particles are trapped in mucus in the pharynx and are then carried back to the intestine.

Water passes through the gill slits into the atrium, a chamber with a ventral opening (the atriopore) just anterior to the anus. Metabolic wastes are excreted by segmentally arranged, ciliated protonephridia that open into the atrium. In contrast to other invertebrates, the blood flows anteriorly in the ventral vessel and posteriorly in the dorsal vessel. This circulatory pattern is similar to that of fishes.

The success of the vertebrates is linked to the evolution of key adaptations

The vertebrates—members of subphylum Vertebrata—are distinguished from other chordates in having a backbone, or **vertebral column,** that forms the skeletal axis of the body. This flexible support develops around the notochord, and in most species it largely replaces the notochord during embryonic development. The vertebral column consists of cartilaginous or bony segments called **vertebrae.** Dorsal projections of the vertebrae enclose the nerve cord along its length. Anterior to the vertebral column, a **cranium,** or braincase, encloses and protects the brain, the enlarged anterior end of the nerve cord.

The cranium and vertebral column are part of the **en-**

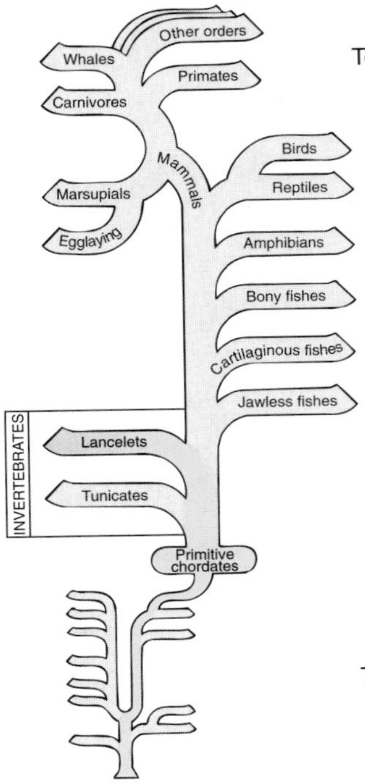

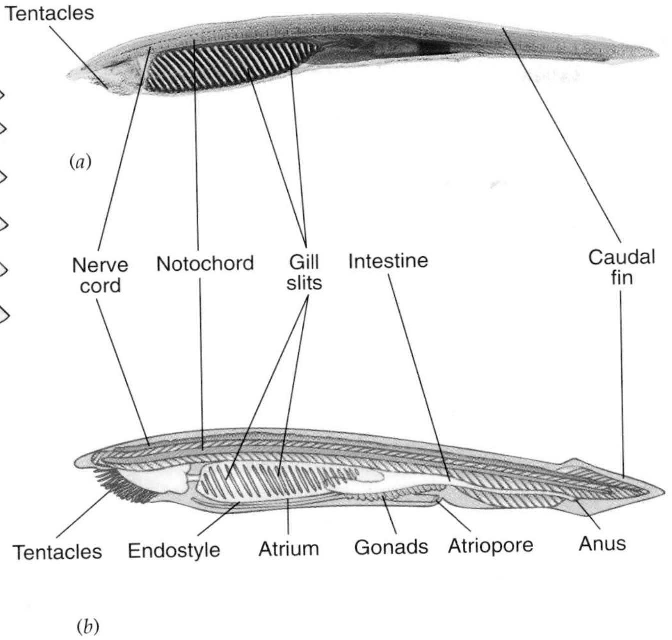

FIGURE 30–7 The lancelet *Branchiostoma* (amphioxus) is a cephalochordate. (***a***) Photograph of a lancelet showing external structure. Note the prominent gill slits. (***b***) Longitudinal section showing internal structure. (Visuals Unlimited/John D. Cunningham)

doskeleton. In contrast to the nonliving exoskeleton of many invertebrates, the vertebrate endoskeleton is a living tissue that grows with the animal. Two pairs of appendages are present in most vertebrates. The fins of fishes stabilize them in the water. As vertebrates moved onto the land, jointed appendages that facilitated locomotion evolved from lobed fins.

Recall that in invertebrates there is an evolutionary trend toward **cephalization,** concentration of nerve cells and sense organs in a definite head. Vertebrate evolution is characterized by *pronounced* cephalization. The brain becomes larger and more elaborate, and its various regions become specialized to perform different functions. Ten or 12 pairs of cranial nerves emerge from the brain and extend to various organs of the body. Vertebrates have well-developed sense organs concentrated in the head: eyes; ears that serve as organs of balance and, in some vertebrates, for hearing as well; and organs of smell and taste.

Vertebrates have two pairs of appendages, a closed circulatory system with a ventral heart, paired kidneys that regulate fluid balance, and a complete digestive tract with large digestive glands (the liver and pancreas). The sexes are almost always separate.

With about 43,000 species, the vertebrates are less diverse and much less numerous than the insects but rival them in their adaptation to an enormous variety of lifestyles. Most vertebrates excel over the insects in their ability to receive and respond appropriately to a wide variety of stimuli.

Living vertebrates are divided into three classes of fishes (sometimes referred to as superclass Pisces) and four classes of land vertebrates (superclass Tetrapoda; Fig. 30–8 and Table 30–3). The classes of fishes include class **Agnatha,** the jawless fishes such as lampreys; class **Chondrichthyes,** the sharks and rays with cartilaginous skeletons; and class **Osteichthyes,** the bony fishes.

The four-legged land vertebrates, or **tetrapods,** are grouped in class **Amphibia,** the frogs, toads, and salamanders; class **Reptilia,** the lizards, snakes, turtles, and alligators; class **Aves,** the birds; and class **Mammalia,** the mammals. Not all the tetrapods have four legs (e.g., the snakes), but all evolved from four-legged ancestors. Nor do all tetrapods now live entirely on land (e.g., sea turtles, penguins, whales, seals), but all of these aquatic forms evolved from terrestrial ancestors.

The jawless fishes are the most primitive vertebrates

The jawless fishes comprise class Agnatha (*a,* "without"; *gnathos,* "jaw"), which includes the extinct **ostracoderms** and the living lampreys and hagfishes. Fragments of ostracoderm scales have been found in rocks from the Ordovician period, indicating that vertebrates evolved in the sea 450 to 500 million years ago. By the Silurian and

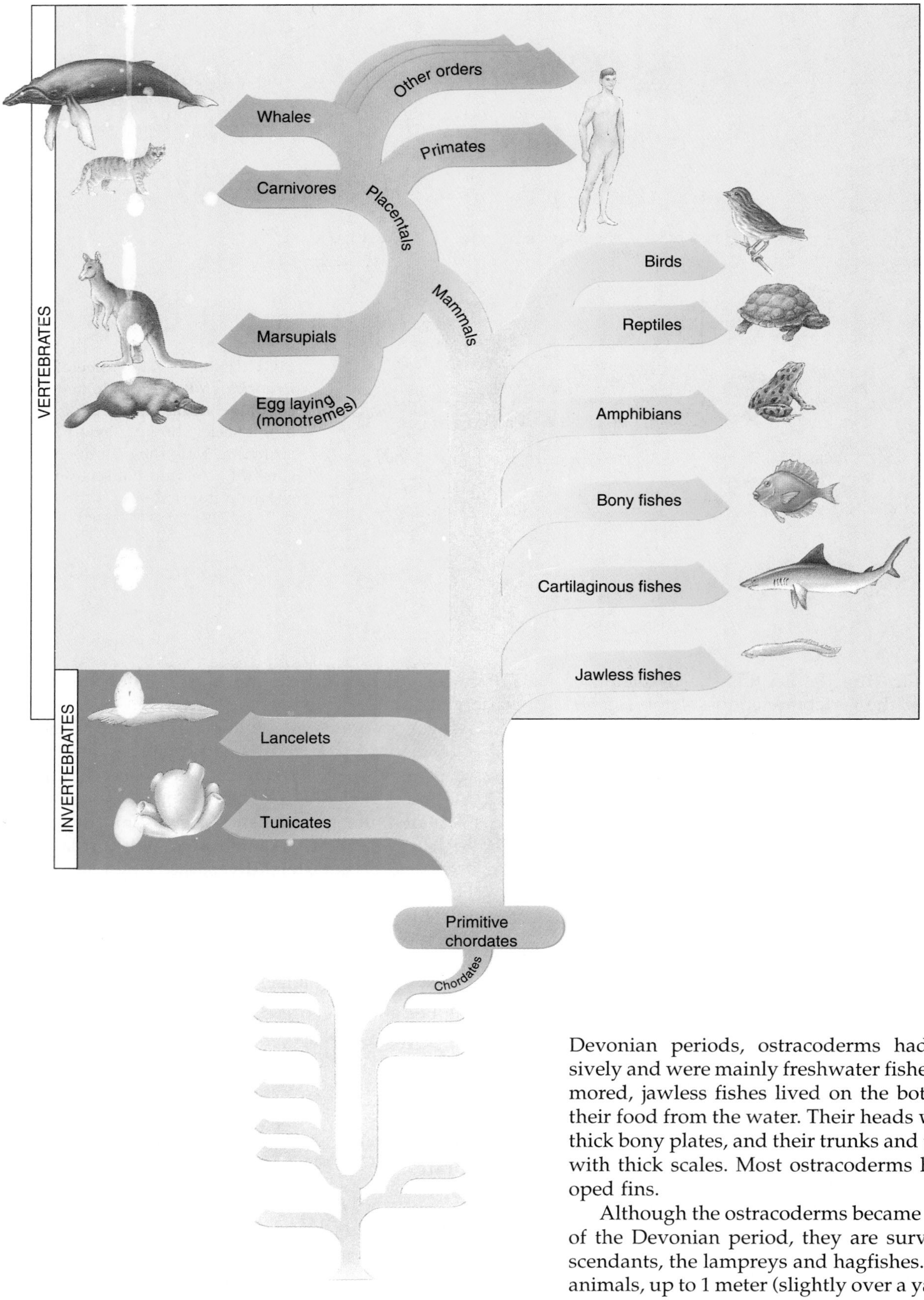

FIGURE 30 This tree illustrates current thinking about the evoluti e vertebrates.

Devonian periods, ostracoderms had radiated extensively and were mainly freshwater fishes. These small, armored, jawless fishes lived on the bottom and strained their food from the water. Their heads were covered with thick bony plates, and their trunks and tails were covered with thick scales. Most ostracoderms had poorly developed fins.

Although the ostracoderms became extinct by the end of the Devonian period, they are survived by their descendants, the lampreys and hagfishes. These eel-shaped animals, up to 1 meter (slightly over a yard) long, are supported by a cartilaginous skeleton. Their smooth skin lacks scales, and they have neither jaws nor paired fins.

Table 30–3 THE VERTEBRATE CLASSES

Class	*Examples*	*Characteristics*
Agnatha	Lampreys, hagfishes	No jaws; skeleton of cartilage; notochord persists throughout life; gills; marine and freshwater.
Chondrichthyes	Sharks, rays, skates	Skeleton of cartilage; gills; placoid scales; jawed marine and freshwater fishes; notochord replaced by vertebrae in adult. Two pairs of fins; oviparous, ovoviviparous, a few species viviparous; well-developed sense organs.
Osteichthyes	Perch, salmon, trout, tuna	Bony fishes; marine and fresh water; gills; swim bladder; generally oviparous.
Amphibia	Salamanders, frogs and toads, caecilians	Aquatic larva undergoes metamorphosis into terrestrial adult; gas exchange through lungs and/or moist skin; heart consists of two atria and single ventricle; systemic and pulmonary circulation.
Reptilia	Turtles, lizards, snakes, alligators	Tetrapods; terrestrial; body covered with hard scales; adapted for reproduction on land (internal fertilization, leathery shell, amnion); lungs; ventricle of heart partially divided.
Aves	Robins, pelicans, eagles, ducks, penguins, ostriches	Tetrapods with feathers; anterior limbs modified as wings; compact, streamlined body; lungs; four-chambered heart; complete separation of oxygen-rich and oxygen-poor blood; endotherms; vocal calls and complex songs.
Mammalia	Monotremes (platypus); marsupials (kangaroo); placentals (whales, humans)	Tetrapods with hair; females nourish young with mammary glands; diaphragm moves air in and out of lungs; highly developed nervous system; endotherms; most viviparous.

The hagfishes are marine scavengers. They burrow for worms and other invertebrates or prey on dead and disabled fishes.

Most lampreys live in fresh water. Some spend their adult lives in the ocean and return to fresh water to reproduce. Many species of adult lampreys are parasites on other fishes (Fig. 30–9). Adult parasitic lampreys have a circular sucking disk around the mouth, which is located on the ventral side of the anterior end of the body. Using this disk to attach to a fish, the lamprey bores through the skin of its host with horny teeth on the disk and tongue. Then the lamprey injects an anticoagulant into its host and sucks blood and soft tissues.

Adult lampreys leave the ocean or lake and swim upstream to spawn. They build a nest, a shallow depression in the gravelly stream bed, into which they shed eggs and sperm. After spawning they die. The fertilized eggs develop into larvae, which drift downstream to a pool and live as suspension feeders in burrows in the muddy bottom for up to seven years. They then undergo a metamorphosis, become adult lampreys, and migrate back to the ocean or lake.

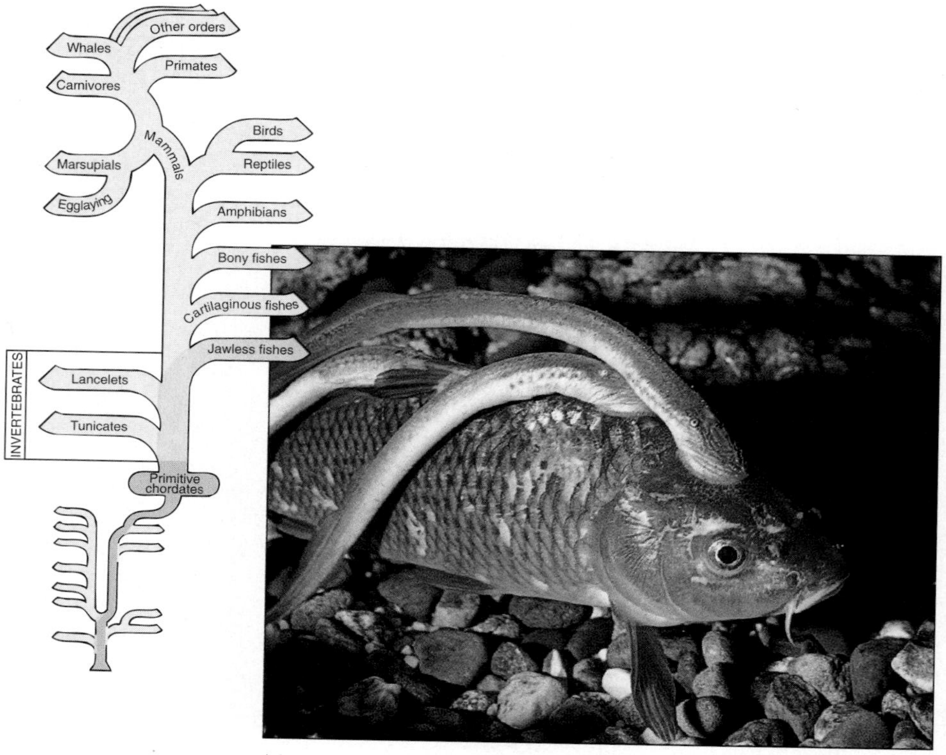

(a)

(b)

FIGURE 30–9 Lampreys are members of class Agnatha. (***a***) Three lampreys attached to a carp by their suction-cup-type mouths. Note the absence of jaws and paired fins. (***b***) Suction-cup-like mouth of adult lamprey (*Estosphenus japonicus*). Note the rasp-like teeth. (*a*, Tom Stack/Tom Stack & Associates; *b*, courtesy of Dr. Kiyoko Uehara)

The earliest jawed fishes are now extinct

Fossil evidence suggests that, during the late Silurian and Devonian periods, fish evolved with jaws and finlike paired appendages. Although they are now extinct, the first two classes of jawed fishes were the Acanthodii and the Placodermi (Fig. 30–10). **Acanthodians** were armored fishes with paired spines and pectoral and pelvic fins. The **placoderms** were armored fishes with paired fins.

The evolution of jaws from a portion of the gill arch skeleton and the development of fins enabled fish to change from suspension-feeding bottom-dwellers to active predators. This change afforded many new opportunities for finding food. The success of the jawed vertebrates may have contributed to the extinction of the ostracoderms.

Class Chondrichthyes includes the sharks, rays, and skates

The ostracoderms and early jawed fishes inhabited mainly fresh water. Only a few ventured into the oceans. Members of class **Chondrichthyes,** the cartilaginous fishes, evolved as successful marine forms in the Devonian period. Most species have remained as ocean dwellers, but a few have secondarily returned to a freshwater habitat.

The chondrichthyes—sharks, rays, and skates (Fig. 30–11)—retain their cartilaginous embryonic skeleton. Although this skeleton is not replaced by bone, it may be strengthened by a deposit of calcium salts. All chondrichthyes have paired jaws and two pairs of fins. The skin contains **placoid scales.** Each scale is a toothlike structure composed of an outer layer of enamel and an inner layer of dentine (Fig. 30–12). The lining of the mouth contains larger, but essentially similar, scales that serve as teeth. The teeth of the higher vertebrates are homologous with these shark scales. Shark teeth are embedded in the flesh and not attached to the jawbones; new teeth develop continuously in rows behind the functional teeth and migrate forward to replace any that are lost.

(*a*) Climatius

(*b*) Dinichthys

FIGURE 30–10 Acanthodians and placoderms flourished in the Devonian period. (***a***) *Climatius,* a spiny-skinned acanthodian with large fin spines and five pairs of accessory fins between the pectoral and pelvic pairs. (***b***) *Dinichthys,* a giant placoderm that grew to a length of nine meters (30 feet). Its head and thorax were covered by bony armor, but the rest of the body and tail were naked.

Because their bodies are denser than water, cartilaginous fishes tend to sink unless they are actively swimming. Most sharks are streamlined predators that swim actively and catch and eat other fishes as well as crustaceans and mollusks. The largest sharks, like the largest whales, feed on plankton. They gulp water through the mouth. Then, as the water passes through the pharynx and out the gill slits, food particles are trapped in a sievelike structure. The whale shark, which may reach a length of more than 12 meters (40 feet), is the largest fish known.

Sharks are attracted to blood, so a wounded swimmer or a skin diver towing speared fish is a target. However, although books and films portray sharks as monstrous enemies, most do not go out of their way to attack humans. In fact, of the approximately 350 known shark species, fewer than 30 have been known to attack humans.

The shark has a complex brain, which is differentiated both structurally and functionally. The spinal cord is protected by vertebrae. Well-developed sense organs enable

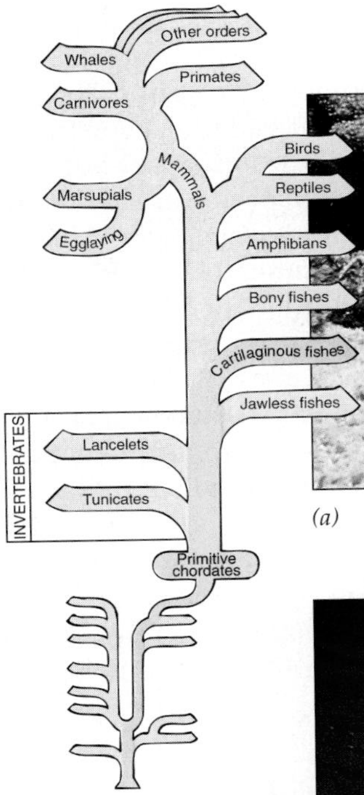

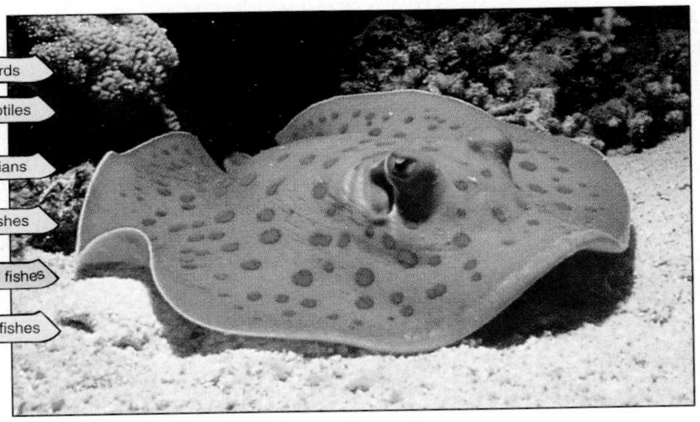

(*a*)

(*b*)

FIGURE 30–11 Sting rays and sharks are members of class Chondrichthyes. (***a***) Blue-spotted sting ray *(Taeniura lymma).* (***b***) Great white shark *(Carcharodon carcharias)* photographed near surface of the water, in Australia. (*a,* Jeffrey L. Rotman/Peter Arnold, Inc.; *b,* Kelvin Aitken/Peter Arnold, Inc.)

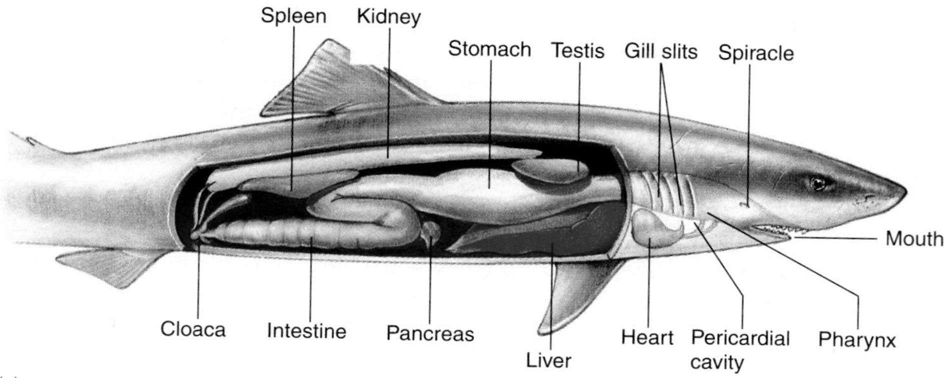

FIGURE 30–12 Sharks have cartilaginous skeletons. (*a*) Internal structure of a shark. (*b*) Structure of a placoid scale.

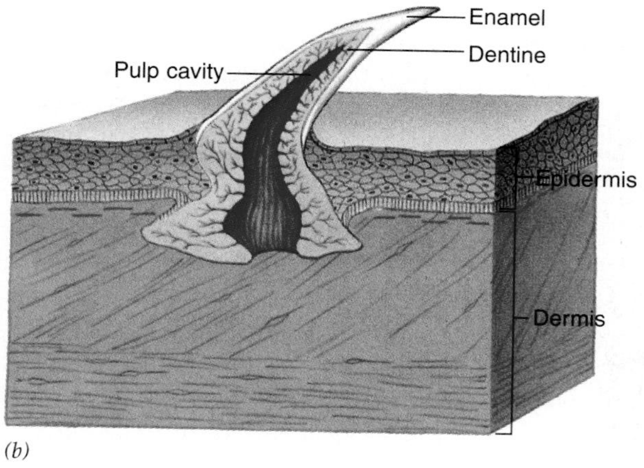

the shark to locate prey in the water. Sharks may detect other animals electrically before sensing them by sight or smell. **Electroreceptors** on the shark's head can sense weak electric currents generated by the muscular activity of animals. The **lateral line organ,** found in all fishes, is a groove along each side of the body with many tiny openings to the outside. Sensory cells in the lateral line organ respond to movement of the water.

Cartilaginous fishes have five to seven pairs of gills. A current of water enters the mouth and passes over the gills and out the gill slits, constantly providing the fish with a fresh supply of dissolved oxygen. However, because they have no mechanism for pumping water over the gills, many sharks depend on their motion for gas exchange.

The digestive tract of the shark consists of the mouth cavity; a long pharynx leading to the stomach; a short, straight intestine; and a **cloaca,** which opens on the underside of the body and is characteristic of many vertebrates (Fig. 30–12). The liver and pancreas discharge digestive juices into the intestine. The cloaca receives digestive wastes as well as urine from the urinary system and, in the female, sperm from the male.

The sexes are separate, and fertilization is internal. In the mature male, each pelvic fin has a slender, grooved section, known as a **clasper,** used to transfer sperm into the female's cloaca. The eggs are fertilized in the upper part of the female's oviducts. Part of the oviduct is modified as a shell gland, which secretes a protective coat around the egg. Skates and some species of sharks are **oviparous;** that is, they lay eggs. Many species of sharks, however, are **ovoviviparous,** meaning that their young are enclosed in eggs that are incubated within the mother's body. During development, the young depend on stored yolk for their nourishment, rather than on transfer of materials from the mother. The young are born after hatching from the eggs. A few species of sharks are **viviparous:** not only do the embryos develop within the uterus, but much of their nourishment is delivered to them by the mother's blood. An intimate relationship is established between the yolk sac surrounding each embryo and the blood vessels in the lining of the uterus.

Most rays and skates are sluggish, flattened creatures that live partly buried in the sand. Their enormous pectoral fins propel them along the bottom where they feed on mussels and clams. The sting ray has a whiplike tail with a barbed spine that can inflict a painful wound. The electric ray has electric organs on either side of the head. These modified muscles can discharge enough electric current (up to 2500 watts) to stun fairly large fishes as well as human swimmers.

Bony fish belong to class Osteichthyes

The class Osteichthyes includes more than 24,000 living species of freshwater and saltwater bony fishes, of many shapes and colors (Fig. 30–13). Bony fishes range in size from the Philippine goby, which is only about 10 millimeters (0.4 inch) long, to the ocean sunfish (or *Mola;* Fig. 19–12*c*), which may reach 1 metric ton (about 2200 lb). Cartilaginous and bony fishes share many characteristics (e.g. continuous tooth replacement), but they also differ in important ways. Although osteichthyes appear earlier in the fossil record, both classes may have evolved about the same time (during the Devonian period).

Most osteichthyes are characterized by a bony skeleton with many vertebrae (Fig. 30–14). Bone has advantages over cartilage: it provides excellent support and serves as a very effective storage site for calcium.

The osteichthyes body is covered with overlapping bony dermal scales. Most species have both median and paired fins, supported by long rays made of cartilage or bone. A lateral bony flap, the **operculum,** extends posteriorly from the head and protects the gills.

Unlike the sharks, bony fishes are generally oviparous. They lay an impressive number of eggs that they fertilize externally. The ocean sun[illegible] example, lays over 300 million eggs! Of course, [illegible]f the eggs and young become food for other ani[illegible]o increase the probability of survival, many speci[illegible]shes build nests for their eggs and even watch ove[illegible]m.

The ray-finned fishes gave rise to modern bony fishes

During the Devonian period, the bony fishes diverged into two major groups: the **ray-finned fishes** (actino-

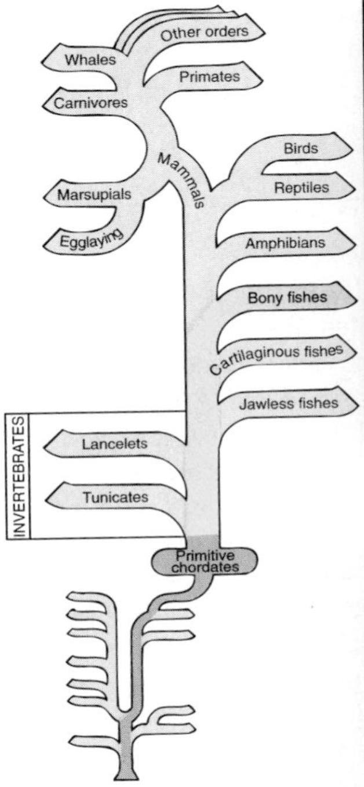

(*a*)

(*b*)

(*c*)

(*d*)

FIGURE 30–13 Bony fishes are a diverse group. (***a***) Spiny puffer fish *(Dictolychthys punctulatus),* in Fiji. (***b***) Schooling goggle-eyes *(Priacanthus hamrur),* Lizard Island, Great Barrier Reef, Australia. (***c***) Parrot fish *(Scarus gibbus)* photographed in the Red Sea. (***d***) The leaf-like processes of this Australian leafy sea dragon *(Phyllapteryx taeniolatus)* help camouflage it in surr[illegible]ding kelp. (*a,* Animals/Animals © 1996 [illegible]ace Watkins; *b,* Fred Bavendam/Peter Arnold, Inc.; *c,* Jeffrey L. Rotman/Peter Arnold, Inc.; *d,* Norbert Wu/Peter Arnold, Inc.)

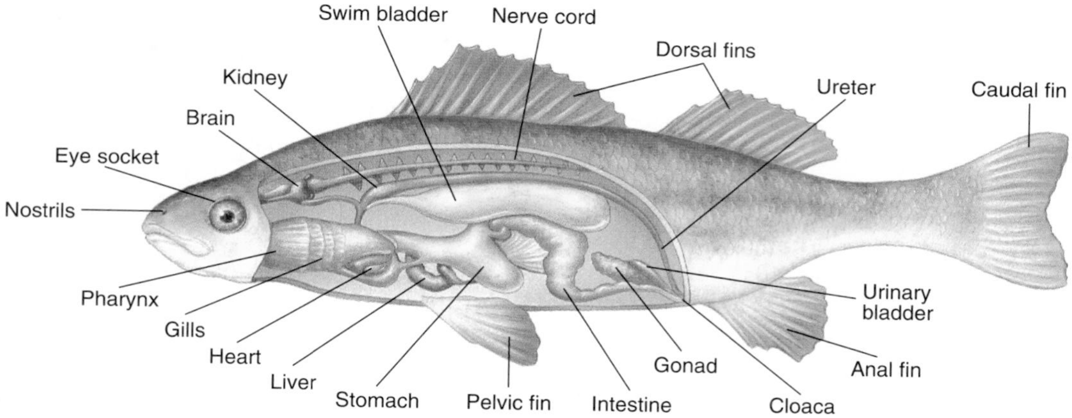

FIGURE 30–14 The perch is a representative bony fish. The swim bladder is a hydrostatic organ that enables the fish to change the density of its body, and remain stationary at a given depth.

pterygians) and the **lobe-finned fishes** (sarcopterygians). The ray-finned fishes gave rise to most modern bony fishes. The ancestors of the ray-finned fishes are thought to have had lungs.

The ray-finned fishes underwent two important adaptive radiations (see Chapter 20). The first gave rise during the late Paleozoic era to a group of fishes that are now mostly extinct. The second began during the early Mesozoic era and gave rise to the modern bony fishes, the **teleosts.** The lungs became modified as a **swim bladder,** an air sac that helps the fish regulate its buoyancy (Fig. 30–14). By secreting gases into the bladder or absorbing gases from it, a fish can change the overall density of its body. This ability allows a fish to hover at a given depth of water without much muscular effort.

Descendants of the lobe-finned fishes moved onto the land

The lobe-finned fishes (sarcopterygians) are characterized by fleshy, lobed fins and lungs. Two lines of sarcopterygians diverged: lungfishes and crossopterygians. During the Devonian period, frequent seasonal droughts caused swamps to become stagnant or even to dry up completely. Fishes with lungs had a tremendous advantage for survival under those conditions. The fleshy fins of lobe-finned fishes could support their weight, enabling them to emerge onto dry land and make their way to another pond or stream. The ability to move about, however awkwardly, on dry land gave these animals access to new food sources—terrestrial plants that were already established and terrestrial insects and arachnids that were rapidly evolving. A vertebrate that could survive on land had less competition for food. Laying eggs on land away from the many ocean predators also increased their chances for successful reproduction. Natural selection favored those individuals best adapted for making their way on land, resulting eventually in the evolution of the amphibians and, later, the reptiles.

Three genera of lungfishes have survived to the present day in the rivers of tropical Africa, Australia, and South America. The lobe-finned crossopterygians, generally considered the ancestors of the land vertebrates, were almost extinct by the end of the Paleozoic era. However, one genus (*Latimeria*), classified as a **coelacanth** (a suborder adapted for marine life), survives in the deep waters off the east coast of Africa near the Comoro Islands (Fig. 30–15). Nearly 2 meters (about 6 feet) long, these giant "living fossils" have neither lungs nor functional swim bladders.

Class Amphibia includes frogs, toads, and salamanders

The first successful **tetrapods,** or land vertebrates, were the **labyrinthodonts** (Figs. 30–16 and 30–17), clumsy, salamander-like animals with short necks and heavy muscular tails. These ancient members of class Amphibia somewhat resembled their ancestors, the lobe-finned fishes. However, they had limbs strong enough to support the weight of their bodies on land. These earliest limbs were five-fingered, a pattern that has largely been retained in vertebrate evolution.

The largest labyrinthodonts were the size of crocodiles. They flourished during the late Paleozoic and early Mesozoic eras, becoming extinct during the Mesozoic era. It is probable that the labyrinthodonts gave rise to other primitive amphibians, to frogs and salamanders, and to the earliest reptiles, the **cotylosaurs.**

Modern amphibians are classified in three orders: Order **Urodela** includes the salamanders, mud puppies, and newts—animals with long tails; order **Anura** is made up

FIGURE 30–15 Ancestors of this coelacanth, a lobe-finned fish, probably gave rise to the amphibians. The paired fins indicate the basic plan of a jointed series of bones that could evolve into the limbs of a terrestrial vertebrate. Living coelacanths (genus *Latimeria*) are difficult to observe because they inhabit deep ocean waters, and when brought to the surface, they do not survive the change in atmospheric pressure. (Dr. J. Metzner/Peter Arnold, Inc.)

of the tailless frogs and toads, with legs adapted for hopping; and order **Apoda** contains the wormlike, legless caecilians (Fig. 30–18). Although some adult amphibians are quite successful as land animals and can live in dry environments, most return to the water to reproduce. Eggs and sperm are generally released in the water.

The embryos of frogs and toads develop into larvae called **tadpoles.** These larvae have tails and gills and feed on aquatic plants. After a time the tadpole undergoes metamorphosis. The gills and gill slits disappear, the tail is resorbed, and limbs emerge. The digestive tract shortens and food preference shifts from plant material to a carnivorous diet, the mouth widens, a tongue develops, the tympanic membrane and eyelids appear, and the eye lens changes shape. Many biochemical changes also accompany the transformation from a completely aquatic life to an amphibious one.

As in insects, amphibian metamorphosis is under hormonal control. It is regulated by hormones secreted by the thyroid gland. Amphibians undergo a single, although complex, change from larva to adult. Several salamanders, such as the mudpuppy *Necturus,* do not undergo complete metamorphosis: they retain many larval characteristics as adults. In this process, known as **neoteny,** animals become sexually mature without completing metamorphosis. Neoteny permits these salamanders to remain aquatic rather than having to compete on land (Fig. 19–13).

Adult amphibians do not depend solely on their primitive lungs for the exchange of respiratory gases. Their moist, glandular skin, which lacks scales and is plentifully supplied with blood vessels, also serves as a respiratory surface. Many mucous glands within the skin help to keep the body surface moist, which is important in gas exchange and helps prevent drying out. The mucus also makes the animal slippery; as a result it can escape from predators. Some amphibians have glands in their skin that secrete poisonous substances harmful to predators.

The coloration of amphibians may conceal them in their habitat or may be very bright and striking. Many of the brightly colored species are poisonous (see Fig. 53–3). Their distinctive colors warn predators that they are not encountering an ordinary amphibian. Some frogs can camouflage themselves by changing color.

The amphibian heart is divided into three chambers: two **atria** receive blood and a single **ventricle** pumps it into the arteries. A double circuit of blood vessels keeps oxygen-rich and oxygen-poor blood partially separate. Blood passes through the **systemic circulation** to the various tissues and organs of the body. Then, after returning to the heart, it is directed through the **pulmonary circulation** to the lungs and skin, where it is recharged with oxygen. The oxygen-rich blood returns to the heart to be pumped out into the systemic circulation again. The comparative anatomy of the heart and circulation of various vertebrate classes is discussed in Chapter 42.

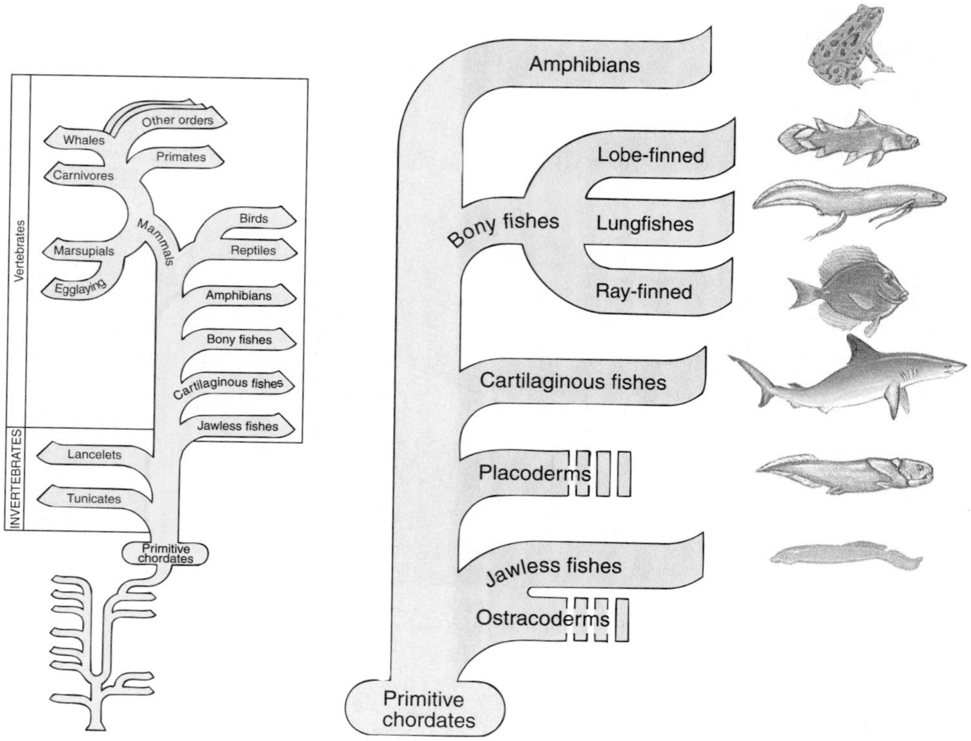

FIGURE 30–16 This phylogenetic tree illustrates the probable evolution of fishes and amphibians.

Class Reptilia includes turtles, lizards, snakes, and alligators

Members of class Reptilia are terrestrial animals that do not need to return to water to reproduce. A variety of adaptations make the reptilian lifestyle possible. The female secretes a protective leathery shell around the egg, which helps prevent the developing embryo from drying out. Because sperm cannot penetrate this shell, fertilization occurs within the body of the female before the shell is added. In this process of internal fertilization, the male uses a copulatory organ to transfer sperm into the female reproductive tract.

As the embryo develops within the protective shell, a membrane called the **amnion** forms and surrounds the embryo. The amnion secretes fluid, providing the embryo with its own private "pond." The amniotic fluid keeps the embryo moist and also serves as a shock absorber should the egg get bounced around. All terrestrial vertebrate embryos have amnions as well as other membranes that protect and support their development (see Chapter 49).

The hard, dry, horny scales that protect the reptilian body from drying are another adaptation to life on land. This scaly protective armor, which also helps protect the reptile from predators, is shed periodically.

The dry reptilian skin cannot serve as an organ for gas exchange. However, reptilian lungs are better-developed than the saclike lungs of amphibians. Divided into many chambers, the reptilian lung provides a greatly increased surface area for gas exchange. Most reptiles have a three-chambered heart that is more efficient than the amphibian heart. The ventricle has a partition, though incomplete, that separates oxygen-rich and oxygen-poor blood. This separation facilitates oxygenation of body tissues.

Another adaptation to life on land is a method of waste disposal that conserves water. Much of the fluid

FIGURE 30–17 An artist's conception of labyrinthodonts (a primitive Paleozoic amphibian). Labyrinthodonts inhabited carboniferous swamps about 345 million years ago. (Photo by Logan, courtesy of Department of Library Services, American Museum of Natural History)

FIGURE 30–18 Modern amphibians include frogs and salamanders. (***a***) This leopard frog *(Rana pipiens)* leaps to evade a predator. (***b***) The red eft newt *(Notophthalmus viridescens)* belongs to the family Salamandridae. Although mainly aquatic, the red eft spends one to three years as a pre-adult in a moist terrestrial environment. (*a*, Stephen Dalton/Photo Researchers, Inc.; *b*, The Stock Market/Roy Morsch)

filtered from the blood by the kidneys is reabsorbed in the kidney tubules and urinary bladder. In aquatic animals, nitrogenous wastes from protein and nucleic acid metabolism are excreted as ammonia. This waste product is quite toxic, and its excretion requires large amounts of water. Reptiles (like birds and most terrestrial arthropods) convert ammonia to uric acid, which is much less toxic and can be excreted as relatively insoluble crystals. These solid crystals do not require much water for their excretion, and so water is conserved.

Like fishes and amphibians, reptiles generally lack metabolic mechanisms for regulating body temperature. They are **ectothermic,** meaning that their body temperature fluctuates with the temperature of the surrounding environment. Some reptiles have behavioral adaptations that enable them to maintain a body temperature higher than that of their environment. You may have observed a lizard basking in the sun, which raises its body temperature and so increases its metabolic rate. This permits the lizard to actively hunt for food. When the body of a reptile is cold, the metabolic rate is low and the animal tends to be sluggish. Ectothermia may explain why reptiles are more successful in warm than in cold climates.

Although a few species of turtles, tortoises, and lizards are herbivores (animals that eat vegetation), most reptiles are carnivores (meat-eaters). Their paired limbs, usually with five toes, are well adapted for running and climbing, and their well-developed sense organs enable them to locate prey.

Modern reptiles belong to three orders

The reptiles living today are assigned to one of three main orders: Order **Chelonia** includes turtles, terrapins, and tortoises; order **Squamata** is composed of lizards, snakes, iguanas, and geckos; and order **Crocodilia** contains crocodiles, alligators, caimans, and gavials (Fig. 30–19).

Members of order Chelonia are enclosed in a protective shell made up of bony plates overlaid by horny scales. Some species can withdraw their heads and legs completely into their shells. The size of adult turtles ranges from about 8 centimeters (3 inches) long to that of the great marine species, which measure more than 2 meters (6.5 feet) in length and may weigh 450 kilograms (almost 1000 pounds). The land species are usually referred to as tortoises, whereas the aquatic forms are called turtles (freshwater types are sometimes called terrapins).

Lizards and snakes are the most common of the modern reptiles. These animals have rows of scales that overlap like shingles on a roof, forming a continuous flexible armor that may be shed periodically. Lizards range in size from certain geckos, which may weigh as little as 1 gram (0.035 ounce) to the Komodo dragon of Indonesia, which may weigh 100 kilograms (220 pounds). Their body sizes and shapes vary greatly. Some—for example, the glass snake, which is really a lizard—are legless. For a discussion of an interesting example of iguana evolution in the Caribbean, see *Making the Connection: The Iguanas of Mona Island, the Galapagos of the Caribbean.*

Snakes are characterized by a flexible, loosely jointed jaw structure that permits them to swallow animals larger than the diameter of their own jaws. Snakes lack legs, and their bodies are elongated. Their eyes, covered by a transparent cuticle, lack movable eyelids. They also lack an external ear opening, a tympanic membrane (eardrum), and a middle ear cavity.

The forked tongue of a snake, which often darts quickly from its mouth, is used as an accessory sensory organ for touch and smell. Chemicals from the ground or air adhere to the tongue, and the tip is then projected into a sense organ located in the roof of the mouth that detects odors. Pit vipers and some boas also have a prominent **sensory pit** on each side of the head that enables them to detect heat. These sense organs permit them to locate and capture small nocturnal mammals.

Some snakes—for example, king snakes, pythons, and boa constrictors—capture their prey by rapidly wrapping themselves around the animal and squeezing it so that it cannot breathe. Others have fangs, which are hollow teeth connected to poison glands in the mouth. When the snake bites, the poison is pumped through the fangs into the prey's body. Some snake poisons cause the breakdown of red blood cells; others, such as that of the coral snake, are neurotoxins that interfere with nerve function. Poisonous snakes of the United States include rattlesnakes, copperheads, cottonmouths, and coral snakes. All except the coral snakes are pit vipers.

The three groups of crocodilians are: (1) the crocodiles of Africa, Asia, and America; (2) the alligators of the southern United States and China, plus the caimans of Central America; and (3) the gavials of Southeast Asia. Most species live in swamps, in rivers, or along sea coasts, burrowing in the mud and feeding on various kinds of animals. Crocodiles are the largest living reptiles; some measure more than 7 meters (23 feet) long.

Reptiles were once the dominant land animals

A labyrinthodont ancestor gave rise to the vertebrates which evolved the reproductive adaptations that permit-

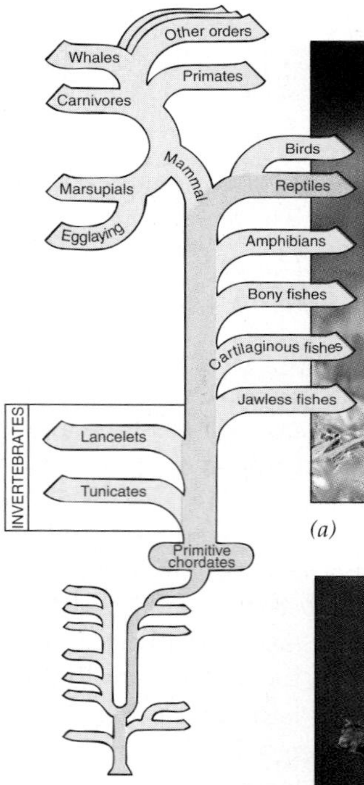

(*a*)

(*b*)

FIGURE 30–19 Modern reptiles include lizards, turtles, and crocodiles. (***a***) This basilisk lizard *(Basiliscus plumifrons)* was photographed in Tortuguero National Park, Costa Rica. (***b***) Like many turtles, this green turtle *(Chelonia mydas),* a native of Hawaii, is a good swimmer. (***c***) This Nile crocodile *(Crocodilus niloticus)* is in the process of hatching. (*a*, Y. Lefevre/Peter Arnold, Inc.; *b, c,* Frans Lanting/Minden Pictures)

(*c*)

ted them to be independent of a watery external environment. Terrestrial vertebrates—reptiles, birds, and mammals—are referred to as **amniotes** because their embryos are enclosed by an amnion. Amniotes are thought to be a monophyletic group; that is, they have a common ancestor (Fig. 30–20).

The earliest reptiles are thought to have resembled lizards. By the late Carboniferous period (about 290 million years ago) amniotes had diverged into three groups. One of these groups evolved into the mammals, a second into the turtles, and a third gave rise to all of the other reptiles and to the birds.

Many more extinct reptiles than living species are known. The Mesozoic era, which ended about 65 million years ago, is known as the Age of Reptiles. During that time reptiles were the dominant terrestrial animals (see Chapter 20). They had radiated into an impressive variety of ecological lifestyles (Fig. 20–11). Some were able to fly, others became marine, and many filled terrestrial habitats. Some of the dinosaurs were among the largest land animals to have ever lived.

The reptiles were the dominant land animals for almost 200 million years. Then, toward the end of the Mesozoic era, many of them—including all of the dinosaurs—disappeared from the fossil record (see Chapter 20). In fact, more than half of all animal species became extinct at that time.

The birds belong to class Aves

Birds, which comprise class Aves, are the only animals with feathers (Fig. 30–21). Thought to have evolved from reptilian scales, feathers are flexible and very strong relative to their light weight. They protect the body, decrease water and heat loss, and aid in flight by presenting a flat surface to the air.

The anterior limbs of birds are wings, usually modified for flight. The posterior limbs are modified for walking, swimming, or perching. Not all birds fly. Some, such as penguins, have small, flipper-like wings used in swimming. Others, such as the ostrich and cassowary, have only vestigial wings but well-developed legs.

In addition to feathers and wings, birds have many other adaptations for flight. Their bodies are compact and streamlined, and the fusion of many bones gives them the rigidity needed for flying. Their bones are strong but very light; many are hollow, containing large air spaces. The jaw is light, and instead of teeth, there is a light, horny beak. Their very efficient lungs have thin-walled extensions, called air sacs, that occupy spaces between the internal organs and within certain bones. Birds, like mammals, have a four-chambered heart and a double circuit of blood flow. Blood delivers oxygen to the tissues and then is recharged with oxygen in the lungs before being pumped out into the systemic circulation again.

The very effective respiratory and circulatory systems provide the cells with enough oxygen to permit a high metabolic rate, which is necessary for the tremendous muscular activity that flying requires. Some of the heat generated by metabolic activities is used to maintain a constant body temperature. Birds and mammals are the only animals that are **endotherms** (sometimes referred to as warm-blooded). This ability to maintain a constant body temperature permits birds to remain active in cold climates.

Birds excrete nitrogenous wastes mainly as semisolid uric acid. Because they lack a urinary bladder, these solid wastes are delivered into the cloaca. They leave the body with the feces, which are dropped frequently. This adaptive mechanism helps to maintain a light body weight.

Birds have become adapted to a variety of environments, and various species have very different types of beaks, feet, wings, tails, and behavioral patterns. Bills are specifically adapted for the type of food the bird eats. Although all birds must eat frequently (because they have a high metabolic rate but do not store much fat), the choice of food varies widely among species. Most birds eat energy-rich foods such as seeds, fruits, worms, mollusks, or arthropods. Warblers and some other species eat mainly insects. Owls and hawks eat rodents, rabbits, and other small mammals. Vultures feed on dead animals. Pelicans, gulls, terns, and kingfishers catch fishes. Some hawks catch snakes and lizards.

An interesting feature of the bird digestive system is the **crop,** an expanded, saclike portion of the digestive tract below the esophagus, in which food is temporarily stored. The stomach is divided into a **proventriculus,** which secretes gastric juices, and a thick, muscular **gizzard,** which grinds food. The bird swallows small bits of gravel that act as "teeth" in the gizzard, mechanically breaking down food.

Birds have a well-developed nervous system with a brain that is proportionately larger than that of reptiles. Birds rely heavily on vision. Their eyes are relatively larger than those of other vertebrates. Hearing is also well developed.

In striking contrast to the relatively silent reptiles, birds are very vocal. Most have short, simple calls that signal danger or influence feeding, flocking, or interaction between parent and young. Songs are usually more complex than calls and are performed mainly by males. Songs are related to reproduction, attracting and keeping a mate, and claiming and defending territory.

One of the most fascinating aspects of bird behavior is the annual migration that many species make. Some birds, such as the golden plover and Arctic tern, fly from Alaska to Patagonia, South America, and back each year, covering perhaps 40,250 kilometers (25,000 miles) en route. Migration and navigation are discussed in Chapter 50.

Making the Connection

The Iguanas of Mona Island, the Galapagos of the Caribbean

In Chapter 17 we discussed the unusual and unique species studied by Charles Darwin on the Galapagos Islands off the Pacific coast of South America. The same evolutionary processes that occurred in the Galapagos Islands have given rise to unique species on other islands.

Consider Mona Island, a small island that is part of Puerto Rico. Mona Island could be considered the "Galapagos of the Caribbean." Located 68 kilometers (42 miles) west of the main island of Puerto Rico and 60 kilometers (37 miles) east of Hispaniola, Mona Island is home to many fascinating animal and plant species.

Mona Island is quite different from the rest of Puerto Rico. The island resembles a great, kidney-shaped white slab of floating limestone, bounded almost completely by towering cliffs rising vertically out of the sea. Although moist tropical trade winds blow over it, its low elevation precludes much precipitation; as a result, it has a semi-arid climate.

In spite of Mona Island's small size (12 kilometers [7.5 miles] long, 6 kilometers [3.75 miles] wide), its distance from large land masses, its lack of a deep, fertile soil, and its low rainfall, it supports 417 species of vascular plants and nearly 700 species of land animals. Many of these species are endemic (found nowhere else). Ancestral species are thought to have colonized the island from nearby land areas. Over countless generations, the colonizers became more and more distinctive as they adapted to the local conditions. In time, they evolved into new species.

Like the Galapagos, Mona Island is now a wildlife sanctuary inhabited by unique plant and animal species, including its most striking endemic, the Mona ground iguana (*Cyclura steinegeri*). The genus *Cyclura* consists of 14 living and fossil species, all of which are confined to the Bahamas and the Greater Antilles. These distinctive reptiles are rapidly becoming extinct. In 1916, 11 species of iguanas were found in the Caribbean, but now only 6 species remain; all of these are confined to remote areas uninhabited by humans.

Mona Island. (Héctor E. Quintero, Ph.D.)

The Mona iguana measures from 1 to 1.3 meters (3–4 feet) in length and has a heavy body, a large head, and a stout, laterally compressed tail. A dorsal crest extends from its head to its tail. Its closest living relative is the rhinoceros iguana (*Cyclura cornuta*) found on Hispaniola. Both species have a distinct series of horns protruding from the snout, but the species can be distinguished by differences in their scale patterns.

Early birds had reptilian characteristics

Birds are thought to have evolved from saurischian dinosaurs, long-tailed animals that moved about on two feet and had forelimbs with three clawed fingers (Chapter 20). Feathers may have first evolved as an adaptation that conserved body heat. Bird ancestors became endothermic and more active. According to one hypothesis, these animals ran along the ground using their feathered forelimbs to swat insects, which they then ate. Another hypothesis holds that bird ancestors climbed trees and used their feathered forelimbs for gliding.

Although the bones of birds are fragile and disintegrate quickly, a few fossils of early birds have been found. The first birds looked very much like reptiles. They had teeth (which modern birds lack), a long tail, and bones with thick walls. Unlike reptiles, their jaws were elongated into beaks, and they had feathers and wings.

One of the earliest known birds, *Archaeopteryx* (meaning "ancient wing"), was about the size of a pigeon and had rather feeble wings. Its jawbones were armed with reptilian-type teeth (so this early bird could get its worm?), and it had a long reptilian tail covered with feathers. Each of its wings was equipped with three claw-bearing digits (Fig. 30–22). Several specimens of this species have been found in the Jurassic limestone of Bavaria, which was laid down about 150 million years ago (see Figure 20–13).

Cretaceous rocks have yielded fossils of other early birds. *Hesperornis,* which lived in North America, was a toothed, aquatic diving bird with powerful hind legs and vestigial wings. *Ichthyornis* was a toothed, flying bird

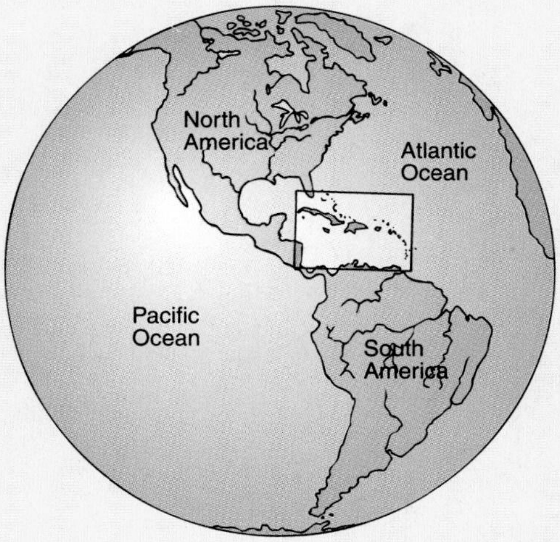

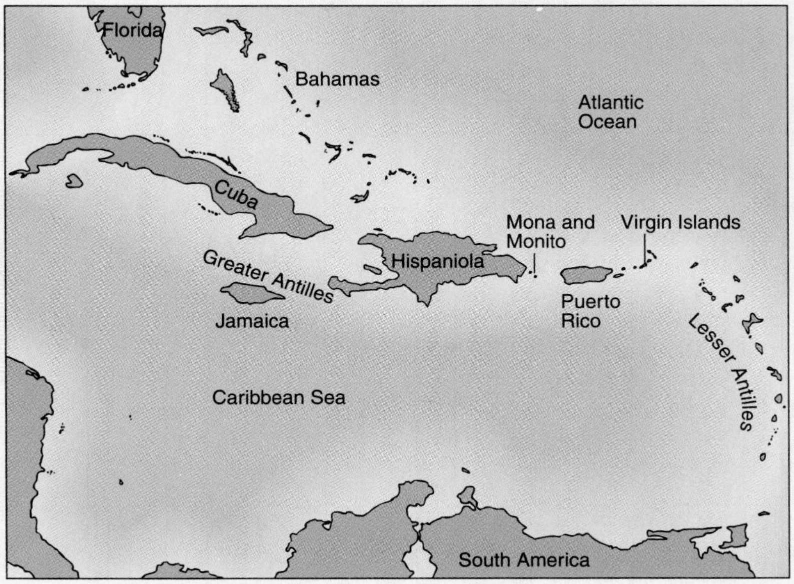

Mona iguanas are found throughout Mona Island, especially along major escarpments, cliff slopes, and sinkholes. In the summer nesting season they congregate on the coastal plains where soils are deeper and more suitable for building nests. Males are usually solitary and strongly territorial.

In 1977 a study estimated the population of iguanas on Mona Island to be 2000. That study also demonstrated a scarcity of young iguanas, which is characteristic of declining populations. The hazards to which Mona iguanas have been subjected are mostly the result of the introduction of other species to the island. Cats, mice, rats, and pigs dig up iguana nests and eat their eggs, whereas goats compete with the iguanas for food. If action is not taken soon to control the introduced species, the Mona island iguana, like many of its nearby relatives, will probably become extinct. Something similar occurred on the Galapagos Islands. Extinction of the Galapagos land iguana first occurred on islands where goats had been living the longest. Currently, stable iguana populations can only be found on those Galapagos islands without goats. ▲

Essay contributed by Hector E. Quintero, Ph.D.

about the size of a seagull. In 1990 a fossil found in the rocky remnants of an ancient lake in China was described as the earliest known example of a bird with modern flying ability. This bird apparently had the adaptations necessary for living in trees. From the Tertiary period onward, the fossil record of birds shows an absence of teeth and progressive changes leading to the modern birds.

Modern birds are a very successful group

Even modern birds possess some characteristics in common with the reptiles. For example, they lay eggs and have reptilian-type scales on their legs.

About 9000 species of birds have been described; these have been classified into 27 orders. Birds inhabit a wide variety of habitats and can be found on all of the continents, most islands, and even the open sea. The largest living birds are the ostriches of Africa, which may be up to 2 meters (6.5 feet) tall and weigh 136 kilograms (300 pounds), and the great condors of the Americas, with wingspans of up to 3 meters (10 feet). The smallest known bird is Helena's hummingbird of Cuba, with a length of less than 6 centimeters (2 inches) and a weight of less than 4 grams (0.14 ounce).

Beautiful and striking colors are found among birds. Color is due partly to pigments deposited during the development of the feathers and partly to reflection and refraction of light of certain wavelengths. Many birds, especially females, are protectively colored by their plumage. Brighter colors are often assumed by the male during the breeding season to help him attract a mate.

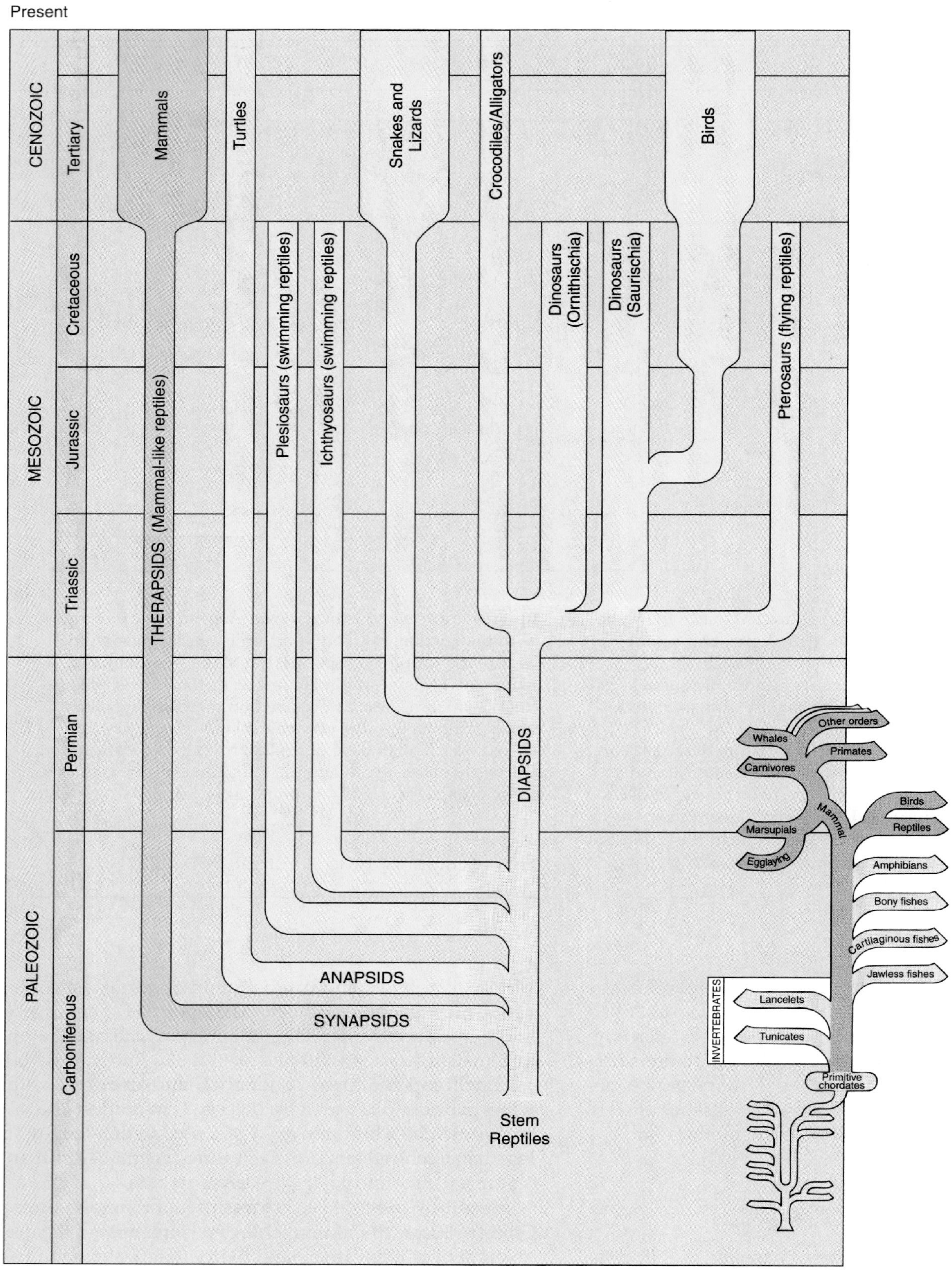

FIGURE 30–20 This hypothetical tree illustrates some proposed evolutionary relationships among extinct and present-day reptiles and some other vertebrates.

FIGURE 30–21 Modern birds are an extremely diverse group. (***a***) A male peacock *(Pavo cristatus)* impressively displaying his feathers. (***b***) This Atlantic puffin *(Fratercula arctica)* has just caught a fish. (***c***) Cattle egrets *(Bubulcus ibis)* photographed in Kenya. (*a*, Frans Lanting/Minden Pictures; *b*, Gary Meszaros/Dembinsky Photo Associates; *c*, D. W. Fawcett)

Mammals have hair and mammary glands

The distinguishing features of mammals are hair; **mammary glands,** which produce milk for the young; and the differentiation of teeth into incisors, canines, premolars, and molars. A muscular **diaphragm** helps to move air into and out of the lungs. Like the birds, mammals are endotherms. The process of maintaining a constant body temperature is enhanced by the covering of hair, which serves as insulation, and by the four-chambered heart and separate pulmonary and systemic circulations. Red blood cells without nuclei serve as excellent oxygen transporters.

Contributing significantly to the success of the mammals is the complex nervous system, which is more highly developed than in any other group of animals. The cerebrum is especially large and complex, with an outer gray region called the cerebral cortex.

Fertilization is always internal, and except for the primitive monotremes that lay eggs, mammals are viviparous. Most mammals develop a **placenta,** an organ of exchange between developing embryo and mother, through which the embryo receives its nourishment and oxygen and rids its blood of wastes.

The limbs of mammals are variously adapted for walking, running, climbing, swimming, burrowing, or flying. In four-legged mammals, the limbs are more directly under the body than they are in reptiles, which contributes to speed and agility. Life processes of mammals are discussed in detail in Part 7.

FIGURE 30–22 ***Archaeopteryx* was a very early bird.** This painting represents the hypothesis that *Archaeopteryx* was a climbing animal that had at least some ability to use its wings and feathers for gliding. Other hypotheses suggest that *Archaeopteryx* remained mainly on the ground and used its wings to trap small insects and its feathers for insulation. (Also see Figure 20–13 for the fossil on which this reconstruction is based.) (From a painting by Rudolph Freund, courtesy of Carnegie Museum of Natural History.)

Early mammals were small, endothermic animals

Mammals are thought to have evolved from a group of reptiles called **therapsids** (Fig. 30–23) during the Triassic period some 200 million years ago. The therapsids were doglike carnivores with differentiated teeth (a mammalian trait) and legs adapted for running. The fossil record indicates that the early mammals were small, about the size of a mouse or shrew. Some may have been endothermic and some may have had fur.

How did the mammals manage to coexist with the reptiles during the 160 million years or so that reptiles ruled Earth? Many adaptations permitted the early mammals to compete for a place on our planet. Perhaps one of the most important was their skill at being inconspicuous. They were **arboreal** (tree-dwelling) and **nocturnal** (active at night), searching for food (mainly insects and plant material and perhaps reptile eggs) at night while the reptiles were inactive. This lifestyle is suggested by the large eye sockets seen in fossil species, indicating that they possessed the large eyes characteristic of present-day nocturnal mammals. By bearing their young alive, mammals avoided the hazards of having their eggs consumed by predators. By nourishing the young and caring for them, the parents could offer both protection and an "education"—which probably focused on how to obtain food and avoid being eaten.

As many reptiles died out, the mammals adapted to their abandoned niches (lifestyles). During this time, the flowering plants, including many trees, underwent adaptive radiation, providing new habitats, sources of food, and protection from predators. Larger forms and numerous varieties of mammals evolved. During the early Cenozoic era (perhaps 50 million years ago), the mammals underwent adaptive radiation, becoming widely distributed and adapted to an impressive variety of ecological lifestyles.

Modern mammals are assigned to three subclasses

By the end of the Cretaceous period, there were three main groups of mammals. Today, mammals inhabit virtually every corner of Earth; they are found on land, in fresh and salt water, and in the air. They range in size from the tiny pigmy shrew, weighing about 25 grams (less than 1 ounce), to the blue whale, which may weigh more than 90,000 kilograms (88 tons) and is thought to be one of the largest animals that has ever lived.

Modern mammals are classified in three subclasses: **Prototheria** includes the egg-laying mammals, also called monotremes; **Metatheria** includes the marsupials, or pouched mammals; and **Eutheria** includes the placental mammals.

Monotremes Are Mammals That Lay Eggs The **monotremes** are the only living order of subclass Prototheria. The two genera include the duck-billed platypus (*Ornithorhynchus*) and the spiny anteater or echidna (*Tachyglossus*) (Fig. 30–24; see also Figs. 1–16 and 22–6). Both are found in Australia and Tasmania; the spiny anteater is also found in New Guinea. The females lay eggs that may be carried in a pouch on the abdomen or kept warm in a nest. When the young hatch, they are nourished with milk from the mammary glands. As its name suggests, the spiny anteater feeds on ants, which it catches with its long sticky tongue. The duck-billed platypus lives in burrows along river banks. It has webbed feet and a flat, beaver-type tail, which aids in swimming. The duck-billed platypus preys on freshwater invertebrates.

FIGURE 30–23 Therapsids were mammal-like reptiles. This therapsid *(Lycaenops)* lived in the late Permian period in South Africa. (Trans. no. 203. Painting by John C. Germann, Department of Library Services, American Museum of Natural History)

Marsupials Are Pouched Mammals **Marsupials** include pouched mammals such as kangaroos and opossums. Embryos begin their development in the mother's uterus, where they are nourished by fluid and yolk. After a few weeks, still in a very undeveloped stage, the young are born. They crawl to the **marsupium** (pouch), where they complete their development. The young marsupial attaches its mouth to a mammary gland nipple and is nourished by its mother's milk (Fig. 30–25).

Like the monotremes, the marsupials are found mainly in Australia. Only the opossum is common in North America. At one time, marsupials probably inhabited much of the world, but they were replaced by the placental mammals. Australia became geographically isolated from the rest of the world before placental mammals reached it, and there the marsupials remained the dominant mammals (see Chapters 17 and 20). They underwent adaptive radiation, paralleling the evolution of placental mammals elsewhere. Thus, in Australia and adjacent islands we find marsupials that correspond to North American placental wolves, bears, rats, moles, flying squirrels, and even cats (Fig. 30–26).

Placental Mammals Complete Embryonic Development within the Mother Most familiar to us are **placental mammals,** characterized by development of a placenta, an organ of exchange between the developing embryo and its mother (Fig. 30–27). The placenta forms from both embryonic membranes and the maternal uterine wall. In it the blood vessels of the embryo come very close to the blood vessels of the mother, so that materials can be exchanged by diffusion. (The two circulations do not normally mix.) The placenta enables the young to remain within the mother's body until embryonic development is complete.

(Text continues on page 682)

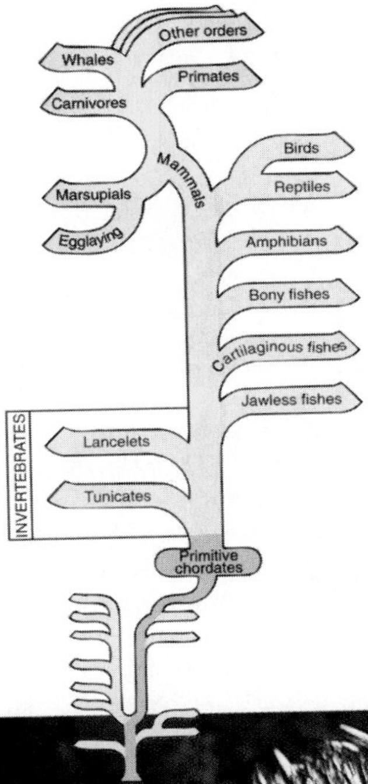

FIGURE 30–24 Monotremes are egg-laying mammals. The spiny anteater (*Tachyglossus* sp.). (Tom McHugh/Photo Researchers, Inc.)

Table 30–4 SOME ORDERS OF LIVING PLACENTAL MAMMALS

Order and Examples		*Some Characteristics*
Insectivora Moles, hedgehogs, and shrews	 Mother African hedgehog with young (*Atelerix albiventris*), Busch Gardens, Florida. (Animals Animals © 1996 Jim Tuten)	Nocturnal; eat insects; considered most primitive placental mammals. Shrew is the smallest living mammal; some weigh less than 5 grams.
Chiroptera Bats	 Lesser long-nosed bat (*Leptony cleris curasoae*) approaching flower. (Merlin D. Tuttle/Bat Conservation International)	Adapted for flying; a fold of skin extends from the elongated fingers to the body and legs, forming a wing. Guided in flight by a type of biological sonar: they emit high-frequency squeaks and are guided by the echoes from obstructions. Eat insects and fruit, or suck blood of other animals.
Carnivora Cats, dogs, wolves, foxes, bears, otters, mink, weasels, skunks	 Gray wolf, black female (*Canis lupus*). (Carl R. Sams II/Peter Arnold, Inc.)	Carnivores with sharp, pointed canine teeth and molars for shearing. Keen sense of smell; complex social interactions. Among fastest, strongest, and smartest animals.
Edentata Sloths, anteaters, armadillos	 Nine-banded armadillo (*Dasypus novemcinctus*), southern Florida. (Animals Animals © 1996 Ken Cole)	Teeth reduced or no teeth. Sloths are sluggish animals that hang upside down from branches; often protectively colored by green algae that grow on their hair. Armadillos are protected by bony plates; eat insects and small invertebrates.
Rodentia Squirrels, beavers, rats, mice, hamsters, porcupines, guinea pigs	 Giant flying squirrel (*Petaurista elegans*), Java, Indonesia. (BIOS/A. Compost/Peter Arnold, Inc.)	Gnawing animals with chisel-like incisors. As they gnaw, teeth are worn down, and so must grow continually.
Lagomorpha Rabbits, hares, pikas	 Pika (*Ochotona princeps*), Banff National Park, Alberta, Canada. (Ed Reschke)	Like rodents, have chisel-like incisors. Long hind legs adapted for jumping. Many have long ears.
Primates Lemurs, monkeys, apes, humans	 Maki catta ring-tailed lemur (*Lemur catta*). (Animals Animals © 1996 Henry Ausloos)	Highly developed brains and eyes. Nails instead of claws. Opposable thumb. Eyes directed forward. Omnivores. Most species arboreal. (Primate evolution is discussed in Chapter 21.)

Order and Examples		Some Characteristics
Perissodactyla Horses, zebras, tapirs, rhinoceroses	 Baird's tapir (*Tapirus bairdii*), Belize. (Kevin Schafer/Peter Arnold, Inc.)	Herbivores. Hoofed with an odd number of digits per foot; one or three toes. Teeth adapted for chewing. Usually large animals with long legs. (Hoofed mammals are referred to as ungulates).
Artiodactyla Cattle, sheep, pigs, deer, giraffes	 American elk (*Cervus elaphus*) browsing on pine in a coniferous forest, Yellowstone National Park. (Ed Reschke)	Hoofed with even number of digits per foot; most have two toes, some have four. Most have antlers or horns. Herbivores; most are ruminants that chew a cud and have a series of stomachs in which bacteria that digest cellulose are incubated; this contributes to their success as herbivores.
Proboscidea Elephants	 African elephant (*Loxodonta africana*) with young, South Africa. (Stan Osolinski/FPG International)	Largest land animals; weigh up to 7 tons; large head; broad ears; long, muscular trunk (proboscis) that is very flexible. Thick, loose skin characteristic. The two upper incisors are elongated as tusks. This order includes the extinct mastodons and wooly mammoths.
Sirenia Sea cows, manatees	 Indian manatee (*Trichechus manatus*), Florida. (Mark J. Thomas/Dembinsky Photo Associates)	Herbivorous, aquatic mammals with finlike forelimbs and no hind limbs. They are probably the basis for most tales about mermaids.
Cetacea Whales, dolphins, porpoises	 Humpback whale calf (*Megaptera novaeangliae*), Harvey Bay, Queensland, Australia. (Kelvin Aitken/Peter Arnold, Inc.)	Adapted for aquatic life with fish-shaped body and broad, paddle-like forelimbs (flippers). Posterior limbs absent. Many have thick layer of fat (blubber) covering the body. Mate and bear their young in the water; young suckled like those of other mammals. Very intelligent. The blue whale is the largest living animal and may be the largest animal that has ever existed.
Pinnipedia Seals, sea lions, walruse	 California sea lions (*Zalophus californianus*), Ano Nuevo Islands. (Frans Lanting/Minden Pictures)	Marine; limbs adapted as flippers for swimming. Carnivores; eat fish.

Placental mammals are born at a more mature stage than are marsupials. Indeed, among some species the young can walk around and begin to interact with other members of the group within a few minutes after birth. Extant (living) placental mammals are classified into about 17 orders. A brief summary of some of these orders is given in Table 30–4.

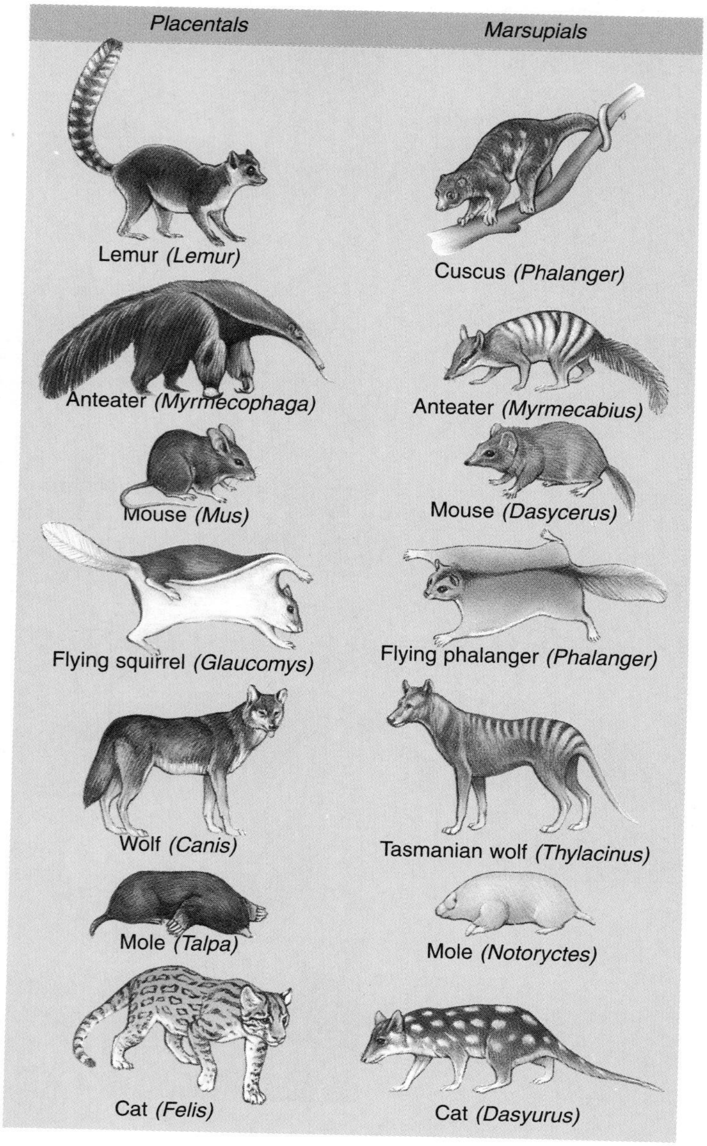

FIGURE 30–26 Placental and marsupial mammals have similar ways of life. For each mammal with a given niche in one group, there is a counterpart in the other group. These similarities include both lifestyles and structural features.

(a)

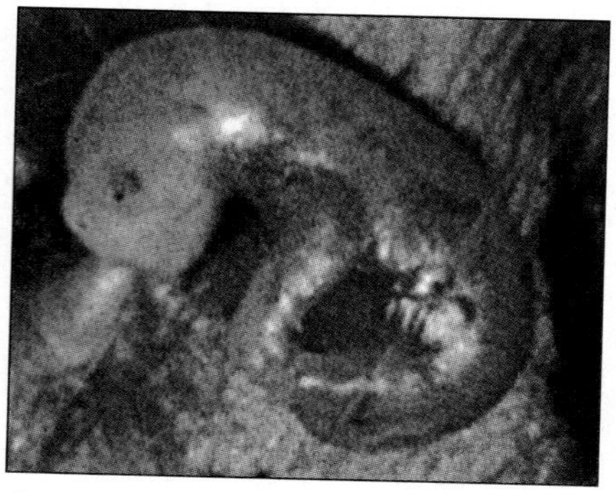

(b)

FIGURE 30–25 Marsupials are pouched mammals. ***(a)*** Eastern gray kangaroo *(Macropus giganteus)* with joey. The kangaroo is native to Australia. ***(b)*** A kangaroo soon after birth. Young marsupials are born in a very immature state. The young continue to develop in the safety of the marsupium (pouch). (*a*, John Cancalosi/Peter Arnold, Inc.; *b*, Robert Anderson, reprinted with permission of Hubbard Scientific Company)

(a)

(b)

(c)

FIGURE 30–27 Placental mammals are nourished during development by nutrients that pass through the placenta, an organ of exchange. (***a***) Bison calves *(Bison bison)* with parent in western USA. (***b***) Humpback whales *(Megaptera novaeangliae)*, exhibiting a rare double breach. (***c***) Polar bears *(Ursus maritimus)* are the largest living land carnivores. They feed mainly on seals. (*a*, Mike Barlow/Dembinsky Photo Associates; *b*, Animals Animals © 1996 James D. Watt; *c*, Michio Hoshino/Minden Pictures)

SUMMARY

I. Animals classified as deuterostomes include echinoderms, chordates, lophophorates, hemichordates, and chaetognaths.

II. Phylum Echinodermata includes marine animals with a spiny skeleton, a water vascular system, and tube feet; the larvae have bilateral symmetry; most of the adults exhibit pentaradial symmetry.
   A. Class Crinoidea includes the sea lilies and feather stars. In these animals, the oral surface is turned upward; some are sessile.
   B. Class Asteroidea consists of the sea stars—animals with a central disk from which radiate five or more arms.
   C. Class Ophiuroidea includes the brittle stars, which resemble asteroids but with longer, more slender arms that are set off more distinctly from the central disk.
   D. Class Echinoidea includes the sea urchins and sand dollars—animals that lack arms; they have a solid shell and are covered with spines.
   E. Class Holothuroidea consists of sea cucumbers—animals with elongated flexible bodies; the mouth is surrounded by a circle of modified tube feet that serve as tentacles.

III. Phylum Chordata consists of three subphyla: Urochordata, Cephalochordata, and Vertebrata. At some time in its life cycle, a chordate has a notochord, a dorsal tubular nerve cord, and pharyngeal gill slits.
   A. The tunicates, which belong to subphylum Urochordata, are sessile, filter-feeding marine animals with tunics.
   B. Subphylum Cephalochordata consists of the lancelets—small, segmented, fishlike animals that exhibit chordate characteristics.
   C. Subphylum Vertebrata includes animals with a vertebral column forming the chief skeletal axis of the body. Vertebrates also have a cranium, pronounced cephalization, a differentiated brain, muscles attached to the endoskeleton for movement, and two pairs of appendages.
      1. Class Agnatha, the jawless fishes, includes the lampreys and hagfishes.
      2. Descendants of the ostracoderms (agnathans that are the earliest known fossil vertebrates) are thought to have evolved jaws and paired appendages and to have given rise to the modern jawed fishes.

3. Class Chondrichthyes, the cartilaginous fishes, consists of the sharks, rays, and skates.
4. Class Osteichthyes, the bony fishes, includes about 20,000 species of freshwater and saltwater fishes. The bony fishes and cartilaginous fishes are thought to have evolved at about the same time. Most modern bony fishes are ray-finned fishes with swim bladders.
5. Modern amphibians include salamanders, frogs and toads, and wormlike caecilians.
   a. Most amphibians return to the water to reproduce; frog embryos develop into tadpoles, which undergo metamorphosis to become adults.
   b. Amphibians use their moist skin as well as lungs for gas exchange; they have a three-chambered heart and systemic and pulmonary circulations.
6. Class Reptilia includes turtles, lizards, snakes, and alligators.
   a. Reptiles are amniotes; most are terrestrial.
   b. Fertilization is internal; most reptiles secrete a leathery protective shell around the egg; the embryo develops an amnion and other membranes, which protect the embryo and keep it moist.
   c. A reptile has a dry skin with horny scales, lungs with many chambers, and a three-chambered heart (with some separation of oxygen-rich and oxygen-poor blood).
   d. Reptiles dominated Earth during the Mesozoic era; then, toward the end of the Cretaceous period, many reptiles, including all of the dinosaurs, became extinct.
7. Birds (class Aves) have many adaptations for flight, including feathers, wings, and light, hollow bones containing air spaces.
   a. Birds have a four-chambered heart, very efficient lungs, a high metabolic rate, and a constant body temperature; they excrete solid wastes (uric acid).
   b. Birds have a well-developed nervous system and excellent vision and hearing.
   c. Birds communicate with simple calls and complex songs.
8. Mammals have hair, mammary glands, and differentiated teeth. They maintain a constant body temperature and have a highly developed nervous system and a muscular diaphragm.
   a. Monotremes — mammals that lay eggs — include the duck-billed platypus and the spiny anteater.
   b. Marsupials are pouched mammals such as kangaroos and opossums. The young are born in an immature stage and complete their development in the marsupium, where they are nourished with milk from the mammary glands.
   c. Placental mammals are characterized by an organ of exchange, the placenta, that develops between the embryo and the mother. Both oxygen and nutrients diffuse across the placenta from mother to embryo, permitting development to take place within the uterus. Living placental mammals are classified into about 17 orders.

SELECTED KEY TERMS

agnathan
amnion
amniote
amphibian
cephalochordate
chondrichthyes
chordate
coelacanth
cotylosaur
cranium
echinoderm
ectotherm
endoskeleton
endotherm
hemichordate
labyrinthodont
lobe-finned fish
lophophorate phyla
mammal
marsupial
monotreme
neoteny
notochord
operculum
osteichthyes
ostracoderm
oviparous
ovoviviparous
pharyngeal gill slit
placenta
placoderm
placoid scales
ray-finned fishes
reptile
swim bladder
teleost
tetrapod
therapsid
tube feet
tunicate
urochordate
vertebrate
viviparous
water vascular system

POST-TEST

1. Adult __________ have pentaradial symmetry.
2. The __________ __________ system and __________ feet are unique to echinoderms.
3. The __________ phyla (e.g. phoronids and brachiopods) are classified as deuterostomes.
4. In addition to a postanal tail and a dorsal, tubular nerve cord, chordates have a(an) __________ and pharyngeal __________ __________.
5. __________ are sessile, marine chordates often mistaken for sponges.
6. __________ are distinguished from all other animals in having a vertebral column; anterior to this structure a(an) __________ encloses and protects the brain.
7. Reptiles, birds, and mammals are referred to as __________ and also as __________ (because they have a fluid-filled membranous sac that keeps the embryo moist).

8. The cranium and vertebral column are part of the vertebrate __________.
9. Sharks have __________ scales.
10. A skeleton composed of cartilage is characteristic of fishes known as __________.
11. Modern bony fishes (Osteichthyes) are thought to have descended from the __________-__________ fishes; the lobe-finned fishes are thought to be the ancestors of the __________.
12. The __________ are thought to have been the first successful tetrapods.
13. The earliest reptiles are known as __________.
14. Animals that can maintain a constant body temperature are known as __________.
15. __________ are mammals that lay eggs.
16. The __________ body is covered with hard, dry, horny scales.
17. The __________, characteristic of most mammals, is an organ of exchange between the developing embryo and its mother.
18. Label the generalized chordate.

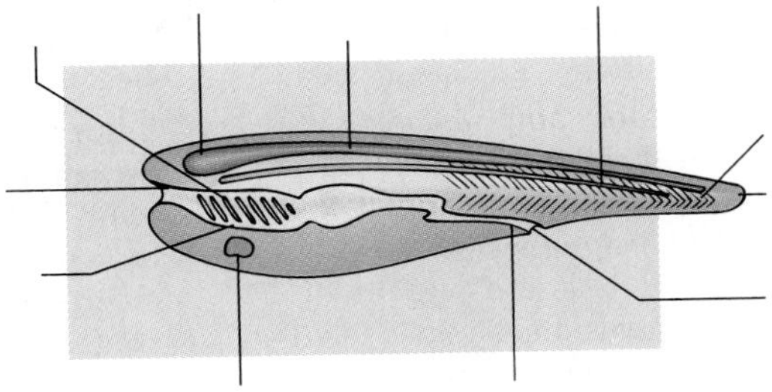

REVIEW QUESTIONS

1. Why are echinoderms thought to be more closely related to chordates than to other phyla? What other phyla are classified as deuterostomes?
2. What are the four principal distinguishing characteristics of a chordate? How are these evident in a tunicate larva? In an adult tunicate? In a lancelet? In a human?
3. What characteristics distinguish the vertebrates from the rest of the chordates?
4. How do lampreys and hagfishes differ from other fish? Of what economic importance are agnathans?
5. Compare the skins of sharks, frogs, snakes, and mammals.
6. Identify organisms that possess each of the following, and give the location and function of each structure: (a) swim bladder; (b) placenta; (c) operculum; (d) amnion; (e) marsupium.
7. Classify each of the following animals, giving the phylum, subphylum, class (and order if you can): (a) human; (b) turtle; (c) lamprey; (d) *Branchiostoma* (Amphioxus); (e) shark; (f) whale; (g) frog; (h) pelican; (i) bat.
8. Which vertebrate groups maintain a constant body temperature? How do they accomplish this? Why is it advantageous?
9. From an evolutionary perspective, what is the significance of each of the following: (a) coelacanths; (b) placoderms; (c) labyrinthodonts; (d) therapsids; (e) *Archaeopteryx*?

YOU MAKE THE CONNECTION

1. What is the function of gills? In general terms, how do they work? Why do you suppose aquatic mammals do not have them?
2. Some paleontologists consider monotremes to be therapsid reptiles rather than mammals. Give arguments for and against this position.
3. Which are more specialized, birds or mammals? Explain your answer.

RECOMMENDED READINGS

Austad, S.N. "The Adaptable Opossum." *Scientific American*, Vol. 258, No. 2, February 1988. The Virginia opossum can adjust the sex ratios of its progeny, an efficient reproductive strategy that helps it adapt quickly to environmental changes.

Griffiths, M. "The Platypus." *Scientific American*, Vol. 258, No. 5, May 1988. Everything you might want to know about this interesting monotreme, including the fact that the platypus has mechanoreceptors and electroreceptors on its beak for detecting prey.

McClanahan, L.L., Ruibal, R. and V.H. Shoemaker. "Frogs and Toads in the Desert," *Scientific American*, Vol. 270, No. 3, March 1994. A discussion of the adaptations that permit certain amphibians to inhabit desert areas.

Monastersky, R. "The Lonely Bird." *Science News*, Vol. 140, No. 7, August 17, 1991. Debate regarding the earliest fossil bird.

Rismiller, P.D., and R. S. Seymour. "The Echidna." *Scientific American*, Vol. 264, No. 2, February 1991. A discussion of the natural history and reproductive behavior of the spiny anteater, a mammal that lays eggs.

Zimmer, C. "Coming on to the Land." *Discover*, Vol. 16, No. 6, June 1995. An interesting discussion of the evolution of terrestrial vertebrates.

Physician Assistant

ROBIN HUNTER-BUSKEY

As an undergraduate at the State University of New York at Stony Brook, Robin Hunter-Buskey took courses in biology, chemistry, anatomy and physiology, among others, maintaining the option to apply to medical school. She ultimately completed a bachelor's degree in physical therapy. After a few years in practice, she decided to take further courses resulting in a second bachelor's degree, qualifying her to become a physician assistant. Now she has the skills to perform many of the tasks in the field of medicine which were once the sole province of doctors, including conducting patient interviews, ordering diagnostic tests, evaluating those tests and developing treatment plans for patients in hospital and clinic settings.

What made you want to pursue a career in healthcare?

I was exposed to alternative medical careers through a work-study program in high school. Students were able to spend one day per week in a particular setting and I chose healthcare, which involved working in hospitals. At the hospital, I worked as a medical assistant, did physical therapy aide work, assisted EKG technicians, and even helped out in the radiology department. So I got exposed to a lot of different career possibilities.

When did you know you wanted to be a physical therapist?

When I first entered Stony Brook I wasn't really sure what I wanted to do, so I began as a general biology major. Once I became aware of the physical therapy program, I realized that I could continue to take the medical school prerequisite courses and have the option of applying to medical school.

Did you have a mentor in college?

Several people were instrumental in helping me pursue my career choices. I spoke with employees in the physical therapy department and I volunteered at local hospitals. Through these efforts I was able to meet a lot of faculty and employees who encouraged me to pursue some type of career in medicine.

How did you become interested in becoming a physician assistant?

Students entering the physical therapy program were advised not to work while obtaining the degree. This is very difficult for people with limited financial resources, and I was one of those students who needed to work while in school. I created a niche for myself as an office assistant in the school of allied health sciences where I was able to work for every program the school offered. I spent a lot of time with the physician assistant program at Stony Brook because they were developing new guidelines for teaching their students. I helped prepare materials and learned more about the profession, which offers the possibility of having a medical background and providing patient care without becoming a doctor.

Did you then decide to become a physician assistant?

Well, I worked as a physical therapist for two years first. During that time I thought I wanted to do more. I went back to Stony Brook and talked with faculty of the physician assistant program, ultimately deciding to apply to the program. Despite the many biology courses I had taken for my first degree, there were still a few prerequisites I needed to obtain before I was successfully accepted into the program.

Describe what you do as a physician assistant.

I work in primary care; this involves seeing patients daily in both a hospital and satellite clinic setting. I interview patients, take their medical histories, examine them, order tests, interpret these tests, and develop treatment plans. I execute and monitor those treatment plans for all of the patients either through office visits or by following up with phone calls. I'm one of the few practitioners here with training in both geriatrics and ob/gyn, so I see all age groups: youth, adult, and elderly.

What do you do that's different from what a doctor does?

I draw my knowledge and skills directly from the experience and training I received in the physical therapy and physician assistant programs. If I have questions, or if there is something I don't understand, I'll ask one of my physician supervisors and we'll follow-up together.

What kind of background would a student need to become a physician assistant?

Most programs have similar prerequisites for a bachelor's degree. Several courses are usually required, including biology, chemistry, college math, psychology, anatomy and physiology with dissection, microbiology, and social sciences and humanities. Actual health care experience is also valuable, and knowledge of first aid and CPR would be helpful.

P A R T 6

Structure and Life Processes in Plants

Crimson clover and lupine add color to a field in Del Norte Redwoods National Park, California. Each terrestrial environment has its own vegetation, determined in part by the area's climate. Humans also greatly affect plant distribution. (Ron Thomas/ FPG International)

CHAPTER 31

Plant Structure, Growth, and Differentiation

This massive Australian gum tree (*Eucalyptus grandis*) is 65.5 meters (215 ft) tall. (Carleton Ray/Photo Researchers, Inc.)

About 90 percent of the approximately 262,000 species of plants are flowering plants, which are characterized by the presence of flowers and of seeds enclosed within fruits. Because flowering plants comprise the largest, most successful group of plants, they are the focus of much of this chapter and Chapters 32 to 36.

Remarkable variety is represented in the 235,000 or so species of flowering plants that live in and are adapted to the many environments offered by our planet. Yet all of these plants—from desert cacti with enormously swollen stems, to cattails partly submerged in marshes, to orchids growing in the uppermost branches of lush tropical rainforest trees—are recognizable as plants. Regardless of size—from *Wolffia microscopica*, the smallest flowering plant known, to Australian gum trees, some of Earth's tallest plants—all plants have the same basic body plan.

Most plants are either herbaceous or woody. *Herbaceous* plants do not develop persistent woody parts above ground, whereas *woody* plants (trees and shrubs) do. **Annuals** are herbaceous plants (such as corn, geranium, and marigold) that grow, reproduce, and die in one year. Other herbaceous plants (such as carrot, cabbage, and beet) are **biennials** and take two years to complete their life cycles before dying. During their first season biennials produce extra food, which they store and use during their second year when they typically form flowers and reproduce. **Perennials** are plants that live for a number of years. In temperate climates, the aerial (above-ground) stems of herbaceous perennials—iris, rhubarb, onion, and asparagus, for example—die back each winter. Their underground parts (roots and underground stems) become dormant during the winter but send out new growth each spring. (In dormancy, an organism reduces its metabolic state to a minimum level to survive unfavorable conditions.) Likewise, in certain tropical climates with pronounced wet and dry seasons, the aerial parts of herbaceous perennials die back and the underground parts become dormant during the dry season. Other tropical plants, such as orchids, are herbaceous perennials that grow year-round.

All woody plants are perennials, and some of them live for hundreds or even thousands of years (see *Making the Connection: Plant Life History Strategies*). In temperate climates, the above-ground stems of woody plants become dormant during the winter. Most temperate woody perennials shed their leaves before winter and produce new stems with new leaves the following spring. Because they have permanent woody stems that are the starting point for new growth the following season, many trees attain massive sizes.

In this chapter we examine the structure of the plant body—its cells, tissues, and tissue systems—and how plants grow.

LEARNING OBJECTIVES

After you have studied this chapter you should be able to

1. Discuss the plant body, including the root system and shoot system.
2. Describe the ground tissue system (parenchyma tissue, collenchyma tissue, and sclerenchyma tissue) of plants.
3. Outline the structure and function of the vascular tissue system (xylem and phloem) of plants.
4. Describe the dermal tissue system (epidermis and periderm) of plants.
5. Discuss what is meant by "growth" in plants, and relate how growth is different in plants and animals.
6. Distinguish between primary and secondary growth.

ROOTS, STEMS, LEAVES, FLOWERS, AND FRUITS MAKE UP THE PLANT BODY

The plant body is organized into a root system and a shoot system (Fig. 31–1). The **root system** is generally the below-ground portion. The above-ground portion, the **shoot system,** consists of a vertical stem that bears leaves, flowers, and fruits containing seeds.

Each plant grows in two different environments: the dark, moist soil and the relatively bright, dry air. Plants must have both roots and shoots because they need resources from *both* environments. Thus, roots branch extensively through the soil, forming a network that anchors the plant firmly in place and absorbs water and dissolved minerals from the soil. Leaves, the flattened organs of photosynthesis, are attached more or less regularly on the stem to capture the sun's light and absorb atmospheric CO_2 needed for photosynthesis.

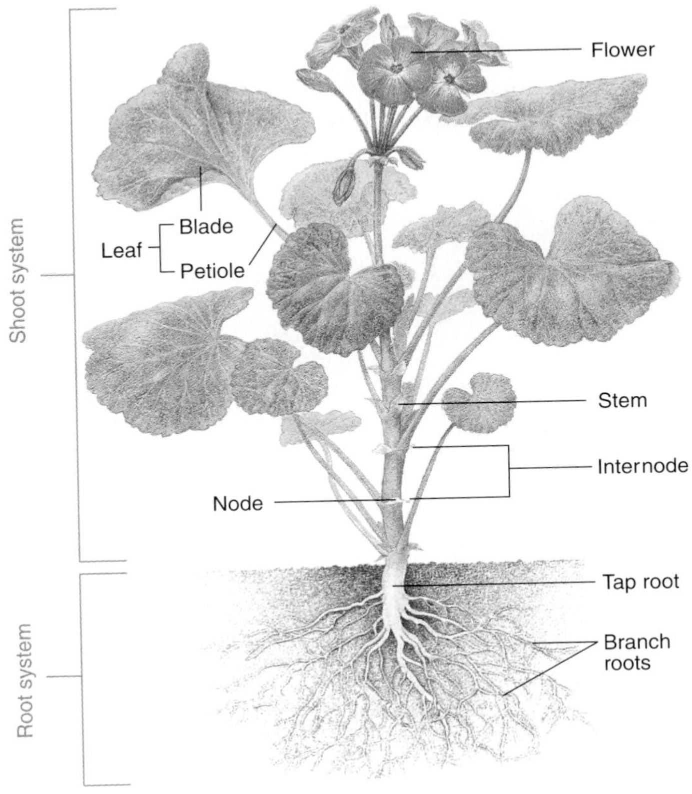

FIGURE 31–1 The body of a geranium is typical of herbaceous plants. The plant body, which consists of a root system and a shoot system, possesses all of the necessary parts to maintain life.

THE PLANT BODY IS COMPOSED OF CELLS AND TISSUES

As in other living organisms, the basic structural and functional unit of plants is the cell. During the course of evolution, plants have developed a diversity of cell types, each specialized for particular functions.

Like animal cells, plant cells are organized into tissues. A **tissue** is a group of cells that form a structural and functional unit. Some plant tissues—known as *simple tissues*—are composed of only one kind of cell, whereas other plant tissues—*complex tissues*—have two or more kinds of cells.

In plants, there are three tissue systems, each of which extends throughout the plant body. Each tissue system contains two or more kinds of tissues (Table 31–1). Most of the plant body is composed of the **ground tissue system,** which consists of three tissues that exhibit a variety of functions, including photosynthesis, storage, and support. The **vascular tissue system,** an intricate plumbing system that extends throughout the plant body, is responsible for conduction of various substances, including water, dissolved minerals, and food (dissolved sugar). It also functions in strengthening and supporting the plant. The **dermal tissue system** provides a covering for the plant body.

Roots, stems, leaves, flower parts, and fruits are referred to as **organs** because each is composed of several different tissues. The tissue systems of different plant organs form an interconnected network throughout the plant. For example, the vascular tissue of a leaf is continuous with the vascular tissue of the stem to which it is attached, and the vascular tissue of the stem is continuous with the vascular tissue of the root.

Making the Connection

Plant Life History Strategies

Under what conditions is it more favorable for a species to be long-lived or short-lived? Woody perennials often live for hundreds of years, whereas some herbaceous annuals may live for only a few weeks or months. Biologists have considered the relative advantages of each life history strategy and have concluded that in some environments a longer life span is advantageous, whereas in others a shorter life span actually increases a species' chances for reproductive success.

When an environment is relatively favorable, it is filled with plants competing for available space. Because such an environment is so crowded, it has few open spots in which new plants can become established. When a plant dies, the empty area is quickly filled by another plant, but not necessarily by the *same* species as before. Thus, an adult perennial survives well in such a habitat, but young plants (perennials or annuals) do not. A plant with a long life span thrives in this type of environment because it can "hold onto" a piece of soil and continue to produce seeds for many years. In a tropical rain forest, for example, competition prevents most young plants from becoming established; hence woody perennials predominate.

In a relatively unfavorable environment, many possible sites are usually available. This type of environment is not crowded, and young plants usually don't have to compete against large, fully established plants. Here, smaller, short-lived plants have the reproductive advantage. These plants are opportunists: they grow and mature quickly during the brief periods when environmental conditions are most favorable. As a result, all of their resources are directed into producing as many seeds as possible before dying. For example, annuals are prevalent in deserts where few woody perennials can survive.

Thus, each organism has its own characteristic life history, with some plants adapted to variable environments and others adapted to stable environments. The longer life span characteristic of woody perennials is just one of several successful life history plans. We return to the matter of different life history strategies in our discussion of population ecology (see Chapter 52). ▲

The ground tissue system is composed of three simple tissues

The bulk of an herbaceous plant is its ground tissue system, which is composed of three tissues: parenchyma, collenchyma, and sclerenchyma (Table 31–2). These tissues can be distinguished by their cell wall structures. Recall that plant cells are surrounded by a cell wall that provides structural support (see Chapter 4). A growing plant cell secretes a thin *primary cell wall,* which stretches and ex-

Table 31–1 TISSUE SYSTEMS, TISSUES, AND CELL TYPES OF FLOWERING PLANTS

Tissue System	*Tissue*	*Cell Types*
Ground tissue system	Parenchyma tissue Collenchyma tissue Sclerenchyma tissue	Parenchyma cells Collenchyma cells Sclerenchyma cells (sclereids, fibers)
Vascular tissue system	Xylem	Tracheids Vessel elements Parenchyma cells Fibers
	Phloem	Sieve tube members Companion cells Parenchyma cells Fibers
Dermal tissue system	Epidermis	Parenchyma cells Guard cells Trichomes
	Periderm	Cork cells Cork cambium cells Cork parenchyma

pands as the cell increases in size. After the cell stops growing, it sometimes secretes a thick, strong *secondary cell wall*, which is deposited *inside* the primary cell wall—that is, between the primary cell wall and the plasma membrane.

Parenchyma cells have thin primary walls

Parenchyma tissue, a simple tissue composed of parenchyma cells, is found throughout the plant body. Parenchyma cells perform a number of important functions for plants, such as photosynthesis, storage, and secretion. Materials stored in parenchyma cells include starch grains, oil droplets, water, and salts (sometimes visible as crystals). The various functions of parenchyma require that they be living, metabolizing cells.

Collenchyma cells have unevenly thickened primary walls

Collenchyma tissue, a simple plant tissue composed of collenchyma cells, is an extremely flexible structural tissue that provides much of the support in soft, nonwoody plant organs. Support is a crucial function in plants, in part because it allows them to grow upward, thus enabling them to compete with other plants for available sunlight in a plant-crowded area. Plants lack the bony skeletal system that is typical of many animals; instead, the plant body is supported by individual cells, including collenchyma cells.

Collenchyma cells, which are usually elongated, are alive at maturity. Their primary cell walls are unevenly thickened and are especially thick in the corners. Collenchyma, which is not found uniformly throughout the plant, often occurs as long strands near stem surfaces and along leaf veins.

Sclerenchyma cells have both primary walls and thick secondary walls

A second simple plant tissue specialized for structural support is **sclerenchyma** tissue, whose cells have both primary and secondary cell walls. The word *sclerenchyma* is derived from a Greek term meaning "hard." The secondary cell walls of sclerenchyma cells become strong and hard due to extreme thickening. At functional maturity, when sclerenchyma tissue is providing support for the plant body, its cells are often dead.

Sclerenchyma tissue may be located in several areas of the plant body. Two types of sclerenchyma cells are sclereids and fibers. **Sclereids** are short, cubical cells common in the shells of nuts and in the pits of stone fruits such as cherries and peaches. **Fibers,** which are long, tapered cells that often occur in patches or clumps, are particularly abundant in the wood and bark of flowering plants.

(Text continues on page 694)

Making the Connection

Relating Cell Wall Chemistry to Different Cell Types

Can parenchyma, collenchyma, and sclerenchyma cells be distinguished by the chemistry of their cell walls? To answer this question, we must first examine the main chemical components of plant cell walls.

Cell walls may contain cellulose, hemicellulose, pectin, and lignin. **Cellulose,** the most abundant polymer in the world, accounts for about 40% to 60% of plant cell walls. As discussed in Chapter 3, cellulose is a polymer of glucose units joined by β-1,4 bonds. Each cellulose molecule consists of thousands of glucose subunits joined to form a flat, ribbon-like chain. From 40 to 70 of these chains lie parallel to one another and connect by hydrogen-bonding to form a **cellulose microfibril,** a strong, tiny strand visible in electron microscopy (see Fig. 3–10*a*).

Cellulose microfibrils are cemented together by hemicelluloses and pectins. **Hemicelluloses** are a group of polysaccharides that vary in their chemical composition from one species to another. Despite their name, the chemical structure of hemicelluloses is in no way related to that of cellulose. **Pectin,** another cementing polysaccharide, is less variable in its monomer composition than are the hemicelluloses.

An important component of secondary plant cell walls, particularly those of wood, is **lignin** (see Chapter 26). Comprising as much as 35% of the secondary cell wall, lignin is a strengthening polymer composed of complex monomers derived from certain amino acids. The chemical structure of lignin has not been completely determined at this time.

Having examined the main chemicals in plant cell walls, we can now generalize about the cell chemistry of parenchyma, collenchyma, and sclerenchyma cells. The thin primary cell walls of parenchyma cells contain predominantly *cellulose.* (Of course, they also contain hemicelluloses and pectin.) Although both parenchyma and collenchyma cells have primary cell walls, their walls are chemically distinct because the thickened areas of collenchyma walls contain large quantities of *pectin* (in addition to cellulose and hemicelluloses). The thick secondary walls of sclerenchyma cells are chemically different because they are rich in *lignin* (in addition to cellulose, hemicelluloses, and pectin). ▲

Table 31–2 CELL TYPES IN THE GROUND TISSUE SYSTEM

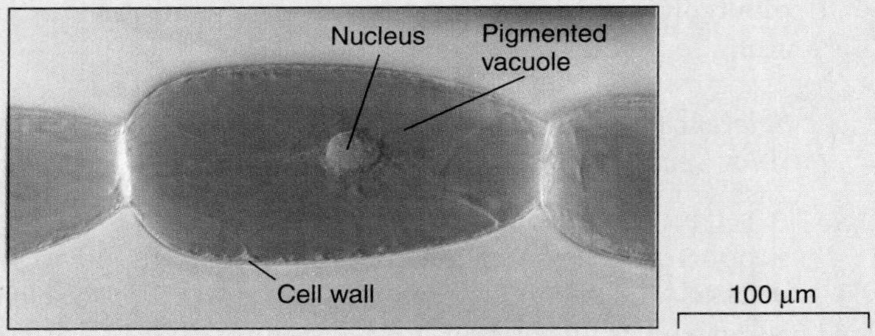

(Phil Gates, University of Durham, Biological Photo Service)

Parenchyma cell

Description
Living, actively metabolizing; thin primary cell walls

Function
Storage; secretion; photosynthesis

Location and comments
Throughout the plant body; shown is a stamen hair of spiderwort (*Tradescantia virginiana*); note the large pigmented vacuole

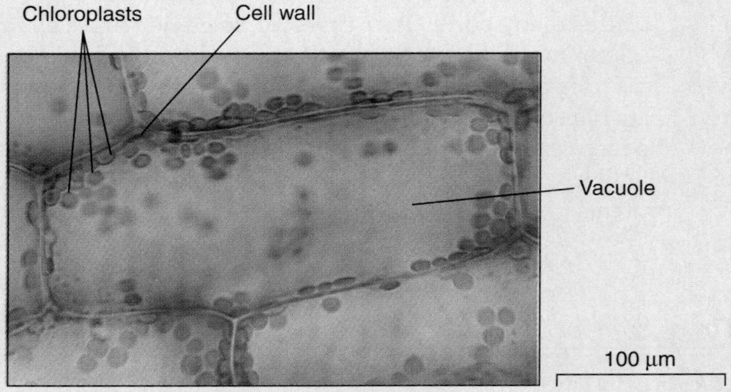

Parenchyma cell

Description
Living, actively metabolizing; thin primary cell walls

Function
Storage; secretion; photosynthesis

Location and comments
Throughout the plant body; shown are leaf cells from an aquatic plant (*Elodea*); note the many chloroplasts surrounding the large, transparent vacuole

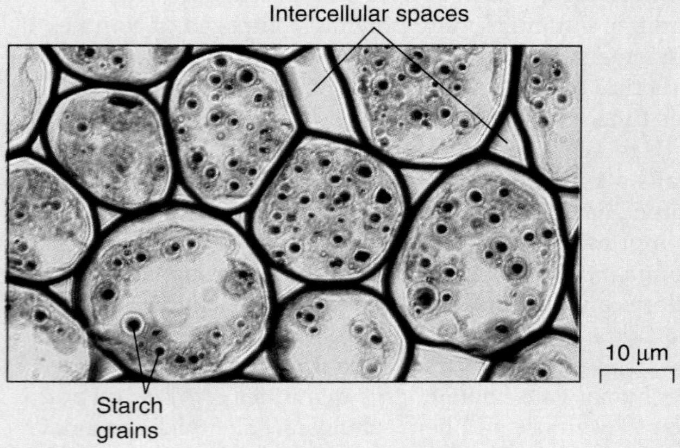

Parenchyma cell

Description
Living, actively metabolizing; thin primary cell walls

Function
Storage; secretion; photosynthesis

Location and comments
Throughout the plant body; shown is a cross section of part of a buttercup (*Ranunculus*) root; note the starch grains filling the cells

Table 31–2 continued

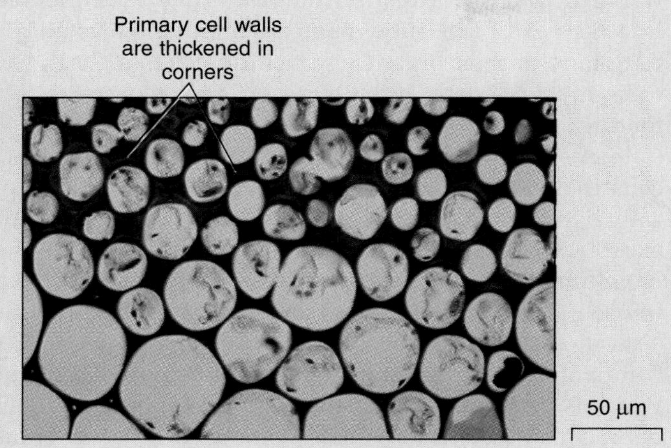

(James Mauseth, University of Texas)

Collenchyma cell

Description
Living; unevenly thickened primary cell walls.

Function
Support

Location and comments
Just under stem epidermis; shown is a cross section of a water lily petiole (leaf stalk); note the unevenly thickened cell walls that are especially thick in the corners, making the cell contents assume a spherical shape

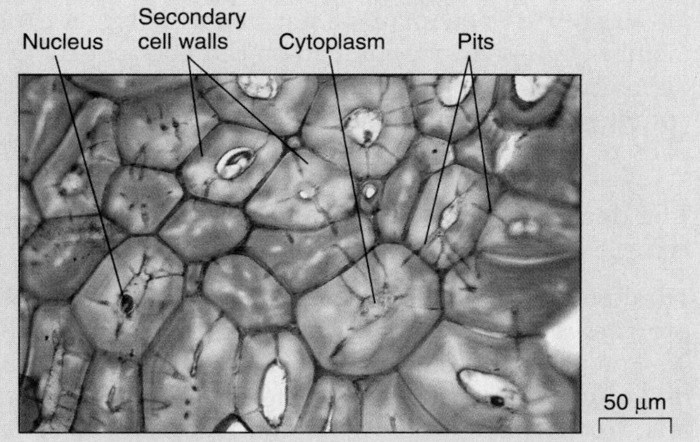

(James Mauseth, University of Texas)

Sclereid

Description
Often living at maturity; thick secondary cell walls; lacks secondary wall at pits

Function
Strength; sclereid-rich tissue is hard and inflexible

Location and comments
Shells of walnuts and coconuts; pits of cherries and peaches

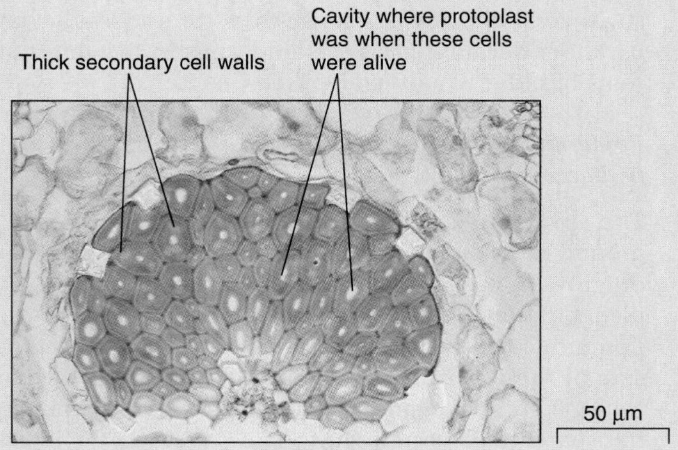

(James Mauseth, University of Texas)

Fiber

Description
Dead at maturity; thick secondary cell walls; fewer pits than in sclereids

Function
Support; provides strength and elasticity

Location and comments
Throughout the plant body; common in stems and certain leaves; shown is a cross section through a clump of fibers from an *Agave* leaf

The vascular tissue system consists of two complex tissues

The vascular tissue system, which is embedded in the ground tissue, transports needed materials throughout the plant via two complex tissues: xylem and phloem (Table 31–3). Both xylem and phloem are continuous throughout the plant body. (Chapter 33 discusses the mechanisms of transport in xylem and phloem.)

The conducting cells in xylem are tracheids and vessel elements

Xylem conducts water and dissolved minerals from the roots to the stems and leaves, and provides structural support. Xylem is a complex tissue composed of four different cell types in flowering plants: tracheids, vessel elements, parenchyma cells, and fibers. Two of the four cell types found in xylem—the **tracheids** and **vessel elements**—actually conduct water and dissolved minerals. In addition to these cells, xylem also contains parenchyma cells that perform storage functions, and fibers that provide support.

Tracheids and vessel elements are highly specialized for conduction. At maturity both cell types are dead and therefore hollow; only their cell walls remain. Tracheids, the chief water-conducting cells in gymnosperms (such as pine) and seedless vascular plants (such as ferns), are long, tapering cells located in patches or clumps. Water is conducted upward, from roots to shoots, passing from one tracheid into another through *pits,* thin areas in their cell walls consisting of a primary wall but no secondary wall.

In addition to tracheids, flowering plants also possess extremely efficient water-conducting cells called vessel elements. The cell diameter of vessel elements is usually wider than that of tracheids. Vessel elements are hollow, but unlike tracheids, the end walls either have holes, known as *perforations,* or they may be entirely dissolved away. Vessel elements are stacked one on top of the other, and water is conducted readily from one vessel element into the next. A stack of vessel elements, called a *vessel,* resembles a miniature water pipe. Vessel elements also have pits that permit the lateral transport (sideways movement) of water from one vessel to another.

Sieve tube members are the conducting cells of phloem

Phloem conducts dissolved sugar throughout the plant and provides structural support. Phloem is a complex tissue composed of four different cell types in flowering plants: sieve tube members, companion cells, fibers, and parenchyma. Fibers are frequently extensive in the phloem of herbaceous plants, providing additional structural support for the plant body.

Sugar is conducted in solution (that is, dissolved in water) through the **sieve tube members,** which are among the most specialized cells in nature. Sieve tube members are stacked end-on-end to form long sieve tubes. The cell's end walls, called *sieve plates,* have a series of holes through which cytoplasm extends from one sieve tube member into the next. Sieve tube members are living at maturity, but many of their organelles, including the nucleus, vacuole, mitochondria, and ribosomes, disintegrate as they mature.

Sieve tube members are among the few eukaryotic cells that can function without nuclei. These cells typically live for less than a year. There are, however, notable exceptions: certain palms have sieve tube members that have remained alive approximately 100 years! Adjacent to each sieve tube member is a **companion cell** that assists in the functioning of the sieve tube member. The companion cell is a living cell, complete with a nucleus, which is thought to direct the activities of both the companion cell and sieve tube member. Numerous **plasmodesmata**—cytoplasmic connections through which cytoplasm extends from one cell to another (see Chapter 5)—occur between a companion cell and its sieve tube member. Although the companion cell does not conduct dissolved sugar, it plays an essential role in moving sugars into the sieve tube members for transport to other parts of the plant (see Chapter 33).

The dermal tissue system consists of two complex tissues

The dermal tissue system—the epidermis and periderm—provides a protective covering over plant parts (Table 31–4 on pages 698–699). In herbaceous plants, the dermal tissue system is a single layer of cells called the epidermis. Woody plants initially produce an epidermis, but it splits apart as the plant increases in diameter due to the production of additional woody tissues. Periderm, a tissue several to many cell layers thick, comprises the outer bark. Periderm replaces the epidermis in the stems and roots of older woody plants (see Chapter 33).

Epidermis is the outermost layer of cells of a herbaceous plant

The **epidermis** is a complex tissue composed mostly of ground parenchyma cells with scattered guard cells and outgrowths called trichomes (discussed shortly). In most plants, the epidermis consists of a single layer of cells. Epidermal cell walls thicken somewhat toward the outside of the plant for protection. Epidermal parenchyma cells generally contain no chloroplasts and are therefore transparent, allowing light to penetrate into the interior tissues of stems and leaves (see Chapter 32). (In both stems and leaves, photosynthetic tissues are *interior to* the epidermis.)

An important requirement of the aerial parts (stems and leaves) of a plant is the ability to control water loss. Epidermal cells of aerial parts secrete a waxy layer called a **cuticle** over the surface of their walls; this wax greatly restricts the loss of water from plant surfaces.

Although the cuticle is extremely efficient at preventing most water loss through epidermal cells, it also prevents the carbon dioxide required for photosynthesis (see Chapter 8) from diffusing from the atmosphere into the leaf or stem. The diffusion of carbon dioxide is facilitated by **stomata** (singular, *stoma*). Stomata are tiny pores formed in the epidermis by two rounded cells called **guard cells.** A number of gases, including carbon dioxide, oxygen, and water vapor, pass through the stomata by diffusion. Stomata are generally open during the day when photosynthesis is occurring, and the loss of water that also takes place when stomata are open helps to cool the leaf.[1] During the night, stomata close to conserve water when photosynthesis is not taking place and cooling is not required. Stomata are discussed in greater detail in Chapter 32.

The epidermis also may contain special outgrowths or hairs, called **trichomes,** which occur in many sizes and shapes and have a variety of functions. Root hairs are simple, unbranched trichomes that increase the surface area of the root epidermis (which comes into contact with the soil) for more effective water and mineral absorption (see Chapter 34). Plants that can tolerate salty environments often have specialized trichomes on their leaves to remove excess salt that accumulates in the plant. The presence of trichomes on the aerial parts of desert plants may increase the reflection of light off the plants, thereby keeping the internal tissues cooler and decreasing water loss. Other trichomes have a protective function. For example, the trichomes on stinging nettle leaves and stems contain irritating substances that discourage herbivorous animals from eating the plant.

Epidermis is replaced by periderm in woody plants

As a woody plant begins to increase in girth, its epidermis is sloughed off and replaced by **periderm.** Periderm forms the outer bark of older stems and roots. It is a complex tissue composed mainly of cork cells and cork parenchyma cells. **Cork cells** are dead at maturity, and their walls are heavily coated with a waterproof substance to help reduce water loss. **Cork parenchyma** cells function primarily in storage.

PLANTS EXHIBIT LOCALIZED GROWTH AT MERISTEMS

Growth is a complex phenomenon involving three different processes: cell division, cell elongation, and cell differentiation. Cell division is an essential part of growth that results in an increase in the number of cells. However, an increase in cell number without an accompanying increase in cell size would contribute little to the overall size increase in a plant. Thus, cell elongation (the expansion of a cell) is an essential part of growth. Cell elongation occurs as the vacuole fills with water, which exerts pressure on the cell wall and causes it to expand. In an onion root cell, the vacuole increases in size some 30 to 150 times during elongation.

Plant cells also **differentiate,** or specialize, into the various cell types just discussed. These cell types comprise the mature plant body and perform the various functions required in a complex, multicellular organism. Although differentiation does not contribute to an increase in size, it is considered an important aspect of growth because it is essential for tissue formation.

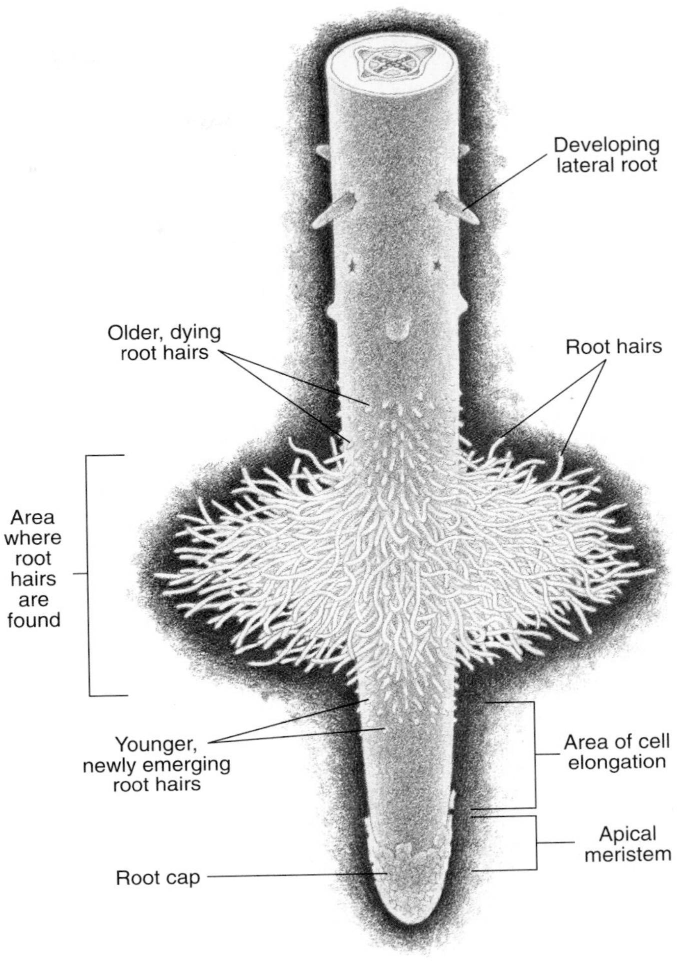

FIGURE 31–2 The root tip is a site of primary growth. Just behind the root cap is the root apical meristem, where cell division occurs. Behind the root apical meristem (farther from the tip) is the area of cell elongation, where cells enlarge and begin to differentiate. Behind this region is the area where root hairs are found.

[1]During drought conditions, the need to conserve water overrides the need to cool the leaves and exchange gases. Thus, in a drought, stomata close during the day.

Table 31–3 SELECTED CELL TYPES IN THE VASCULAR TISSUE SYSTEM

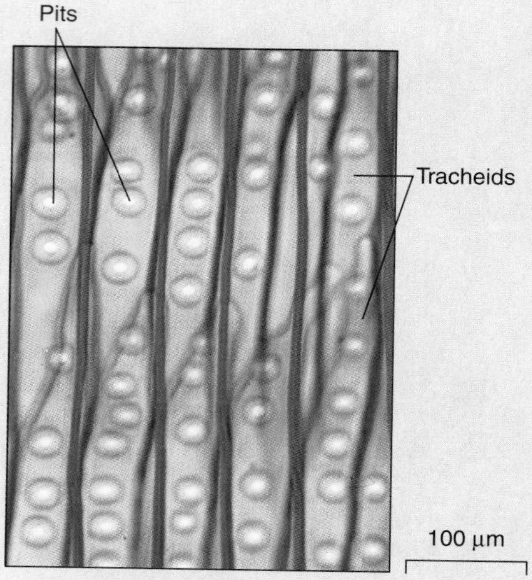

(Visuals Unlimited/John D. Cunningham)

Tracheid

Description
Dead at maturity; lacks secondary wall at pits

Function
Conduction of water and minerals

Location and comments
Occurs in clumps in xylem throughout plant body; shown is a longitudinal section of pine tracheids

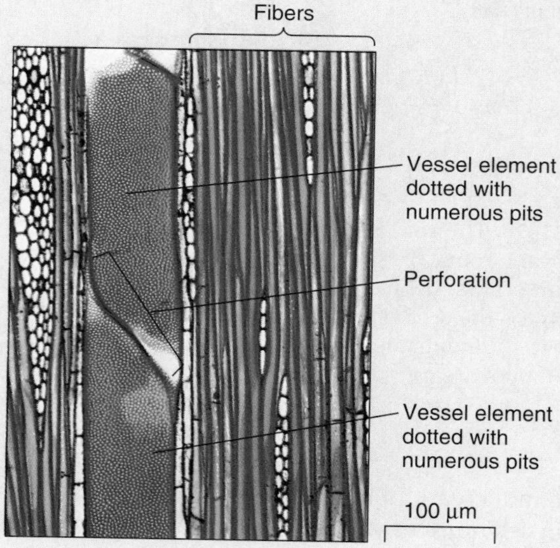

(James Mauseth, University of Texas)

Vessel element

Description
Dead at maturity; end walls have perforations; lacks secondary wall at pits

Function
Conduction of water and minerals

Location and comments
Xylem throughout plant body; vessel elements are more efficient than tracheids in conduction; shown is a longitudinal section of two vessel elements

One difference between plants and animals is the *location* of growth. When a young animal is growing, all parts of its body grow, although not necessarily at the same rate. However, when plants grow, their cells divide only in specific areas, called **meristems,** composed of cells that do not differentiate. Meristematic cells retain the ability to divide by mitosis (see Chapter 9), a trait that differentiated cells lose. The persistence of meristems means

Table 31–3 continued

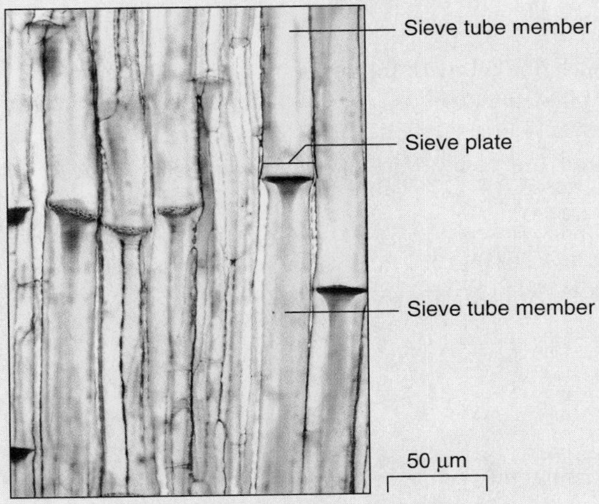

(James Mauseth, University of Texas)

Sieve tube member

Description
Living but lacks nucleus and other organelles at maturity; end walls are sieve plates

Function
Conduction of sugar in solution

Location and comments
Phloem throughout plant body; shown is a longitudinal section through a clump of sieve tube members

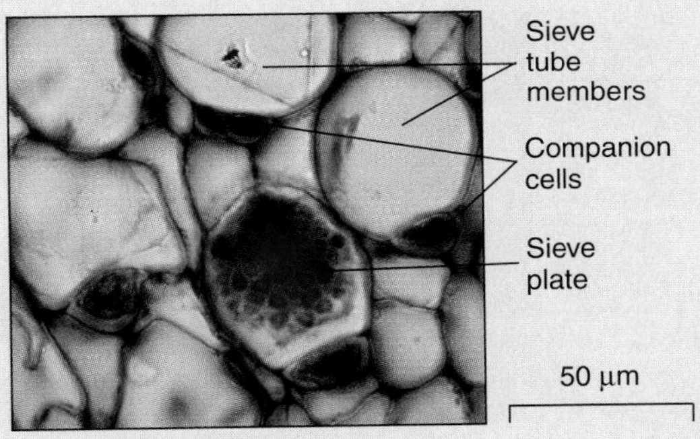

(J. Robert Waaland/University of Washington/Biological Photo Service)

Companion cell

Description
Living; has cytoplasmic connections with sieve tube member

Function
Assists in moving sugars into and out of sieve tube member

Location and comments
Phloem throughout plant body; shown is phloem in cross section

that plants retain the capability for growth throughout their entire life spans.

Two kinds of growth may occur in plants. **Primary growth** refers to an increase in the length of a plant. All plants have primary growth, which forms the entire plant body in herbaceous plants and the young, soft shoots and roots in woody trees and shrubs. **Secondary growth** refers to an increase in the girth of a plant. For the most part, only gymnosperms and woody dicots (see Chapter 27) have secondary growth. Tissues produced by secondary growth comprise the wood and bark, which make up most of the bulk of trees and shrubs. A few annuals—geranium and sunflower, for example—have secondary growth despite the fact that they lack obvious wood and bark tissues. (Monocots, which were introduced in Chapter 27, lack secondary growth, although certain monocots, such as palms, develop thickened trunks by a special form of primary growth.)

Primary growth takes place at apical meristems

Primary growth occurs as a result of the activity of **apical meristems,** areas located at the tips of roots and the buds of stems. Such growth is evident when a root tip (Fig. 31–2) is examined. The root tip is covered by a protective layer of cells called a root cap. Directly behind the root cap is the root apical meristem, which consists of meristematic cells. Meristematic cells, which are quite small and "boxy" in shape, remain small because they are continually dividing.

Table 31–4 SELECTED CELL TYPES IN THE DERMAL TISSUE SYSTEM

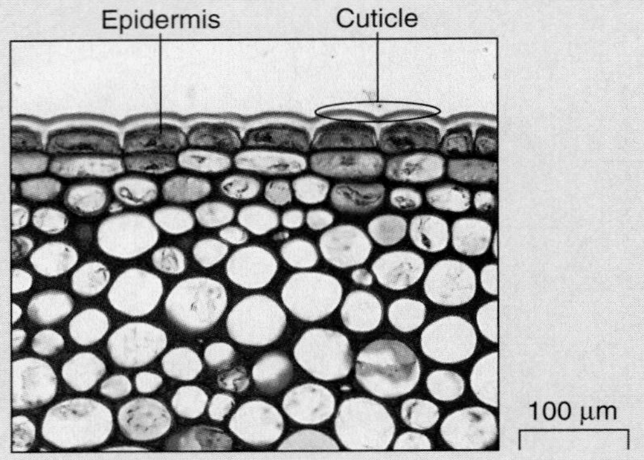

(James Mauseth, University of Texas)

Epidermal cell (ground parenchyma)

Description
Living parenchyma cell with thin primary wall; outer wall often thicker and covered by a noncellular waxy layer (cuticle)

Function
Protective covering over surface of plant body; helps reduce water loss

Location and comments
Epidermis is usually one cell thick; shown is a cross section through epidermis of ivy (*Hedera helix*) stem

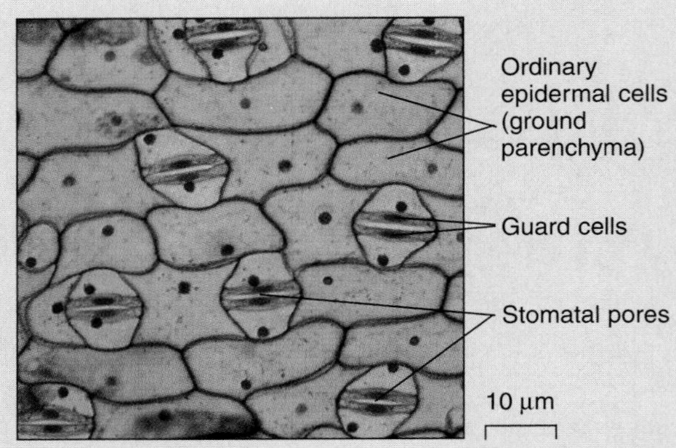

(James Bell/Photo Researchers, Inc.)

Guard cell

Description
Chloroplast-containing cell that occurs in pairs; pair changes shape to open and close stomatal pore

Function
Opens and closes stomatal pore

Location and comments
Epidermis of stems and leaves; shown is epidermis of spiderwort (*Tradescantia*) leaf

Further back from the root tip, just behind the area of cell division, is an area of cell elongation where the cells are no longer dividing but instead are growing larger. Some differentiation also occurs in the area of cell elongation, and immature tissues become evident. The immature tissues continue to develop and differentiate into the mature tissues of the adult plant. Further back from the tip, above the area of cell elongation, the cells have completely differentiated and are fully mature. Root hairs are evident in this area.

A stem bud—the terminal bud, for example—is quite different in appearance from a root tip (Fig. 31–3). A dome of tiny, regularly-arranged meristematic cells—the stem apical meristem—is located within every bud. *Leaf primordia* (embryonic leaves) and *bud primordia* (embryonic buds) arise from the embryonic stem tip. The tiny leaf primordia tend to cover and protect the stem apical meristem. As the cells formed by the stem apical meristem elongate in the area behind the tip, the stem apical meristem is pushed upward. Subsequent cell divisions produce additional stem tissue and cause new leaf primordia to appear. Further back from the stem tip, the immature cells differentiate into the three tissue systems.

Secondary growth takes place at lateral meristems

In addition to primary growth, woody trees and shrubs have secondary growth. These plants increase in length by primary growth and increase in girth by secondary growth. The increase in girth, which occurs in areas that are no longer elongating, is due to cell divisions that take place in **lateral meristems,** areas extending the entire length of the stems and roots, except at the tips. Two lateral meristems are responsible for secondary growth: the vascular cambium and the cork cambium (Fig. 31–4).

The **vascular cambium** is a layer of meristematic cells that forms a thin cylinder around the stem and root trunk. It is located between the wood and bark of a woody plant. Cells of the vascular cambium divide, adding more cells to the *wood* (secondary xylem) and *inner bark* (secondary phloem).

(Text continues on page 700)

Table 31–4 (*continued*)

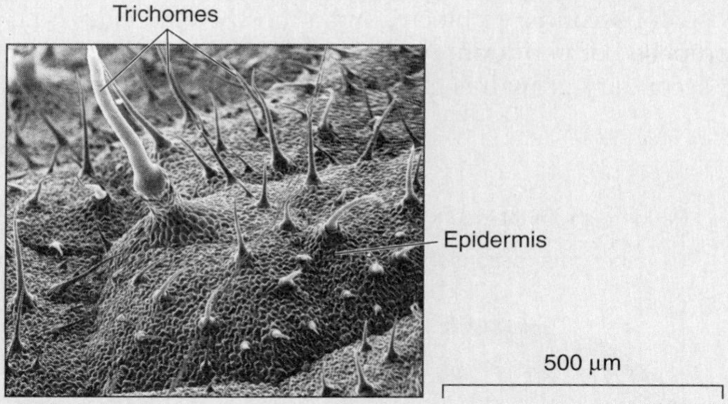

(Biophoto Associates)

Trichome

Description
Hair or other epidermal outgrowth; may be single-celled or multicellular; occurs in variety of sizes and shapes

Function
Varied: absorption; secretion; excretion; protection; reduces water loss

Location and comments
Epidermis; shown is nettle leaf, which has trichomes that break off inside the skin of animals and inject irritating substances that cause stinging

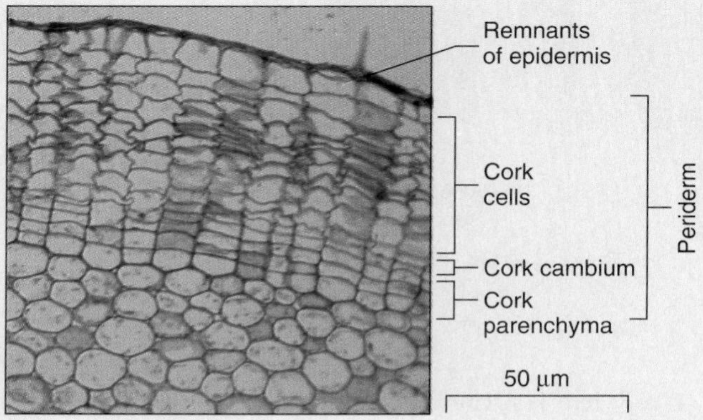

(Dennis Drenner)

Cork cell

Description
Dead at maturity; cell walls are impregnated with waterproof materials

Function
Reduces water loss and prevents disease-causing organisms from penetrating

Location and comments
Produced in large amounts; often forms just under the epidermis; replaces epidermis in older stems and roots; shown is a cross section through periderm of a geranium stem

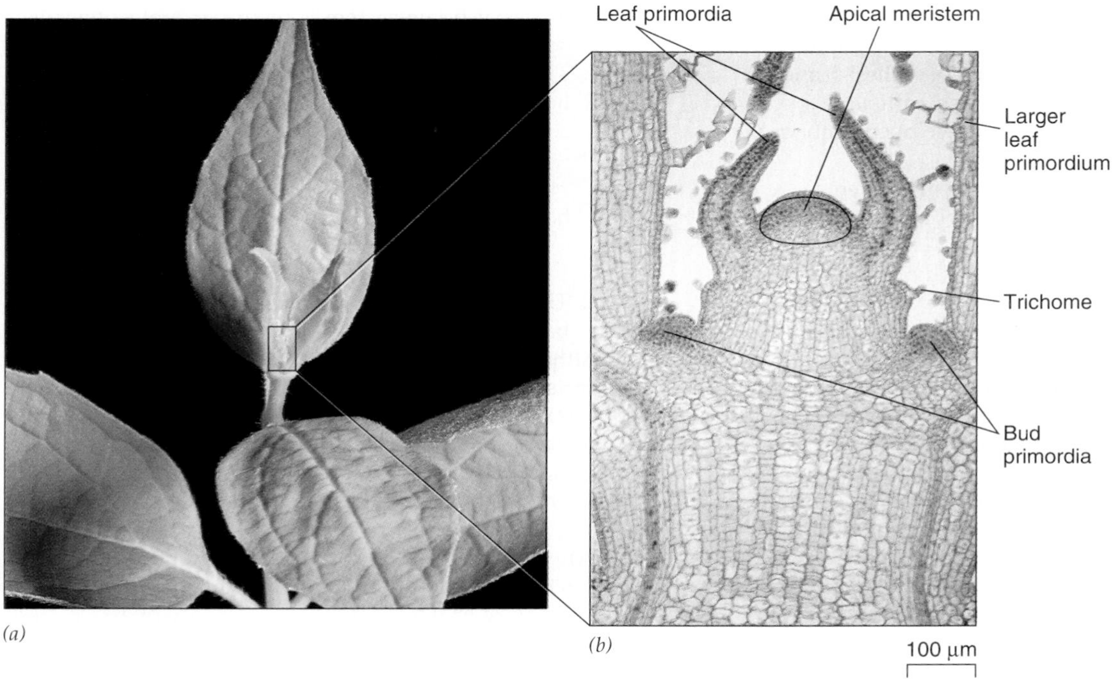

FIGURE 31–3 Buds are sites of primary growth. (*a*) This shoot tip has several expanded leaves as well as two sets of tiny leaves surrounding and protecting the bud. (*b*) A longitudinal section through a bud showing the stem apical meristem, leaf primordia, and bud primordia. (*a, b,* James Mauseth, University of Texas)

The **cork cambium** is composed of a thin cylinder or irregular arrangement of meristematic cells and is located in the *outer bark*. Cells of the cork cambium divide to form the cork cells and cork parenchyma cells that make up the periderm (discussed previously).

We are now ready to give a more precise definition of bark: **Bark** is the outermost covering over woody stems and roots. It consists of two regions, a living *inner bark*, composed of secondary phloem, and a mostly dead *outer bark*, composed of periderm. A more comprehensive discussion of secondary growth is given in Chapters 33 and 34.

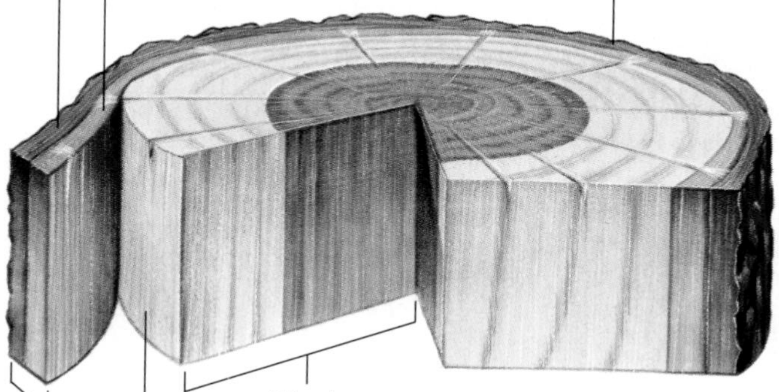

FIGURE 31–4 In secondary growth, plants increase in girth as a result of the activity of two lateral meristems. The vascular cambium (a thin layer of cells sandwiched between the wood and bark) produces the secondary vascular tissues: the wood (secondary xylem), and the inner bark (secondary phloem). The cork cambium produces the outer bark tissues (the periderm) that replace the epidermis in the secondary plant body.

SUMMARY

I. The plant body typically consists of a root system and a shoot system.
   A. The root system is generally underground and obtains water and dissolved minerals for the plant. Roots also anchor the plant firmly in place.
   B. The shoot system is generally aerial and obtains sunlight and carbon dioxide for the plant.
      1. The shoot system consists of a vertical stem that bears leaves (the main organs of photosynthesis) and reproductive structures (flowers and fruits).
      2. Terminal and lateral buds (undeveloped embryonic shoots) develop on stems.
II. The plant body is composed of three tissue systems.
   A. The ground tissue system consists of three tissues with a variety of functions.
      1. Parenchyma tissue is composed of living parenchyma cells that possess thin primary cell walls. Functions of parenchyma tissue include photosynthesis, storage, and secretion.
      2. Collenchyma tissue is composed of collenchyma cells with unevenly thickened primary cell walls. This tissue provides flexible structural support.
      3. Sclerenchyma tissue is composed of sclerenchyma cells that have both primary and secondary cell walls. Sclerenchyma cells are often dead at maturity, but they provide structural support.
   B. The vascular tissue system conducts materials throughout the plant body and provides strength and support.
      1. Xylem is a complex tissue that conducts water and dissolved minerals. The actual conducting cells of xylem are tracheids and vessel elements.
      2. Phloem is a complex tissue that conducts sugar in solution. Sieve tube members are the conducting cells of phloem; they are assisted by companion cells.
   C. The dermal tissue system is the outer protective covering of the plant body.
      1. The epidermis is a complex tissue that covers the herbaceous plant body.
         a. The epidermis that covers aerial parts secretes a layer of wax, called the cuticle, that reduces water loss.
         b. Gas exchange between the interior of the shoot system and the surrounding atmosphere occurs through stomata.
      2. The periderm is a complex tissue that covers the plant body in woody plants.
   D. Although separate organs (roots, stems, leaves, flower parts, and fruits) exist in the plant, many tissues are integrated throughout the plant body, providing continuity from organ to organ.
III. Growth in plants is localized in specific regions, called meristems, and involves cell division, cell elongation, and cell differentiation.
   A. Primary growth is an increase in stem and root length.
      1. Primary growth occurs in all plants.
      2. Primary growth is due to the activity of apical meri-

stems that are localized at the tips of roots and the buds of stems.

B. Secondary growth is an increase in stem and root girth (thickness).
   1. In addition to primary growth, woody plants also have secondary growth.
   2. Secondary growth is localized, typically in long cylinders of active growth throughout the length of older stems and roots.
   3. The two lateral meristems responsible for secondary growth are the vascular cambium and the cork cambium.

SELECTED KEY TERMS

annual
apical meristem
bark
biennial
cellulose
collenchyma
companion cell
cork cambium
cork cell
cork parenchyma
cuticle
dermal tissue system
differentiate
epidermis
fiber
ground tissue system
guard cell
lateral meristem
lignin
meristem
organ
parenchyma
pectin
perennial
periderm
phloem
primary growth
root system
sclereid
sclerenchyma
secondary growth
shoot system
sieve tube member
stoma (stomata)
tissue
tracheid
trichome
vascular cambium
vascular tissue system
vessel element
xylem

POST-TEST

1. Plants that complete their life cycles in one year are called __________; those that complete them in two years are __________; and those that live year after year are __________.
2. Most of the plant body consists of the __________ tissue system.
3. Storage, secretion, and photosynthesis are the functions of __________.
4. The two simple tissues that are specialized for support are __________ and sclerenchyma.
5. Transport of various materials throughout the plant body is performed by the __________ tissue system.
6. Conduction of water and minerals in the xylem occurs in __________ and vessel elements.
7. Conduction of sugar in solution in the sieve tube members of the phloem is aided by __________ cells.
8. The outer covering of plants with primary growth is __________, whereas plants with secondary growth are covered by __________.
9. The epidermis secretes a noncellular layer of wax, known as a __________, over its surface.
10. Tiny pores known as __________ dot the surface of the epidermis of leaves and stems; each pore is bordered by two __________ __________.
11. Outgrowths of the epidermis are called __________.
12. Localized areas within the plant body where cell divisions occur are known as __________.
13. Primary growth, an increase in the length of a plant, occurs at __________ meristems.
14. The two lateral meristems responsible for secondary growth are the __________ __________ and the cork cambium.
15. Cellulose characterizes parenchyma cell walls; __________ characterizes collenchyma walls; and __________ characterizes sclerenchyma walls.

REVIEW QUESTIONS

1. What are some of the functions of roots? Of shoots?
2. What are the three tissue systems in plants? Describe the functions of each.
3. Compare the cellular structures and functions of parenchyma, collenchyma, and sclerenchyma tissues.
4. What are the functions of xylem and phloem?
5. Compare and contrast epidermis and periderm.
6. What is the role of plant meristems?
7. Distinguish between primary and secondary growth. Between apical and lateral meristems.

8. Label the following diagram:

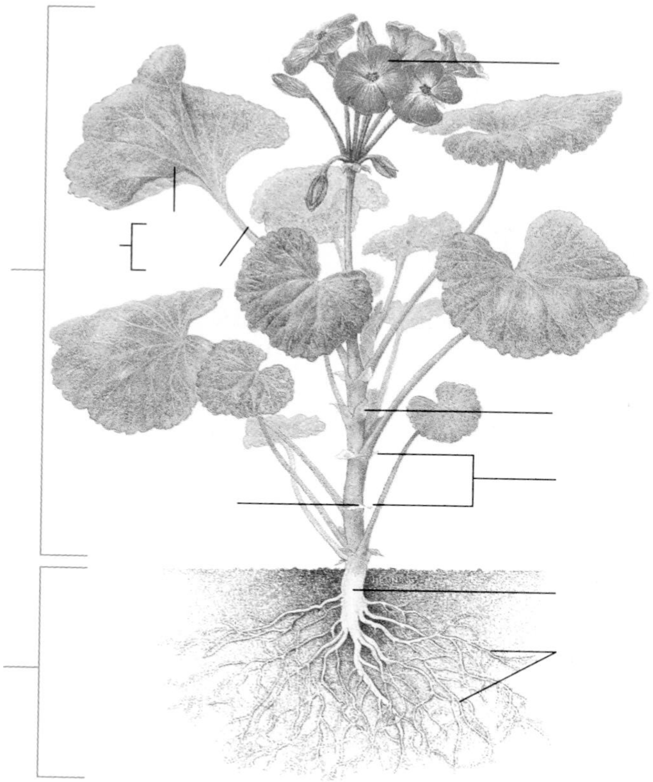

YOU MAKE THE CONNECTION

1. Grasses have a special meristem situated at the base of the leaves. Relate this information to what you know about mowing the lawn.
2. A couple carved a heart with their initials into a tree trunk four feet above ground level; the tree was 25 feet tall at the time. Twenty years later the tree was 50 feet tall. How far above the ground were the initials? Explain your answer.
3. Sclerenchyma in plants is the functional equivalent of bone in more complex animals (both sclerenchyma and bone provide support). However, sclerenchyma is dead, while bone is living tissue. What are some of the advantages of a plant having dead support cells? Can you think of any disadvantage?

RECOMMENDED READINGS

Bell, A.D. *Plant Form: An Illustrated Guide to Flowering Plant Morphology,* Oxford University Press, Oxford, 1991. The external features of plants are examined in this beautifully illustrated book.

Mauseth, J.D. *Botany: An Introduction to Plant Biology,* 2nd ed. Philadelphia, Saunders College Publishing, 1995. A general botany text with a structural emphasis.

Raven, P.H., R.F. Evert, and S.E. Eichhorn. *Biology of Plants,* 5th ed. New York, Worth Publishers, 1992. A general botany text with an evolutionary emphasis.

Rensberger, B. "Getting to the Root of Plant Growth." *Washington Post,* Science Section, 13 July 1992. Why roots generally grow downward and shoots upward.

C H A P T E R 3 2

Leaf Structure and Function

Leaves of a sugar maple tree absorb sunlight for photosynthesis.
(John Mielcarek/Dembinsky Photo Associates)

Suppose for a moment that you are taking a course in engineering and have been asked to design an efficient solar collector capable of converting the radiant energy it collects into chemical energy. Where would you start? It might be helpful to check library books and periodicals to see how solar collectors/energy converters have been designed in the past. In this instance, it would also be wise to ask a biologist, or even a biology student like yourself, whether anything comparable exists in nature. The answer, of course, is yes: plants have organs that are effective solar collectors and energy converters. They are called *leaves.*

Plants allocate many resources into the production of leaves. According to John E. Dale, a University of Edinburgh botanist who specializes in leaf development, a large maple tree annually produces 46.5 square meters (500 square feet) of leaves, which weigh more than 113 kilograms (250 pounds). The metabolic costs of producing so many leaves are high, but leaves are essential to the survival of the tree.

Leaves gather the sunlight necessary for **photosynthesis,** the biological process that converts radiant energy into the chemical energy of sugar and other organic molecules, which plants then use to fuel their metabolic processes (see Chapters 7 and 8). During a single summer, the leaves of the maple tree will fix about 454 kilograms (1000 pounds) of carbon dioxide into sugars and other organic compounds. Leaves are the main photosynthetic organs of most plants, and the structure of a leaf is superbly adapted for its primary function of photosynthesis. Most leaves are thin and flat, a shape that allows maximum absorption of light energy and the efficient internal diffusion of gases. As a result of their ordered arrangement on the stem, leaves efficiently catch the sun's rays. The leaves of plants form an intricate green mosaic, bathed in sunlight and atmospheric gases.

In addition to meeting the requirements of photosynthesis, the structure of foliage leaves is also adapted to prevent excessive water loss. Leaves possess several features that help reduce or control water loss, the most important of which is a thin, transparent layer of wax that covers the leaf surface.

However, some features that optimize photosynthesis actually *promote* water loss. For example, plants have minute pores that allow an exchange of gases for photosynthesis, but these tiny openings also allow water to escape into the atmosphere as water vapor. Thus leaf structure represents a trade-off between photosynthesis and water conservation.

LEARNING OBJECTIVES

After you have studied this chapter you should be able to

1. Describe the major tissues of the leaf: epidermis, mesophyll, xylem, and phloem.
2. Compare leaf anatomy in dicots and monocots.
3. Relate leaf structure to its function of photosynthesis.
4. Outline the physiological changes that accompany stomatal opening and closing.
5. Discuss transpiration and its effects on plants.
6. Define leaf abscission, and explain why it occurs and what physiological and anatomical changes precede it.
7. List five modified leaves and give the function of each.

FOLIAGE LEAVES VARY GREATLY IN EXTERNAL FORM

Leaves are the most variable plant organ—so much so that plant biologists had to develop specific terminology to describe their shapes, margins (edges), vein patterns, and the way they attach to stems. Because each leaf is characteristic of the plant on which it grows, many plants can be identified by their leaves alone. Leaves may be round, needle-like, scalelike, cylindrical, heart-shaped, fan-shaped, or thin and narrow. They vary in size from those of the raffia palm (*Raphia ruffia*), whose leaves often grow to over 20 meters (65 feet) long, to those of duckweeds (*Lemna minor,* for example), whose leaves are so small that 20 of them laid end-to-end equal 2.5 centimeters (one inch).

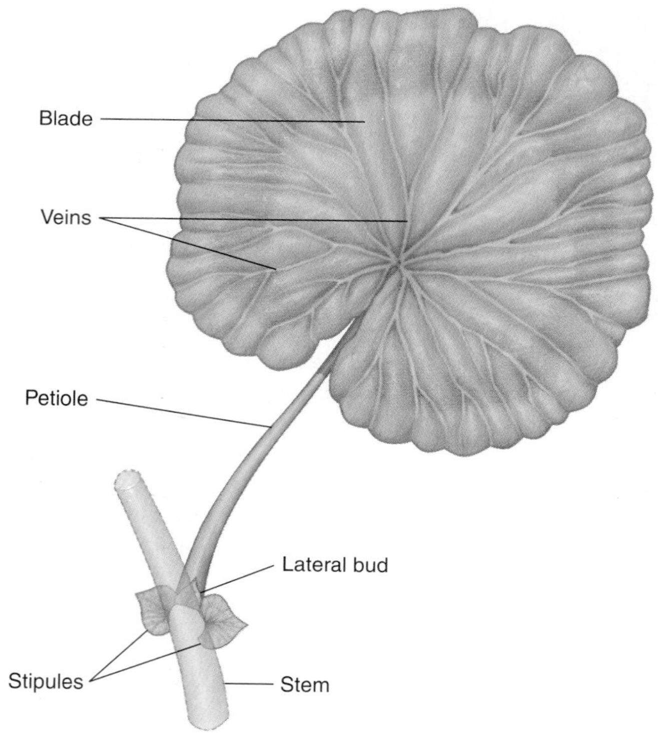

FIGURE 32–1 A geranium leaf consists of a blade, a petiole, and two stipules at the base of the leaf. Note the lateral bud in the leaf axil.

Most leaves are composed of three parts: a blade, a petiole, and a pair of stipules. The broad, flat portion of a leaf is the **blade;** the stalk that attaches the blade to the stem is the **petiole;** and the **stipules** are leaflike outgrowths at the base of the petiole (Fig. 32–1). Some leaves do not have petioles or stipules.

Leaves may be *simple* (having a single blade) or *compound* (having a blade divided into two or more leaflets) (Fig. 32–2). Sometimes it is difficult to tell whether a plant has formed one compound leaf or a small stem bearing several simple leaves. One easy way to determine if a plant has simple or compound leaves is to look for lateral buds, also called axillary buds because each develops in a leaf *axil* (the angle between the stem and petiole; Fig. 32–1). Lateral buds form at the base of a leaf, whether it is simple or compound. However, lateral buds never develop at the base of leaflets. Also, the leaflets of a compound leaf lie in a single plane (you can lay a compound leaf flat on a table), whereas simple leaves are not arranged in one plane on a stem.

Leaves are arranged on a stem in one of three possible ways (Fig. 32–3). Plants such as the walnut have an *alternate leaf arrangement,* with one leaf at each node. (A *node* is the area of the stem where one or more leaves are attached.) In an *opposite leaf arrangement,* as occurs in lilac, two leaves grow at each node. In a *whorled leaf arrangement,* as occurs in catalpa, three or more leaves grow at each node.

Leaf blades may possess *parallel venation* (in which the primary veins run approximately parallel to one another—generally characteristic of monocots[1]) or *netted venation* (veins that are branched in such a way that they resem-

[1]Recall that flowering plants, the focus of this chapter, are divided into two groups, informally called dicots and monocots, based on a number of structural features (see Chapter 27). Dicots include such well-known plants as bean, petunia, oak, cherry, apple, rose, and snapdragon; examples of monocots include corn, lily, grasses, palms, tulip, orchid, and banana.

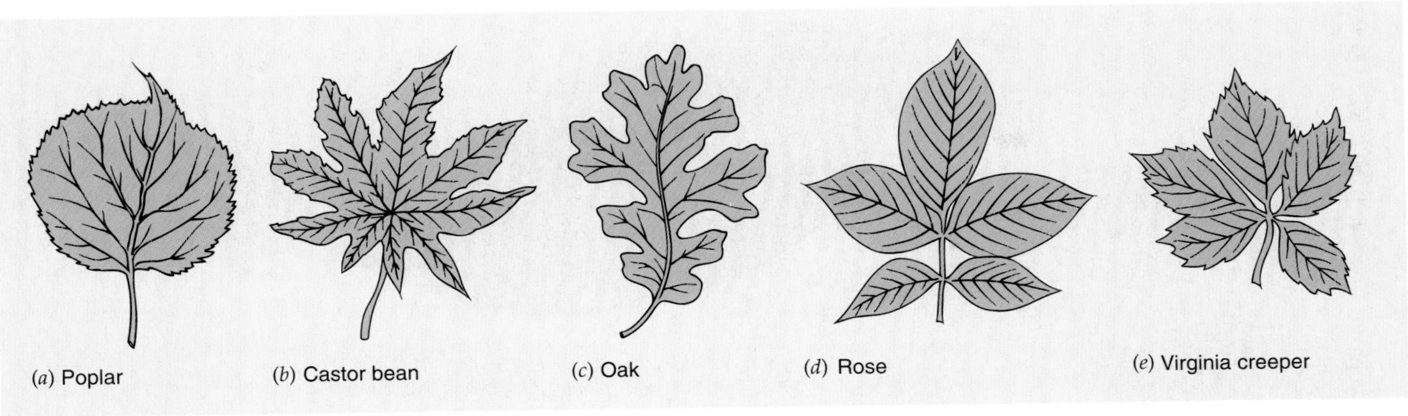

FIGURE 32–2 Leaves may be simple or compound. Simple leaves (***a, b, c***) come in a variety of sizes and shapes. Compound leaves (***d, e***) may be pinnately compound (***d***) or palmately compound (***e***).

ble a net—generally characteristic of dicots; Fig. 32–4). Netted veins can be *palmately netted,* in which several major veins radiate out from one point, or *pinnately netted,* in which major veins branch off along the entire length of the main vein.

EPIDERMIS, MESOPHYLL, XYLEM, AND PHLOEM ARE THE MAJOR TISSUES OF THE LEAF

The leaf is a complex organ composed of several different tissues, all of which are organized to optimize photosynthesis (Fig. 32–5). Because the leaf blade has upper and lower surfaces, it is covered by two epidermal layers, the **upper epidermis** and the **lower epidermis.** Most cells in these layers are living parenchyma cells (see Chapter 31) that lack chloroplasts and are relatively transparent. One interesting feature of leaf epidermal cells is that the cell wall facing outside of the leaf is somewhat thicker than the cell wall facing inward. This extra thickness may provide the plant with additional protection against injury or water loss.

Because leaves have such a large surface area exposed to the atmosphere, water loss by evaporation from the leaf's surface is unavoidable. However, epidermal cells secrete a waxy layer, the **cuticle,** that reduces water loss (see Table 31–4). The cuticle varies in thickness in different plants, in part due to environmental conditions. As one might expect, the leaves of plants adapted to hot, dry

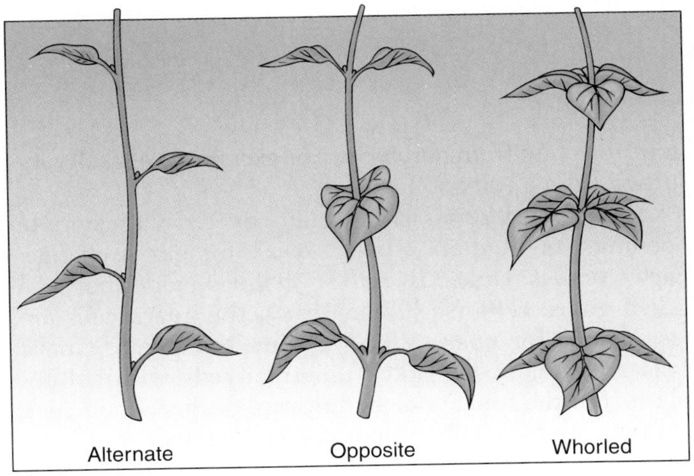

FIGURE 32–3 Leaf arrangement on a stem may be alternate, opposite, or whorled, depending on the number of leaves at each node.

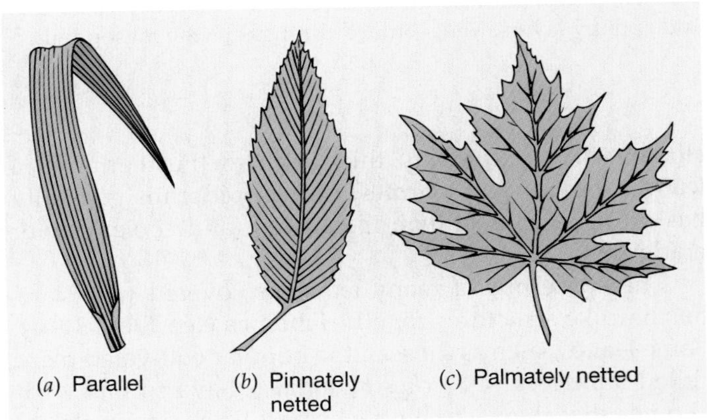

FIGURE 32–4 Venation patterns in leaves may be parallel or netted. (*a*) Kentucky bluegrass has parallel veins, which are characteristic of monocot leaves. Siberian elm (***b***) and silver maple (***c***) have netted veins. Siberian elm is pinnately netted and silver maple is palmately netted.

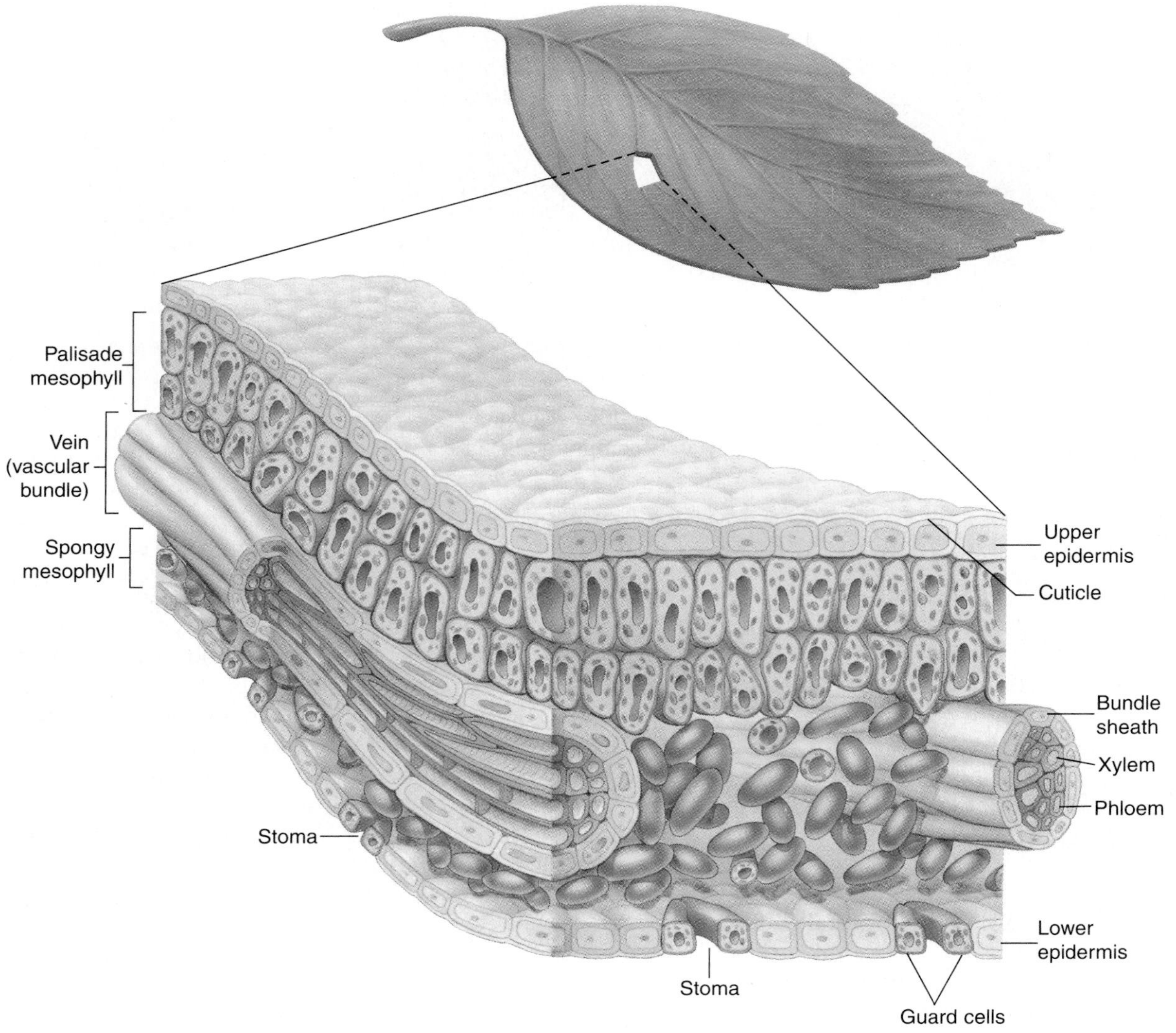

FIGURE 32–5 A magnified view of a block of leaf tissue shows the internal arrangement of tissues in a typical leaf blade. The blade is covered by an upper and lower epidermis. The photosynthetic tissue, called mesophyll, is often arranged into palisade and spongy layers. Veins branch throughout the mesophyll.

climates have extremely thick cuticles. Furthermore, a leaf's exposed (and warmer) upper epidermis generally has a thicker cuticle than its shaded (and cooler) lower epidermis.

The epidermis of many leaves is covered with various hairlike structures called **trichomes** (see Table 31–4). Some leaves, such as those of the popular cultivated plant called lamb's ears (*Stachys byzantina*), have so many trichomes that they actually feel fuzzy. Trichomes help reduce water loss from the leaf surface by retaining a layer of moist air next to the leaf. A leaf covered with trichomes is difficult for an insect to walk over or eat. Trichomes also reflect sunlight, thereby protecting the plant from overheating. Some trichomes secrete stinging irritants (for protection from herbivores) or excrete excess salts absorbed from a salty soil.

The leaf epidermis is typically covered with minute openings, or **stomata** (sing., *stoma*), for gas exchange. Each stoma is flanked by two specialized epidermal cells called **guard cells** (see Table 31–4). The guard cells are responsible for opening and closing the stoma. Guard cells are usually the only epidermal cells with chloroplasts, but the functional significance of this is not clear. Stomata are especially numerous on the lower epidermis (an average of about 100 stomata per square millimeter) and in many cases are located *only* on the lower surface. This adaptation reduces water loss because stomata on the lower epidermis are shielded from direct sunlight.

The photosynthetic tissue of the leaf, called the **mesophyll** (from Greek, meaning "the middle of the leaf"), is sandwiched between the upper epidermis and the lower epidermis. Mesophyll cells, which are parenchyma cells packed with chloroplasts, are loosely arranged with many air spaces between them that facilitate gas exchange. In many plants, the mesophyll is divided into two sub-layers. Toward the upper epidermis, the cells are stacked closely together in a **palisade** layer. In the lower portion, they are more loosely and irregularly arranged in a **spongy** layer. The two layers have different functions. Palisade mesophyll is the main site of photosynthesis in the leaf. Although photosynthesis also occurs in the spongy mesophyll, its primary function is to allow for diffusion of gases (particularly CO_2) within the leaf.

Palisade mesophyll may be further organized into one, two, three, or even more rows of cells. The presence of additional layers of palisade mesophyll is an adaptation due in part to environmental conditions. Leaves exposed to direct sunlight are thicker because they contain more rows of palisade mesophyll than do shaded leaves on the same plant. In direct sunlight, the light is high enough to effectively penetrate multiple layers of palisade mesophyll, allowing all layers to photosynthesize efficiently.

The **veins,** or **vascular bundles,** of a leaf extend through the mesophyll. Branching is extensive, and no mesophyll cell is very far from a vein. Each vein contains two types of vascular tissue: xylem and phloem (see Chapter 31). **Xylem,** which conducts water and dissolved minerals, is usually located on the upper side of a vein, toward the upper epidermis, whereas **phloem,** which conducts dissolved sugars, is usually confined to the lower side of a vein.

Veins are usually surrounded by one or more layers of nonvascular cells that make up the **bundle sheath.** Bundle sheaths are composed of parenchyma or sclerenchyma cells (see Chapter 31). Frequently, the bundle sheath has support columns, called **bundle sheath extensions,** that extend through the mesophyll from the upper epidermis to the lower epidermis (Fig. 32–6). Bundle sheath extensions may be composed of parenchyma, collenchyma, or sclerenchyma cells.

Leaf structure differs in dicots and monocots

The leaves of dicots are usually composed of a broad, flattened blade and a petiole. As mentioned previously, dicot leaves typically have netted venation. In contrast, monocot leaves often lack a petiole; they are narrow and wrap around the stem, forming a sheath. Parallel venation is characteristic of monocot leaves.

Dicots and certain monocots also differ in internal leaf anatomy (Fig. 32–7). The mesophyll of dicots and many monocots typically contains both palisade and spongy layers. In contrast, mesophyll in certain monocot leaves (corn and other grasses) is not differentiated into palisade and spongy tissue. Because dicots have netted veins, a cross section of a dicot blade often shows veins in cross section as well as lengthwise views. In a cross section of a monocot leaf, on the other hand, the parallel venation pattern produces evenly spaced veins, all of which are in cross section.

Differences between the guard cells in dicot and certain monocot leaves also occur (Fig. 32–8). The guard cells of dicots and many monocots are shaped like tiny kidney beans. Other monocot leaves (those of grasses,

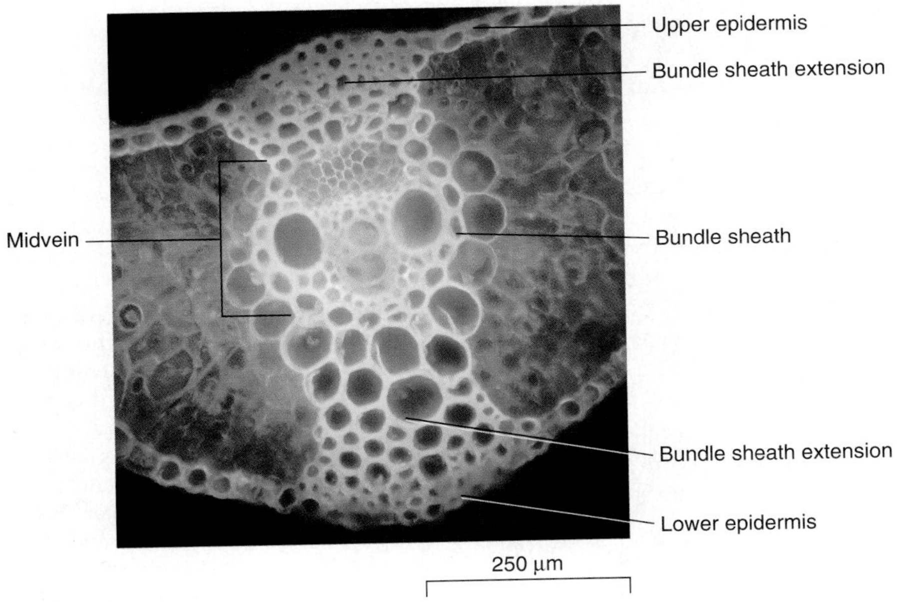

FIGURE 32–6 A bundle sheath extension extends from the bundle sheath to the epidermis. Cross section of a wheat midvein, showing bundle sheath extensions to both the upper epidermis and lower epidermis. (Phil Gates, University of Durham/Biological Photo Service)

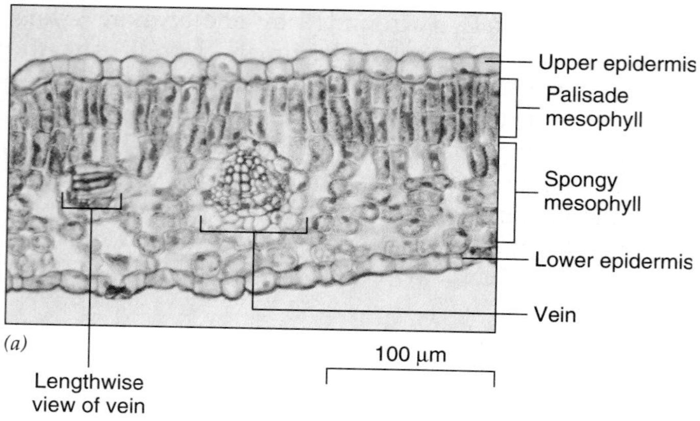

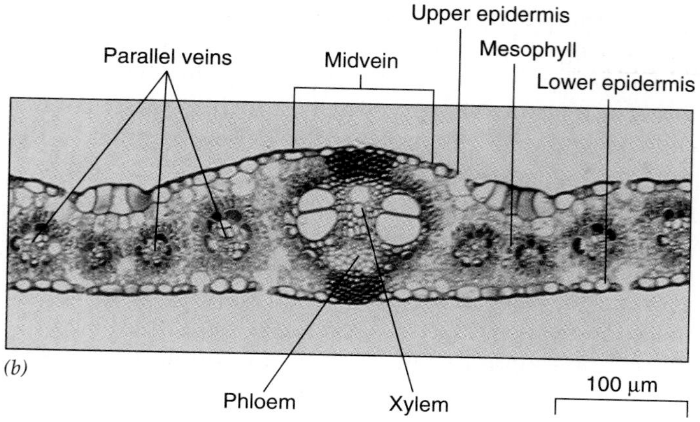

FIGURE 32–7 Dicot and monocot leaves usually can be distinguished by examining cross sections of the blades. ***(a)*** Privet, a dicot, has a mesophyll with distinct palisade and spongy sections. ***(b)*** Corn (*Zea mays*), a monocot. Note the absence of distinct regions of palisade and spongy mesophyll. Also evident is the evenly spaced parallel venation characteristic of monocots. (*a,* James Mauseth, University of Texas; *b,* Carolina Biological Supply Company/Phototake–NYC)

reeds, and sedges) have guard cells shaped like dumbbells. These structural differences affect how the cells swell or shrink to open or close the pore. Guard cells of both monocots and dicots may be associated with special epidermal cells called **subsidiary cells** that are structurally different from all other epidermal cells. Subsidiary cells provide a reservoir of water and ions that move into and out of the guard cells as they change shape (discussed later in this chapter).

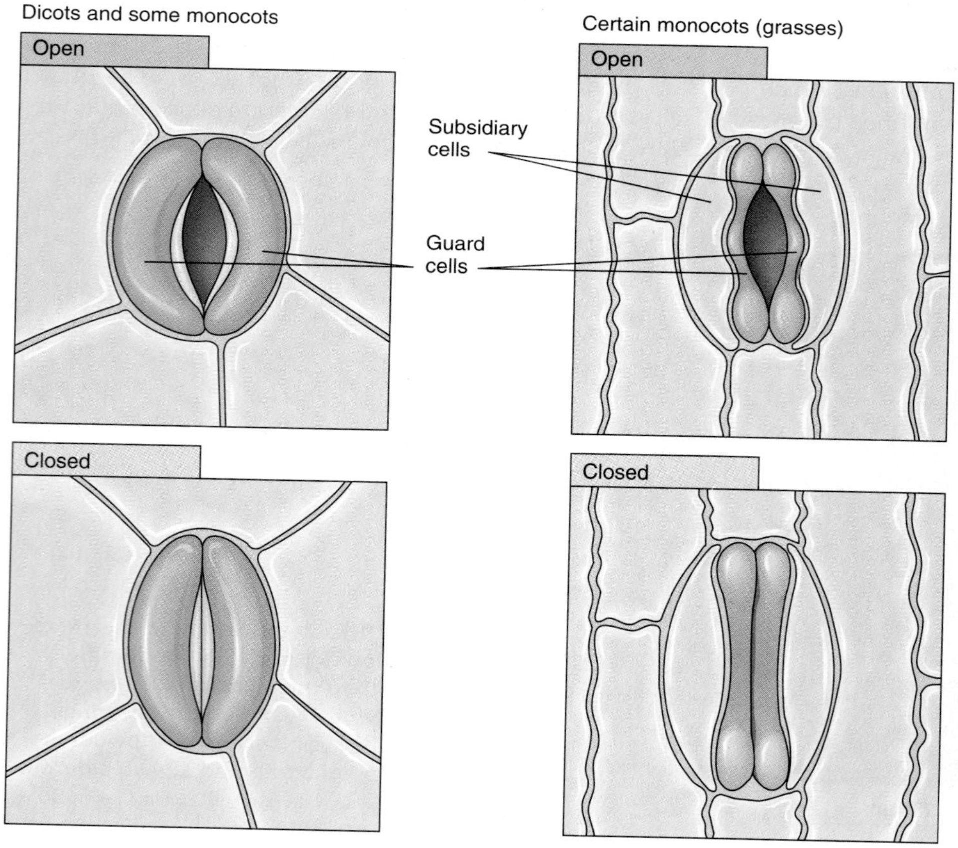

FIGURE 32–8 Although guard cells vary in shape, they all have the same function: control over the opening and closing of stomata. Guard cells of dicots and many monocots are bean-shaped. Other monocot guard cells (those of grasses, reeds, and sedges) are narrow in the center and thicker at each end. Each narrow guard cell is associated with a special epidermal cell called a subsidiary cell.

STRUCTURE IS RELATED TO FUNCTION IN LEAVES

The primary function of leaves is to collect radiant energy and convert it to the chemical energy stored in the bonds of organic molecules such as glucose. This process, called photosynthesis, was examined in detail in Chapter 8. During photosynthesis, plants take relatively simple molecules (carbon dioxide and water) and convert them into sugars. Oxygen is given off as a by-product. The sugar formed during photosynthesis is used by the plant in two ways. First, it is broken down by aerobic respiration to release the chemical energy stored in its bonds for other cellular activities (see Chapter 7). Second, sugar molecules provide the cell with basic building materials. The cell modifies glucose, converting it into all the organic compounds needed by the plant (starch and cellulose, amino acids, lipids, nucleic acids, etc.).

How is leaf structure related to its primary function of photosynthesis? The epidermis of a leaf is relatively transparent and allows light to penetrate to the interior of the leaf where the photosynthetic tissue, the mesophyll, is located. Stomata, which dot the leaf surfaces, permit the exchange of gases between the atmosphere and the leaf's internal tissues. Carbon dioxide, a raw material of photosynthesis, diffuses into the leaf through stomata, and the oxygen produced during photosynthesis diffuses rapidly out of the leaf through stomata. (Stomata, adapted to optimize gas exchange, also permit other gases, including air pollutants, to enter the leaf. See *Focus On: The Effects of Air Pollution on Leaves*.)

Water required for photosynthesis is obtained from the soil and transported in the xylem to the leaf, where it moistens the surfaces of mesophyll cells. The loose arrangement of the mesophyll tissue, with air spaces between cells, allows for rapid diffusion of carbon dioxide to the mesophyll cell surfaces; there it dissolves in a film of water before diffusing into the cells.

The veins not only supply the photosynthetic tissue with water and minerals (from the roots, by way of the xylem), but also carry (in the phloem) the dissolved sugar produced during photosynthesis to other parts of the plant. Bundle sheaths and bundle sheath extensions associated with the veins provide additional support to prevent the leaf, which is structurally weak because of the large amount of air space in the mesophyll, from collapsing under its own weight.

The leaves of each type of plant help it survive in the environment to which it is adapted

The environment to which a particular plant is adapted is reflected in its leaf structure. Although both aquatic plants and those adapted to dry conditions perform photosynthesis and have the same basic leaf anatomy, their leaves are modified to enable them to survive different environmental conditions. The leaves of water lilies, for example, have petioles long enough to allow the blade to float on the water's surface (Fig. 32–9). These petioles and other submerged parts have an internal system of air ducts; oxygen moves through these ducts from the floating leaves to the underwater roots and stems, which live in a poorly aerated environment.

The leaves of conifers, an important group of woody trees and shrubs that include pine, spruce, fir, redwood, and cedar, are waxy needles.[2] Most conifers are ever-

[2]As discussed in Chapter 27, conifers are gymnosperms, one of the two groups of seed plants (the other group is the flowering plants, or angiosperms). Unlike flowering plants, whose seeds are enclosed in fruits, conifers bear their seeds in cones.

FIGURE 32–9 An unusual view of water lily leaves shows their petioles as well as their blades, which float on the water's surface. The length of the petioles depends on the depth of the water. (Frans Lanting/Minden Pictures)

Focus On

The Effects of Air Pollution on Leaves

The air we breathe is often dirty and contaminated with many pollutants, particularly in urban areas. Air pollution consists of gases, liquids, or solids present in the atmosphere in levels high enough to harm humans and other organisms, as well as nonliving materials. Although air pollutants can come from natural sources—as, for example, when lightning causes a forest fire or when a volcano erupts—human activities make a major contribution to global air pollution. Motor vehicles and industry are the two main human sources of air pollution.

All parts of a plant can be damaged by air pollution, but leaves are particularly susceptible because of their structure and function (see figure). The thin blade provides a large surface area that comes into contact with the surrounding air. The thousands of tiny stomatal pores that dot the epidermis and allow gas exchange with the atmosphere also permit pollutants to diffuse into the leaf. Just as lungs, the organs of gas exchange in humans, often are affected by air pollution, so too the leaves of a plant are most affected.

Trees provide a dramatic demonstration of the effect of air pollution on biological longevity. According to American Forests (formerly the American Forestry Association), the average lifespan for a tree living in a rural setting, where air pollution is generally low, is 150 years. Even in the most ideal urban setting, a tree typically lives only 60 years, and in a normal city setting the average lifespan is 32 years. Trees living in a downtown setting, where air quality is lowest, live only seven years on average. It should be noted, however, that air pollution is not the only cause of short lifespan in urban trees. Other factors include inadequate room for root growth, lack of water, and polluted runoff from streets and sidewalks.

Many studies have shown that the overall productivity of crop plants is reduced by high levels of most forms of air pollution. The worst pollutant in terms of yield loss is ozone, a toxic gas produced when sunlight catalyzes a reaction between pollutants emitted by motor vehicles and industries. In plants, ozone inhibits photosynthesis because it damages the mesophyll cells, probably by altering the permeability of their cell membranes. Exposure to low levels of air pollution often causes a decline in photosynthesis without any other symptoms of injury. Lesions on leaves and other obvious symptoms appear at much higher levels of air pollution.

When air pollution, including acid rain, is combined with other environmental stresses (such as low winter temperatures, prolonged droughts, insects, and bacterial, fungal, and viral diseases), it can cause plants to decline and die. More than half of the red spruce trees in the mountains of the northeastern United States have died since the mid-1970s. Other tree species, such as sugar maples, for example, are also dying. Many still-living trees are exhibiting symptoms of **forest decline,** characterized by a gradual deterioration and often eventual death. The general symptoms of forest decline are reduced vigor and growth, but some plants exhibit specific symptoms, such as yellowing of needles in conifers.

Air pollutants may or may not be the *primary* stress that results in forest decline, but the presence of air pollution lowers plant resistance to other stress factors. When one or more stresses weaken a tree, then an additional stress may be enough to cause death. ■

Plants exposed to air pollution exhibit a variety of symptoms, including reduced growth. Compare the size and overall vigor of the radish grown in clean air (*left*) with the radish damaged by air pollution (*right*). Note the extensive leaf damage. (Visuals Unlimited/Science VU)

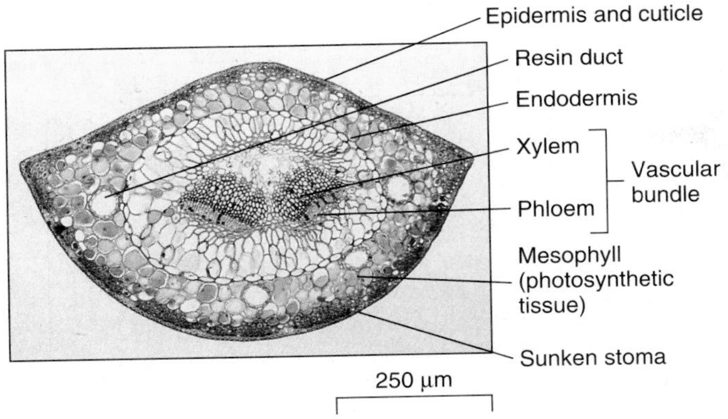

FIGURE 32–10 A pine needle (leaf) in cross section is somewhat rounded and thickened. The thick, waxy cuticle and sunken stomata are two structural adaptations that enable the pine tree to retain its needles throughout the winter. (Dennis Drenner)

green, which means they keep their leaves throughout the year. Conifers dominate a large portion of Earth's land area, particularly in northern forests and mountains. Their needles have structural adaptations that help them survive winter, the driest part of the year. (Winter is arid even in areas of heavy snows because roots cannot absorb water from soil when the soil temperature is low.) Indeed, many of the structural features of needles are also found in many desert plants.

Figure 32–10 shows a cross section of a pine needle. Note that the needle is somewhat thickened rather than thin and bladelike. The needle's relative thickness, which results in less surface area exposed to the air, reduces water loss. Other features that help conserve water include the thick waxy cuticle and sunken stomata; these permit gas exchange while minimizing water loss. Thus, needles help conifers tolerate the dry conditions that exist for part of the year. With the warming of spring, water again becomes available, and the needles quickly resume photosynthesis. (In contrast, *deciduous* plants—those that shed their leaves—have a lag period before resuming growth because they first have to grow new leaves.)

STOMATAL OPENING AND CLOSING ARE DUE TO CHANGES IN GUARD CELL TURGIDITY

Stomata are adjustable pores that are usually open during the day when carbon dioxide is required for photosynthesis, and closed at night when photosynthesis is shut down (see the section on CAM photosynthesis in Chapter 8 for an interesting exception). The opening and closing of stomata are controlled by changes in the shape of the two guard cells that surround each pore. When water moves into guard cells from surrounding cells, they become turgid (swollen) and the inner cell walls bend outward at the center, producing a pore (Fig. 32–11*a*). When water leaves the guard cells, they become flaccid (limp) and collapse against one another, closing the pore (Fig. 32–11*b*).

Although the opening and closing of stomata are triggered by light or darkness, other environmental factors are also involved, including carbon dioxide concentration. A low concentration of CO_2 in the leaf induces stomata to open even in the dark. The effects of light and CO_2 concentration on stomatal opening are interrelated. Photosynthesis, which occurs in the presence of light, reduces the internal concentration of CO_2 in the leaf, triggering stomatal opening. Another environmental factor that affects stomatal opening and closing is dehydration (water

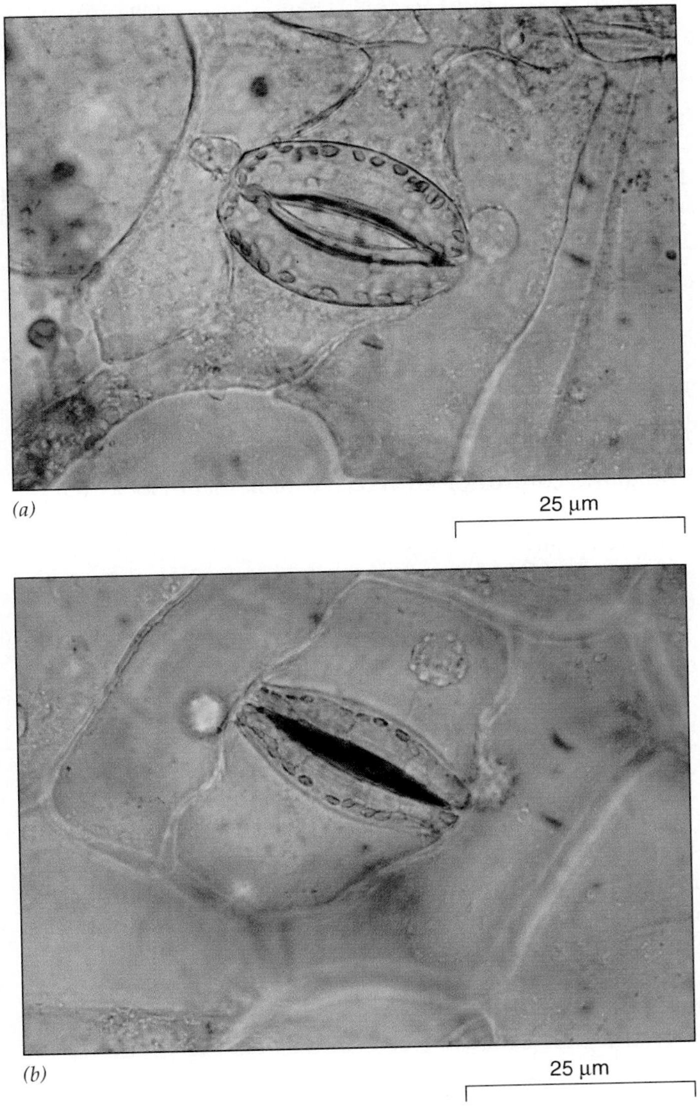

FIGURE 32–11 Stomata are usually open during the day and closed at night. Shown are (***a***) open and (***b***) closed stomata from a *Zebrina* leaf. (*a, b,* Dwight R. Kuhn)

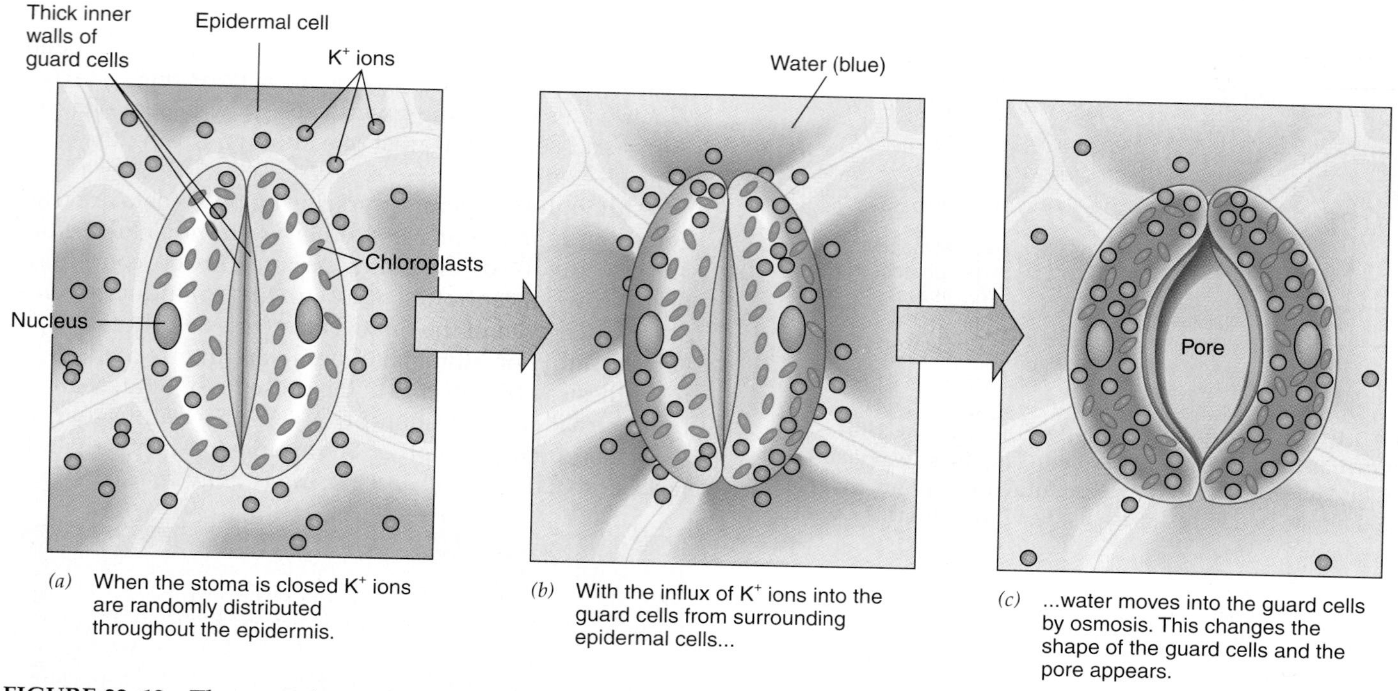
Thick inner walls of guard cells
Epidermal cell
K+ ions
Chloroplasts
Nucleus
Water (blue)
Pore
(a) When the stoma is closed K+ ions are randomly distributed throughout the epidermis.
(b) With the influx of K+ ions into the guard cells from surrounding epidermal cells...
(c) ...water moves into the guard cells by osmosis. This changes the shape of the guard cells and the pore appears.

FIGURE 32–12 The potassium ion mechanism of stomatal opening and closing involves the movement of K^+ ions (followed by water) into and out of the guard cells.

stress). During a prolonged drought, stomata will remain closed even during the day. Stomatal opening and closing are also under hormonal control (see Chapter 36).

The opening and closing of stomata also appear to be under the control of an internal biological clock that in some way measures time. For example, plants placed in continual darkness persist in opening and closing their stomata at more or less the same time each day. Such biological rhythms that follow an approximate 24-hour cycle are known as **circadian rhythms.** Other examples of circadian rhythms are provided in Chapter 36.

The potassium ion mechanism explains stomatal opening and closing

The data from numerous experiments and observations suggest that stomata open and close by the **potassium ion (K^+) mechanism.** Light in some way triggers an influx of potassium ions (K^+) into the guard cells from the surrounding cells of the epidermis (Fig. 32–12). This movement of potassium ions, which has been experimentally measured and verified, occurs by active transport through specific ion channels in the guard cells' plasma membranes, and requires ATP. The ATP supplies energy to pump protons out of the guard cells, producing an electrochemical gradient that can drive the uptake of potassium ions through specific K^+ channels (an example of a linked cotransport system, discussed in Chapter 5). The K^+ accumulates in the vacuoles of the guard cells.

The increased concentration of K^+ in the guard cells lowers the relative concentration of water in those cells. As a result, water passes into the guard cells from surrounding epidermal cells by osmosis. (Recall from the discussion of osmosis in Chapter 5 that when a cell has a solute concentration greater than that of surrounding cells, water flows into the cell.) The increased turgidity of the guard cells changes their shape, and the pore opens.

In the late afternoon or early evening, stomata may close by a reversal of the opening process.[3] The potassium ions are pumped out of the guard cells into the surrounding epidermal cells. Water leaves the guard cells by osmosis, the cells lose their turgidity, and the pore closes as they collapse.

As in other plant responses to light, stomata vary in their sensitivity to different colors—that is, different wavelengths—of light. Stomatal opening is most pronounced in blue and, to a lesser extent, in red light. Also, dim blue light induces stomatal opening, whereas dim red light does not. Recall from Chapter 8 that any plant

[3]Stomatal closure may not be an *exact* reversal of stomatal opening. There is evidence that Ca^{2+} triggers stomatal closure but inhibits stomatal opening. The actual mechanism whereby Ca^{2+} exerts this effect is under investigation.

Making the Connection

Trees, Transpiration, and Climate

Do trees influence an area's climate—that is, the average weather conditions that occur in that area over a period of years? Factors that determine climate include temperature and precipitation, both of which are influenced by transpiration.

Transpiration by trees, for example, influences the local temperature of forests. If you have ever walked into a forest on a hot summer day, you have probably noticed that the air is cooler and moister there than outside the forest. Transpired water has a cooling effect, not only for the plant but also for the local area.

Transpiration is an important part of the **hydrologic cycle** (see Chapter 54), in which water cycles from the oceans and land to the atmosphere, and then back to the oceans and land. As a result of transpiration, water evaporates from the surfaces of leaves and stems to form clouds in the atmosphere. Thus, transpiration eventually results in precipitation. As you might expect, forest trees release substantial amounts of moisture into the air by transpiration (see figure). For this reason, when a large forest is burned or cut down, rainfall declines and droughts become more common in that region. This phenomenon has been observed where large tracts of tropical rain forests have been burned to grow crops; often people have observed droughts in areas that formerly had rainfall almost every day of the year. ▲

Forests play an important role in the hydrologic cycle by returning most of the water that falls as precipitation to the atmosphere by transpiration.

response to light must involve a pigment, a molecule that absorbs the light prior to inducing a particular biological response. The responses of stomata to different colors of light and other data suggest that the pigment involved in stomatal opening and closing is yellow (yellow pigments strongly absorb blue light and weakly absorb red light) and is located in the guard cells.

Leaves lose water by transpiration and guttation

Despite leaf adaptations such as the cuticle, approximately 99% of the water that a plant absorbs from the soil is lost by evaporation from the leaves and, to a lesser extent, the stems. Loss of water vapor by evaporation from aerial plant parts is called **transpiration.**

The cuticle is extremely effective in reducing water loss from transpiration. It is estimated that only 1% to 3% of the water lost from a plant passes directly through the cuticle. Most transpiration occurs through the stomata. The numerous stomatal pores that are so effective in gas exchange for photosynthesis also provide openings through which water vapor escapes. In addition, the loose arrangement of the mesophyll cells provides a large surface area within the leaf from which water can evaporate.

A number of environmental factors influence the rate of transpiration. At higher temperatures more water is

(a)

(b)

FIGURE 32–13 Zucchini leaves wilt temporarily in the late afternoon of a hot day but recover overnight. (*a*) Late afternoon. (*b*) The following morning. Note that wilting helps reduce the surface area from which transpiration occurs. During the night, the plants recover by absorbing water from the soil while transpiration is negligible. (Ken Knott)

lost from plant surfaces. Light increases the transpiration rate, in part because it triggers stomatal opening and in part because it increases the leaf's temperature. Wind and dry air increase transpiration, but humid air *decreases* transpiration because the air is already saturated, or nearly so, with water vapor.

Although transpiration may seem wasteful, particularly to farmers in arid lands, it is an essential process. Transpiration is responsible for water movement in plants, and without it water would not reach the leaves from the soil (see Chapter 33). The large amount of water plants lose by transpiration may provide some additional benefits. Transpiration, like sweating in humans, cools the leaves and stems. When water passes from a liquid state to a vapor, it absorbs a great deal of heat. When the water molecules leave the plant as water vapor, they carry this heat with them. Thus, the cooling effect of transpiration may prevent the plant from overheating, particularly in direct sunlight. On a hot summer day, for example, the internal temperature of leaves is measurably lower than that of the surrounding air.

A second benefit of transpiration is that it provides the plant with sufficient essential minerals. The water a plant transpires is initially absorbed from the soil, where it is not present as pure water but rather as a dilute solution of dissolved mineral salts. Water leaves the plant during transpiration but minerals do not. Many of these minerals are required for the plant's growth. It has been suggested that transpiration enables a plant to take in sufficient water to provide enough essential minerals, and that plants cannot satisfy their mineral requirements if the transpiration rate is not high enough.

There is no doubt, however, that under certain circumstances transpiration can be harmful to a plant. On hot summer days, plants frequently lose more water by transpiration than they can take in from the soil. Their cells experience a loss of turgor, and the plant wilts (Fig. 32–13). If a plant is able to recover overnight, because of the combination of negligible transpiration (recall that stomata are closed) and absorption of water from the soil, the plant is said to have experienced *temporary wilting*. Most plants recover from temporary wilting with no ill effects. In cases of prolonged drought, however, the soil may not contain sufficient moisture to permit recovery from wilting. A plant that cannot recover overnight is said to be *permanently wilted* and will die unless water is supplied immediately.

Some plants exude water as a liquid

Many leaves have special structures through which liquid water is literally forced out. This loss of liquid water, known as **guttation,** occurs when transpiration is negligible and available soil moisture is high. Guttation typically occurs at night because the stomata are closed, but water continues to move into the roots by osmosis. People sometimes incorrectly attribute the early morning droplets of water on leaf margins to dew—water condensation from the air—rather than to guttation (Fig. 32–14).

LEAF ABSCISSION ALLOWS PLANTS IN TEMPERATE CLIMATES TO SURVIVE WINTER

All trees shed leaves; many conifers, for example, shed their needles year-round. The leaves of deciduous plants,

FIGURE 32–14 In guttation, liquid water is exuded from leaves. Guttation in lady's mantle (*Alchemilla vulgaris*). Many people mistake guttation for early morning dew. (Dennis Drenner)

however, turn color and **abscise,** or fall off, once a year—as winter approaches in temperate climates or before the dry period in climates with pronounced wet and dry seasons. In temperate forests, for example, most woody plants with broad leaves shed their leaves in order to survive the low temperatures of winter. During winter, the plant's metabolism, including its photosynthetic machinery, slows down a great deal. As a result, plants have little need for leaves at that time.

Another reason for abscission is related to a plant's water requirements, which become critical during the winter. As mentioned previously, as the ground chills, absorption of water by the roots is inhibited. When the ground freezes, *no* absorption occurs. If the broad leaves were to stay on the plant during the winter, the plant would continue to lose water by transpiration but would be unable to replace it with water absorbed from the soil.

Leaf abscission is a complex process that involves many physiological changes, all of which are initiated and orchestrated by changing levels of plant hormones (see Chapter 36). Briefly, the process is this: As autumn approaches, sugars, amino acids, and many essential minerals (such as nitrogen, phosphorus, and possibly potassium) are transported from the leaves into the woody tissues. Chlorophyll breaks down, allowing the orange carotenes and yellow xanthophylls, some of the accessory pigments in the chloroplasts of leaf cells, to become evident. Recall from Chapter 8 that accessory pigments are always present in the leaf but are masked by the chlorophyll. In addition, red water-soluble pigments may be synthesized and stored in the vacuoles of leaf cells in some species; their function is unknown. The various combinations of these pigments are responsible for the brilliant colors found in autumn landscapes in temperate climates.

In many leaves, abscission occurs at an abscission zone near the base of the petiole

The area where a petiole detaches from the stem of the plant is structurally different from surrounding tissues. This area, called the **abscission zone,** is composed primarily of thin-walled parenchyma cells and, because it contains few fibers, is anatomically weak (Fig. 32–15). A protective layer of cork cells develops on the stem side of the abscission zone. These cells have a waxy waterproof material impregnated in their walls. Enzymes then dissolve the *middle lamella* (the "cement" that holds the primary cell walls of adjacent cells together) in the abscission zone. Once this process is completed, nothing holds the leaf to the stem but a few xylem cells. A sudden breeze is enough to make the final break, and the leaf detaches. The protective layer of cork remains, sealing off the area and forming a leaf scar.

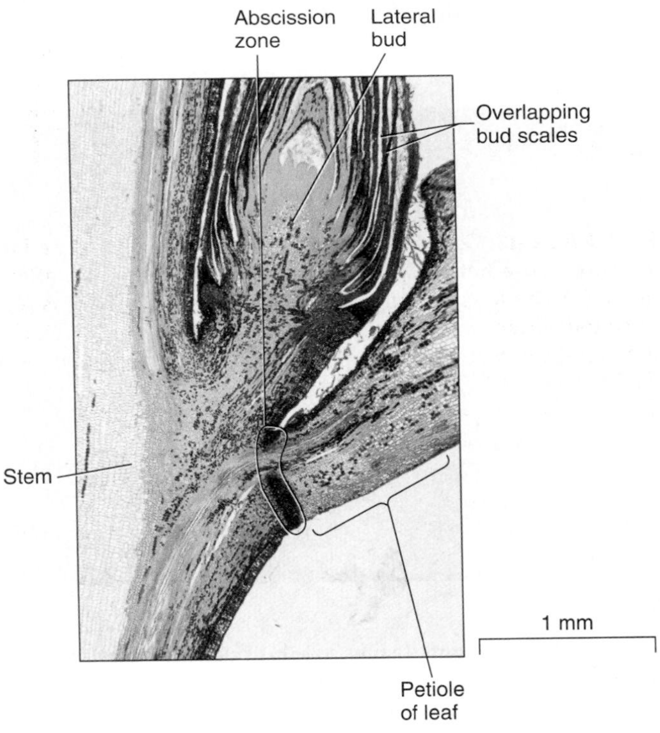

FIGURE 32–15 A leaf's abscission zone is an area of thin-walled cells at the petiole's base where the leaf will separate from the stem. A longitudinal section through a tree branch, showing the abscission zone at the base of a petiole. A lateral bud with its protective bud scales is evident above the petiole. (James Mauseth, University of Texas)

Leaves with Functions Other than Photosynthesis Exhibit Modifications in Structure

Although photosynthesis is the main function of leaves, certain leaves possess special modifications for other functions. Some plants have leaves specialized for protection. **Spines,** modified leaves that are hard and pointed, may be found on many desert plants like cacti (Fig. 32–16*a*). In the cactus, the main organ of photosynthesis is the stem rather than the leaf. Spines discourage animals from eating the succulent (juicy) stem tissue.

(a) (b) (c) (d) (e)

FIGURE 32–16 Some leaves exhibit modifications for functions other than photosynthesis. (*a*) The leaves of cacti such as this barrel cactus are modified as spines for protection. **(*b*)** Tendrils, which wind around objects and aid vines in climbing, may be modified leaves or stems. Shown are tendrils of bur cucumber (*Echinocystis lobata*). **(*c*)** Overlapping bud scales protect buds. Shown here are a terminal bud and two lateral buds. **(*d*)** The leaves of bulbs such as the onion are fleshy for storage of food materials and water. **(*e*)** Some plants have thick succulent leaves modified for water storage as well as photosynthesis. The succulent leaves of the string-of-beads plant are spherical to reduce surface area, thereby conserving water. (*a, e,* James Mauseth, University of Texas; *b,* Stephen P. Parker/Photo Researchers, Inc.; *c, d,* Dennis Drenner)

Vines are climbing plants whose stems cannot support their own weight, and so they often possess **tendrils** that help keep the vine attached to the structure on which it is growing. The tendrils of many vines, such as peas, cucumbers, and squash, are specialized leaves (Fig. 32–16*b*). However, some tendrils, such as those of ivy, Virginia creeper, and grape, are specialized stems.

The winter buds of a dormant woody plant are covered by **bud scales,** modified leaves that protect the delicate meristematic tissue of the bud from being injured and drying out (Fig. 32–16*c*).

Leaves may also be modified for storage of water or food. For example, a **bulb** is a short underground stem to which large, fleshy leaves are attached (Fig. 32–16*d*). Onions and tulips form bulbs. Many plants adapted to arid conditions—for example, jade plant (*Crassula arborescens*), medicinal aloe (*Aloe barbadensis*), and string-of-beads (*Senecio rowleyanus*)—have succulent leaves for water storage (Fig. 32–16*e*). These leaves are usually green and function for photosynthesis as well.

Some of the most remarkable examples of modified leaves are those of insectivorous plants

Insectivorous plants are plants that capture insects. Most insectivorous plants grow in poor soil that is deficient in certain essential minerals, particularly nitrogen. These plants meet some of their mineral requirements by digesting insects and other small animals. The leaves of insectivorous plants are adapted to attract, capture, and digest their animal prey.

Some insectivorous plants have passive traps. The leaves of a pitcher plant, for example, are shaped so that rainwater collects and forms a reservoir that also contains digestive enzymes secreted by the plant (Fig. 32–17). Some pitchers are quite large; in the tropics, for example, pitcher plants may be large enough to hold one liter (approximately one quart) or more of liquid.

An insect attracted by the odor or nectar of the pitcher may lean over the edge and fall in. Although it may make repeated attempts to escape, the insect is prevented from crawling out by a row of stiff spines that point downward around the lip of the pitcher. The insect eventually drowns, and part of its body is digested and absorbed.

FIGURE 32–17 The insectivorous pitcher plant has leaves modified to form a water-collecting pitcher that drowns its prey. Shown are pitchers of *Nepenthes,* which is native to the tropics and has the most elaborate of all pitcher plant leaves. The extreme tip of the leaf forms a lid that partially shields the pitcher from rainfall that would dilute the solution of digestive enzymes inside the pitcher. (James Mauseth, University of Texas)

The Venus flytrap is an insectivorous plant with active traps. Its leaves resemble tiny bear traps (see Fig. 1–5). Each side of the leaf contains three small hairs. If an insect alights and brushes against two of the hairs, the trap springs shut with amazing rapidity (the mechanism whereby the leaves shut is discussed further in Chapter 36). After the insect has died and been digested, the trap reopens and the indigestible remains fall out.

SUMMARY

I. Leaves exhibit variation in shape and form.
   A. Leaves typically consist of a broad, flat blade, a stalklike petiole, and small, leaflike outgrowths from the base called stipules.
   B. Leaves may be simple or compound; have alternate, opposite, or whorled leaf attachment; and have parallel or netted (either pinnately netted or palmately netted) venation.

II. Leaf structure is adapted for its primary function of photosynthesis.
   A. Most leaves have a broad, flattened blade that is quite efficient in collecting the sun's radiant energy.
   B. The transparent epidermis allows light to penetrate into the mesophyll, where photosynthesis occurs.
      1. Stomata are small pores in the epidermis that permit gas exchange needed for photosynthesis and

transpiration needed to obtain water and minerals from the soil.

2. The waxy cuticle coats the epidermis, enabling the plant to survive the dry conditions of a terrestrial existence.

C. The mesophyll tissue contains air spaces that permit rapid diffusion of carbon dioxide and water into, and oxygen out of, mesophyll cells.

D. Leaf veins have xylem to conduct water and essential minerals to the leaf, and phloem to conduct sugar produced by photosynthesis to the rest of the plant.

III. Monocot and dicot leaves can be distinguished based on their external and internal structures.

A. Monocot leaves have parallel venation.

B. Dicot leaves have netted venation.

IV. Stomata generally open during the day and close at night.

A. The potassium ion mechanism explains the opening and closing process.

1. Light triggers an influx of potassium ions (K^+) into the guard cells. The resulting osmotic movement of water into the guard cells causes their shape to change, forming a pore.

2. Stomata close when potassium ions leave the guard cells, which causes water to flow out by osmosis. As the guard cells become less turgid and collapse, the pore closes.

B. A number of factors affect stomatal opening and closing, including light or darkness, CO_2 concentration, water stress, and the plant's circadian rhythm.

V. Transpiration is the loss of water vapor from plants.

A. It occurs primarily through the stomata.

B. The rate of transpiration is affected by environmental factors like temperature, wind, and relative humidity.

C. Transpiration appears to be both beneficial and harmful to the plant.

VI. Guttation is the release of liquid water from leaves that occurs through special structures when transpiration is negligible and available soil moisture is high.

VII. Leaf abscission is a complex process involving physiological and anatomical changes prior to leaf fall.

VIII. Leaves may be modified for functions other than photosynthesis.

A. Spines are leaves adapted to provide protection.

B. Some tendrils are leaves modified for grasping and holding onto other structures (to support weak stems).

C. Bud scales are leaves modified to protect delicate meristematic tissue or dormant buds.

D. Bulbs are short underground stems with fleshy leaves specialized for storage.

E. Some leaves are modified to trap insects.

SELECTED KEY TERMS

abscission
abscission zone
blade
bud scale
bulb
bundle sheath
bundle sheath extension
circadian rhythm
cuticle
epidermis, upper and lower
forest decline
guard cell
guttation
mesophyll
palisade mesophyll
petiole
phloem
photosynthesis
potassium ion mechanism
spine
spongy mesophyll
stipule
stoma (stomata)
subsidiary cell
tendril
transpiration
trichome
vascular bundle
vein
xylem

POST-TEST

1. The __________ is the photosynthetic tissue in the middle of the leaf.
2. Gas exchange occurs through tiny pores formed by two __________ __________.
3. Most stomata are usually located on the __________ epidermis of the leaf.
4. The __________ is a thin, noncellular layer of wax secreted by the epidermis of leaves.
5. The layer of cells that encircles a vein, or vascular bundle, is called a(an) __________ __________.
6. The __________ of a leaf vein transports water and dissolved minerals.
7. Sugars produced in the leaf during photosynthesis are transported to the rest of the plant by the __________.
8. Most of the water that a plant absorbs from the soil is lost by the process of __________.
9. When transpiration is negligible, plants such as grasses exude excess water by __________.
10. Responses based on an internal, 24-hour biological clock are called __________ __________.
11. The __________ __________ mechanism explains the opening and closing of stomata.
12. The fall of leaves in the autumn is known as leaf __________.
13. __________ are modified leaves that enable a stem to climb.
14. Cactus __________ are leaves that are modified for protection against animals.

REVIEW QUESTIONS

1. Give the general equation for photosynthesis (see Chapter 8), and discuss how the leaf is organized to deliver the raw materials and products of photosynthesis.
2. How is leaf structure related to both photosynthesis and transpiration? How is it tied to each physiological process?
3. Relate the series of physiological changes that occur in guard cells during stomatal opening and closing.
4. Discuss at least two ways that, during the course of evolution, certain leaves adapted to conserve water.
5. How do environmental factors (sunlight, temperature, humidity, wind) influence the rate of transpiration? How does the environment influence stomatal opening and closing?
6. Why do many woody plants lose their leaves in autumn?
7. Discuss the specialized features of the leaves of insectivorous plants.
8. Label the diagram to the right.

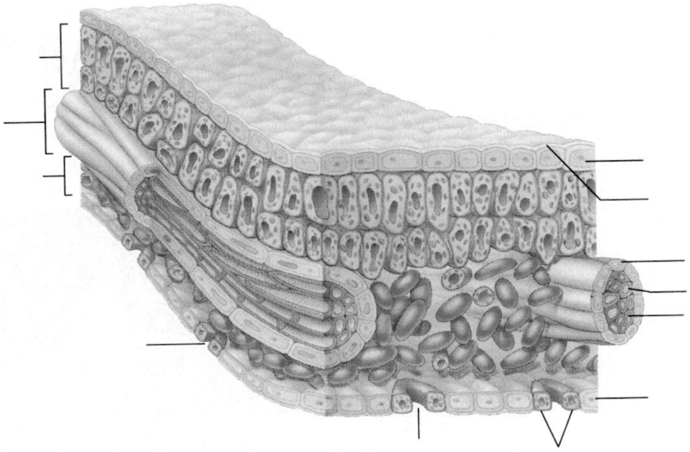

YOU MAKE THE CONNECTION

1. Given that (1) xylem is located toward the upper epidermis in leaf veins while phloem is toward the lower epidermis, and (2) the vascular tissue of a leaf is continuous with that of the stem, suggest one possible arrangement of vascular tissues in the stem that accounts for the arrangement of vascular tissue in the leaf.
2. Suppose you are asked to observe a micrograph of a leaf cross section and determine which side is covered by the upper epidermis and which side by the lower epidermis. How would you make this decision?
3. What might be some of the advantages of a plant with a few very large leaves? Disadvantages? What might be some advantages of having many very small leaves? Disadvantages?

RECOMMENDED READINGS

Dale, J.E. "How Do Leaves Grow?" *BioScience,* Vol. 42, No. 6 June 1992. Leaf research is "branching out" in new directions—away from describing leaf variations and toward understanding changes that occur during leaf development.

Dale, J.E. "Power Plants." *The Sciences,* September/October 1994. What is currently known about how a leaf primordium, which starts out as a bump of cells on the stem apical meristem, develops into a fully expanded leaf.

Lipske, M. "Forget Hollywood: These Bloodthirsty Beauties Are for Real," *Smithsonian,* December 1992. Examines some of the relationships between insectivorous plants and animals.

Mauseth, J.D. *Botany: An Introduction to Plant Biology,* 2nd edition. Philadelphia, Saunders College Publishing, 1995. A comprehensive introduction to plant biology.

Raven, P.H., R.F. Evert, and S.E. Eichhorn. *Biology of Plants,* 5th edition. New York, Worth Publishers, 1992. A general botany text with an evolutionary emphasis.

Vogel, S. "When Leaves Save the Tree," *Natural History,* September 1993. In high winds, leaves roll or clump together to minimize drag.

C H A P T E R 3 3

Stems and Plant Transport

Baobab trees, native to Africa, Madagascar, and Australia, store large volumes of water and starch in their stems. The massive trunks have on occasion been hollowed out and used as prison cells. (Frans Lanting/Minden Pictures)

The vegetative (nonreproductive) body of a vascular plant has three parts: roots, leaves, and stems. Roots serve to anchor the plant and absorb materials from the soil. Leaves are for photosynthesis, converting radiant energy into the chemical energy of carbohydrate molecules.

Stems perform three main functions in plants. First, they support leaves and reproductive structures. The upright position of most stems and the arrangement of the leaves on them allow each leaf to absorb maximum light for use in photosynthesis. Reproductive structures (flowers and fruits) are located on stems in areas accessible to insects, birds, and air currents, which transfer pollen from flower to flower and help disperse seeds and fruits.

Second, stems provide internal transport. They conduct water and dissolved minerals from the roots, where these materials are absorbed from the soil, to leaves and other plant parts. Stems also conduct the sugar produced in leaves by photosynthesis to roots and other parts of the plant. Remember, however, that stems are not the only plant organ that conducts materials. The vascular system is continuous throughout all parts of a plant, and conduction occurs in roots, stems, leaves, and reproductive structures.

Third, stems produce new living tissue. They continue to grow throughout a plant's life, producing buds that develop into stems with new leaves and/or reproductive structures. In addition to the main functions of support, conduction, and production of new stem tissues, a number of stems are modified for asexual reproduction (see Chapter 35). Also, some stems are specialized to store starch or, if green, to manufacture sugar by photosynthesis.

Stems exhibit varied forms ranging from ropelike vines to massive tree trunks. They link a plant's roots to its leaves and are usually located above ground (although many plants have underground stems). Stems can be either herbaceous (consisting of soft, nonwoody tissues) or woody (developing extensive hard tissues of wood and bark). Most stems are more or less circular in cross section, although some—such as those of mint—are square, and others—such as those of sedges—are triangular.

LEARNING OBJECTIVES

After you have studied this chapter you should be able to

1. Describe three functions of stems.
2. Label cross sections of herbaceous dicot and monocot stems, and describe the functions of each tissue.
3. Outline the transition from primary growth to secondary growth in a woody stem, name the two lateral meristems, and describe the tissues that arise from each.
4. Describe the pathway of water movement in plants.
5. Discuss tension-cohesion and root pressure as mechanisms to explain the rise of water and dissolved minerals in xylem.
6. Describe the pathway of sugar translocation in plants.
7. Discuss the pressure-flow hypothesis of sugar translocation in phloem.

A WOODY TWIG DEMONSTRATES THE EXTERNAL STRUCTURE OF STEMS

Although stems exhibit great variation in structure and growth, they all have **buds,** which are undeveloped embryonic shoots. A **terminal bud** is the embryonic shoot located at the tip of a stem. When a terminal bud is dormant (that is, unopened and not actively growing), it is covered and protected by an outer protective layer of **bud scales,** which are modified leaves. **Lateral buds** are located in the *axils* of a plant's leaves. An axil is the upper angle between a leaf and the stem to which it is attached. When terminal and lateral buds grow, they form stems that bear leaves and/or flowers. The area on a stem where each leaf is attached is called a *node,* and the region between two successive nodes is an *internode.*

A woody twig of a deciduous tree that has shed its leaves can be used to demonstrate stem structures (Fig. 33–1). The terminal bud is covered by bud scales that protect its delicate tip during dormancy. When the bud resumes growth, the bud scales covering the terminal bud fall off, leaving a **bud scale scar** on the stem where the bud scales were attached. Because temperate plants form terminal buds at the end of each year's growing season, counting the number of bud scale scars on a twig indicates its age. A **leaf scar** shows where each leaf was attached on the stem; the vascular (conducting) tissue that extends from the stem out into the leaf forms **bundle scars** within a leaf scar. Lateral buds may be found above the leaf scars. Also, the bark of a woody twig has **lenticels,** sites of loosely arranged cells that allow oxygen to diffuse into the interior of the woody stem. Lenticels look like tiny marks, or specks, on the bark of a twig.

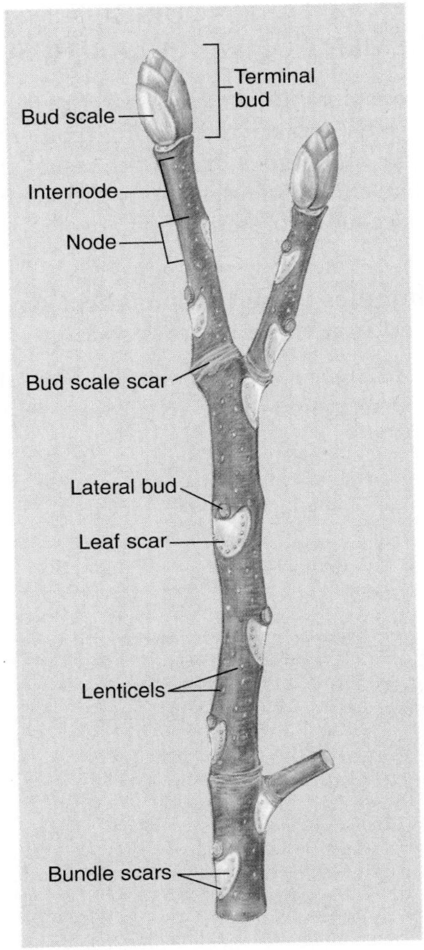

FIGURE 33–1 The external structure of a woody twig of horse chestnut (*Aesculus hippocastanum*) is shown in its winter condition. The age of a woody twig can be determined by the number of bud scale scars (don't count side branches). How old is this twig?

STEMS ORIGINATE AND DEVELOP AT MERISTEMS

Recall from Chapter 31 that plants have two different types of growth. Primary growth is an increase in the length of a plant and occurs at **apical meristems** located at the tips of stems and roots. Secondary growth is an increase in the width of a plant and is due to the activity of **lateral meristems** located along the sides of stems and roots. The new tissues formed by the lateral meristems are called *secondary tissues* to distinguish them from *primary tissues* produced by apical meristems.

All plants have primary growth; some plants have both primary and secondary growth. Recall that stems with only primary growth are herbaceous, while those with both primary and secondary growth are woody.[1] A woody plant increases in length (by primary growth) at the tips of its stems and roots, while its older stems and roots further back from the tips increase in width by secondary growth. In other words, at the same time that secondary growth is adding wood and bark (thereby causing the stem to thicken), primary growth (which increases the length of the stem) continues.

HERBACEOUS DICOT AND MONOCOT STEMS CAN BE DISTINGUISHED BY THE ARRANGEMENT OF VASCULAR TISSUES

Although considerable variation exists in stems, they all possess an outer protective covering (epidermis or periderm), one or more types of ground tissue, and vascular tissues (xylem and phloem). Let us first consider the structure of herbaceous dicot stems and then of monocot stems.

Vascular bundles of herbaceous dicot stems are arranged in a circle in cross section

A young sunflower stem is a representative herbaceous dicot stem that exhibits primary growth (Fig. 33–2). Its outer covering, the **epidermis,** provides protection in herbaceous stems, as it does in leaves and herbaceous roots. The epidermis is covered by a cuticle, a waxy layer that also covers the leaf epidermis (see Chapter 32) and reduces water loss from the stem surface.

Inside the epidermis is a complex tissue called the **cortex.** It is a layer several cells thick that may contain parenchyma, collenchyma, and sclerenchyma cells (see Chapter 31). As might be expected from the various types of cells that it contains, the cortex in herbaceous dicot stems can have three functions: photosynthesis, storage, and support. If a stem is green, photosynthesis occurs in the cortex (in parenchyma cells containing chloroplasts). Parenchyma in the cortex also serves as a storage tissue; starch grains and crystals are frequently evident as stored materials in cortical parenchyma cells. Collenchyma and sclerenchyma in the cortex provide strength and structural support for the stem.

The vascular tissues provide conduction and support. In herbaceous dicot stems, the vascular tissues are located in bundles that, when viewed in cross section, are arranged in a circle. However, viewed lengthwise these bundles extend as long strands throughout the length of a stem and are continuous with vascular tissues of both roots and leaves.

Each vascular bundle contains both **xylem,** which transports water and dissolved minerals from roots to leaves, and **phloem,** which transports dissolved sugar out of the leaves. Xylem is located on the inner side of the vascular bundle, and phloem is found toward the outside. Sandwiched between xylem and phloem is a single

[1]Many herbaceous stems (such as those of geranium and sunflower) also have a limited amount of secondary growth.

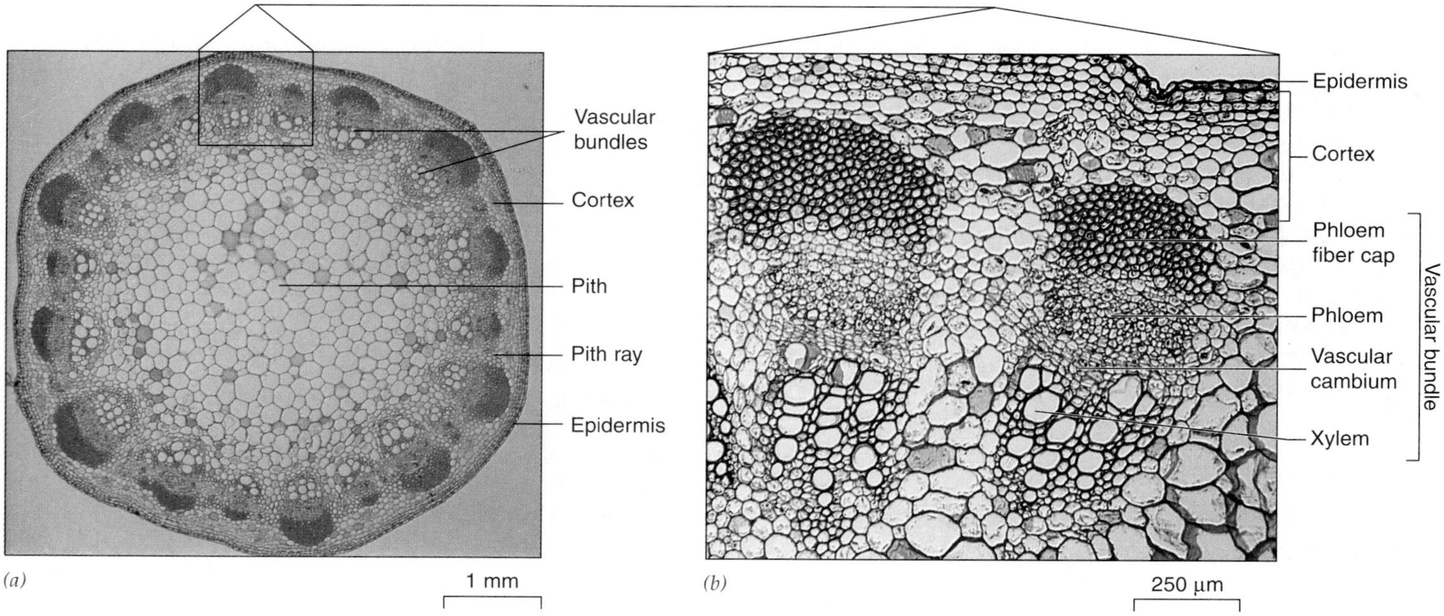

FIGURE 33–2 A cross section of the internal structure of a stem of sunflower, a herbaceous dicot, shows vascular bundles arranged in a circle around a central core of pith. (*a*) Cross section of a sunflower stem. (*b*) Close-up of two vascular bundles. Xylem is located toward a stem's interior, and phloem is toward the exterior. Each vascular bundle is "capped" by a batch of fibers for additional support. (*a*, Dennis Drenner; *b*, Ed Reschke)

Making the Connection

Vines, Evolutionary Adaptations, and Ecology

How are vines adapted to their environment? Vines are weak-stemmed plants that depend on other plants for support. Because vines do not expend many resources to produce structurally strong stems, they are able to grow extremely rapidly, ascending through the shaded understory to the sunlit canopy of forest trees where they produce luxuriant foliage.

Vines have a variety of adaptations that fit their climbing lifestyle. As newly germinated plants, many vine seedlings grow *away* from sunlight rather than toward it. In growing toward the darkest part of its environment, a vine seedling usually encounters a large tree that it then ascends. Along their trunks, woody vines (known as *lianas*) often produce special roots with adhesive pads that stick to the bark of the host tree. Herbaceous vines frequently have tendrils—modified leaves or stems that wrap around supports (see Chapter 32)—while other vines are *twiners,* with stems that spiral around their host as they ascend it.

Vines are most numerous and diverse in tropical forests, particularly tropical rain forests (see Chapter 51), where both herbaceous vines and lianas abound. Lianas grow from the upper branches of one forest tree to another, connecting the tops of the trees and providing a walkway for many of the canopy's animal residents. These and other vines provide nectar and fruit for many tree-dwelling animals.

Temperate forests, boreal (far northern) forests, and island forests have far fewer vines than do tropical rain forests. It is thought that vines are less common in temperate and boreal forests because they are less resistant to droughts and fires than are the trees that grow there. Vines in tropical rain forests generally need not adapt to long droughts or forest fires. Vines are thought to be relatively uncommon on islands because the seeds of most vines, which are dispersed by wind, cannot reach islands to colonize them. ▲

layer of cells called the **vascular cambium,** a lateral meristem responsible for secondary growth (to be discussed shortly).

Because stems grow through the air and must support the plant body, they have to be much stronger than roots. Fibers may be found in both xylem and phloem, although they are usually more extensive in phloem (see Chapter 31). These fibers add considerable strength to the herbaceous stem. In sunflowers and certain other herbaceous dicot stems, phloem contains a cluster of fibers, called a **phloem fiber cap,** that helps strengthen the stem. The phloem fiber cap is not a part of the structure of all herbaceous dicot stems.

At the center of the herbaceous dicot stem is **pith,** a tissue composed of large, thin-walled parenchyma cells that function primarily for storage. Due to the arrangement of the vascular tissues in bundles, there is no distinct separation of cortex and pith between the vascular bundles. The areas between the vascular bundles are often referred to as **pith rays.**

Vascular bundles are scattered throughout monocot stems

Monocot stems, as exemplified by the herbaceous stem of corn, are covered by an epidermis with its waxy cuticle. As in herbaceous dicot stems, the vascular tissues run in strands throughout the length of a stem (Fig. 33–3). In cross section, however, the vascular bundles, which contain xylem toward the inside and phloem toward the outside, are not arranged in a circle as they are in herbaceous dicot stems but instead are scattered throughout the stem. Each vascular bundle is enclosed in a bundle sheath of supporting sclerenchyma cells. The monocot stem does not have distinct areas of cortex and pith. The **ground tissue** in which the vascular tissues are embedded performs the same functions as cortex and pith in herbaceous dicot stems.

The lateral meristems—vascular cambium and cork cambium (see Chapter 31)—that give rise to secondary growth do not occur in monocot stems. Monocots have primary growth only and do not produce the secondary tissues, wood and bark. Although some monocots such as palm trees attain considerable size, they do so by a modified form of primary growth rather than by secondary growth. Stems of monocots like bamboo and palm contain a great deal of sclerenchyma tissue, which makes them extremely hard.

WOODY PLANTS HAVE STEMS WITH SECONDARY GROWTH

Woody plants undergo secondary growth, an increase in the width of stems and roots. Secondary growth occurs as a result of the activity of two lateral meristems: vascular cambium and cork cambium. Among flowering plants, only woody dicots (such as apple, hickory, and maple) have secondary growth. Gymnosperms (such as pine, juniper, and spruce) also have secondary growth. Table 33–1 summarizes the relationships of meristems and tissues in a woody dicot stem.

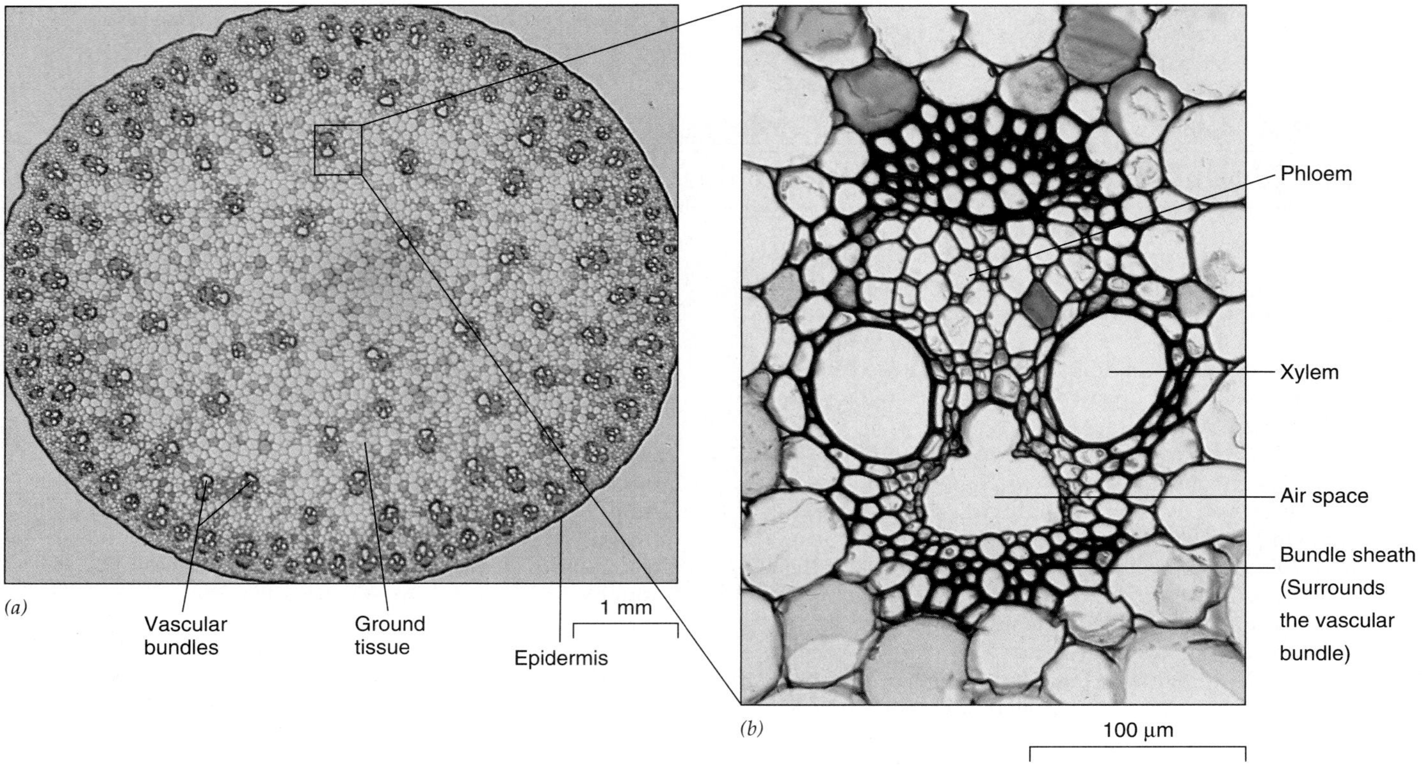

FIGURE 33–3 The interior of a stem of corn, a monocot, has vascular bundles scattered throughout ground tissue. (***a***) Cross section of a corn stem. (***b***) Close-up of a vascular bundle. The air space is the site where the first xylem elements were formed. The entire bundle is enclosed in a bundle sheath of sclerenchyma for additional support. (*a*, Carolina Biological Supply Company/Phototake-NYC; *b*, Ed Reschke)

Cells in **vascular cambium** divide and produce two tissues: secondary xylem (wood) to replace primary xylem, and secondary phloem (inner bark) to replace primary phloem. Primary xylem and primary phloem are not able to transport materials indefinitely and so must be replaced if the plant is to survive long-term. Cells of the second lateral meristem, **cork cambium,** divide to produce cork cells and cork parenchyma. Cork cambium

Table 33–1 ANATOMICAL RELATIONSHIPS IN A DICOT STEM*

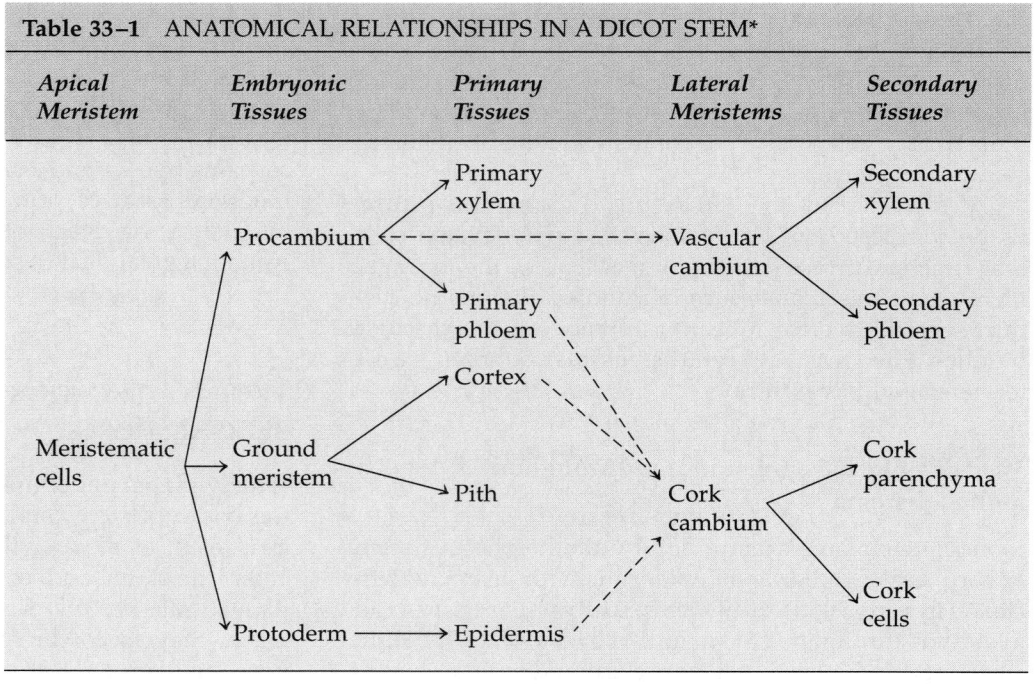

Apical Meristem	*Embryonic Tissues*	*Primary Tissues*	*Lateral Meristems*	*Secondary Tissues*
Meristematic cells	Procambium	Primary xylem; Primary phloem	Vascular cambium	Secondary xylem; Secondary phloem
	Ground meristem	Cortex; Pith	Cork cambium	Cork parenchyma; Cork cells
	Protoderm	Epidermis		

*Dashed lines indicate tissues from which the lateral meristems may arise.

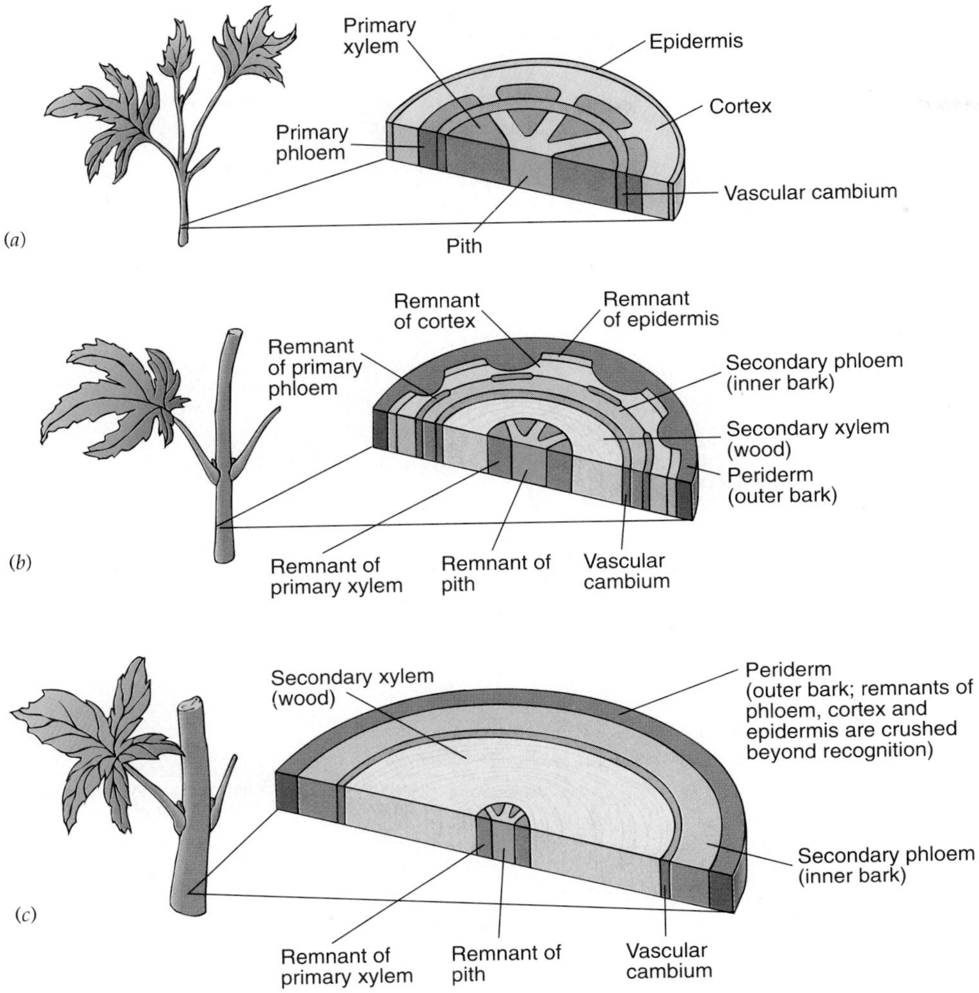

FIGURE 33–4 Development of secondary growth in a dicot stem is shown in cross section. Vascular cambium and the tissues it produces are shown; cork cambium is not depicted. **(*a*)** A herbaceous dicot stem. Vascular cambium is sandwiched between primary xylem and primary phloem in each vascular bundle. At the onset of secondary growth, vascular cambium arises in the parenchyma between the vascular bundles, forming a cylinder of meristematic tissue that appears as a circle in cross section *(shown)*. **(*b*)** Vascular cambium begins to divide, forming secondary xylem on the inside and secondary phloem on the outside. Note that the primary xylem and primary phloem in the original vascular bundles are split apart in this process. **(*c*)** A young woody stem. Vascular cambium produces significantly more secondary xylem than secondary phloem.

and the tissues it produces are collectively referred to as **periderm** (outer bark), which functions as a replacement for the epidermis.

Vascular cambium gives rise to secondary xylem and secondary phloem

Primary tissues in woody stems are organized similarly to those in herbaceous dicot stems, with the vascular cambium a thin layer of cells sandwiched between xylem and phloem in the vascular bundles. Once secondary growth begins, however, the internal structure of a stem changes considerably (Fig. 33–4). Although vascular cambium is not initially a solid cylinder of cells (because the vascular bundles are separated by pith rays), it becomes continuous when production of secondary tissues begins. This continuity develops because certain parenchyma cells in each pith ray retain the ability to divide. These cells connect to vascular cambium cells in each vascular bundle, thus forming a complete ring of vascular cambium.

When a cell in the vascular cambium divides radially (inward or outward), one of the daughter cells remains meristematic; that is, it remains as a part of the vascular cambium. The other cell may divide again several times, but eventually it stops dividing and develops into mature secondary tissue.

Cells in the vascular cambium divide to produce tissues in two directions. The cells formed from the dividing vascular cambium are located either *inside* the ring of vascular cambium (the secondary xylem, or wood), or *outside* it (the secondary phloem, or inner bark; Fig. 33–5). Thus, vascular cambium is a thin layer of cells sandwiched in between the wood and inner bark, the two tissues that it produces (Fig. 33–6).

As the stem increases in circumference, the number of cells in the vascular cambium must also increase. This occurs by an occasional division of a vascular cambium cell at right angles to its normal direction of division. (Both daughter cells remain meristematic.)

What happens to the original primary tissues of a stem once secondary growth develops? As a stem increases in thickness, the orientation of the original primary tissues changes. For example, secondary xylem and secondary phloem are laid down between the primary xylem and primary phloem within each vascular bundle. Therefore, as vascular cambium forms secondary tissues, the pri-

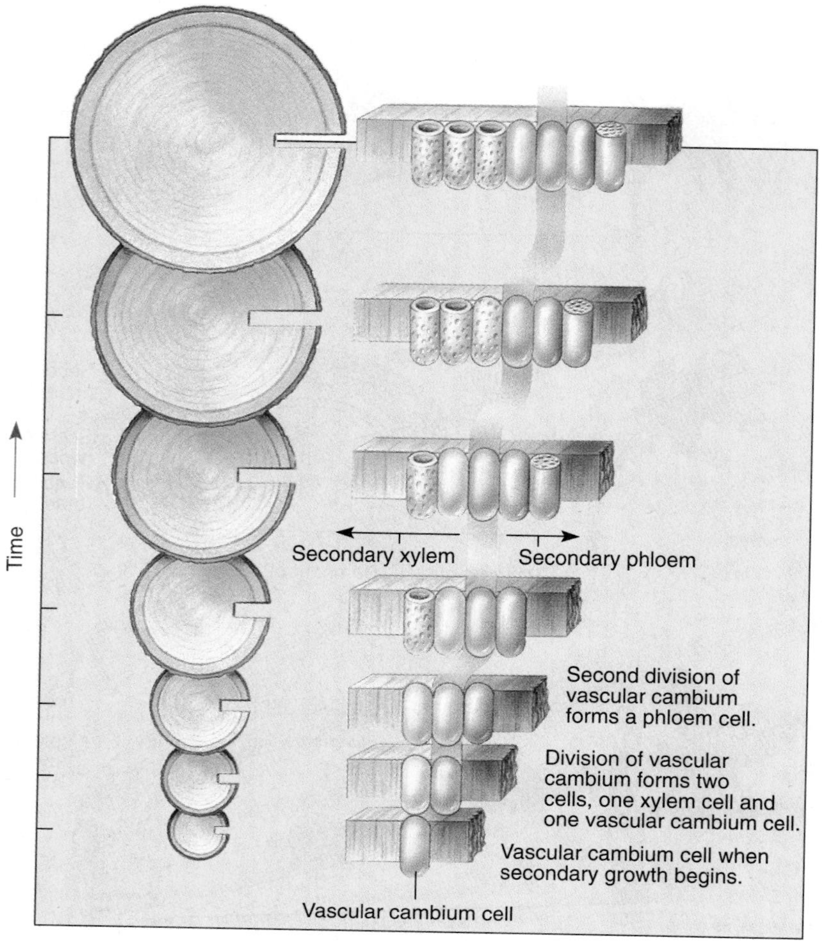
Time
Secondary xylem
Secondary phloem
Second division of vascular cambium forms a phloem cell.
Division of vascular cambium forms two cells, one xylem cell and one vascular cambium cell.
Vascular cambium cell when secondary growth begins.
Vascular cambium cell

FIGURE 33–5 This radial view of a dividing vascular cambium cell shows the stages in the development of secondary xylem and phloem. To study the figure, start at the bottom and move up. Note that vascular cambium divides in two directions, forming secondary xylem to the inside and secondary phloem to the outside. These cells differentiate to form the mature cell types associated with xylem and phloem. As secondary xylem accumulates, vascular cambium moves outward, and the woody stem increases in diameter.

mary xylem and primary phloem in each vascular bundle are separated from one another (Fig. 33–7).

The primary tissues remaining inside the cylinder of secondary tissues (consisting of pith and primary xylem) are under pressure from the changes in growth and soon get crushed almost beyond recognition. The primary tissues located outside the cylinder of secondary growth (that is, primary phloem, cortex, and epidermis) are also subjected to the pressures produced by secondary growth and are split apart and sloughed off.

Secondary tissues replace the primary tissues in function. Secondary xylem conducts water and dissolved minerals in the woody plant from roots to leaves. It contains the same types of cells found in primary xylem: water-conducting tracheids and vessel elements (see Chapter 31) in addition to parenchyma cells and fibers. The arrangement of the different cell types in secondary xylem produces distinctive wood characteristics of each species.

Secondary phloem conducts dissolved sugar, for example, from its place of manufacture (leaves) to a place of storage (roots). The same types of cells found in primary phloem (sieve tube members, companion cells, parenchyma cells, and fibers) are also found in secondary phloem, although there are usually more fibers in secondary phloem than in primary phloem.

Whereas secondary xylem and secondary phloem provide transport of water, minerals, and sugar vertically throughout the plant, materials must also move horizontally—that is, laterally. These materials move laterally through **rays,** which are chains of parenchyma cells that radiate out from the center of the stem. Rays, which are often continuous from the secondary xylem to the secondary phloem, are formed by the vascular cambium. Water and dissolved minerals are transported laterally through rays from the secondary xylem to the secondary phloem. Likewise, rays form pathways for the lateral transport of dissolved sugar from the secondary phloem to the secondary xylem.

Cork cambium produces periderm

Cork cambium, which usually arises from parenchyma cells in the epidermis or outer cortex, produces **periderm,** the functional replacement for the epidermis. Cells of the cork cambium retain their ability to divide. Cork cam-

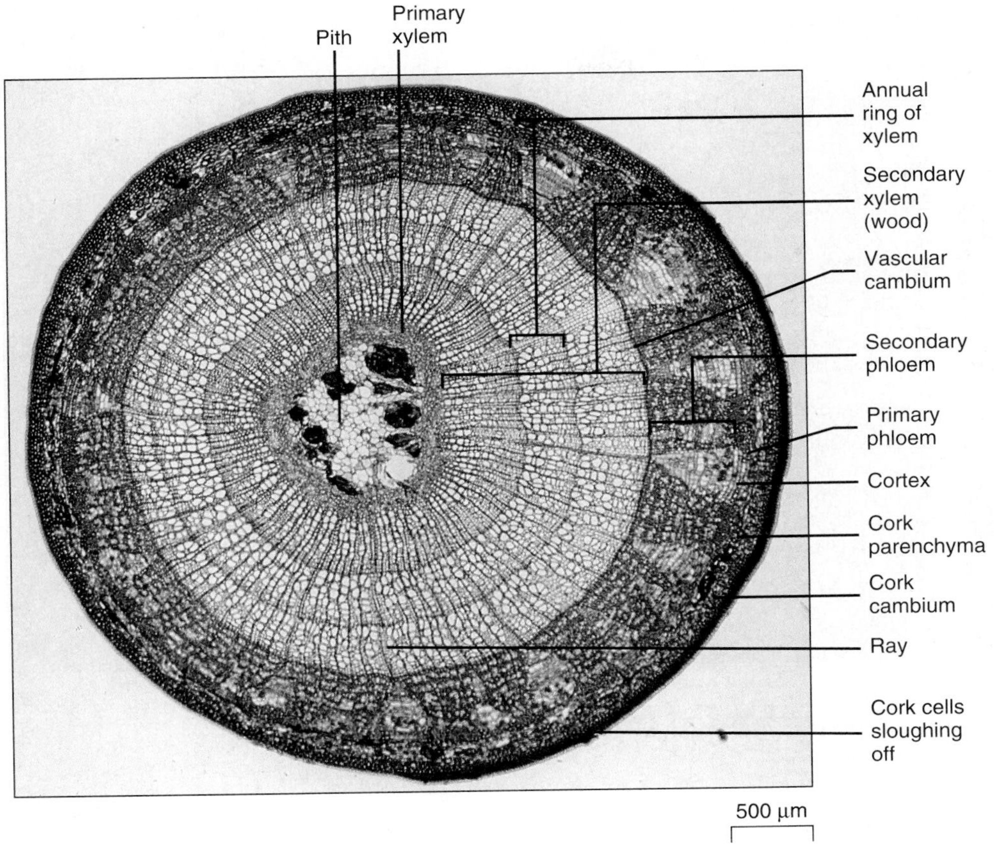

FIGURE 33–6 Secondary tissues are evident in this cross section of a three-year-old *Tilia* (basswood) stem. (Carolina Biological Supply Company/Phototake-NYC)

bium forms either a continuous cylinder of dividing cells (similar to vascular cambium) or a series of small arcs of meristematic cells that cut into successively deeper layers of tissue—that is, into primary phloem and the outer part of secondary phloem. This explains why outer bark is scaly or furrowed. The thickness, patterns, and texture of bark vary considerably from species to species (Fig. 33–8). These differences are due to varying growth rates of the cork cambium.

As is true of vascular cambium, cork cambium may divide to form new tissues in two directions: to its inside and its outside. Cork cells, formed to the outside of cork cambium, are dead at maturity and have heavily *suberized* (waterproofed) walls. Cork cells protect a stem against mechanical injury, mild fires, temperature extremes, and water loss. To its inside, cork cambium sometimes forms cork parenchyma, a tissue that stores water and starch granules. Cork parenchyma is only one to several cells thick, much thinner than the cork cell layer.

Cork is impermeable to water and gases, yet the internal tissues of the woody stem must be able to exchange gases with the surrounding atmosphere. As a stem thickens from secondary growth, the epidermis, including stomata that allowed gas exchange for the herbaceous stem, dies. Stomata are replaced by lenticels, areas in which cork cells are loosely arranged with lots of intercellular spaces to facilitate gas exchange (Fig. 33–9).

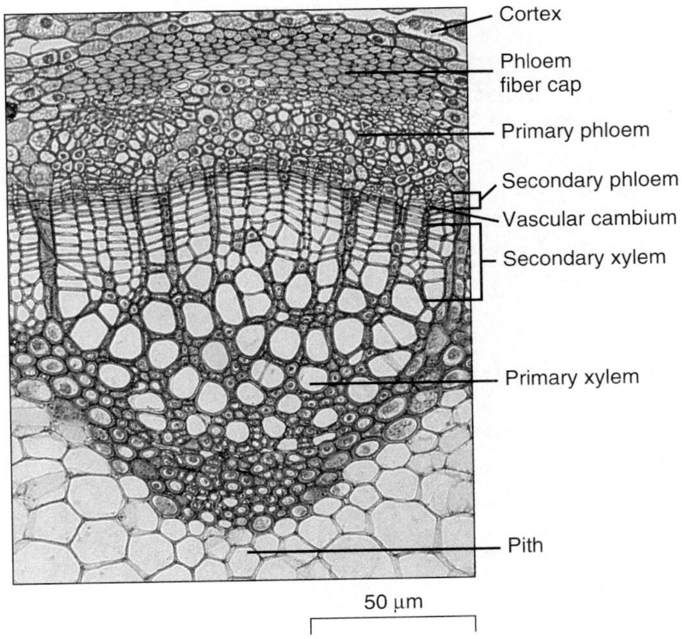

FIGURE 33–7 Cross section through part of a magnolia stem shows a vascular bundle that has been split apart by secondary growth. Compare this vascular bundle with the one in Figure 33–2*b*, which has primary growth only. (Dennis Drenner)

FIGURE 33–8 **Bark varies considerably among species.** (*a*) Pecan. (*b*) Snake-bark maple. (*c*) Sycamore. (*a, b, c,* James Mauseth, University of Texas)

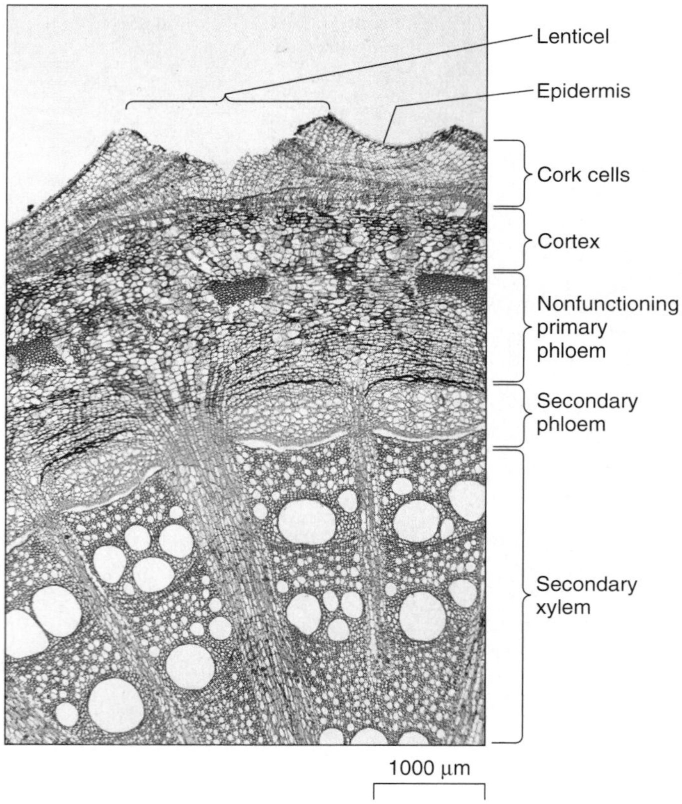

FIGURE 33–9 Lenticels permit gas exchange through the periderm. Cross section through the periderm of a stem, showing a lenticel. The epidermis has ruptured due to the proliferation of loosely arranged cork cells in the lenticel. (James Mauseth, University of Texas)

Common terms associated with wood are based on plant structure

If you've ever examined different types of lumber, you may have noticed that some trees have wood with two different colors (Fig. 33–10). The functional secondary xylem—that is, the part that conducts water and dissolved minerals—is the *sapwood,* a thin layer of younger, lighter-colored wood that is closest to the bark. *Heartwood,* the older wood in the center of the tree, is typically a brownish-red color. A microscopic examination of heartwood reveals that its vessels and tracheids are plugged up with pigments, tannins, gums, resins, and other materials. Therefore, heartwood no longer functions in conduction. Heartwood is denser than sapwood and provides structural support for trees. Some evidence suggests that heartwood is also more resistant to decay.

Almost everyone has heard of hardwood and softwood. Botanically speaking, *hardwood* is the wood of flowering plants and *softwood* is the wood of conifers (cone-bearing gymnosperms). The wood of pine and other conifers typically lacks fibers and vessel elements. The only kind of conducting cell in gymnosperm wood is the tracheid. These structural differences generally make conifer wood softer than the wood of flowering plants, although there is a substantial variation from one species to another. The balsa tree, for example, is a "hardwood" whose extremely light, soft wood is used to fashion airplane models.

Plants that grow in temperate climates where there is a growing period (during spring and summer) and a dor-

FIGURE 33–10 The wood of older trees consists of a dense, central heartwood and an outer layer of sapwood. The sapwood is the functioning xylem that conducts water and dissolved minerals. (Carolyn Iverson)

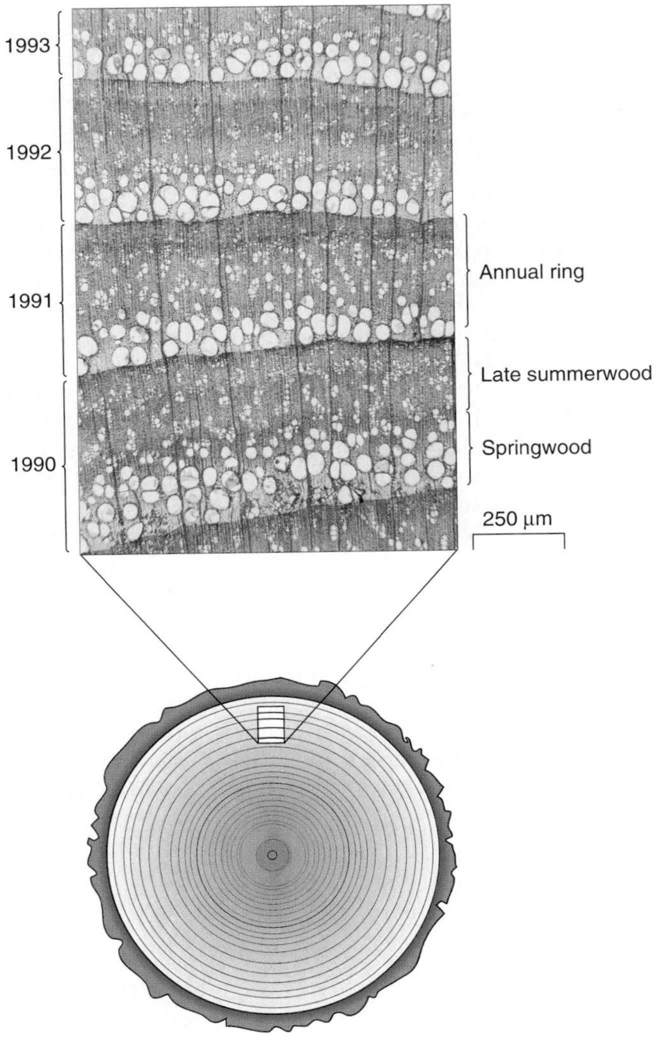

FIGURE 33–11 Portions of several annual rings are visible in this cross section. Note the differences in cell size between the springwood and the late summerwood. (James Mauseth, University of Texas)

mant period (during winter) exhibit *annual rings,* concentric circles found in cross sections of wood (Fig. 33–10). To determine the age of a woody stem in a temperate zone, simply count the annual rings. In the tropics, environmental conditions—particularly precipitation patterns (seasonal or year-round)—determine the presence or absence of rings, and rings are not a reliable method of determining the ages of tropical trees.

Examination of annual rings with a magnifying lens reveals no actual "ring," or line, separating one year's growth from the next. The appearance of a ring in cross section is due to differences in cell size and cell wall thickness between secondary xylem formed at the end of the preceding year's growth and that formed at the beginning of the following year's growth. In the spring, when water is plentiful, wood formed by vascular cambium has thin-walled, large-diameter conducting cells and few fibers, and is appropriately called *springwood.* As summer progresses and water becomes less plentiful, the wood formed, known as *late summerwood,* has thicker-walled, narrower conducting cells and many fibers. It is these differences in cell size between the late summerwood of the preceding year and the springwood of the following year that give the appearance of rings (Fig. 33–11). A great deal of information about climate in past times can be learned from the study of annual rings of ancient trees (see *Focus On: Tree Ring Analysis*).

As a woody stem increases in width over the years, the branches that it bears grow along with it as long as they are alive. If a branch dies, it no longer continues to grow with the stem. In time the stem grows out and surrounds the base of the branch. The basal portion of an embedded dead branch is called a *knot.* It is possible for a knot to contain bark as well as wood. The presence of knots in wood reduces its commercial value, except for ornamental purposes. Some plants, such as knotty pine, are valued for their high production of knots.

Transport in Plants Occurs in Xylem and Phloem

Roots obtain water and dissolved minerals from the soil. Once in roots, these materials are transported upward to

Focus On

Tree Ring Analysis

In temperate climates, the age of a tree can be determined by counting the number of annual rings. Other useful information can be determined by analyzing tree rings as well. For example, the size of each ring varies depending on local weather conditions, including precipitation and temperature. Sometimes the variation in tree rings can be attributed to a single environmental factor, and similar patterns appear in the rings of different tree species over a large geographical area. For example, trees in the Southwestern United States have similar ring patterns due to variations in the amount of annual precipitation. Years with adequate precipitation produce larger rings of growth, whereas years of drought produce much smaller rings.

It is possible to study ring sequences going back several thousand years. First, a *master chronology*, a complete sample of rings dating back as far as possible, is developed (see figure). A small core of wood is bored out of the trunk of an old living tree to obtain a sample of rings. The oldest rings (those toward the center of the tree) are matched with the youngest rings (those toward the outside) of an older tree or even an old piece of wood from a house. A master chronology of the area is obtained by using successively older and older sections of wood, even those found in prehistoric dwellings, and overlapping their matching ring sequences. The longest master chronology is of bristlecone pines in the Western United States; it goes back almost 9,000 years.

Dendrochronology, the study of both visible and microscopic details of tree rings, has been used extensively in several unrelated fields. Tree ring analysis has been extremely useful in dating prehistoric sites of Native Americans in the Southwest. For example, the Cliff Palace in the Mesa Verde National Park dates back to the year 1073 A.D. Tree ring analysis indicates that an extended drought forced the original inhabitants to abandon their homes.

Climatologists use tree ring data to study past climatic patterns. For example, tree ring analysis was used in 1993 to reconstruct annual temperatures in California's Sierra Nevada Mountains for the past 2000 years. Tree ring analysis is also useful in other disciplines, including ecology (to study changes in a forest community over time), environmental science (to study the effects of air pollution), and geology (to locate the epicenters and estimate the magnitudes of earthquakes that occurred in the past). ■

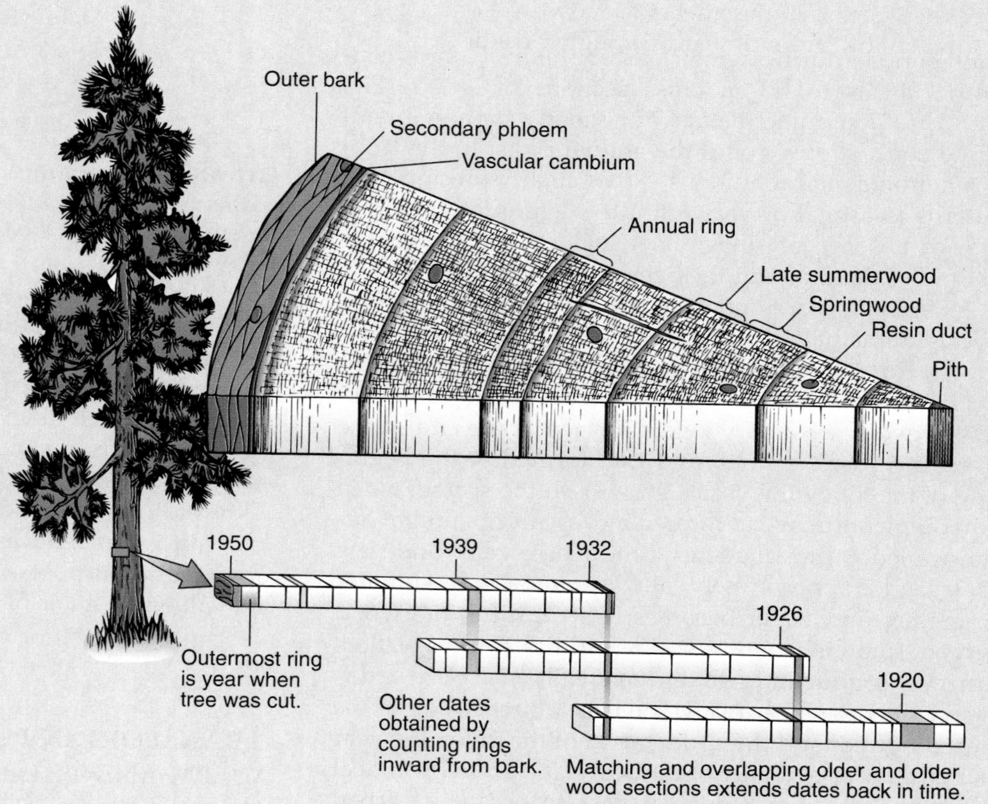

In tree ring dating, a master chronology is developed using progressively older pieces of wood from the same geographical area. By matching the rings of a wood sample of unknown age to the master chronology, the age of the sample can be accurately determined. Ring matching, once a painstakingly tedious job, is usually done by computer today.

stems, leaves, and other structures. Furthermore, sugar molecules manufactured in leaves by photosynthesis are transported in solution (that is, dissolved in water) throughout the plant, including into the subterranean roots. Water and dissolved minerals are transported from roots to other parts of the plant in xylem, whereas dissolved sugar is **translocated** in phloem.

Xylem transport and phloem translocation do not resemble the movement of materials in animals, because in plants nothing *circulates* in a system of vessels. Water and minerals, transported in xylem, travel in one direction only (upward), whereas translocation of dissolved sugar may occur upward or downward in separate phloem cells. Also, xylem transport and phloem translocation differ from internal circulation in animals because plants lack a heart. In plants, movement in both xylem and phloem is driven largely by natural physical processes rather than by a pumping organ.

How, exactly, do materials travel in the continuous system of the plant's vascular tissues? We first examine water and its movement through the plant, and later we discuss the translocation of dissolved sugar.

WATER AND MINERALS ARE TRANSPORTED IN XYLEM

Water initially moves horizontally into roots from the soil, passing through several tissues until it reaches xylem. Once the water is in the tracheids and vessel elements of root xylem, it travels upward through a continuous network of xylem cells from root to stem to leaf. Dissolved minerals are carried along passively in the water. The plant does not expend any energy of its own to transport water, which moves as a result of natural physical processes.

Water and dissolved minerals are transported within tracheids and vessel elements, which are hollow dead cells in xylem tissue. The transport in xylem sap is the most rapid of any movement in plants (Table 33–2).

How does water move to the tops of plants? It is either pushed up from the bottom of the plant or pulled up to the top of the plant. Although both mechanisms exist, current evidence indicates that most water is transported through xylem by being *pulled* to the top of the plant.

Water movement can be explained by the concept of water potential

In order to understand how water moves, it is helpful to introduce **water potential,** which is defined as the free energy of water. Water potential is important in plant physiology because it is a measure of a cell's ability to absorb water by osmosis (see Chapter 5). Water potential also provides a measure of water's tendency to evaporate from cells.

Table 33–2 XYLEM AND PHLOEM TRANSPORT RATES IN SELECTED PLANTS*

Plant	*Maximum Rate in Xylem (cm/hr)*	*Maximum Rate in Phloem (cm/hr)*
Conifer	120	48
Woody dicot	4,400	120
Herbaceous dicot/monocot	6,000	168–660
Herbaceous vine	15,000	72

*Xylem and phloem rates are from different plants within each general group and should be used for comparative purposes only. Adapted from Mauseth, J.D. *Botany: An Introduction to Plant Biology*, 2nd ed., Philadelphia, Saunders College Publishing, 1995.

The water potential of pure water is conventionally set at 0 megapascals[2] because it cannot be measured directly. It is possible to measure differences in the free energy of water molecules in different situations, however. When solutes are dissolved in water, the free energy of water decreases.[3] This means that dissolved solutes lower the water potential to a negative number. *Water moves from a region of higher (less negative) water potential to a region of lower (more negative) water potential.*

The water potential for the soil varies, depending on how much water it contains. When a soil is extremely dry, its water potential is very low (very negative). When a soil is moister, its water potential is higher, although it is still a negative number because dissolved minerals are present in dilute concentrations.

The water potential in root cells is also negative owing to the presence of dissolved minerals, sugars, and other solutes. Roots contain more dissolved materials than does soil water, however, unless the soil is extremely dry. This means that under normal conditions the water potential of the root is more negative than the water potential of the soil. Thus, water moves by osmosis from the soil into the root.

Tension-cohesion pulls water up a stem

According to the **tension-cohesion model,** also known as the transpiration-cohesion model, water is pulled up the

[2]A megapascal (MPa) is a unit of pressure equal to 9.87 atmospheres or 145.1 pounds per square inch.

[3]Solutes induce the formation of hydration layers, in which water molecules surround polar molecules and ions, keeping them in solution by preventing them from coming together. The association of water molecules in hydration layers reduces their motion, thereby decreasing their free energy.

plant as a result of a *tension* produced at the top of the plant (Fig. 33–12). This tension, which resembles that produced by sucking liquid up a straw, is caused by the evaporative pull of transpiration. Recall from Chapter 32 that **transpiration** is the evaporation of water vapor from plants. Most water loss from transpiration takes place through the numerous microscopic pores (stomata) present on leaf and stem surfaces. The tension extends from leaves, where most transpiration occurs, down the stems and into the roots. It draws water up stem xylem to leaf cells that have lost water as a result of transpiration, and pulls water from root xylem into stem xylem. As water is pulled upward, additional water from the soil is drawn into the roots. Thus, the pathway of water movement is as follows:

soil → root xylem → stem xylem
→ leaf xylem → atmosphere

This upward pulling of water is only possible as long as there is a solid, unbroken column of water in xylem throughout the plant. Water forms an unbroken column in xylem because of the *cohesiveness* of water molecules. Recall from Chapter 2 that water molecules are cohesive—that is, strongly attracted to one another—because of *hydrogen-bonding*. In addition, the *adhesion* of water to the walls of xylem cells (also the result of hydrogen-bonding) is an important factor in maintaining an unbroken column of water. Thus, the cohesive and adhesive properties of water enable it to form an unbroken column that can be pulled up through the xylem.

The movement of water in xylem due to the tension-cohesion mechanism can be explained in terms of water potential. The atmosphere has an *extremely negative* water potential. There is a water potential gradient from the least negative (the soil) up through the plant to the most negative (the atmosphere). This gradient literally pulls the water from the soil up through the plant.

Is tension-cohesion powerful enough to explain the rise of water in the tallest plants? Plant biologists have calculated that the tension produced by transpiration is strong enough to pull water up 150 meters (500 feet) in tubes the diameter of xylem vessels. Because the tallest trees are no more than 117 meters (375 feet) high, the tension-cohesion model easily accounts for the transport of water. The tension-cohesion model is considered the dominant mechanism of xylem transport in most plants.

Root pressure pushes water from the root up a stem

In the less important mechanism for water transport, known as **root pressure,** water that moves into a plant's roots from the soil is *pushed* up through xylem toward the

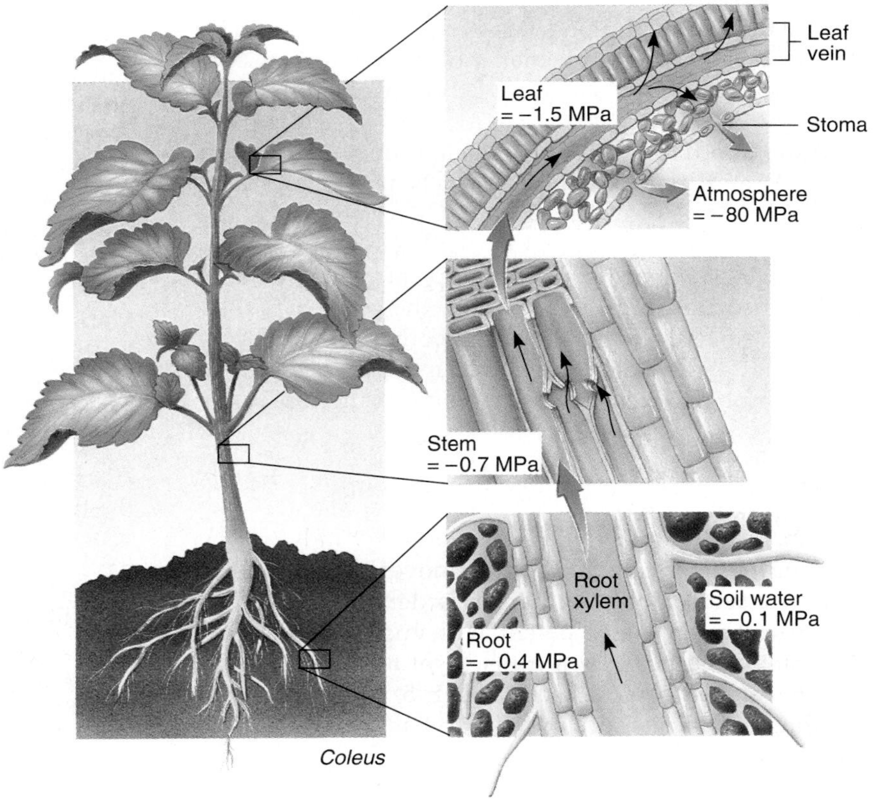

FIGURE 33–12 According to the tension-cohesion model, the evaporative pull of transpiration draws water up the plant from the soil to the atmosphere. (*Top*) Water vapor diffuses from the surfaces of leaf mesophyll cells to the drier atmosphere through stomata. This produces a tension that pulls water out of leaf xylem toward the mesophyll cells. (*Middle*) The cohesion of the water molecules, caused by hydrogen-bonding, allows unbroken columns of water to be pulled up the narrow vessels and tracheids of stem xylem. (*Bottom*) This in turn pulls water up root xylem, forming a continuous column of water from root xylem to stem xylem to leaf xylem. As water moves upward in the root, it produces a pull that causes soil water to diffuse into the root.

top of the plant. Water moves into roots by osmosis because of the difference in water potential between the soil and the cytoplasm of root cells. The accumulation of water in root tissues produces a pressure that forces the water up through the xylem.

Guttation, a phenomenon in which liquid water is forced out through special openings in the leaves (see Chapter 32), is a manifestation of root pressure. However, root pressure is not a strong enough force to explain the rise of water to the tops of coastal redwoods and other tall trees. Root pressure exerts an influence in smaller plants, particularly in the spring when the soil is quite wet, but it clearly does not cause water to rise 100 meters or more in the tallest plants. Furthermore, root pressure does not occur to any appreciable extent in summer (when water is often not plentiful in soil), yet the movement of water is greatest during hot summer days.

SUGAR IN SOLUTION IS TRANSLOCATED IN PHLOEM

The sugar produced during photosynthesis is converted into the disaccharide sucrose (common table sugar) before being loaded into phloem and translocated to the rest of the plant. Sucrose is the predominant photosynthetic product carried in phloem. Translocation in phloem sap is not as swift-moving as xylem transport (Table 33–2).

Fluid within phloem tissue moves both upward and downward. Sucrose is translocated in phloem from a *source,* an area of excess sugar supply (usually a leaf), to the *sink,* an area of storage or metabolism such as roots, apical meristems, fruits, or seeds.

The pressure-flow hypothesis explains translocation in phloem

Translocation of dissolved sugar in phloem is explained by the **pressure-flow hypothesis,** which postulates that dissolved sugar moves in phloem by means of a pressure gradient (that is, a difference in pressure). The pressure gradient exists between the source, where the sugar is loaded into phloem, and the sink, where the sugar is removed from phloem.

At the source, the dissolved sucrose is moved from a leaf's mesophyll cells, where it was manufactured, into the companion cells of phloem. This movement occurs by active transport, a process that requires ATP (Fig. 33–13). Once the sugar is in the companion cell, it readily moves into a sieve tube member through many plasmodesmata, the cytoplasmic connections between the two cells. The increase in dissolved sugars in the sieve tube member at the source decreases (makes more negative) the water potential of that cell. As a result, water moves by osmosis into the sieve tubes, increasing the hydrostatic pressure inside them. This pressure pushes the sugar solution through phloem much as water is forced through a hose.

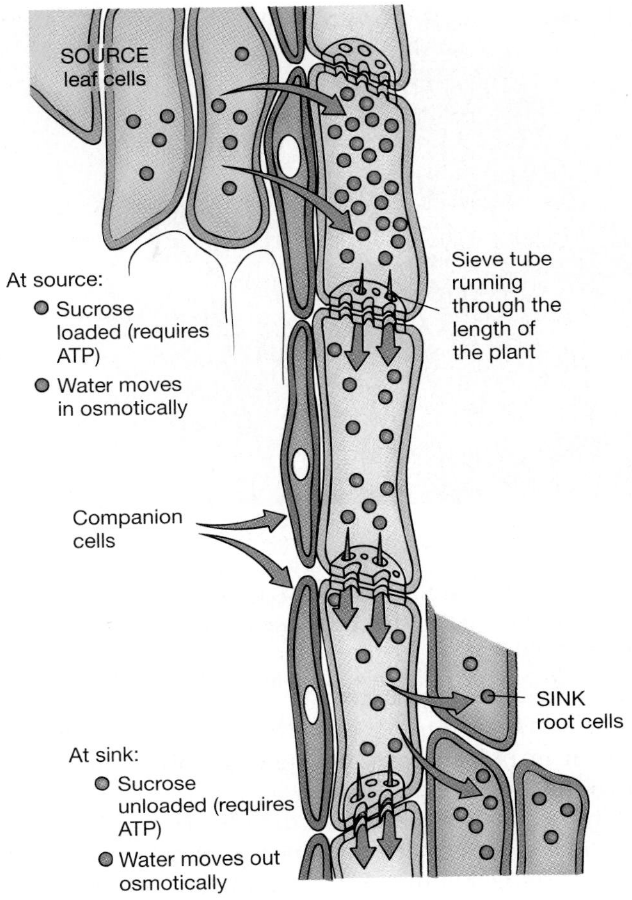

FIGURE 33–13 The pressure-flow hypothesis explains phloem translocation. Sugar is actively loaded into the sieve tube member at the source. As a result, water moves into the sieve tube member by osmosis. At the sink, the sugar is actively unloaded, and water leaves the sieve tube member (again by osmosis). The pressure gradient from source to sink causes translocation from the area of higher hydrostatic pressure (the source) to the area of lower hydrostatic pressure (the sink).

At its destination at the sink, sugar is actively unloaded from the sieve tube members, with ATP again being required. With the loss of sugar, the water potential in the sieve tube members at the sink increases (becomes less negative). Therefore, water moves out of the sieve tubes by osmosis and into surrounding cells. This decreases the hydrostatic pressure inside the sieve tubes at the sink.

Thus, the pressure-flow hypothesis explains the movement of dissolved sugar in phloem by means of a pressure gradient. It is the difference in sugar concentrations between the source (where a high sugar concentration causes a high pressure) and the sink (where a low sugar concentration causes a low pressure) that causes trans-

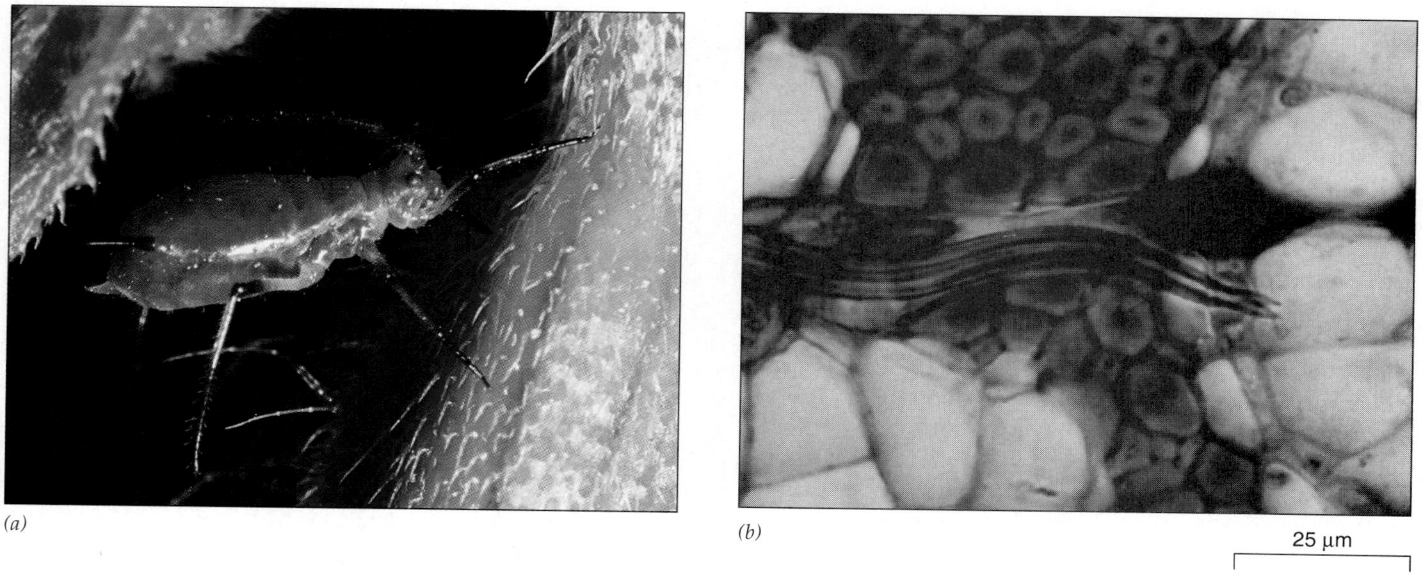

(a) (b)

FIGURE 33–14 Aphids, tiny insects about 6 millimeters (0.24 inch) in length, have been used to help study translocation in phloem. (*a*) Mature aphid feeding on lower side of a stem. (*b*) Microscopic view of phloem, showing the aphid mouthpart that has penetrated a sieve tube member. (*a,* Dwight Kuhn; *b,* M. H. Zimmerman, *Science* 133:73–79 [Fig. 4], January 13, 1961. Copyright 1996 by the American Association for the Advancement of Science)

location in phloem, as water and dissolved sugar flow down the pressure gradient.

The actual translocation of dissolved sugar in phloem does not require metabolic energy. However, both the loading of sugar at the source and the unloading of sugar at the sink require energy derived from ATP to move the sugar across cell membranes by active transport.

Although the pressure-flow hypothesis adequately explains current data on phloem translocation, much remains to be learned about this complex process. Phloem translocation is difficult to study in plants. Because phloem cells are under pressure, cutting into phloem to observe it releases the pressure and causes the contents of the sieve tube members to be sucked against one end wall. Aphids, small wingless insects that insert their mouthparts into phloem sieve tubes for feeding, have been a useful tool in phloem research (Fig. 33–14). The pressure in the punctured phloem drives the sugar solution through the aphid's mouthpart and into its digestive system. When the aphid's mouthpart is severed from its body by a laser beam, the sugar solution continues to flow through the mouthpart at a rate proportional to the pressure in phloem; this rate can be measured.

Much useful information about phloem translocation has also been obtained using radioactive tracers. When a leaf is exposed to CO_2 labelled with carbon-14, the radioactive carbon becomes fixed into the sugars produced by photosynthesis. The translocation of the sugar in phloem can be followed using special techniques. After exposure to the labelled CO_2, the stem tissue is freeze-dried, sliced into thin sections, and placed on photographic film. The part of the film in contact with radioactive substances is exposed. After the film is developed, the exact location of the radioactive sugars in the sieve tubes can be determined.

SUMMARY

I. The main functions of stems are support, conduction, and production of new stem tissues.

II. Woody twigs demonstrate the external structure of stems.
   A. Buds are undeveloped embryonic shoots. A terminal bud is located at the tip of a stem, whereas lateral buds are located in leaf axils.
   B. A dormant bud is covered and protected by bud scales. When the bud resumes growth, bud scales covering the terminal bud fall off, leaving a bud scale scar.
   C. The area on a stem where each leaf is attached is called a node, and the region of a stem between two successive nodes is an internode.
   D. A leaf scar shows where each leaf was attached on the stem.
   E. Lenticels are sites of loosely arranged cells that allow oxygen to diffuse into the interior of a woody stem.

III. Herbaceous stems possess an epidermis, vascular tissue, and either cortex and pith or ground tissue.
   A. The epidermis is a protective layer covered by a water-conserving cuticle. Stomata permit gas exchange.

B. Xylem conducts water and dissolved minerals, and phloem conducts dissolved sugar.
C. The cortex, pith, and ground tissue function primarily for storage.

IV. Although herbaceous stems all have the same basic tissues, the arrangement of the tissues varies considerably.
A. Herbaceous dicot stems have the vascular bundles arranged in a circle (in cross section) and have a distinct cortex and pith.
B. Monocot stems have vascular bundles scattered in ground tissue.

V. Secondary growth occurs in some flowering plants (woody dicots) and in all conifers.
A. Vascular cambium arises between the primary xylem and the primary phloem.
1. Vascular cambium produces secondary xylem (wood) to the inside, and secondary phloem (inner bark) to the outside.
2. Secondary xylem conducts water and dissolved minerals in the woody stem. Secondary phloem conducts dissolved sugar in the woody stem.
B. Cork cambium arises near a stem's surface.
1. Cork cambium produces cork parenchyma to the inside and cork cells to the outside.
2. Cork cells are the functional replacement for epidermis in the woody stem. Cork parenchyma functions primarily for storage in the woody stem.
C. Rays are chains of parenchyma cells that radiate out from the center of a stem and form pathways for the lateral movement of materials between the secondary xylem and the secondary phloem.

VI. Water and dissolved minerals move upward in xylem from roots to stems and leaves.
A. Water potential is a measure of the free energy of water.
1. Pure water has a water potential of 0 megapascals, whereas water with dissolved solutes has a negative water potential.
2. Water moves from an area of higher (less negative) water potential to an area of lower (more negative) water potential.
B. The tension-cohesion model explains the rise of water in even the largest plants.
1. The evaporative pull of transpiration causes tension at the top of the plant. This tension is the result of a water potential gradient that ranges from the slightly negative water potentials in the soil and roots to the very negative water potentials in the leaves and atmosphere.
2. As a result of the gradient in water potentials, water moves from the soil to root xylem to stem xylem to leaf xylem to the atmosphere.
3. As a consequence of the cohesive and adhesive properties of water, the column of water pulled up through the plant remains unbroken.
C. Root pressure, caused by the movement of water into roots from the soil, helps explain the rise of water in small plants, particularly when the soil is wet.

VII. Dissolved sugar is translocated upward or downward in phloem.
A. Sucrose is the predominant sugar translocated in phloem.
B. The movement of materials in phloem is explained by the pressure-flow hypothesis.
1. Sugar is actively loaded (requiring ATP) into the sieve tubes at the source. As a result, water moves into these sieve tubes by osmosis, increasing the hydrostatic pressure inside the sieve tubes at the source.
2. Sugar is actively unloaded (requiring ATP) from the sieve tubes at the sink. As a result, water leaves these sieve tubes by osmosis, decreasing the hydrostatic pressure inside the sieve tubes at the sink.
3. The flow of materials between source and sink is driven by the hydrostatic pressure gradient produced by water entering phloem at the source and water leaving phloem at the sink.

SELECTED KEY TERMS

apical meristem
bud
bud scale
bud scale scar
bundle scar
cork cambium
cortex
dendrochronology
epidermis
ground tissue
lateral bud
lateral meristem
leaf scar
lenticel
periderm
phloem
phloem fiber cap
pith
pith ray
pressure-flow hypothesis
ray
root pressure
tension-cohesion model
terminal bud
translocation
transpiration
vascular cambium
water potential
xylem

POST-TEST

1. All stems have undeveloped embryonic shoots called __________.
2. The __________ is the protective outer layer of cells covering herbaceous stems.
3. The areas between the vascular bundles in herbaceous dicot stems are often referred to as __________ __________.
4. The tissue in monocot stems in which the vascular tissues are embedded is called __________ __________.
5. Ground tissue in monocot stems performs the same functions as __________ and __________ in herbaceous dicot stems.
6. The two lateral meristems responsible for secondary growth are the __________ __________ and the __________ __________.
7. Cork cambium and the tissues it produces are collectively called the __________ and make up the outer region of the bark.
8. Horizontal movement of materials in woody plants occurs in __________.

9. __________ is the science of tree ring analysis.
10. The movement of dissolved sugar in phloem is called __________ .
11. The free energy of water in a particular situation is referred to as its __________ __________ .
12. This mechanism of water movement, __________ __________, is not strong enough to explain the rise of water to the tops of the tallest trees.
13. In the __________-__________ model, a tension is produced at the top of the plant by the evaporative pull of transpiration.
14. According to the __________-__________ hypothesis, dissolved sugar moves in phloem by means of a pressure gradient that exists between the source (where sugar is loaded into phloem) and the sink (where sugar is removed from phloem).

REVIEW QUESTIONS

1. List several functions of stems and describe the tissue(s) responsible for each.
2. If you are examining a cross section of a herbaceous stem of a flowering plant, how would you determine whether it is a dicot or a monocot?
3. What happens to the primary tissues of a stem when secondary growth occurs?
4. When a strip of bark is peeled off of a tree branch, what tissues are actually removed?
5. Distinguish between vascular cambium and cork cambium, and describe the tissues that arise from each.
6. Name and briefly describe the model used to explain the rise of water in the tallest trees.
7. Describe root pressure. What are its limitations?
8. Describe the pressure-flow hypothesis of sugar movement in phloem, including the activities at source and sink.
9. Label the various tissues of this herbaceous dicot stem. Give at least one function for each of the tissues.

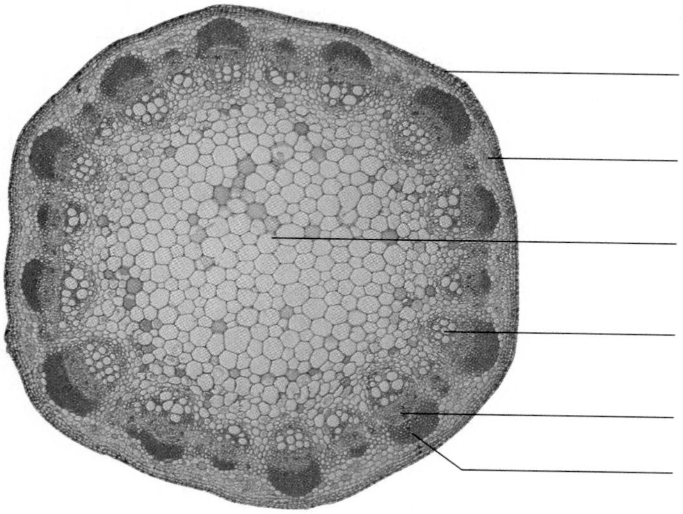

(Dennis Drenner)

YOU MAKE THE CONNECTION

1. When secondary growth is initiated, certain cells become meristematic and begin to divide. Could a mature tracheid ever do this? A sieve tube member? Why or why not?
2. Why does the wood of many tropical trees lack annual rings?
3. Why is hardwood more desirable than softwood for making furniture? Explain your answer based on the structural differences between hardwood and softwood.

RECOMMENDED READINGS

Angier, N. "Warming? Tree Rings Say Not Yet." *The New York Times,* December 1, 1992. How the Laboratory of Tree Ring Research at the University of Arizona deciphers the history of Earth's climate by analyzing tree rings.

Mauseth, J.D. *Botany: An Introduction to Plant Biology,* 2nd ed., Philadelphia, Saunders College Publishing, 1995. A comprehensive introduction to plant biology.

Moffett, M. "These Plants Claw and Strangle Their Way to the Top." *Smithsonian,* September 1993. Spectacular photographs accompany this article on vines that grow in tropical rain forests.

Raven, P.H., R.F. Evert, and S.E. Eichhorn. *Biology of Plants,* 5th ed. New York, Worth Publishers, 1992. A general botany text with an evolutionary emphasis.

Scuderi, L.A. "A 2000-Year Tree Ring Record of Annual Temperatures in the Sierra Nevada Mountains." *Science,* Vol. 259, March 5, 1993. How tree rings were used to reconstruct the mean temperatures in the Sierra Nevadas of California.

CHAPTER 34

Roots and Mineral Nutrition

Beets and other root crops are important sources of human food. (Dennis Drenner)

In Chapters 32 and 33 we discussed the aerial vegetative structures of a plant: the leaves and stems. In this chapter we turn to the third major vegetative organ: the roots. Because roots are underground and out of sight, people do not always appreciate the important functions that they perform. First, as anyone who has ever pulled weeds can attest, roots anchor a plant securely in the soil. Because a plant remains in one location throughout its life and needs a solid foundation from which to grow, firm anchorage is essential to its survival.

Second, roots absorb water and dissolved mineral salts such as nitrate, phosphate, and sulfate from the soil. These materials are then transported throughout the plant in the xylem.

Storage is the third main function performed by many roots. Surplus sugars produced in the leaves by photosynthesis are transported in the phloem to the roots for storage (usually as starch or sucrose) until needed. Carrot roots, for example, have an extensive phloem for this purpose. Although roots use some photosynthetic products for their own respiratory needs, most of them are stored and later transported out of the roots for use by the plant. Root systems may be modified for storage—as both *taproots* (beets, carrots, radishes, and turnips) and *fibrous roots* (sweet potatoes and yams). Plants with storage taproots are usually biennials (see Chapter 31) that store their food reserves in the root during the first year's growth and use it to reproduce during the second year's growth. Other plants, particularly those living in arid regions, possess storage roots adapted to store water.

In certain species, roots are modified for functions other than anchorage, absorption/conduction, and storage. Roots specialized to perform unusual functions are discussed later in this chapter.

LEARNING OBJECTIVES

After you have studied this chapter you should be able to

1. Describe the functions of roots, and how their structure correlates with these functions.
2. Distinguish between the structures of roots and stems.
3. Label cross sections of a primary dicot root and a monocot root, and describe the functions of each tissue.
4. Trace the pathway of water from the soil through the various root tissues.
5. Discuss the structure of roots with secondary growth, and describe how secondary tissues form.
6. Describe several roots that are modified to perform unusual functions.
7. List the five components of soil, and give the ecological significance of each.
8. Describe the factors involved in soil formation.
9. Distinguish between macronutrients and micronutrients, and give a physiological role in plants for each essential element.

PLANTS HAVE ONE OF TWO BASIC TYPES OF ROOT SYSTEMS

Branching underground root systems are often more extensive than a plant's aboveground parts. The roots of a corn plant, for example, may grow to a depth of 2.5 meters (about 8 feet) and spread outward 1.2 meters (4 feet) from the stem. Desert-dwelling tamarisk trees reportedly have roots that grow to a depth of 50 meters (163 feet) to tap underground water. The total root length, not counting root hairs, of a four-month-old rye plant was found to exceed 500 kilometers (310 miles)! The extent of a plant's root depth and spread varies considerably among different species and even among different individuals in the same species. Soil conditions, also discussed in this chapter, greatly affect the extent of root growth.

Two types of root systems—a taproot system and a fibrous root system (Fig. 34–1)—may develop. A **taproot** system consists of one main root (formed from the enlarging *radicle,* or embryonic root) with many smaller lateral roots coming out of it. Lateral roots often initially occur in regular rows along the length of the main root. Taproots are characteristic of dicots and gymnosperms. A dandelion is a good example of a common herbaceous plant with a taproot system. A few woody trees (hickory, for example) retain their taproots, which become quite massive as the plants age. Most mature trees, however, do not retain their taproots and have root systems that consist of large, shallow lateral roots from which other roots branch off and grow downward.

A **fibrous root** system has several to many roots of the same size developing from the end of the stem, with

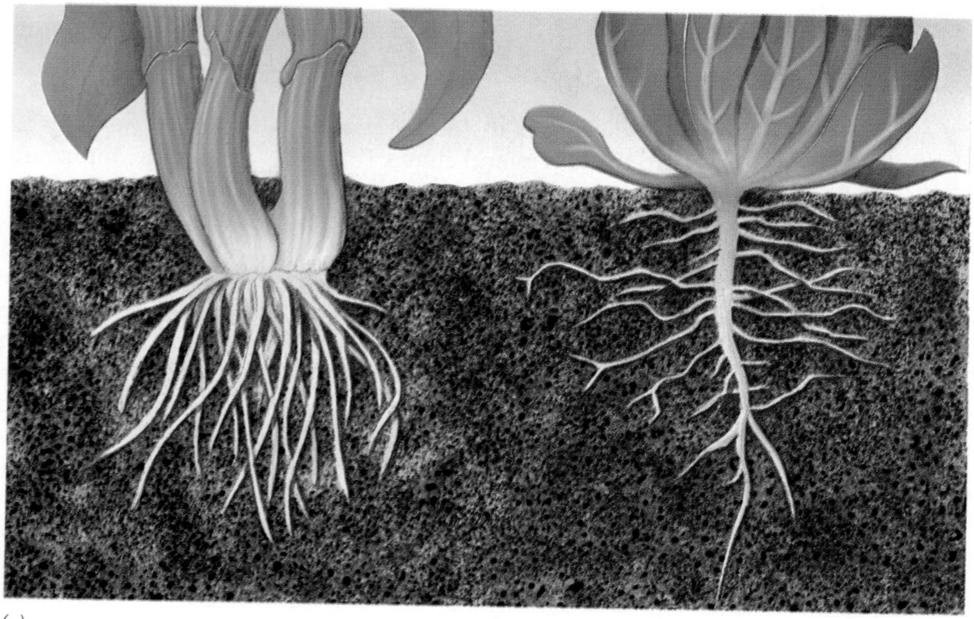

(a) (b)

FIGURE 34–1 Plants have two types of root systems. (*a*) The roots of a fibrous root system are adventitious and develop from stem tissue. (*b*) A taproot system develops from the embryonic root in the seed.

smaller lateral roots branching off of these roots. Fibrous root systems form in plants that have a short-lived embryonic root. The roots initially originate from the base of the embryonic root, and later from stem tissue. Because fibrous roots do not arise from pre-existing roots, but rather from the stem, they are said to be **adventitious.** Adventitious organs occur in an unusual location, such as roots that develop on a stem, or buds that develop on roots. Onions, crabgrass, and other monocots have fibrous root systems.

Taproot and fibrous root systems are adapted to obtain water in different ways. For example, taproot systems often extend down into the soil to obtain water located deep underground, whereas fibrous root systems, which are located relatively close to the surface of the soil, are adapted to obtain rainwater from a larger area as it drains into the soil.

UNLIKE STEMS, ROOTS POSSESS A ROOT CAP AND ROOT HAIRS

Roots have several structures, such as a root cap and root hairs, not found in stems. Although stems and leaves have various types of hairs, they are distinct from root hairs in structure and function.

Each root tip is covered by a **root cap,** a protective thimble-like layer many cells thick that covers the delicate root apical meristem (Fig. 34–2*a*; also see Fig. 31–2). As the root grows, pushing its way through the soil, parenchyma cells of the root cap are sloughed off and replaced by new cells formed by the root apical meristem. Stems do not possess a structure equivalent to the root cap.

The root cap also appears to be involved in orienting the root so that it grows downward. When a root cap is removed, the root apical meristem grows a new cap. However, until the root cap has regenerated, the root grows randomly rather than in the direction of gravity.

Root hairs are short-lived extensions of single epidermal cells located just behind the growing root tip. Root hairs continually form in the area of cell maturation closest to the root tip to replace those that are dying off at the more mature end of the root hair zone (see Fig. 31–2). Each root hair is short (typically less than 1 centimeter, or 0.4 inch, in length), but they are quite numerous. Root hairs greatly increase the absorptive capacity of roots by increasing their surface area in contact with moist soil (Fig. 34–2*b*). Soil particles are coated with a microscopically thin layer of water in which minerals are dissolved. The root hairs establish an intimate contact with soil particles, allowing absorption of much of this water.

Unlike stems, roots lack nodes and internodes and do not usually produce leaves or buds. While herbaceous roots have certain primary tissues (such as epidermis, cortex, xylem, phloem, and pith) found in herbaceous stems, these tissues are arranged quite differently. Table 34–1 summarizes the major differences between roots and stems in herbaceous dicots.

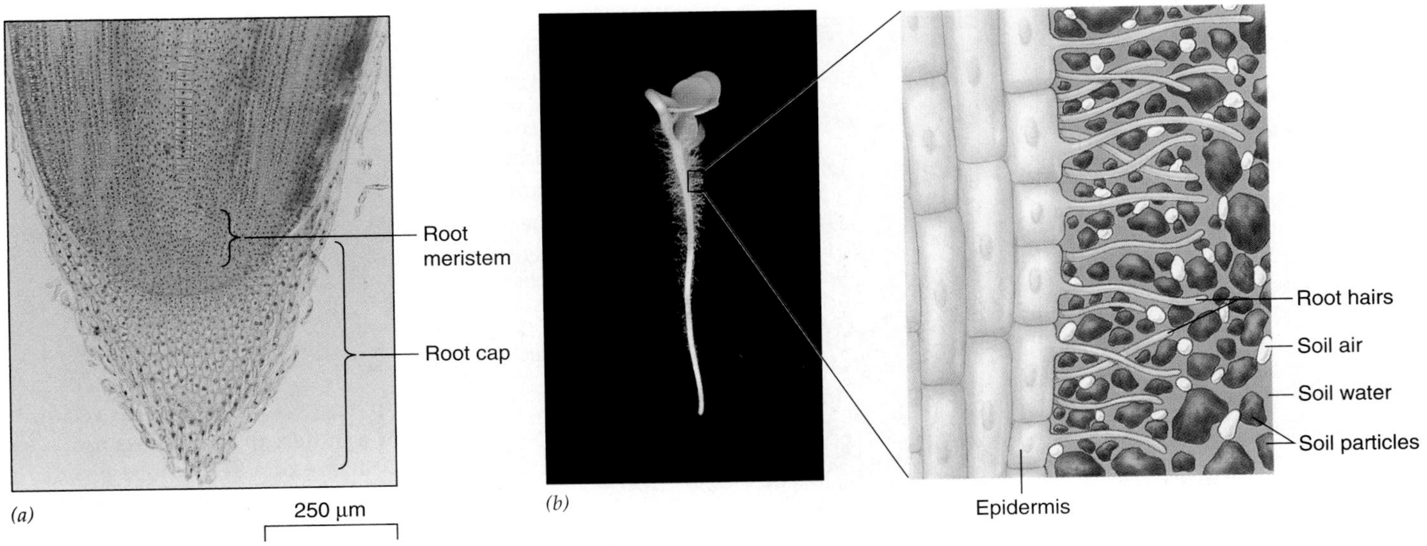

FIGURE 34–2 Root caps and root hairs are structures unique to roots. (***a***) Tip of a corn root showing its root cap. The root apical meristem is protected by the root cap. (***b***) Root hairs on a radish seedling. Each delicate hair is an extension of a single cell of the root epidermis. Root hairs increase the surface area of the root in contact with the soil. (*a*, Bruce Iverson; *b*, Dennis Drenner)

Table 34–1 GENERAL DIFFERENCES BETWEEN HERBACEOUS DICOT ROOTS AND STEMS*

Roots	*Stems*
No nodes or internodes	Nodes and internodes
No leaves or buds	Leaves and buds
Nonphotosynthetic	Photosynthetic
No pith	Pith
No cuticle	Cuticle
Root cap	No cap
Root hairs	Trichomes
Pericycle	No pericycle
Endodermis	No endodermis
Branches form internally from the pericycle	Branches form externally from lateral buds

*Some exceptions to these general differences exist.

THE ARRANGEMENT OF VASCULAR TISSUES DISTINGUISHES HERBACEOUS DICOT ROOTS AND MONOCOT ROOTS

Although considerable variation exists in roots, they all possess an outer protective covering (epidermis or periderm), a cortex for storage of starch and other organic molecules, and vascular tissues for conduction. Let us first consider the structure of herbaceous dicot roots.

Xylem and phloem are not organized into bundles in herbaceous dicot roots

The buttercup root is a representative dicot root with primary growth (Fig. 34–3). Like other parts of this herbaceous dicot, its roots are covered by a single layer of protective tissue, the **epidermis.** The root hairs are a modification of the root epidermis that enables it to absorb more water from the soil. The root epidermis does not secrete a thick waxy cuticle in the region of root hairs because this layer would impede the absorption of water from the soil. Both the lack of a cuticle and the presence of root hairs increase absorption. (*Why* water moves from the soil into the root was explained in Chapter 33.)

Most of the water that enters the root moves along the cell walls rather than entering the cells. One of the major components of cell walls is cellulose, which absorbs wa-

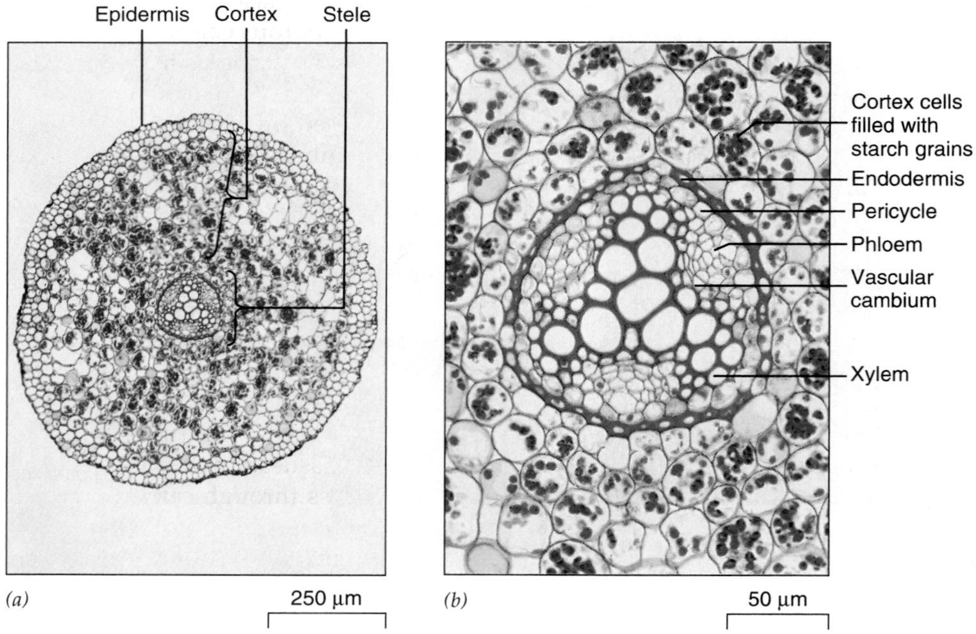

FIGURE 34–3 This cross section of a buttercup root shows the structure of a herbaceous dicot root. (***a***) Cortex comprises the bulk of the root. (***b***) Close-up of the center of the root. Note the solid core of vascular tissues. (*a, b,* James Mauseth, University of Texas)

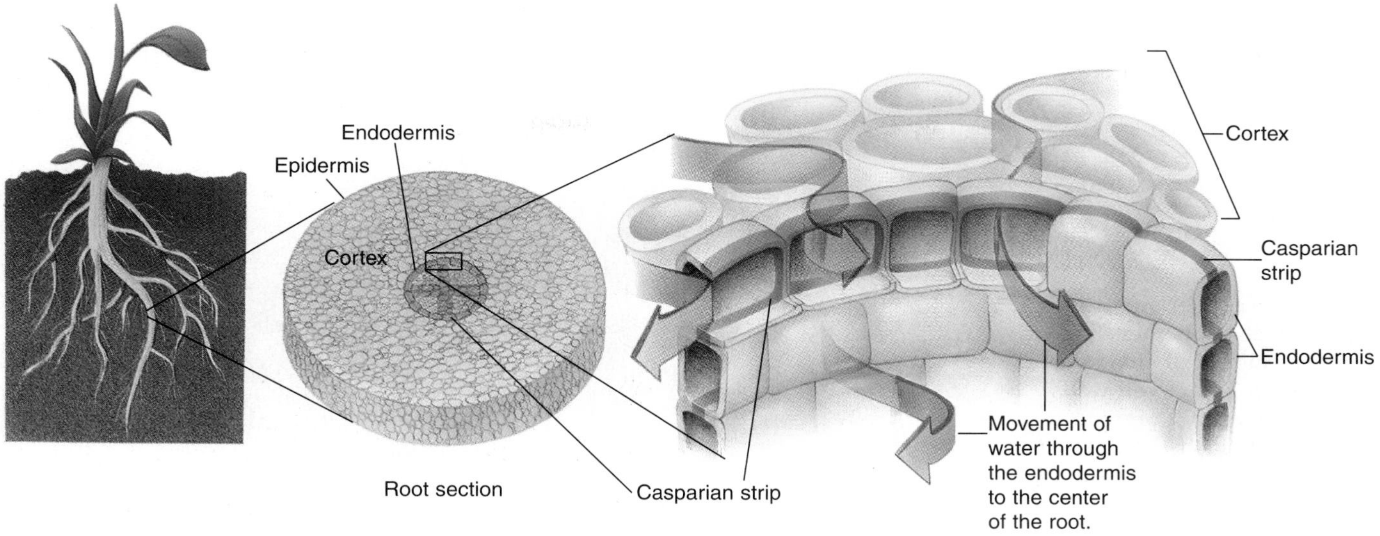

FIGURE 34–4 The endodermis controls mineral uptake by the root. Note the Casparian strip around the radial and transverse walls.

ter like a sponge. An example of the absorptive properties of cellulose is found in cotton balls, which are almost pure cellulose.

The **cortex** of a herbaceous dicot root, which is primarily composed of loosely arranged parenchyma cells with large intercellular (*between* cells) spaces, comprises the bulk of the root. Roots usually lack supporting collenchyma cells, probably because the soil supports the root, although they may develop some sclerenchyma (another supporting tissue; see Chapter 31) as they age. The primary function of the root cortex is storage. A microscopic examination of the parenchyma cells that form the cortex often reveals numerous starch grains. Starch, an insoluble carbohydrate composed of glucose subunits, is the most common form of stored nutrients in plants.

The large intercellular spaces, a common feature of root cortex, provide a pathway for water uptake and allow for aeration of the root. The oxygen that root cells need for aerobic respiration (see Chapter 7) diffuses from air spaces in the soil into the intercellular spaces of the cortex, and from there into the cells of the root.

The inner layer of the cortex, the **endodermis,** controls the amounts and kinds of minerals that enter the xylem in the root's interior. Harmful minerals may be blocked, while beneficial minerals are absorbed into the xylem and thereby transported to the stem, leaves, and other parts of the plant.

Structurally, the endodermis differs from the rest of the cortex. Endodermal cells fit snugly against each other, and each has a special bandlike region, called a **Casparian strip** (Fig. 34–4), on its radial and transverse walls. If you compare the endodermis to a cylinder constructed of bricks, endodermal cells correspond to the bricks, and the Casparian strips correspond to the mortar between them. Casparian strips contain *suberin,* a fatty material that is waterproof. (Recall from Chapter 33 that suberin is also the waterproof material in cork cell walls.)

The water and dissolved minerals that enter the root cortex from the epidermis move in solution along two pathways, the apoplast and symplast. The **apoplast** follows along the interconnected porous cell walls, whereas the **symplast** is from one cell's cytoplasm to the next through plasmodesmata (Fig. 34–5). Up to this point, most of the water and dissolved minerals may not have passed through a plasma membrane or entered the cytoplasm of a root cell (i.e., most may have traveled along the apoplast). However, water and minerals are prevented from continuing along the cell walls by the waterproof Casparian strip on the radial and transverse walls of endodermal cells. Water enters by osmosis, whereas minerals enter the endodermal cells by passing through carrier proteins in their plasma membranes. Thus, the endodermis controls the movement of minerals into the root, even though it is an internal tissue and minerals pass through other tissues to reach it.

Dissolved mineral ions pass through carrier proteins in the plasma membranes of endodermal cells by active transport (see Chapter 5). In active transport, the mineral ions move *against* the concentration gradient—that is, from an area of *low* concentration to an area of *high* con-

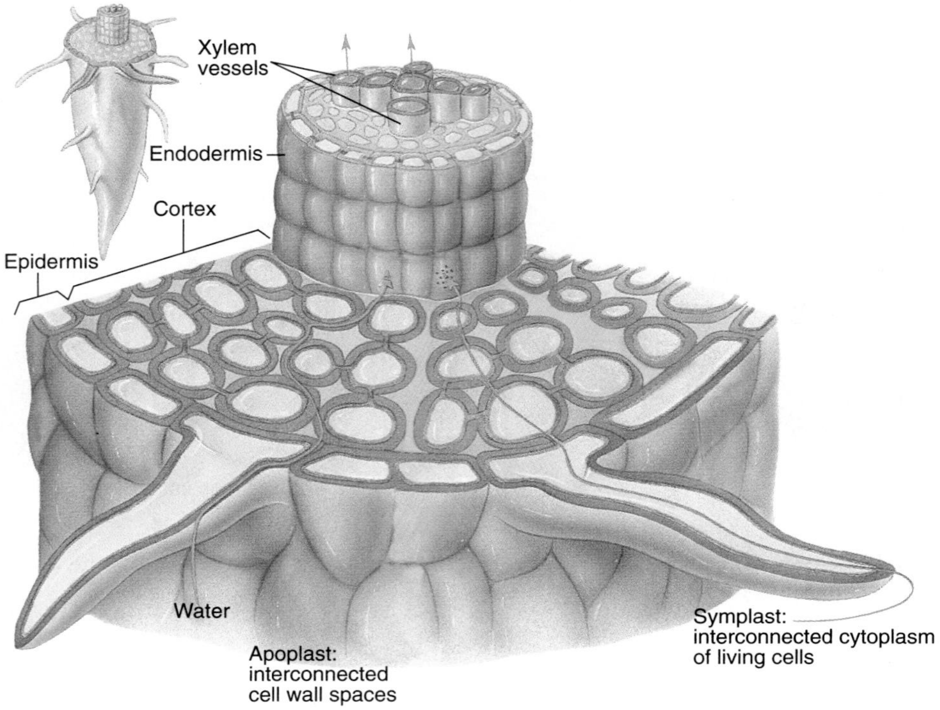

FIGURE 34–5 Water and dissolved minerals that enter the root travel from cell to cell along the interconnected porous cell walls (the apoplast) or from one cell's cytoplasm to another through plasmodesmata (the symplast). On reaching the endodermis, water and minerals can only continue to move into the root's center if they pass through a plasma membrane and enter an endodermal cell. The Casparian strip blocks the passage of water and minerals along the endodermal cell wall.

centration of that mineral. One of many reasons that root cells require sugar and oxygen for aerobic respiration is that active transport of mineral ions across biological membranes requires the expenditure of energy, usually in the form of ATP. From the endodermis, water and mineral ions enter the root xylem (precisely *how* this is done is not known) and are conducted to the rest of the plant.

Just inside the endodermis is a single layer of cells called the **pericycle,** which gives rise to multicellular lateral roots, also called branch roots (Fig. 34–6). The pericycle is composed of parenchyma cells that remain meristematic. Lateral roots originate when cells in a portion of the pericycle, usually at the tip of a xylem arm, start dividing. As it grows, the lateral root breaks through several layers of root tissue (endodermis, cortex, and epidermis) before entering the soil. Each lateral root has all the structures and features—root cap, root hairs, epidermis, cortex, endodermis, pericycle, xylem, and phloem—of the larger root from which it branches. In addition to producing lateral roots, the pericycle is involved in forming the lateral meristems that produce secondary growth in woody roots (discussed in the next section).

At the center of a primary dicot root is the **stele,** a central cylinder of vascular tissues (Fig. 34–3*b*). **Xylem,** the center-most tissue, often has two, three, four, or more extensions, or "xylem arms." **Phloem** is located in patches

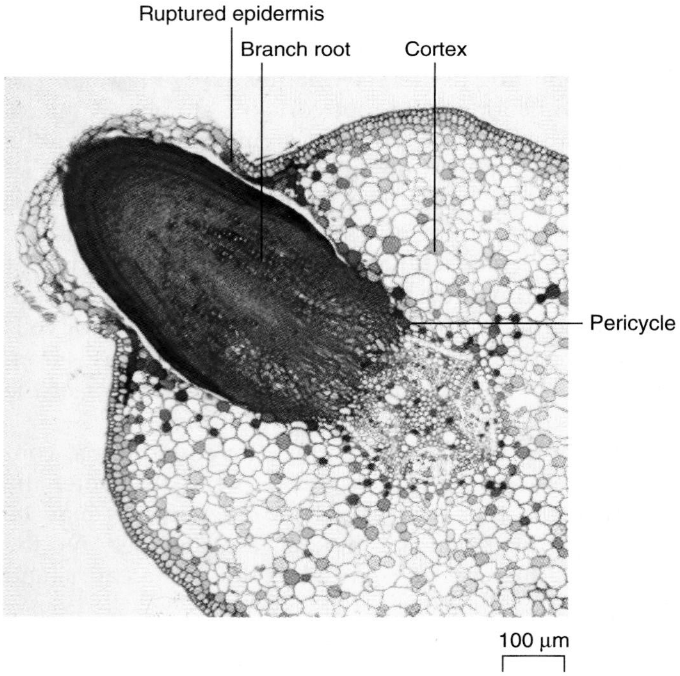

FIGURE 34–6 A multicellular lateral root emerges from the root. Lateral roots originate at the pericycle. (James Mauseth, University of Texas)

between the xylem arms. The xylem and phloem of the root have the same functions as in the rest of the plant: water and dissolved minerals are conducted in xylem, and dissolved sugar is conducted in phloem.

After passing through the endodermal cells, water enters the root xylem, often at one of the xylem arms. Up to this point the pathway of water has been horizontal from the soil into the center of the root:

> root hair → epidermis → cortex → endodermis
> → pericycle → root xylem

Once water enters the xylem, it is transported upward through root xylem into stem xylem and throughout the rest of the plant (see Chapter 33).

Phloem conducts dissolved sugar (sucrose) from the leaves, where it is made by photosynthesis, to the root, where it is stored—usually as starch, or from the root where it is stored, to other parts of the plant, where it is used for growth and maintenance of tissues. The same kinds of cells found in stem xylem and phloem are found in root xylem and phloem, except that roots typically have fewer fibers for support. The **vascular cambium,** which gives rise to secondary tissues, is sandwiched between the xylem and phloem. Because it possesses an inner core of vascular tissue, the primary dicot root lacks a pith.

Xylem does not form the central tissue in some monocot roots

Although monocot roots exhibit considerable variation in structure when compared to dicot roots, they possess the same basic tissues. Starting at the outside of a greenbriar root (a representative monocot), there is epidermis, then cortex, endodermis, and pericycle (Fig. 34–7). Unlike herbaceous dicot roots, the vascular tissues in a monocot root do not form a solid cylinder in the center. Instead, the phloem and xylem are located in separate alternating bundles arranged in a circle around the central **pith.**

Because the vast majority of monocots do not have secondary growth, no vascular cambium exists in monocot roots. Despite their lack of secondary growth, long-lived monocots, such as palms, often have thickened roots because the cortex expands.

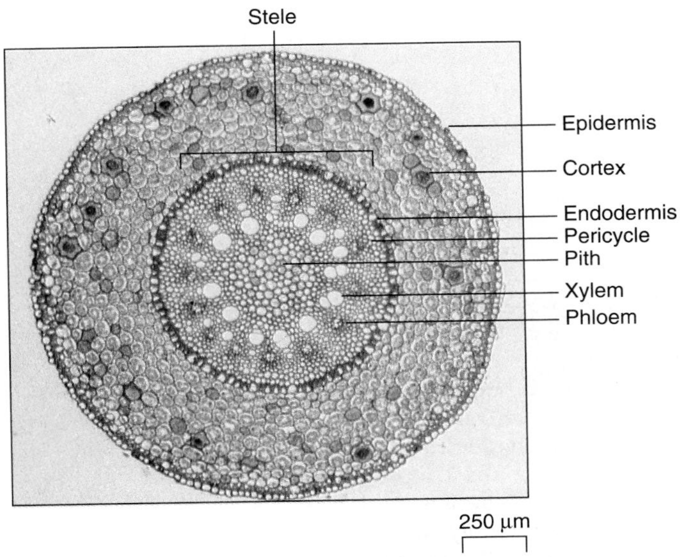

FIGURE 34–7 This cross section of a greenbriar (*Smilax*) root shows the structure of a monocot root. As in herbaceous dicot roots, the cortex is extensive. (Dennis Drenner)

Woody plants have roots with secondary growth

Plants that produce stems with secondary growth also produce roots with secondary growth. Recall from Chapter 33 that these plants—gymnosperms and woody dicots—have primary growth at apical meristems, while secondary growth occurs at lateral meristems. The production of secondary tissues occurs some distance back from the root tips and is the result of the activity of two lateral meristems: the vascular cambium and the cork cambium. Major roots of trees are often massive and possess both wood and bark. In temperate climates, the wood of both roots and stems exhibits annual rings in cross section.

Before secondary growth starts in a root, the vascular cambium is sandwiched between the primary xylem and the primary phloem (Fig. 34–8). At the onset of secondary growth, the vascular cambium extends out to the pericycle, forming a continuous, noncircular loop. As the vascular cambium divides to produce secondary tissues, it eventually forms a cylinder of vascular cambium that continues to divide, producing secondary xylem (wood) to the inside and secondary phloem (inner bark) to the outside. As additional secondary xylem is produced, the root increases in girth and the vascular cambium continues to move outward.

The primary root tissues are gradually crushed as the root increases in girth. The root epidermis is replaced by periderm, composed of cork cells and cork parenchyma produced by the cork cambium. The cork cambium in the root initially arises from regions in the pericycle.

Some roots are specialized for unusual functions

Adventitious roots often arise from the nodes of stems. Many aerial adventitious roots are adapted for functions

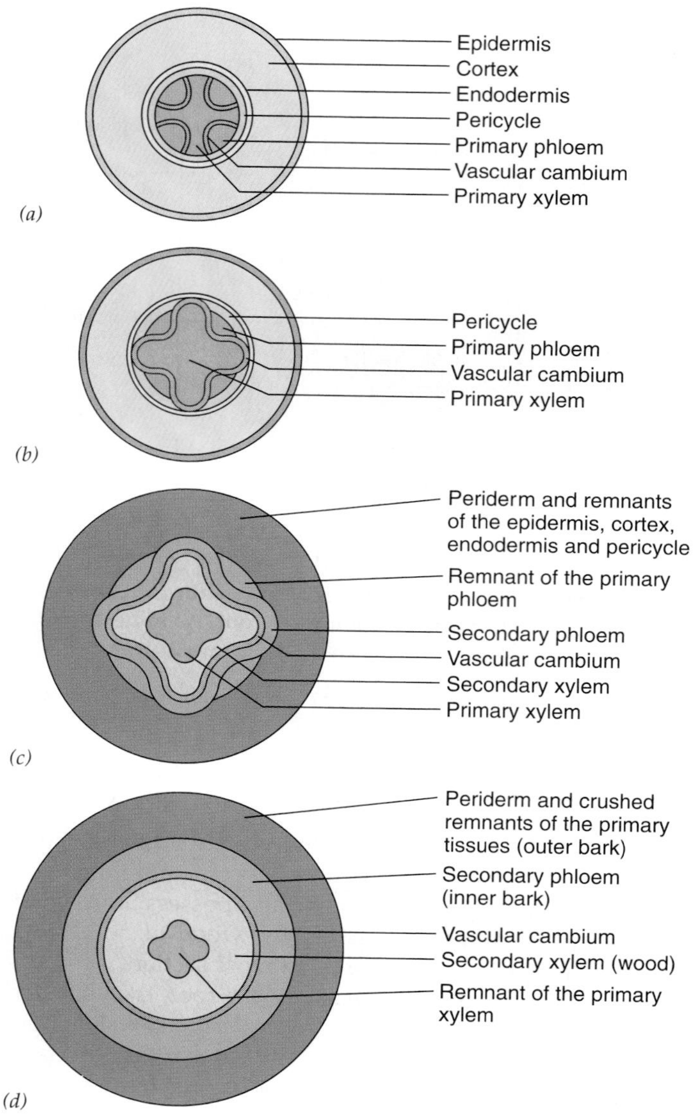

FIGURE 34–8 Development of secondary vascular tissues in a primary root is shown in cross section. (***a***) The tissues in a primary root. (***b***) At the onset of secondary growth, the vascular cambium extends out to the pericycle, forming a continuous, noncircular loop. (***c***) The vascular cambium produces secondary xylem to its inside and secondary phloem to its outside. Note that the primary phloem is pushed outward. (***d***) Over time, the ring of vascular cambium gradually becomes circular. As it continues to divide, the primary tissues found in the outer bark are crushed almost beyond recognition.

other than anchorage, absorption, conduction, or storage. **Prop roots** are adventitious roots that develop from branches or from a vertical stem and grow downward into the soil to help support the plant in an upright position (Fig. 34–9*a,b*). Prop roots are more common in monocots than in dicots. Corn and sorghum, both monocots, are herbaceous plants that produce prop roots. Many tropical dicot trees—such as red mangrove (*Rhizophora mangle*) and banyan (*Ficus benghalensis*)—also produce prop roots.

Other aerial roots have additional functions. In swampy or tidal environments where the soil is flooded or waterlogged, roots often grow upward until they are above the high-tide level. Even though roots live in the soil, they still require oxygen for aerobic respiration. Flooded soils are depleted of oxygen, so these aerial "breathing" roots, known as **pneumatophores,** may assist in getting oxygen to the submerged roots. Pneumatophores have a well-developed system of internal air spaces that is continuous with the submerged parts of the root, presumably allowing gas exchange. Black mangrove, white mangrove, and bald cypress are examples of plants with pneumatophores (Fig. 34–9*c*).

Epiphytes (plants that grow attached to other plants) and climbing plants have aerial roots that anchor the plant to the bark, branch, or other surface on which it grows. Some epiphytes have aerial roots specialized for functions other than anchorage. Certain epiphytic orchids, for example, have photosynthetic roots that make up the bulk of the plant. Epiphytic roots may absorb moisture as well. Some parasitic epiphytes such as mistletoe have roots that penetrate the host plant tissues and absorb nutrients (Fig. 34–9*d*). Another plant that starts its life as an epiphyte is the strangler fig, which produces long roots that eventually reach the ground and anchor the plant (now a tree rather than an epiphyte) in the soil. The original tree that the strangler fig grew on is often killed as the strangler fig grows around it, competing with it for light and other resources.

Plants that produce corms or bulbs (underground stems or buds specialized for asexual reproduction; see Chapter 35) often have wiry **contractile roots** in addition to their "normal" roots. The contractile roots grow into the soil and then contract (the root cells shorten or totally collapse), thus pulling the corm or bulb deeper into the soil (Fig. 34–9*e*). Contractile roots are necessary for bulbs and corms because each succeeding year's growth is *on top of* the preceding year's growth. As a result, bulbs and corms tend to move upward in the soil over time. With-

FIGURE 34–9 Many roots are specialized for unusual functions. (*a, b*) Prop roots are adventitious roots that provide additional support. (***a***) Prop roots in corn (*Zea mays*) arise near the base of the stem. (***b***) The banyan tree has prop roots that develop from branches. (***c***) Black mangrove (*Avicennia germinans*) produces pneumatophores that possibly provide oxygen for roots buried in anaerobic (oxygen-deficient) mud. (***d***) Micrograph of a parasitized juniper branch showing a parasitic mistletoe root that obtains nutrients from the host on which it lives. (***e***) Plants that produce corms or bulbs often have contractile roots. During successive seasons, contractile roots pull the corm or bulb deeper into the soil. ▶ (*a,* Dennis Drenner; *b,* James Mauseth, University of Texas; *c,* Visuals Unlimited/Nada Pecnik; *d,* Courtesy of C. Calvin, Portland State University; *e,* Courtesy of Judith Jernstedt, University of California, Davis)

Stem
Prop roots
(a)
Branch
Prop roots
(b)
Pneumatophores
(c)
Juniper bark
Mistletoe root (parasitic root)
Juniper wood
(d)
100 μm
Contractile roots
(e)

out contractile roots they would eventually be exposed at the soil's surface. Contractile roots are more common in monocots, but certain dicots and ferns also possess them.

ROOTS FORM RELATIONSHIPS WITH OTHER SPECIES

As roots of certain trees grow through the soil, they sometimes encounter roots of other trees of the same or different species. When this occurs, they may grow together to form a natural **graft,** as, for example, between a birch and a maple (Fig. 34–10). Because their vascular tissues are connected in the graft, dissolved sugars and other materials such as hormones pass between the two trees. Root grafts have been observed in over 160 different species of trees.

The roots of most plant species form mutually beneficial relationships with certain soil fungi (see Chapters 25 and 53). These associations, known as **mycorrhizae,** permit the transfer of materials (such as amino acids) from roots to fungus. At the same time, essential minerals such as phosphorus move from fungus to roots. The threadlike body of the fungal partner extends into the soil, extracting minerals well beyond where the plant's roots grow. The relationship is considered mutually beneficial because when mycorrhizae are not present, neither the fungus nor the plant grows as well.

The roots of some plants—clover, peas, and soybeans, for example—form an association with certain nitrogen-fixing bacteria. **Nodules** (swellings) that house millions of the bacteria develop on the roots (see Fig. 54–3*a*). Like mycorrhizae, the association between nitrogen-fixing bacteria and the roots of plants is mutually beneficial. Bacteria receive the photosynthesis products from plants while helping them to meet their nitrogen requirements.

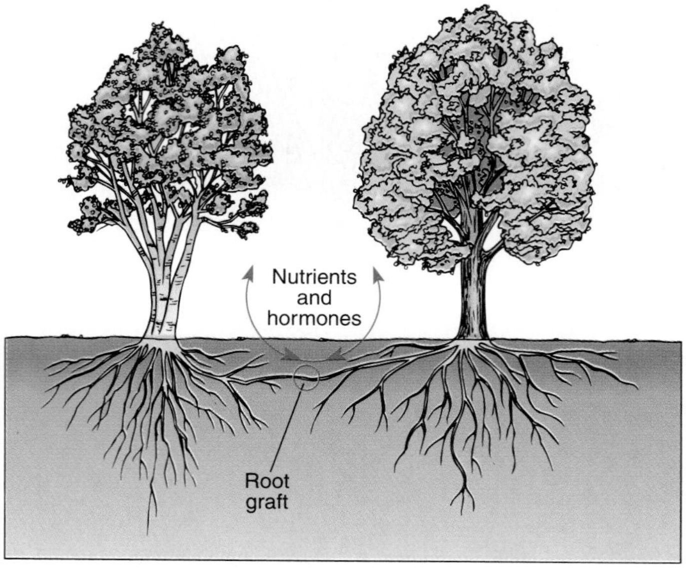

FIGURE 34–10 Root grafts often occur between two different trees, such as a birch (*left*) and a maple (*right*).

SOIL ANCHORS PLANTS AND PROVIDES WATER AND MINERALS

Soil is the ground underfoot, a relatively thin layer of Earth's crust that has been modified by the natural actions of weather, wind, water, and organisms. It is easy to take soil for granted. We walk on and over it throughout our lives, but rarely stop to think about how important it is to our survival. Vast numbers and kinds of organisms colonize soil and depend on it for shelter and food. Plants anchor themselves in soil and from it they receive water and essential minerals. Thirteen of the 16 different elements essential for plant growth are obtained directly from the soil. Plants could not survive on their own without soil, and because we depend on plants for our food, humans could not exist without soil, either.

Soils are formed from rock (called parent rock) that is gradually broken down into smaller and smaller particles by biological, chemical, and physical **weathering.** Two important factors that work together in the weathering of rock are climate and living organisms. Soil organisms such as lichens[1] produce acids that etch tiny cracks, or fissures, in the rock surface; water then seeps into these cracks. If the parent rock is located in a temperate climate, the alternate freezing and thawing during winter causes the cracks to enlarge, breaking off small pieces of rock. Small plants can then become established and send their roots into the larger cracks, fracturing the rock further.

Topography, a region's surface features—including the presence or absence of mountains and valleys—is also involved in soil formation. Steep slopes often have little or no soil on them because the soil and rock are continually transported down the slopes by gravity; runoff from precipitation tends to amplify erosion on steep slopes. Moderate slopes, on the other hand, may encourage the formation of deep soils.

The disintegration of solid rock into finer and finer mineral particles and the accumulation of organic material in the soil take an extremely long time, usually thousands of years. Soil forms continually as the weathering of parent rock beneath already-formed soil continues to add new soil. Soil thickness varies from a thin film on young lands such as those near the poles and on mountaintops, to more than 3 meters (10 feet) deep on very old lands such as tropical rain forests.

[1]A lichen is a dual organism composed of a fungus and an alga or cyanobacterium (see Chapter 25).

Soil is composed of inorganic minerals, organic matter, soil organisms, soil air, and soil water

The five distinct components of soil—inorganic mineral particles, organic matter, soil organisms, water, and air—all interact with one another. Some minerals, for example, continually cycle from the soil to living organisms, which use them in their biological processes. When the organisms die, they are decomposed by bacteria and other soil organisms, returning the minerals to the soil.

The inorganic mineral particles, which come from weathered rock, constitute most of what we call soil. Because different rocks are composed of different minerals, soils vary in composition. Also, soils formed from the same kind of parent rock may not develop in the same way because other factors such as weather, topography, and living organisms differ.

The texture or structural characteristic of a soil is determined by the amounts (measured in percentages by weight) of the different-sized inorganic mineral particles—sand, silt, and clay—that it contains. The size assignments for sand, silt, and clay are arbitrary; they give soil scientists a way to classify soil texture. Any particles larger than 2 millimeters in diameter, called stones, are not considered soil particles because they do not have any direct value to plants. The largest soil particles are called *sand* (0.02 to 2 mm in diameter), the medium-sized particles are called *silt* (0.002 to 0.02 mm in diameter), and the smallest particles are called *clay* (less than 0.002 mm in diameter) (Fig. 34–11). Sand particles are large enough to be seen easily with the eye; silt particles (about the size of flour particles) are barely visible with the eye; and most individual clay particles are too small to be seen with an ordinary light microscope; they can only be seen under an electron microscope.

The clay component of a soil is particularly important in determining many of its characteristics, in part because clay particles have the greatest surface area. If the surface areas of about 450 grams (1 pound) of clay particles were laid out side by side, they would occupy 1 hectare (2.5 acres). Each clay particle has negative electrical charges on its outer surface that attract and bind cations (positively charged mineral ions). Many cations, such as potassium (K^+) and magnesium (Mg^{2+}), are essential for plant growth and are "held" in the soil for plant use by their interactions with clay particles. These ions and the water that forms a film around the soil particles are absorbed by the plant's roots. In contrast, anions (negatively charged mineral ions) are usually not held as tightly in the soil and are often washed out of the root zone.

Soil always contains a mixture of different-sized particles, but the proportions vary from one soil to another. A *loam,* which is an ideal agricultural soil, has an optimum combination of different soil particle sizes: it contains approximately 40% each of sand and silt, and about 20% of clay. Generally speaking, the larger particles provide aeration while the smaller ones bind together into clumps and hold mineral nutrients and water. Soils with larger proportions of sand are not as desirable for most plants because they do not hold water and mineral ions well; plants grown in such soils are more susceptible to drought. Soils with larger proportions of clay are also not desirable for most plants because they provide poor drainage. Clay soils tend to get compacted and rob the soil of spaces that can be filled by water and air.

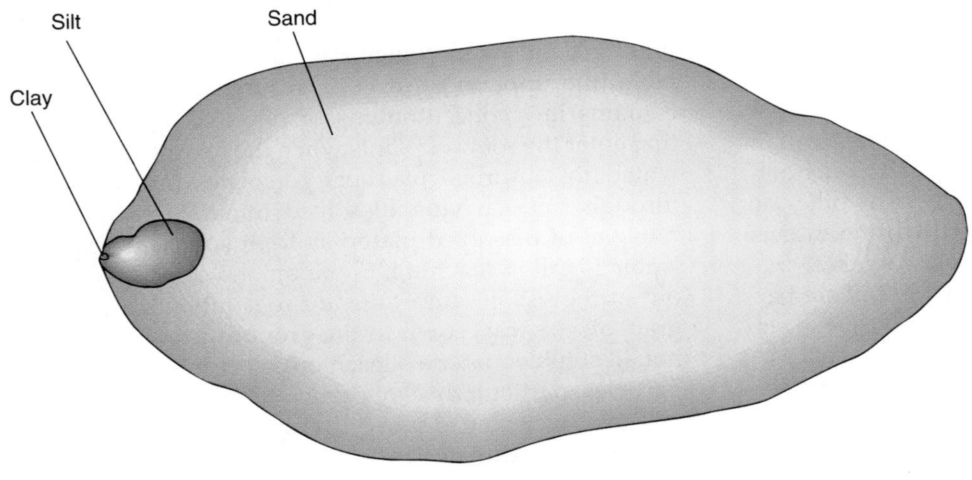

FIGURE 34–11 This diagram compares the relative sizes of sand, silt, and clay particles.

FIGURE 34–12 Humus is partially decomposed organic material that comes primarily from plant and animal remains. Soil rich in humus has a loose, somewhat spongy structure with several properties (such as increased water-holding capacity) that are beneficial for plants and other organisms living in it. (USDA/Soil Conservation Service)

A soil's organic matter consists of the wastes and remains of soil organisms

The organic matter in soil is composed of litter (dead leaves and branches on the soil's surface), droppings (animal dung), and the dead remains of plants, animals, and microorganisms in various stages of decomposition. Organic matter is decomposed by microorganisms, particularly bacteria and fungi, that inhabit the soil. During decomposition, essential minerals are released into the soil, where they may be bound by soil particles or absorbed by new plants. Organic matter increases the soil's water-holding capacity by acting much like a sponge. For this reason gardeners often add organic matter to soils, especially sandy soils, which are naturally low in organic matter. The partly decayed organic portion of the soil is referred to as **humus** (Fig. 34–12).

The organisms living in the soil form a complex community in numbers too vast to comprehend

A single teaspoon of fertile agricultural soil may contain millions of living microorganisms such as bacteria, fungi, algae, protozoa, and microscopic worms. Many other organisms also colonize soil, including earthworms, insects, plant roots, and animals such as moles, snakes, and groundhogs (Fig. 34–13). Most numerous in soil are bacteria, which number in the hundreds of millions per gram of soil.

Worms are some of the most important organisms living in soil. Earthworms, probably one of the most familiar soil inhabitants, eat soil and obtain energy and raw materials by digesting humus (see Chapter 29). *Castings,* bits of soil that have passed through the gut of an earthworm, are deposited on the soil surface. In this way, minerals from deeper layers are brought to upper layers. Earthworm tunnels serve to aerate the soil, and the worms' waste products and corpses add organic material to deeper layers of the soil.

Ants live in the soil in enormous numbers, constructing tunnels and chambers that help to aerate it. Members of soil-dwelling ant colonies forage on the surface for bits of food, which they carry back to their nests. Not all of this food is eaten, however, and its eventual decomposition helps increase the organic matter in the soil.

About 30% to 60% of soil volume is composed of pore spaces between soil particles

A healthy soil has numerous pore spaces of different sizes around and among the soil particles. Pore spaces are filled with varying proportions of soil air and soil water (Fig. 34–14), both of which are necessary to produce a moist but well-drained soil that sustains plants and other soil-dwelling organisms. Generally speaking, water is held in the smaller pores, while air is found in the larger pores. After a prolonged rain, almost all of the pore spaces may be filled with water, but water drains rapidly from the larger pore spaces, drawing air from the atmosphere into those spaces.

Soil air contains the same gases as atmospheric air, although they are usually present in different proportions. As a result of aerobic respiration by soil organisms, there is usually less oxygen and more carbon dioxide in soil air than in atmospheric air. (Recall from Chapter 7 that aerobic respiration uses oxygen and produces carbon dioxide.) Among the important gases in soil air are oxygen (O_2), required by soil organisms for aerobic respiration; nitrogen (N_2), used by nitrogen-fixing soil organisms; and carbon dioxide (CO_2). Carbon dioxide, a product of aerobic respiration, dissolves in water to form carbonic acid, H_2CO_3, a weak acid that helps to weather rock.

Soil water originates as precipitation, which drains downward, or as groundwater (water stored in porous underground rock), which rises upward from the water table (the uppermost level of groundwater). Soil water contains low concentrations of dissolved mineral salts that enter the roots of plants when they absorb water. Soil water not absorbed by roots percolates (moves down) through soil, carrying dissolved minerals with it. The removal of dissolved materials from soil by percolating water is called **leaching.** Some anions completely leach out of the soil because they are so soluble that they migrate all the way down to the groundwater. (Recall that not all soluble materials leach out of the soil because soil particles, particularly clay particles, bind them.) It is also possible for water to move *upward* through the soil and carry dissolved materials with it.

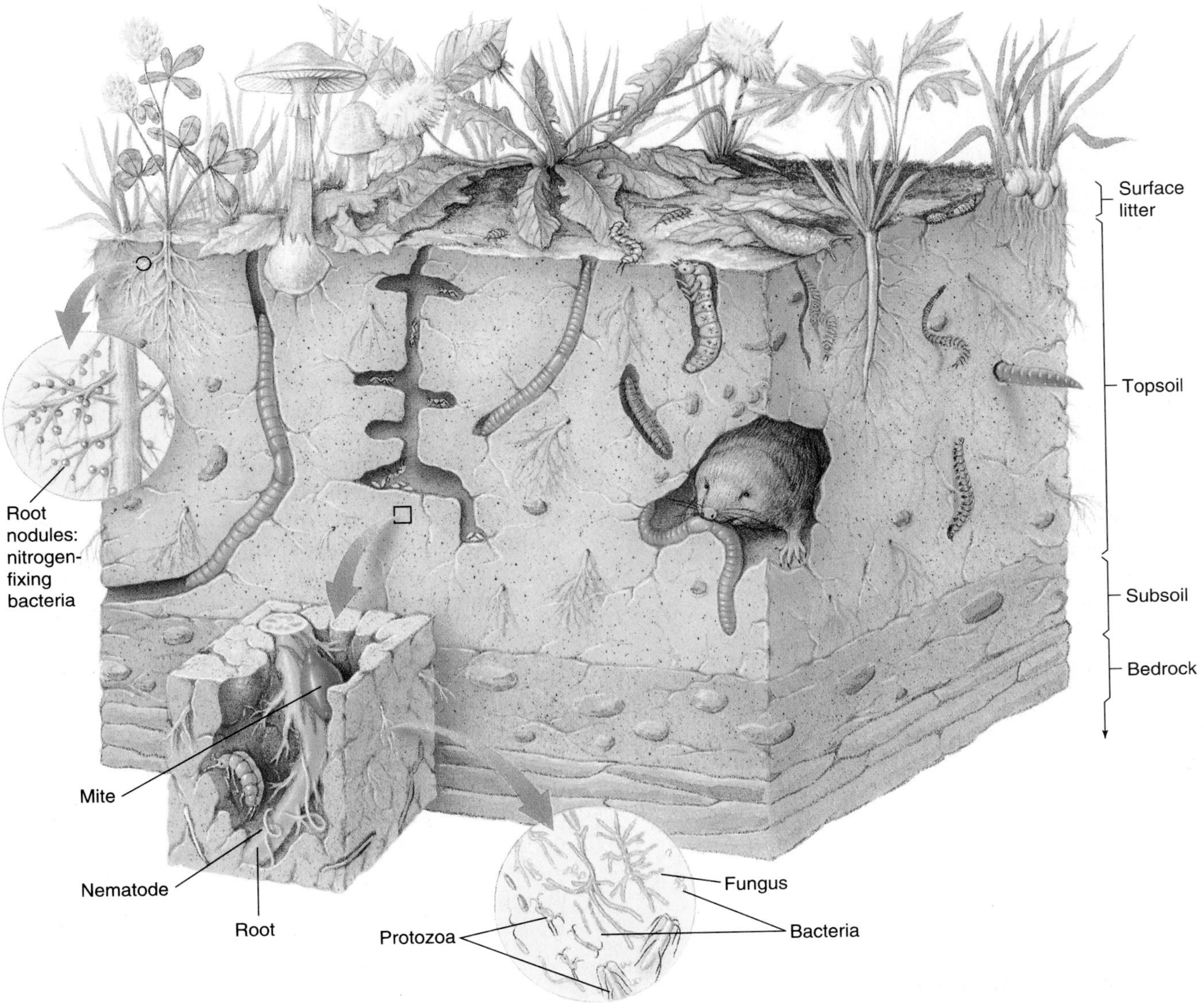

FIGURE 34–13 The diversity of life in fertile soil is remarkable. It includes plants, algae, fungi, earthworms, flatworms, roundworms, insects, spiders and mites, bacteria, and burrowing animals such as moles and groundhogs.

Soil pH affects soil characteristics and plant growth

As discussed in Chapter 2, soil acidity is measured using the pH scale, which runs from 0 (extremely acidic) through 7 (neutral) to 14 (extremely alkaline). The pH of most soils ranges from 4 to 8, but some soils are outside this range. The soil of the Pygmy Forest in Mendocino County, California, is extremely acidic (pH 2.8–3.9). At the other extreme, certain saline soils in Death Valley, California, have a pH of 8.5.

Soil pH affects plants and is, in turn, influenced by plants and other organisms that inhabit it. Plants are affected by soil pH partly because the solubility of certain minerals varies with differences in pH. Soluble mineral elements can be absorbed by the plant, whereas insoluble forms cannot. At a low pH, for example, the aluminum and manganese in soil water are more soluble and are sometimes absorbed by the roots in toxic concentrations. Certain mineral salts essential for plant growth, such as calcium phosphate, become less soluble at a higher pH.

Soil pH also affects the leaching of nutrients. An acidic soil has less ability to bind positively charged ions to it. As a result, certain minerals essential for plant growth, such as potassium (K^+), are leached more readily from

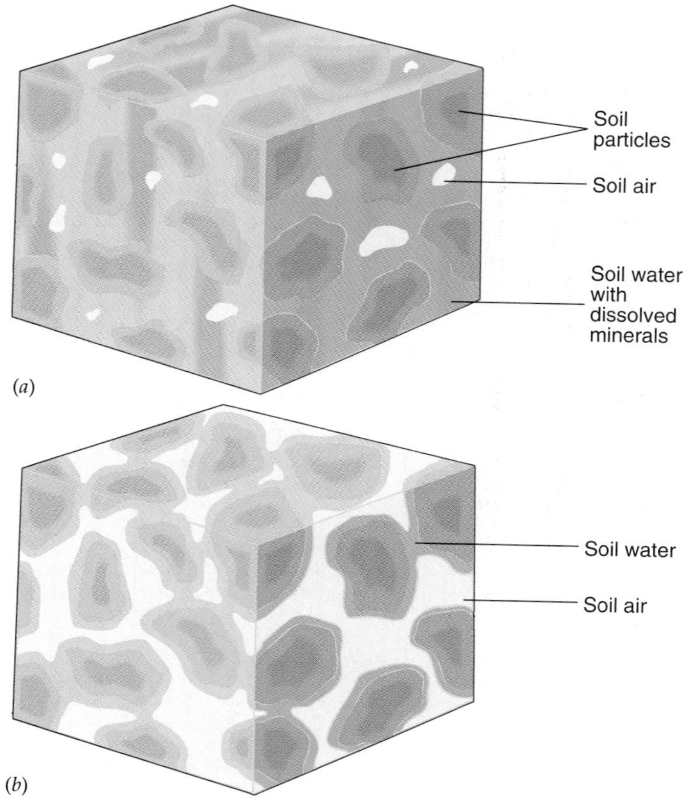

FIGURE 34–14 Pore space is occupied by varying amounts of soil air and soil water. (*a*) In a wet soil, most of the pore space is filled with water. (*b*) In a dry soil, a thin film of water is tightly bound to soil particles, and most of the pore space is occupied by soil air.

acidic soil. The optimum soil pH for most plant growth is 6.5 to 7.5, because most nutrients needed by plants are available to the plants in that pH range.

SOIL PROVIDES MOST OF THE MINERALS FOUND IN PLANTS

More than 90 naturally occurring elements exist on Earth, and more than 60 of these, including elements as common as carbon and as rare as gold, have been found in plant tissues. Not all of these elements are considered essential for plant growth, however.

How do biologists determine whether an element is essential? It is impossible to conduct mineral nutrition experiments by growing plants in soil because soil is too complex and contains too many elements. Thus, one of the most useful methods to test whether or not an element is essential is **hydroponics,** which is the growing of plants in aerated water to which mineral salts have been added (Fig. 34–15). Hydroponics has other applications in addition to its scientific use (see *Focus On: Commercial Hydroponics*).

If biologists suspect that a particular element is essential for plant growth, they grow plants in a nutrient solution that contains all known essential elements *except* the one in question. If plants grown in the absence of that element are unable to develop normally or to complete their life cycle, the element may be essential. Additional criteria are used to confirm whether an element is essential. For example, it must be demonstrated that the element has a direct effect on the plant's metabolism, and that the element is essential for a wide variety of plants.

FIGURE 34–15 In this hydroponics experiment at the University of California at Davis, the plants are grown in a liquid solution of mineral nutrients through which air is bubbled (to allow the roots to respire). (L. Migdale/Science Source/Photo Researchers, Inc.)

Focus On

Commercial Hydroponics

Hydroponics, the practice of growing plants in an aerated solution of mineral salts, has been used by scientists to determine which elements are essential for plant growth. Initially, entrepreneurs hailed hydroponics as the scientific way to grow plants in places where soil was poor or unavailable. The expenses involved in commercially growing produce for human consumption, however, prevented hydroponics from becoming more than a curiosity. Recent technical improvements have revived the interest in commercial hydroponics (see figure).

Hydroponics has great potential in several places. It is being tried experimentally in desert countries in the Middle East, where the soil is too arid to support cultivation and water is unavailable for irrigation. When plants are grown hydroponically in greenhouses, little water is used compared with traditional irrigation methods. Hydroponics is also being tried in temperate climates, particularly to produce crops such as lettuce, tomatoes, and herbs in winter.

Hydroponics has other advantages. First, it is possible to grow crops hydroponically under conditions in which disease-causing microorganisms and insect pests are completely absent. This means that the crops are not exposed to chemical pesticides. Hydroponics can also be used to grow crops near their area of use, thus saving on transportation costs. (Most foods consumed by Americans travel an average of 1,300 miles.)

The main disadvantage of hydroponics is the expense. Plants grown hydroponically must be supplied with a nutrient solution that is continually monitored and adjusted. Heating and lighting costs are high. Aeration of the roots was a major expense until developments like the nutrient-film technique helped cut costs. In this technique, plants are grown in trenches produced by two sheets of plastic through which a nutrient solution trickles. In this way the roots get adequate aeration. The nutrient solution is saved and reused, which cuts down on water and mineral costs.

Although hydroponics will probably never replace traditional agriculture, it has been shown to be a viable alternative in certain situations. As new techniques are developed, hydroponics may become even more common. ■

Lettuce is one of several hydroponic crops now grown commercially.
(Hank Morgan/Photo Researchers, Inc.)

Table 34–2 FUNCTIONS OF ESSENTIAL ELEMENTS

Element	*Major Functions*
Carbon	Component of carbohydrate, lipid, protein, and nucleic acid molecules
Hydrogen	Component of carbohydrate, lipid, protein, and nucleic acid molecules
Oxygen	Component of carbohydrate, lipid, protein, and nucleic acid molecules
Nitrogen	Component of proteins, nucleic acids, chlorophyll, certain coenzymes
Phosphorus	In nucleic acids, phospholipids, ATP (energy transfer compound)
Calcium	In cell walls; involved in membrane permeability; enzyme activator
Magnesium	In chlorophyll; enzyme activator in carbohydrate metabolism
Sulfur	In certain amino acids and vitamins
Potassium	Osmosis and ionic balance; opening and closing of stomata; enzyme activator (for 40+ enzymes)
Chlorine	Ionic balance; involved in photosynthesis
Iron	Part of enzymes involved in photosynthesis, respiration, and nitrogen fixation
Manganese	Part of enzymes involved in respiration and nitrogen metabolism; required for photosynthesis
Copper	Part of enzymes involved in photosynthesis
Zinc	Part of enzymes involved in respiration and nitrogen metabolism
Molybdenum	Part of enzymes involved in nitrogen metabolism
Boron	Exact role unclear; involved in membrane transport and calcium utilization

Sixteen elements are essential for plant growth

Sixteen elements have been found essential for plant growth (Table 34–2). Nine of these are required in fairly large quantities (greater than 0.05% dry weight) and are therefore known as **macronutrients.** These include carbon, hydrogen, oxygen, nitrogen, phosphorus, potassium, sulfur, calcium, and magnesium. The remaining seven **micronutrients** are needed in trace amounts (less than 0.05% dry weight) for normal plant growth and development. These include iron, boron, manganese, copper, molybdenum, chlorine, and zinc.

Four of the 16 elements—carbon, oxygen, hydrogen, and nitrogen—come from water or gases in the atmosphere. Carbon is obtained from carbon dioxide (CO_2) in the atmosphere during photosynthesis. Oxygen is obtained from atmospheric oxygen (O_2) and water. Water also supplies hydrogen to the plant. Plants absorb their nitrogen from the soil as ions of nitrogen salts, but the nitrogen in nitrogen salts ultimately comes from atmospheric nitrogen (see *Making the Connection: Soil Nitrogen and the Interdependence of Living Organisms*). The remaining 12 essential elements are obtained from the soil as dissolved mineral ions. Their ultimate source is the parent rock from which the soil was formed.

SOIL CAN BE DAMAGED BY HUMAN MISMANAGEMENT

Soil is a valuable natural resource on which humans depend for food. Many human activities generate or accentuate soil problems, including mineral depletion, soil erosion, and accumulation of salt.

Mineral depletion occurs in soils that are farmed

In a natural ecosystem, the essential minerals removed from the soil by plants are returned when the plants or the animals that eat them die and decompose. An agricultural system disrupts this pattern. Crops, which contain minerals, are removed from the cycle when they are harvested. As a result, they cannot decay and release their nutrients back into the soil. Thus, over time, soil that is farmed eventually loses its fertility—that is, its ability to produce abundant crops.

The three elements that are most often limiting factors for plants are nitrogen, phosphorus, and potassium. In order to sustain the productivity of agricultural soils, fertilizers are periodically added to replace these minerals.

Making the Connection

Soil Nitrogen and the Interdependence of Living Organisms

How does the requirement of nitrogen by living organisms show the interdependence of life? Because nitrogen is an essential part of biologically important molecules such as proteins and nucleic acids, all living organisms must have nitrogen in order to survive. Animals, including humans, get their nitrogen from proteins and other nitrogen-containing compounds in the foods they eat. Ultimately, this nitrogen is traced back to plant sources.

Plants obtain their nitrogen in much simpler forms—nitrate (NO_3^-) and ammonium (NH_4^+) ions—from the soil. But where do the nitrate and ammonium ions come from? Nitrogen is converted into those forms from atmospheric nitrogen (N_2) by a few species of bacteria (see Chapter 23). Thus, the nitrogen supply of the entire biosphere, including humans, comes from a few, at first glance insignificant, organisms. Without these bacteria, the supply of nitrogen in a form suitable for plants would be rapidly depleted, and plants would die. With the demise of plants, animals, including humans, would be unable to survive. We say more about the crucial role of microorganisms in the nitrogen cycle in Chapter 54. ▲

The two main types of fertilizers are organic and inorganic. *Organic fertilizers* include such natural materials as cow manure, crop residues, bone meal, and compost. Green manure, a special type of organic fertilizer, is actually a crop that is planted in the soil and deliberately plowed under to decompose instead of being harvested. Frequently the plant grown as green manure has nitrogen-fixing bacteria living in root nodules, thereby increasing the amount of nitrogen in the soil. Organic fertilizers are complex and their exact compositions vary. The mineral nutrients in them become available to plants only as the organic material decomposes. For that reason, organic fertilizers are slow-acting and long-lasting.

Inorganic fertilizers are manufactured from chemical compounds, and their exact compositions are known. Because they are soluble, they are immediately available to plants. Inorganic fertilizers are available in the soil for only a short period of time, however, because they quickly leach away. Most inorganic fertilizers contain the three elements (nitrogen, phosphorus, and potassium) that are usually the limiting factors in plant growth. The numbers on fertilizer bags (for example, 10–20–20) tell the relative concentrations of each of the three elements (N,P,K).

Soil erosion is the loss of soil from the land

Water, wind, ice, and other agents cause **soil erosion,** the wearing away or removal of soil from the land. Water and wind are particularly effective in removing soil: rainfall loosens soil particles that can then be transported away by moving water (Fig. 34–16), while wind loosens soil and blows it away, particularly if the soil is exposed and dry.

Soil erosion is a national and international problem that does not make the headlines very often. To get a feeling for how serious the problem is, consider that approximately 2.7 billion metric tons (3.0 billion tons) of topsoil are lost from U.S. farmlands as a result of soil erosion *each year.* The U.S. Department of Agriculture estimates that approximately one-fifth of U.S. cropland is vulnerable to soil erosion damage. Because erosion reduces the amount of soil in an area, it limits the growth of plants. Erosion also causes a loss in soil fertility because essential minerals and organic matter are also removed. As a result of these losses, the productivity of eroded agricultural soils declines.

Humans often accelerate soil erosion through poor soil management practices. Agriculture is not the only culprit, since the removal of natural plant communities

FIGURE 34–16 Precipitation causes soil erosion in an open field. Gullies can grow rapidly because they are prone to greater rates of erosion as they provide channels for the runoff of water when it rains. (USDA/Soil Conservation Service)

during the construction of roads and buildings also accelerates erosion. In addition, unsound logging practices such as clear-cutting large forested areas for lumber and pulpwood cause severe erosion.

Soil erosion has an impact on other natural resources. Sediment that enters streams, rivers, and lakes degrades water quality and fish habitats. If the sediment contains pesticide and fertilizer residues, they further pollute the water.

Sufficient plant cover reduces the amount of soil erosion: leaves and stems cushion the impact of rainfall, and roots help to hold the soil in place. Although soil erosion is a natural process, plant cover makes it negligible in many natural ecosystems.

Salt accumulates in soil that is improperly irrigated

Although irrigation improves the agricultural productivity of arid and semiarid lands, it can also cause salt to accumulate in the soil, a process called **salinization.** In a natural scenario, as a result of precipitation runoff, rivers carry dissolved salts away. Irrigation water, however, normally soaks into the soil and does not run off the land into rivers, so when it evaporates, the salt remains behind and accumulates in the soil. Salty soil results in a decline in productivity and in extreme cases renders the soil completely unfit for crop production.

Most plants cannot obtain all the water they need from salty soil because a water balance problem exists: water moves by osmosis *out* of plant roots and into the salty soil. Obviously, most plants cannot survive under these conditions (see Fig. 5–15). Plant species that thrive in saline soils have special adaptations that enable them to tolerate the high amount of salt. Most crops, unless they have been genetically selected to tolerate high salt, are not productive in saline soil.

SUMMARY

I. Anchorage, absorption, conduction, and storage are the main functions of roots.

II. Two types of root systems exist in plants.
   A. A taproot system has one main root from which many smaller lateral roots extend.
   B. A fibrous root system has several to many adventitious roots of the same size developing from the end of the stem.

III. Roots have several structures not found in stems.
   A. Each root tip is covered by a root cap, a protective layer that covers the delicate root apical meristem and may orient the root so that it grows downward.
   B. Root hairs, short-lived extensions of epidermal cells, increase the surface area of the root in contact with the moist soil.
   C. Unlike stems, roots lack nodes and internodes and do not usually produce leaves or buds.

IV. Primary roots possess an epidermis, ground tissues (cortex and in certain roots, pith), and vascular tissues (xylem and phloem).
   A. Epidermis protects the root, and its root hairs aid in absorption of water and dissolved minerals.
   B. Cortex consists of parenchyma cells that often store starch.
   C. Endodermis, the innermost layer of the cortex, controls mineral uptake into the root xylem.
      1. Cells of the endodermis possess a Casparian strip around their radial and transverse walls that is impermeable to water and dissolved minerals.
      2. Minerals therefore must pass through carrier proteins in the plasma membranes of the endodermal cells.
   D. Pericycle gives rise to lateral roots and lateral meristems.
   E. Xylem conducts water and dissolved minerals; phloem conducts dissolved sugar.

V. Roots vary in internal structure.
   A. Xylem and phloem of herbaceous dicot roots form a solid mass in the center of the root. In contrast, the center of monocot roots often consists of pith.
   B. Monocot roots lack a vascular cambium and therefore do not have secondary growth.
   C. Roots of gymnosperms and woody dicots develop secondary tissues (wood and bark).

VI. Some roots are modified for specialized functions.
   A. Prop roots develop from branches or from a vertical stem and grow downward into the soil to help support the plant in an upright position.
   B. Pneumatophores are aerial "breathing" roots that may assist in getting oxygen to the submerged roots.
   C. Certain epiphytes have roots modified to photosynthesize, absorb moisture, or, if parasitic, penetrate a host's tissues and absorb nutrients.
   D. Corms and bulbs often have contractile roots that grow into the soil and then contract, thereby pulling the corm or bulb deeper into the soil.

VII. Roots often form associations with other species.
   A. A root graft is a natural union between the roots of trees belonging to the same or different species.
   B. Mycorrhizae are mutually beneficial associations between roots and soil fungi.
   C. Root nodules are swellings that develop on certain roots and house millions of nitrogen-fixing bacteria.

VIII. Soil is the complex material in which roots grow.
   A. Factors influencing soil formation include parent rock, climate, living organisms, the passage of time, and topography.

B. Soil is composed of inorganic minerals, organic matter, living organisms, soil air, and soil water.

IX. Plants require 16 essential elements for normal growth.

A. Nine elements are macronutrients: carbon, oxygen, hydrogen, nitrogen, potassium, phosphorus, sulfur, magnesium, and calcium.

B. Seven elements are micronutrients: iron, boron, manganese, copper, zinc, molybdenum, and chlorine.

SELECTED KEY TERMS

adventitious root
apoplast
Casparian strip
contractile root
cortex
endodermis
epidermis
fibrous root
graft
humus
hydroponics
leaching
macronutrient
micronutrient
mycorrhizae
nodule
pericycle
phloem
pith
pneumatophore
prop root
root cap
root hair
soil erosion
stele
symplast
taproot
vascular cambium
weathering
xylem

POST-TEST

1. A __________ system consists of one main root (formed from the enlarging embryonic root) with many smaller lateral roots coming out of it.
2. __________ roots are produced at unusual places on the plant.
3. Plants with bulbs often have __________ roots that pull the bulb deeper into the ground.
4. Certain plants adapted to flooded soil produce aerial "breathing" roots known as __________ .
5. Unlike stems, roots produce a(an) __________ __________ (a protective covering over the delicate root apical meristem) and __________ __________ (short-lived, single-celled extensions of epidermal cells).
6. The waterproof region around the radial and transverse walls of endodermal cells is the __________ __________ .
7. The __________ is the origin of lateral roots.
8. The center of a herbaceous dicot root is composed of __________ , whereas the center of a monocot root is composed of __________ .
9. Sometimes the roots of two different trees form a natural __________ through which materials such as dissolved sugars and hormones pass.
10. __________ are mutually-beneficial associations between certain soil fungi and the roots of most plant species.
11. Soils are formed from parent rock that is gradually broken down into smaller and smaller particles by biological, chemical, and physical __________ .
12. Some precipitation percolates through soil, __________ certain dissolved minerals out of the soil.
13. The technique of growing plants in aerated water with dissolved mineral salts is known as __________ .
14. __________ are essential elements required in fairly large quantities.
15. Much water that enters roots moves from cell to cell along interconnected porous cell walls, known as the __________ .

REVIEW QUESTIONS

1. List several functions of roots, and describe the tissue(s) responsible for each function.
2. How would you distinguish between a root hair and a small lateral root?
3. If you were examining a cross section of a primary root of a flowering plant, how would you determine whether it was a dicot or a monocot?
4. Trace the pathway of water from the soil through the various tissues in a herbaceous dicot root.
5. How does a herbaceous root develop secondary tissues?
6. Distinguish among the following specialized roots: (a) storage root; (b) prop root; (c) aerial "breathing" root; (d) photosynthetic root; and (e) contractile root.
7. List the five components of soil, and tell how each is important to plants.
8. Explain how weathering processes convert rock into soil.
9. Distinguish between macronutrients and micronutrients.
10. Label the various tissues of this herbaceous dicot root:

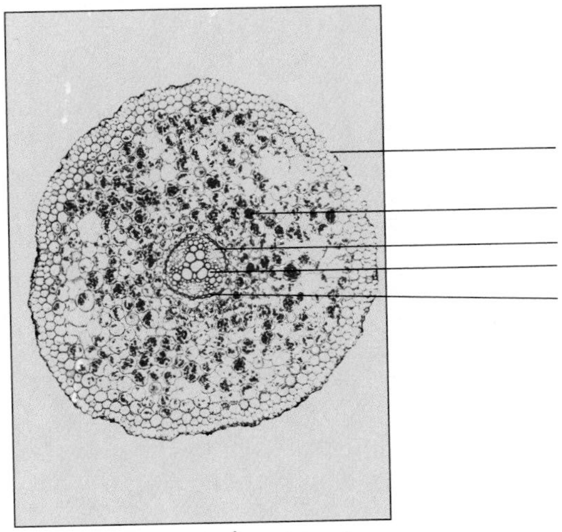

(James Mauseth, University of Texas)

YOU MAKE THE CONNECTION

1. A mesquite root is found penetrating a mine shaft about 46 meters (150 feet) below the surface of the soil. How might you determine *when* the root first grew into the shaft? (*Hint*: mesquite is a woody plant.)
2. A barrel cactus that is two feet tall and one foot in diameter has roots over 10 feet long. However, all of the plant's roots are found in the soil at a depth of two to six inches. What possible adaptive value does such a shallow root system confer on a desert plant?
3. You are given a plant structure that was found growing in the soil and are asked to determine whether it is a root or an underground stem. How would you identify the plant part without a microscope? With a microscope?
4. How would you design an experiment to determine whether gold is essential for plant growth? What would you use for an experimental control?
5. Why does overwatering a plant often kill it?

RECOMMENDED READINGS

Brown, J.C., and V.D. Jolley. "Plant Metabolic Responses to Iron-deficiency Stress." *BioScience,* Vol. 39, No. 8, September 1989. An in-depth discussion of the factors that make iron in the soil available to roots.

Feldman, L.J. "The Habits of Roots." *BioScience,* Vol. 38, No. 9, October 1988. How roots interact with their soil environment.

Mauseth, J.D. *Botany: An Introduction to Plant Biology,* 2nd ed. Philadelphia, Saunders College Publishing, 1995. A comprehensive introduction to plant biology.

Moore, P. "Upwardly Mobile Roots." *Nature,* Vol. 341, September 21, 1989. An essay on unusual root behaviors, including roots that grow out of the soil and up the trunk of a tree to obtain nutrients leaching from the forest canopy.

C H A P T E R 3 5

Reproduction in Flowering Plants

The five petals of each trumpet honeysuckle (*Lonicera sempervirens*) flower are fused to form a tubular corolla.
(James L. Castner)

One of the reasons for the success of flowering plants is their reproductive flexibility. Many flowering plants are able to reproduce both sexually and asexually. Sexual reproduction entails the fusion of reproductive cells (egg and sperm cells, called gametes). The union of gametes, which is called *fertilization*, occurs within the ovary of the flower. After fertilization, flowering plants produce seeds and fruits.

The offspring of plants that reproduce sexually exhibit a great deal of genetic variation. They may resemble one of the parent plants, both of the parents, or neither of the parents. Sexual reproduction offers the advantage of new combinations of genes not found in either parent. These new gene combinations might make an individual plant better suited to its environment. (The details of how this variation occurs are discussed in Chapters 9 and 10.)

In flowering plants asexual reproduction does not usually involve the formation of flowers, seeds, and fruits. Instead, offspring usually form when part of an existing plant becomes separated from the rest of the plant by the death of tissues. This part subsequently grows to form a complete, independent plant.

Flowering plants have many methods of asexual reproduction, most of which involve modified vegetative parts. In particular, a number of stems are modified asexual structures — rhizomes, tubers, bulbs, corms, and stolons. Because asexual reproduction involves only one parent, and no fusion of gametes occurs, the offspring of asexual reproduction are genetically similar to each other and to the parent plant from which they came.

In this chapter we examine various aspects of both sexual and asexual reproduction in flowering plants.

LEARNING OBJECTIVES

After you have studied this chapter you should be able to

1. Label the parts of a flower on a diagram, and describe the functions of each part.
2. Relate where eggs and pollen grains are formed within the flower, and distinguish between pollination and fertilization.
3. Compare the general features that characterize flowers pollinated in different ways (by insects, birds, bats, and wind).
4. Define coevolution and give examples of ways that plants and their animal pollinators have affected one another's evolution.
5. Trace the stages of embryo development in flowering plants, and list and define the main parts of seeds.
6. Distinguish among simple, aggregate, multiple, and accessory fruits, give examples of each type, and cite several different methods of seed and fruit dispersal.
7. State the difference between sexual and asexual reproduction, and distinguish among the following methods of asexual reproduction: rhizomes, tubers, stolons, corms, bulbs, plantlets, suckers, and apomixis.

FLOWERS ARE INVOLVED IN SEXUAL REPRODUCTION

Flowers are reproductive shoots usually composed of four kinds of organs—sepals, petals, stamens, and carpels—arranged in whorls (circles) on the end of a flower stalk (Fig. 35–1*a*). In flowers with all four organs, the normal order of whorls from the flower's periphery to the center is:

sepals → petals → stamens → carpels

The tip of the stalk enlarges to form a **receptacle** on which some or all of the flower parts are borne. Sepals and petals consist of sterile modified leaves, while stamens and carpels consist of fertile modified leaves. All four floral parts are important in the reproductive process, but only the stamens (the "male" organs) and carpels (the "female" organs) participate directly in reproduction.

Sepals, which constitute the lowest and outermost whorl on a floral shoot, cover and protect the flower parts when the flower is a bud. Sepals are leaflike in shape and form, and are often green. Some sepals, such as those in lily flowers, resemble petals. As the blossom opens from a bud, the sepals fold back to reveal the more conspicuous petals. The collective term for all the sepals of a flower is **calyx.**

The whorl just above the sepals consists of **petals,** which are broad, flat, and thin (like sepals and leaves) but tremendously varied in shape and frequently brightly colored. Petals play an important role in ensuring that sexual reproduction will occur. Sometimes petals are fused to form a tube (as in honeysuckle flowers in the chapter opening photograph) or other floral shape (as in snapdragons). The collective term for all the petals of a flower is **corolla.**

Just inside the petals are the **stamens,** the male reproductive organs. Each stamen is composed of a thin stalk, called a **filament,** at the top of which is an **anther,** a saclike structure in which pollen grains form (Fig. 35–1*b*; see also Fig. 27–9). For sexual reproduction to occur, pollen grains must be transferred from the anther to the female reproductive structure (the carpel), usually of another flower of the same species. Each pollen grain produces two cells surrounded by a tough outer wall. One cell divides mitotically to form two nonflagellated male gametes, known as sperm, and the other produces a **pollen tube** through which the sperm travel to reach the ovule.

In the center of most flowers are one or more **carpels,** the female reproductive organs. Carpels bear **ovules,** which are structures with the potential to develop into seeds. The carpels of a flower may be separate or fused together into a single structure. The female part of the flower, often referred to as a **pistil,** may consist of a single carpel (a *simple* pistil) or a group of fused carpels (a *compound* pistil). Each pistil has three sections: a **stigma,** on which the pollen grains land; a **style,** a necklike structure through which the pollen tube grows; and an **ovary,** a juglike structure that contains one or more ovules (Fig. 35–1*c*; see also Fig. 27–9). Each ovule contains an embryo sac that forms one female gamete (an egg) and two **polar nuclei.** The egg and both polar nuclei participate directly in fertilization.

POLLINATION IS THE FIRST STEP TOWARD FERTILIZATION

Before fertilization can occur, pollen grains must travel from the anther (where they form) to the stigma. The transfer of pollen grains from anther to stigma is known

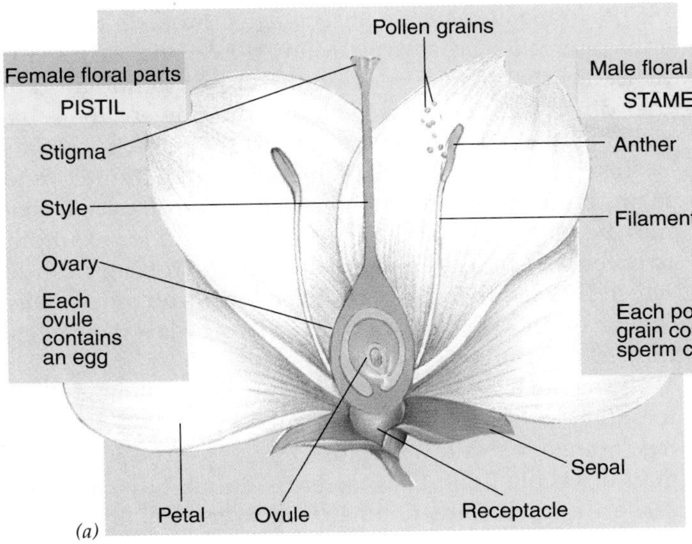

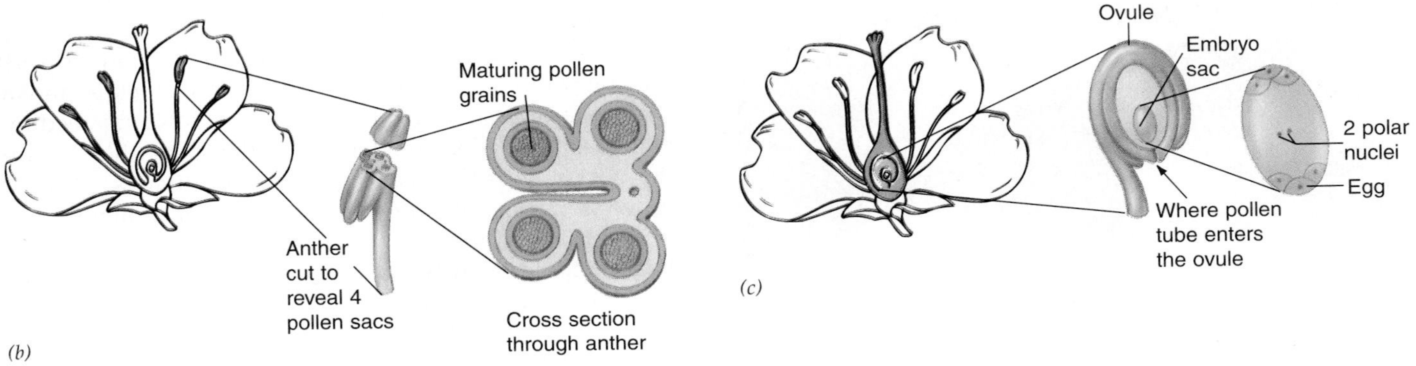

FIGURE 35–1 The flower bears the organs for sexual reproduction in flowering plants. (*a*) Cutaway view of a "typical" flower. (*b*) Pollen grains develop within sacs in the anther. (*c*) An embryo sac, with its egg and polar nuclei, forms within the ovule. The ovary depicted here contains a single ovule.

as **pollination.** Plants are *self-pollinated* if pollination occurs within the same flower or a different flower on the same plant. When pollen grains are transferred to a flower on another plant, we say the plant is *cross-pollinated.* Flowering plants accomplish pollination in a variety of ways. Some flowers are pollinated by beetles, bees, flies, butterflies, moths, wasps, or other insects. Animals such as birds, bats, snails, and small rodents also pollinate plants. Wind is the agent of pollination for certain flowers, while a few aquatic flowers are pollinated by water.

Flowering plants and their animal pollinators have affected one another's evolution

Flowers pollinated by animals have various features to attract them, including showy petals (a visual attractant) and scent (an olfactory attractant) (Fig. 35–2). One of the rewards for the animal pollinator is food. Nectar is a sugary solution produced in special floral glands called **nectaries** and used as an energy-rich food by pollinators. Pollen grains are also a protein-rich food for many animals. As they move from flower to flower searching for

FIGURE 35–2 Devil's tongue produces a disagreeable floral odor that causes a visible reaction. The odor-causing chemicals react with HCl on the filter paper to produce a visible precipitate. Devil's tongue is pollinated by beetles and flies. (Courtesy of J. M. Patt and B. J. D. Meeuse)

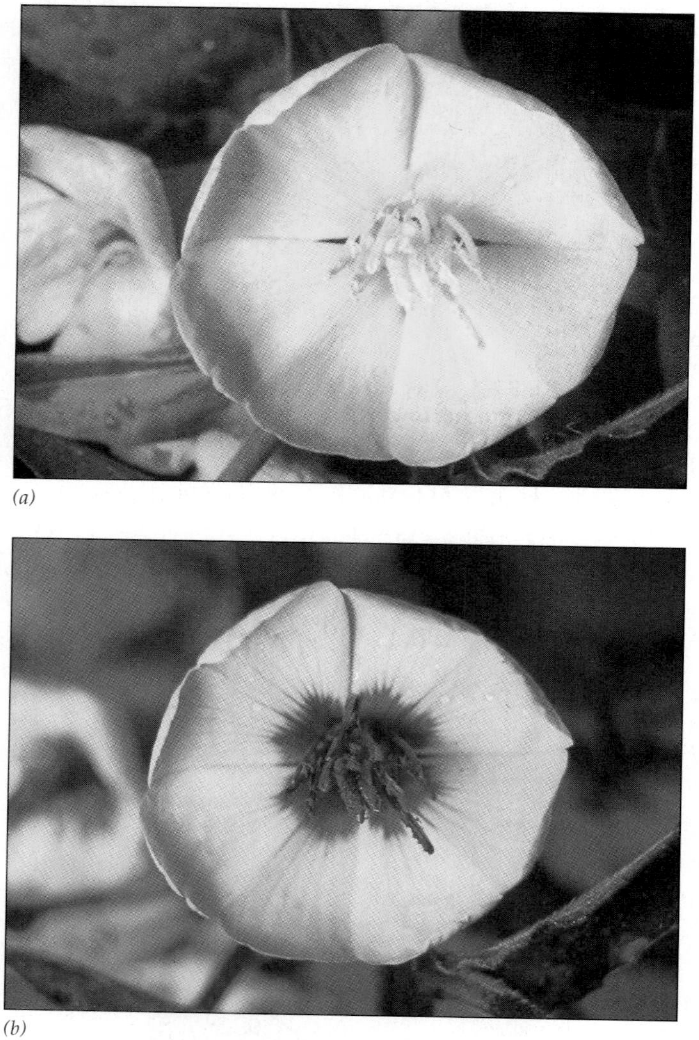

(a)

(b)

FIGURE 35–3 Many insect-pollinated flowers have ultraviolet markings that are invisible to humans but quite conspicuous to insects. **(a)** A flower as seen by the human eye is solid yellow. **(b)** The same flower viewed under ultraviolet radiation provides clues about how the insect eye perceives it. The outer portions of the petals appear purple to a bee whereas the inner parts appear yellow. These differences in coloration draw attention to the center of the flower, where the pollen grains and nectar are located. (*a, b,* Thomas Eisner)

food, they inadvertently carry pollen grains on their body parts, thus facilitating sexual reproduction in plants.

Plants pollinated by insects often have blue or yellow petals. The insect eye does not perceive color in the same way that the human eye does. Most insects see well in the violet, blue, and yellow range of visible light but do not perceive red as a distinct color. Consequently, flowers pollinated by insects are not usually red. Insects can also see in the ultraviolet range, wavelengths that are invisible to the human eye. Insects see ultraviolet radiation as a color called *bee's purple.* Many flowers have dramatic markings that may or may not be visible to humans but that direct insects to the center of the flower where the pollen grains and nectar are located (Fig. 35–3).

Insects have a well-developed sense of smell, and many insect-pollinated flowers have a strong scent that may be pleasant or foul. The carrion plant, for example, is pollinated by flies and smells like the rotting flesh in which flies lay their eggs. As flies move from one smelly flower to another looking for a place to lay their eggs, they transfer pollen grains.

Birds such as hummingbirds are important pollinators (Fig. 35–4*a*). Flowers pollinated by birds are usually red, orange, or yellow because birds see well in this region of visible light. Because birds do not have a strong sense of smell, bird-pollinated flowers usually lack a scent.

Bats, which feed at night and do not see well, are important pollinators, particularly in the tropics where they are most abundant (Fig. 35–4*b*). Bat-pollinated flowers are night-blooming and have dull white petals and a strong scent, usually of fermented fruit. Nectar-feeding bats are attracted to the flowers by the scent; they lap up the nectar with their long, extendible tongues. As they move from flower to flower, they transfer pollen grains.

Animal pollinators and the plants they pollinate have had such close, interdependent relationships over time that they have affected the evolution of certain physical and behavioral features in each other. The term **coevolution** describes such mutual adaptation, in which two different organisms interact so closely that they become increasingly adapted to one another.

During the time that plants were coevolving specialized features such as petals, scent, and nectar to attract pollinators, animal pollinators coevolved specialized body parts and behaviors that adapted them to aid pollination as they obtain nectar and pollen grains as a reward. For example, coevolution is responsible for the hairy bodies of bumblebees, which catch and hold the sticky pollen grains for transport from one flower to another. Coevolution is also responsible for the long, curved beaks of certain honeycreepers, Hawaiian birds that insert their beaks into tubular flowers to obtain nectar (Fig. 35–5). (The long, tubular corolla of the flowers that honeycreepers visit also came about through coevolution.)

Animal behavior has also coevolved, sometimes in bizarre ways. The flowers of certain orchids (*Ophrys*), for example, resemble female wasps in coloring and shape. The resemblance between one *Ophrys* species and female wasps is so strong that male wasps mount the flowers and attempt to copulate with them. During this misdirected activity, a pollen sac usually attaches to the back of the wasp. When the frustrated wasp departs and attempts to copulate with another orchid flower, pollen grains are transferred to that flower.

(a)

(b)

FIGURE 35–4 Animals are pollinating agents for many flowering plants. (*a*) A ruby-throated hummingbird obtains nectar from a trumpet vine flower. The pollen grains on the bird's feathers will be carried to the next plant. (*b*) A lesser long-nosed bat obtains nectar from a cardon cactus flower. Bats, which go from flower to flower seeking nectar, are important pollinators in the tropics. (*a*, Dan Dempster/Dembinsky Photo Associates; *b*, Merlin Tuttle/Bat Conservation International/Photo Researchers, Inc.)

A number of obligate relationships exist between animal pollinators and the flowering plants they pollinate in which each depends on the other for its survival. For example, a yucca plant native to the Southwest can only be pollinated by one species of moth, the female of which can only lay her eggs inside the yucca flower's ovary. Both the yucca and the moth would become extinct if something happened to the other, as neither would be able to reproduce successfully.

Some flowering plants depend on wind to disperse pollen

Flowering plants such as grasses, ragweed, and maples, among others, are pollinated by wind and produce many flowers that are often inconspicuous (Fig. 35–6). Wind-pollinated plants do not produce large, colorful petals, scent, or nectar. Some have large, feathery stigmas, presumably to make it easier to trap wind-borne pollen grains. Because wind pollination is a hit-or-miss affair, the likelihood of pollen grains landing on a stigma of the same species of flower is slim. Wind-pollinated plants therefore produce large quantities of pollen grains.

FIGURE 35–5 The 'i'iwi, one of the Hawaiian honeycreepers, is thought to possess its gracefully curved bill in order to sip nectar from flowers of the lobelia. The 'i'iwi bill fits perfectly into the long, tubular lobelia flowers. (Robert J. Western)

FIGURE 35–6 Wind is a pollination agent for a number of flowering plants. Clusters of male oak flowers, which lack petals, dangle from a tree branch and shed a shower of pollen when the wind blows. (Dr. Jeremy Burgess/Science Photo Library/Photo Researchers, Inc.)

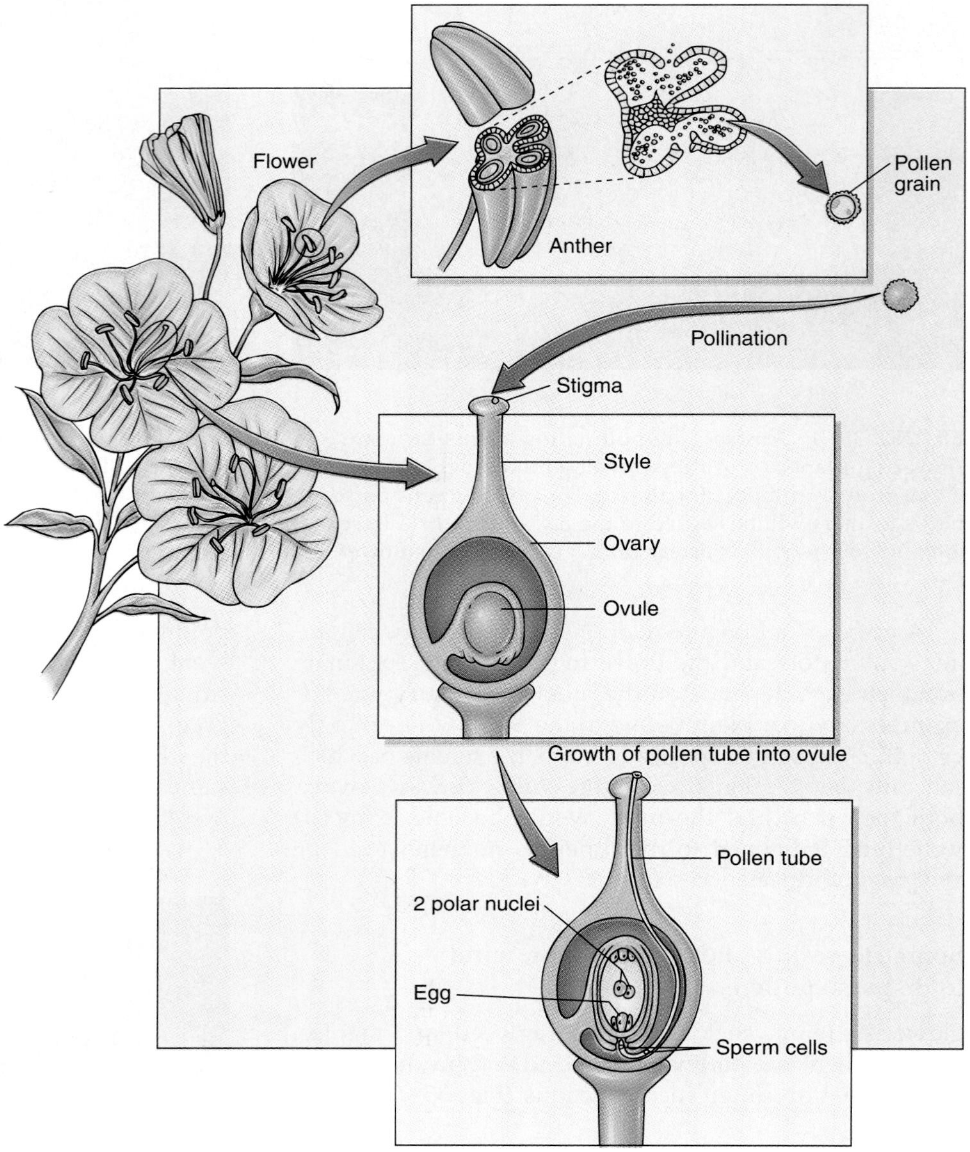

FIGURE 35–7 After pollination, a pollen tube grows through the style to the ovule in the ovary. Two sperm move down the pollen tube and enter the ovule, where they participate in double fertilization.

FERTILIZATION OCCURS, FOLLOWED BY SEED AND FRUIT DEVELOPMENT

Once pollen grains have been transferred from anther to stigma, one of the two cells in the pollen grain grows a thin pollen tube down through the style and into an ovule in the ovary. The second cell within the pollen grain divides to form two male gametes (the sperm), which move down the pollen tube and enter the ovule (Fig. 35–7; see also Fig. 27–9).

A unique double fertilization process occurs in flowering plants

The egg within the ovule unites with one of the sperm, forming a zygote (fertilized egg) that will develop into an embryonic plant contained in a seed. The two polar nuclei in the ovule fuse with the second sperm to form the first cell of the **endosperm,** the nutritive tissue that surrounds the embryonic plant in a seed. This process, in which two separate cell fusions occur, is called **double fertilization.** It is, with one exception, unique to flowering plants. (A type of double fertilization has been reported in *Ephedra,* a gymnosperm.)

After double fertilization has occurred, the ovule develops into a seed and the ovary surrounding it develops into a fruit. (Double fertilization and the flowering plant life cycle are discussed in greater detail in Chapter 27.)

Embryonic development in seeds is orderly and predictable

Flowering plants produce a young plant embryo complete with stored nutrients in a compact package, the

Focus On

Seed Banks

More than 100 seed collections, called *seed banks,* exist around the world and collectively hold more than 3 million samples of thousands of different kinds of plants (see figure). The U.S. Department of Agriculture's seed bank at Fort Collins, Colorado, for example, stores seeds of about 200,000 different species and varieties.

Most seed banks help to preserve the genetic variation within different varieties of crops. Farmers typically discontinue planting local varieties when newer, improved varieties become available. The newer varieties have desirable genetic characteristics—a greater yield, for example—but the local, discarded varieties also contain valuable genes. Maintaining the genetic diversity present in local crop varieties will help provide genes that we may need in the future.

Seed banks offer the advantage of storing a large amount of live plant genetic material in a very small space. Seeds stored in seed banks are safe from habitat destruction, climatic changes, and general neglect. There have even been some instances of using seeds from seed banks to reintroduce a plant species that has become extinct in the wild.

There are some disadvantages to seed banks. Seeds do not remain alive indefinitely and must be germinated periodically so that new ones can be collected. In addition, accidents such as fires or power failures can result in the permanent loss of the genetic diversity represented in a seed bank. Also, many types of plants—avocados and coconuts, for example—cannot be stored as seeds. The seeds of these plants do not tolerate being dried out, which is a necessary step before they can be frozen. (If 20% or more moisture remains in a frozen seed, it will probably die.) Some seeds cannot be stored successfully because they only remain viable for a short period of time—a few months or even just a few days.

Perhaps the most important disadvantage of seed banks is that plants stored in this manner remain stagnant in an evolutionary sense. They do not adapt in response to changes in their natural environments. As a result, they may be less fit for survival when they are reintroduced into the wild.

Despite their shortcomings, seed banks are increasingly viewed as an important way to safeguard seeds for future generations. Other international efforts to preserve plant genetic diversity are also being planned and implemented. For example, some farmers may soon be paid to set aside some of their land for cultivating local varieties of crops, thereby preserving the genetic diversity in agriculturally important plants. ■

These small vials of seeds are stored in the seed bank in Svalbard, Norway. (Courtesy of Nordiska Genbanken, Alnarp, Svenge)

seed, that develops after fertilization occurs. The nutrients in seeds are not only used by germinating plant embryos but also provide food for animals, including humans (see *Focus On: Seed Banks*). Development of the embryo following fertilization is possible because of the constant flow of nutrients into the developing seed from the parent plant.

Cell divisions of the fertilized egg to form a multicellular embryo proceed in a variety of ways in flowering plants. The following description is of dicot embryonic development; monocot embryonic development is identical in the early stages.

The two cells formed as a result of the first division of the fertilized egg establish polarity, or direction, in the embryo. The bottom cell (toward the outside of the ovule) typically develops into a **suspensor,** which is a multicellular structure that anchors the embryo and aids in nutrient uptake from the endosperm. The top cell (toward the inside of the ovule) develops into the actual embryo.

Initially, the top cell divides to form a short chain of cells, called a *proembryo* (Fig. 35–8). As cell division continues, a small ball of cells, often called a *globular embryo,* develops. Cells begin to develop into specialized tissues during this stage. When the dicot embryo starts to develop its two cotyledons (seed leaves), it has two lobes and resembles a heart; this is often called the *heart stage.* (Because monocot embryos develop a single cotyledon, they are more cylindrical at this stage.) During the *torpedo stage,* the embryo continues to grow as the cotyledons elongate. As the embryo enlarges, it often curves back on itself and crushes the suspensor beyond recognition. To summarize, stages of embryonic development in dicots include:

proembryo → globular embryo →
heart stage embryo → torpedo stage embryo

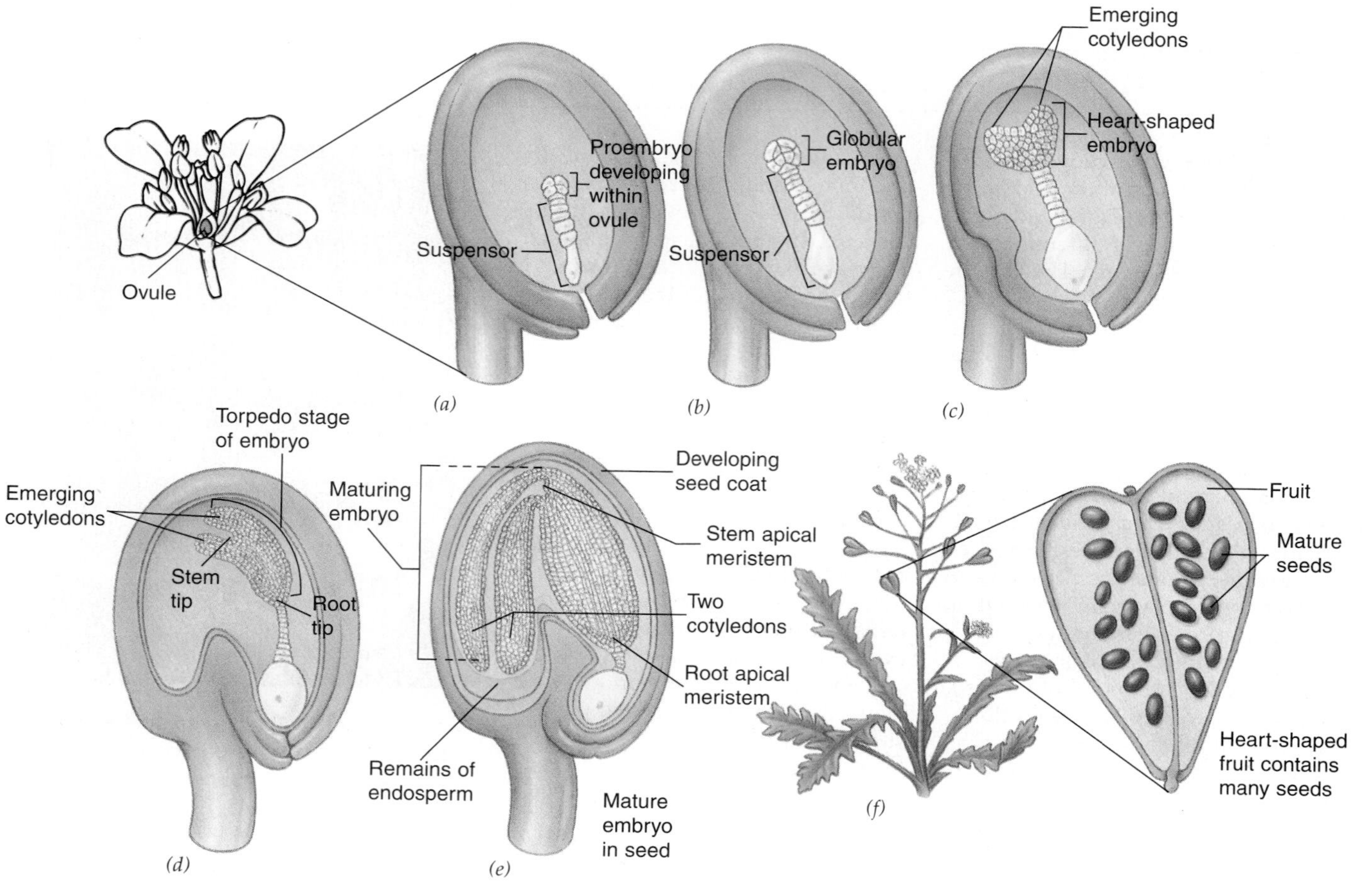

FIGURE 35–8 Shepherd's purse, a common dicot weed, is an excellent model for embryonic development in flowering plants. (*a*) The proembryo is the earliest multicellular stage of the embryo (shown outside the ovule). (*b*) As cell division continues, the embryo becomes a ball of cells, called the globular stage. (*c*) As the two cotyledons begin to emerge, the embryo is shaped like a heart. (*d*) The cotyledons continue to elongate, forming the torpedo stage. (*e*) A mature embryo within the seed. The food originally stored in the endosperm has been almost completely depleted during embryonic growth and development. Most of the food for the embryonic plant is in the two cotyledons. (*f*) A longitudinal section through a heart-shaped fruit of shepherd's purse reveals numerous tiny seeds, each containing a mature embryo.

The mature seed contains an embryonic plant and storage materials

A mature seed contains an embryonic plant and food (stored in either the cotyledons or endosperm), surrounded by a tough, protective **seed coat.** The seeds in turn are enclosed within a fruit.

The mature embryo within the seed consists of a short embryonic root, or **radicle**; an embryonic shoot; and one or two seed leaves, or **cotyledons.** Monocots have a single cotyledon, whereas dicots have two cotyledons (Fig. 35–9; see also Fig. 27–7). The short portion of the embryonic shoot connecting the radicle to one or two cotyledons is known as the **hypocotyl.** The shoot apex, or terminal bud, located above the point of attachment of the cotyledon(s) is the **plumule.**

Because the embryonic plant is nonphotosynthetic, it must have nutrients during germination (very early growth) until it becomes self-sufficient. (Seed germination and early growth are discussed in Chapter 36.) The cotyledons of many plants function as storage organs and become large and thick as they absorb the food reserves (starches, oils, and proteins) initially produced as endosperm. These seeds have little or no endosperm at maturity. Examples of seeds with thick, fleshy cotyledons are peas, beans, squash seeds, sunflower seeds, and peanuts. Other plants—wheat and corn, for example—have thin cotyledons that function primarily to help the young plant digest and absorb food stored in the endosperm.

Fruits are mature, ripened ovaries

After double fertilization takes place within the ovule, the ovule develops into a seed (just described), and the ovary surrounding it develops into a **fruit.** For example, a pea

Making the Connection

Seed Size and Ecology

How large should seeds be? Seed size varies considerably in flowering plants, from the microscopic, dustlike seeds of orchids to the giant seeds of the double coconut *(Lodoicea seychellarum)*, which weigh as much as 27 kilograms (almost 60 pounds). Despite this variation among different species, seed size is a remarkably constant trait within a species.

Assuming that a given plant species invests a fixed amount of its energy in reproduction, is it more advantageous to produce a large number of small seeds, or a few big ones? By observing the seed sizes that predominate in different environments, biologists have concluded that in some environments, a smaller seed appears to be advantageous, whereas in others, a larger seed may be better.

For example, plants that grow in widely scattered open sites (such as old fields) usually produce smaller seeds, perhaps because they can be more easily dispersed over large areas than can larger seeds. On the other hand, wide dispersal is probably less important for plants adapted to densely vegetated areas such as forests. These plants generally produce bigger seeds with an ample food reserve that may confer a greater likelihood of becoming successfully established in a shaded environment.

Other ecological factors are associated with seed size. Larger seeds are typical of plants that live in arid habitats (possibly because the food stored in a large seed allows a young seedling to establish an extensive root system quickly, thereby enabling it to survive the dry climate). Island plants also produce larger seeds than similar species on the nearby mainland. In this case, biologists hypothesize that large seeds are less likely to be widely dispersed (and therefore less likely to fall into the ocean, where chances of survival are slim). On the other hand, seeds that remain dormant in the soil for a long period of time tend to be smaller. (The exact reason for this phenomenon is not known.)

Despite many observations that associate seed size with specific environments, the general principles concerning the adaptive advantages of large seeds versus small seeds remain to be determined. ▲

pod is a fruit, and the peas within it are seeds. A fruit may contain one or more seeds; some orchid fruits contain several thousand to a few million! Fruits provide protection for the enclosed seeds and sometimes aid in their dispersal.

There are several types of fruits; their differences result from variations in the structure or arrangement of the flowers from which they were formed. The four basic types of fruits are simple fruits, aggregate fruits, multiple fruits, and accessory fruits (Table 35–1).

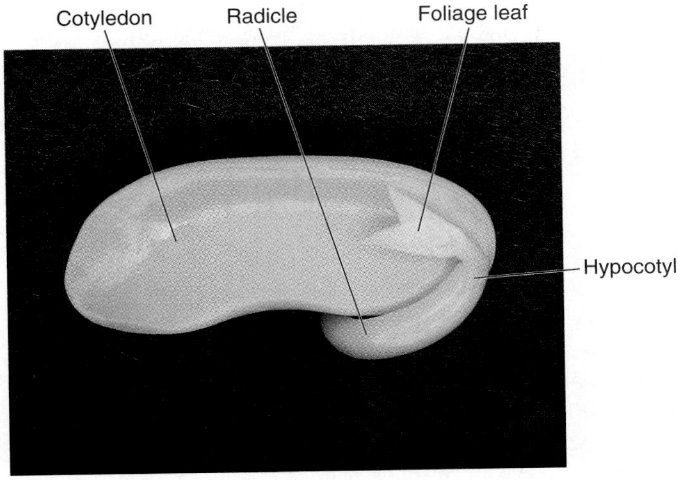

FIGURE 35–9 A bean seed has been dissected to show the radicle, hypocotyl, cotyledon, and foliage leaf produced at the shoot apex. Its seed coat and one of its two cotyledons were removed. This particular seed had begun germinating, so its radicle is larger than in an ungerminated seed. (James Mauseth, University of Texas)

Most fruits are simple fruits (Fig. 35–10). A **simple fruit** develops from a single pistil (which may consist of a single carpel or several fused carpels). At maturity, simple fruits may be fleshy or dry. Two examples of simple, fleshy fruits are berries and drupes. A **berry** is a fleshy fruit that has soft tissues throughout and contains few to many seeds; a blueberry is a berry, as are grapes, cranberries, bananas, and tomatoes (Fig. 35–10*a*). Many so-called berries do not fit the botanical definition. Strawberries, raspberries, and mulberries, for example, are not berries.

A **drupe** is a simple, fleshy (or fibrous) fruit that contains a hard, stony pit surrounding a single seed (Fig. 35–10*b*). Examples of drupes include peaches, plums, olives, avocados, and almonds (the almond shell is actually the stony pit, which remains after the rest of the fruit has been removed).

Many simple fruits are dry at maturity; some of these split open, usually along seams, to release their seeds. A milkweed pod is an example of a **follicle,** a simple, dry fruit that splits open along *one* side to release its seeds. A **legume** is a simple, dry fruit that splits open along *two* sides (the top and bottom). Pea pods are legumes, as are green beans, although both are generally harvested before the fruit has dried out and split open. Pea seeds are usually removed from the fruit and consumed, whereas

Table 35–1 THE MAIN TYPES OF FRUITS

I. Simple fruit—develops from a flower with a single pistil
   - A. Fleshy
     1. Berry—soft and fleshy throughout; usually has many seeds (tomato, grape)
     2. Drupe—hard, stony pit; single-seeded (peach, cherry)
   - B. Dry, splits open to release seeds
     1. Follicle—splits along one side (milkweed, columbine)
     2. Legume—splits along two sides (bean, pea)
     3. Capsule—splits along many sides or pores (poppy, iris)
   - C. Dry, does not split open
     1. Grain—single-seeded; seed fully fused to fruit wall (wheat, corn)
     2. Achene—single-seeded; seed attached to fruit wall at base only (sunflower)
     3. Nut—hard, thick fruit wall (acorn, chestnut)

II. Aggregate fruit—develops from a flower with many separate ovaries (raspberry, blackberry)

III. Multiple fruit—develops from many flowers borne together on a common floral stalk; the ovaries of these flowers fuse together to form a single fruit (pineapple, mulberry)

IV. Accessory fruit—develops from a flower in which the receptacle or floral tube enlarges and becomes part of the mature fruit (apple, strawberry)

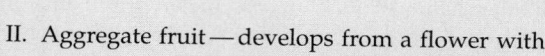

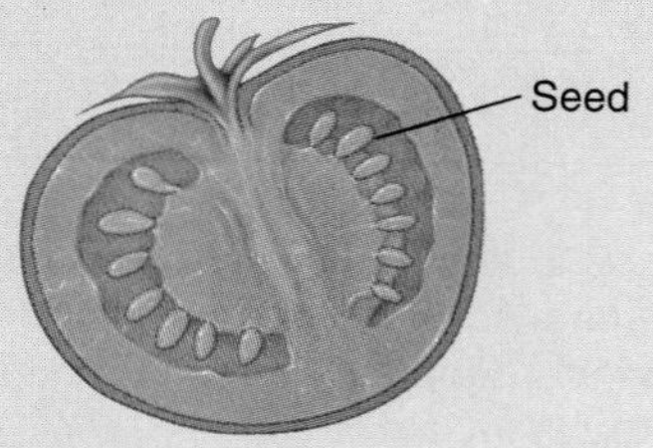

BERRY

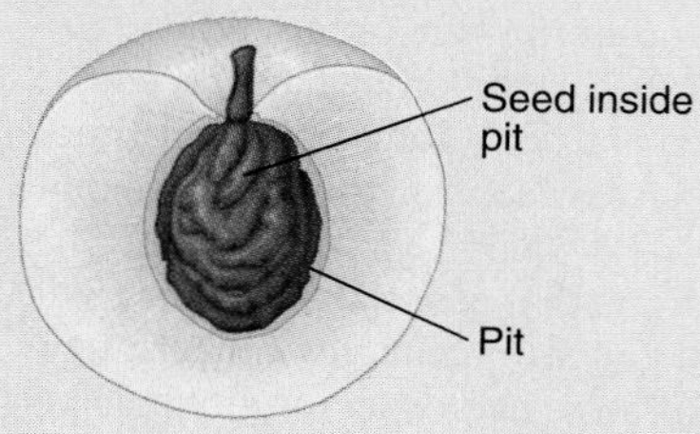

DRUPE

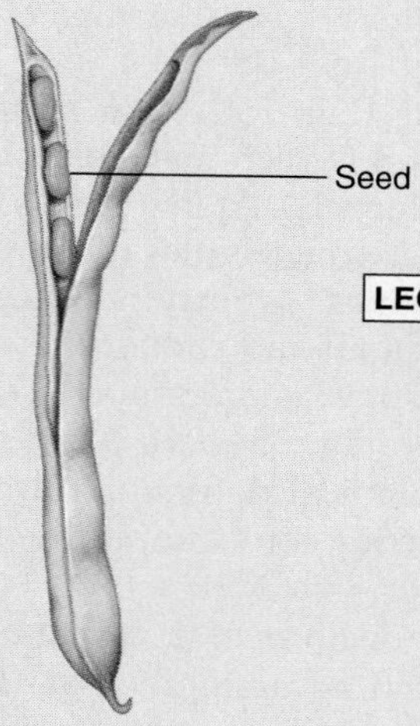

LEGUME

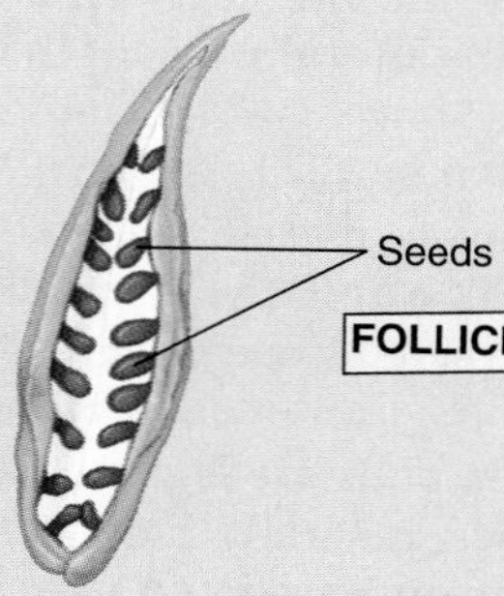

FOLLICLE

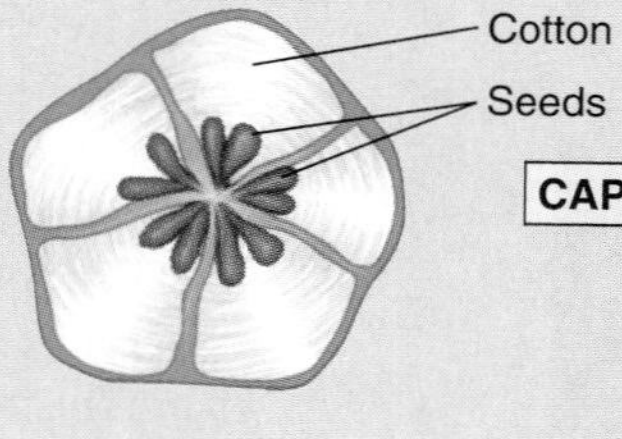

CAPSULE

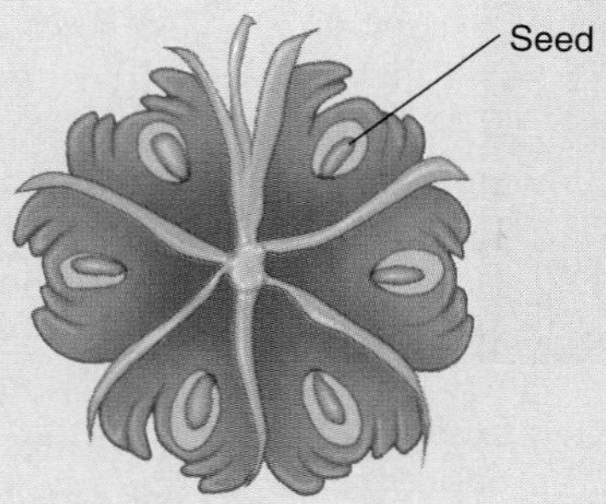

MULTIPLE FRUIT

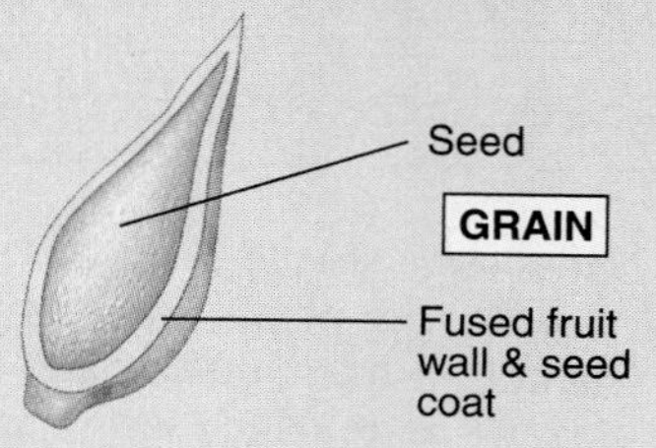

GRAIN

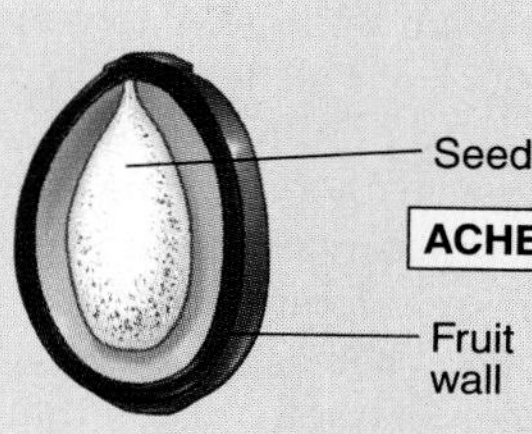

ACHENE

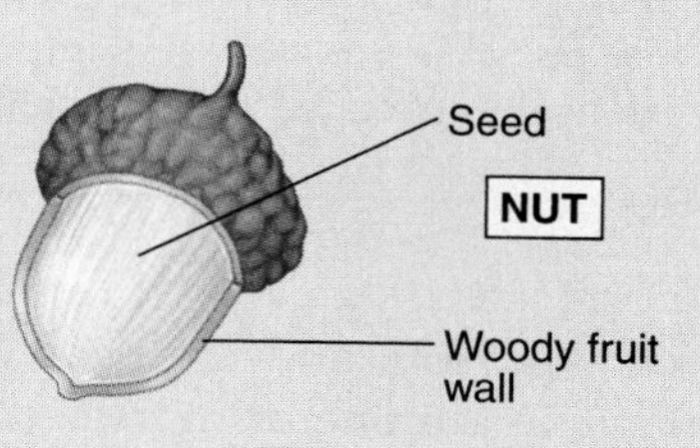

NUT

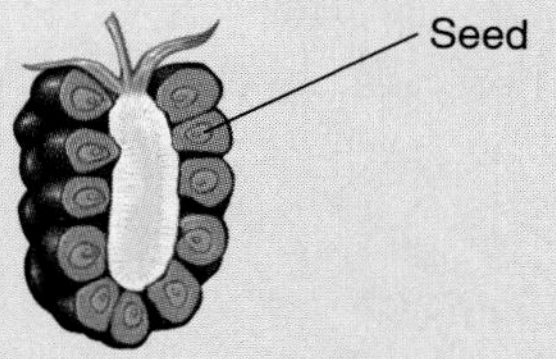

AGGREGATE FRUIT

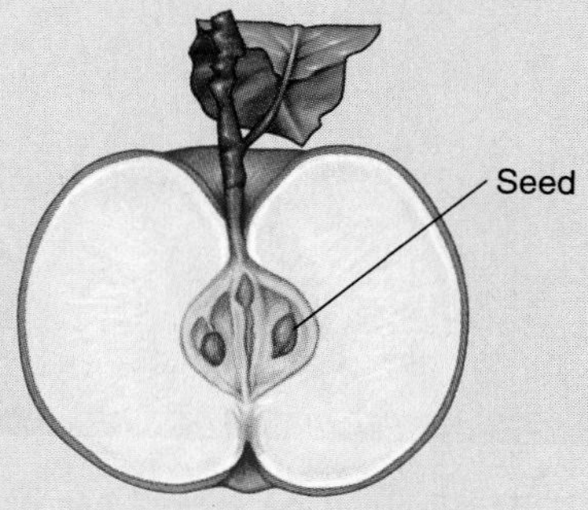

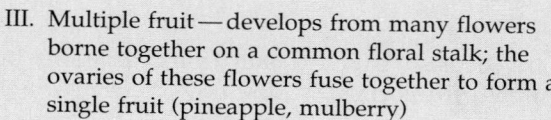

ACCESSORY FRUIT

(a)

(b)

(c)

(d)

FIGURE 35–10 Simple fruits, such as berries, drupes, capsules, and grains, are either fleshy or dry at maturity. (***a***) The tomato, a berry, is a simple fruit composed of soft tissues throughout. (***b***) The peach is a drupe. The hard, stony pit that surrounds the single seed is part of the fruit. (***c***) A capsule is a simple, dry fruit that splits open along multiple sides or pores at maturity. These iris capsules split along three sides. (***d***) Corn fruits are grains. In grains the fruit wall is fused to the seed coat. (*a, d,* Dennis Drenner; *b,* Carlyn Iverson; *c,* G. R. Roberts)

in green beans the entire fruit is eaten. A **capsule** is a simple, dry fruit that splits open along *multiple* sides or pores (Fig. 35–10*c*). Poppy and cotton fruits are capsules.

Other simple, dry fruits—**grains,** for example—do not split open at maturity (Fig. 35–10*d*). Each grain contains a single seed. Because the seed coat is fused to the fruit wall, it appears that a grain is a seed rather than a fruit. Kernels of corn and wheat are fruits of this type.

An **achene** is similar to a grain in that it is simple, dry, does not split open at maturity, and contains a single seed. However, the seed coat of an achene is not fused to the fruit wall. Instead, the single seed is attached to the fruit wall at one point only, permitting an achene to be separated from its seed. The sunflower fruit is an example of an achene. The fruit wall (the shell) can be peeled off to obtain the sunflower seed within.

Nuts are simple, dry fruits that have a stony fruit wall and do not split open at maturity. Unlike achenes, which are small fruits derived from a simple pistil, nuts are usually large and often derived from a compound pistil. Examples of nuts include chestnuts, acorns, and hazelnuts. Many so-called nuts do no fit the botanical definition. Peanuts and Brazil nuts, for example, are seeds, not nuts.

Aggregate fruits are a second main type of fruit. An **aggregate fruit** is formed from a single flower that contains several to many separate (free) carpels (Fig. 35–11). After fertilization, each ovary from each individual carpel enlarges. As they enlarge, the ovaries may fuse together

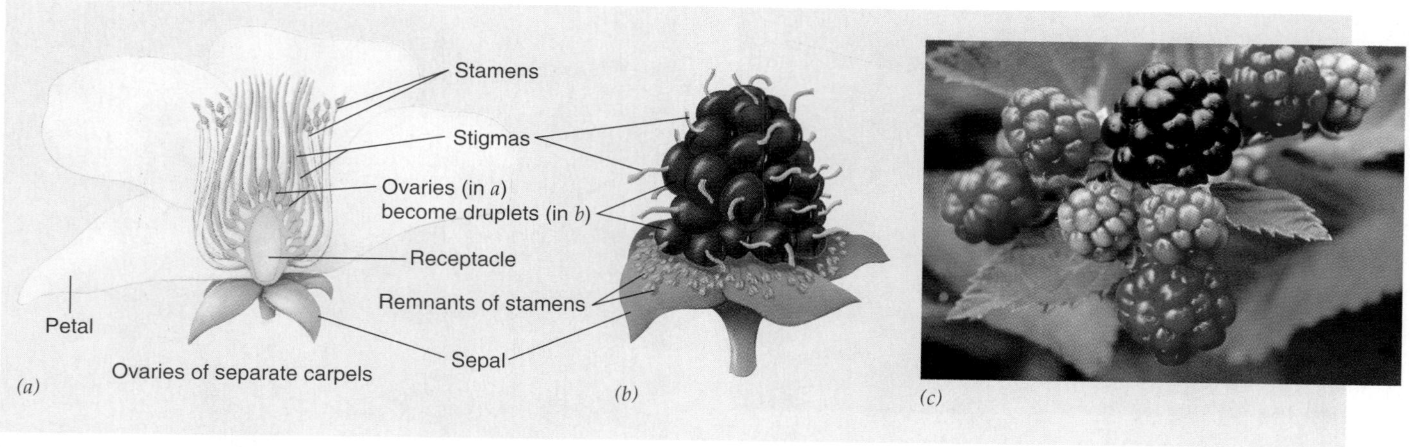

FIGURE 35–11 Blackberries are aggregate fruits. (***a***) Cutaway view of a blackberry flower, showing the many separate carpels in the center of the flower. (***b***) A developing blackberry fruit is an aggregate of tiny drupes. The little "hairs" on the blackberry are remnants of stigmas and styles. (***c***) Developing fruits at various stages of maturity. (*c*, Dennis Drenner)

to form a single fruit. Raspberries, blackberries, and magnolia fruits are examples of aggregate fruits.

A third type is the **multiple fruit,** formed from the ovaries of many flowers that grow in close proximity on a common floral stalk. The ovary from each flower fuses with nearby ovaries as it develops and enlarges after fertilization. Pineapples, figs, and mulberries are multiple fruits (Fig. 35–12).

Accessory fruits are the fourth type. They differ from other fruits in that other plant tissues in addition to ovary tissue make up the fruit. For example, the edible portion of a strawberry is the red, fleshy receptacle. Apples and pears are also accessory fruits; the outer part of each is an enlarged *floral tube* (consisting of receptacle tissue along with portions of the calyx) that surrounds the ovary (Fig. 35–13).

FIGURE 35–12 The pineapple is a multiple fruit formed by fusion of the ovaries of many separate flowers. (Courtesy of Dole Packaged Foods Company)

Seed dispersal is highly varied in flowering plants

The seeds and fruits of flowering plants are dispersed by wind, animals, water, and explosive dehiscence. Effective methods of seed dispersal have made it possible for certain plants to expand their geographical range. In some cases, the seed is the actual agent of dispersal, whereas in others, the fruit performs this role. In tumbleweeds, such as Russian thistle, the *entire plant* is the agent of dispersal because it detaches and blows across the ground, scattering seeds as it bumps along. Tumbleweeds are lightweight but have a great deal of wind resistance and are sometimes blown many kilometers by the wind.

Wind is responsible for seed dispersal in many plants. Plants such as maple have winged fruits adapted for wind dispersal (Fig. 35–14*a*). Light, feathery plumes are other structures that allow seeds or fruits to be transported by wind, often for considerable distances. Both dandelion fruits and milkweed seeds have this type of adaptation.

Some plants have special structures that aid in dispersal of their seeds and fruits by animals. The spines and barbs of burdock burs and similar fruits catch in animal fur and are dispersed as the animal moves about (Fig. 35–14*b*). Fleshy, edible fruits are also adapted for animal dispersal. As these fruits are eaten, the seeds are either discarded or swallowed. Seeds that are swallowed have thick seed coats and are not digested, but instead are passed through the digestive tract and deposited with the animal's feces some distance away from the parent plant. In fact, some seeds will not germinate unless they have passed through an animal's digestive tract. (Interestingly, a 1994 paper published in the journal *Ecology* reported that some edible fruits contain laxatives to *speed* the passage of seeds through the animal's digestive tract. The

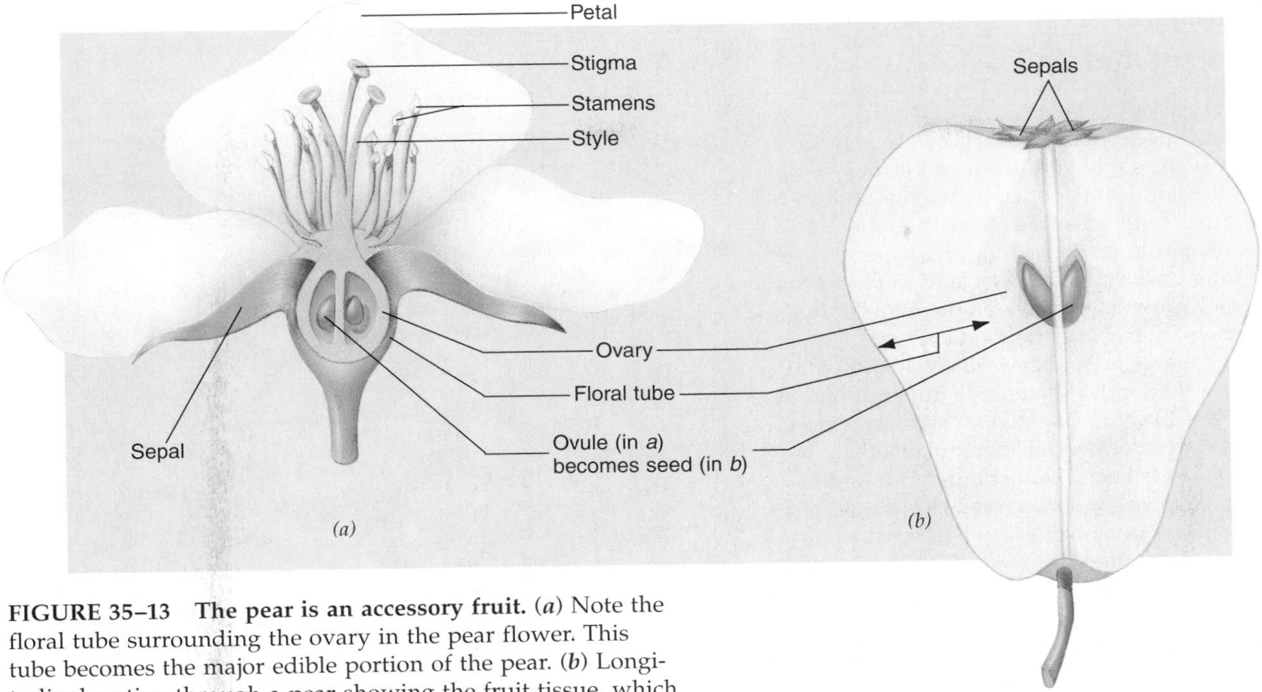

FIGURE 35–13 The pear is an accessory fruit. (*a*) Note the floral tube surrounding the ovary in the pear flower. This tube becomes the major edible portion of the pear. (*b*) Longitudinal section through a pear showing the fruit tissue, which is derived from both the floral tube and the ovary.

(*a*)

(*b*)

FIGURE 35–14 Seeds (and fruits) are dispersed in a variety of ways. (*a*) Maple fruits have wings that rotate and float in the wind, enabling them to travel a short distance from the parent tree. (*b*) Burdock burs (the hooked fruits) are carried from the parent plant after they become matted in bird feathers, mammal fur, or human clothing. (*a*, James Mauseth, University of Texas; *b*, DPA/Dembinsky Photo Associates)

Making the Connection

Seed Dispersal and Ants

How does a plant species disperse its seeds to places where they can successfully germinate and grow? As discussed in this chapter, plants possess a variety of dispersal methods that increase the chances of seeds landing in suitable locations. Regardless of how they are dispersed from the parent plant, however, most seeds land in places that are unsuitable for growth, or are eaten by animals such as mice and squirrels shortly after being dispersed.

Some plants have increased seed survival because of a dispersal method that buries their seeds underground. Such seeds are less likely to be eaten by animals. The role of burying seeds is performed for many plants by ants, which collect the seeds and take them underground to their nests. Ants disperse and bury seeds for hundreds of different plant species in almost every terrestrial environment, from northern coniferous forests to tropical rain forests to deserts.

Both ants and flowering plants benefit from their association. The ants ensure the reproductive success of the plants whose seeds they bury, and the plants supply food to the ants. A seed that is collected and taken underground by ants often contains a special structure called an *elaiosome,* or *oil body,* that protrudes from the seed (see figure). Elaiosomes are a nutritious food for ants, which carry seeds underground before removing the elaiosome. Once an elaiosome is removed from a seed, the ants discard the undamaged seed in an underground refuse pile, which happens to be rich in organic material (such as ant droppings and dead ants) and contains the minerals required by young seedlings. Thus, ants not only bury the seeds away from animals that might eat them, but also place the seeds in rich soil that is ideal for germination and seedling growth. ▲

Bloodroot (*Sanguinaria canadensis*) seeds have elaiosomes, or oil bodies. The brown part is the seed proper, and the white part is the elaiosome. The seeds have been placed on an oak leaf to indicate scale. (Marion Lobstein)

longer these seeds spend in the digestive tract, the less likely they are to germinate.)

Animals such as squirrels and many bird species also help to disperse acorns and other fruits and seeds by burying them for winter use. Many buried seeds are never used by the animal and germinate the following spring.

The coconut is an example of a fruit adapted for dispersal by water. It has air spaces and corky floats that make it buoyant and capable of being carried by ocean currents for thousands of kilometers. When it washes ashore, the seed within it germinates and grows into a coconut palm tree.

Some seeds are dispersed by neither wind, animals, nor water, but instead are found in fruits that use *explosive dehiscence,* in which the fruit bursts open suddenly and quite often violently, forcibly discharging its seeds (Fig. 35–15). Pressures due to differences in turgor or to drying out cause these fruits to burst open. The fruits of plants such as touch-me-not and bitter cress split open so explosively that seeds are scattered a meter or more.

Asexual reproduction in flowering plants may involve modified stems, leaves, or roots

Flowering plants have many kinds of asexual reproduction, a number of which involve modified stems (rhizomes, tubers, bulbs, corms, and stolons). A **rhizome** is a horizontal underground stem; it may be fleshy, indicating that it is used for storing food materials such as starch (Fig. 35–16). Although rhizomes resemble roots, they are really stems, as indicated by the presence of scalelike leaves, buds, nodes, and internodes. Rhizomes frequently

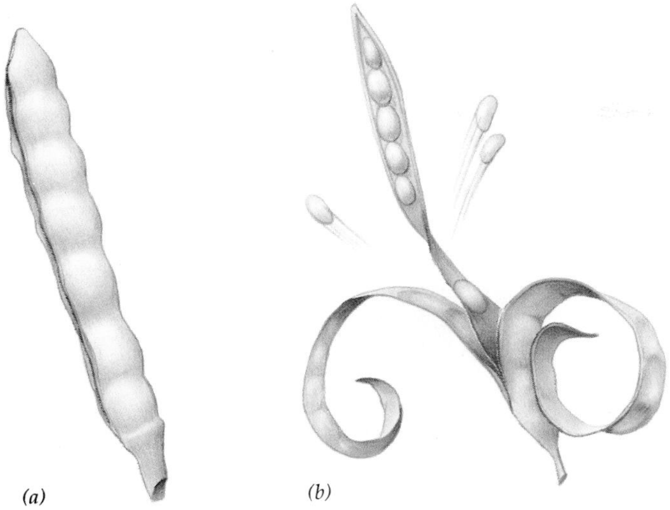

FIGURE 35–15 Bitter cress (*Cardamine*) dehisces explosively. (*a*) An intact fruit before it has opened. (***b***) The fruit splits open with explosive force, propelling the seeds some distance from the plant.

branch in different directions. Over time, the old portion of the rhizome dies, and two branches eventually separate to become two distinct plants. Irises, bamboos, and many grasses are examples of plants that reproduce asexually by forming rhizomes.

Another underground stem is the **tuber,** which is greatly enlarged for food storage. When the attachment between a tuber and its parent plant breaks (often due to the death of the parent), the tuber grows into a separate plant. White potatoes and elephant's ear (*Caladium*) are examples of plants that produce tubers (Fig. 35–17). The "eyes" of a potato are actually lateral buds, evidence that the tuber is an underground stem rather than a storage root like sweet potatoes and carrots.

A **bulb** is a shortened underground bud in which fleshy storage leaves are attached to a short stem (Fig. 35–18*a*). Bulbs are globose (round) and are covered by paper-like bulb scales, which are modified leaves. They frequently form small daughter bulbs that are initially attached to the parent bulb; when the parent bulb dies and rots away, each daughter bulb can become established as a separate plant. Lilies, tulips, onions, and daffodils form bulbs.

A **corm** is an underground stem that superficially resembles a bulb (Fig. 35–18*b*). Unlike the bulb, whose food is stored in underground leaves, the corm's storage organ is a thickened underground stem covered by papery scales (modified leaves). Lateral buds that give rise to new corms frequently arise; the death of the parent corm separates these daughter corms, which then become established as separate plants. Plants that produce corms include crocus, gladiolus, and cyclamen.

Stolons, or runners, are horizontal above-ground stems that run along the ground's surface and typically have long internodes (Fig. 35–19). Buds develop along the stolon, and each bud gives rise to a new plant that

FIGURE 35–16 Irises have horizontal underground stems called rhizomes. New aerial shoots arise from buds that develop on the rhizome.

FIGURE 35–17 Potatoes form rhizomes, which enlarge into tubers at the ends. (Carlyn Iverson)

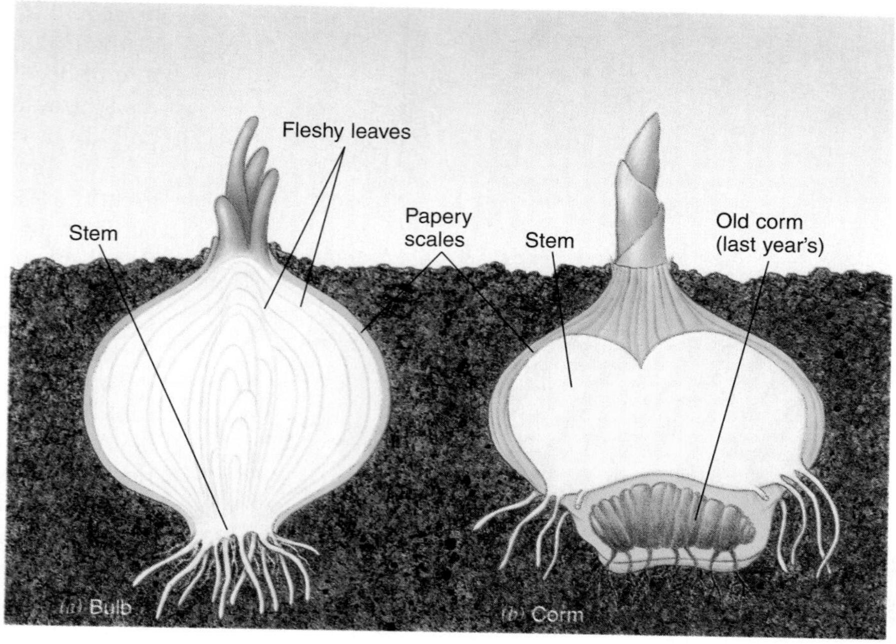

FIGURE 35–18 Bulbs and corms are underground stems that resemble one another. (*a*) A bulb is an underground stem to which overlapping, fleshy leaves are attached; most of the bulb consists of leaves. (*b*) A corm is almost entirely stem tissue surrounded by a few papery scales.

roots in the ground. When the stolon dies, the daughter plants are separated. The strawberry plant produces stolons.

Some plants are capable of forming **plantlets** (small plants) along their leaf margins. *Kalanchoe,* commonly called "mother of thousands," has meristematic tissue in the leaf that gives rise to an individual plantlet at each notch in the leaf (Fig. 35–20). When these plantlets attain a certain size, they drop to the ground, root, and grow.

Some plants reproduce asexually by producing **suckers,** above-ground stems that develop from adventitious buds on the roots. Each sucker grows additional roots and becomes an independent plant when the parent plant dies. Examples of plants that form suckers include black locust, apple, cherry, blackberry, and aspen (Fig. 35–21). Some weeds—field bindweed, for example—are able to produce a large number of suckers. These plants are difficult to control, because pulling the plant out of the soil

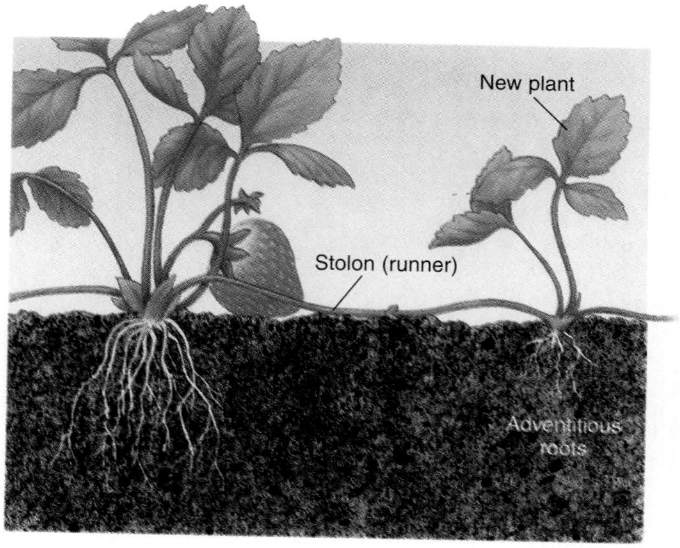

FIGURE 35–19 Strawberries reproduce asexually by forming stolons, or runners.

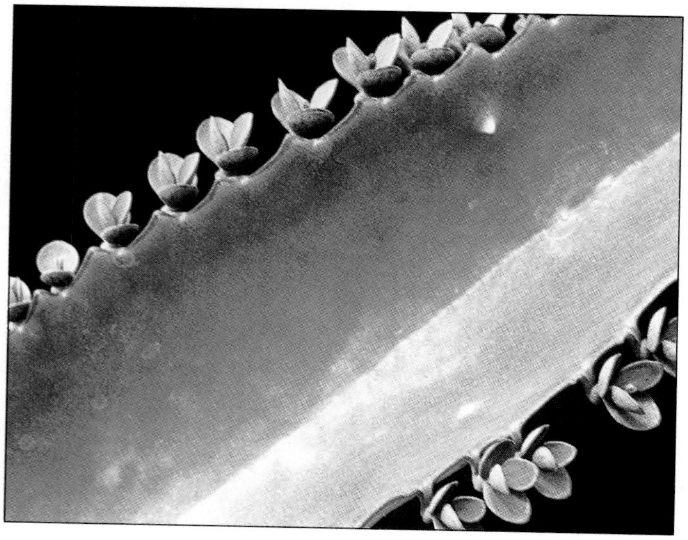

FIGURE 35–20 The "mother of thousands" (*Kalanchoe*) produces plantlets along the margins of its leaves. The young plantlets will drop off and root in the ground.
(Dennis Drenner)

FIGURE 35–21 A grove of aspen trees is often descended from a single tree that reproduced asexually by forming suckers. Because all the trees in the grove are genetically identical, their responses to the environment are uniform. In spring they break dormancy simultaneously, and in the fall their leaves turn color at the same time. (Sharon Cummings/Dembinsky Photo Associates)

seldom removes all of the roots. The roots of field bindweed can grow as deep as 3 meters (10 feet) into the soil. In fact, in response to wounding, the roots produce additional suckers that can be a considerable nuisance.

Apomixis is the production of seeds without the sexual process

Sometimes plants produce embryos in seeds without meiosis and fusion of gametes, a process known as **apomixis.** For example, an embryo may develop from a diploid cell in the ovule rather than from a diploid zygote that forms from the union of two haploid gametes. The seeds produced by apomixis are a form of asexual reproduction because there is no fusion of gametes. The embryo is genetically identical to the original parent. However, the advantage of apomixis over other methods of asexual reproduction is that the seeds and fruits produced can be dispersed by methods associated with sexual reproduction. Examples of plants that reproduce by apomixis include dandelions, citrus trees, blackberries, garlic, and certain grasses.

SUMMARY

I. The offspring produced by sexual reproduction are genetically variable.
  A. Sexual reproduction occurs in the flower.
    1. Sepals cover and protect the flower parts when the flower is a bud.
    2. Petals play an important role in attracting animal pollinators to the flower.
    3. Stamens produce pollen grains.
    4. Each carpel bears one or more ovules.
  B. Each pollen grain contains two cells. One generates two sperm, and the other produces a pollen tube through which the sperm will reach the ovule.
  C. An egg and two polar nuclei are formed in the ovule. Both egg and polar nuclei participate directly in fertilization.

II. Pollination is the transfer of pollen grains from anther to stigma.
  A. Some plants rely on animals to transfer pollen grains.
    1. Coevolution occurs when two different organisms (such as flowering plants and their animal pollinators) form an interdependent relationship and affect the course of one another's evolution.
    2. Flowers pollinated by insects are often yellow or blue and possess a scent.
    3. Bird-pollinated flowers are often yellow, orange, or red and do not have a strong scent.
    4. Bat-pollinated flowers often have dusky white petals and possess a scent.
  B. Plants pollinated by wind often have smaller petals or lack petals altogether, and do not produce a scent or nectar; wind-pollinated flowers make copious amounts of pollen.

III. After pollination, fertilization (fusion of gametes) occurs.
  A. A pollen tube grows down the style into the ovary, and the two sperm travel down it into the ovule.
  B. Flowering plants have a double fertilization.
    1. In the ovule, the egg fuses with one sperm, forming a fertilized egg that eventually develops into a multicellular embryo in the seed.
    2. The two polar nuclei fuse with the second sperm, forming a nutritive tissue called endosperm.

IV. The seed and fruit develop as a result of successful fertilization.
  A. A dicot embryo develops in the seed in an orderly fashion, from proembryo to globular embryo to the heart stage to the torpedo stage. The mature flowering plant embryo consists of a radicle, a hypocotyl, one or two cotyledons, and a plumule.
  B. A mature seed contains both a young embryo and nutritive tissue (stored in the endosperm or cotyledons) for use during germination.
  C. Seeds are enclosed within fruits, which are mature, ripened ovaries.
    1. Simple fruits develop from a single pistil that consists of one carpel or several fused carpels. Some simple fruits (berries, drupes) are fleshy at maturity, whereas

others (follicles, legumes, capsules, grains, achenes, nuts) are dry.
2. Aggregate fruits develop from a single flower with many separate ovaries.
3. Multiple fruits develop from many ovaries of many flowers growing in close proximity on a common axis.
4. In accessory fruits, the major part of the fruit consists of tissue other than ovary tissue.

D. Seeds and fruits are adapted for various means of dispersal, including wind, animals, water, and explosive dehiscence.

V. Asexual reproduction involves the formation of offspring without the fusion of gametes. The offspring are genetically similar to the single parent plant.
A. Rhizomes, tubers, bulbs, corms, and stolons are stems specialized for asexual reproduction.
B. Some leaves have meristematic tissue along their margins and give rise to plantlets.
C. Roots may develop adventitious buds that develop into suckers. Suckers develop additional roots and may give rise to new plants.
D. Apomixis is the production of seeds and fruits without sexual reproduction.

SELECTED KEY TERMS

accessory fruit
achene
aggregate fruit
anther
apomixis
berry
bulb
calyx
capsule
carpel
coevolution
corm
corolla
cotyledon
double fertilization
drupe
endosperm
filament
follicle
fruit
grain
hypocotyl
legume
multiple fruit
nectary
nut
ovary
ovule
petal
pistil
plantlet
plumule
polar nuclei
pollen tube
pollination
radicle
rhizome
seed
seed coat
sepal
simple fruit
stamen
stigma
stolon
style
sucker
suspensor
tuber

POST-TEST

1. The pistil is composed of a(an) __________, style, and ovary.
2. The petals of a flower are collectively called a(an) __________.
3. __________ involves the transfer of pollen grains from anther to stigma.
4. The observation that deep tubular flowers are pollinated by insects with long mouthparts, whereas short flowers are pollinated by insects with short mouthparts, is explained by __________.
5. The process of __________ __________ in flowering plants involves one sperm fusing with an egg cell and one sperm fusing with two polar nuclei.
6. __________, the nutritive tissue in the seeds of flowering plants, is formed from the union of a sperm and two polar nuclei.
7. The __________ is a multicellular structure that anchors the embryo and aids in nutrient uptake from the endosperm.
8. After fertilization an ovule develops into a(an) __________.
9. A(an) __________ may be defined as a mature, ripened ovary.
10. __________ fruits form from many ovaries of a single flower, whereas __________ fruits develop from many ovaries of many separate flowers.
11. Apples, strawberries, and pears are examples of __________ fruits.
12. A(an) __________ is a horizontal, underground stem that is specialized for asexual reproduction.
13. The white potato is an example of an underground storage stem called a(an) __________.
14. Place the following events in the correct order.
    (1) pollen tube grows into ovule
    (2) insect lands on flower to drink nectar
    (3) embryo develops within the seed
    (4) fertilization occurs
    (5) pollen carried by insect drops onto stigma

REVIEW QUESTIONS

1. What is the difference between pollination and fertilization? Which process occurs first?
2. Describe a "typical" insect-pollinated flower.
3. What is coevolution?
4. Put the following stages of embryonic development in order and briefly describe each: torpedo stage, globular stage, proembryo, and heart stage.
5. Distinguish among simple, aggregate, multiple, and accessory fruits, and give examples of each.
6. Explain some of the features possessed by seeds and fruits that are dispersed by animals.
7. Distinguish between sexual and asexual reproduction.
8. List and describe four modified stems that are involved in sexual reproduction.

YOU MAKE THE CONNECTION

1. Draw pictures to show the kinds of flowers that might form simple, aggregate, multiple, and accessory fruits.
2. Could seed dispersal by ants be considered an example of coevolution? Why or why not?
3. Which type of reproduction, sexual or asexual, might be more beneficial in the following circumstances, and why? (a) a perennial (plant that lives more than 2 years) in a stable environment; (b) an annual (plant that lives one year) in a rapidly changing environment; (c) a plant adapted to an extremely narrow climate range.

RECOMMENDED READINGS

Beattie, A.J. "Ant Plantation." *Natural History,* February 1990. Discusses the role that ants play in dispersing certain seeds.

Cox, P.A. "Water-pollinated Plants." *Scientific American,* Vol. 269, No. 4, October 1993. Some aquatic plants rely on water currents to transfer pollen to female flowers.

Fleming, T.H. "Cardon and the Night Visitors." *Natural History,* October 1994. Many cacti are pollinated by bats.

Handel, S.N., and A.J. Beattie. "Seed Dispersal by Ants." *Scientific American,* Vol. 263, No. 2, August 1990. How plants induce ants to disperse seeds without harming them.

Heinrich, B. "Of Bedouins, Beetles, and Blooms." *Natural History,* May 1994. In the Judean Desert, competition among various flowers for animal pollinators has resulted in convergent evolution. Many flower species have red petals (even buttercups, which are yellow elsewhere) and are pollinated by unusual beetles that can see red.

Kaerns, C.A., and D.W. Inouye. "Pistil-packing Flies." *Natural History,* April 1993. Flies are effective pollinators of many kinds of flowering plants.

Meyerowitz, E.M. "The Genetics of Flower Development." *Scientific American,* Vol. 271, No. 5, November 1994. Studies of normal and mutant *Arabidopsis* flowers have shown it possible to predict how mutations will alter the structure of flowers.

Moore, P.D. "Not Only Gone with the Wind." *Nature,* Vol. 35, No. 14, May 1992. Some plant species have *two* methods of seed dispersal instead of one. Violets, for example, have explosive dehiscence as their primary means of dispersal, and ants for secondary dispersal after the seeds have been scatteed by explosive dehiscence.

Wolinksy, C. "Wildflowers of Western Australia." *National Geographic,* Vol. 187, No. 1, January 1995. About 12,000 species of flowering plants bloom in Western Australia, many found nowhere else.

C H A P T E R 3 6

Growth Responses and Regulation of Growth

The growth and development of plants is linked to many environmental cues. Poinsettias form flowers in response to the shortening days and lengthening nights of late summer or early fall.

(David Cavagnaro)

The ultimate control of plant growth and development is genetic (see Chapter 16). If the genes required for development of a particular trait—for example, leaf shape, flower color, or type of root system—are not present, that characteristic does not develop. When a particular gene is present, its *expression,* that is, how the gene exhibits itself as an observable feature in a plant, is determined by a variety of factors. The *location* of a cell in the young plant body, for example, has a profound effect on gene expression during development, causing some genes in that cell to be turned off and others to be turned on. Thus, location helps to determine what each cell ultimately becomes.

Environmental cues (such as changing day length, variation in precipitation, and temperature) also exert an important influence on gene expression, as they do on all aspects of plant growth and development. The initiation of sexual reproduction, for example, is often under environmental control, particularly in temperate latitudes. Such control is important for the plant's survival because the timing of sexual reproduction is critical to reproductive success: all flowering plants in temperate climates must flower and form seeds before dormancy is induced by the onset of winter. A number of plants detect changes in the relative amounts of daylight and darkness that accompany the seasons, and flower in response to these changes. Other plants have temperature requirements that induce sexual reproduction.

In this chapter we consider the role of plant **hormones,** organic chemical compounds produced in one part of a plant and transported to another part, where they elicit some kind of physiological response. The production of hormones, which is under both genetic and environmental control, regulates all aspects of plant growth and development. Growth and development include seed germination and the growth of seedlings (young plants that develop from germinating seeds) into mature plants. A plant's responses to changes in various aspects of its environment, including temperature, light, gravity, and touch, are also a part of growth and development.

LEARNING OBJECTIVES

After you have studied this chapter you should be able to

1. Summarize the influence of environmental factors on the germination of seeds.
2. Discuss genetic and environmental factors that affect plant growth and development.
3. Explain how flowering is induced by varying amounts of light and darkness, and describe the role of phytochrome in flowering.
4. Distinguish between a turgor movement and a tropism, and describe phototropism, gravitropism, and thigmotropism.
5. Define circadian rhythm and give an example.
6. List several different ways each of the following hormones affects plant growth and development: auxins, gibberellins, cytokinins, ethylene, and abscisic acid.
7. Describe the hormone regulation of each of the following: seed dormancy, seed germination, apical dominance, stem elongation, fruit ripening, and leaf abscission.

BOTH EXTERNAL AND INTERNAL FACTORS AFFECT SEED GERMINATION AND EARLY GROWTH

In Chapter 35 we discussed how pollination and fertilization are followed by seed and fruit development. Each seed develops from an ovule and contains an embryonic plant, plus food reserves to provide nourishment for the embryo during germination. A mature seed—one in which the embryo is fully developed—is often dormant and may not germinate immediately even if growing conditions are ideal.

A number of factors influence whether or not a seed germinates. Many of these are environmental cues, including the presence of water and oxygen, proper temperature, and sometimes the presence of light penetrating the soil surface. For example, no seed germinates unless it has absorbed water, because the embryo in a mature seed is dehydrated. When a seed germinates, its metabolic machinery is turned on, and numerous materials are synthesized and degraded. Because cells require water for active metabolism, water is an absolute requirement for germination.

The absorption of water by a dry seed is known as **imbibition.** As a seed imbibes water, it often swells to several times its original size (Fig. 36–1). Cells imbibe water by osmosis (see Chapter 5) and by adsorption of water onto and into colloidal materials such as cellulose, pectin, and starches within the seed. Water is attracted to and bound to these materials by adhesion (see Chapter 2).

Seed germination and subsequent growth also require a great deal of energy. Because plants obtain this energy by converting the energy of food molecules stored in the seed's endosperm or cotyledons (see Chapter 35) into ATP by aerobic respiration (see Chapter 7), oxygen is usually needed during germination. Some plants such as rice are able to respire anaerobically during the early stages of germination and seedling growth. This enables rice plants to grow and become established in flooded soil, an environment that would suffocate most young plants.

Temperature is another environmental factor that affects germination. In a sample of seeds of the same species, some germinate at each temperature in a broad range of temperatures. Each plant species, however, has an optimal, or ideal, temperature at which the largest number of seeds germinate. For most plants, the optimal germination temperature falls between 25° and 30°C (77°–86°F). Some seeds, such as those of apples, require exposure to prolonged periods of cold before their seeds germinate at any temperature. Some plants, especially those with tiny seeds such as lettuce, require light for germination.

Some of the environmental factors needed for seed germination help ensure the survival of the young plant. The requirement of a prolonged cold period ensures that seeds adapted to temperate climates germinate in the spring rather than in the winter. A light requirement ensures that a tiny seed germinates only if it is close to the surface of the soil. If such a seed were to germinate sev-

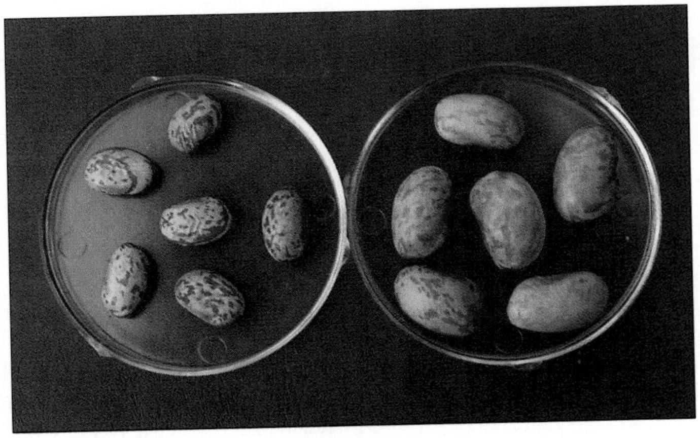

FIGURE 36–1 Dry seeds must imbibe water before they germinate. Pinto bean seeds (*a*) before and (*b*) after imbibition. (Marion Lobstein)

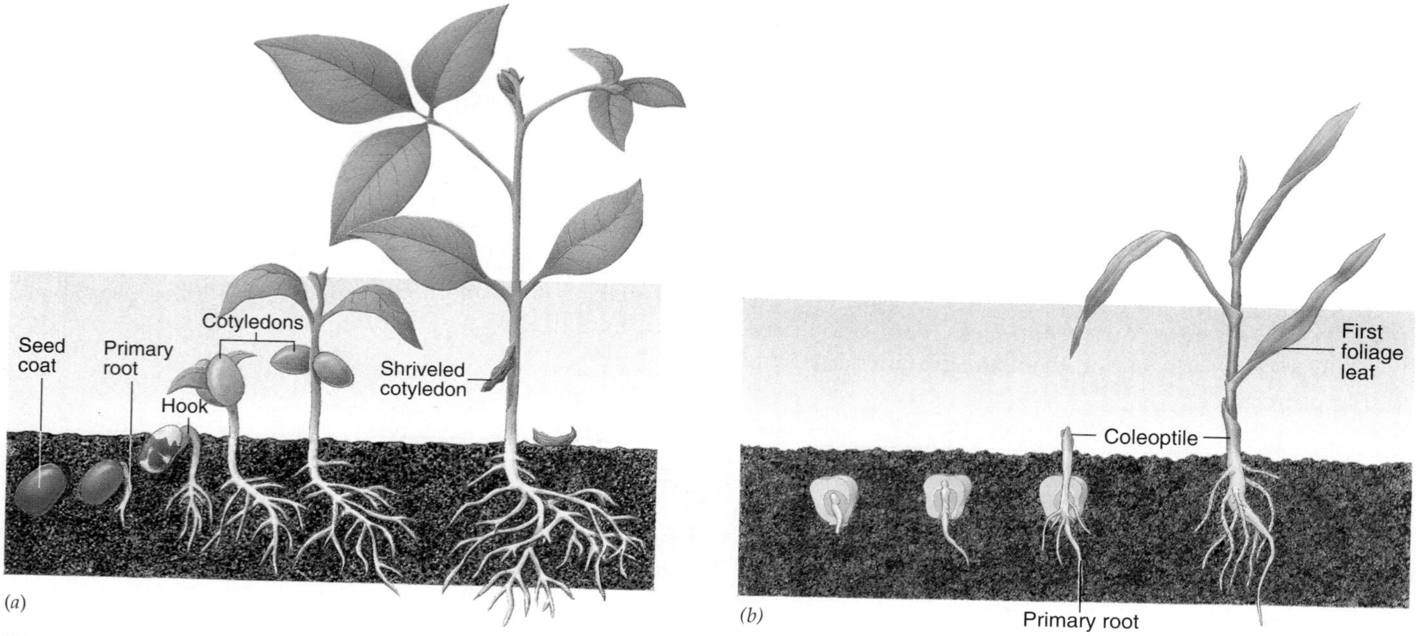

FIGURE 36–2 Young seedlings of dicots and monocots have different patterns of growth. (***a***) Seed germination and growth of a young soybean, a dicot. Note the hook in the stem, which protects the delicate stem tip as it moves up through the soil. Once the shoot has emerged from the soil, the hook straightens. (***b***) Seed germination and growth of a young corn plant. Note the coleoptile, a sheath of cells that emerges first from the soil. The shoot and leaves grow up through the middle of the coleoptile.

eral inches down, it might not have enough food reserves to grow to the surface. On the other hand, if this seed were to remain dormant until a soil disturbance brought it to the surface (where its germination would be triggered by light), it would have a much greater likelihood of survival.

In certain seeds, internal factors, which are under genetic control, prevent germination even when all external conditions are favorable. Many seeds are dormant either because the embryo is immature and must develop further, or because certain chemicals (such as abscisic acid, discussed later in the chapter) are present. Such chemical inhibitors help ensure the survival of the plant. The seeds of many desert plants, for example, often contain high levels of abscisic acid that is washed out only when rainfall is sufficient to support the plant's growth after the seed germinates.

Dicots and monocots exhibit characteristic patterns of early growth

Once conditions are right for seed germination, the first part of the plant to emerge from the seed is the embryonic root. As the root grows and forces its way through the soil, it encounters considerable friction from soil particles. The delicate apical meristem of the root tip is protected by a root cap (see Chapter 34).

Next to emerge from the seed is the shoot. Stem tips are not protected by a structure comparable to anything like a root cap, but plants have different ways to protect the delicate stem tip as it grows up through the soil to the surface. The stem of a bean seedling (a dicot), for instance, curves over to form a hook so that the stem tip and cotyledons are actually *pulled* up through the soil (Fig. 36–2*a*). Corn and other grasses (monocots) have a special sheath of cells called a **coleoptile** that surrounds the young shoot (Fig. 36–2*b*). First, the coleoptile pushes up through the soil, and then the leaves and stem grow up through the middle of the coleoptile.

Certain parts of a plant continue to grow throughout its life. This **indeterminate growth,** the ability to grow indefinitely, is characteristic of stems and roots. Other parts of a plant, such as leaves and flowers, have **determinate growth;** they stop growing after reaching a certain size. The size of these structures varies from species to species and from individual to individual, depending on the plant's genetic programming and on environmental conditions such as availability of sunlight, water, and essential minerals.

Environmental cues affect flowering and other plant responses

Photoperiodism is any response of a plant to the relative lengths of daylight and darkness. Flowering is one of several physiological activities that are photoperiodic in many plants. Plants are classified into three main groups—short-day, long-day, and day-neutral—on the basis of how photoperiodism affects their flowering.

Short-day plants were initially defined as plants that flower when the day length is equal to or less than some critical length. However, later experiments showed that the important factor in the initiation of flowering in short-day plants is *not* the short period of daylight but the long, uninterrupted period of darkness. In other words, *short-day plants flower when the night length is equal to or greater than some critical period* (Fig. 36–3*a*). Although the minimum critical night length varies among different short-day species, it is between 8 and 14 hours for most plants. Examples of short-day plants are chrysanthemum, aster, ragweed, and poinsettia, which flower when they detect the shortening days and lengthening nights of late summer or fall.

Long-day plants were initially defined as plants that flower when the day length is equal to or greater than some critical period. As with short-day plants, however, night length is the determining factor, and so a more accurate definition is that *long-day plants flower when the night length is equal to or less than some critical period* (Fig. 36–3*b*). Although the maximum critical night length varies among different long-day species, it is between 8 and 15 hours for most plants. Plants such as red clover, spinach, black-eyed Susan, and lettuce are long-day plants that flower when they detect the lengthening days and shortening nights of spring and early summer.

Some plants, called **day-neutral plants,** do not initiate flowering in response to seasonal changes in the period of daylight and darkness, but instead respond to some other type of stimulus, external or internal. Tomato, dandelion, string bean, grape, cucumber, corn, and pansy are examples of day-neutral plants. Many of these plants originated in the tropics where day length does not vary appreciably during the year.

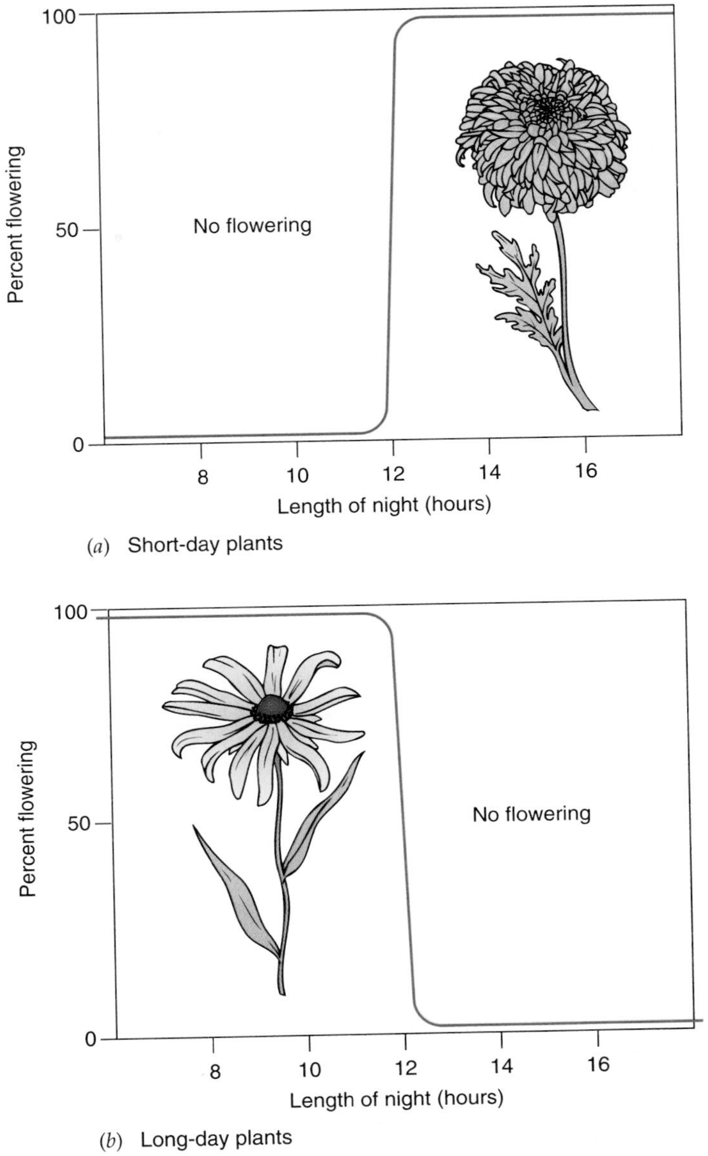

FIGURE 36–3 Short-day and long-day plants have different photoperiodic responses. (***a***) Short-day plants flower when the night length is equal to or exceeds a certain critical length. (***b***) Long-day plants flower when the night length is equal to or less than a certain critical length. The critical length, arbitrarily chosen as 12 hours in both (***a***) and (***b***), varies among different species.

Phytochrome detects varying periods of day length

For a plant or any other living organism to have a biological response to light, it must contain a light-sensitive substance, called a *photoreceptor*, to absorb the light. The photoreceptor for photoperiodism and a number of other light-initiated plant responses is a blue-green pigment called **phytochrome**. Phytochrome, which is present in cells of all vascular plants, occurs in two forms and readily converts from one form to the other after absorption of light of specific wavelengths. One form, designated P_r (for *r*ed-absorbing *p*hytochrome), strongly absorbs red light with a wavelength of 660 nanometers. In the process, the shape of the molecule changes to the second form of phytochrome, P_{fr} (Fig. 36–4). This form is so designated because it absorbs *f*ar-*r*ed light, which is red light with a wavelength of 730 nanometers. When P_{fr} absorbs far-red light, it reverts back to the original form, P_r. The P_{fr} form of phytochrome is less stable than the P_r form, so it slowly reverts to P_r in the dark.

What does a pigment that absorbs red light and far-red light have to do with daylight and darkness? Sunlight is composed of the entire spectrum of visible light, in addition to ultraviolet and infrared radiation. Because sunlight contains more red than far-red light, however, a

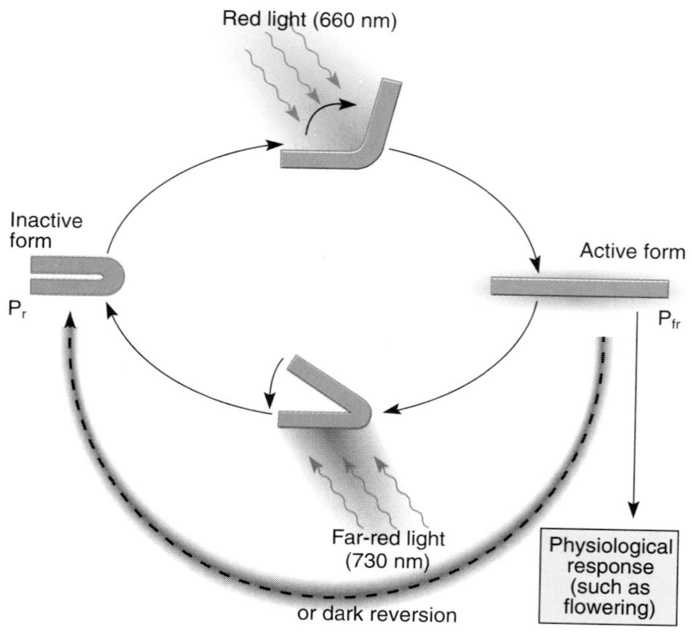

plant's level of P_{fr} increases when it is exposed to sunlight. During the night, its level of P_{fr} slowly decreases (as it reverts back to P_r).

In short-day plants, the P_{fr} form of phytochrome *inhibits* flowering, so that in order to flower, these plants need long nights. The long period of darkness allows the P_{fr} to revert back to P_r, so the plant has some minimum time during the 24-hour period with *no* P_{fr} present. This absence of P_{fr} initiates flowering. Biologists have experimented with short-day plants by growing them under a short-day/long-night regimen that includes a short burst of red light in the middle of the night (Fig. 36–5). Exposure to red light for a few minutes in the middle of the night will prevent flowering in short-day plants. This effect occurs because the brief exposure to red light con-

FIGURE 36–4 Phytochrome, a pigment that occurs in two forms designated P_r and P_{fr}, readily converts from one form to the other. Red light (660 nm) converts P_r to P_{fr}, while far-red light (730 nm) converts P_{fr} to P_r.

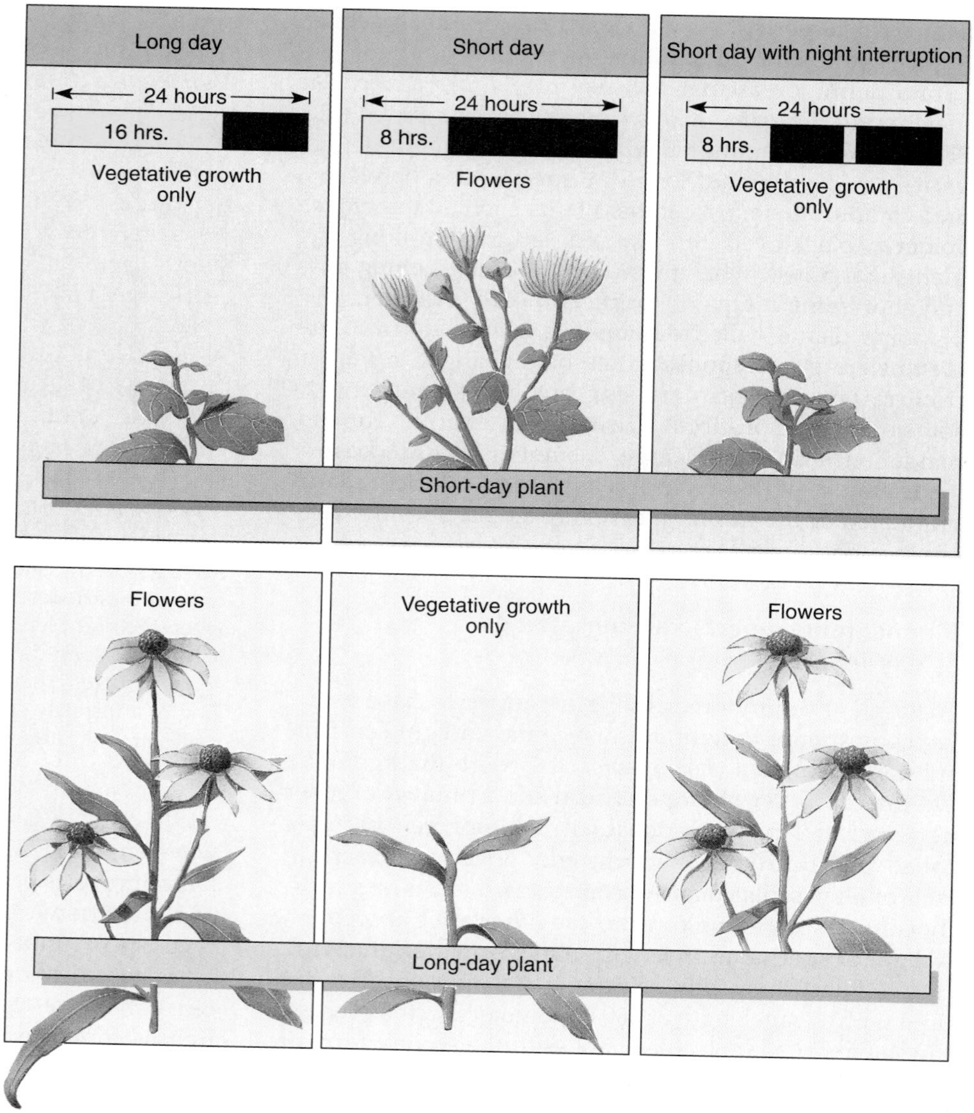

FIGURE 36–5 A short-day plant and a long-day plant differ in their photoperiodic responses to periods of light and dark. Note that the short-day plant does not flower when exposed to a short day (8 hours of daylight) and a long night (16 hours of darkness) interrupted with a brief flash of light. This same treatment induces the long-day plant to flower.

Making the Connection

Phytochrome and Competition Among Shade-Avoiding Plants for Sunlight

Can plants sense the presence of nearby plants? The answer is yes: they not only detect the presence of their neighbors, but they also react to nearby plants by changing the way they grow and develop.

Although plants do not possess real eyes, they are able to use their phytochrome "eye" to perceive changes in the ratio of red to far-red light that result from the presence of nearby plants. The leaves of neighboring plants absorb much more red than far-red light. (Recall from Chapter 8 that the green pigment chlorophyll used in photosynthesis strongly absorbs red light.) Therefore, in a densely plant-populated area, the ratio of red light to far-red light (r/fr) decreases, and far-red light is reflected from leaves of nearby plants. As a result, the phytochrome "eye" absorbs more far-red light, which in turn triggers a series of responses that cause the plant to grow taller.

When a plant is using many of its resources for stem elongation, it has fewer resources to allocate for new leaves and branches. However, for a plant shaded by its neighbors, a rapid increase in stem length is advantageous because once this plant is taller than its neighbors, it obtains a larger share of unfiltered sunlight. We discuss other aspects of competition in Chapters 52 and 53. ▲

verts some of the phytochrome from the P_r form to the P_{fr} form. Therefore, the plant lacks a sufficient nighttime period free of P_{fr}.

In long-day plants, the active form of phytochrome, P_{fr}, *induces* flowering. The long days cause these plants to produce predominantly P_{fr}. During the short nights, some P_{fr} is changed to P_r, but because the night is short, the plant still retains a P_{fr} level sufficient to stimulate flowering.

Plant biologists are puzzled by the observation that P_{fr} inhibits flowering in short-day plants but induces flowering in long-day plants. Why different plants respond so differently to P_{fr} is not known at this time. Nevertheless, the importance of phytochrome to plants cannot be over-emphasized. Timing of day length and darkness is the most reliable way for plants to measure the change from one season to the next. This measurement is crucial for their survival, particularly in environments where the climate goes through a regular, annual pattern of favorable and unfavorable seasons.

Phytochrome is involved in many other responses to light

Phytochrome is involved in the light requirement that some seeds have for germination. Seeds with a light requirement must be exposed to red light. Exposure to red light converts P_r to P_{fr}, and germination occurs. Other physiological functions under the influence of phytochrome include sleep movements in leaves (discussed shortly), shoot dormancy, leaf abscission (see Chapter 32), and pigment formation in flowers, fruits, and leaves.

Temperature may also affect reproduction

Certain plants have a temperature requirement that must be met if they are to flower. The promotion of flowering by exposure to low temperature for a period of time is known as **vernalization.** Although the exact temperature and amount of time required varies among species, most vernalization temperatures occur between 0° and 10°C (32°–50°F). The part of the plant that must be exposed to low temperature varies. For some plants the moist *seeds* must be exposed to several weeks of low temperature in order for flowering to be induced. For other plants, recently germinated *seedlings* have a "cold" requirement.

In some plants the requirement of a low-temperature period is absolute, meaning that they will not flower without vernalization. Other plants will flower sooner if exposed to low temperatures, but will still flower at a later date if not exposed to low temperatures.

Examples of plants with a low-temperature requirement include annuals such as winter wheat and biennials such as carrots. Winter wheat is planted in the fall and germinates at that time. The young seedlings are exposed to cold during the winter and subsequently flower after resuming growth the following spring.

Carrots and other biennials grow vegetatively (their leaves, stems, and roots grow) the first year, storing surplus food in their roots (Fig. 36–6). If the roots are not harvested, the plants flower and reproduce sexually during the second year, after having been exposed to the cold temperatures of winter. Carrots left in a warm environment and not exposed to low temperatures continue vegetative growth indefinitely and do not initiate sexual reproduction.

The external stimulus that a plant responds to (in this case, low temperature) may be moderated and influenced by internal conditions, such as hormone levels in the plant. It is possible, for example, to eliminate the low-temperature requirement for flowering in biennials by treating the plants with gibberellin (discussed later in the chapter).

(a) (b)

FIGURE 36–6 Biennials have a low-temperature requirement for flowering. (*a*) The carrot, a biennial, grows vegetatively the first year, storing surplus food in its root. (*b*) After the plant has been dormant through the winter, the energy stored in the root is used in the second year for flowering and reproduction. (*a*, Carlyn Iverson; *b*, Jack Bostrack/CBR Images)

A BIOLOGICAL CLOCK INFLUENCES MANY PLANT RESPONSES

Plants, animals, and eukaryotic microorganisms appear to have an internal timer, or biological clock, that approximates a 24-hour cycle. These internal cycles, known as **circadian rhythms** (from the Latin *circum,* "around," and *diurn,* "daily"), help the organism determine the time of day. (In contrast, photoperiodism enables a plant to detect time of year.)

In the absence of external cues, circadian rhythms repeat every 20 to 30 hours, although in nature the rising and setting sun resets the biological clock so that the cycle repeats every 24 hours. Interestingly, phytochrome has been implicated as the photoreceptor involved in resetting the biological clock for many plants.

One example of a circadian rhythm in plants is the opening and closing of stomata (see Chapter 32). Plants placed in continual darkness for extended periods continue to open and close their stomata on an approximate 24-hour cycle.

Sleep movements observed in the common bean and other plants provide another example of a circadian rhythm (Fig. 36–7). During the day, bean leaves are extended horizontally for optimal light absorption, but at night they fold together so that they are perpendicular to their daytime position. (The biological significance of sleep movements is unknown at this time.) These movements occur independently of Earth's 24-hour cycle. If bean plants are placed in continual darkness or continual light, sleep movements continue on an approximate 24-hour cycle. The first plant with a mutant biological clock gene was reported in 1995 (see *On the Cutting Edge: Circadian Clock Mutants in* Arabidopsis).

Why do plants and other organisms possess circadian rhythms? A number of predictable environmental changes—sunrise and sunset, for example—occur during the course of each 24-hour period. These predictable changes may be important to an individual organism, causing it to change its behavior (in the case of animals) or its physiological activities. It is thought that circadian rhythms help an organism to synchronize repeated daily activities so that it does them at the same appropriate time each day.

CHANGES IN TURGOR CAN INDUCE TEMPORARY PLANT MOVEMENTS

Mimosa pudica, the sensitive plant, dramatically folds its leaves and droops in response to touch or to an electrical, chemical, or thermal stimulus (Fig. 36–8). The response, which typically occurs in a few seconds, spreads throughout the plant even if only one leaflet is initially aroused. When a *Mimosa* leaf is touched, an electrical impulse moves down the leaf to special cells located in **pulvini** (sing., *pulvinus*), organs at the base of each leaflet, each cluster of leaflets, and each petiole. The pulvinus is

(Text continues on page 785)

(a) (b)

FIGURE 36–7 A bean seedling exhibits sleep movements. (*a*) Leaf position at noon. (*b*) Leaf position at midnight. It is not known why some plants exhibit sleep movements, but they occur on an approximate 24-hour cycle, regardless of the amount of light or darkness to which they are exposed. (*a, b,* Dennis Drenner)

(a)

(b)

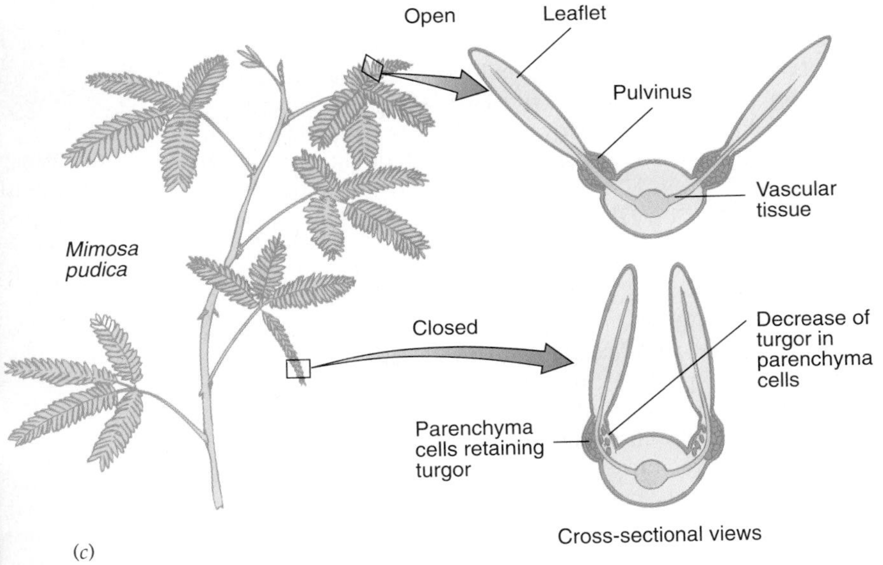

(c)

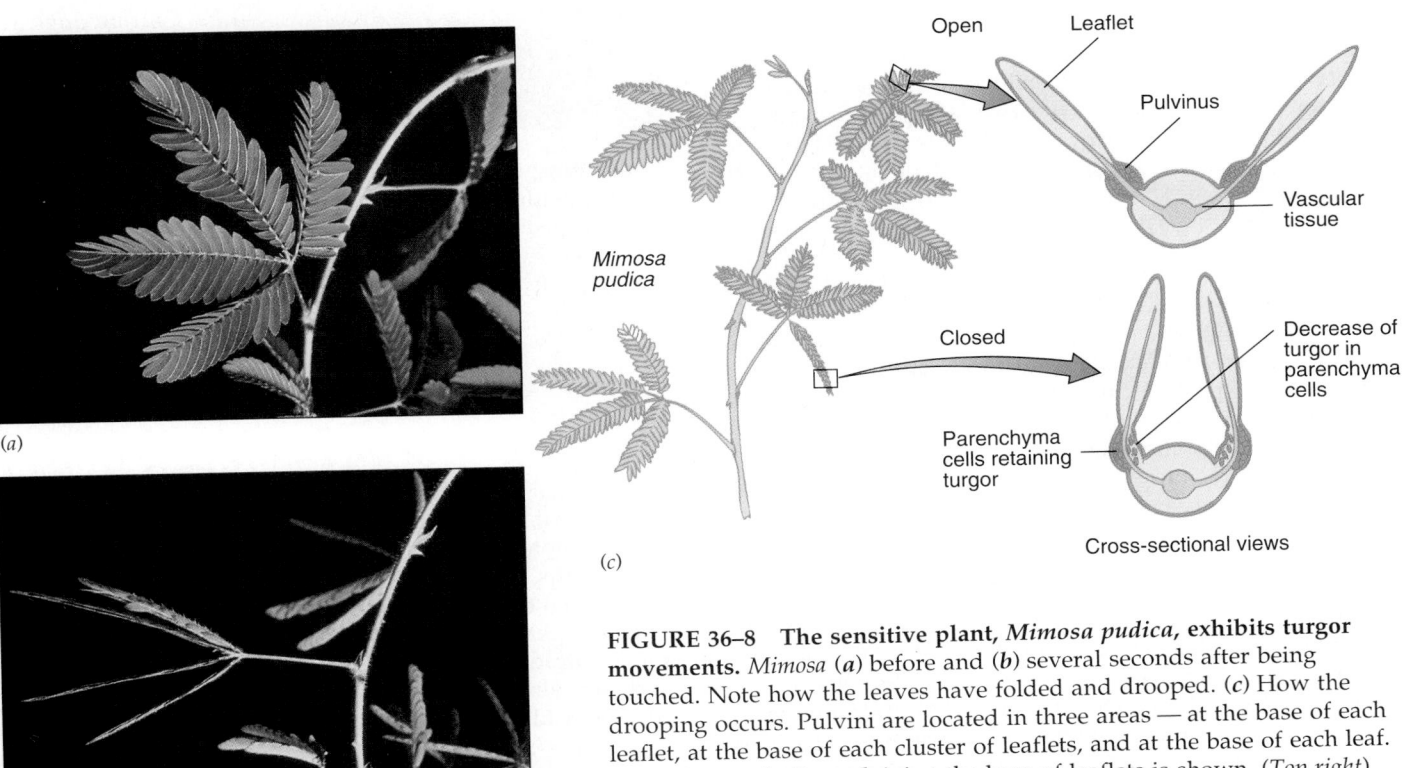

FIGURE 36–8 The sensitive plant, *Mimosa pudica*, exhibits turgor movements. *Mimosa* (*a*) before and (*b*) several seconds after being touched. Note how the leaves have folded and drooped. (*c*) How the drooping occurs. Pulvini are located in three areas — at the base of each leaflet, at the base of each cluster of leaflets, and at the base of each leaf. Only changes in the pulvini at the base of leaflets is shown. (*Top right*) Section through two leaflets, showing their pulvini when the leaf is undisturbed. (*Bottom right*) Section through the two leaflets, showing how a loss of turgor produces the folding of the leaves. (*a, b,* Dennis Drenner)

ON THE CUTTING EDGE

Circadian Clock Mutants in Arabidopsis

Objective: To identify *Arabidopsis* plants with mutant circadian clock genes.

Method: Develop a line of transgenic *Arabidopsis* plants in which the promoter region of an *Arabidopsis* gene known to be influenced by a circadian clock is spliced to the luciferase gene from fireflies.

Results: When exposed to a chemical called luciferin, wild-type (normal) transgenic plants that contain the luciferase gene glow during the day and do not glow at night. Transgenic plants that are circadian clock mutants glow at unpredictable times.

Conclusion: Plants with mutant circadian clocks can be identified and studied.

Circadian clock genes have been identified and cloned in organisms as diverse as cyanobacteria, fungi, fruit flies, and mice, but until recently such genes in plants evaded biologists. To identify circadian clock genes, researchers typically study mutant individuals with abnormal internal rhythms—for example, a fungus that develops reproductive structures at the "wrong" time of day. However, plant rhythms are generally less obvious than those in other organisms, and mutants have therefore been extremely difficult to identify. (Imagine how daunting it would be, for example, to monitor stomata from *thousands* of plants to determine if any are opening or closing at unpredictable times.)

In 1995 a research team led by Steve Kay of the National Science Foundation Center for Biological Timing at the University of Virginia reported that they had identified circadian clock mutants in seedlings of *Arabidopsis*, a diminutive plant from the mustard family (see Chapter 16).* One of the *Arabidopsis* genes known to exhibit a circadian rhythm is *cab*, the gene that codes for chlorophyll-*a*/*b* binding protein. The *cab* gene is actively transcribed during the day but switched off at night. As do other genes that follow a circadian rhythm, the *cab* gene continues this rhythm on an approximate 24-hour cycle when *Arabidopsis* plants are placed in continual light.

A transgenic line of *Arabidopsis* plants was developed in which the *cab* promoter (a regulatory sequence; see Chapter 13) was spliced to the luciferase gene from fireflies. The luciferase gene codes for an enzyme that catalyzes a reaction causing the chemical luciferin to glow. It was hoped that linking the luciferase gene to the promoter of *cab*, which is influenced by as-yet-unidentified circadian clock genes, would cause the luciferase gene to be expressed during the day but not at night.

The anticipated results were obtained. When *Arabidopsis* seedlings containing the recombinant DNA were sprayed with luciferin, they glowed only during the day (see figure). All that remained was to screen the seedlings for individuals that glowed either earlier or later than normal. Altogether, 26 mutant seedlings, representing at least 21 separate mutations, were identified. When various circadian rhythms were examined in these mutants, it was found that their leaf movements were timed differently than those of wild-type plants. Other circadian rhythms were normal, however, which may indicate that *Arabidopsis* contains more than one circadian clock gene.

Biologists are currently making a concerted effort to identify and clone the circadian clock gene in *Arabidopsis*. After that goal is reached, the gene can be compared with known circadian clock genes in other organisms to determine how similar they are. Biologists also want to learn how the wild-type circadian clock gene works.

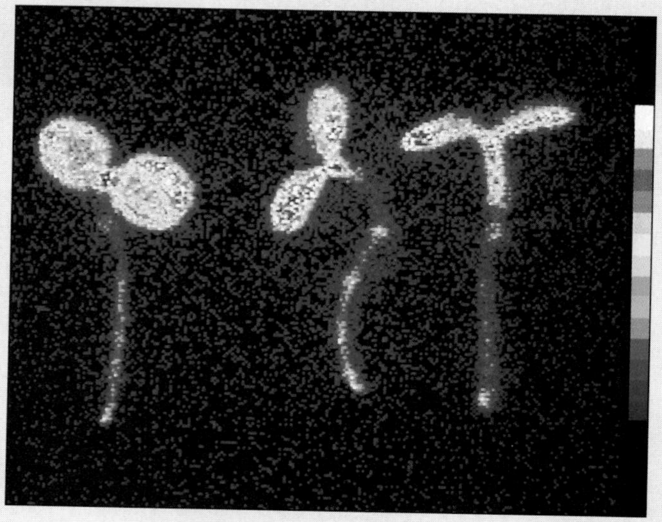

Three transgenic *Arabidopsis* seedlings glow after being sprayed with luciferin. The timing of their bioluminescence helped researchers identify circadian clock mutants. (The intensity of light produced by the glowing seedlings was digitally transformed into color. White, pink, and red indicate the strongest intensities.) (Courtesy of Steve A. Kay, University of Virginia)

*Millar, A.J., I.A. Carre, and S.A. Kay. "Circadian Clock Mutants in *Arabidopsis* Identified by Luciferase Imaging." *Science,* Vol. 267, February 24, 1995.

a somewhat swollen joint that acts as a hinge. When the electrical signal reaches cells in the pulvinus, it triggers a loss of turgor in certain pulvinal cells as potassium ions exit, causing water to leave the cells by osmosis. The sudden change in turgor causes the leaf movement. Such **turgor movements** (also called **nastic movements**) are temporary and reversible; the movement of potassium ions and water back into the pulvinus cells causes the plant part to return to its original position, although recovery takes several to many minutes longer than the original movement.

The mechanism by which the Venus flytrap leaf closes (see Chapter 1) is similar to that of *Mimosa.* An electrical signal, which moves much more rapidly than in *Mimosa,* induces a movement of potassium ions out of certain cells, followed by the exit of water. The loss of turgor causes the leaf to snap shut.

Changes in turgor are also responsible for **solar tracking,** the ability of leaves or flowers of certain plants to follow the sun's movement across the sky. Frequently, the leaves of such plants arrange themselves so that they are perpendicular to the sun's rays, regardless of the time of day or the sun's position in the sky. This allows for maximal light absorption throughout the day. Many solar trackers have pulvini at the bases of their petioles. Turgor changes in the pulvinus cells help position the leaf in its proper orientation relative to the sun. Sunflower, soybean, and cotton are examples of solar trackers.

A TROPISM IS GROWTH IN RESPONSE TO AN EXTERNAL STIMULUS FROM A SPECIFIC DIRECTION

A plant may respond to an external stimulus, such as light, gravity, or touch, by directional growth. Such responses, called **tropisms,** are *permanent* changes in the position of a plant part. Tropisms may be positive or negative, depending on whether the plant grows toward (positive) or away from (negative) the stimulus. Tropisms are under hormonal control (discussed later in the chapter).

Phototropism is the growth of a plant caused by the direction of light. Most growing shoot tips exhibit positive phototropism and bend (grow) toward light, something you may have noticed after placing houseplants near a sunny window. This growth response increases the likelihood that stems and leaves will receive adequate light for photosynthesis. The bending response of phototropism is triggered by blue light with wavelengths less than 500 nanometers. The photoreceptor that absorbs blue light and triggers the phototropic response is thought to be a flavoprotein. However, it has never been isolated and biochemically identified because it is present in such low concentrations in the cell. In 1993 the gene that codes for this pigment was isolated, and ongoing research should determine the signal pathway by which blue light triggers phototropism.

Growth in response to the direction of gravity is called **gravitropism.** Shoot tips generally exhibit negative gravitropism (they grow away from the center of Earth), whereas root tips exhibit positive gravitropism. The root cap is apparently the site of gravity perception in roots. Special cells in the root cap contain large starch grains known as **statoliths** that collect toward the bottoms of the cells in response to gravity (Fig. 36–9). If the root is

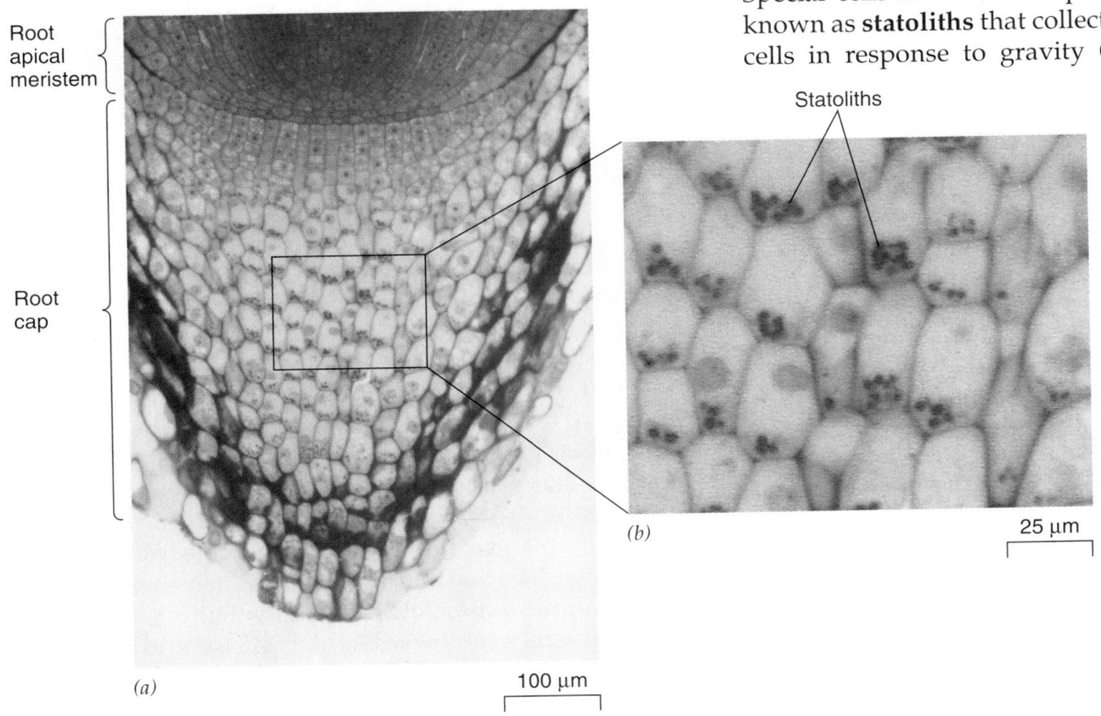

FIGURE 36–9 Cells in the root cap contain starch grains called statoliths. (*a*) A root cap of a root that is growing downward, in the direction of gravity. (*b*) Close-up of statoliths, which have collected on the bottoms of the cells. (*a, b,* Courtesy of R. Moore and E. McClellan, Baylor University)

moved—as when a potted plant is laid on its side—the statoliths tumble to a new position, always settling in the direction of gravity. The side of the root opposite where the statoliths collected begins elongating shortly thereafter.

Thigmotropism is growth in response to a mechanical stimulus, such as contact with a solid object. The twining or curling growth of tendrils, which helps attach a vine to some type of support, is an example of thigmotropism (see Fig. 32–16*b*). Tropisms in plants may also be caused by other stimuli in the environment such as water, temperature, chemicals, and oxygen.

THIGMOMORPHOGENESIS IS A GROWTH RESPONSE TO MECHANICAL STRESS

Plants growing in a natural environment are subjected to rain, hail, wind, and contact with passing animals. All of these forms of mechanical stress alter the growth and development of plants, making them shorter and stockier than plants grown in a greenhouse and therefore more protected from mechanical stress. Developmental response to mechanical stimuli is known as **thigmomorphogenesis.**

HORMONES ARE CHEMICAL MESSENGERS THAT REGULATE GROWTH AND DEVELOPMENT

The study of plant hormones and their effects is quite challenging because they are effective in extremely small amounts, and each hormone elicits *many* different responses. (This contrasts with the action of animal hormones, which are usually very specific.) In addition, the effects of different plant hormones overlap, so it is difficult to determine which hormone, if any, is the primary cause of a particular response. Also, plant hormones may stimulate a certain response at one concentration but inhibit that response at a different concentration.

Biologists have identified five classes of plant hormones thus far: auxins, gibberellins, cytokinins, ethylene, and abscisic acid (Tables 36–1 and 36–2). Together these hormones control the growth and development of a plant.

Auxin promotes cell elongation

Charles Darwin, the 19th century naturalist best known for his theory of evolution by natural selection (see Chapter 17), also provided the first evidence for the existence of auxin. The experiments that Darwin and his son Francis performed on newly germinated canary grass seedlings involved positive phototropism (Fig. 36–10). As with all grasses, the first part of a canary grass seedling to emerge from the soil is the coleoptile, which, when exposed to light from only one direction, bends toward it. The bending occurs close to the tip of the coleoptile.

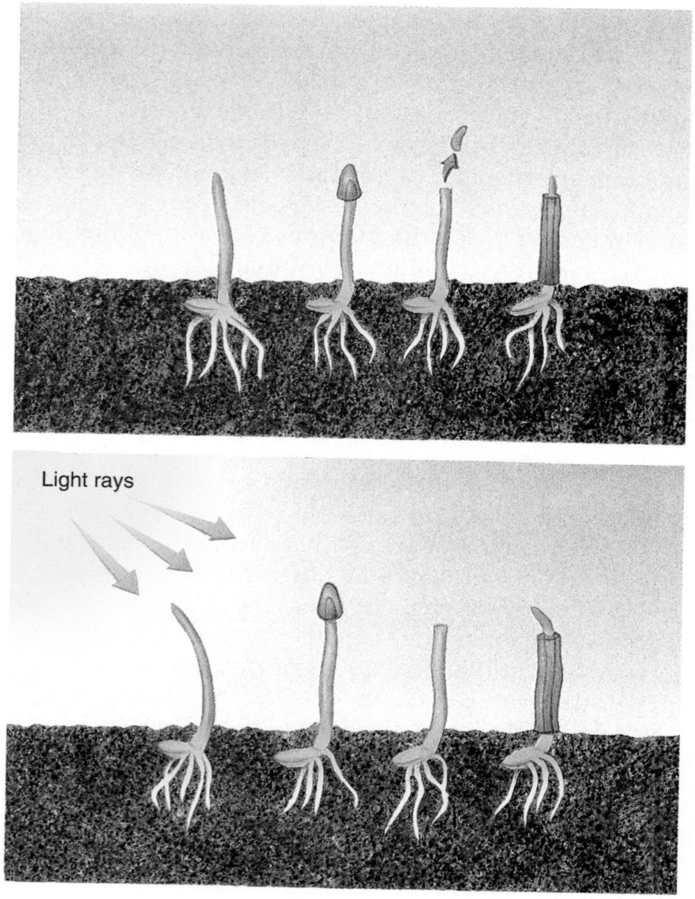

FIGURE 36–10 Charles and Francis Darwin conducted phototropism experiments with coleoptiles of canary grass seedlings. *(Upper row)* Some plants were uncovered *(far left)*, some were covered only at the tip, some had the tip removed, and some were covered everywhere but at the tip. *(Lower row)* After exposure to light coming from one direction, the uncovered plants and the plants with uncovered tips *(far right)* grew toward the light. The plants with covered tips *(center left)* or tips removed *(center right)* did not bend toward light.

The Darwins tried to influence this bending in several ways. For example, they covered the tip of the coleoptile as soon as it emerged from the soil. Even though they covered that part of the coleoptile *above* where the bend occurs, the plants treated in this manner did not bend. On other plants, they removed the coleoptile tip and found that bending did not occur. When the bottom of the coleoptile was shielded from the light, however, the coleoptile still bent toward light. From these experiments, the Darwins concluded that "some influence is transmitted from the upper to the lower part, causing it to bend."

Table 36–1 THE FIVE PLANT HORMONES

Hormone	*Chemical Structure*	*Site of Production*	*Method of Translocation*
Auxin (IAA)	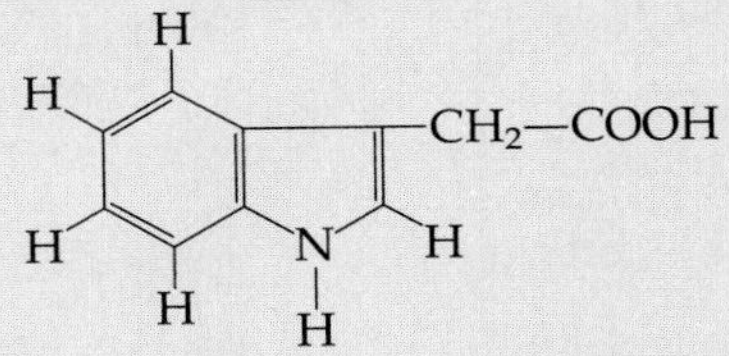	Shoot apical meristem, young leaves, seeds	Polar transport in parenchyma cells
Gibberellin (GA_3)	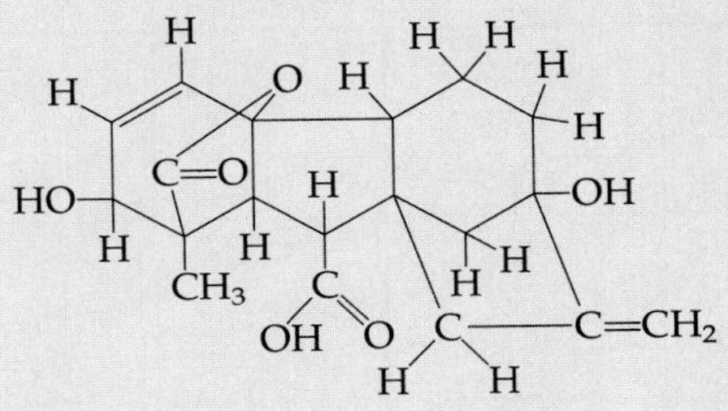	Young leaves, root and shoot apical meristems, embryo in seed	Unknown
Cytokinin (Zeatin)	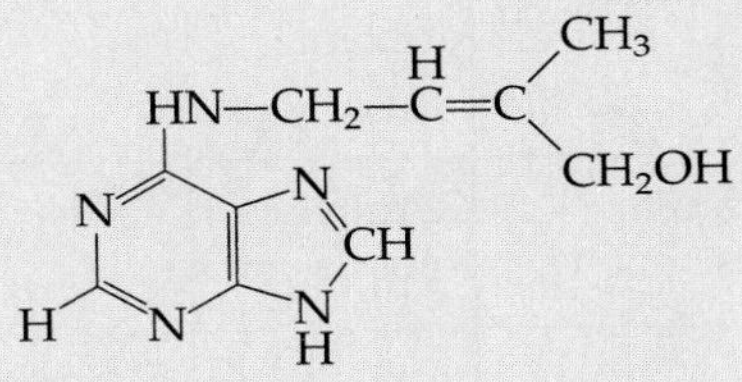	Roots	Xylem
Ethylene	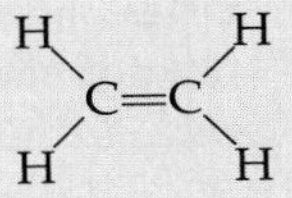	Stem nodes, ripening fruit, senescing tissue	Unknown (diffusion?)
Abscisic acid	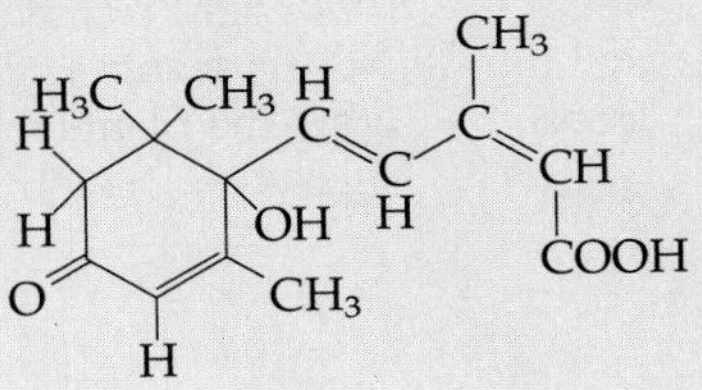	Older leaves, root cap, stem	Vascular tissue

In the 1920s Frits Went, a young Dutch scientist, isolated the phototropic hormone from oat coleoptiles. He removed the coleoptile tips and placed them on tiny blocks of agar for a period of time. When he put one of these agar blocks on the side of a decapitated coleoptile in the dark, bending occurred (Fig. 36–11). This indicated that the substance had diffused from the coleoptile tip into the agar, and later from the agar into the decapitated coleoptile. Went named this substance **auxin** (from the Greek *aux,* "enlarge" or "increase"). The purification and elucidation of auxin's chemical structure were accomplished in the early 1940s by a research team led by Kenneth Thimann at the California Institute of Technology.

Auxin was originally defined as any of a group of compounds that stimulate phototropic curvature in coleoptiles and stems. A broader definition is that auxin is a group of natural and artificial hormones that regulate growth, often by promoting cell elongation. The main auxin found in plants is *indoleacetic acid* (*IAA*).

The movement of auxin in the plant is said to be *polar;* that is, it moves downward from its site of production, usually the shoot apical meristem. Young leaves and seeds are also sites of auxin production.

Auxin promotes cell growth by triggering cell elongation, an effect it apparently exerts by changing the cell

(Text continues on page 789)

Table 36–2 SOME OF THE INTERACTIONS AMONG PLANT HORMONES

Physiological Activity	*Auxin*	*Gibberellin*	*Cytokinin*	*Ethylene*	*Abscisic Acid*	*Other Factors for Some Plants*
Seed germination		Promotes germination	?		Inhibits germination	Cold requirement, light requirement
Growth of seedling into mature plant	Cell elongation, organogenesis*	Cell division and elongation	Cell division and differentiation, organogenesis*			
Apical dominance	Inhibits lateral bud development		Promotes lateral bud development	?		
Initiation of reproduction (flowering)		Stimulates flowering in some plants†	?			Cold requirement, photoperiod requirement
Fruit development and ripening	Fruit development	Fruit development		Promotes ripening		Light requirement (for pigment formation)
Leaf abscission	Inhibits abscission		Inhibits abscission	Promotes abscission		Light requirement
Winter dormancy of plant		Breaks dormancy	?		Promotes dormancy	Light requirement
Seed dormancy		Breaks dormancy	?		Promotes dormancy	

*In plant tissue culture.

†Gibberellin cannot be considered *the* flowering hormone. There is evidence for a flowering hormone that has not yet been isolated and characterized.

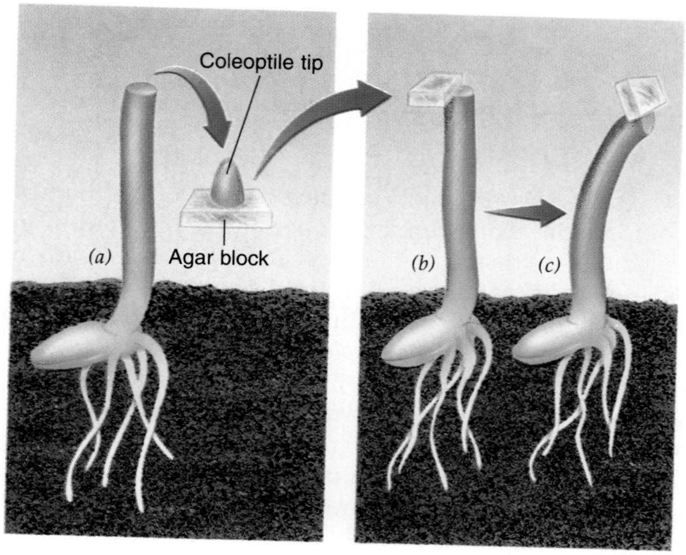

FIGURE 36–11 Frits Went isolated auxin from coleoptiles. (*a*) Coleoptile tips were placed on agar blocks for a period of time. (*b*) The agar block was transferred to a decapitated coleoptile. It was placed off-center, and the coleoptile was left in continual darkness. (*c*) The coleoptile bent, indicating that a chemical had been transferred from the original coleoptile tip to the agar block and from there to one side of the decapitated coleoptile.

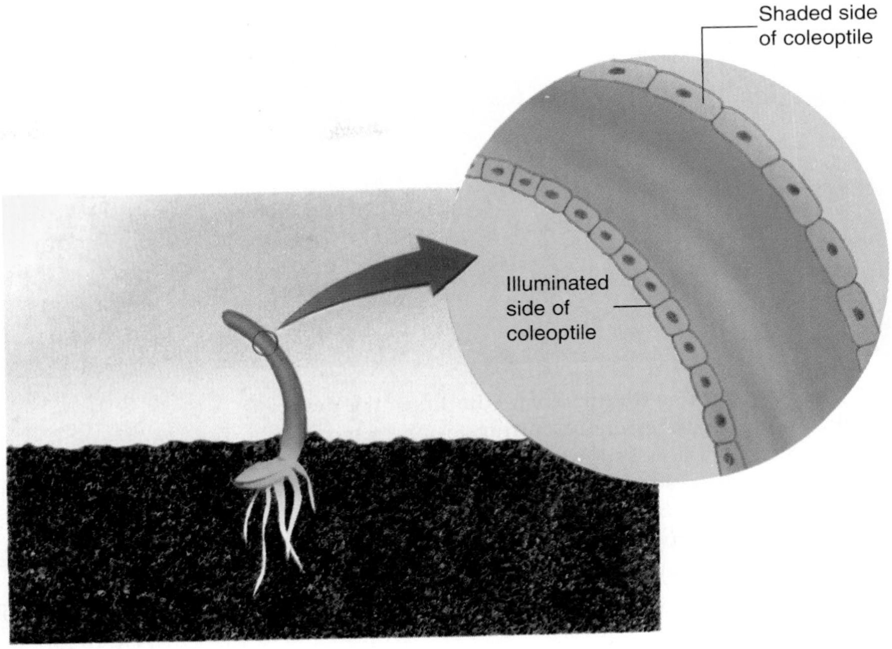

FIGURE 36–12 Phototropism is caused by the unequal distribution of auxin. Auxin travels down the side of the stem or coleoptile *away* from the light, causing cells on the shaded side to elongate. Therefore, the stem or coleoptile bends toward light.

walls so they can expand. Auxin's effect on cell elongation also provides an explanation for phototropism. When a plant is exposed to a light from only one direction, or to a light gradient (stronger on one side), the auxin migrates laterally to the shaded side of the stem before moving down the stem by polar transport. As a result, the cells on the shaded side of the stem elongate more than the cells on the light side, and the stem bends toward the light (Fig. 36–12). Auxin is thought to be involved in gravitropism and thigmotropism as well.

Auxin exerts other effects on plants. For example, certain plants do not "branch out" when they grow. Growth in these plants occurs almost exclusively from the apical meristem rather than from lateral buds. Such plants are said to exhibit **apical dominance** (Fig. 36–13). In plants with strong apical dominance, it appears that auxin produced in the apical meristem inhibits the development of lateral buds into actively growing shoots. When the apical meristem is pinched off, the auxin source is removed and lateral buds grow into branches.

Auxin produced by developing seeds stimulates the development of the fruit. When auxin is applied to certain flowers in which fertilization has not occurred (and, therefore, in which seeds are *not* developing), the ovary enlarges and develops into a seedless fruit. Seedless tomatoes have been produced in this manner.[1] Auxin is not the only hormone involved in fruit development, however.

[1]Not all seedless fruits are produced by treatment with auxin. In Thompson seedless grapes, for example, fertilization occurs, but the embryos abort and thus the seeds fail to develop.

A number of manufactured, or synthetic, auxins with structures similar to indoleacetic acid have been made. Synthetic auxin is used to stimulate root development on stem cuttings for asexual propagation, particularly of woody plants with horticultural importance (Fig. 36–14). Other synthetic auxins, 2,4-D and 2,4,5-T, are used as selective herbicides (weed killers) that kill plants by disrupting their normal hormonal balance. Synthetic auxins are *selective* in that they kill broadleaf dicots (such as dandelions) but do not harm monocots (such as lawn grass or cereal crops); the reason for this selectivity is not known.

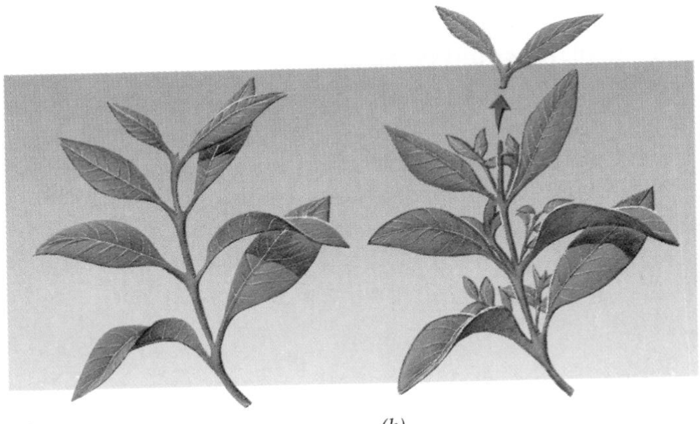

(a) *(b)*

FIGURE 36–13 Auxin inhibits the development of lateral buds. (*a*) When a plant's tip (the source of auxin) is intact, its lateral buds do not develop. (*b*) The stem tip has been removed. Because no auxin is moving down from the tip, lateral buds develop into branches.

FIGURE 36–14 A synthetic auxin stimulates the development of roots on honeysuckle cuttings. Many adventitious roots formed on a honeysuckle cutting placed in a high auxin concentration (*left*); whereas fewer roots formed in a lower auxin concentration (*middle*). The cutting, placed in water (*right*), served as a control and did not form roots in the same time period. (Visuals Unlimited/Joe Eakes, Color Advantage)

A gene that codes for an enzyme that regulates auxin has been isolated

In 1994 biologists at Michigan State University isolated the first gene known to be involved in regulating auxin. Called the *iaglu* gene, it codes for an enzyme that changes auxin from its active to its inactive form. The *iaglu* gene has been found in many different kinds of plants, and biologists suspect it may be present in all plants.

The *iaglu* gene has received a lot of attention because it may make it possible to modify plant growth by genetic engineering. For example, by inserting extra copies of the *iaglu* gene into a plant, researchers hypothesize that smaller amounts of active auxin would be available and, as a result, the plant would be shorter and bushier. By inserting *reversed* copies of the *iaglu* gene, researchers think that larger amounts of active auxin would be available and the plant would be taller:

> reversed copy of *iaglu* gene is inserted → plant makes reversed copy of RNA by transcription → reversed RNA binds by complementary base-pairing to normal mRNA produced by transcription of normal *iaglu* gene → translation is blocked and *iaglu* enzyme is not produced → more active auxin is present in plant → plant grows taller

Ultimately, biologists want to be able to alter growth by manipulating hormones in specific parts of the plant, in order to make larger fruits or larger clusters of flowers, for instance. How easily plant growth can be manipulated using the *iaglu* gene remains to be seen. Other as-yet-unknown genes may also be involved in controlling auxin production and regulation, in which case manipulating the *iaglu* gene by itself will not be particularly effective.

Gibberellins promote stem elongation

In the 1920s a Japanese biologist was studying the "foolish seedling" disease, which makes young rice seedlings grow extremely tall and spindly, fall over, and die. These symptoms were caused by a fungus that produces a chemical substance named **gibberellin.** Not until after World War II did scientists in Europe and North America learn of the exciting work done by the Japanese. During the 1960s gibberellins were isolated from healthy plants and determined to be hormones involved in many normal plant functions. In the case of the "foolish seedling" disease, the symptoms were caused by an abnormally high gibberellin concentration in the plant tissue.

Gibberellins promote stem elongation in many plants. When gibberellin is applied to a plant, this elongation may be spectacular, particularly in plants that normally have very short stems. Some dwarf mutant corn and pea plants will grow to a normal height when treated with gibberellins (Fig. 36–15*a*). Gibberellins are also involved in **bolting,** the rapid elongation of a floral stalk that occurs naturally in many plants when they initiate flowering (Fig. 36–15*b*). In all these cases, gibberellins promote stem elongation by causing cells to divide as well as elongate. The actual mechanism of cell elongation appears to be different from that caused by auxin, however.

Gibberellins are involved in several reproductive processes in plants. They stimulate flowering, particularly in long-day plants. In addition, they can substitute for the low temperature that biennials normally require before initiating flowering. If gibberellins are applied to biennials during their first year of growth, flowering occurs without the period of low temperature. Gibberellins, like auxin, affect the development of fruits. Gibberellins are applied to several varieties of grapes to produce larger fruits.

Gibberellins are also involved in the germination of seeds in many plants. In a classic experiment involving the germination of barley seeds, the release of gibberellin from the embryo was shown to trigger the synthesis of α-amylase, which digests starch in the endosperm. As a result, glucose is available for absorption by the embryo. Although the formation of enzymes to mobilize starch reserves occurs in many types of seeds, control of enzyme formation by gibberellin appears to occur exclusively in seeds of cereals and other grasses. In addition to mobilizing food reserves in newly germinated grass seeds, the application of gibberellins can substitute for light or low-temperature seed germination requirements in other plants.

(a)

(b)

FIGURE 36–15 Stem elongation is triggered by gibberellin. (***a***) An experiment testing the effects of gibberellin on normal and dwarf corn plants shows that dwarf plants respond to gibberellin much more dramatically than normal plants. From left to right: dwarf, untreated; dwarf treated with gibberellin; normal treated with gibberellin; normal, untreated. (This dwarf variety is a mutant with a single recessive gene that impairs gibberellin metabolism.) (***b***) Bolting is regulated by gibberellin. Many biennials grow as a rosette (a circular cluster of leaves close to the ground) during their first year (*left*) and then bolt when they initiate flowering in their second year (*right*). (*a*, Courtesy of B.O. Phinney, University of California, Los Angeles; *b*, Robert E. Lyons)

Cytokinins promote cell division

During the 1940s and 1950s, a number of researchers involved in **tissue culture**—a technique that involves isolating cells from plants and growing them in a nutrient medium—were trying to find substances that might induce plant cells to divide (see *Focus On: Cell and Tissue Culture*). They discovered that cells would not divide without some substance found in both coconut water and herring sperm. Finally, the active substance was isolated from herring sperm and called **cytokinin** because it induces cell division, or cytokinesis. In 1963 the first naturally occurring cytokinin was identified from corn, and since that time several similar molecules have been identified from other plants.

In intact plants, cytokinins promote cell division and differentiation, in which young, relatively unspecialized cells become mature, more specialized cells. They are also a required ingredient in any plant tissue culture medium and must be present in order for cells to divide. In tissue culture, cytokinins interact with auxin during the formation of plant organs like roots and stems (Fig. 36–16). For example, in tobacco tissue culture, a high ratio of cytokinin to auxin induces shoots to form, whereas a low ratio induces roots to form.

Cytokinins and auxin also interact in the control of apical dominance. Here their relationship is antagonistic: auxin inhibits the growth of lateral buds, and cytokinin promotes their growth.

One interesting effect of cytokinins on plant cells is to delay **senescence,** or aging (Fig. 36–17). Plant cells, like all living cells, go through a natural aging process. This process is accelerated in cells of plant parts that are cut, such as cut stems. It is thought that plants must have a continual supply of cytokinins from the roots. Cut stems, of course, lose their source of cytokinins and therefore age and die rapidly. When cytokinins are sprayed on leaves of a cut stem, they remain green, while any unsprayed leaves turn yellow and die.

Ethylene stimulates abscission and fruit ripening

Ethylene, a gaseous hormone produced by plants, was first discovered in 1934. Many diverse plant processes are

Focus On

Cell and Tissue Culture

Cells can be isolated from certain plants and grown in a chemically-defined, sterile nutrient medium. In initial experiments with such cultures, plant cells could be kept alive, but they did not divide. It was later discovered that the addition of certain natural materials such as the liquid endosperm of coconut, also known as coconut water, induced cells to divide in culture. Because coconut water has a complex chemical composition, the division-inducing substance (a cytokinin) was not chemically identified for some time. By the late 1950s, plant cells from a variety of sources could be cultured successfully, dividing to produce a mass of undifferentiated cells, or **callus.**

In 1958, F.C. Steward, a plant physiologist at Cornell University, succeeded in generating an entire carrot plant from a single callus cell derived from a carrot root. This demonstrated conclusively that each plant cell contains a genetic blueprint for all features of an entire organism. His work also showed that an entire plant can be grown from a single cell, provided the proper genes are expressed at the appropriate times.

Since Steward's pioneering work, many plants have been successfully cultured using a variety of cell sources. Plants have been regenerated from different tissues, organ explants (excised organs or parts such as root apical meristems or young embryos), and single cells. Under certain conditions, the genes that control embryonic development are expressed in plant tissue culture, and all of the plant's embryonic stages in the seed can be observed in their normal progression (see figure).

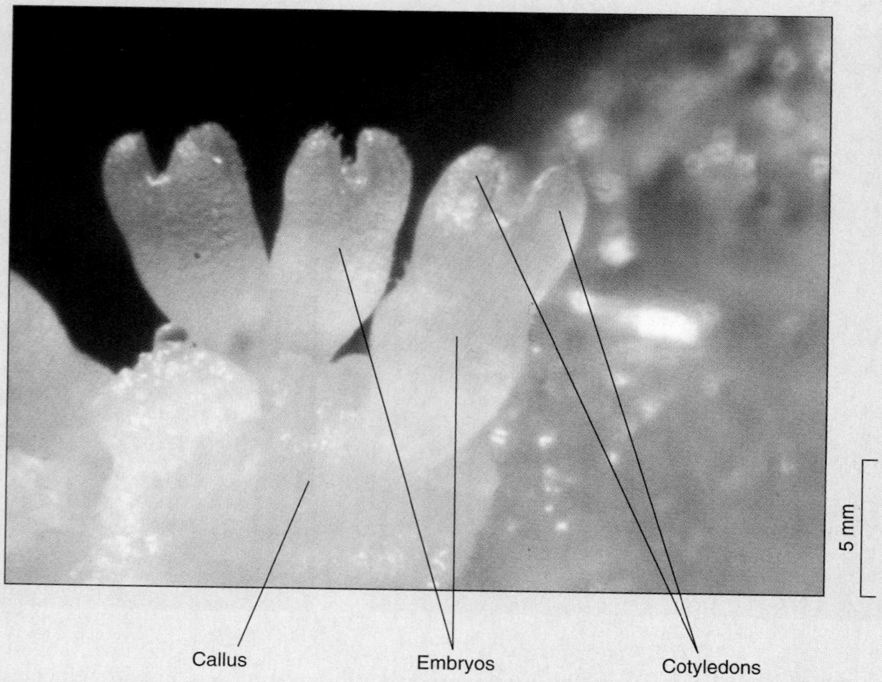

Tiny embryos bud from callus in culture. Compare these embryos with the one depicted in Figure 35–8*d*. (Courtesy of Dennis Gray, University of Florida)

Cell and tissue culture techniques are used to help answer many fundamental questions involving growth and development in plants (see Chapter 16). These techniques also have great practical potential. Using tissue culture, it is possible to regenerate large numbers of genetically identical plants from cells of a single, genetically superior plant. It is also possible to alter the genetic composition of a cell while it is in culture and then have these changes expressed in the whole plant during regeneration. This provides a valuable tool to those scientists who practice genetic engineering in an effort to introduce desirable new traits into crop species (see Chapter 14). Research involving plant cell and tissue culture is one of the exciting and fruitful areas of biological research today. ■

influenced by it. Ethylene inhibits cell elongation, promotes the germination of seeds, and is involved in plant responses to wounding or invasion by disease-causing microorganisms.

Ethylene also plays a major role in many aspects of senescence, including the ripening process in fruits. As a fruit ripens, it produces ethylene, which triggers an acceleration of ripening. This induces the fruit to produce more ethylene, which further accelerates ripening. The expression "one rotten apple spoils the lot" is true. A rotten apple is one that is overripe and produces large amounts of ethylene, which causes a "domino effect" as it diffuses and triggers the ripening process in nearby apples. Ethylene is used commercially to promote the uni-

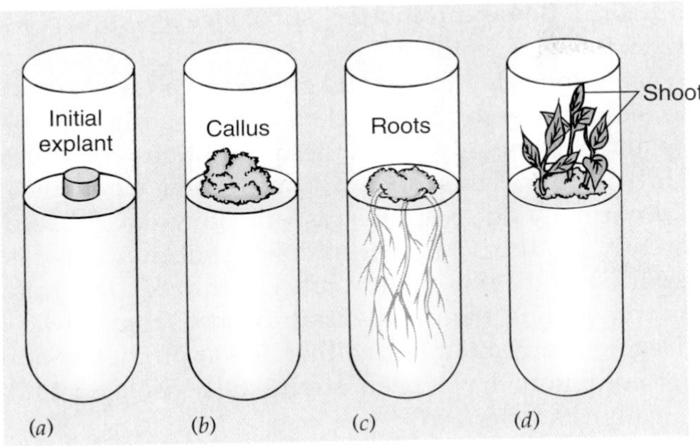

FIGURE 36–16 Tobacco tissue culture responds to varying levels of auxin and cytokinin. (*a*) The initial explant is a small piece of sterile tissue from the pith of a tobacco stem, which is placed on a nutrient agar medium. Varying amounts of auxin and cytokinin in the culture media produce different growth responses. (*b*) Nutrient agar containing 2.0 milligrams of auxin per liter and 0.2 milligrams of cytokinin per liter caused cells to divide and form a clump of undifferentiated tissue called a callus. (*c*) When the callus is transplanted to a medium with a high ratio of auxin to cytokinin, roots differentiate. (*d*) Shoot growth is stimulated by a medium containing a low ratio of auxin to cytokinin.

FIGURE 36–17 Senescence was delayed in the green leaf by repeated applications of cytokinin. The other leaf is aging because it was not treated with cytokinin. (Courtesy of A.C. Leopold, Cornell University)

form ripening of bananas. Bananas are picked while green and shipped to their destination, where they are exposed to ethylene before being delivered to grocery stores.

Ethylene has been implicated as the hormone that induces leaf abscission. However, abscission is affected by two plant hormones that are antagonistic toward one another, ethylene and auxin. As a leaf ages (when autumn approaches), its level of auxin decreases. Concurrently, cells in the abscission zone begin producing ethylene. To further complicate the process, cytokinins are possibly involved in abscission. Cytokinins, like auxin, decrease in concentration as leaf tissue ages.

Abscisic acid promotes bud and seed dormancy

Abscisic acid was discovered simultaneously in 1963 by two independent research teams. Its name is an unfortunate choice because abscisic acid is primarily involved in dormancy and does not induce abscission in most plants. Abscisic acid is sometimes referred to as the plant "stress hormone" because it promotes changes in plant tissues that are stressed, or exposed to unfavorable conditions such as freezing, high salt levels, and droughts. (Recall that ethylene also affects plant responses to certain stresses.)

The effect of abscisic acid on plants suffering from water stress is best understood. The level of abscisic acid increases at least 20-fold in the leaves of plants exposed to severe drought conditions. The high level of abscisic acid in the leaves triggers the closing of stomata, which saves the water normally transpired through the stomata, thereby increasing the plant's likelihood of survival.

The onset of winter could also be considered a type of stress on plants. As winter approaches, woody plants cease growing, and protective coverings of bud scales form over terminal buds. These adaptations are promoted by abscisic acid. Another winter adaptation that involves abscisic acid is dormancy in seeds. As mentioned earlier in this chapter, many seeds have high levels of abscisic acid in their tissues and are therefore unable to germinate until the abscisic acid washes out. In a corn mutant unable to synthesize abscisic acid, the seeds germinate as soon as the embryos are mature, while still attached to the ear (Fig. 36–18).

The evidence that abscisic acid is the only hormone involved in both bud and seed dormancy is not conclusive, particularly because the addition of gibberellin reverses the effects of dormancy. In seeds the level of abscisic acid decreases during the winter, while the level of gibberellin increases. Cytokinins have also been implicated in breaking dormancy. Once again we see that a single physiological activity such as dormancy may be controlled in plants by the interaction of several hormones. The actual response made by a plant may be the result of changing hormone ratios rather than the effects of each individual hormone.

Florigen is a hypothetical hormone that promotes flowering

Grafting experiments in which different tobacco species are grafted together indicate that both flower-promoting and flower-inhibiting substances exist. *Nicotiana silvestris* is a long-day plant, and a variety of *N. tabacum* is a day-neutral plant. When a long-day tobacco is grafted to a day-neutral tobacco and exposed to long days and short nights, both plants flower; that is, the day-neutral tobacco plant flowers sooner than it normally would (Fig. 36–19). It appears that the flower-promoting substance is induced in the long-day tobacco and transported to the day-neutral tobacco through the graft union, causing the day-neutral tobacco to flower sooner than expected. The hypothetical flower-promoting substance is known as **florigen.**

When a long-day tobacco is grafted to a day-neutral tobacco and exposed to short days and long nights, neither plant flowers. As long as these conditions continue, the day-neutral plants do not flower even when they would normally do so. In this case, the long-day tobacco apparently produces a flower-inhibiting substance that is transported to the day-neutral tobacco through the graft union, preventing the day-neutral tobacco from flowering. Despite many attempts, neither flower promoters nor flower inhibitors have been successfully isolated and chemically characterized.

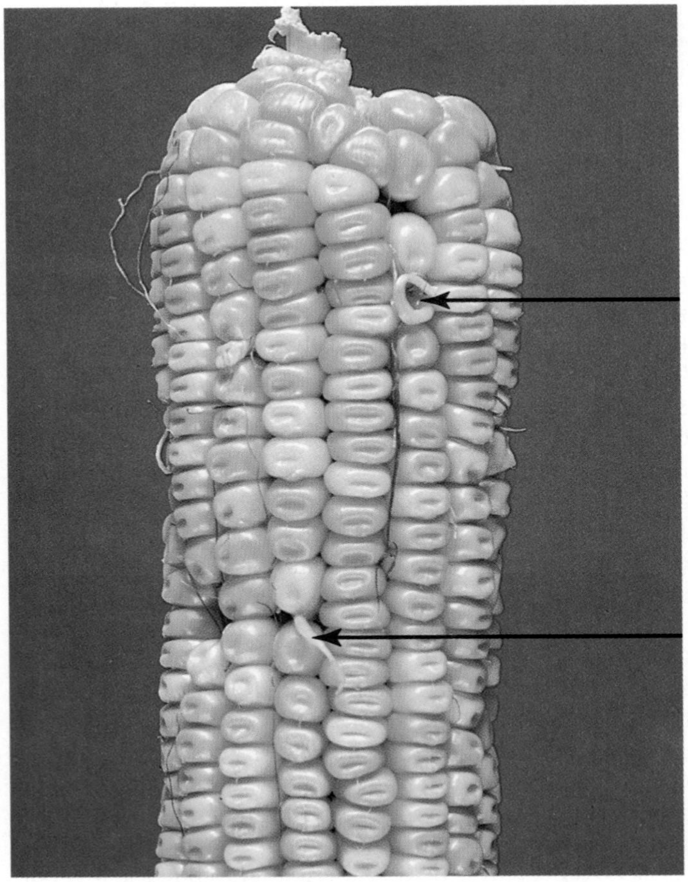

FIGURE 36–18 Inability to produce abscisic acid can prevent seed dormancy in corn. Some of the white kernels have germinated prematurely, while still on the ear, producing white roots (*arrows*). (Courtesy of M.G. Neuffer)

OTHER CHEMICALS HAVE BEEN IMPLICATED IN PLANT GROWTH AND DEVELOPMENT

In addition to the five major plant hormones, a number of chemicals are implicated in certain specific aspects of plant growth and development. Four such chemical regulators are polyamines, polypeptides, oligosaccharins, and salicylic acid. Not much is currently known about how these chemicals function, but each is the focus of intense research.

Polyamines, organic molecules with two or more amino ($-NH_2$) groups, affect a variety of physiological activities such as cell division, fruit development, and senescence. Polyamines are not considered plant hormones because (1) they are present in high concentrations (about 1000 times greater than hormones such as auxin), and (2) they are not transported extensively through the plant. Polyamines may function by influencing gene expression; they appear to increase both transcription of DNA and translation of mRNA.

Although many animal hormones are known to be polypeptides, the first *plant polypeptide* with hormonal properties was not isolated until 1991. Known as *systemin,* this polypeptide is transported systemically throughout the plant in response to wounding by insects. It appears to stimulate a natural defense mechanism at extremely low concentrations — as low as one part per trillion. Systemin may trigger the plant to produce protease inhibitors, molecules that disrupt insect digestion, thereby curbing leaf damage done by caterpillars and other herbivorous insects. The discovery of systemin has caused a flurry of research in search of additional polypeptide regulators in plants.

Oligosaccharins are cell-wall carbohydrate fragments consisting of short, branched chains of sugar residues. They are present in extremely small quantities in cells and active at much lower concentrations (100 to 1000 times lower) than hormones such as auxin. As do many hor-

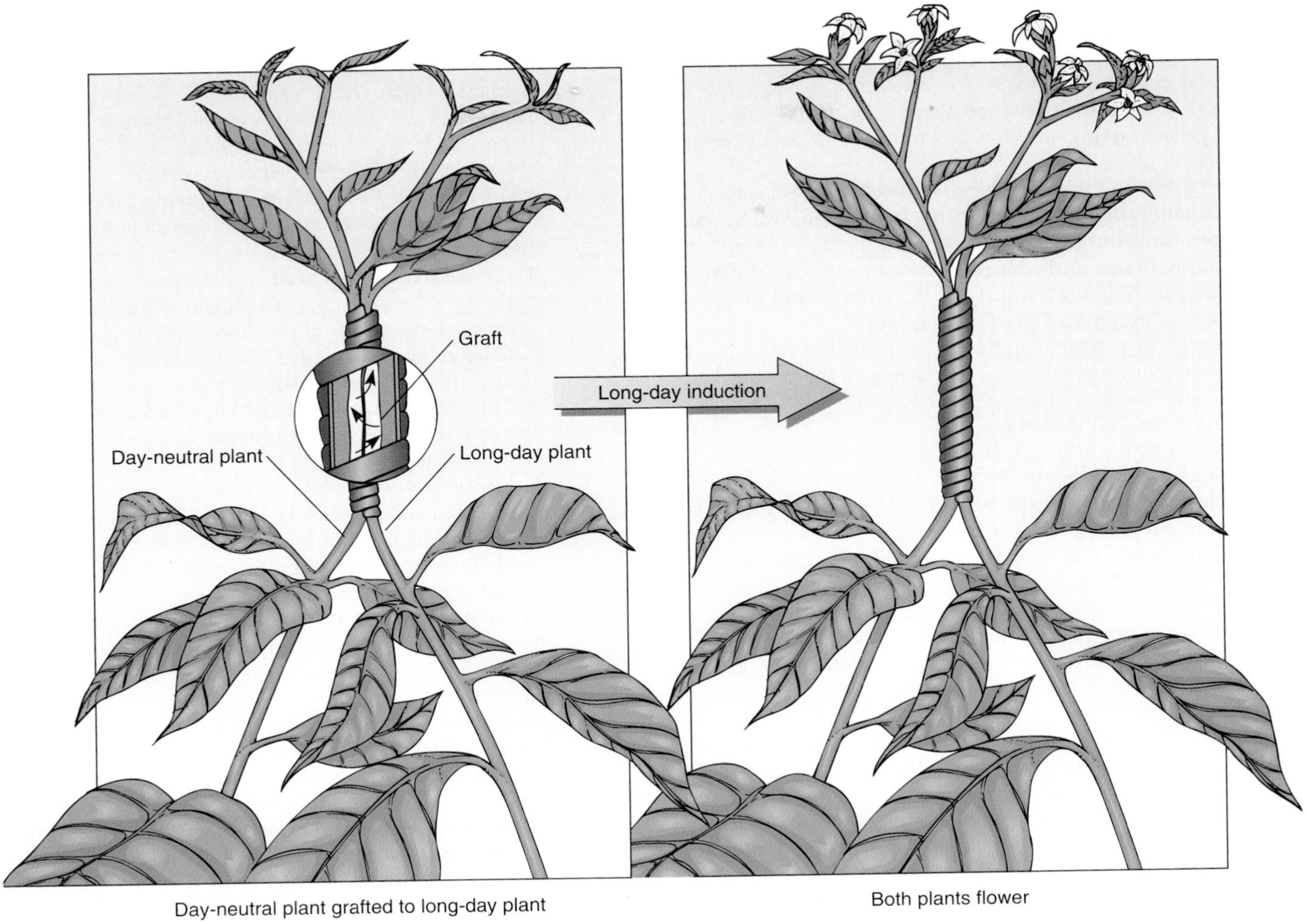

FIGURE 36–19 When a long-day tobacco is grafted to a day-neutral tobacco and both plants are exposed to a long-day/short-night regimen, they both flower. The day-neutral plant flowers sooner than it normally would, presumably because a flower-promoting substance (florigen) passes from the long-day plant to the day-neutral one through the graft.

mones, oligosaccharins exert their effects by binding to membrane receptors and altering gene expression. Different oligosaccharins appear to have distinct functions. Some trigger the production of *phytoalexins* (from the Greek *phyto*, "plant", and *alexi*, "to ward off"), antibiotics that kill plant pathogens, particularly fungi. Other oligosaccharins inhibit flowering and induce vegetative (nonreproductive) growth. One oligosaccharin has been implicated in a negative feedback loop that regulates the effects of auxin; the presence of auxin apparently stimulates synthesis of this oligosaccharin, which in turn lessens the effect of auxin on cell elongation.

Salicylic acid was first extracted from willow (*Salix*) bark and is the active ingredient in aspirin (acetylsalicylic acid). More recently, salicylic acid has been shown to help defend plants against insect pests and pathogens such as viruses. When a plant is under attack, the concentration of salicylic acid increases, and it spreads systemically throughout the plant. It is thought that salicylic acid binds to a cell receptor, thereby switching on genes that code for proteins that fight infection and promote wound healing. In 1993 the receptor to which salicylic acid binds was identified as catalase, an enzyme that catalyzes the degradation of hydrogen peroxide in eukaryotic cells. The binding of salicylic acid to catalase starts a chain of events that, while still hypothetical, is supported by current data:

salicylic acid binds to catalase → catalase is inhibited → hydrogen peroxide level increases in the cell → specific proteins (transcription factors?) are activated → specific genes are turned on and expressed → natural defense response occurs.

SUMMARY

I. The location of a cell in the young plant body affects gene expression during development by causing some genes in that cell to be turned off and others to be turned on.

II. Seed germination is affected by both internal and external factors.
   A. External environmental factors that may affect seed germination include requirements for oxygen, water, temperature, and light.
   B. Internal factors affecting whether or not a seed germinates include the maturity of the embryo and the presence or absence of chemical inhibitors.

III. Plant growth and development is controlled not only by internal genetic factors but also by factors in the physical environment (such as changing day length, variation in precipitation, and temperature).

IV. Many plants flower in response to specific cues from the environment.
   A. Photoperiodism is the response of plants to the duration and timing of light and dark.
      1. In many plants, flowering is a photoperiodic response; some are short-day plants and others are long-day plants. In day-neutral plants, flowering is not affected by photoperiod.
      2. The photoreceptor in photoperiodism is phytochrome, a blue-green pigment with two forms, P_r and P_{fr}.
   B. Vernalization is the promotion of flowering by exposure of seeds or young seedlings to low temperatures.

V. Circadian rhythms are regular rhythms in growth or activities of a plant or other organism that approximate the 24-hour day and are reset by the rising and setting of the sun.

VI. Turgor movements and tropisms are the two kinds of plant movements that occur in response to external stimuli.
   A. Turgor movements are caused by temporary turgor changes in special cells. Solar tracking, the ability of leaves or flowers to follow the sun, is caused by turgor movements.
   B. Tropisms are permanent directional growth responses.
      1. Phototropism is plant growth in response to the direction of light.
      2. Gravitropism is plant growth in response to the influence of gravity.
      3. Thigmotropism is plant growth in response to contact with a solid object.

VII. Plants respond to hormones, chemical messengers that regulate plant growth and development.
   A. Hormones are effective in extremely small concentrations.
   B. The functions of plant hormones overlap.
   C. Many physiological activities of plants may be due to the interactions of several hormones rather than to the effect of a single hormone.
   D. There are five classes of plant hormones.
      1. Auxin is involved in cell elongation, tropisms, apical dominance, and fruit development.
      2. Gibberellins are involved in stem elongation, flowering, and seed germination.
      3. Cytokinins promote cell division and differentiation, delay senescence, and interact with auxin in apical dominance.
      4. Ethylene has a role in the ripening of fruits, leaf abscission, and senescence.
      5. Abscisic acid is the "stress hormone." It is involved in stomatal closure due to water stress and in bud and seed dormancy.
   E. Grafting experiments indicate the existence of both flower-promoting (florigen) and flower-inhibiting substances, neither of which have been successfully isolated.
   F. A number of chemicals are implicated in certain specific aspects of plant growth and development.
      1. Polyamines are organic molecules with two or more amino ($-NH_2$) groups that affect a variety of plant processes, possibly by influencing gene expression.
      2. Systemin, a plant polypeptide with hormonal properties, stimulates a natural defense mechanism in which the plant produces molecules that disrupt insect digestion.
      3. Oligosaccharins, sugar residues that are fragments of cell-wall carbohydrates, have a variety of functions and exert their effects by binding to membrane receptors and altering gene expression.
      4. Salicylic acid helps defend plants against pathogens and insect pests. It may bind to a cell receptor, thereby switching on genes that fight infection and promote wound-healing.

SELECTED KEY TERMS

abscisic acid
apical dominance
auxin
bolting
callus
circadian rhythm
coleoptile
cytokinin
day-neutral plant
determinate growth
ethylene
florigen
gibberellin
gravitropism
hormone
imbibition
indeterminate growth
long-day plant
nastic movements
photoperiodism
phototropism
phytochrome
pulvinus
senescence
short-day plant
solar tracking
statolith
thigmomorphogenesis
thigmotropism
tissue culture
tropism
turgor movements
vernalization

POST-TEST

1. A plant's response to the relative amounts of daylight and darkness is known as __________ .
2. The active form of __________ , P_{fr}, is formed when red light is absorbed.
3. The promotion of flowering by a low-temperature treatment is known as __________ .
4. The __________ is an organ at the base of the petiole in *Mimosa* that can undergo rapid changes in turgor, causing dramatic leaf movements.
5. In __________ __________ , plants use turgor movements to position organs such as leaves optimally in sunlight.
6. Regularly recurring sleep movements observed in beans are an example of a(an) __________ rhythm.
7. __________ is the growth of a plant due to the direction of light.
8. Plant roots generally exhibit positive __________ .
9. The twining of tendrils is an example of __________ .
10. Cell expansion by growth is permanent, whereas cell expansion by __________ __________ is temporary.
11. __________ are organic compounds produced in one part of the plant and transported to another, where they cause a positive or negative effect.
12. A synthetic __________ known as 2,4-D is used as a selective herbicide.
13. Research on a fungal disease of rice provided the first clues about __________ .
14. __________ interact with auxins during the formation of plant organs in tissue culture.
15. __________ delay senescence, whereas __________ promotes it.
16. The only plant hormone that is a gas is __________ .
17. The "stress hormone" is __________ __________ .
18. __________ __________ promotes the dormancy of woody twigs.

REVIEW QUESTIONS

1. What factors influence the germination of seeds? Explain the role that each of these factors plays in germination, and discuss why the germinating seed responds the way it does.
2. Why are plant growth and development so sensitive to environmental cues?
3. What is phytochrome? Describe its role in flowering.
4. Distinguish between photoperiodism and circadian rhythms.
5. Distinguish between turgor movements and tropisms.
6. How is auxin involved in phototropism?
7. Discuss the various hormones involved in each of the following physiological processes: (a) germination of seeds; (b) stem elongation; (c) ripening of fruits; (d) abscission of leaves; (e) dormancy of seeds.
8. Summarize the roles of auxins, gibberellins, cytokinins, ethylene, and abscisic acid.

YOU MAKE THE CONNECTION

1. Predict whether flowering would be expected to occur in the following situations. Explain each answer.
   (a) A short-day plant is exposed to 15 hours of daylight and 9 hours of darkness.
   (b) A short-day plant is exposed to 9 hours of daylight and 15 hours of darkness.
   (c) A short-day plant is exposed to 9 hours of daylight and 15 hours of darkness, with a 10-minute exposure to red light in the middle of the night.
2. What benefits are conferred on a plant when its stems exhibit positive phototropism and its roots exhibit positive gravitropism?

RECOMMENDED READINGS

Chapin, F.S., III. "Integrated Responses of Plants to Stress." *BioScience,* Vol. 41, January 1991. Plants respond to many environmental stresses by altered growth.

Evans, M.L., R. Moore, and K.H. Hasenstein. "How Roots Respond to Gravity." *Scientific American,* Vol. 255, No. 6 December 1986. A closer look at gravitropism.

"Leaves with Clocks," in "Breakthroughs" section of *Discover,* September 1993. Sleep movements are another example of circadian rhythms in plants.

Mores, P.B., and N.H. Chua, "Light Switches and Plant Genes." *Scientific American,* Vol. 258, No. 4 April 1988. How an environmental cue is related to gene expression.

Career Visions

Zoo Instructor

JOHN RESANOVICH

Ever since elementary school John Resanovich has been interested in biology, but he wasn't sure how he would turn that interest into a career. While attending Long Island University, Southampton, he worked in various settings, including research laboratories. However, it was the position he obtained after college at a working farm that proved he could combine his interests in biology, animals, and people into a rewarding career. At The Wildlife Conservation Society of the Bronx Zoo *John works as one of seven instructors who teach courses to students of all ages, conveying the Zoo's message of resource conservation. John is currently working on a Master's degree in computer education at Iona College, New Rochelle so that he can use new technologies in the Zoo's education program.*

What did you plan to do with your undergraduate degree?

I wasn't sure. Beginning in elementary school I was interested in biology. During college I worked at Brookhaven National Laboratories doing lab work, but I discovered that I wasn't interested in pursuing that as a career.

Did you have a mentor who helped you learn about other careers you might pursue?

My father was a big influence; he had always exposed me to the environment and to nature. In college I had a few different professors who pushed me that way as well.

Were any particular courses influential?

Early on I took a seminar in conservation biology and natural history interpretation. The natural history interpretation portion of the class involved getting out and talking to people about what was going on in the local habitats. That course really pushed me into doing things in the environment.

What did you do after college?

I worked at a living history museum—Phillipsburg Manor in North Tarrytown, New York—which is actually a colonial era working farm. That job involved working with people and animals, which I really enjoyed. I gave tours and also worked on the farm. Everybody there helped out in all areas, so I also worked with livestock. That job synthesized two of my interests and helped me focus on what I wanted to do.

Then you went to work at the Bronx Zoo?

Yes, and it entailed passing a pretty rigorous interview that included a teaching demonstration. They were interested in how well I worked with children in addition to my knowledge of biology.

Describe your job. What do you do on a typical day?

The Zoo has seven full-time instructors who teach classes. We provide programs that vary depending on the age level and interests of the children. For a very young class we focus on basic concepts; things like how animal sizes, shapes, and textures help them to survive. Further along, we might focus on habitat ecology, for example, we'll discuss the features and the ecological importance of the rain forest. At all levels we emphasize the Zoo's conservation message, which is that if we use our precious resources indiscriminately we will lose them. Classrooms are located throughout the Zoo and often look out onto an exhibit, so I might be able to show the kids a baboon right outside of the window. We use this resource to make a connection between the students and the animals and guide them to observe behaviors such as why a male baboon is surrounded by a group of females. We ask the students to determine the importance of this male's job in keeping other males away. We might ask them to consider the various adaptations that help animals survive. Does that tail help them? Do their hands help them? What parts of their bodies help them?

Do you teach higher-level age groups?

We teach anyone from pre-kindergarten to adults. For adults we offer a variety of programs including classes on specific animals, such as wolves. For adults or family groups, we try to have role-playing games or an activity that will keep both adults and children active and interested. We try to limit the straight lecture material in adult classes because they can get that anywhere. The Zoo is a phenomenal resource and we want to use it in the most thoughtful and creative way.

What kind of background would a student need to do your job?

When we hire instructors or interns, we look for people who have a good solid background in biology or have been committed environmental educators. The best thing would be a concentration in both. What the Zoo really wants is someone with a strong desire to positively influence and inspire future conservationists.

P A R T 7

Structure and Life Processes in Animals

Bottlenose dolphins (*Tursiops truncatus*), Sea of Cortez, Mexico. (Kevin Schafer/Peter Arnold, Inc.)

C H A P T E R 37

The Animal Body: Introduction to Structure and Function

All animals are multicellular. A large animal such as the elephant (*Loxodonta africana*) is composed of more cells than the much smaller impala (*Aetyceros melampus*), drinking nearby. (Frans Lanting/Minden Pictures)

Why is it that we do not see amoebas the size of whales slithering around? Most cells, in fact, are microscopic in size (see Chapter 4). Recall that when its size approaches the limits of efficiency, a cell divides to form two cells. In unicellular organisms, cell division results in the production of two new individuals. In multicellular organisms, new cells remain associated to form one individual. Animals can grow to large sizes because they are composed of many cells.

The number of cells, not their individual sizes, is responsible for the size of an organism. The cells of an earthworm, a bird, and an elephant are all about the same size. The elephant is larger because its genes are programmed to provide for a larger number of cells. Consider how different an earthworm and an elephant are, not only in size, but also in body form and lifestyle. Such diversity also results from multicellularity.

Multicellularity permits the specialization of cells. In a unicellular organism such as a bacterium or a flagellate, the single cell must carry on all the activities necessary for life. In a multicellular organism, cells specialize to perform specific tasks. Recall that a **tissue** consists of a group of closely associated, similar cells adapted to carry out specific functions. Animal tissues are classified as epithelial, connective, muscle, and nervous. Each kind of tissue (and subtissue) is composed of cells with characteristic sizes, shapes, and arrangements. Some tissues are specialized to transport materials, whereas others contract, enabling the organism to move. Still others secrete hormones that regulate metabolic processes.

You may recall from Chapter 1 that tissues associate to form **organs** such as the heart or stomach. Groups of tissues and organs form the **organ systems** of the body. In this unit we discuss just how cells associate with one another and how tissues, organs, and organ systems perform specialized functions such as regulation of body temperature, hearing, and digestion.

LEARNING OBJECTIVES

After you have studied this chapter you should be able to

1. Discuss the advantages of multicellularity.
2. Define tissue, organ, and organ system.
3. Compare the general structure and function of the four principal kinds of animal tissues: epithelial, connective, muscle, and nervous tissues.
4. Describe the main types of epithelial tissue and give their functions.
5. Compare the main types of connective tissue and summarize their functions.
6. Contrast the three types of muscle tissue and their functions.
7. Identify the cells that make up nervous tissue and give their functions.
8. List and briefly describe the organ systems of a complex animal.
9. Define homeostasis, and discuss how each organ system helps maintain homeostasis.
10. Distinguish between benign and malignant neoplasms, and contrast the cells of a malignant neoplasm with those of normal tissue.

EPITHELIAL TISSUES COVER THE BODY AND LINE ITS CAVITIES

Epithelial tissue (also called **epithelium**) covers body surfaces and lines cavities. It forms the outer layer of the skin and the linings of the digestive, respiratory, excretory, and reproductive tracts. Epithelial tissue consists of cells fitted tightly together to form a continuous layer, or sheet, of cells. One surface of the sheet is typically free because it lines a cavity, such as the lumen (cavity) of the intestine, or covers the body (outer layer of the skin). The other surface of an epithelial layer is attached to the underlying tissue by a noncellular **basement membrane** composed of tiny fibers and nonliving polysaccharide material produced by the epithelial cells. Table 37–1 illustrates the main types of epithelial tissue, indicates their locations in the body, and describes their functions.

Epithelial tissues perform a wide variety of functions, including protection, absorption, secretion, and sensation. The epithelial layer of the skin—the epidermis—covers the entire body and protects it from threats such as mechanical injury, chemicals, bacteria, and fluid loss. The epithelial tissue lining the digestive tract absorbs nutrients and water into the body. Some epithelial cells are organized into **glands** that secrete cell products like hormones, enzymes, or sweat. Other epithelial cells are specialized as sensory receptors that receive information from the environment. For example, epithelial cells in taste buds and in the nose are specialized as chemical receptors.

Everything that enters or leaves the body must cross one or more layers of epithelium. Food taken into the mouth and swallowed is not really "inside" the body until it is absorbed through the epithelium of the gut and enters the blood. To a large extent, the permeabilities of the various epithelia regulate the exchange of substances between the different parts of the body, as well as between the organism and the external environment.

Many epithelial membranes are subjected to continuous wear and tear. As outer cells are sloughed off, they must be replaced by new ones from below. Such epithelial tissues generally have a rapid rate of cell division that continuously produces new cells to replace those lost.

Three types of epithelial cells can be distinguished on the basis of shape (Table 37–1). **Squamous** epithelial cells are thin, flattened cells shaped like pancakes or flagstones. **Cuboidal** epithelial cells are short cylinders that from the side appear cube-shaped, resembling dice. Actually, each cell has a complex shape, usually an eight-sided polyhedron.

Columnar epithelial cells look like tiny columns or cylinders when viewed from the side. The nucleus is usually located near the base of the cell. Viewed from above or in cross section, these cells often appear hexagonal. On its free surface, a columnar epithelial cell may have cilia that beat in a coordinated way, moving materials over the tissue surface. Most of the respiratory tract is lined with ciliated epithelium that moves particles of dust and other foreign material away from the lungs.

Epithelial tissue may be **simple**—that is, composed of one layer of cells—or **stratified,** composed of two or more layers (Table 37–1). Simple epithelium is usually located where materials must diffuse through it or where substances are secreted, excreted, or absorbed. Stratified epithelial tissue is found where protection is required. For example, it makes up the outer layer of the skin and lines the mouth of humans and other vertebrates.

A third arrangement of epithelial cells is **pseudostratified** epithelium, so named because its cells falsely appear to be layered. Although all of its cells rest on a basement membrane, not every cell extends to the free surface of the tissue. This arrangement gives the impression of two or more cell layers. Some of the respiratory passageways are lined with pseudostratified epithelium equipped with cilia.

(Text continues on page 803)

Table 37–1 EPITHELIAL TISSUES

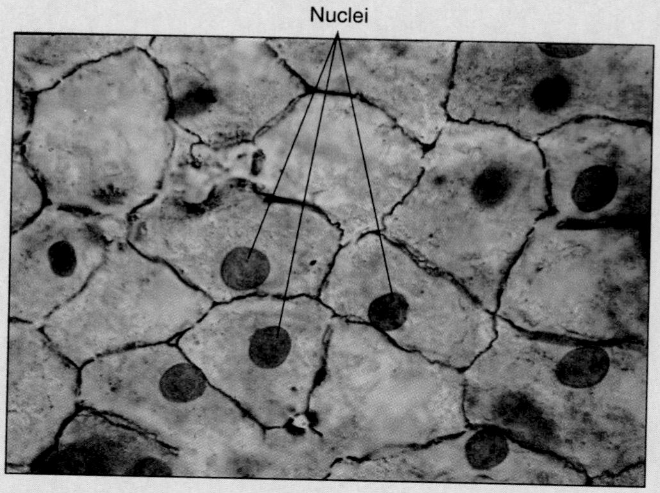

Simple squamous epithelium. (Ed Reschke)

Simple squamous epithelium

Main Locations
Air sacs of lungs; lining of blood vessels

Functions
Passage of materials where little or no protection is needed and where diffusion is major form of transport

Description and Comments
Cells are flat and arranged as single layer

25 μm

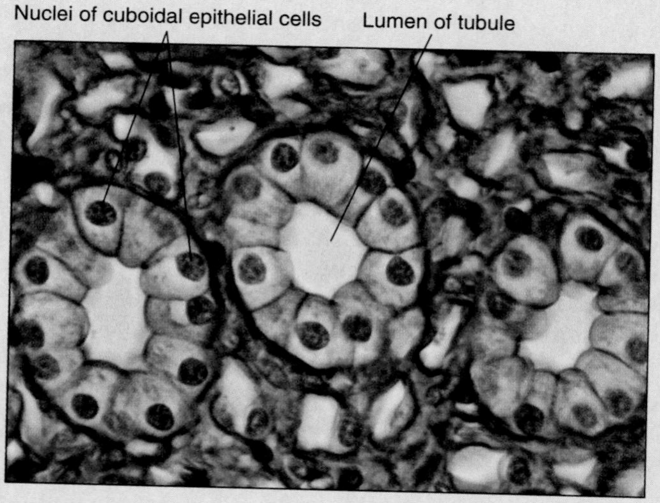

Simple cuboidal epithelium. (Ed Reschke)

Simple cuboidal epithelium

Main Locations
Linings of kidney tubules; gland ducts

Functions
Secretion and absorption

Description and Comments
Single layer of cells; photograph shows cross-section through tubules; from the side each cell looks like short cylinder; sometimes have microvilli for absorption

25 μm

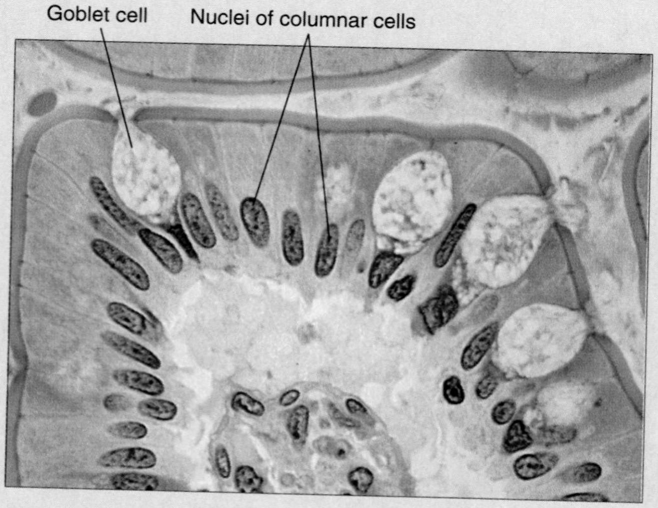

Simple columnar epithelium. (Ed Reschke)

Simple columnar epithelium

Main Locations
Linings of much of digestive tract and upper part of respiratory tract

Functions
Secretion, especially of mucus; absorption; protection; movement of mucous layer

Description and Comments
Single layer of columnar cells; sometimes with enclosed secretory vesicles (in goblet cells); highly developed Golgi complex; often ciliated

25 μm

Table 37–1 continued

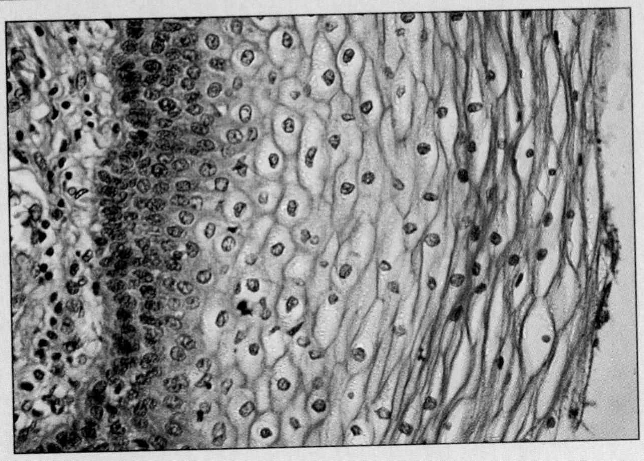

Stratified squamous epithelium. (Ed Reschke)

Stratified squamous epithelium

Main Locations
Skin; mouth lining; vaginal lining

Functions
Protection only; little or no absorption or transit of materials; outer layer continuously sloughed off and replaced from below

Description and Comments
Several layers of cells, with only the lower ones columnar and metabolically active; division of lower cells causes older ones to be pushed upward toward surface, becoming flatter as they move

25 μm

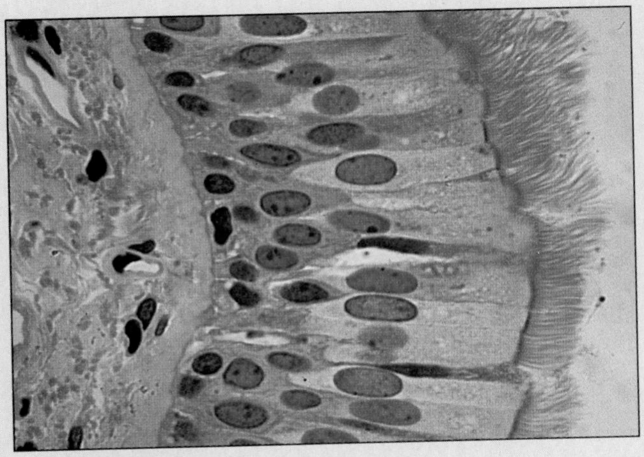

Pseudostratified columnar epithelium, ciliated. (Ed Reschke)

Pseudostratified epithelium

Main Locations
Some respiratory passages; ducts of many glands

Functions
Secretion; protection; movement of mucus

Description and Comments
Ciliated, mucus-secreting, or with microvilli; comparable in many ways to columnar epithelium except that not all cells are the same height—thus, though all cells contact the same basement membrane, the tissue appears stratified

25 μm

Because the linings of blood and lymph vessels have a different embryonic origin than "true" epithelium, they are referred to as **endothelium.** Structurally, the cells of these linings are typical epithelial cells.

A gland consists of one or more epithelial cells specialized to produce and secrete a product such as sweat, milk, mucus, wax, saliva, hormones, or enzymes (Fig. 37–1). The epithelial tissue lining the cavities and passageways of the body typically contains some specialized mucus-secreting cells called **goblet cells.** The mucus lubricates these surfaces and facilitates the movement of materials.

Glands can be classified as exocrine or endocrine. **Exocrine glands,** like goblet cells and sweat glands, secrete their products onto a free epithelial surface, typically through a duct (tube). **Endocrine glands** lack ducts. As described in Chapter 47, endocrine glands release their products, called **hormones,** into the tissue fluid surrounding them. Hormones are then transported in the blood.

An **epithelial membrane** consists of a sheet of epithelial tissue and a layer of underlying connective tissue. Types of epithelial membranes include mucous membranes and serous membranes. A **mucous membrane,** or mucosa, lines a body cavity that opens to the outside of the body, such as the digestive or respiratory tract. The epithelial layer secretes mucus that lubricates the tissue and protects it from drying.

A **serous membrane** lines a body cavity that does not open to the outside of the body. It consists of simple squamous epithelium over a thin layer of loose connective tissue. This type of membrane secretes fluid into the cavity it lines. Some serous membranes that may be familiar to you are the pleural membranes lining the pleural cavities

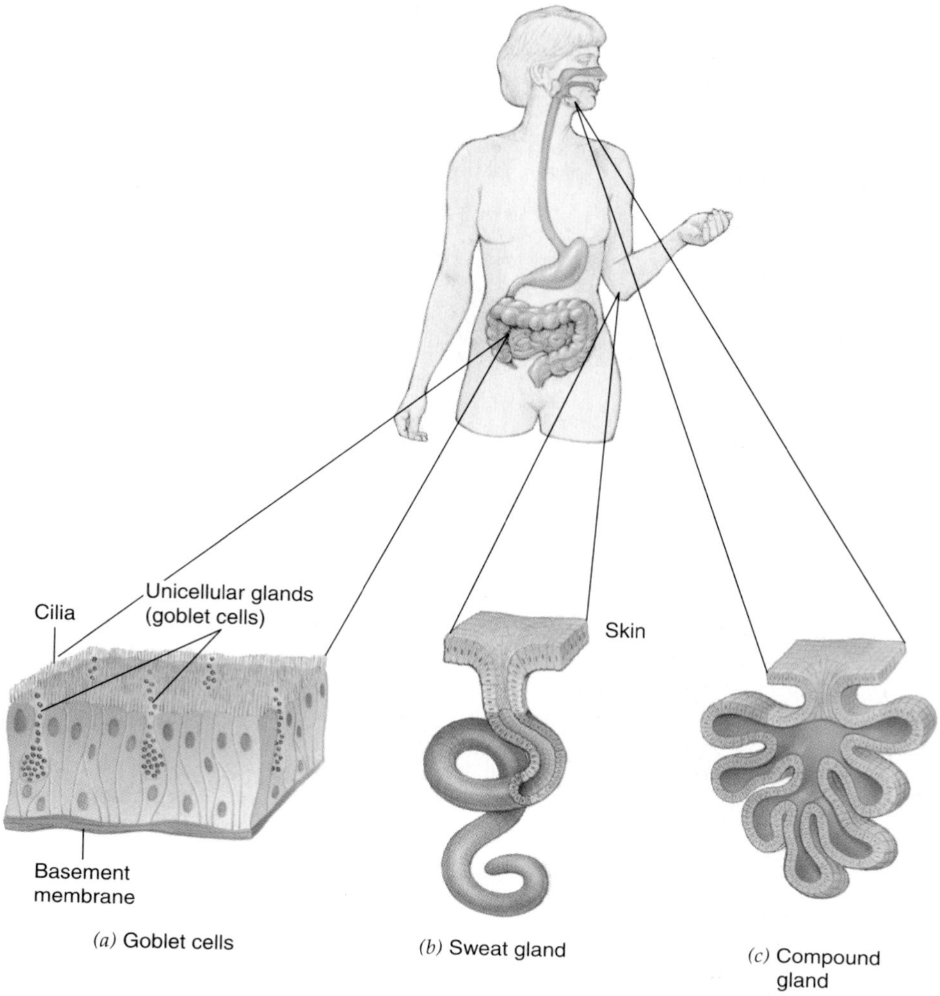

FIGURE 37–1 A gland consists of one or more epithelial cells. (*a*) Goblet cells are unicellular glands that secrete mucus. (*b*) Sweat glands are simple tubular glands with coiled tubes. The walls of the gland are constructed of simple cuboidal epithelium. (*c*) Compound glands, such as the parotid salivary glands, have branched ducts.

around the lungs, the pericardial membranes lining the pericardial cavity around the heart, and the peritoneum lining the abdominal cavity.

CONNECTIVE TISSUES JOIN AND SUPPORT OTHER BODY STRUCTURES

Almost every organ in the body has a framework of **connective tissue** that supports and cushions it. Cartilage and bone are examples of connective tissues that support the vertebrate body and protect organs such as the heart and lungs. Blood is a connective tissue that transports materials, and adipose tissue stores fat.

There are many kinds of connective tissues and many systems for classifying them. Some of the main types of connective tissue are: (1) loose and dense connective tissues; (2) elastic connective tissue; (3) reticular connective tissue; (4) adipose tissue; (5) cartilage; (6) bone; and (7) blood, lymph, and tissues that produce blood cells. The tissues vary widely in their structural details and in the functions they perform (Table 37–2).

Typically, connective tissues contain relatively few cells; these are embedded in an extensive **intercellular substance** consisting of threadlike microscopic **fibers** scattered throughout a **matrix,** a thin gel composed of polysaccharides, secreted by the cells. The cells of different kinds of connective tissues differ in their shapes and structures and in the kinds of fibers and matrices they secrete. The nature and function of each kind of connective tissue are determined in part by the structure and properties of the intercellular substance.

Connective tissue contains collagen, reticular, and elastic fibers

Connective tissue contains three types of fibers: collagen, elastic, and reticular. **Collagen fibers,** the most numerous type, are composed of the protein collagen, the most abundant protein in the body. Collagen is a very tough material. In fact, the tensile strength of collagen fibers has been compared to that of steel. (Meat is tough because of its collagen content.) When treated with hot water, collagen is converted into gelatin, a soluble protein. Collagen fibers are wavy and flexible but resist stretching. Their structure allows them to remain intact when the tissue is stretched.

Elastic fibers branch and fuse to form networks. They can be stretched by a force and then (like a stretched rubber band) return to their original size and shape when the force is removed. Elastic fibers, composed of the protein elastin, are an important component of structures that must stretch.

Reticular fibers are very small branched fibers that form delicate networks not visible in ordinary stained slides. They become apparent when a tissue is stained with silver. Reticular fibers are composed of collagen and some glycoprotein.

Connective tissue contains specialized cells

Fibroblasts are connective tissue cells that produce the fibers as well as the protein and carbohydrate complexes of the matrix. Fibroblasts release protein components that become arranged to form the characteristic fibers. These cells are especially active in developing tissue and are important in healing wounds. As tissues mature, the number of fibroblasts decreases and they become less active.

Macrophages, the scavenger cells of the body, commonly wander through connective tissues, cleaning up cellular debris and phagocytizing foreign matter, including bacteria. Among the other types of cells seen in connective tissues are mast cells, which release histamine during allergic reactions; adipose (fat) cells; and plasma cells, which secrete antibodies.

Loose connective tissue is widely distributed

Loose connective tissue (also called *areolar tissue*) is the most widely distributed connective tissue in the body. Found as a thin filling between body parts, it serves as a reservoir for fluid and salts. Nerves, blood vessels, and muscles are wrapped in this tissue. Together with adipose tissue, loose connective tissue forms the subcutaneous (below-the-skin) layer that attaches skin to the muscles and other structures beneath. Loose connective tissue consists of fibers running in all directions through a semifluid matrix. Its flexibility permits the parts it connects to move.

Dense connective tissue consists mainly of fibers

Dense connective tissue is very strong, though somewhat less flexible than loose connective tissue. Collagen fibers predominate. In **irregular** dense connective tissue, the collagen fibers are arranged in bundles distributed in all directions through the tissue. This type of tissue is found in the lower layer (dermis) of the skin.

In **regular** dense connective tissue, the collagen bundles are arranged in a definite pattern, making the tissue greatly resistant to stress. Tendons, the cable-like cords that connect muscles to bones, consist of dense connective tissue.

Elastic tissue is found in structures that must expand

Elastic connective tissue consists mainly of bundles of parallel elastic fibers. Structures that must expand and then return to their original size, such as lung tissue and the walls of large arteries, contain elastic connective tissue.

Reticular connective tissue provides support

Reticular connective tissue is composed mainly of interlacing reticular fibers. It forms a supporting framework in many organs, including the liver, spleen, and lymph nodes.

Adipose tissue stores energy

Adipose tissue is rich with fat cells, which store fat and release it when fuel is needed for cellular respiration. Adipose tissue is found in the subcutaneous layer and in tissue that cushions internal organs. An immature fat cell is somewhat star-shaped. As fat droplets accumulate within the cytoplasm, the cell assumes a more rounded appearance (Fig. 37–2). Fat droplets merge with one another, eventually forming a single large drop of fat that occupies most of the volume of the mature fat-storing cell. The

(*Text continues on page 808*)

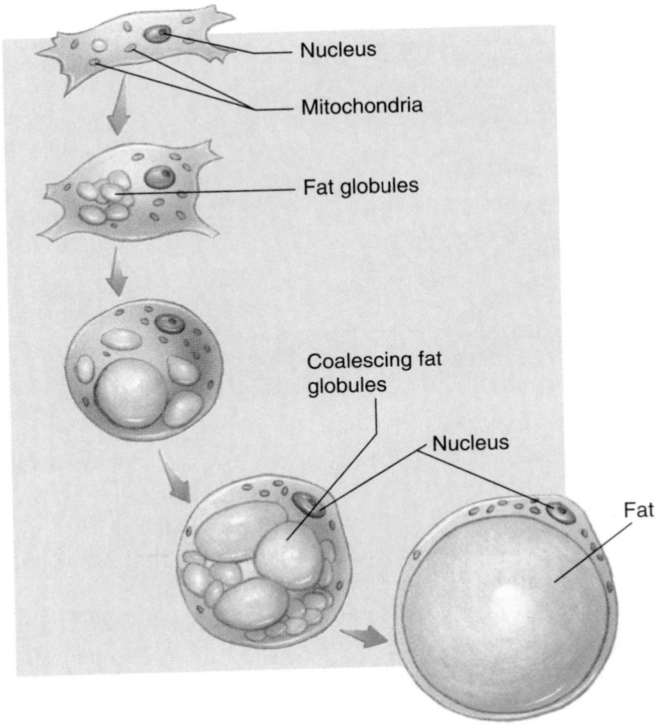

FIGURE 37–2 Fat is stored in fat cells which make up adipose tissue. As fat droplets accumulate in the cytoplasm, they coalesce to form a very large globule of fat. Such a fat globule may occupy most of the cell, pushing the cytoplasm and the organelles to the periphery. See Table 37–2 for a photomicrograph of fat cells filled with fat.

Table 37–2 CONNECTIVE TISSUES

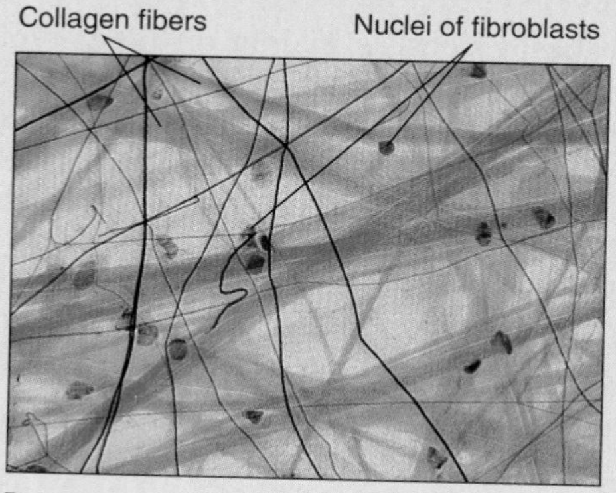

Loose connective tissue. (Ed Reschke)

Loose (areolar) connective tissue

Main Locations
Everywhere that support must be combined with elasticity, such as subcutaneous layer

Functions
Support; reservoir for fluid and salts

Description and Comments
Fibers produced by fibroblast cells embedded in semifluid matrix and mixed with miscellaneous group of other cells

50 µm

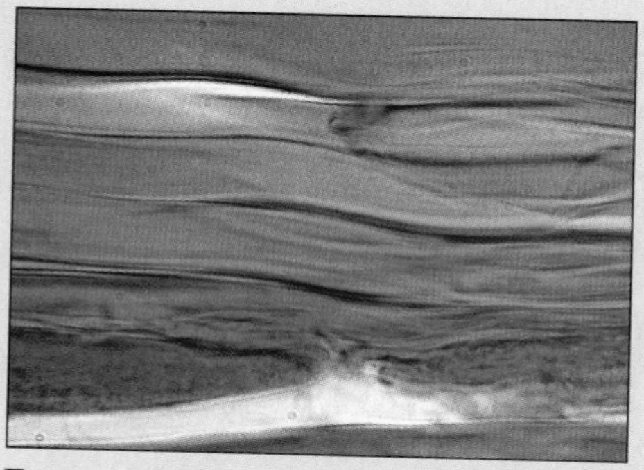

Dense connective tissue. (Dennis Drenner)

Dense connective tissue

Main Locations
Tendons; many ligaments; dermis of skin

Functions
Support; transmission of mechanical forces

Description and Comments
Collagen fibers may be regularly or irregularly arranged

25 µm

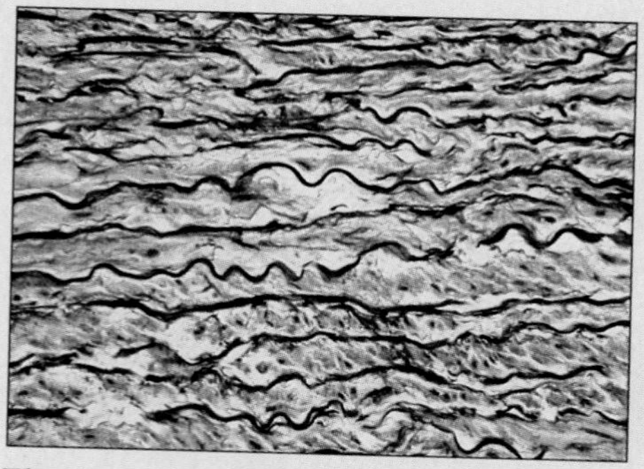

Elastic connective tissue. (Ed Reschke)

Elastic connective tissue

Main Locations
Structures that must both expand and return to their original size, such as lung tissue and large arteries

Functions
Confers elasticity

Description and Comments
Branching elastic fibers interspersed with fibroblasts

50 µm

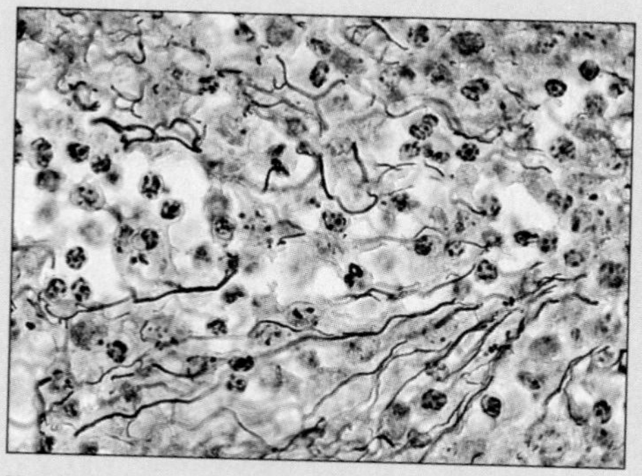

Reticular connective tissue. (Ed Reschke)

Reticular connective tissue

Main Locations
Framework of liver; lymph nodes; spleen

Functions
Support

Description and Comments
Consists of interlacing reticular fibers

50 µm

Table 37–2 continued

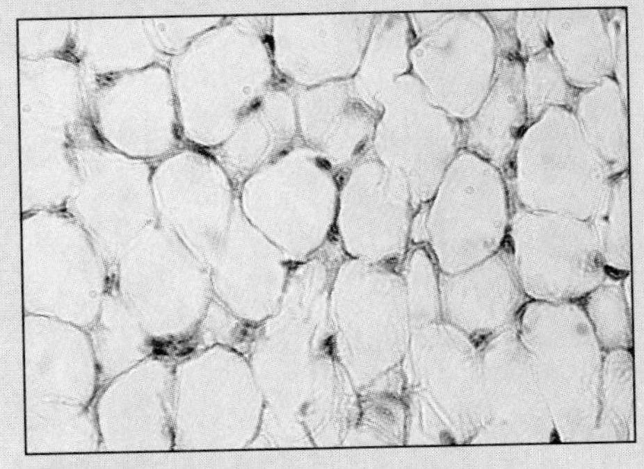

Adipose tissue. (Dennis Drenner)

Adipose tissue

Main Locations
Subcutaneous layer; pads around certain internal organs

Functions
Food storage; insulation; support of such organs as mammary glands, kidneys

Description and Comments
Fat cells are star-shaped at first; fat droplets accumulate until typical ring-shaped cells are produced

50 µm

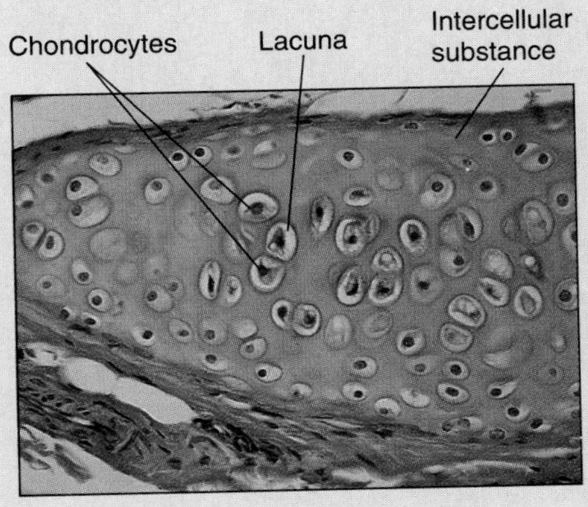

Cartilage. (Ed Reschke)

Cartilage

Main Locations
Supporting skeleton in sharks, rays, and some other vertebrates; ends of bones in other vertebrates; supporting rings in walls of some respiratory tubes; tip of nose; external ear

Functions
Flexible support and reduction of friction in bearing surfaces

Description and Comments
Cells (chondrocytes) separated from one another by intercellular substance; cells occupy lacunae

50 µm

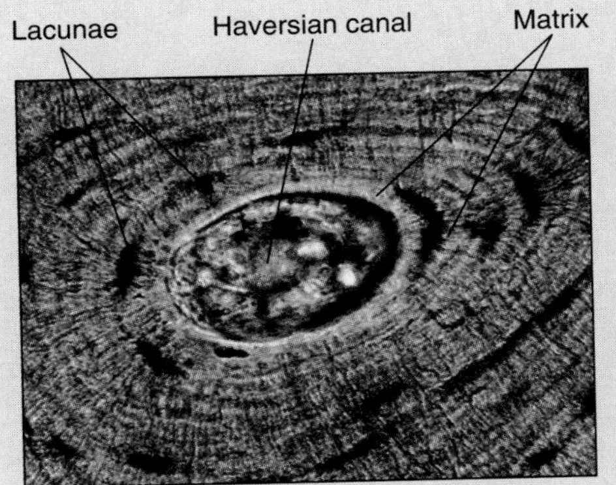

Bone. (Dennis Drenner)

Bone

Main Locations
Forms skeletal structure in most vertebrates

Functions
Support and protection of internal organs; calcium reservoir; skeletal muscles attach to bones

Description and Comments
Osteocytes in lacunae; in compact bone, lacunae arranged in concentric circles about haversian canals

50 µm

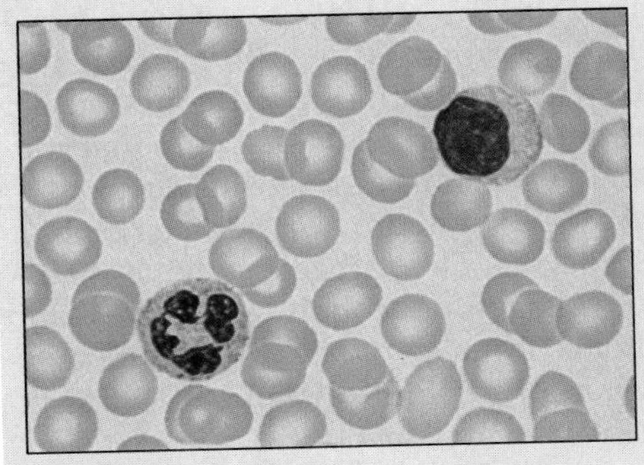

Blood. (Ed Reschke)

Blood

Main Locations
Within heart and blood vessels of circulatory system

Functions
Transports oxygen, nutrients, wastes, and other materials

Description and Comments
Consists of cells dispersed in fluid intercellular substance

25 µm

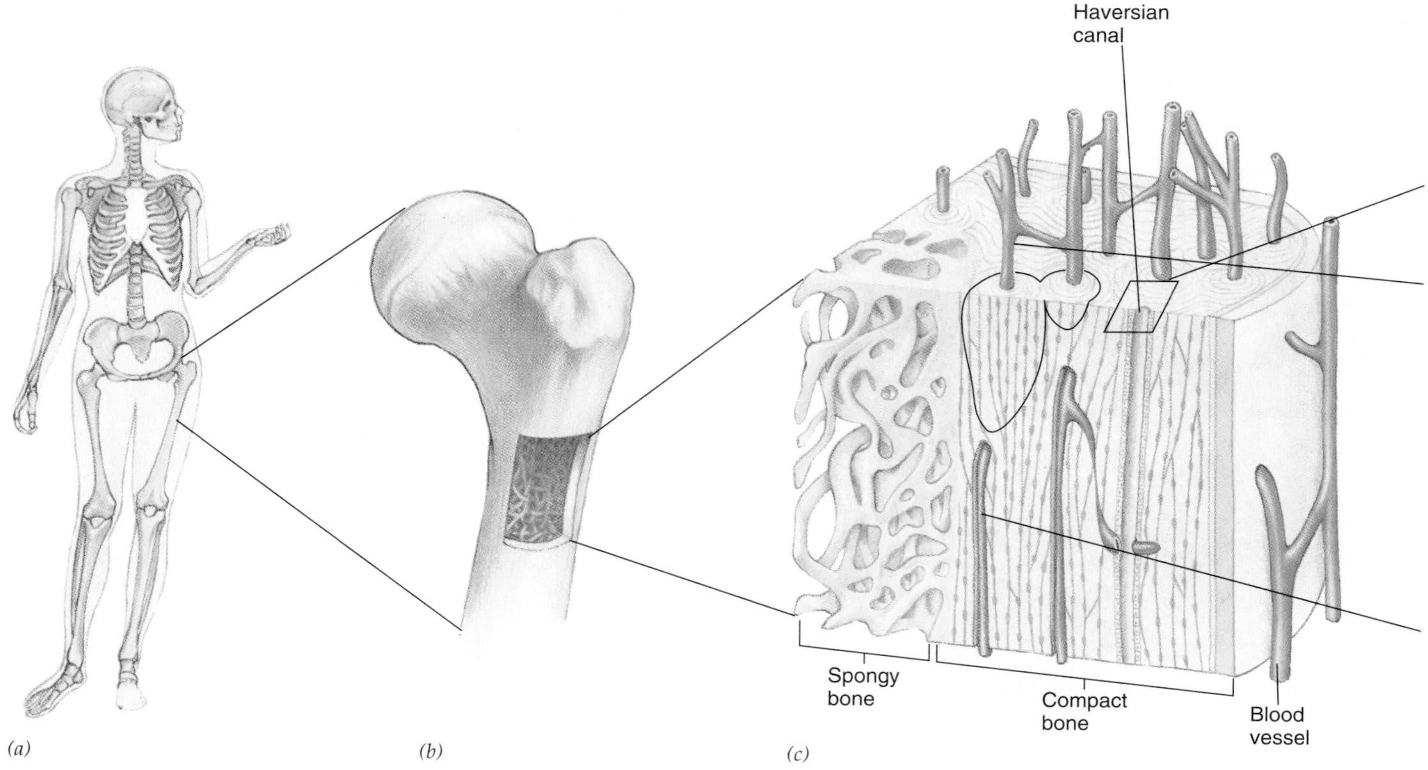

FIGURE 37–3 Compact bone is made up of units called osteons. (*a*) The human skeleton consists mainly of bone. (*b*) A bone is cut open exposing its internal structure. (*c*) Blood vessels and nerves run through the haversian canal within each osteon of compact bone.

cytoplasm and its organelles are pushed to the cell edges, where a bulge is typically formed by the nucleus. A cross section of such a fat cell looks like a ring (the cytoplasm) with a single stone (the nucleus).

When you study a section of adipose tissue through a microscope, it may remind you of chicken wire. The rings of cytoplasm are the "wire," and the large spaces indicate where fat drops existed before they were dissolved by chemicals used to prepare the tissue. These spaces may cause the cells to collapse, giving the tissue a wrinkled appearance.

Cartilage and bone provide support

The supporting skeleton of a vertebrate is composed of cartilage, or of both cartilage and bone. Recall that **cartilage** is the supporting skeleton in the embryonic stages of all vertebrates, but is largely replaced in the adult by bone in all but sharks and rays. In humans, cartilage forms the supporting structure of the external ear, the supporting rings in the walls of the respiratory passageways, the tip of the nose, the ends of some bones, and the discs that serve as cushions between our vertebrae.

Cartilage is firm yet elastic. Its cells, called **chondrocytes,** secrete a hard, rubbery matrix around themselves. Chondrocytes also secrete collagen fibers, which become embedded in the matrix and strengthen it. Chondrocytes eventually come to lie, singly or in groups of two or four, in small cavities in the matrix called **lacunae** (Table 37–2). Chondrocytes remain alive and are nourished by nutrients and oxygen that diffuse through the matrix. Cartilage tissue lacks nerves, lymph vessels, and blood vessels.

Bone is the main vertebrate skeletal tissue. It is similar to cartilage in that it consists mostly of matrix material that contains lacunae; this material is inhabited by the bone cells. These cells, called **osteocytes,** secrete and maintain the matrix (Fig. 37–3). Unlike cartilage, however, bone is a highly vascular tissue with a substantial blood supply. Osteocytes communicate with one another and with capillaries by tiny channels (canaliculi) that contain long cytoplasmic extensions of the osteocytes.

Diffusion alone would not provide sufficient nourishment for the osteocytes. This is because bone matrix consists not only of collagen, mucopolysaccharides, and other organic materials, but also of hydroxyapatite crystals, composed mainly of calcium phosphate. Diffusion through this substance is very slow.

Osteocytes are arranged in concentric layers called *lamellae,* which are formed by the matrix. In turn, the

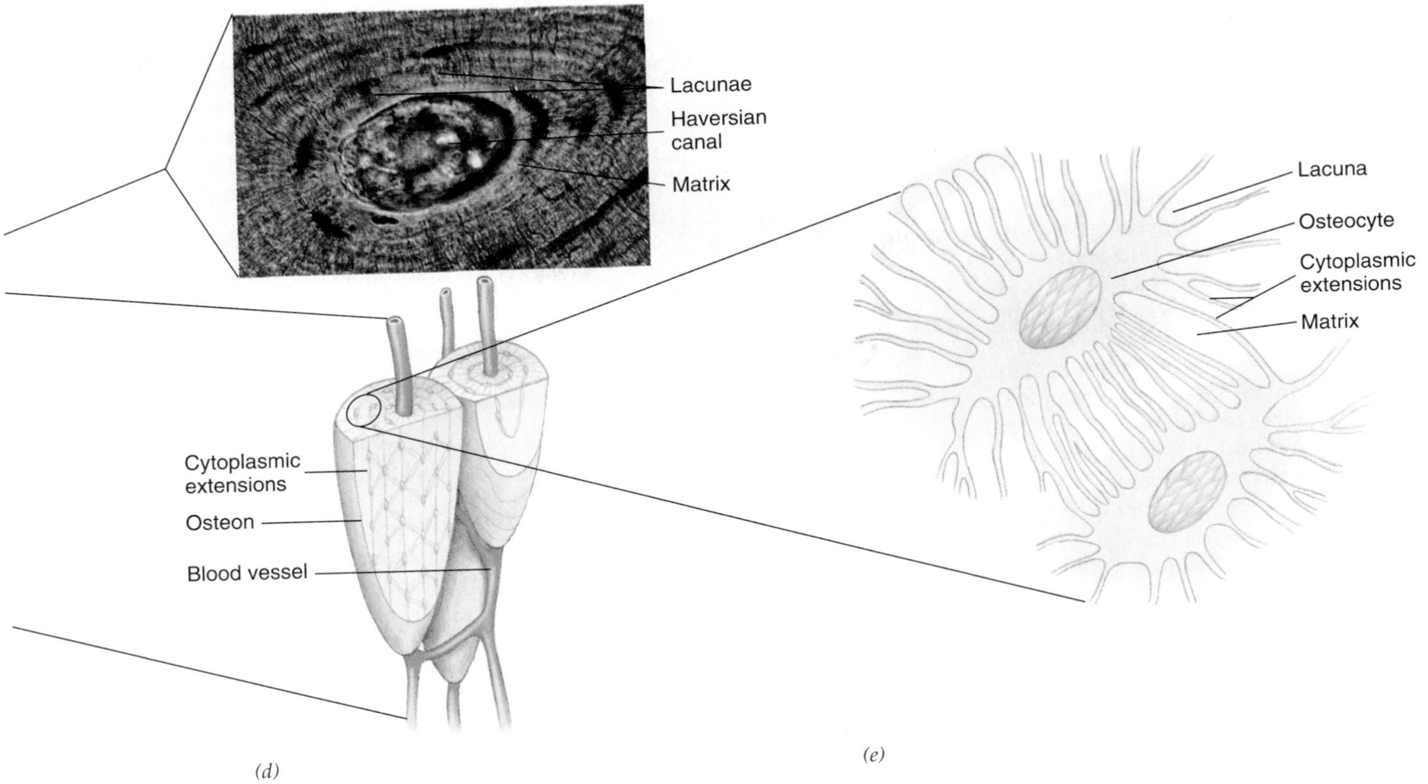

FIGURE 37–3 (continued) (*d*) and (*e*) The bone matrix is rigid and hard. Osteocytes become trapped within lacunae but communicate with one another by way of cytoplasmic extensions that extend through tiny canals. (Dennis Drenner)

lamellae surround central microscopic channels known as **haversian canals,** through which run capillaries and nerves. Each spindle-shaped unit of bone—called an **osteon**—consists of a central blood vessel, surrounding lamellae, and osteocytes.

Bone tissue also contains large multinucleated cells, called **osteoclasts,** that can dissolve and remove the bony substance, as can the osteocytes themselves. The shape and internal architecture of the bone gradually change in response to normal growth processes and physical stress. The calcium salts of bone render the matrix very hard, and collagen prevents the bony matrix from being overly brittle. Bones are amazingly light and strong. Most have a large central **marrow cavity** that may contain a spongy tissue called marrow. Yellow marrow consists mainly of fat. Red marrow is the connective tissue in which blood cells are produced.

Blood and lymph are circulating tissues

Blood and **lymph** are circulating tissues that help other parts of the body communicate and interact. As do all connective tissues, they consist of specialized cells dispersed in an intercellular substance. Blood and lymph are the focus of Chapters 42 and 43.

In mammals, blood is composed of red blood cells, white blood cells, and platelets suspended within **plasma,** the liquid, noncellular part of the blood. Plasma consists of water, proteins, salts, and a variety of substances such as hormones that it transports from one part of the body to another.

The **red blood cells** (*erythrocytes*) of humans and other vertebrates contain the red respiratory pigment hemoglobin, which transports oxygen. The red blood cells of most mammals are flattened, biconcave discs that lack nuclei (Table 37–2). Those of other vertebrates are oval and have nuclei.

Human blood contains five main types of **white blood cells** (*leukocytes*), each with distinct size, shape, structure, and functions. The white blood cells are an important line of defense against disease-causing microorganisms.

Platelets are small fragments broken off from large cells in the bone marrow. In complex vertebrates they play a key role in blood clotting.

MUSCLE TISSUE IS SPECIALIZED TO CONTRACT

The movements of most animals result from the contraction of the elongated, cylindrical, or spindle-shaped cells

Focus On

Unwelcome Tissues: Cancers

In this chapter we have focused on normal tissues. A **neoplasm** ("new growth"), or **tumor,** is an abnormal mass of cells. A neoplasm may be benign or malignant (cancerous). A benign ("kind") tumor tends to grow slowly, and its cells stay together. Because benign tumors form masses with distinct borders, they can usually be removed surgically.

A malignant ("wicked") neoplasm, or **cancer,** usually grows much more rapidly and invasively than a benign tumor. Unlike the cells of benign tumors, cancer cells do not retain normal structural features. Recall that cancer cells lack normal regulatory mechanisms due to the presence of at least one, and perhaps several, oncogenes. Membrane proteins that would normally help regulate cell division and interaction with other cells are replaced by tumor-specific proteins. When a transformed cell multiplies, all the cells derived from it are also abnormal. Cancers that develop from connective tissues or muscle are referred to as **sarcomas.** Those that originate in epithelial tissue are called **carcinomas.**

Two basic defects in behavior that characterize most cancer cells are rapid multiplication and abnormal relations with neighboring cells. Unlike normal cells, which respect one another's boundaries and form tissues in an orderly, organized manner, cancer cells grow helter-skelter upon one another and infiltrate normal tissues. They are apparently no longer able to receive or respond appropriately to signals from surrounding cells; communication is lacking.

Death from cancer almost always results from **metastasis,** a migration of cancer cells through blood or lymph channels to distant parts of the body. Once there, they multiply, forming new malignant neoplasms that can interfere with the normal functions of the tissues being invaded. Cancer often spreads so rapidly and extensively that surgeons are unable to locate or remove all the malignant masses.

Studies suggest that many neoplasms grow to several millimeters in diameter and then enter a dormant stage, which may last for months or even years. At some point, cells of the neoplasm release a chemical substance that stimulates nearby blood vessels to develop new capillaries, that grow into the abnormal mass of cells. Nourished by its new blood supply, the neoplasm may begin to grow rapidly. Newly formed blood vessels have leaky walls, and so are an important route for metastasis. Malignant cells enter the blood through these walls and are transported to new sites.

Cancer cells multiply rapidly and invade normal tissues, interfering with normal function. A healthy bronchial passageway is lined with ciliated cells. The cilia (shown in orange) sweep dust and other foreign particles away from the lungs. Cancer cells (shown in green) invade the bronchial wall and crowd out normal cells lining the bronchial passageway.
(© Boehringer Ingelheim International GmbH, Photo by Lennart Nilsson, *The Incredible Machine,* National Geographic Society.)

Cancer researchers are looking for ways to predict or prevent metastasis and also for treatments for cancer patients whose tumors have already metastasized. Several promising research approaches have been reported. For example, Lance Liotta, Chief of the Laboratory of Pathology at the National Cancer Institute, has researched how cancer cells invade body tissues. He and his colleagues discovered that increased levels of a class of protein-cleaving enzymes (metalloproteinases) correlated with metastasis of several types of human cancers. These investigators found that the enzymes attack collagen, a protein found in connective tissues. When they blocked the activity of the enzymes (by treating tumor cells with antibodies to the enzymes), invasion by the cancer cells did not occur. This research could lead to the development of drugs that prevent metastasis.

Why some persons are more susceptible to cancer than others remains somewhat of a mystery. Apparently, cancer cells occur frequently in everyone, but in most people the immune system destroys them. (The immune system provides protection from disease organisms and other foreign invaders.) According to this theory, cancer is a failure of the immune system.

Several genes have been identified that, when mutated, greatly increase the risk for specific types of cancer. Alleles of some genes appear to affect an individual's level of tolerance to carcinogens (cancer-producing agents). More than 80% of cancer cases are thought to be triggered by carcinogens in the environment.

Cancer is the second greatest cause of death in the United States. One in three persons in the United States gets cancer at some time in his or her life, and two out of three cancer patients die within 5 years of diagnosis. Currently, the key to survival is early diagnosis and treatment with some combination of surgery, hormonal treatment, radiation therapy, and drugs that suppress mitosis, such as chemotherapy. Because cancer is actually an entire family of closely related diseases (there are more than 100 distinct varieties), it is probable that no single cure exists. Most investigators agree, however, that a greater understanding of the control mechanisms and communication systems of cells is necessary before effective cures can be developed.

Risk of developing cancer can be decreased by following these recommendations:

1. Do not smoke or use tobacco. Smoking is responsible for more than 80% of lung cancer cases.
2. Avoid prolonged exposure to the sun. When in the sun, use sunscreen or sun block. Exposure to the sun is responsible for almost all of the 400,000 cases of skin cancer reported each year.
3. Increase the fiber content of your diet and avoid high-fat, smoked, salt-cured, and nitrite-cured foods.
4. Avoid unnecessary exposure to x rays.
5. Women should examine their breasts each month and obtain annual Pap tests. Men should regularly examine their testes, and should have prostate examinations yearly after age 50. ■

Table 37–3 MUSCLE TISSUES

	Skeletal	*Smooth*	*Cardiac*
Location	Attached to skeleton	Walls of stomach, intestines, etc.	Walls of heart
Type of Control	Voluntary	Involuntary	Involuntary
Shape of Fibers	Elongated, cylindrical, blunt ends	Elongated, spindle-shaped, pointed ends	Elongated, cylindrical fibers that branch and fuse
Striations	Present	Absent	Present
Number of Nuclei per Fiber	Many	One	One or two
Position of Nuclei	Peripheral	Central	Central
Speed of Contraction	Most rapid	Slowest	Intermediate
Ability to Remain Contracted	Least	Greatest	Intermediate

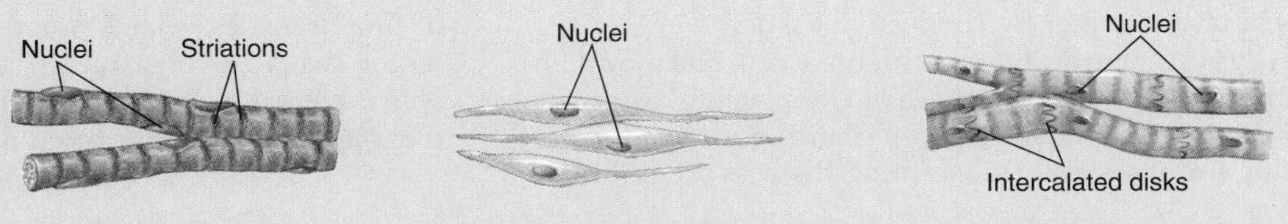

(*a*) Skeletal muscle fibers (*b*) Smooth muscle fibers (*c*) Cardiac muscle fibers

of **muscle tissue.** Each muscle cell is referred to as a **fiber** because of its length. A muscle fiber contains many thin, longitudinal, parallel contractile fibers called **myofibrils.** Two proteins, **myosin** and **actin,** are the chief components of myofibrils, and play a key role in contraction of muscle cells.

Vertebrates have three types of muscle tissue: smooth, skeletal, and cardiac (Table 37–3). **Smooth muscle** occurs in the walls of the digestive tract, uterus, blood vessels, and certain other internal organs. Each spindle-shaped fiber contains a single nucleus.

Skeletal muscle makes up the large muscle masses attached to the bones of the body. Skeletal muscle fibers are very long—up to 2 or 3 centimeters (about 1 inch). Each skeletal muscle fiber has many nuclei, an exception to the generalization that cells contain only one nucleus. The nuclei of skeletal muscle fibers are also unusual in their position. They lie peripherally, just under the plasma membrane, which frees the entire central part of the skeletal muscle fiber for the contractile units, the myofibrils. This arrangement appears to be an adaptation that increases the efficiency of contraction. Whereas skeletal muscle fibers are generally under voluntary control, cardiac and smooth muscle fibers are normally not regulated at will.

Light microscopy reveals that both skeletal and cardiac fibers have alternating light and dark transverse stripes, or **striations,** that change their relative sizes during contraction. Striated muscle fibers can contract rapidly but cannot remain contracted for a long period of time. They must relax and rest momentarily before contracting again.

Cardiac muscle is the main tissue of the heart. The fibers of cardiac muscle are joined end-to-end, and they branch and rejoin, forming complex networks. One or two nuclei are found within each fiber. A characteristic feature of cardiac muscle tissue is the presence of *intercalated discs,* specialized junctions where the fibers join.

NERVOUS TISSUE CONTROLS MUSCLES, GLANDS, AND OTHER ORGANS

Nervous tissue is composed of **neurons,** cells specialized for conducting electrochemical nerve impulses, and **glial cells,** cells that support and nourish the neurons (Fig. 37–4). Certain neurons receive signals from the external or internal environment and transmit them to the spinal cord and brain. Other neurons process and store the information. Still others transmit information from the

Making the Connection

Stress and Homeostasis

How does stress affect the body? **Stressors** are changes in the internal or external environment that disturb homeostasis, causing **stress** to the body. Examples of external stressors are heat, cold, noise, abnormal pressure, and lack of oxygen. Internal stressors include changes in blood pressure, pH, and salt concentration, as well as high and low blood-sugar levels. Many stressors occur routinely. The body responds automatically by activating homeostatic mechanisms that expertly manage the stress. These mechanisms often involve the nervous and endocrine systems.

Other stressors are more severe and may cause serious disruption. When homeostatic mechanisms are unable to restore the steady (normal) state, the stress may cause malfunction that leads to disease or even death. In fact, we can view death as the failure of some homeostatic mechanism. In Chapter 47 we will discuss the role of the adrenal glands in regulating homeostatic adjustments that help the body effectively cope with stress. ▲

brain and spinal cord to the muscles, glands, and other organs of the body. Neurons communicate at junctions called **synapses.** A **nerve** consists of a great many neurons bound together by connective tissue.

A typical neuron has an enlarged **cell body** containing the nucleus, and two types of cytoplasmic extensions (see Chapter 39). **Dendrites** are fibers specialized for receiving impulses and transmitting them to the cell body. The single **axon** transmits impulses away from the cell body. Axons are usually long and smooth but may give off an occasional branch. They typically end in a group of fine branches. Axons range in length from a millimeter or two to over a meter. Those extending from the spinal cord down the arm or leg in a human, for example, may be a meter or more in length.

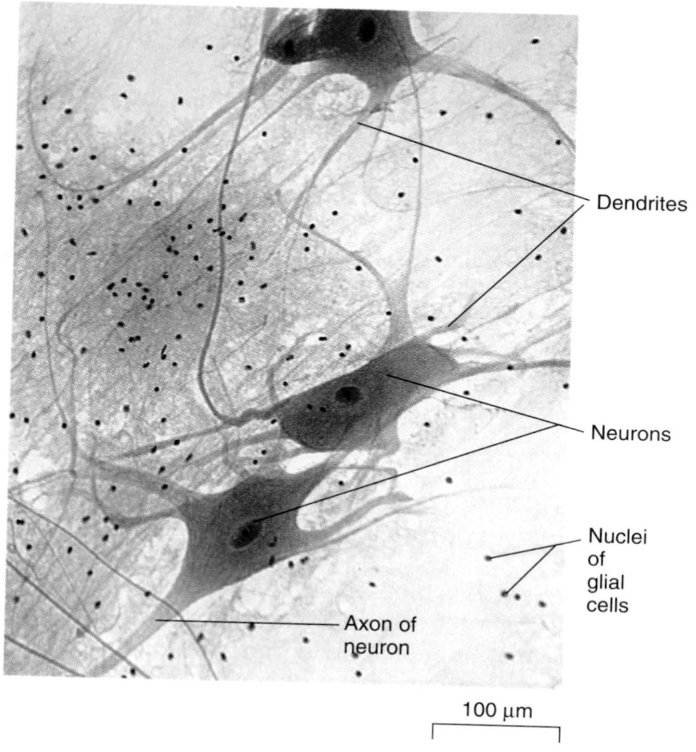

FIGURE 37–4 Nervous tissue consists of neurons and glial cells. (Ed Reschke)

COMPLEX ANIMALS HAVE ORGANS AND ORGAN SYSTEMS

Although an animal organ may be composed predominantly of one type of tissue, other types are needed to provide support, protection, and a blood supply, and to allow transmission of nerve impulses. For example, the heart consists mainly of cardiac muscle tissue, but it is lined and covered by endothelium (a tissue that resembles epithelium), contains blood vessels composed of smooth muscle and connective tissue, and is regulated by nervous tissue.

Several tissues and organs may work together, performing a specialized set of functions. Such an organized group of structures makes up an organ system. In complex animals we can identify ten major organ systems that work together to make up the organism (Fig. 37–5). The main organ systems of complex animals include the **integumentary, skeletal, muscle, nervous, circulatory, digestive, respiratory, urinary, endocrine,** and **reproductive systems.** Table 37–4 summarizes their principal organs and functions.

Consider the digestive system as an example of an organ system. Its organs include the mouth, esophagus, stomach, small and large intestines, liver, pancreas, and salivary glands. The digestive system processes food, reducing it to small molecular components. The products of digestion are absorbed into the blood, which transports them to all of the cells.

Table 37–4 THE MAMMALIAN ORGAN SYSTEMS AND THEIR FUNCTIONS

System	*Components*	*Functions*	*Homeostatic Ability*
Integumentary	Skin, hair, nails, sweat glands	Covers and protects body	Sweat glands help control body temperature; as barrier, the skin helps maintain steady state
Skeletal	Bones, cartilage, ligaments	Supports and protects body; provides for movement and locomotion; stores calcium	Helps maintain constant calcium level in blood
Muscular	Skeletal muscle; cardiac muscle; smooth muscle	Moves parts of skeleton; provides locomotion; moves internal materials	Ensures vital functions requiring movement, e.g., cardiac muscle circulates the blood
Digestive	Mouth, esophagus, stomach, intestines, liver, pancreas	Ingests and digests foods; absorbs them into blood	Maintains adequate supplies of fuel molecules and building materials
Circulatory	Heart, blood vessels, blood; lymph and lymph structures (lymphatic system is a subsystem of the circulatory system)	Transports materials from one part of body to another; defends body against disease	Transports oxygen, nutrients, hormones; removes wastes; maintains water and ionic balance of tissues
Respiratory	Lungs, trachea, and other air passageways	Exchanges gases between blood and external environment	Maintains adequate blood oxygen content and helps regulate blood pH; eliminates carbon dioxide
Urinary	Kidney, bladder, and associated ducts	Excretes metabolic wastes; removes excessive substances from blood	Helps regulate volume and composition of blood and body fluids
Nervous	Nerves and sense organs; brain and spinal cord	Receives stimuli from external and internal environment; conducts impulses; integrates activities of other systems	Principal regulatory system
Endocrine	Ductless glands (e.g., pituitary, adrenal, thyroid) and tissues that secrete hormones	Regulates blood chemistry and many body functions	In conjunction with nervous system, regulates metabolic activities and blood levels of various substances
Reproductive	Testes, ovaries, and associated structures	Provides for continuation of species	Passes on genetic endowment of individuals; maintains secondary sexual characteristics

ORGAN SYSTEMS WORK TOGETHER TO MAINTAIN HOMEOSTASIS

In a complex animal, billions of cells are organized to form tissues, organs, and organ systems. The organism functions effectively, in large part, because very precise control mechanisms maintain an appropriate internal environment. If the organism is to survive and function, the composition of the fluids that bathe its cells must be carefully regulated. An appropriate concentration of nutrients, oxygen and other gases, ions, and compounds needed for metabolism must be available at all times. In

(Text continues on page 816)

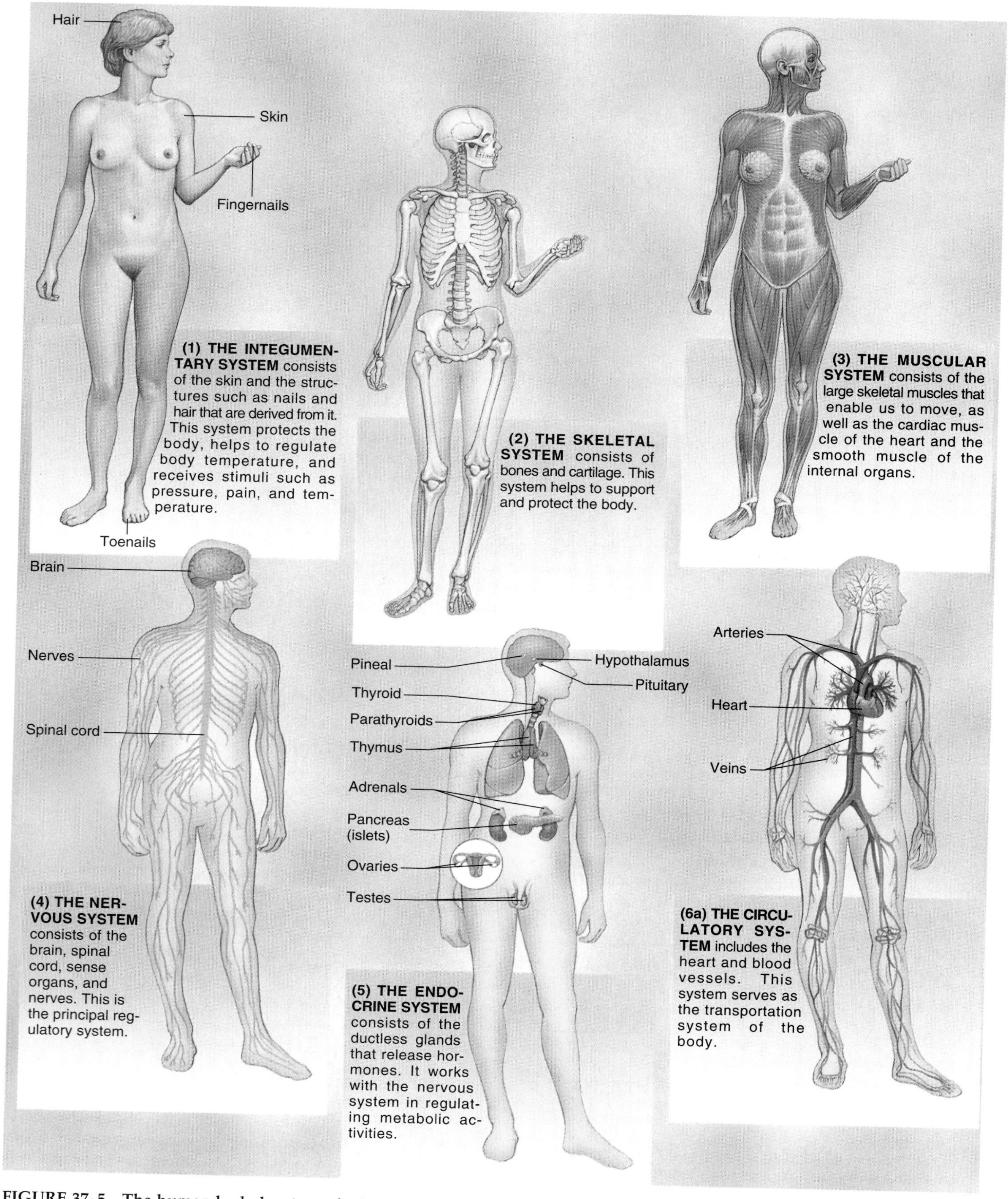
Hair
Skin
Fingernails
Toenails
(1) THE INTEGUMENTARY SYSTEM consists of the skin and the structures such as nails and hair that are derived from it. This system protects the body, helps to regulate body temperature, and receives stimuli such as pressure, pain, and temperature.
(2) THE SKELETAL SYSTEM consists of bones and cartilage. This system helps to support and protect the body.
(3) THE MUSCULAR SYSTEM consists of the large skeletal muscles that enable us to move, as well as the cardiac muscle of the heart and the smooth muscle of the internal organs.
Brain
Nerves
Spinal cord
(4) THE NERVOUS SYSTEM consists of the brain, spinal cord, sense organs, and nerves. This is the principal regulatory system.
Pineal
Hypothalamus
Pituitary
Thyroid
Parathyroids
Thymus
Adrenals
Pancreas (islets)
Ovaries
Testes
(5) THE ENDOCRINE SYSTEM consists of the ductless glands that release hormones. It works with the nervous system in regulating metabolic activities.
Arteries
Heart
Veins
(6a) THE CIRCULATORY SYSTEM includes the heart and blood vessels. This system serves as the transportation system of the body.

FIGURE 37–5 **The human body has ten principal organ systems.**

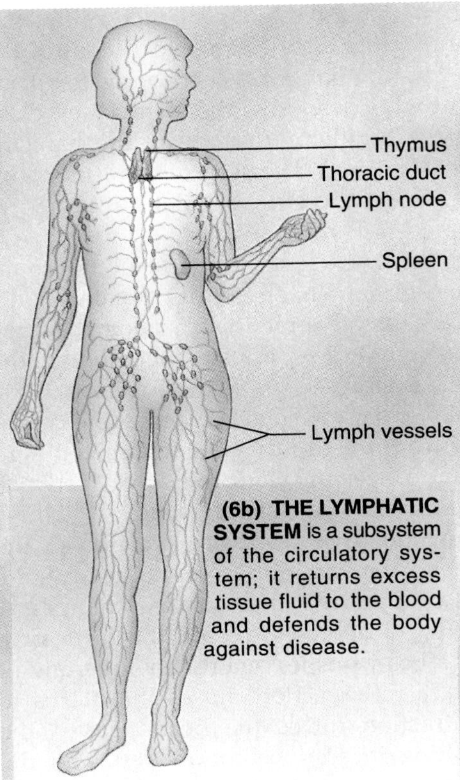

(6b) THE LYMPHATIC SYSTEM is a subsystem of the circulatory system; it returns excess tissue fluid to the blood and defends the body against disease.

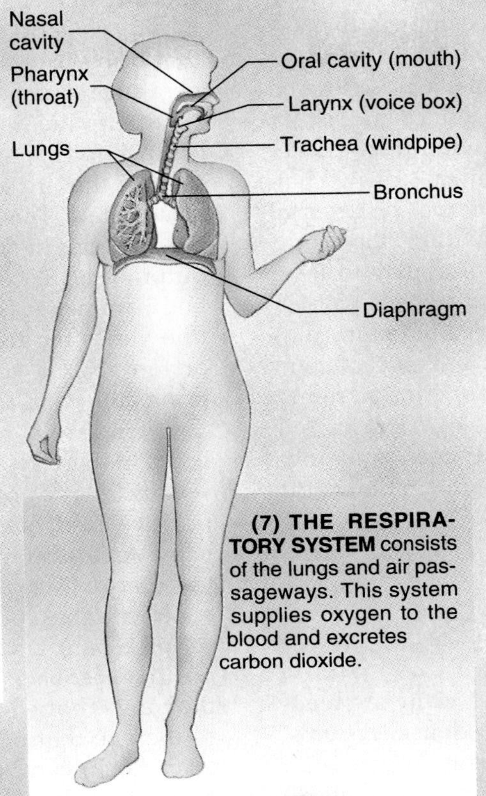

(7) THE RESPIRATORY SYSTEM consists of the lungs and air passageways. This system supplies oxygen to the blood and excretes carbon dioxide.

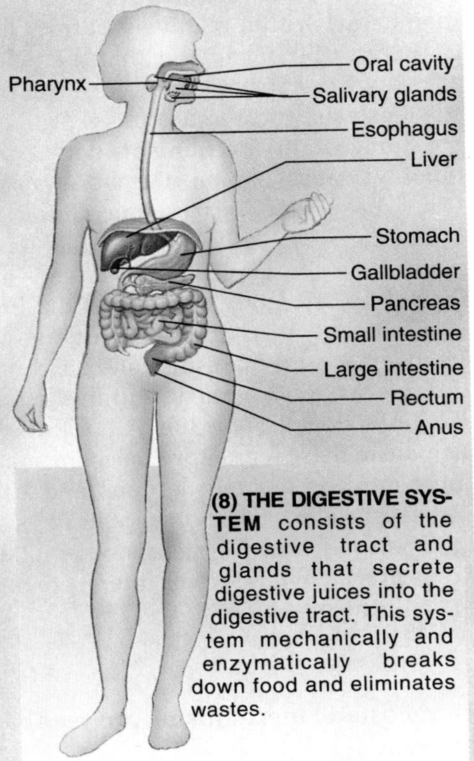

(8) THE DIGESTIVE SYSTEM consists of the digestive tract and glands that secrete digestive juices into the digestive tract. This system mechanically and enzymatically breaks down food and eliminates wastes.

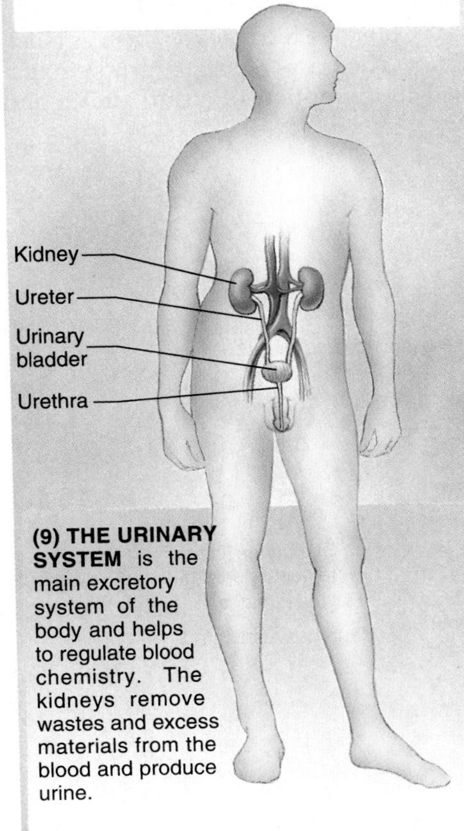

(9) THE URINARY SYSTEM is the main excretory system of the body and helps to regulate blood chemistry. The kidneys remove wastes and excess materials from the blood and produce urine.

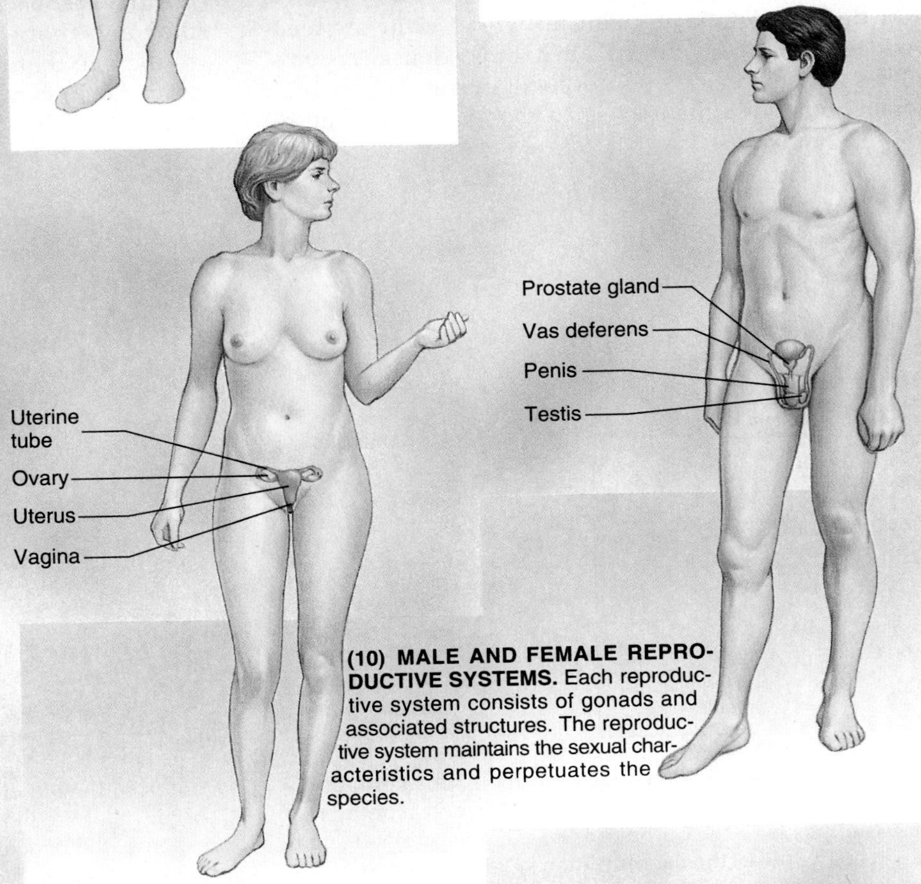

(10) MALE AND FEMALE REPRODUCTIVE SYSTEMS. Each reproductive system consists of gonads and associated structures. The reproductive system maintains the sexual characteristics and perpetuates the species.

addition, internal temperature and pressure must be maintained within relatively narrow limits.

Recall from Chapter 1 that the tendency to maintain a relatively constant internal environment is termed **homeostasis,** and the processes that accomplish the task are **homeostatic mechanisms.** First coined by the physiologist Walter Cannon, the word *homeostasis* is derived from the Greek *homoios,* meaning "same," and *stasis,* "standing." Actually, the internal environment never really stays the same. Homeostasis is always being challenged by **stressors,** changes in the internal or external environment that affect conditions within the body. Homeostatic mechanisms interact continuously to keep the internal environment within the narrow physiological limits that support life. All of the organ systems participate in these regulatory mechanisms, but most of them are controlled by the nervous and endocrine systems. Homeostasis is a basic concept in physiology. During our study of the organ systems, we will discuss numerous ways in which the systems interact to maintain the steady state of the organism.

Homeostatic mechanisms are feedback systems

How do homeostatic mechanisms work? Many are **feedback systems,** sometimes called biofeedback systems. Such a system consists of a cycle of events in which information about a change (e.g., a change in temperature) is fed back into the system so that the regulator (the temperature-regulating center in the brain) can control the process (temperature regulation). When body temperature becomes too high or too low, the change serves as input, triggering the regulator to activate mechanisms that bring it back to normal. The return to normal temperature signals the temperature-regulating center to "shut off" the homeostatic mechanisms.

In this type of feedback system, the response counteracts the inappropriate change, thus restoring the steady state. This is a **negative feedback system,** because the response of the regulator is opposite (negative) to the output (Fig. 37–6). Most homeostatic mechanisms in the body are negative feedback systems. When some condition varies too far from the steady state (either too high or too low), a control system using negative feedback brings the condition back to the steady state.

There are a few **positive feedback systems** in the body. In these systems the variation from the steady state sets off a series of changes that intensify the changes. A positive feedback cycle operates during the birth of a baby. As the baby's head pushes against the opening of the uterus (cervix), a reflex action causes the uterus to contract. The contraction forces the head against the cervix again, resulting in another contraction, and the positive feedback cycle is repeated again and again until the baby is born. Many positive feedback sequences (such as the one that occurs when a person goes into circulatory shock) can lead to disruption of steady states and even to death.

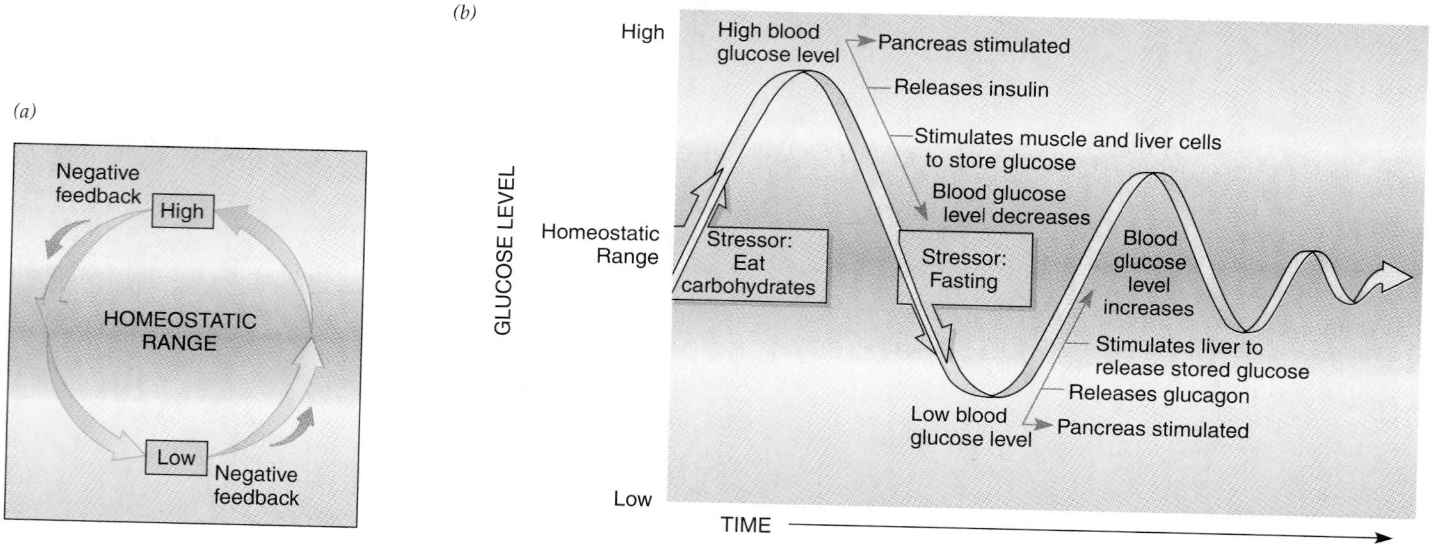

FIGURE 37–6 Negative feedback mechanisms maintain homeostasis. (*a*) In negative feedback the response of the regulator is opposite to the output. (*b*) Regulation of blood-sugar level. When you eat, the concentration of glucose in your blood increases. This increase stimulates the pancreas to release insulin, a hormone that stimulates cells to take up glucose from the blood and store it. After several hours, the glucose level of the blood begins to fall below its normal level. In response, the pancreas releases another hormone, glucagon, that stimulates cells to release stored glucose. This action increases glucose concentration.

Homeostatic mechanisms regulate blood-sugar level

Regulation of blood-sugar level provides a good example of homeostatic mechanisms at work (Fig. 37–6*b*). When you wake up in the morning your blood-sugar (glucose) level is about 90 milligrams of glucose per 100 milliliters of blood. Perhaps you eat a big breakfast that includes pastry or a doughnut. Many of the starches and sugars in your breakfast are digested to glucose. The glucose is then absorbed into the circulatory system, causing the blood-sugar level to rise.

An increase in blood-sugar level stimulates the pancreas to release the hormone insulin. This hormone causes the body cells to remove glucose from the blood, and stimulates the liver and muscle cells to store glucose (as glycogen). As a result, the glucose level in the blood decreases and returns to the normal fasting level of 90 mg/100 mL.

After several hours, when the glucose level of the blood begins to fall below the normal level, the pancreas releases the hormone glucagon. This hormone raises the blood-sugar level by stimulating the liver cells to slowly convert glycogen to glucose and thereby release their stored glucose. In this way, insulin and glucagon act in seesaw fashion to maintain a steady state of glucose in the blood.

SUMMARY

I. Multicellular organisms can be much larger and more diverse than unicellular ones, and their cells can specialize, performing specific functions.

II. A tissue consists of a group of similarly specialized cells that associate to perform one or more functions. Animal tissues are classified as epithelial, connective, muscular, or nervous.
   - A. Epithelial tissue may form a continuous layer, or sheet, of cells covering a body surface or lining a body cavity. Some epithelial tissue is specialized to form glands.
      1. Epithelial tissue functions in protection, absorption, secretion, or sensation.
      2. Epithelial cells may be squamous, cuboidal, or columnar in shape.
      3. Epithelial tissue may be simple, stratified, or pseudostratified (summarized in Table 37–1).
   - B. Connective tissue joins other tissues of the body, supports the body and its organs, and protects underlying organs. Connective tissue consists of relatively few cells; these are separated by intercellular substance composed of fibers scattered through a matrix.
      1. The intercellular substance contains collagen, elastic, and reticular fibers. Connective tissue contains specialized cells such as fibroblasts and macrophages.
      2. Some types of connective tissue are loose and dense connective tissues, elastic connective tissue, reticular connective tissue, adipose tissue, cartilage, bone, and blood (summarized in Table 37–2).
   - C. Muscle tissue is composed of cells specialized to contract. Each cell is an elongated fiber containing many contractile units called myofibrils. The chief components of myofibrils are the proteins actin and myosin.
      1. Skeletal muscle is striated and under voluntary control.
      2. Cardiac muscle is striated; its contraction is involuntary.
      3. Smooth muscle contracts involuntarily. It is responsible for movement of food through the digestive tract and for movement of other body organs.
   - D. Nervous tissue is composed of neurons, which are cells specialized for conducting impulses, and glial cells, which are supporting cells.

III. Organs and tissues work together, forming organ systems. In complex animals, ten principal organ systems work together, making up the living organism. Among these are the digestive system, the nervous system, and the skeletal system (summarized in Table 37–4).

IV. Homeostasis is the body's automatic tendency to maintain a constant internal environment, or steady state. It is maintained by negative feedback mechanisms.

SELECTED KEY TERMS

adipose tissue
blood
bone tissue
cardiac muscle
cartilage
chondrocyte
collagen
connective tissue
dense connective tissue
elastic connective tissue
endocrine gland
endocrine system
epithelial tissue
exocrine gland
fiber
gland
glial cell
goblet cell
homeostasis
integumentary system
intercellular substance
loose connective tissue
macrophage
matrix
muscle tissue
myofibril
negative feedback system
nervous tissue
organ
organ system
osteocyte
osteon
positive feedback system
skeletal muscle
smooth muscle
stressor
tissue

POST-TEST

1. A group of closely associated cells that carry out specific functions forms a(an) __________ .
2. A(an) __________ consists of epithelial cells that produce and secrete a product.
3. Tissues and organs working together can form a(an) __________ __________ .
4. Intercalated discs are characteristic of __________ __________ .
5. __________ muscle is striated and voluntary.
6. Supporting cells found in nervous tissue are called __________ cells.
7. The organ system made up of glands that secrete hormones is the __________ system.
8. The organ system that covers the body is the __________ system.
9. In a(an) __________ feedback system, the response counteracts the inappropriate change.
10. Tissue that covers and lines body surfaces is __________ tissue.
11. Tissue that receives stimuli and transmits impulses is __________ tissue.
12. The contractile fibers in muscle tissue are __________ .
13. Tissue that contains fibroblasts and a great deal of intercellular substance is __________ tissue.
14. A widely distributed connective tissue that along with adipose tissue forms the subcutaneous layer is __________ connective tissue.
15. Tissue that has a hard, rubbery matrix and lacks blood vessels is __________ .
16. Tissue composed of osteons is __________ tissue.
17. Tissue found in walls of large arteries and in lungs is __________ connective tissue.
18. __________ glands secrete their product onto a free epithelial surface.
19. Homeostatic mechanisms in the body depend mainly on __________ __________ systems.
20. Changes in the internal or external environment that disrupt homeostasis are called __________ .

REVIEW QUESTIONS

1. What advantages do multicellular organisms have over unicellular organisms? Can you think of any disadvantages?
2. What are the functions of epithelial tissues? How are these tissues adapted to carry out these functions?
3. What is the structure of bone? Of adipose tissue? Of loose connective tissue? How is each adapted to carry out its special functions?
4. Compare the properties of the three types of muscle.
5. How is the neuron uniquely adapted for its function?
6. What kinds of tissues would you expect to find in the following organs: (a) lung; (b) heart; (c) intestine; (d) salivary glands?
7. List the principal organ systems found in a complex animal and give the functions of each.
8. How do the cells of a malignant neoplasm differ from those of a normal tissue? What is metastasis?
9. Identify each type of tissue in the figure.

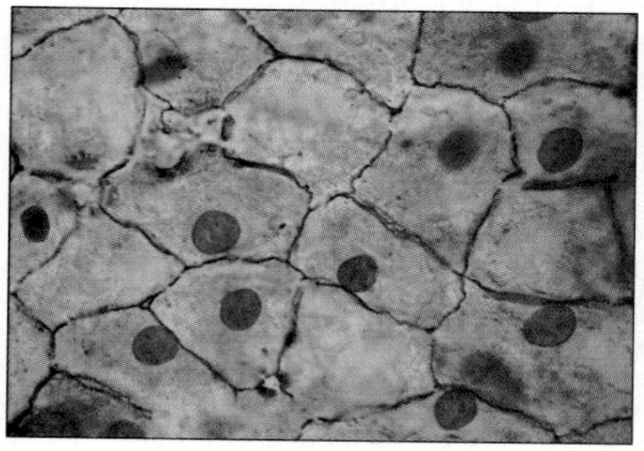

(a)

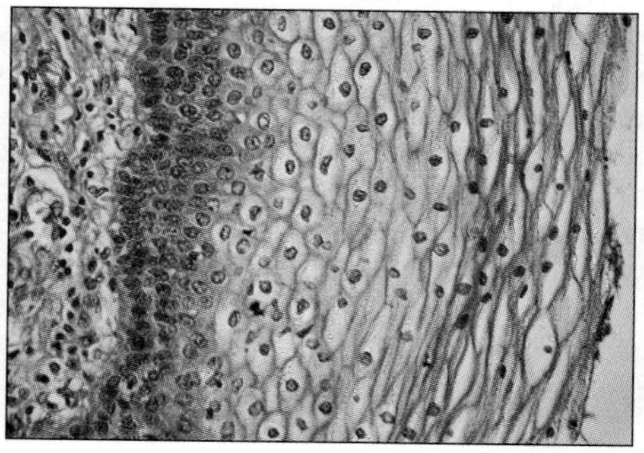

(b)

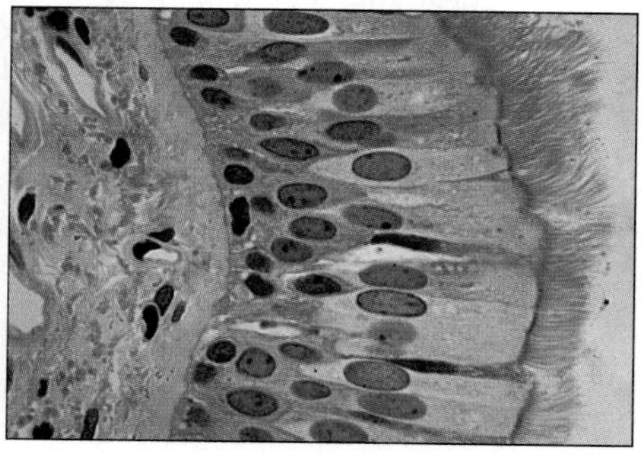

(c)

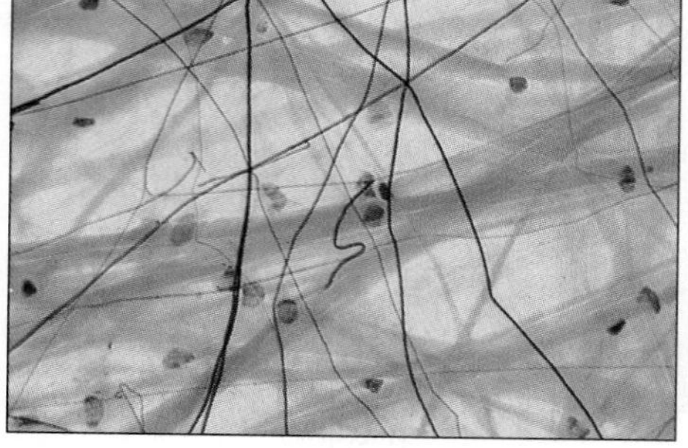

(d)

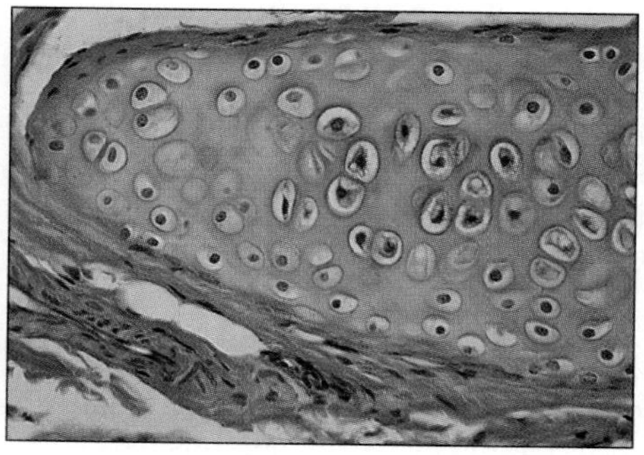

(e)

(Ed Reschke)

YOU MAKE THE CONNECTION

1. Imagine that all of the epithelium in a complex animal, such as a human, suddenly disappeared. What effects might this have on the body and its ability to function?

2. What would connective tissue be like if it had no intercellular substance? What effect would its absence have on the body?

RECOMMENDED READINGS

Fackelman, K. "Variations on a Theme: Interplay of Genes and Environment Elevates Cancer Risk," *Science News,* Vol. 147, No. 18, May 6, 1995. An overview of cancer-causing genes and polymorphisms and how they interact with environmental factors in increasing risk for cancer.

Liotta, L.A. "Cancer Cell Invasion and Metastasis." *Scientific American,* Vol. 266, No. 2, February 1992. A discussion of current research on mechanisms and treatment of metastatic cancer.

National Geographic Society Book Service. *The Incredible Machine.* National Geographic, Washington, D.C., 1986. A beautiful and informative introduction to the human body, featuring the art of renowned photographer Lennart Nilsson.

Solomon, E. P., R. Schmidt, and P. Adragna. *Human Anatomy and Physiology,* 2d ed. Saunders College Publishing, Philadelphia, 1990. A very readable presentation of human anatomy and physiology.

C H A P T E R 38

Protection, Support, and Movement: Skin, Skeleton, and Muscle

Some epithelial coverings secrete mucus. The mucus secreted by the foot of the striped land snail (*Helicella candicans*) produces a slime track through which the animal glides. (Y. Momatiuk/Photo Researchers, Inc.)

Some animals run; some jump; some fly. Others remain rooted to one spot, sweeping their surroundings with tentacles. Many contain internal circulating fluids, pumped by hearts and contained by hollow vessels that maintain their pressure with gentle squeezing. Most have digestive systems that push food along with peristaltic contractions. In all these cases, each action is powered by muscle, a specialized tissue that, however varied its effects, performs only one action: it contracts.

In many animals, the muscle and skeletal systems work together. Muscles responsible for locomotion are anchored to the skeleton, which gives them something firm to pull or push against. In humans and many other animals, bones serve as levers that transmit the force necessary to move body parts. The skeleton also serves to support the body and protect the delicate organs within.

The epithelial coverings of invertebrates and the skin of vertebrates also protect underlying tissues. These coverings may be specialized to perform additional functions such as temperature regulation, gas exchange, excretion of wastes, or secretion of mucus or other substances.

In our survey of the animal kingdom in Chapters 28 to 30, we made many references to protective coverings, movement, and skeletal systems in terms of adaptation of individual animal groups. In Chapter 37 we described animal tissues, organs, and systems, laying the foundation for the more detailed discussions in this and later chapters. Here we focus on skin, skeleton, and muscle—systems that are closely interrelated in function and significance. We compare these systems across several animal groups and then focus on their structures and functions in the human body.

LEARNING OBJECTIVES

After you have studied this chapter you should be able to

1. Compare the structure and function of vertebrate skin with those of the external epithelium of invertebrates, and identify the principal derivatives of vertebrate skin.
2. Compare the advantages and disadvantages of different types of skeletal systems, including the hydrostatic skeleton, the exoskeleton, and the endoskeleton.
3. Identify the main divisions of the vertebrate skeleton and the bones that make up each division.
4. Describe the structure of a typical long bone, and differentiate between endochondral and intramembranous bone development.
5. Describe the gross and microscopic structures of skeletal muscle.
6. List, in sequence, the events that take place during muscle contraction.
7. Compare the roles of glycogen, creatine phosphate, and ATP in providing energy for muscle contraction.
8. Describe the antagonistic action of muscles.
9. Summarize the functional relationship between skeletal and muscle tissues.

OUTER COVERINGS PROTECT THE BODY

Epithelial tissue covers all external and internal surfaces of the animal body. The epithelial covering of invertebrates and the skin of vertebrates form a protective shield around the body.

The epithelium of invertebrates may function in secretion or gas exchange

In invertebrates, the external epithelium protects the body and may also be specialized for secretion, absorption, or gas exchange. Epithelial cells may be modified as sensory cells that are selectively sensitive to light, chemical stimuli, or mechanical stimuli such as contact or pressure.

In many species, the epithelium contains secretory cells that produce a protective cuticle or secrete lubricants or adhesives. In some species, secretory cells release odorous secretions that are used for communication among members of the species or for marking trails. In other species, the cells produce poisonous secretions that are used for offense or defense. In earthworms, a lubricating mucous secretion serves as a moist slime for diffusion of gases across the body wall and for reduction of friction during movement through the soil.

In some species, an epithelial secretion may be limited to a particular region of the body surface. The foot of the gastropod mollusk, for example, releases a mucous secretion, producing a slime track along which the snail glides. Certain insects such as weaver ants release epithelial secretions as fine, very strong threads that are used to construct nests. The spinning glands of many spiders develop from epithelial cells, and lepidopteran insects such as butterflies and moths synthesize silk from amino acids in silk-forming glands.

The vertebrate skin functions in protection and temperature regulation

In many fishes, in the African ant-eating pangolin (a mammal), and in some reptiles, skin has developed into a set of scales formidable enough to be considered armor. Even human skin has considerable strength. Structures derived from skin include fingernails and toenails, hair, sweat glands, oil (sebaceous) glands, and several types of sensory receptors that give us the ability to feel pressure, temperature, and pain (Fig. 38–1).

The skin of mammals contains mammary glands, specialized in females for secretion of milk. Oil glands in human skin empty via short ducts into hair follicles (Fig. 38–2). They secrete a substance called **sebum,** a complex mixture of fats, waxes, and hydrocarbons. In humans these glands are especially numerous on the face and scalp. The oil secreted keeps the hair moist and pliable and prevents the skin from drying and cracking. Sebum also contains substances that inhibit the growth of harmful bacteria. (At puberty, excessive sebum, produced in response to increased levels of sex hormones, may fill the glands and follicles, producing acne.)

In humans, the skin helps maintain body temperature (see Chapter 46). About 2.5 million sweat glands secrete sweat, and its evaporation from the surface of the skin lowers the body temperature. Constriction and dilation of capillaries in the skin are also part of homeostatic mechanisms for regulating body temperature.

The skin in some other vertebrates differs considerably from ours. Instead of hairs, birds have feathers, which form in a manner comparable to hairs and provide even more effective insulation than fur. Among the ectothermic vertebrates, we find epidermal scales (as in reptiles), naked skin covered with mucus (as in many amphibians), and skin with bony or toothlike scales (as in fishes). Some skin, such as that of certain tropical frogs, is equipped with poison glands. Skin and its derivatives

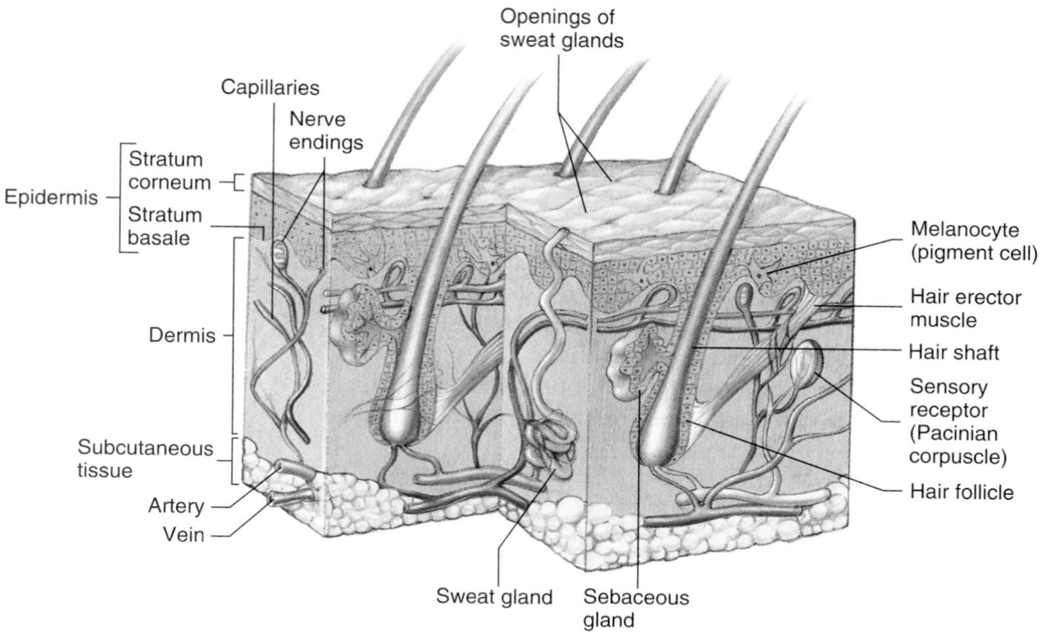

FIGURE 38–1 The integumentary system of mammals is a complex organ system consisting of skin and structures such as hair and nails that are derived from the skin.

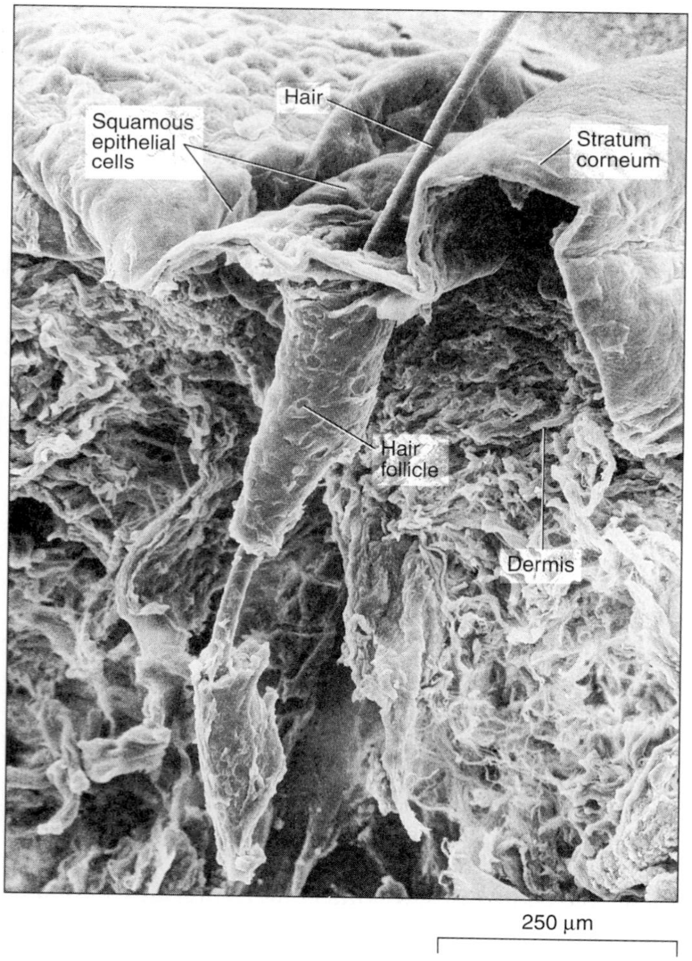

FIGURE 38–2 This scanning electron micrograph shows a hair follicle in human skin. (Courtesy of Dr. Karen A. Holbrook)

are often brilliantly colored in connection with courtship rituals, territorial displays, and other kinds of communication. The human blush pales alongside the spectacular displays of animals such as peacocks.

The epidermis is a waterproof protective barrier

The outer layer of skin, the **epidermis,** is the interface between the delicate tissues within the vertebrate body and the hostile outside environment. The epidermis consists of several strata, or sublayers. The deepest is **stratum basale,** and the most superficial is **stratum corneum** (Fig. 38–1). Pigment cells in stratum basale and in the dermis produce melanin, a pigment that contributes to the color of the skin (see *Making the Connection: Skin Cancer and the Environment*). In stratum basale, cells continuously divide and are pushed outward as other cells are produced below them. As the epidermal cells move toward the skin surface, they mature. Because most vertebrates have no capillaries in the epidermis, the maturing cells receive less and less nourishment, which diminishes their metabolic activity.

As they move toward the body surface, epidermal cells manufacture **keratin,** an elaborately coiled protein that gives the skin considerable mechanical strength and flexibility. Keratin is quite insoluble and serves to waterproof the body surface. As epidermal cells move through the stratum corneum, they die. When they reach the outer surface of the skin, they wear off and must be continuously replaced.

Making the Connection

Skin Cancer and the Environment

Skin cancer is on the rise. Most cases are caused by excessive, chronic exposure to the ultraviolet radiation of sunlight. Ultraviolet radiation damages DNA, causing mutations that lead to malignant transformation (see Chapter 16, *Focus On: Oncogenes and Cancer*). The ozone layer in the stratosphere, which absorbs incoming ultraviolet radiation from the sun, is being destroyed by pollutants such as chlorofluorocarbons. Environmental scientists hypothesize that continuing depletion of the ozone shield around our planet will cause one million extra cases of skin cancer each year, and that about 30,000 of these victims will die. Malignant melanoma is a type of skin cancer that spreads rapidly through the body and may cause death within a few months after diagnosis.

Exposure to ultraviolet rays causes the epidermis to thicken and stimulates pigment cells in the skin to produce melanin at an increased rate. An increase in melanin causes the skin to become darker. Melanin is an important protective screen against the sun because it absorbs some of the harmful ultraviolet rays. The suntan so prized by sun worshippers is actually a sign that the skin has been exposed to too much ultraviolet radiation. When the melanin is not able to absorb all the ultraviolet rays, the skin becomes inflamed, or sunburned. Because dark-skinned people have more melanin, they suffer less sunburn, wrinkling, and skin cancer.

Most skin cancer can be prevented by protecting the body from sunlight and other forms of ultraviolet radiation. Tanning machines that use artificial light to induce tanning also expose the body to ultraviolet rays and so pose the same risks as sunbathing. Those who choose to expose themselves to sunlight can protect their skin by limiting the time they spend in the sun, and by applying effective sunscreens. ▲

The dermis contains blood vessels and other structures

Beneath the epidermis lies the **dermis** (Fig. 38–1), composed of a dense, fibrous connective tissue made up mainly of collagen fibers. Collagen imparts strength and flexibility to the skin. The major part of each sweat gland is embedded in the dermis, and the hair follicles extend into it (Fig. 38–2). The dermis also contains blood vessels, which nourish the skin, and sensory receptors for touch, pain, and temperature. Mammalian skin rests on a layer of **subcutaneous tissue** that is composed mainly of adipose tissue. This fatty tissue insulates the body from outside temperature extremes.

SKELETONS ARE IMPORTANT IN LOCOMOTION, PROTECTION, AND SUPPORT

In some animals without hard skeletons, muscle acts directly on the jelly-like substance of the body itself, or on a fluid-filled body cavity. In more complex animals, a hard skeleton receives, transmits, and transforms the contraction of muscle tissue into the variety of motions animals use. In addition to its function in locomotion, the skeleton supports the body and protects the internal organs.

In echinoderms and chordates, the skeleton is internal, in the form of plates or shafts of calcium-impregnated tissue (such as cartilage or bone). But in most animals, the skeleton is not a living tissue but a lifeless deposit atop the epidermis—a shell or **exoskeleton.**

In hydrostatic skeletons, body fluids are used to transmit force

Imagine an elongated balloon full of water. If you were to pull on it, it would lengthen and become thinner. It would do the same if you squeezed it. Conversely, if you pushed the ends toward the center, it would shorten and thicken. Many animals, including cnidarians and annelids, have a **hydrostatic skeleton** that works something like a balloon filled with water. Contracting muscles push against a tube of water (or other fluid). Because fluids cannot be compressed, the force is transmitted through the fluid and generates movement of the body.

In *Hydra* and other cnidarians, cells of the two body layers are capable of contraction. The contractile cells in the outer epidermal layer are arranged longitudinally, whereas the contractile cells of the inner layer (the gastrodermis) are arranged circularly around the central body axis (Fig. 38–3). The two groups of cells work in **antagonistic** fashion: what one can do, the other can undo. When the epidermal (longitudinal) layer contracts, the hydra shortens. Because of the fluid in the gastrovascular cavity, force is transmitted so that the hydra thickens as well. On the other hand, when the inner (circular) layer contracts, the hydra thins, and its fluid contents force it to lengthen.

Mechanically, the hydra is little more than a simple bag of fluid. The fluid acts as a hydrostatic skeleton because it transmits force when the contractile cells contract against it. Hydrostatic skeletons permit only crude mass movements of the body or its appendages. Delicate move-

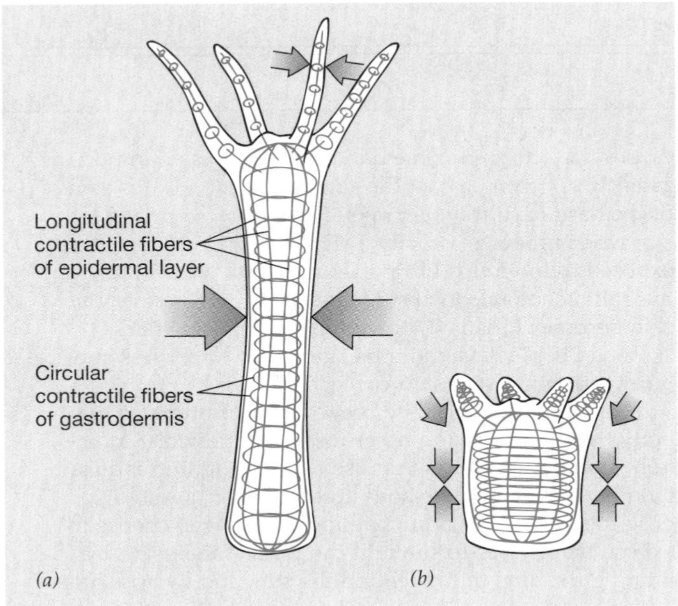

FIGURE 38–3 Movement in *Hydra* depends on its hydrostatic skeleton. The longitudinally arranged cells are antagonistic to the cells arranged in circles around the body axis. (*a*) Contraction of the circular muscles elongates the body. (*b*) Contraction of the longitudinal muscles shortens the body.

ments are difficult, because force tends to be transmitted equally in all directions throughout the entire fluid-filled body of the animal. For example, it is not easy for the hydra to thicken one part of its body while thinning another.

The more sophisticated hydrostatic skeleton of the annelid worm enables it to be more versatile in movement. Recall from Chapter 29 that the earthworm's body consists of a series of transverse partitions, or septa, that isolate portions of the body cavity and its contained fluid. This isolation permits the hydrostatic skeletons of each segment to be largely independent of one another. Thus, the contraction of the circular muscle in the elongating anterior end need not interfere with the action of the longitudinal muscle in the segments of the posterior.

Hydrostatic skeletons work fine for many types of small aquatic animals, and for animals like the burrowing earthworm. However, most animals that move rapidly and in a highly coordinated way have a skeletal framework composed of minerals, proteins, and carbohydrates. Even so, some examples of hydrostatic skeletons occur even in complex invertebrates equipped with shells or endoskeletons, and in vertebrates with endoskeletons of cartilage or bone. Among mollusks, for example, the clam extends and anchors its foot by a hydrostatic blood pressure mechanism similar to that used by earthworms. Sea stars and sea urchins move their tube feet by an ingenious version of the hydrostatic skeleton. And even the human penis becomes erect and stiff because of the turgidity of pressurized blood in its cavernous spaces.

Mollusks and arthropods have nonliving exoskeletons

In both mollusks and arthropods, the exoskeleton is a nonliving product of the epidermal cells. In mollusks, the exoskeleton basically provides protection — a retreat used in emergencies — with the bulk of the naked, tasty body exposed at other times.

The exoskeletons of arthropods serve not only to protect but to transmit forces. In this respect they are comparable to the skeletons of vertebrates. Although the arthropod exoskeleton is a continuous, one-piece sheath covering the entire body, it varies greatly in thickness and flexibility. Large, thick, inflexible plates are separated from one another by thin, flexible joints arranged segmentally. Enough joints are provided to make the arthropod's body as flexible as those of many vertebrates. The exoskeleton is also extensively modified to form specialized tools or weapons, or is otherwise adapted to a vast variety of lifestyles.

A major disadvantage of the rigid arthropod exoskeleton is that it interferes with growth. To accommodate growth, the arthropod must **molt** — that is, shed its exoskeleton and replace it with a new, larger one (Fig. 38–4). During molting the animal is weak and vulnerable to predators.

FIGURE 38–4 A greengrocer cicada is shown molting. This insect requires 13 years to mature. (Judy Davidson/Science Photo Library/Photo Researchers, Inc.)

Internal skeletons are capable of growth

Endoskeletons, or internal skeletons, are best developed in echinoderms and chordates. Composed of living tissue, the endoskeleton grows along with the animal as a whole, eliminating the need for molting. The endoskeleton probably also permits a greater variety of possible motions than does an exoskeleton.

Echinoderms have spicules and plates of calcium salts embedded in the tissues of their body walls, forming what amounts to an internal shell that provides support and protection (Fig. 38–5). Many echinoderm endoskeletons bear spines that project to the outer surface.

The internal skeletons of vertebrates provide support and protection, and transmit forces. Members of class Chondrichthyes (sharks and rays) have skeletons composed of cartilage, but in most vertebrates the skeleton consists mainly of bone. Although the skeletons of adult vertebrates vary, their bones have been shown to be homologous. For example, a number of elements of the human skull originate embryonically in the same way that the gill arches of fishes do. Some of these develop into the tiny middle ear bones (the malleus, incus, and stapes).

FIGURE 38–5 The echinoderm endoskeleton provides support and protection. The endoskeleton of the sea urchin is composed of spicules and plates of nonliving calcium salts embedded in tissues of the body wall. (Brian Parker/Tom Stack & Associates)

The vertebrate skeleton has two main divisions: the axial and appendicular skeletons. The **axial skeleton,** located along the central axis of the body, consists of the skull, vertebral column, ribs, and sternum (breastbone). The **appendicular skeleton** consists of the bones of the limbs (arms and legs) plus the bones making up the girdles—the shoulder (pectoral) girdle and most of the hip (pelvic) girdle—that connect the limbs to the axial skeleton (Fig. 38–6).

The **skull,** the bony framework of the head, consists of the cranial and facial bones. In the human, eight cranial bones enclose the brain, and 14 bones make up the facial portion of the skull. Several cranial bones that are single in the adult human result from the fusion of two or more bones that are separate in the embryo or even in the newborn.

The vertebrate spine, or **vertebral column,** supports the body and bears its weight. In mammals it consists of 24 **vertebrae** and two fused bones, the **sacrum** and **coccyx.** The vertebral column consists of the **cervical** (neck) region, with seven vertebrae; the **thoracic** (chest) region, with 12 vertebrae; the **lumbar** (back) region, with five vertebrae; the **sacral** (pelvic) region, with five fused vertebrae; and the **coccygeal** region, also composed of fused vertebrae.

Although vertebrae differ in size and shape in various regions of the vertebral column, a typical vertebra consists of a bony **centrum** (central portion), which bears most of the body weight, and a dorsal ring of bone called the **neural arch,** which surrounds and protects the delicate spinal cord. Vertebrae may also have projections for the attachment of ribs and muscles and for articulating (joining) with neighboring vertebrae. The first vertebra, the **atlas** (named for the mythical Greek who held the world on his shoulders), has rounded depressions on its upper surface into which fit two projections from the base of the skull.

The **rib cage** is a bony "basket" formed by the **sternum** (breastbone), thoracic vertebrae, and, in mammals, 12 pairs of ribs. The rib cage protects the internal organs of the chest, including the heart and lungs. It also supports the chest wall, preventing it from collapsing as the diaphragm contracts with each breath. Each pair of ribs is attached dorsally to a separate vertebra. Of the 12 pairs of ribs in the human, the first seven are attached ventrally to the sternum; the next three are attached indirectly by cartilages; and the last two, the "floating ribs," have no attachments to the sternum.

The **pectoral girdle** consists of two collarbones, or **clavicles,** and two shoulderblades, or **scapulas.** The **pelvic girdle** consists of a pair of large bones, each composed of three fused hipbones. Whereas the pelvic girdle is securely fused to the vertebral column, the pectoral girdle is loosely and flexibly attached to it by muscles.

Each human limb consists of 30 bones, and terminates in five **digits**—the fingers and toes. The more special-

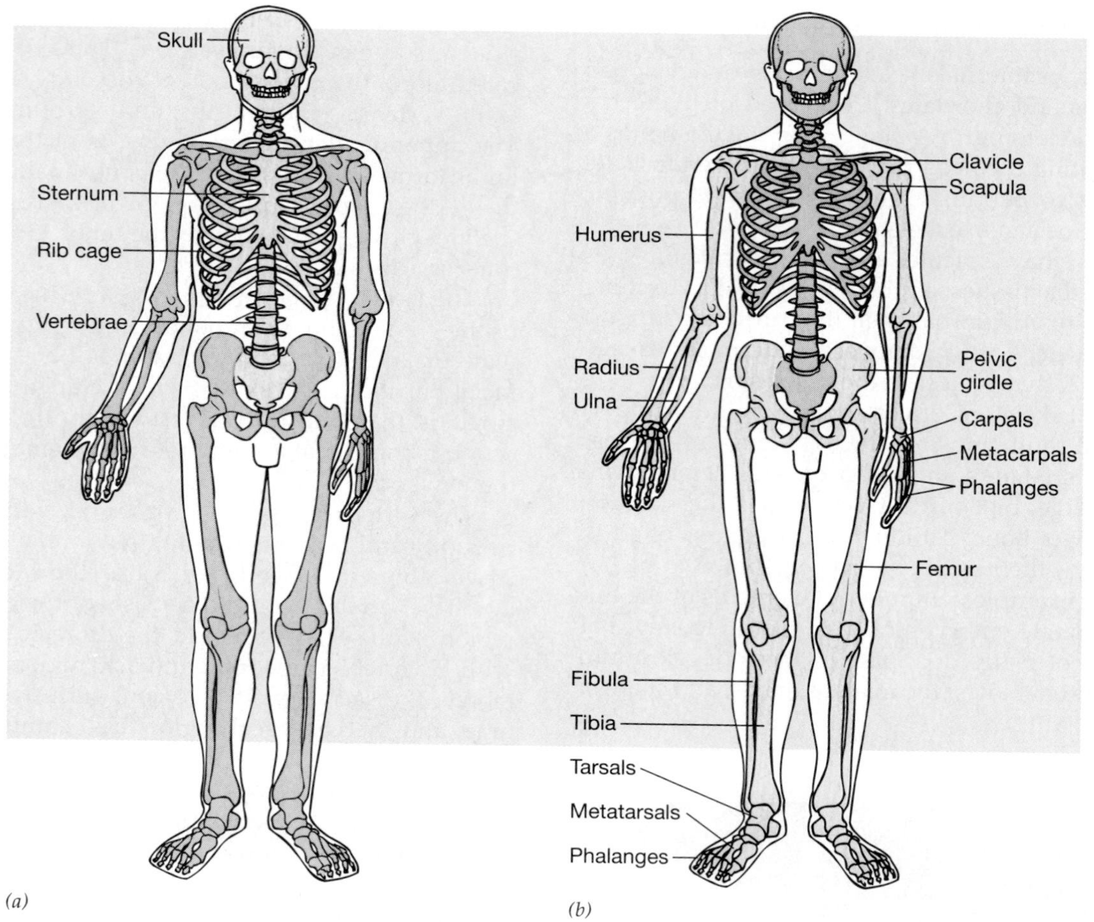

FIGURE 38–6 The skeletal system provides support and protection. (***a***) Bones of the axial skeleton, anterior view. (***b***) Bones of the appendicular skeleton, anterior view.

ized appendages of other animals may be characterized by four digits (as in the pig), three (as in the rhinoceros), two (as in the camel), or one (as in the horse).

Great apes and humans have a highly specialized feature: the opposable thumb. (Great apes also have an opposable big toe. Although the human big toe is not opposable, it is similar enough in structure to the thumb that it can be surgically substituted.) The opposable thumb can be readily wrapped around objects, such as a tree limb, in climbing, and it is especially useful in grasping and manipulating objects.

In humans, the upper limbs are not generally used for locomotion as they are in other mammals, including the great apes. Our hands have been emancipated. The combination of opposable thumbs and upright posture enables us to use our hands to write, shape, build, and use weapons. These abilities have permitted the human species to change its environment to a greater extent than any other organism in the history of the Earth.

A typical long bone consists of compact and spongy bone

The radius, one of the two bones of the forearm, is a typical long bone (Fig. 38–7). Its numerous muscle attachments are arranged in such a way that the bone operates as a lever, amplifying the motion generated by the muscles. By themselves, muscles cannot shorten enough to produce large movements of the body parts to which they are attached.

Like other bones, the radius is covered by a connective tissue membrane, the **periosteum,** capable of producing fresh layers of bone and thus increasing the bone's diameter. The main shaft of a long bone is its **diaphysis;** the expanded ends are called **epiphyses.** In children, a disk of cartilage, the **metaphysis,** is found between the epiphyses and the diaphysis. Metaphyses are growth centers that disappear at maturity, becoming vague **epiphyseal lines.** Long bones contain a central cavity filled with

bone marrow. The **red marrow** found in certain bones produces blood cells. **Yellow marrow** consists mainly of a fatty connective tissue.

The radius has a thin outer shell of **compact bone,** which is very dense and hard. Compact bone is found primarily near the surfaces of a bone, where it provides great strength. Recall from Chapter 37 that compact bone consists of interlocking spindle-shaped units called **osteons,** or **haversian systems** (see Fig. 37–3). Within an osteon, **osteocytes** (bone cells) are found in small cavities called **lacunae** (sing., *lacuna*). The lacunae are arranged in concentric circles around central **haversian canals.** Blood vessels that nourish the bone tissue pass through the haversian canals. Threadlike extensions of the cytoplasm of the osteocytes extend through narrow channels called *canaliculi*. These cellular extensions connect the osteocytes.

Interior to the thin shell of compact bone is a filling of **spongy bone** (also called *cancellous bone*). Spongy bone consists of a mesh of thin strands of bone that, despite its loose structure, provides most of the bone's mechanical strength. The spaces within spongy bone are filled with bone marrow.

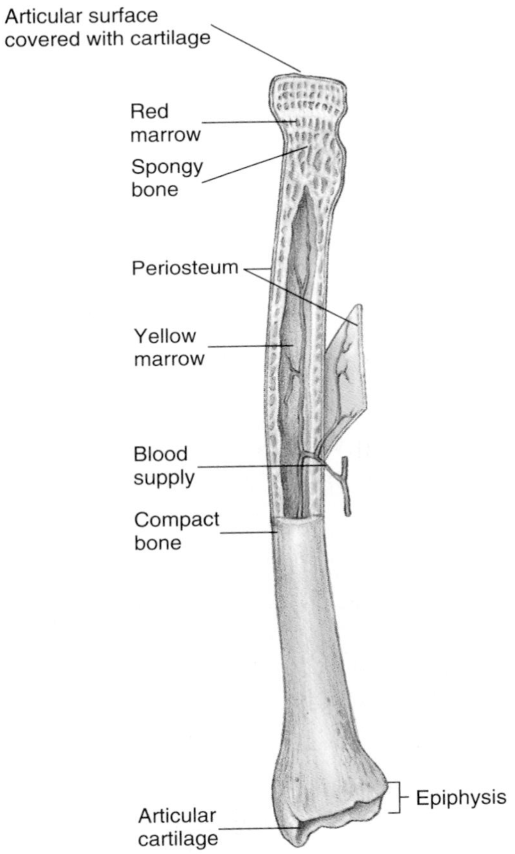

FIGURE 38–7 A typical long bone has a thin shell of compact bone and a filling of spongy bone.

Bones are remodeled throughout life

During fetal development, bones form in two ways. Long bones, such as the radius, develop from cartilage templates in a process called **endochondral bone development.** In contrast, the flat bones of the skull, the vertebrae, and some other bones develop from a noncartilage connective tissue scaffold. This process is known as **intramembranous bone development.**

Osteoblasts are bone-building cells. They secrete the protein collagen, which forms the strong fibers of bone. The compound hydroxyapatite, composed mainly of calcium phosphate, is present in the tissue fluid. It automatically crystallizes around the collagen fibers, forming the hard matrix of bone. As the matrix forms around the osteoblasts, they become isolated within the lacunae. The trapped osteoblasts are then referred to as osteocytes.

Bones are modeled during growth, and remodeled continuously throughout life in response to physical stresses and other changing demands. As muscles develop in response to physical activity, the bones to which they are attached thicken and become stronger. As a bone grows, tissue must be removed from its interior, especially from the walls of the marrow cavity. This process keeps the bone from becoming too heavy. **Osteoclasts** are very large cells that break down bone in a process referred to as bone resorption. The osteoclasts move about, secreting enzymes that digest bone. Osteoclasts and osteoblasts work together to shape bones. Most bone is replaced as many as ten times during the course of an average lifetime.

Joints are junctions between bones

Joints hold bones together, and many of them permit flexibility and movement. At joint surfaces, the outer surface of each bone consists of articular cartilages. One way to classify joints is according to the degree of movement they allow. The sutures found between bones of the skull are **immovable joints.** In a suture, bones are held together by a thin layer of dense fibrous connective tissue, which may be replaced by bone in the adult. **Slightly movable joints,** found between vertebrae, are made of cartilage and help absorb shock.

Most joints are **freely movable joints.** Each is enclosed by a joint capsule composed of connective tissue and lined with a membrane that secretes a lubricant called **synovial fluid.** This thick fluid reduces friction during movement and also absorbs shock. The joint capsule is typically reinforced by **ligaments,** bands of fibrous connective tissue that connect bones and limit movement at the joint.

Joints wear down with time and use. In osteoarthritis, a common joint disorder, cartilage repair does not keep up with degeneration, and the articular cartilage

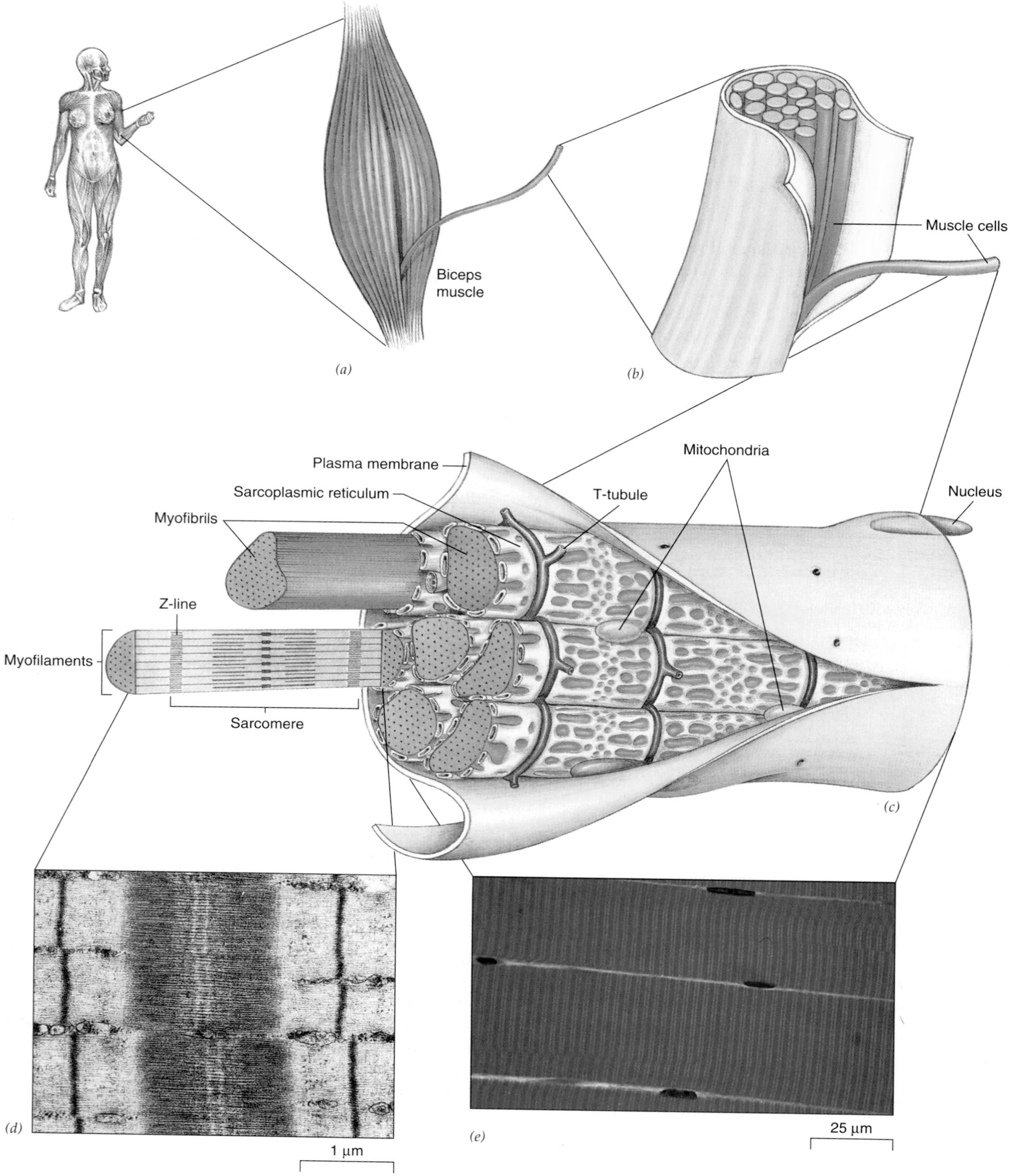

FIGURE 38–8 The muscular system permits complex movement. (*a*) A muscle such as the biceps in the arm consists of many bundles of muscle cells. (*b*) A bundle of muscle cells wrapped in a connective tissue covering. (*c*) Structure of a muscle cell showing the structure of a myofibril. The Z lines mark the ends of the sarcomeres. (*d*) Electron micrograph of striated muscle. (*e*) Light photomicrograph showing striations. (*d*, D. W. Fawcett; *e*, Ed Reschke)

wears out. In rheumatoid arthritis, the synovial membrane thickens and becomes inflamed. Synovial fluid accumulates, causing pressure and pain, and the joints stiffen.

MUSCLE IS THE CONTRACTILE TISSUE THAT PERMITS MOVEMENT IN MOST ANIMALS

All eukaryotic cells contain the contractile protein **actin.** This protein is the major component of microfilaments and is important in many cell processes such as amoeboid movement and attachment of cells to a surface. In most cells, actin is functionally associated with the contractile protein **myosin.** Actin and myosin are most highly organized in **muscle** cells.

We have discussed the contractile cells of the hydrostatic skeleton of *Hydra* and other cnidarians. In flatworms and most other animal groups, muscle occurs as a specialized tissue organized into definite layers, or straplike bands. In complex animals the muscles serve as motors that generate mechanical forces and motion. Muscles permit locomotion, manipulation of objects, circulation of blood, movement of food through the digestive tract, and many other movements. The three types of muscle—skeletal, smooth, and cardiac—each specialized for its particular task, were described in Chapter 37. Recall that both skeletal and cardiac muscle are striated (banded).

Bivalve mollusks have both smooth and striated muscle. The smooth muscle, which is capable of slow, sustained contraction, can be used to keep the two shells tightly closed for long periods, even days. The striated muscle, which contracts rapidly, shuts the shells quickly when the mollusk is threatened. Arthropod muscles are typically striated.

A vertebrate muscle may consist of thousands of muscle fibers

In vertebrates, each skeletal muscle may be considered an organ. Its elongated cells, referred to as fibers, are organized in bundles, called **fascicles,** that are wrapped by connective tissue. The biceps in your arm, for example, consists of thousands of individual muscle fibers and their connective tissue coverings.

Recall that each striated muscle fiber is a long, cylindrical cell with many nuclei (Fig. 38–8). The plasma membrane (known as the *sarcolemma* in a muscle fiber) has multiple inward extensions that form a set of **T tubules** (transverse tubules). The cytoplasm of a muscle fiber is referred to as **sarcoplasm,** and the endoplasmic reticulum as **sarcoplasmic reticulum.**

Threadlike structures called **myofibrils** run lengthwise through the muscle fiber. They are composed of two types of even smaller structures, the **myofilaments.** The thick myofilaments, called **myosin filaments,** consist mainly of the protein myosin. The thin **actin filaments** consist mostly of the protein actin; they also contain the regulatory proteins **tropomyosin** and the **troponin complex.** Myosin and actin filaments overlap lengthwise in the muscle fibers, producing the pattern of transverse bands, or striations, characteristic of striated muscle (Fig. 38–8 and 38–9). The bands are designated by specific letters.

Myosin and actin filaments are organized into structures called **sarcomeres,** the basic units of muscle contraction. Sarcomeres are joined at their ends by an interweaving of filaments called the *Z line.* Hundreds of sarcomeres connected end-to-end make up a myofibril.

Contraction occurs when actin and myosin filaments slide past each other

Muscle contraction occurs when the sarcomeres, and thus the muscle fibers, shorten. This shortening takes place when the actin and myosin filaments slide past each other, increasing their overlap. This concept of muscle contraction is referred to as the **sliding filament model.** It is important to understand that neither the actin nor myosin filaments change in length. The length of the muscle shortens as the filaments increase their overlap. (You might think of an extension ladder. The overall ladder length changes as the ends get closer or farther apart, but the length of each ladder section stays the same.) In the following paragraphs, we describe the current understanding of this model.

In a **motor unit,** a motor neuron (a neuron that stimulates muscle) is functionally connected with an average of about 150 muscle fibers (Fig. 38–10). When a motor neuron is stimulated, it releases the neurotransmitter acetylcholine into the synaptic cleft (the gap) between the motor neuron and each muscle fiber. The acetylcholine binds with receptors on each muscle fiber, causing an electrical change, called **depolarization,** along its sarcolemma. Depolarization may cause an electrical impulse, or **action potential,** to be generated.

An action potential is a wave of depolarization that travels along the sarcolemma, down the T tubule membranes, and through the sarcoplasmic reticulum. The action potential alters the permeability of the sarcoplasmic reticulum. This change opens protein channels in the sarcoplasmic reticulum, allowing calcium ions to move out of storage and flow into the sarcoplasm.

Calcium ions bind to troponin molecules in the actin filaments, which causes the tropomyosin-troponin complex to change shape. This change uncovers active sites on the actin (Fig. 38–11). When the active sites are un-

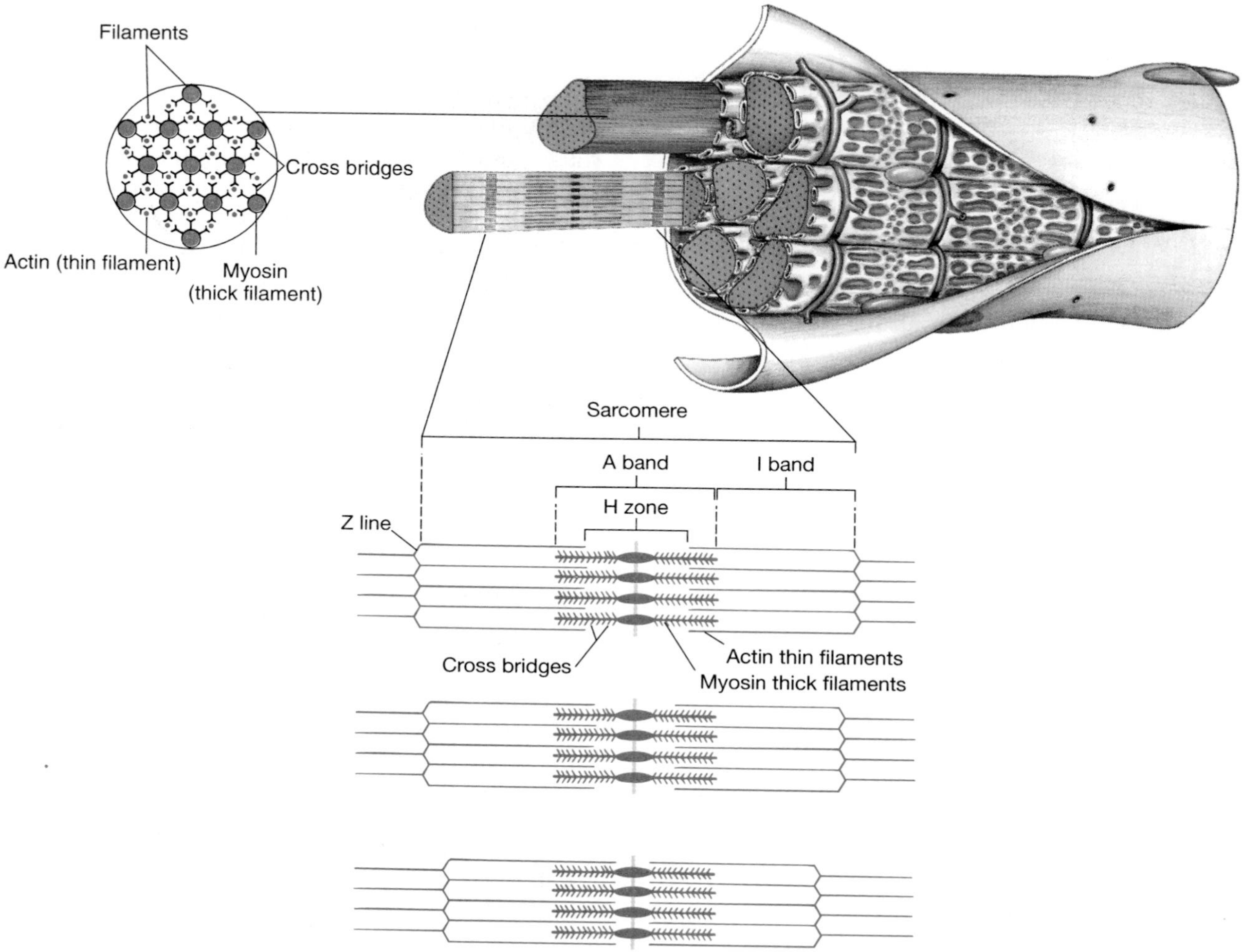

FIGURE 38–9 Myofibrils are threadlike structures that contain actin and myosin filaments. The cross section *(at top left)* of a myofibril shows the arrangement of actin and myosin filaments. Filaments slide past each other during contraction. The regular pattern of overlapping filaments gives skeletal and cardiac muscle their striated appearance. The filaments are shown sliding toward each other, increasing the amount of overlap. The muscle cell is shortened as the sarcomeres become shorter. At bottom, maximum contraction has occurred; the sarcomere has shortened considerably.

covered, myosin binds to them, forming **cross bridges** that link the myosin and actin filaments.

One end of each myosin molecule is folded into two globular structures called heads. The rounded heads of the myosin molecules extend away from the body of the myosin filament. Each myosin molecule also has a long tail that joins other myosin tails to form the thick filament. The myosin head has the ability to break down ATP in the presence of calcium and use the liberated energy for the contraction process.

ATP is bound to the myosin when the muscle fiber is at rest (not contracting). When the muscle fiber is activated, calcium ions activate ATP breakdown; myosin acts as an enzyme in this process. When ATP breaks down to ADP and phosphate, its energy is not released until the actin interacts with myosin. It is thought that after the myosin head attaches to the actin filament, it releases the ADP and flexes (bends) about 45 degrees. This flexing motion is the power stroke that pulls the actin filament along toward the center of the sarcomere. Thus, ATP provides the energy to move the cross bridges.

Each myosin head then picks up a new ATP and releases its hold on the first set of active sites. This binding of ATP to the myosin head must occur before the cross

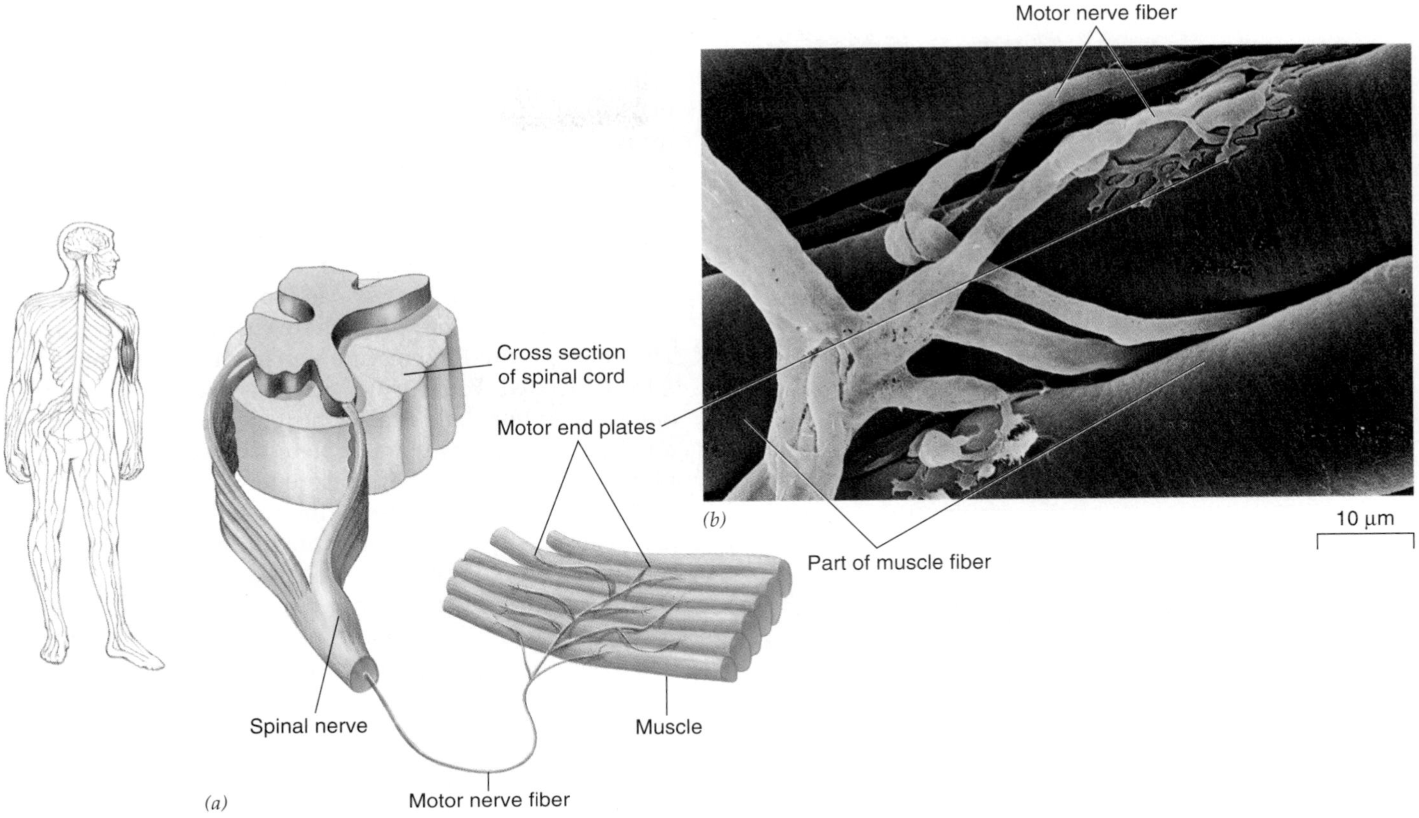

FIGURE 38–10 A motor unit consists of a motor nerve fiber and the muscle fibers that it innervates. A motor unit typically includes about 150 muscle fibers, but some units have less than a dozen fibers, while others have several hundred. (*a*) The motor unit illustrated here shows only a single motor nerve fiber. (*b*) Scanning electron micrograph of some of the cells in a motor unit. Note how the neuron branches to innervate all of the cells in the motor unit. (*b*, Don Fawcett/Science Source/Photo Researchers, Inc.)

bridge can detach itself from the actin and begin a new cycle.

Energized once again, the myosin heads reach for a second set of active sites—those closer to the center of the sarcomere. The process is repeated with a third set, and so on (Fig. 38–11). This series of stepping motions pulls the myosin and actin filaments past one another. Thus, when the cross bridges attach, move 45 degrees, detach, and then reattach farther along the actin filament, the muscle shortens. (One way to visualize this process is to imagine the myosin heads engaging "hand-over-hand" in a kind of tug-of-war on the actin filaments.)

During muscle contraction, the actin filaments move toward the center of the myofibril. As this occurs, the muscle shortens. When many sarcomeres contract simultaneously, they produce the contraction of the muscle as a whole. The process of muscle contraction is summarized in Figure 38–12.

After contracting, muscle fibers return to their resting state. Acetylcholine in the synaptic cleft is inactivated by the enzyme cholinesterase. Calcium ions are then pumped back into the sarcoplasmic reticulum by active transport, a process that requires ATP. Without calcium ions, the active sites on the actin filaments are again covered by tropomyosin-troponin complex. The actin filaments slide back to their original position, and relaxation of the muscle occurs. This entire series of events happens in milliseconds.

Even when we are not moving, our muscles are in a state of partial contraction known as **muscle tone.** At any given moment some muscle fibers are contracted, stimulated by messages from nerve cells. Muscle tone is an unconscious process that helps keep muscles prepared for action. When the motor nerve to a muscle is cut, the muscle becomes limp (completely relaxed), or flaccid.

ATP powers muscle contraction

Muscle cells are often called on to perform strenuously, and so they must be provided with large amounts of energy. The immediate source of energy necessary for muscle contraction is ATP, needed not only for the pull ex-

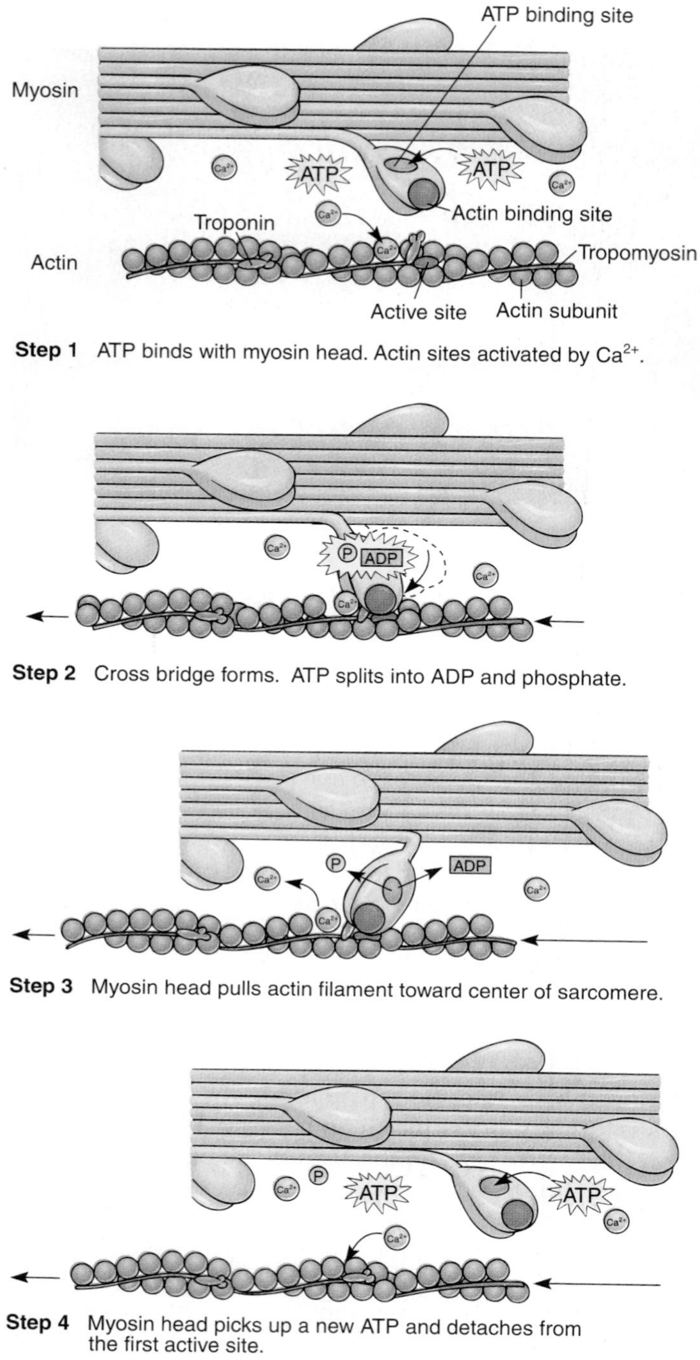
ATP binding site
Myosin
ATP
ATP
Actin binding site
Troponin
Actin
Tropomyosin
Active site
Actin subunit
Step 1 ATP binds with myosin head. Actin sites activated by Ca2+.
P
ADP
Step 2 Cross bridge forms. ATP splits into ADP and phosphate.
P
ADP
Step 3 Myosin head pulls actin filament toward center of sarcomere.
P
ATP
ATP
Step 4 Myosin head picks up a new ATP and detaches from the first active site.

FIGURE 38–11 In this model of muscle contraction, cross bridges move the thin and thick filaments past each other. Parts of this model are still hypothetical.

erted by the cross bridges, but also for their release from each active site as they engage in their tug-of-war on the actin filaments. Rigor mortis, the temporary but very marked muscular rigidity that appears after death, results from ATP depletion following the cessation of cellular respiration upon death.[1] Without ATP, the cross bridges that have been formed are not released.

Sufficient energy can be held in ATP molecules for only the first few seconds of strenuous activity. Fortunately, muscle cells have a backup energy storage compound called **creatine phosphate** that can be stockpiled. The energy stored in creatine phosphate is transferred to ATP as needed. But during vigorous exercise the supply of creatine phosphate is also short-lived. As ATP and creatine phosphate stores are depleted, muscle cells must replenish their supplies of these energy-rich compounds.

Fuel is stored in muscle fibers in the form of glycogen, a large polysaccharide formed from hundreds of glucose units. Stored glycogen is degraded, yielding glucose, which is then broken down in cellular respiration. When sufficient oxygen is available, enough energy is captured from the glucose to produce needed quantities of ATP and creatine phosphate.

During strenuous exercise, oxygen may not be available in sufficient quantity to meet the needs of the rapidly metabolizing muscle cells. Under these conditions, muscle cells are capable of breaking down fuel molecules anaerobically (without oxygen) for short periods. Recall from Chapter 7 that lactic acid fermentation is a method of generating ATP anaerobically, but not in great quantity. ATP depletion results in weaker contractions and muscle fatigue. Accumulation of the waste product lactic acid also contributes to muscle fatigue. The buildup of lactic acid and depletion of glycogen during muscle exertion make up an **oxygen debt.** The debt is paid back during the period of rapid breathing and aerobic metabolism that typically follows strenuous exercise. Lactic acid accumulates at a much slower rate in well-conditioned athletes.

The energy conversion of muscular contraction is not very efficient. Only about 30% of the chemical energy of the glucose fuel is actually converted to mechanical work. The remaining energy is accounted for as heat, produced mainly by frictional forces within the muscle cell. That is

[1]Rigor mortis does not persist indefinitely, however, for the entire contractile apparatus of the muscles eventually decomposes, restoring pliability. The phenomenon is temperature-dependent, so, given the prevailing temperature, a medical examiner can estimate the time of death of a cadaver from its degree of rigor mortis. Perhaps it should be said that rigor mortis by itself is not muscular contraction; it only tends to freeze the corpse in its position at the time of death. Thus, tales of corpses sitting up and pointing to their murderers, and otherwise carrying on posthumously may be entertaining but have no factual basis.

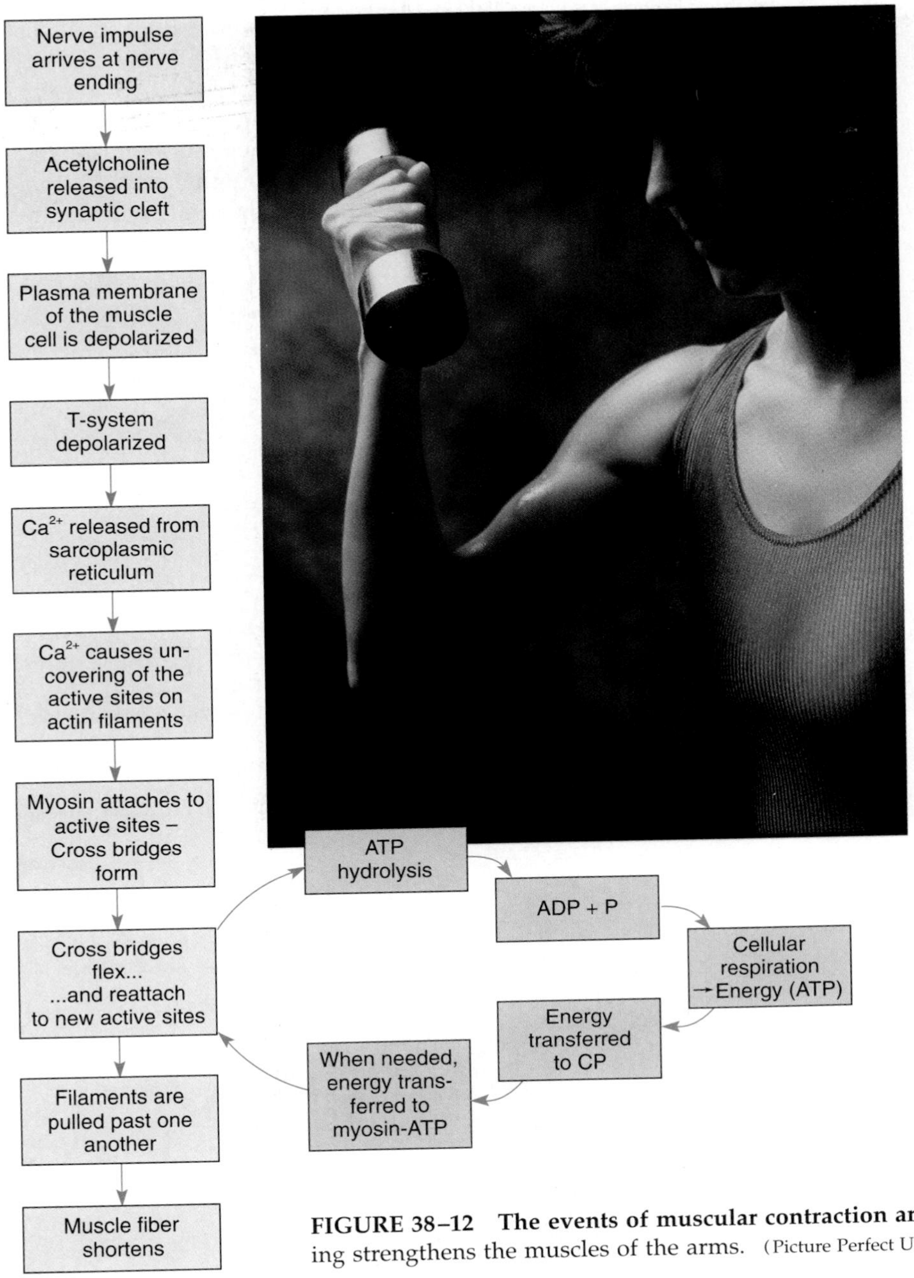
Nerve impulse arrives at nerve ending
Acetylcholine released into synaptic cleft
Plasma membrane of the muscle cell is depolarized
T-system depolarized
Ca^{2+} released from sarcoplasmic reticulum
Ca^{2+} causes uncovering of the active sites on actin filaments
Myosin attaches to active sites – Cross bridges form
Cross bridges flex... ...and reattach to new active sites
Filaments are pulled past one another
Muscle fiber shortens
ATP hydrolysis
ADP + P
Cellular respiration → Energy (ATP)
Energy transferred to CP
When needed, energy transferred to myosin-ATP

FIGURE 38–12 The events of muscular contraction are summarized here. Weightlifting strengthens the muscles of the arms. (Picture Perfect USA/Dembinsky Photo Associates)

why we get hot during hard physical labor, and shiver when we are cold. The muscle contractions involved in shivering are one way of producing heat to warm the body.

Skeletal muscle action depends on muscle pairs that work antagonistically

Skeletal muscles produce movements by pulling on **tendons.** Tendons are tough cords of connective tissue that anchor muscles to bone. Tendons then pull on bones. Most muscles pass across a joint and are attached to the bones that form the joint. When the muscle contracts, it draws one bone toward or away from the bone with which it articulates.

Muscles can only pull; they cannot push. Muscles act **antagonistically** to one another, which means that the movement produced by one can be reversed by another. The biceps muscle, for example, permits you to flex your arm, whereas the triceps muscle allows you to extend it once again (Fig. 38–13). Thus, the biceps and triceps work antagonistically.

The muscle that contracts to produce a particular action is known as the **agonist.** The muscle that produces the opposite movement is the **antagonist.** When the agonist is contracting, the antagonist is relaxed. Generally, movements are accomplished by groups of muscles working together, so there may be several agonists and several antagonists in any action. Note that muscles that are agonists in one movement may be antagonists in another. The superficial muscles of the human body are shown in Figures 38–14 and 38–15.

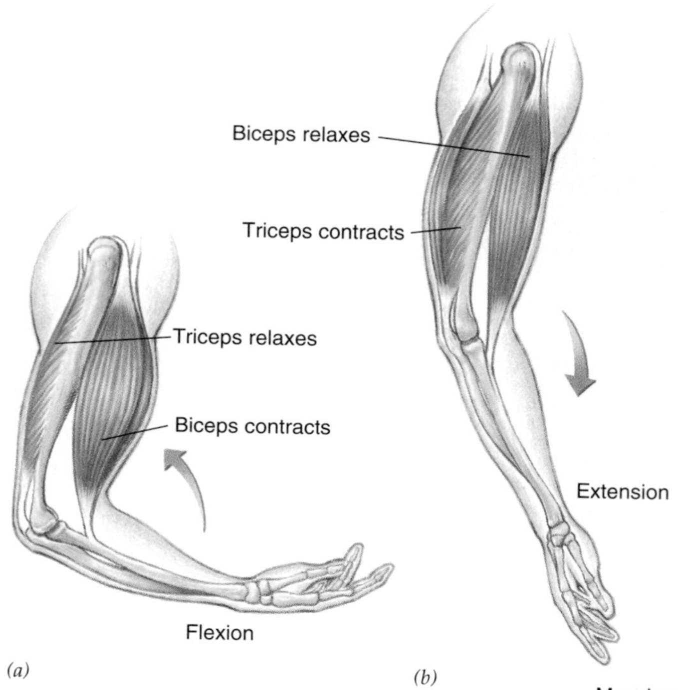

FIGURE 38–13 The biceps and triceps muscles function antagonistically.

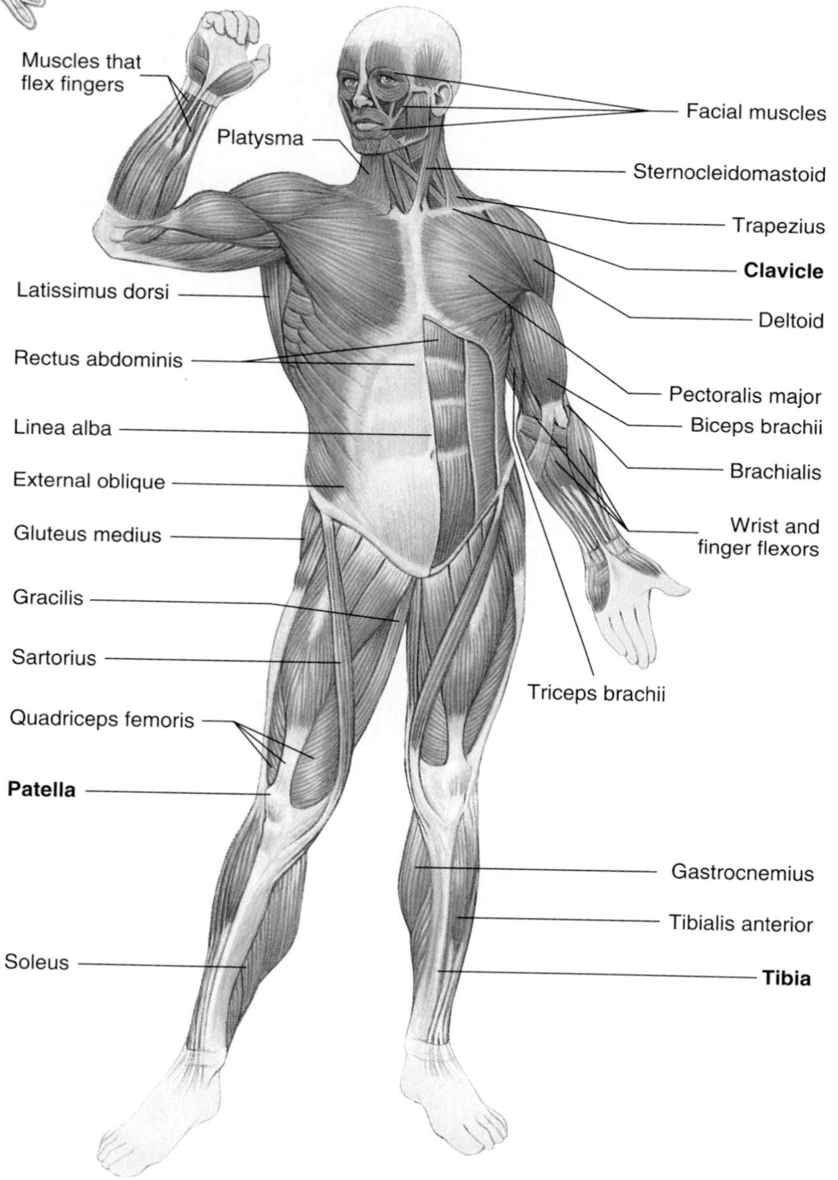

FIGURE 38–14 Some of the prominent superficial muscles of the human body are illustrated in this anterior view.

Smooth, cardiac, and skeletal muscle respond in specific ways

Each type of muscle is specialized for particular types of responses. Smooth muscle often contracts in response to simple stretching, and its contraction tends to be lengthy. It is well adapted to performing such tasks as the regulation of blood pressure by sustained contraction of the walls of the arterioles. Although smooth muscle contracts slowly, it shortens much more than striated muscle does. Though not well suited for running or flying, smooth muscle squeezes superlatively.

Cardiac muscle contracts abruptly and rhythmically, propelling blood with each contraction. Sustained contraction of cardiac muscle would be disastrous!

Skeletal muscle, when stimulated by a single brief stimulus, contracts with a single quick contraction called a **simple twitch.** Simple twitches ordinarily do not occur except in laboratory experiments. In the normal animal, skeletal muscle receives a series of separate stimuli very close together. These produce not a series of simple twitches, however, but a single, smooth, sustained contraction called **tetanus.** Depending upon the identity and number of our muscle cells tetanically contracting, we might thread a needle, haul a rope, or run a mile.

Not all muscular activities are the same. Dancing and, even more so, typing require quick response rather than the long, sustained effort that might be appropriate in hauling a rope. In many animals, entire muscles are specialized for quick or slow responses. In chickens, for

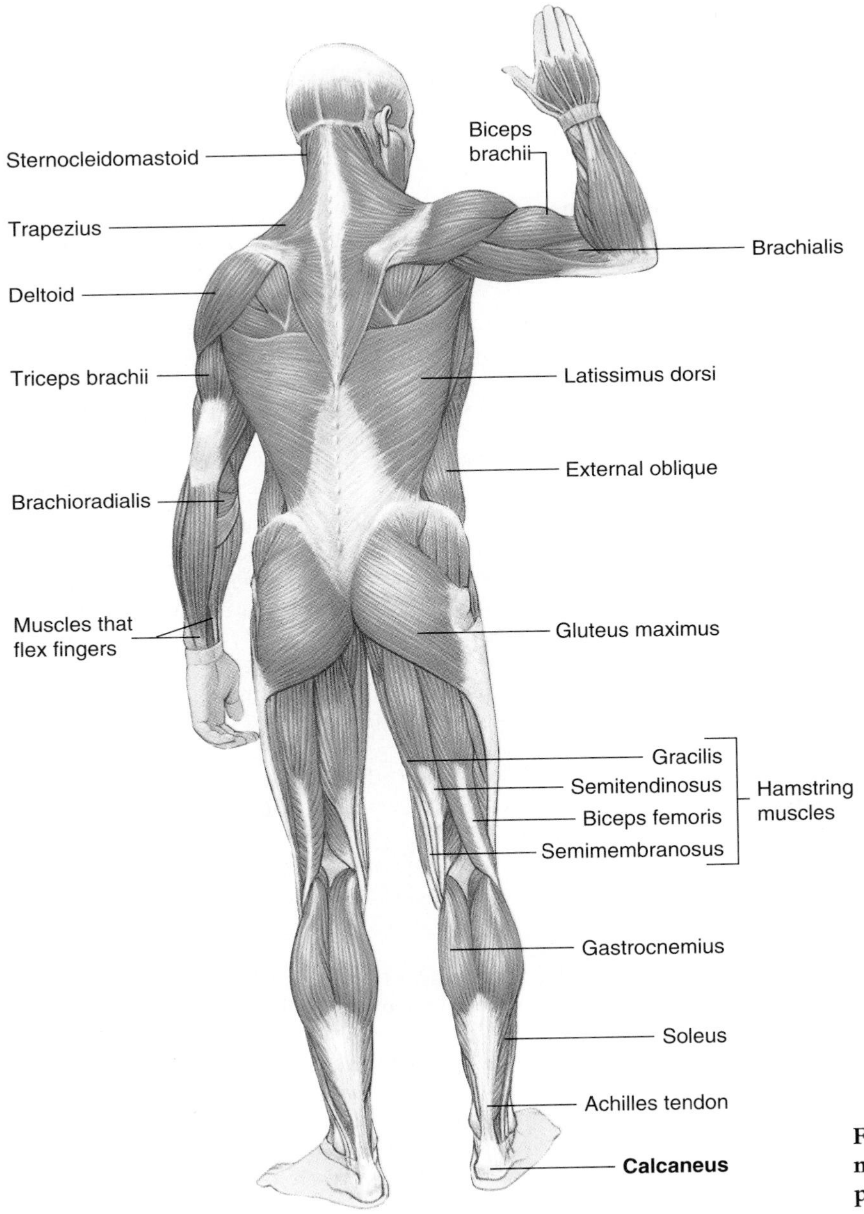

FIGURE 38–15 Some of the prominent superficial muscles of the human body are illustrated in this posterior view.

instance, the white breast muscles are efficient for quick responses, because flight is an escape mechanism for chickens. On the other hand, they walk about on the ground all day, so the dark meat of the leg and thigh is composed of muscle specialized for more sustained activity.

There is no human equivalent of light and dark meat. However, like other vertebrates, we do possess **fast-twitch fibers** specialized for fast response, and **slow-twitch fibers** specialized for slow response. We also have intermediate fibers that can be distinguished microscopically with appropriate staining.

Slow-twitch fibers have less sarcoplasmic reticulum than fast-twitch fibers. They have many mitochondria and a rich supply of capillaries. Slow-twitch fibers also have more **myoglobin,** a red pigment similar to hemoglobin. Myoglobin serves as an intermediary responsible for the transfer of oxygen from hemoglobin to the aerobic metabolic processes of the muscle cell. Myoglobin may also have a modest capacity to store oxygen.

The proportions of slow-twitch and fast-twitch fibers vary from person to person and from muscle to muscle in the same person. The relative proportions of the two appear to be genetically determined, and they influence the kind of athletic activity at which each of us has the greatest potential proficiency. Someone whose leg and thigh muscles contain a high proportion of fast-twitch fibers could, with proper training, become a good sprinter. An athlete with a greater proportion of slow-twitch fibers may be better suited to marathon activities. Recent evidence, however, suggests that the proportions of the two kinds of fibers can be changed by appropriate training.

SUMMARY

I. Invertebrate epithelial tissue may be specialized for sensory or respiratory functions, to produce a protective cuticle, to secrete lubricants or adhesives, to produce odorous or poisonous secretions, or to produce threads for nests or webs.

II. Vertebrate skin protects, may prevent dehydration, may be specialized for secretion or reception of stimuli, and may help regulate body temperature. Human skin includes nails, hair, sweat glands, oil glands, and sensory receptors.
   A. In human skin, cells in the stratum basale of the epidermis continuously divide. As they are pushed upward toward the skin surface, these cells mature, flatten, produce keratin, and eventually die.
   B. The dermis, which consists of dense, fibrous connective tissue, rests on a layer of subcutaneous tissue composed largely of fat.

III. The skeleton transmits mechanical forces generated by muscle and also supports and protects the body.
   A. Many invertebrates have a hydrostatic skeleton in which fluid is used to transmit forces generated by contractile cells or muscle.
   B. Exoskeletons are characteristic of mollusks and arthropods. The arthropod skeleton, composed mainly of chitin, is jointed for flexibility. This nonliving skeleton does not grow, making it necessary for arthropods to molt periodically.

IV. The endoskeletons of echinoderms and chordates are composed of living tissue and therefore are capable of growth.
   A. The vertebrate skeleton consists of an axial portion and an appendicular portion.
   B. A typical long bone consists of a thin outer shell of compact bone surrounding the inner spongy bone and a central marrow cavity.
   C. Long bones develop from cartilage replicas during endochondral bone formation. Other bones, such as the flat bones of the skull, develop from a noncartilage connective tissue model by intramembranous bone development.

V. All animals have the ability to move. Specialized muscle tissue is found in most invertebrate phyla and in all of the vertebrates. As muscle tissue contracts (shortens), it moves body parts by pulling on them.
   A. A muscle such as the biceps is an organ made up of hundreds of muscle fibers.
   B. The striations of skeletal muscle fibers reflect the overlapping of their actin and myosin filaments. A sarcomere is a contractile unit of thin (actin) and thick (myosin) filaments.
   C. During muscle contraction, the thin filaments are pulled between the myosin filaments toward the center of the myofibril.
      1. Muscle contraction begins when a motor neuron releases acetylcholine into the cleft between the neuron and the muscle fiber.
      2. The acetylcholine combines with receptors on the surface of the muscle fiber, resulting in depolarization of the sarcolemma and transmission of an action potential.
      3. The impulse spreads through the T tubules and stimulates calcium ion release from the sarcoplasmic reticulum.
      4. Calcium ions initiate a process that uncovers the active sites of the actin filaments.
      5. Cross bridges form, linking the myosin and actin filaments.
      6. The cross bridges flex and reattach to new active sites so that the filaments are pulled past one another and the muscle shortens.
   D. ATP is the immediate source of energy for muscle contraction. Muscle tissue has an intermediate energy storage compound, creatine phosphate. Glycogen is the fuel stored in muscle fibers.
   E. Muscles act antagonistically to one another.

SELECTED KEY TERMS

actin filament (thin filament)
appendicular skeleton
axial skeleton
compact bone
creatine phosphate
cross bridge
dermis
endoskeleton
epidermis
exoskeleton
hydrostatic skeleton
joint
keratin
ligament
motor unit
muscle
muscle tone
myofibril
myofilament
myosin filament (thick filament)
osteoblast
osteoclast
osteocyte
osteon
oxygen debt
periosteum
red marrow
sarcomere
simple twitch
skeletal muscle
spongy bone
stratum basale
stratum corneum
T tubule
tendon
vertebral column

POST-TEST

1. The vertebrate skin consists of two main layers, the outer ________ and the inner ________ .
2. The cells of the stratum ________ of the epidermis are dead and almost waterproof.
3. The protein ________ contributes mechanical strength, flexibility, and waterproofing to the skin.
4. The principal function of ________ skeletons is transmission of muscular force.
5. The skeleton of an echinoderm or chordate is known as a(an) ________ .
6. A typical long bone has a thin outer shell of ________ bone and a filling of ________ bone.
7. The joint capsule of a freely movable joint is reinforced by ________ .
8. The two types of myofilaments in muscle tissue are ________ filaments and ________ filaments.
9. ________ ________ is an energy storage compound that can be stockpiled in muscle cells.
10. A(an) ________ is a unit of myosin (thick) and actin (thin) filaments in a muscle fiber.
11. The inward extensions of the muscle fiber plasma membrane are known as ________ ________ .
12. When active sites are uncovered, myosin binds to them, forming ________ ________ .
13. Muscles are always in a state of partial contraction called ________ ________ .
14. Lactic acid buildup and glycogen depletion during strenuous exercise can result in ________ debt.

REVIEW QUESTIONS

1. Compare vertebrate skin with the external epithelium of invertebrates.
2. What properties does keratin confer on human skin?
3. What is a hydrostatic skeleton? What functions does it perform?
4. How do the septa in the annelid worm contribute to the flexibility of its hydrostatic skeleton?
5. What are some disadvantages of an exoskeleton? Some advantages?
6. Describe the divisions of the human skeleton.
7. Draw a typical long bone, such as the radius, and label its parts.
8. What are the functions of osteoblasts and osteoclasts? Why is it important that bones be continuously remodeled?
9. Compare the two types of filaments in muscle tissue.
10. Outline a model for the sequence of events that causes a muscle cell to contract, beginning with the stimulation of its nerve and including cross bridge action.
11. What is the role of ATP in muscle contraction? What are the functions of creatine phosphate and glycogen?
12. Label the bones of the skeleton.

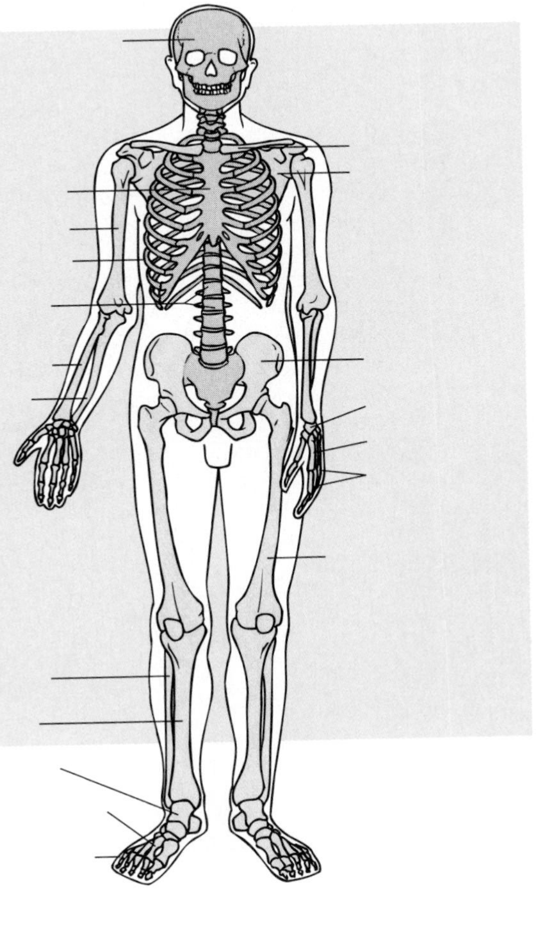

YOU MAKE THE CONNECTION

1. Compare the arthropod exoskeleton to the vertebrate endoskeleton. What are some benefits and some disadvantages of each type of skeleton?
2. What are some examples of hydrostatic support in plants?
3. Why is it important that a muscle be able to switch roles, sometimes acting as an agonist and at other times acting as an antagonist?

RECOMMENDED READINGS

Hadley, N. F. "The Arthropod Cuticle." *Scientific American,* Vol. 255, No. 1, July 1986. The arthropod exoskeleton is largely responsible for the adaptive success of the phylum. The author discusses the properties that enable the exoskeleton to provide protection and support.

Mayor, M.B., and J. Collier. "The Technology of Hip Replacement." *Scientific American Science and Medicine,* May/June 1994. An interesting account of technology for replacing damaged joints.

Zimmer, C. "See How They Run." *Discover,* September 1994. A discussion of animal locomotion.

C H A P T E R 3 9

Neural Control: Neurons

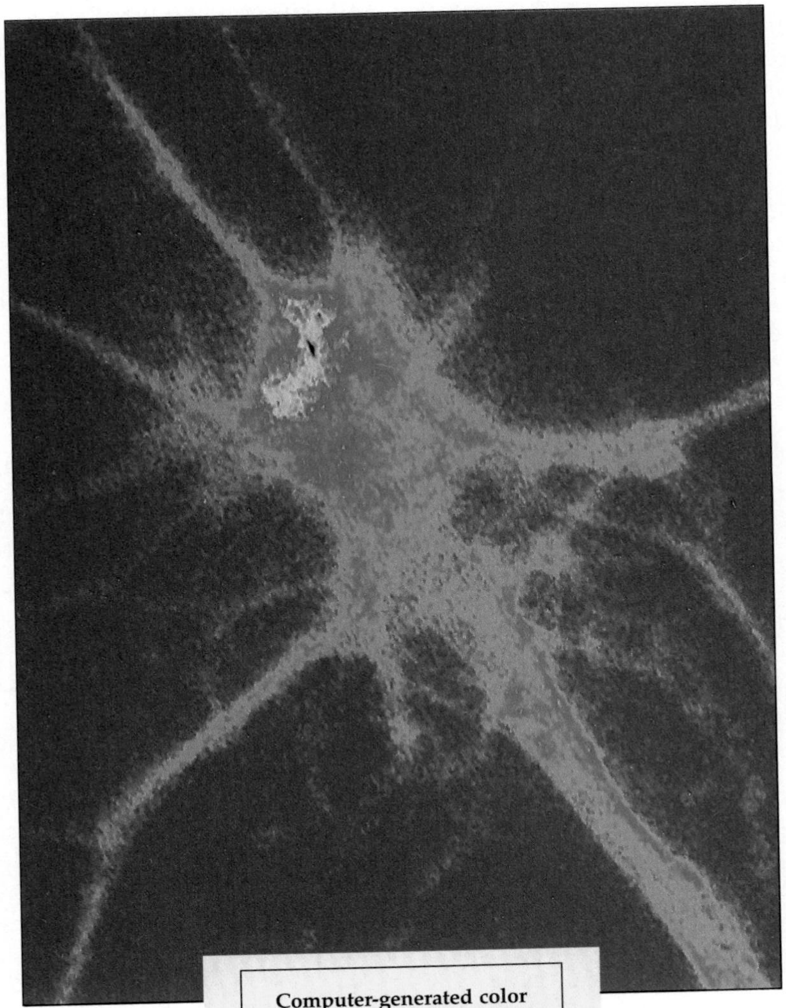

Computer-generated color image of adhesion molecules (CAMs) seen as red and yellow patches on a sensory neuron taken from a sea snail *(Aplysia)*. CAMs may play a role in learning. (Courtesy of Samuel Schacher, Ph.D., Center for Neurobiology and Behavior, College of Physicians and Surgeons of Columbia University)

An earthworm quickly retreats into its burrow when it senses vibrations from approaching footsteps. With lightning speed, a frog darts out its tongue to capture a fly. A college student learns the principles of biology or calculus. All these activities and countless others are made possible by the nervous system. In complex animals the endocrine system works with the nervous system to regulate many behaviors and physiological processes. The endocrine system generally provides relatively slow and long-lasting regulation, whereas the nervous system permits very rapid responses.

Changes within the body or in the outside world that can be detected by an organism are termed stimuli. An organism's ability to survive and to maintain homeostasis depends largely on how effectively it responds to internal and external stimuli.

Thousands of stimuli bombard an organism each day. Appropriate reaction to a stimulus involves four processes: reception, transmission, integration, and response by muscles or glands. **Reception,** the process of detecting a stimulus, is the job of the neurons and of specialized sense organs such as the eyes and ears. **Transmission** is the process of sending messages along a neuron, from one neuron to another, or from a neuron to a muscle or gland. A neural message is transmitted from a receptor to the **central nervous system (CNS),** which consists of the brain and spinal cord. Neurons that transmit information to the CNS are called **afferent neurons**, also called **sensory neurons.** Afferent neurons generally transmit information to **association neurons** (also called **interneurons**). **Integration** involves sorting and interpreting incoming information and determining the appropriate response. In vertebrates, integration is primarily the function of association neurons in the CNS. Neural messages are transmitted from the CNS by **efferent neurons,** or **motor neurons.** The **response** is carried out by **effectors,** the muscles and glands (Fig. 39–1).

LEARNING OBJECTIVES

After you have studied this chapter you should be able to

1. Trace the flow of information through the nervous system. Include the four processes — reception, transmission, integration, and response — involved in responding to a stimulus.
2. Draw a typical neuron, label its parts (including myelin and cellular sheaths), and give the function of each.
3. Summarize the process by which an impulse is transmitted along a neuron.
4. Explain the ionic basis of the resting membrane potential, and relate the conduction of an action potential to changes in ion distribution.
5. Compare continuous conduction with saltatory conduction.
6. Trace the events that take place in synaptic transmission, and draw diagrams to support your description.
7. Identify the neurotransmitters described in the chapter, and describe mechanisms for inactivating them.
8. Identify factors that affect speed of neural transmission.
9. Describe how a postsynaptic neuron integrates incoming stimuli and "decides" whether or not to fire.
10. Draw diagrams of neural circuits showing convergence, divergence, facilitation, and reverberation, and explain why each is important.

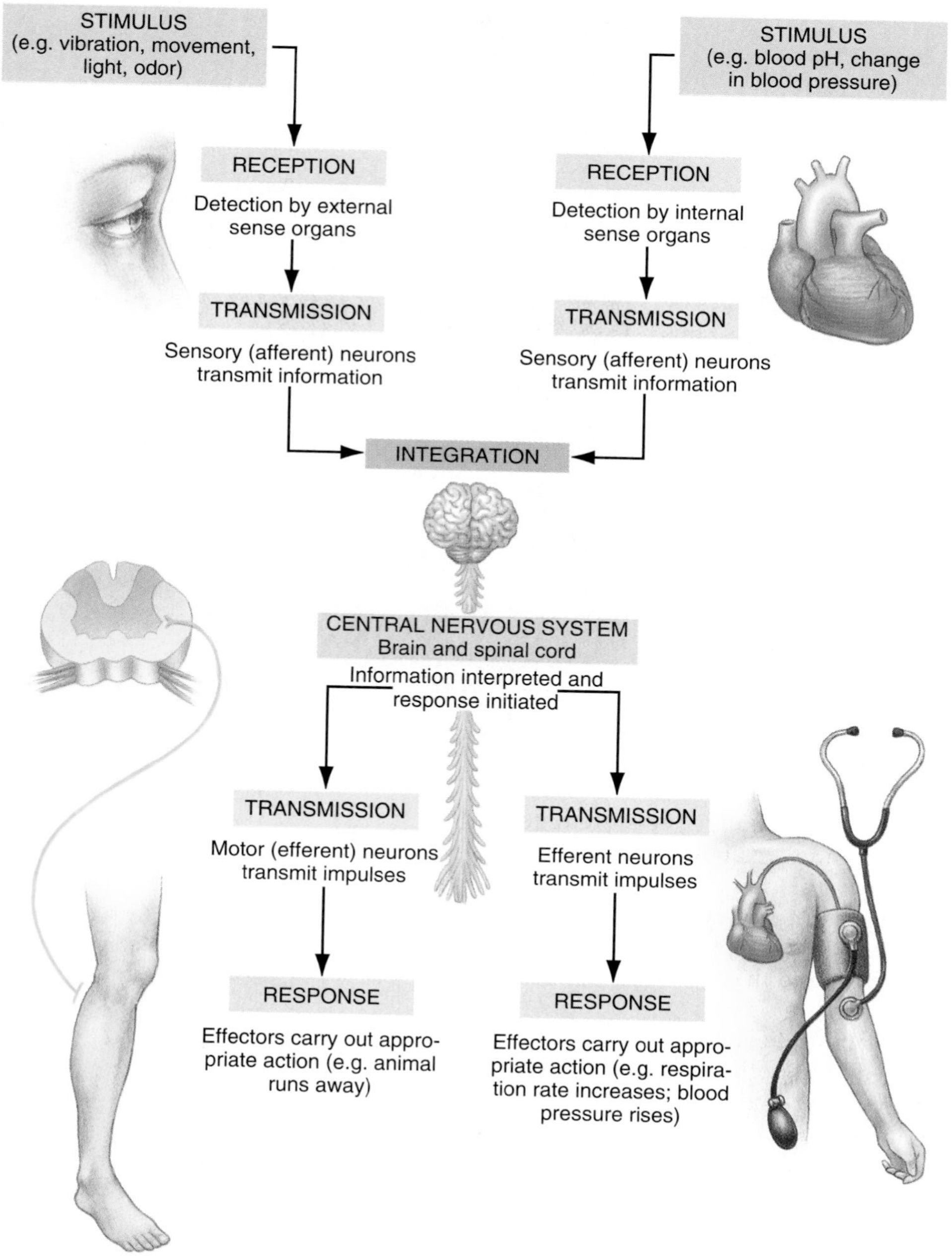

FIGURE 39–1 Whether a stimulus originates in the outside world or inside the body, information must be received, transmitted to the central nervous system, integrated, then transmitted to muscles or glands that carry out the response.

The cell types of the nervous system are neurons and glial cells

In complex animals the functional unit of the nervous system is the **neuron,** a cell specialized to receive and send information. The neuron works by producing and transmitting rapid electrical signals called **nerve impulses.** A second cell type unique to the nervous system is the **glial cell,** which provides support for the neurons.

Glial cells support and protect neurons

Some glial cells envelop neurons and form insulating sheaths around them. Others are phagocytic and remove invading microorganisms and debris from the nervous tissue. A third type of glial cell lines the cavities of the brain and spinal cord. Still others anchor neurons to blood vessels and may play a role in transmission of impulses. **Schwann cells,** glial cells found outside the central nervous system, form sheaths around some neurons. Sometimes glial cells are referred to collectively as the **neuroglia,** which literally means "nerve glue."

A typical neuron consists of a cell body, dendrites, and an axon

Highly specialized to receive stimuli and transmit messages in the form of electrical impulses, the neuron is distinguished from all other cells by its long cytoplasmic extensions, or processes. Examine the structure of a common type of neuron, the multipolar neuron in Figure 39–2.

The largest portion of the neuron, the **cell body,** contains the bulk of the cytoplasm, the nucleus, and most of the other organelles. Two types of cytoplasmic extensions project from the cell body of a multipolar neuron. Numerous dendrites extend from one end, and a long, single axon projects from the opposite end.

Dendrites are typically short, highly branched fibers specialized to receive stimuli and send nerve impulses to the cell body. The cell body integrates incoming signals and can also receive impulses directly.

Although microscopic in diameter, an **axon** may be 1 meter (over 1 yard) or more in length and may branch along its course. It conducts nerve impulses away from the cell body to another neuron, or to a muscle or gland. At its end the axon branches, forming many **axon terminals.** Each terminal ends in a tiny **synaptic knob.** The knobs release **neurotransmitters,** chemicals that transmit signals from one neuron to another, or from neuron to effector.

The axons of many neurons outside the central nervous system have two coverings: an outer **cellular sheath,** or **neurilemma,** and an inner **myelin sheath** (Fig. 39–3). Both are formed by Schwann cells. The cellular sheath is formed by Schwann cells that line up along the axon. The myelin sheath, which lies between the axon and the cellular sheath, is formed when the Schwann cell winds its plasma membrane around the axon several times.

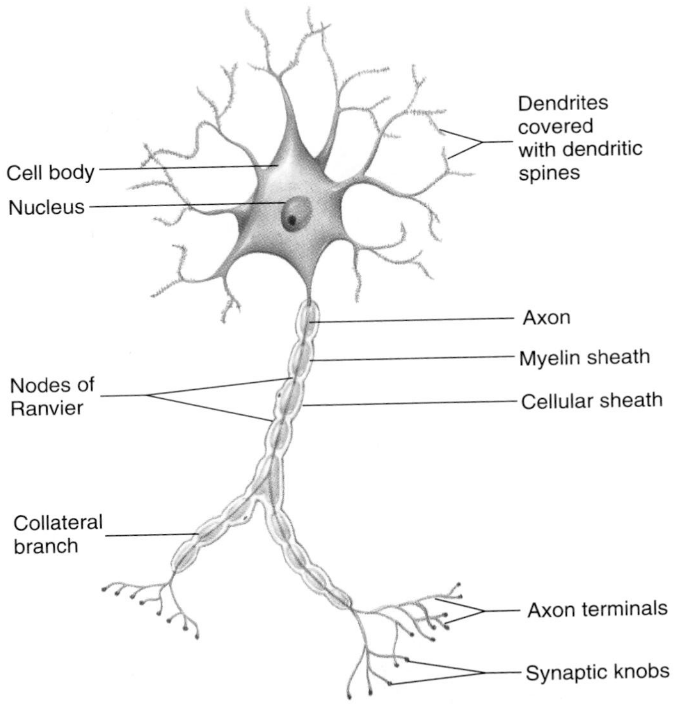

FIGURE 39–2 The structure of a multipolar neuron includes dendrites, a cell body, and an axon. The axon of this neuron is myelinated, so the myelin sheath is shown as well as the cellular sheath.

The plasma membrane of the Schwann cell is rich in **myelin,** a white, fatty substance. Myelin is an excellent electrical insulator, and its presence influences transmission of nerve impulses. Gaps in the myelin sheath, called **nodes of Ranvier,** occur between successive Schwann cells. At these points the axon is not insulated with myelin.

Almost all axons more than 2 micrometers in diameter are myelinated, meaning that they have myelin sheaths. Those of smaller diameter are generally unmyelinated. In the brain and spinal cord, myelin sheaths are formed by certain glial cells (oligodendrocytes) but cellular sheaths are not present.

In **multiple sclerosis,** a neurological disease that affects about 300,000 people in the United States alone, patches of myelin deteriorate at irregular intervals along axons and are replaced by scar tissue. This damage interferes with conduction of neural impulses, and the victim suffers loss of coordination, tremor, and partial or complete paralysis of parts of the body. The cause of multiple sclerosis has been a mystery, but some evidence suggests that it is an autoimmune disease, in which the body attacks its own tissue, in this case neurons and glial cells (see Chapter 43).

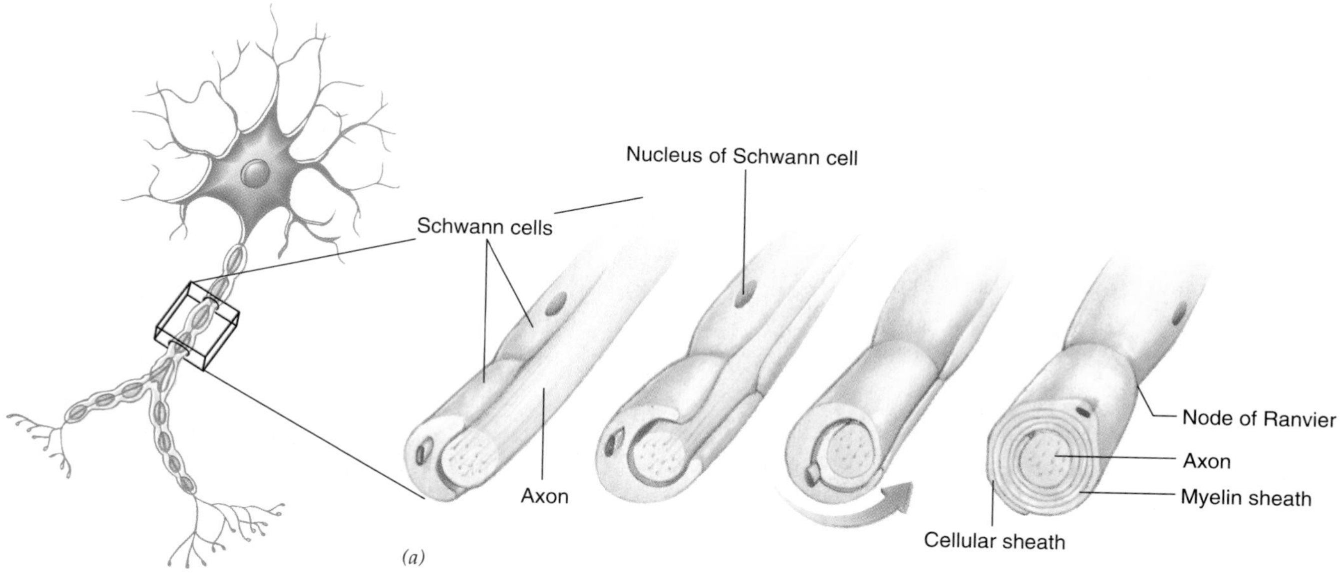

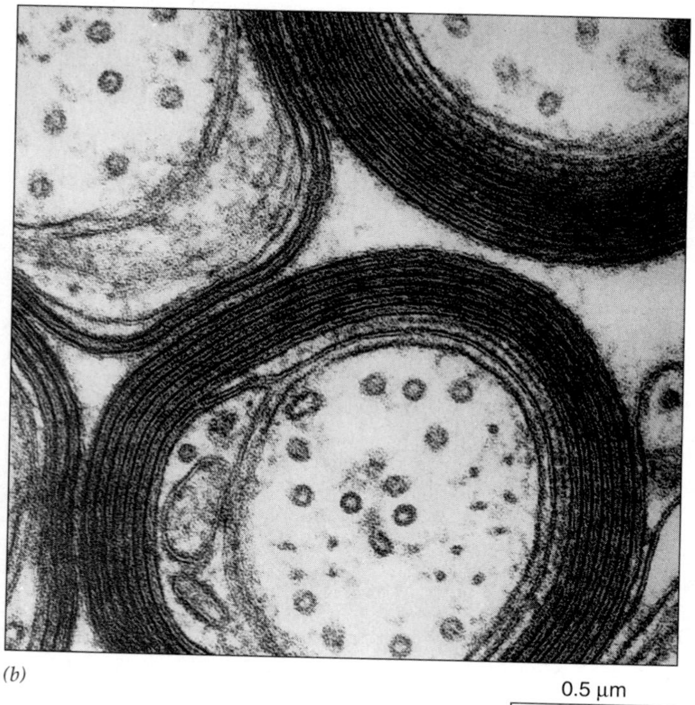

FIGURE 39–3 The myelin sheath provides insulation around the axon. (*a*) A Schwann cell wraps its plasma membrane many times around the axon of a peripheral neuron to form a myelin sheath. The rest of the Schwann cell remains outside the myelin sheath, forming the cellular sheath. (*b*) Electron micrograph of a section through a myelinated axon showing the myelin sheath. (Visuals Unlimited/C. S. Raine)

The cellular sheath is important in the regeneration of injured neurons. When an axon is cut, the portion separated from the cell body deteriorates and is phagocytized by surrounding cells, but the cellular sheath remains intact. The cut end of the axon grows slowly through the empty cellular sheath, and eventually at least partial neural function may be restored.

A **nerve** consists of hundreds or even thousands of axons wrapped together in connective tissue (Fig. 39–4). We can compare a nerve to a telephone cable. The individual axons correspond to the wires that run through the cable, and the sheaths and connective tissue coverings correspond to the insulation. Within the CNS, bundles of axons are referred to as **tracts** or **pathways** rather than nerves. Outside the CNS, the cell bodies of neurons are usually grouped together in masses called **ganglia** (sing., *ganglion*). Inside the CNS, collections of cell bodies are generally referred to as nuclei rather than ganglia.

Neurons Convey Information by Transmitting Rapidly Moving Electrical Impulses

When a neuron receives a stimulus that is strong enough, its axon fires a nerve impulse. This is an electrical current that travels rapidly down the axon into the synaptic knobs. Once initiated, the electrical impulse is self-propagating. For a nerve impulse to be fired, however, the plasma membrane has to maintain what is called a resting potential.

The resting potential is the difference in electrical charge across the plasma membrane

In a resting neuron—one not transmitting an impulse—the inner surface of the plasma membrane has a negative charge compared with the interstitial fluid surrounding it (Fig. 39–5). The plasma membrane is said to be electrically **polarized**, meaning that one side, or pole, has a

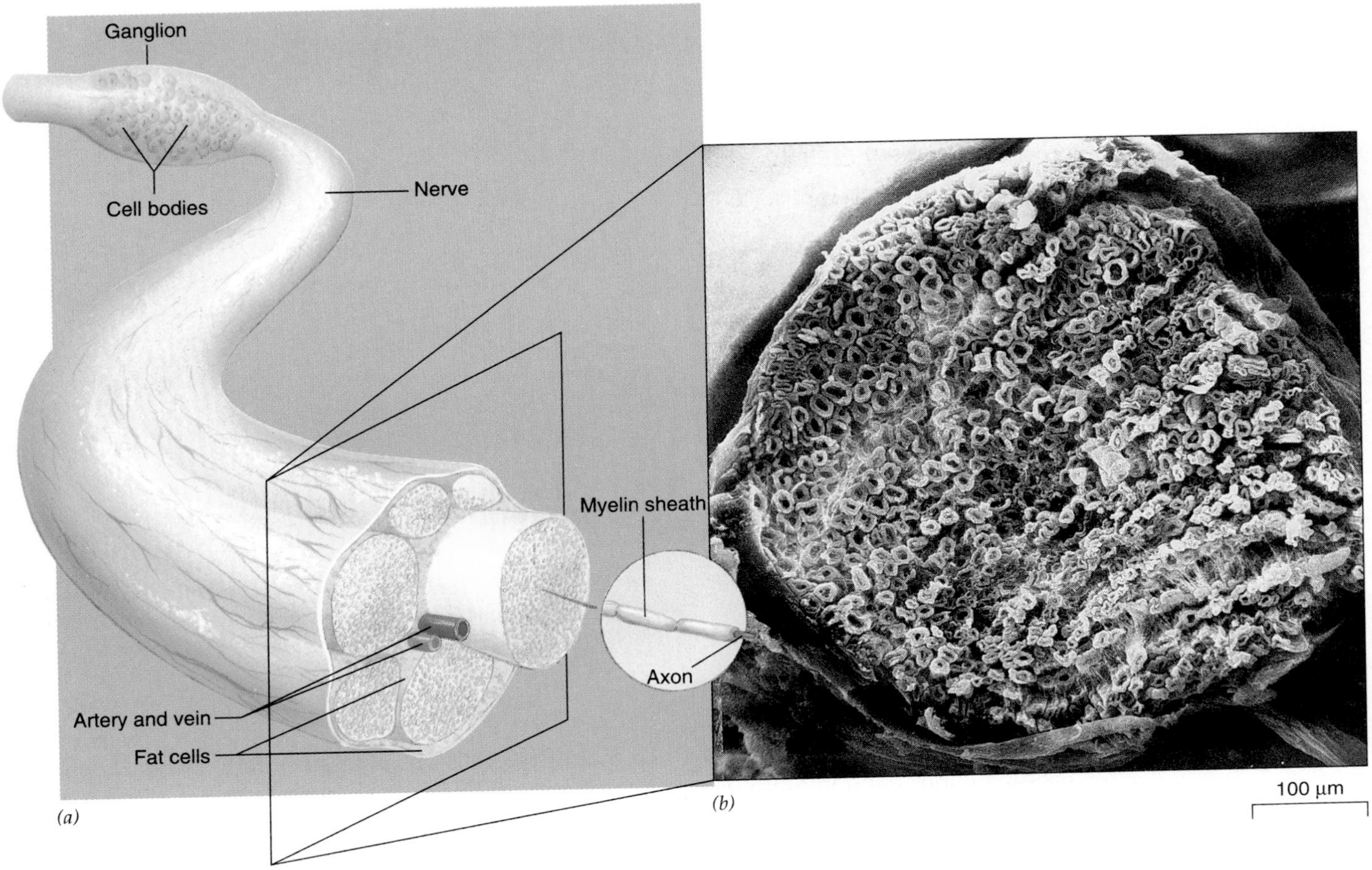

FIGURE 39–4 A nerve consists of bundles of axons held together by connective tissue. (*a*) Structure of a nerve and a ganglion. The cell bodies belonging to the axons of a nerve are grouped together in a ganglion. (*b*) Electron micrograph showing a cross section through a myelinated afferent nerve of a bullfrog. (*b*, E. R. Lewis/University of California/Biological Photo Service)

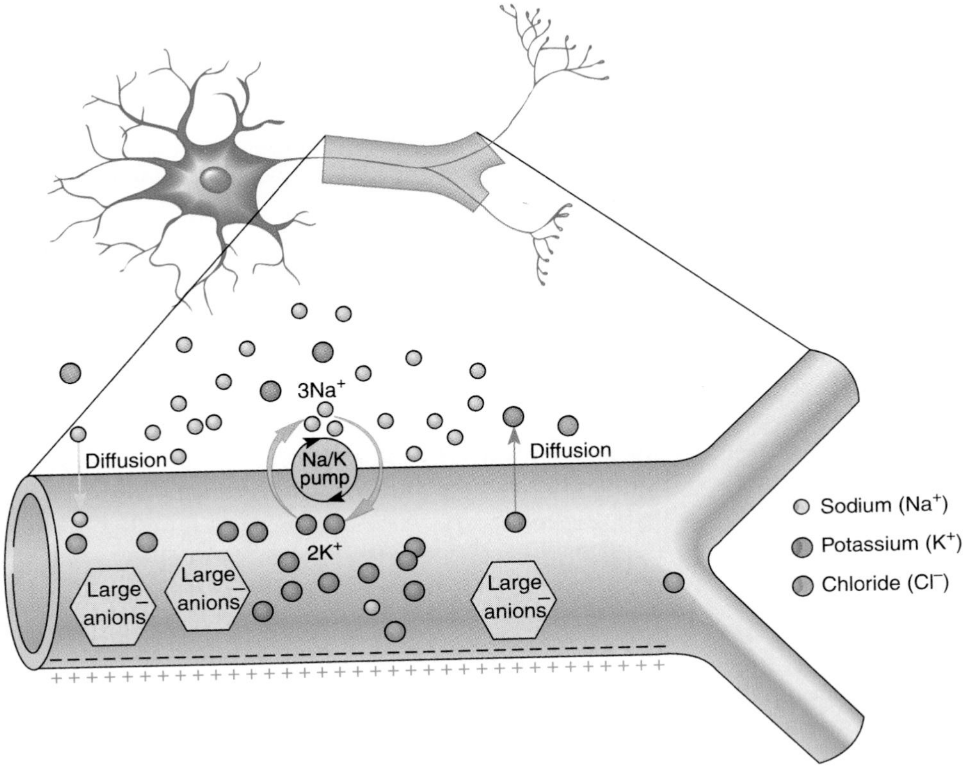

FIGURE 39–5 The axon of a resting neuron is negatively charged compared to the surrounding interstitial fluid. As shown in this segment of an axon of a resting (nonconducting) neuron, sodium-potassium pumps in the plasma membrane actively pump sodium out of the cell and pump potassium in. Sodium is unable to diffuse back to any extent, but potassium does diffuse out along its concentration gradient. Negatively charged proteins and other large anions are present in the cell. Because of the unequal distribution of ions and charged molecules, the inside of the axon is negatively charged relative to the surrounding interstitial fluid.

different charge than the other side. When electrical charges are separated in this way, an electrical potential energy difference exists across the membrane. Should the charges be permitted to come together, they have the *potential* of doing work. In resting neurons, the potential difference across the plasma membrane is called the **membrane potential** or **resting potential.**

The resting potential may be expressed in units called *millivolts* (mV). (A millivolt equals one-thousandth of a volt and is a unit for measuring electrical potential.) The resting potential of a neuron amounts to about 70 mV. By convention this is expressed as −70 mV because the inner surface of the plasma membrane is negatively charged relative to the interstitial fluid. The potential can be measured by placing one electrode, insulated except at the tip, inside the cell, and a second electrode on the outside surface. The two electrodes are connected with an instrument such as a galvanometer, which measures current by electromagnetic action. If both electrodes are placed on the outside surface of the neuron, no potential difference between them is registered. All points on the outside of the membrane are at the same potential. The neuron can be thought of as a biological battery. If its plasma membrane, which is only about one-millionth of a centimeter thick, were 1 centimeter (0.39 inch) thick, the membrane potential would amount to an impressive 70,000 volts!

How does the resting potential develop? It is a direct result of a slight excess of positive ions outside the plasma membrane and a slight excess of negative ions inside the membrane. The distribution of ions inside neurons and in the interstitial fluid surrounding them is similar to that of most other cells in the body. The K^+ concentration is about ten times greater inside a resting neuron than outside the cell. In contrast, the Na^+ concentration is about ten times greater outside than inside the neuron.

This ionic imbalance is brought about by several factors. The neuron plasma membrane has very efficient **sodium-potassium pumps** that actively transport sodium ions out of the cell and potassium ions into the cell (Fig. 5–17). Because the pumps work against a concentration gradient and an electrochemical gradient, ATP is required. For every three sodium ions pumped out of the cell, two potassium ions are pumped in. Thus, more positive ions are pumped out than in.

Ions also cross the membrane by facilitated diffusion through membrane proteins that form ion-specific channels. Net movement of ions occurs from an area of higher concentration to one of lower concentration (Fig. 39–6). However, the ease of passage through an ion channel varies according to the type of ion. Sodium ions move through sodium channels much less easily than potassium ions pass through potassium channels. In fact, in the resting neuron the membrane is up to 100 times more permeable to potassium than to sodium. Consequently, sodium ions pumped out of the neuron cannot easily pass back into the cell, but potassium ions pumped into the neuron can diffuse out.

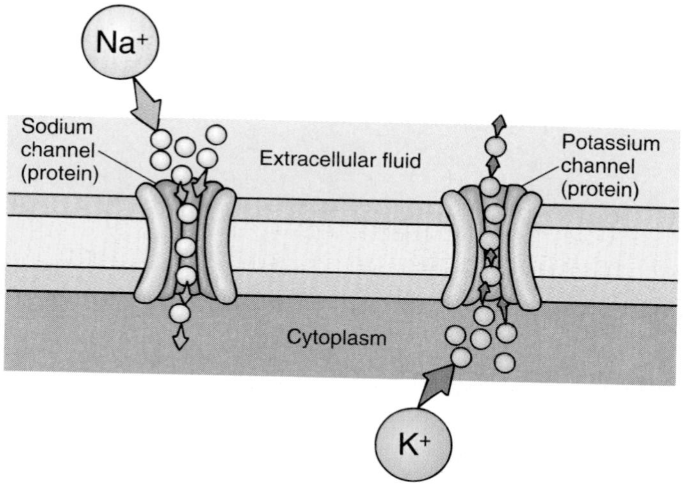

FIGURE 39–6 Proteins in the plasma membrane form ion-specific channels. Ions move through these channels down their concentration gradient from a region of higher to a region of lower concentration.

Potassium ions leak out through the membrane along their concentration gradient until the positive charge outside the membrane reaches a level that repels the outflow of more positively charged potassium ions. A steady state is reached when the potassium outflow equals the inward flow of sodium ions. At this point a potential difference of about −70 mV has developed across the membrane, establishing the resting potential.

Contributing to the overall ion distribution are large numbers of negatively charged proteins and organic phosphates within the neuron that are too large to diffuse out. Some of the potassium ions in the cell help to neutralize these negative charges, but there are not enough potassium ions to neutralize all of them. The plasma membrane is permeable to negatively charged chloride ions, but because of the positively charged ions that accumulate outside the membrane, chloride ions are attracted to the outside and tend to accumulate there.

The resting potential is mainly due to the presence of large protein anions inside the cell and to the outward diffusion of potassium ions along their concentration gradient. However, the conditions for this diffusion must first be established by the action of the sodium-potassium pumps. Remember that the active transport of ions by these pumps is a form of cellular work and therefore requires energy. Both sodium and potassium ions are pumped against a concentration gradient.

The nerve impulse is an action potential

Neurons are highly excitable cells. They have the ability to respond to stimuli and to convert stimuli into neural

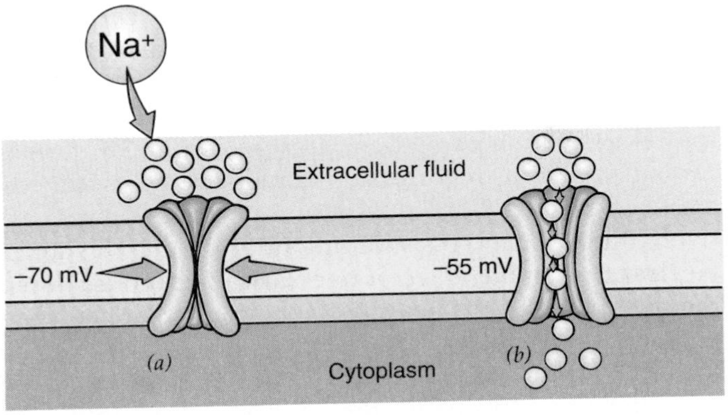

FIGURE 39–7 Voltage-activated ion channels are present in the plasma membrane of the axon and cell body. (*a*) In the resting state, the voltage-activated Na^+ channel is closed. (*b*) When the voltage reaches threshold level, the voltage-activated gate opens, allowing Na^+ to flow into the cell. After a certain amount of time elapses, inactivating gates close the channels.

impulses. An electrical, chemical, or mechanical stimulus may alter the resting potential by increasing the membrane's permeability to sodium. If the neuron membrane is only slightly stimulated, only a local disturbance may occur in the membrane. If the stimulus is sufficiently strong, however, it may result in an **action potential**—the transmission of a **neural impulse.**

In addition to the sodium-potassium pumps and passive ion channels already discussed, the plasma membrane of the axon and cell body contains specific **voltage-activated ion channels,** which open when they detect a change in the membrane potential (Fig. 39–7). When the voltage reaches a certain critical point—the **threshold level**—the protein making up the channel changes shape so that the channel opens. Gating of channels permits regulation. The "gates" are actually extensions of the transport protein molecule that makes up the channel. When the gates open, specific ions can pass through the channel.

The membrane of the neuron can **depolarize** by about 15 mV—that is, to a resting potential of about −55 mV—without actually initiating an impulse. However, when the extent of depolarization is greater than −55 mV, the threshold level is reached. At that point the voltage-activated sodium ion channels open, and sodium ions pass into the cell, moving from an area of higher concentration to an area of lower concentration. After a certain period of time, inactivating gates at the cytoplasmic end of the channel close the channels. Their closing is dependent on time rather than on voltage.

Potassium (K^+) channels also open when the threshold level is reached. They open more slowly and stay open until they sense that the resting potential has been restored.

An almost explosive action occurs as the action potential is produced. The neuron membrane quickly reaches zero potential and even overshoots to about +35 mV, as a momentary reversal in polarity takes place. The sharp rise and fall of the action potential is referred to as a **spike.** Figure 39–8 illustrates an action potential that has been recorded by placing one electrode inside an axon, and one just outside.

The action potential is an electrical current of sufficient strength to induce collapse of the resting potential in the adjacent area of the membrane. The depolarization in one area causes adjacent voltage-gated channels to open, allowing sodium ions to enter. This action causes adjacent voltage-gated channels to open, permitting sodium ions to enter in that area, and the process is repeated until the end of the axon is reached. Thus, the area of depolarization then spreads, like a chain reaction, down the length of the axon (at a constant velocity and amplitude for each type of neuron). A neural impulse is

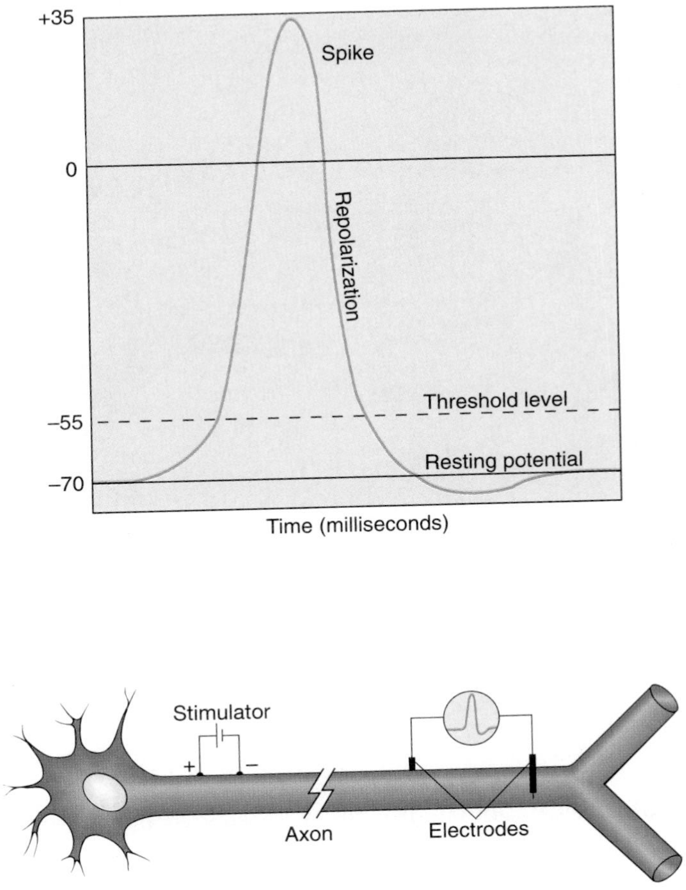

FIGURE 39–8 An action potential can be recorded by placing one electrode inside the cell and one just outside the plasma membrane. When the axon depolarizes to about −55 millivolts, an action potential is generated. (The numerical values are examples. They vary for different nerve cells.)

transmitted as a **wave of depolarization** that travels down the neuron. Conduction of a neural impulse is sometimes compared to burning a trail of gunpowder. Once the gunpowder is ignited at one end of the trail, the flame moves steadily to the other end by igniting the powder particles ahead of it.

By the time the action potential moves a few millimeters down the axon, the membrane over which it has just passed begins to **repolarize** (Fig. 39–9). After a certain time, the sodium gates close, and the membrane again becomes impermeable to sodium. Potassium gates in the membrane open at this time (due to the threshold level of voltage), allowing potassium to leak out at the point of stimulation. This leakage of potassium ions returns the interior of the membrane to its relatively negative state, repolarizing the membrane. This entire mechanism—depolarization and then repolarization—can take place in less than 1 millisecond.

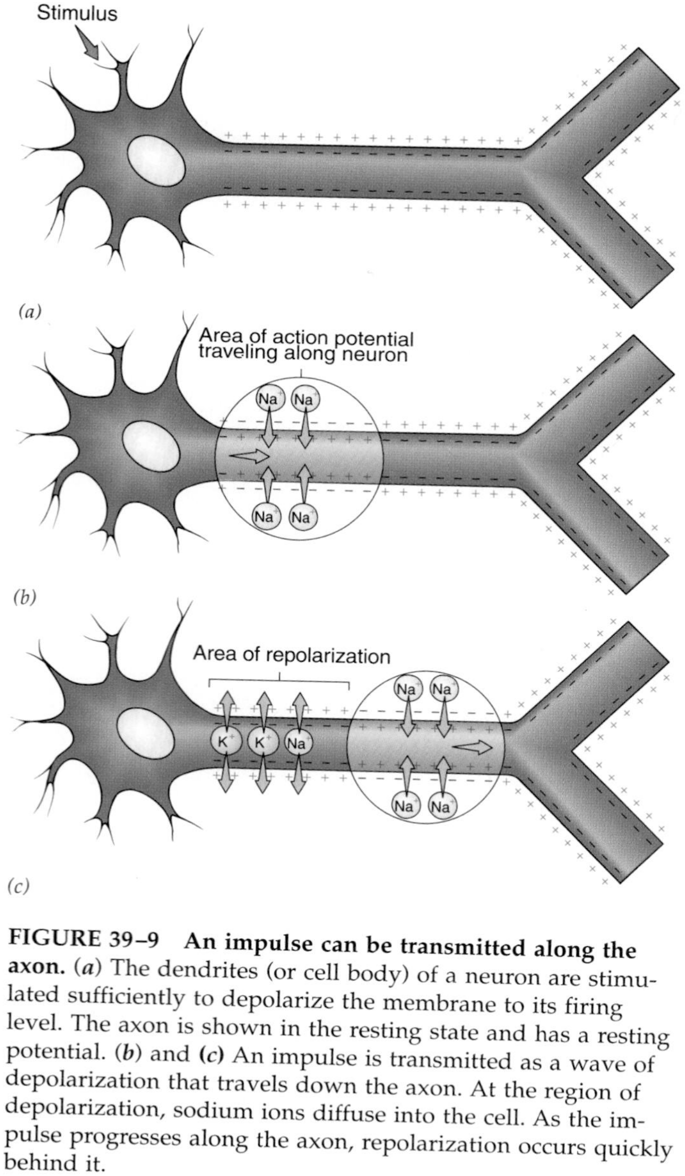

FIGURE 39–9 An impulse can be transmitted along the axon. (*a*) The dendrites (or cell body) of a neuron are stimulated sufficiently to depolarize the membrane to its firing level. The axon is shown in the resting state and has a resting potential. (*b*) and (*c*) An impulse is transmitted as a wave of depolarization that travels down the axon. At the region of depolarization, sodium ions diffuse into the cell. As the impulse progresses along the axon, repolarization occurs quickly behind it.

Even though repolarization takes place very quickly, the redistribution of sodium and potassium to normal resting conditions requires more time. Resting conditions are reestablished when the sodium-potassium pump actively transports excess sodium out of the cell. It should be clear that as the wave of depolarization moves down the membrane of the neuron, the normal polarized state is quickly reestablished behind it. We might imagine the action potential as a ring-shaped zone of negative charge traveling along the axon from one end to the other.

During the millisecond or so in which it is depolarized, the axon membrane is in an **absolute refractory period:** it cannot transmit another action potential no matter how great a stimulus is applied. Then, for a few additional milliseconds, while the resting condition is being reestablished, the axon can transmit impulses only when they are more intense than is normally required. This is the **relative refractory period.** Even with the limits imposed by their refractory periods, most neurons can transmit several hundred impulses per second.

Saltatory conduction is rapid

The smooth, progressive impulse transmission just described, called **continuous conduction,** occurs in unmyelinated neurons. Conduction in myelinated axons depends upon a similar pattern of current flow. However, in myelinated neurons the myelin acts as an effective electrical insulator around the axon except at the nodes of Ranvier. The axon plasma membrane makes direct contact with the surrounding interstitial fluid only at the nodes, and sodium channels are concentrated there. The action potential jumps along the axon from one node of Ranvier to the next (Fig. 39–10). The ion activity at the active node serves to depolarize the next node along the axon. This type of impulse transmission, known as **saltatory conduction,** is faster than the continuous step-by-step type of conduction. A myelinated axon can conduct an impulse up to 50 times faster than the fastest unmyelinated axon.

Saltatory conduction has another advantage over continuous conduction: it requires less energy. Current flows only at the nodes, so fewer sodium and potassium ions are displaced. As a result, the cell does not have to work so hard to reestablish resting conditions each time an impulse is conducted.

The neuron obeys an all-or-none law

Any stimulus too weak to depolarize the neuron to threshold level cannot fire the neuron. It merely sets up a local response that fades and dies within a few millimeters from the point of stimulus. A stimulus strong

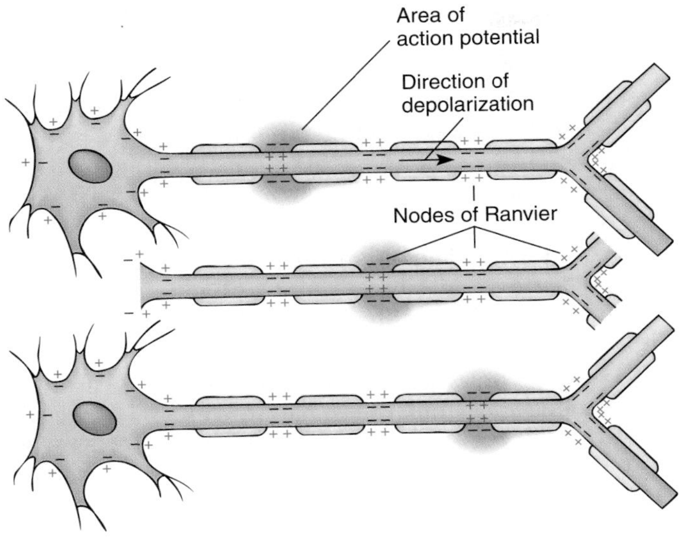

FIGURE 39–10 Saltatory conduction occurs in myelinated axons. The impulse leaps along from one node of Ranvier to the next.

enough to depolarize the neuron to its critical threshold level results in transmission of an impulse along the axon.

The threshold level varies with the type of neuron. For example, receptor neurons have lower thresholds to certain stimuli than do other types of neurons. A stimulus stronger than that necessary to fire a particular neuron results only in the transmission of an identical action potential. The neuron either transmits an action potential or it does not. No variation exists in the strength of a single impulse. Thus, the neuron obeys an **all-or-none law.**

But how can this be? Sensations, after all, do come in different levels of intensity. We have no difficulty distinguishing between the pain of a severe toothache and that of a minor cut on the arm. This apparent inconsistency is explained by the fact that intensity of sensation depends on the number of neurons stimulated and on their frequency of discharge. Suppose you burn your hand. The larger the area burned, the more pain receptors are stimulated and the more neurons are depolarized. Also, the stronger the stimulus, the greater the number of action potentials each neuron transmits per unit of time.

Certain substances affect excitability

Any substance that increases the permeability of the membrane to sodium causes the neuron to become more excitable than normal. Other substances decrease the permeability of the membrane to sodium, making the neuron less excitable.

Calcium balance is essential to normal neural function. Calcium ions are thought to bind to the proteins that make up the sodium channels. The positively charged calcium ions affect the channel proteins, increasing the voltage necessary to open the gates. When calcium ions are too numerous, neurons are less excitable and more difficult to fire.

In contrast, when insufficient numbers of calcium ions are present, the sodium gates apparently fail to close completely between action potentials. As a result, sodium leaks into the cell. This lowers the resting potential, bringing the neuron closer to firing. The neuron fires more easily and sometimes even spontaneously. As a result, the muscle innervated by the neuron may go into spasm, or *tetany.* The condition known as low-calcium tetany can occur when the parathyroid glands do not secrete sufficient hormone.

Many narcotics and anesthetics block conduction of nerve impulses. Local anesthetics such as cocaine, procaine (Novocain), and lidocaine (Xylocaine) affect the gates of the sodium channels, decreasing the permeability of the neuron to sodium. Excitability may be so reduced that the neuron cannot transmit an impulse through the anesthetized region.

DDT and other chlorinated hydrocarbon pesticides interfere with the action of the sodium pump. When nerves are poisoned by such substances, they are unable to transmit impulses. Although the human nervous system can be damaged by these poisons, insects are even more sensitive to them, and may die when exposed to them.

SYNAPTIC TRANSMISSION OCCURS BETWEEN NEURONS

Recall that a synapse is the junction between two neurons or between a neuron and an effector. The synapse between a neuron and a muscle cell is called a **neuromuscular junction** or **motor end plate.** A neuron that *terminates* at a specific synapse is referred to as a **presynaptic neuron**, while a neuron that *begins* at a synapse is known as a **postsynaptic neuron.** Note that these terms are relative to a specific synapse. A neuron that is postsynaptic with respect to one synapse may be presynaptic to the next synapse in the sequence.

Based on how pre- and postsynaptic neurons communicate, two types of synapses have been identified: **electrical synapses** and **chemical synapses.** In electrical synapses, the presynaptic and postsynaptic neurons occur very close together (within 2 nanometers of one another) and form gap junctions (see Chapter 5). The two cells are connected by a protein channel. Electrical synapses allow the passage of ions from one cell to another, permitting an impulse to be directly transmitted from presynaptic to postsynaptic neuron.

Electrical synapses are found between axons and cell body, axons and dendrites, dendrites and dendrites, and between two cell bodies. Such synapses permit rapid communication between neurons and help synchronize the activity of many adjacent neurons.

Although some electrical synapses have been identified in both invertebrates and vertebrates, most synapses are thought to be chemical. In chemical synapses the pre- and postsynaptic cells are separated by a space, the **synaptic cleft,** more than 20 nanometers wide (less than a millionth of an inch). Because depolarization is a property of the plasma membrane, when an impulse reaches the end of the axon it is unable to jump the gap. Chemical messengers called **neurotransmitters,** or transmitter substances, conduct the neural message across the synapse.

Neurotransmitters affect postsynaptic neurons

The synaptic knobs, also known as presynaptic terminals, continuously synthesize neurotransmitter. Mitochondria in the synaptic knobs provide the ATP required for this synthesis (Fig. 39–11). Needed enzymes are produced in the cell body and transported down the axon to the synaptic knobs. After it is produced, the neurotransmitter is stored in small membrane-bounded sacs, which are called **synaptic vesicles,** within the cytoplasm of the synaptic knobs.

Each time an action potential reaches a synaptic knob, the resulting change in membrane potential activates voltage-sensitive calcium channels. Calcium ions from the surrounding interstitial fluid then pass into the synaptic knob. The Ca^{2+} causes synaptic vesicles to fuse with the presynaptic membrane and release neurotransmitter molecules into the synaptic cleft by exocytosis (Fig. 39–11).

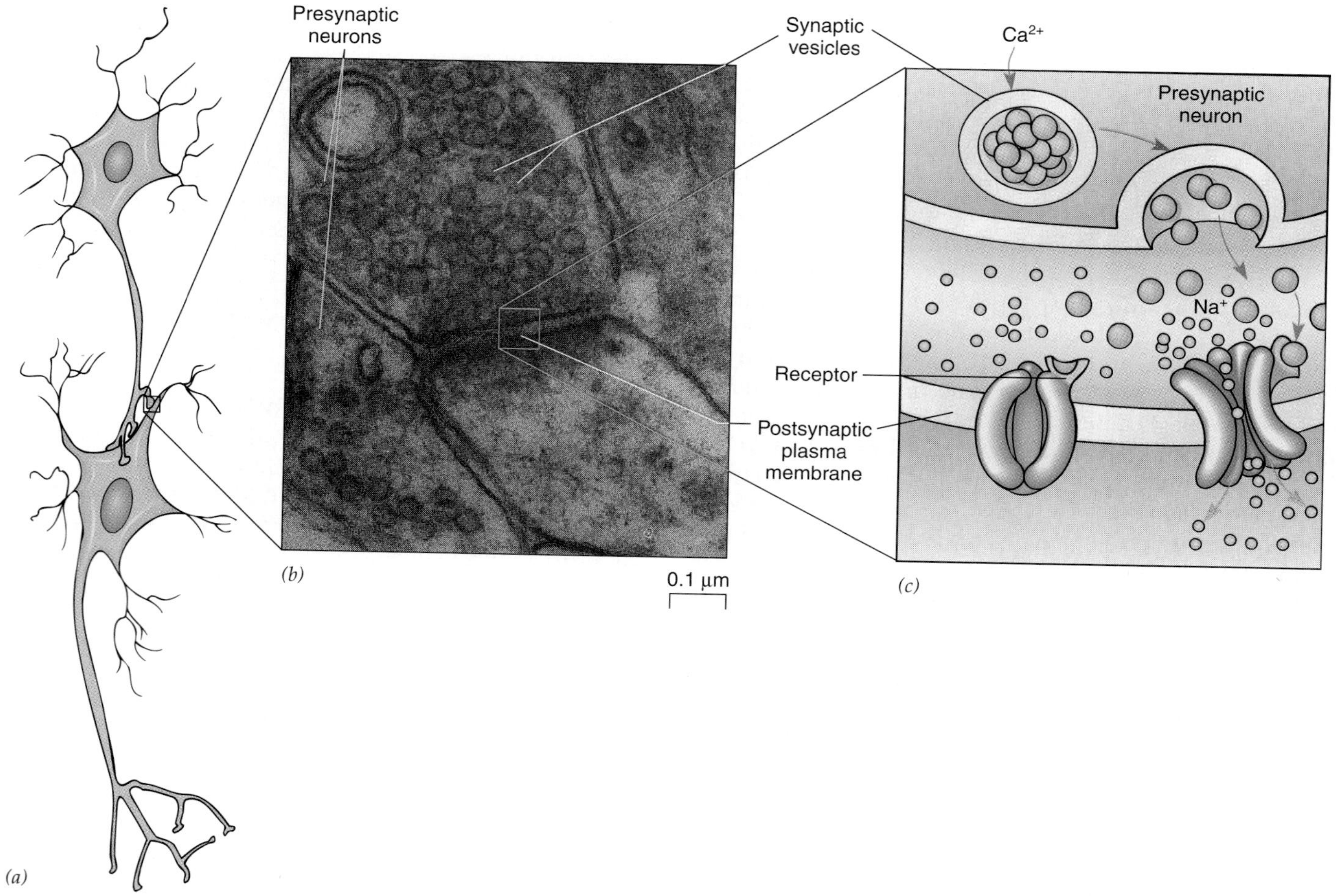

FIGURE 39–11 Impulses are transmitted between neurons, or from a neuron to an effector by neurotransmitters. ***(a)*** In most synapses the wave of depolarization is unable to jump across the synaptic cleft between the two neurons. The problem is solved by chemical transmission across synapses. ***(b)*** The electron micrograph shows synaptic knobs filled with synaptic vesicles. ***(c)*** When an impulse reaches the synaptic knobs at the end of a presynaptic neuron, calcium ions enter the synaptic knob from the interstitial fluid. The calcium ions apparently cause the synaptic vesicles to fuse with the plasma membrane and release a neurotransmitter into the synaptic cleft. The neurotransmitter diffuses across the synaptic cleft and may combine with receptors in the membrane of the postsynaptic neuron. The permeability of the postsynaptic membrane to certain ions may change, resulting in either depolarization or hyperpolarization. Sufficient depolarization can trigger an impulse in the postsynaptic neuron. (*b,* J.F. Gennaro/Photo Researchers, Inc.)

Neurotransmitter molecules then diffuse across the synaptic cleft and combine with specific **receptors** on the dendrites or cell bodies of postsynaptic neurons (or on the membranes of effectors). These receptors are proteins that control **chemically activated ion channels.** When the neurotransmitter binds with the receptor, the channel opens, permitting the passage of specific ions through the membrane. The resulting redistribution of ions affects the electrical potential of the membrane, causing either depolarization or hyperpolarization (making the membrane potential more negative). If sufficiently intense, a local depolarization can set off an action potential.

If repolarization is to occur quickly, any excess neurotransmitter in the synaptic cleft must be removed. Some neurotransmitters are inactivated by enzymes. Others are taken back into the presynaptic axon terminal by a pumping mechanism.

Signals may be excitatory or inhibitory

When neurotransmitter molecules combine with receptors on the membrane of a postsynaptic neuron, the postsynaptic neuron may come closer to firing, or go farther away from firing. For example, if the combination of the neurotransmitter molecules with the receptors opens sodium channels, the resulting influx of sodium ions may partially or completely depolarize the membrane. If sufficient sodium ions enter to change the membrane potential, from −70 mV to, let us say, −60 mV, the membrane would be only −5 mV away from threshold. Thus, a lower stimulus than normal would cause the neuron to fire. A change in membrane potential that brings the neuron closer to firing is called an **excitatory postsynaptic potential (EPSP).**

Another mechanism that may bring about an EPSP involves potassium channels. When the neurotransmitter serotonin combines with its receptor, it activates **adenylyl cyclase,** an enzyme in the postsynaptic membrane. Adenylyl cyclase converts ATP to **cyclic AMP (cAMP)** (Fig. 3–29), which activates a protein kinase. In turn, the protein kinase phosphorylates a protein that closes the K^+ channels. Potassium ions are then unable to diffuse out of the cell, resulting in depolarization.

Some neurotransmitter-receptor combinations *hyperpolarize* the postsynaptic membrane; that is, they make its membrane potential more negative. For example, if the membrane potential changes from −70 mV to −80 mV, the membrane is farther away from threshold and a stronger stimulus will be required to fire the neuron. Because such an action takes the neuron farther away from the firing level, a potential change in this direction is called an **inhibitory postsynaptic potential (IPSP).**

Like an EPSP, an IPSP can be produced in several ways. When the neurotransmitter binds to the receptor, K^+ channels may open. As K^+ leaves the cell, it becomes more negative, and the result is hyperpolarization of the membrane. Other receptors that produce IPSPs permit Cl^- to flow into the cell. This increased membrane permeability to Cl^- also hyperpolarizes the membrane.

When certain types of receptors on the postsynaptic neuron are activated, the reactivity of the neuron can be affected for long periods of time, even years. Neurotransmitters that cause such long-term effects are sometimes referred to as *modulators.* They appear to work by affecting genes or activating enzymes that increase or decrease the number of receptors.

Graded potentials vary in magnitude

Each EPSP or IPSP is a local response in the neuron membrane. Such responses are referred to as **graded potentials** because they vary in magnitude depending on the strength of the stimulus applied. A local change in potential can cause a flow of electrical current. The greater the change in potential, the greater the flow of current. Such a local current flow can function as a signal only over a very short distance, because it fades out within a few millimeters of its point of origin. As we will see, however, graded potentials can be added together, resulting in action potentials.

One EPSP is usually too weak to trigger an action potential by itself. Its effect is subliminal—that is, below threshold level. (Even though subthreshold EPSPs do not produce an action potential, they do affect the membrane potential.) EPSPs may be added together in a process known as **summation. Temporal summation** occurs when repeated stimuli cause new EPSPs to develop before previous EPSPs have decayed. By summation of several EPSPs, the neuron may be brought to the critical firing level. When several synaptic knobs release neurotransmitter simultaneously, the postsynaptic neuron is stimulated at several places at once. This effect, called **spatial summation,** can also bring the postsynaptic neuron to the threshold level.

Even with summation, the postsynaptic neuron may not be depolarized sufficiently to conduct an impulse. However, the neuron is said to be **facilitated,** meaning that its membrane is nearer the threshold for firing than it would normally be in the resting state. Additional stimulation of the neuron can more easily bring it to the firing level. Facilitation is further discussed later in this chapter.

Many types of neurotransmitters are known

More than 40 different chemical compounds are now known or suspected to function as neurotransmitters. Some are small molecules that act rapidly. Others are neuropeptides, larger molecules that appear to act more slowly and that modulate the effect of other neurotransmitters. Each type of neuron is thought to release only one small-molecule type of neurotransmitter. However, the neuron may release one or more neuropeptides at the same time. A postsynaptic neuron may have recep-

Table 39–1 SOME NEUROTRANSMITTERS

Substance	*Where Secreted*	*Comments*
Acetylcholine	Nerve-muscle junctions; autonomic system;* parts of brain	Inactivated by cholinesterase
Biogenic amines		
Norepinephrine	Autonomic system; reticular activating system and other areas of brain and spinal cord	Inactivated slowly by monoamine oxidase (MAO); mainly inactivated by reabsorption by synaptic knob vesicles; norepinephrine level in brain affects mood
Dopamine	Limbic system; cerebral cortex; basal ganglia; hypothalamus	Thought to affect motor function; may be involved in schizophrenia; amount reduced in Parkinson's disease
Serotonin (5-hydroxytryptamine, 5-HT)	Limbic system; hypothalamus; cerebellum; spinal cord	May play role in sleep; LSD antagonizes serotonin; thought to be inhibitory
Amino acid		
GABA (gamma-aminobutyric acid)	Spinal cord; cerebral cortex; cerebellum	Acts as inhibitor; may play role in pain perception
Neuropeptides		
Endorphins	CNS and pituitary gland	Neuropeptides that have morphine-like properties and suppress pain; may help regulate cell growth; linked to learning and memory
Enkephalins	Brain and digestive tract	Neuropeptides that inhibit pain impulses by inhibiting release of substance P; bind to same receptors in brain as morphine
Substance P	Brain and spinal cord; sensory nerves; intestine	Transmits pain impulses from pain receptors into CNS

*These and other structures listed in this table are discussed in Chapter 40.

tors for more than one type of neurotransmitter. Indeed, some of its receptors may be excitatory and some may be inhibitory.

Two extensively studied neurotransmitters are acetylcholine and norepinephrine (NE) (Table 39–1). In Chapter 38 we discussed how **acetylcholine,** released from motor neurons, triggers muscle contraction. Acetylcholine is also released by some neurons in the brain and in the autonomic nervous system (see Chapter 40). Cells that release this neurotransmitter are referred to as **cholinergic neurons.**

Acetylcholine has an excitatory effect on skeletal muscle. It opens sodium channels, which increases the permeability of the muscle fiber membrane to sodium. The influx of sodium depolarizes the membrane, leading to contraction. In contrast, acetylcholine has an inhibitory effect on cardiac muscle, resulting in a decreased heart rate. Whether a neurotransmitter excites or inhibits is apparently a property of the postsynaptic receptors with which it combines.

After acetylcholine is released by a presynaptic neuron and combined with receptors on the postsynaptic neuron, excess acetylcholine must be removed. The enzyme cholinesterase breaks it down into its chemical components, choline and acetate.

Norepinephrine is released by sympathetic neurons (see Chapter 40) and by many neurons in the brain and spinal cord. Neurons that release norepinephrine are called **adrenergic neurons.** Norepinephrine and the neurotransmitters **serotonin** and **dopamine** belong to a class of compounds called **biogenic amines,** or catecholamines.

After release, most of the excess biogenic amines are actively transported into the vesicles in the synaptic knobs, a process known as *re-uptake.* Some are degraded by the enzymes catechol-O-methyl transferase and **monoamine oxidase (MAO).** MAO is also thought to help regulate the concentration of biogenic amines in the synaptic knobs. Biogenic amines affect mood, and their imbalance has been linked to depression, attention deficit disorder, and psychosis (e.g., schizophrenia). Antidepressants and

Making the Connection

Neurons, Genes, and Alzheimer's Disease

How can researchers use their knowledge of gene function, proteins, and neurons to solve the mystery of Alzheimer's disease? This progressive, degenerative brain disorder afflicts more than four million people in the United States alone, making it the fourth leading cause of death. Although Alzheimer's disease strikes individuals in midlife, more than 90% of Alzheimer's patients develop the disease after age 65. In fact, it is the leading cause of senile dementia, which is the loss of memory, judgment, and the ability to reason that we often associate with aging. For reasons still unknown, more women than men suffer from Alzheimer's disease.

In Alzheimer's disease, cells are lost in certain parts of the brain, including the cerebral cortex and hippocampus, areas that are important in thinking and remembering. Neurons that secrete the neurotransmitter acetylcholine are especially affected. Two abnormalities that develop in brain tissue as we age — *senile plaques* and *neurofibrillary tangles* — are especially characteristic of Alzheimer's disease. These abnormal developments appear to damage brain cells, leading to deterioration in general function. Researchers are investigating the biochemistry and genetic basis of both plaques and tangles for clues to the causes and cures of Alzheimer's disease.

Senile plaques are clusters of abnormal neurons and glial cells. Investigators have demonstrated that senile plaques have a central core consisting of a protein fragment called beta amyloid peptide. The precursor (beta-APP) of this protein fragment is a large transmembrane protein coded by a gene located on chromosome 21. Normal brain cells make a soluble form of beta amyloid peptide. Alzheimer's disease may develop when an imbalance in brain metabolism results in an insoluble peptide form, which leads to plaque formation. The plaques are thought to disrupt calcium homeostasis, which in turn leads to neural malfunction and brain cell death.

Neurofibrillary tangles consist of abnormal accumulations of certain cytoskeletal proteins in the neuron cytoplasm. One of the proteins involved, called **tau,** normally stimulates the protein tubulin to form microtubules. When too many phosphates attach to tau, it can no longer adhere to microtubules. Instead, tau molecules join with one another to form fibrous deposits that make up the neurofibrillary tangles. Neurons that have such tangles form fewer synapses with other neurons.

In 1993, neurologist Allen Roses at Duke University Medical Center discovered a new clue to the Alzheimer's disease mystery. We all have a gene that codes for a protein called apolipoprotein E, or **Apo-E,** which helps transport cholesterol in the blood. Three different forms of this protein are known. The common form, Apo-E3, binds to tau and may inhibit phosphate-bonding. Apo-E3 may also be important in transporting beta amyloid to cells for processing. The Apo-E4 form of the protein (which differs in only one amino acid) does not inhibit phosphate-bonding.

Roses reported that individuals who are homozygous for the Apo-E4 allele (that is, have two copies of the gene that codes for the Apo-E4 form of the protein) are eight times more likely to develop Alzheimer's disease than are individuals with the more common form of the gene, Apo-E3. Thus, the Apo-E4 gene may prove to be an important risk factor. Would you want to know if you had this gene?

Clues to untangling the mysteries of Alzheimer's disease are leading researchers in many different directions. By mid-1995 genes on three different chromosomes had been identified that are associated with increased risk of early onset of Alzheimer's disease.

Some investigators are exploring an inherited defect in mitochondria. One current focus of study involves certain nerve growth factors that appear to have a protective role.

More than a dozen drugs designed to slow the progress of Alzheimer's disease are in clinical trials. One of the challenges of developing a drug has been getting the medication through the blood-brain barrier (see Chapter 40). Current research on the causes and cures of Alzheimer's disease is providing new insights into the metabolism of neurons and the function of the nervous system. ▲

many other mood-affecting drugs work by altering the levels of biogenic amines in the brain.

Other neurotransmitters listed in Table 39–1 include serotonin, gamma-aminobutyric acid (GABA), and the endorphins. GABA, an amino acid, is an inhibitory neurotransmitter in the brain and spinal cord. Drugs such as the benzodiazepines and the barbiturates used to treat anxiety enhance the actions of GABA.

NERVE FIBERS MAY BE CLASSIFIED IN TERMS OF SPEED OF CONDUCTION

In the laboratory we can demonstrate that an impulse can move in both directions within a single axon. In the body, however, an impulse generally stops when it reaches the dendrites, because no neurotransmitter is there to conduct it across the synapse. This limitation imposed by neurotransmitter makes neural pathways function as one-way streets. The usual direction of transmission is from the axon of the presynaptic neuron, across the synapse, and then to the dendrite or cell body of the postsynaptic neuron.

Compared with the speed of an electrical current or the speed of light, a nerve impulse travels rather slowly. Its speed varies from about 0.5 meter per second to more than 120 meters (400 feet) per second. What factors affect speed of transmission? In general, the greater the diameter of an axon, the greater its speed of conduction. The largest neurons seem also to be the most heavily myelinated, and the more myelin a neuron has, the faster it transmits impulses. In myelinated neurons, the distance

Making the Connection

Neurotransmitters, Thought Processes, and Schizophrenia

How is the neurotransmitter dopamine linked to the disturbed thought processes and hallucinations characteristic of schizophrenia? Schizophrenia is a disease involving thought disorder and decline in general functioning. According to the dopamine hypothesis, individuals suffering from schizophrenia have an abnormally high number of dopamine receptors in certain areas of the brain. More dopamine receptors result in overactivity of the neurons that respond to dopamine.

When the neurons that respond to dopamine are overactive, they overstimulate areas in the temporal lobes of the brain that can produce hallucinations — perceptual experiences like seeing or hearing things that have no basis in present reality.

The dopamine hypothesis is based on two types of evidence. First, antipsychotic drugs reduce the frequency of hallucinations and delusions by blocking dopamine receptors. Second, amphetamines, cocaine, and other drugs that are biochemically related to dopamine cause symptoms of schizophrenia in normal individuals and increase symptoms in those suffering from schizophrenia.

The neurotransmitter serotonin also plays a role in schizophrenia. Serotonin inhibits activity in some neurons. When serotonin levels are low, dopamine activity is not kept in check. A normal balance of dopamine and serotonin appears to be necessary for normal neural functioning. ▲

between successive nodes of Ranvier is also important: the farther apart the nodes, the faster the axon conducts.

When considering speed of conduction through a sequence of neurons, the number of synapses must be taken into account, because each time an impulse is conducted from one neuron to another there is a slight synaptic delay (about 0.5 millisecond). This delay is due to the time required for the release of neurotransmitter, its diffusion, and its binding to postsynaptic membrane receptors.

NEURAL IMPULSES MUST BE INTEGRATED

Neural integration is the process of sorting and interpreting incoming signals, and then determining an appropriate response. Each neuron may synapse with hundreds of other neurons. Indeed, as much as 40% of a postsynaptic neuron's dendritic surface and cell body may be covered by synaptic knobs of presynaptic neurons. It is the job of the dendrites and the cell body of every neuron to integrate the hundreds of messages that continually bombard them.

EPSPs and IPSPs occur continually in postsynaptic neurons, and IPSPs cancel the effects of some of the EPSPs. The postsynaptic cell body and dendrites continually tabulate such molecular transactions. It is important to remember that each EPSP and IPSP is not an all-or-none response. Rather, each is a local response (it does not travel like an action potential) that may be added to or subtracted from other EPSPs and IPSPs. After the neuron membrane has completed its chemical tabulations, the neuron may be inhibited, facilitated, or brought to threshold level. If sufficient EPSPs have been received to bring the neuron to threshold level, an all-or-none action potential is initiated and travels down the axon. This mechanism provides for integration of hundreds of tiny "messages" (EPSPs and IPSPs) before an impulse is actually transmitted along the axon of a postsynaptic neu-

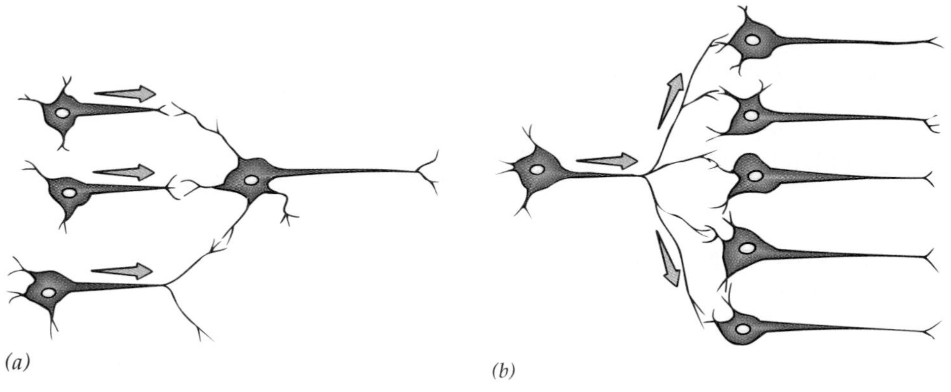

FIGURE 39–12 Neural circuits can be organized in many ways. (*a*) Convergence of neural input. Several presynaptic neurons synapse with one postsynaptic neuron. This organization in a neural circuit permits one neuron to receive signals from many sources. (*b*) Divergence of neural output. A single presynaptic neuron synapses with several postsynaptic neurons. This organization allows one neuron to communicate with many others.

ron. Local responses permit the neuron and the entire nervous system a far greater range of response than would be the case if every EPSP generated an action potential.

Where does neural integration take place? Every neuron acts as a tiny integrator, sorting through (on a molecular level) the hundreds and thousands of bits of information continually bombarding it. Since more than 90% of the neurons in the body are located in the CNS, most neural integration takes place there, within the brain and spinal cord. These neurons are responsible for making most of the "decisions." In the next chapter the brain and spinal cord will be examined in some detail.

NEURONS ARE ORGANIZED INTO CIRCUITS

The CNS contains millions of neurons, but it is not just a tangled mass of nerve cells. Its neurons are organized into separate **neuronal pools,** or networks, and within each pool the neurons are arranged in specific pathways, or **circuits.** Although each pool has some special characteristics, the neural circuits in all of the pools share many organizational features. For example, convergence and divergence are probably characteristic of all of them.

In **convergence,** a single neuron is controlled by converging signals from two or more presynaptic neurons (Fig. 39–12*a*). An association neuron in the spinal cord, for instance, may receive converging information from sensory neurons entering the cord, from neurons bringing information from various parts of the brain, and from neurons coming from different levels of the spinal cord. Information from all these sources has to be integrated before a neural message (action potential) can be sent and an appropriate motor neuron stimulated. Convergence is an important mechanism by which the CNS can integrate the information that impinges on it from various sources.

In **divergence,** a single presynaptic fiber stimulates many postsynaptic neurons (Fig. 39–12*b*). Each presynaptic neuron may branch and synapse with 25,000 or more different postsynaptic neurons. For example, a single neuron transmitting an impulse from the motor area of the brain may synapse with hundreds of association neurons in the spinal cord, and each of these may diverge in turn, so that hundreds of muscle fibers may be stimulated.

The mechanism of **facilitation,** discussed earlier in this chapter, is illustrated in Figure 39–13. Neither Neuron A nor Neuron B by itself can fire Neuron 2 or 3. However, stimulation by either A or B does slightly depolarize the neuron. This facilitates the postsynaptic neuron so that, if the other (or any other) presynaptic neuron stimulates it, threshold level is reached more easily and an action potential may be generated. Many neural interactions within the CNS depend on facilitation.

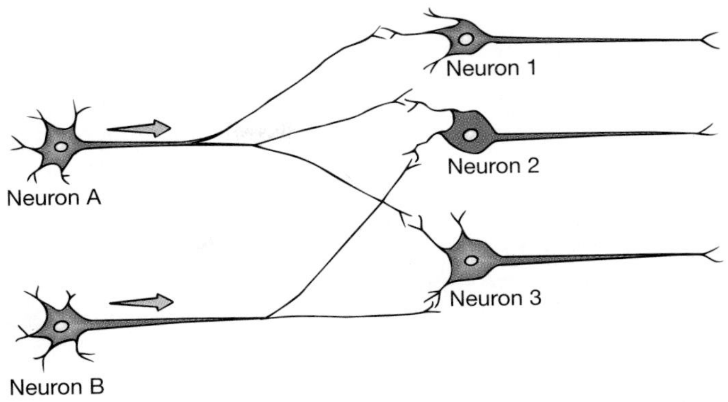

FIGURE 39–13 In facilitation, a neuron becomes more likely to fire as a result of a subthreshold stimulus. Neither neuron A nor neuron B can itself fire neuron 2 or 3. However, stimulation by either A or B does depolarize the neuron toward the threshold level (if the stimulation is excitatory). This facilitates the postsynaptic neuron, so if another presynaptic neuron stimulates it, the threshold level may be reached and an action potential generated.

One of the most important kinds of neural circuits is the **reverberating circuit,** a neural pathway arranged so that a neuron collateral (branch) synapses with an association neuron (Fig. 39–14). The association neuron synapses with a neuron in the sequence that can again send new impulses through the circuit. This is an example of positive feedback. New impulses can be generated again and again until the synapses fatigue (from depletion of neurotransmitter) or until stopped by some sort of inhibition. Reverberating circuits are thought to be important in rhythmic breathing, mental alertness, and perhaps short-term memory.

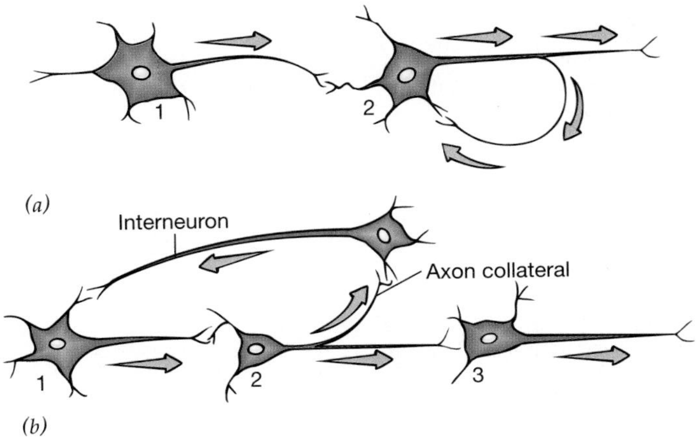

FIGURE 39–14 In a reverberating circuit a neuron may fire multiple times. (***a***) A simple reverberating circuit in which an axon collateral of the second neuron turns back on its own dendrites, so the neuron continues to stimulate itself. (***b***) In this neural circuit an axon collateral of the second neuron synapses with an interneuron. The interneuron synapses with the first neuron in the sequence. New impulses are triggered again and again in the first neuron, causing reverberation.

SUMMARY

I. Information flow through the nervous system begins with reception. Information is transmitted to the CNS, where integration takes place. Appropriate nerve impulses are then transmitted by neurons to the effectors that carry out the response.

II. Glial cells and neurons are the two types of cells characteristic of neural tissue.
   A. Glial cells are supporting cells.
   B. Neurons are specialized to receive stimuli and transmit impulses.
      1. A typical neuron consists of a cell body with many branched dendrites and a single long axon.
      2. Outside the central nervous system, axons are surrounded by a cellular sheath, and many are also enveloped by a myelin sheath.
      3. A nerve consists of several hundred axons wrapped in connective tissue; a ganglion is a mass of cell bodies.

III. Neurons rapidly transmit electrical impulses.
   A. A neuron that is not transmitting an impulse has a resting potential.
      1. The inner surface of the plasma membrane is negatively charged compared to the outside; the membrane is polarized.
      2. Sodium-potassium pumps continuously transport sodium ions out of the neuron and potassium ions in.
      3. Potassium ions can leak out more readily than sodium ions can leak in.
   B. Excitatory stimuli are thought to open sodium channels in the plasma membrane. This permits sodium to enter the cell and alter the membrane potential toward depolarization. Inhibitory stimuli hyperpolarize the membrane.
   C. When the membrane potential reaches threshold level, an action potential may be generated.
      1. The action potential is a wave of depolarization that moves down the axon.
      2. The action potential obeys an all-or-none law.
      3. As the action potential moves down the axon, repolarization occurs very quickly behind it.
   D. Saltatory conduction takes place in myelinated neurons. In this type of transmission, depolarization skips along the axon from one node of Ranvier to the next.
   E. Excitability of a neuron can be affected by calcium concentration and by certain substances such as local anesthetics and pesticides.

IV. Synaptic transmission generally depends on release of a neurotransmitter from vesicles in the synaptic knobs of the presynaptic axon.
   A. Neurotransmitter diffuses across the synaptic cleft and combines with specific receptors on the postsynaptic neuron.
   B. This may cause an excitatory postsynaptic potential (EPSP) or an inhibitory postsynaptic potential (IPSP).
   C. Some neurotransmitters are small molecules that act rapidly. Others are neuropeptides, larger molecules that modulate the effects of the small-molecule neurotransmitters. The biogenic amines are a group of neurotransmitters that includes norepinephrine, serotonin, and dopamine.

V. Neural integration is the process of adding and subtracting incoming signals and determining whether or not to fire an impulse.

VI. Complex neural pathways are possible because of reverberating circuits and neural associations such as convergence, divergence, and facilitation.

SELECTED KEY TERMS

acetylcholine
action potential
all-or-none law
association neuron
axon
biogenic amine
convergence
dendrite
depolarization
divergence
dopamine
excitatory postsynaptic potential (EPSP)
facilitation
ganglion (ganglia)
glial cell
inhibitory postsynaptic potential (IPSP)
integration
motor neuron
myelin sheath
nerve
neuron
neurotransmitter
norepinephrine
postsynaptic neuron
presynaptic neuron
reception
refractory period
resting potential
saltatory conduction
Schwann cell
sensory neuron
serotonin
sodium-potassium pump
summation
synapse
synaptic knob
threshold level
transmission

POST-TEST

1. The process of conducting neural information is termed ________ .
2. A functional connection between neurons is called a(an) ________ .
3. The process of detecting a change in the environment is termed ________ .
4. A nerve cell is properly referred to as a(an) ________ .
5. The supporting cells of nervous tissue are called ________ cells.
6. ________ are neuron extensions specialized to transmit messages toward the cell body; the ________ functions to transmit impulses from the cell body to another neuron.
7. In some neurons outside the central nervous system, Schwann cells produce a(an) ________ ________ surrounding the neuron.
8. The cell bodies of neurons are massed together in ________ .
9. The ________ potential of a neuron is due in large part to the outward diffusion of potassium ions along their concentration gradient.
10. ________ stimuli are thought to open sodium channels, thereby permitting sodium to rush into the cell.

11. The passage of sodium ions into the neuron causes __________ of the plasma membrane.
12. A wave of depolarization that travels down the axon is called a neural impulse or __________ __________ .
13. Because there is no variation in the intensity of an action potential, the neuron is said to obey a(an) __________–__________–__________ law.
14. In __________ conduction, ion activity at one node depolarizes the next node along the axon.
15. When an impulse reaches the synaptic knobs, it stimulates the release of __________ .
16. The neurotransmitter that stimulates muscle contraction is __________ .
17. Serotonin and dopamine belong to a class of neurotransmitter known as biogenic __________ .
18. In __________ , a single presynaptic neuron stimulates many postsynaptic neurons.

REVIEW QUESTIONS

1. Distinguish between a neuron and a nerve.
2. Imagine that a very unfriendly-looking monster suddenly approaches you. What processes must take place within your nervous system before you can make your escape?
3. What happens when a neuron reaches threshold level?
4. Describe the functions of the following: (a) myelin; (b) cellular sheath; (c) glial cells; (d) dendrites; (e) synaptic knobs.
5. What is meant by the resting potential of a neuron? How do sodium-potassium pumps contribute to the resting potential?
6. What is an action potential? What is responsible for it?
7. What is the all-or-none law?
8. Contrast saltatory conduction with conduction in an unmyelinated neuron.
9. How is neural function affected by the presence of too much calcium? Too little calcium?
10. Describe the functions of the following substances: (a) acetylcholine; (b) cholinesterase; (c) norepinephrine.
11. Contrast convergence and divergence.
12. What is summation? Describe facilitation.
13. Label the diagram on the right.

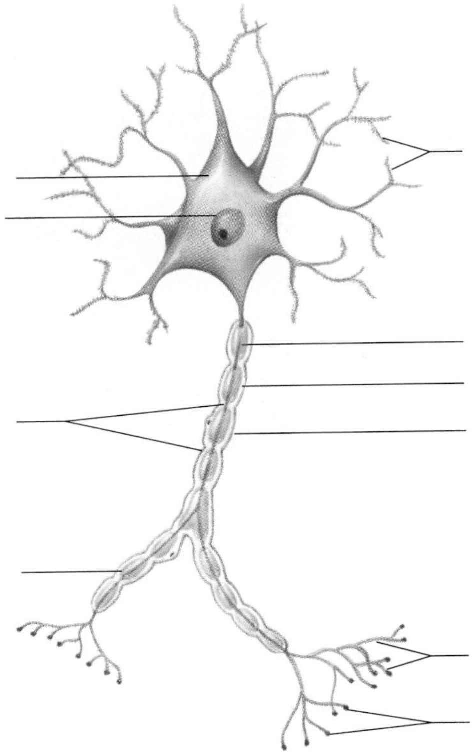

YOU MAKE THE CONNECTION

1. We have no difficulty distinguishing between the pain of a severe toothache and that of a minor cut on the arm. How can you explain the difference in intensity if the all-or-none law is true?
2. Stimulant drugs increase the activity of the nervous system. Propose two mechanisms involving synaptic transmission that could explain the action of a stimulant.

RECOMMENDED READINGS

Gottlieb, D. I. "Gabaergic Neurons." *Scientific American*, Vol. 258, No. 2 February 1988. GABA is an inhibitory neurotransmitter in the brains of all mammals.

Fuller, R. W. "Neural Functions of Serotonin," *Scientific American Science & Medicine*, Vol. 2, No. 4, July/August 1995. The neurotransmitter serotonin is released by neurons in many parts of the brain, and many drugs, including Prozac, work by modifying serotonin levels.

Kimelberg, H. K., and M. D. Norenberg. "Astrocytes." *Scientific American*, Vol. 260, No. 4 April 1989. A discussion of these interesting glial cells.

Neher, E., and B. Sakmann. "The Patch Clamp Technique." *Scientific American*, Vol. 266, No. 3, March 1992. The 1991 Nobel Prize-winning authors discuss their technique for isolating ion channels in cell membranes.

Pennisi, E. "A Molecular Whodunit: New Twists in the Alzheimer's Mystery." *Science News*, Vol. 145, No. 1, January 1, 1994. An overview of current thinking regarding Alzheimer's disease.

Selkoe, D. J. "Amyloid Protein and Alzheimer's Disease." *Scientific American*, Vol. 265, No. 5, November 1991. A discussion of Alzheimer's disease, and research that may lead to treatment.

CHAPTER 40

Neural Regulation: Nervous Systems

The grace and precision of the leap of a European tree frog (*Hyla aborea*) is directed and coordinated by its nervous sytem.

(Animals Animals © 1996 Stephen Dalton)

An animal's lifestyle is closely related to the type and complexity of its nervous system. The sessile sponge, with its very simple lifestyle, has no specialized nerve cells and no nervous system. Its activities are limited to basic life-support processes such as filtering food and reproducing. Whatever responses it makes depends on actions of individual cells.

The simplest organized nervous system is the relatively inefficient **nerve net** found in *Hydra* and other cnidarians (Fig. 40–1). This is an adequate system for an animal that remains rooted in one spot, waiting for dinner to brush by its tentacles. *Hydra* does respond to predators and prey, discharging nematocysts (stinging structures) and coordinating the movements of its tentacles to capture food.

In a nerve net, neurons are scattered throughout the body. Neither a central control organ nor definite pathways are present. In some neurons and across some synapses, impulses are transmitted in more than one direction. The impulses become less intense as they spread from the region of initial stimulation. More neurons receive the message if the stimulus is strong than do if it is weak.

The nerve net produces responses that simultaneously involve large parts of the body. Such a diffuse pattern of transmission is adequate in a radially symmetric animal that moves sluggishly. An advantage of the nerve net is that the cnidarian can respond to dinner approaching from any direction. Some cnidarians have two nerve nets. One, that transmits slowly, coordinates movement of the tentacles, while the other, which is faster, coordinates swimming.

With its more highly developed vertebrate nervous system, a frog can hop about in search of food and eject its tongue with lightning speed to capture a passing fly. However, neither the *Hydra* nor the frog is capable of solving algebra problems or learning about its own biology. An animal's range of possible responses depends in large part on the number of its neurons and how they are organized in the nervous system. As animal groups evolved, nervous systems became increasingly complex.

LEARNING OBJECTIVES

After you have studied this chapter you should be able to

1. Contrast nerve nets and radial nervous systems with bilateral nervous systems.
2. Compare the vertebrate nervous system with a bilateral invertebrate nervous system.
3. Trace the development of the principal vertebrate brain regions from the forebrain, midbrain, and hindbrain, and compare the brains of fish, amphibians, reptiles, birds, and mammals.
4. Describe the functions and structure of the spinal cord.
5. Locate the following parts of the human brain and give the functions of each: medulla, pons, midbrain, thalamus, hypothalamus, cerebellum, and cerebrum. Include a description of the functional areas of the cerebrum.
6. Compare the reticular activating system with the limbic system.
7. Contrast REM and non-REM sleep.
8. Review current theories of information processing as presented in this chapter.
9. Cite experimental evidence linking environmental stimuli with changes in the brain and with learning and motor abilities.
10. Describe the organization of the vertebrate nervous system, comparing the central nervous system with the peripheral nervous system, and comparing the somatic system with the autonomic system.
11. Contrast the sympathetic and parasympathetic divisions of the autonomic system, giving examples of the effects of these systems on specific organs such as the heart and intestine.
12. Discuss the biological actions and effects on mood of the following types of drugs: alcohol, antidepressants, barbiturates, antianxiety drugs, antipsychotic drugs, opiates, stimulants, hallucinogens, and marijuana.

MANY INVERTEBRATES HAVE COMPLEX NERVOUS SYSTEMS

The radial nervous system of the sea star and other echinoderms is a modified nerve net, more complex than the cnidarian nervous system described in the chapter introduction. The echinoderm nervous system consists of a nerve ring around the mouth, from which a large radial nerve extends into each arm. Branches of these nerves, which form a network somewhat similar to the nerve net of *Hydra*, coordinate movement of the animal. In sea stars, a nerve net mediates the responses of the dermal gills to touch.

Bilaterally symmetric animals generally have a more complex nervous system than radially symmetric animals. A bilaterally symmetric animal typically moves forward, head first. With sense organs concentrated at the front of the body, the animal can detect an enemy quickly or sense food in time to capture it. We can identify the following trends in the evolution of bilateral nervous systems:

1. An **increased number of nerve cells.**
2. A **concentration of nerve cells,** forming masses of tissue that become ganglia and **brain,** and thick cords of tissue that become nerve cords and nerves.
3. A **specialization of function.** For example, transmission of nerve impulses in one direction requires both *afferent* nerves, which conduct impulses toward a central nervous system (CNS), and *efferent* nerves, which transmit impulses away from the CNS and to the effectors (muscles and glands). Certain parts of the central nervous system are also usually specialized to perform specific functions; as a result, distinct structural and functional regions can be identified.
4. An **increased number of association neurons and more complex synaptic contacts.** This permits much greater integration of incoming messages, provides a greater range of responses, and allows more precision in responses.
5. **Cephalization, or formation of a head.** A bilaterally symmetric animal generally moves in a forward direction. Concentration of sense organs at the front end of the body is important either for detecting an enemy quickly enough to escape, or for seeing or smelling food in time to capture it. Response can be more rapid if these sense organs are linked by short pathways to

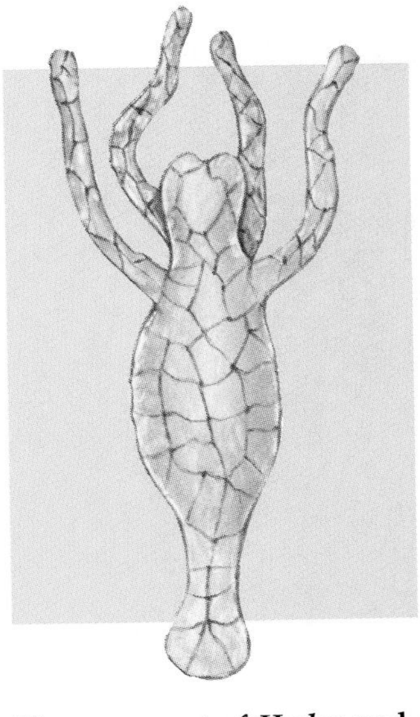

FIGURE 40–1 The nerve net of *Hydra* and other cnidarians is the simplest organized nervous system. No central control organ and no definite neural pathways are present.

decision-making nerve cells nearby. Therefore, nerve cells are usually concentrated in the head region, and organized to form a brain.

In planarian flatworms, the head region contains concentrations of nerve cells referred to as **cerebral ganglia** (Fig. 40–2). These serve as a primitive "brain" and exert some measure of control over the rest of the nervous system. Two solid ventral longitudinal nerve cords extend from the ganglia to the posterior end of the body. Transverse nerves connect the two nerve cords and connect the brain with the eyespots. This arrangement is referred to as a ladder-type nervous system.

In annelids and arthropods, the nervous system also includes a pair of ventrally located longitudinal nerve cords (Fig. 40–3). The cell bodies of the nerve cells are massed into pairs of ganglia located in *each* body segment. Afferent and efferent neurons are located in lateral nerves that link the ganglia with muscles and other body structures.

If an earthworm's brain is removed, the animal can move almost as well as before, but when it bumps into an obstacle, it persists in a futile effort to move forward rather than turning aside. Its brain is necessary for the earthworm to respond adaptively to environmental change.

In some arthropods, the cerebral ganglia are somewhat brainlike in that specific functional regions have been identified in them. Mollusks typically have at least three pairs of ganglia: (1) *cerebral ganglia,* which serve as a coordinating center for complex reflexes and have a motor function, are found dorsal to the esophagus; (2) *visceral ganglia,* which control shell opening and closing, are distributed among the organs; and (3) *pedal ganglia,* which control the movement of the foot, are found in the foot. The visceral and pedal ganglia are connected to the cerebral ganglia by nerve cords.

In cephalopods such as the octopus, the nerve cells are concentrated in a central region (Fig. 40–4). All the ganglia are massed in a ring surrounding the esophagus that contains about 168 million nerve cells. With this complex nervous system, it is no wonder that the octopus is capable of considerable learning and can be taught quite complex tasks. In fact, the octopus is considered to be among the most intelligent invertebrates.

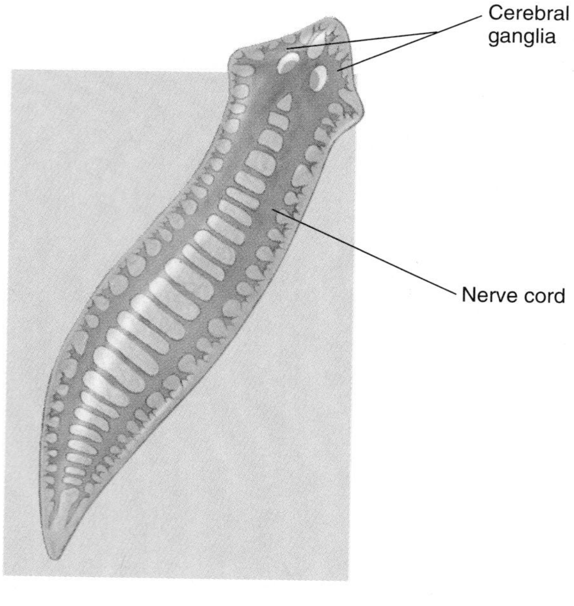

FIGURE 40–2 Planarian flatworms have a ladder-type nervous system. Cerebral ganglia in the head region serve as a simple brain and, to some extent, control the rest of the nervous system.

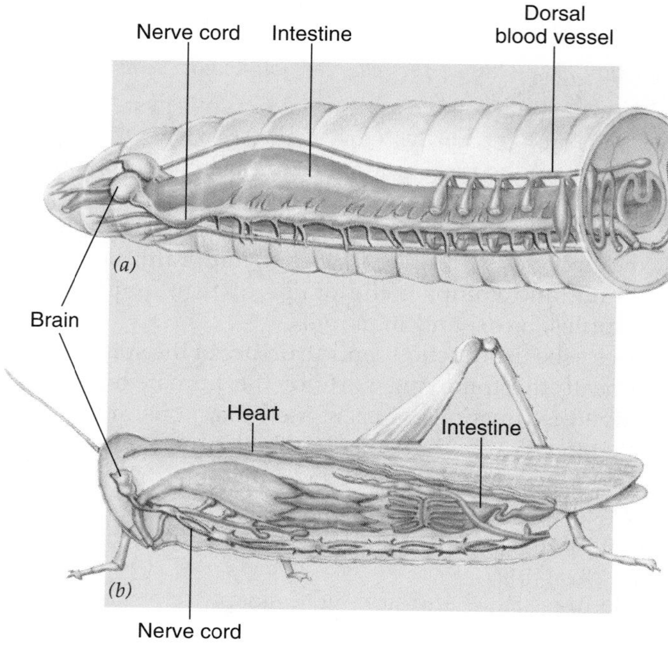

FIGURE 40–3 The nervous systems of annelids and arthropods are somewhat similar, each having a dorsal anterior brain and one or more ventral nerve cords. (*a*) The nervous system of the earthworm is typical of those found in other annelids. The cell bodies of the neurons are located in ganglia found in each body segment. They are connected by the ventral nerve cord. (*b*) In the insect nervous system the cerebral ganglia (a simple brain) are connected to two ventral nerve cords.

KEY FEATURES OF THE VERTEBRATE NERVOUS SYSTEM ARE THE HOLLOW DORSAL NERVE CORD AND THE WELL-DEVELOPED BRAIN

The vertebrate nervous system has two main divisions: the **central nervous system (CNS)** and the **peripheral nervous system (PNS)** (Table 40–1). The CNS consists of a complex tubular brain that is continuous with the single dorsal tubular spinal cord. Serving as central control, these organs integrate incoming information and determine appropriate responses.

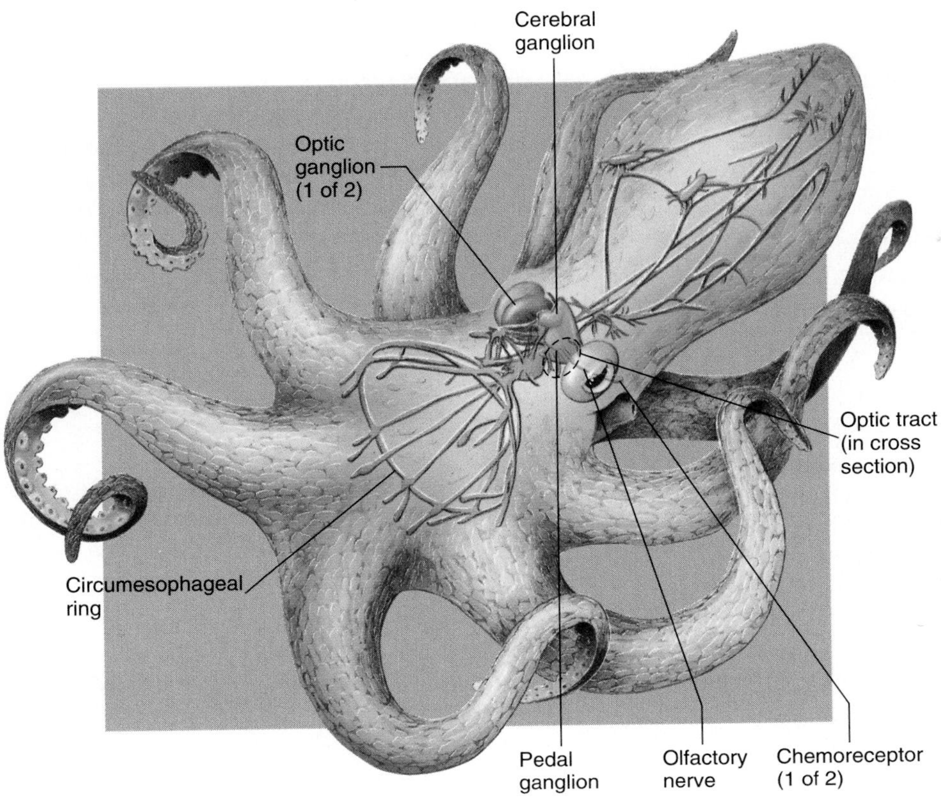

FIGURE 40–4 The cephalopod nervous system contains millions of nerve cells. Several ganglia, including the cerebral, optic, and pedal ganglia, make up the "brain."

The PNS is made up of the sensory receptors (e.g., tactile, auditory, and visual receptors) and the nerves, which are the communication lines. Various parts of the body are linked to the brain by cranial nerves and to the spinal cord by spinal nerves. Afferent neurons in these nerves continuously inform the CNS of changing conditions. Then efferent neurons transmit the "decisions" of the CNS to appropriate muscles and glands, which make the adjustments needed to maintain homeostasis.

For convenience the PNS can be subdivided into **somatic** and **autonomic** portions. Most of the receptors and nerves concerned with changes in the external environment are somatic. Those that regulate the internal environment are autonomic. Both systems have sensory (afferent) nerves, which transmit messages from receptors to the CNS, and motor (efferent) nerves, which transmit information back from the CNS to the structures that must respond. In the autonomic system there are two kinds of efferent pathways: **sympathetic** and **parasympathetic** nerves.

THE EVOLUTION OF THE VERTEBRATE BRAIN IS MARKED BY INCREASING COMPLEXITY

All vertebrates, from fishes to mammals, have the same basic brain structure. Different parts of the brain are spe-

Table 40–1 DIVISIONS OF THE VERTEBRATE NERVOUS SYSTEM

Central nervous system (CNS)

Brain
Spinal cord

Peripheral nervous system (PNS)

Somatic portion
1. Receptors
2. Afferent (sensory) nerves — transmit information from receptors to CNS
3. Efferent nerves — transmit information from CNS to skeletal muscles

Autonomic portion
1. Receptors
2. Afferent (sensory) nerves — transmit information from receptors in internal organs to CNS
3. Efferent nerves — transmit information from CNS to glands and involuntary muscle in organs
   a. Sympathetic nerves — generally stimulate activity that results in mobilization of energy (e.g., speeds heartbeat)
   b. Parasympathetic nerves — action results in energy conservation or restoration (e.g., slows heartbeat)

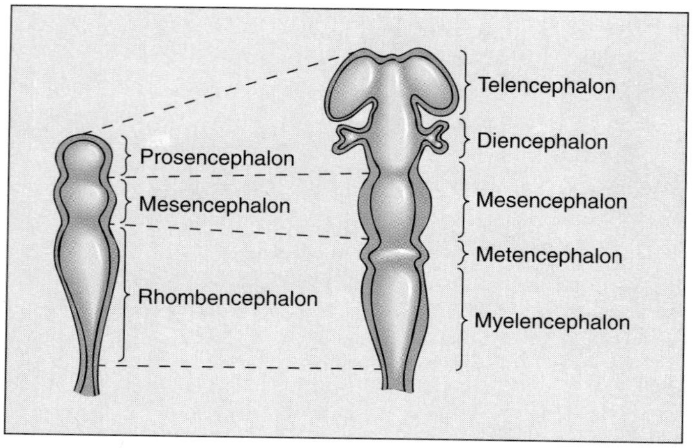

FIGURE 40–5 Early in the development of the vertebrate embryo, the anterior end of the neural tube differentiates into the prosencephalon (forebrain), mesencephalon (midbrain), and rhombencephalon (hindbrain). These primary divisions subdivide and then give rise to specific structures of the adult brain (see Table 40–2).

cialized in the various vertebrate classes, and the evolutionary trend is toward increasing complexity, especially of the cerebrum and cerebellum.

In the early vertebrate embryo, the brain and spinal cord differentiate from a single tube of tissue, the **neural tube.** Anteriorly, the tube expands and develops into the brain. Posteriorly, the tube becomes the spinal cord. Brain and spinal cord remain continuous, and their cavities communicate. As the brain begins to differentiate, three bulges become visible: the hindbrain, midbrain, and forebrain (Fig. 40–5).

The hindbrain develops into the medulla, pons, and cerebellum

The **hindbrain** (rhombencephalon) subdivides to form the metencephalon, which gives rise to the **cerebellum** and **pons,** and the myelencephalon, which gives rise to the **medulla.** The medulla, pons, and midbrain make up

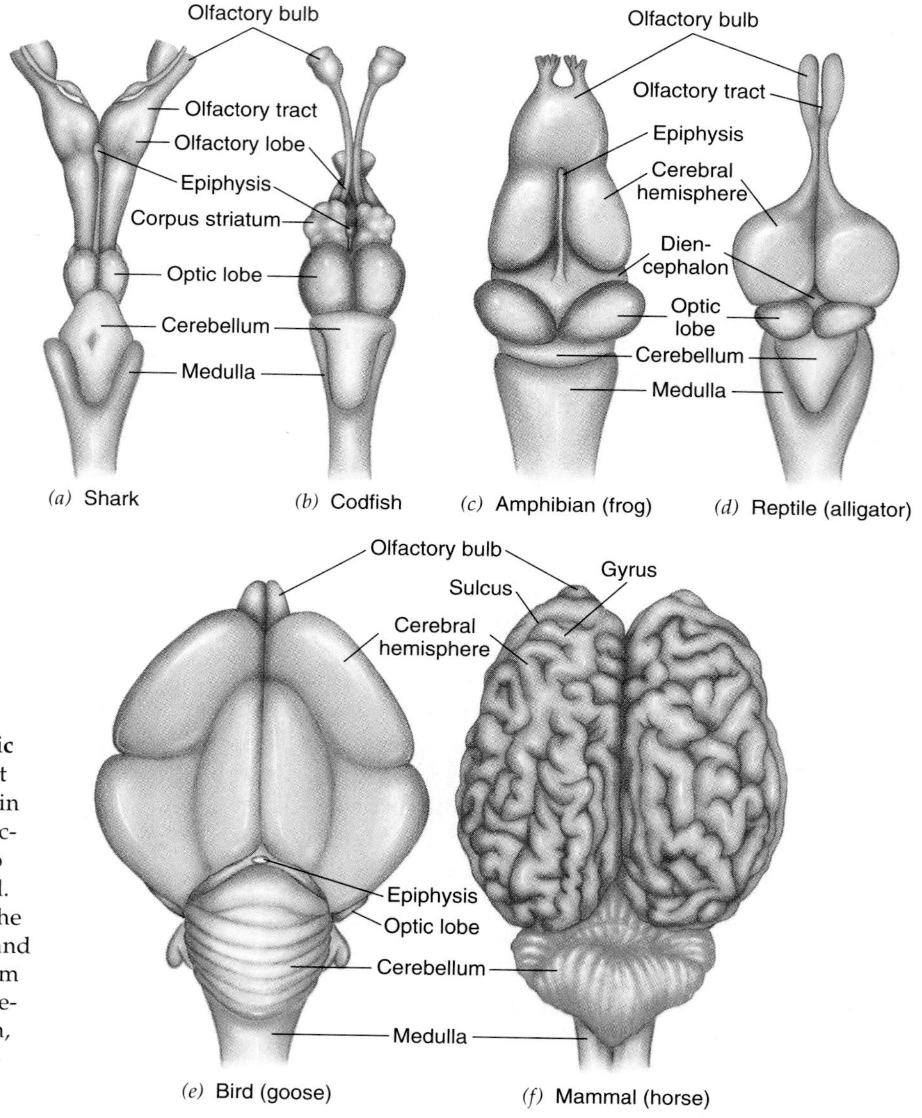

FIGURE 40–6 Comparison of the brains of members of six vertebrate classes reveals basic similarities and evolutionary trends. Note that different parts of the brain may be specialized in the various groups. For example, the large olfactory lobes in the shark brain (*a*) are essential to this predator's highly developed sense of smell. (*b* through *f*) During the course of evolution, the cerebrum and cerebellum have become larger and more complex. In the mammal (*f*), the cerebrum is the most prominent part of the brain; the cerebral cortex, the thin outer layer of the cerebrum, is highly convoluted (folded), which greatly increases its surface area.

the **brain stem,** the elongated portion of the brain that looks like a stalk holding up the cerebrum.

The medulla, the most posterior part of the brain, is continuous with the spinal cord. Its cavity, the **fourth ventricle,** communicates with the central canal of the spinal cord and with a channel (the cerebral aqueduct) that runs through the midbrain. This channel connects the third ventricle (within the diencephalon) with the fourth ventricle.

The walls of the medulla are thick and made up largely of nerve tracts (bundles of axons) that connect the spinal cord with various parts of the brain. In complex vertebrates, the medulla contains discrete nuclei (masses of nerve cell bodies) that serve as vital centers, regulating respiration, heartbeat, and blood pressure. Other reflex centers in the medulla regulate such activities as swallowing, coughing, and vomiting.

The size and shape of the cerebellum vary greatly among the vertebrate classes (Fig. 40–6). Development of the cerebellum in different animals is roughly correlated with the extent and complexity of muscular activity. In some fishes, birds, and mammals the cerebellum is highly developed, whereas it tends to be small in agnathans, amphibians, and reptiles. The cerebellum coordinates muscle activity and is responsible for muscle tone, posture, and equilibrium.

Injury or removal of the cerebellum results in impaired muscle coordination. A bird without a cerebellum cannot fly, and its wings thrash about jerkily. When the human cerebellum is injured by a blow or by disease, muscular movements are uncoordinated. Any activity requiring delicate coordination, such as threading a needle, is very difficult, if not impossible.

In mammals, a large mass of fibers known as the pons forms a bulge on the anterior surface of the brain stem. The pons is a bridge connecting the spinal cord and medulla with upper parts of the brain. It contains centers that help regulate respiration and nuclei that relay impulses from the cerebrum to the cerebellum.

The midbrain is most prominent in fishes and amphibians

In fish and amphibians, the **midbrain,** or mesencephalon, is the most prominent part of the brain, serving as the main association area. It receives incoming sensory information, integrates it, and sends decisions to appropriate motor nerves. The dorsal portion of the midbrain is differentiated to some extent. For example, the **optic lobes** are specialized for visual interpretations.

In reptiles, birds, and mammals, many of the functions of the optic lobes are assumed by the cerebrum, which develops from the forebrain. In mammals, the midbrain consists of the **superior colliculi,** centers for visual reflexes such as pupil constriction, and the **inferior colliculi,** centers for certain auditory reflexes. The mammalian midbrain also contains a center (the red nucleus) that integrates information about muscle tone and posture.

The forebrain gives rise to the thalamus, hypothalamus, and cerebrum

As indicated in Table 40–2, the **forebrain,** or prosencephalon, subdivides to form the **telencephalon** and **diencephalon.** The telencephalon gives rise to the **cere-**

Table 40–2 DIFFERENTIATION OF CNS STRUCTURES

Early Embryonic Divisions	*Subdivisions*	*Derivatives in Adult*	*Cavity*
Brain			
Forebrain (prosencephalon)	Telencephalon	Cerebrum	Lateral ventricles (first and second ventricles)
	Diencephalon	Thalamus, hypothalamus, epiphysis (pineal body)	Third ventricle
Midbrain (mesencephalon)	Midbrain	Optic lobes in fish and amphibians; superior and inferior colliculi	Cerebral aqueduct
Hindbrain (rhombencephalon)	Metencephalon	Cerebellum, pons	
	Myelencephalon	Medulla	Fourth ventricle
Spinal cord		Spinal cord	Central canal

brum, and the diencephalon to the **thalamus** and **hypothalamus.** The lateral ventricles (also called the first and second ventricles) are located within the cerebrum. Each lateral ventricle is connected with the third ventricle (within the diencephalon) by way of a channel.

The diencephalon contains the thalamus and hypothalamus. In all vertebrate classes, the thalamus is a relay center for motor and sensory messages. In mammals, all sensory messages (except those from the olfactory receptors) are delivered to the thalamus before they are relayed to the sensory areas of the cerebrum.

The hypothalamus, which lies below the thalamus, forms the floor of the third ventricle. The hypothalamus contains olfactory centers and is the principal integration center for the regulation of the viscera (internal organs). It provides input to centers in the medulla and spinal cord that regulate activities such as heart rate, respiration, and digestive system function. In reptiles, birds, and mammals, the hypothalamus controls body temperature. It also regulates appetite and water balance and is involved in emotional and sexual responses. As will be discussed in Chapter 47, the hypothalamus links the nervous and endocrine systems, and produces certain hormones.

The telencephalon gives rise to the cerebrum and, in most vertebrate groups, the **olfactory bulbs.** These structures are important in the chemical sense of smell—the dominant sense in most aquatic and terrestrial vertebrates. In fact, much of brain development in vertebrates appears to be focused on the integration of olfactory information. In fish and amphibians, the cerebrum is almost entirely devoted to the integration of olfactory information.

Birds are an exception among the vertebrates in that their sense of smell is generally poorly developed. A part of their cerebrum called the **corpus striatum,** however, is highly developed. This structure is thought to control the innate, stereotypical, yet complex action patterns characteristic of birds. Just above the corpus striatum is a region thought to govern learning in birds.

In most vertebrates, the cerebrum is divided into right and left **hemispheres.** Most of the cerebrum is made of **white matter,** which consists mainly of axons that connect various parts of the brain. In mammals and most reptiles, a layer of **gray matter,** the **cerebral cortex,** makes up the outer portion of the cerebrum. Gray matter contains cell bodies and dendrites.

Certain reptiles and all mammals possess a type of cortex, the **neopallium,** not found in less complex vertebrates. It serves as an association area—a region that links sensory and motor functions and is responsible for higher functions such as learning. The neopallium is very extensive in mammals, making up the bulk of the cerebrum. In humans, about 90% of the cerebral cortex is neopallium, consisting of six distinct cell layers.

In mammals, the cerebrum is the most prominent part of the brain. During embryonic development, it expands and grows backwards, covering many other brain structures. The cerebrum is responsible for many of the functions performed by other parts of the brain in other vertebrates. In addition, it has many complex association functions that are lacking in reptiles, amphibians, and fish.

In small or simple mammals, the cerebral cortex may be smooth. However, in large complex mammals, the surface area is greatly expanded by numerous folds called **convolutions** or **gyri** (sing., *gyrus*). The furrows between them are called **sulci** (sing., *sulcus*) if shallow and **fissures** if deep. The number of folds (not the size of the brain) has been associated with complexity of brain function.

THE HUMAN CENTRAL NERVOUS SYSTEM IS THE MOST COMPLEX BIOLOGICAL MECHANISM KNOWN

The soft, fragile human brain and spinal cord are well protected. Encased within bone, they are covered by three layers of connective tissue, the **meninges.** The three meningeal layers are the tough, outer **dura mater,** the middle **arachnoid,** and the thin, vascular **pia mater,** which adheres closely to the tissue of the brain and spinal cord (Fig. 40–7). Meningitis is a disease in which these coverings become infected and inflamed.

Between the arachnoid and the pia mater is the subarachnoid space, which contains **cerebrospinal fluid (CSF).** The fluid is produced by special networks of capillaries, collectively called the **choroid plexus,** that project from the pia mater into the ventricles. The choroid plexus extracts nutrients from the blood and adds them to the CSF. Together the choroid plexus and the arachnoid serve as a barrier between blood and CSF. They prevent harmful substances in the blood from entering the brain.

CSF is a shock-absorbing fluid that cushions the brain and spinal cord against mechanical injury. CSF also serves as a medium for the exchange of nutrients and waste products between the blood and brain. CSF circulates down through the ventricles and passes into the subarachnoid space surrounding the brain and spinal cord. It is then reabsorbed into large blood sinuses within the dura mater.

The spinal cord transmits impulses to and from the brain and controls many reflex activities

The tubular **spinal cord** extends from the base of the brain to the level of the second lumbar vertebra. A cross section through the spinal cord reveals a small **central canal** surrounded by an area of gray matter shaped somewhat like the letter H (Fig. 40–8). Outside the gray matter, the spinal cord consists of white matter. The gray matter is composed of large masses of cell bodies, dendrites, un-

FIGURE 40–7 The brain is well protected by several coverings and by the cerebrospinal fluid. (*a*) Frontal section through the superior part of the brain. Note the large sinus shown between two layers of the dura mater. Blood leaving the brain flows into such sinuses and then circulates to the large jugular veins in the neck. (*b*) The cerebrospinal fluid circulates through the brain and spinal cord. This cushioning fluid, produced by the choroid plexi in the walls of the ventricles, circulates through the ventricles and subarachnoid space. It is continuously produced and continuously reabsorbed into the blood of the dural sinuses.

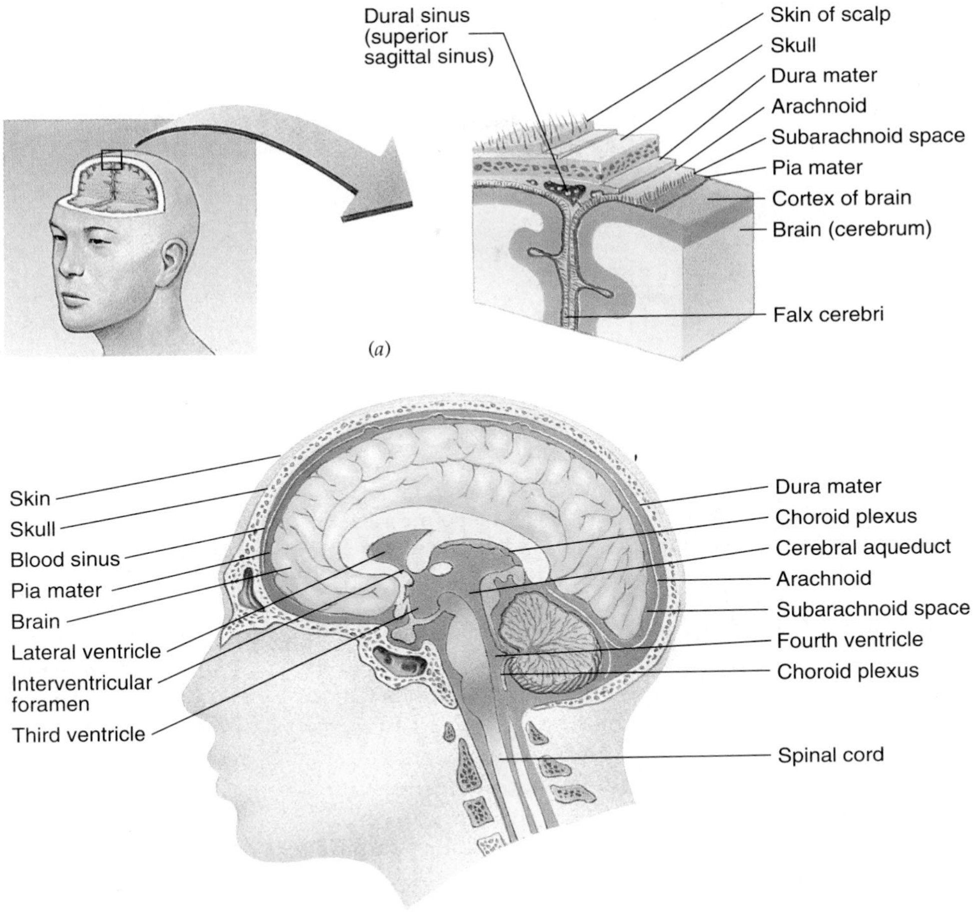

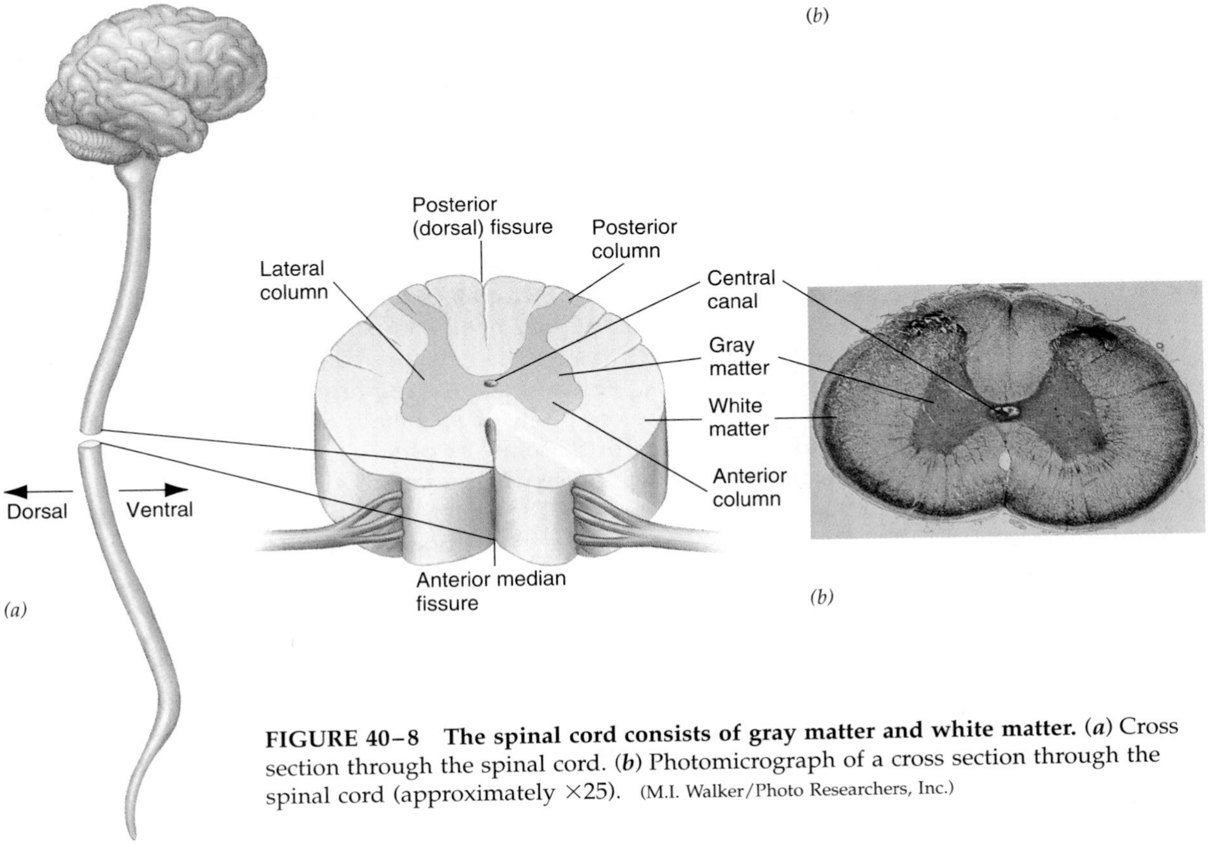

FIGURE 40–8 The spinal cord consists of gray matter and white matter. (*a*) Cross section through the spinal cord. (*b*) Photomicrograph of a cross section through the spinal cord (approximately ×25). (M.I. Walker/Photo Researchers, Inc.)

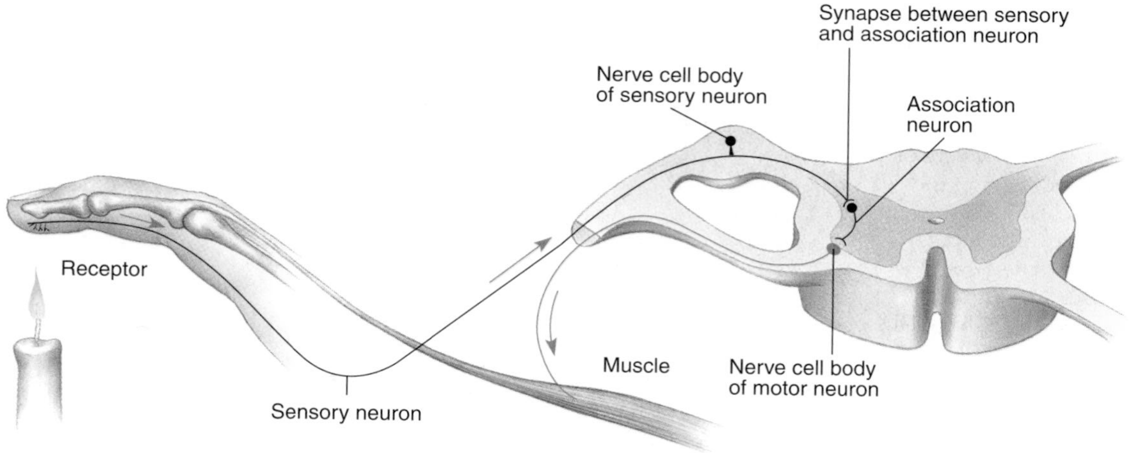

FIGURE 40–9 The withdrawal reflex depends on a chain of three neurons. A sensory neuron transmits the message from the receptor to the central nervous system, where it synapses with an association neuron. Then an appropriate motor neuron (*shown in red*) transmits an impulse to the muscles that move the hand away from the flame (the response).

myelinated axons, glial cells, and blood vessels, and is subdivided into sections called **horns.** The white matter consists of myelinated axons arranged in bundles called **tracts** or **pathways.** Long ascending tracts conduct impulses up the cord to the brain. For example, the spinothalamic tracts in the anterior and lateral columns of the white matter conduct pain and temperature information from sensory neurons in the skin. The pyramidal tracts are descending tracts that convey impulses from the cerebrum to spinal motor nerves at various levels in the cord.

A **reflex action** is a relatively fixed response pattern to a simple stimulus. The response is predictable and automatic, not requiring conscious thought. Many of the activities of the body, such as breathing, are regulated by reflex actions.

Although most reflex actions are much more complex, let us consider a **withdrawal reflex,** in which a neural circuit consisting of only three types of neurons is needed to carry out a response to a stimulus (Fig. 40–9). Suppose you touch a hot stove. Almost instantly, and even before you are consciously aware of what has happened, you jerk your hand away. In this brief instant a message has been carried by a sensory neuron from pain receptors in the skin to the spinal cord. In the tissue of the spinal cord, the message is transmitted from the sensory neuron to an association neuron. Finally, the message is transmitted to an appropriate motor neuron, which conducts it to groups of muscles that respond by contracting and pulling the hand from the stove. Actually, many neurons in sensory, association, and motor nerves participate in such a reaction, and complicated switching is involved. For example, we move our hands *up* from a hot stove but *down* from a hot light bulb. Generally, we are not even consciously aware that all these responding muscles exist.

Quite probably, at the same time that the association neuron sends a message out along a motor neuron, it also sends one up the spinal cord to the conscious areas of the brain. As you withdraw your hand from the hot stove, you become aware of what has happened and feel the pain. This awareness, however, is not part of the reflex response.

The largest, most prominent part of the human brain is the cerebrum

The structure and functions of the main parts of the human brain are summarized in Table 40–3. The human brain is illustrated in Figures 40–10 and 40–11. As in other mammals, the human cerebral cortex is functionally divided into three areas: (1) the **sensory areas,** which receive incoming signals from the sense organs; (2) the **motor areas,** which control voluntary movement; and (3) the **association areas,** which link the sensory and motor areas and are responsible for thought, learning, language, memory, judgment, and personality.

Investigators have been able to map the cerebral cortex, locating the areas responsible for different functions. The **occipital lobes** contain the visual centers. Stimulation of these areas, even by a blow on the back of the head, causes the sensation of light; their removal causes blindness. The centers for hearing are located in the **temporal lobes** of the brain above the ear; stimulation by a blow causes a sensation of noise. Removal of both auditory areas causes deafness. Removal of one does not cause deafness in one ear, but rather produces a decrease in the auditory acuity of both ears.

A groove called the **central sulcus** crosses the top of each hemisphere from medial to lateral edge. This groove partially separates the **frontal lobes** from the **parietal**

Table 40–3 THE BRAIN

Structure	*Description*	*Function*
Brain stem		
Medulla	Continuous with spinal cord; primarily made up of nerves passing from spinal cord to rest of brain	Contains vital centers (clusters of neuron cell bodies) that control heartbeat, respiration, and blood pressure; contains centers that control swallowing, coughing, vomiting
Pons	Forms bulge on anterior surface of brain stem	Connects various parts of brain with one another; contains respiratory center
Midbrain	Just above pons; largest part of brain in lower vertebrates; in humans, most of its functions are assumed by cerebrum	Center for visual and auditory reflexes (e.g., pupil reflex, blinking, adjusting ear to volume of sound)
Thalamus	At top of brain stem	Main sensory relay center for conducting information between spinal cord and cerebrum. Neurons in thalamus sort and interpret all incoming sensory information (except olfaction) before relaying messages to appropriate neurons in cerebrum
Hypothalamus	Just below thalamus; pituitary gland is connected to hypothalamus by stalk of neural tissue	Contains centers for control of body temperature, appetite, fat metabolism, and certain emotions; regulates pituitary gland; link between "mind" (cerebrum) and "body" (physiological mechanisms)
Cerebellum	Second largest division of brain	Reflex center for muscular coordination and refinement of movements; when it is injured, performance of voluntary movements is uncoordinated and clumsy
Cerebrum	Largest, most prominent part of human brain; more than 70% of brain's cells located here; longitudinal fissure divides cerebrum into right and left hemispheres, each divided by shallow sulci (furrows) into six lobes: frontal, parietal, temporal, insular, occipital, and limbic	Center of intellect, memory, consciousness, and language; also controls sensation and motor functions
Cerebral cortex (outer gray matter)	Arranged into convolutions (folds) that increase surface area; functionally, cerebral cortex is divided into:	
	1. Motor cortex	Controls movement of voluntary muscles
	2. Sensory cortex	Receives incoming information from eyes, ears, pressure and touch receptors, etc.
	3. Association cortex	Site of intellect, memory, language, and emotion; interprets incoming sensory information
White matter	Consists of myelinated axons of neurons that connect various regions of brain; these axons are arranged into bundles (tracts)	Connects: 1. Neurons within same hemisphere 2. Right and left hemispheres 3. Cerebrum with other parts of brain and spinal cord

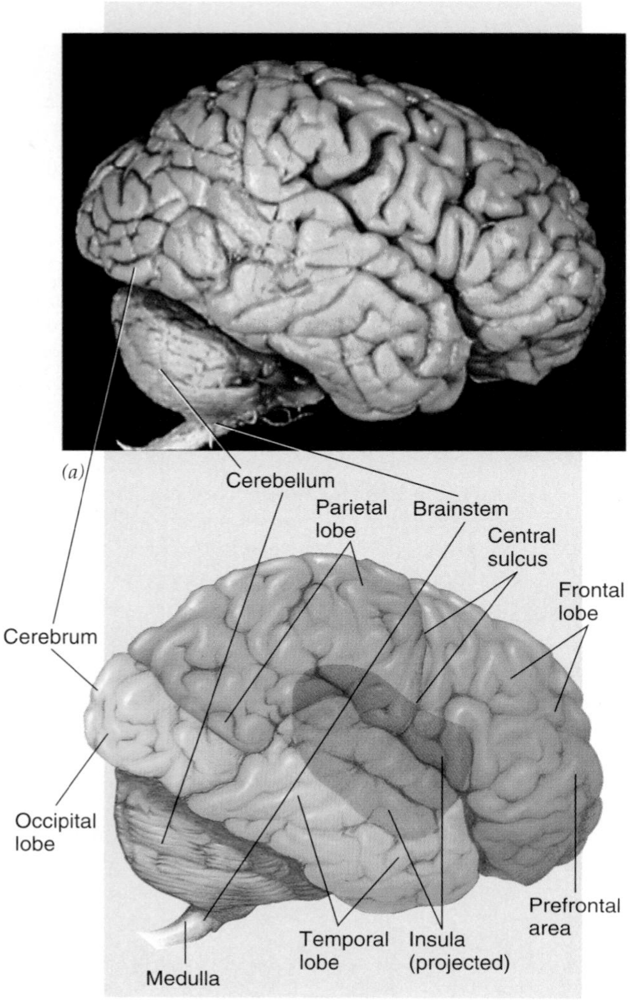

FIGURE 40–10 The human brain is one of the most complex mechanisms known. (*a*) Photograph of the human brain, lateral view. Note that the cerebrum covers the diencephalon and part of the brain stem. (*b*) Lateral view of the human brain showing the lobes of the cerebrum. Part of the brain has been made transparent so that the underlying insular lobe can be located. (Visuals Unlimited/Fred Hossler)

lobes. The **primary motor areas** in the frontal lobes control the skeletal muscles. The **primary sensory areas** in the parietal lobes receive information regarding heat, cold, touch, and pressure from sense organs in the skin.

The size of the motor area in the brain for any given part of the body is proportional not to the amount of muscle but to the complexity of movement involved. Predictably, areas for the control of the hands and face are relatively large (Fig. 40–12). A similar relationship exists between the sensory area and the region of the skin from which it receives impulses. In connections between the body and the brain, not only do the fibers cross so that one side of the brain controls the opposite side of the body, but another "reversal" makes the uppermost part of the cortex control the lower limbs of the body.

When all the areas of known function are plotted, they cover almost all of the rat's cortex, a large part of the dog's, a moderate amount of the monkey's, and only a small part of the total surface of the human cortex. The

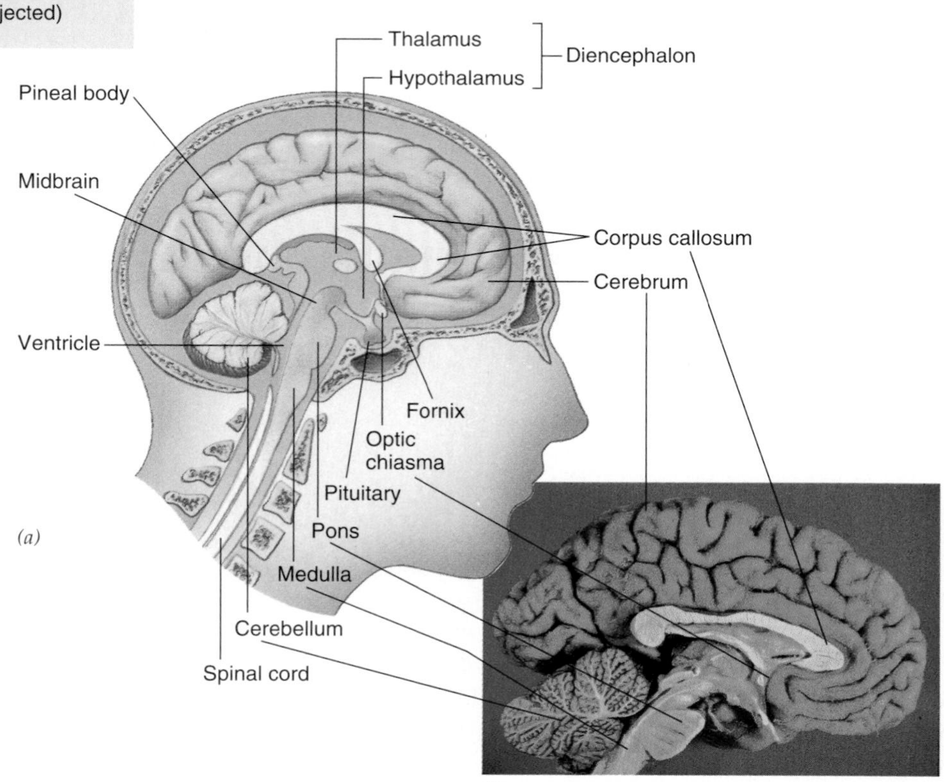

FIGURE 40–11 In this midsagittal section through the human brain, half of the brain has been cut away, exposing structures normally covered by the cerebrum. Compare the diagram (*a*) with the photograph of the human brain (*b*). (*b*, Science Photo Library/Photo Researchers, Inc.)

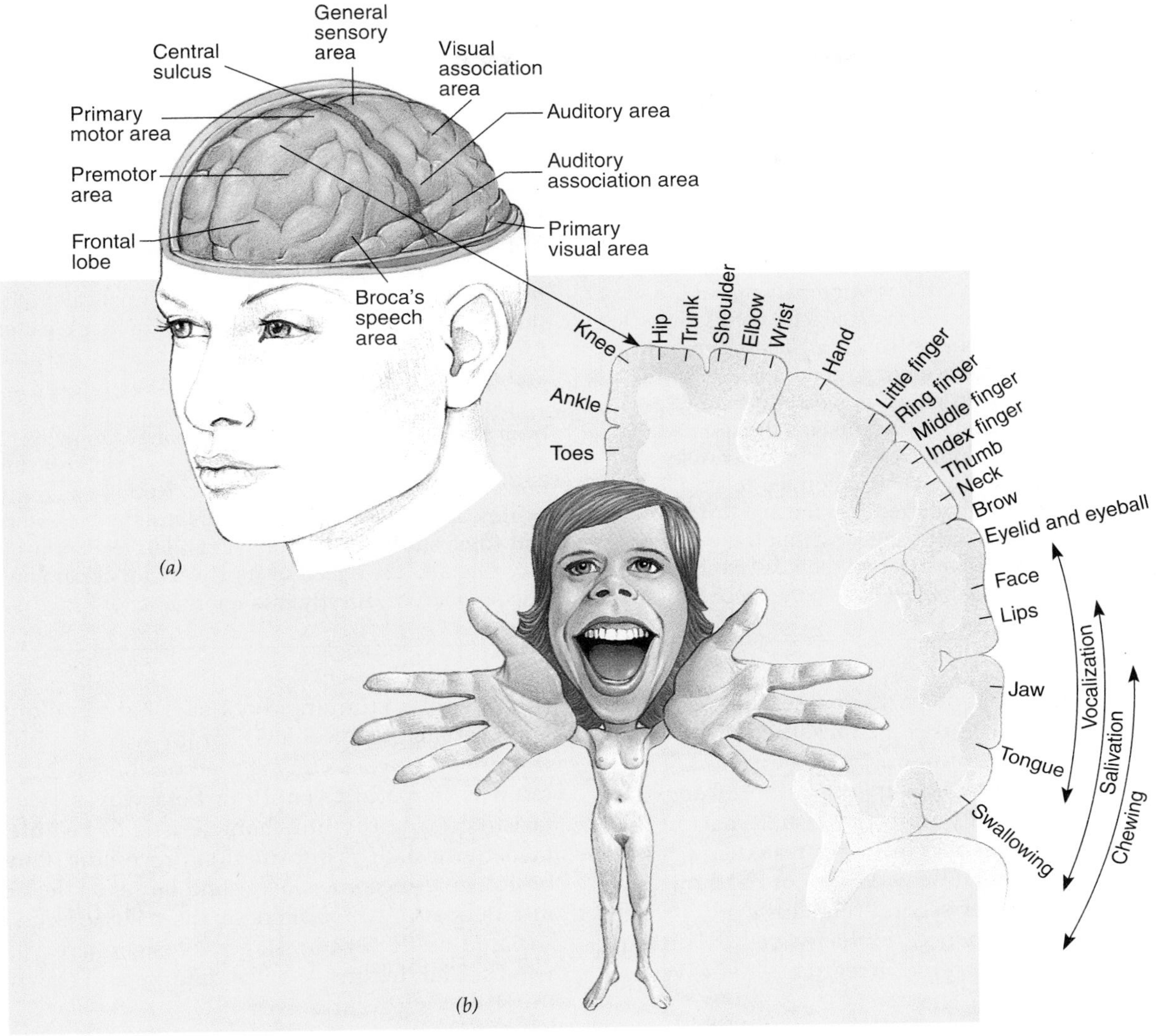

FIGURE 40–12 Many functional areas of the brain have been identified. (*a*) Several sensory, motor, and association areas are shown in the diagram. (*b*) A cross section through the primary motor area (precentral gyrus) showing which area of cerebral cortex controls each body part. The figure shown here, known as a motor homunculus, is proportioned to reflect the amount of cerebral cortex that controls each body part. Note that more cortical tissue is devoted to controlling those body structures capable of skilled, complex movement.

remaining cortical areas are the association areas. Somehow the association regions integrate all the diverse impulses reaching the brain into a meaningful unit so that an appropriate response is made. When disease or accident destroys the functioning of one or more association areas, the ability to recognize certain kinds of symbols may be lost. For example, the names of objects may be forgotten, although their functions are remembered and understood.

The white matter of the cerebrum lies beneath the cerebral cortex. Nerve fibers of the white matter connect the cortical areas with one another and with other parts of the nervous system. A large band of white matter, the **corpus callosum,** connects the right and left hemispheres.

The brain integrates information

Most of the integration of the nervous system takes place within the functional regions of the cerebral cortex. Integrative activities include arousal and sleep, emotion, and information processing.

Brain activity cycles in a sleep-wake pattern

Brain activity can be studied by measuring and recording the electrical potentials, or "brain waves," given off by various parts of the brain when they are active. This electrical activity can be recorded by a device known as an electroencephalograph. To obtain an **electroencephalogram (EEG),** a recording of this electrical activity, elec-

Making the Connection

Dopamine and Motor Function

How does the neurotransmitter dopamine affect motor function? The function of dopamine was discovered through an interesting series of somewhat unrelated events. During the mid-1950s, the drug reserpine became popular as a major tranquilizer for mental patients. Then, in 1959, investigators noticed that some patients taking reserpine experienced distressing side effects, such as muscle rigidity and persistent tremors (shaking). These symptoms were very similar to those seen in patients with Parkinson's disease, a disorder in which movement is slow, shaky, and difficult. Victims of Parkinson's disease have a shuffling gait and suffer from tremors even when they are not attempting to move. This observation led to studies showing that reserpine greatly reduces the amount of dopamine in two of the basal ganglia within the white matter of the cerebrum. Investigators then discovered that patients with Parkinson's disease have only about 50% of the normal amount of dopamine in their basal ganglia.

Attempts to administer dopamine to these patients were not successful because dopamine cannot penetrate the blood-brain barrier. However, L-dopa, a substance from which dopamine is synthesized in the body, does penetrate the blood-brain barrier. Its use has dramatically relieved the symptoms of Parkinson's disease in many patients. Dopamine is thought to restore balance between inhibition and excitation of neurons involved in motor function.

Dopamine depletion may occur with aging. As a result, even healthy persons experience changes in motor abilities as they age. Body movements and even reflexes slow, and movement becomes more difficult. Studies suggest that treatment with L-dopa may be helpful. Investigators have had some success using animal models of Parkinson's disease to study the effects of transplanting dopamine-producing neurons directly into the brain. ▲

trodes are taped to different parts of the scalp, and the activity of the cerebral cortex is measured. The EEG shows that the brain is continuously active.

On the EEG, the most regular indication of activity, **alpha waves,** comes mainly from the visual areas in the occipital lobes when the person being tested is resting quietly with eyes closed. These waves occur rhythmically at the rate of about ten per second (Fig. 40–13).

When the eyes are opened, alpha waves disappear and are replaced by more rapid, irregular waves. When

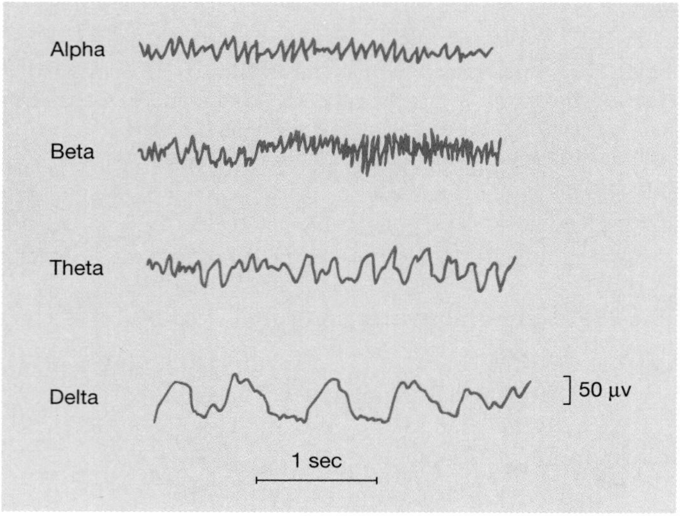

FIGURE 40–13 These electroencephalograms were made while the subject was excited, relaxed, and in various stages of sleep. Recordings made during excitement show brain waves that are rapid and of small amplitude, whereas in sleep the waves are much slower and of greater amplitude. The regular waves characteristic of the relaxed state are called alpha waves.

some regular stimulus, such as a light blinking at regular intervals, is presented, brain waves with a similar rhythm appear. As you are reading this biology text, your brain is (hopefully) emitting **beta waves.** These have a fast-frequency rhythm characteristic of heightened mental activity such as information processing. During sleep, brain waves become slower and larger as the person falls into deeper unconsciousness; these slow, large waves associated with certain stages of sleep are called **delta waves.** The dreams of a sleeping person are mirrored in flurries of irregular waves.

Certain brain diseases change the pattern of brain waves. Individuals with epilepsy, for example, exhibit a distinctive, readily recognizable, abnormal wave pattern. The location of a brain tumor or the site of brain damage caused by a blow to the head can sometimes be determined by noting the part of the brain that shows abnormal waves.

Not much is known about the neurophysiology that underlies changes in brain wave patterns or about sleep-wake cycles. The **reticular activating system (RAS)** is a complex neural pathway within the brain stem and thalamus. Sometimes referred to as an arousal system, the RAS receives messages from neurons in the spinal cord and from many other parts of the nervous system, and communicates with the cerebral cortex by complex neural circuits. The RAS maintains consciousness (awareness), and the extent of its activity helps determine the state of alertness. When the RAS bombards the cerebral cortex with stimuli, you feel alert and are able to focus your attention on specific thoughts. If the RAS is severely damaged, the victim may pass into a deep, permanent coma.

Sleep is a state of unconsciousness during which there is decreased electrical activity of the cerebral cortex, and from which a person can be aroused. Two main stages of sleep are recognized: **non-REM** and **REM.** *REM* is an acronym for *rapid eye movements.* During non-REM sleep, sometimes called normal sleep, metabolic rate decreases, breathing slows, and blood pressure decreases. Delta waves, thought to be generated spontaneously by the cerebral cortex when it is not driven by impulses from other parts of the brain, are characteristic of non-REM sleep.

Every 90 minutes or so, a sleeping person enters the REM stage for a time. During this stage, which accounts for about one fourth of total sleep time, the eyes move about rapidly beneath the closed but fluttering lids. Brain waves change to a desynchronized pattern of beta waves. Sleep researchers claim that everyone dreams, especially during REM sleep. Dreams may result from release of norepinephrine within the RAS, which generates stimulating impulses that are fed into the cerebral cortex.

The raphe nuclei in the brain stem (lower pons and medulla) appear to be important in producing sleep. Many of the neurons that project from the raphe nuclei release serotonin, the neurotransmitter thought to cause sleep. During REM sleep, neurons in the brain stem release norepinephrine. This neurotransmitter is thought to cause the heightened activity in certain regions of the brain that is characteristic of REM sleep.

The sleep-wake cycle may result from fatigue of the RAS after many hours of activity. Sleep centers are then activated and their neurons release serotonin. After sufficient rest, the inhibitory neurons of the sleep centers become less excitable, while the excitatory neurons of the RAS become more excitable.

Why sleep is necessary is not understood. Apparently only vertebrates with fairly well-developed cerebral cortices sleep. When a person stays awake for unusually long periods, fatigue and irritability result, and even routine tasks cannot be performed well.

Not only is non-REM sleep required, but REM sleep is apparently also necessary. In sleep deprivation experiments performed with human volunteers, lack of REM sleep makes subjects anxious and irritable. When the subjects are permitted to sleep normally again, they spend more time than usual in the REM stage for a period. Many types of drugs alter sleep patterns and affect the amount of REM sleep. Certain drugs that induce sleep, for example, may increase total sleeping time but decrease REM time. When a person stops taking such a drug, several weeks may be required before normal sleep patterns are reestablished.

The limbic system affects emotional aspects of behavior

The **limbic system** is an action system of the brain that consists of certain structures of the cerebrum and diencephalon. It affects the emotional aspects of behavior, sexual behavior, biological rhythms, autonomic responses, and motivation, including feelings of pleasure and punishment. Stimulation of certain areas of the limbic system in an experimental animal results in increased general activity and may cause fighting behavior or extreme rage.

When an electrode is implanted in the so-called reward center of the limbic system, a rat may press a lever that stimulates this area as many as 15,000 times per hour. Stimulation of the reward center is apparently so rewarding that an animal foregoes food and drink and may continue to press the lever until it drops from exhaustion. When an electrode is implanted in the punishment center of the limbic system, an experimental animal quickly learns to press a lever to avoid stimulating that area. The reward and punishment centers are thought to be important in influencing motivation and behavior.

Learning involves the storage of information and its retrieval

Learning is a relatively long-lasting adaptive change in behavior resulting from experience. Laboratory experiments have shown that members of every animal phylum can learn. Several types of learning will be discussed in Chapter 50. In order for learning to occur, we must be able to remember what we experience. **Memory** is the storage of information and the ability to recall it. A thought is an awareness of an experience or the development of a concept.

Information processing involves three levels of memory Three levels of memory are recognized: sensory, short-term, and long-term. At any moment you are bombarded with thousands of bits of sensory information. At this very moment, your eyes are receiving information about the words on this page, the objects around you, and the intensity of the light in the room. At the same time you may be hearing a variety of sounds—rock music, your friends talking in the next room, the hum of an air conditioner. Your olfactory epithelium may sense cologne or the smell of coffee. Sensory receptors in your hands may be receiving information regarding the weight and position of your book. The bits of sensory information on which we focus our attention are registered in **sensory memory.** Sensory memory can hold a lot of information, but it is lost within about 1 second. Attention is an important component of sensory memory. When we attend to information in sensory memory, pattern recognition—the process of identifying the stimuli—begins.

In order for a sensory memory to be identified or recognized, we must relate it to past experience or past knowledge, and this requires further processing. As pattern recognition and encoding occur, we become aware of stimuli. **Short-term memory** is the information that we are aware of at the moment. Short-term memory can hold only about seven chunks of information (a chunk is some unit such as a word, syllable, or number).

Short-term memory allows us to recall information for a few minutes. Usually when we look up a phone number, for example, we remember it only long enough to dial. Should we need the same number an hour later, we have to look it up again. Keeping information in short-term memory requires rehearsal. If we redirect our attention, the new stimulus interferes with recall of the information already present, and we forget. Short-term memory is necessary for understanding speech.

One theory of short-term memory suggests that it is based on reverberating circuits (see Chapter 39). A memory circuit may continue to reverberate for several minutes until it fatigues, or until new signals are received that interfere with the old.

Once we have processed information into **long-term memory,** we no longer have to focus attention on it in order to remember. Apparently, when information is selected for long-term storage, the brain rehearses the material and then stores it in association with similar memories.

Retrieval of information stored in long-term memory is of considerable interest, especially to students. Some researchers think that once information is deposited in long-term memory, it remains within the brain permanently. The challenge is to find it when you need it. When you seem to forget something, the problem may be that you have not effectively searched for the memory. Information retrieval can be improved by careful storage. One method is to form strong associations between items when they are being stored.

Sensory memory, short-term memory, and long-term memory are all part of the information processing system. Sensory memory can be viewed as a component of short-term memory, and short-term memory can be thought of as an activated component of long-term memory (Fig. 40–14). Information can be transferred back and forth from one component to another. For example, information from long-term memory can be retrieved and temporarily transferred to short-term memory, at which time we focus on the activated information.

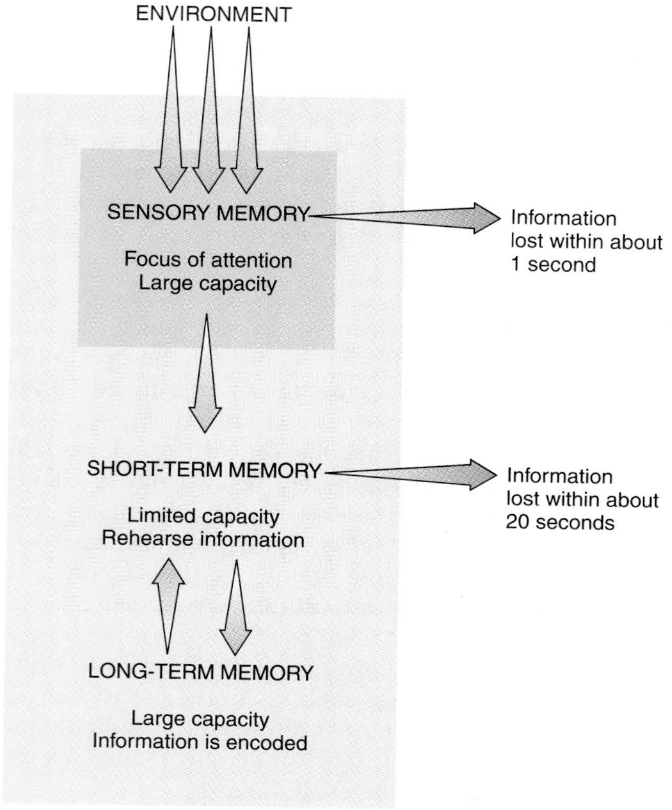

FIGURE 40–14 This simplified model suggests how information is processed in the human brain. Sensory memory is represented here as an activated component of short-term memory. Short-term memory is represented as an activated component of long-term memory.

Memory storage involves new synaptic connections What physiological changes take place during memory storage? Recent studies suggest that when information is processed for long-term memory storage, genes are activated that result in the formation of new synaptic connections. In at least some types of learning, changes take place in the synaptic knobs or postsynaptic neurons that permanently facilitate or inhibit the transmission of impulses within a newly established circuit. In some cases, specific neurons may become more sensitive to neurotransmitter.

Memory circuits are formed throughout the brain Researchers have worked with the brains of experimental animals for years without finding specific regions where information is stored. Some forms of learning take place in association areas within lower brain regions—the thalamus, for example. Even very simple animals like flatworms that completely lack a cerebral cortex are capable of some types of learning.

When large areas of cerebral cortex are destroyed, information is lost somewhat in proportion to the extent of lost tissue. However, no specific area can be labeled the "memory bank."

Memories are thought to be stored within the many sensory areas of the brain. For example, visual memories may be stored in the visual centers of the occipital lobes, and auditory memories may be stored in the temporal lobes. Memories are thought to be integrated in many areas of the brain, including association areas of the cerebral cortex, parts of the limbic system (the amygdala and the hippocampus), and the thalamus and hypothalamus. A very important association area for complex thought processes is Wernicke's area in the temporal lobe. This area is an important center for language function involved in the recognition and interpretation of words.

Neurons within the association areas form highly complex pathways that permit complicated reverberation. Several minutes are required for a memory to become consolidated within long-term memory. Should a person suffer a brain concussion or undergo electroshock therapy, for example, memory of what happened immediately prior to the incident may be completely lost. This is known as *retrograde amnesia*. When the hippocampus is damaged, a person can recall information stored in the past but can no longer convert new short-term memories into long-term memories.

Experience affects the brain

Studies show that environmental experience may cause physical as well as chemical changes in brain structure. When rats are provided with a stimulating environment and given the opportunity to learn, they exhibit increases in the size of cell bodies and nuclei of brain neurons. A great number of glial cells develop, and there is an increased concentration of synaptic contacts. Some investigators have reported that the cerebral cortex becomes thicker and heavier, and biochemical changes take place. Other experiments have indicated that animals reared in a complex environment may be able to process and remember information more quickly than animals not given such advantages. Rats provided with the basic necessities but deprived of stimulation and/or social interaction do not exhibit these changes.

Early environmental stimulation can also enhance the development of motor areas in the brain. Experiments with rats have shown that the brains of animals encouraged to exercise become slightly heavier than those of control animals. Characteristic changes occur within the cerebellum, including the development of larger dendrites. On the basis of this research, investigators have suggested that early experience may help a young child develop his or her physical potential.

Apparently, during early life, certain critical or sensitive periods of nervous system development occur that are influenced by environmental stimuli. For example, when the eyes of young mice first open, neurons in the visual cortex develop large numbers of dendritic spines (structures in which synaptic contact takes place). If the animals are kept in the dark and deprived of visual stimuli, fewer dendritic spines form. If the mice are exposed to light later in life, some new dendritic spines form but never the number that develop in a mouse reared in a normal environment.

Studies linking the development of the brain with environmental experience indicate that early stimulation is important for the sensory, motor, and intellectual development of children. Such studies have led to the rapidly expanding educational toy market and to widespread acceptance of early education programs. Researchers have also suggested that continuing environmental stimulation is needed to maintain the status of the cerebral cortex in later life.

THE PERIPHERAL NERVOUS SYSTEM INCLUDES SOMATIC AND AUTONOMIC SUBDIVISIONS

The **peripheral nervous system (PNS)** consists of the sensory receptors, the nerves that link them with the CNS, and the nerves that link the CNS with the effectors (muscles and glands). The portion of the PNS that helps the body respond to changes in the external environment is the somatic system. The nerves and receptors that maintain homeostasis despite internal changes make up the autonomic nervous system.

The somatic system helps the body adjust to the external environment

The **somatic nervous system** includes the receptors that react to changes in the external environment, the sensory neurons that keep the CNS informed of those changes, and the motor neurons that adjust the positions of the skeletal muscles, maintaining the body's well-being. In mammals, 12 pairs of **cranial nerves** emerge from the brain. They transmit information regarding the senses of smell, sight, hearing, and taste from the special sensory receptors, and information from the general sensory receptors, especially in the head region. The cranial nerves also bring orders from the CNS to the voluntary muscles that control movements of the eyes, face, mouth, tongue, pharynx, and larynx. The names of the cranial nerves and the structures they innervate are given in Table 40–4. For example, cranial nerve X, the vagus nerve (which forms part of the autonomic system), innervates the internal organs of the chest and upper abdomen.

In humans, 31 pairs of spinal nerves emerge from the spinal cord. Named for the general region of the vertebral column from which they originate, they comprise eight pairs of cervical spinal nerves, twelve pairs of thoracic, five pairs of lumbar, five pairs of sacral, and one pair of coccygeal spinal nerves.

Each spinal nerve has a **dorsal root** and a **ventral root** (Fig. 40–15). The dorsal root consists of sensory (afferent) fibers that transmit information from the sensory receptors to the spinal cord. Just before the dorsal root joins with the cord, it is marked by a swelling, the **spinal ganglion,** which consists of the cell bodies of the sensory neurons. The ventral root is a grouping of the motor (efferent) fibers leaving the cord en route to the muscles and glands. Cell bodies of the motor neurons occur within the gray matter of the cord.

Table 40–4 THE CRANIAL NERVES OF MAMMALS

Number	*Name*	*Origin of Sensory Fibers*	*Effector Innervated by Motor Fibers*
I	Olfactory	Olfactory epithelium of nose (smell)	None
II	Optic	Retina of eye (vision)	None
III	Oculomotor	Proprioceptors* of eyeball muscles (muscle sense)	Muscles that move eyeball (with IV and VI); muscles that change shape of lens; muscles that constrict pupil
IV	Trochlear	Proprioceptors of eyeball muscles	Other muscles that move eyeball
V	Trigeminal	Teeth and skin of face	Some of muscles used in chewing
VI	Abducens	Proprioceptors of eyeball muscles	Other muscles that move eyeball
VII	Facial	Taste buds of anterior part of tongue	Muscles used for facial expression; submaxillary and sublingual salivary glands
VIII	Auditory (vestibulocochlear)	Cochlea (hearing) and semicircular canals (senses of movement, balance, and rotation)	None
IX	Glossopharyngeal	Taste buds of posterior third of tongue and lining of pharynx	Parotid salivary gland; muscles of pharynx used in swallowing
X	Vagus	Nerve endings in many of the internal organs; lungs, stomach, aorta, larynx	Parasympathetic fibers to heart, stomach, small intestine, larynx, esophagus, and other organs
XI	Spinal accessory	Muscles of shoulder	Muscles of neck and shoulder
XII	Hypoglossal	Muscles of tongue	Muscles of tongue

*Proprioceptors are receptors located in muscles, tendons, or joints that provide information about body position and movement.

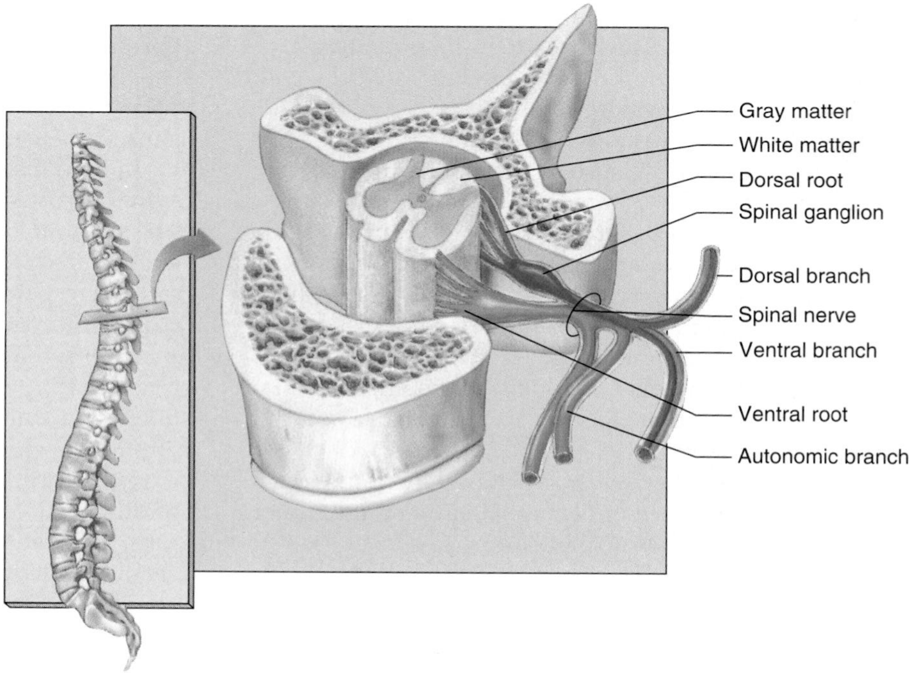

FIGURE 40–15 Dorsal and ventral roots emerge from the spinal cord and join to form a spinal nerve. The spinal nerve divides into several branches.

Table 40–5 COMPARISON OF SYMPATHETIC AND PARASYMPATHETIC SYSTEMS

Characteristic	*Sympathetic System*	*Parasympathetic System*
General effect	Prepares body to cope with stressful situations	Restores body to resting state after stressful situation; actively maintains normal configuration of body functions
Extent of effect	Widespread throughout body	Localized
Neurotransmitter released at synapse with effector	Norepinephrine (usually)	Acetylcholine
Duration of effect	Lasting	Brief
Outflow from CNS	Thoracolumbar levels of spinal cord	Craniosacral levels (from brain and spinal cord)
Location of ganglia	Chain and collateral ganglia	Terminal ganglia
Number of postganglionic fibers with which each preganglionic fiber synapses	Many	Few

Peripheral to the junction of the dorsal and ventral roots, each spinal nerve divides into branches. The dorsal branch serves the skin and muscles of the back. The ventral branch serves the skin and muscles of the sides and ventral part of the body. The autonomic branch innervates the viscera. The ventral branches of several spinal nerves form tangled networks called plexi (sing., *plexus*). Within a plexus, the fibers of a spinal nerve may separate and then regroup with fibers that originated in other spinal nerves. Thus, nerves emerging from a plexus consist of neurons that originated in several different spinal nerves. Among the principal plexi are the cervical plexus, the brachial plexus, the lumbar plexus, and the sacral plexus.

The autonomic system regulates the internal environment

The autonomic system helps to maintain homeostasis in the internal environment. For instance, it regulates the heart rate and maintains a constant body temperature. The autonomic system works automatically and without voluntary input. Its effectors are smooth muscle, cardiac muscle, and glands. Like the somatic system, it is functionally organized into reflex pathways. Receptors within the viscera relay information via afferent nerves to the CNS. The information is integrated at various levels. Then, the decision is transmitted along efferent nerves to the appropriate muscles or glands.

Information from the organs is transmitted to the CNS by afferent neurons. Some of these, found within cranial or spinal nerves, are thought to synapse with association neurons in the CNS. They bring information about blood pressure, respiration, heartbeat, digestive tract contractions, and other visceral activities.

The efferent portion of the autonomic system is subdivided into **sympathetic** and **parasympathetic systems.** In general, the sympathetic nerves operate to stimulate organs and to mobilize energy, especially in response to stress, whereas the parasympathetic nerves influence organs to conserve and restore energy, particularly during quiet, calm activities. Many organs are innervated by both

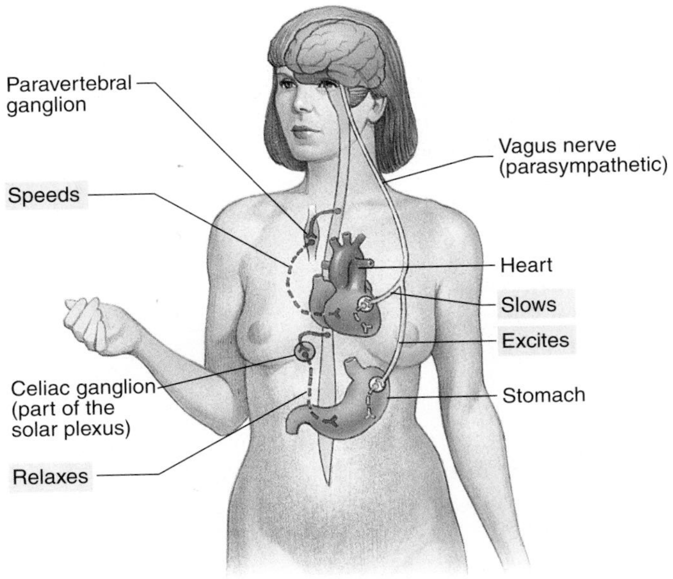

FIGURE 40–16 The heart and stomach, like most organs, are innervated by both sympathetic and parasympathetic nerves. Sympathetic nerves are shown in red; postganglionic fibers are shown as dotted lines.

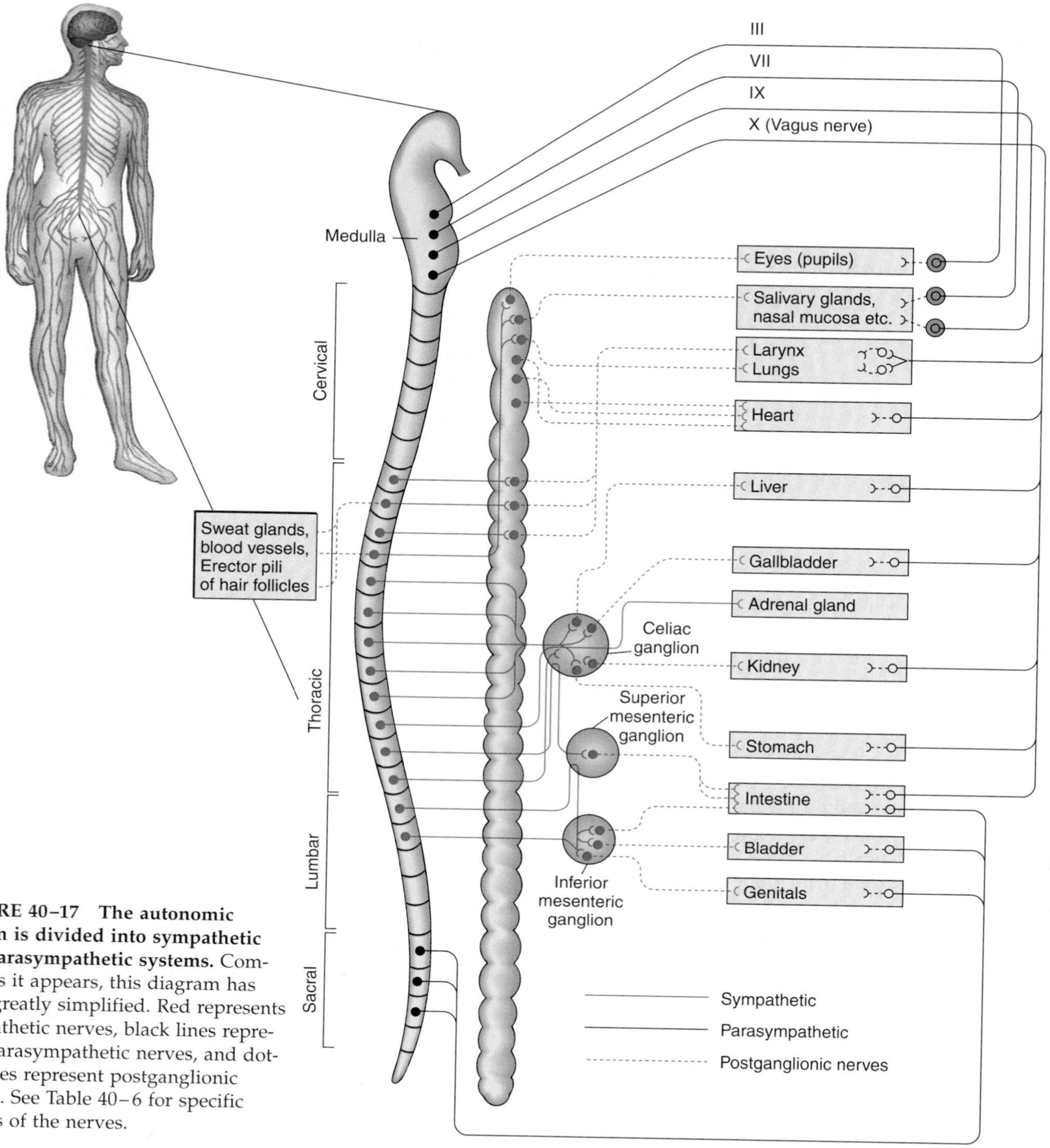

FIGURE 40–17 The autonomic system is divided into sympathetic and parasympathetic systems. Complex as it appears, this diagram has been greatly simplified. Red represents sympathetic nerves, black lines represent parasympathetic nerves, and dotted lines represent postganglionic nerves. See Table 40–6 for specific actions of the nerves.

types of nerves, which act upon the organ in a complementary way (Figs. 40–16 and 40–17; Table 40–5). For example, the heart rate is slowed by impulses from its parasympathetic nerve fibers and speeded up by messages from its sympathetic nerve supply.

Instead of using a single efferent neuron, as in the somatic system, the autonomic system uses a relay of two neurons between the CNS and the effector. The first neuron, called the **preganglionic neuron,** has a cell body and dendrites within the CNS. Its axon, part of a peripheral nerve, ends by synapsing with a **postganglionic neuron.** The dendrites and cell body of the postganglionic neuron are in a ganglion outside the CNS. Its axon terminates near or on the effector. The sympathetic ganglia are paired, and a chain of them — the **paravertebral sympathetic ganglion chain** — occurs on each side of the spinal cord from the neck to the abdomen. Some sympathetic preganglionic neurons do not end in these ganglia but in-

Table 40–6 COMPARISON OF SYMPATHETIC AND PARASYMPATHETIC ACTIONS ON SELECTED EFFECTORS*

Effector	*Sympathetic Action*	*Parasympathetic Action*
Heart	Increases rate and strength of contraction	Decreases rate; no direct effect on strength of contraction
Bronchial tubes	Dilates	Constricts
Iris of eye	Dilates pupil	Constricts pupil
Sex organs (male)	Constricts blood vessels; ejaculation	Dilates blood vessels; erection
Blood vessels	Generally constricts	No innervation for many
Sweat glands	Stimulates	No innervation
Intestine	Inhibits motility	Stimulates motility and secretion
Liver	Stimulates glycogenolysis (conversion of glycogen to glucose)	No effect
Adipose tissue	Stimulates free fatty acid release from fat cells	No effect
Adrenal medulla	Stimulates secretion of epinephrine and norepinephrine	No effect
Salivary glands	Stimulates thick, viscous secretion	Stimulates a profuse, watery secretion

*Refer to Figure 40–17 as you study this table. Notice that many other examples could be added to this list.

stead pass on to ganglia in the abdomen, close to the aorta and its major branches. These ganglia are known as **collateral ganglia.** Parasympathetic preganglionic neurons synapse with postganglionic neurons in **terminal ganglia** near or within the walls of the organs they innervate.

The sympathetic and parasympathetic systems also differ in the neurotransmitters they release at the synapse with the effector. Both preganglionic and postganglionic parasympathetic neurons secrete acetylcholine. Sympathetic postganglionic neurons release norepinephrine (although their preganglionic neurons secrete acetylcholine). Table 40–6 compares the actions of the sympathetic and parasympathetic systems on selected effectors.

The autonomic system got its name from the original belief that it was independent of the CNS—that is, autonomous. Physiologists have shown that this is not so and that the hypothalamus and many other parts of the CNS help to regulate the autonomic system. Although the autonomic system usually functions automatically, its activities can be consciously influenced. Biofeedback provides a person with visual or auditory evidence concerning the status of an autonomic body function. For example, a tone may be sounded when blood pressure reaches a desirable level. Using such techniques, subjects have learned to control autonomic activities such as blood pressure, brain wave pattern, heart rate, and blood sugar level. Even certain abnormal heart rhythms can be consciously modified.

MANY DRUGS AFFECT THE NERVOUS SYSTEM

About 25% of all prescribed drugs are taken to alter psychological conditions, and almost all the commonly abused drugs affect mood. Many of them act by changing the levels of neurotransmitters within the brain. In particular, levels of norepinephrine, serotonin, and dopamine are thought to influence mood. For example, when excessive amounts of norepinephrine are released in the RAS, we feel stimulated and energetic, whereas low concentrations of this neurotransmitter reduce anxiety. Table 40–7 lists several commonly used and abused drugs and gives their effects. Also see *Focus On: Alcohol Abuse* and *Focus On: Crack Cocaine.*

Habitual use of almost any mood-altering drug can result in **psychological dependence,** in which the user becomes emotionally dependent on the drug. When deprived of it, the user craves the feeling of euphoria (well-being) that the drug induces. Some drugs induce **tolerance** after several weeks. This means that response to the drug decreases, and greater amounts are required to obtain the desired effect. Tolerance often occurs because the liver cells are stimulated to produce more of the enzymes that metabolize and inactivate the drug. It can also occur due to a decrease in the number of postsynaptic receptors (a mechanism known as downregulation).

(*Text continues on page 878*)

Table 40–7 EFFECTS OF SOME COMMONLY USED DRUGS

Name of Drug	*Effect on Mood*	*Actions on Body*	*Side Effects/Dangers Associated with Abuse*
Antidepressants			
Tricyclic antidepressants (e.g., Elavil, Anafranil)	Elevate mood; relieve depression; also used to treat obsessive-compulsive behavior	Most block reuptake of biogenic amines (especially norepinephrine) increasing their concentration in the synapses	Can cause sedation, weight gain, sexual dysfunction
Selective serotonin reuptake inhibitors (SSRIs) (Prozac, Zoloft, Effexor)	Relieve depression; used to treat obsessive-compulsive behavior	Block serotonin reuptake	Nausea, headache, insomnia, anxiety
MAO inhibitors	Relieve depression	Block enzymatic breakdown of biogenic amines, increasing their concentration in the synapses	Liver toxicity, excessive CNS stimulation; overdose may affect blood pressure, cause hallucinations
Anti-anxiety drugs			
Benzodiazepines (e.g., Xanax, Valium, Librium)	Sedation; induce sleep	Bind to GABA receptor complex on post-synaptic neuron; chloride channels open, causing hyperpolarization; cause relaxation of skeletal muscles	Drowsiness, confusion; psychological dependence, addiction; effects additive with alcohol
Barbiturates "Downers" (e.g., Phenobarbital)	Sedation; induce sleep	Bind to receptor adjacent to GABA receptor on chloride channels; chloride channels open so more chloride ions move in, causing hyperpolarization	Drowsiness, confusion; psychological dependence, tolerance, addiction; overdose can cause severe CNS depression, resulting in coma and death (especially lethal in combination with alcohol)
Antipsychotic medications			
Phenothiazines (e.g., Thorazine, Mellaril, Stelazine)	Relieve some symptoms of schizophrenia; reduce impulsive and aggressive behavior	Block dopamine receptors	Muscle spasms, shuffling gait, and other symptoms of Parkinson's disease
Newer antipsychotic medications (e.g., Clozaril, Risperdal)	Similar to phenothiazines, but also appear to increase motivation	Block dopamine receptors and several other neurotransmitter receptors	Fewer side effects than with the phenothiazines
Narcotic analgesics Opiates (e.g., morphine, codeine, Demerol, heroin)	Euphoria; sedation; relieve pain	Mimic actions of endorphins; bind to opiate receptors	Depress respiration; constrict pupils; impair coordination; tolerance, psychological dependence, addiction; convulsions, death from overdose

Table 40–7 continued

Name of Drug	*Effect on Mood*	*Actions on Body*	*Side Effects/Dangers Associated with Abuse*
Cocaine	Euphoria; excitation followed by depression	CNS stimulation followed by depression; autonomic stimulation; dilation of pupils; local anesthesia; inhibits reuptake of norepinephrine, dopamine, and serotonin	Mental impairment, convulsions, hallucinations, unconsciousness; death from overdose
Marijuana	Euphoria	Impairs coordination; impairs depth perception and alters sense of timing; impairs short-term memory (probably by decreasing acetylcholine levels in the hippocampus); inflames eyes; causes peripheral vasodilation	In large doses, sensory distortions, hallucinations; evidence of lowered sperm counts and testosterone (male hormone) levels
Amphetamines "Uppers," "pep pills" (e.g., Dexedrine)	Euphoria, stimulation, hyperactivity	Stimulate release of dopamine and norepinephrine; enhance flow of impulses in RAS; increase heart rate; raise blood pressure; dilate pupils	Tolerance, possible physical dependence, hallucinations; increased blood pressure; psychotic episodes; death from overdose
LSD (lysergic acid diethylamide)	Overexcitation; sensory distortions, hallucinations	Alters levels of transmitters in brain; potent CNS stimulator; dilates pupils, sometimes unequally; increases heart rate; raises blood pressure	Irrational behavior
Methaqualone (e.g., Quaalude, Sopor)	Hypnotic	Depresses CNS; depresses certain spinal reflexes	Tolerance, physical dependence; convulsions, death
Caffeine	Increases mental alertness; decreases fatigue and drowsiness	Acts on cerebral cortex; relaxes smooth muscle; stimulates cardiac and skeletal muscle; increases urine volume	Very large doses stimulate centers in the medulla (may slow the heart); toxic doses may cause convulsions
Nicotine	Lessens psychological tension	Stimulates sympathetic nervous system; stimulates synthesis of lipid in arterial wall	Tolerance, physical dependence; stimulates development of atherosclerosis
Alcohol	Euphoria, relaxation, release of inhibitions	Depresses CNS; impairs vision, coordination, judgment; lengthens reaction time	Physical dependence, damage to pancreas, liver cirrhosis, brain damage

Focus On

Alcohol: The Most Abused of All Drugs

Our society is saturated with alcohol. Children see more than 100,000 beer commercials on television before they are legally old enough to drink. According to a national study released by the University of Michigan in 1994, we are experiencing an increase in the use of alcohol and other drugs. For example, 28% of high school seniors acknowledged having five or more drinks in a row during the two weeks prior to a survey.

After tobacco, alcohol is the leading cause of premature death in the United States. It is linked to more than 100,000 deaths every year, and costs our society more than $100 billion annually. According to the pollster Louis Harris, there are 28 million alcoholics in the United States, and about one in three homes includes someone with a serious drinking problem. Alcohol abuse is not limited to adults. About 4.6 million adolescents, or nearly one of every three high school students, experience negative consequences from alcohol use, including difficulty with parents, poor performance at school, and trouble with the law.

Alcohol abuse results in physiological, psychological, and social impairment for the abuser, and has serious negative consequences for family, friends, and society. Alcohol abuse has been linked to the following:

- More than 50% of all traffic fatalities.
- More than 50% of violent crimes.
- More than 50% of suicides.
- More than 60% of child abuse and spouse abuse cases.
- More than 15,000 babies born each year with serious birth defects because their mothers drank alcohol during pregnancy.
- Fetal alcohol syndrome is a leading cause of mental retardation in the U.S.
- More breast cancer. Recent studies suggest that as few as three drinks per week increase breast cancer risk by 50%.
- Greater risk of liver disease and brain impairment for women than men (clinical effects are seen in women consuming about half the alcohol consumed by men).

A single drink — that is, 12 ounces of beer, 5 ounces of wine, or 1.5 ounces of 80-proof liquor — results in a blood alcohol concentration of approximately 20 mg/dL (milligrams per deciliter). This represents about 0.5 ounce of pure alcohol in the blood. Alcohol accumulates in the blood because absorption occurs more rapidly than do oxidation and excretion. Every cell in the body can take in alcohol from the blood. At first the drinker may feel stimulated. But alcohol actually causes depression of the central nervous system, probably by affecting specific neurotransmitters in the brain.

As blood alcohol concentration rises, information processing, judgment, memory, sensory perception, and motor coordination all become progressively impaired. Depression and drowsiness generally occur. Contrary to popular belief, alcohol decreases sexual performance in males. Some individuals become loud, angry, or violent. In most states, a blood alcohol concentration of 100 mg/dL (or 0.10) legally defines driving while intoxicated (DWI). A 150-pound man typically reaches this level by drinking two or three beers in an hour.

In chronic drinkers, cells of the central nervous system adapt to the presence of the drug. This causes tolerance (more and more alcohol is needed to experience the same effect) and physical dependence. Abrupt withdrawal can result in sleep disturbances, severe anxiety, tremors, seizures, hallucinations, and psychoses.

Treatment for alcohol problems includes various forms of psychotherapy, including relapse prevention therapy in which individuals are encouraged not to consider lapses from abstinence as failure. The group support offered by Alcoholics Anonymous (AA) has proved effective for many struggling with alcohol abuse. ■

Use of some drugs, such as heroin, tobacco, alcohol, and barbiturates, may also result in **addiction** (physical dependence), in which physiological changes occur that make the user dependent on the drug. When the drug is withheld, the addict suffers withdrawal symptoms characteristic of that drug. Addiction can also occur because certain drugs, such as morphine, have components similar to substances that body cells normally manufacture on their own. The continued use of such a drug followed by sudden withdrawal causes potentially dangerous physiological effects, because the body's natural production of these substances has been depressed. It may be some time before homeostasis is reestablished.

Focus On

Crack Cocaine

With the exception of alcoholics, the majority of persons seeking treatment for drug abuse are now crack cocaine addicts. Cocaine use by teenagers alone has increased about 400% during the past 10 years, involving an estimated 2 million youngsters. Crack is a very concentrated and extremely powerful form of cocaine—five to ten times more addictive than other forms of cocaine. This drug is produced in illegal, makeshift labs by converting powdered cocaine into small "rocks" that are up to 80% pure cocaine. Crack is smoked in pipes or added to tobacco or marijuana cigarettes.

Use of crack results in an intense, brief high beginning in 4 to 6 seconds and lasting for 5 to 7 minutes. Physiologically, crack stimulates a massive release of catecholamine neurotransmitters (norepinephrine and dopamine) in the brain and blocks reuptake. Excitation of the sympathetic nervous system occurs, and users report experiencing feelings of self-confidence, power, and euphoria. As the neurotransmitters are depleted, the high is followed by a "crash," a period of deep depression. The abuser experiences an intense craving for another crack "hit" in order to get more stimulation.

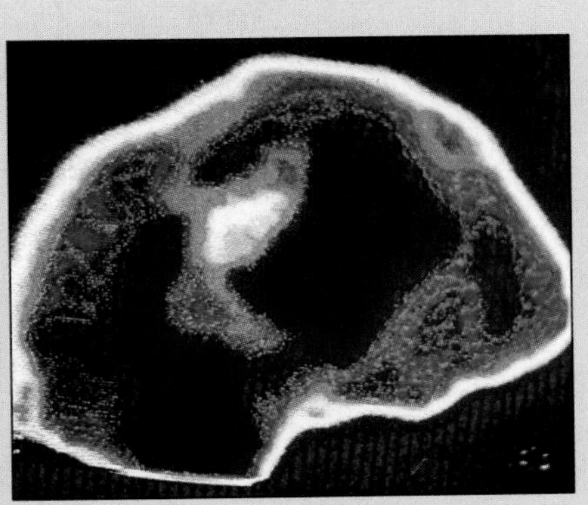

Cocaine use causes neurological damage. This color-enhanced SPECT (single photon emission computer tomography) scan of the brain of a cocaine abuser shows decreased metabolism (white area) near the frontal region. (Howard Sochurek, Courtesy of Brigham and Women's Hospital, Boston)

Some abusers spend days smoking crack without stopping to eat or sleep. Although a vial of rocks can be obtained for about $20, many abusers develop habits that cost hundreds of dollars a week. Supporting an expensive drug habit leads many abusers to prostitution, drug dealing, and other forms of crime.

Cocaine addicts report problems with memory loss, fatigue, depression, insomnia, paranoia, loss of sexual drive, violent behavior, and attempts at suicide. Crack can cause respiratory problems, brain seizures, cardiac arrest, and elevation of blood pressure that leads to stroke. Many users have suffered fatal reactions to impurities in the drug or have died as a result of accidents related to its use.

Researchers at Columbia University are developing a genetically engineered antibody that can bind to cocaine and destroy it. The antibody can apparently remain in the blood for long periods of time. Clinical trials are planned to test this potential treatment for cocaine addiction. ■

SUMMARY

I. Among invertebrates, nerve nets and radial nervous systems are typical of radially symmetrical animals. Bilateral nervous systems are characteristic of bilaterally symmetrical animals.
   A. A nerve net consists of nerve cells scattered throughout the body; no central nervous system is present. Responses are generally slow and imprecise.
   B. Echinoderms typically have a nerve ring and nerves that extend into various parts of the body.
   C. In a bilateral nervous system, nerve cells concentrate to form nerves, nerve cords, ganglia, and (in complex animals) a brain. An increase in number of neurons, especially the association neurons, permits a wide range of responses.

II. In the vertebrate embryo, the brain and spinal cord arise from the neural tube. The anterior end of the tube differentiates into forebrain, midbrain, and hindbrain.
   A. The hindbrain subdivides into the metencephalon and myelencephalon.
      1. The myelencephalon develops into the medulla, which contains the vital centers and other reflex centers.
      2. The metencephalon gives rise to the cerebellum, which is responsible for muscle tone, posture, and equilibrium, and to the pons, which connects various parts of the brain.
   B. The midbrain is the largest part of the brain in fishes and amphibians. It is their main association area, linking sen-

sory input and motor output. In reptiles, birds, and mammals, the midbrain serves as a center for visual and auditory reflexes.

C. The forebrain differentiates to form the diencephalon and telencephalon.
   1. The diencephalon develops into the thalamus and hypothalamus.
      a. The thalamus is a relay center for motor and sensory information.
      b. The hypothalamus controls autonomic functions; links nervous and endocrine systems; controls temperature, appetite, and fluid balance; and is involved in some emotional and sexual responses.
   2. The telencephalon develops into the cerebrum and olfactory bulbs.
      a. In fishes and amphibians, the cerebrum mainly integrates incoming sensory information.
      b. In birds, the corpus striatum controls stereotypical but complex behavior patterns. Another part of the cerebrum is thought to govern learning.
      c. In mammals, the neopallium accounts for a large part of the cerebral cortex, and the cerebrum has complex association functions.

III. The human brain and spinal cord are protected by bone and three meninges and are cushioned by cerebrospinal fluid.
   A. The spinal cord transmits impulses to and from the brain and controls many reflex activities.
      1. The spinal cord consists of ascending tracts, which transmit information to the brain, and descending tracts, which transmit information from the brain. Its gray matter contains nuclei that serve as reflex centers.
      2. A withdrawal reflex involves receptors; sensory, association, and motor neurons; and effectors.
   B. The human cerebral cortex consists of motor areas that control voluntary movement; sensory areas that receive incoming sensory information; and association areas that link sensory and motor areas and are responsible for learning, language, thought, and judgment.
   C. Brain activity cycles in a sleep-wake pattern.
      1. Alpha wave patterns are characteristic of relaxed states, beta wave patterns of heightened mental activity, and delta waves of non-REM sleep.
      2. The reticular activating system (RAS) is an arousal system.
      3. Electrical activity of the cerebral cortex and metabolic rate slows during non-REM sleep. REM sleep is characterized by dreaming.
      4. After many hours of activity, neurons in the RAS are thought to fatigue. Sleep centers are then activated and release serotonin.
   D. The limbic system affects the emotional aspects of behavior, motivation, sexual behavior, autonomic responses, and biological rhythms.
   E. Learning is a change in behavior that results from experience. Memory is the storage of knowledge and the ability to retrieve it.
      1. Information is transferred from sensory memory to short-term memory and then to long-term memory.
      2. Long-term memory storage may involve gene activation and formation of new synaptic connections.
      3. Memories appear to be stored throughout the sensory areas of the cerebrum; they are integrated in many areas of the brain, including the association areas of the cerebral cortex and parts of the limbic system.
   F. The physical structure and chemistry of the brain can be altered by environmental experience.

IV. The peripheral nervous system consists of sensory receptors and nerves, including the cranial and spinal nerves and their branches.

V. The autonomic system regulates the internal activities of the body.
   A. The sympathetic system permits the body to respond to stressful situations.
   B. The parasympathetic system influences organs to conserve and restore energy.

VI. Some of the types of drugs that affect the nervous system are amphetamines, antianxiety drugs, barbiturates, antipsychotic drugs, antidepressants, narcotic analgesics, and alcohol. Many drugs alter mood by increasing or decreasing the concentrations of specific neurotransmitters within the brain.

SELECTED KEY TERMS

addiction
alpha waves
association areas
autonomic system
beta waves
brain
brain stem
central nervous system (CNS)
cerebellum
cerebral cortex
cerebral ganglia
cerebrospinal fluid (CSF)
cerebrum
corpus callosum
cranial nerve
delta waves
dorsal root (of spinal nerve)
fissure
forebrain
frontal lobes
gray matter
gyrus (gyri)
hindbrain
hypothalamus
learning
limbic system
medulla
memory
meninges
midbrain
motor areas
nerve net
non-REM sleep
occipital lobes
olfactory bulbs
parasympathetic system
parietal lobes
peripheral nervous system (PNS)
pons
reflex action
REM sleep
reticular activating system (RAS)
sensory areas
somatic system
spinal cord
spinal ganglion
spinal nerve
sulcus (sulci)
sympathetic system
temporal lobe
thalamus
tolerance
tract
ventral root (of spinal nerve)
ventricle (of brain)
white matter

POST-TEST

1. In a nerve __________, impulses are transmitted in all directions and there is no central control organ.
2. The cerebral __________ in a flatworm serve as a simple brain.
3. The two main divisions of the vertebrate nervous system are the __________ nervous system and the __________ nervous system.
4. In vertebrates, the embryonic neural tube expands anteriorly to form the __________, and develops posteriorly into the __________ __________.
5. The medulla, pons, and midbrain make up the __________ __________.
6. The fourth __________ is the cavity located within the medulla.
7. The most prominent part of the amphibian brain is the __________.
8. The __________ links nervous and endocrine systems.
9. The most prominent part of the mammalian brain is the __________.
10. The __________ coordinates muscle activity.
11. The __________ contains several vital centers that regulate respiration, heart rate, and blood pressure.
12. Descending __________ convey impulses from the brain down the spinal cord.
13. Deep furrows between gyri in the cerebrum are __________.
14. The gray matter of the human cerebrum is the cerebral __________.
15. The primary motor areas are located in the __________ lobes; the visual centers are located in the __________ lobes.
16. An action system of the brain concerned with emotional behavior and with reward and punishment is the __________ __________.
17. As you answer these questions, your brain is emitting __________ waves.
18. Dreaming takes place during __________ sleep.
19. The __________ nervous system mobilizes energy and helps the body respond to stress.
20. Sensory nerves enter the spinal cord through the __________ root.
21. When some drugs are taken for several weeks, increasingly greater amounts are needed to obtain the desired effect. This is called __________.

REVIEW QUESTIONS

1. Contrast the nervous system of a planarian flatworm with that of *Hydra.*
2. What are some characteristics of a bilateral nervous system? What are some advantages of this type of system?
3. Compare the flatworm nervous system with that of a vertebrate. In each case, what is the relationship between the type of nervous system and the animal's lifestyle?
4. Identify the parts of the adult vertebrate brain derived from the embryonic forebrain, midbrain, and hindbrain.
5. Give the function of each of the following structures in the human brain: (a) medulla; (b) midbrain; (c) cerebellum; (d) thalamus; (e) hypothalamus; (f) cerebrum.
6. Compare the midbrain and cerebrum of the fish with those of the mammal.
7. Describe the protective coverings of the human CNS, and give the function of the cerebrospinal fluid.
8. In what sequence is information processed through the levels of memory? How might the various memory levels be adaptive? Where are memories stored?
9. Cite experimental evidence supporting the view that environmental experience can alter the brain.
10. Contrast the functions of the CNS with those of the PNS.
11. Compare the functions of the somatic and autonomic systems.
12. Contrast the structure and function of the sympathetic system with those of the parasympathetic system.
13. Describe how each of the following drugs affects the CNS: (a) alcohol; (b) antipsychotic drugs; (c) antidepressants; (d) amphetamines.
14. Label the diagram.

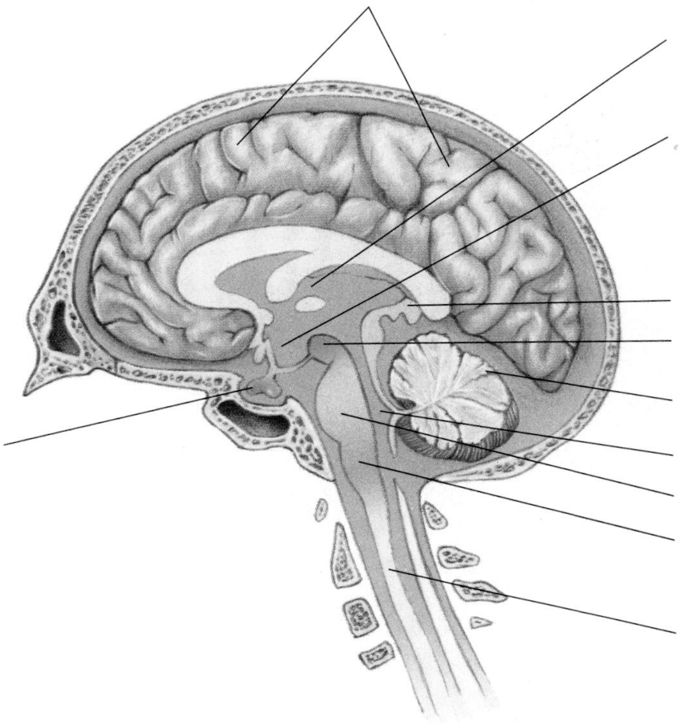

YOU MAKE THE CONNECTION

1. What general trends can you identify in the evolution of the vertebrate brain?
2. In what way does electrical activity of the brain reflect your state of consciousness? Of what use might it be to learn to control your brain wave patterns?
3. Imagine that you have just become a parent. What kind of things could you do to enhance the development of your child's academic abilities?
4. Hypothesize a possible relationship between inheritance and intelligence based on gene activation during information processing.

RECOMMENDED READINGS

Hinton, G.E., D.C. Plaut, and T. Shallice. "Simulating Brain Damage." *Scientific American,* Vol. 269, No. 4, October 1993. When a network of computer-simulated neurons is trained to read and is then damaged, errors made are similar to those made by adults with brain damage.

Kalil, R. "Synapse Formation in the Developing Brain." *Scientific American,* Vol. 261, No. 6, December 1989. Developing neurons generate impulses that modify existing synapses and result in the formation of new connections.

Musto, D.F. "Opium, Cocaine and Marijuana in American History." *Scientific American,* Vol. 265, No. 1, July 1991. A review of the history of drug use in the United States provides perspective on current reaction to drug use.

Selkoe, D.J. "Amyloid Protein and Alzheimer's Disease." *Scientific American,* Vol. 265, No. 5, November 1991. Is the accumulation of a protein fragment in the brain one cause of Alzheimer's disease?

Spector, R., and C.E. Johanson. "The Mammalian Choroid Plexus." *Scientific American,* Vol. 261, No. 5, November 1989. The choroid plexus serves an important purpose in nourishing and protecting the nervous system.

CHAPTER 41

Sensory Reception

Bats navigate by ultrasonic echoes that are inaudible to the human ear. They emit high-pitched clicking sounds and use the echoes to locate objects. California leaf-nosed bat (*Macrotus californicus*) approaching a flower. (Merlin D. Tuttle/Bat Conservation International/Photo Researchers, Inc.)

Sense organs link organisms with the outside world and enable them to receive information about their environment. The kinds of sense organs an animal possesses determine just how it perceives its surroundings. We humans live in a world of rich colors, numerous shapes, and varied sounds. But we cannot hear the high-pitched whistles that are audible to dogs and cats, or the ultrasonic echoes by which bats navigate. Nor do we ordinarily recognize our friends by their distinctive odors. And although vision is our dominant and most refined sense, we are blind to the ultraviolet hues that light up the world for insects.

Sensory receptors are structures that respond to information about changes in the internal or external environment. Sensory receptors consist of neuron endings or specialized cells in close contact with neurons. Human taste buds, for example, are modified epithelial cells connected to neurons.

Sensory receptors, along with other types of cells, make up the familiar complex **sense organs**: eyes, ears, nose, and taste buds. Traditionally, mammals are said to have the five senses of sight, hearing, smell, taste, and touch. Balance is now also recognized as a sense, and touch is viewed as a compound sense that allows us to detect pressure, pain, and temperature. In this chapter, we will also consider receptors that enable us to sense muscle tension and joint position.

LEARNING OBJECTIVES

After you have studied this chapter you should be able to

1. Compare exteroceptors, proprioceptors, and interoceptors, and explain the importance of each group.
2. Distinguish among the five kinds of receptors that are classified according to the types of energy to which they respond. Give examples of sense organs of each type.
3. Describe how a sensory receptor functions: define energy transduction, receptor potential, and adaptation in your answer.
4. Describe the following mechanoreceptors: tactile receptors, statocysts, lateral line organs, and proprioceptors.
5. Compare the function of the saccule and utricle with that of the semicircular canals in maintaining equilibrium.
6. Trace the path taken by sound waves through the structures of the ear, and explain how the organ of Corti functions as an auditory receptor.
7. Describe the receptors of taste and smell.
8. Relate the presence of thermoreceptors to the lifestyles of animals that have them.
9. Contrast simple eyes, compound eyes, and vertebrate eyes.
10. Label the structures of the vertebrate eye on a diagram, and give the functions of each of its accessory structures.
11. Compare the two types of photoreceptors in the human retina; include the role of rhodopsin in your explanation.

SENSORY RECEPTORS CAN BE CLASSIFIED ACCORDING TO THE STIMULI TO WHICH THEY RESPOND

Exteroceptors receive stimuli from the outside environment, enabling an animal to know and explore the world, search for food, find and attract a mate, recognize friends, and detect enemies. **Proprioceptors** are sensory receptors within muscles, tendons, and joints that enable the animal to perceive the positions of its arms, legs, head, and other body parts, along with the orientation of its body as a whole. With the help of proprioceptors, humans can get dressed or eat even in the dark.

Interoceptors are sensory receptors within body organs that detect changes in pH, osmotic pressure, body temperature, and the chemical composition of the blood. We are not usually conscious of messages sent to the central nervous system (CNS) by these receptors as they work continuously to maintain homeostasis. We become aware of their activity when they enable us to perceive such diverse internal conditions as thirst, hunger, nausea, pain, and orgasm.

Another way sensory receptors can be classified is according to the *types* of stimuli to which they respond (Table 41–1). **Mechanoreceptors** respond to mechanical energy—touch, pressure, gravity, stretching, and movement. **Chemoreceptors** respond to certain chemical compounds, and **photoreceptors** detect light energy. **Thermoreceptors** respond to heat and cold. Some fish have well-developed **electroreceptors,** which detect electrical energy.

RECEPTOR CELLS WORK BY PRODUCING RECEPTOR POTENTIALS

Receptor cells absorb energy, transduce (convert) that energy into electrical energy, and produce a **receptor potential.** In its capacity as a detector or sensor, a receptor receives a small amount of energy from the environment. Each kind of receptor is especially sensitive to one particular form of energy. For example, photoreceptor cells (rods and cones) in the eye absorb light energy, while temperature receptors respond to infrared thermal energy (heat).

Receptor cells are remarkably sensitive to appropriate stimuli. The photoreceptors of the eye, for example, are stimulated by an extremely faint beam of light, whereas only a very strong light can directly stimulate the optic nerve. The minute amount of vinegar that can be tasted or vanilla that can be smelled would have no effect if applied directly to a nerve fiber.

Multiple kinds of environmental stimuli trigger receptor cells to perform biological work. When unstimulated, the neuron maintains a resting potential—that is, a potential difference between the inside and the outside of the neuron. When the receptor is stimulated, the permeability of its plasma membrane is altered. If the potential difference increases, the cell becomes hyperpolarized. If it decreases, the cell becomes depolarized.

The receptor potential, the state of depolarization caused by a stimulus, is a graded response that spreads relatively slowly down the dendrite, fading as it goes. If the neuron is sufficiently depolarized to reach its thresh-

Table 41–1 CLASSIFICATION OF RECEPTORS BY STIMULI TO WHICH THEY RESPOND

Type of Receptor	*Examples*	*Effective Stimuli*
Mechanoreceptors	Tactile receptors Pacinian corpuscles Meissner's corpuscles	Touch, pressure
	Proprioceptors Muscle spindles Golgi tendon organs Joint receptors	Movement, body position Muscle contraction Stretch of a tendon Movement in ligaments
	Lateral line organs in fish	Waves, currents in water
	Statocysts in invertebrates	Gravity
	Labyrinth of vertebrate ear Saccule and utricle Semicircular canals Hair cells in the cochlea	 Gravity; linear acceleration Angular acceleration Pressure waves (sound)
Chemoreceptors	Taste buds; olfactory epithelium	Specific chemical compounds
Thermoreceptors	Temperature receptors in blood-sucking insects and ticks; pit organs in pit vipers; nerve endings and receptors in skins and tongues of many animals	Heat
Electroreceptors	Organs in skin of some fishes	Electrical currents in water
Photoreceptors	Eyespots; ommatidia of arthropods; rods and cones in retinas of vertebrates	Light energy

old level, an action potential is generated. The action potential travels along the axon to the central nervous system (CNS). In summary:

stimulus (e.g., light energy) → transduction into electrical energy → receptor potential → action potential

The receptor performs three important functions: It (1) detects a stimulus in the environment by absorbing energy; (2) converts the energy of the stimulus into electrical energy; and (3) produces a receptor potential that may result in the transmission of information to the CNS by an action potential. With minor variations, this is how all receptors operate.

A strong stimulus causes a greater depolarization of the receptor membrane and results in more action potentials than does a weak one. This occurs because the receptor potential is a *graded* response. In contrast, according to the all-or-none law, the amplitude (size) of each *action potential* has no relation to the stimulus; it is characteristic of the particular neuron.

Sensation depends on transmission of a "coded" message

How do you know whether you are seeing a blue sky, tasting a doughnut, or hearing a note played by a piano? All action potentials are qualitatively the same. Light of the wavelength 450 nanometers (blue), sugar molecules (sweet), and sound waves of 440 hertz (A above middle C) all cause transmission of similar action potentials. Our ability to differentiate stimuli depends on both the sensory receptor itself and on the brain. We can distinguish the color of a blue sky from the scent of cologne; a sweet taste from a light breeze; or the sound of a piano from the heat of the sun because cells of each sensory receptor are connected to specific neurons in particular parts of the brain. Because a receptor normally responds to only one type of stimulus—for example, light or sound—the brain interprets a message arriving from a particular receptor as meaning that a certain type of stimulus occurred (e.g., a flash of color).

Sensation takes place in the brain. Interpretation of the message and the type of sensation depends on which association neurons receive the message. The rods and cones of the eye do not see. When stimulated, they send a message to the brain that interprets the signals and translates them into a rainbow, an elephant, or a child. Artificial (e.g., electrical) stimulation of brain centers can also result in sensation. On the other hand, many sensory messages never give rise to sensations at all. For example, certain chemoreceptors sense internal changes in the body but never stir our consciousness.

When stimulated, a sensory receptor initiates what might be considered a "coded" message, composed of action potentials transmitted by nerve fibers. This coded message is later decoded in the brain. Impulses from the sensory receptor may differ in (1) the total number of fibers transmitting, (2) the specific fibers carrying action potentials, (3) the total number of action potentials passing over a given fiber, and (4) the frequency of the action potentials passing over a given fiber. For example, the difference in sound intensity between the gentle rustling of leaves and a clap of thunder depends on the number of neurons transmitting action potentials, as well as the frequency of the action potentials transmitted by each neuron. Just how the sensory receptor initiates different codes and how the brain analyzes and interprets them to produce various sensations are not completely understood.

Receptors adapt to stimuli

Many receptors do not continue to respond at the initial rate, even if the stimulus continues at the same intensity. With time, the frequency of action potentials in the sensory neuron decreases—because the sensory neuron becomes less responsive to stimulation, the receptor produces a smaller receptor potential, or both. This diminishing response to a continued, constant stimulus is called **sensory adaptation.**

Some receptors, such as those for pain or cold, adapt so slowly that they continue to trigger action potentials as long as the stimulus persists. Other receptors adapt rapidly, permitting an animal to ignore persistent unpleasant or unimportant stimuli. For example, when you first pull on a pair of tight jeans, your pressure receptors let you know that you are being squished, and you may feel uncomfortable. Soon, though, these receptors adapt, and you hardly notice the sensation of the tight fit. In the same way, we quickly adapt to odors that at first seem to assault our senses.

MECHANORECEPTORS RESPOND TO TOUCH, PRESSURE, GRAVITY, STRETCH, OR MOVEMENT

Mechanoreceptors are activated when they change shape as a result of being mechanically pushed or pulled. Some mechanoreceptors enable an organism to maintain its body position with respect to gravity (for us, head up and feet down; for a dog, dorsal side up and ventral side down; for a tree sloth, ventral side up and dorsal side down). Other mechanoreceptors are concerned with maintaining postural relations—the position of one part of the body with respect to another. This information is essential for all forms of locomotion and for all coordinated and skilled movements, from spinning a cocoon to completing a reverse one-and-a-half dive with twist.

Mechanoreceptors provide information about the shape, texture, weight, and topographical relations of objects in the external environment. Certain mechanoreceptors affect the operation of internal organs. For example, they supply information about the presence of food in the stomach, feces in the rectum, urine in the bladder, and a fetus in the uterus.

Touch receptors are located in the skin

The simplest mechanoreceptors are free nerve endings in the skin. These nerve endings detect touch, pressure, and pain when stimulated by objects that contact the body surface. Thousands of more specialized touch receptors are also located in the skin (Fig. 41–1). Meissner's cor-

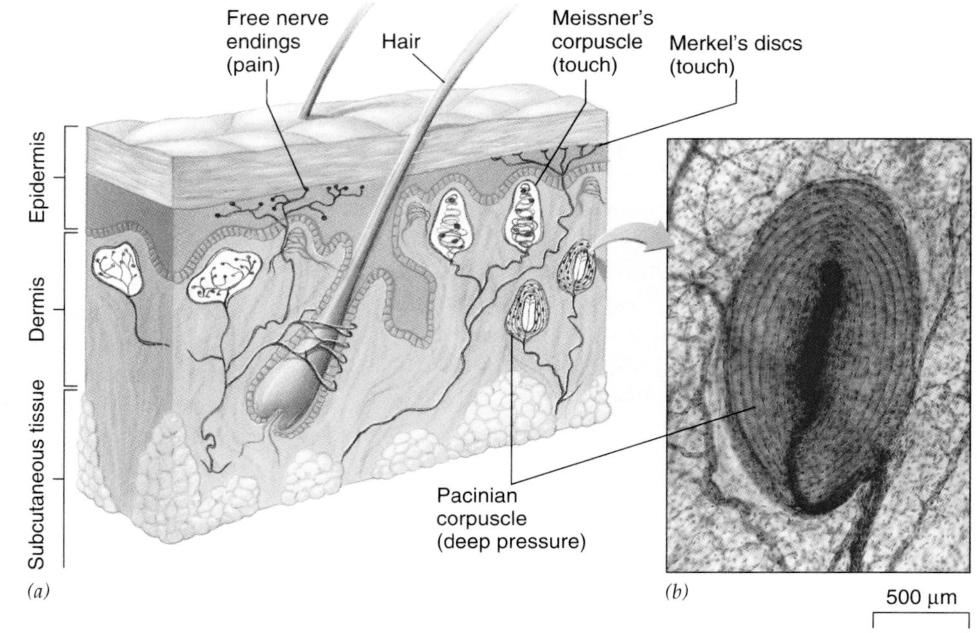

FIGURE 41–1 Several types of sensory receptors are located in the human skin. (*a*) This diagrammatic section through the human skin illustrates the free nerve endings that respond to pain; tactile hairs, Merkel's discs, and Meissner's corpuscles that respond to touch; and Pacinian corpuscles that respond to deep pressure. (*b*) The Pacinian corpuscle, a deep pressure receptor. (Ed Reschke)

puscles are sensitive to light touch. Merkel's discs permit us to know that an object continues to touch the skin. Ruffini's end-organs, which adapt very slowly, inform us of heavy and continuous touch and pressure.

The **Pacinian corpuscle** is especially sensitive to deep pressure that causes rapid movement of the tissues. The nerve ending is surrounded by connective tissue layers interspersed with fluid. Compression causes displacement of the layers, which stimulates the axon. Even though the displacement is maintained under steady compression, the receptor potential rapidly falls to zero and action potentials cease — an excellent example of sensory adaptation. The Pacinian corpuscle is a *phasic receptor,* a receptor that is stimulated only when there is movement of the tissue.

In many invertebrates as well as vertebrates, tactile (touch) receptors lie at the base of a hair or bristle. Tactile hairs may sense the body's orientation in space with respect to gravity. They may also detect air and water vibrations, and contact with other objects. The tactile receptor is stimulated indirectly when the hair is bent or displaced. A receptor potential develops, and a few action potentials may be generated. This type of receptor is a phasic receptor that responds only when the hair is moving. Even though the hair may be maintained in a displaced position, the receptor is not stimulated unless there is motion.

Many invertebrates have gravity receptors called statocysts

Every organism is oriented in a characteristic way with respect to gravity. When displaced from this normal position, it quickly adjusts its body to reassume normal orientation. In complex animals, receptors must continually send information to the CNS regarding the position and movements of the body.

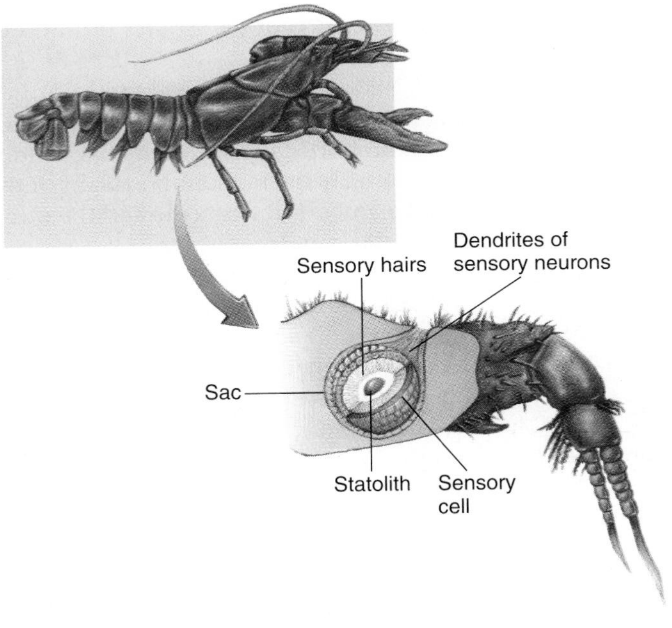

FIGURE 41–2 Many invertebrates have statocysts, receptors that serve as sensors of gravitational force.

Many invertebrates have gravity receptors called **statocysts.** A statocyst is basically an infolding of the epidermis lined with receptor cells that have hairs (Fig. 41–2). The cavity contains one or more **statoliths,** tiny granules of loose sand grains or calcium carbonate. The particles are held together by an adhesive material secreted by cells of the statocyst. Normally the particles are pulled downward by gravity and stimulate the hair cells. When the position of the statolith changes, the hairs of the receptor cells are bent. This mechanical displacement results in receptor potentials and action potentials that inform the CNS of the change in position. By "knowing" which hair cells are firing, the animal knows where "down" is and so can correct any abnormal orientation.

In a classic experiment, the function of the statocyst was demonstrated by substituting iron filings for sand grains in the statocysts of crayfish. The force of gravity was overcome by holding magnets above the animals: the iron filings were attracted upward toward the magnets, and the crayfish began to swim upside down in response to the new information provided by their gravity receptors.

Lateral line organs supplement vision in fish

Lateral line organs are found in fish and in aquatic and larval amphibians. They are thought to supplement vision by informing the animal of obstacles in its way and of moving objects such as prey, enemies, and schooling mates.

Typically this sensory receptor consists of a long canal running the length of the body on each side and continuing into the head (Fig. 41–3). The canals are lined with receptor cells that have hairs. The tips of the hairs are enclosed by a **cupula,** a mass of gelatinous material secreted by the receptor cells. The receptor cells are thought to respond to waves, currents, and disturbances in the water. The water moves the cupula and causes the hairs to bend. This results in messages being dispatched to the CNS.

Proprioceptors help coordinate muscle movement

Proprioceptors are sensory receptors that continually respond to tension and movement in muscles and joints. Vertebrates have three main types of proprioceptors: **muscle spindles,** which detect muscle movement (Fig. 41–4); **Golgi tendon organs,** which determine stretch in the tendons that attach muscle to bone; and **joint receptors,** which detect movement in ligaments. These are *tonic* (static) sensory receptors rather than phasic receptors. The receptor potential is maintained (though not at constant magnitude) as long as the stimulus is present, and action potentials continue to be generated. Thus, information about position is continuously supplied.

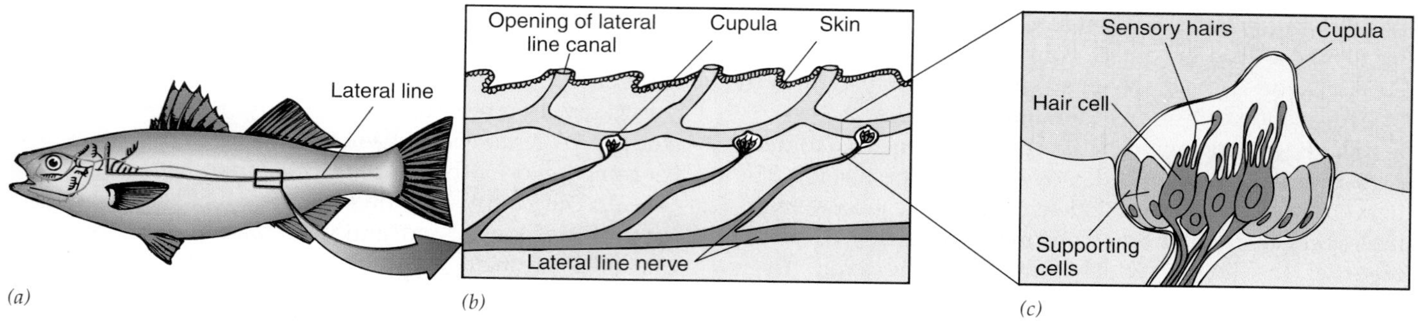

FIGURE 41–3 The receptor cells of the lateral line organ respond to waves, currents, or disturbances in the water. They inform the fish of obstacles or moving objects.

By means of these sensory receptors, we can carry out activities such as dressing or playing the piano even with our eyes closed. Impulses from the proprioceptors are important in ensuring the harmonious contractions of the distinct muscles involved in a single movement; without such receptors, complicated, skillful acts would be impossible. These organs are also important in maintaining balance.

Proprioceptors were discovered only a little more than 100 years ago. They are probably more numerous and more continually active than any of the other sensory receptors, although we are less aware of them than of most of the others. We obtain some idea of what life without proprioceptors would be like when a leg or arm "goes to sleep" and, partly due to lack of proprioception, grows numb.

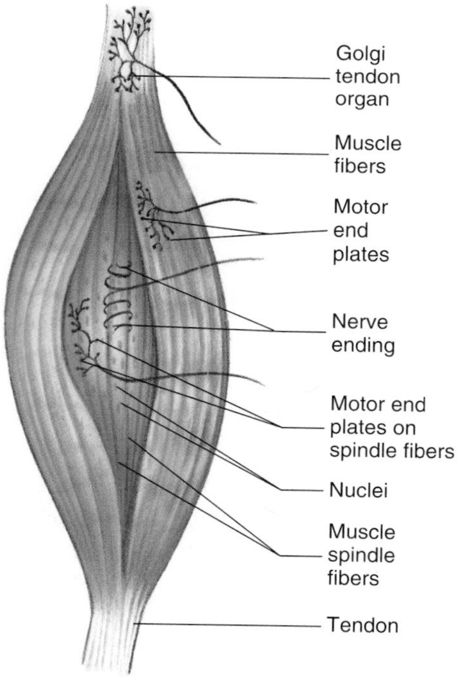

FIGURE 41–4 Proprioceptors respond to changes in movement, tension, and position in muscles and joints. Muscle spindles detect muscle movement; Golgi tendon organs determine stretch in tendons.

The mammalian muscle spindle, one of the more versatile stretch receptors, helps maintain muscle tone. It consists of a bundle of specialized muscle fibers, in the center of which is a region encircled by sensory nerve endings. These neurons respond statically and continue to transmit signals for a prolonged period of time, in proportion to the degree of stretch. Some of the nerve endings also exhibit a strong dynamic response to an increase in length, but only *while* the length is actually increasing.

The vestibular apparatus of the vertebrate ear maintains equilibrium

When we think of the ear, we think of hearing. However, in vertebrates its main function is to help maintain equilibrium. Typically the ear also contains gravity receptors. Although many vertebrates do not have outer or middle ears, all have inner ears.

The inner ear consists of a complicated group of interconnected canals and sacs, referred to as the **labyrinth,** which includes a membranous labyrinth that fits inside a bony labyrinth. In mammals the membranous labyrinth consists of two saclike chambers, the **saccule** and the **utricle,** and three **semicircular canals.**

Collectively, the saccule, utricle, and semicircular canals are referred to as the **vestibular apparatus** (Fig. 41–5). Destruction of these organs leads to a considerable loss of the sense of equilibrium. A pigeon whose vestibular apparatus has been destroyed cannot fly but in time can relearn how to maintain equilibrium using visual stimuli. Equilibrium in the human depends not only on stimuli from the organs in the inner ear but also on the sense of vision, stimuli from the proprioceptors, and stimuli from cells sensitive to pressure in the soles of the feet.

The saccule and utricle house gravity detectors in the form of small calcium carbonate ear stones called **otoliths** (Fig. 41–6). The sensory cells of these structures are sim-

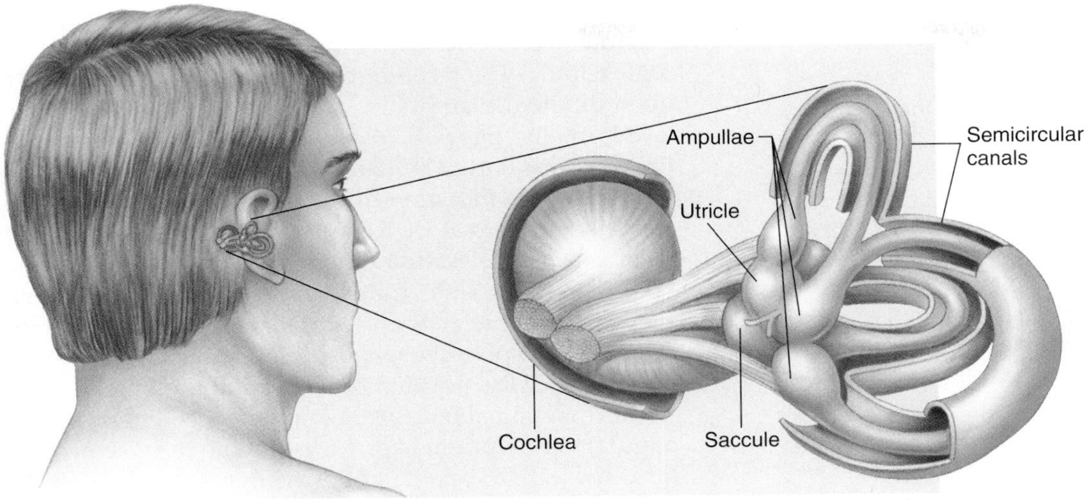

FIGURE 41–5 The human inner ear is illustrated here with the membranous labyrinth exposed. Because this is a posterior view, the utricle and saccule can be seen.

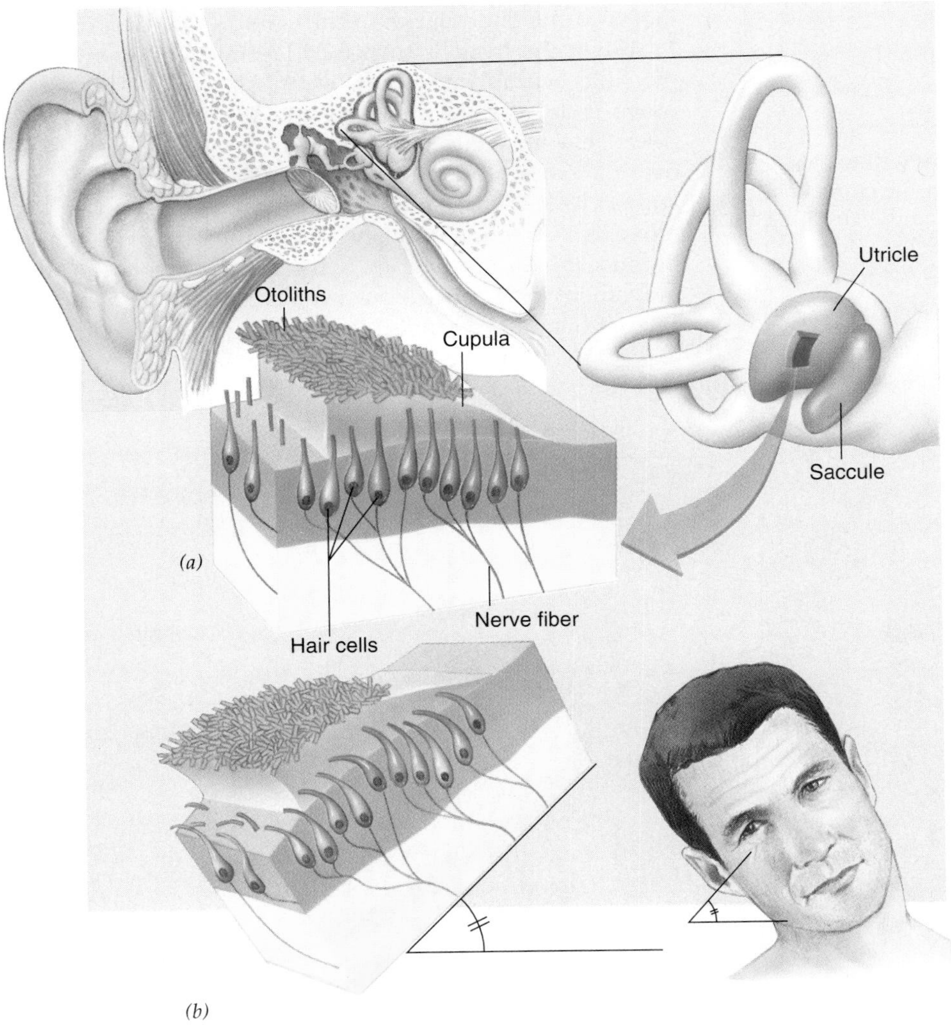

FIGURE 41–6 The saccule and utricle are part of the labyrinth. Compare the positions of the otoliths and hairs in (***a***) with those in (***b***). Changes in head position cause the force of gravity to distort the cupula, which in turn distorts the hairs of the hair cells. The hair cells respond by sending impulses down the vestibular nerve (part of the auditory nerve) to the brain.

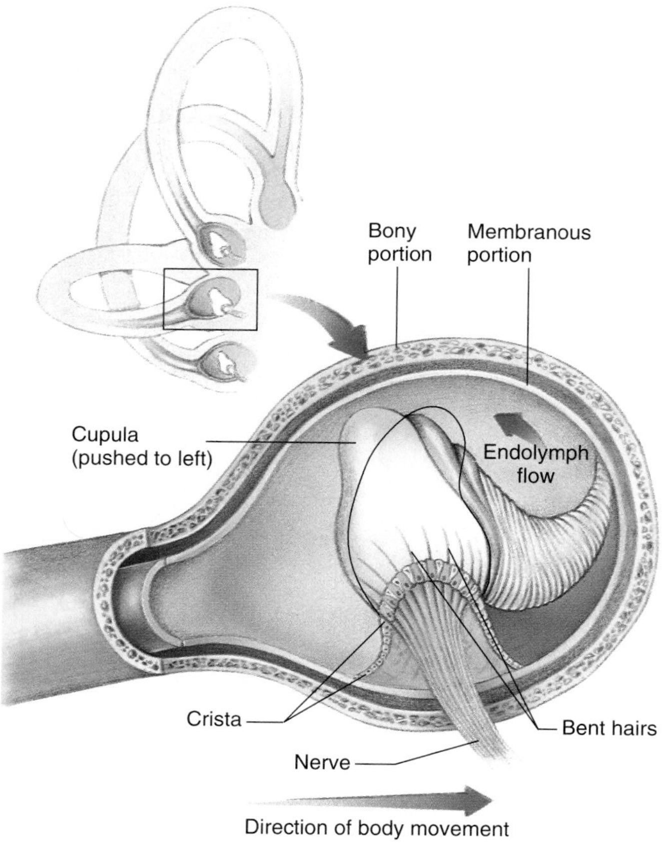

FIGURE 41–7 Movement of the endolymph within the ampulla distorts the cupula. The hair cells of the cupula are bent, and changes are reported to the brain via the vestibular nerve.

ilar to those of the lateral line organ. Each consists of a group of **hair cells** surrounded at their tips by a gelatinous cupula. The receptor cells in the saccule and utricle lie in different planes.

Normally, the pull of gravity causes the otoliths to press against particular hair cells, stimulating them to initiate impulses that are sent to the brain by way of sensory nerve fibers at their bases. When the head is tilted or in **linear acceleration** (an increase in speed when the body is moving in a straight line), the otoliths press on the hairs of other cells and stimulate them. This enables the animal to perceive its position relative to the ground regardless of the position of its head at the time.

Information about turning movements, referred to as **angular acceleration,** is provided by the three semicircular canals. Each canal, a hollow ring connected with the utricle, lies at right angles to the other two and is filled with fluid called **endolymph.** At one of the openings of each canal into the utricle is a small, bulblike enlargement, the **ampulla.** Within each ampulla lies a clump of hair cells called a **crista** (pl., *cristae*), similar to the groups of hair cells in the utricle and saccule but lacking otoliths. The receptor cells of the cristae are stimulated by movements of the endolymph in the canals (Fig. 41–7).

When the head is turned, a lag in the movement of the fluid within the canals occurs so that the hair cells move in relation to the fluid and are stimulated by its flow. This stimulation produces not only the consciousness of rotation but also certain reflex movements in response to it. These reflexes cause the eyes and head to move in a direction opposite that of the original rotation. Because the three canals are in three different planes, head movement in any direction stimulates fluid movement in at least one of the canals.

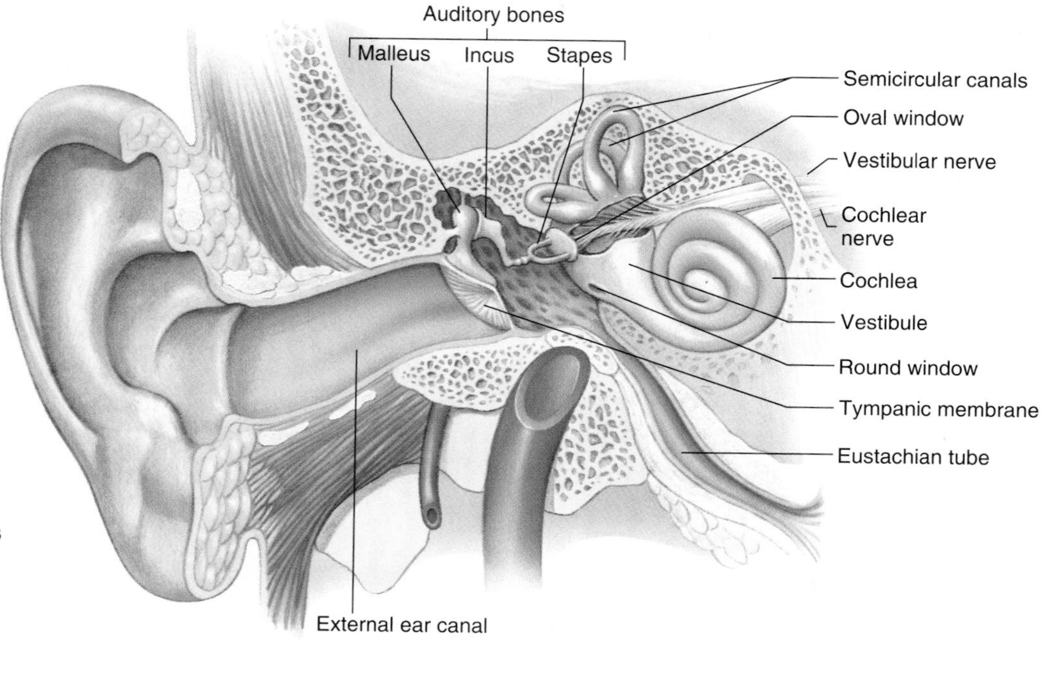

FIGURE 41–8 The human ear is structured to direct sound waves from outside the body to receptors in the inner ear.

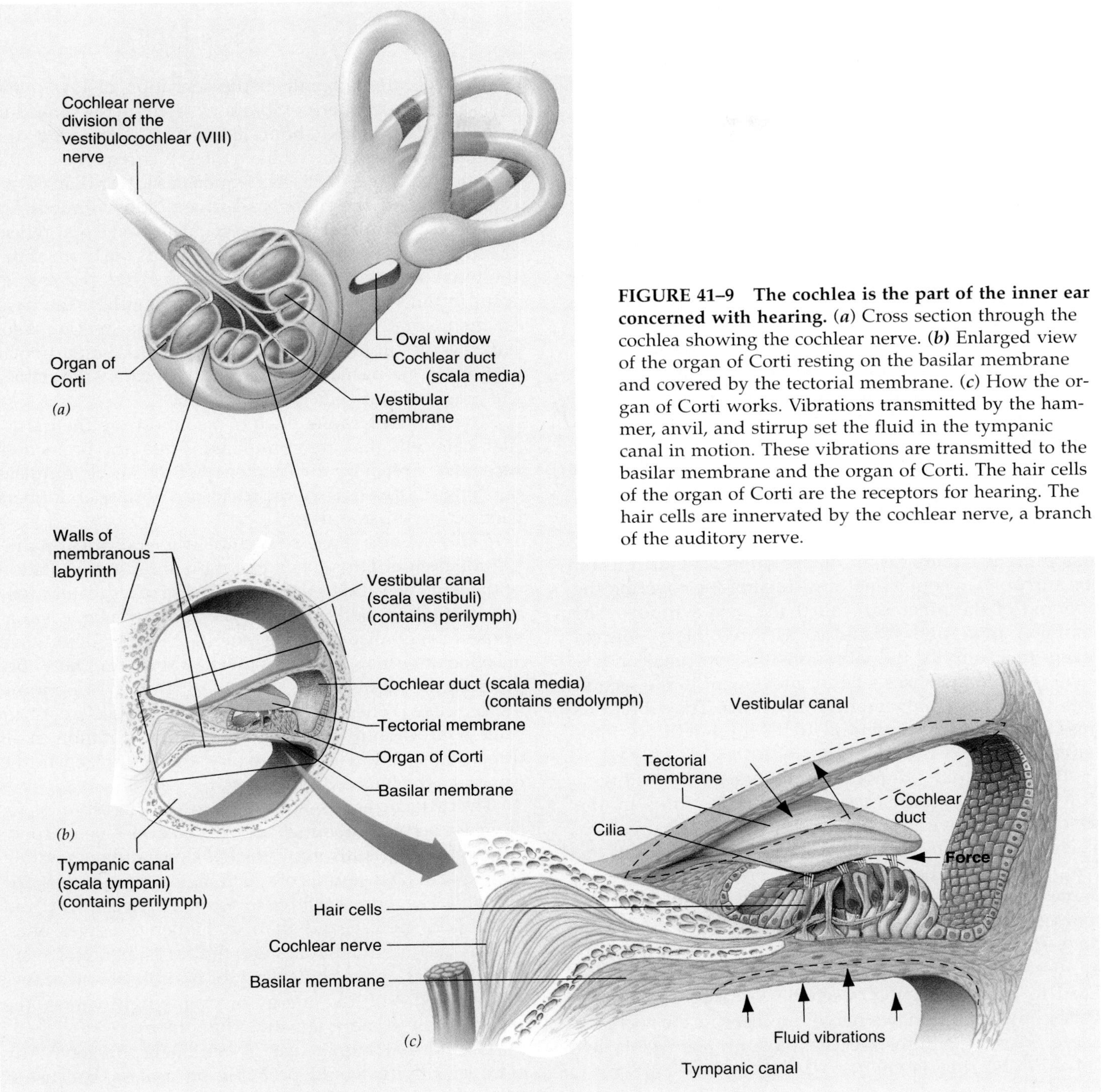

FIGURE 41–9 The cochlea is the part of the inner ear concerned with hearing. (*a*) Cross section through the cochlea showing the cochlear nerve. (***b***) Enlarged view of the organ of Corti resting on the basilar membrane and covered by the tectorial membrane. (*c*) How the organ of Corti works. Vibrations transmitted by the hammer, anvil, and stirrup set the fluid in the tympanic canal in motion. These vibrations are transmitted to the basilar membrane and the organ of Corti. The hair cells of the organ of Corti are the receptors for hearing. The hair cells are innervated by the cochlear nerve, a branch of the auditory nerve.

We humans are used to movements in the horizontal plane, but not to vertical movements (parallel to the long axis of the upright body). The motion of an elevator or of a ship pitching in a rough sea stimulates the semicircular canals in an unusual way and may cause seasickness or motion sickness, with resultant nausea or vomiting. When a seasick person lies down, the movement stimulates the semicircular canals in a more familiar way, and nausea is less likely to occur.

Auditory receptors are located in the cochlea

Many arthropods and most vertebrates have sound receptors, but for many of them hearing does not seem to be important. It is important in tetrapods, however, and both birds and mammals have a highly developed sense of hearing. Their auditory receptors, located in the **cochlea** of the inner ear, contain mechanoreceptor hair cells that detect pressure waves.

The cochlea is a spiral tube that resembles a snail's shell (Figs. 41–8 and 41–9). If the cochlea were uncoiled, it would be seen to consist of three canals separated from each other by thin membranes and coming almost to a point at the apex. Two of these canals, or ducts—the **vestibular canal** (also known as the *scala vestibuli*) and the **tympanic canal** (*scala tympani*)—are connected with one another at the apex of the cochlea and are filled with a fluid known as **perilymph.** The middle canal, the

cochlear duct (*scala media*), is filled with endolymph and contains the auditory organ, the **organ of Corti.**

Each organ of Corti contains about 18,000 hair cells arranged in rows that extend the length of the coiled cochlea. Each cell is equipped with hairlike projections that extend into the cochlear duct. The hair cells rest on the **basilar membrane,** which separates the cochlea from the tympanic canal. Overhanging and in contact with the hair cells of the organ of Corti is another membrane, the **tectorial membrane.** Movement of the hair cells initiates impulses in the fibers of the cochlear nerve.

In terrestrial vertebrates, accessory structures transform sound waves in the air into pressure waves in the cochlear fluid. In the human ear, for example, sound waves pass through the **external auditory canal** and set the **tympanic membrane,** or eardrum (the membrane separating the outer ear and the middle ear), vibrating. The vibrations are transmitted across the middle ear by three tiny bones, the **malleus, incus,** and **stapes** (or hammer, anvil, and stirrup—so called because of their shapes). The hammer is in contact with the eardrum, and the stirrup is in contact with the membrane covering the opening of the inner ear, called the **oval window.** The hammer, anvil, and stirrup act as three interconnected levers that amplify the vibrations. A very small movement in the first causes a larger movement in the second and a very large movement in the third. The vibrations pass through the oval window to the fluid in the vestibular canal.

Because liquids cannot be compressed, the oval window could not cause movement of the fluid in the vestibular duct without an escape valve for the pressure. This is provided by the **round window** at the end of the tympanic canal. The pressure wave presses on the membranes separating the three ducts, is transmitted to the tympanic canal, and causes a bulging of the round window. The movements of the basilar membrane produced by these pulsations cause the hair cells of the organ of Corti to rub against the overlying tectorial membrane. This stimulation initiates nerve impulses in the cochlear nerve. We can summarize the sequence of events involved in hearing as follows:

> sound waves enter external auditory canal → tympanic membrane vibrates → malleus, incus, and stapes vibrate → intensity of the vibrations is amplified → oval window vibrates → vibrations are conducted through fluid → basilar membrane vibrates → hair cells in the organ of Corti are stimulated → cochlear nerve transmits impulses to brain

Sounds differ in pitch, loudness, and tone quality. Pitch depends on frequency of sound waves. Low-frequency vibrations result in the sensation of low pitch, whereas high-frequency vibrations result in the sensation of high pitch. Up to about 4000 cycles per second, the nerve impulses produced have the same frequency as the sounds that cause them. At frequencies lower than 60 cycles per second, the entire basilar membrane vibrates. Action potentials are set up in the cochlear nerve that reflect the rhythm of the sound, and this informs the brain about the pitch. Frequencies greater than 60 cycles per second result in unequal vibration along the length of the basilar membrane. Sounds of a given frequency set up resonance waves in the cochlear fluid that cause a particular section of the basilar membrane to vibrate, stimulating the group of hair cells in that section.

The brain infers the pitch of a sound from the particular hair cells that are stimulated. Thus, the brain may recognize a pitch by the frequency of the nerve impulses reaching it as well as by the identities of the nerve fibers conducting the impulses.

Loud sounds cause resonance waves of greater amplitude (height); these lead to a more intense stimulation of the hair cells and to the transmission of a greater number of impulses per second by the auditory nerve. Variations in the quality of sound, such as those evident when an oboe, a cornet, and a violin play the same note, depend on the number and kinds of overtones, or harmonics, produced. These provide stimulation to different hair cells in addition to the main stimulation common to all three. Thus, differences in quality are recognized in the *pattern* of the hair cells stimulated.

The human ear is equipped to register sound frequencies between about 20 and 20,000 cycles per second, although individuals vary greatly. Dogs and some other animals can hear sounds of much higher frequencies. The human ear is more sensitive to sounds between 1000 and 4000 cycles per second than to higher or lower ones. Within this intermediate range, the ear is extremely sensitive. In fact, when the energy of audible sound waves is compared with the energy of visible light waves, the ear is ten times more sensitive than the eye.

The normal human ear is extremely efficient. Any more sensitivity would probably be useless, because it would enable us to pick up the random movement of air molecules, which we would perceive as a constant hiss or buzz. Similarly, if the eye were more sensitive, a steady light would appear to flicker because the eye would perceive the individual photons (light particles) striking it.

Deafness may be caused by injury to or malformation of either the sound-transmitting mechanism of the outer, middle, or inner ear, or the sound-perceiving mechanism of the inner ear. Intense sound can injure the organ of Corti. Members of rock bands and workers subjected to loud, high-pitched noises over a period of years frequently become deaf to high tones because the cells near the base of the organ of Corti become damaged.

Making the Connection

Chemoreception and Animal Behavior

How do chemical compounds affect the interaction of animals? Many species use chemoreception to communicate information such as danger, ownership of territory, and availability for mating. Insects, for example, use specific chemical compounds to communicate, defend themselves against predators, and recognize specific foods. Many vertebrates employ chemical secretions to mark territory, attract sexual partners, track prey, and defend themselves.

The chemical messengers are **pheromones,** small volatile molecules that are dispersed into the environment. For example, an ovulating female (one physiologically ready to mate) may release pheromones as part of her vaginal secretion. These chemical odors are detected by males, and information is transmitted from olfactory neurons to the brain. The effect may be increased sexual interest. Whether or not humans rely on pheromones is not known, but the importance of chemoreception in communication is evident by the use of colognes and perfumes to attract the attention of the opposite sex. ▲

CHEMORECEPTORS ARE ASSOCIATED WITH THE SENSES OF TASTE AND SMELL

Two highly sensitive chemoreceptive systems are the senses of **taste** (gustation) and **smell** (olfaction).

Taste buds detect dissolved food molecules

The organs of taste in mammals are **taste buds,** located in the mouth, mostly on the tongue. In humans, they are found mainly in tiny elevations, or papillae. Each of the 3000 or so taste buds on the human tongue is an oval epithelial capsule containing several taste receptors (Fig. 41–10). The tips of the taste receptor cells extend into a pore on the surface of the tongue. The taste receptors detect food molecules dissolved in saliva, setting up receptor potentials that cause action potentials in associated neurons.

Traditionally, four basic tastes have been recognized: sweet, sour, salty, and bitter. Although the greatest sensitivity to each of these tastes occurs in a given area of the human tongue (Fig. 41–10), not all papillae are restricted to a single category of taste. Flavor depends on the four basic tastes in combination with smell, texture, and tem-

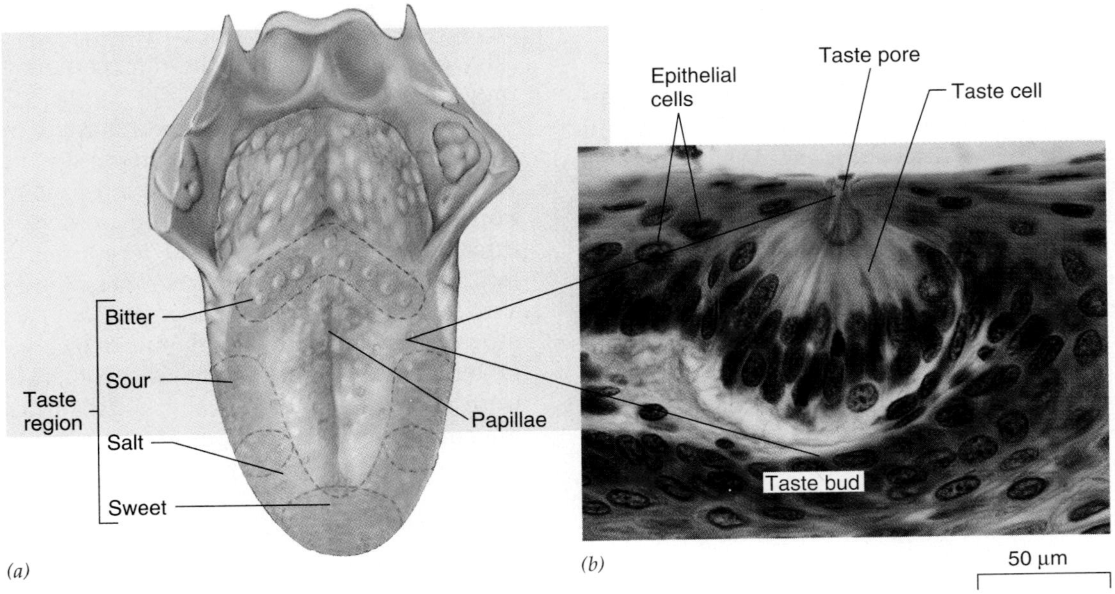

FIGURE 41–10 Taste buds (receptors) are located mainly on the surface of the tongue. (*a*) The surface of the tongue, showing distribution of taste buds which are sensitive to sweet, bitter, sour, and salt. A single taste receptor may respond to more than one category of taste. (***b***) A taste bud consists of an epithelial capsule containing several taste receptors. (*b*, Ed Reschke)

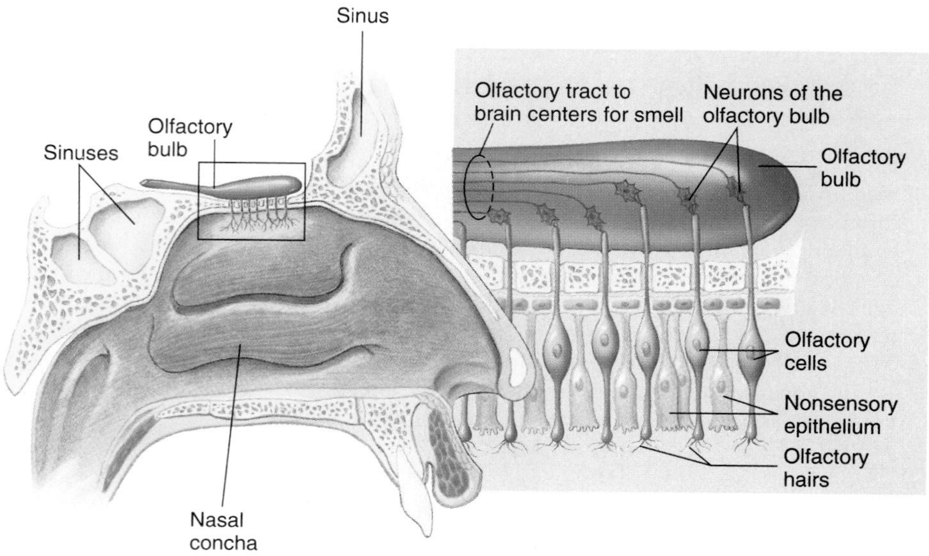

FIGURE 41–11 Smell depends on thousands of chemoreceptors located in the upper wall of the nasal cavities. The olfactory cells are neurons located in the olfactory epithelium.

perature. Smell affects flavor because odors pass from the mouth to the nasal chamber via the internal nares. No doubt you have observed that when your nose is congested, food seems to have little "taste." The taste buds are not affected, but the blockage of nasal passages severely reduces the participation of olfactory reception in the composite sensation of flavor.

The olfactory epithelium is responsible for the sense of smell

In terrestrial vertebrates, olfaction occurs in the nasal epithelium. In humans, the **olfactory epithelium** is found in the roof of the nasal cavity (Fig. 41–11). It contains about 100 million specialized olfactory cells with axons that extend upward as the fibers of the olfactory nerves. The end of each olfactory cell on the epithelial surface bears several olfactory hairs that are thought to react to odors (chemicals) in the air.

Mixtures of primary smell sensations (from 7 to 50 different olfactory stimulants have been hypothesized) are thought to produce the broad spectrum of odors that we are capable of perceiving. The olfactory organs respond to remarkably small amounts of a substance. For example, ionone, the synthetic substitute for the odor of violets, can be detected by most people when it is present in a concentration of only one part to more than 30 billion parts of air.

Despite its sensitivity, smell is perhaps the sense that adapts most quickly. The olfactory receptors adapt about 50% in the first second or so after stimulation, so even offensively odorous air may seem odorless after only a few minutes.

Thermoreceptors are sensitive to heat

Heat is another form of radiant energy to which organisms respond. Although not much is known about their specific thermoreceptors, many invertebrates are sensitive to changes in temperature. Mosquitoes, ticks, and other blood-sucking arthropods use thermoreception in their search for an endothermic host. Some have temperature receptors on their antennae that are sensitive to changes of less than 0.5°C. At least two types of snakes, pit vipers and boas, use thermoreceptors to locate their prey (Fig. 41–12).

In mammals, which are endothermic, free nerve endings and specialized receptors in the skin and tongue detect temperature changes in the outside environment. In humans, changes in temperature are detected by at least three types of receptors: cold receptors, warmth receptors, and pain receptors (for temperature extremes). (For a discussion on pain, see *Focus On*: *Pain Perception*.) Thermoreceptors in the hypothalamus detect internal changes in temperature and receive and integrate information from thermoreceptors on the body surface. The hypothalamus then initiates homeostatic mechanisms that ensure a constant body temperature.

Electroreceptors detect electrical currents in water

Electric organs are found in a few species of rays and bony fishes. Some of these animals have electroreceptors on the body that are linked with the neurons supplying the lateral line organs. Electroreceptors are used to detect

FIGURE 41–12 The pit organ of this bamboo viper (*Trimersurus stejnegeri*) is a sense organ located between each eye and nostril. The pit organ of many snakes can detect the heat from a warmblooded animal up to a distance of 1 to 2 meters. (Animals Animals © 1996 Zig Leszczynski)

electrical currents in the water. In species that produce a weak current, the electric organs help in orientation. This is particularly useful in murky water, where visibility and olfaction are poor. Some animals with electric organs can deliver powerful shocks that stun prey or enemies.

PHOTORECEPTORS USE PIGMENTS TO ABSORB LIGHT

All organisms respond to light. Most animals have photoreceptors that use pigments to absorb light energy. **Rhodopsins** are the photosensitive pigments found in the eyes of cephalopod mollusks, arthropods, and vertebrates. Light energy striking a light-sensitive receptor cell containing these pigments triggers chemical changes in the pigment molecules. As a result, the receptor cell may transmit a nerve impulse.

Eyespots, simple eyes, and compound eyes are found among invertebrates

The simplest light-sensitive structures in animals are found in certain cnidarians and flatworms (Fig. 41–13). They are **eyespots,** called **ocelli,** that detect light but do not see objects. Eyespots are often bowl-shaped clusters of light-sensitive cells within the epidermis. They may detect the direction of the source of light and distinguish light intensity.

Effective image formation requires a more complex **eye,** usually with a lens. A lens is a structure that concentrates light on a group of photoreceptors. Vision also requires a brain that can interpret the action potentials generated by the photoreceptors. The brain must integrate information about movement, brightness, location, position, and shape of the visual stimulus.

Two fundamentally different types of eyes evolved: the camera eye of vertebrates and some mollusks (squids and octopods), and the compound eye of the arthropods. (Vertebrate and mollusk eyes are analogous structures. They evolved independently of one another.)

Compound eyes found in crustaceans and insects are structurally and functionally different from vertebrate

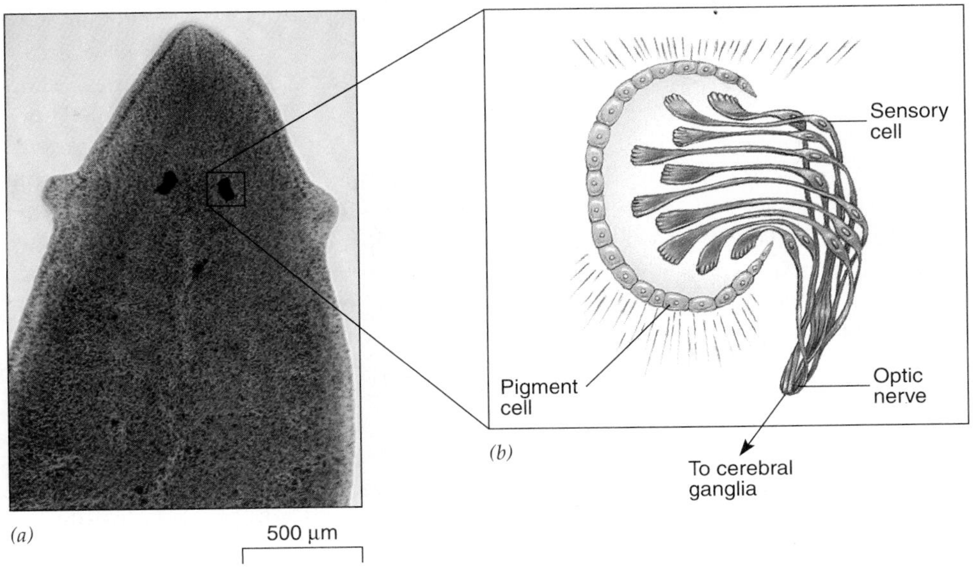

FIGURE 41–13 The simplest light-sensitive structures are the ocelli, or eyespots, found in certain invertebrates. (***a***) Planarian worm (*Planaria agilis*) showing eyespots. (***b***) Structure of the eyespot of a planarian worm. (Terry Ashley/Tom Stack & Associates)

Focus On

Pain Perception

Pain is a signal that protects us from harmful stimuli. An excess of any type of stimulus—pressure, heat, cold, excessive mechanical stretch, specific chemical compounds—stimulates pain receptors. In the human body, pain receptors are dendrites of certain sensory neurons found in almost every tissue.

When stimulated, pain receptors send a message through sensory neurons to the spinal cord. The message is transmitted to the opposite side of the spinal cord and then sent upward to the thalamus, where pain perception begins (see figure). From the thalamus, impulses are sent into the parietal lobes of the cerebrum. At that time the individual becomes fully aware of the pain and can evaluate the situation. How threatening is the stimulus? How intense is the pain? What can be done about it? From the thalamus, messages are also sent to a region in the limbic system, which is the brain's emotional center. Pain perception is colored by emotional experience.

Pain can be initiated or facilitated at many levels. The intensity of one's pain perception depends on the particular situation and on how one has learned to deal with pain. A child with a bruised knee may emotionally heighten the feeling of pain, whereas a professional fighter may virtually ignore a long series of well-delivered blows.

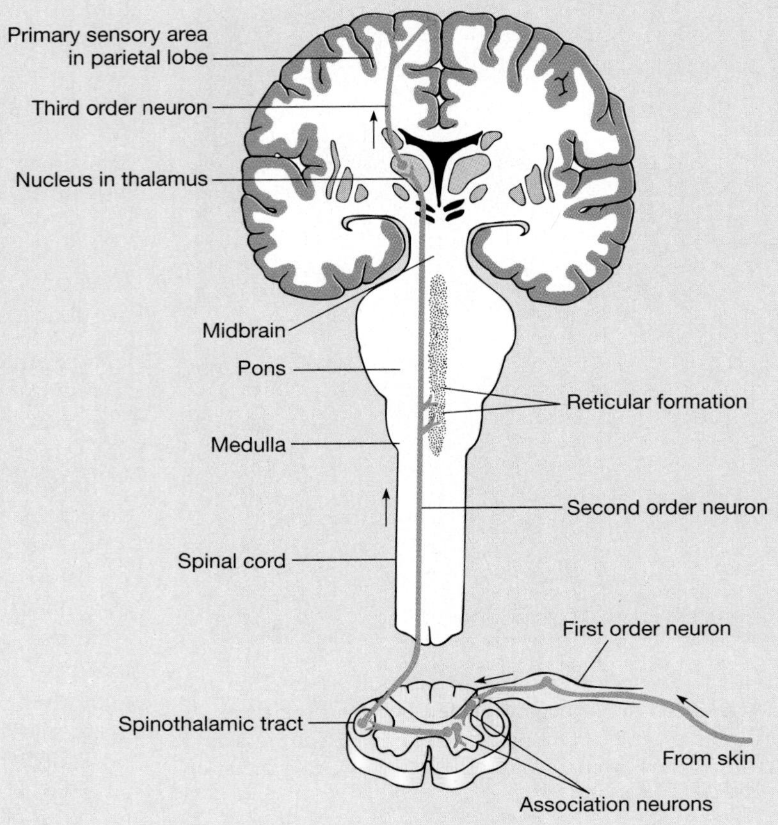

The brain locates pain on the basis of past experience. Generally, pain at the body surface is accurately projected back to the injured area. For example, when you step on a nail, your brain perceives the pain and then projects it back to the injured foot, so that you feel pain at the site of puncture. Artificial stimulation of the leg nerves may produce a sensation of pain in the foot even though the foot is untouched. In fact, for years after an amputation, a patient may feel **phantom pain** in the missing limb. This occurs

eyes (Fig. 41–14). The surface of a compound eye appears faceted, which means having many faces, like a diamond. Each **facet** is the convex cornea of one of the eye's visual units, called **ommatidia.** The number of ommatidia varies with the species. For example, the eyes of certain crustaceans have only 20 ommatidia apiece, whereas the eye of a dragonfly has as many as 28,000.

Each ommatidium has a transparent covering, the **cornea,** and a lens that focuses light onto a light-sensitive receptor, the **rhabdome.** Each receptor receives light from only a small portion of the visual field and forms a small, inverted image. In this way, the receptor samples the average light intensity from that area. All of the ommatidia together produce a composite image, or **mosaic** picture. A sheath of pigmented cells envelops each ommatidium.

Arthropod eyes usually adapt to different intensities of light. In nocturnal and crepuscular (dark- or dusk-loving) insects and many crustaceans, pigment is capable of migrating proximally and distally. When the pigment is in the proximal position, each ommatidium is shielded from its neighbor, and only light entering directly along its axis can stimulate the receptors. When the pigment is in the distal position, light striking at any angle can pass through several ommatidia and stimulate many retinal because, when the severed nerve is stimulated and sends a message to the brain, the brain "remembers" the nerve as it originally was—connected to the missing limb.

Most internal organs are poorly supplied with pain receptors. For this reason pain from internal structures is often difficult to locate. In fact, pain is often not projected back to the organ that is stimulated. Instead it is *referred* to an area just under the skin that may be some distance from the organ involved. The area to which the pain is referred generally is connected to nerve fibers from the same level of the spinal cord as the organ involved.

A person with angina who feels heart pain in the left arm is experiencing **referred pain.** The pain originates in the heart as a result of ischemia (insufficient blood in the blood vessels of the heart muscle) but is actually felt in the arm. One explanation is that neurons from both the heart and the arm converge on the same neurons in the central nervous system. The brain interprets the incoming message as coming from the body surface because somatic pain is far more common than pain from internal organs; the brain acts on the basis of past experience. When pain is felt both at the site of the distress and as referred pain, it may seem to spread, or *radiate,* from the organ to the superficial area.

The physiology of pain is not completely understood. Within the body, several different compounds stimulate pain receptors. These include histamine, serotonin, bradykinin, acids, and enzymes that break down proteins. The sensory neurons that transmit pain impulses to the spinal cord release the peptide neurotransmitter **substance P.**

The body has a variety of mechanisms for analgesia (pain control). Neurophysiologists have long known that opiates, such as morphine, relieve pain. They work by blocking the release of substance P. More than ten opiates have now been discovered in the brain, spinal cord, and pituitary gland. Among the more important ones are beta-endorphin (for "endogenous morphine-like"), enkephalins, and dynorphin. The opiates work by blocking the release of substance P. These compounds are more powerful analgesics than morphine, and several are currently being investigated as potential analgesic (pain-killing) drugs.

In the spinal cord, certain neurons release serotonin, which acts on a set of interneurons that synapse with pain neurons. These interneurons release enkephalin, which is thought to cause presynaptic inhibition of incoming pain-transmitting neurons. Enkephalin may block calcium channels, preventing release of substance P. In this way the body can block pain signals when they first enter the spinal cord.

Many areas of the brain have opiate receptors, so pain signals that reach the brain can be blocked by beta-endorphin, enkephalins, and dynorphins. The neurotransmitter GABA (gamma aminobutyric acid) is also thought to inhibit release of substance P in some areas of the brain.

Some neurobiologists think that endorphins may explain the mechanism of action of acupuncture. For thousands of years acupuncture has been used to relieve pain, but how it works has remained a mystery. There is now some evidence that acupuncture needles stimulate nerves deep within the muscles, which in turn stimulate the pituitary gland and parts of the brain to release endorphins.

You may have observed that rubbing the skin around an injury can alleviate pain. Activating certain large sensory neurons that deliver messages about touch or pressure can stimulate the interneurons in the spinal cord that release enkephalin. Acupuncture may also stimulate these large sensory neurons.

Clinical methods have been developed for relieving pain by electrical stimulation of large sensory nerve fibers. Stimulation of the skin over a painful area with electrodes has successfully relieved pain in some patients. This procedure is called transcutaneous electrical nerve stimulation. In a few patients, electrodes have been implanted in the brain itself, allowing the patient to control chronic pain by stimulating the release of endorphins. ■

units. As a result, sensitivity is increased in dim light, and the eye is protected from excessive stimulation in bright light. Pigment migration is under neural control in insects and under hormonal control in crustaceans. In some species it follows a daily rhythm.

Compound eyes do not perceive form well. Although the lens system of each ommatidium is adequate to focus a small inverted image, there is little evidence that they are actually perceived as images by the organism. However, all the ommatidia together do produce a composite image. Each ommatidium, in gathering a point of light from a narrow sector of the visual field, is in fact sampling a mean intensity from that sector. All of these points of light taken together form a mosaic picture.

To appreciate the nature of this mosaic picture, we need only look at a newspaper photograph through a magnifying glass; it is a mosaic of many dots of different intensities. The clarity and definition of the picture depend on how many dots there are per unit area—the more dots, the better the picture. So it is with the compound eye. The image perceived by the animal is probably much better in quality than might be suspected from the structure of the compound eye. The nervous system of an insect is apparently capable of image processing

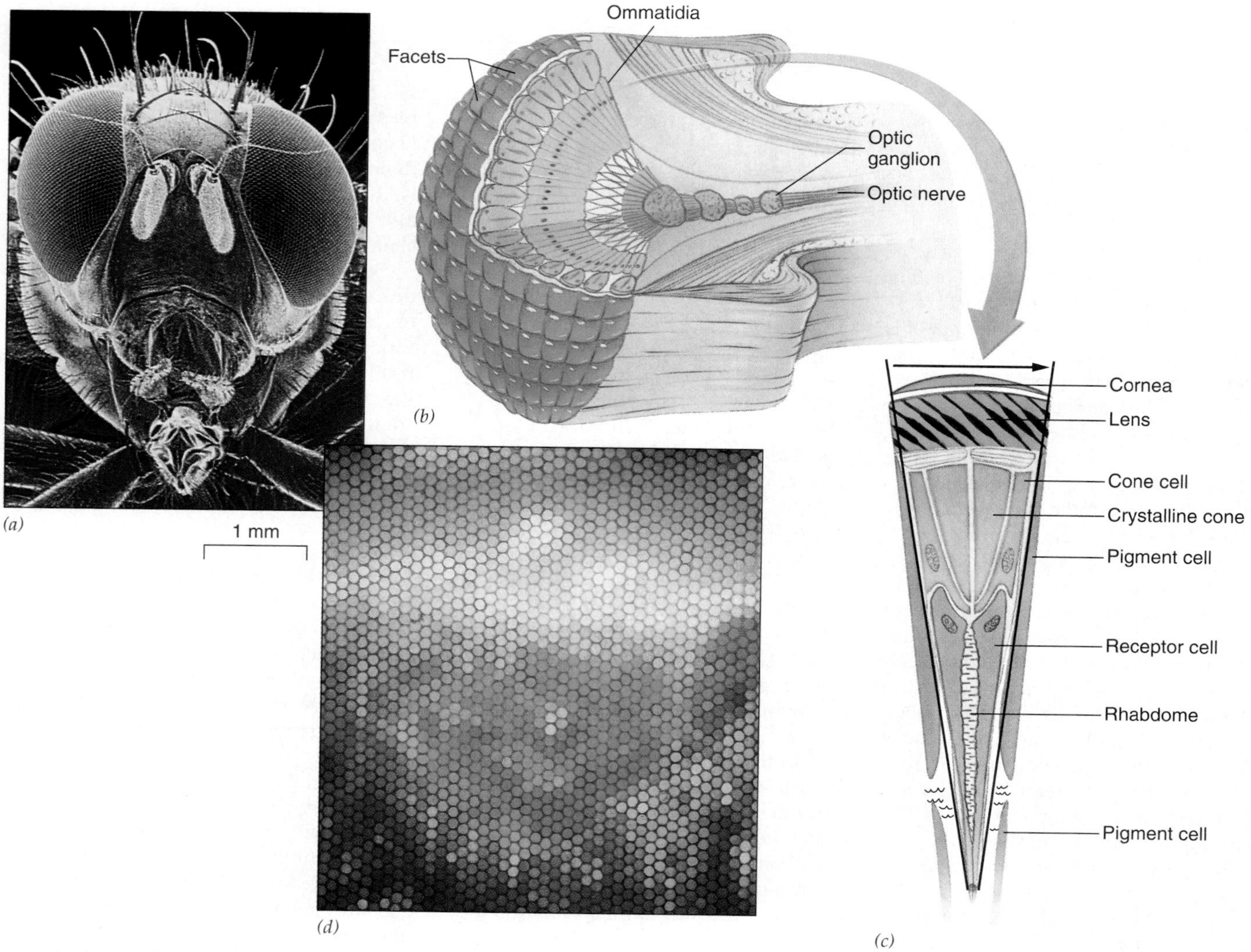

FIGURE 41–14 Compound eyes are found in crustaceans and insects. (***a***) The Mediterranean fruit fly (*Ceratitis capitata*) has prominent compound eyes. (***b***) Structure of the compound eye showing several ommatidia. This type of eye registers changes in light and shade, permitting the animal to detect movement. (***c***) Structure of an ommatidium. The rhabdome is the light-sensitive core of the ommatidium. (***d***) A bee eye's view of poppies. (*a,* David Scharff/Peter Arnold, Inc.; *d,* John Lithgoe/Seaphot, Ltd.)

similar to that employed to improve the quality of photographs sent to Earth by robot spacecraft.

Although the compound eye can form only coarse images, it compensates by being able to follow flickers to higher frequencies. Flies are able to detect up to about 265 flickers per second. In contrast, the human eye can detect only 45 to 53 flickers per second; for us, flickering lights fuse above these values, so we see motion pictures as smooth movement and light provided by an ordinary bulb as steady. To an insect, both motion pictures and room lighting must flicker horribly. The advantage of the insect's high critical flicker fusion threshold is that it permits immediate detection of any movement by prey or enemy. The compound eye is an important adaptation to the arthropod's way of life.

Compound eyes differ from our eyes in another respect. They are sensitive to wavelengths of light in the range from red to ultraviolet (UV). Accordingly, an insect can see UV light well, and its world of color is very different from ours. Because flowers deflect UV light to various degrees, two flowers that appear identically colored to us may appear strikingly dissimilar to insects (see Fig. 35–3).

Vertebrate eyes form sharp images

The position of the eyes in the head of humans and certain other higher vertebrates permits both to be focused on the same object (Fig. 41–15). This **binocular vision** is an important factor in judging distance and depth. Vari-

(*a*)

(*b*)

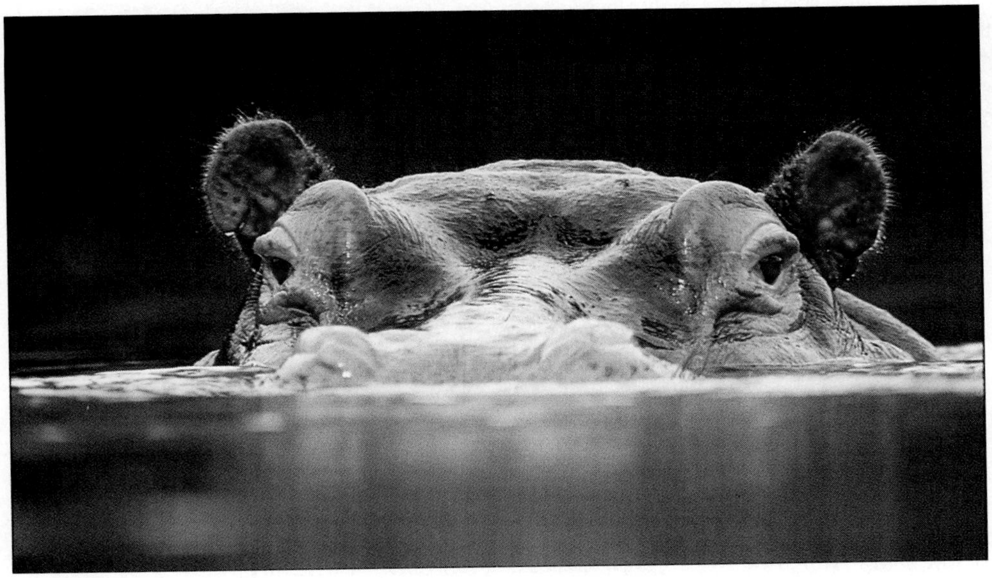

(*c*)

FIGURE 41–15 The position of the eyes varies in different vertebrates, resulting in differences in vision. (***a***) The eyes of the zebra are positioned laterally, enabling the animal to see on both sides. Even while grazing, it can spot a predator approaching from behind. (***b***) Like many other nocturnal animals, the owl monkey (*Aotus evingatus*) has large eyes. Its eyes are positioned at the front of the head, and it has binocular vision, permitting it to judge distances. (***c***) The orbits (bony cavities that contain the eyeballs) of the hippopotamus are elevated, enabling the animal to see even when most of its head is under water. (*a*, Diane Blell/Peter Arnold, Inc.; *b*, Animals Animals © 1996 Stephen Dalton; *c*, Frans Lanting/Minden Pictures)

ations of eye position in other vertebrates offer different advantages. For example, the eyes of grazing animals are positioned laterally, enabling them to spot a predator approaching from behind.

The vertebrate eye can be compared to a camera. An adjustable lens can be focused for different distances, and a diaphragm, called the **iris,** regulates the size of the light opening, called the **pupil** (Fig. 41–16). The **retina** corresponds to the light-sensitive film used in a camera. Outside the retina is the **choroid layer,** a sheet of cells filled with black pigment that absorbs extra light and prevents internally reflected light from blurring the image. (Cameras are also black on the inside.)

The outer coat of the eyeball, called the **sclera,** is a tough, opaque, curved sheet of connective tissue that protects the inner structures and helps to maintain the rigidity of the eyeball. On the front surface of the eye, this sheet becomes the thinner, transparent **cornea,** through which light enters. The cornea serves as a fixed lens.

The lens of the eye is a transparent, elastic ball just behind the iris. It bends the light rays coming in and brings them to a focus on the retina. The lens is aided by the curved surface of the cornea and by the refractive properties (ability to bend light rays) of the liquids inside the eyeball. The **anterior cavity** between the cornea and the lens is filled with a watery substance, the **aqueous fluid.**

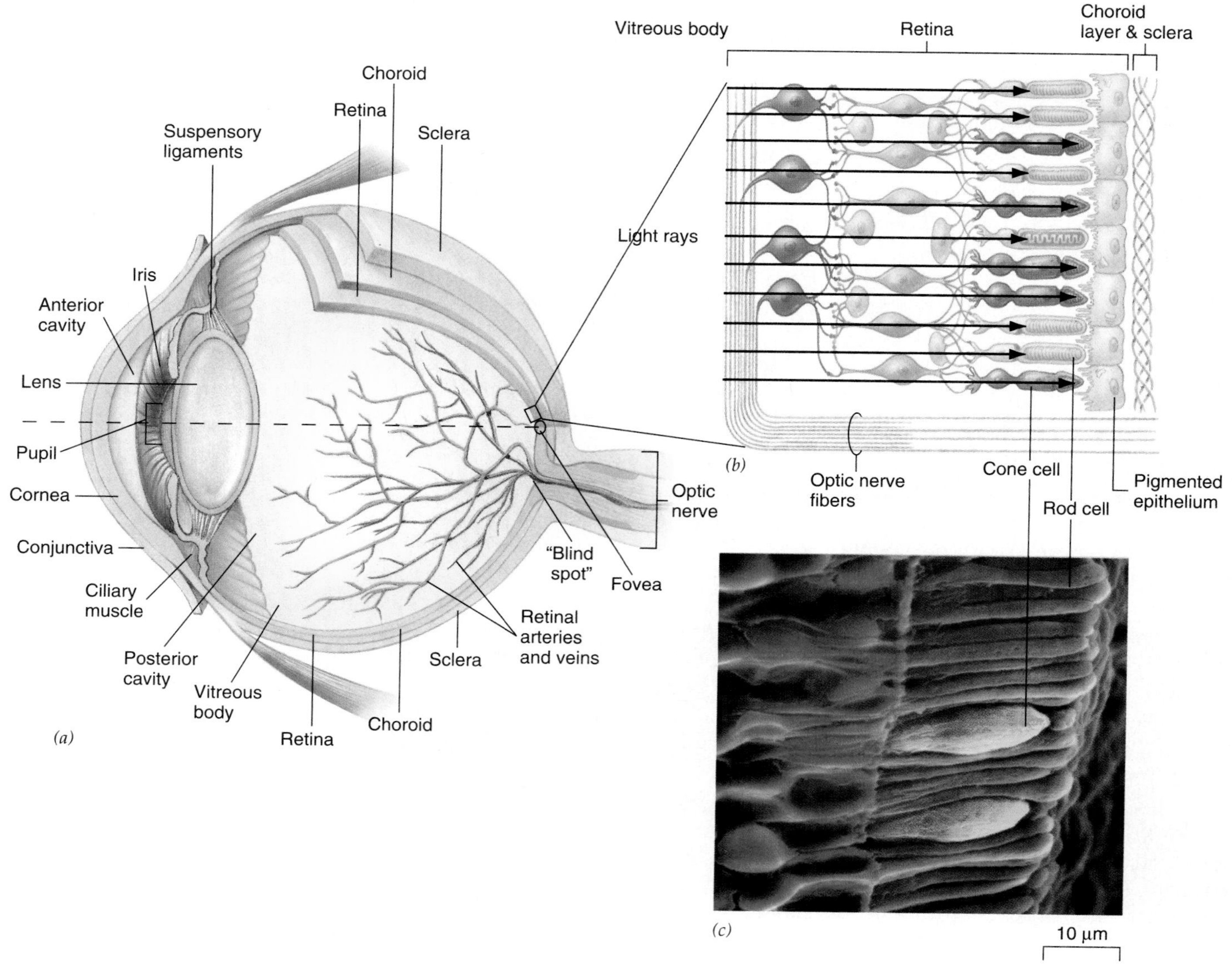

FIGURE 41–16 Light passes through the human eye to photoreceptor cells in the retina. (*a*) In this lateral view, the eye is shown partly sectioned along the sagittal plane to show its internal structures. (*b*) The elaborate interconnections among the various layers of neurons in the retina allow them to interact and to influence one another in a number of ways. The rods and cones are the photoreceptor cells. (*c*) Rods (red elongated structures) and two cones (shorter, thicker, yellow structures) are seen in this micrograph. The elongated rods permit us to see shape and movement, whereas the shorter cones allow us to view our world in color. (Lennart Nilsson, from *The Incredible Machine*, p. 279)

The larger **posterior cavity** between the lens and the retina is filled with a more viscous fluid, the **vitreous body.** Both fluids are important in maintaining the shape of the eyeball by providing an internal fluid pressure.

At its anterior margin, the choroid is thick and projects medially into the eyeball to form the **ciliary body,** which consists of ciliary processes and the ciliary muscle. The ciliary processes are glandlike folds that project toward the lens and secrete the aqueous fluid.

The eye has the power of **accommodation,** meaning it can change focus for near or far vision by changing the curvature of the lens. This is accomplished by the **ciliary muscle,** a part of the ciliary body that is attached to the lens by the suspensory ligaments. When the eye is at rest, the ciliary muscle is relaxed and the ligaments are tense. This pulls the lens into a flattened (ovoid) shape, focusing the eye for far vision. When the ciliary muscle contracts, tension on the suspensory ligaments lessens. Then the elastic lens assumes a rounder shape for near vision.

In *nearsightedness* (*myopia*), the eyeball is elongated. The light rays converge at a point in front of the retina and are diverging again when they reach it. This results in a blurred image. Concave lenses correct for the nearsighted condition by bringing the light rays to a focus at a point farther back. In a *farsighted* eye, the eyeball is too short and the retina too close to the lens. Light rays strike the retina before they have converged, again resulting in a blurred image. Convex lenses correct for the farsighted condition by causing the light rays to converge farther forward.

The amount of light entering the eye is regulated by the iris, a ring of smooth muscle that appears as blue, green, gray, or brown depending on the amount and nature of pigment present. The iris is composed of two mutually antagonistic sets of muscle fibers. One set is arranged circularly and contracts to *decrease* the size of the pupil. The other is arranged radially and contracts to *increase* the size of the pupil. The response of these muscles to a change in light intensity requires from 10 to 30 seconds. Thus, when a person steps from a light to a dark area, some time is needed for the eyes to adapt, and when a person steps from a dark room to a brightly lighted area, the eyes are dazzled until the size of the pupil is decreased. The retina of the eye can also adapt to changes in light intensity.

Each eye has six muscles that extend from the surface of the eyeball to various points in the bony socket. These muscles enable the eye as a whole to move and be oriented in a given direction. Cranial nerves innervate the muscles in such a way that the eyes normally move together and focus on the same area.

The retina contains light-sensitive rods and cones

The light-sensitive structure in the vertebrate eye is the retina, which lines the posterior two thirds of the eyeball, covering the choroid. The retina contains abundant photoreceptor cells called, according to their shapes, **rods** and **cones.** The human eye has about 125 million rods and 6.5 million cones. Rods function in dim light, allowing us to detect shape and movement. They are not sensitive to colors. Because the rods are more numerous in the periphery of the retina, you can see an object better in dim light if you look slightly to one side of it (allowing the image to fall on the rods).

Cones are responsible for bright-light vision, for the perception of fine detail, and for color vision. They permit color vision by being differentially sensitive to different frequencies (colors) of light. The cones are concentrated in the **fovea,** a small depressed area in the center of the retina. The fovea is the region of sharpest vision.

Only those objects more or less directly in front of our eyes can be perceived in color. This can be demonstrated by a simple experiment. Close one eye and focus the other on some point straight ahead. As a colored object is gradually brought into view from the side, you will be aware of its presence and of its size and shape before you are aware of its color. Only when the object is brought closer to the direct line of vision, so that its image falls on a part of the retina containing cones, can its color be determined.

Light must pass through several layers of connecting neurons in the retina to reach the rods and cones (Fig. 41–16). The axons of the sensory neurons extend across the surface of the retina and unite to form the **optic nerve,** which then passes out of the eyeball. This area is called the "blind spot"; because it lacks rods and cones, images falling on it cannot be perceived.

In summary, vision involves the following sequence of events:

> light passes through cornea → through aqueous fluid → through lens → through vitreous body → image forms on retina → optic nerve transmits nerve impulses to visual areas of the cerebral cortex

A chemical change in rhodopsin leads to the response of a rod to light

Rhodopsin in the rod cells and some very closely related pigments in the cone cells are responsible for the ability to see. Rhodopsin consists of opsin, a polypeptide that is chemically joined with **retinal,** a pigment made from vitamin A. Two isomers of retinal exist: the *cis* form and the *trans* form.

When light strikes rhodopsin, it transforms *cis*-retinal to *trans*-retinal (Fig. 41–17). This change in shape causes rhodopsin to break down into its components, opsin and retinal. During this process the rhodopsin is converted into a series of intermediate compounds. These reactions result in a voltage change across the plasma membrane of the rod or cone cell. The photoreceptors determine whether action potentials occur in neighboring cells. Neural signals are transmitted through various types of neurons in the retina, eventually resulting in transmission of impulses to the brain by the optic nerves.

Just how visual images are processed is not certain. The size, intensity, and location of light stimuli determine initial processing in the retina. The pattern of neuron firing in the retina appears to be very important. The optic nerves are thought to transmit information to the brain by way of complex, encoded signals. The optic nerves cross in the floor of the hypothalamus, forming an X-shaped structure, the *optic chiasma.* Many of the axons of the optic nerves cross over and then extend to the opposite side of the brain. Axons of the optic nerves end in the thalamus. From there, neurons convey information to the visual cortex in the cerebrum.

Color vision depends on three different types of cones

Three different types of cones, each containing a different cone pigment, function in color vision. They are commonly referred to as blue, green, or red cones. Each type of cone can respond to light within a considerable range

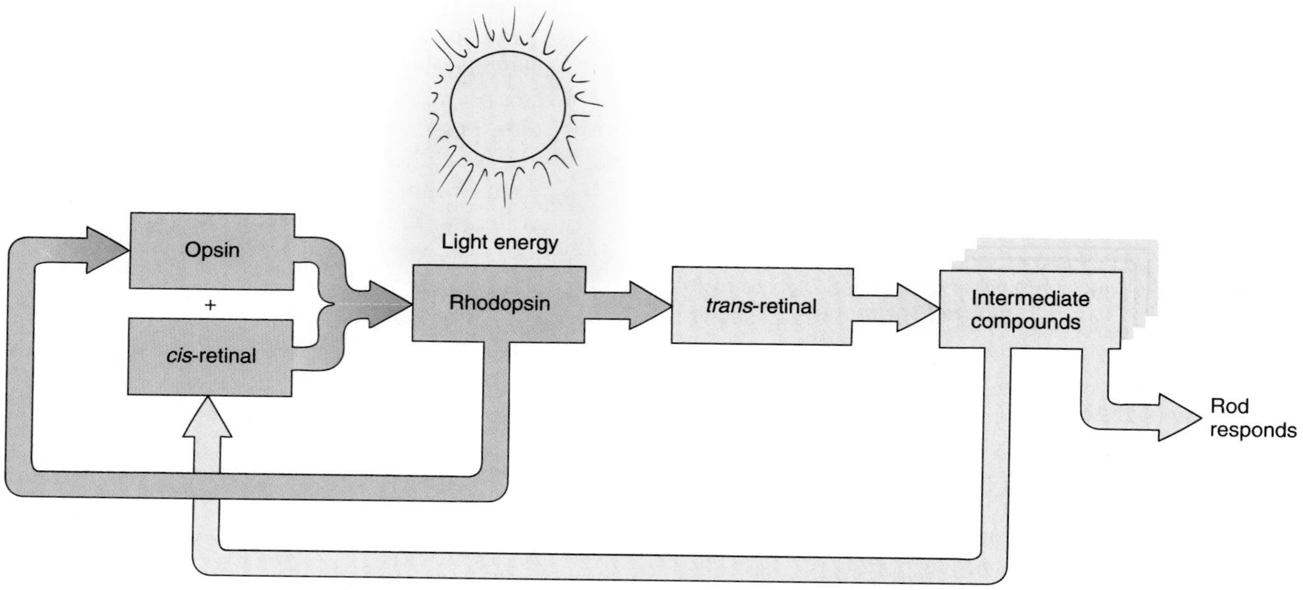

FIGURE 41–17 When light strikes rhodopsin, it breaks down, activating a series of chemical reactions. These events lead to response by the rod. See text for further explanation.

of wavelengths, but is named for the wavelength its pigment responds to most strongly. For example, red light can be absorbed by all three types of cones, but those cones most sensitive to red act as red receptors. By comparing the relative responses of the three types of cones, the brain can detect colors of intermediate wavelengths.

Color blindness occurs when one or more of the three types of cones is absent. This is usually an inherited X-linked condition (see Chapter 10, Fig. 10–13).

SUMMARY

I. Sensory receptors are specialized to respond to specific energy stimuli in the environment.
   A. Sensory receptors may be neuron endings or specialized cells in close contact with neurons.
   B. Sense organs are composed of receptor cells and accessory cells.

II. Exteroceptors are sensory receptors that receive information from the outside world. Proprioceptors are sensory receptors within muscles, tendons, and joints; they enable the animal to perceive orientation of the body and the positions of its parts. Interoceptors are sensory receptors within body organs.

III. Based on the types of stimuli to which they respond, sensory receptors also may be classified as mechanoreceptors, chemoreceptors, photoreceptors, thermoreceptors, or electroreceptors.

IV. Receptor cells absorb energy, transduce that energy into electrical energy, and produce receptor potentials.

V. Adaptation of a receptor to a continuous stimulus results in diminished perception.

VI. Mechanoreceptors respond to touch, pressure, gravity, stretch, or movement.
   A. The tactile receptors in the skin are mechanoreceptors that respond to mechanical displacement of hairs or of the receptor cells themselves.
   B. Statocysts are gravity receptors found in many invertebrates.
   C. Lateral line organs supplement vision in fish and some amphibians by informing the animal of moving objects or objects in its path.
   D. Muscle spindles, Golgi tendon organs, and joint receptors are proprioceptors that continually respond to tension and movement in the muscles and joints.
   E. The saccule and utricle of the vertebrate ear contain otoliths that change position when the head is tilted or when the body is moving in a straight line. The hair cells stimulated by otoliths send impulses to the brain, enabling the animal to perceive the direction of gravity.
   F. The semicircular canals of the vertebrate ear inform the brain about turning movements. Their cristae are stimulated by movements of the endolymph.
   G. In birds and mammals, the organ of Corti within the cochlea contains auditory receptors.

1. Sound waves pass through the external auditory canal, cause the eardrum to vibrate, and are transmitted through the middle ear by the hammer, anvil, and stirrup.
2. Vibrations pass through the oval window to fluid within the vestibular duct. Pressure waves press on the membranes that separate the three ducts of the cochlea.
3. Movements of the basilar membrane stimulate the hair cells of the organ of Corti by rubbing them against the overlying tectorial membrane.
4. Nerve impulses are initiated in the dendrites of the auditory neurons that lie at the base of each hair cell.

VII. Chemoreceptors include receptors for taste and smell.
   A. Taste receptors are specialized epithelial cells in taste buds.
   B. The olfactory epithelium contains specialized olfactory cells with axons that extend upward as fibers of the olfactory nerves.

VIII. Thermoreceptors are important in endothermic animals because they provide cues about body temperature. In some invertebrates they are used to locate an endothermic host.

IX. Photoreceptors in very simple eyes detect light, but such eyes do not form images effectively. Effective image formation and interpretation are called vision.
   A. The compound eye found in insects and crustaceans consists of ommatidia, which collectively produce a mosaic image.
   B. In the human eye, light enters through the cornea, is focused by the lens, and produces an image on the retina. The iris regulates the amount of light that can enter.
   C. When light strikes rhodopsin in the rod cells, a chemical change in retinal breaks down the rhodopsin, leading to a change in the rod cell.
   D. Rods form images in black and white, whereas cones permit color vision.

SELECTED KEY TERMS

ampulla
basilar membrane
chemoreceptor
cochlea
compound eye
cones
cornea
electroreceptor
endolymph
exteroceptor
eyespot
fovea
Golgi tendon organ
hair cell
interoceptor
iris
joint receptor
labyrinth
lateral line organ
lens
mechanoreceptor
muscle spindle
olfactory epithelium
ommatidium (ommatidia)
organ of Corti
otolith
Pacinian corpuscle
photoreceptor
proprioceptor
receptor potential
retina
rhodopsin
rods
saccule
semicircular canal
sensory adaptation
sensory receptor
statocyst
taste bud
tectorial membrane
thermoreceptor
utricle
vestibular apparatus

POST-TEST

1. A(an) __________ __________ detects changes in the environment and transmits that information into the nervous system.
2. __________ are sensory receptors that receive stimuli from the outside environment; __________ enable an animal to perceive orientation of the body.
3. __________-receptors detect light energy; __________-receptors respond to touch, gravity, or movement.
4. Receptor cells absorb and transduce __________ and produce a(an) __________ __________.
5. The diminishing response of a receptor to a continued, constant stimulus is called sensory __________.
6. In many invertebrates, __________ serve as gravity receptors; their action depends on mechanical displacement of receptor cell hairs by change in position of a statolith.
7. The __________ __________ organ of fishes is thought to supplement vision.
8. Three main types of vertebrate proprioceptors are __________ __________; which detect muscle movement; __________ tendon organs, which determine tendon stretch; and __________ receptors, which detect ligament movement.
9. The Pacinian __________ responds to deep pressure.
10. The inner ear of a vertebrate consists of interconnected canals and sacs called the __________; in jawed vertebrates this structure consists of two saclike chambers, the __________ and __________, and three __________ canals.
11. The rocks in your head (within the saccule and utricle), called __________, are actually gravity detectors.
12. Each semicircular canal is filled with fluid called __________; at one of the openings of each canal into the utricle is a small enlargement, the __________.
13. The __________, located in the inner ear, contains mechanoreceptor hair cells that detect sound waves.
14. The auditory receptors are located within the organ of __________.
15. The senses of taste and smell depend on __________-receptors.
16. The principal photosensitive pigment in the eyes of vertebrates is __________.
17. The light-sensitive part of the human eye is the __________.
18. The __________ regulates the size of the pupil.
19. __________ are responsible for color vision.
20. The region of keenest vision is the __________.
21. The visual units of a compound eye are __________.

REVIEW QUESTIONS

1. Contrast mechanoreceptors with chemoreceptors.
2. Which sensory receptors permit us to perform actions such as getting dressed or finding our way into bed when our eyes are closed? Explain how they work.
3. What is the physiological basis of seasickness?
4. Draw a diagram of the human eye, labeling all parts. How are rods and cones distributed in the retina?
5. Discuss the mechanism by which photoreceptors are stimulated by light. What is the function of rhodopsin? How is it regenerated?
6. How does the human eye adjust to near and far vision and to bright and dim light?
7. Draw a diagram of the human ear, labeling all parts.
8. Discuss the mechanism by which the sensory cells of the ear are stimulated by sound waves.
9. What are otoliths, and what is their role in maintaining equilibrium?
10. Contrast the function of the insect's compound eye with that of the vertebrate eye.
11. Explain the statement "Vision happens mainly in the brain."
12. Label the following diagrams.

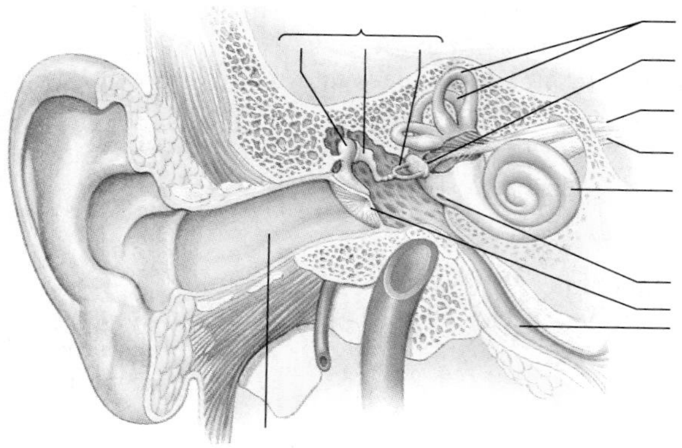

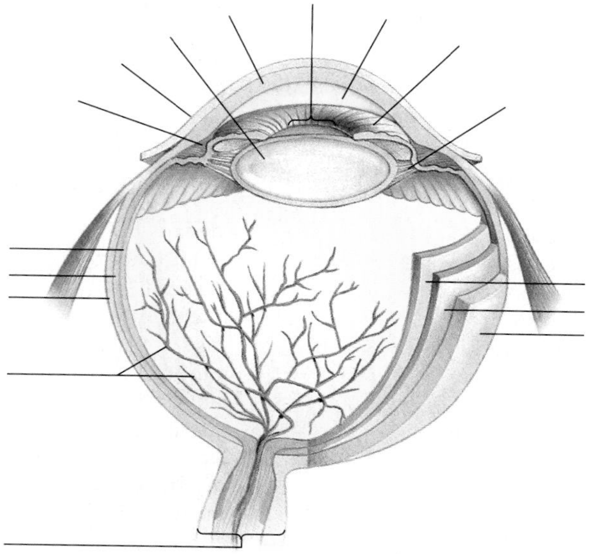

YOU MAKE THE CONNECTION

1. If all neurons transmit the same type of message, how do we know the difference between sound and light? How are we able to distinguish between an intense pain and a mild one? How are these discriminations adaptive?
2. Connoisseurs can recognize many varieties of cheese or wine by tasting. How can this be, when there are only four types of taste receptors? How might it be adaptive for organisms to have only four or five basic tastes?

RECOMMENDED READINGS

Borg, E., and S.A. Counter. "The Middle-Ear Muscles." *Scientific American,* Vol. 261, No. 2, August 1989. The neuromuscular control system of the middle ear prevents sensory overload and enhances sound discrimination.

DISCOVER Special Issue, "The Mystery of Sense," Vol. 14, No. 6, June 1993. Several articles discussing vision, hearing, touch, smell, and taste.

Kirchner, W.H., and W.F. Towne. "The Sensory Basis of the Honeybee's Dance Language." *Scientific American,* Vol. 270, No. 6, June 1994. Bees can hear the sounds produced by the dances they use to communicate.

Long, M.E. "The Sense of Sight." *National Geographic,* Vol. 182, No. 5, November 1992. A beautifully illustrated discussion of vision and new technology for restoring sight.

Nathans, J. "The Genes for Color Vision." *Scientific American,* Vol. 260, No. 2, February 1989. The genes that code for the color-detecting proteins of the eye offer clues about the evolution of normal color vision.

CHAPTER 42

Internal Transport

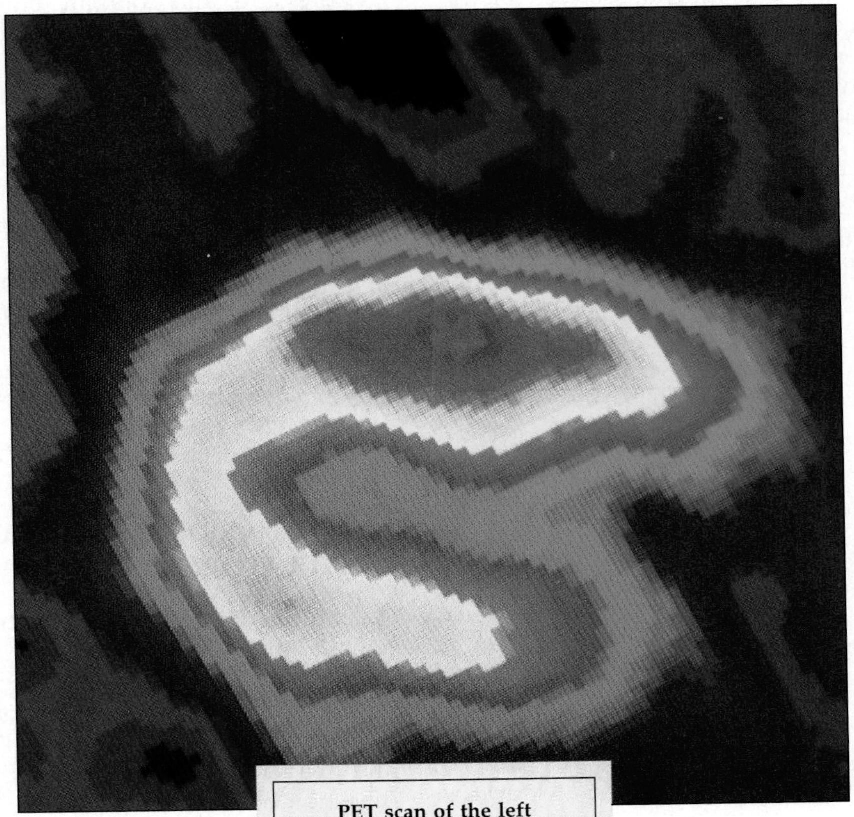

PET scan of the left ventricle of a human heart. The scan measures glucose levels in the tissue, providing information about the condition of heart muscle. (The Stock Market/ Ted Horowitz © 1992)

Most cells require a continuous supply of nutrients and oxygen, and removal of waste products. In very small animals, these metabolic needs can be met by simple diffusion. No specialized circulatory structures are present in sponges, cnidarians, ctenophores, flatworms, or nematodes. Because the bodies of these invertebrates are only a few cells thick, diffusion is an effective mechanism for distributing materials to and from their cells.

In larger complex animals, diffusion cannot supply enough raw materials to all of the cells. Specialized structures are required to transport materials. These structures make up the **circulatory system.**

A circulatory system typically consists of the following:

1. Blood, a connective tissue consisting of cells and cell fragments dispersed in fluid known as plasma.
2. A pumping organ, generally a heart.
3. A system of blood vessels or spaces through which the blood circulates.

Arthropods and most mollusks have an **open circulatory system,** in which the heart pumps blood into vessels that have open ends. Blood spills out of them, filling large spaces that make up the **hemocoel** (blood cavity) and bathing the cells of the body directly. Blood reenters the circulatory system through openings in the heart (in arthropods) or through open-ended vessels that lead to the gills (in mollusks).

Annelids, some mollusks (cephalopods), echinoderms, and vertebrates have a **closed circulatory system.** In them, blood flows through a continuous circuit of blood vessels. The walls of the smallest blood vessels are thin enough to permit diffusion of gases, nutrients, and wastes between blood in the vessels and the extracellular fluid that bathes the cells.

LEARNING OBJECTIVES

After you have studied this chapter you should be able to

1. Compare internal transport in animals with no circulatory system, those with an open circulatory system, and those with a closed circulatory system.
2. Relate structural adaptations of the vertebrate circulatory system to each function it performs.
3. Compare the structure and functions of red blood cells, white blood cells, and platelets.
4. Summarize the events involved in blood clotting.
5. Compare the structure and function of different types of blood vessels, including arteries, arterioles, capillaries, and veins.
6. Trace the evolution of the vertebrate heart from fish to mammal.
7. Describe the structure and function of the human heart, and label a diagram of the heart. Include a description of cardiac muscle and of the heart's conduction system.
8. Trace the events of the cardiac cycle, and relate normal heart sounds to the events of this cycle.
9. Define cardiac output, describe how it is regulated, and identify factors that affect it.
10. Identify factors that determine and regulate blood pressure, and compare blood pressure in different types of blood vessels.
11. Trace a drop of blood through the pulmonary and systemic circulations, naming in sequence each structure through which it passes.
12. Identify the risk factors for atherosclerosis, trace the progress of the disorder, and summarize its possible complications (including angina pectoris and myocardial infarction).
13. List the functions of the lymphatic system, and describe how it operates to maintain fluid balance.

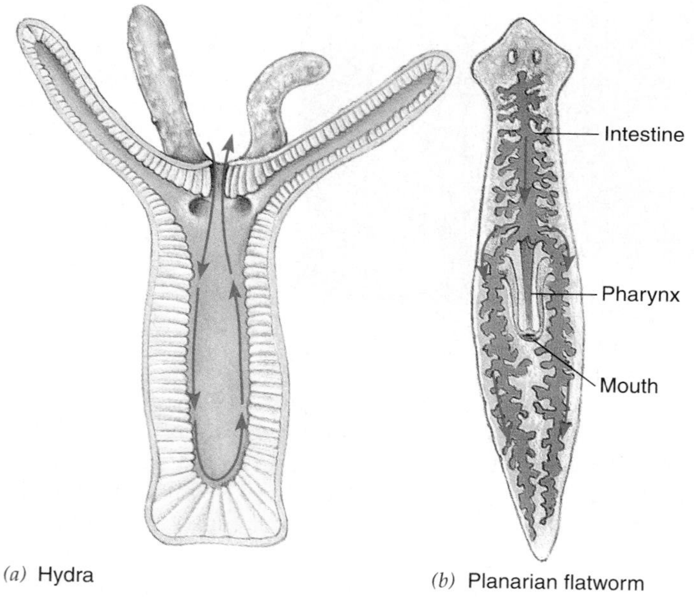

FIGURE 42–1 Some invertebrates have no circulatory system. (*a*) In *Hydra* and other cnidarians, the gastrovascular cavity serves a circulatory function, permitting nutrients to come in contact with the body cells. (*b*) In planarian flatworms, the branched intestine conducts food to all regions of the body.

SOME INVERTEBRATES HAVE NO CIRCULATORY SYSTEM

As indicated in the chapter introduction, many small, aquatic invertebrates have no circulatory system. In cnidarians, the central gastrovascular cavity (as its name implies) serves as a circulatory organ as well as a digestive organ (Fig. 42–1). The animal's tentacles capture prey and deliver it through the mouth into the cavity, where digestion occurs. The digested nutrients then pass into the cells lining the cavity, and through them to cells of the outer layer. As the animal stretches and contracts, movements of the body stir up the contents of the gastrovascular cavity and help distribute nutrients to all parts of the body.

The flattened body of the flatworm permits effective gas exchange by diffusion. Its branched intestine brings nutrients to all regions of the body. As in cnidarians, circulation is aided by contractions of the muscles of the body wall, which agitate the intestinal fluid and the tissue fluid. Metabolic rate tends to be higher than in cnidarians, allowing a more active lifestyle. The branching excretory system of planarians provides for internal transport of wastes that are then expelled from the body.

Fluid in the body cavity of nematodes and other pseudocoelomate animals helps to circulate materials. Nutrients, oxygen, and wastes dissolve in this fluid and diffuse through it to and from the individual cells of the body. Body movements of the animal result in movement of the fluid, facilitating distribution of these materials to all parts of the body.

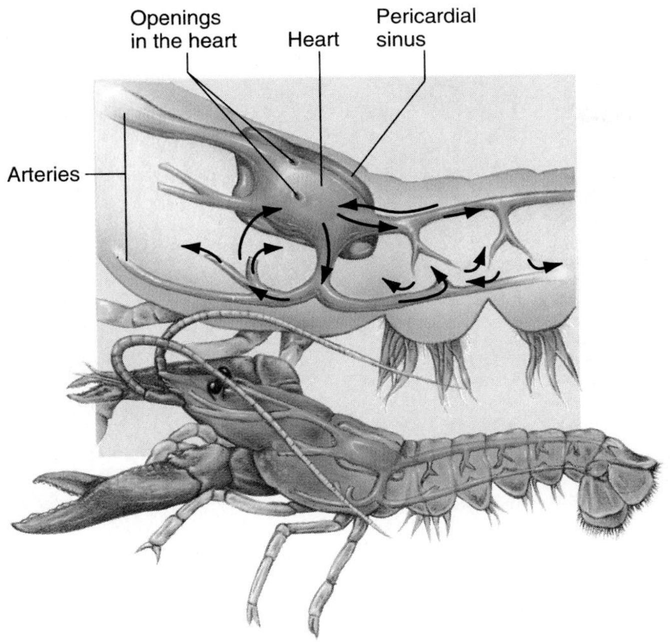

FIGURE 42–2 The crayfish, like other arthropods, has an open circulatory system. Lateral view.

MANY INVERTEBRATES HAVE AN OPEN CIRCULATORY SYSTEM

In animals with open circulatory systems, blood and **interstitial fluid** (tissue fluid) are not distinguishable, and are collectively referred to as **hemolymph.** In the open circulatory system of most mollusks, the heart typically consists of three chambers: two atria and a ventricle. The atria receive hemolymph from the gills. The ventricle pumps oxygen-rich hemolymph to the tissues. Blood vessels conducting hemolymph from the heart open into large spaces called **sinuses,** enabling the hemolymph to bathe the body cells. Such hemolymph-filled spaces form a **hemocoel.** From the hemocoel, hemolymph passes into vessels that lead to the gills. There it is recharged with oxygen and passes into blood vessels that return it to the heart.

Some mollusks, as well as arthropods, have a hemolymph pigment, **hemocyanin,** that contains copper. Hemocyanin transports oxygen and imparts a bluish color to the hemolymph of these animals (the original bluebloods!).

In arthropods, a tubular heart pumps hemolymph into blood vessels (arteries) that deliver it to the sinuses of the hemocoel (Fig. 42–2). Hemolymph then circulates through the hemocoel, eventually returning to the pericardial cavity surrounding the heart. Hemolymph enters the heart through tiny openings that are equipped with valves to prevent backflow. Some insects have accessory "hearts" that help pump hemolymph through the extremities, particularly the wings. Circulation of the hemolymph is faster during muscular movement. Thus, when an animal is active and most in need of nutrients for fuel, its own movement ensures effective circulation.

An open circulatory system cannot provide enough oxygen to maintain the active lifestyle of insects. Indeed, insect hemolymph mainly distributes nutrients and hormones. Oxygen diffuses directly to the cells through a system of air tubes (tracheae) that make up the respiratory system.

SOME INVERTEBRATES HAVE A CLOSED CIRCULATORY SYSTEM

A rudimentary closed circulatory system is found in the proboscis worms (phylum Nemertea). This system consists of a complete network of blood vessels but no heart. Blood flow depends on movements of the animal and on contractions in the walls of the large blood vessels.

Earthworms and other annelids have a complex closed circulatory system (Fig. 42–3). Two main blood vessels extend lengthwise in the body. The ventral vessel conducts blood posteriorly, and the dorsal vessel conducts blood anteriorly. Dorsal and ventral vessels are connected by lateral vessels in every segment. Branches of the lateral vessels deliver blood to the surface, where it is oxygenated. In the anterior part of the worm, five pairs of contractile blood vessels (sometimes referred to as "hearts") connect dorsal and ventral vessels. Contractions of these paired vessels and of the dorsal vessel, as well as contraction of the muscles of the body wall, circulate the blood. Earthworms have **hemoglobin,** the same red pigment that transports oxygen in vertebrate

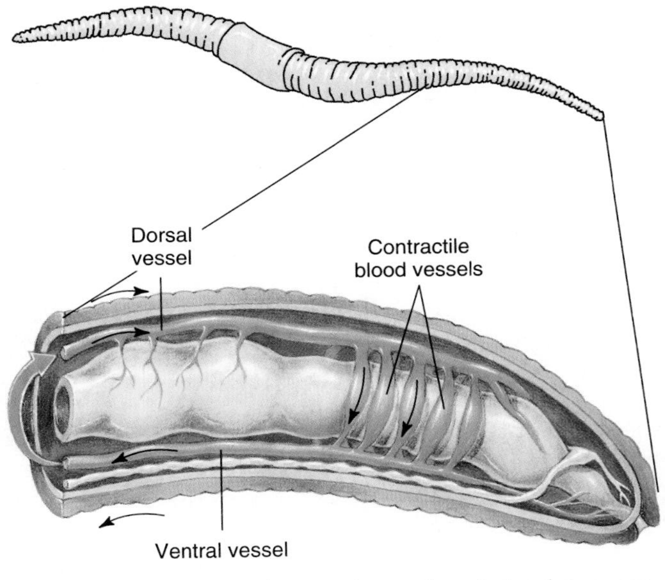

FIGURE 42–3 The earthworm has a closed circulatory system. Five pairs of contractile blood vessels deliver blood from the dorsal vessel to the ventral vessel.

Table 42–1 CELLULAR COMPONENTS OF BLOOD

	Normal Range	*Function*	*Pathology*
Red blood cells	Male: 4.2–5.4 million/μL Female: 3.6–5.0 million/μL	Oxygen transport; carbon dioxide transport	Too few: anemia Too many: polycythemia
Platelets	150,000–400,000/μL	Essential for clotting	Clotting malfunctions; bleeding; easy bruising
White blood cells (total)	5000–10,000/μL		
Neutrophils	About 60% of WBCs	Phagocytosis	Too many: may be due to bacterial infection, inflammation, leukemia (myelogenous)
Eosinophils	1–3% of WBCs	Play some role in allergic response	Too many: may result from allergic reaction, parasitic infestation
Basophils	1% of WBCs	May play role in prevention of clotting in body	
Lymphocytes	25–35% of WBCs	Produce antibodies; destroy foreign cells	Atypical lymphocytes present in infectious mononucleosis; too many may be due to leukemia (lymphocytic), certain viral infections
Monocytes	6% of WBCs	Differentiate to form macrophages	May increase in monocytic leukemia and fungal infections

blood. However, their hemoglobin is not contained within red blood cells but is dissolved in the blood plasma.

Although other mollusks have an open circulatory system, the fast-moving cephalopods, such as the squid and octopus, require a more efficient means of internal transport. They have a closed system made even more effective by the presence of accessory "hearts" at the base of the gills, which speed the passage of blood through the gills.

The Closed Circulatory System of Vertebrates Is Adapted to Carry Out a Variety of Functions

The circulatory system is basically similar in all vertebrates, from fishes, frogs, and reptiles to birds and mammals. All have a ventral, muscular heart that pumps blood into a closed system of blood vessels. The vertebrate circulatory system consists of heart, blood vessels, blood, lymph, lymph vessels, and associated organs such as the thymus, spleen, and liver. The tiniest blood vessels, **capillaries,** have very thin walls that permit exchange of materials between blood and extracellular fluid.

The vertebrate circulatory system performs several functions:

1. Transports nutrients from the digestive system and from storage depots to each cell of the body.
2. Transports oxygen from respiratory structures (gills or lungs) to the cells of the body.
3. Transports metabolic wastes from each cell to organs that excrete them.
4. Transports hormones from endocrine glands to target tissues.
5. Helps maintain fluid balance.
6. Defends the body against invading microorganisms.
7. Helps distribute metabolic heat within the body and helps maintain normal body temperature in endothermic (warm-blooded) animals.

Vertebrate Blood Consists of Plasma, Blood Cells, and Platelets

In vertebrates, **blood** consists of a pale yellowish fluid, known as **plasma,** in which red blood cells, white blood cells, and platelets are suspended (Fig. 42–4; Table 42–1). In humans the total circulating blood volume is about 8% of the body weight—5.6 liters (6 quarts) in a 70-kilogram

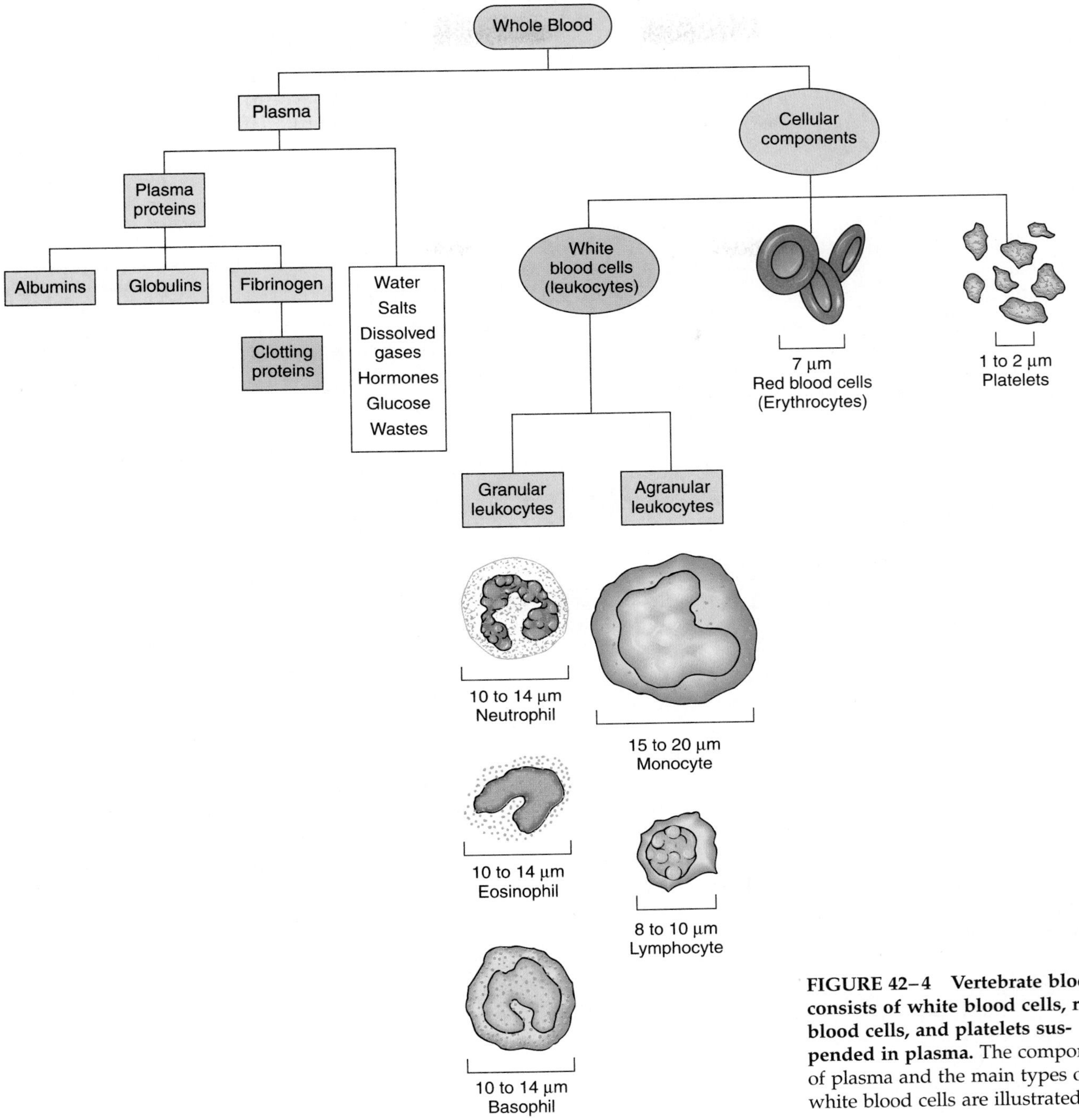

FIGURE 42–4 Vertebrate blood consists of white blood cells, red blood cells, and platelets suspended in plasma. The components of plasma and the main types of white blood cells are illustrated.

(154-pound) person. This is about the amount of oil in the crankcase of most cars!

About 55% of the blood is plasma. The remaining 45% is made up of blood cells and platelets. Because cells and platelets are heavier than plasma, they can be separated from it by the process of centrifugation. Plasma does not separate from blood cells in the body because the blood is constantly mixed as it circulates in the blood vessels.

Plasma is the fluid component of blood

Plasma is composed of water (about 92%), proteins (about 7%), salts, and a variety of materials being transported, such as dissolved gases, nutrients, wastes, and hormones. Plasma is in dynamic equilibrium with the interstitial fluid bathing the cells and with the intracellular fluid within the cells. As blood passes through the capillaries, substances constantly move into and out of the plasma. Changes in its composition initiate responses on the part of one or more organs of the body to restore its normal steady state.

Plasma contains several kinds of **plasma proteins,** each with specific properties and functions: fibrinogen; alpha, beta, and gamma globulins; albumin; and lipoproteins. Fibrinogen is one of the proteins involved in the clotting process. When the proteins involved in blood clotting have been removed from the plasma, the remaining liquid is called **serum.** The gamma globulin fraction contains many types of antibodies that provide immunity to diseases such as measles and infectious hepatitis. Purified human gamma globulin is sometimes used to treat certain diseases, or to reduce the possibility of contracting a disease. **Lipoproteins** in the blood transport triacylglycerols (triglycerides) and cholesterol (see Chapter 45).

Plasma proteins, especially albumins and globulins, help regulate the distribution of fluid between plasma and tissue fluid. Too large to pass readily through the walls of blood vessels, these proteins contribute to the blood's osmotic pressure, important in maintaining an appropriate blood volume. Plasma proteins (along with the hemoglobin in the red blood cells) are also important acid-base buffers. They help keep the pH of the blood within a narrow range—near its normal, slightly alkaline pH of 7.4.

Red blood cells transport oxygen

Erythrocytes, also called **red blood cells (RBCs),** are highly specialized for transporting oxygen. In all vertebrates except mammals, erythrocytes have nuclei. In mammals, the nucleus is ejected from the erythrocyte as the cell develops. Each mammalian erythrocyte is a flexible, biconcave disc, 7 to 8 micrometers (μm) in diameter and 1 to 2 micrometers thick. An internal elastic framework maintains the disc shape and permits the cell to bend and twist as it passes through blood vessels even smaller than its own diameter. Its biconcave shape provides a high ratio of surface area to volume, allowing efficient diffusion of oxygen and carbon dioxide into and out of the cell. In a human, about 30 trillion erythrocytes circulate in the blood—approximately 5.4 million per microliter in an adult male, 5 million per microliter in an adult female.

Erythrocytes are produced within the red bone marrow of certain bones: vertebrae, ribs, breast bone, skull bones, and long bones. As an erythrocyte develops, it produces great quantities of **hemoglobin,** the oxygen-transporting pigment that gives vertebrate blood its red color. (Oxygen transport is discussed in Chapter 44.) The life span of a human erythrocyte is about 120 days. As blood circulates through the liver and spleen, phagocytic cells remove worn-out erythrocytes from the circulation. These erythrocytes are then disassembled, and some of their components are recycled. In the human body, more than 2.4 million RBCs are destroyed every second, so an equal number must be produced in the bone marrow to replace them.

Anemia is a deficiency in hemoglobin (often accompanied by a decrease in the number of RBCs). When hemoglobin is insufficient, the amount of oxygen transported is inadequate to supply the body's needs. An anemic person may complain of feeling weak and may become easily fatigued. Three general causes of anemia are (1) loss of blood due to hemorrhage or internal bleeding; (2) decreased production of hemoglobin or red blood cells as in iron-deficiency anemia; and (3) increased rate of RBC destruction—the **hemolytic anemias** such as sickle cell anemia.

White blood cells defend the body against disease organisms

The **leukocytes,** or **white blood cells (WBCs),** are specialized to defend the body against harmful bacteria and other microorganisms. Leukocytes are amoeba-like cells capable of independent movement. Some types routinely slip through the walls of blood vessels and enter the tissues. Human blood contains five kinds of leukocytes, which may be classified as either granular or agranular (Fig. 42–4).

The **granular leukocytes** are manufactured in the red bone marrow. These cells are characterized by large lobed nuclei and distinctive granules in their cytoplasm. The three varieties of granular leukocytes are the neutrophils, eosinophils, and basophils. **Neutrophils,** the principal phagocytic cells in the blood, are especially adept at seeking out and ingesting bacteria. They also phagocytize the remains of dead tissue cells, a clean-up task that must be done after injury or infection. Most of the granules in neutrophils contain enzymes that digest ingested material.

Eosinophils have large granules that stain bright red with eosin, an acidic dye. These cells increase in number during allergic reactions and during parasitic (e.g., tapeworm) infestations. **Basophils** exhibit deep blue granules when stained with basic dyes. Like eosinophils, these cells are thought to play a role in allergic reactions. Basophils contain large amounts of **histamine,** which they release in injured tissues and in allergic responses. Because they contain the anticlotting chemical **heparin,** basophils may play a role in preventing blood from clotting inappropriately within the blood vessels.

Agranular leukocytes lack large distinctive granules, and their nuclei are rounded or kidney-shaped. Two types of agranular leukocytes are lymphocytes and monocytes. Some **lymphocytes** are specialized to produce antibodies, whereas others attack foreign invaders such as bacteria or viruses directly. Just how they manage these feats is discussed in the next chapter.

Monocytes are the largest WBCs, reaching 20 micrometers in diameter. They are manufactured in the bone marrow. After circulating in the blood for about 24 hours, a monocyte leaves the circulation and completes its development in the tissues. The monocyte greatly enlarges and becomes a **macrophage,** a giant scavenger cell. All of the macrophages found in the tissues develop in this way. Macrophages voraciously engulf bacteria, dead cells, and debris.

In human blood there are normally about 7000 WBCs per microliter (μL) of blood (only one for every 700 RBCs). During bacterial infections the number may rise sharply, so that a WBC count is a useful diagnostic tool. The proportion of each kind of WBC is determined by a differential WBC count. The normal distribution of leukocytes is indicated in Table 42–1.

Leukemia is a form of cancer in which any one of the various kinds of white cells multiplies rapidly within the bone marrow. Many of these cells do not mature, and their large numbers crowd out developing RBCs and platelets, leading to anemia and impaired clotting. A common cause of death from leukemia is internal hemorrhaging, especially in the brain. Another frequent cause of death is infection; although the white cell count may rise dramatically, the cells are immature and abnormal and cannot defend the body against disease organisms. No cure for leukemia has been discovered, but radiation treatment and therapy with antimitotic drugs can induce partial or complete remissions lasting 15 years or longer in some patients.

Platelets function in blood clotting

In most vertebrates other than mammals, the blood contains small, oval cells called **thrombocytes,** which have

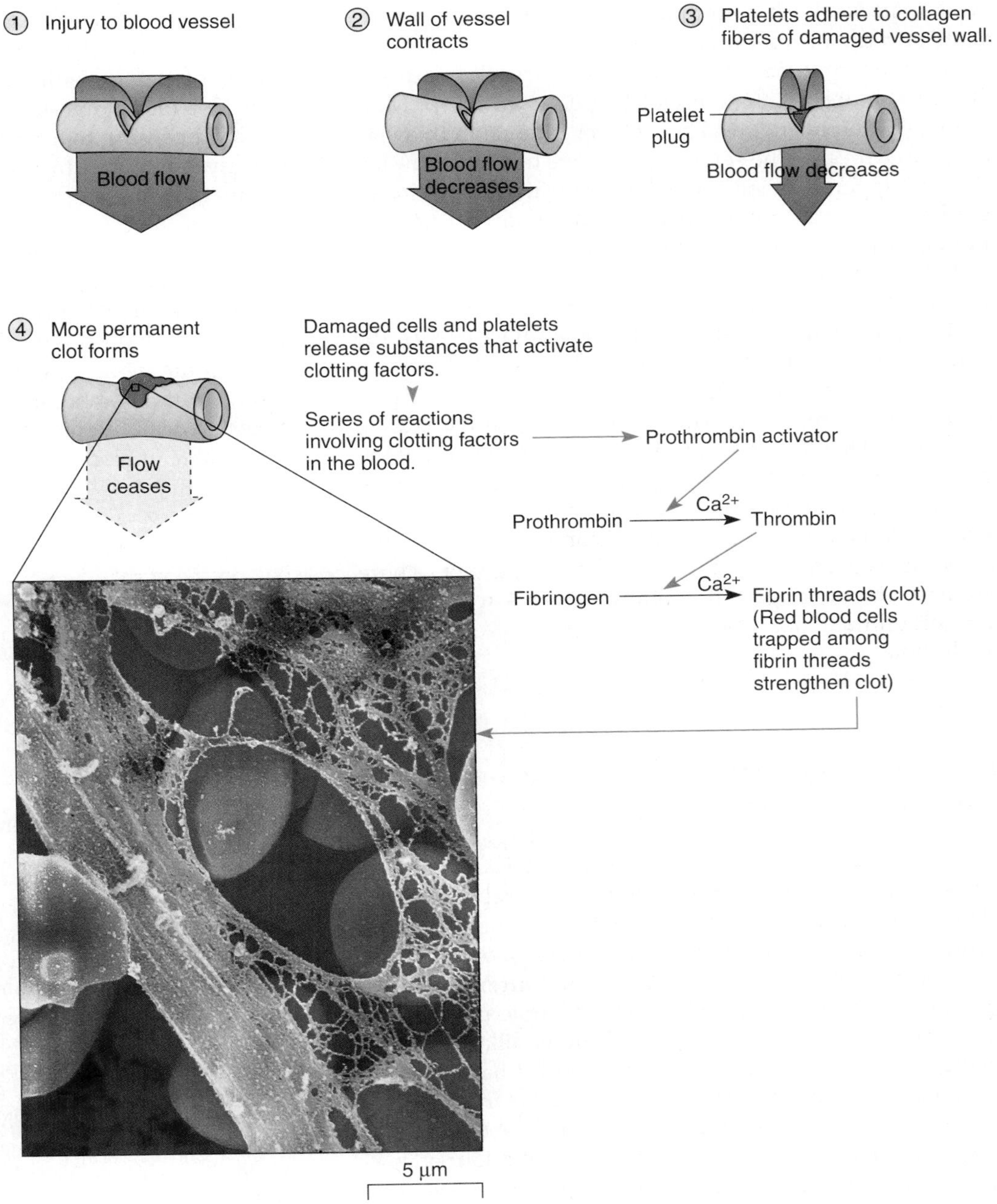

FIGURE 42–5 Platelets and a variety of clotting factors are important in blood clotting. Scanning electron micrograph (false-color) of part of a blood clot, showing red blood cells enmeshed in a network of fibrin. (Lennart Nilsson © Boehringer Ingelheim International, GmbH.)

nuclei. In mammals, thrombocytes are tiny spherical or disc-shaped bits of cytoplasm that lack nuclei. They are usually referred to as **platelets.** About 300,000 platelets per microliter are present in human blood. Platelets are formed from bits of cytoplasm that are pinched off from very large cells (megakaryocytes) in the bone marrow. Thus, a platelet is not a whole cell but a fragment of cytoplasm enclosed by a membrane.

Platelets play an important role in **hemostasis** (the control of bleeding). When a blood vessel is cut, it constricts, reducing loss of blood. Platelets stick to the rough, cut edges of the vessel, physically patching the break in the wall. As platelets begin to gather, they release substances that attract other platelets. The platelets become sticky and adhere to collagen fibers in the blood vessel wall. Within about 5 minutes after injury they form a platelet plug, or temporary clot.

At the same time that the temporary clot forms, a stronger, more permanent clot begins to develop. More than 30 different chemical substances interact in this very complex process. The series of reactions that leads to clotting is triggered when one of the clotting factors in the blood is activated by contact with the injured tissue. In **hemophilia,** one of the clotting factors is absent as a result of an inherited genetic mutation (see Chapter 15). The clotting process is summarized in Figure 42–5.

Prothrombin, a plasma protein manufactured in the liver, requires vitamin K for its production. In the presence of clotting factors, calcium ions, and compounds released from platelets, prothrombin is converted to **thrombin.** Then thrombin catalyzes the conversion of the soluble plasma protein **fibrinogen** to an insoluble protein, **fibrin.** Once formed, fibrin polymerizes, producing long threads that stick to the damaged surface of the blood vessel and form the webbing of the clot. These threads trap blood cells and platelets, which help strengthen the clot.

VERTEBRATES HAVE THREE MAIN TYPES OF BLOOD VESSELS

The vertebrate circulatory system includes three main types of blood vessels: arteries, capillaries, and veins (Fig. 42–6). An **artery** carries blood away from the heart, toward other tissues. When an artery enters an organ, it divides into many smaller branches called **arterioles.** The arterioles deliver blood into the microscopic **capillaries.** After blood courses through an organ, capillaries eventually merge to form **veins** that transport it back toward the heart. Because this basic plan is similar in all vertebrates, much can be learned about the human circulatory system by dissecting an animal such as a shark or a frog.

A blood vessel wall has three layers (Fig. 42–6). The innermost layer (tunica intima), which lines the blood vessel, consists mainly of endothelium, a tissue that resembles squamous epithelium. The middle layer (tunica media) consists of connective tissue and smooth muscle cells, and the outer coat (tunica adventitia) is composed of connective tissue rich in elastic and collagen fibers.

The thick walls of arteries and veins prevent gases and nutrients from passing through. Materials are exchanged between the blood and interstitial fluid bathing the cells through the capillary walls, which are only one cell thick (Fig. 42–6). Capillary networks in the body are so extensive that at least one of these tiny vessels is located close to almost every cell in the body. The total length of all capillaries in the body has been estimated to be more than 60,000 miles.

Smooth muscle in the arteriole wall can constrict **(vasoconstriction)** or relax **(vasodilation),** changing the radius of the arteriole. Such changes help maintain appropriate blood pressure and can help control the volume of blood passing to a particular tissue. Changes in blood flow are regulated by the nervous system in response to the metabolic needs of the tissue, as well as by the demands of the body as a whole. For example, when a tissue is metabolizing rapidly, it needs a greater supply of nutrients and oxygen. During exercise, arterioles within the muscles dilate, increasing by more than tenfold the amount of blood flowing to the muscle cells.

If all the blood vessels were dilated at the same time, there would not be sufficient blood to fill them completely. Normally the liver, kidneys, and brain receive the lion's share of blood. However, if an emergency suddenly

FIGURE 42–6 Blood generally flows through a sequence of blood vessels and lymph vessels. (*a*) The heart pumps blood into arteries. It then passes through arterioles, capillaries, and veins, which deliver it back to the heart. Some plasma leaves the capillaries and becomes interstitial fluid. Lymphatic vessels return excess interstitial fluid to the blood by way of ducts that lead into large veins in the shoulder region. (*b*) Comparison of the structure of the wall of an artery, vein, and capillary. All three vessels are lined with endothelium, a tissue similar to simple squamous epithelium. Notice that the wall of the capillary is composed of a single cell formed into a tube. A series of such cells makes up the entire capillary. (*c*) The photomicrograph illustrates that red blood cells must pass through capillaries in almost single file. (Lennart Nilsson © Boehringer Ingelheim International, GmbH) ▶

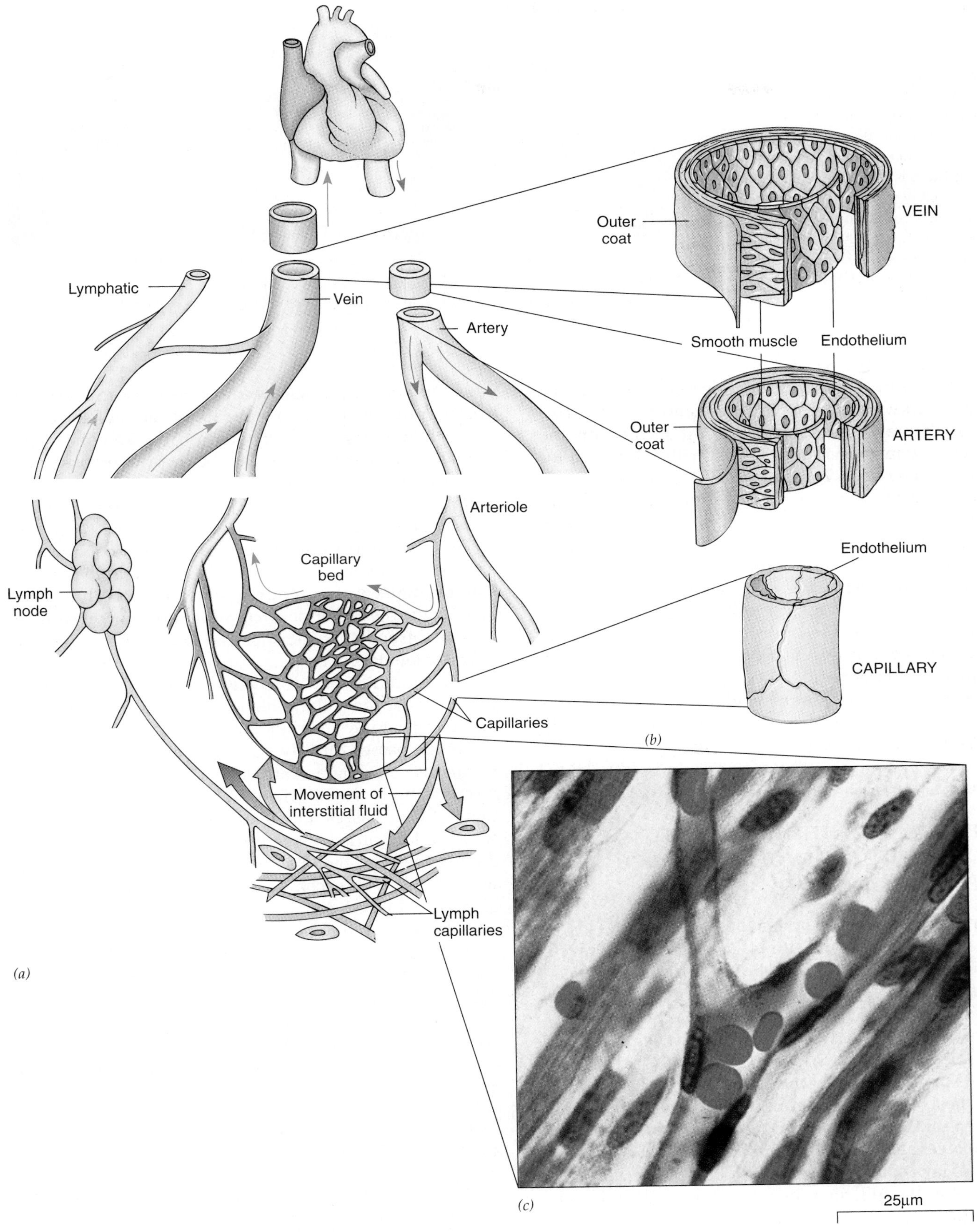
Lymphatic
Vein
Artery
Outer coat
VEIN
Smooth muscle
Endothelium
Outer coat
ARTERY
Arteriole
Capillary bed
Lymph node
Endothelium
CAPILLARY
Capillaries
Movement of interstitial fluid
Lymph capillaries
(a)
(b)
(c)
25μm

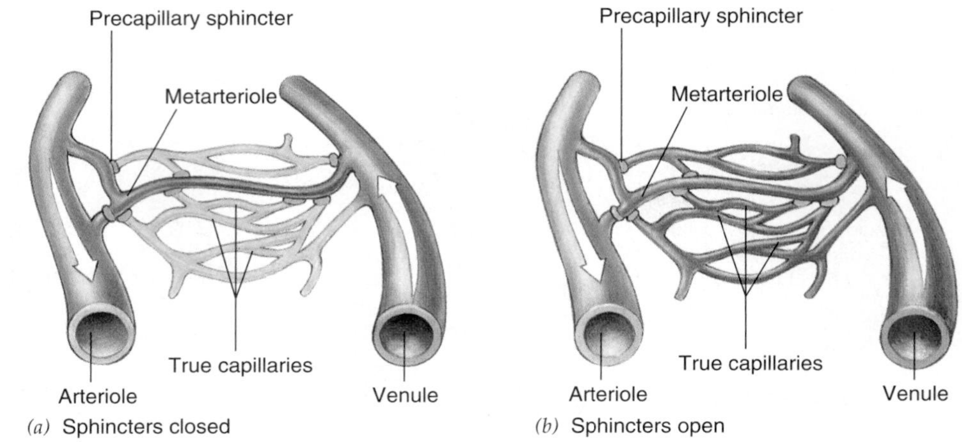

FIGURE 42–7 As a tissue becomes active, the pattern of blood flow through its capillary networks changes. (*a*) When the tissue is inactive, only the metarterioles are open. (*b*) When the tissue becomes active, the decreased oxygen tension in the tissue brings about a relaxation of the precapillary sphincters, and the capillaries open. This increases the blood supply and the delivery of oxygen to the active tissue.

occurred requiring rapid action, the blood would be rerouted quickly in favor of the heart and muscles. This would enable rapid, effective action. At such a time the digestive system and kidneys can do with less blood, for they are not critical in responding to the crisis.

The small vessels that directly link arterioles with venules (small veins) are **metarterioles.** The so-called true capillaries branch off from the metarterioles and then rejoin them (Fig. 42–7). True capillaries also interconnect with one another. Wherever a capillary branches from a metarteriole, a smooth muscle cell called a precapillary sphincter is present. Precapillary sphincters open and close continuously, directing blood first to one and then to another section of tissue. These sphincters also (along with the smooth muscle in the walls of arteries and arterioles) regulate the blood supply to each organ and its subdivisions (Fig. 42–7).

ADAPTATIONS OF THE HEART AND CIRCULATION HAVE EVOLVED IN EACH VERTEBRATE CLASS

The vertebrate circulatory system became modified in the course of evolution as the site of gas exchange changed from gills to lungs and as vertebrates became active, endothermic animals. The vertebrate heart has one or two **atria** (sing., *atrium*), chambers that receive blood returning from the tissues, and a single or divided **ventricle** that pumps blood into the arteries (Fig. 42–8). Additional chambers are present in some classes.

In fishes, blood flows through a single circuit passing through a capillary network in the gills, where blood becomes oxygenated, and then through capillaries located in other organs of the body. After blood circulates through capillaries in the gills its pressure is low, so blood leaving the gills passes very slowly to the other organs. Circulation is helped by the movements of the fish while it is swimming. This single-circuit, low-pressure circulatory system permits only a low metabolic rate in fishes.

Although it contains only one atrium and one ventricle, the fish heart has four chambers. A thin-walled sinus venosus receives blood returning from the tissues and pumps it into the atrium. The atrium then pumps blood into the ventricle. Next, the ventricle pumps blood into an elastic conus arteriosus, which does not contract.

In amphibians, blood flows through a double circuit. The **pulmonary circulation** delivers it to the lungs and skin, while the **systemic circulation** transports it to all organs of the body. Oxygen-rich and oxygen-poor blood are kept somewhat separate.

The amphibian heart has two atria and one ventricle. A sinus venosus collects blood returning from the veins and pumps it into the right atrium. Blood returning from the lungs passes directly into the left atrium. Both atria pump into the single ventricle, but oxygen-poor blood is pumped out of the ventricle before oxygen-rich blood enters it. Blood passes into an artery (the conus arteriosus) with a fold that helps keep the blood separate. Much of the oxygen-poor blood is directed to the lungs and skin, where it can be recharged with oxygen. Oxygen-rich blood is delivered into arteries that conduct it to the various tissues of the body.

Reptiles also have a double circuit of blood flow, made more efficient by a wall that partly divides the ventricle. Although some mixing of oxygen-rich and oxygen-poor blood occurs, it is minimized by the timing of contractions of the left and right sides of the heart and by pressure differences. (In crocodiles, the wall between the ventricles is complete, so the heart consists of two atria and two ventricles.)

Unlike birds and mammals, amphibians and reptiles do not ventilate their lungs continuously. Therefore, it would be inefficient to pump blood through the pulmonary circulation continuously. The shunts between the two sides of the heart allow the blood to be distributed to the lungs as needed.

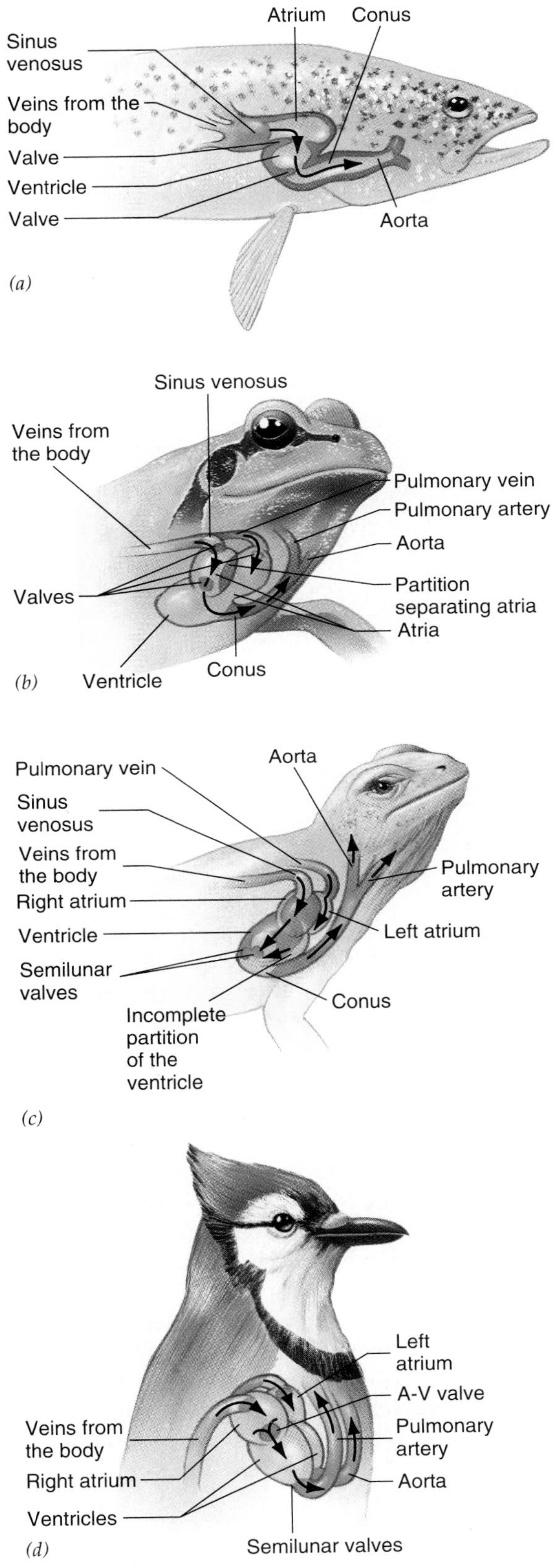

FIGURE 42–8 Through evolution, the vertebrate heart has become adapted for the way of life of each class of vertebrates. (***a***) The single atrium and ventricle of the fish heart is part of a single circuit of blood flow. (***b***) In amphibians blood flows through a double circuit and the heart consists of two atria and one ventricle. (***c***) The reptilian heart has two atria and two ventricles, but the wall separating the ventricles is incomplete so that blood from the right and left chambers mixes to some extent. (***d***) Birds and mammals have two atria and two ventricles, and blood rich in oxygen is kept completely separate from oxygen-poor blood.

In birds and mammals, the wall between the ventricles is complete, preventing the mixing of oxygen-rich blood in the left chamber with oxygen-poor blood in the right chamber. The conus has split and become the base of the **aorta** (the largest artery) and pulmonary artery. No sinus venosus is present as a separate chamber, but a vestige remains as the sinoatrial node (the pacemaker).

Complete separation of the right and left sides of the heart makes it necessary for blood to pass through the heart twice each time it tours the body. As a result, it is possible to maintain higher blood pressures, and materials are delivered to the tissues rapidly and efficiently. Because the blood of birds and mammals contains more oxygen per unit volume and circulates more rapidly than in other vertebrates, the tissues receive more oxygen. As a result, birds and mammals can maintain a higher metabolic rate and a constant, high body temperature even in cold surroundings.

The pattern of blood circulation in birds and mammals may be summarized as follows:

> veins (conduct blood from organs) → right atrium → right ventricle → pulmonary arteries → capillaries in the lungs → pulmonary veins → left atrium → left ventricle → aorta → arteries (conduct blood to organs) → capillaries

THE HUMAN HEART IS WELL ADAPTED FOR PUMPING BLOOD

Not much bigger than a fist and weighing less than a pound, the human **heart** is a remarkable organ that beats about 2.5 billion times in an average lifetime, pumping about 300 million liters (80 million gallons) of blood (Fig. 42–9). To meet the body's changing needs, it can vary its output from 5 to more than 20 liters of blood per minute.

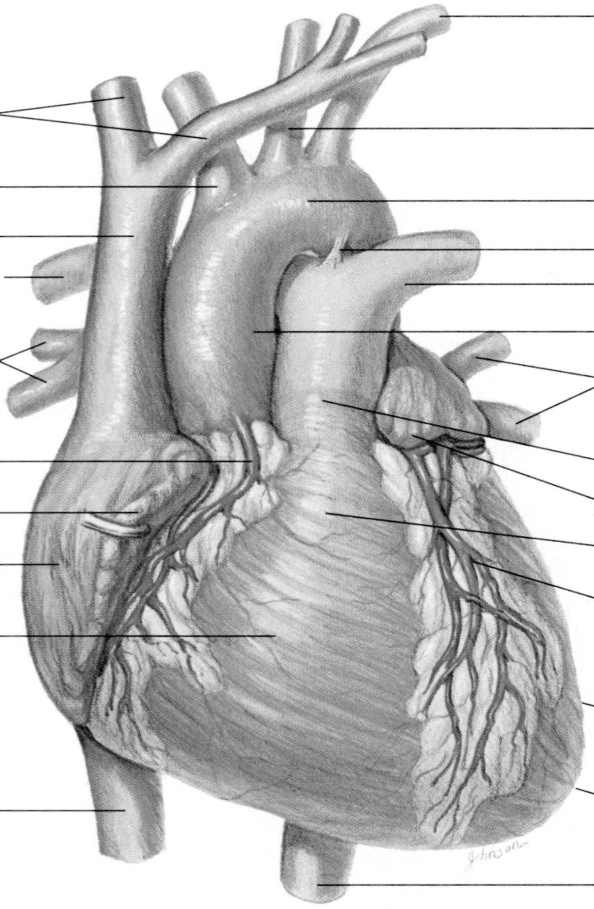

(a)

The heart is a hollow, muscular organ located in the chest cavity directly under the breastbone. Enclosing it is a tough connective tissue sac, the **pericardium.** The inner surface of the pericardium and the outer surface of the heart are covered by a smooth layer of epithelial-like cells. Between these two surfaces is a small **pericardial cavity** filled with fluid, which reduces friction to a minimum as the heart beats.

The wall of the heart is composed mainly of cardiac muscle attached to a framework of collagen fibers. The right atrium and ventricle are separated from the left atrium and ventricle by a wall, or **septum** (Fig. 42–10). Between the atria the wall is known as the **interatrial septum;** between the ventricles it is the **interventricular septum.** On the interatrial septum a shallow depression, the **fossa ovalis,** marks the place where an opening, the **foramen ovale,** was located in the fetal heart. In the fetus, the foramen ovale permits the blood to flow directly from right to left atrium so that very little passes to the nonfunctional lungs. At the upper surface of each atrium lies a small muscular pouch called an **auricle.**

To prevent blood from flowing backward, the heart is equipped with valves that close automatically. The valve between the right atrium and right ventricle is called the **right atrioventricular (AV) valve** (also known as the *tricuspid valve*). The **left AV valve** (between the left atrium and left ventricle) is referred to as the **mitral valve.** The AV valves are held in place by stout cords, or "heartstrings," the **chordae tendineae.** These cords attach the valves to the **papillary muscles** that project from the walls of the ventricles.

When blood returning from the tissues fills the atria, blood pressure on the AV valves forces them to open into the ventricles, which then fill with blood. As the ventricles contract, blood is forced back against the AV valves, pushing them closed. Contraction of the papillary muscles and tensing of the chordae tendineae prevent them from opening backward into the atria. These valves are like swinging doors that can open in only one direction.

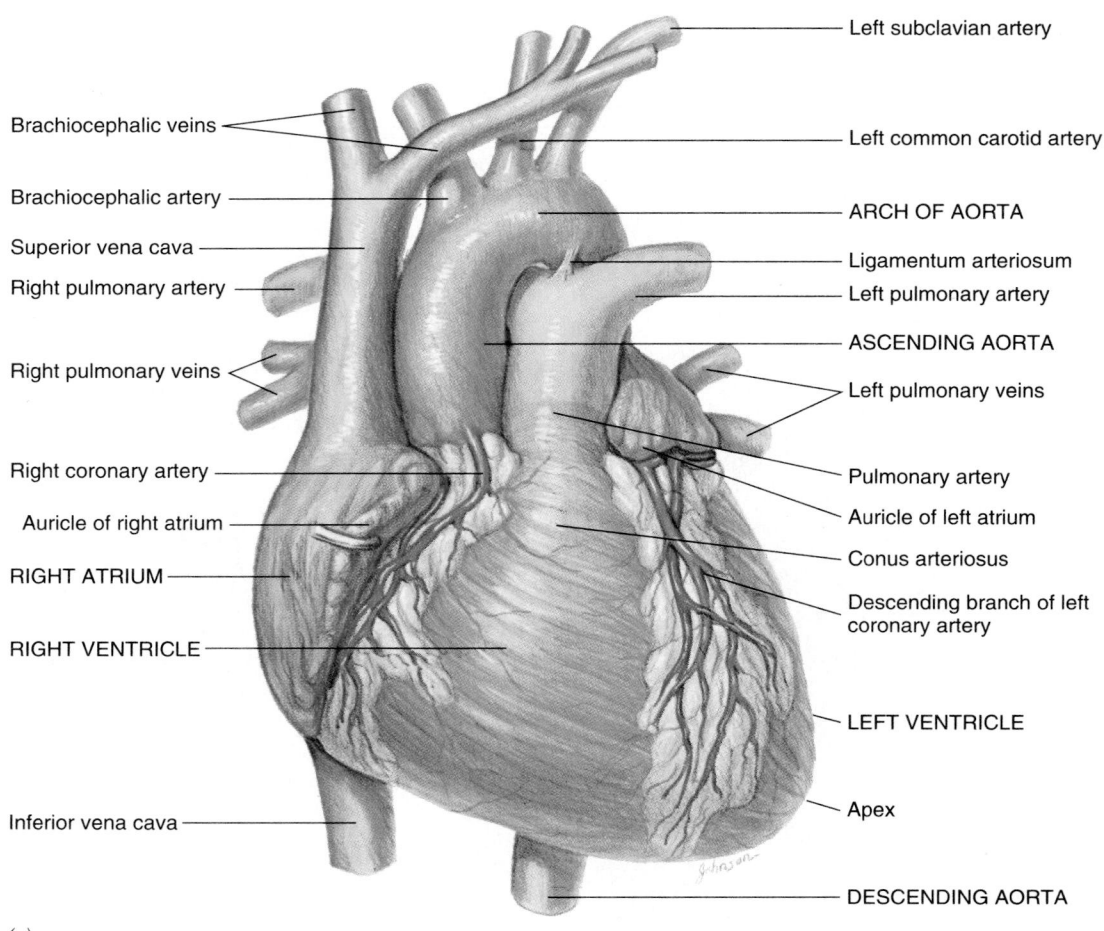

Semilunar valves (named for their flaps, which are shaped like half-moons) guard the exits from the heart. The semilunar valve between the left ventricle and the aorta is the **aortic valve,** and the one between the right ventricle and the pulmonary artery is the **pulmonary valve.** When blood passes out of the ventricles, the flaps of the semilunar valves are pushed aside and offer no resistance to blood flow. But when the ventricles are relaxing and filling with blood from the atria, the blood pres-

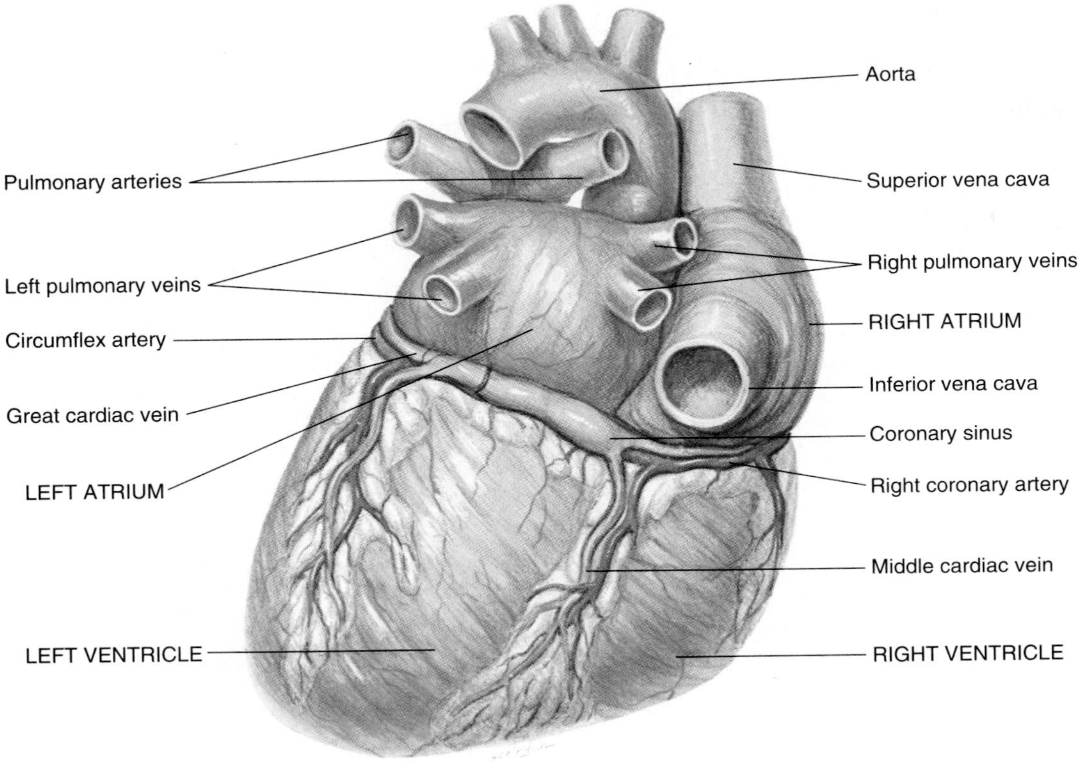

(*b*)

FIGURE 42–9 The human heart has a left and right atrium and a left and right ventricle. (*a*) Anterior external view of the human heart. Note the coronary blood vessels that bring blood to and from the heart muscle itself. (*b*) Posterior external view.

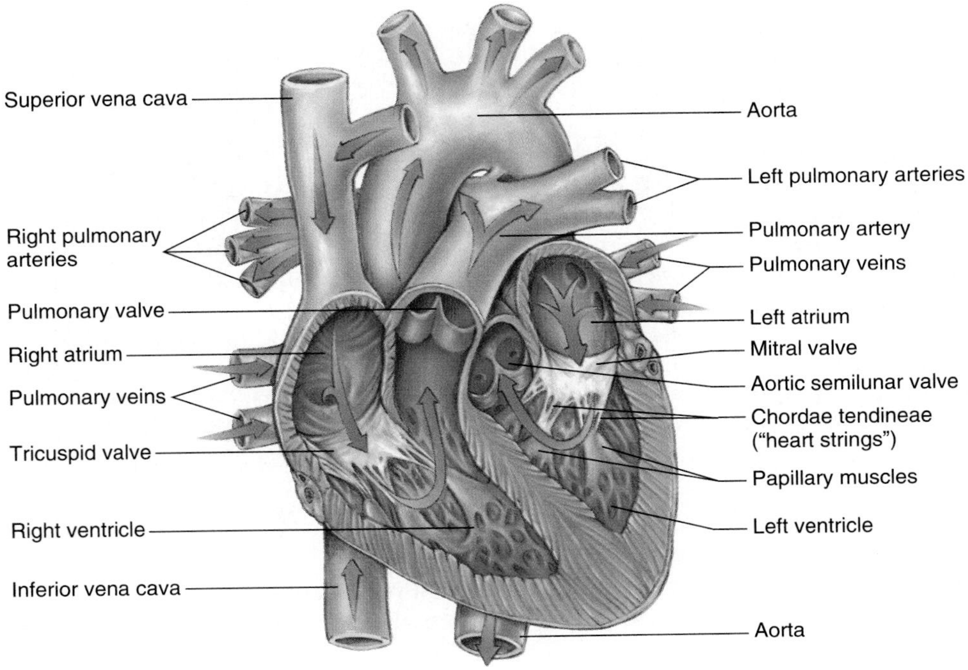

FIGURE 42–10 The atria receive blood; the ventricles pump blood into the arteries. Section through the human heart showing chambers, valves, connecting blood vessels, and pattern of blood flow.

sure in the arteries is higher than that in the ventricles. Blood then fills the pouches of the valves, stretching them across the artery so that blood cannot flow back into the ventricle.

Valve deformities are sometimes present at birth, or they may result from certain diseases such as rheumatic fever or syphilis. Inflammation and scarring may narrow the blood passageway by thickening the valves. Sometimes erosion of the valve tissues prevents the flaps from closing tightly, causing blood to leak backward and reducing the efficiency of the heartbeat. Diseased valves can be surgically replaced with artificial valves.

Each heartbeat is initiated by a pacemaker

Horror films frequently feature a scene in which a heart cut out of the body of its owner continues to beat. Scriptwriters of these tales actually have some factual basis for their gruesome fantasies, for when carefully removed from the body, the heart does continue to beat for many hours if kept in a nutritive, oxygenated fluid. This is possible because the contractions of cardiac muscle begin within the muscle itself and can occur independently of any nerve supply.

A specialized conduction system ensures that the heart beats in a regular and effective rhythm. Each beat is initiated by the **pacemaker,** called the **sinoatrial (SA) node** (Fig. 42–11). The SA node is a small mass of specialized cardiac muscle in the posterior wall of the right atrium near the opening of a large vein, the superior vena cava. Ends of the SA node fibers fuse with surrounding ordinary atrial muscle fibers so that each action potential spreads through both atria, producing atrial contraction.

One group of atrial muscle fibers conducts the action potential directly to the **atrioventricular (AV) node,** located in the right atrium along the lower part of the septum. Here transmission is delayed briefly, permitting the atria to complete their contraction before the ventricles begin to contract. From the AV node the action potential spreads into specialized muscle fibers called **Purkinje fibers.** These large fibers make up the **atrioventricular (AV) bundle.** The AV bundle divides, sending branches into each ventricle. When an impulse reaches the ends of the Purkinje fibers, it spreads through the ordinary cardiac muscle fibers of the ventricles.

At their ends cardiac muscle cells are joined by dense bands called **intercalated discs** (Fig. 42–11*b* and *c*). Each disc is a type of gap junction (see Chapter 5) in which two cells are connected through pores. This type of junction is of great physiological importance because it offers very little resistance to the passage of an action potential. Ions move easily through the gap junctions, allowing the entire atrial (or ventricular) muscle mass to contract as one giant cell.

Each minute the heart beats about 70 times. One complete heartbeat takes about 0.8 second and is referred to as a **cardiac cycle.** That portion of the cycle in which contraction occurs is known as **systole;** the period of relaxation is **diastole.** Figure 42–12 shows the sequence of events that occur during one cardiac cycle.

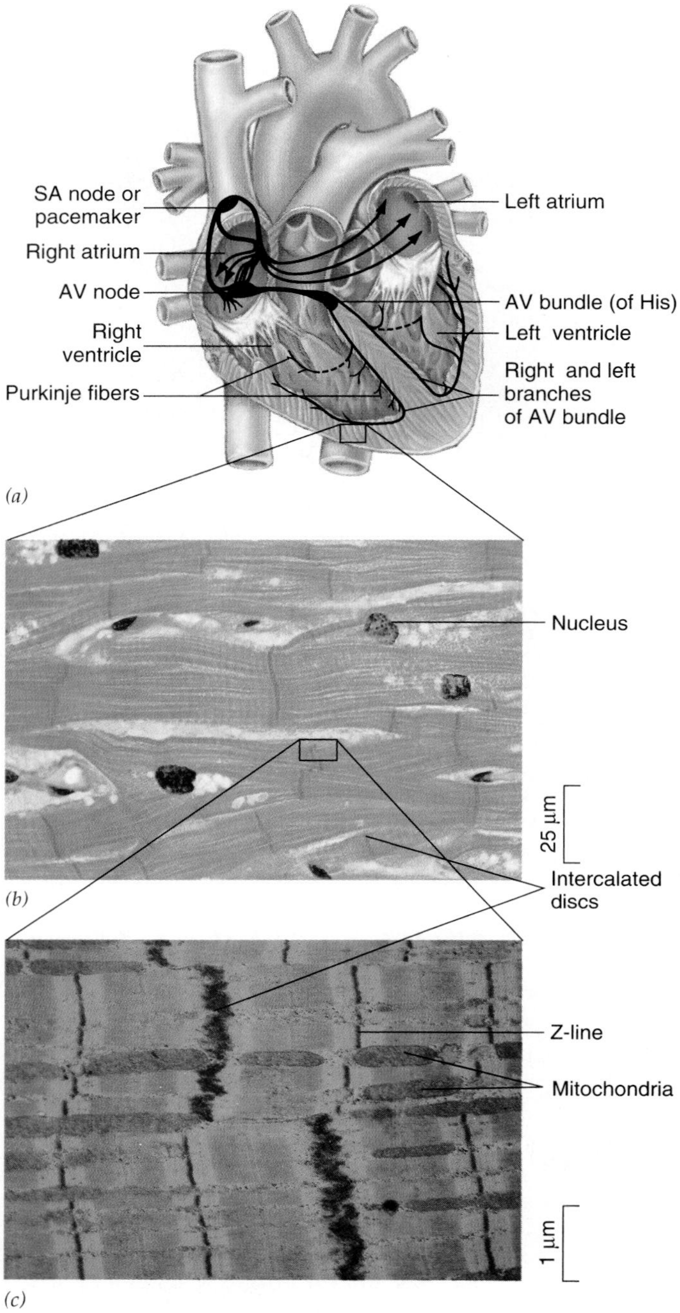

FIGURE 42–11 The conduction system of the heart maintains rhythmic contraction. (***a***) The SA node initiates each heartbeat. The action potential spreads through the muscle fibers of the atria, producing atrial contraction. Transmission is briefly delayed at the AV node. Then, the action potential spreads through specialized muscle fibers into the ventricles. (***b***) Cardiac muscle viewed with the light microscope (approximately ×250). (***c***) Electron micrograph of cardiac muscle. (*b*, Ed Reschke; *c*, Visuals Unlimited/Don Fawcett)

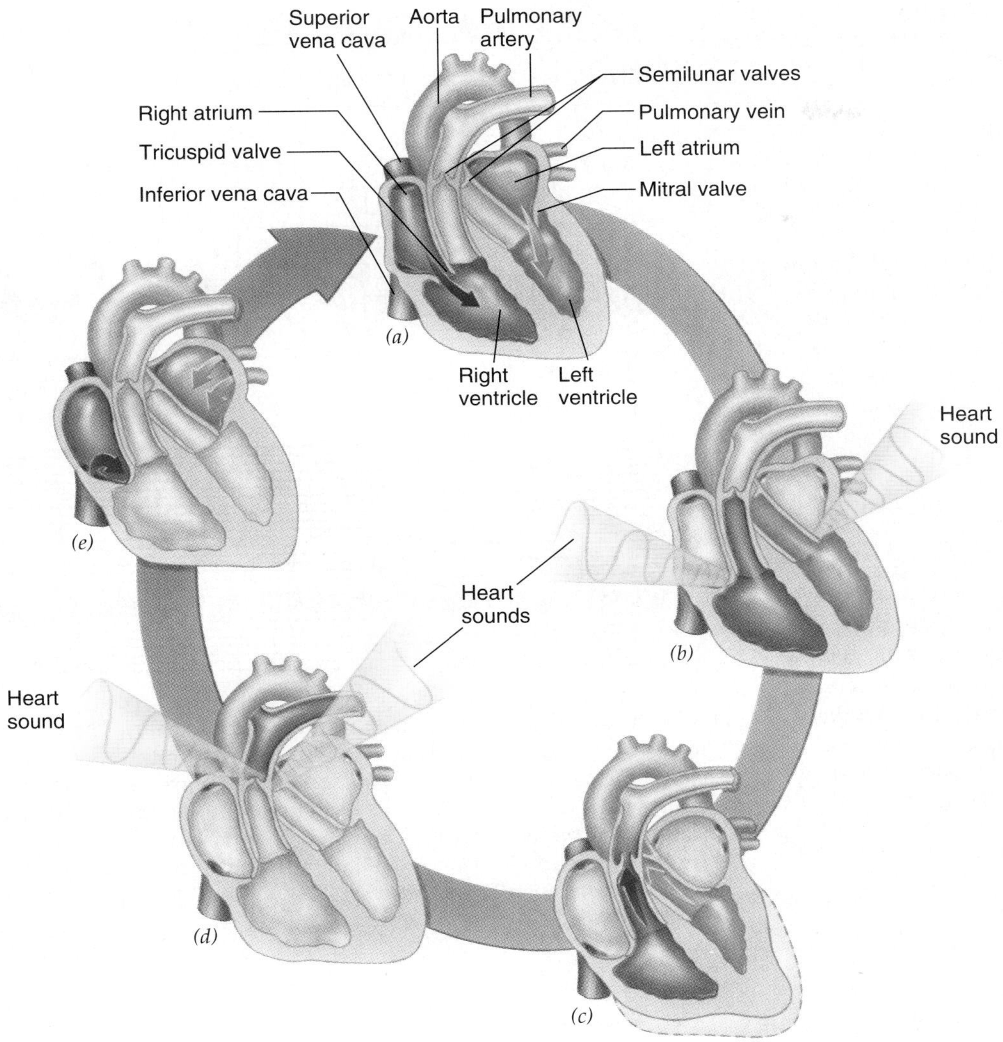

FIGURE 42–12 The contraction of both atria followed by both ventricles makes a cardiac cycle. Arrows indicate the direction of blood flow; dotted lines indicate the change in size as contraction occurs. (*a*) Atrial systole: Atria contract, and blood is pushed through the open tricuspid (right AV) and mitral valves into the ventricles. The semilunar valves are closed. (*b*) Beginning of ventricular systole: Ventricles begin to contract, and pressure within the ventricles increases and closes the tricuspid and mitral valves, causing the first heart sound. (*c*) Period of rising pressure. The semilunar valves open when the pressure within the ventricle exceeds that in the arteries, and blood spurts into the aorta and pulmonary artery. (*d*) At the beginning of ventricular diastole, the pressure in the relaxing ventricles drops below that in the arteries. The semilunar valves snap shut, causing the second heart sound. (*e*) Period of falling pressure: Blood flows from the veins into the relaxed atria. The tricuspid and mitral valves open when the pressure in the ventricles falls below that in the atria; blood then flows into the ventricles.

You can measure your heart rate by placing a finger over the radial artery in your wrist or the carotid artery in your neck and counting the pulsations. Arterial **pulse** is the alternate expansion and recoil of an artery. Each time the left ventricle pumps blood into the aorta, the elastic wall of the aorta expands to accommodate the blood. This expansion moves in a wave down the aorta and the arteries that branch from it. When the wave passes, the elastic arterial wall snaps back to its normal size.

Two main heart sounds can be distinguished

When you listen to the heartbeat with a stethoscope you can hear two main heart sounds, "lub-dup," which repeat rhythmically. The first heart sound, lub, is low-pitched, not very loud, and fairly long-lasting. It is caused mainly by the closing of the AV valves and marks the beginning of ventricular systole. The lub sound is quickly followed by the higher-pitched, louder, sharper, and shorter dup sound. Heard almost as a quick snap, the dup

Focus On

The Electrical Activity of the Heart

An electrocardiogram, or ECG, begins with a **P wave,** which represents the spread of an impulse through the atria just before atrial contraction (see figure). Then a **QRS complex** appears, reflecting the spread of an impulse through the ventricles just before they contract. As the ventricles recover, currents generated are reflected on the graph as a **T wave.** The heart then repeats its pattern of electrical impulses, generating a new P wave, QRS complex, and T wave.

Abnormalities in the ECG indicate disorders in the heart or its rhythm. One class of disorders that can be diagnosed with the help of the ECG is **heart block.** In this condition, transmission of an impulse is delayed or blocked at some point in the conduction system. **Artificial pacemakers** can be implanted in patients with severe heart block. A pacemaker is implanted beneath the skin, and its electrodes are connected to the heart. This device provides continuous rhythmic impulses that avoid the block and drive the heartbeat. ■

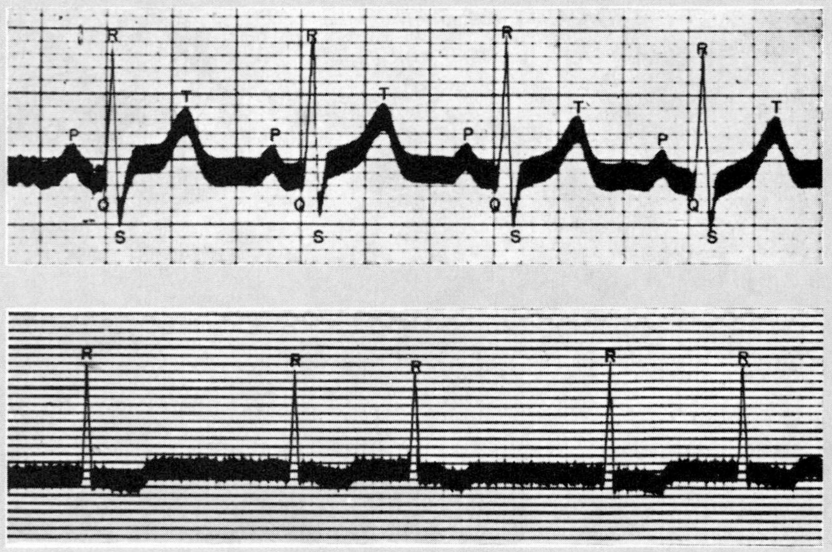

(a)

(b)

Electrocardiograms. (*a*) Tracing from a normal heart. The P wave corresponds to the contraction of the atria, the QRS complex to the contraction of the ventricles, and the T wave to the relaxation of the ventricles. (*b*) Tracing from a patient with atrial fibrillation. The individual muscle fibers of the atria twitch rapidly and independently. There is no regular atrial contraction and no P wave. The ventricles beat independently and irregularly, causing the QRS wave to appear at irregular intervals. (Courtesy of Dr. Lewis Dexter and Peter Bent Brigham Hospital, Boston, Mass.)

marks the closing of the semilunar valves and the beginning of ventricular diastole.

The quality of these sounds tells a discerning physician much about the state of the valves. When the semilunar valves are injured, a soft hissing noise ("lub-shhh") is heard in place of the normal sound. This is known as a **heart murmur** and may be caused by any condition that prevents the valves from closing tightly, permitting blood to flow backward into the ventricles during diastole.

The electrical activity of the heart can be recorded

As each wave of contraction spreads through the heart, electrical currents flow into the tissues surrounding the heart and onto the body surface. By placing electrodes on the body surface on opposite sides of the heart, the electrical activity can be amplified and recorded by either an oscilloscope or an electrocardiograph. The written record produced is called an **electrocardiogram** (**ECG** or **EKG**). (See *Focus On: The Electrical Activity of the Heart.*)

Cardiac output varies with the body's need

The volume of blood pumped by one ventricle during one beat is called the **stroke volume.** By multiplying the stroke volume by the number of times the ventricle beats per minute, the **cardiac output** can be computed. In other words, the cardiac output is the volume of blood pumped by one ventricle in 1 minute. For example, in a resting adult the heart may beat about 72 times per minute and pump about 70 milliliters of blood with each contraction.

$$\begin{aligned}\text{cardiac output} &= \text{stroke volume} \times \text{heart rate (number of ventricular contractions per minute)} \\ &= 70\ \text{ml/stroke} \times 72\ \text{strokes/minute} \\ &= 5040\ \text{ml/min (about 5 liters/min)}\end{aligned}$$

The cardiac output varies dramatically with the changing needs of the body. During stress or heavy exercise, the normal heart can increase its cardiac output fourfold to fivefold, so that 20 to 30 liters of blood can be pumped per minute. Cardiac output varies with changes in either stroke volume or heart rate.

Stroke volume depends on venous return

Stroke volume depends mainly on venous return, the amount of blood delivered to the heart by the veins. According to **Starling's law of the heart,** the greater the amount of blood delivered to the heart by the veins, the more blood the heart pumps (within physiological limits). When extra amounts of blood fill the heart chambers, the cardiac muscle fibers are stretched to a greater extent and contract with greater force, pumping a larger volume of blood into the arteries. This increase in stroke volume increases the cardiac output (Fig. 42–13).

The release of norepinephrine by sympathetic nerves also increases the force of contraction of the cardiac muscle fibers. Epinephrine released by the adrenal glands during stress has a similar effect on the heart muscle. When the force of contraction increases, the stroke volume increases, and this in turn increases cardiac output.

Heart rate is regulated by the nervous system

Although the heart is capable of beating independently of its control system, its rate is, in fact, carefully regulated by the nervous system (Fig. 42–13). Sensory receptors in the walls of certain blood vessels and heart chambers are sensitive to changes in blood pressure. When stimulated, they send messages to **cardiac centers** in the medulla of the brain. These cardiac centers maintain control over two sets of autonomic nerves that pass to the SA node. Sympathetic nerves release norepinephrine, which speeds the heart rate and increases the strength of contraction. Parasympathetic nerves release acetylcholine, which slows the heart.

Hormones also influence heart rate. During stress, the adrenal glands release epinephrine and norepinephrine, which speed the heart. An elevated body temperature can greatly increase heart rate. During fever, the heart may beat more than 100 times per minute. As you might expect, heart rate decreases when body temperature is lowered. This is why a patient's temperature may be deliberately lowered during heart surgery.

BLOOD PRESSURE DEPENDS ON BLOOD FLOW AND RESISTANCE TO BLOOD FLOW

Blood pressure is the force exerted by the blood against the inner walls of the blood vessels. It is determined by the blood flow and the resistance to that flow. Blood flow depends directly on the pumping action of the heart.

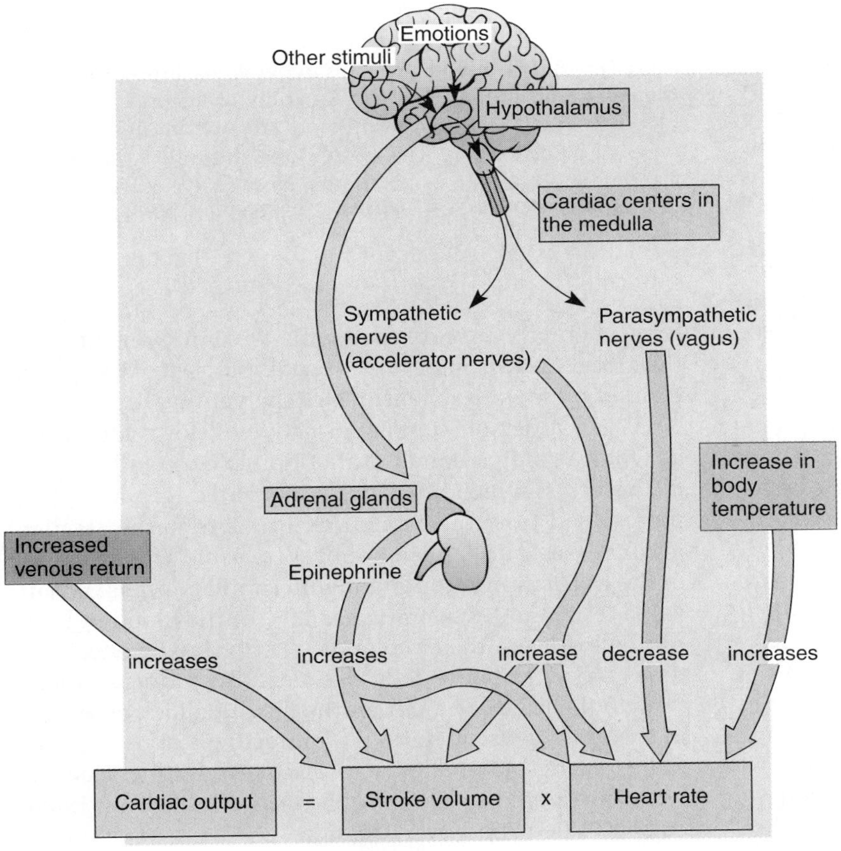

FIGURE 42–13 Multiple factors influence cardiac output.

When cardiac output increases, blood flow increases, causing a rise in blood pressure. When cardiac output decreases, blood flow decreases, causing a fall in blood pressure. The volume of blood flowing through the system also affects blood pressure. If blood volume is reduced by hemorrhage or by chronic bleeding, the blood pressure drops. On the other hand, an increase in blood volume results in an increase in blood pressure. For example, a high dietary intake of salt causes water retention. This results in an increase of blood volume and leads to higher blood pressure.

Blood flow is impeded by resistance; when the resistance to flow increases, blood pressure rises. **Peripheral resistance** is the resistance to blood flow caused by viscosity of the blood and by friction between the blood and the wall of the blood vessel. In the blood of a healthy person, viscosity remains fairly constant and is only a minor factor influencing changes in blood pressure. More important is the friction between the blood and the wall of the blood vessel. The length and diameter of a blood vessel determine the amount of surface area in contact with the blood. The length of a blood vessel does not change, but the diameter, especially of an arteriole, does. A small change in the diameter of a blood vessel causes a big change in blood pressure.

Blood pressure in arteries rises during systole and falls during diastole. Normal blood pressure (measured in the upper arm) for a young male adult is about 120/80 millimeters of mercury, abbreviated mm Hg, as measured by the sphygmomanometer. Systolic pressure is indicated by the numerator, diastolic by the denominator.

When the diastolic pressure consistently measures more than 95 mm Hg, the patient may be suffering from high blood pressure, or **hypertension.** In hypertension, there is usually increased vascular resistance, especially in the arterioles and small arteries. The heart's workload increases because it must pump against this greater resistance. As a result, the left ventricle enlarges and may begin to deteriorate in function. Heredity, obesity, and possibly high dietary salt intake are thought to be important in the development of hypertension.

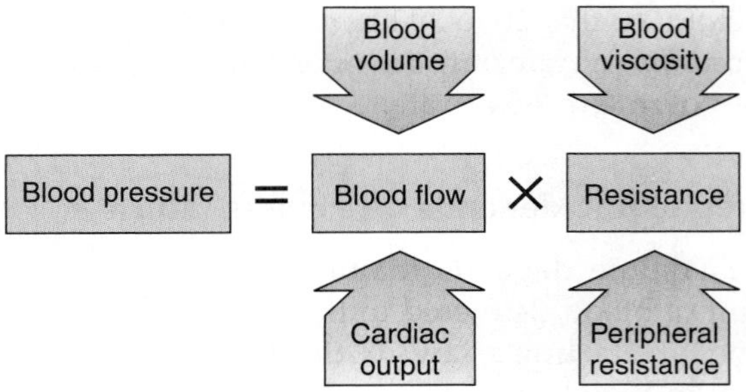

(a)

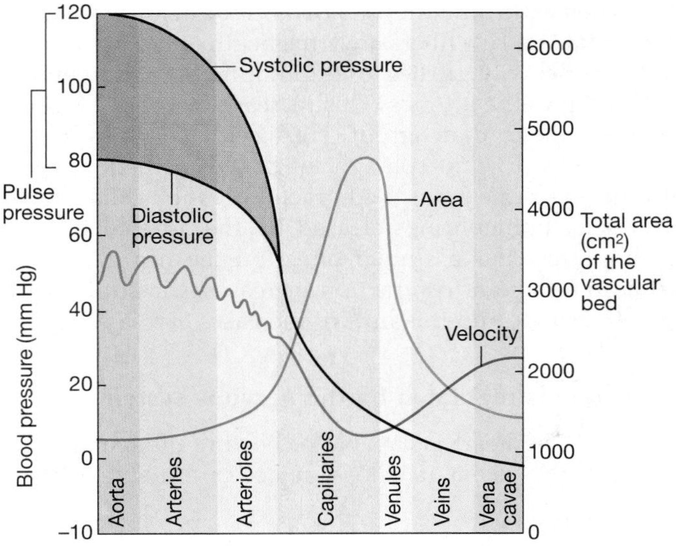

(b)

FIGURE 42–14 Blood pressure depends on blood flow and the resistance to that flow. (*a*) A variety of factors can affect blood flow and resistance. (*b*) Blood pressure in different types of blood vessels. The systolic and diastolic variations in arterial blood pressures are shown. Note that the venous pressure drops below zero (below atmospheric pressure) near the heart.

Blood pressure is highest in arteries

As you might guess, blood pressure is greatest in the large arteries, decreasing as blood flows away from the heart and through the smaller arteries and capillaries (Fig. 42–14). By the time blood enters the veins, its pressure is very low, even approaching zero. Flow rate can be maintained in veins at low pressure because they are low-resistance vessels. Their diameter is larger than that of corresponding arteries, and there is little smooth muscle in their walls. Flow of blood through veins depends on several factors, including muscular movement, which compresses veins. Most veins larger than 2 millimeters (0.08 inch) in diameter that conduct blood against the force of gravity are equipped with valves to prevent backflow (Fig. 42–15). Such valves usually consist of two cusps formed by inward extensions of the vein wall.

When a person stands perfectly still for a long time, as when a soldier stands at attention, blood tends to pool in the veins, which when fully distended, can accept no more blood from the capillaries. Pressure in the capillaries increases, and large amounts of plasma are forced out of the circulation through the thin capillary walls. Within just a few minutes, as much as 20% of the blood volume can be lost from the circulation—with drastic effect. Arterial blood pressure falls dramatically, reducing blood flow to the brain. Sometimes the resulting lack of oxygen in the brain causes fainting, a protective response aimed at increasing blood supply to the brain. Lifting a person who has fainted to an upright position can result in circulatory shock and even death.

Making the Connection

Cardiovascular System, Exercise, and Human Physiology

What are the benefits of exercise to the cardiovascular system and the rest of the body? Exercise improves fitness. Specifically, exercise improves muscle tone and strength, and enhances muscular and cardiovascular endurance. Recent studies suggest that exercise helps reduce hypertension (high blood pressure) and increases high-density lipoproteins (the good lipoproteins that help lower cholesterol levels) in the blood. Exercise also helps reduce stress.

Your endurance determines how long you can continue physical activities, such as running, swimming, and dancing. Endurance requires cardiovascular fitness as well as muscle strength. Endurance can be improved by aerobic exercise, which is any activity that increases the heart rate and requires oxygen. Aerobic exercise includes such activities as rapid walking, in-line skating, and bicycling.

Exercise results in marked physiological changes in the body. For example, it alters the rate and distribution of blood flow to the organs. Contracting muscles need more oxygen, so blood flow to the muscles increases dramatically. During aerobic exercise, it can increase more than 20 times. This increase helps meet not only the greater demand for oxygen, but also the greater need for removal of CO_2 and other wastes generated during physical activity. Thus, exercise has a purifying effect on body systems. Although the rate of blood flow to the brain is maintained, blood flow to the digestive organs and the kidneys is markedly reduced, so more blood is available for the muscles.

Regular aerobic exercise strengthens the heart muscle and decreases the resting heart rate. The heart does not have to work as hard. Athletic training often results in heart enlargement (hypertrophy), including thickening of the heart walls and an increase in the size of its chambers. This leads to an increase in the amount of blood the heart can pump with each contraction, and increases the strength of cardiac muscle contraction. The heart becomes more efficient in its pumping capacity both at rest and during physical activity.

Another benefit of exercise is increased metabolic rate, which makes it easier to maintain desired body weight and amount of body fat. As a result, more food can be eaten without gaining weight.

The American College of Sports Medicine recommends that healthy adults perform aerobic exercise for about 20 to 60 minutes three to five days each week. The activity should be intense enough to increase the heart rate to 70% to 90% of maximum. You can estimate maximum heart rate by subtracting your age from 220. For example, if you are 20 years old, your maximum heart rate is 220 minus 20, or 200 beats per minute. If you are in good health, and exercise at a level that raises your heart rate to 70% to 90% of maximum, your heart rate will increase to about 160 beats per minute.

If you have not been exercising and are just beginning a program, mild exercise such as walking can raise your heart rate to aerobic levels. As you improve your fitness, you will need to intensify your exercise to reach aerobic range. For example, you might speed up your walking or skating. ▲

Blood pressure is carefully regulated

Each time you get up from a horizontal position, changes occur in your blood pressure. Several complex mechanisms interact to maintain normal blood pressure so that you do not faint when you get out of bed each morning or change position during the day. When blood pressure falls, sympathetic nerves to the blood vessels stimulate vasoconstriction so that pressure rises again.

The **baroreceptors** present in the walls of certain arteries and in the heart wall are sensitive to changes in

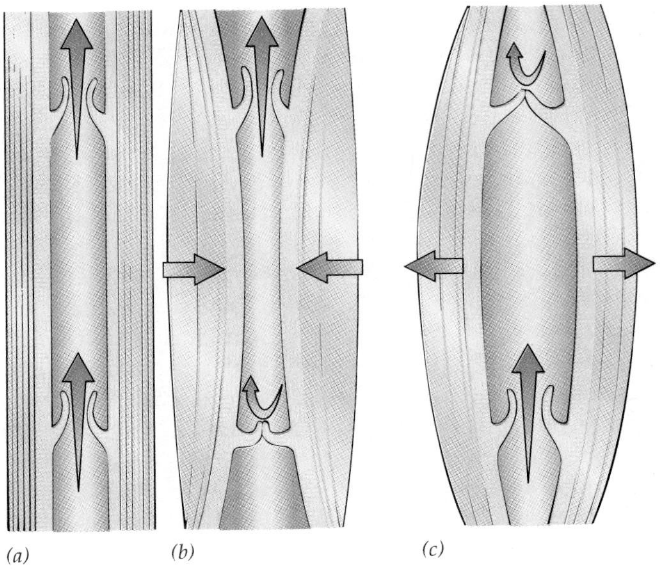

FIGURE 42–15 The contraction of skeletal muscles helps move blood through the veins. (*a*) Resting condition. (*b*) Muscles contract and bulge, compressing veins and forcing blood toward the heart. The lower valve prevents backflow. (*c*) Muscles relax, and the vein expands and fills with blood from below. The upper valve prevents backflow.

blood pressure. When an increase in blood pressure stretches the baroreceptors, messages are sent to the cardiac and vasomotor centers in the medulla of the brain. The cardiac center stimulates parasympathetic nerves that slow the heart, lowering blood pressure. The vasomotor center inhibits sympathetic nerves that constrict arterioles; this action lowers blood pressure. These neural reflexes continuously work to maintain blood pressure within normal limits.

Hormones are also involved in regulating blood pressure. The **angiotensins** are a group of hormones that act as powerful vasoconstrictors. The enzyme **renin** stimulates formation of angiotensins from a plasma protein. The kidneys release renin in response to low blood pressure. They also act indirectly to maintain blood pressure by influencing blood volume. This is accomplished by hormonal regulation of the rate at which salt and water are excreted.

In birds and mammals blood is pumped through a pulmonary and a systemic circuit

One of the main jobs of the circulation is to bring oxygen to all the cells of the body. In humans, as in other mammals and in birds, blood is charged with oxygen in the lungs. Then it is returned to the heart to be pumped out into the arteries that deliver it to the other tissues and organs of the body. There is a double circuit of blood vessels: (1) the **pulmonary circulation,** which connects the heart and lungs; and (2) the **systemic circulation,** which connects the heart with all of the tissues of the body. This general pattern of circulation may be traced in Figure 42–16.

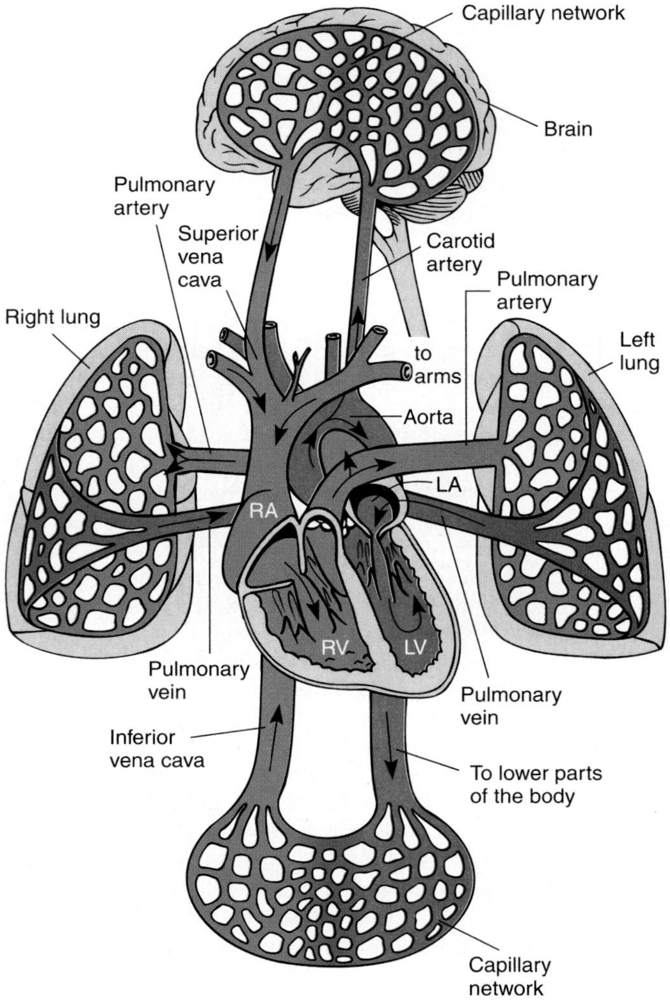

FIGURE 42–16 This highly simplified diagram illustrates the pattern of circulation through the systemic and pulmonary circuits. Red represents oxygen-rich blood; blue represents oxygen-poor blood.

The pulmonary circulation oxygenates the blood

Blood from the tissues returns to the right atrium of the heart partly depleted of its oxygen supply. This oxygen-poor blood, loaded with carbon dioxide, is pumped by the right ventricle into the pulmonary circulation. As it emerges from the heart, the large pulmonary trunk branches to form the two pulmonary arteries, one going to each lung. These are the only arteries in the body that carry oxygen-poor blood.

In the lungs the pulmonary arteries branch into smaller and smaller vessels, which finally give rise to extensive networks of pulmonary capillaries that bring blood to all of the air sacs of the lungs. As blood circulates through the pulmonary capillaries, carbon dioxide diffuses out of the blood and into the air sacs. Oxygen from the air sacs diffuses into the blood so that, by the time blood enters the pulmonary veins leading back to the left atrium of the heart, it is charged with oxygen. Pulmonary veins are the only veins in the body that carry blood rich in oxygen.

In summary, blood flows through the pulmonary circulation in the following sequence:

right atrium → right ventricle → pulmonary arteries → pulmonary capillaries (in lungs) → pulmonary veins → left atrium

The systemic circulation delivers blood to the tissues

Blood entering the systemic circulation is pumped by the left ventricle into the **aorta,** the largest artery of the body. Arteries that branch off from the aorta conduct blood to all regions of the body. Some of the principal branches include the **coronary arteries** to the heart wall itself, the **carotid arteries** to the brain, the **subclavian arteries** to

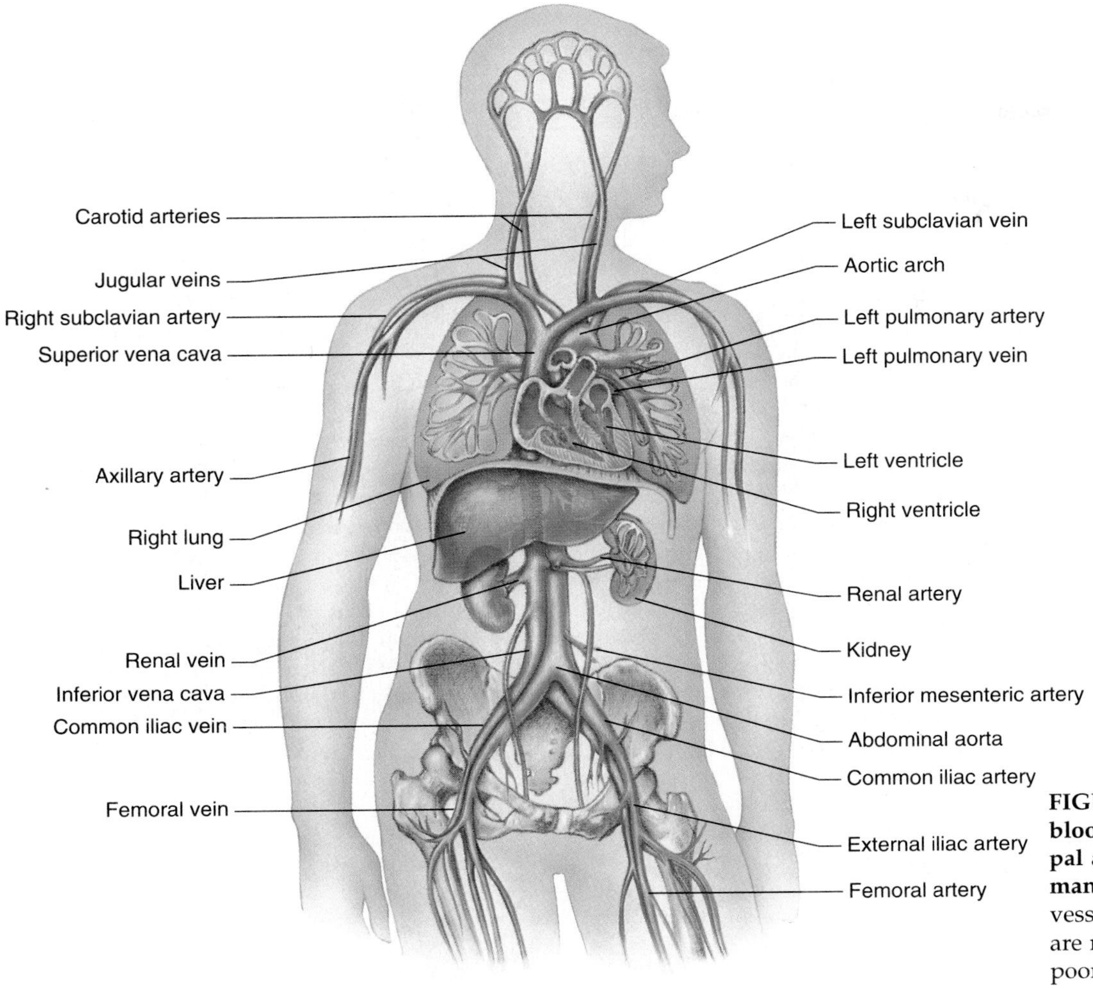

FIGURE 42–17 Circulation of blood through some of the principal arteries and veins of the human body is illustrated. Blood vessels carrying oxygen-rich blood are red; those carrying oxygen-poor blood are blue.

the shoulder region, the **mesenteric artery** to the intestine, the **renal arteries** to the kidneys, and the **iliac arteries** to the legs (Fig. 42–17). Each of these arteries gives rise to smaller branches, which in turn give rise to smaller and smaller vessels, somewhat like branches of a tree that divide until they form tiny twigs. Eventually blood flows into the capillary network within each tissue or organ.

Blood returning from the capillary networks within the brain passes through the **jugular veins.** Blood from the shoulders and arms drains into the **subclavin veins.** These veins and others returning blood from the upper portion of the body merge to form a very large vein that empties blood into the right atrium. In humans this vein is called the **superior vena cava. Renal veins** from the kidneys, **iliac veins** from the lower limbs, **hepatic veins** from the liver, and other veins from the lower portion of the body return blood to the **inferior vena cava,** which delivers blood to the right atrium.

As an example of blood circulation through the systemic circuit, let us trace a drop of blood from the heart to the right leg and back to the heart:

> left atrium → left ventricle → aorta → right common iliac artery → smaller arteries in leg → capillaries in leg → small vein in leg → common iliac vein → inferior vena cava → right atrium

The coronary circulation delivers blood to the heart

The heart muscle is not nourished by the blood within its chambers because its walls are too thick for nutrients and oxygen to diffuse through them to reach all of the muscle fibers. Instead, the cardiac muscle is supplied by *coronary arteries* branching from the aorta at the point where that vessel leaves the heart. These arteries branch, giving rise to a network of blood vessels within the wall of the heart. Nutrients and gases are exchanged through the *coronary capillaries.* Blood from these capillaries flows into *coronary veins,* which join to form a large vein, the *coronary sinus.* The coronary sinus empties directly into the right atrium; it does not join either of the venae cavae.

When one of the coronary arteries is blocked, the cells in the area of the heart muscle served by that artery die due to deprivation of oxygen and nutrients, and the affected muscle stops contracting. If sufficient cardiac muscle is affected, the heart may stop beating entirely. This is a common cause of "heart attack" and is often the result of atherosclerosis (see *Focus On: Cardiovascular Disease*).

Four arteries deliver blood to the brain

Four arteries—two internal carotid arteries and two vertebral arteries (branches of the subclavian arteries)—

Focus On

Cardiovascular Disease

Cardiovascular disease is the number-one cause of death in the United States and in most other industrial societies. Most often death results from some complication of **atherosclerosis** (hardening of the arteries as a result of lipid and calcium deposition). Although it can affect almost any artery, the disease most often develops in the aorta and in the coronary and cerebral arteries. When it occurs in the cerebral arteries, it can lead to a **cerebrovascular accident (CVA),** commonly referred to as a stroke.

Although there is apparently no single cause of atherosclerosis, several major risk factors have been identified:

1. **Elevated cholesterol levels** in the blood, often associated with diets rich in total calories, total fats, saturated fats, and cholesterol.
2. **Hypertension.** The higher the blood pressure, the greater the risk.
3. **Cigarette smoking.** The risk of developing atherosclerosis is two to six times greater in smokers than in nonsmokers and is directly proportional to the number of cigarettes smoked daily.
4. **Diabetes mellitus,** an endocrine disorder in which glucose is not metabolized normally.

The risk of developing atherosclerosis also increases with age. Estrogen hormones are thought to offer some protection in women until after menopause, when the concentration of these hormones decreases. Other suggested risk factors currently being studied are obesity, hereditary predisposition, lack of exercise, stress and behavior patterns, and dietary factors.

In atherosclerosis, lipids are deposited in the smooth muscle cells of the arterial wall. These cells proliferate, and the inner lining thickens. More lipid, especially cholesterol from low-density lipoproteins, accumulates in the wall. Eventually calcium is deposited there, contributing to the slow formation of hard plaque. As the plaque develops, arteries lose their ability to stretch when filled with blood, and they become progressively occluded (blocked), as shown in the figure. As the artery narrows, less blood can pass through to reach the tissues served by that vessel, and the tissue may become **ischemic** (lacking in blood). Under these conditions the tissue is deprived of an adequate oxygen supply.

When a coronary artery becomes narrowed, **ischemic heart disease** can develop. Sufficient oxygen may reach the heart tissue during normal activity, but the increased need for oxygen during exercise or emotional stress results in the pain known as **angina pectoris.** Persons with this condition often carry nitroglycerin pills for use during an attack. This drug dilates veins, reducing venous return. Cardiac output is lowered so that the heart is not working so hard and requires less oxygen. Nitroglycerin also dilates the coronary arteries slightly, allowing more blood to reach the heart muscle.

Myocardial infarction (MI), popularly referred to as heart attack, is a very serious, often fatal, consequence of ischemic heart disease. MI often results from a sudden decrease in coronary blood supply. The portion of cardiac muscle deprived of oxygen dies within a few minutes and is then referred to as an **infarct.** MI is the leading cause of death and disability in the United States. Just what triggers the sudden decrease in blood supply is a matter of some debate. It is thought that in some cases an episode of ischemia triggers a fatal arrhythmia such as **ventricular fibrillation,** a condition in which the ventricles contract very rapidly without actually pumping blood. In other cases, a *thrombus* (clot) may form in a diseased coronary artery. Because the arterial wall is roughened, platelets may adhere to it and initiate clotting.

If the thrombus blocks a sizable branch of a coronary artery, blood flow to a portion of heart muscle is impeded or completely halted. When this condition, referred to as a *coronary occlusion,* prevents blood flow to a large region of cardiac muscle, the heart may stop beating—that is, cardiac arrest may occur—and death can follow within moments. If only a small region of the heart is affected, however, the heart may continue to function. Cells in the region deprived of oxygen die and are replaced by scar tissue. ■

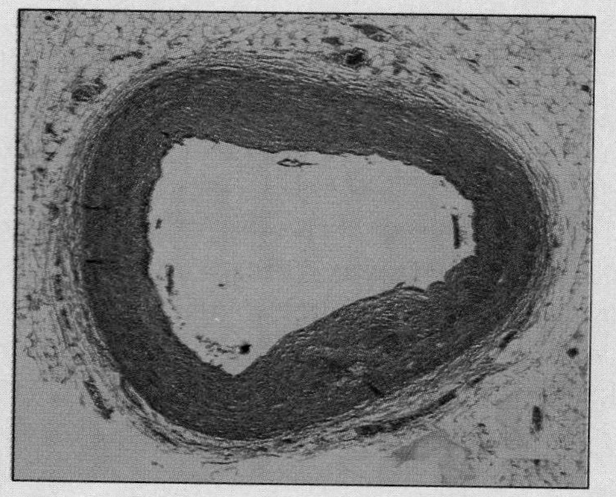

(*a*) 500 μm

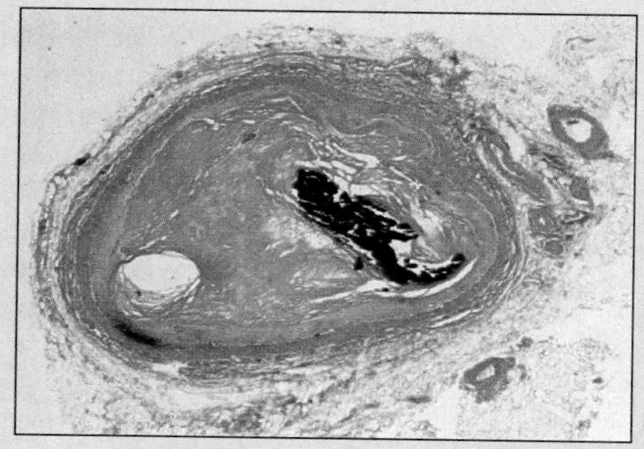

(*b*) 500 μm

Progression of atherosclerosis. Cross sections through two arteries showing changes that take place in atherosclerosis. (***a***) Normal coronary artery. (***b***) This artery is almost completely blocked with atherosclerotic plaque. (*a,* Visuals Unlimited/Cabisco; *b,* Visuals Unlimited/Sloop-Ober) deliver blood to the brain. At the base of the brain, branches of these arteries form an arterial circuit called the **circle of Willis.** In the event that one of the arteries serving the brain becomes blocked or injured in some way, this arterial circuit helps ensure blood delivery to the brain cells via other vessels. Blood from the brain returns to the superior vena cava by way of the internal jugular veins at either side of the neck.

The hepatic portal system delivers nutrients to the liver

Blood almost always travels from artery to capillary to vein. An exception to this sequence occurs in the **hepatic portal system,** which delivers blood rich in nutrients to the liver. Blood is conducted to the small intestine by the superior mesenteric artery. Then, as it flows through capillaries within the wall of the intestine, blood picks up glucose, amino acids, and other nutrients. This blood passes into the mesenteric vein and then into the **hepatic portal vein.** Instead of going directly back to the heart (as most veins would), the hepatic portal vein delivers nutrients to the liver.

Within the liver, the hepatic portal vein gives rise to an extensive network of tiny blood sinuses. As blood courses through the hepatic sinuses, liver cells remove nutrients and store them. Eventually liver sinuses merge to form hepatic veins, which deliver blood to the inferior vena cava. The hepatic portal vein contains blood that, although laden with food materials, has already given up some of its oxygen to the cells of the intestinal wall. Oxygen-rich blood is supplied to the liver by the hepatic artery.

THE LYMPHATIC SYSTEM IS AN ACCESSORY CIRCULATORY SYSTEM

In addition to the blood circulatory system, vertebrates have an accessory circulatory system, the **lymphatic system** (Fig. 42–18), which has three important functions: (1) to collect and return interstitial fluid to the blood; (2) to defend the body against disease organisms by way of immune mechanisms; and (3) to absorb lipids from the digestive tract. In this section we focus on the first function. Immunity is discussed in Chapter 43, and lipid absorption is discussed in Chapter 45.

The lymphatic system consists of lymphatic vessels and lymph tissue

The lymphatic system consists of (1) an extensive network of **lymphatic vessels** that conduct **lymph,** the clear, watery fluid formed from interstitial fluid, and (2) **lymph tissue,** a type of connective tissue with large numbers of lymphocytes. Lymph tissue is organized into small masses of tissue called **lymph nodes** and **lymph nodules.** The tonsils, thymus gland, and spleen, which consist mainly of lymph tissue, are also part of the lymphatic system.

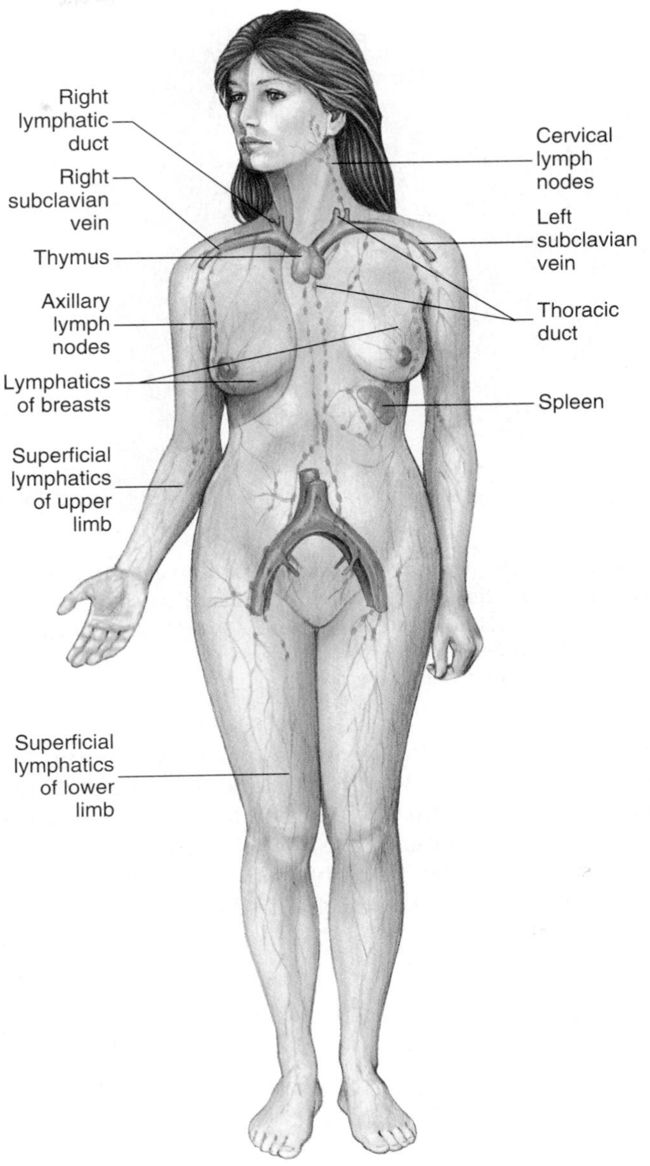

FIGURE 42–18 The lymphatic system extends into most tissues of the body. Note that while the lymphatic vessels extend throughout the body, the lymph nodes are clustered in certain regions. The right lymphatic duct drains lymph from the upper right quadrant of the body. The thoracic duct drains lymph from other regions of the body.

Tiny "dead-end" capillaries of the lymphatic system extend into almost all of the tissues of the body (Fig. 42–19). Lymph capillaries join to form larger **lymphatics** (lymph veins). There are no lymph arteries.

Interstitial fluid enters lymph capillaries and then is referred to as lymph. The lymph is conveyed into lym-

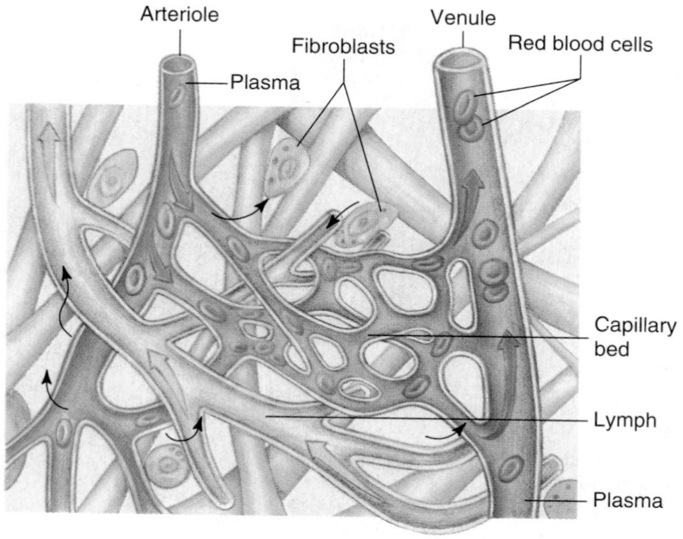

FIGURE 42–19 Lymph capillaries drain excess interstitial fluid from the tissues. Note that blood capillaries are connected to vessels at both ends, whereas lymph capillaries, shown in yellow, are "dead-end streets." The arrows indicate direction of flow.

phatics. At certain locations the lymphatics empty into lymph nodes, where lymph is filtered. Bacteria and other harmful materials are filtered from the lymph. The lymph then flows into lymphatics leaving the lymph node. Lymphatics from all over the body conduct lymph toward the shoulder region. These vessels join the circulatory system at the base of the subclavian veins by way of ducts: the **thoracic duct** on the left side and the **right lymphatic duct** on the right.

Tonsils are masses of lymph tissue under the lining of the oral cavity and throat. (When enlarged, the pharyngeal tonsils in back of the nose are called **adenoids.**) Tonsils help protect the respiratory system from infection by destroying bacteria and other foreign matter that enter the body through the mouth or nose. Unfortunately, tonsils are sometimes overcome by invading bacteria, become the site of frequent infection themselves, and then become targets for surgical removal.

Some nonmammalian vertebrates such as the frog have lymph "hearts," which pulsate and squeeze lymph along. However, in mammals the walls of the lymph vessels themselves pulsate. Valves within the lymph vessels prevent the lymph from flowing backward. When muscles contract or when arteries pulsate, pressure on the lymph vessels increases lymph flow. The rate at which lymph flows is slow and variable, and the total lymph flow is about 100 milliliters per hour—very much slower than the 5 liters per minute of blood flowing in the vascular system.

The lymphatic system plays an important role in fluid homeostasis

When blood enters a capillary network it is under rather high pressure, so some plasma is forced out of the capillaries and into the tissues. Once it leaves the blood ves-

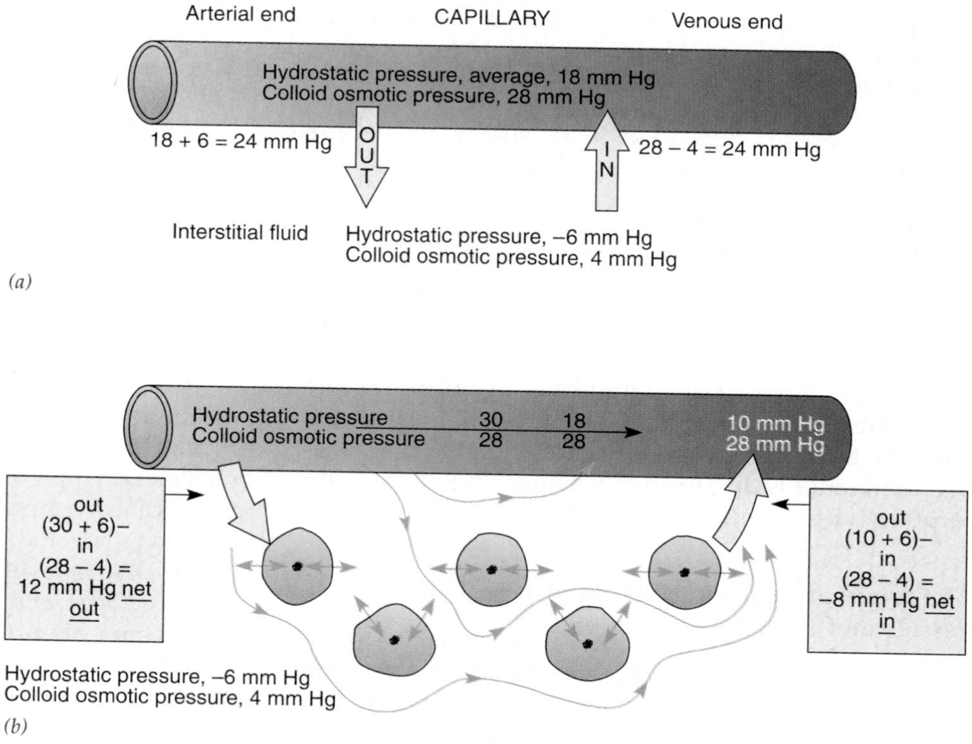

FIGURE 42–20 Materials are exchanged between blood and tissue fluid through the thin walls of the capillaries. (*a*) Hydrostatic and osmotic pressures are responsible for the exchange of materials between blood and interstitial fluid. (*b*) The path by which water flows in and out of capillaries to the interstitial fluid.

sels, this fluid is called interstitial fluid, or tissue fluid. It is somewhat similar to plasma but contains no red blood cells or platelets and only a few white blood cells. Its protein content is about one fourth of that found in plasma. This is because proteins are too large to pass easily through capillary walls. Smaller molecules dissolved in the plasma do pass out with the fluid leaving the blood vessels. Thus, interstitial fluid contains glucose, amino acids, other nutrients, and oxygen, as well as a variety of salts. This nourishing fluid bathes all the cells of the body.

The main force pushing plasma out of the blood is hydrostatic pressure—that is, the pressure exerted by the blood on the capillary wall, caused by the beating of the heart (Fig. 42–20). At the venous ends of the capillaries the hydrostatic pressure is much lower. Here the principal force is the osmotic pressure of the blood, which acts to draw fluid back into the capillary. However, osmotic pressure is not entirely effective. Not as much fluid is returned to the circulation as escapes. Furthermore, protein does not return effectively into the venous capillaries and instead tends to accumulate in the interstitial fluid. These potential problems are so serious that without the lymphatic system, fluid balance in the body would be significantly disturbed within a few hours, and death would occur within about 24 hours. The lymphatic system preserves fluid balance by collecting about 10% of the interstitial fluid and the protein that accumulates in it.

The walls of the lymph capillaries are composed of endothelial cells that overlap slightly. When interstitial fluid accumulates, it presses against these cells, pushing them inward like tiny swinging doors that can swing in only one direction. As fluid accumulates within the lymph capillary, these cell doors are pushed closed.

Obstruction of the lymphatic vessels can lead to **edema,** a swelling that results from excessive accumulation of interstitial fluid. Lymphatic vessels can be blocked as a result of injury, inflammation, surgery, or parasitic infection. For example, when a breast is removed (mastectomy) because of cancer, lymph nodes in the underarm region are often also removed in an effort to prevent the spread of cancer cells. The disrupted lymph circulation may cause the patient's arm to swell greatly. Fortunately, new lymph vessels develop within a few months and the swelling slowly subsides.

Filariasis, a parasitic infection caused by a larval nematode that is transmitted to humans by mosquitoes, also disrupts lymph flow. The adult worms live in the lymph veins, blocking lymph flow. Interstitial fluid then accumulates, causing tremendous swelling. The term elephantiasis has been used to describe the swollen legs sometimes caused by this disease, because they resemble the huge limbs of an elephant (Fig. 42–21).

FIGURE 42–21 Lymphatic drainage is blocked in the limbs of this individual because of the parasitic infection known as filariasis. The condition characterized by such swollen limbs is elephantiasis. (Lawrence S. Burr)

SUMMARY

I. Small, simple invertebrates, such as sponges, cnidarians, and flatworms, depend on diffusion for internal transport. More complex animals require a specialized circulatory system.

II. Arthropods and most mollusks have an open circulatory system in which blood flows into a hemocoel, bathing the tissues directly.

III. Some invertebrates and all vertebrates have a closed circulatory system in which blood flows through a continuous circuit of blood vessels.

IV. The vertebrate circulatory system consists of a muscular heart that pumps blood into a system of arteries, capillaries, and veins. This system transports nutrients, oxygen, wastes, and hormones; helps maintain fluid balance and body temperature; and defends the body against disease.

V. Vertebrate blood consists of liquid plasma in which red blood cells, white blood cells, and platelets are suspended.
   A. Plasma consists of water, salts, substances in transport, and plasma proteins, including albumins, globulins, and fibrinogen.
   B. Red blood cells transport oxygen and carbon dioxide.
   C. White blood cells defend the body against disease organisms. Lymphocytes and monocytes are agranular white blood cells; neutrophils, eosinophils, and basophils are granular white cells.
   D. Platelets patch damaged blood vessels and release substances essential for blood clotting.

VI. Arteries carry blood away from the heart chambers; veins return blood to the heart chambers.
   A. Arterioles constrict and dilate to regulate blood pressure and distribution of blood to the tissues.
   B. Capillaries are the thin-walled exchange vessels through which materials pass back and forth between the blood and tissues.

VII. The vertebrate heart consists of one or two atria, which receive blood, and one or two ventricles, which pump blood into the arteries.
   A. The four-chambered hearts of birds and mammals separate oxygen-rich blood from oxygen-poor blood.
   B. The heart is enclosed by a pericardium and is equipped with valves that prevent backflow of blood.
   C. The sinoatrial node initiates each heartbeat. A specialized conduction system ensures that the heart beats in a coordinated manner.
   D. Cardiac output equals stroke volume times heart rate.
      1. Stroke volume depends on venous return and on neural messages and hormones.
      2. Heart rate is regulated mainly by the nervous system but is influenced by hormones, body temperature, and certain other factors.
      3. The heart rate can be measured by counting the pulse. Arterial pulse is the alternate expansion and recoil of an artery.

VIII. Blood pressure is the force exerted by the blood against the inner walls of the blood vessel.
   A. Blood pressure is greatest in the arteries, and decreases as blood flows through the capillaries.
   B. Baroreceptors sensitive to changes in blood pressure send messages to the cardiac and vasomotor centers in the medulla of the brain. When informed of an increase in blood pressure, the cardiac center stimulates parasympathetic nerves that slow the heart rate, and the vasomotor center inhibits sympathetic nerves that constrict blood vessels. These actions reduce blood pressure.
   C. Angiotensins raise blood pressure.

IX. The pulmonary circulation connects heart and lungs; the systemic circulation connects the heart with the tissues.
   A. In the pulmonary circulation, the right ventricle pumps blood into the pulmonary arteries, one going to each lung. Blood circulates through pulmonary capillaries in the lung, and then is conducted to the left atrium by a pulmonary vein.
   B. In the systemic circulation, the left ventricle pumps blood into the aorta, which branches into arteries leading to the body organs. After flowing through capillary networks within various organs, blood flows into veins that conduct it to the right atrium.
   C. The coronary circulation supplies the heart muscle with blood.
   D. The hepatic portal system circulates nutrient-rich blood through the liver.

X. Atherosclerosis leads to ischemic heart disease, in which the heart muscle does not receive sufficient blood. Myocardial infarction is a very serious consequence of ischemic heart disease.

XI. The lymphatic system collects interstitial fluid and returns it to the blood. It plays an important role in homeostasis of fluids.

SELECTED KEY TERMS

anemia
aorta
arteriole
artery
atherosclerosis
atrioventricular valve
atrium (atria)
baroreceptors
blood
blood pressure
capillary
cardiac cycle
cardiac output
cardiovascular disease
circulatory system
closed circulatory system
diastole
fibrin
hemoglobin
hepatic portal system
interstitial fluid
lymph
lymphatic system
lymph node
mitral valve
myocardial infarction (MI)
open circulatory system
pacemaker
plasma
plasma protein
platelets
pulmonary circulation
pulse
red blood cell (erythrocyte)
semilunar valve
systemic circulation
systole
tissue fluid
vasoconstriction
vasodilation
vein
ventricle
white blood cell (leukocyte)

POST-TEST

1. In a(an) __________ circulatory system, the heart pumps blood into a hemocoel.
2. The fluid component of blood is __________ .
3. A deficiency in hemoglobin is referred to as __________ .

4. During clotting, fibrinogen is converted to __________ by the action of thrombin.
5. The force exerted by the blood against the inner walls of the blood vessels is known as __________ __________ .
6. The pulmonary vein delivers blood to the left __________ .
7. The __________ __________ system (vein) delivers blood to the liver.
8. Blood pressure is sensed by __________ within certain arteries.
9. The __________ valve is located between the left atrium and left ventricle.
10. __________ is the pigment in vertebrate blood that transports oxygen.
11. __________ proteins include albumins and globulins.
12. __________ are blood vessels important in maintaining appropriate blood pressure.
13. Nutrients and other materials are exchanged through the walls of __________ .
14. In birds and mammals, blood leaving the right __________ enters one of the pulmonary arteries.
15. The __________ valves guard the exits of the heart.
16. The portion of the cardiac cycle during which the heart contracts is referred to as __________ ; the relaxation phase is __________ .
17. The __________ __________ is the volume of blood that the heart pumps in one minute.
18. The largest artery in the human body is the __________ .
19. In __________ , the arterial wall thickens and may block the passage of blood.
20. __________ is interstitial fluid that enters lymph vessels.

REVIEW QUESTIONS

1. List five functions of the vertebrate circulatory system.
2. Compare how nutrients and oxygen are transported to the body cells in a hydra, planarian, earthworm, insect, and frog.
3. Contrast an open with a closed circulatory system.
4. List the functions of the main groups of plasma proteins.
5. Contrast the structure and functions of red and white blood cells.
6. Summarize the process by which blood clots.
7. Compare the functions of arteries, capillaries, and veins. Why are arterioles important in maintaining homeostasis?
8. Compare the heart of a fish, amphibian, reptile, and bird.
9. Define cardiac output, and describe factors that influence it.
10. What is the relationship between blood pressure and peripheral resistance? What mechanisms regulate blood pressure?
11. Trace the path of a red blood cell: (a) from the inferior vena cava to the aorta; and (b) from the jugular vein to the kidney.
12. What is the function of the hepatic portal system? How does its sequence of blood vessels differ from that in most other circulatory routes?
13. List four risk factors associated with the development of atherosclerosis. How is atherosclerosis linked to ischemic heart disease? Describe myocardial infarction.
14. How does the lymph system help maintain fluid balance?
15. What is the relationship between plasma, interstitial fluid, and lymph?
16. Blood pressure is low in capillaries. Explain how this helps to retain fluid in the circulation.
17. Label the diagram.

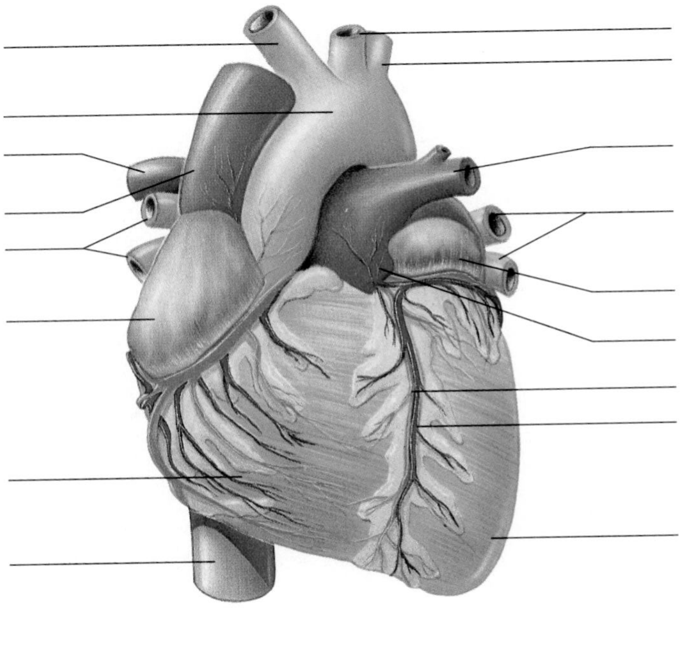

YOU MAKE THE CONNECTION

1. How do the changes that take place during exercise affect the heart? Compare the heart function of an adult who does not exercise with that of an athlete.
2. Cartilage lacks blood and lymph vessels. How do you imagine its cells are nourished? What effect, if any, would a lack of blood vessels have on cartilage healing after an injury?
3. How is the heart of the fish specifically adapted to its lifestyle?
4. When the nerves to the heart are cut, the heart rate increases to about 100 contractions per minute. What does this indicate about the regulation of the heart rate?

RECOMMENDED READINGS

Cantin, M., and J. Genest. "The Heart as an Endocrine Gland." *Scientific American,* Vol. 254, No. 2, February 1986. In addition to its pumping function, the heart produces a hormone that helps regulate blood pressure and volume.

Golde, D. W. "The Stem Cell." *Scientific American,* Vol. 265, No. 6, December 1991. A discussion of the cells that give rise to the various types of blood cells, with emphasis on implications in the development of new treatments for cancer and immune diseases.

Harken, A. H. "Surgical Treatment of Cardiac Arrhythmias." *Scientific American,* Vol. 269, No. 1, July 1993. A surgical procedure is used to correct dangerously rapid heartbeats.

Lawn, R. M. "Lipoprotein(a) in Heart Disease." *Scientific American,* Vol. 266, No. 6, June 1992. Lipoprotein(a) transports cholesterol; this protein can raise the risk of a heart attack.

Stamler, J. and Neaton, J.D. "Benefits of Lower Cholesterol," *Scientific American: Science & Medicine,* Vol. 1, No. 2, May/June 1994. Recent declines in death rate from heart disease appear to be related to lifestyle changes.

Zivin, J. A., and D. W. Choi. "Stroke Therapy." *Scientific American,* Vol. 265, No. 1, July 1991. New strategies are being evaluated to limit the brain damage that occurs in these cardiovascular accidents.

CHAPTER 43

Internal Defense

The T lymphocyte (*green*) shown in this color scanning EM is infected with HIV (human immunodeficiency virus; *red*). (NIBSC/Science Photo Library/Photo Researchers, Inc.)

Animals have internal defense mechanisms to protect them against disease-causing organisms that enter the body in air, food, and water, and through wounds in the skin. Disease-causing microorganisms including certain viruses, bacteria, fungi, and protozoa are referred to as **pathogens.** Internal defense depends on the animal's ability to distinguish between *self* and *nonself.* Such recognition is possible because organisms are biochemically unique. Cells have surface proteins different from those on the cells of other species or even other members of the same species. An organism "knows" its own cells and "recognizes" those of other organisms as foreign.

Pathogens have macromolecules on their cell surfaces that the body recognizes as foreign. A single bacterium may have from 10 to more than 1000 distinct macromolecules on its surface. When a bacterium invades an animal, its distinctive surface macromolecules stimulate the animal's defense mechanisms. A substance capable of stimulating an immune response is called an **antigen.** Many macromolecules, including proteins, RNA, DNA, and some carbohydrates, are antigenic.

Nonspecific defense mechanisms provide general protection against pathogens. These mechanisms prevent most pathogens from entering the body, and rapidly destroy those pathogens that do penetrate the outer defenses. Phagocytosis of invading bacteria is an example of a nonspecific defense mechanism.

Specific defense mechanisms are tailor-made to combat specific antigens associated with each pathogen. Specific defense mechanisms are collectively referred to as **immune responses.** The term *immune* is derived from a Latin word meaning "safe." **Immunology,** the study of specific defense mechanisms, is one of the most exciting fields of biomedical research today.

Immune responses can be directed to the particular type of pathogen infecting the body. One of the most important specific defense mechanisms is the production of **antibodies,** highly specific proteins that bind to antigens. In complex animals, internal defense includes immunological memory, the capacity to respond more effectively the second time foreign molecules invade the body.

LEARNING OBJECTIVES

After you have studied this chapter you should be able to:

1. Compare in general terms the type of internal defense mechanisms in invertebrates and vertebrates.
2. Distinguish between specific and nonspecific defense mechanisms, and describe nonspecific mechanisms.
3. Define the terms *antigen* and *antibody*, describe how antigens stimulate immune responses, and draw the basic structure of an antibody.
4. Identify the cells of the immune system, and contrast T and B lymphocytes (T cells and B cells) with respect to their development and function. (Include the role of the thymus.)
5. Summarize the mechanisms of antibody-mediated immunity, including the effects of antigen-antibody complexes on pathogens; include a discussion of the complement system.
6. Describe the mechanisms of cell-mediated immunity, including development of memory cells.
7. Contrast a secondary with a primary immune response.
8. Compare active and passive immunity, giving examples of each.
9. Describe how the body destroys cancer cells.
10. Summarize the immunological basis of graft rejection, and explain how the effects of graft rejection can be minimized.
11. Describe the immunological basis of autoimmune diseases, give two examples, and list possible causes.
12. Explain the immunological basis of allergy, and briefly describe the events that occur during (a) a hayfever response, and (b) systemic anaphylaxis.
13. Describe the cause of AIDS, the risk factors and the progress of the disease, and the difficulties encountered in developing a vaccine.

INVERTEBRATES HAVE MAINLY NONSPECIFIC INTERNAL DEFENSE MECHANISMS

All invertebrate species that have been studied demonstrate the ability to distinguish between self and nonself. Invertebrates make nonspecific immune responses such as phagocytosis and the inflammatory response. Most invertebrates can also demonstrate some specificity in their immune responses.

Sponge cells have specific glycoproteins on their surfaces that enable them to distinguish between self and nonself. When cells of two different species are mixed together, they reassort according to species. When two different species of sponges are forced to grow in contact with each other, tissue is destroyed along the region of contact. Cnidarians also reject grafted tissue and destroy foreign tissue.

Coelomate invertebrates (those with a coelom) have two types of amoeba-like phagocytes that engulf and destroy bacteria and other foreign matter. Many coelomate invertebrates also have nonspecific substances in the hemolymph that kill bacteria, inactivate cells of some pathogens, and cause some foreign cells to clump. In mollusks, these hemolymph substances enhance phagocytosis by the phagocytes.

Certain cnidarians (e.g., corals) and annelids (e.g., earthworms) have specific immune mechanisms and immunological memory. In them, and in some echinoderms and simple chordates, the body appears to remember antigens for a short period of time; thus it can respond more effectively when they are encountered again. Echinoderms and tunicates are the simplest animals known to have differentiated white blood cells that perform immune functions.

VERTEBRATES LAUNCH BOTH NONSPECIFIC AND SPECIFIC IMMUNE RESPONSES

Like invertebrates, vertebrates protect themselves against pathogens with both nonspecific and specific defense mechanisms (Fig. 43–1). More sophisticated specific defense mechanisms (immune responses) are possible because vertebrates have a specialized lymphatic system (Chapter 42). Only vertebrates have true lymphocytes. In the discussion that follows, we will focus on the human immune system, with references to those of other vertebrates.

VERTEBRATE NONSPECIFIC DEFENSE MECHANISMS INCLUDE MECHANICAL AND CHEMICAL BARRIERS

An animal's first line of defense against pathogens is its outer covering. For example, the *intact* human skin presents both a mechanical and a chemical barrier to microorganisms. Sweat and sebum contain chemicals that destroy certain types of bacteria. **Lysozyme,** an enzyme found in many tissues, and in tears and other body fluids, attacks the cell walls of many gram-positive bacteria.

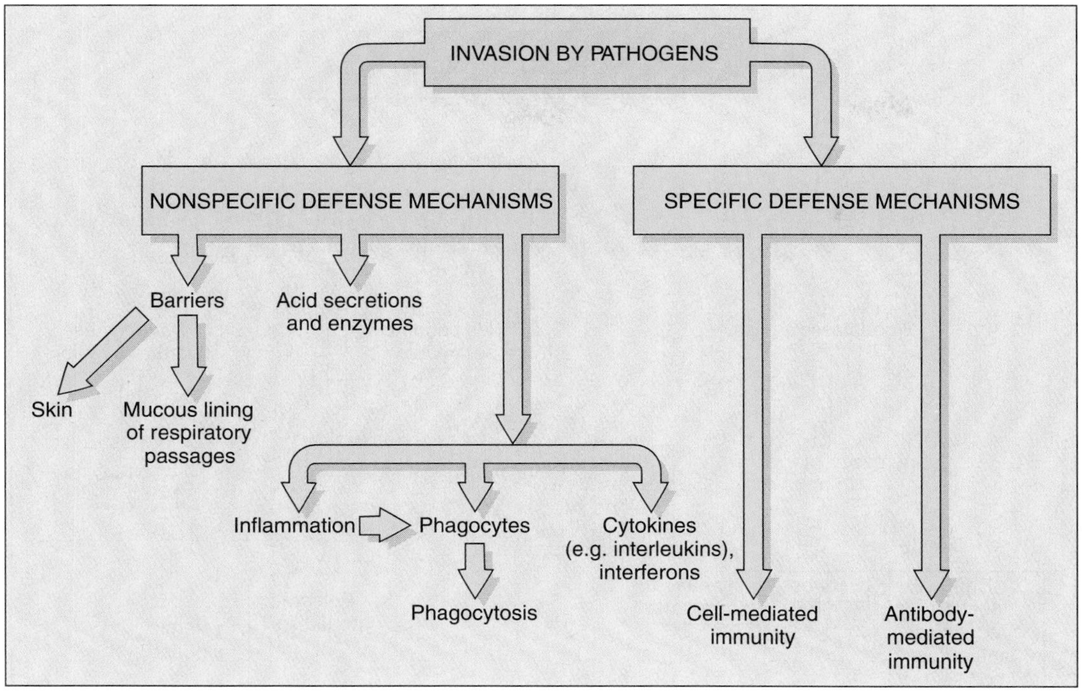

FIGURE 43–1 Most animals have both nonspecific and specific defense mechanisms. Nonspecific mechanisms prevent entrance of pathogens and act rapidly to destroy those that manage to cross the barriers. Specific defense mechanisms take longer to mobilize but are highly effective in destroying invaders.

Microorganisms that enter with food are usually destroyed by the acid secretions and enzymes of the stomach. Pathogens that enter the body with inhaled air may be filtered out by hairs in the nose or trapped in the sticky mucous lining of the respiratory passageways. There, they may be destroyed by phagocytes.

Cytokines are important in both nonspecific and specific immune responses

Cells of the immune system secrete a remarkable number of regulatory proteins known as **cytokines.** Some of the many kinds of cytokines known are discussed in this chapter. Cytokines can act on the cells that produce them, can regulate the activity of nearby cells, and in some cases can modify the functions of cells at a considerable distance.

When infected by viruses or other intracellular parasites (some types of bacteria, fungi, and protozoa), cells respond by secreting cytokines called **interferons.** One group of interferons is produced by either white blood cells (interferon alpha) or by fibroblasts (interferon beta). These interferons signal neighboring cells to produce proteins that inhibit viral replication. Viruses produced in cells exposed to interferon are not very effective at infecting cells. Another group of interferons (interferon gamma), also produced by white blood cells, has a variety of actions, including the ability to enhance the activities of other immune cells. This group of interferons can stimulate macrophages to destroy tumor cells and host cells that have been infected by viruses.

Since their discovery in 1957, interferons have been the focus of much research. Recombinant DNA techniques are now used to produce large quantities of some interferons. The U.S. Food and Drug Administration (FDA) has approved interferons for treating several diseases, including hepatitis B and hepatitis C, genital warts, a type of leukemia, a type of multiple sclerosis, and AIDS-related Kaposi's sarcoma. Interferons are being tested in clinical trials for treatment of HIV infection and several types of cancer.

Interleukins are cytokines secreted mainly by macrophages and lymphocytes. They are numbered according to their order of discovery. This diverse group of proteins regulates interactions between lymphocytes and other cells of the body. Therefore, some interleukins have widespread effects. For example, during infection, **interleukin-1 (IL-1)** can reset the body's thermostat in the hypothalamus, resulting in fever and its symptoms.

Tumor necrosis factors (TNF) are secreted by macrophages (TNF alpha) and by lymphocytes called T cells (TNF beta). TNF can kill tumor cells, offering promise in terms of immunotherapy for cancer patients. In addition, TNF can stimulate immune cells to initiate an inflammatory response.

Complement leads to destruction of pathogens

Complement, so-named because it *complements* the action of other defense mechanisms, consists of a complex of more than 20 proteins present in plasma and other body

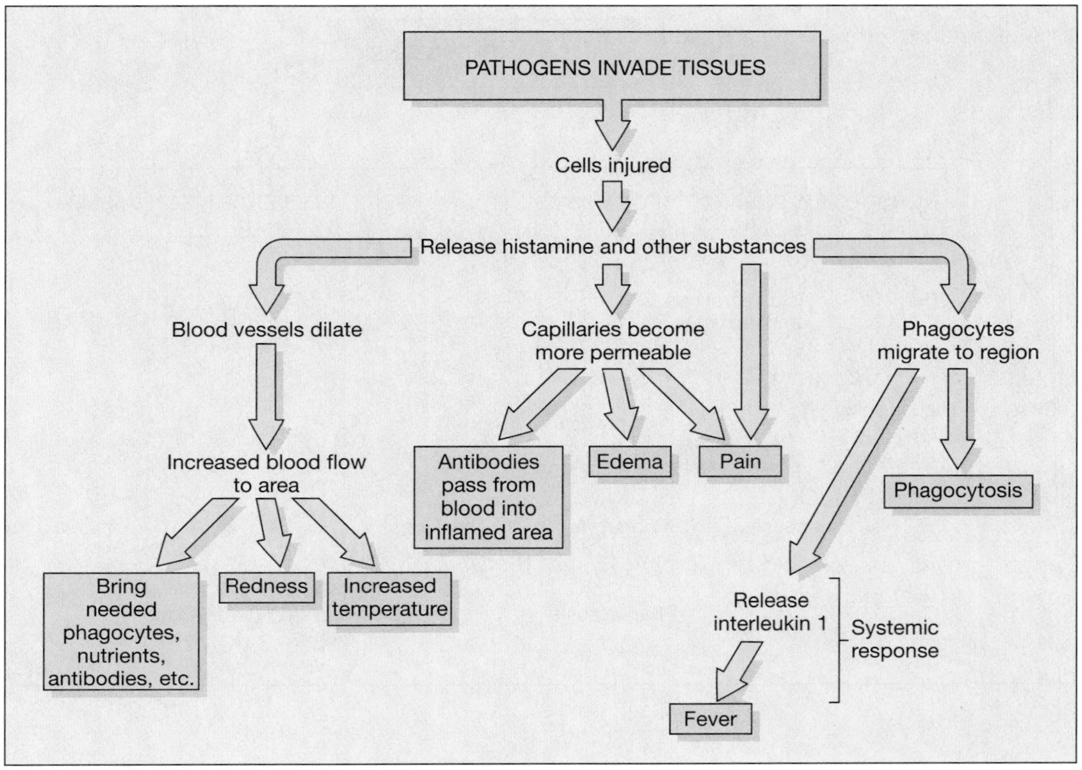

FIGURE 43–2 Inflammation is a vital process that localizes protective immune mechanisms at the area of infection. It permits phagocytic cells, antibodies, and other needed compounds to enter the tissue where microbial invasion is taking place.

fluids. Normally, complement proteins are inactive until the body is exposed to an antigen. Certain pathogens activate the complement system directly. In other cases, when antigen and antibody combine, they stimulate a series of reactions that activate the complement system. The antibody is said to "fix" complement. Proteins of the complement system then work to destroy pathogens.

Complement proteins, unlike antigens, are not specific. They act against any antigen. Some complement proteins increase inflammation by stimulating histamine release. Others attract white blood cells to the site of infection. Certain complement proteins digest portions of the pathogen cell. Others coat the pathogens, making them less "slippery" so that phagocytes (macrophages and neutrophils) can phagocytize them.

Inflammation is a protective mechanism

When pathogens invade tissues, they trigger an **inflammatory response** (Fig. 43–2). Injured cells and certain white blood cells (basophils) release **histamine** and other compounds that dilate blood vessels in the affected area. Blood flow increases to the infected region, bringing great numbers of phagocytic cells. The increased blood flow makes the skin feel warm, and makes skin that contains little pigment appear red. Phagocytes migrate out of the capillaries and into the infected tissues.

Histamine and other chemicals released also make capillaries in the inflamed area more permeable, allowing fluid and antibodies to leave the circulation and enter the tissues. As the volume of tissue fluid increases, **edema** (swelling) occurs. The edema (and also certain substances released by the injured cells) causes the pain characteristic of inflammation. Thus, the clinical characteristics of inflammation are *heat, edema, pain,* and possible *redness.*

Although inflammation is often a local response, sometimes the entire body is involved. **Fever** is a common clinical symptom of widespread inflammatory response. **Macrophages** and certain other cells release compounds, such as the cytokine interleukin-1 (IL-1), that reset the body's thermostat in the hypothalamus, resulting in fever. Prostaglandins (an important group of compounds derived from fatty acids) are also involved in this resetting process.

Fever interferes with the growth of some microorganisms, perhaps by decreasing circulating levels of the iron that they require. Fever also promotes the activity of certain lymphocytes (T cells) and production of antibodies, and may increase phagocytosis. As a result, a short-term low fever may help speed recovery.

Sometimes infection by gram-negative bacteria, such as *Salmonella typhi,* results in the release of large amounts of TNF and other cytokines. This can lead to the potentially lethal condition known as *septic shock.* Thus, cytokines can sometimes have harmful effects.

Phagocytes destroy pathogens

One of the main functions of inflammation appears to be increased phagocytosis (see Chapter 5). A neutrophil can phagocytize 20 or so bacteria before it becomes inactivated (perhaps by leaking lysosomal enzymes) and dies. A macrophage can phagocytize about 100 bacteria during its lifespan.

Can bacteria counteract the phagocyte's attack? Certain bacteria release enzymes that destroy the membranes of the phagocyte's lysosomes. The powerful lysosomal enzymes then spill out into the cytoplasm and may destroy the phagocyte. Some bacteria have cell walls or capsules that resist the action of lysosomal enzymes.

Some macrophages wander through the tissue, phagocytizing foreign matter (including bacteria) and, when appropriate, releasing antiviral agents (Fig. 43–3). Others stay in one place and destroy bacteria that pass by. For example, air sacs in the lungs contain large numbers of macrophages that destroy foreign matter entering with inhaled air.

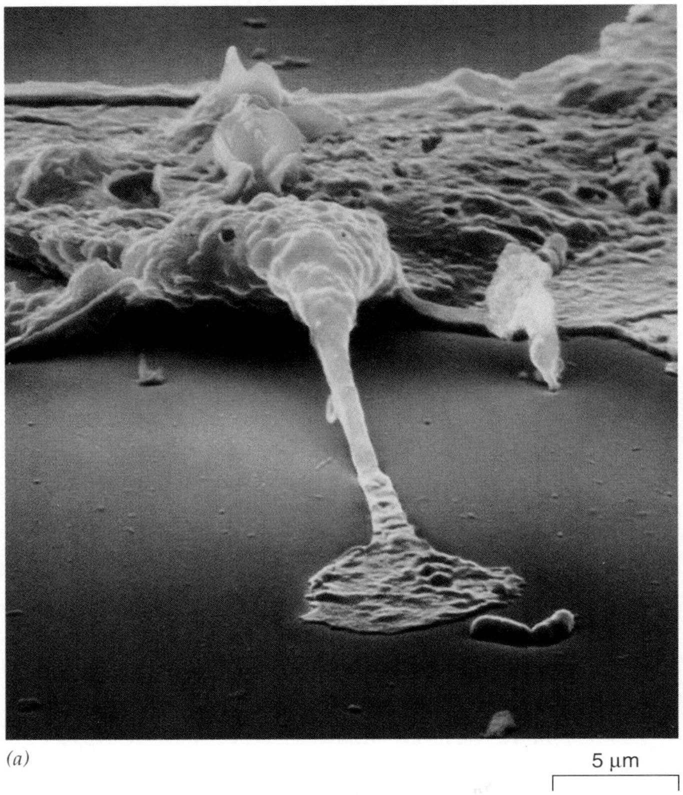

(a)

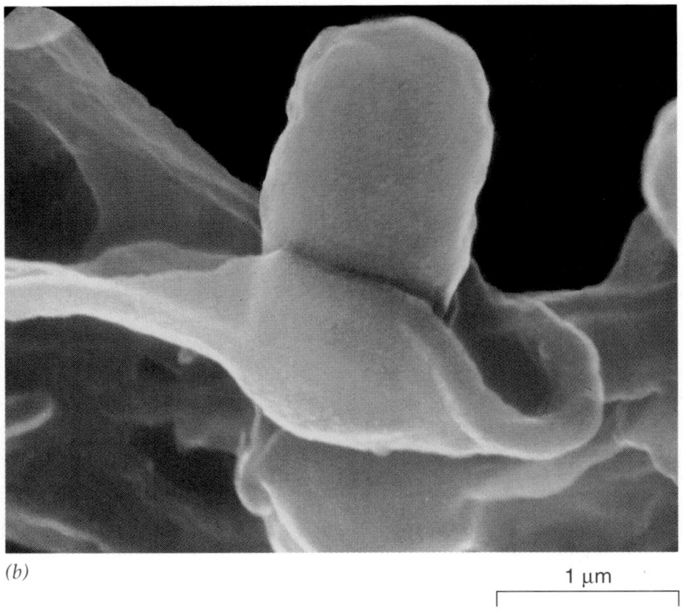

(b)

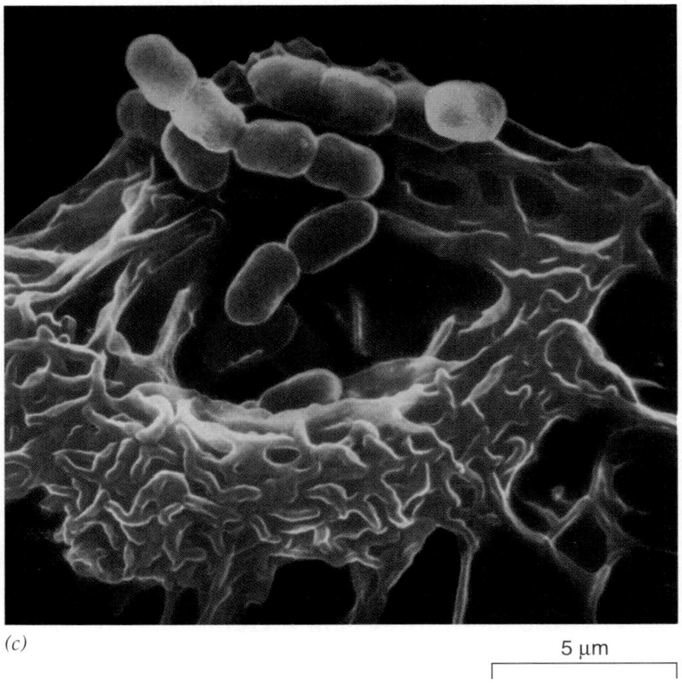

(c)

FIGURE 43–3 The macrophage is a remarkably efficient warrior. (***a***) A macrophage extends a pseudopod toward an invading *E. coli* bacterium that is already multiplying. (***b***) The bacterium is trapped within the engulfing pseudopod. (***c***) The macrophage sucks the trapped bacteria in along with its own plasma membrane. The macrophage's plasma membrane will seal over the bacteria, and powerful lysosomal enzymes will destroy them. (Lennart Nilsson, © Boehringer Ingelheim International GmbH)

Specific Defense Mechanisms Include Antibody-Mediated Immunity and Cell-Mediated Immunity

While nonspecific defense mechanisms are destroying pathogens and preventing the spread of infection, the body is also mobilizing its specific defense mechanisms.

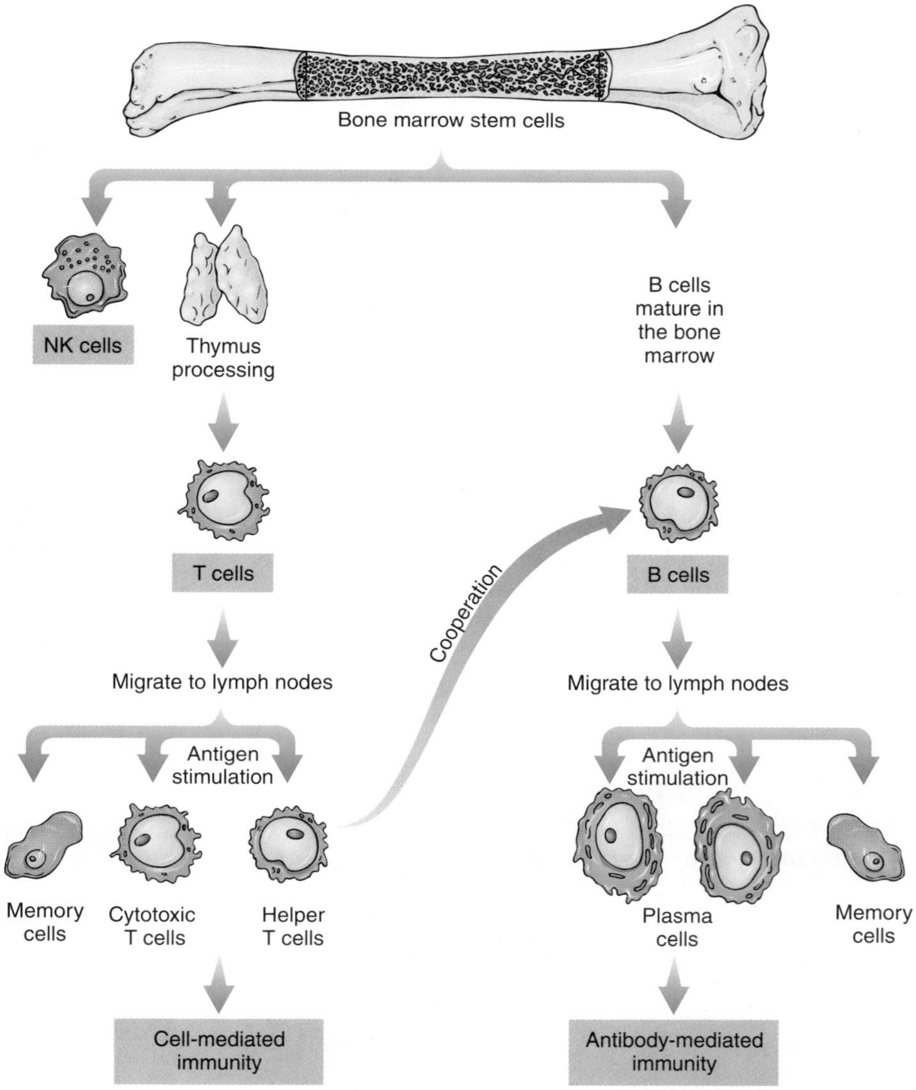

FIGURE 43–4 T cells, B cells, and NK cells have different developmental histories and different functions.

Several days are required to activate specific immune responses, but once in gear, these mechanisms are extremely effective.

Specific defense mechanisms depend on the lymphatic system and its cells (see Chapter 42). Lymphocytes develop and mature in primary lymph organs, including the bone marrow and the thymus gland. Secondary lymph organs, where large numbers of lymphocytes mature and reside, include the spleen and the lymph tissues strategically positioned throughout the body.

Two main types of specific immunity are antibody-mediated immunity and cell-mediated immunity. In **antibody-mediated immunity,** certain lymphocytes, called B cells, mature into plasma cells that produce specific antibodies. In **cell-mediated immunity,** lymphocytes called T cells attack tumor cells and cells infected by invading pathogens. Specific immunity depends on several types of cells.

Cells of the immune system include lymphocytes and phagocytes

Immune responses depend on two main groups of white blood cells: lymphocytes and phagocytes. The millions of lymphocytes are the main warriors in specific immune responses. Lymphocytes are stationed strategically in the lymphatic tissue throughout the body. Three main types of lymphocytes are **T lymphocytes,** or **T cells; B lymphocytes,** or **B cells;** and **natural killer (NK) cells.** NK cells kill virally infected cells and tumor cells. Phagocytes include neutrophils and macrophages.

T cells are responsible for cellular immunity

T cells originate from stem cells in the bone marrow (Fig. 43–4). On their way to the lymph tissues, the future T cells stop off in the **thymus gland** for processing. (The "T" in T cells stands for *thymus-derived.*) The thymus makes T cells immunocompetent—that is, capable of immunological response (discussed in a later section).

Two main categories of T cells have been identified. The first group, known as *CD8 cells* because they have a surface marker designated CD8, includes cytotoxic T cells and suppressor T cells. **Cytotoxic T cells (T_C),** also known as *killer T cells,* recognize and destroy cells with foreign antigens on their surfaces. Among their target cells are virus-infected cells, cancer cells, and foreign tissue grafts. T cells kill their target cells by releasing a variety of cytokines and enzymes (including **perforins** and **granzymes**).

It is thought that some T cells, called **suppressor T cells (T_S),** release cytokines that inhibit the activity of other T cells and of B cells.

The second category of T cells is composed of **helper T cells,** also known as *CD4 cells* because they have a surface marker designated CD4. Helper T cells secrete substances that activate or enhance the immune response. Two subsets of helper T cells are known: helper 1 (Th 1) cells and helper 2 (Th 2) cells. They secrete different types of cytokines and have different functions. Helper 1 cells mainly promote cell-mediated immune responses. Helper 2 cells stimulate specific antibody production. The balance of helper 1 and helper 2 cells appears to be important in effective response to infection.

B cells are responsible for antibody-mediated immunity

Millions of *B cells* are produced in the bone marrow daily. Unlike T cells, which must develop in the thymus, B cells mature in the bone marrow. Each B cell carries receptors needed to bind with a specific type of antigen. When a B cell comes into contact with the specific type of antigen that binds to its receptors, it divides rapidly, forming a clone of identical cells. These B cells develop into **plasma cells,** the cells that are specialized to secrete antibodies. A plasma cell can produce more than 10 million molecules of antibody per hour!

Although T and B lymphocytes have different functions and life histories, they are similar in appearance when viewed with a light microscope. Sophisticated techniques such as fluorescence microscopy, however, demonstrate that the B and T cells can be differentiated by their unique cell surface macromolecules. They also tend to locate in (or "home" to) separate regions of the spleen, lymph nodes, and other lymph tissues.

Natural killer cells attack cancer cells

Natural killer cells are large, granular lymphocytes closely related to T cells. They originate in the bone marrow and differentiate outside the thymus. At first, immunologists thought that NK cells functioned only against tumor cells. However, studies have shown that these cells are active against a wide variety of targets, including cells infected with viruses, some bacteria, and some fungi.

NK cells function without prior sensitization and do not require antigen presentation (discussed in the next section). They kill target cells by both nonspecific and specific (antibody requiring) processes. Like cytotoxic T cells, NK cells release cytokines (including interferon gamma, IL-2, IL-12, and TNF) as well as enzymes such as perforins and granzymes that destroy target cells.

NK activity is stimulated by several cytokines, including interferon gamma. When NK levels are high, resistance to certain cancers may be increased. Certain stressors have been shown to decrease NK cell activity and thus to enhance tumor growth.

Macrophages are important in both nonspecific and specific defense responses

When a macrophage ingests a bacterium, it digests most, but not all, of the bacterial antigens. Fragments of the antigens are displayed on the surface of the macrophage. The macrophage can be described as an **antigen-presenting cell (APC)** that displays bacterial antigens as well as its own surface proteins. The displayed antigen activates helper T cells.

Macrophages secrete about 100 different compounds, including interferons and enzymes that destroy bacteria. When macrophages are stimulated by bacteria, they secrete interleukins, which activate B cells and helper T cells. Interleukins also promote a general response to injury, causing fever and activating other mechanisms that defend the body against invasion.

The thymus "instructs" T cells and produces hormones

Present in all vertebrates, the **thymus gland** has at least two functions. The first is to confer immunological competence on T cells. As T cells move through the thymus, they divide many times and develop specific surface proteins with distinctive receptor sites. Only the T cells possessing specific receptors are selected to divide. This is a form of positive selection.

T cells that react to self-antigens undergo programmed cell death. This is a form of negative selection. Immunologists estimate that more than 90% of developing T cells are negatively selected. The remaining T cells differentiate and leave the thymus to take up residence in other lymph tissues, or to launch immune responses in infected tissues. By selecting only appropriate T cells, the thymus gland insures that T cells can distinguish between the body's own antigens and foreign antigens.

The differentiation of the majority of T cells within the thymus is thought to take place just before birth and during the first few months of postnatal life. If the thymus is removed before this processing takes place, an animal is not able to develop cellular immunity. If the thymus is removed after that time, cellular immunity does not seem to be seriously impaired.

The second function of the thymus is that of an endocrine gland. **Thymosin,** one of several hormones secreted by the thymus, is thought to stimulate T cells after they leave the thymus, causing them to become immunologically active.

The major histocompatibility complex permits recognition of self

The ability of the vertebrate immune system to distinguish self from nonself depends largely on a group of membrane proteins known as the **major histocompatibility complex (MHC).** In humans, the MHC is called the **HLA** (human leukocyte antigen) **group.** These markers are present on the surface of most cells and are slightly different in each individual. A group of closely linked genes codes for the MHC proteins. These genes are polymorphic; that is, within the population there are multiple alleles for each locus (sometimes more than 40 alleles for a given gene).

With so many possible combinations, no two people, except identical twins, are likely to have the same MHC proteins on their cells. Thus, the MHC gives a biochemical "fingerprint." The more closely related two individuals are, the more HLA antigens they have in common.

The MHC is divided into three groups of genes that code for distinct sets of proteins. These proteins differ in terms of tissue distribution and chemical structure. MHC class I antigens are found on most nucleated cells, and are important in distinguishing between self and nonself. They bind with foreign antigens produced within cells (for example, by viruses or by foreign tissue grafts), forming a molecular complex that is displayed on the cell surface. This MHC antigen-foreign antigen complex is recognized by cytotoxic T cells.

MHC class II antigens are found only on cells of the immune system, particularly B cells, macrophages, some T cells, and dendritic cells (specialized cells located throughout the body, especially in the spleen and lymph nodes). Class II antigens regulate the interactions among T cells, B cells, and antigen-presenting cells. MHC class II antigens bind with peptides from proteins that have entered the cell via foreign sources such as bacteria. The MHC class II and foreign antigen complex is displayed

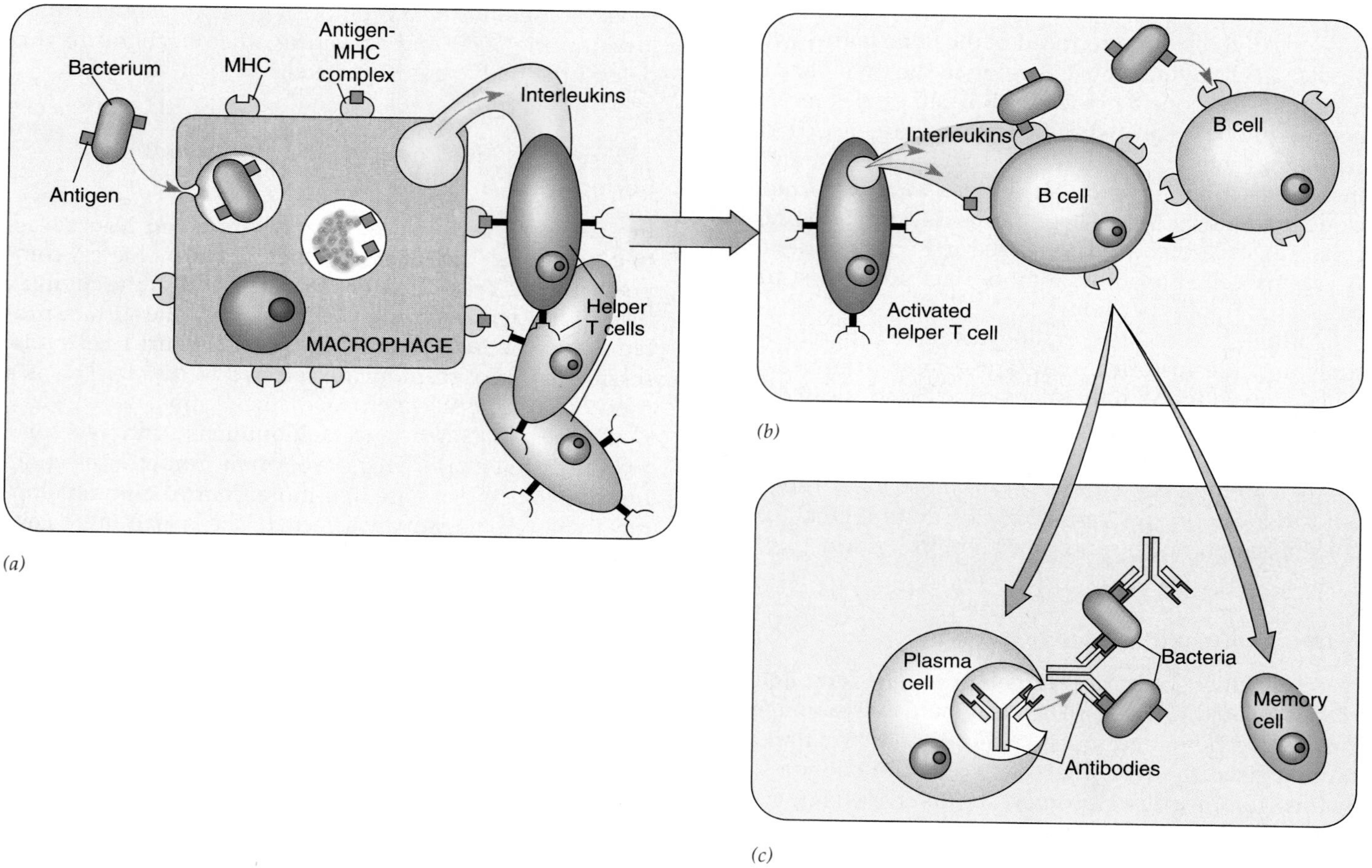

FIGURE 43–5 B cell activation requires antigen presentation and helper T cells. (*a*) The macrophage engulfs a bacterium, breaks it down, and presents bacterial antigens on its surface in combination with MHC. Helper T cells are activated when their receptors combine with this foreign antigen-MHC complex and they are stimulated by interleukins secreted by the macrophage. (*b*) B cells can combine with specific antigens. However, B cells are generally activated by interleukins secreted by helper T cells. (*c*) Once activated, the B cell divides, forming a clone of cells. Some of these differentiate into plasma cells that secrete antibodies. Others develop into memory B cells.

on the cell surface and stimulates helper T cells. MHC class III proteins include components of the complement system.

Antibody-mediated immunity is a chemical warfare mechanism

B cells are responsible for antibody-mediated immunity (also called humoral immunity). A given B cell can produce one specific antibody. The antibody molecules serve as receptors that combine with antigens. Only the variety of B cell with a matching receptor can bind with a particular antigen. This binding activates the B cell.

In most cases, activation of B cells is a complex process that involves macrophages and helper T cells (Fig. 43–5). Recall that a macrophage displays fragments of antigen from a pathogen it has engulfed. The foreign antigen forms a complex with MHC antigens of the macrophage. This foreign antigen-MHC complex is then displayed on the cell surface.

When a macrophage displaying a foreign antigen-MHC complex contacts a helper T cell with complementary receptors, a complicated interaction occurs. One result of this interaction is that the macrophage secretes interleukins such as IL-1, which activates helper T cells. A T cell does not recognize an antigen that is presented alone; the antigen must be presented to the T cell as part of a foreign antigen-MHC complex on the surface of an antigen-presenting cell.

A B cell binds with complementary antigen and displays it with MHC protein (class II) on its surface. An activated helper T cells binds with the foreign antigen-MHC complex on the B cell. The activated helper T cell then releases interleukins, which, together with antigen, stimulate the B cells to divide and differentiate.

Once activated, B cells increase in size. Then they divide by mitosis, each giving rise to a clone of identical cells (Fig. 43–6). This cell division in response to a specific antigen is known as **clonal selection.** Some of these B cells mature into plasma cells that secrete the type of antibody specific to the antigen. Unlike T cells, most plasma cells do not leave the lymph nodes. Only the antibodies they secrete pass out of the lymph tissues and make their way via the lymph and blood to the infected area. This sequence is summarized as follows:

> pathogen (bearing foreign antigens) invades body → macrophage phagocytizes pathogen → foreign antigen-MHC antigen complex displayed on macrophage cell surface → helper T cell binds with foreign antigen-MHC complex → activated helper T cell interacts with a B cell that displays the same complex → B cell activated → clone of B cells produced → B cells differentiate → plasma cells → secrete antibody

Some activated B cells do not differentiate into plasma cells, but instead become **memory cells.** In these cells, a "survival gene" is activated that permits them to prevent the programmed death that is the fate of other B cells. Memory cells continue to live and produce small amounts of antibody long after an infection has been overcome. This antibody, part of the gamma globulin fraction of the plasma, becomes part of the body's arsenal of chemical weapons. Should the same pathogen enter the body again, this circulating antibody immediately targets it for destruction. At the same time, memory cells are stimulated to divide, producing new clones of the appropriate plasma cells.

Antibodies are grouped in five classes

Antibodies, known more formally as **immunoglobulins,** or **Ig,** are highly specific proteins produced in response to specific antigens. The function of an antibody is to bind to an antigen. The antibody does not destroy the antigen directly. Rather, it *labels* the antigen for destruction.

Antibodies are grouped in five classes. Using the abbreviation Ig for immunoglobulin, the classes are designated IgG, IgM, IgA, IgD, and IgE. In humans, about 75% of the antibodies in the blood belong to the **IgG group**; these are part of the gamma globulin fraction of the plasma. IgG and **IgM** defend against many pathogens carried in the blood, including bacteria, viruses, and some fungi. IgG and IgM antibodies stimulate macrophages and activate the complement system.

IgA, present in mucus, tears, saliva, and milk, prevents viruses and bacteria from attaching to epithelial surfaces. This immunoglobulin defends against inhaled or ingested pathogens. IgA is strategically located in the respiratory passageways, digestive tract, urinary tract, and reproductive tract.

IgD has a low concentration in the plasma. Along with IgM, it is an important immunoglobulin on the B cell surface. IgD helps activate B cells following antigen binding. Although **IgE** has a low serum concentration, it is important because it can bind to cells containing potent pharmacologic mediators. When an antigen binds to IgE, histamine is released. Histamine is responsible for many allergy symptoms. IgE is also responsible for immunity to invading parasitic worms.

A typical antibody is a Y-shaped molecule consisting of four polypeptide chains

How does an antibody "recognize" a particular antigen? In a protein antigen, specific sequences of amino acids make up an **antigenic determinant** (Fig. 43–7). These amino acids give part of the antigen molecule a specific shape that can be recognized by an antibody or T-cell receptor. Usually, an antigen has 5 to 10 antigenic determinants on its surface. Some have 200 or even more. These antigenic determinants may differ from one another, so several different kinds of antibodies can combine with a single complex antigen.

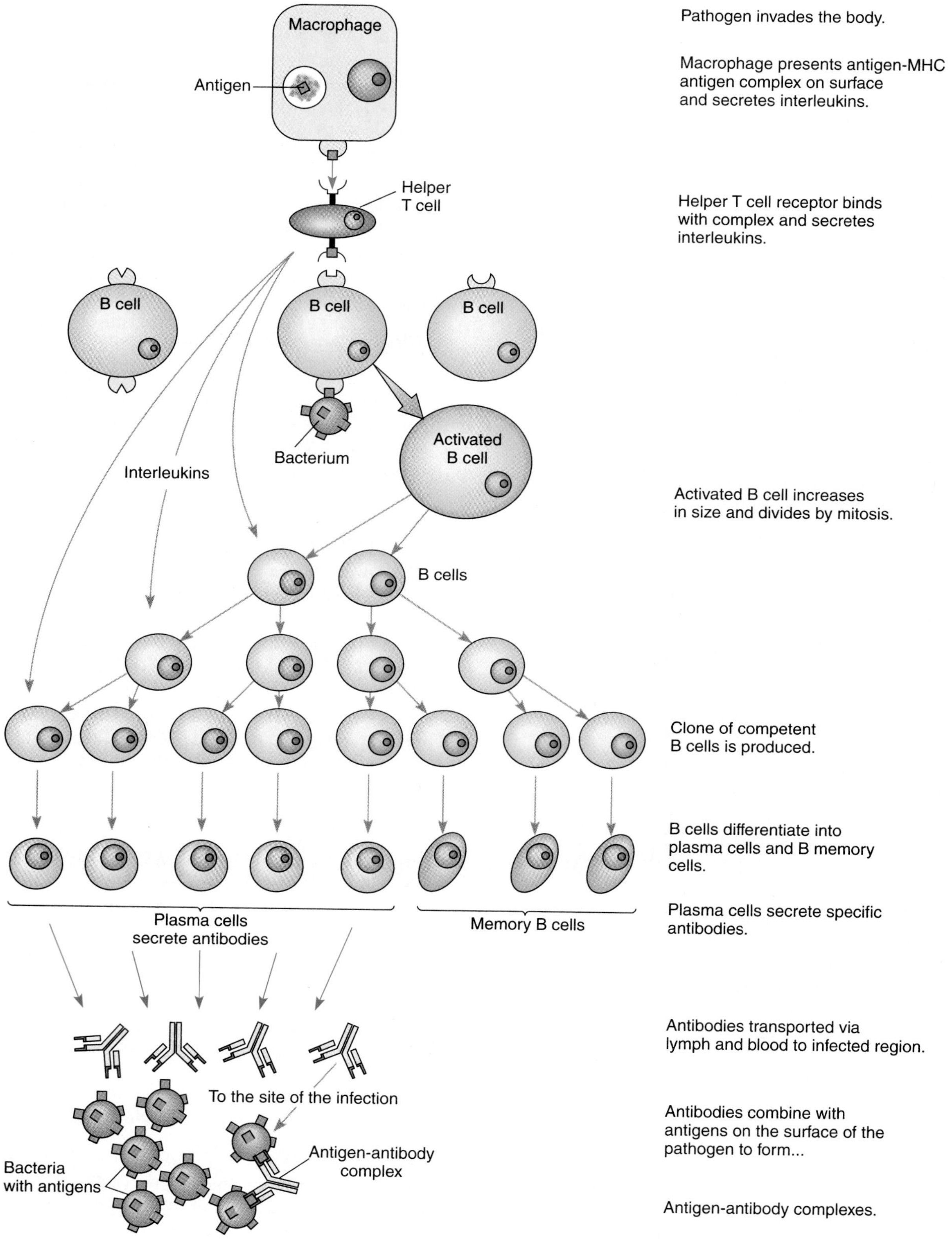
Pathogen invades the body.
Macrophage
Antigen
Macrophage presents antigen-MHC antigen complex on surface and secretes interleukins.
Helper T cell
Helper T cell receptor binds with complex and secretes interleukins.
B cell
B cell
B cell
Bacterium
Activated B cell
Interleukins
Activated B cell increases in size and divides by mitosis.
B cells
Clone of competent B cells is produced.
B cells differentiate into plasma cells and B memory cells.
Plasma cells secrete antibodies
Memory B cells
Plasma cells secrete specific antibodies.
Antibodies transported via lymph and blood to infected region.
To the site of the infection
Antibodies combine with antigens on the surface of the pathogen to form...
Antigen-antibody complex
Bacteria with antigens
Antigen-antibody complexes.

FIGURE 43–6 B cells are responsible for antibody-mediated immunity. When a B cell binds with a specific antigen and a helper T cell releases interleukins, the B cell becomes activated. Once activated, the B cell multiplies, producing a large clone of cells. Many of these differentiate and become plasma cells, which secrete antibodies. The plasma cells remain in the lymph tissues, but the antibodies are transported to the site of infection by the blood or lymph. The antigen-antibody complexes that form destroy pathogens. Some of the B cells become memory cells that continue to secrete small amounts of antibody for years after the infection is over.

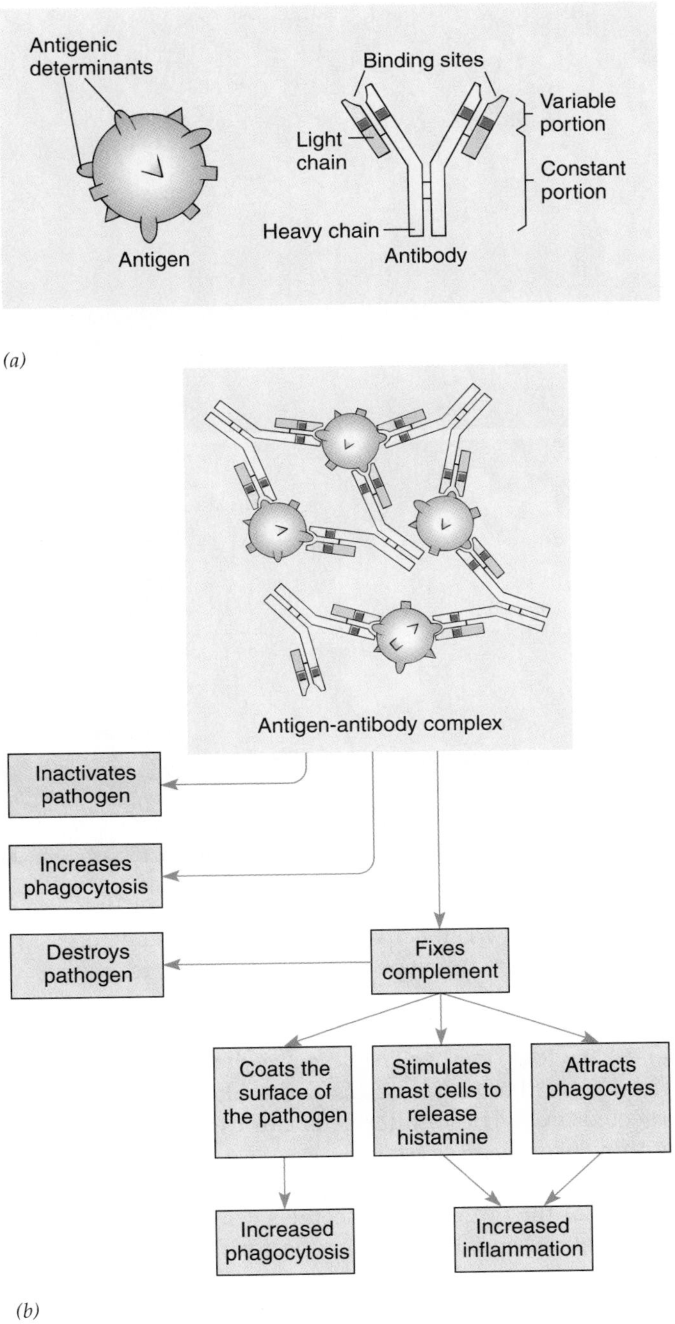

FIGURE 43–7 Antigen and antibody combine, forming antigen-antibody complexes. (*a*) The antibody molecule is composed of two light chains and two heavy chains, joined together by disulfide bonds. The constant and variable regions of the chains are indicated. (*b*) Antigen-antibody complexes directly inactivate pathogens and increase phagocytosis. They also activate the complement system.

Certain drugs and some molecules found in dust are too small to be antigenic, yet they do stimulate immune responses. These substances, called **haptens,** become antigenic by attaching to the surface of a protein. For example, the antibiotic penicillin is a small molecule that cannot by itself cause an immune response. However, once penicillin is degraded in the body, its components can bind to serum proteins, forming an antigenic determinant. In that form, penicillin is recognized as a foreign substance and can stimulate an immune response. Antibodies produced against it can participate in an allergic response to penicillin.

A typical antibody is a Y-shaped molecule in which the two arms of the Y are binding sites (Fig. 43–7). This shape permits the antibody to combine with two antigen molecules, and allows formation of **antigen-antibody complexes.** The tail of the Y performs functions such as binding to cells or activating the complement system.

The antibody molecule consists of four polypeptide chains: two identical long chains called heavy chains, and two identical short chains called light chains (Fig. 43–7). Each chain has a constant segment, a junctional segment, and a variable segment. In the **constant segment,** or **C region,** of the heavy chains, the amino acid sequence is constant within a particular immunoglobulin class. Thus, five types of heavy chain C regions are known, each defining one of the five immunoglobulin classes. The C region may be thought of as the handle portion of a door key. Like the elongated part of a key that slides into a lock, the amino acid sequence of the **junctional segment,** or **J region,** is somewhat variable. Finally, like the pattern of bumps and notches at the end of a key, the **variable segment,** or **V region,** has a unique amino acid sequence. In B-cell receptors, the variable region of the immunoglobulin protrudes from the B cell, whereas the constant region anchors the molecule to the cell.

The V region is the part of the key that is unique for a specific antigen (the lock). At its variable regions, the antibody folds three-dimensionally, assuming a shape that enables it to combine with a specific antigen. When they meet, antigen and antibody fit together *somewhat* like a lock and key. They must fit in just the right way for the antibody to be effective (Fig. 43–8). A given antibody can bind with different strengths, or **affinities,** to different antigens. In the course of an immune response, stronger (higher affinity) antibodies are generated.

An antibody molecule has two main functions. It combines with antigen and it activates processes that destroy the antigen that binds to it. For example, the antibody may stimulate phagocytosis.

The binding of antibody to antigen activates other defense mechanisms

Antibodies mark a pathogen as foreign by combining with an antigen on its surface. Generally, several antibodies bind with several antigens, creating a mass of clumped antigen-antibody complexes. The combination of antigen and antibody activates several defense mechanisms:

1. The antigen-antibody complex may inactivate the pathogen or its toxin. For example, when an antibody

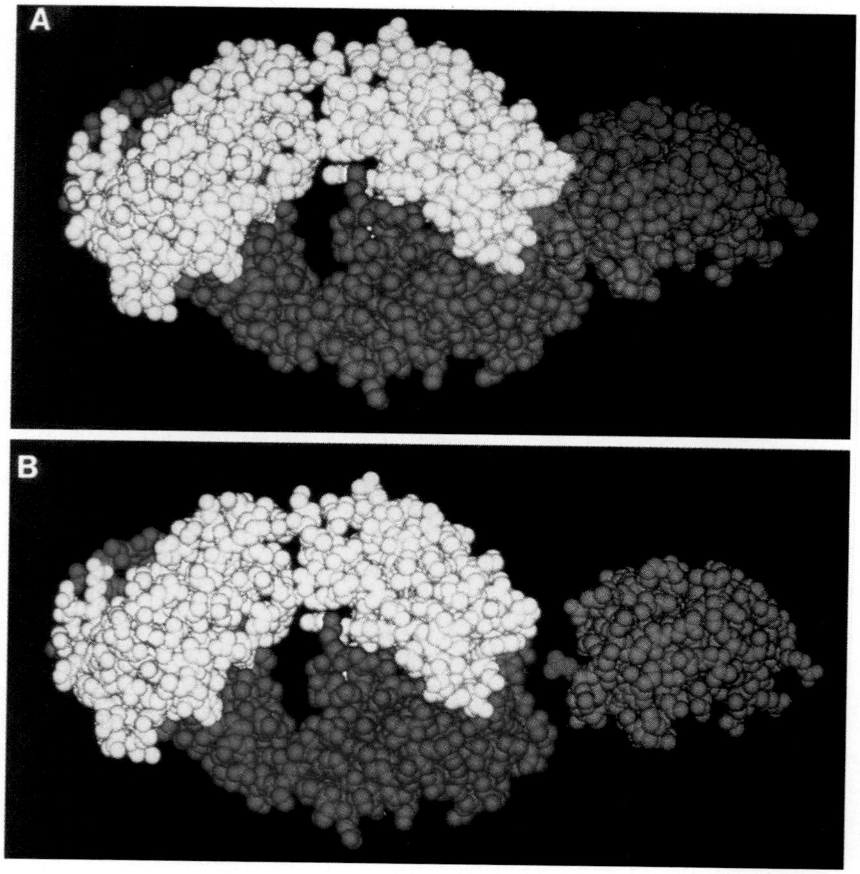

FIGURE 43–8 The components of an antigen-antibody complex fit together as shown in this computer-simulation of their molecular structure. The antigen lysozyme is shown in green, the heavy chain of the antibody is shown in blue, and the light chain in yellow. (*a*) The antigenic determinant, shown in red, fits into a groove in the antibody molecule. (*b*) The antigen-antibody complex has been pulled apart.

attaches to the surface of a virus, the virus may lose its ability to attach to a host cell.

2. The antigen-antibody complex stimulates phagocytic cells to ingest the pathogen.
3. Antibodies of the IgG and IgM groups work mainly through the complement system (discussed earlier in this chapter). When antibodies combine with a specific antigen on a pathogen, complement proteins *complement* their action by destroying the pathogens.

Cell-mediated immunity provides cellular warriors

The T cells and macrophages are responsible for cell-mediated immunity (Fig. 43–9). They destroy cells infected with viruses and cells that have been altered in some way, such as cancer cells. They also destroy foreign grafts such as transplanted kidneys. There are thousands of different populations of T cells. Each T cell has more than 50,000 identical receptors (TCRs) that bind a specific antigen.

How do the T cells know which cells to attack? T cells do not recognize antigens unless they are presented properly. For example, when a virus infects a cell, some of the viral protein is broken down to peptides and displayed with MHC on the cell surface. T cell receptors can react with such antigen-MHC complexes. Only T cells with receptors that bind to the specific antigen presented become activated. Generally, fewer than 1 in 10,000 T cells can respond. Helper T cell activation is thought to be necessary for activating many aspects of the immune response.

Once activated, a T cell increases in size and gives rise to a clone of cytotoxic T cells and memory cells (Fig. 43–9). Cytotoxic T cells make up the cellular infantry. They leave the lymph nodes and make their way to the infected area. These killer cells can destroy a target cell within seconds after contact.

After a cytotoxic T cell combines with antigen on the surface of the target cell, it secretes granules. These granules contain enzymes, such as perforins and granzymes, that can destroy the cell. Under some conditions, cytotoxic T cells release proteins, called **lymphotoxins,** that are especially toxic for cancer cells. After releasing cytotoxic substances, the T cell disengages itself from its victim cell and seeks out a new target. This sequence is summarized as follows:

virus invades body cell → foreign antigen-MHC complex displayed on cell surface → specific T cell activated by this complex → clone of T cells produced → some become cytotoxic T cells → cytotoxic T cells migrate to area of infection → proteins released that destroy target cells

(*Text continues on page 948*)

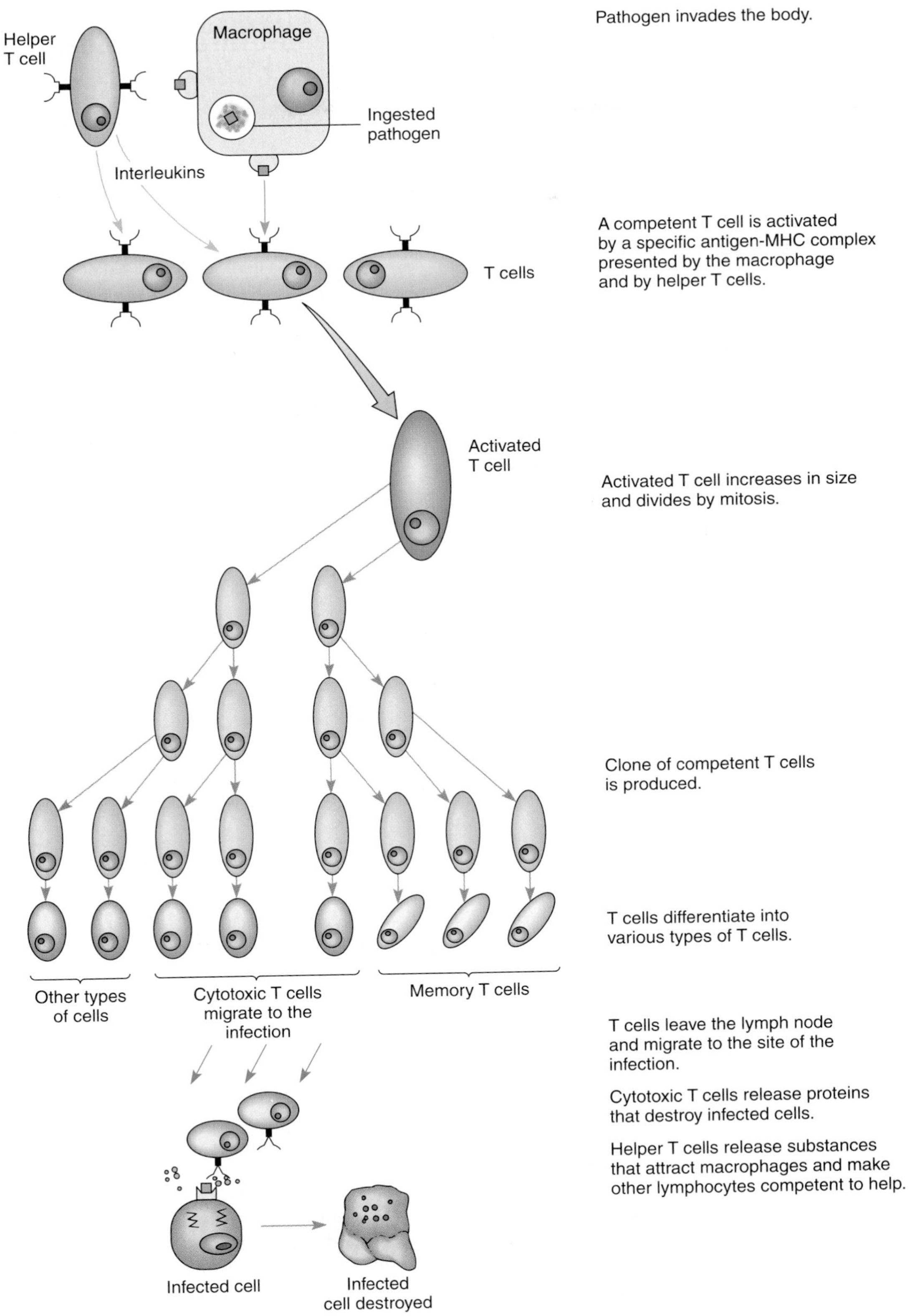
Pathogen invades the body.
Helper T cell
Macrophage
Ingested pathogen
Interleukins
A competent T cell is activated by a specific antigen-MHC complex presented by the macrophage and by helper T cells.
T cells
Activated T cell
Activated T cell increases in size and divides by mitosis.
Clone of competent T cells is produced.
T cells differentiate into various types of T cells.
Other types of cells
Cytotoxic T cells migrate to the infection
Memory T cells
T cells leave the lymph node and migrate to the site of the infection.
Cytotoxic T cells release proteins that destroy infected cells.
Helper T cells release substances that attract macrophages and make other lymphocytes competent to help.
Infected cell
Infected cell destroyed

FIGURE 43–9 T cells are responsible for cell-mediated immunity. When activated by the foreign antigen-MHC complex presented by a macrophage and by interleukins, a T cell gives rise to a clone of cells. Some of these cells become cytotoxic T cells, which migrate to the site of infection, and release proteins that destroy invading pathogens. Other cells become helper T cells, suppressor cells, or memory T cells.

Making the Connection

Immunity and Genetics

How can the immune system recognize every possible antigen, even those produced by newly mutated viruses never before encountered during the evolution of our species? Do our cells contain millions of separate antibody genes, each coding for an antibody with a different specificity? Although each human cell has a large amount of DNA, it is not enough to provide a different gene to code for each specific antibody molecule.

Recombinant DNA technology (see Chapter 14) has allowed researchers to make direct comparisons between the DNA of the B cells that produce antibodies and of other cells of the body. They have found that separate DNA segments code for different regions of the heavy and light chains of an antibody (see Fig. 43–7). During the development of a B cell, the DNA segments are shuffled and then joined, making one combined gene. By shuffling the gene segments in this way, the number of potential combinations is enormous! Millions of different types of B (and T) cells are produced. By chance, one of those cells may produce just the right antibody to destroy a virus that invades the body. You create diverse combinations of things in your everyday life. A familiar example is when you make your own ice cream sundae. Imagine the varied combinations that are possible using ten types of ice cream, six types of sauce, and 15 kinds of toppings.

Additional sources of antibody diversity are known. For example, the DNA of the mature B lymphocytes that codes for the variable regions of immunoglobulins mutates very readily. These mutations produce genes that code for slightly different antibodies.

We have used antibody diversity here as an example, but similar genetic mechanisms account for the diversity of T cell receptors. Although we may actually use only a relatively few types of antibodies or T cells in a lifetime, the remarkable diversity of the immune system prepares it to attack most potentially harmful antigens that may invade the body. ▲

DNA rearrangement occurs during production of an antibody. In undifferentiated cells, gene segments are present for a number of different variable (V) regions, for one or more junction (J) regions, and for one or more different constant (C) regions. During differentiation, the segments are rearranged. A gene segment extending from the end of one of the V segments to the beginning of one of the J segments may be deleted. This produces a gene that can be transcribed. The RNA transcript is processed to remove introns and the mRNA produced is translated. Each of the polypeptide chains of an antibody molecule is produced in this way. This diagram is greatly simplifed.

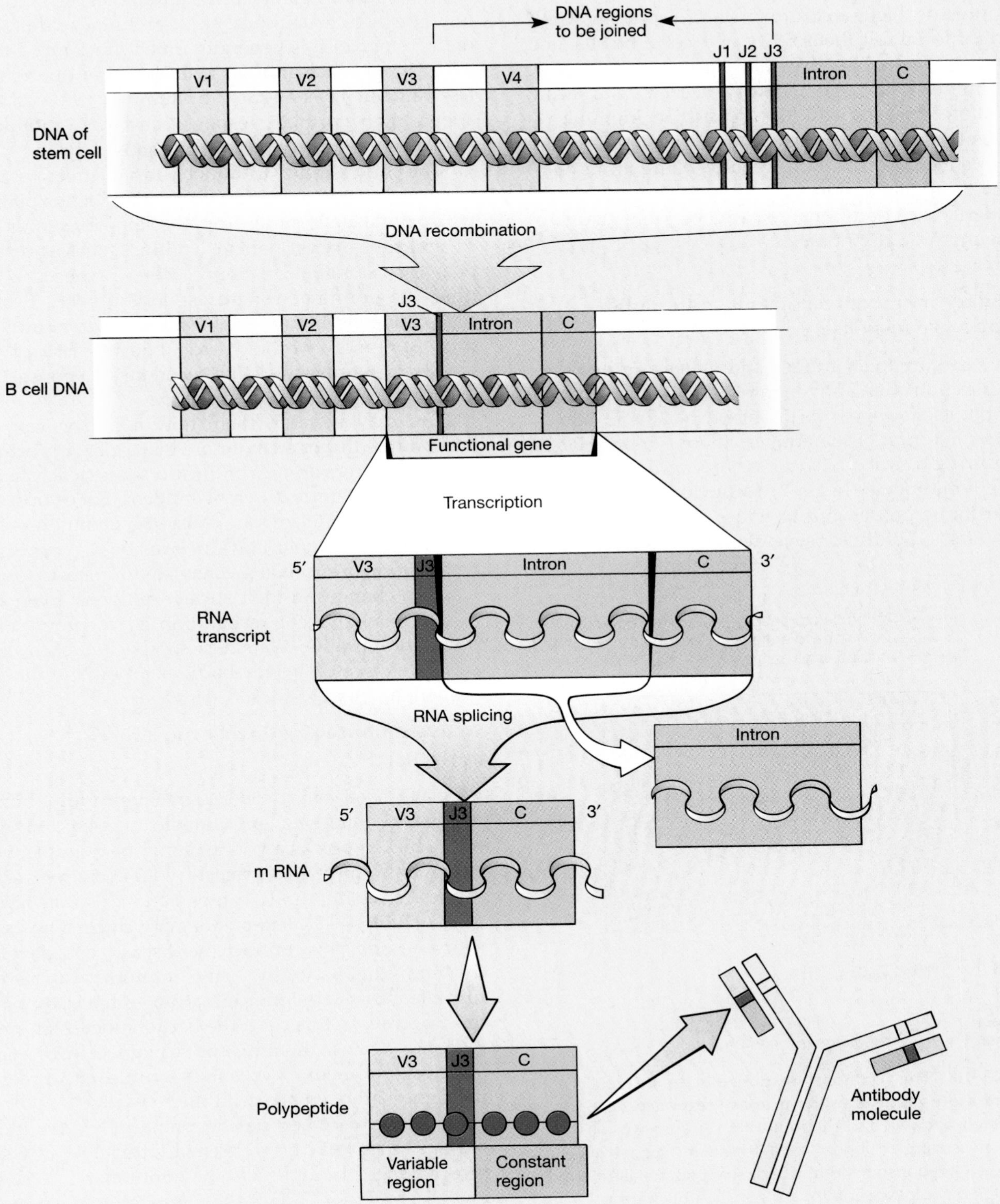
DNA regions
to be joined
J1 J2 J3
V1
V2
V3
V4
Intron
C
DNA of
stem cell
DNA recombination
J3
V1
V2
V3
Intron
C
B cell DNA
Functional gene
Transcription
5′
V3
J3
Intron
C
3′
RNA
transcript
RNA splicing
Intron
5′
V3
J3
C
3′
m RNA
V3
J3
C
Polypeptide
Variable
region
Constant
region
Antibody
molecule

Helper T cells and macrophages at the site of infection secrete substances such as interleukins that help regulate immune function. For example, some interleukins stimulate activated T cells to divide. Other interleukins enhance inflammation, attracting great numbers of macrophages to the site of infection.

The existence of a separate group of CD8 cells referred to as suppressor T cells is controversial. Suppressor T cells are thought to inhibit the activity of T cells, B cells, and macrophages. If they are a separate group of cells, they multiply more slowly than cytotoxic T cells. As a result, it takes more than a week to suppress an immune response. This allows sufficient time for the immune response to effectively defend the body. Some immunologists now think that suppressor cells are cytotoxic T cells that, under some conditions, release cytokines that suppress the immune response.

A secondary immune response is more rapid than a primary response

The first exposure to an antigen stimulates a **primary response.** Injection of an antigen into an animal causes specific antibodies to appear in the blood plasma in 3 to 14 days. After injection of the antigen, there is a brief latent period, during which the antigen is recognized and appropriate lymphocytes begin to form clones. Then there is a logarithmic phase, during which the antibody concentration rises rapidly for several days until it reaches a peak (Fig. 43–10). IgM is the principal antibody synthesized during the primary response. Finally, there is a decline phase, during which the antibody concentration decreases to a very low level.

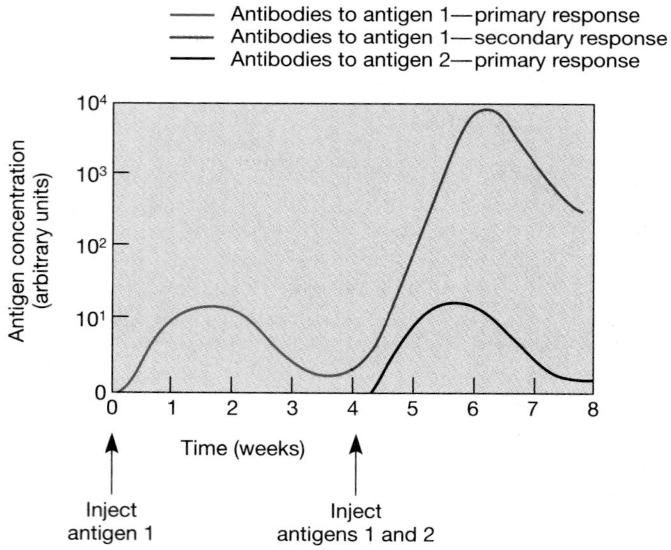

FIGURE 43–10 The secondary response to an antigen is greater and more rapid than the primary response. Antigen 1 was injected at day 0 and the immune response was assessed by measuring antibody levels to the antigen. At week 4, the primary response had subsided. Antigen 1 was injected again along with a new protein, antigen 2. A primary response was made to the newly encountered antigen 2.

A second injection of the same antigen, even years later, results in a **secondary response** (Fig. 43–10). Because memory cells bearing antibodies to that antigen may persist throughout an individual's life, the secondary response is generally much more rapid than the primary response and has a shorter latent period. Much less antigen is necessary to stimulate a secondary response than a primary response, and more antibodies are produced. The predominant antibody is IgG.

The body's ability to launch a rapid, effective response during a second encounter with an antigen explains why we do not usually suffer from the same infectious disease several times. Most persons contract measles or chicken pox, for example, only once. When exposed a second time, the immune system responds quickly, destroying the pathogens before they have time to multiply and cause symptoms of the disease. Booster shots of vaccine are given in order to elicit a secondary response, thus reinforcing immunological memory.

You may wonder, then, how a person can get influenza (the flu) or a cold more than once. Unfortunately, there are many varieties of these diseases, each caused by a virus with slightly different antigens. For example, more than 100 different viruses cause the common cold, and new varieties of cold and flu virus evolve continuously by mutation (a survival mechanism for them), which may result in changes in their surface antigens. Even a slight change may prevent recognition by memory cells. Because the immune system is so specific, each different antigen is treated by the body as a new immunological challenge.

Active immunity follows exposure to antigens

We have been considering **active immunity,** immunity developed following exposure to antigens. When you have chicken pox as a young child, for example, you develop immunity that protects you from contracting it again. Active immunity can be *naturally* or *artificially* induced (Table 43–1). If someone with chicken pox sneezes near you and you contract the disease, you develop active immunity naturally. Active immunity can also be artificially induced by **immunization** — that is, by exposure to a **vaccine.** In this case, the body launches an immune response against the antigens contained in the vaccine, and develops memory cells so that future encounters with the same pathogen are dealt with swiftly.

Effective vaccines can be prepared in a number of ways. A virus may be weakened (attenuated) by successive passage through cells of nonhuman hosts. In the process, mutations adapt the pathogen to the nonhuman host so that it can no longer cause disease in humans.

Table 43–1 ACTIVE AND PASSIVE IMMUNITY

Type of Immunity	*When Developed*	*Development of Memory Cells*	*Duration of Immunity*
Active			
Naturally induced	After pathogens enter the body through natural encounter (e.g., person with measles sneezes on you)	Yes	Many years
Artificially induced	After immunization with a vaccine	Yes	Many years
Passive			
Naturally induced	After transfer of antibodies from mother to developing baby	No	Few months
Artificially induced	After injection with gamma globin	No	Few months

This is how Sabin polio vaccine and measles vaccine are produced.

Whooping cough and typhoid fever vaccines are made from killed pathogens that still have the necessary antigens to stimulate an immune response. Tetanus and botulism vaccines are made from toxins secreted by the respective pathogens. The toxin is altered so that it can no longer destroy tissues, but its antigenic determinants are still intact.

Most vaccines consist of the entire pathogen, attenuated or killed, or of a protein from the pathogen. Researchers are investigating a number of approaches that would reduce potential side effects. For example, they are now designing vaccines consisting of synthetic peptides that are only a small part of the antigen. When any vaccine is introduced into the body, the immune system actively develops clones, produces antibodies, and develops more memory cells.

Passive immunity is borrowed immunity

In **passive immunity,** an individual is given antibodies actively produced by another organism. The serum or gamma globulin that contains these antibodies can be obtained from humans or animals. Animal sera are less desirable because their nonhuman proteins can themselves act as antigens, stimulating an immune response that may result in an illness known as serum sickness.

Passive immunity is borrowed immunity, and its effects are not lasting. It is used to boost the body's defense temporarily against a particular disease. For example, when someone is diagnosed with hepatitis A, a form of viral hepatitis that can be spread through contaminated food or water, persons at risk of infection can be injected with gamma globulin containing antibodies to the hepatitis pathogen. However, such gamma globulin injections offer protection for only a few weeks. Because the body has not actively launched an immune response, it has no memory cells and cannot produce antibodies to the pathogen. Once the injected antibodies are broken down, the immunity disappears.

Pregnant women confer natural passive immunity on their developing babies by manufacturing antibodies for them. These maternal antibodies, of the IgG class, pass through the placenta (the organ of exchange between mother and developing child) and provide the fetus and newborn infant with a defense system until its own immune system matures. Babies who are breast-fed continue to receive immunoglobulins, particularly IgA, in their milk.

Normally the body effectively defends itself against cancer

A few normal cells may be transformed into precancer cells every day in each of us in response to radiation, certain viruses, or chemical carcinogens in the environment. Because they are abnormal cells, some of their surface proteins are different from those of normal body cells. Such proteins act as antigens, stimulating an immune response that generally destroys these abnormal cells. In this way, the immune system provides surveillance against malignancy.

Although many components of the immune system help defend against cancer cells, NK cells and cytotoxic T cells appear to be most critical (Fig. 43–11). T cells produce interleukins, which attract macrophages and NK cells and activate them. The T cells also produce interferons, which have an antitumor effect. The macrophages themselves produce factors, including TNF (tumor necrosis factor), that inhibit tumor growth.

What are the mechanisms by which tumor cells prevent the body from launching an immune response? The cancer cells may remain so antigenetically similar to nor-

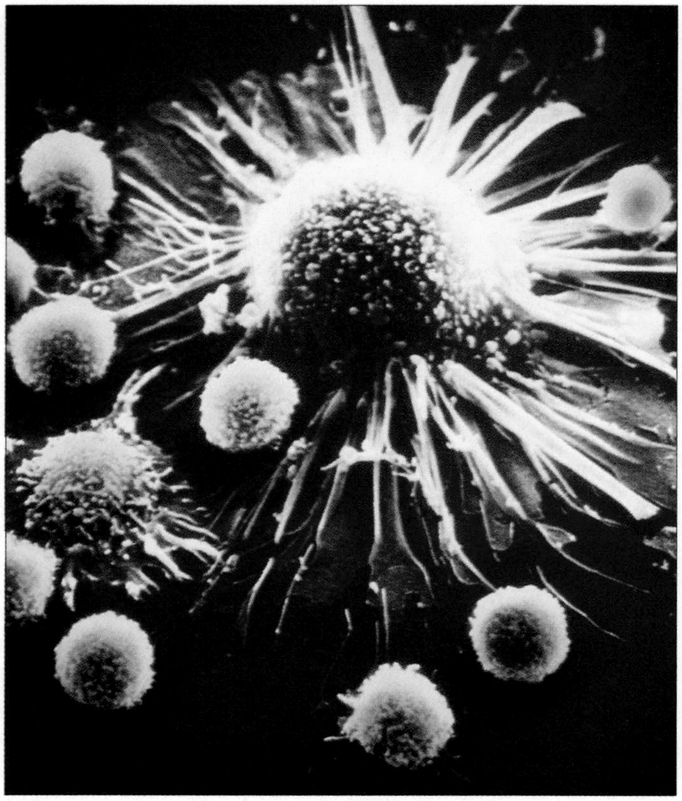

(a) 10 μm

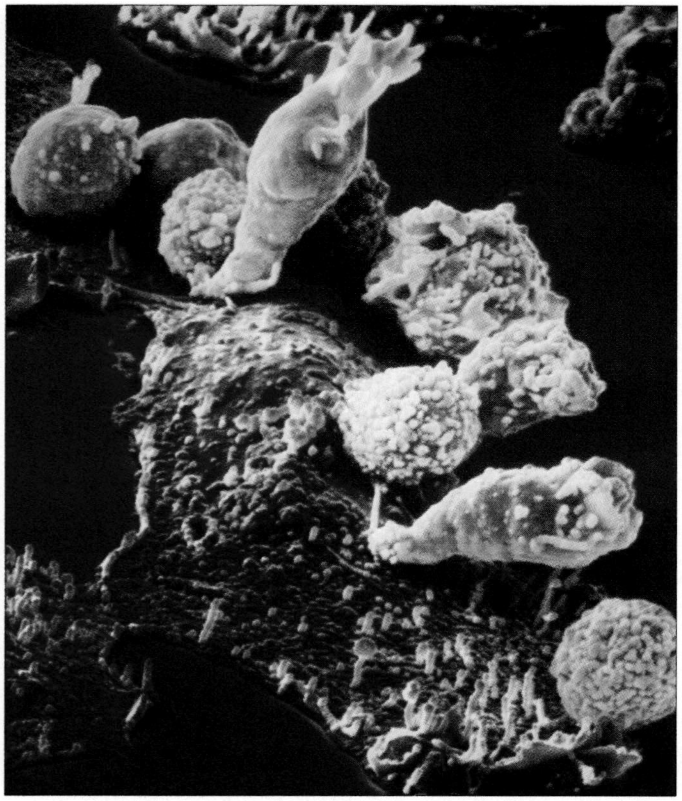

(b) 10 μm

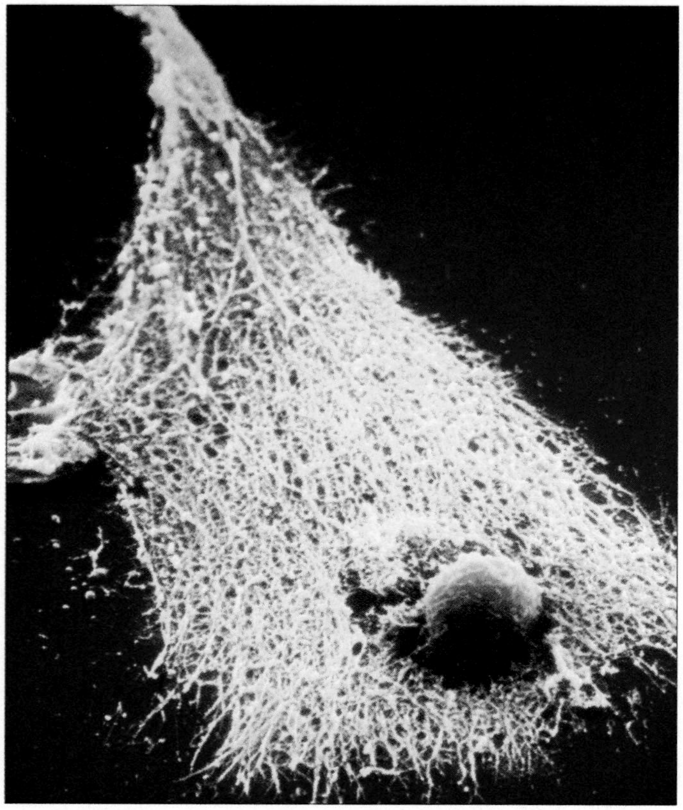

(c) 10 μm

FIGURE 43–11 Cytotoxic T cells defend the body against cancer cells. (*a*) An army of cytotoxic T cells surrounds a large cancer cell. The T cells recognize the cancer cell as nonself because it displays altered antigens on its surface. (*b*) Some of the cytotoxic T cells elongate as they chemically attack the cancer cell, breaking down its plasma membrane. (*c*) The cancer cell has been destroyed. Only a collapsed fibrous cytoskeleton remains. (Lennart Nilsson, © Boehringer Ingelheim International GmbH)

mal cells that the immune system cells may fail to recognize them as foreign. By maintaining a "low profile" in this way, cancer cells can multiply until a tumor gets so large that the host is unable to destroy it.

Sometimes, cells of the immune system do recognize cancer cells but are unable to destroy them. Cancer cells can stimulate B cells to produce IgG antibodies that combine with antigens on the surfaces of the cancer cells. These **blocking antibodies** may block the T cells so that they are unable to adhere to the surface of the cancer cells and destroy them. For some unknown reason, the blocking antibodies are not able to activate the complement system that would destroy the cancer cells. Interestingly, in this case the presence of antibodies is harmful.

An exciting approach in cancer research involves the laboratory production of **monoclonal antibodies** (see *Making the Connection: Antibodies and Biomedical Research*). In this procedure, mice are injected with antigens from human cancer cells. Cells from the mice that produce an-

Making the Connection

Antibodies and Biomedical Research

What are **monoclonal antibodies** and why are they important? Monoclonal antibodies are identical antibodies produced by cells cloned from a single cell. These antibodies, important in immunological research, hold promise as therapeutic agents.

To produce monoclonal antibodies in the laboratory, mice are injected with the antigen of interest. After the mice have produced antibodies to the antigen, B cells containing the antibodies are collected. These cells are suspended in a culture medium together with myeloma (a type of cancer) cells from other mice. Unlike normal cells, these cancer cells can live and divide in tissue culture indefinitely. Many of the cancer cells fuse to the antibody producing cells. The hybrid cells, called hybridomas, have properties of the two "parent" cells, and they continue to secrete antibodies.

Researchers select hybrid cells that are manufacturing the specific antibody needed and then clone them in a separate cell culture. Cells of this clone secrete large amounts of the specific antibody needed—thus, the name monoclonal antibodies. Monoclonal antibodies specific for a single antigenic determinant can now be produced.

Monoclonal antibodies are being used in a variety of clinical diagnostic tests, procedures, and treatments including cancer treatment and pregnancy tests. In several very sensitive pregnancy tests, antibodies are prepared by injecting mice with human chorionic gonadotropin (hCG), a hormone present in the blood and urine of pregnant women. An antigen-antibody response in the urine of the woman being tested indicates that she is pregnant. ▲

tibodies to the cancer cell antigens are used to make hybridomas. Researchers select hybrid cells that are manufacturing the specific antibody needed and then clone them in a separate cell culture. When injected into the same patient whose cancer cells were used to stimulate their production, these antibodies are highly specific for destroying the cancer cells.

In trial studies, monoclonal antibodies are being tagged with toxic drugs that are then delivered specifically to the cancer cells. Such targeted drug therapy would cause fewer side effects than current types of chemotherapy, which damage normal cells as well as cancer cells.

Graft rejection is an immune response against transplanted tissue

Skin can be successfully transplanted from one part of the same body to another or from one identical twin to another. However, when skin is taken from one person and grafted onto the body of a non-twin, it is rejected and sloughs off. Why? Recall that tissues from the same individual or from identical twins have identical MHC alleles and thus the same MHC antigens. Such tissues are compatible.

Because there are many alleles for each of the MHC genes, it is difficult to find identical matches. When a tissue or organ is taken from a donor and transplanted to the body of a non-twin host, several of the MHC antigens are likely to be different. The host's immune system regards the graft as foreign and launches an immune response called **graft rejection.** T cells attack the transplanted tissue and can destroy it within a week (Fig. 43–12).

Before transplants are performed, tissues from the patient and from potential donors must be typed and matched as closely as possible. Cell typing is somewhat similar to blood typing but is more complex. If all of the MHC antigens are matched, the graft has about a 95% chance of surviving the first year. Unfortunately, not many persons are lucky enough to have an identical twin to supply spare parts, so perfect matches are difficult to find. Furthermore, some parts such as the heart cannot

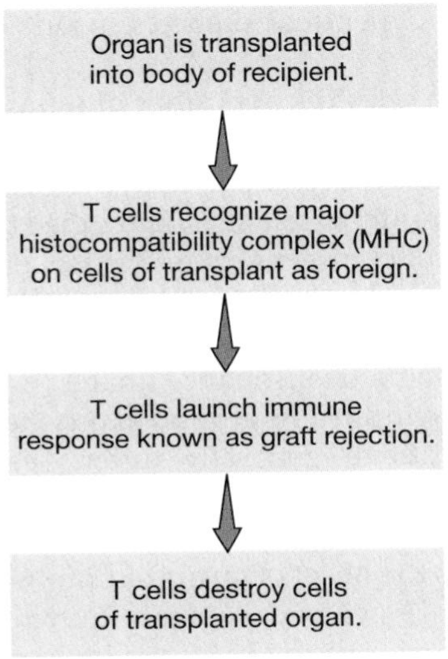

FIGURE 43–12 T cells are responsible for graft rejection.

be spared. Most organs to be transplanted, therefore, are removed from unrelated donors, often from patients who have just died.

To prevent graft rejection in less compatible matches, physicians use drugs that suppress the immune system. Unfortunately, while these immunosuppressive drugs reduce graft rejection, they also make the transplant patient more vulnerable to pneumonia or other infections, and increase the risk of certain types of tumor growths. If the patient can survive the first few months, dosages of these drugs can be reduced. New, powerful immunosuppressive drugs are increasing the success rates for transplants of kidneys, heart, bone marrow, liver, pancreas, and intestine.

Certain sites in the body are immunologically privileged

A few immunologically privileged locations exist in which foreign tissue is accepted by a host. The brain and cornea are examples. Corneal transplants are highly successful because the cornea has almost no blood or lymphatic vessels associated with it and so is out of reach of most lymphocytes. Furthermore, antigens in the corneal graft probably would not find their way into the circulatory system, and so would not stimulate an immune response.

In an autoimmune disease the body attacks its own tissues

Hypersensitivity refers to a damaging immunological response to an antigen that is normally harmless. **Autoimmune disease** is a form of hypersensitivity that occurs when the body reacts immunologically against its own tissues. Some of the diseases that result from such failure in recognizing self (self-tolerance) are rheumatoid arthritis, multiple sclerosis, systemic lupus erythematosus (SLE), insulin-dependent diabetes, psoriasis, and scleroderma.

In multiple sclerosis and other autoimmune diseases, the body produces abnormal antibodies. Genetic factors are known to play a role. For example, when one identical twin has multiple sclerosis, the other twin has about a 30% chance of contracting the disease also. In multiple sclerosis, antibodies attack the glial cells that produce the myelin sheath surrounding neurons in the brain and spinal cord. Magnetic resonance images (MRIs) show the absence of myelin sheaths around axons that are normally myelinated. Patients generally suffer weakness and visual problems, and become progressively more disabled.

Studies also indicate that viral or bacterial infection often precedes the onset of an autoimmune disease. Some pathogens have evolved a tactic known as molecular mimicry. They trick the body by producing molecules that look like self-molecules. For example, an adenovirus that causes respiratory and intestinal illness produces a peptide that mimics myelin protein. When the body launches responses to the adenovirus peptide, it may also begin to attack the similar self-molecule, myelin.

Allergic reactions are abnormal immune responses

In **allergic reactions,** individuals manufacture antibodies against mild antigens, called **allergens,** that do not stimulate a response in nonallergic individuals. Allergic reactions are referred to as type I hypersensitivity. In many kinds of allergic reactions, distinctive IgE immunoglobulins are produced. About 20% of the population of the United States is plagued by an allergic disorder such as allergic asthma or hayfever. A predisposition toward these disorders appears to be inherited.

Let us examine a common allergic reaction—a hayfever response to ragweed pollen (Fig. 43–13). The first step is *sensitization.* Macrophages degrade the allergen and display fragments of it to T cells. The activated T cells then stimulate B cells to become plasma cells and produce IgE. These antibodies attach to receptors on **mast cells,** which are large connective tissue cells filled with distinctive granules. Each IgE molecule attaches to a receptor by its C region end, leaving the V region end free to combine with the ragweed pollen allergen.

The second step is *activation of mast cells.* When a sensitized, allergic person inhales the microscopic pollen, allergen molecules rapidly attach to the IgE on mast cells. This binding of allergen with IgE antibody stimulates the mast cell to release granules filled with chemicals like histamine and serotonin that cause inflammation (Fig. 43–14). These substances cause blood vessels to dilate and capillaries to become more permeable, leading to edema and redness. Such responses cause the victims' nasal passages to become swollen and irritated. Their noses run, they sneeze, their eyes water, and they feel generally uncomfortable.

A third step may occur in which *the allergic response is prolonged.* Chemical compounds released by the mast cells lure certain white blood cells to leave the circulation and migrate to the inflamed area. These cells then release compounds that damage tissue and prolong the allergic reaction.

allergens on pollen → plasma cells sensitized → IgE released → IgE combines with mast cell receptors → bound IgE combines with allergen → mast cell releases granules → histamine and other chemicals released → allergic symptoms

In **allergic asthma,** an allergen-IgE response occurs in the bronchioles of the lungs. Mast cells release *Slow-Reacting Substance of Anaphylaxis* (SRS-A), which causes

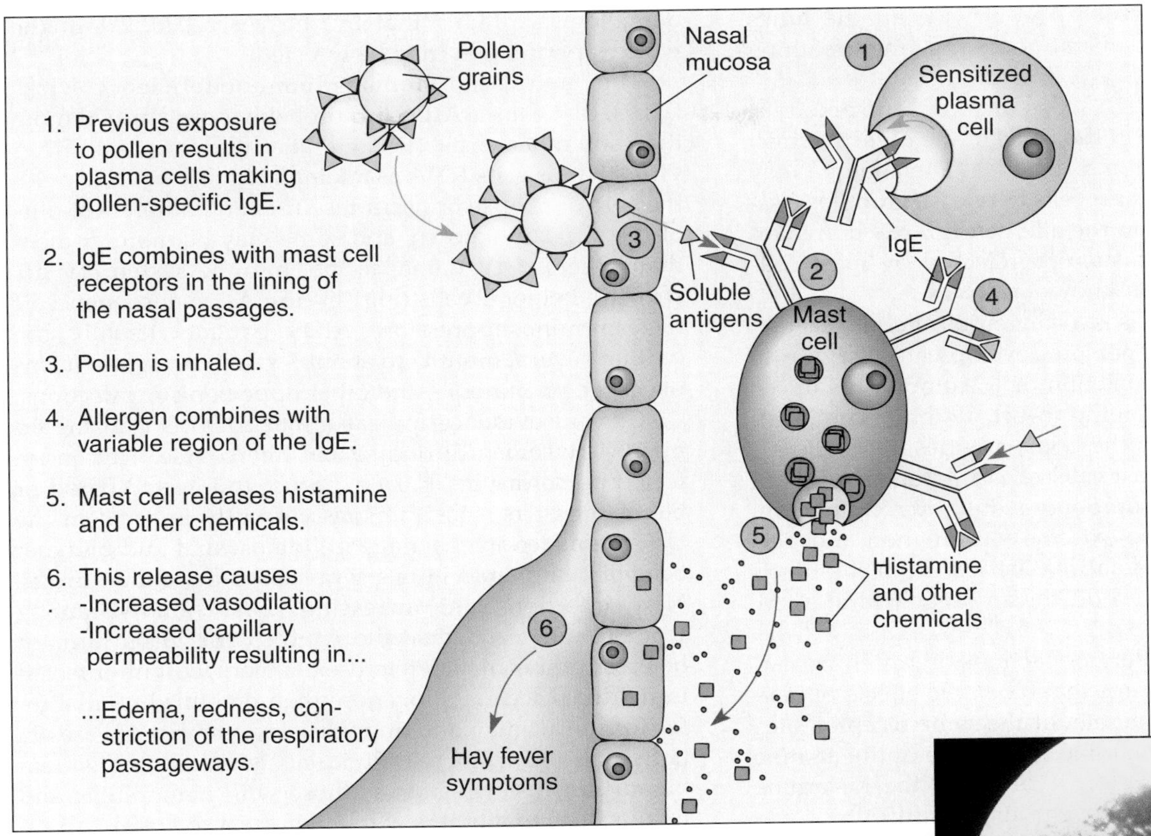
Pollen grains
Nasal mucosa
1
Sensitized plasma cell
1. Previous exposure to pollen results in plasma cells making pollen-specific IgE.
2. IgE combines with mast cell receptors in the lining of the nasal passages.
3. Pollen is inhaled.
4. Allergen combines with variable region of the IgE.
5. Mast cell releases histamine and other chemicals.
6. This release causes
-Increased vasodilation
-Increased capillary permeability resulting in...
...Edema, redness, constriction of the respiratory passageways.
3
IgE
2
4
Soluble antigens
Mast cell
5
Histamine and other chemicals
6
Hay fever symptoms

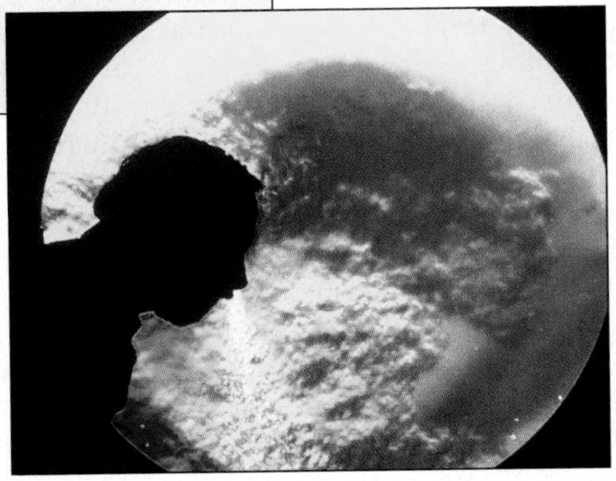

FIGURE 43–13 Pollen causes a common type of allergic response in many people. The photograph shows the expulsion of air and droplets of mucus in a sneeze. (Schlieren photo by Dr. Gary Settles/Photo Researchers, Inc.)

5 μm

FIGURE 43–14 Mast cells release granules that cause allergy symptoms. When an allergen combines with IgE bound to a mast cell receptor, the mast cell explodes, releasing granules filled with histamine and other chemicals that cause the symptoms of an allergic response. (Lennart Nilsson, © Boehringer Ingelheim International GmbH)

smooth muscle to constrict. The airways in the lungs sometimes contract for several hours, making breathing difficult.

Certain foods or drugs act as allergens in some persons, causing a reaction in the walls of the digestive tract that leads to discomfort and diarrhea. The allergen may be absorbed and cause mast cells to release granules elsewhere in the body. When the allergen-IgE reaction takes place in the skin, the histamine released by mast cells causes swollen red welts known as **hives.**

Systemic anaphylaxis is a dangerous allergic reaction that can occur when a person develops an allergy to a specific drug such as penicillin or to compounds in the venom injected by a stinging insect. Within minutes after the substance enters the body, a widespread allergic reaction takes place. Mast cells release large amounts of histamine and other compounds into the circulation. These compounds cause extreme vasodilation and permeability. So much plasma may be lost from the blood that circulatory shock and death can occur within a few minutes.

The symptoms of allergic reactions are often treated with **antihistamines,** drugs that block the effects of histamines. These drugs compete for the same receptor sites on cells targeted by histamine. When the antihistamine combines with the receptor, it prevents the histamine from combining and thus prevents its harmful effects. Antihistamines are useful clinically in relieving the symptoms of hives and hayfever. They are not completely effective because mast cells release substances other than histamines that also cause allergic symptoms.

In serious allergic disorders, patients are sometimes given a form of immunotherapy known as desensitization. Small amounts of the allergen are injected weekly over a period of months or years. Just how this treatment works is not completely understood, but production of IgG antibody to the allergen, or inducing T cell tolerance may be involved.

AIDS is an immune disease caused by a retrovirus

Acquired immune deficiency syndrome, or simply **AIDS,** is a deadly disease currently spreading throughout the world's human population at an alarming rate. First recognized in 1981, epidemiologists estimate that more than 20 million people worldwide are now infected with the AIDS virus. These numbers may reflect only a small percentage of the actual number of individuals infected. More than three million people have developed the disease and most of them have already died. In the United States alone, more than 400,000 individuals have been diagnosed with AIDS, and more than one million more are estimated to be infected with the AIDS virus even though they are presently symptom-free. Some epidemiologists predict that by the year 2000, 2% of the world's population may be infected.

The retrovirus, **human immunodeficiency virus (HIV),** that causes AIDS has probably been studied more than any other virus. (Recall that a retrovirus is an RNA virus that uses its RNA as a template to make DNA with the help of reverse transcriptase.) Several different strains of the virus are known, and some may be more virulent than others. HIV damages the immune system by destroying helper T cells (Fig. 43–15). As a consequence of this immunosuppression, AIDS patients usually die (within several months to about 5 years) from rare forms of cancer, pneumonia, and other opportunistic infections.

Current evidence indicates that AIDS is transmitted mainly by semen during sexual intercourse with an infected person or by direct exposure to infected blood or blood products. Most persons currently infected in the U.S. are males who engage in homosexual and bisexual behavior and individuals who use intravenous drugs. New infections are increasing most rapidly among women and teenagers who contract the virus through heterosexual contact. Heterosexual contact with infected individuals accounts for more than one-third of HIV infections in women and an increasing number of cases in both men and women. Individuals who require frequent blood transfusions, such as those with hemophilia, and infants born to mothers with AIDS are also at risk. An estimated 10% of AIDS patients are children born to infected mothers.

Effective blood-screening procedures have been developed to safeguard blood bank supplies, markedly reducing the risk of infection from blood transfusion in developed countries. Use of a latex condom during sexual intercourse provides some protection against the virus. Use of a spermicide containing nonoxynol-9 is thought to provide additional protection.

HIV is not spread by casual contact. People do not contract the virus by hugging, kissing, sharing a drink, or using the same bathroom facilities. Friends and family members who live with AIDS patients are not more likely to get the virus.

Progression of AIDS may depend on a combination of genetic and environmental factors. Some evidence exists that susceptibility is also affected by psychosocial factors, including personality variables and coping styles that intensify environmental stressors.

When HIV infects the body, an immune response may be launched. The infected individual may experience mild flu-like symptoms (fever and aching muscles) for a week or so. Some cells infected by HIV are not destroyed and the virus may continue to replicate slowly for many years.

After a time, a progression of symptoms occurs, including swollen lymph glands, night sweats, fever, and weight loss (Fig. 43–16). In about one-third of AIDS patients, the virus infects the nervous system, causing AIDS

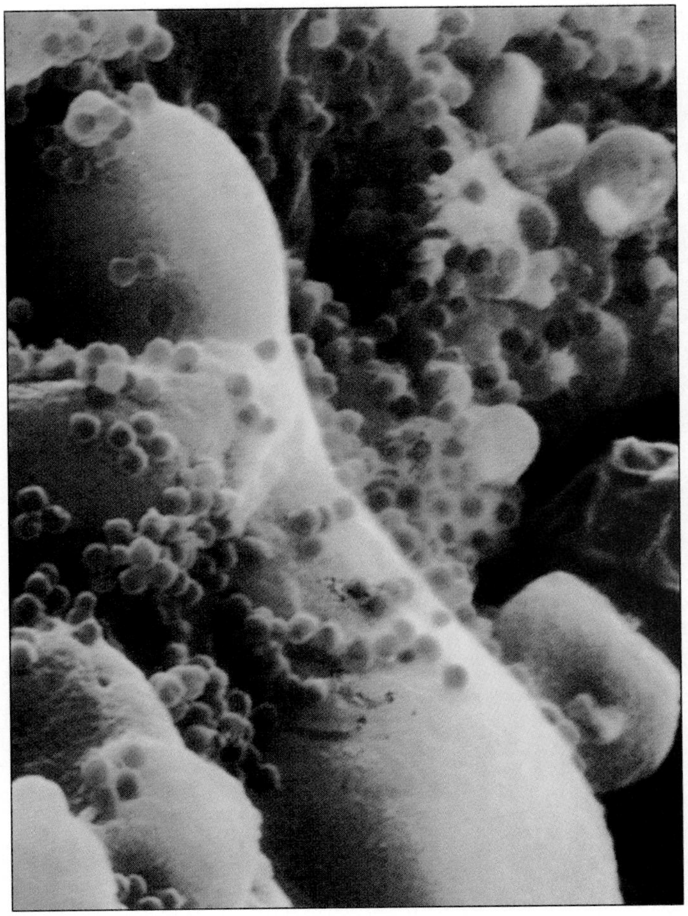

(a) 1 μm

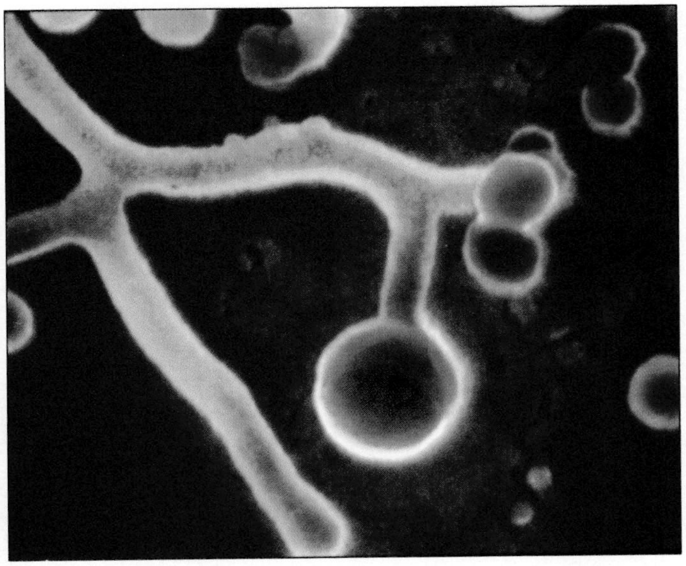

(b) 0.25 μm

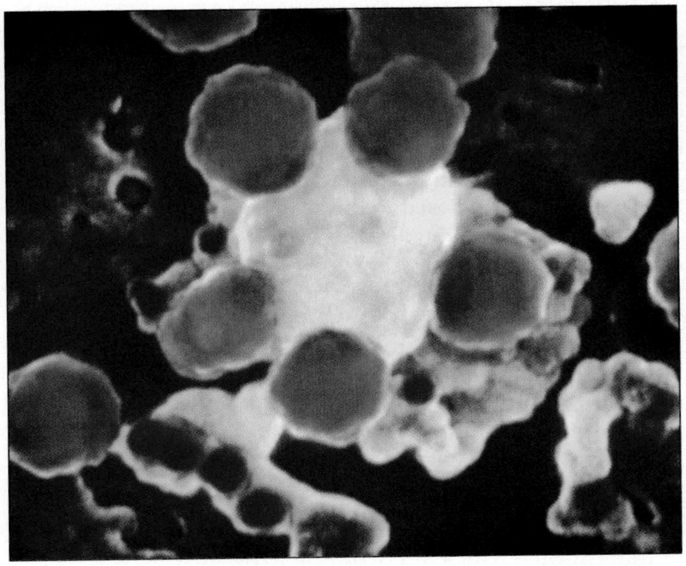

(c) 0.25 μm

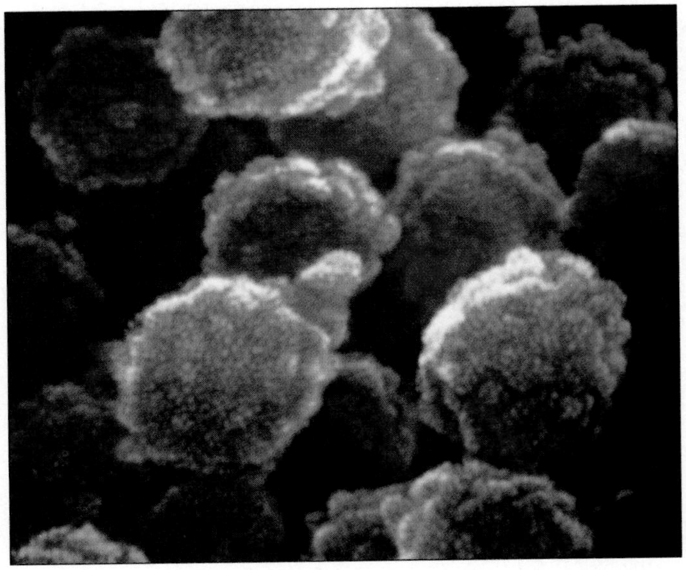

(d) 0.1 μm

FIGURE 43–15 HIV seriously impairs the immune system by rapidly destroying helper T cells. (***a***) HIV virus particles (*blue*), which cause AIDS, attack a helper T cell *(light blue).* (***b***) HIV-1 virus particles budding from the ends of the branched microvilli of a T cell. (***c***) An even higher magnification of virus particles budding from a "bleb" (a cytoplasmic extension broader than a microvillus). Note the pentagonal symmetry often evident in other biological branching and flowering structures. (***d***) HIV-1 virus particles at extremely high magnification. The surfaces are grainy and the outlines slightly blurred because the preparation was coated with coarse-grained heavy metal (palladium) salt. (Lennart Nilsson, © Boehringer Ingelheim International GmbH)

dementia complex. These patients exhibit progressive cognitive, motor, and behavioral dysfunction that typically ends in coma and death.

HIV enters a host cell by attaching a protein on its outer envelope to CD4, a protein present on the surface of several types of immune system cells. Helper T cells appear to be the main target of the virus (chapter opening figure and Fig. 43–15). HIV destroys helper T cells and over time causes a dramatic decrease in their numbers. Some evidence suggests that the virus gradually destroys the lymph nodes. When the helper T cell population is depressed, the ability to resist infection is severely impaired, making the AIDS patients more susceptible to cancer and other opportunistic infections (Fig. 43–15).

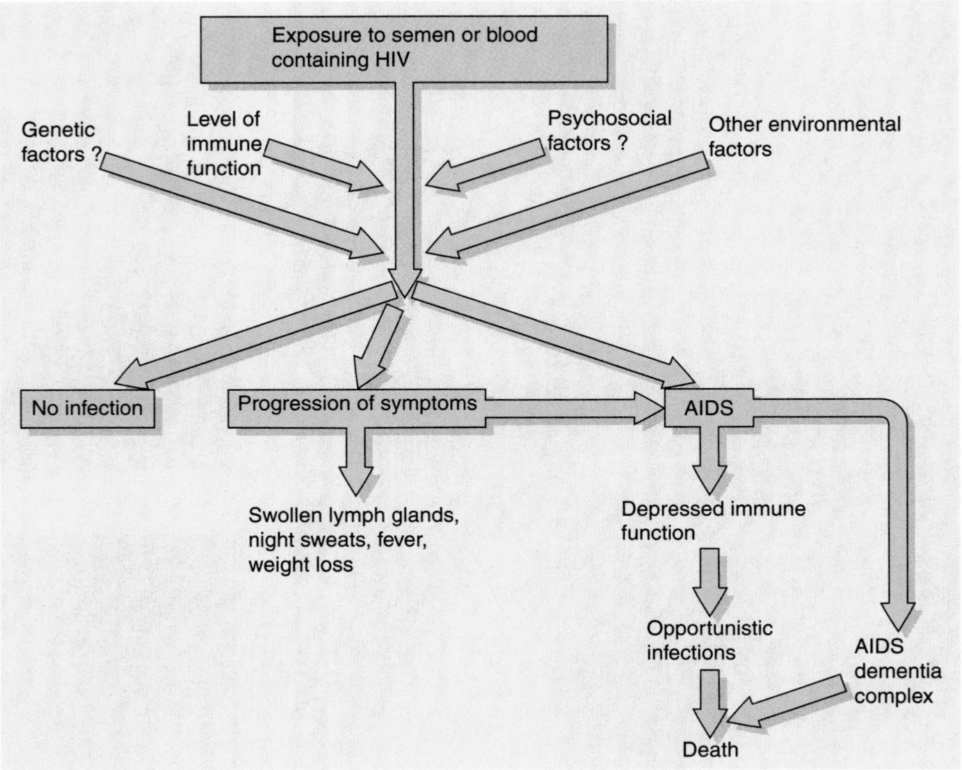

FIGURE 43–16 Exposure to semen or blood that contains HIV can lead to AIDS. Although some exposed individuals apparently do not become infected, the risk increases with multiple exposures. Many factors apparently determine whether a person exposed to the AIDS virus contracts the disease.

Researchers throughout the world are searching for drugs that will successfully combat the AIDS virus. Because HIV often infects the central nervous system, an effective drug must cross the blood-brain barrier. AZT (azidothymidine), the first drug developed to treat HIV infection, can prolong the period prior to the onset of AIDS symptoms. AZT blocks HIV replication by blocking the action of reverse transcriptase, the enzyme used by the retrovirus to synthesize DNA. This process is essential for the virus to incorporate itself into the host cell's DNA.

Unfortunately, the reverse transcriptase used by the virus to synthesize DNA makes many mistakes: about one in every 2000 nucleotides it incorporates is incorrect. By mutating in this way, the virus has developed strains that are resistant to AZT. As other drugs effective against HIV are developed and approved, a combination of treatments aimed at different parts of the HIV life cycle may prove more effective.

Effective vaccination is the simplest way of preventing a disease. However, developing a vaccine against HIV has been a most daunting challenge for virologists. Because of its high rate of mutations, new viral strains with new antigens evolve quickly. A vaccine would not be effective against new antigens, and so would quickly become obsolete. Other barriers to the development of a vaccine include the absence of an effective animal model for AIDS and the ethical and practical difficulties associated with finding human volunteers in whom to test the vaccine.

While immunologists work to develop a successful vaccine and effective drugs to treat infected patients, massive educational programs are being developed to impede the proliferation of AIDS. Educating the public that multiple sexual partners increase the risk of AIDS and teaching sexually active individuals the importance of "safe" sex may help to slow the epidemic. Some have suggested that public health facilities offer free condoms to those who are sexually active, and free sterile hypodermic needles to those addicted to drugs. The cost of these measures would be far less than the cost of medical care for increasing numbers of AIDS patients and the toll in human suffering.

SUMMARY

I. Internal defense depends on the ability of an organism to distinguish between self and nonself.

II. Many invertebrates depend mainly on nonspecific responses such as phagocytosis.

III. Vertebrates have nonspecific defense mechanisms, such as the skin and mucous lining of the respiratory tract, that prevent the entrance of pathogens. Should pathogens break through these first line defenses, other nonspecific defense mechanisms are activated.

A. Cytokines are regulatory proteins, such as interferons and interleukins, secreted by cells of the immune system. Interferons prevent viral replication.

B. Complement proteins increase the inflammatory response and phagocytosis; some complement proteins digest portions of the pathogen cell.
C. When pathogens invade tissues, they trigger an inflammatory response, which brings needed phagocytic cells and antibodies to the infected area.
D. Neutrophils and macrophages phagocytize and destroy bacteria.

IV. Vertebrates have specific defense mechanisms, called immune responses, that include antibody-mediated immunity and cell-mediated immunity.
A. Three main types of lymphocytes are T cells, B cells, and natural killer cells.
B. The thymus gland confers immunological competence on T cells, and also secretes hormones.
C. In antibody-mediated immunity, B cells are activated when they combine with antigen. Activation requires an antigen-presenting cell (such as a macrophage) that has a foreign antigen-MHC complex displayed on its surface. A helper T cell that secretes interleukins is also needed.
   1. B cells multiply, giving rise to clones of cells.
   2. Some B cells differentiate to become plasma cells, which secrete specific antibodies. Others become memory cells.
D. Antibodies are highly specific proteins called immunoglobulins. They are produced in response to specific antigens. Antibodies are grouped in five classes according to the constant regions of their heavy polypeptide chains.
E. An antibody combines with a specific antigen to form an antigen-antibody complex, which may inactivate the pathogen, stimulate phagocytosis, or activate the complement system.
F. In cell-mediated immunity, specific T cells are activated by helper T cells and by a foreign antigen-MHC complex on a cell surface. Activated T cells multiply, giving rise to a clone.
   1. Some T cells differentiate to become cytotoxic T cells, which migrate to the site of infection and chemically destroy cells infected with viruses.
   2. Some activated T cells remain in the lymph nodes as memory cells; others become helper T cells.
G. Second exposure to an antigen evokes a secondary immune response, which is more rapid and more intense than the primary response.
H. Active immunity develops as a result of exposure to antigens; it may occur naturally after recovery from a disease or be artificially induced by immunization.
I. Passive immunity is a temporary condition that develops when an individual receives antibodies produced by another person or animal.
J. Normally, the immune system destroys precancer cells when they arise; diseases such as cancer develop when this immune mechanism fails to operate effectively.
K. Transplanted tissues have MHC antigens that stimulate graft rejection, an immune response that can destroy the transplant.
L. In autoimmune diseases, the body reacts immunologically against its own tissues.
M. In an allergic response, an allergen can stimulate production of IgE antibody, which combines with the receptors on mast cells. When allergen combines with the IgE, the mast cells release histamine and other substances that cause symptoms of allergy such as inflammation.
N. Acquired immune deficiency syndrome (AIDS) is caused by a retrovirus known as human immunodeficiency virus (HIV). HIV damages the immune system by destroying helper T cells.

SELECTED KEY TERMS

active immunity
AIDS (acquired immune deficiency syndrome)
allergen
allergic reaction
antibody
antibody-mediated immunity
antigen
antigen-antibody complex
autoimmune disease
B cell (B lymphocyte)
cell-mediated immunity
complement
cytokine
cytotoxic T cell
graft rejection
helper T cell
histamine
HIV (human immunodeficiency virus)
immune response
immunization
immunoglobulin
inflammatory response
interferon
interleukin
macrophage
mast cell
memory cell
MHC (major histocompatibility complex)
natural killer cell (NK cell)
passive immunity
pathogen
plasma cell
primary response
secondary response
suppressor T cell
T cell (T lymphocyte)
thymus gland

POST-TEST

1. An antigen is a substance capable of stimulating a(an) __________ __________.
2. Specific proteins produced in response to specific antigens are called __________.
3. When infected by viruses, some cells respond by producing proteins called __________.
4. The clinical characteristics of a(an) __________ __________ include warmth, redness, pain, and swelling.
5. T lymphocytes originate in the __________ __________ and are processed in the __________ __________.
6. __________ T cells release perforins, and other potent enzymes.
7. When the body is invaded by the same pathogen a second time, the immune response can be launched more rapidly due to the presence of __________ cells.
8. The cells that produce antibodies are __________ cells.
9. An antigenic determinant gives the antigen molecule a specific configuration that can be "recognized" by a(an) __________.
10. The __________ gland confers immunological competence upon T cells.
11. The complement system is activated when a(an) __________-__________ complex is formed.
12. After a macrophage ingests a pathogen, it may display foreign antigen on its surface, along with __________ proteins.
13. Although artificially induced, immunization is a form of __________ immunity.
14. An individual receiving injections of antibodies produced by another organism is receiving __________ immunity.
15. In __________ __________, the host launches an effective attack against foreign tissue, such as a transplanted kidney.
16. __________ __________ cells are lymphocytes especially adapted to destroy cancer cells.
17. A(An) __________ is a mild antigen that does not stimulate a response in an individual who is not __________.
18. In a typical allergic reaction, __________ cells secrete histamine and other compounds that cause inflammation.
19. __________ is the retrovirus that causes AIDS.

REVIEW QUESTIONS

1. How does the body distinguish between self and nonself? Are invertebrates capable of making this distinction?
2. Contrast specific and nonspecific defense mechanisms. Which type confronts invading pathogens immediately? How do the two types of processes work together?
3. How does inflammation help to restore homeostasis?
4. Give two specific ways in which cell-mediated and antibody-mediated immune responses are similar, and two ways in which they are different.
5. Describe how antibodies destroy pathogens.
6. John is immunized against measles. Jack contracts measles from a playmate in nursery school because his parents failed to have him immunized. Compare the immune responses of the two children. Five years later, John and Jack are playing together when Judy, who is coming down with measles, sneezes on both of them. Compare their immune responses.
7. Why is passive immunity temporary?
8. What is graft rejection? What is the immunological basis for it?
9. List the immunological events that take place in a common type of allergic reaction such as hayfever.
10. What is an autoimmune disease? Give two examples.
11. Specificity, diversity, and memory are key features of the immune system. Giving specific examples, explain how each of these features is important.

YOU MAKE THE CONNECTION

1. Imagine that you are a researcher developing new HIV treatments. What approaches might you take? What public policy decisions would you recommend that might help slow the spread of AIDS while new treatments or vaccines are being developed?
2. Macrophages can be selectively destroyed in the body by the administration of a certain chemical. What would be the effects of such a loss of macrophages? Which do you think would have a greater effect on the immune system, loss of macrophages or loss of B cells?
3. What are the advantages of having MHC antigens? Disadvantages? What do you think would be the consequences of not having them?

RECOMMENDED READINGS

Black, P.H. "Psychoneuroimmunology: Brain and Immunity," *Scientific American: Science & Medicine,* Vol. 2, No. 6, November/December 1995. A discussion of what is known about the relationship between stress and immunosuppression.

Boon, T. "Teaching the Immune System to Fight Cancer." *Scientific American,* Vol. 266, No. 3, March 1993. Certain molecules on cancer cells serve as targets for cells of the immune system. These tumor-rejection antigens are the focus of research and may be useful in treating cancer.

Collins, T. "Adhesion Molecules in Leukocyte Emigration," *Scientific American: Science & Medicine,* Vol. 2, No. 6, November/December 1995. How white blood cells leave the circulation and enter the tissues in search of foreign molecules.

Eisenbarth, G.S. and D. Bellgrau; "Autoimmunity," *Scientific American: Science & Medicine,* Vol. 1, No. 2, May/June 1994. A discussion of what is known about autoimmune disease.

Fischetti, V.A. "Streptococcal M Protein." *Scientific American,* Vol. 264, No. 6, June 1991. The bacteria that cause strep throat and rheumatic fever depend on a surface protein to evade the body's defenses.

Johnson, H.M., F.W. Bazer, B.E. Szente, and M.A. Jarpe. "How Interferons Fight Disease." *Scientific American,* Vol. 270, No. 5, May 1994. An interesting article that discusses how interferons protect against disease and how they are used clinically.

Johnson, H.M., J.K. Russell, and C.H. Pontzer. "Superantigens in Human Disease." *Scientific American,* Vol. 266, No. 4, April 1992. The article discusses how superantigens cause disease such as toxic shock and certain types of food poisoning.

Langer, R. and J. P .Vacanti, "Artificial Organs," *Scientific American,* Vol. 273, No. 3, September 1995. A preview of possible 21st century technologies for fabricating artificial organs.

Levine, A.J. "Tumor Suppressor Genes." *Scientific American: Science & Medicine,* Jan./Feb. 1995. An article that makes important connections between genetics, immunology, virology, molecular biology, and epidemiology.

Nowak, M.A. and A.J. McMichael. "How HIV Defeats the Immune System." *Scientific American,* Vol. 273, No. 2, August 1995. A discussion of the hypothesis that continued mutation of the HIV virus allows it to eventually overcome the body's defenses.

Scientific American, Vol. 269, No. 3, September 1993. A single-topic issue on life, death, and the immune system.

Smith, K.A. "Interleukin-2." *Scientific American,* Vol. 262, No. 3, March 1990. A discussion of the regulatory functions of interleukin-2.

Streit, W.J. and C.A. Kincaid-Colton, "The Brain's Immune System," *Scientific American,* Vol. 273, No. 5, November 1995. Microglia, a type of glial cells, have immunological properties; however, sometimes they damage neurons.

Tizard, I.R. *Immunology: An Introduction,* 4th ed. Saunders College Publishing, Philadelphia, 1995. A basic introduction to immunology.

C H A P T E R 44

Gas Exchange

White-tail deer (*Odocoileus virginianus*) in fog. (Skip Moody/Dembinsky Photo Associates)

Most animal cells require a continuous supply of oxygen. Some cells, such as mammalian brain cells, may be damaged beyond repair if their supply is cut off for only a few minutes. Animal cells must also rid themselves of carbon dioxide. The exchange of gases between an organism and its environment is known as **respiration.**

The exchange of gases is a fairly simple process in small, aquatic organisms such as sponges, hydras, and flatworms. Dissolved oxygen from the surrounding water diffuses into the cells, while carbon dioxide diffuses out of the cells and into the water. No specialized respiratory structures are needed.

Oxygen diffuses slowly through tissues. In an organism more than 1 millimeter thick, diffusion alone does not provide a satisfactory method of gas exchange. Specialized respiratory structures such as gills or lungs are required to ensure adequate exchange of oxygen and carbon dioxide.

Gases move into and out of cells by diffusion. If the air or water supplying oxygen to the cells can be continuously renewed, more oxygen will be available. For this reason animals carry on **ventilation;** that is, they actively move air or water over their respiratory surfaces. Sponges do this by setting up a current of water through the channels of their bodies by means of flagella. Most fishes gulp water, which then passes over their gills. Terrestrial vertebrates breathe air.

LEARNING OBJECTIVES

After you have studied this chapter you should be able to

1. Compare the advantages and disadvantages of gas exchange in air with those in water.
2. Describe the following adaptations for gas exchange: the body surface; tracheal tubes; gills; and lungs.
3. Describe the function of respiratory pigments.
4. Trace a breath of air through the human respiratory system from external nares to air sacs.
5. Summarize the mechanics and the regulation of breathing.
6. Describe how oxygen and carbon dioxide are exchanged in the lungs and in the tissues.
7. Explain the role of hemoglobin in oxygen transport, and identify factors that determine and influence the oxygen-hemoglobin dissociation curve.
8. Outline the mechanisms by which carbon dioxide is transported in the blood.
9. Describe the physiological effects of each of the following: hyperventilation; sudden decompression at 12,000 meters altitude; and surfacing too quickly from a deep-sea dive.
10. Describe the defense mechanisms that protect the lungs, and the effects on the respiratory system of breathing polluted air.

RESPIRATORY STRUCTURES ARE ADAPTED FOR GAS EXCHANGE IN AIR OR WATER

Gills are adapted for gas exchange in water, while tracheal tubes and lungs are respiratory structures adapted for gas exchange in air. In order for oxygen and carbon dioxide to diffuse across cell membranes, they must be dissolved in water. For this reason, whether an animal makes its home on land or water, gas exchange can take place only across a moist surface.

Because gas molecules must be dissolved in water to pass through plasma membranes, an aquatic animal may seem to have an advantage. However, because water has much greater density and viscosity (resistance to flow) than does air, a large aquatic animal must expend more energy to move water over its respiratory surface than it would to move air. For example, a fish uses up to 20% of its total energy expenditure to perform the muscular work needed to move water over its gills. An air-breather uses much less energy, only 1% or 2% of its total, to ventilate its lungs.

Gas exchange in air has other advantages over gas exchange in water. Compared to water, air contains a much higher concentration of molecular oxygen. Oxygen also diffuses much faster through air than through water. Another advantage is that air is not salty like seawater. Air-breathers do not have to cope with diffusion of ions into their body fluids along with oxygen. As a result, they have an easier time maintaining appropriate concentrations of ions in their body fluids.

On the other hand, organisms that respire in air struggle continuously with water loss. Because respiratory surfaces must be kept moist so that oxygen and carbon dioxide can pass through the plasma membranes, terrestrial animals require adaptations to minimize drying out. The lungs of air-breathing vertebrates are located deep within the body, not exposed like gills. Air must pass through a long sequence of passageways before reaching the moist respiratory surfaces of the lungs, and expired air must again pass through these airways before leaving the body. These adaptations help protect the lungs from the drying effects of air.

FOUR MAIN TYPES OF SURFACES FOR GAS EXCHANGE HAVE EVOLVED

Not only must respiratory structures be moist, they must also have thin walls through which diffusion can easily occur. In addition, respiratory structures are generally richly supplied with blood vessels to facilitate transport and exchange of respiratory gases.

Four main types of respiratory surfaces used by animals are the body surface, tracheal tubes, gills, and lungs (Fig. 44–1; Table 44–1).

The body surface may be adapted for gas exchange

Gas exchange occurs through the entire body surface in many animals, including nudibranch mollusks, most annelids, and a few vertebrates (Fig. 44–2). All of these animals are small, with a high surface-to-volume ratio. They also have a low metabolic rate that requires smaller quantities of oxygen per cell. In aquatic animals, the body surface is kept moist by the surrounding water. In terrestrial animals, the body secretes fluids that keep its surface moist.

How does an animal such as an earthworm exchange gases across its body surface? Gland cells in the epider-

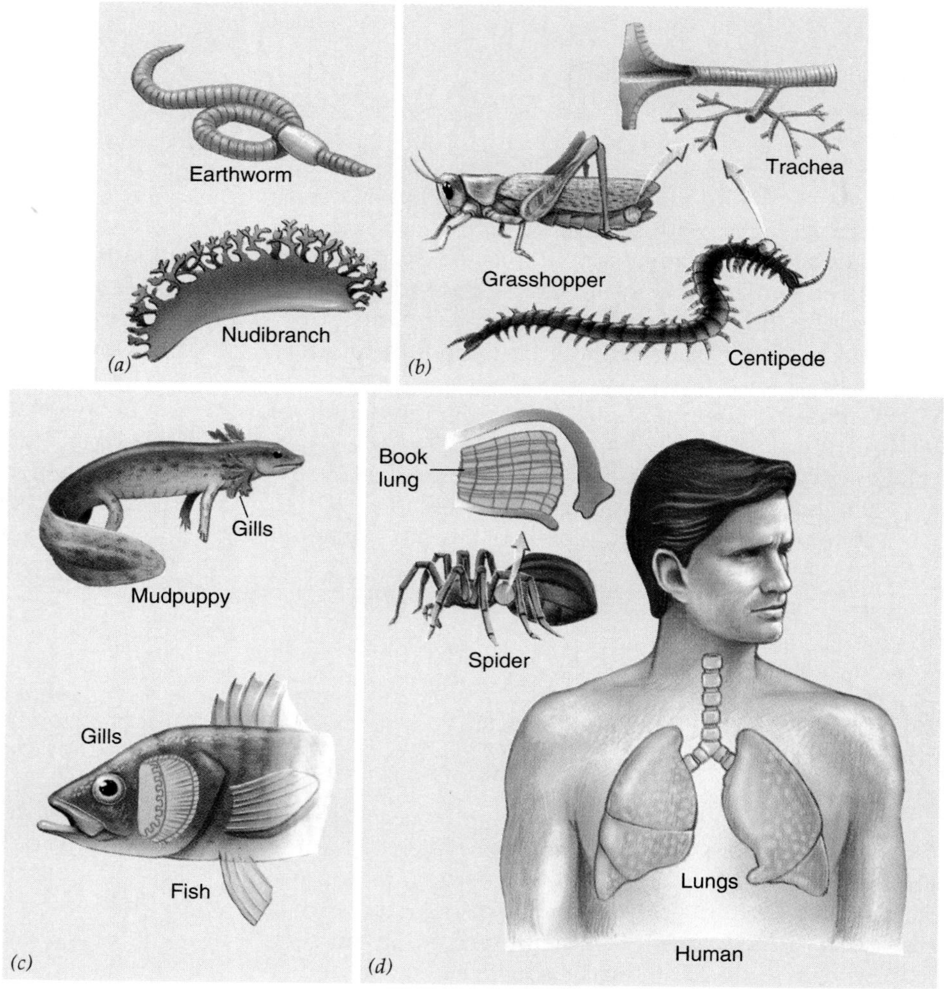

FIGURE 44–1 Animals have a variety of adaptations for exchanging gases. (***a***) Some animals exchange gases through the body surface. (Nudibranchs are mollusks). (***b***) Insects and some other arthropods exchange gases through a system of tracheal tubes. (***c***) Many aquatic animals have gills for gas exchange. (***d***) Lungs are adaptations for terrestrial gas exchange.

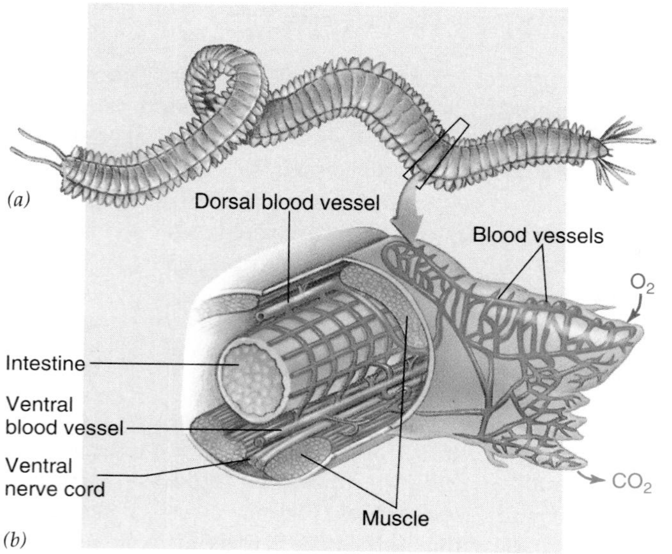

FIGURE 44–2 Some animals exchange gases across the body surface. (***a***) The clamworm *Nereis virens* has parapodia (limblike extensions of each body segment) typical of the polychaete worms. (***b***) The parapodium acts as an extension of the body wall in gas exchange with the surrounding water. Note the rich supply of blood vessels for transport of carbon dioxide and oxygen to and from the body surface.

mis secrete mucus, which keeps the body surface moist while also offering protection. Oxygen, present in air pockets in the loose soil that the earthworm inhabits, dissolves in the mucus. Then, the oxygen diffuses through the body wall and into blood circulating in a network of capillaries just beneath the outer cell layer. Carbon dioxide is transported by the blood to the body surface; from there it diffuses out into the environment.

Tracheal tube systems of arthropods deliver air directly to the cells

In insects and some other arthropods (e.g., chilopods, diplopods, some mites, and some spiders), the respiratory system consists of a network of **tracheal tubes** (Fig. 44–3). Air enters the tracheal tubes through a series of up to 20 tiny openings called **spiracles** along the body surface. In large and active insects, air moves into and out of the spiracles by body movements or by rhythmic movements of the tracheal tubes. For example, the grasshopper draws air in through the first four pairs of spiracles when the abdomen expands. Then the abdomen contracts, forcing air out through the last six pairs of spiracles.

(Text continues on page 964)

Table 44–1 PRINCIPAL TYPES OF RESPIRATORY STRUCTURES

Structure	*Animals that Use Such Structures*	*Description*	*Comment*
Body surface	Nudibranch mollusks, most annelids, small arthropods, some vertebrates	In terrestrial forms the body surface is kept moist by mucus secretion. Blood vessels or coelomic fluid present just below the surface receive oxygen and transport it to other body regions.	These animals have high surface-to-volume ratio and low metabolic rate. In some animals (e.g., frogs), other respiratory structures are present.
Tracheal tubes	Insects, some mites, some spiders, millipedes, centipedes	Air enters through spiracles and passes through the branching tracheal tubes to all parts of the body, terminating in fluid-filled tracheoles. Gas exchange occurs by diffusion between tracheoles and body cells.	In large or active animals ventilation occurs by movements of the body or of tracheal tubes.
Gills	Found mainly in aquatic animals, some annelids, mollusks, crustaceans, fishes, and amphibians	Moist, thin structures that grow out from the body surface. In gills of bony fishes, each gill consists of many filaments containing blood vessels.	A great deal of energy must be expended in ventilation of the gills.
Lungs	Arachnids, some mollusks, most vertebrates	Respiratory structures that develop as ingrowths of the body surface or from the wall of a body cavity. In vertebrates a series of air passageways may terminate in thin-walled air sacs within the lungs.	Most modern fishes lack lungs.

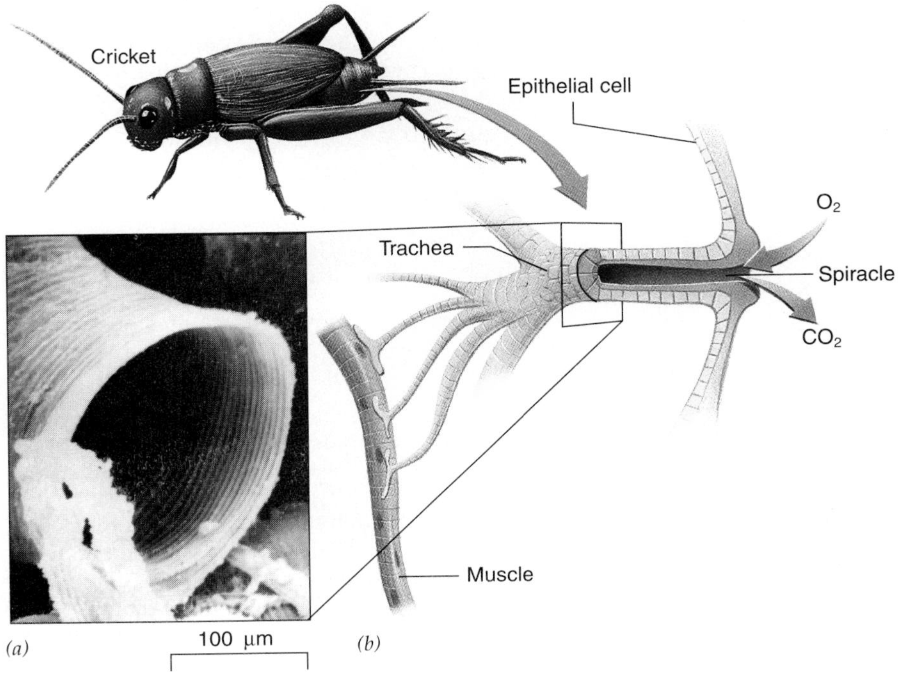

FIGURE 44–3 Tracheal tubes are characteristic of insects and some other invertebrates. (***a***) A scanning electron micrograph of a mole cricket trachea (approximately × 1300). The wrinkles are a part of a long spiral that wraps around the tube, which may strengthen the tracheal wall somewhat like the spring strengthens the plastic hoses of many vacuum cleaners. The tracheal wall is composed of chitin. (***b***) Each tracheal tube and its branches conduct oxygen to the body cells of the insect. (*a*, Courtesy of Dr. James L. Nation and *Stain Technology*, Vol. 58, 1983)

Once inside the body, the air passes through the branching tracheal tubes, which extend to all parts of the animal. The tracheal tubes terminate in microscopic, fluid-filled tracheoles. Gases are exchanged between this fluid and the body cells. Oxygen is not transported by the blood in insects.

Gills of aquatic animals are specialized respiratory surfaces

Found mainly in aquatic animals, **gills** are moist, thin structures that extend from the body surface. They are supported by the buoyancy of water but tend to collapse in air. In many animals, the outer surface of the gills is exposed to water, whereas the inner side is in close contact with networks of blood vessels.

Sea stars and sea urchins have **dermal gills** that project from the body wall. Their ciliated epidermal cells ventilate the gills by beating a stream of water over them. Gases are exchanged through the gills between the water and the coelomic fluid inside the body.

Various types of gills are found in some annelids, aquatic mollusks, crustaceans, fishes, and amphibians. Molluskan gills are folded, providing a large surface for respiration. In clams and other bivalve mollusks and in simple chordates, the gills are adapted for trapping and sorting food. Rhythmic beating of cilia draws water over the gill area, and food is filtered out of the water as gases are exchanged. In mollusks gas exchange also takes place through the mantle.

In chordates, the gills are usually internal. A series of slits perforates the pharynx, and the gills are located along the edges of these gill slits. In bony fishes, the fragile gills are protected by an external bony plate, the **operculum.** Movements of the operculum help to pump oxygenated water in through the mouth. The water flows across the gills and then exits through the gill slits.

Each gill in the bony fish consists of many **filaments,** which provide an extensive surface for gas exchange (Fig. 44–4). The filaments extend out into the water, which continuously flows over them. A capillary network delivers blood to the gill filaments, facilitating diffusion of oxygen and carbon dioxide between blood and water. The very impressive efficiency of this system is possible because blood flows in a direction opposite to the move-

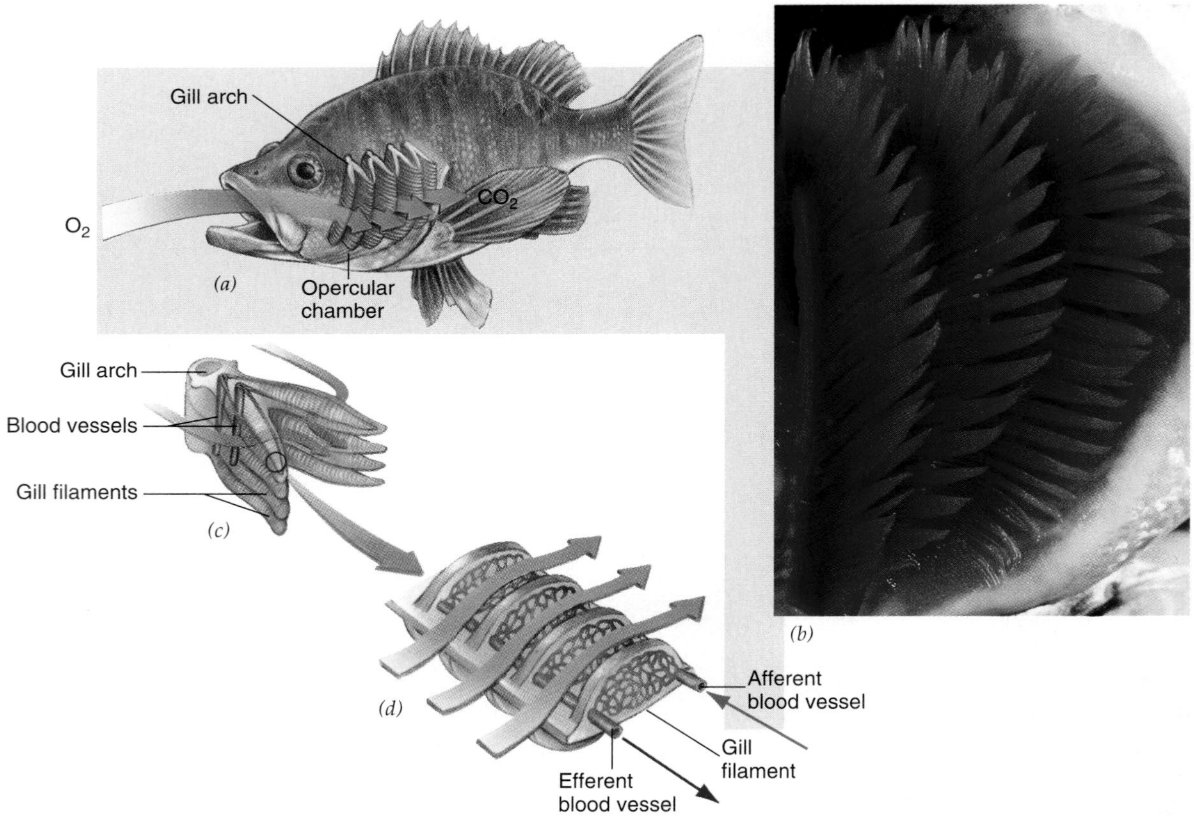

FIGURE 44–4 Gills provide an extensive surface for gas exchange. (*a*) The gills are located under a bony plate, the operculum, which has been removed in this side view. The gills occupy the opercular chamber and form the lateral wall of the pharyngeal cavity. (*b*) Gills of the salmon. (*c*) Each gill consists of a cartilaginous gill arch to which two rows of leaf-like gill filaments are attached. As water passes among the gill filaments, blood circulates through them. (*d*) Each gill filament has many smaller extensions rich in capillaries. Blood entering the capillaries is deficient in oxygen. The blood flows through the capillaries in a direction opposite to that taken by the water. This countercurrent exchange system efficiently charges the blood with oxygen. (*b*. Animals Animals © 1996 G.I. Bernard)

ment of the water. This arrangement, referred to as a **countercurrent exchange system,** maximizes the difference in oxygen concentration between blood and water throughout the area where the two remain in contact.

If blood and water flowed in the *same* direction, the difference between the oxygen concentrations in blood (low) and water (high) would be very large initially and very small at the end. The oxygen concentration in the water would decrease as that in the blood increased. When the concentrations in the two fluids became equal, an equilibrium would be reached and the net diffusion of oxygen would stop. A great deal of oxygen would remain in the water.

In the countercurrent exchange system, however, blood low in oxygen comes in contact with water that is partly oxygen-depleted. Then, as the blood becomes more and more oxygen-rich, it comes in contact with water with a progressively higher concentration of oxygen. In this way a high rate of diffusion is maintained, ensuring that a very high percentage (more than 80%) of the available oxygen in the water diffuses into the blood.

Oxygen and carbon dioxide do not interfere with one another's diffusion, and they simultaneously diffuse in opposite directions. This is because oxygen is more concentrated outside the gills than within, but carbon dioxide is more concentrated inside the gills than outside. Thus, the same countercurrent exchange mechanism that ensures efficient inflow of oxygen also results in equally efficient outflow of carbon dioxide.

Terrestrial vertebrates exchange gases through lungs

Lungs are respiratory structures that develop as ingrowths of the body surface or from the wall of a body cavity such as the pharynx. For example, the book lungs of spiders are enclosed in an inpocketing of the abdominal wall. These lungs consist of a series of thin parallel plates of tissue (like the pages of a book) filled with hemolymph (Fig. 44–1). The plates of tissue are separated by air spaces that receive oxygen from the outside environment through a spiracle.

Larger mollusks have some means of forcefully moving air across the respiratory surface. This helps ensure adequate oxygenation in active animals.

Fossil evidence suggests that crossopterygian fishes, thought to be the ancestors of the amphibians, had lungs somewhat similar to those of modern lungfish. Three genera of lungfish are known today. They live in the headwaters of the Nile, in the Amazon, and in certain Australian rivers. Streams inhabited by the lungfish dry up during seasonal droughts. During these dry periods lungfish remain in the mud of the stream, exchanging gases by means of their lungs. These fishes are also equipped with gills, which they use when swimming. The African lungfish uses both its gills and its lungs throughout the year and cannot survive if deprived of air.

Remains of crossopterygian fishes occur extensively in the fossil record. Those found in Devonian strata are thought to be similar to the ancestors of amphibians. Numerous amphibian fossils also occur in adjacent ancient strata. The geological evidence suggests that periodic droughts occurred in Devonian times. Some paleontologists hypothesize that lungs may have evolved originally as an adaptation to these conditions, and that lungs or lunglike structures may have been present in all early bony fishes.

Most modern fishes have no lungs, but nearly all of them do possess homologous swim bladders (see Chapter 30). The swim bladder permits a fish to control its buoyancy by adjusting its density to the surrounding water. When the fish swims toward the surface where the pressure is lower, its swim bladder expands. Gases are removed, and it returns to its normal size. When the fish swims to greater depths where the pressure is higher, gases are added to the swim bladder.

Some amphibians do not have lungs. Among plethodontid (lungless) salamanders, for example, all gas exchange takes place in the pharynx, or across the thin, wet skin. However, even though they depend mainly on their body surface for gas exchange, most amphibians have lungs (Fig. 44–5). The lungs of mud puppies are two long, simple sacs richly supplied by capillaries. Frogs and toads

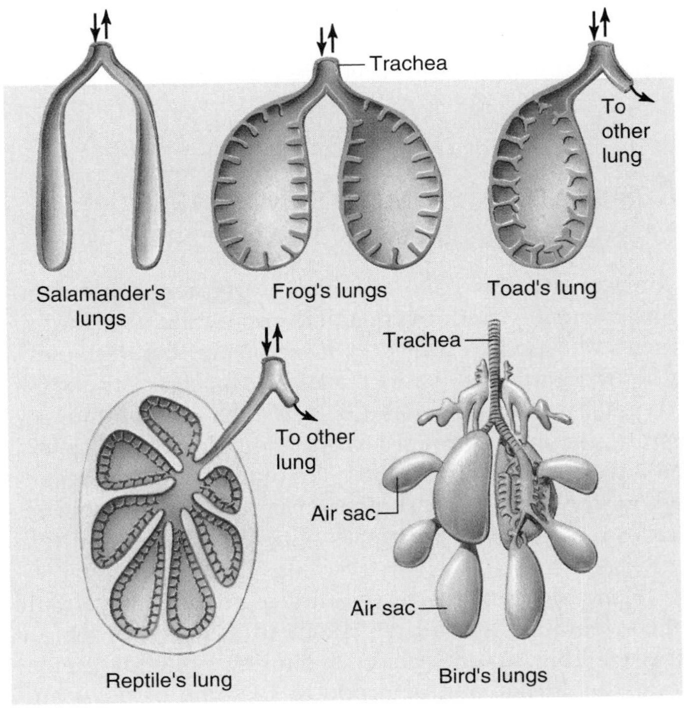

FIGURE 44–5 The surface area of the lung has increased during the evolution of the vertebrates. Salamander lungs are simple sacs, frog lungs have small ridges in the lung wall that help increase surface area, and birds have an elaborate system of lungs and air sacs. Mammalian lungs (Fig. 44-6) have millions of air sacs that increase the surface available for gas exchange.

have ridges containing connective tissue on the inside of the lungs, thereby increasing the respiratory surface somewhat.

The lungs of most reptiles are rather simple, with only some folding of the wall to increase the surface for gas exchange. Because gas exchange is not very effective, these reptiles are unable to sustain long periods of activity. In some lizards, turtles, and crocodiles, the lungs are more complex, having many subdivisions that give them a spongy texture.

Birds are very active animals with high metabolic rates. They require large amounts of oxygen to sustain their activities and have evolved highly efficient respiratory systems. Their lungs have extensions (usually nine) called **air sacs,** which reach into all parts of the body and even penetrate some of the bones. The air sacs act as bellows drawing air into the system. Collapse of the air sacs during expiration (exhalation) forces air out.

The respiratory system is arranged so that air flows in one direction through the lungs and is renewed during each inspiration. The lungs have tiny, thin-walled ducts, the **parabronchi,** which are open at both ends. Gas exchange takes place across the walls of these ducts. The direction of blood flow in the lungs is opposite that of air flow through the parabronchi. This countercurrent flow increases the amount of oxygen that enters the blood.

The lungs of mammals are very complex and have an enormous surface area. Using the human respiratory system as an example, we examine gas exchange in mammals in later sections of this chapter.

RESPIRATORY PIGMENTS INCREASE CAPACITY FOR OXYGEN TRANSPORT

Complex animals have **respiratory pigments** that combine reversibly with oxygen and greatly increase the capacity of blood to transport it. For example, the hemoglobin in human blood increases its capacity to transport oxygen by about 75 times. Oxygen enters the pulmonary capillaries and combines with hemoglobin in the red blood cells. Then, as blood circulates through tissues where the oxygen concentration is low, hemoglobin releases oxygen, which diffuses out of the blood and into the tissue cells.

Hemoglobin is the respiratory pigment characteristic of vertebrates. It is also present in many invertebrate species from several phyla, including annelids, nematodes, mollusks, and arthropods. In some of these animals the hemoglobin is dispersed in the plasma rather than confined to blood cells.

Hemoglobin is actually a general name for a group of related compounds, all of which consist of an iron-porphyrin, or heme, group bound to a protein known as a globin (see Fig. 3–24). The protein portion of the molecule varies in size, amino acid composition, and physical properties among various species.

Hemocyanins, another type of respiratory pigment, are blue, copper-containing proteins found in many species of mollusks and arthropods. These do not have a heme (porphyrin) group. When oxygen is combined with the copper, the compound appears blue. Without oxygen, it is colorless. Hemocyanins are dispersed in the blood rather than confined within cells.

THE HUMAN RESPIRATORY SYSTEM IS TYPICAL OF AIR-BREATHING VERTEBRATES

The respiratory system in humans and other air-breathing vertebrates consists of a series of tubes through which air passes on its journey from the nostrils to the air sacs of the lungs and back (Fig. 44–6).

The airway conducts air into the lungs

A breath of air enters the body through the **nostrils** and flows through the **nasal cavities.** As air passes through the nose, it is filtered, moistened, and brought to body temperature. The nasal cavities are lined with a moist, ciliated epithelium that is rich in blood vessels. Inhaled dirt, bacteria, and other foreign particles are trapped in the stream of mucus that is produced by cells within the epithelium, and pushed along toward the throat by the cilia. In this way, foreign particles are delivered to the digestive system, which can more effectively dispose of such materials than can the delicate lungs. A person normally swallows more than a pint of nasal mucus each day, more during an infection or allergic reaction.

The back of the nasal cavities is continuous with the throat region, or **pharynx.** Air finds its way into the pharynx whether one breathes through the nose or mouth. An opening in the floor of the pharynx leads into the **larynx,** sometimes called the "Adam's apple." Because the larynx contains the vocal cords, it is also referred to as the voice box. Cartilage embedded in its wall prevents the larynx from collapsing and makes it hard to the touch when felt through the neck.

During swallowing, a flap of tissue called the **epiglottis** automatically closes off the larynx so that food and liquid enter the esophagus rather than the lower airway. Should this defense mechanism fail and foreign matter enter the sensitive larynx, a **cough reflex** is initiated, expelling the material. Despite these mechanisms, choking sometimes occurs. (See *Focus On: Choking.*)

From the larynx, air passes into the **trachea,** or windpipe, which is kept from collapsing by C-shaped rings of cartilage in its wall. The trachea divides into two branches, the **bronchi** (sing., *bronchus*), each of which connects to a lung.

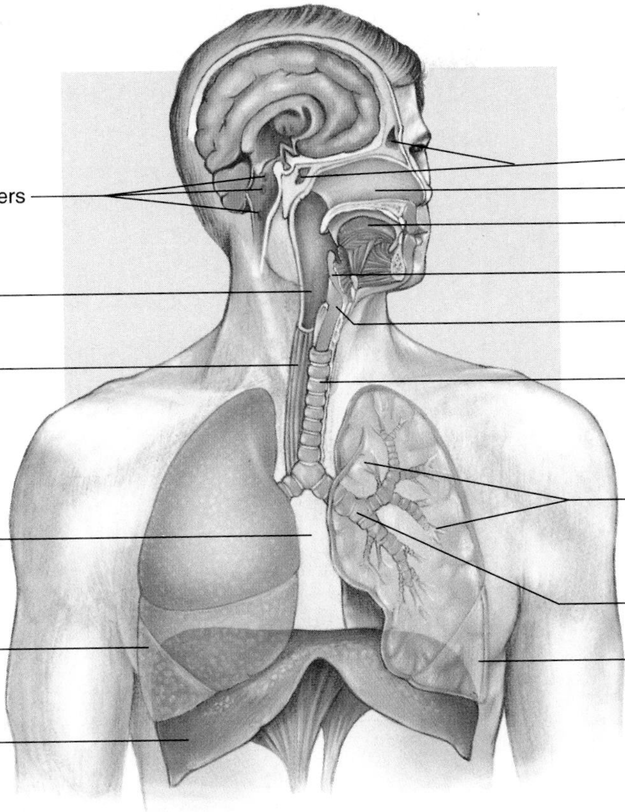

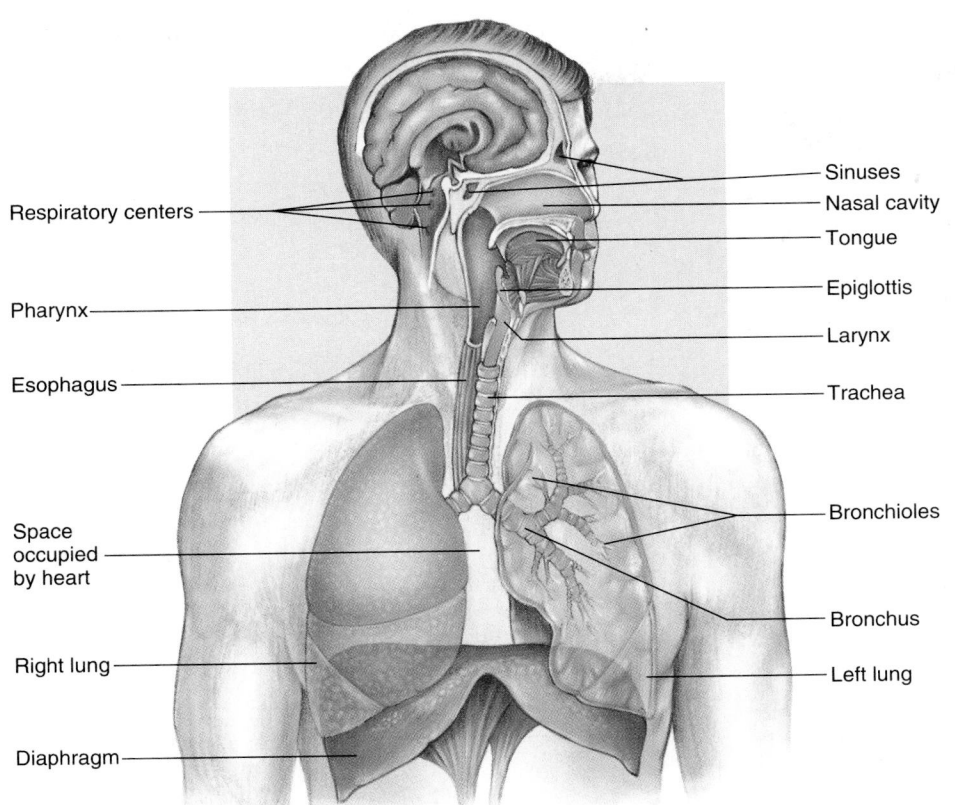

FIGURE 44–6 In the human respiratory system, the paired lungs are located in the thoracic cavity. The muscular diaphragm forms the floor of the thoracic cavity, separating it from the abdominal cavity beneath. An internal view of one lung illustrates its extensive system of air passageways. The microscopic air sacs are shown in Figure 44–7.

Both trachea and bronchi are lined by a mucous membrane containing ciliated cells. Many medium-sized particles that have escaped the cleansing mechanisms of nose and larynx are trapped here. Mucus that contains these particles is constantly beaten upward by the cilia to the pharynx, where it is periodically swallowed. This mechanism, functioning as a cilia-propelled mucous elevator, helps keep foreign material out of the lungs.

Gas exchange occurs in the alveoli of the lungs

The lungs are large, paired, spongy organs occupying the thoracic (chest) cavity. The right lung is divided into three lobes, the left lung into two lobes. Each lung is covered with a **pleural membrane,** which forms a continuous sac that encloses the lung and becomes the lining of the thoracic cavity. The space between the pleural membranes covering the lung and the pleural membrane lining the thoracic cavity is called the **pleural cavity.** A film of fluid in the pleural cavity provides lubrication between the lungs and the chest wall.

Because the lung consists largely of air tubes and elastic tissue, it is a spongy, elastic organ with a very large internal surface area for gas exchange. In normal adults this surface area is estimated to be approximately the size of a tennis court.

Inside the lungs the bronchi branch, becoming smaller and more numerous. These branches give rise to more than one million tiny **bronchioles** in each lung. Each bronchiole ends in a cluster of tiny air sacs, the **alveoli** (sing., *alveolus*) (Fig. 44–7). The alveoli are lined by an extremely thin, single layer of epithelial cells. Gases diffuse freely through the wall of each alveolus and into the capillaries that surround it. Only two thin cells—the epithelia of the alveolar wall and the capillary wall—separate the air in the alveolus from the blood.

In summary, the sequence of structures through which air passes after it enters the body is:

nostrils → nasal cavities → pharynx → larynx → trachea → bronchus → bronchiole → alveolus

Ventilation is accomplished by breathing

Breathing is the mechanical process of moving air from the environment into the lungs and of expelling air from the lungs. Inhaling air is referred to as **inspiration;** exhaling air is **expiration.** A resting adult breathes about 14 times each minute.

The thoracic cavity is closed so that no air can enter except through the trachea. (When the chest wall is punc-

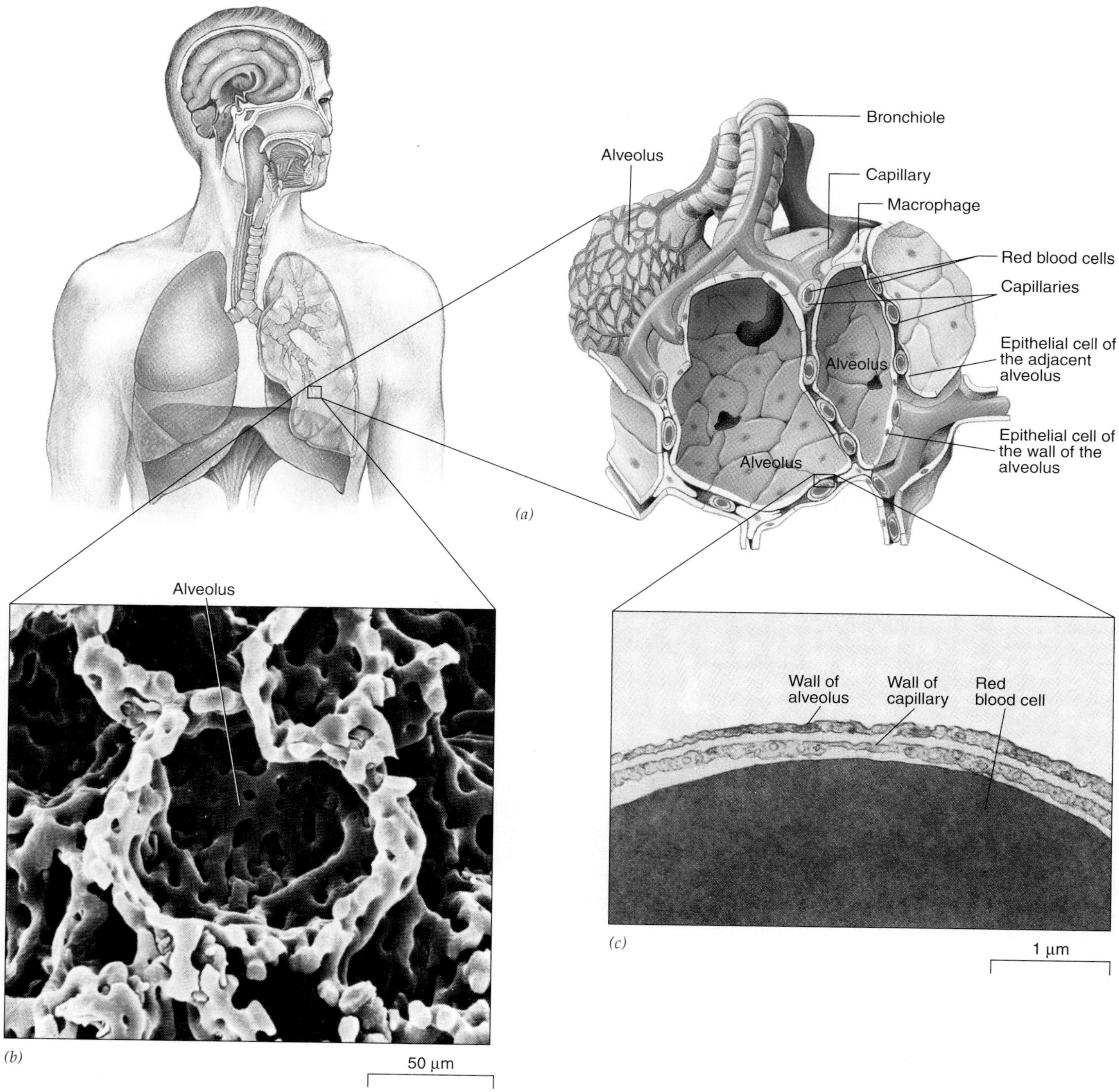

FIGURE 44–7 Gas exchange takes place across the thin wall of the alveolus. (*a*) Structure of the alveolus. Note that the alveolar wall consists of extremely thin squamous epithelium. Between the walls of the alveoli lie extensive capillary networks. (*b*) Scanning electron micrograph showing the capillary network surrounding a portion of several alveoli. (*c*) An enlargement of a portion of a capillary similar to that shown in (*b*). The dark structure extending through the capillary is a portion of a red blood cell. The wall of the alveolus is visible just above the wall of the capillary. Notice the very short distance oxygen must travel to get from the air within the alveolus to the red blood cell that will transport it to the body tissues. (*b*, Kessel, R. G. and R. H. Kardon, *Tissues and Organs: A Text-Atlas of Scanning Electron Microscopy*. San Fransicso, W. H. Freeman Co., 1979; *c*, Courtesy of Drs. Peter Gehr, Marianne Bachofen, and Ewald R. Wiebel)

tured, for example by a gunshot wound, air enters the pleural space and the lung collapses.)

During inspiration, the volume of the thoracic cavity is increased by the contraction of the **diaphragm,** the dome-shaped muscle that forms the floor of the thoracic cavity. When the diaphragm contracts, it moves downward, increasing the volume of the thoracic cavity (Fig. 44–8). During forced inspiration, when a large volume of air is inspired, the rib muscles contract as well. This action moves the ribs upward, which also increases the vol-

Focus On

Choking

Choking kills an estimated 8000 to 10,000 people per year in the United States. Many victims have long suffered from some degree of paralysis or other malfunction of the muscles involved in swallowing, often without consciously realizing it. Swallowing is a very complex process in which the mouth, pharynx, esophagus, and vocal cords must be coordinated with great precision. Functional muscular disorders of this mechanism can originate in a variety of ways — as birth defects, for example, or from brain tumors or vascular accidents involving the swallowing center of the brain stem.

Choking is more likely to occur in restaurants, where social interactions and unfamiliar surroundings are likely to distract a person's attention from swallowing, and where alcohol is more likely to be taken with the meal. A large number of choking victims have a substantial blood alcohol content upon autopsy, which suggests the possibility that in them a marginally effective swallowing reflex was further and fatally compromised by the effects of alcohol on the brain.

Anyone who begins to choke and gasp during a meal should be asked if he or she can speak. If not, as indicated by shaking the head or other gestures, the person is probably suffering a laryngeal obstruction rather than a coronary heart attack. If you are present during such an episode, be aware that you can take certain steps that can save the person's life. Stand behind the victim, bring your arms around the person's waist, and clasp your hands just above the beltline. Your thumbs should be facing inward against the victim's body. Then squeeze abruptly and strongly in an upward direction. Repeat the upward thrusts five times. In most instances the residual air in the lungs will pop the obstruction out like a cork from a bottle. If the obstruction is not relieved, reassess the victim's status and reattempt multiple thrusts. This procedure is called the abdominal thrust, or **Heimlich maneuver.**

If the Heimlich maneuver must be performed with the victim lying down, place the person face up. Kneel astride the victim's, hips and, with one of your hands on top of the other, place the heel of your bottom hand on the abdomen slightly above the navel but below the rib cage. Press into the victim's abdomen with a quick upward thrust. This may be repeated if necessary. ■

ume of the thoracic cavity. The lungs adhere to the walls of the thoracic cavity, so when the volume of the thoracic cavity increases, the space within each lung also increases. The air in the lungs now has more space in which to move about, and the pressure of the air in the lungs falls 2 or 3 mm Hg below the pressure of the air outside the body. As a result of this pressure difference, air from the outside rushes in through the respiratory passageways and fills the lungs until the two pressures are equal once again.

Expiration occurs when the diaphragm and rib muscles relax. The volume of the chest cavity decreases, increasing the pressure in the lungs (to 2–3 mm Hg above atmospheric pressure). The millions of distended air sacs deflate, expelling the air that was inhaled. The pressure returns to normal and the lung is ready for another change of air. Thus, in inspiration, the millions of tiny alveoli fill with air like so many balloons. Then during expiration, the air rushes out of the alveoli, partially deflating them.

The quantity of air respired can be measured

The amount of air moved into and out of the lungs with each normal resting breath is called the **tidal volume.** The normal tidal volume is about 500 milliliters. The **vital capacity** is the maximum amount of air a person can exhale after filling the lungs to the maximum extent.

Vital capacity is greater than tidal volume because the lungs are not completely emptied of stale air and filled with fresh air with each breath. Table 44–2 shows the percentages of oxygen and carbon dioxide present in exhaled

Table 44–2 COMPOSITION OF INHALED AIR COMPARED WITH THAT OF EXHALED AIR

	% Oxygen (O_2)	*% Carbon Dioxide (CO_2)*	*% Nitrogen (N_2)*
Inhaled air (atmospheric air)	20.9	0.04	79
Exhaled air (alveolar air)	14.0	5.60	79

As indicated, the body uses up about one third of the inhaled oxygen. The percentage of CO_2 increases more than 100-fold because it is produced during cellular respiration.

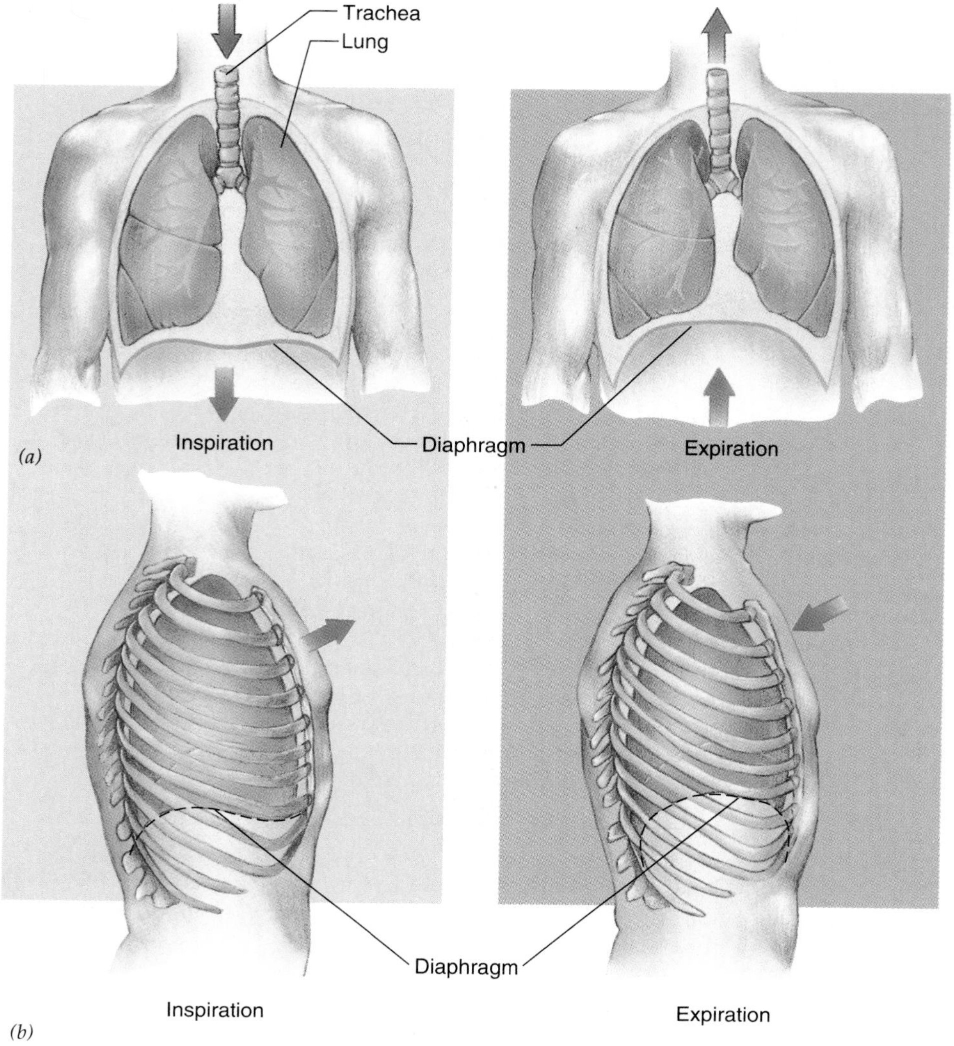

FIGURE 44–8 Breathing requires contraction of the diaphragm and intercostal (chest) muscles. (*a*) Changes in the position of the diaphragm in expiration and inspiration result in changes in the volume of the chest cavity. (*b*) Changes in the position of the rib cage and diaphragm in expiration and inspiration. The elevation of the front ends of the ribs by the intercostal muscles causes an increase in the front-to-back dimension of the chest, and a corresponding increase in the volume of the chest cavity. When the volume of the chest cavity is increased, air moves into the lungs.

air compared with inhaled air. Because carbon dioxide is produced during cellular respiration, more of this gas—100 times more, in fact—enters the alveoli from the blood than is present in air inhaled from the environment.

Gas exchange takes place in the air sacs

The respiratory system delivers oxygen to the air sacs, but if oxygen were to remain in the lungs, all the other body cells would soon die. The vital link between air sac and body cell is the circulatory system. Each air sac serves as a tiny depot from which oxygen is loaded into blood brought close to the alveolar air by capillaries (Fig. 44–9).

Oxygen molecules diffuse from the air sacs, where they are more concentrated, into the blood in the pulmonary capillaries, where they are less concentrated. At the same time, carbon dioxide moves from the blood, where it is more concentrated, to the air sacs, where it is less concentrated. Each gas diffuses through the cells lining the alveoli and the cells lining the capillaries.

Cellular respiration results in the continuous production of carbon dioxide and utilization of oxygen. Consequently, the concentration of oxygen in the cells is lower than in the capillaries entering the tissues, and the concentration of carbon dioxide is higher in the cells than in the capillaries. Thus, as blood circulates through capillaries of a tissue such as brain or muscle, oxygen moves by diffusion from the blood to the cells, while carbon dioxide moves from the cells into the blood.

The factor that determines the direction and rate of diffusion is the pressure or tension of the particular gas. According to Dalton's law of partial pressures, in a mixture of gases the total pressure of the mixture is the sum of the pressures of the individual gases. Each gas exerts, independently of the others, a partial pressure—the same pressure it would exert if it were present alone. At sea level, the barometric pressure (the pressure of Earth's atmosphere) is able to support a column of mercury (Hg) 760 millimeters high. Because the atmosphere is made up of about 21% oxygen, oxygen's share of that pressure is

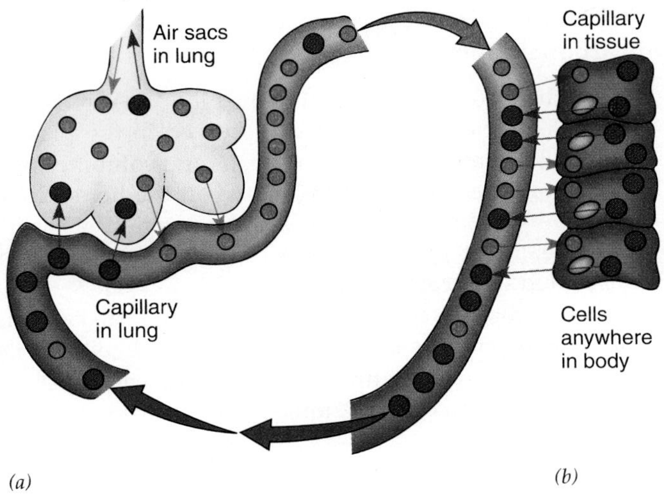

FIGURE 44–9 Gases are exchanged in the lungs and in the tissues. (*a*) Exchange of gases between alveoli (air sacs) and capillaries in the lung. The concentration of oxygen is greater in the alveoli than in the pulmonary capillaries, so oxygen moves from the alveoli into the blood. Carbon dioxide is more concentrated in the blood than in the alveoli, so it moves out of the capillaries and into the alveoli. (*b*) Exchange of gases between the capillary and body cells. Here, oxygen is more concentrated in the blood than in the cells, so it moves out of the capillary into the cells. Carbon dioxide is more concentrated in the cells, so it diffuses out of the cells and moves into the blood.

$0.21 \times 760 = 160$ mm Hg. Thus, 160 mm Hg is the partial pressure of atmospheric oxygen, abbreviated P_{O_2}. The partial pressure of atmospheric CO_2 is 0.3 mm Hg.

Table 44–2 shows the composition of inhaled air compared with exhaled air. **Fick's law** explains that the amount of oxygen or carbon dioxide that diffuses across the membrane of the air sac depends on the differences in partial pressure on the two sides of the membrane and also on the surface area of the membrane. The greater the difference in pressure and the larger the surface area, the faster the gas will diffuse.

Because blood passes through the lung capillaries too rapidly to reach equilibrium with the alveolar air, the partial pressure of oxygen in arterial blood is only about 100 mm Hg. The P_{O_2} in the tissues is still lower, ranging from 0 to 40 mm Hg. Consequently, oxygen diffuses out of the capillaries and into the tissues. Not all of the oxygen leaves the blood, however. The blood passes through the tissue capillaries too rapidly for equilibrium to be reached. As a result, the partial pressure of oxygen in venous blood returning to the lungs is about 40 mm Hg. Thus, expired air has had only about one third of its oxygen removed and can be breathed over again—a good thing for those in need of mouth-to-mouth resuscitation!

Oxygen is transported in combination with hemoglobin

At rest, the cells of the human body use about 250 milliliters of oxygen per minute, or about 300 liters every 24 hours. With exercise or work this rate may increase as much as 10- or 15-fold. If oxygen were simply dissolved in plasma, blood would have to circulate through the body at a rate of 180 liters per minute to supply enough oxygen to the cells at rest. This is because, as we have seen, oxygen is not very soluble in blood plasma. Actually, the blood of a human at rest circulates at a rate of about 5 liters per minute and supplies all of the oxygen the cells need. Why are only 5 liters per minute rather than 180 liters per minute required?

The answer is hemoglobin, the respiratory pigment in red blood cells. Hemoglobin transports about 97% of the oxygen. Only about 3% is dissolved in the plasma. Plasma in equilibrium with alveolar air can take up only 0.25 milliliter of oxygen per 100 milliliters, but the properties of hemoglobin permit whole blood to carry some 20 milliliters of oxygen per 100 milliliters. The protein portion of hemoglobin is composed of four peptide chains, typically two α and two β chains, each attached to a heme (porphyrin) ring. An iron atom is bound in the center of each heme ring.

Hemoglobin has the remarkable property of forming a weak chemical bond with oxygen. An oxygen molecule can attach to the iron atom in each heme. In the lung (or gill), oxygen diffuses into the red blood cells and combines with hemoglobin (Hb) to form **oxyhemoglobin (HbO_2).**

$$Hb + O_2 \rightleftarrows HbO_2$$

Because the chemical bond formed between the oxygen and the hemoglobin is weak, the reaction is readily reversible. In the body tissues, the reaction proceeds to the left, releasing oxygen. Oxyhemoglobin is bright scarlet, giving arterial blood its color; deoxygenated hemoglobin is purple, giving venous blood a darker hue.

The ability of oxygen to combine with hemoglobin and be released from oxyhemoglobin is influenced by several factors, including pH, concentrations of carbon dioxide and oxygen, and temperature. The **oxygen-hemoglobin dissociation curves** shown in Figure 44–10 illustrate that as oxygen concentration increases, there is a progressive increase in the amount of hemoglobin that is combined with oxygen. This value is known as the **percent saturation** of the hemoglobin. The percent saturation is highest in the pulmonary capillaries, where the concentration of oxygen is greatest. In the capillaries of the tissues, where there is less oxygen, the oxyhemoglobin dissociates, releasing oxygen. There, the percent saturation of hemoglobin is correspondingly lower.

Carbon dioxide reacts with water in the plasma to form carbonic acid, H_2CO_3. An increase in the carbon dioxide concentration increases the acidity (lowers the pH) of the blood. Oxyhemoglobin dissociates more readily in a more acidic environment. Lactic acid released from active muscles also lowers the pH of the blood and has a similar effect on the oxygen-hemoglobin dissociation curve. Displacement of the oxygen-hemoglobin dissociation curve by a change in pH is known as the **Bohr effect.**

Some carbon dioxide is transported by the hemoglobin molecule. Although it attaches to the hemoglobin molecule in a different way and at a different site than oxygen, the attachment of a carbon dioxide molecule causes the release of an oxygen molecule from the hemoglobin. The effect of carbon dioxide concentration on the oxygen-hemoglobin dissociation curve is important. In the capillaries of the lungs (or gills in fishes), carbon dioxide concentration is relatively low and oxygen concentration is high, so oxygen saturates a very high percentage of hemoglobin. In the capillaries of the tissues, carbon dioxide concentration is high and oxygen concentration is low, so oxygen is readily released from the hemoglobin.

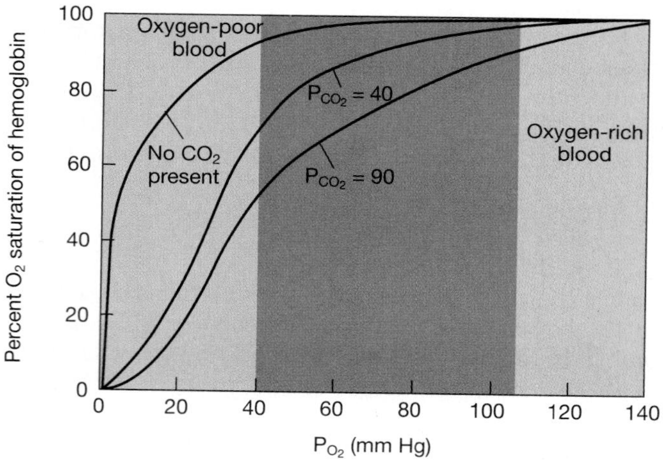

FIGURE 44–10 Oxygen dissociation curves show what happens when oxygen concentration increases. Note the progressive increase in the amount of hemoglobin that is combined with oxygen. The curves also show how carbon dioxide affects the dissociation of oxyhemoglobin. Look at the vertical axis (labeled *percent saturation*). If the blood contains 20% O_2 by volume, which is one fourth of the amount it could contain, it is said to be 25% saturated. The left curve shows what happens if no carbon dioxide is present. The middle curve shows the situation when the P_{CO_2} is 40, which is typical of arterial blood. The right curve indicates P_{CO_2} of 90, an unhealthy concentration. Now find the location on the horizontal axis where the partial pressure of oxygen is 40, and follow the line up through the curves. Notice how the saturation of hemoglobin with oxygen differs among the three curves, even though the partial pressure of oxygen is the same.

Carbon dioxide is transported mainly as bicarbonate ions

Carbon dioxide is transported in the blood in three ways. About 7% of carbon dioxide is dissolved in the plasma. Another 20% enters the red blood cells and combines with hemoglobin. Because the bond between the hemoglobin and carbon dioxide is very weak, the reaction is readily reversible. Most of the carbon dioxide (about 70%) is transported as **bicarbonate ions (HCO_3^-).**

Carbon dioxide combines with water to form carbonic acid. This reaction is catalyzed in red blood cells by the enzyme **carbonic anhydrase.** The carbonic acid dissociates, forming hydrogen ions and bicarbonate ions (HCO_3^-).

$$\underset{\text{Carbon dioxide}}{CO_2} + \underset{\text{Water}}{H_2O} \xrightarrow{\text{Carbonic anhydrase}} \underset{\text{Carbonic acid}}{H_2CO_3} \rightarrow \underset{\text{Hydrogen ion}}{H^+} + \underset{\text{Bicarbonate ion}}{HCO_3^-}$$

Most of the hydrogen ions released from the carbonic acid combine with hemoglobin, which is a very effective buffer. Many of the bicarbonate ions diffuse into the plasma. Chloride ions diffuse into the red blood cells to replace the negative charges of bicarbonate ions, a process known as the **chloride shift.** As CO_2 diffuses out of the alveolar capillaries, the resulting lower CO_2 concentration causes the reversal of the reaction sequence.

Any condition (such as pneumonia) that interferes with the removal of carbon dioxide by the lungs can lead to respiratory acidosis. In this condition there is an increased concentration of carbonic acid and bicarbonate ions in the blood. Although the pH of the blood is not actually acidic, it is lower than normal.

Breathing is regulated by respiratory centers in the brain

Respiratory centers in the brain stem regulate breathing, which is a rhythmic, involuntary process (Fig. 44–6). Neurons in the dorsal region of the medulla regulate the basic rhythm of breathing. These neurons send a burst of impulses to the diaphragm and external intercostal (chest) muscles, causing them to contract. After several seconds, these neurons become inactive, and expiration occurs. Respiratory centers in the pons help control the transition from inspiration to expiration. These centers can stimulate or inhibit the medullary respiratory centers.

The cycle of activity and inactivity repeats itself so that at rest we breathe about 14 times per minute. A group of neurons in the ventral region of the medulla becomes active only when we need to breath forcefully. Overdose of certain medications such as barbiturates depresses the respiratory centers and may lead to respiratory failure.

The basic rhythm of respiration can be altered in response to changing needs of the body. When you are engaged in a strenuous game of tennis, you require more

Making the Connection

Gas Exchange and Cellular Respiration

How does gas exchange between an animal and its environment relate to cellular respiration? Gas exchange between cells and their environment is referred to as **respiration.** We can divide respiration in animals into two phases: organismic and cellular. During **organismic respiration** oxygen from the environment is taken up by the animal and delivered to its individual cells. At the same time, carbon dioxide generated during cellular respiration is excreted into the environment.

In this chapter we discussed how animals exchange gases on an organismic level, and how oxygen is delivered to each cell of the body. The oxygen supplied to the cells by organismic respiration is used in cellular respiration. Recall from Chapter 7 that **cellular respiration** is the complex series of reactions by which cells oxidize organic compounds such as glucose, releasing carbon dioxide and energy. Most cells are dependent on oxygen because it is essential for aerobic cellular respiration. Oxygen serves as the final electron acceptor in the electron transport system, and without it most cells quickly die.

Another important biological concept discussed in this book is the interdependence of producers and consumers, which includes gas exchange. Gas exchange replenishes the oxygen and carbon dioxide in the atmosphere. Recall from Chapter 1 that most organisms depend on producers for oxygen. Plants, algae, and certain bacteria carry on photosynthesis, during which water is split to produce molecular oxygen. On the other hand, carbon dioxide produced during cellular respiration by consumers and decomposers, as well as producers, is used by producers for photosynthesis. ▲

oxygen than when studying biology. During exercise the rate of cellular respiration increases, producing more carbon dioxide. The body must dispose of this carbon dioxide through increased ventilation. Carbon dioxide concentration is the most important chemical stimulus for regulating the rate of respiration. Specialized nerve endings called **chemoreceptors** in the medulla and in the walls of the aorta and carotid arteries are sensitive to changes in arterial carbon dioxide concentration. When stimulated they send impulses to the respiratory centers, which leads to an increase in breathing rate.

The chemoreceptors in the walls of the aorta and carotid arteries are sensitive to changes in hydrogen ion concentration and oxygen concentration, as well as to carbon dioxide levels. Recall that an increase in carbon dioxide concentration results in an increase in hydrogen ions from carbonic acid, lowering pH. Even a slight decrease in pH stimulates these chemoreceptors, leading to a faster breathing rate. As carbon dioxide is removed by the lungs, the hydrogen ion concentration in the blood and other body fluids decreases, and homeostasis is restored.

Interestingly, oxygen concentration generally does not play an important role in regulating respiration. Only if the partial pressure of oxygen falls markedly do the chemoreceptors in the aorta and carotid arteries become stimulated to send messages to the respiratory centers.

Although breathing is an involuntary process, the action of the respiratory centers can be consciously influenced for a short time by stimulating or inhibiting them. For example, you can inhibit respiration by holding your breath. You cannot hold your breath indefinitely, however, because eventually you feel a strong impulse to breathe. Even if you were able to ignore this, you would eventually pass out and breathing would resume.

Individuals who have stopped breathing because of drowning, smoke inhalation, electric shock, or cardiac arrest can sometimes be sustained by mouth-to-mouth resuscitation until their own breathing reflexes are initiated again. Cardiopulmonary resuscitation (CPR) is a method for aiding victims who have suffered respiratory and/or cardiac arrest. For an overview of the procedure, see *Focus On: Cardiopulmonary Resuscitation (CPR).*

Hyperventilation reduces carbon dioxide concentration

Underwater swimmers and some Asiatic pearl divers voluntarily hyperventilate before going under water. By taking a series of deep inhalations and exhalations, one can markedly reduce the carbon dioxide content of the alveolar air and of the blood. As a result, it takes longer before the impulse to breathe becomes irresistible.

When hyperventilation is continued for a long period, dizziness and sometimes unconsciousness may occur. This is because a certain concentration of carbon dioxide is needed in the blood to maintain normal blood pressure. (This mechanism operates by way of the vasoconstrictor center in the brain, which maintains the muscle tone of blood vessel walls.) Furthermore, if divers hold their breath too long, the low concentration of oxygen may result in unconsciousness and drowning.

High flying or deep diving can disrupt homeostasis

The barometric pressure decreases at progressively higher altitudes. Because the concentration of oxygen in the air remains at 21%, the partial pressure of oxygen de-

Focus On

Cardiopulmonary Resuscitation (CPR)

Cardiopulmonary resuscitation, or **CPR,** is a method for aiding victims of accidents or heart attacks who have suffered cardiac and/or respiratory arrest. Taking a course in CPR could prepare you to save someone's life. CPR should not be used if the victim has a pulse or is able to breathe. It must be started immediately, because irreversible brain damage may occur within about 4 minutes of respiratory arrest. Here are its ABCs:

Airway Clear airway by extending victim's neck. This is sometimes sufficient to permit breathing to begin again.

Breathing Use mouth-to-mouth resuscitation.

Circulation Attempt to restore circulation by using external cardiac compression.

Although not intended to substitute for a CPR course, the procedure is summarized below:

I. Establish the unresponsiveness of the victim.

II. Procedure for mouth-to-mouth resuscitation.

1. Place the victim on his or her back on a firm surface.
2. Clear the throat and mouth, and tilt the head back so that the chin points outward. Make sure the tongue is not blocking the airway. Pull the tongue forward if necessary.
3. Pinch the nostrils shut and forcefully exhale into the victim's mouth. Be careful, especially in children, not to overinflate the lungs.
4. Remove your mouth and listen for air rushing out of the lungs.
5. Repeat about 12 times per minute. Do not wait more than 5 seconds between breaths.

III. Procedure for external cardiac compression:

1. Place the heel of one hand on the lower third of the victim's breastbone. Keep your fingertips lifted off the chest. (In infants, two fingers should be used for cardiac compression; in children, use only the heel of the hand.)
2. Place the heel of the other hand at a right angle to and on top of the first hand.
3. Apply firm pressure downward so that the breastbone moves about 4 to 5 cm (1.6 to 2 in) toward the spine. Downward pressure must be about 5.4 to 9 kg (12 to 20 lb) with adults (less with children). Excessive pressure can fracture the sternum or ribs, resulting in punctured lungs or a lacerated liver. This rhythmic pressure can often keep blood moving through the heart and great vessels of the thoracic cavity in sufficient quantities to sustain life.
4. Relax your hands between compressions to allow the chest to expand.
5. Repeat at the rate of at least 80–100 compressions per minute. (For infants or young children, 100 compressions per minute are appropriate.) Fifteen compressions should be applied, then two breaths, in a ratio of 15:2 (5:1 for children). ■

creases along with the barometric pressure. At an altitude of 6000 meters (19,500 feet), the barometric pressure is about 350 mm Hg, the partial pressure of oxygen is about 75 mm Hg, and the hemoglobin in arterial blood is about 70% saturated with oxygen. At 10,000 meters (33,000 feet) the barometric pressure is about 225 mm Hg, the partial pressure of oxygen is 50 mm Hg, and arterial oxygen saturation is only 20%. Thus, getting sufficient oxygen from the air becomes an ever-increasing problem at higher altitudes.

When a person moves to a high altitude, the body adjusts over a period of time by producing a greater number of red blood cells. In a person breathing pure oxygen at 10,000 meters, the oxygen would have a partial pressure of 225 mm Hg, and the hemoglobin would be almost fully saturated with oxygen. Above 13,000 meters, however, barometric pressure is so low that even breathing pure oxygen does not permit complete oxygen saturation of arterial hemoglobin.

A person can remain conscious only until the arterial oxygen saturation falls to between 40% and 50%. This level is reached at about 7000 meters (23,000 feet) when the person is breathing air, or 14,500 meters (47,100 feet) when pure oxygen is used. All high-flying jets have airtight cabins pressurized to the equivalent of the barometric pressure at an altitude of about 2000 meters.

Hypoxia, a deficiency of oxygen, results in drowsiness, mental fatigue, headache, and sometimes euphoria. The ability to think and make judgments is impaired, as is the ability to perform tasks requiring coordination. If a jet flying at 11,700 meters (over 38,000 feet) underwent sudden decompression, the pilot would lose consciousness in about 30 seconds and become comatose in about 1 minute.

In addition to the problems of hypoxia, a rapid decrease in barometric pressure can cause **decompression sickness** (commonly known as the "bends" because those suffering from it bend over in pain). Whenever the barometric pressure drops below the total pressure of all gases dissolved in the blood and other body fluids, the dissolved gases tend to come out of solution and form gas bubbles. A familiar example of such bubbling occurs each time you uncap a bottle of soda, thus reducing the pressure in the bottle. The carbon dioxide is released from so-

Focus On

Adaptations of Diving Mammals

The Weddell seal can swim under the ice at a depth of 500 meters (1640 feet) for more than an hour without coming up for air. The bottle-nosed whale can remain in the ocean depths for as long as 2 hours. Porpoises, whales, seals, beavers, mink, and several other air-breathing mammals have adaptations that permit them to dive for food or disappear below the water surface for several minutes to elude their enemies.

Diving mammals have about twice the volume of blood, relative to their body weight, as nondivers. Many have high concentrations of myoglobin, an oxygen-binding pigment similar to hemoglobin, in muscles. They do not, however, have larger lungs than nondiving mammals.

When a mammal dives to its limit, a group of physiological mechanisms known collectively as the **diving reflex** are activated. Breathing stops. Bradycardia (slowing of the heart rate) occurs. The heart rate may decrease to one tenth of the normal, reducing the body's consumption of oxygen and energy. Blood is redistributed, with the lion's share going to the brain and heart, the organs that can least withstand anoxia. Skin, muscles, digestive organs, and other internal organs can survive with less oxygen, and so receive less blood while an animal is submerged. Muscles shift from aerobic to anaerobic metabolism.

Diving mammals do not take in extra air before a dive. In fact, seals exhale before they dive, and the lungs of whales are compressed during diving. These adaptations are thought to reduce the chance of decompression sickness, because with less air in the lungs there is less nitrogen in the blood to dissolve during the dive.

Atlantic spotted dolphins *(Stenella frontalis).* (François Gohier/Photo Researchers, Inc.)

The diving reflex is present to some extent in humans, where it may act as a protective mechanism during birth when an infant may be deprived of oxygen for several minutes. Many cases of near-drownings, especially of young children, have been documented in which the victim had been submerged for as long as 45 minutes in very cold water before being rescued and resuscitated. In many of these survivors there was no apparent brain damage. The shock of the icy water slows the heart rate, increases blood pressure, and shunts the blood to the internal organs of the body that most need oxygen (blood flow in the arms and legs decreases). Metabolic rate decreases so that less oxygen is required. ■

lution and bubbles out into the air. In the body, nitrogen has a low solubility in blood and tissues. When it comes out of solution, the bubbles formed may damage tissues and block capillaries, interfering with blood flow. The clinical effects of decompression sickness are pain, dizziness, paralysis, unconsciousness, and even death.

Decompression sickness is even more common in deep-sea diving than in high-altitude flying. As a diver descends, the surrounding pressure increases tremendously — 1 atmosphere for each 10 meters. To prevent the collapse of the lungs, a diver must be supplied with air under pressure, thereby exposing the lungs to very high alveolar gas pressures.

At sea level an adult human has about 1 liter of nitrogen dissolved in the body, with about half in the fat and half in the body fluids. After a diver's body has been saturated with nitrogen at a depth of 100 meters (325 feet) the body fluids contain about 10 liters of nitrogen. To prevent this nitrogen from rapidly bubbling out of solution and causing decompression sickness, the diver must be brought to the surface gradually, with stops at certain levels on the way up. This allows the nitrogen to be expelled slowly through the lungs.

Some mammals can spend rather long periods of time in the ocean depths without coming up for air (see *Focus On: Adaptations of Diving Mammals*).

Focus On

The Effects of Smoking

Tobacco smoke is responsible for the premature deaths of nearly half a million individuals every year in the U.S. alone. Cigarette smoke is a "portable" air pollutant. Smokers move about, exhaling tobacco smoke into the air we all must breathe. This environmental tobacco smoke has been linked to death from lung cancer of about 3000 nonsmokers each year.

Almost every American who takes up smoking is a child — 3000 every day, more than one million every year. Former Surgeon General Dr. Antonia C. Novello reported in the August 18, 1993 edition of the Journal of the American Medical Association that 10% of the children who begin smoking start by the fourth grade, and nearly two thirds start by the tenth grade.

Some facts about smoking:

- Smoking is the single most preventable cause of death in our society.
- In the U.S. alone, smoking costs us more than a billion dollars every week in health costs and lost productivity.
- The life of a 30-year-old who smokes 15 cigarettes a day is shortened by an average of more than five years.
- If you smoke more than one pack per day, you are about 20 times more likely to develop lung cancer than is a nonsmoker. According to the American Cancer Society, cigarette smoking causes more than 75% of all lung cancer deaths.
- If you smoke, you are more likely to develop atherosclerosis, and you double your chances of dying from cardiovascular disease.
- If you smoke, you are 20 times more likely to develop chronic bronchitis and emphysema than is a nonsmoker.
- If you smoke, you are seven times more likely to develop peptic ulcers (especially malignant ulcers) than is a nonsmoker.
- If you smoke, you have about 5% less oxygen circulating in your blood (because carbon monoxide binds to hemoglobin) than does a nonsmoker.
- If you smoke when you are pregnant, your baby will weigh about 6 ounces less at birth, and there is double the risk of miscarriage, stillbirth, and infant death.
- Workers who smoke one or more packs of cigarettes per day are absent from their jobs because of illness 33% more often than are nonsmokers.
- Nonsmokers confined in living rooms, offices, automobiles, or other places with smokers are adversely affected by the smoke. For example, when parents of infants smoke, the infant has double the risk of contracting pneumonia or bronchitis in its first year of life.
- When smokers quit smoking, their risk of dying from chronic pulmonary disease, cardiovascular disease, or cancer decreases. (Precise changes in risk depend on the number of years the person smoked, the number of cigarettes smoked per day, the age of starting to smoke, and the number of years since quitting.)
- If everyone in the United States stopped smoking, nearly half a million lives would be saved each year. ■

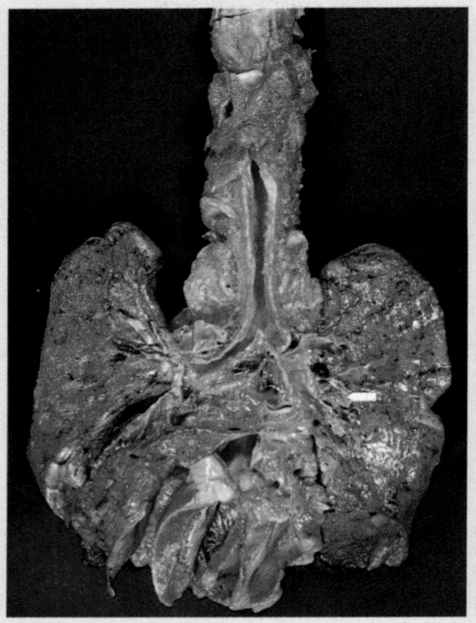

(a)

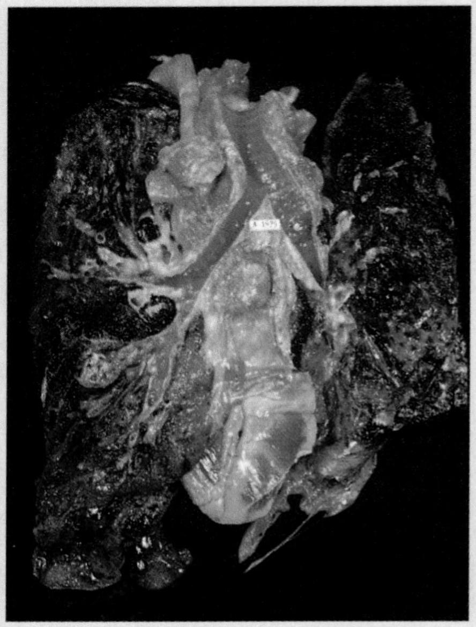

(b)

A diseased lung compared with a healthy lung. (*a*) Normal human lungs and major bronchi. (*b*) Human lungs and heart showing effects of cigarette smoking. (*a, b,* Martin Rotker/Taurus Photos)

FIGURE 44–11 Air pollution contributes to respiratory disorders. Industry spews tons of pollutants into the atmosphere. Air pollution in Copsa Mica (built by the Romanian dictator Nicolae Ceausescu as a "model" industrial city) contributes to the population's short life expectancy of 55 years. (Florent Flipper/Unicorn Stock Photos)

BREATHING POLLUTED AIR DAMAGES THE RESPIRATORY SYSTEM

Several defense mechanisms protect the delicate lungs from the harmful substances we breathe (Fig. 44–11). The hair around the nostrils, the ciliated mucous lining in the nose and pharynx, and the cilia-mucous elevator serve to trap foreign particles in inspired air. One of the body's most rapid defense responses to breathing dirty air is **bronchial constriction.** In this process, the bronchial tubes narrow, increasing the chance that inhaled particles will land on the sticky mucous lining. Unfortunately, bronchial constriction increases airway constriction so that less air can pass through to the lungs, thus decreasing the amount of oxygen available to body cells. Fifteen puffs on a cigarette during a five-minute period increases airway resistance as much as threefold, and this added resistance to breathing lasts more than 30 minutes. Chain smokers and those who breathe heavily polluted air may remain in a state of chronic bronchial constriction.

Neither the smallest bronchioles nor the alveoli are equipped with mucus or ciliated cells. Foreign particles that get through other respiratory defenses and find their way into the alveoli may be engulfed by macrophages. The macrophages may then accumulate in the lymph tissue of the lungs. Lung tissues of chronic smokers and those who work in dirty fossil-fuel-burning industries contain large blackened areas where carbon particles have been deposited (Fig. 44–12).

Continued insult to the respiratory system results in disease. Chronic bronchitis and emphysema are **chronic obstructive pulmonary diseases** that have been linked to smoking and breathing polluted air. More than 75% of patients with **chronic bronchitis** have a history of heavy cigarette smoking (see *Focus On: The Effects of Smoking*). In chronic bronchitis, irritation from inhaled pollutants

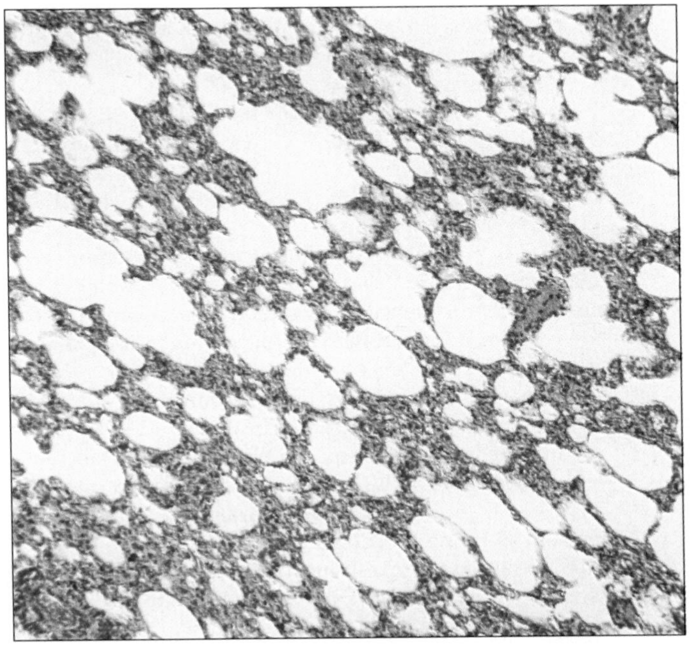

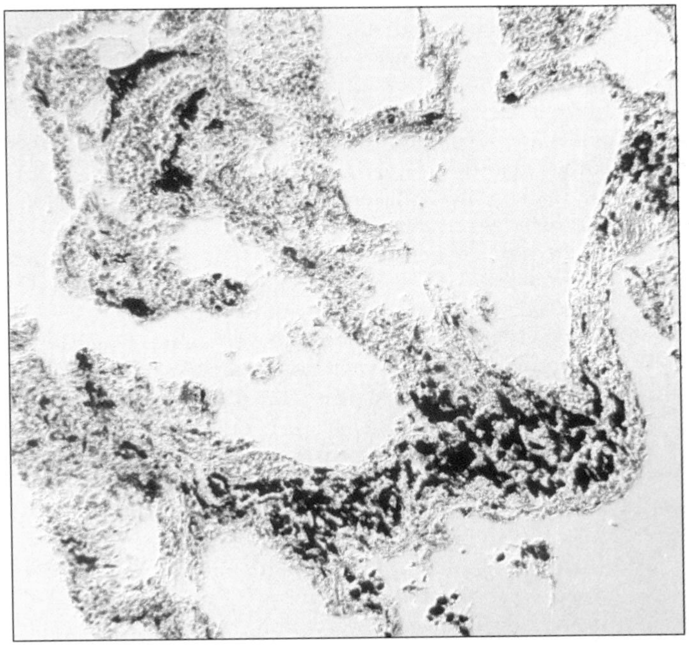

FIGURE 44–12 Diseased lung tissue contrasts starkly with normal tissue. (*a*) Normal lung tissue. (*b*) Lung tissue with accumulated carbon particles. Despite the body's defenses, when we inhale smoky, polluted air, especially over a long period of time, dirt particles do enter the lung tissue and some remain lodged there. (*a*, *b*, Alfred Paseika/Taurus Photos)

causes the bronchial tubes to secrete too much mucus. Ciliated cells, damaged by the pollutants, cannot effectively clear the mucus and trapped particles from the airways. The body resorts to coughing in an attempt to clear the airways. The bronchioles become constricted and inflamed, and the patient is short of breath.

Victims of chronic bronchitis often develop **pulmonary emphysema,** a disease most common in cigarette smokers. In this disorder, alveoli lose their elasticity, and walls between adjacent alveoli are destroyed. The surface area of the lung is so reduced that gas exchange is seriously impaired. Air is not expelled effectively and stale air accumulates in the lungs. The emphysema victim struggles for every breath and still the body does not get enough oxygen. To compensate, the right ventricle of the heart pumps harder and becomes enlarged. Emphysema patients frequently die of heart failure.

Cigarette smoking is also the main cause of lung cancer. More than ten of the compounds in the tar of tobacco smoke have been shown to cause cancer. These carcinogenic substances irritate the cells lining the respiratory passages and alter their metabolic balance. Normal cells are transformed into cancer cells, which may multiply rapidly and invade surrounding tissues.

SUMMARY

I. Gas exchange in air has advantages and disadvantages compared with gas exchange in water. Air contains more molecular oxygen than does water; oxygen diffuses more rapidly through air than through water; and less energy is required for ventilation. However, terrestrial animals must have adaptations that prevent drying.

II. In very small organisms, gas exchange can occur effectively by diffusion, but in larger and more complex forms, specialized respiratory structures are required.
   - A. Many invertebrates, including nudibranch mollusks, most annelids and small arthropods, and some vertebrates, exchange gases across the body surface.
   - B. In insects and some other arthropods, the respiratory system consists of a network of tracheal tubes.
   - C. Gills are moist, thin projections of the body surface that occur mainly in aquatic animals.
     1. In chordates, gills are usually internal, located along the edges of the gill slits.
     2. In bony fishes, a countercurrent exchange system maximizes diffusion of oxygen into the blood and diffusion of carbon dioxide out of the blood.
   - D. Some mollusks and terrestrial vertebrates have lungs with some means of ventilating them.

III. Respiratory pigments increase the capacity of blood to transport oxygen.

IV. The human respiratory system includes the lungs and a system of airways. A breath of air passes in sequence through the nose, pharynx, larynx, trachea, bronchi, bronchioles, and alveoli.
   - A. During breathing, the diaphragm and rib muscles contract, expanding the chest cavity. The membranous walls of the lungs move outward along with the chest walls, decreasing the pressure within the lungs. Air from outside the body rushes in through the air passageways and fills the lungs until the pressure once more equals atmospheric pressure.
   - B. Tidal volume is the amount of air moved into and out of the lungs with each normal breath. Vital capacity is the maximum volume that can be exhaled after the lungs are filled to the maximum extent.
   - C. Oxygen and carbon dioxide are exchanged between alveoli and blood by diffusion.
   - D. About 97% of the oxygen in the blood is transported as oxyhemoglobin.
     1. As oxygen concentration increases, there is a progressive increase in the amount of hemoglobin that combines with oxygen.
     2. Owing to lowered pH, oxyhemoglobin dissociates more readily as carbon dioxide concentration increases (Bohr effect).
   - E. About 70% of the carbon dioxide in the blood is transported as bicarbonate ions.
   - F. Respiratory centers located in the medulla and pons regulate respiration. These centers are stimulated by chemoreceptors sensitive to an increase in carbon dioxide concentration, an increase in hydrogen ions, and to very low oxygen concentration.
   - G. Hyperventilation reduces the concentration of carbon dioxide in the alveolar air and in the blood.
   - H. As altitude increases, barometric pressure decreases and less oxygen enters the blood. This situation can lead to hypoxia. A rapid decrease in barometric pressure can cause decompression sickness.

V. Inhaling polluted air results in bronchial constriction, increased mucous secretion, damage to ciliated cells, and coughing. It can eventually lead to chronic bronchitis, emphysema, and lung cancer.

SELECTED KEY TERMS

air sac
alveolus (alveoli)
bicarbonate ion
bronchial constriction
bronchus (bronchi)
countercurrent exchange system
diaphragm
epiglottis
expiration
Fick's law
gill
inspiration
larynx
lung
oxygen-hemoglobin dissociation curve
oxyhemoglobin
parabronchus (parabronchi)
pharynx
pleural cavity
pleural membrane
respiration
respiratory center
tidal volume
trachea
tracheal tube
ventilation
vital capacity

POST-TEST

1. Gas exchange between an animal and its environment is known as ________ .
2. The process of actively moving air or water over a respiratory surface is called ________ .
3. In insects, the respiratory system consists of a network of ________ tubes.
4. The operculum is a bony plate that protects the ________ in bony fishes.
5. Respiratory structures that develop as ingrowths of the body surface are ________ .
6. In birds, the lungs have several extensions referred to as ________ ________ .
7. In the mammalian respiratory system, inhaled air passing through the larynx next enters the ________ and then passes into a(an) ________ .
8. In the mammalian respiratory system, gas exchange takes place through the thin walls of the ________ .
9. In mammals, the floor of the thoracic cavity is formed by the ________ .
10. In humans, the ________ seals off the larynx during swallowing.
11. The ________ membrane covers the lung.
12. The maximum amount of air a person can exhale after filling the lungs to the maximum extent is the ________ ________ .
13. Oxygen is mainly transported as ________ .
14. Most carbon dioxide is transported in the blood as ________ ________ .
15. Breathing is regulated by ________ ________ in the medulla and pons.
16. Bronchial ________ is one of the body's most rapid responses to breathing dirty air.
17. Label the diagram.

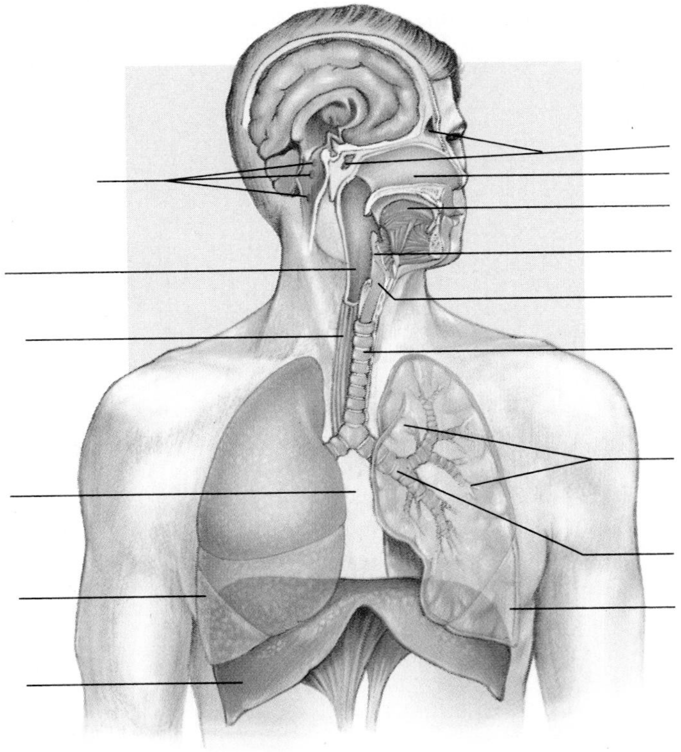

REVIEW QUESTIONS

1. Why are specialized respiratory structures necessary in a tadpole but not in a flatworm?
2. Compare ventilation in a sponge with ventilation in a human.
3. Compare gas exchange in the following animals: (a) earthworm; (b) grasshopper; (c) fish; (d) frog.
4. What challenge is solved by respiratory pigments? How?
5. Why does alveolar air differ in composition from atmospheric air? Explain.
6. What physiological mechanisms bring about an increase in rate and depth of breathing during exercise? Why is such an increase necessary?
7. In what way might it be advantageous for a fish to have lungs as well as gills? What function do the "lungs" of modern fishes serve?
8. How does the countercurrent exchange system increase the efficiency of gas exchange between gills and blood?
9. What mechanisms does the human respiratory system have for getting rid of inhaled dirt? What happens when so much dirty air is inhaled that these mechanisms cannot function effectively?
10. What factors affect the dissociation of oxyhemoglobin?
11. What happens to a deep-sea diver who surfaces too quickly? Why?

YOU MAKE THE CONNECTION

1. What problems would be faced by a terrestrial animal that had gills instead of lungs?
2. Some aquatic mammals such as whales and dolphins use lungs rather than gills for gas exchange. Propose a hypothesis to explain this.
3. What are the advantages of having millions of alveoli rather than a pair of simple, balloon-like lungs?

RECOMMENDED READINGS

Feder, M. E., and W. W. Burggren. "Skin Breathing in Vertebrates." *Scientific American,* Vol. 253, No. 5, November 1985. An interesting account describing how skin functions in gas exchange in some vertebrates.

Moon, R. E., R. D. Vann and P. B. Bennett. "The Physiology of Decompression Illness." *Scientific American,* Vol. 272, No. 2, August 1995. A discussion of recent findings that explain some aspects of decompression sickness.

Zapol, W. M. "Diving Adaptations of the Weddell Seal." *Scientific American,* Vol. 256, No. 6, June 1987. Physiological adaptations enable the seal to swim deeper and hold its breath longer than most other mammals.

CHAPTER 45

Processing Food and Nutrition

Giant water bug eating salamander larva.

(Gary Meszaros, Dembinsky Photo Associates)

Nutrients are substances in food that are used as energy sources to run the systems of the body, as ingredients to make compounds for metabolic processes, and as building blocks in the growth and repair of tissues. Obtaining nutrients is of such vital importance that both individual organisms and ecosystems are structured around the central theme of **nutrition,** the process of taking in and using food. An organism's body plan and its lifestyle are adapted to its particular mode of obtaining food.

All animals are **heterotrophs,** organisms that must obtain their energy and nourishment from the organic molecules manufactured by other organisms. With only slight variations, all animals require the same basic nutrients: minerals, vitamins, carbohydrates, lipids, and proteins. Carbohydrates, lipids, and proteins can all be used as energy sources. Eating too much of any of these nutrients can result in weight gain, whereas eating too few nutrients or an unbalanced diet can result in malnutrition and death.

Most animals have a digestive system that processes the food they eat. After foods are selected and obtained, they are taken into the digestive cavity. **Ingestion** generally includes taking the food into the mouth and swallowing it. The process of breaking down food is called **digestion.** Because animals eat the macromolecules tailor-made by and for other organisms, they must break down these molecules and refashion them for their own needs. We cannot incorporate the proteins in steak directly into our own muscles, for example. The body digests the steak, mechanically breaking down the large bites of meat into smaller ones and then enzymatically hydrolyzing (breaking down with the addition of water) the proteins into their component amino acids.

The amino acids in our example can then be passed through the lining of the digestive tract and into the blood by **absorption.** Like other nutrients, amino acids are transported to the body cells. In muscle cells, they can be used to make human muscle proteins. Food that is not digested and absorbed is discharged from the body. This process is called **egestion** in simple animals and **elimination** in more complex animals.

LEARNING OBJECTIVES

After you have studied this chapter you should be able to

1. Compare food processing—including ingestion, digestion, absorption, and elimination—in an animal (such as *Hydra*) that has a single-opening digestive system, with that in an animal (such as a vertebrate) that has a digestive system with two openings.
2. Identify on a diagram or model each of the structures of the human digestive system described in this chapter, and give the function of each.
3. Trace the pathway traveled by an ingested meal in the human digestive system, and describe the step-by-step digestion of carbohydrate, protein, and lipid.
4. Summarize the functions of the accessory digestive glands of humans and other terrestrial vertebrates.
5. Draw and label a diagram of an intestinal villus, and explain how its structure is adapted to its function.
6. Trace the fate of glucose, lipids, and amino acids after their absorption, and discuss their roles in the body.
7. Discuss the roles of vitamins and minerals, and distinguish between water-soluble and fat-soluble vitamins.
8. Contrast basal metabolic rate with total metabolic rate; write the basic energy equation for maintaining body weight, and describe the consequences of altering it in either direction.
9. In general terms, describe the problem of world food supply relative to world population, and describe the effects of malnutrition.
10. Summarize the challenges encountered in obtaining adequate amounts of amino acids in a vegetarian diet, and describe how a nutritionally balanced vegetarian diet could be planned.

ANIMALS ARE ADAPTED TO THEIR MODE OF NUTRITION

Some animals are **herbivores,** or primary consumers, which eat mainly plant materials (Fig. 45–1). Because animals cannot digest cellulose of plant cell walls, herbivores have evolved many adaptations for extracting nutrients from the plant material they eat. For example, many herbivores have a symbiotic relationship with bacteria that inhabit their digestive tracts. In exchange for food and shelter, the bacteria break down cellulose cell walls, allowing the host's digestive enzymes access to the nutrients within the plant cells. Termites, cows, and horses are among the herbivores that enjoy such symbiotic relationships.

In the cud-chewing ruminants (cattle, sheep, deer), the stomach is divided into four chambers. Bacteria inhabiting the first two chambers digest cellulose, splitting some of it into sugars, which are then used by the bacteria themselves. The bacteria produce fatty acids during their metabolism, some of which are absorbed by the animal and serve as an important energy source. Food that is not sufficiently chewed, called cud, is regurgitated into the animal's mouth and chewed again.

Many herbivores eat great quantities of food. Grasshoppers, locusts, elephants, and cattle, for example, all spend a major part of their lives eating. Most of what they eat is not efficiently digested and is eliminated from the body almost unchanged as waste. However, they eat large enough quantities to provide the nourishment necessary to sustain their life processes.

Herbivores are sometimes eaten by flesh-eating **carnivores,** which may also eat one another. Carnivores (secondary and higher level consumers in ecosystems) are adapted for capturing and killing prey. Some carnivores seize their victims and swallow them alive and whole (Fig. 45–1). Others paralyze, crush, or shred their prey before ingesting it. Carnivorous mammals have well-developed canine teeth for stabbing during combat. The digestive juice of the stomach breaks down proteins, and because meat is more easily digested than plant food, their digestive tracts are shorter than those of herbivores.

Omnivores, such as bears and humans, consume both plants and animals. Earthworms ingest large amounts of soil containing both animal and plant material. The blue whale, the largest animal, is a filter feeder that strains out tiny algae and animals as it swims. Omnivores often possess adaptations that permit them to distinguish among a wide range of smells and tastes and thereby select a variety of foods.

SOME INVERTEBRATES HAVE DIGESTIVE SYSTEMS WITH A SINGLE OPENING

The simplest invertebrates, sponges, have no digestive system at all. Sponges obtain food by filtering microscopic organisms from the surrounding water. Individual cells phagocytize the food particles, and digestion is *intracellular* within food vacuoles. Wastes are egested into

FIGURE 45–1 Animals are adapted for obtaining and processing food. (***a***) The acorn weevil is herbivorous. This little beetle uses its impressively long "snout" both for feeding and for making a hole in the acorn through which it deposits an egg. When it has hatched, the larva feeds on the contents of the acorn seed. (***b***) Giant panda (*Ailuropoda melanoleuca*) eating bamboo. (***c***) The mouth of the carnivorous long-nose butterfly fish (*Forcipiger longirostris*) is adapted for extracting small worms and crustaceans from tight spots in coral reefs. (***d***) This carnivorous snake (*Dromicus* sp.) is strangling a lava lizard (*Tropiduris* sp.). (*a*, Darwin Dale/Photo Researchers, Inc.; *b*, Tom McHugh/Photo Researchers, Inc.; *c*, Zig Leszczynski © 1996 Animals Animals; *d*, Frans Lanting/Minden Pictures)

the water that continuously circulates through the sponge body.

Cnidarians (such as hydras and jellyfish) and flatworms have a digestive system with only a single opening. Cnidarians capture small aquatic animals with the help of their stinging cells and tentacles (Fig. 45–2*a*). The mouth opens into a large gastrovascular cavity. Cells lining this digestive cavity secrete enzymes that break down proteins. Digestion continues *intracellularly* within food vacuoles, and digested nutrients diffuse into other cells.

The **mucosa,** a layer of epithelial tissue and underlying connective tissue, lines the *lumen* (inner space) of the digestive tract. In the stomach and intestine, the mucosa is greatly folded to increase the secreting and absorbing surface. Surrounding the mucosa is the **submucosa,** a connective tissue layer rich in blood vessels, lymphatic vessels, and nerves.

Surrounding the submucosa is a **muscle layer** consisting of two sublayers of smooth muscle. In the inner sublayer the muscle fibers are arranged circularly around the digestive tube. In the outer sublayer the muscle fibers are arranged longitudinally. The outer connective tissue coat of the digestive tract is the **adventitia.** Below the level of the diaphragm, the adventitia becomes the **visceral peritoneum.** By various folds it is connected to the **parietal peritoneum,** a sheet of connective tissue that lines the walls of the abdominal and pelvic cavities. The visceral and parietal peritoneums enclose part of the coelom called the **peritoneal cavity.** Inflammation of the peritoneum, called **peritonitis,** can be very serious because infection can spread along the peritoneum to most of the abdominal organs.

Food processing begins in the mouth

Imagine that you have just taken a big bite of a hamburger. The mouth is specialized for ingestion and for beginning the digestive process. Mechanical digestion begins as you bite, grind, and chew the meat and bun with your teeth. Unlike the simple, pointed teeth of fish, amphibians, and reptiles, the teeth of mammals vary in size and shape and are specialized to perform specific functions. The chisel-shaped **incisors** are used for biting, while the long, pointed **canines** are adapted for tearing food (Fig. 45–6). The flattened surfaces of the **premolars** and **molars** are specialized for crushing and grinding.

Each tooth is covered by **enamel,** the hardest substance in the body (Fig. 45–7). Most of the tooth consists of **dentin,** which resembles bone in composition and hardness. Beneath the dentin is the **pulp cavity,** a soft connective tissue containing blood and lymph vessels, and nerves.

While the food is being mechanically disassembled by the teeth, it is also moistened by saliva. Some of its molecules dissolve, enabling you to taste the food. Recall from Chapter 41 that taste buds are located on the tongue and other surfaces of the mouth. Three pairs of **salivary glands** secrete about a liter of saliva into the mouth cavity each day. Saliva contains an enzyme, **salivary amylase,** which initiates the digestion of starch into sugar.

The pharynx and esophagus conduct food to the stomach

After the bite of food has been chewed and fashioned into a lump called a **bolus,** it is swallowed—that is, moved through the **pharynx** and into the **esophagus.** The pharynx, or throat, is a muscular tube that serves as the hallway of the respiratory system as well as the digestive system. During swallowing, the opening to the airway is closed by a small flap of tissue, the **epiglottis.**

Waves of muscular contraction, called **peristalsis,** sweep the bolus through the pharynx and esophagus toward the stomach (Fig. 45–8). Circular muscle fibers in the wall of the esophagus contract around the top of the bolus, pushing it downward. Almost at the same time, longitudinal muscles around the bottom of the bolus and below it contract, shortening the tube.

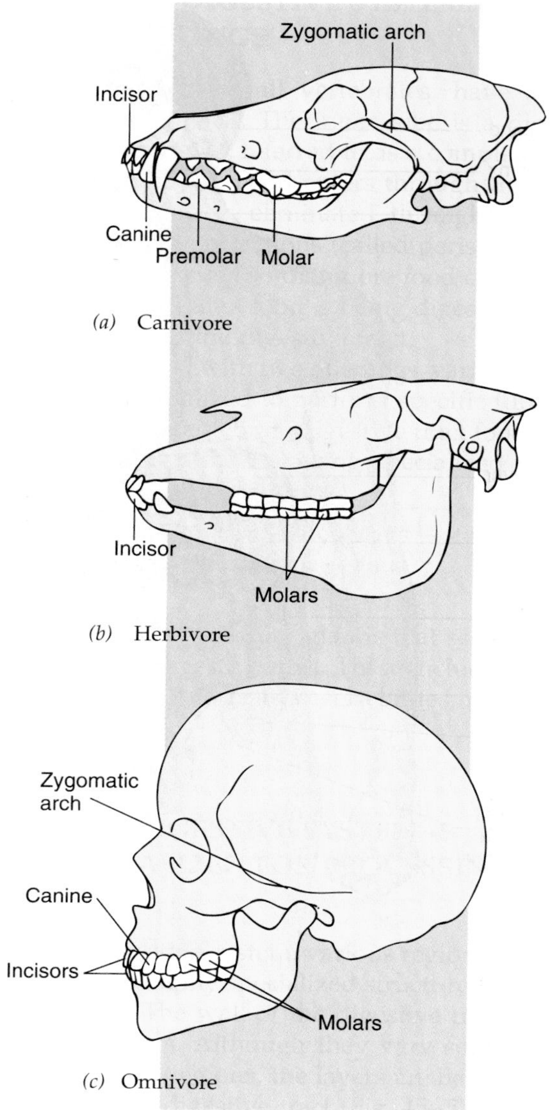

FIGURE 45–6 Teeth are adapted to diet. (*a*) Skull of a coyote, a carnivore, showing the pointed incisors and canines used to rip flesh. (*b*) In contrast, herbivores, such as cows and horses, have incisors and canines adapted for cutting off bits of vegetation. The broad ridge surfaces of their molars are adapted for grinding plant material. (*c*) Omnivores, such as humans, have relatively unspecialized teeth.

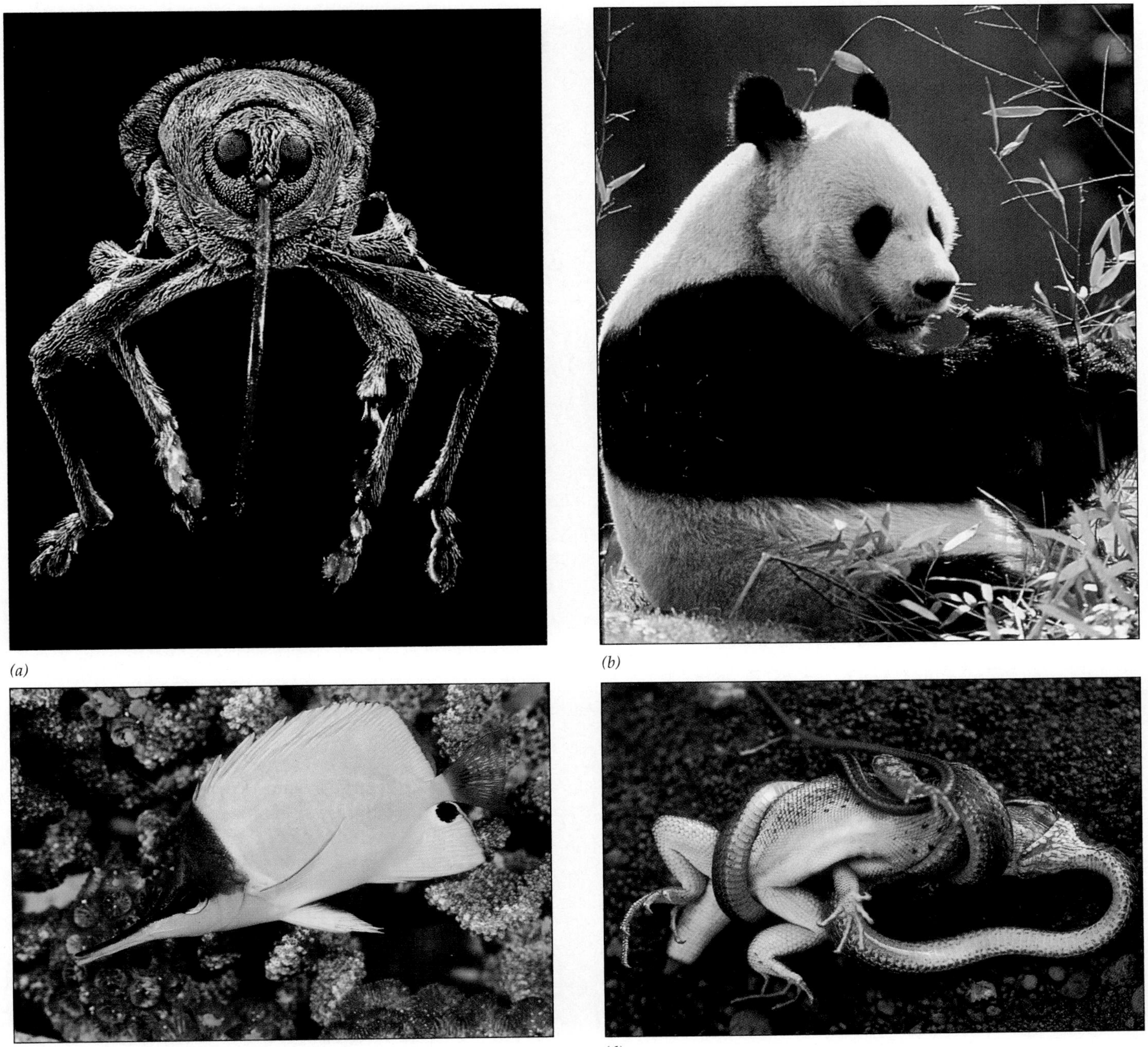

(a) (b) (c) (d)

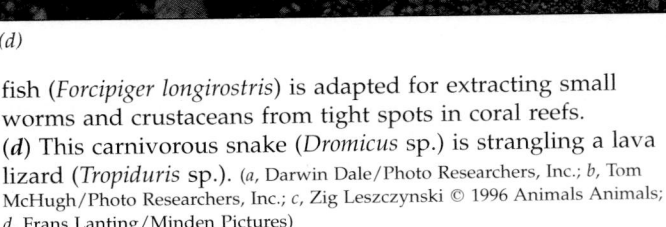

FIGURE 45–1 Animals are adapted for obtaining and processing food. (**a**) The acorn weevil is herbivorous. This little beetle uses its impressively long "snout" both for feeding and for making a hole in the acorn through which it deposits an egg. When it has hatched, the larva feeds on the contents of the acorn seed. (**b**) Giant panda (*Ailuropoda melanoleuca*) eating bamboo. (**c**) The mouth of the carnivorous long-nose butterfly fish (*Forcipiger longirostris*) is adapted for extracting small worms and crustaceans from tight spots in coral reefs. (**d**) This carnivorous snake (*Dromicus* sp.) is strangling a lava lizard (*Tropiduris* sp.). (*a*, Darwin Dale/Photo Researchers, Inc.; *b*, Tom McHugh/Photo Researchers, Inc.; *c*, Zig Leszczynski © 1996 Animals Animals; *d*, Frans Lanting/Minden Pictures)

the water that continuously circulates through the sponge body.

Cnidarians (such as hydras and jellyfish) and flatworms have a digestive system with only a single opening. Cnidarians capture small aquatic animals with the help of their stinging cells and tentacles (Fig. 45–2*a*). The mouth opens into a large gastrovascular cavity. Cells lining this digestive cavity secrete enzymes that break down proteins. Digestion continues *intracellularly* within food vacuoles, and digested nutrients diffuse into other cells.

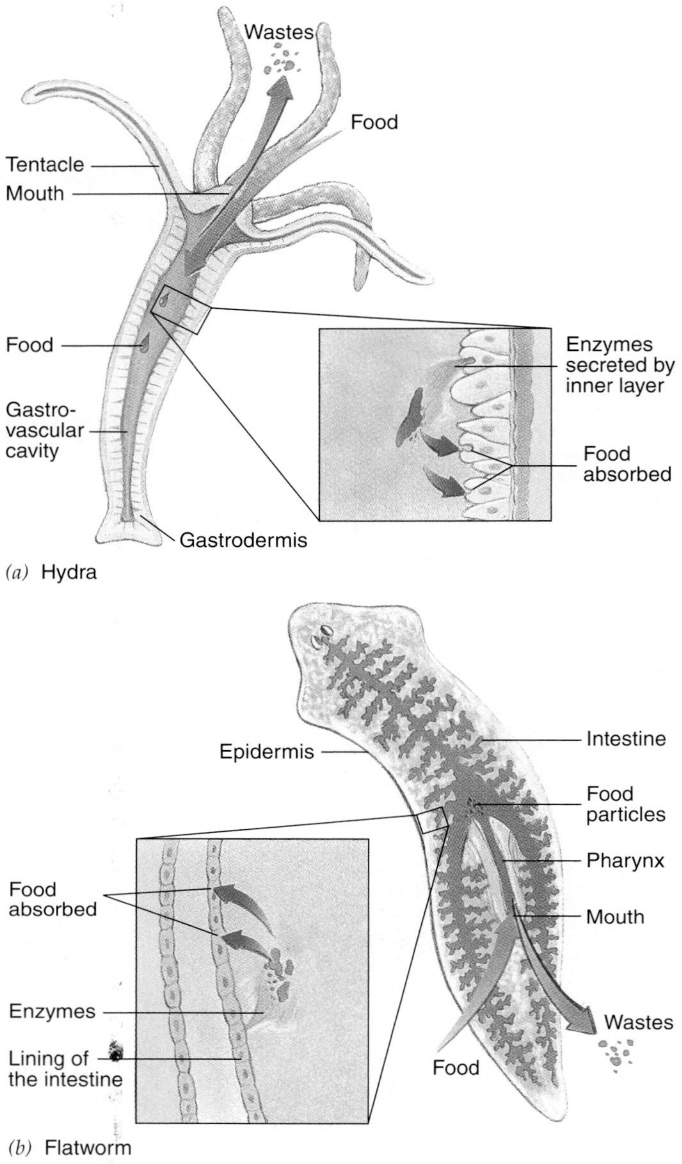

FIGURE 45–2 Some invertebrates have simple digestive systems. The hydra (*a*) and the flatworm (*b*) have digestive tracts with a single opening that serves as both mouth and anus.

Undigested food particles are egested through the mouth by contraction of the body.

Free-living flatworms begin to digest their prey even before ingesting it. They extend the pharynx out through their mouth and secrete digestive enzymes onto the prey (Fig. 45–2*b*). When ingested, the food enters the branched gastrovascular cavity where enzymes continue to digest it. Partly digested food fragments are then phagocytized by cells lining the gastrovascular cavity, and digestion is completed within food vacuoles. As in cnidarians, the flatworm digestive system has only one opening, so undigested wastes are egested through the mouth.

MOST ANIMAL DIGESTIVE SYSTEMS HAVE TWO OPENINGS

Most invertebrates, and all vertebrates, have a tube-within-a-tube body plan. The inner tube is a digestive tract with two openings, referred to as a complete digestive system (Fig. 45–3). Food enters through the mouth and undigested food is eliminated through the anus. Waves of muscular contractions (called peristalsis) push the food in one direction, so that more food can be taken in while previously eaten food is being digested and absorbed farther down the digestive tract.

In a digestive tract with two openings, various regions of the tube are specialized to perform specific functions. For example, in the vertebrate digestive tract food passes in sequence through the following specialized regions:

mouth → pharynx (throat) → esophagus → stomach → small intestine → large intestine → anus

All vertebrates have accessory glands that secrete digestive juices into the digestive tract. These include the liver, the pancreas, and, in terrestrial vertebrates, the salivary glands.

THE HUMAN DIGESTIVE SYSTEM IS HIGHLY SPECIALIZED FOR PROCESSING FOOD

In the human digestive system, various regions of the digestive tract have highly specialized structures and functions (Fig. 45–4). The wall of the digestive tract is composed of four layers. Although they vary somewhat in structure in various regions, the layers are basically similar throughout the digestive tract (Fig. 45–5).

(Text continues on page 986)

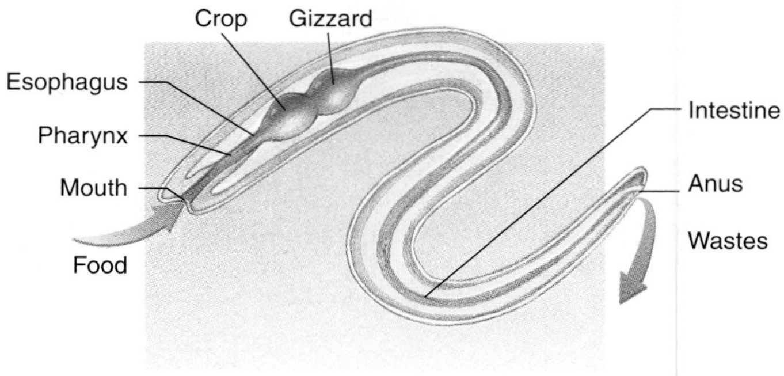

FIGURE 45–3 The earthworm, like most complex animals, has a complete digestive tract extending from mouth to anus. Various regions of the digestive tract are specialized to perform different food processing functions.

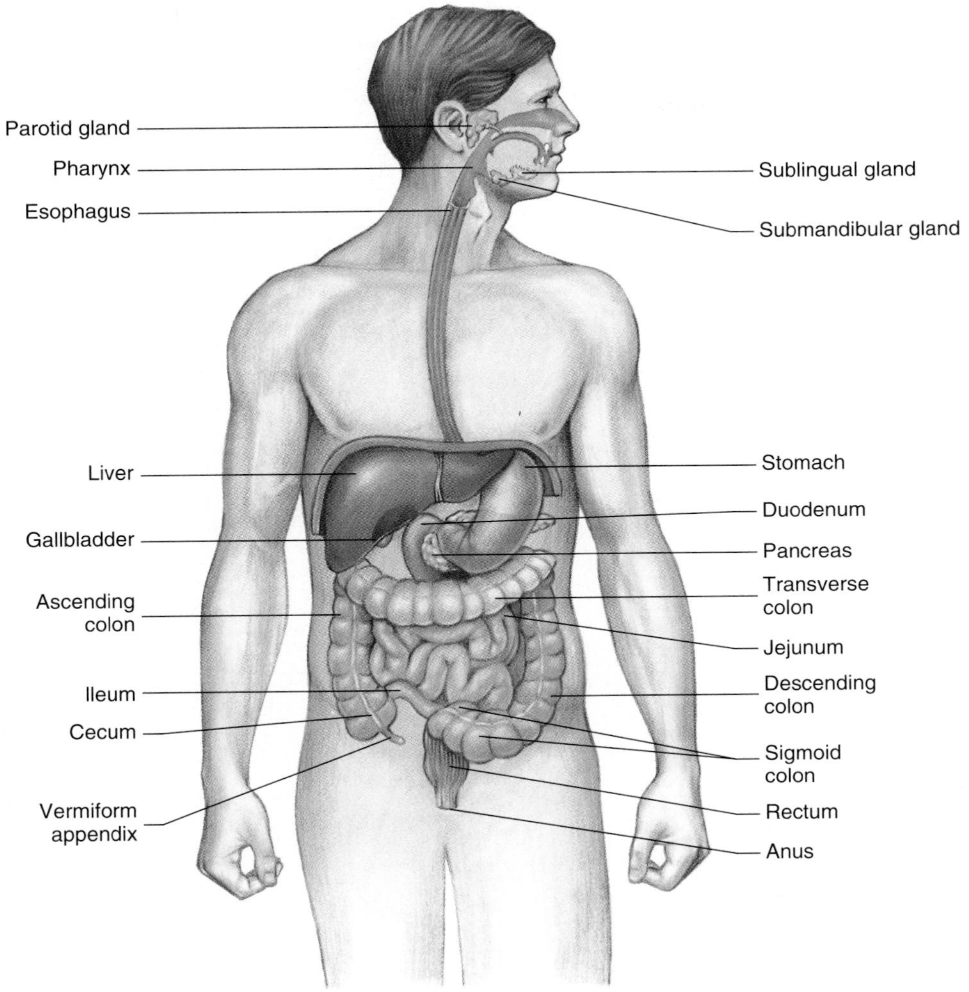

FIGURE 45–4 The human digestive tract is a long, coiled tube extending from mouth to anus. Locate the three types of accessory glands.

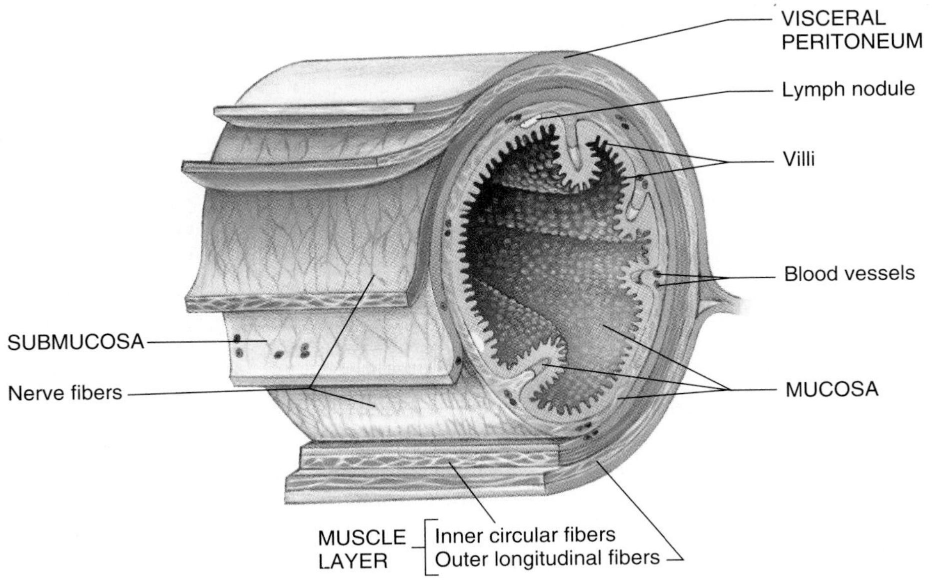

FIGURE 45–5 The wall of the digestive tract consists of four layers.

The **mucosa,** a layer of epithelial tissue and underlying connective tissue, lines the *lumen* (inner space) of the digestive tract. In the stomach and intestine, the mucosa is greatly folded to increase the secreting and absorbing surface. Surrounding the mucosa is the **submucosa,** a connective tissue layer rich in blood vessels, lymphatic vessels, and nerves.

Surrounding the submucosa is a **muscle layer** consisting of two sublayers of smooth muscle. In the inner sublayer the muscle fibers are arranged circularly around the digestive tube. In the outer sublayer the muscle fibers are arranged longitudinally. The outer connective tissue coat of the digestive tract is the **adventitia.** Below the level of the diaphragm, the adventitia becomes the **visceral peritoneum.** By various folds it is connected to the **parietal peritoneum,** a sheet of connective tissue that lines the walls of the abdominal and pelvic cavities. The visceral and parietal peritoneums enclose part of the coelom called the **peritoneal cavity.** Inflammation of the peritoneum, called **peritonitis,** can be very serious because infection can spread along the peritoneum to most of the abdominal organs.

Food processing begins in the mouth

Imagine that you have just taken a big bite of a hamburger. The mouth is specialized for ingestion and for beginning the digestive process. Mechanical digestion begins as you bite, grind, and chew the meat and bun with your teeth. Unlike the simple, pointed teeth of fish, amphibians, and reptiles, the teeth of mammals vary in size and shape and are specialized to perform specific functions. The chisel-shaped **incisors** are used for biting, while the long, pointed **canines** are adapted for tearing food (Fig. 45–6). The flattened surfaces of the **premolars** and **molars** are specialized for crushing and grinding.

Each tooth is covered by **enamel,** the hardest substance in the body (Fig. 45–7). Most of the tooth consists of **dentin,** which resembles bone in composition and hardness. Beneath the dentin is the **pulp cavity,** a soft connective tissue containing blood and lymph vessels, and nerves.

While the food is being mechanically disassembled by the teeth, it is also moistened by saliva. Some of its molecules dissolve, enabling you to taste the food. Recall from Chapter 41 that taste buds are located on the tongue and other surfaces of the mouth. Three pairs of **salivary glands** secrete about a liter of saliva into the mouth cavity each day. Saliva contains an enzyme, **salivary amylase,** which initiates the digestion of starch into sugar.

The pharynx and esophagus conduct food to the stomach

After the bite of food has been chewed and fashioned into a lump called a **bolus,** it is swallowed—that is, moved through the **pharynx** and into the **esophagus.** The pharynx, or throat, is a muscular tube that serves as the hallway of the respiratory system as well as the digestive system. During swallowing, the opening to the airway is closed by a small flap of tissue, the **epiglottis.**

Waves of muscular contraction, called **peristalsis,** sweep the bolus through the pharynx and esophagus toward the stomach (Fig. 45–8). Circular muscle fibers in the wall of the esophagus contract around the top of the bolus, pushing it downward. Almost at the same time, longitudinal muscles around the bottom of the bolus and below it contract, shortening the tube.

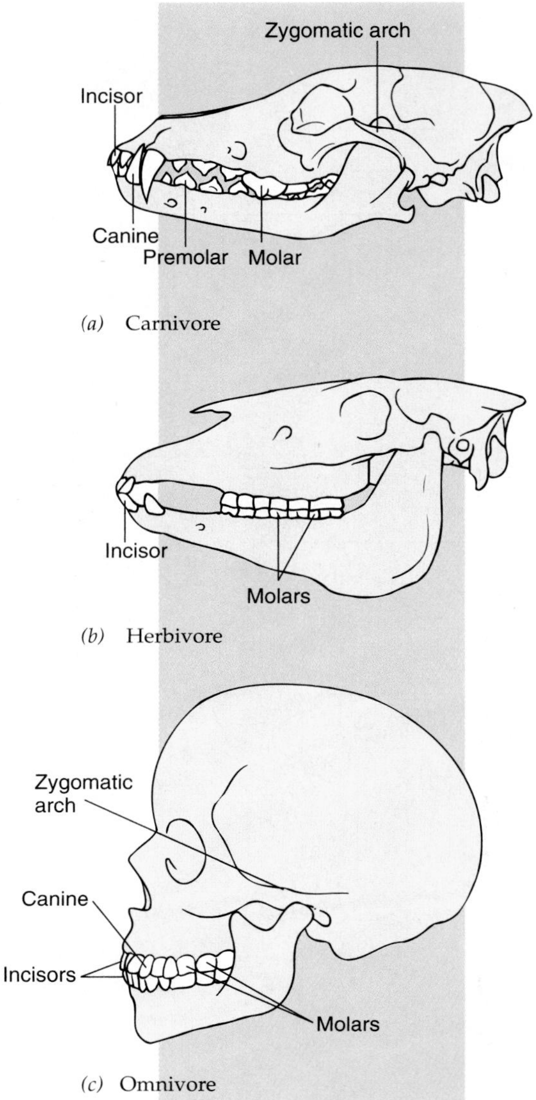

FIGURE 45–6 Teeth are adapted to diet. (***a***) Skull of a coyote, a carnivore, showing the pointed incisors and canines used to rip flesh. (***b***) In contrast, herbivores, such as cows and horses, have incisors and canines adapted for cutting off bits of vegetation. The broad ridge surfaces of their molars are adapted for grinding plant material. (***c***) Omnivores, such as humans, have relatively unspecialized teeth.

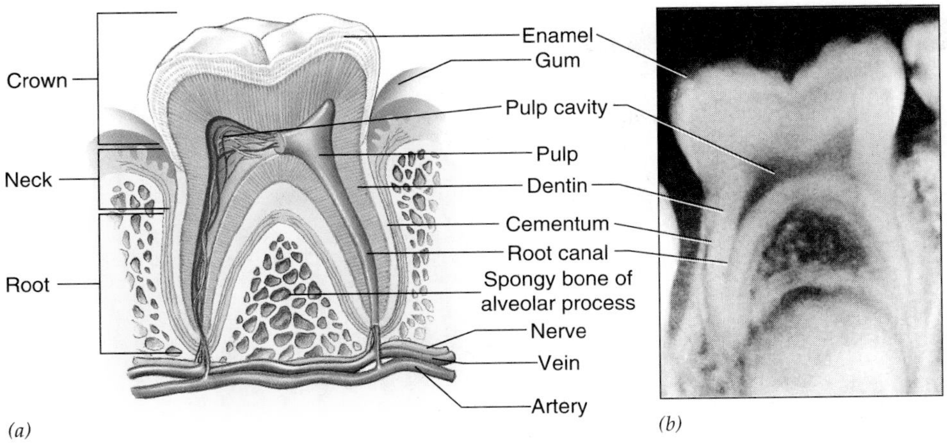

FIGURE 45–7 A tooth consists of a crown, neck, and root. (*a*) Sagittal section through a human lower molar. (*b*) X ray of a healthy tooth.

When the body is in an upright position, gravity helps to move the food through the esophagus, but gravity is not essential. Astronauts are able to eat in its absence, and even if you are standing on your head, food will reach your stomach.

Food is mechanically and enzymatically digested in the stomach

The entrance to the large, muscular **stomach** is normally closed by a ring of muscle at the lower end of the esophagus. When a peristaltic wave passes down the esophagus, the muscle relaxes, permitting the bolus to enter the stomach (Fig. 45–9). When empty, the stomach is collapsed and shaped almost like a hotdog. Folds of the stomach wall, called **rugae,** give the inner lining a wrinkled appearance. As food enters, the rugae gradually smooth out, expanding the capacity of the stomach to more than a liter.

The stomach is lined with a simple columnar epithelium that secretes large amounts of mucus. Tiny pits mark the entrances to the millions of **gastric glands,** which extend deep into the stomach wall. **Parietal cells** in the gastric glands secrete hydrochloric acid and a substance known as **intrinsic factor,** needed for adequate absorption of vitamin B. **Chief cells** in the gastric glands secrete **pepsinogen.** When pepsinogen comes in contact with the acidic gastric juice in the stomach, it is converted to **pepsin,** the main digestive enzyme of the stomach. Pepsin hydrolyzes proteins, converting them to short polypeptides (see *Focus On: Peptic Ulcers*).

What changes occur in our bite of hamburger during its three- to four-hour stay in the stomach? The stomach churns and chemically degrades the food so that it assumes the consistency of a thick soup; this partially digested food is called **chyme.** Protein digestion then begins, and much of the hamburger protein is degraded to polypeptides. Digestion of the starch in the bun to small polysaccharides and maltose continues until salivary amylase is inactivated by the acidic pH of the stomach. When digestion in the stomach is complete, peristaltic waves propel the chyme through the stomach exit, the **pylorus,** and into the small intestine.

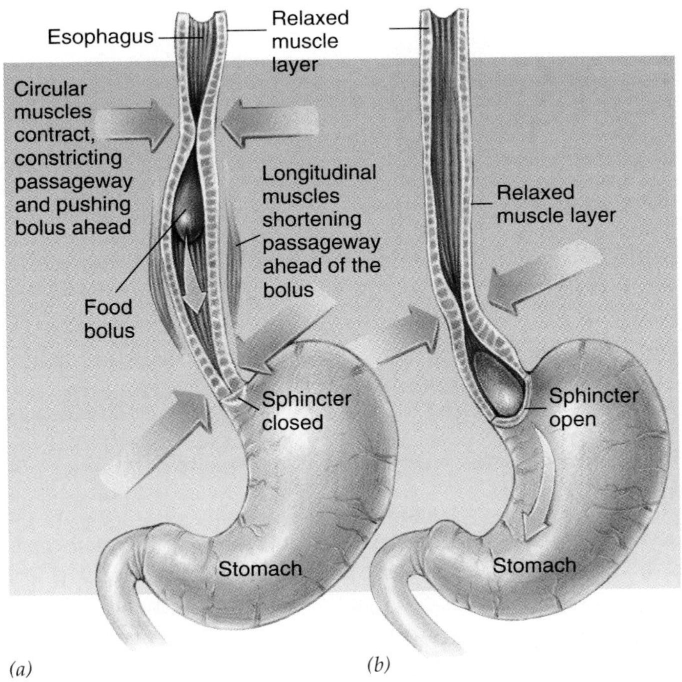

FIGURE 45–8 Food is moved through the digestive tract by waves of muscular contractions known as peristalsis. **(*a*)** A bolus is moved down through the esophagus by peristaltic contractions. **(*b*)** When the sphincter (ring of muscle) at the entrance of the stomach opens, food enters the stomach.

Most enzymatic digestion takes place inside the small intestine

Digestion of food is completed in the **small intestine,** and nutrients are absorbed through its wall. The small intestine has three regions: the **duodenum,** the **jejunum,** and the **ileum.** Most chemical digestion takes place in the duo-

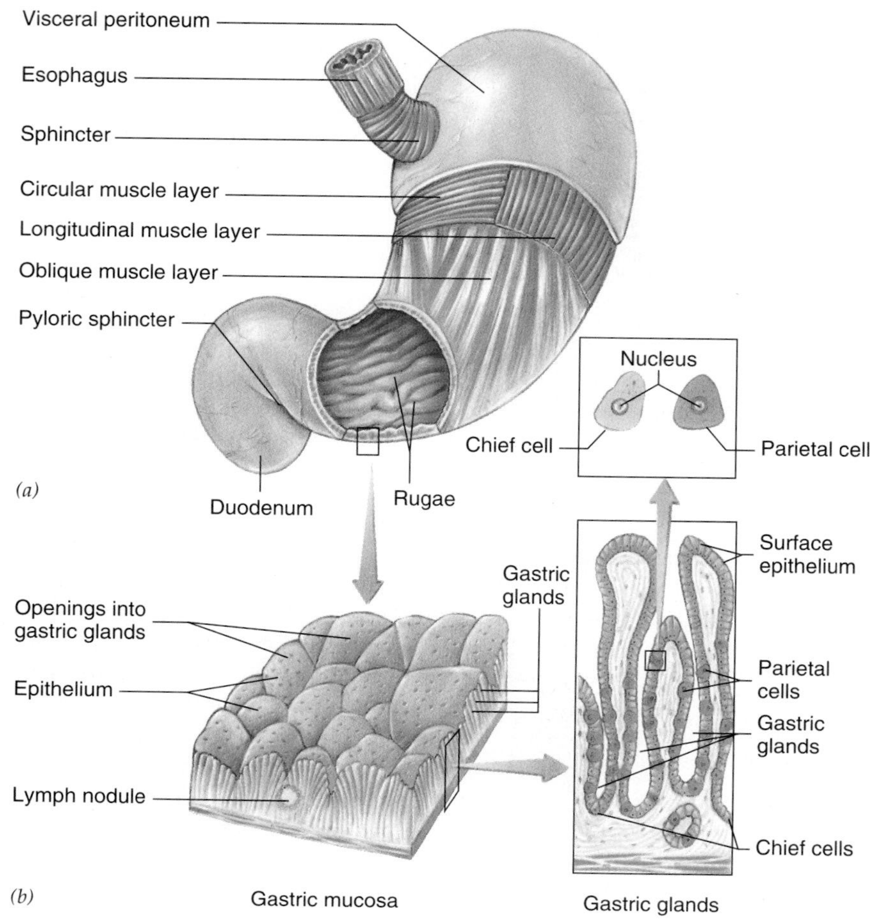

FIGURE 45–9 From the esophagus, food enters the stomach, where it is mechanically and enzymatically digested. (***a***) Structure of the stomach. (***b***) The stomach lining and gastric glands.

denum, the first portion of the small intestine, not in the stomach as is commonly believed. Bile from the liver and enzymes from the pancreas are released into the duodenum and act on the chyme. Then enzymes produced by the epithelial cells lining the duodenum catalyze the final steps in the digestion of the major types of nutrients.

The lining of the small intestine appears velvety because of its millions of tiny finger-like projections, the intestinal **villi** (sing., *villus*) (Fig. 45–10). The villi increase the surface area of the small intestine for digestion and absorption of nutrients. The intestinal surface is further expanded by thousands of microvilli, folds of cytoplasm on the exposed surface of the epithelial cells of the villi. About 600 microvilli protrude from the surface of each cell, giving the epithelial lining a fuzzy appearance when viewed with the electron microscope.

If the intestinal lining were smooth like the inside of a water pipe, food would zip right through the intestine and many valuable nutrients would not be absorbed. Folds in the wall of the intestine, the villi, and microvilli together increase the surface area of the small intestine by about 600 times. If we could unfold and spread out the lining of the small intestine of an adult human, its surface would approximate the size of a tennis court.

The liver secretes bile, which mechanically digests fats

Just under the diaphragm lies the **liver,** the largest and also one of the most complex organs in the body (Fig. 45–11). A single liver cell can carry on more than 500 separate metabolic activities. The liver's food processing functions include the following:

1. Secretes **bile,** which is important in the mechanical digestion of fats.
2. Helps maintain homeostasis by removing or adding nutrients to the blood.

Focus On

Peptic Ulcers

Several protective mechanisms prevent the gastric juice from digesting the wall of the stomach. Cells of the gastric mucosa secrete an alkaline mucus that coats the stomach wall and neutralizes the acidity of the gastric juice along the lining. In addition, the epithelial cells of the lining fit tightly together, preventing gastric juice from leaking between them and onto the tissue beneath. If some of the epithelial cells become damaged, they are quickly replaced. In fact, about a half million of these cells are shed and replaced every minute!

These mechanisms sometimes malfunction and a small bit of the stomach lining is digested, leaving an open sore or **peptic ulcer.** Peptic ulcers often occur in the duodenum, and sometimes occur in the lower part of the esophagus. A bacterium, *Helicobacter pylori,* has been implicated as a causative factor in ulcers. *H. pylori* infects the mucus-secreting cells of the stomach lining. The resulting decrease in protective mucus may lead to peptic ulcers or cancer.

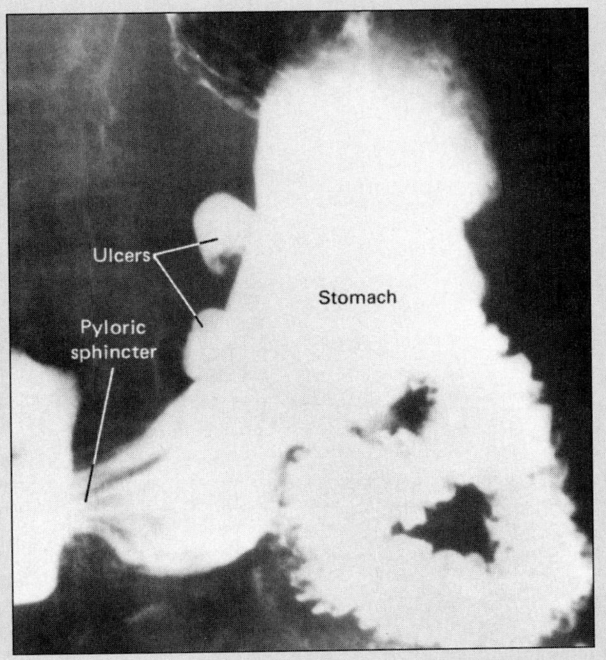

Ulcers in the wall of the stomach are visible in this x ray. The stomach and intestine have been filled with a contrast medium, making them appear white. This fluid also fills the cavities of the ulcers. (Courtesy of Dr. Jon Ehringer)

Peptic ulcers may bleed, which may lead to anemia. If the ulcer extends into the muscle layer, large blood vessels may also be damaged, resulting in hemorrhage. A **perforated ulcer** is one that extends all the way through the wall of the stomach or other affected organ. The opening created may allow bacteria and food to pass through to the peritoneum, leading to peritonitis and shock. Perforation is the main cause of death from ulcers. ■

3. Converts excess glucose to glycogen, and stores it.
4. Converts excess amino acids to fatty acids and urea.
5. Stores iron and certain vitamins.
6. Detoxifies alcohol, drugs, and poisons that enter the body.

Bile consists of water, bile salts, bile pigments, cholesterol, salts, and lecithin (a phospholipid). Bile produced in the liver is stored in the pear-shaped **gallbladder.** The gallbladder concentrates the bile and releases it into the duodenum as needed. Bile mechanically digests fats by a detergent-like action that decreases the surface tension of fat particles. This action permits the fat molecules to disperse so they can be attacked by lipase (fat-digesting enzymes). The dispersion of fat globules by bile is called **emulsification.** Bile contains no digestive enzymes and so does not enzymatically digest food.

The pancreas secretes digestive enzymes

The **pancreas** is an elongated gland that secretes both digestive enzymes and hormones that help regulate the level of glucose in the blood. Among its enzymes are **trypsin** and **chymotrypsin,** which digest polypeptides to dipeptides; **pancreatic lipase,** which degrades neutral fats; **pancreatic amylase,** which breaks down almost all types of carbohydrates, except cellulose, to disaccharides; and **ribonuclease** and **deoxyribonuclease,** which split the nucleic acids ribonucleic acid (RNA) and deoxyribonucleic acid (DNA) to free nucleotides.

Enzymatic digestion occurs as food moves through the digestive tract

As chyme moves through the digestive tract by peristalsis, mixing contractions, and motions of the villi, enzymes

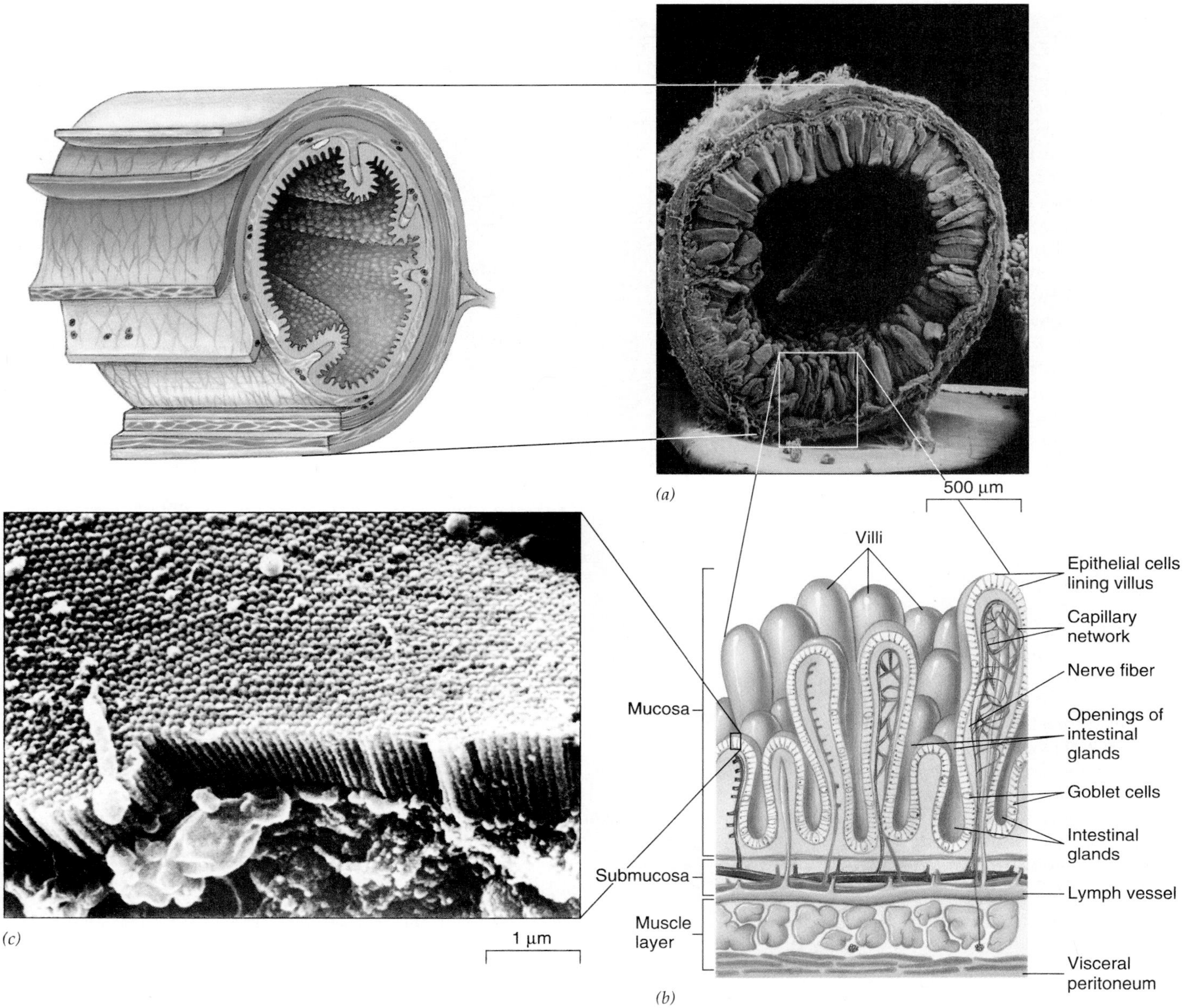

FIGURE 45–10 The inner wall of the small intestine is studded with villi and tiny openings that lead into the intestinal glands. (*a*) Scanning electron micrograph of a cross section of the small intestine. (*b*) Microscopic view of a small portion of the intestinal wall. Some of the villi have been opened to show the blood and lymph vessels within. (*c*) Scanning electron micrograph of the surface of an epithelial cell from the lining of the small intestine, showing microvilli. The epithelium has been cut vertically, allowing the microvilli to be viewed from the side as well as from above. (*a*, Visuals Unlimited/G. Shih-R. Kessel; *c*, Courtesy of J.D. Hoskings, W.G. Henk, and Y.Z. Abdelbaki, from the *American Journal of Veterinary Research*, Vol. 43, No. 10)

come into contact with the nutrients and digest them. The various digestive agents are summarized in Table 45–1.

Carbohydrates are digested to monosaccharides

Polysaccharides, such as starch and glycogen, are important components of the food ingested by humans and most other animals. The glucose units of these large molecules are connected by glycosidic bonds linking carbon 4 (or 6) of one glucose molecule with carbon 1 of the adjacent glucose molecule. These bonds are hydrolyzed by **amylases,** enzymes that digest polysaccharides to the disaccharide maltose (Table 45–2). Although the amylase of the digestive tract can split the α-glycosidic bonds

(Text continues on page 993)

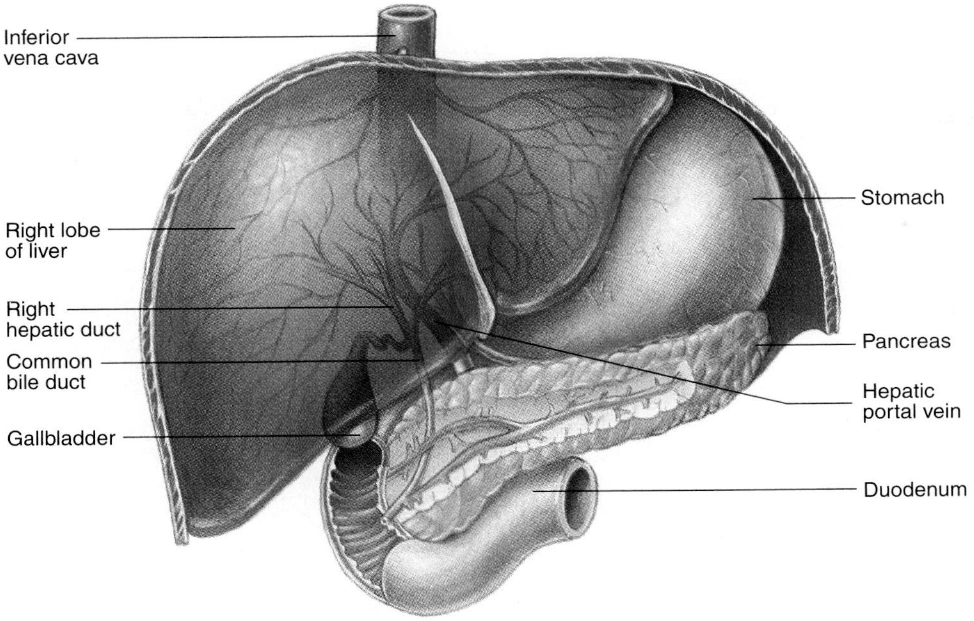

FIGURE 45–11 The liver and pancreas are accessory digestive glands. The gallbladder stores bile from the liver. Note the ducts.

Table 45–1 SUMMARY OF DIGESTIVE AGENTS

Agent (Active Form)	*Optimal pH*	*Substrate*	*Result*
Oral cavity			
Salivary amylase	6.9	Carbohydrates	Maltose and small polysaccharides
Mucus		Food	Moistening and lubrication
Stomach			
Hydrochloric acid (HCl)		Pepsinogen	Pepsin
Pepsin	2.0	Proteins	Polypeptides
Mucus		Food and gastric mucosa	Protection
Liver			
Bile salts		Fats	Emulsification
Pancreas			
Pancreatic amylase	7.1	Polysaccharides	Disaccharides
Pancreatic lipase	8.0	Emulsified fats	Fatty acids and glycerol
Trypsin	8.0	Polypeptides	Dipeptides
Chymotrypsin	8.0	Polypeptides	Dipeptides
Carboxypeptidase	8.0	Peptides	Dipeptides and amino acids
Ribonuclease	8.0	RNA	Nucleotides
Deoxyribonuclease	8.0	DNA	Nucleotides
Sodium bicarbonate		Acidic chyme	Neutralizes acidity
Small intestine			
Maltase	8.0	Maltose	Glucose
Sucrase	8.0	Sucrose	Glucose and fructose
Lactase	8.0	Lactose	Glucose and galactose
Aminopeptidase	8.0	Peptides	Amino acids
Dipeptidase	8.0	Dipeptides	Amino acids
Enterokinase	6.9	Trypsinogen	Trypsin (active)
Mucus		Intestinal mucosa	Protection
Intestinal juice		Chyme	Medium for digestion and absorption

Table 45–2 SUMMARY OF DIGESTION

Location	*Source of Enzyme*	*Digestive Process**
Carbohydrate digestion		
Mouth	Salivary glands	Polysaccharides (e.g., starch) —salivary amylase→ Maltose + Small polysaccharides 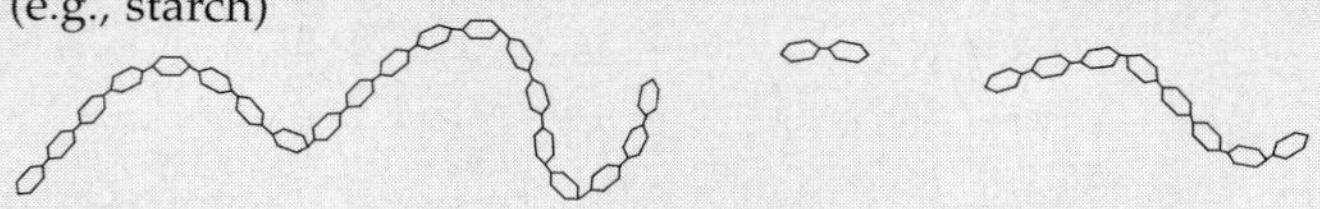
Stomach		Action continues until salivary amylase is inactivated by acidic pH
Small intestine	Pancreas	Undigested polysaccharides and small polysaccharides —pancreatic amylase→ Maltose
	Intestine	Disaccharides hydrolyzed to monosaccharides as follows:
		Maltose (malt sugar) —maltase→ Glucose + Glucose
		Sucrose (table sugar) —sucrase→ Glucose + Fructose
		Lactose (milk sugar) —lactase→ Glucose + Galactose
Protein digestion		
Stomach	Stomach (gastric glands)	Protein —pepsin→ Short polypeptides
Small intestine	Pancreas	Polypeptides (A—A—A—A—A / A—A—A—A—A) —trypsin, chymotrypsin→ Tripeptides + Dipeptides (A—A—A A—A)
		Dipeptides (A—A) —carboxypeptidase, aminopeptidase→ Free amino acids (A A A)
	Small intestine	Tripeptides + Dipeptides (A—A—A A—A) —peptidases→ Free amino acids (A A A)
Lipid digestion		
Small intestine	Liver	Glob of fat —bile salts→ Emulsified fat (individual triacylglycerols) 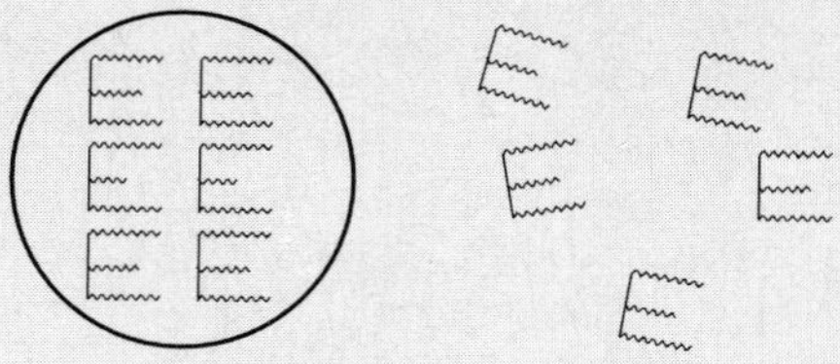
	Pancreas	Triacylglycerol —lipase→ Fatty acids + Glycerol 

*⬭ = monosaccharide; [triacylglycerol symbol] = triacylglycerol; E = glycerol; ~~~ = fatty acid; A = amino acid.

present in starch and glycogen, they cannot split the β-glycosidic bonds present in cellulose.

Amylase cannot split the bond between the two glucose units of maltose. Enzymes produced by the cells lining the small intestine break down disaccharides such as maltose to monosaccharides. **Maltase,** for example, splits maltose into two glucose molecules. Hydrolysis occurs while the disaccharides are being absorbed through the epithelium.

Proteins are digested to amino acids

Several kinds of proteolytic enzymes are secreted into the digestive tract (Table 45–1). Each breaks peptide bonds at one or more specific locations in a polypeptide chain.

Trypsin, secreted in an inactive form by the pancreas, is activated by an enzyme called enterokinase. The trypsin then activates chymotrypsin and carboxypeptidase, as well as additional trypsin. Pepsin, trypsin, and chymotrypsin break certain internal peptide bonds of proteins and polypeptides. Carboxypeptidase removes amino acids with free terminal carboxyl groups from the end of polypeptide chains. Aminopeptidase works at the other end of the chain, splitting off the amino acid with a free terminal amino group. The combined action of these enzymes results in the breakdown of the protein molecules to dipeptides and amino acids. **Dipeptidases** released by the duodenum then split the remaining small peptides to amino acids.

Fats are digested to fatty acids and monoacylglycerols

Lipids are usually ingested as large masses of triacylglycerols (also called triglycerides). They are digested mainly within the duodenum by pancreatic lipase (Table 45–2). Like many other proteins, lipase is water-soluble, but its substrates are not. Thus, the enzyme can attack only the fat molecules at the surface of a mass of fat. The bile salts act like detergents to reduce the surface tension of fats. Their action breaks large masses of fat into smaller droplets. This greatly increases the surface area of fat exposed to the action of lipase and so increases the rate of lipid digestion.

Conditions in the intestine are usually not optimal for the complete hydrolysis of lipids to glycerol and fatty acids. The products of lipid digestion therefore include monoacylglycerols (monoglycerides) and diacylglycerols (diglycerides) as well as glycerol and fatty acids. Undigested triacylglycerols remain as well, and some of these are absorbed without digestion.

Nerves and hormones regulate digestion

Most digestive enzymes are produced only when food is present in the digestive tract. Salivary gland secretion is controlled entirely by the nervous system, but secretion of other digestive juices is regulated by both nerves and hormones. At least four hormones—**gastrin, secretin, cholecystokinin (CCK),** and **gastric inhibitory peptide (GIP)**—regulate the digestive system (Table 45–3). All of these hormones are polypeptides secreted by endocrine cells in the mucosa of certain regions of the digestive tract. Several other messenger peptides have been identified in the digestive tract, and some of these may be important in regulating digestive activity.

As an example of the regulation of the digestive system, consider the secretion of gastric juice. Seeing, smelling, tasting, or even thinking about food causes the brain to send neural messages to the gastric glands in the stomach, stimulating them to secrete. In addition, when food distends the stomach, stretch receptors send neural messages to the medulla. The medulla then sends messages to endocrine cells in the stomach wall that secrete the hormone gastrin. Although gastrin is absorbed into the blood, it returns to the stomach, where it stimulates gastric juice release, gastric emptying, and intestinal motility.

Absorption takes place mainly through the villi of the small intestine

Only a few substances—water, simple sugars, salts, alcohol, and certain drugs—are small enough to be absorbed through the stomach wall. Absorption of nutrients is primarily the job of the intestinal villi. As illustrated in Figure 45–10, the wall of a villus consists of a single layer of epithelial cells. Inside each villus is a network of capillaries and a central lymph vessel, called a lacteal.

To reach the blood (or lymph), a nutrient molecule must pass through an epithelial cell of the intestinal lining and through a cell of the blood or lymph vessel lining. Absorption occurs by a combination of simple diffusion, facilitated diffusion, and active transport. Because glucose and amino acids cannot diffuse through the intestinal lining, they must be absorbed by active transport. Absorption of these nutrients is coupled with the active transport of sodium (see Chapter 5). Fructose is absorbed by facilitated diffusion.

Amino acids and glucose are transported to the liver by the **hepatic portal vein.** In the liver this vein divides into a vast network of sinusoids (tiny blood vessels similar to capillaries). As the nutrient-rich blood moves slowly through the liver, nutrients and certain toxic substances are removed from the circulation.

The products of lipid digestion are absorbed by a different process and different route (Fig. 45–12). Fatty acids and monoacylglycerols combine with bile salts to form soluble complexes called **micelles.** The micelles serve as a reservoir for fatty acids and monoacylglycerols. Because their solubility is low, very small amounts of free fatty acids and monoacylglycerols are present in the intestine. As these products of fat digestion are absorbed by diffusion (they are soluble in the lipid of the plasma membrane),

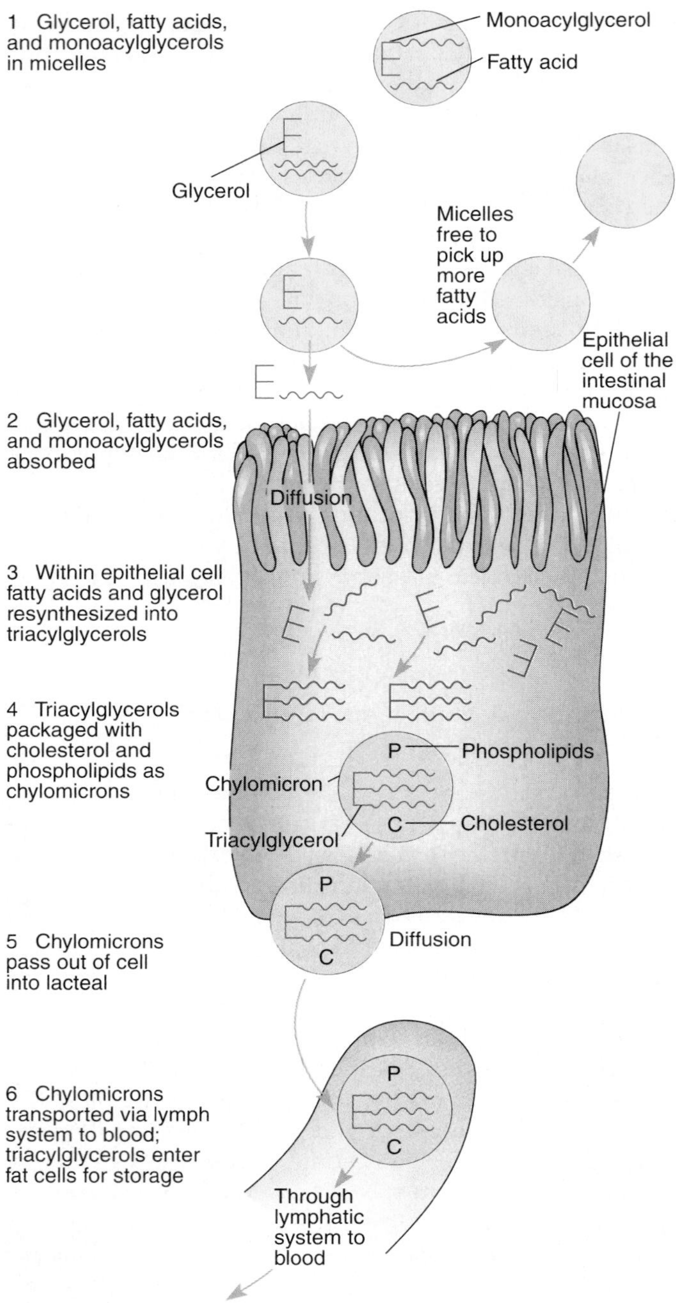
1 Glycerol, fatty acids, and monoacylglycerols in micelles
Monoacylglycerol
Fatty acid
Glycerol
Micelles free to pick up more fatty acids
Epithelial cell of the intestinal mucosa
2 Glycerol, fatty acids, and monoacylglycerols absorbed
Diffusion
3 Within epithelial cell fatty acids and glycerol resynthesized into triacylglycerols
4 Triacylglycerols packaged with cholesterol and phospholipids as chylomicrons
P
Phospholipids
Chylomicron
C
Cholesterol
Triacylglycerol
P
C
Diffusion
5 Chylomicrons pass out of cell into lacteal
P
C
6 Chylomicrons transported via lymph system to blood; triacylglycerols enter fat cells for storage
Through lymphatic system to blood

FIGURE 45–12 Epithelial cells lining the intestine absorb lipids, which then enter the lymphatic system. Fatty acids and monoacylglycerols combine with bile salts to form micelles, which transport the fatty substances to the epithelial cells. In the epithelial cells, fatty acids and glycerol combine to form triacylglycerols. These molecules, along with phospholipids and cholesterol, form protein-covered globules called chylomicrons, which are transported to the blood by the lymphatic system.

they are immediately replaced by molecules from the micelles. This process greatly facilitates absorption of fats. As monoacylglycerols and fatty acids leave a micelle, new fatty acids and monoacylglycerols combine with it. Micelles also dissolve cholesterol and fat-soluble vitamins, facilitating their absorption.

Once free fatty acids and monoacylglycerols enter an epithelial cell in the intestinal lining, they are reassembled as triacylglycerols in the smooth endoplasmic reticulum. The triacylglycerols are then packaged into droplets, which become larger as absorbed cholesterol and phospholipids are added. The droplets are then covered with a thin coat of protein. These protein-covered fat droplets, called **chylomicrons,** diffuse out of the epithelial cell and into the lacteal (lymph vessel) of the villus.

Chylomicrons are transported by the lymph to the subclavian veins where the lymph and its contents enter the blood. About 90% of absorbed fat enters the blood circulation in this indirect way. The rest, mainly short-chain fatty acids such as those in butter, are absorbed directly into the blood. After a meal rich in fats, the great number of chylomicrons in the blood may give the plasma a turbid, milky appearance for a few hours.

Most of the nutrients in the chyme are absorbed by the time it reaches the end of the small intestine. What is left (mainly waste) passes through a sphincter, the **ileocecal valve,** into the large intestine.

The large intestine eliminates waste

Undigested material, such as the cellulose of plant foods, along with unabsorbed chyme, passes into the **large intestine** (Fig. 45–4). Although only about 1.3 meters (about 4 feet) long, this organ is referred to as "large" because its diameter is greater than that of the small intestine. The small intestine joins the large intestine about 7 centimeters (2.8 inches) from the end of the large intestine, forming a blind pouch, the **cecum.** The **vermiform appendix** projects from the end of the cecum. (Appendicitis is an inflammation of the appendix.) The functions of the cecum and appendix in humans are not known. They are generally considered vestigial organs. Herbivores such as rabbits have a large, functional cecum that holds food while bacteria digest its cellulose.

From the cecum to the **rectum** (the last portion of the large intestine) the large intestine is known as the **colon.** The regions of the large intestine are the cecum; ascending colon; transverse colon; descending colon; sigmoid colon; rectum; and anus, the opening for the elimination of wastes.

As the chyme passes slowly through the large intestine, water and sodium are absorbed from it, and it gradually assumes the consistency of normal feces. Bacteria inhabiting the large intestine enjoy the last remnants of the meal and return the favor by producing vitamin K

Table 45–3 HORMONAL CONTROL OF DIGESTION

Hormone	*Source*	*Target Tissue*	*Actions*	*Factors that Stimulate Release*
Gastrin	Stomach (mucosa)	Stomach (gastric glands)	Stimulates gastric glands to secrete pepsinogen	Distention of the stomach by food; certain substances such as partially digested proteins and caffeine
Secretin	Duodenum (mucosa)	Pancreas	Signals secretion of sodium bicarbonate	Acidic chyme acting on mucosa of duodenum
		Liver	Stimulates bile secretion	
Cholecystokinin (CCK)	Duodenum (mucosa)	Pancreas	Stimulates release of digestive enzymes	Presence of fatty acids and partially digested proteins in duodenum
		Gallbladder	Stimulates emptying of bile	
Gastric inhibitory peptide (GIP)	Duodenum (mucosa)	Stomach	Decreases stomach churning, thus slowing emptying	Presence of fatty acids or glucose in duodenum

and certain B vitamins that can be absorbed and used.

A distinction should be made between elimination and excretion. *Elimination* is the process of getting rid of digestive wastes—materials that have not been absorbed from the digestive tract and did not participate in metabolic activities. *Excretion* refers to the process of getting rid of metabolic wastes, and in mammals is mainly the function of the kidneys and lungs. The large intestine, however, does excrete bile pigments.

When chyme passes through the intestine too rapidly, **defecation** (expulsion of feces) becomes more frequent, and the feces are watery. This condition, called diarrhea, may be caused by anxiety, certain foods, or by disease organisms that irritate the intestinal lining. Prolonged diarrhea results in loss of water and salts, and leads to dehydration—a serious condition, especially in infants.

Constipation results when chyme passes through the intestine too slowly. Because more water than usual is removed from the chyme, the feces may be hard and dry. Constipation is often caused by a diet containing insufficient fiber.

In Western countries, cancer of the colon and rectum accounts for more new cases each year than do any other cancers except lung cancer (Fig. 45–13). Studies indicate that a high incidence of this disease is found in populations whose diets are very low in fiber and high in animal protein, fat, and refined carbohydrate. It has been suggested that diets low in fiber result in less frequent defecation, allowing prolonged contact between the mucous membrane of the colon and such carcinogens as nitrites (used as preservatives) in foods.

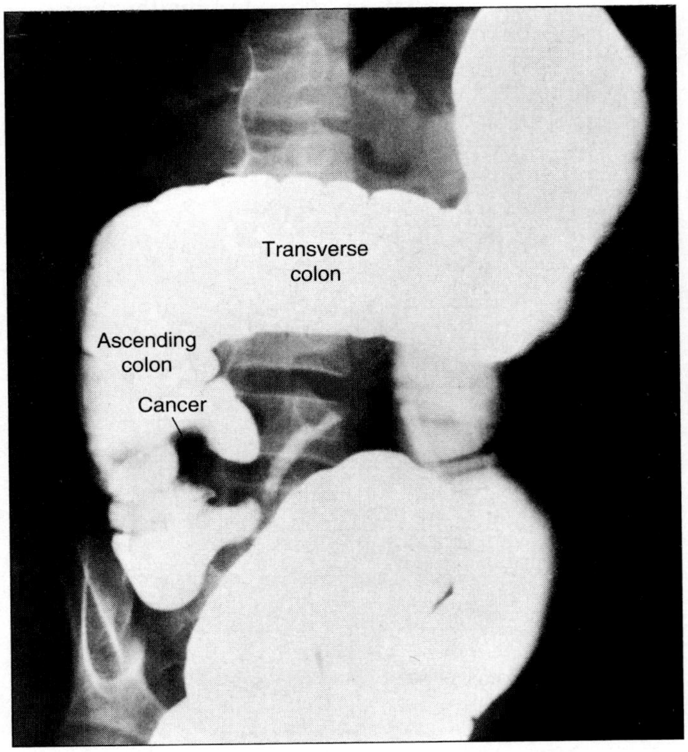

FIGURE 45–13 Colon cancer is evident as a mass that projects into the lumen in this radiographic view of the large intestine. The large intestine has been filled with a suspension of barium sulfate, which makes irregularities in the wall visible.

ADEQUATE AMOUNTS OF REQUIRED NUTRIENTS ARE NECESSARY TO SUPPORT METABOLIC PROCESSES

With only slight variation, all animals require the same basic nutrients: carbohydrates, lipids, proteins, vitamins, and minerals. Although not considered a nutrient in a strict sense, water is a necessary dietary component. Sufficient fluid must be ingested to replace that lost in urine, sweat, feces, and breath.

Adequate amounts of essential nutrients are necessary for metabolic processes. Recall from Chapter 1 that metabolism refers to all of the chemical processes that take place in the body. Metabolic processes include anabolism and catabolism. Anabolism refers to synthetic processes such as the production of proteins. Catabolism includes breakdown processes such as hydrolysis. Nutritionists measure the energy value of food in Calories. (A Calorie, spelled with a capital C, is actually a kilocalorie, defined as the amount of heat required to raise the temperature of a kilogram of water one degree centigrade.)

Once nutrients are absorbed from the digestive tract, they are transported by the blood or lymph. Surplus nutrients in the blood are taken up by the liver cells, where they are either stored or converted into other materials. Under normal circumstances, blood leaving the liver carries sufficient nutrients to meet the requirements of all of the cells of the body. The blood has been appropriately described as a traveling smorgasbord from which each cell selects whatever nutrients it needs to carry on its metabolic processes.

Carbohydrates are a major energy source in the human diet

Sugars and starches are the principal sources of energy in the ordinary human diet. However, they are not considered essential nutrients because the body can obtain sufficient energy from a mixture of proteins and fats. In the average American diet, carbohydrates provide about 50% of the Calories ingested daily.

Most carbohydrates are ingested in the form of starch and cellulose, both polysaccharides. (You may want to review the discussion of carbohydrates in Chapter 3.) Nutritionists refer to polysaccharides as **complex carbohydrates.** Foods rich in complex carbohydrates include rice, potatoes, corn, and other cereal grains. These are the least expensive foods, and for this reason the proportion of carbohydrate in a family's diet often reflects economic status. Very poor people subsist on diets that are almost exclusively carbohydrate, while the more affluent enjoy the costlier protein-rich foods such as meat and dairy products.

Nutritionists suggest that we increase our consumption of complex carbohydrates and fiber by eating more fruits, vegetables, and whole grains. **Fiber** is mainly a complex mixture of cellulose and other indigestible carbohydrates. The U.S. diet is low in fiber due to low in-

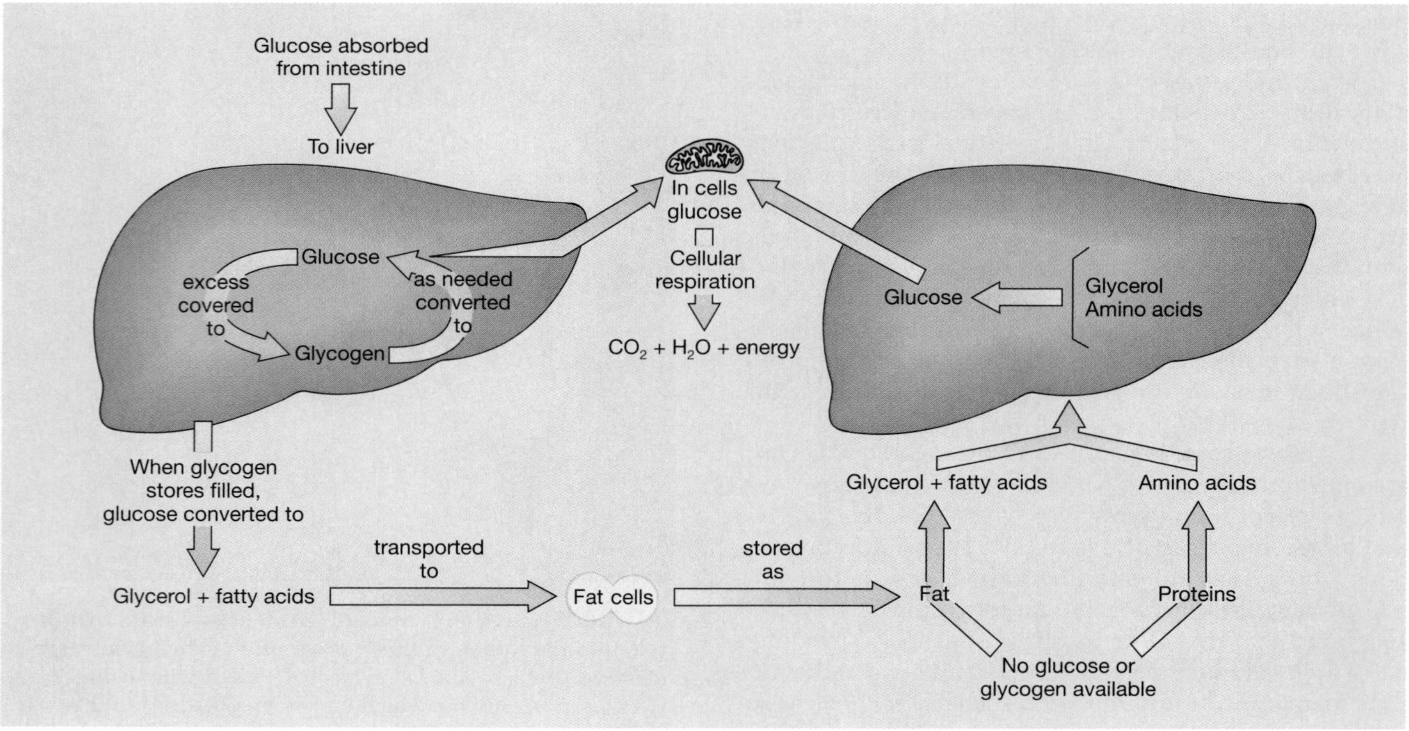

FIGURE 45–14 Glucose is used as an energy source by cells of the body. Glucose is stored as glycogen, and excess glucose is converted to fat. The liver plays a central role in maintaining an appropriate level of glucose in the blood.

take of fruits and vegetables and use of refined flour. Increasing fiber in the diet may decrease the risk of colon cancer. Fiber may also stimulate the feeling of being satisfied (satiety) with the amount of food intake, and thus could be useful in treating obesity.

Monosaccharides, the products of carbohydrate digestion, are converted to glucose in the liver. Excess glucose is stored as glycogen (Fig. 45–14). Hormones secreted by the pancreas act on the liver to regulate the concentration of glucose in the blood (blood sugar level).

When an excess of carbohydrate-rich food is eaten, the liver cells may become fully packed with glycogen and still have more glucose coming in. Liver cells then convert excess glucose to fatty acids and glycerol. These compounds are converted to triacylglycerols and sent to the fat depots of the body for storage.

Lipids are used as an energy source and to make biological molecules

Cells use ingested lipids as fuel, as components of cell membranes, and to make lipid compounds such as steroid hormones and bile salts. Lipids account for about 40% of the Calories in the average U.S. diet. Three polyunsaturated fatty acids (linoleic, linolenic, and arachidonic acids) are essential fatty acids that must be obtained in the human diet. Given these and sufficient nonlipid nutrients, the body can make all of the lipid compounds (including fats, cholesterol, phospholipids, and prostaglandins) that it needs.

About 98% of the lipids in the diet are ingested in the form of triacylglycerols (triglycerides). (Recall from Chapter 3 that a triacylglycerol is a glycerol molecule chemically combined with three fatty acids; see Fig. 3–12.) Triacylglycerols may be saturated—that is, fully loaded with hydrogens or they may be monounsaturated (containing one double bond in the carbon chain of a fatty acid, which allows two fewer hydrogen atoms to be present), or polyunsaturated (containing two or more double bonds).

Generally, animal foods are rich in both saturated fats and cholesterol, while plant foods contain unsaturated fats and no cholesterol. Commonly used polyunsaturated vegetable oils are corn, soy, cottonseed, and safflower oils. Olive, canola, and peanut oils contain large amounts of monounsaturated fats. Butter contains mainly saturated fats.

The average U.S. diet provides about 700 milligrams of cholesterol each day, whereas only about 300 milligrams is the recommended maximum. High cholesterol sources include egg yolks, butter, and meat. The body is not dependent on dietary sources for cholesterol because it is able to synthesize cholesterol from other nutrients. In fact, dietary intake of saturated fats can increase cholesterol level markedly.

When needed, stored fats are hydrolyzed to fatty acids and released into the blood. Before these fatty acids can be used by cells as fuel, they must be broken down into smaller compounds and combined with coenzyme A to form molecules of acetyl coenzyme A (Fig. 45–15). Acetyl-

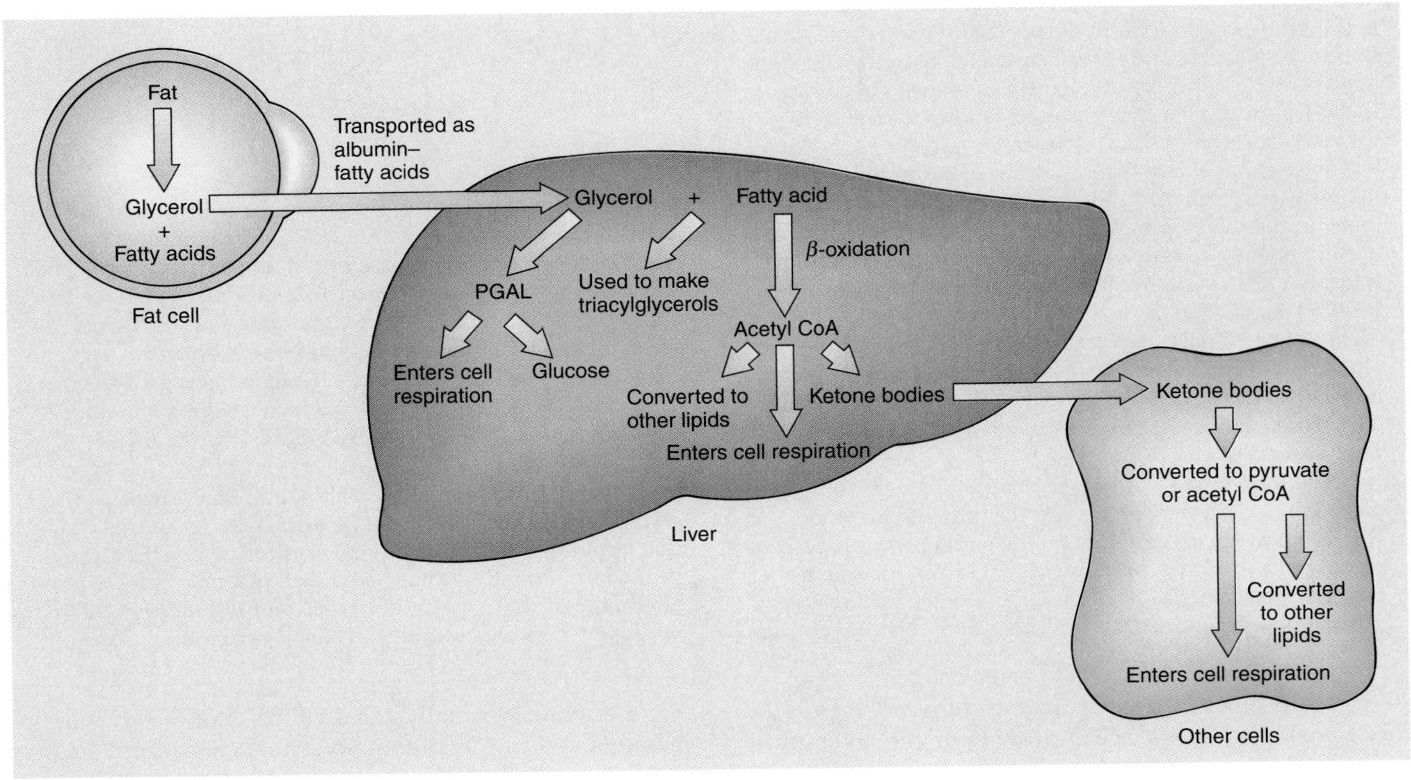

FIGURE 45–15 The liver plays an important role in converting glycerol and fatty acids to compounds that can be used as fuel in cellular respiration.

Making the Connection

Nutrition and Heart Disease

Just how important is it to count fat grams? Lipids have been the focus of much research because of their role in atherosclerosis, a progressive disease in which the arteries become clogged with fatty material. As discussed in Chapter 42, atherosclerosis leads to circulatory impairment and heart disease. Cholesterol and triacylglycerols are not transported freely in the blood plasma but are bound to proteins and transported as large molecular complexes, called lipoproteins. Some plasma cholesterol is transported on high-density lipoproteins (HDLs), but most is transported on low-density lipoproteins (LDLs). A protein on the LDL surface binds with a protein, an LDL receptor, on the plasma membrane of body cells (see Fig. 5–22).

High levels of LDL have been linked to increased risk for coronary artery and heart disease. LDL receptors help regulate cholesterol level in the blood by combining with LDL so that it can be taken up by body cells. If the blood contains too much LDL, there will not be enough LDL receptors to bind it; the excess may then be deposited in the arterial wall. This process is thought to involve highly reactive molecules in the arterial wall that oxidize the LDL cholesterol.

When LDL levels are high, HDL may play a protective role and decrease risk for coronary heart diseases. HDL may collect cholesterol and transport it to the liver. A healthy proportion of HDL to LDL can apparently be promoted by a regular exercise program, by diet (reducing intake of animal fats and cholesterol), by maintaining appropriate body weight (obesity has a negative effect on lipoprotein levels), and by not smoking cigarettes. Omega-3 fatty acids, found in fish oils, are thought to decrease LDL levels and play other protective roles in decreasing the risk for coronary heart disease.

Genetics appears to be important in determining risk for cardiovascular disease. Some individuals with high-fat diets do not develop high blood cholesterol levels. However, in about one-third of the population, a diet high in saturated fats and cholesterol raises the blood cholesterol level by as much as 25%.

Ingestion of polyunsaturated fats may decrease the blood cholesterol level. For these reasons, many people now cook with vegetable oils rather than with butter and lard, drink skim milk rather than whole milk, and eat ice milk instead of ice cream.

Some individuals with high cholesterol levels do not develop coronary heart disease. On the other hand, others who seem to have a safe risk profile suddenly drop dead of coronary heart disease. Lipoprotein(a)—a particle composed of proteins, plus cholesterol and other lipids—may prove a missing piece to this puzzle. Lipoprotein(a) appears to be highly concentrated in the blood of individuals whose coronary heart disease cannot be explained. Its concentration appears to be genetically determined; it does not appear to be affected by changes in diet. Just how this particle affects the cardiovascular system is not fully understood, but researchers are currently studying its actions. ▲

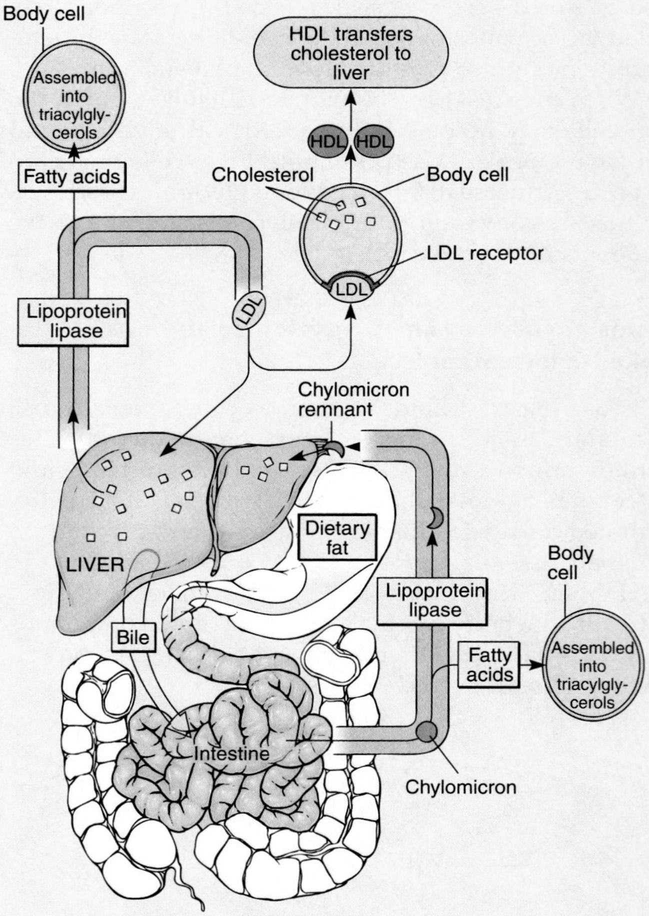

Chylomicrons transport lipids from the intestine to the cells of the body. As chylomicrons circulate in the blood, the enzyme lipoprotein lipase breaks down triacylglycerols. The fatty acids and glycerol can then be taken up by the cells. What is left of the chylomicron, a chylomicron remnant made up mainly of cholesterol and protein, is taken up by the liver. Cholesterol is then packaged in LDLs (low-density lipoproteins) which can be taken up by cells. Cholesterol can be returned to the liver in HDLs.

coenzyme A enters the citric acid cycle (see Chapter 7). This transformation is accomplished in the liver by a process known as beta-oxidation.

For transport to the cells, acetyl coenzyme A is converted into one of three types of *ketone bodies* (four-carbon ketones). Normally, the level of ketone bodies in the blood is low, but in certain abnormal conditions, such as starvation and diabetes mellitus, fat metabolism is tremendously increased. Ketone bodies are then produced so rapidly that their level in the blood becomes ex-

Focus On

Vegetarian Diets

Most of the world's population depends almost entirely on plants, especially cereal grains (usually rice, wheat, or corn) as the staple food. None of these foods contains adequate amounts of all of the essential amino acids. Besides being deficient in some of the essential amino acids, plant foods contain a lower percentage of protein than do animal foods. Meat contains about 25% protein, whereas even the new high-yield grains contain only 5% to 13% protein. What protein is available in plant food is also less digestible than that found in animal foods. Because most of the protein is encased within indigestible cellulose cell walls, much of it passes right through the digestive tract.

Despite these potential nutritional problems, more and more people are turning to vegetarian diets. One reason for the shift is that in meat-based diets we ingest a great deal of fat along with the protein, leading to health problems. Another reason is that meats are becoming increasingly expensive because they are ecologically expensive to produce. About twenty-one kilograms of protein in grain, for example, is required to produce just one kilogram of beef protein. If the human population of our planet continues to expand at a much greater rate than does our food production, more grain will be diverted for human food and less for animal feed. The price of meat will continue to soar and may become unaffordable for many of us.

A vegetarian diet can be nutritionally balanced provided one takes into account the nutritional risks (especially in growing children) associated with non-meat diets. An important guideline is to select foods that complement one another. This requires knowledge of which amino acids are deficient in each kind of food. Since the body cannot store amino acids, all of the essential amino acids must be ingested at the same meal. For example, if rice is eaten for dinner, and beans for lunch the next day, the body will not have all the essential amino acids needed at the same time to manufacture proteins. If beans and rice are eaten together, however, all of the needed amino acids are provided, because one food contributes what the other lacks. Similarly, if dairy products are not excluded from the vegetarian diet, then macaroni can be paired with cheese, or cereal with milk, and all of the essential amino acids can be obtained.

Other concerns regarding vegetarian diets include deficiency of iron, calcium, and certain vitamins. Dairy products, which are a major source of calcium and vitamin D (in fortified milk), are excluded in certain (vegan) vegetarian diets. Vitamin B_{12} is found almost exclusively in animal products.

Balanced vegetarian diets offer many health benefits. Increases in legume, grain, vegetable, and fruit intake provide more fiber, as well as more of certain vitamins and minerals. Studies of Seventh Day Adventists, a religious group that excludes animal products from their diets, compared with non-Seventh Day Adventists living in the same area indicate that the incidence of cardiovascular disease is reduced by about 50%. ■

cessive, which may cause the blood to become too acidic. Such disruption of normal pH balance can lead to death. (How Eskimos, who live on extremely high-fat diets, manage to maintain acid-base homeostasis is something of a mystery.)

Proteins serve as enzymes and as structural components of cells

Proteins are essential building blocks of cells, serve as enzymes, and are also used to make needed substances such as hemoglobin and myosin. Protein consumption is an index of a country's (or an individual's) economic status, because high-quality protein tends to be the most expensive and least available of the nutrients.

The recommended daily adult intake of protein is about 56 grams—only about an eighth of a pound (dry weight). In the United States and other developed countries, most people eat about twice as much protein as they require. In other parts of the world, protein poverty is one of the most pressing health problems. Millions of humans suffer from poor health, disease, and even death as a consequence of protein malnutrition.

Ingested proteins are degraded in the digestive tract to amino acids. Of the 20 or so amino acids important in nutrition, approximately eight (nine in children) cannot be synthesized by humans at all, or at least not in sufficient quantities to meet the body's needs. These, which must be provided in the diet, are referred to as **essential amino acids.**

Complete proteins, those that contain the most appropriate distribution of amino acids for human nutrition, are found in eggs, milk, meat, and fish. Some foods, such as gelatin or soybeans, contain a high proportion of protein. However, they either do not contain all of the essential amino acids, or they do not contain them in proper nutritional proportions. Most plant proteins are deficient in one or more essential amino acids (See *Focus On: Vegetarian Diets*).

Amino acids circulating in the blood can be taken up by cells and used for the synthesis of proteins. Excess amino acids are removed from circulation by the liver. In the liver cells, these are deaminated; that is, the amino group is removed (Fig. 45–16). During deamination, ammonia forms from the amino group. Ammonia, which is toxic at high concentrations, is converted to urea and excreted from the body.

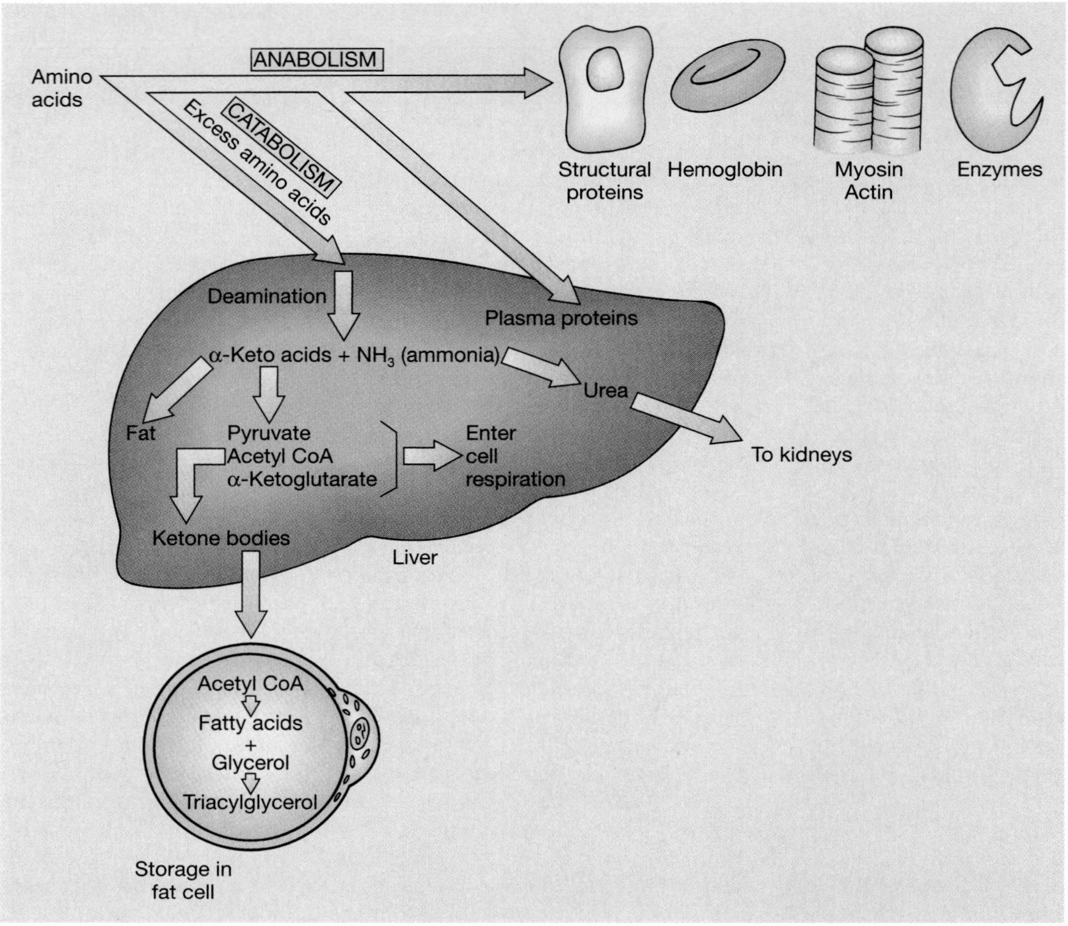

FIGURE 45–16 The liver plays a central role in protein metabolism. Deamination of amino acids and conversion of the amino groups to urea take place there. In addition, many proteins are synthesized in the liver. Note that you can gain weight on a high protein diet because excess amino acids can be converted to fat.

The remaining carbon chain of the amino acid (called a keto acid) may be converted into carbohydrate or lipid and used as fuel or stored. Thus, even people who eat high-protein diets can gain weight if they eat too much.

Vitamins are organic compounds essential for normal metabolism

Vitamins are organic compounds required in the diet in relatively small amounts for normal biochemical functioning. Many are components of coenzymes (see Chapter 6). Vitamins may be divided into two main groups. **Fat-soluble vitamins** are those that can be dissolved in fat and include vitamins A, D, E, and K. **Water-soluble vitamins** are the B and C vitamins. Table 45–4 provides the sources, functions, and consequences of deficiency for most of the vitamins.

Health professionals debate the advisability of taking large amounts of certain specific vitamins, such as vitamin C to prevent colds, or vitamin E to protect against vascular disease. Some studies suggest that vitamin A (found in yellow and green vegetables) and vitamin C (found in citrus fruit and tomatoes) may help protect against certain forms of cancer. We do not yet understand all of the biochemical roles played by vitamins, or the interactions among various vitamins and other nutrients.

We do know that large overdoses of vitamins, like vitamin deficiency, can be harmful. Moderate overdoses of the B and C vitamins are excreted in the urine, but surpluses of the fat-soluble vitamins are not easily excreted and can accumulate to harmful levels.

Minerals are inorganic nutrients required by cells

Minerals are inorganic nutrients ingested in the form of salts dissolved in food and water (Table 45–5). Essential minerals required in amounts of 100 milligrams or more daily include sodium, chloride, potassium, calcium, phosphorus, magnesium, and sulfur. Several others, such as iron, copper, iodine, fluoride, and selenium, are **trace**

Making the Connection

Nutrition and Cell Function

Why are antioxidants an important part of the diet? Normal cellular processes that require oxygen produce highly reactive molecules such as peroxides and **free radicals.** These compounds snatch electrons from other molecules such as DNA. Free radicals are also generated by tobacco smoke and other forms of air pollution.

Cells have **antioxidants** that destroy free radicals and other reactive molecules. However, when too many free radicals are present, they damage DNA, proteins, and unsaturated fatty acids. Damage to DNA can cause mutations that can lead to cancer. Injury to unsaturated fatty acids can result in damage to cell membranes. Free radicals are thought to contribute to atherosclerosis by causing oxidation of LDL cholesterol. The effects of oxidative damage to the body over the years are thought to contribute to the aging process.

Certain vitamins, including A, C, and E, function as antioxidants. For example, vitamins A and E protect cell membranes from free radicals. Selenium, zinc, copper, manganese, and iron are minerals that are parts of antioxidant enzyme systems. Nutritionists recommend that we increase the antioxidants in our diet by eating fruit, vegetables, and other foods high in antioxidant vitamins and minerals. ▲

elements, minerals that are required in amounts of less than 100 milligrams per day.

Minerals are needed as components of body tissues and fluids. Salt content (about 0.9%) is vital in maintaining the fluid balance of the body, and since salts are lost from the body daily in sweat, urine, and feces, they must be replaced by dietary intake. Sodium chloride (common table salt) is the salt needed in largest quantity in blood and other body fluids. A deficiency results in dehydration.

Iron is the mineral most likely to be deficient in the diet. In fact, iron deficiency is one of the most widespread nutritional problems in the world. In most developing countries, an estimated two-thirds of children and women of childbearing age suffer from iron deficiency. In the U.S., Europe, and Japan, 10% to 20% of women of childbearing age have this deficiency.

Energy Metabolism Is Balanced When Energy Input Equals Energy Output

The amount of energy liberated by the body per unit time is a measure of the **metabolic rate.** Much of the energy expended by the body is ultimately converted to heat. Metabolic rate may be expressed either in Calories of heat energy expended per day or as a percentage above or below a standard normal level.

The **basal metabolic rate (BMR)** is the rate at which the body releases heat as a result of breaking down fuel molecules. BMR is the body's basic cost of living—that is, the rate of energy used during resting conditions. This energy is required to maintain body functions such as heart contraction, breathing, and kidney function. An individual's **total metabolic rate** is the sum of his or her BMR and the energy used to carry on all daily activities. For example, a laborer has a greater total metabolic rate than does an executive whose job requirements do not include a substantial amount of movement and who does not exercise regularly.

An average-sized man who does not engage in any exercise program and who sits at a desk all day expends about 2000 Calories daily. If the food he eats each day also contains about 2000 Calories, he will be in a state of energy balance; that is, his energy input will equal his energy output. This is an extremely important concept, because body weight remains constant when:

$$\text{Energy (Calorie) input} = \text{Energy output}$$

When energy output is greater than energy input, stored fat is burned and body weight decreases. On the other hand, people gain weight when they take in more energy (Calories) in food than they expend in daily activity—in other words, when:

$$\text{Energy (Calorie) input} > \text{Energy output}$$

Obesity is a serious nutritional problem

Obesity, the excess accumulation of body fat, is a serious form of malnutrition that has become a problem of epidemic proportions in affluent societies. An overweight person places an extra burden on the heart and is susceptible to heart disease and other ailments. Obese persons generally die at a younger age than do people of normal weight. Yet one-third of our working population is 25% or more overweight.

Recent research suggests that we inherit the risk for obesity. Some people do appear to gain weight more readily than others (see *On the Cutting Edge: In Search of a Cure for Obesity*). Obesity can result from an increase in the

(Text continues on page 1004)

Table 45–4 THE VITAMINS

*Vitamins and U.S. RDA**	*Actions*	*Effect of Deficiency*
Fat-soluble		
Vitamin A, retinol 5000 IU†	Converted to retinal; essential for normal vision; essential for normal growth and differentiation of cells; promotes normal growth, reproduction, and immunity	Growth retardation; night blindness; worldwide 500,000 preschool children are blinded by vitamin A deficiency each year.
Vitamin D, calciferol 400 IU	Promotes calcium absorption from digestive tract; essential to normal growth and maintenance of bone	Bone deformities; rickets in children; osteomalacia in adults
Vitamin E, tocopherols 30 IU	Antioxidant; protects unsaturated fatty acids and cell membranes	Increased catabolism of unsaturated fatty acids, so that not enough are available for maintenance of cell membranes; prevention of normal growth
Vitamin K, probably about 1 mg	Essential for blood clotting	Prolonged blood clotting time
Water-soluble		
Vitamin C, ascorbic acid 60 mg	Needed for synthesis of collagen and other intercellular substances; formation of bone matrix and tooth dentin, intercellular cement; needed for metabolism of several amino acids; may help body withstand injury from burns and bacterial toxins	Scurvy (wounds heal very slowly and scars become weak and split open; capillaries become fragile; bone does not grow or heal properly)
B-complex vitamins		
Vitamin B_1, Thiamine 1.5 mg	Active form is a coenzyme in many enzyme systems; important in carbohydrate and amino acid metabolism	Beriberi (weakened heart muscle, enlarged right side of heart, nervous system and digestive tract disorders)
Vitamin B_2, riboflavin 1.7 mg	Used to make coenzymes (e.g., FAD) essential in cellular respiration	Dermatitis, inflammation and cracking at corners of mouth; confusion
Niacin 20 mg	Component of important coenzymes (NAD^+ and $NADP^+$); essential to cellular respiration	Pellagra (dermatitis, diarrhea, mental symptoms, muscular weakness, fatigue)
Vitamin B_6, pyridoxine 2 mg	Derivative is coenzyme in many reactions in amino acid metabolism	Dermatitis; digestive tract disturbances; convulsions
Pantothenic acid 10 mg	Constituent of coenzyme A (important in cellular metabolism)	Deficiency extremely rare
Folic acid 0.4 mg	Coenzyme needed for reactions involved in nucleic acid synthesis and for maturation of red blood cells	A type of anemia; certain birth defects; increased risk of cardiovascular disease
Biotin 0.3 mg	Coenzyme important in metabolism	———
Vitamin B_{12} 6 mg	Coenzyme important in metabolism	A type of anemia

*RDA is the recommended dietary allowance, established by the Food and Nutrition Board of the National Research Council, to maintain good nutrition for healthy persons.

†International Unit: the amount that produces a specific biological effect and is internationally accepted as a measure of the activity of the substance.

Table 45–4 continued

Sources	Comments
Liver, fortified milk, yellow and green vegetables such as carrots and broccoli	Can be formed from provitamin carotene (a yellow or orange pigment); sometimes called anti-infection vitamin because helps maintain epithelial membranes; excessive amounts harmful
Fish oils, egg yolk, fortified milk, butter, margarine	Two types: D_2, a synthetic form; D_3, formed by action of ultraviolet rays from sun on a cholesterol compound in the skin; excessive amounts harmful
Oils made from cereals; nuts; leafy greens	———
Normally supplied by intestinal bacteria; leafy greens, legumes	Antibiotics may kill bacteria; then supplements needed
Citrus fruits, strawberries, tomatoes	Possible role in preventing common cold or in the development of acquired immunity (?); harmful in very excessive doses
Liver, yeast, cereals, meat, green leafy vegetables	Deficiency common in alcoholics
Liver, cheese, milk, eggs, green leafy vegetables	———
Liver, cheese, milk, eggs, green leafy vegetables	———
Liver, meat, cereals, legumes	Many common drugs interfere with vitamin B_6 metabolism
Widespread in foods	———
Produced by intestinal bacteria; liver, cereals, dark green leafy vegetables	———
Produced by intestinal bacteria; liver, chocolate, egg yolk	———
Liver, meat, fish	Contains cobalt; intrinsic factor secreted by gastric mucosa needed for absorption

Table 45–5 SOME IMPORTANT MINERALS AND THEIR FUNCTIONS

Mineral	*Functions*	*Sources; Comments*
Calcium	Component of bones and teeth; essential for normal blood clotting; needed for normal muscle and nerve function	Milk and other dairy products, fish, green leafy vegetables; bones serve as calcium reservoir
Phosphorus	Performs more functions than any other mineral. Structural component of bone; component of ATP, DNA, RNA, and phospholipids	Meat, dairy products, cereals
Sulfur	Component of many proteins and vitamins	High-protein foods such as meat, fish, legumes, nuts
Potassium	Principal positive ion within cells; influences muscle contraction and nerve excitability	Occurs in many foods
Sodium	Principal positive ion in interstitial fluid; important in fluid balance; essential for conduction of nerve impulses	Many foods, table salt; too much ingested in average American diet; excessive amounts may contribute to high blood pressure
Chloride	Principal negative ion of interstitial fluid; important in fluid balance and in acid-base balance	Many foods; table salt
Magnesium	Needed for normal muscle and nerve function	Nuts, whole grains, green, leafy vegetables
Copper	Component of enzyme needed for melanin synthesis; component of many other enzymes; essential for hemoglobin synthesis	Liver, eggs, fish, whole wheat flour, beans
Iodine	Component of thyroid hormones (hormones that increase metabolic rate). Deficiency results in goiter (abnormal enlargement of thyroid gland)	Seafood, iodized salt, vegetables grown in iodine-rich soils
Manganese	Necessary to activate arginase, an enzyme essential for urea formation; activates many other enzymes	Whole-grain cereals, egg yolks, green vegetables; poorly absorbed from intestine
Iron	Component of hemoglobin, myoglobin, important respiratory enzymes (cytochromes), and other enzymes essential to oxygen transport and cellular respiration. Deficiency results in anemia and may impair cognitive function	Mineral most likely to be deficient in diet. Good sources: meat (especially liver), nuts, egg yolk, legumes
Fluoride	Component of bones and teeth; makes teeth resistant to decay; excess causes tooth mottling	In areas where it does not occur naturally, fluoride may be added to municipal water supplies (fluoridation)
Zinc	Cofactor for at least 70 enzymes; helps regulate synthesis of certain proteins; needed for growth and repair of tissues; deficiency may impair cognitive function	Meat, milk, yogurt, some seafood
Selenium	Antioxidant (part of a peroxidase that breaks down peroxides)	Seafood, eggs, liver

size of fat cells or from an increase in the number of fat cells, or both. The number of fat cells in the adult is apparently determined mainly by the amount of fat stored during infancy and childhood. When we are overfed early in life, abnormally large numbers of fat cells are formed. Later in life, these fat cells may be fully stocked

ON THE CUTTING EDGE

In Search of a Cure for Obesity

Objective: To characterize the normal *ob* gene product, known as Ob protein, in the hope that this knowledge will lead to the development of a pharmacological treatment for obesity.

Method: The normal allele of the *ob* gene was transferred into the genome of bacteria, which then produced large amounts of Ob protein. This protein was then injected, in various doses and intervals, into obese and non-obese mice.

Results: When injected into obese mice, the Ob protein reduced appetite and increased energy use, resulting in weight loss, mainly from loss of body fat.

Conclusions: The Ob protein is a hormone that appears to act on receptors in the brain; it may serve an endocrine function to help regulate fat storage in the body.

In the 1950s a spontaneous recessive mutation occurred in a strain of laboratory mice that causes the mutant mice to become grossly obese. The increase in adipose tissue in these mice is part of a syndrome that parallels morbid obesity in humans, a condition in which an individual's body weight is 100 or more pounds (45.5 kilograms) above normal. In 1994, Jeffrey M. Friedman's research team at Rockefeller University isolated the gene (*ob* gene) that, when mutated, was responsible for the obese phenotype. Mice with the mutated gene apparently lack some weight regulating substance.

In 1995, three different research teams reported in *Science* that injections of the *ob* gene product, known as Ob protein, resulted in weight loss by obese and non-obese mice. All three groups transferred the *ob* gene into bacteria, which then made large quantities of the protein. When the Ob protein was injected into grossly obese mice, their appetites decreased and they lost weight, almost exclusively from loss of body fat.

Friedman's team fed the same diet to obese mice who were injected with the Ob protein and to a control group of obese mice who did not receive the Ob protein. The mice in the treated group lost 50% more weight than the untreated animals. This research group suggested the name *leptin* for the Ob protein.

Frank Collins' research group at Amgen showed that treated mutant mice became more active and that their metabolic rate increased. Arthur Campfield's team at Hoffman-La Roche Inc. reported that Ob protein acts directly on neural networks in the brain. They suggested that Ob protein is a circulating protein (a hormone?) produced in adipose tissue. This protein acts on centers in the brain responsible for regulating feeding behavior and energy metabolism.

Campfield suggests that human obesity is more complex, probably a polygenetic disease. Humans with certain alleles may be at higher risk for obesity and gain weight when on a high-fat diet. Researchers are looking for receptors in the brain that might combine with leptin or a related hormone. The problem in overweight humans could be lack of a hormone or some problem with the receptors (in the brain?) that bind it.

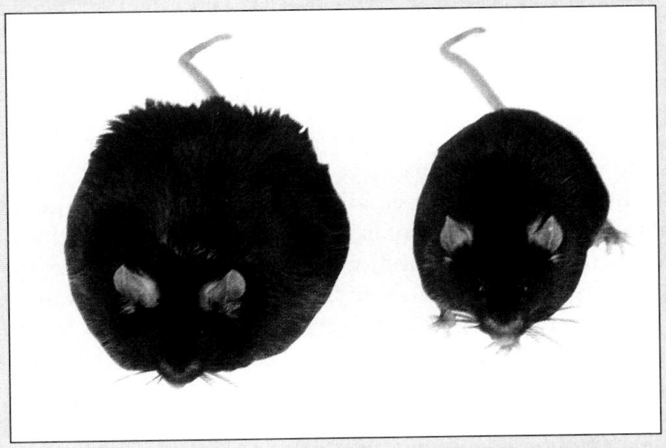

Mutant mice before and after treatment with Ob protein.
(Courtesy of John Sholtis/The Rockefeller University, New York, NY)

The mechanisms of energy balance and body weight are complex and likely involve many regulatory mechanisms. At the time of this writing it is not known whether Ob protein will turn out to be a miracle drug for weight control, as suggested by popular media. However, investigators continue to uncover clues to the puzzle of human obesity.

Science, Vol. 269, 28 July 1995. This issue includes a Research News summary by Marcia Baringa and three articles on weight regulation.

with excess lipids or may be shrunken, but they are always there. People with such increased numbers of fat cells are thought to be more susceptible to obesity than are those with normal numbers.

Most overweight people overeat due to a combination of poor eating habits and psychological factors. Whatever the underlying causes, overeating and/or underexercising appear to be the routes to obesity. For every 9.3 Calories of excess food taken into the body, about 1 gram of fat is stored. (An excess of about 140 Calories per day for a month results in a 1-pound weight gain.)

Because so many people are overweight, dieting has generated a multimillion-dollar industry of diet foods, formulas, pills, books, clubs, slenderizing devices, and even surgical procedures such as gastric stapling. Unfortunately, there is no magical cure for obesity. The only sure (and healthful) way to lose weight is to adjust food consumption and activity level so that energy intake is less than energy output. Then the body will have to draw on its fat stores for the missing calories. As fat is mobilized and burned, body weight decreases. This can generally be accomplished by a combination of increased exercise and decreased caloric intake (1000 to 1500 Calories daily for the mildly obese). Most nutritionists agree that the best reducing diet is well-balanced and provides the bulk of the calories in the form of complex carbohydrates.

Malnutrition can cause serious health problems

While millions of people eat too much, more than a billion humans do not have enough to eat, or do not eat a balanced diet. Individuals suffering from malnutrition are weak, easily fatigued, and highly susceptible to infection. Essential amino acids, iron, calcium, and vitamin A are commonly deficient nutrients.

Of all the required nutrients, essential amino acids are the one most often deficient in the diet. Millions of people suffer from poor health and a lowered resistance to disease because of protein deficiency. Children's physical and mental development are retarded when the essential building blocks of cells are not provided in the diet. Because their bodies cannot manufacture antibodies (which are proteins) and cells needed to fight infection, common childhood diseases, such as measles, whooping cough, and chicken pox, are often fatal in children suffering from protein malnutrition.

In young children, severe protein malnutrition results in the condition known as **kwashiorkor.** This term is an African word that means "first-second." It refers to the situation in which a first child is displaced from its mother's breast when a younger sibling is born. The older child is placed on a diet of starchy cereal or cassava that is deficient in protein. Growth becomes stunted, muscles are wasted, edema develops (as displayed by a swollen belly), the child becomes apathetic and anemic, and metabolism is impaired (Fig. 45–17). Without essential amino acids, the digestive enzymes themselves cannot be manufactured, so what little protein is ingested cannot be digested. Dehydration and diarrhea occur and often lead to death.

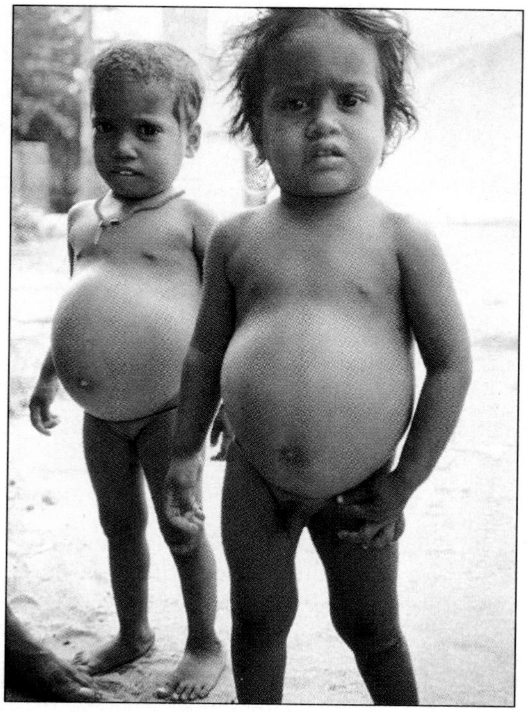

FIGURE 45–17 Millions of children suffer from *kwashiorkor,* a disease caused by severe protein deficiency. Note the characteristic swollen belly, which results from fluid imbalance. (United Nations Food and Agricultural Organization, photo by P. Pittet)

SUMMARY

I. Processing food includes ingestion, digestion, absorption of nutrients, and elimination of wastes.
II. An organism's body plan and lifestyle are adapted to its mode of nutrition.
   A. The simplest invertebrates, the sponges, have no digestive system; they digest food intracellularly.
   B. Cnidarians and flatworms have digestive systems with only one opening, which serves as both mouth and anus. Cnidarians have intracellular and extracellular digestion.
   C. In more complex invertebrates and in all vertebrates, the digestive tract is a complete tube with an opening at each end. Digestion is extracellular.
III. In vertebrates, various parts of the digestive tract are specialized to perform specific functions.
   A. Mechanical digestion and enzymatic digestion of carbohydrates begin in the mouth.
   B. As food is swallowed, it is propelled through the pharynx and esophagus. A bolus of food is moved along through the digestive tract by peristaltic action.

C. In the stomach, food is mechanically digested by vigorous churning, and proteins are enzymatically digested by the action of pepsin in the gastric juice.
D. Most enzymatic digestion takes place in the duodenum, which receives secretions from the liver and pancreas and produces several digestive enzymes of its own.
E. The liver produces bile, which emulsifies fats.
F. The pancreas releases enzymes that digest protein, lipid, and carbohydrate, as well as RNA and DNA.
G. Activities of the digestive system are regulated by both nerves and hormones.
H. Chyme is enzymatically digested as it moves through the digestive tract.
  1. Polysaccharides are digested to maltose by salivary and pancreatic amylases. Maltase in the small intestine splits maltose into glucose, the main product of carbohydrate digestion.
  2. Proteins are split by pepsin in the stomach and by proteolytic enzymes in the pancreatic juice. The dipeptides produced are then split by dipeptidases. The end products of protein digestion are amino acids.
  3. Lipids are emulsified by bile salts and then hydrolyzed by lipase in the pancreatic juice.
I. Most nutrients are absorbed through the thin walls of the intestine.
J. The large intestine is responsible for the elimination of undigested wastes. It also incubates bacteria that produce vitamin K and certain B vitamins.

IV. For a balanced diet, humans and other animals require carbohydrates, lipids, proteins, vitamins, and minerals.
A. Most carbohydrates are ingested in the form of polysaccharides—starch and cellulose.
  1. Carbohydrates are used mainly as fuel.
  2. Glucose concentration in the blood is carefully regulated. Excess glucose is stored as glycogen and can also be converted to fat.
B. Lipids are used as fuel, as components of cell membranes, and to synthesize steroid hormones and other lipid substances.
  1. Most lipids are ingested in the form of triacylglycerols.
  2. Cholesterol is transported by low-density lipoproteins (LDLs) and high-density lipoproteins (HDLs). High levels of LDL are associated with increased risk for heart disease.
  3. Fatty acids are converted to molecules of acetyl coenzyme A and used as fuel. Excess fatty acids are stored as fat.
C. Proteins serve as enzymes and are essential structural components of cells.
  1. The best distribution of essential amino acids is found in the complete proteins of animal foods.
  2. Excess amino acids are deaminated by liver cells. Amino groups are converted to urea and excreted in urine; the remaining keto acids are converted to carbohydrate and used as fuel, or converted to lipid and stored in fat cells.
D. Vitamins are organic compounds required in small amounts for many biochemical processes. Many serve as components of coenzymes.
E. Minerals are inorganic nutrients ingested as salts dissolved in food and water.

V. Basal metabolic rate (BMR) is the body's cost of metabolic living.
A. Total metabolic rate is the BMR plus the energy used to carry on daily activities.
B. When energy (Calorie) input equals energy output, body weight remains constant.

VI. Obesity is a serious nutritional problem in which an excess amount of fat accumulates in the adipose tissues. A person gains weight by taking in more energy, in the form of Calories, than is expended in activity.

VII. Millions of people suffer from malnutrition. Essential amino acids are the nutrients most often deficient in the diet.

SELECTED KEY TERMS

absorption
adventitia
amylase
basal metabolic rate (BMR)
bile
chylomicron
colon
dentin
digestion
duodenum
elimination
emulsification
enamel
essential amino acid
gallbladder
gastrin
ingestion
large intestine
liver
metabolic rate
mineral
molar
mucosa
nutrients
nutrition
pancreas
pepsin
peristalsis
peritoneum
pharynx
premolar
pulp cavity
rugae
salivary gland
small intestine
stomach
submucosa
villus (villi)
vitamin

POST-TEST

1. The process of taking food into the body is called __________ .
2. __________ consists of mechanically and enzymatically breaking down food into molecules small enough to be absorbed.
3. __________ is the process of getting rid of undigested and unabsorbed food.
4. The layer of epithelial and connective tissue that lines the lumen of the digestive tract in vertebrates is the __________ .

5. Teeth with flattened surfaces for crushing and grinding food are __________ and __________ .
6. Salivary __________ is an enzyme that initiates the digestion of starch to sugar.
7. A mammalian tooth consists mainly of __________ .
8. Protein digestion begins in the __________ .
9. Bacteria that produce vitamin K inhabit the __________ __________ .
10. Bile is produced by the __________ .
11. __________ is a hormone secreted by the stomach; it stimulates the gastric glands.
12. The surface area of the stomach is increased by the presence of __________ .
13. Food leaving the stomach next enters the __________ .
14. Absorption takes place mainly through finger-like __________ in the lining of the small intestine.
15. Many __________ function as components of coenzymes.
16. __________ are generally ingested in the form of salts.
17. __________ amino acids must be provided in the diet.
18. The body's rate of energy use during resting conditions is the __________ __________ __________ .

REVIEW QUESTIONS

1. If you were presented with an unfamiliar animal and asked to determine its nutritional lifestyle, how would you go about this task?
2. Compare the advantages of a two-opening digestive tract with those of a single-opening digestive tract.
3. How are digestive structures and methods of processing food in sponges, hydras, and flatworms adapted to each group's lifestyle? Give specific examples.
4. Why must food be digested?
5. Trace a bite of food through the human digestive tract, listing each structure through which it passes.
6. Give the functions of the three types of accessory glands that secrete digestive juices in vertebrates. Identify their secretions.
7. The inner lining of the digestive tract is not smooth like the inside of a water pipe. Why is this advantageous? What structures increase its surface area?
8. Summarize the step-by-step digestion of (a) carbohydrates; (b) lipids; (c) proteins.
9. What happens to ingested cellulose in humans? Why?
10. Draw and label an intestinal villus.
11. How does the absorption of fat differ from the absorption of glucose?
12. List the nutrients that must be included in a balanced diet. What are some of the difficulties in planning a nutritionally balanced vegetarian diet?
13. Why, specifically, is each of the following essential? (a) iron; (b) calcium; (c) iodine; (d) vitamin A; (e) vitamin K; (f) essential amino acids.
14. Draw a diagram to illustrate the fate of each of the following in the body: (a) glucose; (b) absorbed amino acids; (c) absorbed fat.
15. Write an equation to describe energy balance and explain what happens when the equation is altered in either direction.
16. Summarize the relationship between diet and heart disease.
17. Label the diagram.

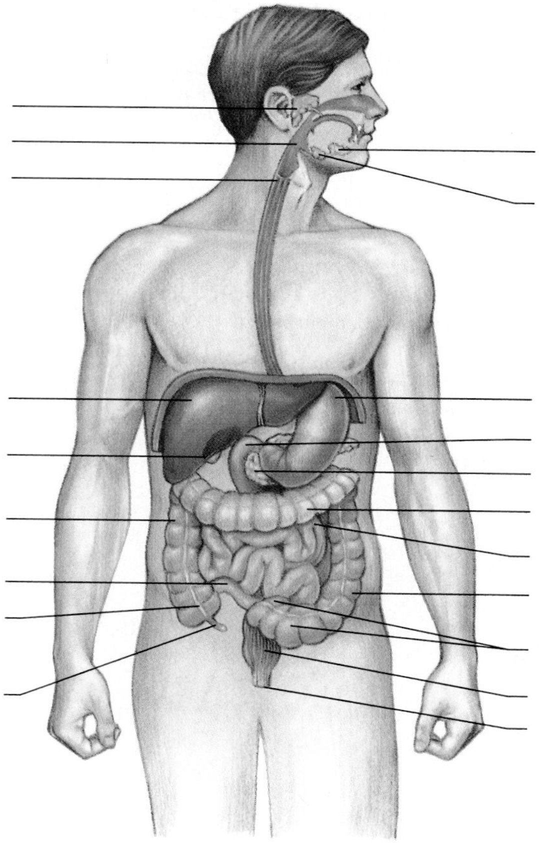

YOU MAKE THE CONNECTION

1. What is the adaptive advantage of the specialization of different regions in a complete digestive tract?
2. A high percentage of adults suffer from gastrointestinal discomfort, including cramps and diarrhea, when they drink milk. What do you think could cause this condition, known as lactose intolerance?
3. Design an experiment to demonstrate the nutritional need for the B vitamin pyridoxine.
4. Why are proteolytic enzymes produced in an inactive form?

RECOMMENDED READINGS

Brown, L. R. et al. *State of the World 1994*. W. W. Norton & Company, New York, 1994. A Worldwatch Institute Report on environmental problems and progress. Includes an analysis of our ability to provide adequate food for our increasing population.

Lawn, R. M. "Lipoprotein(a) in Heart Disease." *Scientific American*, Vol. 266, No. 6, June 1992. Lipoprotein(a) may increase heart attack risk by binding to arterial plaques and to blood clots.

Sanderson, S. L., and R. Wassersug. "Suspension-Feeding Vertebrates." *Scientific American*, Vol. 262, No. 3, March 1990. Animals that filter food out of the surrounding water can grow to large sizes.

Scrimshaw, N. S. "Iron Deficiency." *Scientific American*, Vol. 265, No. 4, October 1991. Iron deficiency is one of the most common and most treatable health problems.

Smolin, L. A., and M. B. Grosvenor. *Nutrition: Science and Applications*. Saunders College Publishing, Philadelphia, 1994. A readable introduction to nutritional science.

Willett, W. C. "Diet and Health: What Should We Eat?" *Science*, April 22, 1994. This article summarizes what is known about the connection between diet and disease.

C H A P T E R 4 6

Osmoregulation, Disposal of Metabolic Wastes, and Temperature Regulation

Organisms must regulate fluid intake in order to maintain fluid balance.
(Frans Lanting/Minden Pictures)

Water shapes life and the distribution of organisms on Earth. Most organisms consist mainly of water, the medium in which most metabolic reactions take place. The simplest animals are small organisms that live in the ocean. They obtain their food and oxygen directly from the seawater that surrounds them, and they release waste products into it. Larger, more complex animals and most terrestrial animals have their own internal sea—the blood and interstitial fluid—to bathe their cells and to dissolve and transport nutrients, gases, and waste products.

Terrestrial animals have a continuous need to conserve water. Water must be taken into the animal body and its loss must be carefully regulated. Water is ingested with food and drink and is also formed in metabolic reactions. Some of this water becomes part of the blood plasma, which transports materials throughout the organism. Then, interstitial fluid forms from the blood plasma and bathes all of the cells of the body. Excess water evaporates from the body surface or is excreted by specialized structures.

The composition of the body fluids must also be kept within limits that the animal can tolerate. Two processes that maintain homeostasis of fluids are osmoregulation and excretion of metabolic wastes. **Osmoregulation** is the active regulation of osmotic pressure of body fluids to keep them from becoming too dilute or too concentrated. **Excretion** is the process of ridding the body of metabolic wastes, including water. Many animals have evolved efficient **excretory systems** that handle these processes and also rid the body of excess water from dietary intake; excess ions; and harmful substances.

Like water, climate is important in determining the distribution of organisms on our planet. Organisms can survive only within certain temperature ranges. Body temperature in *ectotherms* depends largely on the temperature of the environment. In *endotherms*, body temperature is determined more by metabolic activity and by homeostatic mechanisms that regulate heat production and heat loss.

LEARNING OBJECTIVES

After you have studied this chapter you should be able to

1. Relate the principal functions of excretory systems to specific osmoregulatory challenges posed by various environments.
2. Contrast the advantages and disadvantages of excreting ammonia, uric acid, or urea.
3. Compare nephridial organs and Malpighian tubules as osmoregulatory organs.
4. Relate the function of the vertebrate kidney to the success of vertebrates in a wide variety of habitats.
5. Compare the adaptations that freshwater fishes have evolved to solve their challenges of osmoregulation with those of marine bony fishes, sharks, and marine mammals.
6. Label on a diagram the organs of the mammalian urinary system, and give the functions of each.
7. Identify on a diagram the principal parts of a nephron (including associated blood vessels), and give the functions of each structure.
8. Trace a drop of filtrate from Bowman's capsule to its release from the body as urine.
9. Describe the regulatory effects of ADH and aldosterone.
10. Describe the strategies used by ectotherms and endotherms to maintain body temperature.

EXCRETORY SYSTEMS HELP MAINTAIN HOMEOSTASIS

Excretory systems maintain homeostasis by selectively adjusting the concentrations of salts and other substances in blood and other body fluids. Typically, an excretory system collects fluid, generally from the blood or interstitial fluid. It then adjusts the composition of this fluid by selectively returning needed substances to the body fluid. Finally, the adjusted excretory product (urine, for example) containing excess or potentially toxic substances is released from the body.

The terms excretion and elimination are sometimes confused (Fig. 46–1). Food materials that have not been absorbed are eliminated from the body in the feces. Such substances never participated in the organism's metabolism or entered body cells, but merely passed through the digestive system.

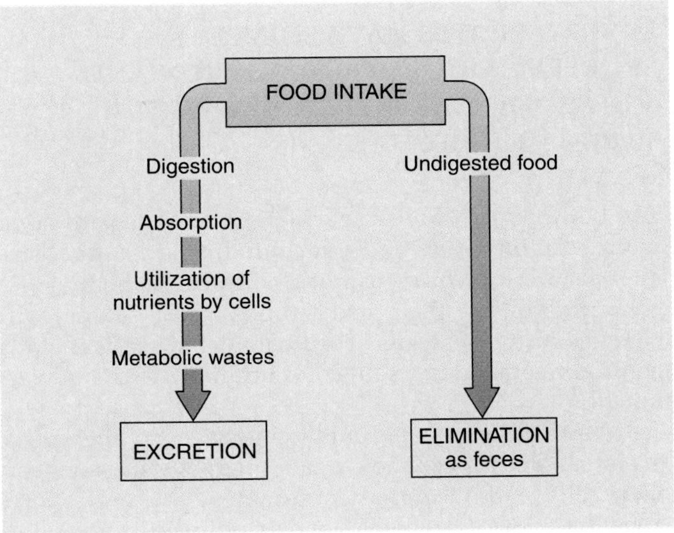

FIGURE 46–1 Excretion is the disposal of metabolic wastes. Excretion can be contrasted with elimination.

THE PRINCIPAL METABOLIC WASTE PRODUCTS ARE WATER, CARBON DIOXIDE, AND NITROGENOUS WASTES

Metabolic wastes must be excreted so that they do not accumulate and reach concentrations that would disrupt homeostasis. The principal metabolic waste products in most animals are water, carbon dioxide, and nitrogenous (nitrogen-containing) wastes. Carbon dioxide is excreted mainly by respiratory structures (see Chapter 44). Excess water is also lost from respiratory surfaces in terrestrial animals. Excretory organs, such as kidneys, remove and excrete most of the water and nitrogenous wastes.

Nitrogenous wastes include ammonia, uric acid, and urea. Recall that amino acids and nucleic acids contain nitrogen. During the breakdown of amino acids, the nitrogen-containing amino group is removed (in a process known as deamination) and converted to **ammonia** (Fig. 46–2). However, ammonia is highly toxic. Some aquatic animals excrete it into the surrounding water before it can build up to toxic concentrations in their tissues. A few terrestrial animals vent it directly into the air. But in many organisms, humans included, ammonia is converted to some less toxic nitrogenous waste such as uric acid or urea.

Uric acid is produced both from ammonia and by the breakdown of nucleotides from nucleic acids. Uric acid forms crystals and can be excreted as a crystalline paste with little fluid loss. This is an important water-

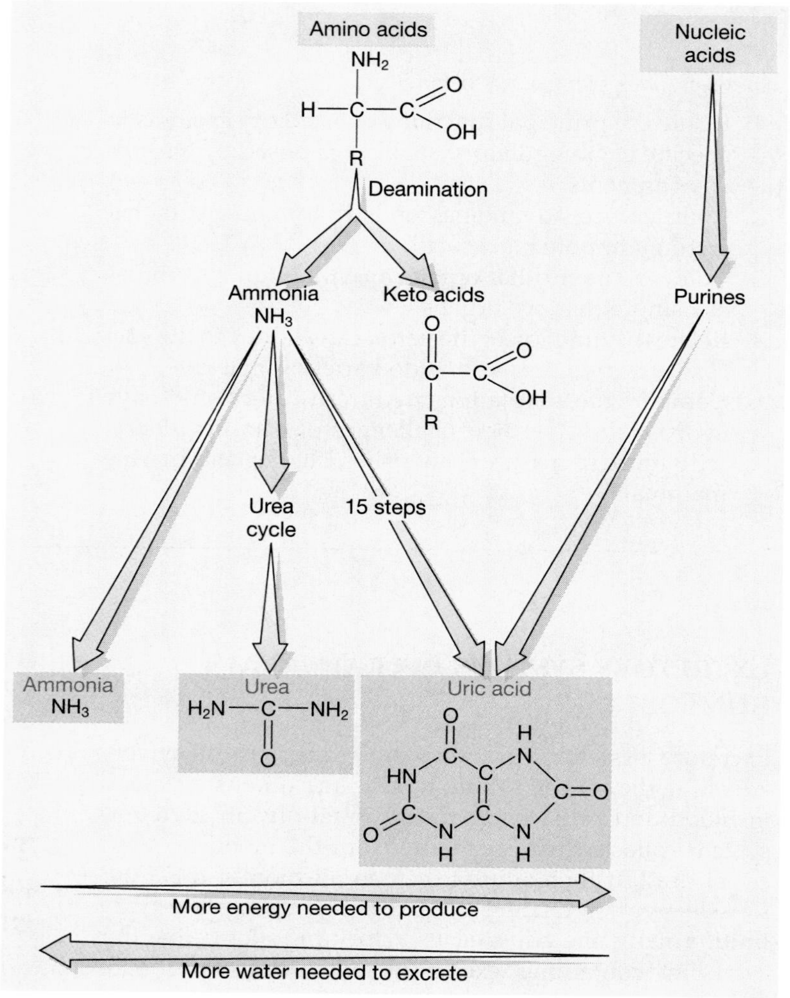

FIGURE 46–2 Nitrogenous wastes are formed by deamination of amino acids and by metabolism of nucleic acids. Ammonia is the first metabolic product of deamination. In many animals, ammonia is converted to urea via the urea cycle. Other animals convert ammonia to uric acid. Energy is required to convert ammonia to urea and uric acid, but less water is required to excrete these wastes.

conserving adaptation in many terrestrial animals, including insects, certain reptiles, and birds. In birds, the absence of a urinary bladder and the frequent excretion of uric acid as part of the feces contribute to the light body weight essential for flight. Also, because uric acid is not toxic and can be safely stored, its excretion is an adaptive advantage for species whose young begin their development enclosed in eggs.

Urea is the principal nitrogenous waste product of amphibians and mammals. It is produced in the liver from ammonia. The sequence of reactions by which urea is synthesized from ammonia and carbon dioxide is known as the **urea cycle.** Like the formation of uric acid, these reactions require specific enzymes and the input of energy by the cells. Urea has the advantage of being far less toxic than ammonia and can accumulate in higher concentrations without causing tissue damage; thus, it can be excreted in more concentrated form. Because urea is highly soluble, it is dissolved in water. More water is needed to excrete urea than to excrete uric acid.

INVERTEBRATES HAVE SOLVED PROBLEMS OF OSMOREGULATION AND METABOLIC WASTE DISPOSAL WITH A VARIETY OF MECHANISMS

The body fluids of most marine invertebrates are in osmotic equilibrium with the surrounding seawater. These animals are known as **osmotic conformers** because the concentration of their body fluids varies along with changes in the seawater. Fortunately, the sea is a stable environment and its salt concentration does not vary much.

Marine sponges and cnidarians need no specialized excretory structures. Their wastes pass by diffusion from their cells to the external environment and are washed away by water currents. They expend little or no energy to excrete wastes. When water becomes stagnant and currents do not wash away wastes, aquatic environments, such as coral reefs, are damaged by their accumulation.

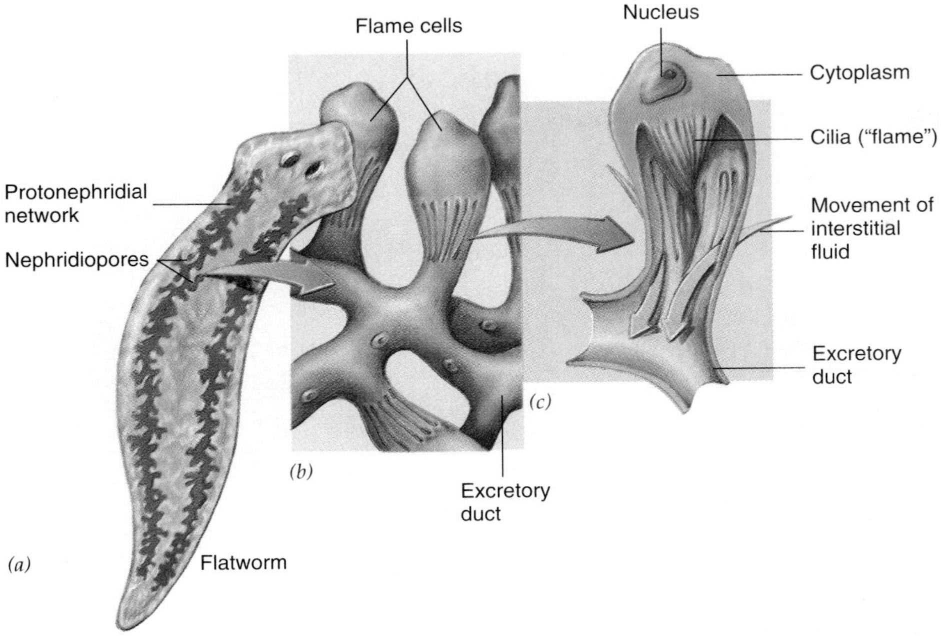

FIGURE 46–3 Nephridial organs are common among invertebrates. (*a*) The excretory organs of a typical flatworm. (*b*) Wastes collected by flame cells pass through excretory ducts of the animal through nephridiopores. (*c*) A single flame cell.

Coastal habitats, such as estuaries that contain brackish water, are much less stable environments than is the open sea. Salt concentrations change frequently with shifting tides. Animals that inhabit these environments are **osmotic regulators.** In a coastal environment where fresh water enters the sea, the water may have a lower salt concentration than the body fluids of the animal. Water osmotically moves into the body, and salt diffuses out. An animal adapted to this environment has excretory structures that actively remove the excess water. Many also have cells in their gills that remove salts from the surrounding water and transport them into the body fluids.

Terrestrial animals have a higher fluid concentration than does the air around them. They tend to lose water by evaporation from the body surface and from respiratory surfaces. They may also lose water as wastes are excreted. Adaptation to life on land has required the evolution of structures and processes that conserve water.

Nephridial organs are specialized for osmoregulation and/or excretion

A **nephridial organ** is a specialized excretory structure that has evolved in many invertebrates. It consists of simple or branching tubes that usually open to the outside of the body through excretory pores (nephridiopores).

In flatworms and nemerteans, metabolic wastes pass through the body surface by diffusion. However, these animals have specialized osmoregulatory nephridial organs, called **protonephridia,** composed of tubules with no internal openings. Their enlarged blind ends consist of **flame cells** with brushes of cilia (whose constant motion reminded early biologists of flickering flames—hence the name) (Fig. 46–3). The flame cells lie in the fluid that bathes the body cells. Fluid enters the flame cells, passes through a system of tubules and excretory ducts, and leaves the body through excretory pores. A variety of other invertebrates, including rotifers, some annelids, and lancelets, have protonephridia.

Most annelids, as well as mollusks, have more complex nephridial organs called **metanephridia** (sing. *metanephridium*). Every segment of an earthworm has a pair of metanephridia. Each metanephridium is a tubule open at both ends. The inner end opens into the coelom as a ciliated funnel (Fig. 46–4), and the outer end opens

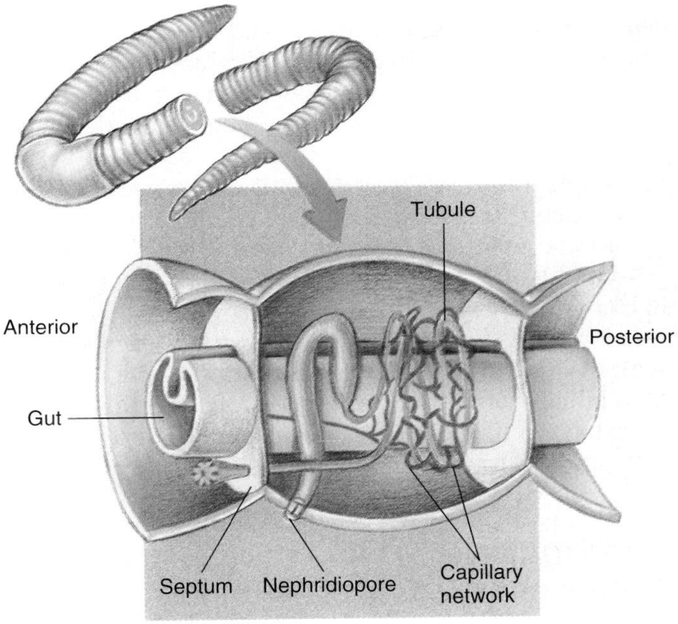

FIGURE 46–4 The earthworm has a series of paired metanephridia. Each consists of a ciliated funnel opening into the coelom, a coiled tubule, and a nephridiopore opening to the outside. This three-dimensional internal view shows parts of three segments.

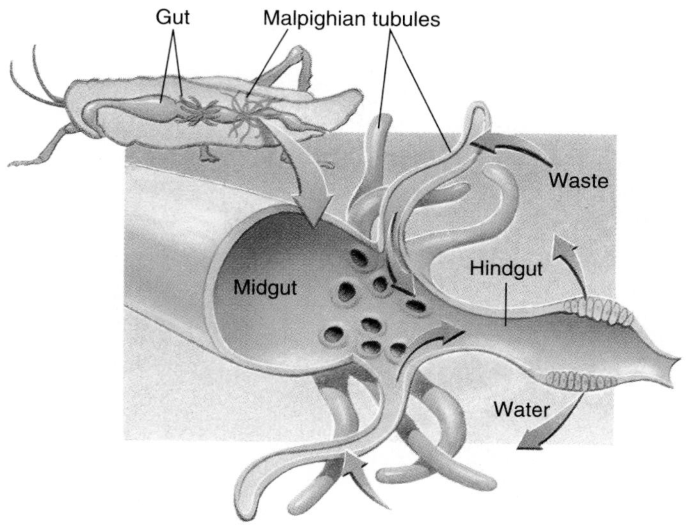

FIGURE 46–5 The slender Malpighian tubules of insects have blind ends located in the hemocoel. Their cells transfer wastes from the hemolymph to the cavity of the tubule. Uric acid, the major waste product, is discharged into the gut.

to the outside through an excretory pore. Around each tubule is a network of capillaries. Fluid from the coelom (body cavity) passes into the tubule, bringing with it whatever it contains—glucose, salts, or wastes. As the fluid moves through the tubule, needed materials (such as water and glucose) are reabsorbed by the capillaries, leaving the wastes behind. In this way, urine is produced that contains concentrated wastes.

Malpighian tubules are an important adaptation for conserving water in insects

The excretory system of insects and spiders consists of **Malpighian tubules** (Fig. 46–5). Two hundred to several hundred tubules may be present, depending on the species. Malpighian tubules have blind ends that lie in the hemocoel (blood cavity), where they are bathed in hemolymph. Their cells transfer salts and wastes by diffusion or active transport from the hemolymph to the cavity of the tubule. The Malpighian tubules empty into the intestine. Water and some salts are reabsorbed into the hemolymph by specialized rectal glands.

Uric acid, the major waste product, is excreted as a semi-dry paste with a minimum of water. Because Malpighian tubules conserve body fluids, these structures have contributed significantly to the success of insects in terrestrial environments.

THE KIDNEY IS THE KEY VERTEBRATE ORGAN OF OSMOREGULATION AND EXCRETION

Vertebrates live successfully in a wide range of habitats—in fresh water, the ocean, tidal regions, and on land, even in extreme environments such as deserts. In response to the requirements of these diverse environments, vertebrates have evolved adaptations for regulating their salt and water content and for excreting wastes. An extreme example is the desert-dwelling kangaroo rat, which must carefully conserve water. It obtains most of its water from its own metabolism, and its kidneys are so efficient that it loses little fluid as urine.

The main osmoregulatory and excretory organ in vertebrates is the **kidney.** In most vertebrates, the skin, lungs or gills, and digestive system also help maintain fluid balance and dispose of metabolic wastes. Some reptiles and birds have salt glands in their heads to excrete the salt that enters their bodies with the seawater they drink.

Freshwater vertebrates must rid themselves of excess water

As fishes began to move into freshwater habitats about 460 million years ago, there was strong selection for the evolution of adaptations for effective osmoregulation. Because the body fluids of freshwater animals have higher salt concentrations than (are hypertonic to) their surroundings, water passes into them osmotically. As a result, they are in constant danger of becoming waterlogged. Freshwater fishes are covered by scales and a mucous secretion that retards the passage of water into the body. However, water enters through the gills. The kidneys of these fishes have become adapted to filter out excess water, and they excrete large amounts of dilute urine (Fig. 46–6).

Water entry, though, is only part of the challenge of osmoregulation in freshwater fishes. These animals also tend to lose salts to the surrounding fresh water. To compensate, special gill cells have evolved that actively trans-

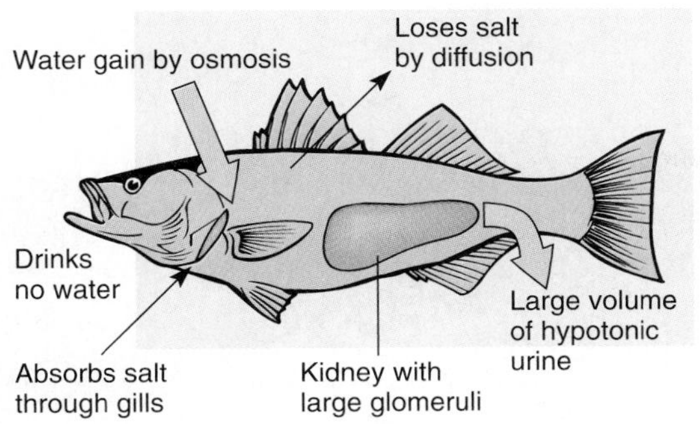

(*a*) Freshwater fish

FIGURE 46–6 Osmoregulation in fishes involves a variety of mechanisms. (*a*) Freshwater fishes live in a hypotonic medium, so water continuously enters the body by osmosis, while salts leave by diffusion. These fishes excrete large quantities of dilute urine, and actively absorb salts through the gills. (*b*) Marine fishes live in a hypertonic medium, and therefore lose water by osmosis. They gain salts from the water they drink and also by diffusion. To compensate, the fish drinks the water, excretes the salt, and produces a small volume of urine. (*c*) The shark solves its osmotic problem differently. It accumulates urea in high enough concentration so that it becomes hypertonic to the surrounding medium. As a result, water enters its body by osmosis. A large quantity of hypotonic urine is excreted.

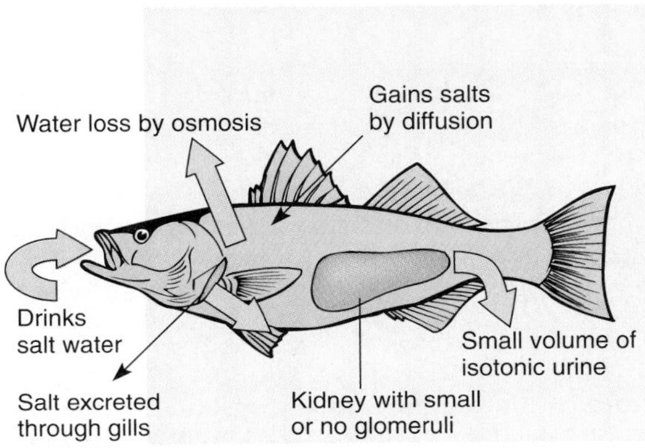

(*b*) Marine bony fish

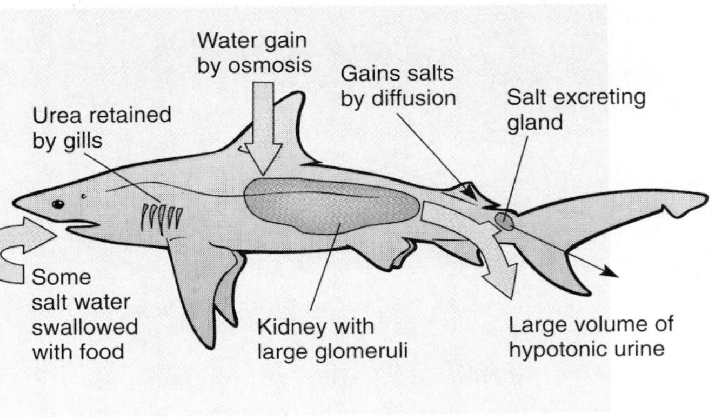

(*c*) Shark

port salts (mainly sodium chloride) from the water into the body.

Most amphibians are at least semiaquatic, and their mechanisms of osmoregulation are similar to those of freshwater fishes. They, too, produce large amounts of dilute urine. For example, through its urine and skin, a frog can lose an amount of water equivalent to one third of its body weight in one day. Active transport of salt inward by special cells in the skin compensates for loss of salt through skin and urine.

Marine vertebrates must replace lost fluid

Freshwater fishes have adapted very successfully to their aquatic habitats. Evolution of body fluids more dilute than seawater is one of their chief adaptations. Thus, when some freshwater fishes returned to the sea about 200 million years ago, their blood and body fluids were less salty than (hypotonic to) their surroundings. They tended to lose water osmotically and to take in salt. To compensate for fluid loss, many marine bony fishes drink seawater (Fig. 46–6). They retain the water and excrete salt by the action of specialized cells in their gills. Very little urine is excreted by the kidneys; the nephrons (microscopic units of the kidney) have only small or no capillary clusters (glomeruli), which filter the blood in other vertebrates.

Marine chondrichthyes (sharks and rays) have different osmoregulatory adaptations that allow them to tolerate the salt concentrations of their environment. These animals accumulate and tolerate urea (Fig. 46–6). Their tissues are adapted to function at concentrations of urea that would be toxic to most other animals. The high urea concentration makes the body fluids slightly hypertonic to seawater. This results in a net inflow of water into their bodies. Their well-developed kidneys excrete a large volume of urine. Excess salt is excreted by the kidneys and, in many species, by a rectal gland.

The heads of certain reptiles and marine birds have salt glands, which can excrete the salt that has entered the body with ingested seawater (Fig. 46–7). Salt glands are usually inactive; they function only in response to osmotic stress. Thus, only when seawater or salty food is ingested do the salt glands excrete a fluid laden with sodium and chloride.

Whales, dolphins, and other marine mammals ingest seawater along with their food. Their kidneys produce a very concentrated urine, much saltier than seawater. This

(a)

(b)

FIGURE 46–7 Marine animals ingest seawater along with their food. (***a***) Marine birds have salt glands that excrete excess salt. This brown pelican (*Pelecanus occidentalis*) was photographed on Sanibel Island, Florida. (***b***) The killer whale (*Orcinus orca*) rids its body of excess salt by producing a very concentrated urine, which is much more salty than seawater. (*a*, Stan Osolinski/Dembinsky Photo Associates; *b*, Ken Lucas/Biological Photo Service)

is an important physiological adaptation, especially for marine carnivores (Fig. 46–7). The high-protein diet of these animals results in production of large amounts of urea, which must be excreted in the urine or, in some cases, by special accessory salt glands.

The mammalian kidney is important in maintaining homeostasis

In mammals, the kidneys are the principal excretory organs. They are responsible for the excretion of most nitrogenous wastes, and for helping to maintain fluid balance by adjusting the salt and water content of the urine. As in other terrestrial vertebrates, the lungs, skin, and digestive system are also important in mammalian osmoregulation and waste disposal (Fig. 46–8). Most carbon dioxide and a great deal of water are excreted by the lungs. Although primarily concerned with the regulation of body temperature, the sweat glands excrete 5% to 10% of all metabolic wastes.

Most of the bile pigments produced by the breakdown of red blood cells are normally excreted by the liver into the intestine. From the intestine they then pass out of the body with the feces. The liver also produces both urea and uric acid.

THE KIDNEYS, URINARY BLADDER, AND THEIR DUCTS MAKE UP THE URINARY SYSTEM

The mammalian **urinary system** consists of the kidneys, the urinary bladder, and associated ducts. The overall structure of the human urinary system is shown in Figure 46–9. Located just below the diaphragm in the "small of the back," the kidneys look like a pair of giant, dark-red lima beans, each about the size of a fist. The outer portion of the kidney is called the **renal cortex;** the inner portion is the **renal medulla** (Fig. 46–10). The renal medulla contains a number (usually eight to ten) of cone-shaped structures called **renal pyramids.** The tip of each pyramid is a renal papilla. Each papilla has several pores, the openings of collecting ducts.

As urine is produced, it flows from collecting ducts through a renal papilla, and into the **renal pelvis,** a funnel-shaped chamber. Urine then flows into one of the paired **ureters,** ducts that connect the kidney with the **urinary bladder.** The urinary bladder is a remarkable organ capable of holding (with practice) up to 800 milliliters (about a pint and a half) of urine. Emptying the bladder changes it from the size of a small melon to that of a pecan. This feat is made possible by the smooth muscle and special epithelium of the bladder wall, which is capable of great shrinkage and stretching.

During **urination,** urine is released from the bladder and flows through the **urethra,** a duct leading to the outside of the body. In the male, the urethra is lengthy and passes through the penis. Semen, as well as urine, is transported through the male urethra. In the female, the urethra is short and transports only urine. Its opening to the outside is just above the opening of the vagina. The length of the male urethra discourages bacterial invasions of the bladder. This length difference helps explain why bladder infections are more common in females than in males.

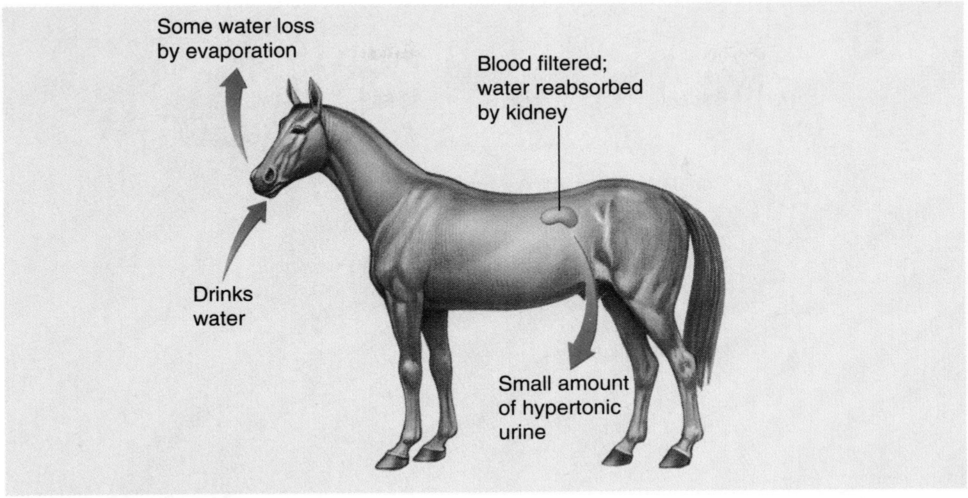

(a) Terrestrial mammal

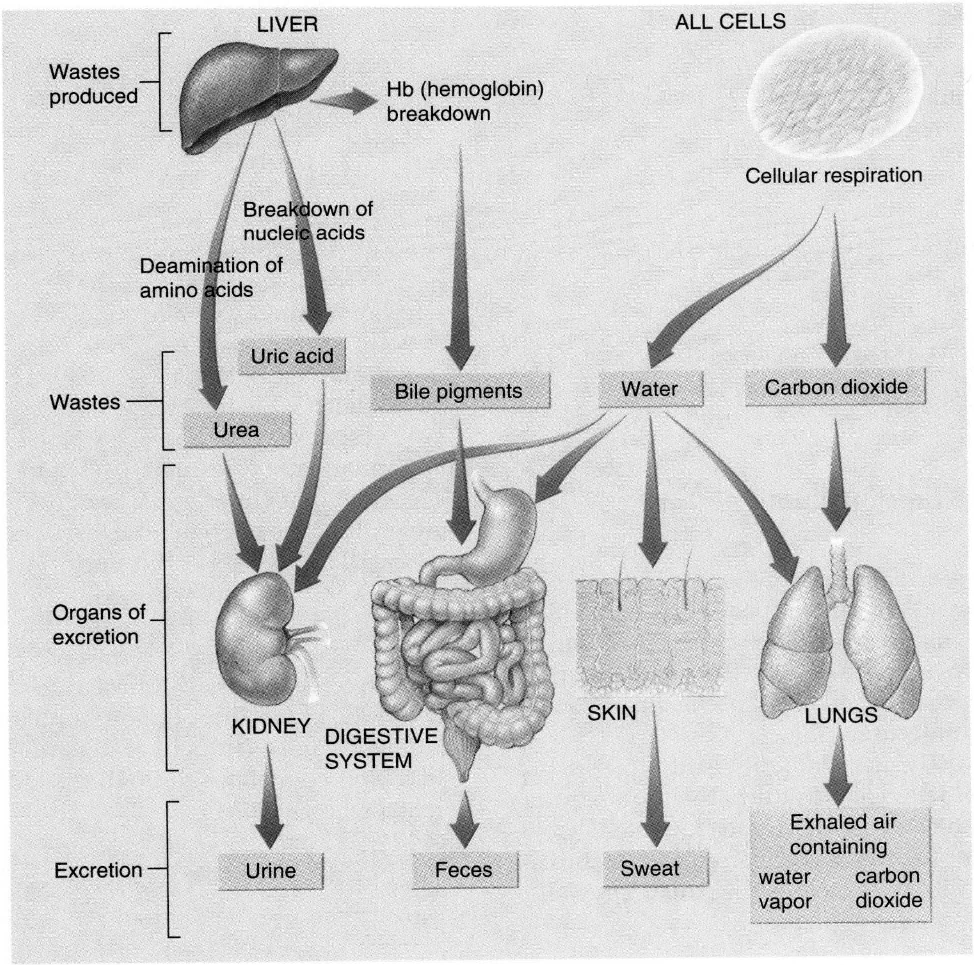

(b)

FIGURE 46–8 In many terrestrial vertebrates, the kidneys, lungs, skin, and digestive system all participate in the disposal of metabolic wastes. ***(a)*** The vertebrate kidney conserves water by reabsorbing it. ***(b)*** Disposal of metabolic wastes in humans and other terrestrial mammals. To conserve water, a small amount of hypertonic urine is usually produced. Nitrogenous wastes are produced by the liver and transported to the kidneys. All cells produce carbon dioxide and some water during cellular respiration.

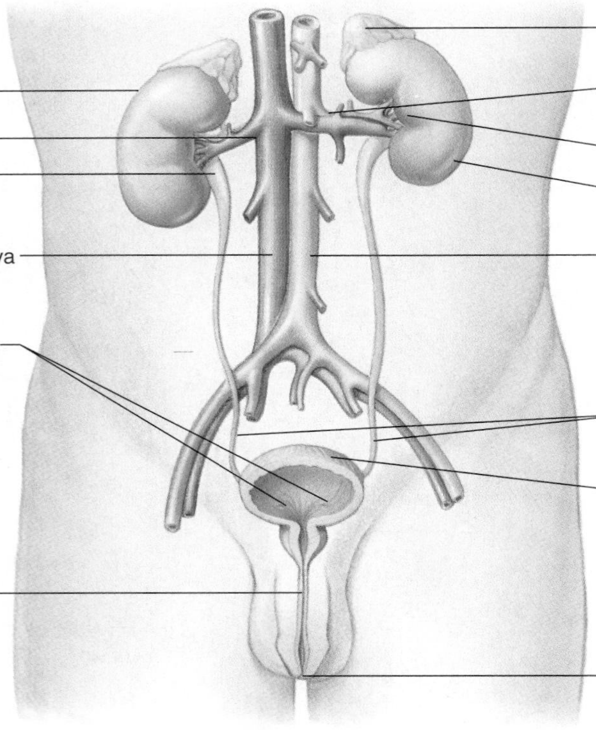

FIGURE 46–9 The human urinary system includes the kidneys, urinary bladder, and associated ducts. The kidneys produce urine, which is conveyed by the ureters to the urinary bladder for temporary storage. The urethra then conducts urine from the bladder to the outside of the body.

In summary, urine flows through the following structures:

> kidney (through renal pelvis) → ureter → urinary bladder → urethra

The nephron is the functional unit of the kidney

Each kidney consists of more than one million functional units called **nephrons.** A nephron consists of a cup-like **Bowman's capsule** connected to a long, partially coiled **renal tubule** (Figs. 46–10 and 46–11). Positioned within the cup-shaped Bowman's capsule is a cluster of capillaries known as a **glomerulus.**

Three main regions of the renal tubule are the **proximal convoluted tubule,** which conducts the filtrate from Bowman's capsule; the **loop of Henle,** an elongated, hairpin-shaped portion; and the **distal convoluted tubule,** which conducts the filtrate to a **collecting duct.** Thus, filtrate passes through the following structures:

> Bowman's capsule → proximal convoluted tubule → loop of Henle → distal convoluted tubule → collecting duct

Blood is delivered to the kidney by the renal artery. Small branches of the renal artery give rise to **afferent arterioles.** (Afferent means "to carry toward.") An afferent arteriole conducts blood into the capillaries that make up each glomerulus. As blood flows through the glomerulus, some of the plasma is forced into Bowman's capsule.

You may recall that in a usual circulatory route, capillaries deliver blood into veins. Circulation in the kidneys is an exception because blood flowing from the glomerular capillaries next passes into an **efferent arteriole.** Each efferent arteriole conducts blood *away* from a glomerulus. The efferent arteriole delivers blood to a second capillary network, the **peritubular capillaries** surrounding the renal tubule.

As blood flows through the first set of capillaries, those of the glomerulus, it is filtered. The peritubular capillaries receive materials returned to the blood by the renal tubule. Blood from the peritubular capillaries enters small veins that eventually lead to the renal vein. In summary, blood circulates through the kidney in the following sequence:

> renal artery → afferent arteriole → capillaries of glomerulus → efferent arteriole → peritubular capillaries → small veins → renal vein

Urine is produced by filtration, reabsorption, and secretion

Urine is produced by a combination of three processes: filtration, reabsorption, and tubular secretion.

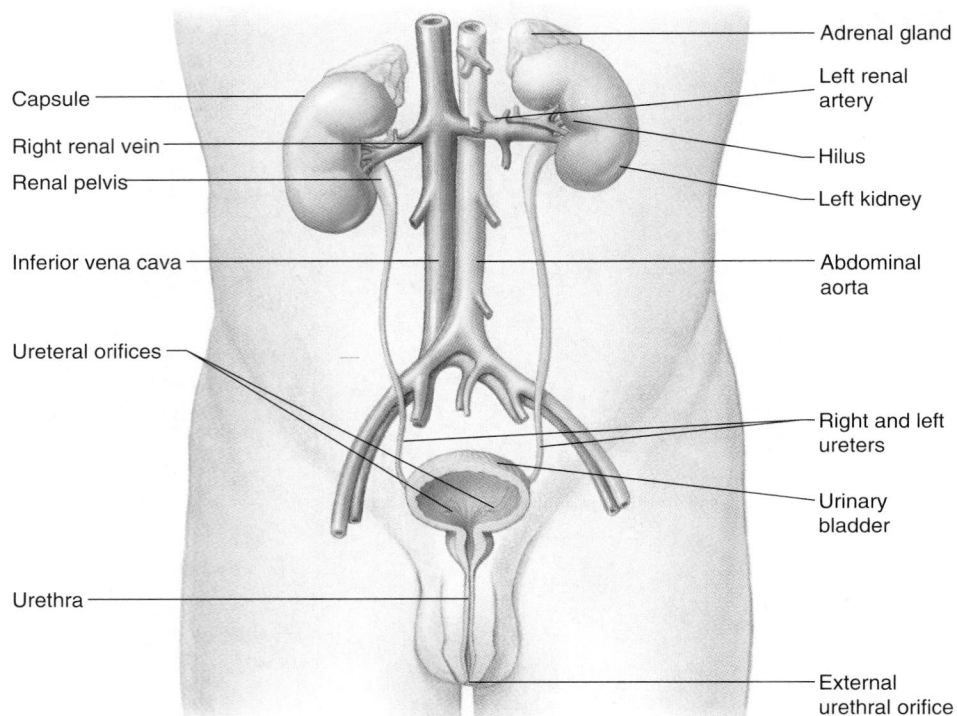

Making the Connection

Disposal of Metabolic Wastes and Neural Control

How does the nervous system affect the body's disposal of metabolic wastes? In mammals, the release of metabolic wastes by the urinary system is regulated by the nervous system. Urination, or the release of urine from the body, is a reflex action. In humans, two sphincter valves (rings of muscle) keep the opening of the urethra closed. When the volume of urine in the bladder reaches about 300 mL, receptors in the bladder wall are stretched sufficiently to stimulate the urination reflex. The sphincter muscles relax, allowing urination to occur.

The sphincter muscles are not under voluntary nervous control in the sense that our skeletal muscles are voluntary. What we call bladder control depends on the ability to facilitate or inhibit this reflex voluntarily. For example, we can voluntarily empty the bladder at a convenient time before it is full. On the other hand, even when the volume of urine in the bladder has exceeded 300 mL, we can inhibit urination until a more convenient time. Such voluntary control cannot be exerted by an immature nervous system. Most babies cannot develop urinary control until at least age two, no matter how hard anxious parents try to teach them. ▲

Filtration is not selective with regard to ions and small molecules

Blood flows through the glomerular capillaries under high pressure, forcing more than 10% of the plasma out of the capillaries and into Bowman's capsule (Fig. 46–12). **Filtration** is somewhat similar to the mechanism whereby tissue fluid is formed as blood flows through other capillary networks in the body. However, much more plasma is filtered in the kidney.

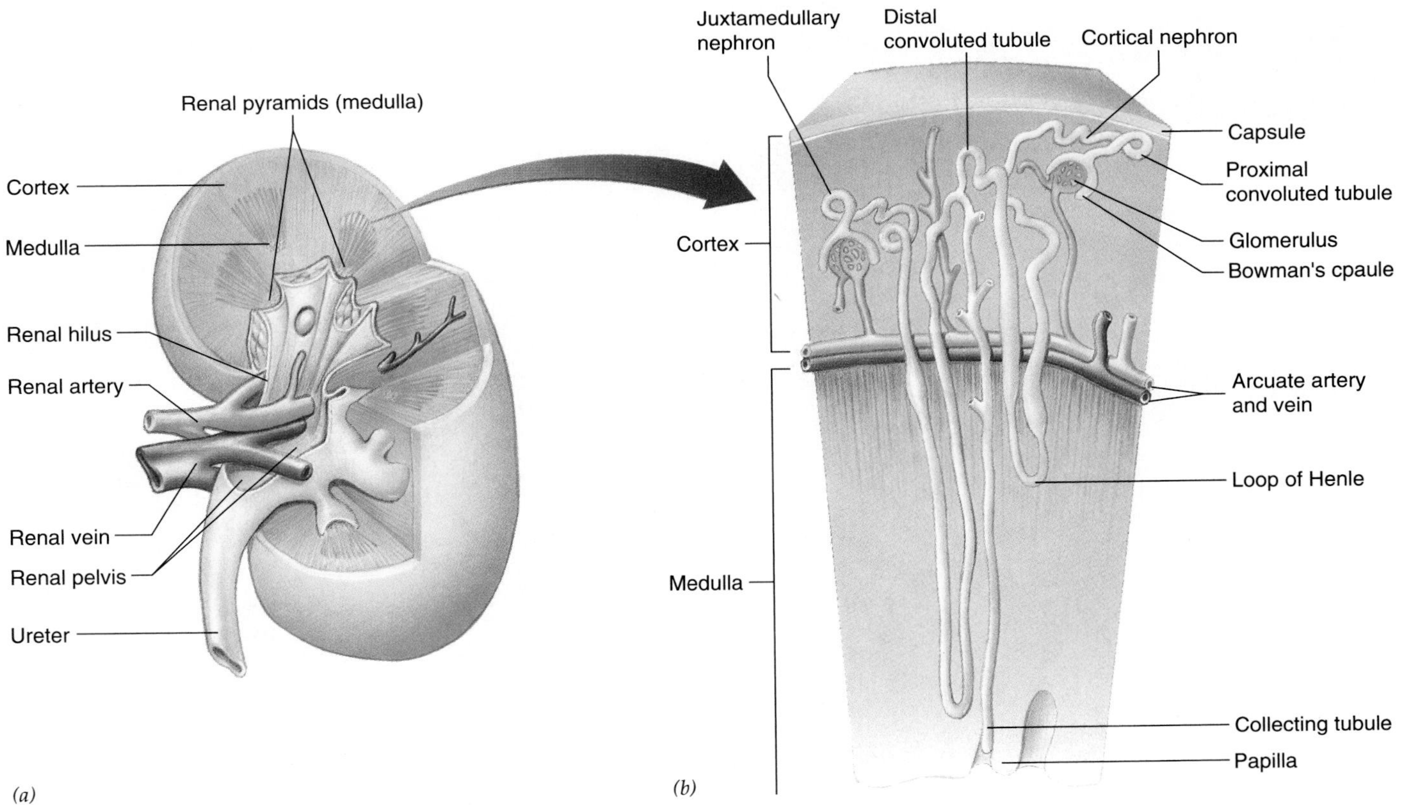

FIGURE 46–10 The kidney has a complex structure. (*a*) The outer region of the kidney is the cortex; the inner region is the medulla. When urine is produced, it flows into the renal pelvis and leaves the kidney through the ureter. The renal artery delivers blood to the kidney; the renal vein drains blood away from the kidney. (*b*) The location of the two main types of nephrons. Longitudinal view.

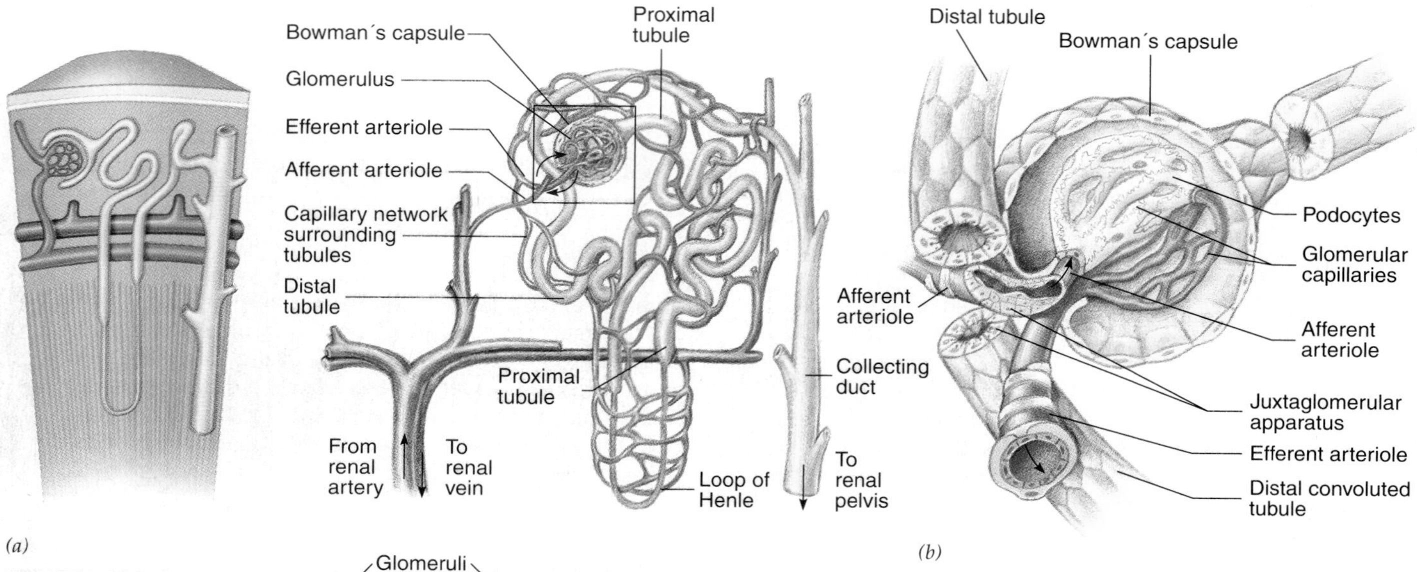

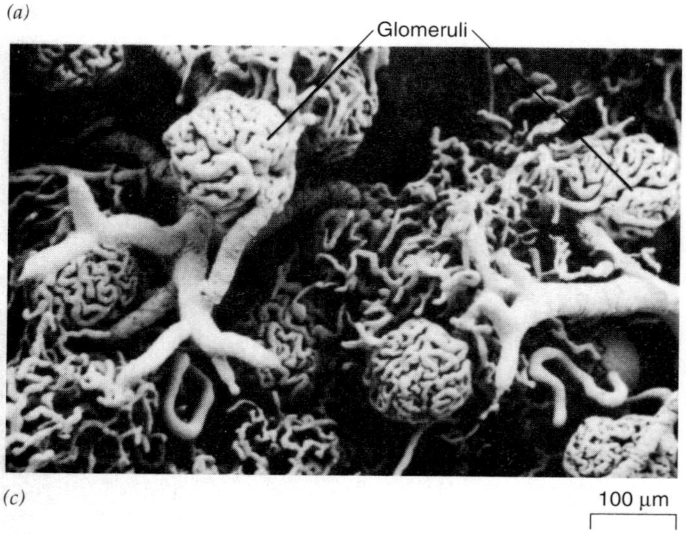

FIGURE 46–11 Each kidney is composed of more than a million microscopic nephrons. (*a*) The location and the basic structure of a nephron. (*b*) Detailed view of Bowman's capsule. Note that the distal convoluted tubule is actually adjacent to the afferent arteriole. The juxtaglomerular apparatus is a small group of cells located in the walls of the tubule and arterioles. (*c*) Low-power scanning electron micrograph of a portion of the kidney cortex, showing glomeruli and associated blood vessels. Urine forms by filtration from the blood in the glomeruli and by adjustment of the filtrate as it passes through the tubules that drain the glomeruli. (*c*, CNRI/Science Photo Library/Photo Researchers, Inc.)

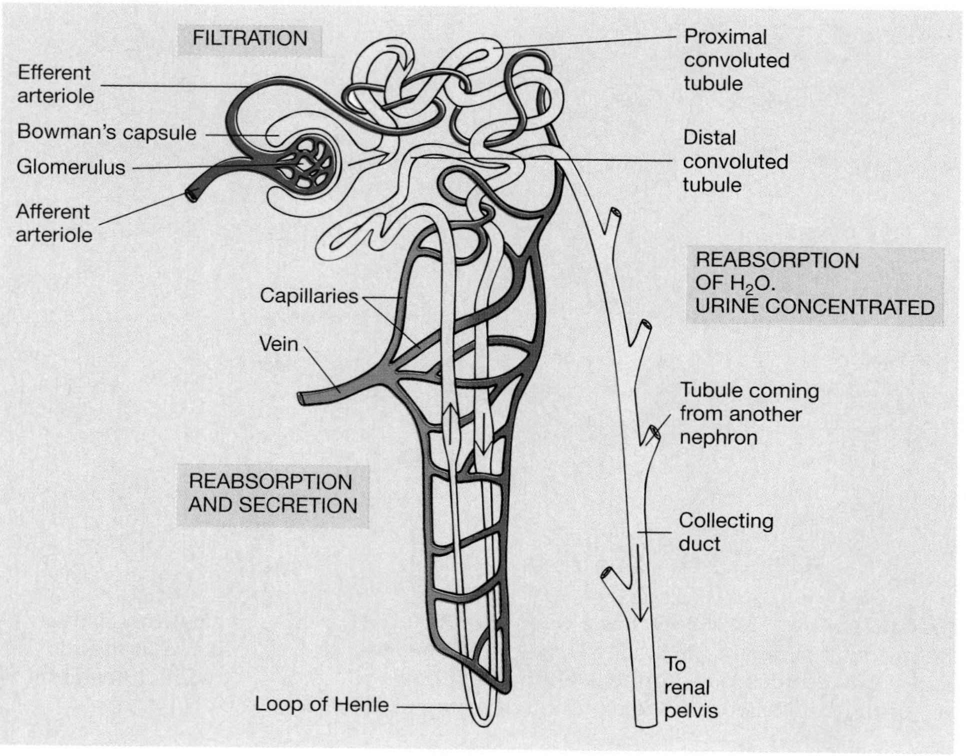

FIGURE 46–12 Filtration, reabsorption, and secretion take place in the nephron.

Several factors contribute to filtration. First, the hydrostatic pressure in the glomerular capillaries is higher than in other capillaries. This high pressure is mainly due to the high resistance to outflow presented by the efferent arteriole, which is smaller in diameter than the afferent arteriole. A second factor that contributes to the large amount of **glomerular filtrate** is the large surface area for filtration provided by the highly coiled glomerular capillaries. A third factor is the great permeability of the glomerular capillaries. Numerous small pores (fenestrations) are present between the squamous cells that form the walls of the glomerular capillaries, making these vessels more porous than typical capillaries.

The permeable glomerular capillary walls and specialized epithelial cells called **podocytes** make up a **filtration membrane.** The podocytes, which make up the wall of Bowman's capsule in contact with the capillaries, have numerous elongated branches called foot processes. These branches cover the surfaces of most of the glomerular capillaries (Fig. 46–13). Foot processes of adjacent podocytes are separated by narrow gaps called **filtration slits (slit pores).**

The filtration membrane holds back blood cells, platelets, and most of the plasma proteins. Fluid and small solutes dissolved in the plasma—such as glucose, amino acids, sodium, potassium, chloride, bicarbonate, other salts, and urea—pass through the barrier and become part of the filtrate.

The total volume of blood passing through the kidneys is about 1200 mL per minute, or about one fourth of the entire cardiac output. The plasma passing through the glomerulus loses more than 10% of its volume to the glomerular filtrate; the rest leaves the glomerulus through the efferent arteriole. The normal glomerular filtration rate amounts to about 180 liters (about 45 gallons) each 24 hours. This is four and a half times the amount of fluid in the entire body! Common sense tells us that urine could not be excreted at that rate. Within a few moments, dehydration would become a life-threatening problem.

Reabsorption is highly selective

The threat to homeostasis posed by the vast amounts of fluid filtered by the kidneys is avoided by **reabsorption.** About 99% of the filtrate is reabsorbed into the blood through the renal tubules, leaving only about 1.5 liters to be excreted as urine during a 24 hour period.

Reabsorption permits precise regulation of blood chemistry by the kidneys. Wastes, excess salts, and other materials remain in the filtrate and are excreted in the urine, while needed substances such as glucose and amino acids are returned to the blood. Each day the tubules reabsorb more than 178 liters of water, 1200 grams (2.6 pounds) of salt, and about 250 grams (0.5 pound) of glucose. Most of this, of course, is reabsorbed many times over.

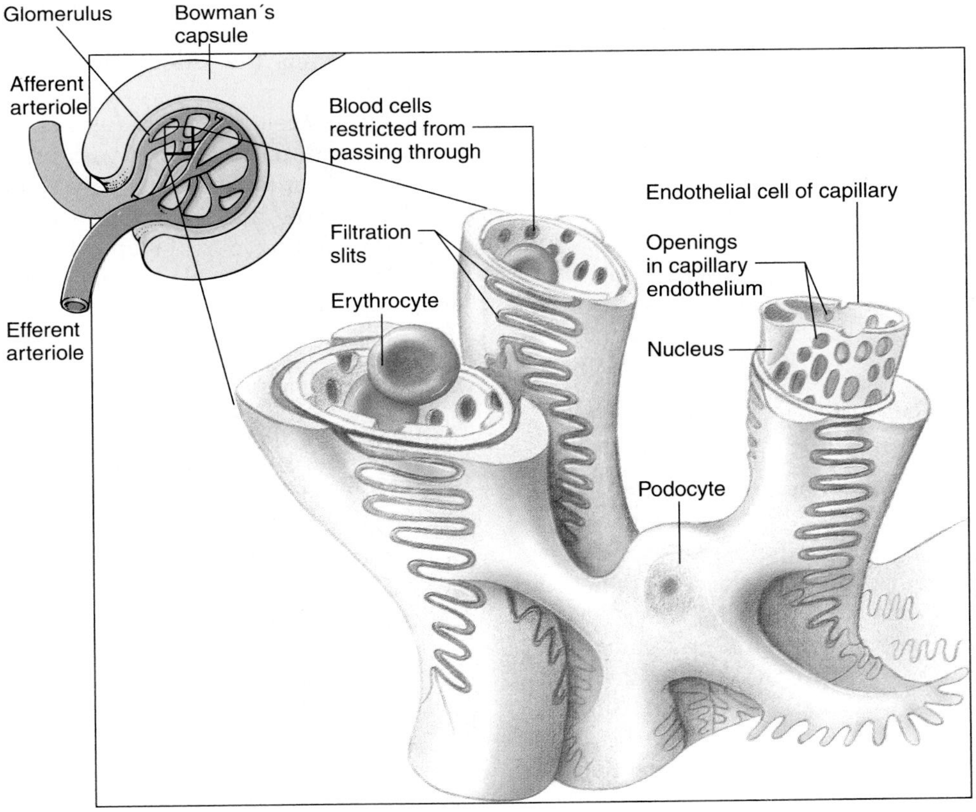

FIGURE 46–13 The podocytes and the glomerular wall make up the filtration membrane of the kidney. This barrier separates the blood of the glomerulus from the lumen of Bowman's capsule.

The simple epithelial cells lining the renal tubule are well adapted for reabsorbing materials. They have abundant microvilli, which increase the surface area for reabsorption (and give the inner lining a "brush border" appearance). These cells also contain numerous mitochondria, which provide the energy for running the cellular "pumps" that actively transport materials.

About 65% of the filtrate is reabsorbed as it passes through the proximal convoluted tubule. Glucose, amino acids, vitamins, and other substances of nutritional value are reabsorbed there, as are many ions, including sodium, chloride, bicarbonate, and potassium. Some of these ions are actively transported; others follow by diffusion. Reabsorption continues as the filtrate passes through the loop of Henle and the distal convoluted tubule. Then the filtrate is further concentrated as it passes through the collecting duct that leads to the renal pelvis.

Normally, substances that are useful to the body, such as glucose or amino acids, are reabsorbed from the renal tubules. If the concentration of a particular substance in the blood is high, however, the tubules may not be able to reabsorb all of it. The maximum rate at which a substance can be reabsorbed is called its **tubular transport maximum (Tm).** For example, the Tm for glucose averages 320 milligrams per minute for an adult human. Normally, the tubular load of glucose is only about 125 milligrams per minute, so almost all of it is reabsorbed. However, if glucose is filtered in excess of the Tm, that excess will not be reabsorbed but will instead pass into the urine.

Each substance that has a Tm also has a **renal threshold** concentration in the plasma. When a substance exceeds its renal threshold, the portion not reabsorbed is excreted in the urine. In a person with uncontrolled diabetes mellitus, the concentration of glucose in the blood exceeds its threshold level (about 150 milligrams of glucose per 100 mL of blood), so glucose is excreted in the urine. Its presence there is evidence of this disorder (Chapter 47).

Some substances are actively secreted from the blood into the filtrate

Some substances, especially potassium, hydrogen, and ammonium ions, are secreted from the blood into the filtrate. Certain drugs, such as penicillin, are also removed from the blood by secretion. Secretion occurs mainly in the region of the distal convoluted tubule.

Secretion of hydrogen ions, an important homeostatic mechanism for regulating the pH of the blood, takes place through the formation of carbonic acid. Carbon dioxide, which diffuses from the blood into the cells of the distal tubules and collecting ducts, combines with water to form carbonic acid. This acid then dissociates, forming hydrogen ions and bicarbonate ions. When the blood becomes too acidic, more hydrogen ions are secreted into the urine.

The secretion of potassium is also very important. When potassium concentration is too high, nerve impulses are not effectively transmitted and the strength of muscle contraction decreases. The heart can be weakened and even fail. When potassium ions are too highly concentrated, they are secreted from the blood into the renal tubules, and then are excreted in the urine. Secretion results partly from a direct effect of the potassium ions on the tubules. In addition, a high potassium ion concentration in the blood stimulates the adrenal cortex to increase its output of the hormone aldosterone. This hormone further stimulates secretion of potassium.

Urine is concentrated as it passes through the renal tubule

In the human kidney, two types of nephrons have been identified: the more numerous cortical nephrons and the more internal juxtamedullary nephrons. **Cortical nephrons** have relatively small glomeruli, and are located almost entirely within the cortex or outer medulla. **Juxtamedullary nephrons** have large glomeruli and their very long loops of Henle extend deep into the medulla. This loop consists of a descending loop, into which filtrate flows, and an ascending loop, through which the filtrate passes on its way to the distal convoluted tubule.

The loop of Henle is specialized to produce a high concentration of sodium chloride in the medulla. This is important in maintaining a highly hypertonic interstitial fluid in the medulla near the bottom of the loop, which in turn permits the kidneys to produce a concentrated urine.

A salt concentration gradient is established, in part, by reabsorption of salt from various regions of the renal tubule. Sodium ions are actively transported out of the proximal tubule, and water follows osmotically. The walls of the descending loop of Henle (the first part of the loop) are relatively permeable to water but relatively impermeable to sodium and urea. There is a high concentration of sodium in the interstitial fluid, so as the filtrate passes down the loop of Henle, water moves out by osmosis. This concentrates the filtrate inside the loop of Henle (Fig. 46–14).

At the turn of the loop of Henle, the walls become more permeable to salt and less permeable to water. As the concentrated filtrate moves up the ascending portion of the loop of Henle, salt diffuses out into the interstitial fluid. This contributes to the high salt concentration in the interstitial fluid in the medulla of the kidney surrounding the loop of Henle. Higher in the ascending part of the loop of Henle, sodium is actively transported out of the tubule.

Because water passes out of the descending portion of the loop of Henle, the filtrate at the bottom of the loop has a high salt concentration. However, because salt (but not water) is removed in the ascending portion, by the time the filtrate moves through the distal tubule, it is iso-

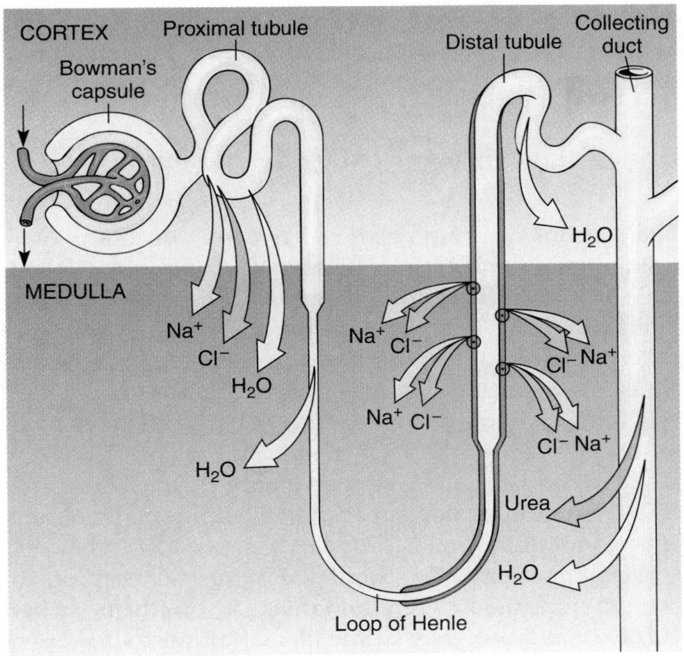

FIGURE 46–14 Various regions of the renal tubule function differently in urine formation. Water passes out of the descending loop of Henle, leaving a more concentrated filtrate inside. In the ascending loop, salt moves out. Chloride pumps transport chloride out into the interstitial fluid, and sodium follows. The saltier the interstitial fluid becomes, the more water moves out of the descending loop. This leaves a concentrated filtrate inside, so more salt passes out. Note that this is a positive feedback system. Urea also moves out into the interstitial fluid through the collecting ducts. Water from the collecting ducts moves out osmotically into this hypertonic tissue fluid.

tonic (or even hypotonic) to blood. The filtrate passes from the renal tubule into a larger **collecting duct** that eventually empties into the renal pelvis.

Note that there is a *counterflow* of fluid through the two limbs of the loop of Henle. Filtrate passing down through the descending portion of the loop is flowing in a direction opposite the filtrate moving upward through the ascending loop. The filtrate is concentrated as it moves down the descending portion of the loop and diluted as it moves up the ascending part of the loop. This countercurrent mechanism helps maintain a high salt concentration in the interstitial fluid of the medulla. The hypertonic interstitial fluid draws water osmotically from the filtrate in the collecting ducts.

The inner medullary collecting ducts are permeable to urea, allowing the concentrated urea in the filtrate to dif-

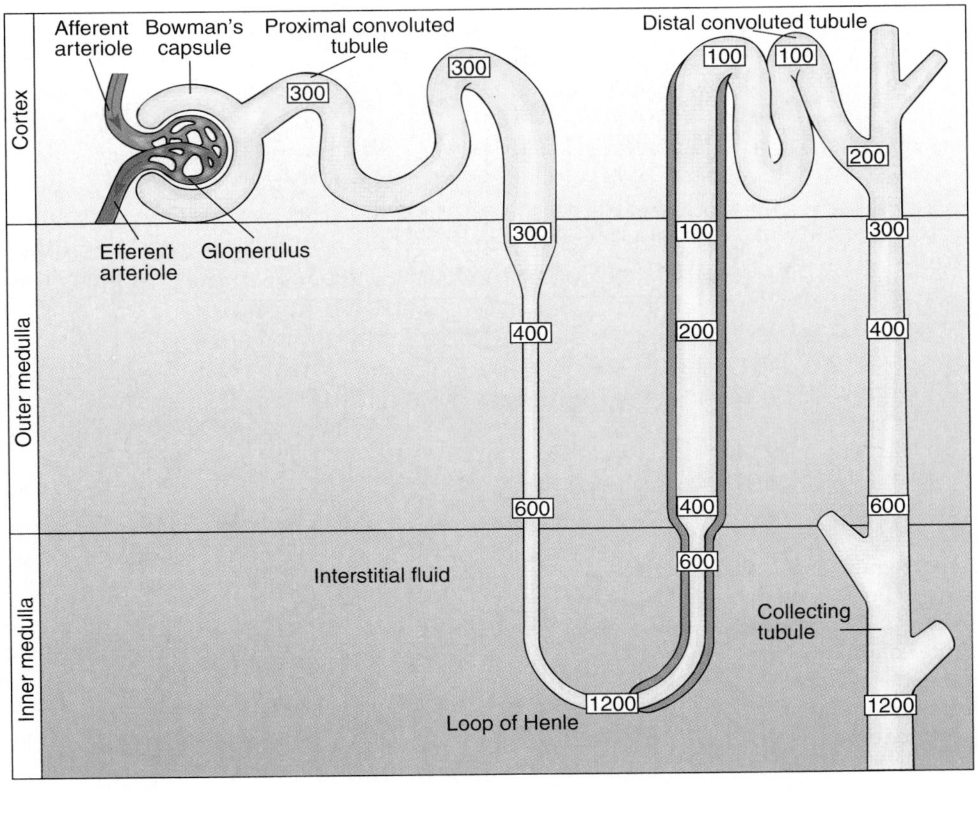

FIGURE 46–15 Filtrate is concentrated as it moves through the nephron. Figure 46–14 illustrates the movement of water and ions through various parts of the nephron. This figure shows the relative concentration of ions, mainly Na^+ and Cl^-, during formation of a very concentrated urine. Numbers indicate the concentration of salt in the filtrate in milliosmols per liter. (A milliosmol is a unit of osmotic pressure equal to one-thousandth gram molecular weight of a substance divided by the number of ions into which a substance dissociates in one liter of solution.) Recall that the more hypertonic a solution is, the higher its osmotic pressure. This mechanism establishes a very hypertonic interstitial fluid near the renal pelvis that draws water osmotically from the filtrate in the collecting ducts. The heavy outline along the thin ascending limb of Henle's loop indicates that this region is relatively impermeable to water.

fuse out into the interstitial fluid. This urea contributes to the high solute concentration of the inner medulla, and so helps in the process of concentrating urine (Fig. 46–15).

The collecting ducts are routed to pass through the zone of very salty interstitial fluid. As the filtrate moves down the collecting duct, water passes osmotically into the interstitial fluid, where it is collected by capillaries. So much water may leave the collecting ducts that highly concentrated urine can be produced. *A hypertonic urine has a low concentration of water and so conserves water.*

Some of the water that diffuses from the filtrate into the interstitial fluid is removed by capillaries known as the **vasa recta** and carried off in the venous drainage of the kidney. The vasa recta are long, looped extensions of the efferent arterioles of the juxtamedullary nephrons. They extend deep into the medulla, only to negotiate a hairpin curve and return to the cortical venous drainage of the kidney.

Blood flows in opposite directions in the ascending and descending regions of the vasa recta, just as filtrate flows in opposite directions in the ascending and descending portions of the loop of Henle. As a consequence of this countercurrent flow, much of the salt and urea that enter the blood leave again; the solute concentration of the blood leaving the vasa recta is only slightly higher than that of the blood entering. This mechanism helps maintain the high solute concentration of the interstitial fluid.

Urine volume is regulated by the hormone ADH

The amount of urine produced depends on the body's need to retain or rid itself of water. We have seen that salt reabsorption in the loops of Henle establishes a very salty interstitial fluid that draws water osmotically from the collecting ducts. Permeability of the collecting ducts to water is regulated by **antidiuretic hormone (ADH).** When the body needs to conserve water, ADH is released from the posterior pituitary gland (Fig. 46–16). This hormone makes the collecting ducts more permeable to water so that more water is reabsorbed, and a small volume of concentrated urine is produced.

Secretion of ADH is stimulated by special receptors in the hypothalamus. When fluid intake is low, the body begins to dehydrate, causing the blood volume to decrease. As blood volume decreases, the *concentration* of salts dissolved in the blood becomes greater, causing an increase in osmotic pressure. Receptors in the hypothalamus are sensitive to this osmotic change and stimulate the poster-

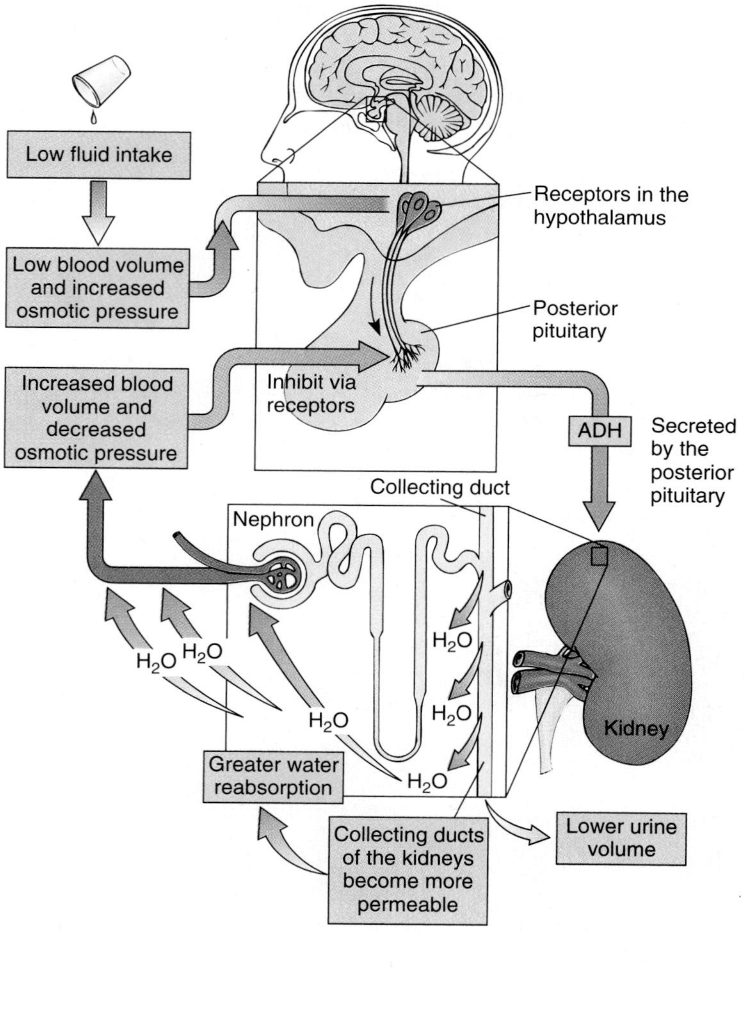

FIGURE 46–16 ADH regulates urine volume. When the body is dehydrated, the hormone ADH increases the permeability of the collecting ducts to water. More water is reabsorbed, and only a small volume of concentrated urine is produced.

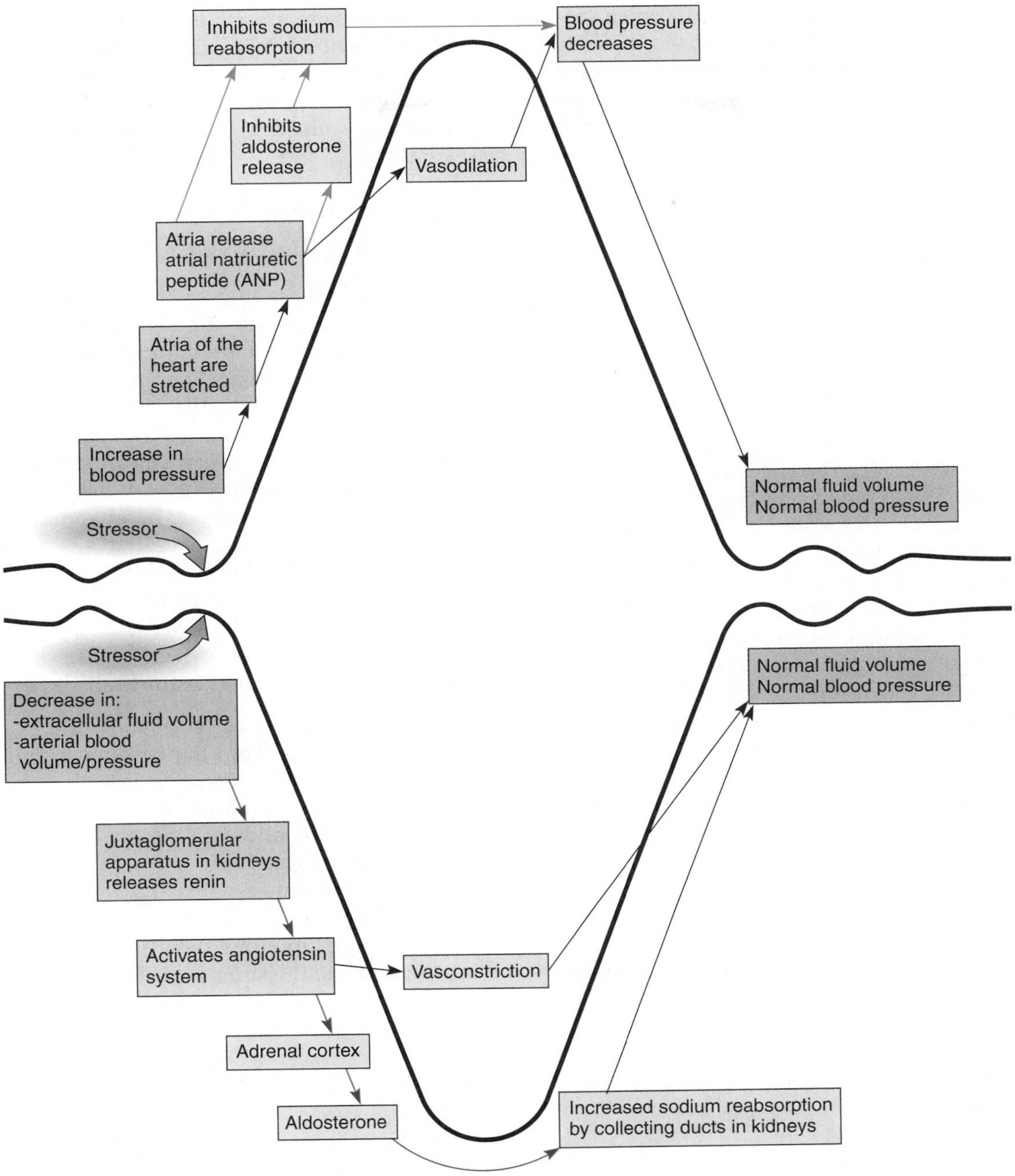

FIGURE 46–17 Aldosterone and ANP help regulate the fluid balance, salt balance, and blood pressure.

ior lobe of the pituitary to release ADH. A thirst center in the hypothalamus also responds to dehydration, stimulating an increase in fluid intake.

When one drinks a great deal of water, the blood becomes diluted and its osmotic pressure falls. Release of ADH by the pituitary gland decreases, lessening the amount of water reabsorbed from the collecting ducts. A large volume of dilute urine is produced.

Occasionally, the pituitary gland malfunctions and does not produce sufficient ADH. The resulting condition is called diabetes insipidus (not to be confused with the more common disorder, diabetes mellitus). This condition can also result from an acquired insensitivity of the kidney to ADH. In diabetes insipidus, water is not efficiently reabsorbed from the ducts, so a large volume of urine is produced. A person with severe diabetes insipidus may excrete up to 25 quarts of urine each day, a serious loss of water to the body. The affected individual becomes dehydrated and must drink almost continually to offset fluid loss. Diabetes insipidus can often be controlled by injections of ADH or by use of an ADH nasal spray.

Sodium reabsorption is regulated by the hormone aldosterone

Sodium is the most abundant extracellular ion, accounting for about 90% of all positive ions outside cells. Sodium concentration is precisely regulated by the hormone **aldosterone,** which is secreted by the cortex of the adrenal glands. Aldosterone stimulates the distal tubules and collecting ducts to increase sodium reabsorption.

Aldosterone secretion can be stimulated by a decrease in blood pressure. When blood pressure falls, cells of the **juxtaglomerular apparatus** secrete the enzyme **renin** and activate the renin-angiotensin pathway. The juxtaglomerular apparatus is a small group of cells located in the region where the renal tubule contacts the afferent and efferent arterioles (Fig. 46–11*b*). Renin acts on a plasma protein, converting it to angiotensin. The active form of angiotensin, called angiotensin II, is a hormone that stimulates aldosterone secretion. Angiotensin II also raises blood pressure directly by constricting the blood vessels.

Another hormone, **atrial natriuretic peptide (ANP),** inhibits sodium reabsorption. Released by the walls of the atria of the heart, ANP acts directly on the adrenal cortex to inhibit aldosterone secretion. ANP also inhibits renin release. The actions of ANP lower blood volume and blood pressure. The renin-angiotensin system and ANP work antagonistically in regulating fluid balance, salt (electrolyte) balance, and blood pressure (Fig. 46–17).

Urine is composed of water, nitrogenous wastes, and salts

By the time the filtrate reaches the renal pelvis, its composition has been precisely adjusted. Useful materials have been returned to the blood by reabsorption. Wastes and excess materials that entered by filtration or secretion have been retained by the tubules. The adjusted filtrate, called **urine,** is composed of about 96% water, 2.5% nitrogenous wastes (mainly urea), 1.5% salts, and traces of other substances such as bile pigments, which may contribute to the characteristic color and odor.

Healthy urine is sterile and has been used to wash battlefield wounds when clean water was not available. However, urine swiftly decomposes when exposed to bacterial action, forming ammonia and other products. It is the ammonia that produces the diaper rash of infants.

The composition of urine yields many clues to body function and malfunction. **Urinalysis,** the physical, chemical, and microscopic examination of urine, is a very important diagnostic tool and is used to monitor many disorders such as diabetes mellitus. Urinalysis is also extensively used in drug testing because breakdown products of some drugs can be identified in the urine for several weeks. (See *Focus On: Treating Kidney Disease.*)

ANIMALS HAVE OPTIMAL TEMPERATURE RANGES

Snowshoe hares, snowy owls, weasels, and a few other animals inhabit cold arctic regions. Others, such as the kangaroo rat, are adapted to desert life. Each animal species has an optimal temperature range and has evolved strategies for regulating body temperature.

Animals produce heat as a byproduct of metabolic activities. Heat can also be gained from the environment or lost to the environment. One way that heat can be transferred to the environment is by **evaporation**—the conversion of a liquid, such as sweat, to a vapor. When sweat evaporates from the surface of the body or when a dog pants, the vaporizing water molecules transfer heat from the body to the surroundings.

Ectotherms absorb heat from their surroundings

As discussed in earlier chapters, most animals are **ectotherms,** which means that their body temperature depends to a large extent on heat from the surrounding environment. Metabolic rate of an ectotherm tends to change with the weather.

Many ectotherms use structural and behavioral strategies to adjust body temperature. For example, lizards bask in the sun, orienting their bodies to expose the maximum surface to the sun's rays (Fig. 46–18). Some insects use a combination of behavior and metabolic heat production to regulate body temperature. The "furry" body of the moth conserves body heat. When a moth prepares for flight, it contracts its flight muscles with little movement of its wings. The heat generated enables the moth to sustain the intense metabolic activity needed for flight. Other behavioral strategies for regulating temperature include hibernation and migration.

An advantage of ectothermy is the absence of direct metabolic cost. Most of the heat for thermoregulation comes from the sun. As a result, ectotherms have a much lower daily energy expenditure than do endotherms, and more of the energy in their food can be converted to growth and reproduction. One of the disadvantages of ectothermy is that activity is limited by daily and seasonal temperature conditions.

Endotherms derive heat from metabolic processes

Birds and mammals are **endotherms,** which means that they have homeostatic mechanisms that maintain a con-

Treating Kidney Disease

Our kidneys excrete metabolic wastes and help regulate the volume and composition of body fluids. Their vital function is compromised in more than 13 million people in the United States who suffer from kidney disease. In fact, kidney disease ranks fourth in prevalence among major human diseases in the United States. Kidney function can be impaired by infections, poisoning by substances such as mercury or carbon tetrachloride, lesions, tumors, kidney stones, shock, or circulatory disease. One of the most common kidney diseases, *glomerulonephritis*, is actually a large number of related chronic diseases in which the filtering units of the kidney (the glomeruli) are damaged. The damage is thought to result from an autoimmune response.

In chronic kidney disease, a progressive loss of function may eventually reach the stage of kidney failure. A decrease in the rate of filtration occurs, causing loss of homeostatic water and salt balance and an inability to effectively excrete urea and other nitrogenous wastes. Retention of water causes edema (swelling). As the concentration of hydrogen ions increases, body fluids become too acidic. Nitrogenous wastes accumulate in the blood and tissues, causing a condition called *uremia*. If untreated, the acidosis and uremia can cause coma and, eventually, death.

Chronic kidney failure is treated by kidney dialysis or by kidney transplant. In dialysis, solutes are exchanged by diffusion across a semi permeable membrane between solutions of different compositions. A kidney dialysis machine can be used to temporarily restore appropriate blood solute balance to a patient whose kidneys are not functioning.

In one type of dialysis (hemodialysis), blood circulates from one of the patient's arteries through the dialysis machine and is then returned through a vein. A plastic tube is surgically inserted into both an artery and a vein in the patient's arm or leg. These tubes are then connected to a circuit of plastic tubing from a dialysis machine. The patient's blood flows through the tubing, which is permeable to certain substances such as nitrogenous wastes and ions, but not permeable to proteins and blood cells.

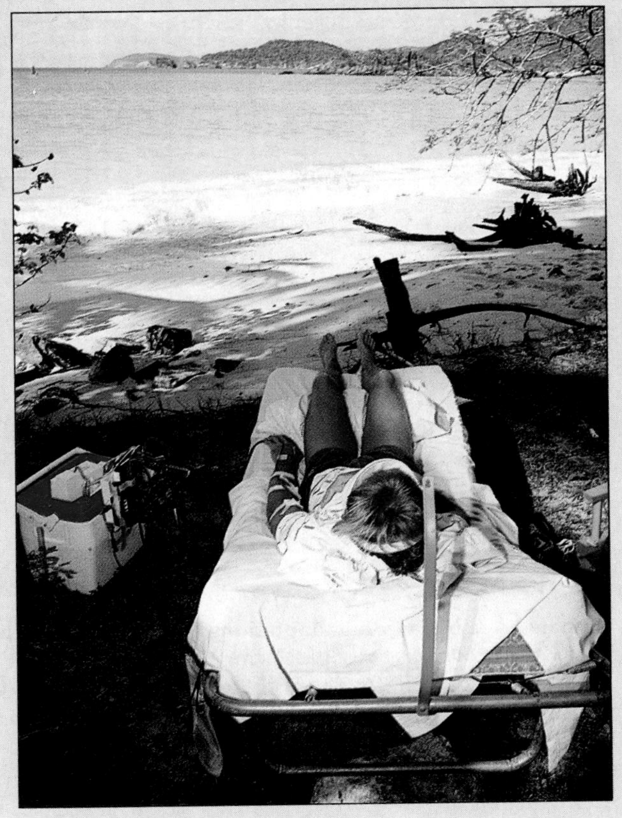

Portable kidney dialyzers give patients the freedom to undergo dialysis even if they are far from a hospital. (©Dan McCoy/Rainbow)

The tubing is immersed in a solution containing most of the normal blood plasma components in their normal proportions. Nitrogenous wastes diffuse down their concentration gradient from the patient's blood through minute pores in the tubing and into the surrounding solution in the dialysis machine. As the blood circulates repeatedly through the tubing in the machine, dialysis continues, eventually adjusting most of the values of the patient's blood chemistry to normal ranges.

Although much improved by recent engineering advances, mechanical dialysis is very expensive, clumsy, and inconvenient. Patients generally undergo dialysis three times a week, and the procedure takes four to six hours. This type of dialysis can also produce serious side effects such as osteoporosis ("brittle bone syndrome," caused by bone calcium loss).

Another dialysis technique, continuous ambulatory peritoneal dialysis (CAPD), uses the patient's own peritoneum (the lining of the abdominal cavity), which is a selectively permeable multicellular membrane. A plastic bag containing dialysis fluid is attached to the patient's abdominal cavity, and the fluid is allowed to run into the abdominal cavity. After about 30 minutes, the fluid is withdrawn into the bag and discarded. This process is repeated about three times each day. This type of dialysis is much more convenient but poses the threat of peritonitis if bacteria enter the body cavity with the dialysis fluid.

Long-term dialysis is not as desirable for the patient as a functioning kidney. With a successful kidney transplant, a patient can live a more normal life with far less long-term expense. At present, more than two-thirds of kidney transplants are successful for several years, although physicians must routinely treat the problems of graft rejection (see Chapter 43). Several recipients of kidney transplants have survived for more than 20 years. ■

(a)

(b)

FIGURE 46–18 Animals have behavioral adaptations for thermoregulation. (***a***) This marine iguana (*Amblyrhynochos cristatus*) is an ectotherm that increases its body temperature by sunning itself. (***b***) The body temperature of this sooty tern (an endotherm) of Hawaii is reduced as heat leaves the body through its open mouth. (*a*, Animals Animals, © 1996, Breck P. Kent; *b*, Frans Lanting/Minden Pictures)

stant body temperature despite changes in the external temperature. Endotherms obtain most of their body heat from their own metabolic processes (see *Making the Connection: Electron Transport and Heat*, Chapter 7). They have homeostatic mechanisms for regulating metabolic rate and for regulating heat exchange with the environment.

Mammals have several homeostatic mechanisms for maintaining a constant body temperature. Mammals can also increase their rate of heat production by contracting muscles or by the action of hormones that increase metabolic rate (e.g., thyroid hormones).

In humans, receptors in the skin, the hypothalamus, and certain other areas are sensitive to body temperature changes (Fig. 46–19). Information about body temperature is sent to the temperature-regulating center in the hypothalamus. Nerves that are part of the somatic system signal muscles to shiver, or allow us to move muscles voluntarily to increase body temperature. When body temperature increases, sympathetic nerves increase the activity of sweat glands.

The autonomic system also helps regulate body temperature by dilating blood vessels in the skin when we are hot. Increased blood flow to the skin brings body heat to the surface. The skin acts as a heat radiator, allowing heat to leave the body. When we are cold, the autonomic system constricts blood vessels in the skin, reducing heat loss.

Important advantages of endothermy include increased rate of enzyme activity and constant body temperature. Endotherms can carry out their activities even in low winter temperatures. However, endotherms are disadvantaged by the high energy cost of thermoregulation during times when they are inactive. We maintain our body temperature even when we are asleep.

FIGURE 46–19 Temperature regulation in the human body is carefully regulated. (***a***) Mechanisms that restore homeostasis when body temperature increases. (***b***) Mechanisms that restore homeostasis when body temperature decreases. (*a*, Merrit Vincent/PhotoEdit; *b*, Jim Brandenburg/Minden Pictures)

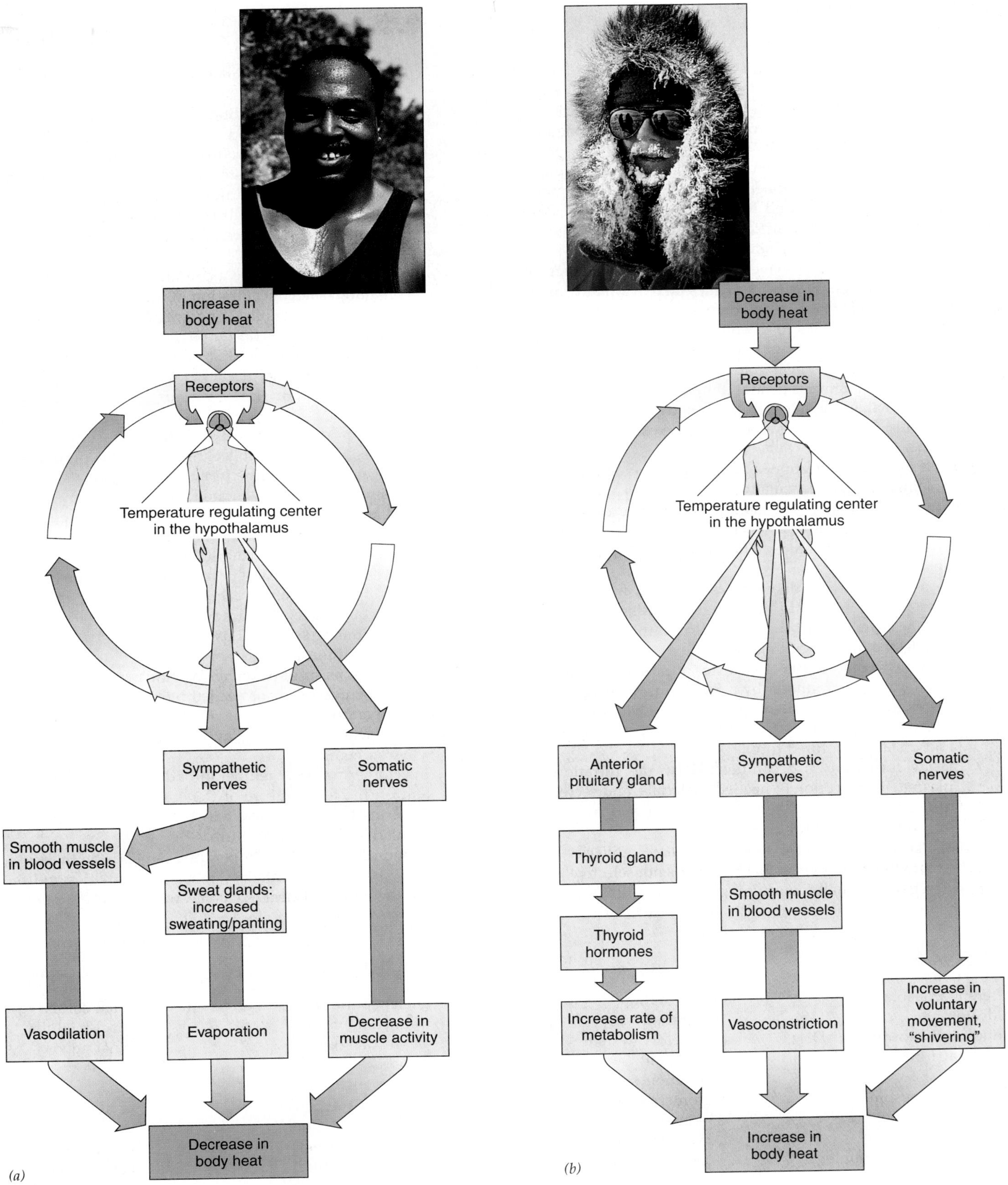
Increase in body heat
Receptors
Temperature regulating center in the hypothalamus
Sympathetic nerves
Somatic nerves
Smooth muscle in blood vessels
Sweat glands: increased sweating/panting
Vasodilation
Evaporation
Decrease in muscle activity
Decrease in body heat
(a)
Decrease in body heat
Receptors
Temperature regulating center in the hypothalamus
Anterior pituitary gland
Sympathetic nerves
Somatic nerves
Thyroid gland
Smooth muscle in blood vessels
Thyroid hormones
Increase rate of metabolism
Vasoconstriction
Increase in voluntary movement, "shivering"
Increase in body heat
(b)

SUMMARY

I. Excretory systems help regulate the concentration of body fluids by osmoregulation and excretion of metabolic wastes.

II. The principal waste products of animal metabolism are water; carbon dioxide; and nitrogenous wastes, including ammonia, urea, and uric acid.

III. Invertebrate mechanisms of osmoregulation and waste disposal include the protonephridia of flatworms, metanephridia of annelids, and Malpighian tubules of insects.

IV. The vertebrate kidney functions in excretion and osmoregulation and is vital in maintaining homeostasis.
   - A. Aquatic vertebrates have the continuous challenge of osmoregulation. Freshwater fishes take in water osmotically; they excrete a large volume of dilute urine.
   - B. Marine bony fishes lose water osmotically. They compensate by drinking seawater and excreting salt through their gills; only a small volume of urine is produced.
   - C. Marine cartilaginous fishes retain large amounts of urea, which enables them to take in water osmotically through the gills. This water can be used to excrete a hypotonic urine.
   - D. Other aquatic vertebrates have specific adaptations for dealing with osmoregulation. Marine birds and reptiles, for example, have salt glands that excrete excess salt.

V. The urinary system is the principal excretory system in humans and other vertebrates. The kidney is the key organ of the urinary system.
   - A. In mammals, the kidneys produce urine, which passes through the ureters to the urinary bladder for storage. During urination, the urine is released through the urethra to the outside of the body.
   - B. Each nephron consists of a cluster of capillaries, called a glomerulus, surrounded by a Bowman's capsule which opens into a long, coiled renal tubule. The renal tubule consists of a proximal convoluted tubule, loop of Henle, and distal convoluted tubule.
   - C. Urine formation is accomplished by the filtration of plasma, reabsorption of needed materials, and secretion of a few substances such as potassium and hydrogen ions into the renal tubule.
      1. Plasma filters out of the glomerular capillaries and into Bowman's capsule. Filtration is nonselective with regard to small molecules; glucose and other needed materials, as well as metabolic wastes, become part of the filtrate.
      2. About 99% of the filtrate is reabsorbed from the renal tubules into the blood. Reabsorption is a highly selective process that returns usable materials to the blood but leaves wastes and excesses of other substances to be excreted in the urine.
      3. In secretion, certain substances and drugs are actively transported into the renal tubule to become part of the urine.
      4. The ability to produce a concentrated urine depends on a high salt concentration in the interstitial fluid of the kidney medulla. The interstitial fluid in the medulla has a salt concentration gradient in which the salt is most concentrated around the bottom of the loop of Henle. This gradient is maintained, in part, by salt reabsorption from various parts of the renal tubule.
      5. Water is drawn by osmosis from the filtrate as it passes through the collecting ducts. This permits the concentration of urine in the collecting ducts.
      6. Urine volume is regulated by the hormone ADH, which is released by the posterior lobe of the pituitary gland in response to an increase in osmotic concentration of the blood (caused by dehydration). ADH increases the permeability of the collecting ducts. As a result, more water is reabsorbed and only a small volume of urine is produced.
      7. Aldosterone and atrial natriuretic peptide (ANP) regulate sodium reabsorption.
      8. Urine consists of water, nitrogenous wastes, salts, and other substances not needed by the body.

VI. Body temperature depends on a balance between heat produced by metabolic activities and the exchange of heat with the environment.
   - A. In ectotherms, body temperature depends to a large extent on the temperature of the environment. Many ectotherms have structural and behavioral strategies for regulating body temperature.
   - B. In endotherms, body temperature depends on the regulation of heat generated by metabolic activities and on the regulation of heat exchange with the environment.

SELECTED KEY TERMS

afferent arteriole
aldosterone
antidiuretic hormone (ADH)
atrial natriuretic peptide (ANP)
Bowman's capsule
collecting duct
cortical nephron
distal convoluted tubule
ectotherm
efferent arteriole
endotherm
evaporation
excretion
excretory system
filtration
filtration membrane
glomerular filtrate
glomerulus
juxtaglomerular apparatus
juxtamedullary nephron
kidney
loop of Henle
Malpighian tubule
metanephridium (metanephridia)
nephridial organ
nephron
osmoregulation
osmotic conformer
osmotic regulator
podocyte
protonephridium (protonephridia)
proximal convoluted tubule
reabsorption
renal cortex
renal medulla
renal pelvis
secretion
urea
ureter
urethra
uric acid
urinary bladder
urinary system
urine

POST-TEST

1. The process of removing metabolic wastes from the body is called ________ .
2. ________ is the process by which an animal regulates its fluid content.
3. The principal nitrogenous waste product of insects and birds is ________ ________ .
4. The principal nitrogenous waste product of amphibians and mammals is ________ .
5. Flatworms have excretory structures called ________ , and earthworms have excretory structures known as ________ .
6. The excretory structures of insects are ________ ________ .
7. Animals whose body fluid concentration varies along with changes in surrounding seawater are referred to as osmotic ________ .
8. The outer portion of the human kidney is the renal ________ .
9. Urine is delivered to the urinary bladder by the ________ .
10. The part of the kidney that receives urine from collecting ducts is the ________ ________ .
11. The vertebrate kidney consists of functional units called ________ .
12. The glomerulus consists of a cluster of ________ surrounded by ________ ________ .
13. Blood is delivered to the glomerulus by a(an) ________ arteriole and leaves the glomerulus in a(an) ________ arteriole.
14. Fluid that leaves the glomerular capillaries and enters Bowman's capsule is called glomerular ________ .
15. When a substance exceeds its renal threshold, the portion not reabsorbed is ________ .
16. Antidiuretic hormone (ADH) increases the permeability of the ________ ________ so that more water is ________ .
17. In ________ , most of the heat for temperature regulation comes from the environment.
18. Sweating cools the body by the process of ________ .

REVIEW QUESTIONS

1. Compare osmoregulation in flatworms with that in insects.
2. What type of osmoregulatory challenge is faced by marine fishes? By freshwater fishes? How are these problems met in each case?
3. What are the principal types of nitrogenous wastes?
4. Name the structure in the mammalian body associated with each of the following: (a) urea formation; (b) urine formation; (c) temporary storage of urine; (d) conduction of urine out of the body.
5. Which part of the nephron is associated with the following: (a) filtration; (b) reabsorption; (c) secretion.
6. Contrast reabsorption with secretion, and filtration with secretion.
7. List the sequence of blood vessels through which a drop of blood passes as it is conducted to and from a nephron.
8. How is urine volume regulated? Explain. Why must victims of untreated diabetes insipidus drink great quantities of water?
9. Describe two treatment strategies for kidney failure. What are the advantages of each?
10. What are some benefits of being an endotherm? What are some disadvantages?
11. Label the diagram.

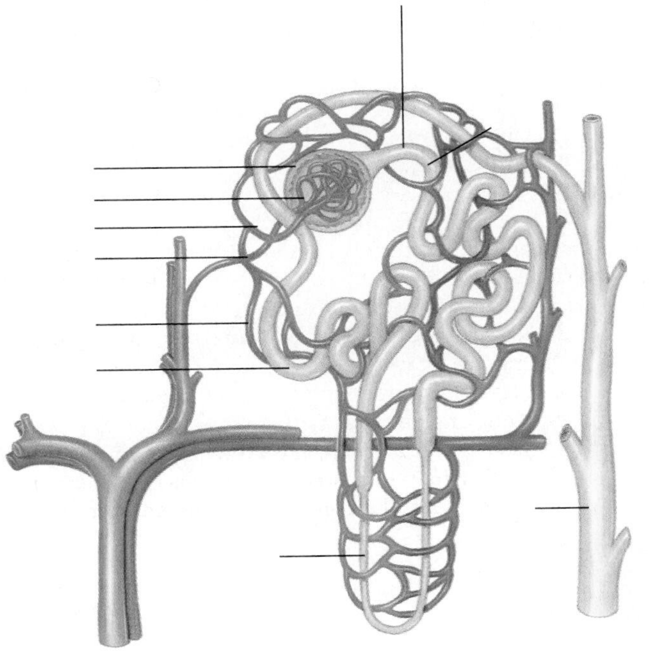

YOU MAKE THE CONNECTION

1. The number of protonephridia in a planarian is related to the salinity of its environment. Planaria inhabiting slightly salty water develop fewer protonephridia, but the number quickly increases when the concentration of salt in the environment is lowered. Explain why.
2. What types of osmoregulatory challenges do humans experience? Explain. What mechanisms do we have to solve these challenges?
3. Why is glucose normally not present in urine? Why is it present in diabetes mellitus? Why do you suppose diabetics experience an increased output of urine?
4. The kangaroo rat's diet consists of dry seeds, and it drinks no water. Speculate about the adaptations this animal would need in order to survive.

RECOMMENDED READINGS

McClanahan, L. L., R. Ruibal, and V. H. Shoemaker. "Frogs and Toads in Deserts." *Scientific American,* Vol. 270, No. 3, March 1994. A discussion of some unusual adaptations for life in the desert.

Solomon, E. P., R. Schmidt, and P. Adragna. *Human Anatomy and Physiology.* Saunders College Publishing, Philadelphia, 1990. Chapters 27 and 28 focus on the human urinary system and fluid balance.

CHAPTER 47

Endocrine Regulation

Many crustaceans can change color to blend with their background. Their pigment cells are controlled by hormones. This juvenile Puget Sound king crab (*Lopholithodes mandtii*) was photographed along the Pacific northwest coast. (SharkSong/M. Kazmers/Dembinsky Photo Associates)

A caterpillar becomes a butterfly. A crustacean changes color. A young girl develops into a woman. An adult copes with chronic stress. These physiological processes and many other adjustments of metabolism, growth, and reproduction are regulated by the **endocrine system.** This system works closely with the nervous system to maintain the steady state of the body. **Endocrinology,** the study of endocrine activity, is a very active and exciting field of biomedical research.

The endocrine system is a diverse collection of cells, tissues, and organs (including specialized glands) that secrete **hormones,** chemical messengers responsible for the regulation of many body processes. The term hormone is derived from a Greek word meaning "to excite." Hormones do indeed affect specific tissues, usually by stimulating a change in some metabolic activity.

Endocrine glands produce hormones and secrete them into the surrounding tissue fluid. These hormones typically diffuse into capillaries and are then transported by the blood. They stimulate responses only in their **target tissues,** which may be another endocrine gland or an entirely different type of organ, such as a bone or the kidney. Often the target tissue is located far from the endocrine gland. Endocrine glands lack ducts; they differ from **exocrine glands** (such as sweat glands and gastric glands), which release their secretions into ducts.

Traditionally, endocrinology focused on the activity of about ten discrete endocrine glands. In recent years investigators have identified specialized cells in the digestive tract, heart, kidneys, and many other parts of the body that also release hormones. In addition, some neurons, known as **neuroendocrine cells,** release hormones called **neurohormones** (Fig. 47–1).

Some cells secrete hormones that diffuse into the interstitial fluid and act on nearby target cells. This type of local regulation, known as **paracrine** regulation, is carried out by the interleukins that regulate immune responses. Other paracrine regulators include growth factors and prostaglandins (discussed in a later section). The scope of endocrinology has been broadened to include the study of chemical messengers produced by a wide variety of organs, tissues, and cells.

Pheromones, another type of chemical messenger, are produced by animals for communication with other animals of the same species. Since pheromones are generally produced by exocrine glands and do not regulate metabolic activities within the animal that produces them, most biologists do not classify them as hormones. Their role in regulating behavior is discussed in Chapter 50.

In this chapter we will discuss and describe the actions of a variety of hormones. We will also examine how overproduction or deficiency of various hormones interferes with normal functioning.

LEARNING OBJECTIVES

After you have studied this chapter you should be able to

1. Define the terms hormone and endocrine gland, and identify sources of hormones other than endocrine glands.
2. Compare the mechanisms of action of steroid and protein-type hormones; include the role of second messengers, such as cyclic AMP.
3. Summarize the roles of hormones in invertebrates.
4. Identify the principal vertebrate endocrine glands, locate them in the body, and list the hormones secreted by each.
5. Summarize the regulation of endocrine glands by negative feedback mechanisms, and give specific examples.
6. Justify the description of the hypothalamus as the link between nervous and endocrine systems, and describe the mechanisms by which the hypothalamus exerts its control.
7. Compare the functions of the posterior and anterior lobes of the pituitary; describe the actions of their hormones.
8. Describe the actions of growth hormone on growth and metabolism, and contrast the consequences of hyposecretion and hypersecretion.
9. Define the actions of the thyroid hormones, their regulation, and the thyroid disorders discussed in this chapter.
10. Contrast the actions of insulin and glucagon, and describe the disorders associated with the malfunction of the islets of the pancreas.
11. Describe the actions of the adrenal glands, including their role in helping the body adapt to stress.

HORMONES CAN BE ASSIGNED TO FOUR CHEMICAL GROUPS

Although hormones are chemically diverse, they generally belong to one of four different chemical groups: steroids; amino acid derivatives; peptides or proteins; or fatty acid derivatives (Figs. 47–2 and 47–3).

In vertebrates, the adrenal cortex, testis, ovary, and placenta secrete steroids synthesized from cholesterol. Progesterone, one of the female sex hormones, is the precursor of the **steroid hormones** produced by the adrenal cortex, as well as of the male hormone testosterone and the female hormone estradiol. The molting hormone (also called ecdysone) of insects is also a steroid.

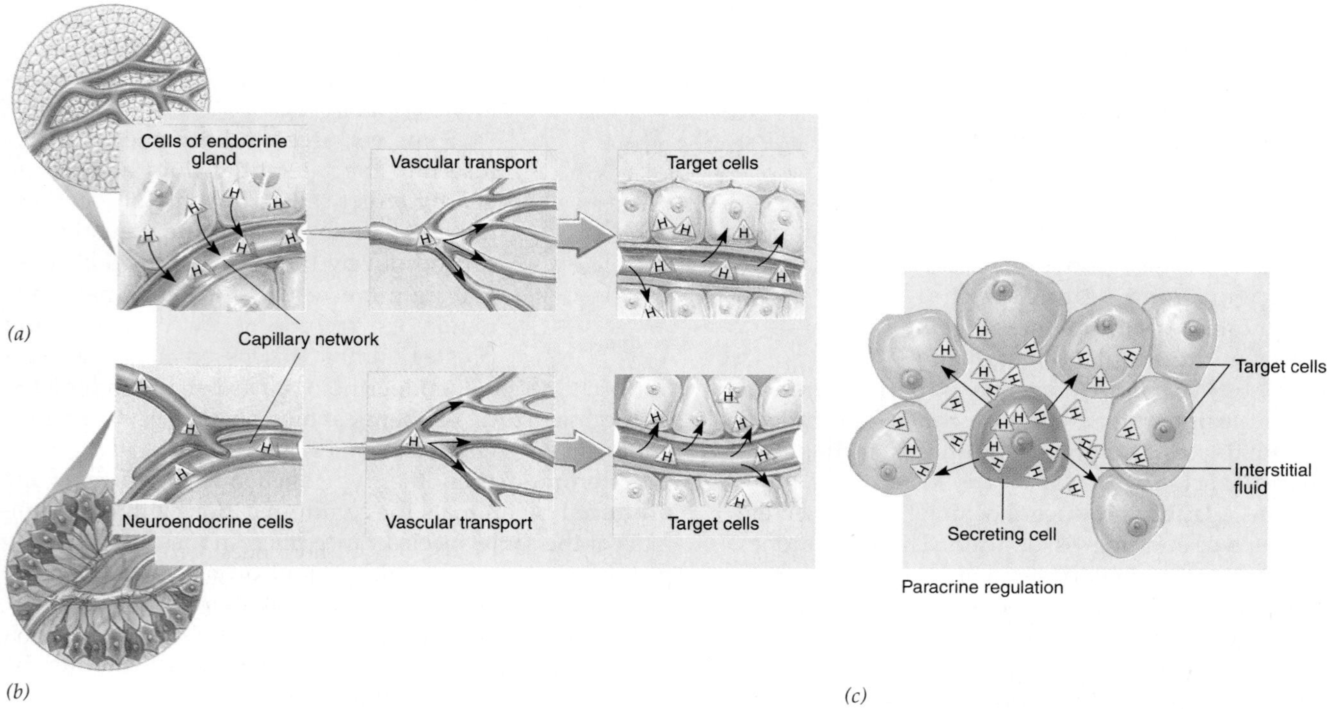

FIGURE 47–1 Hormones can be secreted by the cells of endocrine tissues or glands, by neuroendocrine cells, or by some other cell types. Hormones may be transported to target cells by the blood as in (***a***) and (***b***) or may reach their target cells by diffusion through the interstitial fluid as in the case of paracrine regulation (***c***).

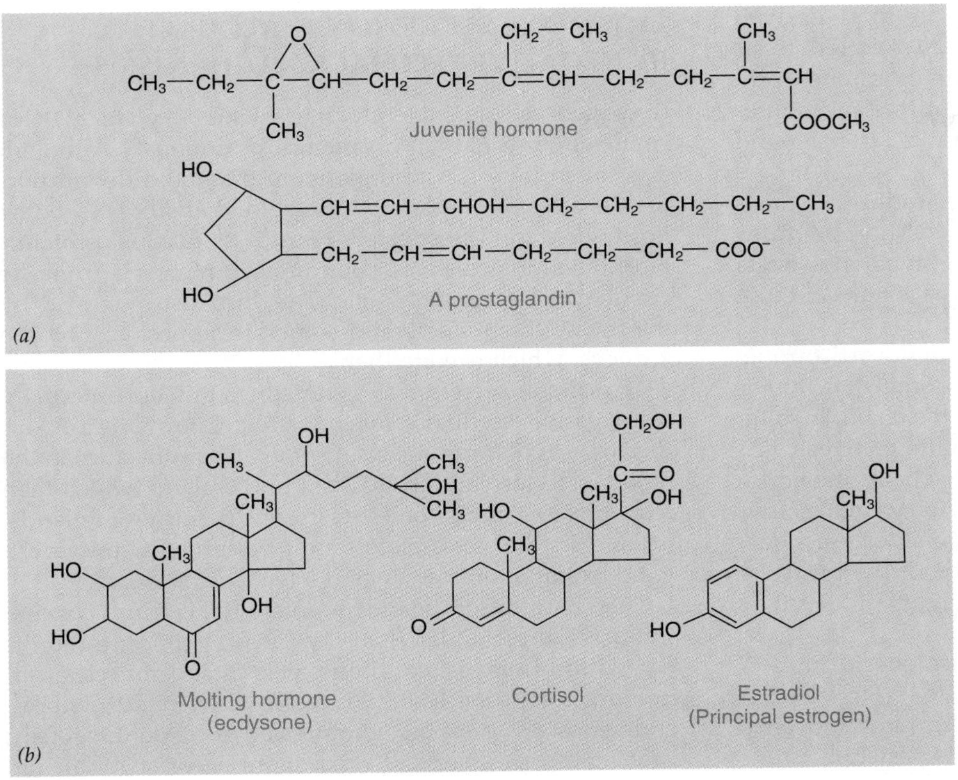

FIGURE 47–2 Major groups of hormones include steroid hormones and hormones derived from fatty acids. (***a***) The juvenile hormone of insects and prostaglandins are derived from fatty acids. (***b***) Cortisol and estradiol, steroid hormones secreted by the adrenal cortex.

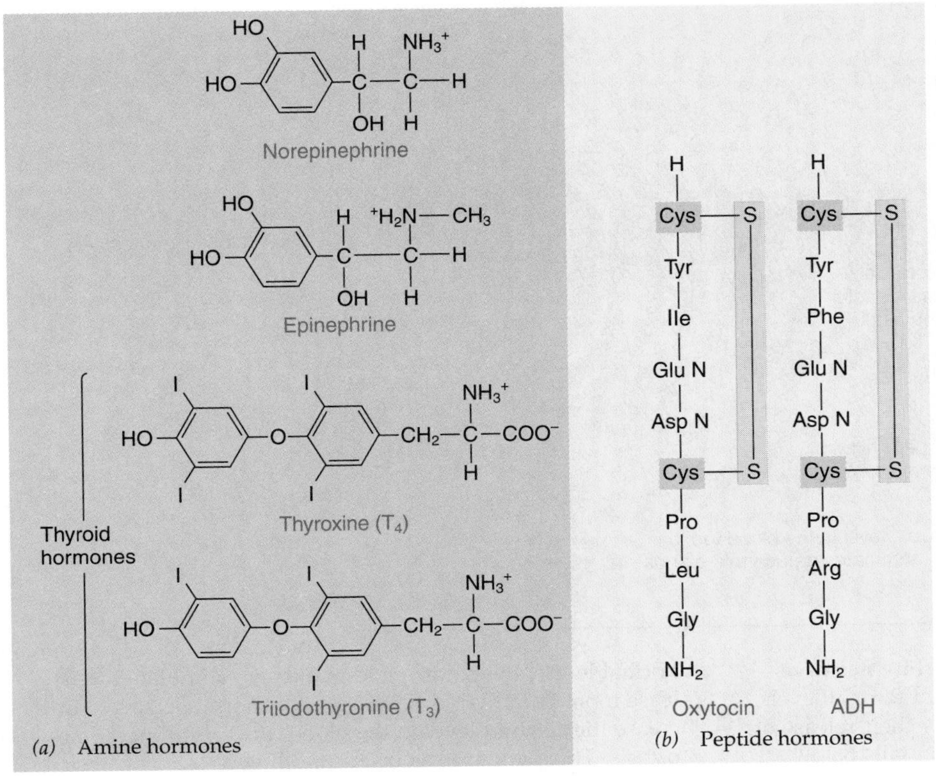

FIGURE 47–3 A variety of hormones belong to the protein family. (***a***) Some hormones derived from amino acids. Note the presence of iodine in the thyroid hormones. (***b***) Peptide hormones produced in the hypothalamus and secreted by the posterior lobe of the pituitary gland. Oxytocin and antidiuretic hormone are both small peptides containing nine amino acids. Note that the structures of these hormones differ by only two amino acids.

Chemically, the simplest hormones are those derived from the amino acid tyrosine; they are referred to as **amines.** Examples include the thyroid hormones produced by the thyroid gland, and epinephrine and norepinephrine produced by the medulla of the adrenal gland (Fig. 47–3*a*).

Oxytocin and antidiuretic hormone, produced by neuroendocrine cells in the hypothalamus, are short **peptides** composed of nine amino acids. Seven of the amino acids are identical in the two hormones, but the actions of these hormones are quite different.

The hormones glucagon, secretin, adrenocorticotropic hormone (ACTH), and calcitonin are somewhat longer peptides with about 30 amino acids in the chain. Insulin, secreted by the islets of Langerhans in the pancreas, is a small protein consisting of two peptide chains joined by disulfide bonds (see Fig. 3–21). Growth hormone, thyroid-stimulating hormone, and the gonadotropic hormones, all secreted by the anterior lobe of the pituitary gland, are proteins with molecular weights of 25,000 or more.

Prostaglandins and the juvenile hormones of insects are hormones derived from fatty acids (Fig. 47–2*a*). Prostaglandins are derivatives of polyunsaturated fatty acids, each with a five-carbon ring in its structure.

HORMONE SECRETION IS REGULATED BY NEGATIVE FEEDBACK MECHANISMS

In vertebrates, most endocrine glands secrete at least small amounts of their hormones continuously. Although present in minute amounts, more than 50 different hormones may be circulating in the blood at all times. Some hormones are transported bound to plasma proteins. Hormone molecules are continuously removed from the circulation by target tissues. They are also removed by the liver, which inactivates some hormones, and by the kidneys, which excrete them.

Hormone secretion is generally regulated internally by **negative feedback mechanisms.** Information about the amount of hormone or of some other substance in the blood or tissue fluid is fed back to the gland, which then responds to restore homeostasis. The parathyroid glands, located in the neck of tetrapod vertebrates, provide a good example of how negative feedback works.

The parathyroid glands regulate the calcium concentration of the blood. When the calcium concentration is not within homeostatic limits, nerves and muscles cannot function properly. For example, when insufficient calcium ions are present, neurons can fire spontaneously, causing muscle spasms. Even a slight decrease in calcium

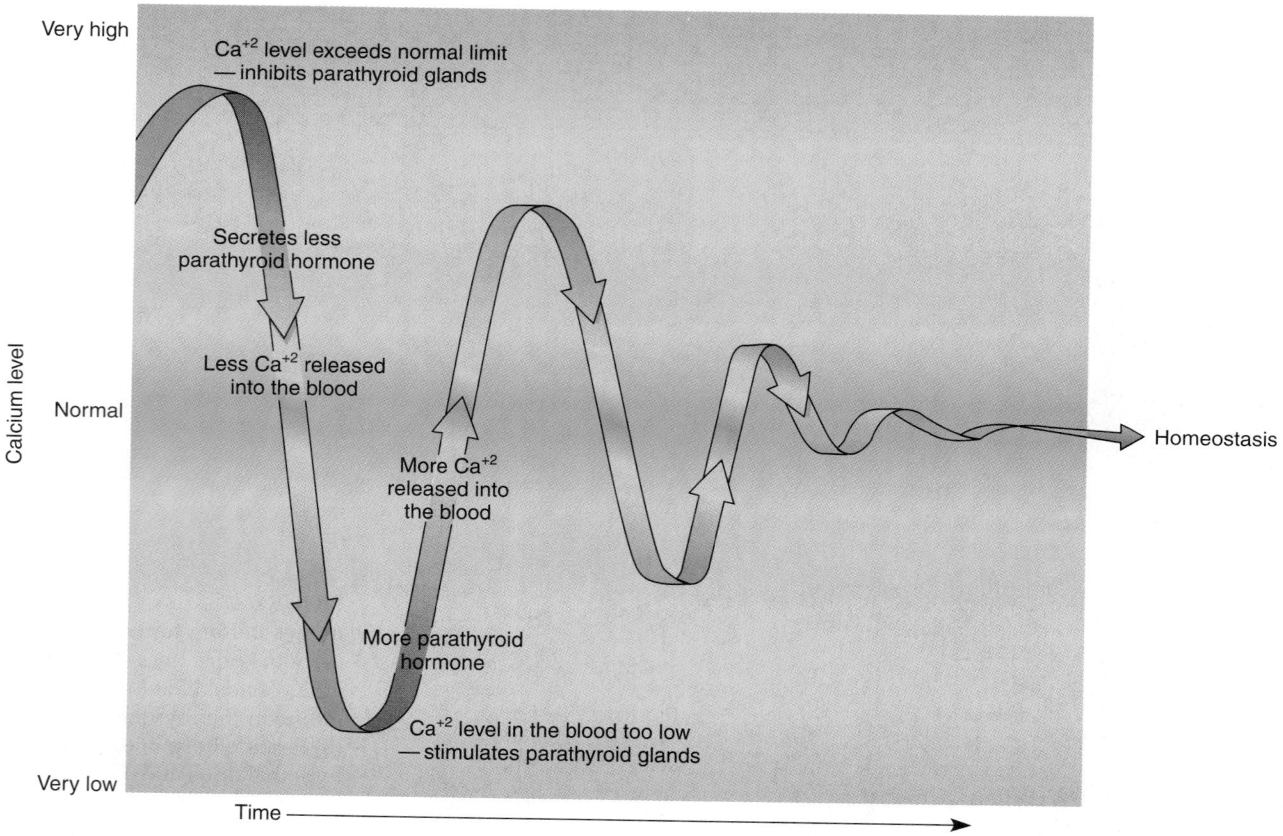

FIGURE 47–4 Most hormones are regulated by negative feedback mechanisms. Should calcium exceed normal levels, the parathyroid glands are inhibited and slow their release of hormone. When the calcium level in the blood falls below normal levels, the parathyroid glands are stimulated to release more parathyroid hormone. This hormone acts to increase the calcium level in the blood, thus restoring homeostasis. This diagram has been simplified.

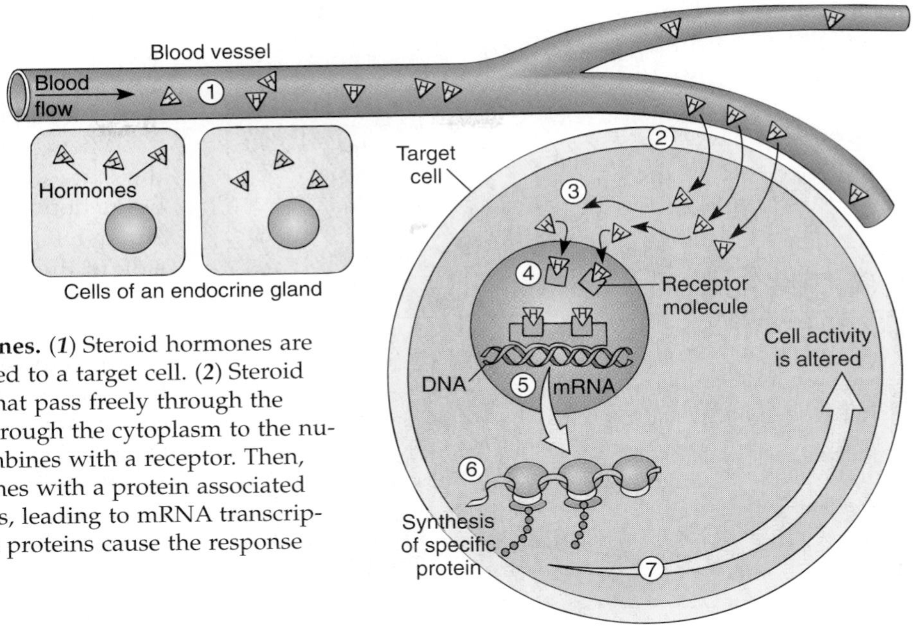

FIGURE 47–5 Steroid hormones activate genes. (*1*) Steroid hormones are secreted by an endocrine gland and transported to a target cell. (*2*) Steroid hormones are small, lipid-soluble molecules that pass freely through the plasma membrane. (*3*) The hormone passes through the cytoplasm to the nucleus. (*4*) Inside the nucleus, the hormone combines with a receptor. Then, the steroid hormone–receptor complex combines with a protein associated with the DNA. (*5*) This activates specific genes, leading to mRNA transcription and (*6*) synthesis of specific proteins. The proteins cause the response recognized as the hormone's action (*7*).

concentration signals the parathyroid glands to release more parathyroid hormone (Fig. 47–4). This hormone increases the concentration of calcium in the blood by stimulating the release of calcium from the bones and by increasing calcium reabsorption by the kidney tubules.

When the calcium concentration rises above normal limits, the parathyroid glands slow their output of hormone. Both responses are negative feedback mechanisms, because in both cases the effects are *opposite* (negative) to the stimulus; that is, more calcium leads to less hormone, and less calcium leads to more hormone. Negative feedback is the basis of most hormone regulation.

HORMONES COMBINE WITH SPECIFIC RECEPTOR PROTEINS IN TARGET CELLS

A hormone may pass through many tissues seemingly "unnoticed" until it reaches its target tissue. How does the target tissue recognize its hormone? Specific receptor proteins in or on the cells of the target tissues bind the hormone. This is a highly specific process. The receptor site is similar to a lock, and the hormones are similar to different keys. Only the hormone that fits the lock—the specific receptor—can influence the metabolic machinery of the cell.

Many hormones do not act directly on enzymes or other cellular compounds. Instead, the receptors serve as intermediaries, acting as **signal transducers.** The receptor converts an extracellular hormone signal into an intracellular signal that affects the function of the cell.

Several types of hormones may be involved in regulating the metabolic activities of a particular type of cell. In fact, many hormones produce a synergistic effect in which the presence of one hormone enhances the effects of another.

Some hormones enter the cell and activate genes

Steroid hormones and thyroid hormones (amino acid derivatives) are relatively small, lipid-soluble (hydrophobic) molecules that easily pass through the plasma membrane of the target cell. Specific protein receptors on the nuclear envelope or in the nucleus bind with the hormone to form a hormone-receptor complex (Fig. 47–5). The complex combines with a protein associated with the DNA. This combination activates certain genes and leads to the synthesis of messenger RNAs coding for specific proteins. These proteins produce the changes in structure or metabolic activity that are the actual effect of the hormone.

Some hormones work through second messengers

Many hormones do not enter the cell, but combine with receptors on the plasma membrane of the target cell. The hormonal message is then relayed to the appropriate site

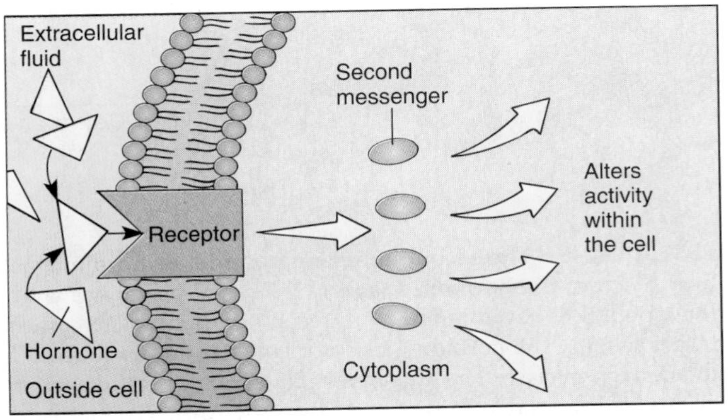

FIGURE 47–6 Peptide hormones combine with receptors on the plasma membrane of a target cell. The hormonal message is relayed by a second messenger.

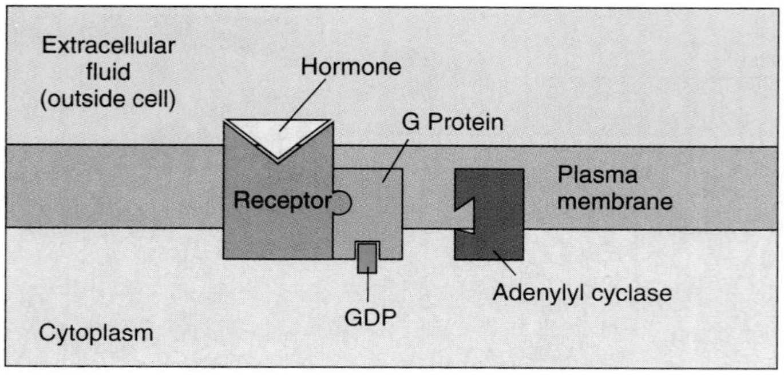

(a)

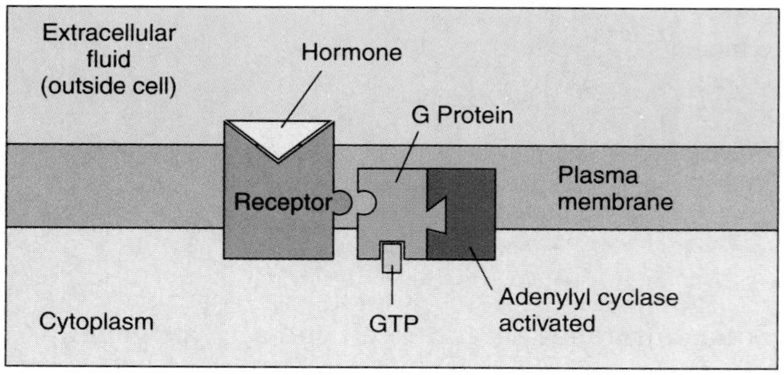

(b)

FIGURE 47–7 G proteins couple adenylyl cyclase activity with membrane receptors. (*a*) A hormone binds to a receptor on the plasma membrane. The hormone–receptor complex binds to a G protein. (*b*) GDP on the G protein is replaced by GTP. G protein undergoes a conformational change (change in shape) allowing it to bind with adenylyl cyclase. The adenylyl cyclase is activated and catalyzes the conversion of ATP to cAMP.

within the cell by a **second messenger** (Fig. 47–6). In the 1960s, Earl Sutherland identified **cyclic AMP (cAMP)** as a second messenger, and for his pioneering work, he received a Nobel prize in 1971.

When certain hormones combine with receptors on the extracellular face of the plasma membrane, a membrane-bound enzyme, **adenylyl cyclase,** is activated. Adenylyl cyclase, however, is located on the cytoplasmic side of the plasma membrane. How then does this activation occur?

A third protein, known as **G protein,** is also located on the cytoplasmic side of the plasma membrane. G protein acts as a shuttle for communication between the receptor and adenylyl cyclase (Fig. 47–7). Two types of G protein exist. One type, G_s, stimulates adenylyl cyclase, and the other, G_i, inhibits it. When the system is inactive, G protein binds to **guanosine diphosphate,** or **GDP,** which is similar to ADP, the hydrolyzed form of ATP.

Most hormones bind to a stimulatory receptor, resulting in activation of adenylyl cyclase. After the receptor-hormone complex is formed, the G protein releases GDP and then binds to **guanosine triphosphate (GTP).** GTP, like ATP, is an important molecule in energy transfers. The binding of G protein with GTP produces a conformational change in the G protein that enables it to bind with, and thereby activate, adenylyl cyclase.

Once activated, adenylyl cyclase catalyzes the conversion of ATP to cyclic AMP (Fig. 47–8). cAMP then activates one or more enzymes known as **protein kinases.** Each type of protein kinase catalyzes the phosphorylation of (addition of a phosphate group to) a specific protein. When the protein is phosphorylated, its function is altered, and it triggers a chain of reactions leading to some

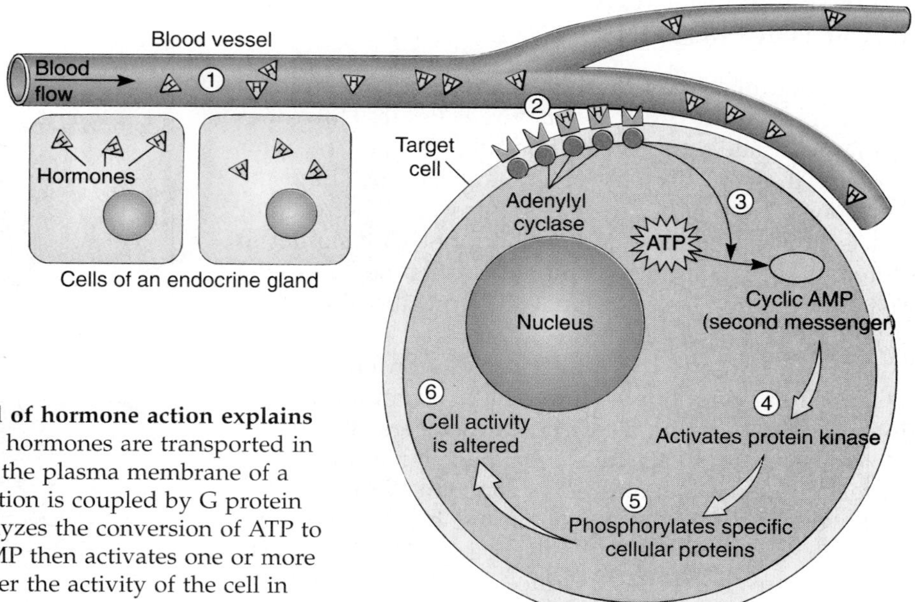

FIGURE 47–8 The second messenger model of hormone action explains how hydrophilic hormones work. (*1*) Peptide hormones are transported in the blood. (*2*) They combine with receptors in the plasma membrane of a target cell. (*3*) The hormone–receptor combination is coupled by G protein to adenylyl cyclase. The adenylyl cyclase catalyzes the conversion of ATP to cyclic AMP, a second messenger. (*4*) Cyclic AMP then activates one or more enzymes which (*5*) change proteins that (*6*) alter the activity of the cell in some way.

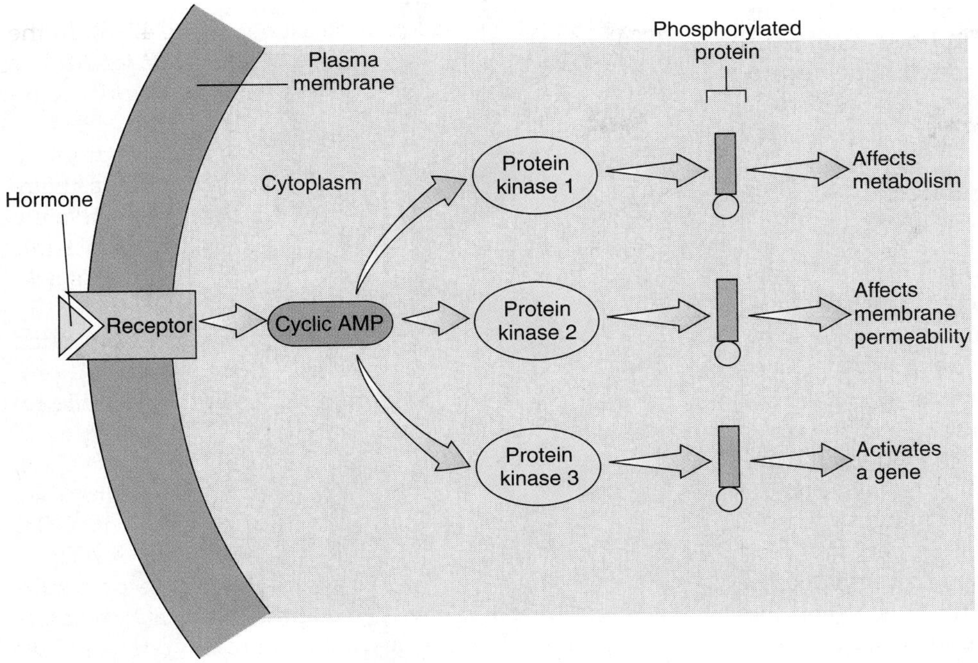

FIGURE 47–9 When hormone action increases the amount of cyclic AMP in the cell, protein kinases are activated. Protein kinases phosphorylate proteins, leading to a variety of effects.

metabolic effect. Any increase in cAMP is temporary. cAMP is rapidly inactivated by enzymes known as phosphodiesterases, which convert it to AMP.

Protein kinases inhibit or activate other enzymes. Each protein kinase acts on a different type of enzyme. Because different types of protein kinases exist in each kind of target cell (and even within different organelles of the same target cell), protein kinases are able to produce a wide variety of responses. For example, activation of one type of protein kinase may have a metabolic effect on the cell, a second protein kinase may affect membrane permeability, and a third may activate genes (Fig. 47–9).

Ions can also serve as second messengers. Certain receptors are linked to gated ion channels, for example calcium channels. When a hormone binds to the receptor, the calcium channel opens, allowing calcium ions to move into the cell. Cyclic AMP can also increase the cellular concentration of calcium by releasing calcium stored in the smooth endoplasmic reticulum. Once the calcium concentration in the cell is increased, calcium ions bind to the protein **calmodulin.** The calcium-calmodulin complex can then activate certain enzymes. Some cellular processes regulated by this complex include membrane phosphorylation, neurotransmitter release, and microtubule disassembly.

Prostaglandins are local chemical mediators

Prostaglandins are modified fatty acids released by many different organs, including the lungs, liver, digestive tract, and certain reproductive organs. Although present in very small quantities, prostaglandins affect a wide range of body processes. They are often referred to as **local hormones** because they act on cells in their immediate vicinity. Prostaglandins, which mimic many of the actions of cyclic AMP, interact with other hormones to regulate various metabolic activities.

About 16 prostaglandins are known, and they have different actions on different tissues. Some reduce blood pressure, whereas others raise it. Various prostaglandins dilate the bronchial passageways, inhibit gastric secretion, stimulate contraction of the uterus, affect nerve function, cause inflammation, and affect blood clotting. Those synthesized in the temperature-regulating center of the hypothalamus cause fever. In fact, the ability of aspirin and acetaminophen to reduce fever and decrease pain depends on the inhibition of prostaglandin synthesis.

Because prostaglandins are involved in the regulation of so many metabolic processes, they have great potential for a variety of clinical uses. At present, prostaglandins are used to induce labor in pregnant women, to induce abortion, and to promote the healing of ulcers in the stomach and duodenum. Their use as a birth control drug is being investigated. Prostaglandins may someday be used to treat a wide variety of illnesses, including asthma, arthritis, kidney disease, certain cardiovascular disorders, and some forms of cancer.

INVERTEBRATE HORMONES REGULATE GROWTH, DEVELOPMENT, METABOLISM, REPRODUCTION, MOLTING, AND PIGMENTATION

Among invertebrates, hormones are secreted mainly by neurons rather than by endocrine glands. These neurohormones regulate regeneration in hydras, flatworms, and annelids; molting and metamorphosis in insects; color changes in crustaceans; and growth, gamete production, reproductive behavior, and metabolic rate in other groups.

Color change in crustaceans is regulated by hormones

Crustaceans possess true (nonneural) endocrine glands, as well as masses of neuroendocrine cells. Their hormonal regulation is complex and affects many activities, including molting, reproduction, heart rate, and metabolism. One of the most interesting—and novel—activities regulated by hormones is color change.

Pigment cells of crustaceans are located in the integument beneath the exoskeleton. Pigment may be black, yellow, red, white, or even blue. Color changes are produced by changes in the distribution of pigment granules within the cells. Distribution of pigment is regulated by hormones produced by neuroendocrine cells. When the pigment is concentrated near the center of a cell, its color is only minimally visible. However, when it is dispersed throughout the cell, the color shows to advantage. These pigments provide protective coloration. By appropriate condensation or dispersal of specific types of pigments, a crustacean can approximate the color of its background.

Insect development is regulated by hormones

Like crustaceans, insects have endocrine glands as well as neuroendocrine cells. The various hormones interact with one another to regulate reproduction, metabolism, growth, and development, including molting and morphogenesis.

Hormonal control of development in insects is complex and varies among the many species. Generally, some environmental factor (e.g., temperature change) affects neuroendocrine cells in the brain. Once activated, these cells produce a hormone referred to as **brain hormone** (**BH,** or ecdysiotropin), which is transported down axons and stored in the paired **corpora cardiaca** (Fig. 47–10). When released from the corpora cardiaca, BH stimulates the **prothoracic glands,** endocrine glands in the prothorax, to produce **molting hormone,** also called **ecdysone.** Molting hormone stimulates growth and molting.

In the immature insect, paired endocrine glands called **corpora allata** secrete **juvenile hormone.** This hormone suppresses metamorphosis at each larval molt so that the insect increases in size while remaining in its immature state (Fig. 47–10); after the molt, the insect is still in a larval stage. When the concentration of juvenile hormone decreases, metamorphosis occurs, and the insect is transformed into a pupa (see Chapter 30). In the absence of juvenile hormone, the pupa molts and becomes an adult.

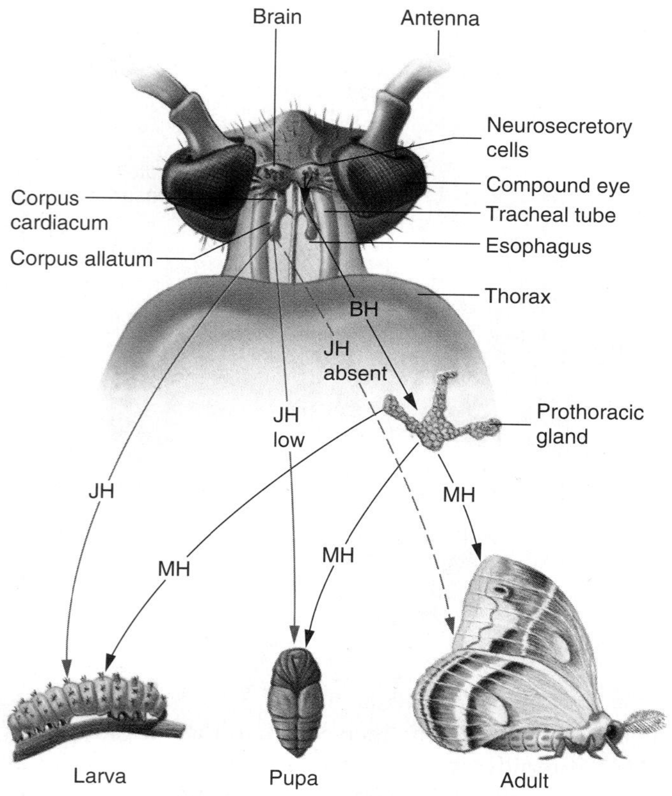

FIGURE 47–10 Neural and endocrine mechanisms regulate growth and molting in insects. The dissection of an insect head shows the paired corpora allata and corpora cardiaca, as well as the brain. Neurosecretory cells in the brain secrete a hormone, BH, which stimulates the prothoracic glands to secrete molting hormone (MH). In the immature insect, the corpora allata secrete juvenile hormone (JH), which suppresses metamorphosis at each larval molt. Metamorphosis to the adult form occurs when molting hormone acts in the absence of juvenile hormone.

The secretory activity of the corpora allata is regulated by the nervous system, and the amount of juvenile hormone decreases with successive molts.

VERTEBRATE HORMONES REGULATE GROWTH, DEVELOPMENT, FLUID BALANCE, METABOLISM, AND REPRODUCTION

Vertebrate hormones regulate such diverse activities as metabolic rate, reproduction, and blood homeostasis. They also help the body adjust to stress. The principal human endocrine glands are illustrated in Figure 47–11. Most vertebrates have similar endocrine glands. Table 47–1 gives the sources, target tissues, and physiological actions of some of the major vertebrate hormones.

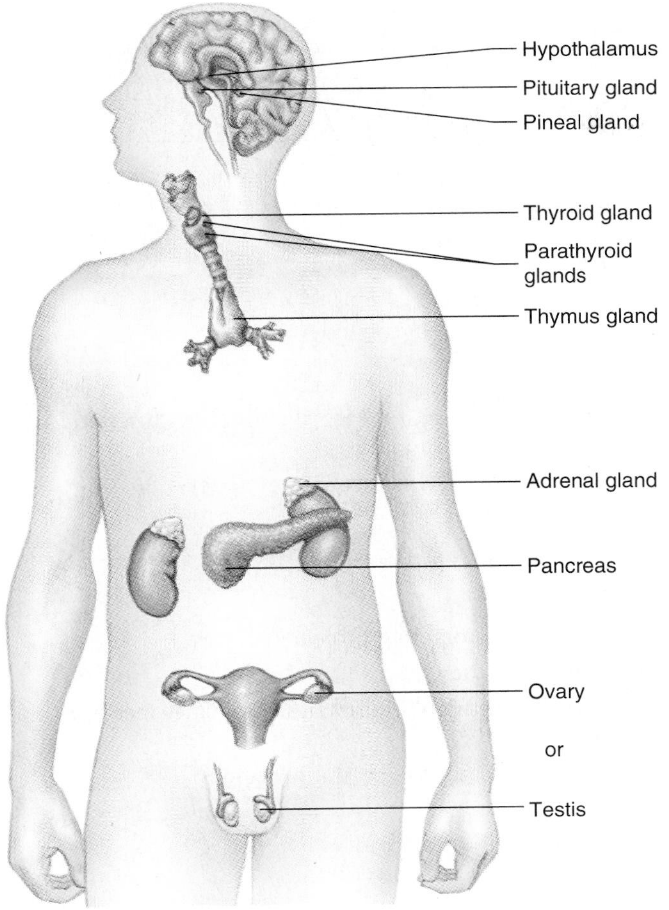

FIGURE 47–11 The principal human endocrine glands are illustrated.

Endocrine disorders may involve too little or too much hormone

When a disorder or disease process affects an endocrine gland, the rate of secretion may become abnormal. If **hyposecretion** (abnormally reduced output) occurs, target cells are deprived of needed stimulation. If **hypersecretion** (abnormally increased output) occurs, the target cells may be overstimulated. In some endocrine disorders, an appropriate amount of hormone is secreted, but the target cell receptors do not function properly. As a result, the target cells may not be able to take up the hormone. Any of these abnormalities leads to predictable metabolic malfunctions and clinical symptoms (Table 47–2).

The hypothalamus integrates neural and endocrine regulation

Most hormone activity is controlled directly or indirectly by the hypothalamus, which links the nervous and endocrine systems. In response to input from other areas of the brain and from hormones in the blood, neurons of the hypothalamus secrete neurohormones that regulate the release of hormones by the pituitary gland.

Because its secretions control the activities of several other endocrine glands, the **pituitary gland** is known as the master gland of the body. Truly a biological marvel, the pituitary gland is the size of a large pea and weighs only about 0.5 gram (0.02 ounce), yet it secretes at least nine distinct hormones that exert far-reaching influence over body activities. Connected to the hypothalamus by a stalk of nervous tissue, the pituitary gland consists of two main lobes, the **anterior lobe** and the **posterior lobe.** In some vertebrates, an intermediate lobe secretes hormones that regulate skin color.

The hypothalamus secretes several **releasing hormones** and **release-inhibiting hormones** that regulate the anterior lobe of the pituitary gland. These neurohormones enter capillaries and pass through special portal veins that connect the hypothalamus with the anterior lobe of the pituitary (Fig. 47–12). (These portal veins, like the hepatic portal vein, do not deliver blood to a larger vein directly but connect two sets of capillaries.) Within the anterior lobe of the pituitary, the portal veins divide into a second set of capillaries. The releasing and release-inhibiting hormones pass through the walls of these capillaries into the tissue of the anterior lobe, where each regulates the production and secretion of specific pituitary hormones.

The posterior lobe of the pituitary gland releases two hormones

Two peptide hormones, **oxytocin** and **antidiuretic hormone (ADH;** see Chapter 46), are secreted by the posterior lobe of the pituitary gland. These hormones are actually produced by specialized nerve cells in the hypothalamus. They reach the posterior lobe of the pituitary by flowing through axons extending from the hypothalamus (Fig. 47–13). Enclosed within tiny vesicles, the hormones accumulate in the axon endings until the neuron is stimulated. Then they are released and diffuse into surrounding capillaries.

In a female, oxytocin levels rise toward the end of pregnancy, stimulating the strong contractions of the uterus needed to expel the baby. Oxytocin is sometimes administered clinically (under the trade name Pitocin) to initiate or speed labor. After birth, when an infant sucks at its mother's breast, sensory neurons signal the release of oxytocin. The hormone stimulates contraction of cells surrounding the milk glands so that milk is let down into the ducts, from which it can be sucked by the infant. Because oxytocin also stimulates the uterus to contract, breast feeding promotes rapid recovery of the uterus to nonpregnant size.

Table 47–1 SOME ENDOCRINE GLANDS AND THEIR HORMONES†

Endocrine Gland and Hormone	*Target Tissue*	*Principal Actions*
Hypothalamus		
Releasing and release-inhibiting hormones	Anterior lobe of pituitary	Stimulates or inhibits secretion of specific hormones
Hypothalamus (production) **Posterior lobe of pituitary** (storage and release)		
Oxytocin	Uterus Mammary glands	Stimulates contraction Stimulates ejection of milk into ducts
Antidiuretic hormone (ADH)	Kidneys (collecting ducts)	Stimulates reabsorption of water; conserves water
Anterior lobe of pituitary		
Growth hormone (GH)	General	Stimulates growth by promoting protein synthesis
Prolactin	Mammary glands	Stimulates milk production
Thyroid-stimulating hormone (TSH)	Thyroid gland	Stimulates secretion of thyroid hormones; stimulates increase in size of thyroid gland
Adrenocorticotropic hormone (ACTH)	Adrenal cortex	Stimulates secretion of adrenal cortical hormones
Gonadotropic hormones* (follicle-stimulating hormone, FSH; luteinizing hormone, LH)	Gonads	Stimulate gonad function and growth
Thyroid gland		
Thyroxine (T_4) and triiodothyronine (T_3)	General	Stimulate metabolic rate; essential to normal growth and development
Calcitonin	Bone	Lowers blood-calcium level by inhibiting bone breakdown
Parathyroid glands		
Parathyroid hormone	Bone, kidneys, digestive tract	Increases blood calcium level by stimulating bone breakdown; stimulates calcium reabsorption by kidneys; activates vitamin D

*The gonadotropic hormones (FSH and LH), and the ovaries and testes and their hormones, are discussed in Chapter 48.
†For more detailed description see Table 48–2.

The anterior lobe of the pituitary gland regulates growth and other endocrine glands

The anterior lobe of the pituitary secretes prolactin, growth hormone, and several **tropic hormones,** which are hormones that stimulate other endocrine glands (Fig. 47–14). **Prolactin** stimulates the cells of the mammary glands to produce milk during lactation.

Growth hormone stimulates protein synthesis

Small children measure themselves periodically against their parents, eagerly awaiting that time when they, too, will be "big." Whether one will be tall or short depends on many factors, including genes, diet, and hormonal balance.

Growth hormone (**GH,** also called **somatotropin**) stimulates body growth by increasing uptake of amino

Table 47–1 continued

Endocrine Gland and Hormone	*Target Tissue*	*Principal Actions*
Islets of Langerhans of pancreas		
Insulin	General	Lowers glucose concentration in the blood by facilitating glucose uptake and utilization by cells; stimulates glycogen production; stimulates fat storage and protein synthesis
Glucagon	Liver; adipose tissue	Raises glucose concentration in the blood; mobilizes fat
Adrenal medulla		
Epinephrine and norepinephrine	Muscle; cardiac muscle; blood vessels; liver; adipose tissue	Help body cope with stress; increase heart rate, blood pressure, metabolic rate; reroute blood; mobilize fat; raise blood-sugar level
Adrenal cortex		
Mineralocorticoids (aldosterone)	Kidney tubules	Maintain sodium and potassium balance; increase sodium reabsorption; increase potassium excretion
Glucocorticoids (cortisol)	General	Help body adapt to long-term stress; raise blood-glucose level; mobilize fat
Pineal gland		
Melatonin	Gonads; pigment cells; other tissues(?)	Influences reproductive processes in hamsters and other animals; pigmentation in some vertebrates; may control biorhythms in some animals; may help control onset of puberty in humans
Ovary†		
Estrogens (estradiol)	General; uterus	Develop and maintain sex characteristics in female; stimulate growth of uterine lining
Progesterone	Uterus; breast	Stimulates development of uterine lining
Testis†		
Testosterone	General; reproductive structures	Develops and maintains sex characteristics of males; promotes spermatogenesis; responsible for adolescent growth spurt
Inhibin	Anterior lobe of pituitary	Inhibits FSH release in male

acids by the cells and by stimulating protein synthesis. Some of the effects of GH on skeletal growth are indirect. GH stimulates the liver to produce peptide growth factors called **somatomedins.** These growth factors (1) promote the linear growth of the skeleton by stimulating growth of cartilage in the epiphyseal plates, and (2) stimulate general tissue growth and increase in size of organs; this results from stimulation of protein synthesis and other anabolic processes.

GH also affects fat and carbohydrate metabolism. It promotes mobilization of fat from adipose tissues, raising the level of free fatty acids in the blood. In this protein-sparing operation, fatty acids become available for cells to use as fuel. How does this help to promote growth?

Fat mobilization by GH is also important during fasting or periods of prolonged stress, when the blood sugar level is low. Can you explain why? Human GH also has several actions that raise the blood sugar level.

Growth is affected by many factors

In adults as well as in growing children, GH is secreted in pulses throughout the day. Secretion of GH is regu-

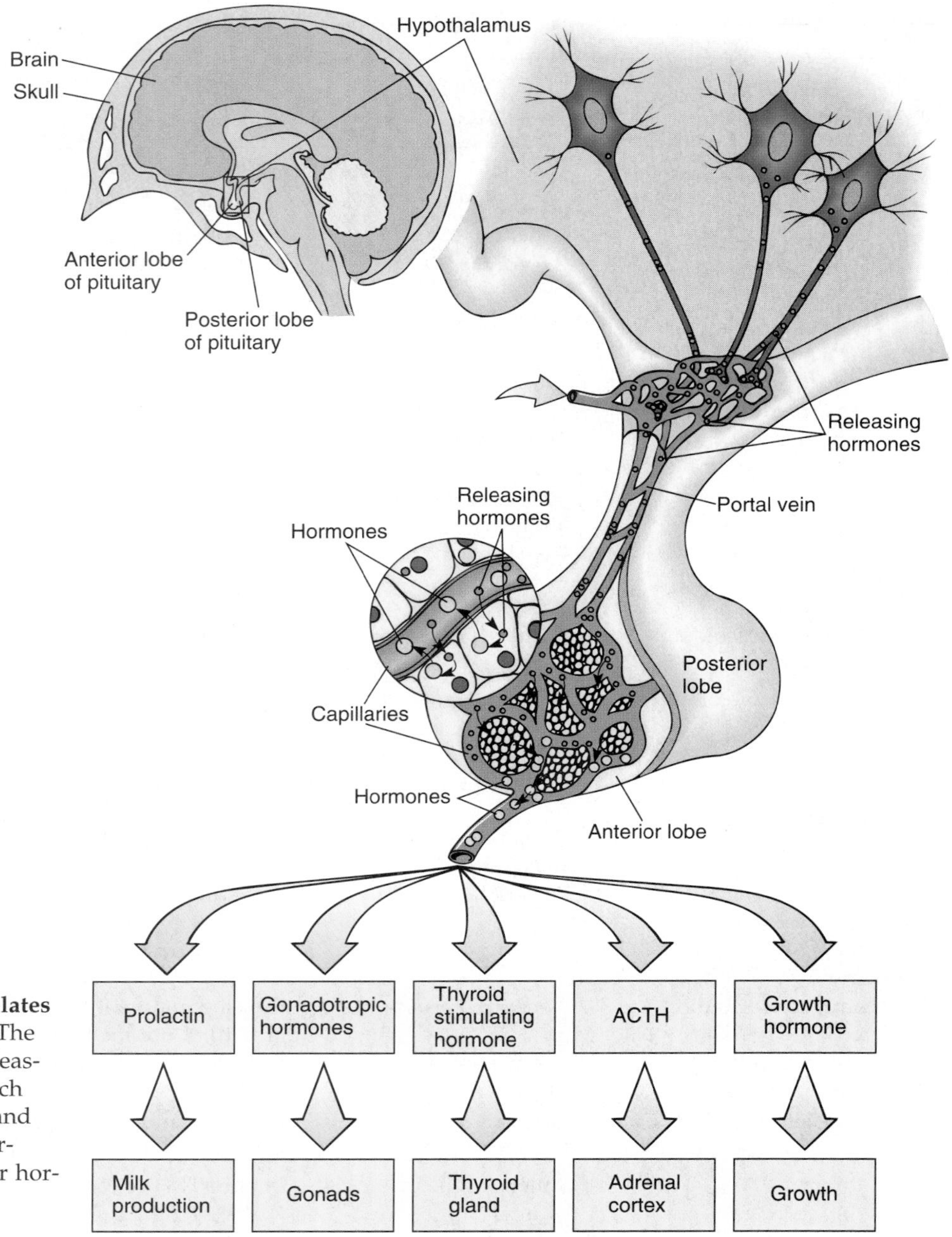

FIGURE 47–12 The hypothalamus regulates the anterior lobe of the pituitary gland. The hypothalamus secretes several specific releasing and release–inhibiting hormones, which reach the anterior lobe of the pituitary gland by way of portal veins. Each releasing hormone stimulates the release of a particular hormone by cells of the anterior lobe.

lated by both a **growth hormone–releasing hormone (GHRH)** and a **growth hormone–inhibiting hormone (GHIH,** also called **somatostatin**) released by the hypothalamus. A high level of GH in the blood signals the hypothalamus to secrete the inhibiting hormone, and the pituitary release of GH slows. A low level of GH in the blood stimulates the hypothalamus to secrete the releasing hormone, which in turn stimulates the pituitary gland to release more GH. Many other factors—including blood sugar level, decrease in amino acid concentration, and stress—influence secretion.

Remember your parents telling you to get plenty of sleep, eat properly, and exercise in order to grow? These age-old notions are supported by research. Secretion of GH does increase during exercise, probably because rapid metabolism by muscle cells lowers the blood sugar level. GH is secreted in a series of pulses 2 to 4 hours after a meal, and also about 1 hour after the onset of deep sleep.

Emotional support is also necessary for proper growth. Growth may be retarded in children who are deprived of cuddling, playing, and other forms of nurture, even when their needs for food and shelter are met. In

Table 47–2 CONSEQUENCES OF ENDOCRINE MALFUNCTION		
Hormone	*Hyposecretion*	*Hypersecretion*
Growth hormone	Pituitary dwarfism	Gigantism if malfunction occurs in childhood; acromegaly in adult
Thyroid hormones	Cretinism (in children); myxedema, a condition of pronounced adult hypothyroidism (metabolic rate is reduced by about 40%; patient feels tired all of the time and may be mentally slow); enlargement of the thyroid gland, goiter (see figure)	Hyperthyroidism; increased metabolic rate, nervousness, irritability; goiter
Parathyroid hormone	Spontaneous discharge of nerves; spasms; tetany; death	Weak, brittle bones; kidney stones
Insulin	Diabetes mellitus	Hypoglycemia
Hormones of adrenal cortex	Addison's disease (inability to cope with stress; sodium loss in urine may lead to shock)	Cushing's disease (edema gives face a full-moon appearance; fat is deposited about trunk; blood glucose level rises; immune responses are depressed)

Goiter commonly results from iodine deficiency. (John Paul Kay/Peter Arnold, Inc.)

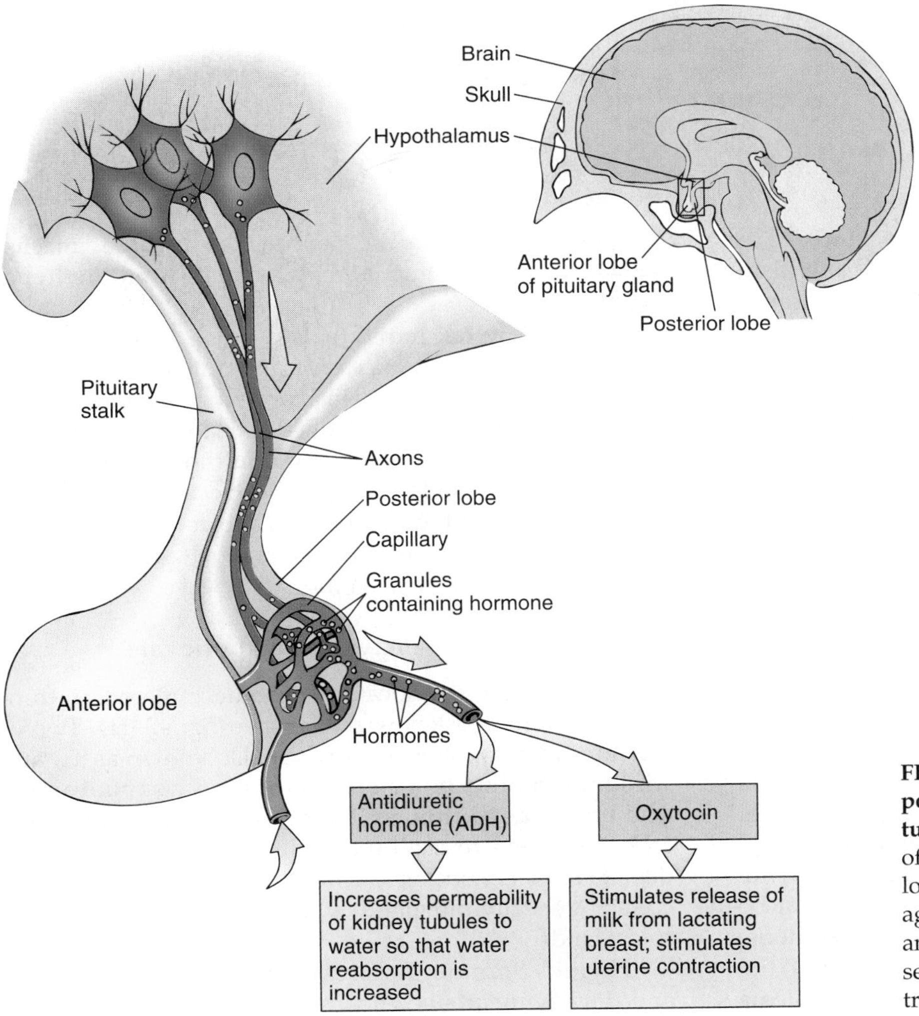

FIGURE 47–13 **The hormones secreted by the posterior lobe of the pituitary are manufactured in cells of the hypothalamus.** The axons of these neurons extend down into the posterior lobe of the pituitary. The hormones are packaged in granules that flow through these axons and are stored in their ends. The hormones are secreted into the interstitial fluid as needed and transported by the circulatory system.

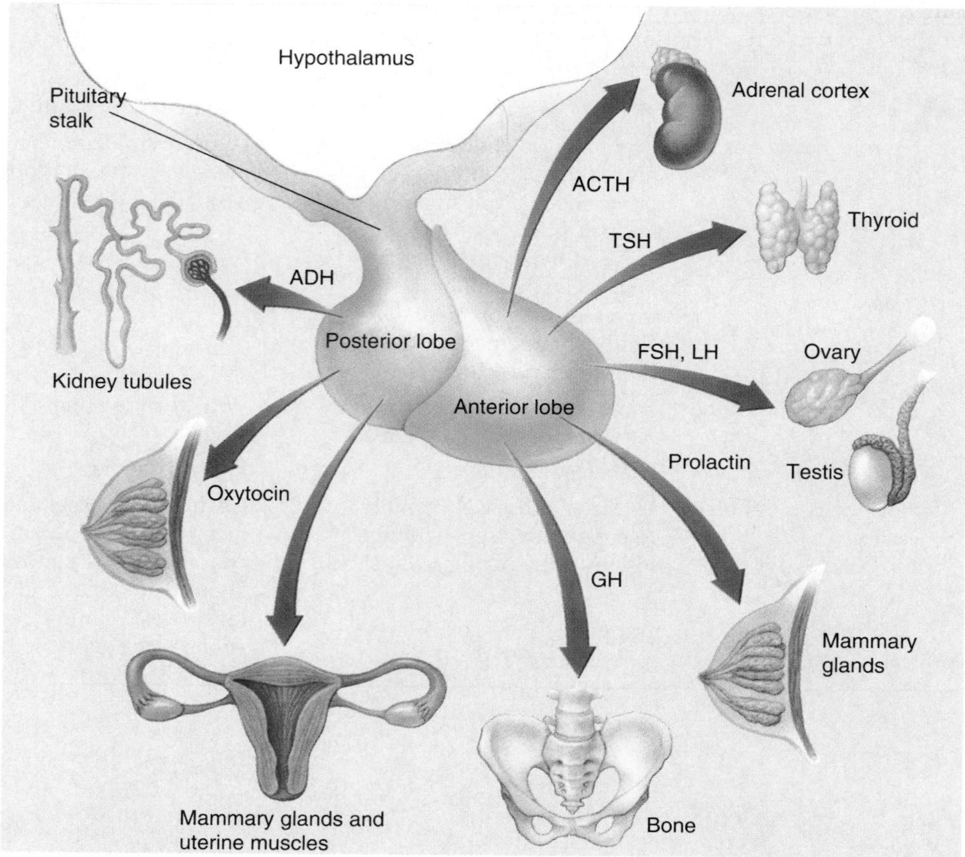

FIGURE 47–14 The anterior and posterior lobes of the pituitary gland release hormones that act on a wide variety of target tissues. The pituitary gland is suspended from the hypothalamus by a stalk of neural tissue.

extreme cases, childhood stress can produce a form of retarded development known as psychosocial dwarfism. Some emotionally deprived children exhibit abnormal sleep patterns, which may be the basis for decreased secretion of GH.

Other hormones also influence growth. Thyroid hormones appear to be necessary for normal GH secretion and function. Sex hormones must be present for the adolescent growth spurt to occur. However, the presence of sex hormones eventually causes the growth centers within the long bones to ossify, so that further increase in height is impossible even when GH is present.

Inappropriate amounts of growth hormone secretion result in abnormal growth

Have you ever wondered why some people fail to grow normally? Some may be **pituitary dwarfs**—individuals whose pituitary glands do not produce sufficient growth hormone during childhood. Although miniature, a pituitary dwarf has normal intelligence and is usually well proportioned. If the growth centers in the long bones are still active when this condition is diagnosed, it can be treated clinically by injection with growth hormone, which can now be synthesized commercially through the use of recombinant DNA technology. Growth problems may also result from the malfunction of other mechanisms, such as the regulating hormones from the hypothalamus.

Abnormally tall individuals develop when the anterior pituitary secretes excessive amounts of growth hormone during childhood. This condition is referred to as **gigantism.** If pituitary malfunction leads to hypersecretion of growth hormone during adulthood, the individual cannot grow taller. Instead, connective tissue thickens, and bones in the hands, feet, and face may increase in diameter. This condition is known as **acromegaly,** which means "large extremities."

Thyroid hormones increase metabolic rate

The **thyroid gland** is located in the neck region, in front of the trachea and below the larynx (Fig. 47–11). Two of the **thyroid hormones, thyroxine,** also known as $\mathbf{T_4}$, and **triiodothyronine,** or $\mathbf{T_3}$, are synthesized from the amino acid tyrosine and from iodine. Thyroxine has four iodine atoms attached to each molecule; T_3 has three. We will describe the actions of calcitonin, another hormone secreted by the thyroid gland, when we discuss the parathyroid glands.

Thyroid hormones are essential for normal growth and development, and they increase the rate of metabolism in most body tissues. They appear to regulate the synthesis of proteins necessary for cellular differentiation. Tadpoles cannot develop into adult frogs without thyroxine.

Thyroid secretion is regulated by negative feedback mechanisms

The regulation of thyroid hormone secretion depends on a negative feedback loop between the anterior pituitary and the thyroid gland (Fig. 47–15). When the concentration of thyroid hormones in the blood rises above normal, the anterior pituitary secretes less thyroid stimulating hormone (TSH):

> concentration of thyroid hormones increases → inhibits anterior pituitary → secretes less TSH → thyroid gland secretes less hormone → thyroid hormone concentrations decrease

Too much thyroid hormone in the blood also affects the hypothalamus, inhibiting secretion of TSH-releasing hormone. However, the hypothalamus is thought to exert its regulatory effects mainly in certain stressful situations, such as extreme weather change. Exposure to very cold weather may stimulate the hypothalamus to increase secretion of TSH-releasing hormone. This action raises body temperature through increased metabolic heat production.

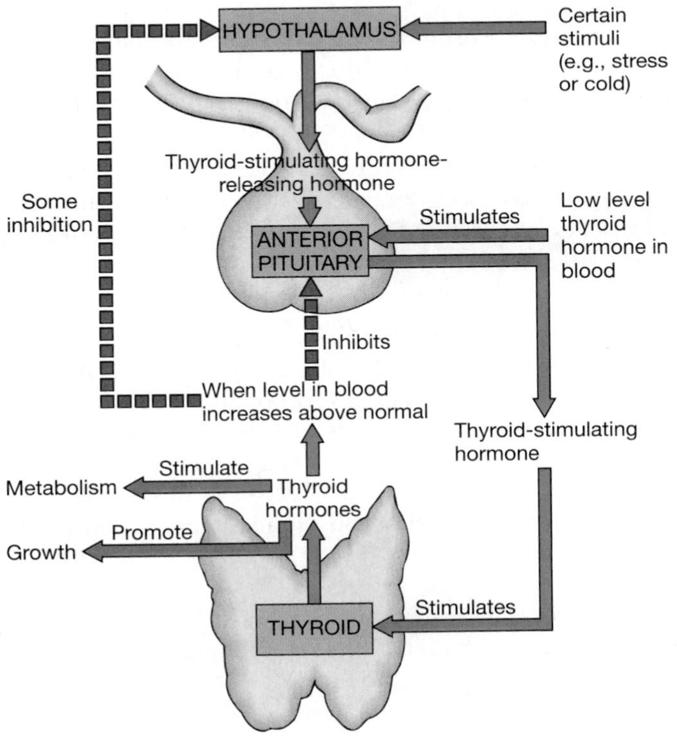

FIGURE 47–15 Thyroid hormone secretion is regulated by negative feedback mechanisms. Green arrows indicate stimulation; red arrows indicate inhibition.

When the concentration of thyroid hormones decreases, the pituitary secretes more TSH, as follows:

> low concentration of thyroid hormones → stimulates anterior pituitary → secretes more TSH → stimulates thyroid gland → secretes more thyroid hormones → thyroid hormone concentrations increase

TSH acts by way of cyclic AMP to promote synthesis and secretion of thyroid hormones and also to promote increased size of the thyroid gland itself.

Malfunction of the thyroid gland leads to specific disorders

Extreme hypothyroidism during infancy and childhood results in low metabolic rate and can lead to **cretinism,** a condition of retarded mental and physical development. When diagnosed early enough and treated with thyroid hormones, the effects of cretinism can be prevented.

An adult who feels like sleeping all of the time, has little energy, and is mentally slow or confused may also be suffering from hypothyroidism. When there is almost no thyroid function, the basal metabolic rate is reduced by about 40% and the patient develops **myxedema,** characterized by a slowing down of physical and mental activity. Hypothyroidism can be treated by oral administration of the missing hormone.

Hyperthyroidism does not cause abnormal growth but does increase metabolic rate by 60% or even more. This increase in metabolism results in the rapid use of nutrients, causing the individual to be hungry and to eat more. But this is not sufficient to meet the demands of the rapidly metabolizing cells, so individuals with this condition often lose weight. They also tend to be nervous, irritable, and emotionally unstable.

Any abnormal enlargement of the thyroid gland is termed a **goiter** and may be associated with either hyposecretion or hypersecretion (see figure in Table 47–2). One cause of hyposecretion is dietary iodine deficiency. Without iodine the gland cannot make thyroid hormones, so their concentration in the blood decreases. In compensation, the anterior pituitary secretes large amounts of TSH. The thyroid gland enlarges, sometimes to gigantic proportions. However, enlargement of the gland cannot increase production of the hormones, because the needed ingredient is still missing. Seafood is a rich source of iodine, and iodine is also added to table salt. In fact, thanks to iodized salt, goiter is no longer common in the United States and other developed countries. In other parts of the world, however, hundreds of thousands still suffer from this easily preventable disorder.

The parathyroid glands regulate calcium concentration

The **parathyroid glands** are embedded in the connective tissue surrounding the thyroid gland. These glands se-

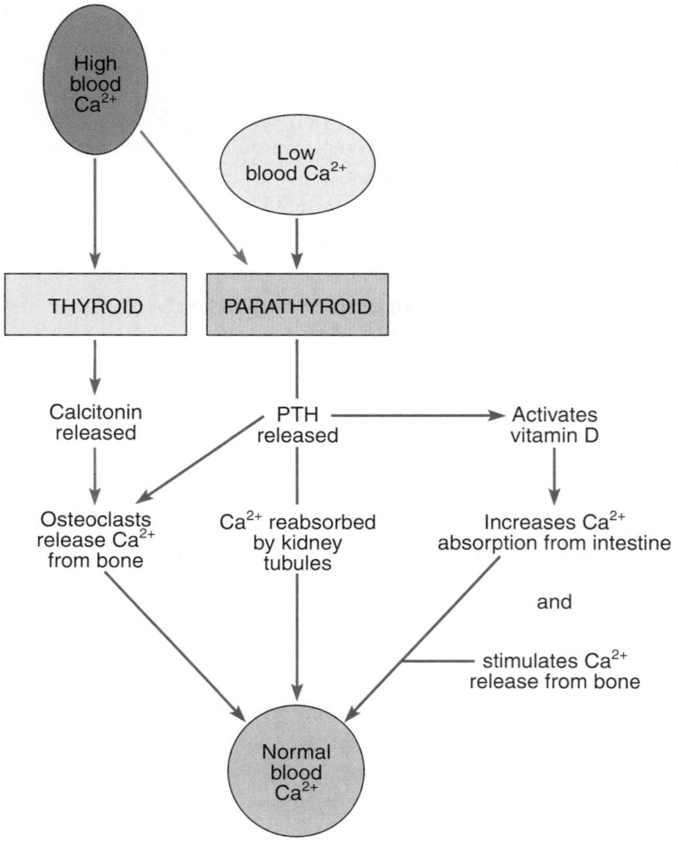

FIGURE 47–16 Hormones from the parathyroid glands and the thyroid gland regulate calcium metabolism. Red arrows indicate inhibition.

crete **parathyroid hormone,** which helps regulate the calcium level of the blood and tissue fluid (Fig. 47–16). Parathyroid hormone stimulates calcium release from bones and calcium reabsorption from the kidney tubules. It also activates vitamin D, which then increases the amount of calcium absorbed from the intestine.

Calcitonin, secreted by the thyroid gland, works antagonistically to parathyroid hormone. When the concentration of calcium rises above homeostatic levels, calcitonin is released and rapidly inhibits removal of calcium from bone.

The islets of the pancreas regulate glucose concentration

In addition to secreting digestive enzymes (see Chapter 45), the pancreas is an important endocrine gland. Its hormones, **insulin** and **glucagon,** are secreted by cells that form little clusters, the **islets of Langerhans,** throughout the pancreas (Fig. 47–17). About one million islets are present in the human pancreas. They are composed of **beta cells,** which secrete insulin, and **alpha cells,** which secrete glucagon.

Insulin lowers the concentration of glucose in the blood

Insulin stimulates cells of many tissues, including muscle and fat cells, to take up glucose from the blood. Once glucose enters muscle cells, it is either used immediately as fuel or stored as glycogen. Insulin also inhibits liver cells from releasing glucose. Thus, insulin activity results in *lowering* the glucose level in the blood.

Insulin also influences fat and protein metabolism. This hormone reduces the use of fatty acids as fuel and instead stimulates their storage in adipose tissue. It also inhibits the use of amino acids as fuel, thus promoting protein synthesis.

Glucagon raises the concentration of glucose in the blood

The actions of **glucagon** are opposite to those of insulin. The main effect of glucagon is to *raise* blood sugar level. It does this by stimulating liver cells to both convert glycogen to glucose, and to make glucose from other metabolites. Glucagon mobilizes fatty acids and amino acids as well as glucose.

Insulin and glucagon secretion are regulated by glucose concentration

Insulin and glucagon secretion are directly controlled by the concentration of glucose in the blood (Fig. 47–18). After a meal, when the blood glucose level rises as a result of intestinal absorption, beta cells are stimulated to increase insulin secretion. Then, as cells remove glucose from the blood, decreasing its concentration, insulin secretion decreases accordingly, as follows:

> blood glucose level too high → stimulates beta cells → insulin secretion increases → blood glucose concentration decreases

When one has not eaten for several hours, the concentration of glucose in the blood begins to fall. When it falls from its normal fasting level of about 90 milligrams of glucose per 100 milliliters of blood to about 70 milligrams of glucose, the alpha cells of the islets increase their secretion of glucagon. Glucose is mobilized from storage in the liver cells, and blood sugar concentration returns to normal:

> blood glucose concentration too low → stimulates alpha cells → glucagon secretion increases → blood glucose concentration increases

The alpha cells respond to the glucose concentration within their own cytoplasm, which reflects the blood

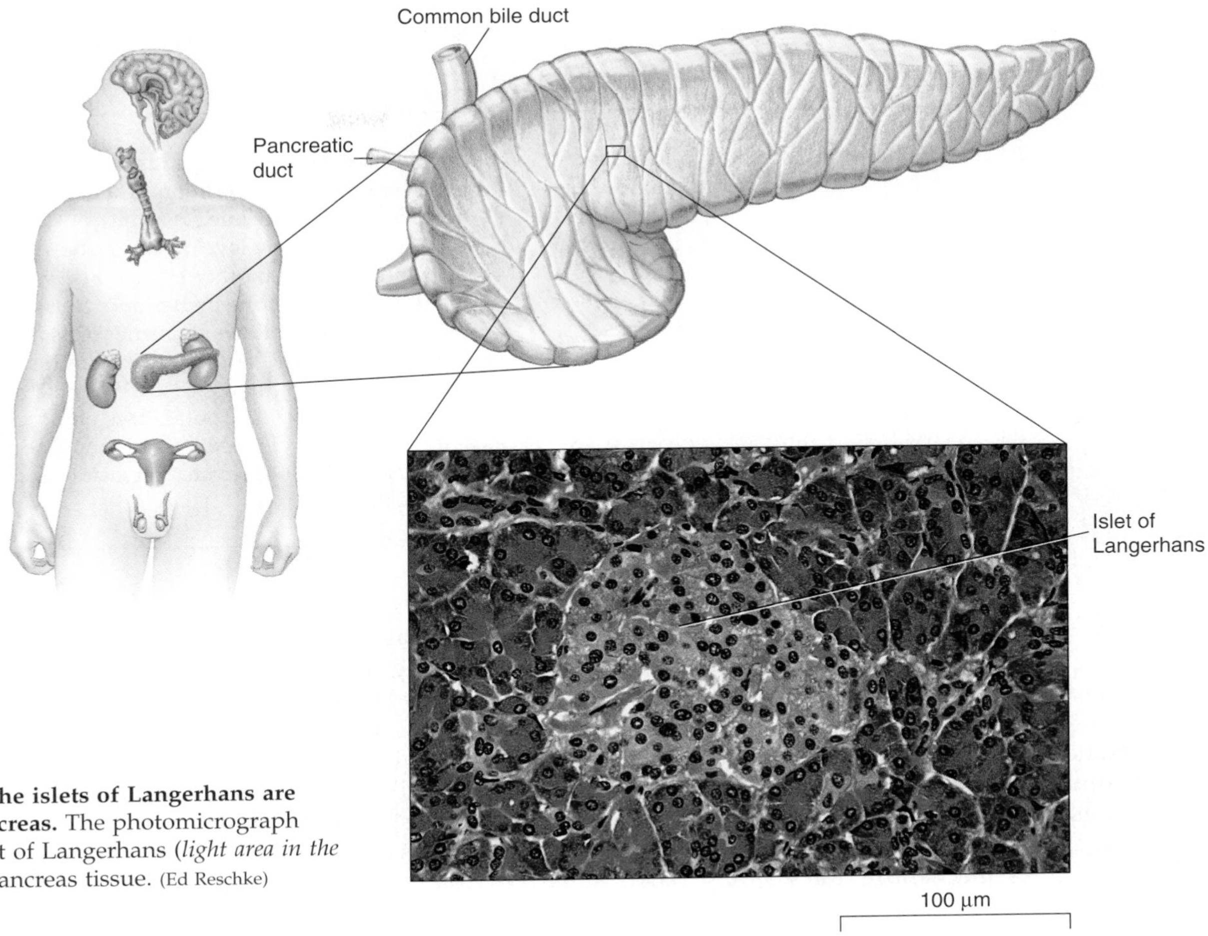

FIGURE 47–17 The islets of Langerhans are located in the pancreas. The photomicrograph shows a single islet of Langerhans (*light area in the center*) in human pancreas tissue. (Ed Reschke)

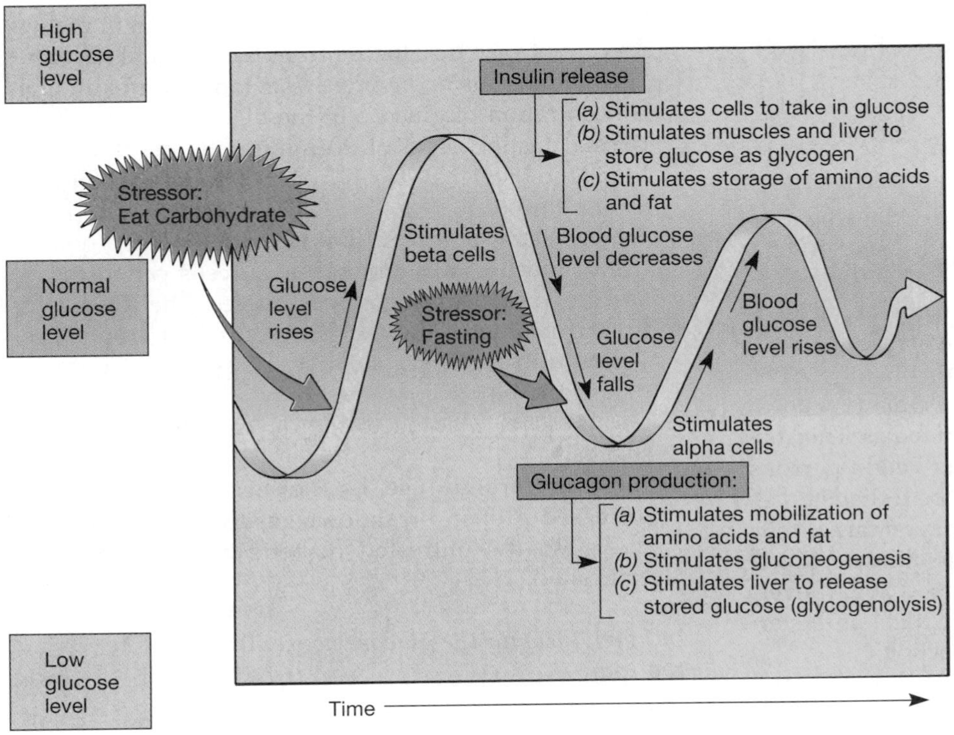

FIGURE 47–18 Insulin and glucagon regulate the concentration of glucose in the blood.

Making the Connection

Hormones, Metabolism, and Kidney Function

How does urine composition provide clues to insulin function? Insulin deficiency leads to disruption of carbohydrate, fat, and protein metabolism, and to electrolyte imbalance. The concentration of glucose is so high that it exceeds the renal threshold: the tubules in the kidneys are unable to return all the glucose in the filtrate to the blood. As a result, glucose is excreted in the urine. The presence of glucose in the urine is a simple screening test for diabetes.

Insulin-dependent cells in a diabetic can take in only about 25% of the glucose they require for fuel. The body turns to fat and protein for energy. Increased fat metabolism increases the formation of compounds known as ketone bodies. These compounds build up in the blood, causing **ketoacidosis,** a condition in which the body fluids and blood become too acidic.

When the ketone level in the blood rises, ketones appear in the urine, another clinical indication of diabetes mellitus. When ketone bodies and glucose are excreted in the urine, water follows by osmosis; as a result, urine volume increases. The resulting dehydration causes the diabetic to feel continually thirsty. When ketones are excreted, they also take sodium, potassium, and some other cations with them. Loss of these ions in the urine causes electrolyte imbalance. If severe, ketoacidosis can lead to coma and death. ▲

sugar level. When blood sugar level is high, there is generally a high level of glucose within the alpha cells, and glucagon secretion is inhibited.

It should be clear that insulin and glucagon work antagonistically to keep blood sugar concentration within normal limits. When the glucose level rises, insulin release brings it back to normal; when it falls, glucagon acts to raise it again. The insulin-glucagon system is a powerful, fast-acting mechanism for keeping blood sugar level within normal limits. Why do you think it is important to maintain a constant blood sugar level? Recall that brain cells depend on a continuous supply of glucose; they ordinarily are unable to use other nutrients as fuel. As we will discuss, several other hormones also affect blood sugar concentration.

Diabetes mellitus is a serious disorder of carbohydrate metabolism

Diabetes mellitus, the most common endocrine disorder, is a worldwide health problem. Of the estimated 10 million diabetics in the United States, about 40,000 die each year as a result of this disorder, making it one of the leading causes of death. Diabetes is a leading cause of blindness, kidney disorders, disease of small blood vessels, gangrene of the limbs, and various other malfunctions.

About 90% of diabetics have type II diabetes. This disorder develops gradually, usually in overweight persons over the age of 40. Some patients with type II diabetes secrete enough insulin, but insulin receptors on target cells cannot bind it, a condition known as insulin resistance. Many type II diabetics can keep their blood sugar levels within normal range by managing their diets and exercising, and so do not require insulin injections.

Insulin-dependent diabetes, referred to as type I (and formerly known as juvenile-onset diabetes), usually develops before age 30. In type I diabetes there is a marked decrease in the number of beta cells in the pancreas, resulting in insulin deficiency. Daily insulin injections are needed to correct the carbohydrate imbalance that results. Type I diabetes is thought to be an autoimmune disease in which antibodies mark the beta cells for destruction. This disorder may be caused by a combination of genetic predisposition and infection by a virus. Patients with diabetes have a shortened life expectancy because impaired lipid metabolism can cause atherosclerotic disease (see Chapter 42).

Similar metabolic disturbances occur in both types of diabetes mellitus:

1. **Decreased use of glucose.** Because the cells of diabetics cannot take up glucose from the blood, it accumulates there, causing *hyperglycemia* (an abnormally high concentration of glucose in the blood). Instead of the normal fasting level of 90 milligrams per 100 milliliters, the level may reach from 300 to more than 1000 milligrams.
2. **Increased fat mobilization.** Despite the large quantities of glucose in the blood, most cells cannot use it and must turn to other fuel sources. The absence of insulin promotes the mobilization of fat stores, providing nutrients for cellular respiration. But unfortunately, the blood lipid level may reach five times the normal level, leading to the development of atherosclerosis.
3. **Increased protein use.** Lack of insulin also results in increased protein breakdown relative to protein synthesis, so the untreated diabetic becomes thin and emaciated.

In hypoglycemia the glucose concentration is too low

Hypoglycemia, low blood glucose concentration, sometimes occurs in people who later develop diabetes. It may be an overreaction by the islets to glucose challenge. Too much insulin is secreted in response to carbohydrate ingestion. About 3 hours after a meal the blood sugar concentration falls below normal, making the individual feel very drowsy. If this reaction is severe enough, the patient may become uncoordinated or even unconscious.

Serious hypoglycemia can develop if diabetics receive injections of too much insulin or if the islets, because of a tumor, secrete too much insulin. The blood sugar concentration may then fall drastically, depriving the brain cells of their needed fuel supply. **Insulin shock** may result, a condition in which the patient may appear drunk, or may become unconscious, suffer convulsions, and even die.

The adrenal glands help the body adapt to stress

The paired **adrenal glands** are small, yellow masses of tissue that lie in contact with the upper ends of the kidneys (Fig. 47–19). Each gland consists of a central portion, the **adrenal medulla,** and a larger outer section, the **adrenal cortex.** Although joined anatomically, the adrenal medulla and cortex develop from different types of tissue in the embryo and function as distinct glands. Both secrete hormones that help to regulate metabolism, and both help the body adjust to stress.

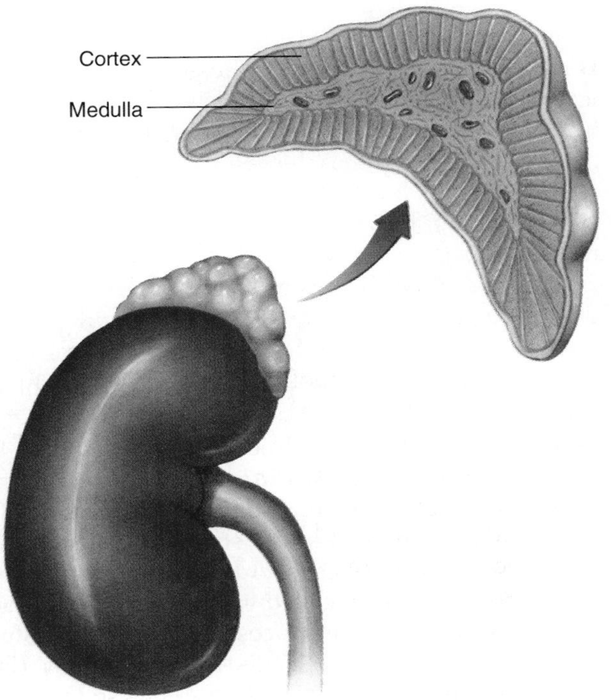

FIGURE 47–19 The adrenal glands are located above the kidneys. Each gland consists of a central adrenal medulla surrounded by an adrenal cortex.

The adrenal medulla initiates an alarm reaction

The adrenal medulla develops from neural tissue, and its secretion is controlled by sympathetic nerves. This endocrine gland secretes **epinephrine** (sometimes called adrenaline) and **norepinephrine** (noradrenaline). Chemically, these hormones are very similar; they belong to the chemical group known as **catecholamines** (derived from amino acids). Recall that norepinephrine also serves as a neurotransmitter released by sympathetic neurons and by some neurons in the central nervous system. When secreted by the adrenal medulla, its effects on the body are similar, but last about ten times longer because the hormone is removed from the blood slowly. About 80% of the hormone output of the adrenal medulla is epinephrine.

Under normal conditions, both epinephrine and norepinephrine are secreted continuously in small amounts. Their secretion is under neural control. When anxiety is aroused, messages are sent from the brain through sympathetic nerves to the adrenal medulla. Acetylcholine released by these neurons triggers the release of epinephrine and norepinephrine.

During a stressful situation, hormone secretion from this gland initiates an alarm reaction enabling you to think more quickly, fight harder, or run faster than usual. Metabolic rate increases by as much as 100%.

Hormones of the adrenal medulla cause blood to be rerouted to those organs essential for emergency action (Fig. 47–20). Blood vessels going to the brain, muscles, and heart are dilated, while those to the skin and kidneys are constricted. Constriction of blood vessels serving the skin has the added advantage of decreasing blood loss in case of hemorrhage (and explains the sudden paling that comes with fear or rage). At the same time, the heart beats faster. Thresholds in the reticular activating system of the brain are lowered, so you become more alert. Strength of muscle contraction increases. The adrenal medullary hormones also raise fatty acid and glucose levels in the blood, ensuring needed fuel for extra energy.

The adrenal cortex helps the body deal with chronic stress

All the hormones of the adrenal cortex are steroids synthesized from cholesterol, which in turn is made from acetyl coenzyme A. Although more than 30 types of steroids have been isolated from the adrenal cortex, this gland produces only three types of hormones in significant amounts: androgen, mineralocorticoids, and glucocorticoids.

An androgen (a masculinizing hormone) known as DHEA (dehydroepiandrosterone) is secreted by the adrenal cortex in both sexes. In the tissues it is converted to **testosterone,** a more powerful androgen and the principal male sex hormone. In males, androgen production by the adrenal cortex is not significant because much more androgen is produced by the testes. However, in fe-

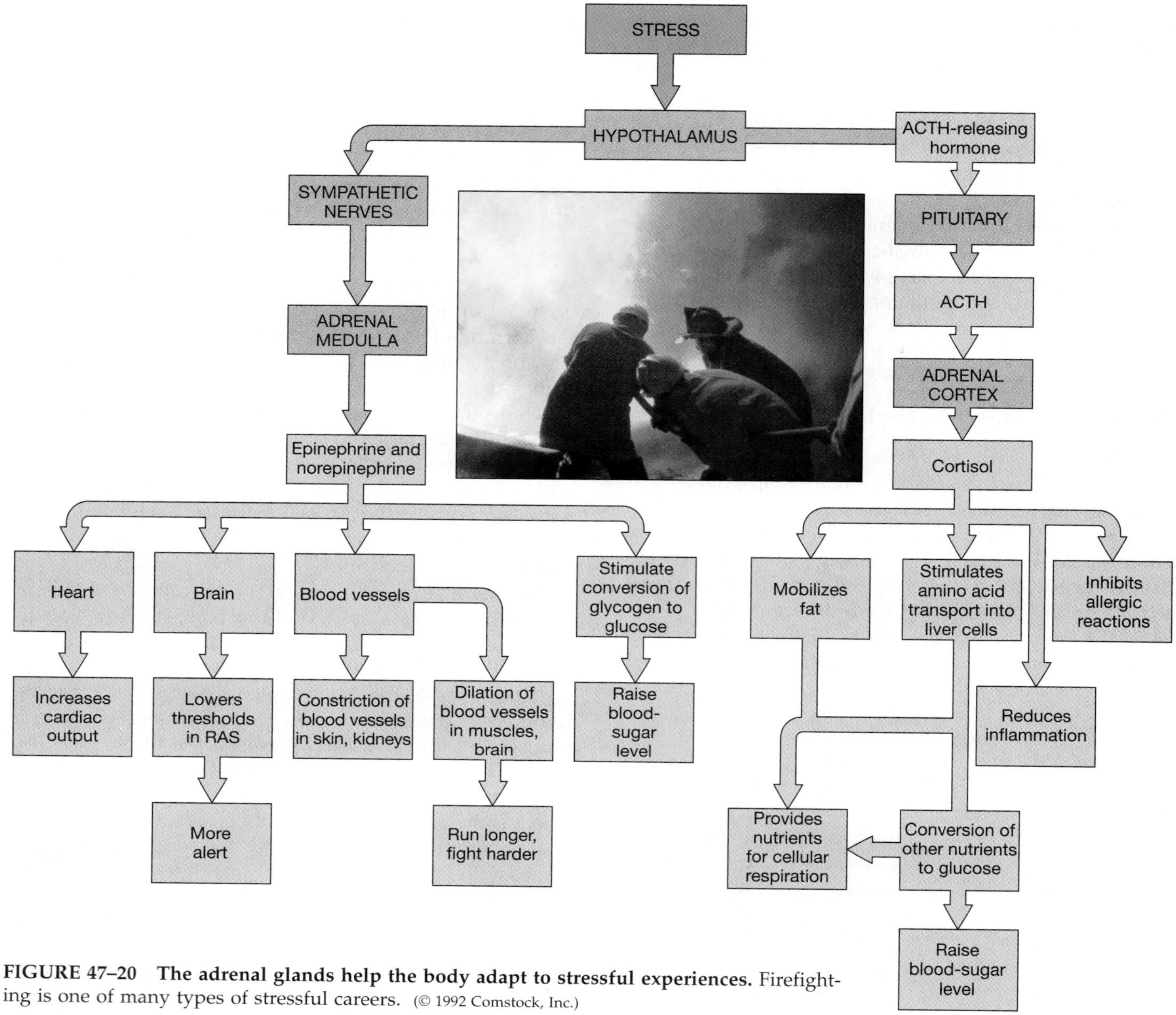

FIGURE 47–20 The adrenal glands help the body adapt to stressful experiences. Firefighting is one of many types of stressful careers. (© 1992 Comstock, Inc.)

males, the androgen produced by the adrenal cortex accounts for most of that circulating in the blood.

The principal **mineralocorticoid** is **aldosterone.** Recall from Chapter 46 that this hormone helps regulate fluid balance by regulating salt balance. In response to aldosterone, the kidneys reabsorb more sodium and excrete more potassium. As a result of increased sodium, extracellular fluid volume increases, which results in greater blood volume and elevated blood pressure.

When the adrenal glands do not produce enough aldosterone, large amounts of sodium are excreted in the urine. Water leaves the body with the sodium (because of osmotic pressure), and the blood volume may be so markedly reduced that the patient dies of low blood pressure.

Cortisol, also called hydrocortisone, accounts for about 95% of the **glucocorticoid** activity of the adrenal cortex. Cortisol helps ensure adequate fuel supplies for the cells when the body is under stress. Its principal action is to stimulate liver cells to produce glucose from other nutrients, a process called *gluconeogenesis.* Cortisol helps provide nutrients for glucose production by stimulating transport of amino acids into liver cells (Fig. 47–20). It also promotes mobilization of fats so that fatty acids are available for conversion to glucose. These actions ensure that glucose and glycogen are produced in

Focus On

Coping with Stress

The breakup of a relationship, an infection, or the anxiety of taking a test when you are not prepared are all stressors that arouse the body to action. The brain and the adrenal glands work together to help the body cope effectively. Information is transferred by nerves and hormones to many tissues and organs. Neural messages from the brain stimulate the adrenal medulla to release epinephrine and norepinephrine that prepare the body for fight or flight.

During stress, the hypothalamus secretes corticotropin-releasing hormone, which signals the anterior pituitary to secrete ACTH. The release of ACTH increases the secretion of glucocorticoids such as cortisol. These hormones adjust metabolism to meet the increased demands of the stressful situation (see Fig. 47–20).

Some forms of stress are short-lived. We react, quickly resolving the situation. Chronic stress, such as that from an unhappy marriage or psychologically toxic work situation, is harmful because of the side effects of long-term elevated levels of glucocorticoids.

Chronic stress has also been shown to damage the brain. Studies indicate that when rodents and monkeys are subjected to prolonged stress, elevation of glucocorticoids leads to degeneration of neurons, especially in the hippocampus (a part of the brain involved in learning and remembering). Elevated concentration of glucocorticoids also impairs the capacity of neurons in the hippocampus to withstand physiological insult such as reduced blood flow or oxygen to the brain.

Individuals approach stressful situations in their lives differently. A stressor that may result in chronic, damaging levels of glucocorticoids in one person may be viewed as a challenge by another. One strategy for reducing psychological and physiological response to stressors is to learn relaxation techniques. A variety of techniques are effective, including meditation, visual imagery, progressive muscle relaxation, and self-hypnosis. Practicing a relaxation technique can result in decreased activity of the sympathetic nervous system and reduced response to norepinephrine. For example, relaxation training has been shown to lower blood pressure in hypertensive patients, decrease the frequency of migraine headaches, and reduce chronic pain. ■

the liver, and the concentration of glucose in the blood rises. Thus, the adrenal cortex provides an important backup system for the adrenal medulla, ensuring glucose supplies when the body is under stress and in need of extra energy (see *Focus On: Coping with Stress*).

Almost any type of stress stimulates the hypothalamus to secrete **corticotropin-releasing factor, CRF.** This hormone stimulates the anterior pituitary to secrete **adrenocorticotropic hormone, ACTH.** ACTH regulates both glucocorticoid and aldosterone secretion. ACTH is so potent that it can result in up to a 20-fold increase in cortisol secretion within minutes. When the body is not under stress, high levels of cortisol in the blood inhibit both CRF secretion by the hypothalamus and ACTH secretion by the pituitary.

Destruction of the adrenal cortex and the resulting decrease in aldosterone and cortisol secretion cause **Addison's disease.** Reduction in cortisol prevents the body from regulating the concentration of glucose in the blood because it cannot synthesize enough glucose. The cortisol-deficient patient also loses the ability to cope with stress. If cortisol levels are significantly depressed, even the stress of mild infections can cause death.

Glucocorticoids are used clinically to reduce inflammation in allergic reactions, infections, arthritis, and certain types of cancer. These hormones help stabilize lysosome membranes so that they do not destroy tissues with their potent enzymes. Glucocorticoids also reduce inflammation by decreasing the permeability of capillary membranes, thereby reducing swelling. In addition, they reduce the effects of histamine and so are used to treat allergic symptoms.

When used in large amounts over long periods of time, glucocorticoids can cause serious side effects. They decrease the number of lymphocytes in the body, reducing the patient's ability to fight infections. Other side effects include ulcers, hypertension, diabetes mellitus, and atherosclerosis.

Abnormally large amounts of glucocorticoids, whether due to disease or drugs, can result in **Cushing's disease.** In this condition, fat is mobilized from the lower part of the body and deposited about the trunk. Edema gives the patient's face a full-moon appearance. Blood sugar level rises to as much as 50% above normal, causing adrenal diabetes. If this condition persists for several months, the beta cells in the pancreas may "burn out" from secreting excessive amounts of insulin. This can result in permanent diabetes mellitus. Reduction in protein synthesis causes weakness and decreases immune responses, so the patient often dies of infection.

Many other hormones are known

Many other tissues of the body secrete hormones. Several hormones secreted by the digestive tract regulate digestive processes. The **thymus gland** produces thymosin, a

Making the Connection

Anabolic Steroids, Physical Endurance, and Drug Abuse

Why do an estimated one million individuals in the U.S., one-half of them adolescents, abuse a group of synthetic hormones known as **anabolic steroids**? These synthetic androgens (male reproductive hormones) were developed in the 1930s to prevent muscle atrophy in patients with diseases that prevented them from moving about. In the 1950s, anabolic steroids became popular with professional athletes, who used them to increase muscle mass, physical strength, endurance, and aggressiveness. In truth, their athletic performance was probably enhanced, at least in part, by drug-induced euphoria and increased enthusiasm for training.

As with other hormones, the amount of steroids circulating in the body is precisely regulated. Use of anabolic steroids interferes with normal physiological processes, and has been linked with a variety of very serious physical and psychological side effects. "Steroid rage" refers to the mood swings, unpredictable rage, increased aggressiveness, and irrational behavior exhibited by many users. Anabolic steroids damage the liver and increase LDL concentration, raising the risk of cardiovascular disease.

In adolescents, the synthetic hormones stunt growth by prematurely closing the growth plates in bones. Abuse of these hormones can shrink the testes, leading to sterility. When their dangerous side effects came to light in the 1960s, steroid use became controversial. In 1973, the Olympic Committee banned the use of anabolic steroids. They are now prohibited worldwide by amateur and professional sports organizations.

Today, the typical anabolic steroid user is a male (95%) athlete (65%), most often a football player, weight lifter, or wrestler. However, about 10% of male high school students have used anabolic steroids and about one-third of these students are not even on a high school team; they use the hormone only to change their physical appearance—to pump up their muscles ("bulk up")—and increase endurance. Many anabolic steroid users have difficulty realistically perceiving their body images. They remain unhappy even after dramatic increases in muscle mass.

According to the U.S. Drug Enforcement Administration, a multimillion-dollar black market exists for anabolic steroids. They are both injected and taken in pill form. A group of researchers reported in the June 2, 1993 issue of the *Journal of the American Medical Association* that even during short-term, relatively low-dose trials, anabolic steroids have a significant effect on mood and behavior. At higher doses, subjects experience radical mood swings, including violent feelings, anger, and hostility. Users experience disturbed thought processes, forgetfulness, and confusion, and often find themselves easily distracted. Anabolic steroids remain in the body for a long time. Their metabolites (breakdown products) can be detected in the urine for up to six months. ▲

hormone that plays a role in immune responses, and the kidneys release hormones, one of which, renin, helps regulate blood pressure. Atrial natriuretic factor (ANF), secreted by the heart, promotes sodium excretion and lowers blood pressure. ANF acts, in part, by inhibiting aldosterone and renin release. The **pineal gland,** located in the brain, produces a hormone called **melatonin,** which influences the onset of sexual maturity. In Chapter 48 we discuss the principal reproductive hormones.

SUMMARY

I. The endocrine system consists of endocrine glands, cells, and tissues that secrete hormones. This system helps regulate many aspects of metabolism, growth, and reproduction. Hormones are transported to their target tissues via the blood.

II. Hormones are steroids, amino acid derivatives, peptides or proteins, or fatty acid derivatives.

III. Hormone secretion is regulated by negative feedback control mechanisms.

IV. Hormones combine with receptors of target cells.
   - A. Some hormones combine with receptors on the plasma membrane of their target cells and act by way of a second messenger, such as cyclic AMP or calcium.
   - B. Steroid hormones combine with receptors within the target cell; the hormone-receptor complex may stimulate a particular gene to initiate protein synthesis.
   - C. Prostaglandins may help regulate hormone action by mimicking cyclic AMP.

V. Many invertebrate hormones are secreted by neurons rather than by endocrine glands. They help to regulate regeneration, molting, metamorphosis, reproduction, and metabolism.
   - A. Pigment distribution in crustaceans is regulated by hormones secreted by neuroendocrine cells.
   - B. Hormones control development in insects.
      1. When stimulated by some environmental factor,

neuroendocrine cells in the insect brain secrete brain hormone (BH).
2. BH stimulates the prothoracic glands to produce molting hormone (ecdysone), which stimulates growth and molting.
3. In the immature insect, the corpora allata secrete juvenile hormone, which suppresses metamorphosis at each larval molt. The amount of juvenile hormone decreases with successive molts.

VI. In vertebrates, hormones help regulate growth, reproduction, salt and fluid balance, and many aspects of metabolism.

VII. Nervous and endocrine system regulation are integrated in the hypothalamus, which regulates the activity of the pituitary gland.
   A. The hormones oxytocin and ADH are produced by the hypothalamus and released by the posterior lobe of the pituitary.
   B. Secretion of anterior pituitary hormones is regulated by releasing and release-inhibiting hormones secreted by the hypothalamus.
   C. The anterior pituitary secretes growth hormone, prolactin, and several tropic hormones.
      1. Growth hormone (GH) stimulates body growth by promoting protein synthesis.
      2. Malfunctions in GH secretion can lead to pituitary dwarfism, gigantism, and acromegaly.

VIII. Thyroid hormones stimulate the rate of metabolism.
   A. Regulation of thyroid secretion depends mainly on a feedback system between the anterior pituitary and the thyroid gland.
   B. Hyposecretion of thyroxine during childhood may lead to cretinism; during adulthood it may result in myxedema. Goiter is associated with both hyposecretion and hypersecretion.

IX. The parathyroid glands regulate the calcium level in the blood.

X. The islets of the pancreas secrete insulin and glucagon.
   A. Insulin stimulates cells to take up glucose from the blood and so lowers blood sugar concentration.
   B. Glucagon raises blood glucose concentration by stimulating conversion of glycogen to glucose and production of glucose from other nutrients.
   C. Insulin and glucagon secretion are regulated directly by blood glucose levels.
   D. In diabetes mellitus, insulin deficiency or insulin resistance results in decreased utilization of glucose, increased fat mobilization, and increased protein utilization.

XI. The adrenal glands secrete hormones that help the body cope with stress.
   A. The adrenal medulla secretes epinephrine and norepinephrine.
   B. The adrenal cortex secretes sex hormones; mineralocorticoids (such as aldosterone, which increases the rate of sodium reabsorption and potassium excretion by the kidneys); and glucocorticoids (such as cortisol, which promotes gluconeogenesis).
   C. The hormones of the adrenal medulla help the body respond to stress by increasing the heart and metabolic rates and the strength of muscle contraction, and also by rerouting blood to those organs needed for fight or flight. The adrenal cortex acts as a backup system, ensuring adequate fuel supplies for the rapidly metabolizing cells.

SELECTED KEY TERMS

adrenal cortex
adrenal medulla
aldosterone
anabolic steroid
anterior lobe of pituitary
calcitonin
cortisol
cretinism
cyclic AMP (cAMP)
diabetes mellitus
endocrine gland
endocrine system
G protein
glucagon
glucocorticoid
goiter
growth hormone (GH)
hormone
hypersecretion
hyposecretion
hypothalamus
insulin
islets of Langerhans
juvenile hormone
mineralocorticoid
molting hormone
myxedema
negative feedback mechanism
norepinephrine
oxytocin
parathyroid gland
parathyroid hormone
pineal gland
pituitary dwarf
posterior lobe of pituitary
prostaglandin
protein kinase
release-inhibiting hormone
releasing hormone
second messenger
somatomedin
target tissue
thyroid gland
thyroid hormones
tropic hormone

POST-TEST

1. __________ glands lack ducts.
2. Endocrine glands produce chemical messengers called __________.
3. A second messenger important in the action of many hormones is __________ AMP.
4. Protein __________ are enzymes that catalyze the phosphorylation of specific proteins.
5. In insects, __________ hormone suppresses metamorphosis at each larval molt.
6. Hormone secretion is generally regulated by __________ __________ mechanisms.
7. In vertebrates, the __________ serves as a link between the nervous and endocrine systems.
8. Hormones bind to specific receptor proteins in their __________ tissues.
9. Cretinism is caused by __________-secretion of __________ hormones during childhood.
10. An abnormal enlargement of the thyroid gland is a(an) __________.
11. The principal action of __________ is to promote the production of glucose from other nutrients.
12. Calcitonin is secreted by the __________ gland(s).

13. In untreated __________ __________, glucose utilization is decreased and the body turns to fat and protein for fuel.
14. The __________ gland(s) increase(s) the concentration of calcium in the blood.
15. The __________ lobe of the __________ gland releases ADH.
16. __________ hormone(s) increase(s) metabolism.
17. The __________ __________ secretes cortisol.
18. Hyposecretion of __________ hormone during childhood can result in pituitary dwarfism.

REVIEW QUESTIONS

1. What is a hormone? What are some of the important functions of hormones?
2. How are hormones transported? How do they "recognize" their target tissues? What is the role of cyclic AMP in hormone action?
3. What are the functions of prostaglandins?
4. Why is the hypothalamus considered the link between the nervous and endocrine systems?
5. Describe the actions of (a) prolactin; (b) oxytocin; and (c) thyroid-stimulating hormone.
6. Draw a diagram to illustrate the regulation of thyroid hormone secretion by the anterior pituitary gland.
7. Explain the hormonal basis for (a) acromegaly; (b) pituitary dwarfism; (c) cretinism; (d) hypoglycemia; and (e) Cushing's disease.
8. Explain the antagonistic actions of insulin and glucagon in regulating blood glucose level.
9. Describe several physiological disturbances that result from diabetes mellitus.
10. What are the actions of epinephrine and norepinephrine?
11. How is the adrenal medulla regulated?
12. What types of hormones are released by the adrenal cortex, and what are the actions of each type?
13. Explain how the adrenal glands help the body deal with stress.
14. An injection of too much insulin may cause a diabetic to go into insulin shock, a condition in which the patient may appear drunk, or may become unconscious, suffer convulsions, and even die. From what you know about the actions of insulin, explain the physiological causes of insulin shock.
15. Label the diagram.

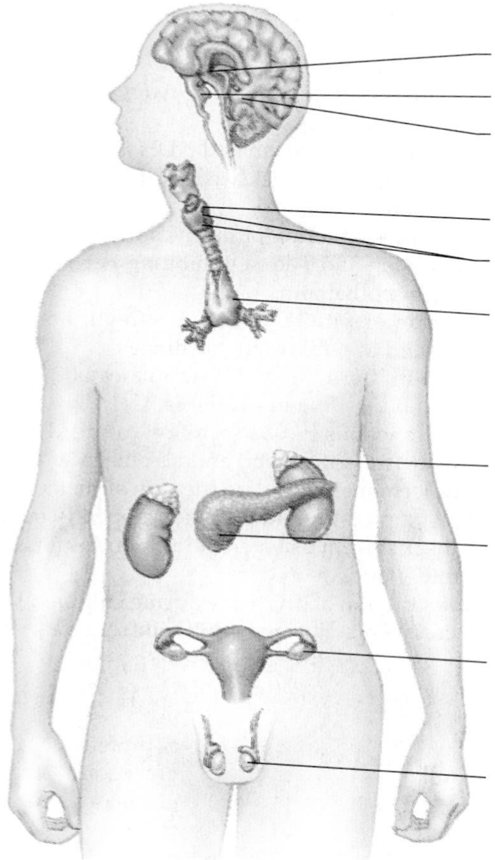

YOU MAKE THE CONNECTION

1. Human males have about the same amount of oxytocin circulating in their blood as do females, but its function in males is unknown. Hypothesize its function and design an ethical experiment to test your hypothesis. (*Hint*: Based on its effects on animal behavior, oxytocin has been referred to as the "nurturing hormone.")
2. How do receptors impart specificity within the endocrine system? What might be the advantages of having such complex mechanisms for hormone action as second messengers?
3. Why do you think it is important to maintain a constant blood sugar level? Several hormones discussed in this chapter affect carbohydrate metabolism. Why is it important to have more than one? How do they interact?

RECOMMENDED READINGS

Atkinson, M., and N. Maclaren. "What Causes Diabetes?" *Scientific American*, Vol. 263, No. 1, July 1990. An informative discussion of this common disorder.

Lienhard, G.E., J. Slot, D.E. James, and M.M. Mueckler. "How Cells Absorb Glucose." *Scientific American*, Vol. 266, No. 1, Jan. 1992. A discussion of recent findings on how insulin helps cells transport glucose.

Linder, M.E., and A.G. Gilman. "G Proteins." *Scientific American*, Vol. 267, No. 1, July 1992. G proteins play an important role in the mechanism of hormone action.

C H A P T E R 4 8

Reproduction

Nudibranchs *(Chromodoris* sp.) mating. Photographed in Indonesia. (Animals Animals © 1996, Bruce Watkins)

If there is one feature of a living organism that is unique, it is its ability to reproduce and perpetuate the species. The survival of each species requires that its members produce new individuals to replace those that die. At the molecular level, reproduction is a function of the capacity of nucleic acids to replicate themselves.

Many invertebrates can reproduce asexually. In **asexual reproduction** a single parent gives rise to two or more offspring that are genetically identical (except for mutation) to the parent. Asexual reproduction is an adaptation of sessile animals that cannot move about to search for mates. For animals that do move about, this method of reproduction is advantageous when the population density is low and mates are not readily available.

Sexual reproduction in animals involves the production and fusion of two types of **gametes**—sperm and eggs. Generally, a male parent contributes the sperm and a female parent contributes the egg, or **ovum** (plural, *ova*). The sperm provides genes coding for some of the male parent's traits, and the egg contributes genes coding for some of the female parent's traits. When sperm and egg unite, a fertilized egg, or **zygote,** forms. The zygote develops into a new organism, similar to both parents but not identical to either. Sexual reproduction has the biological advantage of promoting genetic variety among the members of a species. The offspring is the product of a particular combination of genes contributed by both parents, rather than a genetic copy of a single individual. By making possible the recombination of the inherited traits of two parents, sexual reproduction gives rise to offspring that may be better able to survive than either parent.

In many animals, reproduction involves remarkably complex structural, functional, and behavioral processes. In vertebrates, these processes are regulated by hormones secreted by the anterior pituitary gland and the gonads. This chapter summarizes the major features of animal reproductive processes and then focuses on human reproduction.

LEARNING OBJECTIVES

After you have studied this chapter you should be able to

1. Compare asexual and sexual reproduction, and compare external and internal fertilization.
2. Label the structures of the human male reproductive system on a diagram, and describe the functions of each.
3. Trace the passage of sperm cells through the human male reproductive system from their origin in the seminiferous tubules to their expulsion from the body in the semen.
4. Describe the actions of testosterone and of the gonadotropic hormones in the male.
5. Label the structures of the human female reproductive system on a diagram, and describe the functions of each.
6. Trace the development of an ovum (egg) and its passage through the female reproductive system until it is fertilized.
7. Identify the important events of the menstrual cycle, such as ovulation and menstruation; explain the actions of each of the hormones involved; and describe the hormonal regulation of the menstrual cycle.
8. Summarize the process of human fertilization.
9. Compare the modes of action, effectiveness, advantages, and disadvantages of the methods of birth control in Table 48–3.
10. Identify common sexually transmitted diseases, and describe their symptoms, effects, and treatments.

ASEXUAL REPRODUCTION IS COMMON AMONG SOME ANIMAL GROUPS

In asexual reproduction, a single parent may split, bud, or fragment to give rise to two or more offspring. Except for mutations, the offspring have hereditary traits identical to those of the parent. Sponges and cnidarians are among the animals that can reproduce by **budding.** A small part of the parent's body separates from the rest and develops into a new individual (Fig. 48–1). Sometimes the buds remain attached and become more-or-less independent members of a colony.

Oyster farmers learned long ago that when they tried to kill sea stars by chopping them in half and throwing the pieces back into the sea, the number of sea stars preying on the oyster bed doubled! A sea star can, in fact, regenerate an entire individual from a single arm. In flatworms, this ability to regenerate a part has become a method of reproduction known as **fragmentation.** The body of the parent may break into several pieces. Each piece then regenerates the missing parts and develops into a whole animal.

Parthenogenesis ("virgin development") is a form of asexual reproduction in which an unfertilized egg develops into an adult animal. It is common among some mollusks, some crustaceans, insects (especially honeybees and wasps), and some reptiles. Parthenogenesis may occur for several generations, followed at some point by sexual reproduction in which males develop, produce sperm, and mate with the females to fertilize their eggs. In some species, parthenogenesis appears to be an adaptation for survival in times of stress or serious population decline.

FIGURE 48–1 ***Hydra*** **reproduces asexually by budding.** A part of the body grows outward and then separates and develops into a new individual. The portion of the parent body that buds is not specialized exclusively for performing a reproductive function. The *Hydra* shown here is also reproducing sexually as evidenced by the egg (*at left*). (Richard Campbell/Biological Photo Service)

Sexual reproduction is the most common type of animal reproduction

Most animals reproduce sexually by fusion of sperm and egg. The egg is typically large and nonmotile, with a store of nutrients that supports the development of the embryo. The sperm is usually small and motile, adapted to propel itself by beating its long, whiplike flagellum.

Many aquatic animals practice **external fertilization** in which the gametes meet outside the body (Fig. 48–2). Mating partners usually release eggs and sperm into the water simultaneously. Many gametes are lost; some are eaten by predators. However, so many gametes are released that sufficient numbers of sperm and egg cells do meet to perpetuate the species.

In **internal fertilization,** matters are left less to chance. The male generally delivers sperm cells directly into the body of the female. Her moist tissues provide the watery medium required for the movement of sperm, and the gametes fuse inside the body. Most terrestrial animals, as well as aquatic reptiles, birds, and mammals, practice internal fertilization.

Hermaphroditism is a form of sexual reproduction in which a single individual produces both eggs and sperm. A few hermaphrodites, such as the tapeworm, are capable of self-fertilization. Earthworms are more typical hermaphrodites. Two animals copulate, and each inseminates the other. In some hermaphroditic species, self-fertilization is prevented by the development of testes and ovaries at different times.

Human reproduction: the male provides sperm

The human male, like other male mammals, has the reproductive role of producing sperm cells and delivering them into the female reproductive tract. When a sperm combines with an egg, it contributes its genes and determines the sex of the offspring. The male reproductive system is illustrated in Figure 48–3.

(a)

(b)

FIGURE 48–2 External fertilization is practiced by many aquatic animals, whereas most terrestrial animals, as well as aquatic reptiles, birds, and mammals, practice internal fertilization. (***a***) External fertilization is illustrated by these spawning frogs (*Rana temporaria*). Most amphibians must return to water for mating. The female lays a mass of eggs, while the male mounts her and simultaneously deposits his sperm in the water. (***b***) Internal fertilization is practiced by mammals, such as these lions (*Pantera leo*). (*a*, Animals Animals © 1996 Zig Leszczynski; *b*, Fritz Polking/Dembinsky Photo Associates)

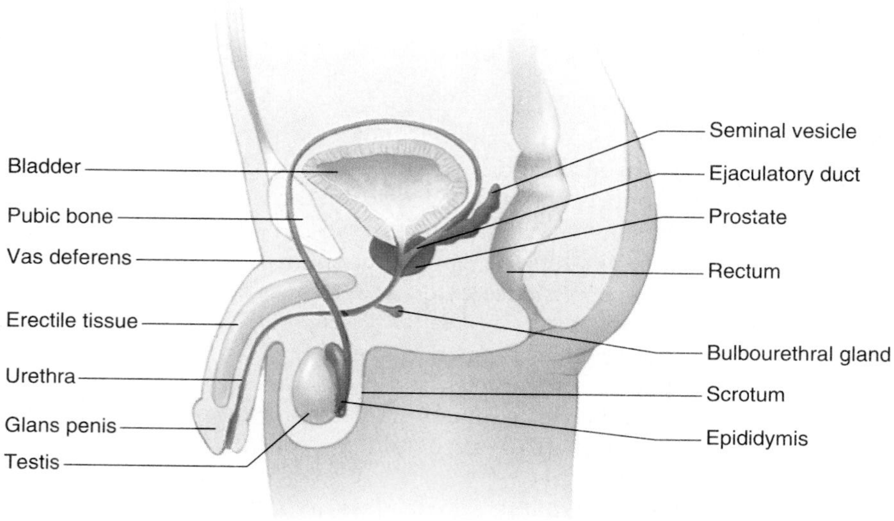

FIGURE 48–3 The scrotum, penis, and pelvic region of the human male are shown in sagittal section to illustrate their internal structure.

The testes produce sperm

In humans and other vertebrates, **spermatogenesis,** the process of sperm cell production, occurs in the paired male gonads, or **testes** (sing., *testis*). Spermatogenesis takes place within a vast tangle of hollow tubules, called the **seminiferous tubules,** within each testis (Fig. 48–4).

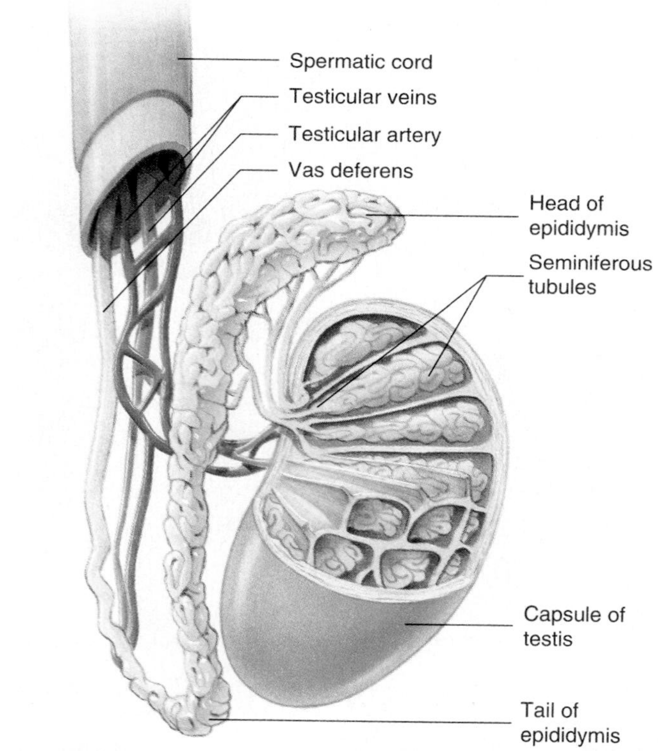

FIGURE 48–4 The testis, epididymis, and spermatic cord are partly dissected and exposed to show their structure. The testis is shown in sagittal section to illustrate the arrangement of the seminiferous tubules.

Spermatogenesis begins with undifferentiated cells, the **spermatogonia** in the walls of the tubules (Fig. 48–5).

The spermatogonia divide by mitosis, producing more spermatogonia. Some enlarge and become **primary spermatocytes,** which undergo meiosis. (You may want to review the discussion of meiosis in Chapter 9.) In many animals, gamete production occurs only in the spring or fall, but humans have no special breeding season. In the human adult male, spermatogenesis proceeds continuously, and millions of sperm are produced each day.

Each primary spermatocyte undergoes a first meiotic division producing **secondary spermatocytes** (Fig. 48–6). In the second meiotic division, each secondary spermatocyte gives rise to two **spermatids.** Four spermatids are produced from the original primary spermatocyte. Each haploid spermatid differentiates into a mature **sperm.** The sequence is as follows:

> spermatogonium (diploid) → primary spermatocyte (diploid) → two secondary spermatocytes (haploid) → four spermatids (haploid) → four mature sperm (haploid)

Each mature sperm consists of a head, a midpiece, and a flagellum (Fig. 48–7). The head consists of the nucleus and a cap, called an **acrosome,** that differentiates from the Golgi complex. The acrosome helps the sperm penetrate the egg. Mitochondria, located in the midpiece of the sperm, provide the energy for movement of the flagellum. The sperm flagellum has the typical eukaryotic 9 + 2 arrangement of microtubules. During its development, most of the sperm's cytoplasm is discarded and is phagocytized by the large nutritive **Sertoli cells** present within the seminiferous tubules.

Making the Connection

Reproduction and Tight Junctions

Of what importance are tight junctions in the cells of the reproductive system? The Sertoli cells in the seminiferous tubules form a ring around the fluid-filled lumen of the tubule. They are joined to one another by tight junctions (see Chapter 5) and together form a blood-testis barrier. This barrier prevents the entrance into the tubule of harmful substances that could interfere with spermatogenesis. It also stops sperm from passing out of the tubule and into the blood, where they could stimulate an immune response.

In the female, the cells (known as granulosa cells) surrounding the developing follicle perform a corresponding function. They are connected by tight junctions that form a protective barrier around the developing ovum. ▲

Human sperm cells cannot develop at body temperature. Although the testes develop within the abdominal cavity of the male embryo, about two months before birth they descend into the **scrotum,** a skin-covered sac suspended from the groin. The scrotum serves as a cooling unit, maintaining sperm below body temperature. In rare cases, the testes do not descend. If this condition is not surgically corrected, the seminiferous tubules eventually degenerate and the male becomes **sterile.**

The scrotum is an outpocketing of the pelvic cavity and is connected to it by the **inguinal canals.** As they descend, the testes pull their blood vessels, nerves, and conducting tubes after them. The inguinal region is a weak place in the abdominal wall. Straining the abdominal

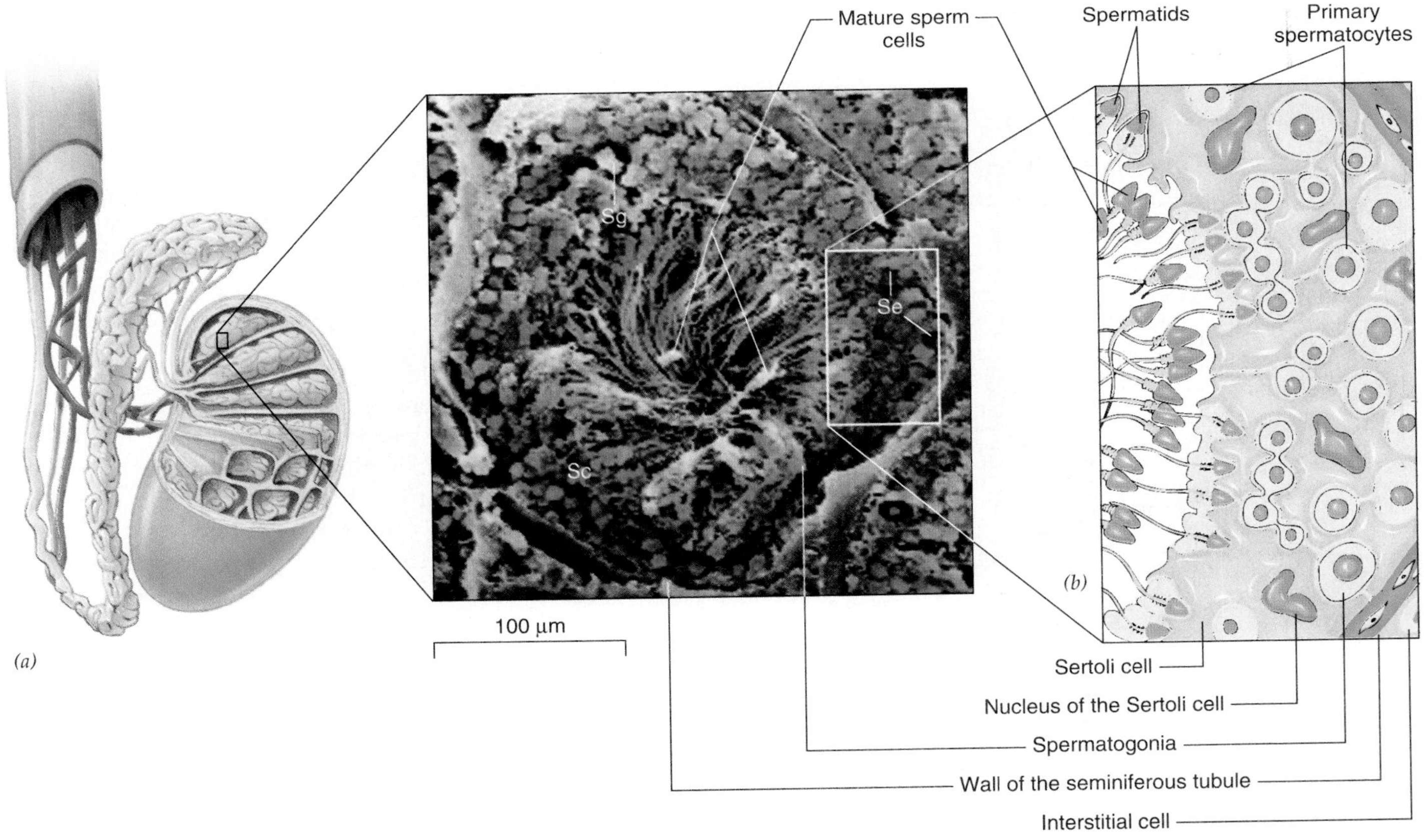

FIGURE 48–5 Spermatogenesis takes place in the seminiferous tubules. (*a*) A scanning electron micrograph of a transverse section through a seminiferous tubule. Se, Sertoli cell; Sc, primary spermatocyte; Sg, spermatogonium. (*b*) Sperm cells can be seen in various stages of development. Can you identify the sequence of sperm cell differentiation? Note the interstitial cells and the large nutritive Sertoli cells. (*a*, Custom Medical Stock Photo)

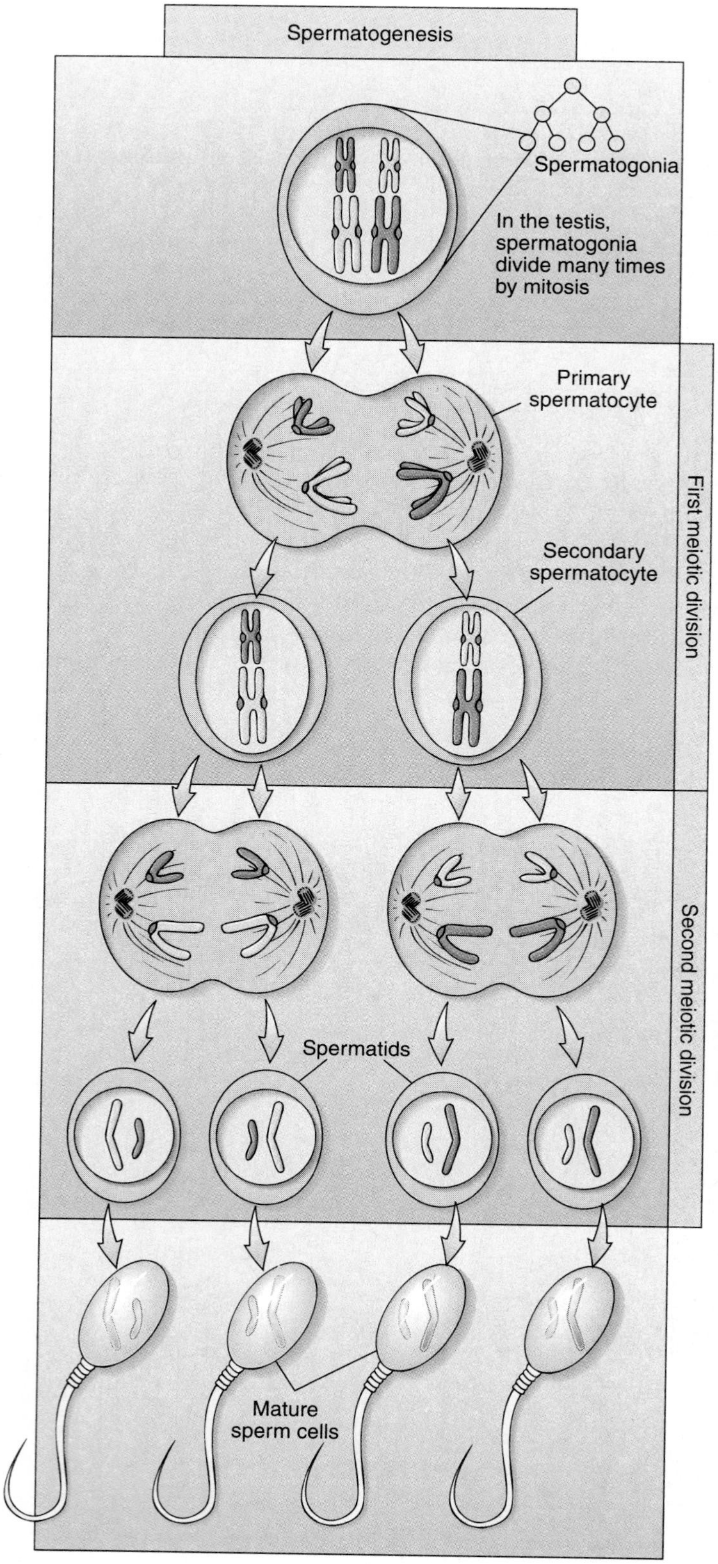

FIGURE 48–6 In spermatogenesis, a primary spermatocyte undergoes meiosis, giving rise to four spermatids. The spermatids differentiate, becoming mature sperm cells.

muscles by lifting heavy objects sometimes results in tearing the inguinal tissue. A loop of intestine can then bulge into the scrotum through the tear, a condition known as an inguinal hernia.

A series of ducts store and transport sperm

Sperm cells leave the seminiferous tubules of each testis through small tubules that empty into a larger coiled tube, the **epididymis.** There sperm complete their maturation and are stored.

During ejaculation, sperm pass from each epididymis into a sperm duct, the **vas deferens** (pl., *vasa deferentia*). The vas deferens extends from the scrotum through the inguinal canal and into the pelvic cavity. Each vas deferens empties into a short **ejaculatory duct,** which passes through the prostate gland and then opens into the **urethra.** The single urethra, which at different times conducts urine and semen, passes through the penis to the outside of the body. Thus, sperm passes in sequence through the following structures:

seminiferous tubules → epididymis → vas deferens → ejaculatory duct → urethra → released from body

The accessory glands produce the fluid portion of semen

As sperm are transported through the conducting tubes, they are mixed with secretions from three types of accessory glands. The 3.5 or so mL of **semen** ejaculated during sexual climax consists of about 400 million sperm cells suspended in the secretions of these glands.

The paired **seminal vesicles** secrete a nutritive fluid rich in fructose and prostaglandins into the vasa deferentia (Fig. 48–3). Nutrients in the secretion of the seminal vesicles provide energy for the sperm after they are ejaculated. The single **prostate gland** secretes an alkaline fluid containing prostaglandins. This secretion is thought to be important in neutralizing the acidic environment of the vagina and in increasing sperm cell motility. Prostaglandins in the semen stimulate contractions of the female uterus, which help move sperm up the female reproductive tract. The prostate gland is a common site of cancer in men over age 50. Its cause is not known, but prostate cancer is thought to be hormone-related.

During sexual arousal, the paired **bulbourethral glands,** located on each side of the urethra, release a mucous secretion. This fluid lubricates the penis, facilitating its penetration into the vagina.

The penis transfers sperm to the female

The **penis** is an erectile copulatory organ that delivers sperm into the female reproductive tract. It consists of a

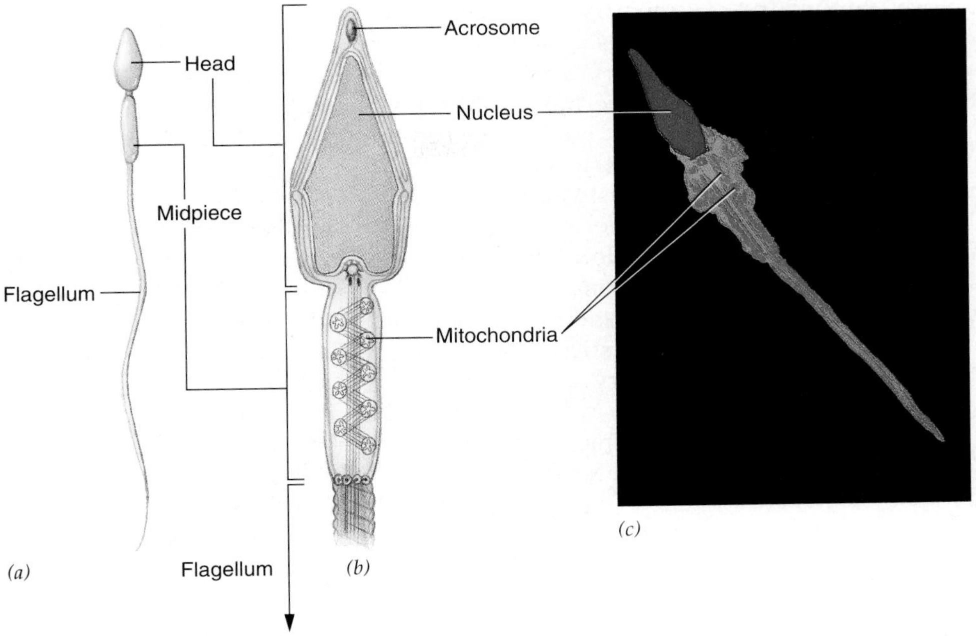

FIGURE 48–7 A mature sperm has a head, midpiece, and flagellum. (*a, b*) Structure of a mature sperm. (*c*) False-color TEM of a human sperm cell. Mitochondria (*colored red*) are visible in the midpiece. (c, Dr. Tony Brain/Science Photo Library/Photo Researchers, Inc.)

long shaft that enlarges to form an expanded tip, the **glans.** Part of the loose-fitting skin of the penis folds down and covers the proximal portion of the glans, forming a cuff called the **prepuce,** or foreskin. In the operation termed circumcision (commonly performed on male babies for either hygienic or religious reasons), the foreskin is removed.

Under the skin, the penis consists of three parallel columns of **erectile tissue,** sometimes called the **cavernous bodies** (Fig. 48–8). One of these columns sur-

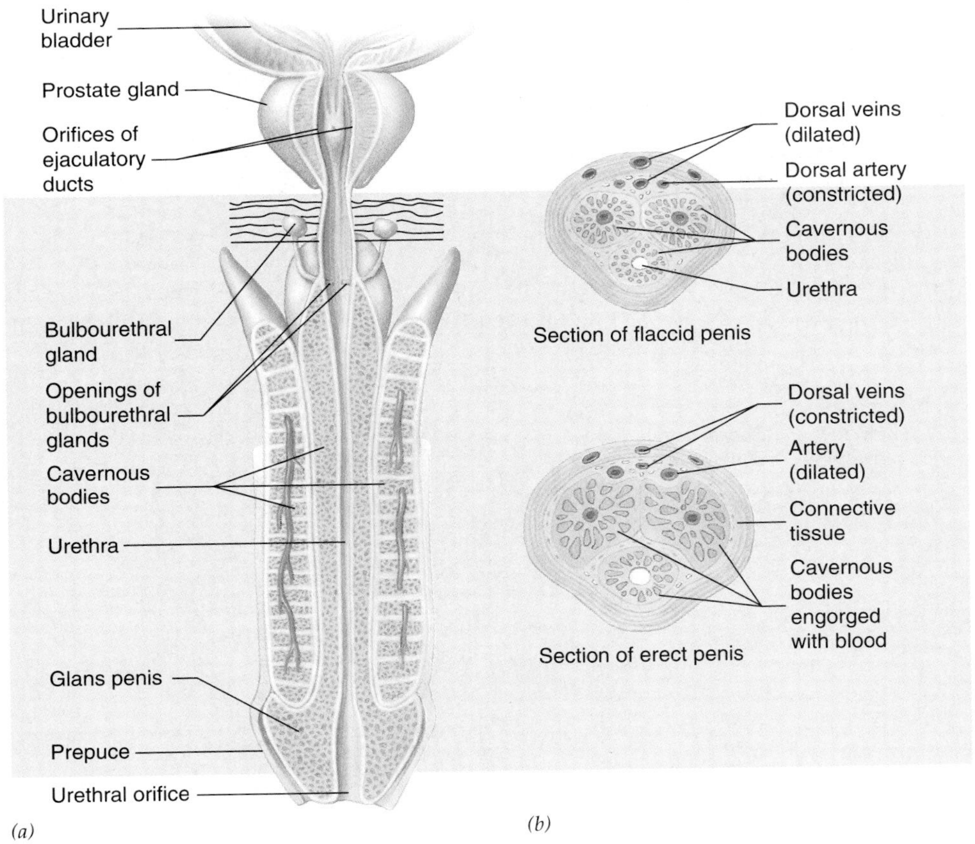

FIGURE 48–8 The internal structure of the penis includes three parallel columns of erectile tissue. (*a*) Longitudinal section through the prostate gland and penis. (*b*) Cross section through flaccid and erect penis. Note that the erectile tissues of the cavernous bodies are engorged with blood in the erect penis.

rounds the portion of the urethra that passes through the penis. When the male is sexually stimulated, nerve impulses cause the arteries of the penis to dilate. Blood rushes into the numerous blood vessels of the erectile tissue, causing the tissue to swell. This compresses veins that conduct blood away from the penis, slowing the outflow of blood. Thus, more blood enters the penis than can leave, causing the erectile tissue to become further engorged with blood. The penis becomes erect—that is, longer, larger in circumference, and firm. Although the human penis contains no bone, penis bones do occur in some other mammals, such as bats, rodents, and some primates.

Reproductive hormones promote sperm production and maintain masculinity

When a boy is about 10 years old the hypothalamus begins to secrete **gonadotropin-releasing hormone (GnRH;** Table 48–1). This hormone stimulates the anterior pituitary to secrete the gonadotropic hormones **follicle-stimulating hormone (FSH)** and **luteinizing hormone (LH).** FSH stimulates development of the seminiferous tubules and may promote spermatogenesis. LH stimulates the **interstitial cells,** lying between the seminiferous tubules in the testes, to secrete the hormone **testosterone,** which is the principal **androgen,** or male sex hormone. The peptide hormone **inhibin,** secreted by the Sertoli cells in the testes, inhibits the secretion of FSH by the anterior pituitary (Fig. 48–9).

Testosterone stimulates spermatogenesis; too little of this hormone can result in sterility. Testosterone causes the adolescent growth spurt in males at about age 13. This hormone stimulates growth of the male reproductive organs and so is responsible for the **primary male sex characteristics.** Testosterone is also responsible for the **secondary sexual characteristics** that develop at puberty. These include growth of facial and body hair, muscle development, and the increase in vocal cord length and thickness that causes the voice to deepen.

What happens when testosterone is absent? If the testes are removed (castration) before puberty, the male is deprived of testosterone and becomes a eunuch. He retains childlike sex organs and does not develop secondary sexual characteristics. If castration occurs after puberty, increased secretion of male hormones by the adrenal glands helps to maintain masculinity.

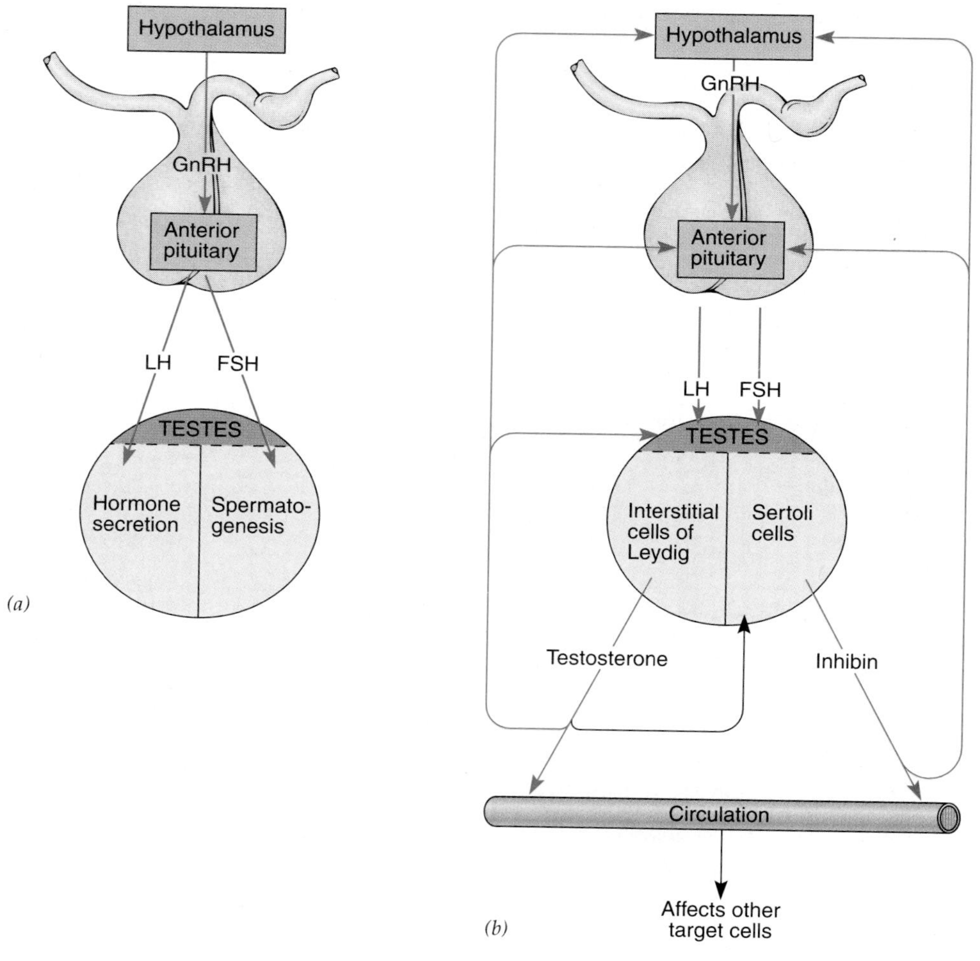

FIGURE 48–9 The hypothalamus, anterior pituitary, and testes interact to regulate the male reproductive system. (***a***) Overview of hormonal regulation. (***b***) Several negative feedback systems operate to regulate hormone level.

Table 48–1 PRINCIPAL MALE REPRODUCTIVE HORMONES		
Endocrine Gland and Hormones	*Principal Target Tissue*	*Principal Actions*
Hypothalamus		
Gonadotropin-releasing hormone (GnRH)	Anterior pituitary	Stimulates release of FSH and LH
Anterior pituitary		
Follicle-stimulating hormone (FSH)	Testes	Stimulates development of seminiferous tubules; may stimulate spermatogenesis
Luteinizing hormone (LH); also called interstitial cell–stimulating hormone (ICSH)	Testes	Stimulates interstitial cells to secrete testosterone
Testes		
Testosterone	General	*Before birth:* stimulates development of primary sex organs and descent of testes into scrotum *At puberty:* responsible for growth spurt; stimulates development of reproductive structures and secondary sex characteristics (male body build, growth of beard, deep voice, etc.) *In adult:* responsible for maintaining secondary sex characteristics; stimulates spermatogenesis
Inhibin	Anterior pituitary	Inhibits secretion of FSH

HUMAN REPRODUCTION: THE FEMALE PRODUCES OVA AND INCUBATES THE EMBRYO

The female reproductive system produces ova, receives the penis and sperm released from it during sexual intercourse, houses and nourishes the embryo during prenatal development, gives birth, and produces milk for the young (lactation). These processes are regulated and coordinated by the interaction of hormones secreted by the hypothalamus, the anterior lobe of the pituitary gland, and the ovaries. The principal organs of the female reproductive system are illustrated in Figures 48–10 and 48–11.

The ovaries produce ova and sex hormones

Like the male gonads, the female gonads, or **ovaries,** produce both gametes and sex hormones. About the size and

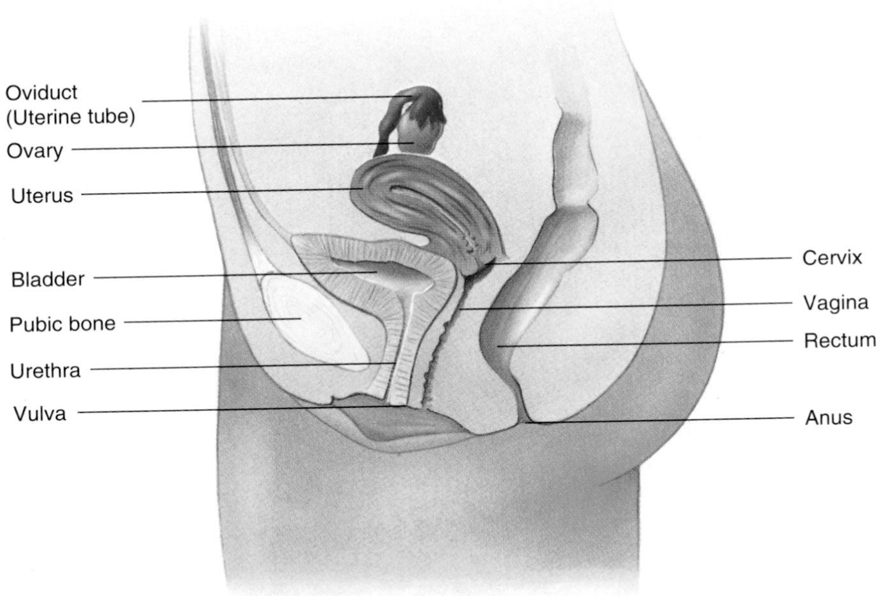

FIGURE 48–10 **This midsagittal section through the female pelvis illustrates the reproductive organs.** Note the position of the uterus relative to the vagina.

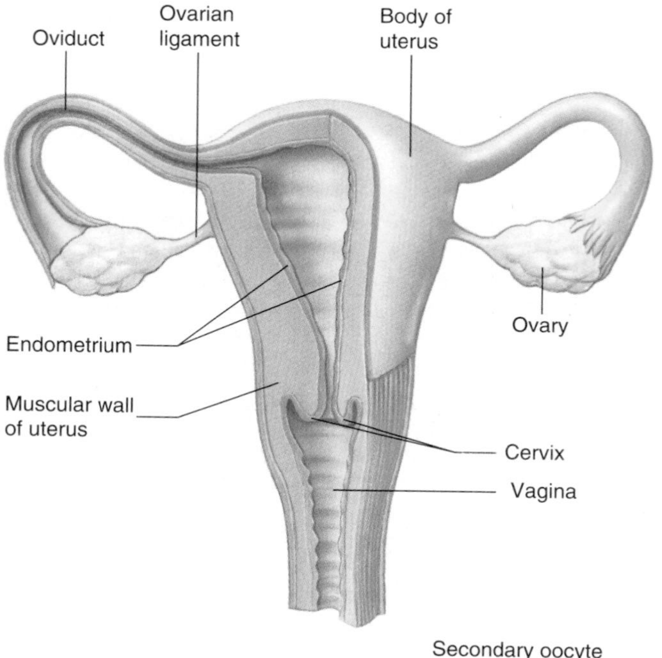

FIGURE 48–11 In this anterior view of the female reproductive system, some organs have been cut open to expose the internal structure. Connective tissue ligaments help hold the reproductive organs in place.

shape of large almonds, the ovaries are located close to the lateral walls of the pelvic cavity, and are held in position by several connective tissue ligaments. Internally, the ovary consists mainly of connective tissue containing scattered ova in various stages of maturation (Fig. 48–12).

The process of ovum production, called **oogenesis,** begins in the ovaries. Before birth, hundreds of thousands of **oogonia** are present in the ovaries. All of a female's gametes originate during embryonic development. No new oogonia are formed after birth. During prenatal development, the oogonia increase in size and become **primary oocytes**. By the time of birth, they are in the prophase of the first meiotic division. At this stage, they enter a rest-

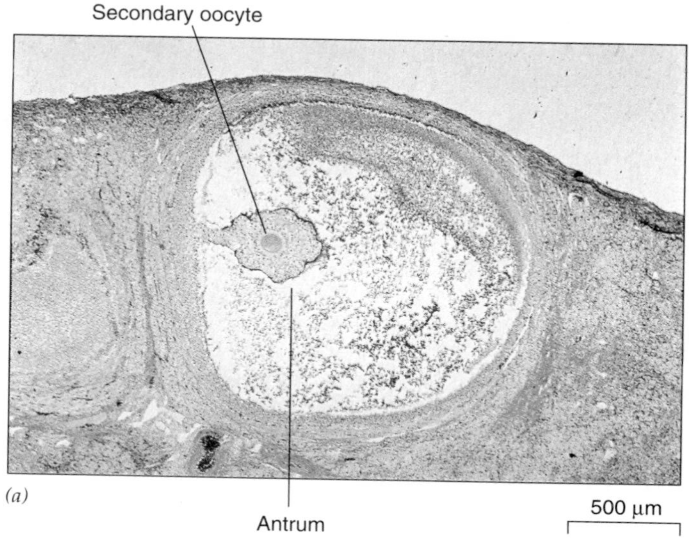

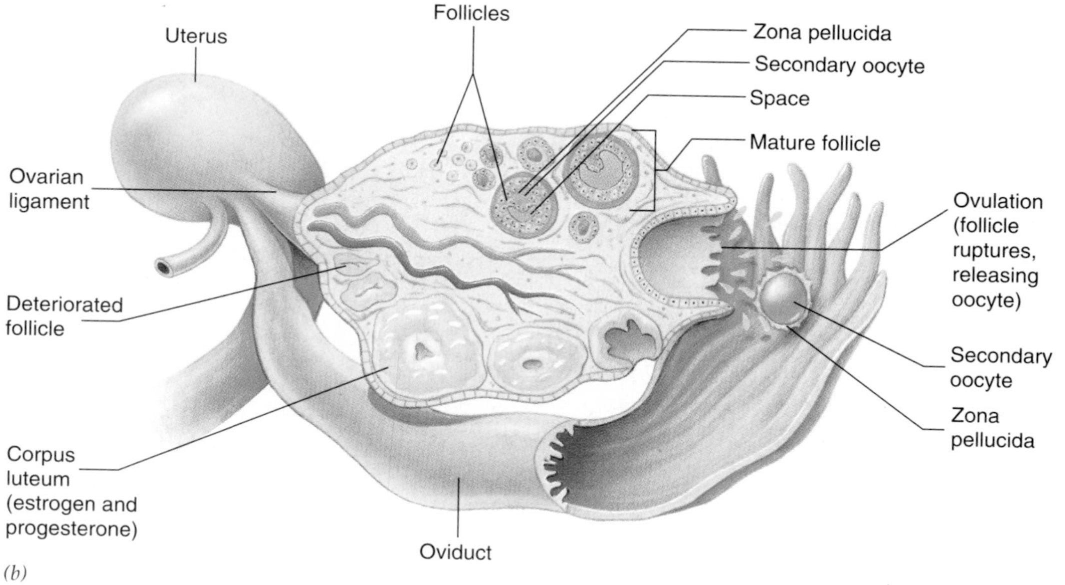

FIGURE 48–12 A microscopic view of the ovary reveals follicles in various stages of development. (***a***) A stained section through a developing follicle. The ovum is surrounded by a layer of follicle cells that will be released along with it. These follicle cells become the corona radiata, a layer that acts as a barrier to sperm cells. (***b***) Ova in different stages of development are scattered throughout the ovary. (*a*, Biophoto Associates)

ing phase that lasts throughout childhood and into adult life.

A primary oocyte and the cluster of cells surrounding it together make up a **follicle.** With the onset of puberty, a few follicles develop each month in response to FSH secreted by the anterior pituitary gland. As the follicle grows, the primary oocyte completes its first meiotic division.

The two cells produced are different in size (Fig. 48–13). The smaller one, the first **polar body,** may later divide, forming two polar bodies, but these eventually disintegrate. The larger cell, the **secondary oocyte,** proceeds to the second meiotic division but remains in metaphase until it is fertilized. When meiosis does continue, the second meiotic division gives rise to a single ovum and a second polar body. The polar bodies are small and apparently serve to dispose of unneeded chromosomes with a minimal amount of cytoplasm. The sequence is as follows:

> oogonium (diploid) → primary oocyte (diploid) →
> secondary oocyte + 1 polar body (both haploid) →
> ovum + 1 polar body (both haploid)

In the male, each primary spermatocyte gives rise to four sperm. In contrast, each primary oocyte generates only one ovum.

As an oocyte develops, it becomes separated from its surrounding follicle cells by a thick membrane, the **zona pellucida.** As the follicle develops, follicle cells secrete fluid, which collects in the space between them (Fig. 48–12). The follicle cells also secrete **estrogens,** female sex hormones.

As a follicle matures, it moves closer to the surface of the ovary, eventually resembling a fluid-filled bulge on the ovarian surface. Typically, only one follicle fully matures each month. Several others may develop for about a week, and then deteriorate.

During **ovulation,** the secondary oocyte is ejected through the wall of the ovary and into the pelvic cavity. The portion of the follicle that remains in the ovary develops into the **corpus luteum,** a temporary endocrine gland. The corpus luteum secretes estrogens and **progesterone.**

The oviducts transport the secondary oocyte

Almost immediately after ovulation, the secondary oocyte passes into the funnel-shaped opening of the **oviduct,** or **uterine tube** (also called the fallopian tube). Peristaltic contractions of the muscular wall of the oviduct and beating of the cilia in its lining help to move the secondary oocyte along toward the uterus. Fertilization takes place within the oviduct. If fertilization does not occur, the secondary oocyte degenerates there.

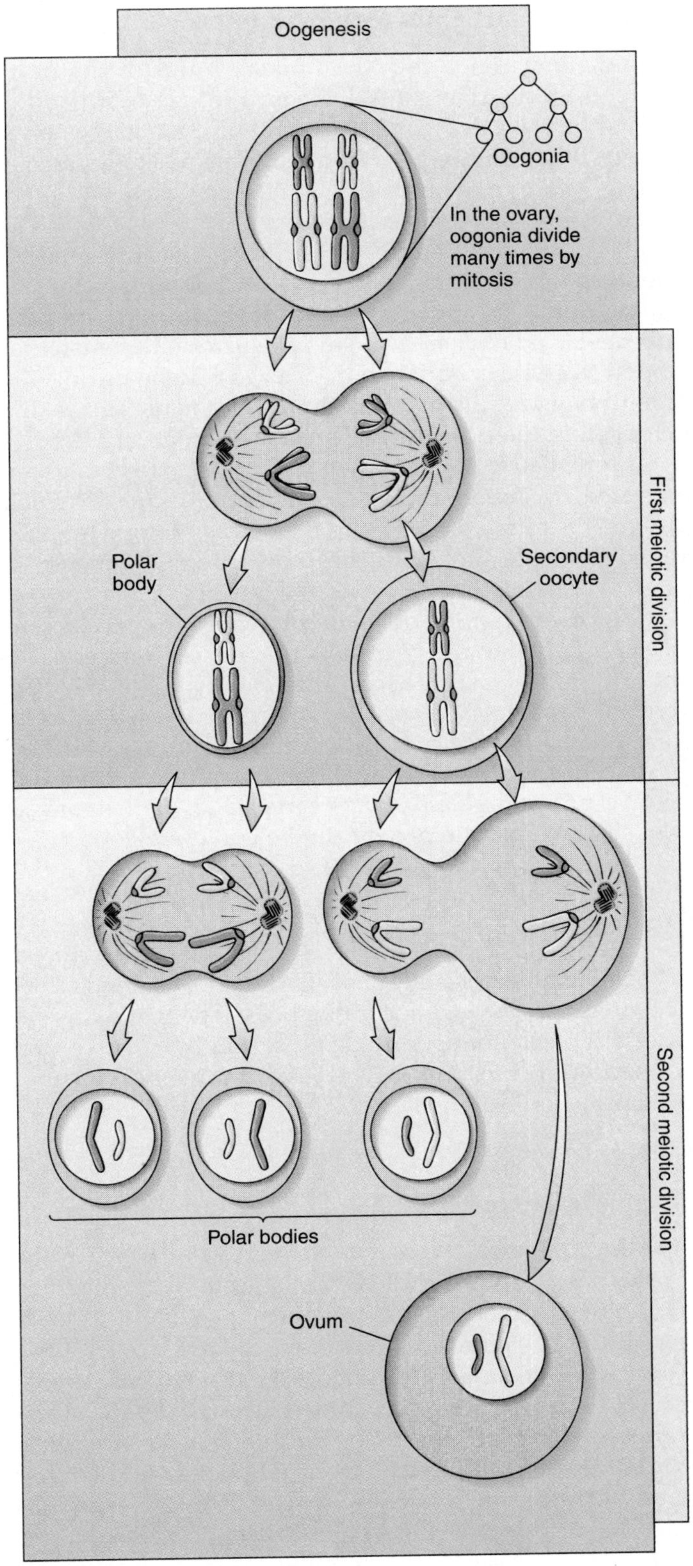

FIGURE 48–13 In oogenesis, only one functional ovum is produced from each primary oocyte. The other cells produced are polar bodies that degenerate. The second meiotic division is completed after fertilization.

The uterus incubates the embryo

The oviducts open into the upper corners of the pear-shaped **uterus** (Fig. 48–11). About the size of a fist, the uterus (or womb) occupies a central position in the pelvic cavity. It has thick walls of smooth muscle and a mucous lining, the **endometrium,** that thickens each month in preparation for possible pregnancy.

If a secondary oocyte is fertilized, the tiny embryo finds its way into the uterus and is implanted in the endometrium. There it grows and develops, sustained by nutrients and oxygen delivered by surrounding maternal blood vessels. If fertilization does not occur during the monthly cycle, the endometrium sloughs off and is discharged in the process known as menstruation.

More than five million women in the United States are affected by **endometriosis,** a painful disorder in which fragments of the tissue lining the uterus migrate to other areas such as the oviducts or ovaries. Endometriosis causes scarring that can lead to infertility.

The lower portion of the uterus, called the **cervix,** projects slightly into the vagina. The cervix is a common site of cancer in women. Detection is usually possible by the routine Papanicolaou test (Pap smear) in which a few cells are scraped from the cervix during a regular gynecological examination and studied microscopically. When cervical cancer is detected at very early stages of malignancy, the chances that the patient can be cured are good.

The vagina receives sperm

The **vagina** is an elastic, muscular tube that extends from the uterus to the exterior of the body. The vagina serves as a receptacle for sperm during sexual intercourse, and as part of the birth canal when development of the fetus is complete (Fig. 48–11).

The vulva are external genital structures

The external female sex organs, collectively known as the **vulva,** include liplike folds, the **labia minora,** that surround the vaginal and urethral openings (Fig. 48–14). External to the delicate labia minora are the thicker, hair-covered **labia majora.** Anteriorly, the labia minora merge to form the prepuce of the **clitoris,** a small erectile structure comparable to the male glans penis. Like the penis, the clitoris contains erectile tissue that becomes engorged with blood during sexual excitement. Rich in nerve endings, the clitoris is highly sensitive to touch, pressure, and temperature, and serves as a center of sexual sensation in the female.

The **mons pubis** is the mound of fatty tissue just above the clitoris at the junction of the thighs and torso. At puberty it becomes covered by coarse pubic hair. The **hymen** is a thin ring of tissue that forms a border around the entrance to the vagina.

The breasts function in lactation

Each breast is composed of 15 to 20 lobes of glandular tissue. The amount of adipose tissue around these lobes determines the size of the breasts and accounts for their

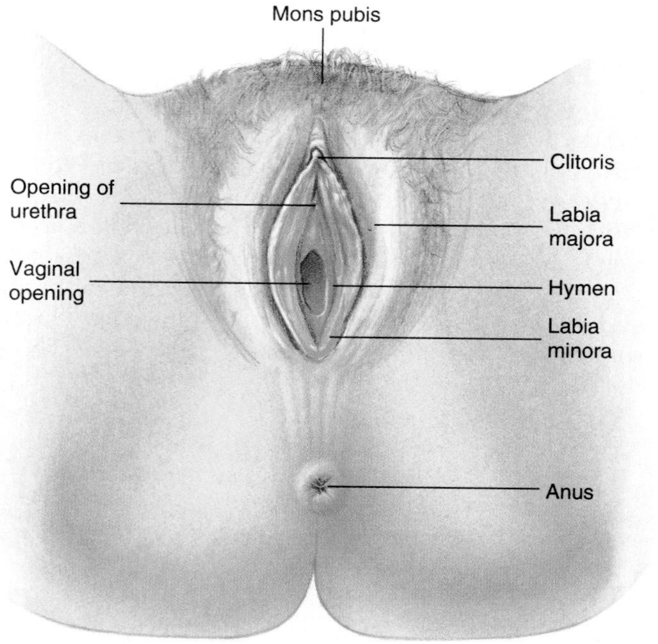

FIGURE 48–14 The external genital structures of the female are referred to as the vulva.

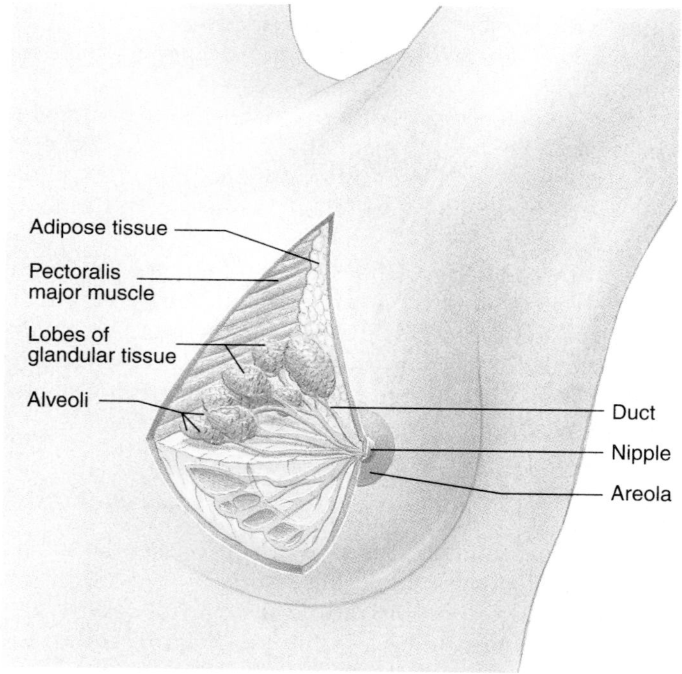

FIGURE 48–15 The mature human female breast contains lobes of glandular tissue. The lobes consist of alveoli, clusters of gland cells.

Focus On

Breast Cancer

Breast cancer is the most common type of cancer among women. Its incidence has increased in recent years, and now the disease strikes about one in every nine women. Next to lung cancer it is the leading cause of cancer deaths in women. The increased incidence of breast cancer is partly due to increased life expectancy.

The causes of breast cancer are not known, but there appears to be a higher risk in women with a family history of the disease. An estimated 5% of breast cancers are familial. Researchers have suggested that mutations of two tumor suppressor genes (BRCA1 and BRCA2) may predispose a woman to develop breast cancer.

Although no conclusive evidence yet exists, some investigators think that other risk factors include a high-fat diet, obesity, exposure to radiation, and exposure to certain chemicals. Smoking cigarettes increases a woman's risk of dying from breast cancer by at least 25%. Women who smoke two packs or more of cigarettes a day have a 75% greater risk. A recent study suggests that taking estrogen for five or more years as hormone replacement therapy after menopause increases the risk of breast cancer by more than 30%.

About 50% of breast cancers begin in the upper, outer quadrant of the breast (see photograph). As a malignant tumor grows, it may adhere to the deep tissue of the chest wall. Sometimes it extends to the skin,

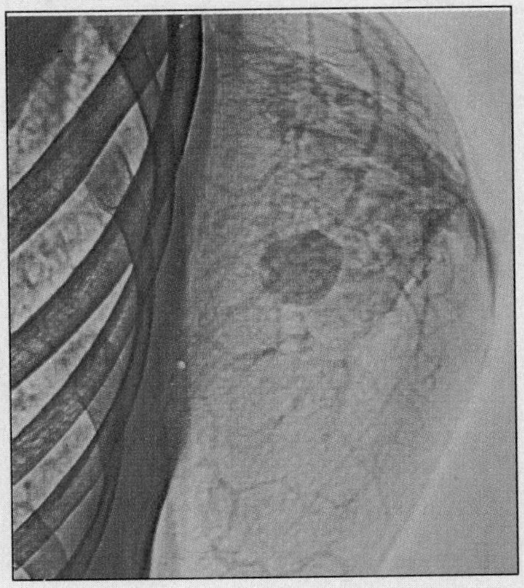

Breast cancer is evident in this mammogram. Note the extensive vascularization. (Visuals Unlimited/SIU)

causing dimpling. Eventually the cancer spreads to the lymphatic system. About two-thirds of breast cancers have metastasized (spread) to the lymph nodes by the time they are first diagnosed. When diagnosis and treatment begin early, 80% of patients survive for five years, and 62% survive for ten years or longer. Untreated patients have a five-year survival rate of only 20%.

Mastectomy (surgical removal of the breast) and radiation treatment are common methods of treating breast cancer. Lumpectomy (surgical removal of only the affected portion of the breast) in conjunction with radiation treatment is thought to be as effective as mastectomy in some cases. Chemotherapy is useful in preventing metastasis, especially in premenopausal patients. A recent development in cancer treatment is the use of biological response modifiers, which include substances such as interferons, interleukins, and monoclonal antibodies (see Chapter 43).

About one-third of breast cancers are estrogen-dependent; that is, their growth depends on circulating estrogens. Removing the ovaries in patients with these tumors relieves the symptoms and may cause remission of the disease for months or even years. The synthetic drug tamoxifen, which blocks estrogen receptors, has been heralded as the most effective drug for preventing new cancers in women who have had breast cancer. However, recent studies suggest that tamoxifen may increase risk of uterine cancer.

Because early detection of breast cancer greatly increases the chances of cure and survival, campaigns have been launched to educate women on the importance of self-examination. Mammography, a soft-tissue radiological study of the breast, is helpful in detecting very small lesions that might not be identified by routine examination. In mammography, lesions show on an x-ray plate as areas of increased density. ■

softness. Gland cells are arranged in grapelike clusters called **alveoli** (Fig. 48–15). Ducts from each cluster join to form a single duct from each lobe, producing 15 to 20 tiny openings on the surface of each nipple. The breasts are the most common site of cancer in women (see *Focus On: Breast Cancer*).

Lactation is the production of milk for the nourishment of the young. During pregnancy, high concentrations of female hormones—estrogens and progesterone—stimulate the breasts to increase in size. For the first couple of days after childbirth, the mammary glands produce a fluid called **colostrum,** which contains protein and lactose but little fat. After birth the hormone **prolactin** stimulates milk production.

Breastfeeding promotes recovery of the uterus, because **oxytocin** released during breastfeeding stimulates the uterus to contract to nonpregnant size. Breastfeeding offers advantages to the baby as well. It promotes a close bond between mother and child, and provides milk tailored to the nutritional needs of the human infant. Breast milk contains antibodies, and breastfed infants have a lower incidence of diarrhea, ear and respiratory infections, and hospital admissions than do bottlefed babies.

Table 48–2 PRINCIPAL FEMALE REPRODUCTIVE HORMONES

Endocrine Gland and Hormones	*Principal Target Tissue*	*Principal Actions*
Hypothalamus		
Gonadotropin-releasing hormone (GnRH)	Anterior pituitary	Stimulates release of FSH and LH
Anterior pituitary		
Follicle-stimulating hormone (FSH)	Ovary	Stimulates development of follicles and secretion of estrogen
Luteinizing hormone (LH)	Ovary	Stimulates ovulation and development of corpus luteum
Prolactin	Breast	Stimulates milk production (after breast has been prepared by estrogen and progesterone)
Ovary		
Estrogens (estradiol)	General	Responsible for growth of sex organs at puberty; stimulate development of secondary sex characteristics (breast development, broadening of pelvis, distribution of fat and muscle)
	Reproductive structures	Induce maturation; stimulate monthly preparation of the endometrium for pregnancy; make cervical mucus thinner and more alkaline
Progesterone (secreted mainly by corpus luteum)	Uterus	Completes preparation of endometrium for pregnancy

Reproductive hormones maintain female characteristics and regulate the menstrual cycle

Table 48–2 lists the actions of the principal female reproductive hormones. Like testosterone in the male, estrogens are responsible for the growth of the sex organs at puberty, for body growth, and for the development of secondary sexual characteristics. In the female, these include the development of the breasts, the broadening of the pelvis, and the characteristic development and distribution of muscle and fat responsible for the shape of the female body.

Hormones of the hypothalamus, anterior pituitary, and ovaries regulate the **menstrual cycle,** the monthly sequence of events that prepares the body for possible pregnancy. The menstrual cycle runs its course every month from puberty until menopause occurs at about age 50.

Although wide variations exist, a typical menstrual cycle is 28 days long (Fig. 48–16). The first day of the cycle is marked by the onset of menstruation, the monthly discharge through the vagina of blood and tissue from the endometrium. Ovulation occurs on about the 14th day of the cycle.

During the menstrual phase of the cycle, which lasts about five days, gonadotropin releasing hormone (GnRH) is released from the hypothalamus. GnRH stimulates the anterior pituitary to release follicle stimulating hormone (FSH) and luteinizing hormone (LH) (Fig. 48–17). FSH stimulates the development of a few follicles in the ovary. After a few days, only one follicle continues to develop.

During the **preovulatory phase** (also called follicular phase) of the menstrual cycle, the developing follicles secrete estrogens. Estrogens stimulate growth of the endometrium, which thickens and develops new blood vessels and glands. The rise in the concentration of estrogens in the blood signals the anterior pituitary to secrete LH.

FSH and LH together stimulate the final maturation of the follicle. The rise in estrogens secreted by the developing follicle stimulates the anterior pituitary to secrete a surge of LH. This stimulating effect is a positive feedback mechanism. The surge of LH is necessary for the final maturation of the follicle and stimulates ovulation to occur at about day 14 of the cycle.

After the secondary oocyte has been ejected from the ovary, the **postovulatory phase** (also called luteal phase) begins. LH stimulates development of the corpus luteum, which secretes a large amount of progesterone and a small amount of estrogens (Fig. 48–17*b*). These hormones stimulate the uterus to continue its preparation for pregnancy. Progesterone stimulates tiny glands in the endometrium to secrete a fluid rich in nutrients.

Although estrogens stimulate secretion of GnRH, FSH, and LH during the preovulatory phase, they inhibit

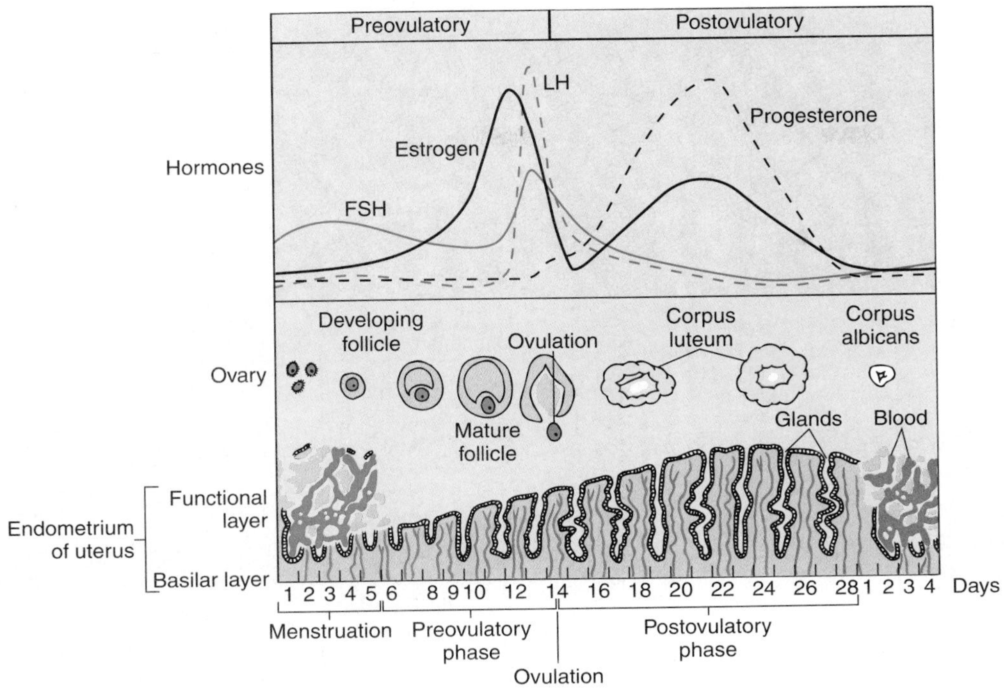

FIGURE 48–16 When fertilization does not occur, the menstrual cycle repeats itself about every 28 days. The events that take place within the ovary and uterus are precisely coordinated by hormones. Compare this illustration with Figure 48–17.

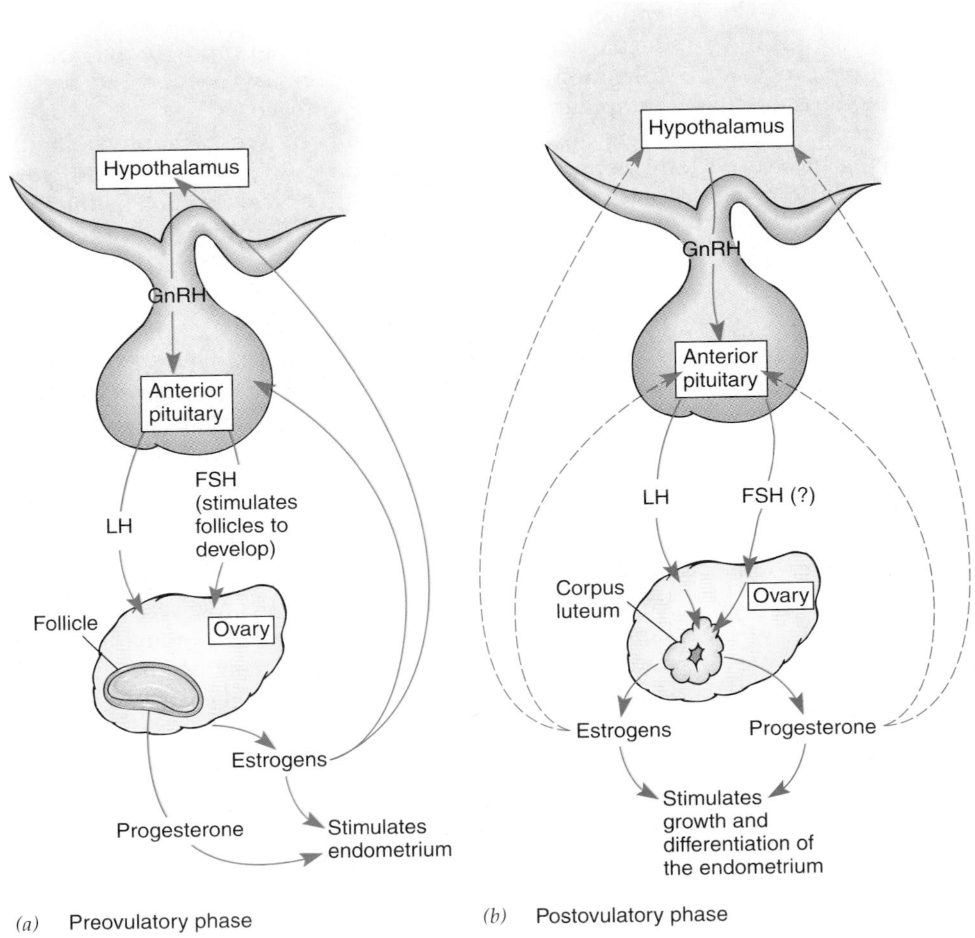

FIGURE 48–17 Hormones from the hypothalamus, anterior pituitary, and ovary interact to regulate the menstrual cycle. (*a*) Hormonal interactions during the preovulatory phase. (*b*) Hormonal interactions during the postovulatory (luteal) phase. Red arrows indicate inhibition.

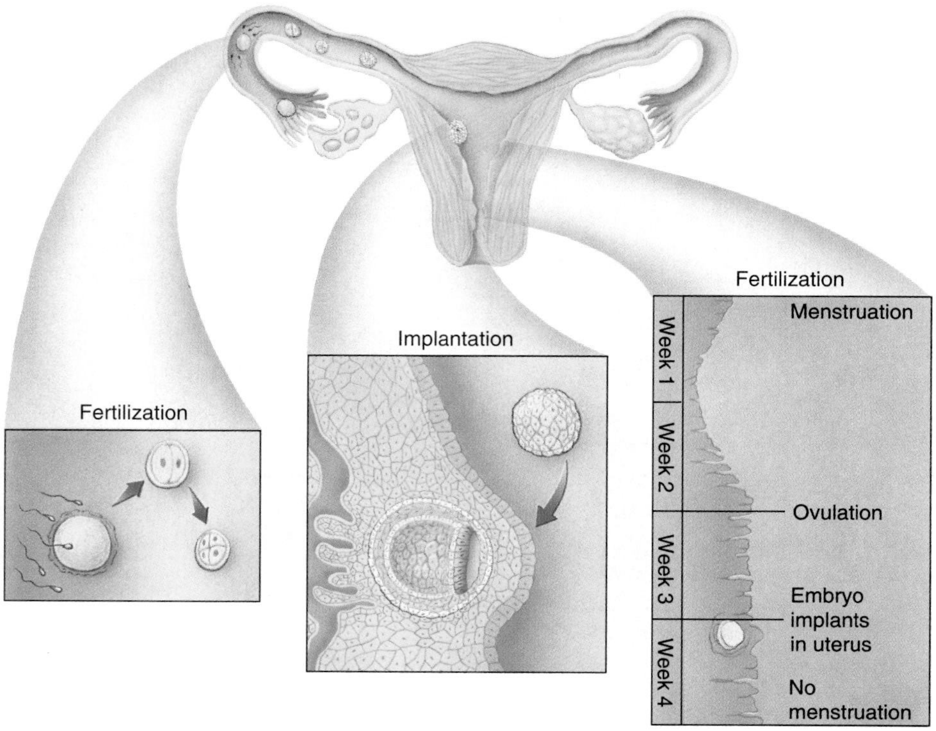

FIGURE 48–18 The menstrual cycle is interrupted when pregnancy occurs. The corpus luteum does not degenerate, and menstruation does not take place. Instead, the wall of the uterus thickens even more so that the embryo can develop within it.

secretion of these hormones during the postovulatory phase. The different effect is attributed to a change in the sensitivity of the hypothalamus to these hormones.

If the secondary oocyte is not fertilized, the corpus luteum begins to degenerate and stops secreting progesterone and estrogens. The concentrations of these hormones in the blood fall markedly. As a result, small arteries in the endometrium constrict, reducing the oxygen supply. Menstruation begins again as cells die and damaged arteries rupture and bleed. Low levels of estrogens and progesterone trigger secretion of FSH and LH by the anterior pituitary once again.

Premenstrual syndrome (PMS) is a condition experienced by some women starting several hours to 10 days before menstruation and ending a few hours after onset of menstruation. Symptoms include fatigue, anxiety, depression, irritability, headache, edema, and skin eruptions. Its cause is not known.

If the secondary oocyte is fertilized, it makes its way to the uterus and begins to implant in the thick endometrium on about the seventh day after fertilization (Fig. 48–18). Membranes that develop around the embryo secrete **human chorionic gonadotropin (hCG),** a hormone that signals the mother's corpus luteum to continue to function.

SEXUAL RESPONSE INVOLVES PHYSIOLOGICAL CHANGES

During copulation, also called **coitus** or sexual intercourse in humans, the male deposits semen into the upper end of the vagina. The complex structures of the male and female reproductive systems, and the physiological, endocrine, and psychological processes associated with sexual activity, are adaptations that promote the successful union of sperm with secondary oocyte, and the development of the resulting embryo.

Sexual stimulation results in two basic physiological responses: increased blood flow (vasocongestion) to reproductive structures and other tissues such as the skin, and increased muscle tension. During vasocongestion, erectile tissues within the penis and clitoris, as well as in other areas of the body, become engorged with blood.

Sexual response includes four phases: sexual desire, excitement, orgasm, and resolution. The *desire* to have sexual activity may be motivated by fantasies or thoughts about sex. This anticipation can lead to (physical) sexual excitement and a sense of sexual pleasure. The *excitement phase* involves vasocongestion and increased muscle tension. Before the penis can enter the vagina and function in coitus, it must be erect. Penile erection is the first male response to sexual excitement. In the female, vaginal lubrication is the first response to effective sexual stimulation. During the excitement phase, the vagina lengthens and expands in preparation for receiving the penis, the clitoris and breasts become vasocongested, and the nipples become erect.

If erotic stimulation continues, sexual excitement heightens. Vasocongestion and muscle tension increase markedly. In both sexes, blood pressure increases and heart rate and breathing accelerate.

Coitus may be initiated during the excitement phase. During coitus the penis creates friction as it is moved inward and outward in the vagina in actions referred to as pelvic thrusts. Physical and psychological sensations resulting from this friction (and from the entire intimate experience between the partners) may lead to **orgasm,** the climax of sexual excitement. In the female, stimulation of the clitoris is important in heightening the sexual excitement that leads to orgasm.

Although it lasts only a few seconds, orgasm is the phase of maximum sexual tension and its release. In both sexes, orgasm is marked by rhythmic contractions of the muscles of the pelvic floor and reproductive structures. These muscular contractions continue at about 0.8-second intervals for several seconds. After the first few contractions, their intensity decreases, and they become less regular and less frequent. Heart rate and respiration more than double, and blood pressure rises markedly, just before and during orgasm. In the male, orgasm is marked by the **ejaculation** of semen from the penis. No fluid ejaculation accompanies orgasm in the female. Orgasm is followed by the **resolution phase,** a state of well-being during which the body is restored to its normal state.

Sexual dysfunction may be caused by psychological or biological factors. For example, chronic inability to sustain an erection, termed *erectile dysfunction* (formerly called impotence), is often associated with psychological issues. This disorder prevents effective coitus. *Vaginismus* is a condition in which, during sexual intercourse, a woman experiences painful involuntary spasms of the outer third of the vaginal muscles. Vaginismus is often associated with a history of sexual abuse.

Fertilization Is the Fusion of Sperm and Egg to Produce a Zygote

In the process of **fertilization,** sperm and ovum fuse to produce a zygote. Fertilization and the subsequent establishment of pregnancy together are referred to as **conception.** When conditions in the vagina and cervix are favorable, sperm begin to arrive at the site of fertilization in the upper uterine tube within 5 minutes after ejaculation. Contractions of the uterus and uterine tubes help transport the sperm. The sperms' own motility is probably most important in approaching and fertilizing the ovum.

If only one sperm is needed to fertilize an ovum, why are millions involved in each act of coitus? For one thing, sperm movement is undirected, so that many lose their way. Others die as a result of unfavorable pH or phagocytosis by leukocytes and macrophages in the female tract. Only a few thousand succeed in traversing the correct uterine tube and reaching the vicinity of the ovum. Additionally, large numbers of sperm may be necessary to penetrate the covering of follicle cells (the corona radiata) that surrounds the ovum. Each sperm is thought

Making the Connection

Some Environmental Effects on Reproduction

What are some environmental factors that can affect the ability to reproduce? **Infertility** is the inability of a couple to achieve conception after not using contraception for at least one year. About 15% of married couples in the United States are affected by infertility. About 30% of cases involve both male and female factors.

A major cause of male infertility is **sterility,** lack of sperm production. When sperm counts drop below 35 million per milliliter, fertility is impaired, and males with fewer than 20 million sperm per milliliter of semen are usually considered sterile. When a couple's attempts to produce a child are unsuccessful, a sperm count and analysis may be performed in a clinical laboratory. Sometimes semen is found to contain large numbers of abnormal sperm, or occasionally no sperm at all.

In the U.S. in the 1970s an average healthy young man produced about 100 million sperm per milliliter of semen. Today that average has dropped to about 60 million. Although the cause of this decrease is not known, low sperm counts have been linked to a variety of environmental factors, including chronic marijuana use, alcohol abuse, and cigarette smoking. Studies show that men who smoke tobacco are more likely than nonsmokers to produce abnormal sperm. Exposure to industrial and environmental toxins such as DDT and PCBs may also result in low sperm count and sterility. The use of anabolic steroids by athletes to accelerate muscle development can cause sterility in both males and females.

One cause of female infertility is scarring of the uterine tubes. Inflammation of the uterine tubes, sometimes caused by sexually transmitted disease such as gonorrhea or chlamydial infections, may result in scarring, which blocks the tubes so that the fertilized ovum cannot pass to the uterus. Sometimes partial constriction of the uterine tube results in tubal pregnancy, in which the embryo begins to develop in the wall of the uterine tube because it cannot progress to the uterus. Uterine tubes are not adapted to bear the burden of a developing embryo; thus, the uterine tube and the embryo it contains must be surgically removed before it ruptures and endangers the life of the mother.

Women with blocked uterine tubes usually can produce ova and can incubate an embryo. However, they need clinical assistance in getting the ovum from the ovary to the uterus (see *Focus On: Novel Origins* in Chapter 49). ▲

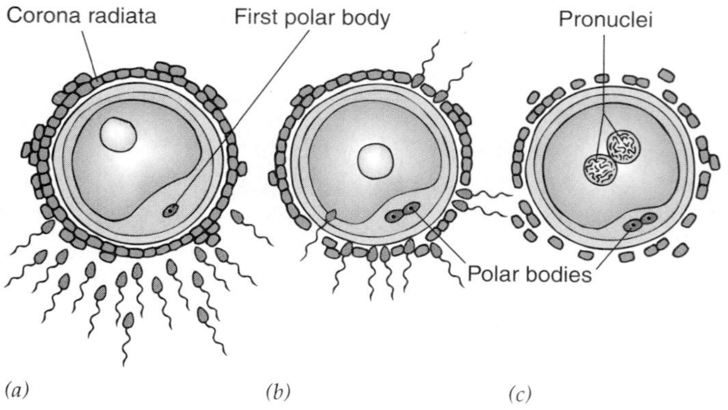

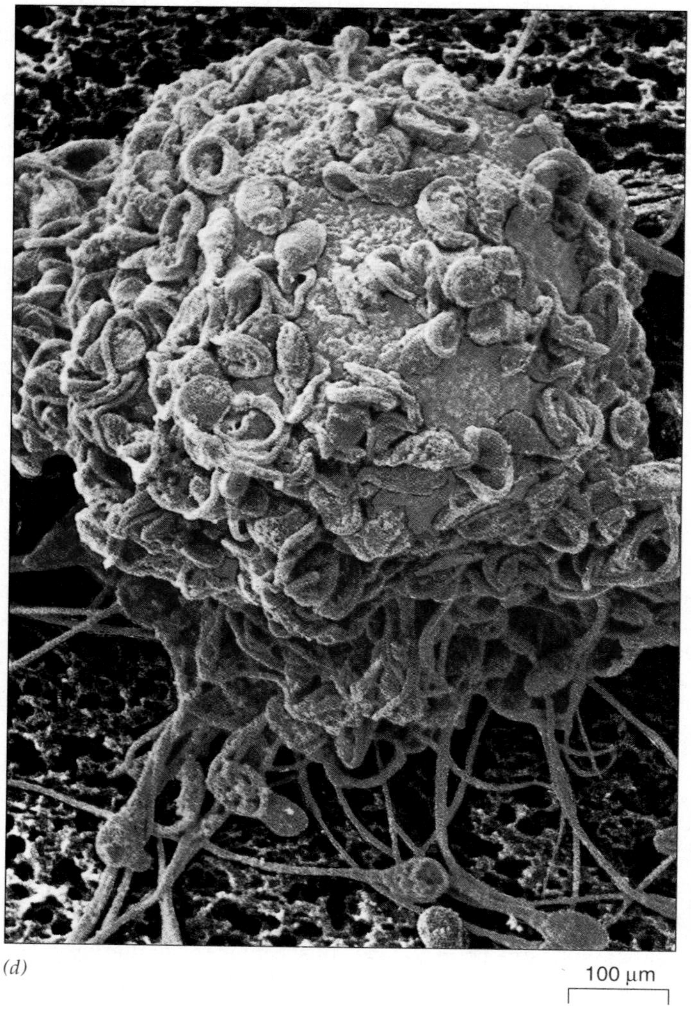

FIGURE 48–19 Fertilization is the fusion of egg and sperm. (***a***) Each sperm is thought to release a small amount of an enzyme that helps to disperse the follicle cells surrounding the ovum. (***b***) After a sperm cell enters, the secondary oocyte completes its second meiotic division, producing an ovum and a polar body. (***c***) Pronuclei of sperm and ovum combine, producing a zygote with the diploid number of chromosomes. (***d***) A scanning electron micrograph of human sperm cells on a test ovum. Sperm are being tested for viability. (*d,* David Scharf/Peter Arnold, Inc.)

to release small amounts of enzymes from its acrosome that help break down the cement-like substance holding the follicle cells together.

As soon as one sperm enters the ovum, a rapid electrical change occurs, followed by a slower chemical change in the plasma membrane of the ovum. These changes prevent the entrance of other sperm. As the fertilizing sperm enters the ovum, it usually loses its flagellum (Fig. 48–19). Sperm entry stimulates the ovum to complete its second meiotic division. The head of the haploid sperm then swells to form the male **pronucleus** and fuses with the female pronucleus, forming the diploid nucleus of the zygote. The process of fertilization is described in more detail in Chapter 49.

After ejaculation into the female reproductive tract, sperm remain alive and retain their ability to fertilize an ovum for only a few days. The ovum itself remains fertile for only about 24 hours after ovulation. Thus, in a very regular 28-day menstrual cycle, sexual intercourse on days 9 to 15 is most likely to result in fertilization. However, many women do not have regular menstrual cycles, and many factors can cause irregular cycles even in women who are generally regular.

In view of the many factors working against fertilization, it may seem remarkable that it ever occurs! Yet the frequency of coitus and the large number of sperm deposited at each ejaculation enable the human species not only to maintain itself, but to increase its numbers at an alarming rate.

BIRTH CONTROL METHODS ALLOW INDIVIDUALS TO CHOOSE

When a fertile, heterosexually active woman uses no form of birth control, her chances of becoming pregnant during the course of a year are about 90%. Any method for deliberately separating sexual intercourse from reproduction is considered **contraception** (literally, "against conception"). Since ancient times, humans have searched for effective contraceptive methods.

Although many couples worldwide agree that it is best to have babies by choice rather than by chance, the majority either are not educated about contraceptive methods or do not have contraceptives available to them. In underdeveloped countries, an estimated 67% of married women lack the means to limit family size. Studies indicate that many of these women would use modern birth control methods if they were available, and affordable, and if someone showed them how.

More than half of the 6.4 million women who become pregnant each year in the U.S. did not intend to do so. Young people particularly lack the knowledge and means of protecting themselves from unwanted pregnancy. Every year more than 1 million teenagers in the United States become pregnant, and thousands of girls aged 14

Novel Origins

Techniques developed through reproduction and human embryo research have been used clinically for several years to help couples who are having difficulty conceiving children. As stated in Chapter 48, about 15% of married couples in the United States experience **infertility,** the inability to achieve conception. In the United States, more than three million infertile couples consult health care professionals each year. Some couples are helped with conventional treatment, for example, with hormone therapy that regulates ovulation. But more than 40,000 couples need more sophisticated clinical help and turn to high-tech assisted reproductive techniques. At present these techniques are expensive and the success rate is low, less than one-third.

A newborn bongo is shown with its surrogate mother, an eland. As a young embryo, the bongo was transplanted into the eland's uterus, where it implanted and developed. Bongos are a rare and elusive species inhabiting dense forests in Africa. The larger and more common elands, members of the same genus, inhabit open areas.

The most common assisted reproductive procedure is **artificial insemination,** the transfer of semen into the female reproductive tract. Artificial insemination is indicated when the male partner of a couple desiring a child is sterile or carries a genetic defect. If the male is infertile due to a low sperm count, his sperm can be concentrated, or alternatively, sperm from a donor can be used. Although the sperm donor usually remains anonymous to the couple involved, his genetic qualifications are screened by physicians.

Intrauterine Insemination (IUI) is a high-tech type of artificial insemination in which a catheter is used to inject sperm directly into the uterus. More than 600,000 IUI procedures are performed annually with a success rate of about 10%.

To assist men with tubal blockages, sperm can be removed from the testes and a single sperm injected into an egg. The zygote is then transferred to the uterus.

Among the common causes of female infertility are failure to ovulate, production of infertile eggs (common in older women), and oviduct scarring that blocks the passage of the secondary oocyte to the uterus. One cause of scarring is the inflammation of the oviducts caused by pelvic inflammatory disease (PID). Women with blocked oviducts can usually produce ova and incubate an embryo normally. However, they need clinical assistance in getting the ovum from the ovary to the uterus.

In **In Vitro Fertilization (IVF),** an ovum is fertilized with sperm in the laboratory and then implanted in the woman's uterus. This procedure was first used in England in 1978 to help a couple who had tried unsuccessfully for several years to have a child. Since that time, thousands of "test-tube babies" have been conceived in this way and born to previously infertile women. Typically, the woman takes a fertility drug that induces ovulation of several eggs. The eggs can be removed and fertilized with sperm. Embryos can be screened for chromosome or gene abnormalities, and healthy ones can be implanted in the uterus. In 1995, the success rate of in vitro fertilization was only about 20%.

Gamete Intrafallopian Transfer (GIFT) is a technique in which eggs and sperm are transferred into a woman's oviduct (fallopian tube). More than 4000 of these expensive procedures are performed each year with a success rate of about 28%. In **Zygote Intrafallopian Transfer (ZIFT),** eggs are fertilized in the laboratory, and the resulting zygotes are inserted into the oviduct. In these procedures, the patient's own eggs may be used, or if she does not produce fertile eggs, they can be contributed by a donor **(oocyte donation).**

Another novel procedure is **host mothering.** A tiny embryo is removed from its natural mother and implanted into a female substitute. The foster mother can support the developing embryo either until birth, or temporarily until it is implanted again into the original mother or another host. This technique has proved useful to animal breeders. For example, embryos from prize sheep can be temporarily implanted into rabbits for easy shipping by air, and then implanted into a host mother sheep, perhaps of inferior quality. Host mothering has the advantage of allowing an animal with superior genetic traits to produce more offspring than would be naturally possible. Host mothering may someday be popular with women who can produce embryos but are either unable or unwilling to carry them to term.

Technology is available to freeze the embryos of many species, including humans, and then successfully transplant them into host mothers. In the future, this procedure may be popular with young women not yet ready to become parents, but who want to preserve embryos for reimplantation at a later time. ■

cause the zygote cytoplasm is not homogeneous, the cytoplasm portioned out to each new cell during cleavage may be different. Such differences help determine the course of development.

In the amphibian egg, fertilization causes rearrangement of some of the superficial cytoplasm that contains dark granules. This shift exposes the underlying crescent lying near the cell equator on the opposite side of where the sperm penetrated the egg. The gray crescent region is thought to contain growth factors from both ends of the egg. Normally, the first cell division separates the gray crescent into the first two cells of the embryo. As cleavage continues, the gray crescent becomes partitioned into certain cells. The cells that contain the parts of the gray crescent eventually develop into the dorsal region of the embryo.

Experiments have confirmed the importance of the gray crescent material to development. When the first two cells of the frog embryo are separated experimentally, each cell develops into a complete tadpole (Fig. 49–3). When the plane of the first division is altered so that the gray crescent is completely absent from one of the cells, that cell does not develop normally.

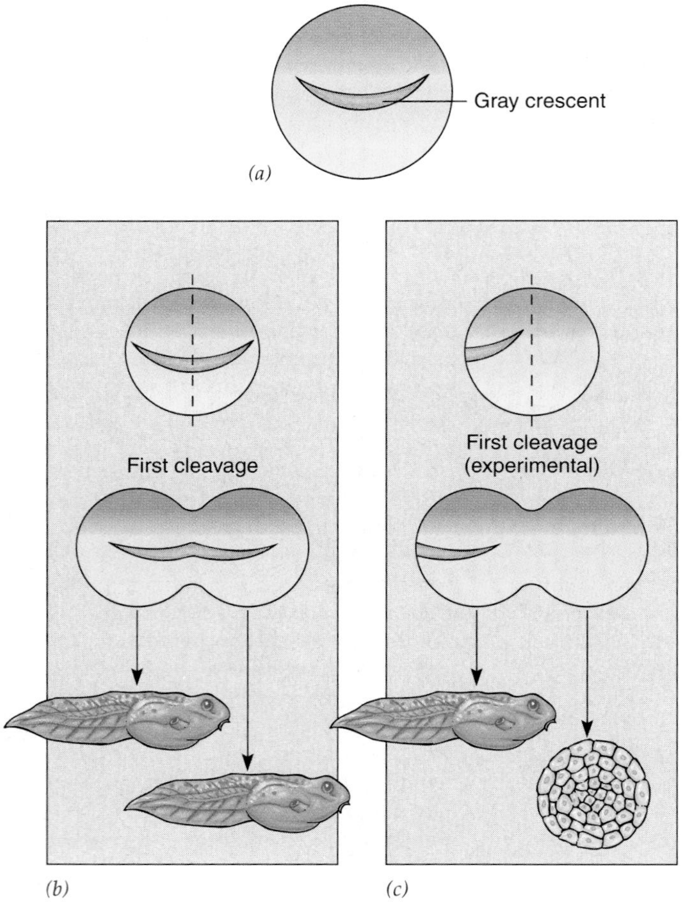

FIGURE 49–3 Cytoplasmic factors affect development. (***a***) The position of the gray crescent in the frog zygote determines the main axes of the body. (***b***) The first division of the zygote partitions the gray crescent into the two daughter cells. If these cells are separated, each can develop into a tadpole. (***c***) If the plane of cleavage is changed experimentally so that only one cell receives the gray crescent, only that cell can develop into a tadpole.

The amount of yolk determines the pattern of cleavage

Many animal eggs contain **yolk,** a mixture of proteins, phospholipids, and fats that serves as food for the developing embryo. The amount and distribution of yolk vary among different animal groups. Most invertebrates and simple chordates have **isolecithal** eggs with relatively small amounts of yolk uniformly distributed through the cytoplasm. Many vertebrate eggs are **telolecithal,** meaning that they have large amounts of yolk concentrated at one end of the cell, known as the **vegetal pole.** The opposite, more metabolically active pole is the **animal pole.**

The amount of yolk in the egg affects the pattern of cleavage. In eggs with evenly distributed yolk, the entire egg divides, producing cells of roughly the same size (holoblastic cleavage). Cleavage of these eggs can be radial or spiral. Radial cleavage is characteristic of deutero-

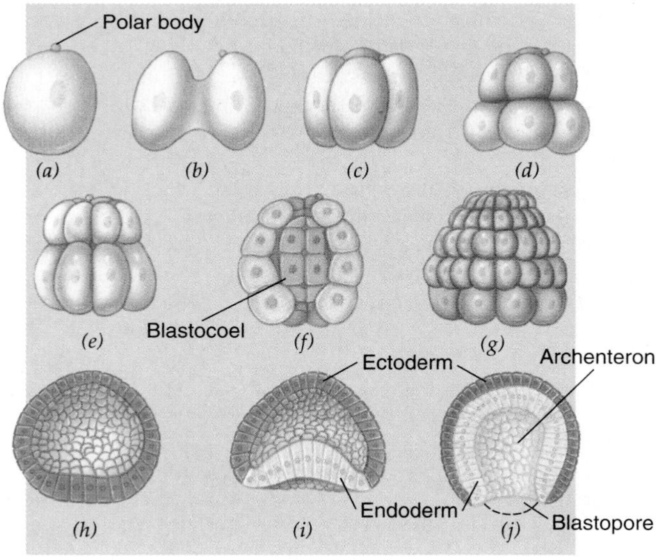

FIGURE 49–4 The pattern of early development in amphioxus is similar to that of the sea star. Yolk is evenly distributed so the entire egg divides, and cleavage is radial. The embryos shown are viewed from the side. (***a***) Mature egg with polar body. (***b* to *e***) Two-, four-, eight-, and 16-cell stages. (***f***) Embryo cut open to show the blastocoel. (***g***) Blastula. (***h***) Blastula cut open. (***i***) Early gastrula showing beginning of invagination at vegetal pole. (***j***) Late gastrula. Invagination is completed and the blastopore has formed.

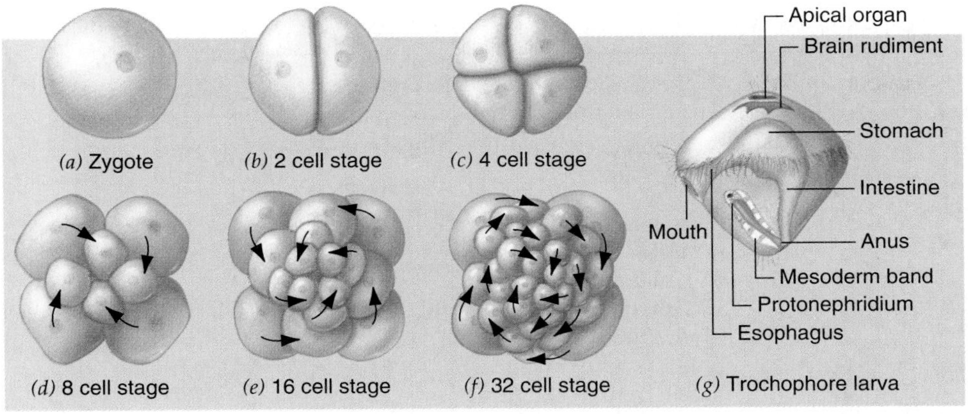

FIGURE 49–5 Annelid embryos undergo spiral cleavage. ***a–f*** are top views of the animal pole. The successive cleavage divisions occur in a spiral pattern as illustrated. (***g***) A typical trochophore larva. The upper half of the trochophore develops into the extreme anterior end of the adult worm; all the rest of the adult body develops from the lower half.

stomes, whereas spiral cleavage is characteristic of protostomes (see Fig. 28–4).

In **radial cleavage,** the first division passes through both animal and vegetal poles and splits the egg into two equal cells. The second cleavage division passes through both poles of the egg at right angles to the first and separates the two cells into four equal cells. The third division is horizontal, at right angles to the other two, and separates the four cells into eight cells: four above and four below the third line of cleavage. This pattern of radial cleavage occurs in the sea star and in the simple chordate amphioxus (Figs. 49–2 and 49–4).

In **spiral cleavage,** after the first two divisions, the plane of cytokinesis is diagonal to the polar axis (Fig. 49–5). This results in a spiral arrangement of cells, with each cell located above and in between the two underlying cells.

The eggs of bony fish and amphibians contain a large amount of yolk concentrated toward the vegetal pole. The cleavage divisions in the vegetal hemisphere are slowed by the presence of the inert yolk. As a result, the blastula consists of many small cells in the animal hemisphere, and fewer but larger cells in the vegetal hemisphere (Fig. 49–6). The blastocoel is displaced toward the animal pole.

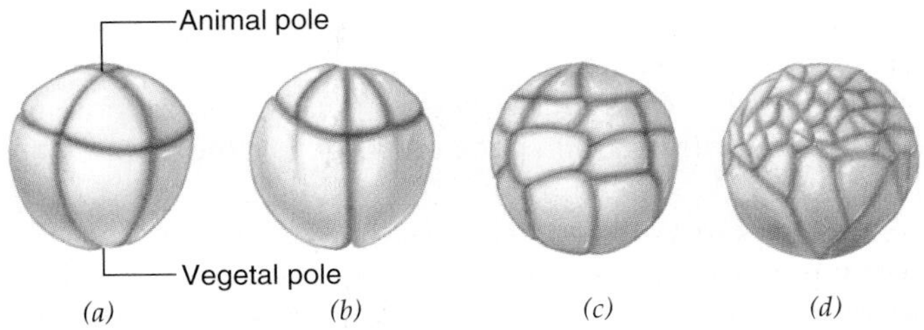

FIGURE 49–6 The large amount of yolk concentrated in the vegetal hemisphere of the frog egg slows cleavage. As a result, fewer cells develop at the vegetal hemisphere than at the animal hemisphere. These embryos are viewed from the side.

The telolecithal eggs of reptiles, birds, and some fish have a very large amount of yolk and only a small amount of cytoplasm concentrated at the animal pole. In such eggs, cell division takes place only in the **blastodisc,** the small disc of cytoplasm at the animal pole (Fig. 49–7); this type of cleavage is termed meroblastic.

In birds and some reptiles, the blastomeres form two layers, an upper epiblast and below that a thin layer of flat cells, the hypoblast. The epiblast and hypoblast are separated from each other by the blastocoel. They are separated from the underlying yolk by another cavity called the subgerminal space. This space appears only under the central portion of the blastodisc.

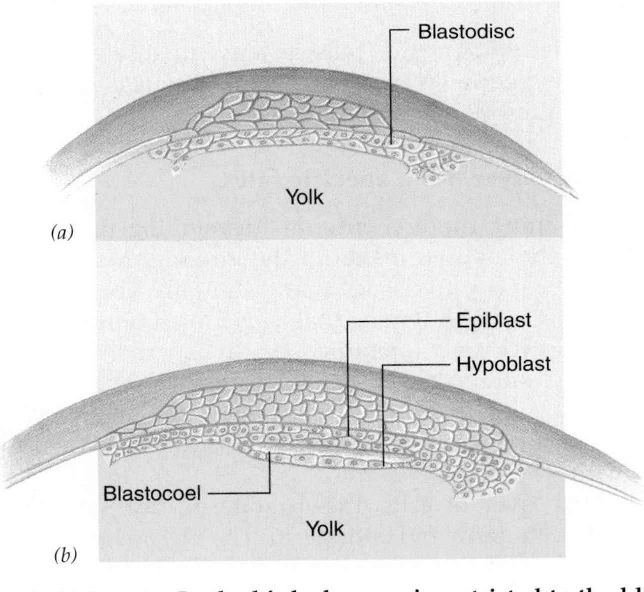

FIGURE 49–7 In the bird, cleavage is restricted to the blastodisc, a small disk of cytoplasm on the upper surface of the egg yolk. (***a***) Note that a space is present beneath the blastodisc, separating it from the unsegmented yolk. (***b***) The blastodisc splits into two tissue layers, an upper epiblast and a lower hypoblast, separated by the blastocoel.

Table 49–1 FATE OF THE GERM LAYERS

GASTRULA	
Ectoderm	Nervous system and sense organs Outer layer of skin (epidermis) and its associated structures (nails, hair, etc.) Pituitary gland
Mesoderm	Skeleton (bone and cartilage) Muscles Circulatory system Excretory system Reproductive system Inner layer of skin (dermis) Outer layers of digestive tube and of structures that develop from it, such as part of respiratory system
Endoderm	Lining of digestive tube and of structures that develop from it, such as lining of the respiratory system

THE GERM LAYERS FORM DURING GASTRULATION

The process by which the blastula becomes a three-layered embryo, or **gastrula,** is called **gastrulation.** Thus, early development proceeds through the following sequence of events:

zygote → cleavage → morula → blastula → gastrulation

During gastrulation, the cells arrange themselves into three distinct **germ layers,** or embryonic tissue layers: the ectoderm, mesoderm, and endoderm.

Each germ layer has a specific fate

The cells lining the **archenteron** (developing digestive cavity) of the embryo make up the **endoderm.** The endoderm gives rise to tissues that eventually line the digestive tract, and to organs that develop as outgrowths of the digestive tract (including the liver, pancreas, and lungs). The outer wall of the gastrula consists of **ectoderm.** This germ layer eventually forms the outer layer of the skin and gives rise to the nervous system and sense organs.

A third layer of cells, the **mesoderm,** develops between the ectoderm and endoderm. The mesoderm gives rise to skeletal tissue, muscle, and the circulatory, excretory, and reproductive systems (Table 49–1).

The pattern of gastrulation is affected by the amount of yolk

Gastrulation in the sea star and in amphioxus is illustrated in Figures 49–2 and 49–4, respectively. Gastrulation begins when a group of cells at the vegetal pole flattens and then bends inward. The cells of a section of the vegetal wall of the blastula move inward (invaginate). The invaginated wall eventually meets the opposite wall, obliterating the blastocoel.

We can roughly demonstrate gastrulation by pushing inward on the wall of a partly deflated rubber ball until it rests against the opposite wall. In a similar way the embryo is converted into a double-walled, cup-shaped structure. The cavity of the cup communicates with the exterior on the side that was originally the vegetal pole of the embryo. The internal wall lines the archenteron, the newly formed cavity of the developing gut. The opening of the archenteron, the **blastopore,** becomes the anus in deuterostomes.

In the amphibian, the large yolk-laden cells in the vegetal half of the blastula obstruct the inward movement at the vegetal pole. Instead, cells from the animal pole move down toward the yolk-rich cells and then inward and away from them, forming the dorsal lip of the blastopore (Fig. 49–8). As the process continues, the blastopore becomes ring-shaped as cells lateral, and then ventral, to the blastopore become involved in the same movements. The yolk-filled cells of the vegetal hemisphere remain as a yolk plug, filling the space enclosed by the lips of the blastopore.

The archenteron forms and is lined on all sides by cells that have moved in from the surface. At first, the archenteron is a narrow slit, but it gradually expands at the anterior end, and eventually obliterates the blastocoel.

Although the details differ somewhat, gastrulation in the bird is basically similar to amphibian gastrulation. Epiblast cells migrate toward the midline to form a thickened cellular region, known as the **primitive streak,** that elongates and narrows as it develops. At its center is a

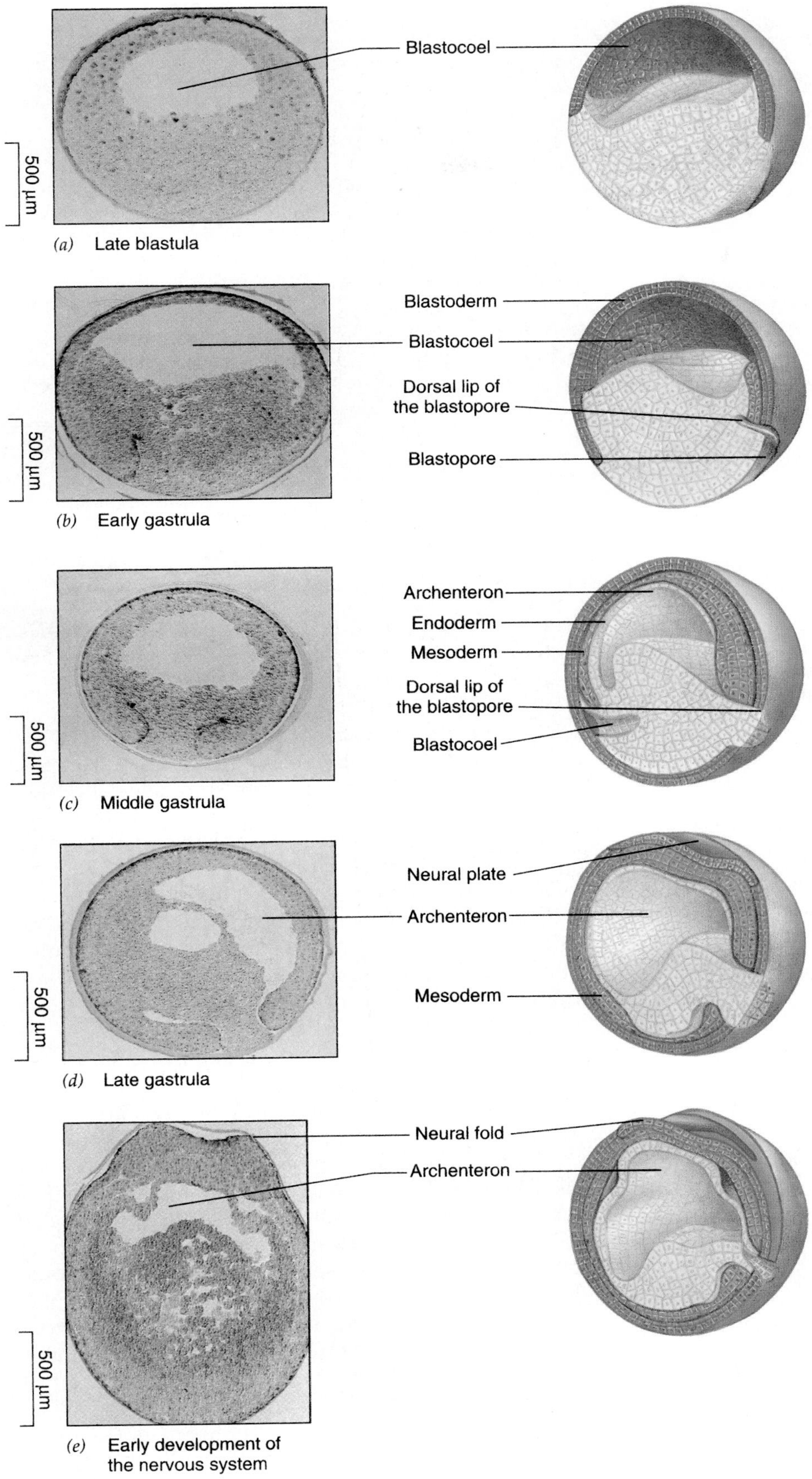

FIGURE 49–8 Gastrulation in a frog embryo shows the formation of the germ layers. The diagrams show the embryo cut in half so you can see inside. (***a***) Late blastula. (***b***) Early gastrula. (***c***) Middle gastrula. (***d***) Late gastrula. (***d***) and (***e***) Nervous system development begins with the formation of the **neural plate.** (Carolina Biological Supply Company/Phototake—NYC)

Making the Connection

Development and Genetics

How important are genes in controlling development? Although nongenetic influences are important in determining cell differentiation, the genes exercise ultimate control over development. Consider a specific experimental example. When ectoderm from the mouth region of an early frog embryo is transplanted to the mouth region of a salamander, the surrounding salamander tissue induces the transplanted cells to develop into mouth structures. However, the mouth that forms is not the characteristic mouth of a salamander. Instead it bears the horny rows of teeth and horny jaws of a frog. Why? As you might guess, the frog cells do not have the capacity to form structures specific to salamanders.

Cellular differentiation is an expression of changes in the activity of specific genes, and genetic activity in turn is influenced by a variety of factors inside and outside the cell. Although each cell of an organism has the same set of genes, differential gene expression results in variations in chemistry, behavior, and structure among cells. Through this process, an embryo develops into an organism composed of more than 200 types of cells, each specialized to perform specific functions. ▲

narrow furrow, the primitive groove. The primitive streak is a dynamic structure. The cells composing it constantly change as they migrate in from the epiblast, sink down at the primitive streak, then move out laterally and anteriorly in the interior (Fig. 49–9). These cells form the mesoderm and also contribute to the endoderm. The primitive streak is homologous to the blastopore of the sea star, amphioxus, and amphibian embryos. However, the embryo of the chick contains no cavity that is homologous to the archenteron.

At the anterior end of the primitive streak is a thickened knot of cells, **Hensen's node.** Cells destined to form the notochord (supporting rod of cartilage-like cells) become concentrated in Hensen's node and then grow anteriorly as a narrow process from the node just beneath the epiblast. Tissue that will form mesoderm moves laterally and anteriorly from the primitive streak between the epiblast (which becomes ectoderm) and the hypoblast (which becomes endoderm).

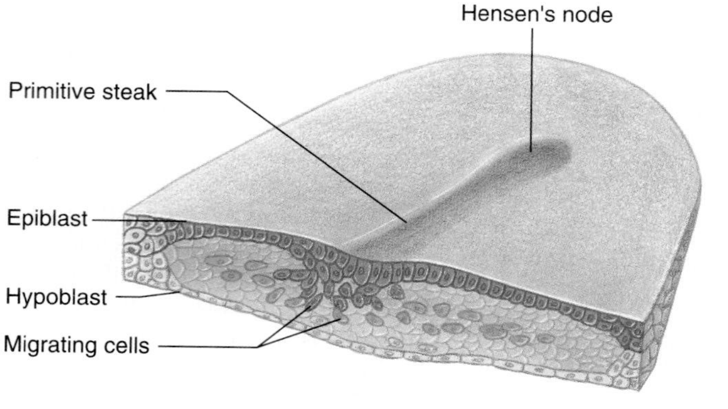

FIGURE 49–9 Gastrulation in birds begins with the development of the primitive streak. The migration of cells during gastrulation results in a three-layered embryo.

ORGANOGENESIS BEGINS WITH THE DEVELOPMENT OF THE NERVOUS SYSTEM

The process of organ formation is called **organogenesis.** The brain, notochord, and spinal cord are among the first organs to develop in the early vertebrate embryo (Figs. 49–8*d* and *e*, and 49–10). First the notochord, the flexible skeletal axis in all chordate embryos, grows forward along the length of the embryo as a cylindrical rod of cells. The developing notochord **induces** (stimulates) the overlying ectoderm to thicken, forming the **neural plate.** Central cells of the neural plate move downward and form a depression called the **neural groove;** the cells flanking the groove on each side form **neural folds.**

Continued cell movement brings the folds closer together until they meet and fuse, forming the **neural tube.** In this process, the neural tube comes to lie beneath the surface. The ectoderm overlying it will form the outer layer of skin. The anterior portion of the neural tube grows and differentiates into the brain; the remainder of the tube develops into the spinal cord.

The induction of the neural plate by the notochord is a good example of the importance of the position of a cell in relation to its cellular neighbors. That position is often critical in determining the fate of a cell because it determines its exposure to substances released from other cells.

Various motor nerves grow out of the developing brain and spinal cord, but sensory nerves have a separate origin. When the neural folds fuse to form the neural tube, bits of nervous tissue known as the **neural crest** arise from the approximate region of fusion on each side of the tube. Neural crest cells migrate downward from their original position and form the dorsal root ganglia of the spinal nerves and the postganglionic sympathetic neurons. From sensory cells in the dorsal root ganglia, dendrites grow out to the sense organ, and axons grow in to the spinal cord. Neural crest cells migrate to various locations

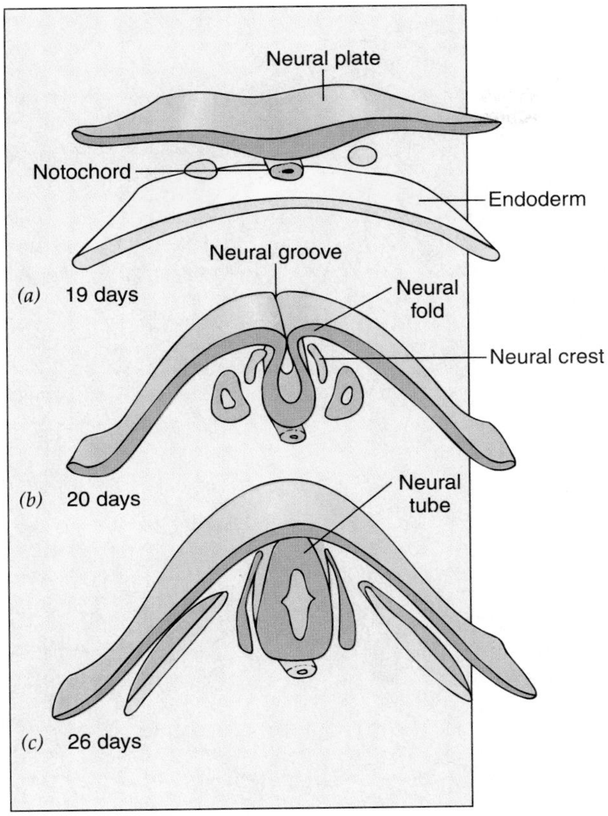

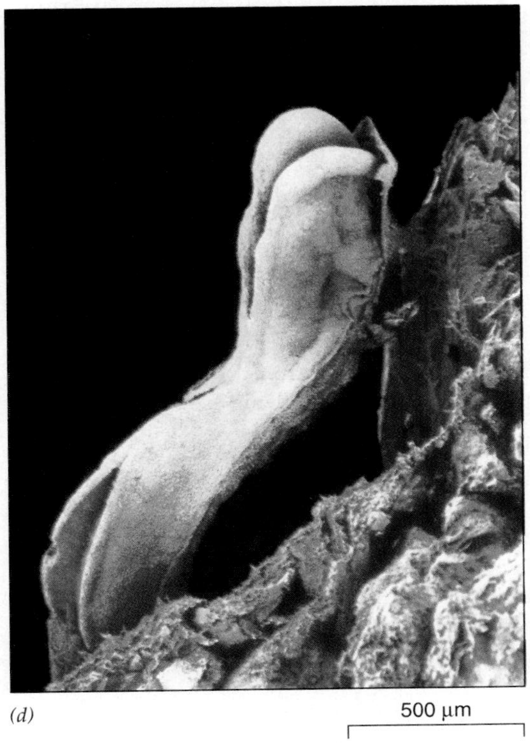

FIGURE 49–10 Development of the nervous system begins with formation of the neural plate. Cross sections of human embryos at successively later stages, illustrating early nervous system development. The neural crest cells give rise to sensory neurons. (***a***) Approximately 19 days. The neural plate has indented to form a shallow groove flanked by neural folds. (***b***) Approximately 20 days. The neural folds approach one another. (***c***) Approximately 26 days. The neural folds have formed the neural tube, which will give rise to the brain at the anterior end of the embryo and the spinal cord posteriorly. (***d***) Photograph of a 20-day-old human embryo shows the developing nervous system. (*d*, Lennart Nilsson, © Boehringer Ingelheim International GmbH, from *A Child Is Born*, p. 76, Dell Publishing Co., Inc.)

in the embryo. They give rise to parts of certain sense organs and nearly all pigment-forming cells in the body.

As the nervous system develops, other organs also begin to take shape. The digestive tract is first formed as a separate foregut and hindgut by the growth and folding of the body wall, which cuts them off as two simple tubes from the original yolk sac (Fig. 49–11*a*). As the embryo grows, these tubes, which are lined with endoderm, grow and become greatly elongated. The liver, pancreas, and trachea originate as hollow, tubular outgrowths from the gut. As the trachea grows downward, it gives rise to the paired lung buds, which develop into lungs.

The most anterior part of the foregut becomes the pharynx. A series of small outpocketings of the pharynx, the **pharyngeal pouches,** bud out laterally (Fig. 49–11*b*). These pouches meet a corresponding set of inpocketings from the overlying ectoderm, the **branchial grooves**. The arches of tissue formed between the grooves are called **branchial arches**. They contain the rudimentary skeletal, neural, and vascular elements of the face, jaws, and neck.

In aquatic vertebrates, the pharyngeal grooves and branchial grooves meet and form a continuous passage from the pharynx to the outside; these are the gills slits, which function as respiratory organs. In fishes and some amphibians, gills develop on the branchial arches. In terrestrial vertebrates, each branchial groove remains separated from the corresponding pharyngeal pouch by a thin membrane of tissue, and these structures give rise to organs more appropriate for life on land.

The heart and circulatory system are among the first structures to take form, and they must function while still developing. The circulatory organs develop from the mesoderm. Other mesodermal structures include the skeleton, muscles, kidneys, and reproductive structures.

EXTRAEMBRYONIC MEMBRANES PROTECT AND NOURISH THE EMBRYO

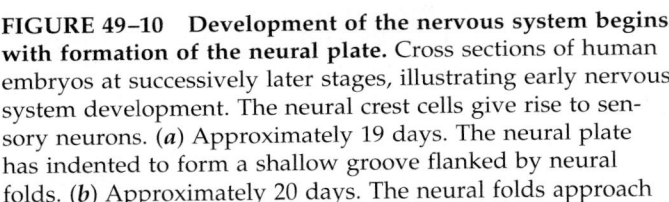

Terrestrial vertebrates have four **extraembryonic membranes:** the chorion, allantois, yolk sac, and amnion (Fig. 49–12). Although they develop from the germ layers, these membranes are not part of the embryo itself and

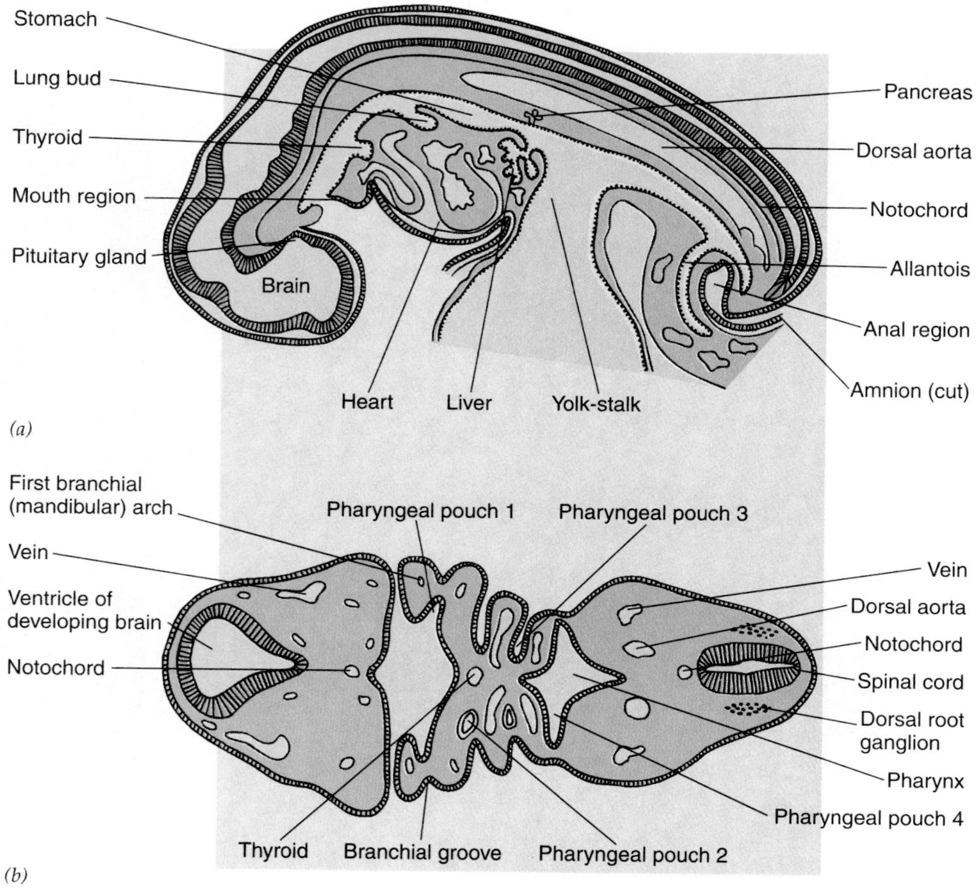

FIGURE 49–11 After gastrulation, the rudiments of the organs develop. (***a***) Sagittal section through human embryo showing some of the structures that develop during the fifth week. Note that the liver, pancreas, and respiratory systems develop as outpocketings from the digestive tract. (***b***) Cross section through a human embryo during the fifth week of development. Note the branchial grooves, pharyngeal pouches, and branchial arches. The embryo is flexed so both the brain and spinal cord are present in the cross section.

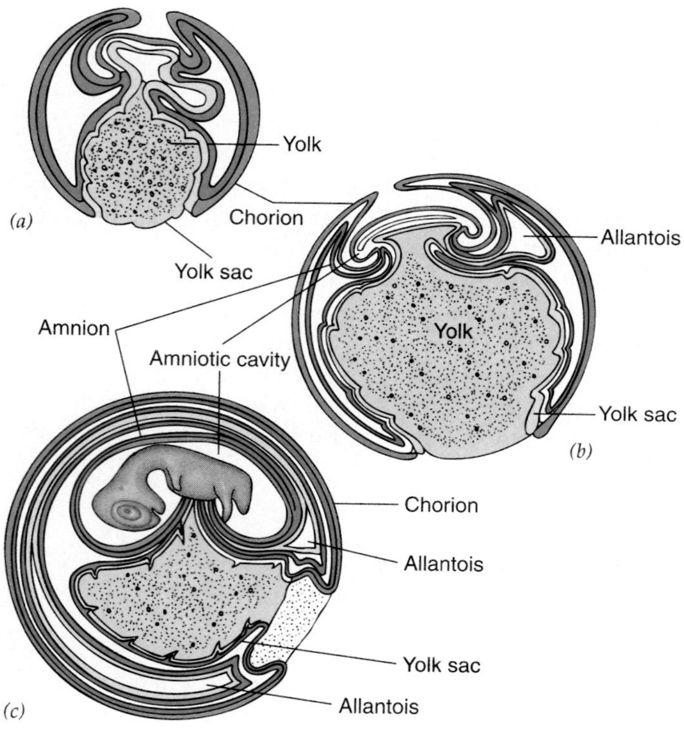

are discarded at birth. The extraembryonic membranes are adaptations to the challenges of embryonic development on land. They protect the embryo, prevent it from drying out, and help in obtaining food and oxygen and eliminating wastes.

The **chorion** and **amnion** enclose the entire embryo (Fig. 49–12*c*). The amniotic cavity, the space between the embryo and the amnion, becomes filled with amniotic fluid secreted by the membrane. Embryos of terrestrial vertebrates develop within this pool of fluid. The amniotic fluid prevents the embryo from drying out and permits it a certain freedom of motion. The fluid also serves as a protective cushion that absorbs shocks and prevents the amniotic membrane from sticking to the embryo.

FIGURE 49–12 Successive stages are shown in the development of the extraembryonic membranes of the chick (*a, b, c*). Each of the membranes develops from a combination of two germ layers. The chorion and amnion form from lateral folds of the ectoderm and mesoderm that extend over the embryo and fuse. The allantois develops from an outpocketing of the gut. The allantois, an elongated sac, and the yolk sac develop from endoderm and mesoderm.

Table 49–2 SOME IMPORTANT DEVELOPMENTAL EVENTS IN THE HUMAN EMBRYO

Time from Fertilization	*Event*
24 hours	Embryo reaches two-cell stage
3 days	Morula reaches uterus
7 days	Blastocyst begins to implant
2.5 weeks	Notochord and neural plate are formed; tissue that will give rise to heart is differentiating; blood cells are forming in yolk sac and chorion
3.5 weeks	Neural tube forming; primordial eye and ear visible; pharyngeal pouches forming; liver bud differentiating; respiratory system and thyroid gland just beginning to develop; heart tubes fuse, bend, and begin to beat; blood vessels are laid down
4 weeks	Limb buds appear; three primary divisions of brain forming
2 months	Muscles differentiating; embryo capable of movement; gonad distinguishable as testis or ovary; bones begin to ossify; cerebral cortex differentiating; principal blood vessels assume final positions
3 months	Sex can be determined by external inspection; notochord degenerates; lymph glands develop
4 months	Face begins to look human; lobes of cerebrum differentiate; eyes, ears, and nose look more "normal"
Third trimester	A covering of downy hair covers the fetus, then later is shed; neuron myelination begins; tremendous growth of body
266 days (from conception)	Birth

The **allantois** is an outgrowth of the developing digestive tract. In reptiles and birds, it stores nitrogenous wastes. In humans, the allantois is small and nonfunctional, except that its blood vessels contribute to the formation of umbilical vessels joining the embryo to the placenta.

In vertebrates with yolk-rich eggs, the **yolk sac** encloses the yolk, slowly digests it, and makes it available to the embryo. A yolk sac develops even in vertebrate embryos with little or no yolk. Its walls serve as temporary centers for the formation of blood cells.

HUMAN PRENATAL DEVELOPMENT REQUIRES ABOUT 266 DAYS

The human **gestation period,** the duration of pregnancy, averages 280 days (40 weeks) from the time of the mother's last menstrual period to the birth of the baby, or 266 days (about 9 months) from the time of conception (Table 49–2).

Development begins in the oviduct

By about 24 hours after fertilization, the human zygote has divided to become a two-celled embryo (Fig. 49–13). Cleavage continues as the embryo is pushed along the oviduct by ciliary action and muscular contraction.

When the embryo enters the uterus on about the fifth day of development, the zona pellucida (its surrounding coat) is dissolved. During the next few days, the embryo floats free in the uterine cavity, nourished by a nutritive fluid secreted by the glands of the uterus. Its cells arrange themselves, forming a blastula, which in mammals is called a **blastocyst** (Fig. 49–14). The outer layer of cells, the **trophoblast,** eventually forms the chorion and amnion that surround the embryo. A little cluster of cells, the **inner cell mass,** projects into the cavity of the blastocyst. The inner cell mass gives rise to the embryo proper.

The embryo implants in the wall of the uterus

On about the seventh day of development the embryo begins to **implant** in the endometrium of the uterus (Figs. 49–14*a* and 49–15). The trophoblast cells in contact with the uterine lining secrete enzymes that erode an area just large enough to accommodate the tiny embryo. Slowly the embryo works its way down into the underlying connective and vascular tissues, and the opening in the lining of the uterus repairs itself. All further development of the embryo takes place *within* the endometrium of the uterus.

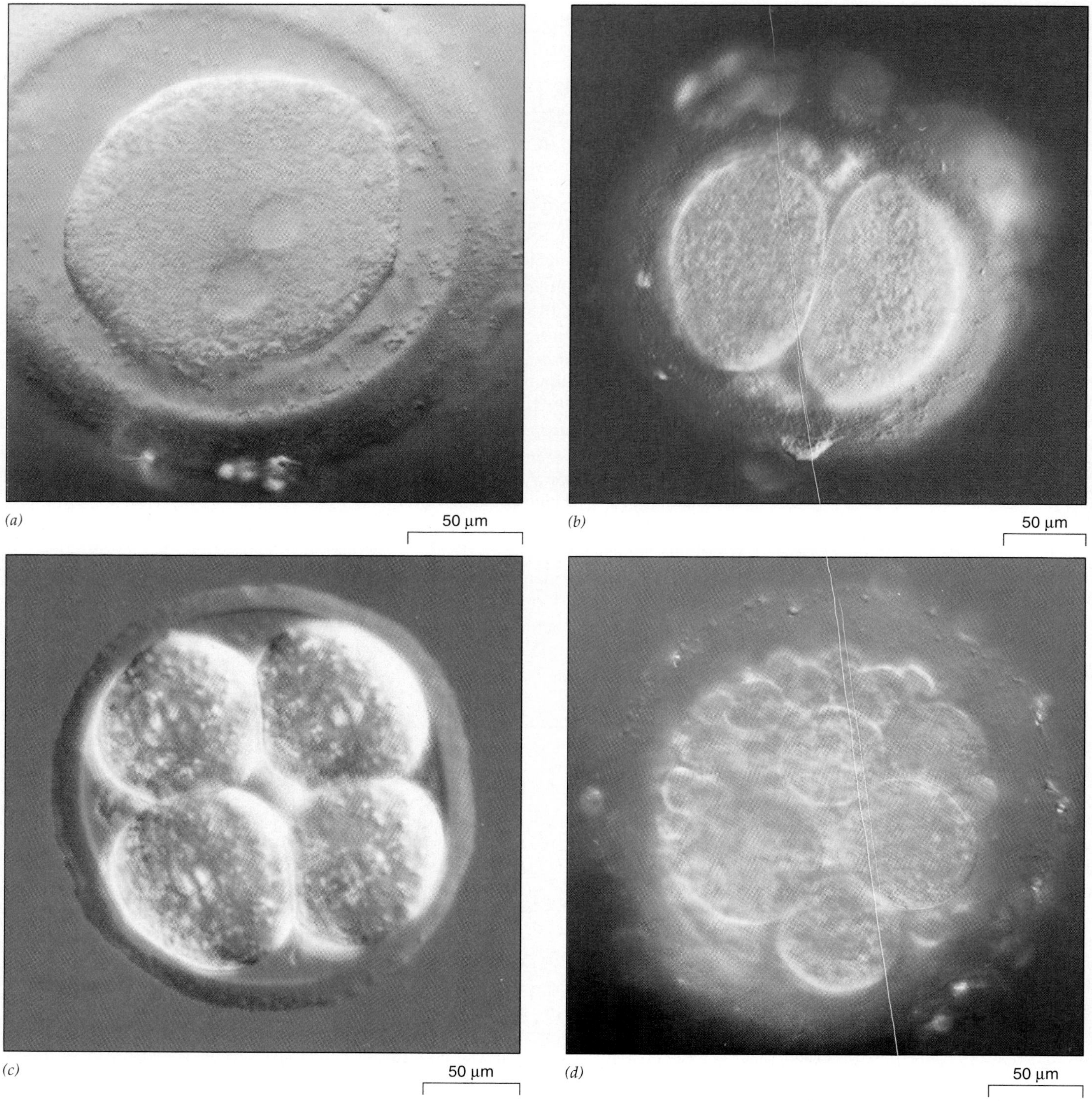

FIGURE 49–13 Cleavage divides the human zygote into many cells. (***a***) The human zygote contains the genetic instructions for producing a complete human. (***b***) Two-cell stage. (***c***) Eight-cell stage. (***d***) Cleavage continues, giving rise to a cluster of cells called the morula. (Lennart Nilsson, from *Being Born,* pp. 14, 15, 17)

The placenta is an organ of exchange

In placental mammals, the **placenta** is the organ of exchange between mother and embryo. The placenta provides nutrients and oxygen for the fetus, and removes wastes, which the mother then excretes (Fig. 49–14*d*). In addition, the placenta is an endocrine organ that secretes hormones to maintain pregnancy.

The placenta develops from both the chorion of the embryo and the uterine tissue of the mother. In early development, the chorion grows rapidly, invading the en-

(Text continues on page 1098)

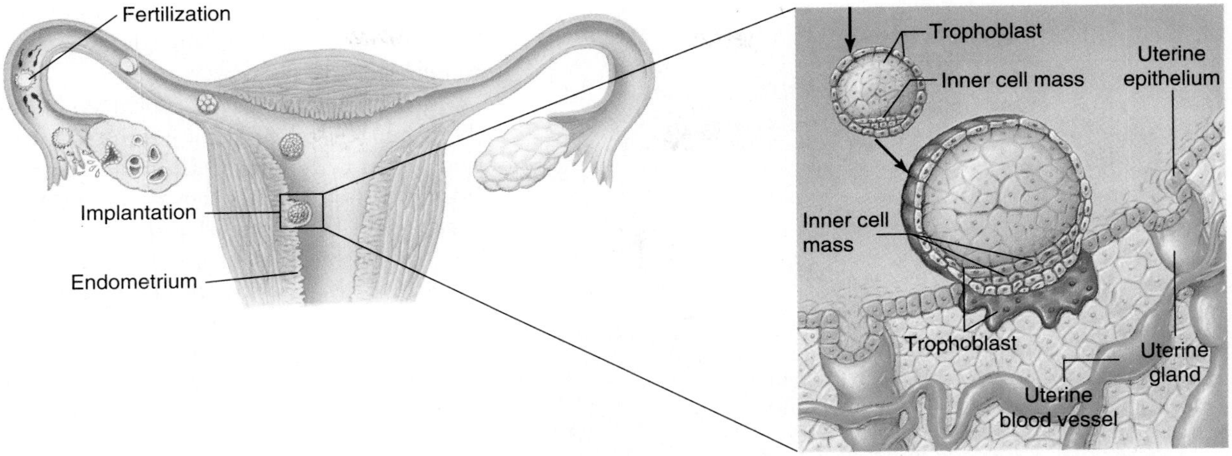

(a)

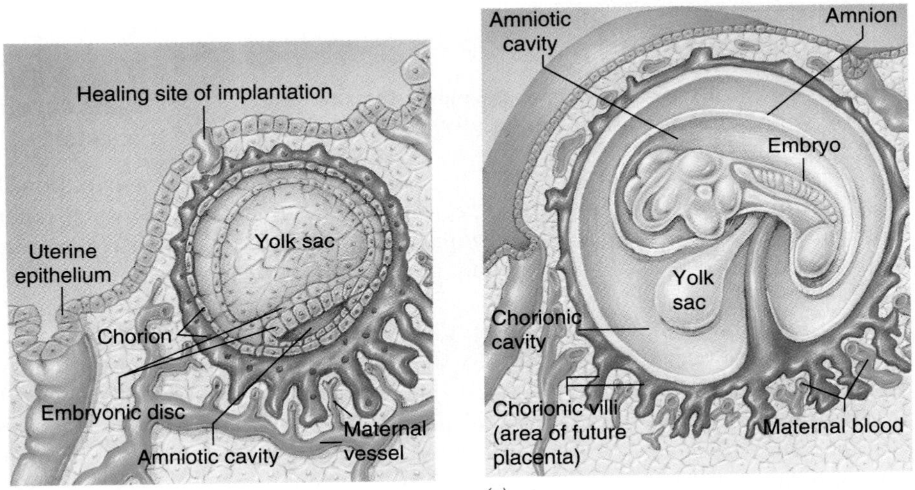

(b)

(c)

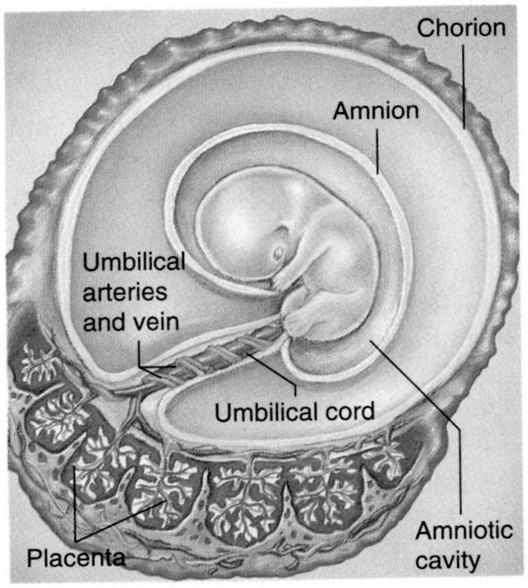

(d)

FIGURE 49–14 After the embryo has completed cleavage on its journey through the oviduct to the uterus, it is implanted in the uterine wall. (*a*) About seven days after fertilization, the blastocyst drifts to an appropriate site along the uterine wall and begins to implant itself. The cells of the trophoblast proliferate and invade the endometrium. (*b*) About 10 days after fertilization, the chorion has formed from the trophoblast. (*c*) After 25 days, intimate relationships have been established between the embryo and the maternal blood vessels. Oxygen and nutrients from the maternal blood are now satisfying the embryo's needs. Note the specialized region of the chorion that will soon become the placenta. (*d*) At about 45 days the embryo and its membranes together are about the size of a ping-pong ball, and the mother still may be unaware of her pregnancy. The amnion filled with amniotic fluid surrounds and cushions the embryo. The yolk sac has been incorporated into the umbilical cord. Blood circulation has been established through the umbilical cord to the placenta.

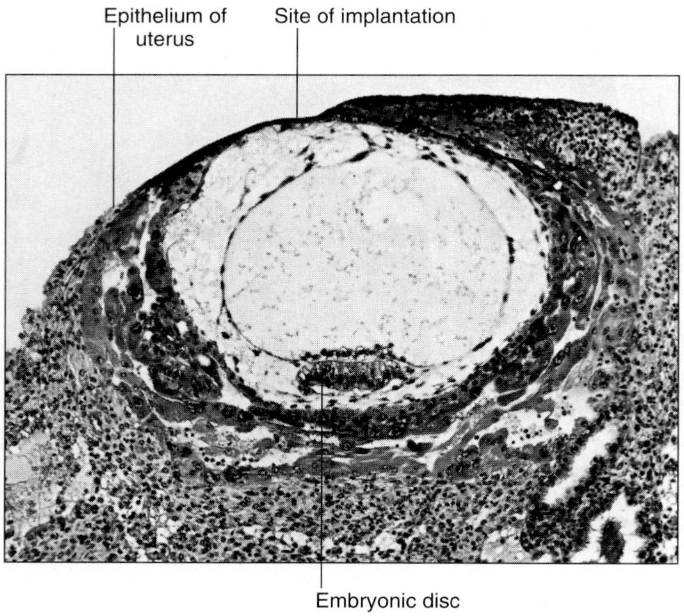

FIGURE 49–15 By twelve days of development, the human blastocyst has implanted itself in the uterine wall. (Courtesy of Carnegie Institute of Washington)

dometrium and forming finger-like projections known as chorionic villi. The villi become vascularized (infiltrated with blood vessels) as the embryonic circulation develops.

As the human embryo grows, the **umbilical cord** develops and connects the embryo to the placenta (Fig. 49–14*d*). The umbilical cord contains the two umbilical arteries and the umbilical vein. The **umbilical arteries** connect the embryo to a vast network of capillaries developing within the villi. Blood from the villi returns to the embryo through the **umbilical vein.**

The placenta consists of the portion of the chorion that develops villi, together with the uterine tissue underlying the villi that contains maternal capillaries and small pools of maternal blood. The fetal blood in the capillaries of the chorionic villi comes in close contact with the mother's blood in the tissues between the villi. However, they are always separated by a membrane through which substances may diffuse or be actively transported. *Maternal and fetal blood do not normally mix in the placenta or any other place.*

Several hormones are produced by the placenta. From the time the embryo first begins to implant itself, its trophoblastic cells release **human chorionic gonadotropin (hCG),** which signals the corpus luteum that pregnancy has begun. In response, the corpus luteum increases in size and releases large amounts of progesterone and estrogens. These hormones stimulate continued development of the endometrium and placenta.

Without hCG, the corpus luteum would degenerate and the embryo would be aborted and flushed out with the menstrual flow. The woman would probably not even know that she had been pregnant. When the corpus luteum is removed before about the 11th week of pregnancy, the embryo is spontaneously aborted. After that time, the placenta itself produces enough progesterone and estrogens to maintain pregnancy.

Organ development begins during the first trimester

Gastrulation occurs during the second and third weeks of development. Then the notochord begins to form and induces formation of the neural plate. The neural tube develops and the forebrain, midbrain, and hindbrain are evident by the fifth week of development. A week or so later the forebrain begins to grow outward, forming the rudiments of the cerebral hemispheres.

The heart begins to develop, and after 3.5 weeks begins to beat spontaneously (Figs. 49–16 and 49–17). Pharyngeal pouches, branchial grooves, and branchial arches form in the region of the developing pharynx. In the floor of the pharynx, a tube of cells grows downward to form the primordial trachea, which gives rise to the lung buds. The digestive system also gives rise to outgrowths that will develop into the liver, gallbladder, and pancreas. Near the end of the first month, the limb buds begin to differentiate; these eventually give rise to arms and legs.

All of the organs continue to develop during the second month (Fig. 49–18). A thin tail becomes evident dur-

FIGURE 49–16 The heart forms from the fusion of two blood vessels. These ventral views of successive stages in the development of the heart show that at first the heart is upside down; the end that receives blood from the veins is at the bottom. As it develops, the heart twists and turns to carry the atrium to a position above the ventricle, and the conus and sinus disappear. The chambers divide to form the four-chambered heart.

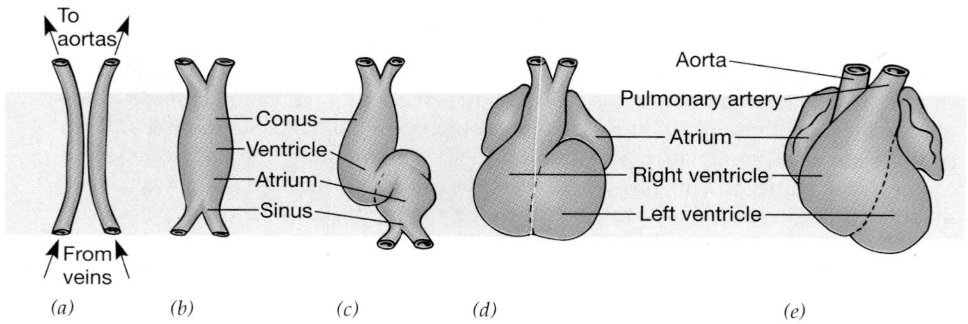

Focus On

Multiple Births

Occasionally, the cells of the two-celled embryo separate and each cell develops into a complete organism. Or sometimes the inner cell mass subdivides, forming two groups of cells, each of which develops independently. Because these cells have identical sets of genes, the individuals formed are exactly alike—**monozygotic,** or **identical, twins.** Very rarely, the two inner cell masses do not completely separate and so give rise to **conjoined (Siamese) twins** (see figure).

Fraternal twins, also called **dizygotic twins,** develop when two eggs are ovulated and each is fertilized by a different sperm. Each zygote has its own distinctive genetic endowment, so the individuals produced are not identical. They may not even be of the same sex. Similarly, triplets (and other multiple births) may be either identical or fraternal.

In the United States, twins are born once in about 80 births (about 30% of twins are monozygotic), triplets once in 80^2 (or 1 in 6400), and quadruplets once in 80^3 (or 1 in 512,000). These statistics appear to be changing. Multiple births have been increasing with increased use of fertility-inducing agents.

A family history of twinning increases the probability of having dizygotic twins. However, giving birth to monozygotic twins does not appear to be influenced by heredity, age of the mother, or other known factors. An estimated two thirds of multiple pregnancies end in the birth of a single baby; the other embryo may be absorbed within the first ten weeks of pregnancy, or spontaneously aborted. Ultrasonic techniques used early in pregnancy can give valuable information regarding the presence of multiple embryos (see Fig. 49–22). Multiple births are associated with increased risk of perinatal mortality and morbidity. An important factor contributing to this risk is low birth weight. ■

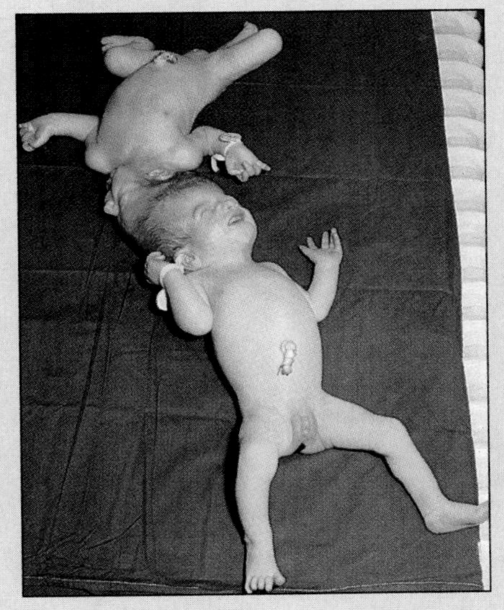

These conjoined (Siamese) twins are joined at the head. (Biophoto Associates)

ing the fifth week but does not grow as rapidly as the rest of the body and so becomes inconspicuous by the end of the second month. Muscles develop, and the embryo becomes capable of movement. The brain begins to send impulses that regulate the functions of some organs, and a few simple reflexes are evident. After the first two months of development, the embryo is referred to as a **fetus.**

By the end of the **first trimester** (the first three months of development) the fetus is recognizably human (Fig. 49–19). The external genital structures have differentiated, indicating the sex of the fetus. Ears and eyes approach their final positions. Some of the skeleton becomes distinct, and the notochord has been replaced by the developing vertebral column. The fetus performs breathing movements, pumping amniotic fluid into and out of its

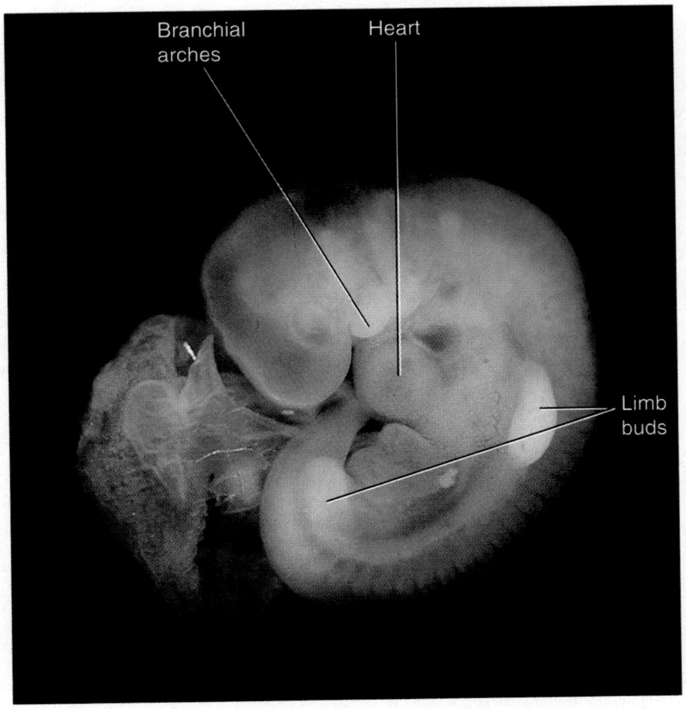

FIGURE 49–17 This photograph of a human embryo at 29 days shows its tail and developing limb buds. The tail will regress during later development. The heart can be seen below the head near the mouth of the embryo. Branchial arches appear as "double chins." At this stage the embryo is about 7 mm (0.3 inch) long. (Lennart Nilsson, from *A Child Is Born,* Dell Publishing Co., Inc.)

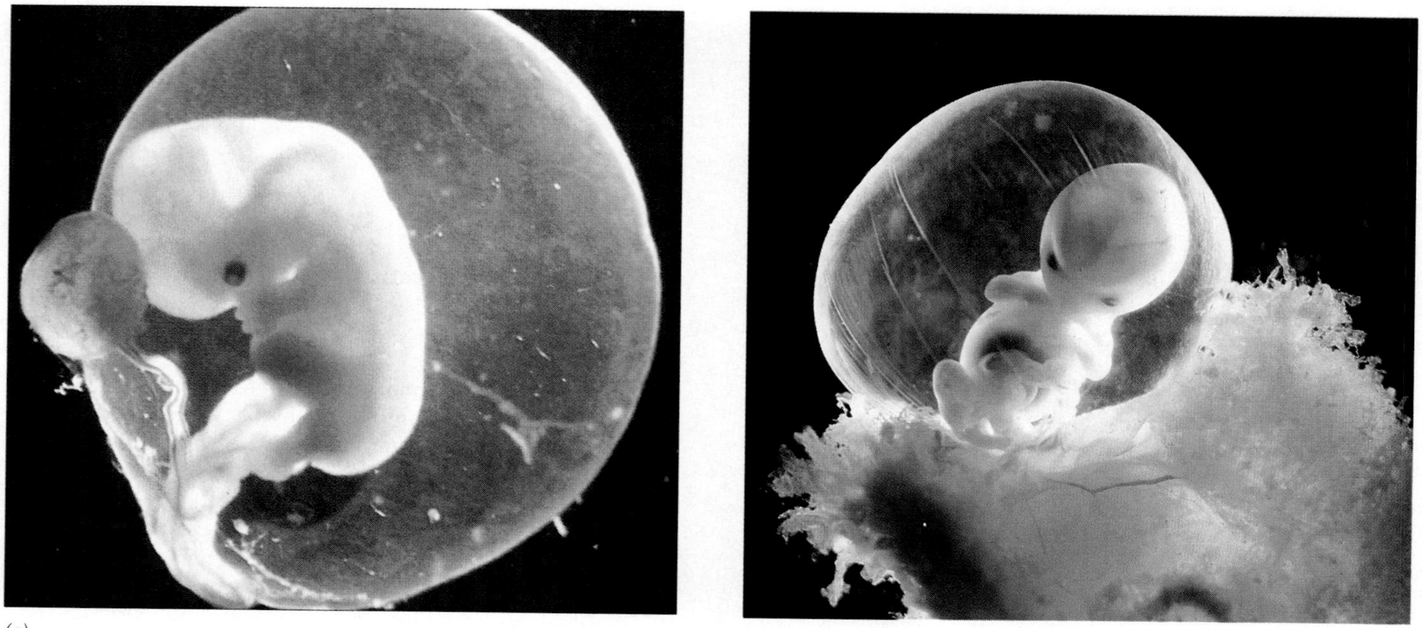

(a) (b)

FIGURE 49–18 The amnion is prominent in these photographs of human embryos during their second month of development. (*a*) Human embryo at 5 1/2 weeks, 1 cm (0.4 inch) long. Limb buds have lengthened and the eyes have become prominent. (*b*) In its 7th week of development, the embryo is 2 cm (0.8 inch) long. The dark red object inside the embryo is the liver. (*a*, Petit Format/Nestle/Photo Researchers, Inc.; *b*, Lennart Nilsson, from *A Child Is Born*, Dell Publishing Co., Inc.)

lungs, and even makes sucking movements. By the end of the third month, the fetus is almost 56 millimeters (about 2.2 inches) long and weighs about 14 grams (0.5 ounce).

Development continues during the second and third trimesters

During the second trimester (months four through six), the fetal heart can be heard with a stethoscope. The fetus moves freely through the amniotic cavity, and during the fifth month the mother usually becomes aware of fetal movements ("quickening").

The fetus grows rapidly during the final trimester (months seven through nine), and final differentiation of tissues and organs occurs. If born at 24 weeks, the fetus has only about a 50% chance, even with the best of medical care, of surviving because its brain is not sufficiently developed to sustain vital functions such as rhythmic breathing and the regulation of body temperature.

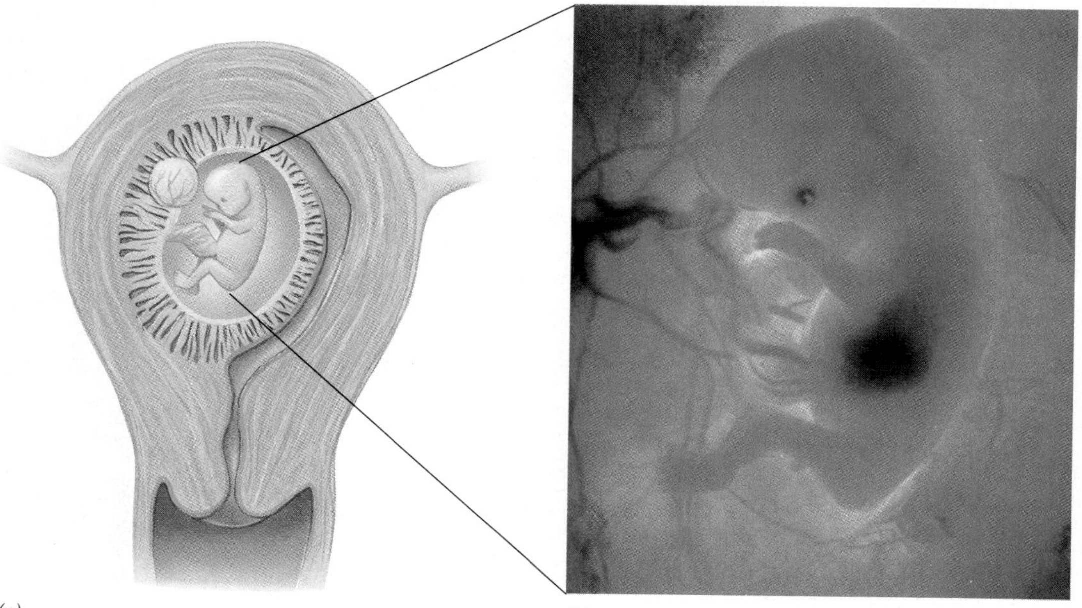

(a) (b)

FIGURE 49–19 The human fetus is shown here at 10 weeks of development. (*a*) Note the position of the fetus within the uterine wall. (*b*) Photograph of fetus enveloped by its extraembryonic membranes. (Nestle/Petit Format/Photo Researchers, Inc.)

During the seventh month the cerebrum grows rapidly and develops convolutions. The grasping and sucking reflexes are evident, and the fetus may suck its thumb. Any infant born before 37 weeks of gestation is considered premature. However, if born after 30 weeks, the baby has a good chance of surviving. At birth the average full-term baby weighs about 3000 grams (6.6 pounds) and measures about 52 centimeters (20 inches) in total length.

The birth process may be divided into three stages

Several hormones are involved in initiating the birth process, which is called **parturition.** Toward the end of pregnancy the muscle fibers of the uterine wall develop additional receptors for oxytocin. As a result the uterus becomes more sensitive to oxytocin, and uterine contractions become stronger. In response to oxytocin, uterine muscle fibers secrete prostaglandins, which also stimulate uterine contraction. A long series of involuntary contractions of the uterus are experienced as **labor.**

Labor may be divided into three stages. During the first stage, which typically lasts about 12 hours, the contractions of the uterus move the fetus toward the cervix, causing the cervix to dilate (open). The cervix also becomes effaced; that is, it loses its normal shape and flattens so that the fetal head can pass through. During the first stage of labor the amnion usually ruptures, releasing about a liter of amniotic fluid, which flows out through the vagina.

During the second stage, which normally lasts between 20 minutes and an hour, the fetus passes through the cervix and the vagina and is born, or "delivered" (Fig. 49–20). With each uterine contraction the woman holds her breath and bears down so that the fetus is expelled from the uterus by the combined forces of uterine contractions and contractions of abdominal wall muscles.

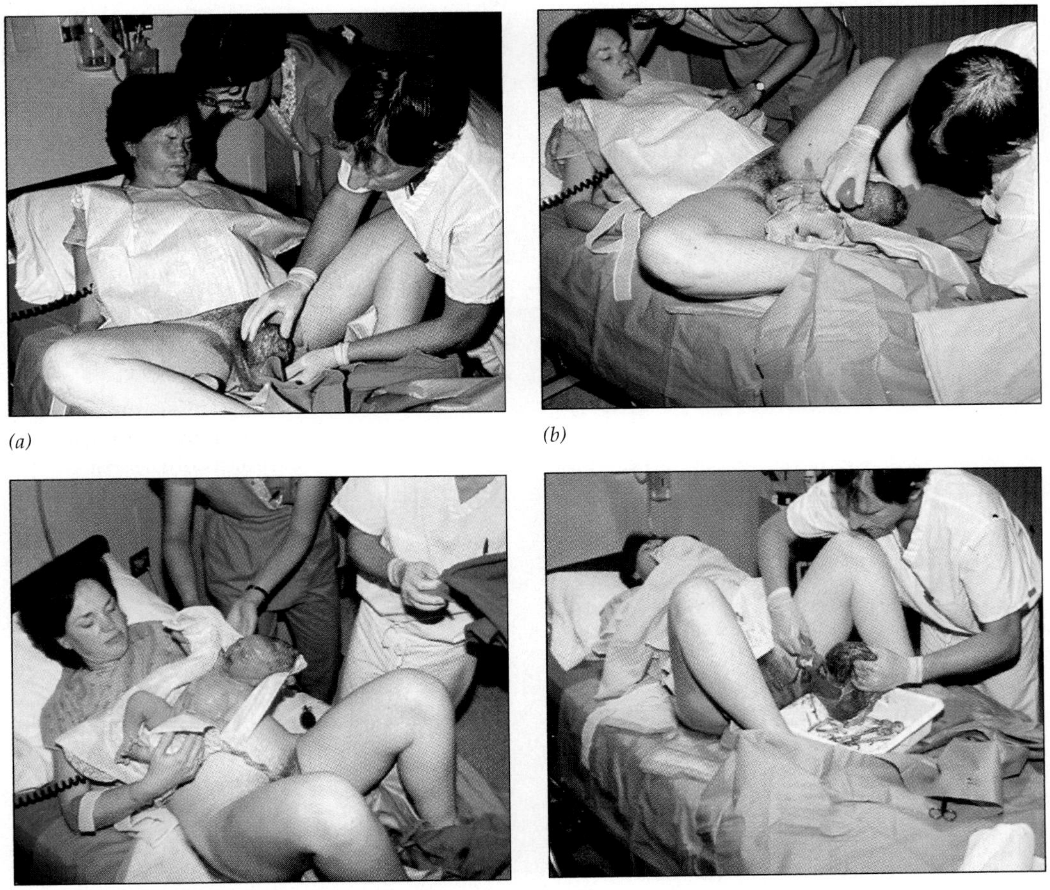

(*a*) (*b*) (*c*) (*d*)

FIGURE 49–20 In about 95% of all human births the baby descends through the cervix and vagina in the head-down position. (***a***) The mother bears down hard with her abdominal muscles, helping to push the baby out. When the head fully appears, the physician or midwife can gently grasp it and guide the baby's entrance into the outside world. (***b***) Once the head has emerged, the rest of the body usually follows readily. The physician gently aspirates the mouth and pharynx to clear the upper airway of any amniotic fluid, mucus, or blood. At this time the neonate usually takes its first breath. (***c***) The baby, still attached to the placenta by its umbilical cord, is presented to its mother. (***d***) During the third stage of labor the placenta is delivered. (Courtesy of Dan Atchinson)

Table 49–3 ENVIRONMENTAL INFLUENCES ON THE EMBRYO

Factor	*Example and Effect*	*Comment*
Nutrition	Severe protein malnutrition doubles number of defects; fewer brain cells are produced, and learning ability may be permanently affected; vitamin deficiencies linked to CNS defects	Growth rate mainly determined by rate of net protein synthesis by embryo's cells; low birth weight
Excessive amounts of vitamins	Vitamin D essential, but excessive amounts may result in form of mental retardation; an excess of vitamins A and K may also be harmful	Vitamin supplements are normally prescribed for pregnant women, but some women mistakenly reason that if one vitamin pill is beneficial, four or five might be even better
Drugs	Many drugs affect development of fetus: even aspirin has been shown to inhibit growth of human fetal cells (especially kidney cells) cultured in laboratory; it may also inhibit prostaglandins, which are concentrated in growing tissue	
Alcohol	When a woman drinks heavily during pregnancy, the baby may be born with fetal alcohol syndrome—that is, deformed and mentally and physically retarded; low birth weight and structural abnormalities have been associated with as little as two drinks a day; some cases of hyperactivity and learning disabilities may result	Fetal alcohol syndrome is one of the leading causes of mental retardation in the United States
Cocaine	Causes extreme constriction of fetal arteries, resulting in retarded development; severe cases may be mentally retarded, have heart defects and other medical problems	Thousands of cocaine-addicted babies are being born to mothers who use cocaine during pregnancy; low birth weight
Heroin	High mortality rate and high prematurity rate	Infants that survive are born addicted and must be treated for weeks or months; low birth weight

After the baby is born, the contractions of the uterus squeeze much of the fetal blood from the placenta back into the infant. The cord is tied and cut, separating the child from the mother. (The stump of the cord gradually shrivels until nothing remains but the scar, the **navel.**)

During the third stage of labor, which lasts 10 or 15 minutes after the birth, the placenta and the fetal membranes are loosened from the lining of the uterus by another series of contractions, and expelled. At this stage they are collectively called the **afterbirth.**

During labor an obstetrician may administer drugs such as oxytocin or prostaglandins to increase the contractions of the uterus, or may assist with special forceps or other techniques. In some women, the opening between the pelvic bones through which the vagina passes is too small to permit the passage of the baby. The baby must then be delivered by **Cesarean section,** an operation in which an incision is made in the abdominal wall and uterus.

The neonate must adapt to its new environment

Important changes take place after birth. During prenatal (before birth) life, the fetus received both food and oxygen from the mother through the placenta. Now the newborn's own digestive and respiratory systems must

Table 49–3 continued

Factor	*Example and Effect*	*Comment*
Thalidomide	Thalidomide, marketed as a mild sedative was responsible for more than 7000 grossly deformed babies born in the late 1950s in 20 countries; principal defect was phocomelia, a condition in which babies are born with extremely short limbs, often with no fingers or toes	This drug interferes with cellular metabolism; most hazardous when taken during fourth to sixth weeks, when limbs are developing
Cigarette smoking	Cigarette smoking reduces the amount of oxygen available to the fetus because some of the maternal hemoglobin is combined with carbon monoxide; may slow growth and can cause subtle forms of damage; in extreme form, carbon monoxide poisoning causes such gross defects as hydrocephaly	Mothers who smoke deliver babies with lower-than-average birth weights and have a higher incidence of spontaneous abortions, stillbirths, and neonatal deaths; studies also indicate a possible link between maternal smoking and slower intellectual development in offspring
Pathogens	Rubella (German measles) virus crosses placenta and infects embryo; interferes with normal metabolism and cell movements; causes syndrome that involves blinding cataracts, deafness, heart malformations, and mental retardation; risk is greatest (about 50%) when rubella is contracted during first month of pregnancy; risk declines with each succeeding month	Rubella epidemic in the United States in 1963–1965 resulted in about 20,000 fetal deaths and 30,000 infants born with gross defects
	HIV can be transmitted from mother to baby before birth, during birth, or postpartum through breastfeeding	See discussion of AIDS in Chapter 43
	Syphilis is transmitted to fetus in about 40% of infected women; fetus may die or be born with defects and congenital syphilis	Pregnant women are routinely tested for syphilis during prenatal examinations
Ionizing radiation	When mother is subjected to x rays or other forms of radiation during pregnancy, infant has higher risk of birth defects and leukemia	Radiation was one of the earliest causes of birth defects to be recognized

function. Correlated with these changes are several major changes in the circulatory system.

Normally, the **neonate** (newborn infant) begins to breathe within a few seconds of birth and cries within half a minute. If anesthetics have been given to the mother, however, the fetus may also have been anesthetized, and breathing and other activities may be depressed. Some infants may not begin breathing until several minutes have passed. This is one of the reasons that many women request childbirth methods that minimize the use of medication.

The neonate's first breath is thought to be initiated by the accumulation of carbon dioxide in the blood after the umbilical cord is cut. The carbon dioxide stimulates the respiratory centers in the medulla. The resulting expansion of the lungs enlarges its blood vessels (which previously were partially collapsed). Blood from the right ventricle flows in increasing amounts through the pulmonary vessels. (During fetal life, blood bypassed the lungs by flowing through an arterial duct connecting the pulmonary artery and aorta.)

Environmental factors affect the embryo

We all know that the growth and development of babies are influenced by the food they eat, the air they breathe, the disease organisms that infect them, and the chemicals or drugs to which they are exposed. *Prenatal* development

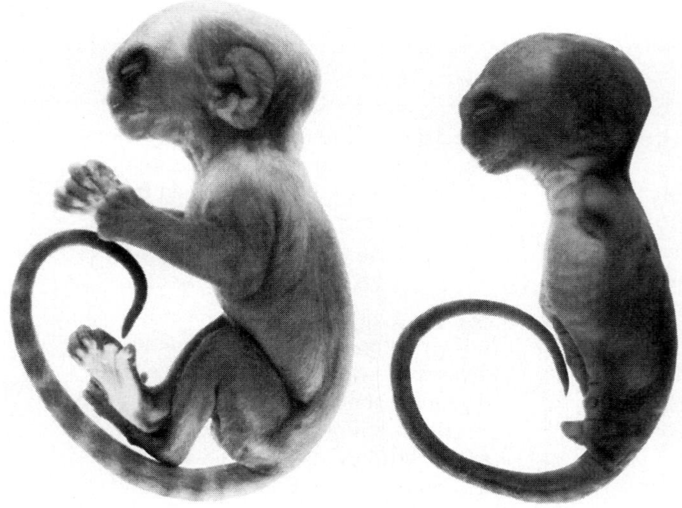

(a) (b)

FIGURE 49–21 Many kinds of drugs affect development. Thalidomide administered to the marmoset (*Callithrix jacchus*) produces a pattern of developmental defects similar to those found in humans. (***a***) Control marmoset fetus obtained from an untreated mother on day 125 of gestation. (***b***) Fetus (same age as control) of marmoset treated with 25 mg/kg thalidomide from days 38 to 52 of gestation. The drug suppresses limb formation, perhaps by interfering with the function of certain nerves. (Courtesy of Dr. W.G. McBride and P.H. Vardy, Foundation 41; from *Development, Growth and Differentiation* 25[4]:361–373, 1983)

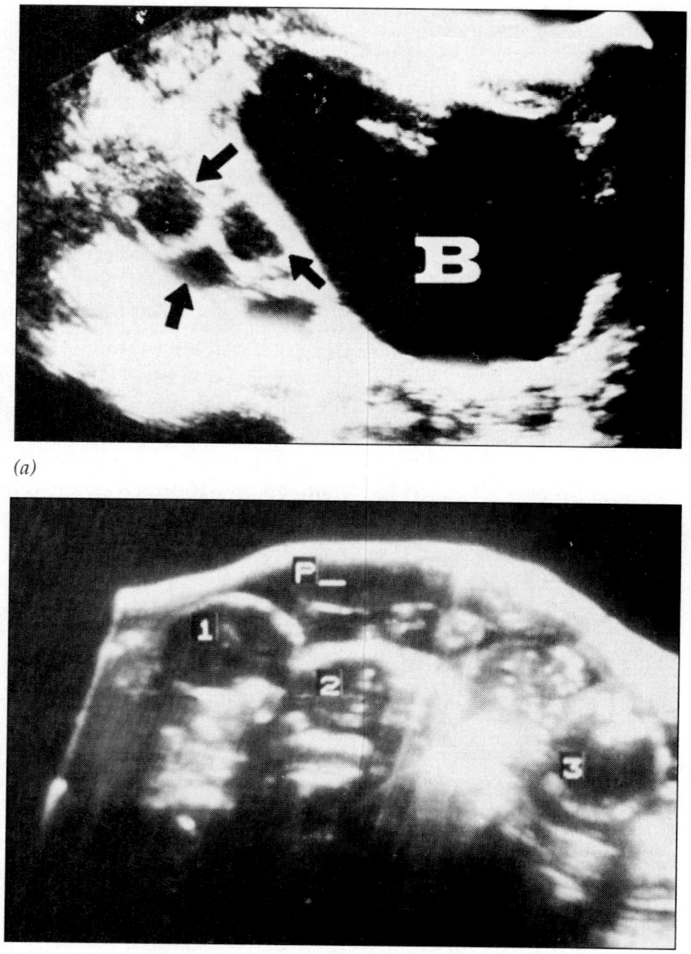

(a)

(b)

FIGURE 49–22 Ultrasonic techniques are used to give the physician information about the fetus. They are also used to monitor follicle maturation and ovulation. (***a***) Sonogram taken with ultrasound techniques, showing three follicles of equal maturation in the left ovary of a human. (***b***) Triplets in the same patient at 16 weeks of pregnancy. Such previews are valuable to the physician in diagnosing defects and predicting multiple births. *P* = placenta. (Also see Figure 15–9.) (Courtesy of Biserka Funduk-Kurjak and Asim Kurjak, from *Acta Obstetrics and Gynecology Scan.* 61:1982)

is also affected by these environmental influences. Life before birth is even more sensitive to environmental changes than it is in the fully formed baby. Anything that circulates in the maternal blood—nutrients, drugs, pathogens, or even gases—may find their way into the blood of the fetus (Fig. 49–21). Table 49–3 on page 1102 describes some of the environmental influences on development. Many of these factors contribute to low birth weight, a condition responsible for a great number of infant deaths.

About 5% of newborns (more than 150,000 babies per year) in the United States have a defect of clinical significance. Such birth defects account for about 15% of deaths among newborns. Birth defects may be caused by genetic or environmental factors, or a combination of the two. Genetic factors were discussed in Chapter 15. In this section, we examine some environmental conditions that affect the well-being of the embryo.

Timing is important. Each developing structure has a critical period during which it is most susceptible to unfavorable conditions. Generally, this critical period occurs early in the development of the structure, when interference with cell movements or divisions may prevent formation of normal shape or size, resulting in permanent malformation. Because most structures form during the first three months of embryonic life, the embryo is most susceptible to environmental factors during this early period. During a substantial portion of this time, the mother may not even realize that she is pregnant and so may not take special precautions to minimize potentially dangerous influences.

Physicians are now able to diagnose some defects while the embryo is in the uterus. In some cases, treatment is possible before birth. Amniocentesis and chorionic villus sampling, discussed in Chapter 15, are techniques used to detect certain defects. Figure 49–22 shows a **sonogram,** a photograph taken of the embryo using ultrasound. Such previews are helpful in diagnosing defects and also in determining the position of the fetus and whether a multiple birth is pending.

Table 49–4 STAGES IN THE HUMAN LIFE CYCLE

Stage	*Time Period*	*Characteristics*
Embryo	Conception to end of eighth week of prenatal development	Development proceeds from single-celled zygote to embryo that is about 30 mm long, weighs 1 g, and has rudiments of all its organs
Fetus	Beginning of ninth week of prenatal development to birth	Period of rapid growth, morphogenesis, and cellular differentiation, changing tiny parasite to physiologically independent organism
Neonate	Birth to 4 weeks of age	Neonate must make vital physiological adjustments to independent life: it must now process its own food, excrete its wastes, obtain oxygen, and make appropriate circulatory changes
Infant	End of fourth week to 2 years of age (sometimes, ability to walk is considered end of infancy)	Rapid growth; deciduous teeth begin to erupt; nervous system develops (myelinization), making coordinated activities possible; language skills begin to develop
Child	Two years to puberty	Rapid growth; deciduous teeth erupt, are slowly shed and replaced by permanent teeth; development of muscular coordination; development of language skills and other intellectual abilities
Adolescent	Puberty (approximately ages 11–14) to adult	Growth spurt; primary and secondary sexual characteristics develop; motor skills and intellectual abilities develop; psychological changes as adolescent approaches adulthood
Young adult	End of adolescence (approximately age 20) to about age 40	Peak of physical development reached; individual assumes adult responsibilities (marriage, fulfilling reproductive potential, and establishing career); after age 30, physiological changes associated with aging begin
Middle-aged adult	Age 40 to about age 65	Physiological aging continues, leading to menopause in women and physical changes associated with aging in both sexes (e.g., graying hair, decline in athletic abilities, wrinkling skin); this is period of adjustment for many as they begin to face their own mortality
Old adult	Age 65 to death	Period of senescence (growing old); physiological aging continues; maintaining homeostasis more difficult when body is stressed; death often results from failure of cardiovascular or immune system

THE HUMAN LIFE CYCLE EXTENDS FROM FERTILIZATION TO DEATH

Development begins at fertilization and continues through the stages of the human life cycle until death (Table 49–4). We have examined briefly the development of the embryo and fetus, the birth process, and the adjustments required of the neonate. The human life cycle then proceeds through the stages of infant, child, adolescent, young adult, middle-aged adult, and old adult.

Homeostatic response to stress decreases during aging

Development encompasses any biological change that takes place within an organism over time, including the changes commonly called **aging.** Changes during the aging process result in decreased function in the older organism. The declining capacities of the various systems in the human body, although most apparent in the elderly, may begin much earlier in life—for example, during childhood or even during prenatal life.

The aging process is far from uniform among different individuals or in various parts of the body. The systems of the body generally decline at different times and rates. On the average, between the ages of 30 and 75 a man loses 64% of his taste buds, 44% of the glomeruli in his kidneys, and 37% of the axons in his spinal nerves. His nerve impulses are propagated at a 10% slower rate, the blood supply to his brain is 20% less, his glomerular filtration rate has decreased 31%, and the vital capacity of his lungs has declined 44%. The aging process is also marked by a progressive decrease in the body's homeostatic ability to respond to stress.

Although relatively little is known about the aging process itself, this is now an active field of scientific investigation. While marked improvements in medicine and public health have led to survival to an advanced age of a larger fraction of the total human population, there has been no corresponding increase in the maximum life expectancy.

Is the aging process genetically programmed like other aspects of human development? Or is it the result of wear and tear on the body? Or is aging a *combination* of inheritance and environmental wear and tear? Cells that are genetically programmed to differentiate and stop dividing appear to be more subject to the changes of aging than are those that continue to divide throughout life. For example, nerve and muscle cells, which lose the capacity for cell division at an earlier age, show a decline in their functional capacities at an earlier age than do tissues such as liver and spleen, which retain the capacity to undergo cell division.

Several hypotheses have been advanced regarding the nature of the aging process: that it is affected by hormonal changes; that it involves the development of autoimmune responses; that it involves the accumulation of specific waste products within the cells; that it involves changes in the molecular structure of macromolecules such as collagen; that the elastic properties of connective tissues decrease owing to an accumulation of calcium, resulting in stiffening of the joints and hardening of the arteries; that it results from the oxidation of certain lipids by free radicals; and that cells are destroyed by hydrolases released by the breaking of lysosomes.

Other hypotheses suggest that continued exposure to cosmic radiation and x radiation leads to the accumulation of somatic mutations that decrease the ability of the cell to carry out its normal functions. The aging process may be part of the program of timed development built into the genome. Like other developmental processes, aging may be accelerated by certain environmental influences, and may occur at different rates in different individuals because of inherited differences. Experimental evidence suggests that aging—at least in rats—can be delayed by caloric restriction: thin rats live longer than fat rats. For now, genetic predisposition may be the best guarantee of a long life.

SUMMARY

I. Development proceeds as a balanced combination of cell division, cell growth, morphogenesis, and cellular differentiation.

II. Fertilization involves four processes: contact and recognition; regulation of sperm entry; fusion of sperm and egg nuclei; activation of the egg.

III. The zygote undergoes cleavage and forms a hollow ball of cells called a blastula.
   - A. The main effect of cleavage is to partition the zygote into many small cells. As cells divide, the distribution of materials in the cytoplasm influences development.
   - B. Yolk is evenly distributed in the eggs of most invertebrates and simple chordates. Cleavage involves division of the entire egg to form cells that are about equal in size.
   - C. In bony fish and amphibians, a concentration of yolk at the vegetal pole slows cleavage so that only a few large cells form there, compared to a large number of smaller cells at the animal pole.
   - D. The meroblastic cleavage that occurs in the telolecithal eggs of reptiles and birds is restricted to the blastodisc.

IV. During gastrulation, the ectoderm, mesoderm, and endoderm form; each of these embryonic tissues gives rise to specific structures.
   - A. In the sea star and in amphioxus, cells from the blastula wall invaginate and eventually meet the opposite wall; the new cavity formed is the archenteron.
   - B. In the amphibian, invagination at the vegetal pole is obstructed by large yolk-laden cells; instead, cells from the animal pole move down over the yolk-rich cells and invaginate, forming the dorsal lip of the blastopore.
   - C. In the bird, invagination occurs at the primitive streak and no archenteron forms.

V. Organogenesis is the process of organ development. The developing notochord induces nervous system development. The brain and spinal cord develop from the neural tube.

VI. Development is regulated by the interaction of genes with cytoplasmic factors and environmental factors.

VII. In terrestrial vertebrates, four extraembryonic membranes—chorion, amnion, allantois, and yolk sac—protect the embryo and help it obtain food and oxygen and eliminate wastes. The amnion is a fluid-filled sac that surrounds the embryo and keeps it moist; it also acts as a shock absorber.

VIII. Early human development follows a fairly typical vertebrate pattern.

A. Cleavage takes place as the embryo is moved toward the uterus.
B. In the uterus, the embryo develops into a blastocyst and implants itself in the endometrium.
C. After the first two months of development, the embryo is referred to as a fetus.
D. In placental mammals, the embryonic chorion and maternal tissue give rise to the placenta, the organ of exchange between mother and developing child.
E. Parturition takes place after about 280 days from the time of the mother's last menstrual period. During the first stage of labor, the cervix becomes dilated and effaced. During the second stage, the baby is delivered. The placenta is delivered during the third stage.

IX. By controlling environmental factors such as diet, smoking, and the intake of alcohol and drugs, a pregnant woman can help ensure the well-being of her unborn child.
X. The human life cycle includes embryo, fetus, neonate, infant, child, adolescent, young adult, middle-aged adult, and old adult.
XI. The aging process is marked by a decrease in homeostatic response to stress.

SELECTED KEY TERMS

acrosome reaction
aging
amnion
animal pole
archenteron
blastocyst
blastula
cellular differentiation
chorion
cleavage
cortical reaction
ectoderm
embryo
endoderm
extraembryonic membranes
fertilization cone
fetus
gastrula
gastrulation
germ layers
gestation period
gray crescent
host mothering
implantation
inner cell mass
labor
mesoderm
morphogenesis
morula
neural plate
neural tube
placenta
umbilical cord
vegetal pole
yolk

POST-TEST

1. Migration and organization of cells to form a structure such as the neural tube is an example of __________.
2. Specialization of cells to form neurons or some other cell type is called __________ __________.
3. The rapid series of mitoses that converts the zygote to a morula is referred to as __________.
4. The process by which the blastula becomes a three-layered embryo is called __________.
5. The germ layer that gives rise to the nervous system is the __________; the germ layer that gives rise to the lining of the digestive tract is the __________.
6. The notochord induces the overlying ectoderm to form the __________ __________.
7. The brain and spinal cord develop from the __________ __________.
8. The __________ secretes fluid that prevents the embryo from drying out and acts as a shock absorber.
9. In humans, the __________ is the organ of exchange between mother and fetus.
10. The cluster of cells that projects into the cavity of the mammalian blastocyst is the __________ __________ __________; it gives rise to the __________.
11. On about the seventh day of development, the human embryo begins to __________ in the uterus.
12. After the first two months of development, the human embryo is referred to as a(an) __________.
13. A human baby is delivered during the second stage of __________.
14. The activation of the sperm by the egg is known as the __________ __________.

REVIEW QUESTIONS

1. Trace the development of a sea star or amphioxus embryo from zygote to gastrula; draw and label diagrams to illustrate your description.
2. How do the mechanisms of fertilization ensure control of both quality (fertilization by a sperm of the same species) and quantity (fertilization by only one sperm)?
3. Contrast cleavage and gastrulation in the sea star or amphioxus, the amphibian, and the bird.
4. Trace some developmental process, such as the formation of the neural tube, and explain how cell division, growth, morphogenesis, and cellular differentiation interact in the process.
5. Give examples of adult structures that develop from each of the germ layers.
6. Why do terrestrial vertebrate embryos develop an amnion? What are its functions?
7. Describe human blastocyst formation and implantation.
8. What kinds of adaptations must the neonate make immediately after birth?
9. What are some of the nongenetic factors that influence development?
10. What steps can the pregnant woman take to help ensure the safety and well-being of her developing child?
11. Describe some of the changes that take place during the aging process.

YOU MAKE THE CONNECTION

1. Construct arguments for and against human embryo research. Consider such techniques as in vitro fertilization or freezing embryos for later implantation in light of concerns about human population growth.
2. What is the adaptive value of developing a placenta?
3. For almost 200 years, scientists debated the preformation theory, which held that the egg cell or the sperm cell contains a completely formed, miniature human. As research techniques improved, the theory of epigenesis was developed. This theory held that the embryo develops from a formless zygote, taking form as development proceeds. Relate these theories to current concepts of development. Can you find any truth in the preformationist view?

RECOMMENDED READINGS

Browder, L.W., C.A. Erickson, and W.R. Jeffery. *Developmental Biology,* 3rd ed. Saunders College Publishing, Philadelphia, 1991. An excellent introduction to animal development.

Rhoades, R., and R. Pflanzer. *Human Physiology,* 3rd ed. Saunders College Publishing, Philadelphia, 1996. Several chapters in this very readable introduction to physiology deal with reproduction and development.

Wasserman, P.M. "Fertilization in Mammals." *Scientific American,* Vol. 259, No. 6, December 1988. A glycoprotein governs many of the events of fertilization.

CHAPTER 50

Animal Behavior

These Grevy's zebras (*Equus grevyi*) are displaying courtship behavior. What part of this behavior is innate? What part is learned? (Photographed in Africa by Stan Osolinski/Dembinsky Photo Associates)

Suppose your professor provided you with a hypodermic syringe full of poison and demanded that you find a particular type of insect, one that you had never seen before and that was armed with active defenses. You then had to inject the ganglia of your victim's nervous system (about which you had been taught nothing) with just enough poison to paralyze but not kill it. You would have difficulty accomplishing these tasks, but a solitary wasp no larger than the first joint of your thumb does it all with elegance and surgical precision, without instruction.

The sand wasp *Philanthus* captures a bee, stings it, and places the paralyzed insect in a burrow excavated in the sand. She then lays an egg on her victim, which is devoured alive by the larva that hatches from that egg. From time to time, *Philanthus* returns to her hidden nest to reprovision it until the larva becomes a hibernating pupa in the fall. Her offspring will repeat this, doing it to perfection without ever having seen it done.

Behavior refers to the responses of an organism to signals from its environment, including those from other organisms. Much of what organisms do can be analyzed in terms of specific behavior patterns that occur in response to stimuli (changes) in the environment. A dog may wag its tail, a bird may sing, a butterfly may release a volatile sex attractant. Behavior is just as diverse as biological structure, and just as characteristic of a given species as its structure and biochemistry. Structure, function, and behavior are all adaptations that help define an organism and equip it for survival.

The capacity for behavior is inherited, but much inherited behavior can be modified by experience. Learning involves persistent changes in behavior that result from experiences.

LEARNING OBJECTIVES

After you have studied this chapter you should be able to

1. Explain why behavior is (a) adaptive, (b) homeostatic, and (c) flexible.
2. Cite examples of biological rhythms, and suggest some of the mechanisms responsible for them.
3. Summarize the contributions of heredity, environment, and maturation to behavior.
4. Classify a learned behavior as an example of (a) classical conditioning, (b) operant conditioning, (c) habituation, or (d) insight learning.
5. Discuss the adaptive significance of imprinting.
6. Postulate biological advantages and disadvantages of migration.
7. Discuss the hypothesis that optimal foraging behavior is adaptive.
8. Give a description of an animal society, and identify the adaptive advantages of cooperative behavior.
9. Summarize the modes of communication that animals employ.
10. Present the concept of a dominance hierarchy, giving at least one example, and propose a possible adaptive significance and social function for this hierarchy.
11. Distinguish between home range and territory, and give three theories about the adaptive significance of territoriality.
12. Discuss the adaptive value of courtship behavior, and describe a pair bond.
13. Compare the society of a social insect with human society.
14. Define kin selection and summarize its proposed role in the maintenance of insect and other animal societies.

BEHAVIOR IS ADAPTIVE

Biologists study behavior in the laboratory and in natural environments, always keeping in mind that what an animal does cannot be isolated from the way in which it lives. **Ethology** is the study of behavior in natural environments from the point of view of adaptation. Thus, a particular behavior may help an organism obtain food or water, acquire and maintain territory in which to live, protect itself, or reproduce.

Behavior tends to be homeostatic for the individual organism, as well as adaptive in the evolutionary sense. Each organism has a set of adaptive behaviors that fit its lifestyle. Certain behavioral responses, however, may lead to the death of the individual while increasing the chance that copies of its genes will survive through the enhanced production or survival of the offspring. In this chapter, we consider how behavior fits into the total life of the organism, enabling it to live and pass on its heritage to those that follow.

FIGURE 50–1 The behavior of these fiddler crabs is synchronized with the tides. The fiddler crabs run about on the surface of the sand at low tide, but must return to their burrows before the tide returns. The males have enlarged claws. (Tom Walker/Photo Researchers, Inc.)

BIOLOGICAL RHYTHMS ANTICIPATE ENVIRONMENTAL CHANGES

An organism adapts by synchronizing its metabolic processes and behavior with the cyclic changes in the external environment. Its behavior can *anticipate* these regular changes. The little fiddler crabs (Fig. 50–1) of marine beaches often emerge from their burrows at low tide to engage in social activities such as territorial disputes.

They must return to their burrows *before* the tide returns to avoid being washed away. How do the crabs "know" that high tide is about to occur?

One might guess that the crabs recognize clues present in the seashore. However, when the crabs are isolated in the laboratory away from any known stimulus that could relate to time and tide, their characteristic behavioral rhythms persist.

A variety of behavioral cycles occur among organisms

In addition to the biological rhythms oriented around tides, other types of biological rhythms occur, including daily, monthly, and annual rhythms. Among many animals, periods of activity and sleep, feeding and drinking, body temperature fluctuations, and secretion of some hormones have cycles that are 24 hours long. Human physiological processes also seem to follow an intrinsic rhythm. Human body temperature, for example, follows a typical daily curve. These daily activities are **circadian** (meaning "approximately one day") **rhythms.** Such biological rhythms suggest that animals have **biological clocks** that are precisely adjusted or reset by environmental cues (see Chapter 30).

The behavior of many animals appears to be organized around circadian rhythms. **Diurnal** animals, like honeybees and pigeons, are most active during the day. **Nocturnal** animals are most active during the hours of darkness. **Crepuscular** animals, like the fiddler crab, are busiest during the twilight hours, at dawn, or both. As in the case of the fiddler crab, there are ecological reasons for these adaptations. If an animal's food is most plentiful in the early morning, for example, its cycle of activity must be regulated so that it becomes active shortly before dawn.

Some biological rhythms of animals reflect the **lunar** (moon) **cycle.** The most striking rhythms are those in marine organisms that are tuned to the changes in the tides caused by the phases of the moon. For instance, a combination of tidal, lunar, and annual rhythms governs the reproductive behavior of the grunion, a small fish that lives off the Pacific coast of the United States.

The grunion swarms from April through June on those three or four nights when the highest tides of the year occur. At precisely the high point of the tide, the fishes squirm onto the beach, deposit eggs and sperm in the sand, and return to the sea in the next wave. By the time the next tide reaches that portion of the beach 15 days later, the young fishes have hatched and are ready to enter the sea.

Biological rhythms are controlled by an internal clock

Biological rhythms are regulated by internal timing mechanisms referred to as biological clocks. Current evidence suggests that most organisms have no single biological clock. Instead, the interaction of a number of biochemical processes (possibly involving cellular membranes) might be responsible for governing physiological and behavioral rhythms. The pineal gland is thought to play a role in the timing systems of birds, rats, humans, and some other vertebrates. Regions of the hypothalamus are also part of the biological clock in mammals. Although some parts of an organism may coordinate or dominate the function of the biological clock, it is likely that every cell has some type of timing mechanism.

BEHAVIOR CAPACITY IS INHERITED AND IS MODIFIED BY ENVIRONMENTAL FACTORS

All behavior has a genetic basis. In other words, the *capacity* for behavior is inherited. Even the capacity to learn is inherited. However, behavior can be modified by the environment; it is a product of the interaction between genetic capacity and environmental influences. Thus, behavior begins with an inherited framework that experience can modify.

The instructions for the honeybee society are inherited and preprogrammed, and the size and structure of the bee's nervous system permit only a limited range of behavior. Yet these insects are not automatons. Within those limits, the complex bee society can respond flexibly to food and other stimuli in the environment.

Some behavior of vertebrates is also largely predetermined by genes. Several species of the lovebird (*Agapornis*) differ not only in appearance but in behavior. One species transports nest-building materials in its bill. Another species tucks materials into its feathers. Studies indicate that the method of transporting materials is inherited but somewhat flexible.

Behavior develops

Behavior involves all body systems, but it is influenced mainly by the nervous and endocrine systems. The capacity for behavior depends on the genetic characteristics that govern the function of these systems. One may think of a continuous scale of behaviors that range from the most rigidly programmed, genetically inherited types, through those that are somewhat modifiable, to those that, although containing a genetic component, are extensively developed through experience.

Before an organism can exhibit any pattern of behavior, it must be physiologically ready to produce the behavior. For example, breeding behavior does not ordinarily occur among birds or most mammals unless steroid sex hormones are present in their blood at certain concentrations. A human baby cannot walk unless its reflex and muscular development permit it to. These states

FIGURE 50–2 Egg-rolling behavior in the European graylag goose is a fixed action pattern (FAP).

of physiological readiness are themselves produced by a continuous interaction with the environment. The level of sex hormones in a bird's blood may be determined by seasonal variations in day length. The baby's muscles develop in response to exercise. Without the trial and error involved in preparing to walk, the ability to walk would be delayed.

A good example of such interaction between readiness and environment is afforded by the white-crowned sparrow, which exhibits considerable regional variation in its song. This bird, even if kept in isolation, eventually sings a very poorly developed but recognizable white-crowned sparrow song. However, if it is allowed to grow up under the care of its parents, it learns the local "dialect" from them. When mature, it can sing the more highly developed local song characteristic of the region. If such learning does not take place early in life, it never will, and if the sparrow later consorts with birds of other species, it does not learn their songs. It appears that the white-crowned sparrow is hatched equipped with a rough genetic pattern of its song. The regional details are filled in by learning.

Some behavior patterns have strong genetic components

The term **instinct** refers to **innate** (inborn) **behavior**—that is, genetically programmed behavior. A classic example of innate behavior in vertebrates is egg-rolling in the European graylag goose. When an egg is removed from the nest of a goose and placed a few inches in front of her, she reaches out with her neck and pulls the egg back into the nest (Fig. 50–2). If, while the goose is rolling the egg back toward the nest, the egg veers off to the side, the goose steers it back on course. Because this type of behavior appears to have a strong genetic component, ethologists refer to it as a **fixed action pattern (FAP).**

A stimulus that elicits an FAP—such as the egg to the graylag goose or the red-colored "belly" to the stickleback fish (Fig. 50–3)—is called a **sign stimulus.** A goose will retrieve a wooden egg instead of a real egg, so the wooden egg is also by definition a sign stimulus.

Behavior is modified by learning

We can define learning as a change in behavior due to experience. The capacity to learn appropriate responses to new situations is adaptive, because **learned behavior** can be shaped to meet the needs of the changing environment

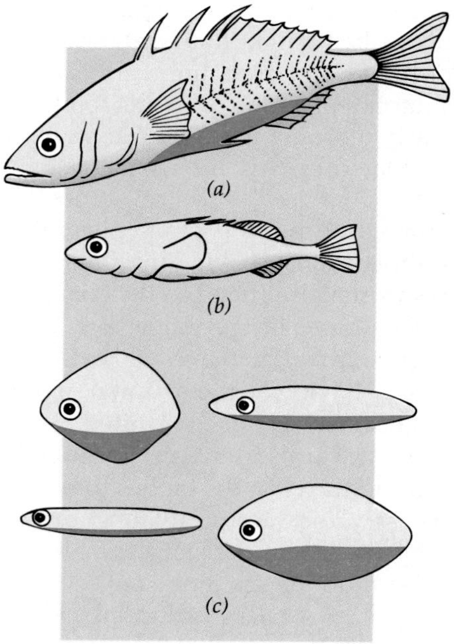

FIGURE 50–3 A sign stimulus triggers a fixed action pattern. A male stickleback fish (*a*) will not attack a realistic model of another male stickleback if it lacks a red belly (*b*), but it will attack another model, however unrealistic, that has a red "belly" (*c*). Therefore, it is the specific red sign stimulus, rather than recognition based on a combination of features, that triggers the aggressive behavior.

Making the Connection

Learned Behavior and Nervous Systems

What determines the level of complexity of learned behavior that an animal is capable of? The simpler the nervous system, the simpler the range of behavior. Even innate behavior depends on the properties of individual neurons, and of their interconnections. For an organism to learn, its neurons must have a large number of potential interactions with one another — not just the few required by stereotyped preprogrammed behavior (such as the sand wasp's ability to sting its victims).

Without the preexistence of the necessary neural circuitry, learned behavior would be impossible. The kind of learned behavior that an animal typically and most easily develops depends on the complexity and layout of its neural circuitry. Learning also depends on changes in the readiness of neurons to form circuits with one another. ▲

that most animals experience. The wasp *Philanthus,* discussed in the chapter introduction, efficiently carries out a complex, although largely genetically programmed, sequence of behaviors. Yet some of her behavior is learned. When *Philanthus* covers a nest with sand, she takes precise bearings on the location of the burrow before flying off again to hunt. There is no way that her ability to locate the burrow could be genetically programmed. How to dig it, how to cover it, how to kill the bees — these behaviors appear to be genetically programmed. But because a burrow can be dug only in a suitable spot, its location must be learned *after* it is dug.

This distinction between innate behavior and behavior modified by experience in the wasp *Philanthus* was determined by the Dutch investigator Niko Tinbergen. Tinbergen surrounded the wasp's burrow with a circle of pine cones, on which the wasp took her bearings (Fig. 50–4). Before she returned with another bee, Tinbergen moved the pine cones. The wasp could not find her burrow without them. Only when the experimenter restored the cones to their original location could the wasp find her burrow. In contrast, the existence of complex programmed behavior is hard to demonstrate in humans. We owe the complexity of our behavior to a *generalized* abil-

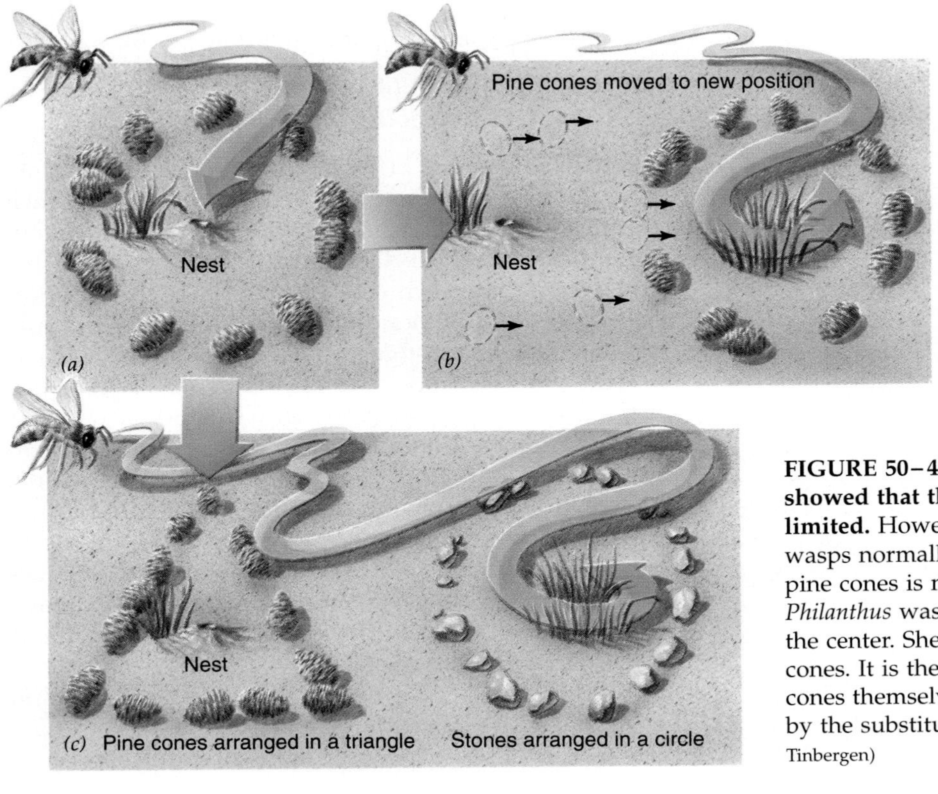

FIGURE 50–4 Tinbergen's sand wasp experiment showed that the ability of *Philanthus* to learn is quite limited. However, it is adequate for situations that these wasps normally encounter in nature. When the ring of pine cones is moved from position (***a***) to position (***b***), the *Philanthus* wasp behaves as if her nest was still located at the center. She learned its position in relation to the cones. It is the arrangement of the cones, rather than the cones themselves, that the wasp responds to, as shown by the substitution of a ring of stones for cones (***c***). (After Tinbergen)

ity to learn. The *Philanthus* wasp's intelligence is as narrowly specialized as her stinger.

In classical conditioning, a reflex becomes associated with a new stimulus

In a type of learning called **classical conditioning,** an association is formed between some normal body function and a stimulus. Ivan Pavlov, a Russian physiologist, studied dogs in the laboratory early in this century. He discovered that when a bell was rung at the same time a dog was fed, an association formed between the sound of the bell and the secretion of saliva. Eventually (Fig. 50–5), when the bell was rung by itself, the dog salivated. Pavlov called the physiologically meaningful stimulus (food, in this case) the *unconditioned stimulus.* The normally irrelevant stimulus (the bell) that became a substitute for it was the *conditioned stimulus.* Because a dog does not normally salivate at the sound of a bell, the association was clearly a learned one. It could also be forgotten. If the bell no longer signaled food, the dog eventually stopped responding to it. Pavlov called this process **extinction.**

In operant conditioning, spontaneous behavior is reinforced

In **operant conditioning** (also called instrumental conditioning), the animal must do something in order to gain a reward (**positive reinforcement**) or avoid a punishment. In a typical experiment, a rat is placed in a cage containing a movable bar. When random actions of the rat result in pressing the bar, a pellet of food rolls down a chute and is delivered to the rat. Thus, the rat is positively reinforced for pressing the bar. Eventually, the rat learns the association and presses the bar whenever it wants food.

In **negative reinforcement,** removal of a stimulus increases the probability that a behavior will occur. For example, a rat may be subjected to an unpleasant stimulus, such as a low-level electric shock. When the animal presses a bar, this negative reinforcer is removed and the animal experiences relief.

Many variations of these techniques have been developed. A pigeon might be trained to peck at a lighted circle to obtain food, a chimpanzee might learn to perform some task in order to get tokens that can be exchanged for food, or children might learn to stay quietly in their seats at school to obtain the teacher's praise. Operant conditioning is probably the way that animals learn to perform complex tasks like walking or perfecting feeding skills.

Operant conditioning plays a role in the development of behaviors that appear to be preprogrammed. An example is the feeding behavior of gull chicks. Herring gull chicks peck the beaks of the parents, which regurgitate partially digested food for them. The chicks are attracted by two stimuli: the general appearance of the parent's beak with its elongated shape and distinctive red spot, and its downward movement as the parent lowers its head. Like the rat's chance pressing of the bar, this behavior is sufficiently functional to get the chicks their first meal, but they waste a lot of energy in pecking. Some pecks are off target and are therefore not rewarded. However, the begging behavior becomes more efficient over time. Thus, a behavior that might appear to be entirely instinctive is perfected by learning (Fig. 50–6).

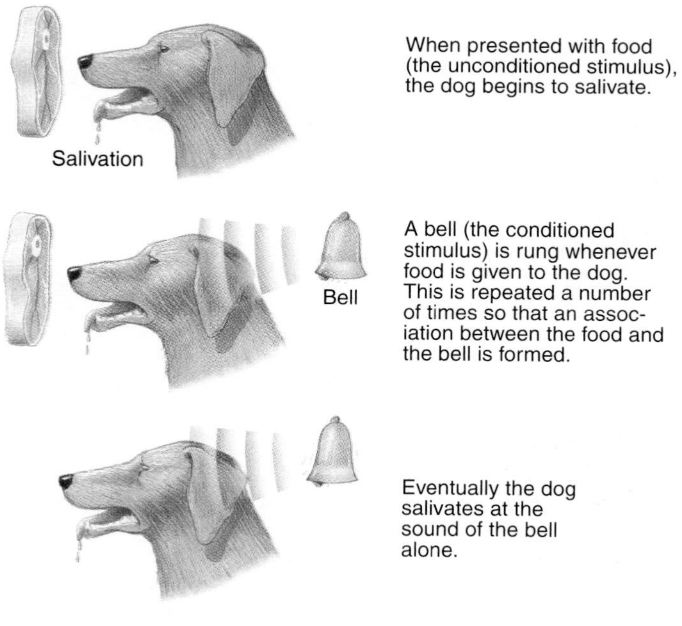

When presented with food (the unconditioned stimulus), the dog begins to salivate.
Salivation
A bell (the conditioned stimulus) is rung whenever food is given to the dog. This is repeated a number of times so that an association between the food and the bell is formed.
Bell
Eventually the dog salivates at the sound of the bell alone.

FIGURE 50–5 Pavlov's experiment showed that dogs can be classically conditioned to salivate at the sound of a bell.

Imprinting is a form of learning that occurs during a critical period

Anyone who has watched a mother duck with her ducklings must have wondered how she can keep track of such a horde of almost identical little creatures tumbling about in the grass, let alone distinguish them from those belonging to another hen (Fig. 50–7). Although she is capable of recognizing her offspring to an extent, basically they have the responsibility of keeping track of her. The survival of the duckling requires that it quickly establish a behavioral bond with its parent. This bond, which is usually formed within a few hours after birth (or hatching), forms by a type of learning known as **imprinting.** An early investigator of imprinting, ethologist Konrad Lorenz, discovered that a newly hatched bird imprints on the first moving object it sees—even a human or an inanimate object such as a colored sphere or light. Although imprinting itself is genetically determined, the bird *learns* the particular animal or object.

Among many types of birds, especially ducks and geese, the older embryos are able to exchange calls with their nest mates and parents right through the porous

FIGURE 50–6 Operant conditioning plays a role in the development of some innate behaviors. The pelican chick learns to be more accurate in begging for food from a parent. (H. Cruickshank/VIREO)

eggshell. When they hatch, at least one parent is normally on hand, emitting the characteristic sounds with which the hatchlings are already familiar. If the parent moves, the chicks follow. This movement plus the sounds produces imprinting. During a brief critical period after hatching, the chicks learn the appearance of the parent.

FIGURE 50–7 Parent-offspring bonds generally form very early. Through imprinting, some young animals follow the first moving object they encounter. Usually, the object is their mother, although it is possible experimentally to imprint many such infants upon unnatural objects. (J. H. Dick/VIREO)

Imprinting establishes the bond between mother and offspring among many mammals, as well as among birds. In many species, the mother also establishes a bond with her offspring during a critical period. The mother in some species of hoofed mammals, such as sheep, will accept her offspring for only a few hours after its birth. If they are kept apart past that time, the young are rejected. Normally, this behavior enables the mother to distinguish her own offspring from those of others, evidently by olfactory cues.

Habituation enables an animal to ignore irrelevant stimuli

In **habituation,** an animal learns to ignore a repeated, irrelevant stimulus. In Figure 50–8, we see the familiar

FIGURE 50–8 After repeated safe encounters with humans, these pigeons are unperturbed by people. This is an example of habituation. (Janet Goldwater)

gathering of pigeons in a city street. These birds have learned by repeated harmless encounters that humans are no more dangerous to them than are cows to crows, and behave accordingly. This is to their advantage. A pigeon intolerant of people might not get enough to eat.

Insight learning uses recalled events to solve new problems

The most complex learning is **insight learning,** which is the ability to adapt past experiences that may involve different stimuli to solve a new problem. A dog can be placed in a blind alley that it must *circumvent* in order to reach a reward. The difficulty lies in the fact that the animal must move *away* from the reward in order to get *to* it. Typically, the dog flings itself at the barrier nearest the food. Eventually, by trial-and-error, the frustrated dog may find its way around the barrier and reach the reward.

In contrast to the dog, a chimpanzee placed in a similar situation is likely to make new associations between tasks it has learned before in order to solve the problem (Fig. 50–9). Primates are especially skilled at insight learning, but some other mammals and a few birds seem to have this ability to some degree.

Learning abilities are biased

Animals learn some things more easily than others. For instance, one does not really need to teach a baby bird to fly. Learning language comes very naturally to humans.

FIGURE 50–9 Insight learning is most developed in primates. Confronted with the problem of reaching food hanging from the ceiling, the chimpanzee stacks boxes until it can climb and reach the food.

A child will learn the speech of its caretakers, even if no one deliberately instructs it. In general, learning biases reflect an animal's specialized mode of life.

The information most important to survival appears to be most easily learned. The same rat that may have taken a dozen trials to perfect the artificial task of pushing a lever to get a reward learns in a *single* trial to avoid a food that has made it ill. Those who poison rats to get rid of them can readily appreciate the adaptive value of this learning ability. Such quick learning in response to an unpleasant experience forms the basis of warning coloration, which is found in many poisonous insects and brilliantly colored, but distasteful, bird eggs. Once made ill by such an egg, predators quickly learn to avoid them.

BEHAVIORAL ECOLOGY EXAMINES INTERACTION OF ANIMALS WITH THEIR ENVIRONMENTS

Behavioral ecologists focus on how animals interact with their environments and on the survival value of their behavior. Two of the many activities that relate behavior and ecology are migration and foraging for food.

Migration is triggered by environmental changes

Migration is the periodic long-distance movement from one location to another with a subsequent return to the first location. Some migrations involve astonishing feats of endurance and navigation. Ruby-throated hummingbirds cross the vast reaches of the Gulf of Mexico twice each year, and the sooty tern travels across the entire South Atlantic from Africa to reach its tiny island breeding grounds south of Florida. (See *Focus On: Migration: Costs and Benefits.*)

The behavioral trigger that sets off migratory behavior varies. Some animals migrate when they mature. In others, certain environmental cues trigger the process. In migratory birds, for example, the pineal gland senses changes in day length and then releases hormones that cause restless behavior. The bird shows an increased readiness to fly and flies for longer periods of time (Fig. 50–10).

The *direction* of travel is obviously important, and this raises the general problem of animal navigation. Honeybees, some fishes, amphibians, birds, sea turtles, and some other animals appear to be sensitive to Earth's magnetic field, and use it as a guide. Adult salmon use the unique odors of different streams to find their way back (sometimes across thousands of miles) to the same stream from which they hatched. Birds appear to navigate by a combination of celestial (sun- and star-related), geographical, and climatic cues.

Working in the 1950s, Franz and Eleonore Sauer hand-reared a number of whitethroats, a species of small Eu-

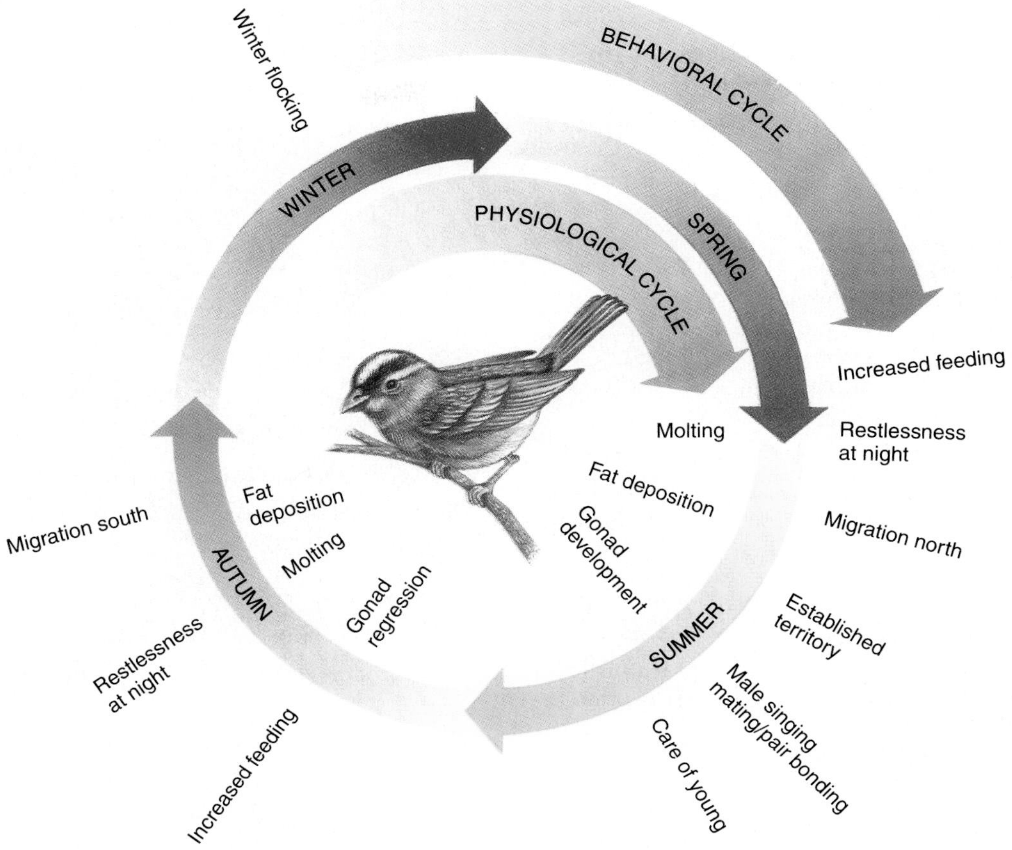

FIGURE 50–10 Seasonal changes occur in the physiology and behavior of the white-crowned sparrow. Note the increased rate of feeding and then restlessness that precedes each period of migration. (From Alcock, J., *Animal Behavior: An Evolutionary Approach,* 2nd ed., Sunderland, Mass., Sinauer Associates, 1979)

ropean warbler. This ruled out the possibility that the parents had transmitted any information to their offspring. When (and only when) the birds could see the star patterns of the night sky, they attempted to fly in the normal migration direction for their species, a direction they had had no opportunity to learn. When the birds were brought into a planetarium that simulated the night sky of a different locale, they attempted to fly in a direction that would have taken them to their normal wintering grounds from that locality. The conclusion seemed inescapable: although the direction of migration was not learned, the birds were able to find their way instinctively by means of celestial navigation.

Efficient foraging behavior contributes to survival

Natural selection produces animals that are efficient at propagating their genes. Such efficiency depends on making the best choices in mate selection, territorial defense, and food foraging. A topic of interest to many behavioral ecologists is **optimal foraging,** which refers to the most efficient way for an animal to obtain food. According to current hypotheses, when animals maximize energy obtained per unit of foraging time, they maximize reproductive success. Many factors—such as avoiding predators while foraging—must be considered. Grizzly bears that spend hours digging Arctic ground squirrels out of burrows provide an example of optimal foraging. Why do these bears ignore larger prey such as caribou? It is energetically more efficient to dig for squirrels because the bears' efforts most probably will be rewarded, whereas caribou might escape, leaving the bears hungry.

Optimal foraging can be illustrated by the following example, which is fictitious for the sake of simplicity. Suppose there exists a monkey that eats only one species of fruit, and that this fruit comes in three kinds of husks: soft, leathery, and hard. It takes 1 minute to remove a soft husk, 5 minutes to remove a leathery husk, and 10 minutes to remove a hard husk. If the forest contains equal

Focus On

Migration: Costs and Benefits

Birds, butterflies, fishes, and sea turtles are among the many animals that travel long distances and then return home to reproduce. Recent DNA tests confirm that loggerhead sea turtles hatch on Florida beaches along the Atlantic, then swim hundreds of miles across the ocean to the Mediterranean Sea. Several years later, those that survive to become adults navigate back, often to the same beach, to lay their eggs.

Green turtles (*Chelonia mydas*) migrate more than 2000 kilometers (1200 miles) across open ocean between their feeding area off the coast of Brazil and their nesting place on Ascension Island, a tiny island less than 10 kilometers (6 miles) wide. There, each turtle will make three to seven trips, about 12 days apart, from the water to an isolated beach, laying about 100 eggs in the sand each time.

A green turtle hatchling (*Chelonia mydas*) makes its way back to the sea. (Kelvin Aitken/Peter Arnold, Inc.)

Why do animals migrate? Migration is a very precise evolutionary adaptation to seasonal changes in climate, availability of food resources, and safe nesting sites. The dramatic annual migration of millions of monarch butterflies from Eastern Canada and the United States to Mexico — a 2500-kilometer journey (about 1500 miles) for some — appears to be related to the availability of milkweed plants on which females lay their eggs. In cold regions these plants die in late autumn and grow again with the warmer weather of spring. The wintering destinations of monarchs also seem to be determined by temperature and humidity.

The benefits of migration are not without cost. Many weeks may be spent each year on energy-demanding journeys. Some animals may become lost or die along the way. And migrating individuals are often at greater risk from predators in unfamiliar areas. In recent years, human activities have interfered with migrations of many kinds of animals. For example, after millions of years of biological success, survival of the sea turtles is threatened by the fishing and shrimping industries. As they migrate, thousands of turtles drown each year in the nets of shrimpers and fishermen. ■

numbers of the three forms, if they are randomly distributed, and if the monkey moves randomly and eats every fruit it encounters, then at the end of a foraging period it should have eaten about the same number of each husk type.

But is it necessarily the best use of the monkey's time and energy to eat all three husk types? Because removing hard husks takes so long, perhaps it is more efficient to pass up these fruits and eat only the soft-husked and leathery-husked fruits. Or maybe the best procedure is to eat only the soft-husked fruits.

These three possible ways of feeding are examples of **foraging strategies.** Two other foraging strategies are to select only the hard-husked fruits, or to select the hard-husked and leathery-husked fruits and pass up the soft-husked fruits. These two strategies are clearly less efficient ways to forage, because they require a greater expenditure of time and energy for the same amount of food. The term *strategy* used in this context does not imply planning as it does in human behavior, but instead refers to the way an animal locates, procures, and handles food.

Assuming that once husked, all three forms can be eaten in the same brief time, we may ask which of the following strategies is most adaptive, that is, which one permits the monkey to forage optimally:

Strategy I: Select the soft-husked fruits whenever encountered, and pass up the other two kinds

Strategy II: Select the soft-husked and leathery-husked fruits whenever encountered, and pass up the hard-husked fruits

Strategy III: Select all three kinds whenever encountered

Strategy I has its trade-offs. Because only every third fruit is soft-husked, the monkey must travel three times as far to get the same number of fruits it would get if it selected leathery-husked and hard-husked fruits as well. Maybe it would be worth traveling farther if it meant spending less time husking, but of course that would depend on *how much* farther.

What do we need to know in order to identify the best foraging strategy? Obviously we need to know how long it takes to get from one fruit tree to the next, and this depends on the density of the fruit trees. If the fruit tree density is so low that it takes an hour to get from one fruit tree to the next, then common sense tells us not to pass up a hard-husked fruit just because it takes 10 minutes to husk. On the other hand, if the fruit trees are so dense that it takes only 2 minutes to get to the next tree, then we would certainly pass up all the hard-husked forms, and would perhaps even consider passing up the leathery-husked forms as well. After all, the next soft-husked fruit is only 6 minutes away, and it takes longer than that to husk a hard-husked fruit.

This relationship between travel time, handling time, and net food intake can be expressed graphically (Fig. 50–11). This graph, based on eight rather than three food items, indicates eight possible foraging strategies. Assuming that the caloric value (expressed as *E*, for energy) of all items is equal, the best item would be the one requiring the least handling time, the worst would be the one requiring the most handling time, and all others would be ranked accordingly.

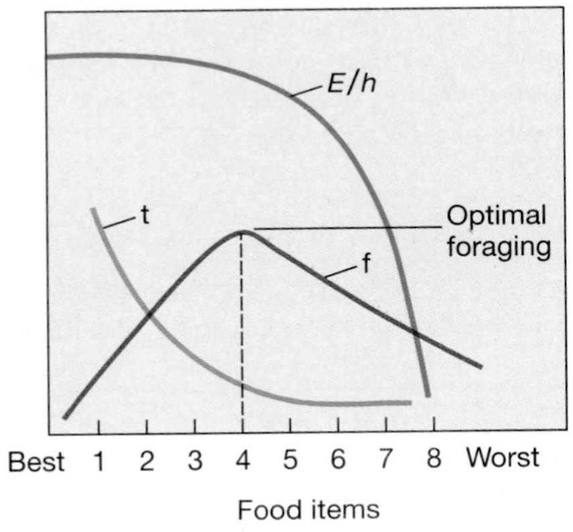

FIGURE 50–11 This graph illustrates optimal foraging, based on food intake (expressed as energy, *E*). In any habitat there is a tradeoff between travel time (*t*) and handling time (*h*). Once *h* and *t* are determined, the optimal foraging strategy can be calculated. Numbers on the horizontal axis refer to foraging strategies concerning eight species of food items, arranged from best to worst. Number 1 refers to selection of only the best food item, number 2 to selection of the best food item plus the second best food item, number 3 to selection of the three best food items, and so forth. In this hypothetical case, an animal that forages optimally (as indicated by the peak of the red curve) would select the four best food items. In better habitats the peak is to the left of 4; in poorer habitats, it is to the right of 4.

The numbers on the horizontal axis do not refer to the specific items, but rather to foraging strategies. In Strategy I the animal selects only the best item, the one requiring the shortest handling time (*h*) per unit of food, and hence the highest ratio of energy to handling time (*E*/*h*). By being so selective, however, the animal has a long traveling time (*t*), because it has chosen to pass up the other seven items. In Strategy II the animal selects the best item plus the second best item, thereby decreasing traveling time but increasing handling time (and consequently decreasing *E*/*h*). As the animal includes more and more unfavorable food items in its diet, travel time decreases but handling time increases. Thus, the animal spends the least amount of time traveling if it selects all eight food items, but by adding more unfavorable items to its diet to reduce travel time, it must also spend more time handling the items.

In good habitats, where good food items are abundant and the animal does not have to travel far to obtain them, the optimal foraging strategy tends toward the left of the graph (Strategy I, II, or III, depending on how good the habitat is). In contrast, the optimal strategy in poor habitats, where it takes longer to find the best food items, is to select more items even though doing so requires more handling time.

How do these theoretical considerations apply to the actual behavior of animals? By measuring travelling, handling, and eating times, and relating these values to the theoretical values calculated for optimal foraging, one can test the hypothesis that a particular animal does, in fact, forage optimally. If the actual values do not fit the theoretical values, it may mean that all relevant variables have not been accounted for. The monkey may choose to forage very inefficiently (in terms of consuming a certain number of Calories each day) by selecting only hard-husked fruits; but it may do this because it prefers the taste of hard-husked fruits or because they contain a critical nutrient missing from the other two forms. Thus, other hypotheses concerning foraging behavior may come to mind.

If the monkey is indeed observed foraging optimally, according to predicted values, it may have learned to do so through trial and error (operant conditioning)—that is, by randomly trying all three strategies and selecting the one associated with the fewest hunger pangs at the end of the day!

SOCIAL BEHAVIOR HAS BOTH BENEFITS AND COSTS

The mere presence of more than one individual does not mean that a behavior is social. Many factors of the physical environment bring animals together in **aggregations,** but whatever interaction they experience may be cir-

Making the Connection

Behavior and Ecology: Cave Bats in Puerto Rico

Among mammals, only rodents have more species than Chiroptera (bats). Bats constitute one of the most diverse groups in terms of feeding habits and reproductive strategies, and are the only mammals capable of active flight as opposed to gliding. The mythical stories often associated with bats have resulted in a poor image, which is unwarranted. Their role as insect predators, plant pollinators, and fruit dispersal agents leaves no question about their important position within ecosystems.

Approximately 31% of Puerto Rico's caves are known to host bat colonies varying in size from a few to hundreds of thousands of individuals. Ten of the 13 bat species known to the island use these caves as preferred roosting sites. Feeding habits vary widely among these cave-dwelling bats: fruit- and insect-eating bats are the most common, followed by the nectar-feeding and fish-eating bats.

It is common for the layperson to associate bats with caves. This relationship is often attributed to the alleged aversion of bats to light. However, cave-dwelling bats often make use of caves because they provide reliable shelter against climatic variations and predators. Foliage and even trees are perishable, but caves remain almost unchanged for very long periods of time. Factors that are potentially important in the selection of roosts include shelter from wind, local increase in air temperature and/or humidity, and a reduction of heat loss.

Some bats are known to select roosting sites that result in entrapment of the metabolic heat they produce. Others may reduce their energy expenditure and at the same time stay warm by clustering, or they may resort to torpor (a state of dormancy that results in conservation of energy). In the neotropics, so-called hot caves are used year-round by several species of bats. These caves are characterized by a single, reduced entrance, a minimum circulation of air, a high population density of bats, a temperature ranging from 28° to 40°C, and relative humidity exceeding 90%.

Antillean bats that use hot caves exhibit high degrees of gregariousness and roost fidelity. At least one species (probably two) is known to occur exclusively in these caves. Although up to seven species may occupy a single cave in Puerto Rico, different species often maintain spatial separation within the roost. It has been suggested that interspecific competition for access to the roost is what limits the species composition and population sizes in these caves. When several species occupy the same cave, they may compete for roosting sites and access to the exit. Narrow cave mouths may physically restrict the flow of bats during periods of activity and limit the number of individuals in the cave. For example, at Cucaracha cave in western Puerto Rico, three species of bats with a total population of 700,000 individuals share a hot cave with a 1.5 m^2 opening.

Many species of bats inhabiting hot caves are prone to rapid dehydration. These species may roost in large groups because of the benefits derived from a thermoneutral environment (one with an ambient temperature at which energy expenditure is minimal) and reduced dehydration. Also, the development of large colonies may increase both foraging success, by functioning as information centers, and reproductive success, by reducing the exposure of newborns to predation and weather. These benefits are opposed by costs associated with permanent use of caves. For example, large numbers of exiting bats attract concentrations of predators to the cave mouth.

Interspecies differences in foraging patterns, associated with diet differences, result in spacing of peak exit times. Such spacing may allow larger numbers of cave-warming bodies to be present than would be likely in either a single-species colony or a random assemblage of species, in which peak exit times might coincide. Multispecies colonies of cave-dwelling bats present opportunities for the study of many behavior patterns. ▲

Contributed by Armando Rodriguez-Durán, Department of Biology, Inter American University, Bayamón, Puerto Rico.

cumstantial. A light shining in the dark is a stimulus that draws large numbers of moths. The high humidity under a log may attract wood lice. Although these aggregations may have adaptive value, they are not truly social, because the organisms are not responding to one another.

We can define **social behavior** as the interaction of two or more animals, usually of the same species. Many animals benefit from living in groups. By cooperation and division of labor, some insects construct elaborate nests and raise young by mass-production methods. Schools of fishes are less vulnerable to predators than are solitary fish. Their large numbers tend to confuse predators, and by working together they can often repel attacks. Antelope can more effectively detect or discourage predators when some individuals in the herd are on watch, and drive off predators by collective action.

Social foraging is an adaptive, efficient strategy, and is used routinely by many animal species. A pack of wolves and a pride of lions have greater success in hunting than the individual wolves or lions would have if hunting alone. Humpback whales engage in cooperative foraging behavior. They surround a school of fishes and corral their prey by blowing bubbles around them. Among some birds of prey, such as ospreys, cooperative hunting results in locating prey more quickly.

Some species that engage in social behavior form societies. A **society** is an actively cooperating group of individuals belonging to the same species. A hive of bees, a flock of birds, a pack of wolves, and a school of fishes are examples of societies. Some societies are loosely organized, whereas others have complex structures. Characteristics of a well-organized society include cooperation and division of labor among animals of different sexes, age groups, or castes. A complex system of communication reinforces the organization of the society. The members tend to remain together and to resist attempts by outsiders to enter the group.

Social behavior offers benefits that increase the chances of perpetuating the genes that produce such behavior. However, social behavior also has certain costs. Living together means increased competition for food and habitats. Social interaction also increases the risk of transmitting disease.

Communication is necessary for social behavior

One animal can influence the behavior of another only if the two of them can exchange mutually recognizable signals (Fig. 50–12). **Communication** is evident when an animal performs an act that changes the behavior of another organism. Communication may be important in finding food, as in the elaborate dances of the honeybees. Animals may communicate to hold a group together, warn of danger, indicate social status, ask for or indicate willingness to provide care, identify members of the same species, or indicate sexual maturity.

Animals communicate in a wide variety of ways

Methods of animal communication are extremely varied. The singing of many birds announces the presence of a territorial male. Killer whales (*Orcinus orca*) communicate by sounds and songs. Members of a given pod have an average of 12 different calls, which vary in pitch and duration and reflect their "emotional" state. Some animals communicate by scent rather than sound. Antelopes rub facial gland secretions on conspicuous objects in their vicinity. Dogs mark territory by frequent urination. Certain fishes, the gymnotids, use electric pulses for navigation and communication, including territorial threat, in a fashion similar to bird vocalization. As sociobiologist Edward O. Wilson said, "The fish, in effect, sing electrical songs."

Pheromones are chemical signals used in communication

Pheromones are chemical signals that convey information between members of a species. They are a simple, widespread means of communication. Most pheromones elicit a very specific, immediate, but transitory type of behavior. Others trigger hormonal activities that result in slow but long-lasting responses. Some pheromones may act in both ways.

An advantage to pheromone communication is that little energy must be expended to synthesize the simple, but distinctive, organic compounds involved. Members of the same species have receptors that fit the molecular configuration of the pheromone; other species usually ignore it. Pheromones are effective in the dark, they can pass around obstacles, and they last for several hours or longer. Major disadvantages of pheromone communication are slow transmission and limited information content. Some animals compensate for the latter disadvantage by secreting different pheromones with different meanings.

Pheromones are important in attracting the opposite sex and in sex recognition in many species. Many female

(a)

(b)

FIGURE 50–12 Communication is necessary for social behavior. (*a*) A male hylid frog of Costa Rica calling to locate a mate. (*b*) Communicating with language. Chimpanzee Tatu (*top*) is signing "food" to Washoe (*below*). Washoe was the first chimp to learn American Sign Language from a human. She then taught other chimps how to sign. (*a*, L.E. Gilbert, University of Texas at Austin/Biological Photo Service; *b*, April Ottey, Chimpanzee and Human Communication Institute, Central Washington University) insects produce pheromones that attract males. We have taken advantage of some sex-attractant pheromones to help control such pests as gypsy moths by luring the males to traps baited with synthetic versions of female pheromones.

Some aspects of the vertebrate sexual cycle are affected by pheromones. When the odor of a male mouse is introduced among a group of females, the reproductive cycles of the female mice become synchronized. In some species of mice, the odor of a strange male, a sign of high population density, causes a newly impregnated female to abort. Among humans, it appears that some unconsciously perceived body odor is capable of synchronizing the menstrual cycles of women who associate closely (for instance, college roommates or cellmates in prison). As we will see, pheromones much more strictly govern the reproduction of many social insects.

FIGURE 50–13 Social animals use many signals to convey messages relating to social dominance. This baboon bares his teeth and screams in an unmistakable show of aggression. (Gerald Lacz/Peter Arnold, Inc.)

Animals often form dominance hierarchies

In the spring, female paper-wasps awaken from hibernation and begin to build a nest cooperatively. During the early course of construction, a series of squabbles among the females takes place in which the combatants bite one another's bodies or legs. Finally, one of the wasps emerges as dominant. After that, she is rarely ever challenged. This queen wasp spends more and more time tending the nest, and less and less time out foraging for herself. She takes the food she needs from the others as they return.

The queen then begins to take an interest in raising a family—her family. Because she is almost always at hand, she is able to prevent other wasps from laying eggs in the brood cells by rushing at them, jaws agape.

The queen can bite any other wasp without serious fear of retaliation. The other wasps are not equal in their relationships with one another. Indeed, the wasps organize into a **dominance hierarchy,** an arrangement of status that regulates aggressive behavior within the society:

Queen > Wasp A > Wasp B > . . . . Wasp J > Wasp K

Dominance hierarchies suppress aggression

Once a dominance hierarchy is established, little or no time is wasted in fighting (Fig. 50–13). When challenged, subordinate wasps exhibit submissive poses that, in turn, inhibit the queen's aggressive behavior. Consequently, few or no colony members are lost through wounds sustained in fighting one another. This ensures greater reproductive success for the colony.

Many factors affect dominance

In some animals, dominance is a simple function of aggressiveness, which is itself often influenced directly by sex hormones. Among chickens, the rooster is the most dominant; as with most vertebrates, the hormone testosterone increases aggressiveness. If a hen receives testosterone injections, her place in the dominance hierarchy shifts upward. When male rhesus monkeys are dominant, their testosterone levels are much higher than when they have been defeated. Not only can estrogen sometimes reduce dominance, and testosterone increase dominance, but dominance may even increase testosterone. It is not always easy to determine cause and effect. (Furthermore, recent studies suggest that the effects of testosterone on human mood and behavior may be different.)

In many species, males and females have separate dominance systems. However, in many monogamous animals, especially birds, the female takes on the dominance status of her mate by virtue of their relationship. This is not always the case, however. Like many fishes, some coral reef fishes (labrids) are capable of sex reversal. The most dominant individual is always male, and the remaining fishes within his territory are all female. If the male dies or is removed, the most dominant female will become the new male. Should any harm come to "him," the next-ranking female will take charge. Other types of fishes exhibit the reverse behavior; in which the most dominant fish is always female.

Many animals defend territory

Most animals have a **home range,** a geographical area that they seldom leave (Fig. 50–14). Because the animal has the opportunity to become familiar with everything in that range, it has an advantage over both its predators and its prey in negotiating cover and finding food.

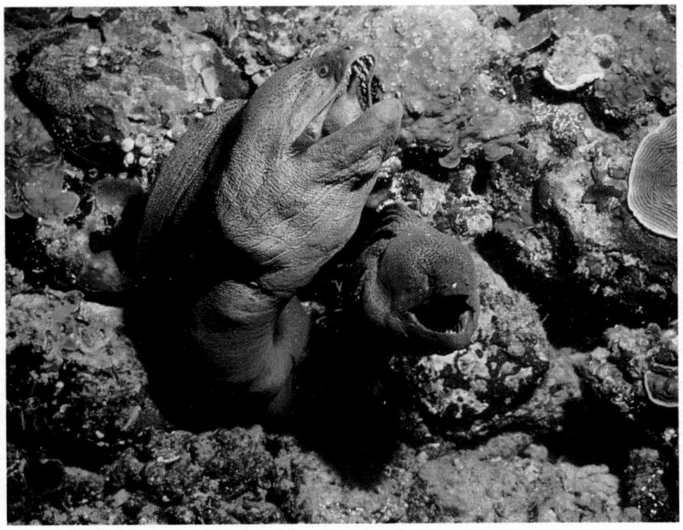

FIGURE 50–14 A coral reef has many secluded areas in which a territorial animal can establish a home range. Among the most territorial of coral reef fishes is the moray eel, which will attack any animal (including a human diver) that comes too close to its shelter. (IFA-Bilderteam-D. Eichler/Peter Arnold, Inc.)

Some, but not all, animals exhibit **territoriality.** They defend a **territory,** a portion of the home range, against other individuals of their species and sometimes against individuals of other species (Fig. 50–15).

Territoriality is easily studied in birds. Typically, the male chooses a territory at the beginning of the breeding season. This behavior results from high sex hormone concentrations in the blood. The males of adjacent territories fight until territorial boundaries become fixed. Generally, the dominance of a male is directly associated with his nearness to the center of his territory. Thus, close to home he is a lion, but when invading some other bird's territory, he is likely to be a lamb. The interplay of dominance values among territorial males eventually produces a neutral line at which neither is dominant. That line is the territorial boundary.

Bird songs announce the existence of a territory and often serve as a substitute for violence. Furthermore, they announce to eligible females that a propertied male resides in the territory. Typically, male birds take up a conspicuous station, sing, and sometimes display striking patterns of coloration to their neighbors, their rivals, and sometimes their mates (Fig. 50–16).

Territoriality among animals appears to be adaptive. It tends to reduce conflict among members of the same species and also helps control population growth. Animals that fail to establish territories fail to reproduce. Territoriality also ensures the most efficient use of environmental resources by encouraging individuals to spread throughout a habitat.

Usually, territorial behavior is related to the specific lifestyle of the animal and to whatever aspect of its ecology is most critical to its reproductive success. For in-

FIGURE 50–15 These sea birds, blue-eyed shags, often defend the territory directly surrounding their nests. This results in a regular spacing of the nests. (Frans Lanting/Minden Pictures)

(a) (b)

FIGURE 50–16 Courtship displays are specific to each species. (*a*) The male great frigatebird (*Frigata minor*) displays himself as part of a courtship ritual. Photographed on Christmas Island, Pacific. (*b*) Courtship signals by male fiddler crabs are specific to each species. This particular sequence of the motion of the large right claw is characteristic of the species *Uca lactae*. (*a*, Sid Bart/Photo Researchers, Inc.)

stance, sea birds may range over hundreds of square miles of open water but exhibit territorial behavior only at crowded nesting sites on an island. The nesting sites are their scarcest resource and the one for which competition is keenest.

Sexual behavior is generally social

For some animals, mating and perhaps the rearing of young are the only forms of social behavior (Fig. 50–17). Let us consider the sex act as a basic example of social

(a) (b)

FIGURE 50–17 Jumping spiders (*Phidippus princeps*) engage in complex courtship behavior before mating. (*a*) The male performs an elaborate dance that inhibits the female's natural aggression toward him, allowing him to get close enough to inseminate her (*b*). (Visuals Unlimited/P. Starborn) conduct. The sex act is obviously adaptive in that in most animals it is necessary for reproduction. It requires *cooperation,* the *temporary suppression of aggressive behavior,* and a *system of communication.*

In many species, success of a male in dominance encounters with other males indicates his quality to the female, and she allows the victorious male to court her. Courtship ensures that the male is a member of the same species, and also provides the female further opportunity to evaluate him. Courtship may also be necessary as a signal to trigger nest building or ovulation.

Courtship rituals can last seconds or hours and often involve complex behaviors. The first display of the male releases a counter behavior by the female. This, in turn, releases additional male behavior, and so on until the pair is ready for copulation.

Courtship is often strenuous and dangerous, especially for males. Tremendous energy is often needed for the fights engaged in by some birds and many mammals, including rams and bull seals. Males may expend additional energy in wallowing, roaring, and leaping about. These behaviors appear to be a test of male fitness. The male with the greatest endurance has the opportunity to mate and to perpetuate his genes.

Among some jumping spiders, mating is preceded by a ritual courtship in which the male temporarily paralyzes the female. While she is thus enthralled, the male inseminates her. Should she recover before he makes his escape, he becomes the main course at his own wedding feast. Even so, he still makes the ultimate material contribution to the eggs the female will produce and, therefore, to the perpetuation of his genes. She would otherwise have to bear the metabolic burden of their production by herself.

FIGURE 50–18 Most birds establish pair bonds. A pair of nesting blue herons (*Ardea herodias*). In many species pair bonds are maintained by grooming or other displays of attraction. (SharkSong/M. Kazmers/Dembinsky Photo Associates)

Pair bonds establish reproductive cooperation

A **pair bond** is a stable relationship between animals of the opposite sex that ensures cooperative behavior in mating and the rearing of the young (Fig. 50–18). In an estimated 90% of bird species, a male pair-bonds with a female during the breeding season. The mechanisms involved in establishing and maintaining the pair bond are often remarkably detailed. Specific cues enable courtship rituals to function as reproductive isolating mechanisms among species (see Chapter 19).

Many organisms care for their young

Care of the young, an important part of successful reproduction in many species, requires parental investment (Fig. 50–19). The benefit of parental care is the increased likelihood that the offspring will survive. The cost is a reduction in the number of offspring that can be produced. Because of the time spent carrying the developing embryo, the female has more to lose than the male if the young do not develop. Thus, females are more likely than males to brood eggs and young, and usually the females invest more in parental care.

Investing time and effort in care of the young is usually less advantageous to a male (assuming that the female can handle the job by herself), for time spent in parenting is time lost from inseminating other females. In some species, it may not even be certain who fathered the offspring. Raising some other male's offspring is a genetic disadvantage. This is probably why male lions kill cubs of a male whose position they have seized.

In some situations, however, a male can benefit by helping to rear his own young or even those of a genetic relative. Receptive females may be scarce, and gathering sufficient food may require more effort than one parent can provide. In some habitats, the young may need protection against predators and sometimes against cannibalistic males of the same species.

Play is often practice behavior

Perhaps you have watched a kitten pouncing on a dead leaf—or practicing a carnivore neck bite or a hind-claw

(a)

(b)

(c)

FIGURE 50–19 Parental investment increases the probability that offspring will survive. (*a*) Adult robins invest energy in feeding their young. (*b*) African elephants (*Loxidonta africana*) protecting young. (*c*) A baby orangutan rides on its mother's back and gains some of her social status. (*a*, Dominique Braud/Dembinsky Photo Associates; *b*, Animals Animals © 1996 David Fritts; *c*, Schafer and Hill/Peter Arnold, Inc.)

disemboweling stroke on a littermate without injury. Many animals, especially young birds and mammals, use play to practice adult patterns of behavior (Fig. 50–20). They may perfect means of escape, the killing of prey, or sexual behavior. In true play, the behavior may not be actually consummated.

Elaborate societies are found among the social insects

Although many insects—such as tent caterpillars, which spin a communal nest—cooperate socially, the most elaborate insect societies are found among the bees, ants, wasps, and termites. (The first three of these all belong, not coincidentally, to the order Hymenoptera.) Insect societies are held together by a complex system of sign stimuli that are keyed to social interaction. As a result, they tend to be quite rigid.

The social organization of honeybees has been studied more extensively than that of any other social insect.

FIGURE 50–20 Cheetah cubs engage in play behavior. Play often serves as a means of practicing behavior that will be used in earnest later in life, possibly in hunting, fighting for territory, or competing for mates. Photographed in Kenya. (BIOS [M&C Denis-Huot]/Peter Arnold, Inc.)

FIGURE 50–21 A queen honeybee inspects wax cells. Note the numerous workers that surround her. They constantly lick secretions from the queen bee. These secretions are transmitted throughout the hive and act to suppress the activity of the workers' ovaries. (Treat Davidson/Photo Researchers, Inc.)

A honeybee society generally consists of a single adult queen, up to 80,000 worker bees (all female), and, at certain times, a few males called drones that fertilize newly developed queens. The queen's job is reproduction: she deposits about 1000 fertilized eggs per day in the wax cells of a comb.

Division of labor in the bee society is mostly determined by age. The youngest worker bees serve as nurses that nourish larval bees. After about a week as nurse bees, workers begin to produce wax and build and maintain the wax cells. Older workers are foragers, bringing home nectar and pollen. Most worker bees die at the ripe old age of 42—days, that is.

The composition of a bee society is controlled by an antiqueen pheromone secreted by the queen. It inhibits the workers from raising a new queen, and prevents development of the ovaries in the workers (Fig. 50–21). If the queen dies, or if the colony becomes so large that the inhibiting effect of the pheromone is dissipated, the workers begin to feed some larvae special food that promotes their development into new queens.

The most sophisticated known mode of communication among bees is a stereotyped series of body movements known as a **dance.** When a honeybee scout locates a rich source of nectar, it communicates the direction and distance of the food source relative to the hive to the other bees. If the food supply is nearby, the scout performs a round dance, which generally excites the other bees and causes them to fly about in all directions (but within a certain distance from the hive) until they have found the nectar. If the source is distant, however, the scout performs a waggle dance.

Karl von Frisch, a pioneer in the study of communication in bees, showed that the orientation of the circular movements indicates the direction of the food source. The frequency of the waggle indicates the distance. Bees use the angle of the sun and light polarization to orient themselves. The dance locates the food in reference to the position of the sun (Fig. 50–22).

In social insects such as bees, males develop from unfertilized eggs and are haploid. Reproductive females

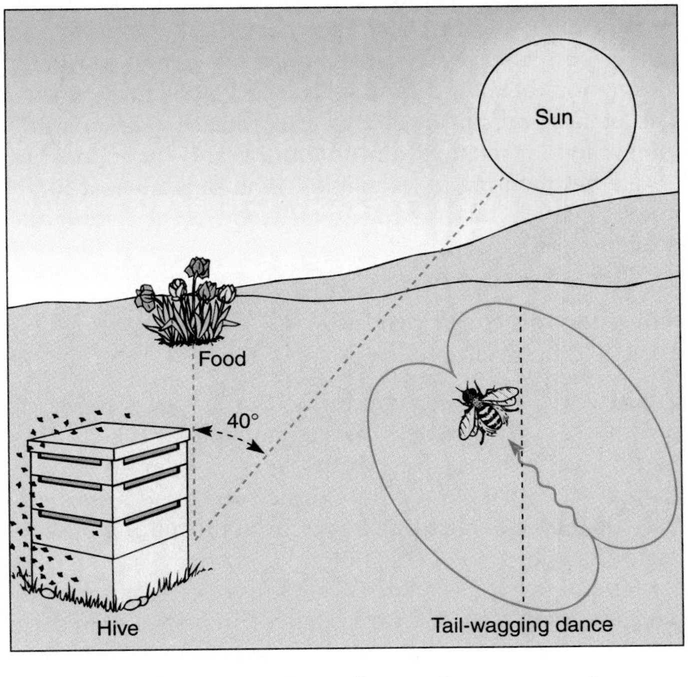

FIGURE 50–22 A scout honeybee performs a waggle dance. The waggle is upward, indicating that the food is toward the sun, and inclined 40 degrees to the left, which reveals the angle of the food source to the sun.

store sperm cells from previous matings in a seminal receptacle. If they permit a sperm cell to contact the egg as it is laid, the resulting insect is female; otherwise, it is male.

Because a drone is haploid, each one of its sperm cells has *all* of his chromosomes; that is, meiosis does not occur during sperm production. The queen bee stores this sperm throughout her lifetime, and uses it to produce worker bees. Thus, the worker bees of a hive are more closely related to one another than would be sisters born of a diploid father. Indeed, they have three-quarters of their genes in common (they share half of the queen's chromosomes and all of the drone's). As a consequence, they are more closely related to one another than they would be to their own offspring, if they could have any. Because new queens will also be their sisters, they are actually more likely to pass on copies of their genes to the next generation by raising these individuals than they would be by producing their own offspring. A worker bee's offspring would have only one-half of its genes in common with its worker mother.

Vertebrate societies tend to be relatively flexible

Vertebrate societies usually contain nothing comparable to the physically and behaviorally specialized castes of honeybees or ants. What is more, except for humans, individual members of vertebrate societies are not as specialized in their tasks as are the social insects. Although in some ways vertebrate societies seem simpler than insect societies, they are also more flexible.

Vertebrate societies share a great range and plasticity of potential behaviors, and can adapt to changing needs. The behavioral plasticity of vertebrates makes possible the symbolic transmission of culture, and this opens up whole new worlds inaccessible to the social insects. Human society is based to a great extent on the symbolic transmission of culture.

Kin selection could produce altruistic behavior

In **altruistic behavior,** one individual appears to act in such a way as to benefit others rather than itself. This type of behavior is frequently seen in complex social groups. Biologists Watts and Stokes observed a particularly clear case of altruistic behavior in the mating of wild turkeys.

Several groups of male turkeys, each of which has an internal dominance hierarchy, gather in a special mating territory. They go through their displays of tail spreading, wing dragging, and gobbling in front of females who come to the area to copulate. One group attains dominance over other groups as a result of cooperation among the males within the group. The dominant member of the dominant group is the one to copulate most frequently with the females.

FIGURE 50–23 Kin selection is evident in prairie dogs. Low-ranking members of this social rodent group act as sentries, risking their own lives by exposing themselves outside their burrows. However, in this way they protect their siblings and by so doing ensure that the genes they share in common will be perpetuated in the population. (Tina Waisman)

Seemingly, the males that have low status within the group they helped to establish gain nothing. Close analysis has shown that members of a group are brothers from the same brood. Because they share many genes with the successful male, they are indirectly perpetuating many of their genes. In this case, altruism is closely related to **kin selection,** the evolutionary effect caused by individuals who favor the survival and reproduction of their relatives (kin) (Fig. 50–23). By promoting the breeding success of a close relative, the nonbreeding turkeys perpetuate copies of many of their own genes.

Among some birds (Florida jays and others), nonreproducing individuals aid in the rearing of the young. Nests tended by these additional helpers as well as parents produce more young than do nests with the same number of eggs overseen only by parents. The nonreproducing helpers, siblings of the parents, are apparently increasing their own biological success by ensuring the successful perpetuation of their genes via their siblings.

Sociobiology explains altruism by kin selection

Sociobiology focuses on the evolution of social behavior through natural selection. It represents a synthesis of population genetics, evolution, and ethology. Like many biologists of the past (such as Darwin), E. O. Wilson and other sociobiologists stress the animal roots of human behavior, emphasizing the effect of kin selection on patterns of inheritance. Many of the concepts discussed in this chapter, such as altruism and paternal investment in care of the young, are based on contributions made by sociobiologists.

From the sociobiological perspective, an organism and its adaptations, including its behavior, ensure that genes make more copies of themselves. The cells and tissues of the body support the functions of the reproductive system, and the reproductive system functions to transmit genetic information to succeeding generations.

Most of the controversy that has been triggered by sociobiology seems related to its possible ethical implications. Sociobiology is often viewed as denying that human behavior is flexible enough to permit substantial improvements in the quality of our social lives. Yet sociobiologists do not disagree with their critics that human behavior is flexible. The debate therefore seems to rest on the *degree* to which human behavior is genetic and the *extent* to which it can be modified.

As sociobiologists acknowledge, people can, through **culture,** change their way of life far more profoundly in a few years than could a hive of bees or a troop of baboons in hundreds of generations of genetic evolution. This capacity to make changes is indeed genetically determined, and that is a great gift. How we use it and what we accomplish with it is not a gift but a responsibility upon which our own well-being and the well-being of other species depend.

SUMMARY

I. Behavior consists of the responses of an organism to signals from its environment.

II. Ethology is the scientific study of behavior in natural environments from the point of view of adaptation.
- A. Behavior tends to be adaptive.
- B. Behavior tends to be homeostatic.

III. It is adaptive for an organism's metabolic processes and behavior to be synchronized with the cyclical changes in the environment.
- A. Some biological rhythms reflect the lunar cycle or the changes in tides due to phases of the moon.
- B. In many species, physiological processes and activity follow circadian rhythms.
- C. No single biological clock has been found. Biological rhythms are thought to be regulated by both internal and external factors.

IV. The capacity for behavior is inherited, and behavior is modified by environmental stimuli.
- A. Before an organism can show any pattern of behavior, it must be physiologically ready to produce the behavior.
- B. Some behavior patterns have strong genetic components. Innate behavior may be triggered by a specific unlearned sign stimulus.
- C. Learning is a change in behavior resulting from experience.
- D. Simpler forms of learning are conditioning (both classical and operant) and habituation.
- E. Imprinting establishes a parent-offspring bond during a critical period early in development.
- F. Insight learning, characteristic of the more "intelligent" animals, involves the ability to see through a problem.

V. Behavioral ecology focuses on interactions between animals and their environment, and on the survival value of the behavior.
- A. In some birds, the need to migrate and the direction of migration appear to be genetically programmed, but navigation may be learned.
- B. Optimal foraging, the most efficient strategy for an animal to obtain food, enhances reproductive success.

VI. Social behavior is adaptive interaction, usually among members of the same species.
- A. A society is a group of individuals of the same species that cooperate in an adaptive manner. In a society, there is a means of communication, cooperation, division of labor, and a tendency to stay together.
- B. Animal communication involves the transmission of signals, but as far as we know it does not use symbolic language in the human sense. Pheromones are chemical signals that convey information between members of a species.
- C. Dominance hierarchies result in the suppression of aggressive behavior.
- D. Organisms often inhabit a home range, from which they seldom or never depart. This range, or some portion of it, may be defended from members of the same (or occasionally different) species.
  1. Defended areas are called territories, and the defensive behavior is territoriality.
  2. Territorial defense is often carried out by display behavior rather than by actual fighting.
- E. Courtship behavior ensures that the male is a member of the same species, and it permits the female to assess the quality of the male.
- F. A pair bond is a stable relationship between a male and a female that ensures cooperative behavior in mating and in rearing the young.
- G. Parental care increases the probability that the offspring will survive. A high investment in parenting is often less advantageous to the male than to the female.
- H. Play gives the young animal a chance to practice adult patterns of behavior.
- I. Insect societies tend to be rigid, with the role of the individual narrowly defined.
- J. Vertebrate societies are far less rigid than are insect societies. Although innate behavior is important, the role of the individual is generally learned.
- K. In altruistic behavior, one individual appears to behave in a way that benefits others rather than itself. Altruism may be closely related to kin selection.
- L. Sociobiology focuses on the evolution of social behavior through natural selection.

SELECTED KEY TERMS

altruistic behavior
behavior
biological clock
circadian rhythm
classical conditioning
crepuscular animal
diurnal animal
dominance hierarchy
ethology
fixed action pattern (FAP)
foraging strategy
habituation
home range
imprinting
innate behavior
insight
kin selection
learned behavior
lunar cycle
migration
nocturnal animal
operant conditioning
optimal foraging
pair bond
pheromone
sign stimulus
social behavior
society
sociobiology
territoriality

POST-TEST

1. __________ may be defined as responses of an organism to signals from its environment.
2. __________ is the study of behavior in natural environments from the point of view of adaptation.
3. A biological rhythm with approximately a 24-hour cycle is a(an) __________ rhythm.
4. Animals that are most active at dawn or twilight are described as __________.
5. A stimulus that elicits a fixed action pattern is a(an) __________ __________.
6. __________ behavior is mainly genetic; __________ behavior develops as a result of experience.
7. __________ is a form of learning in which a young animal forms a strong attachment to a moving object (usually its parent) within a few hours of birth.
8. In __________ an organism learns to ignore a repeated, irrelevant stimulus.
9. Salivation by a student when the noon bell rings is an example of __________ conditioning.
10. Adaptive interactions among members of a population are referred to as __________ behavior.
11. A(An) __________ is a group of individuals belonging to the same species that cooperate in an adaptive manner and have a means of communicating with one another.
12. __________ are chemical signals that convey information between members of a species.
13. An arrangement of members of a population by status is called a(an) __________ __________.
14. The geographical area that members of a population seldom leave is the __________ __________.
15. ____________________ tends to reduce conflict and control population growth.
16. A(An) __________ __________ is a stable relationship between two animals of the opposite sex that ensures cooperative behavior in mating and rearing the young.
17. The extensive behavioral repertoire of the bee is almost entirely __________ behavior.
18. In __________ behavior, one individual appears to act to benefit others rather than itself.
19. __________ selection favors the indirect perpetuation of an animal's genes by a relative.

REVIEW QUESTIONS

1. In what ways are the behaviors of *Philanthus*, the sand wasp, adaptive?
2. Why is it adaptive for some species to be diurnal, but for others to be nocturnal or crepuscular?
3. Behavior capacity is inherited and is modified by learning. Give an example.
4. When Konrad Lorenz kept a greylag goose isolated from other geese for the first week of its life, the goose persisted in following him about in preference to other geese. How can this behavior be explained?
5. How does physiological readiness affect instinctual behavior? How does it affect learned behavior?
6. What distinguishes an organized society from a mere aggregation of organisms? Cite an example of an organized society, and describe characteristics that qualify the society as organized.
7. How does an organism learn its place in a dominance hierarchy? What determines this place? What are the advantages of a dominance hierarchy?
8. What is territoriality? What functions does it seem to serve?
9. What is kin selection? How is kin selection used by sociobiologists to explain the evolution of altruistic behavior?
10. Do animals play just for the fun of it?
11. What are some advantages of courtship rituals?
12. Males and females have different reproductive strategies. What is meant by this statement? (Base your answer on evolutionary success.)
13. Contrast the "language" of bees with human language.

YOU MAKE THE CONNECTION

1. What might be the adaptive value of sea turtle migration? Consider how this adaptive value has changed if, as a result of human activities, migration now puts sea turtles at greater risk than if they restricted their habitat to a single location. Discuss the possible evolutionary mechanisms by which the turtles' behavior may (or may not) adapt to these environmental pressures.
2. Using what you know about animal learning, propose experiments to test whether it might be possible to teach sea turtles "safer" reproductive behavior. As you do, discuss the

following questions: What might be the broader, long-term ecological consequences of your proposals? Is it advisable for humans to attempt to alter the behavior of another species? Is this an ethical way of preserving a species that has become endangered due to human activities?

3. Behavioral ecologists have demonstrated that young tiger salamanders can become cannibals and devour other salamanders. Investigators observed that the cannibal salamanders preferred eating unrelated salamanders to their own cousins, and preferred eating cousins to their siblings. Explain this behavior based on what you have learned in this chapter.
4. What similarities between the transmission of information by symbolic language and by heredity can you think of? What are some differences?

RECOMMENDED READINGS

Ellis, D.H., J.C. Bednarz, D.G. Smith, and S.P. Flemming. "Social Foraging Classes in Raptorial Birds." *BioScience,* Vol. 43, No. 1, January 1993. Many bird species have developed cooperative hunting.

Greenspan, R.J. "Understanding the Genetic Construction of Behavior." *Scientific American,* Vol. 272, No. 4, April 1995. Studies of courtship and mating in the fruit fly suggest how genes might affect complex behavior.

Harbrecht, D. "Games Animals Play." *International Wildlife,* Vol. 23, No. 5, September/October 1993. An interesting discussion on why animals play.

Kirchner, W.H. and W.F. Towne, "The Sensory Basis of the Honeybee's Dance Language," *Scientific American,* Vol. 270, No. 6, June 1994. A discussion of honeybee communication.

Lohmann, K.J. "How Sea Turtles Navigate," *Scientific American,* Vol. 266, No. 1, January 1992. Research has begun to identify the biological compasses and maps that guide sea turtles as they migrate across hundreds of miles of ocean.

Martin, G. "Killer Culture." *Discover,* December 1993. An interesting account of killer whale societies.

Narins, P.M. "Frog Communication," *Scientific American,* Vol. 273, No. 2, August 1995. Some frogs croak loudly to ensure that they are heard; others transmit vibrations by tapping on the ground or on plants.

Pfennig, D.W. and P.W. Sherman, "Kin Recognition," *Scientific American,* Vol. 272, No. 6, June 1995. The ability of many animals to recognize their relatives may have a variety of survival advantages.

Pool, R. "Putting Game Theory to the Test." *Science,* March 17, 1995. A discussion of how game theory can be applied to animal behavior, and of the applications of this data.

Sherman, P.W., J. Jarvis, and S.H. Braude. "Naked Mole Rats." *Scientific American,* Vol. 267, No. 2, August 1992. These animals have a social structure that resembles that of some insects.

Career Visions

Science Journalist

GINA KOLATA

Aspiring journalists can take a lesson in tenacity from The New York Times *reporter Gina Kolata. Although she always knew she liked to write, her success as a journalist came after studies that might have resulted in a career as a research scientist. After obtaining a bachelor's degree in microbiology from the University of Maryland she proceeded to the Massachusetts Institute of Technology for some graduate work in that discipline and ultimately received a Master's degree in mathematics from the University of Maryland. However, she still wanted to write and her perseverance led to a rewarding career reporting on new scientific discoveries, fascinating scientists, and other topics that capture her imagination.*

What did you plan to do after receiving a degree in microbiology?

I thought I would become a researcher. Ironically, I always hated lab work because you were demonstrating something for which you knew the outcome. Research seemed much more interesting since the outcome is unknown. My first year at M.I.T. was wonderful—I attended seminars and wrote papers all the time, but the following summer I was expected to choose a research advisor and work in his/her lab. I found that I disliked research, and decided to try to become a writer instead.

How did you go about working to become a writer?

I met a writer on the beach while I was on a summer fellowship at the Woods Hole Biological Laboratory. He told me he got his start by writing a book review, so I wrote to the same publication and offered to write a book review for them. They accepted my offer and published the review verbatim. I think I got $25.00 for it, but I was so excited to be published! I took a leave of absence from school to try to become a writer, but it's really hard to get a job when you have one publication with your name on it. After some effort I decided I would never make it, and took a job as a lab technician.

And it was at this time that you decided to get a degree in applied mathematics?

I wanted to go into a field without labs. I entered the University of Maryland's math program, but after a short time I decided I just didn't have the kind of genius required to really make a difference in mathematics. It's not just a matter of being able to do problems or take courses on a graduate level. You really have to be born with a talent for mathematics, and I don't have that kind of talent. I decided to leave school and become a writer, but my advisor counseled me to get my degree anyway, and I did so.

But you still wanted to write. How did you become a journalist?

I called and wrote to just about every place in the Washington, D.C. yellow pages. I knew I had to create opportunities for myself. I read books on writing and how to write—always working to improve my skills. *Science* magazine offered a job, and while it wasn't a writing position, I took it anyway and offered to write articles on my own time without compensation. They eventually promoted me to the staff writer position I coveted, but after I developed expertise writing those pieces it wasn't very challenging to do them anymore. I began taking freelance assignments from all sorts of magazines. I wrote a monthly column in the *Journal of Investigative Dermatology,* and for a German science magazine as well.

When did you join *The New York Times*?

I wrote a letter to the science editor there. He called and said he didn't have a position to offer right away, but he did want to meet with me. He flew down to Washington and we had the interview, but when I didn't hear from him afterwards I thought, "Well, you tried and it didn't work out." Three years later, he called and offered me a job, and I've been at *The New York Times* ever since.

Now you write about topics in science for *The New York Times*?

I write about biology, math, computers, medicine, and ethics. I've written about subjects that weren't science at all, but I am in the science section and that's what I really like to write about.

How do you get ideas for stories?

Sometimes people will suggest things, or I'll see something in the science journals that I think is really important. I pay attention to what's going on in science, and there are so many things going on that I could never quite get to all the stories that sound really good. Often, I'll profile someone I think is a really wonderful scientist. That's fun too—it tells you how scientists live their lives, how they think, and what they think is important.

What kind of educational background would be helpful to an aspiring science journalist?

Understanding biology would be really helpful because you won't be afraid of it. You learn how to think. Not every science reporter has a background similar to my own, but it does help to have special expertise. It's not just a matter of being able to write and ask good questions, you need to know the subject you're covering really well.

P A R T V I I I

The Interactions of Life: Ecology

Energy flows through an ecosystem as one organism feeds on another: a grasshopper consumes a spotted knap weed but in turn is food for a praying mantis. (Skip Moody/ Dembinsky Photo Associates)

C H A P T E R 5 1

Ecology and the Geography of Life

Tube coral polyps (*Tubastraea*) extend for feeding. A complex multitude of relationships and interactions occur at coral reefs. (Visuals Unlimited/Dave B. Fleetham)

Living organisms and the physical environment interact in an immense and complicated web of relationships. **Ecology** is the study of interactions among organisms, and between organisms and their **abiotic** (nonliving) **environment.** Ecology is the broadest field in biology, with explicit links to every other biological discipline. Its universality also brings subjects into view that are not traditionally part of biology. Geology and earth science are extremely important to ecology, especially when ecologists examine the abiotic environment of planet Earth. Because humans are living organisms, all of our activities have a bearing on ecology; even economics and politics have profound ecological implications.

The levels of biological organization that interest ecologists are those above the level of the individual organism. Organisms are arranged into **populations** in which members of the same species live together in the same area at the same time. A population ecologist might study a population of polar bears, for example, or of marsh grasses, to see how it interacts with and is adapted to its environment.

Populations are organized into communities. A **community** consists of all the populations of all of the different species that live and interact together within an area. A community ecologist might study how organisms interact with one another —including who eats whom—in a coral reef or in an alpine meadow community. **Ecosystem** is an even more inclusive term because it is a community together with its abiotic environment. Thus, an ecosystem includes not only all the interactions among the living organisms of a community, but also the interactions between the organisms and their abiotic environment. An ecosystem ecologist, for example, might examine how temperature, light, precipitation, and soil factors affect the organisms living in a desert or in a coastal bay.

All of Earth's communities of organisms comprise the **biosphere.** The organisms of the biosphere depend on one another and on the divisions of Earth's physical environment: the atmosphere, hydrosphere, and lithosphere. The *atmosphere* is the gaseous envelope surrounding Earth, the *hydrosphere* is Earth's supply of water (liquid and frozen, fresh and salty), and the *lithosphere* is the soil and rock of Earth's crust. The **ecosphere** encompasses the biosphere and its interactions with the atmosphere, hydrosphere, and lithosphere. Ecologists who study the biosphere or ecosphere examine the complex interrelationships among Earth's atmosphere, land, water, and living organisms.

LEARNING OBJECTIVES

After you have studied this chapter you should be able to

1. Define ecology, and distinguish among the following terms: population, community, ecosystem, biosphere, and ecosphere.
2. Define biome, and briefly describe the nine major terrestrial biomes, giving attention to the climate, soil, and characteristic plants and animals of each.
3. Relate at least one human effect on each of the biomes described.
4. Explain the important environmental factors that affect aquatic ecosystems.
5. Distinguish among plankton, nekton, and benthos.
6. Briefly describe the various freshwater, estuarine, and marine ecosystems, giving attention to the environmental characteristics and representative organisms of each.
7. Define ecotone and describe some of its features.

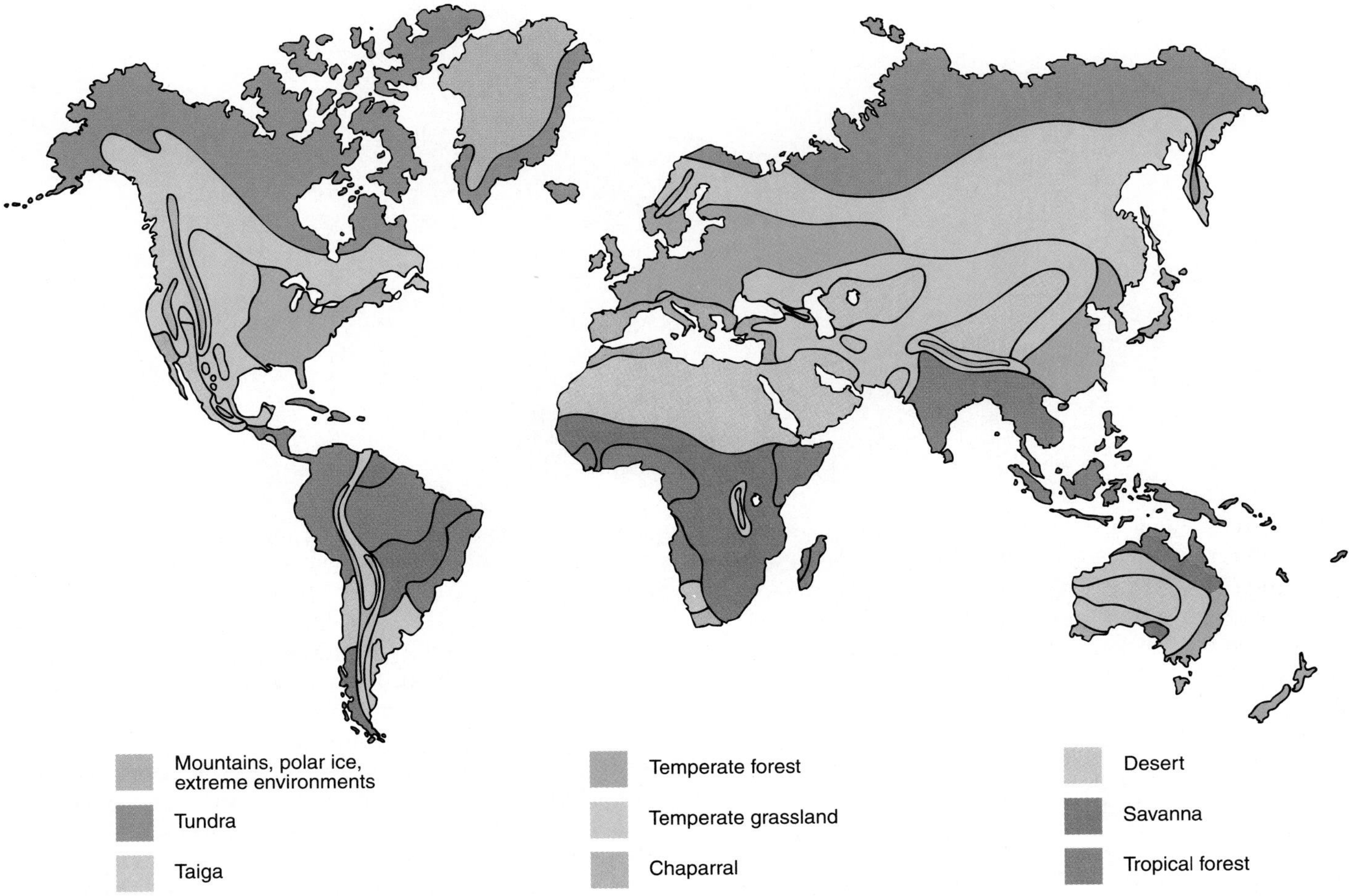

FIGURE 51–1 A highly simplified representation of the world's biomes shows sharp boundaries between biomes. Biomes actually grade together at their boundaries.

BIOMES ARE LARGELY DETERMINED BY CLIMATE

A **biome** is a large, relatively distinct terrestrial region characterized by similar climate, soil, plants, and animals regardless of where it occurs in the world. Because it is so large in area, a biome encompasses a number of interacting ecosystems. In terrestrial ecology, a biome is considered the next level of ecological organization above that of community and ecosystem. A biome's boundaries are largely determined by invisible climate barriers, with temperature and precipitation being most important (Figs. 51–1 and 51–2; climate is discussed further in Chapter 54). Near the poles, temperature is generally the overriding climate factor, whereas in temperate and tropical regions, precipitation becomes more significant (Fig.

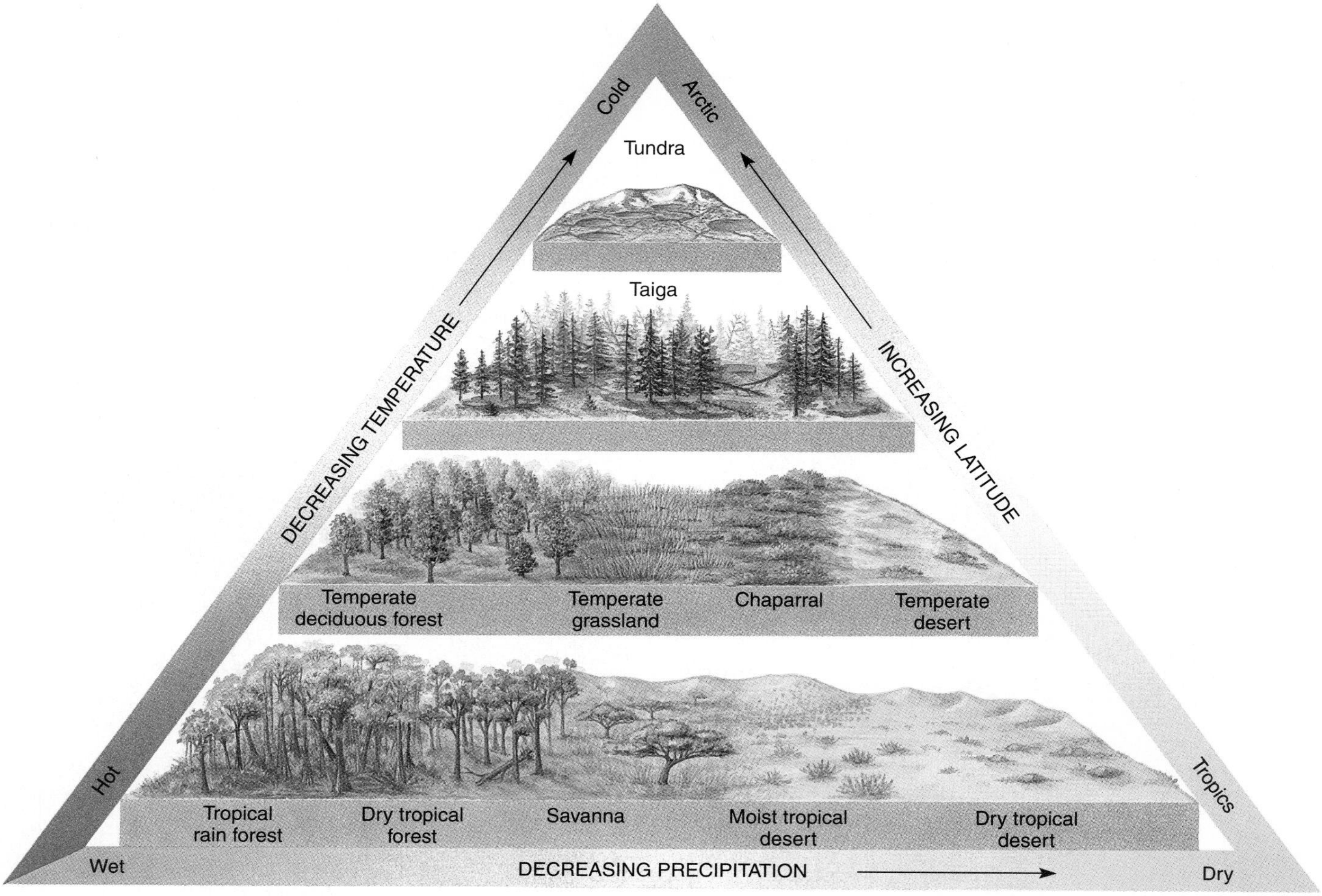

FIGURE 51–2 Biomes are distributed primarily in accordance with two climate factors: temperature and precipitation. In the higher latitudes, temperature is the more important of the two. In temperate and tropical zones, precipitation is a significant determinant of community composition.

51–3). Nine biomes are discussed: tundra, taiga, temperate rain forest, temperate deciduous forest, temperate grassland, chaparral, desert, savanna, and tropical rain forest.

Cold, boggy plains of the far north are called tundra

Tundra (also called **arctic tundra**) occurs in extreme northern latitudes wherever snow melts seasonally (Fig. 51–4). The Southern Hemisphere has no equivalent of the arctic tundra because it has no land in the proper latitudes. A similar ecosystem located in the higher elevations of mountains, above the tree line, is called **alpine tundra** to distinguish it from arctic tundra (see *Making the Connection: Comparing Altitudes and Latitudes*).

Arctic tundra has long, harsh winters and extremely short summers. Although the growing season, with its warmer temperatures, is short (from 50 to 160 days depending on location), the days are long. Above the Arctic Circle the sun does not set at all for many days in midsummer, although the amount of light at midnight is only one-tenth that at noon. There is little precipitation (10–25 centimeters, or 4–10 inches, per year) over much of the tundra, with most of it falling during summer months.

Tundra soils tend to be geologically young, since most were formed only after the last Ice Age.[1] These soils are usually nutrient-poor and have little organic litter. Although the soil surface melts during the summer, tundra has a layer of permanently frozen ground called **permafrost** that varies in depth and thickness. Because permafrost interferes with drainage, the soil is usually waterlogged during the summer. Permafrost also prevents roots of larger plants from becoming established. Limited precipitation, combined with low temperatures, flat topography (surface features), and permafrost, produces a landscape of broad shallow lakes, sluggish streams, and bogs.

[1]Glacier ice, which occupied about 29% of Earth's land during the last Ice Age, began retreating about 17,000 years ago. Today, glacier ice occupies about 10% of the land surface.

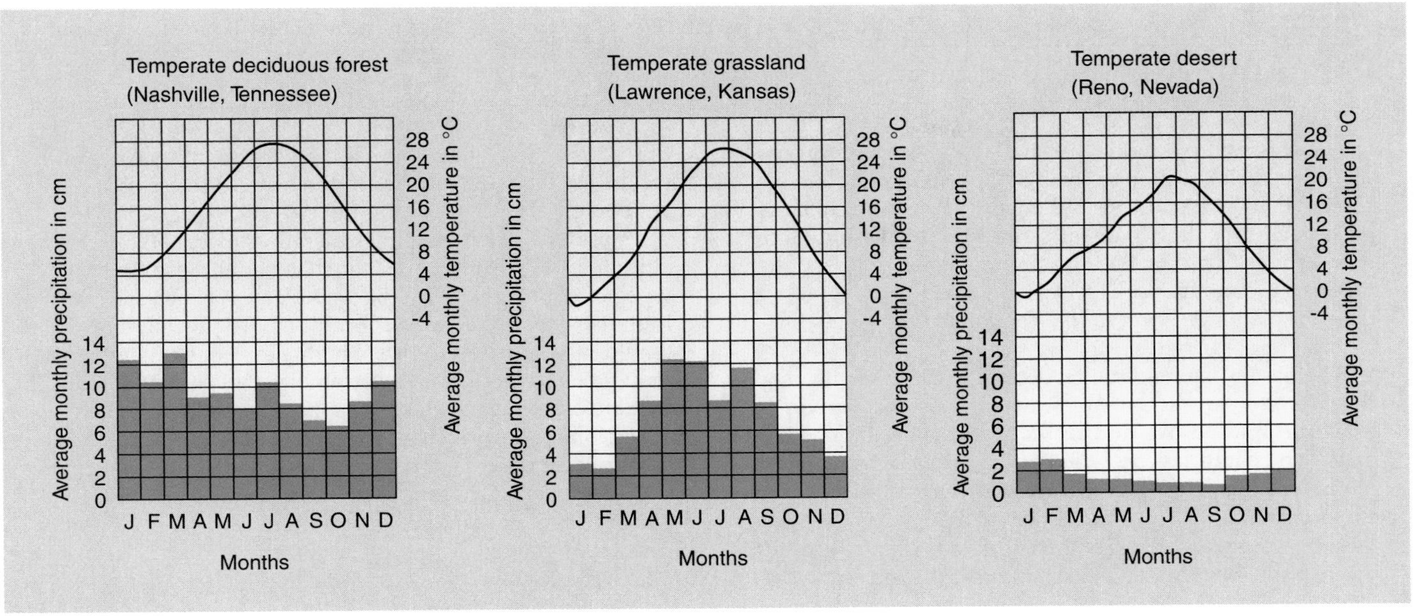

FIGURE 51–3 Average monthly temperature (*black*) and precipitation (*blue*) are given for three temperate biomes in North America. Although temperature is approximately the same in all three locations, precipitation varies a great deal, resulting in deciduous forest where precipitation is plentiful, grassland where it is less plentiful and more seasonal, and desert where it is quite low.

Few species are found in the tundra, but individual species often exist in great numbers. Tundra is dominated by mosses, lichens (such as reindeer moss), grasses, and grasslike sedges. No readily recognizable trees or shrubs grow except in sheltered locations, although dwarf willows and other dwarf trees are common. Tundra plants seldom grow taller than 30 centimeters (12 inches) in open areas.

Year-round animal life of the tundra includes lemmings, voles, weasels, arctic foxes, snowshoe hares, ptarmigan, snowy owls, and musk-oxen. In the summer, caribou migrate north to the tundra to graze on sedges, grasses, and dwarf willow. Dozens of bird species also migrate north in summer to nest and feed on abundant insects. Mosquitoes, blackflies, and deerflies survive the winter as eggs or pupae and occur in great numbers during summer weeks. There are no reptiles or amphibians.

Tundra regenerates quite slowly after it has been disturbed. Even casual use by hikers can cause damage. Long-lasting injury, likely to persist for hundreds of years, has been done to large portions of the arctic tundra as a result of oil exploration and military use.

Taiga is an evergreen forest of the north

Just south of the tundra is the **taiga,** or **boreal forest,** which stretches across both North America and Eurasia and covers approximately 11% of Earth's land (Fig. 51–5). A biome comparable to the taiga is not found in the South-

FIGURE 51–4 During its short growing season, small, hardy plants grow in tundra, the northernmost biome that encircles the Arctic Ocean. (Michio Hoshino/Minden Pictures)

Making the Connection

Comparing Altitudes and Latitudes

How is hiking up a mountain similar to traveling toward the North Pole? Although the distance up a mountain is shorter than an excursion to the North Pole, the same major vegetation patterns are encountered. This altitude-latitude similarity occurs because the temperature drops going up a mountain just as it does on a trip north. The types of plants growing on the mountain change as the temperatures change.

The base of a mountain in Colorado, for example, might be covered by deciduous trees, which shed their leaves every autumn. Above that altitude, where the climate is colder and more severe, a coniferous *subalpine forest* resembling northern taiga grows. Higher still, where the climate is quite cold, a kind of tundra occurs, with vegetation composed of grasses, sedges, and small tufted plants. This tundra is called *alpine tundra* to distinguish it from arctic tundra. At the top of the mountain, a permanent ice or snow cap might be found, similar to the nearly lifeless polar land areas.

Important environmental differences exist between high altitudes and high latitudes, however, and these affect the types of organisms found in each place. Alpine tundra typically lacks permafrost and receives more precipitation than does arctic tundra. Also, high elevations of temperate mountains do not have the great extremes of day length that are associated with the changing seasons in high-latitude biomes. Furthermore, the intensity of solar radiation is greater at high elevations than at high latitudes. At high elevations, the sun's rays pass through less atmosphere, which results in greater exposure to ultraviolet radiation (less UV is filtered out by the atmosphere) than occurs at high latitudes. ▲

ern Hemisphere. Winters are extremely cold and severe, although not as harsh as in the tundra. The growing season of the boreal forest is somewhat longer than that of the tundra. Taiga receives little precipitation, perhaps 50 centimeters (20 inches) per year, and its soil is typically acidic, mineral-poor, and characterized by a deep layer of partly decomposed pine and spruce needles at the surface. Permafrost is patchy and, where found, is often deep underneath the soil. Taiga contains numerous ponds and lakes in water-filled depressions that were dug in the ground by grinding ice sheets during the last Ice Age.

FIGURE 51–5 Taiga is coniferous forest that occurs in cold regions of the Northern Hemisphere adjacent to the tundra. (Charlie Ott/Photo Researchers, Inc.)

Deciduous trees such as aspen or birch, which shed their leaves in autumn, may form striking stands in taiga, but overall, spruce, fir, and other conifers (cone-bearing evergreens) clearly dominate. Conifers have many drought-resistant adaptations such as needle-like leaves with a minimal surface area for water loss. Such an adaptation enables conifers to withstand the "drought" of the northern winter months, when roots cannot absorb water because the ground is frozen.

Animal life of the boreal forest includes some larger species such as caribou (which migrate from the tundra to the taiga for winter), wolves, bears, and moose. However, most animal life is medium-sized to small, and includes rodents, rabbits, and fur-bearing predators such as lynx, sable, and mink. Most species of birds are seasonally abundant but migrate to warmer climates for winter. Insects are abundant, but there are few amphibians and reptiles except in the southern taiga.

Most of the taiga is not well suited to agriculture because of its short growing season and mineral-poor soil. However, the boreal forest yields vast quantities of lumber and pulpwood (for making paper products), plus animal furs and other forest products.

Temperate rain forest is characterized by cool weather, dense fog, and high precipitation

A coniferous **temperate rain forest** occurs on the northwest coast of North America. Similar vegetation exists in southeastern Australia and in southern South America. Annual precipitation in this biome is high, from 200 to 380 centimeters (80–152 inches), and this is augmented by condensation of water from dense coastal fogs. The proximity of temperate rain forest to the coastline moderates the temperature so that seasonal fluctuation is narrow; winters are mild and summers are cool. Temperate rain forest has a relatively nutrient-poor soil, although its organic content may be high. (Needles and large fallen branches and trunks accumulate on the ground as litter that takes many years to decay and release nutrients back to the soil.)

The dominant vegetation in the North American temperate rain forest is large evergreen trees such as western hemlock, Douglas fir, Sitka spruce, and western arborvitae. Temperate rain forest is rich in epiphytic vegetation, smaller plants that grow nonparasitically on the large trees (Fig. 51–6). Epiphytes in this biome are mainly mosses, lichens, and ferns. Squirrels, deer, and numerous bird species are common temperate rain forest animals.

Temperate rain forest is one of the richest wood producers in the world, supplying us with lumber and pulpwood. It is also one of the most complex ecosystems. Care must be taken to avoid overharvesting original old-growth forest, however, because such an ecosystem takes hundreds of years to develop. The logging industry typically harvests old-growth forest and replants the area with a monoculture (a single species) of trees that it can harvest in 40- to 60-year cycles. Thus, the old-growth forest ecosystem, once harvested, never has a chance to redevelop.

Temperate deciduous forest has a dense canopy of broad-leaf trees that overlie saplings and shrubs

Seasonality (hot summers and cold winters) is characteristic of **temperate deciduous forest,** which occurs in temperate areas where precipitation ranges from about 75 to 126 centimeters (30–50 inches) annually. Typically, the soil of a temperate deciduous forest consists of a topsoil rich in organic material, and a deep, clay-rich lower layer. As organic materials decay, mineral ions are released. If they are not immediately absorbed by the roots of living trees, these ions leach into the clay, where they may be retained.

Temperate deciduous forests of the northeastern and middle eastern United States are dominated by broad-leaf hardwood trees, such as oak, hickory, and beech, that lose their foliage annually (Fig. 51–7). In southern reaches, the

FIGURE 51–6 Temperate rain forest is characterized by high amounts of precipitation. Note the epiphytes hanging from the branches of coniferous trees. (Terry Donnelly/Dembinsky Photo Associates)

FIGURE 51–7 The broad-leaf trees that dominate the temperate deciduous forest will shed their leaves before winter. (Dennis Drenner)

number of broad-leaf evergreen trees, such as magnolia, increases.

Temperate deciduous forest originally contained a variety of large mammals such as puma, wolves, bison, and other species now extinct, plus deer, bears, many small mammals, and birds. Both reptiles and amphibians abounded, together with a denser and more varied insect life than exists today.

As in Europe, much of the original temperate deciduous forest of North America was removed by logging and land clearing. Where it has been allowed to regenerate, temperate deciduous forest is often in a seminatural state—that is, highly modified by humans.

Worldwide, temperate deciduous forest was among the first biomes to be converted to agricultural use. In Europe and Asia, for example, many soils that originally supported temperate deciduous forest have been cultivated by traditional agricultural methods for thousands of years without substantial loss in fertility. During the 20th century, however, intensive agricultural practices resulting in the degradation of perhaps 15% to 20% of Earth's total agricultural land have been widely adopted. Most damage to farmland has been done since the end of World War II.

Grasslands occur in temperate areas of moderate precipitation

Summers are hot, winters are cold, and rainfall is often uncertain in **temperate grasslands.** Annual precipitation averages 25 to 75 centimeters (10–30 inches). In grasslands with less precipitation, minerals tend to accumulate in a marked layer just below the topsoil. These minerals tend to wash out of the soil in areas with more precipitation. Grassland soil contains considerable organic material because above-ground portions of many grasses die off each winter and contribute to the organic content of the soil, while the roots and rhizomes (underground stems) survive underground. Also, many grasses are sod formers: their roots and rhizomes form a thick, continuous underground mat.

The North American Midwest is an excellent example of a moist temperate grassland. Although few trees grow except near rivers and streams, grasses grow in great profusion in the thick, rich soil (Fig. 51–8). Formerly, certain species of grass grew as tall as a person on horseback, and the land was covered with herds of grazing animals, particularly bison. The principal predators were wolves, although in sparser, drier areas their place was taken by coyotes. Smaller fauna included prairie dogs and their predators (foxes, black-footed ferrets, and various birds of prey), grouse, reptiles (such as snakes and lizards), and great numbers of insects.

The North American grassland was so well suited to agriculture that little of it remains. Almost nowhere can we see even an approximation of what our ancestors experienced in the Midwest. It is not surprising that the American Midwest, the Ukraine, and other moist temperate grasslands became the breadbaskets of the world because they provide ideal growing conditions for crops such as corn and wheat, which are also grasses.

FIGURE 51–8 Moist temperate grasslands are mostly treeless but contain a profusion of grasses and other herbaceous flowering plants. The photo shows a relatively moist, tall-grass prairie in Minnesota. (Visuals Unlimited/Ann B. Swengel)

Chaparral is a thicket of evergreen shrubs and small trees

Some temperate environments have mild winters with abundant rainfall, combined with extremely dry summers. Such Mediterranean climates, as they are called, occur not only in the area around the Mediterranean Sea but also in California, western Australia, portions of Chile, and South Africa. In the North American Southwest, this environment is known as **chaparral.** Chaparral soil is thin and infertile. Frequent fires occur naturally in this environment, particularly in late summer and autumn (see *Focus On: Wildfires* in Chapter 54).

Chaparral vegetation looks strikingly similar in different areas of the world, even though the individual species are quite different. Chaparral is usually dominated by a dense growth of evergreen shrubs and may contain drought-resistant pine or scrub oak trees (Fig. 51–9). During the rainy winter season the environment may be lush and green, but the plants lie dormant during the hot, dry summer. Trees and shrubs often have hard, small, leathery leaves that resist water loss. Many plants are also specifically fire-adapted and grow best in the months following a fire. Such growth is possible because fire releases minerals that were tied up in above-ground parts of plants that burned. The underground parts of some plants and the seeds of many others are not destroyed by the fire, however, and with the new availability of essential minerals, plants sprout vigorously during winter rains. Mule deer, wood rats, chipmunks,

FIGURE 51–9 Chaparral, which consists primarily of drought-resistant evergreen shrubs, develops where hot, dry summers alternate with mild, rainy winters. Shown is chaparral in the Santa Monica Mountains, California. (Visuals Unlimited/John Cunningham)

lizards, and many species of birds are common animals of the chaparral.

Fires that occur at irregular intervals in California chaparral vegetation are often quite costly because they consume expensive homes built on the hilly chaparral landscape. Unfortunately, efforts to control naturally occurring fires sometimes backfire. Denser, thicker vegetation tends to accumulate when periodic fires are prevented; then, when a fire does occur, it is much more severe. Removing the chaparral vegetation, whose roots hold the soil in place, can also cause problems: witness the mud slides that sometimes occur during winter rains in these areas.

Deserts are arid ecosystems

Deserts are dry areas found in both temperate and tropical regions. The low water content of the desert atmosphere leads to daily temperature extremes of heat and cold, so that a major change in temperature occurs in a single 24-hour period. Deserts vary greatly depending on the amount of precipitation they receive, which is generally less than 25 centimeters (10 inches) per year. A few deserts are so dry that virtually no plant life occurs in them. As a result of sparse vegetation, desert soil is low in organic material but often high in mineral content. In some regions, the concentration of certain soil minerals reaches toxic levels.

Desert vegetation includes both perennials and, after a rain, annuals. Plants in North American deserts include cacti, yuccas, Joshua trees, and widely scattered bunchgrass (Fig. 51–10). Desert plants tend to have reduced leaves or no leaves, an adaptation that conserves water.

FIGURE 51–10 Desert inhabitants are strikingly adapted to the demands of their environment. The warmer deserts of North America characterized by summer rainfall frequently contain large cacti such as the giant saguaro, which grows 15 to 18 meters (50–60 feet) tall. Photosynthesis is carried out by the stem, which also expands accordion-style to store water. Leaves are modified into spines, which discourage herbivores. (Willard Clay/Dembinsky Photo Associates)

Other desert plants shed their leaves for most of the year, growing only during the brief moist season. Desert plants are noted for **allelopathy,** an adaptation in which toxic substances secreted by roots or shed leaves inhibit the establishment of competing plants nearby.[2] Many desert plants possess defensive spines, thorns, or toxins to resist the heavy grazing pressure often experienced in this food- and water-deficient environment.

Desert animals tend to be small. During the heat of the day they remain under cover or return to shelter periodically, whereas at night they come out to forage or

[2]Allelopathy causes uniform dispersion; see Figure 52–2.

hunt. In addition to desert-adapted insects, there are many specialized desert reptiles (such as lizards, tortoises, and snakes) and a few desert-adapted amphibians (frogs and toads). Mammals include such rodents as the American kangaroo rat, which does not need to drink water but can subsist solely on the water content of its food (primarily seeds and insects). American deserts are also home to jack-rabbits, while Australian deserts have kangaroos. Carnivores like the African fennec fox and some birds of prey, especially owls, live on rodents and rabbits. During the driest months of the year, many desert insects, amphibians, reptiles, and mammals tunnel underground, where they remain inactive; this period of dormancy is known as *aestivation.*

American deserts have been altered by humans in several ways. Off-road vehicles damage desert vegetation, which sometimes takes years to recover, and certain cacti and desert tortoises are rare as a result of poaching. Also, houses, factories, and farms built in desert areas require vast quantities of water, which must be imported from distant areas.

Savanna is a tropical grassland with scattered trees

The **savanna** biome is a tropical grassland with widely scattered clumps of low trees (Fig. 51–11). Savanna is found in areas of low or seasonal rainfall with prolonged dry periods. Temperatures in savannas vary little throughout the year, and seasons are regulated by precipitation, not by temperature as they are in temperate grasslands. Annual precipitation is 85 to 150 centimeters (34–60 inches). Savanna soil is low in essential mineral nutrients, in part because the parent rock from which it is formed is infertile. Savanna soil is often rich in aluminum, and in places the aluminum reaches levels that are toxic to many plants. Although the African savanna is best known, savanna also occurs in South America, western India, and northern Australia.

Savanna is characterized by wide expanses of grasses interrupted by occasional trees such as *Acacia,* which bristles with thorns that provide protection against herbivores. Both trees and grasses have fire-adapted features such as extensive underground root systems; these enable them to survive the periodic fires that sweep through savanna.

The world's greatest assemblage of hoofed mammals occurs in the African savanna. Here live great herds of herbivores, including wildebeests, antelopes, giraffes, zebras, and elephants. Large predators, such as lions and hyenas, kill and scavenge the herds. In areas of seasonally varying rainfall, the herds and their predators may migrate annually.

Savanna is rapidly being converted to rangeland for cattle and other domesticated animals, which are replacing the big herds of wild animals. In some places severe overgrazing has converted marginal savanna to desert.

Tropical rain forests are lush equatorial forests

Tropical rain forests occur where temperatures are warm throughout the year and precipitation occurs almost daily. The annual precipitation of tropical rain forests is 200 to 450 centimeters (80–180 inches). Much of this precipitation comes from locally recycled water that enters the atmosphere by transpiration (see Chapter 32) of the forest's own trees.

Tropical rain forests are often located in areas with ancient, highly weathered, mineral-poor soil. Little organic

FIGURE 51–11 Savanna is tropical grassland with widely scattered trees. Savanna such as that found in eastern Africa formerly supported large herds of grazing animals and their predators; these are swiftly vanishing under pressure from pastoral and agricultural land use. (Frans Lanting/Minden Pictures)

(a)

(b)

FIGURE 51–12 Except at riverbanks, tropical rain forest has a closed canopy that admits little light to the forest floor. (***a***) A broad view of tropical rainforest vegetation along a riverbank. (***b***) Tropical rainforest trees typically possess buttress roots that support them in the shallow, often wet soil. (*a*, Frans Lanting/Minden Pictures; *b*, Richard W. Pippen)

matter accumulates in such soils. Because temperatures are high year round, decay organisms and detritus-feeding ants and termites decompose organic litter quite rapidly. Mineral nutrients from decomposing materials are quickly absorbed by roots. Thus, minerals of tropical rain forests are tied up in the vegetation rather than in the soil.

Tropical rain forest is very productive despite the scarcity of mineral nutrients in the soil. Its plants capture a lot of energy by photosynthesis, stimulated by abundant solar energy and precipitation.

Of all the biomes, the tropical rain forest is unexcelled in species diversity. No single species dominates. A person can often travel for 0.4 kilometer (0.25 mile) without encountering two members of the same tree species.

Trees of tropical rain forests are usually evergreen flowering plants (Fig. 51–12*a*). Their roots are often shallow and concentrated near the surface in a mat only a few centimeters (an inch or so) thick. The root mat catches and absorbs almost all mineral nutrients released from leaves and litter by decay processes. Swollen bases or braces called *buttresses* hold the trees upright and aid in the extensive distribution of the shallow roots (Fig. 51–12*b*).

A fully developed rain forest has at least three distinct stories of vegetation. The topmost story consists of the crowns of occasional very tall trees, some 50 meters (164 feet) or more in height, that are exposed to direct sunlight. The middle story, which reaches a height of 30 to 40 meters (100–130 feet), forms a continuous canopy of leaves overhead that lets in little sunlight for the support of the sparse understory. The understory itself consists of both smaller plants specialized for life in shade, and seedlings of taller trees. Vegetation of tropical rain forests is not dense at ground level except near stream banks or where a fallen tree has opened the canopy.

Tropical rainforest trees support extensive epiphytic communities of smaller plants such as orchids and bromeliads. Although epiphytes grow in crotches of branches, on bark, or even on the leaves of their hosts, they only use their host trees for physical support, not for nourishment.

Because little light penetrates to the understory, many plants living there are adapted to climb already-established host trees rather than to invest their meager photosynthetic resources in the dead cellulose tissues of their own trunks. Lianas (woody tropical vines), some as thick as a human thigh, twist up through the branches of tropical rainforest trees (see *Making the Connection: Vines, Evolutionary Adaptations, and Ecology* in Chapter 33).

Rainforest animals include the most abundant and varied insect, reptile, and amphibian fauna on Earth. Birds, too, are varied and often brilliantly colored. Most rainforest mammals, such as sloths and monkeys, live only in the trees, although some large ground-dwelling mammals, including Asian elephants, are also found in rain forests.

Unless strong conservation measures are initiated soon, human population growth and industrialization in tropical countries will spell the end of tropical rain forests by the middle of the next century. Biologists know that many rainforest organisms will become extinct before they have even been scientifically described. The ecological impacts of tropical rainforest destruction are discussed in Chapter 55.

AQUATIC ECOSYSTEMS OCCUPY MOST OF EARTH'S SURFACE

Not surprisingly, aquatic life zones are different in almost all respects from terrestrial ecosystems. For example, in biomes, temperature and precipitation are major determinants of plant and animal inhabitants, and light is relatively plentiful (except in certain environments such as the rainforest floor). Significant environmental factors in aquatic ecosystems are quite different. Temperature is often less important in watery environments, and water is obviously never an important limiting factor.

Salinity (the concentration of dissolved salts, such as sodium chloride, in a body of water) affects the kinds of organisms present in aquatic ecosystems, as does the amount of dissolved oxygen. Water greatly interferes with the penetration of light, so floating aquatic organisms that photosynthesize must remain near the water's surface, and vegetation attached to the bottom can grow only in shallow water. In addition, low levels of essential mineral nutrients often limit the number and distribution of organisms in certain aquatic environments.

Aquatic ecosystems contain three main ecological categories of organisms: free-floating plankton, strongly swimming nekton, and bottom-dwelling benthos. **Plankton** are usually small or microscopic organisms that are relatively feeble swimmers. For the most part, they are carried about at the mercy of currents and waves. They are unable to swim far horizontally, but some species are capable of large vertical migrations, and are found at different depths of water at different times of the day or at different seasons. Plankton are generally subdivided into two major categories: phytoplankton and zooplankton. **Phytoplankton** (photosynthetic cyanobacteria and free-floating algae) are producers that form the base of most aquatic food webs. **Zooplankton** are nonphotosynthetic organisms that include protozoa and the larval stages of many animals.

Nekton are larger, more strongly swimming organisms such as fishes, turtles, and whales. **Benthos** are bottom-dwelling organisms that fix themselves to one spot (oysters and barnacles), burrow into the sand (many worms and echinoderms), or simply walk about on the bottom (lobsters and brittle stars).

Freshwater Ecosystems Are Important Ecologically

Freshwater ecosystems include rivers and streams (flowing-water ecosystems), lakes and ponds (standing-water ecosystems), and marshes and swamps (freshwater wetlands). Each type of freshwater ecosystem is distinguished by its own specific environmental conditions and characteristic organisms.

Although freshwater ecosystems occupy a relatively small portion of Earth's surface, they have an important role in the hydrologic cycle: they assist in recycling precipitation that flows as surface runoff to the ocean (see Chapter 54). Large bodies of fresh water also help moderate daily and seasonal temperature fluctuations on nearby land.

Rivers and smaller streams are flowing-water ecosystems

Many different conditions exist along the length of a river or stream. The nature of a **flowing-water ecosystem** changes greatly from its source (where it begins) to its mouth (where it empties into another body of water). For example, headwater streams (small streams that are the sources of a river) are usually shallow, cold, swiftly flowing, and therefore highly oxygenated. In contrast, rivers downstream from the headwaters are wider and deeper,

FIGURE 51–13 As a river's course levels out downstream from the headwaters, it flows more slowly and winds from side to side. An aerial view of the Mara River in Kenya shows many bends, called meanders. (Altitude [Y. Arthus-B.]/ Peter Arnold, Inc.)

not as cold, slower-flowing, and therefore less oxygenated (Fig. 51–13).

The kinds of organisms found in flowing-water ecosystems vary greatly from one stream to another, depending primarily on the strength of the current. In streams with fast currents, inhabitants may have adaptations such as suckers to attach themselves to rocks so they are not swept away, or they may have flattened bodies that enable them to slip under or between rocks. Organisms in large, slow-moving streams and rivers do not need such adaptations, although they are typically streamlined (as are most aquatic organisms) to lessen resistance during movement through water. Where current is slow, plants and animals of the headwaters are replaced by those characteristic of ponds and lakes.

Unlike other freshwater ecosystems, streams and rivers depend on land for much of their energy. In headwater streams, for example, up to 99% of the energy input comes from detritus (dead organic material such as leaves) carried from the land into streams and rivers by wind or surface drainage. Downstream, rivers contain more producers, and therefore have a slightly lower dependence on detritus as a source of energy than do the headwaters.

Lakes and ponds are standing-water ecosystems

Standing-water ecosystems are characterized by zonation. A large lake has three basic layers or zones: the littoral, limnetic, and profundal zones (Fig. 51–14). Smaller lakes and ponds typically lack a profundal zone. The **littoral zone** is a shallow water area along the shore of a lake or pond. It includes emergent vegetation, such as cattails and burreeds, plus several deeper-dwelling aquatic plants and algae. The littoral zone is the most highly pro-

(a)

FIGURE 51–14 A lake is a standing-water ecosystem surrounded by land. (***a***) A satellite mosaic of part of northeastern North America shows the five Great Lakes. From left to right, they are Lake Superior, Lake Michigan, Lake Huron, Lake Erie, and Lake Ontario. (***b***) Zonation in a large lake. (*a*, U. S. Geological Survey)

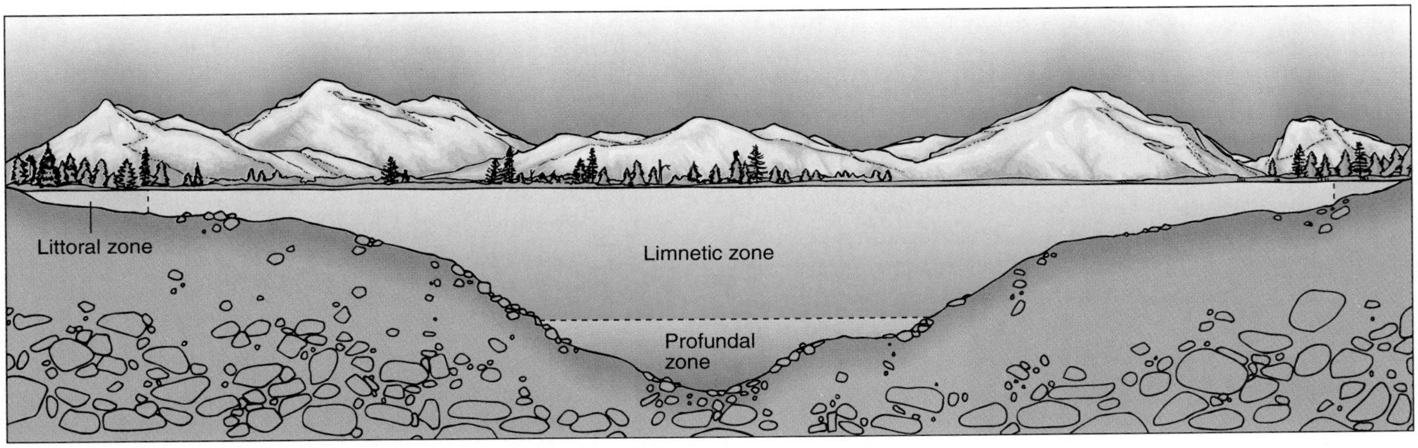

(b)

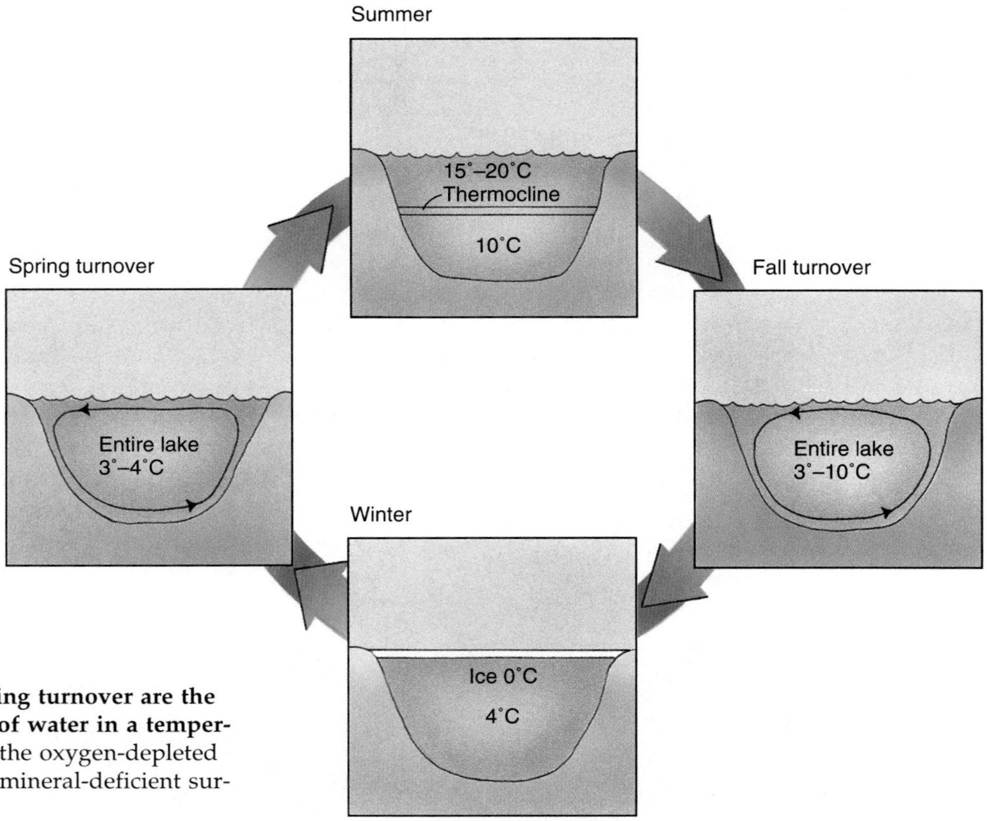

FIGURE 51–15 Fall turnover and spring turnover are the mixing of the upper and lower layers of water in a temperate lake. This mixing brings oxygen to the oxygen-depleted depths of the lake, and minerals to the mineral-deficient surface waters.

ductive zone of the lake—that is, photosynthesis is greatest here—in part because it receives nutrient inputs from surrounding land that stimulate the growth of plants and algae. Animals of the littoral zone include frogs and their tadpoles, turtles, worms, crayfish and other crustaceans, insect larvae, and many fishes such as perch, carp, and bass.

The limnetic zone is the open water away from the shore; it extends down as far as sunlight penetrates. The main organisms of the limnetic zone are microscopic phytoplankton and zooplankton. Larger fishes also spend some of their time in the limnetic zone, although they may visit the littoral zone to feed and reproduce. Owing to its depth, less vegetation grows here than in the littoral zone.

Beneath the limnetic zone of a large lake is the profundal zone. Because no light penetrates to this depth, plants and algae do not live here. Food drifts into the profundal zone from the littoral and limnetic zones. Bacteria decompose dead plants and animals that reach the profundal zone, liberating minerals. These minerals are not effectively recycled because no photosynthetic organisms are present to absorb them and incorporate them into the food web. As a result, the profundal zone tends to be both mineral-rich and oxygen-deficient, with few forms of life occupying it other than anaerobic bacteria.

Thermal stratification occurs in temperate lakes

The marked layering of large temperate lakes caused by light penetration is accentuated by **thermal stratification,** in which the temperature changes sharply with depth. Thermal stratification occurs because the summer sunlight penetrates and warms surface waters, making them less dense.[3] In summer, cool (and therefore more dense) water remains at the lake bottom and is separated from warm (and therefore less dense) water above by an abrupt temperature transition called the **thermocline.**

In temperate lakes, falling temperatures in autumn cause a mixing of the lake waters called the **fall turnover** (Fig. 51–15). (Since little seasonal temperature variation occurs in the tropics, such turnovers are uncommon there.) As surface water cools, its density increases and eventually it displaces the less dense, warmer, mineral-rich water beneath. Warmer water then rises to the surface where it, in turn, cools and sinks. This process of cooling and sinking continues until the lake reaches a uniform temperature throughout.

When winter comes, surface water cools to below 4°C, its temperature of greatest density, and, if it is cold

[3]Recall that the density of water is greatest at 4°C; both above and below this temperature, water is less dense.

enough, ice forms. Ice, which forms at 0°C, is less dense than cold water; thus, ice forms on the surface, and the water on the lake bottom is warmer than the ice.

In the spring, a **spring turnover** occurs as ice melts and surface water reaches 4°C. Surface water again sinks to the bottom, and bottom water returns to the surface. As summer arrives, thermal stratification occurs once again.

The mixing of deeper, nutrient-rich water with surface, nutrient-poor water during the fall and spring turnovers brings essential minerals to the surface. The sudden presence of large amounts of essential minerals in surface waters encourages the development of large algal populations, which form temporary blooms in the fall and spring.

Freshwater wetlands are transitional between aquatic and terrestrial ecosystems

Freshwater wetlands, usually covered by shallow water for at least part of the year, have characteristic soils and water-tolerant vegetation. They include marshes, in which grasslike plants dominate, and swamps, in which woody trees or shrubs dominate (Fig. 51–16). Freshwater wetlands also include hardwood bottomland forests (lowlands along streams and rivers that are periodically flooded), prairie potholes (small, shallow ponds that formed when glacial ice melted at the end of the last Ice Age), and peat moss bogs (peat-accumulating wetlands where mosses dominate).

Wetland plants, which are highly productive, provide enough food to support a wide variety of organisms. Wetlands are valued as a wildlife habitat for migratory waterfowl and many other bird species, beaver, otters, muskrats, and game fishes. Wetlands help control flooding by acting as holding areas for excess water when rivers flood their banks. The floodwater stored in wetlands then drains slowly back into the rivers, providing a steady flow of water throughout the year. Wetlands also serve as groundwater recharging areas. One of their most important roles is to help cleanse and purify water by trapping and holding pollutants in the flooded soil.

At one time wetlands were thought of as wastelands, areas that needed to be filled in or drained so that farms, housing developments, and industrial plants could be built on them. Wetlands are also breeding places for mosquitoes and therefore were viewed as a menace to public health. Today, however, the crucial environmental services that wetlands provide are widely recognized, and wetlands are somewhat protected by law.

Estuaries occur where fresh water and saltwater meet

Where the ocean meets the land there may be one of several kinds of ecosystems: a rocky shore, a sandy beach, an intertidal mud flat, or a tidal estuary. An **estuary** is a coastal body of water, partly surrounded by land, with access to the open ocean and a large supply of fresh water from rivers. It usually contains **salt marshes,** shallow swampy areas dominated by grasses in which tides and precipitation cause salinity to fluctuate between that of seawater and that of fresh water (Fig. 51–17). Many estuaries undergo marked variations in temperature, salinity, and other physical properties in the course of a year. To survive there, estuarine organisms must have a wide tolerance to such changes.

FIGURE 51–16 Freshwater swamps are inland areas permanently saturated or even covered by water and dominated by trees. Shown is a cypress swamp in Water Spring, South Carolina. (Visuals Unlimited/Milton H. Tierney, Jr.)

FIGURE 51–17 Cordgrass (*Spartina*) is the dominant vegetation in a salt marsh at Assateague Island National Seashore, Maryland. (Connie Toops)

Estuaries are among the most fertile ecosystems in the world, often having much greater productivity than the adjacent ocean or freshwater river. This high productivity is brought about by (1) the action of tides, which promote a rapid circulation of nutrients and help remove waste products; (2) the transport of nutrients from land into rivers and creeks that empty into the estuary; and (3) the presence of many plants, which provide an extensive photosynthetic carpet and whose roots and stems also mechanically trap much potential food material. Plants decay after they die, forming the basis of detritus food webs. Most commercially important fin fishes and shellfish spend their larval stages in estuaries among the protective tangle of decaying stems.

Salt marshes have often appeared to be worthless, empty stretches of land to uninformed people. As a result, they have been used as dumps, becoming severely polluted, or have been filled with dredged bottom material to form artificial land for residential and industrial development. A large part of the estuarine environment has been lost in this way, along with many of its benefits: wildlife habitat, sediment and pollution trapping, and flood control.

Mangrove forests, the tropical equivalent of salt marshes, cover perhaps 70% of tropical coastlines. Like salt marshes, mangrove forests provide valuable environmental services. They are breeding grounds and nurseries for several commercially important fish species and their roots stabilize the soil, thereby preventing coastal erosion.

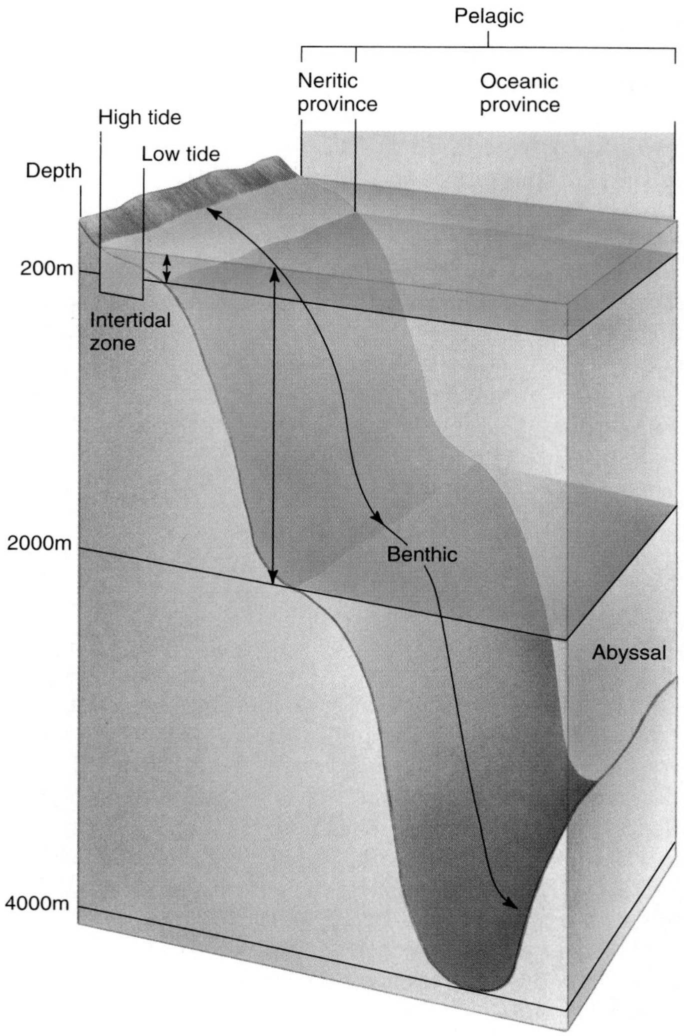

FIGURE 51–18 The ocean has three main life zones: the intertidal zone, the benthic environment, and the pelagic environment. The pelagic environment consists of the neritic province and the oceanic province.

MARINE ECOSYSTEMS DOMINATE EARTH'S SURFACE

Although oceans and lakes are comparable in many ways, they have many differences. Depths of even the deepest lakes do not approach those of the oceanic abysses, which are extremely deep areas that extend more than 6 kilometers (3.6 miles) below the sunlit surface. Oceans are profoundly influenced by tides and currents. Gravitational pulls of both sun and moon produce two tides a day throughout the oceans, but the height of those tides varies with the phases of the moon (a full moon causes the highest tides), the season, and the local topography.

The immense marine environment is subdivided into several zones: the intertidal zone, the benthic (ocean floor) environment, and the pelagic (ocean water) environment (Fig. 51–18). The pelagic environment is in turn divided into two provinces—the neritic province and the oceanic province.

The intertidal zone is transitional between land and ocean

The shoreline area between low and high tide is called the **intertidal zone.** Although high levels of light and nutrients, together with an abundance of oxygen, make the intertidal zone a biologically productive environment, it is also a stressful one. If an intertidal beach is sandy, inhabitants must contend with a constantly shifting environment that threatens to engulf them and gives them scant protection against wave action. Consequently, most sand-dwelling organisms, such as mole crabs, are continual and active burrowers. Because they are able to follow the tides up and down the beach, most do not have any notable adaptations to survive drying or exposure.

A rocky shore provides a fine anchorage for seaweeds and animals. However, it is exposed to constant wave action when immersed during high tides, and to drying and temperature changes when exposed to the air during low tides. A typical rocky-shore inhabitant has some way of sealing in moisture, perhaps by closing its shell, if it has one, plus a powerful means of anchoring itself to rocks. Mussels, for example, have horny, threadlike anchors,

FIGURE 51–19 Sea palms (*Postelsia*), sturdy seaweeds adapted to life in a rocky intertidal zone, are exposed at low tide. The sea palm is common on the rocky Pacific coast from Vancouver Island to California. Its base is firmly attached to the rocky substrate, enabling it to withstand pounding wave action. (François Gohier/Photo Researchers, Inc.)

and barnacles have special cement glands. Rocky-shore intertidal algae (seaweeds) usually have thick, gummy polysaccharide coats, which dry out slowly when exposed, and flexible bodies not easily broken by wave action (Fig. 51–19). Some rocky-shore community inhabitants hide in burrows or crevices at low tide.

Coral reefs are an important part of the benthic environment

The benthic environment is the ocean floor. It consists mostly of sand and mud in which many marine animals, such as worms and clams, burrow. Bacteria are common in marine sediments, but until recently it was assumed that bacteria did not extend very far into the sediments. In 1994 in the journal *Nature,* however, living bacteria were reported found in ocean sediments more than 500 meters (1625 feet) below the ocean floor at several different sites in the Pacific Ocean.

Particularly productive benthic communities in shallow ocean waters include kelp forests, seaweed beds, and coral reefs. We restrict our discussion here to coral reefs.

The living portions of coral reefs must grow in shallow waters where light penetrates. Many coral reefs are composed principally of red coralline algae that require light for photosynthesis. Coral animals also require light for the large number of symbiotic dinoflagellates, known as **zooxanthellae,** that live and photosynthesize in their tissues (see Chapters 28 and 53). (Although species of coral without zooxanthellae exist, only those with zooxanthellae build reefs.) In addition to obtaining food from the zooxanthellae living inside them, coral animals capture food at night by using their stinging tentacles to paralyze small animals that drift nearby (see chapter introduction photo).

Coral reefs grow slowly, as coral organisms build on the calcareous remains of countless organisms before them. The basic pattern of coral reef development was first outlined by the famous naturalist Charles Darwin (best known for his theory of natural selection; see Chapter 17). Coral reefs start in warm, shallow water—for example, directly attached to the shore of a volcanic island—to form a **fringing reef** (Fig. 51–20). After the volcano becomes inactive, it slowly erodes and sinks back into the ocean. As this occurs, the channel between the island and the reef widens, and the reef becomes a **barrier reef.** A barrier reef is therefore separated from the nearby land by open water. If the volcano erodes or sinks completely below the water level, the outcome is an **atoll,** a circular coral reef surrounding a central lagoon of quiet water. An atoll is the final stage in the development of coral reefs associated with ocean volcanoes.

fringing reef → barrier reef → atoll

The waters in which coral reefs are found are often poor in nutrients. Other factors are favorable for high productivity, however, including the presence of zooxanthellae, favorable temperature, and plenty of sunlight.

Coral reef ecosystems contain hundreds of species of fishes and invertebrates, such as giant clams, sea urchins, sea stars, sponges, brittle stars, sea fans, and shrimp. Many are brightly colored to advertise the fact that they are poisonous. The conspicuous lion fish, for example, defends itself from potential predators with spiny projections that contain a powerful poison (Fig. 51–21*a*).

The complex multitude of relationships and interactions that occur at coral reefs seems comparable only to the tropical rain forest among terrestrial ecosystems. As in the rain forest, however, competition is intense, particularly for light and space to grow.

Many unusual relationships occur at coral reefs. Certain tiny fishes, for example, swim inside the mouths of larger fishes and remove potentially harmful parasites. Fishes sometimes line up at these cleaning stations, waiting their turn to be serviced. As another example, *Podillopora* corals have a remarkable defense mechanism that protects them from coral browsers such as the crown-of-thorns sea star (Fig. 51–21*b*). Tiny crabs live among the branches of the coral. When a sea star crawls onto the coral, the crabs swarm over it, nipping off its tube feet and eventually killing it if it does not promptly retreat! (Since the late 1960s, population explosions of the crown-of-thorns sea stars have occurred on many coral reefs, causing extensive devastation of the coral. It is not known what causes these outbreaks, nor is it known if human factors such as pollution play a part.)

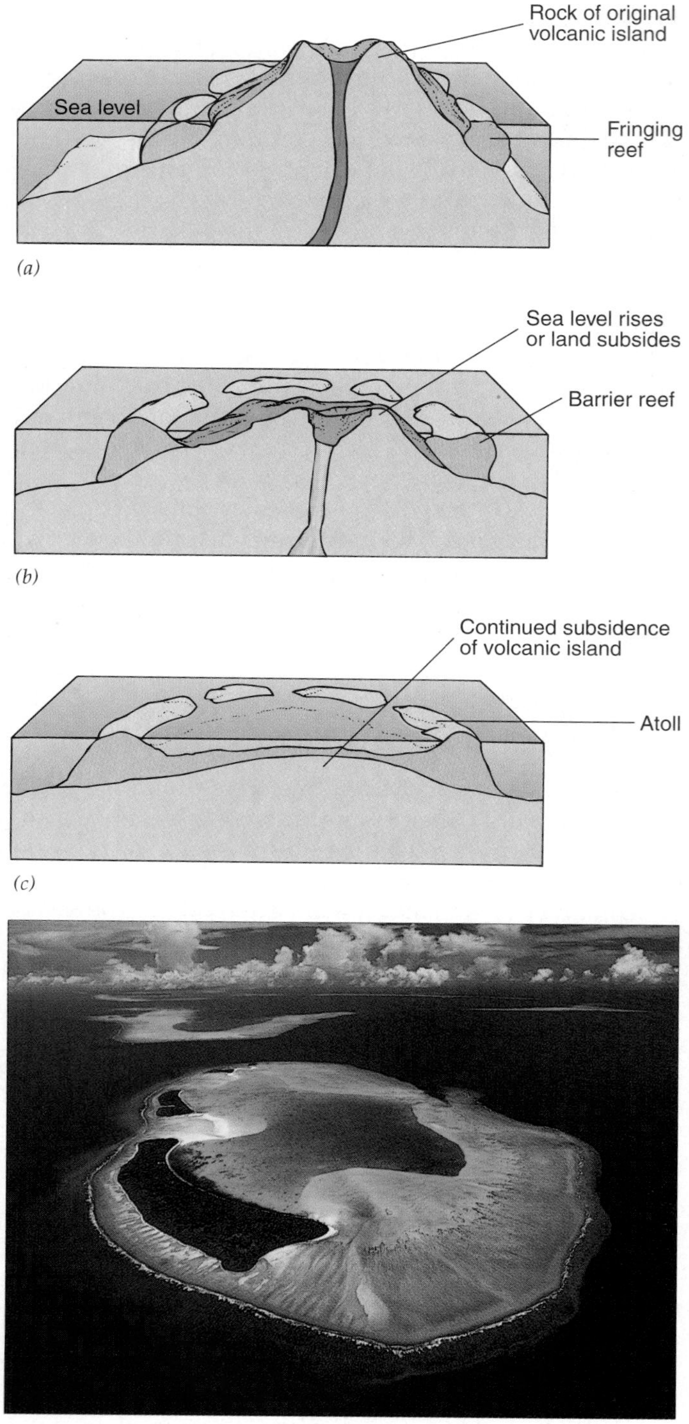

FIGURE 51–20 Coral colonies form reefs that change over time. (***a***) A fringing reef grows in the shallow water around the shores of a volcanic island. (***b***) Over time, as the coral continues to grow, the island sinks or the sea level rises, forming a barrier reef. (***c***) More time passes, and the island sinks below the surface, leaving an atoll. (***d***) The Kayangel Atoll (Belau) in the western Pacific Ocean is a classic example of an atoll. Note the ring-shaped coral island that surrounds a central lagoon. (Douglas Faulkner/Photo Researchers, Inc.)

FIGURE 51–21 Many animals live around coral reefs. (***a***) The bold candy-stripe pattern of the lion fish (*Pterois volitans*) advertises its poisonous dorsal spines to potential predators. (***b***) In recent decades, the crown-of-thorns sea star (*Acanthaster planci*) has killed large sections of coral reefs. The crown-of-thorns sea star is widely distributed in the Indian and Pacific Oceans. Photo was taken in the Red Sea. (*a*, Karl & Jill Wallin/FPG International; *b*, SharkSong/M. Kazmers/Dembinsky Photo Associates)

The neritic province consists of shallow waters close to shore

The **neritic province** is open ocean that overlies the ocean floor from the shoreline to a depth of 200 meters (650 feet). Organisms that live in the neritic province are all floaters or swimmers. The upper reaches of the neritic province make up the **euphotic region,** which extends from the surface to a depth of approximately 100 meters. Light penetrates the euphotic zone in sufficient amounts to support photosynthesis. Large numbers of phytoplankton, particularly diatoms in cooler waters and dinoflagellates in warmer waters, produce food by photosynthesis and are thus the base of food webs. Small zooplankton (in-

FIGURE 51–22 The huge manta ray, which makes its home in the neritic province, looks more fearsome than it is. It swims in the open ocean, peacefully grazing on plankton and small fishes. A full-grown manta ray can have a "wingspan" of 5.5 meters (more than 18 feet) and weigh 1.4 metric tons (3000 pounds). This manta ray was photographed in the Pacific Ocean off the coast of Mexico. Note the two remoras clinging to its body. Remoras feed on scraps of food left by their host. (Visuals Unlimited/Alex Kerstitch)

cluding tiny crustaceans, jellyfish, comb jellies, protists such as foraminiferans, and larvae of barnacles, sea urchins, worms, and crabs) feed on phytoplankton.

Zooplankton are consumed by plankton-eating nekton such as herring, sardines, squid, baleen whales, and manta rays (Fig. 51–22); these in turn become prey for carnivorous nekton such as sharks, tuna, porpoises, and toothed whales. Nekton are mostly confined to the shallower neritic waters (fewer than 60 meters, or 195 feet, deep) because that is where their food is.

The oceanic province comprises most of the ocean

The average depth of the world's oceans is 4000 meters (2.4 miles). The **oceanic province** is that part of the open ocean that overlies the ocean floor at depths greater than 200 meters. It is the largest marine environment, comprising about 75% of the ocean's waters. The oceanic province is characterized by cold temperatures, high hydrostatic pressure, and an absence of sunlight; these environmental conditions are uniform throughout the year.

Most organisms of the oceanic province depend on **marine snow,** organic debris that drifts down into the **aphotic** ("without light") **region** from the upper, lighted regions. Organisms of this little-known realm are filter-feeders, scavengers, or predators. Many are gelatinous invertebrates, some of which attain great sizes. Fishes of the oceanic province are strikingly adapted to darkness and food scarcity. For example, the rare gulper eel has huge jaws that enable it to swallow large prey (Fig. 51–23). (An organism that encounters food infrequently needs to eat as much as possible when it has the chance.) Many animals of the oceanic province have illuminated organs that enable them to see one another for mating or to capture food (see Fig. 6–13).

Ecosystems interact with one another

We have discussed the various terrestrial biomes and aquatic ecosystems as distinct and separate entities, but they intergrade with one another at their boundaries. The transition zone where two ecosystems or biomes meet and intergrade is called an **ecotone.** Ecotones range in size from quite small, such as the area where an agricultural field meets a woodland, to continental in scope. For example, at the border between tundra and taiga, an extensive ecotone occurs that consists of tundra vegetation interspersed with small, scattered conifers. Such ecotones provide habitat diversity and are often populated by a greater variety and density of organisms than either adjacent ecosystem.

Ecologists who study ecotones look for adaptations that enable organisms to survive there. They also examine the relationship between biological diversity and ecotones, and how ecotones change over time. Long-term studies of ecotones have revealed that they are far from static. The ecotone boundary between desert and semiarid grassland in southern New Mexico, for example, has moved northward during the past 50 years as the desert ecosystem has expanded into the grassland.

FIGURE 51–23 The gulper eel, a resident of the oceanic province, uses its "trap-door" jaws to swallow prey larger than itself! The tail (not visible) makes up most of the length of a gulper eel's body. Gulper eels grow to 1.8 meters (6 feet). (Courtesy of Bruce H. Robison, Monterey Bay Aquarium)

SUMMARY

I. Ecology is the study of all relationships involving organisms and their abiotic environment.
   A. A population is all the members of a particular species that live together in the same area.
   B. A community is all the populations of different species living in the same area. An ecosystem is a community and its environment.
   C. The biosphere is all the communities on Earth. The ecosphere, the largest ecosystem on Earth, comprises the interactions among the biosphere, atmosphere, lithosphere, and hydrosphere.

II. A biome is a large, relatively distinct terrestrial region with characteristic climate, soil, plants, and animals. Because it is so large in area, a biome encompasses a number of interacting ecosystems.
   A. Tundra, the northernmost biome, is characterized by a frozen layer of subsoil (permafrost), and low-growing vegetation that is adapted to extreme cold and a short growing season.
   B. The taiga, or boreal forest, lies south of the tundra and is dominated by coniferous trees.
   C. Temperate rain forest, such as occurs on the northwest coast of North America, receives high precipitation and is dominated by large conifers.
   D. Temperate deciduous forest, which occurs where precipitation is relatively high, is dominated by broad-leaf trees that lose their leaves seasonally.
   E. Temperate grassland typically possesses a deep, mineral-rich soil and has moderate but uncertain precipitation.
   F. The chaparral is characterized by thickets of small-leaf shrubs and trees and a climate of wet, mild winters and dry summers.
   G. Deserts, found in both temperate and tropical areas with low levels of precipitation, possess communities whose organisms have specialized water-conserving adaptations.
   H. Tropical grassland, called savanna, has widely scattered trees interspersed with grassy areas. It occurs in tropical areas with low or seasonal rainfall.
   I. Tropical rain forest is characterized by mineral-poor soil and high rainfall that is evenly distributed throughout the year. Tropical rain forest has a high species diversity.

III. In aquatic ecosystems, important environmental factors include salinity, the amount of dissolved oxygen, and the availability of light for photosynthesis.
   A. Aquatic life is ecologically divided into plankton (free-floating), nekton (strongly swimming), and benthos (bottom-dwelling).
   B. Phytoplankton are photosynthetic algae and cyanobacteria that form the base of the food web in most aquatic communities.

IV. Freshwater ecosystems include flowing water (rivers and smaller streams), standing water (lakes and ponds), and freshwater wetlands.
   A. In flowing-water ecosystems the water flows in a current. Flowing-water ecosystems have few phytoplankton and depend on detritus from the land for much of their energy. The organisms in flowing-water ecosystems vary greatly depending on the current, which is swifter in headwaters than downstream.
   B. Freshwater lakes are divided into zones on the basis of water depth. The marginal littoral zone contains emergent vegetation and heavy growths of algae. The limnetic zone is open water away from the shore. The deep, dark profundal zone holds little life other than decomposers.
   C. Freshwater wetlands, lands that are transitional between freshwater and terrestrial ecosystems, are usually covered by shallow water and have characteristic soils and vegetation.

V. An estuary is a coastal body of water, partly surrounded by land, with access to the ocean and a large supply of fresh water from rivers.
   A. Estuaries are very productive, in part because they receive a high input of nutrients from the adjacent land.
   B. One of the important roles of estuaries is to provide a nursery for the young of many aquatic organisms.

VI. Four important marine environments are the intertidal zone, the benthic environment, the neritic province, and the oceanic province.
   A. The intertidal zone is the shoreline area between low and high tide. Organisms of the intertidal zone possess adaptations to resist wave action and the extremes of being covered by water (high tide) and exposed to air (low tide).
   B. The benthic environment is the ocean floor. Coral reefs are particularly important benthic communities in shallow ocean waters. Coral reefs, which have high species diversity and high productivity, are classified as fringing reefs, barrier reefs, and atolls.
   C. The neritic province is open ocean from the shoreline to a depth of 200 meters (650 feet). Organisms that live in the neritic province are all floaters or swimmers. Phytoplankton in the euphotic zone are the base of the food web.
   D. The oceanic province is that part of the open ocean that is deeper than 200 meters. The uniform environment is one of darkness, cold temperature, and high pressure. Animal inhabitants of the oceanic province are either predators, or scavengers that subsist on detritus that drifts in from other areas of the ocean.

SELECTED KEY TERMS

abiotic environment
allelopathy
aphotic region
atoll
barrier reef
benthos
biome
biosphere
boreal forest
chaparral
community
deciduous
desert
ecology
ecosphere
ecosystem

(Selected Key Terms continued on next page)

ecotone
estuary
euphotic region
fall turnover
flowing-water ecosystem
freshwater wetlands
fringing reef
intertidal zone
limnetic zone
littoral zone
mangrove forest
marine snow
nekton
neritic province
oceanic province
permafrost
phytoplankton
plankton
population
profundal zone
salinity
salt marsh
savanna
spring turnover
standing-water ecosystem
taiga
temperate deciduous forest
temperate grassland
temperate rain forest
thermal stratification
thermocline
tropical rain forest
tundra
zooplankton
zooxanthellae

POST-TEST

1. ________ is the branch of biology that studies the relationships among organisms, and between organisms and their environment.
2. All members of the same species that live together in the same place are called a(an) ________.
3. All the populations of different species interacting in an area compose a(an) ________.
4. A(an) ________ encompasses the interactions between a community and its environment.
5. The northernmost biome, ________, typically has little precipitation, a short growing season, and permafrost.
6. South of tundra is the ________, which consists of coniferous forests with many lakes.
7. ________ ________ forests of the northeastern and middle eastern United States are dominated by broad-leaf hardwood trees that lose their foliage annually.
8. The thickest, richest soil in the world occurs in temperate ________.
9. ________ is a thicket of evergreen shrubs and small trees that is found in areas with Mediterranean climates.
10. ________ are arid ecosystems found in both temperate and tropical regions.
11. The ________ is a tropical grassland interspersed with widely spaced trees.
12. The ________ ________ forest is the biome with the highest species diversity.
13. Organisms in aquatic environments fall into three categories: free-floating ________, strongly swimming ________, and bottom-dwelling ________.
14. Temperate-zone lakes are thermally stratified, with warm and cold layers separated by a transitional ________.
15. Emergent vegetation grows in the ________ zone of freshwater lakes.
16. A(an) ________ is a coastal body of water with access to the open ocean and a large supply of fresh water from rivers.
17. The ________ zone is the ocean shoreline between low and high tide.
18. The first stage in coral reef development, the ________ ________, is directly attached to the shore of a volcanic island.
19. The ________ province is open ocean from the shoreline to a depth of 200 meters.
20. The inhabitants of the dark ________ province live on an input of dead organisms from other marine environments.
21. The transition zone where two ecosystems or biomes meet and intergrade is called a(an) ________.

REVIEW QUESTIONS

1. List and define the levels of biological organization that ecologists study.
2. What climate and soil factors produce the major biomes?
3. Describe representative organisms of these forest biomes: (a) taiga; (b) temperate deciduous forest; (c) temperate rain forest; (d) tropical rain forest.
4. In which biome do you live? If your biome does not match the description given in this text, explain the discrepancy.
5. Compare tundra with desert, and temperate grassland with savanna.
6. Distinguish among plankton, nekton, and benthos, giving examples of each.
7. What environmental factors are most important in determining the adaptations of organisms that live in aquatic environments?
8. Distinguish between freshwater wetlands and estuaries, and between flowing-water and standing-water ecosystems.
9. List and briefly describe the four main marine environments.
10. Which aquatic ecosystem is often compared to tropical rain forests? Why?

YOU MAKE THE CONNECTION

1. In which biomes would migration be most common? Hibernation? Aestivation? Explain your answers.
2. Develop a hypothesis to explain why there are no amphibians or reptiles in the tundra. How would you test your hypothesis?
3. Develop a hypothesis to explain why animals adapted to the desert are usually small. How would you test your hypothesis?
4. Why do most of the animals of the tropical rain forest live in trees?
5. What would happen to the organisms in a river with a fast current if a dam were built? Explain your answer.

RECOMMENDED READINGS

Broad, W.J. "Fantastic World Found in Sea's Middle Depths." *New York Times,* September 26, 1994. The Monterey Bay Aquarium Research Institute is exploring the little-known realm of life in the ocean's dark middle depths.

Chadwick, D.H. "Roots of the Sky." *National Geographic,* Vol. 184, No. 4, October 1993. Wildlife flourishes in the patchy remnants of the North American prairie.

Goulding, M. "Flooded Forests of the Amazon." *Scientific American,* Vol. 268, No. 3, March 1993. During the rainy season, vast sections of the Amazonian rain forest are inundated, making them a uniquely aquatic as well as terrestrial ecosystem. Many of the organisms living here possess special adaptations that enable them to survive in such a changing environment.

Johnson, B.L., W.B. Richardson, and T.J. Naimo. "Past, Present, and Future Concepts in Large River Ecology." *BioScience,* Vol. 45, No. 3, March 1995. Provides an overview of what is known about large river ecology. This issue of *BioScience* contains a number of articles about flowing-water ecosystems.

Kappel-Smith, D. "Fickle Desert Blooms: Opulent One Year, No-shows the Next." *Smithsonian,* March 1995. Vegetation varies widely in deserts from year to year, depending on local conditions.

Levine, J.S. "Dusk and Dawn Are Rush Hours on the Coral Reef." *Smithsonian,* October 1993. Examines the night life of a coral reef.

"Life in the Deep." *Discover,* November 1993. A photo essay of several unusual organisms adapted to various parts of the marine environment.

McClanahan, L.L., R. Ruibal, and V.H. Shoemaker. "Frogs and Toads in the Desert." *Scientific American,* Vol. 270, No. 3, March 1994. A few amphibians have adapted to survive dry climates.

Moffett, M. "These Plants Claw and Strangle their Way to the Top." *Smithsonian,* September 1993. Many unusual vines are found in tropical forests.

CHAPTER 52

Population Ecology

A population of Mexican poppies blooms in the desert after the winter rains. (Jim Brandenburg/Minden Pictures)

Individuals are part of a larger organization—a group composed of members of the same species that live together in the same area at the same time. Such a group is called a **population.** Populations exhibit characteristics distinctive from those of the individuals of which they are composed. Some of the traits discussed in this chapter that are unique to populations are birth and death rates, growth rates, population density, sex ratios, and age structure.

Although communities are composed of all the populations of all the different species that live together within an area, populations have properties that communities lack. Communities, for example, do not share a common gene pool (see Chapter 18). As a result, there is no obvious way for natural selection to produce adaptive change at the community level. Instead, gene frequency changes resulting from natural selection occur in the populations that make up communities.

Population ecology deals with the number of individuals of a particular species that are found in an area, and how and why those numbers change (or remain fixed) over time. Population ecologists try to determine the population processes common to all populations. They study how a population responds to its environment—that is, to competition for resources, predation, and other environmental pressures. Population growth cannot increase indefinitely because of such environmental controls.

Additional aspects of populations that interest biologists are their biological success or failure (extinction), their evolution, their population genetics, and how they affect the normal functioning of communities and ecosystems. Biologists in applied disciplines must understand populations in order to manage forests, field crops, game, fishes, and other populations of economic importance.

In this chapter we focus on the dynamics of population ecology characteristic of all organisms. We then extend our discussion to humans because the same parameters that affect other populations also affect the human population.

LEARNING OBJECTIVES

After you have studied this chapter you should be able to

1. Define what is meant by a population, and discuss what population ecology entails.
2. Define population density and dispersion, and describe the main types of population dispersion.
3. List and explain the four factors that produce changes in population size.
4. Define biotic potential and carrying capacity, and explain the differences between J-shaped and S-shaped growth curves.
5. Distinguish between density-dependent and density-independent factors, and give examples of each.
6. Distinguish between *K* strategies and *r* strategies, and give an example of each.
7. Describe Type I, Type II, and Type III survivorship curves, and explain which curve corresponds to *r* selection and which to *K* selection.
8. Summarize the history of human population growth.
9. Explain how developed and developing countries differ in population characteristics such as infant mortality rate, total fertility rate, replacement-level fertility, and age structure.

DENSITY AND DISPERSION ARE IMPORTANT FEATURES OF POPULATIONS

By itself, the size of a population tells us relatively little. Population size is more meaningful when the boundaries of that population are defined. Consider, for example, the difference between a thousand mice in 100 hectares (250 acres) and a thousand mice in one hectare (2.5 acres).

Sometimes a population is too large to study in its entirety. Such a population is examined by sampling a part of it and then expressing the population in terms of density—as, for example, the number of grass plants per square meter or the number of cabbage aphids per square inch of cabbage leaf. **Population density,** then, is the number of individuals of a species per unit of area or volume at a given time.

Different environments vary in the population density of any species of organism that they can support. This density may also vary in a single habitat from season to season or year to year. As an example, consider red grouse populations in northwest Scotland at two locations only 2.5 kilometers (1.5 miles) apart. At one location the population density remained stable during a three-year period, but at the other it almost doubled in two years and then declined to its initial density once again. The reason was likely a difference in habitat. The area where the population density increased had been experimentally burned, and young plant growth produced after the burn was beneficial to the red grouse. So population density is determined in large part by external factors in the environment.

The individuals comprising the population often exhibit characteristic patterns of **dispersion,** or spacing. Individuals may be dispersed uniformly (evenly spaced), randomly, or clumped into clusters.

Uniform dispersion occurs when individuals are more evenly spaced than would be expected from a random occupation of a given habitat (Fig. 52–1*a*). A nesting colony of seabirds, in which the birds place their nests at a more-or-less equal distance from each other, is an ex-

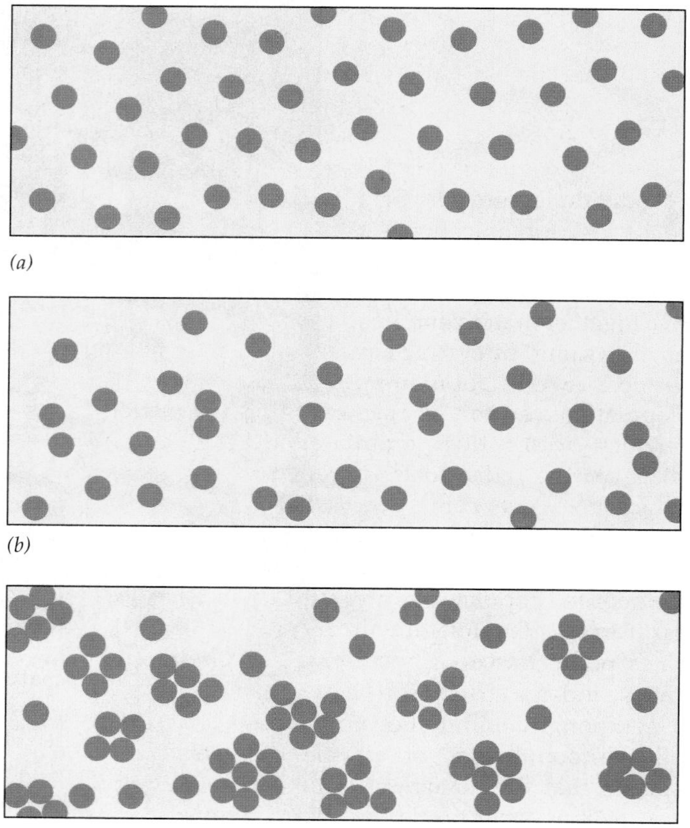

(*a*) (*b*) (*c*)

FIGURE 52–1 Individuals of a population may have different patterns of dispersion in the habitat. (*a*) Uniform dispersion is usually caused by negative, or antagonistic, interactions among individuals of a population. (*b*) Random dispersion, which is not common in nature, occurs when individuals distribute themselves with no regard for the positions of other individuals in the population. (*c*) Clumped dispersion is usually the result of mutual attractions among individuals within a population.

FIGURE 52–2 The uniform dispersion of certain desert plants such as the creosote bush and the saltbush is probably due to allelopathy, the production of toxic substances that inhibit the growth of plants nearby. (William E. Ferguson)

ample of uniform dispersion. In this case, uniform dispersion occurs as a result of antagonistic interactions among the nesting birds. (Each pair places its nest just beyond the reach of nearby nesting birds.) Uniform dispersion also occurs when competition among individuals is severe, when they wage chemical warfare on one another by allelopathy (Fig. 52–2), or when they exhibit territorial behavior.

Random dispersion occurs when individuals in a population are spaced in an unpredictable or random way that is unrelated to the presence of others (Fig. 52–1*b*). Of the three major types of dispersion, random dispersion is least common and hardest to observe. Random dispersion may occur infrequently because important environmental factors affecting dispersion usually do not occur at random. Flour beetle larvae in a container of flour are randomly dispersed, for example, but their environment is unusually homogeneous.

Perhaps the most common spacing is **clumped dispersion** (also called **aggregated distribution**), which occurs when individuals are concentrated in specific portions of the habitat (Fig. 52–1*c*). Clumped dispersion often results from the patchy distribution of resources. It is also caused in animals by the presence of family groups and pairs, and in plants by limited seed dispersal or by asexual reproduction. An entire grove of aspen trees, for example, may originate asexually from a single plant (see Fig. 35–21). Clumped dispersion may sometimes be advantageous, as social animals derive many benefits from their association.

MATHEMATICAL MODELS DESCRIBE POPULATION GROWTH

Part of the progression of scientific knowledge is the discovery of common threads or patterns among separate observations. As mentioned previously, population ecologists wish to understand general processes that are shared by many different populations, so they develop models, mathematical equations that represent population dynamics. No population model is a perfect representation of all populations, but models do serve to illuminate complex processes. Moreover, mathematical modeling enhances the scientific process by providing a framework with which experimental population studies can be compared. As more knowledge accumulates, the model is refined and made more precise.

Populations of organisms, whether they are sunflowers, elephants, or humans, change over time. On a *global* scale, this change is due to two factors: **natality,** the rate at which organisms produce offspring (the birth rate), and **mortality,** the rate at which organisms die (the death rate). In humans the birth rate is usually expressed as the number of births per 1000 people per year, and the death rate as the number of deaths per 1000 people per year.

Population growth is a special case of population change. To determine the rate of change, we must also take into account the time interval involved—that is, the change in time. To express these changes, we employ the Greek letter *delta* (Δ):

$$\Delta N/\Delta t = b - d$$

where ΔN is the change in the number of individuals in the population, Δt the change in time, b the natality, and d the mortality. Thus, the rate of change, or **growth rate** (r) of a population is the birth rate minus the death rate ($r = b - d$).

A modification of this equation tells us the rate at which the population is growing at a particular instant in time (rather than the average growth rate during the entire period of study)—that is, its instantaneous growth rate ($dN/\Delta d$).[1] Using differential calculus, this growth rate can be expressed as

$$dN/dt = rN$$

where N is the number of individuals in the existing population, t the time, and r the growth rate.

Because $r = b - d$, if organisms in the population are born faster than they die, r is a positive value and population size increases. If organisms in the population die faster than they are born, r is a negative value and population size decreases. If r is equal to zero, births and deaths match, and population size is stable despite continued reproduction.

Migration affects the growth rate in some populations

In addition to the number of births and deaths, migration (movement from one region or country to another)

[1]The symbol dN is the differential of N; it is not a product, nor should the d in dN be confused with the death rate, d. The same applies to dt.

Time	Number of bacteria
0	1
20 min	2
40 min	4
1 hour	8
1 hr, 20 min	16
1 hr, 40 min	32
2 hours	64
2 hr, 20 min	128
2 hr, 40 min	256
3 hours	512
3 hr, 20 min	1,024
3 hr, 40 min	2,048
4 hours	4,096
4 hr, 20 min	8,192
4 hr, 40 min	16,384
5 hours	32,768
5 hr, 20 min	65,536
5 hr, 40 min	131,072
6 hours	262,144
6 hr, 20 min	524,288
6 hr, 40 min	1,048,576
7 hours	2,097,152
7 hr, 20 min	4,194,304
7 hr, 40 min	8,388,608
8 hours	16,777,216
8 hr, 20 min	33,554,432
8 hr, 40 min	67,108,864
9 hours	134,217,728
9 hr, 20 min	268,435,456
9 hr, 40 min	536,870,912
10 hours	1,073,741,824

(a)

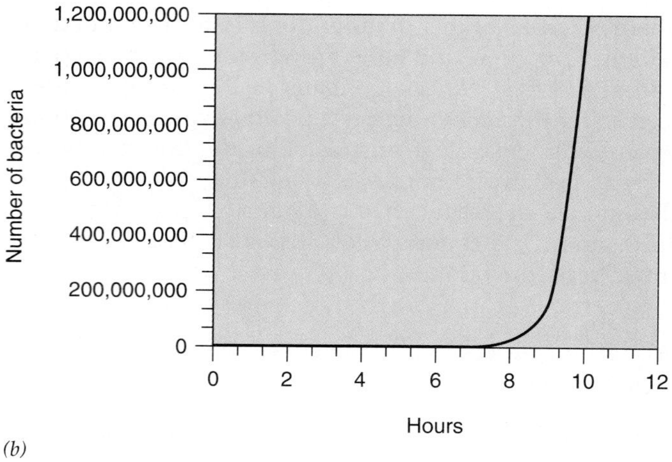

(b)

FIGURE 52–3 Exponential growth has a characteristic J-shaped curve when plotted. (***a***) When bacteria divide every 20 minutes, their numbers increase exponentially. This projection assumes a zero death rate, but even if a certain percentage of each generation of bacteria died, exponential growth would still occur; it would just take longer to reach the very high numbers. (***b***) When these data are graphed, the curve of exponential growth has a J shape. The ideal conditions under which bacteria or other organisms reproduce exponentially rarely occur in nature, and when they do, they are of short duration.

must be considered when examining changes in populations on a *local* scale. There are two types of migration: immigration and emigration. **Immigration** occurs when individuals enter a population and thus increase its size. **Emigration** occurs when individuals leave a population and thus decrease its size. The growth rate of a local population must take into account birth rate (b), death rate (d), immigration (i), and emigration (e). The growth rate is equal to the value of the birth rate minus the death rate, plus the value of immigration minus emigration:

$$r = (b - d) + (i - e)$$

Each population has a characteristic biotic potential

The maximum rate at which an organism or population could increase under ideal conditions is known as its **biotic potential** (r_{max}). Different species have different biotic potentials. A particular species' biotic potential is influenced by several factors, including the age at which reproduction begins, the percentage of the lifespan during which the organism is capable of reproducing, and the number of offspring produced during each period of reproduction. These factors, called life history characteristics, determine whether a particular species has a large or small biotic potential.

Generally, larger organisms such as blue whales and elephants have the smallest biotic potentials, whereas microorganisms have the greatest biotic potentials. Under ideal conditions (an environment with unlimited resources), certain bacteria can reproduce by splitting in half every 20 minutes. At this rate of growth, a single bacterium would increase to a population of more than one billion in just 10 hours!

If we plot the population number versus time, the graph has a J shape that is characteristic of **exponential growth,** the constant growth rate that occurs under optimal conditions (Fig. 52–3). When a population grows exponentially, the larger that population gets, the faster it grows.

Regardless of which organism we are considering, whenever the population is growing at its biotic potential, population number plotted versus time gives a curve of the same shape; the only variable is time. That is, it may take longer for an elephant population than for a bacterial population to reach a certain size, but both populations will always increase exponentially under ideal conditions.

No population can increase exponentially indefinitely

Certain populations may exhibit exponential growth for short periods of time. Exponential growth has been experimentally demonstrated in bacterial and protist cultures and in certain insects. However, organisms cannot reproduce indefinitely at their biotic potential because the environment sets limits, which are collectively called **environmental resistance.** Using the preceding example, bacteria would never be able to reproduce unchecked for an extended period of time because they would run out of food and living space, and poisonous wastes would accumulate in their vicinity. With crowding, they would also become more susceptible to parasites and predators. As their environment changed, their birth rate (b) would decline and their death rate (d) would increase due to shortages of food, increased predation, increased competition, and other environmental stresses. Conditions might worsen to a point where d would exceed b and the population would decrease. The number of organisms in a population, then, is controlled by the ability of the environment to support them. As N increases, so does environmental resistance.

Over longer periods of time, the rate of population growth for most organisms decreases to around zero. This leveling out occurs at or near the limits of the environment to support the population. The **carrying capacity (*K*)** represents the largest population that can be maintained for an indefinite period of time by a particular environment.

When a population regulated by environmental resistance is graphed over longer periods of time, the curve has a characteristic S shape that shows the population's initial exponential increase (note the curve's J shape at the start, when environmental resistance is low), followed by a leveling out as the carrying capacity of the environment is approached (Fig. 52–4). The S-shaped growth curve can be modeled by a modified growth equation called a logistic equation. It describes a population increasing from a small number of individuals to a larger number of individuals, limited by a finite environment. The logistic equation takes the carrying capacity of the environment into account.[2]

$$dN/dt = rN\,[(K - N)/K]$$

[2]The logistic model of population growth was developed to explain population growth in continually breeding populations. Similar models exist for populations that have specific breeding seasons.

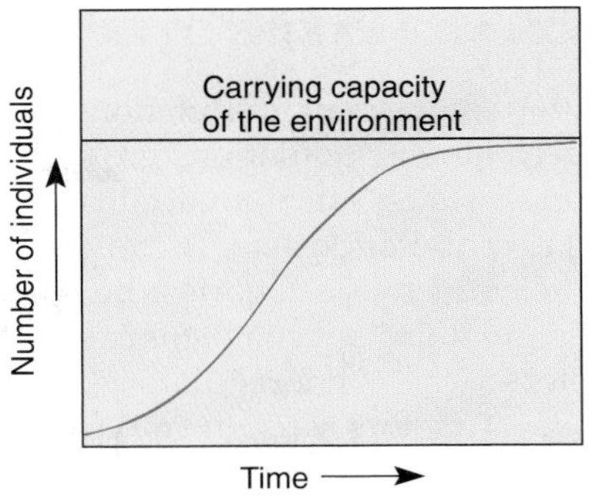

FIGURE 52–4 In many laboratory studies, exponential population growth slows as the carrying capacity (*K*) of the environment is approached. The logistic model of population growth, when graphed, has a characteristic S-shaped curve.

When the number of organisms (N) is small, the rate of population growth is unchecked by the environment, because the expression $[(K - N)/K]$ has a value of almost 1. But as the population (N) begins to approach the carrying capacity (K), the growth rate falls nearly to zero, despite high population density, because $[(K - N)/K]$ approaches 0 in value.

Although the S curve is a simplification of actual population changes over time, it appears to fit many populations that have been studied in the laboratory, as well as a few studied in nature. For example, G.F. Gause, a Russian ecologist who conducted experiments during the 1930s, grew a population of *Paramecium caudatum* in a test tube. He supplied a constant but limited amount of food daily and replenished the growth medium occasionally to eliminate the buildup of metabolic wastes. Under these conditions, the population of *P. caudatum* increased exponentially at first. The paramecia became so numerous that the water was cloudy with them. But then their growth rate declined, and the population leveled off.

A population does not always stabilize at K, but may temporarily rise far higher. It may then experience a **population crash,** an abrupt decline from high to low population density. Such an abrupt change is commonly observed in bacterial cultures and other populations whose resources are nonrenewable or inflexible. When the vital resources are consumed, the organisms die. In 1910 a small herd of 26 reindeer was introduced on one of the Pribilof Islands of Alaska. The herd's population increased exponentially for about 25 years until there were approximately 2000 reindeer, many more than the island could support. The reindeer overgrazed the vegetation until the plant life was almost wiped out. Then, in slightly over a decade, as reindeer died from starvation, the number of reindeer plunged to eight, one third the size of the original introduced population.

Population size varies over time

The observation that small populations tend to grow rapidly, whereas larger populations grow less rapidly or even decline in size, suggests that certain mechanisms operate to regulate population size. Factors that affect population size fall into two categories: density-dependent factors (the "regulatory" mechanisms) and density-independent factors (the "nonregulatory" mechanisms). These two sets of factors vary in importance from one species to another, and in most cases probably interact in complex ways to determine the size of a population.

Density-dependent factors regulate population size

If a change in population density alters how an environmental factor affects a population, then the environmental factor is said to be a **density-dependent factor.** As population density increases, density-dependent factors tend to slow population growth by causing an increase in death rate and/or a decrease in birth rate. The effect of these density-dependent factors on population growth becomes stronger as the population density increases; that is, *density-dependent factors affect a larger proportion (not just a larger number) of the population.* Density-dependent factors can also exert an effect on population growth when population density declines; a decrease in population density results in enhancement of population growth by density-dependent factors. These factors either decrease the death rate and/or increase the birth rate. Density-dependent factors, then, tend to cause a population to maintain itself at a relatively constant size that is near the carrying capacity of the environment.

Predation, disease, and competition are examples of density-dependent factors. As the density of a population increases, for example, predators are more likely to find an individual of a given prey species. Also, when population density is high, the members of a population encounter one another more frequently, and the chance of their transmitting infectious disease organisms increases.

Most studies of density dependence have been conducted in laboratory settings where all density-dependent factors except one are controlled experimentally. But populations in natural settings are exposed to a complex set of variables that continually change. As a result, in natural communities it can be difficult to evaluate the relative effects of different density-dependent factors. For example, it has been noted that on tropical islands inhabited by lizards, few spiders occur, whereas more spiders and more species of spiders are found on lizard-free islands. Deciding to study these observations experimentally, researchers staked out plots of vegetation (mainly sea grape shrubs) and enclosed some of them with lizard-proof screens (Fig. 52–5). Some of the plots were emptied of all lizards, and web-building spiders were introduced into all the plots. At the end of a two-year study period, spider population densities averaged 2.5 times higher in the lizard-free enclosures than in enclosures with lizards. Moreover, the species diversity of spiders was higher in the lizard-free areas. Even the results of this relatively simple experiment may be explained by a combination of two density-dependent factors: predation (lizards eat spiders) and competition (lizards compete with spiders for insect prey). In this case, the effects of the two factors in determining spider population size cannot be evaluated separately.

FIGURE 52–5 Spiller and Schoener, two ecologists at the University of California at Davis, tested the effect of the presence of lizards on spider population density on Bahamian islands. Shown is one of the enclosures used for their two-year study. (Courtesy of Thomas W. Schoener)

Competition is an important density-dependent factor

As population density increases, so does competition for resources such as living space, food, cover, water, minerals, and sunlight. Eventually, it may reach the point where many members of a population fail to obtain the minimum of whatever resource is in shortest supply. This raises the death rate and/or lowers the birth rate, inhibiting further population growth.

Competition occurs both *within* a given population (**intraspecific competition**) and *among* populations of different species (**interspecific competition**). We consider the effects of intraspecific competition here; interspecific competition is discussed in Chapter 53.

Individuals of the same species compete for a resource in limited supply by scramble competition or by contest competition. In **scramble competition** all the individuals in a population "share" the limited resource equally, so that at high population densities none of them obtains an adequate amount. In **contest competition** certain dominant individuals obtain an adequate supply of the limited resource at the expense of other individuals in the population. The populations of species in which scram-

Making the Connection

Predator-Prey Dynamics and Population Genetics

Are there any real-life examples of population genetics influencing predator-prey dynamics? As discussed in this chapter, during the early 1900s a small herd of moose wandered across the ice of frozen Lake Superior to an island, Isle Royale. In the ensuing years, until 1949, moose became successfully established there, although the population experienced oscillations. The story continues, beginning in 1949, when a few Canadian wolves wandered across the frozen lake and discovered abundant moose prey on the island. The wolves also remained and became established.

Moose have been the wolves' primary winter prey on Isle Royale. Wolves hunt in packs by encircling a moose and trying to get it to run so they can attack it from behind. (The moose is the wolf's largest, most dangerous prey. A standing moose is more dangerous than one that is running because when standing, it can kick and slash its attackers with its hooves.)

Since 1958, wildlife biologists have studied the populations of moose and wolves on Isle Royale (see figure). Not unexpectedly, they found that the two populations fluctuated over the years. Generally, as the population of wolves increased, the population of moose declined. Wolves tended to reduce the population of moose, but they did not eliminate them from the island. Studies indicate that wolves primarily feed on the very old and very young in the moose population. Healthy moose in their peak reproductive years are not eaten.

Despite the fact that both populations appeared to be in a state of dynamic equilibrium, a new episode recently began in the Isle Royale story. The wolf population has plunged, from 50 animals in 1980 to 16 animals in 1995. As expected, the moose population is increasing; in 1995 there were 2424 moose on Isle Royale. Biologists think that the extreme genetic uniformity of the wolf population is one of the reasons for its decline. Genetic studies of the Isle Royale wolves indicate that they are all descended from the same female.

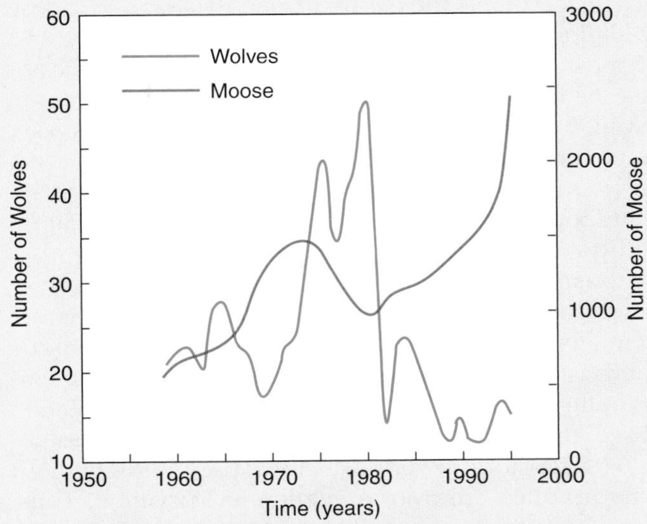

Fluctuations in the populations of wolves and moose on Isle Royale, 1959 to 1995. (Rolf O. Peterson, Michigan Technological University)

Populations that lack genetic variability cannot adapt well to changing environmental conditions. Such populations often have low reproductive success. In 1994, for example, only two wolf pups were born on the island. A genetically uniform population is also more susceptible to disease. Analysis of their blood revealed the presence of antibodies to canine parvovirus, indicating that the wolves have been exposed to this deadly disease.

It is not known if the wolf population on Isle Royale will recover or disappear altogether. Given their genetic uniformity, however, the outlook for their long-term survival is poor. ▲

ble competition operates often oscillate over time, and there is always a risk that the population size will drop to zero. In contrast, those species in which contest competition operates experience a relatively small drop in population size caused by the death of individuals that are unable to compete successfully.

The moose population on Isle Royale, the largest island in Lake Superior, provides a vivid example of scramble competition (Fig. 52–6). Isle Royale differs from most

FIGURE 52–6 A moose browses on vegetation along the shoreline of Isle Royale in Lake Superior. Individual moose compete with one another for food during the long, cold winter months. (Rolf O. Peterson, Michigan Technological University)

islands in that large mammals can walk to it when the lake freezes over in winter. The minimum distance to be walked is 24 kilometers (15 miles), however, so this has happened infrequently. Around 1900, some moose reached the island for the first time. By 1935 the moose population on the island had increased to about 3000 and had consumed almost all the edible vegetation. In the absence of this food resource, nearly 90% of the moose starved. By 1948 the population had recovered to its peak number once again, and again most of the moose starved. Thus, scramble competition for scarce resources can result in population oscillations rather than in population stability.

Intraspecific competition among red grouse involves contest competition. When red grouse populations are low, the birds are less aggressive, and most young birds establish a territory (an area defended against other members of the same species). However, when the population is large, the birds are much more aggressive, and establishing a territory is very difficult. Those birds without territories often die from predation or starvation. Thus, birds with territories use a larger share of the limited resource (the territory with its associated food and cover), whereas birds without territories cannot compete successfully.

Density-independent factors influence populations

Any environmental factor that affects the size of a population but is not influenced by changes in population density is called a **density-independent factor.** Random weather events that reduce population size serve as density-independent factors. These often affect population density in unpredictable ways. A severe blizzard, hurricane, or fire, for example, may cause extreme and irregular reductions in a vulnerable population and thus might be considered largely density-independent (Fig. 52–7).

Consider a density-independent factor that influences mosquito populations in arctic environments. These insects produce several generations per summer and achieve high population densities by the end of the season. A shortage of food does not seem to be a limiting factor for mosquitoes, nor is there any shortage of ponds in which to breed. What puts a stop to the skyrocketing mosquito population is winter. Not a single adult mosquito survives winter, and the entire population must grow afresh the next summer from the few eggs and hibernating larvae that survive. Thus, severe winter weather is a density-independent factor that affects arctic mosquito populations.

It is difficult to think of many density-independent limiting factors that bear absolutely *no* relationship to population density. Social animals, for example, are often able to resist dangerous weather conditions by means of their collective behavior, as in the case of sheep huddling together in a snowstorm. In this case it appears that the greater the population density, the better their ability to resist the environmental pressure of a snowstorm.

FIGURE 52–7 Forest fires are an example of a density-independent factor. Density-independent factors are not influenced by changes in population density. (Doug Sokell/Tom Stack & Associates)

Natural populations may fluctuate chaotically

An additional reason that populations fluctuate over time may be their inherent instability. Ecologists in recent years have tried to understand why, in the absence of any obvious density-dependent and density-independent factors, certain populations undergo wild and unpredictable changes. Some ecologists have turned to the mathematical theory of chaos to help explain the state of flux displayed by some populations. Chaos is the tendency of a simple system to exhibit complex, erratic dynamics. Researchers from the University of California at Davis published the results of a computer model of Dungeness crab populations in the journal *Science* in 1994. The assumptions that they built into the model were simple: (1) Adults produce many larvae, most of which die from competition and predation. (2) After reproducing, the adults die. (3) Most of the surviving larvae do not migrate far away. The computer model predicted the crab population for the next 20,000 years. Despite the simple assumptions, the model showed occasional wild fluctuations that were independent of any environmental factors (Fig. 52–8). It therefore appears that a certain amount of chaos may be inherent in population dynamics.

DIFFERENT SPECIES HAVE DIFFERENT REPRODUCTIVE STRATEGIES

Imagine an organism possessing the "perfect" life history strategy that ensures reproduction at its biotic potential. In other words, this hypothetical organism produces the

Table 52–1 A COMPARISON OF SELECTED FEATURES OF *r*-SELECTED AND *K*-SELECTED SPECIES

Feature	*r-Selected Species*	*K-Selected Species*
Development time to maturity	Short	Long
Body size	Small	Large
Lifespan	Short	Long
Number of offspring per reproductive event	Many	Few
Age at first reproductive event	Early	Late
Number of reproductive events per lifetime	Often only one	Often more than one
Parental care invested in offspring	No	Yes

maximum number of offspring, and the majority of these offspring survive to reproductive maturity. Such an organism would have to mature soon after it was born so that it could begin reproducing at an early age. It would have to reproduce frequently throughout its life and produce large numbers of offspring each time. Further, it would have to be able to care for its young in order to assure their survival.

In nature, such an organism does not exist because if an organism were to put all its energy into being a perfect "reproductive machine," it would not expend any energy toward ensuring its own survival. Animals must use energy to hunt for food, and plants need energy to grow taller than surrounding plants to obtain adequate sunlight. Natural selection, then, requires organisms to make trade-offs in the expenditure of energy. Organisms, if they are to be successful, must do what is required to survive, both as individuals and as a species.

Thus, each species has its own life history strategy—its own reproductive characteristics, body size, habitat requirements, migration patterns, and so on—that is a series of trade-offs reflecting this energy compromise. Although many different life histories exist, organisms often fall into two main groups with respect to their reproductive strategies: *r*-selected species and *K*-selected species.

Populations described by the concept of *r* **selection** have a strategy that has evolved through the natural selection of traits that lead to a high growth rate. (Recall that *r* designates the growth rate. Because such organisms have a high *r*, they are known as *r* **strategies** or ***r*-selected species.**) Small body size and large broods are typical of many *r* strategists (Table 52–1), which are usually opportunists found in variable, temporary, or unpredictable environments where the probability of long-term survival is minimal.

Some of the best examples of *r* strategists are common weeds such as the dandelion (Fig. 52–9*a*). Dandelions are propagated by parachute-like fruits containing little seeds that are scattered widely by any strong breeze or wind. In temperate climates, any suitable habitat can be colonized by dandelions as soon as it becomes available.

The seeds of some varieties of dandelions are produced by the asexual process of apomixis (see Chapter

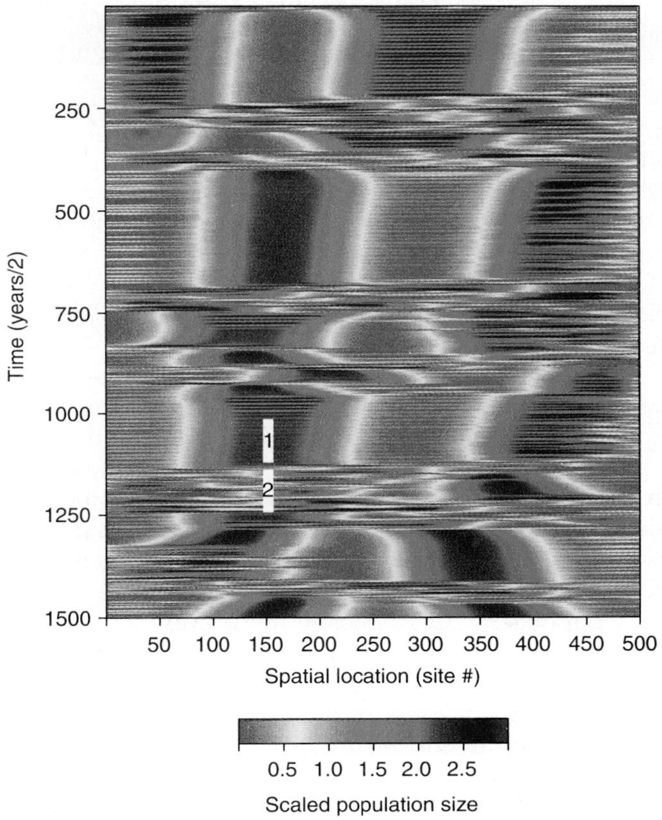

FIGURE 52–8 A computer model of population dynamics in a Dungeness crab population projected over 3,000 years shows periods of stability as well as occasional wild fluctuations. For example, examine site #150 over time by visually scanning from the top to the bottom of the figure above the #150 site location. Two short vertical lines have been designated 1 and 2 on the figure. The area of the first short line shows one time period during the simulation when the population at site #150 did not fluctuate appreciably. The second short line shows the same location during a later time period when the population fluctuated wildly. (Courtesy of Kevin Higgins and Alan Hastings)

35). After germinating, the young plant uses another form of asexual reproduction. The dandelion has a long taproot that splits longitudinally to form two roots, then four roots, and so on, so that natural clones of plants spread

(a)

(b)

FIGURE 52–9 Most organisms have one of two reproductive strategies, *r* selection or *K* selection. (***a***) Dandelions (*Taraxacum officinale*) are *r* strategists—annuals that mature early and produce many small seeds. A dandelion population fluctuates from year to year but rarely approaches the carrying capacity of its environment. (***b***) Tawny owls (*Strix aluco*) are *K* strategists: they maintain a fairly constant population size at or near the carrying capacity. They mature slowly, delay reproduction, and have a relatively large body size. (*a*, Marion Lobstein; *b*, Stephen Dalton/Photo Researchers, Inc.)

out from the original plant in genetically identical clumps.

Asexual reproduction has two key advantages for dandelions and some other *r* strategists. First, asexual reproduction in plants does not require insect pollinators or even other plants, both of which may be in short supply in unpredictable habitats. Second, as long as the particular kind of habitat for which an *r* strategist is adapted is common, asexual reproduction preserves a genotype adapted to that habitat.

In populations described by the concept of ***K* selection,** evolution has selected traits that maximize the chance of surviving in an environment where *N* is near *K*. (Recall that *K* is the carrying capacity of the environment. The population of a species with *K* selection is maintained at or near the carrying capacity, and as a result, its individuals have little need for a high reproductive rate.) These organisms, called ***K* strategists** or ***K*-selected species,** do not produce large numbers of offspring. They characteristically have long lifespans with slow development, late reproduction, large body size, and low reproductive rate (Table 52–1). Animals that are *K* strategists also invest in parental care of their young. *K* strategists are found in relatively constant or stable environments, where they have a high competitive ability.

Tawny owls (*Strix aluco*) are *K* strategists that pair-bond for life, with both members of a pair living and hunting in adjacent, well-defined territories (Fig. 52–9*b*). They regulate their reproduction in accordance with the resources, especially food, present in their territories. In an average year, 30% of the birds do not breed. If food supplies are more limited than initially indicated, many of those that do breed fail to incubate their eggs. Rarely do the owls lay the maximum number of eggs, and often they delay breeding until late in the season, when the rodent populations on which they depend have become large. Thus, tawny owls regulate their population behaviorally so that its number stays at or near the carrying capacity of the environment. Starvation, an indication that the tawny owl population has exceeded the carrying capacity, rarely occurs.

Survivorship is related to *r* and *K* selection

The various reproductive strategies that organisms possess, from *r* selection to *K* selection, are associated with different patterns of **survivorship,** the proportion of individuals in a population that survive to a particular age. Figure 52–10 is a graph of the three main survivorship

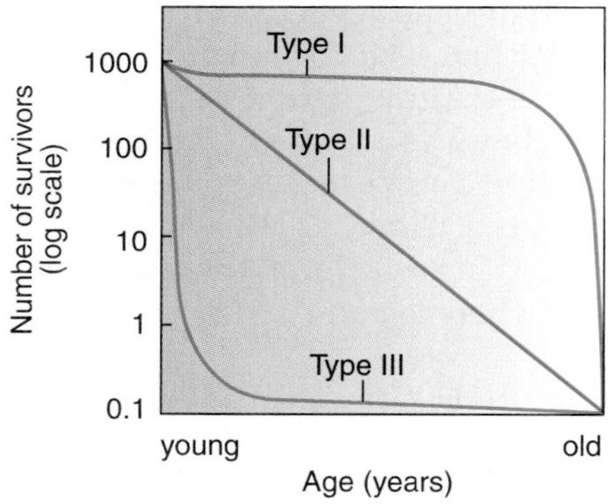

FIGURE 52–10 Survivorship curves represent the ideal survivorships of organisms in which mortality is greatest in old age (Type I), spread evenly across all age groups (Type II), and greatest among the young (Type III). The survivorship of most organisms can be compared to these curves.

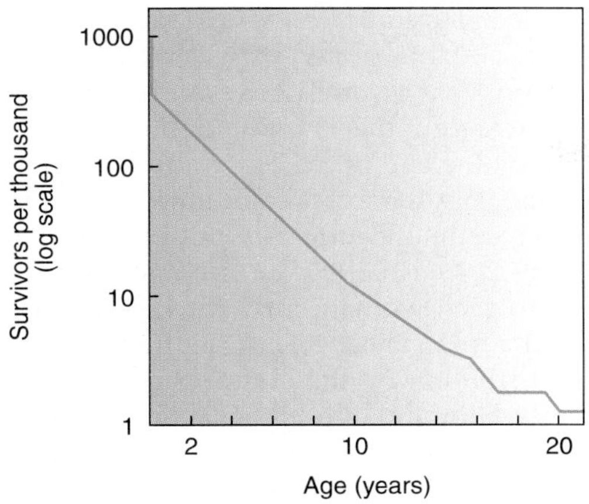

FIGURE 52–11 The survivorship curve for a herring gull population reveals Type III survivorship as chicks and Type II survivorship as adults. Data collected from Kent Island, Maine, 1934 to 1939. Baby gulls were banded to establish identity.

curves that ecologists recognize. In Type I survivorship, as exemplified by humans, the probability of survival decreases with age; that is, mortality is concentrated later in life. *K*-selected species are characterized by Type I survivorship. Humans and other *K*-selected species are long-lived organisms whose young have a high likelihood of survival.

In Type III survivorship, the probability of survival increases with age; that is, mortality is concentrated early in life, and those that survive early mortality subsequently have a high rate of survival. Type III survivorship is characteristic of oysters and other *r*-selected species. Young oysters have three free-swimming larval stages before settling down and secreting a shell. These larvae are vulnerable to predation, and few survive to adulthood.

In Type II survivorship, intermediate between Types I and III, the probability of survival does not change with age. Mortality is spread evenly across all age groups, resulting in a linear decline in survivorship. This constancy probably results from essentially random events that cause death with little age bias. Certain annual plants and some lizards have Type II survivorship.

The three survivorship curves are generalizations, and not all organisms fit one of the three. Many organisms have one type of survivorship curve early in life and another type as adults. Herring gulls, for example, start out with a Type III survivorship curve but develop a Type II curve as adults (Fig. 52–11). Note that most mortality occurs almost immediately despite the protection and care given to the chicks by the parent bird. Herring gull chicks die from predation or attack by other herring gulls, inclement weather, infectious diseases, or starvation following death of the parent. Once the chicks become independent, their survivorship increases dramatically, and mortality occurs at about the same rate throughout their remaining lives. As a result, few or no herring gulls die from the degenerative diseases of "old age" that cause death in most humans.

THE PRINCIPLES OF POPULATION ECOLOGY APPLY TO HUMAN POPULATIONS

Now that we have examined some of the basic concepts of population ecology, we can apply those concepts to the human population. Examine Figure 52–12, which shows the world increase in the human population since the development of agriculture approximately 10,000 years ago. Observe how the human population is increasing exponentially. The characteristic J curve of exponential growth reflects the decreasing amount of time it has taken to add each additional billion people to our numbers. It took thousands of years for the "human population to reach 1 billion, a milestone that took place in 1800. It took 130 years to reach 2 billion (in 1930), 30 years to reach 3 billion (in 1960), 15 years to reach 4 billion (in 1975), and 12 years to reach 5 billion (in 1987). The population is projected to reach 6 billion in 1998.

Thomas Malthus (1766–1834), a British economist, was one of the first to recognize that the human popula-

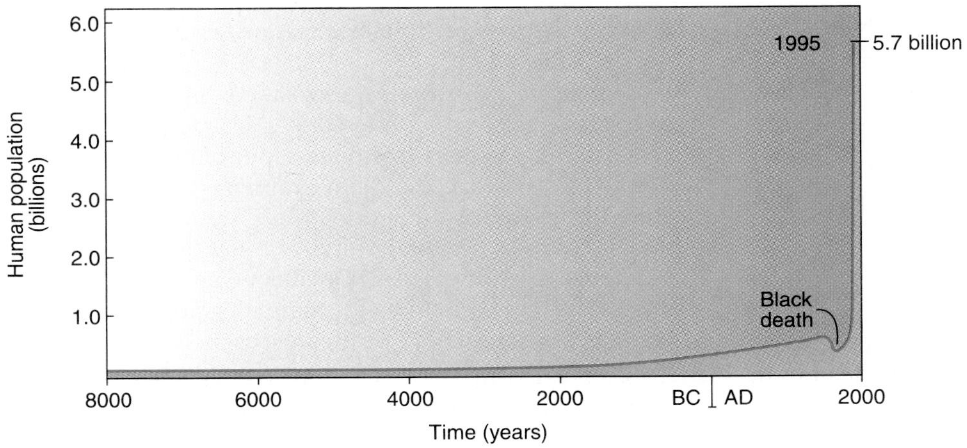

FIGURE 52–12 The human population has been increasing exponentially since the beginning of agriculture (10,000 years ago) to the present.

tion cannot continue to increase indefinitely (see Chapter 17). He pointed out that human population growth was not always desirable (a view contrary to the beliefs of his day) and that the human population was capable of increasing faster than the food supply. He maintained that the inevitable consequences of population growth were famine, disease, and war.

The overall growth rate for the human population has increased dramatically over the past two centuries. This change was not caused by an increase in the birth rate (*b*). In fact, the world birth rate has actually declined slightly during the past 200 years. The increase in growth rate is due instead to a large *decrease in the death rate (d),* which has occurred primarily because of greater food production, better medical care, and improved sanitation practices. For example, from 1920 to 1995, the death rate in Mexico fell from approximately 40 to 5 per 1000 individuals, whereas the birth rate dropped from approximately 40 to 27 per 1000 individuals (Fig. 52–13).

The human population has reached a turning point, however. Although our numbers continue to increase, the growth rate has declined over the past several years. Population experts at the United Nations and the World Bank have projected that the growth rate will continue to slowly decrease until **zero population growth**—the point at which the birth rate equals the death rate ($r = 0$)—is attained. They anticipate that toward the end of the 21st century, the human population will level off at approximately 10.4 billion. This number, which is significantly greater than the 1995 world population of 5.7 billion, is based on many assumptions and will be affected by the actions that individuals and nations take between now and then.

Population projections are "what if" exercises: given certain assumptions about future tendencies in natality, mortality, and migration, an area's population can be calculated for a given number of years into the future. Population projections indicate that changes are upcoming, but they must be interpreted with care because they vary depending on what assumptions have been made. For example, in projecting that the population will stabilize at 10.4 billion by the end of the 21st century, population experts assume that the average number of children born to each woman in all countries will have declined to just about 2.0 by 2040. (In 1995 the average number of children born to each woman on Earth was 3.1.) If that decline does not occur by 2040, our population will not stabilize at 10.4 billion by the end of the 21st century, but will stabilize later and at a higher number. For example, if the population continues to grow at its 1995 rate, there will be almost 23 billion humans toward the end of the 21st century. This number is almost five times larger than our current population.

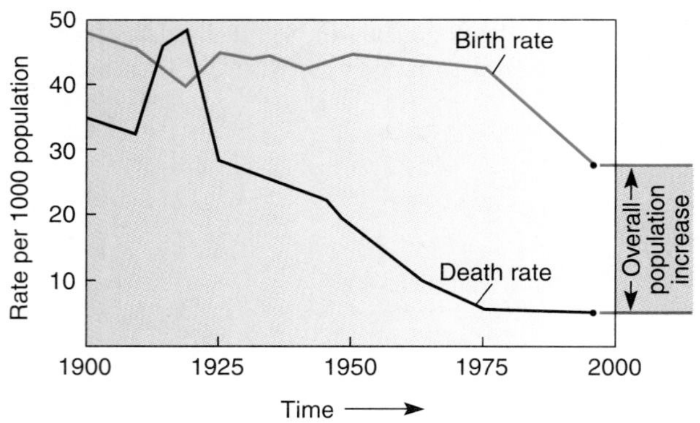

FIGURE 52–13 In Mexico, both birth and death rates have declined in this century. Because the death rate has declined much more than the birth rate, however, Mexico has experienced a high growth rate. (The high death rate prior to 1920 was caused by the Mexican Revolution.)

The main unknown factor in this population growth scenario is the carrying capacity of the environment. No one knows how many humans can be supported by Earth, and projections and estimates vary widely depending on what assumptions are made. It is also not clear what will happen to the human population if or when the carrying capacity is approached. Optimists suggest that the human population will stabilize because of a decrease in the birth rate. Some experts take a more pessimistic view and predict that the widespread degradation of our environment caused by our ever-expanding numbers will result in a massive wave of human deaths. Some experts think human population has already exceeded the carrying capacity of the environment.

Not all countries have the same growth rates

While world population numbers illustrate overall trends, they do not describe other important aspects of the human population story, such as population differences from country to country. **Demographics,** the branch of sociology that deals with population statistics, provides interesting information about the populations of various countries. As you probably know, not all parts of the world have the same rates of population increase. Countries can be classified into two groups, developed and developing, based on their rates of population growth, degrees of industrialization, and relative prosperity (Table 52–2).

Developed countries (also called highly developed countries) such as the United States, Canada, France, Australia, and Japan have low rates of population growth, are highly industrialized, and have high per capita GNPs relative to the rest of the world. (GNP stands for gross national product, the total value of a nation's annual output in goods and services.) Developed countries have the lowest birth rates. Indeed, some developed countries (such as Germany) have birth rates just below that needed to sustain the population and are thus declining slightly in numbers. Highly developed countries also have low

Table 52–2 COMPARISON OF 1995 POPULATION DATA IN DEVELOPED AND DEVELOPING COUNTRIES

	Developed	*Developing*	
	(Highly Developed) United States	**(Moderately Developed) Brazil**	**(Less Developed) Ethiopia**
Fertility rate	2.0	2.9	7.0
Doubling time at current rate	105 years	41 years	23 years
Infant mortality rate	8.0 per 1000	58 per 1000	120 per 1000
Life expectancy at birth	76 years	66 years	52 years
Per capita GNP (U.S. $; 1995)	$24,750	$3020	$100
Women using modern contraception	65%	56%	3%

infant mortality rates (the number of infant deaths per 1000 live births). The infant mortality rate of the United States, for example, was 8.0 in 1995, compared with a 1995 worldwide rate of 62. In addition, highly developed countries have longer life expectancies (76 years in the U.S. versus 66 years worldwide) and higher average per capita GNPs ($24,750 in the U.S. versus $4,500 worldwide in 1995).

Developing countries fall into two subcategories: moderately developed and less developed. Mexico, Turkey, Thailand, and most countries of South America are *moderately developed.* When compared with highly developed countries, the birth rates and infant mortality rates of moderately developed countries are higher. Moderately developed countries have a medium level of industrialization, and their average per capita GNPs are lower than those of highly developed countries. *Less developed* countries, which include Bangladesh, Niger, Ethiopia, and Laos, have the highest birth rates, the highest infant mortality rates, the lowest life expectancies, and the lowest average per capita GNPs in the world.

One way to represent the population growth of a country is to determine the **doubling time,** the amount of time it would take for its population to double in size, assuming that its current growth rate did not change. A look at a country's doubling time can identify it as a highly, moderately, or less developed country: the shorter the doubling time, the less developed the country. At 1995 growth rates, the doubling time is 19 years for Togo, 23 years for Ethiopia, 34 years for Mexico, 105 years for the United States, and 578 years for Belgium.

It is also instructive to examine **replacement-level fertility,** the number of children a couple must produce in order to "replace" themselves. Replacement-level fertility is usually given as 2.1 children in developed countries and 2.7 children in developing countries. The number is greater than 2.0 because some children die before they reach reproductive age. Thus, higher infant mortality rates are the main reason that replacement levels in developing countries are greater than in developed countries. The **total fertility rate,** the average number of children born to a woman during her lifetime, is 3.1 worldwide, well above replacement levels.

Table 52–3 FERTILITY CHANGES IN SELECTED DEVELOPING COUNTRIES

	*Total Fertility Rate**	
Country	*1960–65*	*1995*
Afghanistan	7.0	6.9
Bangladesh	6.7	4.3
Brazil	6.2	2.9
China	5.9	1.9
Egypt	7.1	3.9
Guatemala	6.9	5.4
India	5.8	3.4
Kenya	8.1	5.7
Mexico	6.8	3.1
Nepal	5.9	5.8
Nigeria	6.9	6.3
Thailand	6.4	2.2

*Total fertility rate = average number of children born to each woman during her lifetime.

The population in many developing countries is beginning to approach stabilization. Fertility rates must decline in order for the population to stabilize (see Table 52–3, and note the general decline in total fertility rate from the 1960s to 1995 in selected developing countries). The total fertility rate in developing countries has decreased from an average of 6.1 children per woman in 1970 to 3.5 in 1995. In countries such as Brazil, Indonesia, and Mexico, fertility rates have declined by at least 25% in the past decade.[3]

[3]Although the fertility rates in these countries have declined, it should be remembered that they still exceed replacement-level fertility. Consequently, the populations in these countries are still increasing.

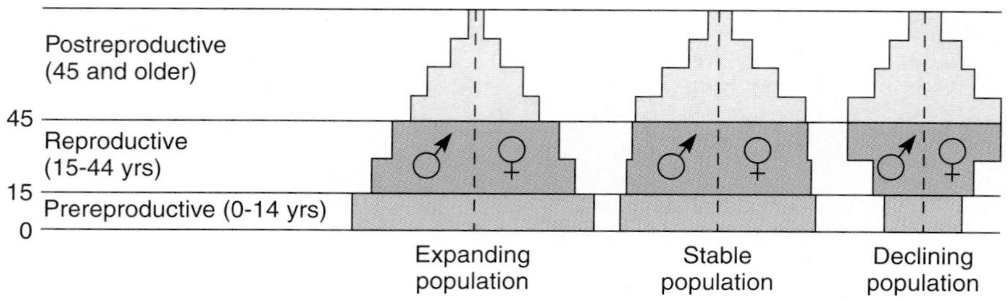

FIGURE 52–14 Generalized age structure diagrams for expanding, stable, and declining populations can be identified by their characteristic shapes. In each, the left half of the diagram represents the males in the population, and the right half represents the females. Each diagram is divided horizontally into age groups, and the width of each segment represents the population size of that group.

The age structure of a country can be used to predict future population growth

In order to predict the future growth of a population, it is important to know its **age structure,** which is the percentages of the population at different ages. The number of males and females at each age, from birth to death, is represented in age structure diagrams (Fig. 52–14). Age structure diagrams are divided vertically to show sex ratios: the left side of the diagram represents the males in a population, and the right side represents the females. The bottom one-third (*blue*) of the diagram represents prereproductive humans (0–14 years of age); the middle one-third (*green*) corresponds to the reproductive ages (15–44 years); and the top one-third (*gray*) depicts postreproductive humans (45 years and older). The width of the individual diagrams at any given point is proportional to the population size; a broader width implies a larger population.

The overall shape of an age structure diagram indicates whether the population is increasing, stable, or shrinking. The age structure diagram of a country with a high growth rate (for example, Nigeria or Venezuela) is shaped like a pyramid (Fig. 52–15*a*). The largest percentage of the population is in the prereproductive age group, providing the momentum for future population growth. When these children mature, they will become the parents of the next generation. Thus, *even if the fertility rate of such a country is at replacement levels, the population will continue to grow.* In contrast, the more tapered bases of the age structure diagrams of countries with stable or declining populations indicate a smaller ratio of children who will become the parents of the next generation.

The age structure diagram of a stable population, one that is neither growing nor shrinking, demonstrates that the number of people at the prereproductive age and at the reproductive age is approximately the same (Fig. 52–15*b*). Also, a larger percentage of the population is older (postreproductive) than in a rapidly increasing population. Many countries in Europe have stable populations.

In a shrinking population, the prereproductive age group is *smaller* than either the reproductive or postre-

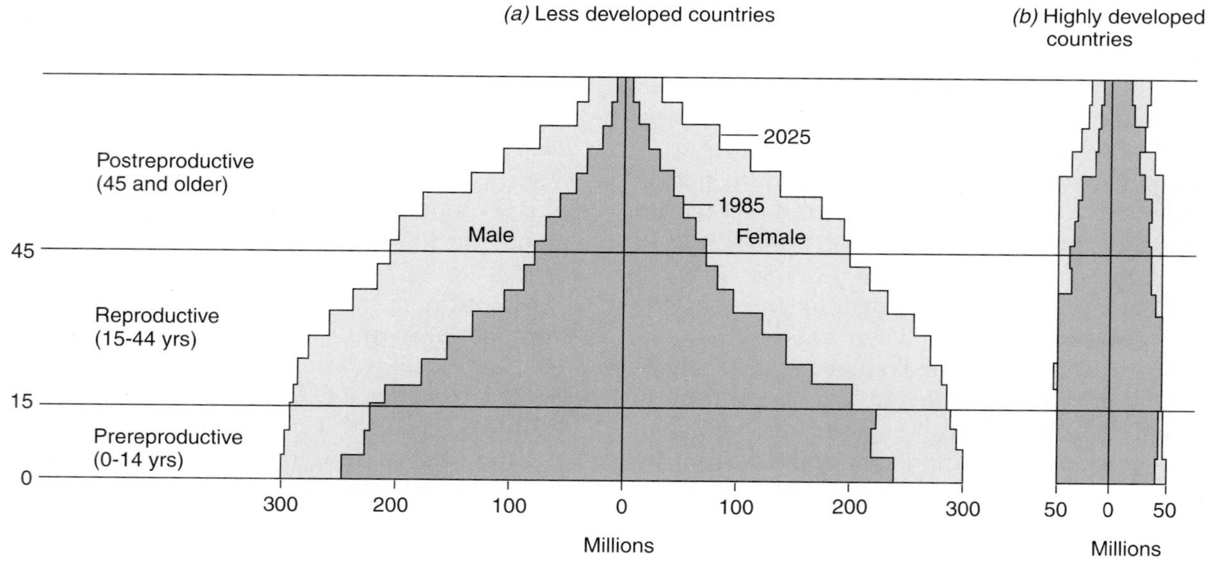

FIGURE 52–15 Less developed countries have a much higher percentage of young people than do highly developed countries. As a result, less developed countries (***a***) are projected to have greater population growth than are highly developed countries (***b***). The dark blue region represents actual age distribution in 1985. The light blue region represents projected age distribution in 2025.

productive groups. Russia, Bulgaria, and Hungary are examples of countries with shrinking populations.

Worldwide, 32% of the human population is under age 15. When these people enter their reproductive years, they have the potential to cause a large increase in the growth rate. Even if the birth rate does not increase, the growth rate will increase simply because there are more people reproducing.

Most of the world population increase that has occurred since 1950 has taken place in developing countries (as a result of the younger age structure and the higher-than-replacement-level fertility rates of their populations). In 1950, 66.8% of the world's population was in the developing countries in Africa, Asia (minus Japan), and Latin America. Between 1950 and 1995, the world's population more than doubled, but most of that growth occurred in developing countries. As a reflection of this, in 1995 the number of people in developing countries had increased to 79.5% of the world's population. Most of the population increase that will occur during the next century will also take place in developing countries, those least able to support such growth. By 2020 it is estimated that about 85% of the people in the world will live in developing countries.

SUMMARY

I. Populations have certain properties, such as birth rates and death rates, that individual organisms lack.
   A. Population density is the number of individuals of a species per unit of area or volume at a given time.
   B. Population dispersion may be uniform, random, or clumped.

II. Population size is affected by the number of births, deaths, immigrants, and emigrants.
   A. The growth rate (r) of a population is its rate of change.
      1. $r = b - d$ when migration is not a factor.
      2. $r = (b - d) + (i - e)$ for a local population (where migration is a factor).
   B. Biotic potential is the maximum rate at which an organism or population could increase in number under ideal conditions.
   C. Population size is modified by environmental resistance (the effects of crowding).
   D. The carrying capacity (K) of the environment is the largest population that can be maintained for an indefinite time by a particular environment.

III. The logistic equation of population growth describes an S-shaped curve when graphed. It shows an initial lag phase (when the population is small), followed by an exponential phase, followed by a leveling phase as the carrying capacity of the environment is reached.

IV. Environmental factors limit population growth.
   A. Density-dependent factors are most effective at limiting population growth when the population density is high. Predation, disease, and competition are examples.
   B. Density-independent factors limit population growth but are not influenced by changes in population density. Hurricanes and fires are examples.

V. Each organism has its own life history strategy.
   A. An r strategy emphasizes a high growth rate. These organisms often have small body sizes, high reproductive rates, and short lifespans, and they inhabit variable environments.
   B. A K strategy emphasizes maintenance of a population near the carrying capacity of the environment. These organisms often have large body sizes, low reproductive rates, and long lifespans, and they inhabit stable environments.
   C. There are three general types of survivorship curves. Type I survivorship, in which mortality is greatest in old age, is typical of K-selected species. Type III survivorship, in which mortality is greatest among the young, is typical of r-selected species. In Type II survivorship, mortality is spread evenly across all age groups.

VI. The principles of population ecology apply to humans.
   A. Currently, human population is increasing exponentially.
      1. The rate of population increase has declined slightly over the past several years, however.
      2. Demographers project that the world population will stabilize ($r = 0$) by the end of the 21st century.
   B. Highly developed countries have the lowest birth rates, lowest infant mortality rates, longest life expectancies, and highest per capita GNPs. Developing countries have the highest birth rates, highest infant mortality rates, shortest life expectancies, and lowest per capita GNPs.
   C. The age structure of a population greatly influences population dynamics. It is possible for a country to have replacement-level fertility and still experience population growth if the largest percentage of the population is in the prereproductive years.

SELECTED KEY TERMS

age structure
biotic potential
carrying capacity
clumped dispersion
contest competition
demographics
density-dependent factor
density-independent factor
developed country
developing country
dispersion
doubling time
emigration
environmental resistance
exponential growth
growth rate
immigration
infant mortality rate
interspecific competition
intraspecific competition
K selection
mortality
natality
population
population crash
population density
population ecology
random dispersion
replacement-level fertility
r selection
scramble competition
survivorship
total fertility rate
uniform dispersion
zero population growth

POST-TEST

1. A(An) ________ is a group of organisms of a single species living in a particular area at a given time.
2. Population ________ is the number of individuals of a species per unit of habitat area at a given time.
3. ________ ________ occurs when individuals are aggregated, or concentrated, in certain portions of the habitat.
4. The birth rate, or ________, increases population size, whereas the death rate, or ________, decreases population size.
5. The maximum rate at which a population could increase under ideal conditions is known as its ________ ________.
6. In a graph of population number versus time, a J-shaped curve is characteristic of ________ growth.
7. Populations never remain at their biotic potential indefinitely because population size is modified by the effects of crowding, known as ________ ________.
8. The ________ ________ is the largest population that can be maintained by a particular environment for an indefinite period of time.
9. An abrupt decline from high to low population density is called a population ________.
10. Predation, disease, and competition are examples of density-________ factors.
11. ________-competition occurs within a population and ________-competition occurs among populations of different species.
12. The effect of density-________ factors on population growth intensifies as population density increases.
13. Density-________ factors regulate the size of a population but are not influenced by changes in population density.
14. Organisms with ________ selection have traits that allow them to have high rates of natural increase.
15. Organisms that are ________ strategists do not use environmental resources and energy to produce large numbers of offspring.
16. The proportion of individuals in a population that survive to a particular age, known as ________, is related to an organism's reproductive strategy.
17. Population experts project that ________ ________ ________, the point at which the human birth rate equals the death rate, will occur toward the end of the 21st century.
18. A developed country has a low ________ ________ rate, which is the number of infant deaths per 1000 live births.
19. The percentages of a population at different ages make up that population's ________ ________.

REVIEW QUESTIONS

1. What kinds of questions do population ecologists study?
2. Give several biological advantages for organisms having a clumped dispersion. What are the disadvantages?
3. Define each of the following and explain its effect on population size: (a) natality; (b) mortality; (c) immigration; (d) emigration.
4. Explain the J-shaped and S-shaped population growth curves in terms of biotic potential and carrying capacity.
5. Draw a graph to represent the long-term growth of a bacterial population cultured in a test tube containing a nutrient medium.
6. Give three examples each of density-dependent and density-independent factors that affect population growth.
7. Draw the three main types of survivorship curves, and relate each to *r* selection and *K* selection.
8. What is replacement-level fertility? Why is it higher in developing countries than in developed countries?
9. Which of the following age structure diagrams indicates a population that is growing? Explain your answer.

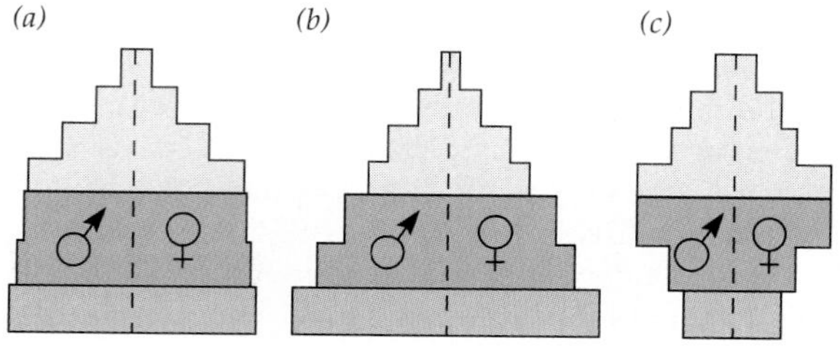

YOU MAKE THE CONNECTION

1. How might pigs at a trough be an example of scramble competition? Of contest competition?
2. A female elephant bears a single offspring every two to four years. Based on this information, do you think elephants are *r*-selected or *K*-selected species? Which survivorship curve do you think is representative of elephants? Explain your answers.
3. In Bolivia, 42% of the population is younger than age 15, and 4% is older than 65. In Austria, 18% of the population is younger than 15, and 15% is older than 65. Which country will have the highest growth rate over the next two decades? Why?

RECOMMENDED READINGS

Bongaarts, J. "Population Policy Options in the Developing World." *Science,* Vol. 263, February 11, 1994. Discusses various policy options that could be implemented during the next century to reduce population growth in developing countries.

Brown, L.R., N. Lenssen, and H. Kane. *Vital Signs 1995: The Trends That Are Shaping Our Future.* W.W. Norton & Company, New York, 1995. Contains a short article ("Population Growth Steady") on world population that summarizes the 1994 Conference on Population and Development.

Cohen, J.E. "Population Growth and Earth's Human Carrying Capacity." *Science,* Vol. 269, July 21, 1995. Explores the question of how many people Earth can support.

Haub, C. "Understanding Population Projections." *Population Bulletin,* Vol. 42, No. 4, December 1987. A comprehensive evaluation of the challenges facing demographers as they calculate population projections.

Sherman, D. "Institutions: Zero Population Growth." *Environment,* Vol. 35, No. 9, November 1993. Examines ZPG, an organization founded to bring Earth's population into balance with its resources and the environment.

World Population Data Sheet, Population Reference Bureau, Washington, D.C., 1995. A chart that provides current population data for all countries. Includes birth rates, death rates, infant mortality rates, total fertility rates, and life expectancies as well as other pertinent information.

CHAPTER 53

Community Ecology

Raccoons (*Procyon lotor*) are common inhabitants of North American forests, particularly near streams. (Gary Meszaros/Dembinsky Photo Associates)

The term *community* has a far broader meaning in ecology than it does in everyday speech. A **community** consists of populations of different species of organisms that live and interact together in a given environment. The organisms that comprise a community interact with one another and are interdependent in a variety of ways. Species compete with one another for food, water, living space, and other resources. Organisms kill and eat other organisms. Species form symbiotic associations with one another. Each organism plays one of three main roles in community life: producer, consumer, or decomposer. The unraveling of the many interactions and interdependencies of organisms living together as a community is one of the goals of community ecologists.

Communities vary greatly in size, do not have precise boundaries, and are rarely completely isolated. They interact with and influence one another in countless ways, even when the interaction is not readily apparent. Furthermore, communities are nested within one another like Chinese boxes: there are communities within communities. A forest is a community, but so is a rotting log in that forest. The log contains bacteria, fungi, slime molds, worms, insects, and perhaps even mice. The microorganisms living within the gut of a termite in the rotting log also form a community.

Organisms occupy an abiotic (nonliving) environment that is as essential to their lives as their interactions with other organisms. Minerals, air, water, and sunlight are just as much a part of a honeybee's environment, for example, as are the flowers that it pollinates and from which it takes nectar. A biological community and its abiotic environment together comprise an **ecosystem.**

This chapter is concerned with making sense of community structure and diversity by finding common patterns and processes in a wide variety of communities. Although the living community is emphasized in this chapter, communities and their physical environments are inseparably linked; the abiotic components of ecosystems are considered in Chapter 54.

LEARNING OBJECTIVES

After you have studied this chapter you should be able to

1. Distinguish between a community and an ecosystem.
2. Define predation, and describe the effects of natural selection on predator-prey relationships.
3. Explain symbiosis, and distinguish among mutualism, commensalism, and parasitism.
4. Define ecological niche, and distinguish between an organism's fundamental niche and its realized niche.
5. Summarize the concept of competitive exclusion.
6. Give several examples of limiting factors, and discuss how they might affect an organism's ecological niche.
7. Summarize the main determinants of species diversity in a community.
8. Define ecological succession, and distinguish between primary succession and secondary succession.
9. Discuss the two views of the nature of communities: the organismic model and the individualistic model.

COMMUNITIES CONTAIN AUTOTROPHS AND HETEROTROPHS

Sunlight is the source of energy that powers almost all life processes on Earth. **Primary producers,** also called *autotrophs* or, simply, *producers,* are organisms that make complex organic molecules from simple inorganic substances (generally carbon dioxide and water), usually using the energy of sunlight to do so. In other words, most producers perform the process of photosynthesis. By incorporating the chemicals they manufacture into their own biomass (living material), producers become potential food resources for other organisms. While plants are the most significant producers on land, algae and cyanobacteria are important producers in aquatic environments. In rift-vent hot spring communities deep in the ocean, nonphotosynthetic bacteria are the producers (see *Focus On: Life Without the Sun;* see also *Making the Connection: Nutrition and Metabolic Diversity* in Chapter 8).

All other organisms in a community are *heterotrophs* that extract energy from organic molecules produced by other organisms. **Consumers** are heterotrophs that obtain energy and body-building materials by feeding on organic molecules formed by other organisms. Consumers that eat producers are called **primary consumers,** which usually means that they are exclusively **herbivores** (plant eaters). Deer and grasshoppers are examples of primary consumers. **Secondary consumers** eat primary consumers and include flesh-eating **carnivores,** which eat other animals. Lions and spiders are examples of carnivores. Other consumers, called **omnivores,** eat a variety of organisms, both plant and animal. Bears, pigs, and humans are examples of omnivores.

Some consumers, called **detritus feeders,** or **detritivores,** consume **detritus,** which is dead organic matter that includes animal carcasses, leaf litter, and feces. Detritus feeders, such as snails, crabs, clams, and worms, are especially abundant in aquatic environments, where they burrow in the bottom muck and consume the organic matter that collects there. Earthworms are terrestrial detritus feeders, as are termites, beetles, snails, and millipedes. Detritus feeders work together with microbial decomposers to destroy dead organisms and waste products. An earthworm, for example, actually eats its way through the soil, digesting much of the organic matter contained there.

Many consumers do not fit readily into a single category of herbivore, carnivore, omnivore, or detritivore. To some degree these organisms modify their food preferences as the need arises. Furthermore, some organisms change food preferences over their lifetime. Tadpoles are primary consumers, but adult frogs are secondary consumers.

Decomposers (also called **saprotrophs**) are microbial heterotrophs that break down organic material and use the decomposition products to supply themselves with energy. They typically release simple inorganic molecules, such as carbon dioxide and mineral salts, that can then be reused by producers. Bacteria and fungi are important examples of decomposers. Dead wood, for example, is invaded first by sugar-metabolizing fungi that consume the wood's simple carbohydrates, such as glucose and maltose. When these carbohydrates are exhausted, fungi, often aided by termites (with symbiotic bacteria in their guts) complete the digestion of the wood by breaking down cellulose, a complex carbohydrate that is the main component of wood.

Producers and decomposers have indispensable roles in ecosystems. Producers provide both food and oxygen for the rest of the community. Decomposers are also necessary for the long-term survival of any community, because without them, dead organisms and waste products would accumulate indefinitely. Without decomposers, essential elements such as potassium, nitrogen, and phosphorus would remain permanently in dead organisms and hence be unavailable for use by new generations of living organisms.

LIVING ORGANISMS INTERACT IN A VARIETY OF WAYS

No species exists independently of other organisms. The primary producers, consumers, and decomposers of a

Focus On

Life Without the Sun

In 1977 an oceanographic expedition aboard the research submersible *Alvin* studied the Galapagos Rift, a deep cleft in the ocean floor off the coast of Ecuador. The expedition revealed, on the floor of the abyss, a series of hydrothermal vents where seawater apparently had penetrated and been heated by the hot rocks below. During its time within the Earth, the water had also been charged with mineral compounds, including hydrogen sulfide, H_2S.

At the tremendous depths of greater than 2500 meters (8200 feet) found in the Galapagos Rift, there is no light for photosynthesis. But the hydrothermal vents support a rich and bizarre variety of life forms (see figure), in contrast with the surrounding "desert" of the abyssal floor. Many of the species in these oases of life were new to science. For example, giant blood-red tube worms almost 3 meters (10 feet) in length cluster in great numbers around the vents. Other animals around the hydrothermal vents include clams, crabs, barnacles, and mussels.

The mystery is, what is the ultimate source of energy for producers in this dark environment? Most deep-sea communities depend on the organic matter that drifts down from the surface waters; in other words, they rely on energy derived from photosynthesis. The Galapagos Rift community is too densely clustered and too productive to be dependent on chance encounters with organic material from surface waters, however.

The base of the food web in these aquatic oases is occupied by chemosynthetic autotrophic bacteria that function as producers but do not photosynthesize. Instead, they possess enzymes that catalyze the oxidation of hydrogen sulfide to produce water and sulfur or sulfate. Such chemical reactions are exergonic (see Chapter 6) and provide the energy required to fix CO_2, which is dissolved in the water, into organic compounds. Thus these bacteria support life in deep-ocean hydrothermal vents. Many of the Galapagos Rift animals consume the bacteria directly by filter feeding; others, such as the giant tube worms, get their energy from bacterial endosymbionts that live inside their bodies.

Therefore, it is accurate to say that *almost* all organisms on Earth depend on the sun for energy. The organisms in abyssal hydrothermal vents are an interesting and unique exception. ■

Chemosynthetic autotrophic bacteria living in the tissues of these tube worms extract energy from hydrogen sulfide to manufacture organic compounds. Because these worms lack digestive systems, they depend on the organic compounds provided by the endosymbiotic bacteria, along with materials filtered from the surrounding water and digested extracellularly. Also visible in the photograph are some filter-feeding mollusks (*yellow*) and a crab (*white*). (Visuals Unlimited/ Science VU-WHOI, D. Foster)

community interact with one another in a variety of complex ways, and each forms associations with other organisms. Three main types of interactions occur among species in a community: predation, symbiosis, and competition. We now examine how predation and symbiosis lead to diverse adaptations when species interact with one another. *Inter*specific competition (competition between species) will be discussed later in the chapter. *Intra*specific competition was discussed in Chapter 52.

Natural selection shapes both prey and predator

Predation is the consumption of one species, the *prey*, by another, the *predator*. It includes animals eating other animals, as well as animals eating plants. Predation has resulted in an evolutionary "arms race," with both the evolution of predator strategies—more efficient ways to catch prey—and prey strategies—better ways to escape the predator. A predator that is more efficient at catching prey exerts a strong selective force on its prey, which over time may evolve some sort of countermeasure. The countermeasure acquired by the prey in turn acts as a strong selective force on the predator. This interdependent evolution of two interacting species is known as **coevolution** (see Chapter 35).

Pursuit and ambush are two predator strategies

A brown pelican sights a fish while in flight. Less than two seconds after diving into the water, it has its catch (Fig. 53–1). Killer whales, which hunt in packs, often herd salmon or tuna into a cove so that they are easier to catch. Any trait that increases hunting efficiency, such as the speed of a brown pelican or the cunning of killer whales, favors predators that pursue their prey. Because these car-

FIGURE 53–1 Pursuit of prey, an effective predator strategy, requires speed. A brown pelican alights in the water, having just caught a fish. (SharkSong/M. Kazmers/Dembinsky Photo Associates)

nivores must be able to process information quickly during the pursuit, their brains are generally larger (relative to body size) than those of the prey they pursue.

Ambush is another effective way to catch prey. The spider pictured on the cover is the same color as the white flower on which it hides. This camouflage keeps unwary insects that visit the flower for nectar from noticing the spider until it is too late. Predators that are able to *attract* prey are particularly effective at ambushing. For example, a diverse group of deep-sea fishes known as angler fish possess rodlike luminescent lures to attract food (see Fig. 6–13).

Chemical protection is an effective plant defense against herbivores

Plants cannot escape predators by fleeing, but they possess a number of adaptations that protect them from being eaten. The presence of spines, thorns, tough leathery leaves, or even thick wax on leaves discourages foraging herbivores from grazing.

Other plants produce an array of protective chemicals that are unpalatable or even toxic to herbivores. The active ingredients in such plants as marijuana, opium poppy, tobacco, and peyote cactus are thought to discourage the foraging of herbivores.

Milkweeds are an excellent example of the biochemical coevolution between plants and herbivores (Fig. 53–2). Milkweeds produce alkaloids and cardiac glycosides, chemicals that are poisonous to all animals except for a small group of insects. During the course of evolution, these insects acquired the ability to either tolerate or metabolize the milkweed toxins. As a result, they can eat milkweeds without being poisoned. These insects avoid competition from other herbivorous insects because few are able to tolerate milkweed toxins. Predators also learn to avoid these insects, which accumulate the toxins in their tissues and are usually brightly colored to announce that fact. The black, white, and yellow banded caterpillar of the monarch butterfly (discussed later in this chapter) is an example of a milkweed feeder.

Animals possess a variety of defensive adaptations

Many animals, such as meadow voles and woodchucks, flee from predators by rapidly running home. Others have mechanical defenses: the barbed quills of a porcupine and the shell of a pond turtle are examples. Some animals live in groups, such as herds of antelope, colonies of honeybees, schools of anchovies, and flocks of pigeons. Because a group has so many eyes, ears, and noses watching, listening, and smelling for predators, this social behavior decreases the likelihood of a predator catching one member unaware.

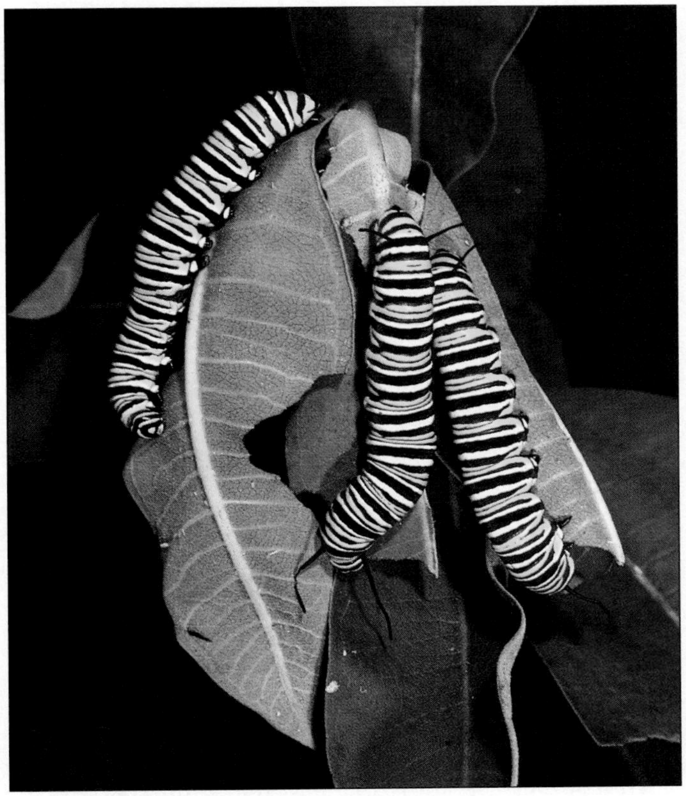

FIGURE 53–2 The common milkweed (*Asclepias syriaca*) is protected by its toxic chemicals. Its leaves are poisonous to most herbivores except monarch caterpillars (*shown*) and a few other insects. (Patti Murray)

FIGURE 53–3 The poison arrow frog (*Dendrobates tinctorius*) advertises its poisonous nature with its conspicuous coloring, warning away would-be predators. (Animals Animals ©1995 Michael Fogden)

Chemical defenses are also common among animal prey. The South American poison arrow frog (*Dendrobates*) has poison glands in its skin. Its bright yellow and black **warning coloration** prompts avoidance by experienced predators (Fig. 53–3). Snakes and other animals that have tried once to eat a poisonous frog do not repeat their mistake! Other examples of warning coloration occur in the striped skunk, which sprays acrid chemicals from its anal glands, and the bombardier beetle, which spews harsh chemicals at potential predators (see Fig. 6–19).

Some animals hide from predators by blending into their surroundings. Such camouflage is often enhanced by the animal's behavior. There are many examples of camouflage (Fig. 53–4). Certain caterpillars resemble twigs so closely that you would never guess they were animals unless they moved. Some katydids resemble leaves not only in color but also in the pattern of veins in their wings. Pipefish are almost perfectly camouflaged in green eel grass. Such camouflage has been preserved and accentuated by means of natural selection.

Sometimes a harmless or edible species (called a *mimic*) is protected from predation by its resemblance to a species that is dangerous in some way (called a *model*). Such a strategy is known as **batesian mimicry.** Numerous cases of this phenomenon exist. For example, a harmless moth may look so much like a bee or wasp that predators avoid it (Fig. 53–5). Even a biologist would hesitate to pick it up!

In **müllerian mimicry,** different species, all of which are poisonous, harmful, or distasteful, resemble one another. Although their harmfulness protects them as individual species, their similarity in appearance works as an added advantage. Potential predators can more easily learn one common warning coloration. Viceroy and monarch butterflies are currently viewed as an example of müllerian mimicry (see *Focus On: Batesian Butterflies Disproved*).

Symbiosis involves close associations in which one species usually lives on or in the other species

Symbiosis is any intimate, long-term relationship or association between two or more species. The partners of a symbiotic relationship, called *symbionts,* may benefit from, be unaffected by, or be harmed by the relationship. The thousands, or perhaps even millions, of symbiotic associations are all products of coevolution. Symbiosis takes three forms: mutualism, commensalism, and parasitism.

In mutualism, benefits are shared

Mutualism is a symbiotic relationship in which both partners benefit. The association between nitrogen-fixing bacteria of the genus *Rhizobium* and legumes (plants such as peas, beans, and clover) is an example of mutualism. Nitrogen-fixing bacteria, which live inside nodules on the roots of legumes, supply the plants with all of the nitrogen they need (see Fig. 54–3*a*). The legumes supply organic molecules to their bacterial symbionts.

Another example of mutualism is the association between reef-building coral animals and dinoflagellates called zooxanthellae (see Chapters 28 and 51). These symbiotic algae live inside cells of the coral, where they photosynthesize and provide the animal with carbon and nitrogen compounds as well as oxygen. Zooxanthellae have a stimulatory effect on corals, causing calcium carbonate skeletons to form around their bodies much faster when the algae are present. The coral, in turn, supplies its zooxanthellae with waste products such as ammonia, which the algae use to make nitrogen compounds for both partners.

Mycorrhizae are mutualistic associations between fungi and the roots of about 90% of all plant families. The fungus absorbs essential minerals, especially phosphorus, from the soil and provides them to the plant. In return, the plant provides the fungus with organic molecules produced by photosynthesis. Plants grow more vigorously in the presence of mycorrhizae (see Fig. 25–8), and they are better able to tolerate environmental stresses such as drought and high soil temperatures.

A remarkable example of mutualism occurs in the rain forests of South and Central America between ants or

(Text continues on page 1178)

(a)

(b)

(c)

FIGURE 53–4 Camouflage helps animals hide from predators. (***a***) Geometrid larvae are caterpillars that resemble twigs. Can you find the caterpillar? (***b***) The wings of this katydid resemble dead leaves. (***c***) The bay pipefish is thin, narrow, and green like the eel grass or algae in which it hides. Note its habit of holding its body in a position that resembles waving eel grass or algae. (*a, b,* James L. Castner; *c,* Doug Wechsler)

(a)

(b)

FIGURE 53–5 In this example of batesian mimicry, a fly (*a*) is the mimic and a yellowjacket wasp (*b*) is the model. (*a, b,* James L. Castner)

Focus On

Batesian Butterflies Disproved

The monarch butterfly (*Danaus plexippus*) is an attractive insect found throughout much of North America (see Fig. *a*). As a caterpillar, it feeds exclusively on milkweed leaves. The milky white liquid produced by the milkweed plant contains poisons that apparently do not harm the insect, but remain in its tissues for life. When a young bird encounters and eats its first monarch butterfly, it gags until the insect is regurgitated. Thereafter, the bird avoids eating the distinctively marked insect.

Many people confuse the viceroy butterfly (*Limenitis archippus*) (see Fig. *b*) with the monarch. The viceroy, which is found throughout most of North America, is approximately the same size, and the color and markings of its wings are almost identical to those of the monarch. As caterpillars, viceroys eat willow and poplar leaves, which do not contain poisonous substances.

During the past century, it was thought that the viceroy butterfly was a tasty food for birds, but that its close resemblance to monarchs gave it some protection against being eaten. In other words, birds that had learned to associate the distinctive markings and coloration of the monarch butterfly with its bad taste tended to avoid viceroys because they were similarly marked. The viceroy butterfly was therefore considered a classic example of batesian mimicry.

(*a*)

(*b*)

Müllerian mimicry is the resemblance between two or more poisonous, harmful, or distasteful organisms. Monarch (***a***) and viceroy (***b***) butterflies are an example of müllerian mimicry. (*a*, John Gerlach/Dembinsky Photo Associates; *b*, Sharon Cummings/Dembinsky Photo Associates)

In the journal *Nature* in 1991, biologists David Ritland and Lincoln Brower of the University of Florida reported the results of an experiment that tested the long-held notion that birds like the taste of viceroys but avoid eating them because of their resemblance to monarchs. They pulled the wings off of different kinds of butterflies — monarchs, viceroys, and several tasty species — and fed the seemingly identical wingless bodies to red-winged blackbirds. The results were surprising: monarchs and viceroys were *equally* distasteful to the birds.

As a result of this work, biologists are reevaluating the evolutionary significance of different types of mimicry. Monarchs and viceroys appear to be an example of müllerian mimicry, in which two or more different species that are distasteful or poisonous have come to resemble one another during the course of evolution. This likeness provides an adaptive advantage because predators learn quickly to avoid all butterflies with the coloration and markings of monarchs and viceroys. As a result, fewer butterflies of either species die, and more individuals survive to reproduce.

The butterfly study provides us with a useful reminder about the scientific method. Expansion of knowledge in science is an ongoing enterprise, and newly acquired evidence always takes precedence over older ideas. Thus, scientific knowledge is not static, but continually changing. ■

wasps and flowering vines. Those vines that produce bright red flowers are usually pollinated by hummingbirds, which are rewarded with nectar from the blossoms. Some insects, however, try to rob the flower of its nectar without performing the important task of pollination. This is where the ants or wasps come into the picture. These insects patrol the plant and attack any would-be nectar robber (Fig. 53–6). In exchange for their vigilance,

FIGURE 53–6 An ant guards the nectar of a passion flower blossom against nectar-robbing thieves. The nectar is thus saved for its intended recipients, pollen-transporting hummingbirds. The ant benefits from this association because it is rewarded with nectar produced by special glands in the leaves. (Michael & Patricia Fogden)

FIGURE 53–7 Epiphytes, small plants that grow attached to the body of a tree, are an example of commensalism. (Carlyn Iverson)

the insect guards are rewarded with nectar produced by the leaves.

Commensalism is taking without harming

Commensalism is a type of symbiosis in which one organism benefits and the other one is neither harmed nor helped. One example of commensalism is the relationship between two kinds of insects—silverfish and army ants. Certain types of silverfish move along in permanent association with the marching columns of army ants and share the food caught in their raids. The army ants derive no apparent benefit or harm from the silverfish. Another example is the relationship between a tropical tree and its epiphytes, which are smaller plants that live attached to the bark of its branches (Fig. 53–7). The epiphyte anchors itself to the tree but does not obtain nutrients or water directly from it. Living on the tree enables it to obtain adequate light, water (as rainfall dripping down the branches), and required minerals (washed out of the tree's leaves by rainfall). Thus, the epiphyte benefits from the association, whereas the tree is apparently unaffected.

Parasitism is taking at another's expense

Parasitism is a symbiotic relationship in which one member, the *parasite,* benefits, while the other, the *host,* is adversely affected. The parasite lives in or on its host, from which it obtains nourishment, but although a parasite may weaken its host, it rarely kills it. (A parasite would have a difficult life if it kept killing off its hosts!) Parasitism is a successful lifestyle; more than 100 parasites, from ticks to tapeworms, live in or on the human species alone!

When a parasite causes disease and sometimes the death of a host, it is called a **pathogen.** For example, humans can become infected with histoplasmosis, a serious, often fatal disease caused by breathing the spores of a certain fungus. The spores grow in the lungs, causing chronic coughing and fever. Eventually the disease progresses to other organs of the body. The spores of the fungus are common in soils with a high concentration of bird droppings, and the disease is more prevalent in warm tropical regions of the world.

Crown gall disease, which is caused by a bacterium, occurs in many different kinds of plants, and results in millions of dollars of damage each year to ornamental and agricultural plants. Crown gall bacteria, which also live on organic debris in the soil, enter plants through small wounds such as those caused by insects. They cause galls (tumor-like growths), often at a plant's crown—that is, the area between the stem and roots at or near the surface of the ground. Although plants seldom die from crown gall disease, they weaken, grow more slowly, and often succumb to other pathogens.

Many parasites do not cause disease. For example, humans can become infected with the beef tapeworm by eating poorly cooked beef infested with immature tapeworms. Once it is inside the human digestive system, the tapeworm attaches itself to the wall of the small intestine and grows rapidly by absorbing nutrients that pass through. A single beef tapeworm living in a human digestive tract does not usually cause any serious symptoms.

THE NICHE DESCRIBES AN ORGANISM'S ROLE IN THE COMMUNITY

We have seen that a diverse assortment of organisms inhabit each community, and that these organisms obtain nourishment in a variety of ways. We have also examined some of the ways that species interact to form interdependent relationships within the community. A biological description of an organism includes: (1) whether it is a producer, consumer, or decomposer; (2) whether it is a predator and/or a prey; and (3) the kinds of associations it forms. Other details are needed, however, to provide a complete picture.

Every organism is thought to have its own role within the structure and function of a community; we call this role its **ecological niche.** An ecological niche, which is difficult to define precisely, takes into account *all* aspects of the organism's existence: all of the physical, chemical, and biological factors that the organism needs to survive, remain healthy, and reproduce. Among other things, the niche includes the physical surroundings in which an organism lives (its **habitat**). An organism's niche also encompasses what it consumes, what consumes it, what or-

(a)

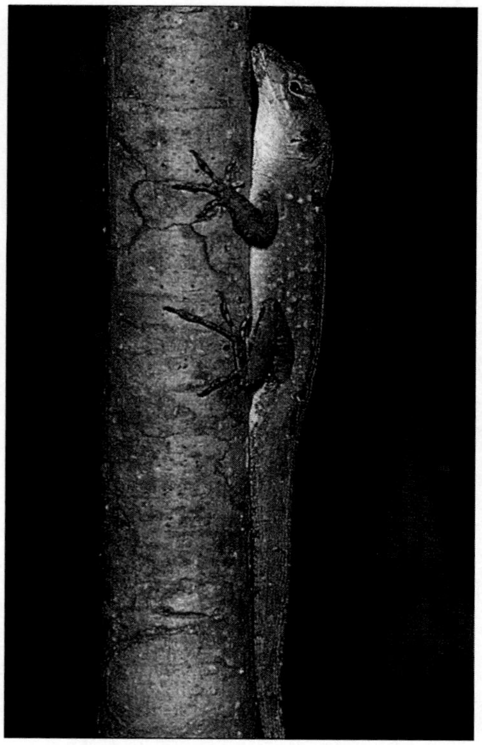

(b)

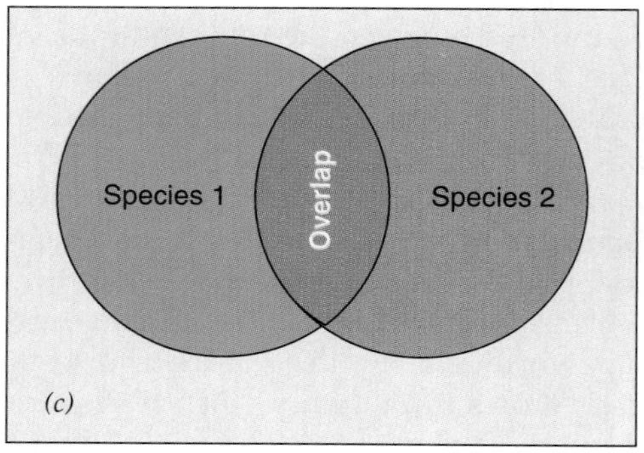

(c)

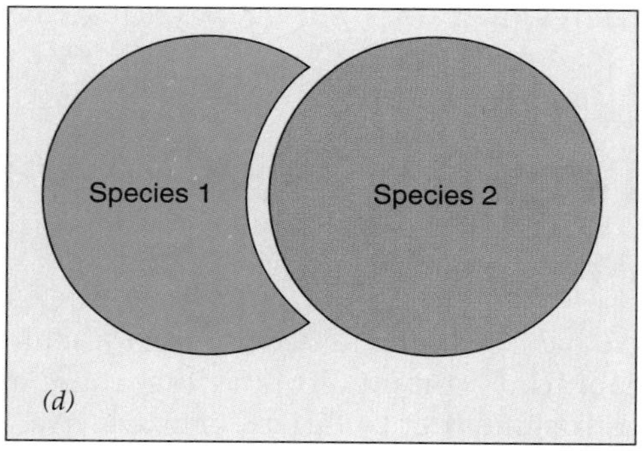

(d)

FIGURE 53–8 Competition can restrict an organism's realized niche. (***a***) The green anole is native to Florida. (***b***) The brown anole was introduced to Florida. (***c***) The fundamental niches of the two lizards overlap. Species 1 represents the green anole, and Species 2 represents the brown anole. (***d***) The brown anole is more competitive in the area where their niches overlap, restricting the niche of the green anole. (*a,* Ed Kanze/Dembinsky Photo Associates; *b,* Connie Toops)

ganisms it competes with, and how it interacts with and is influenced by the abiotic components of its environment (e.g., light, temperature, and moisture). The niche, then, represents the totality of an organism's adaptations, its use of resources, and the lifestyle to which it is suited. Thus, a complete description of an organism's ecological niche involves numerous dimensions.

The ecological niche of an organism may be far broader in potential than in actuality. Put differently, an organism is usually capable of using much more of its environment's resources or of living in a wider assortment of habitats than it actually does. The potential ecological niche of an organism is its **fundamental niche,** but various factors such as competition with other species may exclude it from part of this fundamental niche. Thus, the lifestyle that an organism actually pursues and the resources that it actually uses make up its **realized niche.**

An example may help to make this distinction clear. The green anole (*Anolis carolinensis*), a lizard native to Florida and other southeastern states, perches on trees, shrubs, walls, or fences during the day waiting for insect and spider prey (Fig. 53–8*a*). In the past these little lizards were widespread in Florida. Several years ago, however, a related species, the brown anole (*Anolis sagrei*), was introduced from Cuba into southern Florida and quickly became common (Fig. 53–8*b*). Suddenly the green anoles became rare, apparently driven out of their habitat by competition from the slightly larger brown anoles. Careful investigation disclosed, however, that green anoles were still around. They were now confined largely to the

Making the Connection

Competition and Natural Selection

Is it possible for familiar species with overlapping niches and ranges to coexist? Sometimes two similar species overlap in their geographical distribution. Where they overlap, however, the two species tend to differ more in their structural, ecological, and behavioral characteristics than they do where each occurs in separate geographical areas. Such divergence in traits in two species living in the same geographical area is known as character displacement. It is thought that character displacement prevents the two species from directly competing, since their differences give them different ecological niches in the same environment.

There are several well-documented examples of character displacement between two closely related species. For example, the flowers of two species of *Solanum* in Mexico are quite similar in areas where either one or the other occurs, but in areas where their distributions overlap, there is a noticeable difference in flower size. Because of this difference, the flowers of the two species are pollinated by different kinds of bees. In other words, character displacement reduces interspecific competition, in this case for the same animal pollinator.

The bill sizes of Darwin's finches provide another example of character displacement (see figure). On large islands in the Galapagos where *Geospiza fortis* and *G. fuliginosa* occur together, their bill depths are distinctive: *G. fuliginosa* has a smaller bill depth that enables it to crack

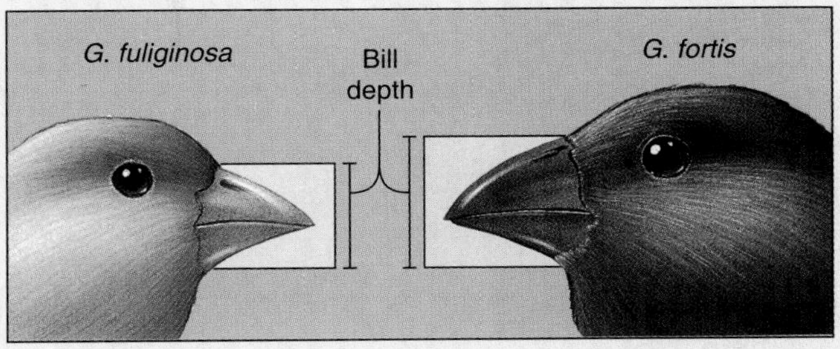

Bill depth exhibits character displacement in two species of Galapagos Island finches, *Geospiza fuliginosa* and *G. fortis*. Where the two species are found on the same island, *G. fuliginosa* has a smaller average bill depth than *G. fortis*.

small seeds, whereas *G. fortis* has a larger bill depth that enables it to crack medium-sized seeds. However, *G. fortis* and *G. fuliginosa* are also found separately on smaller islands. Where this occurs their bill depths tend to be the same intermediate size, perhaps because there is no competition from the other species.

Although these examples can be explained in terms of character displacement, it was only recently, in 1994, that an experiment was conducted to test the character displacement hypothesis. See *On the Cutting Edge: Experimental Evidence of Character Displacement.* ▲

vegetation in wetlands and to the foliated crowns of trees, where they were less easily seen.

The habitat portion of the green anole's fundamental niche includes the trunks and crowns of trees, exterior house walls, and many other locations. Where they became established, brown anoles were able to drive green anoles out from all but wetlands and tree crowns, so that the green anole's realized niche became much smaller as a result of interspecific competition (Fig. 53–8*c*, *d*). Because communities consist of numerous species, many of which compete to some extent, the complex interactions among them produce each one's realized niche.

Competition between two species with identical or similar niches leads to competitive exclusion

When two species are similar, as are the green and brown anoles, their fundamental niches may overlap. However, according to one school of thought, no two species can indefinitely occupy the same niche in the same community because competitive exclusion eventually occurs. In **competitive exclusion,** one species is excluded from a niche by another as a result of interspecific competition. Although it is possible for different species to compete for some necessary resource without being total competitors, two species with absolutely identical ecological niches cannot coexist. Coexistence *can* occur, however, if the overlap between the two niches is reduced. In the lizard example, direct competition between the two species was reduced as the brown anole excluded the green anole from most of its former physical habitat.

The initial evidence that interspecific competition determines a species' realized niche came from a series of laboratory experiments by the Russian biologist G. F. Gause in 1934 (see descriptions of other experiments by Gause in Chapter 52). In one study Gause grew populations of two species of *Paramecium* (a type of protozoon): *P. aurelia* and the larger *P. caudatum* (Fig. 53–9). When grown in separate test tubes, each species quickly increased its population to a high level, which it maintained for some time thereafter. When grown together, however, only *P. aurelia* thrived, while *P. caudatum* dwindled and eventually died out. Under different sets of culture conditions, *P. caudatum* prevailed over *P. aurelia*. Gause interpreted this to mean that although one set of conditions

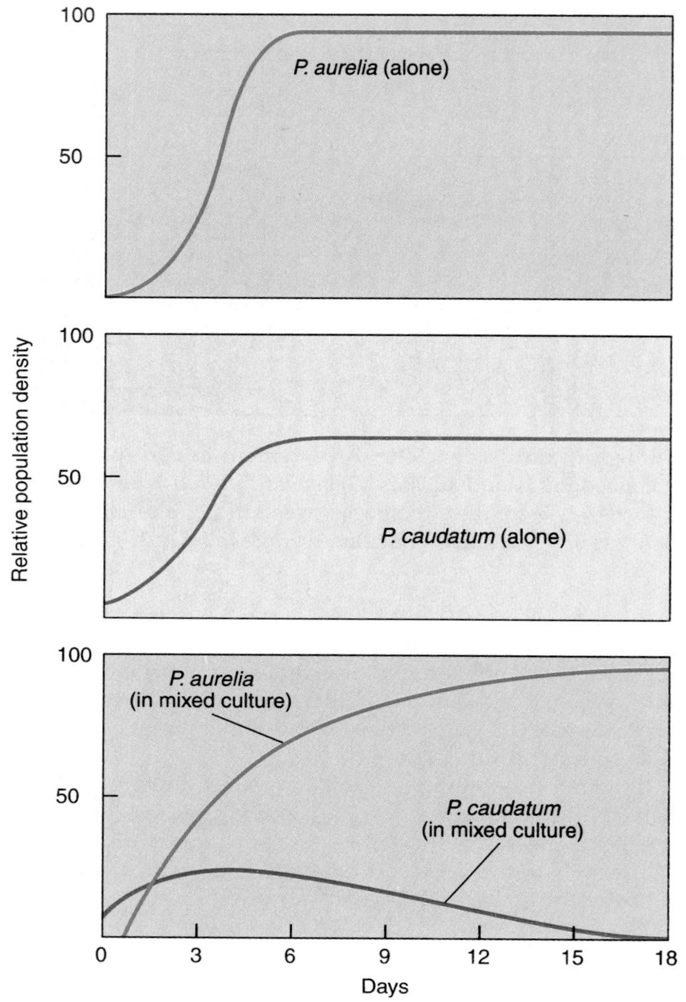

FIGURE 53–9 Competition between two species of *Paramecium* was studied by G. F. Gause. The top two graphs show how a population of each species of *Paramecium* flourishes when grown alone; the bottom graph shows how they grow together, in competition with each other. (After Gause)

favored one species, a different set favored the other. Nonetheless, because both species were similar, in time one or the other would eventually triumph at the other's complete cost.

Sometimes there are apparent contradictions to the competitive exclusion principle. In Florida, for example, native and introduced (nonnative) fish seem to coexist in identical niches. Similarly, botanists have observed closely competitive plant species in the same location. Although such situations seem to contradict the concept of competitive exclusion, the realized niches of these organisms may differ in some way that biologists do not yet understand.

Limiting factors restrict an organism's ecological niche

The environmental factors that actually determine a species' ecological niche can be extremely difficult to identify. For this reason, the concept of the ecological niche is largely abstract, although some of its dimensions can be experimentally determined. Any environmental variable that tends to restrict the ecological niche of an organism is called a **limiting factor.**

What factors actually determine an ecological niche? A niche is basically determined by the total of a species' structural, physiological, and behavioral adaptations. Such adaptations determine, for example, the tolerance of an organism for environmental extremes. If any feature of its environment lies outside the bounds of its tolerance, then the organism cannot live there. Just as you would not expect to find a cactus living in a pond, you would not expect water lilies in a desert.

Most of the limiting factors that have been studied are simple variables such as the soil mineral content, temperature extremes, and precipitation amounts. Such investigations have disclosed that any factor exceeding an organism's tolerance, or present in quantities smaller than the required minimum, limits the occurrence of that organism in a community. By their interaction, such factors help to define an organism's ecological niche.

Limiting factors may affect only part of an organism's life cycle. For instance, although adult blue crabs can live in almost fresh water, they cannot become permanently established in such areas because their larvae cannot tolerate fresh water. Similarly, the ring-necked pheasant, a popular game bird, has been widely introduced in North America but has not become established in the southern United States. The adult birds do well, but the eggs cannot develop properly in the high temperatures found there.

COMMUNITIES VARY IN SPECIES RICHNESS

Species richness, the number of species within a community, varies greatly from one community to another. Tropical rain forests and coral reefs are examples of communities with extremely high species diversity. In contrast, geographically isolated islands and mountaintops exhibit low species diversity.

What determines the number of species in a community? There seems to be no single answer, but several factors appear to be significant. Species diversity is related to the abundance of potential ecological niches. An already complex community offers a greater variety of potential ecological niches than does a simple community (Fig. 53–10). It may become even more complex if organisms potentially capable of filling those niches evolve or migrate into the community. Thus, it appears that species richness is self-reinforcing.

Species diversity is inversely related to the geographical isolation of a community. Isolated island communities tend to be much less diverse than are communities

(Text continues on page 1184)

Experimental Evidence of Character Displacement

Objective: To test the hypothesis that interspecific competition in stickleback fish promotes character displacement.

Method: Place a species of stickleback fish that includes a range of phenotypes (the Cranby species) into two divided experimental ponds. Add individuals of a potential competitor species (the limnetic species) to one separated half of each pond.

Results: In the presence of the limnetic species, the Cranby individuals that were *least* like the limnetic species were more likely to survive and grow.

Conclusion: Evidence was obtained of natural selection against phenotypes most similar to the competitor species. Such natural selection is consistent with the hypothesis of character displacement.

There are many examples in nature of similar species that are markedly different where they are found together, but very alike (yet distinguishable, at least by experts) where each is found alone. For example, two species of threespine stickleback fish, a limnetic species and a benthic species, occur in many lakes of coastal British Columbia. Where they are found together, individuals belonging to the limnetic species are small, have narrow mouths, and eat plankton in the open water. Individuals belonging to the benthic species, on the other hand, are large, have wide mouths, and feed on invertebrates along the shoreline. In lakes where each species occurs alone, the fish are "generalists": they are intermediate in form and eat both plankton and invertebrates.

The hypothesis of character displacement, which has been advanced to explain these observations in threespine sticklebacks and many other organisms, had never been experimentally tested until recently. Dolph Schluter, a researcher at the University of British Columbia who had previously made extensive observations on natural threespine stickleback populations, designed just such an experiment.* He chose to study two species.† One was the limnetic species from a two-species lake. The other, referred to as the Cranby species, was from a nearby lake where it occurs alone. Prior to beginning his experiments, Schluter performed a series of genetic crosses in the laboratory to make the Cranby species, which exhibits the intermediate phenotype, more variable. After the crosses, the Cranby species contained a range of phenotypes, from benthic-like to intermediate to limnetic-like.

Schluter began his experiments by dividing two ponds into separate halves. (Two ponds were used to provide a replication of the experiment.) He placed 1800 young Cranby individuals into each half-pond. To one of the halves of each pond he also added 1200 young limnetic individuals; these were the experimental halves. The half-ponds with only Cranby individuals were the control halves.

After 90 days, the fish were harvested and preserved. About 65 randomly chosen Cranby individuals from each group were sized (measured for length), and their phenotypes were noted. There were no significant differences in survival and growth among the three types of Cranby individuals when they were grown alone. However, in the presence of the limnetic species, Cranby individuals with a limnetic phenotype exhibited significantly lower rates of growth and survival than did the benthic and intermediate types (see figure).

Because only one generation was studied, this experiment investigated only the first step (natural selection) of

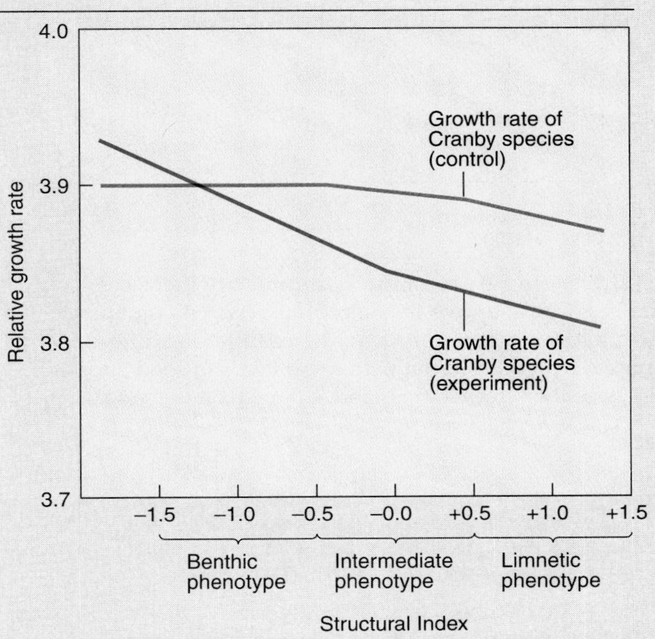

A comparison of growth rates for the Cranby species alone (*control*), and in the presence of the limnetic species (*experiment*). The structural index reflects the range of Cranby phenotypes, from those most benthic-like (−1.5) to those most limnetic-like (+1.5). Values for relative growth rates were obtained statistically. (Courtesy of Dolph Schluter)

the process of character displacement. The next step, an evolutionary change in gene frequencies over successive generations, will require a long-term evaluation of several generations.

This experiment has been both hailed as a landmark and criticized for its design and analysis. For example, some have argued that the experimental half-ponds had much higher population densities (1800 plus 1200 fish) than the control half-ponds (only 1800 fish) and that this might have affected the results. Further refinements of the experimental design, including studies on multiple generations, are expected to resolve some of these concerns, and to raise many interesting new questions.

*Schluter, D. "Experimental Evidence That Competition Promotes Divergence in Adaptive Radiation." *Science,* November 1994.

†Threespine stickleback fish were once thought to be a single species, *Gasterosteus aculeatus.* Although several species are now known to exist, all are currently lumped together in the *Gasterosteus aculeatus* complex because their taxonomic relationships have not been determined.

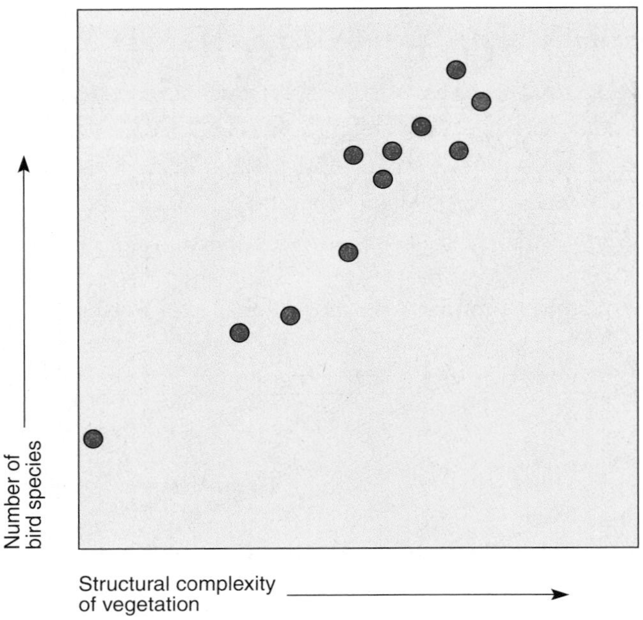

FIGURE 53–10 As community complexity increases, more species are able to live in the community. A community in which the vegetation is structurally complex—a forest, for example—provides birds with more kinds of food and hiding places than does a grassland. (After MacArthur and MacArthur, 1961)

in similar environments on continents. This is due partly to the difficulty encountered by many species in reaching and successfully colonizing the island. Also, sometimes species become locally extinct as a result of random events. In isolated habitats such as islands or mountaintops, extinct species cannot be readily replaced. Isolated areas are likely to be small and to possess fewer potential ecological niches.

Generally, species diversity is inversely related to the environmental stress of a habitat. Only those species capable of tolerating extreme conditions can live in an environmentally stressed community. Thus, the species diversity of a polluted stream is low compared to that of a nearby pristine stream.

Similarly, the species diversity of high-latitude (further from the equator) communities with harsh climates is less than that of lower-latitude (closer to the equator) communities with milder climates. Although the equatorial countries of Colombia, Ecuador, and Peru occupy only 2% of Earth's land, they contain an astonishing 40,000 native plant species. The United States and Canada, with a significantly larger land area, possess a total of 700 native plant species. Ecuador alone contains more than 1300 native species of birds—twice as many as the United States and Canada combined.

Species diversity is usually greater at the margins of distinct communities than it is in their centers. This is because **ecotones**—transitional zones where two or more communities meet—contain all or most of the ecological niches of the adjacent communities as well as some unique to the ecotone (see Chapter 51). This change in species composition produced at ecotones is known as the **edge effect.**

Species diversity is reduced when any one species enjoys a decided position of dominance within a community that enables it to appropriate a disproportionate share of available resources, thus crowding out, or outcompeting, other species. Ecologist James H. Brown of the University of New Mexico addressed species composition and diversity in experiments conducted from 1977 to 1994 in the Chihuahuan desert of southeastern Arizona. In one experiment, the removal of a single dominant species, the kangaroo rat, from several plots resulted in an increased diversity of rodent species. This was ascribed both to less competition for food and also to an altered habitat, because the number of plants increased dramatically after the removal of the kangaroo rats.

Species diversity is greatly affected by geological history. Tropical rain forests are very old, stable communities that have undergone few climatic changes through the entire history of Earth. During this time, myriad species evolved in tropical rain forests, having experienced few or no abrupt climatic changes that might have led to their extinction. In contrast, glaciers have repeatedly altered temperate and arctic regions during Earth's history. An area recently vacated by glaciers has a low species diversity because few species have as yet had a chance to enter it and become established.

Species richness causes community stability—or does it?

For a long time ecologists thought that community stability—the ability of a community to withstand disturbances—was a consequence of community complexity; that is, a community with considerable species diversity was thought to be more stable than a community with less species diversity. According to this view, the greater the species diversity, the less critically important any one species should be. With a multitude of possible interactions within the community, it appeared unlikely that any single disturbance could affect enough components of the system to make a significant difference in its functioning. Evidence for this view can be found in the fact that destructive outbreaks of pests are more common in cultivated fields, which are low-diversity communities, than in natural communities with greater species diversity. As another example, the almost complete loss of the American chestnut tree to the chestnut blight fungus had little ecological impact on the highly diverse Appalachian woodlands of which it used to be a part.

The greatest objection of some ecologists to the apparent relationship between community stability and species richness arises from mathematical models. Such models indicate that more complex communities are actually less stable than less complex ones, at least if the assemblage of species in them is random. (According to modern views of community structure, the random assemblage of species is a reasonable assumption.)

A study reported in the journal *Nature* in 1994, however, strengthened the link between species richness and community stability. Researchers from the Universities of Minnesota and Montreal studied species composition in 207 plots of Minnesota grasslands for seven years. During the study period, Minnesota's worst drought in 50 years occurred (1987–88). The biologists found that those plots with the greatest number of plant species lost less ground cover and recovered faster than species-poor plots. Additional research will have to be done, however, before any firm conclusions can be made about the relationship between species richness and community stability.

SUCCESSION IS COMMUNITY CHANGE OVER TIME

A mature community does not spring into existence full-blown but develops gradually, through a series of stages. The process of community development over time, which involves species in one stage being replaced by different species, is called **succession.** An area is initially colonized by certain species that are replaced over time by others, which themselves may be replaced much later by still others.

Ecologists initially thought that succession inevitably led to a stable and persistent community, known as a *climax community.* But more recently, the traditional view has fallen out of favor. The apparent stability of a climax forest, for example, is probably the result of how long trees live relative to the human lifespan. It is now recognized that mature climax communities are not in a state of permanent equilibrium, but rather in a state of continual flux. A mature community changes in species composition and in the relative abundance of each species despite the fact that it retains a uniform appearance.

Succession is usually described in terms of the changes in the species composition of an area's vegetation, although each successional stage also has its own characteristic kinds of animals and other organisms. The time involved in ecological succession is on the order of tens, hundreds, or thousands or years, not the millions of years involved in the evolutionary time scale.

Sometimes a community develops in a "lifeless" environment

Primary succession is the change in species composition over time in a habitat that was not previously inhabited by organisms. No soil exists when primary succession begins. Bare rock surfaces, such as recently formed volcanic lava and rock scraped clean by glaciers, are examples of sites where primary succession might occur.

During primary succession on bare rock, the rock is eventually transformed into soil

Although the details vary from one site to another, in primary succession on bare rock, one might first observe a community of lichens, which are dual organisms composed of a fungus and an alga or cyanobacterium (Fig. 53–11*a*). Lichens are often the most important element in the **pioneer** community, which is the first community to appear in a succession. Lichens secrete acids that help to break the rock apart, which is how soil starts to form. Lichens also add valuable organic materials to the young soil. Over time, the lichen community may be replaced by mosses and drought-resistant ferns, followed in turn by tough grasses and herbs. Once sufficient soil has formed, grasses and herbs may be replaced by low shrubs (Fig. 53–11*b*), which in turn may be replaced by forest trees (Fig. 53–11*c*) in several distinct stages. Primary succession on bare rock, which may take hundreds or thousands of years, might proceed from a pioneer community to a forest community as follows:

lichens → mosses → grasses → shrubs → trees

Primary succession occurs on sand dunes

Lake and ocean shores have extensive sand dunes, which are deposited by wind and water. At first these dunes are blown about by the wind. The sand dune environment is severe, with high temperatures during the day and low temperatures at night. The sand may also be deficient in certain mineral nutrients needed by plants. As a result, few plants can tolerate the environmental conditions of a sand dune.

Grasses are common pioneer plants on sand dunes around the Great Lakes. As the grasses extend over the surface of a dune, their roots hold it in place, helping to stabilize it (Fig. 53–12). At this point, mat-forming shrubs may invade, further stabilizing the dune. Much later, shrubs may be replaced by pines, which years later are replaced by oaks. Because the soil fertility remains low, oaks are rarely replaced by other forest trees. A summary of how primary succession on sand dunes around the Great Lakes might proceed is

grasses → shrubs → pine trees → oak trees

Sometimes communities develop in an environment where a previous community existed

Secondary succession is the change in species composition over time in a habitat already substantially modified by a preexisting community. Soil is already present at these sites. An open area caused by a forest fire and an

(a)

(b)

(c)

FIGURE 53–11 Primary succession occurs after the retreat of glaciers. Although these photos were not taken in the same area, they show some of the stages of primary succession on glacial moraine (rocks, gravel, and sand deposited by a glacier). (***a***) The barren landscape exposed after the retreat of the glacier is initially colonized by lichens, then mosses. (***b***) At a later time, dwarf trees and shrubs colonize the area. (***c***) Still later, spruces dominate the community. (*a, c,* Wolfgang Kaehler; *b,* Visuals Unlimited/Glenn N. Oliver)

FIGURE 53–12 Empire Bluffs, Sleeping Bear Sand Dunes National Lakeshore, Michigan, shows several stages of primary succession on sand dunes. The first plants to colonize a sand dune are grasses (*foreground*), followed by low-growing shrubs (*midground on left*), and later by trees (*background*). (Joe Sroka/Dembinsky Photo Associates)

abandoned agricultural field are common examples of sites where secondary succession occurs.

Secondary succession has been studied extensively in abandoned farmland, particularly in North Carolina. Although it takes more than 100 years for secondary succession to occur at a single site, it is possible for a single researcher to study old field succession in its entirety by observing different sites in the same area that have been undergoing succession for different amounts of time. By examining court records, the biologist can accurately determine when each field was abandoned.

Abandoned farmland in North Carolina is colonized by a predictable succession of communities. The first year after cultivation ceases, the field is dominated by crabgrass. During the second year, horseweed, a larger plant that outgrows crabgrass, is the dominant species. Horseweed does not dominate for more than one year, however, because decaying horseweed roots inhibit the growth of young horseweed seedlings. In addition, horseweed does not compete well with the other plants—broomsedge, ragweed, and aster—that appear and become established during the third year. Typically, broomsedge, which is drought-tolerant, outcompetes aster, which is not.

In 5 to 15 years, the dominant plants in an abandoned field are pines such as shortleaf pine and loblolly pine. Through the buildup of litter (pine needles and branches) on the soil, pines produce conditions that cause the earlier dominant plants to decline in importance.

Over time, pines give up their dominance to hardwoods such as oaks. The replacement of pines by oaks depends primarily on the environmental alterations produced by the pines. The pine litter causes soil changes, such as an increase in water-holding capacity, that are necessary in order for young hardwood seedlings to become established. In addition, hardwood seedlings are more tolerant of shade than are pine seedlings. Secondary succession on abandoned farmland might proceed as follows:

crabgrass → horseweed → perennial weeds → pine trees → hardwood trees

Animal life also changes during secondary succession

As secondary succession proceeds, a progression of animal life follows the changes in vegetation. Although a few animals — the short-tailed shrew, for example — are found in all stages of abandoned farmland succession in North Carolina, most animals appear with certain stages and disappear with others. During the crabgrass and weed stages of secondary succession, the habitat is characterized by open fields that support grasshoppers, meadow mice, cottontail rabbits, and birds such as grasshopper sparrows and meadowlarks. As young pine seedlings become established, animals of open fields give way to animals common in mixed herbaceous and shrubby habitats. Now white-tailed deer, white-footed mice, ruffed grouse, robins, and song sparrows are common, whereas grasshoppers, meadow mice, grasshopper sparrows, and meadowlarks disappear. As the pine seedlings grow into trees, animals of the forest replace those common in mixed herbaceous and shrubby habitats. Cottontail rabbits give way to red squirrels; and ruffed grouse, robins, and song sparrows are replaced by warblers and veeries. Thus, each stage of succession supports its own characteristic animal life.

ECOLOGISTS CONTINUE TO STUDY THE NATURE OF COMMUNITIES

F. E. Clements, an ecologist who was professionally active during the first third of this century, was struck by the worldwide uniformity of large tracts of vegetation — for example, tropical rain forests in South America, Africa, and Southeast Asia (see Chapter 51). He also noted that even though the species composition of a community in a particular habitat might be different from that of a community in a habitat with a similar climate elsewhere in the world, overall the components of the two communities were usually analogous. He viewed communities as something like compound organisms — "superorganisms" whose member species cooperated with one another in a manner that resembled the cooperation of the parts of an individual organism's body. Clements' view was that a community went through certain stages of development, like those of an organism, and eventually reached an adult state; the developmental process was succession, and the adult state was the climax community. This cooperative view of the community, called the *organismic model,* stresses the interaction of the members, which tend to cluster in groups within discrete community boundaries.

Opponents of the organismic model, particularly H. A. Gleason, have held that biological interactions are less important in the production of communities than are climate, soil, and even chance; indeed, the concept of a community is questionable. It might be a classification category with no reality outside the minds of ecologists, reflecting little more than the tendency of organisms with similar environmental requirements to live in similar places. This school of thought, called the *individualistic model,* emphasizes species individuality, with each species having its own particular living requirements. It holds that communities are therefore not interdependent associations of organisms; rather, each species is spread across a continuum of areas that meets its own individual requirements.

Studies testing the organismic and individualistic hypotheses of communities do not seem to support Clements' interactive concept. Instead, most studies favor the individualistic model. However, this does not mean that species do not interact or form important associations within a community.

SUMMARY

I. A biological community consists of a group of organisms of different species that interact and live together. A community and its abiotic environment comprise an ecosystem.

II. The major roles of organisms in communities are those of autotrophs and heterotrophs.
   A. Producers are the photosynthetic autotrophs at the base of most food webs. They include plants and algae.
   B. Consumers are heterotrophs that feed on other organisms.
   C. Decomposers are microbial heterotrophs (bacteria and fungi) that recycle the components of dead organisms and organic wastes.

III. Predation is the consumption of one species (the prey) by another (the predator).
   A. During coevolution between predator and prey, the predator evolves more efficient ways to catch prey, and the prey evolves better ways to escape the predator.
   B. Two effective predator strategies are pursuit and ambush.
   C. Plants possess a number of adaptations that protect

them from being eaten, including spines, thorns, tough leathery leaves, and protective chemicals that are unpalatable or toxic to herbivores.

D. Animals possess many strategies that help them avoid being killed and eaten.
1. Many animals flee from predators, some have mechanical defenses, and some associate in groups.
2. Animals that possess chemical defenses also exhibit warning coloration.
3. Some animals hide from predators by blending into their surroundings.
4. In batesian mimicry, a harmless or edible species resembles another species that is dangerous in some way. Predators avoid the mimic as well as the model.
5. In müllerian mimicry, several different species, all of which are poisonous, harmful, or distasteful, resemble one another. Predators easily learn to avoid their common warning coloration.

IV. Symbiosis is any intimate association between two or more species.
A. In mutualism, both partners benefit.
B. In commensalism, one organism benefits and the other is unaffected.
C. In parasitism, one organism (the parasite) benefits while the other (the host) is harmed.

V. The distinctive lifestyle and role of an organism in a community is its ecological niche. An organism's ecological niche takes into account all aspects of the organism's existence.
A. Organisms are potentially able to exploit more resources and play a broader role in the life of their community than they actually do.
1. The potential ecological niche for an organism is its fundamental niche, whereas the niche it actually occupies is its realized niche.
2. Interspecific competition is one of the chief biological determinants of a species' ecological niche.
B. It is thought that no two species can occupy the same niche in the same community for an indefinite period of time because competitive exclusion occurs. In competitive exclusion, one species is excluded by another as a result of competition for a resource in limited supply.
C. An organism's limiting factors (such as the mineral content of soil, temperature extremes, and amount of precipitation) tend to restrict its realized niche.

VI. Community complexity is related to a variety of factors.
A. Community complexity is expressed in terms of species richness, or diversity.
B. Species diversity is often great when there are many potential ecological niches, when a community is not isolated or severely stressed, in ecotones, and in communities with long histories.

VII. Succession is the orderly replacement of one community by another.
A. Primary succession occurs in an area that has not previously been inhabited (for example, bare rock or shifting sand dunes).
B. Secondary succession begins in an area where there was a preexisting community and well-formed soil (for example, abandoned farmland).

VIII. There are two views of the nature of communities.
A. The organismic model views a community as a "superorganism" that goes through certain stages of development (succession) toward adulthood (climax). In this view, biological interactions are primarily responsible for species composition, and organisms are highly interdependent.
B. Most ecologists support the individualistic model, which challenges the concept of a highly interdependent community. According to this model, factors other than biological interactions are the primary determinants of species composition in a community.

SELECTED KEY TERMS

batesian mimicry
carnivore
character displacement
coevolution
commensalism
community
competitive exclusion
consumer
decomposer
detritivore
detritus
detritus feeder
ecological niche
ecosystem
ecotone
edge effect
fundamental niche
habitat
herbivore
limiting factor
müllerian mimicry
mutualism
niche
omnivore
parasitism
pathogen
pioneer
predation
primary consumer
primary producer
primary succession
realized niche
saprotroph
secondary consumer
secondary succession
species richness
succession
symbiosis
warning coloration

POST-TEST

1. A __________ is an association of populations of different species living together in one area.
2. Ecologically speaking, grasses are classified as primary __________ and foxes are classified as consumers.
3. __________ eat meat, __________ eat plants, and __________ eat a variety of organisms.
4. Monarch and viceroy butterflies are an example of __________ mimicry.
5. A symbiotic association in which organisms are beneficial to one another is known as __________ .
6. The symbiotic association exemplified by silverfish and army ants that live together and share the food caught by the army ants is __________ .
7. An organism's ecological __________ is the totality of its adaptations, its use of resources, and the lifestyle to which it is fitted.

8. The __________ __________ is the ideal ecological niche that an organism could potentially occupy.
9. Competition from other species helps to determine an organism's __________ niche.
10. "Complete competitors cannot coexist" is a statement of the principle of __________ __________.
11. The __________ __________ signifies that species diversity is greater where two communities meet than at the center of either community.
12. The change in composition of species in a community over time is called __________.
13. A succession of communities on bare rock is an example of __________ succession.
14. Lichens are called __________ because they are the first inhabitants of bare rock.
15. An abandoned field is a site of __________ succession.
16. The tendency for two similar species to differ from one another more markedly in areas where they occur together is known as __________ __________.
17. An unpalatable species demonstrates its threat to potential predators by displaying conspicuous colors, known as __________ __________.

REVIEW QUESTIONS

1. Distinguish a community from an ecosystem.
2. Distinguish between a decomposer and a detritus feeder.
3. Describe how natural selection has affected predator-prey relationships.
4. List the three kinds of symbiosis and give an example of each.
5. Why is an organism's realized niche usually narrower, or more restricted, than its fundamental niche?
6. What is the principle of competitive exclusion?
7. Describe at least two of the determinants of species diversity in a community.
8. Distinguish between primary and secondary succession and give an example of each.
9. Contrast the organismic and individualistic models of the nature of communities.

YOU MAKE THE CONNECTION

1. In what symbiotic relationships are humans involved?
2. Describe the ecological niche of humans. Do you think our realized niche has changed during the past 1000 years? Why or why not?
3. Which part of the stickleback experiment described in the *Cutting Edge* box exhibited directional selection? Explain your answer.
4. How is interspecific competition related to secondary succession on old fields?

RECOMMENDED READINGS

Beardsley, T. "Recovery Drill." *Scientific American,* Vol. 263, No. 5, November 1990. Some conventional ideas about how communities respond to environmental catastrophes are being challenged by the recovery of Mount St. Helens.

Boucher, D. H. "Growing Back after Hurricanes." *BioScience,* Vol. 40, No. 3, March 1990. The significance of periodic environmental catastrophes, such as hurricanes, for communities is causing ecologists to reconsider the notion of a climax community.

Conniff, R. "Yellowstone's 'Rebirth' Amid the Ashes Is Not Neat or Simple, But It's Real." *Smithsonian,* September 1989. Secondary succession of the forests that were burned during the September 1988 fire in Yellowstone.

Lutz, R.A., and R.M. Haymon. "Rebirth of a Deep-Sea Vent." *National Geographic,* Vol. 186, No. 5, November 1994. Recently, undersea lava flows buried some hydrothermal vent communities off the west coast of Mexico. The authors follow their quick rebirth as nearby vent species migrated into the area.

Mohlenbrock, R.H. "Mount St. Helens, Washington." *Natural History,* June 1990. Secondary succession in areas devastated by the eruption of Mount St. Helens.

Moore, P.D. "Vegetation's Place in History." *Nature,* Vol. 347, October 25, 1990. A brief discussion of differing views on the nature of communities.

Science, Vol. 269, July 21, 1995. This issue contains a number of news and research articles that highlight the science of ecology.

CHAPTER 54

Ecosystems and the Ecosphere

Planet Earth, with its living organisms, hydrosphere, lithosphere, and atmosphere, is the largest ecosystem. (NASA)

Almost completely isolated from everything in the Universe but sunlight, our planet Earth has often been compared to a vast spaceship whose life support system consists of the organisms that inhabit it, plus energy from the sun. Earth's organisms produce oxygen, cleanse the air, adjust gases, transfer energy, and recycle waste products with great efficiency. Yet none of these processes would be possible without the abiotic (nonliving) environment of our spaceship Earth. Much of the climate to which organisms have adapted is determined by the sun, which warms the planet, powers the hydrologic cycle (causes precipitation), and drives ocean currents and atmospheric circulation patterns.

The science of ecology deals with the abiotic environment as well as with living organisms. Individual communities and their abiotic environments are **ecosystems.** An ecosystem encompasses all the interactions among organisms living together in a particular place, and among those organisms and their abiotic environment. Earth, which encompasses the biosphere and its interactions with the hydrosphere, lithosphere, and atmosphere (see Chapter 51), is the **ecosphere,** the largest ecosystem.

This chapter will develop three key concepts about ecosystems. First, matter, the material of which organisms are composed, moves in numerous cycles from one part of an ecosystem to another—that is, from one organism to another, and from organisms to the abiotic environment and back again. All materials vital to life are continually recycled through ecosystems and so become available to new generations of organisms.

Second, unlike matter, which is cyclic in ecosystems, energy flow is linear. Energy moves through food webs of ecosystems in a one-way direction from the environment to organisms and back to the environment. Once energy has been used to do biological work for an organism, it is unavailable to other organisms because, as work is performed, the energy is changed to heat and given off into the cooler surroundings. Energy cannot be recycled and reused.

Third, the abiotic environment causes conditions that determine where and how successfully species live. We will consider four aspects of the abiotic environment that have helped shape the biotic component of ecosystems: solar radiation, the atmosphere, the ocean, and weather and climate.

LEARNING OBJECTIVES

After you have studied this chapter you should be able to

1. Compare how matter and energy operate in ecosystems.
2. Diagram the carbon, nitrogen, phosphorus, and hydrologic cycles.
3. Summarize the concept of energy flow through a food web.
4. Draw and explain typical pyramids of numbers, biomass, and energy.
5. Distinguish between gross primary productivity and net primary productivity.
6. Summarize the effects of solar energy on Earth's temperatures.
7. Discuss the roles of solar energy and the Coriolis effect in the production of global air and water flow patterns.
8. Define El Niño-Southern Oscillation (ENSO) and describe some of its effects.
9. Give three causes of regional precipitation differences.

MATTER CYCLES THROUGH ECOSYSTEMS

Matter cycles from the living world to the abiotic environment and back again. We call these cycles **biogeochemical cycles.** Earth is essentially a closed system (one into which matter does not enter and from which matter cannot escape). The materials used by organisms cannot be "lost," although they can end up in locations outside the reach of organisms. Usually, however, materials are reused and often recycled both within and among ecosystems.

Four different biogeochemical cycles of matter—carbon, nitrogen, phosphorus, and water—are representative of all biogeochemical cycles. These four cycles are particularly important to organisms, as they involve materials used to make the chemical compounds of cells. Carbon, nitrogen, and water have gaseous components and so cycle over large distances of the atmosphere with relative ease. Phosphorus, however, is an element that is completely nongaseous, and as a result, only local cycling of phosphorus occurs easily.

Carbon dioxide is the pivotal molecule of the carbon cycle

Carbon must be available to organisms because proteins, nucleic acids, lipids, carbohydrates, and other molecules essential to life contain carbon. Carbon is present in the atmosphere as the gas carbon dioxide (CO_2), which makes up approximately 0.03% of the atmosphere. It is also present in the ocean as dissolved carbon dioxide, carbonate (CO_3^{2-}) and bicarbonate (HCO_3^-); and in rocks such as limestone. The global movement of carbon between the abiotic environment, including the atmosphere, and organisms is known as the **carbon cycle** (Fig. 54–1).

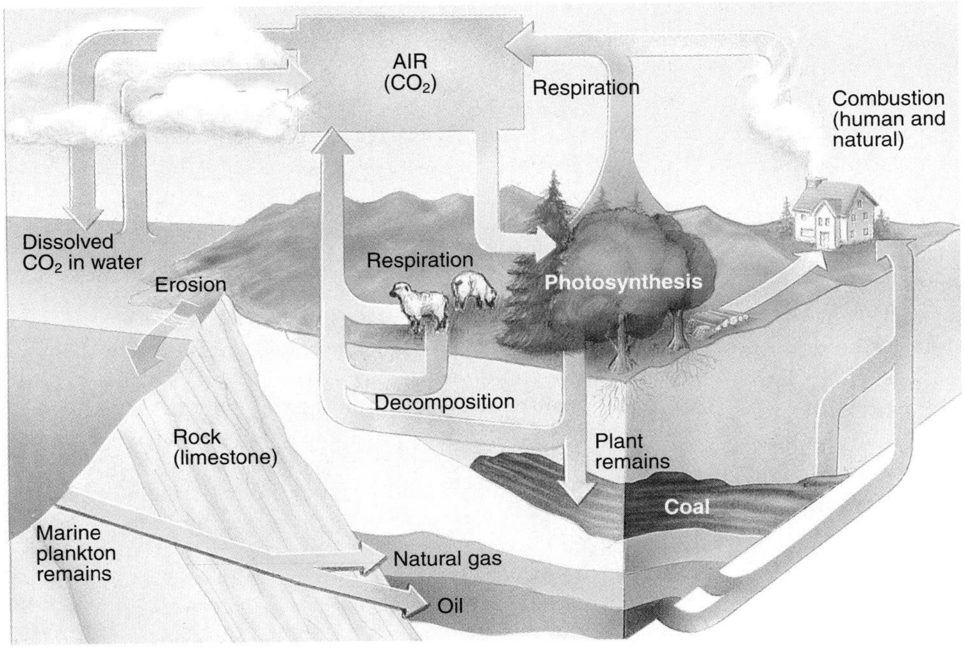

FIGURE 54–1 In the carbon cycle, carbon cycles between the abiotic environment and organisms. Carbon (as carbon dioxide) enters organisms when primary producers photosynthesize. Carbon returns to the environment by respiration, combustion, and erosion. The carbon in fossil fuels, limestone rock, and marine animal shells may take millions of years to cycle back to the biotic world.

During photosynthesis, plants, algae, and cyanobacteria remove carbon dioxide from the air and *fix,* or incorporate, it into complex chemical compounds such as glucose. Plants use much of the glucose to make cellulose, amino acids, nucleic acids, and other compounds. Thus, photosynthesis incorporates carbon from the abiotic environment into the biological compounds of primary producers. Many of these compounds are used as fuel for cellular respiration by the producer that made them, by a consumer that eats the producer, or by a decomposer that breaks down the remains of the producer or consumer. Thus, carbon dioxide is returned to the atmosphere by the process of cellular respiration. A similar carbon cycle occurs in aquatic ecosystems between aquatic organisms and dissolved carbon dioxide in the water.

Sometimes the carbon in biological molecules is not recycled back to the abiotic environment for some time. For example, a large amount of carbon is stored in the wood of trees, where it may stay for several hundred years or even longer. In addition, millions of years ago vast coal beds formed from the bodies of ancient trees that were buried under anaerobic conditions before they had fully decayed. Similarly, the oils of unicellular marine organisms probably gave rise to the underground deposits of oil and natural gas that accumulated in the geological past. Coal, oil, and natural gas, called **fossil fuels** because they formed from the remains of ancient organisms, are vast depositories of carbon compounds—the end products of photosynthesis that occurred millions of years ago.

The carbon in coal, oil, natural gas, and wood may be returned to the atmosphere by the process of burning, or combustion. In combustion, organic molecules are rapidly oxidized (combined with oxygen), converting them into carbon dioxide and water with an accompanying release of light and heat.

An even greater amount of carbon that leaves the carbon cycle for millions of years is incorporated into the shells of marine organisms. When these organisms die, their shells sink to the ocean floor and are covered by sediments. The shells form seabed deposits thousands of meters thick that are eventually cemented together to form a sedimentary rock called limestone. Earth's crust is dynamically active, and over millions of years, sedimentary rock on the bottom of the sea floor may be lifted up to form land surfaces. (The summit of Mount Everest, for example, is composed of sedimentary rock.) When limestone is exposed by the process of geological uplift, it is slowly eroded away by chemical and physical weathering processes. This returns the carbon to the water and atmosphere, where it is available to participate in the carbon cycle once again.

Thus photosynthesis removes carbon from the abiotic environment and incorporates it into biological molecules, while cellular respiration, combustion, and erosion return carbon to the water and atmosphere of the abiotic environment.

Bacteria are essential to the nitrogen cycle

Nitrogen is crucial for all organisms because it is an essential part of proteins and nucleic acids. Because Earth's atmosphere is about 80% nitrogen gas (N_2), it would appear that there could be no possible shortage of nitrogen for organisms. However, molecular nitrogen is so stable that it does not readily combine with other elements. Therefore, N_2 must be broken apart before the nitrogen can combine with other elements to form proteins and nucleic acids. The overall reaction that breaks up N_2 and combines nitrogen with such elements as oxygen and hydrogen requires a great deal of energy.

The **nitrogen cycle,** in which nitrogen cycles between the abiotic environment and organisms, has five steps: nitrogen fixation, nitrification, assimilation, ammonification, and denitrification (Fig. 54–2). All of these steps except assimilation are performed by bacteria.

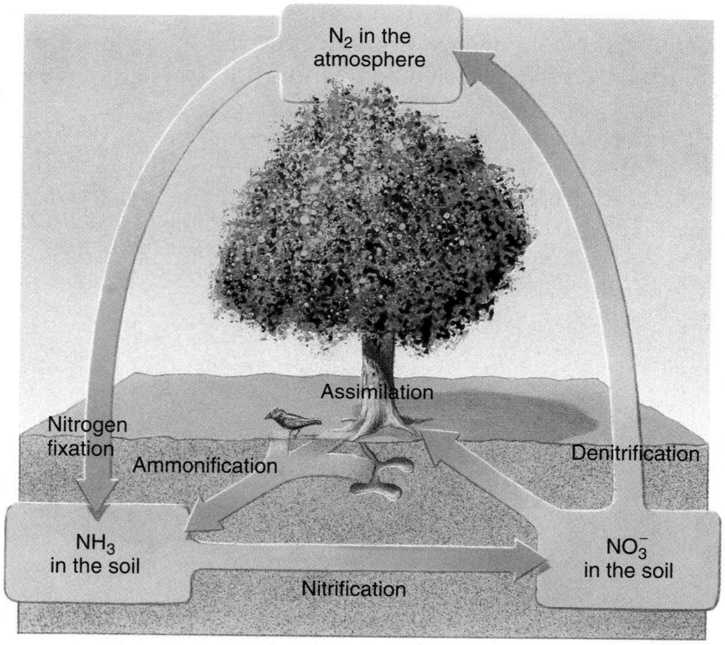

FIGURE 54–2 The nitrogen cycle has five steps that involve organisms. One, nitrogen-fixing bacteria convert atmospheric nitrogen into ammonia. Two, ammonia is converted to nitrates by nitrifying bacteria in the soil. Nitrate is the main form of nitrogen absorbed by plants. Three, plants assimilate nitrate and use it to produce proteins and nucleic acids; animals eat plant proteins and produce animal proteins as a part of assimilation. Four, when organisms die, their nitrogen compounds are broken down by ammonifying bacteria, and ammonia is released. Five, some nitrogen is returned to the atmosphere by denitrifying bacteria, which convert nitrate to molecular nitrogen.

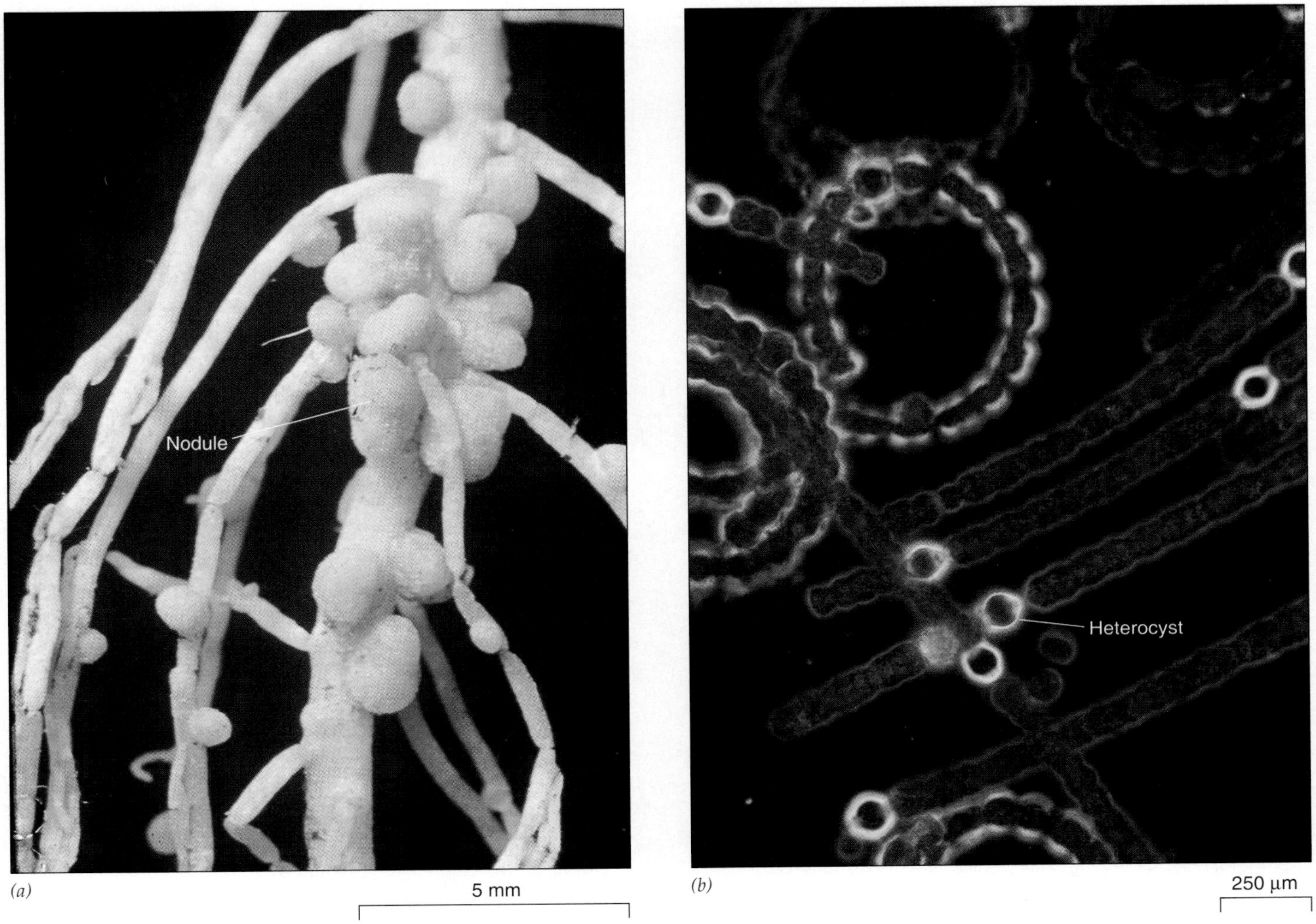

FIGURE 54–3 Nitrogenase, the enzyme involved in biological nitrogen fixation, only functions in the absence of oxygen. (*a*) Root nodules of a pea plant provide an oxygen-free environment for nitrogen-fixing *Rhizobium* bacteria that live in them. (*b*) *Anabaena,* a cyanobacterium, has distinctive oxygen-excluding heterocysts in which nitrogen fixation occurs. (*a,* Hugh Spencer/Photo Researchers, Inc.; *b,* Dennis Drenner)

The first step in the nitrogen cycle, biological **nitrogen fixation,** involves the conversion of gaseous nitrogen (N_2) to ammonia (NH_3). This process is called nitrogen fixation because nitrogen is *fixed* into a form that organisms can use. Although nitrogen can also be fixed as nitrate by combustion, volcanic action, lightning discharges, and industrial means (each of these processes supplies enough energy to break up molecular nitrogen), most nitrogen fixation is biological. It is carried out by nitrogen-fixing bacteria, including cyanobacteria, in soil and aquatic environments. Nitrogen-fixing bacteria employ an enzyme called **nitrogenase** to break up molecular nitrogen and combine it with hydrogen.

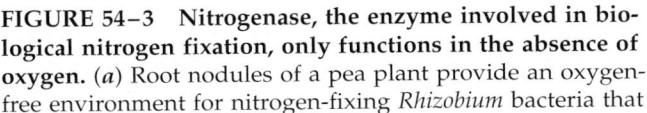

Because nitrogenase functions only in the absence of oxygen, the bacteria that fix nitrogen must insulate the enzyme from oxygen by some means. Some nitrogen-fixing bacteria live beneath layers of oxygen-excluding slime on the roots of a number of plants. Other important nitrogen-fixing bacteria, in the genus *Rhizobium,* live in special swellings, or **nodules** (Fig. 54–3*a*), on the roots of legumes such as beans and peas and some woody plants.

In aquatic environments, most nitrogen fixation is performed by cyanobacteria. Filamentous cyanobacteria have special oxygen-excluding cells called **heterocysts** that function to fix nitrogen (Fig. 54–3*b*). Some water ferns have cavities in which cyanobacteria live, in a manner comparable to the way *Rhizobium* lives in root nodules of legumes. Other cyanobacteria fix nitrogen in symbiotic association with cycads or some other terrestrial plants, or as the photosynthetic partner of certain lichens.

The reduction of nitrogen gas to ammonia by nitrogenase is a remarkable accomplishment by living organisms, achieved without the tremendous heat, pressure, and energy required to do the same thing during the manufacture of commercial fertilizers. Even so, nitrogen-fixing bacteria must consume the energy in 12 grams of glucose (or the equivalent) to fix a single gram of nitrogen biologically.

The second step of the nitrogen cycle is **nitrification,** the conversion of ammonia (NH_3) to nitrate (NO_3^-). Nitrification, a two-step process, is accomplished by soil bacteria. First, the soil bacteria *Nitrosomonas* and *Nitrococcus* convert ammonia to nitrite (NO_2^-). Then the soil bacterium *Nitrobacter* oxidizes nitrite to nitrate. The process of nitrification furnishes these bacteria, called nitrifying bacteria, with energy.

In the third step, called **assimilation,** roots absorb either nitrate (NO_3^-) or ammonia (NH_3) that was formed by nitrogen fixation and nitrification, and incorporate the nitrogen into plant proteins and nucleic acids. When animals consume plant tissues they assimilate nitrogen as well, by taking in plant nitrogen compounds and converting them to animal compounds.

The fourth step is **ammonification,** which is the conversion of organic nitrogen compounds into ammonia. Ammonification begins when organisms produce nitrogen-containing wastes such as urea (in urine) and uric acid (in the wastes of birds). These substances, along with the nitrogen compounds in dead organisms, are decomposed, releasing the nitrogen into the abiotic environment as ammonia (NH_3). The bacteria that perform ammonification in both the soil and aquatic environments are called ammonifying bacteria. The ammonia produced by ammonification is available for the processes of nitrification and assimilation; as a matter of fact, most available nitrogen in the soil derives from the recycling of organic nitrogen by ammonification.

The fifth, and final, step of the nitrogen cycle is **denitrification,** which is the reduction of nitrate (NO_3^-) to gaseous nitrogen (N_2). Denitrifying bacteria reverse the action of nitrogen-fixing and nitrifying bacteria by returning nitrogen to the atmosphere as nitrogen gas. Denitrifying bacteria are anaerobic and therefore live and grow best where there is little or no free oxygen. For example, they are found deep in the soil near the water table, an environment that is nearly oxygen-free.

The phosphorus cycle lacks a gaseous component

In the **phosphorus cycle,** phosphorus, which does not exist in a gaseous state and therefore does not enter the atmosphere, cycles from the land to sediments in the ocean and back to the land (Fig. 54–4). As water runs over rocks containing phosphorus, it gradually wears away the surface and carries off inorganic phosphate (PO_4^{3-}) molecules.

The erosion of phosphorus rocks releases phosphate into the soil, where it is taken up by roots. Once in plant cells, phosphate is incorporated into a variety of biolog-

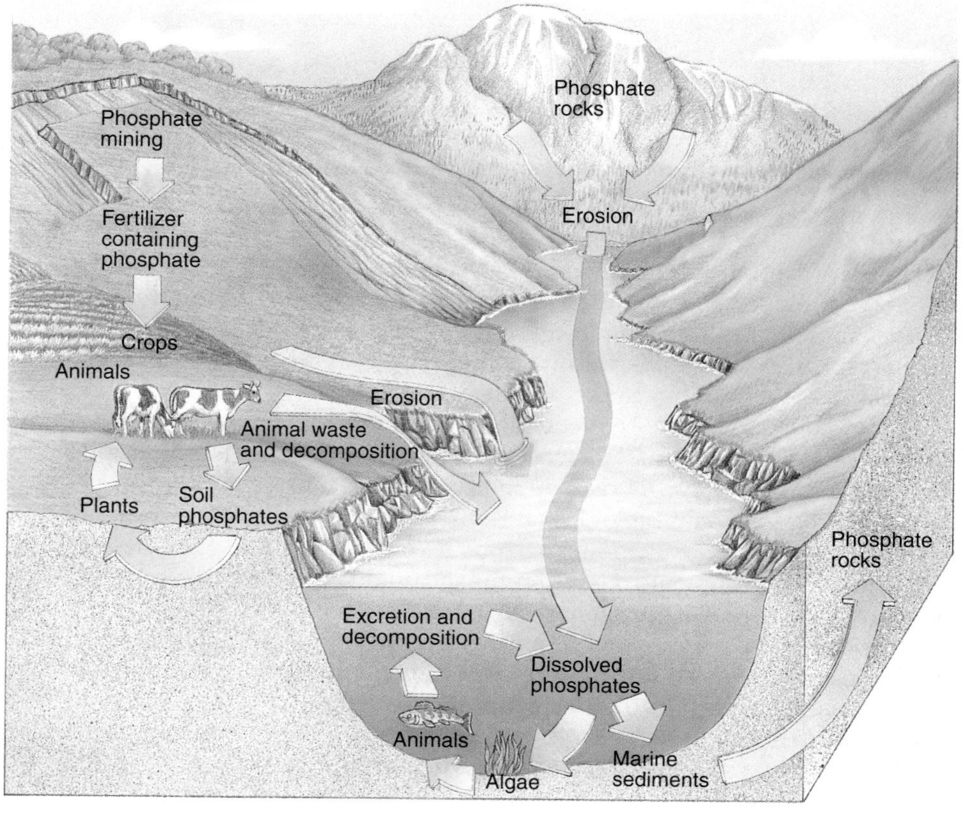

FIGURE 54–4 In the phosphorus cycle, phosphorus cycles between the abiotic environment and organisms. Recycling of phosphorus (as phosphate) is slow because no biologically important form of phosphorus is gaseous. Phosphate that becomes part of marine sediments may take millions of years to solidify into rock, uplift as mountains, and erode again to become available to organisms.

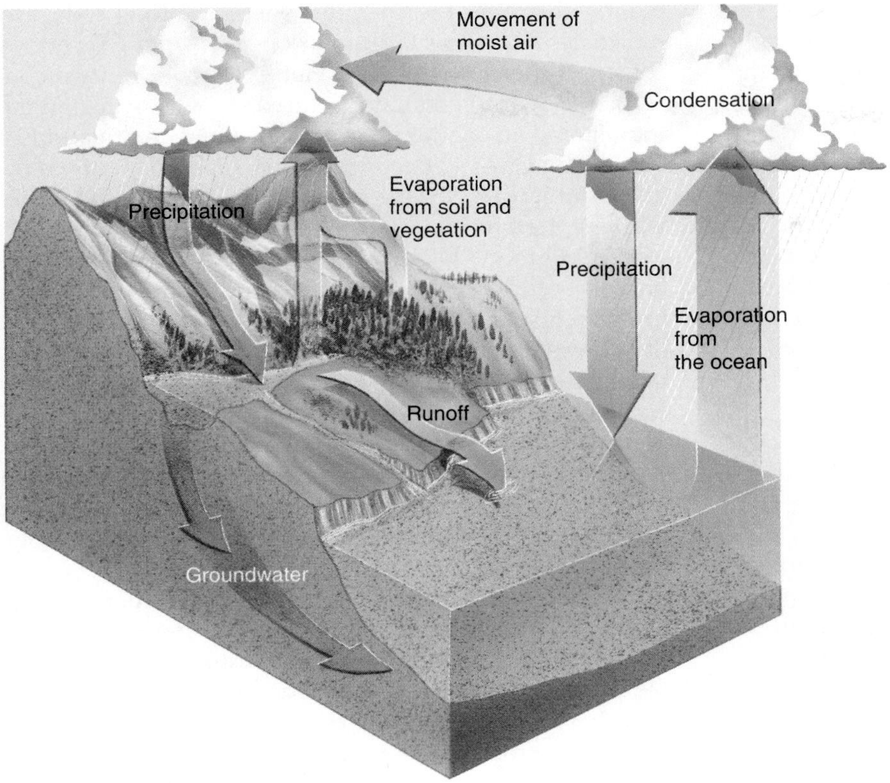

FIGURE 54–5 Water cycles from the ocean to the atmosphere to the land and back to the ocean in the hydrologic cycle. Although some water molecules are unavailable for thousands of years (locked up in polar ice, for example), all water molecules eventually cycle through the hydrologic cycle.

ical molecules, including nucleic acids. Animals obtain most of their required phosphorus from the food they eat, although in some places drinking water may contain a substantial amount of inorganic phosphate. Phosphate released by decomposers becomes part of the pool of inorganic phosphate in the soil that can be reused by plants.

Phosphorus cycles through aquatic ecosystems in much the same way that it does through terrestrial ecosystems. Dissolved phosphate enters aquatic ecosystems through absorption by algae and aquatic plants, which are consumed by plankton and larger organisms. These are in turn eaten by a variety of fin and shellfish. Ultimately, decomposers that break down wastes and dead organisms release inorganic phosphate into the water, making it available for use by aquatic producers.

Phosphate can be lost from biological cycles. Some phosphate is carried from the land by streams and rivers to the ocean, where it can be deposited on the sea floor and remain for millions of years. The geological process of uplift may someday expose these sea floor sediments as new land surfaces, from which phosphate will be once again eroded. Phosphate is also mined for agricultural use in phosphate fertilizers.

Some phosphate in the aquatic food web finds its way back to the land. A tiny portion of fish and aquatic invertebrates are eaten by sea birds, which may defecate on the land where they roost. Guano, the manure of sea birds, contains large amounts of phosphate and nitrate. Once on land, these minerals may be absorbed by plant roots. The phosphate contained in guano may enter terrestrial food webs in this way, although the amounts involved are quite small.

Water circulates in the hydrologic cycle

Life on planet Earth would be impossible without water, because all life forms, from bacteria to plants and animals, contain it. Water continuously circulates from the ocean to the atmosphere to the land and back to the ocean, providing us with a renewable supply of purified water on land. This complex cycle, known as the **hydrologic cycle,** results in a balance between water in the ocean, on the land, and in the atmosphere (Fig. 54–5). Water moves from the atmosphere to the land and ocean in the form of precipitation (rain, sleet, snow, or hail). When water evaporates from the ocean surface and from soil, streams, rivers, and lakes on land, it forms clouds in the atmosphere. In addition, transpiration, the loss of water vapor from land plants, adds more water to the atmosphere.

Water may evaporate from land and reenter the atmosphere directly. Alternatively, it may flow in rivers and streams to coastal estuaries, where fresh water meets the ocean. The movement of surface water from land to ocean is called *runoff,* and the area of land being drained by runoff is called a *watershed.* Water also percolates (seeps) downward in the soil to become *groundwater,* where it is trapped and held for a time. Groundwater supplies water to the soil, to streams and rivers, and to plants.

Regardless of its physical form (solid, liquid, or vapor) or location, every molecule of water eventually moves through the hydrologic cycle. As is true of the other cycles, water (in the form of glaciers and icecaps) can be lost from the cycle for thousands of years.

Tremendous quantities of water are cycled annually between Earth and its atmosphere. An estimated 396,000 cubic kilometers (95,000 cubic miles), an amount that is difficult to visualize, enters the atmosphere each year. Approximately three-fourths of this water reenters the ocean directly as precipitation over water; the remainder falls on land.

THE FLOW OF ENERGY THROUGH ECOSYSTEMS IS LINEAR

The passage of energy in a one-way direction through an ecosystem is known as **energy flow.** Energy enters an ecosystem as radiant energy (sunlight), a tiny portion (far less than 1%) of which is trapped by primary producers during photosynthesis. The energy, now in chemical form, is stored in the bonds of organic molecules such as glucose. When these molecules are broken apart by cellular respiration, energy becomes available (in the form of ATP) to do work such as repairing tissues, producing body heat, or reproducing. As the work is accomplished, energy escapes the organisms and dissipates into the environment as heat. Ultimately, this heat energy radiates into space. Thus, once energy has been used by an organism, it is unavailable for reuse (Fig. 54–6; see also the discussion of the second law of thermodynamics in Chapter 6).

Energy flow describes who eats whom in ecosystems

In an ecosystem, energy flow occurs in **food chains,** in which energy from food passes from one organism to the next in a sequence. Primary producers form the beginning of the food chain by capturing the sun's energy through photosynthesis. Herbivores (and omnivores) eat plants, obtaining the chemical energy of the producers' molecules as well as building materials from which they construct their own tissues. Herbivores are in turn consumed by carnivores and omnivores, who reap the energy stored in the herbivores' molecules. At the end of food chains are the decomposers, which break down organic molecules in the remains (carcasses and body wastes) of all members of the food chain.

Simple food chains as just described rarely occur in nature, because few organisms eat just one kind, or are eaten by just one kind, of other organism. More typically, the flow of energy and materials through ecosystems takes place in accordance with a range of food choices for each organism involved. In an ecosystem of average complexity, hundreds of alternative pathways are possible. Thus, a **food web,** which is a complex of interconnected food chains in an ecosystem, is a more realistic model of the flow of energy and materials through ecosystems (Fig. 54–7).

The most important thing to remember about energy flow in ecosystems is that it is linear, or one-way. That is, energy can move along a food web from one organism to the next as long as it is not used. Once energy has been used by an organism, however, it is lost as heat and is unavailable for any other organism in the ecosystem.

Ecological pyramids illustrate how ecosystems work

Each level in a food web is called a **trophic level.** The first trophic level is formed by primary producers (organisms that photosynthesize), the second by primary consumers (herbivores), the third by secondary consumers (carnivores and omnivores), and so on (Fig. 54–8).

Ecologists sometimes compare trophic levels by determining the number of organisms, the biomass, or the relative energy found at each level. This information is presented graphically as **ecological pyramids.**

A **pyramid of numbers** shows the number of organisms at each trophic level in a given ecosystem, with greater numbers illustrated by a wider pyramid (Fig. 54–9). In most pyramids of numbers, each successive

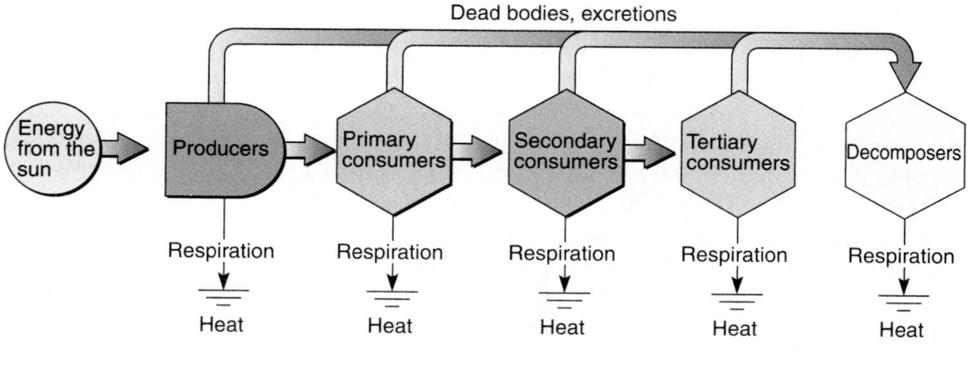

FIGURE 54–6 Energy flow occurs linearly (in a one-way direction) through ecosystems. Energy enters ecosystems from an external source (the sun) and exits as heat loss. Much of the energy acquired by a given trophic level is used for metabolic purposes at that level and is therefore unavailable to the next trophic level.

FIGURE 54–7 This deciduous forest food web is greatly simplified compared to what actually happens in nature. Groups of species are lumped into single categories such as "Spiders," many other species are not included, and numerous links in the web are not shown.

trophic level is occupied by fewer organisms. Thus, the number of herbivores (such as zebras and wildebeests) is greater than the number of carnivores (such as lions). Inverted pyramids of numbers, in which higher trophic levels have *more* organisms than lower trophic levels, are often observed among decomposers, parasites, tree-dwelling herbivorous insects, and similar organisms. One tree can provide food for thousands of leaf-eating insects, for example.

A **pyramid of biomass** illustrates the total biomass at each successive trophic level. **Biomass** is a quantitative estimate of the total mass, or amount, of living material;

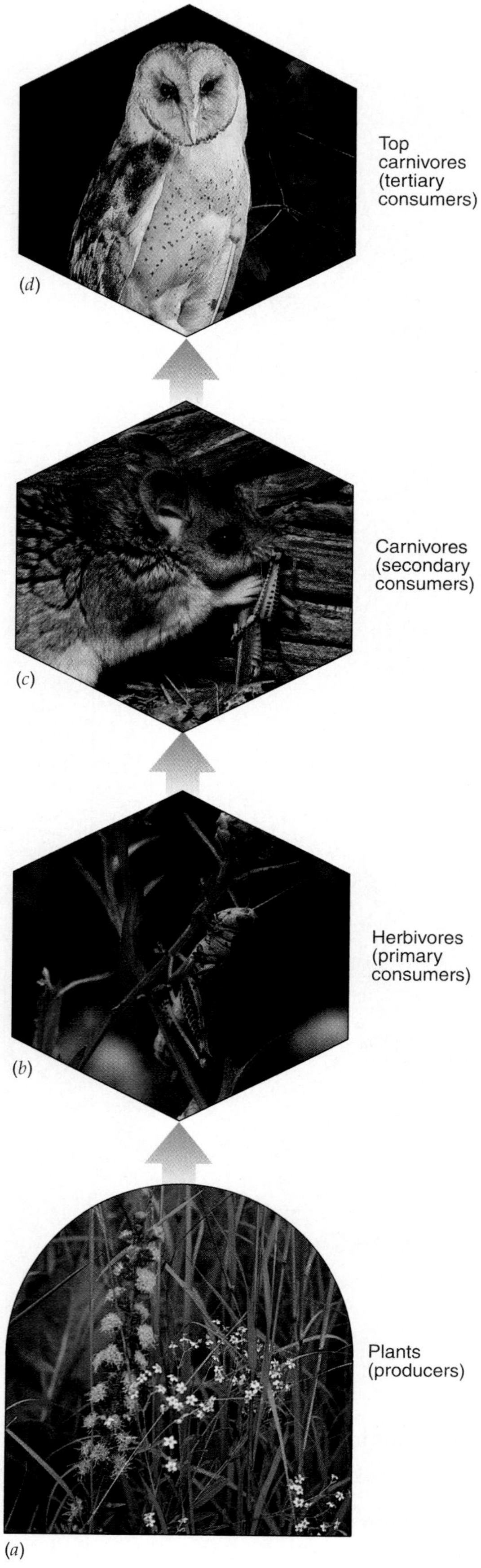

FIGURE 54–8 Trophic levels help us simplify food webs by grouping organisms based on their position in the web. (***a***) Plants occupy the first trophic level in an ecosystem. (***b***) The grasshopper is a primary consumer and occupies the second trophic level. (***c***) The third trophic level is represented by the grasshopper mouse, which eats grasshoppers. (***d***) The barn owl is a top carnivore at the fourth trophic level. (*a,* Courtesy of Virginia Kline, University of Wisconsin; *b,* Animals Animals © 1996 Richard Kolar; *c,* Tom McHugh/Photo Researchers, Inc.; *d,* John Mielcarek/Dembinsky Photo Associates)

it indicates the amount of fixed energy at a particular time. Units of measure vary; biomass may be represented as total volume, dry weight, or live weight. Typically, the pyramids illustrate a progressive reduction of biomass in succeeding trophic levels (Fig. 54–10*a*). Assuming an average biomass reduction of about 90% for each trophic level,[1] 10,000 kilograms of grass should be able to support 1000 kilograms of crickets, which in turn support 100 kilograms of frogs. By this logic, the biomass of frog-eaters (such as herons) could only weigh, at most, about 10 kilograms. From this brief exercise, you can see that although carnivores may eat no vegetation, a great deal of vegetation is required to support them.

Occasionally we find an inverted pyramid of biomass in which the primary consumers outweigh the producers

[1]The 90% reduction in biomass is an approximation; actual biomass reduction varies widely in nature.

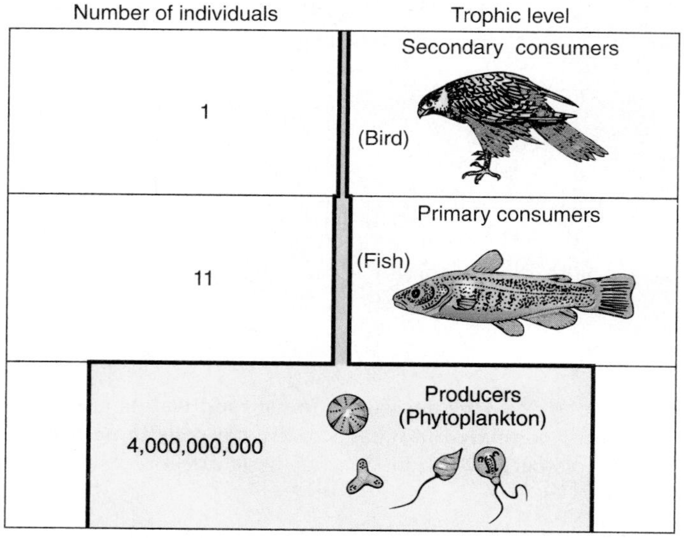

FIGURE 54–9 A pyramid of numbers is based on the number of organisms at each trophic level. Typically, there are more primary producers than primary consumers, more primary consumers than secondary consumers, and so on.

Making the Connection

Interactions Among Trophic Levels in a Food Web

Do food webs indicate relationships other than predator-prey interactions? It turns out that both negative and positive interactions occur in a food web. Because food webs are descriptions of "who eats whom," they indicate the negative effects that predators have on their prey. For example, consider a simple food chain: grass → field mouse → owl. The owl, which kills and eats mice, obviously exerts a negative effect on the mouse population; in like manner, field mice, which eat grass seeds, reduce the grass population.

One trophic level in a food web also influences other trophic levels to which it is *not* directly linked. Primary producers and top carnivores do not exert direct effects on one another, yet each is indirectly affected by the other. In our example, the owls help the producers by keeping the population of seed-eating mice under control. Likewise, the grasses benefit owls by supporting a population of mice on which the owl population feeds. These indirect interactions are often mutualistic in nature and may be as important as direct, predator-prey interactions (see Chapter 53) in food web dynamics. ▲

(Fig. 54–10*b*). In these instances, the producers—usually algae—are highly productive and reproduce quickly despite being rapidly consumed by efficient herbivores such as zooplankton and small fish. Thus, although at any point in time relatively few algae are present, the rate of biomass production of the primary consumers is much less than that of the producers.

A **pyramid of energy** indicates the energy content (usually expressed in calories) in the biomass of each trophic level. These pyramids show that most energy dissipates into the environment when going from one trophic level to another (Fig. 54–11). Less energy reaches each successive trophic level from the level beneath it because some of the energy at the lower level is used by those organisms to perform work, while some of it is lost (no biological process is 100% efficient). Energy pyramids explain why there are few trophic levels: *Food webs are short because of the dramatic reduction in energy content that occurs at each successive trophic level.*

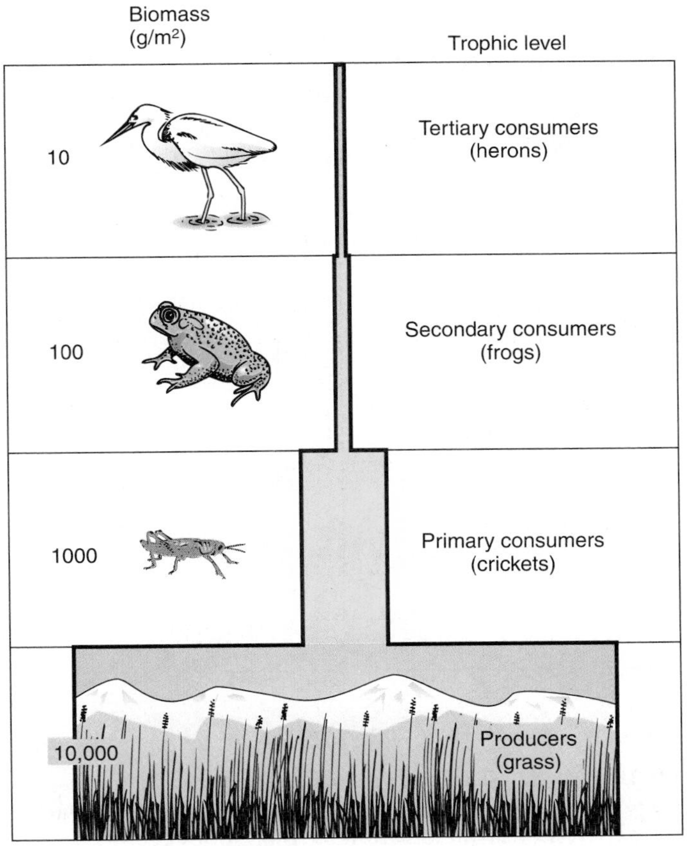

(*a*)

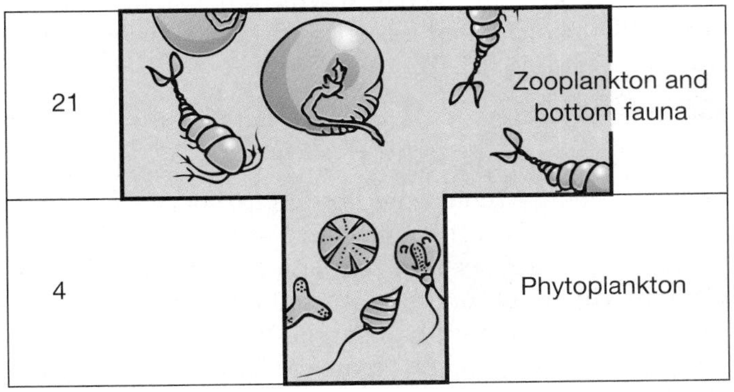

(*b*)

FIGURE 54–10 Pyramids of biomass are based on the biomass at each trophic level and typically resemble pyramids of numbers. (*a*) A pyramid of biomass for a hypothetical area of a temperate grassland. (*b*) An inverted biomass pyramid occurs when highly productive lower trophic levels experience high rates of turnover.

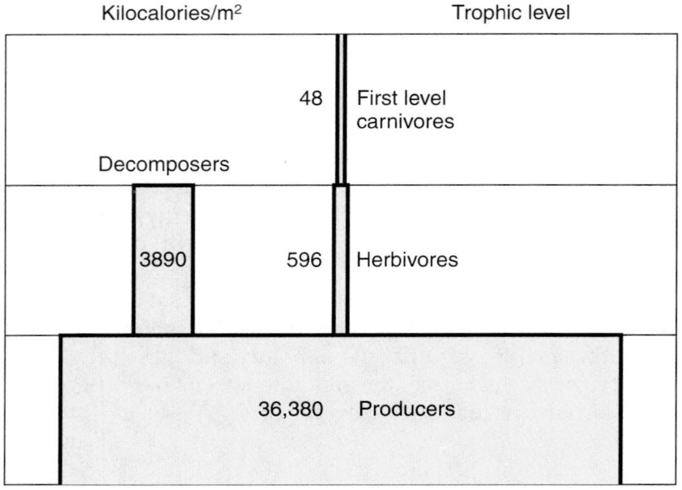

FIGURE 54–11 Energy flow, the functional basis of ecosystem structure, is represented by a pyramid of energy. Note the relatively large role played by bacterial decomposers, which actually operate on all trophic levels.

Ecosystems vary in productivity

The **gross primary productivity** (GPP) of an ecosystem is the *rate* at which energy is captured during photosynthesis.[2] Thus, gross primary productivity is the total amount of photosynthetic energy captured in a given period of time. Of course, plants must respire to provide energy for their life processes, and cellular respiration acts as a drain on photosynthetic output. Energy that remains in plant tissues after cellular respiration has occurred is called **net primary productivity (NPP).** That is, net primary productivity is the amount of biomass (the energy stored in plant tissues) found in excess of that broken down by a plant's cellular respiration. Net primary productivity represents the *rate* at which this organic matter is actually incorporated into plant tissues in order to produce growth (Fig. 54–12).

Net primary productivity = Gross primary productivity − Plant respiration
(Plant growth) (Total photosynthesis)

Only the energy represented by net primary productivity is available for consumers, and of this energy only a portion is actually used by them. Both gross and net primary productivity can be expressed in terms of kilocalories (energy fixed by photosynthesis) per square meter per year, or of dry weight (grams of carbon incorporated into tissue) per square meter per year.

A number of factors may interact to determine productivity. Some plants are more efficient than others in fixing carbon. Environmental factors are also important. The influx of solar energy, availability of mineral nutrients, availability of water, and other climatic factors are important, as are the degree of maturity of the community, the severity of human modification, and other factors that are difficult to assess. For example, the high productivity of intertidal communities along an ocean shoreline is due largely to wave action. Many intertidal organisms (such as mussels) are sedentary detritus filter feeders whose food is carried to them by wave action that allows them to expend less energy obtaining food.

Ecologists use different methods to measure primary productivity, depending on whether gross or net primary productivity is being assessed. Methods also vary from one type of ecosystem to another. On land, for example, ecologists might cut, dry, and weigh plants at the end of a growing season to measure NPP. This method would not be useful to measure NPP in the ocean, however, where microscopic algae are the main producers.

Although different methods are employed, it is possible to compare NPPs of diverse ecosystems or biomes (Table 54–1). Ecosystems differ strikingly in their productivity. On land, tropical rain forests (see Chapter 51) have the highest productivity, probably due to their abundant rainfall, warm temperatures, and intense sunlight. As you might expect, tundra with its harsh, cold winters and deserts with their lack of precipitation are the least productive biomes.

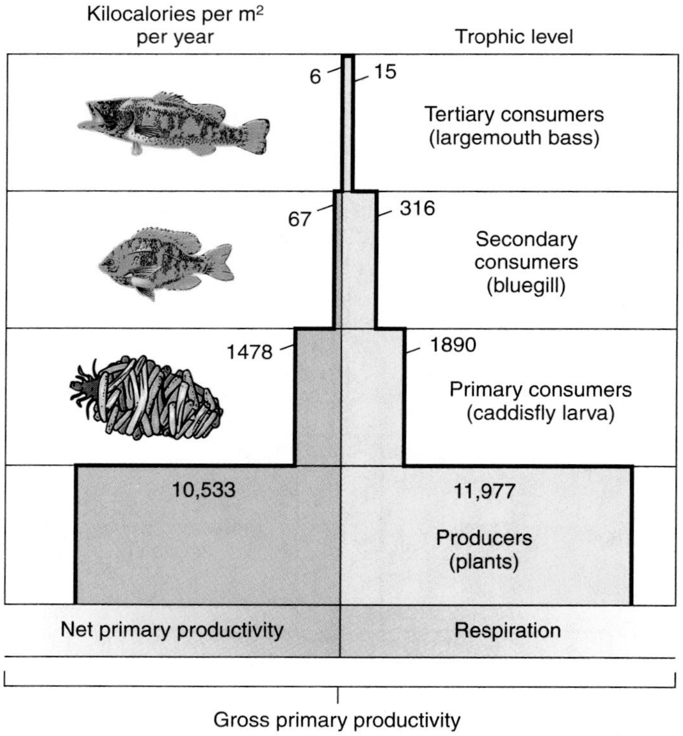

FIGURE 54–12 This pyramid of energy for a river ecosystem illustrates gross primary productivity and net primary productivity. Measurements are in kilocalories per square meter per year.

[2]Gross and net primary productivities are referred to as *primary* because plants occupy the first position in food webs.

Focus On

The Gaia Hypothesis

One of the most unusual and controversial hypotheses suggested in recent years is the Gaia hypothesis, which states that Earth can be viewed as a single living organism. (Gaia is derived from the Greek *gaea,* which means "the Earth.") According to the Gaia hypothesis, planet Earth is alive in the sense that it is capable of self-maintenance. Organisms interact with the abiotic environment to produce and maintain Earth's chemical composition, temperature, and other characteristics. Thus, the environment and organisms depend on one another and work together as a homeostatic mechanism.

As an example of the Gaia mechanism, consider Earth's temperature. It is generally accepted that the temperature has remained relatively constant at a level suitable for life over the 3.5 to 4 billion years that life has existed. Yet there is evidence that the sun has been heating up during that time. Why hasn't Earth warmed up as well? Gaia proponents think that Earth's temperature has remained the same because the level of atmosphere-warming CO_2 has dropped. This happened because living Earth compensated for increased sunlight by "fixing" CO_2 into calcium carbonate shells for countless billions of marine plankton. As the plankton died, their shells sank to the ocean floor, thus removing CO_2 from the system. This Gaia planetary temperature mechanism is an example of a feedback loop between the abiotic environment and organisms, which mutually interact to regulate Earth's temperature.

Another example of interactions between the abiotic and living components of Gaia's Earth involves the salinity of the ocean. As terrestrial rocks are weathered, the ocean tends to get saltier and saltier, in time becoming too saline to support life. However, geological evidence indicates that the salinity of the ocean has remained constant for millions of years. Gaia proponents suggest that a feedback loop exists in which bacteria remove excess salt from the ocean in salt flats, shallow bays along tropical and subtropical parts of the ocean where bacteria grow in such numbers that they form great mats.

Many scientists are reluctant to accept the Gaia hypothesis, although some consider it a useful metaphor. Almost everyone agrees that the environment modifies organisms and that organisms modify the environment to some extent, especially on a local scale. However, the idea that Earth's organisms *adjust* the abiotic environment to meet their needs has few backers, in part because it is difficult to test. ■

In the three forest biomes listed in Table 54–1, water availability does not affect NPP as much as does temperature, which depends on light energy. (Water is still an important factor in forests, however. In warmer climates, a greater annual precipitation is required to support forest vegetation.) In biomes with comparable annual temperatures — temperate deciduous forest, temperate grassland, and temperate desert, for example — water availability (as annual precipitation) directly affects NPP. Availability of essential minerals such as nitrogen and phosphorus can also affect NPP.

Table 54–1 NET PRIMARY PRODUCTIVITIES FOR SELECTED BIOMES

Biome	*NPP (grams of tissue/ m^2 of land/year)*
Tropical rain forest	1800
Temperate deciduous forest	1250
Boreal forest	800
Savanna	700
Temperate grassland	500
Tundra	140
Desert	70

Wetlands (swamps and marshes), which connect terrestrial and aquatic environments, are extremely productive. The most productive aquatic ecosystems are algal beds, coral reefs, and estuaries. The unavailability of mineral nutrients in the sunlit region of the open ocean makes this area an extremely unproductive aquatic desert.

ABIOTIC FACTORS INFLUENCE WHERE AND HOW SUCCESSFULLY AN ORGANISM CAN SURVIVE

We have seen how organisms depend on the abiotic environment to supply essential materials (in biogeochemical cycles) and energy. Abiotic factors such as solar radiation, the atmosphere, the ocean, and weather and climate also affect organisms (see *Focus On: The Gaia Hypothesis* for an interesting view of organisms and their abiotic environment).

THE SUN WARMS EARTH

The sun makes life on Earth possible. It warms the planet to habitable temperatures. Without the sun's energy, the temperature on planet Earth would approach absolute zero (−273°C) and all water would be frozen, even in the ocean. The hydrologic cycle, carbon cycle, and other biogeochemical cycles are powered by the sun, which also determines climate to a great extent. The sun's energy is captured by photosynthetic organisms and used to make organic compounds that are required by almost all forms of life. Most of our fuels—wood, oil, coal, and natural gas, for example—represent solar energy captured by photosynthetic organisms. Without the sun, life would cease.

The sun's energy, which is the product of a massive nuclear fusion reaction, is emitted into space in the form of electromagnetic radiation, especially visible light, infrared, and ultraviolet radiation. An infinitesimal portion of this energy—one billionth of the sun's total production—strikes the atmosphere, and a minute part of this tiny trickle of energy operates the ecosphere.

On average, 30% of the solar radiation that falls on Earth is immediately reflected away by clouds and surfaces, especially snow, ice, and ocean. The remaining 70% is absorbed by Earth and runs the water cycle, drives winds and ocean currents, powers photosynthesis, and warms the planet. Ultimately, however, all of this energy is lost by the continual radiation of long-wave infrared (heat) energy into space. (If heat gains were not balanced by losses, the Earth would heat up or cool down.)

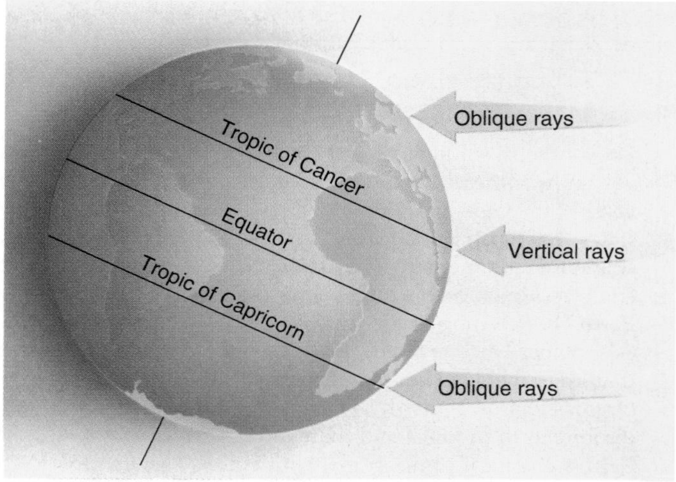

(a)

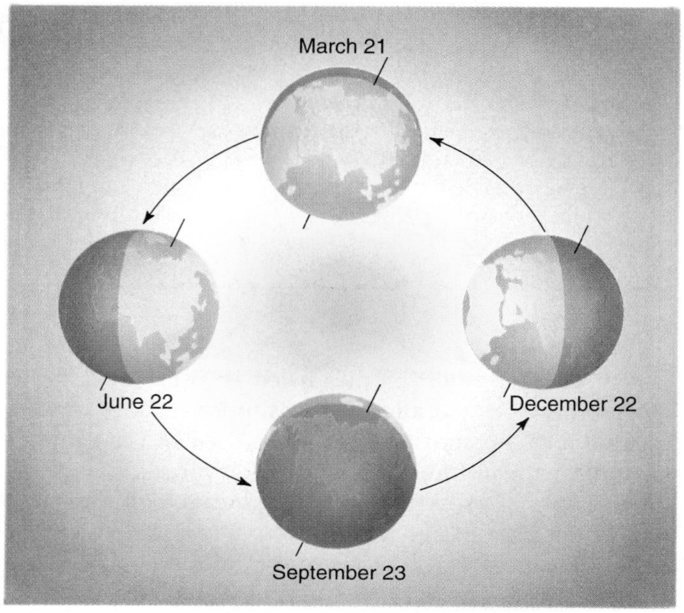

(b)

FIGURE 54–13 The amount of energy per unit area that the surface receives is determined by the angle at which the sun's rays strike Earth. (*a*) This angle varies from one geographical location to another due to the inclination of the planet on its axis. The month of June is represented here. (*b*) As Earth travels around the sun, the inclination of its axis remains the same. Thus, the sun's rays hit the Northern Hemisphere obliquely during winter months and more directly during summer months.

Temperature changes with latitude

The most significant local variations in Earth's temperature are produced because the sun's energy does not uniformly reach all places. A combination of our planet's roughly spherical shape and the tilted angle of its axis produces a great deal of variation in the exposure of the surface to the energy delivered by sunlight.

The principal difference this tilting makes is on the angles at which the sun's rays strike different areas at any one time (Fig. 54–13*a*). They hit nearly vertically near the equator, concentrating the energy and producing warmer temperatures. Near the poles the sun's rays strike more obliquely, and, as a result, are spread over a larger surface area. Also, rays of light entering the atmosphere obliquely near the poles must pass through a deeper envelope of air than those entering near the equator. This causes more of the sun's energy to be scattered and reflected back into space, which further lowers temperatures near the poles. Thus, the solar energy that reaches polar regions is less concentrated and produces lower temperatures.

Temperature changes with season

Seasons are determined by two main factors: Earth's inclination on its axis (the more important factor), and distance from the sun, which varies during the year. Since Earth's inclination on its axis is always the same (23.5°), during half of the year (March 21 to September 22) the Northern Hemisphere tilts *toward* the sun, concentrating the sunlight and making the days longer (Fig. 54–13*b*). Thus, the Northern Hemisphere is warm at this time. During the other half of the year (September 22 to March 21) the Northern Hemisphere tilts *away* from the sun, giving it a lower concentration of sunlight and shorter days. (The orientation of the Southern Hemisphere is just the opposite.)

ORGANISMS DEPEND ON THE ATMOSPHERE

The atmosphere is an invisible layer of gases that envelops the Earth. Oxygen (21%) and nitrogen (78%) are the predominant gases in the atmosphere, accounting for about 99% of dry air; other gases, including argon, carbon dioxide, neon, and helium, make up the remaining 1%. In addition, water vapor and dust particles are present. The atmosphere becomes less dense as it extends outward into space; most of its mass is found near Earth's surface.

The atmosphere performs several ecologically important functions. It protects the surface from most of the sun's ultraviolet radiation and x rays, and from lethal amounts of cosmic rays from space. Without this shielding by the atmosphere, life as we know it would cease. While the atmosphere protects Earth from high-energy radiations, it allows visible light and some infrared radiation to penetrate, and they warm the surface and the lower atmosphere. This interaction between the atmosphere and solar energy is responsible for weather and climate.

Organisms depend on the atmosphere, but they also maintain and, in certain instances, modify its composition. For example, atmospheric oxygen is thought to have increased to its present level as a result of millions of years of photosynthesis. The level is maintained by a balance between oxygen-producing photosynthesis and oxygen-using aerobic respiration.

Atmospheric circulation is driven by the sun

In large measure, differences in temperature caused by variations in the amount of solar energy at different locations drive the circulation of the atmosphere. The warm surface near the equator heats the air with which it comes into contact, causing this air to expand and rise. As the warm air rises, it flows away from the equator, cools, and sinks again. Much of it recirculates back to the same areas it left, but the remainder flows toward the poles, where eventually it is chilled. Similar upward movements of warm air and its subsequent flow toward the poles occur at higher latitudes (further from the equator) as well (Fig. 54–14). As air cools, it sinks and flows toward the equator, generally beneath the sheets of warm air flowing toward the pole at the same time. The constant motion of air transfers heat from the equator toward the poles, and as the air returns, it cools the land over which it passes. This constant turnover does not equalize temperatures over Earth's surface, but it does moderate them.

The atmosphere exhibits complex horizontal movements called winds

In addition to global circulation patterns, the atmosphere exhibits complex horizontal movements that are commonly referred to as **winds.** The nature of wind, with its turbulent gusts, eddies, and lulls, is complex and difficult to understand or predict. It results in part from differences in atmospheric pressure and from the rotation of Earth.

The gases that constitute the atmosphere have weight and exert a pressure that is, at sea level, about 1013 millibars (millibars are units of atmospheric pressure; 1013 millibars is equivalent to 14.7 pounds per square inch). Air pressure is variable, however, and changes with altitude, temperature, and humidity. Winds tend to blow from areas of high atmospheric pressure to areas of low pressure; the greater the difference between the high and low pressure areas, the stronger the wind.

Earth's rotation influences the direction that wind blows. Because Earth rotates from west to east, wind swerves to the right in the Northern Hemisphere and to the left in the Southern Hemisphere. This tendency of moving air to be deflected from its path by Earth's rotation is known as the **Coriolis effect.**

The Coriolis effect can be visualized by imagining that you and a friend are standing about 10 feet apart on a

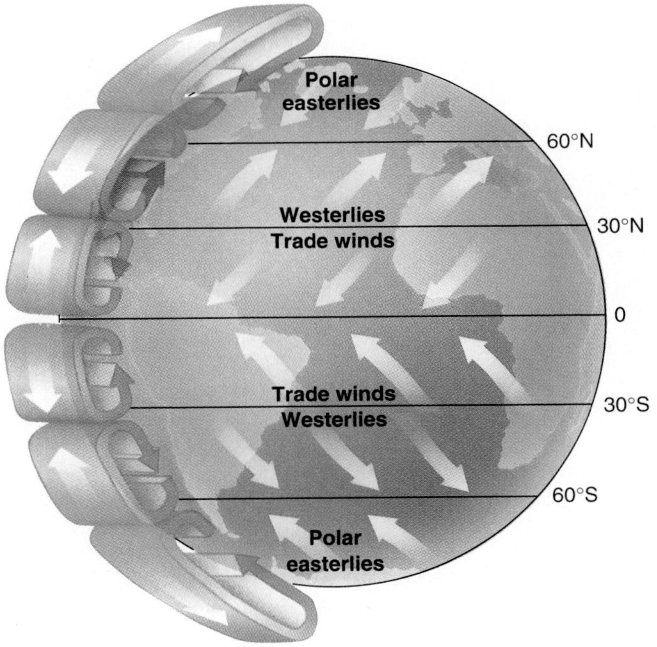

FIGURE 54–14 Atmospheric circulation transports heat from the equator to the poles. The greatest solar energy input occurs at the equator, heating air most strongly in that area. The air rises and travels poleward, but is cooled in the process so that much of it descends again around 30 degrees latitude in both hemispheres. At higher latitudes the patterns of air movement are more complex.

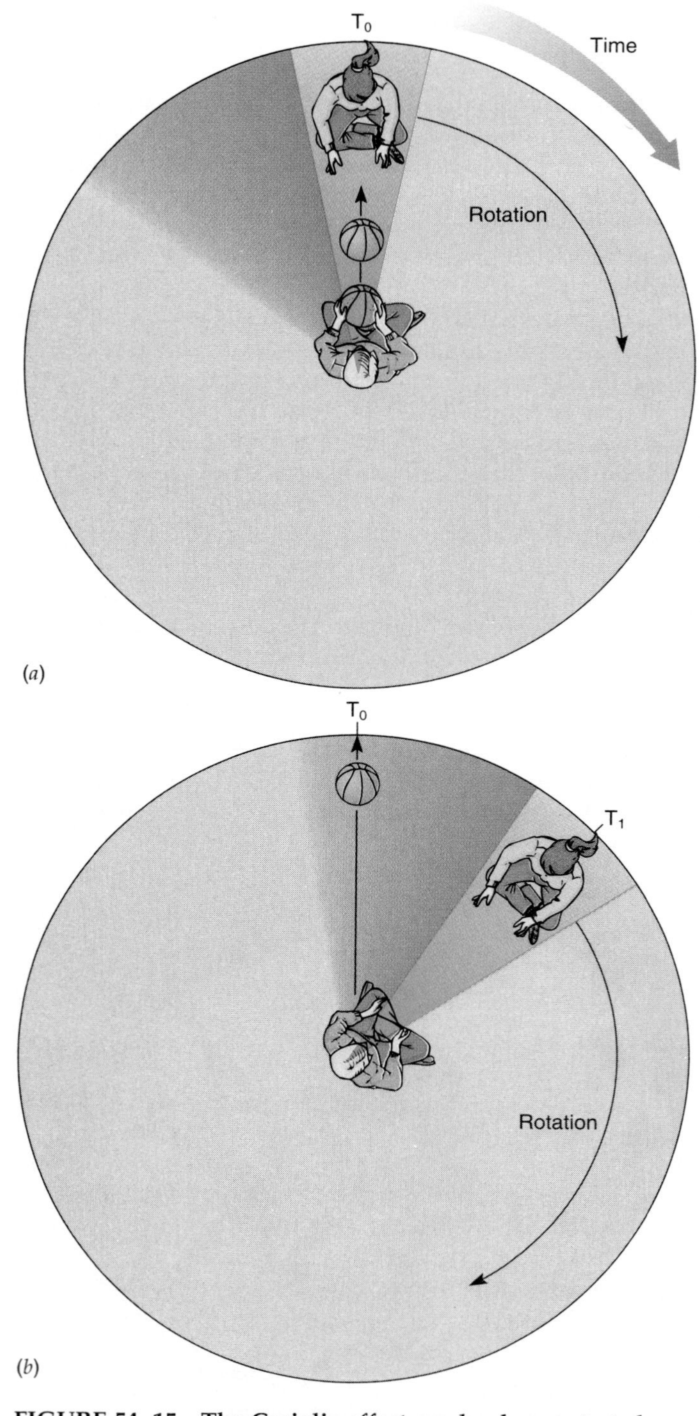

FIGURE 54–15 The Coriolis effect can be demonstrated on a playground merry-go-round (shown from above). The center of the merry-go-round is analogous to the South Pole, and the outer edge to the equator. (*a*) If you throw a ball to a friend at time zero (T_0), when the merry-go-round is rotating clockwise, the ball at time T_1 (*b*) appears to curve to the left instead of going straight. Because of Earth's rotation, winds appear to curve to the left in the Southern Hemisphere (and to the right in the Northern Hemisphere).

merry-go-round that is turning in a clockwise direction (Fig. 54–15). Suppose you throw a ball directly at your friend. By the time the ball arrives, your friend is no longer in that spot. The ball will have apparently swerved far to the left. This is how the Coriolis effect works in the Southern Hemisphere.

To visualize how the Coriolis effect works in the Northern Hemisphere, imagine that you and your friend are standing on the same merry-go-round, only this time it is moving counterclockwise. Now when you throw the ball, it will appear to swerve far to the right of your friend.

The atmosphere has three **prevailing winds,** major surface winds that blow more or less continually (Fig. 54–14). Prevailing winds that blow from the northeast near the North Pole or the southeast near the South Pole are *polar easterlies.* Winds that blow in the mid-latitudes from the southwest in the Northern Hemisphere or the northwest in the Southern Hemisphere are *westerlies.* Tropical winds that blow from the northeast in the Northern Hemisphere or the southeast in the Southern Hemisphere are *trade winds.*

THE GLOBAL OCEAN COVERS MOST OF EARTH'S SURFACE

The global ocean is a huge body of salt water that surrounds the continents and covers almost three-fourths of Earth's surface. It is a single, continuous body of water, but geographers divide it into four sections — the Pacific, Atlantic, Indian, and Arctic Oceans — separated by the continents. The Pacific Ocean is the largest by far: it covers one-third of Earth's surface and contains more than half of Earth's water.

Surface ocean currents are driven by winds and by the Coriolis effect

The persistent prevailing winds blowing over the ocean produce mass movements of surface ocean water known as **ocean currents** (Fig. 54–16); circular ocean currents generated by the prevailing winds are called *gyres.* For example, in the North Atlantic, the tropical trade winds tend to blow toward the west, whereas the westerlies in the mid-latitudes blow toward the east. This helps establish a clockwise gyre in the North Atlantic. Thus, surface ocean currents and winds tend to move in the same direction, although there are many variations to this general rule. Other factors that contribute to ocean currents include the Coriolis effect, varying densities of water, and the position of land masses.

The paths that surface ocean currents travel are partly caused by the Coriolis effect. Earth's rotation from west to east causes surface ocean currents to swerve to the right in the Northern Hemisphere, producing a clockwise gyre of water currents. In the Southern Hemisphere, ocean cur-

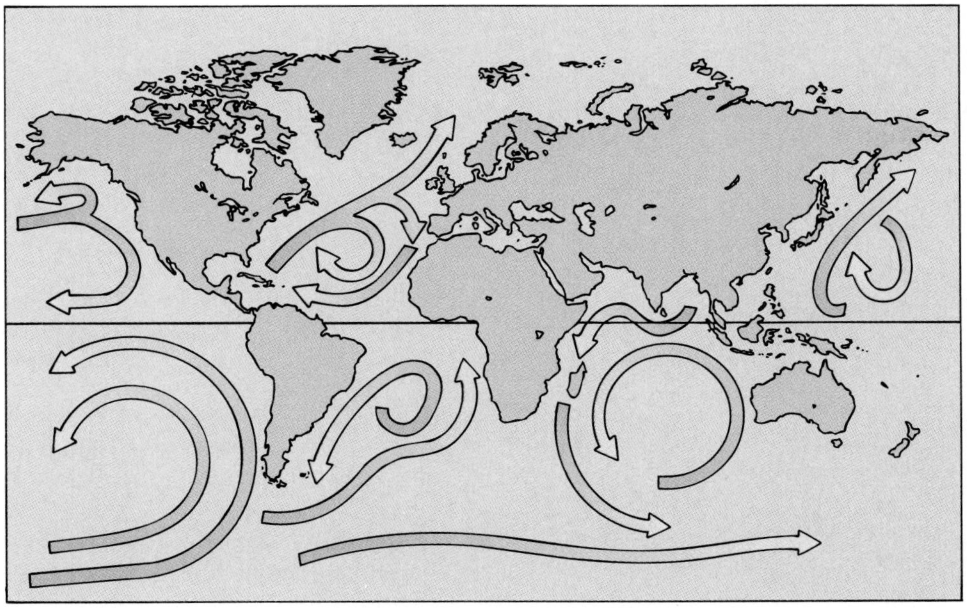

FIGURE 54–16 The basic pattern of surface ocean currents is caused largely by the action of winds. The clockwise flow in the Northern Hemisphere and counterclockwise flow in the Southern Hemisphere result partly from the Coriolis effect.

rents swerve to the left, producing a counterclockwise gyre.

The varying density (mass per unit volume) of sea water affects deep ocean currents. Colder or saltier sea water is denser than warmer or more dilute sea water. (The density of water increases with decreasing temperature down to 4°C.) Thus, colder ocean water sinks and flows under warmer water, producing currents far below the surface. Deep ocean currents often travel in different directions and at different speeds than do surface currents, in part because the Coriolis effect is more pronounced at greater depths.

The position of land masses also affects oceanic circulation. The ocean is not uniformly distributed over the globe (Fig. 54–17). The Southern Hemisphere contains much more water than the Northern Hemisphere. Therefore, the circumpolar ("around the pole") flow of water in the Southern Hemisphere is almost unimpeded by land masses.

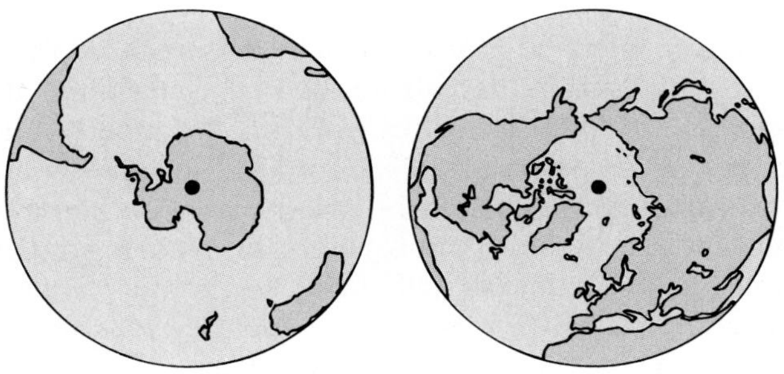

(*a*) (*b*)

FIGURE 54–17 The Southern and Northern Hemisphere have greatly differing proportions of land and water, with far more water in the Southern Hemisphere. (***a***) Southern Hemisphere. (***b***) Northern Hemisphere.

The ocean interacts with the atmosphere

The ocean and the atmosphere are strongly linked, with wind from the atmosphere affecting the ocean currents, and heat from the ocean affecting atmospheric circulation. One of the best examples of the interaction between ocean and atmosphere is the **El Niño-Southern Oscillation (ENSO)** event. ENSO is a periodic warming of surface waters of the tropical East Pacific that alters both oceanic and atmospheric circulation patterns, and results in unusual weather in areas far from the tropical Pacific (Fig. 54–18). Normally, westward-blowing trade winds restrict the warmest waters to the western Pacific (near Australia). Every three to seven years, however, the trade winds weaken and the warm mass of water expands eastward to South America, raising surface temperatures in the East Pacific to 3 to 4 degrees above normal. Ocean currents, which normally flow westward in this area, slow down, stop altogether, or even reverse and go eastward. The phenomena is called *El Niño* (Spanish for "the child") because the warming usually reaches the fishing grounds off Peru just before Christmas. Most ENSOs last about two years.

ENSO has a devastating effect on the fisheries off the west coast of South America. The warmer temperatures and accompanying changes in ocean circulation patterns prevent nutrient-laden deeper waters from **upwelling** (coming to the surface). This results in a severe drop in the populations of anchovies and many other marine fishes. For example, during the 1982–83 ENSO, the worst ever recorded, the anchovy population decreased by 99%. (Some other species, such as shrimp and scallops, thrive during an ENSO event.)

ENSO also alters global air currents, directing unusual weather to areas far from the tropical Pacific. The 1994–95

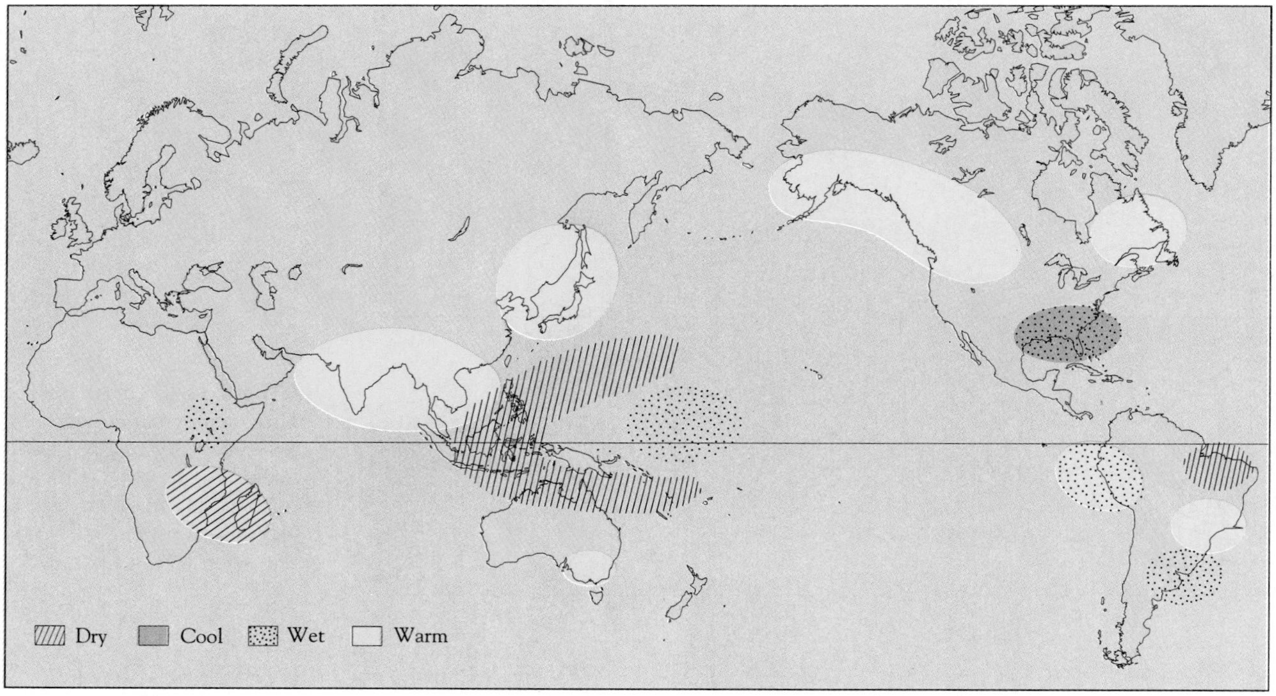

FIGURE 54–18 An El Niño-Southern Oscillation can drastically alter the climate in many areas remote from the Pacific Ocean. During ENSO events, widely separated regions can experience drier, wetter, cooler, or warmer conditions than usual.

ENSO, for example, resulted in heavy snows in parts of the western United States. This ENSO was also responsible for the unusually warm winter in northeastern states and for the torrential rains that flooded California, Arizona, and western Europe. In addition, the effects of this ENSO have been linked to droughts in Australia and Indonesia.

Temperature and Precipitation Are Important Aspects of Climate

Weather refers to the conditions in the atmosphere at a given place and time; it includes temperature, atmospheric pressure, precipitation, cloudiness, humidity, and wind. Weather changes from one hour to the next and from one day to the next.

Climate comprises the average weather conditions that occur in a given place over a period of years. The two most important factors that help determine an area's climate are temperature (both average temperature and temperature extremes) and precipitation (both average precipitation and seasonal distribution). Other climatic factors include wind, humidity, fog, and cloud cover. Wildfires (those started by lightning) are an important aspect of climate in some areas (see *Focus On: Wildfires*).

Day-to-day variations, day-to-night variations, and seasonal variations in climatic factors are also important dimensions of climate that affect organisms. Temperature, precipitation, and other aspects of climate are influenced by latitude, elevation, topography, vegetation, distance from the ocean, and location on a continent or other land mass. Unlike weather, which changes rapidly, climate changes slowly—over hundreds or thousands of years.

Earth has many different climates, and because they are relatively constant for many years, organisms have adapted to them. The wide variety of organisms on Earth are here in part because of the large number of different climates, ranging from cold, snow-covered polar climates to hot tropical climates where it rains almost every day.

Precipitation patterns are affected by air and water movements, and by surface features of the land

Precipitation refers to any form of water, such as rain, snow, sleet, and hail, that falls to the surface from the atmosphere. Precipitation varies from one location to another and has a profound effect on the distribution and kinds of organisms. One of the driest places on Earth is in the Atacama Desert in Chile, where the average annual rainfall is 0.05 centimeter (0.02 inch). In contrast, Mount Waialeale in Hawaii, Earth's wettest spot, receives an average annual precipitation of 1200 centimeters (472 inches).

Differences in precipitation depend on several factors. The heavy-rainfall areas of the tropics result mainly from the equatorial upwelling of moisture-laden air. High surface-water temperatures (recall the enormous amount of solar energy striking the equator) cause the evaporation of vast quantities of water from tropical parts of the ocean.

Focus On

Wildfires

Wildfires are an important environmental force in many geographical areas. Those areas most prone to wildfires have wet seasons followed by dry seasons. Vegetation that grows and accumulates during the wet season dries out enough during the dry season to burn easily. When lightning hits the ground, it ignites the dry organic material, and a fire spreads through the area.

Fires have several effects on the environment. First, combustion frees the minerals that were locked in dry organic matter. The ashes remaining after a fire are rich in potassium, phosphorus, calcium, and other minerals essential for plant growth. Thus, vegetation flourishes following a fire. Second, fire removes plant cover and exposes the soil, which stimulates the germination of seeds requiring bare soil, and encourages the growth of shade-intolerant plants. Third, fire can cause increased soil erosion because it removes plant cover, leaving the soil more vulnerable to wind and water.

Grasses adapted to wildfire have underground stems and buds that are unaffected by a fire sweeping over them. After the aerial parts have been killed by fire, the underground parts send up new sprouts. Fire-adapted trees such as bur oak and ponderosa pine have a thick bark that is resistant to fire. (In contrast, fire-sensitive trees such as many hardwoods have a thin bark.) Certain pines such as jack and lodgepole pine depend on fire for successful reproduction, because the heat of the fire opens the cones so that the seeds can be released.

Fires were a part of the natural environment long before humans appeared, and many terrestrial ecosystems have adapted to it. African savanna, California chaparral, North American grasslands, and pine forests of the southern United States are some of the fire-adapted ecosystems (see Chapter 51). For example, fire helps maintain grasses as the dominant vegetation in grasslands by removing fire-sensitive hardwood trees.

Fire is a necessary tool of ecological managment. Here, a controlled burn helps maintain a ponderosa pine (*Pinus ponderosa*) stand in Oregon. (Joan Landsberg, USDA Forest Service)

The influence of fire on plants became even more pronounced once humans appeared. Because humans deliberately and accidentally set fires, fire became a more common occurrence. Humans set fires for many reasons: to provide the grasses and shrubs that many game animals require; to clear the land for agriculture and human development; and to reduce enemy cover in times of war.

Humans also try to prevent fires, and sometimes this effort can have disastrous consequences. When fire is excluded from a fire-adapted ecosystem, organic litter accumulates. As a result, when a fire does occur, it is much more destructive. The deadly fire in Colorado during the summer of 1994, which claimed the lives of 14 firefighters, was blamed in part on decades of suppressing fires in the region. Prevention of fire also converts grassland to woody vegetation and facilitates the invasion of fire-sensitive trees into fire-adapted forests.

Humans sometimes conduct *controlled burns*, a tool of ecological management in which the organic litter is deliberately burned before it has accumulated to dangerous levels. Controlled burns are also used to suppress fire-sensitive trees, thereby maintaining the natural fire-adapted ecosystem (see figure). ■

Prevailing winds blow the resulting moist air over land masses.

Heating of air by land surfaces warmed by the sun causes moist air to rise. As it rises, the air cools and its moisture content decreases (cool air holds less water vapor than warm air). When air reaches its saturation point (when it cannot hold any additional water vapor), clouds form and water is released as precipitation. The air eventually returns to the surface on both sides of the equator between the Tropics of Cancer and Capricorn (latitudes

23.5 degrees north and south, respectively). By then, most of the moisture has precipitated so that dry air returns to the equator. This dry air makes little biological difference over the ocean, but its lack of moisture produces some of the great tropical deserts, such as the Sahara.

Air is also dried from long journeys over land masses. Near the windward (side from which the prevailing wind blows) coasts of continents, rainfall may be heavy. However, in the temperate zones—the areas between the tropical and polar zones—continental interiors are usually dry, because they are far from the ocean that replenishes water in the air passing over it.

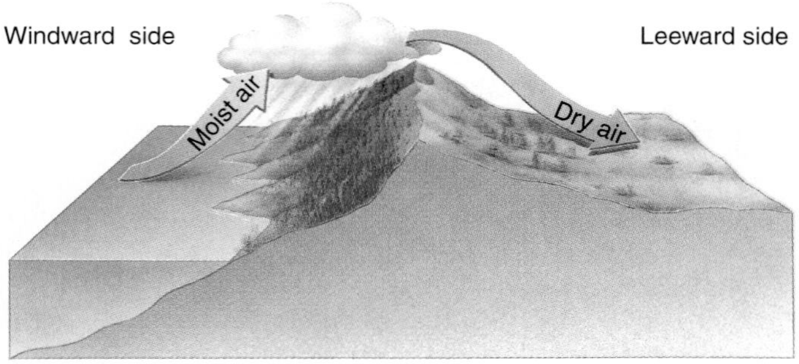

FIGURE 54–19 When wind blows moist air over a mountain range, precipitation occurs on the windward side of the mountain, causing a dry rain shadow on the leeward side. Such a rain shadow occurs in Washington state east of the Cascade Mountains.

Some mountains cause rain shadows

Moisture is removed from humid air by mountains, which force air to rise. As it gains altitude, the air cools, clouds form, and precipitation occurs—primarily on the windward slopes of the mountains. As the air mass moves down on the other side of the mountain, it is warmed, thereby lessening the chance of precipitation of any remaining moisture. This situation exists on the west coast of North America, where precipitation falls on the western slopes of mountains that are close to the coast. The dry lands on the sides of the mountains away from the prevailing wind (in this case, east of the mountain range) are called **rain shadows** (Fig. 54–19).

Microclimates are local variations in climate

Differences in elevation, in the steepness and direction of slopes, and in exposure to sunlight and prevailing winds may produce local variations in climate known as **microclimates,** which can be quite different from their overall surroundings. The microclimate of its habitat is of primary importance to an organism, because that is the climate an organism actually experiences and must cope with.

Sometimes organisms modify their own microclimate. For instance, trees modify the local climate within a forest so that in summer the temperature is usually lower, and the relative humidity greater, than outside the forest. The temperature and humidity beneath the litter of the forest floor differ still more; in the summer this area is considerably cooler and moister than the surrounding forest. As another example, many desert-dwelling animals burrow to avoid surface climatic conditions that would kill them in minutes. The cooler daytime microclimate in their burrows permits them to survive until night, when the surface cools off and they can come out to forage or hunt.

SUMMARY

I. Biogeochemical cycles are the cycling of matter from the abiotic environment to organisms and then back to the environment.
   A. Carbon enters plants, algae, and cyanobacteria as CO_2, which is incorporated into organic molecules by photosynthesis. Cellular respiration, combustion, and erosion return CO_2 to the atmosphere, making it available for producers again.
   B. The nitrogen cycle has five steps.
      1. Biological nitrogen fixation is the conversion of nitrogen gas to ammonia.
      2. Nitrification is the biological conversion of ammonia to nitrate, one of the main forms of nitrogen used by plants.
      3. Assimilation is the biological conversion of nitrates or ammonia to proteins and other nitrogen-containing compounds by plants. The conversion of plant proteins into animal proteins is also considered assimilation.
      4. Ammonification is the biological conversion of organic nitrogen to ammonia.
      5. Denitrification is the biological conversion of nitrate to nitrogen gas.
   C. The phosphorus cycle has no biologically important gaseous compounds.
      1. Phosphorus erodes from rock as inorganic phosphate, which is absorbed from the soil by the roots of plants.
      2. Animals obtain the phosphorus they need from their diets. Decomposers release inorganic phosphate into the environment.
      3. Phosphorus can be lost from biological cycles for millions of years when it washes into the ocean and is deposited in sea beds.
   D. The hydrologic cycle, which continually renews the supply of water that is so essential to life, involves an exchange of water between the land, the atmosphere, and organisms.
      1. Water enters the atmosphere by evaporation and

transpiration, and leaves the atmosphere as precipitation.
2. On land, water filters through the ground or runs off to lakes, rivers, and the ocean.

II. Energy flows through an ecosystem in a linear direction, from the sun to producer to consumer to decomposer. Much of this energy is converted to less-useful heat as it moves from one organism to another.
  A. Trophic relationships may be expressed as food chains or, more realistically, as food webs, which show the many alternative pathways that energy may take among the producers, consumers, and decomposers of an ecosystem.
  B. Ecological pyramids express the progressive reduction in numbers of organisms, biomass, and energy found in successively higher trophic levels.
  C. Gross primary productivity of an ecosystem is the rate at which energy accumulates as biomass during photosynthesis. Net primary productivity is the energy that remains (as biomass) after plants respire.

III. The unique planetary environment of Earth makes life possible. Sunlight is the primary (almost the sole) source of energy available to the biosphere.
  A. Of the solar energy that reaches Earth, 30% is immediately reflected away and the remaining 70% is absorbed.
  B. Ultimately, all absorbed solar energy is reradiated into space as infrared (heat) radiation.
  C. A combination of Earth's roughly spherical shape and the tilted angle of its axis concentrates solar energy at the equator and dilutes it at the poles.
    1. The tropics are therefore hotter and less variable in climate than are temperate and polar areas.
    2. Seasons are determined by two main factors: the inclination of Earth's axis (the more important factor) and the distance from the sun, which varies during the year.

IV. The atmosphere protects the surface from most of the sun's ultraviolet radiation and x rays, and from lethal amounts of cosmic rays from space. At the same time, visible light and some infrared radiation penetrate to warm the surface and the lower part of the atmosphere.
  A. Atmospheric heat transferred from the equator to the poles produces both a movement of warm air toward the poles and of cool air toward the equator, thus moderating the climate.
  B. Winds result in part from differences in atmospheric pressure and from the Coriolis effect.

V. The global ocean is a single, continuous body of water that surrounds and covers almost three-fourths of Earth's surface.
  A. Surface ocean currents result from prevailing winds, the Coriolis effect, the varying water densities, and the land mass positions.
  B. The El Niño-Southern Oscillation (ENSO) event alters both ocean and atmospheric currents and results in unusual weather patterns.

VI. An area's climate comprises the average weather conditions that occur there over a period of years.
  A. The two most important climatic factors are temperature (both average temperature and temperature extremes) and precipitation (both average precipitation and seasonal distribution).
  B. Precipitation is greatest where warm air passes over the ocean, absorbing moisture, and is then cooled, such as when humid air is forced upward by mountains.
  C. Deserts develop in the rain shadows of mountain ranges or in continental interiors.

SELECTED KEY TERMS

ammonification
assimilation
biogeochemical cycle
biomass
carbon cycle
climate
Coriolis effect
denitrification
ecological pyramid
ecosphere
ecosystem
El Niño-Southern Oscillation (ENSO)
energy flow
food chain
food web
fossil fuel
gross primary productivity (GPP)
heterocyst
hydrologic cycle
microclimate
net primary productivity (NPP)
nitrification
nitrogen cycle
nitrogen fixation
nitrogenase
nodule
ocean current
phosphorus cycle
prevailing wind
pyramid of biomass
pyramid of energy
pyramid of numbers
rain shadow
trophic level
upwelling

POST-TEST

1. A community and its abiotic environment best defines a(an) __________ .
2. Unlike the movement of matter, which is cyclic in ecosystems, __________ __________ is linear.
3. The global movement of carbon between the atmosphere and organisms is known as the __________ __________ .
4. In biological __________ __________ , gaseous nitrogen is converted to ammonia.
5. The conversion of ammonia to nitrate, known as __________ , is a two-step process performed by soil bacteria.
6. The __________ __________ does not have a gaseous component, but cycles from the land to sediments in the ocean and back to the land.
7. The global cycling of water is called the __________ __________ .

8. Each level in a food web is called a(an) __________ level.
9. A(An) __________ __________ is a complex of interconnected food chains in an ecosystem.
10. The three types of ecological pyramids are the pyramids of __________, __________, and __________ (any order).
11. The quantitative estimate of the total mass, or amount, of living material is called __________.
12. __________ primary productivity equals __________ primary productivity minus plant respiration.
13. An area's __________ refers to its average weather conditions over a period of years.
14. Surface __________ __________ and winds tend to move in the same general direction.
15. The __________ __________, which results from the rotation of Earth, displaces the paths of atmospheric and oceanic currents to the right (Northern Hemisphere) and the left (Southern Hemisphere).
16. The periodic warming of surface waters of the tropical East Pacific that alters both oceanic and atmospheric circulation patterns, and results in unusual weather in areas far from the tropical Pacific, is known as __________ (abbreviated form).
17. Mountain ranges may produce downwind arid __________ __________.
18. Local variation in climate is called a(an) __________.

REVIEW QUESTIONS

1. Why is the cycling of matter essential to the continuance of life?
2. Diagram the carbon cycle, including the following processes: photosynthesis, cellular respiration, combustion, and erosion.
3. List and describe the five steps in the nitrogen cycle.
4. Describe energy flow through a food web.
5. What are trophic levels and how are they related to ecological pyramids?
6. Define gross primary productivity and net primary productivity.
7. What basic forces determine the circulation of the atmosphere? Describe the general directions of atmospheric circulation.
8. What basic forces produce the main ocean currents? Describe the general directions of main ocean currents.
9. What are some of the factors that produce regional differences in precipitation?
10. Label the following diagram.

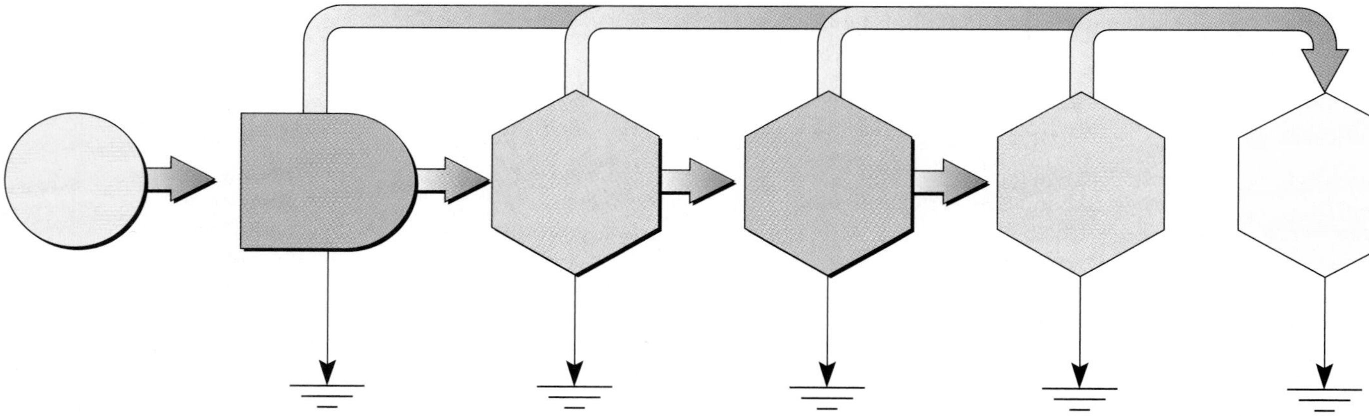

YOU MAKE THE CONNECTION

1. Describe the simplest stable ecosystem that you can imagine.
2. The connection box on food webs focused on interactions among trophic levels. Food webs also show us much about interactions *within* a given trophic level. What kind of interactions would these be?
3. Why might industrial polluters think that the Gaia hypothesis gives them permission to pollute the air, water, and soil indefinitely? How would you respond to such an opinion?
4. Would the microclimate of an ant be the same as that of an elephant living in the same area? Why or why not?

RECOMMENDED READINGS

Barlow, C., and T. Volk. "Gaia and Evolutionary Biology." *BioScience,* Vol. 42, No. 9, October 1992. The Gaia hypothesis has stimulated researchers to look for connections between organisms and the global environment.

Brewer, R. *The Science of Ecology,* 2nd edition. Saunders College Publishing, Philadelphia, 1994. A readable general textbook on the principles of ecology, including ecosystem ecology.

Caraco, N.F. "Disturbance of the Phosphorus Cycle: A Case of Indirect Effects of Human Activity." *Tree,* Vol. 8, No. 2, February 1993. Humans have affected the phosphorus cycle both directly and indirectly.

Chahine, M.T. "The Hydrologic Cycle and Its Influence on Climate." *Nature,* Vol. 359, October 1, 1992. A review of our current understanding of how the hydrologic cycle affects global climate.

Golnaraghi, M., and R. Kaul. "The Science of Policymaking: Responding to ENSO." *Environment,* Vol. 37, No. 1, January/February 1995. Peru and Brazil have adopted strategies to cope with ENSO events, which scientists can now forecast about one year in advance.

Gribbin, J. "Climate Now." *New Scientist,* Vol. 130, March 16, 1991. A discussion of the main forces that control climate.

Pimm, S. L., J. H. Lawton, and J. E. Cohen. "Food Web Patterns and Their Consequences." *Nature,* Vol. 350, April 25, 1991. A review article on current ecological knowledge of food webs.

CHAPTER 55

Humans in the Environment

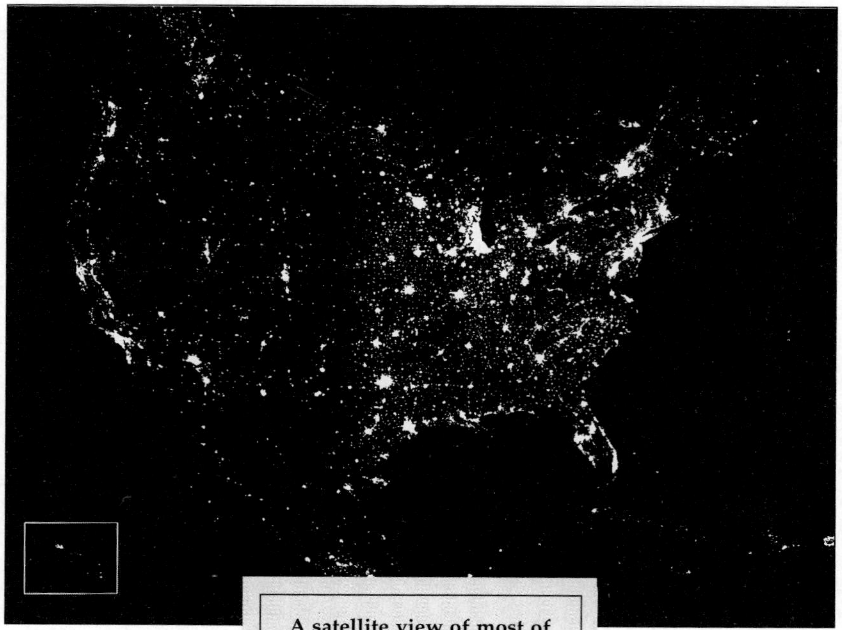

A satellite view of most of North America at night shows cities and major metropolitan areas. The people in these cities are completely dependent on natural ecosystems for survival. (© Hansen Planetarium, SLC, UT, USA and W. T. Sullivan III, Univ. of Washington)

Humans have been present on Earth for about 200,000 years (see Chapter 21), which is a brief span of time compared with the age of our planet (some 4.6 billion years). Despite our relatively short tenure on Earth, our biological success has been unparalleled. Our numbers have increased dramatically—the human population is expected to surpass 6 billion by 1998—and we have expanded our biological range, moving into almost every habitat on Earth.

Wherever we have gone, we have altered the environment and shaped it to meet our needs. In only a few generations we have transformed the face of Earth, placed a great strain on Earth's resources and resilience, and profoundly affected other organisms. Thus, the impact of humans on the environment merits special study in biology, not merely because we ourselves are humans but also because our impact on the rest of the biosphere has been so extensive.

Humans do not live alone on Earth, nor are we above the laws of nature: our actions do have consequences. We have many partners who share Earth with us, and we would not live long without them. Thus, one of our principal goals is to identify ways to avoid upsetting the biological systems that support us.

Many environmental concerns exist today, too many to be considered in a single chapter. We therefore focus our attention on four serious environmental issues: declining biological diversity, deforestation, global warming, and ozone depletion in the stratosphere.

Living organisms are important natural resources that are not fully appreciated. Biological diversity contributes toward a sustainable environment—one that provides a life support system that enables humans as well as other species to survive. The current reduction in biological diversity caused by the extinction of many species results in lost opportunities and lost solutions to future problems.

The world's forests provide many environmental services (habitat for wildlife, watershed protection, and prevention of soil erosion, to name a few) as well as commercially important timber. The greatest problem facing forests today is deforestation, the permanent removal of forest.

Production of atmospheric pollutants that trap solar heat in the atmosphere will probably affect Earth's climate during the 21st century. Global warming could alter food production, destroy forests, reduce biological diversity, submerge coastal areas, and change precipitation patterns.

The chemical destruction of stratospheric ozone by gaseous air pollutants may permit large amounts of solar ultraviolet radiation to penetrate to Earth's surface. Living organisms, including humans, will be harmed by increased exposure to ultraviolet radiation.

LEARNING OBJECTIVES

After you have studied this chapter you should be able to

1. Distinguish among threatened species, endangered species, and extinct species.
2. Discuss at least four causes of declining biological diversity, and identify the most important cause.
3. Compare in situ and ex situ conservation measures.
4. Discuss the ecological benefits of forests.
5. State at least three reasons that tropical forests are disappearing today.
6. Name at least three greenhouse gases, and explain how greenhouse gases contribute to global warming.
7. Describe how global warming may affect sea level, precipitation patterns, living organisms (including humans), and food production.
8. Give examples of ways to prevent, mitigate, and adapt to global warming.
9. Distinguish between surface ozone and stratospheric ozone.
10. Cite the potential effects of ozone destruction in the stratosphere.

SPECIES ARE DISAPPEARING AT AN ALARMING RATE

Extinction, the death of a species, occurs when the last individual member of a species dies (see Chapter 19). It is an irreversible loss, because once a species becomes extinct, it can never reappear. Biological extinction is the eventual fate of all species, much as death is the eventual fate of all individual organisms.

Although extinction is a natural biological process, it has been greatly accelerated by human activities. Burgeoning human populations have spread into almost all areas of Earth, and whenever humans invade an area, the habitats of many plants and animals are disrupted or destroyed, which can lead to their extinction. For example, the dusky seaside sparrow became extinct in 1987 largely because of the destruction of its habitat (see Fig. 19–16).

Earth's **biological diversity,** which is the variety of living organisms and of the ecosystems in which they live, is currently decreasing at an alarming rate. It is likely that thousands of species will be eliminated within the next few decades. As many as one-fourth of Earth's plant families[1] may be extinct by the end of the 21st century, and countless animal species that depend on those plants for food and habitat will probably also become extinct.

Some biologists fear that we are entering the greatest period of mass extinction in Earth's history, but the current situation differs from previous periods of mass extinction in several respects. First, its cause is directly attributable to human activities. Second, it is occurring in a tremendously compressed period of time (just a few decades as opposed to millions of years), much faster than rates of speciation (or replacement). Perhaps even more sobering, larger numbers of plant species are becoming extinct today than in previous mass extinctions. Because plants are the base of terrestrial food webs, extinction of animals that depend on plants cannot be far behind.

According to the United States' Endangered Species Act, a species is designated as **endangered** when its numbers are so severely reduced that it is in danger of becoming extinct throughout all or a significant part of its range. When extinction is less imminent but the population of a particular species is still quite small, the species is said to be **threatened.** A threatened species is likely to become endangered in the foreseeable future. Endangered or threatened status for a species represents a decline in biological diversity, since it implies severely diminished genetic variability. Endangered and threatened species are at greater risk of extinction than species with greater genetic variability, because long-term survival and evolution depend on **genetic diversity,** the genetic variation *within* a species.

Human activities contribute to declining biological diversity

Species become endangered and extinct for a variety of reasons, many caused by human activities. These include environmental pollution and the destruction or modification of habitats. Humans also upset the delicate balance of living organisms in a given area by introducing new, exotic species or by controlling pests or predators. Uncontrolled hunting and commercial harvest are also factors.

The majority of species facing extinction today are endangered because of *destruction of natural habitats* (Fig. 55–1). Building roads, parking lots, and buildings; clearing forests to grow crops or graze domestic animals; and logging forests for timber all take their toll on natural habitats. Draining marshes converts aquatic habitats to terrestrial ones, while building dams floods terrestrial habitats. Habitat destruction threatens the survival of species because most organisms require a particular type of environment.

[1]Recall that a family consists of a number of related genera, each of which consists of a number of related species. When a family becomes extinct, all the species of all the genera comprising that family cease to exist.

FIGURE 55–1 Many species are endangered as a result of habitat destruction. The California condor is a scavenger bird that lives off carrion and requires a large (hundreds of square kilometers) undisturbed territory in order to find enough food. On the brink of extinction, this highly endangered species was not found in the wild from 1987 to 1992. Zoo-bred California condors were released into the wild beginning in 1992. (Tom McHugh/Photo Researchers, Inc.)

Humans often leave small patches of natural landscape, which may be completely surrounded by roads, fields, buildings, and such. These habitat fragments may be able to support only a fraction of the species found in the original, unaltered environment (see *Making the Connection: Forest Fragmentation and Declining Bird Populations*).

Even habitats left undisturbed and in their natural state are modified by human activities that produce acid precipitation and other forms of *pollution*. Acid precipitation is thought to have contributed to the decline of large stands of forest trees and to the biological death of many freshwater lakes. The production of other types of pollutants also adversely affects wildlife. Such pollutants include industrial and agricultural chemicals, organic pollutants from sewage, acid wastes seeping from mines, and thermal pollution from the heated waste water of industrial plants.

The *introduction of a foreign, or exotic, species* into an area where it is not native often harms local organisms. The foreign species may compete with native species for food or habitat, or may prey on them. An introduced competitor or predator causes a greater negative effect on local organisms than do native competitors or predators. (Most introduced species lack natural agents, such as diseases, predators, and competitors, that would otherwise control them.) Although exotic species may be introduced into new areas by natural means, humans are usually responsible for such introductions, either knowingly or accidentally.

Many examples exist of introduced exotic species causing local organisms to become endangered or extinct. In 1958 a carnivorous snail was deliberately introduced in Oahu, an island in Hawaii, as a way to control another snail species that had been introduced by humans and had become a pest. The newly introduced species started consuming native snails that were already in jeopardy from human collectors and human-introduced rats (Fig. 55–2). As a result, *Achatinella*, a Hawaiian snail genus with numerous species known for their rainbow-colored shells, has all but disappeared from Oahu.

Islands are particularly susceptible to the introduction of exotic species (see *Focus On: The Lesson of New Zealand*). For example, Abingdon Island, one of the Galapagos Islands off the coast of South America, was home to an endemic (found nowhere else) giant tortoise. In 1957 sev-

FIGURE 55–2 The introduction of exotic species often threatens native species. *Euglandia rosea*, an exotic species in Oahu, consumes a smaller native *Achatinella* species. (Robert J. Western)

Making the Connection

Forest Fragmentation and Declining Bird Populations

How is habitat fragmentation related to declining bird populations? Evidence is mounting that the populations of many birds that were once abundant have been dropping steadily for more than a decade. Across the North American continent, population declines have been noted in every major group of birds. In the eastern United States, for example, 70% of the 300 to 400 species of neotropical birds show declining numbers. Neotropical birds spend the winter in Central and South America and the Caribbean and then migrate north to breed in the United States and Canada during the summer. Cerulean warblers, olive-sided flycatchers, yellow-billed cuckoos, and rose-breasted grosbeaks are examples of declining neotropical birds.

Neotropical birds are faced with changing environments in both their winter and summer homes, and loss of habitat is the main reason for their decline. These migratory birds are under stress from the burning of tropical rain forests in Central and South America, as well as the fragmenting of forested areas in North America to accommodate suburban development, agriculture, and logging.

Several studies have shown that fragmentation of forests increases the likelihood of reproductive failure for neotropical birds, many of which are insect-eaters that nest only in forests. As a result of forest fragmentation, the nests are more likely to be located near a *forest edge,* the boundary between the forest and surrounding farmlands or residential neighborhoods, rather than deep in the forest. Nests that are within 100 meters (328 feet) of a forest edge are more vulnerable to predation by small mammalian predators such as raccoons and feral cats (domestic cats that have gone wild). Blue jays, American crows, and other egg-eating birds do most of their hunting along the forest edge.

Nest parasitism by brown-headed cowbirds is also more significant along the forest edge. Female cowbirds lay their eggs in the nests of other bird species. In a 1989-to-1993 study that monitored 5000 nests of neotropical birds, the majority of nests were parasitized in areas with less than 55% forest cover, and in many cases the nests contained more cowbird eggs than eggs that belonged there. Nest parasitism causes reproductive failure because the host parents provide food for the larger, more aggressive cowbird babies, while their own young starve. ▲

eral fishermen introduced goats to Abingdon Island, and within five years the Abingdon tortoise was extinct. The goats, with no natural predators on the island, had greatly increased in number and had eaten the tortoises' food. In Hawaii, the introduction of mouplan sheep has imperiled both the mamane tree (because the sheep eat it) and a species of honeycreeper, an endemic bird that relies on the tree for food.

Sometimes species become endangered or extinct as a result of deliberate *efforts to eradicate or control their numbers,* often because they prey on game animals or livestock. Populations of large predators such as the wolf, mountain lion, and grizzly bear have been decimated by ranchers, hunters, and government agents. Predators of game animals and livestock are not the only animals vulnerable to human control efforts. Some animals are killed because their lifestyles cause problems for humans. The Carolina parakeet, a beautiful green, red, and yellow bird endemic to the southeastern United States, was extinct by 1920, exterminated by farmers because it ate fruit from their trees. Prairie dogs and pocket gophers were poisoned and trapped by ranchers because their burrows weakened the ground on which unwary cattle grazed. As a result of sharply decreased numbers of prairie dogs and pocket gophers, the black-footed ferret, a natural predator of these animals, became endangered. None was found in the wild in the United States between 1986 and 1991. A small number had survived in zoos, however, and a successful captive breeding program enabled biologists to release 50 black-footed ferrets to the Wyoming prairie in 1991.

In addition to the control of predators and pests, humans hunt for three other reasons. *Commercial hunting* is the killing of animals for profit, by selling their fur, for example; *sport hunting* is the killing of animals for recreation; and *subsistence hunting* is the killing of animals for food. Subsistence hunting has caused the extinction of certain species in the past but is not a major cause today, mainly because so few human groups still rely on subsistence hunting for their food supply. Sport hunting was also a major factor in the extinction of animals in the past. The passenger pigeon, for example, was one of the most common birds in North America in the early 1800s, but a century of overhunting resulted in its extinction in the early 1900s. Sport hunting is now strictly controlled in most countries.

Commercial hunting, however, continues to endanger a number of larger animals such as the tiger, cheetah, and snow leopard, whose beautiful furs are quite valuable. Rhinoceroses are slaughtered for their horns (used for dagger handles in the Middle East and as a medicine and an aphrodisiac in Asia) and bears for their gallbladders (used in Asian medicine to treat ailments ranging from indigestion to hemorrhoids). Although these animals are

Focus On

The Lesson of New Zealand

Although the effect of humans on the environment is acknowledged to be considerable, few places exist in the world where human occupation is recent and where the history and impact of our occupation are well known. New Zealand provides an interesting illustration of these rare circumstances. This large archipelago was first occupied by the Maoris about 1000 years ago, and their invasion was followed some 800 years later by European colonization. As a result, we know much about the impact of human settlement on the organisms of New Zealand.

New Zealand separated from the ancient southern continent of Gondwana about 70 million years ago, before it could be invaded by snakes and placental mammals (see Chapter 20). As a result, the animal life of New Zealand became dominated by birds. During their evolution in this isolated environment, the birds evolved trusting natures, and many became weak fliers while others, such as the moas and kiwis, became large and flightless. Thus, when the first Maori settlers arrived, they found a land containing a unique assemblage of animals that were easily hunted.

These first humans had a dramatic impact on New Zealand. Not only did they introduce hunting to the islands, but they also brought dogs, rats, and agriculture. For the first time the birds had to contend with mammalian predation, against which they were poorly adapted, and human-induced habitat modification. The birds had little chance to adapt. By the time Europeans arrived, at least 35 species (39%) of New Zealand's unique birds had become extinct.

With active European settlement, which began in the 1840s, forest clearance accelerated so that now only 23% of New Zealand's original forest remains. The newly cleared land was given over to a pastoral, European style of agriculture. In addition, a huge range of exotic organisms was purposely introduced to New Zealand; at present some 800 species of nonnative plants, 39 species of birds, and 33 species of mammals are established. Extinction of endemic birds has continued in the face of increased hunting pressure and the efficiency of European firearms, the further reduction of the forests, the damaging impact of browsing mammals on the surviving native vegetation, and the introduction of a larger array of predatory mammals. During the period of European settlement, an additional nine species (10%) of New Zealand's endemic land birds have become extinct.

New Zealand has not been alone in its experience. But because its history is so well known, it provides a dramatic example of how completely and rapidly humans can alter major ecosystems. The biological lessons of New Zealand are clear. Ecosystems are complex, dynamic entities with their own unique evolutionary histories. However, major disruption can often lead to the unraveling of their fabric. Increasingly, humans have become the primary cause of such disruptions so that now our species has become a major force in shaping the future of Earth and its ecosystems. Although we have the necessary knowledge to safeguard the ecological integrity of our planet, do we have the wisdom and will to use it responsibly? Only time will tell. ■

Contributed by Alex L. A. Middleton, Department of Zoology, University of Guelph, Guelph, Ontario, Canada.

protected by law, demand for their products on the black market has promoted illegal hunting. When illegally obtained wildlife trophies are confiscated by police or park rangers, they are destroyed (Fig. 55–3).

In contrast to commercial hunting, in which target organisms are killed, *commercial harvesting* removes live organisms from the wild. Organisms that are commercially harvested end up in zoos, aquaria, research laboratories, and pet stores. Several million birds are commercially harvested each year for the pet trade, but unfortunately, many die in transit and many more die from improper treatment in their owners' homes. At least nine bird species are now threatened or endangered because of commercial harvest. Although it is illegal to capture endangered animals from the wild, a thriving black market exists, mainly because collectors in the United States, Europe, and Japan will pay extremely large sums for rare tropical birds. Imperial Amazon macaws, for example, fetch up to $30,000 each.

Animals are not the only organisms threatened by commercial harvest. A number of unique or rare plants have been collected from the wild so extensively that they are now classified as endangered. These include carnivorous plants, certain cacti, and orchids.

There are two types of conservation efforts to save wildlife

The two strategies to protect wildlife are in situ conservation and ex situ conservation. **In situ conservation,** which includes the establishment of parks and reserves, concentrates on preserving biological diversity in the wild. A high priority of in situ conservation is the identification and protection of sites with a great deal of diversity. With increasing demands on land, however, in situ conservation cannot guarantee the preservation of all types of biological diversity. **Ex situ conservation** involves conserving biological diversity in

FIGURE 55–3 The Kenyan government destroys elephant tusks confiscated from poachers. Burning ensures that such illegally obtained wildlife trophies will never be sold on the black market. (Animals Animals/© 1996 Steve Turner)

human-controlled settings. Breeding captive species in zoos and seed storage of genetically diverse crops (see Chapter 35) are examples of ex situ conservation.

In situ conservation is the best way to preserve biological diversity

Many nations appreciate the need to protect their biological heritage, and for this reason have set aside areas for wildlife habitats. Such natural ecosystems offer the best strategy for the long-term protection and preservation of biological diversity. There are currently more than 3000 national parks, marine sanctuaries, refuges, forests, and other protected areas throughout the world. Some of these areas have been chosen to protect specific endangered species. The first such refuge, established in 1903 at Pelican Island, Florida, was set aside to protect the brown pelican. Today the National Wildlife Refuge System of the United States includes more than 400 refuges.

Many protected areas have multiple uses. National parks may serve recreational needs, for example, whereas national forests may be open for logging, grazing, and farming operations. The mineral rights to many refuges are privately owned, and some wildlife refuges have had oil, gas, and other mineral development. For instance, the D'Arbonne Wildlife Refuge in Louisiana, which is a sanctuary for 145 species of birds, has soil and water pollution from natural gas wells. Hunting is allowed in more than half of the wildlife refuges in the United States, and military exercises are conducted in several of them. The Air Force, for example, conducts low-flying jet exercises and live fire exercises over portions of the Prieta Wildlife Refuge, an Arizona refuge established for bighorn sheep.

Certain parts of the world are critically short of protected areas. In addition to tropical rain forests, protected areas are needed in the tropical grasslands and savannas of Brazil and Australia, and in the dry forests widely scattered around the world. Tropical desert communities are underprotected in northern Africa and Argentina, and the wildlife of many islands and lakes also need protection.

Restoring damaged or destroyed habitats is the goal of restoration ecology

Disturbed lands can be reclaimed and converted into natural ecosystems with high levels of biological diversity. In certain cases, an area is *restored* to its original species composition, whereas in other cases, the area is *rehabilitated* by establishing at least some of the original species.

One of the most famous examples of ecological restoration has been carried out since 1934 by the University of Wisconsin–Madison Arboretum (Fig. 55–4). Several different communities, including a tallgrass prairie, a xeric (dry) prairie, and several types of pine and maple forests native to Wisconsin, have been carefully developed on damaged agricultural land.

Restoration ecology establishes wildlife habitats and also provides additional benefits, such as the regeneration of soil damaged by agriculture or mining. The disadvantages of restoration ecology include the time and expense required to restore an area. Nonetheless, restoration ecology is an important aspect of wildlife conservation.

Ex situ conservation is used in an attempt to save species on the brink of extinction

Zoos, aquaria, and botanical gardens attempt to save certain endangered species from extinction. Eggs may be collected from the wild, for example, or the remaining few animals may be captured and bred in zoos and other research facilities. Special techniques such as artificial insemination and host mothering are used to increase the

(a) (b)

FIGURE 55–4 The University of Wisconsin–Madison Arboretum has pioneered restoration ecology. (*a*) The restoration of the prairie was at an early stage in November, 1935. The men are digging holes to plant prairie grass sod. (*b*) The prairie as it looks today. This picture was taken at approximately the same location as the 1935 photograph. (Courtesy of Virginia Kline, University of Wisconsin–Madison Arboretum)

number of offspring. In **artificial insemination,** sperm collected from a suitable male of a rare species is used to artificially impregnate a female (perhaps located in another zoo in a different city or even in another country). In **host mothering,** a female of a rare species is treated with fertility drugs, which cause her to produce multiple eggs. Some of these eggs are collected, fertilized with sperm, and surgically implanted into females of a related but less rare species, who later give birth to offspring of the rare species. (Host mothering is also discussed in *Focus On: Novel Origins* in Chapter 49.)

A few spectacular successes have occurred in captive breeding programs, in which large enough numbers of a species have been produced to reestablish small populations in the wild. Conservation efforts, for example, have helped the bald eagle make a remarkable comeback. In the mid-1970s, the first eagles bred in captivity were released to the wild. In addition to captive breeding programs, biologists also remove eagle eggs from their nests in the wild, raise the baby eagles in wildlife refuges, and then return them to the wild. (Removal of eggs actually helps increase the number of eagles because nesting eagles often lay more eggs to replace those that were removed.) As a result of continuing efforts, the number of nesting pairs in the contiguous United States increased from 1000 to 4000 between 1975 and 1993. In 1994 the bald eagle was removed from the endangered list and transferred to the less critical threatened list.

Attempting to save a species on the brink of extinction is extremely expensive. Moreover, zoos, aquaria, and botanical gardens do not have the space to try to save *all* endangered species. For example, zoos have traditionally focused on large, charismatic animals because the public is more interested in them. Such ex situ conservation efforts ignore the millions of less glamorous but ecologically important species. However, they do serve a useful purpose by educating the public about the value of wildlife conservation. Although ex situ conservation is a valuable strategy, protecting natural habitats so that species do not become endangered is clearly more effective.

DEFORESTATION IS OCCURRING AT AN UNPRECEDENTED RATE

The permanent destruction of all tree cover in an area is known as **deforestation.** When forests are destroyed, soil fertility decreases and soil erosion increases. Increased sedimentation of waterways caused by soil erosion can harm aquatic ecosystems. In drier areas, deforestation can lead to the formation of deserts.

Making the Connection

Keystone Species and Conservation Biology

Within an ecosystem, are certain species more important than others? Such species, called **keystone species,** do exist and are crucial in determining the nature of the entire ecosystem — that is, the ecosystem's species composition and its functioning. Keystone species are not necessarily the most abundant species in the ecosystem. Even when present in relatively small numbers, the members of a keystone species influence the entire ecosystem, usually by affecting the amount of available food, water, or some other resource.

Identifying and protecting keystone species is a crucial goal of conservation biology because if one of them disappears from an ecosystem, many other organisms in that ecosystem may also go extinct. One example of a keystone species is a top predator such as the gray wolf. Where wolves were hunted to extinction, for example, the populations of deer and other herbivores increased explosively. As these herbivores overgrazed the vegetation, many plant species that could not tolerate such grazing pressure disappeared. Many smaller animals such as insects were also lost from the ecosystem because the plants that they depended on for food became less abundant. Thus, the disappearance of the wolf resulted in an ecosystem with considerably less biological diversity.

Because fig trees produce a continuous crop of fruits, they appear to be keystone species in tropical rain forests of Peru. Monkeys, fruit-eating birds, and other vertebrates of the forest do not normally consume large quantities of figs in their diets. During that time of the year when other fruits are less plentiful, however, fig trees become important in sustaining fruit-eating vertebrates. Should the fig trees disappear, most of the fruit-eating vertebrates would also be eliminated. Thus, protecting fig trees in such tropical rainforest ecosystems increases the likelihood that monkeys, birds, and other vertebrates will survive. ▲

Deforestation also contributes to the loss of biological diversity. In particular, tropical species often have limited ranges within a forest, so they are quite vulnerable to habitat destruction or modification. Wildlife in temperate areas, including migratory birds and butterflies, also suffer because of tropical deforestation.

Forests, particularly on hillsides and mountains, provide nearby lowlands with some protection from floods by trapping and absorbing precipitation. When a forest is cut down, the watershed cannot hold water nearly as well, and the total amount of surface runoff flowing into rivers and streams increases. This not only causes soil erosion, but puts lowland areas at extreme risk of flooding.

Regional and global climate changes are induced by deforestation. Transpiring trees release substantial amounts of moisture into the air. This moisture falls back to the surface in the hydrologic cycle. When forest is removed, rainfall declines and droughts become common in that region. Deforestation contributes to an increase in global temperature (discussed shortly) because deforestation results in a release of stored carbon into the atmosphere as carbon dioxide, which causes air to retain solar heat.

Where and why are forests disappearing?

During the past 1000 years, forests in temperate areas were largely cleared for housing and agriculture. Today, however, deforestation in the tropics is occurring much more rapidly and over a much larger area. Most remaining undisturbed tropical rain forest, which occurs in the Amazon and Congo River basins of South America and Africa, is being cleared and burned at a rate unprecedented in human history. Tropical rain forests are also being destroyed at an extremely rapid rate in southern Asia, Indonesia, Central America, and the Philippines. Ten countries — Brazil, Indonesia, Zaire, Burma, Colombia, India, Malaysia, Mexico, Nigeria, and Thailand — account for 76% of tropical deforestation.

The three main causes of deforestation in tropical rain forests are subsistence agriculture, commercial logging, and cattle ranching, whereas the main cause of deforestation in dry tropical forests is fuel wood consumption. Other reasons for the destruction of tropical forests include hydroelectric dams and mining.

Subsistence agriculture is the most important cause of deforestation

Subsistence agriculture, in which just enough food is produced to feed oneself and one's family, accounts for 60% of tropical deforestation. In many developing countries where tropical rain forests are located, the majority of people do not own the land on which they live and work. Most subsistence farmers were displaced from traditional farmlands because of an inequitable distribution of land ownership. They have no place to go except into the forest, which they clear to grow food.

Subsistence farmers often follow logging access roads. These poor rural people first cut down the forest and allow it to dry; then they burn the area (Fig. 55–5) and plant crops immediately after burning. This is known as **slash-and-burn agriculture.** Yields from the first crop are often quite high because the nutrients that had been in the trees

Making the Connection

Deforestation and Biogeochemical Cycles

Does deforestation of tropical rain forests disrupt nutrient cycling by which various materials move from the abiotic environment to living organisms? In a natural ecosystem, essential minerals cycle from the soil to living organisms, and then back again to the soil when those organisms die and decay (recall the biogeochemical cycles discussed in Chapter 54). An agricultural system disrupts this pattern when crops are harvested. Much plant material, containing minerals, is removed from the cycle, sent to market, turned into waste and sewage, and deposited in another ecosystem. As a result, it fails to decay and release its nutrients back to the soil from which it came. Thus, over time, soil that is farmed inevitably loses its fertility.

When tropical rain forests are removed to produce agricultural land, mineral depletion is particularly severe. Recall from Chapter 52 that tropical rainforest soils are nutrient-poor because nutrients are stored in vegetation. Any minerals released as dead organisms decay are promptly reabsorbed by plant roots and their mutualistic fungi. If this did not occur, heavy rainfall would quickly wash the nutrients away. Nutrient reabsorption by vegetation is so effective that tropical rainforest soils can support luxuriant plant growth despite the relative infertility of the soil, *as long as the forest remains intact.*

When a rain forest is cleared, its efficient nutrient cycling is disrupted. Removing vegetation that so effectively stores nutrients allows minerals to wash out of the system. Crops will grow in these soils for only a few years before the small mineral reserves in the soil are depleted. ▲

are available in the soil. However, soil productivity declines at a rapid rate, so that subsequent crops are poor. In a short time the farmers must move to new forests and repeat the process. Often, cattle ranchers then claim the land for grazing because land that is not rich enough to support crops can support livestock.

Slash-and-burn agriculture done on a small scale with long periods (20 to 100 years) between cycles is actually sustainable. But when millions of people try to obtain a living in this way, the land is not allowed to lie uncultivated for an adequate recovery period.

FIGURE 55–5 Tropical rain forest in Brazil is burned to provide agricultural land. This type of cultivation is known as slash-and-burn agriculture. (Earth Scenes/© 1996 Dr. Nigel Smith)

Commercial loggers remove vast tracts of tropical rain forests

Commercial logging, mostly for export abroad, accounts for 21% of tropical deforestation. Most tropical countries allow commercial logging to proceed at a much faster rate than is sustainable. For example, in parts of Malaysia, current logging practices remove the forest almost twice as fast as the sustainable rate. If this continues, Malaysia will soon experience shortages of timber and will have to start importing logs. When that happens, Malaysia will have lost future revenues, both from logging and from harvesting other forest products, from its newly vanished forests.

Cattle ranching also causes deforestation

Approximately 12% of tropical rainforest destruction is done to provide open rangeland for cattle. After the forests are cleared, cattle can be raised on the land for six to ten years, after which time shrubby plants, known as scrub savanna, take over the range. Much of the beef raised on these ranches, which are often owned by foreign companies, is exported to fast-food restaurants.

Fuel wood consumption destroys dry tropical forests

Perhaps half of the wood consumed worldwide is used in developing countries for heating and cooking fuel (Fig. 55–6). Fuel wood consumption is a greater concern in dry tropical forests—tropical areas subjected to a wet season and a prolonged dry season—than it is in humid forests.

(a) (b)

FIGURE 55–6 Half of the people in the world use wood fires to cook food. (*a*) A woman in India lights a wood cooking fire. (*b*) A woman in Kenya prepares her first meal using a solar oven that she has just built. In average sunlight, food will cook in two to four hours. Solar ovens, which can be built from inexpensive, easily available materials, reduce the number of trees destroyed for fuel. (*a*, Inga Spence/Tom Stack & Associates; *b*, © 1993 John Reader, courtesy of Earthwatch Expeditions)

CERTAIN ATMOSPHERIC POLLUTANTS MAY AFFECT EARTH'S CLIMATE

Imagine a world in which beautiful tropical islands, such as the Maldives in the Indian Ocean, disappear forever under the waves. Closer to home, consider the permanent flooding of Louisiana's bayous and the Florida Keys. Imagine forests in the southern United States dying from high temperatures and drought. Think of the effects of an annual hurricane season that is one month longer than it is now. Consider the consequences of tropical pests and diseases spreading northward into the United States and remaining throughout the winter.

Although none of these events has yet occurred, there is a chance that some or all of them will, possibly even during your lifetime. This is because Earth may become warmer during the next century than it has been for hundreds of thousands of years. **Global warming** is the long-term increase in Earth's average temperature. Unlike past climate changes, which occurred over thousands of years or longer, the global warming that faces us now would take place in a matter of decades.

Although most scientists agree that the planet will continue to warm, they disagree over how rapidly the warming will proceed, how severe it will be, whether or not it has already started, and where it will be most pronounced. One of the complications that makes global warming difficult to predict is that certain air pollutants from both natural and human sources actually exert a cooling influence. As a result of these uncertainties, many people, including policy-makers, are confused about what should be done.

Greenhouse gases cause global warming

Carbon dioxide (CO_2) and certain other trace gases, including methane (CH_4), surface ozone (O_3),[2] nitrous oxide (N_2O), and chlorofluorocarbons (CFCs), are accumulating in the atmosphere as a result of human activities (Table 55–1). The concentration of atmospheric CO_2 has increased from about 280 parts per million (ppm) approximately 200 years ago (before the Industrial Revolution began) to 360 ppm today (Fig. 55–7). And atmospheric CO_2 is still increasing, as are levels of the other trace gases associated with global warming.

[2]Surface ozone (more precisely, ozone in the troposphere) is a greenhouse gas as well as a component of photochemical smog. Ozone in the upper atmosphere, the stratosphere, provides an important planetary service that is discussed later in this chapter.

Table 55–1 INCREASE IN SELECTED ATMOSPHERIC GREENHOUSE GASES, PRE-INDUSTRIAL REVOLUTION TO 1992

	Concentration in Air	
Gas	*Preindustrial*	*1992*
Carbon dioxide	274 ppm*	356.2 ppm
Methane	0.7 ppm	1.7 ppm
Chlorofluorocarbon-12	0 ppb†	0.4 ppb
Chlorofluorocarbon-11	0 ppb	0.3 ppb
Nitrous oxide	280 ppb	306.0 ppb

*ppm = parts per million.
†ppb = parts per billion.

For example, every time you drive your car, the combustion of gasoline releases CO_2 and N_2O, and triggers the production of surface O_3. Every day CO_2 is released by the burning of rain forest in the Amazon. One way that CFCs get into the atmosphere is from old, leaking refrigerators and air conditioners. Decomposition in landfills is a major source of CH_4.

Global warming occurs because these gases are able to retain heat that normally would dissipate into space (Fig. 55–8). Some heat from the atmosphere is transferred to the ocean and raises its temperature as well. As the atmosphere and ocean warm, Earth's overall temperature increases. Because CO_2 and other gases trap the sun's radiation in much the same way that glass does in a greenhouse, the global warming produced in this manner is known as the **greenhouse effect,** and gases that retain heat in the atmosphere are called **greenhouse gases.**

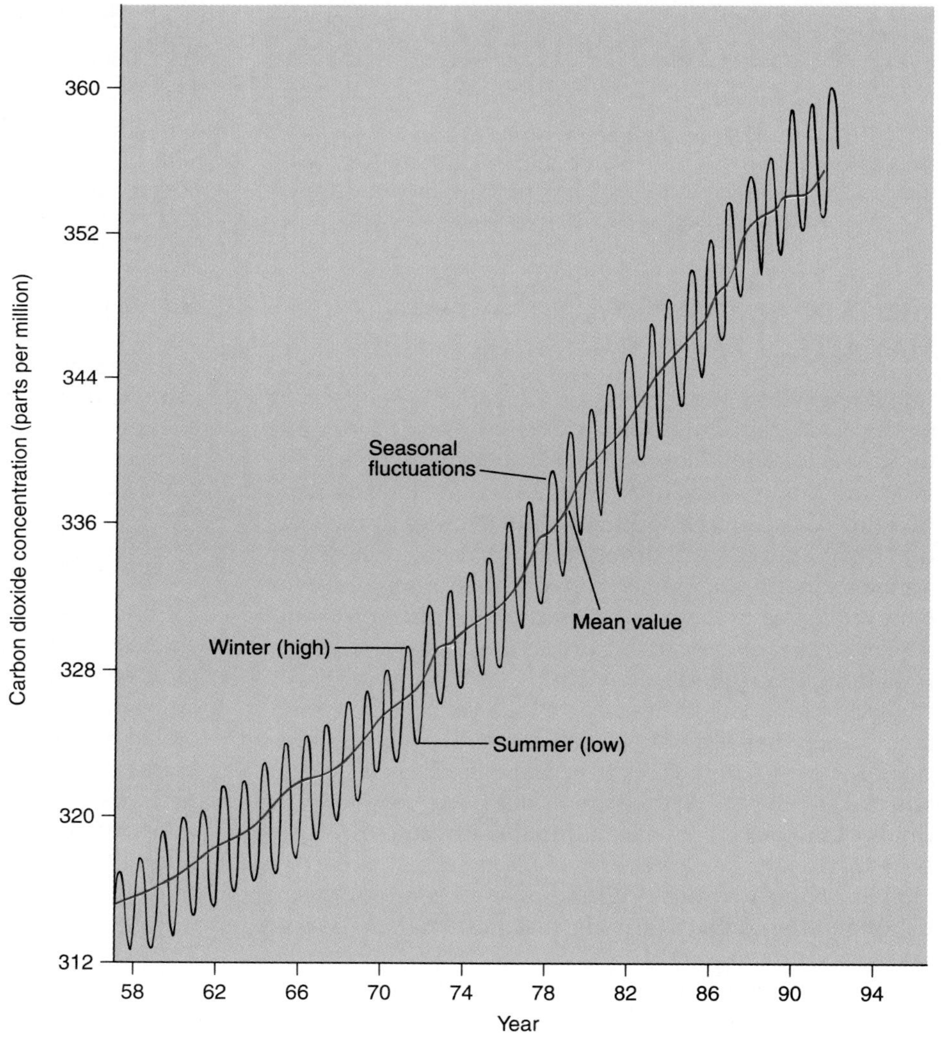

FIGURE 55–7 The concentration of carbon dioxide in the atmosphere has shown a slow but steady increase for many years. These measurements were taken at the Mauna Loa Observatory, far from urban areas where CO_2 is emitted by factories, power plants, and motor vehicles. The seasonal fluctuation each year corresponds to winter (a high level of CO_2) and summer (a lower level of CO_2) and is caused by greater photosynthesis in summer.

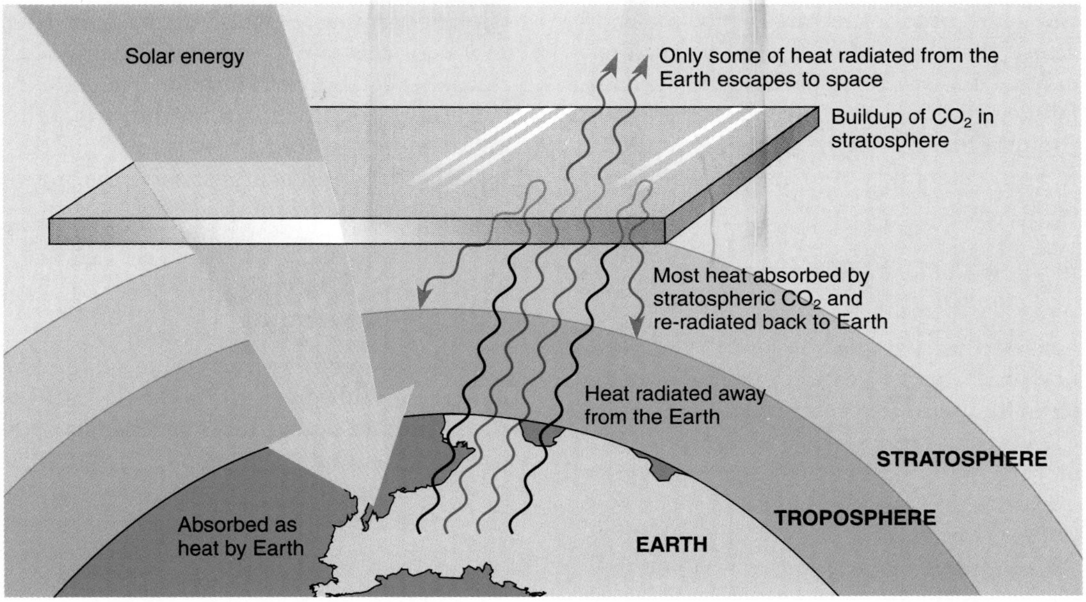

FIGURE 55–8 The buildup of carbon dioxide and other greenhouse gases causes the greenhouse effect, which results in global warming. The increased level of carbon dioxide in the stratosphere is depicted in this diagram as a sheet of glass to emphasize that in this situation CO_2 functions much like the glass walls and roof of a greenhouse.

Global warming could profoundly affect all life

Because the interactions among the atmosphere, the ocean, and the land are too complex and too large to study in the laboratory, climatologists develop models using computer simulations. A model, however, is only as good as the data and assumptions on which it is based, and a number of uncertainties are built into global warming models. For example, if global warming causes more low-lying clouds to form, they will block some sunlight and decrease warming. On the other hand, global warming may cause more high, thin cirrus clouds to form, which would actually increase the greenhouse effect. As new data about these uncertainties become available, they are used to make the predictions more precise.

Current models predict that a doubling of the CO_2 concentration in the atmosphere will cause the average temperature of Earth to increase by between 2° and 5°C before the end of the 21st century. Warming, however, will not be uniform from region to region. At current rates of fossil fuel combustion and deforestation, scientists expect a doubling of CO_2 to occur within the next 50 years. However, the warming trend will probably be slower than the increasing CO_2 might indicate because the ocean takes longer than the atmosphere to absorb heat. The second half of the next century will probably experience greater warming than will the first half.

With global warming, the sea level is expected to rise

If the overall temperature of Earth increases by just a few degrees, there could be a major thawing of glaciers and the polar icecaps. In addition, thermal expansion of the ocean will probably occur because water, like other substances, expands when it is warmer. These two changes may cause the sea level to rise by as little as 0.2 meter (8 inches) or as much as 2.2 meters (7 feet) by 2050, flooding low-lying coastal areas. Coastal areas that are not inundated will be more likely to suffer damage from hurricanes and typhoons.

Countries most vulnerable to a rise in the sea level have dense populations that inhabit low-lying river deltas—countries such as Bangladesh, Egypt, Vietnam, and Mozambique. A rising sea level, for example, could cause Bangladesh to lose as much as 25% of its land. Since 1970, at least 300,000 people in Bangladesh have been killed by tropical storms that caused increased flooding and high waves. A rising sea level caused by global warming would put even more people at risk in this densely populated nation.

With global warming, precipitation patterns will change

Global warming is also expected to change precipitation patterns, causing some areas to have more frequent droughts. At the same time, other areas may be more likely to experience flooding. The frequency and intensity of storms may increase also. All of these factors could affect the availability and quality of fresh water in many locations. It is thought that currently arid or semi-arid areas will become even drier and suffer the greatest water-shortage problems.

With global warming, the habitable ranges of organisms will change

Biologists have started to examine some of the potential consequences of global warming on wildlife. Each species reacts to changes in temperature differently. Some species will undoubtedly become extinct, particularly those with narrow temperature requirements, those confined to small reserves or parks, and those living in fragile ecosystems, whereas other species may survive in greatly reduced numbers and ranges. Ecosystems considered most vulnerable to species loss in the short term are polar seas, coral reefs, mountains, coastal wetlands, tundra, taiga, and temperate forests.

Some species may be able to migrate to new environments or adapt to the changing conditions in their present habitats. Also, some species may be unaffected by global warming, whereas others may emerge from it as winners, with greatly expanded numbers and ranges. Those considered most likely to prosper include weeds, pests, and disease-carrying organisms that are already common in many different environments.

Biologists generally agree that global warming will have an unusually severe impact on plants because they are unable to move about when conditions change. Although seeds may be dispersed over long distances, seed dispersal has definite limits in terms of how fast a plant species can migrate. Moreover, soil characteristics, water availability, competition with other species, and human alterations of natural habitats all affect the rate at which plants can move into a new area.

The increase in atmospheric CO_2 and the resulting increase in temperature will probably not harm human health directly. Human health will be indirectly affected, however, as such disease carriers as mosquitoes, which transmit malaria and encephalitis, expand their ranges into the newly warm areas. People can also expect more frequent and severe heat waves during summer months.

With global warming, agriculture will be affected

Global warming will increase problems for agriculture, which is already beset with the challenge of providing enough food for a hungry world without doing irreparable damage to the environment. The rising sea level will inundate river deltas, which are some of the world's best agricultural lands. Certain agricultural pests and disease-causing organisms will probably proliferate. Global warming may also increase the frequency and duration of droughts. To get an idea of how devastating droughts can be, consider the drought experienced in the American grain belt during the summer of 1988. Wheat yields were reduced by 40%, causing the United States, usually the world's largest exporter of grain, to consume more grain than it produced.

Many actions have been suggested to deal with global warming

Even if we were to immediately stop polluting the atmosphere with greenhouse gases, Earth would still undergo climate change to some extent because of the greenhouse gases that have already accumulated during the past 100 years. The amount and severity of global warming depend on how much additional greenhouse gas we add to the atmosphere.

There are three different strategies to manage global warming: prevention, mitigation, and adaptation. *Prevention* of global warming involves developing ways to prevent the buildup of greenhouse gases in the atmosphere. It is the ultimate and best solution because it is permanent. Prevention is also the least likely response because it requires technological "fixes" that have not yet been developed. *Mitigation,* which involves ways to moderate or postpone global warming, gives us time to pursue other, more permanent solutions. Further, a mitigative strategy gives us time to understand more fully how global warming operates so we can avoid some of its worst consequences. *Adaptation* is responding to changes brought about by global warming. Developing adaptive strategies to climate change assumes that global warming is unavoidable.

We can respond by preventing the buildup of greenhouse gases in the atmosphere

The development of alternatives (such as solar energy) to fossil fuels offers a permanent solution to the global warming challenge caused by increased CO_2 emissions.[3] Solar energy, which can be used directly to heat water and buildings and generate electricity, is inexhaustible and has fewer environmental problems than do fossil fuels. Solar and other alternative energy will not be practical for widespread adoption, however, until their technologies have been improved. Governments must increase available funding for research into renewable energy technologies. In addition, the invention of technological innovations to trap CO_2 being emitted from smokestacks (currently not possible) would help prevent global warming and yet allow us to use fossil fuels for energy.

[3]Alternatives to the other greenhouse gases will also have to be developed, but we focus on CO_2 here because it is produced in the greatest quantity and has the largest total effect of all the greenhouse gases.

We can respond by mitigating global warming

One of the most effective ways to mitigate global warming involves forests. As you know, atmospheric CO_2 is removed from the air by actively growing forests, which incorporate carbon into leaves, stems, and roots by photosynthesis. On the other hand, deforestation releases CO_2 into the atmosphere as trees are burned or decomposed. We can mitigate global warming by planting and maintaining new forests and by protecting existing forests. In addition, we can help mitigate global warming by increasing the energy efficiency of automobiles and appliances, which would reduce the output of CO_2.

We can respond by adapting to the reality of global warming

Government planners and social scientists from many countries are developing a number of strategies to help us adapt to global warming. For example, what should people living in coastal areas do? They can move inland away from the dangers of storm surges, although this solution has high societal and economic costs. An alternative plan, which is also extremely expensive, is to build dikes and levees to protect coastal land. The Dutch, who have been doing this sort of thing for several hundred years, have offered their technical expertise to several developing nations threatened by a rise in sea level.

We also have to adapt to shifting agricultural zones. Many temperate countries are in the process of evaluating semitropical crops to determine the best ones to substitute for traditional crops if or when the climate warms. Drought-resistant strains of trees are being developed by large lumber companies now, because the trees planted today will be harvested in the middle of the next century when global warming may already be in progress.

THE AMOUNT OF STRATOSPHERIC OZONE IS DROPPING

Ozone (O_3) is a form of oxygen that is a human-made pollutant in the lower atmosphere, but a naturally produced, essential part of the stratosphere (see Fig. 20–6). The **stratosphere,** which encircles our planet some 10 to 45 kilometers (6 to 28 miles) above Earth's surface, contains a layer of ozone that shields the surface from much of the ultraviolet radiation coming from the sun (Fig. 55–9). If ozone disappeared from the stratosphere, Earth would become unlivable for most forms of life. (Ozone in the lower atmosphere does not replenish the ozone that has been depleted from the stratosphere because it is converted back to oxygen in a few days.)

The problem of **ozone depletion,** which is the thinning of the ozone layer in the stratosphere, was demonstrated in 1984 with the discovery of a large hole—that is, a thin spot—in the ozone layer over Antarctica (Fig. 55–10). Ozone levels decrease there by as much as 67% each year. A smaller ozone hole also exists in the stratospheric ozone layer over the Arctic, but probably the most disquieting news is that worldwide levels of stratospheric

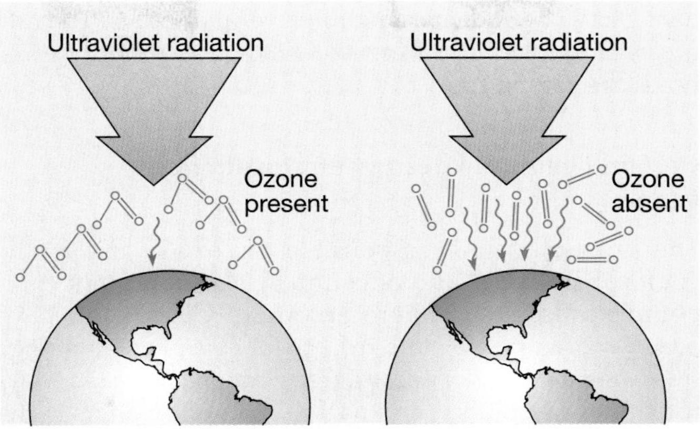

(*a*) (*b*)

FIGURE 55–9 Ultraviolet radiation is largely blocked by the ozone layer. (***a***) Stratospheric ozone absorbs ultraviolet radiation, effectively shielding Earth's surface. (***b***) When stratospheric ozone is absent, more high-energy ultraviolet radiation penetrates the atmosphere to Earth's surface, where its presence harms living organisms.

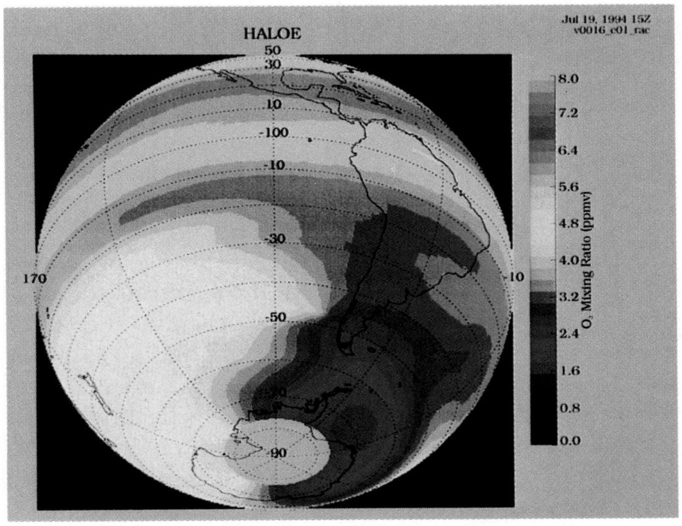

FIGURE 55–10 A computer-generated image of part of the Southern Hemisphere on July 19, 1994, reveals the ozone "hole" (*blue area*) over Antarctica. Relatively low ozone levels (*green areas*) extend into much of South America as well as Central America. Normal ozone levels are shown in yellow; no measurements were taken for the white area over the South Pole. The ozone hole is not stationary but moves about as a result of air currents. (NASA)

ozone have been decreasing for several decades. The rate of ozone depletion during the 1980s was roughly three times the rate in the 1970s.

Certain chemicals destroy stratospheric ozone

The primary culprit responsible for ozone loss in the stratosphere is a group of commercially important compounds called chlorofluorocarbons, or CFCs. CFCs have been used as propellants in aerosol cans, coolants (e.g., freon) in air conditioners and refrigerators, foam (e.g., styrofoam) for insulation and packaging, and cleaners in the electronics industry. Additional compounds that also attack ozone include halons (found in many fire extinguishers); methyl bromide (used as a pesticide in agriculture); methyl chloroform (used to degrease metals); and carbon tetrachloride (used in many industrial processes, including the manufacture of pesticides and dyes).

CFCs and similar compounds drift up to the stratosphere and become widely dispersed. Once there, ultraviolet radiation breaks them down into chlorine, fluorine, and carbon. Under certain stratospheric conditions, chlorine is capable of reacting with ozone, converting it into molecular oxygen. The chlorine is not altered by this process, and as a result, a single chlorine atom can break down many thousands of ozone molecules.

Ozone depletion harms living organisms

Ozone molecules in the stratosphere absorb incoming solar ultraviolet radiation. With depletion of the ozone layer, more ultraviolet radiation reaches Earth's surface. Excessive exposure to ultraviolet radiation is linked to a number of human health problems, including cataracts, skin cancer, and a weakened immune system. Much scientific evidence also documents crop damage from exposure to high levels of ultraviolet radiation.

Biologists are concerned that the ozone hole over Antarctica could damage plankton that forms the base of the food web for the surrounding ocean. A 1992 study confirmed that increased ultraviolet radiation is penetrating surface waters around Antarctica and that productivity of Antarctic phytoplankton has declined by at least 6% to 12% as a result. If the productivity of phytoplankton continues to decline, the complex food web of Antarctica, which includes fishes, seals, penguins, whales, and vast populations of birds, will be at risk.

International cooperation will prevent significant depletion of the ozone layer

In 1987, representatives from a number of countries met in Montreal to sign the Montreal Protocol, an agreement to reduce CFC production by 50% by 1998. Then, in 1990, more than 90 countries signed an agreement to phase out all use of CFCs by 2000. Despite these agreements, the news about the ozone layer has continually worsened since 1990. As a result, some nations have taken even stricter measures to limit CFC production. At an international meeting in Copenhagen in 1992, representatives agreed to phase out CFCs, carbon tetrachloride, halons, methyl chloroform, and similar chemicals by early 1996. They also agreed to freeze production of other ozone-destroying chemicals.

The ozone depletion problem is far from over. Unfortunately, CFCs are extremely stable and will continue to deplete stratospheric ozone well into the 21st century. In the final analysis, however, ozone depletion may become one of the 20th century's most significant environmental success stories. When faced with a serious environmental crisis, governments, scientists, industrialists, and citizens of many nations worked together to help solve the problem.

Environmental problems are interrelated

Several connections have been pointed out among the four environmental problems discussed in this chapter. For example, deforestation, global warming, and ozone depletion will likely cause future reductions in biological diversity. Similarly, any environmental problem you can think of—even if not discussed in this chapter—is related to other environmental concerns.

Most of all, however, environmental problems are connected to overpopulation. We have seen that the rate of human population growth is greatest in developing countries, but that developed nations are overpopulated as well, in the sense that they have a high per capita consumption of resources (see Chapter 52). Many natural resources are needed to provide the air conditioners, disposable diapers, cars, video cassette recorders, and other "comforts" of life in developed nations. Thus, *consumption overpopulation* (the disproportionately large consumption of resources by people in developed countries) affects the environment as much as the population explosion in the developing world does.

As living organisms, we share much in common with the fate of other life forms on the planet. We are not immune to the environmental damage we have produced. We differ from other organisms, however, in our capacity to reflect on the consequences of our actions and to alter our behavior accordingly. Humans, both individually and collectively, are able to bring about change. This talent is the key to any hope of ensuring the biosphere's survival.

SUMMARY

I. A reduction in biological diversity is occurring wordwide.
   A. A species becomes extinct when its last individual member dies.
      1. A species whose severely reduced numbers put it in danger of extinction is called an endangered species.
      2. When extinction is less imminent but the population is quite low, a species is called a threatened species.
   B. Human activities that contribute to a reduction in biological diversity include habitat loss and disturbance, pollution, introduction of foreign species, pest and predator control, hunting, and commercial harvest. Of these, habitat loss is the most significant.
   C. Efforts to preserve biological diversity in the wild are known as in situ conservation. Ex situ conservation, such as captive breeding, occurs in human-controlled settings.

II. Forests provide us with many ecological benefits, including watershed protection, soil erosion prevention, climate moderation, and wildlife habitat.
   A. The greatest problem facing forests today is tropical deforestation.
   B. In the tropics, forests are destroyed to provide colonizers with temporary agricultural land; produce timber, particularly for developed nations; provide open rangeland for cattle; and supply people with fuel wood.

III. Carbon dioxide and other greenhouse gases cause the air to retain heat, which warms the Earth.
   A. The increased CO_2 and other greenhouse gases in the atmosphere are causing concerns about a major global warming that may occur during the next century.
      1. The combustion of fossil fuels produces pollutants, especially CO_2.
      2. Other greenhouse gases are methane, nitrous oxide, chlorofluorocarbons, and surface ozone.
   B. Global warming may cause a rise in sea level, changes in precipitation patterns, the death of forests, the extinction of many living organisms, and problems for agriculture. It could result in the displacement of thousands or even millions of people.
   C. The challenges of global warming can be met by prevention (stop polluting the air with greenhouse gases), mitigation (slow down the rate of global warming), and/or adaptation (make adjustments to live with global warming).

IV. The ozone layer in the stratosphere helps to shield Earth's surface from damaging ultraviolet radiation.
   A. The total amount of ozone in the stratosphere is slowly declining, and large ozone holes develop over Antarctica and the Arctic each year.
   B. The attack on the ozone layer is caused by chlorofluorocarbons and similar chlorine-containing compounds.

SELECTED KEY TERMS

artificial insemination
biological diversity
deforestation
endangered species
ex situ conservation
extinction
genetic diversity
global warming
greenhouse effect
greenhouse gases
host mothering
in situ conservation
keystone species
ozone (O_3)
ozone depletion
slash-and-burn agriculture
stratosphere
subsistence agriculture
threatened species

POST-TEST

1. ________ is the permanent loss of a species.
2. A(an) ________ species is severely reduced in number and in imminent danger of becoming extinct.
3. A(an) ________ species is not in imminent danger of extinction, but its numbers are low enough for concern.
4. Habitat destruction or modification is the most important reason for declining ________ ________.
5. ________ ________ conservation concentrates on preserving biological diversity in the wild.
6. Breeding captive species in zoos is an example of ________ ________ conservation.
7. In ________ ________, sperm from a male of a rare species is used to artificially impregnate a female (perhaps located in another zoo).
8. In ________ ________, a female of a rare species is treated with fertility drugs, her eggs are collected and fertilized with sperm, and each embryo is surgically implanted into a female of a similar species.
9. A(an) ________ ________ greatly influences the species composition of the ecosystem to which it belongs.
10. The permanent loss of forests is known as ________.
11. Poor rural people living in the tropics practice a type of subsistence agriculture known as ________ and ________.
12. ________ ________, which is the long-term increase in Earth's average temperature, may occur over the next few decades.
13. The ________ effect causes warming when certain atmospheric gases retain heat near the surface.
14. Important ________ ________ include carbon dioxide, methane, surface ozone, nitrous oxide, and chlorofluorocarbons.
15. ________ is a human-made pollutant in the lower (surface) atmosphere but a natural and beneficial gas in the stratosphere.
16. The ozone layer in the ________ absorbs incoming ultraviolet radiation from the sun.
17. ________ ________ is most pronounced in the atmosphere over Antarctica.

REVIEW QUESTIONS

1. Which organism is more likely to become extinct, an endangered species or a threatened species? Explain your answer.
2. How does habitat destruction contribute to declining biological diversity?
3. Which type of conservation measure, in situ or ex situ, will help the greatest number of species? Why?
4. State at least three environmental benefits that forests provide.
5. Discuss two reasons for tropical deforestation.
6. Describe the greenhouse effect, and list three or more greenhouse gases.
7. What are some of the significant problems that may be caused by global warming during the next century?
8. Discuss and give an example of each of the three approaches (prevention, mitigation, and adaptation) for dealing with global warming.
9. What environmental service does the stratospheric ozone layer provide?
10. Describe some of the consequences of the thinning ozone layer.

YOU MAKE THE CONNECTION

1. If most of the world's biological diversity were to disappear, how would it affect your life?
2. Because new species will eventually evolve to replace those that humans are driving to extinction, why is declining biological diversity such a threat to us?
3. Would controlling the growth of the human population have any effect on biological diversity? Why or why not?
4. Why might captive breeding programs that reintroduce species into natural environments be doomed to failure?
5. If all of the world's tropical rain forests were destroyed, how would it affect your life?
6. If you were given the task of developing a policy for the U.S. to deal with global climate change during the next 100 years, would you stress prevention, mitigation, or adaptation? Explain your answer. What do you think the U.S. government is currently stressing?
7. It has been suggested that the wisest way to "use" fossil fuels would be to leave them in the ground. How would this affect global warming? Energy supplies?
8. This statement was overheard in an elevator: "CFCs cannot be the cause of ozone depletion over Antarctica because there are no refrigerators in Antarctica." Criticize the reasoning behind this statement.
9. It has been suggested that a more descriptive name for *Homo sapiens* is "*Homo dangerous*." Explain the specific epithet, based on what you have learned in this chapter.

RECOMMENDED READINGS

Cohn, J.P. "The Flight of the California Condor." *Bioscience,* Vol. 43, No. 4, April 1993. Captive breeding of this endangered species has been so successful that it has been reintroduced into the wild.

Conniff, R. "From *Jaws* to Laws: Now the Big, Bad Shark Needs Protection from Us." *Smithsonian,* May 1993. New conservation measures have been passed to help shark populations recover from human predation.

Hansen, J. et al. "How Sensitive Is the World's Climate?" *Research & Exploration* (a scholarly publication of the National Geographic Society), Spring 1993. Although interpreting climate data is difficult, current data indicate that the climate is warming as a result of human activities.

Line, L. "Silence of the Songbirds." *National Geographic,* Vol. 183, No. 6, June 1993. Populations of migratory songbirds that spend their summers in North America and their winters in Central and South America have declined dramatically during the past few years.

Potten, C.J. "America's Illegal Wildlife Trade: A Shameful Harvest." *National Geographic,* Vol. 180, No. 3, September 1991. Illegal hunting and the black market for animal products are decimating America's wildlife.

Repetto, R. "Deforestation in the Tropics." *Scientific American,* Vol. 262, No. 4, April 1990. Examines government policies that encourage the destruction of tropical forests.

Ryan, J.C. "When Nature Loses Its Cool." *World Watch,* Vol. 5, No. 5, September/October 1992. A sobering discussion of the effects that global warming will probably have on wildlife.

Tattersall, I. "Madagascar's Lemurs." *Scientific American,* Vol. 268, No. 1, January 1993. Habitat destruction in Madagascar has caused the extinction of hundreds of species of lemurs and threatened many others.

Vincent, J.R. "The Tropical Timber Trade and Sustainable Development." *Science,* Vol. 256, June 19, 1992. Tropical countries are exporting wood from their forests faster than can be sustained. This paper examines the root causes of this problem.

Youth, H. "Flying into Trouble." *World Watch* Vol. 7, No. 1, January/February, 1994. Examines the reasons why bird populations are declining worldwide.

Zimmer, C. "Son of Ozone Hole." *Discover,* October 1993. A new computer model of atmospheric circulation presents some disquieting news about the ozone hole over Antarctica.

Career Visions

Genetic Counselor

ANDY FAUCETT

Andy Faucett, M.S. obtained a Bachelor's degree in biology from Baptist College at Charleston and began medical school. During his first year of medical school, he realized his uncertainty about the physician role and his desire for a career with more opportunity to care for a patient's emotional needs. He went on to work as a teacher and counselor of emotionally disturbed high-school students and, through some research, eventually discovered the career that would enable him to combine his aptitude for science with his affinity for working with people. After completing a Master's degree program in human genetics at Sarah Lawrence College, Andy became a genetic counselor. Working primarily with prenatal patients, he evaluates family histories, tests patients to assess the genetic risks of disease to their children, and provides counseling to parents.

What do genetic counselors do?

There are different types of genetic counselors. Some work in prenatal, others in pediatrics, and many now work with adult conditions as a result of recent findings from the Human Genome Project. I work primarily with couples who are either considering a pregnancy or are expecting and have a family history of birth defects. Many patients come in due to an age-related risk. I also counsel individuals whose pregnancy testing or ultrasound indicates a higher risk for birth defects. For these patients, we conduct a much more thorough ultrasound along with other tests. Once or twice a week we find something that's pretty serious and my role is to explain it to the parents, talk about what it means, what the chances are of a good outcome, what the chances are of a bad outcome, if we can do anything to change it, and what other tests should be considered. So part of genetic counseling is science, but there's also a strong patient care component. Discovering a problem during your pregnancy is pretty devastating, so genetic counselors are there to help them deal with it and to make sure they have support. I describe the counseling aspect as crisis counseling. I'm involved in people's lives for a very short time—usually one or two visits. I have to get to know them quickly, help determine what kind of support they need, and see that they get it.

What do you mean by age-related risk?

The risk for a chromosome problem such as Down syndrome and the other trisomies—the extra chromosome syndromes—increases with a mother's age. Women over thirty-five in the United States are routinely offered testing. Perhaps twenty-five percent of the people I see have concerns related to the mother's age.

Is age-related risk testing routine?

Genetic counselors use a term called "non-directive" which means we stay in the middle. I would say that anyone who is thirty-five or older should be offered testing. They should be given the chance to come in and discuss the advantages and disadvantages of testing. When it comes to testing in a pregnancy there isn't a "should or a shouldn't," it has more to do with your perception of the risk. Is this important information to you, is this something you would want to know? For example, some thirty-nine year olds say, "My risk is high and I want every test you have," while other people who are the same age are less concerned. My role is to help make sure they have all the information needed to make a decision. Many patients have pre-conceived ideas that are quite incorrect. They may think that if it's your first pregnancy the risk of defects is greater—that's not true; that if you have ten children the risk is lower—that's not true; that if your family history is completely clean and there's nothing for five generations your risk is lower—and that's not true. Consequently, a lot of my job involves removing misconceptions and providing people with the right information.

What kinds of courses did you take to become a genetic counselor?

I would divide the coursework into three tracks. The first is primarily human genetics with a concentration of birth defects, inheritance patterns, calculations of risk, and tests that are available—DNA testing for example. There is a counseling component which instructs students in how to perform individual counseling, how to relate to people, and how to respond to body language. In my program there was a third component which entailed eight-hundred hours of practical rotation in which students worked with genetic counselors and gained hands-on experience.

Does a student need to be in a pre-med program to become a genetic counselor?

Actually there are a number of counselors who began with undergraduate degrees in psychology or a social science. There is usually some adjustment in the graduate degree program so that a student with a strong psychology background might take fewer counseling courses and more genetics. You need to be comfortable with science and working with people. I would encourage students to contact the National Society of Genetic Counselors.* They can provide a list of training programs in which students might arrange to visit and talk to a genetic counselor.

*For further information about a career in genetic counseling contact the National Society of Genetic Counselors at 610-872-7608.

APPENDIX A

The Classification of Organisms

The system of cataloging organisms used in this book is described in Chapter 1 and in Part 5. Recall that although we use a five-kingdom classification, some biologists now recognize six kingdoms; they divide kingdom Prokaryotae into kingdom Archaebacteria and kingdom Eubacteria. We have omitted many groups, especially extinct ones, in order to simplify the vast number of diverse categories of living organisms and their relationships to one another. Note that we have omitted the viruses from this survey, because they do not fit into any of the five kingdoms.

KINGDOM PROKARYOTAE: BACTERIA

Prokaryotic organisms that lack nuclear envelopes, mitochondria, and other membranous organelles. Typically unicellular, but some form colonies or filaments. Mainly heterotrophic, but some groups are photosynthetic or chemosynthetic. Reproduction is primarily asexual by fission. Bacteria are nonmotile or move by beating flagella. When present, flagella are solid (rather than the 9 + 2 type typical of eukaryotes). More than 10,000 species.

Bacterial nomenclature and taxonomic practices do not currently correspond to those of other organisms. As explained in Chapter 23, the editors of *Bergey's Manual of Systematic Bacteriology*[1] have divided prokaryotes into four groups based on their cell wall type. Representative types of each group are described in Table 23–2.

The four groups are: (1) bacteria with a gram-negative type of cell wall; (2) bacteria with a gram-positive type of cell wall; (3) bacteria that lack cell walls (mycoplasmas); and (4) bacteria with cell walls that lack peptidoglycan (archaebacteria).

Subkingdom Archaebacteria

Prokaryotes with cell walls lacking peptidoglycan. Three main groups (considered phyla by some microbiologists) are *methanogens, halobacteria,* and *thermoacidophiles.* Distinguished by their ribosomal RNA, lipid structure, and specific enzymes. Archaebacteria are found in extreme environments—hot springs, sea vents, dry and salty seashores, boiling muds, and near ash-ejecting volcanoes.

Subkingdom Eubacteria

Very large, diverse group. Some taxonomists divide the eubacteria into three phyla based on their cell walls.

Gram-negative bacteria. *Nitrogen-fixing aerobic bacteria, enterobacteria, spirochetes, cyanobacteria, rickettsias, chlamydias,* and *myxobacteria.* Thin cell wall containing peptidoglycan.

Gram-positive bacteria. *Lactic acid bacteria, streptococci, staphylococci, clostridia,* and *actinomycetes.* Thick cell wall of peptidoglycan. All are nonphotosynthetic, and many produce spores.

Bacteria that lack a rigid cell wall. *Mycoplasmas.* Extremely small bacteria bounded by a plasma membrane.

[1]*Bergey's Manual of Systematic Bacteriology, 9th ed.,* Williams & Wilkins, Baltimore, Philadelphia, 1994.

KINGDOM PROTISTA

Primarily unicellular or simple multicellular eukaryotic organisms that do not form tissues and that exhibit relatively little division of labor. Most modes of nutrition occur in this kingdom. Life cycles may include both sexually and asexually reproducing phases and may be extremely complex, especially in parasitic forms. Locomotion is by cilia, flagella, amoeboid movement, or by other means. Flagella and cilia have 9 + 2 structure.

Animal-like Protists

Phylum Zoomastigina. *Flagellates.* Single cells that move by means of flagella. Some free-living; many symbiotic; some pathogenic. Reproduction usually asexual by binary fission.

Phylum Rhizopoda. *Amoebas.* Shelled or naked single-celled protists whose movement is associated with pseudopods.

Phylum Foraminifera. *Foraminiferans.* Unicellular protists that produce calcareous tests (shells) with pores through which cytoplasmic projections extend, forming a sticky net to entangle prey.

Phylum Actinopoda. *Actinopods.* Unicellular protists that produce axopods (long, filamentous cytoplasmic projections) that protrude through pores in their siliceous shells.

Phylum Ciliophora. *Ciliates.* Unicellular organisms that move by means of cilia. Reproduction is asexual by binary fission or sexual by conjugation. About 7200 species.

Phylum Apicomplexa. *Sporozoa.* Parasitic single-celled protists that reproduce by spores and have no means of locomotion. Reproduce by an unusual type of multiple fission. Some pathogenic. About 3900 species.

Plant-like Protists

Phylum Dinoflagellata. *Dinoflagellates.* Unicellular (some colonial), photosynthetic, biflagellate. Cell walls, composed of overlapping cell plates, contain cellulose. Contain chlorophylls *a* and *c* and carotenoids, including fucoxanthin. About 2100 species.

Phylum Bacillariophyta. *Diatoms.* Unicellular (some colonial), photosynthetic. Most nonmotile, but some move by gliding. Cell walls composed of silica rather than cellulose. Contain chlorophylls *a* and *c* and carotenoids, including fucoxanthin. About 5600 species.

Phylum Euglenophyta. *Euglenoids.* Unicellular, photosynthetic, two flagella (one of them very short). Flexible outer covering. Contain chlorophylls *a* and *b* and carotenoids. About 1000 species.

Phylum Chlorophyta. *Green algae.* Mainly aquatic. Unicellular, colonial, siphonous, and multicellular forms. Some motile and flagellated. Photosynthetic; contain chlorophylls *a* and *b* and carotenoids. About 7000 species.

Phylum Rhodophyta. *Red algae.* Most multicellular (some unicellular), mainly marine. Some (coralline algae) have bodies impregnated with calcium carbonate. No motile cells. Photosynthetic; contain chlorophyll *a*, carotenoids, phycocyanin, and phycoerythrin. About 4000 species.

Phylum Phaeophyta. *Brown algae.* Multicellular, often quite large (kelps). Photosynthetic; contain chlorophylls *a* and *c* and carotenoids, including fucoxanthin. Biflagellate reproductive cells. About 1500 species.

Fungal-like Protists

Phylum Myxomycota. *Plasmodial slime molds.* Spend part of life cycle as a thin, streaming, multinucleate plasmodium that creeps along on decaying leaves or wood. Flagellated or amoeboid reproductive cells; form spores in sporangia. About 500 species.

Phylum Acrasiomycota. *Cellular slime molds.* Vegetative (nonreproductive) form unicellular; move by pseudopods. Amoeba-like cells aggregate to form a multicellular pseudoplasmodium that eventually develops into a fruiting body that bears spores. About 70 species.

Phylum Oomycota. *Water molds.* Consist of branched, coenocytic mycelia. Cellulose and/or chitin in cell walls. Produce biflagellate asexual spores. Sexual stage involves production of oospores. Some parasitic. About 580 species.

KINGDOM FUNGI

All eukaryotic, mainly multicellular organisms that are heterotrophic with saprotrophic or parasitic nutrition. Body form often a mycelium, and cell walls consist of chitin. No flagellated stages. Reproduce by means of spores, which may be produced sexually or asexually. Cells usually haploid or dikaryotic with brief diploid period following fertilization.

Phylum Zygomycota. Produce sexual resting spores called zygospores, and asexual spores in a sporangium. Hyphae are coenocytic. Many are heterothallic (two mating types). About 765 species.

Phylum Ascomycota. *Sac fungi.* Sexual reproduction involves formation of ascospores in little sacs called asci. Asexual reproduction involves production of spores called conidia, which pinch off from conidiophores. Hyphae usually have perforated septa. About 30,000 species.

Phylum Basidiomycota. *Club fungi.* Sexual reproduction involves formation of basidiospores on a basidium. Asexual reproduction uncommon. Heterothallic. Hyphae usually have perforated septa. About 16,000 species.

Phylum Deuteromycota. *Imperfect fungi.* Sexual stage has not been observed. Most reproduce only by conidia. About 17,000 species.

KINGDOM PLANTAE

Multicellular eukaryotic organisms with differentiated tissues and organs. Cell walls contain cellulose. Cells frequently contain large vacuoles, photosynthetic pigments in plastids. Photosynthetic pigments are chlorophylls *a* and *b* and carotenoids. Nonmotile. Reproduce both asexually and sexually, with alternation of gametophyte and sporophyte generations.

Phylum Bryophyta. *Mosses.* Nonvascular plants that lack xylem and phloem. Marked alternation of generations with dominant gametophyte generation. Motile sperm. The gametophytes generally form a dense green mat consisting of individual plants. About 9000 species.

Phylum Hepatophyta. *Liverworts.* Nonvascular plants that lack xylem and phloem. Marked alternation of generations with dominant gametophyte generation. Motile sperm. The gametophytes of certain species have a flat, liver-like thallus; other species are more mosslike in appearance. About 6000 species.

Phylum Anthocerophyta. *Hornworts.* Nonvascular plants that lack xylem and phloem. Marked alternation of generations with dominant gametophyte generation. Motile sperm. The gametophyte is a small, flat green thallus with scalloped edges. Spores are produced on an erect, hornlike stalk. About 100 species.

Phylum Pterophyta. *Ferns.* Vascular plants with a dominant sporophyte generation. Generally homosporous. Gametophyte is free-living and photosynthetic. Reproduce by spores. Motile sperm. About 11,000 species.

Phylum Psilotophyta. *Whisk ferns.* Vascular plants with a dominant sporophyte generation. Homosporous. Stem is distinctive because it branches dichotomously; plant lacks true roots and leaves. The gametophyte is subterranean and nonphotosynthetic and forms a mycorrhizal relationship with a fungus. Motile sperm. About 12 species.

Phylum Sphenophyta. *Horsetails.* Vascular plants with hollow, jointed stems and reduced scalelike leaves. Although modern representatives are small, some extinct species were treelike. Homosporous. Gametophyte a tiny photosynthetic plant. Motile sperm. About 15 species.

Phylum Lycophyta. *Club mosses.* Sporophyte plants are vascular with branching rhizomes and upright stems that bear microphylls. Although modern representatives are small, some extinct species were treelike. Some homosporous, others heterosporous. Motile sperm. About 1000 species.

Phylum Coniferophyta. *Conifers.* Heterosporous vascular plants with woody tissues (trees and shrubs) and needle-shaped leaves. Most are evergreen. Seeds are usually borne naked on the surface of cone scales. Nutritive tissue in the seed is haploid female gametophyte tissue. Nonmotile sperm. About 550 species.

Phylum Cycadophyta. *Cycads.* Heterosporous, vascular, dioecious plants that are small and shrubby or larger and palmlike. Produce naked seeds in conspicuous cones. Flagellated sperm. About 140 species.

Phylum Ginkgophyta. *Ginkgo.* Broad-leaved deciduous trees that bear naked seeds directly on branches. Dioecious. Contain vascular tissues. Flagellated sperm. The ginkgo tree is the only living representative. One species.

Phylum Gnetophyta. *Gnetophytes.* Woody shrubs, vines, or small trees that bear naked seeds in cones. Contain vascular tissues. Possess many features similar to flowering plants. About 70 species.

Phylum Anthophyta. *Flowering plants.* Largest, most successful group of plants. Heterosporous; dominant sporophytes with extremely reduced gametophytes. Contain vascular tissues. Bear flowers, fruits, and seeds (enclosed in a fruit; seeds contain endosperm as nutritive tissue). Double fertilization. About 235,000 species.

KINGDOM ANIMALIA: ANIMALS

Multicellular eukaryotic heterotrophs with differentiated cells. In most animals, cells are organized to form tissues, and tissues are organized to form organs; in complex forms, specialized body systems carry on specific functions. Most animals have a well-developed nervous system and can respond rapidly to changes in their environment. Most are capable of locomotion during some time in their life cycle. Most animals reproduce sexually with large nonmotile eggs and flagellated sperm.

Phylum Porifera. *Sponges.* Mainly marine. Body bears many pores through which water circulates. Food is filtered from the water by collar cells (choanocytes). Solitary or form colonies. Asexual reproduction by budding; external sexual reproduction in which sperm are released and swim to internal egg. Larva is motile. About 9000 species.

Phylum Cnidaria. *Hydras, jellyfish, sea anemones, corals.* Tentacles surrounding mouth. Stinging cells (cnidocytes) that contain stinging structures called nematocysts. Polyp and medusa forms. Planula larva. Solitary or colonial. Marine, with a few freshwater forms. About 10,000 species.

Phylum Ctenophora. *Comb jellies.* Biradial symmetry. Free-swimming; marine. Two tentacles and eight longitudinal rows of cilia resembling combs; animal moves by means of these bands of cilia. About 100 species.

ACOELOMATES

No body cavity; region between body wall and internal organs filled with tissue.

Phylum Platyhelminthes. *Flatworms.* Planarians are free-living; flukes and tapeworms are parasitic. Body dorsoventrally flattened; cephalization; three tissue layers. Simple nervous system with ganglia in head region. Excretory organs are protonephridia with flame cells. About 20,000 species.

Phylum Nemertea. *Proboscis worms* (also called ribbon worms). Long, dorsoventrally flattened body with complex proboscis used for defense and for capturing prey. Simplest animal to have definite organ systems. Complete digestive tract. Circulatory system with blood. About 900 species.

Pseudocoelomates

Body cavity not completely lined with mesoderm. Complete digestive tract extending from mouth to anus.

Phylum Nematoda. *Roundworms. Ascaris,* hookworms, pinworms. Slender, elongated, cylindrical worms; covered with cuticle. Free-living and parasitic forms. About 12,000 species.

Phylum Rotifera. *Wheel animals.* Microscopic, wormlike animals. Anterior end has ciliated crown that looks like a wheel when the cilia beat. Posterior end tapers to a foot. Constant number of cells. About 1800 species.

COELOMATES

These animals have a true coelom, a body cavity completely lined with mesoderm.

Protostomes

Coelomates with spiral, determinate cleavage; mouth develops from blastopore.

Phylum Mollusca. *Snails, clams, squids, octopods.* Unsegmented, soft-bodied animals usually covered by a dorsal shell. Have a ventral, muscular foot. Most organs located above foot in visceral mass. A shell-secreting mantle covers the visceral mass and forms a mantle cavity, which contains gills. Trochophore and/or veliger larva. About 50,000 species.

Phylum Annelida. *Segmented worms.* Polychaetes, earthworms, leeches. Both body wall and internal organs are segmented. Body segments separated by septa. Some have nonjointed appendages. Setae used in locomotion. Closed circulatory system; metanephridia; specialized regions of digestive tract. Trochophore larva. About 15,000 species.

Phylum Arthropoda. *Crabs, shrimp, insects, spiders, mites, ticks, centipedes, millipedes.* Segmented animals with paired, jointed appendages and a hard exoskeleton made of chitin. Open circulatory system with dorsal heart. Hemocoel occupies most of body cavity, and coelom is reduced. About one million species.

Deuterostomes

Coelomates with radial, indeterminate cleavage. Blastopore develops into anus, and mouth forms from a second opening.

Phylum Phoronida. One of the lophophorate phyla. Tube-dwelling marine worms.

Phylum Ectoprocta. *Bryozoans.* One of the lophophorate phyla. Mainly marine; sessile colonies produced by asexual budding.

Phylum Brachiopoda. *Lamp shells.* One of the lophophorate phyla. Marine, solitary; body enclosed between two shells. About 300 species.

Phylum Echinodermata. *Sea stars, sea urchins, sand dollars, sea cucumbers.* Marine animals that have pentaradial symmetry as adults but bilateral symmetry as larvae. Endoskeleton of small, calcareous plates. Water vascular system; tube feet for locomotion. About 7000 species.

Phylum Chaetognatha. *Arrow worms.* Marine, make up part of plankton. Ciliated. Pharyngeal gill slits. About 100 species.

Phylum Hemichordata. *Acorn worms.* Marine animals with an anterior muscular proboscis, connected by a collar region to a long wormlike body. The larval form resembles an echinoderm larva. About 85 species.

Phylum Chordata. *Tunicates, lancelets, vertebrates.* Notochord; pharyngeal gill slits; dorsal, tubular nerve cord; and postanal tail present at some time in life cycle. About 50,000 species.

APPENDIX B

Careers in Biology

The following organizations and professional associations will provide career information upon request.

General

American Association for the Advancement of Science
Office of Opportunity in Sciences
1333H, N.W.
Washington, D.C. 20005
202-326-6400
email: info@aas.org

American Institute of Biological Science
730 11th Street, N.W.
Washington, D.C. 20001-4521
202-628-1500
email: cgabriel@aibs.org

Association for Tropical Biology
Dr. Julie Denslow
Louisiana State University
Department of Plant Biology
502 Life Sciences Building
Baton Rouge, LA 70803-1705
504-388-8485

Association for Women in Science
1346 Connecticut Avenue, N.W., Suite 1122
Washington, D.C. 20036
202-408-0742

Biosciences Information Service
2100 Arch Street
Philadelphia, PA 19103-1399
215-587-4800
email: info@mail.biosis.org

Federation of American Societies for Experimental Biology
9650 Rockville Pike
Bethesda, MD 20814
301-530-7000
email: gmoore@dp.faseb.org

Agronomy

American Society of Agronomy
667 South Segoe Road
Madison, WI 53711
608-273-8080
email: lmalicon@agronomy.org

Allied Health

American Nurses' Association
600 Maryland Avenue, S.W.
Suite 100 West
Washington, D.C. 20024
202-651-7000
email: rmurray@ana.com

National Association for Practical Nurse Education and Service
1400 Spring Street, Suite 300
Silver Springs, MD 20910
301-588-2491

Art and Communications

American Medical Writers Association
9650 Rockville Pike
Bethesda, MD 20814
301-493-0003

Association of Medical Illustrators
1819 Peachtree Street, N.E., Suite 620
Atlanta, GA 30309
404-350-7900
email: amwa@amwa.org

Biological Photographic Association
115 Stoneridge Drive
Chapel Hill, NC 27514
919-967-8247

Medical Library Association
6 North Michigan Avenue, Suite 300
Chicago, IL 60602
312-419-9094
email: info@mlahg.org

National Association of Science Writers
P.O. Box 294
Greenlawn, NY 11740
516-757-5664

Biochemistry

American Society for Biochemistry and Molecular Biology
9650 Rockville Pike
Bethesda, MD 20814
301-530-7145
email: asbmb@asbmb.faseb.org

Biomedical Engineering

Biomedical Engineering Society
P.O. Box 2399
Culver City, CA 90231
310-618-9322

Biophysics

Biophysical Society
9650 Rockville Pike, Room 512
Bethesda, MD 20814
301-530-7114
email: society@biophysics.faseb.org

Botany

American Phytopathological Society
3340 Pilot Knob Road
St. Paul, MN 55121
612-454-7250
email: aps@scisoc.org

American Society of Plant Physiologists
15501 Monona Drive
Rockville, MD 20855
301-251-0560

Botanical Society of America
c/o Christopher Haufler
Department of Botany
University of Kansas
Lawrence, KS 66045-2106
913-864-4301

Mycological Society of America
c/o Dr. Charles Mins
Department of Plant Pathology
University of Georgia
Athens, GA 30602
706-542-1291

Phycological Society of America
c/o Dr. Robert Sheath
Department of Botany
Memorial University
St. John's, NF, Canada A1B 3XB

Cell Biology

American Society for Cell Biology
9650 Rockville Pike
Bethesda, MD 20814
301-530-7153
email: ascbinfo@ascb.faseb.org

International Society of Differentiation
Department of Genetics and Cell Biology
University of Minnesota
750 Biological Sciences Center
St. Paul, MN 55108
612-624-2285

Dentistry

American Association of Dental Schools
1625 Massachusetts Avenue, 6th Floor, N.W.
Washington, D.C. 20036
202-667-9433
email: aads@aads.jhu.edu

American Dental Association
211 E. Chicago Avenue
Chicago, IL 60611
312-440-2500

Education

National Association of Biology Teachers
11250 Roger Bacon Drive, No. 19
Reston, VA 22090
703-471-1134
email: nabter@aol.com

Society for Public Health Education
2001 Addison Street, Suite 220
Berkeley, CA 94704
510-644-9242

Entomology

Entomological Society of America
9301 Annapolis Road
Lanham, MD 20706
301-731-4535
email: info@entsoc.org

Environment-Ecology

Ecological Society of America
2010 Massachusetts Avenue, N.W.
Suite 400
Washington, D.C. 20036
202-833-8773

National Wildlife Federation
1400 16th Street, N.W.
Washington, D.C. 20036
202-797-6800
web site: www.nwf.org/nwf

The Nature Conservancy
1815 N. Lynn Street
Arlington, VA 22209
703-841-5300
nadine@esa.org

Western Society of Naturalists
c/o Dr. Michael Foster
Moss Landing Marine Labs
P.O. Box 450
Moss Landing, CA 95039
408-755-8658

Environmental Law

Environmental Defense Fund
257 Park Avenue South, 16th Floor
New York, NY 10016
212-686-4191

Environmental Law Institute
1616 P Street, N.W., 2nd Floor
Washington, D.C. 20036
202-939-3800

Forensics

American Academy of Forensic Sciences
410 North 21st Street
Suite 203
Colorado Springs, CO 80904
719-636-1100
email: memship@aafs.org

Genetics

Genetics Society of America
9650 Rockville Pike
Bethesda, MD 20814-3998
301-571-1825
email: society@genetics.faseb.org

National Society of Genetic Counselors
233 Canterbury Drive
Walingford, PA 19086
610-872-7608

Marine Biology

Virginia Institute of Marine Science
P.O. Box 134b
Gloucester Point, VA 23062
804-642-7000

Woods Hole Oceanographic Institute
86 Water Street
Woods Hole, MA 02543
508-548-1400

Medicine

American Academy of Family Physicians
8880 Ward Parkway
Kansas City, MO 64114
816-333-9700

American Academy of Pediatrics
141 Northwest Point Boulevard
P.O. Box 927
Elk Grove Village, IL 60009-0927
708-228-5005

American Medical Association
515 North State Street
Chicago, IL 60610
312-464-5000

International Association of Environmental Mutagen Societies
c/o Herbert S. Rosenkrantz
Department of Environmental Health Science
Case Western University
Cleveland, OH 44106

National Association of Healthcare Recruitment
P.O. Box 5769
Akron, OH 44372
216-867-3088

National Environmental Health Association
720 S. Colorado Boulevard, Suite 970, S. Tower
Denver, CO 80222
303-756-9090
email: staffrus@neha.org

Research!America (Medical Research)
1522 King Street
Alexandria, VA 22314
703-739-2577
email: resamer@capaccess.org

Microbiology

American Society of Microbiology
1325 Massachusetts Avenue, N.W.
Washington, D.C. 20005
202-737-3600

Foundation for Microbiology
c/o Bruce H. Waksman
300 E. 54th Street, Suite 5K
New York, NY 10022
212-759-8729

Society for Industrial Microbiology
3929 Old Lee Highway
Suite 92A
Fairfax, VA 22030
703-691-3357

Microscopy

Microscopy Society of America
Box MSA
Woods Hole, MA 02453
800-538-3672
email: lmaser@mbl.edu

Nutrition

American Dietetic Association
216 West Jackson Boulevard, Suite 800
Chicago, IL 60606
312-899-0040

Institute of Food Technologists
221 N. LaSalle Street, Suite 800
Chicago, IL 60601
312-782-8424

Pharmaceuticals

American Association of Colleges of Pharmacy
1426 Prince Street
Alexandria, VA 22314
703-739-2330

American Pharmaceutical Association
2215 Constitution Avenue, N.W.
Washington, D.C. 20037
202-628-4410

Physical Therapy

American Physical Therapy Association
Department of Information Services
1111 N. Fairfax Street
Alexandria, VA 22314
703-684-2782

Physiology

American Physiological Society
c/o Dr. Martin Frank
9560 Rockville Pike
Bethesda, MD 20814
301-530-7160
email: mfrank@aps.faseb.org

Psychiatry/Psychology

American Psychiatric Association
Division of Public Affairs
1400 K Street, N.W.
Washington, D.C. 20005
202-682-6325
email: mbennett@psych.org

American Psychological Association
750 First Street, N.E.
Washington, D.C. 20002
202-336-5500
email: education@apa.org

Taxonomy

American Society of Plant Taxonomists
David M. Johnson, Secretary
Department of Botany, Microbiology
Ohio Wesleyan University
Delaware, OH 43015

Zoology

American Association of Zoo Keepers
Topeka Zoological Park
635 S.W. Gage Boulevard
Topeka, KS 66606
913-272-5821 or 800-242-4519

American Society of Mammalogists
c/o Dr. H. Duane Smith
290 MLB Museum
Brigham Young University
Provo, UT 84602
801-378-2492
email: duane@museumbyu.edu

American Society of Zoologists
401 N. Michigan Avenue
Chicago, IL 60611-4267
312-527-6640

The previous edition of this textbook featured career vision interviews which have been replaced in this edition. The following list contains brief descriptions of the careers featured in the last edition. If you want to know more about these careers contact the appropriate professional societies listed above.

Electron Microscopist: Electron microscopy is used to test the composition and integrity of products, among many other industrial and academic uses. Positions are available in universities, museums, and industry.
Environmental Lawyer: An environmental lawyer advises clients on the legal issues concerning the use and/or protection of resources. They may defend clients who are not in compliance with environmental laws, or they may work for environmental advocacy groups.
Forensic Scientist (Criminalist): A criminalist conducts examinations of forensic evidence gathered at a crime scene.
Medical Illustrator: A medical illustrator uses an understanding of anatomy and physiology to draw artwork that appears in textbooks and medical journals, among other publications.
Naturalist: Naturalist is a term that covers a variety of positions. A park ranger who takes people through a natural site and explains how plants adapt or what animals are there is a naturalist. He/she may also work in a natural history museum educating the public through tours and classes.
Pharmaceutical Sales Representative: A biology degree enables a pharmaceutical sales representative to explain the advantages of his/her products effectively to medical professionals.
Science Writer: Science writers can work for magazines, newspapers, and may even write textbooks. They may also work in public information offices at universities or research facilities explaining scientific information to the media.
Teacher: Teachers with biology degrees can apply their education in various settings from primary schools to universities.

APPENDIX C

Post-Test Answers

Chapter 1

1. metabolism 2. homeostasis 3. stimulus 4. Cilia 5. asexual reproduction 6. adapt 7. cells 8. molecules 9. DNA 10. Hormones 11. tissue 12. producers; decomposers; consumers 13. cellular respiration 14. genus 15. Fungi 16. Prokaryotae 17. Mutations 18. natural selection 19. hypothesis 20. theory

Chapter 2

1. elements 2. electrons 3. atomic number 4. atomic mass 5. isotopes 6. orbital 7. ions; cations, anions 8. hydrophilic; hydrophobic 9. Valence 10. molecule 11. ionic 12. covalent 13. hydrogen 14. heat of vaporization 15. acid; base 16. pH 17. buffer 18. electrolytes

Chapter 3

1. monosaccharide 2. fatty acids 3. amino acids; peptide 4. Cellulose 5. amphipathic 6. amino, carboxyl (either order) 7. glycogen 8. sugar, phosphate (either order) 9. condensation synthesis 10. enantiomers 11. unsaturated 12. tertiary 13. aldehyde 14. secondary 15. ATP

Chapter 4

1. resolving power 2. rough ER 3. chromatin 4. lysosomes 5. perioxisomes 6. Mitochondria 7. plastids 8. thylakoid; photosynthesis 9. microtubules 10. microtubules, microfilaments (either order); intermediate 11. cilia, flagella (either order); microtubules 12. cell wall 13. Vacuoles; lysosomes 14. nucleolar; nucleus 15. glycoproteins

Chapter 5

1. fluid mosaic 2. amphipathic 3. Integral; peripheral 4. diffusion 5. osmosis 6. isotonic (or isosmotic) 7. hypertonic (or hyperosmotic) 8. hypotonic (or hypo-osmotic) 9. facilitated diffusion 10. active transport 11. active 12. cotransport 13. phagocytosis 14. receptor-mediated 15. Desmosomes 16. Gap junctions 17. Tight junctions 18. Plasmodesmata

Chapter 6

1. Work 2. kinetic 3. Thermodynamics 4. first law of thermodynamics 5. entropy 6. second law of thermodynamics 7. free 8. endergonic 9. exergonic (or spontaneous) 10. free 11. coupled 12. activation 13. catalyst 14. Enzymes 15. enzyme-substrate 16. active site 17. coenzyme 18. competitive

Chapter 7

1. catabolism 2. anabolism 3. reduction 4. glycolysis 5. glycolysis 6. NADH 7. citrate 8. citric acid cycle 9. NAD^+, FAD (either order) 10. Cytochromes 11. chemiosmosis 12. ATP synthase 13. glycolysis 14. fermentation; aerobic respiration 15. facultative anaerobes 16. ethyl alcohol 17. lactate 18. NAD^+

Chapter 8

1. autotrophs 2. thylakoid 3. photolysis 4. photons 5. action spectrum 6. $NADP^+$ 7. carotenoids 8. photosystem 9. photophosphorylation 10. noncyclic phosphorylation 11. chemiosmosis 12. thylakoid 13. carbon fixation 14. Calvin 15. C_4 (or Hatch-Slack) 16. bundle sheath 17. photorespiration

Chapter 9

1. chromatin 2. karyotype 3. generation time 4. S (or interphase) 5. prophase; metaphase; anaphase; telophase 6. sister chromatids 7. kinetochore 8. metaphase 9. cytokinesis 10. cell plate 11. colchicine 12. asexual 13. clone 14. homologous 15. diploid; polyploid 16. $2n$; n 17. synapsis 18. bivalent, tetrad (either order) 19. crossing over

Chapter 10

1. locus 2. alleles 3. monohybrid 4. genotype 5. phenotype 6. dominant 7. recessive 8. homozygous; heterozygous 9. F_1 10. product 11. sum 12. dihybrid 13. linked 14. polygenic 15. multiple 16. inbreeding 17. hybrid vigor

Genetics Problems

1. **a.** all yellow **b.** $\frac{1}{2}$ yellow: $\frac{1}{2}$ green **c.** all yellow **d.** $\frac{3}{4}$ yellow: $\frac{1}{4}$ green
2. 150
3. The short-winged condition is recessive. Both parents are heterozygous.
4. The blue-eyed man is homozygous recessive; his brown-eyed wife is heterozygous.

5. Repeated matings of the roan bull and the white cow will yield an approximate 1:1 ratio of roan to white offspring. Repeated matings among roan offspring will yield red, roan, and white offspring in an approximate ratio of 1:2:1. The mating of two red individuals will yield only red offspring.
6. There are 36 possible outcomes when a pair of dice is rolled. There are six ways of obtaining a seven: 1,6; 6,1; 2,5; 5,2; 3,4; and 4,3. There are five ways of rolling a six: 1,5; 5,1; 2,4; 4,2; and 3,3. There are also five ways of rolling an eight: 2,6; 6,2; 3,5; 5,3; and 4,4.
7. The genotype of the brown, spotted rabbit is *bbSS*. The genotype of the black, solid rabbit is *BBss*. An F_1 rabbit would be black, spotted (*BbSs*). The F_2 is expected to be $\frac{9}{16}$ black, spotted (*B_S_*), $\frac{3}{16}$ black, solid (*B_ss*), $\frac{3}{16}$ brown, spotted (*bbS_*), and $\frac{1}{16}$ brown, solid (*bbss*).
8. The F_1 offspring are expected to be black, short-haired. There is a $\frac{1}{16}$ chance of a brown and tan, long-haired offspring in the F_2 generation.
9. pleiotropic
10. The rooster is *PpRr*; hen A is *PpRR*; hen B is *Pprr*; and hen C is *PPRR*.
11. **a.** Both parents are heterozygous (for a single locus).
    b. One parent is heterozygous and the other homozygous recessive (for a single locus).
    c. Both parents are heterozygous (for two loci).
    d. One parent is heterozygous and the other homozygous (for two loci).
12. The expected types of F_2 plants are: $\frac{1}{16}$ of the plants produce 4-pound fruits; $\frac{1}{4}$ of the plants produce 3.5-pound fruits; $\frac{3}{8}$ of the plants produce 3-pound fruits; $\frac{1}{4}$ of the plants produce 2.5-pound fruits; and $\frac{1}{16}$ of the plants produce 2-pound fruits.
13. All females are barred and all males are nonbarred, thus allowing the sex of the chicks to be determined from their phenotypes.
14. This is a two-point test cross involving linked loci. The parental classes of offspring are *Aabb* and *aaBb*; the recombinant classes are *AaBb* and *aabb*. There is 4.6% recombination, which corresponds to 4.6 map units between the loci.
15. If genes B and C are 10 map units apart, gene A is in the middle. If genes B and C are 2 map units apart, gene C is in the middle.

Chapter 11

1. transformation; DNA 2. deoxyribose 3. purines; pyrimidines 4. deoxyribose; phosphodiester 5. adenine, thymine (either order); guanine, cytosine (either order) 6. x-ray diffraction 7. parallel 8. replication 9. semi-conservative replication 10. replication 11. 5′; 3′ 12. nucleoside triphosphates 13. Okazaki fragments; DNA ligase 14. leading 15. DNA polymerase 16. bidirectional 17. nucleosomes 18. chromatin; scaffolding proteins

Chapter 12

1. transcription 2. translation 3. codon 4. transfer 5. ribosome 6. RNA polymerase 7. uracil 8. promoter 9. anticodon 10. transfer RNA 11. ribosomal RNA 12. initiation; initiation 13. elongation; translocation 14. termination 15. polyribosome (or polysome) 16. introns; exon 17. capped 18. polyadenylated (or poly-A) 19. nonsense; frame-shift

Chapter 13

1. operons 2. operator 3. repressor; negative 4. Inducible; allosteric 5. Repressible; corepressor 6. activator 7. Catabolite activator 8. regulon 9. constitutive 10. repressor 11. Transcription factors 12. feedback inhibition 13. upstream promoter 14. enhancers 15. amplification 16. heterochromatin 17. euchromatin

Chapter 14

1. vector 2. recombinant 3. palindromes 4. plasmids; bacteriophages 5. restriction enzyme; DNA ligase 6. clone 7. genomic 8. cDNA; reverse transcriptase 9. genomic; cDNA 10. polymerase chain reaction 11. restriction map 12. sequencing 13. gel electrophoresis 14. probes 15. transgenic

Chapter 15

1. isogenic 2. karyotype 3. birth 4. aneuploidy 5. trisomic 6. monosomic 7. nondisjunction 8. translocation 9. Down 10. Klinefelter; Turner 11. metabolism 12. sickle cell 13. cystic fibrosis 14. amniocentesis 15. Chorionic villus 16. genetic counseling

Chapter 16

1. differentiated; determination 2. morphogenesis; pattern formation 3. nuclear equivalence 4. totipotent 5. gene expression 6. polytene 7. maternal effect 8. Zygotic 9. morphogens 10. Homeotic 11. Homeodomains 12. mosaic 13. induction 14. regulative 15. Transgenic 16. homeotic 17. genomic rearrangements 18. Gene amplification

Chapter 17

1. evolution 2. gene pool 3. natural selection 4. natural selection 5. artificial selection 6. synthetic theory 7. neo-Darwinism 8. index fossils 9. radioisotopes 10. vestigial 11. analogous 12. convergent evolution 13. biogeography 14. range 15. DNA sequencing

Chapter 18

1. population 2. microevolution 3. Hardy-Weinberg 4. genetic drift 5. gene flow 6. mutation 7. Natural selection 8. stabilizing 9. directional

10. Disruptive 11. genetic polymorphism
12. heterozygote advantage 13. frequency-dependent
14. neutral mutation

Chapter 19

1. biological species concept 2. temporal isolation
3. mechanical isolation 4. hybrid inviability
5. allopatric 6. sympatric 7. allopolyploidy
8. punctuated equilibrium 9. gradualism
10. macroevolution 11. preadaptation 12. allometric growth 13. extinction 14. adaptive radiation

Chapter 20

1. chemical evolution 2. hydrothermal vents
3. Protobionts 4. Heterotrophs; autotrophs
5. autotroph 6. stromatolites 7. anaerobes
8. endosymbiont theory 9. Precambrian time
10. Paleozoic; Mesozoic; Cenozoic 11. periods; epochs
12. Paleozoic 13. Mesozoic 14. Cenozoic
15. continental drift 16. plate tectonics

Chapter 21

1. anthropoids 2. prosimians 3. prehensile
4. hominoids 5. brachiates 6. supraorbital
7. foramen magnum 8. hominids 9. *Australopithecus*
10. *habilis* 11. *erectus* 12. *sapiens* 13. Neandertal
14. cultural evolution

Chapter 22

1. taxonomy 2. genus; specific 3. genus 4. family
5. phylum 6. Protista 7. Fungi 8. monophyletic
9. homologous 10. derived 11. clocks 12. Phenetics
13. cladistic 14. classical evolutionary 15. species
16. polyphyletic

Chapter 23

1. pathogens 2. virus 3. capsid 4. bacteria
5. Lytic 6. temperate 7. transduction
8. Peptidoglycan 9. saprobes or saprotrophic
10. cocci; bacilli; spirilla 11. gram positive
12. chemoautotrophs 13. conjugation
14. archaebacteria 15. exotoxins

Chapter 24

1. protists 2. plankton 3. protozoa 4. pseudopodia
5. tests 6. axopods 7. flagellates 8. ciliates
9. conjugation 10. sporozoans 11. Dinoflagellates
12. red tide 13. diatoms 14. euglenoids
15. red algae 16. red algae 17. brown algae
18. plasmodium 19. cellular slime molds
20. zoospores; oospores

Chapter 25

1. hyphae; mycelium 2. septa; coenocytic 3. spores
4. zygospores 5. heterothallic 6. conidia 7. budding
8. ascospores; asci 9. basidiocarp 10. basidiospore
11. gills 12. imperfect fungi 13. lichen 14. saprotrophs 15. mycorrhizae 16. haustoria

Chapter 26

1. bryophytes 2. cuticle 3. lignin 4. stomata
5. archegonium; antheridium 6. alternation of generations
7. mosses 8. gametophyte 9. thallus 10. Hornworts
11. xylem; phloem 12. ferns 13. sori 14. megaphyll
15. Whisk ferns 16. Horsetails 17. microphyll
18. heterospory

Chapter 27

1. gymnosperms 2. monoecious 3. pollen grain
4. pollination 5. cycads, *Ginkgo* (any order)
6. angiosperms; flowering plants 7. monocots
8. endosperm 9. stigma, style, ovary (any order)
10. carpel 11. incomplete, imperfect (any order)
12. ovary; ovule 13. embryo sac 14. progymnosperms

Chapter 28

1. invertebrates 2. sessile 3. Acoelomate
4. ectoderm 5. mesoderm 6. protostome 7. Radial
8. protostome 9. ventral 10. spicules 11. collar
12. polyp, medusa 13. Cnidaria 14. Platyhelminthes (flatworms) 15. auricles 16. Nemertea 17. Rotifera
18. Ctenophora 19. Platyhelminthes 20. Nematoda

Chapter 29

1. exoskeleton 2. radula 3. open 4. trochophore
5. metamerism 6. mantle 7. Nudibranchs 8. Setae
9. arthropods 10. mandibles 11. chelipods
12. Malpighian 13. tracheated 14. Uniramia
15. Chelicerata

Chapter 30

1. echinoderms 2. water vascular; tube 3. lophophorate
4. notochord; gill slits 5. Tunicates 6. Vertebrates; cranium 7. tetrapods; amniotes 8. endoskeleton
9. placoid 10. Chondrichthyes 11. ray-finned; amphibians (tetrapods) 12. labyrinthodonts
13. cotylosaurs 14. endotherms 15. Monotremes
16. reptile 17. placenta

Chapter 31

1. annuals; biennials; perennials 2. ground
3. parenchyma 4. collenchyma 5. vascular

6. tracheids 7. companion 8. epidermis; periderm 9. cuticle 10. stomata; guard cells 11. trichomes 12. meristems 13. apical 14. vascular cambium 15. pectin; lignin

CHAPTER 32

1. mesophyll 2. guard cells 3. lower 4. cuticle 5. bundle sheath 6. xylem 7. phloem 8. transpiration 9. guttation 10. circadian rhythms 11. potassium ion 12. abscission 13. Tendrils 14. spines

CHAPTER 33

1. buds 2. epidermis 3. pith rays 4. ground tissue 5. cortex, pith (any order) 6. cork cambium, vascular cambium (any order) 7. periderm 8. rays 9. Dendrochronology 10. translocation 11. water potential 12. root pressure 13. tension-cohesion 14. pressure-flow

CHAPTER 34

1. taproot 2. Adventitious 3. contractile 4. pneumatophores 5. root cap; root hairs 6. Casparian strip 7. pericycle 8. xylem; pith 9. graft 10. Mycorrhizae 11. weathering 12. leaching 13. hydroponics 14. Macronutrients 15. apoplast

CHAPTER 35

1. stigma 2. corolla 3. Pollination 4. coevolution 5. double fertilization 6. Endosperm 7. suspensor 8. seed 9. fruit 10. Aggregate; multiple 11. accessory 12. rhizome 13. tuber 14. 2-5-1-4-3

CHAPTER 36

1. photoperiodism 2. phytochrome 3. vernalization 4. pulvinus 5. solar tracking 6. circadian 7. Phototropism 8. gravitropism 9. thigmotropism 10. turgor movements 11. Hormones 12. auxin 13. gibberellins 14. Cytokinins 15. Cytokinins; ethylene 16. ethylene 17. abscisic acid 18. Abscisic acid

CHAPTER 37

1. tissue 2. gland 3. organ system 4. cardiac muscle 5. Skeletal 6. glial 7. endocrine 8. integumentary 9. negative 10. epithelial 11. nervous 12. myofibrils 13. connective 14. loose 15. cartilage 16. bone 17. elastic 18. Exocrine 19. negative feedback 20. stressors

CHAPTER 38

1. epidermis; dermis 2. corneum 3. keratin 4. hydrostatic 5. endoskeleton 6. compact; spongy (cancellous) 7. ligaments 8. actin (thin), myosin (thick) 9. creatine phosphate 10. sarcomere 11. T tubules 12. cross bridges 13. muscle tone 14. oxygen

CHAPTER 39

1. transmission 2. synapse 3. reception 4. neuron 5. glial 6. dendrites; axon 7. myelin sheath (or cellular sheath) 8. ganglia 9. resting 10. Excitatory 11. depolarization 12. action potential 13. all or none 14. saltatory 15. neurotransmitter 16. acetylcholine 17. amines 18. divergence

CHAPTER 40

1. nerve net 2. ganglia 3. central, peripheral 4. brain; spinal cord 5. brain stem 6. ventricle 7. midbrain 8. hypothalamus 9. cerebrum 10. cerebellum 11. medulla 12. tracts 13. fissures 14. cortex 15. frontal; occipital 16. limbic system 17. beta 18. REM 19. sympathetic 20. dorsal 21. tolerance

CHAPTER 41

1. sensory receptor 2. Exteroceptors; proprioceptors 3. Photo-; mechano- 4. energy; receptor potential 5. adaptation 6. statocysts 7. lateral line 8. muscle spindles; Golgi; joint 9. corpuscle 10. labyrinth; saccule, utricle; semicircular 11. otoliths 12. endolymph; ampulla 13. cochlea 14. Corti 15. chemoreceptors 16. rhodopsins 17. retina 18. iris 19. Cones 20. fovea 21. ommatidia

CHAPTER 42

1. open 2. plasma 3. anemia 4. fibrin 5. blood pressure 6. atrium 7. hepatic portal 8. baroreceptors 9. mitral 10. hemoglobin 11. Plasma 12. Arterioles 13. capillaries 14. ventricle 15. semilunar 16. systole; diastole 17. cardiac output 18. aorta 19. atherosclerosis 20. lymph

CHAPTER 43

1. immune response 2. antibodies (or immunoglobulins) 3. interferons 4. inflammation 5. bone marrow; thymus gland 6. Cytotoxic 7. memory 8. plasma (differentiated B cells) 9. antibody 10. thymus 11. antigen-antibody 12. MHC 13. active 14. passive 15. graft rejection 16. natural killer 17. allergen; allergic 18. mast 19. HIV (human immunodeficiency virus)

Chapter 44

1. respiration 2. ventilation 3. tracheal 4. gills 5. lungs 6. air sacs 7. trachea; bronchus 8. alveoli 9. diaphragm 10. epiglottis 11. pleural 12. vital capacity 13. oxyhemoglobin 14. bicarbonate ions 15. respiratory centers 16. constriction

Chapter 45

1. ingestion 2. digestion 3. Elimination 4. mucosa 5. molars, premolars 6. amylase 7. dentin 8. stomach 9. large intestine 10. liver 11. gastrin 12. rugae 13. duodenum (small intestine) 14. villi 15. vitamins 16. Minerals 17. Essential 18. basal metabolic rate (BMR)

Chapter 46

1. excretion 2. Osmoregulation 3. uric acid 4. urea 5. protonephridia; metanephridia 6. Malpighian tubules 7. conformers 8. cortex 9. ureter 10. renal pelvis 11. nephrons 12. capillaries; Bowman's capsule 13. afferent; efferent 14. filtrate 15. excreted (in the urine) 16. collecting ducts; reabsorbed 17. ectotherms 18. evaporation

Chapter 47

1. Endocrine 2. hormones 3. cyclic 4. kinases 5. juvenile 6. negative feedback 7. hypothalamus 8. target 9. hypo-; thyroid 10. goiter 11. cortisol 12. thyroid 13. diabetes mellitus 14. parathyroid 15. posterior; pituitary 16. thyroid 17. adrenal cortex 18. growth

Chapter 48

1. fragmentation 2. Parthenogenesis 3. hermaphroditic 4. gamete; zygote 5. sterile 6. scrotum 7. seminiferous; testes 8. Testosterone 9. prostate, seminal vesicles 10. vas deferens 11. ovaries 12. endometrium 13. corpus luteum 14. follicles 15. oviduct (uterine tube) 16. Estrogens 17. uterus 18. therapeutic 19. sterilization 20. menstrual; ovulation 21. syphilis

Chapter 49

1. morphogenesis 2. cellular differentiation 3. cleavage 4. gastrulation 5. ectoderm; endoderm 6. neural plate 7. neural tube 8. amnion 9. placenta 10. inner cell mass; embryo 11. implant 12. fetus 13. labor 14. acrosome reaction

Chapter 50

1. Behavior 2. Ethology 3. circadian 4. crepuscular 5. releaser 6. Innate; learned 7. Imprinting 8. habituation 9. classical (Pavlovian) 10. social 11. society 12. Pheromones 13. dominance hierarchy 14. home range 15. Territorial behavior 16. pair bond 17. innate 18. altruistic 19. Kin

Chapter 51

1. Ecology 2. population 3. community 4. ecosystem 5. tundra 6. taiga 7. Temperate deciduous 8. grasslands 9. Chaparral 10. Deserts 11. savanna 12. tropical rain 13. plankton; nekton; benthos 14. thermocline 15. littoral 16. estuary 17. intertidal 18. fringing reef 19. neritic 20. oceanic 21. ecotone

Chapter 52

1. population 2. density 3. Clumped dispersion 4. natality; mortality 5. biotic potential 6. exponential 7. environmental resistance 8. carrying capacity 9. crash 10. dependent 11. Intra-; inter- 12. dependent 13. independent 14. *r* 15. *K* 16. survivorship 17. zero population growth 18. infant mortality 19. age structure

Chapter 53

1. community 2. producers 3. Carnivores; herbivores; omnivores 4. müllerian 5. mutualism 6. commensalism 7. niche 8. fundamental niche 9. realized 10. competitive exclusion 11. edge effect 12. succession 13. primary 14. pioneers 15. secondary 16. character displacement 17. warning coloration

Chapter 54

1. ecosystem 2. energy flow 3. carbon cycle 4. nitrogen fixation 5. nitrification 6. phosphorus cycle 7. hydrologic cycle 8. trophic 9. food web 10. numbers, biomass, energy (any order) 11. biomass 12. Net; gross 13. climate 14. ocean currents 15. Coriolis effect 16. ENSO 17. rain shadows 18. microclimate

Chapter 55

1. Extinction 2. endangered 3. threatened 4. biological diversity 5. In situ 6. ex situ 7. artificial insemination 8. host mothering 9. keystone species 10. deforestation 11. slash and burn 12. Global warming 13. greenhouse 14. greenhouse gases 15. Ozone 16. stratosphere 17. Ozone depletion

APPENDIX D

Understanding Biological Terms

Your task of mastering new terms will be greatly simplified if you learn to dissect each new word. Many terms can be divided into a prefix, the part of the word that precedes the main root, the word root itself, and often a suffix, a word ending that may add to or modify the meaning of the root. As you progress in your study of biology, you will learn to recognize the more common prefixes, word roots, and suffixes. Such recognition will help you analyze new terms so that you can more readily determine their meaning and will also help you remember them.

Prefixes

a-, ab- from, away, apart (abduct, move away from the midline of the body)
a-, an-, un- less, lack, not (asymmetrical, not symmetrical)
ad- (also **af-, ag-, an-, ap-**) to, toward (adduct, move toward the midline of the body)
allo- different (allometric growth, different rates of growth for different parts of the body during development)
ambi- both sides (ambidextrous, able to use either hand)
andro- a man (androecium, the male portion of a flower)
anis- unequal (anisogamy, sexual reproduction in which the gametes are of unequal sizes)
ante- forward, before (anteflexion, bending forward)
anti- against (antibody, proteins that have the capacity to react against foreign substances in the body)
auto- self (autotroph, organism that manufactures its own food)
bi- two (biennial, a plant that takes two years to complete its life cycle)
bio- life (biology, the study of life)
circum-, circ- around (circumcision, a cutting around)
co-, con- with, together (congenital, existing with or before birth)
contra- against (contraception, against conception)
cyt- cell (cytology, the study of cells)
di- two (disaccharide, a compound made of two sugar molecules chemically combined)
dis- apart (dissect, cut apart)
ecto- outside (ectoplasm, outer layer of cytoplasm)
end-, endo- within, inner (endoplasmic reticulum, a network of membranes found within the cytoplasm)
epi- on, upon (epidermis, upon the dermis)
ex-, e-, ef- out from, out of (extension, a straightening out)
extra- outside, beyond (extraembryonic membrane, a membrane that encircles and protects the embryo)
gravi- heavy (gravitropism, growth of a plant in response to gravity)
hemi- half (cerebral hemisphere, lateral half of the cerebrum)
hetero- other, different (heterozygous, having unlike members of a gene pair)
homo-, hom- same (homologous, corresponding in structure; homozygous, having identical members of a gene pair)
hyper- excessive, above normal (hypersecretion, excessive secretion)
hypo- under, below, deficient (hypotonic, a solution whose osmotic pressure is less than that of a solution with which it is compared)
in-, im- not (incomplete flower, a flower that does not have one or more of the four main parts)
inter- between, among (interstitial, situated between parts)
intra- within (intracellular, within the cell)
iso- equal, like (isotonic, equal osmotic concentration)
macro- large (macronucleus, a large, polyploid nucleus found in ciliates)
mal- bad, abnormal (malnutrition, poor nutrition)
mega- large, great (megakaryocyte, giant cell of bone marrow)
meso- middle (mesoderm, middle tissue layer of the animal embryo)
meta- after, beyond (metaphase, the stage of mitosis after prophase)
micro- small (microscope, instrument for viewing small objects)
mono- one (monocot, a group of flowering plants with one cotyledon, or seed leaf, in the seed)
oligo- small, few, scant (oligotrophic lake, a lake deficient in nutrients and organisms)
oo- egg (oocyte, developing egg cell)
paedo- a child (paedomorphosis, the preservation of a juvenile characteristic in an adult)
para- near, beside, beyond (paracentral, near the center)
peri- around (pericardial membrane, membrane that surrounds the heart)
photo- light (phototropism, growth of a plant in response to the direction of light)
poly- many, much, multiple, complex (polysaccharide, a carbohydrate composed of many simple sugars)
post- after, behind (postnatal, after birth)
pre- before (prenatal, before birth)
pseudo- false (pseudopod, a temporary protrusion of a cell, i.e., "false foot")
retro- backward (retroperitoneal, located behind the peritoneum)
semi- half (semilunar, half-moon)
sub- under (subcutaneous tissue, tissue immediately under the skin)
super, supra- above (suprarenal, above the kidney)
sym- with, together (sympatric speciation, evolution of a new species within the same geographical region as the parent species)
syn- with, together (syndrome, a group of symptoms that occur together and characterize a disease)
trans- across, beyond (transport, carry across)

Suffixes

-able, -ible able (viable, able to live)
-ad used in anatomy to form adverbs of direction (cephalad, toward the head)
-asis, -asia, -esis condition or state of (euthanasia, state of "good death")
-cide kill, destroy (biocide, substance that kills living things)
-emia condition of blood (anemia, a blood condition in which there is a lack of red blood cells)
-gen something produced or generated or something that produces or generates (pathogen, an organism that produces disease)
-gram record, write (electrocardiogram, a record of the electrical activity of the heart)
-graph record, write (electrocardiograph, an instrument for recording the electrical activity of the heart)
-ic adjective-forming suffix which means *of* or *pertaining to* (ophthalmic, of or pertaining to the eye)
-itis inflammation of (appendicitis, inflammation of the appendix)
-logy study or science of (cytology, study of cells)
-oid like, in the form of (thyroid, in the form of a shield)
-oma tumor (carcinoma, a malignant tumor)
-osis indicates disease (psychosis, a mental disease)
-pathy disease (dermopathy, disease of the skin)
-phyll leaf (mesophyll, the middle tissue of the leaf)
-scope instrument for viewing or observing (microscope, instrument for viewing small objects)

Some Common Word Roots

abscis cut off (abscission, the falling off of leaves or other plant parts)
angi, angio vessel (angiosperm, plants that produce seeds enclosed within a fruit or "vessel")
apic tip, apex (apical meristem, area of cell division located at the tips of plant stems and roots)
arthr joint (arthropods, invertebrate animals with jointed legs and segmented bodies)
aux grow, enlarge (auxin, a plant hormone involved in growth and development)
bi, bio life (biology, study of life)
blast a formative cell, germ layer (osteoblast, cell that gives rise to bone cells)
brachi arm (brachial artery, blood vessel that supplies the arm)
bry grow, swell (embryo, an organism in the early stages of development)
cardi heart (cardiac, pertaining to the heart)
carot carrot (carotene, a yellow, orange, or red pigment in plants)
cephal head (cephalad, toward the head)
cerebr brain (cerebral, pertaining to the brain)
cervic, cervix neck (cervical, pertaining to the neck)
chlor green (chlorophyll, a green pigment found in plants)
chondr cartilage (chondrocyte, a cartilage cell)
chrom color (chromosome, deeply staining body in nucleus)
cili small hair (cilium, a short, fine cytoplasmic hair projecting from the surface of a cell)
coleo a sheath (coleoptile, a protective sheath that encircles the stem in grass seedlings)
conjug joined together (conjugation, a sexual phenomenon in certain protists)
cran skull (cranial, pertaining to the skull)
cyt cell (cytology, study of cells)
decid falling off (deciduous, a plant that sheds its leaves at the end of the growing season)
dehis split (dehiscent fruit, a fruit that splits open at maturity)
derm skin (dermatology, study of the skin)
ecol dwelling, house (ecology, the study of organisms in relation to their environment, i.e., "their house")
enter intestine (enterobacteria, a group of bacteria that include species that inhabit the intestines of humans and other animals)
evol to unroll (evolution, descent with modification, or gradual directional change)
fil a thread (filament, the thin stalk of the stamen in flowers)
gamet a wife or husband (gametangium, the part of a plant or protist that produces reproductive cells)
gastr stomach (gastrointestinal tract, the digestive tract)
glyc, glyco sweet, sugar (glycogen, storage form of glucose)
gon seed (gonad, an organ that produces gametes)
gutt a drop (guttation, loss of water as liquid "drops" from plants)
gymn naked (gymnosperm, a plant that produces seeds that are not enclosed with a fruit, i.e., "naked")
hem blood (hemoglobin, the pigment of red blood cells)
hepat liver (hepatic, of or pertaining to the liver)
hist tissue (histology, study of tissues)
hom, homeo same, unchanging, steady (homeostasis, reaching a steady state)
hydr water (hydrolysis, a breakdown reaction involving water)
leuk white (leukocyte, white blood cell)
menin membrane (meninges, the three membranes that envelop the brain and spinal cord)
morph form (morphogenesis, development of body form)
my, myo muscle (myocardium, muscle layer of the heart)
myc a fungus (mycelium, the vegetative body of a fungus)
nephr kidney (nephron, microscopic unit of the kidney)
neur, nerv nerve (neuromuscular, involving both the nerves and muscles)
occiput back part of the head (occipital, back region of the head)
ost bone (osteology, study of bones)
path disease (pathologist, one who studies disease processes)
ped, pod foot (bipedal, walking on two feet)
pell skin (pellicle, a flexible covering over the body of certain protists)
phag eat (phagocytosis, process by which certain cells ingest particles and foreign matter)
phil love (hydrophilic, a substance that attracts, i.e., "loves," water)
phloe bark of a tree (phloem, food-conducting tissue in plants which corresponds to bark in woody plants)
phyt plant (xerophyte, a plant adapted to xeric, or dry, conditions)
plankt wandering (plankton, microscopic aquatic protists that float or drift passively)
rhiz root (rhizome, a horizontal, underground stem that superficially resembles a root)
scler hard (sclerenchyma, cells that provide strength and support in the plant body)

sipho a tube (siphonous, a type of tubular body form found in certain algae)
som body (chromosome, deeply staining body in the nucleus)
sor heap (sorus, a cluster or "heap" of sporangia in a fern)
spor seed (spore, a reproductive cell that gives rise to individual offspring in plants and protists)
stom a mouth (stoma, a small pore, i.e., "mouth," in the epidermis of plants)
thigm a touch (thigmotropism, plant growth in response to touch)
thromb clot (thrombus, a clot within a blood vessel)
tropi turn (thigmotropism, growth of a plant in response to contact with a solid object, as when a tendril "turns" or wraps around a wire fence)
visc pertaining to an internal organ or body cavity (viscera, internal organs)
xanth yellow (xanthophyll, a yellowish pigment found in plants)
xyl wood (xylem, water-conducting tissue in plant, the "wood" of woody plants)
zoo an animal (zoology, the science of animals)

APPENDIX E

Abbreviations

The biological sciences employ a great many abbreviations, and with good reason. Many technical terms in biology and biological chemistry are both long and difficult to pronounce. Yet it can be difficult for beginners, when confronted with something like NADPH or EPSP, to understand the reference. Here are some of the common abbreviations employed in biology for your ready reference.

ABA Abscisic acid
ACTH Adrenocorticotropic hormone
ADA Adenosine deaminase
ADH Antidiuretic hormone
ADP Adenosine monophosphate
AIDS Acquired immune deficiency syndrome
AMP Adenosine diphosphate
amu Atomic mass unit (dalton)
ATP Adenosine triphosphate
AV node or valve Atrioventricular node or valve (of heart)
B lymphocyte or B cell Lymphocyte responsible for antibody-mediated immunity
BH Brain hormone (of insects)
BMR Basal metabolic rate
C_3 Three-carbon pathway for carbon fixation (Calvin cycle)
C_4 Four-carbon pathway for carbon fixation (Hatch-Slack pathway)
CAM Crassulacean acid metabolism
cAMP Cyclic adenosine monophosphate
CAP Catabolite activator protein
cDNA Complementary deoxyribonucleic acid
CFTR Cystic fibrosis transmembrane conductance regulator
CNS Central nervous system
CoA Coenzyme A
COPD Chronic obstructive pulmonary disease
CP Creatine phosphate
CPR Cardiopulmonary resuscitation
CSF Cerebrospinal fluid
CVS Cardiovascular system
DNA Deoxyribonucleic acid
ECG Electrocardiogram
EEG Electroencephalogram
EKG Electrocardiogram
EM Electron microscope or micrograph
ENSO El Niño-Southern Oscillation
EPSP Excitatory postsynaptic potential (of a neuron)
ER Endoplasmic reticulum
F_1 First filial generation
F_2 Second filial generation
Factor VIII Blood clotting factor (absent in hemophiliacs)
$FAD/FADH_2$ Flavin adenine dinucleotide (oxidized and reduced forms, respectively)
FAP Fixed action pattern
FSH Follicle-stimulating hormone
G_1 phase First gap phase (of the cell cycle)
G_2 phase Second gap phase (of the cell cycle)
G protein Cell signaling molecule that requires GTP
GABA Gaba-aminobutyric acid
GH Growth hormone (somatotropin)
GnRH Gonadotropin-releasing hormone
GTP Guanosine triphosphate
hCG Human chorionic gonadotropin
HDL High density lipoprotein
hGH Human growth hormone
HIV Human immunodeficiency virus
HLA Human leukocyte antigen
IAA Indole acetic acid (natural auxin)
Ig Immunoglobulin, as in IgA, IgG, etc.
IPSP Inhibitory postsynaptic potential
IUD Intrauterine device
JH Juvenile hormone (of insects)
kb Kilobase
LDH Lactic dehydrogenase enzyme
LDL Low density lipoprotein
LH Luteinizing hormone
LM Light microscope or micrograph
LSD Lysergic acid diethylamide
MAO Monoamine oxidase
MHC Major histocompatibility complex
MI Myocardial infarction
MPF Mitosis-promoting factor
mRNA Messenger RNA
mtDNA Mitochondrial DNA
MTOC Microtubule organizing center
9 + 2 structure Cilium or flagellum (of a eukaryote)
9 × 3 structure Centriole or basal body (of a eukaryote)
$NAD^+/NADH$ Nicotinamide adenine dinucleotide (oxidized and reduced forms, respectively)
$NADP^+/NADPH$ Nicotinamide adenine dinucleotide phosphate (oxidized and reduced forms, respectively)
NK cell Natural killer cell
P Parental generation
P680 Reaction center of photosystem II
P700 Reaction center of photosystem I
PABA Para-aminobenzoic acid
PCR Polymerase chain reaction
PEP Phosphoenolpyruvate
P_{fr} Phytochrome (form that absorbs far red light)
PGA Phosphoglycerate
PGAL Glyceraldehyde-3-phosphate (formerly known as phos-
phoglyceraldehyde)
PID Pelvic inflammatory disease
PKU Phenylketonuria

PNS Peripheral nervous system
pre-mRNA Precursor messenger RNA (in eukaryotes)
P_r Phytochrome (form that absorbs red light)
PTH Parathyroid hormone
RAS Reticular activating system
RBC Red blood cell (erythrocyte)
REM sleep Rapid eye movement sleep
RFLP Restriction fragment length polymorphism
RNA Ribonucleic acid
rRNA Ribosomal RNA
Rubisco Ribulose bisphosphate carboxylase/oxygenase
RuBP Ribulose bisphosphate
S phase Synthetic phase (of the cell cycle)
SA node Sinoatrial node (of heart)
SCID Severe combined immune deficiency
SEM Scanning electron microscope or micrograph
snRNP Small nuclear ribonucleoprotein complex
STD Sexually transmitted disease
T Generation time (in cell cycle)
T lymphocyte or T cell Lymphocyte responsible for cell-mediated immunity
TATA box Base sequence in eukaryotic promoter
TCA cycle Tricarboxylic acid cycle (synonym for citric acid cycle)
TEM Transmission electron microscope or micrograph
tRNA Transfer RNA
UPE Upstream promoter element
UV light Ultraviolet light
WBC White blood cell (leukocyte)

Glossary

abiotic environment The nonliving, physical environment.

abortion (uh-bor'shun) Expulsion of an embryo or fetus before it is capable of surviving.

abscisic acid (ab-sis'ik) A plant hormone involved in dormancy and responses to stress.

abscission (ab-sizh'en) The normal (usually seasonal) fall of leaves or other plant parts, such as fruits or flowers.

abscission layer The area at the base of the petiole where the leaf will break away from the stem. Also known as abscission zone.

absorption (ab-sorp'shun) The movement of nutrients and other substances through the wall of the digestive tract and into the blood or lymph.

absorption spectrum A graph of the amount of light at specific wavelengths that has been absorbed as light passes through a substance. Each type of molecule has a characteristic absorption spectrum. Compare with action spectrum.

accessory fruit A fruit composed primarily of tissue other than ovary tissue, e.g., apple, pear. Compare with aggregate, simple, and multiple fruits.

acetyl coenzyme A (acetyl CoA) (ah-see'til) A key intermediate compound in metabolism; consists of a two-carbon acetyl group covalently bonded to coenzyme A.

acetyl group ` A two-carbon group derived from acetic acid (acetate).

acetylcholine (ah"see-til-koh'leen) A common neurotransmitter released by cholinergic neurons, including motor neurons.

achene (a-keen') A simple, dry fruit with one seed in which the fruit wall is separate from the seed coat, e.g., sunflower fruit.

acid A substance that is a hydrogen ion (proton) donor; acids unite with bases to form salts. An acidic solution has a pH less than 7 and turns blue litmus paper red.

acoelomate (a-seel'oh-mate) Animal lacking a body cavity (coelom). Compare with coelomate and pseudocoelomate.

acquired immune deficiency syndrome (AIDS) A fatal disease caused by the human immunodeficiency virus (HIV).

acromegaly (ak"roh-meg'ah-lee) A condition characterized by overgrowth of the extremities of the skeleton, fingers, toes, jaws, and nose. It may be produced by excessive secretion of growth hormone by the anterior pituitary gland.

acrosome (ak'roh-sohm) A caplike structure covering the head of a sperm cell.

actin (ak'tin) The protein of which microfilaments are composed. Actin, together with the protein myosin, is responsible for muscle contraction.

actinopods (ak-tin'o-podz) Protozoa characterized by axopods that protrude through pores in their shells.

action potential The brief change in electrical activity developed across the plasma membrane of a muscle or nerve cell during activity; a neural impulse.

action spectrum A graph of the effectiveness of light at specific wavelengths in promoting a light-requiring reaction. Compare with absorption spectrum.

activation energy The energy required to initiate a chemical reaction.

activator protein A transcription factor that acts as a positive regulator, stimulating transcription when bound to DNA. Compare with repressor protein.

active site Specific region of an enzyme (generally near the surface) that accepts one or more substrates and catalyzes a chemical reaction. Compare with allosteric site.

active transport All forms of transport of a substance across a membrane that do not rely on the potential energy of a concentration gradient for the substance being transported, and therefore require an additional energy source (often ATP); includes carrier-mediated active transport, endocytosis, and exocytosis.

adaptation (1) Evolutionary modification that improves the chances of survival and reproductive success; (2) A decline in the response of a receptor subjected to repeated or prolonged stimulation.

adaptive radiation The evolution of a large number of related species from an unspecialized ancestral organism.

addiction Physical dependence on a drug, generally based on physiological changes that take place in response to the drug; when the drug is withheld, the addict suffers characteristic withdrawal symptoms.

adenine (ad'eh-neen) A nitrogenous purine base that is a component of nucleic acids and ATP.

adenosine triphosphate (ATP) (a-den'oh-seen) An organic compound containing adenine, ribose, and three phosphate groups; of prime importance for energy transfers in cells.

adhesive Having the property of sticking to some other substance.

adipose tissue (ad'i-pohs) Tissue in which fat is stored.

adrenal cortex (ah-dree'-nul kor'-teks) The outer region of each adrenal gland; secretes steroid hormones, including mineralocorticoids and glucocorticoids.

adrenal glands (ah-dree'nul) Paired endocrine glands, one located just superior to each kidney.

adrenal medulla (ah-dree'nul meh-dull'uh) The inner region of each adrenal gland; secretes epinephrine and norepinephrine.

adrenergic neuron (ad-ren-er'jik) A neuron that releases norepinephrine or epinephrine as a neurotransmitter. Compare with cholinergic neuron.

adventitious root (ad"ven-tish'us) A root that arises in an unusual position on a plant, such as from a stem or leaf.

aerobe Organism that grows or metabolizes only in the presence of molecular oxygen. Compare with anaerobe.

aerobic (air-oh'bik) Growing or metabolizing only in the presence of molecular oxygen. Compare with anaerobic.

aerobic respiration See respiration.
afferent (af′fer-ent) Leading toward some point of reference. Compare with efferent.
age structure The percentage of a population, including the number of people of each sex, at different ages. Age structure diagrams represent the number of males and females at each age in the population.
aggregate fruit A fruit that develops from a single flower with many separate carpels, e.g., raspberry. Compare with simple, accessory, and multiple fruits.
aggregated dispersion See clumped dispersion.
agnathans (ag-na′thanz) Jawless fishes; class of vertebrates, including lampreys, hagfishes, and many extinct forms.
albinism (al′bih-niz-em) A hereditary inability to form melanin pigment, resulting in light coloration.
AIDS See acquired immune deficiency syndrome.
albumin (al-bew′min) A class of protein found in most animal tissues; a fraction of plasma proteins.
aldehyde An organic molecule containing a carbonyl group bonded to at least one hydrogen atom.
aldosterone (al-dos′tur-ohn) Steroid hormone produced by the vertebrate adrenal cortex; regulates excretion of sodium and other ions.
algae (al′gee) (sing. *alga*) An informal group of unicellular or simple multicellular, photosynthetic protists that are important producers in aquatic ecosystems; includes dinoflagellates, diatoms, euglenoids, green algae, red algae, and brown algae.
alkaptonuria Genetic disease in which one of the enzymes required in the breakdown of the amino acids phenylalanine and tyrosine is missing; homogentisic acid excreted in the urine of affected persons oxidizes and turns black when exposed to air.
allantois (a-lan′toe-iss) One of the extraembryonic membranes of reptiles, birds, and mammals; most of the allantois is detached at hatching or birth.
alleles (al-leels′) Genes governing variation of the same character that occupy corresponding positions (loci) on homologous chromosomes; alternative forms of a gene.
allelopathy (uh-leel′uh-path″ee) An adaptation in which toxic substances secreted by roots or shed leaves inhibit the establishment of competing plants nearby.
allergen A substance that stimulates an allergic reaction.
allergy A hypersensitivity to some substance in the environment, manifested as hay fever, skin rash, asthma, food allergies, etc.
allometric growth Varied rates of growth for different parts of the body during development, resulting in a change of shape or proportion.
all-or-none law The principle that neurons transmit an impulse in a similar way no matter how weak or strong the stimulus is. The neuron either transmits an action potential (all) or does not (none).
allopatric speciation (al-oh-pa′trik) Speciation that occurs when one population becomes geographically separated from the rest of the species and subsequently evolves. Compare with sympatric speciation.
allopolyploid (al″oh-pol′ee-ploid) A polyploid whose chromosomes are derived from two different species.
allosteric site (al-oh-steer′ik) A site located on an enzyme molecule that is separate from the active site; enables a substance other than the normal substrate to bind to the molecule, thereby changing the shape of the molecule and thus its activity.
alpha carbon The asymmetric carbon in an amino acid to which the carboxyl group, the amino group, and the side chain are attached.
alpha helix A type of secondary structure of a polypeptide chain consisting of a regular coiled structure, maintained by hydrogen bonds.
alpine tundra An ecosystem located in the higher elevations of mountains, above the tree line and below the snow line. Compare with arctic tundra. See tundra.
alternation of generations A type of life cycle characteristic of plants in which they spend part of their life in a multicellular n gametophyte stage and part in a multicellular $2n$ sporophyte stage. The gametophyte develops from a spore and produces gametes; the sporophyte develops from a zygote and produces spores.
altruistic behavior Behavior in which one organism helps another, seemingly at its own risk or expense.
alveolus (al-vee′o-lus) (pl. *alveoli*) (1) An air sac of the lung through which gas exchange with the blood takes place; (2) A saclike unit of some glands.
amino acid (uh-mee′no) An organic compound containing an amino group ($—NH_2$) and a carboxyl group ($—COOH$); may be joined by peptide bonds to form the polypeptide chains of protein molecules.
amino group A weakly basic functional group; abbreviated $—NH_2$
aminoacyl-tRNA (uh mee″no-as′seel) Molecule consisting of an amino acid covalently linked to a transfer RNA.
aminoacyl-tRNA synthetase One of a family of enzymes, each responsible for covalently linking an amino acid to its specific transfer RNA.
ammonification (uh-moe″nuh-fah-kay′shun) The conversion of nitrogen-containing organic compounds to ammonia (NH_3) by certain bacteria (ammonifying bacteria) in the soil; part of the nitrogen cycle.
amniocentesis (am″nee-oh-sen-tee′sis) Sampling of the amniotic fluid surrounding a fetus in order to obtain information about its development and genetic makeup. Compare with chorionic villus sampling.
amnion (am′nee-on) An extraembryonic membrane that forms a fluid-filled sac for the protection of the developing embryo.
amoeba (a-mee′ba) A unicellular protozoon that moves by means of pseudopodia.
amphibians Vertebrate class that includes salamanders, frogs, and caecilians.
amphipathic molecule (am″fih-pa′thik) Molecule containing both hydrophobic and hydrophilic regions.
ampulla Any small sac-like extension, for example, the expanded structure at the end of each semicircular canal of the ear.
amylase (am′-uh-laze) Starch-digesting enzyme, e.g., human salivary amylase or pancreatic amylase.
amyloplasts Colorless plastids (leukoplasts) specialized for the storage of large amounts of starch in cells of roots and tubers.
anabolic steroids Synthetic androgens that increase muscle mass, physical strength, endurance, and aggressiveness.
anabolism (an-ab′oh-lizm) The aspect of metabolism in which simpler substances are combined to form more complex substances, resulting in the storage of energy, the production of new cellular materials, and growth. Compare with catabolism.

anaerobe Organism that grows or metabolizes only in the absence of molecular oxygen. Compare with aerobe.

anaerobic (an″air-oh′bik) Growing or metabolizing only in the absence of molecular oxygen. Compare with aerobic.

analogous features (uh-nal′uh-gus) Features similar in function or appearance in different species as a result of convergent evolution rather than shared ancestry. Compare with homologous features.

anaphase (an′uh-faze) The stage of mitosis, and of meiosis I and II, in which the chromosomes move to opposite poles of the cell; anaphase occurs after metaphase and before telophase.

anaphylaxis (an″uh-fih-lak′sis) An acute allergic reaction following sensitization to a foreign substance or other substance.

ancestral characters Traits present in an ancestral species that have remained essentially unchanged in the group being considered.

androgen (an′dro-jen) Any substance that possesses masculizing properties, such as a sex hormone. See testosterone.

anemia (uh-nee′mee-uh) A deficiency of hemoglobin or red blood cells.

aneuploidy (an′you-ploy-dee) Any chromosomal aberration in which there are either extra or missing copies of certain chromosomes.

angiosperms (an′jee-oh-spermz″) The traditional name for flowering plants. A very large (about 235,000 species), very diverse phylum of plants that form flowers for sexual reproduction and produce seeds enclosed in fruits. Include monocots and dicots.

angiotensin (an-jee-o-ten′sin) A compound formed by the action of renin and found in blood; stimulates aldosterone secretion by the renal cortex.

animal pole The non-yolky, metabolically active, pole of a vertebrate or echinoderm egg. Compare with vegetal pole.

anion (an′eye-on) A particle with one or more units of negative charge, such as a chloride ion (Cl^-) or hydroxide ion (OH^-). Compare with cation.

anisogamy (an″eye-sog′uh-mee) Sexual reproduction involving motile gametes of similar form but dissimilar size. Compare with isogamy and oogamy.

annelids (an′eh-lids) Phylum of segmented worms with true coeloms, such as earthworms.

annual plant A plant that completes its entire life cycle in one year or less. Compare with perennial and biennial.

antennae (sing. *antenna*) Sensory structures characteristic of some arthropod groups.

anterior Toward the head end of a bilaterally symmetrical animal. Compare with posterior.

anther (an′thur) The part of the stamen in flowers that produces microspores and, ultimately, pollen grains.

antheridium (an″thur-id′ee-im) (pl. *antheridia*) In plants, the multicellular male gametangium (sex organ) that produces sperm. Compare with archegonium.

anthropoid (an′thra-poid) A member of a suborder of primates that includes monkeys, apes, and humans.

antibody (an-tih-bod′ee) Protein compound (immunoglobulin) produced by plasma cells in response to specific antigens and having the capacity to react against the antigens.

antibody-mediated immunity A type of internal defense mechanism in which B cells differentiate into plasma cells and produce antibodies that combine with foreign antigen, leading to the destruction of pathogens.

anticodon (an′ty-koh″don) A sequence of three nucleotides in transfer RNA that is complementary to, and combines with, the three nucleotide codon on messenger RNA, thus helping to specify the addition of a particular amino acid to the end of a growing polypeptide.

antidiuretic hormone (ADH) (an″ty-dy-uh-ret′ik) A hormone secreted by the posterior lobe of the pituitary that controls the rate of water reabsorption by the kidney.

antigen (an′tih-jen) Any substance capable of stimulating an immune response; usually a protein or large carbohydrate that is foreign to the body.

antigen-antibody complex The combination of antigen and antibody molecules.

anti-oncogene A gene (also known as a tumor suppressor gene) whose normal role is to block cell division in response to certain growth inhibiting factors; when mutated, may contribute to the formation of a cancer cell. Compare with oncogene.

anus (ay′nus) The distal end and outlet of the digestive tract.

aorta (ay-or′tah) The largest and main systemic artery of the vertebrate body; arises from the left ventricle and branches to distribute blood to all parts of the body except the lungs.

aphotic region (ay-fote′ik) The lower layer of the ocean (deeper than 100 meters or so) where light does not penetrate.

apical dominance (ape′ih-kl) The inhibition of lateral buds by a shoot tip.

apical meristem (mehr′ih-stem) An area of dividing tissue located at the tip of a shoot or root; apical meristems cause an increase in the length of the plant body. Compare with lateral meristem.

apoenzyme (ap″oh-en′zime) Protein portion of an enzyme; requires the presence of a specific coenzyme to become a complete functional enzyme.

apomixis (ap″uh-mix′us) A type of reproduction in which fruits and seeds are formed asexually.

apoplast A continuum consisting of the interconnected, porous plant cell walls. Compare with symplast.

arachnids (ah-rack′nids) Eight-legged arthropods such as spiders, scorpions, ticks, and mites.

arboreal (are-bore′ee-uhl) Adapted to living in trees.

archaebacteria (ar″kuh-bak-teer′ee-uh) Anaerobic prokaryotic organisms with a number of features that set them apart from the rest of the bacteria. Compare with eubacteria.

archegonium (ar″ke-go′nee-um) (pl. *archegonia*) In plants, the multicellular female gametangium (sex organ) that contains an egg. Compare with antheridium.

archenteron (ar-ken′ter-on) The central cavity of the gastrula stage of embryonic development, which is lined with endoderm; primitive digestive system.

arctic tundra See tundra.

arteriole (ar-teer′ee-ole) A very small artery. Vasoconstriction and vasodilation of arterioles help regulate blood pressure.

artery A thick-walled blood vessel that carries blood away from the heart and toward the body organs.

arthropods (ar′-throh-podz) Invertebrate phylum characterized by a hard exoskeleton, a segmented body, and paired, jointed appendages.

artificial insemination The impregnation of a female by artificially introducing sperm from a male.

artificial selection Selection by humans of traits that are desirable in plants or animals, and breeding only those individuals that possess the desired traits.
ascocarp (ass'koh-karp) The fruiting body of an ascomycete.
ascomycete (ass"koh-my'seat) Member of a phylum of fungi characterized by the production of nonmotile asexual conidia and sexual ascospores.
ascospore (ass'koh-spor) One of a set of sexual spores, usually eight, contained in a special spore case (an ascus) of an ascomycete.
ascus (ass'kus) A sac-like spore case in ascomycetes that contains sexual spores called ascospores.
asexual reproduction Reproduction in which there is no fusion of gametes and in which the genetic makeup of parent and of offspring is usually identical. Compare with sexual reproduction.
assimilation (of nitrogen) The conversion of inorganic nitrogen (nitrate, NO_3^-, or ammonia, NH_3) to the organic molecules of living things; part of the nitrogen cycle.
association neuron See interneuron.
association areas Areas of the brain that link sensory and motor areas; responsible for thought, learning, memory, language abilities, judgment, and personality.
asters Clusters of microtubules radiating out from the poles in dividing cells that have centrioles; present in animal cells, but not in the cells of flowering plants or of most gymnosperms.
atherosclerosis (ath"ur-oh-skle-row'sis) A progressive disease in which lipid deposits accumulate in the inner lining of arteries, leading eventually to impaired circulation and heart disease.
atoll (a'tol) A circular coral reef with a central lagoon.
atom The smallest quantity of an element that can retain the chemical properties of that element.
atomic mass The total number of protons and electrons in an atom; expressed in atomic mass units or daltons.
atomic mass unit (amu) The approximate mass of a proton or neutron. Also called dalton.
atomic number The number of protons in the atomic nucleus of an atom, which uniquely identifies the element to which the atom corresponds.
atomic weight A value that is numerically equal to the total number of protons and neutrons in an atom; atomic weight has no units.
ATP synthase Large enzyme complex that catalyzes the formation of ATP from ADP and inorganic phosphate by chemiosmosis; contains a transmembrane channel through which protons diffuse down a concentration gradient; located in the inner mitochondrial membrane, the thylakoid membrane of chloroplasts, and the plasma membrane of bacteria.
atrial natriuretic peptide (ANP) A hormone released by the atrium of the heart; helps regulate sodium excretion.
atrioventricular (AV) node (ay"tree-oh-ven-trik'you-lur) Mass of specialized cardiac tissue that receives an impulse from the sinoatrial node (pacemaker) and conducts it to the ventricles.
atrioventricular (AV) valve (of the heart) A valve between each atrium and its ventricle that prevents backflow of blood. The right AV valve is the tricuspid valve, the left AV valve is the mitral valve.
atrium (of the heart) (ay'tree-um) A chamber that receives blood from the veins.
Australopithecus (os"tra-lo-pith'uh-cus) A genus of extinct primitive hominids that originated in Africa. Australopithecines walked erectly.
autoimmune disease (aw"toh-ih-mune') A disease in which the body produces antibodies against its own cells or tissues.
autonomic nervous system (aw-tuh-nom'ik) The portion of the peripheral nervous system that controls the visceral functions of the body, e.g., regulates smooth muscle, cardiac muscle, and glands, thereby helping to maintain homeostasis. Its divisions are the sympathetic and parasympathetic nervous systems. Compare with somatic nervous system.
autosome (aw'toh-sohm) A chromosome other than the sex (X and Y) chromosomes.
autotroph (aw'toh-trof) Organism that synthesizes organic compounds from inorganic raw materials. Also called producer. Compare with heterotroph.
auxin (awk'sin) A plant hormone involved in various aspects of growth and development, such as stem elongation, apical dominance, and root formation on cuttings.
avirulent Unable to cause disease in a host. Compare with virulent.
Avogadro's number The number of units (6.02×10^{23}) present in one mole of any substance.
axon (aks'on) The long extension of the neuron that transmits nerve impulses away from the cell body.
axopods (aks'o-podz) Long, filamentous cytoplasmic projections characteristic of actinopods.
B cell (B lymphocyte) A type of white blood cell responsible for antibody-mediated immunity. When stimulated, B lymphocytes differentiate to become plasma cells that produce antibodies. Compare with T lymphocyte.
bacilli (bah-sill'eye) (sing. *bacillus*) Rod-shaped bacteria.
bacteria (bak-teer'ee-uh) Unicellular, prokaryotic microorganisms. Most bacteria are decomposers, but some are parasites and others are autotrophs.
bacteriophage (bak-teer'ee-oh-fayj) Virus that can infect a bacterium (literally, "bacteria eater"). Also called phage.
balanced polymorphism (pol"ee-mor'fizm) The presence in a population of two or more morphs (genetic variants) that are maintained in a stable frequency over several generations.
bark The outermost covering over woody stems and roots; consists of two regions, a living inner bark, composed of secondary phloem, and a mostly dead outer bark, composed of periderm.
baroreceptors (bare"oh-ree-sep'torz) Receptors within certain blood vessels that are stimulated by changes in blood pressure.
Barr body A condensed and inactivated X-chromosome appearing as a distinctive dense spot in the nucleus of certain cells of female mammals.
barrier reef A coral reef that is separated from the nearby land.
basal body (bay'sl) Structure involved in the organization and anchorage of a cilium or flagellum. Structurally similar to a centriole; each is in the form of a cylinder composed of 9 triplets of microtubules (9×3 structure).
basal metabolic rate (BMR) The amount of energy expended by the body at resting conditions, when no food is being digested and no voluntary muscular work is being performed.
base A hydrogen ion (proton) acceptor; bases unite with acids to form salts. A basic solution has a pH greater than 7 and turns red litmus paper blue.

base-substitution mutation Also known as a point mutation; a change in one base pair in DNA.

basidiocarp (ba-sid′e-o-karp) The fruiting body of a basidiomycete; a mushroom is an example of a basidiocarp.

basidiomycete (ba-sid″e-o-my′seat) Member of a phylum of fungi characterized by the production of sexual basidiospores.

basidiospore (ba-sid′e-o-spor) One of a set of sexual spores, usually four, borne on a basidium of a basidiomycete.

basidium (ba-sid′ee-um) The club-like spore-producing organ of basidiomycetes that bears sexual spores called basidiospores.

basilar membrane The multicellular tissue in the inner ear that separates the cochlear duct from the tympanic canal; the sensory cells of the organ of Corti rest on this membrane.

batesian mimicry (bate′see-un mim′ih-kree) The resemblance of a harmless or palatable species to one that is dangerous, unpalatable, or poisonous. Compare with müllerian mimicry.

behavioral isolation A reproductive isolating mechanism (prezygotic) in which reproduction between similar species is prevented because each group possesses its own characteristic courtship behavior. Also called sexual isolation.

benthos (ben′thos) Bottom-dwelling sea organisms that fix themselves to one spot, burrow into the sediment, or simply walk about on the ocean floor.

berry A simple, fleshy fruit in which the fruit wall is soft throughout, e.g., tomato, banana, grape.

beta-galactosidase (β-galactosidase) Enzyme responsible for cleaving the disaccharide galactose into its component monosaccharides, glucose, and galactose.

beta oxidation Process by which fatty acids are converted to acetyl CoA prior to entry into the citric acid cycle.

beta pleated sheet A regular folded sheet-like structure resulting from hydrogen bonding between two different polypeptide chains or two regions of the same polypeptide chain.

biennial plant (by-en′ee-ul) A plant that takes two years to complete its life cycle. Compare with annual and perennial.

bilateral symmetry A body shape with right and left halves that are approximately mirror images of one another. Compare with radial symmetry.

bile The fluid secreted by the liver; emulsifies fats.

binary fission (by′nare-ee fish′un) Equal division of a cell or organism into two; a type of asexual reproduction.

binomial system of nomenclature (by-nome′ee-ul) System of naming a species by the combination of the genus name and a specific epithet.

biogenic amines A class of neurotransmitters that includes norepinephrine, serotonin, and dopamine.

biogeochemical cycle (bye″o-jee″o-kem′e-kl) Process by which matter cycles from the living world to the nonliving physical environment and back again. Examples of biogeochemical cycles include the carbon cycle, the nitrogen cycle, and the phosphorus cycle.

biogeography The study of the geographical distribution of organisms.

biological clocks Mechanisms by which activities of organisms are adapted to regularly recurring changes in the environment.

biological diversity The number and variety of living organisms; biological diversity is considered at all levels, including genetic diversity within a species, species diversity, and ecosystem diversity.

biological species concept See species.

biomass (bye′o-mas) A quantitative estimate of the total mass, or amount, of living material in a particular ecosystem.

biome (by′ohm) A large, relatively distinct terrestrial region characterized by a similar climate, soil, plants, and animals, regardless of where it occurs on Earth.

biosphere All of Earth's living organisms. Compare with ecosphere.

biotic potential The maximum rate at which an organism or population could increase when environmental conditions are optimal.

bipedal Walking on two feet.

biramous appendages Appendages with two jointed branches at their ends; characteristic of crustaceans.

bivalent (by-vale′ent or biv′ah-lent) See tetrad.

blade (1) The thin, expanded part of a leaf; (2) The flat, leaf-like structure of certain multicellular algae.

blastocoel (blas′toh-seel) The fluid-filled cavity of a blastula.

blastocyst See blastula.

blastopore (blas′toh-pore) Primitive opening into the body cavity of an early embryo that may become the mouth (in protostomes) or anus (in deuterostomes) of the adult organism.

blastula (blas′tew-lah) In animal development, a hollow ball of cells produced by cleavage of a fertilized ovum. Known as a blastocyst in mammalian development.

blood A fluid, circulating connective tissue that transports nutrients and other materials through the body of many types of animals. In vertebrates, consists of red blood cells, white blood cells, and platelets suspended in a fluid component called plasma.

blood pressure The force exerted by blood against the inner walls of the blood vessels.

bolting Production of a tall flower stalk by a plant that grows vegetatively as a rosette (growth habit with a short stem and a circular cluster of leaves).

bond energy The energy required to break a particular chemical bond.

bone tissue Principal vertebrate skeletal tissue; a type of connective tissue.

boreal forest (bor′ee-uhl) See taiga.

bottleneck A sudden decrease in a population size due to adverse environmental factors; may result in genetic drift.

Bowman's capsule Double-walled sac of cells that surrounds the glomerulus of each nephron.

brachiate (brake′ee-ate) To swing, arm to arm, from one branch to another.

brachiopods (bray′kee-oh-pods) Phylum of marine invertebrate deuterostomes possessing a pair of shells, and internally, a pair of coiled arms with ciliated tentacles; one of the lophophorate phyla.

brain A concentration of nervous tissue that controls neural function; in vertebrates, the anterior, enlarged portion of the central nervous system.

brain stem The part of the vertebrate brain that includes the medulla, pons, midbrain, thalamus, and hypothalamus.

branchial Pertaining to the gills or gill region.

bronchial constriction Narrowing of the bronchial tubes in response to breathing dirty air.

bronchiole (bronk′ee-ole) Tiny air duct in the lung that branches from a bronchus; divides to form air sacs (alveoli).

bronchus (bronk'us) (pl. *bronchi*) One of the branches of the trachea and its immediate branches within the lung.

brown alga One of a phylum of predominantly marine algae that are multicellular and contain the pigments chlorophyll *a* and *c*, and carotenoids, including fucoxanthin.

bryophytes (bry'oh-fites) Nonvascular plants including mosses, liverworts, and hornworts.

bud An undeveloped shoot that can develop into flowers, stems, or leaves. Buds are enclosed in bud scales.

bud scale A modified leaf that covers and protects a dormant bud.

bud scale scar Scar on a twig left when a bud scale abscises from the terminal bud.

budding Asexual reproduction in which a small part of the parent's body separates from the rest and develops into a new individual. Characteristic of yeasts and certain other organisms.

buffer A substance in a solution that tends to lessen the change in hydrogen ion concentration (pH) that otherwise would be produced by adding an acid or base.

bulb A globose, fleshy, underground bud; a bulb is a short stem with fleshy leaves, e.g., onion.

bundle scar Marks on a leaf scar left when vascular bundles of the petiole break during leaf abscission.

bundle sheath cells Tightly packed cells that form a sheath around the veins of a leaf.

bundle sheath extension Support cells that extend from the bundle sheath of a vein toward the upper and/or lower epidermis.

C_3 plant Plant that carries out carbon fixation solely by the Calvin cycle. Compare with C_4 plant and CAM plant.

C_4 plant Plant that fixes carbon initially by the Hatch-Slack pathway, in which the reaction of CO_2 with phosphoenolpyruvate is catalyzed by PEP carboxylase in leaf mesophyll cells; the products are transferred to the bundle sheath cells, where the Calvin cycle takes place. Compare with C_3 plant and CAM plant.

calcitonin (kal-sih-toh'nin) A hormone secreted by the thyroid gland that rapidly lowers the calcium content in the blood.

callus (kal'us) Undifferentiated tissue formed on an explant (excised tissue or organ) in plant tissue culture.

calorie The amount of heat energy required to raise the temperature of 1 gram of water 1° C; equivalent to 4.184 joules. The Calorie used in metabolism, the kilocalorie, is the heat required to raise the temperature of 1 kilogram of water 1° C.

Calvin cycle Cyclic series of reactions in the chloroplast stroma in photosynthesis; fixes carbon dioxide and produces carbohydrate.

calyx (kay'liks) The collective term for the sepals of a flower.

cambium See lateral meristems.

CAM plant Plant that carries out crassulacean acid metabolism; carbon is initially fixed into organic acids at night in the reaction of CO_2 and phosphoenolpyruvate, catalyzed by PEP carboxylase; during the day the acids break down to yield CO_2, which enters the Calvin cycle. Compare with C_3 plant and C_4 plant.

capillaries (kap'i-lare-eez) Microscopic blood vessels in the tissues that permit exchange of materials between tissues and blood.

capillary action The ability of water to move in small diameter tubes as a consequence of its cohesive and adhesive properties.

capsid Protein coat surrounding the nucleic acid of a virus.

capsule (1) The portion of the moss sporophyte that contains spores; (2) A simple, dry, dehiscent fruit that opens along many seams or pores to release seeds, such as cotton fruits; (3) A gelatinous coat that surrounds some bacteria.

carbohydrate Compound containing carbon, hydrogen, and oxygen, in the approximate ratio of C:2H:O; examples include sugars, starch, and cellulose.

carbon cycle The worldwide circulation of carbon from the abiotic environment into living things and back into the abiotic environment.

carbon fixation reactions Reduction reactions of photosynthesis in which carbon from carbon dioxide becomes incorporated into organic molecules, leading to the production of carbohydrate; requires ATP and NADPH.

carbonyl group A polar functional group consisting of a carbon attached to an oxygen by a double bond; found in aldehydes and ketones.

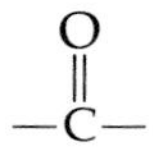

carboxyl group A weakly acidic functional group; abbreviated —COOH.

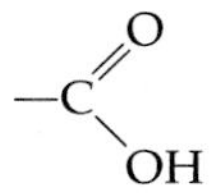

carcinogen (kar-sin'oh-gen, kar'sin-oh-jen) An agent that causes cancer or accelerates its development.

cardiac (kar'dee-ak) Pertaining to the heart.

cardiac cycle One complete heart beat.

cardiac muscle Distinctive involuntary, striated type of muscle found in the vertebrate heart. Compare with smooth muscle and skeletal muscle.

cardiac output The volume of blood pumped by the heart per unit of time.

cardiovascular disease Disease of the heart or blood vessels; the leading cause of death in most industrial societies.

carnivore (kar'ni-vor) An animal that feeds on other animals; flesh-eater.

carotenoids (ka-rot'n-oidz) A group of yellow to orange plant pigments synthesized from isoprene subunits; include carotenes and xanthophylls.

carpel (kar'pul) The female reproductive unit of a flower; carpels bear ovules.

carrier-mediated active transport Transport across a membrane of a substance from a region of low concentration to a region of high concentration; requires both a transport protein with a binding site for the specific substance and an energy source (often ATP).

carrier-mediated transport Any form of transport across a membrane that utilizes a membrane-bound transport protein with a binding site for a specific substance; includes both facilitated diffusion and carrier-mediated active transport.

carrying capacity The maximum population that a particular habitat can support and sustain (long-term).

cartilage Flexible skeletal tissue of vertebrates; a type of connective tissue.

Casparian strip (kas-pare'ee-un) A band of waterproof material around the radial and transverse walls of endodermal root cells.

catabolism The aspect of metabolism in which complex substances are broken down to form simpler substances; catabolic reactions are particularly important in releasing chemical energy stored by the cell. Compare with anabolism.

catabolite activator protein (CAP) A positively acting regulator that becomes active when bound to cAMP; active CAP stimulates transcription of the lactose operon and a number of other operons that code for enzymes used in catabolic pathways.

catalase Iron-containing enzyme responsible for breaking down hydrogen perioxide (H_2O_2) to oxygen and water.

catalyst (kat'ah-list) A substance that increases the speed at which a chemical reaction occurs without being used up in the reaction. Enzymes are biological catalysts.

cation Particle bearing one or more units of positive charge, such as a hydrogen ion (H^+) or calcium ion (Ca^{2+}). Compare with anion.

cDNA library A collection of recombinant plasmids that contain complementary DNA (cDNA) copies of mRNA templates. The cDNA, which lacks introns, is synthesized by reverse transcriptase.

cell The basic structural and functional unit of life, which consists of living material bounded by a membrane.

cell cycle Cyclic series of events in the life of a dividing eukaryotic cell; consists of mitosis, cytokinesis, and the stages of interphase. The time required to complete one cell cycle is the generation time (T).

cell fractionation Technique used to separate the components of cells by subjecting them to centrifugal force.

cell plate Structure that forms during cytokinesis in plants, separating the two daughter cells produced by mitosis.

cell theory The theory that the cell is the basic unit of life, of which all living things are composed, and that all cells are derived from preexisting cells.

cell wall Structure outside the plasma membrane of certain cells; may contain cellulose (plant cells), chitin (most fungal cells), peptidoglycan and/or lipopolysaccharide (most bacterial cells), or other material.

cellular differentiation Development toward a more mature state; a process changing a relatively unspecialized cell to a more specialized cell.

cellular respiration See respiration.

cellular slime mold A phylum of fungus-like protists whose feeding stage consists of unicellular, amoeboid organisms that aggregate to form a pseudoplasmodium during reproduction.

cellulose (sel'yoo-lohs) A structural polysaccharide composed of beta glucose subunits; the main constituent of plant primary cell walls.

Cenozoic era A geological era that began 65 million years ago and extends to the present time.

center of origin The geographical area where a given species originated.

central nervous system (CNS) In vertebrates, the brain and spinal cord. Compare with peripheral nervous system (PNS).

centrifuge Device used to separate cells or their components by subjecting them to centrifugal force.

centriole (sen'tree-ohl) One of a pair of small, cylindrical organelles lying at right angles to each other near the nucleus in the cytoplasm of animal cells and certain protist and plant cells; each centriole is in the form of a cylinder composed of 9 triplets of microtubules (9×3 structure).

centromere (sen' tro-meer) Specialized constricted region of a chromatid; contains the kinetochore. In cells at prophase and metaphase, sister chromatids are joined in the vicinity of their centromeres.

cephalization The evolution of a head; the concentration of nervous tissue and sense organs at the front end of the animal.

cephalochordates Members of the chordate subphylum that includes the lancelets.

cerebellum (ser-eh-bel'um) A convoluted subdivision of the vertebrate brain concerned with the coordination of muscular movements, muscle tone, and balance.

cerebral cortex (ser-ee'brul kor'tex) The outer layer of the cerebrum composed of gray matter and consisting mainly of nerve cell bodies.

cerebrospinal fluid (CSF) The fluid that bathes the CNS of vertebrates.

cerebrum (ser-ee'brum) Large subdivision of the vertebrate brain; in humans, it functions as the center for learning, voluntary movement, and interpretation of sensation.

chaparral (shap"uh-ral') A biome with a Mediterranean climate (mild, moist winters and hot, dry summers). Chaparral vegetation is characterized by drought-resistant, small-leaved evergreen shrubs and small trees.

character displacement The tendency for two similar species to diverge (become more different) in areas where their ranges overlap, presumably to reduce interspecific competition.

chelicerae (keh-lis'er-ee) Pincers; the first pair of appendages in certain arthropods; clawlike appendages located immediately anterior to the mouth, and used to manipulate food into the mouth.

chemical bond A force of attraction between atoms in a compound.

chemical compound Two or more elements combined in a fixed ratio.

chemical equilibrium State of a chemical reaction in which the rate of the forward reaction is equal to the rate of the reverse reaction and the concentrations of the reactants and products are not changing.

chemical evolution The origin of life from nonliving matter.

chemical formula A representation of the composition of a compound; the elements are indicated by chemical symbols with subscripts to indicate their ratios.

chemical symbol Abbreviation for an element; usually the first letter (or first and second letters) of the English or Latin name.

chemiosmosis Process by which phosphorylation of ADP to form ATP is coupled to the transfer of electrons down a membrane-bound electron transport chain; the electron transport chain powers proton pumps that produce a proton gradient across the membrane; ATP is formed as protons diffuse through transmembrane channels in ATP synthase complexes; chemiosmosis is the mechanism of oxidative phosphorylation in aerobic respiration and of photophosphorylation in photosynthesis.

chemoautotroph (kee"moh-aw'toh-trof) Chemosynthetic autotroph; organism that obtains energy from inorganic compounds and synthesizes organic compounds from inorganic raw materials; includes some bacteria.

chemoheterotroph (kee"moh-het'ur-oh-trof) Chemosynthetic heterotroph; organism that uses organic compounds as a source of energy and carbon; includes animals, fungi, and many bacteria.

chemoreceptor (kee″moh-ree-sep′tor) A sensory receptor that responds to chemical stimuli.

chiasma (ky-az′muh) (pl. *chiasmata*) An X-shaped site in a tetrad (bivalent) marking the location where homologous (nonsister) chromatids previously underwent crossing over.

chimera Organism composed of two or more kinds of genetically dissimilar cells.

chitin (ky′tin) A nitrogen-containing structural polysaccharide that forms the exoskeleton of insects and the cell walls of many fungi.

chlorophyll (klor′oh-fil) A group of light-trapping green pigments found in most photosynthetic organisms.

chloroplasts (klor′oh-plastz) Membranous organelles that are the sites of photosynthesis in eukaryotes; occur in some plant and algal cells.

choanocyte (koh-an′oh-sight) A unique cell having a flagellum surrounded by a thin cytoplasmic collar; characteristic of sponges and one group of protists.

cholinergic neuron (kohl″in-air′jik) A nerve cell that secretes acetylcholine as a neurotransmitter. Compare with adrenergic neuron.

chondrichthyes A class of cartilaginous fishes, including the sharks, rays, and skates.

chondrocytes Cartilage cells.

chordates (kor′dates) Phylum of deuterostome animals that possess, at some time in their lives, a cartilaginous dorsal skeletal structure called a notochord; a dorsal, tubular nerve cord; pharyngeal gill grooves; and a postanal tail.

chorion (kor′ee-on) An extraembryonic membrane in reptiles, birds, and mammals that forms an outer cover around the embryo, and in mammals contributes to the formation of the placenta.

chorionic villus sampling (CVS) (kor″ee-on′ik) Study of extraembryonic cells that are genetically identical to the cells of an embryo, making it possible to assess its genetic makeup. Compare with amniocentesis.

chromatid (kroh′ mah-tid) One of the two identical halves of a duplicated chromosome; the two chromatids that make up a chromosome are referred to as sister chromatids.

chromatin (kro′mah-tin) The complex of DNA, protein, and RNA that makes up eukaryotic chromosomes.

chromoplasts Pigment-containing plastids; usually found in flowers and fruits.

chromosomes (kro′moh-soms) Structures in the cell nucleus, composed of chromatin and containing the genes. The chromosomes become visible with the microscope as distinct structures when the cell divides.

chylomicrons (kie-low-my′kronz) Protein-covered fat droplets produced in the intestinal cells; enter the lymphatic system and are transported to the blood.

ciliate (sil′e-ate) A unicellular protozoon covered by many short cilia.

cilium (sil′ee-um) (pl. *cilia*) One of many short, hairlike structures that project from the surface of some eukaryotic cells and are used for locomotion or movement of materials across the cell surface; structurally like flagella, including two central single microtubules surrounded by a cylinder of nine double microtubules (9 + 2 structure), covered by a plasma membrane.

circadian rhythm (sir-kay′dee-un) An internal rhythm that approximates the 24-hour day. Circadian rhythms are found in plants, animals, and eukaryotic microorganisms.

circulatory system The body system that moves blood through the body; functions in internal transport and protects the body from disease.

citrate (citric acid) A 6-carbon organic acid.

citric acid cycle Series of chemical reactions in aerobic cellular respiration in which acetyl coenzyme A is completely degraded to carbon dioxide and water with the release of metabolic energy, which is used to produce ATP; also known as the Krebs cycle and the tricarboxylic acid (TCA) cycle.

clade A taxon containing a common ancestor and all the taxa descended from it; a monophyletic group.

cladistics An approach to classification based on recency of common ancestry, rather than degree of structural similarity. Compare with phenetics and classical evolutionary taxonomy.

cladogram A branching diagram that illustrates taxonomic relationships based on the principles of cladistics.

class A taxonomic category made up of related orders.

classical conditioning A type of learning in which an association is formed between some normal response to a stimulus and a new stimulus, after which the new stimulus elicits the response.

classical evolutionary taxonomy An approach to classification that considers both evolutionary branching and the extent of divergence that has occurred since a group branched from an ancestral group. Compare with cladistics and phenetics.

cleavage Series of mitotic cell divisions that converts the zygote to a multicellular blastula.

climate The average weather conditions that occur in a place over a period of years. Includes temperature and precipitation.

clitoris (klit′o-ris) A small, erectile structure at the anterior part of the vulva in female mammals; homologous to the male penis.

cloaca (klow-a′ka) An exit chamber in less complex vertebrates that receives digestive wastes and urine; may also serve as an exit for gametes (the reproductive system).

clone A population of cells descended by mitotic division from a single ancestral cell, or a population of genetically identical organisms asexually propagated from a single individual.

closed circulatory system A type of circulatory system in which the blood flows through a continuous circuit of blood vessels; characteristic of annelids, cephalopods, and vertebrates. Compare with open circulatory system.

club mosses A phylum of seedless vascular plants with a life cycle similar to ferns.

clumped dispersion The spatial distribution pattern in which individuals are more clustered than in a random distribution. Compare with random dispersion and uniform dispersion.

cnidarians (cny-dare′ee-anz) Phylum of animals that have stinging cells called cnidocytes, two tissue layers, and radial symmetry; include hydras and jellyfish.

cnidocytes Stinging cells characteristic of cnidarians.

cochlea (koke′lee-ah) The structure of the inner ear of mammals that contains the auditory receptors (organ of Corti).

coccus (kok′us) (pl. *cocci*) A bacterium with a spherical shape.

codon (koh′don) A triplet of mRNA bases that specifies an amino acid, a start signal, or a signal to terminate the polypeptide.

codominance (koh″dom′in-ants) Condition in which two alleles of a locus are expressed in a heterozygote.

coelacanths A genus of lobe-finned fish that have survived to the present day.
coelom (see′lum) The main body cavity of most animals; a true coelom is lined with mesoderm. Compare with pseudocoelom.
coelomate (seel′oh-mate) Animal possessing a true coelom. Compare with acoelomate and pseudocoelomate.
coenocyte (see′no-site) An organism consisting of a multinucleated cell; an organism in which the nuclei are not separated from one another by septa.
coenzyme (koh-en′zime) An organic cofactor for an enzyme; generally participates in the reaction by transferring some component.
coenzyme A (CoA) Organic cofactor responsible for transferring groups derived from organic acids.
coevolution The interdependent evolution of two or more species that occurs as a result of their interactions over a long period of time.
cofactor A non-protein substance needed by an enzyme for normal activity; some cofactors are inorganic (usually metal ions); others are organic cofactors, known as coenzymes.
cohesive Having the property of sticking together.
colchicine A drug that blocks the division of eukaryotic cells by binding to tubulin subunits, which make up the microtubules that comprise the major component of the mitotic spindle.
coleoptile (kol-ee-op′tile) A protective sheath that encloses the young stem in certain monocots.
collagen (kol′ah-gen) Protein in connective tissues.
collecting duct A tube in the kidney that receives filtrate from several nephrons and conducts it to the renal pelvis.
collenchyma (kol-en′kih-mah) Living cells with moderately but unevenly thickened primary cell walls; collenchyma cells help support the herbaceous plant body.
colon (koe′lon) Portion of the large intestine between the cecum and rectum.
commensalism (kuh-men′sul-iz-m) A type of symbiosis in which one organism benefits and the other one is neither harmed nor helped. Compare with mutualism and parasitism.
community An association of different species living together in a defined habitat with some degree of interdependence. Compare with ecosystem.
compact bone Dense, hard bone tissue found mainly near the surfaces of a bone.
companion cell A cell in the phloem of flowering plants that is responsible for loading and unloading sugar into the sieve tube member for translocation.
competitive exclusion The concept that no two species with identical living requirements can occupy the same ecological niche indefinitely. Eventually, one species will be excluded by the other as a result of interspecific competition for a resource in limited supply.
competitive inhibitor A substance that binds to the active site of an enzyme, thus lowering the rate of the reaction catalyzed by the enzyme. Compare with noncompetitive inhibitor.
complement A group of proteins in blood and other body fluids that are activated by an antigen-antibody complex.
complementary DNA (cDNA) DNA synthesized by reverse transcriptase, using RNA as a template.
complete flower A flower that possesses all four parts: sepals, petals, stamens, and carpels. Compare with incomplete flower.
compound eye An eye, such as that of an insect, composed of many light-sensitive units called ommatidia.
concentration gradient A change in the concentration of a substance from one point to another.
condensation synthesis Reaction in which two monomers are combined covalently through the removal of the equivalent of a water molecule. Compare with hydrolysis.
cone (1) In botany, a reproductive structure in many gymnosperms that produces either microspores or megaspores. (2) In zoology, one of the conical photoreceptive cells of the retina that is particularly sensitive to bright light, and, by distinguishing light of various wavelengths, mediates color vision. Compare with rod.
conidiophore (kah-nid′e-o-for″) A specialized hypha that bears conidia.
conidium (kah-nid′e-um) (pl. *conidia*) An asexual spore that is usually formed at the tip of a specialized hypha called a conidiophore.
conifer (kon′ih-fur) Any of a large phylum of gymnosperms that are woody trees and shrubs with needle-like, mostly evergreen, leaves and with seeds in cones.
conjugation (kon″jew-gay′shun) (1) A sexual phenomenon in certain protists that involves exchange or fusion of a cell with another cell; (2) A mechanism for DNA exchange in bacteria that involves cell to cell contact.
connective tissue Animal tissue consisting mostly of intercellular substance (fibers scattered through a matrix) in which the cells are embedded, e.g., bone.
consanguineous mating A mating between close relatives.
constitutive gene A gene that is constantly transcribed.
consumers Organisms that cannot synthesize their own food from inorganic materials and therefore must use the bodies of other organisms as a source of energy and body-building materials. Also called heterotrophs. Compare with producers.
contest competition Intraspecific competition in which certain dominant individuals obtain an adequate supply of the limited resource at the expense of other individuals in the population. Compare with scramble competition.
continental drift The theory that continents were once joined together and later split and drifted apart.
contraception Any method used to intentionally prevent pregnancy.
contractile root (kun-trak′til) A specialized type of root that contracts and pulls a bulb or corm deeper into the soil.
control group In a scientific experiment, a group in which the experimental variable is kept constant. The control provides a standard of comparison used to verify the results of the experiment.
controlled mating A mating in which the genotypes of the parents are known.
convergent circuit A neural pathway in which a postsynaptic neuron is controlled by signals coming from two or more presynaptic neurons.
convergent evolution (kun-vur′jent) The independent evolution of structural or functional similarity in two or more organisms of widely different, unrelated ancestry.
convoluted tubule See proximal convoluted tubule and distal convoluted tubule.
corepressor Substance that binds to a repressor protein, converting it to its active form, which is capable of preventing transcription.

Coriolis effect (kor″e-o′lis) The tendency of moving air or water to be deflected from its path to the right in the Northern Hemisphere and to the left in the Southern Hemisphere. Caused by the direction of Earth's rotation.

cork cambium (kam′bee-um) A lateral meristem that produces cork cells and cork parenchyma; cork cambium and the tissues it produces make up the outer bark on a woody plant. Compare with vascular cambium.

cork cell A cell in the bark that is produced outwardly by the cork cambium; cork cells are dead at maturity and function for protection and reduction of water loss.

cork parenchyma (par-en′kih-mah) One or more layers of parenchyma cells produced inwardly by the cork cambium.

corm A short, thickened underground stem specialized for food storage and asexual reproduction, e.g., crocus, gladiolus.

cornea (kor′nee-ah) Transparent covering of an eye.

corolla (kor-ohl′ah) A collective term for the petals of a flower.

corpus callosum (kah-loh′sum) In mammals, a large bundle of nerve fibers interconnecting the two cerebral hemispheres.

corpus luteum (loo′tee′um) The temporary endocrine tissue in the ovary that develops from the ruptured follicle after ovulation; secretes progesterone and estrogen.

cortex (kor′tex) (1) The outer part of an organ, such as the cortex of the kidney (compare with medulla); (2) The tissue between the epidermis and vascular tissue in the stems and roots of many herbaceous plants.

cotransport The active transport of a substance from a region of low concentration to a region of high concentration by coupling its transport to the transport of a substance down its concentration gradient.

cotyledon (kot″i-lee′dun) The seed leaf of a plant embryo, which may contain food stored for germination.

cotylosaurs The first reptiles; also known as stem reptiles.

countercurrent exchange system A biological mechanism that enables maximum exchange between two fluids. The two fluids must be flowing in opposite directions and have a concentration gradient between them.

coupled reactions Set of reactions in which an exergonic reaction provides the free energy required to drive an endergonic reaction; energy coupling generally occurs through a common intermediate.

covalent bond Chemical bond involving shared pairs of electrons; may be single, double, or triple (with one, two, or three shared pairs of electrons respectively).

covalent compound A compound in which atoms are held together by covalent bonds.

cranial nerves Ten to twelve pairs of nerves in vertebrates that emerge directly from the brain.

cranium The bony framework that protects the brain in vertebrates.

crassulacean acid metabolism See CAM plant.

creatine phosphate Energy storing compound found in muscle cells.

crenated Term used to describe the shriveled appearance of an animal cell, such as a red blood cell, that has lost water due to osmosis. Compare with plasmolysis.

cretinism (kree′tin-izm) A chronic condition due to lack of thyroid secretion during fetal development and early childhood; results in retarded physical and mental development if untreated.

cristae (kris′tee) (sing. *crista*) Shelflike or finger-like inward projections of the inner membrane of a mitochondrion.

Cro-Magnons Prehistoric humans with modern features (tall, erect, lacking a heavy brow) who lived in Europe some 40,000 years ago.

cross bridges The connections of myosin with actin that link thick and thin filaments in muscle fibers.

crossing over The breaking and rejoining of homologous (nonsister) chromatids during early meiotic prophase I, resulting in an exchange of genetic material.

ctenophores (teen′oh-forz) Phylum of marine animals ("comb jellies") whose bodies consist of two layers of cells enclosing a gelatinous mass. The outer surface is covered with comblike rows of cilia, by which the animal moves.

cultural evolution The progressive addition of knowledge to the human experience.

cuticle (kew′tih-kl) (1) A noncellular waxy covering over the epidermis of the above-ground portion of plants; reduces water loss; (2) The outer covering of some animals.

cyanobacteria (sy-an″oh-bak-teer′ee-uh) Prokaryotic photosynthetic microorganisms that possess chlorophyll and produce oxygen during photosynthesis. Formerly known as blue-green algae.

cycad (sih′kad) Any of a phylum of gymnosperms that live mainly in tropical and semitropical regions and have stout stems (to 20 meters in height) and fern-like leaves.

cyclic AMP (cAMP) A form of adenosine monophosphate in which the phosphate is part of a ring-shaped structure; acts as a regulatory molecule and second messenger in organisms ranging from bacteria to humans.

cyclic phosphorylation In photosynthesis, a series of reactions involving Photosystem I; ATP is formed by chemiosmosis, but O_2 and NADPH are not produced. Compare with noncyclic phosphorylation.

cyclins Regulatory proteins whose levels oscillate during the cell cycle; activate cyclin-dependent protein kinases.

cystic fibrosis A genetic disease with an autosomal recessive inheritance pattern; characterized by secretion of abnormally thick mucus, particularly in the respiratory and digestive systems.

cytochromes (sy′toh-kromz) Iron-containing heme proteins of an electron transport system.

cytogenetics The branch of genetics that relates inheritance to cell structure and function, with particular emphasis on the chromosomes.

cytokines Regulatory proteins released by cells of the immune system; interferons and interleukins are examples.

cytokinesis (sy″toh-kih-nee′sis) Stage of cell division in which the cytoplasm is divided to form two daughter cells.

cytokinin (sy″toh-kih′nin) A plant hormone involved in various aspects of plant growth and development, such as cell division and delay of senescence.

cytoplasm Term referring to the cell contents with the exception of the nucleus.

cytosine A nitrogenous pyrimidine base that is a component of nucleic acids.

cytoskeleton Dynamic internal network of protein fibers; includes microfilaments, intermediate filaments, and microtubules.

cytosol Fluid component of the cytoplasm in which the organelles are suspended.

cytotoxic T cell T lymphocyte that destroys cancer cells and other pathogenic cells on contact. Also known as killer T cell.
dalton See atomic mass unit (amu).
day-neutral plant A plant whose flowering is not controlled by variations in day length that occur with changing seasons.
deamination (dee-am-ih-nay'shun) Removal of an amino group ($—NH_2$) from an amino acid or other organic compound.
decarboxylation Reaction in which a molecule of CO_2 is removed from a carboxyl group of an organic acid.
deciduous Term describing leaves or other structures that are shed at regular intervals; e.g., leaves of deciduous trees fall off every autumn.
decomposers Microbial heterotrophs that break down dead organic material and use the decomposition products as a source of energy. Also called saprotrophs or saprobes.
deductive reasoning Reasoning that operates from generalities to specifics and can make relationships among data more apparent. Compare with inductive reasoning.
deforestation The permanent removal of forest without adequate replanting.
dehydrogenation (dee-hy"dro-jen-ay'shun) A form of oxidation in which hydrogen atoms are removed from a molecule.
demographics The branch of sociology that deals with human population statistics (such as density and distribution).
denature (dee-nay'ture) To alter the physical properties and three-dimensional structure of a protein, nucleic acid, or other macromolecule by treating it with excess heat, strong acids, or strong bases.
dendrite (den'drite) A short branch of a neuron that receives and conducts nerve impulses toward the cell body.
dendrochronology (den"dro-kruh-naal'uh-gee) A method of dating using the annual rings of trees.
denitrification (dee-nie"tra-fuh-kay'shun) The conversion of nitrate (NO_3^-) to nitrogen gas (N_2) by certain bacteria (denitrifying bacteria) in the soil; part of the nitrogen cycle.
dense connective tissue A type of tissue that may be irregular as in the dermis of the skin or regular as in tendons.
density-dependent factor An environmental factor whose effects on a population change as population density changes; density-dependent factors tend to retard population growth as population density increases and enhance population growth as population density decreases.
density gradient centrifugation A technique whereby organelles or large molecules can be separated on the basis of differences in density.
density-independent factor An environmental factor that affects the size of a population but is not influenced by changes in population density.
deoxyribonucleic acid (DNA) Double stranded nucleic acid; contains genetic information coded in specific sequences of its constituent nucleotides.
deoxyribose Pentose sugar lacking a hydroxyl ($—OH$) group on carbon-2'; a constituent of DNA.
depolarization (dee-pol"ar-ih-zay'shun) A decrease in the charge difference across a plasma membrane; may result in an action potential in a neuron or muscle cell.
derived characters Traits that evolved relatively recently and so are not present in the ancestral species being considered.
dermal tissue system The tissue that forms the outer covering over a plant; the epidermis or periderm.
dermis (dur'mis) The layer of dense connective tissue beneath the epidermis in the skin of vertebrates.
desert A temperate or tropical biome in which lack of precipitation limits plant growth.
desmosomes (dez'moh-somz) Button-like plaques, present on two opposing cell surfaces, that hold the cells together by means of protein filaments that span the intercellular space.
determinate growth Growth of limited duration, as for example, in flowers and leaves. Compare with indeterminate growth.
determination The developmental process by which one or more cells become progressively committed to a particular fate.
detritivore (duh-try'tuh-vore) An organism (such as an earthworm or crab) that consumes fragments of dead organisms; also called detritus feeder.
detritus (duh-try'tus) Organic debris from decomposing organisms.
detritus feeder See detritivore.
deuteromycetes (doot"er-o-my'seats) Imperfect fungi. An artificial grouping of fungi characterized by the absence of sexual reproduction but usually having other traits similar to ascomycetes.
deuterostome (doo'ter-oh-stome) Major division of the animal kingdom in which the anus develops from the blastopore; this group is characterized by radial cleavage; group of coelomate animals that includes the echinoderms and chordates. Compare with protostome.
developed country An industrialized country; characterized by a low fertility rate, low infant mortality rate, and high per capita income. Developed countries include the United States, Canada, Japan, and European countries; also called highly developed country. Compare with developing country.
developing country A country not highly industrialized; characterized by a high fertility rate, high infant mortality rate, and low per capita income. Compare with developed country.
diabetes mellitus (mel'i-tus) An endocrine disorder caused by insulin deficiency.
diacylglycerol (di"as-il-glis'-er-ol) A neutral fat consisting of a glycerol combined chemically with two fatty acids; also called diglyceride.
dialysis The diffusion of certain solutes across a selectively permeable membrane.
diaphragm In mammals, the muscular floor of the thorax (chest cavity); contracts during inspiration, expanding the chest cavity.
diastole (di-as'toh-lee) Phase of the cardiac cycle in which the heart is relaxed. Compare with systole.
diatom (die'eh-tom") A usually unicellular alga that is covered by an ornate, siliceous shell consisting of two overlapping halves; an important component of plankton in both marine and fresh waters.
dichotomous (di-kaut'uh-mus) A type of branching in which one part always divides into two more or less equal parts.
dicot (dy'kot) One of the two classes of flowering plants; dicot seeds contain two cotyledons, or seed leaves; compare with monocot.
differentiation (dif"ah-ren-she-ay'shun) Development toward a more mature state; a process changing a young, relatively unspecialized cell to a more specialized cell.

diffusion Net movement of particles (atoms, molecules, or ions) from a region of higher concentration to a region of lower concentration (i.e., down a concentration gradient), resulting from random motion.
digestion The breakdown of food to smaller molecules.
diglyceride See diacylglycerol.
dihybrid cross (dy-hy'brid) A genetic cross that takes into account the behavior of alleles of two loci. Compare with monohybrid cross.
dikaryotic (dy-kare-ee-ot'ik) Condition of having two nuclei per cell (i.e., $n + n$), characteristic of certain fungal hyphae. Compare with monokaryotic.
dimer An association of two monomers.
dinoflagellate (dy"noh-flaj'eh-late) A unicellular biflagellate, typically marine, alga that is an important component of plankton; usually photosynthetic.
dioecious (dy-ee'shus) Having male and female reproductive structures on separate plants; compare with monoecious.
dipeptide A compound consisting of two amino acids linked by a peptide bond.
diploid (dip'loyd) The condition of having two sets of chromosomes per nucleus. Compare with haploid and polyploid.
directional selection The gradual replacement of one phenotype with another due to environmental change that favors those phenotypes at one of the extremes of the normal distribution. Compare with stabilizing selection and disruptive selection.
disaccharide (dy-sak'ah-ride) A sugar produced by covalently linking two monosaccharides.
disomy Condition in which both members of a chromosome pair are present. Compare with monosomy and trisomy.
dispersion The pattern of distribution in space of the individuals of a population; may be clumped, random, or uniform.
disruptive selection A special type of directional selection in which changes in the environment favor two or more variant phenotypes at the expense of the mean. Compare with stabilizing selection and directional selection.
distal Remote, farther from the point of reference. Compare with proximal.
distal covoluted tubule The part of the renal tubule that extends from the loop of Henle to the collecting duct. Compare with proximal convoluted tubule.
divergent circuit A neural pathway in which a presynaptic neuron stimulates many postsynaptic neurons.
division A taxonomic category below that of kingdom, comparable to a phylum, formerly used in classifying plants, fungi, and certain protists.
DNA See deoxyribonucleic acid.
DNA-dependent RNA polymerase See RNA polymerase.
DNA ligase Enzyme capable of joining the 5' and 3' ends of two DNA fragments; essential in DNA replication and used in recombinant DNA technology.
DNA polymerases Family of enzymes that catalyze the synthesis of DNA from a DNA template, by adding nucleotides to a growing 3' end.
DNA replication The process by which DNA is duplicated; ordinarily a semiconservative process in which a double helix gives rise to two double helices, each with an "old" strand and a newly synthesized strand.
DNA sequencing Procedure by which the sequence of nucleotides in DNA is determined.
domain (1) A structural and functional region of a protein; (2) A taxonomic category that includes one or more kingdoms.
dominance hierarchy A linear "pecking order" into which animals in a population may organize according to status; regulates aggressive behavior within the population.
dominance, principle of Genetic principle that states that in an F_1 hybrid the allele contributed by one parent (the dominant allele) masks expression of the allele contributed by the other parent (the recessive allele). First noted by Gregor Mendel, the principle of dominance has many exceptions.
dominant allele (al-leel') An allele that is always expressed when it is present, regardless of whether it is homozygous or heterozygous. Compare with recessive allele.
dopamine A neurotransmitter of the biogenic amine group.
dorsal (dor'sl) Toward the uppermost surface or back of an animal. Compare with ventral.
dosage compensation Genetic mechanism by which the expression of X-linked genes is made equivalent in XX females and XY males.
double fertilization A process in the flowering plant life cycle in which there are two fertilizations; one fertilization results in the formation of a zygote that develops into a young plant, while the second results in the formation of endosperm (nutritive tissue).
doubling time The amount of time it takes for a population to double in size, assuming that its current rate of increase does not change.
Down syndrome An inherited condition in which individuals have abnormalities of the face, eyelids, tongue, and other parts of the body, and are physically and mentally retarded; usually results from trisomy of chromosome 21.
drupe (droop) A simple, fleshy fruit in which the inner wall of the fruit is hard and stony, e.g., peach, cherry.
duodenum (doo"o-dee'num) The portion of the small intestine into which the contents of the stomach first enter.
ecdysone (ek'dih-sone) Steroid hormone that induces molting (ecdysis) in insects.
echinoderms (eh-kine'oh-derms) Phylum of spiny-skinned marine deuterostome invertebrates that possess a water vascular system; include sea stars, sea urchins, and sea cucumbers.
ecological niche See niche.
ecological pyramid A graphic representation of the relative energy value at each trophic level. See pyramid of biomass, pyramid of energy, and pyramid of numbers.
ecology (ee-kol'uh-jee) A discipline of biology that studies the interrelations among living things and their environments.
ecosphere (ee'koh-sfeer) The interactions among and between all the Earth's living organisms and the air (atmosphere), land (lithosphere), and water (hydrosphere) that they occupy. Compare with biosphere.
ecosystem (ee'koh-sis-tem) The interacting system that encompasses a community and its nonliving, physical environment. Compare with community.
ecotone The transition zone where two ecosystems or biomes meet and intergrade.
ectoderm (ek'toh-derm) The outer of the three embryonic germ layers; gives rise to the skin and nervous system. Compare with mesoderm and endoderm.

ectotherm An animal whose temperature fluctuates with that of the environment; may use behavioral adaptations to regulate temperature. Sometimes referred to as cold-blooded. Compare with endotherm.

edge effect The ecological phenomenon in which ecotones between adjacent communities often contain a greater number of species or greater population densities of certain species than either adjacent community.

effector A muscle or gland that contracts or secretes in direct response to nerve impulses.

efferent (ef'fur-ent) Leading away from some point of reference. Compare with afferent.

ejaculation (ee-jak"yoo-lay'shun) A sudden expulsion, as in the ejection of semen from the penis.

electrolyte Substance that dissociates into ions when dissolved in water; the resulting solution can conduct an electrical current.

electron A particle with one unit of negative charge and negligible mass, located outside the atomic nucleus. See also neutron and proton.

electron configuration The arrangement of the electrons around the atom. In a Bohr model the electron configuration is depicted as a series of concentric circles.

electron microscope Microscope capable of producing high resolution, highly magnified images through the use of an electron beam (rather than light). Transmission electron microscopes (TEM) produce images of thin sections; scanning electron microscopes (SEM) produce images of surfaces.

electron shell Group of orbitals of electrons with similar energies.

electron transport system A series of chemical reactions during which hydrogens or their electrons are passed along from one acceptor molecule to another (the electron transport chain), with the release of energy.

electronegativity A measure of an atom's attraction for electrons.

electroreceptor A receptor that responds to electrical stimuli.

element A substance that cannot be changed to a simpler substance by a chemical reaction.

elimination Ejection of undigested food from the body.

El Niño-Southern Oscillation (ENSO) (el nee'nyo) A recurring climatic phenomenon that involves a surge of warm water in the Pacific Ocean and unusual weather patterns elsewhere in the world.

embryo (em'bree-oh) (1) A young organism before it emerges from the egg, seed, or body of its mother; (2) Developing human until the end of the second month, after which it is referred to as a fetus; (3) Young sporophyte plant produced following fertilization of an egg and subsequent development of the zygote.

embryo sac The female gametophyte generation in flowering plants.

emigration A type of migration in which individuals leave a population and thus decrease its size. Compare with immigration.

enantiomers Two chemical compounds that are mirror images.

endangered species A species whose numbers are so severely reduced that it is in imminent danger of extinction throughout all or part of its range. Compare with threatened species.

endergonic reaction (end"er-gon'ik) Nonspontaneous reaction; a reaction requiring a net input of free energy. Compare with exergonic reaction.

endocrine gland (en'doh-crin) Gland that secretes hormones directly into the blood or tissue fluid instead of into ducts. Compare with exocrine gland.

endocrine system Body system that helps regulate metabolic activities; consists of ductless glands and tissues that secrete hormones.

endocytosis (en"doh-sy-toh'sis) The active transport of substances into the cell by the formation of invaginated regions of the plasma membrane that pinch off and become cytoplasmic vesicles. Compare with exocytosis.

endoderm (en'doh-derm) The inner germ layer of the early embryo; becomes the lining of the digestive tract and the structures that develop from the digestive tract — liver, lungs, and pancreas. Compare with ectoderm and mesoderm.

endodermis (en"doh-der'mis) The innermost layer of the plant root cortex. Endodermal cells have a waterproof Casparian strip around their radial and transverse walls that ensures that water and minerals can enter the root xylem only by passing through the endodermal cells.

endolymph (en'doh-limf) The fluid of the membranous labyrinth of the ear.

endomembrane system See internal membrane system.

endometrium (en"doh-mee'tree-um) Uterine lining.

endoplasmic reticulum (ER) (en'doh-plaz"mik reh-tik'yoo-lum) Interconnected network of internal membranes in eukaryotic cells enclosing a compartment, the ER lumen. Rough ER has ribosomes attached to the cytosolic surface; smooth ER, a site of lipid biosynthesis, lacks ribosomes.

endorphins (en-dor'finz) Polypeptide transmitters of certain brain and visceral neurons whose action is mimicked by opiate alkaloids. Compare with enkephalins.

endoskeleton (en"doh-skel'eh-ton) Bony and/or cartilaginous supporting structures within the body that provide support. Compare with exoskeleton.

endosperm (en'doh-sperm) The $3n$ nutritive tissue that is formed at some point in the development of all angiosperm seeds.

endospore A resting cell formed by certain bacteria; highly resistant to heat, radiation, and disinfectants.

endosymbiont (en"doe-sim'bee-ont) An organism that lives inside the body of another kind of organism. Endosymbionts may benefit their host (mutualism) or harm their host (parasitism).

endosymbiont theory Theory that certain organelles such as mitochondria and chloroplasts originated as symbiotic prokaryotes that lived inside other, free-living, prokaryotic cells.

endothelium (en-doh-theel'ee-um) The tissue that lines the cavities of the heart, blood, and lymph vessels.

endotherm (en'doh-therm) An animal that uses metabolic energy to maintain a constant body temperature despite variations in environmental temperature; e.g., birds and mammals. Compare with ectotherm.

endotoxin A poisonous substance in the cell walls of gram-negative bacteria. Compare with exotoxin.

energy The capacity to do work; can be expressed in kilojoules or kilocalories.

energy flow The passage of energy in a one-way direction through an ecosystem.

enhancers Regulatory elements that can be located long distances away from the actual coding regions of a gene.

enkephalins (en-kef'ah-linz) Polypeptide transmitter substances employed by certain brain neurons that seem to function in pain perception and whose action is mimicked by opiate alkaloids. Compare with endorphins.

enterocoely Process by which the coelom forms as a cavity within mesoderm produced by outpocketings of the primitive gut (archenteron); characteristic of many deuterostomes. Compare with schizocoely.

enthalpy The total potential energy of a system; sometimes referred to as the heat content of the system.

entropy (en'trop-ee) Disorderliness; a quantitative measure of the amount of the random, disordered energy that is unavailable to do work.

environmental resistance Unfavorable environmental conditions, such as crowding, that prevent organisms from reproducing indefinitely at their biotic potential.

enzyme (en'zime) An organic catalyst (usually a protein) that accelerates a specific chemical reaction by lowering the activation energy required for that reaction.

eosinophil (ee-oh-sin'oh-fil) A type of white blood cell whose cytoplasmic granules absorb acidic stains.

epidermis (ep-ih-dur'mis) (1) An outer layer of cells covering the body of plants and animals; functions primarily for protection; (2) The outer layer of vertebrate skin.

epididymis (ep-ih-did'ih-mis) (pl. *epididymides*) A coiled tube that receives sperm from the testis and conveys it to the vas deferens.

epiglottis A thin, flexible structure that guards the entrance to the larynx, preventing food from entering the airway during swallowing.

epinephrine (ep-ih-nef'rin) The chief hormone produced by the adrenal medulla; stimulates the sympathetic nervous system.

epistasis (ep'ih-sta-sis) Condition in which certain alleles of one locus can alter the expression of alleles of a different locus.

epithelial tissue (ep-ih-theel'ee-al) The type of animal tissue that covers body surfaces, lines body cavities, and forms glands; also called epithelium.

epoch The smallest unit of geological time; a subdivision of a period.

era In geology, one of the main divisions of geological time. Eras are subdivided into periods.

erythroblastosis fetalis Serious condition in which Rh^+ red blood cells of a fetus are destroyed by maternal anti-Rh antibodies.

erythrocyte (er-eeth'roh-site) Vertebrate red blood cell.

essential amino acids Amino acids that cannot be synthesized by the organism, and so must be ingested in food.

ester linkage Covalent linkage formed by the reaction of a carboxyl group and a hydroxyl group, with the removal of the equivalent of a water molecule; the linkage includes an oxygen atom bonded to a carbonyl group.

estrogens (es'troh-jens) Female sex hormones produced by the ovary; promote the development and maintenance of female reproductive structures and of secondary sexual characteristics.

estuary (es'choo-wear-ee) A coastal body of water that connects to an ocean, in which fresh water from the land mixes with salt water from the oceans.

ethology (ee-thol'oh-jee) The study of animal behavior under natural conditions from the point of view of adaptation.

ethyl alcohol A 2-carbon alcohol.

ethylene (eth'ih-leen) A plant hormone involved in various aspects of plant growth and development, such as leaf abscission and fruit ripening.

eubacteria (yoo"bak-teer'ee-ah) Prokaryotes other than the archaebacteria.

euchromatin (yoo-croh'mah-tin) Loosely coiled chromatin that is generally capable of transcription. Compare with heterochromatin.

euglenoids (yoo-glee'noids) A group of mostly freshwater, flagellated algae that move by means of an anterior flagellum and are usually photosynthetic.

eukaryote (yoo"kar'ee-ote) Organism whose cells possess nuclei and other membrane-bounded organelles. Compare with prokaryote.

euphotic region (yoo-fote'ik) The surface layer of the ocean (the top 100 meters or so), where light penetrates and photosynthesis occurs.

eustachian tube (yoo-stay'shee-un) The auditory tube passing between the middle ear cavity and the pharynx in vertebrates; permits the equalization of pressure on the tympanic membrane.

evolution Any cumulative genetic change in a population of organisms from generation to generation. Evolution leads to differences in populations and explains the origin of all of the organisms that exist today or have ever existed.

excitatory postsynaptic potential (EPSP) A change in membrane potential that brings a neuron closer to the firing level. Compare with inhibitory postsynaptic potential (IPSP).

excretion (ek-skree'shun) The discharge from the body of a waste product of metabolism (not to be confused with the elimination of undigested food materials).

excretory system The body system in animals that functions to discharge metabolic wastes.

exergonic reaction (ex"er-gon'ik) A reaction characterized by a release of free energy. Also called spontaneous reaction. Compare with endergonic reaction.

exocrine gland (ex'oh-crin) Gland that excretes its products through a duct that opens onto a free surface such as the skin (e.g., sweat glands). Compare with endocrine gland.

exocytosis (ex"oh-sy-toh'sis) The active transport of materials out of the cell by fusion of cytoplasmic vesicles with the plasma membrane. Compare with endocytosis.

exon (1) A protein-coding region of a eukaryotic gene; (2) The RNA transcribed from such a region. Compare with intron.

exoskeleton (ex"oh-skel'eh-ton) An external skeleton, such as the shell of mollusks or outer covering of arthropods; provides protection and sites of attachment for muscles. Compare with endoskeleton.

exotoxin A poisonous substance released by certain bacteria. Compare with endotoxin.

expiration Breathing out; expelling air from the lungs to the outside environment. Compare with inspiration.

exponential growth Population growth that occurs at a constant rate of increase over a period of time. When the increase in number versus time is plotted on a graph, exponential growth produces a characteristic J-shaped curve.

ex situ conservation Conservation efforts that involve conserving biological diversity in human-controlled settings. Compare with in situ conservation.

exteroceptor (ex′tur-oh-sep″tor) One of the sense organs that receives sensory stimuli from the outside world, such as the eyes or touch receptors. Compare with interoceptor.

extinction The elimination of a species; occurs when the last individual member of a species dies.

F_1 generation (first filial generation) The first generation of hybrid offspring resulting from a cross between parents from two different true-breeding lines.

F_2 generation (second filial generation) The offspring of the F_1 generation.

facilitated diffusion The passive transport of ions or molecules by a specific carrier protein in a membrane. As in simple diffusion, net transport is down a concentration gradient, and no additional energy has to be supplied.

facilitation A process in which a neuron is brought close to threshold level by stimulation from various presynaptic neurons.

facultative anaerobe Organism capable of carrying out aerobic respiration, but able to switch to fermentation when oxygen is unavailable; e.g. yeast. Compare with obligate anaerobe.

$FAD/FADH_2$ Oxidized and reduced forms, respectively, of flavin adenine dinucleotide; coenzyme that transfers electrons (as hydrogen) in cellular metabolism, including cellular respiration.

fall turnover A mixing of water in temperate lakes caused by falling temperatures in autumn. Compare with spring turnover.

Fallopian tube See oviduct.

family A taxonomic category made up of related genera.

fatty acid An organic acid containing a long hydrocarbon chain, with no double bonds (saturated fatty acid), one double bond (monounsaturated fatty acid), or two or more double bonds (polyunsaturated fatty acid); fatty acids are components of neutral fats and phospholipids.

feedback control System in which the accumulation of the product of a reaction affects its rate of production.

fermentation Anaerobic process by which ATP is produced by a series of redox reactions in which organic compounds serve as electron donors and as electron acceptors.

fern One of a phylum of seedless vascular plants that reproduce by spores produced in sporangia usually borne on leaves in sori; ferns have an alternation of generations between the dominant sporophyte (a fern plant) and the gametophyte (a prothallus).

fertilization Fusion of n gametes; results in the formation of a $2n$ zygote.

fetus The unborn human offspring from the third month of pregnancy to birth.

fiber (1) In plants, a type of sclerenchyma. Fibers are long, tapered cells with thick walls. Compare with sclereid; (2) In animals, an elongated cell such as a muscle or nerve cell.

fibrin An insoluble protein formed from the plasma protein fibrinogen during blood clotting.

fibrous root system A root system in plants that has many roots that are similar in length and thickness. Compare with taproot system.

filament In flowering plants, the thin stalk of a stamen; the filament bears an anther at its tip.

first law of thermodynamics The law of conservation of energy, which states that the total energy of any closed system (any object plus its surroundings, i.e., the universe) remains constant.

fitness A measure of the ability of an organism, owing to its genotype, to compete successfully and make a genetic contribution to subsequent generations.

fixed action pattern (FAP) An innate behavior triggered by a sign stimulus.

flagellate (flaj″eh-late″) A unicellular, nonphotosynthetic protozoon that possesses one or more long, whip-like flagella.

flagellum (flah-jel′um) (pl. *flagella*) Long, whip-like movable structure extending from the cell and used in locomotion. (1) Eukaryote flagella are composed of two central single microtubules surrounded by a cylinder of nine double microtubules (9 + 2 structure), all covered by a plasma membrane. (2) Prokaryote flagella are filaments rotated by special structures located in the plasma membrane and cell wall.

flame cells Collecting cells that have cilia; part of the osmoregulatory system of flatworms.

flavin adenine dinucleotide See $FAD/FADH_2$.

florigen (flor′uh-jen) A hypothetical plant hormone that promotes flowering.

flowering plants See angiosperms.

flowing-water ecosystem A river or stream ecosystem.

fluid-mosaic model The modern picture of the plasma membrane and other cellular membranes in which protein molecules float in a phospholipid bilayer.

fluorescence Emission of light of a longer wavelength (lower energy) than the light originally absorbed.

follicle (fol′i-kl) (1) A simple, dry, dehiscent fruit that splits open at maturity along one seam to liberate the seeds; (2) A small sac of cells in the mammalian ovary that contains a maturing egg; (3) The pocket in the skin from which a hair grows.

follicle-stimulating hormone (FSH) A gonadotropic hormone secreted by the anterior lobe of the pituitary gland; stimulates follicle development in the ovaries of females and sperm production in the testes of males.

food chain The series of organisms through which energy flows in an ecosystem. Each organism in the series eats or decomposes the preceding organism in the chain. See food web.

food web A complex interconnection of all of the food chains in an ecosystem.

foramen magnum The opening in the vertebrate skull through which the spinal cord passes.

foraminiferan (for″am-in-if′er-an) A marine protozoon that produces a shell, or test, that encloses an amoeboid body.

forest decline A gradual deterioration (and often death) of many trees in a forest. The cause of forest decline is unclear at present, and it may involve a combination of factors.

fossil Parts or traces of an ancient organism usually preserved in rock.

fossil fuel Combustible deposits in Earth's crust; composed of the remnants of prehistoric organisms that existed millions of years ago. Examples include oil, natural gas, and coal.

founder effect Genetic drift that results from a small population colonizing a new area.

fovea (foe′vee-ah) The area of sharpest vision in the retina; cone cells are concentrated here.

frameshift mutation Mutation that results when nucleotides are inserted into or deleted from the DNA, thereby altering the reading frame so that all of the codons downstream from the mutation are changed.

free energy The maximum amount of energy in a system that is available to do work (without a change in temperature).
frequency-dependent selection Selection in which the relative fitness of different genotypes is related to how frequently they occur in the population.
freshwater wetlands Land that is transitional between freshwater and terrestrial ecosystems and is covered with water for at least part of the year; marshes and swamps are examples.
fringing reef A coral reef that is attached directly to the shore.
frontal lobes In mammals, the anterior part of the cerebrum.
fruit In flowering plants, a mature, ripened ovary. Fruits contain seeds and usually provide seed protection and dispersal.
fruiting body A multicellular structure that contains the sexual spores of certain fungi; refers to the ascocarp of an ascomycete and the basidiocarp of a basidiomycete.
FSH See follicle-stimulating hormone.
fucoxanthin (few″koh-zan′thin) The brown carotenoid pigment found in brown algae, diatoms, and dinoflagellates.
functional group A group of atoms that confers distinctive properties on an organic molecule (or region of a molecule) to which it is attached; examples include hydroxyl, carbonyl, carboxyl, amino, phosphate, and sulfhydryl groups.
fundamental niche The potential ecological niche that an organism could have if there were no competition from other species. Compare with realized niche.
fungus (pl. *fungi*) A heterotrophic eukaryote with chitinous cell walls and a body usually in the form of a mycelium of branched, thread-like hyphae, or unicellular. Most fungi are decomposers; some are parasitic.
G protein One of a group of proteins that bind GTP and are involved in the transfer of signals across the plasma membrane.
G_1 phase The first gap phase within the interphase stage of the cell cycle; G_1 occurs before DNA synthesis (S phase) begins.
G_2 phase Second gap phase within the interphase stage of the cell cycle; G_2 occurs after DNA synthesis (S phase) and before mitosis.
gallbladder A small sac that stores bile.
gametangium (gam″uh-tan′gee-um) Special multicellular or unicellular structure of plants, protists, and fungi in which gametes are formed.
gamete (gam′eet) A sex cell; in plants and animals, an egg or sperm. In sexual reproduction, the union of gametes results in the formation of a zygote. The chromosome number of a gamete is designated n. Species that are not polyploid have haploid gametes and diploid zygotes.
gametic isolation (gam-ee′tik) A reproductive isolating mechanism (prezygotic) in which sexual reproduction between two closely related species cannot occur because of chemical differences in the gametes.
gametophyte generation (gam-ee′toh-fite) The n, gamete-producing stage in the life cycle of a plant. Compare with sporophyte generation.
ganglion (gang′glee-on) (pl. *ganglia*) A mass of cell bodies of neurons located outside the central nervous system.
gap junction Structure consisting of specialized regions of the plasma membrane of two adjacent cells; contains numerous pores that allow passage of certain small molecules and ions between them.
gastrin (gas′trin) A hormone released by the stomach mucosa; stimulates the gastric glands to secrete pepsinogen.
gastrovascular cavity A central digestive cavity with a single opening that functions as both mouth and anus; characteristic of cnidarians and flatworms.
gastrula (gas′troo-lah) Early stage of embryonic development during which the three germ layers form.
gel electrophoresis Procedure by which proteins and nucleic acids can be separated on the basis of size and charge as they migrate through a gel in an electrical field.
gene A segment of DNA that serves as a unit of hereditary information. Includes a transcribable DNA sequence (plus associated sequences regulating its transcription) that yields a protein or RNA product with a specific function. Most eukaryotic genes are in chromosomes.
gene amplification Process by which multiple copies of a gene are produced by selective replication, thus allowing for increased synthesis of the gene product.
gene flow The movement of alleles between local populations due to the migration of individuals. Gene flow can have significant evolutionary consequences.
gene pool All the alleles of all the genes present in a freely interbreeding population.
gene therapy Any one of a variety of methods designed to correct a disease or alleviate its symptoms through the introduction of genes into the affected person's cells.
generation time (T) The time required for the completion of one cell cycle.
genetic bottleneck See bottleneck.
genetic drift A random change in allele frequency in a small breeding population.
genetic engineering Manipulation of genes, often through recombinant DNA technology.
genetic polymorphism (pol″ee-mor′fizm) The presence in a population of two or more alleles for a given gene.
genetic probe A single stranded nucleic acid (either DNA or RNA) that can be used to identify a complementary sequence by hydrogen-bonding to it.
genome (jee′nome) Originally, all the genetic material in a cell or organism. The term is used more than one way, depending on context: for example, an organism's haploid genome is all the DNA contained in one haploid set of its chromosomes, and its mitochondrial genome is all the DNA in a mitochondrion.
genomic DNA library A collection of recombinant plasmids in which all the DNA in the genome is represented.
genotype (jeen′oh-type, jen′oh-type) The genetic makeup of an individual. Compare with phenotype.
genus (jee′nus) A taxonomic category made up of related species.
geometric isomer One of two chemical compounds that differ in the arrangement of their atoms with respect to a double bond. A molecule with two large groups on the same side of a double bond is the *cis* isomer; a *trans* isomer has two large groups on opposite sides of the bond.
germ layers Primitive embryonic tissue layers; endoderm, mesoderm, or ectoderm.
germ line In animals, the line of cells that will ultimately undergo meiosis to form gametes.
gibberellin (jib″ur-el′lin) A plant hormone involved in many aspects of plant growth and development, such as stem elongation, flowering, and seed germination.

gills (1) The respiratory organs of aquatic animals, usually thin-walled projections from the body surface or from some part of the digestive tract; (2) The spore-bearing plate-like structures under the caps of mushrooms.

ginkgo (ging'ko) Member of an ancient gymnosperm group that consists of a single living representative *(Ginkgo biloba),* which is a hardy, deciduous tree with broad, fan-shaped leaves and naked, fleshy seeds (on female trees).

gland Body structure specialized for secretion.

glial cells In nervous tissue, cells that support and nourish neurons.

global warming The long-term increase in Earth's average temperature.

globulin (glob'yoo-lin) One of a class of proteins in blood plasma, some of which (gamma globulins) function as antibodies.

glomerulus (glom-air'yoo-lus) The cluster of capillaries at the proximal end of a nephron; the glomerulus is surrounded by Bowman's capsule.

glucagon (gloo'kah-gahn) A pancreatic hormone that stimulates glycogen breakdown, thereby increasing the concentration of glucose in the blood. Compare with insulin.

glucose A very common hexose aldehyde sugar.

glyceraldehyde-3-phosphate (PGAL) Phosphorylated 3-carbon compound that is an important intermediate in glycolysis and in the Calvin cycle.

glycerol A three-carbon alcohol, with a hydroxyl group on each carbon; a component of neutral fats and phospholipids.

glycocalyx (gly"koh-kay'lix) A coating on the outside of an animal cell, formed by the polysaccharide portions of glycoproteins and glycolipids associated with the plasma membrane.

glycogen (gly'koh-jen) The principal storage polysaccharide in animal cells; formed from glucose and stored primarily in the liver and (to a lesser extent) in muscle cells.

glycolipid A lipid with covalently attached carbohydrates.

glycolysis (gly-kol'ih-sis) The first stage of cellular respiration, literally the "splitting of sugar." The metabolic conversion of glucose into pyruvate, accompanied by the production of ATP.

glycoprotein (gly'koh-proh-teen) A protein with covalently attached carbohydrates.

glycosidic linkage Covalent linkage joining two sugars; includes an oxygen atom bonded to a carbon of each sugar.

glyoxysomes (gly-ox'ih-somz) Membrane-bounded structures in cells of certain plant seeds; contain a large array of enzymes that convert stored fat to sugar.

gnetophyte (nee'toe-fite) One of a small phylum of unusual gymnosperms that possess many features similar to flowering plants.

GnRH See gonadotropin-releasing hormone.

goblet cells Unicellular glands that secrete mucus.

goiter (goy'ter) An enlargement of the thyroid gland.

Golgi complex (goal'jee) Organelle composed of stacks of flattened membranous sacs. Mainly responsible for modifying, packaging, and sorting proteins that will be secreted or targeted to other organelles of the internal membrane system or to the plasma membrane. Also called Golgi body or Golgi apparatus.

gonad (goh'nad) A gamete-producing gland; an ovary or a testis.

gonadotropin-releasing hormone (GnRH) A hormone secreted by the hypothalamus that stimulates the anterior pituitary to secrete FSH or LH.

gonadotropins (go-nad"oh-troh'pins) Hormones produced by the anterior pituitary gland and the embryo that stimulate the function of the testes and ovaries; include follicle-stimulating hormone, luteinizing hormone, and human chorionic gonadotropin.

gradualism The idea that evolutionary change of a species is due to a slow, steady accumulation of changes over time. Compare with punctuated equilibrium.

graft rejection An immune response directed against a transplanted tissue or organ.

grain A simple, dry one-seeded fruit in which the fruit wall is fused to the seed coat, making it impossible to separate the fruit from the seed, e.g., corn and wheat kernels.

granum (pl. *grana*) A stack of thylakoids within a chloroplast.

gravitropism (grav"ih-troh'pizm) Growth of a plant in response to gravity.

gray crescent The grayish area of cytoplasm that marks the region where gastrulation begins in an amphibian embryo.

gray matter Nerve tissue in the brain and spinal cord that contains cell bodies, dendrites, and unmyelinated axons. Compare with white matter.

green alga A member of a diverse phylum of algae that contain the same pigments as plants (chlorophylls *a* and *b* and carotenoids).

greenhouse effect The global warming of Earth's atmosphere produced by the buildup of carbon dioxide and other greenhouse gases, which trap the sun's radiation in much the same way that glass does in a greenhouse.

greenhouse gases Trace gases in the atmosphere that allow the sun's energy to penetrate to the Earth's surface but do not allow as much of it to escape as heat.

gross primary productivity The rate at which energy accumulates in an ecosystem (as biomass) during photosynthesis. Compare with net primary productivity.

ground state The lowest energy state of an atom.

ground tissue Tissue other than epidermis and vascular tissues in monocot stems; equivalent to cortex and pith in herbaceous dicot stems.

ground tissue system All tissues in the plant body other than the dermal tissue system and vascular tissue system; consists of parenchyma, collenchyma, and sclerenchyma.

growth factor An extracellular protein that signals certain cells to grow and divide.

growth hormone (GH) A hormone secreted by the anterior lobe of the pituitary gland; stimulates growth of body tissues. Also called somatotropin.

growth rate The rate of change of a population; calculated by subtracting the death rate from the birth rate (in populations with little or no migration).

guanine (gwan'een) A nitrogenous purine base that is a component of nucleic acids.

guard cell One of a pair of epidermal cells that adjust their shape to form a stomatal pore for gas exchange.

guttation (gut-tay'shun) The appearance of water droplets on leaves, forced out through leaf pores by root pressure.

gymnosperm (jim'noh-sperm) Any of a group of seed plants in which the seeds are not enclosed in an ovary; gymnosperms frequently bear their seeds in cones. Includes four phyla: conifers, cycads, ginkgoes, and gnetophytes.

habitat The natural environment or place where an organism, population, or species lives.

habituation (hab-it″yoo-ay′shun) The process by which organisms become accustomed to a stimulus and no longer respond to it.

half-life The period of time required for a radioisotope (i.e., radioactive isotope) to change into a different material.

haploid (hap′loyd) The condition of having one set of chromosomes per nucleus. Compare with diploid and polyploid.

Hardy-Weinberg principle The theorem that allele frequencies do not change from generation to generation in a large population in the absence of microevolutionary processes (mutation, genetic drift, gene flow, natural selection).

Hatch-Slack pathway See C_4 plant.

haustorium (hah-stor′e-um) (pl. *haustoria*) A specialized hypha of a parasitic fungus that penetrates a host cell to absorb food and other materials.

Haversian canals (ha-vur′zee-un) Channels extending through the matrix of bone; contain blood vessels and nerves.

hCG See human chorionic gonadotropin.

heat The total amount of kinetic energy in a sample of a substance.

heat energy Energy that flows between two objects that differ in temperature; unit of measurement is the calorie or kilocalorie.

heat of vaporization The amount of heat energy that must be supplied to change one gram of a substance from the liquid phase to the vapor phase.

helicases Enzymes that unwind the two strands of a DNA double helix.

helper T cell T lymphocyte that facilitates the ability of B lymphocytes to form an antibody-producing clone in response to an antigen.

hemichordates A phylum of sedentary, wormlike deuterostomes.

hemizygous (hem″ih-zy′gus) Possessing only one allele for a particular locus; a human male is hemizygous for all X-linked genes. Compare with homozygous and heterozygous.

hemocoel Blood cavity characteristic of animals with an open circulatory system.

hemoglobin (hee′moh-gloh″bin) The red, iron-containing protein pigment of erythrocytes that transports oxygen and carbon dioxide and aids in regulation of pH.

hemophilia (hee″-moh-feel′ee-ah) A hereditary disease in which blood does not clot properly; the form known as hemophila A has an X-linked recessive inheritance pattern.

hepatic (heh-pat′ik) Pertaining to the liver.

hepatic portal system The portion of the circulatory system that carries blood from the intestine through the liver.

herbivore (erb′i-vore) An animal that feeds on plants or algae. Also called primary consumer.

hermaphrodite (her-maf′roh-dite) An organism that possesses both male and female sex organs.

heterochromatin (het′ur-oh-kroh′mah-tin) Highly coiled and compacted chromatin in an inactive state. Compare with euchromatin.

heterocyst (het′ur-oh-sist″) An oxygen-excluding cell of cyanobacteria that is the site of nitrogen fixation.

heterogametic Term describing an individual that produces two classes of gametes with respect to their sex chromosome constitutions. Human males (XY) are heterogametic, producing X and Y sperm. Compare with homogametic.

heterospory (het″ur-os′pur-ee) Production of two types of n spores, microspores (male) and megaspores (female). Compare with homospory.

heterothallic (het″ur-oh-thal′ik) Pertaining to certain algae and fungi that have two mating types; only by combining a plus strain and a minus strain can sexual reproduction occur. Compare with homothallic.

heterotroph (het′ur-oh-trof) Organism that cannot synthesize its own food from inorganic raw materials and therefore must live either at the expense of other organisms or upon decaying matter. Also called consumer. Compare with autotroph.

heterozygote advantage A phenomenon in which the heterozygous condition confers some special advantage on an individual that either homozygous condition does not (i.e., *Aa* has a higher degree of fitness than does *AA* or *aa*).

heterozygous (het-ur′oh-zye′gus) Possessing a pair of unlike alleles for a particular locus. Compare with homozygous.

hexose A monosaccharide containing six carbon atoms.

highly developed country See developed country.

histamine (his′tah-meen) Substance released from mast cells that is involved in allergic and inflammatory reactions.

histones (his′tones) Small, positively charged (basic) proteins in the cell nucleus that bind to the negatively charged DNA.

holdfast The basal structure for attachment to solid surfaces found in multicellular algae.

homeobox A DNA sequence of approximately 180 base pairs found in many homeotic genes and some other genes that are important in development; genes containing homeobox sequences code for certain transcription factors.

home range A geographical area that an animal seldom or never leaves.

homeostasis (home″ee-oh-stay′sis) The balanced internal environment of the body; the automatic tendency of an organism to maintain such a steady state.

homeotic gene (home″ee-o′tik) A gene that controls the formation of specific structures during development. Such genes were originally identified through insect mutants in which one body part is substituted for another.

hominid (hah′min-id) Any of a group of extinct and living humans.

hominoid (hah′min-oid) The apes and hominids.

Homo erectus An extinct hominid that made stone tools, wore clothing, built fires, and lived in caves.

homogametic Term referring to an individual that produces gametes with identical sex chromosome constitutions. Human females (XX) are homogametic, producing all X eggs. Compare with heterogametic.

Homo habilis An extinct hominid that lived in Africa and fashioned crude stone tools.

homologous chromosomes (hom-ol′ah-gus) Chromosomes that are similar in morphology and genetic constitution. In humans there are 23 pairs of homologous chromosomes; one member of each pair is inherited from the mother and the other from the father.

homologous features Features similar in underlying structure in different species as a result of a common evolutionary origin. Compare with analogous features.

Homo sapiens The modern human species.

homospory (hoh″mos′pur-ee) Production of one type of n spore that gives rise to a bisexual gametophyte. Compare with heterospory.

homothallic (ho″meh-thal′ik) Term referring to certain algae and fungi that are self-fertile. Compare with heterothallic.

homozygous (hoh″moh-zy′gous) Possessing a pair of identical alleles for a particular locus. Compare with heterozygous.

hormone An organic chemical messenger in multicellular organisms produced in one part of the body and transported to another part where it affects some aspect of metabolism.

hornwort A type of spore-producing, nonvascular, thallose plant with a life cycle similar to mosses.

horsetail A phylum of seedless vascular plants with a life cycle similar to ferns.

host mothering The introduction of an embryo from one species into the uterus of another species, where it implants and develops. The host mother subsequently gives birth and may raise the offspring as her own.

human chorionic gonadotropin (hCG) A hormone secreted by cells surrounding the early embryo; signals the mother's corpus luteum to continue to function.

human immunodeficiency virus (HIV) The retrovirus that causes AIDS (acquired immune deficiency syndrome).

human leukocyte antigen (HLA) See major histocompatibility complex.

humus (hew′mus) Organic matter in various stages of decomposition in the soil; gives soil a dark brown or black color.

Huntington's disease A genetic disease that has an autosomal dominant inheritance pattern and causes mental and physical deterioration.

hybrid Offspring of two genetically dissimilar parents.

hybrid breakdown A reproductive isolating mechanism (postzygotic) in which, although an interspecific hybrid is fertile and produces a second (F_2) generation, the F_2 has defects that prevent it from successfully reproducing.

hybrid inviability A reproductive isolating mechanism (postzygotic) in which the embryonic development of an interspecific hybrid is aborted.

hybrid sterility A reproductive isolating mechanism (postzygotic) in which an interspecific hybrid cannot reproduce successfully.

hybrid vigor Genetic superiority of an F_1 hybrid over either parent, due to the presence of heterozygosity for a number of different loci.

hybridization (1) Interbreeding between members of two different taxa; (2) Interbreeding between genetically dissimilar parents; (3) In molecular biology, complementary base pairing between nucleic acid (DNA or RNA) strands from different sources.

hybridized orbitals Orbitals that have become rearranged and have taken on new shapes as a consequence of covalent bond formation.

hydration Process of association of a substance with the partial positive and/or negative charges of water molecules.

hydrocarbon An organic compound composed solely of hydrogen and carbon atoms.

hydrogen bond A weak attractive force existing between a hydrogen atom with a partial positive charge and an electronegative atom (usually oxygen or nitrogen) with a partial negative charge.

hydrologic cycle The water cycle, which includes evaporation, precipitation, and flow to the seas. The hydrologic cycle supplies terrestrial organisms with a continual supply of fresh water.

hydrolysis Reaction in which a covalent bond between two subunits is broken through the addition of the equivalent of a water molecule; a hydrogen atom is added to one subunit and a hydroxyl group to the other. Compare with condensation synthesis.

hydrophilic Attracted to water. Compare with hydrophobic.

hydrophobic Repelled by water. Compare with hydrophilic.

hydroponics (hy″dra-paun′iks) Growing plants in an aerated solution of dissolved inorganic minerals (that is, without soil).

hydrostatic skeleton A type of skeleton found in some invertebrates in which contracting muscles push against a tube of fluid.

hydrothermal vent A crack in the ocean floor through which mineral-laden hot water seeps.

hydroxide ion Anion (negatively charged particle) consisting of oxygen and hydrogen; usually written OH^-.

hydroxyl group (hy-drok′sil) Polar functional group; abbreviated $-OH$.

hyperosmotic See hypertonic.

hypertonic Term referring to a solution having an osmotic pressure (or solute concentration) greater than that of the solution with which it is compared. Also called hyperosmotic. Compare with hypotonic and isotonic.

hypha (hy′fah) (pl. *hyphae*) One of the thread-like filaments composing the mycelium of a water mold or fungus.

hypocotyl (hy′poh-kah″tl) The part of the axis of a plant embryo or seedling below the point of attachment of the cotyledons.

hypoosmotic See hypotonic.

hypothalamus (hy-poh-thal′uh-mus) Part of the mammalian brain that regulates the pituitary gland, the autonomic system, emotional responses, body temperature, water balance, and appetite; located below the thalamus.

hypothesis An educated guess testable by observation and experimentation. Compare with theory.

hypotonic Term referring to a solution having an osmotic pressure (or solute concentration) less than that of the solution with which it is compared. Also called hypoosmotic. Compare with hypertonic and isotonic.

imaginal discs Paired structures in an insect larva that develop into specific adult structures during complete metamorphosis.

imago The adult form of an insect.

imbibition (im″bi-bish′en) The absorption of water by a seed prior to germination.

immigration A type of migration in which individuals enter a population and thus increase the population size. Compare with emigration.

immune response The production of antibodies or T cells in response to foreign antigens.

immunoglobulin (im-yoon″oh-glob′yoo-lin) See antibody.

imperfect flower A flower that lacks either stamens or carpels. Compare with perfect flower.

imperfect fungi See deuteromycetes.

implantation The embedding of a developing embryo in the wall (endometrium) of the uterus.

imprinting A form of learning by which a young bird or mammal forms a strong social attachment to an individual (usually a parent) or object within a few hours after hatching or birth.

inbreeding Mating of genetically similar individuals. Homozygosity increases with each successive generation of inbreeding.

incomplete dominance Condition in which neither member of a pair of contrasting alleles is completely expressed when the other is present.

incomplete flower A flower that lacks one or more of the four parts: sepals, petals, stamens, and/or carpels. Compare with complete flower.

independent assortment, principle of Genetic principle, first noted by Gregor Mendel, that states that the alleles of unlinked loci are randomly distributed to gametes.

indeterminate growth Unrestricted growth, as for example, in stems and roots. Compare with determinate growth.

index fossils Fossils restricted to a narrow unit of time and found in the same sedimentary layers in different geographical areas.

induced-fit model Hypothesis that the active site of certain enzymes becomes conformed to the shape of the substrate molecule.

inducer molecule Substance that binds to a repressor protein, converting it to its inactive form, which is unable to prevent transcription.

induction Phenomenon in embryology in which the differentiation of a cell or group of cells is influenced by interactions with neighboring cells.

inductive reasoning Reasoning that uses specific examples to draw a general conclusion or discover a general principle. Compare with deductive reasoning.

infant mortality rate The number of infant deaths per 1000 live births. (An infant is a child in its first two years of life.)

inflammatory response The response of body tissues to injury or infection, characterized clinically by heat, swelling, redness, and pain, and physiologically by increased dilation of blood vessels and increased phagocytosis.

ingestion Process of taking food (or other material) into the body.

inhibitory postsynaptic potential (IPSP) A change in membrane potential that brings a neuron farther from the firing level. Compare with excitatory postsynaptic potential (EPSP).

initiation codon The codon AUG, which serves as the signal to begin translation of messenger RNA. Also called a start codon.

innate behavior Behavior that is inherited and typical of the species.

inner cell mass The cluster of cells in the early mammalian embryo that gives rise to the embryo proper.

inorganic compound A simple substance that does not contain a carbon backbone.

insight learning Complex learning process in which an animal adapts past experience to solve a new problem that may involve different stimuli.

in situ conservation Conservation efforts that concentrate on preserving biological diversity in the wild. Compare with ex situ conservation.

inspiration Breathing in; drawing air into the lungs. Compare with expiration.

instinct A genetically determined pattern of behavior or responses that is not based on the individual's previous experience but appears to be innate behavior.

insulin (in'suh-lin) A hormone secreted by the pancreas that lowers blood glucose concentration. Compare with glucagon.

integral membrane protein A protein that is tightly associated with the lipid bilayer of a biological membrane; a transmembrane integral protein spans the bilayer. Compare with peripheral membrane protein.

integration Sorting and interpreting of neural impulses.

integumentary system (in-teg"yoo-men'tar-ee) The body's covering, including the skin and its nails, glands, hair, and other associated structures.

intercellular substance In connective tissues, the combination of matrix and fibers in which the cells are embedded.

interferon (in"tur-feer'on) A protein (cytokine) produced by animal cells when challenged by a virus. Important in immune responses, it prevents viral reproduction and enables cells to resist a variety of viruses.

interkinesis The stage between meiosis I and meiosis II. Interkinesis is usually brief; the chromosomes may decondense, reverting at least partially to an interphase-like state, but DNA synthesis and chromosome duplication do not occur.

interleukins A diverse group of regulatory proteins (cytokines) produced mainly by macrophages and lymphocytes.

intermediate filaments Cytoplasmic fibers that are part of the cytoskeletal network and are intermediate in size between microtubules and microfilaments.

internal membrane system Group of membranous structures that interact extensively through direct connections or by means of vesicles; includes the endoplasmic reticulum, outer membrane of the nuclear envelope, Golgi complex, lysosomes, and the plasma membrane. Also called the endomembrane system.

interneuron (in"tur-noor'on) A nerve cell that carries impulses from one nerve cell to another and is not directly associated with either an effector or a sensory receptor. Also known as an association neuron.

interoceptor (in'tur-oh-sep"tor) A sense organ within a body organ that transmits information regarding chemical composition, pH, osmotic pressure, or temperature. Compare with exteroceptor.

interphase Stage of the cell cycle between successive mitotic divisions; its subdivisions are the G_1 (first gap), S (DNA synthesis), and G_2 (second gap) phases.

interspecific competition The interaction between members of different species that vie for the same resource in an ecosystem (such as food, living space, or other resources). Compare with intraspecific competition.

interstitial cells (of testis) The cells between the seminiferous tubules that secrete testosterone.

interstitial fluid The fluid that bathes the tissues of the body. Also called tissue fluid.

intertidal zone The marine shoreline area between the high tide mark and the low tide mark.

intraspecific competition The interaction between members of the same species that vie for the same resource in an ecosystem (such as food, living space, or other resources). Compare with interspecific competition.

intron A non-protein-coding region of a eukaryotic gene and also of the pre-mRNA transcribed from such a region. Introns do not appear in mRNA. Compare with exon.

invertebrate An animal without a backbone (vertebral column); invertebrates account for about 95% of animal species.

in vitro Occurring outside a living organism (literally "in glass").

in vivo Occurring in a living organism.

ion An atom or group of atoms bearing one or more units of electrical charge, either positive (cation) or negative (anion).
ionic bond Chemical attraction between a cation and an anion.
ionic compound Substance consisting of cations and anions, which are attracted by their opposite charges; ionic compounds do not consist of molecules.
iris The pigmented portion of the vertebrate eye.
irreversible inhibitor A substance that permanently inactivates an enzyme. Compare with reversible inhibitor.
islets of Langerhans (eye'lets of Lahng'er-hanz) The endocrine portion of the pancreas that secretes glucagon and insulin, hormones that regulate the concentration of glucose in the blood.
isogamy (eye-sog'uh-mee) Sexual reproduction involving motile gametes of similar form and size. Compare with anisogamy and oogamy.
isogenic Term describing a strain of organisms in which all individuals are genetically identical and homozygous at all loci.
isomer (eye'soh-mer) One of two or more chemical compounds having the same chemical formula, but different structural formulas; examples include structural and geometric isomers and enantiomers.
isoprene units Five-carbon hydrocarbon monomers that make up certain lipids such as carotenoids and steroids.
isosmotic See isotonic.
isotonic (eye'soh-ton'ik) Term applied to solutions that have identical concentrations of solute molecules and hence the same osmotic pressure. Also called isosmotic. Compare with hypertonic and hypotonic.
isotope (eye"suh-tope) An alternate form of an element with a different number of neutrons, but the same number of protons and electrons. See also radioisotopes.
joint Junction between two or more bones of the skeleton.
joule A unit of energy, equivalent to 0.239 calories.
juvenile hormone (JH) An arthropod hormone that preserves juvenile structure during a molt. Without it, metamorphosis toward the adult form takes place.
juxtaglomerular apparatus (juks"tah-glo-mer'yoo-lar) A structure in the kidney that secretes renin in response to a decrease in blood pressure.
***K* selection** A reproductive strategy in which a species (known as a *K*-selected species or *K* strategist) typically has a large body size, slow development, long life-span, and does not devote a large proportion of its metabolic energy to the production of offspring. Compare with *r* selection.
karyotype (kare'ee-oh-type) The chromosomal constitution of an individual. Representations of the karyotype are generally prepared by photographing the chromosomes and arranging the homologous pairs according to size, centromere position and pattern of bands.
keratin (kare'ah-tin) A horny, water-insoluble protein found in the epidermis of vertebrates and in nails, feathers, hair, and horns.
ketone An organic molecule containing a carbonyl group bonded to two carbon atoms.
keystone species A species whose presence largely determines the species composition and functioning of the ecosystem of which it is a part.
kidney The paired vertebrate organ important in excretion of metabolic wastes and in osmoregulation.
killer T cell See cytotoxic T cells.
kilobase (kb) 1000 bases or base pairs of a nucleic acid.
kilocalorie 1000 calories. See calorie.
kilojoule 1000 joules. See joule.
kin selection The evolutionary effect caused by animals who behave in ways that favor the survival and reproduction of their own relatives (kin).
kinases Enzymes that catalyze the transfer of phosphate groups from ATP to acceptor molecules. Protein kinases activate or inactivate proteins through the addition of phosphates at specific locations.
kinetic energy Energy of motion. Compare with potential energy.
kinetochore (kin-eh'toh-kore) Portion of the chromosome centromere to which the mitotic spindle fibers attach.
kingdom A broad taxonomic category made up of related phyla; most biologists currently recognize five kingdoms of living organisms.
Klinefelter syndrome Inherited condition in which the affected individual is a sterile male with an XXY karyotype.
knuckle-walking Quadrupedal walking in which the digits of the arms are folded underneath. Characteristic of gorillas and chimpanzees.
Krebs cycle See citric acid cycle.
labyrinth The system of interconnecting canals of the inner ear of vertebrates.
labyrinthodonts First successful group of tetrapods.
lactation (lak-tay'shun) The production or release of milk from the breast.
lacteal (lak'tee-al) One of the many lymphatic vessels in the intestinal villi that absorb fat.
lactic acid A 3-carbon organic acid; also known as lactate.
lagging strand Strand of DNA that is synthesized as a series of short segments, called Okazaki fragments, which are then covalently joined by DNA ligase. Compare with leading strand.
lamins Proteins attached to the inner surface of the nuclear envelope that provide a type of skeletal framework.
large intestine The portion of the digestive tract of humans (and other vertebrates) consisting of the cecum, colon, rectum, and anus.
larva (pl. *larvae*) An immature form in the life history of some animals; may be unlike the parent.
larynx (lare'inks) The organ at the upper end of the trachea that contains the vocal cords.
lateral bud A bud in the axil of a leaf. Compare with terminal bud.
lateral meristems Areas of localized cell division on the side of a plant that give rise to secondary tissues. Lateral meristems, including the vascular cambium and the cork cambium; cause an increase in the girth of the plant body. Compare with apical meristem.
leaching The process by which dissolved materials are washed away or carried with water down through the various layers of the soil.
leader sequence Noncoding sequence of nucleotides in mRNA that is transcribed from the region that precedes (is upstream to) the coding region.
leading strand Strand of DNA that is synthesized continuously. Compare with lagging strand.
leaf scar Scar on a stem, left when a leaf abscises.
learning A change in the behavior of an animal that results from experiences during its lifetime.

lectins Chemicals derived from certain plants that stimulate cells to undergo mitosis.

legume (leg'yoom) (1) A simple, dry fruit that splits open at maturity along two seams to release its seeds; (2) Any member of the pea family, e.g., pea, bean, peanut, alfalfa.

lens The oval, transparent structure located behind the iris of the vertebrate eye; bends incoming light rays and brings them to a focus on the retina.

lenticels (len'tih-sels) Porous swellings of cork cells in the stems of woody plants; facilitate the exchange of gases.

leukocytes (loo'koh-sites) White blood cells; colorless cells that defend the body against disease-causing organisms; exhibit amoeboid movement.

leukoplasts Colorless plastids; include amyloplasts, which are used for starch storage in cells of roots and tubers.

LH See luteinizing hormone.

lichen (ly'ken) A compound organism composed of a symbiotic alga (or cyanobacterium) and fungus.

ligament (lig'uh-ment) A connective tissue cable or strap that connects bones to each other or holds other organs in place.

light dependent reactions Reactions of photosynthesis in which light energy absorbed by chlorophyll is used to synthesize ATP and usually NADPH.

lignin (lig'nin) The substance found in many plant cell walls that confers rigidity and strength, particularly in woody tissues.

limbic system In vertebrates, an action system of the brain. In humans, plays a role in emotional responses, motivation, autonomic function, and sexual response.

limiting factor An environmental factor that restricts the growth, distribution, or abundance of a particular population.

limnetic zone (lim-net'ik) The open water area away from the shore of a lake or pond extending down as far as sunlight penetrates. Compare with littoral zone and profundal zone.

linkage The tendency for a group of genes carried by the same chromosome to be inherited together in successive generations.

lipase (lip'ase) Fat-digesting enzyme.

lipid Any of a group of organic compounds that are insoluble in water, but soluble in nonpolar solvents; lipids serve as energy storage and are important components of cell membranes.

littoral zone (lit'or-ul) The region of shallow water along the shore of a lake or pond. Compare with limnetic zone and profundal zone.

liver A large, complex organ that secretes bile, helps maintain homeostasis by removing or adding nutrients to the blood, and performs many other metabolic functions.

liverworts A phylum of spore-producing, nonvascular, thallose or leafy plant with a life cycle similar to mosses.

locus (gene locus) The place on the chromosome at which the gene for a given trait occurs; the locus is a segment of the chromosomal DNA containing information that controls some character of the organism.

long-day plant A plant that flowers in response to lengthening days (shortening nights); compare with short-day plant.

loop of Henle (Hen'lee) The U-shaped loop of a mammalian kidney tubule which extends down into the renal medulla.

loose connective tissue A type of connective tissue that is widely distributed in the body; consists of fibers strewn through a semifluid matrix.

lophoporate phyla Three related invertebrate deuterostome phyla. See brachiopods.

lumen (loo'men) (1) The space enclosed by a membrane, such as the lumen of the endoplasmic reticulum; (2) The cavity or channel within a tube or tubular organ, such as a blood vessel or the digestive tract.

lung A respiratory organ that functions in gas exchange; enables an animal to breathe air.

luteinizing hormone (LH) (loot'-eh-nigh-zing) Gonadotropic hormone secreted by the anterior pituitary; stimulates ovulation and maintains the corpus luteum in the ovaries of females; stimulates testosterone production by interstitial cells in the testes of males.

lymph (limf) The colorless fluid within the lymphatic vessels that is derived from blood plasma and resembles it closely in composition; contains white cells; ultimately, returned to the blood.

lymph nodes A mass of lymph tissue surrounded by a connective tissue capsule; manufactures lymphocytes and filters lymph.

lymphatic system A subsystem of the cardiovascular system; returns excess interstitial fluid to the circulation; defends the body against disease organisms.

lymphocyte (limf'oh-site) White blood cell with nongranular cytoplasm that is responsible for immune responses. See B cell and T cell.

lysis (ly'sis) The process of disintegration of a cell or some other structure.

lysogenic conversion The change in properties of bacteria that result from the presence of a prophage.

lysosomes (ly'soh-somes) Intracellular organelles present in many animal cells; contain a variety of hydrolytic enzymes.

macroevolution (mak"roh-eh-voh-loo'shun) Large-scale evolutionary change. Compare with microevolution.

macromolecule A very large organic molecule, such as a protein or nucleic acid.

macronucleus A large nucleus found, along with one or several micronuclei, in ciliates. The macronucleus regulates cell metabolism and growth.

macronutrient An essential element that is required in fairly large amounts for normal plant growth. Compare with micronutrient.

macrophage (mak'roh-faje) A large phagocytic cell capable of ingesting and digesting bacteria and cellular debris. Macrophages are also antigen-presenting cells.

major histocompatibility complex (MHC) A group of membrane proteins, present on the surface of most cells, that are slightly different in each individual. In humans, the MHC is called the HLA (human leukocyte antigen) group.

malignant Term used to describe cancer cells (tumor cells that are able to invade tissue and metastasize).

Malpighian tubules (mal-pig'ee-an) The excretory organs of many arthropods.

mammals Class of vertebrates characterized by hair, mammary glands, and differentiation of teeth.

mandible (man'dih-bl) (1) The lower jaw of vertebrates; (2) Jawlike, external mouthparts of insects.

mangrove forest A wetland dominated by mangrove trees in which the salinity fluctuates between that of sea water and fresh water; mangrove trees grow on protected shores in the intertidal zone.

mantle In the mollusk, a fold of tissue that covers the visceral mass and that usually produces a shell.

marine snow The organic debris (plankton, dead organisms, fecal material, etc.) that "rains" into the dark area of the oceanic province from the lighted euphotic region above; marine snow is the primary food of organisms that live in the ocean's depths.

marsupials (mar-soo'pee-uls) A subclass of mammals, characterized by the presence of an abdominal pouch in which the young are carried for some time after they are born; the young are born in a very undeveloped condition.

mast cell A type of cell found in connective tissue; contains histamine and is important in allergic reactions.

matrix (may'triks or mat'riks) (1) The interior of the compartment enclosed by the mitochondrial inner membrane; (2) Nonliving material secreted by and surrounding connective tissue cells; contains a network of microscopic fibers.

matter Anything that has mass and takes up space.

maxillae Appendages used for manipulating food; characteristic of crustaceans.

mechanical isolation A reproductive isolating mechanism (prezygotic) in which fusion of the gametes of two species is prevented by morphological or anatomical differences.

mechanoreceptor (meh-kan'-oh-ree-sep"tor) A sensory cell or organ that perceives mechanical stimuli, for example, touch, pressure, hearing, and balance.

medulla (meh-dul'uh) (1) The inner part of an organ, such as the medulla of the kidney (compare with cortex); (2) The most posterior part of the vertebrate brain, lying next to the spinal cord.

medusa A jellyfish-like animal; a free-swimming, umbrella-shaped stage in the life cycle of certain cnidarians. Compare with polyp.

megaphyll (meg'uh-fil) Type of leaf found in horsetails, ferns, gymnosperms, and angiosperms that contains multiple vascular strands (i.e., complex venation). Compare with microphyll.

megaspore (meg'uh-spor) The *n* spore in heterosporous plants that gives rise to a female gametophyte. Compare with microspore.

meiosis (my-oh'sis) Process in which a $2n$ cell undergoes two successive nuclear divisions (meiosis I and meiosis II), potentially producing four *n* nuclei; leads to the formation of gametes in animals and spores in plants.

memory cell B or T lymphocyte that permits rapid mobilization of immune response on second or subsequent exposure to a particular antigen.

meninges (meh-nin'jeez) (sing. *meninx*) The three membranes that envelop the brain and spinal cord: the dura mater, arachnoid, and pia mater.

menopause The period (usually from 45 to 55 years of age) in women when the recurring menstrual cycle ceases.

menstruation (men-stroo-ay'shun) The monthly discharge of blood and degenerated uterine lining in the human female; marks the beginning of each menstrual cycle.

meristem (mer'ih-stem) A localized area of mitotic cell division in the plant body. See apical meristem and lateral meristem.

mesenchyme (mes'en-kime) A loose, often jelly-like connective tissue containing undifferentiated cells; found in the embryos of vertebrates and the adults of some invertebrates.

mesoderm (mez'oh-derm) The middle layer of the three basic tissue layers that develop in the early embryo; gives rise to connective tissue, muscle, bone, blood vessels, kidneys, and many other structures. Compare with ectoderm and endoderm.

mesophyll (mez'oh-fil) Photosynthetic tissue in the interior of a leaf; sometimes differentiated into palisade mesophyll and spongy mesophyll.

Mesozoic era That part of geological time extending from roughly 248 to 65 million years ago.

messenger RNA (mRNA) RNA that specifies the amino acid sequence of a protein; transcribed from DNA.

metabolic rate Energy production of an organism per unit time. See basal metabolic rate.

metabolism The sum of all the chemical processes that occur within a cell or organism: the transformations by which energy and matter are made available for use by the organism. See anabolism and catabolism.

metamerism Condition in which the body is divided into a series of similar segments; characteristic of annelids and arthropods.

metamorphosis (met"ah-mor'fuh-sis) Transition from one developmental stage to another, such as from a larva to an adult.

metanephridia (sing. *metanephridium*) The excretory organs of annelids and mollusks; each metanephridium consists of a tubule open at both ends; at one end a ciliated funnel opens into the coelom and the other end of the tube opens to the outside of the body.

metaphase (met'ah-faze) The stage of mitosis, and of meiosis I and II, in which the chromosomes line up on the equatorial plane of the cell. Occurs after prophase and before anaphase.

metastasis (met-tas'tuh-sis) The spreading of cancer cells from one organ or part of the body to another.

MHC See major histocompatibility complex.

microbodies Membrane-bounded structures in eukaryotic cells that contain enzymes; include peroxisomes and glyoxisomes.

microclimate Local variations in climate produced by differences in elevation, in the steepness and direction of slopes, and in exposure to prevailing winds.

microevolution Changes in allele frequencies that occur within a population over successive generations. Compare with macroevolution.

microfilaments Thin fibers composed of actin protein subunits; form part of the cytoskeleton.

microfossils Ancient traces (fossils) of microscopic life.

micronucleus One or more smaller nuclei found, along with the macronucleus, in ciliates. The micronucleus is involved in sexual reproduction.

micronutrient An essential element that is required in trace amounts for normal plant growth. Compare with macronutrient.

microphyll (mi'kro-fil) Type of leaf found in club mosses; contains one vascular strand (i.e., simple venation). Compare with megaphyll.

microsphere A protobiont formed by adding water to abiotically-formed polypeptides.

microspore (mi'kro-spor) The *n* spore in heterosporous plants that gives rise to a male gametophyte. Compare with megaspore.

microtubule (my-kroh-too'bewl) Hollow cylindrical fiber composed of tubulin protein subunits; microtubules are major components of the cytoskeleton and found in mitotic spindles, cilia, flagella, centrioles, and basal bodies.

microtubule organizing center (MTOC) Region of the cell from which microtubules extend and which appears to serve as the major site of assembly of microtubules from tubulin subunits. The MTOCs of many organisms (including animals, but not flowering plants nor most gymnosperms) contain a pair of centrioles.

microvilli (sing. *microvillus*) Minute projections of the plasma membrane that increase the surface area of the cell; found mainly in cells concerned with absorption or secretion, such as those lining the intestine or the kidney tubules.

middle lamella Layer composed of pectin polysaccharides that serves to cement together the primary cell walls of adjacent plant cells.

migration Movement of an organism (individual or population) from one place to another. See immigration and emigration.

mineralocorticoids (min"ur-al-oh-kor'tih-koidz) Hormones produced by the adrenal cortex that regulate mineral metabolism and, indirectly, fluid balance. The principal mineralocorticoid is aldosterone.

minerals Inorganic nutrients ingested in the form of salts dissolved in food and water.

missense mutation Mutation that causes one amino acid to be substituted for another in the resulting protein product.

mitochondria (my"toh-kon'dree-ah) (sing. *mitochondrion*) Spherical or elongate intracellular organelles that are the sites of oxidative phosphorylation in eukaryotes; include an outer membrane and an inner membrane.

mitosis (my-toh'sis) Division of the cell nucleus, resulting in two daughter nuclei, each with the same number of chromosomes as parent nucleus. Mitosis consists of four phases: prophase, metaphase, anaphase, and telophase. Cytokinesis (division of the cytoplasm to form two separate cells) usually overlaps the telophase stage.

mitosis promoting factor (MPF) A cyclin dependent protein kinase that controls the transition of a cell from the G_2 stage of interphase to mitosis; formerly known as maturation promoting factor.

mitotic spindle See spindle.

mitral valve Bicuspid valve between the left atrium and left ventricle of the heart. Also known as the left atrioventricular valve.

mobile genetic element See transposon.

mole The atomic mass of an element or the molecular mass of a compound, expressed in grams; one mole of any substance has 6.02×10^{23} units (Avogadro's number).

molecular clock analysis A comparison of the DNA nucleotide sequences of related organisms in order to estimate when they diverged from one another during the course of evolution.

molecular mass The sum of the atomic masses of the atoms that make up a single molecule of a compound; expressed in atomic mass units (amu) or daltons.

molecular weight The sum of the atomic weights of the atoms that make up a single molecule of a compound; atomic weight has no units.

molecule The smallest particle of a covalently bonded element or compound that has the composition and properties of a larger part of the substance.

mollusks A phylum of coelomate protostome animals characterized by a soft body, visceral mass, mantle, and foot.

molting The shedding and replacement of an outer covering such as an exoskeleton.

monoacylglycerol (mon"o-as"-il-glis'er-ol) A neutral fat consisting of glycerol combined chemically with a single fatty acid. Also called monoglyceride.

monocot (mon'oh-kot) One of the two classes of flowering plants; monocot seeds contain a single cotyledon, or seed leaf. Compare with dicot.

monocyte (mon'oh-site) A type of white blood cell; a large phagocytic, nongranular leukocyte that enters the tissues and differentiates into a macrophage.

monoecious (mon-ee'shus) Having male and female reproductive parts in separate flowers on the same plant; compare with dioecious.

monoglyceride See monoacylglycerol.

monohybrid cross A genetic cross that takes into account the behavior of alleles of a single locus. Compare with dihybrid cross.

monokaryotic (mon"o-kare-ee-ot'ik) Condition of having a single n nucleus per cell, characteristic of certain fungal hyphae. Compare with dikaryotic.

monomer (mon'oh-mer) A molecule of a compound of relatively low molecular weight that can be linked with other similar molecules to form a polymer.

monophyletic group (mon"oh-fye-let'ik) A group made up of organisms that evolved from a common ancestor. Compare with polyphyletic group.

monosaccharide (mon-oh-sak'ah-ride) A simple sugar that cannot be degraded by hydrolysis to a simpler sugar.

monosomy Condition in which only one member of a chromosome pair is present. Compare with disomy and trisomy.

monotremes (mon'oh-treems) Egg-laying mammals such as the duck-billed platypus of Australia.

monounsaturated fatty acid See fatty acid.

morphogen Any chemical agent thought to be responsible for the processes of cellular differentiation and pattern formation that lead to morphogenesis.

morphogenesis (mor-foh-jen'eh-sis) The development of the form and structures of an organism and its parts.

mortality The number of deaths per 1000 people per year.

morula (mor'yoo-lah) An early embryo consisting of a solid ball of cells.

mosaic development Highly invariant developmental pattern in which the fate of each blastomere becomes determined at a very early stage. Compare with regulative development.

mosses A phylum of nonvascular plants with an alternation of generations in which the dominant n gametophyte (a leafy moss plant) alternates with a $2n$ sporophyte that remains attached to the gametophyte.

motor neuron Efferent neuron; a neuron that transmits impulses away from the central nervous system.

motor unit All the skeletal muscle fibers that are stimulated by a single motor neuron.

mRNA cap An unusual nucleotide, 7-methylguanylate, that is added to the 5′ end of a eukaryotic messenger RNA. Capping enables eukaryotic ribosomes to bind to the message.

mucosa (mew-koh'suh) A mucous membrane, especially in the lining of the digestive and respiratory tracts.

mucus (mew'cus) A sticky secretion composed of covalently linked protein and carbohydrate; serves to lubricate body parts and trap particles of dirt and other contaminants. (The adjectival form is spelled mucous.)

müllerian mimicry (mul-ler'ee-un mim'ih-kree) The resemblance of dangerous, unpalatable, or poisonous species to one another so that they are more easily recognized by potential predators, which learn to avoid all of them after tasting one. Compare with batesian mimicry.

multiple alleles (al-leels') Three or more alleles of single locus (in a population), such as the alleles governing the ABO series of blood types.

multiple fruit A fruit that develops from many ovaries of many separate flowers, e.g., pineapple. Compare with simple, aggregate, and accessory fruits.

muscle A tissue specialized for contraction, or an organ that produces movement by contraction.

mutagen (mew'tah-jen) Any agent capable of producing mutations.

mutation Any change in DNA; may include a change in the nucleotide base pairs of a gene, a rearrangement of genes within the chromosomes so that their interactions produce different effects, or a change in the chromosomes themselves.

mutualism A symbiotic relationship in which both partners benefit from the association. Compare with parasitism and commensalism.

mycelium (my-seel'ee-um) (pl. *mycelia*) The vegetative body of fungi and certain protists (water molds); consists of a branched network of hyphae.

mycorrhizae (my"kor-rye'zee) Mutualistic associations of fungi and plant roots that aid in the plant's absorption of essential minerals from the soil.

myelin sheath (my'eh-lin) The white fatty material that forms a sheath around the axons of certain nerve cells, which are then called myelinated fibers.

myocardial infarction (MI) Heart attack; serious consequence occurring when the heart muscle receives insufficient oxygen.

myofibrils (my-oh-fy'brilz) Tiny threadlike structures in the cytoplasm of striated and cardiac muscle that are responsible for contractions of the cell; contain myofilaments.

myofilament (my-oh-fil'uh-ment) One of the filaments making up a myofibril; the structural unit of muscle proteins in a muscle cell. See thick (myosin) filament and thin (actin) filament.

myoglobin (my'oh-gloh"bin) A hemoglobin-like oxygen transferring protein found in muscle.

myosin (my'oh-sin) A protein that, together with actin, is responsible for muscle contraction.

n The chromosome number of a gamete. The chromosome number of a zygote is $2n$. If an organism is not polyploid, the n gametes are haploid and the $2n$ zygotes are diploid.

NAD^+/NADH Oxidized and reduced forms, respectively, of nicotinamide adenine dinucleotide; coenzyme that transfers electrons (as hydrogen), particularly in catabolic pathways, including cellular respiration.

$NADP^+$/NADPH Oxidized and reduced forms, respectively, of nicotinamide adenine dinucleotide phosphate; coenzyme that acts as an electron (hydrogen) transfer agent, particularly in anabolic pathways, including photosynthesis.

nastic movement See turgor movement.

natality The number of births per 1000 people per year.

natural selection The mechanism of evolution proposed by Charles Darwin: The tendency of organisms that possess favorable adaptations to their environment to survive and become the parents of the next generation. Evolution occurs when natural selection results in changes in allele frequencies in a population.

natural killer cell (NK cell) A large, granular lymphocyte that functions in both nonspecific and specific defense; releases cytokines and proteolytic enzymes that target tumor cells and cells infected with viruses and other pathogens.

Neandertal Primitive humans who lived in Europe and Asia from about 200,000 to 27,000 years ago.

nectary (nek'ter-ee) A gland or other structure that secretes nectar.

needle The leaf of a conifer such as pine.

negative feedback system A homeostatic mechanism in which a change in some condition triggers a response that counteracts, or reverses, the changed condition; examples include (1) the regulation of many biochemical processes and (2) how mammals maintain body temperature. Compare with positive feedback system.

nekton (nek'ton) Free-swimming aquatic organisms such as fish and turtles. Compare with plankton.

nematocyst (nem-at'oh-sist) A stinging structure found within cnidocytes (stinging cells) in cnidarians; used for anchorage, defense, and capturing prey.

nematodes Phylum of pseudocoelomate animals commonly known as roundworms.

nemerteans Phylum of acoelomate animals commonly known as ribbon worms; possess a complete digestive tract.

neo-Darwinism See synthetic theory of evolution.

neoplasm See tumor.

neoteny (nee-ot'eh-nee) Retention of juvenile or larval features in an otherwise adult (sexually mature) animal. Also known as paedomorphosis.

nephridial organ (neh-frid'ee-al) The excretory organ of many invertebrates; consists of simple or branching tubes that usually open to the outside of the body through pores.

nephron (nef'ron) The functional, microscopic unit of the vertebrate kidney.

neritic province (ner-ih'tik) Ocean water that extends from the shoreline to where the bottom reaches a depth of 200 m. Compare with oceanic province.

nerve A bundle of axons (or dendrites) wrapped in connective tissue that conveys impulses between the central nervous system and some other part of the body.

nerve net A system of interconnecting nerve cells found in cnidarians and echinoderms.

nervous tissue A type of animal tissue specialized for conducting electrochemical impulses.

net primary productivity Energy that remains in an ecosystem (as biomass) after cellular respiration has occurred. Compare with gross primary productivity.

neural crest (noor'ul) Group of cells along the neural tube that migrate and form structures of the peripheral nervous system and certain other structures.

neural transmission See transmission, neural.

neural tube The structure in the early vertebrate embryo that gives rise to the brain and spinal cord.

neuron (noor'on) A nerve cell; a conducting cell of the nervous system that typically consists of a cell body, dendrites, and an axon.

neurotransmitter A chemical messenger used by neurons to transmit impulses across a synapse.
neutral fat A lipid used for energy storage, consisting of a glycerol covalently bonded to one, two or three fatty acids. See monoacylglycerol, diacylglycerol, and triacylglycerol.
neutral mutation A mutation that does not confer any selective advantage or disadvantage to the organism.
neutron (noo' tron) An electrically neutral particle with a mass of one atomic mass unit (amu) found in the atomic nucleus. See also proton and electron.
neutrophil (new'truh-fil) A type of granular leukocyte.
niche (nich) The totality of an organism's adaptations, its use of resources, and the life-style to which it is fitted in its community. The niche describes how an organism uses materials in its environment as well as how it interacts with other organisms; also called ecological niche.
nicotinamide adenine dinucleotide See NAD^+/NADH.
nicotinamide adenine dinucleotide phosphate See $NADP^+$/NADPH.
nitrification (nie"tra-fuh-kay'shun) The conversion of ammonia to nitrate by certain bacteria (nitrifying bacteria) in the soil; part of the nitrogen cycle.
nitrogen cycle The worldwide circulation of nitrogen from the abiotic environment into living things and back into the abiotic environment.
nitrogen fixation The conversion of atmospheric nitrogen (N_2) to ammonia (NH_3) by certain bacteria; part of the nitrogen cycle.
nitrogenase (nie-traa'jen-ase) The enzyme responsible for nitrogen fixation under anaerobic conditions.
nodule Swellings on the roots of plants, such as legumes, in which symbiotic nitrogen-fixing bacteria (*Rhizobium*) live.
noncompetitive inhibitor A substance that lowers the rate at which an enzyme catalyzes a reaction, but does not bind to the active site. Compare with competitive inhibitor.
noncyclic phosphorylation In photosynthesis, a series of reactions involving both Photosystem II and Photosystem I. Results in the formation of ATP (by chemiosomosis), NADPH, and O_2. Compare with cyclic phosphorylation.
nondisjunction Abnormal separation of sister chromatids or of homologous chromosomes caused by their failure to disjoin (move apart) properly during mitosis or meiosis.
nonsense mutation Mutation that results in an amino-acid-specifying codon being changed to a termination codon. When the abnormal mRNA is translated, the resulting protein is usually truncated and nonfunctional.
norepinephrine (nor-ep-ih-nef'rin) A neurotransmitter that is also a hormone secreted by the adrenal medulla.
notochord (no'toe-kord) The flexible, longitudinal rod in the anteroposterior axis that serves as an internal skeleton in the embryos of all chordates and in the adults of some.
nuclear area Region of a bacterial cell that contains DNA, but is not enclosed by a membrane.
nuclear envelope The double membrane system that encloses the cell nucleus of eukaryotes.
nuclear pores Structures in the nuclear envelope that allow passage of certain materials between the cell nucleus and the cytoplasm.
nucleolus (new-klee'oh-lus) (pl. *nucleoli*) Specialized structure in the cell nucleus formed from regions of several chromosomes; site of assembly of the ribosomal subunits.
nucleoplasm The contents of the cell nucleus.
nucleoside (new'klee-oh-side) Molecule consisting of a nitrogenous base (purine or pyrimidine) and a pentose sugar.
nucleoside triphosphate Molecule consisting of a nitrogenous base, a pentose sugar, and three phosphate groups; adenosine triphosphate (ATP) is an example.
nucleosomes (new'klee-oh-somz) Repeating units of chromatin structure, each consisting of a length of DNA wound around a complex of eight histone molecules (two of four different types) plus a DNA linker region associated with a fifth histone protein.
nucleotide (noo'klee-oh-tide) A molecule composed of one or more phosphate groups, a 5-carbon sugar (ribose or deoxyribose) and a nitrogenous base (purine or pyrimidine).
nucleus (new'klee-us) (pl. *nuclei*) (1) The central region of an atom, containing the protons and neutrons; (2) A cellular organelle in eukaryotes that contains the DNA and serves as the control center of the cell; (3) A mass of nerve cell bodies in the central nervous system.
nut A simple, dry fruit that contains a single seed and is surrounded by a hard fruit wall.
nutrients The chemical substances in food that are used as components for synthesizing needed materials and/or as energy sources.
nutrition The process of taking in and using food (nutrients).
obligate anaerobe Organism that grows only in the absence of oxygen. Compare with facultative anaerobe.
occipital lobes Posterior areas of the mammalian cerebrum; interpret visual stimuli from the retina of the eye.
ocean current A pathway of flowing water in the ocean. Ocean currents flow both at the surface and far beneath the surface.
oceanic province That part of the open ocean that overlies an ocean bottom deeper than 200 meters. Compare with neritic province.
Okazaki fragment One of many short segments of DNA, each 100 to 1000 nucleotides long, that must be joined by DNA ligase to form the lagging strand in DNA replication.
olfactory epithelium Tissue containing the smell-sensing neurons.
ommatidium (om"ah-tid'ee-um) (*pl. ommatidia*) One of the light-detecting units of a compound eye, consisting of a cornea, lens, and rhabdome.
omnivore (om'nih-vore) An animal that eats a variety of plant and animal materials.
oncogene (on'koh-jeen) An abnormally functioning gene implicated in causing cancer. Compare with proto-oncogene and anti-oncogene.
onycophorans (on"ih-kof'or-anz) Phylum of rare, tropical, caterpillar-like animals, structurally intermediate between annelids and arthropods, possessing an annelid-like excretory system, and claw-tipped short legs.
oocytes (oh'uh-sites) Meiotic cells that give rise to egg cells (ova).
oogamy (oh-og'uh-me) The fertilization of a large, nonmotile female gamete by a small, motile male gamete. Compare with isogamy and anisogamy.
oogenesis (oh"oh-jen'eh-sis) Production of female gametes (eggs). Compare with spermatogenesis.
oospore A thick-walled, resistant spore formed from a zygote during sexual reproduction in water molds and certain algae.

open circulatory system A type of circulatory system in which the blood bathes the tissues directly; characteristic of arthropods and many mollusks. Compare with closed circulatory system.

operant (instrumental) conditioning A type of learning in which an animal is rewarded or punished for performing a behavior it discovers by chance.

operator site One of the control regions of an operon. The DNA segment to which a repressor binds, thereby inhibiting the transcription of the adjacent structural genes of the operon.

operculum In bony fishes, a protective flap of the body wall that covers the gills.

operon (op'er-on) In prokaryotes, a group of structural genes that are coordinately controlled and transcribed as a single message, plus their adjacent regulatory elements.

optimal foraging The process of obtaining food in a manner that maximizes benefits and/or minimizes costs.

orbital Region in which electrons occur in an atom or molecule.

order A taxonomic category made up of related families.

organ A specialized structure, such as the heart or liver, made up of tissues and adapted to perform a specific function or group of functions.

organ of Corti (kor'-tie) The structure within the inner ear of vertebrates that contains receptor cells that sense sound vibrations.

organ system Body system; an organized group of tissues and organs that work together to perform a specialized set of functions.

organelle One of the specialized structures within the cell, such as the mitochondria, Golgi apparatus, ribosomes, or contractile vacuole.

organic compound A compound composed of a backbone made up of carbon atoms.

organism Any living system composed of one or more cells.

orgasm (or'gazm) The climax of sexual excitement.

origin of replication A specific site on the DNA where replication can begin.

osmoregulation (oz"moh-reg-yoo-lay'shun) The active regulation of the osmotic pressure of body fluids so that they do not become excessively dilute or excessively concentrated.

osmosis (oz-moh'sis) Net movement of water (the principal solvent in biological systems) by diffusion through a selectively permeable membrane from a region of higher concentration of water (a hypotonic solution) to a region of lower concentration of water (a hypertonic solution).

osmotic pressure The pressure that must be exerted on the hypertonic side of a selectively permeable membrane to prevent diffusion of water (by osmosis) from the side containing pure water.

osteichthyes The vertebrate class of bony fishes.

osteocyte (os'tee-oh-site) A mature bone cell; an osteoblast that has become embedded within the bone matrix and occupies a lacuna.

osteon (os'tee-on) Spindle-shaped unit of bone composed of concentric layers of osteocytes; Haversian system of bone.

otoliths (oh'toe-liths) Small calcium carbonate crystals in the saccule and utricle of the inner ear; sense gravity and are important in static equilibrium.

outbreeding The mating of individuals of unrelated strains.

ovary (oh'var-ee) (1) In animals, one of the paired female gonads; responsible for producing eggs and sex hormones; (2) In flowering plants, the base of the carpel that contains ovules; ovaries develop into fruits after fertilization.

overdominance Condition in which the phenotype of a heterozygous individual is more extreme than that of either of its homozygous parents.

oviduct (oh'vih-dukt) Tube that carries ova from the ovary to the uterus, cloaca, or body exterior. Also called fallopian tube or uterine tube.

oviparous (oh-vip'ur-us) Bearing young in the egg stage of development; egg-laying.

ovoviviparous (oh'voh-vih-vip"ur-us) A type of development in which the young hatch from eggs incubated inside the mother's body.

ovulation (ov-u-lay'shun) The release of an ovum from the ovary.

ovule (ov'yool) The structure (i.e., megasporangium) in the ovary that develops into the seed following fertilization.

ovum (pl. *ova*) Female gamete of an animal.

oxaloacetate Four-carbon compound; important intermediate in the citric acid cycle and in the C_4 and CAM pathways of carbon fixation in photosynthesis.

oxidation The loss of one or more electrons (or hydrogen atoms) by an atom, ion, or molecule. Compare with reduction.

oxidative phosphorylation (fos"for-ih-lay'shun) The production of ATP using energy derived from the transfer of electrons in the electron transport system of mitochondria; occurs by chemiosmosis.

oxygen debt The oxygen necessary to metabolize the lactic acid produced during strenuous exercise.

oxygen dissociation curve A curve depicting the percentage saturation of hemoglobin with oxygen, as a function of certain variables such as oxygen concentration, carbon dioxide concentration, or pH.

oxyhemoglobin Hemoglobin that has combined with oxygen.

oxytocin (ok"see-tow'sin) Hormone secreted by the hypothalamus and released by the posterior lobe of the pituitary gland; stimulates contraction of the pregnant uterus and the ducts of mammary glands.

ozone A blue gas, O_3, that has a distinctive odor. Ozone is a human-made pollutant near Earth's surface (in the troposphere) but a natural and essential component of the stratosphere.

ozone depletion The thinning of the ozone layer in the stratosphere.

P generation (parental generation) Members of two different true breeding lines that are crossed to produce the F_1 generation.

P680 Chlorophyll *a* molecules that serve as the reaction center of Photosystem II, transferring photoexcited electrons to a primary acceptor; named by their absorption peak at 680 nm.

P700 Chlorophyll *a* molecules that serve as the reaction center of Photosystem I, transferring photoexcited electrons to a primary acceptor; named by their absorption peak at 700 nm.

pacemaker (of the heart) The sinoatrial (SA) node of the heart; specialized cardiac muscle where each heartbeat begins.

Pacinian corpuscle (pah-sin'-ee-an kor'pus-el) A receptor located in the dermis of the skin that responds to pressure.

paedomorphosis See neoteny.

pair bond A stable relationship between animals of opposite sex that ensures cooperative behavior in mating and rearing the young.
paleoanthropology (pay″lee-o-an-thro-pol′uh-gee) The study of human evolution.
Paleozoic era That part of geological time extending from 570 to 248 million years ago.
palindromic Reading the same forward and backward. Some DNA sequences are said to be palindromic because the base sequence of one strand is the reverse of the base sequence in its complement; for example, the complement of 5′-GAATTC-3′ is 3′-CTTAAG-5′.
palisade meosphyll (mez′oh-fil) The vertically stacked, columnar mesophyll cells near the upper epidermis in certain leaves. Compare with spongy mesophyll.
pancreas (pan′kree-us) Large gland located in the vertebrate abdominal cavity. The pancreas produces pancreatic juice containing digestive enzymes; also serves as an endocrine gland, secreting the hormones insulin and glucagon.
parabronchi (sing. *parabronchus*) Thin-walled ducts in the lungs of birds; gases are exchanged across their walls.
parapodia (par″uh-poh′dee-ah) (pl. *parapodium*) Paired, thickly bristled paddle-like appendages extending laterally from each segment of polychaete worms.
parasite Any heterotrophic organism that obtains nourishment from the living tissue of another organism (the host).
parasitism (par′uh-si-tiz″m) A symbiotic relationship in which one member (the parasite) benefits and the other (the host) is adversely affected. Compare with commensalism and mutualism.
parasympathetic nervous system A division of the autonomic nervous system concerned with the control of the internal organs; functions to conserve or restore energy. Compare with sympathetic nervous system.
parathyroid glands Small, pea-sized glands closely adjacent to the thyroid gland; their secretion regulates calcium and phosphate metabolism.
parathyroid hormone (PTH) Hormone secreted by the parathyroid glands; regulates calcium and phosphate metabolism.
parenchyma (par-en′kih-mah) Highly variable living plant cells that have thin primary walls; function in photosynthesis, the storage of nutrients, and/or secretion.
parthenogenesis (par″theh-noh-jen′eh-sis) The development of an unfertilized egg into an adult organism; common among honey bees, wasps, and certain other arthropods.
passive immunity Temporary immunity that depends on the presence of immunoglobulins produced by another organism.
patch clamp technique A method that allows researchers to study the ion channels of a tiny patch of membrane by tightly sealing a micropipette to the patch and measuring the flow of ions through the channels.
pathogen (path′oh-gen) An organism, usually a microorganism, capable of producing disease.
pattern formation The developmental process by which cells become organized into recognizable structures.
pedigree analysis Technique by which an inheritance pattern is traced through multiple generations.
penis The male sexual organ of copulation in reptiles and mammals.
pentose A sugar molecule containing five carbons.
pepsin (pep′sin) An enzyme produced in the stomach that initiates digestion of protein.
peptide (pep′ tide) A compound consisting of a chain of amino acid groups. A dipeptide consists of two amino acids, a polypeptide of many.
peptide bond A distinctive covalent carbon-to-nitrogen bond that links amino acids in peptides and proteins.
peptidoglycan (pep″tid-oh-gly′kan) A modified protein or peptide possessing an attached carbohydrate; component of the bacterial cell wall.
perennial plant (pur-en′ee-ul) A woody or herbaceous plant that grows year after year (that is, lives more than 2 years). Compare with annual and biennial.
perfect flower A flower that has both stamens and carpels. Compare with imperfect flower.
pericentriolar material Rather amorphous, somewhat fibrillar matter surrounding the centrioles in the microtubule organizing centers in cells of animals and other organisms possessing centrioles; chemically similar to the material in the microtubule organizing centers of organisms lacking centrioles.
pericycle (pehr′eh-sy″kl) A layer of meristematic cells typically found between the endodermis and phloem in roots.
periderm (pehr′ih-durm) The outer bark of woody stems and roots; composed of cork cells, cork cambium, and cork parenchyma, along with traces of primary tissues.
period An interval of geological time that is a subdivision of an era. Each period is divided into epochs.
peripheral membrane protein A protein associated with one of the surfaces of a biological membrane. Compare with integral membrane protein.
peripheral nervous system (PNS) In vertebrates, the nerves and receptors that lie outside the central nervous system. Compare with central nervous system (CNS).
peristalsis (pehr″ih-stal′sis) Rhythmic waves of muscular contraction and relaxation in the walls of hollow tubular organs, such as the ureter or parts of the digestive tract, that serve to move the contents through the tube.
permafrost Permanently frozen subsoil characteristic of frigid areas such as the tundra.
peroxisomes (pehr-ox′ ih-somz) Membrane-bounded organelles in eukaryotic cells containing enzymes that produce or degrade hydrogen peroxide.
petal One of the colored cluster of modified leaves that constitute the next-to-outermost whorl of a flower.
petiole (pet′ee-ohl) The part of a leaf that attaches to a stem.
pH The negative logarithm of the hydrogen ion concentration of a solution (expressed as moles per liter). Neutral pH is 7; values less than 7 are acidic, and those greater than 7 are basic.
phage See bacteriophage.
phagocytosis (fag″oh-sy-toh′sis) Literally, "cell eating"; a type of endocytosis by which certain cells engulf food particles, microorganisms, foreign matter, or other cells.
pharynx (far′inks) Part of the digestive tract. In complex vertebrates it is bounded anteriorly by the mouth and nasal cavities and posteriorly by the esophagus and larynx; the throat region in humans.
phenetics (feh-neh′tiks) An approach to classification based on measurable similarities in phenotypic characters, without consideration of homology, analogy, or other evolutionary relationship. Compare with cladistics and classical evolutionary taxonomy.

phenotype (fee'noh-type) The physical or chemical expression of an organism's genes. Compare with genotype.

phenylketonuria (PKU) (fee"nl-kee"toh-noor'ee-ah) An inherited disease in which there is a deficiency of the enzyme that normally converts phenylalanine to tyrosine; results in mental retardation if untreated.

pheromone (feer'oh-mone) A substance secreted by an organism to the external environment that influences the development or behavior of other members of the same species.

phloem (floh'em) The vascular tissue that conducts dissolved sugar and other organic compounds in plants.

phloem fiber cap A cluster of fibers formed just outside the phloem in vascular bundles of many herbaceous dicot stems.

phosphate group A weakly acidic functional group that can release one or two hydrogen ions.

phosphodiester linkage Covalent linkage between two nucleotides in a strand of DNA or RNA; includes a phosphate group bonded to the sugars of two adjacent nucleotides.

phosphoenolpyruvate (PEP) 3-carbon phosphorylated compound that is an important intermediate in glycolysis and is a reactant in the initial carbon fixation step in the C_4 and CAM pathways of carbon fixation in photosynthesis.

phosphoglycerate (PGA) Phosphorylated 3-carbon compound that is an important metabolic intermediate.

phospholipids (fos"foh-lip'idz) Fatlike substances in which there are two fatty acids and a phosphorus-containing group attached to glycerol; major components of cellular membranes.

phosphorus cycle The worldwide circulation of phosphorus from the abiotic environment into living things and back into the abiotic environment.

phosphorylation (fos"for-ih-lay'shun) The introduction of a phosphate group into an organic molecule. Kinases are enzymes that catalyze certain phosphorylation reactions.

photoautotroph Photosynthetic autotroph; organism that obtains energy from light and synthesizes organic compounds from inorganic raw materials; includes plants, algae, and some bacteria.

photoheterotroph Photosynthetic heterotroph; organism that is able to carry out photosynthesis, but unable to fix carbon dioxide and therefore requires organic compounds as a carbon source; include some bacteria.

photolysis (foh-tol'uh-sis) The photochemical splitting of water in the light-dependent reactions of photosynthesis, catalyzed by a specific enzyme.

photon (foh'ton) A particle of electromagnetic radiation; one quantum of radiant energy.

photoperiodism (foh"toh-peer'ee-od-izm) The physiological response (such as flowering) of plants to variations in the length of daylight and darkness.

photophosphorylation (foh"toh-fos-for-ih-lay'shun) The production of ATP in photosynthesis.

photoreceptor (foh"toh-ree-sep'tor) (1) A sense organ specialized to detect light; (2) A pigment that absorbs light before triggering a physiological response.

photorespiration (foh"toh-res-pur-ay'shun) Process that reduces the efficiency of photosynthesis in C_3 plants at high light intensities; consumes oxygen and produces carbon dioxide through the degradation of Calvin cycle intermediates.

photosynthesis (foh"toh-sin'thuh-sis) The biological process that captures light energy and transforms it into the chemical energy of organic molecules (such as carbohydrates), which are manufactured from carbon dioxide and water. Photosynthesis is performed by plants, algae, and certain bacteria.

photosystem A group of chlorophyll molecules, accessory pigments, and associated electron acceptors that emits electrons in response to light; located in the thylakoid membrane (in photoautotrophic eukaryotes).

phototropism (foh"toh-troh'pizm) The growth of a plant in response to the direction of light.

phycocyanin (fy"koh-sy'ah-nin) A blue pigment found in cyanobacteria and red algae.

phycoerythrin (fy"koh-ee-rih'thrin) A red pigment found in cyanobacteria and red algae.

phylogenetic tree A branching diagram that shows lines of descent among a group of related species.

phylogeny (fy-loj'en-ee) The complete evolutionary history of a group of organisms.

phylum (fy'lum) A taxonomic grouping of related, similar classes; a category beneath the kingdom and above the class.

phytochrome (fy'toh-krome) A blue-green, proteinaceous pigment involved in a wide variety of physiological responses to light; occurs in two interchangeable forms depending on the ratio of red to far-red light.

phytoplankton (fy"toh-plank'tun) Microscopic floating algae and cyanobacteria that are the base of most aquatic food webs. Compare with zooplankton. See plankton.

pia mater (pee'a may'ter) The inner membrane covering the brain and spinal cord; the innermost of the meninges.

pili (pil'eye) (sing. *pilus*) Hairlike structures on the surface of many bacteria. Function in conjugation or attachment.

pineal gland (pie-nee'al) Endocrine gland located in the brain.

pinocytosis (pin"oh-sy-toh'sis) Cell drinking; a type of endocytosis by which cells engulf and absorb droplets of liquids.

pioneer The first organism to colonize an area and begin the first stage of succession.

pistil The female reproductive organ of a flower; consists of either a single carpel or two or more fused carpels.

pith The innermost tissue in the stems and roots of many herbaceous plants; primarily a storage tissue.

pituitary gland (pit-oo'ih-tehr"ee) Endocrine gland located below the hypothalamus; secretes several hormones influencing a wide range of physiological processes.

placenta (plah-sen'tah) The partly fetal and partly maternal organ whereby materials are exchanged between fetus and mother in the uterus of eutherian (placental) mammals.

placoderms (plak'oh-durms) A group of extinct jawed fishes.

plankton Free-floating, mainly microscopic aquatic organisms found in the upper layers of the water; composed of phytoplankton and zooplankton. Compare with nekton.

plantlet A small, immature plant.

planula larva (plan'yoo-lah) A ciliated larval form found in cnidarians.

plasma membrane The selectively permeable surface membrane that encloses the cell contents and through which all materials entering or leaving the cell must pass.

plasma cell Cell that secretes antibodies; differentiated B lymphocyte.

plasma proteins Proteins such as albumins, globulins, and fibrinogen that circulate in the blood plasma.

plasmids (plaz'midz) Small circular DNA molecules that carry genes separate from the main bacterial DNA.

plasmodesmata (sing. *plasmodesma*) Cytoplasmic channels connecting adjacent plant cells and allowing for the movement of molecules and ions between cells.
plasmodial slime mold (plaz-moh'dee-uhl) A fungus-like protist whose feeding stage consists of a plasmodium.
plasmodium (plaz-moh'dee-um) (1) A multinucleate, ameboid mass of living matter; a coenocyte. (2) The feeding phase of the life cycle of plasmodial slime molds.
plasmolysis (plaz-mol"ih-sis) The shrinkage of cytoplasm and the pulling away of the plasma membrane from the cell wall when a plant cell (or other walled cell) loses water, usually in a hypertonic environment. Compare with crenated.
plastids (plas'tidz) A family of membrane-bounded organelles occurring in photosynthetic eukaryotic cells; include chloroplasts, chromoplasts, and amyloplasts and other leukoplasts.
plate tectonics Theory that Earth's crust is divided into several plates that move about relative to one another.
platelets (playt'lets) Cell fragments in the blood that function in clotting; also called thrombocytes.
platyhelminthes Phylum of acoelomate animals commonly known as flatworms.
pleiotropic (ply"oh-troh'pik) Term referring to an allele that affects a number of characteristics of an individual.
pleural membrane (ploor'ul) The membrane that lines the thoracic cavity and envelops each lung.
ploidy (ploy'dee) Relating to the number of sets of chromosomes in a cell. See haploid, diploid, and polyploid.
plumule (ploom'yool) The embryonic shoot apex, or terminal bud, located above the point of attachment of the cotyledon(s).
pneumatophore (noo-mat'uh-for") Roots that extend up out of the water in swampy areas and are thought to provide aeration between the atmosphere and submerged roots.
point mutation See base-substitution mutation.
polar body A small *n* cell produced during oogenesis in female animals that does not develop into a functional ovum.
polar covalent bond Chemical bond formed by the sharing of electrons between atoms that differ in electronegativity; the end of the bond near the more electronegative atom has a partial negative charge, the other end has a partial positive charge.
polar molecule Molecule that has one end with a partial positive charge and the other with a partial negative charge.
polar nucleus In flowering plants, one of two *n* cells in the embryo sac that fuse with a sperm during double fertilization to form the 3*n* endosperm.
pollen grain The immature male gametophyte of seed plants (i.e., all gymnosperms and angiosperms) that produces sperm capable of fertilization.
pollen tube A tube that forms after germination of the pollen grain and that grows from the stigma through the style into the ovary; two sperm pass through the pollen tube into the ovule.
pollination (paul"uh-nay'shen) In seed plants, the transfer of pollen from the male to the female part of the plant.
polyadenylation (pol"ee-a-den-uh-lay'shun) That part of eukaryotic mRNA processing in which multiple adenine-containing nucleotides (a poly A tail) are added to the 3′ end of the molecule.
polygenic inheritance (pol"ee-jen'ik) Inheritance in which several independently assorting or loosely linked non-allelic genes modify the intensity of a trait or contribute to the phenotype in additive fashion.
polymer (pol' ih-mer) A molecule built up from repeating subunits of the same general type (monomers), such as a protein, nucleic acid, or polysaccharide.
polymerase chain reaction (PCR) A method by which a targeted DNA fragment can be amplified in vitro to produce millions of copies.
polymorphism (pol"ee-mor'fizm) (1) The existence of two or more phenotypically different individuals within the same species; (2) The presence of more than one allele for a given locus in a population.
polyp (pol'ip) Hydra-like animal; the sessile stage of the life cycle of certain cnidarians. Compare with medusa.
polypeptide A compound consisting of many amino acids linked by peptide bonds.
polyphyletic group (pol"ee-fye-let'ik) A group made up of organisms that evolved from two or more different ancestors. Compare with monophyletic group.
polyploid (pol'ee-ployd) Possessing more than two sets of chromosomes per nucleus. Compare with diploid and haploid.
polyribosome A complex consisting of a number of ribosomes attached to an mRNA during translation; also known as a polysome.
polysaccharide (pol-ee-sak'ah-ride) A carbohydrate consisting of many monosaccharide subunits; examples are starch, glycogen, and cellulose.
polysome See polyribosome.
polytene Term describing a giant chromosome consisting of many (usually more than 1000) parallel DNA double helices. Polytene chromosomes are typically found in cells of the salivary glands and some other tissues of certain insects, such as the fruit fly, *Drosophila.*
polyunsaturated fatty acid See fatty acid.
pons (ponz) The white bulge that is the part of the brainstem between the medulla and the midbrain; connects various parts of the brain.
population A group of organisms of the same species that live in the same geographical area at the same time.
population bottleneck See bottleneck.
population crash An abrupt decline in the size of a population.
population density The number of individuals of a species per unit of area or volume at a given time.
population ecology That branch of biology that deals with the numbers of a particular species that are found in an area and how and why those numbers change (or remain fixed) over time.
poriferans Animal phylum that comprises the sponges.
positive feedback system A homeostatic mechanism in which a change in some condition triggers a response that intensifies the changing condition. Compare with negative feedback system.
posterior Toward the tail-end of a bilaterally symmetrical animal. Compare with anterior.
postsynaptic neuron A neuron that transmits an impulse away from a synapse. Compare with presynaptic neuron.
potassium (K^+) ion mechanism Mechanism by which plants open and close their stomata. The influx of potassium ions into guard cells causes water to move in by osmosis, changing the shape of the guard cells and opening the pore.

potential energy Stored energy; energy that can do work as a consequence of its position or state. Compare with kinetic energy.

preadaptation An evolutionary change in an existing biological structure that enables it to have a different function.

Precambrian time All of geological time before the Paleozoic era, encompassing approximately the first 4 billion years of Earth's history.

predation Relationship in which one organism (the predator) kills and devours another organism (the prey).

prehensile (pree-hen'sil) Adapted for grasping by wrapping around an object, as in a prehensile tail.

pre-mRNA RNA precursor to mRNA in eukaryotes; contains both introns and exons.

pressure-flow hypothesis The mechanism by which dissolved sugar is thought to be transported in phloem.

presynaptic neuron A neuron that transmits an impulse to a synapse. Compare with postsynaptic neuron.

prevailing wind One of three winds that blow more or less continually: polar easterlies, westerlies, and trade winds.

primary consumer A consumer that eats producers; also called herbivore.

primary growth An increase in the length of a plant that occurs at the tips of the shoots and roots due to the activity of apical meristems. Compare with secondary growth.

primary mycelium A mycelium in which the cells are monokaryotic and haploid; a mycelium that grows from either an ascospore or a basidiospore. Compare with secondary mycelium.

primary producer An organism that produces complex organic molecules from simple inorganic substances. In most ecosystems, primary producers are photosynthetic organisms (plants, algae, and cyanobacteria). Also called autotroph and producer.

primary response The response of the immune system to first exposure to an antigen. Compare with secondary response.

primary structure (of a protein) The complete sequence of amino acids in a polypeptide chain, beginning at the amino end and ending at the carboxyl end. Compare with secondary, tertiary, and quaternary protein structure.

primary succession An ecological succession that occurs on land that has not previously been inhabited by plants; no soil is present initially. Compare with secondary succession.

primosome A complex of proteins responsible for synthesizing the RNA primers required in DNA synthesis.

principal energy level The energy level of the electrons making up a particular electron shell. Electrons at the same principal energy level have similar, but not necessarily identical, energies.

prion An infectious agent that consists only of protein.

producer See primary producer.

product Substance formed by a chemical reaction.

product law Rule for combining the probabilities of independent events by multiplying their individual probabilities.

profundal zone (pro-fun'dl) The deepest zone of a large lake. Compare with littoral zone and limnetic zone.

progesterone (pro-jes'-ter-own) A steroid hormone secreted by the corpus luteum and placenta; stimulates the uterus and breasts.

progymnosperm (pro-jim'noh-sperm) An extinct group of plants that are thought to have been the ancestors of gymnosperms.

prokaryote (pro-kar'ee-ote) Cell that lacks a nucleus and other membrane-bounded organelles; include the bacteria, members of Kingdom Prokaryotae. Compare with eukaryote.

promoter Site on DNA to which RNA polymerase attaches to begin transcription.

prop root An adventitious root that arises from the stem and provides additional support for a plant such as corn.

prophage (pro'faj) A latent state of a bacteriophage in which the viral genome is inserted into the bacterial host chromosome.

prophase The first stage of mitosis, and of meiosis I and meiosis II. During prophase the chromosomes become visible as distinct structures, the nuclear envelope breaks down, and a spindle forms. (Meiotic prophase I is more complex, and includes synapsis of homologous chromosomes and crossing over.)

proplastids Organelles that are plastid precursors; may mature into various specialized plastids, including chloroplasts, chromoplasts, leukoplasts, or amyloplasts.

proprioceptor (pro"pree-oh-sep-tor) Receptors in muscles, tendons, and joints that respond to changes in movement, tension, and position; enable an animal to perceive the position of its body.

prosimian (pro"sim'ee-un) A suborder of primates that includes the lemurs, lorises, and tarsiers.

prostaglandins (pros"tah-glan'dinz) Derivatives of unsaturated fatty acids that produce a wide variety of hormone-like effects; synthesized by most cells of the body; sometimes called local hormones.

prostate gland A gland in male animals that produces an alkaline secretion that is part of the semen.

protein A large, complex organic compound composed of covalently linked amino acid subunits; contains carbon, hydrogen, oxygen, nitrogen, and sulfur.

prothallus (pro-thal'us) (pl. *prothalli*) The free-living, n gametophyte in ferns and other seedless vascular plants.

protist (proh'tist) One of a vast kingdom of eukaryotic organisms, primarily single-celled or simple multicellular; mostly aquatic.

protobionts (proh"toh-by'ontz) Assemblages of organic polymers that spontaneously form under certain conditions. Protobionts may have been involved in chemical evolution.

proton A particle present in the nuclei of all atoms that has one unit of positive charge and a mass of one atomic mass unit (amu). See electron and neutron.

proto-oncogene A gene (also known as a cellular oncogene) that normally promotes cell division in response to the presence of certain growth factors; when mutated it may become an oncogene, possibly leading to the formation of a cancer cell. Compare with oncogene.

protonema (proh"toh-nee'mah) (pl. *protonemata*) In mosses, a filament of n cells that grows from a spore and develops into leafy moss gametophytes.

protonephridia (proh"toh-nef-rid'ee-ah) (sing. *protonephridium*) The flame-cell excretory organs of flatworms and some other simple invertebrates.

protostome (proh'toh-stome) Major division of the animal kingdom in which the blastopore develops into the mouth, and the anus forms secondarily; includes the annelids, arthropods, and mollusks. Compare with deuterostome.

protozoa (proh″toh-zoh′a) (sing. *protozoon*) An informal group of unicellular, animal-like protists, including amoebas, foraminiferans, actinopods, ciliates, flagellates, and sporozoans.

provirus (pro′vy′rus) A part of a virus, consisting of nucleic acid only, that has been inserted into a host genome.

proximal Closer to the point of reference. Compare with distal.

proximal convoluted tubule The part of the renal tubule that extends from Bowman's capsule to the loop of Henle. Compare with distal convoluted tubule.

pseudocoelom (soo″doh-see′lom) A body cavity between the mesoderm and endoderm; derived from the blastocoel. Compare with coelom.

pseudocoelomate (soo″doh-seel′oh-mate) Animal possessing a pseudocoelom. Compare with coelomate and acoelomate.

pseudogene A DNA segment resembling a gene, but containing mutations that prevent it from functioning.

pseudoplasmodium (soo″doe-plaz-moh′dee-um) In cellular slime molds, an aggregation of amoeboid cells to form a spore-producing fruiting body during reproduction.

pseudopodium (sou″doe-poe′dee-um) (pl. *pseudopodia*) A temporary extension of an amoeboid cell that is used for feeding and locomotion.

puff In a polytene chromosome, a decondensed region that is a site of intense RNA synthesis.

pulmonary circulation The part of the circulatory system that delivers blood to and from the lungs for oxygenation. Compare with systemic circulation.

pulse, arterial Alternate expansion and recoil of an artery.

pulvinus (pul-vy′nus) A special structure, often located at the base of the petiole, that functions in leaf movement by changes in turgor.

punctuated equilibrium The idea that evolution proceeds with periods of inactivity (i.e., periods of little or no change within a species) followed by very active phases, so that major adaptations or clusters of adaptations appear suddenly in the fossil record. Compare with gradualism.

Punnett square Grid structure, first developed by Reginald Punnett, that allows direct calculation of the probabilities of occurrence of all possible offspring of a genetic cross.

pupa (pew′pah) (pl. *pupae*) A stage in the development of an insect, between the larva and the imago (adult); a form that neither moves nor feeds, and may be in a cocoon.

pure line See true breeding line.

purines (pure′eenz) Nitrogenous bases with carbon and nitrogen atoms in two attached rings; components of nucleic acids, ATP, NAD^+, and certain other biologically active substances. Examples are adenine and guanine.

pyramid of biomass An ecological pyramid that illustrates the total biomass (for example, the total dry weight of all living organisms in a community) at each successive trophic level.

pyramid of energy An ecological pyramid that shows the energy flow through each trophic level of an ecosystem.

pyramid of numbers An ecological pyramid that shows the number of organisms at each successive trophic level in a given ecosystem.

pyrimidines (pyr-im′ih-deenz) Nitrogenous bases, each composed of a single ring of carbon and nitrogen atoms; components of nucleic acids. Examples are thymine, cytosine, and uracil.

pyruvic acid (pyruvate) A 3-carbon organic acid; the end product of glycolysis.

quadrupedal (kwad′roo-ped″ul) Walking on all fours.

quantitative trait A trait that shows continuous variation in a population (e.g., human height). Quantitative traits usually have polygenic inheritance patterns.

quantum A small, discrete quantity of energy.

quaternary structure (of a protein) The overall conformation of a protein produced by the interaction of two or more polypeptide chains. Compare with primary, secondary, and tertiary protein structure.

***r* selection** A reproductive strategy in which a species (known as an *r*-selected species or *r* strategist) typically has a small body size, rapid development, short life-span, and devotes a large proportion of its metabolic energy to the production of offspring. Compare with *K* selection.

radial cleavage Pattern of blastomere production in which the cells are located directly above or below one another; characteristic of early deuterostome embryos. Compare with spiral cleavage.

radial symmetry A body plan in which any section through the mouth and down the length of the body divides the body into similar halves. Jellyfish and other cnidarians have radial symmetry. Compare with bilateral symmetry.

radicle (rad′ih-kl) The embryonic root of a seed plant.

radioactive decay The process in which a radioactive element emits radiation and, as a result, its nucleus changes into the nucleus of a different element.

radioisotopes Unstable isotopes that spontaneously emit radiation. Also called radioactive isotopes.

radula (rad′yoo-lah) A rasplike structure in the digestive tract of chitons, snails, squids, and certain other mollusks.

rain shadow An area on the downwind side of a mountain range with very little precipitation. Deserts often occur in rain shadows.

random dispersion The spatial distribution pattern in which the presence of one individual has no effect on the distribution of other individuals. Compare with clumped dispersion and uniform dispersion.

range The area where a particular species occurs.

ray A chain of parenchyma cells (one to many cells thick) that functions for lateral transport in stems and roots of woody plants.

ray-finned fishes Group of bony fishes that gave rise to most modern osteichthyes.

reabsorption The selective removal of certain substances from the glomerular filtrate by the renal tubules and collecting ducts of the kidney, and their return into the blood.

reactant Substance that participates in a chemical reaction.

reaction center Portion of an antenna complex within a photosystem that includes chlorophyll *a* molecules capable of transferring electrons to a primary acceptor; the reaction center of Photosystem I is P700 and of Photosystem II is P680.

realized niche The life-style that an organism actually pursues, including the resources that it actually uses. An organism's realized niche is narrower than its fundamental niche because of interspecific competition. Compare with fundamental niche.

receptacle The end of a flower stalk where the flower parts (sepals, petals, stamens, and carpels) are attached.

reception Process of detecting a stimulus.

receptor-mediated endocytosis A type of endocytosis in which extracellular molecules become bound to specific receptors on the cell surface and then enter the cytoplasm enclosed in vesicles.

recessive allele (al-leel′) Allele that is not expressed in the heterozygous state. Compare with dominant allele.

recombinant DNA Any DNA molecule made by combining genes from different organisms.

recombination, genetic The appearance of new gene combinations. Recombination in eukaryotes generally results from meiotic events, either crossing over or shuffling of chromosomes.

red alga Member of a diverse phylum of algae that contain the pigments chlorophyll *a*, carotenoids, phycocyanin, and phycoerythrin.

red blood cell (RBC) See erythrocyte.

red tide A red or brown coloration of ocean water caused by a population explosion, or bloom, of dinoflagellates.

redox reaction (ree′dox) Chemical reaction in which one or more electrons are transferred from one substance (the substance that becomes oxidized) to another (the substance that becomes reduced). See oxidation and reduction.

reduction The gain of one or more electrons (or hydrogen atoms) by an atom, ion, or molecule. Compare with oxidation.

reflex action An automatic, involuntary response to a given stimulus that generally functions to restore homeostasis.

refractory period The brief period of time that must elapse after the response of a neuron or muscle fiber, during which it cannot respond to another stimulus.

regulative development Very plastic developmental pattern in which each individual blastomere retains totipotency. Compare with mosaic development.

regulon A group of operons that are coordinately controlled.

renal (ree′nl) Pertaining to the kidney.

renal pelvis The funnel-shaped chamber of the kidney that receives urine from the collecting ducts; urine then moves into the ureters.

renin (reh′nin) A hormone released by the kidney in response to ischemia or lowered pulse pressure, which changes angiotensinogen into angiotesin, leading to an increase in blood pressure.

replacement-level fertility The number of children a couple must produce in order to "replace" themselves. The average number is greater than 2 because some children die before reaching reproductive age.

replication fork Y-shaped structure produced during semiconservative replication of DNA.

replication See DNA replication.

repressor protein A transcription factor that acts as a negative regulator, inhibiting transcription when bound to DNA; some repressors require a corepressor to be active; some other repressors become inactive when bound to an inducer molecule. Compare with activator protein.

reproduction Process by which new individuals are produced. See asexual reproduction and sexual reproduction.

reproductive isolating mechanisms The reproductive barriers that prevent a species from interbreeding with another species. As a result, each species' gene pool is isolated from other species.

reptile Class of vertebrate animals characterized by dry skin with horny scales and adaptations for terrestrial reproduction; includes turtles, snakes, and alligators.

resolution See resolving power.

resolving power The ability of a microscope to show fine detail, defined as the minimum distance between two points at which they can be seen as separate images. Also referred to as resolution.

respiration (1) Cellular respiration is the process by which cells generate ATP through a series of redox reactions in which the terminal electron acceptor is an inorganic compound. In aerobic cellular respiration the terminal electron acceptor is molecular oxygen; in anaerobic cellular respiration the terminal acceptor is an inorganic molecule other than oxygen. (2) Organismic respiration is the act or function of gas exchange in a complex animal, generally through a specialized respiratory surface, such as a lung or gill.

respiratory centers Centers in the medulla and pons that regulate breathing.

resting potential The membrane potential (difference in electrical charge) of a neuron in which no action potential is occurring. The typical resting potential is about −70 millivolts.

restriction enzyme One of a class of enzymes that cleave DNA at specific base sequences; produced by bacteria to degrade foreign DNA; used in recombinant DNA technology.

restriction fragment length polymorphism (RFLP) analysis A technique that permits assessment of the degree of relatedness among individuals within a population. Individuals can be compared on the basis of the different patterns of DNA fragments generated when their DNA is cut with the same restriction enzyme.

restriction map A physical map of DNA in which sites cut by specific restriction enzymes serve as landmarks.

reticular activating system (RAS) (reh-tik′yoo-lur) A diffuse network of neurons in the brain stem responsible for maintaining consciousness.

reticulum (reh-tik′yoo-lum) A general term referring to any network (consisting of fibrils, filaments, or membranes).

retina (ret′ih-nah) The innermost of the three layers of the eyeball, which is continuous with the optic nerve and contains the light-sensitive rod and cone cells.

retrovirus (ret′roh-vy″rus) An RNA virus that uses reverse transcriptase to produce a DNA intermediate in the host cell.

reverse transcriptase Enzyme produced by retroviruses to enable the transcription of DNA from the viral RNA in the host cell.

reversible inhibitor A substance that forms weak bonds with an enzyme, temporarily interfering with its function; a reversible inhibitor can be competitive or noncompetitive. Compare with irreversible inhibitor.

Rh factors Red blood cell antigens, first identified in *Rh*esus monkeys. Persons possessing these antigens are Rh^+, those lacking them are Rh^-. See erythroblastosis fetalis.

rhizome (ry′zome) A horizontal underground stem.

rhodopsin (roh-dop′sin) Visual purple; a light-sensitive pigment found in the rod cells of the vertebrate eye and also employed for photosynthesis by certain bacteria.

ribonucleic acid (RNA) A family of single-stranded nucleic acids that function mainly in protein synthesis.

ribosomal RNA (rRNA) See ribosomes.

ribosomes (ry′boh-sohms) Organelles that are part of the protein synthesis machinery of both prokaryotic and eukaryotic cells; consist of a larger and smaller subunit, each composed of ribosomal RNA (rRNA) and ribosomal proteins.

ribozyme (ry'boh-zime) A molecule of RNA that has catalytic properties.

ribulose bisphosphate (RuBP) A 5-carbon phosphorylated compound with a high energy potential reacts with carbon dioxide in the initial step of the Calvin cycle.

ribulose bisphosphate carboxylase See Rubisco.

RNA polymerase Enzyme that catalyzes the synthesis of RNA from a DNA template. Also referred to as DNA-dependent RNA polymerase.

RNA primer Sequence of about five RNA nucleotides that are synthesized during DNA replication to provide a 3′ end to which DNA polymerase can add nucleotides. The RNA primer is later degraded and replaced with DNA.

rod One of the rod-shaped, light-sensitive cells of the retina, which are particularly sensitive to dim light and mediate black and white vision. Compare with cone.

root cap A covering of cells over the root tip that protects the delicate meristematic tissue directly behind it.

root graft The process of roots from two different plants growing together and becoming permanently attached to one another.

root hair An extension, or outgrowth, of a root epidermal cell. Root hairs increase the absorptive capacity of roots.

root pressure The pressure in root xylem that occurs as a result of water moving into roots from the soil.

root system The underground portion of a plant that anchors it in the soil and absorbs water and dissolved minerals.

rough ER See endoplasmic reticulum.

Rubisco Common name of ribulose bisphosphate carboxylase, the enzyme that catalyzes the reaction of carbon dioxide with ribulose bisphosphate in the Calvin cycle.

rugae (roo'jee) Folds, such as those in the lining of the stomach.

S phase Stage in interphase of the cell cycle during which DNA and other chromosomal constituents are synthesized. Compare with G_1 and G_2 phases.

saccule The structure within the vestibule of the inner vertebrate ear that along with the utricle houses the receptors of static equilibrium.

salinity The concentration of dissolved salts (such as sodium chloride) in a body of water.

salivary glands Accessory digestive glands found in vertebrates; in humans there are three pairs; also present in some invertebrates.

salt Ionic compound consisting of an anion other than a hydroxide ion and a cation other than a hydrogen ion. A salt can be formed by the reaction of an acid and a base.

salt marsh A wetland dominated by grasses in which the salinity fluctuates between that of sea water and fresh water; salt marshes are usually located in estuaries.

saltatory conduction Transmission of a neural impulse along a myelinated neuron; ion activity at one node depolarizes the next node along the axon.

saprobe See decomposer.

saprotroph (sap'roh-trof) See decomposer.

sarcolemma (sar″koh-lem'mah) The muscle cell plasma membrane.

sarcomere (sar'koh-meer) A segment of a striated muscle cell located between adjacent Z-lines that serves as a unit of contraction.

sarcoplasmic reticulum System of vesicles in a muscle cell that surrounds the myofibrils and releases calcium in muscle contraction; a modified endoplasmic reticulum.

saturated fatty acid See fatty acid.

savanna (suh-van'uh) A tropical grassland containing scattered trees; found in areas of low rainfall or seasonal rainfall with prolonged dry periods.

scaffolding proteins Nonhistone proteins that help maintain the structure of a chromosome.

schizocoely (skiz'oh-seely) Process of coelom formation in which the mesoderm splits into two layers, forming a cavity between them; characteristic of protostomes. Compare with enterocoely.

Schwann cells Supporting cells found in nervous tissue outside the central nervous system.

sclereid (skler'id) In plants, a sclerenchyma cell that is variable in shape but typically not long and tapered. Compare with fiber.

sclerenchyma (skler-en'kim-uh) Cells that provide strength and support in the plant body, are often dead at maturity, and have extremely thick walls; includes fibers and sclereids.

scramble competition Intraspecific competition in which all of the individuals in a population "share" the limited resource equally, so that at high population densities none of them obtains an adequate amount. Compare with contest competition.

scrotum (skroh'tum) The external sac of skin found in most male mammals that contains the testes and their accessory organs.

second law of thermodynamics The physical law that states that the total amount of entropy in the universe continually increases.

second messenger A substance within a cell that relays a message and (usually) triggers a response to a hormone combined with a receptor at the cell's surface; cyclic AMP and calcium are examples.

secondary consumer An organism that consumes primary consumers; also called carnivore.

secondary growth An increase in the girth of a plant due to the activity of the vascular cambium and cork cambium; secondary growth results in the production of secondary tissues, i.e., wood and bark. Compare with primary growth.

secondary mycelium A dikaryotic mycelium formed by the fusion of two primary hyphae. Compare with primary mycelium.

secondary response Rapid production of antibodies induced by a second exposure to antigen several days, weeks, or even months after the initial exposure. Compare with primary response.

secondary structure (of a protein) A regular geometric shape produced by hydrogen bonding between the atoms of the uniform polypeptide backbone; includes the alpha helix and the beta pleated sheet. Compare with primary, tertiary, and quaternary protein structure.

secondary succession An ecological succession that takes place after some disturbance destroys the existing vegetation; soil is already present. Compare with primary succession.

secretory vesicles Small cytoplasmic vesicles that move substances from an internal membrane system to the plasma membrane.

seed A plant reproductive body composed of a young, multicellular plant and nutritive tissue (food reserves), enclosed by a seed coat.

seed coat The outer protective covering of a seed.

seed fern An extinct group of seed-bearing woody plants with fern-like leaves; seed ferns probably descended from progymnosperms and gave rise to cycads and possibly ginkgoes.

segregation, principle of Genetic principle, first formulated by Gregor Mendel, that states that two alleles of a locus become separated into different gametes.

selectively permeable membrane A membrane that allows some substances to cross it more easily than others. Biological membranes are generally permeable to water, but restrict the passage of many solutes.

semen Fluid composed of sperm suspended in various glandular secretions that is ejaculated from the penis during orgasm.

semicircular canals The passages in the vertebrate inner ear containing structures that control the sense of equilibrium (balance).

semiconservative replication See DNA replication.

semilunar valves Valves between the ventricles of the heart and the arteries that carry blood away from the heart.

seminal vesicles (1) In mammals, glandular sacs that secrete a component of seminal fluid; (2) In some invertebrates, structures that store sperm.

seminiferous tubules (sem-ih-nif′er-ous) Coiled tubules in the testes in which spermatogenesis takes place in male vertebrates.

senescence (seh-nes′ns) The aging process.

sensory neuron A neuron that transmits an impulse from a receptor to the central nervous system.

sepal (see′pul) One of the outermost parts of a flower, usually leaf-like in appearance, that protect the flower as a bud.

septum (pl. *septa*) A cross-wall or partition; for example, the walls that divide a hypha into cells.

serotonin A neurotransmitter of the biogenic amine group.

Sertoli cells (sur-tole′ee) Supporting cells of the tubules of the testis.

sessile (ses′sile) Permanently attached to one location. Coral animals, for example, are sessile.

setae (sing. *seta*) Bristle-like structures that aid in annelid locomotion.

sex influenced trait Genetic trait that is expressed differently in males and females.

sex-linked gene Gene carried on a sex chromosome. In mammals almost all sex-linked genes are borne on the X chromosome (are X-linked).

sexual isolation See behavioral isolation.

sexual reproduction Type of reproduction in which two gametes (usually, but not necessarily, contributed by two different parents) fuse to form a zygote. Compare with asexual reproduction.

shoot system The above-ground portion of a plant, such as the stem and leaves.

short-day plant A plant that flowers in response to shortening days (lengthening nights); compare with long-day plant.

sickle cell anemia An inherited form of anemia in which there is abnormality in the hemoglobin beta chains; the inheritance pattern is autosomal recessive.

sieve tube members Cells that conduct dissolved sugar in the phloem of flowering plants.

sign stimulus Any stimulus that elicits an innate response in an animal.

signal transduction Process by which a signal is relayed in the cell and produces a response.

simple fruit A fruit that develops from a single ovary. Compare with aggregate, accessory, and multiple fruits.

sinoatrial (SA) node Mass of specialized cardiac tissue in which the impulse triggering the heartbeat originates; the pacemaker of the heart.

skeletal muscle Voluntary (striated) muscle of vertebrates, so-called because it usually is directly or indirectly attached to some part of the skeleton. Compare with cardiac muscle and smooth muscle.

slash-and-burn agriculture A type of agriculture in which tropical rain forest is cut down, allowed to dry, and burned. The crops that are planted immediately afterwards thrive because the ashes provide nutrients; in a few years, however, the soil is depleted and the land must be abandoned.

small intestine Portion of the vertebrate digestive tract that extends from the stomach to the large intestine.

small nuclear ribonucleoprotein complexes (snRNP) Aggregations of RNA and protein responsible for binding to pre-mRNA in eukaryotes and catalyzing the excision of introns and the splicing of exons.

smooth ER See endoplasmic reticulum.

smooth muscle Vertebrate muscle tissue that lacks transverse striations; found mainly in sheets surrounding hollow organs, such as the intestine. Also called involuntary muscle because it is controlled by the autonomic nervous system. Compare with cardiac muscle and skeletal muscle.

snRNP See small nuclear ribonucleoprotein complexes.

social behavior Interaction of two or more animals, usually of the same species.

sociobiology A branch of biology that focuses on the evolution of social behavior through natural selection.

sodium-potassium pump Cellular active transport system that transports sodium ions out of, and potassium ions into, cells.

soil erosion The wearing away or removal of soil from the land. Although soil erosion occurs naturally from precipitation and runoff, human activities (such as clearing the land) accelerate it.

solar tracking The ability of leaves or flowers of certain plants to follow the sun by aligning themselves either perpendicular or parallel to the sun's rays.

solute A dissolved substance. Compare with solvent.

solvent Substance capable of dissolving other substances. Compare with solute.

somatic cell A cell of the body not involved in formation of gametes.

somatic nervous system That part of the vertebrate peripheral nervous system that keeps the body in adjustment with the external environment; includes the sensory receptors on the body surface and within the muscles, and the nerves that link them with the central nervous system. Compare with autonomic nervous system.

somatotropin See growth hormone.

soredium (sor-id′e-um) (pl. *soredia*) In lichens, a type of asexual reproductive structure that consists of a cluster of algal cells surrounded by fungal hyphae.

sorus (soh′rus) (pl. *sori*) In ferns, a cluster of spore-producing sporangia.

Southern blot hybridization A technique in which DNA fragments, previously separated by gel electrophoresis, are transferred to a nitrocellulose membrane. A specific radioactive genetic probe is then allowed to hybridize to complementary fragments, thus marking their locations.

speciation Evolution of a new species.
species A group of organisms with similar structural and functional characteristics that in nature breed only with each other and have a close common ancestry; a group of organisms with a common gene pool. Also known as biological species concept.
species richness The number of species in a community.
specific epithet The second part of the name of a species.
specific heat The amount of heat energy that must be supplied to raise the temperature of 1 gram of a substance 1 degree Celsius.
sperm The motile, *n* male reproductive cell of animals and some plants and protists; spermatozoan.
spermatid (spur′ma-tid) An immature sperm cell.
spermatocyte (spur-mah′toh-site) A meiotic cell that gives rise to spermatids, and ultimately to mature sperm cells.
spermatogenesis (spur″mah-toh-jen′eh-sis) The production of sperm by meiosis. Compare with oogenesis.
spermatozoa (spur-mah-toh-zoh′uh) Mature sperm cells.
sphincter (sfink′tur) A group of circularly arranged muscle fibers, the contractions of which close an opening, such as the pyloric sphincter at the exit of the stomach.
spinal cord In vertebrates, the dorsal, tubular nerve cord.
spinal nerves In vertebrates, the nerves that emerge from the spinal cord.
spindle Structure consisting mainly of microtubules that provides the framework for chromosome movement during cell division. Also known as mitotic spindle.
spine A leaf that is modified for protection, such as a cactus spine.
spiracle (speer′ih-kl) An opening for gas exchange, such as the opening on the body surface of a trachea in insects.
spiral cleavage Distinctive spiral pattern of blastomere production in an early protostome embryo. Compare with radial cleavage.
spleen Abdominal organ located just below the diaphragm that removes worn-out blood cells and bacteria from the blood and plays a role in immunity.
spongy mesophyll (mez′oh-fil) The loosely arranged mesophyll cells near the lower epidermis in certain leaves. Compare with palisade mesophyll.
spontaneous reaction See exergonic reaction.
sporangium (spor-an′jee-um) (pl. *sporangia*) A spore case, found in plants, certain protists, and fungi.
spore A reproductive cell that gives rise to individual offspring in plants, fungi, and certain algae and protozoa.
sporophyll (spor′oh-fil) A leaf-like structure that bears spores.
sporophyte generation (spor′oh-fite) The $2n$ spore-producing stage in the life cycle of a plant. Compare with gametophyte generation.
sporozoa (spor″uh-zoe′uh) A group of parasitic protozoa that lack structures for locomotion and produce spores as infective agents. Malaria is caused by a sporozoan.
spring turnover A mixing of lake water in temperate lakes that occurs in spring as ice melts and the surface water reaches 4°C, its temperature of greatest density. Compare with fall turnover.
stabilizing selection Natural selection that acts against extreme phenotypes and favors intermediate variants; associated with a population well-adapted to its environment. Compare with directional selection and disruptive selection.
stamen (stay′men) The male part of flower; consists of a filament and anther.
standing-water ecosystem A lake or pond ecosystem.
starch A polysaccharide composed of alpha glucose subunits; made by plants for energy storage.
start codon See initiation codon.
statocyst (stat′oh-sist) An invertebrate sense organ containing one or more granules; used in a variety of animals to sense gravity and motion.
statoliths (stat′uh-liths) (1) Starch grains in certain root cap cells that sense the direction of gravity and are involved in the gravitropism response; (2) Granules of loose sand or calcium carbonate found in statocysts.
stele The cylinder in the center of roots and stems that contains the vascular tissue.
sterilization A procedure that renders an individual incapable of producing offspring; the most common surgical procedures are vasectomy in the male and tubal ligation in the female.
steroids (steer′oids) Complex molecules containing carbon atoms arranged in four attached rings, three of which contain six carbon atoms each and the fourth of which contains five. Cholesterol and certain hormones, including the male and female sex hormones of vertebrates, are examples.
stigma Portion of the carpel where pollen grains land during pollination (and prior to fertilization).
stipe A short stalk or stem-like structure that is a part of the body of certain multicellular algae.
stipule (stip′yule) One of a pair of scalelike or leaflike structures found at the base of certain leaves.
stolon (stow′lon) An above-ground, horizontal stem with long internodes; stolons often form buds that develop into separate plants.
stomach Muscular region of the vertebrate digestive tract extending from the esophagus to the small intestine.
stomata (sing. *stoma*) Small pores located in the epidermis of plants that provide for gas exchange for photosynthesis; each stoma is flanked by two guard cells, which are responsible for its opening and closing.
stop codon See termination codon.
stratosphere The layer of the atmosphere between the troposphere and the mesosphere. It contains a thin ozone layer that protects life by filtering out much of the sun's ultraviolet radiation.
stratum basale (strat′um bah-say′lee) The deepest sublayer of the human epidermis, consisting of cells that continuously divide. Compare with stratum corneum.
stratum corneum The most superficial sublayer of the human epidermis. Compare with stratum basale.
strobilus (stroh′bil-us) (pl. *strobili*) In certain plants, a cone-like structure that bears spore-producing sporangia.
stroke volume The volume of blood pumped by one ventricle during one contraction.
stroma A fluid space of the chloroplast, enclosed by the chloroplast inner membrane and surrounding the thylakoids; site of the reactions of the Calvin cycle.
stromatolite (stroh-mat′oh-lite) A column-like rock that is composed of many minute layers of prokaryotic cells, usually cyanobacteria.
structural formula Representation of a molecule that shows the spatial arrangement of the atoms.

structural isomer One of two or more chemical compounds having the same chemical formula, but differing in the covalent arrangement of their atoms; examples include glucose and fructose.

style The neck connecting the stigma to the ovary of a carpel.

subsidiary cell A structurally distinct epidermal cell associated with a guard cell.

subsistence agriculture The production of enough food to feed oneself and one's family with little left over to sell or reserve for times of famine.

substrate A substance on which an enzyme acts; a reactant in an enzymatically catalyzed reaction.

succession The sequence of vegetation changes in a community over time.

sucker A shoot formed from an adventitious bud that develops on certain roots or underground stems. Suckers give rise to new plants.

sulcus (sul'kus) (pl. *sulci*) A groove, trench, or depression, especially one occurring on the surface of the brain, separating the gyri.

sulfhydryl group Functional group abbreviated -SH; found in organic compounds called thiols.

sum law Rule for combining the probabilities of mutually exclusive events by adding their individual probabilities.

summation The process of adding together excitatory postsynaptic potentials (EPSPs).

suppressor T cell T lymphoctye that suppresses the immune response.

supraorbital ridge (soop"rah-or'bit-ul) Prominent bony ridges above the eye sockets. Ape skulls have prominent supraorbital ridges.

surface tension The attraction that the molecules at the surface of a liquid may have for each other.

survivorship The proportion of individuals in a population that survive to a particular age; usually presented as a survivorship curve.

suspensor (suh-spen'sur) In plant embryo development, a multicellular structure that anchors the embryo and aids in nutrient absorption from the endosperm.

swim bladder Hydrostatic organ in bony fishes that permits the fishes to hover at a given depth.

symbiosis (sim-bee-oh'sis) An intimate relationship between two or more organisms of different species. See commensalism, mutualism, and parasitism.

sympathetic nervous system A division of the autonomic nervous system; its general effect is to mobilize energy, especially during stress situations; prepares the body for fight-or-flight response. Compare with parasympathetic nervous system.

sympatric speciation (sim-pa'trik) The evolution of a new species within the same geographical region as the parental species. Compare with allopatric speciation.

symplast A continuum consisting of the cytoplasm of many plant cells, connected by plasmodesmata. Compare with apoplast.

synapse (sin'aps) The junction between two neurons or between a neuron and an effector (muscle or gland).

synapsis (sin-ap'sis) The process of physical association of homologous chromosomes during prophase I of meiosis.

synaptic knobs The ends of axon terminals that contain neurotransmitters.

synaptonemal complex Structure, visible with the electron microscope, that forms between synapsed homologous chromosomes.

syngamy (sin'gah-mee) The union of the gametes in sexual reproduction.

synthetic theory of evolution The synthesis of previous theories, especially of Mendelian genetics with Darwin's theory of evolution by natural selection, to formulate a comprehensive explanation of evolution; also called neo-Darwinism.

systematics Modern taxonomy, based on evolutionary relationships.

systemic circulation That part of the circulatory system that delivers blood to and from the tissues and organs of the body. Compare with pulmonary circulation.

systole The part of the cardiac cycle when the heart is contracting. Compare with diastole.

T cell (T lymphocyte) Lymphocyte that is processed in the thymus. T cells have a wide variety of immune functions but are primarily responsible for cell-mediated immunity. Compare with B cell.

T tubules Transverse tubules; system of inward extensions of the muscle fiber plasma membrane.

taiga (tie'gah) Northern coniferous forest biome found primarily in Canada, northern Europe, and Siberia.

taproot system A root system in plants that has one main root with smaller roots branching off it. Compare with fibrous root system.

target tissue or cell A cell or tissue that is affected by a hormone.

TATA box Component of a eukaryotic promoter region; consists of a sequence of bases located about 30 base pairs upstream from the transcription initiation site.

taxon A formal taxonomic group at any level, for example phylum or genus.

taxonomy (tax-on'ah-mee) The science of naming, describing, and classifying organisms; see systematics.

Tay-Sachs disease A serious genetic disease in which abnormal lipid metabolism in the brain causes mental deterioration in affected infants and young children; inheritance pattern is autosomal recessive.

tectorial membrane (tek-tor'ee-ul) The roof membrane of the organ of Corti in the cochlea of the ear.

teleosts Modern bony fishes.

telophase (teel'oh-faze or tel'oh-faze) The last stage of mitosis, and of meiosis I and II, when, having reached the poles, chromosomes become decondensed, and a nuclear envelope forms around each group.

temperate deciduous forest A forest biome that occurs in temperate areas where annual precipitation ranges from about 75 cm to 125 cm.

temperate grassland A grassland characterized by hot summers, cold winters, and less rainfall than is found in a temperate deciduous forest biome.

temperate rain forest A coniferous biome characterized by cool weather, dense fog, and high precipitation. Found on the north Pacific coast of North America.

temperate virus A virus that can become integrated into the host DNA as a prophage.

temperature The average kinetic energy of the particles in a sample of a substance.

temporal isolation A reproductive isolating mechanism (prezygotic) in which genetic exchange is prevented between similar species because they reproduce at different times of the day, season, or year.

tendon A connective tissue structure that joins a muscle to another muscle, or a muscle to a bone. Tendons transmit the force generated by a muscle.

tendril A leaf or stem that is modified for holding or attaching onto objects.

tension-cohesion model The mechanism by which water and dissolved minerals are thought to be transported in xylem; water is pulled upward under tension due to transpiration while maintaining an unbroken column in xylem due to cohesion.

terminal bud A bud at the tip of a stem. Compare with lateral bud.

termination codon Also known as a stop codon. Any codon in mRNA that does not code for an amino acid. This stops translation at that point.

territoriality Behavior pattern in which one organism (usually a male) stakes out a territory of its own and defends it against intrusion by other members of the same species and sex.

tertiary structure (of a protein) (tur'she-air"ee) The overall three-dimensional shape of a polypeptide that is determined by interactions involving the amino acid side chains. Compare with primary, secondary, and quaternary protein structure.

test A shell.

test cross Genetic cross in which either an F_1 individual, or an individual of unknown genotype, is mated to a homozygous recessive individual.

testis (tes'tis) (pl. *testes*) The male gonad that produces sperm and the male hormone testosterone; in humans and certain other mammals, the testes are situated in the scrotum.

testosterone (tes-tos'ter-own) The principal male sex hormone (androgen); a steroid hormone produced by the interstitial cells of the testes; stimulates spermatogenesis and is responsible for primary and secondary sex characteristics in the male.

tetrad Chromosome complex formed by the synapsis of homologous chromosomes during meiotic prophase I; also known as a bivalent. A tetrad contains four chromatids.

tetrapods (tet'rah-podz) Four-limbed vertebrates: the amphibians, reptiles, birds, and mammals.

thalamus (thal'uh-mus) The part of the vertebrate brain that serves as a main relay center, transmitting information between the spinal cord and the cerebrum.

thallus (thal'us) (pl. *thalli*) The simple body of an alga, fungus, or nonvascular plant that lacks roots, stems, or leaves; for example, a liverwort thallus or a lichen thallus.

theory A widely accepted idea supported by a large body of observations and experiments. Compare with hypothesis.

therapsids (ther-ap'sids) A group of mammal-like reptiles of the Permian period; gave rise to the mammals.

thermal stratification The marked layering (separation into warm and cold layers) of temperate lakes during the summer. See thermocline.

thermocline (thur'moe-kline) A marked and abrupt temperature transition in temperate lakes between warm surface water and cold deeper water. See thermal stratification.

thermodynamics (thurm"oh-dy-nam'iks) Principles governing energy transfer (often expressed in terms of heat transfer). See first law of thermodynamics and second law of thermodynamics.

thermoreceptor A sensory receptor that responds to heat.

thick filaments Filaments composed mainly of myosin; found in muscle fibers. Compare with thin filaments.

thigmomorphogenesis A plant developmental response to mechanical stresses such as wind, rain, hail, and contact with passing animals.

thigmotropism (thig"moh-troh'pizm) Plant growth in response to contact with a solid object, such as the twining of plant tendrils.

thin filaments Filaments composed mainly of actin; found in muscle fibers. Compare with thick filaments.

threatened species A species in which the population is small enough for it to be at risk of becoming extinct throughout all or part of its range, but not so small that it is in imminent danger of extinction. Compare with endangered species.

threshold level The potential that a neuron or other excitable cell must reach for an action potential to be initiated.

thrombocytes See platelets.

thrombus (throm'bus) A blood clot formed within a blood vessel or within the heart.

thylakoids (thy'lah-koidz) Interconnected system of flattened, sac-like membranous structures inside the chloroplast; the thylakoid membranes contain chlorophyll and the electron transport chain, and enclose a compartment, the thylakoid space.

thymine (thy'meen) A nitrogenous pyrimidine base found in DNA.

thymus gland (thy'mus) An endocrine gland that functions as part of the lymphatic system; important in the development of the ability to make immune responses.

thyroid gland An endocrine gland that lies anterior to the trachea and releases hormones that regulate the rate of metabolism.

thyroid hormones Hormones, including thyroxin, secreted by the thyroid gland; stimulate rate of metabolism.

tidal volume The volume of air moved into and out of the lungs with each normal resting breath.

tight junctions Specialized structures that form between some animal cells, producing a tight seal that prevents materials from passing through the spaces between the cells.

tissue A group of closely associated, similar cells that work together to carry out specific functions.

tissue culture The growth of tissue or cells in a synthetic growth medium under sterile conditions.

tissue fluid See interstitial fluid.

tolerance Decreased response to a drug over time.

tonoplast The membrane surrounding a vacuole.

topoisomerases Enzymes that relieve twists and kinks in a DNA molecule by breaking and rejoining the strands.

torsion Twisting of the visceral mass characteristic of gastropod mollusks.

total fertility rate The average number of children born to a woman during her lifetime.

totipotency (toh-ti-poh'tun-cee) Ability of a cell (or nucleus) to provide information for the development of an entire organism.

trace element An element required by an organism, but in very small amounts.

trachea (tray'kee-uh) (pl. *tracheae*) (1) Principal thoracic air duct of terrestrial vertebrates; windpipe; (2) One of the microscopic air ducts branching throughout the body of most terrestrial arthropods and some terrestrial mollusks.

tracheal tubes See trachea.
tracheid (tray'kee-id) A type of water-conducting and supporting cell in the xylem of vascular plants.
tract A bundle of nerve fibers within the CNS.
transcription The synthesis of RNA from a DNA template.
transcription factors DNA-binding proteins that regulate transcription; include positively acting activators and negatively acting repressors.
transduction The transfer of a genetic fragment from one cell to another, e.g., from one bacterium to another, by a virus.
transfer RNA (tRNA) RNA molecules that bind to specific amino acids and serve as adapter molecules in protein synthesis. The tRNA anticodons bind to complementary mRNA codons.
transformation (1) The incorporation of genetic material into a cell, thereby changing its phenotype; (2) The conversion of a normal cell to a malignant cell.
transgenic organism A plant or animal that has incorporated foreign DNA into its genome.
translation Conversion of information provided by mRNA into a specific sequence of amino acids in a polypeptide chain; process also requires transfer RNA and ribosomes.
translocation (1) The movement of organic materials (dissolved food) in the phloem of a plant; (2) Chromosome abnormality in which part of one chromosome has become attached to another; (3) Part of the elongation cycle of protein synthesis in which a transfer RNA attached to the growing polypeptide chain is transferred from the A site to the P site.
transmembrane protein An integral membrane protein that spans the lipid bilayer.
transmission, neural Conduction of a neural impulse along a neuron or from one neuron to another.
transpiration Loss of water vapor from the aerial surfaces of a plant (i.e., leaves and stems).
transpiration-cohesion model See tension-cohesion model.
transport vesicles Small cytoplasmic vesicles that move substances from one membrane system to another.
transposon (tranz-poze'on) A DNA segment that is capable of moving from one chromosome to another, or to different sites within the same chromosome; also referred to as a transposable element or mobile genetic element.
transverse tubules See T tubules.
triacylglycerol (try-as"il-glis'er-ol) A neutral fat consisting of a glycerol combined chemically with three fatty acids; also called triglyceride.
tricarboxylic acid (TCA) cycle See citric acid cycle.
trichocyst (trik'oh-sist) A cellular organelle found in certain ciliates that can discharge a thread-like structure that may aid in trapping and holding prey.
trichome (try'kohm) A hair or other appendage growing out from the epidermis of a plant.
triglyceride See triacylglycerol.
triose A sugar molecule containing three carbons.
triplet A sequence of three nucleotides that serves as the basic unit of genetic information.
triplet code The sequences of three nucleotides that compose the codons, the units of genetic information in mRNA that specify the order of amino acids in a polypeptide chain.
trisomy (try'sohm-ee) Condition in which a chromosome is present in triplicate, instead of the normal pair (disomy). Compare with monosomy and disomy.
trochophore larva (troh'koh-for) A larval form found in mollusks and many polychaetes.
trophic level (trow'fik) Each level in a food web. All producers belong to the first trophic level, all herbivores belong to the second trophic level, and so on.
trophoblast (troh'foh-blast) The outer cell layer of a late blastocyst which, in placental mammals, gives rise to the chorion and to the fetal contribution to the placenta.
tropic hormone (trow'pic) A hormone, produced by one endocrine gland, that targets another endocrine gland.
tropical rain forest A lush, species-rich forest biome that occurs in tropical areas where the climate is very moist throughout the year. Tropical rain forests are also characterized by old, infertile soils.
tropism (troh'pizm) In plants, a directional growth response that is elicited by an environmental stimulus.
tropomyosin (troh-poh-my'oh-sin) A regulatory muscle protein involved in the control of contraction.
true breeding line A genetically pure strain of organism, i.e. one in which all individuals are homozygous for the traits under consideration. Also called a pure line.
tube feet Structures characteristic of echinoderms; function in locomotion.
tuber A thickened underground stem adapted for food storage, e.g., white potato.
tumor Mass of tissue that grows in an uncontrolled manner; a neoplasm.
tumor suppressor gene See anti-oncogene.
tundra (tun'dra) A treeless biome (between the taiga in the south and the polar ice cap in the north) that consists of boggy plains covered by lichens and small plants such as mosses. The tundra is characterized by harsh, very cold winters and extremely short summers. Also called arctic tundra. Compare with alpine tundra.
tunicates Chordates belonging to subphylum Urochordata; sea squirts.
turgor movement A temporary movement of a plant organ that results from changes in turgor in certain of its cells. Also called nastic movement.
turgor pressure (tur'gor) Hydrostatic pressure that develops within a walled cell, such as a plant cell, when the osmotic pressure of the cell's contents is greater than the osmotic pressure of the surrounding fluid.
Turner syndrome An inherited condition is which only one sex chromosome (an X chromosome) is present in cells; karyotype is designated XO. Affected individuals are sterile females.
ultrasound imaging Technique in which high frequency sound waves (ultrasound) are used to provide an image of an internal structure.
ultrastructure The fine detail of a cell, generally only observable by use of an extremely sophisticated microscope, such as an electron microscope.
uniform dispersion The spatial distribution pattern in which individuals are more regularly spaced than in a random distribution. Compare with random dispersion and clumped dispersion.
unsaturated fatty acid See fatty acid.
upstream promoter elements (UPE) Components of a eukaryotic promoter, found upstream of the RNA polymerase-binding site; the strength of a promoter is affected by the number and type of UPEs present.

upwelling An upward movement of water that brings nutrients from the ocean depths to the surface. Where upwelling occurs, the ocean is very productive.

uracil (yur'ah-sil) A nitrogenous pyrimidine base found in RNA.

urea (yur-ee'ah) The principal nitrogenous excretory product of mammals; one of the water-soluble end products of protein metabolism.

ureter (yoo-ree'tur) One of the paired tubular structures that conducts urine from the kidney to the bladder.

urethra (yoo-ree'thruh) The tube that conducts urine from the bladder to the outside of the body.

uric acid (yoor'ik) The principal nitrogenous excretory product of insects, birds, and reptiles; a relatively insoluble end product of protein metabolism; also occurs in mammals as an end product of purine metabolism.

urinary bladder An organ that receives urine from the ureters and temporarily stores it.

urinary system Body system in mammals that consists of kidneys, urinary bladder, and associated ducts.

urochordates Subphylum of chordates; includes the tunicates.

uterine tube (yoo'tur-in) See oviduct.

uterus (yoo'tur-us) The womb; the hollow, muscular organ of the female reproductive tract in which the fetus undergoes development.

utricle The structure within the vestibule of the vertebrate inner ear that along with the saccule houses the receptors of static equilibrium.

vaccine (vak-seen') A commercially produced weakened or killed antigen of a particular disease that stimulates the body to make antibodies.

vacuole (vak'yoo-ole) A fluid-filled, membrane-bounded sac found within the cytoplasm; may function in storage, digestion, or water elimination.

vagina The elastic, muscular tube, extending from the cervix to its orifice, that receives the penis during sexual intercourse and serves as the birth canal.

valence electrons The electrons in the outer electron shell, known as the valence shell, of an atom; in the formation of a chemical bond an atom can accept electrons into its valence shell, or donate or share valence electrons.

van der Waals forces Weak attractive forces between the atoms of nonpolar molecules; caused by interactions among fluctuating charges.

vas deferens (def'ur-enz) One of the paired sperm ducts that connects the epididymis of the testes to the ejaculatory duct.

vascular cambium A lateral meristem in plants that produces secondary xylem (wood) and secondary phloem (inner bark). Compare with cork cambium.

vascular tissue system The tissues specialized for translocation of materials throughout the plant body, i.e., the xylem and phloem.

vasoconstriction Narrowing of blood vessels.

vasodilation Expansion of the diameter of blood vessels.

vector (1) Any carrier or means of transfer; (2) Agent, such as a plasmid or virus, that transfers genetic information; (3) Agent that transfers a parasite from one host to another.

vegetal pole The yolky pole of a vertebrate or echinoderm egg. Compare with animal pole.

vein (1) A blood vessel that carries blood from the tissues toward the heart; (2) A strand of vascular tissue that is part of the network of conducting tissue in a leaf.

ventilation The process of actively moving air or water over a respiratory surface.

ventral Toward the lowermost surface or belly of an animal. Compare with dorsal.

ventricle (1) A cavity in an organ; (2) One of the several cavities of the brain; (3) One of the chambers of the heart that receives blood from an atrium.

vernalization (vur″nul-i-za'shun) The induction of flowering by a low temperature treatment.

vertebrates A chordate subphylum of animals that possess a bony vertebral column; include fishes, amphibians, reptiles, birds, and mammals.

vesicle (ves'ih-kl) Any small sac, especially a small spherical membrane-bounded compartment, within the cytoplasm.

vessel element A type of water-conducting cell in the xylem of vascular plants.

vestibular apparatus Collectively, the saccule, utricle, and semicircular canals of the inner ear.

vestigial (ves-tij'ee-ul) Rudimentary; an evolutionary remnant of a formerly functional structure.

villus (pl. *villi*) A multicellular, minute, elongated projection from the surface of a membrane, e.g., villi of the mucosa of the small intestine.

viroid (vy'roid) Tiny, naked virus consisting only of nucleic acid.

virulent Able to cause disease in a host. Compare with avirulent.

virus A tiny pathogen composed of a core of nucleic acid usually encased in protein and capable of infecting living cells. A virus is characterized by total dependence upon a living host.

viscera (vis'ur-uh) The internal body organs, especially those located in the abdominal or thoracic cavities.

visceral mass Concentration of body organs (viscera) located above the foot; characteristic of mollusks.

vital capacity The maximum volume of air a person can expire after filling the lungs to the maximum extent.

vitamin A complex organic molecule required in very small amounts for normal metabolic functioning.

viviparous (vih-vip'er-us) Bearing living young that develop within the body of the mother.

vulva The liplike external genital structures of the female.

warning coloration The conspicuous coloring of a poisonous or distasteful organism that enables potential predators to easily see and recognize it.

water mold A fungus-like protist with a body consisting of a coenocytic mycelium that reproduces asexually by forming motile zoospores and sexually by forming oospores.

water potential Free energy of water; the water potential of pure water is zero, and that of solutions is a negative value. Differences in water potential are used to predict the direction of water movement (always from a region of less negative water potential to a region of more negative water potential).

water vascular system Unique hydraulic system of echinoderms; functions in locomotion and feeding.

wavelength The distance from one wave peak to the next; the energy of electromagnetic radiation is inversely proportional to its wavelength.

weathering A chemical or physical process that helps form soil from rock; during weathering, the rock is gradually broken down into smaller and smaller pieces.

whisk ferns A phylum of seedless vascular plants with a life cycle similar to ferns.

white matter Central nervous system tissue containing myelinated axons. Compare with gray matter.

wild type The phenotypically normal (naturally occurring) form of a gene or organism.

wobble The ability of some tRNA anticodons to associate with more than one mRNA codon; in these cases the 5′ base of the anticodon is capable of forming hydrogen bonds with more than one kind of base in the 3′ position of the codon.

work Any change in the state or motion of matter.

X-linked gene Gene carried on an X chromosome.

x-ray diffraction A technique for determining the spatial arrangement of the components of a crystal. A beam of x rays directed toward a sample becomes diffracted (bent) by the ordered components of the crystal. The resulting pattern of spots on a photographic plate can be analyzed mathematically to determine the distances between the atoms and molecules.

xylem (zy′lem) The vascular tissue that conducts water and dissolved minerals in plants.

XYY karyotype Chromosome constitution that causes affected individuals (who are fertile males) to be unusually tall, with severe acne.

yeast A unicellular fungus (ascomycete) that reproduces asexually by budding or fission and sexually by ascospores.

yolk sac One of the extraembryonic membranes; a pouch-like outgrowth of the digestive tract of certain vertebrate embryos that grows around the yolk, digests it, and makes it available to the rest of the developing embryo.

zero population growth Point at which the birth rate equals the death rate. A population with zero population growth does not change in size.

zona pellucida (pel-loo′sih-duh) The thick, transparent covering that surrounds the plasma membrane of an ovum.

zooplankton (zoh″oh-plank′tun) The nonphotosynthetic organisms present in plankton. See plankton. Compare with phytoplankton.

zoospore (zoh′oh-spore) A flagellated motile spore produced asexually by certain algae, water molds, and other protists.

zooxanthellae (zoh″oh-zan-thel′ee) (sing. *zooxanthella*) Endosymbiotic, photosynthetic dinoflagellates found in certain marine invertebrates; their mutualistic relationship with corals enhances the corals' reef-building ability.

zygomycetes (zy″gah-my′seat) Fungi characterized by the production of nonmotile asexual spores and sexual zygospores.

zygospore (zy′gah-spor) A thick-walled sexual spore produced by a zygomycete.

zygote The $2n$ cell that results from the union of n gametes in sexual reproduction. Species that are not polyploid have haploid gametes and diploid zygotes.

Index

Boldface page numbers indicate pages on which the index term is defined; "il" following the page number indicates an illustration, "t" a table, "f" a focus, and "n" a footnote.

SCIENTIFIC MEASUREMENT

SOME COMMON PREFIXES

		Examples
kilo	1,000	a kilogram is 1,000 grams
centi	0.01	a centimeter is 0.01 meter
milli	0.001	a milliliter is 0.001 liter
micro (μ)	one millionth	a micrometer is 0.000001 (one millionth) of a meter
nano (n)	one billionth	a nanogram is 10^{-9} (one billionth) of a gram
pico (p)	one trillionth	a picogram is 10^{-12} (one trillionth) of a gram

The relationship between mass and volume of water (at 20°C)

$1\ g = 1\ cm^3 = 1\ mL$

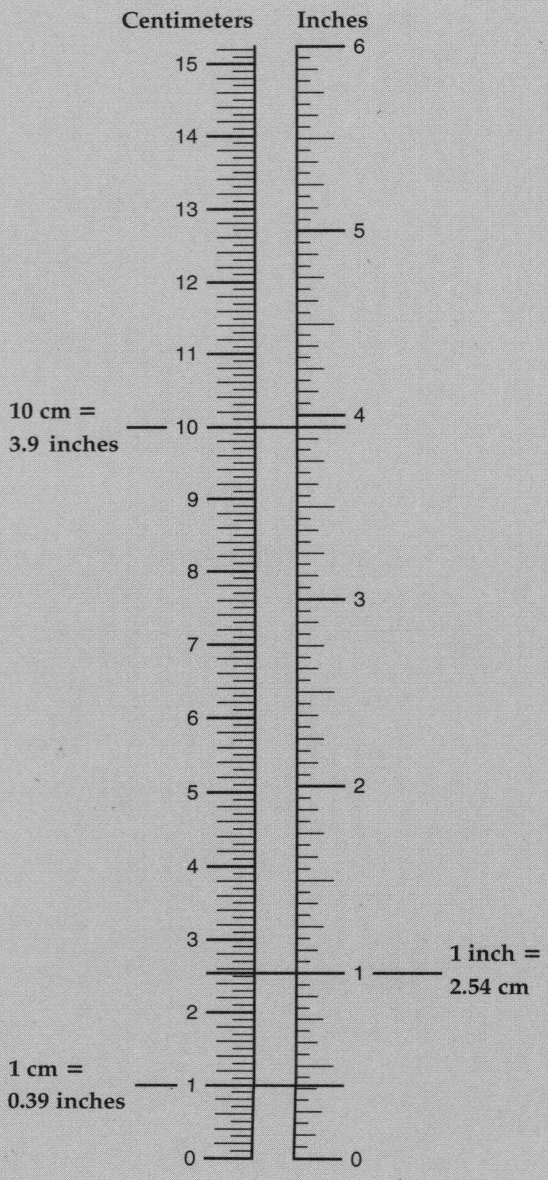

SOME COMMON UNITS OF LENGTH

Unit	Abbreviation	Equivalent
meter	m	approximately 39 in
centimeter	cm	10^{-2} m
millimeter	mm	10^{-3} m
micrometer	μm	10^{-6} m
nanometer	nm	10^{-9} m
angstrom	Å	10^{-10} m

Length Conversions

1 in = 2.5 cm	1 mm = 0.039 in
1 ft = 30 cm	1 cm = 0.39 in
1 yd = 0.9 m	1 m = 39 in
1 mi = 1.6 km	1 m = 1.094 yd
	1 km = 0.6 mi

To convert	*Multiply by*	*To obtain*
inches	2.54	centimeters
feet	30	centimeters
centimeters	0.39	inches
millimeters	0.039	inches

STANDARD METRIC UNITS

		Abbreviation
Standard unit of mass	gram	g
Standard unit of length	meter	m
Standard unit of volume	liter	L